Contents—1999 HVAC APPLICATIONS

Contents—1998 REFRIGERATION

2001 ASHRAE® HANDBOOK

FUNDAMENTALS

Inch-Pound Edition

American Society of Heating, Refrigerating and Air-Conditioning Engineers, Inc.

1791 Tullie Circle, N.E., Atlanta, GA 30329

(404) 636-8400 http://www.ashrae.org

DEDICATED TO THE ADVANCEMENT OF

THE PROFESSION AND ITS ALLIED INDUSTRIES

Volunteer members of ASHRAE Technical Committees and others compiled the information in this handbook, and it is generally reviewed and updated every four years. Comments, criticisms, and suggestions regarding the subject matter are invited. Any errors or omissions in the data should be brought to the attention of the Editor. Additions and corrections to Handbook volumes in print will be published in the Handbook published the year following their verification and as soon as verified on the ASHRAE Internet website.

DISCLAIMER

ISBN 1-883413-87-7

CONTENTS

LOAD AND ENERGY CALCULATIONS

SPACE AIR DESIGN

GENERAL

ADDITIONS AND CORRECTIONS

INDEX

Composite index to the 1998 Refrigeration, 1999 HVAC Applications, 2000 HVAC Systems and Equipment, and 2001 Fundamentals volumes.

CONTRIBUTORS

In addition to the Technical Committees, the following individuals contributed significantly to this volume. The appropriate chapter numbers follow each contributor's name.

Charles H. Bemisderfer (1)
York International

Donald C. Erickson (1)
Energy Concepts Co.

Hans-Martin Hellmann (1)
Zent-Frenger

Thomas H. Kuehn (1, 6)
University of Minnesota

Christopher P. Serpente (1)
Carrier Corp.

Robert M. Tozer (1)
Waterman-Gore M&E Consulting Engineers

Anthony M. Jacobi (2)
University of Illinois at Urbana-Champaign

Dr. Arthur E. Bergles (3)
Rensselaer Polytechnic Institute

Michael M. Ohadi (3, 5)
University of Maryland

Steven J. Eckels (4)
Kansas State University

Rick J. Couvillion (5)
University of Arkansas

Albert C. Kent (5)
Southern Illinois University

Ray Rite (5)
Trane Company

Jason T. LeRoy (6)
Trane Company

Warren E. Blazier, Jr. (7)
Warren Blazier Associates, Inc.

Alfred C. C. Warnock (7)
National Research Council Canada

Larry G. Berglund (8)
U.S. Army Research Institute for Environmental Medicine

Gemma Kerr (9, 12)
InAir Environmental Ltd.

D.J. Marsick (9)
U.S. Department of Energy

D.J. Moschandreas (9)
The Institute for Science, Law, and Technology

Kenneth M. Wallingford (9)
NIOSH

Richard S. Gates (10)
University of Kentucky

Albert J. Heber (10)
Purdue University

Farhad Memarzadeh (10)
National Institutes of Health

Gerald L. Riskowski (10, 11)
University of Illinois

Yuanhui Zhang (10)
University of Illinois

Roger C. Brook (11)
Michigan State University

Joe F. Pedelty (12, 13)
Holcomb Environmental Services

Pamela Dalton (13)
Monell Chemical Senses Center

Martin Kendal-Reed (13)
Florida State University Sensory Research Institute

James C. Walker (13)
Florida State University Research Institute

Rick Stonier (14)
Gray Wolf Sensing Solutions

Monica Y. Amalfitano (15)
P2S Engineering

James J. Coogan (15)
Siemens Building Technologies

David M. Underwood (15)
USA-CERL

John J. Carter (16)
CPP Inc.

Dr. David J. Wilson (16)
University of Alberta

William J. Coad (17)
McClure Engineering Assoc.

David Grumman (17)
Grumman/Butkus Associates

Douglas W. DeWerth (18)

Robert G. Doerr (19)
The Trane Company

Eric W. Lemmon (20)
National Institute of Standards and Technology

Mark O. McLinden (20)
National Institute of Standards and Technology

Steven G. Penoncello (20)
University of Idaho

Sherry K. Emmrich (21)
Dow Chemical

Lewis G. Harriman III (22)
Mason-Grant Company

Hugo L.S.C. Hens (23)
Katholieke Universiteit

Achilles N. Karagiozis (23)
Oak Ridge National Laboratory

Hartwig M. Kuenzel (23)
Fraunhofer Institut Bauphysik

Anton TenWolde (23)
Forest Products Laboratory

William Brown (24)
Morrison Herschfield

Andre Desjarlais (24)
Oak Ridge National Laboratory

David Roodvoets (24)
DLR Consultants

William B. Rose (24, 25)
University of Illinois

William P. Goss (25)
University of Massachusetts

Malcolm S. Orme (26)
Air Infiltration and Ventilation Centre

Andrew K. Persily (26)
National Institute of Standards and Technology

Brian A. Rock (26)
University of Kansas

Armin F. Rudd (26)
Building Science Corporation

Max H. Sherman (26)
Lawrence Berkeley National Laboratory

Iain S. Walker (26)
Lawrence Berkeley National Laboratory

Craig P. Wray (26)
Lawrence Berkeley National Laboratory
University of Nebraska-Omaha

Grenville K. Yuill (26)
University of Nebraska-Omaha

Robert Morris (27)
Environment Canada
Climate and Water Products Division

Raymond G. Alvine (28)

Jim Norman (28)
AAA Enterprises

Charles F. Turdik (28)

Lynn Bellenger (29)
Pathfinder Engineers LLP

Steven Bruning (29)
Newcomb & Boyd

Curt Petersen (29)

Thomas Romine (29)
Romine, Romine & Burgess

Christopher Wilkins (29)
Hallam Associates

D. Charlie Curcija (30)
University of Massachusetts

William C. duPont (30)
Lawrence Berkeley National Laboratory

John F. Hogan (30)

Joseph H. Klems (30)
Lawrence Berkeley National Laboratory

W. Ross McCluney (30)
Florida Solar Energy Center

M. Susan Reilly (30)
Enermodal Engineering

Eleanor S. Lee (30)
Lawrence Berkeley National Laboratory

David Tait (30)
Tait Solar

Leslie Norford (31)
Massachusetts Institute of Technology

Fred S. Bauman (32)
University of California

Mohammad H. Hosni (32)
Kansas State University

Leon Kloostra (32)
TITUS, Division of Tomkins

Raymond H. Horstman (33)
Boeing Commercial Airplanes

Herman F. Behls (34)
Behls & Associates

Albert W. Black (35)
McClure Engineering Associates

ASHRAE HANDBOOK COMMITTEE

Dennis J. Wessel, Chair

2001 Fundamentals Volume Subcommittee: **George Reeves,** Chair

David E. Claridge **Frederick H. Kohloss** **Brian A. Rock** **T. David Underwood** **Michael W. Woodford**

ASHRAE HANDBOOK STAFF

Jeanne Baird, Associate Editor

Scott A. Zeh, Nancy F. Thysell, and **Jayne E. Jackson,** Publishing Services

W. Stephen Comstock,
Director, Communications and Publications
Publisher

ASHRAE TECHNICAL COMMITTEES AND TASK GROUPS

SECTION 1.0—FUNDAMENTALS AND GENERAL

1.1 Thermodynamics and Psychrometrics
1.2 Instruments and Measurements
1.3 Heat Transfer and Fluid Flow
1.4 Control Theory and Application
1.5 Computer Applications
1.6 Terminology
1.7 Operation and Maintenance Management
1.8 Owning and Operating Costs
1.9 Electrical Systems
1.10 Energy Resources

SECTION 2.0—ENVIRONMENTAL QUALITY

2.1 Physiology and Human Environment
2.2 Plant and Animal Environment
2.3 Gaseous Air Contaminants and Gas Contaminant Removal Equipment
2.4 Particulate Air Contaminants and Particulate Contaminant Removal Equipment
2.6 Sound and Vibration Control
2.7 Seismic and Wind Restraint Design
TG Buildings' Impacts on the Environment
TG Global Climate Change

SECTION 3.0—MATERIALS AND PROCESSES

3.1 Refrigerants and Secondary Coolants
3.2 Refrigerant System Chemistry
3.3 Refrigerant Contaminant Control
3.4 Lubrication
3.5 Desiccant and Sorption Technology
3.6 Water Treatment
3.8 Refrigerant Containment

SECTION 4.0—LOAD CALCULATIONS AND ENERGY REQUIREMENTS

4.1 Load Calculation Data and Procedures
4.2 Weather Information
4.3 Ventilation Requirements and Infiltration
4.4 Building Materials and Building Envelope Performance
4.5 Fenestration
4.6 Building Operation Dynamics
4.7 Energy Calculations
4.10 Indoor Environmental Modeling
4.11 Smart Building Systems
4.12 Integrated Building Design
TG Mechanical Systems Insulation

SECTION 5.0—VENTILATION AND AIR DISTRIBUTION

5.1 Fans
5.2 Duct Design
5.3 Room Air Distribution
5.4 Industrial Process Air Cleaning (Air Pollution Control)
5.5 Air-to-Air Energy Recovery
5.6 Control of Fire and Smoke
5.7 Evaporative Cooling
5.8 Industrial Ventilation
5.9 Enclosed Vehicular Facilities
5.10 Kitchen Ventilation

SECTION 6.0—HEATING EQUIPMENT, HEATING AND COOLING SYSTEMS AND APPLICATIONS

6.1 Hydronic and Steam Equipment and Systems
6.2 District Energy
6.3 Central Forced Air Heating and Cooling Systems
6.4 In Space Convection Heating
6.5 Radiant Space Heating and Cooling
6.6 Service Water Heating
6.7 Solar Energy Utilization
6.8 Geothermal Energy Utilization
6.9 Thermal Storage
6.10 Fuels and Combustion

SECTION 7.0—PACKAGED AIR-CONDITIONING AND REFRIGERATION EQUIPMENT

7.1 Residential Refrigerators and Food Freezers
7.4 Combustion Engine Driven Heating and Cooling Equipment
7.5 Mechanical Dehumidification Equipment and Heat Pipes
7.6 Unitary and Room Air Conditioners and Heat Pumps

SECTION 8.0—AIR-CONDITIONING AND REFRIGERATION SYSTEM COMPONENTS

8.1 Positive Displacement Compressors
8.2 Centrifugal Machines
8.3 Absorption and Heat Operated Machines
8.4 Air-to-Refrigerant Heat Transfer Equipment
8.5 Liquid-to-Refrigerant Heat Exchangers
8.6 Cooling Towers and Evaporative Condensers
8.7 Humidifying Equipment
8.8 Refrigerant System Controls and Accessories
8.10 Pumps and Hydronic Piping
8.11 Electric Motors and Motor Control

SECTION 9.0—AIR-CONDITIONING SYSTEMS AND APPLICATIONS

9.1 Large Building Air-Conditioning Systems
9.2 Industrial Air Conditioning
9.3 Transportation Air Conditioning
9.4 Applied Heat Pump/Heat Recovery Systems
9.5 Cogeneration Systems
9.6 Systems Energy Utilization
9.7 Testing and Balancing
9.8 Large Building Air-Conditioning Applications
9.9 Building Commissioning
9.10 Laboratory Systems
9.11 Clean Spaces
9.12 Tall Buildings
TG Combustion Gas Turbine Inlet Air Cooling Systems

SECTION 10.0—REFRIGERATION SYSTEMS

10.1 Custom Engineered Refrigeration Systems
10.2 Automatic Icemaking Plants and Skating Rinks
10.3 Refrigerant Piping, Controls, and Accessories
10.4 Ultra-Low Temperature Systems and Cryogenics
10.5 Refrigerated Distribution and Storage Facilities
10.6 Transport Refrigeration
10.7 Commercial Food and Beverage Cooling Display and Storage
10.8 Refrigeration Load Calculations
10.9 Refrigeration Application for Foods and Beverages
TG Mineral Oil Circulation

ASHRAE Research: Improving the Quality of Life

The American Society of Heating, Refrigerating and Air-Conditioning Engineers is the world's foremost technical society in the fields of heating, ventilation, air conditioning, and refrigeration. Its members worldwide ideas, identify needs, support research, and write the industry's standards for testing and practice. The result is that engineers are better able to keep indoor environments safe and productive while protecting and preserving the outdoors for generations to come.

One of the ways that ASHRAE supports its members' and industry's need for information is through ASHRAE Research. Thousands of individuals and companies support ASHRAE Research annually, enabling ASHRAE to report new data about material properties and building physics and to promote the application of innovative technologies.

The chapters in ASHRAE Handbooks are updated through the experience of members of ASHRAE technical committees and through results of ASHRAE Research reported at ASHRAE meetings and published in ASHRAE special publications and in *ASHRAE Transactions*.

For information about ASHRAE Research or to become a member contact, ASHRAE, 1791 Tullie Circle, Atlanta, GA 30329; telephone: 404-636-8400; www.ashrae.org.

The 2001 *ASHRAE Handbook*

The Fundamentals volume covers basic principles and provides data for the practice of HVAC&R technology. Although design data change little over time, research sponsored by ASHRAE and others continues to generate new information that meets the evolving needs of the people and industries that rely on HVAC&R technology to improve the quality of life. The ASHRAE technical committees that prepare chapters strive to provide new information, clarify existing information, delete obsolete materials and reorganize chapters to make the Handbook more understandable and easier to use. In this volume, some of the changes and additions are as follows:

- Chapter 1, Thermodynamics and Refrigeration Cycles, includes new sections on ideal thermal and absorption cycles, multiple stage cycles, and thermodynamic representation of absorption cycles. The section on ammonia water cycles has been expanded.
- Chapter 12, Air Contaminants, has undergone major revisions. Material has been added from the 1999 *ASHRAE Handbook*, Chapter 44, Control of Gaseous Indoor Air Contaminants. Health-related material with standards and guidelines for exposure has been moved to Chapter 9, Indoor Environmental Health.
- Chapter 15, Fundamentals of Control now includes new or revised figures on discharge air temperature control, step input process, and pilot positioners. New are sections on networking and fuzzy logic, revised descriptions on dampers and modulating control, and text on chilled mirror humidity sensors and dispersive infrared technology.
- Chapter 17, Energy Resources, contains new sections on sustainability and designing for effective energy resource use.
- Chapter 19, Refrigerants, provides information on phaseout of CFC and HCFC refrigerants and includes new data on R-143a and R-404A, R-407C, R-410A, R-507, R-508A, and R-508B blends.
- Chapter 20, Thermophysical Properties of Refrigerants, has new data on R-143a and R-245fa. Though most CFC Refrigerants have been removed from the chapter, R-12 has been retained to assist in making comparisons. Revised formulations have been used for many of the HFC refrigerants, conforming to international standards where applicable.
- Chapter 23, Thermal and Moisture Control in Insulated Assemblies—Fundamentals, now has a reorganized section on economic insulation thickness, a revised surface condensation section, and a new section on moisture analysis models.
- Chapter 26, Ventilation and Infiltration, includes rewritten stack pressure and wind pressure sections. New residential sections discuss averaging time variant ventilation, superposition methods, the enhanced (AIM-2) model, air leakage through automatic doors, and central air handler blowers in ventilation systems. The nonresidential ventilation section has also been rewritten, and now includes a commercial building envelope leakage measurements summary.
- Chapter 27, Climatic Design Information, now contains new monthly, warm-season design values for some United States locations. These values aid in consideration of seasonal variations in solar geometry and intensity, building occupancy, and use patterns.
- Chapter 29, Nonresidential Cooling and Heating Load Calculations, now contains enhanced data on internal loads, an expanded description of the heat balance method, and the new, simplified radiant time series (RTS) method.
- Chapter 30, Fenestration, now has revised solar heat gain and visible transmittance sections, including information on the solar heat gain coefficients (SHGC) method. The chapter now also has a rewritten section on solar-optical properties of glazings, an expanded daylighting section, and a new section on occupant comfort and acceptance.
- Chapter 31, Energy Estimating and Modeling Methods, now contains improved model forms for both design and existing building performance analysis. A new section describes a simplified method for calculating heat flow through building foundations and basements. Sections on secondary equipment and bin-energy method calculations have added information, while the section on data-driven models has been rewritten and now illustrates the variable-base degree-day method.
- Chapter 32, Space Air Diffusion, has been reorganized to be more user-friendly. The section on principles of jet behavior now includes simpler equations with clearer tables and figures. Temperature profiles now accompany characteristics of different outlets, with stagnant regions identified. The section on underfloor air distribution and task/ambient conditioning includes updates from recent ASHRAE-sponsored research projects.
- Chapter 33, HVAC Computational Fluid Dynamics, is a new chapter that provides an introduction to computational methods in flow modeling, including a description of computational fluid dynamics (CFD) with discussion of theory and capabilities.
- Chapter 34, Duct Design, includes revisions to duct sealing requirements from ASHRAE *Standard* 90.1, and has been expanded to include additional common fittings, previously included in electronic form in ASHRAE's Duct Fitting Database.

This Handbook is published both as a bound print volume and in electronic format on a CD-ROM. It is available in two editions—one contains inch-pound (I-P) units of measurement, and the other contains the International System of Units (SI).

Look for corrections to the 1998, 1999, and 2000 Handbooks on the Internet at **http://www.ashrae.org**. Any changes in this volume will be reported in the 2002 *ASHRAE Handbook* and on the ASHRAE web site.

If you have suggestions for improving a chapter or you would like more information on how you can help revise a chapter, e-mail ashrae@ashrae.org; write to Handbook Editor, ASHRAE, 1791 Tullie Circle, Atlanta, GA 30329; or fax 404-321-5478.

CHAPTER 1

THERMODYNAMICS AND REFRIGERATION CYCLES

THERMODYNAMICS is the study of energy, its transformations, and its relation to states of matter. This chapter covers the application of thermodynamics to refrigeration cycles. The first part reviews the first and second laws of thermodynamics and presents methods for calculating thermodynamic properties. The second and third parts address compression and absorption refrigeration cycles, the two most common methods of thermal energy transfer.

THERMODYNAMICS

A **thermodynamic system** is a region in space or a quantity of matter bounded by a closed surface. The surroundings include everything external to the system, and the system is separated from the surroundings by the system boundaries. These boundaries can be movable or fixed, real or imaginary.

The concepts that operate in any thermodynamic system are **entropy** and **energy**. Entropy measures the molecular disorder of a system. The more mixed a system, the greater its entropy; conversely, an orderly or unmixed configuration is one of low entropy. Energy has the capacity for producing an effect and can be categorized into either stored or transient forms as described in the following sections.

Stored Energy

Thermal (internal) energy is the energy possessed by a system caused by the motion of the molecules and/or intermolecular forces.

Potential energy is the energy possessed by a system caused by the attractive forces existing between molecules, or the elevation of the system.

$$\mathrm{PE} = mgz \qquad (1)$$

where

m = mass
g = local acceleration of gravity
z = elevation above horizontal reference plane

Kinetic energy is the energy possessed by a system caused by the velocity of the molecules and is expressed as

$$\mathrm{KE} = mV^2/2 \qquad (2)$$

where V is the velocity of a fluid stream crossing the system boundary.

Chemical energy is energy possessed by the system caused by the arrangement of atoms composing the molecules.

Nuclear (atomic) energy is energy possessed by the system from the cohesive forces holding protons and neutrons together as the atom's nucleus.

Energy in Transition

Heat (*Q*) is the mechanism that transfers energy across the boundary of systems with differing temperatures, always toward the lower temperature. Heat is positive when energy is added to the system (see Figure 1).

Work is the mechanism that transfers energy across the boundary of systems with differing pressures (or force of any kind), always toward the lower pressure. If the total effect produced in the system can be reduced to the raising of a weight, then nothing but work has crossed the boundary. Work is positive when energy is removed from the system (see Figure 1).

Mechanical or **shaft work** (*W*) is the energy delivered or absorbed by a mechanism, such as a turbine, air compressor, or internal combustion engine.

Flow work is energy carried into or transmitted across the system boundary because a pumping process occurs somewhere outside the system, causing fluid to enter the system. It can be more easily understood as the work done by the fluid just outside the

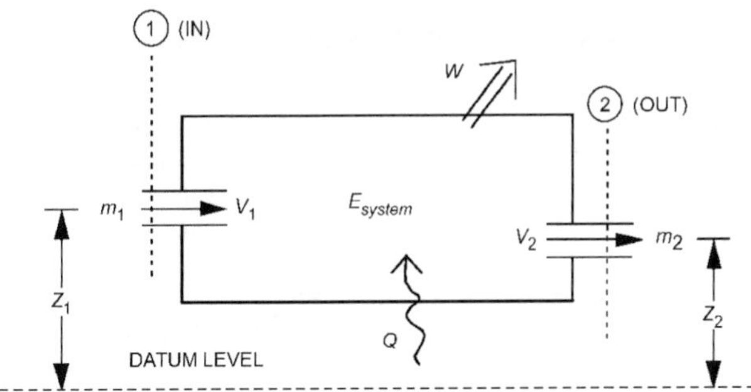

Fig. 1 Energy Flows in General Thermodynamic System

The preparation of the first and second parts of this chapter is assigned to TC 1.1, Thermodynamics and Psychrometrics. The third part is assigned to TC 8.3, Absorption and Heat-Operated Machines.

system on the adjacent fluid entering the system to force or push it into the system. Flow work also occurs as fluid leaves the system.

$$\text{Flow Work (per unit mass)} = pv \tag{3}$$

where p is the pressure and v is the specific volume, or the volume displaced per unit mass.

A **property** of a system is any observable characteristic of the system. The **state** of a system is defined by listing its properties. The most common thermodynamic properties are temperature T, pressure p, and specific volume v or density ρ. Additional thermodynamic properties include entropy, stored forms of energy, and enthalpy.

Frequently, thermodynamic properties combine to form other properties. **Enthalpy (*h*)**, a result of combining properties, is defined as

$$h \equiv u + pv \tag{4}$$

where u is internal energy per unit mass.

Each property in a given state has only one definite value, and any property always has the same value for a given state, regardless of how the substance arrived at that state.

A **process** is a change in state that can be defined as any change in the properties of a system. A process is described by specifying the initial and final equilibrium states, the path (if identifiable), and the interactions that take place across system boundaries during the process.

A **cycle** is a process or a series of processes wherein the initial and final states of the system are identical. Therefore, at the conclusion of a cycle, all the properties have the same value they had at the beginning.

A **pure substance** has a homogeneous and invariable chemical composition. It can exist in more than one phase, but the chemical composition is the same in all phases.

If a substance exists as liquid at the saturation temperature and pressure, it is called **saturated liquid**. If the temperature of the liquid is lower than the saturation temperature for the existing pressure, it is called either a **subcooled liquid** (the temperature is lower than the saturation temperature for the given pressure) or a **compressed liquid** (the pressure is greater than the saturation pressure for the given temperature).

When a substance exists as part liquid and part vapor at the saturation temperature, its quality is defined as the ratio of the mass of vapor to the total mass. Quality has meaning only when the substance is in a saturated state; i.e., at saturation pressure and temperature.

If a substance exists as vapor at the saturation temperature, it is called **saturated vapor**. (Sometimes the term **dry saturated vapor** is used to emphasize that the quality is 100%.) When the vapor is at a temperature greater than the saturation temperature, it is **superheated vapor**. The pressure and temperature of superheated vapor are independent properties, since the temperature can increase while the pressure remains constant. Gases are highly superheated vapors.

FIRST LAW OF THERMODYNAMICS

The first law of thermodynamics is often called the **law of the conservation of energy**. The following form of the first law equation is valid only in the absence of a nuclear or chemical reaction.

Based on the first law or the law of conservation of energy for any system, open or closed, there is an energy balance as

$$\begin{bmatrix}\text{Net Amount of Energy}\\\text{Added to System}\end{bmatrix} = \begin{bmatrix}\text{Net Increase in Stored}\\\text{Energy of System}\end{bmatrix}$$

or

$$\text{Energy In} - \text{Energy Out} = \text{Increase in Energy in System}$$

Figure 1 illustrates energy flows into and out of a thermodynamic system. For the general case of multiple mass flows in and out of the system, the energy balance can be written

$$\begin{aligned}&\sum m_{in}\left(u + pv + \frac{V^2}{2} + gz\right)_{in}\\&- \sum m_{out}\left(u + pv + \frac{V^2}{2} + gz\right)_{out} + Q - W\\&= \left[m_f\left(u + \frac{V^2}{2} + gz\right)_f - m_i\left(u + \frac{V^2}{2} + gz\right)_i\right]_{system}\end{aligned} \tag{5}$$

where subscripts i and f refer to the initial and final states, respectively.

The steady-flow process is important in engineering applications. Steady flow signifies that all quantities associated with the system do not vary with time. Consequently,

$$\begin{aligned}&\sum_{\substack{\text{all streams}\\\text{leaving}}} \dot{m}\left(h + \frac{V^2}{2} + gz\right)\\&- \sum_{\substack{\text{all streams}\\\text{entering}}} \dot{m}\left(h + \frac{V^2}{2} + gz\right) + \dot{Q} - \dot{W} = 0\end{aligned} \tag{6}$$

where $h = u + pv$ as described in Equation (4).

A second common application is the closed stationary system for which the first law equation reduces to

$$Q - W = [m(u_f - u_i)]_{system} \tag{7}$$

SECOND LAW OF THERMODYNAMICS

The second law of thermodynamics differentiates and quantifies processes that only proceed in a certain direction (irreversible) from those that are reversible. The second law may be described in several ways. One method uses the concept of entropy flow in an open system and the irreversibility associated with the process. The concept of irreversibility provides added insight into the operation of cycles. For example, the larger the irreversibility in a refrigeration cycle operating with a given refrigeration load between two fixed temperature levels, the larger the amount of work required to operate the cycle. Irreversibilities include pressure drops in lines and heat exchangers, heat transfer between fluids of different temperature, and mechanical friction. Reducing total irreversibility in a cycle improves the cycle performance. In the limit of no irreversibilities, a cycle will attain its maximum ideal efficiency.

In an open system, the second law of thermodynamics can be described in terms of entropy as

$$dS_{system} = \frac{\delta Q}{T} + \delta m_i s_i - \delta m_e s_e + dI \tag{8}$$

where

dS_{system} = total change within system in time dt during process
$\delta m_i s_i$ = entropy increase caused by mass entering (incoming)
$\delta m_e s_e$ = entropy decrease caused by mass leaving (exiting)
$\delta Q/T$ = entropy change caused by reversible heat transfer between system and surroundings
dI = entropy caused by irreversibilities (always positive)

Equation (8) accounts for all entropy changes in the system. Rearranged, this equation becomes

$$\delta Q = T[(\delta m_e s_e - \delta m_i s_i) + dS_{sys} - dI] \tag{9}$$

In integrated form, if inlet and outlet properties, mass flow, and interactions with the surroundings do not vary with time, the general equation for the second law is

$$(S_f - S_i)_{system} = \int_{rev} \frac{\delta Q}{T} + \sum (ms)_{in} - \sum (ms)_{out} + I \quad (10)$$

In many applications the process can be considered to be operating steadily with no change in time. The change in entropy of the system is therefore zero. The irreversibility rate, which is the rate of entropy production caused by irreversibilities in the process, can be determined by rearranging Equation (10)

$$\dot{I} = \sum (\dot{m}s)_{out} - \sum (\dot{m}s)_{in} - \int \frac{\dot{Q}}{T_{surr}} \quad (11)$$

Equation (6) can be used to replace the heat transfer quantity. Note that the absolute temperature of the surroundings with which the system is exchanging heat is used in the last term. If the temperature of the surroundings is equal to the temperature of the system, the heat is transferred reversibly and Equation (11) becomes equal to zero.

Equation (11) is commonly applied to a system with one mass flow in, the same mass flow out, no work, and negligible kinetic or potential energy flows. Combining Equations (6) and (11) yields

$$\dot{I} = \dot{m}\left[(s_{out} - s_{in}) - \frac{h_{out} - h_{in}}{T_{surr}}\right] \quad (12)$$

In a cycle, the reduction of work produced by a power cycle or the increase in work required by a refrigeration cycle is equal to the absolute ambient temperature multiplied by the sum of the irreversibilities in all the processes in the cycle. Thus the difference in the reversible work and the actual work for any refrigeration cycle, theoretical or real, operating under the same conditions becomes

$$\dot{W}_{actual} = \dot{W}_{reversible} + T_0 \sum \dot{I} \quad (13)$$

THERMODYNAMIC ANALYSIS OF REFRIGERATION CYCLES

Refrigeration cycles transfer thermal energy from a region of low temperature T_R to one of higher temperature. Usually the higher temperature heat sink is the ambient air or cooling water. This temperature is designated as T_0, the temperature of the surroundings.

The first and second laws of thermodynamics can be applied to individual components to determine mass and energy balances and the irreversibility of the components. This procedure is illustrated in later sections in this chapter.

Performance of a refrigeration cycle is usually described by a **coefficient of performance**. COP is defined as the benefit of the cycle (amount of heat removed) divided by the required energy input to operate the cycle, or

$$\text{COP} \equiv \frac{\text{Useful refrigerating effect}}{\text{Net energy supplied from external sources}} \quad (14)$$

For a mechanical vapor compression system, the net energy supplied is usually in the form of work, mechanical or electrical, and may include work to the compressor and fans or pumps. Thus

$$\text{COP} = \frac{Q_{evap}}{W_{net}} \quad (15)$$

In an absorption refrigeration cycle, the net energy supplied is usually in the form of heat into the generator and work into the pumps and fans, or

$$\text{COP} = \frac{Q_{evap}}{Q_{gen} + W_{net}} \quad (16)$$

In many cases the work supplied to an absorption system is very small compared to the amount of heat supplied to the generator, so the work term is often neglected.

Application of the second law to an entire refrigeration cycle shows that a completely reversible cycle operating under the same conditions has the maximum possible Coefficient of Performance. A measure of the departure of the actual cycle from an ideal reversible cycle is given by the **refrigerating efficiency**:

$$\eta_R = \frac{\text{COP}}{(\text{COP})_{rev}} \quad (17)$$

The Carnot cycle usually serves as the ideal reversible refrigeration cycle. For multistage cycles, each stage is described by a reversible cycle.

EQUATIONS OF STATE

The equation of state of a pure substance is a mathematical relation between pressure, specific volume, and temperature. When the system is in thermodynamic equilibrium,

$$f(p, v, T) = 0 \quad (18)$$

The principles of statistical mechanics are used to (1) explore the fundamental properties of matter, (2) predict an equation of state based on the statistical nature of a particular system, or (3) propose a functional form for an equation of state with unknown parameters that are determined by measuring thermodynamic properties of a substance. A fundamental equation with this basis is the **virial equation**. The virial equation is expressed as an expansion in pressure p or in reciprocal values of volume per unit mass v as

$$\frac{pv}{RT} = 1 + B'p + C'p^2 + D'p^3 + \ldots \quad (19)$$

$$\frac{pv}{RT} = 1 + (B/v) + (C/v^2) + (D/v^3) + \ldots \quad (20)$$

where coefficients B', C', D', etc., and B, C, D, etc., are the virial coefficients. B' and B are second virial coefficients; C' and C are third virial coefficients, etc. The virial coefficients are functions of temperature only, and values of the respective coefficients in Equations (19) and (20) are related. For example, $B' = B/RT$ and $C' = (C - B^2)/(RT)^2$.

The ideal gas constant R is defined as

$$R = \lim_{p \to 0} \frac{(pv)_T}{T_{tp}} \quad (21)$$

where $(pv)_T$ is the product of the pressure and the volume along an isotherm, and T_{tp} is the defined temperature of the triple point of water, which is 491.69°R. The current best value of R is 1545.32 ft·lb$_f$/(lb mole·°R).

The quantity pv/RT is also called the **compressibility factor**; i.e., $Z = pv/RT$ or

$$Z = 1 + (B/v) + (C/v^2) + (D/v^3) + \ldots \quad (22)$$

An advantage of the virial form is that statistical mechanics can be used to predict the lower order coefficients and provide physical significance to the virial coefficients. For example, in Equation

(22), the term B/v is a function of interactions between two molecules, C/v^2 between three molecules, etc. Since the lower order interactions are common, the contributions of the higher order terms are successively less. Thermodynamicists use the partition or distribution function to determine virial coefficients; however, experimental values of the second and third coefficients are preferred. For dense fluids, many higher order terms are necessary that can neither be satisfactorily predicted from theory nor determined from experimental measurements. In general, a truncated virial expansion of four terms is valid for densities of less than one-half the value at the critical point. For higher densities, additional terms can be used and determined empirically.

Digital computers allow the use of very complex equations of state in calculating p-v-T values, even to high densities. The Benedict-Webb-Rubin (B-W-R) equation of state (Benedict et al. 1940) and the Martin-Hou equation (1955) have had considerable use, but should generally be limited to densities less than the critical value. Strobridge (1962) suggested a modified Benedict-Webb-Rubin relation that gives excellent results at higher densities and can be used for a p-v-T surface that extends into the liquid phase.

The B-W-R equation has been used extensively for hydrocarbons (Cooper and Goldfrank 1967):

$$P = (RT/v) + (B_oRT - A_o - C_o/T^2)/v^2 + (bRT - a)/v^3$$

$$+ (a\alpha)/v^6 + [c(1 + \gamma/v^2)e^{(-\gamma/v^2)}]/v^3T^2 \qquad (23)$$

where the constant coefficients are A_o, B_o, C_o, a, b, c, α, γ.

The Martin-Hou equation, developed for fluorinated hydrocarbon properties, has been used to calculate the thermodynamic property tables in Chapter 20 and in ASHRAE *Thermodynamic Properties of Refrigerants* (Stewart et al. 1986). The Martin-Hou equation is as follows:

$$p = \frac{RT}{v-b} + \frac{A_2 + B_2T + C_2e^{(-kT/T_c)}}{(v-b)^2} + \frac{A_3 + B_3T + C_3e^{(-kT/T_c)}}{(v-b)^3}$$

$$+ \frac{A_4 + B_4T}{(v-b)^4} + \frac{A_5 + B_5T + C_5e^{(-kT/T_c)}}{(v-b)^5} + (A_6 + B_6T)e^{av} \qquad (24)$$

where the constant coefficients are A_i, B_i, C_i, k, b, and α.

Strobridge (1962) suggested an equation of state that was developed for nitrogen properties and used for most cryogenic fluids. This equation combines the B-W-R equation of state with an equation for high density nitrogen suggested by Benedict (1937). These equations have been used successfully for liquid and vapor phases, extending in the liquid phase to the triple-point temperature and the freezing line, and in the vapor phase from 18 to 1800°R, with pressures to 150,000 psi. The equation suggested by Strobridge is accurate within the uncertainty of the measured p-v-T data. This equation, as originally reported by Strobridge, is

$$p = RT\rho + \left[Rn_1T + n_2 + \frac{n_3}{T} + \frac{n_4}{T^2} + \frac{n_5}{T^4}\right]\rho^2$$

$$+ (Rn_6T + n_7)\rho^3 + n_8T\rho^4$$

$$+ \rho^3\left[\frac{n_9}{T^2} + \frac{n_{10}}{T^3} + \frac{n_{11}}{T^4}\right]\exp(-n_{16}\rho^2)$$

$$+ \rho^5\left[\frac{n_{12}}{T^2} + \frac{n_{13}}{T^3} + \frac{n_{14}}{T^4}\right]\exp(-n_{16}\rho^2) + n_{15}\rho^6 \qquad (25)$$

The 15 coefficients of this equation's linear terms are determined by a least-square fit to experimental data. Hust and Stewart (1966) and Hust and McCarty (1967) give further information on methods and techniques for determining equations of state.

In the absence of experimental data, Van der Waals' principle of corresponding states can predict fluid properties. This principle relates properties of similar substances by suitable reducing factors; i.e., the p-v-T surfaces of similar fluids in a given region are assumed to be of similar shape. The critical point can be used to define reducing parameters to scale the surface of one fluid to the dimensions of another. Modifications of this principle, as suggested by Kamerlingh Onnes, a Dutch cryogenic researcher, have been used to improve correspondence at low pressures. The principle of corresponding states provides useful approximations, and numerous modifications have been reported. More complex treatments for predicting property values, which recognize similarity of fluid properties, are by generalized equations of state. These equations ordinarily allow for adjustment of the p-v-T surface by introduction of parameters. One example (Hirschfelder et al. 1958) allows for departures from the principle of corresponding states by adding two correlating parameters.

CALCULATING THERMODYNAMIC PROPERTIES

While equations of state provide p-v-T relations, a thermodynamic analysis usually requires values for internal energy, enthalpy, and entropy. These properties have been tabulated for many substances, including refrigerants (See Chapters 6, 20, and 38) and can be extracted from such tables by interpolating manually or with a suitable computer program. This approach is appropriate for hand calculations and for relatively simple computer models; however, for many computer simulations, the overhead in memory or input and output required to use tabulated data can make this approach unacceptable. For large thermal system simulations or complex analyses, it may be more efficient to determine internal energy, enthalpy, and entropy using fundamental thermodynamic relations or curves fit to experimental data. Some of these relations are discussed in the following sections. Also, the thermodynamic relations discussed in those sections are the basis for constructing tables of thermodynamic property data. Further information on the topic may be found in references covering system modeling and thermodynamics (Stoecker 1989, Howell and Buckius 1992).

At least two intensive properties must be known to determine the remaining properties. If two known properties are either p, v, or T (these are relatively easy to measure and are commonly used in simulations), the third can be determined throughout the range of interest using an equation of state. Furthermore, if the specific heats at zero pressure are known, specific heat can be accurately determined from spectroscopic measurements using statistical mechanics (NASA 1971). Entropy may be considered a function of T and p, and from calculus an infinitesimal change in entropy can be written as follows:

$$ds = \left(\frac{\partial s}{\partial T}\right)_p dT + \left(\frac{\partial s}{\partial p}\right)_T dp \qquad (26)$$

Likewise, a change in enthalpy can be written as

$$dh = \left(\frac{\partial h}{\partial T}\right)_p dT + \left(\frac{\partial h}{\partial p}\right)_T dp \qquad (27)$$

Using the relation $Tds = dh - vdp$ and the definition of specific heat at constant pressure, $c_p \equiv (\partial h/\partial T)_p$, Equation (27) can be rearranged to yield

$$ds = \frac{c_p}{T}dT + \left[\left(\frac{\partial h}{\partial p}\right)_T - v\right]\frac{dp}{T} \tag{28}$$

Equations (26) and (28) combine to yield $(\partial s/\partial T)_p = c_p/T$. Then, using the Maxwell relation $(\partial s/\partial p)_T = -(\partial v/\partial T)_p$, Equation (26) may be rewritten as

$$ds = \frac{c_p}{T}dT - \left(\frac{\partial v}{\partial T}\right)_p dp \tag{29}$$

This is an expression for an exact derivative, so it follows that

$$\left(\frac{\partial c_p}{\partial p}\right)_T = -T\left(\frac{\partial^2 v}{\partial T^2}\right)_p \tag{30}$$

Integrating this expression at a fixed temperature yields

$$c_p = c_{po} - \int_0^p T\left(\frac{\partial^2 v}{\partial T^2}\right) dp_T \tag{31}$$

where c_{p0} is the known zero pressure specific heat, and dp_T is used to indicate that the integration is performed at a fixed temperature. The second partial derivative of specific volume with respect to temperature can be determined from the equation of state. Thus, Equation (31) can be used to determine the specific heat at any pressure.

Using $Tds = dh - vdp$, Equation (29) can be written as

$$dh = c_p dT + \left[v - T\left(\frac{\partial v}{\partial T}\right)_p\right]dp \tag{32}$$

Equations (28) and (32) may be integrated at constant pressure to obtain

$$s(T_1,p_0) = s(T_0,p_0) + \int_{T_0}^{T_1} \frac{c_p}{T} dT_p \tag{33}$$

and
$$h(T_1,p_0) = h(T_0,p_0) + \int_{T_0}^{T_1} c_p dT \tag{34}$$

Integrating the Maxwell relation $(\partial s/\partial p)_T = -(\partial v/\partial T)_p$ gives an equation for entropy changes at a constant temperature as

$$s(T_0,p_1) = s(T_0,p_0) - \int_{p_0}^{p_1} \left(\frac{\partial v}{\partial T}\right)_p dp_T \tag{35}$$

Likewise, integrating Equation (32) along an isotherm yields the following equation for enthalpy changes at a constant temperature

$$h(T_0,p_1) = h(T_0,p_0) + \int_{p_0}^{p_1} \left[v - T\left(\frac{\partial v}{\partial T}\right)_p\right]dp \tag{36}$$

Internal energy can be calculated from $u = h - pv$.

Combinations (or variations) of Equations (33) through (36) can be incorporated directly into computer subroutines to calculate properties with improved accuracy and efficiency. However, these equations are restricted to situations where the equation of state is valid and the properties vary continuously. These restrictions are violated by a change of phase such as evaporation and condensation, which are essential processes in air-conditioning and refrigerating devices. Therefore, the Clapeyron equation is of particular value; for evaporation or condensation it gives

$$\left(\frac{dp}{dT}\right)_{sat} = \frac{s_{fg}}{v_{fg}} = \frac{h_{fg}}{Tv_{fg}} \tag{37}$$

where

s_{fg} = entropy of vaporization
h_{fg} = enthalpy of vaporization
v_{fg} = specific volume difference between vapor and liquid phases

If vapor pressure and liquid and vapor density data are known at saturation, and these are relatively easy measurements to obtain, then changes in enthalpy and entropy can be calculated using Equation (37).

Phase Equilibria for Multicomponent Systems

To understand phase equilibria, consider a container full of a liquid made of two components; the more volatile component is designated i and the less volatile component j (Figure 2A). This mixture is all liquid because the temperature is low—but not so low that a solid appears. Heat added at a constant pressure raises the temperature of the mixture, and a sufficient increase causes vapor to form, as shown in Figure 2B. If heat at constant pressure continues to be added, eventually the temperature will become so high that only vapor remains in the container (Figure 2C). A temperature-concentration (T-x) diagram is useful for exploring details of this situation.

Figure 3 is a typical T-x diagram valid at a fixed pressure. The case shown in Figure 2A, a container full of liquid mixture with mole fraction $x_{i,0}$ at temperature T_0, is point 0 on the T-x diagram. When heat is added, the temperature of the mixture increases. The point at which vapor begins to form is the **bubble point**. Starting at point 0, the first bubble will form at temperature T_1, designated by point 1 on the diagram. The locus of bubble points is the **bubble point curve**, which provides bubble points for various liquid mole fractions x_i.

When the first bubble begins to form, the vapor in the bubble may not have the i mole fraction found in the liquid mixture. Rather, the mole fraction of the more volatile species is higher in the vapor than in the liquid. Boiling prefers the more volatile species, and the T-x diagram shows this behavior. At T_1, the vapor-forming bubbles have an i mole fraction of $y_{i,1}$. If heat continues to be added, this preferential boiling will deplete the liquid of species

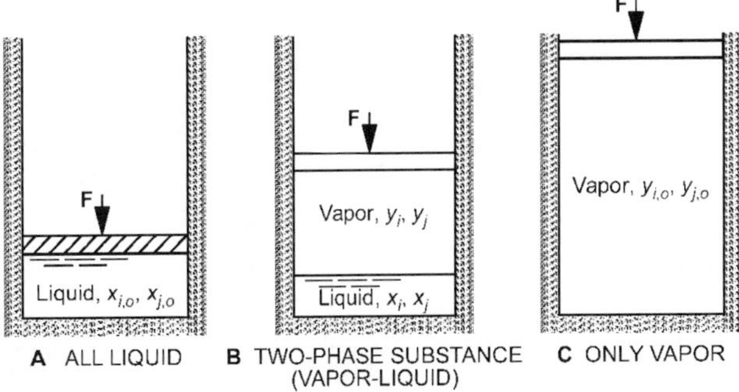

Fig. 2 Mixture of *i* and *j* Components in Constant Pressure Container

i and the temperature required to continue the process will increase. Again, the T-x diagram reflects this fact; at point 2 the i mole fraction in the liquid is reduced to $x_{i,2}$ and the vapor has a mole fraction of $y_{i,2}$. The temperature required to boil the mixture is increased to T_2. Position 2 on the T-x diagram could correspond to the physical situation shown in Figure 2B.

If the constant-pressure heating continues, all the liquid eventually becomes vapor at temperature T_3. At this point the i mole fraction in the vapor $y_{i,3}$ equals the starting mole fraction in the all-liquid mixture $x_{i,1}$. This equality is required for mass and species conservation. Further addition of heat simply raises the vapor temperature. The final position 4 corresponds to the physical situation shown in Figure 2C.

Starting at position 4 in Figure 3, the removal of heat leads to 3, and further heat removal would cause droplets rich in the less volatile species to form. This point is called the **dew point**, and the locus of dew points is called the **dew-point curve**. The removal of heat will cause the mixture to reverse through points 3, 2, 1, and to starting point 0. Because the composition shifts, the temperature required to boil (or condense) this mixture changes as the process proceeds. This mixture is therefore called **zeotropic**.

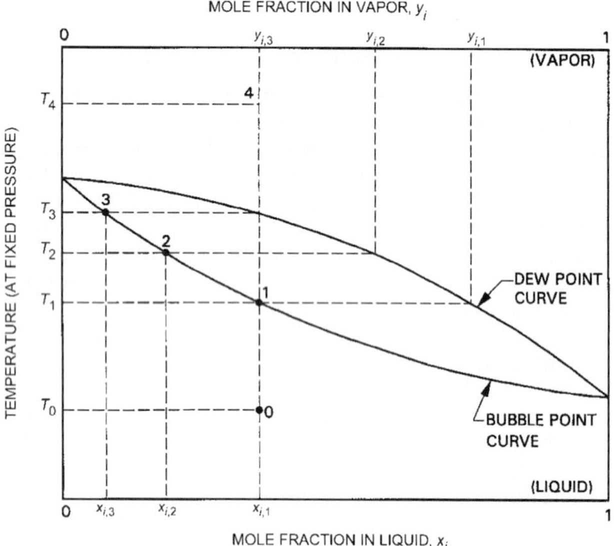

Fig. 3 Temperature-Concentration (T-x) Diagram for Zeotropic Mixture

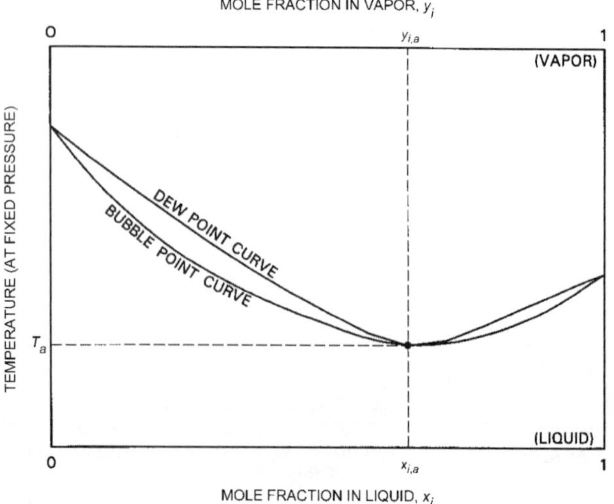

Fig. 4 Azeotropic Behavior Shown on T-x Diagram

Most mixtures have T-x diagrams that behave as previously described, but some have a markedly different feature. If the dew point and bubble point curves intersect at any point other than at their ends, the mixture exhibits what is called **azeotropic** behavior at that composition. This case is shown as position a in the T-x diagram of Figure 4. If a container of liquid with a mole fraction x_a were boiled, vapor would be formed with an identical mole fraction y_a. The addition of heat at constant pressure would continue with no shift in composition and no temperature glide.

Perfect azeotropic behavior is uncommon, while near azeotropic behavior is fairly common. The azeotropic composition is pressure dependent, so operating pressures should be considered for their impact on mixture behavior. Azeotropic and near-azeotropic refrigerant mixtures find wide application. The properties of an azeotropic mixture are such that they may be conveniently treated as pure substance properties. Zeotropic mixtures, however, require special treatment, using an equation-of-state approach with appropriate mixing rules or using the fugacities with the standard state method (Tassios 1993). Refrigerant and lubricant blends are a zeotropic mixture and can be treated by these methods (see Thome 1995 and Martz et al. 1996a, b).

COMPRESSION REFRIGERATION CYCLES

CARNOT CYCLE

The Carnot cycle, which is completely reversible, is a perfect model for a refrigeration cycle operating between two fixed temperatures, or between two fluids at different temperatures and each with infinite heat capacity. Reversible cycles have two important properties: (1) no refrigerating cycle may have a coefficient of performance higher than that for a reversible cycle operated between the same temperature limits, and (2) all reversible cycles, when operated between the same temperature limits, have the same coefficient of performance. Proof of both statements may be found in almost any textbook on elementary engineering thermodynamics.

Figure 5 shows the Carnot cycle on temperature-entropy coordinates. Heat is withdrawn at the constant temperature T_R from the region to be refrigerated. Heat is rejected at the constant ambient temperature T_0. The cycle is completed by an isentropic expansion and an isentropic compression. The energy transfers are given by

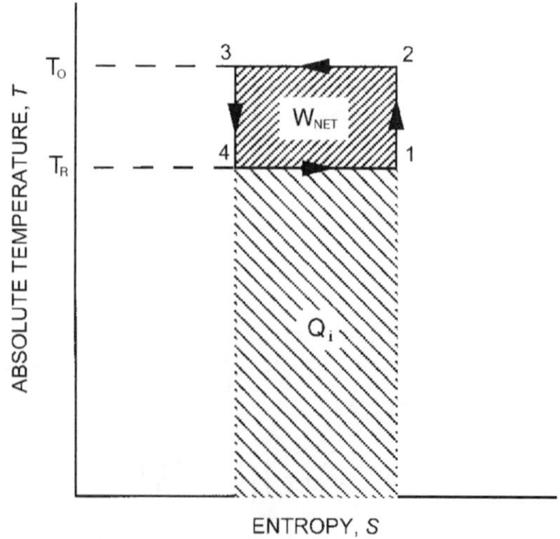

Fig. 5 Carnot Refrigeration Cycle

$$Q_0 = T_0(S_2 - S_3)$$
$$Q_i = T_R(S_1 - S_4) = T_R(S_2 - S_3)$$
$$W_{net} = Q_o - Q_i$$

Thus, by Equation (15),

$$\text{COP} = \frac{T_R}{T_0 - T_R} \quad (38)$$

Example 1. Determine entropy change, work, and coefficient of performance for the cycle shown in Figure 6. Temperature of the refrigerated space T_R is 400°R and that of the atmosphere T_0 is 500°R. Refrigeration load is 200 Btu.

Solution:

$$\Delta S = S_1 - S_4 = Q_i/T_R = 200/400 = 0.500 \text{ Btu/°R}$$
$$W = \Delta S(T_0 - T_R) = 0.5(500 - 400) = 50 \text{ Btu}$$
$$\text{COP} = Q_i/(Q_o - Q_i) = Q_i/W = 200/50 = 4$$

Flow of energy and its area representation in Figure 6 is:

Energy	Btu	Area
Q_i	200	b
Q_o	250	$a + b$
W	50	a

The net change of entropy of any refrigerant in any cycle is always zero. In Example 1 the change in entropy of the refrigerated space is $\Delta S_R = -200/400 = -0.5$ Btu/°R and that of the atmosphere is $\Delta S_o = 250/500 = 0.5$ Btu/°R. The net change in entropy of the isolated system is $\Delta S_{total} = \Delta S_R + \Delta S_o = 0$.

The Carnot cycle in Figure 7 shows a process in which heat is added and rejected at constant pressure in a two-phase region of a refrigerant. Saturated liquid at state 3 expands isentropically to the low temperature and pressure of the cycle at state d. Heat is added isothermally and isobarically by evaporating the liquid phase refrigerant from state d to state 1. The cold saturated vapor at state 1 is compressed isentropically to the high temperature in the cycle at state b. However the pressure at state b is below the saturation pressure corresponding to the high temperature in the cycle. The compression process is completed by an isothermal compression process from state b to state c. The cycle is completed by an isothermal and isobaric heat rejection or condensing process from state c to state 3.

Applying the energy equation for a mass of refrigerant m yields (all work and heat transfer are positive)

$${}_3W_d = m(h_3 - h_d)$$
$${}_1W_b = m(h_b - h_1)$$
$${}_bW_c = T_0(S_b - S_c) - m(h_b - h_c)$$
$${}_dQ_1 = m(h_1 - h_d) = \text{Area def1d}$$

The net work for the cycle is

$$W_{net} = {}_1W_b + {}_bW_c - {}_3W_d = \text{Area d1bc3d}$$

and

$$\text{COP} = \frac{{}_dQ_1}{W_{net}} = \frac{T_R}{T_0 - T_R}$$

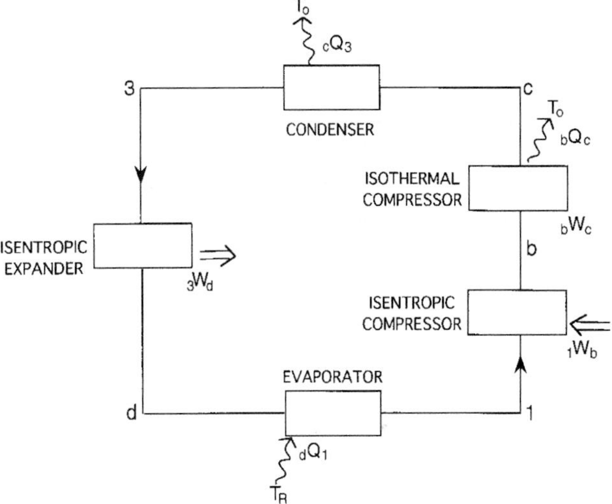

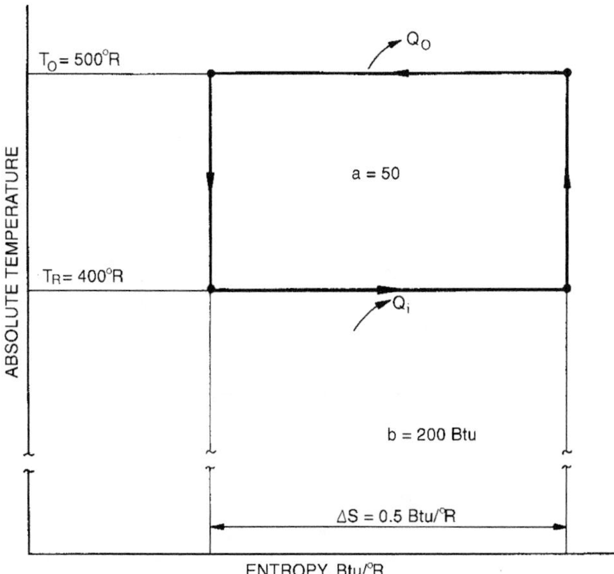

Fig. 6 Temperature-Entropy Diagram for Carnot Refrigeration Cycle of Example 1

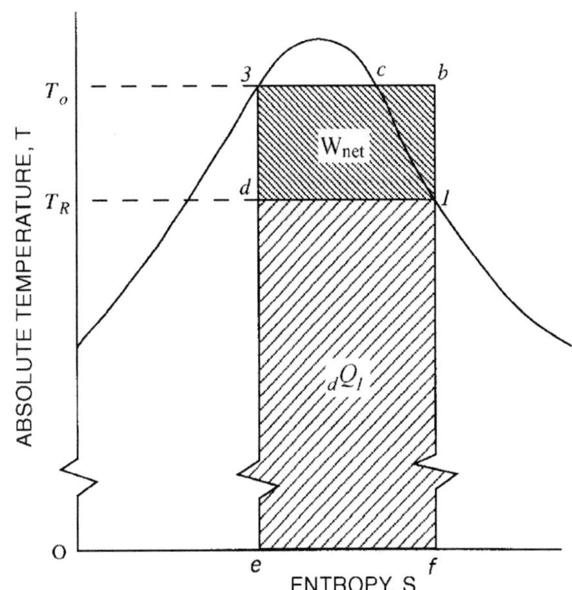

Fig. 7 Carnot Vapor Compression Cycle

THEORETICAL SINGLE-STAGE CYCLE USING A PURE REFRIGERANT OR AZEOTROPIC MIXTURE

A system designed to approach the ideal model shown in Figure 7 is desirable. A pure refrigerant or an azeotropic mixture can be used to maintain constant temperature during the phase changes by maintaining a constant pressure. Because of such concerns as high initial cost and increased maintenance requirements, a practical machine has one compressor instead of two and the expander (engine or turbine) is replaced by a simple expansion valve. The valve throttles the refrigerant from high pressure to low pressure.

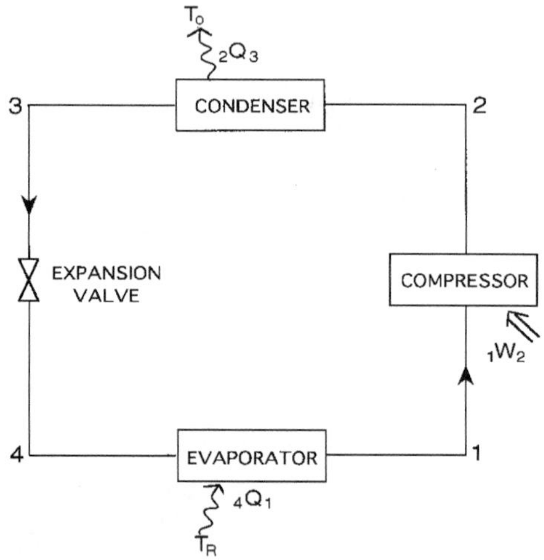

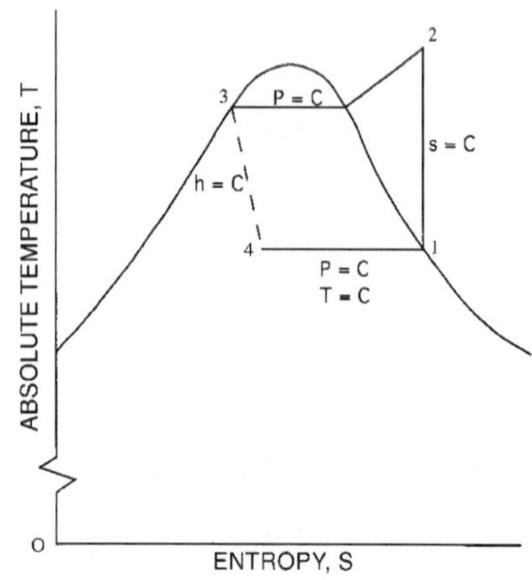

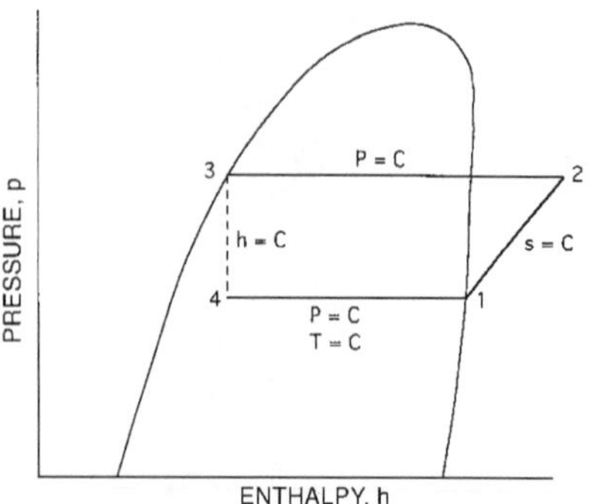

Fig. 8 Theoretical Single-Stage Vapor Compression Refrigeration Cycle

Figure 8 shows the theoretical single-stage cycle used as a model for actual systems.

Applying the energy equation for a mass of refrigerant m yields

$$
\begin{aligned}
{}_4Q_1 &= m(h_1 - h_4) \\
{}_1W_2 &= m(h_2 - h_1) \\
{}_2Q_3 &= m(h_2 - h_3) \\
h_3 &= h_4
\end{aligned}
\tag{39}
$$

The constant enthalpy throttling process assumes no heat transfer or change in potential or kinetic energy through the expansion valve.

The coefficient of performance is

$$
\mathrm{COP} = \frac{{}_4Q_1}{{}_1W_2} = \frac{h_1 - h_4}{h_2 - h_1} \tag{40}
$$

The theoretical compressor displacement CD (at 100% volumetric efficiency), is

$$
\mathrm{CD} = \dot{m}v_3 \tag{41}
$$

which is a measure of the physical size or speed of the compressor required to handle the prescribed refrigeration load.

Example 2. A theoretical single-stage cycle using R-134a as the refrigerant operates with a condensing temperature of 90°F and an evaporating temperature of 0°F. The system produces 15 tons of refrigeration. Determine (a) the thermodynamic property values at the four main state points of the cycle, (b) the coefficient of performance of the cycle, (c) the cycle refrigerating efficiency, and (d) rate of refrigerant flow.

Solution:

(a) Figure 9 shows a schematic p-h diagram for the problem with numerical property data. Saturated vapor and saturated liquid properties for states 1 and 3 are obtained from the saturation table for R-134a in Chapter 20. Properties for superheated vapor at state 2 are obtained by linear interpolation of the superheat tables for R-134a in Chapter 20. Specific volume and specific entropy values for state 4 are obtained by determining the quality of the liquid-vapor mixture from the enthalpy.

$$
x_4 = \frac{h_4 - h_f}{h_g - h_f} = \frac{41.430 - 12.090}{103.015 - 12.090} = 0.3227
$$

$$
\begin{aligned}
v_4 &= v_f + x_4(v_g - v_f) = 0.011187 + 0.3227(2.1587 - 0.011187) \\
&= 0.7047 \text{ ft}^3/\text{lb}
\end{aligned}
$$

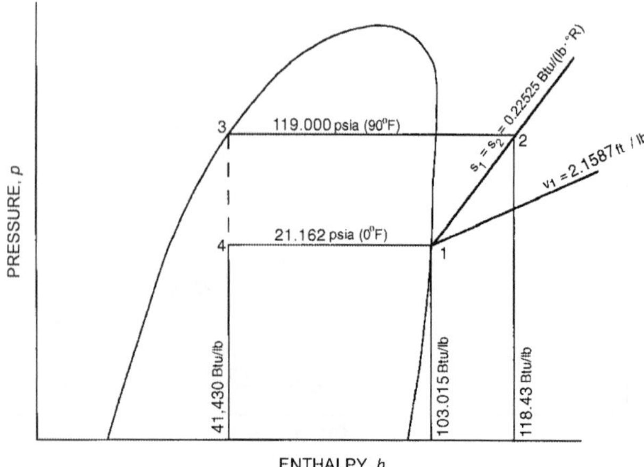

Fig. 9 Schematic *p-h* Diagram for Example 2

$$s_4 = s_f + x_4(s_g - s_f) = 0.02744 + 0.3227(0.22525 - 0.02744)$$
$$= 0.09127\,\text{Btu/(lb} \cdot {}^\circ\text{R)}$$

The property data are tabulated in Table 1.

(b) By Equation (40)

$$\text{COP} = \frac{103.015 - 41.430}{118.43 - 103.015} = 4.00$$

(c) By Equation (17)

$$\eta_R = \frac{\text{COP}(T_3 - T_1)}{T_1} = \frac{(4.00)(90)}{459.6} = 0.78 \text{ or } 78\%$$

(d) The mass flow of refrigerant is obtained from an energy balance on the evaporator. Thus

$$\dot{m}(h_1 - h_4) = q_i = 15 \text{ tons}$$

$$\text{and } \dot{m} = \frac{(15\text{ tons})(200\,\text{Btu/min ton})}{(103.015 - 41.430)\,\text{Btu/lb}} = 48.7 \text{ lb/min}$$

The saturation temperatures of the single-stage cycle have a strong influence on the magnitude of the coefficient of performance. This influence may be readily appreciated by an area analysis on a temperature-entropy (*T-s*) diagram. The area under a reversible process line on a *T-s* diagram is directly proportional to the thermal energy added or removed from the working fluid. This observation follows directly from the definition of entropy [see Equation (8)].

In Figure 10 the area representing Q_o is the total area under the constant pressure curve between states 2 and 3. The area representing the refrigerating capacity Q_i is the area under the constant pressure line connecting states 4 and 1. The net work required W_{net} equals the difference $(Q_o - Q_i)$, which is represented by the shaded area shown on Figure 10.

Table 1 Thermodynamic Property Data for Example 2

State	*t*, °F	*p*, psia	*v*, ft³/lb	*h*, Btu/lb	*s*, Btu/lb °R
1	0	21.162	2.1587	103.015	0.22525
2	103.6	119.00	0.4238	118.43	0.22525
3	90.0	119.00	0.01359	41.430	0.08518
4	0	21.162	0.7047	41.430	0.09127

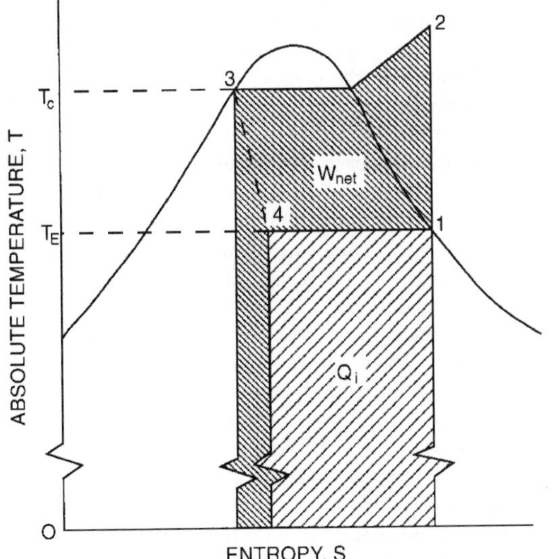

Fig. 10 Areas on *T-s* Diagram Representing Refrigerating Effect and Work Supplied for Theoretical Single-Stage Cycle

Because COP = Q_i/W_{net}, the effect on the COP of changes in evaporating temperature and condensing temperature may be observed. For example, a decrease in evaporating temperature T_E significantly increases W_{net} and slightly decreases Q_i. An increase in condensing temperature T_C produces the same results but with less effect on W_{net}. Therefore, for maximum coefficient of performance, the cycle should operate at the lowest possible condensing temperature and at the maximum possible evaporating temperature.

LORENZ REFRIGERATION CYCLE

The Carnot refrigeration cycle includes two assumptions which make it impractical. The heat transfer capacity of the two external fluids are assumed to be infinitely large so the external fluid temperatures remain fixed at T_0 and T_R (they become infinitely large thermal reservoirs). The Carnot cycle also has no thermal resistance between the working refrigerant and the external fluids in the two heat exchange processes. As a result, the refrigerant must remain fixed at T_0 in the condenser and at T_R in the evaporator.

The Lorenz cycle eliminates the first restriction in the Carnot cycle and allows the temperature of the two external fluids to vary during the heat exchange. The second assumption of negligible thermal resistance between the working refrigerant and the two external fluids remains. Therefore the refrigerant temperature must change during the two heat exchange processes to equal the changing temperature of the external fluids. This cycle is completely reversible when operating between two fluids, each of which has a finite but constant heat capacity.

Figure 11 is a schematic of a Lorenz cycle. Note that this cycle does not operate between two fixed temperature limits. Heat is added to the refrigerant from state 4 to state 1. This process is assumed to be linear on *T-s* coordinates, which represents a fluid with constant heat capacity. The temperature of the refrigerant is increased in an isentropic compression process from state 1 to state 2. Process 2-3 is a heat rejection process in which the refrigerant temperature decreases linearly with heat transfer. The cycle is concluded with an isentropic expansion process between states 3 and 4.

The heat addition and heat rejection processes are parallel so the entire cycle is drawn as a parallelogram on *T-s* coordinates. A Carnot refrigeration cycle operating between T_0 and T_R would lie between states 1, a, 3, and b. The Lorenz cycle has a smaller refrigerating effect than the Carnot cycle and more work is required. However this cycle is a more practical reference to use than the Carnot cycle when a refrigeration system operates between two single-phase fluids such as air or water.

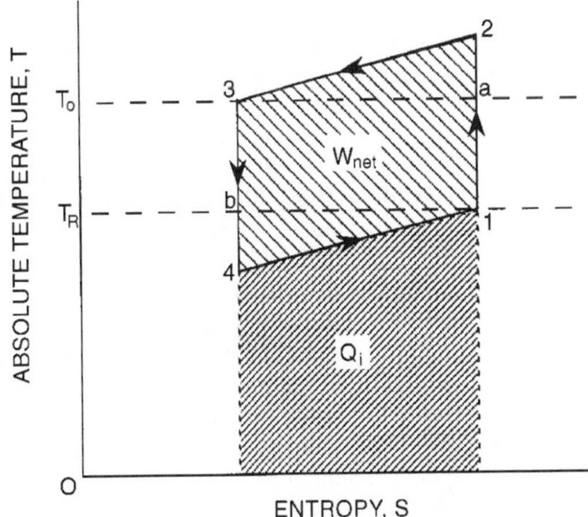

Fig. 11 Processes of Lorenz Refrigeration Cycle

The energy transfers in a Lorenz refrigeration cycle are as follows, where ΔT is the temperature change of the refrigerant during each of the two heat exchange processes.

$$Q_0 = (T_0 + \Delta T/2)(S_2 - S_3)$$

$$Q_i = (T_R - \Delta T/2)(S_1 - S_4) = (T_R - \Delta T/2)(S_2 - S_3)$$

$$W_{net} = Q_0 - Q_R$$

Thus by Equation (15),

$$\text{COP} = \frac{T_R - (\Delta T/2)}{T_O - T_R + \Delta T} \qquad (42)$$

Example 3. Determine the entropy change, the work required, and the coefficient of performance for the Lorenz cycle shown in Figure 11 when the temperature of the refrigerated space is $T_R = 400$°R, the ambient temperature is $T_0 = 500$°R, the ΔT of the refrigerant is 10°R and the refrigeration load is 200 Btu.

Solution:

$$\Delta S = \int_4^1 \frac{Q_i}{T} = \frac{Q_i}{T_R - (\Delta T/2)} = \frac{200}{395} = 0.5063 \text{ Btu/°R}$$

$$Q_O = [T_O + (\Delta T/2)]\Delta S = (500 + 5)0.5063 = 255.68 \text{ Btu}$$

$$W_{net} = Q_O - Q_R = 255.68 - 200 = 55.68 \text{ Btu}$$

$$\text{COP} = \frac{T_R - (\Delta T/2)}{T_O - T_R + \Delta T} = \frac{400 - (10/2)}{500 - 400 + 10} = \frac{395}{200} = 3.591$$

Note that the entropy change for the Lorenz cycle is larger than for the Carnot cycle at the same temperature levels and the same capacity (see Example 1). That is, the heat rejection is larger and the work requirement is also larger for the Lorenz cycle. This difference is caused by the finite temperature difference between the working fluid in the cycle compared to the bounding temperature reservoirs. However, as discussed previously, the assumption of constant temperature heat reservoirs is not necessarily a good representation of an actual refrigeration system because of the temperature changes that occur in the heat exchangers.

THEORETICAL SINGLE-STAGE CYCLE USING ZEOTROPIC REFRIGERANT MIXTURE

A practical method to approximate the Lorenz refrigeration cycle is to use a fluid mixture as the refrigerant and the four system components shown in Figure 8. When the mixture is not azeotropic and the phase change processes occur at constant pressure, the temperatures change during the evaporation and condensation processes and the theoretical single-stage cycle can be shown on *T-s* coordinates as in Figure 12. This can be compared with Figure 10 in which the system is shown operating with a pure simple substance or an azeotropic mixture as the refrigerant. Equations (14), (15), (39), (40), and (41) apply to this cycle and to conventional cycles with constant phase change temperatures. Equation (42) should be used as the reversible cycle COP in Equation (17).

For zeotropic mixtures, the concept of constant saturation temperatures does not exist. For example, in the evaporator, the refrigerant enters at T_4 and exits at a higher temperature T_1. The temperature of saturated liquid at a given pressure is the **bubble point** and the temperature of saturated vapor at a given pressure is called the **dew point**. The temperature T_3 on Figure 12 is at the bubble point at the condensing pressure and T_1 is at the dew point at the evaporating pressure.

An analysis of areas on a *T-s* diagram representing additional work and reduced refrigerating effect from a Lorenz cycle operating

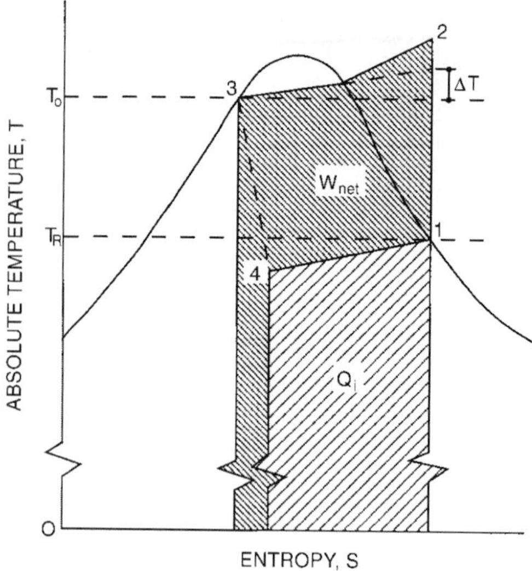

Fig. 12 Areas on *T-s* Diagram Representing Refrigerating Effect and Work Supplied for Theoretical Single-Stage Cycle Using Zeotropic Mixture as Refrigerant

between the same two temperatures T_1 and T_3 with the same value for ΔT can be performed. The cycle matches the Lorenz cycle most closely when counterflow heat exchangers are used for both the condenser and the evaporator.

In a cycle that has heat exchangers with finite thermal resistances and finite external fluid capacity rates, Kuehn and Gronseth (1986) showed that a cycle which uses a refrigerant mixture has a higher coefficient of performance than a cycle that uses a simple pure substance as a refrigerant. However, the improvement in COP is usually small. The performance of the cycle that uses a mixture can be improved further by reducing the thermal resistance of the heat exchangers and passing the fluids through them in a counterflow arrangement.

MULTISTAGE VAPOR COMPRESSION REFRIGERATION CYCLES

Multistage vapor compression refrigeration is used when several evaporators are needed at various temperatures such as in a supermarket or when the temperature of the evaporator becomes very low. Low evaporator temperature indicates low evaporator pressure and low refrigerant density into the compressor. Two small compressors in series have a smaller displacement and usually operate more efficiently than one large compressor that covers the entire pressure range from the evaporator to the condenser. This is especially true in refrigeration systems that use ammonia because of the large amount of superheating that occurs during the compression process.

The thermodynamic analysis of multistage cycles is similar to the analysis of single-stage cycles. The main difference is that the mass flow differs through various components of the system. A careful mass balance and energy balance performed on individual components or groups of components ensures the correct application of the first law of thermodynamics. Care must also be exercised when performing second law calculations. Often the refrigerating load is comprised of more than one evaporator, so the total system capacity is the sum of the loads from all evaporators. Likewise the total energy input is the sum of the work into all compressors. For multistage cycles, the expression for the coefficient of performance given in Equation 15 should be written as

$$\text{COP} = \sum Q_i / W_{net} \qquad (43)$$

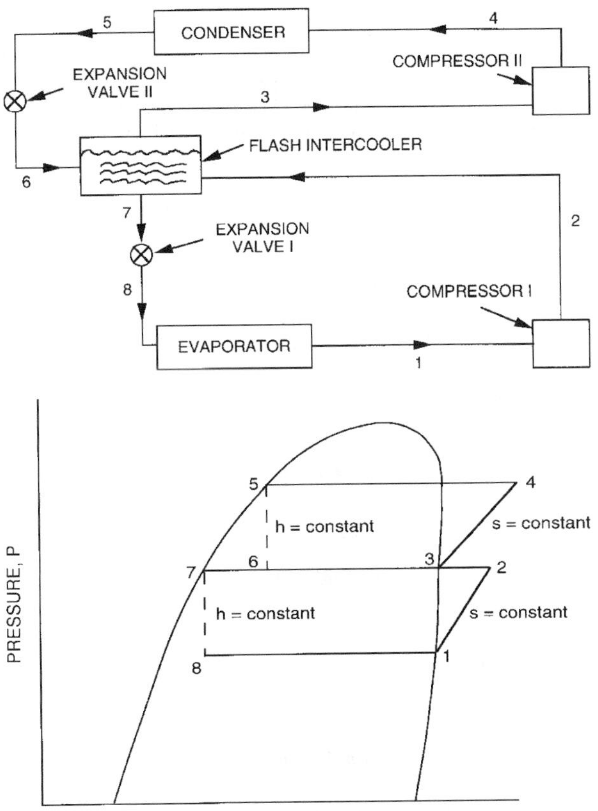

Fig. 13 Schematic and Pressure-Enthalpy Diagram for Dual-Compression, Dual-Expansion Cycle of Example 4

When compressors are connected in series, the vapor between stages should be cooled to bring the vapor to saturated conditions before proceeding to the next stage of compression. Intercooling usually minimizes the displacement of the compressors, reduces the work requirement, and increases the COP of the cycle. If the refrigerant temperature between stages is above ambient, a simple intercooler that removes heat from the refrigerant can be used. If the temperature is below ambient, which is the usual case, the refrigerant itself must be used to cool the vapor. This is accomplished with a flash intercooler. Figure 13 shows a cycle with a flash intercooler installed.

The superheated vapor from compressor I is bubbled through saturated liquid refrigerant at the intermediate pressure of the cycle. Some of this liquid is evaporated when heat is added from the superheated refrigerant. The result is that only saturated vapor at the intermediate pressure is fed to compressor II. A common assumption is to operate the intercooler at about the geometric mean of the evaporating and condensing pressures. This operating point provides the same pressure ratio and nearly equal volumetric efficiencies for the two compressors. Example 4 illustrates the thermodynamic analysis of this cycle

Example 4. Determine the thermodynamic properties of the eight state points shown in Figure 13, the mass flows, and the COP of this theoretical multistage refrigeration cycle when R-134a is the refrigerant. The saturated evaporator temperature is 0°F, the saturated condensing temperature is 90°F, and the refrigeration load is 15 tons. The saturation temperature of the refrigerant in the intercooler is 40°F, which is nearly at the geometric mean pressure of the cycle.

Solution:

Thermodynamic property data are obtained from the saturation and superheat tables for R-134a in Chapter 20. States 1, 3, 5, and 7 are obtained directly from the saturation table. State 6 is a mixture of liquid and vapor. The quality is calculated by

Table 2 Thermodynamic Property Values for Example 4

State	Temperature, °F	Pressure, psia	Specific Volume, ft³/lb	Specific Enthalpy, Btu/lb	Specific Entropy, Btu/lb·°R
1	0.00	21.162	2.1587	103.015	0.22525
2	48.296	49.724	0.9785	110.49	0.22525
3	40.00	49.724	0.9534	108.705	0.22172
4	95.666	119.000	0.4110	116.47	0.22172
5	90.00	119.000	0.01359	41.430	0.08518
6	40.00	49.724	0.2000	41.430	0.08708
7	40.00	49.724	0.01252	24.694	0.05359
8	0.00	21.162	0.3095	24.694	0.05486

$$x_6 = \frac{h_6 - h_7}{h_3 - h_7} = \frac{41.430 - 24.694}{108.705 - 24.694} = 0.19921$$

Then,

$$v_6 = v_7 + x_6(v_3 - v_7) = 0.01252 + 0.19921(0.9534 - 0.012521)$$
$$= 0.2000 \text{ ft}^3/\text{lb}$$
$$s_6 = s_7 + x_6(s_3 - s_7) = 0.05359 + 0.19921(0.22172 - 0.05359)$$
$$= 0.08708 \text{ Btu/lb} \cdot {}^\circ\text{R}$$

Similarly for state 8,

$$x_8 = 0.13862,\ v_8 = 0.3095 \text{ ft}^3/\text{lb},\ s_8 = 0.05486 \text{ Btu/lb} \cdot {}^\circ\text{R}$$

States 2 and 4 are obtained from the superheat tables by linear interpolation. The thermodynamic property data are summarized in Table 2.

The mass flow through the lower circuit of the cycle is determined from an energy balance on the evaporator.

$$\dot{m}_1 = \frac{\dot{Q}_i}{h_1 - h_8} = \frac{15 \text{ tons (200 Btu/min ton)}}{(103.015 - 24.696)\text{Btu/lb}} = 38.30 \text{ lb/min}$$

$$\dot{m}_1 = \dot{m}_2 = \dot{m}_7 = \dot{m}_8$$

For the upper circuit of the cycle,

$$\dot{m}_3 = \dot{m}_4 = \dot{m}_5 = \dot{m}_6$$

Assuming the intercooler has perfect external insulation, an energy balance on it is used to compute $\dot{m}_3$.

$$\dot{m}_6 h_6 + \dot{m}_2 h_2 = \dot{m}_7 h_7 + \dot{m}_3 h_3$$

Rearranging and solving for $\dot{m}_3$,

$$\dot{m}_3 = \dot{m}_2 \frac{h_7 - h_2}{h_6 - h_3} = 38.30 \text{ lb/min} \frac{24.694 - 110.49}{41.430 - 108.705} = 48.84 \text{ lb/min}$$

$$\dot{W}_I = \dot{m}_1(h_2 - h_1) = 38.30 \text{ lb/min}(110.49 - 103.015)\text{Btu/lb}$$
$$= 286.3 \text{ Btu/min}$$

$$\dot{W}_{II} = \dot{m}_3(h_4 - h_3) = 48.84 \text{ lb/min}(116.47 - 108.705)\text{Btu/lb}$$
$$= 379.2 \text{ Btu/min}$$

$$\text{COP} = \frac{\dot{Q}_i}{\dot{W}_I + \dot{W}_2} = \frac{15 \text{ tons(200 Btu/min ton)}}{(286.3 + 379.2)\text{Btu/min}} = 4.51$$

Examples 2 and 4 have the same refrigeration load and operate with the same evaporating and condensing temperatures. The two-stage cycle in Example 4 has a higher COP and less work input than the single-stage cycle. Also the highest refrigerant temperature leaving the compressor is about 96°F for the two-stage cycle versus about 104°F for the single-stage cycle. These differences are more pronounced for cycles operating at larger pressure ratios.

ACTUAL REFRIGERATION SYSTEMS

Actual systems operating steadily differ from the ideal cycles considered in the previous sections in many respects. Pressure drops occur everywhere in the system except in the compression process. Heat transfers occur between the refrigerant and its environment in all components. The actual compression process differs substantially from the isentropic compression assumed above. The working fluid is not a pure substance but a mixture of refrigerant and oil. All of these deviations from a theoretical cycle cause irreversibilities within the system. Each irreversibility requires additional power into the compressor. It is useful to understand how these irreversibilities are distributed throughout a real system. Insight is gained that can be useful when design changes are contemplated or operating conditions are modified. Example 5 illustrates how the irreversibilities can be computed in a real system and how they require additional compressor power to overcome. The input data have been rounded off for ease of computation.

Example 5. An air-cooled, direct-expansion, single-stage mechanical vapor-compression refrigerator uses R-22 and operates under steady conditions. A schematic drawing of this system is shown in Figure 14. Pressure drops occur in all piping and heat gains or losses occur as indicated. Power input includes compressor power and the power required to operate both fans. The following performance data are obtained:

Ambient air temperature,	t_O	= 90°F
Refrigerated space temperature,	t_R	= 20°F
Refrigeration load,	$\dot{Q}_{evap}$	= 2 tons
Compressor power input,	$\dot{W}_{comp}$	= 3.0 hp
Condenser fan input,	$\dot{W}_{CF}$	= 0.2 hp
Evaporator fan input,	$\dot{W}_{EF}$	= 0.15 hp

Refrigerant pressures and temperatures are measured at the seven locations shown on Figure 14. Table 3 lists the measured and computed thermodynamic properties of the refrigerant neglecting the dissolved oil. A pressure-enthalpy diagram of this cycle is shown in Figure 15 and is compared with a theoretical single-stage cycle operating between the air temperatures t_R and t_O.

Compute the energy transfers to the refrigerant in each component of the system and determine the second law irreversibility rate in each component. Show that the total irreversibility rate multiplied by the absolute ambient temperature is equal to the difference between the actual power input and the power required by a Carnot cycle operating between t_R and t_O with the same refrigerating load.

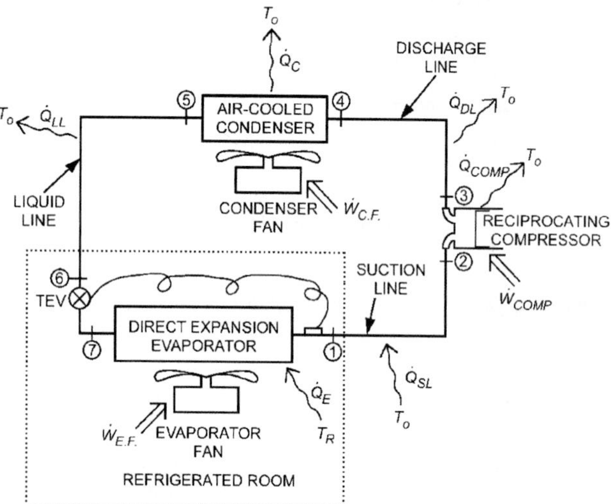

Fig. 14 Schematic of Real, Direct-Expansion, Single-Stage Mechanical Vapor-Compression Refrigeration System

Solution: The mass flow of refrigerant is the same through all components, so it is only computed once through the evaporator. Each component in the system is analyzed sequentially beginning with the evaporator. Equation (6) is used to perform a first law energy balance on each component and Equation (13) is used for the second law analysis. Note that the temperature used in the second law analysis is the absolute temperature.

Evaporator:

Energy balance

$$_7\dot{Q}_1 = \dot{m}(h_1 - h_7) = 24{,}000 \text{ Btu/h}$$

$$\dot{m} = \frac{24000}{(106.4 - 36.8)} = 345 \text{ lb/h}$$

Second law

$$_7\dot{I}_1 = \dot{m}(s_1 - s_7) - \frac{_7\dot{Q}_1}{T_R}$$

$$= 345(0.2291 - 0.0800) - \frac{24000}{479.67}$$

$$= 1.405 \text{ Btu/h}\cdot{}^\circ\text{R}$$

Suction Line:

Energy balance

$$_1\dot{Q}_2 = \dot{m}(h_2 - h_1)$$

$$= 345(108.1 - 106.4) = 586 \text{ Btu/h}$$

Table 3 Measured and Computed Thermodynamic Properties of Refrigerant 22 for Example 5

	Measured		Computed		
State	Pressure, psia	Temperature, °F	Specific Enthalpy, Btu/lb	Specific Entropy, Btu/lb·°R	Specific Volume, ft³/lb
1	45.0	15.0	106.4	0.2291	1.213
2	44.0	25.0	108.1	0.2330	1.276
3	210.0	180.0	128.8	0.2374	0.331
4	208.0	160.0	124.8	0.2314	0.318
5	205.0	94.0	37.4	0.0761	0.014
6	204.0	92.0	36.8	0.0750	0.014
7	46.5	9.0	36.8	0.0800	0.308

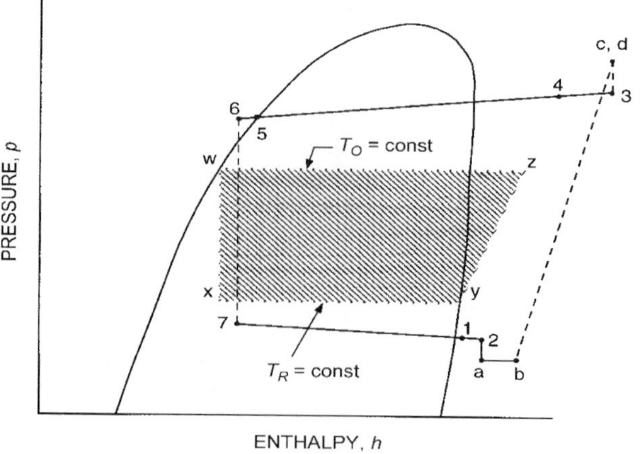

Fig. 15 Pressure-Enthalpy Diagram of Actual System and Theoretical Single-Stage System Operating Between Same Inlet Air Temperatures T_R and T_O

Second law

$$ {}_1\dot{I}_2 = \dot{m}(s_2 - s_1) - \frac{{}_1\dot{Q}_2}{T_O} $$

$$ = 345(0.2330 - 0.2291) - 586/549.67 $$

$$ = 0.279 \text{ Btu/h} \cdot {}^\circ\text{R} $$

***Compressor*:**
Energy balance

$$ {}_2\dot{Q}_3 = \dot{m}(h_3 - h_2) + {}_2\dot{W}_3 $$

$$ = 345(128.8 - 108.1) - 3.0(2545) $$

$$ = -494 \text{ Btu/h} $$

Second law

$$ {}_2\dot{I}_3 = \dot{m}(s_3 - s_2) - \frac{{}_2\dot{Q}_3}{T_O} $$

$$ = 345(0.2374 - 0.2330) - (-494/549.67) $$

$$ = 2.417 \text{ Btu/h} \cdot {}^\circ\text{R} $$

***Discharge Line*:**
Energy balance

$$ {}_3\dot{Q}_4 = \dot{m}(h_4 - h_3) $$

$$ = 345(124.8 - 128.8) = -1380 \text{ Btu/h} $$

Second law

$$ {}_3\dot{I}_4 = \dot{m}(s_4 - s_3) - \frac{{}_3\dot{Q}_4}{T_O} $$

$$ = 345(0.2314 - 0.2374) - (-1380/549.67) $$

$$ = 0.441 \text{ Btu/h} \cdot {}^\circ\text{R} $$

Condenser:
Energy balance

$$ {}_4\dot{Q}_5 = \dot{m}(h_5 - h_4) $$

$$ = 345(37.4 - 124.8) = -30{,}153 \text{ Btu/h} $$

Second law

$$ {}_4\dot{I}_5 = \dot{m}(s_5 - s_4) - \frac{{}_4\dot{Q}_5}{T_O} $$

$$ = 345(0.0761 - 0.2314) - (-30{,}153/549.67) $$

$$ = 1.278 \text{ Btu/h} \cdot {}^\circ\text{R} $$

Liquid Line:
Energy balance

$$ {}_5\dot{Q}_6 = \dot{m}(h_6 - h_5) $$

$$ = 345(36.8 - 37.4) = -207 \text{ Btu/h} $$

Second law

$$ {}_5\dot{I}_6 = \dot{m}(s_6 - s_5) - \frac{{}_5\dot{Q}_6}{T_O} $$

$$ = 345(0.0750 - 0.0761) - (-207/549.67) $$

$$ = 0 \text{ Btu/h} \cdot {}^\circ\text{R} $$

Expansion Device:
Energy balance

$$ {}_6\dot{Q}_7 = \dot{m}(h_7 - h_6) = 0 $$

Second law

$$ {}_6\dot{I}_7 = \dot{m}(s_7 - s_6) $$

$$ = 345(0.0800 - 0.0750) = 1.725 \text{ Btu/h} \cdot {}^\circ\text{R} $$

These results are summarized in Table 4. For the Carnot cycle,

Table 4 Energy Transfers and Irreversibility Rates for Refrigeration System in Example 5

Component	$\dot{Q}$, Btu/h	$\dot{W}$, Btu/h	$\dot{I}$, Btu/h·°R	$\dot{I}/\dot{I}_{total}$, %
Evaporator	24000	0	1.405	19
Suction line	586	0	0.279	4
Compressor	–494	7635	2.417	32
Discharge line	–1380	0	0.441	6
Condenser	–30153	0	1.278	17
Liquid line	–207	0	≈ 0	≈ 0
Expansion device	0	0	1.725	23
Totals	–7648	7635	7.545	

$$ \text{COP}_{Carnot} = \frac{T_R}{T_o - T_R} = \frac{479.67}{70} = 6.852 $$

The Carnot power requirement for the 2 ton load is

$$ \dot{W}_{Carnot} = \frac{\dot{Q}_{evap}}{\text{COP}_{Carnot}} = \frac{24000}{6.852} = 3502 \text{ Btu/h} $$

The actual power requirement for the compressor is

$$ \dot{W}_{comp} = \dot{W}_{Carnot} + \dot{I}_{total}T_o $$

$$ = 3502 + 7.545 \times 549.67 = 7649 \text{Btu/h} $$

This result is within computational error of the measured power input to the compressor of 7635 Btu/h.

The analysis demonstrated in Example 5 can be applied to any actual vapor compression refrigeration system. The only required information for the second law analysis is the refrigerant thermodynamic state points and mass flow rates and the temperatures in which the system is exchanging heat. In this example, the extra compressor power required to overcome the irreversibility in each component is determined. The component with the largest loss is the compressor. This loss is due to motor inefficiency, friction losses, and irreversibilities due to pressure drops, mixing, and heat transfer between the compressor and the surroundings. The unrestrained expansion in the expansion device is the next largest loss. This loss could be reduced by using an expander rather than a throttling process. An expander may be economical on large machines.

All heat transfer irreversibilities on both the refrigerant side and the air side of the condenser and evaporator are included in the analysis. The refrigerant pressure drop is also included. The only items not included are the air-side pressure drop irreversibilities of the two heat exchangers. However these are equal to the fan power requirements as all the fan power is dissipated as heat.

An overall second law analysis, such as in Example 5, shows the designer those components with the most losses, and it helps determine which components should be replaced or redesigned to improve performance. However, this type of analysis does not identify the nature of the losses. A more detailed second law analysis in which the actual processes are analyzed in terms of fluid flow and heat transfer is required to identify the nature of the losses (Liang and Kuehn 1991). A detailed analysis will show that most irreversibilities associated with heat exchangers are due to heat transfer, while pressure drop on the air side causes a very small loss and the refrigerant pressure drop causes a negligible loss. This finding indicates that promoting refrigerant heat transfer at the expense of increasing the pressure drop usually improves performance. This analysis does not provide the cost/benefits associated with reducing component irreversibilities. The use of a thermo-economic technique is required.

ABSORPTION REFRIGERATION CYCLES

An absorption cycle is a heat-activated thermal cycle. It exchanges only thermal energy with its surroundings—no appreciable mechanical energy is exchanged. Furthermore, no appreciable conversion of heat to work or work to heat occurs in the cycle.

Absorption cycles find use in applications where one or more of the heat exchanges with the surroundings is the useful product. This includes refrigeration, air conditioning, and heat pumping. The two great advantages of this type of cycle in comparison to other cycles with similar product are

- No large rotating mechanical equipment is required
- Any source of heat can be used, including low-temperature sources (e.g., waste heat)

IDEAL THERMAL CYCLE

All absorption cycles include at least three thermal energy exchanges with their surroundings; that is, energy exchange at three different temperatures. The highest temperature and lowest temperature heat flows are in one direction, and the mid-temperature one (or two) is in the opposite direction. In the **forward cycle**, the extreme temperature (hottest and coldest) heat flows are into the cycle. This cycle is also called the heat amplifier, heat pump, conventional cycle, or Type I cycle. When the extreme temperature heat flows are out of the cycle, it is called a **reverse cycle**, heat transformer, temperature amplifier, temperature booster, or Type II cycle. Figure 16 illustrates both types of thermal cycles.

This fundamental constraint of heat flow into or out of the cycle at three or more different temperatures establishes the first limitation on cycle performance. By the first law of thermodynamics (at steady state),

$$Q_{hot} + Q_{cold} = -Q_{mid} \tag{44}$$

(positive heat quantities are into the cycle)

The second law requires that

$$\frac{Q_{hot}}{T_{hot}} + \frac{Q_{cold}}{T_{cold}} + \frac{Q_{mid}}{T_{mid}} \geq 0 \tag{45}$$

with equality holding in the ideal case.

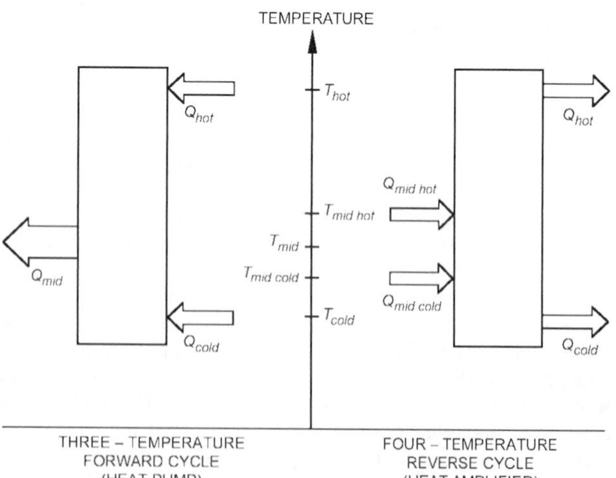

Fig. 16 Thermal Cycles

From these two laws alone (i.e., without invoking any further assumptions) it follows that, for the ideal forward cycle,

$$COP_{ideal} = \frac{Q_{cold}}{Q_{hot}} = \frac{T_{hot} - T_{mid}}{T_{hot}} \frac{T_{cold}}{T_{mid} - T_{cold}} \tag{46}$$

The heat ratio Q_{cold}/Q_{hot} is commonly called the **coefficient of performance (COP)**, which is the cooling realized divided by the driving heat supplied.

Heat that is rejected to ambient may be at two different temperatures, creating a **four-temperature cycle**. The ideal COP of the four-temperature cycle is also expressed by Equation (46), with T_{mid} signifying the entropic mean heat rejection temperature. In that case, T_{mid} is calculated as follows:

$$T_{mid} = \frac{Q_{mid\ hot} + Q_{mid\ cold}}{\frac{Q_{mid\ hot}}{T_{mid\ hot}} + \frac{Q_{mid\ cold}}{T_{mid\ cold}}} \tag{47}$$

This expression results from assigning all the entropy flow to the single temperature T_{mid}.

The ideal COP for the four-temperature cycle requires additional assumptions, such as the relationship between the various heat quantities. Under the assumptions that $Q_{cold} = Q_{mid\ cold}$ and $Q_{hot} = Q_{mid\ hot}$, the following expression results:

$$COP_{ideal} = \frac{T_{hot} - T_{mid\ hot}}{T_{hot}} \frac{T_{cold}}{T_{mid\ cold}} \frac{T_{cold}}{T_{mid\ hot}} \tag{48}$$

WORKING FLUID PHASE CHANGE CONSTRAINTS

Absorption cycles require at least two working substances—a sorbent and a fluid refrigerant; and each substance achieves its cycle function with a phase change. Given this constraint, many combinations are not achievable. The first result of invoking the phase change constraints is that the various heat flows assume known identities. As illustrated in Figure 17, the refrigerant phase changes occur in an evaporator and a condenser, and those of the sorbent in an absorber and a desorber (generator). For the **forward absorption cycle**, the highest temperature heat is always supplied to the generator,

$$Q_{hot} \equiv Q_{gen} \tag{49}$$

and the coldest heat is supplied to the evaporator:

$$Q_{cold} \equiv Q_{evap} \tag{50}$$

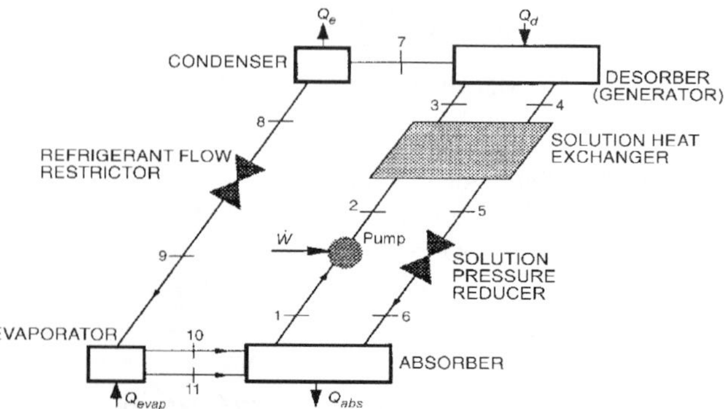

Fig. 17 Single-Effect Absorption Cycle

For the **reverse absorption cycle**, the highest temperature heat is rejected from the absorber, and the lowest temperature heat is rejected from the condenser.

The second result of the phase change constraint is that for all known refrigerants and sorbents over pressure ranges of interest,

$$Q_{evap} \approx Q_{cond} \tag{51}$$

and

$$Q_{gen} \approx Q_{abs} \tag{52}$$

These two relations are true because the latent heat of phase change (vapor ↔ condensed phase) is relatively constant when far removed from the critical point. Thus, each heat input can not be independently adjusted.

The ideal single-effect forward cycle COP expression is

$$\mathrm{COP}_{ideal} \le \frac{T_{gen} - T_{abs}}{T_{gen}} \frac{T_{evap}}{T_{cond} - T_{evap}} \frac{T_{cond}}{T_{abs}} \tag{53}$$

Equality holds only if the heat quantities at each temperature may be adjusted to specific values, which as shown below is not possible.

The third result of invoking the phase change constraint is that only three of the four temperatures T_{evap}, T_{cond}, T_{gen}, and T_{abs} may be independently selected.

Practical liquid absorbents for absorption cycles have a significant negative deviation from behavior predicted by Raoult's law. This has the beneficial effect of reducing the required amount of absorbent recirculation, at the expense of reduced lift and increased sorption duty. The practical effect of the negative deviation is that for most absorbents,

$$\frac{Q_{abs}}{Q_{cond}} \approx 1.2 \text{ to } 1.3 \tag{54}$$

and

$$T_{gen} - T_{abs} \approx 1.2(T_{cond} - T_{evap}) \tag{55}$$

The net result of applying the above approximations and constraints to the ideal cycle COP for the single-effect forward cycle is

$$\mathrm{COP}_{ideal} \approx 1.2 \frac{T_{evap} T_{cond}}{T_{gen} T_{abs}} \approx \frac{Q_{cond}}{Q_{abs}} \approx 0.8 \tag{56}$$

In practical terms, the temperature constraint reduces the ideal COP to about 0.9, and the heat quantity constraint further reduces it to about 0.8.

Another useful result is

$$T_{gen\ min} = T_{cond} + T_{abs} - T_{evap} \tag{57}$$

where $T_{gen\ min}$ is the minimum generator temperature necessary to achieve a given evaporator temperature.

Alternative approaches are available that lead to nearly the same upper limit on ideal cycle COP. For example, one approach equates the exergy production from a "driving" portion of the cycle to the exergy consumption in a "cooling" portion of the cycle (Tozer et al. 1997). This leads to the expression

$$\mathrm{COP}_{ideal} \le \frac{T_{evap}}{T_{abs}} = \frac{T_{cond}}{T_{gen}} \tag{58}$$

Another approach derives the idealized relationship between the cycle **lift** ($T_{cond} - T_{evap}$) and **drop** ($T_{gen} - T_{abs}$), i.e., between the two temperature differences that define the cycle.

WORKING FLUIDS

The working fluids for absorption cycles naturally fall into four categories, each requiring a different approach to cycle modeling and thermodynamic analysis. For the liquid absorbents, the important distinction is whether the absorbent is volatile or nonvolatile. In the latter case, the vapor phase is always pure refrigerant (neglecting noncondensables), and analysis is relatively straightforward. For volatile absorbents, wherein vapor concentration is variable, the cycle and component modeling techniques must keep track of vapor concentration as well as liquid concentration.

The sorbent may be either liquid phase or solid phase. For the solid sorbents, the important distinction is whether the solid is a physisorbent (also known as adsorbent) or a chemisorbent. With the physisorbent, the sorbent temperature depends on both pressure and refrigerant loading (bivariance), the same as for the liquid absorbents. In contrast, the chemisorbent temperature does not vary with loading, at least over small ranges, and hence a different modeling approach is required.

Beyond these distinctions, various other characteristics are either necessary or desirable for suitable liquid absorbent-refrigerant pairs, as follows:

Absence of Solid Phase (Solubility Field). The refrigerant-absorbent pair should not form a solid over the expected range of composition and temperature. If a solid forms, it will stop flow and cause equipment to shut down. Controls must prevent operation beyond the acceptable solubility range for the pair.

Relative Volatility. The refrigerant should be much more volatile than the absorbent so the two can be separated easily. Otherwise, cost and heat requirements may be excessive. Many of the absorbents are effectively nonvolatile.

Affinity. The absorbent should have a strong affinity for the refrigerant under conditions in which absorption takes place. Affinity means a negative deviation from Raoult's law and results in an activity coefficient of less than unity for the refrigerant. Strong affinity allows less absorbent to be circulated for the same refrigeration effect, reducing sensible heat losses. A smaller liquid heat exchanger to transfer heat from the absorbent to the pressurized refrigerant-absorption solution is also a benefit of affinity. On the other hand, as affinity increases, extra heat is required in the generators to separate refrigerant from the absorbent, and the COP suffers.

Pressure. Operating pressures, established by the thermodynamic properties of the refrigerant, should be moderate. High pressure requires the use of heavy-walled equipment, and significant electrical power may be required to pump the fluids from the low-pressure side to the high-pressure side. Vacuum requires the use of large-volume equipment and special means of reducing pressure drop in the refrigerant vapor paths.

Stability. High chemical stability is required because fluids are subjected to severe conditions over many years of service. Instability can cause undesirable formation of gases, solids, or corrosive substances. The purity of all components charged into the system is critical for high performance and corrosion prevention.

Corrosion. Most absorption fluids corrode materials used in construction. Therefore, corrosion inhibitors are used.

Safety. Precautions as dictated by code are followed in the cases where fluids are toxic, inflammable, or at high pressure. Codes vary according to country and region.

Transport Properties. Viscosity, surface tension, thermal diffusivity, and mass diffusivity are important characteristics of the refrigerant-absorbent pair. For example, low viscosity promotes heat and mass transfer and reduces pumping power.

Latent Heat. The refrigerant latent heat should be high, so the circulation rate of the refrigerant and absorbent can be minimized.

Environmental Soundness. The two parameters of greatest concern are the global warming potential and the ozone depletion potential.

No refrigerant-absorbent pair meets all requirements. Unfortunately, many requirements work at cross-purposes. For example, a greater solubility field goes hand-in-hand with reduced relative volatility. Thus, selection of a working pair is inherently a compromise.

Water-lithium bromide and ammonia-water offer the best compromises of thermodynamic performance and have no known detrimental environmental effect (zero ozone depletion potential and zero global warming potential).

The ammonia-water pair meets most requirements, but its volatility ratio is low and it requires high operating pressures. Ammonia is also a Safety Code Group 2 fluid (ASHRAE *Standard* 15), which restricts its use indoors.

Advantages of the water-lithium bromide pair include high safety, high volatility ratio, high affinity, high stability, and high latent heat. However, this pair tends to form solids and operates at deep vacuum. Because the refrigerant turns to ice at 32°F, the pair cannot be used for low-temperature refrigeration. Lithium bromide (LiBr) crystallizes at moderate concentrations, as would be encountered in air-cooled chillers, which ordinarily limits the pair to applications where the absorber is water-cooled and the concentrations are lower. However, using a combination of salts as the absorbent can reduce this crystallization tendency enough to permit air cooling (Macriss 1968). Other disadvantages of the water-lithium bromide pair include the low operating pressures and high viscosity. This is particularly detrimental to the absorption step; however, alcohols with a high relative molecular mass enhance LiBr absorption. Proper equipment design and additives can overcome these disadvantages.

Other refrigerant-absorbent pairs are listed in Table 5 (Macriss and Zawacki 1989, ISHPC 1999). Several refrigerant-absorbent pairs appear suitable for certain cycles and may solve some problems associated with traditional pairs. However, stability, corrosion, and property information on several is limited. Also, some of the fluids are somewhat hazardous.

Table 5 Refrigerant-Absorbent Pairs

Refrigerant	Absorbents
H_2O	Salts
	Alkali halides
	LiBr
	$LiClO_3$
	$CaCl_2$
	$ZnCl_2$
	ZnBr
	Alkali nitrates
	Alkali thiocyanates
	Bases
	Alkali hydroxides
	Acids
	H_2SO_4
	H_3PO_4
NH_3	H_2O
	Alkali thiocyanates
TFE	NMP
(Organic)	E181
	DMF
	Pyrrolidone
SO_2	Organic solvents

ABSORPTION CYCLE REPRESENTATIONS

The quantities of interest to absorption cycle designers are temperature, concentration, pressure, and enthalpy. The most useful plots are those with linear scales and in which the key properties plot as straight lines. Some of the following plots are used:

- **Absorption plots** embody the vapor-liquid equilibrium of both the refrigerant and the sorbent. Plots of vapor-liquid equilibrium on linear pressure-temperature coordinates have a logarithmic shape and hence are little used.
- In the **van't Hoff plot** (ln P versus $-1/T$), the constant concentration contours plot as nearly straight lines. Thus, it is more readily constructed (e.g., from sparse data) in spite of the awkward coordinates.
- The **Dühring diagram** (solution temperature versus reference temperature) retains the linearity of the van't Hoff plot, while eliminating the complexity of nonlinear coordinates. Thus, it has found extensive use (see Figure 20). The primary drawback is the need for a reference substance.
- The **Gibbs plot** (solution temperature versus T ln P) retains most of the advantages of the Dühring plot (linear temperature coordinates, concentration contours are straight lines) while eliminating the recourse to a reference substance.
- The **Merkel plot** (enthalpy versus concentration) is used to assist thermodynamic calculations and to solve the distillation problems that arise with volatile absorbents. It has also been used for basic cycle analysis.
- **Temperature-entropy coordinates** are occasionally used to relate absorption cycles to their mechanical vapor compression counterparts.

CONCEPTUALIZING THE CYCLE

The basic absorption cycle shown in Figure 17 must be altered in many cases to take advantage of the available energy. Examples include the following: (1) the driving heat is much hotter than the minimum required $T_{gen\,min}$: a multistage cycle boosts the COP; and (2) the driving heat temperature is below $T_{gen\,min}$: a different multistage cycle (half-effect cycle) can reduce the $T_{gen\,min}$.

Multistage means that one or more of the four basic exchangers (generator, absorber, condenser, evaporator) are present at two or more places in the cycle at different pressures or concentrations. **Multieffect** is a special case of multistaging, signifying the number of times the driving heat is used as it transits the cycle. Thus there are several types of two-stage cycles: the double-effect cycle, the half-effect cycle, and the two-stage, triple-effect cycle.

Two or more single-effect absorption cycles, such as shown in Figure 17, can be combined to form a multistage cycle by coupling any of the components. **Coupling** implies either (1) sharing of component(s) between the cycles to form an integrated single hermetic cycle or (2) alternatively exchanging heat between components belonging to two hermetically separate cycles that operate at (nearly) the same temperature level.

Figure 18 shows a **double-effect absorption cycle** formed by coupling the absorbers and evaporators of two single-effect cycles into an integrated, single hermetic cycle. Heat is transferred between the high-pressure condenser and intermediate-pressure generator. The heat of condensation of the refrigerant (generated in the high-temperature generator) generates additional refrigerant in the lower temperature generator. Thus, the prime energy provided to the high-temperature generator is **cascaded** (used) twice in the cycle, making it a double-effect cycle. With the generation of additional refrigerant from a given heat input, the cycle COP increases. Commercial water-lithium bromide chillers normally use this cycle.

The cycle COP can be further increased by coupling additional components and by increasing the number of cycles that are combined. This way, several different multiple-effect cycles can be combined by pressure-staging and/or concentration-staging. The double-effect cycle, for example, is formed by pressure-staging two single-effect cycles.

Figure 19 shows twelve generic triple-effect cycles identified by Alefeld and Radermacher (1994). Cycle 5 is a pressure-staged cycle, and Cycle 10 is a concentration-staged cycle. All other cycles are pressure- and concentration-staged. Cycle 1, which is called a dual loop cycle, is the only cycle consisting of two loops that doesn't circulate absorbent in the low-temperature portion of the cycle.

Each of the cycles shown in Figure 19 can be made with one, two, or sometimes three separate **hermetic loops**. Dividing a cycle into separate hermetic loops allows the use of a different working fluid in each loop. Thus, a corrosive and/or high-lift absorbent can be restricted to the loop where it is required, and a conventional additive-enhanced absorbent can be used in other loops to reduce the system cost significantly. As many as 78 hermetic loop configurations can be synthesized from the twelve triple-effect cycles shown in Figure 19. For each hermetic loop configuration, further variations are possible according to the absorbent flow pattern (e.g., series or parallel), the absorption working pairs selected, and various other hardware details. Thus, literally thousands of distinct variations of the triple-effect cycle are possible.

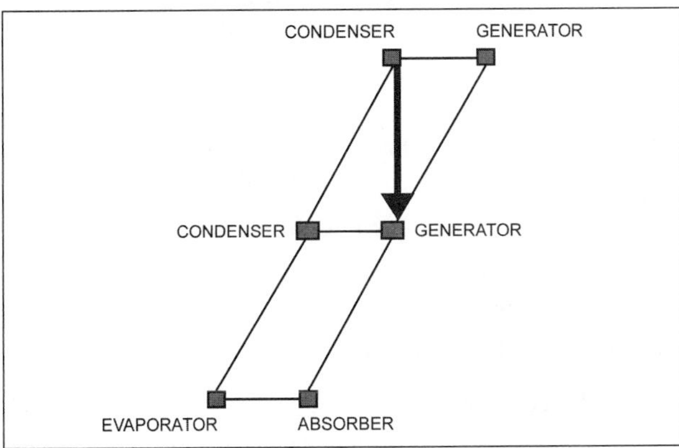

Fig. 18 Double-Effect Absorption Cycle

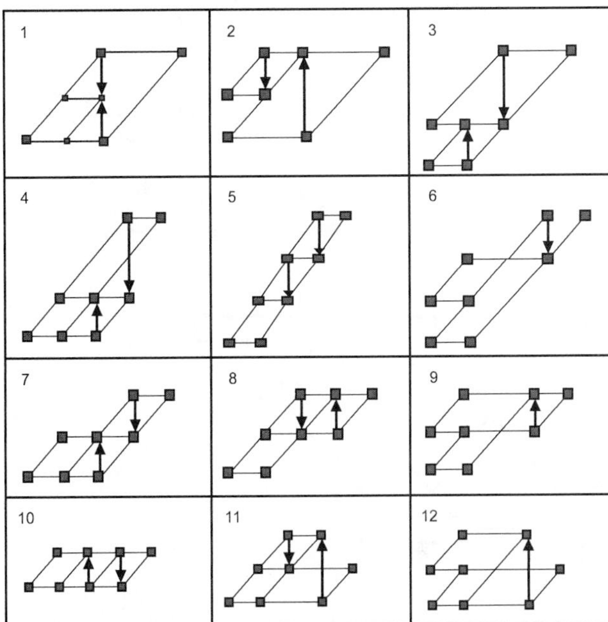

Fig. 19 Generic Triple-Effect Cycles

The ideal analysis can be extended to these multistage cycles (Alefeld and Radermacher 1994). A similar range of cycle variants is possible for situations calling for the half-effect cycle, in which the available heat source temperature is below $t_{gen\ min}$.

ABSORPTION CYCLE MODELING

Analysis and Performance Simulation

A physical-mathematical model of an absorption cycle consists of four types of thermodynamic equations: mass balances, energy balances, relations describing the heat and mass transfer, and equations for the thermophysical properties of the working fluids.

As an example of simulation, Figure 20 shows a Dühring plot of a single-effect water-lithium bromide absorption chiller. The chiller is hot water driven, rejects waste heat from the absorber and the condenser to a stream of cooling water, and produces chilled water. A simulation of this chiller starts by specifying the assumptions (Table 6) and the design parameters and the operating conditions at the design point (Table 7). Design parameters are the specified *UA* values and the flow regime (co/counter/crosscurrent, pool, or film) of all heat exchangers (evaporator, condenser, generator, absorber, solution heat exchanger) and the flow rate of weak solution through the solution pump.

One complete set of input operating parameters could be the design point values of the chilled water and cooling water temperatures $t_{chill\ in}$, $t_{chill\ out}$, $t_{cool\ in}$, $t_{cool\ out}$, the hot water flow rate $\dot{m}_{hot}$, and the total cooling capacity Q_e. With this information, a cycle simulation calculates the required hot water temperatures; the cooling water flow rate; and the temperatures, pressures, and concentrations at all internal state points. Some additional assumptions are made that reduce the number of unknown parameters.

Table 6 Assumptions for Single-Effect Water-Lithium Bromide Model ()

Assumptions
• Generator and condenser as well as evaporator and absorber are under same pressure
• Refrigerant vapor leaving the evaporator is saturated pure water
• Liquid refrigerant leaving the condenser is saturated
• Strong solution leaving the generator is boiling
• Refrigerant vapor leaving the generator has the equilibrium temperature of the weak solution at generator pressure
• Weak solution leaving the absorber is saturated
• No liquid carryover from evaporator
• Flow restrictors are adiabatic
• Pump is isentropic
• No jacket heat losses
• The lmtd (log mean temperature difference) expression adequately estimates the latent changes

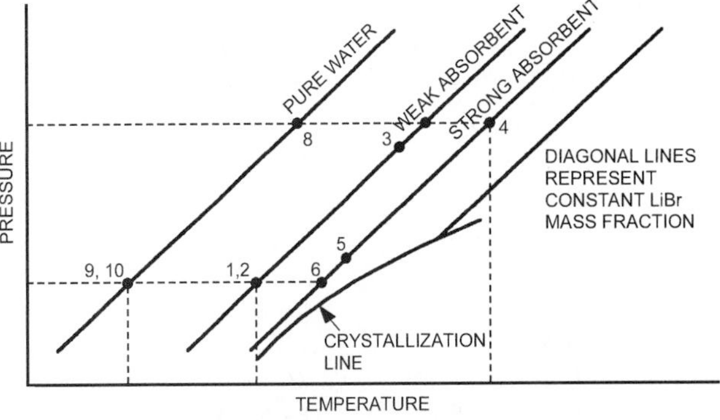

Fig. 20 Single-Effect Water-Lithium Bromide Absorption Cycle Dühring Plot

With these assumptions and the design parameters and operating conditions as specified in Table 7, the cycle simulation can be conducted by solving the following set of equations:

Mass Balances

$$\dot{m}_{refr} + \dot{m}_{strong} = \dot{m}_{weak} \quad (59)$$

$$\dot{m}_{strong}\xi_{strong} = \dot{m}_{weak}\xi_{weak} \quad (60)$$

Energy Balances

$$\begin{aligned}\dot{Q}_{evap} &= \dot{m}_{refr}(h_{vapor,\,evap} - h_{liq,\,cond})\\ &= \dot{m}_{chill}(h_{chill\ in} - h_{chill\ out})\end{aligned} \quad (61)$$

$$\begin{aligned}\dot{Q}_{evap} &= \dot{m}_{refr}(h_{vapor,\,gen} - h_{liq,\,cond})\\ &= \dot{m}_{cool}(h_{cool\ out} - h_{cool\ mean})\end{aligned} \quad (62)$$

$$\begin{aligned}\dot{Q}_{abs} &= \dot{m}_{refr}h_{vapor,\,evap} + \dot{m}_{strong}h_{strong,\,gen}\\ &- \dot{m}_{weak}h_{weak,\,abs} - \dot{Q}_{sol} = \dot{m}_{cool}(h_{cool\ mean} - h_{cool\ in})\end{aligned} \quad (63)$$

$$\begin{aligned}\dot{Q}_{gen} &= \dot{m}_{refr}h_{vapor,\,gen} + \dot{m}_{strong}h_{strong,\,gen}\\ &- \dot{m}_{weak}h_{weak,\,abs} - \dot{Q}_{sol} = \dot{m}_{hot}(h_{hot\ in} - h_{hot\ out})\end{aligned} \quad (64)$$

Table 7 Design Parameters and Operating Conditions for Single-Effect Water-Lithium Bromide Absorption Chiller

	Design Parameters	Operating Conditions
Evaporator	UA_{evap} = 605,000 Btu/h·°F, countercurrent film	$t_{chill\ in}$ = 53.6°F $t_{chill\ out}$ = 42.8°F
Condenser	UA_{cond} = 342,300 Btu/h·°F, countercurrent film	$t_{cool\ out}$ = 95°F
Absorber	UA_{abs} = 354,300 Btu/h·°F, countercurrent film-absorber	$t_{cool\ in}$ = 80.6°F
Generator	UA_{gen} = 271,800 Btu/h·°F, pool-generator	$\dot{m}_{hot}$ = 590,000 lb/h
Solution	UA_{sol} = 64,100 Btu/h·°F, countercurrent	
General	$\dot{m}_{weak}$ = 95,200 lb/h	$\dot{Q}_{evap}$ = 7.33 × 10^6 Btu/h

Table 8 Simulation Results for Single-Effect Water-Lithium Bromide Absorption Chiller

	Internal Parameters	Performance Parameters
Evaporator	$t_{vapor,evap}$ = 35.2°F $p_{sat,evap}$ = 0.1 psia	$\dot{Q}_{evap}$ = 7.33 × 10^6 Btu/h $\dot{m}_{chill}$ = 677,000 lb/h
Condenser	$T_{liq,cond}$ = 115.2°F $p_{sat,cond}$ = 1.48 psia	$\dot{Q}_{cond}$ = 7.92 × 10^6 Btu/h $\dot{m}_{cool}$ = 1.260 × 10^6lb/h
Absorber	ξ_{weak} = 59.6% t_{weak} = 105.3°F $t_{strong,abs}$ = 121.8°F	$\dot{Q}_{abs}$ = 10.18 × 10^6 Btu/h $t_{cool,mean}$ = 88.7°F
Generator	ξ_{strong} = 64.6% $t_{strong,gen}$ = 218.3°F $t_{weak,gen}$ = 198.3°F $t_{weak,sol}$ = 169°F	$\dot{Q}_{gen}$ = 10.78 × 10^6Btu/h $t_{hot\ in}$ = 257°F $t_{hot\ out}$ = 239°F
Solution	$t_{strong,sol}$ = 144.3°F $t_{weak,sol}$ = 169°F	$\dot{Q}_{sol}$ = 2.815 × 10^6 Btu/h ε = 65.4%
General	$\dot{m}_{vapor}$ = 7380 lb/h $\dot{m}_{strong}$ = 87,800 lb/h	COP = 0.68

$$\begin{aligned}\dot{Q}_{sol} &= \dot{m}_{strong}(h_{strong,\,gen} - h_{strong,\,sol})\\ &= \dot{m}_{weak}(h_{weak,\,sol} - h_{weak,\,abs})\end{aligned} \quad (65)$$

Heat Transfer Equations

$$\dot{Q}_{evap} = UA_{evap}\frac{t_{chill\ in} - t_{chill\ out}}{\ln\left(\dfrac{t_{chill\ in} - t_{vapor,\,evap}}{t_{chill\ out} - t_{vapor,\,evap}}\right)} \quad (66)$$

$$\dot{Q}_{cond} = UA_{cond}\frac{t_{cool\ out} - t_{cool\ mean}}{\ln\left(\dfrac{t_{liq,\,cond} - t_{cool\ mean}}{t_{liq,\,cond} - t_{cool\ out}}\right)} \quad (67)$$

$$\dot{Q}_{abs} = UA_{abs}\frac{(t_{strong,\,abs} - t_{cool\ mean}) - (t_{weak,\,abs} - t_{cool\ in})}{\ln\left(\dfrac{t_{strong,\,abs} - t_{cool\ mean}}{t_{weak,\,abs} - t_{cool\ in}}\right)} \quad (68)$$

$$\dot{Q}_{gen} = UA_{gen}\frac{(t_{hot\ in} - t_{strong,\,gen}) - (t_{hot\ out} - t_{weak,\,gen})}{\ln\left(\dfrac{t_{hot\ in} - t_{strong,\,gen}}{t_{hot\ out} - t_{weak,\,gen}}\right)} \quad (69)$$

$$\dot{Q}_{sol} = UA_{sol}\frac{(t_{strong,\,gen} - t_{weak,\,sol}) - (t_{strong,\,sol} - t_{weak,\,abs})}{\ln\left(\dfrac{t_{strong,\,gen} - t_{weak,\,sol}}{t_{strong,\,sol} - t_{weak,\,abs}}\right)} \quad (70)$$

Fluid Property Equations at each state point

Thermal Equations of State: $h_{water}(t,p)$, $h_{sol}(t,p,\xi)$
Two-Phase Equilibrium: $t_{water,sat}(p)$, $t_{sol,sat}(p,\xi)$

The results are listed in Table 8.

Double-Effect Cycle

Double-effect cycle calculations can be performed in a manner similar to that illustrated for the single-effect cycle. Mass and energy balances of the model shown in Figure 21 were calculated using the inputs and assumptions listed in Table 9. The results are shown in Table 10. The COP is quite sensitive to several inputs and

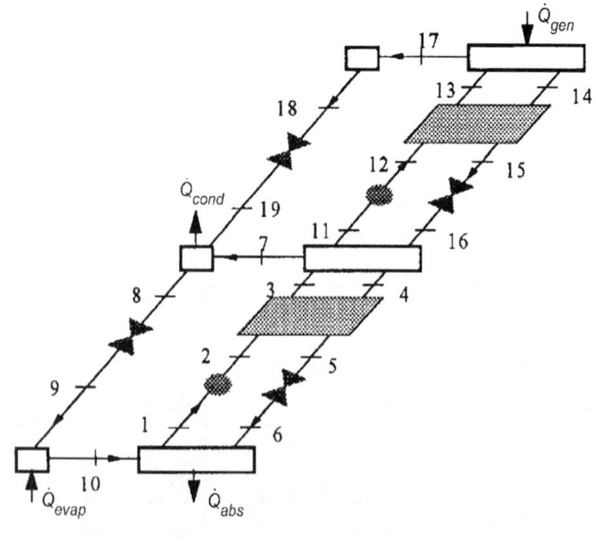

Fig. 21 Double-Effect Water-Lithium Bromide Absorption Cycle with State Points

assumptions. In particular, the effectiveness of the solution heat exchangers and the driving temperature difference between the high-temperature condenser and the low-temperature generator influence the COP strongly.

Table 9 Inputs and Assumptions for Double-Effect Water-Lithium Bromide Model

Inputs		
Capacity	Q_{evap}	500 tons (refrig.)
Evaporator temperature	t_{10}	41.1°F
Desorber solution exit temperature	t_{14}	339.3°F
Condenser/absorber low temperature	t_1, t_8	108.3°F
Solution heat exchanger effectiveness	ε	0.6

Assumptions

- Steady state
- Refrigerant is pure water
- No pressure changes except through flow restrictors and pump
- State points at 1, 4, 8, 11, 14, and 18 are saturated liquid
- State point 10 is saturated vapor
- Temperature difference between high-temperature condenser and low-temperature generator is 9°F
- Parallel flow
- Both solution heat exchangers have same effectiveness
- Upper loop solution flow rate is selected such that upper condenser heat exactly matches lower generator heat requirement
- Flow restrictors are adiabatic
- Pumps are isentropic
- No jacket heat losses
- No liquid carryover from evaporator to absorber
- Vapor leaving both generators is at equilibrium temperature of entering solution stream

Table 10 State Point Data for Double-Effect Water-Lithium Bromide Cycle of Figure 21

No.	h, Btu/lb	$\dot{m}$, lb/min	p, psia	Q, Fraction	t, °F	x, % LiBr
1	50.6	1263.4	0.13	0.0	108.3	
2	50.6	1263.4	1.21		108.3	59.5
3	78.3	1263.4	1.21		168.1	59.5
4	106.2	1163.7	1.21	0.0	208.0	64.6
5	76.1	1163.7	1.21		137.9	64.6
6	76.1	1163.7	0.13	0.004	127.8	64.6
7	1143.2	42.3	1.21		186.2	0.0
8	76.2	99.8	1.21	0.0	108.3	0.0
9	76.2	99.8	0.13	0.063	41.1	0.0
10	1078.6	99.8	0.13	1.0	41.1	0.0
11	86.7	727.3	1.21	0.0	186.2	59.5
12	86.7	727.3	16.21		186.2	59.5
13	129.4	727.3	16.21		278.0	59.5
14	162.7	669.9	16.21	0.0	339.3	64.6
15	116.4	669.9	16.21		231.6	64.6
16	116.4	669.9	1.21	0.008	210.3	64.6
17	1197.4	57.4	16.21		312.2	0.0
18	185.0	57.4	16.21	0.0	217.0	0.0
19	185.0	57.4	1.21	0.105	108.3	0.0

COP = 1.195	$\dot{Q}_{evap}$ = 6.000 × 10^6 Btu/h
Δt = 9.0°F	$\dot{Q}_{gen}$ = 5.019 × 10^6 Btu/h
ε = 0.600	$\dot{Q}_{shx1}$ = 2.103 × 10^6 Btu/h
$\dot{Q}_{abs}$ = 7.936 × 10^6 Btu/h	$\dot{Q}_{shx2}$ = 1.862 × 10^6 Btu/h
$\dot{Q}_{gen}$ = 3.488 × 10^6 Btu/h	$\dot{W}_{p1}$ = 0.032 hp
$\dot{Q}_{cond}$ = 3.085 × 10^6 Btu/h	$\dot{W}_{p2}$ = 0.258 hp

AMMONIA-WATER ABSORPTION CYCLES

Ammonia-water absorption cycles are similar to the water-lithium bromide cycles, but with some important differences. The differences arise due to the lower latent heat of ammonia compared to water, the volatility of the absorbent, and the different pressure and solubility ranges. The latent heat of ammonia is only about half that of water, so, for the same duty, the refrigerant and absorbent mass circulation rates are roughly double that of water-lithium bromide. As a result, the sensible heat loss associated with heat exchanger approaches is greater. Accordingly, ammonia-water cycles incorporate more techniques to reclaim sensible heat described in Hanna et al. (1995). The refrigerant heat exchanger (RHX), also known as refrigerant subcooler, which improves COP by about 8%, is the most important (Holldorff 1979). Next is the absorber heat exchanger (AHX), accompanied by a generator heat exchanger (GHX) (Phillips 1976). These either replace or supplement the traditional solution heat exchanger (SHX). These components would also benefit the water-lithium bromide cycle, except that the deep vacuum in that cycle makes them impractical there.

The volatility of the water absorbent is also key. It makes the distinction between crosscurrent, cocurrent, and countercurrent mass exchange more important in all of the latent heat exchangers (Briggs 1971). It also requires a distillation column on the high-pressure side. When improperly implemented, this column can impose both cost and COP penalties. Those penalties are avoided by refluxing the column from an internal diabatic section (e.g., solution cooled rectifier [SCR]) rather than with an external reflux pump.

The high-pressure operating regime makes it impractical to achieve multieffect performance via pressure-staging. On the other hand, the exceptionally wide solubility field facilitates concentration-staging. The generator-absorber heat exchange (GAX) cycle is an especially advantageous embodiment of concentration-staging (Modahl and Hayes 1988).

Ammonia-water cycles can equal the performance of water-lithium bromide cycles. The single-effect or basic GAX cycle yields the same performance as a single-effect water-lithium bromide cycle; the branched GAX cycle (Herold et al. 1991) yields the same performance as a water-lithium bromide double-effect cycle; and the VX GAX cycle (Erickson and Rane 1994) yields the same performance as a water-lithium bromide triple-effect cycle. Additional advantages of the ammonia-water cycle include refrigeration capability, air-cooling capability, all mild steel construction, extreme compactness, and capability of direct integration into industrial processes. Between heat-activated refrigerators, gas-fired residential air conditioners, and large industrial refrigeration plants, this technology has accounted for the vast majority of absorption activity over the past century.

Figure 22 shows the diagram of a typical single-effect ammonia-water absorption cycle. The inputs and assumptions in Table 11 are used to calculate a single-cycle solution, which is summarized in Table 12.

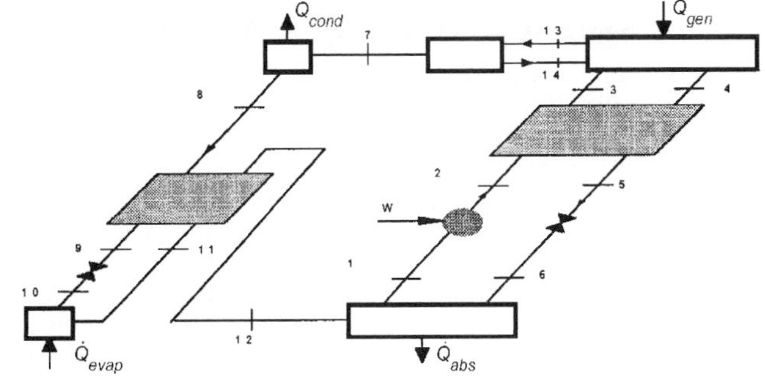

Fig. 22 Single-Effect Ammonia-Water Absorption Cycle

Table 11 Inputs and Assumptions for Single-Effect Ammonia/Water Cycle of Figure 22

Inputs		
Capacity	$\dot{Q}_{evap}$	500 tons (refrig.)
High-side pressure	p_{high}	211.8 psia
Low-side pressure	p_{low}	74.7 psia
Absorber exit temperature	t_1	105°F
Generator exit temperature	t_4	203°F
Rectifier vapor exit temperature	t_7	131°F
Solution heat exchanger eff.	ε_{shx}	0.692
Refrigerant heat exchanger eff.	ε_{rhx}	0.629

Assumptions

- Steady state
- No pressure changes except through flow restrictors and pump
- States at points 1, 4, 8, 11, and 14 are saturated liquid
- States at point 12 and 13 are saturated vapor
- Flow restrictors are adiabatic
- Pump is isentropic
- No jacket heat losses
- No liquid carryover from evaporator to absorber
- Vapor leaving generator is at equilibrium temperature of entering solution stream

Table 12 State Point Data for Single-Effect Ammonia/Water Cycle of Figure 22

No.	h, Btu/lb	$\dot{m}$, lb/min	p, psia	Q, Fraction	t, °F	x, Fraction NH_3
1	–24.55	1408.2	74.7	0.0	105.0	0.50094
2	–24.05	1408.2	211.8		105.5	0.50094
3	38.47	1408.2	211.8		163.0	0.50094
4	83.81	1203.0	211.8	0.0	203.0	0.41612
5	10.61	1203.0	211.8		135.6	0.41612
6	10.61	1203.0	74.7	0.006	132.0	0.41612
7	579.51	205.2	211.8	1.0	131.0	0.99809
8	76.61	205.2	211.8	0.0	100.1	0.99809
9	35.28	205.2	211.8		64.1	0.99809
10	35.28	205.2	74.7	0.049	41.1	0.99809
11	522.55	205.2	74.7	0.953	42.8	0.99809
12	563.88	205.2	74.7	1.0	87.0	0.99809
13	613.91	209.9	211.8	1.0	174.5	0.98708
14	51.72	4.6	211.8	0.0	174.5	0.50094

$COP_C = 0.571$	$\dot{Q}_{evap} = 6.00 \times 10^6$ Btu/h
$\Delta t_{rhx} = 36.00$°F	$\dot{Q}_{gen} = 1.051 \times 10^7$ Btu/h
$\Delta t_{shx} = 30.1$°F	$\dot{Q}_{rhx} = 5.089 \times 10^5$ Btu/h
$\varepsilon_{rhx} = 0.629$	$\dot{Q}_r = 5.805 \times 10^5$ Btu/h
$\varepsilon_{shx} = 0.692$	$\dot{Q}_{shx} = 5.283 \times 10^6$ Btu/h
$\dot{Q}_{abs} = 9.784 \times 10^6$ Btu/h	$\dot{W} = 9.22$ hp
$\dot{Q}_{cond} = 6.192 \times 10^6$ Btu/h	

NOMENCLATURE FOR EXAMPLES

c_p = specific heat at constant pressure
COP = coefficient of performance
g = local acceleration of gravity
h = enthalpy, Btu/lb
I = irreversibility
$\dot{I}$ = irreversibility rate
m = mass
$\dot{m}$ = mass flow, lb/min
p = pressure
Q = heat energy, Btu
$\dot{Q}$ = rate of heat flow, Btu/h
R = ideal gas constant
s = entropy, Btu/lb·°R
S = total entropy
t = temperature, °F
T = absolute temperature, °R
u = internal energy
W = mechanical or shaft work
$\dot{W}$ = rate of work, power
v = specific volume, ft³/lb
V = velocity of fluid
x = mass fraction (of either lithium bromide or ammonia)
x = vapor quality (fraction)
z = elevation above horizontal reference plane
Z = compressibility factor
ε = heat exchanger effectiveness
η = efficiency
ρ = density, lb/ft³

Subscripts

abs = absorber
cond = condenser or cooling mode
cg = condenser to generator
evap = evaporator
fg = fluid to vapor
gen = generator
gh = high-temperature generator
o, 0 = reference conditions, usually ambient
p = pump
R = refrigerating or evaporator conditions
sol = solution
rhx = refrigerant heat exchanger
shx = solution heat exchanger

REFERENCES

Alefeld, G. and R. Radermacher. 1994. *Heat conversion systems*. CRC Press, Boca Raton.

Benedict, M., G.B. Webb, and L.C. Rubin. 1940. An empirical equation for thermodynamic properties of light hydrocarbons and their mixtures. *Journal of Chemistry and Physics* 4:334.

Benedict, M. 1937. Pressure, volume, temperature properties of nitrogen at high density, I and II. *Journal of American Chemists Society* 59(11): 2224.

Briggs, S.W. 1971. Concurrent, crosscurrent, and countercurrent absorption in ammonia-water absorption refrigeration. *ASHRAE Transactions* 77(1):171.

Cooper, H.W. and J.C. Goldfrank. 1967. B-W-R Constants and new correlations. *Hydrocarbon Processing* 46(12):141.

Erickson, D.C. and M. Rane. 1994. Advanced absorption cycle: Vapor exchange GAX. *Proceedings of the International Absorption Heat Pump Conference*. Chicago.

Hanna, W.T., et al. 1995. Pinch-point analysis: An aid to understanding the GAX absorption cycle. *ASHRAE Technical Data Bulletin* 11(2).

Herold, K.E., et al. 1991. The branched GAX absorption heat pump cycle, *Proceedings of Absorption Heat Pump Conference.* Tokyo.

Hirschfelder, J.O. et al. 1958. Generalized equation of state for gases and liquids. *Industrial and Engineering Chemistry* 50:375.

Holldorff, G. 1979. Revisions up absorption refrigeration efficiency. *Hydrocarbon Processing* 58(7):149.

Howell, J.R. and R.O. Buckius. 1992. *Fundamentals of engineering thermodynamics*, 2nd ed. McGraw-Hill, New York.

Hust, J.G. and R.D. McCarty. 1967. Curve-fitting techniques and applications to thermodynamics. *Cryogenics* 8:200.

Hust, J.G. and R.B. Stewart. 1966. Thermodynamic property computations for system analysis. *ASHRAE Journal* 2:64.

Kuehn, T.H. and R.E. Gronseth. 1986. The effect of a nonazeotropic binary refrigerant mixture on the performance of a single stage refrigeration cycle. Proceedings *International Institute of Refrigeration Conference,* Purdue University, p. 119.

Liang, H. and T.H. Kuehn. 1991. Irreversibility analysis of a water to water mechanical compression heat pump. *Energy* 16(6):883.

Macriss, R.A. 1968. Physical properties of modified LiBr solutions. AGA Symposium on Absorption Air-Conditioning Systems, February.

Macriss, R.A. and T.S. Zawacki. 1989. Absorption fluid data survey: 1989 update. Oak Ridge National Laboratories *Report* ORNL/Sub84-47989/4.

Martin, J.J. and Y. Hou. 1955. Development of an equation of state for gases. *AIChE Journal* 1:142.

Martz, W.L., C.M. Burton, and A.M. Jacobi. 1996a. Liquid-vapor equilibria for R-22, R-134a, R-125, and R-32/125 with a polyol ester lubricant: Measurements and departure from ideality. *ASHRAE Transactions* 102(1):367-74.

Martz, W.L., C.M. Burton and A.M. Jacobi. 1996b. Local composition modeling of the thermodynamic properties of refrigerant and oil mixtures. *International Journal of Refrigeration* 19(1):25-33.

Modahl, R.J. and F.C. Hayes. 1988. Evaluation of commercial advanced absorption heat pump. *Proceedings of the 2nd DOE/ORNL Heat Pump Conference.* Washington.

NASA. 1971. SP-273. US Government Printing Office, Washington, D.C.

Phillips, B. 1976. Absorption cycles for air-cooled solar air conditioning. *ASHRAE Transactions* 82(1):966. Dallas.

Stewart, R.B., R.T. Jacobsen, and S.G. Penoncello. 1986. *ASHRAE Thermodynamic properties of refrigerants.* ASHRAE, Atlanta, GA.

Strobridge, T.R. 1962. The thermodynamic properties of nitrogen from 64 to 300 K, between 0.1 and 200 atmospheres. National Bureau of Standards *Technical Note* 129.

Stoecker, W.F. 1989. *Design of thermal systems*, 3rd ed. McGraw-Hill, New York.

Tassios, D.P. 1993. *Applied chemical engineering thermodynamics.* Springer-Verlag, New York.

Thome, J.R. 1995. Comprehensive thermodynamic approach to modeling refrigerant-lubricant oil mixtures. *International Journal of Heating, Ventilating, Air Conditioning and Refrigeration Research* 1(2):110.

Tozer, R.M. and R.W. James. 1997. Fundamental thermodynamics of ideal absorption cycles. *International Journal of Refrigeration* 20 (2):123-135.

BIBLIOGRAPHY

Bogart, M. 1981. *Ammonia absorption refrigeration in industrial processes.* Gulf Publishing Co., Houston, TX.

Herold, K.E., R. Radermacher, and S.A. Klein. 1996. *Absorption chillers and heat pumps.* CRC Press, Boca Raton.

Jain, P.C. and G.K. Gable. 1971. Equilibrium property data for aqua-ammonia mixture. *ASHRAE Transactions* 77(1):149.

Moran, M.J. and Shapiro, H. 1995. *Fundamentals of engineering thermodynamics*, 3rd Ed. John Wiley and Sons, Inc. New York.

Van Wylen, C.J. and R.E. Sonntag. 1985. *Fundamentals of classical thermodynamics*, 3rd ed. John Wiley and Sons, New York.

Zawacki, T.S. 1999. Effect of ammonia-water mixture database on cycle calculations. *Proceedings of the International Sorption Heat Pump Conference.* Munich.

CHAPTER 2

FLUID FLOW

FLOWING fluids in heating, ventilating, air-conditioning, and refrigeration systems transfer heat and mass. This chapter introduces the basics of fluid mechanics that are related to HVAC processes, reviews pertinent flow processes, and presents a general discussion of single-phase fluid flow analysis.

FLUID PROPERTIES

Fluids differ from solids in their reaction to shearing. When placed under shear stress, a solid deforms only a finite amount, whereas a fluid deforms continuously for as long as the shear is applied. Both liquids and gases are fluids. Although liquids and gases differ strongly in the nature of molecular actions, their primary mechanical differences are in the degree of compressibility and liquid formation of a free surface (interface).

Fluid motion can usually be described by one of several simplified modes of action or models. The simplest is the ideal-fluid model, which assumes no resistance to shearing. Ideal flow analysis is well developed (Baker 1983, Schlichting 1979, Streeter and Wylie 1979), and when properly interpreted is valid for a wide range of applications. Nevertheless, the effects of viscous action may need to be considered. Most fluids in HVAC applications can be treated as Newtonian, where the rate of deformation is directly proportional to the shearing stress. Turbulence complicates fluid behavior, and viscosity influences the nature of the turbulent flow.

Density

The density ρ of a fluid is its mass per unit volume. The densities of air and water at standard indoor conditions of 68°F and 14.696 psi (sea level atmospheric pressure) are

$$\rho_{water} = 62.3\ \text{lb}_m/\text{ft}^3$$

$$\rho_{air} = 0.075\ \text{lb}_m/\text{ft}^3$$

Viscosity

Viscosity is the resistance of adjacent fluid layers to shear. For shearing between two parallel plates, each of area A and separated by distance Y, the tangential force F per unit area required to slide one plate with velocity V parallel to the other is proportional to V/Y:

$$F/A = \mu(V/Y)$$

where the proportionality factor μ is the **absolute viscosity** or **dynamic viscosity** of the fluid. The ratio of the tangential force F to area A is the **shearing stress** τ, and V/Y is the **lateral velocity gradient** (Figure 1A). In complex flows, velocity and shear stress may vary across the flow field; this is expressed by the following differential equation:

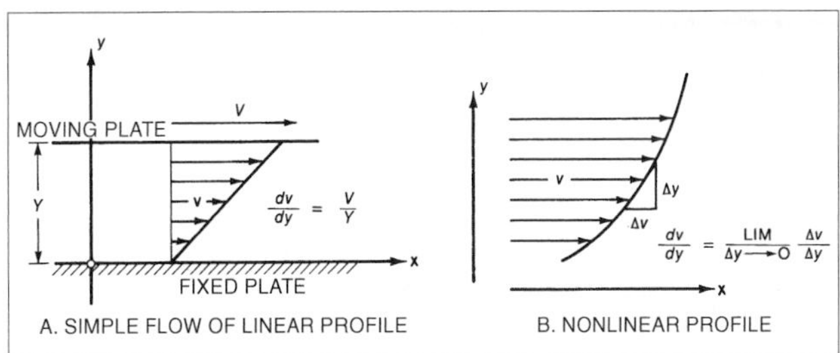

Fig. 1 Velocity Profiles and Gradients in Shear Flows

$$\tau = \mu \frac{dv}{dy} \qquad (1)$$

The velocity gradient associated with viscous shear for a simple case involving flow velocity in the x direction but of varying magnitude in the y direction is illustrated in Figure 1B.

Absolute viscosity μ depends primarily on temperature. For gases (except near the critical point), viscosity increases with the square root of the absolute temperature, as predicted by the kinetic theory. Liquid viscosity decreases with increasing temperature. Viscosities of various fluids are given in Chapter 38.

Absolute viscosity has dimensions of force · time/length^2. At standard indoor conditions, the absolute viscosities of water and dry air are

$$\mu_{water} = 6.7 \times 10^{-4}\ \text{lb}_m/\text{ft}\cdot\text{s} = 2.1 \times 10^{-5}\ \text{lb}_f\cdot\text{s}/\text{ft}^2$$

$$\mu_{air} = 1.2 \times 10^{-5}\ \text{lb}_m/\text{ft}\cdot\text{s} = 3.7 \times 10^{-7}\ \text{lb}_f\cdot\text{s}/\text{ft}^2$$

The centipoise is another common unit of viscosity (1 centipoise = 1 g/(s·m) = 1 mPa·s). At standard conditions, water has a viscosity close to 1.0 centipoise.

In fluid dynamics, **kinematic viscosity** ν is the ratio of absolute viscosity to density:

$$\nu = \mu/\rho$$

The stoke (1 cm^2/s) and centistoke (1 mm^2/s) are common units for kinematic viscosity. At standard indoor conditions, the kinematic viscosities of water and dry air are

$$\nu_{water} = 1.08\times10^{-5}\ \text{ft}^2/\text{s}$$

$$\nu_{air} = 1.7\times10^{-4}\ \text{ft}^2/\text{s}$$

Note that the inch-pound system of units requires the conversion factor $g_c = 32.1740\ \text{lb}_m\cdot\text{ft/s}^2\cdot\text{lb}_f$ to make some equations containing lb_f and lb_m dimensionally consistent. The conversion factor g_c is not shown in the equations, but is included as needed.

The preparation of this chapter is assigned to TC 1.3, Heat Transfer and Fluid Flow.

BASIC RELATIONS OF FLUID DYNAMICS

This section considers homogeneous, constant-property, incompressible fluids and introduces fluid dynamic considerations used in most analyses.

Continuity

Conservation of matter applied to fluid flow in a conduit requires that

$$\int \rho v \, dA = \text{constant}$$

where

v = velocity normal to the differential area dA
ρ = fluid density

Both ρ and v may vary over the cross section A of the conduit. If both ρ and v are constant over the cross-sectional area normal to the flow, then

$$\dot{m} = \rho V A = \text{constant} \tag{2a}$$

where $\dot{m}$ is the mass flow rate across the area normal to the flow. When flow is effectively incompressible, ρ = constant; in pipeline and duct flow analyses, the average velocity is then $V = (1/A)\int v dA$. The continuity relation is

$$Q = AV = \text{constant} \tag{2b}$$

where Q is the volumetric flow rate. Except when branches occur, Q is the same at all sections along the conduit.

For the ideal-fluid model, flow patterns around bodies (or in conduit section changes) result from displacement effects. An obstruction in a fluid stream, such as a strut in a flow or a bump on the conduit wall, pushes the flow smoothly out of the way, so that behind the obstruction, the flow becomes uniform again. The effect of fluid inertia (density) appears only in pressure changes.

Pressure Variation Across Flow

Pressure variation in fluid flow is important and can be easily measured. Variation across streamlines involves fluid rotation (vorticity). Lateral pressure variation across streamlines is given by the following relation (Bober and Kenyon 1980, Olson 1980, Robertson 1965):

$$\frac{\partial}{\partial r}\left(\frac{p}{\rho} + gz\right) = \frac{v^2}{r} \tag{3}$$

where

r = radius of curvature of the streamline
z = elevation

This relation explains the pressure difference found between the inside and outside walls of a bend and near other regions of conduit section change. It also states that pressure variation is hydrostatic ($p + \rho gz$ = constant) across any conduit where streamlines are parallel.

Bernoulli Equation and Pressure Variation along Flow

A basic tool of fluid flow analysis is the Bernoulli relation, which involves the principle of energy conservation along a streamline. Generally, the Bernoulli equation is not applicable across streamlines. The first law of thermodynamics can be applied to mechanical flow energies (kinetic and potential) and thermal energies: heat is a form of energy and energy is conserved.

The change in energy content ΔE per unit mass of flowing material is a result from the work W done on the system plus the heat Q absorbed:

$$\Delta E = W + Q$$

Fluid energy is composed of kinetic, potential (due to elevation z), and internal (u) energies. Per unit mass of fluid, the above energy change relation between two sections of the system is

$$\Delta\left(\frac{v^2}{2} + gz + u\right) = E_M - \Delta\left(\frac{p}{\rho}\right) + Q$$

where the work terms are (1) the external work E_M from a fluid machine (E_M is positive for a pump or blower) and (2) the pressure or flow work p/ρ. Rearranging, the energy equation can be written as the **generalized Bernoulli equation**:

$$\Delta\left(\frac{v^2}{2} + gz + \frac{p}{\rho}\right) + \Delta u = E_M + Q \tag{4}$$

The term in parentheses in Equation (4) is the **Bernoulli constant**:

$$\frac{p}{\rho} + \frac{v^2}{2} + gz = B \tag{5a}$$

In cases with no viscous action and no work interaction, B is constant; more generally its change (or lack thereof) is considered in applying the Bernoulli equation. The terms making up B are fluid energies (pressure, kinetic, and potential) per mass rate of fluid flow. Alternative forms of this relation are obtained through multiplication by ρ or division by g:

$$p + \frac{\rho v^2}{2} + \rho gz = p + \frac{\rho v^2}{2} + \gamma z = \rho B \tag{5b}$$

$$\frac{p}{\rho g} + \frac{v^2}{2g} + z = \frac{p}{\gamma} + \frac{v^2}{2g} + z = \frac{B}{g} \tag{5c}$$

where $\gamma = \rho g$ is the specific weight.

The first form involves energies per volume flow rate, or pressures; the second involves energies per mass flow rate, or heads. In gas flow analysis, Equation (5b) is often used with the γz term dropped as negligible. Equation (5a) should be used when density variations occur. For liquid flows, Equation (5c) is commonly used. Identical results are obtained with the three forms if the units are consistent and the fluids are homogeneous.

Many systems of pipes or ducts and pumps or blowers can be considered as one-dimensional flow. The Bernoulli equation is then considered as velocity and pressure vary along the conduit. Analysis is adequate in terms of the section-average velocity V of Equation (2a) or (2b). In the Bernoulli relation [Equations (4) and (5)], v is replaced by V, and variation across streamlines can be ignored; the whole conduit is now taken as one streamline. Two- and three-dimensional details of local flow occurrences are still significant, but their effect is combined and accounted for in factors.

The kinetic energy term of the Bernoulli constant B is expressed as $\alpha V^2/2$, where the **kinetic energy factor** ($\alpha > 1$) expresses the ratio of the true kinetic energy of the velocity profile to that of the mean flow velocity.

For laminar flow in a wide rectangular channel, $\alpha = 1.54$, and for laminar flow in a pipe, $\alpha = 2.0$. For turbulent flow in a duct $\alpha \approx 1$.

Heat transfer Q may often be ignored. The change of mechanical energy into internal energy Δu may be expressed as E_L. Flow analysis involves the change in the Bernoulli constant ($\Delta B = B_2 - B_1$) between stations 1 and 2 along the conduit, and the Bernoulli equation can be expressed as

$$\left(\frac{p}{\rho} + \alpha\frac{V^2}{2} + gz\right)_1 + E_M = \left(\frac{p}{\rho} + \alpha\frac{V^2}{2} + gz\right)_2 + E_L \qquad (6a)$$

or, dividing by g, in head form as

$$\left(\frac{p}{\gamma} + \alpha\frac{V^2}{2g} + z\right)_1 + H_M = \left(\frac{p}{\gamma} + \alpha\frac{V^2}{2g} + z\right)_2 + H_L \qquad (6b)$$

The factors E_M and E_L are defined as positive, where $gH_M = E_M$ represents energy added to the conduit flow by pumps or blowers, and $gH_L = E_L$ represents energy dissipated, that is, converted into heat as mechanically nonrecoverable energy. A turbine or fluid motor thus has a negative H_M or E_M. For conduit systems with branches involving inflow or outflow, the total energies must be treated, and analysis is in terms of $\dot{m}B$ and not B.

When real-fluid effects of viscosity or turbulence are included, the continuity relation in Equation (2b) is not changed, but V must be evaluated from the integral of the velocity profile, using time-averaged local velocities.

In fluid flow past fixed boundaries, the velocity at the boundary is zero and shear stresses are produced. The equations of motion then become complex and exact solutions are difficult to find, except in simple cases.

Laminar Flow

For steady, fully developed laminar flow in a parallel-walled conduit, the shear stress τ varies linearly with distance y from the centerline. For a wide rectangular channel,

$$\tau = \left(\frac{y}{b}\right)\tau_w = \mu\frac{dv}{dy}$$

where

τ_w = wall shear stress = $b\,(dp/ds)$
$2b$ = wall spacing
s = flow direction

Because the velocity is zero at the wall ($y = b$), the integrated result is

$$v = \left(\frac{b^2 - y^2}{2\mu}\right)\frac{dp}{ds}$$

This is the **Poiseuille-flow parabolic velocity profile** for a wide rectangular channel. The average velocity V is two-thirds the maximum velocity (at $y = 0$), and the longitudinal pressure drop in terms of conduit flow velocity is

$$\frac{dp}{ds} = -\left(\frac{3\mu V}{b^2}\right) \qquad (7)$$

The parabolic velocity profile can also be derived for the axisymmetric conduit (pipe) of radius R but with a different constant. The average velocity is then half the maximum, and the pressure drop relation is

$$\frac{dp}{ds} = -\left(\frac{8\mu V}{R^2}\right) \qquad (8)$$

Turbulence

Fluid flows are generally turbulent, involving random perturbations or fluctuations of the flow (velocity and pressure), characterized by an extensive hierarchy of scales or frequencies (Robertson 1963). Flow disturbances that are not random, but have some degree of periodicity, such as the oscillating vortex trail behind bodies, have been erroneously identified as turbulence. Only flows involving random perturbations without any order or periodicity are turbulent; the velocity in such a flow varies with time or locale of measurement (Figure 2).

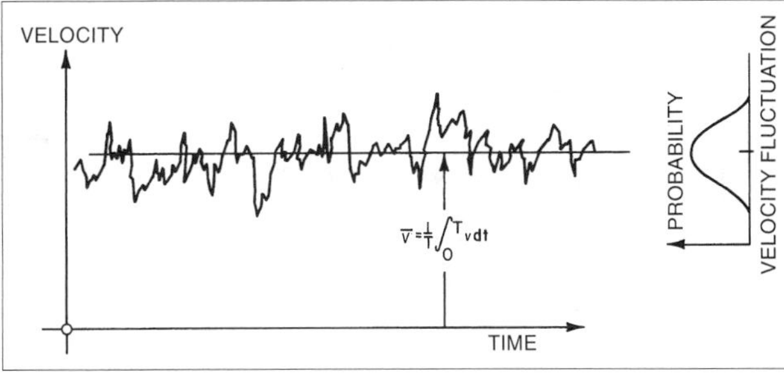

Fig. 2 Velocity Fluctuation at Point in Turbulent Flow

Turbulence can be quantified by statistical factors. Thus, the velocity most often used in velocity profiles is the temporal average velocity $\bar{v}$, and the strength of the turbulence is characterized by the root-mean-square of the instantaneous variation in velocity about this mean. The effects of turbulence cause the fluid to diffuse momentum, heat, and mass very rapidly across the flow.

The **Reynolds number** Re, a dimensionless quantity, gives the relative ratio of inertial to viscous forces:

$$\text{Re} = VL/\nu$$

where

L = characteristic length
ν = kinematic viscosity

In flow through round pipes and tubes, the characteristic length is the diameter D. Generally, laminar flow in pipes can be expected if the Reynolds number, which is based on the pipe diameter, is less than about 2300. Fully turbulent flow exists when $\text{Re}_D > 10{,}000$. Between 2300 and 10,000, the flow is in a transition state and predictions are unreliable. In other geometries, different criteria for the Reynolds number exist.

BASIC FLOW PROCESSES

Wall Friction

At the boundary of real-fluid flow, the relative tangential velocity at the fluid surface is zero. Sometimes in turbulent flow studies, velocity at the wall may appear finite, implying a fluid slip at the wall. However, this is not the case; the difficulty is in velocity measurement (Goldstein 1938). Zero wall velocity leads to a high shear stress near the wall boundary and a slowing down of adjacent fluid layers. A velocity profile develops near a wall, with the velocity increasing from zero at the wall to an exterior value within a finite lateral distance.

Laminar and turbulent flow differ significantly in their velocity profiles. Turbulent flow profiles are flat compared to the more pointed profiles of laminar flow (Figure 3). Near the wall, velocities of the turbulent profile must drop to zero more rapidly than those of the laminar profile, so the shear stress and friction are much greater in the turbulent flow case. Fully developed conduit flow may be characterized by the **pipe factor**, which is the ratio of average to maximum (centerline) velocity. Viscous velocity profiles result in pipe factors of 0.667 and 0.50 for wide rectangular and axisymmetric conduits. Figure 4 indicates much higher values for rectangular and circular conduits for turbulent flow. Due to the flat velocity profiles, the kinetic energy factor α in Equation (6) ranges from 1.01 to 1.10 for fully developed turbulent pipe flow.

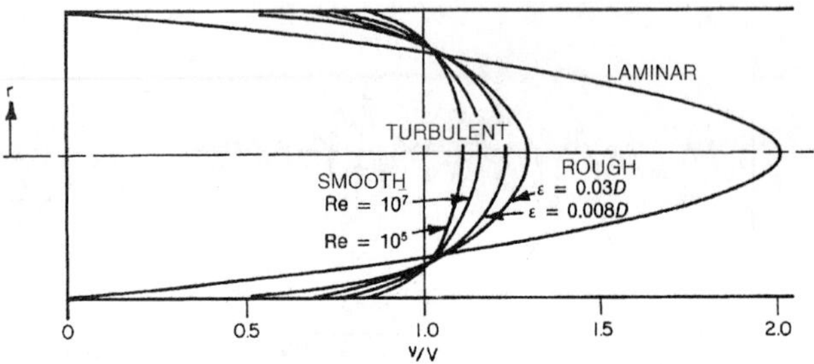

Fig. 3 Velocity Profiles of Flow in Pipes

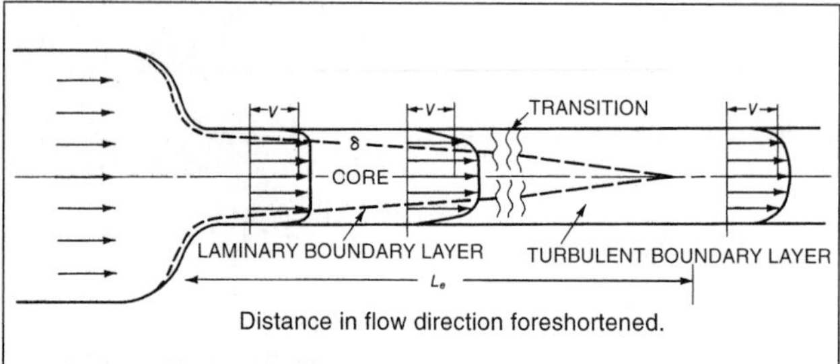

Fig. 5 Flow in Conduit Entrance Region

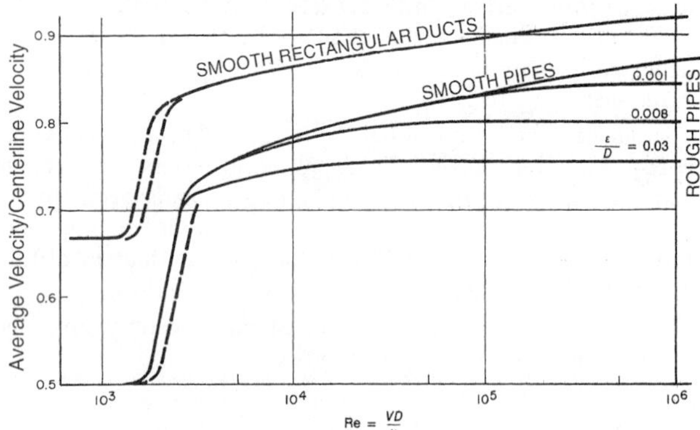

Fig. 4 Pipe Factor for Flow in Conduits

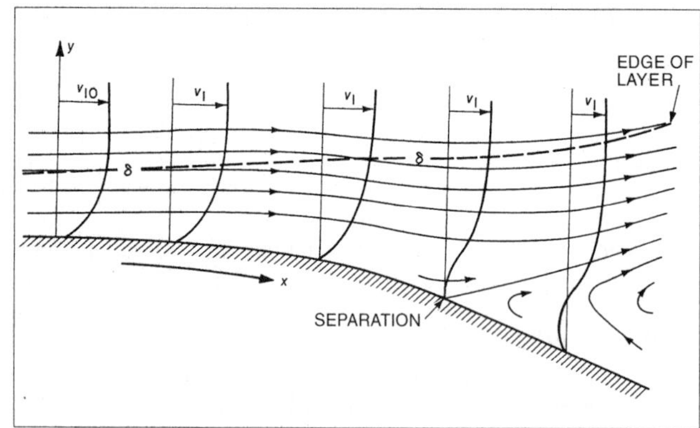

Fig. 6 Boundary Layer Flow to Separation

Boundary Layer

In most flows, the friction of a bounding wall on the fluid flow is evidenced by a boundary layer. For flow around bodies, this layer (which is quite thin relative to distances in the flow direction) encompasses all viscous or turbulent actions, causing the velocity in it to vary rapidly from zero at the wall to that of the outer flow at its edge. Boundary layers are generally laminar near the start of their formation but may become turbulent downstream of the transition point (Figure 5). For conduit flows, spacing between adjacent walls is generally small compared with distances in the flow direction. As a result, layers from the walls meet at the centerline to fill the conduit.

A significant boundary-layer occurrence exists in a pipeline or conduit following a well-rounded entrance (Figure 5). Layers grow from the walls until they meet at the center of the pipe. Near the start of the straight conduit, the layer is very thin (and laminar in all probability), so the uniform velocity core outside has a velocity only slightly greater than the average velocity. As the layer grows in thickness, the slower velocity near the wall requires a velocity increase in the uniform core to satisfy continuity. As the flow proceeds, the wall layers grow (and the centerline velocity increases) until they join, after an entrance length L_e. Application of the Bernoulli relation of Equation (5) to the core flow indicates a decrease in pressure along the layer. Ross (1956) shows that although the entrance length L_e is many diameters, the length in which the pressure drop and head loss rates significantly exceed those for fully developed flow is on the order of 10 diameters for turbulent flow in smooth pipes.

In more general boundary-layer flows, as with wall layer development in a diffuser or for the layer developing along the surface of a strut or turning vane, pressure gradient effects can be severe and may even lead to separation. The development of a layer in an adverse-pressure gradient situation (velocity v_l at edge $y = \delta$ of layer decreasing in flow direction) with separation is shown in Figure 6. Downstream from the separation point, fluid backflows near the wall. Separation is due to frictional velocity (thus local kinetic energy) reduction near the wall. Flow near the wall no longer has energy to move into the higher pressure imposed by the decrease in v_l at the edge of the layer. The locale of this separation is difficult to predict, especially for the turbulent boundary layer. Analyses verify the experimental observation that a turbulent boundary layer is less subject to separation than a laminar one because of its greater kinetic energy.

Flow Patterns with Separation

In technical applications, flow with separation is common and often accepted if it is too expensive to avoid. Flow separation may be geometric or dynamic. Dynamic separation is shown in Figure 6. Geometric separation (Figures 7 and 8) results when a fluid stream passes over a very sharp corner, as with an orifice; the fluid generally leaves the corner irrespective of how much its velocity has been reduced by friction.

For geometric separation in orifice flow (Figure 7), the outer streamlines separate from the sharp corners and, because of fluid inertia, contract to a section smaller than the orifice opening, the **vena contracta**, with a limiting area of about six-tenths of the orifice opening. After the vena contracta, the fluid stream expands rather slowly through turbulent or laminar interaction with the fluid along its sides. Outside the jet, fluid velocity is small compared to that in the jet. Turbulence helps spread out the jet, increases the losses, and brings the velocity distribution back to a more uniform profile. Finally, at a considerable distance downstream, the velocity profile returns to the fully developed flow of Figure 3.

Other geometric separations (Figure 8) occur at a sharp entrance to a conduit, at an inclined plate or damper in a conduit, and at a sudden expansion. For these, a vena contracta can be identified; for sudden expansion, its area is that of the upstream contraction. Ideal-fluid theory, using free streamlines, provides insight and predicts contraction coefficients for valves, orifices, and vanes (Robertson 1965). These geometric flow separations are large loss-producing devices. To expand a flow efficiently or to have an entrance with

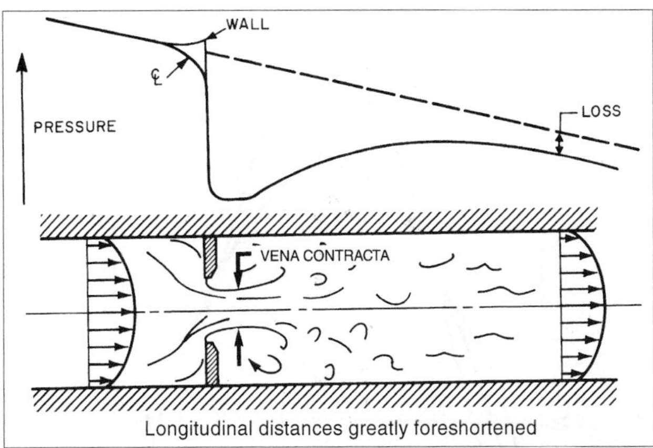

Fig. 7 Geometric Separation, Flow Development, and Loss in Flow Through Orifice

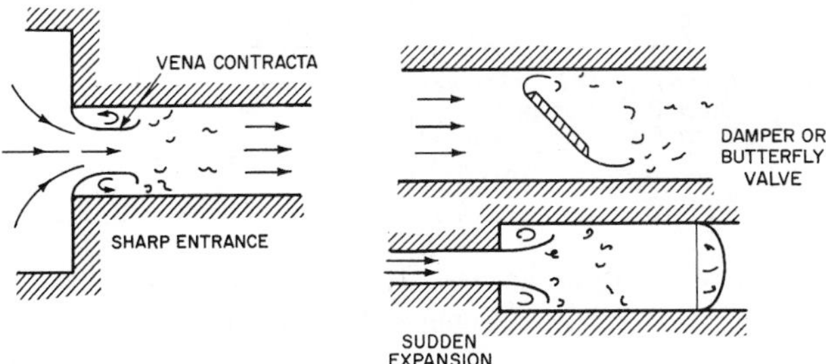

Fig. 8 Examples of Geometric Separation Encountered in Flows in Conduits

minimum losses, the device should be designed with gradual contours, a diffuser, or a rounded entrance.

Flow devices with gradual contours are subject to separation that is more difficult to predict, because it involves the dynamics of boundary layer growth under an adverse pressure gradient rather than flow over a sharp corner. In a diffuser, which is used to reduce the loss in expansion, it is possible to expand the fluid some distance at a gentle angle without difficulty (particularly if the boundary layer is turbulent). Eventually, separation may occur (Figure 9), which is frequently asymmetrical because of irregularities. Downstream flow involves flow reversal (backflow) and excess losses exist. Such separation is termed **stall** (Kline 1959). Larger area expansions may use splitters that divide the diffuser into smaller divisions less likely to have separations (Moore and Kline 1958). Another technique for controlling separation is to bleed some low-velocity fluid near the wall (Furuya et al. 1976). Alternatively, Heskested (1965, 1970) shows that suction at the corner of a sudden expansion has a strong positive effect on geometric separation.

Drag Forces on Bodies or Struts

Bodies in moving fluid streams are subjected to appreciable fluid forces or drag. Conventionally expressed in coefficient form, drag forces on bodies can be expressed as

$$D = C_D \rho A V^2/2 \tag{9}$$

where A is the projected (normal to flow) area of the body. The **drag coefficient** C_D depends on the body's shape and angularity and on the Reynolds number of the relative flow in terms of the body's characteristic dimension.

For Reynolds numbers of 10^3 to above 10^5, the C_D of most bodies is constant due to flow separation, but above 10^5, the C_D of

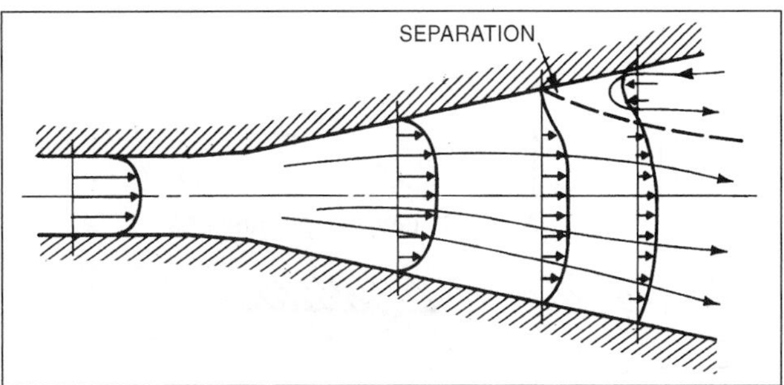

Fig. 9 Separation in Flow in Diffuser

Table 1 Drag Coefficients

Body Shape	$10^3 < \text{Re} < 2 \times 10^5$	$\text{Re} > 3 \times 10^5$
Sphere	0.36 to 0.47	~0.1
Disk	1.12	1.12
Streamlined strut	0.1 to 0.3	< 0.1
Circular cylinder	1.0 to 1.1	0.35
Elongated rectangular strut	1.0 to 1.2	1.0 to 1.2
Square strut	~2.0	~2.0

rounded bodies drops suddenly as the surface boundary layer undergoes transition to turbulence. Typical C_D values are given in Table 1; Hoerner (1965) gives expanded values.

For a strut crossing a conduit, the contribution to the head loss of Equation (6b) is

$$H_L = C_D \left(\frac{A}{A_c}\right)\left(\frac{V^2}{2g}\right) \tag{10}$$

where

A_c = conduit cross-sectional area
A = area of the strut facing the flow

Cavitation

Liquid flow with gas- or vapor-filled pockets can occur if the absolute pressure is reduced to vapor pressure or less. In this case, a cavity or series of cavities forms, because liquids are rarely pure enough to withstand any tensile stressing or pressures less than vapor pressure for any length of time (John and Haberman 1980, Knapp et al. 1970, Robertson and Wislicenus 1969). Robertson and Wislicenus (1969) indicate significant occurrences in various technical fields, chiefly in hydraulic equipment and turbomachines.

Initial evidence of cavitation is the collapse noise of many small bubbles that appear initially as they are carried by the flow into regions of higher pressure. The noise is not deleterious and serves as a warning of the occurrence. As flow velocity further increases or pressure decreases, the severity of cavitation increases. More bubbles appear and may join to form large fixed cavities. The space they occupy becomes large enough to modify the flow pattern and alter performance of the flow device. Collapse of the cavities on or near solid boundaries becomes so frequent that the cumulative impact in time results in damage in the form of cavitational erosion of the surface or excessive vibration. As a result, pumps can lose efficiency or their parts may erode locally. Control valves may be noisy or seriously damaged by cavitation.

Cavitation in orifice and valve flow is indicated in Figure 10. With high upstream pressure and a low flow rate, no cavitation occurs. As pressure is reduced or flow rate increased, the minimum pressure in the flow (in the shear layer leaving the edge of the orifice) eventually approaches vapor pressure. Turbulence in this layer causes fluctuating pressures below the mean (as in vortex cores) and small bubble-like cavities. These are carried downstream into the

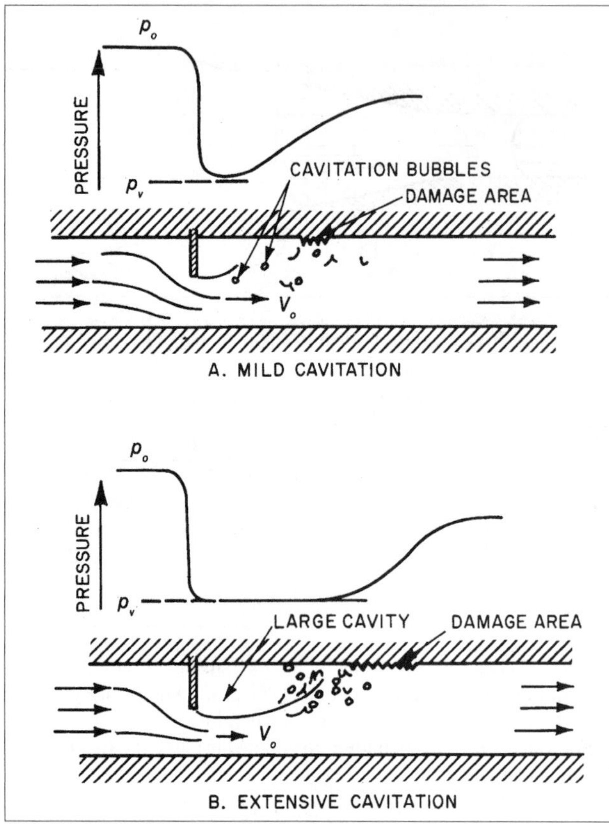

Fig. 10 Cavitation in Flows in Orifice or Valve

region of pressure regain where they collapse, either in the fluid or on the wall (Figure 10A). As the pressure is reduced, more vapor- or gas-filled bubbles result and coalesce into larger ones. Eventually, a single large cavity results that collapses further downstream (Figure 10B). The region of wall damage is then as many as 20 diameters downstream from the valve or orifice plate.

Sensitivity of a device to cavitation occurrence is measured by the **cavitation index** or **cavitation number**, which is the ratio of the available pressure above vapor pressure to the dynamic pressure of the reference flow:

$$\sigma = \frac{2(p_o - p_v)}{\rho V_o^2} \tag{11}$$

where p_v is the vapor pressure, and the subscript o refers to appropriate reference conditions. Valve analyses use such an index in order to determine when cavitation will affect the discharge coefficient (Ball 1957).With flow-metering devices such as orifices, venturis, and flow nozzles, there is little cavitation, because it occurs mostly downstream of the flow regions involved in establishing the metering action.

The detrimental effects of cavitation can be avoided by operating the liquid-flow device at high enough pressures. When this is not possible, the flow must be changed or the device must be built to withstand cavitation effects. Some materials or surface coatings are more resistant to cavitation erosion than others, but none is immune. Surface contours can be designed to delay the onset of cavitation.

Nonisothermal Effects

When appreciable temperature variations exist, the primary fluid properties (density and viscosity) are no longer constant, as usually assumed, but vary across or along the flow. The Bernoulli equation in the form of Equations (5a) through (5c) must be used, because volumetric flow is not constant. With gas flows, the thermodynamic process involved must be considered. In general, this is assessed in applying Equation (5a), written in the following form:

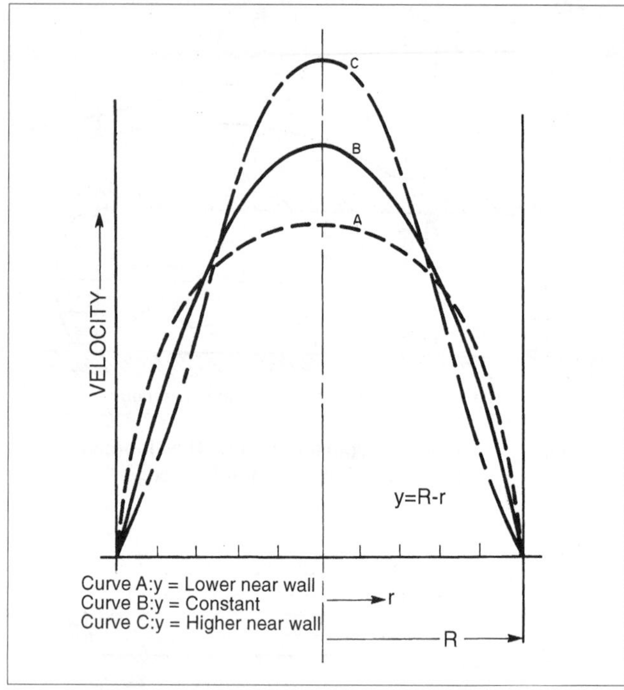

Fig. 11 Effect of Viscosity Variation on Velocity Profile of Laminar Flow in Pipe

$$\int \frac{dp}{\rho} + \frac{V^2}{2} + gz = B \tag{12}$$

Effects of viscosity variations also appear. With nonisothermal laminar flow, the parabolic velocity profile (Figure 3) is no longer valid. For gases, viscosity increases as the square root of absolute temperature, and for liquids, it decreases with increasing temperature. This results in opposite effects.

For fully developed pipe flow, the linear variation in shear stress from the wall value τ_w to zero at the centerline is independent of the temperature gradient. In the section on Laminar Flow, τ is defined as $\tau = (y/b)\tau_w$, where y is the distance from the centerline and $2b$ is the wall spacing. For pipe radius $R = D/2$ and distance from the wall $y = R - r$ (see Figure 11), then $\tau = \tau_w\ (R - y)/R$. Then, solving Equation (1) for the change in velocity gives

$$dv = \left[\frac{\tau_w(R-y)}{R\mu}\right] dy = -\left(\frac{\tau_w}{R\mu}\right) r\, dr \tag{13}$$

When the fluid has a lower viscosity near the wall than at the center (due to external heating of liquid or cooling of gas via heat transfer through the pipe wall), the velocity gradient is steeper near the wall and flatter near the center, so the profile is generally flattened. When liquid is cooled or gas is heated, the velocity profile becomes more pointed for laminar flow (Figure 11). Calculations were made for such flows of gases and liquid metals in pipes (Deissler 1951). Occurrences in turbulent flow are less apparent. If enough heating is applied to gaseous flows, the viscosity increase can cause reversion to laminar flow.

Buoyancy effects and gradual approach of the fluid temperature to equilibrium with that outside the pipe can cause considerable variation in the velocity profile along the conduit. Thus, Colborne and Drobitch (1966) found the pipe factor for upward vertical flow of hot air at a Reynolds number less than 2000 reduced to about 0.6 at 40 diameters from the entrance, then increased to about 0.8 at 210 diameters, and finally decreased to the isothermal value of 0.5 at the end of 320 diameters.

Compressibility

All fluids are compressible to some degree; their density depends on the pressure. Steady liquid flow may ordinarily be treated as incompressible, and incompressible flow analysis is satisfactory for gases and vapors at velocities below about 4000 to 8000 fpm, except in long conduits.

For liquids in pipelines, if flow is suddenly stopped, a severe pressure surge or water hammer is produced that travels along the pipe at the speed of sound in the liquid. This pressure surge alternately compresses and decompresses the liquid. For steady gas flows in long conduits, a decrease in pressure along the conduit can reduce the density of the gas significantly enough to cause the velocity to increase. If the conduit is long enough, velocities approaching the speed of sound are possible at the discharge end, and the Mach number (the ratio of the flow velocity to the speed of sound) must be considered.

Some compressible flows occur without heat gain or loss (adiabatically). If there is no friction (conversion of flow mechanical energy into internal energy), the process is reversible as well. Such a reversible adiabatic process is called isentropic, and follows the relationship

$$p/\rho^k = \text{constant}$$
$$k = c_p/c_v$$

where k, the ratio of specific heats at constant pressure and volume, has a value of 1.4 for air and diatomic gases.

The Bernoulli equation of steady flow, Equation (12), as an integral of the ideal-fluid equation of motion along a streamline, then becomes

$$\int \frac{dp}{\rho} + \frac{V^2}{2} = \text{constant} \tag{14}$$

where, as in most compressible flow analyses, the elevation terms involving z are insignificant and are dropped.

For a frictionless adiabatic process, the pressure term has the form

$$\int_1^2 \frac{dp}{\rho} = \frac{k}{k-1}\left(\frac{p_2}{\rho_2} - \frac{p_1}{\rho_1}\right) \tag{15}$$

Then, between stations 1 and 2 for the isentropic process,

$$\frac{p_1}{\rho_1}\left(\frac{k}{k-1}\right)\left[\left(\frac{p_2}{p_1}\right)^{(k-1)/k} - 1\right] + \frac{V_2^2 - V_1^2}{2} = 0 \tag{16}$$

Equation (16) replaces the Bernoulli equation for compressible flows and may be applied to the stagnation point at the front of a body. With this point as station 2 and the upstream reference flow ahead of the influence of the body as station 1, $V_2 = 0$. Solving Equation (16) for p_2 gives

$$p_s = p_2 = p_1\left[1 + \left(\frac{k-1}{2}\right)\frac{\rho_1 V_1^2}{kp_1}\right]^{k/(k-1)} \tag{17}$$

where p_s is the stagnation pressure.

Because kp/ρ is the square of the acoustic velocity a and the Mach number $M = V/a$, the stagnation pressure relation becomes

$$p_s = p_1\left[1 + \left(\frac{k-1}{2}\right)M_1^2\right]^{k/(k-1)} \tag{18}$$

For Mach numbers less than one,

$$p_s = p_1 + \frac{\rho_1 V_1^2}{2}\left[1 + \frac{M_1}{4} + \left(\frac{2-k}{24}\right)M_1^4 + \dots\right] \tag{19}$$

When $M = 0$, Equation (19) reduces to the incompressible flow result obtained from Equation (5a). Appreciable differences appear when the Mach number of the approaching flow exceeds 0.2. Thus a pitot tube in air is influenced by compressibility at velocities over about 13,000 fpm.

Flows through a converging conduit, as in a flow nozzle, venturi, or orifice meter, also may be considered isentropic. Velocity at the upstream station 1 is negligible. From Equation (16), velocity at the downstream station is

$$V_2 = \sqrt{\frac{2k}{k-1}\left(\frac{p_1}{\rho_1}\right)\left[1 - \left(\frac{p_2}{p_1}\right)^{(k-1)/k}\right]} \tag{20}$$

The mass flow rate is

$$\dot{m} = V_2 A_2 \rho_2 =$$
$$A_2\sqrt{\frac{2k}{k-1}(p_1\rho_1)\left[\left(\frac{p_2}{p_1}\right)^{2/k} - \left(\frac{p_2}{p_1}\right)^{(k+1)/k}\right]} \tag{21}$$

The corresponding incompressible flow relation is

$$\dot{m}_{in} = A_2\rho\sqrt{2\Delta p/\rho} = A_2\sqrt{2\rho(p_1 - p_2)} \tag{22}$$

The compressibility effect is often accounted for in the **expansion factor** Y:

$$\dot{m} = Y\dot{m}_{in} = A_2 Y\sqrt{2\rho(p_1 - p_2)} \tag{23}$$

Y is 1.00 for the incompressible case. For air ($k = 1.4$), a Y value of 0.95 is reached with orifices at $p_2/p_1 = 0.83$ and with venturis at about 0.90, when these devices are of relatively small diameter (D_2/D_1 less than 0.5).

As p_2/p_1 decreases, the flow rate increases, but more slowly than for the incompressible case because of the nearly linear decrease in Y. However, the downstream velocity reaches the local acoustic value and the discharge levels off at a value fixed by upstream pressure and density at the critical ratio:

$$\left.\frac{p_2}{p_1}\right|_c = \left(\frac{2}{k+1}\right)^{k/(k-1)} = 0.53 \text{ for air} \tag{24}$$

At higher pressure ratios than critical, **choking** (no increase in flow with decrease in downstream pressure) occurs and is used in some flow control devices to avoid flow dependence on downstream conditions.

FLOW ANALYSIS

Fluid flow analysis is used to correlate pressure changes with flow rates and the nature of the conduit. For a given pipeline, either the pressure drop for a certain flow rate, or the flow rate for a certain pressure difference between the ends of the conduit, is needed. Flow analysis ultimately involves comparing a pump or blower to a conduit piping system for evaluating the expected flow rate.

Generalized Bernoulli Equation

Internal energy differences are generally small and usually the only significant effect of heat transfer is to change the density ρ. For gas or vapor flows, use the generalized Bernoulli equation in the pressure-over-density form of Equation (6a), allowing for the thermodynamic process in the pressure-density relation:

$$\int_1^2 \frac{dp}{\rho} + \alpha_1 \frac{V_1^2}{2} + E_M = \alpha_2 \frac{V_2^2}{2} + E_L \quad (25a)$$

The elevation changes involving z are often negligible and are dropped. The pressure form of Equation (5b) is generally unacceptable when appreciable density variations occur, because the volumetric flow rate differs at the two stations. This is particularly serious in friction-loss evaluations where the density usually varies over considerable lengths of conduit (Benedict and Carlucci 1966). When the flow is essentially incompressible, Equation (25a) is satisfactory.

Example 1. Specify the blower to produce an isothermal airflow of 400 cfm through a ducting system (Figure 12). Accounting for intake and fitting losses, the equivalent conduit lengths are 60 and 165 ft and the flow is isothermal. The head at the inlet (station 1) and following the discharge (station 4), where the velocity is zero, is the same. The frictional losses H_L are evaluated as 24.5 ft of air between stations 1 and 2, and 237 ft between stations 3 and 4.

Solution: The head form of the generalized Bernoulli relation is used in place of Equation (25a), which also could be used:

$$(p_1/\rho_1 g) + \alpha_1(V_1^2/2g) + z_1 + H_M$$
$$= (p_2/\rho_2 g) + \alpha_2(V_2^2/2g) + z_2 + H_L \quad (25b)$$

The term $V_1^2/2g$ can be calculated as follows:

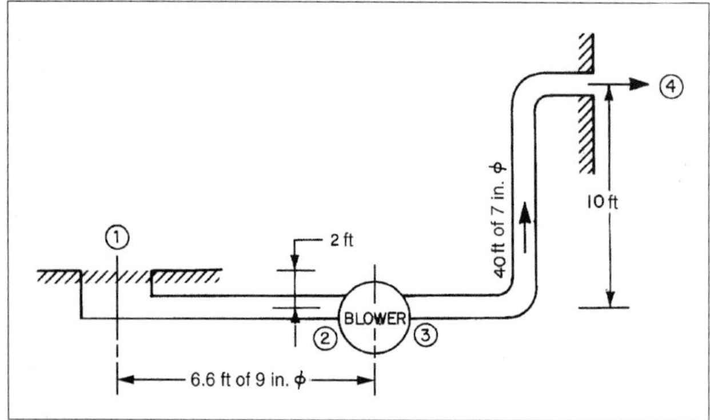

Fig. 12 Blower and Duct System for Example 1

$$A_1 = \pi\left(\frac{D}{2}\right)^2 = \pi\left(\frac{9/12}{2}\right)^2 = 0.44 \text{ ft}^2$$

$$V_1 = Q/A_1 = \left(\frac{400 \text{ ft}^3}{\text{min}}\right)\left(\frac{1 \text{ min}}{60 \text{s}}\right)/0.44 \text{ ft}^2 = 15.1 \text{ ft/s}$$

$$V_1^2/2g = (15.1)^2/2(32) = 3.56 \text{ ft}$$

The term $V_2^2/2g$ can be calculated in a similar manner.

In Equation (25b), H_M is evaluated by applying the relation between any two points on opposite sides of the blower. Because conditions at stations 1 and 4 are known, they are used, and the location-specifying subscripts on the right side of Equation (25b) are changed to 4. Note that $p_1 = p_4 = p$, $\rho_1 = \rho_4 = \rho$, and $V_1 = V_4 = 0$. Thus,

$$(p/\rho g) + 0 + 2 + H_M = (p/\rho g) + 0 + 10 + (24.5 + 237)$$

so H_M = 269.5 ft of air. For standard air, this corresponds to 3.89 in. of water.

The pressure (or head) difference measured across the blower (between stations 2 and 3), is often taken as the head H_M. It can be obtained by calculating the static pressure at stations 2 and 3. Applying Equation (25b) successively between stations 1 and 2 and between 3 and 4 gives

$$(p_1/\rho g) + 0 + 2 + 0 = (p_2/\rho g) + (1.06 \times 3.56) + 0 + 24.5$$
$$(p_3/\rho g) + (1.03 \times 9.70) + 0 + 0 = (p_4/\rho g) + 0 + 10 + 237$$

where α just ahead of the blower is taken as 1.06, and just after the blower as 1.03; the latter value is uncertain because of possible uneven discharge from the blower. Static pressures p_1 and p_4 may be taken as zero gage. Thus,

$$p_2/\rho g = -26.2 \text{ ft of air}$$
$$p_3/\rho g = 237 \text{ ft of air}$$

The static head across the blower, the difference between these two numbers, is 263.2 ft, which is *not* the H_M calculated after Equation (25b) as 269.5 ft. The apparent discrepancy results from ignoring the velocity heads at stations 2 and 3. Actually, H_M is the change in total head (pressure head plus true kinetic energy) across the machine:

$$H_M = (p_3/\rho g) + \alpha_3(V_3^2/2g) - [(p_2/\rho g) + \alpha_2(V_2^2/2g)]$$
$$= 237 + (1.03 \times 9.70) - [-26.2 + (1.06 \times 3.54)]$$
$$= 247 - (-22.5) = 269.5 \text{ ft of air}$$

The required blower head is the same, no matter how it is evaluated. It is the specific energy added to the system by the machine. Only when the conduit size and velocity profiles on both sides of the machine are the same is E_M or H_M simply found from $\Delta p = p_3 - p_2$.

Conduit Friction

The loss term E_L or H_L of Equation (6a) or (6b) accounts for friction caused by conduit-wall shearing stresses and losses from conduit-section changes. H_L is the loss of energy (ft·lb$_f$) per unit weight (lb$_f$) of flowing fluid.

In real-fluid flow, a frictional shear occurs at bounding walls, gradually influencing the flow further away from the boundary. A lateral velocity profile is produced and flow energy is converted into heat (fluid internal energy), which is generally unrecoverable (a loss). This loss in fully developed conduit flow is evaluated through the **Darcy-Weisbach equation**:

$$(H_L)_f = f\left(\frac{L}{D}\right)\left(\frac{V^2}{2g}\right) \quad (26)$$

where L is the length of conduit of diameter D and f is the **friction factor**. Sometimes a numerically different relation is used with the **Fanning friction factor** (one-quarter of f). The value of f is nearly constant for turbulent flow, varying only from about 0.01 to 0.05.

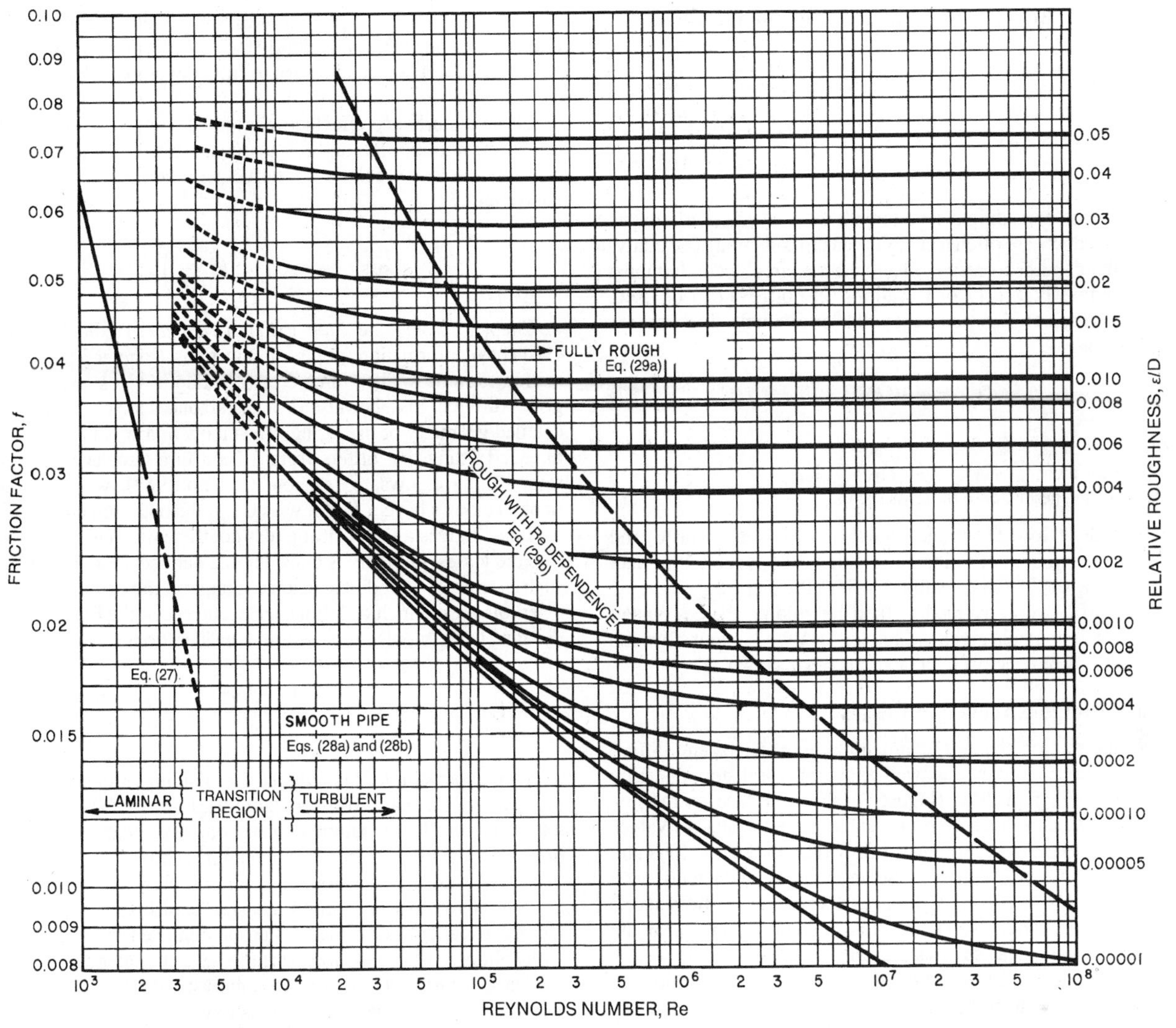

Fig. 13 Relation Between Friction Factor and Reynolds Number
(Moody 1944)

For fully developed laminar-viscous flow in a pipe, the loss is evaluated from Equation (8) as follows:

$$(H_L)_f = \frac{L}{\rho g}\left(\frac{8\mu V}{R^2}\right) = \frac{32L\nu V}{D^2 g} = \frac{64}{VD/\nu}\left(\frac{L}{D}\right)\left(\frac{V^2}{2g}\right) \qquad (27)$$

where $\mathrm{Re} = VD/\nu$ and $f = 64/\mathrm{Re}$. Thus, for laminar flow, the friction factor varies inversely with the Reynolds number.

With turbulent flow, friction loss depends not only on flow conditions, as characterized by the Reynolds number, but also on the nature of the conduit wall surface. For smooth conduit walls, empirical correlations give

$$f = \frac{0.3164}{\mathrm{Re}^{0.25}} \qquad \text{for Re} < 10^5 \qquad (28a)$$

$$f = 0.0032 + \frac{0.221}{\mathrm{Re}^{0.237}} \qquad \text{for } 10^5 < \mathrm{Re} < 3 \times 10^6 \qquad (28b)$$

Generally, f also depends on the wall roughness ε. The variation is complex and best expressed in chart form (Moody 1944) as shown in Figure 13. Inspection indicates that, for high Reynolds numbers and relative roughness, the friction factor becomes independent of the Reynolds number in a **fully-rough flow regime**. Then

$$\frac{1}{\sqrt{f}} = 1.14 + 2\log(D/\varepsilon) \qquad (29a)$$

Values of f between the values for smooth tubes and those for the fully-rough regime are represented by **Colebrook's natural roughness function**:

$$\frac{1}{\sqrt{f}} = 1.14 + 2\log(D/\varepsilon) - 2\log\left[1 + \frac{9.3}{\mathrm{Re}(\varepsilon/D)\sqrt{f}}\right] \qquad (29b)$$

A transition region appears in Figure 13 for Reynolds numbers between 2000 and 10,000. Below this critical condition, for smooth walls, Equation (27) is used to determine f; above the critical condition, Equation (28b) is used. For rough walls, Figure 13 or Equation (29b) must be used to assess the friction factor in turbulent flow. To do this, the roughness height ε, which may increase with conduit use or aging, must be evaluated from the conduit surface (Table 2).

Table 2 Effective Roughness of Conduit Surfaces

Material	ε, ft
Commercially smooth brass, lead, copper, or plastic pipe	0.000005
Steel and wrought iron	0.00015
Galvanized iron or steel	0.0005
Cast iron	0.00085

Although the preceding discussion has focused on circular pipes and ducts, air ducts are often rectangular in cross section. The equivalent circular conduit corresponding to the noncircular conduit must be found before Figure 13 or Equations (28) or (29) can be used. Based on turbulent flow concepts, the **equivalent diameter** is determined by

$$D_{eq} = 4A/P_w \tag{30}$$

where

A = flow area
P_w = wetted perimeter of the cross section

For turbulent flow, D_{eq} is substituted for D in Equation (26) and the Reynolds number definition in Equation (27). Noncircular duct friction can be evaluated to within 5% for all except very extreme cross sections. A more refined method for finding the equivalent circular duct diameter is given in Chapter 34. With laminar flow, the loss predictions may be off by a factor as large as two.

Section Change Effects and Losses

Valve and section changes (contractions, expansions and diffusers, elbows or bends, tees), as well as entrances, distort the fully developed velocity profiles (Figure 3) and introduce extra flow losses (dissipated as heat) into pipelines or duct systems. Valves produce such extra losses to control flow rate. In contractions and expansions, flow separation as shown in Figures 8 and 9 causes the extra loss. The loss at rounded entrances develops as the flow accelerates to higher velocities. The resulting higher velocity near the wall leads to wall shear stresses greater than those of fully developed flow (Figure 5). In flow around bends, the velocity increases along the inner wall near the start of the bend. This increased velocity creates a secondary motion, which is a double helical vortex pattern of flow downstream from the bend. In all these devices, the disturbance produced locally is converted into turbulence and appears as a loss in the downstream region.

The return of disturbed flow to a fully developed velocity profile is quite slow. Ito (1962) showed that the secondary motion following a bend takes up to 100 diameters of conduit to die out but the pressure gradient settles out after 50 diameters.

With laminar flow following a rounded entrance, the entrance length depends on the Reynolds number:

$$L_e/D \approx 0.06\ \mathrm{Re} \tag{31}$$

At Re = 2000, a length of 120 diameters is needed to establish the parabolic profile. The pressure gradient reaches the developed value of Equation (26) much sooner. The extra head drop is 1.2 velocity head; the change in profile from uniform to parabolic results in a drop of 1.0 velocity head (since $\alpha = 2.0$), and the rest is due to excess friction. With turbulent flow, 80 to 100 diameters following the rounded entrance are needed for the velocity profile to become fully developed, but the friction loss per unit length reaches a value close to that of the fully developed flow value more quickly. After six diameters, the loss rate at a Reynolds number of 10^5 is only 14% above that of fully developed flow in the same length, while at 10^7, it is only 10% higher (Robertson 1963). For a sharp entrance, the flow separation (Figure 8) causes a greater disturbance, but fully developed flow is achieved in about half the length required for a rounded entrance. With sudden expansion, the pressure change settles out in about eight times the diameter change ($D_2 = D_1$), while the velocity profile takes at least a 50% greater distance to return to fully developed pipe flow (Lipstein 1962).

These disturbance effects are assumed compressed (in the flow direction) into a point, and the losses are treated as locally occurring. Such a loss is related to the velocity head by the **fitting loss coefficient** K:

$$\text{Loss of section} = K\left(\frac{V^2}{2g}\right) \tag{32}$$

Chapter 35 and the *Pipe Friction Manual* (Hydraulic Institute 1961) have information for pipe applications. Chapter 34 gives information for airflow. The same type of fitting in pipes and ducts may give a different loss, because flow disturbances are controlled by the detailed geometry of the fitting. The elbow of a small pipe may be a threaded fitting that differs from a bend in a circular duct. For 90 screw-fitting elbows, K is about 0.8 (Ito 1962), whereas smooth flanged elbows have a K as low as 0.2 at the optimum curvature.

Table 3 gives a list of fitting loss coefficients. These values indicate the losses, but there is considerable variance. Expansion flows, such as from one conduit size to another or at the exit into a room or reservoir, are not included. For such occurrences, the **Borda loss prediction** (from impulse-momentum considerations) is appropriate:

$$\text{Loss at expansion} = \frac{(V_1 - V_2)^2}{2g} = \frac{V_1^2}{2g}\left(1 - \frac{A_1}{A_2}\right)^2 \tag{33}$$

Such expansion loss is reduced by avoiding or delaying separation using a gradual diffuser (Figure 9). For a diffuser of about 7° total angle, the loss is minimal, about one-sixth that given by Equation (33). The diffuser loss for total angles above 45 to 60° exceeds that of the sudden expansion, depending somewhat on the diameter ratio of the expansion. Optimum design of diffusers involves many factors; excellent performance can be achieved in short diffusers with splitter vanes or suction. Turning vanes in miter bends produce the least disturbance and loss for elbows; with careful design, the loss coefficient can be reduced to as low as 0.1.

For losses in smooth elbows, Ito (1962) found a Reynolds number effect (K slowly decreasing with increasing Re) and a minimum loss at a bend curvature (bend radius to diameter ratio) of 2.5. At this optimum curvature, a 45° turn had 63%, and a 180° turn approximately 120%, of the loss of a 90° bend. The loss does not vary linearly with the turning angle because secondary motion occurs.

Use of coefficient K presumes its independence of the Reynolds number. Crane Co. (1976) found a variation with the Reynolds number similar to that of the friction factor; Kittridge and Rowley (1957) observed it only with laminar flow. Assuming that K varies with Re similarly to f, it is convenient to represent fitting losses as adding to the effective length of uniform conduit. The **effective length** of a fitting is then

$$L_{eff}/D = K/f_{ref} \tag{34}$$

where f_{ref} is an appropriate reference value of the friction factor. Deissler (1951) uses 0.028, and the air duct values in Chapter 34 are based on an f_{ref} of about 0.02. For rough conduits, appreciable errors can occur if the relative roughness does not correspond to that used when f_{ref} was fixed. It is unlikely that the fitting losses involving separation are affected by pipe roughness. The effective length method for fitting loss evaluation is still useful.

When a conduit contains a number of section changes or fittings, the values of K are added to the fL/D friction loss, or the L_{eff}/D of the fittings are added to the conduit length L/D for evaluating the total loss H_L. This assumes that each fitting loss is fully developed and its disturbance fully smoothed out before the next section change. Such an assumption is frequently wrong, and the total loss can be overestimated. For elbow flows, the total loss of adjacent bends may be over or underestimated. The secondary flow pattern following an elbow is such that when one follows another, perhaps in a different plane, the secondary flow of the second elbow may reinforce or partially cancel that of the first. Moving the second elbow a few diameters can reduce the total loss (from more than twice the amount) to less than the loss from one elbow. Screens or perforated plates can be used for smoothing velocity profiles (Wile 1947) and flow spreading. Their effectiveness and loss coefficients depend on their amount of open area (Baines and Peterson 1951).

Compressible Conduit Flow

When friction loss is included, as it must be except for a very short conduit, the incompressible flow analysis previously considered applies until the pressure drop exceeds about 10% of the initial pressure. The possibility of sonic velocities at the end of relatively long conduits limits the amount of pressure reduction achieved. For an inlet Mach number of 0.2, the discharge pressure can be reduced to about 0.2 of the initial pressure; for an inflow at $M = 0.5$, the discharge pressure cannot be less than about $0.45p_1$ in the adiabatic case and about $0.6p_1$ in isothermal flow.

Analysis of such conduit flow must treat density change, as evaluated from the continuity relation in Equation (2), with the frictional occurrences evaluated from wall roughness and Reynolds number correlations of incompressible flow (Binder 1944). In evaluating valve and fitting losses, consider the reduction in K caused by compressibility (Benedict and Carlucci 1966). Although the analysis differs significantly, isothermal and adiabatic flows involve essentially the same pressure variation along the conduit, up to the limiting conditions.

Control Valve Characterization

Control valves are characterized by a discharge coefficient C_d. As long as the Reynolds number is greater than 250, the orifice equation holds for liquids:

$$Q = C_d A_o \sqrt{2\Delta P/\rho} \qquad (35)$$

where

A_o = area of orifice opening
P = absolute pressure

The discharge coefficient is about 0.63 for sharp-edged configurations and 0.8 to 0.9 for chamfered or rounded configurations. For gas flows at pressure ratios below the choking critical [Equation (24)], the mass rate of flow is

$$\dot{m} = C_d A_o C_1 \left(\frac{P_u}{\sqrt{T_u}}\right) \sqrt{\frac{P_d}{P_u}} \sqrt{1 - \left(\frac{P_d}{P_u}\right)^{(k-1)/k}} \qquad (36)$$

where

$C_1 = \sqrt{2k/R(k-1)}$
k = ratio of specific heats at constant pressure and volume
R = gas constant
T = absolute temperature
u, d = subscripts referring to upstream and downstream positions

Incompressible Flow in Systems

Flow devices must be evaluated in terms of their interaction with other elements of the system, for example, the action of valves in modifying flow rate and in matching the flow-producing device (pump or blower) with the system loss. Analysis is via the general Bernoulli equation and the loss evaluations noted previously.

A valve regulates or stops the flow of fluid by throttling. The change in flow is not proportional to the change in area of the valve opening. Figures 14 and 15 indicate the nonlinear action of valves in controlling flow. Figure 14 shows a flow in a pipe discharging water from a tank that is controlled by a gate valve. The fitting loss coefficient K values are those of Table 3; the friction factor f is 0.027. The degree of control also depends on the conduit L/D ratio. For a relatively long conduit, the valve must be nearly closed before its high K value becomes a significant portion of the loss. Figure 15 shows a control damper (essentially a butterfly valve) in a duct discharging air from a plenum held at constant pressure. With a long duct, the damper does not affect the flow rate until it is about one-quarter closed. Duct length has little effect when the damper is more than half closed. The damper closes the duct totally at the 90° position ($K = \infty$).

Flow in a system (pump or blower and conduit with fittings) involves interaction between the characteristics of the flow-producing device (pump or blower) and the loss characteristics of the pipeline or duct system. Often the devices are centrifugal, in which case the head produced decreases as the flow increases, except for the lowest flow rates. System head required to overcome losses increases roughly as the square of the flow rate. The flow rate of a given system is that for which the two heads match, where the two curves of head versus flow rate intersect (point 1 in Figure 16).

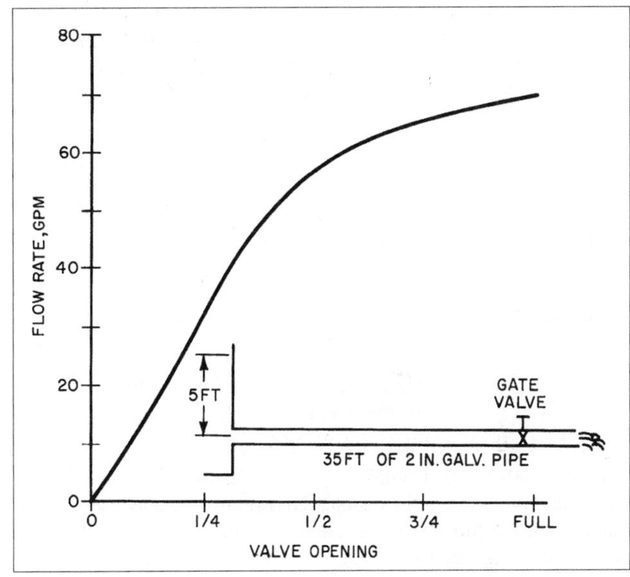

Fig. 14 Valve Action in Pipeline

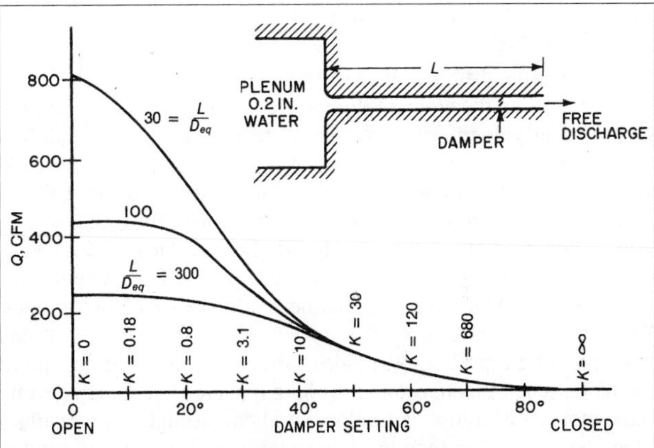

Fig. 15 Effect of Duct Length on Damper Action

Table 3 Fitting Loss Coefficients of Turbulent Flow

Fitting	Geometry	$K = \frac{\Delta p/\rho g}{V^2/2g}$
Entrance	Sharp	0.50
	Well-rounded	0.05
Contraction	Sharp ($D_2/D_1 = 0.5$)	0.38
90° Elbow	Miter	1.3
	Short radius	0.90
	Long radius	0.60
	Miter with turning vanes	0.2
Globe valve	Open	10
Angle valve	Open	5
Gate valve	Open	0.19 to 0.22
	75% open	1.10
	50% open	3.6
	25% open	28.8
Any valve	Closed	∞
Tee	Straight through flow	0.5
	Flow through branch	1.8

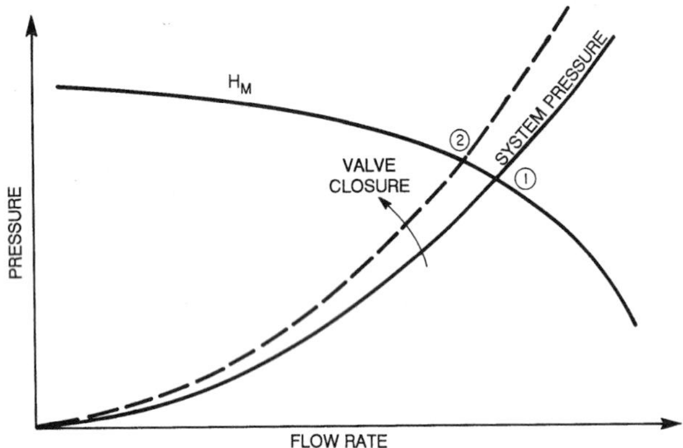

Fig. 16 Matching of Pump or Blower to System Characteristics

When a control valve (or damper) is partially closed, it increases the losses and reduces the flow (point 2 in Figure 16). For cases of constant head, the flow decrease due to valving is not as great as that indicated in Figures 14 and 15.

Flow Measurement

The general principles noted (the continuity and Bernoulli equations) are basic to most fluid-metering devices. Chapter 14 has further details.

The pressure difference between the stagnation point (total pressure) and that in the ambient fluid stream (static pressure) is used to give a point velocity measurement. The flow rate in a conduit is measured by placing a pitot device at various locations in the cross section and spatially integrating the velocity profile found. A single point measurement may be used for approximate flow rate evaluation. When the flow is fully developed, the pipe-factor information of Figure 4 can be used to estimate the flow rate from a centerline measurement. Measurements can be made in one of two modes. With the pitot-static tube, the ambient (static) pressure is found from pressure taps along the side of the forward-facing portion of the tube. When this portion is not long and slender, static pressure indication will be low and velocity indication high; as a result, a tube coefficient less than unity must be used. For parallel conduit flow, wall piezometers (taps) may take the ambient pressure, and the pitot tube indicates the impact (total pressure).

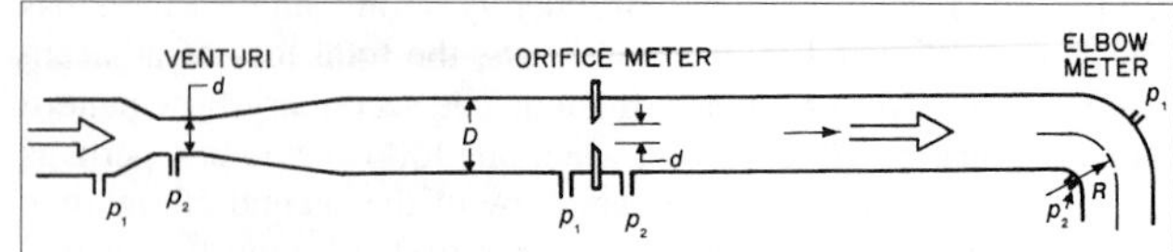

Fig. 17 Differential Pressure Flowmeters

The venturi meter, flow nozzle, and orifice meter are flow rate metering devices based on the pressure change associated with relatively sudden changes in conduit section area (Figure 17). The elbow meter (also shown in Figure 17) is another differential pressure flowmeter. The flow nozzle is similar to the venturi in action, but does not have the downstream diffuser. For all these, the flow rate is proportional to the square root of the pressure difference resulting from fluid flow. With the area change devices (venturi, flow nozzle, and orifice meter), a theoretical flow rate relation is found by applying the Bernoulli and continuity equations in Equations (6) and (2) between stations 1 and 2:

$$Q_{theor} = \frac{\pi d^2}{4}\sqrt{\frac{2g\Delta h}{1-\beta^4}} \tag{37}$$

where

$\Delta h = h_1 - h_2 = \Delta p/\rho g$ (h = static pressure head)
$\beta = d/D$ = ratio of throat (or orifice) diameter to conduit diameter

The actual flow rate through the device can differ because the approach flow kinetic energy factor α deviates from unity and because of small losses. More significantly, the jet contraction of orifice flow is neglected in deriving Equation (37), to the extent that it can reduce the effective flow area by a factor of 0.6. The effect of all these factors can be combined into the discharge coefficient C_d:

$$Q = C_d Q_{theor} = C_d\left(\frac{\pi d^2}{4}\right)\sqrt{\frac{2g\Delta p}{1-\beta^4}} \tag{38}$$

Sometimes an alternate coefficient is used of the form

$$\frac{C_d}{\sqrt{1-\beta^4}}$$

For compressible fluid metering, the expansion factor Y as described by Equation (23) must be included, and the mass flow rate is

$$\dot{m} = C_d Y \rho Q_{theor} = C_d Y\left(\frac{\pi d^2}{4}\right)\sqrt{\frac{2\rho\Delta p}{1-\beta^4}} \tag{39}$$

Values of Y depend primarily on the pressure ratio p_2/p_1, and also on the metering device and k value of the particular gas.

The general mode of variation in C_d for orifices and venturis is indicated in Figure 18 as a function of Reynolds number and, to a lesser extent, diameter ratio β. For Reynolds numbers less than 10, the coefficient varies as $\sqrt{\text{Re}}$.

The elbow meter employs the pressure difference between inside and outside the bend as the metering signal (Murdock et al. 1964). A momentum analysis gives the flow rate as

$$Q_{theor} = \frac{\pi D^2}{4}\sqrt{\frac{R}{2D}(2g\Delta h)} \tag{40}$$

where R is the radius of curvature of the bend. Again, a discharge coefficient C_d is needed; as in Figure 18, this drops off for the lower Reynolds numbers (below 10^5). These devices are calibrated in pipes

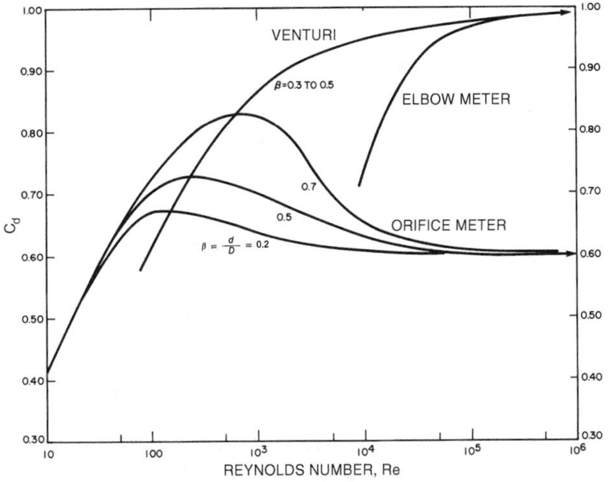

Fig. 18 Flowmeter Coefficients

with fully developed velocity profiles, so they must be located far enough downstream of sections that modify the approach velocity.

Unsteady Flow

Conduit flows are not always steady. In a compressible fluid, the acoustic velocity is usually high and the conduit length is rather short, so the time of signal travel is negligibly small. Even in the incompressible approximation, system response is not instantaneous. If a pressure difference Δp is applied between the conduit ends, the fluid mass must be accelerated and wall friction overcome, so a finite time passes before the steady flow rate corresponding to the pressure drop is achieved.

The time it takes for an incompressible fluid in a horizontal constant-area conduit of length L to achieve steady flow may be estimated by using the unsteady flow equation of motion with wall friction effects included. On the quasi-steady assumption, friction is given by Equation (26); also by continuity, V is constant along the conduit. The occurrences are characterized by the relation

$$\frac{dV}{d\theta} + \left(\frac{1}{\rho}\right)\frac{dp}{ds} + \frac{fV^2}{2D} = 0 \tag{41}$$

where

θ = time
s = distance in the flow direction

Since a certain Δp is applied over the conduit length L,

$$\frac{dV}{d\theta} = \frac{\Delta p}{\rho L} - \frac{fV^2}{2D} \tag{42}$$

For laminar flow, f is given by Equation (27), and

$$\frac{dV}{d\theta} = \frac{\Delta p}{\rho L} - \frac{32\mu V}{\rho D^2} = A - BV \tag{43}$$

Equation (43) can be rearranged and integrated to yield the time to reach a certain velocity:

$$\theta = \int d\theta = \int \frac{dV}{A - BV} = -\frac{1}{B}\ln(A - BV) \tag{44}$$

and

$$V = \frac{\Delta p}{L}\left(\frac{D^2}{32\mu}\right)\left[1 - \frac{\rho L}{\Delta p}\exp\left(\frac{-32\nu\theta}{D^2}\right)\right] \tag{45a}$$

For long times ($\theta \to \infty$), this indicates steady velocity as

$$V_\infty = \frac{\Delta p}{L}\left(\frac{D^2}{32\mu}\right) = \frac{\Delta p}{L}\left(\frac{R^2}{8\mu}\right) \tag{45b}$$

as by Equation (8). Then, Equation (45a) becomes

$$V = V_\infty\left[1 - \frac{\rho L}{\Delta p}\exp\left(\frac{-f_\infty V_\infty \theta}{2D}\right)\right] \tag{46}$$

where

$$f_\infty = \frac{64\nu}{V_\infty D}$$

The general nature of velocity development for starting-up flow is derived by more complex techniques; however, the temporal variation is as given above. For shutdown flow (steady flow with $\Delta p = 0$ at $\theta > 0$), the flow decays exponentially as $e^{-\theta}$.

Turbulent flow analysis of Equation (41) also must be based on the quasi-steady approximation, with less justification. Daily et al. (1956) indicate that the frictional resistance is slightly greater than the steady-state result for accelerating flows, but appreciably less for decelerating flows. If the friction factor is approximated as constant,

$$\frac{dV}{d\theta} = \frac{\Delta p}{\rho L} - \frac{fV^2}{2D} = A - BV^2$$

and, for the accelerating flow,

$$\theta = \frac{1}{\sqrt{AB}}\tanh^{-1}\left(V\sqrt{\frac{B}{A}}\right)$$

or

$$V = \sqrt{\frac{A}{B}}\tanh(\theta\sqrt{AB})$$

Because the hyperbolic tangent is zero when the independent variable is zero and unity when the variable is infinity, the initial ($V = 0$ at $\theta = 0$) and final conditions are verified. Thus, for long times ($\theta \to \infty$),

$$V_\infty = \sqrt{\frac{A}{B}} = \sqrt{\frac{\Delta p/\rho L}{f_\infty/2D}} = \sqrt{\frac{\Delta p}{\rho L}\left(\frac{2D}{f_\infty}\right)}$$

which is in accord with Equation (26) when f is constant (the flow regime is the fully rough one of Figure 13). The temporal velocity variation is then

$$V = V_\infty \tanh(f_\infty V_\infty \theta/2D) \tag{47}$$

In Figure 19, the turbulent velocity start-up result is compared with the laminar one in Figure 19, where initially the turbulent is steeper but of the same general form, increasing rapidly at the start but reaching V_∞ asymptotically.

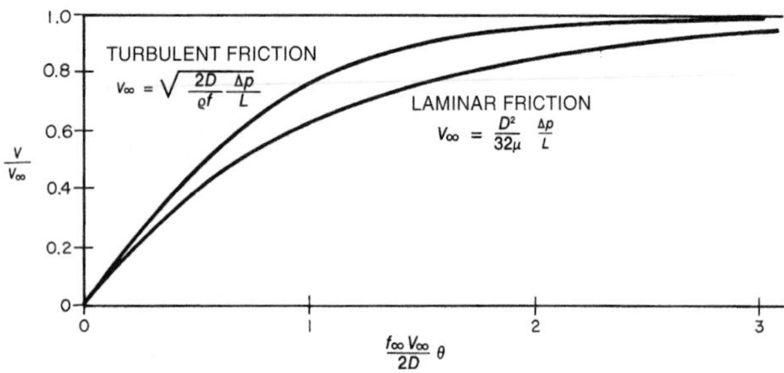

Fig. 19 Temporal Increase in Velocity Following Sudden Application of Pressure

NOISE FROM FLUID FLOW

Noise in flowing fluids results from unsteady flow fields and can be at discrete frequencies or broadly distributed over the audible range. With liquid flow, cavitation results in noise through the collapse of vapor bubbles. The noise in pumps or fittings (such as valves) can be a rattling or sharp hissing sound. It is easily eliminated by raising the system pressure. With severe cavitation, the resulting unsteady flow can produce indirect noise from induced vibration of adjacent parts. See Chapter 46 of the 1999 *ASHRAE Handbook—Applications* for more information on sound control.

The disturbed laminar flow behind cylinders can be an oscillating motion. The shedding frequency f of these vortexes is characterized by a Strouhal number $St = fd/V$ of about 0.21 for a circular cylinder of diameter d, over a considerable range of Reynolds numbers. This oscillating flow can be a powerful noise source, particularly when f is close to the natural frequency of the cylinder or some nearby structural member so that resonance occurs. With cylinders of another shape, such as impeller blades of a pump or blower, the characterizing Strouhal number involves the trailing edge thickness of the member. The strength of the vortex wake, with its resulting vibrations and noise potential, can be reduced by breaking up the flow with downstream splitter plates or boundary-layer trip devices (wires) on the cylinder surface.

Noise produced in pipes and ducts, especially from valves and fittings, is associated with the loss through such elements. The sound pressure of noise in water pipe flow increases linearly with the head loss; the broad-band noise increases, but only in the lower frequency range. Fitting-produced noise levels also increase with fitting loss (even without cavitation) and significantly exceed noise levels of the pipe flow. The relation between noise and loss is not surprising because both involve excessive flow perturbations. A valve's pressure-flow characteristics and structural elasticity may be such that for some operating point it oscillates, perhaps in resonance with part of the piping system, to produce excessive noise. A change in the operating point conditions or details of the valve geometry can result in significant noise reduction.

Pumps and blowers are strong potential noise sources. Turbomachinery noise is associated with blade-flow occurrences. Broadband noise appears from vortex and turbulence interaction with walls and is primarily a function of the operating point of the machine. For blowers, it has a minimum at the peak efficiency point (Groff et al. 1967). Narrow-band noise also appears at the blade-crossing frequency and its harmonics. Such noise can be very annoying because it stands out from the background. To reduce this noise, increase clearances between impeller and housing, and space impeller blades unevenly around the circumference.

REFERENCES

Baines, W.D. and E.G. Peterson. 1951. An investigation of flow through screens. *ASME Transactions* 73:467.

Baker, A.J. 1983. *Finite element computational fluid mechanics*. McGraw-Hill, New York.

Ball, J.W. 1957. Cavitation characteristics of gate valves and globe values used as flow regulators under heads up to about 125 ft. *ASME Transactions* 79:1275.

Benedict, R.P. and N.A. Carlucci. 1966. *Handbook of specific losses in flow systems*. Plenum Press Data Division, New York.

Binder, R.C. 1944. Limiting isothermal flow in pipes. *ASME Transactions* 66:221.

Bober, W. and R.A. Kenyon. 1980. *Fluid mechanics*. John Wiley and Sons, New York.

Colborne, W.G. and A.J. Drobitch. 1966. An experimental study of non-isothermal flow in a vertical circular tube. *ASHRAE Transactions* 72(4):5.

Crane Co. 1976. Flow of fluids. Technical *Paper* No. 410. New York.

Daily, J.W., et al. 1956. Resistance coefficients for accelerated and decelerated flows through smooth tubes and orifices. *ASME Transactions* 78:1071.

Deissler, R.G. 1951. Laminar flow in tubes with heat transfer. National Advisory Technical *Note* 2410, Committee for Aeronautics.

Furuya, Y., T. Sate, and T. Kushida. 1976. The loss of flow in the conical with suction at the entrance. *Bulletin of the Japan Society of Mechanical Engineers* 19:131.

Goldstein, S., ed. 1938. *Modern developments in fluid mechanics*. Oxford University Press, London. Reprinted by Dover Publications, New York.

Groff, G.C., J.R. Schreiner, and C.E. Bullock. 1967. Centrifugal fan sound power level prediction. *ASHRAE Transactions* 73(II): V.4.1.

Heskested, G. 1965. An edge suction effect. *AIAA Journal* 3:1958.

Heskested, G. 1970. Further experiments with suction at a sudden enlargement. *Journal of Basic Engineering, ASME Transactions* 92D:437.

Hoerner, S.F. 1965. *Fluid dynamic drag*, 3rd ed. Published by author, Midland Park, NJ.

Hydraulic Institute. 1961. *Pipe friction manual*. New York.

Ito, H. 1962. Pressure losses in smooth pipe bends. *Journal of Basic Engineering, ASME Transactions* 4(7):43.

John, J.E.A. and W.L. Haberman. 1980. *Introduction to fluid mechanics*, 2nd ed. Prentice Hall, Englewood Cliffs, NJ.

Kittridge, C.P. and D.S. Rowley. 1957. Resistance coefficients for laminar and turbulent flow through one-half inch valves and fittings. *ASME Transactions* 79:759.

Kline, S.J. 1959. On the nature of stall. *Journal of Basic Engineering, ASME Transactions* 81D:305.

Knapp, R.T., J.W. Daily, and F.G. Hammitt. 1970. *Cavitation*. McGraw-Hill, New York.

Lipstein, N.J. 1962. Low velocity sudden expansion pipe flow. *ASHRAE Journal* 4(7):43.

Moody, L.F. 1944. Friction factors for pipe flow. *ASME Transactions* 66:672.

Moore, C.A. and S.J. Kline. 1958. Some effects of vanes and turbulence in two-dimensional wide-angle subsonic diffusers. National Advisory Committee for Aeronautics, Technical *Memo* 4080.

Murdock, J.W., C.J. Foltz, and C. Gregory. 1964. Performance characteristics of elbow flow meters. *Journal of Basic Engineering, ASME Transactions* 86D:498.

Olson, R.M. 1980. *Essentials of engineering fluid mechanics*, 4th ed. Harper and Row, New York.

Robertson, J.M. 1963. A turbulence primer. University of Illinois (Urbana, IL), Engineering Experiment Station *Circular* 79.

Robertson, J.M. 1965. *Hydrodynamics in theory and application.* Prentice-Hall, Englewood Cliffs, NJ.

Robertson, J.M. and G.F. Wislicenus, ed. 1969 (discussion 1970). *Cavitation state of knowledge.* American Society of Mechanical Engineers, New York.

Ross, D. 1956. Turbulent flow in the entrance region of a pipe. *ASME Transactions* 78:915.

Schlichting, H. 1979. *Boundary layer theory*, 7th ed. McGraw-Hill, New York.

Streeter, V.L. and E.B. Wylie. 1979. *Fluid mechanics*, 7th ed. McGraw-Hill, New York.

Wile, D.D. 1947. Air flow measurement in the laboratory. *Refrigerating Engineering*: 515.

CHAPTER 3

HEAT TRANSFER

HEAT is energy in transit due to a temperature difference. The thermal energy is transferred from one region to another by three modes of **heat transfer**: conduction, convection, and radiation. Heat transfer is among a group of energy transport phenomena that includes mass transfer (see Chapter 5), momentum transfer or fluid friction (see Chapter 2), and electrical conduction. Transport phenomena have similar rate equations, in which flux is proportional to a potential difference. In heat transfer by conduction and convection, the potential difference is the temperature difference. Heat, mass, and momentum transfer are often considered together because of their similarities and interrelationship in many common physical processes.

This chapter presents the elementary principles of single-phase heat transfer with emphasis on heating, refrigerating, and air conditioning. Boiling and condensation are discussed in Chapter 4. More specific information on heat transfer to or from buildings or refrigerated spaces can be found in Chapters 25 through 31 of this volume and in Chapter 12 of the 1998 *ASHRAE Handbook—Refrigeration*. Physical properties of substances can be found in Chapters 18, 22, 24, and 36 of this volume and in Chapter 8 of the 1998 *ASHRAE Handbook—Refrigeration*. Heat transfer equipment, including evaporators, condensers, heating and cooling coils, furnaces, and radiators, is covered in the 2000 *ASHRAE Handbook—Systems and Equipment*. For further information on heat transfer, see the section on Bibliography.

HEAT TRANSFER PROCESSES

Thermal Conduction. This is the mechanism of heat transfer whereby energy is transported between parts of a continuum by the transfer of kinetic energy between particles or groups of particles at the atomic level. In gases, conduction is caused by elastic collision of molecules; in liquids and electrically nonconducting solids, it is believed to be caused by longitudinal oscillations of the lattice structure. Thermal conduction in metals occurs, like electrical conduction, through the motion of free electrons. Thermal energy transfer occurs in the direction of decreasing temperature, a consequence of the second law of thermodynamics. In solid opaque bodies, thermal conduction is the significant heat transfer mechanism because no net material flows in the process and radiation is not a factor. With flowing fluids, thermal conduction dominates in the region very close to a solid boundary, where the flow is **laminar** and parallel to the surface and where there is no eddy motion.

Thermal Convection. This form of heat transfer involves energy transfer by fluid movement and molecular conduction (Burmeister 1983, Kays and Crawford 1980). Consider heat transfer to a fluid flowing inside a pipe. If the Reynolds number is large enough, three different flow regions exist. Immediately adjacent to the wall is a **laminar sublayer** where heat transfer occurs by thermal conduction; outside the laminar sublayer is a transition region called the **buffer layer**, where both eddy mixing and conduction effects are significant; beyond the buffer layer and extending to the center of the pipe is the **turbulent region**, where the dominant mechanism of transfer is eddy mixing.

In most equipment, the main body of fluid is in turbulent flow, and the laminar layer exists at the solid walls only. In cases of low-velocity flow in small tubes, or with viscous liquids such as glycol (i.e., at low Reynolds numbers), the entire flow may be laminar with no transition or turbulent region.

When fluid currents are produced by external sources (for example, a blower or pump), the solid-to-fluid heat transfer is termed **forced convection**. If the fluid flow is generated internally by nonhomogeneous densities caused by temperature variation, the heat transfer is termed **free convection** or **natural convection**.

Thermal Radiation. In conduction and convection, heat transfer takes place through matter. In thermal radiation, there is a change in energy form from internal energy at the source to electromagnetic energy for transmission, then back to internal energy at the receiver. Whereas conduction and convection heat transfer rates are driven primarily by temperature difference and somewhat by temperature level, radiative heat transfer rates increase rapidly with temperature levels (for the same temperature difference).

Although some generalized heat transfer equations have been mathematically derived from fundamentals, they are usually obtained from correlations of experimental data. Normally, the correlations employ certain dimensionless numbers, shown in Table 1, that are derived from dimensional analysis or analogy.

STEADY-STATE CONDUCTION

For steady-state heat conduction in one dimension, the Fourier law is

$$q = -(kA)\frac{dt}{dx} \tag{1}$$

where

q = heat flow rate, Btu/h
k = thermal conductivity, Btu·ft/h·ft^2·°F
A = cross-sectional area normal to flow, ft^2
dt/dx = temperature gradient, °F/ft

Equation (1) states that the heat flow rate q in the x direction is directly proportional to the temperature gradient dt/dx and the cross-sectional area A normal to the heat flow. The proportionality factor is the thermal conductivity k. The minus sign indicates that the heat flow is positive in the direction of decreasing temperature. Conductivity values are sometimes given in other units, but consistent units must be used in Equation (1).

The preparation of this chapter is assigned to TC 1.3, Heat Transfer and Fluid Flow.

Table 1 Dimensionless Numbers Commonly Used in Heat Transfer

Name	Symbol	Value[a]	Application
Nusselt number	Nu	hD/k, hL/k, $q''D/\Delta tk$, or $q''L/\Delta tk$	Natural or forced convection, boiling or condensing
Reynolds number	Re	GD/μ or $\rho VL/\mu$	Forced convection, boiling or condensing
Prandtl number	Pr	$\mu c_p/k$	Natural or forced convection, boiling or condensing
Stanton number	St	h/Gc_p	Forced convection
Grashof number	Gr	$L^3\rho^2\beta g\Delta t/\mu^2$ or $L^3\rho^2 g\Delta t/T\mu^2$	Natural convection (for ideal gases)
Fourier number	Fo	$\alpha\tau/L^2$	Unsteady-state conduction
Peclet number	Pe	GDc_p/k or Re Pr	Forced convection (small Pr)
Graetz number	Gz	GD^2c_p/kL or Re Pr D/L	Laminar forced convection

[a]A list of the other symbols used in this chapter appears in the section on Symbols.

Equation (1) may be integrated along a path of constant heat flow rate to obtain

$$q = k\left(\frac{A_m}{L_m}\right)\Delta t = \frac{\Delta t}{R} \qquad (2)$$

where

A_m = mean cross-sectional area normal to flow, ft^2
L_m = mean length of heat flow path, ft
Δt = overall temperature difference, °F
R = thermal resistance, °F·h/Btu

Thermal resistance R is directly proportional to the mean length L_m of the heat flow path and inversely proportional to the conductivity k and the mean cross-sectional area A_m normal to the flow. Equations for thermal resistances of a few common shapes are given in Table 2. Mathematical solutions to many heat conduction problems are addressed by Carslaw and Jaeger (1959). Complicated problems can be solved by graphical or numerical methods such as described by Croft and Lilley (1977), Adams and Rogers (1973), and Patankar (1980).

Analogy to Electrical Conduction. Equation (2) is analogous to Ohm's law for electrical circuits: thermal current (heat flow) in a **thermal circuit** is directly proportional to the thermal potential (temperature difference) and inversely proportional to the thermal resistance. This electrical-thermal analogy can be used for heat conduction in complex shapes that resist solution by exact analytical means. The thermal circuit concept is also useful for problems involving combined conduction, convection, and radiation.

OVERALL HEAT TRANSFER

In most steady-state heat transfer problems, more than one heat transfer mode is involved. The various heat transfer coefficients may be combined into an overall coefficient so that the total heat transfer can be calculated from the terminal temperatures. The solution to this problem is much simpler if the concept of a thermal circuit is employed.

Local Overall Heat Transfer Coefficient—Resistance Method

Consider heat transfer from one fluid to another by a three-step steady-state process: from a warmer fluid to a solid wall, through the wall, then to a colder fluid. An **overall heat transfer coefficient** U based on the difference between the bulk temperatures $t_1 - t_2$ of the two fluids is defined as follows:

$$q = UA(t_1 - t_2) \qquad (3)$$

where A is the surface area. Because Equation (3) is a definition of U, the surface area A on which U is based is arbitrary; it should always be specified in referring to U.

Table 2 Solutions for Some Steady-State Thermal Conduction Problems

System	*R* in Equation $q = \Delta t/R$
Flat wall or curved wall if curvature is small (wall thickness less than 0.1 of inside diameter) 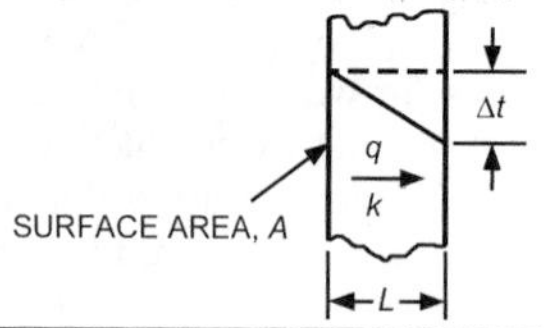	$R = \frac{L}{kA}$
Radial flow through a right circular cylinder 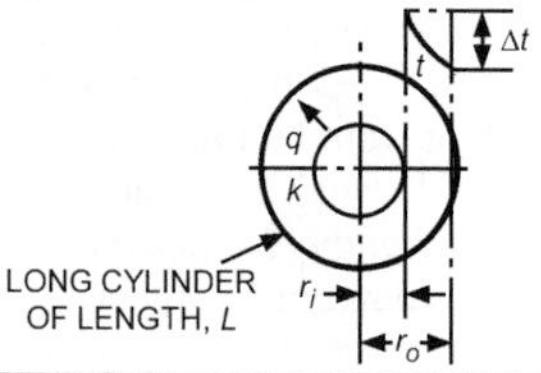	$R = \frac{\ln(r_o/r_i)}{2\pi kL}$
Buried cylinder 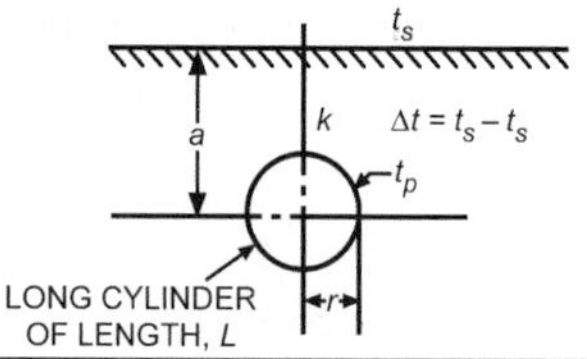	$R = \frac{\ln\left[(a + \sqrt{a^2 - r^2})/r\right]}{2\pi kL}$ $= \frac{\cosh^{-1}(a/r)}{2\pi kL} \quad (L \gg 2r)$
Radial flow in a hollow sphere 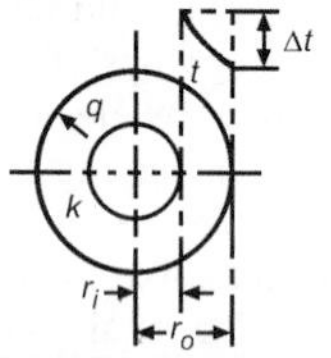	$R = \frac{(1/r_i - 1/r_o)}{4\pi k}$

L, r, a = dimensions, ft
k = thermal conductivity at average material temperature, Btu·ft/h·ft^2·°F
A = surface area, ft^2

The temperature drops across each part of the heat flow path are

$$t_1 - t_{s1} = qR_1$$
$$t_{s1} - t_{s2} = qR_2$$
$$t_{s2} - t_2 = qR_3$$

where t_{s1}, and t_{s2} are the warm and cold surface temperatures of the wall, respectively, and R_1, R_2, and R_3 are the thermal resistances. Because the same quantity of heat flows through each thermal resistance, these equations combined yield the following:

$$\frac{t_1 - t_2}{q} = \frac{1}{UA} = R_1 + R_2 + R_3 \tag{4}$$

As shown above, the equations are analogous to those for electrical circuits; for thermal current flowing through n resistances in *series*, the resistances are additive.

$$R_o = R_1 + R_2 + R_3 + \ldots + R_n \tag{5}$$

Similarly, **conductance** is the reciprocal of resistance, and for heat flow through resistances in *parallel*, the conductances are additive:

$$C = \frac{1}{R_o} = \frac{1}{R_1} + \frac{1}{R_2} + \frac{1}{R_3} + \ldots + \frac{1}{R_n} \tag{6}$$

For convection, the thermal resistance is inversely proportional to the **convection coefficient** h_c and the applicable surface area:

$$R_c = \frac{1}{h_c A} \tag{7}$$

The thermal resistance for radiation is written similarly to that for convection:

$$R_r = \frac{1}{h_r A} \tag{8}$$

The **radiation coefficient** h_r is a function of the temperatures, radiation properties, and geometrical arrangement of the enclosure and the body in question.

Resistance Method Analysis. Analysis by the resistance method can be illustrated by considering heat transfer from air outside to cold water inside an insulated pipe. The temperature gradients and the nature of the resistance analysis are shown in Figure 1.

Because air is sensibly transparent to radiation, some heat transfer occurs by both radiation and convection to the outer insulation surface. The mechanisms act in parallel on the air side. The total transfer then passes through the insulating layer and the pipe wall by thermal conduction, and then by convection and radiation into the cold water stream. (Radiation is not significant on the water side because liquids are sensibly opaque to radiation, although water transmits energy in the visible region.) The contact resistance between the insulation and the pipe wall is assumed negligible.

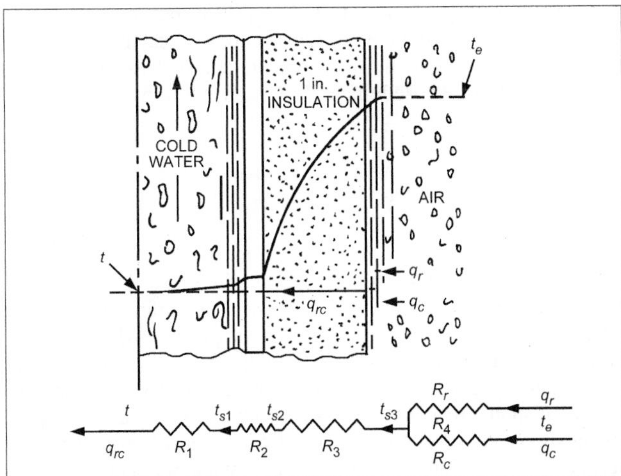

Fig. 1 Thermal Circuit Diagram for Insulated Cold Water Line

The heat transfer rate q_{rc} for a given length L of pipe may be thought of as the sum of the rates q_r and q_c flowing through the parallel resistances R_r and R_c associated with the surface radiation and convection coefficients. The total flow then proceeds through the resistance R_3 offered to thermal conduction by the insulation, through the pipe wall resistance R_2, and into the water stream through the convection resistance R_1. Note the analogy to direct-current electricity. A temperature (potential) drop is required to overcome resistances to the flow of thermal current. The total resistance to heat transfer R_o is the sum of the individual resistances:

$$R_o = R_1 + R_2 + R_3 + R_4 \tag{9}$$

where the resultant parallel resistance R_4 is obtained from

$$\frac{1}{R_4} = \frac{1}{R_r} + \frac{1}{R_c} \tag{10}$$

If the individual resistances can be evaluated, the total resistance can be obtained from this relation. The heat transfer rate for the length of pipe L can be established by

$$q_{rc} = \frac{t_e - t}{R_o} \tag{11}$$

For a unit length of the pipe, the heat transfer rate is

$$\frac{q_{rc}}{L} = \frac{t_e - t}{R_o L} \tag{12}$$

The temperature drop Δt through each individual resistance may then be calculated from the relation:

$$\Delta t_n = R_n q_{rc} \tag{13}$$

where n = 1, 2, and 3.

Mean Temperature Difference

When heat is exchanged between two fluids flowing through a heat exchanger, the local temperature difference Δt varies along the flow path. Heat transfer may be calculated using

$$q = UA\Delta t_m \tag{14}$$

where U is the overall uniform coefficient of heat transfer from fluid to fluid, A is the area associated with the coefficient U, and Δt_m is the appropriate mean temperature difference.

For parallel flow or counterflow exchangers and for any exchanger in which one fluid temperature is substantially constant, the mean temperature difference is

$$\Delta t_m = \frac{\Delta t_1 - \Delta t_2}{\ln(\Delta t_1/\Delta t_2)} = \frac{\Delta t_1 - \Delta t_2}{2.3\log(\Delta t_1/\Delta t_2)} \tag{15}$$

where Δt_1, and Δt_2 are the temperature differences between the fluids at each end of the heat exchanger. Δt_m is called the **logarithmic mean temperature difference**. For the special case of $\Delta t_1 = \Delta t_2$ (possible only with a counterflow heat exchanger with equal capacities), which leads to an indeterminate form of Equation (15), $\Delta t_m = \Delta t_1 = \Delta t_2$.

Equation (15) for Δt_m is true only if the overall coefficient and the specific heat of the fluids are constant through the heat exchanger, and no heat losses occur (often well-approximated in practice). Parker et al. (1969) give a procedure for cases with variable overall coefficient U.

Calculations using Equation (14) and Δt_m are convenient when terminal temperatures are known. In many cases, however, the temperatures of the fluids leaving the exchanger are not known. To avoid trial-and-error calculations, an alternate method involves the use of three nondimensional parameters, defined as follows:

1. Exchanger Heat Transfer Effectiveness ε

$$\varepsilon = \frac{(t_{hi} - t_{ho})}{(t_{hi} - t_{ci})} \quad \text{when } C_h = C_{min}$$

$$\varepsilon = \frac{(t_{co} - t_{ci})}{(t_{hi} - t_{ci})} \quad \text{when } C_c = C_{min} \tag{16}$$

where

$C_h = (\dot{m}\, c_p)_h$ = hot fluid capacity rate, Btu/h·°F
$C_c = (\dot{m}\, c_p)_c$ = cold fluid capacity rate, Btu/h·°F
C_{min} = smaller of capacity rates C_h and C_c
t_h = terminal temperature of hot fluid, °F. Subscript *i* indicates entering condition; subscript *o* indicates leaving condition.
t_c = terminal temperature of cold fluid, °F. Subscripts *i* and *o* are the same as for t_h.

2. Number of Exchanger Heat Transfer Units (NTU)

$$\text{NTU} = \frac{AU_{avg}}{C_{min}} = \frac{1}{C_{min}} \int_A U\, dA \tag{17}$$

where *A* is the area used to define overall coefficient *U*.

3. Capacity Rate Ratio Z

$$Z = \frac{C_{min}}{C_{max}} \tag{18}$$

For a given exchanger, the heat transfer effectiveness can generally be expressed for a given exchanger as a function of the number of transfer units and the capacity rate ratio:

$$\varepsilon = f\,(\text{NTU, Z, flow arrangement}) \tag{19}$$

The effectiveness is independent of the temperatures in the exchanger. For any exchanger in which the capacity rate ratio *Z* is zero (where one fluid undergoes a phase change; e.g., in a condenser or evaporator), the effectiveness is

$$\varepsilon = 1 - \exp(-\text{NTU}) \tag{20}$$

Heat transferred can be determined from

$$q = C_h(t_{hi} - t_{ho}) = C_c(t_{co} - t_{ci}) \tag{21}$$

Combining Equations (16) and (21) produces an expression for heat transfer rate in terms of entering fluid temperatures:

$$q = \varepsilon C_{min}(t_{hi} - t_{ci}) \tag{22}$$

The proper mean temperature difference for Equation (14) is then given by

$$\Delta t_m = \frac{(t_{hi} - t_{ci})\varepsilon}{\text{NTU}} \tag{23}$$

The effectiveness for **parallel flow exchangers** is

$$\varepsilon = \frac{1 - \exp[-\text{NTU}(1 + Z)]}{1 + Z} \tag{24}$$

For $Z = 1$,

$$\varepsilon = \frac{1 - \exp(-2\ \text{NTU})}{2} \tag{25}$$

The effectiveness for **counterflow exchangers** is

$$\varepsilon = \frac{1 - \exp[-\text{NTU}(1 - Z)]}{1 - Z\exp[-\text{NTU}(1 - Z)]} \tag{26}$$

$$\varepsilon = \frac{\text{NTU}}{1 + \text{NTU}} \quad \text{for } Z = 1 \tag{27}$$

Incropera and DeWitt (1996) and Kays and London (1984) show the relations of ε, NTU, and *Z* for other flow arrangements. These authors and Afgan and Schlunder (1974) present graphical representations for convenience.

TRANSIENT HEAT FLOW

Often, the heat transfer and temperature distribution under unsteady-state (varying with time) conditions must be known. Examples are (1) cold storage temperature variations on starting or stopping a refrigeration unit; (2) variation of external air temperature and solar irradiation affecting the heat load of a cold storage room or wall temperatures; (3) the time required to freeze a given material under certain conditions in a storage room; (4) quick freezing of objects by direct immersion in brines; and (5) sudden heating or cooling of fluids and solids from one temperature to a different temperature.

The equations describing transient temperature distribution and heat transfer are presented in this section. Numerical methods are the simplest means of solving these equations because numerical data are easy to obtain. However, with some numerical solutions and off-the-shelf software, the physics that drives the energy transport can be lost. Thus, analytical solution techniques are also included in this section.

The fundamental equation for unsteady-state conduction in solids or fluids in which there is no substantial motion is

$$\frac{\partial t}{\partial \tau} = \alpha \left(\frac{\partial^2 t}{\partial x^2} + \frac{\partial^2 t}{\partial y^2} + \frac{\partial^2 t}{\partial z^2} \right) \tag{28}$$

where thermal diffusivity α is the ratio $k/\rho c_p$; *k* is thermal conductivity; ρ, density; and c_p, specific heat. If α is large (high conductivity, low density and specific heat, or both), heat will diffuse faster.

One of the most elementary transient heat transfer models predicts the rate of temperature change of a body or material being held at constant volume with uniform temperature, such as a well-stirred reservoir of fluid whose temperature is changing because of a net rate of heat gain or loss:

$$q_{net} = (Mc_p)\frac{dt}{d\tau} \tag{29}$$

where *M* is the mass of the body, c_p is the specific heat at constant pressure, and q_{net} is the net heat transfer rate to the substance (heat transfer into the substance is positive, and heat transfer out of the substance is negative). Equation (29) is applicable when the pressure around the substance is constant; if the volume of the substance is constant, c_p should be replaced by the constant volume specific heat c_v. It should be noted that with the density of solids and liquids being almost constant, the two specific heats are almost equal. The term q_{net} may include heat transfer by conduction, convection, or radiation and is the difference between the heat transfer rates into and out of the body.

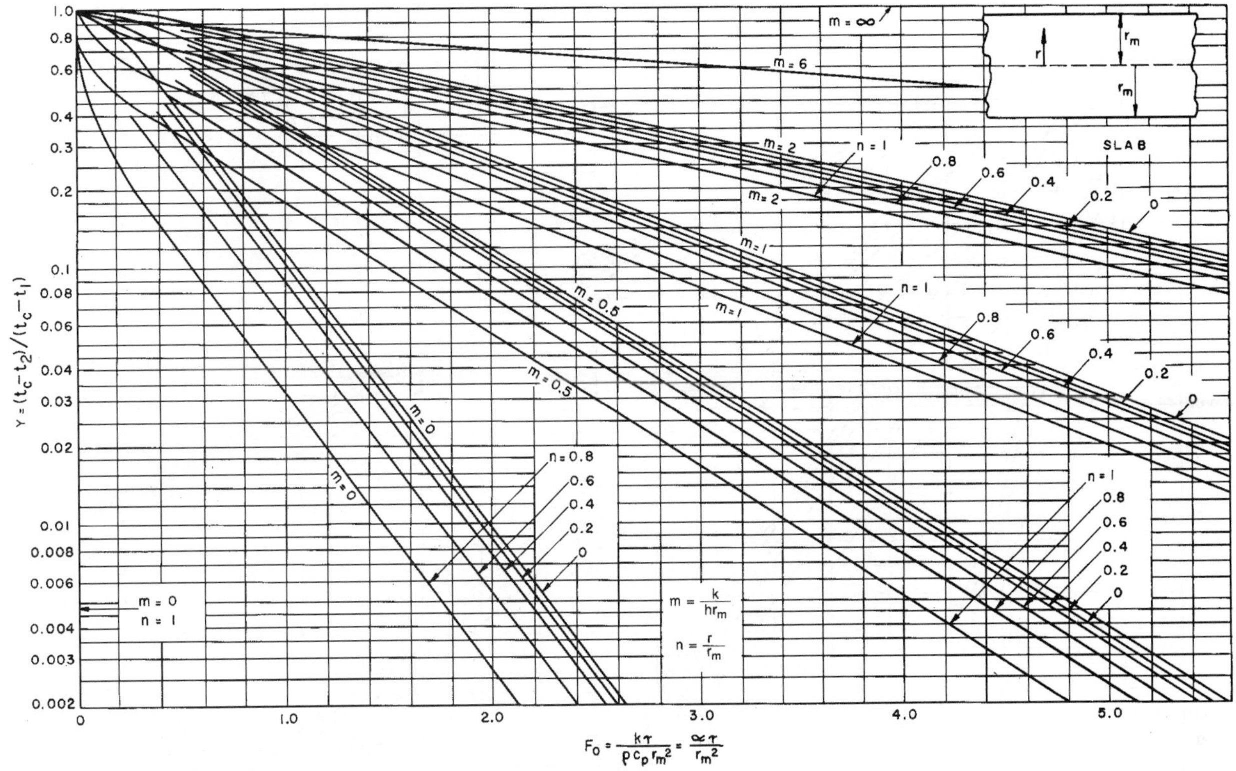

Fig. 2 Transient Temperatures for Infinite Slab

From Equations (28) and (29), it is possible to derive expressions for temperature and heat flow variations at different instants and different locations. Most common cases have been solved and presented in graphical forms (Jakob 1957, Schneider 1964, Myers 1971). In other cases, it is simpler to use numerical methods (Croft and Lilley 1977, Patankar 1980). When convective boundary conditions are required in the solution of Equations (28) and (29), h values based on steady-state correlations are often used. However, this approach may not be valid when rapid transients are involved.

Estimating Cooling Times

Cooling times for materials can be estimated (McAdams 1954) by Gurnie-Lurie charts (Figures 2, 3, and 4), which are graphical solutions for the heating or cooling of infinite slabs, infinite cylinders, and spheres. These charts assume an initial uniform temperature distribution and no change of phase. They apply to a body exposed to a constant temperature fluid with a constant surface convection coefficient of h.

Using Figures 2, 3, and 4, it is possible to estimate both the temperature at any point and the average temperature in a homogeneous mass of material as a function of time in a cooling process. It is possible to estimate cooling times for rectangular-shaped solids, cubes, cylinders, and spheres.

From the point of view of heat transfer, a cylinder insulated on its ends behaves like a cylinder of infinite length, and a rectangular solid insulated so that only two parallel faces allow heat transfer behaves like an infinite slab. A thin slab or a long, thin cylinder may be also considered infinite objects.

Consider a slab having insulated edges being cooled. If the cooling time is the time required for the center of the slab to reach a temperature of t_2, the cooling time can be calculated as follows:

1. Evaluate the temperature ratio $(t_c - t_2)/(t_c - t_1)$.

   where

   t_c = temperature of cooling medium
   t_1 = initial temperature of product
   t_2 = final temperature of product at center

   Note that in Figures 2, 3, and 4, the temperature ratio $(t_c - t_2)/(t_c - t_1)$ is designated as Y to simplify the equations.

2. Determine the radius ratio r/r_m designated as n in Figures 2, 3, and 4.

   where

   r = distance from centerline
   r_m = half thickness of slab

3. Evaluate the resistance ratio k/hr_m designated as m in Figures 2, 3, and 4.

   where

   k = thermal conductivity of material
   h = heat transfer coefficient

4. From Figure 2 for infinite slabs, select the appropriate value of $k\tau/\rho c_p r_m^2$ designated as F_o in Figures 2, 3, and 4.

   where

   τ = time elapsed
   c_p = specific heat
   ρ = density

5. Determine τ from the value of $k\tau/\rho c_p r_m^2$.

Multidimensional Temperature Distribution

The solution for semi-infinite slabs and cylinders (shown in Figures 2, 3, and 4) can be used to find the temperatures in finite rectangular solids or cylinders.

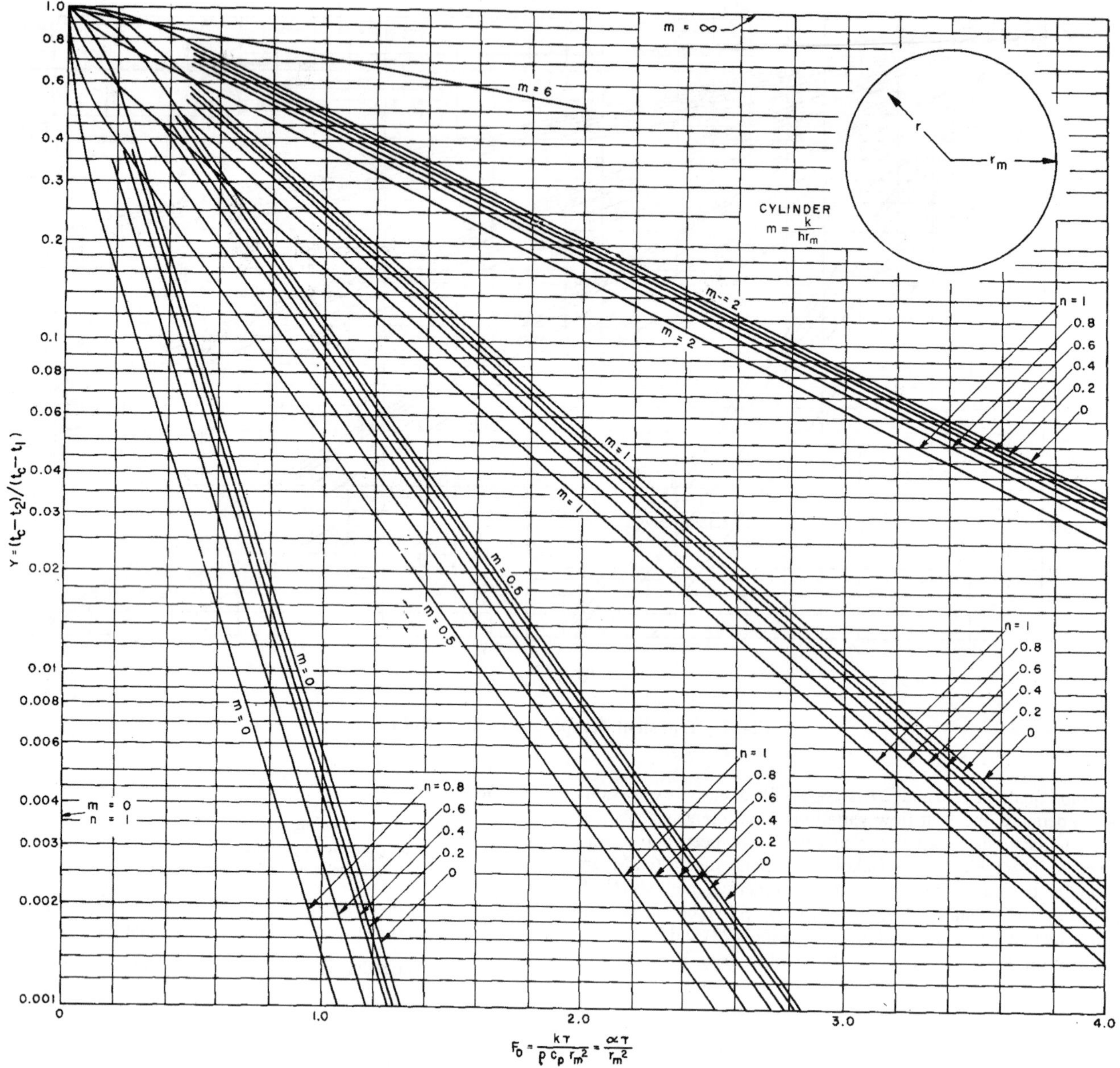

Fig. 3 Transient Temperatures for Infinite Cylinder

The temperature in the finite object can be calculated from the temperature ratio Y of the infinite objects that intersect to form the finite object. The product of the temperature ratios of the infinite objects is the temperature ratio of the finite object; for example, for the finite cylinder of Figure 5,

$$Y_{fc} = Y_{is} Y_{ic} \tag{30}$$

where

Y_{fc} = temperature ratio of finite cylinder
Y_{is} = temperature ratio of infinite slab
Y_{ic} = temperature ratio of infinite cylinder

For a finite rectangular solid,

$$Y_{frs} = (Y_{is})_1 (Y_{is})_2 (Y_{is})_3 \tag{31}$$

where Y_{frs} = temperature ratio of finite rectangular solid, and subscripts 1, 2, and 3 designate three infinite slabs. The convective heat transfer coefficients associated with one pair of parallel surfaces need not be equal to the coefficient associated with another pair. However, the temperature of the fluid adjacent to every surface should be the same. In evaluating the resistance ratio and the Fourier number Fo, the appropriate values of the heat transfer coefficient and the characteristic dimension should be used.

Heat Exchanger Transients

Determination of the transient behavior of heat exchangers is becoming increasingly important in evaluating the dynamic behavior of heating and air-conditioning systems. Many studies of the transient behavior of counterflow and parallel flow heat exchangers have been conducted; some are listed in the section on Bibliography.

THERMAL RADIATION

Radiation, one of the basic mechanisms for energy transfer between different temperature regions, is distinguished from conduction and convection in that it does not depend on an intermediate

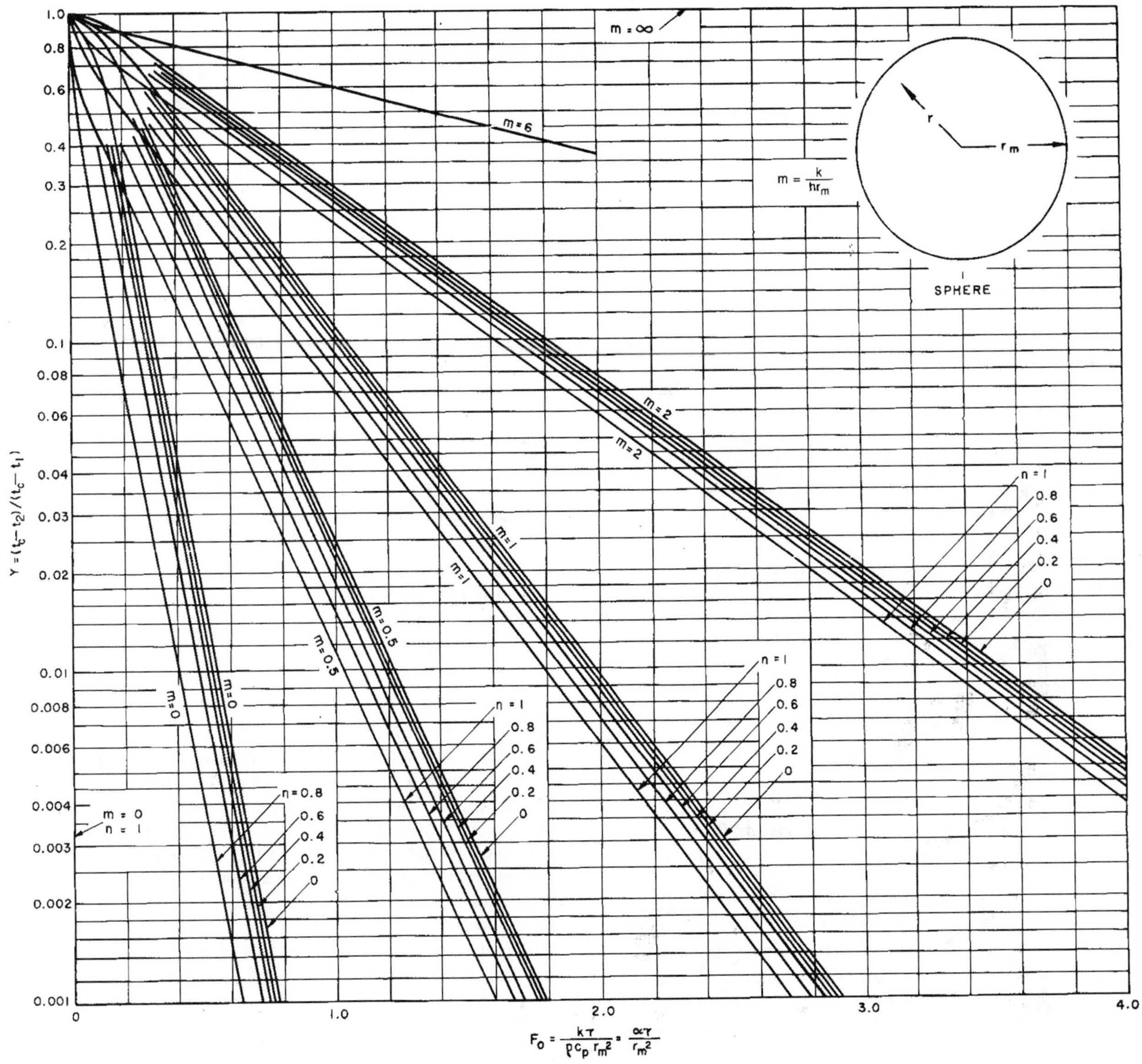

Fig. 4 Transient Temperatures for Spheres

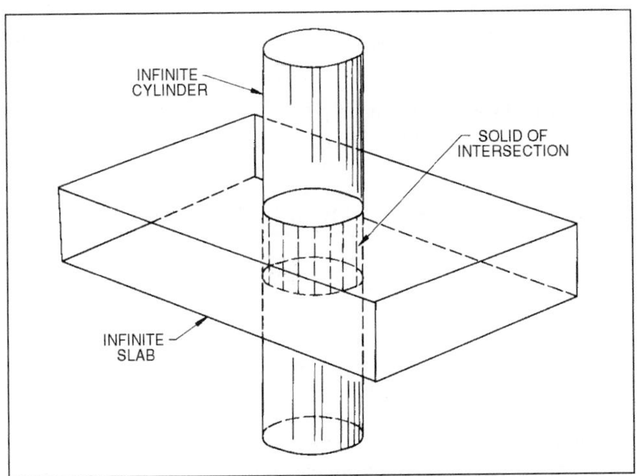

Fig. 5 Finite Cylinder of Intersection from Intersection of Infinite Cylinder and Infinite Slab

material as a carrier of energy but rather is impeded by the presence of material between the regions. The radiation energy transfer process is the consequence of energy-carrying electromagnetic waves that are emitted by atoms and molecules due to changes in their energy content. The amount and characteristics of radiant energy emitted by a quantity of material depend on the nature of the material, its microscopic arrangement, and its absolute temperature. Although rate of energy emission is independent of the surroundings, the **net** energy transfer rate depends on the temperatures and spatial relationships of the surface and its surroundings.

Blackbody Radiation

The rate of thermal radiant energy emitted by a surface depends on its absolute temperature. A surface is called **black** if it can absorb all incident radiation. The total energy emitted per unit time per unit area of black surface W_b to the hemispherical region above it is given by the **Stefan-Boltzmann law**.

$$W_b = \sigma T^4 \tag{32}$$

Table 3 Emittances and Absorptances for Some Surfaces[a]

Class	Surfaces	Total Normal Emittance[b] At 50 to 100 °F	Total Normal Emittance[b] At 1000 °F	Absorptance for Solar Radiation
1	A small hole in a large box, sphere, furnace, or enclosure	0.97 to 0.99	0.97 to 0.99	0.97 to 0.99
2	Black nonmetallic surfaces such as asphalt, carbon, slate, paint, paper	0.90 to 0.98	0.90 to 0.98	0.85 to 0.98
3	Red brick and tile, concrete and stone, rusty steel and iron, dark paints (red, brown, green, etc.)	0.85 to 0.95	0.75 to 0.90	0.65 to 0.80
4	Yellow and buff brick and stone, firebrick, fireclay	0.85 to 0.95	0.70 to 0.85	0.50 to 0.70
5	White or light cream brick, tile, paint or paper, plaster, whitewash	0.85 to 0.95	0.60 to 0.75	0.30 to 0.50
6	Window glass	0.90	—	c
7	Bright aluminum paint; gilt or bronze paint	0.40 to 0.60	—	0.30 to 0.50
8	Dull brass, copper, or aluminum; galvanized steel; polished iron	0.20 to 0.30	0.30 to 0.50	0.40 to 0.65
9	Polished brass, copper, monel metal	0.02 to 0.05	0.05 to 0.15	0.30 to 0.50
10	Highly polished aluminum, tin plate, nickel, chromium	0.02 to 0.04	0.05 to 0.10	0.10 to 0.40
11	Selective surfaces			
	Stainless steel wire mesh	0.23 to 0.28	—	0.63 to 0.86
	White painted surface	0.92	—	0.23 to 0.49
	Copper treated with solution of $NaClO_2$ and NaOH	0.13	—	0.87
	Copper, nickel, and aluminum plate with CuO coating	0.09 to 0.21	—	0.08 to 0.93

[a]See also Chapter 36, McAdams (1954), and Siegel and Howell (1981).

[b]Hemispherical and normal emittance are unequal in many cases. The hemispherical emittance may vary from up to 30% greater for polished reflectors to 7% lower for nonconductors.

[c]Absorbs 4 to 40% depending on its transmittance.

where W_b is the total rate of energy emission per unit area, and σ is the Stefan-Boltzmann constant (0.1712×10^{-8} Btu/h·ft^2·°R^4).

The heat radiated by a body comprises electromagnetic waves of many different frequencies or wavelengths. Planck showed that the spectral distribution of the energy radiated by a blackbody is

$$W_{b\lambda} = \frac{C_1 \lambda^{-5}}{e^{C_2/\lambda T} - 1} \tag{33}$$

where

$W_{b\lambda}$ = monochromatic emissive power of blackbody, Btu/h·ft^2·μm
λ = wavelength, μm
T = temperature, °R
C_1 = first Planck's law constant = 1.1870×10^8 Btu·μm^4/h·ft^2
C_2 = second Planck's law constant = 2.5896×10^4 μm·°R

$W_{b\lambda}$ is the **monochromatic emissive power**, defined as the energy emitted per unit time per unit surface area at wavelength λ per unit wavelength interval around λ; that is, the energy emitted per unit time per unit surface area in the interval $d\lambda$ is equal to $W_{b\lambda} d\lambda$.

The Stefan-Boltzmann equation can be obtained by integrating Planck's equation:

$$W_b = \sigma T^4 = \int_0^\infty W_{b\lambda} d\lambda \tag{34}$$

Wien showed that the wavelength of maximum emissive power multiplied by the absolute temperature is a constant:

$$\lambda_{max} T = 5216 \text{ μm} \cdot \text{°R} \tag{35}$$

where λ_{max} is the wavelength at which the monochromatic emissive power is a maximum and not the maximum wavelength. Equation (35) is known as **Wien's displacement law**. According to this law, the maximum spectral emissive power is displaced to shorter wavelengths with increasing temperature, such that significant emission eventually occurs over the entire visible spectrum as shorter wavelengths become more prominent. For additional details, see Incropera and DeWitt (1996).

Actual Radiation

Substances and surfaces diverge variously from the Stefan-Boltzmann and Planck laws. W_b and $W_{b\lambda}$ are the maximum emissive powers at a surface temperature. Actual surfaces emit and absorb less than these maximums and are called **nonblack**. The emissive power of a nonblack surface at temperature T radiating to the hemispherical region above it is written as

$$W = \varepsilon W_b = \varepsilon \sigma T^4 \tag{36}$$

where ε is known as the **hemispherical emittance**. The term emittance conforms to physical and electrical terminology; the suffix "ance" denotes a property of a piece of material as it exists. The ending "ivity" denotes a property of the bulk material independent of geometry or surface condition. Thus, emittance, reflectance, absorptance, and transmittance refer to actual pieces of material. Emissivity, reflectivity, absorptivity, and transmissivity refer to the properties of materials that are optically smooth and thick enough to be opaque.

The emittance is a function of the material, the condition of its surface, and the temperature of the surface. Table 3 lists selected values; Siegel and Howell (1981) and Modest (1993) have more extensive lists.

The monochromatic emissive power of a nonblack surface is similarly written as

$$W_\lambda = \varepsilon_\lambda W_{b\lambda} = \varepsilon_\lambda \left(\frac{C_1 \lambda^{-5}}{e^{C_2/\lambda T} - 1} \right) \tag{37}$$

where ε_λ is the monochromatic hemispherical emittance. The relationship between ε and ε_λ is given by

$$W = \varepsilon \sigma T^4 = \int_0^\infty W_\lambda d\lambda = \int_0^\infty \varepsilon_\lambda W_{b\lambda} d\lambda$$

or

$$\varepsilon = \frac{1}{\sigma T^4} \int_0^\infty \varepsilon_\lambda W_{b\lambda} d\lambda \tag{38}$$

If ε_λ does not depend on λ, then, from Equation (38), $\varepsilon = \varepsilon_\lambda$. Surfaces with this characteristic are called **gray**. Gray surface characteristics are often assumed in calculations. Several classes of surfaces approximate this condition in some regions of the spectrum. The simplicity is desirable, but care must be exercised, especially if temperatures are high. Assumption of grayness is sometimes made because of the absence of information relating ε_λ and λ.

When radiant energy falls on a surface, it can be absorbed, reflected, or transmitted through the material. Therefore, from the first law of thermodynamics,

$$\alpha + \tau + \rho = 1 \tag{39}$$

where

- α = fraction of incident radiation absorbed or **absorptance**
- τ = fraction of incident radiation transmitted or **transmittance**
- ρ = fraction of incident radiation reflected or **reflectance**

If the material is opaque, as most solids are in the infrared, $\tau = 0$ and $\alpha + \rho = 1$. For a black surface, $\alpha = 1$, $\rho = 0$, and $\tau = 0$. Platinum black and gold black are as black as any actual surface and have absorptances of about 98% in the infrared. Any desired degree of blackness can be simulated by a small hole in a large enclosure. Consider a ray of radiant energy entering the opening. It will undergo many internal reflections and be almost completely absorbed before it has a reasonable probability of passing back out of the opening.

Certain flat black paints also exhibit emittances of 98% over a wide range of conditions. They provide a much more durable surface than gold or platinum black and are frequently used on radiation instruments and as standard reference in emittance or reflectance measurements.

Kirchhoff's law relates emittance and absorptance of any opaque surface from thermodynamic considerations; it states that for any surface where the incident radiation is independent of angle or where the surface is diffuse, $\varepsilon_\lambda = \alpha_\lambda$. If the surface is gray, or the incident radiation is from a black surface at the same temperature, then $\varepsilon = \alpha$ as well, but many surfaces are not gray. For most surfaces listed in Table 3, absorptance for solar radiation is different from emittance for low-temperature radiation. This is because the wavelength distributions are different in the two cases, and ε_λ varies with wavelength.

The foregoing discussion relates to total hemispherical radiation from surfaces. Energy distribution over the hemispherical region above the surface also has an important effect on the rate of heat transfer in various geometric arrangements.

Lambert's law states that the emissive power of radiant energy over a hemispherical surface above the emitting surface varies as the cosine of the angle between the normal to the radiating surface and the line joining the radiating surface to the point of the hemispherical surface. This radiation is **diffuse radiation**. The Lambert emissive power variation is equivalent to assuming that radiation from a surface in a direction other than normal occurs as if it came from an equivalent area with the same emissive power (per unit area) as the original surface. The equivalent area is obtained by projecting the original area onto a plane normal to the direction of radiation. Black surfaces obey the Lambert law exactly. The law is approximate for many actual radiation and reflection processes, especially those involving rough surfaces and nonmetallic materials. Most radiation analyses are based on the assumption of gray diffuse radiation and reflection.

In estimating heat transfer rates between surfaces of different geometries, radiation characteristics, and orientations, it is usually assumed that

- All surfaces are gray or black
- Radiation and reflection are diffuse
- Properties are uniform over the surfaces
- Absorptance equals emittance and is independent of the temperature of the source of incident radiation
- The material in the space between the radiating surfaces neither emits nor absorbs radiation

These assumptions greatly simplify problems, although results must be considered approximate.

Angle Factor

The distribution of radiation from a surface among the surfaces it irradiates is indicated by a quantity variously called an interception, a view, a configuration, a shape factor, or an angle factor. In terms of two surfaces i and j, the **angle factor** F_{ij} from surface i to surface j is the ratio of the radiant energy leaving surface i and directly reaching surface j to the total radiant energy leaving surface i. The angle factor from j to i is similarly defined, merely by interchanging the roles of i and j. This second angle factor is not, in general, numerically equal to the first. However, the reciprocity relation $F_{ij}A_i = F_{ji}A_j$, where A is the surface area, is always valid. Note that a concave surface may "see itself" ($F_{ii} \neq 0$), and that if n surfaces form an enclosure,

$$\sum_{j=1}^{n} F_{ij} = 1 \tag{40}$$

The angle factor F_{12} between two surfaces is

$$F_{12} = \frac{1}{A_1}\int_{A_1}\int_{A_2} \frac{\cos\phi_1 \cos\phi_2}{\pi r^2}\, dA_1\, dA_2 \tag{41}$$

where dA_1, and dA_2 are elemental areas of the two surfaces, r is the distance between dA_1 and dA_2, and ϕ_1 and ϕ_2 are the angles between the respective normals to dA_1 and dA_2 and the connecting line r. Numerical, graphical, and mechanical techniques can solve this equation (Siegel and Howell 1981, Modest 1993). Numerical values of the angle factor for common geometries are given in Figure 6.

Calculation of Radiant Exchange Between Surfaces Separated by Nonabsorbing Media

A surface radiates energy at a rate independent of its surroundings and absorbs and reflects incident energy at a rate dependent on its surface condition. The net energy exchange per unit area is denoted by q or q_j for unit area A_j. It is the rate of emission of the surface minus the total rate of absorption at the surface from all radiant effects in its surroundings, possibly including the return of some of its own emission by reflection off its surroundings. The rate at which energy must be supplied to the surface by other exchange processes if its temperature is to remain constant is q; therefore, to define q, the total radiant surroundings (in effect, an enclosure) must be specified.

Several methods have been developed to solve certain problems. To calculate the radiation exchange at each surface of an enclosure of n opaque surfaces by simple, general equations convenient for machine calculation, two terms must be defined:

- G = irradiation; total radiation incident on surface per unit time and per unit area
- J = radiosity; total radiation that leaves surface per unit time and per unit area

The radiosity is the sum of the energy emitted and the energy reflected:

$$J = \varepsilon W_b + \rho G \tag{42}$$

Because the transmittance is zero, the reflectance is

$$\rho = 1 - \alpha = 1 - \varepsilon$$

Thus,

$$J = \varepsilon W_b + (1 - \varepsilon)G \tag{43}$$

The net energy lost by a surface is the difference between the radiosity and the irradiation:

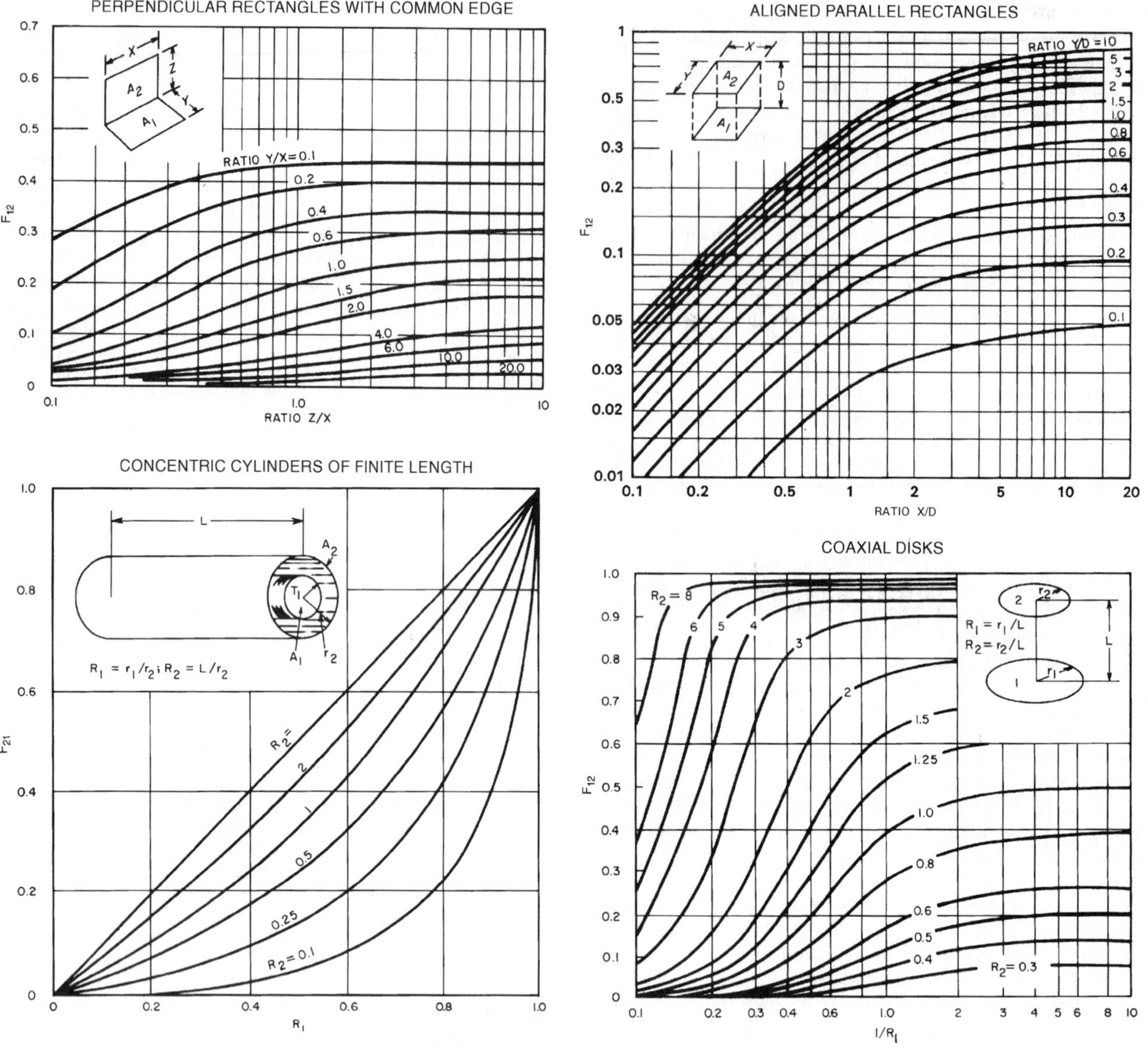

Fig. 6 Radiation Angle Factor for Various Geometries

$$q/A = J - G = \varepsilon W_b + (1 - \varepsilon)G - G \quad (44)$$

Substituting for G in terms of J from Equation (43),

$$q = \frac{W_b - J}{(1 - \varepsilon)/\varepsilon A} \quad (45)$$

Consider an enclosure of n isothermal surfaces with areas of $A_1, A_2, \ldots, A_n$, emittances of $\varepsilon_1, \varepsilon_2, \ldots, \varepsilon_n$, and reflectances of $\rho_1, \rho_2, \ldots, \rho_n$, respectively.

The irradiation of surface i is the sum of the radiation incident on it from all n surfaces:

$$G_i A_i = \sum_{j=1}^{n} F_{ji} J_j A_j = \sum_{j=1}^{n} F_{ij} J_j A_i$$

or

$$G_i = \sum_{j=1}^{n} F_{ij} J_j$$

Substituting in Equation (44) yields the following simultaneous equations when each of the n surfaces is considered:

$$J_i = \varepsilon_i W_{bi} + (1 - \varepsilon_i) \sum_{j=1}^{n} F_{ij} J_j \qquad i = 1, 2, \ldots, n \quad (46)$$

Equation (46) can be solved manually for the unknown Js if the number of surfaces is small. The solution for more complex enclosures requires a computer.

Once the radiosities (Js) are known, the net radiant energy lost by each surface is determined from Equation (45) as

$$q_i = \frac{W_{bi} - J_i}{(1 - \varepsilon_i)/\varepsilon_i A_i}$$

If the surface is black, Equation (45) becomes indeterminate, and an alternate expression must be used, such as

$$q_i = \sum_{j=1}^{n} J_i A_i F_{ij} - J_j A_j F_{ji}$$

or

$$q_i = \sum_{j=1}^{n} F_{ij} A_i (J_i - J_j) \qquad (47)$$

since

$$F_{ij} A_i = F_{ji} A_j$$

All diffuse radiation processes are included in the aforementioned enclosure method, and surfaces with special characteristics are assigned consistent properties. An opening is treated as an equivalent surface area A_e with a reflectance of zero. If energy enters the enclosure diffusely through the opening, A_e is assigned an equivalent temperature; otherwise, its temperature is taken as zero. If the loss through the opening is desired, q_2 is found. A window in the enclosure is assigned its actual properties.

A surface in **radiant balance** is one for which radiant emission is balanced by radiant absorption; heat is neither removed from nor supplied to the surface. Reradiating surfaces (insulated surfaces with $q_{net} = 0$), can be treated in Equation (46) as being perfectly reflective (i.e., $\varepsilon = 0$). The equilibrium temperature of such a surface can be found from

$$T_k = \left(\frac{J_k}{\sigma}\right)^{0.25}$$

once Equation (46) has been solved for the radiosities.

Use of angle factors and radiation properties as defined assumes that the surfaces are diffuse radiators—a good assumption for most nonmetals in the infrared region, but a poor assumption for highly polished metals. Subdividing the surfaces and considering the variation of radiation properties with angle of incidence improves the approximation but increases the work required for a solution.

Radiation in Gases

Elementary gases such as oxygen, nitrogen, hydrogen, and helium are essentially transparent to thermal radiation. Their absorption and emission bands are confined mainly to the ultraviolet region of the spectrum. The gaseous vapors of most compounds, however, have absorption bands in the infrared region. Carbon monoxide, carbon dioxide, water vapor, sulfur dioxide, ammonia, acid vapors, and organic vapors absorb and emit significant amounts of energy.

Radiation exchange by opaque solids is considered a surface phenomenon. Radiant energy does, however, penetrate the surface of all materials. The absorption coefficient gives the rate of exponential attenuation of the energy. Metals have large absorption coefficients, and radiant energy penetrates only a few hundred angstroms at most. Absorption coefficients for nonmetals are lower. Radiation may be considered a surface phenomenon unless the material is transparent. Gases have small absorption coefficients, so the path length of radiation through gas becomes very significant.

Beer's law states that the attenuation of radiant energy in a gas is a function of the product $p_g L$ of the partial pressure of the gas and the path length. The monochromatic absorptance of a body of gas of thickness L is then given by

$$\alpha_{\lambda L} = 1 - e^{-\alpha \lambda L} \qquad (48)$$

Because absorption occurs in discrete wavelengths, the absorptances must be summed over the spectral region corresponding to the temperature of the blackbody radiation passing through the gas.

Table 4 Emittance of CO_2 and Water Vapor in Air at 75 °F

Path Length, ft	CO_2, % by Volume		Relative Humidity, %			
	0.1	0.3	1.0	10	50	100
10	0.03	0.06	0.09	0.06	0.17	0.22
100	0.09	0.12	0.16	0.22	0.39	0.47
1000	0.16	0.19	0.23	0.47	0.64	0.70

The monochromatic absorption coefficient α_λ is also a function of temperature and pressure of the gas; therefore, detailed treatment of gas radiation is quite complex.

Estimated emittance for carbon dioxide and water vapor in air at 75°F is a function of concentration and path length (Table 4). The values are for a hemispherically shaped body of gas radiating to an element of area at the center of the hemisphere. Among others, Modest (1993), Siegel and Howell (1981), and Hottel and Sarofim (1967) describe geometrical calculations in their texts on radiation transfer. Generally, at low values of $p_g L$, the mean path length L (or equivalent hemispherical radius for a gas body radiating to its surrounding surfaces) is four times the mean hydraulic radius of the enclosure. A room with a dimensional ratio of 1:1:4 has a mean path length of 0.89 times the shortest dimension when considering radiation to all walls. For a room with a dimensional ratio of 1:2:6, the mean path length for the gas radiating to all surfaces is 1.2 times the shortest dimension. The mean path length for radiation to the 2 by 6 face is 1.18 times the shortest dimension. These values are for cases where the partial pressure of the gas times the mean path length approaches zero ($p_g L \approx 0$). The factor decreases with increasing values of $p_g L$. For average rooms with approximately 8 ft ceilings and relative humidity ranging from 10 to 75% at 75°F, the effective path length for carbon dioxide radiation is about 85% of the ceiling height, or 6.8 ft. The effective path length for water vapor is about 93% of the ceiling height, or 7.4 ft. The effective emittance of the water vapor and carbon dioxide radiating to the walls, ceiling, and floor of a room 16 ft by 48 ft with 8 ft ceilings is in the following tabulation.

Relative Humidity, %	ε_g
10	0.10
50	0.19
75	0.22

The radiation heat transfer from the gas to the walls is then

$$q = \sigma A_w \varepsilon_g (T_g^4 - T_w^4) \qquad (49)$$

The examples in Table 4 and the preceding text indicate the importance of gas radiation in environmental heat transfer problems. Gas radiation in large furnaces is the dominant mode of heat transfer, and many additional factors must be considered. Increased pressure broadens the spectral bands, and interaction of different radiating species prohibits simple summation of the emittance factors for the individual species. Departures from blackbody conditions necessitate separate calculations of the emittance and absorptance. McAdams (1954) and Hottel and Sarofim (1967) give more complete treatments of gas radiation.

NATURAL CONVECTION

Heat transfer involving motion in a fluid due to the difference in density and the action of gravity is called **natural convection** or **free convection**. Heat transfer coefficients associated with gases for natural convection are generally much lower than those for forced convection, and it is therefore important not to ignore radiation in calculating the total heat loss or gain. Radiant transfer may be of the same order of magnitude as natural convection, even at room

Table 5 Natural Convection Heat Transfer Coefficients

I. General relationships	$\text{Nu} = c(\text{Gr Pr})^n$	(1)
	$h = c\frac{k}{L}\left(\frac{L^3\rho^2\beta g\Delta t}{\mu^2}\right)_f^n\left(\frac{\mu c_p}{k}\right)_f^n$	(2)
Characteristic length L		
Vertical plates or pipes	L = height	
Horizontal plates	L = length	
Horizontal pipes	L = diameter	
Spheres	$L = 0.5 \times$ diameter	
Rectangular block, with horizontal length L_h and vertical length L_v	$1/L = (1/L_h) + (1/L_v)$	
II. Planes and pipes		
Horizontal or vertical planes, pipes, rectangular blocks, and spheres (excluding horizontal plates facing downward for heating and upward for cooling)		
(a) Laminar range, when Gr Pr is between 10^4 and 10^8	$\text{Nu} = 0.56(\text{Gr Pr})^{0.25}$	(3)
(b) Turbulent range, when Gr Pr is between 10^8 and 10^{12}	$\text{Nu} = 0.13(\text{Gr Pr})^{0.33}$	(4)
III. Wires		
For horizontal or vertical wires, use L = diameter, for Gr Pr between 10^{-7} and 1	$\text{Nu} = (\text{Gr Pr})^{0.1}$	(5)
IV. With air		
$\text{Gr Pr} = 1.6 \times 10^6 L^3 \Delta t$ (at 70°F, L in ft, Δt in °F)		
(a) Horizontal cylinders		
Small cylinder, laminar range	$h = 0.27(\Delta t/L)^{0.25}$	(6)
Large cylinder, turbulent range	$h = 0.18(\Delta t)^{0.33}$	(7)
(b) Vertical plates		
Small plates, laminar range	$h = 0.29(\Delta t/L)^{0.25}$	(8)
Large plates, turbulent range	$h = 0.19(\Delta t)^{0.33}$	(9)
(c) Horizontal plates, facing upward when heated or downward when cooled		
Small plates, laminar range	$h = 0.27(\Delta t/L)^{0.25}$	(10)
Large plates, turbulent range	$h = 0.22(\Delta t)^{0.33}$	(11)
(d) Horizontal plates, facing downward when heated or upward when cooled		
Small plates	$h = 0.12(\Delta t/L)^{0.25}$	(12)

temperatures, because wall temperatures in a room can affect human comfort (see Chapter 8).

Natural convection is important in a variety of heating and refrigeration equipment: (1) gravity coils used in high-humidity cold storage rooms and in roof-mounted refrigerant condensers, (2) the evaporator and condenser of household refrigerators, (3) baseboard radiators and convectors for space heating, and (4) cooling panels for air conditioning. Natural convection is also involved in heat loss or gain to equipment casings and interconnecting ducts and pipes.

Consider heat transfer by natural convection between a cold fluid and a hot surface. The fluid in immediate contact with the surface is heated by conduction, becomes lighter, and rises because of the difference in density of the adjacent fluid. The viscosity of the fluid resists this motion. The heat transfer is influenced by (1) gravitational force due to thermal expansion, (2) viscous drag, and (3) thermal diffusion. Gravitational acceleration g, coefficient of thermal expansion β, kinematic viscosity $\nu = \mu/\rho$, and thermal diffusivity $\alpha = k/\rho c_p$ affect natural convection. These variables are included in the dimensionless numbers given in Equation (1) in Table 5. The Nusselt number Nu is a function of the product of the Prandtl number Pr and the Grashof number Gr. These numbers, when combined, depend on the fluid properties, the temperature difference Δt between the surface and the fluid, and the characteristic length L of the surface. The constant c and the exponent n depend on the physical configuration and the nature of flow.

Natural convection cannot be represented by a single value of exponent n, but it can be divided into three regions:

1. **Turbulent** natural convection, for which n equals 0.33
2. **Laminar** natural convection, for which n equals 0.25
3. A region that has GrPr less than for laminar natural convection, for which the exponent n gradually diminishes from 0.25 to lower values

Note that for wires, the GrPr is likely to be very small, so that the exponent n is 0.1 [Equation (5) in Table 5].

To calculate the natural-convection heat transfer coefficient, determine GrPr to find whether the boundary layer is laminar or turbulent; then apply the appropriate equation from Table 5. The correct characteristic length indicated in the table must be used. Because the exponent n is 0.33 for a turbulent boundary layer, the characteristic length cancels out in Equation (2) in Table 5, and the heat transfer coefficient is independent of the characteristic length, as seen in Equations (7), (9), and (11) in Table 5. Turbulence occurs when length or temperature difference is large. Because the length of a pipe is generally greater than its diameter, the heat transfer coefficient for vertical pipes is larger than for horizontal pipes.

Convection from horizontal plates facing downward when heated (or upward when cooled) is a special case. Because the hot air is above the colder air, theoretically no convection should occur. Some convection is caused, however, by secondary influences such as temperature differences on the edges of the plate. As an approximation, a coefficient of somewhat less than half the coefficient for a heated horizontal plate facing upward can be used.

Because air is often the heat transport fluid, simplified equations for air are given in Table 5. Other information on natural convection is available in the section on Bibliography under Heat Transfer, General.

Observed differences in the comparison of recent experimental and numerical results with existing correlations for natural convective heat transfer coefficients indicate that caution should be used when applying coefficients for (isolated) vertical plates to vertical surfaces in enclosed spaces (buildings). Bauman et al. (1983) and

Altmayer et al. (1983) developed improved correlations for calculating natural convective heat transfer from vertical surfaces in rooms under certain temperature boundary conditions.

Natural convection can affect the heat transfer coefficient in the presence of weak forced convection. As the forced-convection effect (i.e., the Reynolds number) increases, "mixed convection" (superimposed forced-on-free convection) gives way to the pure forced-convection regime. In these cases, other sources describing combined free and forced convection should be consulted, since the heat transfer coefficient in the mixed-convection region is often larger than that calculated based on the natural- or forced-convection calculation alone. Metais and Eckert (1964) summarize natural-, mixed-, and forced-convection regimes for vertical and horizontal tubes. Figure 7 shows the approximate limits for horizontal tubes. Other studies are described by Grigull et al. (1982).

FORCED CONVECTION

Forced air coolers and heaters, forced air- or water-cooled condensers and evaporators, and liquid suction heat exchangers are examples of equipment that transfer heat primarily by forced convection.

When fluid flows over a flat plate, a **boundary layer** forms adjacent to the plate. The velocity of the fluid at the plate surface is zero and increases to its maximum free stream value just past the edge of the boundary layer (Figure 8). Boundary layer formation is important because the temperature change from plate to fluid (thermal resistance) is concentrated here. Where the boundary layer is thick, thermal resistance is great and the heat transfer coefficient is small. Flow within the boundary layer immediately downstream from the leading edge is laminar and is known as **laminar forced convection**. As flow proceeds along the plate, the laminar boundary layer increases in thickness to a critical value. Then, turbulent eddies develop within the boundary layer, except for a thin **laminar sublayer** adjacent to the plate.

The boundary layer beyond this point is a **turbulent boundary layer**, and the flow is **turbulent forced convection**. The region between the breakdown of the laminar boundary layer and the establishment of the turbulent boundary layer is the **transition region**. Because the turbulent eddies greatly enhance heat transport into the main stream, the heat transfer coefficient begins to increase rapidly through the transition region. For a flat plate with a smooth leading edge, the turbulent boundary layer starts at Reynolds numbers, based on distance from the leading edge, of about 300,000 to 500,000. In blunt-edged plates, it can start at much smaller Reynolds numbers.

For long tubes or channels of small hydraulic diameter, at sufficiently low flow velocity, the laminar boundary layers on each wall grow until they meet. Beyond this point, the velocity distribution does not change, and no transition to turbulent flow takes place. This is called **fully developed laminar flow**. For tubes of large diameter or at higher velocities, transition to turbulence takes place and **fully developed turbulent flow** is established (Figure 9). Therefore, the length dimension that determines the critical Reynolds number is the hydraulic diameter of the channel. For smooth circular tubes, flow is laminar for Reynolds numbers below 2100 and turbulent above 10,000.

Table 6 lists various forced-convection correlations. In the generalized, dimensionless formula of Equation (1) in Table 6, heat transfer is determined by flow conditions and by the fluid properties, as indicated by the Reynolds number and the Prandtl number. This equation can be modified to Equation (4) in Table 6 to get the **heat transfer factor** j. The heat transfer factor is related to the **friction factor** f by the interrelationship of the transport of momentum and heat; it is approximately $f/2$ for turbulent flow in straight ducts. These factors are plotted in Figure 10.

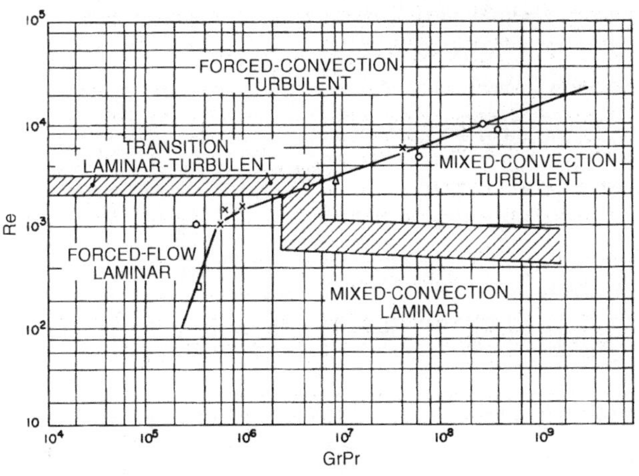

Fig. 7 Regimes of Free, Forced, and Mixed Convection for Flow-Through Horizontal Tubes

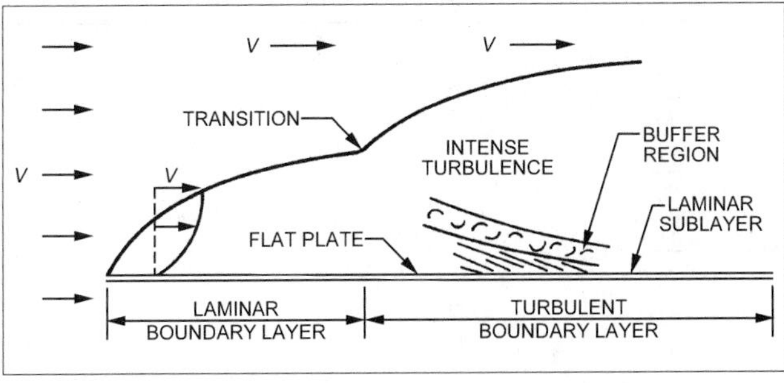

Fig. 8 Boundary Layer Buildup on Flat Plate (Vertical Scale Magnified)

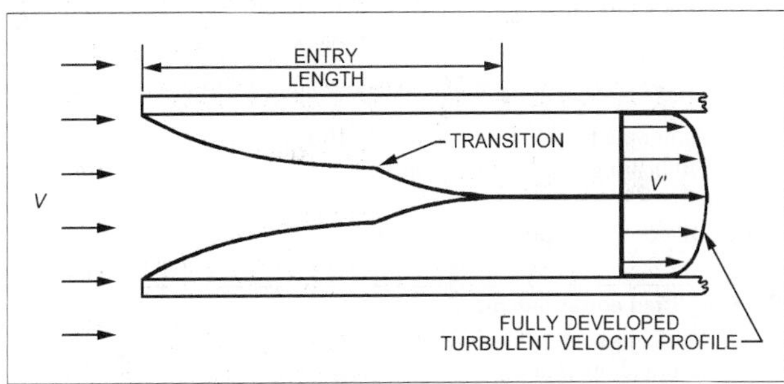

Fig. 9 Boundary Layer Buildup in Entry Length of Tube or Channel

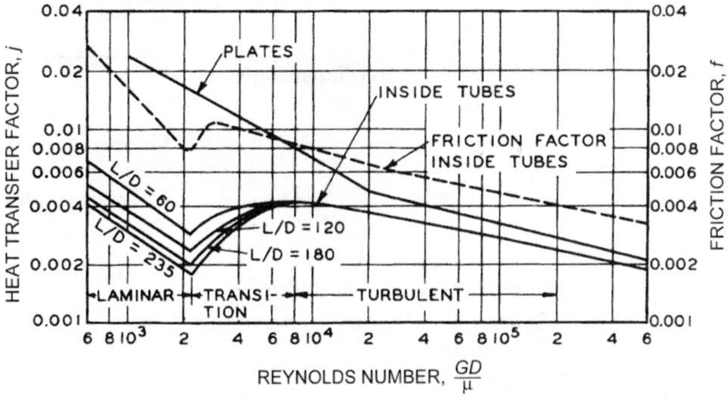

Fig. 10 Typical Dimensionless Representation of Forced-Convection Heat Transfer

Table 6 Equations for Forced Convection

Description	Reference Author	Page	Eq. No.	Equation		
I. Generalized correlations						
(a) Turbulent flow inside tubes	Jakob	491	(23-36)	$\frac{hD}{k} = c\left(\frac{GD}{\mu}\right)^m\left(\frac{\mu c_p}{k}\right)^n$		(1)
(1) Using fluid properties based on bulk temperature t	McAdams	219	(9-10a)	$\frac{hD}{k} = 0.023\left(\frac{GD}{\mu}\right)^{0.8}\left(\frac{\mu c_p}{k}\right)^{0.4}$	(See Note a)	(2)
(2) Same as (1), except μ at surface temperature t_s	McAdams	219	(9-10c)	$\frac{h}{c_p G}\left(\frac{c_p\mu}{k}\right)^{2/3}\left(\frac{\mu_s}{\mu}\right)^{0.14} = \frac{0.023}{(GD/\mu)^{0.2}}$		(3)
(3) Using fluid properties based on film temperature $t_f = 0.5(t_s + t)$, except c_p in Stanton modulus	McAdams	219	(9-10b)	$\frac{h}{c_p G}\left(\frac{c_p\mu}{k}\right)_f^{2/3} = \frac{0.023}{(GD/\mu_f)^{0.2}} = j$		(4)
(4) For viscous fluids (viscosities higher than twice water), using viscosity μ at bulk temperature t and μ_s at surface temperature t_s	Jakob	547	(26-12)	$\frac{hD}{k} = 0.027\left(\frac{GD}{\mu}\right)^{0.8}\left(\frac{\mu c_p}{k}\right)^{1/3}\left(\frac{\mu}{\mu_s}\right)^{0.14}$		(5)
(b) Laminar flow inside tubes						
(1) For large D or high Δt, the effect of natural convection should be included	Jakob	544	(26-5)	$\frac{hD}{k} = 1.86\left[\left(\frac{GD}{\mu}\right)\left(\frac{c_p\mu}{k}\right)\left(\frac{D}{L}\right)\right]^{1/3}\left(\frac{\mu}{\mu_s}\right)^{0.14}$		(6)
(2) For very long tubes				When $\left(\frac{GD}{\mu}\right)\left(\frac{c_p\mu}{k}\right)\left(\frac{D}{L}\right) < 20$, Eq. (6) should not be used		
(c) Annular spaces, turbulent flow. All fluid properties at bulk temperature except μ_s at surface temperature t_s	McAdams	242	(9-32c)	$\frac{h}{c_p G}\left(\frac{c_p\mu}{k}\right)^{2/3}\left(\frac{\mu_s}{\mu}\right)^{0.14} = \frac{0.023}{(DeG/\mu)^{0.2}}$		(7)
II. Simplified equations for gases, turbulent flow inside tubes (Units are in lb_m, h, ft, °F, and Btu.)						
(a) Most common gases, turbulent flow (assuming $\mu = 0.0455$ lb_m/ft · h and $\mu c_p/k = 0.78$)	Obtained from Eq. (2)			$h = 0.0144(c_p G^{0.8}/D^{0.2})$		(8)
(b) Air at ordinary temperatures	Obtained from Eq. (2)			$h = c(G^{0.8}/D^{0.2})$	(See Note b)	(9)
(c) Fluorinated hydrocarbon refrigerant gas at ordinary pressures	Obtained from Eq. (2)			$h = c(G^{0.8}/D^{0.2})$	(See Note b)	(10)
(d) Ammonia gas at approximately 150°F, 300 psi	Obtained from Eq. (2)			$h = 0.00756(G^{0.8}/D^{0.2})$		(11)
At 0°F, 24 psi	Obtained from Eq. (2)			$h = 0.00604(G^{0.8}/D^{0.2})$		(12)
III. Simplified equations for liquids, turbulent flow inside tubes (Units are in lb_m, h, ft, °F, and Btu.)						
(a) Water at ordinary temperatures, 40 to 200°F. V is velocity in fps, D is tube ID in inches.	McAdams	228	(9-19)	$h = \frac{150(1 + 0.011t)V^{0.8}}{D^{0.2}}$		(13)
(b) Fluorinated hydrocarbon refrigerant liquid	Obtained from Eq. (2)			$h = c(G^{0.8}/D^{0.2})$	(See Note b)	(14)
(c) Ammonia liquid at approximately 100°F	Obtained from Eq. (2)			$h = 0.0156(G^{0.8}/D^{0.2})$		(15)
(d) Oil heating, approximate equation	Brown and Marco	146	(7-15)	$h = 0.034V/\mu_f^{0.63}$		(16)
(e) Oil cooling, approximate equation	Brown and Marco	146	(7-15)	$h = 0.0255V/\mu_f^{0.63}$		(17)
IV. Simplified equations for air						
(a) Vertical plane surfaces, V of 16 to 100 fps (room temperature)[c]	McAdams	249	(9-42)	$h' = 0.5V^{0.78}$		(18)
(b) Vertical plane surfaces, $V < 16$ fps (room temperature)[c]	McAdams	249	(9-42)	$h' = 0.99 + 0.21V$		(19)
(c) Single cylinder cross flow (film temperature = 200°F) $1000 < GD/\mu_f < 50{,}000$	McAdams	261	(10-3c)	$h = 0.026(G^{0.6}/D^{0.4})$		(20)
(d) Single sphere $17 < GD/\mu_f < 70{,}000$	McAdams	265	(10-6)	$h = 0.37\frac{k_f}{D}\left(\frac{GD}{\mu_f}\right)^{0.6}$		(21)
V. Gases flowing normal to pipes (dimensionless)						
(a) Single cylinder Re from 0.1 to 1000	McAdams	260	(10-3)	$\frac{hD}{k_f} = 0.32 + 0.43\left(\frac{GD}{\mu}\right)^{0.52}$		(22)
Re from 1000 to 50,000	McAdams	260	(10-3)	$\frac{hD}{k_f} = 0.24\left(\frac{GD}{\mu_f}\right)^{0.6}$		(23)
(b) Unbaffled staggered tubes, 10 rows. Approximate equation for turbulent flow[d]	McAdams	272	(10-11a)	$\frac{hD}{k_f} = 0.33\left(\frac{G_{max}D}{\mu_f}\right)^{0.6}\left(\frac{\mu c_p}{k}\right)_f^{1/3}$		(24)
(c) Unbaffled in-line tubes, 10 rows. Approximate equation for turbulent flow[d] ($G_{max}D/\mu_f$) from 2000 to 32,000	McAdams	272	(10-11a)	$\frac{hD}{k_f} = 0.26\left(\frac{G_{max}D}{\mu_f}\right)^{0.6}\left(\frac{\mu c_p}{k}\right)_f^{1/3}$		(25)

[a] McAdams (1954) recommends this equation for heating and cooling. Others recommend exponents of 0.4 for heating and 0.3 for cooling, with a change in constant.
[b] Table 7 in Chapter 2 of the 1981 *ASHRAE Handbook—Fundamentals* lists values for c.
[c] h' is expressed in Btu/h·ft²·°F based on initial temperature difference.
[d] G_{max} is based on *minimum* free area. Coefficients for tube banks depend greatly on geometrical details. These values approximate only.

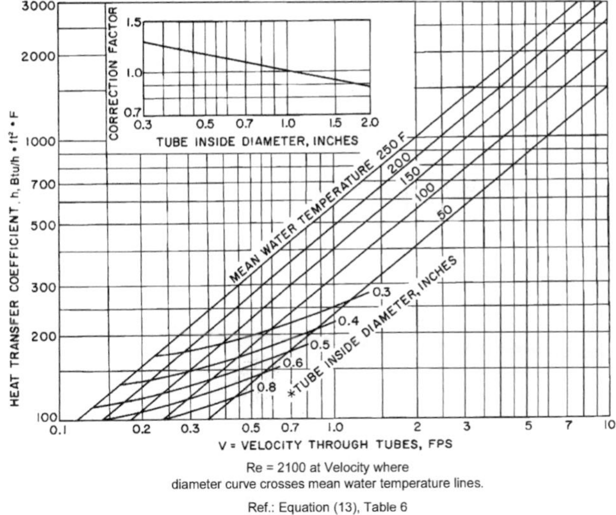

Fig. 11 Heat Transfer Coefficient for Turbulent Flow of Water Inside Tubes

The characteristic length D is the diameter of the tube, outside or inside, or the length of the plane plate. For other shapes, the hydraulic diameter D_h is used. With a uniform surface temperature and assuming uniform heat transfer coefficient the inlet and exit temperatures are related by:

$$D_h = 2r_h = 4 \times \frac{\text{Cross-sectional area for flow}}{\text{Total wetted perimeter}}$$

This reduces to twice the distance between surfaces for parallel plates or an annulus.

Simplified equations applicable to common fluids under normal operating conditions appear in Equations (8) through (25) of Table 6. Figure 11 gives graphical solutions for water. However, the value of the convective heat transfer coefficient with internal flows varies in the direction of the flow because of the temperature dependence of the properties of the fluids. In such a case a representative value of the heat transfer coefficient evaluated at the inlet and exit temperatures can be used in the above equation.

HEAT TRANSFER AUGMENTATION TECHNIQUES

As discussed by Bergles (1998), techniques applied to augment (enhance) heat transfer can be classified as passive methods, which require no direct application of external power, or as active schemes, which require external power. Examples of passive techniques include rough surfaces, extended surfaces, displaced promoters, and vortex flow devices. Examples of active techniques include mechanical aids, surface vibration, fluid vibration, and electrostatic fields. The effectiveness of a given augmentation technique depends largely on the mode of heat transfer or the type of heat exchanger to which it is applied.

When augmentation is used, the dominant thermal resistances in Equation (9) should be considered; that is, do not invest in reducing an already low thermal resistance (increasing an already high heat transfer coefficient). Additionally, heat exchangers with a large number of heat transfer units (NTU) show relatively small gains in effectiveness with augmentation [see Equations (24) and (26)]. Finally, the increased friction factor that usually accompanies the heat transfer augmentation must be considered.

Passive Techniques

Several examples of tubes with internal roughness or fins are shown in Figure 12. Rough surfaces of the spiral repeated rib variety are widely used to improve in-tube heat transfer with water, as in flooded chillers. The roughness may be produced by spirally indenting the outer wall, forming the inner wall, or inserting coils. Longitudinal or spiral internal fins in tubes can be produced by extrusion or forming and give a substantial increase in the surface area. The fin efficiency (see the section on Fin Efficiency on p. 3.20) can usually be taken as unity. Twisted strips (vortex flow devices) can be inserted as original equipment or as retrofit devices. From a practical point of view, the twisted tape width should be such that the tape can be easily inserted or removed. Ayub and Al-Fahed (1993) addressed the issue of the clearance between the twisted tape and tube inside dimension.

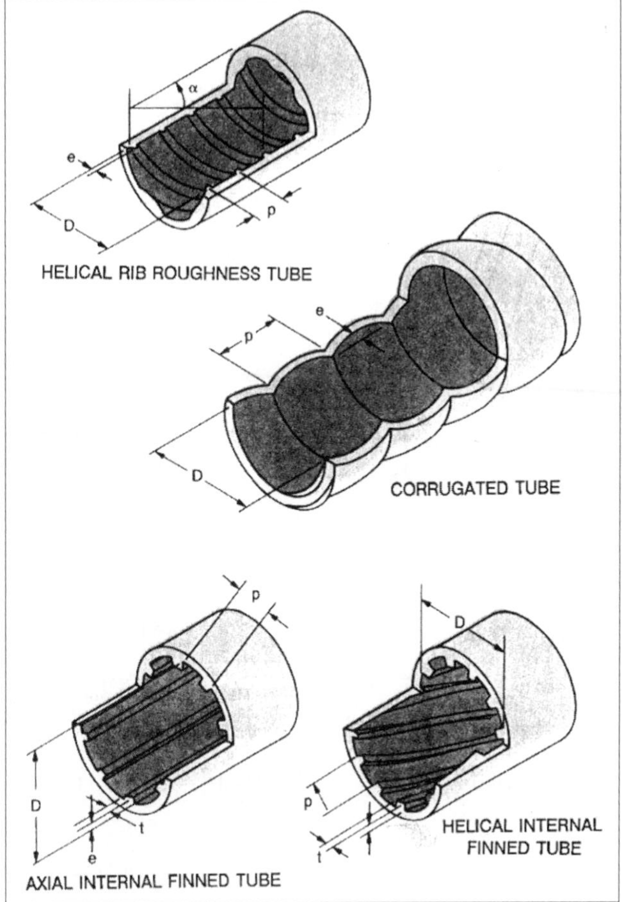

Fig. 12 Typical Tube-Side Enhancements

Microfin tubes (internally finned tubes with about 60 short fins around the circumference) are widely used in refrigerant evaporation and condensers. Since the gas entering the condenser in vapor-compression refrigeration is superheated, a considerable portion of the condenser acts to desuperheat the flow (i.e., it is single phase). Some data on the single-phase performance of microfin tubes, showing considerably higher heat transfer coefficients than for plain tubes, are available (e.g., Khanpara et al. 1986, Al-Fahed et al. 1993), but the upper Reynolds numbers of about 10,000 are lower than those found in practice. This deficiency is being addressed in a current ASHRAE research project.

The increased friction factor may not require increased pumping power if the flow rate can be adjusted or if the length of the heat exchanger can be reduced. Nelson and Bergles (1986) discuss this issue of performance evaluation criteria, especially for HVAC applications.

Of concern in chilled water systems is the fouling that in some cases may seriously reduce the overall heat transfer coefficient U. In general, fouled enhanced tubes perform better than fouled plain tubes,

Table 7 Equations for Augmented Forced Convection (Single Phase)

Description	Equation	
I. Turbulent in-tube flow of liquids (a) Spiral repeated rib[a]	$\frac{h_a}{h_s} = \left\{\left[1 + 2.64\left(\frac{GD}{\mu}\right)^{0.036}\left(\frac{e}{d}\right)^{0.212}\left(\frac{p}{d}\right)^{-0.21}\left(\frac{\alpha}{90}\right)^{0.29}\left(\frac{c_p\mu}{k}\right)^{-0.024}\right]^7\right\}^{1/7}$ $\frac{f_a}{f_s} = \left\{1 + \left[29.1\left(\frac{GD}{\mu}\right)^{w}\left(\frac{e}{d}\right)^{x}\left(\frac{p}{d}\right)^{y}\left(\frac{\alpha}{90}\right)^{z}\left(1 + \frac{2.94}{n}\right)\sin\beta\right]^{15/16}\right\}^{16/15}$ *where* $w = 0.67 - 0.06(p/d) - 0.49(\alpha/90)$ $x = 0.37 - 0.157(p/d)$ $y = -1.66 \times 10^{-6}(GD/\mu) - 0.33(\alpha/90)$ $z = 4.59 + 4.11 \times 10^{-6}(GD/\mu) - 0.15(p/d)$ $h_s = \frac{(k/D)(f_s/2)\left(\frac{GD}{\mu}\right)\left(\frac{c_p\mu}{k}\right)}{1 + 12.7(f_s/2)^{0.5}\left[\left(\frac{c_p\mu}{k}\right)^{2/3} - 1\right]}$ $f_s = \left[1.58\ln\left(\frac{GD}{\mu}\right) - 3.28\right]^{-2}$	
(b) Fins[b]	$\frac{hD_h}{k} = 0.023\left(\frac{c_p\mu}{k}\right)^{0.4}\left(\frac{GD_h}{\mu}\right)^{0.8}\left(\frac{A_F}{A_{Fi}}\right)^{0.1}\left(\frac{A_i}{A}\right)^{0.5}(\sec\alpha)^3$ $f_h = 0.046\left(\frac{GD_h}{\mu}\right)^{-0.2}\left(\frac{A_F}{A_{Fi}}\right)^{0.5}(\sec\alpha)^{0.75}$	Note that in computing the Reynolds number for (b) and (c) there is allowance for the reduced cross-sectional area.
(c) Twisted-strip inserts[c]	$\left(\frac{hd}{k}\right)\Big/\left(\frac{hd}{k}\right)_{y=\infty} = [1 + 0.769/y]$ $\left(\frac{hd}{k}\right)_{y=\infty} = 0.023\left(\frac{GD}{\mu}\right)^{0.8}\left(\frac{c_p\mu}{k}\right)^{0.4}\left(\frac{\pi}{\pi - 4\delta/d}\right)^{0.8}\left(\frac{\pi + 2 - 2\delta/d}{\pi - 4\delta/d}\right)^{0.2}\phi$ *where* $\phi = (\mu_b/\mu_w)^n$ $n = \begin{cases} 0.18 \text{ for liquid heating} \\ 0.30 \text{ for liquid cooling} \end{cases}$ $f = \frac{0.0791}{(GD/\mu)^{0.25}}\left(\frac{\pi}{\pi - 4\delta/d}\right)^{1.75}\left(\frac{\pi + 2 - 2\delta/d}{\pi - 4\delta/d}\right)^{1.25}\left(1 + \frac{2.752}{y^{1.29}}\right)$	
II. Turbulent in-tube flow of gases (a), (b) Bent-strip inserts[d]	$\frac{hD}{k}\left(\frac{T_w}{T_b}\right)^{0.45} = 0.258\left(\frac{GD}{\mu}\right)^{0.6}$ $\frac{hD}{k}\left(\frac{T_w}{T_b}\right)^{0.45} = 0.208\left(\frac{GD}{\mu}\right)^{0.63}$	
(c) Twisted-strip inserts[d]	$\frac{hD}{k}\left(\frac{T_w}{T_b}\right)^{0.45} = 0.122\left(\frac{GD}{\mu}\right)^{0.65}$	
(d) Bent-tab inserts[d]	$\frac{hD}{k}\left(\frac{T_w}{T_b}\right)^{0.45} = 0.406\left(\frac{GD}{\mu}\right)^{0.54}$	Note that in computing the Reynolds number there is no allowance for the flow blockage of the insert.
III. Offset strip fins for plate-fin heat exchangers[e]	$\frac{h}{c_pG} = 0.6522\left(\frac{GD_h}{\mu}\right)^{-0.5403}\alpha^{-0.1541}\delta^{0.1499}\gamma^{-0.0678}\left[1 + 5.269 \times 10^{-5}\left(\frac{GD_h}{\mu}\right)^{1.340}\alpha^{0.504}\delta^{0.456}\gamma^{-1.055}\right]^{0.1}$ $f_h = 9.6243\left(\frac{GD_h}{\mu}\right)^{-0.7422}\alpha^{-0.1856}\delta^{-0.3053}\gamma^{-0.2659}\left[1 + 7.669 \times 10^{-8}\left(\frac{GD_h}{\mu}\right)^{4.429}\alpha^{0.920}\delta^{3.767}\gamma^{0.236}\right]^{0.1}$ where h/c_pG, f_h, and GD_h/μ are based on the hydraulic diameter, given by $D_h = 4shl/[2(sl + hl + th) + ts]$	

References:
[a]Ravigururajan and Bergles (1985)
[b]Carnavos (1979)
[c]Manglik and Bergles (1993)
[d]Junkhan et al. (1985)
[e]Manglik and Bergles (1990)

as shown in studies of scaling due to cooling tower water (Knudsen and Roy 1983) and particulate fouling (Somerscales et al. 1991). A comprehensive review of fouling with enhanced surfaces is presented by Somerscales and Bergles (1997).

Fire-tube boilers are frequently fitted with **turbulators** to improve the turbulent convective heat transfer coefficient constituting the dominant thermal resistance. Also, due to the high gas temperatures, radiation from the convectively heated insert to the tube wall can represent as much as 50% of the total heat transfer. (Note, however, that the magnitude of the convective contribution decreases as the radiative contribution increases because of the reduced temperature difference.) Two commercial bent-strip inserts, a twisted-strip insert, and a simple bent-tab insert are depicted in Figure 13. Design equations, for convection only, are included in Table 7. Beckermann and Goldschmidt (1986) present procedures to include radiation, and Junkhan et al. (1985, 1988) give friction factor data and performance evaluations.

Several enhanced surfaces for gases are depicted in Figure 14. The offset strip fin is an example of an interrupted fin that is often found in compact plate fin heat exchangers used for heat recovery from exhaust air. Design equations are included in Table 7. These equations are comprehensive in that they apply to laminar and transitional flow as well as to turbulent flow, which is a necessary feature because the small hydraulic diameter of these surfaces drives the Reynolds number down. Data for other surfaces (wavy, spine, louvered, etc.) are given in the section on Bibliography.

Plastic heat exchangers have been suggested for HVAC applications (Pescod 1980) and are being manufactured for refrigerated sea water (RSW) applications. They could be made of materials impervious to corrosion, say from acidic condensate when cooling a gaseous stream (flue gas heat recovery), and could easily be manufactured with enhanced surfaces. Several companies now offer heat exchangers in plastic, including various enhancements.

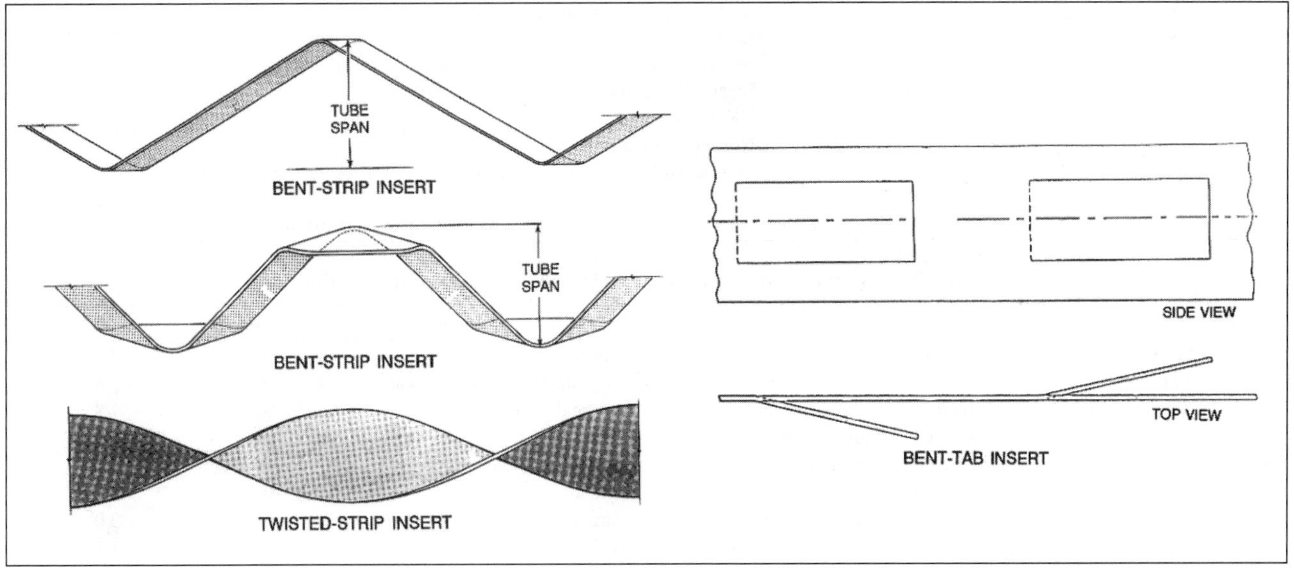

Fig. 13 Turbulators for Fire-Tube Boilers

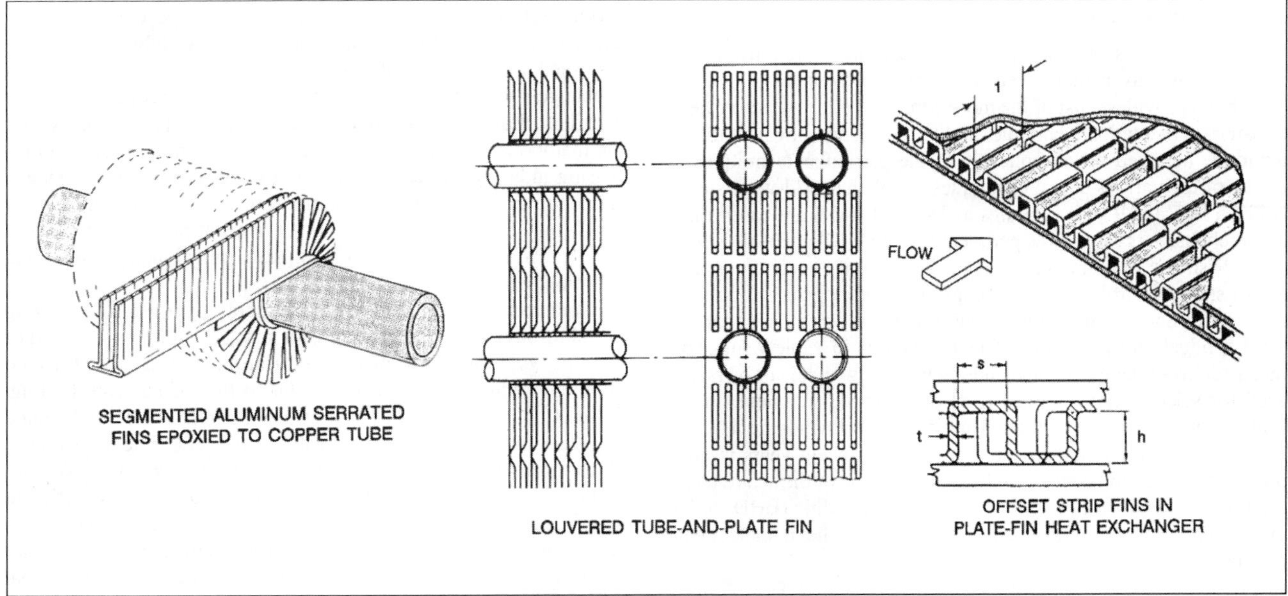

Fig. 14 Enhanced Surfaces for Gases

Table 8 Active Heat Transfer Augmentation Techniques and the Most Relevant Heat Transfer Modes

	Heat Transfer Mode					
Technique	Forced Convection (Gases)	Forced Convection (Liquids)	Boiling	Evaporation	Condensation	Mass Transfer
Mechanical aids	NA	B	C	C	NA	B
Surface vibration	B	B	B	B	B	A
Fluid vibration	C	B	B	B	D	B
Electrostatic/Electrohydrodynamic	B	B	A	A	A	A
Suction/Injection	C	B	NA	NA	B	B
Jet impingement	B	B	NA	B	NA	C
Rotation	C	C	A	A	A	A
Induced flow	B	B	NA	NA	NA	C

A = Most significant B = Significant C = Somewhat significant D = Not significant NA = Not believed to be applicable

Table 9 World-Wide Status of Active Techniques

Technique	Country or Countries
Mechanical aids	Universally used in selected applications
Surface vibration	USA; not significant
Fluid vibration	Sweden; mostly used for sonic cleaning
Electrostatic/Electrohydrodynamic	Japan, USA and UK; successful prototypes in operation in Japan
Other electrical methods	UK, France, and USA
Suction/Injection	No significant activity
Jet impingement	France and USA; high temperature units and aerospace applications
Rotation	US industry and R&D-based in UK
Induced flow	USA leader in the technology, particularly in combustion

Table 10 Selected Studies on Mechanical Aids, Suction, and Injection

Source	Process	Heat Transfer Surface	Fluid	Maximum α
Valencia et al. (1996)	Natural convection	Fin type tube	Air	0.5
Jeng et al. (1995)	Natural convection/Suction	Asymmetric isothermal wall	Air	1.4
Inagaki and Komori (1993)	Turbulent natural convection/Suction	Vertical plate	Air	1.8
Dhir et al. (1992)	Forced convection/Injection	Tube	Air	1.45
Duignan et al. (1993)	Forced convection/Film boiling	Horizontal plate	Air	2.0
Son and Dhir (1993)	Forced convection/Injection	Annuli	Air	1.85
Malhotra and Mujumdar (1991)	Water to bed/Stirring	Granular bed	Air	3.0
Aksan and Borak (1987)	Pool of water/Stirring	Tube coils	Water	1.7
Hagge and Junkhan (1975)	Forced convection/Scraping	Cylindrical wall	Air	11.0
Hu and Shen (1996)	Turbulent natural convection	Converging ribbed tube	Air	1.0

Active Techniques

Unlike the passive techniques, active techniques require use of external power to sustain the enhancement mechanism.

Table 8 provides a list of the more commonly known active heat transfer augmentation techniques and the corresponding heat transfer mode believed most applicable to the particular technique. A listing of the various active techniques and their world-wide status is given in Table 9. The rankings in Tables 8 and 9 are based on a comprehensive review of the pertinent literature (Ohadi et al. 1996). Table 9 shows that among the eight listed active techniques, the electrostatic/electrohydrodynamic and rotation appear to apply to almost all important heat transfer modes, at least from an enhancement applicability view point. The information in Table 9 suggests that aside from the mechanical aids technique, which is universally used for selected applications, most other active techniques have limited commercial use so far and are still in the development stage. However, the significant research progress in recent years is now promising expedited commercialization for some of the active techniques, such as the electrostatic/electrohydrodynamic (EHD) technique. The following is a brief summary of the enhancement techniques listed in Table 9.

Mechanical Aids. Augmentation by mechanical aids involves stirring the fluid by mechanical means, such as mixers, stirrers, or surface scrapers. Stirrers and mixers that scrape the surface are extensively used in the chemical processing of highly viscous fluids, such as the flow of highly viscous plastic with air. Heat exchangers that employ mechanical aids for enhancement are often called **mechanically assisted heat exchangers**. The surface scraping method is widely used for viscous liquids in the chemical processing industry and can be applied to duct flow of gases. Hagge and Junkhan (1974) reported tenfold improvement in heat transfer coefficient for laminar flow of air over a flat plate. Table 10 provides a listing of selected works on mechanical aids, suction, and injection.

Injection. This method involves supplying a gas to a flowing liquid through a porous heat transfer surface or injecting a fluid of a similar type upstream of the heat transfer test section. Injected bubbles produce an agitation similar to that of nucleate boiling. Gose et al. (1957) bubbled gas through sintered or drilled heated surfaces and found that the heat transfer coefficient increased 500% in laminar flow and about 50% in turbulent flow. Wayner and Bankoff (1965) demonstrated that the heat transfer coefficient could be increased by 150% if a porous block was placed on the surface to stabilize the flow of liquid toward the surface. Tauscher et al. (1970) demonstrated up to a five-fold increase in local heat transfer coefficients by injecting a similar fluid into a turbulent tube flow, but the effect dies out at a length-to-diameter ratio of 10.

Table 11 Selected Studies on Rotation

Source	Process	Heat Transfer Surface	Fluids	Rotation Speed, rpm	Max. α
Prakash and Zerle (1995)	Natural convection	Ribbed duct	Air	Given as a function	1.3
Mochizuki et al. (1994)	Natural convection	Serpentine duct	Air	Given as a function	3.0
Lan (1991)	Solidification	Vertical tube	Water	400	NA
McElhiney and Preckshot (1977)	External condensation	Horizontal tube	Steam	40	1.7
Nichol and Gacesa (1970)	External condensation	Vertical cylinder	Steam	2700	4.5
Astaf'ev and Baklastov (1970)	External condensation	Circular disc	Steam	2500	3.4
Tang and McDonald (1971)	Nucleate boiling	Horizontal heated circular cylinder	R-113	1400	<1.2
Marto and Gray (1971)	In-tube boiling	Vertical heated circular cylinder	Water	2660	1.6

NA = Not Available

Table 12 Selected Studies on EHD Technique

Source	Process	Heat Transfer Surface/Electrode	Fluid	P/Q, %	Max. α
Poulter and Allen (1986)	Internal flow	Tube/Wire	Aviation fuel-hexane	NA	20
Fernandez and Poulter(1987)	Internal flow	Tube/Wire	Transformer oil	NA	23
Ohadi et al. (1996)	Internal flow	Smooth surface/Rod	PAO (oil-based fluid)	1.2	3.2
Ohadi et al. (1991)	Internal flow	Tube/Wire	Air	15	3.2
Owsenek and Seyed-Yagoobi(1995)	Forced convection	Horizontal flat plate	Air	NA	25
Ohadi et al. (1995)	In-tube boiling	Microfin tube/Helical	R-134a	0.1	6.5
Ohadi et al. (1995)	In-tube condensation	Smooth tube/Rod wire	R-134a	0.1	7.0
Wawzyniak and Seyed-Yagoobi (1996)	External condensation	Enhanced and smooth tubes/ Circular rods	R-113	0.08	6.2
Seyed-Yagoobi et al. (1996)	Pool boiling	Horizontal tube/ Straight and circular wires	R-123	0.1	12.6

NA = Not Available; P = EHD Power; Q = Heat Exchanger Capacity

The practical application of injection appears to be rather limited because of difficulty of cost-effective supplying and removing of the injection fluid.

Suction. The suction method involves fluid removal through a porous heated surface leading to reduced heat/mass transfer resistance at the surface. Kinney (1968) and Kinney and Sparrow (1970) reported that applying suction at the surface increased heat transfer coefficients for laminar film and turbulent flows, respectively. For laminar film condensation, Antonir and Tamir (1977) and Lienhard and Dhir (1972) indicated that heat transfer coefficient can be improved by several hundred percent when film thickening is reduced by suction. Jeng et al. (1995) conducted experiments on a vertical parallel channel with asymmetric, isothermal walls. A porous wall segment was embedded into a segment of the test section wall, and enhancement occurred as the hot air was sucked from the channel. The local heat transfer coefficient increased with increasing porosity. The maximum heat transfer enhancement obtained was 140%.

Fluid or Surface Vibration. Fluid or surface vibrations are natural processes that occur in most heat exchangers; however, naturally occurring vibration is rarely factored into thermal design. Vibration equipment is expensive, and use of this technique for heat transfer enhancement does not have industrial applications at this stage of development. Vibrations of a wire in a forced convecting airflow enhanced the heat transfer to air up to 300% (Nesis et al. 1994), depending on the amplitude of vibrations and frequency. By using standing waves in a fluid, the input power was reduced by 75% compared with a fan that provides the same heat transfer rate (Woods 1992). Lower frequencies are preferable because they are less harmful to those who use this method of augmenting heat transfer.

Rotation. Rotation is a type of heat transfer enhancement that occurs naturally in rotating electrical machinery, gas turbine blades, and some other equipment. The rotating evaporator, the rotating heat pipe, the Higee distillation column, and the Rotex absorption cycle heat pump are typical examples of previous work in this area. In rotating evaporators, the rotation effectively distributes the liquid on the outside surface of the rotating surface. Rotation of the heat transfer surface also seems to be a promising method for effectively removing the condensate and decreasing liquid film thickness. Substantial increases in heat transfer coefficients have been demonstrated by using centrifugal force, which may be several times greater than the gravity force.

As shown in Table 11, the heat transfer enhancement obtained in the various studies varies from slight improvement up to 450%, depending on the system and rotation speed. The rotation technique is of particular interest for use in two-phase flows, particularly in boiling and condensation. It has been demonstrated that this technique is not effective in the gas-to-gas heat recovery mode in laminar flow, but its application is more likely in turbulent flow. High power consumption, sealing and vibration problems, moving parts, and the expensive equipment required for rotation are some drawbacks of the rotation technique.

Electrohydrodynamics. The electrohydrodynamic (EHD) enhancement of heat transfer refers to the coupling of an electric field with the fluid field in a dielectric fluid medium. The net effect is the production of secondary motions that destabilize the thermal boundary layer near the heat transfer surface, leading to heat transfer coefficients that are often an order of magnitude higher than those achievable by any of the conventional enhancement techniques. Among the various active augmentation techniques, EHD has benefited the most from substantial research in the past two decades in Japan, the United States, and the United Kingdom (Ohadi 1991). Its applicability for heat transfer enhancement in many applications has already been demonstrated, including for refrigeration/HVAC systems, process heat exchangers, waste heat recovery devices, cryogenics, aircraft environmental control systems, avionic cooling systems, and space thermal systems.

Selected work on EHD heat transfer enhancement is listed in Table 12. The work has involved studies on phase change processes as well as the indicated single-phase process. In fact, EHD-enhanced boiling and condensation have attracted the most attention from industrial and academic researchers.

The EHD effect is generally applied by placing wire or plate electrodes parallel and adjacent to the heat transfer surface. Figure 15 presents four electrode configurations for augmentation of forced-convection heat/mass transfer in-tube flows. A high-voltage, low-current electric field charges the electrode and establishes the electrical body force required to initiate and sustain augmentation.

When compared with other active techniques, a number of important advantages have contributed to the fast progress of the

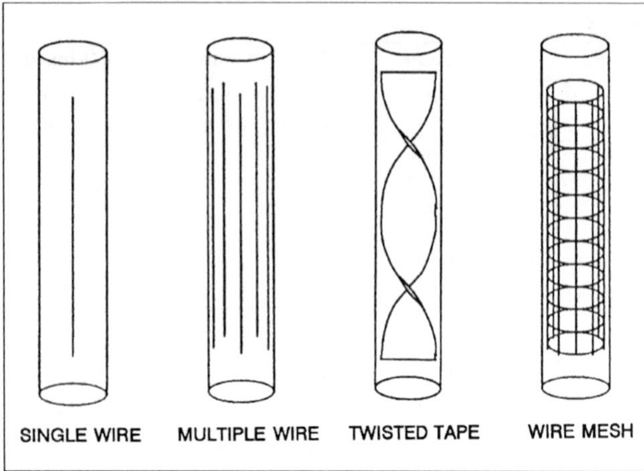

Fig. 15 Electrode Configurations for Internal Forced-Convection Flow

EHD technique in recent years. Rotation, injection, and vibration are generally mechanically complex and somewhat cumbersome to manufacture. Furthermore, the energy required to operate these systems can be a significant fraction of the power employed in pumping the fluid. In these respects, the EHD method of enhancement is superior.

The safety aspect of the EHD technique may be misjudged if careful attention is not paid to the manner in which these systems operate. Although EHD systems work at high voltages, the fact that very small currents are employed reduces the hazard of EHD fields to well below that of conventional low-voltage household appliances and ensures that the electrical power consumed by the EHD process is extremely small (less than 1% in most cases). Although the EHD technique appears to be well ahead of other active techniques, a number of issues remain to be addressed before successful implementation of this technique in practical heat exchangers can be realized. The most important issues include (1) any long-term effects the electric field may have on the heat exchanger, working fluid, and components, (2) the development of low-cost, high-voltage power supplies, and (3) identifying manufacturing processes that can lead to inexpensive mass production of EHD-enhanced heat exchangers. The encouraging news is that research work addressing some of the issues has already been initiated by researchers in the academic and government sectors and in private companies and research institutions. Additional details on the principles, applicability, and limitations of the EHD technique can be found elsewhere (Ohadi et al. 1991; Ohadi et al. 1999; Yabe 1991).

EXTENDED SURFACE

Heat transfer from a prime surface can be increased by attaching **fins** or **extended surfaces** to increase the area available for heat transfer. Fins provide a more compact heat exchanger with lower material costs for a given performance. To achieve optimum design, fins are generally located on the side of the heat exchanger where the heat transfer coefficients are low (such as the air side of an air-to-water coil). Equipment with an extended surface includes natural- and forced-convection coils and shell-and-tube evaporators and condensers. Fins are also used inside tubes in condensers and dry expansion evaporators.

Fin Efficiency

As heat flows from the root of a fin to its tip, temperature drops because of the thermal resistance of the fin material. The temperature difference between the fin and the surrounding fluid is therefore greater at the root than at the tip, causing a corresponding variation in the heat flux. Therefore, increases in fin length result in proportionately less additional heat transfer. To account for this effect, **fin efficiency** ϕ is defined as the ratio of the actual heat transferred from the fin to the heat that would be transferred if the entire fin were at its root or base temperature:

$$\phi = \frac{\int h(t - t_e)dA}{\int h(t_r - t_e)dA} \tag{50}$$

where ϕ is the fin efficiency, t_e is the temperature of the surrounding environment, and t_r is the temperature at the fin root. Fin efficiency is low for long fins, thin fins, or fins made of low thermal conductivity material. Fin efficiency decreases as the heat transfer coefficient increases because of the increased heat flow. For natural convection in air-cooled condensers and evaporators, where h for the air side is low, fins can be fairly large and fabricated from low-conductivity materials such as steel instead of from copper or aluminum. For condensing and boiling, where large heat transfer coefficients are involved, fins must be very short for optimum use of material.

The heat transfer from a finned surface, such as a tube, which includes both finned or secondary area A_s and unfinned or prime area A_p is given by the following equation:

$$q = (h_p A_p + \phi h_s A_s)(t_r - t_e) \tag{51}$$

Assuming the heat transfer coefficients for the finned surface and prime surface are equal, a *surface efficiency* ϕ_s can be derived for use in Equation (52).

$$\phi_s = 1 - \left(\frac{A_s}{A}\right)(1 - \phi) \tag{52}$$

$$q = \phi_s hA(t_r - t_e) \tag{53}$$

where A is the total surface area, equal to the sum of the finned and prime areas ($A = A_s + A_p$).

Temperature distribution and fin efficiencies for various fin shapes are derived in most heat transfer texts. Figures 16 through 19 show curves and equations for annular fins, straight fins, and spines. For constant thickness square fins, the efficiency of a constant thickness annular fin of the same area can be used. More accuracy, particularly with rectangular fins of large aspect ratio, can be obtained by dividing the fin into circular sectors (Rich 1966).

Rich (1966) presents results for a wide range of geometries in a compact form for equipment designers by defining a dimensionless thermal resistance Φ:

$$\Phi = \frac{R_f t_o k}{l^2} \tag{54}$$

$$R_f = \left(\frac{1}{h}\right)\left(\frac{1}{\phi} - 1\right) \tag{55}$$

where

Φ = dimensionless thermal resistance
ϕ = fin efficiency
t_o = fin thickness at fin base
l = length dimension = $r_t - r_o$ for annular fins
W = for rectangular fins

Rich (1966) also developed expressions for Φ_{max}, the maximum limiting value of Φ. Figure 20 gives Φ_{max} for annular fins of constant and tapered cross section as a function of $R = r_t/r_o$ (i.e., the

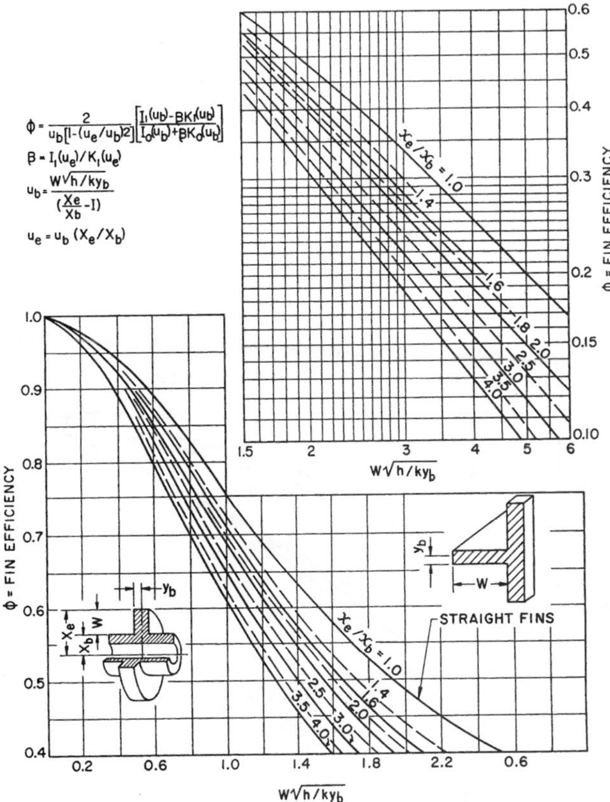

Fig. 16 Efficiency of Annular Fins of Constant Thickness

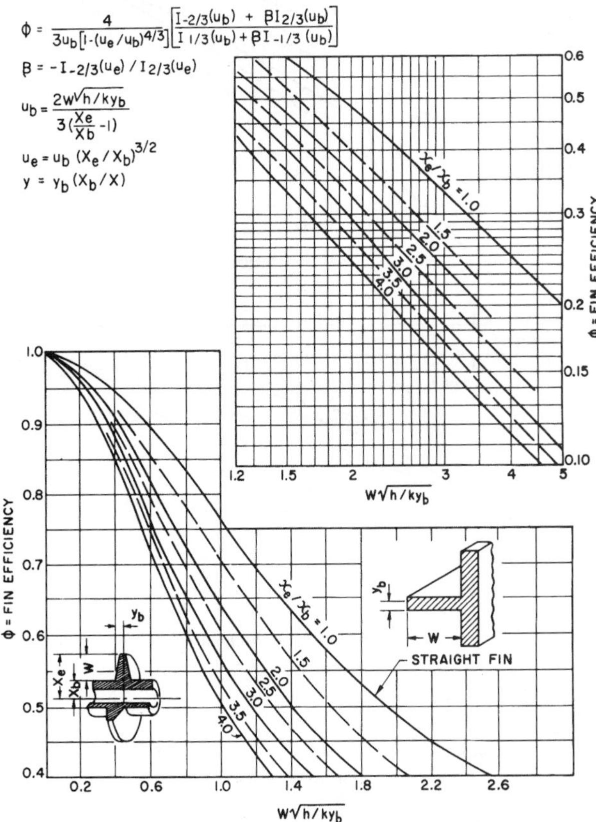

Fig. 17 Efficiency of Annular Fins with Constant Metal Area for Heat Flow

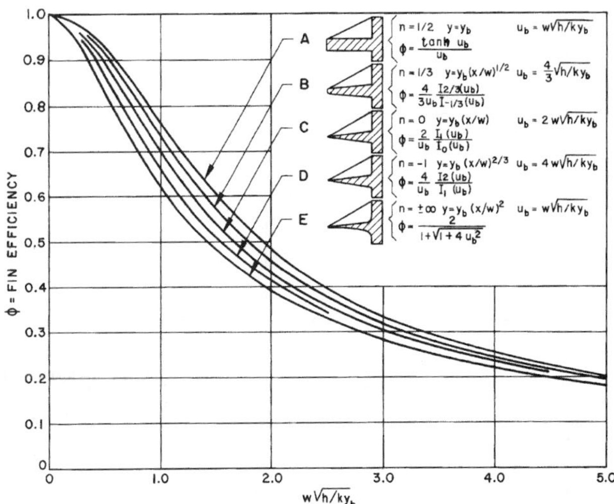

Fig. 18 Efficiency of Several Types of Straight Fin

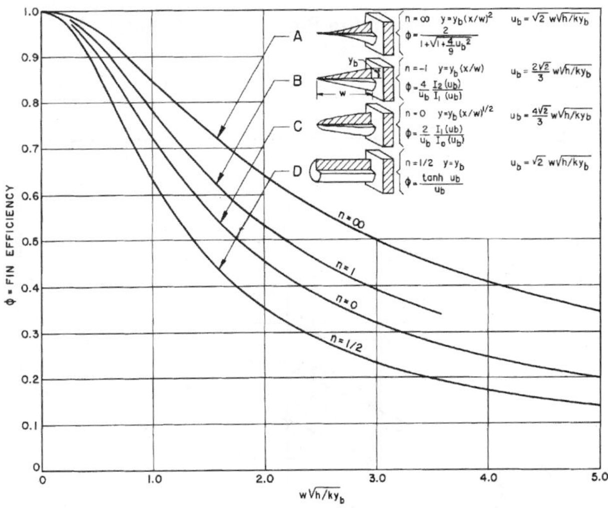

Fig. 19 Efficiency of Four Types of Spine

ratio of the fin tip-to-root radii). Figure 21 gives Φ_{max} for rectangular fins of a given geometry as determined by the sector method. Figure 22 gives correction factors (Φ/Φ_{max}) for the determination of Φ from Φ_{max} for both annular and rectangular fins.

Example. This example illustrates the use of the fin resistance for a rectangular fin typical of that for an air-conditioning coil.

Given: L = 0.75 in. $\quad t_o$ = 0.006 in.
W = 0.50 in. $\quad h$ = 10 Btu/h·ft²·°F
r_o = 0.25 in. $\quad k$ = 100 Btu·ft/h·ft²·°F

Solution: From Figure 18 at $W/r_o = 2.0$ and $L/W = 1.5$,

$$\Phi_{max} = R_{f(max)} t_o k/W^2 = 1.12$$

$$R_{f(max)} = \frac{1.12 \times 0.50^2}{0.006 \times 100 \times 12} = 0.0389 \text{ ft}^2\cdot°\text{F}\cdot\text{h/Btu}$$

The correction factor Φ/Φ_{max}, which is multiplied by $R_{f(max)}$ to give R_f, is given in Figure 22 as a function of the fin efficiency. As a first approximation, the fin efficiency is calculated from Equation (54a) assuming $R_f = R_{f(max)}$.

$$\phi = 1/(1 + hR_f) \approx 0.72$$

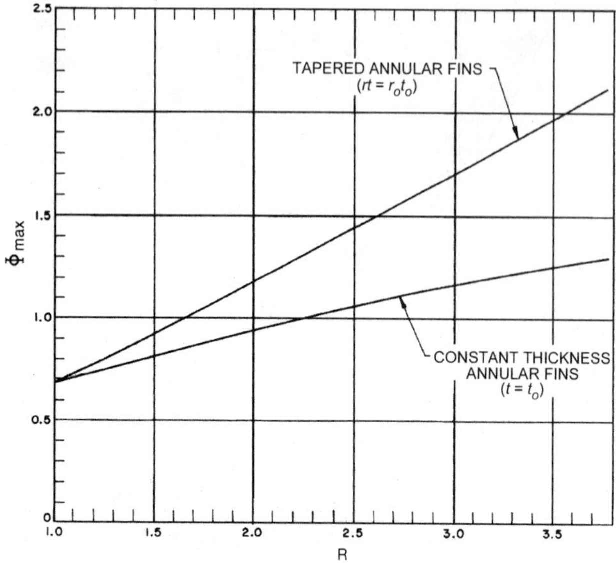

Fig. 20 Maximum Fin Resistance of Annular Fins
(Gardner 1945)

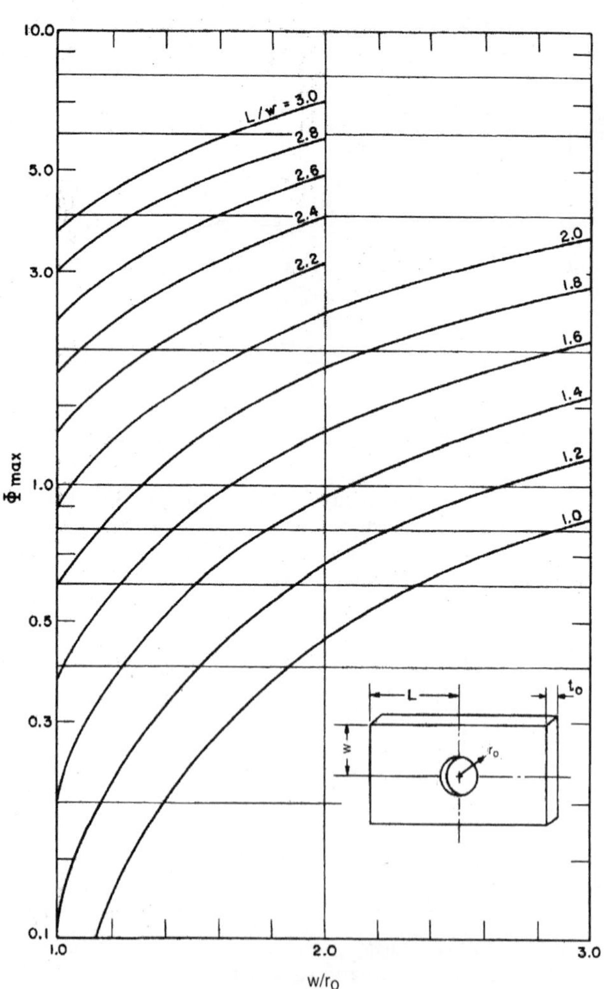

Fig. 21 Maximum Fin Resistance of Rectangular Fins Determined by Sector Method

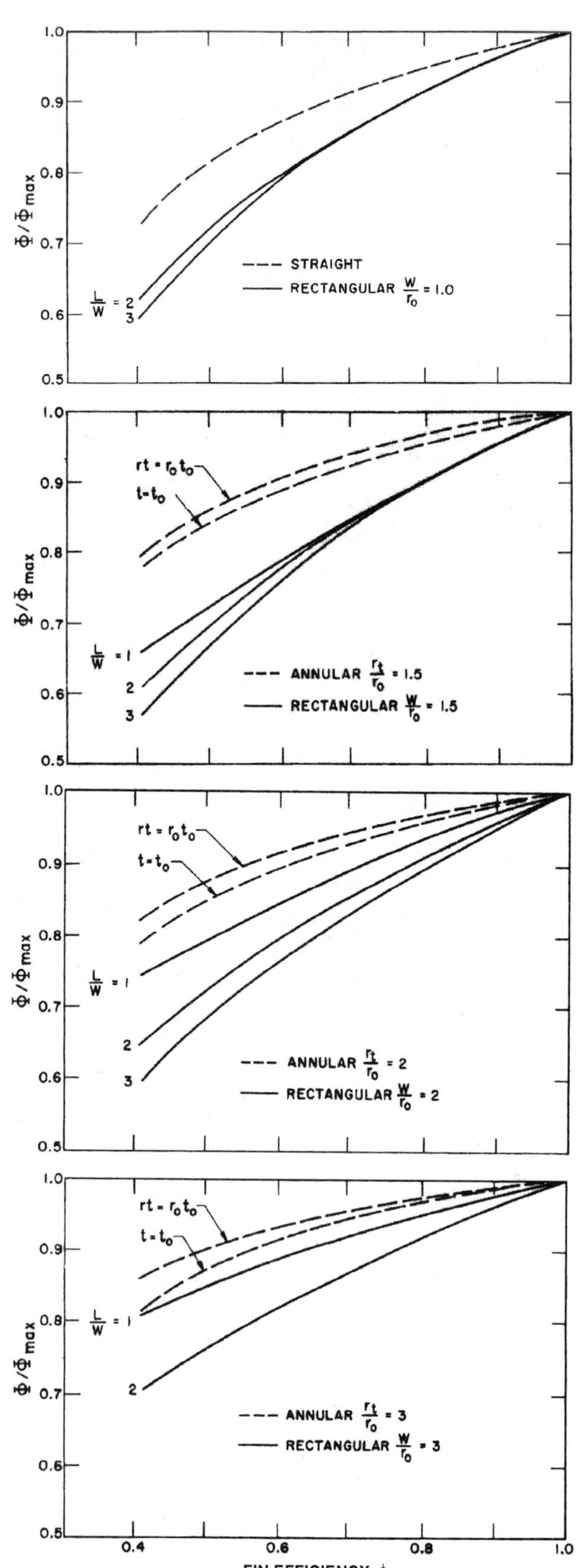

Fig. 22 Variation of Fin Resistance with Efficiency for Annular and Rectangular Fins
(Gardner 1945)

Interpolating between $L/W = 1$ and $L/W = 2$ at $W/r_o = 2$ gives

$$\Phi/\Phi_{max} = 0.88$$

Therefore,

$$R_f = 0.88 \times 0.0389 = 0.0342 \text{ ft}^2 \cdot {}^\circ\text{F} \cdot \text{h/Btu}$$

The above steps may now be repeated using the corrected value of fin resistance.

$$\phi = 0.745$$
$$\Phi/\Phi_{max} = 0.90.75$$
$$R_f = 0.035 \text{ ft}^2 \cdot {}^\circ\text{F} \cdot \text{h/Btu}$$

Note that the improvement in accuracy by reevaluating Φ/Φ_{max} is less than 1% of the overall thermal resistance (environment to fin base). The error produced by using $R_{f(max)}$ without correction is less than 3%. For many practical cases where greater accuracy is not warranted, a single value of R_f, obtained by estimating Φ/Φ_{max}, can be used over a range of heat transfer coefficients for a given fin. For approximate calculations, the fin resistance for other values of k and t_o can be obtained by simple proportion if the range covered is not excessive.

Schmidt (1949) presents approximate, but reasonably accurate, analytical expressions (for computer use) for circular, rectangular, and hexagonal fins. Hexagonal fins are the representative fin shape for the common staggered tube arrangement in finned-tube heat exchangers.

Schmidt's empirical solution is given by

$$\phi = \frac{\tanh(mr_i\Phi)}{mr_i\Phi}$$

where $m = \sqrt{2h/kt}$ and Φ is given by

$$\Phi = [(r_e/r_i) - 1][1 + 0.35\ln(r_e/r_i)]$$

For **circular fins,**

$$r_e/r_i = r_o/r_i$$

For **rectangular fins,**

$$r_e/r_i = 1.28\psi\sqrt{\beta - 0.2}, \quad \psi = M/r_i, \quad \beta = L/M \geq 1$$

where M and L are defined by Figure 23 as $a/2$ or $b/2$, depending on which is greater.

For **hexagonal fins,**

$$r_e/r_i = 1.27\psi\sqrt{\beta - 0.3}$$

where ψ and β are defined as above and M and L are defined by Figure 24 as $a/2$ or b (whichever is less) and $0.5\sqrt{(a^2/2)^2 + b^2}$, respectively.

The section on Bibliography lists other sources of information on finned surfaces.

Thermal Contact Resistance

Fins can be extruded from the prime surface (e.g., the short fins on the tubes in flooded evaporators or water-cooled condensers) or they can be fabricated separately, sometimes of a different material, and bonded to the prime surface. Metallurgical bonds are achieved by furnace-brazing, dip-brazing, or soldering. Nonmetallic bonding materials, such as epoxy resin, are also used. Mechanical bonds are obtained by tension-winding fins around tubes (spiral fins) or expanding the tubes into the fins (plate fins). Metallurgical bonding, properly done, leaves negligible thermal resistance at the joint but is not always economical. Thermal resistance of a mechanical bond

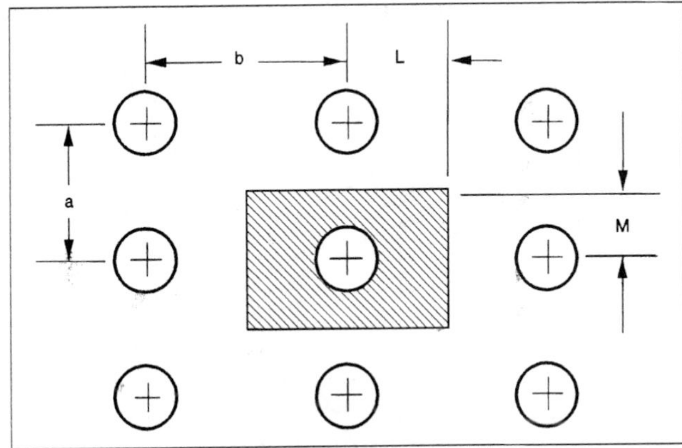

Fig. 23 Rectangular Tube Array

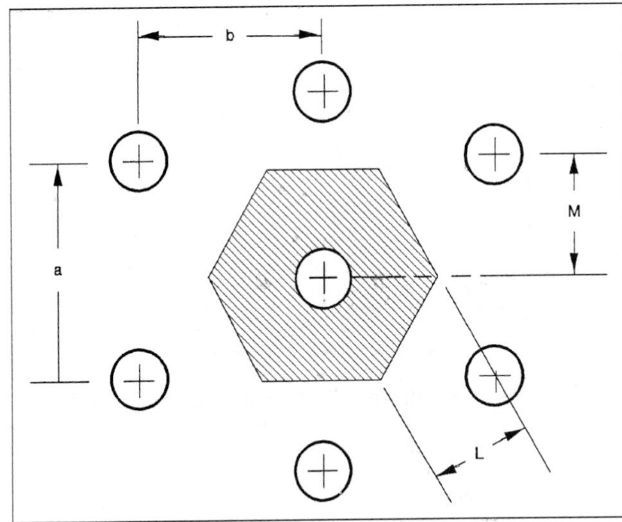

Fig. 24 Hexagonal Tube Array

may or may not be negligible, depending on the application, quality of manufacture, materials, and temperatures involved. Tests of plate fin coils with expanded tubes have indicated that substantial losses in performance can occur with fins that have cracked collars; but negligible thermal resistance was found in coils with continuous collars and properly expanded tubes (Dart 1959).

Thermal resistance at an interface between two solid materials is largely a function of the surface properties and characteristics of the solids, the contact pressure, and the fluid in the interface, if any. Eckels (1977) models the influence of fin density, fin thickness, and tube diameter on contact pressure and compared it to data for wet and dry coils. Shlykov (1964) shows that the range of attainable contact resistances is large. Sonokama (1964) presents data on the effects of contact pressure, surface roughness, hardness, void material, and the pressure of the gas in the voids. Lewis and Sauer (1965) show the resistance of adhesive bonds, and Kaspareck (1964) and Clausing (1964) give data on the contact resistance in a vacuum environment.

Finned-Tube Heat Transfer

The heat transfer coefficients for finned coils follow the basic equations of convection, condensation, and evaporation. The arrangement of the fins affects the values of constants and the exponential powers in the equations. It is generally necessary to refer to test data for the exact coefficients.

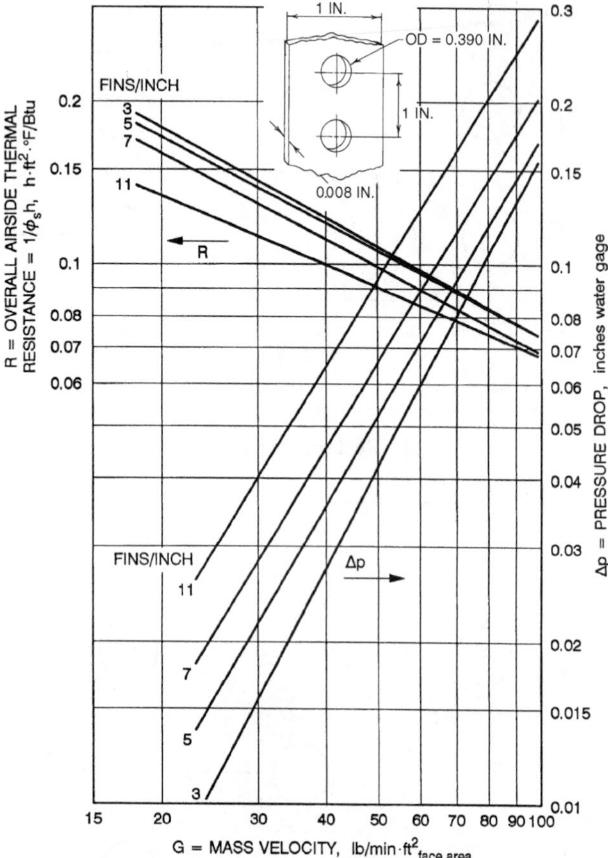

Fig. 25 Overall Air-Side Thermal Resistance and Pressure Drop for 1-Row Coils
(Shepherd 1946)

For natural-convection finned coils (gravity coils), approximate coefficients can be obtained by considering the coil to be made of tubular and vertical fin surfaces at different temperatures and then applying the natural-convection equations to each. This calculation is difficult because the natural-convection coefficient depends on the temperature difference, which varies at different points on the fin.

Fin efficiency should be high (80 to 90%) for optimum natural-convection heat transfer. A low fin efficiency reduces the temperature near the tip. This reduces Δt near the tip and also the coefficient h, which in natural convection depends on Δt. The coefficient of heat transfer also decreases as the fin spacing decreases because of interfering convection currents from adjacent fins and reduced free-flow passage; 2 to 4 in. spacing is common. Generally, high coefficients result from large temperature differences and small flow restriction.

Edwards and Chaddock (1963) give coefficients for several circular fin-on-tube arrangements, using fin spacing δ as the characteristic length and in the form $\mathrm{Nu} = f(\mathrm{Gr}\,\mathrm{Pr}\,\delta/D_o)$, where D_o is the fin diameter.

Forced-convection finned coils are used extensively in a wide variety of equipment. The fin efficiency for optimum performance is smaller than that for gravity coils because the forced-convection coefficient is almost independent of the temperature difference between the surface and the fluid. Very low fin efficiencies should be avoided because an inefficient surface gives a high (uneconomical) pressure drop. An efficiency of 70 to 90% is often used.

As fin spacing is decreased to obtain a large surface area for heat transfer, the coefficient generally increases because of higher air velocity between fins at the same face velocity and reduced equivalent diameter. The limit is reached when the boundary layer formed on one fin surface (Figure 8) begins to interfere with the boundary layer formed on the adjacent fin surface, resulting in a decrease of the heat transfer coefficient, which may offset the advantage of larger surface area.

Selection of the fin spacing for forced-convection finned coils usually depends on economic and practical considerations, such as fouling, frost formation, condensate drainage, cost, weight, and volume. Fins for conventional coils generally are spaced 14 to 6 per inch, except where factors such as frost formation necessitate wider spacing.

Several means are used to obtain higher coefficients with a given air velocity and surface, usually by creating air turbulence, generally with a higher pressure drop: (1) staggered tubes instead of in-line tubes for multiple-row coils; (2) artificial additional tubes, or collars or fingers made by suitably forming the fin materials; (3) corrugated fins instead of plane fins; and (4) louvered or interrupted fins.

Figure 25 shows data for one-row coils. The thermal resistances plotted include the temperature drop through the fins, based on one square foot of total external surface area.

The section on Bibliography lists other sources of information on fins.

SYMBOLS

A = surface area for heat transfer
A_F = cross-sectional flow area
C = conductance; or fluid capacity rate
C_1, C_2 = Planck's law constants [see Equation (33)]
c = coefficient or constant
c_p = specific heat at constant pressure
c_v = specific heat at constant volume
D = tube (inside) or rod diameter; or diameter of the vessel
d = diameter; or prefix meaning differential
E = electric field
e = emissivity; or protuberance height
F = angle factor [see Equations (40) and (41)]
Fo = Fourier number (see Table 1 and Figures 2, 3, and 4)
f = Fanning friction factor for single-phase flow; or electric body force
G = mass velocity; or irradiation
Gr = Grashof number
g = gravitational acceleration
h = heat transfer coefficient; or offset strip fin height
I = modified Bessel function
ID = inside diameter
J = mechanical equivalent of heat; or radiosity
j = heat transfer factor [see Equation (4), Table 6]
k = thermal conductivity
L = length; or height of liquid film
l = length; or length of one module of offset strip fins
M = mass; or molecular weight
m = general exponent
$\dot{m}$ = mass rate of flow
n = general number [see Equation (2) in Table 5 or Table 7 (c)]; or ratio r/r_m (see Figures 2, 3, and 4); or number of blades
NTU = number of exchanger heat transfer units [see Equation (17)]
Nu = Nusselt number
p = pressure; or fin pitch; or repeated rib pitch
Pr = Prandtl number
q = rate of heat transfer
q'' = heat flux
R = thermal resistance
Re = pipe Reynolds number (GD/μ); or film Reynolds number ($4\Gamma/\eta$)
Re* = rotary Reynolds number (D^2N_p/η)
r = radius
s = lateral spacing of offset fin strips
T = absolute temperature
t = temperature; or fin thickness at base
U = overall heat transfer coefficient
V = linear velocity
W = work; or total rate of energy emission; or fin dimension
W_λ = monochromatic emissive power
x, y, z = lengths along principal coordinate axes
Y = temperature ratio (see Figures 2, 3, and 4)

y = one-half diametrical pitch of a twisted tape: length of 180° revolution/tube diameter
Z = ratio of fluid capacity rates [see Equation (18)]
α = thermal diffusivity = $k/\rho c_p$ [Equation (28)]; or absorptance; or spiral angle for helical fins; or aspect ratio of offset strip fins, s/h; or enhancement factor: ratio of enhanced to unenhanced heat transfer coefficient—conditions remaining the same.
β = coefficient of thermal expansion; or contact angle of rib profile
Γ = mass flow of liquid per unit length
γ = ratio, t/s
Δ = difference between values
δ = distance between fins; or ratio t/l; or thickness of twisted tape
ε = hemispherical emittance; or exchanger heat transfer effectiveness [see Equation (16)]; or dielectric constant
λ = wavelength
μ = absolute viscosity
ν = kinematic viscosity
ρ = density; or reflectance
σ = Stefan-Boltzmann constant
τ = time; or transmittance [see Equation (39)]
Φ = fin resistance defined by Equation (54); Φ_{max} is maximum limiting value of Φ
ϕ = fin efficiency [see Equation (50)]; or angle [see Equation (41)]; or temperature correction factor [see Table 7(c)]

Subscripts

a = augmented
b = blackbody; or based on bulk fluid temperature
c = convection; or critical; or cold (fluid)
e = equivalent; or environment
f = film; or fin
fc = finite cylinder
frs = finite rectangular solid
g = gas
h = horizontal; or hot (fluid); or hydraulic
i = inlet; or inside; or particular surface (radiation); or based on maximum inside (envelope) diameter
ic = infinite cylinder
if = interface
is = infinite slab
iso = isothermal conditions
j = particular surface (radiation)
k = particular surface (radiation)
L = thickness
l = liquid
m = mean
max = maximum
min = minimum
n = counter variable
o = outside; or outlet; or overall; or at base of fin
p = prime heat transfer surface
r = radiation; or root (fin); or reduced
s = surface; or secondary heat transfer surface; or straight or plain; or accounting for flow blockage of twisted tape
st = static (pressure)
t = temperature; or terminal temperature; or tip (fin)
v = vapor; or vertical
w = wall; or wafer
λ = monochromatic
∞ = bulk

REFERENCES

Adams, J.A. and D.F. Rogers. 1973. *Computer aided heat transfer analysis*. McGraw-Hill, New York.

Afgan, N.H. and E.U. Schlunder. 1974. *Heat exchangers: Design and theory sourcebook*. McGraw-Hill, New York.

Aksan, D. and F. Borak. 1987. Heat transfer coefficients in coiled stirred tank systems. *Can. J. Chem. Eng.* 65:1013-14.

Al-Fahed, S.F., Z.H. Ayub, A.M. Al-Marafie, and B.M. Soliman. 1993. Heat transfer and pressure drop in a tube with internal microfins under turbulent water flow conditions. *Exp. Thermal and Fluid Science* 7:249-53.

Altmayer, E.F., A.J. Gadgil, F.S. Bauman, and R.C. Kammerud. 1983. Correlations for convective heat transfer from room surfaces. *ASHRAE Transactions* 89(2A):61-77.

Antonir, I. and A. Tamir. 1977. The effect of surface suction on condensation in the presence of a noncondensible gas. *International Journal of Heat and Mass Transfer* 99:496-99.

Astaf'ev, B.F. and A.M. Baklastov. 1970. Condensation of steam on a horizontal rotating disc. *Teploenergetika* 17:55-57.

Ayub, Z.H. and S.F. Al-Fahed. 1993. The effect of gap width between horizontal tube and twisted tape on the pressure drop in turbulent water flow, *International Journal of Heat and Fluid Flow* 14(1):64-67.

Bauman, F., A. Gadgil, R. Kammerud, E. Altmayer, and M. Nansteel. 1983. Convective heat transfer in buildings. *ASHRAE Transactions* 89(1A): 215-33.

Beckermann, C. and V. Goldschmidt. 1986. Heat transfer augmentation in the flueway of a water heater. *ASHRAE Transactions* 92(2B):485-95.

Bergles, A.E. 1998. Techniques to enhance heat transfer. In *Handbook of heat transfer,* 3rd ed., 11.1-11.76. McGraw-Hill, New York.

Brown, A.I. and S.M. Marco. 1958. *Introduction to heat transfer*, 3rd ed. McGraw-Hill, New York.

Burmeister, L.C. 1983. *Convective heat transfer*. John Wiley and Sons, New York.

Carnavos, T.C. 1979. Heat transfer performance of internally finned tubes in turbulent flow. In *Advances in enhanced heat transfer*, pp. 61-67. American Society of Mechanical Engineers, New York.

Carslaw, H.S. and J.C. Jaeger. 1959. *Conduction of heat in solids*. Oxford University Press, England.

Clausing, A.M. 1964. Thermal contact resistance in a vacuum environment. ASME *Paper* 64-HT-16, Seventh National Heat Transfer Conference.

Croft, D.R. and D.G. Lilley. 1977. *Heat transfer calculations using finite difference equations*. Applied Science Publishers, London.

Dart, D.M. 1959. Effect of fin bond on heat transfer. *ASHRAE Journal* 5:67.

Dhir, V.K., F. Chang, and G. Son. 1992. Enhancement of single-phase forced convection heat transfer in tubes and ducts using tangential flow injection. *Annual Report*. Contract No. 5087-260135. Gas Research Institute.

Duignan, M., G. Greene, and T. Irvine. 1993. The effect of surface gas injection on film boiling heat transfer. *Journal of Heat Transfer* 115:986-92.

Eckels, P.W. 1977. Contact conductance of mechanically expanded plate finned tube heat exchangers. AIChE-ASME Heat Transfer Conference, Salt Lake City, UT.

Edwards, J.A. and J.B. Chaddock. 1963. An experimental investigation of the radiation and free-convection heat transfer from a cylindrical disk extended surface. *ASHRAE Transactions* 69:313.

Fernandez, J. and R. Poulter. 1987. Radial mass flow in electrohydrodynamically-enhanced forced heat transfer in tubes. *International Journal of Heat and Mass Transfer* 80:2125-36.

Gardner, K.A. 1945. Efficiency of extended surface. *ASME Transactions* 67:621.

Gose, E.E., E.E. Peterson, and A. Acrivos. 1957. On the rate of heat transfer in liquids with gas injection through the boundary layer. *Journal of Applied Physics* 28:1509.

Grigull, U. et al. 1982. Heat transfer. *Proceedings of the Seventh International Heat Transfer Conference*, Munich 3. Hemisphere Publishing, New York.

Hagge, J.K. and G.H. Junkhan. 1974. Experimental study of a method of mechanical augmentation of convective heat transfer in air. *HTL3*, ISU-ERI-Ames-74158, Nov. 1975. Iowa State University, Ames, Iowa.

Hottel, H.C. and A.F. Sarofim. 1967. *Radiation transfer*. McGraw-Hill, New York.

Hu, Z. and J. Shen. 1996. Heat transfer enhancement in a converging passage with discrete ribs. *Int. J. Heat Mass Trans.* 39:1719.

Inagaki, T. and I. Komori. 1993. Experimental study of heat transfer enhancement in tubulent natural convection along a vertical flat plate, Part 1: The effect of injection and suction. *Heat Transactions—Japanese Research* 22:387.

Incropera, F.P. and D.P. DeWitt. 1996. *Fundamentals of heat transfer*. John Wiley and Sons, New York.

Jakob, M. 1949, 1957. *Heat transfer*, Vols. I and II. John Wiley and Sons, New York.

Jeng, Y., J. Chen, and W. Aung. 1995. Heat transfer enhancement in a vertical channel with asymmetric isothermal walls by local blowing or suction. *International Journal of Heat and Fluid Flow* 16:25.

Junkhan, G.H. et al. 1985. Investigation of turbulators for fire tube boilers. *Journal of Heat Transfer* 107:354-60.

Junkhan, G.H., A.E. Bergles, V. Nirmalan, and W. Hanno. 1988. Performance evaluation of the effects of a group of turbulator inserts on heat transfer from gases in tubes. *ASHRAE Transactions* 94(2):1195-1212.

Kaspareck, W.E. 1964. Measurement of thermal contact conductance between dissimilar metals in a vacuum. ASME *Paper* 64-HT-38, Seventh National Heat Transfer Conference.

Kays, W.M. and A.L. London. 1984. *Compact heat exchangers*, 3rd ed. McGraw-Hill, New York.

Kays, W.M. and M. Crawford. 1980. *Convective heat and mass transfer*, 2nd ed. McGraw-Hill, New York.

Khanpara, J.C., A.E. Bergles, and M.B. Pate. 1986. Augmentation of R-113 in-tube condensation with micro-fin tubes. HTD 65:21-32.

Kinney, R.B. 1968. Fully developed frictional and heat transfer characteristics of laminar flow in porous tubes. *International Journal of Heat and Mass Transfer* 11:1393-1401.

Kinney, R.B. and E.M. Sparrow. 1970. Turbulent flow: Heat transfer and mass transfer in a tube with surface suction. *Journal of Heat Transfer* 92:117-25.

Knudsen, J.G. and B.V. Roy. 1983. Studies on scaling of cooling tower water. *Fouling of heat enhancement surfaces*, pp. 517-30. Engineering Foundation, New York.

Lan, C.W. 1991. Effects of rotation on heat transfer fluid flow and interface in normal gravity floating zone crystal growth. *Journal of Crystal Growth* 114:517.

Lewis, D.M. and H.J. Sauer, Jr. 1965. The thermal resistance of adhesive bonds. ASME *Journal of Heat Transfer* 5:310.

Lienhard, J. and V.A. Dhir. 1972. Simple analysis of laminar film condensation with suction. *Journal of Heat Transfer* 94:334-36.

Malhotra, K. and A.S. Mujumdar. 1991. Wall to bed contact heat transfer rates in mechanically stirred granular beds. *International Journal of Heat and Mass Transfer* 34:724-35.

Manglik, R.M. and A.E. Bergles. 1990. The thermal-hydraulic design of the rectangular offset-strip-fin-compact heat exchanger. In *Compact heat exchangers*, pp. 123-49. Hemisphere Publishing, New York.

Manglik, R.M. and A.E. Bergles. 1993. Heat transfer and pressure drop correlation for twisted-tape insert in isothermal tubes: Part II —Transition and turbulent flows. *Journal of Heat Transfer* 115:890-96.

Marto, P.J. and V.H. Gray. 1971. Effects of high accelerations and heat fluxes on nucleate boiling of water in an axisymmetric rotating boiler. NASA *Technical Note* TN. D-6307. Washington, DC.

McAdams, W.H. 1954. *Heat transmission*, 3rd ed. McGraw-Hill, New York.

McElhiney, J.E. and G.W. Preckshot. 1977. Heat transfer in the entrance length of a horizontal rotating tube. *International Journal of Heat and Mass Transfer* 20:847-54.

Metais, B. and E.R.G. Eckert. 1964. Forced, mixed and free convection regimes. ASME *Journal of Heat Transfer* 86(C2)(5):295.

Mochizuki, S., J. Takamura, and S. Yamawaki. 1994. Heat transfer in serpentine flow passages with rotation. *J. Turbomachinery* 116:133.

Modest, M.F. 1993. *Radiation heat transfer*. McGraw-Hill, New York.

Myers, G.E. 1971. *Analytical methods in conduction heat transfer*. McGraw-Hill, New York.

Nelson, R.M. and A.E. Bergles. 1986. Performance evaluation for tubeside heat transfer enhancement of a flooded evaporative water chiller. *ASHRAE Transactions* 92(1B):739-55.

Nesis, E.I., A.F. Shatalov, and N.P. Karmatskii. 1994. Dependence of the heat transfer coefficient on the vibration amplitude and frequency of a vertical thin heater. *J. Eng. Physics and Thermophysics* 67(No. 1-2).

Nichol, A.A. and M. Gacesa. 1970. Condensation of steam on a rotating vertical cyclinder. *Journal of Heat Transfer* 144-52.

Ohadi, M.M. 1991. Heat transfer enhancement in heat exchangers. *ASHRAE Journal* (December):42-50.

Ohadi, M.M., S. Dessiatoun, A. Singh, K. Cheung, and M. Salehi. 1995a. EHD-Enhancement of boiling/condensation heat transfer of alternate refrigerants, *Progress Report* No. 6. Presented to the U.S. Department of Energy and the EHD Consortium Members, under DOE Grant No. DE-FG02-93CE23803.A000, Chicago, January 1995.

Ohadi, M.M., S.V. Dessiatoun, J. Darabi, and M. Salehi. 1996. Active augmentation of single-phase and phase-change heat transfer—An overview. In *Process enhanced and multiphase heat transfer: A festschrift for A.E. Bergles*, R.M. Manglik and A.D. Kraus, eds. Begell House Publishers, pp. 277-86.

Ohadi, M.M., J. Darabi, and B. Roget. 2000. Electrode design, materials, and fabrication fundamentals. In *Advances in Heat Transfer*,C.L. Tien, ed. Begell House Publishers.

Ohadi, M.M. et al. 1991. Electrohydrodynamic enhancement of heat transfer in a shell-and-tube heat exchanger. *Enhanced Heat Transfer* 4(1):19-39.

Ohadi, M.M., R. Papar, T.L. Ng, M. Faani, and R. Radermacher. 1992. EHD enhancement of shell-side boiling heat transfer coefficients of R-123/oil mixture. *ASHRAE Transactions* 98(2):427-34.

Owsenek, B. and J. Seyed-Yagoobi. 1995. Experimental investigation of corona wind heat transfer enhancement with a heated horizontal flat plate. *Journal of Heat Transfer* 117:309.

Parker, J.D., J.H. Boggs, and E.F. Blick. 1969. *Introduction to fluid mechanics and heat transfer*. Addison Wesley Publishing, Reading, MA.

Patankar, S.V. 1980. *Numerical heat transfer and fluid flow*. McGraw-Hill, New York.

Pescod, D. 1980. An advance in plate heat exchanger geometry giving increased heat transfer. HTD 10:73-77, American Society of Mechanical Engineers, New York.

Poulter, R. and P.H.G. Allen. 1986. Electrohydrodynamincally augmented heat and mass transfer in the shell tube heat exchanger. *Proceedings of the Eighth International Heat Transfer Conference* 6:2963-68.

Prakash, C. and R. Zerle. 1995. Prediction of turbulent flow and heat transfer in a ribbed rectangular duct with and without rotation. *J. Turbomachinery* 117:255.

Ravigururajan, T.S. and A.E. Bergles. 1985. General correlations for pressure drop and heat transfer for single-phase turbulent flow in internally ribbed tubes. *Augmentation of heat transfer in energy systems*, 52:9-20. American Society of Mechanical Engineers, New York.

Rich, D.G. 1966. The efficiency and thermal resistance of annular and rectangular fins. *Proceedings of the Third International Heat Transfer Conference*, AIChE 111:281-89.

Schmidt, T.E. 1949. Heat transfer calculations for extended surfaces. *Refrigerating Engineering* 4:351-57.

Schneider, P.J. 1964. *Temperature response charts*. John Wiley and Sons, New York.

Seyed-Yagoobi, J., C.A. Geppert, and L.M. Geppert. 1996. Electrohydrodynamically enhanced heat transfer in pool boiling. *ASME Journal of Heat Transfer* 118:233-37.

Shepherd, D.G. 1946. Performance of one-row tube coils with thin plate fins, low velocity forced convection. *Heating, Piping, and Air Conditioning* (April).

Shlykov, Y.P. 1964. Thermal resistance of metallic contacts. *International Journal of Heat and Mass Transfer* 7(8):921.

Siegel, R. and J.R. Howell. 1981. *Thermal radiation heat transfer*. McGraw-Hill, New York.

Somerscales, E.F.C. et al. 1991. Particulate fouling of heat transfer tubes enhanced on their inner surface. *Fouling and enhancement interactions*. HTD 164:17-28. American Society of Mechanical Engineers, New York.

Somerscales, E.F.C. and A.E. Bergles. 1997. Enhancement of heat transfer and fouling mitigation. *Advances in Heat Transfer* 30:197-253. Academic Press, San Diego.

Son, G. and V.K. Dhir. 1993. Enhancement of heat transfer in annulus using tangential flow injection. HTD-Vol. 246, *Heat Transfer in Turbulent Flows*, ASME.

Sonokama, K. 1964. Contact thermal resistance. *Journal of the Japan Society of Mechanical Engineers* 63(505):240. English translation in RSIC-215, AD-443429.

Sunada, K., A. Yabe, T. Taketani, and Y. Yoshizawa. 1991. Experimental study of EHD pseudo-dropwise condensation. *Proceedings of the ASME-JSME Thermal Engineering Joint Conference* 3:47-53.

Tang, S. and T.W. McDonald. 1971. A study of boiling heat transfer from a rotating horizontal cylinder. *International Journal of Heat and Mass Transfer* 14:1643-57.

Tauscher, W.A., E.M. Sparrow, and J.R. Lloyd. 1970. Amplification of heat transfer by local injection of fluid into a turbulent tube flow. *International Journal of Heat and Mass Transfer* 13:681-88.

Uemura, M., S. Nishio, and I. Tanasawa. 1990. Enhancement of pool boiling heat transfer by static electric field. *Ninth International Heat Transfer Conference*, 75-80.

Valencia, A., M. Fiebig, and V.K. Mitra. 1996. Heat transfer enhancement by longitudinal vortices in a fin tube heat exchanger. *Journal of Heat Transfer* 118:209.

Wawzyniak, M. and J. Seyed-Yagoobi. 1996. Augmentation of condensation heat transfer with electrohydrodynamics on vertical enhanced tubes. *ASME Journal of Heat Transfer* 118:499-502.

Wayner, P.C., Jr. and S.G. Bankoff. 1965. Film boiling of nitrogen with suction on an electrically heated porous plate. *AIChE Journal* 11:59-64.

Woods, B.G. 1992. Sonically enhanced heat transfer from a cylinder in cross flow and its impact on process power consumption. *International Journal of Heat and Mass Transfer* 35:2367-76.

Yabe, A. and H. Maki. 1988. Augmentation of convective and boiling heat transfer by applying an electrohydrodynamical liquid jet. *International Journal of Heat Mass Transfer* 31(2):407-17.

BIBLIOGRAPHY

Fins

General

Gunter, A.Y. and A.W. Shaw. 1945. A general correlation of friction factors for various types of surfaces in cross flow. *ASME Transactions* 11:643.

Shah, R.K. and R.L. Webb. 1981. *Compact and enhanced heat exchangers, heat exchangers, theory and practice*, pp. 425-68. J. Taborek et al., eds. Hemisphere Publishing, New York.

Webb, R.L. 1980. Air-side heat transfer in finned tube heat exchangers. *Heat Transfer Engineering* 1(3):33-49.

Smooth

Clarke, L. and R.E. Winston. 1955. Calculation of finside coefficients in longitudinal finned heat exchangers. *Chemical Engineering Progress* 3:147.

Elmahdy, A.H. and R.C. Biggs. 1979. Finned tube heat exchanger: Correlation of dry surface heat transfer data. ASHRAE *Transactions* 85:2.

Ghai, M.L. 1951. Heat transfer in straight fins. General discussion on heat transfer. London Conference, September.

Gray, D.L. and R.L. Webb. 1986. Heat transfer and friction correlations for plate finned-tube heat exchangers having plain fins. Proceedings of Eighth International Heat Transfer Conference, San Francisco.

Wavy

Beecher, D.T. and T.J. Fagan. 1987. Fin patternization effects in plate finned tube heat exchangers. ASHRAE *Transactions* 93:2.

Yashu, T. 1972. Transient testing technique for heat exchanger fin. *Reito* 47(531):23-29.

Spine

Abbott, R.W., R.H. Norris, and W.A. Spofford. 1980. Compact heat exchangers for General Electric products—Sixty years of advances in design and manufacturing technologies. *Compact heat exchangers—History, technological advancement and mechanical design problems.* R.K. Shah, C.F. McDonald, and C.P. Howard, eds. Book No. G00183, pp. 37-55. American Society of Mechanical Engineers, New York.

Moore, F.K. 1975. Analysis of large dry cooling towers with spine-fin heat exchanger elements. ASME *Paper* No. 75-WA/HT-46.

Rabas, T.J. and P.W. Eckels. 1975. Heat transfer and pressure drop performance of segmented surface tube bundles. ASME *Paper* No. 75-HT-45.

Weierman, C. 1976. Correlations ease the selection of finned tubes. *Oil and Gas Journal* 9:94-100.

Louvered

Hosoda, T. et al. 1977. Louver fin type heat exchangers. *Heat Transfer Japanese Research* 6(2):69-77.

Mahaymam, W. and L.P. Xu. 1983. Enhanced fins for air-cooled heat exchangers—Heat transfer and friction factor correlations. Y. Mori and W. Yang, eds. Proceedings of the ASME-JSME Thermal Engineering Joint Conference, Hawaii.

Senshu, T. et al. 1979. Surface heat transfer coefficient of fins utilized in air-cooled heat exchangers. *Reito* 54(615):11-17.

Circular

Jameson, S.L. 1945. Tube spacing in finned tube banks. *ASME Transactions* 11:633.

Katz, D.L. 1954-55. Finned tubes in heat exchangers; Cooling liquids with finned coils; Condensing vapors on finned coils; and Boiling outside finned tubes. Bulletin reprinted from *Petroleum Refiner.*

Heat Exchangers

Gartner, J.R. and H.L. Harrison. 1963. Frequency response transfer functions for a tube in crossflow. ASHRAE *Transactions* 69:323.

Gartner, J.R. and H.L. Harrison. 1965. Dynamic characteristics of water-to-air crossflow heat exchangers. ASHRAE *Transactions* 71:212.

McQuiston, F.C. 1981. Finned tube heat exchangers: State of the art for the air side. ASHRAE *Transactions* 87:1.

Myers, G.E., J.W. Mitchell, and R. Nagaoka. 1965. A method of estimating crossflow heat exchanger transients. ASHRAE *Transactions* 71:225.

Stermole, F.J. and M.H. Carson. 1964. Dynamics of forced flow distributed parameter heat exchangers. *AIChE Journal* 10(5):9.

Thomasson, R.K. 1964. Frequency response of linear counterflow heat exchangers. *Journal of Mechanical Engineering Science* 6(1):3.

Wyngaard, J.C. and F.W. Schmidt. Comparison of methods for determining transient response of shell and tube heat exchangers. ASME *Paper* 64-WA/HT-20

Yang, W.J. Frequency response of multipass shell and tube heat exchangers to timewise variant flow perturbance. ASME *Paper* 64-HT-18.

Heat Transfer, General

Bennet, C.O. and J.E. Myers. 1984. *Momentum, heat and mass transfer*, 3rd ed. McGraw-Hill, New York.

Chapman, A.J. 1981. *Heat transfer*, 4th ed. Macmillan, New York.

Holman, J.D. 1981. *Heat transfer*, 5th ed. McGraw-Hill, New York.

Kern, D.Q. and A.D. Kraus. 1972. *Extended surface heat transfer.* McGraw-Hill, New York.

Kreith, F. and W.Z. Black. 1980. *Basic heat transfer.* Harper and Row, New York.

Lienhard, J.H. 1981. *A heat transfer textbook.* Prentice Hall, Englewood Cliffs, NJ.

McQuiston, F.C. and J.D. Parker. 1988. *Heating, ventilating and air-conditioning, analysis and design*, 4th ed. John Wiley and Sons, New York.

Rohsenow, W.M. and J.P. Hartnett, eds. 1973. *Handbook of heat transfer.* McGraw-Hill, New York.

Sissom, L.E. and D.R. Pitts. 1972. *Elements of transport phenomena.* McGraw-Hill, New York.

Smith, E.M. 1997. *Thermal design of heat exchangers.* John Wiley and Sons, Chichester, UK.

Todd, J.P. and H.B. Ellis. 1982. *Applied heat transfer.* Harper and Row, New York.

Webb, R.L. and A.E. Bergles. 1983. Heat transfer enhancement, second generation technology. *Mechanical Engineering* 6:60-67.

Welty, J.R. 1974. *Engineering heat transfer.* John Wiley and Sons, New York.

Welty, J.R., C.E. Wicks, and R.E. Wilson. 1972. *Fundamentals of momentum, heat and mass transfer.* John Wiley and Sons, New York.

Wolf, H. 1983. *Heat transfer.* Harper and Row, New York.

CHAPTER 4

TWO-PHASE FLOW

TWO-PHASE FLOW is encountered extensively in the air-conditioning, heating, and refrigeration industries. A combination of liquid and vapor refrigerant exists in flooded coolers, direct-expansion coolers, thermosiphon coolers, brazed and gasketed plate evaporators and condensers, and tube-in-tube evaporators and condensers, as well as in air-cooled evaporators and condensers. In the pipes of heating systems, steam and liquid water may both be present. Because the hydrodynamic and heat transfer aspects of two-phase flow are not as well understood as those of single-phase flow, no single set of correlations can be used to predict pressure drops or heat transfer rates. Instead, the correlations are for specific thermal and hydrodynamic operating conditions.

This chapter presents the basic principles of two-phase flow and provides information on the vast number of correlations that have been developed to predict heat transfer coefficients and pressure drops in these systems.

BOILING

Commonly used refrigeration evaporators are (1) flooded evaporators, where refrigerants at low fluid velocities boil outside or inside tubes; and (2) dry expansion shell-and-tube evaporators, where refrigerants at substantial fluid velocities boil outside or inside tubes.

Two-phase heat and mass transport are characterized by various flow and thermal regimes, whether vaporization takes place under natural convection or in forced flow. Unlike single-phase flow systems, the heat transfer coefficient for a two-phase mixture depends on the flow regime, the thermodynamic and transport properties of both the vapor and the liquid, the roughness of the heating surface, the wetting characteristics of the surface-liquid pair, and other parameters. Therefore, it is necessary to consider each flow and boiling regime separately to determine the heat transfer coefficient.

Accurate data defining limits of regimes and determining the effects of various parameters are not available. The accuracy of correlations in predicting the heat transfer coefficient for two-phase flow is in most cases not known beyond the range of the test data.

Boiling and Pool Boiling in Natural Convection Systems

Regimes of Boiling. The different regimes of pool boiling described by Farber and Scorah (1948) verified those suggested by Nukiyama (1934). The regimes are illustrated in Figure 1. When the temperature of the heating surface is near the fluid saturation temperature, heat is transferred by convection currents to the free surface where evaporation occurs (Region I). Transition to nucleate boiling occurs when the surface temperature exceeds saturation by a few degrees (Region II).

In **nucleate boiling** (Region III), a thin layer of superheated liquid is formed adjacent to the heating surface. In this layer, bubbles nucleate and grow from spots on the surface. The thermal resistance

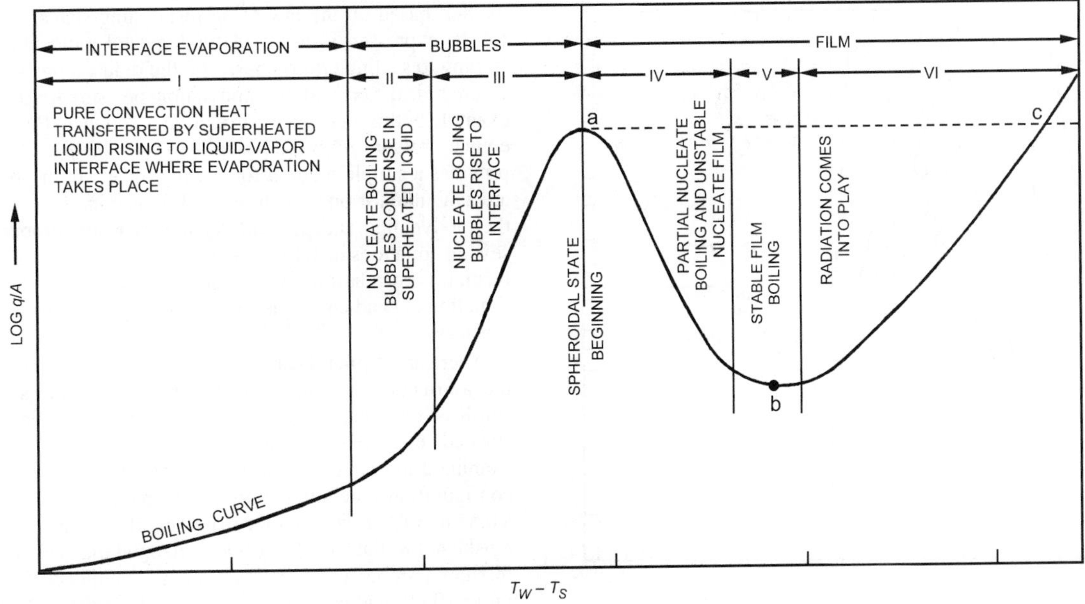

Fig. 1 Characteristic Pool Boiling Curve

The preparation of this chapter is assigned to TC 1.3, Heat Transfer and Fluid Flow.

of the superheated liquid film is greatly reduced by bubble-induced agitation and vaporization. Increased wall temperature increases bubble population, causing a large increase in heat flux.

As heat flux or temperature difference increases further and as more vapor forms, the flow of the liquid toward the surface is interrupted, and a vapor blanket forms. This gives the **maximum heat flux**, which is at the departure from nucleate boiling (DNB) at point a, Figure 1. This flux is often termed the **burnout heat flux** or **boiling crisis** because, for constant power-generating systems, an increase of heat flux beyond this point results in a jump of the heater temperature (to point c, Figure 1), often beyond the melting point of a metal heating surface.

In systems with controllable surface temperature, an increase beyond the temperature for DNB causes a decrease of heat flux density. This is the **transition boiling regime** (Region IV); liquid alternately falls onto the surface and is repulsed by an explosive burst of vapor.

At sufficiently high surface temperatures, a stable vapor film forms at the heater surface; this is the **film boiling regime** (Regions V and VI). Because heat transfer is by conduction (and some radiation) across the vapor film, the heater temperature is much higher than for comparable heat flux densities in the nucleate boiling regime. The minimum film boiling (MFB) heat flux (point b) is the lower end of the film boiling curve.

Free Surface Evaporation. In Region I, where surface temperature exceeds liquid saturation temperature by less than a few degrees, no bubbles form. Evaporation occurs at the free surface by convection of superheated liquid from the heated surface. Correlations of heat transfer coefficients for this region are similar to those for fluids under ordinary natural convection [Equations (1) through (4) in Table 1].

Nucleate Boiling. Much information is available on boiling heat transfer coefficients, but no universally reliable method is available for correlating the data. In the nucleate boiling regime, heat flux density is not a single, valued function of the temperature but depends also on the nucleating characteristics of the surface, as illustrated by Figure 2 (Berenson 1962).

The equations proposed for correlating nucleate boiling data can be put in a form that relates heat transfer coefficient h to temperature difference $(t_w - t_{sat})$:

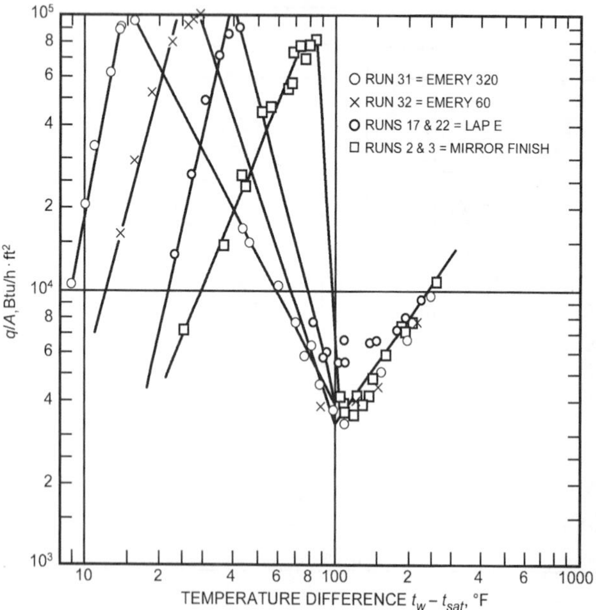

Fig. 2 Effect of Surface Roughness on Temperature in Pool Boiling of Pentane

$$h = \text{constant}\ (t_w - t_{sat})^a \quad (1)$$

Exponent a is normally about 3 for a plain, smooth surface; its value depends on the thermodynamic and transport properties of the vapor and the liquid. Nucleating characteristics of the surface, including the size distribution of surface cavities and the wetting characteristics of the surface-liquid pair, affect the value of the multiplying constant and the value of the exponent a in Equation (1).

A generalized correlation cannot be expected without consideration of the nucleating characteristics of the heating surface. A statistical analysis of data for 25 liquids by Hughmark (1962) shows that in a correlation not considering surface condition, deviations of more than 100% are common.

In the following sections, correlations and nomographs for prediction of nucleate and flow boiling of various refrigerants are given. For most cases, these correlations have been tested for refrigerants, such as R-11, R-12, R-113, and R-114, that have now been identified as environmentally harmful and are no longer being used in new equipment. Extensive research on the thermal and fluid characteristics of alternative refrigerants/refrigerant mixtures has taken place in recent years, and some correlations have been suggested.

Stephan and Abdelsalam (1980) developed a statistical approach for estimating the heat transfer during nucleate boiling. The correlation [Equation (5) in Table 1] should be used with a fixed contact angle θ regardless of the fluid. Cooper (1984) proposed a dimensional correlation for nucleate boiling, shown as Equation (6) in Table 1. The dimensions required are listed in Table 1. This correlation is recommended for fluids with poorly defined physical properties.

Gorenflo (1993) proposed a nucleate boiling correlation based on a set of reference conditions and a base heat transfer coefficient as shown in Equation (7) in Table 1. The correlation was developed for a reduced pressure p_r of 0.1 and the reference conditions given in Table 1. Base heat transfer coefficients are given for three fluids in Table 1, and Gorenflo (1993) should be consulted for additional fluids.

In addition to correlations dependent on thermodynamic and transport properties of the vapor and the liquid, Borishansky et al. (1962) and Lienhard and Schrock (1963) documented a correlating method based on the law of corresponding states. The properties can be expressed in terms of fundamental molecular parameters, leading to scaling criteria based on the reduced pressure, $p_r = p/p_c$, where p_c is the critical thermodynamic pressure for the coolant. An example of this method of correlation is shown in Figure 3. The reference pressure p^* was chosen as $p^* = 0.029p_c$. This correlation provides a simple method for scaling the effect of pressure if data are available for one pressure level. It also has an advantage if the thermodynamic and particularly the transport properties used in several equations in Table 1 are not accurately known. In its present form, this correlation gives a value of $a = 2.33$ for the exponent in Equation (1) and consequently should apply for typical aged metal surfaces.

There are explicit heat transfer coefficient correlations based on the law of corresponding states for various substances (Borishansky and Kosyrev 1966), halogenated refrigerants (Danilova 1965), and flooded evaporators (Starczewski 1965). Other investigations examined the effects of oil on boiling heat transfer from diverse configurations, including boiling from a flat plate (Stephan 1963b); a 0.55 in. OD horizontal tube using an oil-R-12 mixture (Tschernobyiski and Ratiani 1955); inside horizontal tubes using an oil-R-12 mixture (Breber et al. 1980, Worsoe-Schmidt 1959, Green and Furse 1963); and commercial copper tubing using R-11 and R-113 with oil content to 10% (Dougherty and Sauer 1974). Additionally, Furse (1965) examined R-11 and R-12 boiling over a flat horizontal copper surface.

Table 1 Equations for Boiling Heat Transfer

Description	References	Equations	
Free convection Free convection boiling, or boiling without bubbles for low Δt and GrPr < 10^8 (all properties based on liquid state)	Jakob (1949 and 1957)	$\mathrm{Nu} = C(\mathrm{Gr})^m(\mathrm{Pr})^n$	(1)
Vertical submerged surface		$\mathrm{Nu} = 0.61(\mathrm{Gr})^{0.25}(\mathrm{Pr})^{0.25}$	(2)
Horizontal submerged surface		$\mathrm{Nu} = 0.16(\mathrm{Gr})^{1/3}(\mathrm{Pr})^{1/3}$	(3)
Simplified equation for water		$h \sim 80(\Delta t)^{1/3}$, where h is in Btu/h·ft²·°F, Δt in °F	(4)
Nucleate boiling	Stephan and Abdelsalam (1980)	$\frac{hD_d}{k_L} = 0.0546\left[\left(\frac{\rho_v}{\rho_l}\right)^{0.5}\left(\frac{qD_d}{Ak_lT_{sat}}\right)\right]^{0.67}\left(\frac{h_{fg}D_d^2}{a_L^2}\right)^{0.248}\left(\frac{\rho_l-\rho_v}{\rho_l}\right)^{-4.33}$ where $D_d = 0.0208\theta\left[\frac{\sigma}{g(\rho_l-\rho_v)}\right]^{1/2}$ with $\theta = 35°$	(5)
	Cooper (1984)	$h = 55P_r^{0.12-0.4343\ln(R_p)}(-0.4343\ln P_r)M^{-0.5}\left(\frac{q}{A}\right)^{0.67}$ where h is in W/(m²·K), q/A is in W/m², and R_p is surface roughness in μm	(6)
	Gorenflo (1993)	$h = h_oF_{PF}\left(\frac{q/A}{(q/A)_o}\right)^{nf}\left(\frac{R_p}{R_{po}}\right)^{0.133}$ where reference conditions are $(q/A)_o$ = 6300 Btu/h·ft², R_{po} = 0.4 μm	(7)
		$F_{PF} = 1.2P_r^{0.27} + 2.5P_r + \frac{P_r}{1-P_r}$	(8)
		$nf = 0.9 - 0.3P_r^{0.3}$ For all fluids except water and helium.	(9)

Fluid	h_o, Btu/h·ft²·°F
R-134a	790
R-22	690
Ammonia	1230

Description	References	Equations	
Critical heat flux	Kutateladze (1951) Zuber et al. (1962)	$\frac{q/A}{\rho_vh_{fg}}\left[\frac{\rho_v^2}{\sigma_tg(\rho_l-\rho_v)}\right]^{0.25} = K_D$ For many liquids, K_D varies from 0.12 to 0.16. Recommended average value is 0.13.	(10)
Minimum heat flux in film boiling from horizontal plate	Zuber (1959)	$\frac{q/A}{\rho_vh_{fg}}\left[\frac{(\rho_l+\rho_v)}{\sigma_tg(\rho_l-\rho_v)}\right]^{0.25} = 0.09$	(11)
Minimum heat flux in film boiling from horizontal cylinders	Lienhard and Wong (1963)	$\frac{q/A}{\rho_vh_{fg}}\left[\frac{(\rho_l+\rho_v)^2}{\sigma_tg(\rho_l-\rho_v)}\right]^{0.25} = 0.114\frac{\left[\frac{2\sigma_t}{g(\rho_l-\rho_v)D^2}\right]^{0.5}}{\left[1+\frac{2\sigma_t}{g(\rho_l-\rho_v)D^2}\right]^{0.25}}$	(12)
Minimum temperature difference for film boiling from horizontal plate	Berenson (1961)	$(t_w-t_{sat}) = 0.127\frac{\rho_vh_{fg}}{k_v}\left[\frac{g(\rho_l-\rho_v)}{\rho_l+\rho_v}\right]^{2/3} \times\left[\frac{\sigma_t}{g(\rho_l-\rho_v)}\right]^{0.5}\left[\frac{\mu_v}{\rho_l-\rho_v}\right]^{1/3}$	(13)
Film boiling from horizontal plate	Berenson (1961)	$h = 0.425\left[\frac{k_v^3\rho_vh_{fg}g(\rho_l-\rho_v)}{\mu_v(t_w-t_{sat})\sqrt{\phi_t/g(\rho_l-\rho_v)}}\right]^{0.25}$	(14)
Film boiling from horizontal cylinders	Anderson et al. (1966)	$h = 0.62\left[\frac{k_v^3\rho_vg(\rho_l-\rho_v)h_{fg}}{D\mu_v(t_w-t_{sat})}\right]^{0.25}$	(15)
Effect of radiation	Anderson et al. (1966)	Substitute $h'_{fg} = h_{fg}\left[1+0.4c_p\frac{t_w-t_b}{h_{fg}}\right]$	

Table 1 Equations for Boiling Heat Transfer (*Continued*)

Description	References	Equations	
Effect of surface tension and of pipe diameter	Breen and Westwater (1962)	$\Lambda/D < 0.8$: $h(\Lambda)^{0.25}/F = 0.60$	(16)
		$0.8 < \Lambda/D < 8$: $hD^{0.25}/F = 0.62$	(17)
		$8 < \Lambda/D$: $h(\Lambda)^{0.25}/F = 0.016\,(\Lambda/D)^{0.83}$	(18)
		where $\Lambda = 2\pi\left[\frac{\sigma_t}{g(\rho_l-\rho_v)}\right]^{0.25}$	
		$F = \left[\frac{\rho_v h_{fg} g(\rho_l-\rho_v)k_v^3}{\mu_v(t_w-t_{sat})}\right]^{0.25}$	
Turbulent film	Frederking and Clark (1962)	$\mathrm{Nu} = 0.15\,(\mathrm{Ra})^{1/3}$ for $\mathrm{Ra} > 5\times10^7$	(19)
		$\mathrm{Ra} = \left[\frac{D^3 g(\rho_l-\rho_v)}{v_v^2\rho_v}\left(\frac{c_p\mu}{k}\right)_v\left(\frac{h_{fg}}{c_p(t_w-t_{sat})}+0.4\right)\frac{a}{g}\right]^{1/3}$	
		a = local acceleration	

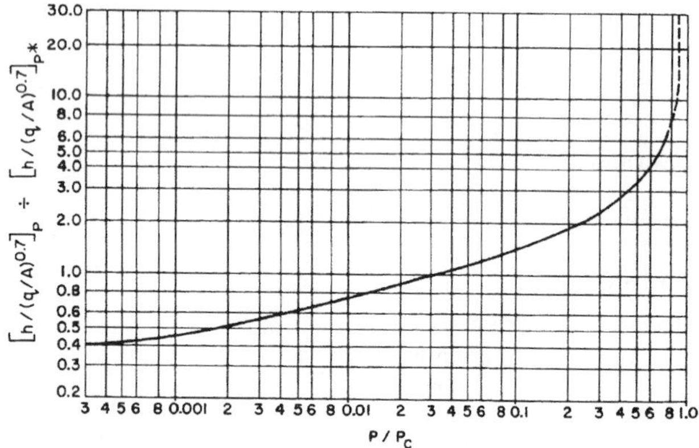

Fig. 3 Correlation of Pool Boiling Data in Terms of Reduced Pressure

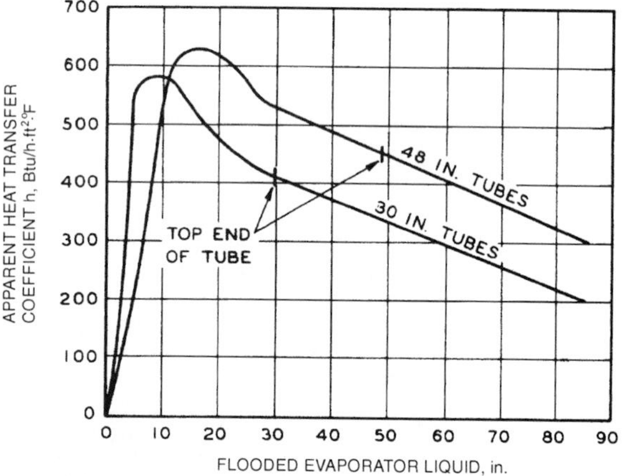

Fig. 4 Boiling Heat Transfer Coefficients for Flooded Evaporator

Maximum Heat Flux and Film Boiling

Maximum or critical heat flux and the film boiling region are not as strongly affected by conditions of the heating surface as the heat flux in the nucleate boiling region, making analysis of DNB and of film boiling more tractable.

Carey (1992) provides a review of the mechanisms that have been postulated to cause the DNB phenomenon in pool boiling. Each model is based on the scenario that vapor blankets, which lead to an increased thermal resistance, exist on portions of the heat transfer surface. It has been proposed that these blankets may result from Helmholtz instabilities.

When DNB (point *a*, Figure 1) is assumed to be a hydrodynamic instability phenomenon, a simple relation, Equation (10) in Table 1, can be derived to predict this flux for pure, wetting liquids (Kutateladze 1951, Zuber et al. 1962). The dimensionless constant K varies from approximately 0.12 to 0.16 for a large variety of liquids. The effect of wettability is still in question. Van Stralen (1959) found that for liquid mixtures, DNB is a function of the concentration.

The minimum heat flux density (point *b*, Figure 1) in film boiling from a horizontal surface and a horizontal cylinder can be predicted by Equations (11) and (12) in Table 1. The numerical factors 0.09 and 0.114 were adjusted to fit experimental data; values predicted by two analyses were approximately 30% higher. Equation (13) in Table 1 predicts the temperature difference at minimum heat flux of film boiling.

The heat transfer coefficient in film boiling from a horizontal surface can be predicted by Equation (14) in Table 1; and from a horizontal cylinder by Equation (15) in Table 1 (Bromley 1950), which has been generalized to include the effect of surface tension and cylinder diameter, as shown in Equations (16), (17), and (18) in Table 1 (Breen and Westwater 1962).

Frederking and Clark (1962) found that for turbulent film boiling, Equation (19) in Table 1 agrees with data from experiments at reduced gravity (Rohsenow 1963, Westwater 1963, Kutateladze 1963, Jakob 1949 and 1957).

Flooded Evaporators

Equations in Table 1 merely approximate heat transfer rates in flooded evaporators. One reason is that vapor entering the evaporator combined with vapor generated within the evaporator can produce significant forced convection effects superimposed on those caused by nucleation. Nonuniform distribution of the two-phase, vapor-liquid flow within the tube bundle of shell-and-tube evaporators or the tubes of vertical-tube flooded evaporators is also important.

Bundle data and design methods for plain, low fin, and enhanced tubes have been reviewed in Thome (1990) and Collier and Thome (1996).

Typical performance of vertical tube natural circulation evaporators, based on data for water, is shown in Figure 4 (Perry 1950). Low coefficients are at low liquid levels because insufficient liquid covers the heating surface. The lower coefficient at high levels is the result of an adverse effect of hydrostatic head on temperature difference and circulation rate. Perry (1950) noted similar effects in horizontal shell-and-tube evaporators.

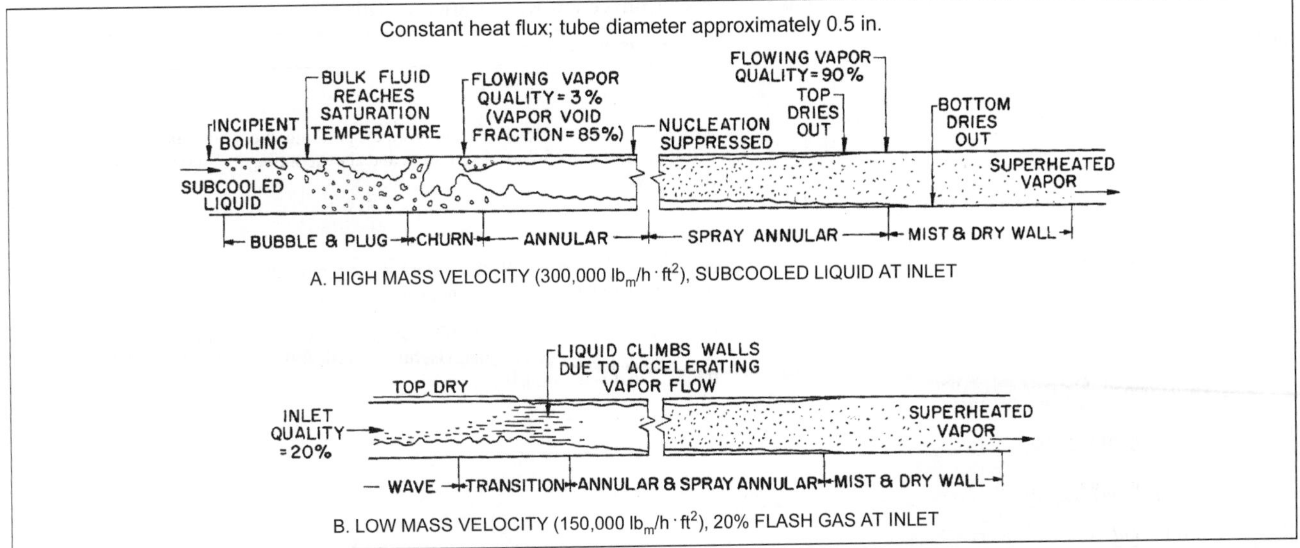

Fig. 5 Flow Regimes in Typical Smooth Horizontal Tube Evaporator

Forced-Convection Evaporation in Tubes

Flow Mechanics. When a mixture of liquid and vapor flows inside a tube, a number of flow patterns occur, depending on the mass fraction of liquid, the fluid properties of each phase, and the flow rate. In an evaporator tube, the mass fraction of liquid decreases along the circuit length, resulting in a series of changing vapor-liquid flow patterns. If the fluid enters as a subcooled liquid, the first indications of vapor generation are bubbles forming at the heated tube wall (nucleation). Subsequently, bubble, plug, churn (or semiannular), annular, spray annular, and mist flows can occur as the vapor content increases for two-phase flows in horizontal tubes. Idealized flow patterns are illustrated in Figure 5A for a horizontal tube evaporator.

Because nucleation occurs at the heated surface in a thin sublayer of superheated liquid, boiling in forced convection may begin while the bulk of the liquid is subcooled. Depending on the nature of the fluid and the amount of subcooling, the bubbles formed can either collapse or continue to grow and coalesce (Figure 5A), as Gouse and Coumou (1965) observed for R-113. Bergles and Rohsenow (1964) developed a method to determine the point of incipient surface boiling.

After nucleation begins, bubbles quickly agglomerate to form vapor plugs at the center of a vertical tube, or, as shown in Figure 5A, vapor plugs form along the top surface of a horizontal tube. At the point where the bulk of the fluid reaches saturation temperature, which corresponds to local static pressure, there will be up to 1% vapor quality because of the preceding surface boiling (Guerrieri and Talty 1956).

Further coalescence of vapor bubbles and plugs results in churn, or semiannular flow. If the fluid velocity is high enough, a continuous vapor core surrounded by a liquid annulus at the tube wall soon forms. This annular flow occurs when the ratio of the tube cross section filled with vapor to the total cross section is approximately 85%. With common refrigerants, this equals a vapor quality of about 0% to 30%. **Vapor quality** is the ratio of mass (or mass flow rate) of vapor to total mass (or mass flow rate) of the mixture. The usual flowing vapor quality or vapor fraction is referred to throughout this discussion. Static vapor quality is smaller because the vapor in the core flows at a higher average velocity than the liquid at the walls (see Chapter 2).

If two-phase mass velocity is high (greater than 150,000 $lb_m/h \cdot ft^2$ for a 0.5 in. tube), annular flow with small drops of entrained liquid in the vapor core (spray) can persist over a vapor quality range from about 10% to more than 90%. Refrigerant evaporators are fed from an expansion device at vapor qualities of approximately 20%, so that annular and spray annular flow predominate in most tube lengths. In a vertical tube, the liquid annulus is distributed uniformly over the periphery, but it is somewhat asymmetric in a horizontal tube (Figure 5A). As vapor quality reaches about 80%, the surface dries out. Chaddock and Noerager (1966) found that in a horizontal tube, dryout occurs first at the top of the tube and progresses toward the bottom with increasing vapor quality (Figure 5A).

If two-phase mass velocity is low (less than 150,000 $lb_m/h \cdot ft^2$ for a 0.5 in. horizontal tube), liquid occupies only the lower cross section of the tube. This causes a wavy type of flow at vapor qualities above about 5%. As the vapor accelerates with increasing evaporation, the interface is disturbed sufficiently to develop annular flow (Figure 5B). Liquid slugging can be superimposed on the flow configurations illustrated; the liquid forms a continuous, or nearly continuous, sheet over the tube cross section. The slugs move rapidly and at irregular intervals. Kattan et al. (1998a) presented a general method for prediction of flow pattern transitions (i.e., a flow pattern map) based on observations for R-134a, R-125, R-502, R-402A, R-404A, R-407C, and ammonia.

Heat Transfer. It is difficult to develop a single relation to describe the heat transfer performance for evaporation in a tube over the full quality range. For refrigerant evaporators with several percentage points of flash gas at entrance, it is less difficult because annular flow occurs in most of the tube length. The reported data are accurate only within geometry, flow, and refrigerant conditions tested; therefore, a large number of methods for calculating heat transfer coefficients for evaporation in tubes is presented in Table 2 (also see Figures 6 through 8).

Figure 6 gives heat transfer data for R-22 evaporating in a 0.772 in. tube (Gouse and Coumou 1965). At low mass velocities (below 150,000 $lb_m/h \cdot ft^2$), the wavy flow regime shown in Figure 5B probably exists, and the heat transfer coefficient is nearly constant along the tube length, dropping at the tube exit as complete vaporization occurs. At higher mass velocities, the flow pattern is usually annular, and the coefficient increases as vapor accelerates. As the surface dries and the flow reaches a 90% vapor quality, the coefficient drops sharply.

Equation (1) in Table 2 is used to estimate average heat transfer coefficients for refrigerants evaporating in horizontal tubes (Pierre

Table 2 Equations for Forced Convection Evaporation in Tubes

Equations		Comments and References
Horizontal Tubes		
$h = C_1\left(\frac{k_l}{d}\right)\left[\left(\frac{GD}{\mu_l}\right)^2\left(\frac{J\Delta x h_{fg}}{L}\right)\right]^n$ where $C_1 = 0.0009$ and $n = 0.5$ for exit qualities < 90%; $C_1 = 0.0082$ and $n = 0.4$ for 11°F superheat at exit	(1)	Average coefficients for R-12 and R-22 evaporating in copper tubes of 0.472 and 0.709 in. ID, from 13.4 to 31.2 ft long, and at evaporating temperatures from –4 to 32°F (Pierre 1955, 1957).
Equation (1) with $C_1 = 0.0225$ and $n = 0.375$		Average coefficients for R-22 evaporating at temperatures from 40 to 80°F in a 0.343 in. ID tube, 8 ft long (Altman et al. 1960b).
$h = Eh_f + Sh_{ncb}$ where	(2)	Compiled from a data base of 3693 data points including data for R-11, R-12, R-22, R-113, R-114, and water. Useful for vertical flows and horizontal flows (Gungor and Winterton 1986).
$E = 1 + 24{,}000B_o^{1.16} + 1.37(1/X_{tt})^{0.86}$	(3)	
$S = [1 + 0.00000115E^2\text{Re}_l^{1.17}]^{-1}$	(4)	
$h_f = 0.023\text{Re}_l^{0.8}\text{Pr}_l^{0.4}(k_l/d)$	(5)	
$\text{Re}_l = \frac{G(1-x)d}{\mu_l}$	(6)	
h_{ncb} is found from Equation (6) in Table 1, which has units W/m². If tube is horizontal and Fr < 0.05, use following multipliers: $E_2 = \text{Fr}_L^{(0.1-2\text{Fr}_l)}$ $S_2 = \text{Fr}_l^{1/2}$		
$h = (C_1C_o^{C_2}(25\text{Fr}_L) + C_3B_o^{C_4}F_{fl})h_f$ where h_f is found from Equations (5) and (6) shown above.	(7)	Compiled from a database of 5246 data points including data for R-11, R-12, R-22, R-113, R-114, R-152a, nitrogen, neon, and water. Tube sizes ranged from 0.25 in. to 1.25 in. (Kandlikar 1990).

Constant	Convective	Nucleate Boiling
C_1	1.136	0.6683
C_2	–0.9	–0.2
C_3	667.2	1058
C_4	0.7	0.7
C_5	0.3	0.3

Use convective constants if $C_o < 0.65$ and nucleate if $C_o > 0.65$. $C_5 = 0$ for vertical tubes and horizontal with $\text{Fr}_l > 0.04$.

Fluid	F_{fl}
R-22	2.2
R-12	1.5
R-152a	1.1

Equations		Comments and References
Vertical Tubes		
$h = 3.4h_l(1/X_{tt})^{0.45}$	(8)	Equations (8) and (9) were fitted to experimental data for vertical upflow in tubes. Both relate to forced-convection evaporation regions where nucleate boiling is suppressed (Guerrieri and Talty 1956, Dengler and Addoms 1956). A multiplying factor is recommended when nucleation is present.
$h = 3.5h_L(1/X_{tt})^{0.5}$ where h_l is from (3), X_{tt} from (7), h_L from (6)	(9)	
$h = 0.74h_L[B_o \times 10^4 + (1/X_{tt})^{0.67}]$ where B_o is from (5), h_L from (6), X_{tt} from (7)	(10)	Local coefficients for water in vertical upflow in tubes with diameters from 0.1162 to 0.4317 in. and lengths of 15 to 40 in. The boiling number B_o accounts for nucleation effects, and the Martinelli parameter X_{tt}, for forced-convection effects (Schrock and Grossman 1962).
$h = h_{mic} + h_{mac}$ where	(11)	Chen developed this correlation reasoning that the nucleation transfer mechanism (represented by h_{mic}) and the convective transfer mechanism (represented by h_{mac}) are additive. h_{mac} is expressed as a function of the two-phase Reynolds number after Martinelli, and h_{mic} is obtained from the nucleate boiling correlation of Forster and Zuber (1955). S_c is a suppression factor for nucleate boiling (Chen 1963).
$h_{mac} = h_lF_c$	(12)	
$h_{mic} = 0.00122\,(S_c)(E)(\Delta t)^{0.24}(\Delta p)^{0.75}$	(13)	
F_c and S_c from Figures 7 and 8; $E = \frac{k_l^{0.79}(c_p)_l^{0.45}\rho_l^{0.49}g_c^{0.25}}{\sigma_l^{0.50}\mu_l^{0.29}h_{fg}^{0.24}\rho_v^{0.24}}$		

Note: Except for dimensionless equations, units are lb_m, h, ft, °F, and Btu.

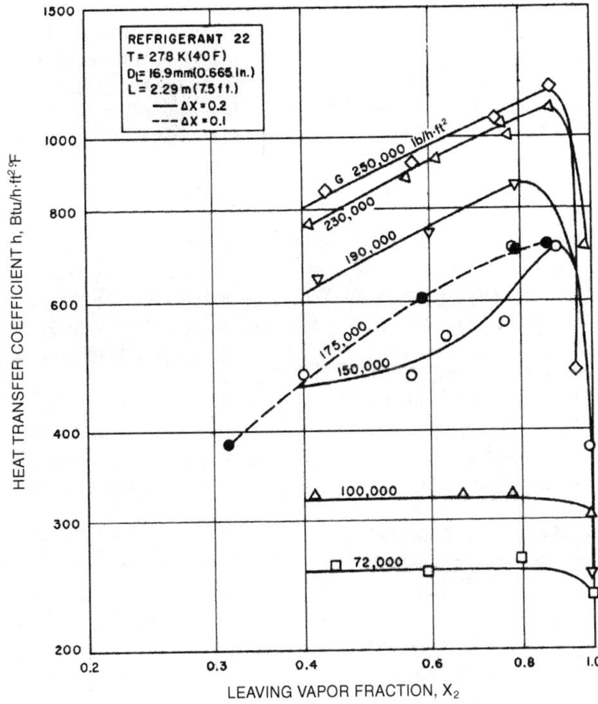

Fig. 6 Heat Transfer Coefficient Versus Vapor Fraction for Partial Evaporation

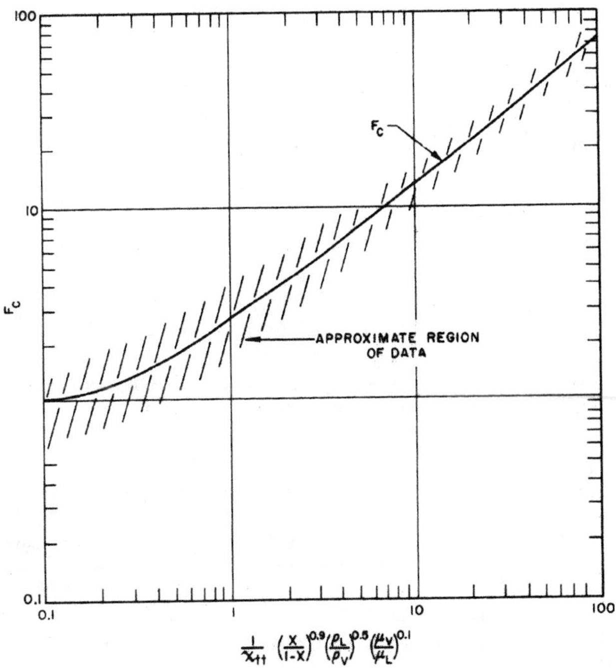

Fig. 7 Reynolds Number Factor F_c

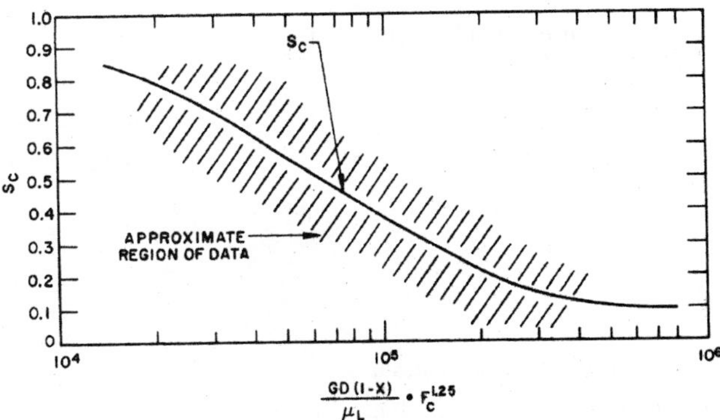

Fig. 8 Suppression Factor S_c

1955, 1957). A number of methods are also available to estimate local heat transfer coefficients during evaporation. Equations (2) through (6) in Table 2 summarize the Gungor and Winterton (1986) model, while Equation (7) gives the Kandlikar (1990) model. In addition, the Shah (1982) model is also recommended for estimating local heat transfer coefficients during annular flow. These local models have been found accurate for a wide range of refrigerants but do not include mechanisms to model dryout. The flow pattern based model of Kattan et al. (1998b) includes specific models for each flow pattern type and has been tested with the newer refrigerants such as R-134a and R-407C.

Heat transfer coefficients during flow in vertical tubes can be estimated with Equations (8), (9), and (10) in Table 2. The Chen correlation shown in Equations (11) through (13) in Table 2 includes terms for the velocity effect (convection) and heat flux (nucleation) and produces local heat transfer coefficients as a function of local vapor quality. The method developed by Steiner and Taborek (1992) includes an asymptotic model for the convection and nucleation component of heat transfer and is recommended for most situations.

The effect of lubricant on the evaporation heat transfer coefficients has been studied by a number of authors (Schalger et al. 1988, Eckels et al. 1993, and Zeurcher et al. 1999). Schalger et al. and Eckels et al. showed that the average heat transfer coefficients during evaporation of R-22 and R-134a in smooth and enhanced tubes are in general decreased by presence of lubricant (up to a 20% reduction at a 5% lubricant concentration by mass). Slight enhancements at lubricant concentrations under 3% are observed with some refrigerant lubricant mixtures. Zeurcher et al. (1999) studied local heat transfer coefficients of refrigerant/lubricant mixtures in the dry wall region of the evaporator (see Figure 5) and proposed prediction methods. The effect of lubricant concentration on local heat transfer coefficients was shown to be dependent on the mass flux and vapor quality. At low mass fluxes, the oil sharply decreased performance, while at higher mass fluxes, enhancements at certain vapor qualities were seen.

CONDENSING

In most applications that use the condensation process, condensation is initiated by removing heat at a solid-vapor interface, either through the walls of the vessel containing the saturated vapor or through the solid surface of a cooling mechanism placed within the saturated vapor. If a sufficient amount of energy is removed, the local temperature of the vapor near the interface will drop below its equilibrium saturation temperature. Because the heat removal process creates a temperature gradient with the lowest temperature near the interface, vapor droplets most likely form at this location. This defines one type of heterogeneous nucleation that can result in either dropwise condensation or film condensation, depending on the physical characteristics of the solid surface and the working fluid.

Dropwise condensation occurs on the cooling solid surface when its surface free energy is relatively low compared to that of the liquid. Examples of this type of interface include highly polished or fatty acid-impregnated surfaces in contact with steam. **Film condensation** occurs when a cooling surface having relatively high surface free energy contacts a fluid having lower surface free energy (see Isrealachvili 1991). This is the type of condensation that occurs in most systems.

The rate of heat transport depends on the condensate film thickness, which depends on the rate of vapor condensation and the rate of condensate removal. At high reduced pressures, the heat transfer coefficients for dropwise condensation are higher than those available in the presence of film condensation at the same surface loading. At low reduced pressures, the reverse is true. For example, there is a reduction of 6 to 1 in the dropwise condensation coefficient of steam when saturation pressure is decreased from 0.9 to 0.16 atm. One method for correlating the dropwise condensation heat transfer coefficient employs nondimensional parameters, including the effect of surface tension gradient, temperature difference, and fluid properties.

When condensation occurs on horizontal tubes and short vertical plates, the condensate film motion is laminar. On vertical tubes and long vertical plates, the film motion can become turbulent. Grober et al. (1961) suggest using a Reynolds number (Re) of 1600 as the critical point at which the flow pattern changes from laminar to turbulent. This Reynolds number is based on condensate flow rate divided by the breadth of the condensing surface. For a vertical tube, the breadth is the circumference of the tube; for a horizontal tube, the breadth is twice the length of the tube. Re = $4\Gamma/\mu_f$, where Γ is the mass flow of condensate per unit of breadth, and μ_f is the absolute (dynamic) viscosity of the condensate at the film temperature t_f. In practice, condensation is usually laminar in shell-and-tube condensers with the vapor outside horizontal tubes.

Vapor velocity also affects the condensing coefficient. When this is small, condensate flows primarily by gravity and is resisted by the viscosity of the liquid. When vapor velocity is high relative to the condensate film, there is appreciable drag at the vapor-liquid interface. The thickness of the condensate film, and hence the heat transfer coefficient, is affected. When vapor flow is upward, a retarding force is added to the viscous shear, increasing the film thickness. When vapor flow is downward, the film thickness decreases and the heat transfer coefficient increases. For condensation inside horizontal tubes, the force of the vapor velocity causes

Table 3 Heat Transfer Coefficients for Film-Type Condensation

Description	References	Equations	
1. **Vertical surfaces**, height L			
Laminar condensate flow, Re = $4\Gamma/\mu_f < 1800$	McAdams (1954)	$h = 1.13F_1(h_{fg}/L\Delta t)^{0.25}$	(1)
	McAdams (1954)	$h = 1.11F_2(b/w_l)^{1/3}$	(2)
	Grigull (1952)	$h = 0.003(F_1)^2(\Delta t L/\mu_f^2 h_{fg})^{0.5}$	(3)
Turbulent flow, Re = $4\Gamma/\mu_f > 1800$	McAdams (1954)	$h = 0.0077F_2(\mathrm{Re})^{0.4}(1/\mu_f)^{1/3}$	(4)
2. **Outside horizontal tubes**, N rows in a vertical plane, length L, laminar flow			
	McAdams (1954)	$h = 0.79F_1(h_{fg}/Nd\Delta t)^{0.25}$	(5)
	McAdams (1954)	$h = 1.05F_2(L/w_l)^{1/3}$	(6)
Finned tubes	Beatty and Katz (1948)	$h = 0.689F_1(h_{fg}/\Delta t D_e)^{0.25}$ where D_e is determined from $\frac{1}{(D_e)^{0.25}} = 1.30\frac{A_s\phi}{A_{eff}(L_{mf})^{0.25}} + \frac{A_p}{A_{eff}(D)^{0.25}}$ with $A_{eff} = A_s\phi + A_p$ and $L_{mf} = a_f/D_o$	(7)
3. **Simplified equations for steam**			
Outside vertical tubes, Re = $4\Gamma/\mu_f < 2100$	McAdams (1954)	$h = 4000/(L)^{0.25}(\Delta t)^{1/3}$	(8)
Outside horizontal tubes, Re = $4\Gamma/\mu_f < 1800$	McAdams (1954)		
Single tube		$h = 3100/(d')^{0.25}(\Delta t)^{1/3}$	(9a)
Multiple tubes		$h = 3100/(Nd')^{0.25}(\Delta t)^{1/3}$	(9b)
4. **Inside vertical tubes**	Carpenter and Colburn (1949)	$h = 0.065\left(\frac{c_{pf}k_f\rho_f f'}{2\mu_f\rho_v}\right)^{0.5}G_m$ *where* $G_m = \left(\frac{G_i^2 + G_iG_o + G_o^2}{3}\right)^{0.5}$	(10)
5. **Inside horizontal tubes**, $\frac{DG_l}{\mu_l} < 5000$			
$1000 < \frac{DG_v}{\mu_l}\left(\frac{\rho_l}{\rho_v}\right)^{0.5} < 20{,}000$	Ackers and Rosson (1960)	$\frac{hD}{k_l} = 13.8\left(\frac{c_p\mu_l}{k_l}\right)^{1/3}\left(\frac{h_{fg}}{c_p\Delta t}\right)^{1/6}\left[\frac{DG_v}{\mu_l}\left(\frac{\rho_l}{\rho_v}\right)^{0.5}\right]^{0.2}$	(11)
$20{,}000 < \frac{DG_v}{\mu_l}\left(\frac{\rho_l}{\rho_v}\right)^{0.5} < 100{,}000$	Ackers and Rosson (1960)	$\frac{hD}{k_l} = 0.1\left(\frac{c_p\mu_l}{k_l}\right)^{1/3}\left(\frac{h_{fg}}{c_p\Delta t}\right)^{1/6}\left[\frac{DG_v}{\mu_l}\left(\frac{\rho_l}{\rho_v}\right)^{0.5}\right]^{2/3}$	(12)
For $\frac{DG_l}{\mu_l} > 5000$ $\frac{DG_v}{\mu_l}\left(\frac{\rho_l}{\rho_v}\right)^{0.5} > 20{,}000$	Ackers et al. (1959)	$\frac{hD}{k_l} = 0.026\left(\frac{c_p\mu_l}{k_l}\right)^{1/3}\left(\frac{DG_e}{\mu_l}\right)^{0.8}$ *where* $G_e = G_v(\rho_l/\rho_v)^{0.5} + G_l$	(13)

Notes: 1. Equations (1) through (10) are dimensional with units of Btu, h, ft, °F, and $\mathrm{lb_m}$. 2. t_f = liquid film temperature = $t_{sat} - 0.75\Delta t$

the condensate to flow. When the vapor velocity is high, the transition from laminar to turbulent flow occurs at Reynolds numbers lower than previously described [i.e., 1600 according to Grober et al. (1961)].

When **superheated** vapor is condensed, the heat transfer coefficient depends on the surface temperature. When the surface temperature is *below* saturation temperature, using the value of h for condensation of saturated vapor that incorporates the difference between the *saturation* temperature and the surface temperature leads to insignificant error (McAdams 1954). If the surface temperature is *above* the saturation temperature, there is no condensation and the equations for gas convection apply.

Correlation equations for condensing heat transfer are given in Table 3. Factors F_1 and F_2, which depend only on the physical properties of the working fluid and which occur often in these equations, have been computed for some commonly used refrigerants in Table 4. Refrigerant properties used in the calculations may be found in Chapter 19.

In some cases, the equations are given in two forms: one is convenient when the amount of refrigerant to be condensed or the condensing load is known; the second is useful when the difference between the vapor temperature and the condensing surface temperature is known.

Condensation on Outside Surface of Vertical Tubes

For film-type condensation on the outside surface of vertical tubes and on vertical surfaces, Equations (1) and (2) in Table 3 are recommended when $4\Gamma/\mu_f$ is less than 1800 (McAdams 1954). For these equations, fluid properties are evaluated at the mean film temperature. When $4\Gamma/\mu_f$ is greater than 1800 (tall vertical plates or tubes), use Equation (3) or (4) in Table 3. Equations (2) and (4) in Table 3 are plotted in Figure 9. The theoretical curve for laminar film-type condensation is shown for comparison. A semitheoretical relationship for turbulent film-type condensation is also shown for Pr values of 1.0 and 5.0 (Colburn 1933-34).

Condensation on Outside Surface of Horizontal Tubes

For a bank of N tubes, Nusselt's equations, increased by 10% (Jakob 1949 and 1957), are given in Equations (5) and (6) in Table 3. Experiments by Short and Brown (1951) with R-11 suggest that drops of condensation falling from row to row cause local turbulence and increase heat transfer.

For condensation on the outside surface of horizontal finned tubes, Equation (7) in Table 3 is used for liquids that drain readily from the surface (Beatty and Katz 1948). For condensing steam outside finned tubes, where liquid is retained in the spaces between the tubes, coefficients substantially lower than those given by Equation (7) in Table 3 were reported. For additional data on condensation outside finned tubes, see Katz et al. (1947). For more on this topic, refer to Webb (1994).

Table 4 Values of Condensing Coefficient Factors for Different Refrigerants (from Chapter 19)

Refrigerant	Film Temperature, °F $t_f = t_{sat} - 0.75\Delta t$	F_1	F_2
Refrigerant 11	75	154	822
	100	153	815
	125	151	803
Refrigerant 12	75	133	672
	100	122	608
	125	112	538
Refrigerant 22	75	153	822
	100	144	755
	125	132	675
Sulfur Dioxide	75	290	1920
	100	299	2000
	125	318	2170
Ammonia	75	409	3040
	100	408	3035
	125	408	3030
Propane	75	159	850
	100	157	845
	125	154	836
Butane	75	156	840
	100	156	843
	125	157	845

$$F_1 = \left(\frac{k_f^3 \rho_f^2 g}{\mu_f}\right)^{0.25} \qquad \text{Units: } \left[\frac{(\text{Btu})^3(\text{lb}_m)}{(\text{h})^4(\text{ft})^7(°\text{F})^3}\right]^{0.25}$$

$$F_2 = \left(\frac{k_f^3 \rho_f^2 g}{\mu_f}\right)^{1/3} \qquad \text{Units: } \left[\frac{(\text{Btu})^3(\text{lb}_m)}{(\text{h})^4(\text{ft})^7(°\text{F})^3}\right]^{1/3}$$

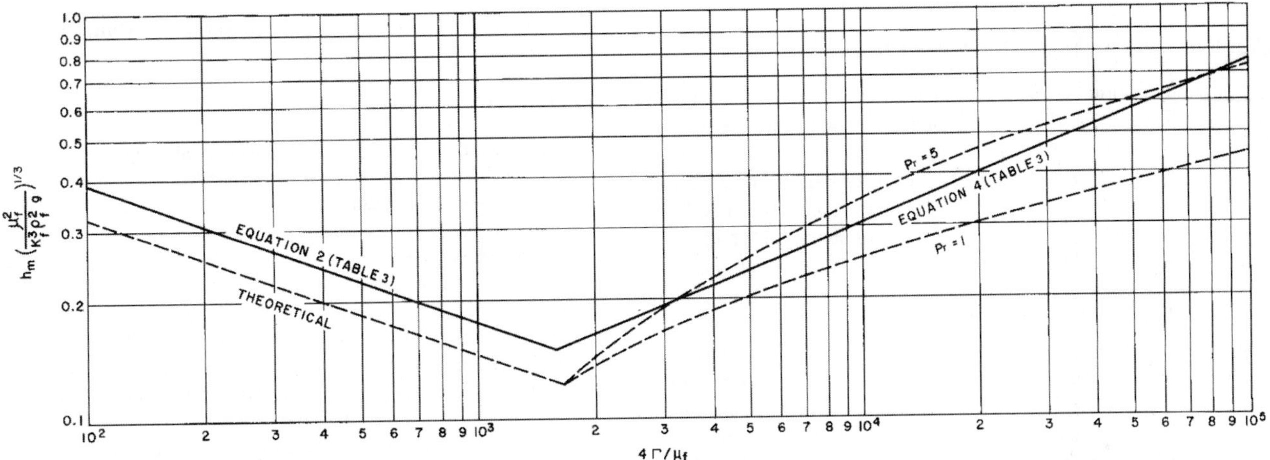

Fig. 9 Film-Type Condensation

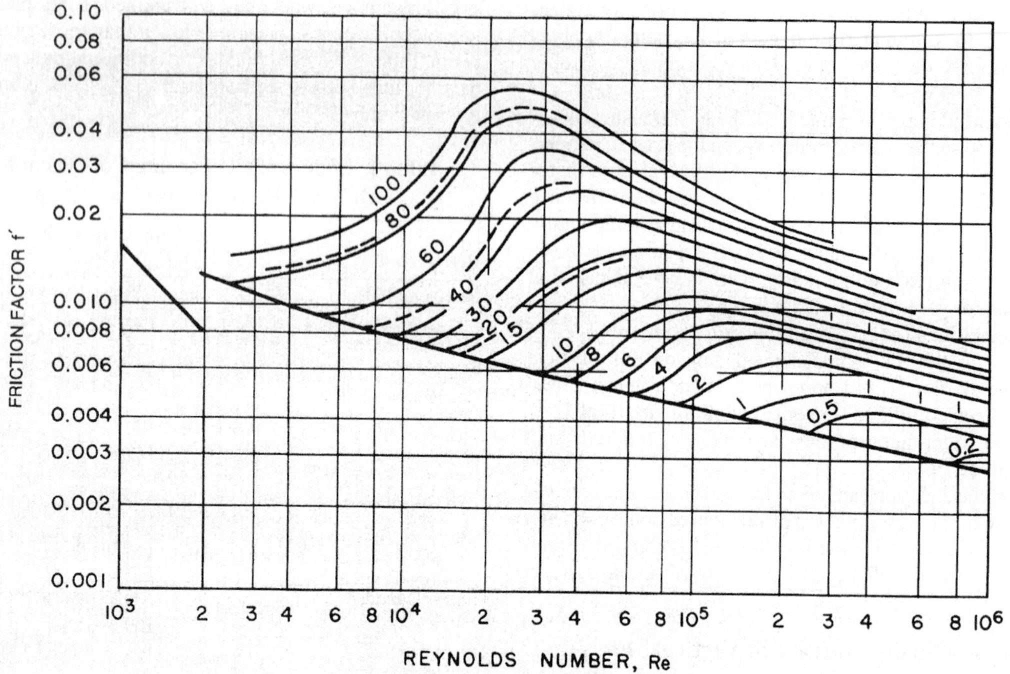

Curve parameter = Γ/ρs, where Γ = liquid flow rate, ρ = liquid density, and s = surface tension of liquid relative to water; values of gas velocity used in calculating *f* and Re are calculated as though no liquid were present.

Fig. 10 Friction Factors for Gas Flow Inside Pipes with Wetted Walls

Simplified Equations for Steam

For film-type steam condensation at atmospheric pressure and film temperature drops of 10 to 150°F, McAdams (1954) recommends Equations (8) and (9) in Table 3.

Condensation on Inside Surface of Vertical Tubes

Condensation on the inside surface of tubes is generally affected by appreciable vapor velocity. The measured heat transfer coefficients are as much as 10 times those predicted by Equation (4) in Table 3. For vertical tubes, Jakob (1949 and 1957) gives theoretical derivations for upward and downward vapor flow. For downward vapor flow, Carpenter and Colburn (1949) suggest Equation (10) in Table 3. The friction factor f' for vapor in a pipe containing condensate should be taken from Figure 10.

Condensation on Inside Surface of Horizontal Tubes

For condensation on the inside surface of horizontal tubes (as in air-cooled condensers, evaporative condensers, and some shell-and-tube condensers), the vapor velocity and resulting shear at the vapor-liquid interface are major factors in analyzing heat transfer. Hoogendoorn (1959) identified seven types of two-phase flow patterns. For semistratified and laminar annular flow, use Equations (11) and (12) in Table 3 (Ackers and Rosson 1960). Ackers et al. (1959) recommend Equation (13) in Table 3 for turbulent annular flow (vapor Reynolds number greater than 20,000 and liquid Reynolds number greater than 5000). For high mass flux (> 40 lb/s·ft^2), the method of Shah (1979) is recommended for predicting local heat transfer coefficients during condensation. A method for using a flow regime map to predict the heat transfer coefficient for condensation of pure components in a horizontal tube is presented in Breber et al. (1980). More recently, the flow regime dependent method of Dobson and Chato (1998) provides a more accurate design approach.

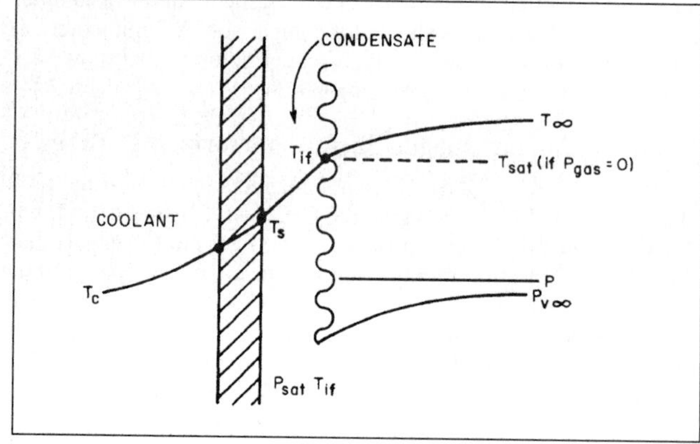

Fig. 11 Origin of Noncondensable Resistance

Noncondensable Gases

Condensation heat transfer rates reduce drastically if one or more noncondensable gases are present in the condensing vapor/gas mixture. In mixtures, the condensable component is termed **vapor** and the noncondensable component is called **gas**. As the mass fraction of gas increases, the heat transfer coefficient decreases in an approximately linear manner. In a steam chest with 2.89% air by volume, Othmer (1929) found that the heat transfer coefficient dropped from about 2000 to about 600 Btu/h·ft^2·°F. Consider a surface cooled to some temperature t_s below the saturation temperature of the vapor (Figure 11). In this system, accumulated condensate falls or is driven across the condenser surface. At a finite heat transfer rate, a temperature profile develops across the condensate that can be estimated from Table 3; the interface of the condensate is at a temperature $t_{if} > t_s$. In the absence of gas, the interface temperature is the vapor saturation temperature at the pressure of the condenser.

The presence of noncondensable gas lowers the vapor partial pressure and hence the saturation temperature of the vapor in equilibrium with the condensate. Further, the movement of the vapor toward the cooled surface implies similar bulk motion of the gas. At the condensing interface, the vapor is condensed at temperature t_{if} and is then swept out of the system as a liquid. The gas concentration rises to ultimately diffuse away from the cooled surface at the same rate as it is convected toward the surface (Figure 11). If gas (mole fraction) concentration is Y_g and total pressure of the system is p, the partial pressure of the bulk gas is

$$p_{g\infty} = Y_{g\infty}p \tag{2}$$

The partial pressure of the bulk vapor is

$$p_{v\infty} = (1 - Y_{g\infty})p = Y_{v\infty}p \tag{3}$$

As opposing fluxes of convection and diffusion of the gas increase, the partial pressure of gas at the condensing interface is $p_{gif} > p_{g\infty}$. By Dalton's law, assuming isobaric condition,

$$p_{gif} + p_{vif} = p \tag{4}$$

Hence, $p_{vif} < p_{v\infty}$.

Sparrow et al. (1967) noted that thermodynamic equilibrium exists at the interface, except in the case of very low pressures or liquid metal condensation, so that

$$p_{vif} = p_{sat}(t_{if}) \tag{5}$$

where $p_{sat}(t)$ is the saturation pressure of the vapor at temperature t. The available Δt for condensation across the condensate film is reduced from $(t_\infty - t_s)$ to $(t_{if} - t_s)$, where t_∞ is the bulk temperature of the condensing vapor-gas mixture, caused by the additional noncondensable resistance.

The equations in Table 3 are still valid for the condensate resistance, but the interface temperature t_{if} must be found. The noncondensable resistance, which accounts for the temperature difference $(t_\infty - t_{if})$, depends on the heat flux (through the convecting flow to the interface) and the diffusion of gas away from the interface.

In simple cases, Sparrow et al. (1967), Rose (1969), and Sparrow and Lin (1964) found solutions to the combined energy, diffusion, and momentum problem of noncondensables, but they are cumbersome.

A general method given by Colburn and Hougen (1934) can be used over a wide range if correct expressions are provided for the rate equations—add the contributions of the sensible heat transport through the noncondensable gas film and the latent heat transport via condensation:

$$h_g(t_\infty - t_{if}) + K_D M_v h_{lv}(p_{v\infty} - p_{vif}) = h(t_{if} - t_s) = U(t_{if} - t_c) \tag{6}$$

where h is from the appropriate equation in Table 3.

The value of the heat transfer coefficient for the stagnant gas depends on the geometry and flow conditions. For flow parallel to a condenser tube, for example,

$$j = \left(\frac{h_g}{(c_p)_g G}\right)\left(\frac{(c_p)_g \mu_{gv}}{K_{Dg}}\right)^{2/3} \tag{7}$$

where j is a known function of Re $= GD/\mu_{gv}$.

The mass transfer coefficient K_D is

$$\frac{K_D}{M_m}\left[\frac{p_{g\infty} - p_{gif}}{\ln(p_{g\infty}/p_{gif})}\right]\left(\frac{\mu_{gv}}{\rho_g D}\right)^{2/3} = j \tag{8}$$

The calculation method requires substitution of Equation (8) into Equation (6). For a given flow condition, G, Re, j, M_m, $p_{g\infty}$, h_g, and h (or U) are known. Assume values of t_{if}; calculate $p_{sat}(t_{if}) = p_{vif}$ and hence p_{gif}. If t_s is not known, use the overall coefficient U to the coolant and t_c in place of h and t_s in Equation (6). For either case, at each location in the condenser, iterate Equation (6) until it balances, giving the condensing interface temperature and, hence, the thermal load to that point (Colburn and Hougen 1934, Colburn 1951). For more detail, refer to Chapter 10 in Collier and Thome (1996).

Other Impurities

Vapor entering the condenser often contains a small percentage of impurities such as oil. Oil forms a film on the condensing surfaces, creating additional resistance to heat transfer. Some allowance should be made for this, especially in the absence of an oil separator or when the discharge line from the compressor to the condenser is short.

PRESSURE DROP

Total pressure drop for two-phase flow in tubes consists of friction, acceleration, and gravitational components. It is necessary to know the **void fraction** (the ratio of gas flow area to total flow area) to compute the acceleration and gravitational components. To compute the frictional component of pressure drop, either the **two-phase friction factor** or the **two-phase frictional multiplier** must be determined.

The homogeneous model provides a simple method for computing the acceleration and gravitational components of pressure drop. The homogeneous model assumes that the flow can be characterized by average fluid properties and that the velocities of the liquid and vapor phases are equal (Collier and Thome 1996, Wallis 1969).

Martinelli and Nelson (1948) developed a method for predicting the void fraction and two-phase frictional multiplier to use with a separated flow model. This method predicts the pressure drops of boiling refrigerants reasonably well. Other methods of computing the void fraction and two-phase frictional multiplier used in a separated flow model are given in (Collier and Thome 1996, Wallis 1969).

The general nature of annular gas-liquid flow in vertical, and to some extent horizontal, pipe is indicated in Figure 12 (Wallis 1970), which plots the effective gas friction factor versus the liquid fraction $(1 - a)$. Here a is the void fraction, or fraction of the pipe cross section taken up by the gas or vapor.

The effective gas friction factor is defined as

$$f_{eff} = \left[\frac{a^{5/2}D}{2\rho_g(4Q_g/\pi D^2)^2}\right]\left(-\frac{dp}{ds}\right) \tag{9}$$

where D is the pipe diameter, ρ_g the gas density, and Q_g the gas volumetric flow rate. The friction factor of gas flowing by itself in the pipe (presumed smooth) is denoted by f_g. Wallis' analysis of the flow occurrences is based on interfacial friction between the gas and liquid. The wavy film corresponds to a conduit of relative roughness ε/D, about four times the liquid film thickness. Thus, the pressure drop relation of vertical flow is

$$-\frac{dp}{ds} + \rho_g g = 0.01\left(\frac{\rho_g}{D^5}\right)\left(\frac{4Q_g}{\pi}\right)^2 \frac{1 + 75(1 - a)}{a^{5/2}} \tag{10}$$

This corresponds to the Martinelli-type analysis with

$$f_{two\text{-}phase} = \phi_g^2 f_g$$

when

$$\phi_g^2 = \frac{1 + 75(1 - a)}{a^{5/2}} \tag{11}$$

The friction factor f_g (of the gas alone) is taken as 0.02, an appropriate turbulent flow value. This calculation can be modified for more detailed consideration of factors such as Reynolds number variation in friction, gas compressibility, and entrainment (Wallis 1970).

In two-phase flow inside horizontal tubes, the pressure gradient is written as the sum of frictional and momentum terms. Thus,

$$\frac{dp}{dz} = \left(\frac{dp}{dz}\right)_f + \left(\frac{dp}{dz}\right)_m \tag{12}$$

In adiabatic two-phase flow, the contribution of the momentum transfer to the overall pressure drop is negligibly small; theoretically, it is nonexistent if the flow is fully developed. In condensation heat transfer, the momentum transfer term contributes to the overall pressure drop due to the mass transfer that occurs at the liquid-vapor interface.

Two basic models were used in developing frictional pressure drop correlations for two-phase adiabatic flow. In the first, the flow of both phases is assumed to be homogeneous; the gas and liquid velocities are assumed equal. The frictional pressure drop is computed as if the flow were single phase, except for introducing modifiers to the single-phase friction coefficient. In the second model, the two phases are considered separate, and the velocities may differ. Two correlations used to predict the frictional pressure drop are those of Lockhart and Martinelli (1949) and Dukler et al. (1964).

In the Lockhart-Martinelli correlation, a parameter X was defined as

$$X = \left[\left(\frac{dp}{dz}\right)_l \div \left(\frac{dp}{dz}\right)_v\right]^{0.5} \tag{13}$$

where

$\left(\frac{dp}{dz}\right)_l$ = frictional pressure gradient, assuming that liquid alone flows in pipe

$\left(\frac{dp}{dz}\right)_v$ = frictional pressure gradient, assuming that gas (or vapor in case of condensation) alone flows in pipe

The frictional pressure gradient due to the single-phase flow of the liquid or vapor depends on the type of flow of each phase, laminar or turbulent. For turbulent flow during condensation, replace X by X_{tt}. Thus,

$$X_{tt} = \left(\frac{1-x}{x}\right)^{0.9}\left(\frac{\mu_l}{\mu_v}\right)^{0.1}\left(\frac{\rho_v}{\rho_l}\right)^{0.5} \tag{14}$$

Lockhart and Martinelli (1949) also defined ϕ_v as

$$\phi_v = \left[\left(\frac{dp}{dz}\right)_f \div \left(\frac{dp}{dz}\right)_v\right]^{0.5} \tag{15}$$

For condensation,

$$\left(\frac{dp}{dz}\right)_v = -\frac{2f_o(xG)^2}{\rho_v D_i} \tag{16}$$

where

$$f_o = \frac{0.045}{(GxD_i/\mu_v)^{0.2}} \tag{17}$$

Here f_o is the friction factor for adiabatic two-phase flow.

By analyzing the pressure drop data of simultaneous adiabatic flow of air and various liquids, Lockhart and Martinelli (1949) correlated the parameters ϕ_v and X and reported the results graphically. Soliman et al. (1968) approximated the graphical results of ϕ_v versus X_{tt} by

$$\phi_v = 1 + 2.85X_{tt}^{0.523} \tag{18}$$

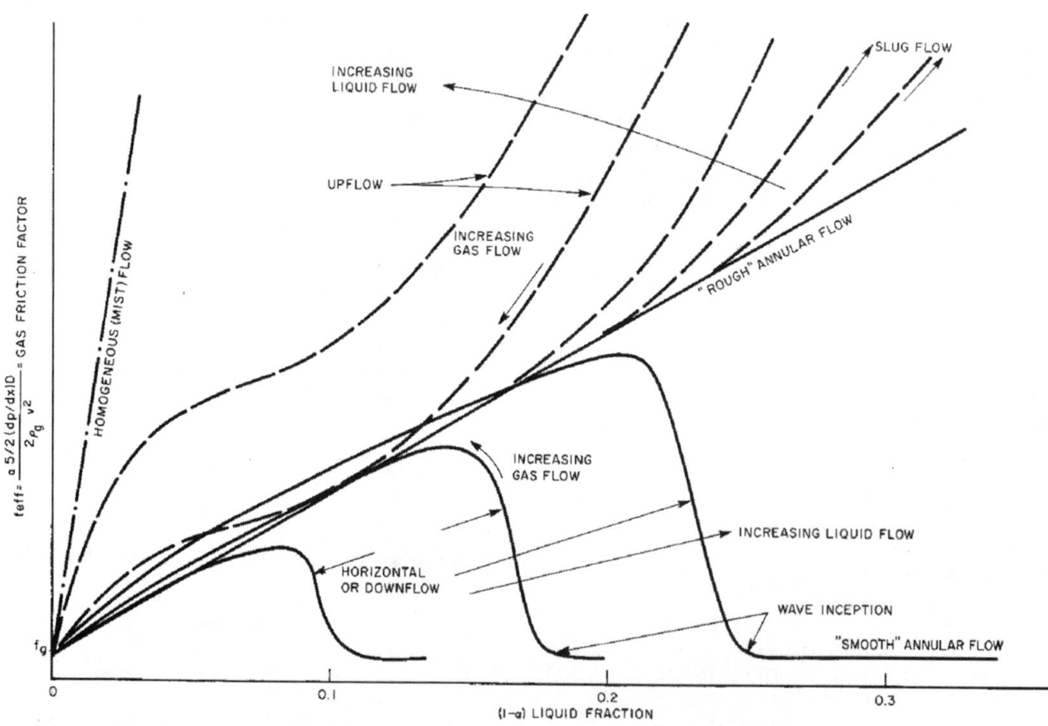

Fig. 12 Qualitative Pressure Drop Characteristics of Two-Phase Flow Regime

In the correlation of Dukler et al. (1964), the frictional pressure gradient is given by

$$\left(\frac{dp}{dz}\right)_f = -\frac{2G^2 f_o \alpha(\lambda)\beta}{D_i \rho_{NS}} \tag{19}$$

where

f_o = single-phase friction coefficient evaluated at two-phase Reynolds number

$$= 0.0014 + 0.125\left(\frac{4\dot{m}_t\beta}{\pi D_i \mu_{NS}}\right)^{-0.32} \tag{20}$$

$$\alpha(\lambda) = 1 - (\ln\lambda)/[1.281 + 0.478\ln\lambda + 0.444(\ln\lambda)^2 + 0.094(\ln\lambda)^3 + 0.00843(\ln\lambda)^4] \tag{21}$$

$$\beta = \left(\frac{\rho_l}{\rho_{NS}}\right)\frac{\lambda^2}{(1-\psi)} + \left(\frac{\rho_v}{\rho_{NS}}\right)\frac{(1-\lambda)^2}{\psi} \tag{22}$$

$$\rho_{NS} = \rho_l\lambda + \rho_v(1-\lambda) \tag{23}$$

$$\mu_{NS} = \mu_l\lambda + \mu_v(1-\lambda) \tag{24}$$

$$\lambda = 1/\left(1 + \frac{x}{(1-x)}\frac{\rho_v}{\rho_e}\right) \tag{25}$$

Because the correlations mentioned here were originally developed for adiabatic two-phase flow, Luu and Bergles (1980) modified the friction coefficients in Equations (16) and (19), using the modifier suggested by Silver and Wallis (1965-66). The modification replaced the friction coefficient f_o with the friction coefficient f_{co}. These terms are related by

$$\left(\frac{f_{co}}{f_o}\right) = \exp\left(\frac{\varepsilon}{2f_o}\right) - \left(\frac{\xi}{f_o}\right) \tag{26}$$

where

$$\xi = \left(\frac{D_i\psi}{2x}\right)\frac{dx}{dz} \tag{27}$$

Because the Lockhart-Martinelli and Dukler correlations for the frictional pressure gradient were based on the separated flow model, the momentum pressure gradient should be as well. Thus,

$$\left(\frac{dp}{dz}\right)_m = -G^2\left(\frac{dx}{dz}\right)\left\{\frac{2x}{\rho_v\psi} - \frac{2(1-x)}{\rho_l(1-\psi)} + q_l\left[\frac{\psi}{x}\frac{1-x}{1-\psi}\frac{1}{\rho_l} - \frac{x(1-\psi)}{\psi(1-x)\rho_v}\right]\right\} \tag{28}$$

To determine $(dp/dz)_m$, the void fraction ψ and the quality gradient must be known. A generalized expression for ψ was suggested by Butterworth (1975):

$$\psi = \frac{1}{1 + A_l[(1-x)/x]^{q_l}(\rho_v/\rho_l)^{r_l}(\mu_l/\mu_v)^{S_l}} \tag{29}$$

Table 5 Constants in Equation (29) for Different Void Fraction Correlations

Model	A_l	q_l	r_l	S_l
Homogeneous (Collier 1972)	1.0	1.0	1.0	0
Lockhart-Martinelli (1949)	0.28	0.64	0.36	0.07
Baroczy (1963)	1.0	0.74	0.65	0.13
Thom (1964)	1.0	1.0	0.89	0.18
Zivi (1964)	1.0	1.0	0.67	0
Turner-Wallis (1965)	1.0	0.72	0.40	0.08

where A_l, q_l, r_l, and S_l are constants and are listed for the various correlations in Table 5.

The quality gradient dx/dz in Equation (28) can be estimated by assuming a constant rate of cooling. In the case of complete condensation, its value is $-1/L$, where L is the length of the condenser tube.

Evaporators and condensers often have valves, tees, bends, and other fittings that contribute to the overall pressure drop of the heat exchanger. Collier and Thome (1996) summarize methods predicting the two-phase pressure drop in these fittings.

ENHANCED SURFACES

Enhanced heat transfer surfaces are used in heat exchangers to improve performance and decrease cost. Condensing heat transfer is often enhanced with circular fins attached to the external surfaces of tubes to increase the heat transfer area. Other enhancement methods, such as porous coatings, integral fins, and reentrant cavities, are used to augment boiling heat transfer on the external surfaces of evaporator tubes. Webb (1981) surveys external boiling surfaces and compares the performances of several enhanced surfaces with the performance of smooth tubes. For heat exchangers, the heat transfer coefficient for the refrigerant side is often smaller than the coefficient for the water side. Thus, enhancing the refrigerant-side surface can reduce the size of the heat exchanger and improve its performance.

Internal fins increase the heat transfer coefficients during evaporation or condensation in tubes. However, internal fins increase the refrigerant pressure drop and reduce the heat transfer rate by decreasing the available temperature difference between hot and cold fluids. Designers should carefully determine the number of parallel refrigerant passes that give optimum loading for best overall heat transfer.

For additional information on enhancement methods in two-phase flow, consult Bergles (1976, 1985), Thome (1990), and Webb (1994).

SYMBOLS

A = area
A_{eff} = total effective area [Equation (7) in Table 3]
a = local acceleration [Equation (19) in Table 1]; void fraction [Equations (9) and (10)]
a_f = area of one side of one film
B_o = boiling number [Equations (3) and (7) in Table 2]
b = breadth of a condensing surface. For vertical tube, $b = \pi d$; for horizontal tube, $b = 2L$
C = a coefficient or constant
C_o = convection number
$C_1 \ldots C_5$ = special constants (see Table 2)
c_p = specific heat at constant pressure
c_v = specific heat at constant volume
D = diameter
D_d = bubble departure diameter [Equation (5) in Table 1]
D_i = inside tube diameter
D_o = outside tube diameter
d = diameter; or prefix meaning differential
(dp/dz) = pressure gradient

$(dp/dz)_f$ = frictional pressure gradient
$(dp/dz)_l$ = frictional pressure gradient, assuming that liquid alone is flowing in pipe
$(dp/dz)_m$ = momentum pressure gradient
$(dp/dz)_v$ = frictional pressure gradient, assuming that gas (or vapor) alone is flowing in pipe
E = special coefficient (Table 2)
F_c = Reynolds number factor [Equation (12) in Table 2 and Figure 7]
F_{PF} = special coefficient [Equation (7) in Table 1]
Fr = Froude number
F_1, F_2 = condensing coefficient factors (Table 4)
f = friction factor for single-phase flow
f' = friction factor for gas flow inside pipes with wetted walls (Figure 10)
f_{co} = friction factor in presence of condensation [Equation (26)]
f_o = friction factor [Equations (17) and (19)]
G = mass velocity
Gr = Grashof number
g = gravitational acceleration
g_c = gravitational constant
h = heat transfer coefficient
h_f = special coefficient [Equation (7) in Table 1]
h_{fg} = latent heat of vaporization or of condensation
j = Colburn *j*-factor
K_D = mass transfer coefficient
k = thermal conductivity
L = length
L_{mf} = mean length of fin [Equation (7) in Table 3]
ln = natural logarithm
M = mass; or molecular weight
M_m = mean molecular weight of vapor-gas mixture
M_v = molecular weight of condensing vapor
m = general exponent [Equations (1) and (6) in Table 1]
$\dot{m}$ = mass rate of flow
N = number of tubes in vertical tier
Nu = Nusselt number
n = general exponent [Equations (1) and (6) in Table 1]
Pr = Prandtl number
p = pressure
p_c = critical thermodynamic pressure for coolant
p_r = reduced pressure
Q = total heat transfer
q = rate of heat transfer
r = radius
Ra = Rayleigh number
Re = Reynolds number
R_p = surface roughness, μm
S = distance along flow direction
S_c = suppression factor (Table 2 and Figure 8)
t = temperature
U = overall heat transfer coefficient
V = linear velocity
x = quality (i.e., vapor fraction = M_v/M); or distance in dt/dx
X_{tt} = Martinelli parameter [Figure 7, Table 2, and Equation (14)]
x,y,z = lengths along principal coordinate axes
Y_g = mole fraction of gas [Equations (2) and (3)]
Y_v = mole fraction of vapor [Equation (3)]
α = thermal diffusivity = $k/\rho c_p$
$\alpha(\lambda)$ = ratio of two-phase friction factor to single-phase friction factor at two-phase Reynolds number [Equation (21)]
β = ratio of two-phase density to no-slip density [Equation (22)]
Γ = mass rate of flow of condensate per unit of breadth (see section on Condensing)
Δ = difference between values
ε = roughness of interface
Λ = special coefficient [Equations (16) through (19) in Table 1]
λ = ratio of liquid volumetric flow rate to total volumetric flow rate [Equation (25)]
μ = absolute (dynamic) viscosity
μ_l = dynamic viscosity of saturated liquid
μ_{NS} = dynamic viscosity of two-phase homogeneous mixture [Equation (24)]
μ_v = dynamic viscosity of saturated vapor
ν = kinematic viscosity
ρ = density
ρ_l = density of saturated liquid
ρ_{NS} = density of two-phase homogeneous mixture [Equation (23)]
ρ_v = density of saturated vapor phase
σ = surface tension
ϕ_g = fin efficiency, Martinelli factor [Equation (11)]
ϕ_v = Lockhart-Martinelli parameter [Equation (15)]
ψ = void fraction

Subscripts and Superscripts

a = exponent in Equation (1)
b = bubble
c = critical or cold (fluid)
cg = condensing
e = equivalent
eff = effective
f = film or fin
g = gas
h = horizontal or hot (fluid) or hydraulic
i = inlet or inside
if = interface
L = liquid
l = liquid
m = mean
mac = convective mechanism [Equations (11) through (13) in Table 2]
max = maximum
mic = nucleation mechanism [Equations (11) through (13) in Table 2]
min = minimum
ncb = nucleate boiling
o = outside or outlet or overall
r = root (fin) or reduced pressure
s = surface or secondary heat transfer surface
sat = saturation (pressure)
t = temperature or terminal temperature of tip (fin)
v = vapor or vertical
w = wall
∞ = bulk
$*$ = reference

REFERENCES

Ackers, W.W., H.A. Deans, and O.K. Crosser. 1959. Condensing heat transfer within horizontal tubes. *Chemical Engineering Progress Symposium Series* 55(29):171-76.

Ackers, W.W. and H.F. Rosson. 1960. Condensation inside a horizontal tube. *Chemical Engineering Progress Symposium Series* 56(30):145-50.

Altman, M., F.W. Staub, and R.H. Norris. 1960b. Local heat transfer and pressure drop for Refrigerant-22 condensing to horizontal tubes. *Chemical Engineering Progress Symposium Series* 56(30):151-60.

Anderson, W., D.G. Rich, and D.F. Geary. 1966. Evaporation of Refrigerant 22 in a horizontal 3/4-in. OD tube. *ASHRAE Transactions* 72(1):28.

Baroczy, C.J. 1963. Correlation of liquid fraction in two-phase flow with application to liquid metals. *North American Aviation Report* SR-8171, El Segundo, CA.

Beatty, K.O. and D.L. Katz. 1948. Condensation of vapors on outside of finned tubes. *Chemical Engineering Progress* 44(1):55.

Berenson, P.J. 1961. Film boiling heat transfer from a horizontal surface. *ASME Journal of Heat Transfer* 85:351.

Berenson, P.J. 1962. Experiments on pool boiling heat transfer. *International Journal of Heat and Mass Transfer* 5:985.

Bergles, A.E. 1976. Survey and augmentation of two-phase heat transfer. *ASHRAE Transactions* 82(1):891-905.

Bergles, A.E. 1985. Techniques to augment heat transfer. In *Handbook of heat transfer application*, 2nd ed. McGraw-Hill, New York.

Bergles, A.E. and W.M. Rohsenow. 1964. The determination of forced convection surface-boiling heat transfer. *ASME Journal of Heat Transfer*, Series C, 86(August):365.

Borishansky, W. and A. Kosyrev. 1966. Generalization of experimental data for the heat transfer coefficient in nucleate boiling. *ASHRAE Journal* (May):74.

Borishansky, V.M., I.I. Novikov, and S.S. Kutateladze. 1962. Use of thermodynamic similarity in generalizing experimental data on heat transfer. Proceedings of the International Heat Transfer Conference.

Breber, G., J.W. Palen, and J. Taborek. 1980. Prediction of the horizontal tubeside condensation of pure components using flow regime criteria. *ASME Journal of Heat Transfer* 102(3):471-76.

Breen, B.P. and J.W. Westwater. 1962. Effects of diameter of horizontal tubes on film boiling heat transfer. AIChE Preprint No. 19, Fifth National Heat Transfer Conference, Houston, TX. *Chemical Engineering Progress* 58(7):67-72.

Bromley, L.A. 1950. Heat transfer in stable film boiling. *Chemical Engineering Progress* (46):221.

Butterworth, D. 1975. A comparison of some void-fraction relationships for co-current gas-liquid flow. *International Journal of Multiphase Flow* 1:845-50.

Carey, V.P. 1992. *Liquid-vapor phase change phenomena: An introduction to the thermophysics of vaporization and condensation processes in heat transfer equipment.* Hemisphere Publishing Corporation, Washington, D.C.

Carpenter, E.F. and A.P. Colburn. 1949. The effect of vapor velocity on condensation inside tubes. General discussion on Heat Transfer and Fluid Mechanics Institute, American Society of Mechanical Engineers, New York.

Chaddock, J.B. and J.A. Noerager. 1966. Evaporation of Refrigerant 12 in a horizontal tube with constant wall heat flux. *ASHRAE Transactions* 72(1):90.

Chen, J.C. 1963. A correlation for boiling heat transfer to saturated fluids on convective flow. ASME *Paper* 63-HT-34. American Society of Mechanical Engineers, New York.

Colburn, A.P. 1933-34. Note on the calculation of condensation when a portion of the condensate layer is in turbulent motion. *AIChE Transactions* No. 30.

Colburn, A.P. 1951. Problems in design and research on condensers of vapours and vapour mixtures. Proceedings of the Institute of Mechanical Engineers, 164:448, London.

Colburn, A.P. and O.A. Hougen. 1934. Design of cooler condensers for mixtures of vapors with noncondensing gases. *Industrial and Engineering Chemistry* 26 (November):1178.

Collier, J.G. and J.R. Thome. 1996. *Convective boiling and condensation*, 3rd ed. Oxford University Press.

Cooper, M.G. 1984. Heat flow rates in saturated nucleate pool boiling-a wide-ranging examination using reduced properties. *Advances in Heat Transfer.* Academic Press, Orlando, 16, 157-239.

Danilova, G. 1965. Influence of pressure and temperature on heat exchange in the boiling of halogenated hydrocarbons. *Kholodilnaya Teknika*, No. 2. English abstract, *Modern Refrigeration* (December).

Dengler, C.E. and J.N. Addoms. 1956. Heat transfer mechanism for vaporization of water in a vertical tube. *Chemical Engineering Progress Symposium Series* 52(18):95.

Dobson, M.K., and J.C. Chato. 1998. Condensation in smooth horizontal tubes. *Journal of Heat Transfer* 120:193-213.

Dougherty, R.L. and H.J. Sauer, Jr. 1974. Nucleate pool boiling of refrigerant-oil mixtures from tubes. *ASHRAE Transactions* 80(2):175.

Dukler, A.E., M. Wicks, III, and R.G. Cleveland. 1964. Frictional pressure drop in two-phase flow: An approach through similarity analysis. *AIChE Journal* 10(January):44-51.

Eckels, S.J., T.M. Doer, and M.B. Pate. 1994. In-tube heat transfer and pressure drop of R-134a and ester lubricant mixtures in a smooth tube and a micro-fin tube: Part 1 evaporation. *ASHRAE Transactions* 100(2)(1994): 265-82.

Farber, E.A. and R.L. Scorah. 1948. Heat transfer to water boiling under pressure. ASME *Transactions* (May):373.

Forster, H.K. and N. Zuber. 1955. Dynamics of vapor bubbles and boiling heat transfer. *AIChE Journal* 1(4):531-35.

Frederking, T.H.K. and J.A. Clark. 1962. Natural convection film boiling on a sphere. In *Advances in cryogenic engineering*, ed. K.D. Timmerhouse, Plenum Press, New York.

Furse, F.G. 1965. Heat transfer to Refrigerants 11 and 12 boiling over a horizontal copper surface. *ASHRAE Transactions* 71(1):231.

Gorenflo, D. 1993. Pool boiling. *VDI-Heat Atlas*. VDI-Verlag, Düsseldorf.

Gouse, S.W., Jr. and K.G. Coumou. 1965. Heat transfer and fluid flow inside a horizontal tube evaporator, Phase I. *ASHRAE Transactions* 71(2):152.

Green, G.H. and F.G. Furse. 1963. Effect of oil on heat transfer from a horizontal tube to boiling Refrigerant 12-oil mixtures. *ASHRAE Journal* (October):63.

Grigull, U. 1952. Wärmeübergang bei Filmkondensation. *Forsch, Gebiete Ingenieurw.* 18.

Grober, H., S. Erk, and U. Grigull. 1961. *Fundamentals of heat transfer.* McGraw-Hill, New York.

Guerrieri, S.A. and R.D. Talty. 1956. A study of heat transfer to organic liquids in single tube boilers. *Chemical Engineering Progress Symposium Series* 52(18):69.

Gungor, K.E., and R.H.S. Winterton. 1986. A general correlation for flow boiling in tubes and annuli. *International Journal Heat Mass Transfer* 29:351-58.

Hoogendoorn, C.J. 1959. Gas-liquid flow in horizontal pipes. *Chemical Engineering Sciences* IX(1).

Hughmark, G.A. 1962. A statistical analysis of nucleate pool boiling data. *International Journal of Heat and Mass Transfer* 5:667.

Isrealachvili, J.N. 1991. *Intermolecular surface forces.* Academic Press, New York.

Jakob, M. 1949 and 1957. *Heat transfer*, Vols. I and II. John Wiley and Sons, New York.

Kandlikar, S.G. 1990. A general correlation for saturated two-phase flow and boiling heat transfer inside horizontal and vertical tubes. *Journal of Heat Transfer* 112:219-28.

Kattan, N., J.R. Thome, and D. Favrat. 1998a. Flow boiling in horizontal tubes Part 1: Development of diabatic two-phase flow pattern map. *Journal of Heat Transfer* 120(1):140-47.

Kattan, N., J.R. Thome, and D. Favrat. 1998b. Flow boiling in horizontal tubes Part 3: Development of new heat transfer model based on flow patterns. *Journal of Heat Transfer* 120(1):156-65.

Katz, D.L., P.E. Hope, S.C. Datsko, and D.B. Robinson. 1947. Condensation of Freon-12 with finned tubes. Part I, Single horizontal tubes; Part II, Multitube condensers. *Refrigerating Engineering* (March):211, (April): 315.

Kutateladze, S.S. 1951. A hydrodynamic theory of changes in the boiling process under free convection. Izvestia Akademii Nauk, USSR, *Otdelenie Tekhnicheski Nauk* 4:529.

Kutateladze, S.S. 1963. *Fundamentals of heat transfer.* E. Arnold Press, London.

Lienhard, J.H. and V.E. Schrock. 1963. The effect of pressure, geometry and the equation of state upon peak and minimum boiling heat flux. *ASME Journal of Heat Transfer* 85:261.

Lienhard, J.H. and P.T.Y. Wong. 1963. The dominant unstable wave length and minimum heat flux during film boiling on a horizontal cylinder. ASME *Paper* No. 63-HT-3. ASME-AIChE Heat Transfer Conference, Boston, August.

Lockhart, R.W. and R.C. Martinelli. 1949. Proposed correlation of data for isothermal two-phase, two-component flow in pipes. *Chemical Engineering Progress* 45(1):39-48.

Luu, M. and A.E. Bergles. 1980. Augmentation of in-tube condensation of R-113. ASHRAE *Research Project* RP-219.

Martinelli, R.C. and D.B. Nelson. 1948. Prediction of pressure drops during forced circulation boiling of water. *ASME Transactions* 70:695.

McAdams, W.H. 1954. *Heat transmission*, 3rd ed. McGraw-Hill, New York.

Nukiyama, S. 1934. The maximum and minimum values of heat transmitted from metal to boiling water under atmospheric pressure. *Journal of the Japanese Society of Mechanical Engineers* 37:367.

Othmer, D.F. 1929. The condensation of steam. *Industrial and Engineering Chemistry* 21(June):576.

Perry, J.H. 1950. *Chemical engineers handbook*, 3rd ed. McGraw-Hill, New York.

Pierre, B. 1955. S.F. Review. *A.B. Svenska Flaktafabriken*, Stockholm, Sweden 2(1):55.

Pierre, B. 1957. *Kylteknisk Tidskrift* 3 (May):129.

Pierre, B. 1964. Flow resistance with boiling refrigerant. *ASHRAE Journal* (September through October).

Rohsenow, W.M. 1963. Boiling heat transfer. In *Modern developments in heat transfer*, ed. W. Ibele. Academic Press, New York.

Rose, J.W. 1969. Condensation of a vapour in the presence of a noncondensable gas. *International Journal of Heat and Mass Transfer* 12:233.

Schlager, L.M., M.B. Pate, and A.E. Bergles. 1987. Evaporation and condensation of refrigerant-oil mixtures in a smooth tube and micro-fin tube. *ASHRAE Transactions* 93:293-316.

Schrock, V.E. and L.M. Grossman. 1962. Forced convection boiling in tubes. *Nuclear Science and Engineering* 12:474.

Shah, M.M. 1979. A general correlation for heat transfer during film condensation inside pipes. *International Journal Heat Mass Transfer* 22:547-56.

Shah, M.M. 1982. A new correlation for saturated boiling heat transfer: equations and further study. *ASHRAE Transactions* 88(1):185-96.

Short, B.E. and H.E. Brown. 1951. Condensation of vapors on vertical banks of horizontal tubes. American Society of Mechanical Engineers, New York.

Silver, R.S. and G.B. Wallis. 1965-66. A simple theory for longitudinal pressure drop in the presence of lateral condensation. Proceedings of Institute of Mechanical Engineering, 180 Part I(1):36-42.

Soliman, M., J.R. Schuster, and P.J. Berenson. 1968. A general heat transfer correlation for annular flow condensation. *Journal of Heat Transfer* 90:267-76.

Sparrow, E.M. and S.H. Lin. 1964. Condensation in the presence of a noncondensable gas. ASME *Transactions*, *Journal of Heat Transfer* 86C:430.

Sparrow, E.M., W.J. Minkowycz, and M. Saddy. 1967. Forced convection condensation in the presence of noncondensables and interfacial resistance. *International Journal of Heat and Mass Transfer* 10:1829.

Starczewski, J. 1965. Generalized design of evaporation heat transfer to nucleate boiling liquids. *British Chemical Engineering* (August).

Steiner, D., and J. Taborek. 1992. Flow boiling heat transfer in vertical tubes correlated by an asymptotic model. *Heat Transfer Engineering* 13(2), 43-69.

Stephan, K. 1963a. The computation of heat transfer to boiling refrigerants. *Kältetechnik* 15:231.

Stephan, K. 1963b. Influence of oil on heat transfer of boiling Freon-12 and Freon-22. Eleventh International Congress of Refrigeration, I.I.R. *Bulletin* No. 3.

Stephan, K. 1963c. A mechanism and picture of the processes involved in heat transfer during bubble evaporation. *Chemic. Ingenieur Technik* 35:775.

Stephan, K. and M. Abdelsalam. 1980. Heat transfer correlations for natural convection boiling. *International Journal of Heat and Mass Transfer* 23:73-87.

Thom, J.R.S. 1964. Prediction of pressure drop during forced circulation boiling water. *International Journal of Heat and Mass Transfer* 7:709-24.

Thome, J.R. 1990. *Enhanced boiling heat transfer.* Hemisphere (Taylor and Francis), New York.

Tschernobyiski, I. and G. Ratiani. 1955. *Kholodilnaya Teknika* 32.

Turner, J.M. and G.B. Wallis. 1965. The separate-cylinders model of two-phase flow. *Report* No. NYO-3114-6. Thayer's School of Engineering, Dartmouth College, Hanover, NH.

Van Stralen, S.J. 1959. Heat transfer to boiling binary liquid mixtures. *Chemical Engineering* (British) 4(January):78.

Wallis, G.B. 1969. *One-dimensional two-phase flow.* McGraw-Hill, New York.

Wallis, G.C. 1970. Annular two-phase flow, Part I: A simple theory, Part II: Additional effect. *ASME Transactions*, *Journal of Basic Engineering* 92D:59 and 73.

Webb, R.L. 1981. The evolution of enhanced surface geometrics for nucleate boiling. *Heat Transfer Engineering* 2(3-4):46-69.

Webb, J.R. 1994. *Enhanced boiling heat transfer.* John Wiley and Sons, New York.

Westwater, J.W. 1963. Things we don't know about boiling. In *Research in Heat Transfer*, ed. J. Clark. Pergamon Press, New York.

Worsoe-Schmidt, P. 1959. Some characteristics of flow-pattern and heat transfer of Freon-12 evaporating in horizontal tubes. *Ingenieren*, International edition, 3(3).

Zeurcher O., J.R. Thome, and D. Favrat. 1998. In-tube flow boiling of R-407C and R-407C/oil mixtures, part II: Plain tube results and predictions. *Int. J. HVAC&R Research* 4(4):373-399.

Zivi, S.M. 1964. Estimation of steady-state steam void-fraction by means of the principle of minimum entropy production. *Journal of Heat Transfer* 86:247-52.

Zuber, N. 1959. Hydrodynamic aspects of boiling heat transfer. U.S. Atomic Energy Commission, Technical Information Service, *Report* AECU 4439. Oak Ridge, TN.

Zuber, N., M. Tribus, and J.W. Westwater. 1962. The hydrodynamic crisis in pool boiling of saturated and subcooled liquids. Proceedings of the International Heat Transfer Conference 2:230, and discussion of the papers, Vol. 6.

CHAPTER 5

MASS TRANSFER

MASS transfer by either molecular diffusion or convection is the transport of one component of a mixture relative to the motion of the mixture and is the result of a **concentration gradient**. In an air-conditioning process, water vapor is added or removed from the air by a simultaneous transfer of heat and mass (water vapor) between the airstream and a wetted surface. The wetted surface can be water droplets in an air washer, wetted slats of a cooling tower, condensate on the surface of a dehumidifying coil, surface presented by a spray of liquid absorbent, or wetted surfaces of an evaporative condenser. The performance of equipment with these phenomena must be calculated carefully because of the simultaneous heat and mass transfer.

This chapter addresses the principles of mass transfer and provides methods of solving a simultaneous heat and mass transfer problem involving air and water vapor. Emphasis is on air-conditioning processes involving mass transfer. The formulations presented can help in analyzing the performance of specific equipment. For a discussion on the performance of air washers, cooling coils, evaporative condensers, and cooling towers, see Chapters 19, 21, 35, and 36, respectively, of the 2000 *ASHRAE Handbook—Systems and Equipment.*

This chapter is divided into (1) the principles of molecular diffusion, (2) a discussion on the convection of mass, and (3) simultaneous heat and mass transfer and its application to specific equipment.

MOLECULAR DIFFUSION

Most mass transfer problems can be analyzed by considering the diffusion of a gas into a second gas, a liquid, or a solid. In this chapter, the diffusing or dilute component is designated as component B, and the other component as component A. For example, when water vapor diffuses into air, the water vapor is component B and dry air is component A. Properties with subscripts *A* or *B* are local properties of that component. Properties without subscripts are local properties of the mixture.

The primary mechanism of mass diffusion at ordinary temperature and pressure conditions is **molecular diffusion**, a result of density gradient. In a binary gas mixture, the presence of a concentration gradient causes transport of matter by molecular diffusion; that is, because of random molecular motion, gas B diffuses through the mixture of gases A and B in a direction that reduces the concentration gradient.

Fick's Law

The basic equation for molecular diffusion is Fick's law. Expressing the concentration of component B of a binary mixture of components A and B in terms of the mass fraction ρ_B/ρ or mole fraction C_B/C, Fick's law is

$$J_B = -\rho D_v \frac{d(\rho_B/\rho)}{di} = -J_A \tag{1a}$$

$$J_B^* = -CD_v \frac{d(C_B/C)}{di} = -J_A^* \tag{1b}$$

where $\rho = \rho_A + \rho_B$ and $C = C_A + C_B$.

The minus sign indicates that the concentration gradient is negative in the direction of diffusion. The proportionality factor D_v is the **mass diffusivity** or the **diffusion coefficient**. The total mass flux $\dot{m}_B''$ and molar flux $\dot{m}_B''^*$ are due to the average velocity of the mixture plus the diffusive flux:

$$\dot{m}_B'' = \rho_B v - \rho D_v \frac{d(\rho_B/\rho)}{dy} \tag{2a}$$

$$\dot{m}_B''^* = C_B v^* - CD_v \frac{d(C_B/C)}{dy} \tag{2b}$$

where v is the mass average velocity of the mixture and v^* is the molar average velocity.

Bird et al. (1960) present an analysis of Equations (1a) and (1b). Equations (1a) and (1b) are equivalent forms of Fick's law. The equation used depends on the problem and individual preference. This chapter emphasizes mass analysis rather than molar analysis. However, all results can be converted to the molar form using the relation $C_B \equiv \rho_B / M_B$.

Fick's Law for Dilute Mixtures

In many mass diffusion problems, component B is dilute; the density of component B is small compared to the density of the mixture. In this case, Equation (1a) can be written as

$$J_B = -D_v \frac{d\rho_B}{dy} \tag{3}$$

when $\rho_B << \rho$ and $\rho_A \approx \rho$.

Equation (3) can be used without significant error for water vapor diffusing through air at atmospheric pressure and a temperature less than 80°F. In this case, $\rho_B < 0.02\rho$, where ρ_B is the density of water vapor and ρ is the density of moist air (air and water vapor mixture). The error in J_B caused by replacing $\rho[d(\rho_B/\rho)/dy]$ with $d\rho_B/dy$ is less than 2%. At temperatures below 140°F where $\rho_B < 0.10\rho$, Equation (3) can still be used if errors in J_B as great as 10% are tolerable.

Fick's Law for Mass Diffusion Through Solids or Stagnant Fluids (Stationary Media)

Fick's law can be simplified for cases of dilute mass diffusion in solids, stagnant liquids, or stagnant gases. In these cases, $\rho_B \ll \rho$ and $v \approx 0$, which yields the following approximate result:

$$\dot{m}_B'' = J_B = -D_v \frac{d\rho_B}{dy} \tag{4}$$

The preparation of this chapter is assigned to TC 1.3, Heat Transfer and Fluid Flow.

Fick's Law for Ideal Gases with Negligible Temperature Gradient

For cases of dilute mass diffusion, Fick's law can be written in terms of partial pressure gradient instead of concentration gradient. When gas B can be approximated as ideal,

$$p_B = \frac{\rho_B R_u T}{M_B} = C_B R_u T \tag{5}$$

and when the gradient in T is small, Equation (3) can be written as

$$J_B = -\left(\frac{M_B D_v}{R_u T}\right)\frac{dp_B}{dy} \tag{6a}$$

or

$$J_B^* = -\left(\frac{D_v}{R_u T}\right)\frac{dp_B}{dy} \tag{6b}$$

If v ≈ 0, Equation (4) may be written as

$$\dot{m}_B'' = J_B = -\left(\frac{M_B D_v}{R_u T}\right)\frac{dp_B}{dy} \tag{7a}$$

or

$$\dot{m}_B''^{*} = J_B^* = -\left(\frac{D_v}{R_u T}\right)\frac{dp_B}{dy} \tag{7b}$$

The partial pressure gradient formulation for mass transfer analysis has been used extensively; this is unfortunate because the pressure formulation [Equations (6) and (7)] applies only to cases where one component is dilute, the fluid closely approximates an ideal gas, and the temperature gradient has a negligible effect. The density (or concentration) gradient formulation expressed in Equations (1) through (4) is more general and can be applied to a wider range of mass transfer problems, including cases where neither component is dilute [Equation (1)]. The gases need not be ideal, nor the temperature gradient negligible. Consequently, this chapter emphasizes the density formulation.

Diffusion Coefficient

For a binary mixture, the diffusion coefficient D_v is a function of temperature, pressure, and composition. Experimental measurements of D_v for most binary mixtures are limited in range and accuracy. Table 1 gives a few experimental values for diffusion of some gases in air. For more detailed tables, see the section on Bibliography at the end of this chapter.

In the absence of data, use equations developed from (1) theory or (2) theory with constants adjusted from limited experimental data. For binary gas mixtures at low pressure, D_v is inversely proportional to pressure, increases with increasing temperature, and is almost independent of composition for a given gas pair. Bird et al. (1960) present the following equation, developed from kinetic theory and corresponding states arguments, for estimating D_v at pressures less than $0.1p_{c\,min}$:

Table 1 Mass Diffusivities for Gases in Air[a]

Gas	D_v, ft²/h
Ammonia	1.08
Benzene	0.34
Carbon dioxide	0.64
Ethanol	0.46
Hydrogen	1.60
Oxygen	0.80
Water vapor	0.99

[a]Gases at 77°F and 14.696 psi.

$$D_v = a\left(\frac{T}{\sqrt{T_{cA}+T_{cB}}}\right)^b \sqrt{\frac{1}{M_A}+\frac{1}{M_B}} \times \frac{(p_{cA}p_{cB})^{1/3}(T_{cA}T_{cB})^{5/12}}{p} \tag{8}$$

where

D_v = diffusion coefficient, ft²/h
a = constant, dimensionless
b = constant, dimensionless
T = absolute temperature, K
p = pressure, psi
M = molecular weight, lb_m/lb mol

The subscripts cA and cB refer to the critical states of the two gases. Analysis of experimental data gives the following values of the constants a and b:

For nonpolar gas pairs,

$$a = 7.266 \times 10^{-3} \text{ and } b = 1.823$$

For water vapor with a nonpolar gas,

$$a = 9.635 \times 10^{-3} \text{ and } b = 2.334$$

A **nonpolar gas** is one for which the intermolecular forces are independent of the relative orientation of molecules, depending only on the separation distance from each other. Air, composed of nonpolar gases O_2 and N_2, is nonpolar.

Equation (8) is stated to agree with experimental data at atmospheric pressure to within about 8% (Bird et al. 1960).

The mass diffusivity D_v for binary mixtures at low pressure is predictable within about 10% by kinetic theory (Reid et al. 1987).

$$D_v = 0.0072\frac{T^{1.5}}{p(\sigma_{AB})^2\Omega_{D,AB}}\sqrt{\frac{1}{M_A}+\frac{1}{M_B}} \tag{9}$$

where

σ_{AB} = characteristic molecular diameter, nm
$\Omega_{D,AB}$ = temperature function, dimensionless

D_v is in ft²/h, p in atmospheres, and T in kelvins. If the gas molecules of A and B are considered rigid spheres having diameters σ_A and σ_B [and $\sigma_{AB} = (\sigma_A/2) + (\sigma_B/2)$], all expressed in nanometers, the dimensionless function $\Omega_{D,AB}$ equals unity. More realistic models for the molecules having intermolecular forces of attraction and repulsion lead to values of $\Omega_{D,AB}$ that are functions of temperature. Reid et al. (1987) present tabulations of this quantity. These results show that D_v increases as the 2.0 power of T at low temperatures and as the 1.65 power of T at very high temperatures.

The diffusion coefficient of moist air has been calculated for Equation (8) using a simplified intermolecular potential field function for water vapor and air (Mason and Monchick 1965).

The following is an empirical equation for mass diffusivity of water vapor in air up to 2000°F (Sherwood and Pigford 1952):

$$D_v = \frac{0.00215}{p}\left(\frac{T^{2.5}}{T+441}\right) \tag{10}$$

where D_v is in ft²/h, p in psi, and T in °R.

Diffusion of One Gas Through a Second Stagnant Gas

Figure 1 shows diffusion of one gas through a second stagnant gas. Water vapor diffuses from the liquid surface into surrounding stationary air. It is assumed that local equilibrium exists through the gas mixture, that the gases are ideal, and that the Gibbs-Dalton law is valid, which implies that the temperature gradient has a negligible effect. Diffusion of water vapor is due to concentration gradient and is given by Equation (6a). There is a continuous gas phase, so the mixture pressure p is constant, and the Gibbs-Dalton law yields

$$p_A + p_B = p = \text{constant} \tag{11a}$$

or

$$\frac{\rho_A}{M_A} + \frac{\rho_B}{M_B} = \frac{p}{R_u T} = \text{constant} \tag{11b}$$

The partial pressure gradient of the water vapor causes a partial pressure gradient of the air such that

$$\frac{dp_A}{dy} = -\frac{dp_B}{dy}$$

or

$$\left(\frac{1}{M_A}\right)\frac{d\rho_A}{dy} = -\left(\frac{1}{M_B}\right)\frac{d\rho_B}{dy} \tag{12}$$

Air, then, diffuses toward the liquid water interface. Because it cannot be absorbed there, a bulk velocity v of the gas mixture is established in a direction away from the liquid surface, so that the net transport of air is zero (i.e., the air is stagnant):

$$\dot{m}''_A = -D_v\frac{d\rho_A}{dy} + \rho_A v = 0 \tag{13}$$

The bulk velocity v transports not only air but also water vapor away from the interface. Therefore, the total rate of water vapor diffusion is

$$\dot{m}''_B = -D_v\frac{d\rho_B}{dy} + \rho_B v \tag{14}$$

Substituting for the velocity v from Equation (13) and using Equations (11b) and (12) gives

$$\dot{m}''_B = \left(\frac{D_v M_B p}{\rho_A R_u T}\right)\frac{d\rho_A}{dy} \tag{15}$$

Integration yields

$$\dot{m}''_B = \frac{D_v M_B p}{R_u T}\left[\frac{\ln(\rho_{AL}/\rho_{A0})}{y_L - y_0}\right] \tag{16a}$$

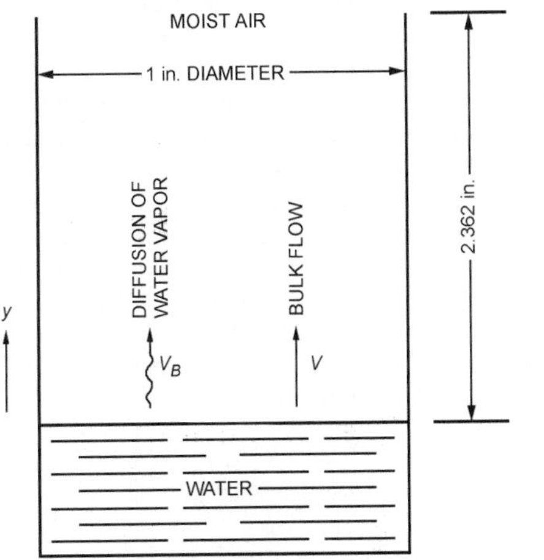

Fig. 1 Diffusion of Water Vapor Through Stagnant Air

or

$$\dot{m}''_B = -D_v P_{Am}\left(\frac{\rho_{BL} - \rho_{B0}}{y_L - y_0}\right) \tag{16b}$$

where

$$P_{Am} \equiv \frac{p}{p_{AL}}\rho_{AL}\left[\frac{\ln(\rho_{AL}/\rho_{A0})}{\rho_{AL} - \rho_{A0}}\right] \tag{17}$$

P_{Am} is the logarithmic mean density factor of the stagnant air. The pressure distribution for this type of diffusion is illustrated in Figure 2. **Stagnant** refers to the net behavior of the air; it does not move because the bulk flow exactly offsets diffusion. The term P_{Am} in Equation (16b) approximately equals unity for dilute mixtures such as water vapor in air at near atmospheric conditions. This condition makes it possible to simplify Equations (16) and implies that in the case of dilute mixtures, the partial pressure distribution curves in Figure 2 are straight lines.

Example 1. A vertical tube of 1 in. diameter is partially filled with water so that the distance from the water surface to the open end of the tube is 2.362 in., as shown in Figure 1. Perfectly dried air is blown over the open tube end, and the complete system is at a constant temperature of 59°F. In 200 h of steady operation, 0.00474 lb of water evaporates from the tube. The total pressure of the system is 14.696 psia (1 atm). Using these data, (a) calculate the mass diffusivity of water vapor in air, and (b) compare this experimental result with that from Equation (10).

Solution:

(a) The mass diffusion flux of water vapor from the water surface is

$$\dot{m}_B = 0.00474/200 = 0.0000237 \text{ lb/h}$$

The cross-sectional area of a 1 in. diameter tube is $\pi(0.5)^2/144$ = 0.005454 ft². Therefore, $\dot{m}''_B$ = 0.004345 lb/ft²·h. The partial densities are determined with the aid of the psychrometric tables.

$$\rho_{BL} = 0; \quad \rho_{B0} = 0.000801 \text{ lb/ft}^3$$

$$\rho_{AL} = 0.0765 \text{ lb/ft}^3; \quad \rho_{A0} = 0.0752 \text{ lb/ft}^3$$

Because $p = p_{AL}$ = 1 atm, the logarithmic mean density factor [Equation (17)] is

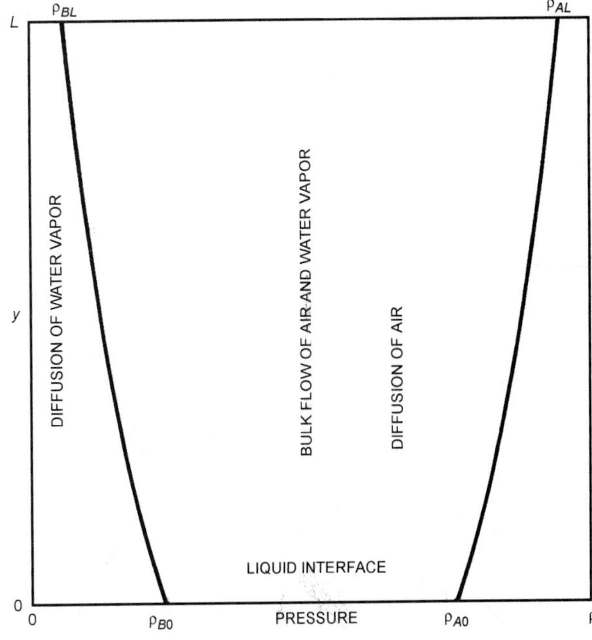

Fig. 2 Pressure Profiles for Diffusion of Water Vapor Through Stagnant Air

$$P_{Am} = 0.0765\left[\frac{\ln(0.0765/0.0752)}{0.0765 - 0.0752}\right] = 1.009$$

The mass diffusivity is now computed from Equation (16b) as

$$D_v = \frac{-\dot{m}_B''(y_L - y_0)}{P_{Am}(\rho_{BL} - \rho_{B0})} = \frac{-(0.004345)(2.362)}{(1.009)(0 - 0.000801)(12)}$$

$$= 1.058 \text{ ft}^2/\text{h}$$

(b) By Equation (10), with $p = 14.696$ psi and $T = 59 + 460 = 519$°R,

$$D_v = \frac{0.00215}{14.696}\left(\frac{519^{2.5}}{519 + 441}\right) = 0.935 \text{ ft}^2/\text{h}$$

Neglecting the correction factor P_{Am} for this example gives a difference of less than 1% between the calculated experimental and empirically predicted values of D_v.

Equimolar Counterdiffusion

Figure 3 shows two large chambers, both containing an ideal gas mixture of two components A and B (e.g., air and water vapor) at the same total pressure p and temperature T. The two chambers are connected by a duct of length L and cross-sectional area A_{cs}. Partial pressure p_B is higher in the left chamber, and partial pressure p_A is higher in the right chamber. The partial pressure differences cause component B to migrate to the right and component A to migrate to the left.

At steady state, the molar flows of A and B must be equal, but in the opposite direction, or

$$\dot{m}_A''^* + \dot{m}_B''^* = 0 \tag{18}$$

because the total molar concentration C must stay the same in both chambers if p and T remain constant. Since the molar fluxes are the same in both directions, the molar average velocity $v^* = 0$. Thus, Equation (7b) can be used to calculate the molar flux of B (or A):

$$\dot{m}_B''^* = \frac{-D_v}{R_u T}\frac{dp_B}{dy} \tag{19}$$

or

$$\dot{m}_B^* = \frac{A_{cs} D_v}{R_u T}\frac{p_{B0} - p_{BL}}{L} \tag{20}$$

or

$$\dot{m}_B = \frac{M_B A_{cs} D_v}{R_u T}\frac{p_{B0} - p_{BL}}{L} \tag{21}$$

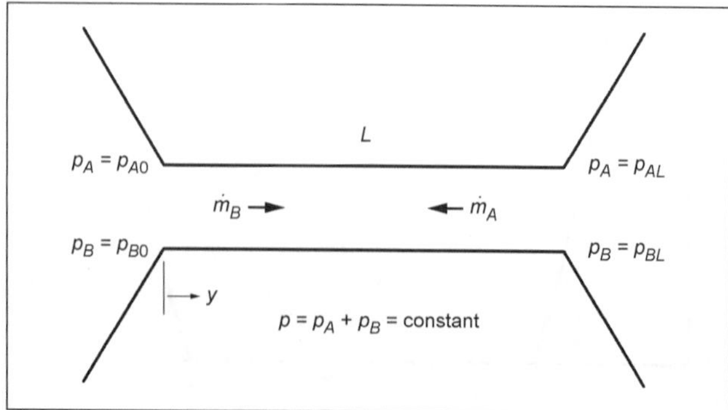

Fig. 3 Equimolar Counterdiffusion

Example 2. One large room is maintained at 70°F (530°R), 14.7 psia, 80% rh. A 60 ft long duct with cross-sectional area of 1.5 ft^2 connects the room to another large room at 70°F, 14.7 psia, 10% rh. What is the rate of water vapor diffusion between the two rooms?

Solution: Let air be component A and water vapor be component B. Equation (21) can be used to calculate the mass flow of water vapor B. Equation (10) can be used to calculate the diffusivity.

$$D_v = \frac{0.00215}{14.7}\left(\frac{530^{2.5}}{530 + 441}\right) = 0.964 \text{ ft}^2/\text{h}$$

From a psychrometric table (Table 3, Chapter 6), the saturated vapor pressure at 70°F is 0.363 psi. The vapor pressure difference $p_{B0} - p_{BL}$ is

$$p_{B0} - p_{BL} = (0.8 - 0.1)0.363 \text{ psi} = 0.254 \text{ psi}$$

Then, Equation (21) gives

$$\dot{m}_b = 18\,\frac{\text{lb}_\text{m}}{\text{lb mol}} \times \frac{\text{lb mol·°R}}{1545 \text{ ft·lb}} \times \frac{144 \text{ in}^2}{\text{ft}^2} \times \frac{1}{530\text{°R}} \times 1.5 \text{ ft}^2$$

$$\times 0.964 \text{ ft}^2/\text{h} \times \frac{0.254 \text{ psi}}{60 \text{ ft}} = 1.94 \times 10^{-5} \text{ lb}_\text{m}/\text{h}$$

Molecular Diffusion in Liquids and Solids

Because of the greater density, diffusion is slower in liquids than in gases. No satisfactory molecular theories have been developed for calculating diffusion coefficients. The limited measured values of D_v show that, unlike for gas mixtures at low pressures, the diffusion coefficient for liquids varies appreciably with concentration.

Reasoning largely from analogy to the case of one-dimensional diffusion in gases and employing Fick's law as expressed by Equation (4) gives

$$\dot{m}_B'' = D_v\left(\frac{\rho_{B1} - \rho_{B2}}{y_2 - y_1}\right) \tag{22}$$

Equation (22) expresses the steady-state diffusion of the solute B through the solvent A in terms of the molal concentration difference of the solute at two locations separated by the distance $\Delta y = y_2 - y_1$. Bird et al. (1960), Hirschfelder et al. (1954), Sherwood and Pigford (1952), Reid and Sherwood (1966), Treybal (1980), and Eckert and Drake (1972) provide equations and tables for evaluating D_v. Hirschfelder et al. (1954) provide comprehensive treatment of the molecular developments.

Diffusion through a solid when the solute is dissolved to form a homogeneous solid solution is known as **structure-insensitive diffusion** (Treybal 1980). This solid diffusion closely parallels diffusion through fluids, and Equation (22) can be applied to one-dimensional steady-state problems. Values of mass diffusivity are generally lower than they are for liquids and vary with temperature.

The diffusion of a gas mixture through a porous medium is common (e.g., the diffusion of an air-vapor mixture through porous insulation). The vapor diffuses through the air along the tortuous narrow passages within the porous medium. The mass flux is a function of the vapor pressure gradient and diffusivity as indicated in Equation (7a). It is also a function of the structure of the pathways within the porous medium and is therefore called **structure-sensitive diffusion**. All of these factors are taken into account in the following version of Equation (7a):

$$\dot{m}_B'' = -\bar{\mu}\,\frac{dp_B}{dy} \tag{23}$$

where μ is called the permeability of the porous medium. Chapter 23 presents this topic in more depth.

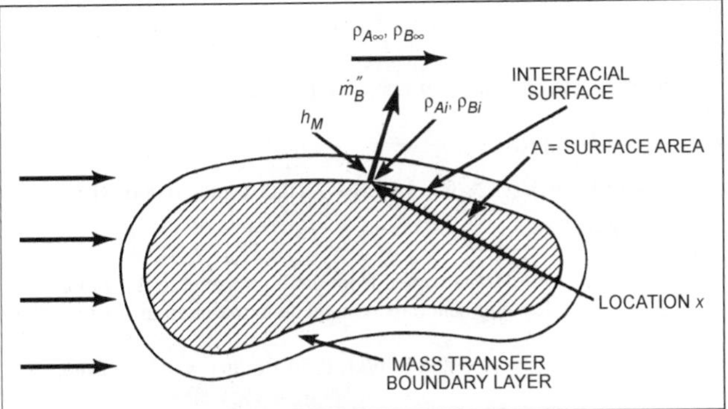

Fig. 4 Nomenclature for Convective Mass Transfer from External Surface at Location *x* Where Surface is Impermeable to Gas A

CONVECTION OF MASS

Convection of mass involves the mass transfer mechanisms of molecular diffusion and bulk fluid motion. Fluid motion in the region adjacent to a mass transfer surface may be laminar or turbulent, depending on geometry and flow conditions.

Mass Transfer Coefficient

Convective mass transfer is analogous to convective heat transfer where geometry and boundary conditions are similar. The analogy holds for both laminar and turbulent flows and applies to both external and internal flow problems.

Mass Transfer Coefficients for External Flows. Most external convective mass transfer problems can be solved with an appropriate formulation that relates the mass transfer flux (to or from an interfacial surface) to the concentration difference across the boundary layer illustrated in Figure 4. This formulation gives rise to the convective mass transfer coefficient, defined as

$$h_M \equiv \frac{\dot{m}_B''}{\rho_{Bi} - \rho_{B\infty}} \quad (24)$$

where

h_M = local external mass transfer coefficient, ft/h
$\dot{m}_B''$ = mass flux of gas B from surface, $lb_m/ft^2 \cdot h$
ρ_{Bi} = density of gas B at interface (saturation density), lb_m/ft^3
$\rho_{B\infty}$ = density of component B outside boundary layer, lb_m/ft^3

If ρ_{Bi} and $\rho_{B\infty}$ are constant over the entire interfacial surface, the mass transfer rate from the surface can be expressed as

$$\dot{m}_B'' = \bar{h}_M(\rho_{Bi} - \rho_{B\infty}) \quad (25)$$

where $\bar{h}_M$ is the average mass transfer coefficient, defined as

$$\bar{h}_M \equiv \frac{1}{A}\int_A h_m\,dA \quad (26)$$

Mass Transfer Coefficients for Internal Flows. Most internal convective mass transfer problems, such as those that occur in channels or in the cores of dehumidification coils, can be solved if an appropriate expression is available to relate the mass transfer flux (to or from the interfacial surface) to the difference between the concentration at the surface and the bulk concentration in the channel, as illustrated in Figure 5. This formulation leads to the definition of the mass transfer coefficient for internal flows:

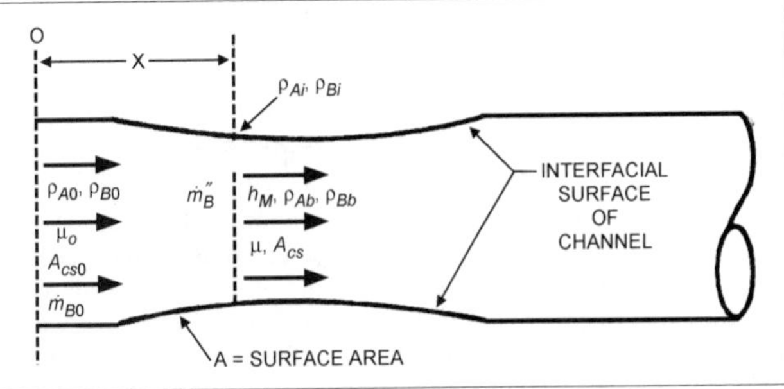

Fig. 5 Nomenclature for Convective Mass Transfer from Internal Surface Impermeable to Gas A

$$h_M \equiv \frac{\dot{m}_B''}{\rho_{Bi} - \rho_{Bb}} \quad (27)$$

where

h_M = internal mass transfer coefficient, ft/h
$\dot{m}_B''$ = mass flux of gas B at interfacial surface, $lb_m/ft^2 \cdot h$
ρ_{Bi} = density of gas B at interfacial surface, lb_m/ft^3
$\rho_{Bb} \equiv (1/\bar{u}_B A_{cs})\int_{A_{cs}} u_B \rho_B\,dA_{cs}$ = bulk density of gas B at location x
$\bar{u}_B \equiv (1/A_{cs})\int_A u_B\,dA_{cs}$ = average velocity of gas B at location x, fpm
A_{cs} = cross-sectional area of channel at station x, ft^2
u_B = velocity of component B in x direction, fpm
ρ_B = density distribution of component B at station x, lb_m/ft^3

Often, it is easier to obtain the bulk density of gas B from

$$\rho_{Bb} = \frac{\dot{m}_{Bo} + \int_A \dot{m}_B''\,dA}{\bar{u}_B A_{cs}} \quad (28)$$

where

$\dot{m}_{Bo}$ = mass flow rate of component B at station $x = 0$, lb_m/h
A = interfacial area of channel between station $x = 0$ and station $x = x$, ft^2

Equation (28) can be derived from the preceding definitions. The major problem is the determination of $\bar{u}_B$. If, however, the analysis is restricted to cases where B is dilute and concentration gradients of B in the x direction are negligibly small, $\bar{u}_B \approx \bar{u}$. Component B is swept along in the x direction with an average velocity equal to the average velocity of the dilute mixture.

Analogy Between Convective Heat and Mass Transfer

Most expressions for the convective mass transfer coefficient h_M are determined from expressions for the convective heat transfer coefficient h.

For problems in internal and external flow where mass transfer occurs at the convective surface and where component B is dilute, it is shown by Bird et al. (1960) and Incropera and DeWitt (1996) that the Nusselt and Sherwood numbers are defined as follows:

$$\text{Nu} = f(X, Y, Z, \text{Pr}, \text{Re}) \quad (29)$$

$$\text{Sh} = f(X, Y, Z, \text{Sc}, \text{Re}) \quad (30)$$

and

$$\overline{\text{Nu}} = g(\text{Pr}, \text{Re}) \quad (31)$$

$$\overline{\text{Sh}} = g(\text{Sc}, \text{Re}) \quad (32)$$

where the function f is the same in Equations (29) and (30), and the function g is the same in Equations (31) and (32). The quantities Pr and Sc are dimensionless Prandtl and Schmidt numbers, respectively, as defined in the section on Symbols. The primary restrictions on the analogy are that the surface shapes are the same and that the temperature boundary conditions are analogous to the density distribution boundary conditions for component B when cast in dimensionless form. Several primary factors prevent the analogy from being perfect. In some cases, the Nusselt number was derived for smooth surfaces. Many mass transfer problems involve wavy, droplet-like, or roughened surfaces. Many Nusselt number relations are obtained for constant temperature surfaces. Sometimes ρ_{Bi} is not constant over the entire surface because of varying saturation conditions and the possibility of surface dryout.

In all mass transfer problems, there is some blowing or suction at the surface because of the condensation, evaporation, or transpiration of component B. In most cases, this blowing/suction phenomenon has little effect on the Sherwood number, but the analogy should be examined closely if $v_i/u_\infty > 0.01$ or $v_i/\bar{u} > 0.01$, especially if the Reynolds number is large.

Example 3. Use the analogy expressed in Equations (31) and (32) to solve the following problem. An expression for heat transfer from a constant temperature flat plate in laminar flow is

$$\overline{\mathrm{Nu}}_L = 0.664\mathrm{Pr}^{1/3}\mathrm{Re}_L^{1/2} \tag{33}$$

$\mathrm{Sc} = 0.35$, $D_v = 3.87 \times 10^{-4}$ ft²/s, and $\mathrm{Pr} = 0.708$ for the given conditions; determine the mass transfer rate and temperature of the water-wetted flat plate in Figure 6 using the heat/mass transfer analogy.

Solution: To solve the problem, properties should be evaluated at film conditions. However, since the plate temperature and the interfacial water vapor density are not known, a first estimate will be obtained assuming the plate t_{i1} to be at 77°F. The plate Reynolds number is

$$\mathrm{Re}_{L1} = \frac{\rho u_\infty L}{\mu} = \frac{(0.0728\ \mathrm{lb/ft^3})(1970\ \mathrm{fpm})(0.328\ \mathrm{ft})}{(1.32\times 10^{-5}\mathrm{lb/ft\cdot s})(60\ \mathrm{s/min})} = 59{,}340$$

The plate is entirely in laminar flow, since the transitional Reynolds number is about 5×10^5. Using the mass transfer analogy given by Equations (31) and (32), Equation (33) yields

$$\overline{\mathrm{Sh}}_{L1} = 0.664\ \mathrm{Sc}^{1/3}\mathrm{Re}_L^{1/2}$$

$$= 0.664(0.35)^{1/3}(59{,}340)^{1/2} = 114$$

From the definition of the Sherwood number,

$$\bar{h}_{M1} = \overline{\mathrm{Sh}}_{L1}D_v/L = (114)(3.87\times 10^{-4}\ \mathrm{ft^2/s})/0.328\ \mathrm{ft} = 0.135\ \mathrm{fps}$$

The psychrometric tables give a humidity ratio W of 0.0121 at 77°F and 60% rh. Therefore,

$$\rho_{B\infty} = 0.0121\rho_{A\infty} = (0.0121)(0.0728\ \mathrm{lb/ft^3}) = 0.000881\ \mathrm{lb/ft^3}$$

Psychrometric tables give the saturation density for water at 77°F as

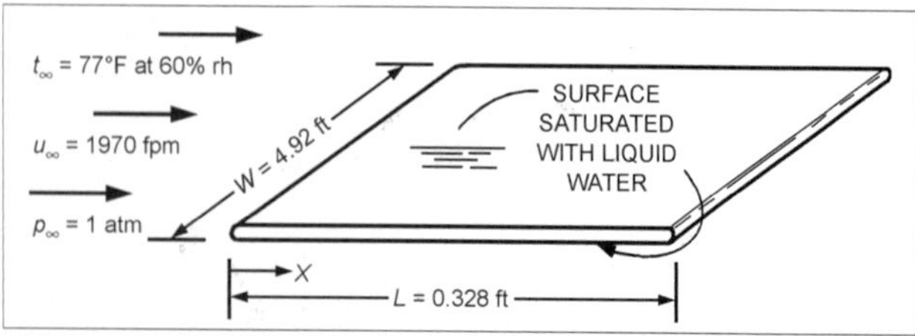

Fig. 6 Water-Saturated Flat Plate in Flowing Airstream

$$\rho_{Bi1} = 0.02017\rho_{A\infty} = (0.02017)(0.0728\ \mathrm{lb/ft^3}) = 0.001468\ \mathrm{lb/ft^3}$$

Therefore, the mass transfer rate from the double-sided plate is

$$\dot{m}_{B1} = \bar{h}_{M1}A(\rho_{Bi} - \rho_{B\infty})$$

$$= (0.135\ \mathrm{fps})(0.328\ \mathrm{ft}\times 4.92\ \mathrm{ft}\times 2)(0.001468 - 0.000881\ \mathrm{lb/ft^3})$$

$$= 0.000256\ \mathrm{lb/s}$$

This mass rate, transformed from the liquid state to the vapor state, requires the following heat rate to the plate to maintain the evaporation:

$$q_{i1} = \dot{m}_{B1}h_{fg} = (0.000256\ \mathrm{lb/s})(1050\ \mathrm{Btu/lb}) = 0.269\ \mathrm{Btu/s}$$

To obtain a second estimate of the wetted plate temperature in this type of problem, the following criteria are used. Calculate the t_i necessary to provide a heat rate of q_{i1}. If this temperature t_{iq1} is above the dew-point temperature t_{id}, set the second estimate at $t_{i2} = (t_{iq1} + t_{i1})/2$. If t_{iq1} is below the dew-point temperature, set $t_{i2} = (t_{id} + t_{i1})/2$. For this problem, the dew point is $t_{id} = 57$°F.

Obtaining the second estimate of the plate temperature requires an approximate value of the heat transfer coefficient.

$$\overline{\mathrm{Nu}}_{L1} = 0.664\mathrm{Pr}^{1/3}\mathrm{Re}_L^{1/2} = 0.664(0.708)^{1/3}(59{,}340)^{1/2}$$

$$= 144.2$$

From the definition of the Nusselt number,

$$\bar{h}_1 = \overline{\mathrm{Nu}}_{L1}k/L = (144.2)(0.0151\ \mathrm{Btu/h\cdot ft\cdot {}^\circ F})/(0.328\ \mathrm{ft})$$

$$= 6.64\ \mathrm{Btu/h\cdot ft^2\cdot {}^\circ F}$$

Therefore, the second estimate for the plate temperature is

$$t_{iq1} = t_\infty - q_{i1}/(\bar{h}_1 A)$$

$$= 77^\circ\mathrm{F} - [(0.269\ \mathrm{Btu/s})(3600\ \mathrm{s/h})/(6.64\ \mathrm{Btu/h\cdot ft^2\cdot {}^\circ F})]$$

$$(2\times 0.328\ \mathrm{ft}\times 4.92\ \mathrm{ft}) = 77^\circ\mathrm{F} - 45^\circ\mathrm{F} = 32^\circ\mathrm{F}$$

This temperature is below the dew-point temperature; therefore,

$$t_{i2} = (57^\circ\mathrm{F} + 77^\circ\mathrm{F})/2 = 67^\circ\mathrm{F}$$

The second estimate of the film temperature is

$$t_{f2} = (t_{i2} + t_\infty)/2 = (67^\circ\mathrm{F} + 77^\circ\mathrm{F})/2 = 72^\circ\mathrm{F}$$

The next iteration on the solution is as follows:

$$\mathrm{Re}_{L2} = 61{,}010$$

$$\overline{\mathrm{Sh}}_{L2} = 0.664(0.393)^{1/3}(61{,}010)^{1/2} = 120$$

$$\bar{h}_{M2} = (120)(3.63\times 10^{-4})/0.328 = 0.133\ \mathrm{fps}$$

The free stream density of the water vapor has been evaluated. The density of the water vapor at the plate surface is the saturation density at 67°F.

$$\rho_{Bi2} = (0.01374)(0.0739\ \mathrm{lb/ft^3}) = 0.00101\ \mathrm{lb/ft^3}$$

$$A = 2\times 4.92\times 0.328 = 3.228\ \mathrm{ft^2}$$

$$\dot{m}_{B2} = (0.133\ \mathrm{fps})(3.228\ \mathrm{ft^2})(0.00101\ \mathrm{lb/ft^3} - 0.000881\ \mathrm{lb/ft^3})$$

$$= 0.0000554\ \mathrm{lb/s}$$

$$q_{i2} = (0.0000554)(1056) = 0.0585\ \mathrm{Btu/s}$$

$$\overline{\mathrm{Nu}}_{L2} = 0.664(0.709)^{1/3}(61{,}010)^{1/2} = 146$$

$$\bar{h}_2 = (146)(0.01493)/0.328 = 6.65\ \mathrm{Btu/h\cdot ft^2\cdot {}^\circ F}$$

$$t_{iq2} = 77^\circ\mathrm{F} - (0.0585)(3600)/(6.65\times 3.228) = 67.2^\circ\mathrm{F}$$

This temperature is above the dew-point temperature; therefore,

$$t_{i3} = (t_{i2} + t_{iq2})/2 = (67 + 67.2)/2 = 67.1°F$$

This is approximately the same result as that obtained in the previous iteration. Therefore, the problem solution is

$$t_i = 67°F$$

$$\dot{m}_B = 0.199 \text{ lb/h}$$

The kind of similarity between heat and mass transfer that results in Equation (29) through Equation (32) can also be shown to exist between heat and momentum transfer. Chilton and Colburn (1934) used this similarity to relate Nusselt number to friction factor by the analogy

$$j_H = \frac{\text{Nu}}{\text{Re Pr}^{(1-n)}} = \text{St Pr}^n = \frac{f}{2} \tag{34}$$

where $n = 2/3$, St = Nu/(Re Pr) is the Stanton number, and j_H is the Chilton-Colburn j-factor for heat transfer. Substituting Sh for Nu and Sc for Pr in Equations (31) and (32) gives the Chilton-Colburn j-factor for mass transfer, j_D:

$$j_D = \frac{\text{Sh}}{\text{Re Sc}^{(1-n)}} = \text{St}_m\text{Sc}^n = \frac{f}{2} \tag{35}$$

where $\text{St}_m = \text{Sh}P_{AM}/(\text{Re Sc})$ is the Stanton number for mass transfer. Equations (34) and (35) are called the **Chilton-Colburn *j*-factor analogy**.

The power of the Chilton-Colburn j-factor analogy is represented in Figures 7 through 10. Figure 7 plots various experimental values of j_D from a flat plate with flow parallel to the plate surface. The solid line, which represents the data to near perfection, is actually $f/2$ from Blasius' solution of laminar flow on a flat plate (left-hand portion of the solid line) and Goldstein's solution for a turbulent boundary layer (right-hand portion). The right-hand portion of the solid line also represents McAdams' (1954) correlation of turbulent flow heat transfer coefficient for a flat plate.

A **wetted-wall column** is a vertical tube in which a thin liquid film adheres to the tube surface and exchanges mass by evaporation or absorption with a gas flowing through the tube. Figure 8 illustrates typical data on vaporization in wetted-wall columns, plotted as j_D versus Re. The spread of the points with variation in $\mu/\rho D_v$ results from Gilliland's finding of an exponent of 0.56, not 2/3, representing the effect of the Schmidt number. Gilliland's equation can be written as follows:

$$j_D = 0.023\text{Re}^{-0.17}\left(\frac{\mu}{\rho D_v}\right)^{-0.56} \tag{36}$$

Similarly, McAdams' (1954) equation for heat transfer in pipes can be expressed as

$$j_H = 0.023\text{Re}^{-0.20}\left(\frac{c_p\mu}{k}\right)^{-0.7} \tag{37}$$

This is represented by the dash-dot curve in Figure 8, which falls below the mass transfer data. The curve $f/2$ representing friction in smooth tubes is the upper, solid curve.

Data for the evaporation of liquids from single cylinders into gas streams flowing transversely to the cylinders' axes are shown in Figure 9. Although the dash-dot line on Figure 9 represents the data, it is actually taken from McAdams (1954) as representative of a large collection of data on heat transfer to single cylinders placed transverse to airstreams. To compare these data with friction, it is necessary to distinguish between total drag and skin friction. Since the analogies are based on skin friction, the normal pressure drag must be subtracted from the measured total drag. At Re = 1000, the skin friction is 12.6% of the total drag; at Re = 31,600, it is only 1.9%. Consequently, the values of $f/2$ at a high Reynolds number, obtained by the difference, are subject to considerable error.

In Figure 10, data on the evaporation of water into air for single spheres are presented. The solid line, which best represents these data, agrees with the dashed line representing McAdams' correlation for heat transfer to spheres. These results cannot be compared with friction or momentum transfer because total drag has not been allocated to skin friction and normal pressure drag. Application of these data to air-water contacting devices such as air washers and spray cooling towers is well substantiated.

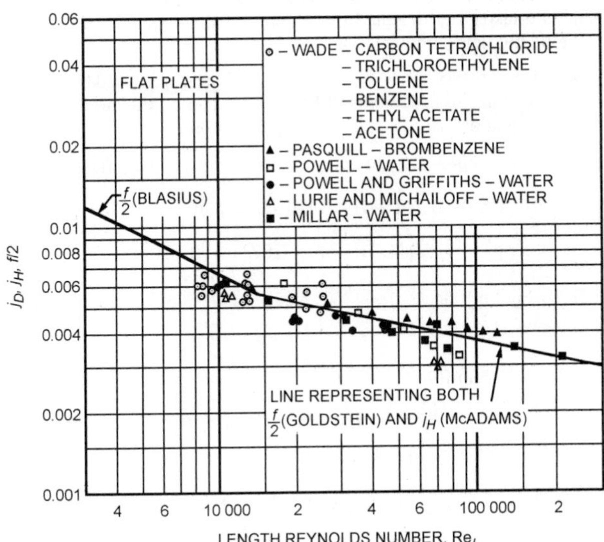

Fig. 7 Mass Transfer from Flat Plate

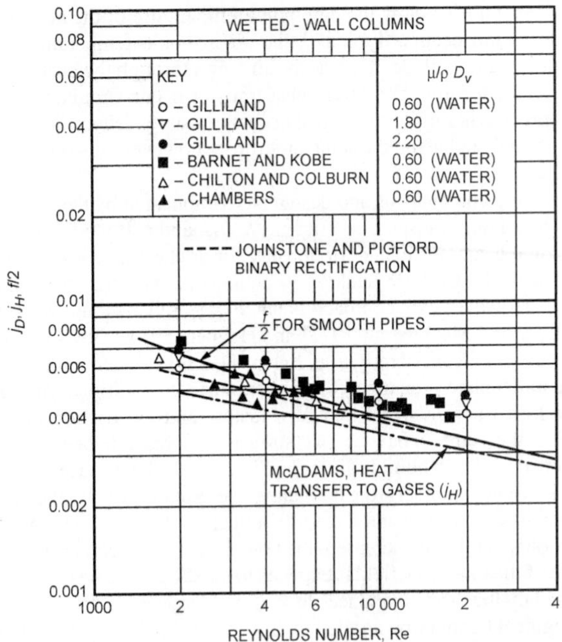

Fig. 8 Vaporization and Absorption in Wetted-Wall Column

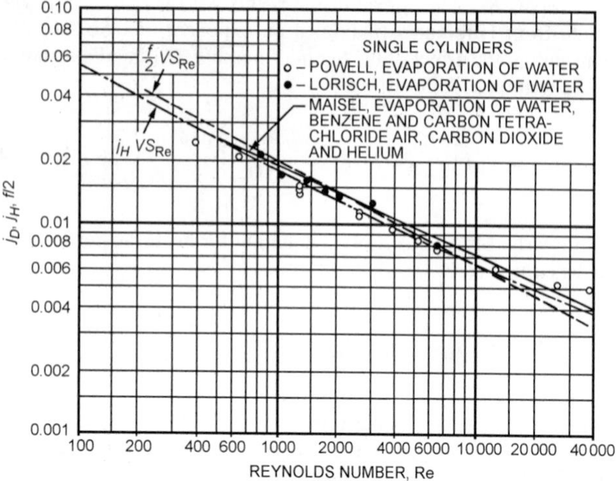

Fig. 9 Mass Transfer from Single Cylinders in Crossflow

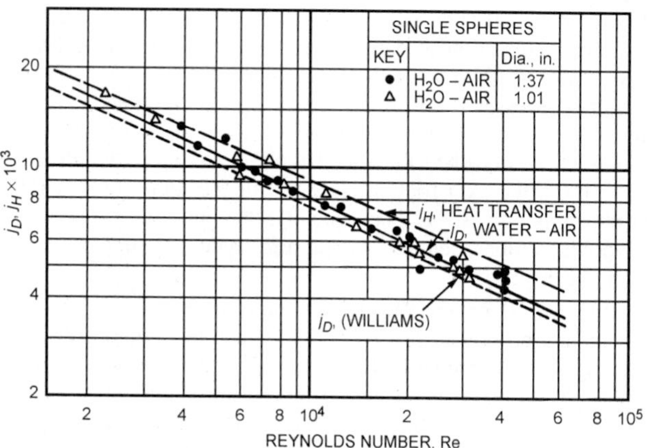

Fig. 10 Mass Transfer from Single Spheres

When the temperature of the heat exchanger surface in contact with moist air is below the dew-point temperature of the air, vapor condensation occurs. Typically, the air dry-bulb temperature and humidity ratio both decrease as the air flows through the exchanger. Therefore, sensible and latent heat transfer occur simultaneously. This process is similar to one that occurs in a spray dehumidifier and can be analyzed using the same procedure; however, this is not generally done.

Cooling coil analysis and design are complicated by the problem of determining transport coefficients h, h_M, and f. It would be convenient if heat transfer and friction data for dry heating coils could be used with the Colburn analogy to obtain the mass transfer coefficients. However, this approach is not always reliable, and work by Guillory and McQuiston (1973) and Helmer (1974) shows that the analogy is not consistently true. Figure 11 shows j-factors for a simple parallel plate exchanger for different surface conditions with sensible heat transfer. Mass transfer j-factors and the friction factors exhibit the same behavior. Dry surface j-factors fall below those obtained under dehumidifying conditions with the surface wet. At low Reynolds numbers, the boundary layer grows quickly; the droplets are soon covered and have little effect on the flow field. As the Reynolds number is increased, the boundary layer becomes thin and more of the total flow field is exposed to the droplets. The roughness caused by the droplets induces mixing and larger j-factors. The data in Figure 11 cannot be applied to all surfaces because the length of the flow channel is also an important variable. However, the water collecting on the surface is mainly responsible for breakdown of the

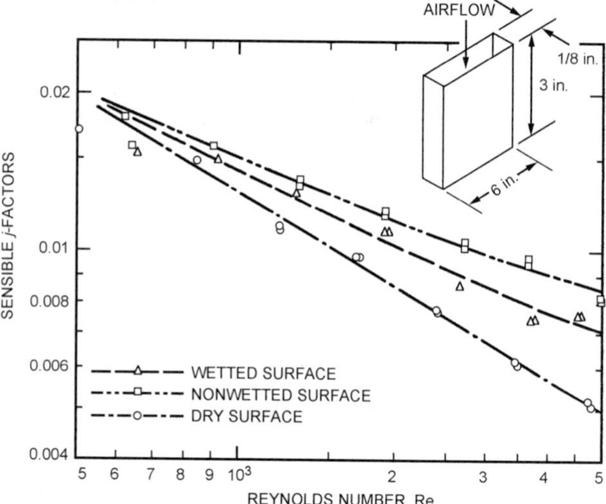

Fig. 11 Sensible Heat Transfer j-Factors for Parallel Plate Exchanger

j-factor analogy. The j-factor analogy is approximately true when the surface conditions are identical. Under some conditions, it is possible to obtain a film of condensate on the surface instead of droplets. Guillory and McQuiston (1973) and Helmer (1974) related dry sensible j- and f-factors to those for wetted dehumidifying surfaces.

The equality of j_H, j_D, and $f/2$ for certain streamline shapes at low mass transfer rates has experimental verification. For flow past bluff objects, j_H and j_D are much smaller than $f/2$, based on total pressure drag. The heat and mass transfer, however, still relate in a useful way by equating j_H and j_D.

Example 4. Using solid cylinders of volatile solids (e.g., naphthalene, camphor, dichlorobenzene) with airflow normal to these cylinders, Bedingfield and Drew (1950) found that the ratio between the heat and mass transfer coefficients could be closely correlated by the following relation:

$$\frac{h}{\rho h_M} = (0.294\ \text{Btu/lb}_m \cdot °\text{F})\left(\frac{\mu}{\rho D_v}\right)^{0.56}$$

For completely dry air at 70°F flowing at a velocity of 31 fps over a wet-bulb thermometer of diameter $d = 0.300$ in., determine the heat and mass transfer coefficients from Figure 9 and compare their ratio with the Bedingfield-Drew relation.

Solution: For dry air at 70°F and standard pressure, $\rho = 0.075$ lb$_m$/ft^3, $\mu = 0.044$ lb$_m$/h·ft, $k = 0.0149$ Btu/h·ft·°F, and $c_p = 0.240$ Btu/lb$_m$·°F. From Equation (10), $D_v = 0.973$ ft^2/h. Therefore,

$$Re_{da} = \rho u_\infty d/\mu = 0.0749 \times 31 \times 3600 \times 0.300/(12 \times 0.044) = 4750$$

$$Pr = c_p\mu/k = 0.240 \times 0.044/0.0149 = 0.709$$

$$Sc = \mu/\rho D_v = 0.044/(0.0749 \times 0.973) = 0.604$$

From Figure 9 at $Re_{da} = 4750$, read $j_H = 0.0088$, $j_D = 0.0099$. From Equations (34) and (35),

$$h = j_H \rho c_p u_\infty/(Pr)^{2/3}$$
$$= 0.0088 \times 0.0749 \times 0.240 \times 31 \times 3600/(0.709)^{2/3}$$
$$= 22.2\ \text{Btu/h} \cdot \text{ft}^2 \cdot °\text{F}$$

$$h_M = j_D u_\infty/(Sc)^{2/3} = 0.0099 \times 31 \times 3600/(0.604)^{2/3}$$
$$= 1550\ \text{ft/h}$$

$$h/\rho h_M = 22.2/(0.0749 \times 1550) = 0.191\ \text{Btu/lb}_m \cdot °\text{F}$$

From the Bedingfield-Drew relation,

$$h/\rho h_M = 0.294(0.604)^{0.56} = 0.222\ \text{Btu/lb}_m \cdot {}^\circ\text{F}$$

Equations (34) and (35) are call the Reynolds analogy when Pr = Sc = 1. This suggests that $h/\rho h_M = c_p = 0.240$ Btu/lb$_m$·°F. This close agreement is because the ratio Sc/Pr is 0.604/0.709 or 0.85, so that the exponent of these numbers has little effect on the ratio of the transfer coefficients.

The extensive developments for calculating heat transfer coefficients can be applied to calculate mass transfer coefficients under similar geometrical and flow conditions using the *j*-factor analogy. For example, Table 6 of Chapter 3 lists equations for calculating heat transfer coefficients for flow inside and normal to pipes. Each equation can be used for mass transfer coefficient calculations by equating j_H and j_D and imposing the same restriction to each stated in Table 6 of Chapter 3. Similarly, mass transfer experiments often replace corresponding heat transfer experiments with complex geometries where exact boundary conditions are difficult to model (Sparrow and Ohadi 1987a, 1987b).

The *j*-factor analogy is useful only at low mass transfer rates. As the rate of mass transfer increases, the movement of matter normal to the transfer surface increases the convective velocity. For example, if a gas is blown from many small holes in a flat plate placed parallel to an airstream, the boundary layer thickens, and resistance to both mass and heat transfer increases with increasing blowing rate. Heat transfer data are usually collected at zero or, at least, insignificant mass transfer rates. Therefore, if such data are to be valid for a mass transfer process, the mass transfer rate (i.e., the blowing) must be low.

The *j*-factor relationship $j_H = j_D$ can still be valid at high mass transfer rates, but neither j_H nor j_D can be represented by data at zero mass transfer conditions. Eckert and Drake (1972) and Chapter 24 of Bird et al. (1960) have detailed information on high mass transfer rates.

Lewis Relation

Heat and mass transfer coefficients are satisfactorily related at the same Reynolds number by equating the Chilton-Colburn *j*-factors. Comparing Equations (34) and (35) gives

$$\text{St Pr}^n = \frac{f}{2} = \text{St}_m \text{Sc}^n$$

Inserting the definitions of St, Pr, St$_m$, and Sc gives

$$\frac{h}{\rho c_p \bar{u}}\left(\frac{c_p \mu}{k}\right)^{2/3} = \frac{h_M P_{Am}}{\bar{u}}\left(\frac{\mu}{\rho D_v}\right)^{2/3}$$

or

$$\frac{h}{h_M \rho c_p} = P_{Am}\left[\frac{(\mu/\rho D_v)}{(c_p \mu/k)}\right]^{2/3}$$

$$= P_{Am}(\alpha/D_v)^{2/3} \qquad (38)$$

The quantity α/D_v is the **Lewis number** Le. Its magnitude expresses relative rates of propagation of energy and mass within a system. It is fairly insensitive to temperature variation. For air and water vapor mixtures, the ratio is (0.60/0.71) or 0.845, and $(0.845)^{2/3}$ is 0.894. At low diffusion rates, where the heat-mass transfer analogy is valid, P_{Am} is essentially unity. Therefore, for air and water vapor mixtures,

$$\frac{h}{h_M \rho c_p} \approx 1 \qquad (39)$$

The ratio of the heat transfer coefficient to the mass transfer coefficient is equal to the specific heat per unit volume of the mixture at constant pressure. This relation [Equation (39)] is usually called the Lewis relation and is nearly true for air and water vapor at low mass transfer rates. It is generally not true for other gas mixtures because the ratio Le of thermal to vapor diffusivity can differ from unity. The agreement between wet-bulb temperature and adiabatic saturation temperature is a direct result of the nearness of the Lewis number to unity for air and water vapor.

The Lewis relation is valid in turbulent flow whether or not α/D_v equals 1 because eddy diffusion in turbulent flow involves the same mixing action for heat exchange as for mass exchange, and this action overwhelms any molecular diffusion. Deviations from the Lewis relation are, therefore, due to a laminar boundary layer or a laminar sublayer and buffer zone where molecular transport phenomena are the controlling factors.

SIMULTANEOUS HEAT AND MASS TRANSFER BETWEEN WATER-WETTED SURFACES AND AIR

A simplified method used to solve simultaneous heat and mass transfer problems was developed using the Lewis relation, and it gives satisfactory results for most air-conditioning processes. Extrapolation to very high mass transfer rates, where the simple heat-mass transfer analogy is not valid, will lead to erroneous results.

Enthalpy Potential

The water vapor concentration in the air is the humidity ratio *W*, defined as

$$W \equiv \frac{\rho_B}{\rho_A} \qquad (40)$$

A mass transfer coefficient is defined using *W* as the driving potential:

$$\dot{m}_B'' = K_M(W_i - W_\infty) \qquad (41)$$

where the coefficient K_M is in lb$_m$/h·ft^2. For dilute mixtures, $\rho_{Ai} \cong \rho_{A\infty}$; that is, the partial mass density of dry air changes by only a small percentage between interface and free stream conditions. Therefore,

$$\dot{m}_B'' = \frac{K_M}{\rho_{Am}}(\rho_{Bi} - \rho_\infty) \qquad (42)$$

where ρ_{Am} = mean density of dry air, lb$_m$/ft^3. Comparing Equation (42) with Equation (24) shows that

$$h_M = \frac{K_M}{\rho_{Am}} \qquad (43)$$

The **humid specific heat** c_{pm} of the airstream is, by definition (Mason and Monchick 1965),

$$c_{pm} = (1 + W_\infty)c_p \qquad (44a)$$

or

$$c_{pm} = \left(\frac{\rho}{\rho_{A\infty}}\right)c_p \qquad (44b)$$

where c_{pm} is in Btu/lb$_{da}$·°F.

Substituting from Equations (43) and (44b) into Equation (39) gives

$$\frac{h\rho_{Am}}{K_M\rho_{A\infty}c_{pm}} = 1 \approx \frac{h}{K_M c_{pm}} \quad (45)$$

since $\rho_{Am} \cong \rho_{A\infty}$ because of the small change in dry-air density. Using a mass transfer coefficient with the humidity ratio as the driving force, the Lewis relation becomes ratio of heat to mass transfer coefficient equals humid specific heat.

For the plate humidifier illustrated in Figure 6, the total heat transfer from liquid to interface is

$$q'' = q''_A + \dot{m}''_B h_{fg} \quad (46)$$

Using the definitions of the heat and mass transfer coefficients gives

$$q'' = h(t_i - t_\infty) + K_M(W_i - W_\infty)h_{fg} \quad (47)$$

Assuming Equation (45) is valid gives

$$q'' = K_M\left[c_{pm}(t_i - t_\infty) + (W_i - W_\infty)h_{fg}\right] \quad (48)$$

The enthalpy of the air is approximately

$$h = c_{pa}t + Wh_s \quad (49)$$

The enthalpy h_s of the water vapor can be expressed by the ideal gas law as

$$h_s = c_{ps}(t - t_o) + h_{fgo} \quad (50)$$

where the base of enthalpy is taken as saturated water at temperature t_o. Choosing t_o = 0°F to correspond with the base of the dry-air enthalpy gives

$$h = (c_{pa} + Wc_{ps})t + Wh_{fgo} = c_{pm}t + Wh_{fgo} \quad (51)$$

If small changes in the latent heat of vaporization of water with temperature are neglected when comparing Equations (49) and (51), the total heat transfer can be written as

$$q'' = K_M(h_i - h_\infty) \quad (52)$$

Where the driving potential for heat transfer is temperature difference and the driving potential for mass transfer is mass concentration or partial pressure, the driving potential for simultaneous transfer of heat and mass in an air water-vapor mixture is, to a close approximation, enthalpy.

Basic Equations for Direct-Contact Equipment

Air-conditioning equipment can be classified as (1) having direct contact between air and water used as a cooling or heating fluid or (2) having the heating or cooling fluid separated from the airstream by a solid wall. Examples of the former are air washers and cooling towers; an example of the latter is a direct-expansion refrigerant (or water) cooling and dehumidifying coil. In both cases, the airstream is in contact with a water surface. Direct contact implies contact directly with the cooling (or heating) fluid. In the dehumidifying coil, the contact with the condensate removed from the airstream is direct, but it is indirect with the refrigerant flowing inside the tubes of the coil. These two cases are treated separately because the surface areas of direct-contact equipment cannot be evaluated.

For the direct-contact spray chamber air washer of cross-sectional area A_{cs} and length l (Figure 12), the steady mass flow rate of dry air per unit cross-sectional area is

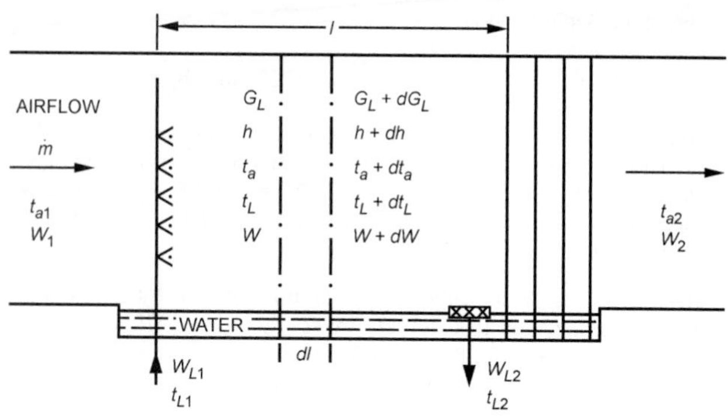

Fig. 12 Air Washer Spray Chamber

$$\dot{m}_a/A_{cs} = G_a \quad (53)$$

and the corresponding mass flux of water flowing parallel with the air is

$$\dot{m}_L/A_{cs} = G_L \quad (54)$$

where

$\dot{m}_a$ = mass flow rate of air, lb/h
G_a = mass flux or flow rate per unit cross-sectional area for air, lb/h·ft²
$\dot{m}_L$ = mass flow rate of liquid, lb/h
G_L = mass flux or flow rate per unit cross-sectional area for liquid, lb/h·ft²

Because water is evaporating or condensing, G_L changes by an amount dG_L in a differential length dl of the chamber. Similar changes occur in temperature, humidity ratio, enthalpy, and other properties.

Because evaluating the true surface area in direct-contact equipment is difficult, it is common to work on a unit volume basis. If a_H and a_M are the area of heat transfer and mass transfer surface per unit of chamber volume, respectively, the total surface areas for heat and mass transfer are

$$A_H = a_H A_{cs} l \quad \text{and} \quad A_M = a_M A_{cs} l \quad (55)$$

The basic equations for the process occurring in the differential length dl can be written for

1. *Mass transfer*

$$-dG_L = G_a dW = K_M a_M(W_i - W)dl \quad (56)$$

That is, the water evaporated, the moisture increase of the air, and the mass transfer rate are all equal.

2. *Heat transfer to air*

$$G_a c_{pm} dt_a = h_a a_H(t_i - t_a)dl \quad (57)$$

3. *Total energy transfer to air*

$$G_a(c_{pm} dt_a + h_{fgo} dW) = [K_M a_M(W_i - W)h_{fg} + h_a a_H(t_i - t_a)]\, dl \quad (58)$$

Assuming $a_H = a_M$ and Le = 1, and neglecting small variations in h_{fg}, Equation (58) reduces to

$$G_a dh = K_M a_M(h_i - h)\, dl \quad (59)$$

The heat and mass transfer areas of spray chambers are assumed to be identical ($a_H = a_M$). Where packing materials, such as wood slats or Raschig rings, are used, the two areas may be considerably different because the packing may not be wet uniformly. The validity of the Lewis relation was discussed previously. It is not necessary to account for the small changes in latent heat h_{fg} after making the two previous assumptions.

4. *Energy balance*

$$G_a dh = \pm G_L c_L dt_L \tag{60}$$

A minus sign refers to parallel flow of air and water; a plus sign refers to counterflow (water flow in the opposite direction from airflow).

The water flow rate changes between inlet and outlet as a result of the mass transfer. For exact energy balance, the term ($c_L t_L dG_L$) should be added to the right side of Equation (60). The percentage change in G_L is quite small in usual applications of air-conditioning equipment and, therefore, can be ignored.

5. *Heat transfer to water*

$$\pm G_L c_L dt_L = h_L a_H (t_L - t_i) dl \tag{61}$$

Equations (56) to (61) are the basic relations for solution of simultaneous heat and mass transfer processes in direct-contact air-conditioning equipment.

To facilitate the use of these relations in equipment design or performance, three other equations can be extracted from the above set. Combining Equations (59), (60), and (61) gives

$$\frac{h - h_i}{t_L - t_i} = -\frac{h_L a_H}{K_M a_M} = -\frac{h_L}{K_M} \tag{62}$$

Equation (62) relates the enthalpy potential for the total heat transfer through the gas film to the temperature potential for this same transfer through the liquid film. Physical reasoning leads to the conclusion that this ratio is proportional to the ratio of gas film resistance ($1/K_M$) to liquid film resistance ($1/h_L$). Combining Equations (57), (59), and (45) gives

$$\frac{dh}{dt_a} = \frac{h - h_i}{t_a - t_i} \tag{63}$$

Similarly, combining Equations (56), (57), and (45) gives

$$\frac{dW}{dt_a} = \frac{W - W_i}{t_a - t_i} \tag{64}$$

Equation (64) indicates that at any cross section in the spray chamber, the instantaneous slope of the air path dW/dt_a on a psychrometric chart is determined by a straight line connecting the air state with the interface saturation state at that cross section. In Figure 13, state 1 represents the state of the air entering the parallel flow air washer chamber of Figure 12. The washer is operating as a heating and humidifying apparatus so that the interface saturation state of the water at air inlet is the state designated 1_i. Therefore, the initial slope of the air path is along a line directed from state 1 to state 1_i. As the air is heated, the water cools and the interface temperature drops. Corresponding air states and interface saturation states are indicated by the letters *a*, *b*, *c*, and *d* in Figure 13. In each instance, the air path is directed toward the associated interface state. The interface states are derived from Equations (60) and (62). Equation (60) describes how the air enthalpy changes with water temperature; Equation (62) describes how the interface saturation state

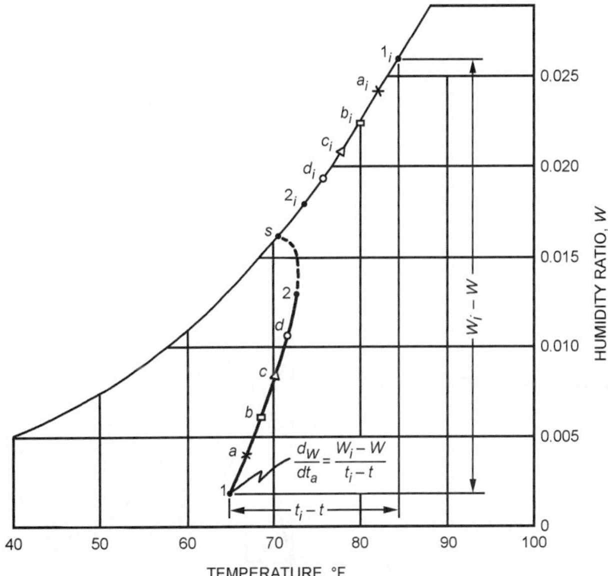

Fig. 13 Air Washer Humidification Process on Psychrometric Chart

changes to accommodate this change in air and water conditions. The solution for the interface state on the normal psychrometric chart of Figure 13 can be determined either by trial and error from Equations (60) and (62) or by a complex graphical procedure (Kusuda 1957).

Air Washers

Air washers are direct-contact apparatus used to (1) simultaneously change the temperature and humidity content of air passing through the chamber and (2) remove air contaminants such as dust and odors. Adiabatic spray washers, which have no external heating or chilling source, are used to cool and humidify air. Chilled spray air washers have an external chiller to cool and dehumidify air. Heated spray air washers, whose external heating source provides additional energy for evaporation of water, are used to humidify and possibly heat air.

Example 5. A parallel flow air washer with the following design conditions is to be designed (see Figure 12).

Water temperature at inlet t_{L1} = 95°F
Water temperature at outlet t_{L2} = 75°F
Air temperature at inlet t_{a1} = 65°F
Air wet-bulb at inlet t'_{a1} = 45°F
Air mass flow rate per unit area G_a = 1200 lb/h·ft²
Spray ratio G_L/G_a = 0.70
Air heat transfer coefficient per cubic foot of chamber volume $h_a a_H$ = 72 Btu/h·°F·ft³
Liquid heat transfer coefficient per cubic foot of chamber volume $h_L a_H$ = 900 Btu/h·°F·ft³
Air volumetric flow rate Q = 6500 cfm

Solution: The air mass flow rate $\dot{m}_a$ = 6500 × 0.075 = 490 lb/min; the required spray chamber cross-sectional area is, then, $A_{cs} = \dot{m}_a/G_a$ = 490 × 60/1200 = 24.5 ft². The mass transfer coefficient is given by the Lewis relation [Equation (45)] as

$$K_M a_M = (h_a a_H)/c_{pm} = 72/0.24 = 300 \text{ lb/h} \cdot \text{ft}^3$$

Figure 14 shows the enthalpy-temperature psychrometric chart with the graphical solution for the interface states and the air path through the washer spray chamber. The solution proceeds as follows:

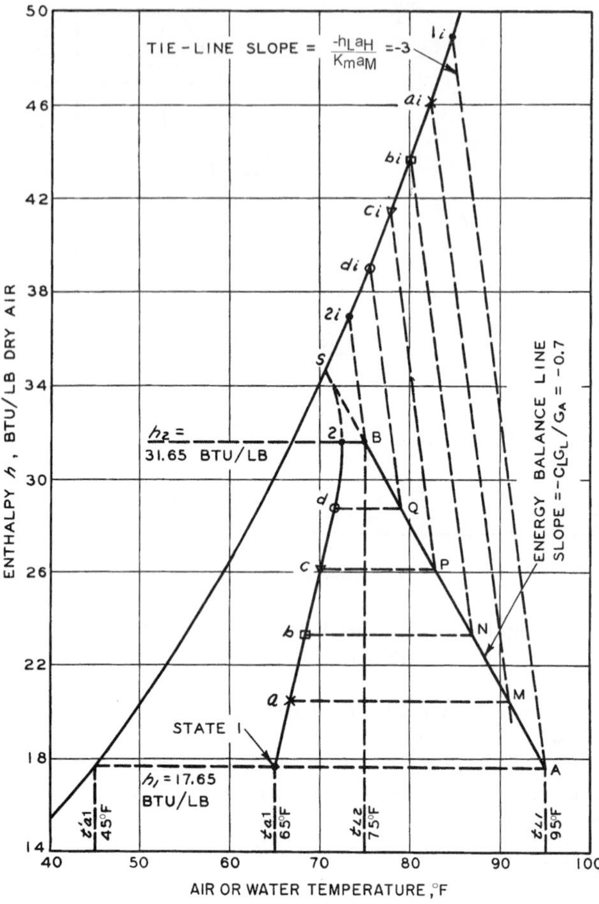

Fig. 14 Graphical Solution for Air-State Path in Parallel Flow Air Washer

1. Enter bottom of chart with t'_{a1} of 45°F, and follow up to saturation curve to establish air enthalpy h_1 of 17.65 Btu/lb. Extend this enthalpy line to intersect initial air temperature t_{a1} of 65°F (state 1 of air) and initial water temperature t_{L1} of 95°F at point A. (Note that the temperature scale is used for both air and water temperatures.)
2. Through point A, construct the *energy balance* line A-B with a slope of

$$\frac{dh}{dt_L} = -\frac{c_L G_L}{G_a} = -0.7$$

Point B is determined by intersection with the leaving water temperature t_{L2} = 75°F. The negative slope here is a consequence of the parallel flow, which results in the air-water mixture's approaching, but not reaching, the common saturation state *s*. (The line A-B has no physical significance in representing any *air state* on the psychrometric chart. It is merely a construction line in the graphical solution.)

3. Through point A, construct the *tie-line* A-1_i having a slope of

$$\frac{h-h_i}{t_L-t_i} = -\frac{h_L a_H}{K_M a_M} = -\frac{900}{300} = -3$$

The intersection of this line with the saturation curve gives the initial interface state 1_i at the chamber inlet. (Note how the energy balance line and tie-line, representing Equations (60) and (62), combine for a simple graphical solution on Figure 14 for the interface state.)

4. The initial slope of the air path can now be constructed, according to Equation (63), drawing line 1-*a* toward the initial interface state 1_i. (The length of the line 1-*a* will depend on the degree of accuracy required in the solution and the rate at which the slope of the air path is changing.)

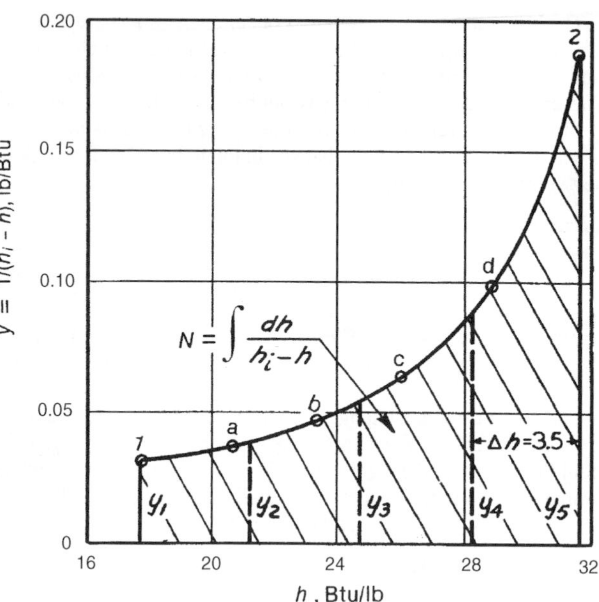

Fig. 15 Graphical Solution of $\int dh/(h_i - h)$

5. Construct the horizontal line *a*-M locating the point M on the energy-balance line. Draw a new tie-line (slope of −3 as before) from M to a_i locating interface state a_i. Continue the air path from *a* to *b* by directing it toward the new interface state a_i. (Note that the change in slope of the air path from 1-*a* to *a*-*b* is quite small, justifying the path incremental lengths used.)
6. Continue in the manner of step 5 until point 2, the final state of the air leaving the chamber, is reached. In this example, six steps are used in the graphical construction with the following results:

State	1	*a*	*b*	*c*	*d*	2
t_L	95.0	91.0	87.0	83.0	79.0	75.0
h	17.65	20.45	23.25	26.05	28.85	31.65
t_i	84.5	82.3	80.1	77.8	75.6	73.2
h_i	49.00	46.25	43.80	41.50	39.10	37.00
t_a	65.0	66.8	68.5	70.0	71.4	72.4

The final state of the air leaving the washer is t_{a2} = 72.4°F and h_2 = 31.65 Btu/lb (wet-bulb temperature t'_{a2} = 67°F).

7. The final step involves calculating the required length of the spray chamber. From Equation (59),

$$l = \frac{G_a}{K_M a_M}\int_1^2 \frac{dh}{(h_i - h)}$$

The integral is evaluated graphically by plotting $1/(h_i - h)$ versus h as shown in Figure 15. Any satisfactory graphical method can be used to evaluate the area under the curve. Simpson's rule with four equal increments of Δh equal to 3.5 gives

$$N = \int_1^2 \frac{dh}{(h_i - h)} \approx (\Delta h/3)(y_1 + 4y_2 + 2y_3 + 4y_4 + y_5)$$

$$N = (3.5/3)[0.0319 + (4\times 0.0400) + (2\times 0.0553) + (4\times 0.0865) + 0.1870] = 0.975$$

The design length is, therefore, l = (1200/300)(0.975) = 3.9 ft.

The method used in Example 5 can also be used to predict the performance of existing direct-contact equipment and can determine the transfer coefficients when performance data from test runs are available. By knowing the water and air temperatures entering and leaving the chamber and the spray ratio, it is possible, by trial and error, to determine the proper slope of the tie-line necessary to achieve the measured final air state. The tie-line slope gives the ratio $h_L a_H / K_M a_M$; $K_M a_M$ is found from the integral relationship in Example 5 from the known chamber length l.

Additional descriptions of air spray washers and general performance criteria are given in Chapter 19 of the 2000 *ASHRAE Handbook—Systems and Equipment.*

Cooling Towers

A cooling tower is a direct-contact heat exchanger in which waste heat picked up by the cooling water from a refrigerator, air conditioner, or industrial process is transferred to atmospheric air by cooling the water. Cooling is achieved by breaking up the water flow to provide a large water surface for air, moving by natural or forced convection through the tower, to contact the water. Cooling towers may be counterflow, crossflow, or a combination of both.

The temperature of the water leaving the tower and the packing depth needed to achieve the desired leaving water temperature are of primary interest for design. Therefore, the mass and energy balance equations are based on an overall coefficient K, which is based on (1) the enthalpy driving force due to h at the bulk water temperature and (2) neglecting the film resistance. Combining Equations (59) and (60) and using the parameters described above yields

$$G_L c_L dt = K_M a_M (h_i - h) dl = G_a dh$$

$$= \frac{K_a dV(h' - h_a)}{A_{cs}} \tag{65}$$

or

$$\frac{K_a V}{\dot{m}_L} = \int_{t_1}^{t_2} \frac{c_L \, dt}{(h' - h_a)} \tag{66}$$

Chapter 36 of the 2000 *ASHRAE Handbook—Systems and Equipment* covers cooling tower design in detail.

Cooling and Dehumidifying Coils

When water vapor is condensed out of an airstream onto an extended surface (finned) cooling coil, the simultaneous heat and mass transfer problem can be solved by the same procedure set forth for direct-contact equipment. The basic equations are the same, except that the true surface area of the coil A is known and the problem does not have to be solved on a unit volume basis. Therefore, if in Equations (56), (57), and (59) $a_M dl$ or $a_H dl$ is replaced by dA/A_{cs}, these equations become the basic heat, mass, and total energy transfer equations for indirect-contact equipment such as dehumidifying coils. The energy balance shown by Equation (60) remains unchanged. The heat transfer from the interface to the refrigerant now encounters the combined resistances of the condensate film ($R_L = 1/h_L$); the metal wall and fins, if any (R_m); and the refrigerant film ($R_r = A/h_r A_r$). If this combined resistance is designated as $R_i = R_L + R_m + R_r = 1/U_i$, Equation (61) becomes, for a coil dehumidifier,

$$\pm \dot{m}_L c_L dt_L = U_i (t_L - t_i) dA \tag{67}$$

(plus sign for counterflow, minus sign for parallel flow).

The tie-line slope is then

$$\frac{h - h_i}{t_L - t_i} = \mp \frac{U_i}{K_M} \tag{68}$$

Figure 16 illustrates the graphical solution on a psychrometric chart for the air path through a dehumidifying coil with a constant refrigerant temperature. Because the tie-line slope is infinite in this case, the energy balance line is vertical. The corresponding interface states and air states are denoted by the same letter symbols, and the solution follows the same procedure as in Example 5.

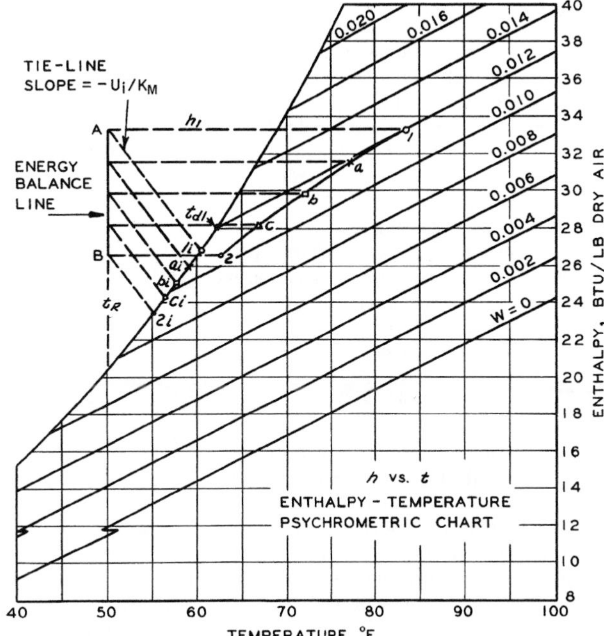

Fig. 16 Graphical Solution for Air-State Path in Dehumidifying Coil with Constant Refrigerant Temperature

If the problem is to determine the required coil surface area for a given performance, the area is computed by the following relation:

$$A = \frac{\dot{m}_a}{K_M} \int_1^2 \frac{dh}{(h_i - h)} \tag{69}$$

This graphical solution on the psychrometric chart automatically determines whether any part of the coil is dry. Thus, in the example illustrated in Figure 16, the entering air at state 1 initially encounters an interface saturation state 1_i, clearly below its dew-point temperature t_{d1}, so the coil immediately becomes wet. Had the graphical technique resulted in an initial interface state above the dew-point temperature of the entering air, the coil would be initially dry. The air would then follow a constant humidity ratio line (the sloping W = constant lines on the chart) until the interface state reached the air dew-point temperature.

Mizushina et al. (1959) developed this method not only for water vapor and air, but also for other vapor-gas mixtures. Chapter 21 of the 2000 *ASHRAE Handbook—Systems and Equipment* shows another related method, based on ARI *Standard* 410, of determining air-cooling and dehumidifying coil performance.

SYMBOLS

a = constant, dimensionless; or surface area per unit volume, ft²/ft³
A = surface area, ft²
A_{cs} = cross-sectional area, ft²
b = exponent, dimensionless
c_L = specific heat of liquid, Btu/lb·°F
c_p = specific heat at constant pressure, Btu/lb·°F
c_{pm} = specific heat of moist air at constant pressure, Btu/lb$_{da}$·°F
C = molal concentration of solute in solvent, lb mol/ft²
d = diameter, ft
D_v = diffusion coefficient (mass diffusivity), ft²/h
f = Fanning friction factor, dimensionless
g_c = gravitational constant, ft·lb$_m$/h²·lb$_f$
G = mass flux, flow rate per unit of cross-sectional area, lb$_m$/h·ft²
h = enthalpy, Btu/lb; or heat transfer coefficient, Btu/h·ft²·°F

h_{fg} = enthalpy of vaporization, Btu/lb_m
h_M = mass transfer coefficient, ft/h
j_D = Colburn mass transfer group = Sh/(Re·$Sc^{1/3}$), dimensionless
j_H = Colburn heat transfer group = Nu/(Re·$Pr^{1/3}$), dimensionless
J = diffusive mass flux, lb_m/h·ft^2
J^* = diffusive molar flux, lb mol/h·ft^2
k = thermal conductivity, Btu/h·ft·°F
K_M = mass transfer coefficient, lb/h·ft^2
l = length, ft
L = characteristic length, ft
L/G = liquid-to-air mass flow ratio
Le = Lewis number = α/D_v, dimensionless
$\dot{m}$ = rate of mass transfer, lb/h
$\dot{m}''$ = mass flux, lb/h·ft^2
$\dot{m}''^*$ = molar flux, lb mol/h·ft^2
M = molecular weight, lb_m/lb mol
Nu = Nusselt number = hL/k, dimensionless
p = pressure, atmospheres or psi
P_{Am} = logarithmic mean density factor
Pr = Prandtl number = $c_p\mu/k$, dimensionless
q = rate of heat transfer, Btu/h
q'' = heat flux per unit area, Btu/h·ft^2
Q = volumetric flow rate, cfm
R_i = combined thermal resistance, ft^2·°F·h/Btu
R_L = thermal resistance of condensate film, ft^2·°F·h/Btu
R_m = thermal resistance across metal wall and fins, ft^2·°F·h/Btu
R_r = thermal resistance of refrigerant film, ft^2·°F·h/Btu
R_u = universal gas constant = 1545 lb_f·ft/lb mol·°R
Re = Reynolds number = $\rho uL/\mu$, dimensionless
Sc = Schmidt number = $\mu/\rho D_v$, dimensionless
Sh = Sherwood number = $h_M L/D_v$, dimensionless
St = Stanton number = $h/\rho c_p \bar{u}$, dimensionless
St_m = mass transfer Stanton number = $h_M P_{Am}/\bar{u}$, dimensionless
t = temperature, °F
T = absolute temperature, °R
u = velocity in x direction, fpm
U_i = overall conductance from refrigerant to air-water interface for dehumidifying coil, Btu/h·ft^2·°F
v = velocity in y direction, fpm
v_i = velocity normal to mass transfer surface for component i, ft/h
V = fluid stream velocity, fpm
W = humidity ratio, lb of water vapor per lb of dry air, lb(w)/lb(da)
x, y, z = coordinate direction, ft
X, Y, Z = coordinate direction, dimensionless
α = thermal diffusivity = $k/\rho c_p$, ft^2/h
ε_D = eddy mass diffusivity, ft^2/h
θ = dimensionless time parameter
μ = absolute (dynamic) viscosity, lb_m/ft·h
$\bar{\mu}$ = permeability, grains·in/h·ft^2·in. Hg
ν = kinematic viscosity, ft^2/h
ρ = mass density or concentration, lb_m/ft^3
σ = characteristic molecular diameter, nm
τ = time
τ_i = shear stress in the x-y coordinate plane, lb_f/ft^2
ω = mass fraction, lb/lb
$\Omega_{D,AB}$ = temperature function in Equation (9)

Subscripts

a = air property
Am = logarithmic mean
A = gas component of binary mixture
B = the more dilute gas component of binary mixture
c = critical state
da = dry air property or air-side transfer quantity
H = heat transfer quantity
i = air-water interface value
L = liquid
m = mean value or metal
M = mass transfer quantity
o = property evaluated at 0°F
s = water vapor property or transport quantity
w = water vapor
∞ = property of main fluid stream

Superscripts

* = on molar basis
$^-$ = average value
$'$ = wet bulb

REFERENCES

Bedingfield, G.H., Jr. and T.B. Drew. 1950. Analogy between heat transfer and mass transfer—A psychrometric study. *Industrial and Engineering Chemistry* 42:1164.

Bird, R.B., W.E. Stewart, and E.N. Lightfoot. 1960. *Transport phenomena.* John Wiley and Sons, New York.

Chilton, T.H. and A.P. Colburn. 1934. Mass transfer (absorption) coefficients—Prediction from data on heat transfer and fluid friction. *Industrial and Engineering Chemistry* 26 (November):1183.

Guillory, J.L. and F.C. McQuiston. 1973. An experimental investigation of air dehumidification in a parallel plate heat exchanger. *ASHRAE Transactions* 79(2):146.

Helmer, W.A. 1974. *Condensing water vapor—Airflow in a parallel plate heat exchanger.* Ph.D. thesis, Purdue University, West Lafayette, IN.

Hirschfelder, J.O., C.F. Curtiss, and R.B. Bird. 1954. *Molecular theory of gases and liquids.* John Wiley and Sons, New York.

Incropera, F.P. and D.P. DeWitt. 1996. *Fundamentals of heat and mass transfer*, 4th ed. John Wiley and Sons, New York.

Kusuda, T. 1957. Graphical method simplifies determination of aircoil, wet-heat-transfer surface temperature. *Refrigerating Engineering* 65:41.

Mason, E.A. and L. Monchick. 1965. Survey of the equation of state and transport properties of moist gases. *Humidity and Moisture* 3. Reinhold Publishing Corporation, New York.

McAdams, W.H. 1954. *Heat transmission*, 3rd ed. McGraw-Hill, New York.

Mizushina, T., N. Hashimoto, and M. Nakajima. 1959. Design of cooler condensers for gas-vapour mixtures. *Chemical Engineering Science* 9:195.

Ohadi, M.M. and E.M. Sparrow. 1989. Heat transfer in a straight tube situated downstream of a bend. *International Journal of Heat and Mass Transfer* 32(2):201-12.

Reid, R.C. and T.K. Sherwood. 1966. *The properties of gases and liquids: Their estimation and correlation*, 2nd ed. McGraw-Hill, New York, pp. 520-43.

Reid, R.C., J.M. Prausnitz, and B.E. Poling. 1987. *The properties of gases and liquids*, 4th ed. McGraw-Hill, New York, pp.21-78.

Sherwood, T.K. and R.L. Pigford. 1952. *Absorption and extraction.* McGraw-Hill, New York, pp. 1-28.

Sparrow, E.M. and M.M. Ohadi. 1987a. Comparison of turbulent thermal entrance regions for pipe flows with developed velocity and velocity developing from a sharp-edged inlet. *ASME Transactions, Journal of Heat Transfer* 109:1028-30.

Sparrow, E.M. and M.M. Ohadi. 1987b. Numerical and experimental studies of turbulent flow in a tube. *Numerical Heat Transfer* 11:461-76.

Treybal, R.E. 1980. *Mass transfer operations*, 3rd ed. McGraw-Hill, New York.

BIBLIOGRAPHY

Bennett, C.O. and J.E. Myers. 1982. *Momentum, heat and mass transfer*, 3rd ed. McGraw-Hill, New York.

DeWitt, D.P. and E.L. Cussler. 1984. *Diffusion, mass transfer in fluid systems.* Cambridge University Press, UK.

Eckert, E.R.G. and R.M. Drake, Jr. 1972. *Analysis of heat and mass transfer.* McGraw-Hill, New York.

Geankopolis, C.J. 1993. *Transport processes and unit operations*, 3rd ed. Prentice Hall, Englewood Cliffs, NJ.

Kays, W.M. and M.E. Crawford. 1993. *Convective heat and mass transfer.* McGraw-Hill, New York.

Mikielviez, J. and A.M.A. Rageb. 1995. Simple theroretical approach to direct-contact condensation on subcooled liquid film. *International Journal of Heat and Mass Transfer* 38(3):557.

CHAPTER 6

PSYCHROMETRICS

PSYCHROMETRICS deals with the thermodynamic properties of moist air and uses these properties to analyze conditions and processes involving moist air.

Hyland and Wexler (1983a,b) developed formulas for thermodynamic properties of moist air and water. However, perfect gas relations can be used instead of these formulas in most air-conditioning problems. Kuehn et al. (1998) showed that errors are less than 0.7% in calculating humidity ratio, enthalpy, and specific volume of saturated air at standard atmospheric pressure for a temperature range of −60 to 120°F. Furthermore, these errors decrease with decreasing pressure.

This chapter discusses perfect gas relations and describes their use in common air-conditioning problems. The formulas developed by Hyland and Wexler (1983a) and discussed by Olivieri (1996) may be used where greater precision is required.

COMPOSITION OF DRY AND MOIST AIR

Atmospheric air contains many gaseous components as well as water vapor and miscellaneous contaminants (e.g., smoke, pollen, and gaseous pollutants not normally present in free air far from pollution sources).

Dry air exists when all water vapor and contaminants have been removed from atmospheric air. The composition of dry air is relatively constant, but small variations in the amounts of individual components occur with time, geographic location, and altitude. Harrison (1965) lists the approximate percentage composition of dry air by volume as: nitrogen, 78.084; oxygen, 20.9476; argon, 0.934; carbon dioxide, 0.0314; neon, 0.001818; helium, 0.000524; methane, 0.00015; sulfur dioxide, 0 to 0.0001; hydrogen, 0.00005; and minor components such as krypton, xenon, and ozone, 0.0002. The apparent molecular mass (weighted average molecular weight of all components), for dry air is 28.9645, based on the carbon-12 scale (Harrison 1965). The gas constant for dry air, based on the carbon-12 scale, is

$$R_{da} = 1545.32/28.9645 = 53.352 \text{ ft}\cdot\text{lb}_f/\text{lb}_m\cdot{}^\circ\text{R} \qquad (1)$$

Moist air is a binary (two-component) mixture of dry air and water vapor. The amount of water vapor in moist air varies from zero (dry air) to a maximum that depends on temperature and pressure. The latter condition refers to **saturation**, a state of neutral equilibrium between moist air and the condensed water phase (liquid or solid). Unless otherwise stated, saturation refers to a flat interface surface between the moist air and the condensed phase. Saturation conditions will change when the interface radius is very small such as with ultrafine water droplets. The relative molecular mass of water is 18.01528 on the carbon-12 scale. The gas constant for water vapor is

$$R_w = 1545.32/18.01528 = 85.778 \text{ ft}\cdot\text{lb}_f/\text{lb}_m\cdot{}^\circ\text{R} \qquad (2)$$

UNITED STATES STANDARD ATMOSPHERE

The temperature and barometric pressure of atmospheric air vary considerably with altitude as well as with local geographic and weather conditions. The standard atmosphere gives a standard of reference for estimating properties at various altitudes. At sea level, standard temperature is 59°F; standard barometric pressure is 29.921 in. Hg. The temperature is assumed to decrease linearly with increasing altitude throughout the troposphere (lower atmosphere), and to be constant in the lower reaches of the stratosphere. The lower atmosphere is assumed to consist of dry air that behaves as a perfect gas. Gravity is also assumed constant at the standard value, 32.1740 ft/s². Table 1 summarizes property data for altitudes to 60,000 ft.

Table 1 Standard Atmospheric Data for Altitudes to 60,000 ft

Altitude, ft	Temperature, °F	Pressure, in. Hg	Pressure, psia
−1000	62.6	31.02	15.236
−500	60.8	30.47	14.966
0	59.0	29.921	14.696
500	57.2	29.38	14.430
1,000	55.4	28.86	14.175
2,000	51.9	27.82	13.664
3,000	48.3	26.82	13.173
4,000	44.7	25.82	12.682
5,000	41.2	24.90	12.230
6,000	37.6	23.98	11.778
7,000	34.0	23.09	11.341
8,000	30.5	22.22	10.914
9,000	26.9	21.39	10.506
10,000	23.4	20.58	10.108
15,000	5.5	16.89	8.296
20,000	−12.3	13.76	6.758
30,000	−47.8	8.90	4.371
40,000	−69.7	5.56	2.731
50,000	−69.7	3.44	1.690
60,000	−69.7	2.14	1.051

Source: Adapted from NASA (1976).

The preparation of this chapter is assigned to TC 1.1, Thermodynamics and Psychrometrics.

The pressure values in Table 1 may be calculated from

$$p = 29.92(1 - 6.8753 \times 10^{-6} Z)^{5.2559} \quad (3)$$

The equation for temperature as a function of altitude is given as

$$t = 59 - 0.00356616Z \quad (4)$$

where

Z = altitude, ft
p = barometric pressure, in. Hg.
t = temperature, °F

Equations (3) and (4) are accurate from –16,500 ft to 36,000 ft. For higher altitudes, comprehensive tables of barometric pressure and other physical properties of the standard atmosphere, in both SI and inch-pound units, can be found in NASA (1976).

THERMODYNAMIC PROPERTIES OF MOIST AIR

Table 2, developed from formulas by Hyland and Wexler (1983a,b), shows values of thermodynamic properties of **moist air** based on the **thermodynamic temperature scale**. This ideal scale differs slightly from practical temperature scales used for physical measurements. For example, the standard boiling point for water (at 14.696 psia or 29.921 in. Hg) occurs at 211.95°F on this scale rather than at the traditional value of 212°F. Most measurements are currently based on the International Temperature Scale of 1990 (ITS-90) (Preston-Thomas 1990).

The following paragraphs briefly describe each column of Table 2:

t = Fahrenheit temperature, based on thermodynamic temperature scale and expressed relative to absolute temperature T in degrees Rankine (°R) by the following relation:

$$T = t + 459.67$$

W_s = humidity ratio at saturation, condition at which gaseous phase (moist air) exists in equilibrium with condensed phase (liquid or solid) at given temperature and pressure (standard atmospheric pressure). At given values of temperature and pressure, humidity ratio W can have any value from zero to W_s.

v_{da} = specific volume of dry air, ft^3/lb_{da}.

v_{as} = $v_s - v_{da}$, difference between specific volume of moist air at saturation and that of dry air itself, ft^3/lb_{da}, at same pressure and temperature.

v_s = specific volume of moist air at saturation, ft^3/lb_{da}.

h_{da} = specific enthalpy of dry air, Btu/lb_{da}. In Table 2, h_{da} has been assigned a value of 0 at 0°F and standard atmospheric pressure.

h_{as} = $h_s - h_{da}$, difference between specific enthalpy of moist air at saturation and that of dry air itself, Btu/lb_{da}, at same pressure and temperature.

h_s = specific enthalpy of moist air at saturation, Btu/lb_{da}.

s_{da} = specific entropy of dry air, $Btu/lb_{da} \cdot °R$. In Table 2, s_{da} has been assigned a value of 0 at 0°F and standard atmospheric pressure.

s_{as} = $s_s - s_{da}$, difference between specific entropy of moist air at saturation and that of dry air itself, $Btu/lb_{da} \cdot °R$, at same pressure and temperature.

s_s = specific entropy of moist air at saturation $Btu/lb_{da} \cdot °R$.

h_w = specific enthalpy of condensed water (liquid or solid) in equilibrium with saturated moist air at specified temperature and pressure, Btu/lb_w. In Table 2, h_w is assigned a value of 0 at its triple point (32.018°F) and saturation pressure.

Note that h_w is greater than the steam-table enthalpy of saturated pure condensed phase by the amount of enthalpy increase governed by the pressure increase from saturation pressure to 1 atm. plus influences from presence of air.

s_w = specific entropy of condensed water (liquid or solid) in equilibrium with saturated air, $Btu/lb_w \cdot °R$; s_w differs from entropy of pure water at saturation pressure, similar to h_w.

p_s = vapor pressure of water in saturated moist air, in. Hg. Pressure p_s differs negligibly from saturation vapor pressure of pure water p_{ws} for conditions shown. Consequently, values of p_s can be used at same pressure and temperature in equations where p_{ws} appears. Pressure p_s is defined as $p_s = x_{ws}p$, where x_{ws} is mole fraction of water vapor in moist air saturated with water at temperature t and pressure p, and where p is total barometric pressure of moist air.

THERMODYNAMIC PROPERTIES OF WATER AT SATURATION

Table 3 shows thermodynamic properties of **water at saturation** for temperatures from –80 to 300°F, calculated by the formulations described by Hyland and Wexler (1983b). Symbols in the table follow standard steam table nomenclature. These properties are based on the thermodynamic temperature scale. The enthalpy and entropy of saturated liquid water are both assigned the value zero at the triple point, 32.018°F. Between the triple-point and critical-point temperatures of water, two states—liquid and vapor—may coexist in equilibrium. These states are called **saturated liquid** and **saturated vapor**.

The **water vapor saturation pressure** is required to determine a number of moist air properties, principally the saturation humidity ratio. Values may be obtained from Table 3 or calculated from the following formulas (Hyland and Wexler 1983b).

The saturation pressure over **ice** for the temperature range of –148 to 32°F is given by

$$\ln p_{ws} = C_1/T + C_2 + C_3T + C_4T^2 + C_5T^3 + C_6T^4 + C_7 \ln T \quad (5)$$

where

C_1 = –1.021 416 5 E+04
C_2 = –4.893 242 8 E+00
C_3 = –5.376 579 4 E–03
C_4 = 1.920 237 7 E–07
C_5 = 3.557 583 2 E–10
C_6 = –9.034 468 8 E–14
C_7 = 4.163 501 9 E+00

The saturation pressure over **liquid water** for the temperature range of 32 to 392°F is given by

$$\ln p_{ws} = C_8/T + C_9 + C_{10}T + C_{11}T^2 + C_{12}T^3 + C_{13} \ln T \quad (6)$$

where

C_8 = –1.044 039 7 E+04
C_9 = –1.129 465 0 E+01
C_{10} = –2.702 235 5 E–02
C_{11} = 1.289 036 0 E–05
C_{12} = –2.478 068 1 E–09
C_{13} = 6.545 967 3 E+00

In both Equations (5) and (6),

ln = natural logarithm
p_{ws} = saturation pressure, psia
T = absolute temperature, °R = °F + 459.67

The coefficients of Equations (5) and (6) have been derived from the Hyland-Wexler equations, which are given in SI units. Due to rounding errors in the derivations and in some computers' calculating precision, the results obtained from Equations (5) and (6) may not agree precisely with Table 3 values.

Table 2 Thermodynamic Properties of Moist Air at Standard Atmospheric Pressure, 14.696 psi (29.921 in. Hg)

Temp., °F	Humidity Ratio, lb_w/lb_{da}	Specific Volume, ft^3/lb_{da}			Specific Enthalpy, Btu/lb_{da}			Specific Entropy, $Btu/lb_{da}\cdot°F$			Condensed Water: Specific Enthalpy, Btu/lb_w	Condensed Water: Specific Entropy, $Btu/lb_w\cdot°F$	Condensed Water: Vapor Pressure, in. Hg	Temp., °F
t	W_s	v_{da}	v_{as}	v_s	h_{da}	h_{as}	h_s	s_{da}	s_{as}	s_s	h_w	s_w	p_s	t
−80	0.0000049	9.553	0.000	9.553	−19.221	0.005	−19.215	−0.04594	0.00001	−0.04592	−193.45	−0.4067	0.000236	−80
−79	0.0000053	9.579	0.000	9.579	−18.980	0.005	−18.975	−0.04531	0.00002	−0.04529	−193.06	−0.4056	0.000255	−79
−78	0.0000057	9.604	0.000	9.604	−18.740	0.006	−18.734	−0.04468	0.00002	−0.04466	−192.66	−0.4046	0.000275	−78
−77	0.0000062	9.629	0.000	9.629	−18.500	0.007	−18.493	−0.04405	0.00002	−0.04403	−192.27	−0.4036	0.000296	−77
−76	0.0000067	9.655	0.000	9.655	−18.259	0.007	−18.252	−0.04342	0.00002	−0.04340	−191.87	−0.4025	0.000319	−76
−75	0.0000072	9.680	0.000	9.680	−18.019	0.007	−18.011	−0.04279	0.00002	−0.04277	−191.47	−0.4015	0.000344	−75
−74	0.0000078	9.705	0.000	9.705	−17.778	0.008	−17.770	−0.04217	0.00002	−0.04215	−191.07	−0.4005	0.000371	−74
−73	0.0000084	9.731	0.000	9.731	−17.538	0.009	−17.529	−0.04155	0.00002	−0.04152	−190.68	−0.3994	0.000400	−73
−72	0.0000090	9.756	0.000	9.756	−17.298	0.010	−17.288	−0.04093	0.00003	−0.04090	−190.27	−0.3984	0.000430	−72
−71	0.0000097	9.781	0.000	9.782	−17.057	0.010	−17.047	−0.04031	0.00003	−0.04028	−189.87	−0.3974	0.000463	−71
−70	0.0000104	9.807	0.000	9.807	−16.806	0.011	−16.817	−0.03969	0.00003	−0.03966	−189.47	−0.3963	0.000498	−70
−69	0.0000112	9.832	0.000	9.832	−16.577	0.012	−16.565	−0.03907	0.00003	−0.03904	−189.07	−0.3953	0.000536	−69
−68	0.0000120	9.857	0.000	9.858	−16.336	0.013	−16.324	−0.03846	0.00003	−0.03843	−188.66	−0.3943	0.000576	−68
−67	0.0000129	9.883	0.000	9.883	−16.096	0.013	−16.083	−0.03785	0.00004	−0.03781	−188.26	−0.3932	0.000619	−67
−66	0.0000139	9.908	0.000	9.908	−15.856	0.015	−15.841	−0.03724	0.00004	−0.03720	−187.85	−0.3922	0.000665	−66
−65	0.0000149	9.933	0.000	9.934	−15.616	0.015	−15.600	−0.03663	0.00004	−0.03659	−187.44	−0.3912	0.000714	−65
−64	0.0000160	9.959	0.000	9.959	−15.375	0.017	−15.359	−0.03602	0.00005	−0.03597	−187.04	−0.3901	0.000766	−64
−63	0.0000172	9.984	0.000	9.984	−15.117	0.018	−15.135	−0.03541	0.00005	−0.03536	−186.63	−0.3891	0.000822	−63
−62	0.0000184	10.009	0.000	10.010	−14.895	0.019	−14.876	−0.03481	0.00005	−0.03476	−186.22	−0.3881	0.000882	−62
−61	0.0000198	10.035	0.000	10.035	−14.654	0.021	−14.634	−0.03420	0.00006	−0.03415	−185.81	−0.3870	0.000945	−61
−60	0.0000212	10.060	0.000	10.060	−14.414	0.022	−14.392	−0.03360	0.00006	−0.03354	−185.39	−0.3860	0.001013	−60
−59	0.0000227	10.085	0.000	10.086	−14.174	0.024	−14.150	−0.03300	0.00006	−0.03294	−184.98	−0.3850	0.001086	−59
−58	0.0000243	10.111	0.000	10.111	−13.933	0.025	−13.908	−0.03240	0.00007	−0.03233	−184.57	−0.3839	0.001163	−58
−57	0.0000260	10.136	0.000	10.137	−13.693	0.027	−13.666	−0.03180	0.00007	−0.03173	−184.15	−0.3829	0.001246	−57
−56	0.0000279	10.161	0.000	10.162	−13.453	0.029	−13.424	−0.03121	0.00008	−0.03113	−183.74	−0.3819	0.001333	−56
−55	0.0000298	10.187	0.000	10.187	−13.213	0.031	−13.182	−0.03061	0.00008	−0.03053	−183.32	−0.3808	0.001427	−55
−54	0.0000319	10.212	0.001	10.213	−12.972	0.033	−12.939	−0.03002	0.00009	−0.02993	−182.90	−0.3798	0.001526	−54
−53	0.0000341	10.237	0.001	10.238	−12.732	0.035	−12.697	−0.02943	0.00009	−0.02934	−182.48	−0.3788	0.001632	−53
−52	0.0000365	10.263	0.001	10.263	−12.492	0.038	−12.454	−0.02884	0.00010	−0.02874	−182.06	−0.3778	0.001745	−52
−51	0.0000390	10.288	0.001	10.289	−12.251	0.041	−12.211	−0.02825	0.00011	−0.02814	−181.64	−0.3767	0.001865	−51
−50	0.0000416	10.313	0.001	10.314	−12.011	0.043	−11.968	−0.02766	0.00011	−0.02755	−181.22	−0.3757	0.001992	−50
−49	0.0000445	10.339	0.001	10.340	−11.771	0.046	−11.725	−0.02708	0.00012	−0.02696	−180.80	−0.3747	0.002128	−49
−48	0.0000475	10.364	0.001	10.365	−11.531	0.050	−11.481	−0.02649	0.00013	−0.02636	−180.37	−0.3736	0.002272	−48
−47	0.0000507	10.389	0.001	10.390	−11.290	0.053	−11.237	−0.02591	0.00014	−0.02577	−179.95	−0.3726	0.002425	−47
−46	0.0000541	10.415	0.001	10.416	−11.050	0.056	−10.994	−0.02533	0.00015	−0.02518	−179.52	−0.3716	0.002587	−46
−45	0.0000577	10.440	0.001	10.441	−10.810	0.060	−10.750	−0.02475	0.00016	−0.02459	−179.10	−0.3705	0.002760	−45
−44	0.0000615	10.465	0.001	10.466	−10.570	0.064	−10.505	−0.02417	0.00017	−0.02400	−178.67	−0.3695	0.002943	−44
−43	0.0000656	10.491	0.001	10.492	−10.329	0.068	−10.261	−0.02359	0.00018	−0.02342	−178.24	−0.3685	0.003137	−43
−42	0.0000699	10.516	0.001	10.517	−10.089	0.073	−10.016	−0.02302	0.00019	−0.02283	−177.81	−0.3675	0.003343	−42
−41	0.0000744	10.541	0.001	10.543	−9.849	0.078	−9.771	−0.02244	0.00020	−0.02224	−177.38	−0.3664	0.003562	−41
−40	0.0000793	10.567	0.001	10.568	−9.609	0.083	−9.526	−0.02187	0.00021	−0.02166	−176.95	−0.3654	0.003793	−40
−39	0.0000844	10.592	0.001	10.593	−9.368	0.088	−9.280	−0.02130	0.00022	−0.02107	−176.52	−0.3644	0.004039	−39
−38	0.0000898	10.617	0.002	10.619	−9.128	0.094	−9.034	−0.02073	0.00024	−0.02049	−176.08	−0.3633	0.004299	−38
−37	0.0000956	10.643	0.002	10.644	−8.888	0.100	−8.788	−0.02016	0.00025	−0.01991	−175.65	−0.3623	0.004575	−37
−36	0.0001017	10.668	0.002	10.670	−8.648	0.106	−8.541	−0.01959	0.00027	−0.01932	−175.21	−0.3613	0.004866	−36
−35	0.0001081	10.693	0.002	10.695	−8.407	0.113	−8.294	−0.01902	0.00028	−0.01874	−174.78	−0.3603	0.005175	−35
−34	0.0001150	10.719	0.002	10.721	−8.167	0.120	−8.047	−0.01846	0.00030	−0.01816	−174.34	−0.3529	0.005502	−34
−33	0.0001222	10.744	0.002	10.746	−7.927	0.128	−7.799	−0.01790	0.00032	−0.01758	−173.90	−0.3582	0.005848	−33
−32	0.0001298	10.769	0.002	10.772	−7.687	0.136	−7.551	−0.01733	0.00034	−0.01699	−173.46	−0.3572	0.006214	−32
−31	0.0001379	10.795	0.002	10.797	−7.447	0.145	−7.302	−0.01677	0.00036	−0.01641	−173.02	−0.3561	0.006601	−31
−30	0.0001465	10.820	0.003	10.822	−7.206	0.154	−7.053	−0.01621	0.00038	−0.01583	−172.58	−0.3551	0.007009	−30
−29	0.0001555	10.845	0.003	10.848	−6.966	0.163	−6.803	−0.01565	0.00040	−0.01525	−172.14	−0.3541	0.007442	−29
−28	0.0001650	10.871	0.003	10.873	−6.726	0.173	−6.553	−0.01510	0.00043	−0.01467	−171.70	−0.3531	0.007898	−28
−27	0.0001751	10.896	0.003	10.899	−6.486	0.184	−6.302	−0.01454	0.00045	−0.01409	−171.25	−0.3520	0.008381	−27
−26	0.0001858	10.921	0.003	10.924	−6.245	0.195	−6.051	−0.01399	0.00048	−0.01351	−170.81	−0.3510	0.008890	−26
−25	0.0001970	10.947	0.003	10.950	−6.005	0.207	−5.798	−0.01343	0.00051	−0.01293	−170.36	−0.3500	0.009428	−25
−24	0.0002088	10.972	0.004	10.976	−5.765	0.220	−5.545	−0.01288	0.00054	−0.01235	−169.92	−0.3489	0.009995	−24
−23	0.0002214	10.997	0.004	11.001	−5.525	0.233	−5.292	−0.01233	0.00057	−0.01176	−169.47	−0.3479	0.010594	−23
−22	0.0002346	11.022	0.004	11.027	−5.284	0.247	−5.038	−0.01178	0.00060	−0.01118	−169.02	−0.3469	0.011226	−22
−21	0.0002485	11.048	0.004	11.052	−5.044	0.261	−4.783	−0.01123	0.00063	−0.01060	−168.57	−0.3459	0.011893	−21
−20	0.0002632	11.073	0.005	11.078	−4.804	0.277	−4.527	−0.01069	0.00067	−0.01002	−168.12	−0.3448	0.012595	−20
−19	0.0002786	11.098	0.005	11.103	−4.564	0.293	−4.271	−0.01014	0.00071	−0.00943	−167.67	−0.3438	0.013336	−19
−18	0.0002950	11.124	0.005	11.129	−4.324	0.311	−4.013	−0.00960	0.00075	−0.00885	−167.21	−0.3428	0.014117	−18
−17	0.0003121	11.149	0.006	11.155	−4.084	0.329	−3.754	−0.00905	0.00079	−0.00826	−166.76	−0.3418	0.014939	−17
−16	0.0003303	11.174	0.006	11.180	−3.843	0.348	−3.495	−0.00851	0.00083	−0.00768	−166.30	−0.3407	0.015806	−16
−15	0.0003493	11.200	0.006	11.206	−3.603	0.368	−3.235	−0.00797	0.00088	−0.00709	−165.85	−0.3397	0.016718	−15
−14	0.0003694	11.225	0.007	11.232	−3.363	0.390	−2.973	−0.00743	0.00093	−0.00650	−165.39	−0.3387	0.017679	−14
−13	0.0003905	11.250	0.007	11.257	−3.123	0.412	−2.710	−0.00689	0.00098	−0.00591	−164.93	−0.3377	0.018690	−13
−12	0.0004128	11.276	0.007	11.283	−2.882	0.436	−2.447	−0.00635	0.00103	−0.00532	−164.47	−0.3366	0.019754	−12
−11	0.0004362	11.301	0.008	11.309	−2.642	0.460	−2.182	−0.00582	0.00109	−0.00473	−164.01	−0.3356	0.020873	−11

Table 2 Thermodynamic Properties of Moist Air at Standard Atmospheric Pressure, 14.696 psi (29.921 in. Hg) (*Continued*)

Temp., °F	Humidity Ratio, lb_w/lb_{da}	Specific Volume, ft^3/lb_{da}			Specific Enthalpy, Btu/lb_{da}			Specific Entropy, $Btu/lb_{da}\cdot°F$			Condensed Water Specific Enthalpy, Btu/lb_w	Condensed Water Specific Entropy, $Btu/lb_w\cdot°F$	Condensed Water Vapor Pressure, in. Hg	Temp., °F
t	W_s	v_{da}	v_{as}	v_s	h_{da}	h_{as}	h_s	s_{da}	s_{as}	s_s	h_w	s_w	p_s	t
−10	0.0004608	11.326	0.008	11.335	−2.402	0.487	−1.915	−0.00528	0.00115	−0.00414	−163.55	−0.3346	0.022050	−10
−9	0.0004867	11.351	0.009	11.360	−2.162	0.514	−1.647	−0.00475	0.00121	−0.00354	−163.09	−0.3335	0.023289	−9
−8	0.0005139	11.377	0.009	11.386	−1.922	0.543	−1.378	−0.00422	0.00127	−0.00294	−162.63	−0.3325	0.024591	−8
−7	0.0005425	11.402	0.010	11.412	−1.681	0.574	−1.108	−0.00369	0.00134	−0.00234	−162.17	−0.3315	0.025959	−7
−6	0.0005726	11.427	0.010	11.438	−1.441	0.606	−0.835	−0.00316	0.00141	−0.00174	−161.70	−0.3305	0.027397	−6
−5	0.0006041	11.453	0.011	11.464	−1.201	0.640	−0.561	−0.00263	0.00149	−0.00114	−161.23	−0.3294	0.028907	−5
−4	0.0006373	11.478	0.012	11.490	−0.961	0.675	−0.286	−0.00210	0.00157	−0.00053	−160.77	−0.3284	0.030494	−4
−3	0.0006722	11.503	0.012	11.516	−0.721	0.712	−0.008	−0.00157	0.00165	0.00008	−160.30	−0.3274	0.032160	−3
−2	0.0007088	11.529	0.013	11.542	−0.480	0.751	0.271	−0.00105	0.00174	0.00069	−159.83	−0.3264	0.033909	−2
−1	0.0007472	11.554	0.014	11.568	−0.240	0.792	0.552	−0.00052	0.00183	0.00130	−159.36	−0.3253	0.035744	−1
0	0.0007875	11.579	0.015	11.594	0.0	0.835	0.835	0.00000	0.00192	0.00192	−158.89	−0.3243	0.037671	0
1	0.0008298	11.604	0.015	11.620	0.240	0.880	1.121	0.00052	0.00202	0.00254	−158.42	−0.3233	0.039694	1
2	0.0008742	11.630	0.016	11.646	0.480	0.928	1.408	0.00104	0.00212	0.00317	−157.95	−0.3223	0.041814	2
3	0.0009207	11.655	0.017	11.672	0.721	0.978	1.699	0.00156	0.00223	0.00380	−157.47	−0.3212	0.044037	3
4	0.0009695	11.680	0.018	11.699	0.961	1.030	1.991	0.00208	0.00235	0.00443	−157.00	−0.3202	0.046370	4
5	0.0010207	11.706	0.019	11.725	1.201	1.085	2.286	0.00260	0.00247	0.00506	−156.52	−0.3192	0.048814	5
6	0.0010743	11.731	0.020	11.751	1.441	1.143	2.584	0.00311	0.00259	0.00570	−156.05	−0.3182	0.051375	6
7	0.0011306	11.756	0.021	11.778	1.681	1.203	2.884	0.00363	0.00635	0.00272	−155.57	−0.3171	0.054060	7
8	0.0011895	11.782	0.022	11.804	1.922	1.266	3.188	0.00414	0.00286	0.00700	−155.09	−0.3161	0.056872	8
9	0.0012512	11.807	0.024	11.831	2.162	1.332	3.494	0.00466	0.00300	0.00766	−154.61	−0.3151	0.059819	9
10	0.0013158	11.832	0.025	11.857	2.402	1.402	3.804	0.00517	0.00315	0.00832	−154.13	−0.3141	0.062901	10
11	0.0013835	11.857	0.026	11.884	2.642	1.474	4.117	0.00568	0.00330	0.00898	−153.65	−0.3130	0.066131	11
12	0.0014544	11.883	0.028	11.910	2.882	1.550	4.433	0.00619	0.00347	0.00966	−153.17	−0.3120	0.069511	12
13	0.0015286	11.908	0.029	11.937	3.123	1.630	4.753	0.00670	0.00364	0.01033	−152.68	−0.3110	0.073049	13
14	0.0016062	11.933	0.031	11.964	3.363	1.714	5.077	0.00721	0.00381	0.01102	−152.20	−0.3100	0.076751	14
15	0.0016874	11.959	0.032	11.991	3.603	1.801	5.404	0.00771	0.00400	0.01171	−151.71	−0.3089	0.080623	15
16	0.0017724	11.984	0.034	12.018	3.843	1.892	5.736	0.00822	0.00419	0.01241	−151.22	−0.3079	0.084673	16
17	0.0018613	12.009	0.036	12.045	4.084	1.988	6.072	0.00872	0.00439	0.01312	−150.74	−0.3069	0.088907	17
18	0.0019543	12.035	0.038	12.072	4.324	2.088	6.412	0.00923	0.00460	0.01383	−150.25	−0.3059	0.093334	18
19	0.0020515	12.060	0.040	12.099	4.564	2.193	6.757	0.00973	0.00482	0.01455	−149.76	−0.3049	0.097962	19
20	0.0021531	12.085	0.042	12.127	4.804	2.303	7.107	0.01023	0.00505	0.01528	−149.27	−0.3038	0.102798	20
21	0.0022592	12.110	0.044	12.154	5.044	2.417	7.462	0.01073	0.00529	0.01602	−148.78	−0.3028	0.107849	21
22	0.0023703	12.136	0.046	12.182	5.285	2.537	7.822	0.01123	0.00554	0.01677	−148.28	−0.3018	0.113130	22
23	0.0024863	12.161	0.048	12.209	5.525	2.662	8.187	0.01173	0.00580	0.01753	−147.79	−0.3008	0.118645	23
24	0.0026073	12.186	0.051	12.237	5.765	2.793	8.558	0.01223	0.00607	0.01830	−147.30	−0.2997	0.124396	24
25	0.0027339	12.212	0.054	12.265	6.005	2.930	8.935	0.01272	0.00636	0.01908	−146.80	−0.2987	0.130413	25
26	0.0028660	12.237	0.056	12.293	6.246	3.073	9.318	0.01322	0.00665	0.01987	−146.30	−0.2977	0.136684	26
27	0.0030039	12.262	0.059	12.321	6.486	3.222	9.708	0.01371	0.00696	0.02067	−145.81	−0.2967	0.143233	27
28	0.0031480	12.287	0.062	12.349	6.726	3.378	10.104	0.01420	0.00728	0.02148	−145.31	−0.2956	0.150066	28
29	0.0032984	12.313	0.065	12.378	6.966	3.541	10.507	0.01470	0.00761	0.02231	−144.81	−0.2946	0.157198	29
30	0.0034552	12.338	0.068	12.406	7.206	3.711	10.917	0.01519	0.00796	0.02315	−144.31	−0.2936	0.164631	30
31	0.0036190	12.363	0.072	12.435	7.447	3.888	11.335	0.01568	0.00832	0.02400	−143.80	−0.2926	0.172390	31
32	0.0037895	12.389	0.075	12.464	7.687	4.073	11.760	0.01617	0.00870	0.02487	−143.30	−0.2915	0.180479	32
32*	0.003790	12.389	0.075	12.464	7.687	4.073	11.760	0.01617	0.00870	0.02487	0.02	0.0000	0.18050	32
33	0.003947	12.414	0.079	12.492	7.927	4.243	12.170	0.01665	0.00905	0.02570	1.03	0.0020	0.18791	33
34	0.004109	12.439	0.082	12.521	8.167	4.420	12.587	0.01714	0.00940	0.02655	2.04	0.0041	0.19559	34
35	0.004277	12.464	0.085	12.550	8.408	4.603	13.010	0.01763	0.00977	0.02740	3.05	0.0061	0.20356	35
36	0.004452	12.490	0.089	12.579	8.648	4.793	13.441	0.01811	0.01016	0.02827	4.05	0.0081	0.21181	36
37	0.004633	12.515	0.093	12.608	8.888	4.990	13.878	0.01860	0.01055	0.02915	5.06	0.0102	0.22035	37
38	0.004820	12.540	0.097	12.637	9.128	5.194	14.322	0.01908	0.01096	0.03004	6.06	0.0122	0.22920	38
39	0.005014	12.566	0.101	12.667	9.369	5.405	14.773	0.01956	0.01139	0.03095	7.07	0.0142	0.23835	39
40	0.005216	12.591	0.105	12.696	9.609	5.624	15.233	0.02004	0.01183	0.03187	8.07	0.0162	0.24784	40
41	0.005424	12.616	0.110	12.726	9.849	5.851	15.700	0.02052	0.01228	0.03281	9.08	0.0182	0.25765	41
42	0.005640	12.641	0.114	12.756	10.089	6.086	16.175	0.02100	0.01275	0.03375	10.08	0.0202	0.26781	42
43	0.005863	12.667	0.119	12.786	10.330	6.330	16.660	0.02148	0.01324	0.03472	11.09	0.0222	0.27831	43
44	0.006094	12.692	0.124	12.816	10.570	6.582	17.152	0.02196	0.01374	0.03570	12.09	0.0242	0.28918	44
45	0.006334	12.717	0.129	12.846	10.810	6.843	17.653	0.02244	0.01426	0.03669	13.09	0.0262	0.30042	45
46	0.006581	12.743	0.134	12.877	11.050	7.114	18.164	0.02291	0.01479	0.03770	14.10	0.0282	0.31206	46
47	0.006838	12.768	0.140	12.908	11.291	7.394	18.685	0.02339	0.01534	0.03873	15.10	0.0302	0.32408	47
48	0.007103	12.793	0.146	12.939	11.531	7.684	19.215	0.02386	0.01592	0.03978	16.10	0.0321	0.33651	48
49	0.007378	12.818	0.152	12.970	11.771	7.984	19.756	0.02433	0.01651	0.04084	17.10	0.0341	0.34937	49
50	0.007661	12.844	0.158	13.001	12.012	8.295	20.306	0.02480	0.01712	0.04192	18.11	0.0361	0.36264	50
51	0.007955	12.869	0.164	13.033	12.252	8.616	20.868	0.02528	0.01775	0.04302	19.11	0.0381	0.37636	51
52	0.008259	12.894	0.171	13.065	12.492	8.949	21.441	0.02575	0.01840	0.04415	20.11	0.0400	0.39054	52
53	0.008573	12.920	0.178	13.097	12.732	9.293	22.025	0.02622	0.01907	0.04529	21.11	0.0420	0.40518	53
54	0.008897	12.945	0.185	13.129	12.973	9.648	22.621	0.02668	0.01976	0.04645	22.11	0.0439	0.42030	54
55	0.009233	12.970	0.192	13.162	13.213	10.016	23.229	0.02715	0.02048	0.04763	23.11	0.0459	0.43592	55
56	0.009580	12.995	0.200	13.195	13.453	10.397	23.850	0.02762	0.02122	0.04884	24.11	0.0478	0.45205	56
57	0.009938	13.021	0.207	13.228	13.694	10.790	24.484	0.02808	0.02198	0.05006	25.11	0.0497	0.46870	57
58	0.010309	13.046	0.216	13.262	13.934	11.197	25.131	0.02855	0.02277	0.05132	26.11	0.0517	0.48589	58
59	0.010692	13.071	0.224	13.295	14.174	11.618	25.792	0.02901	0.02358	0.05259	27.11	0.0536	0.50363	59

*Extrapolated to represent metastable equilibrium with undercooled liquid.

Table 2 Thermodynamic Properties of Moist Air at Standard Atmospheric Pressure, 14.696 psi (29.921 in. Hg) (*Continued*)

Temp., °F	Humidity Ratio, lb_w/lb_{da}	Specific Volume, ft^3/lb_{da}			Specific Enthalpy, Btu/lb_{da}			Specific Entropy, $Btu/lb_{da} \cdot °F$			Condensed Water: Specific Enthalpy, Btu/lb_w	Condensed Water: Specific Entropy, $Btu/lb_w \cdot °F$	Condensed Water: Vapor Pressure, in. Hg	Temp., °F
t	W_s	v_{da}	v_{as}	v_s	h_{da}	h_{as}	h_s	s_{da}	s_{as}	s_s	h_w	s_w	p_s	t
60	0.011087	13.096	0.233	13.329	14.415	12.052	26.467	0.02947	0.02442	0.05389	28.11	0.0555	0.52193	60
61	0.011496	13.122	0.242	13.364	14.655	12.502	27.157	0.02994	0.02528	0.05522	29.12	0.0575	0.54082	61
62	0.011919	13.147	0.251	13.398	14.895	12.966	27.862	0.03040	0.02617	0.05657	30.11	0.0594	0.56032	62
63	0.012355	13.172	0.261	13.433	15.135	13.446	28.582	0.03086	0.02709	0.05795	31.11	0.0613	0.58041	63
64	0.012805	13.198	0.271	13.468	15.376	13.942	29.318	0.03132	0.02804	0.05936	32.11	0.0632	0.60113	64
65	0.013270	13.223	0.281	13.504	15.616	14.454	30.071	0.03178	0.02902	0.06080	33.11	0.0651	0.62252	65
66	0.013750	13.248	0.292	13.540	15.856	14.983	30.840	0.03223	0.03003	0.06226	34.11	0.0670	0.64454	66
67	0.014246	13.273	0.303	13.577	16.097	15.530	31.626	0.03269	0.03107	0.06376	35.11	0.0689	0.66725	67
68	0.014758	13.299	0.315	13.613	16.337	16.094	32.431	0.03315	0.03214	0.06529	36.11	0.0708	0.69065	68
69	0.015286	13.324	0.326	13.650	16.577	16.677	33.254	0.03360	0.03325	0.06685	37.11	0.0727	0.71479	69
70	0.015832	13.349	0.339	13.688	16.818	17.279	34.097	0.03406	0.03438	0.06844	38.11	0.0746	0.73966	70
71	0.016395	13.375	0.351	13.726	17.058	17.901	34.959	0.03451	0.03556	0.07007	39.11	0.0765	0.76567	71
72	0.016976	13.400	0.365	13.764	17.299	18.543	35.841	0.03496	0.03677	0.07173	40.11	0.0783	0.79167	72
73	0.017575	13.425	0.378	13.803	17.539	19.204	36.743	0.03541	0.03801	0.07343	41.11	0.0802	0.81882	73
74	0.018194	13.450	0.392	13.843	17.779	19.889	37.668	0.03586	0.03930	0.07516	42.11	0.0821	0.84684	74
75	0.018833	13.476	0.407	13.882	18.020	20.595	38.615	0.03631	0.04062	0.07694	43.11	0.0840	0.87567	75
76	0.019491	13.501	0.422	13.923	18.260	21.323	39.583	0.03676	0.04199	0.07875	44.10	0.0858	0.90533	76
77	0.020170	13.526	0.437	13.963	18.500	22.075	40.576	0.03721	0.04339	0.08060	45.10	0.0877	0.93589	77
78	0.020871	13.551	0.453	14.005	18.741	22.851	41.592	0.03766	0.04484	0.08250	46.10	0.0896	0.96733	78
79	0.021594	13.577	0.470	14.046	18.981	23.652	42.633	0.03811	0.04633	0.08444	47.10	0.0914	0.99970	79
80	0.022340	13.602	0.487	14.089	19.222	24.479	43.701	0.03855	0.04787	0.08642	48.10	0.0933	1.03302	80
81	0.023109	13.627	0.505	14.132	19.462	25.332	44.794	0.03900	0.04945	0.08844	49.10	0.0951	1.06728	81
82	0.023902	13.653	0.523	14.175	19.702	26.211	45.913	0.03944	0.05108	0.09052	50.10	0.0970	1.10252	82
83	0.024720	13.678	0.542	14.220	19.943	27.120	47.062	0.03988	0.05276	0.09264	51.09	0.0988	1.13882	83
84	0.025563	13.703	0.561	14.264	20.183	28.055	48.238	0.04033	0.05448	0.09481	52.09	0.1006	1.17608	84
85	0.026433	13.728	0.581	14.310	20.424	29.021	49.445	0.04077	0.05626	0.09703	53.09	0.1025	1.21445	85
86	0.027329	13.754	0.602	14.356	20.664	30.017	50.681	0.04121	0.05809	0.09930	54.09	0.1043	1.25388	86
87	0.028254	13.779	0.624	14.403	20.905	31.045	51.949	0.04165	0.05998	0.10163	55.09	0.1061	1.29443	87
88	0.029208	13.804	0.646	14.450	21.145	32.105	53.250	0.04209	0.06192	0.10401	56.09	0.1080	1.33613	88
89	0.030189	13.829	0.669	14.498	21.385	33.197	54.582	0.04253	0.06392	0.10645	57.09	0.1098	1.37893	89
90	0.031203	13.855	0.692	14.547	21.626	34.325	55.951	0.04297	0.06598	0.10895	58.08	0.1116	1.42298	90
91	0.032247	13.880	0.717	14.597	21.866	35.489	57.355	0.06810	0.04340	0.11150	59.08	0.1134	1.46824	91
92	0.033323	13.905	0.742	14.647	22.107	36.687	58.794	0.04384	0.07028	0.11412	60.08	0.1152	1.51471	92
93	0.034433	13.930	0.768	14.699	22.347	37.924	60.271	0.04427	0.07253	0.11680	61.08	0.1170	1.56248	93
94	0.035577	13.956	0.795	14.751	22.588	39.199	61.787	0.04471	0.07484	0.11955	62.08	0.1188	1.61154	94
95	0.036757	13.981	0.823	14.804	22.828	40.515	63.343	0.04514	0.07722	0.12237	63.08	0.1206	1.66196	95
96	0.037972	14.006	0.852	14.858	23.069	41.871	64.940	0.04558	0.07968	0.12525	64.07	0.1224	1.71372	96
97	0.039225	14.032	0.881	14.913	23.309	43.269	66.578	0.04601	0.08220	0.12821	65.07	0.1242	1.76685	97
98	0.040516	14.057	0.912	14.969	23.550	44.711	68.260	0.04644	0.08480	0.13124	66.07	0.1260	1.82141	98
99	0.041848	14.082	0.944	15.026	23.790	46.198	69.988	0.04687	0.08747	0.13434	67.07	0.1278	1.87745	99
100	0.043219	14.107	0.976	15.084	24.031	47.730	71.761	0.04730	0.09022	0.13752	68.07	0.1296	1.93492	100
101	0.044634	14.133	1.010	15.143	24.271	49.312	73.583	0.04773	0.09306	0.14079	69.07	0.1314	1.99396	101
102	0.046090	14.158	1.045	15.203	24.512	50.940	75.452	0.04816	0.09597	0.14413	70.06	0.1332	2.05447	102
103	0.047592	14.183	1.081	15.264	24.752	52.621	77.373	0.04859	0.09897	0.14756	71.06	0.1349	2.11661	103
104	0.049140	14.208	1.118	15.326	24.993	54.354	79.346	0.04901	0.10206	0.15108	72.06	0.1367	2.18037	104
105	0.050737	14.234	1.156	15.390	25.233	56.142	81.375	0.04944	0.10525	0.15469	73.06	0.1385	2.24581	105
106	0.052383	14.259	1.196	15.455	25.474	57.986	83.460	0.04987	0.10852	0.15839	74.06	0.1402	2.31297	106
107	0.054077	14.284	1.236	15.521	25.714	59.884	85.599	0.05029	0.11189	0.16218	75.06	0.1420	2.38173	107
108	0.055826	14.309	1.279	15.588	25.955	61.844	87.799	0.05071	0.11537	0.16608	76.05	0.1438	2.45232	108
109	0.057628	14.335	1.322	15.657	26.195	63.866	90.061	0.05114	0.11894	0.17008	77.05	0.1455	2.52473	109
110	0.059486	14.360	1.367	15.727	26.436	65.950	92.386	0.05156	0.12262	0.17418	78.05	0.1473	2.59891	110
111	0.061401	14.385	1.414	15.799	26.677	68.099	94.776	0.05198	0.12641	0.17839	79.05	0.1490	2.67500	111
112	0.063378	14.411	1.462	15.872	26.917	70.319	97.237	0.05240	0.13032	0.18272	80.05	0.1508	2.75310	112
113	0.065411	14.436	1.511	15.947	27.158	72.603	99.760	0.05282	0.13434	0.18716	81.05	0.1525	2.83291	113
114	0.067512	14.461	1.562	16.023	27.398	74.964	102.362	0.05324	0.13847	0.19172	82.04	0.1543	2.91491	114
115	0.069676	14.486	1.615	16.101	27.639	77.396	105.035	0.05366	0.14274	0.19640	83.04	0.1560	2.99883	115
116	0.071908	14.512	1.670	16.181	27.879	79.906	107.786	0.05408	0.14713	0.20121	84.04	0.1577	3.08488	116
117	0.074211	14.537	1.726	16.263	28.120	82.497	110.617	0.05450	0.15165	0.20615	85.04	0.1595	3.17305	117
118	0.076586	14.562	1.784	16.346	28.361	85.169	113.530	0.05492	0.15631	0.21122	86.04	0.1612	3.26335	118
119	0.079036	14.587	1.844	16.432	28.601	87.927	116.528	0.05533	0.16111	0.21644	87.04	0.1629	3.35586	119
120	0.081560	14.613	1.906	16.519	28.842	90.770	119.612	0.05575	0.16605	0.22180	88.04	0.1647	3.45052	120
121	0.084169	14.638	1.971	16.609	29.083	93.709	122.792	0.05616	0.17115	0.22731	89.04	0.1664	3.54764	121
122	0.086860	14.663	2.037	16.700	29.323	96.742	126.065	0.05658	0.17640	0.23298	90.03	0.1681	3.64704	122
123	0.089633	14.688	2.106	16.794	29.564	99.868	129.432	0.05699	0.18181	0.23880	91.03	0.1698	3.74871	123
124	0.092500	14.714	2.176	16.890	29.805	103.102	132.907	0.05740	0.18739	0.24480	92.03	0.1715	3.85298	124
125	0.095456	14.739	2.250	16.989	30.045	106.437	136.482	0.05781	0.19314	0.25096	93.03	0.1732	3.95961	125
126	0.098504	14.764	2.325	17.090	30.286	109.877	140.163	0.05823	0.19907	0.25729	94.03	0.1749	4.06863	126
127	0.101657	14.789	2.404	17.193	30.527	113.438	143.965	0.05864	0.20519	0.26382	95.03	0.1766	4.18046	127
128	0.104910	14.815	2.485	17.299	30.767	117.111	147.878	0.05905	0.21149	0.27054	96.03	0.1783	4.29477	128
129	0.108270	14.840	2.569	17.409	31.008	120.908	151.916	0.21800	0.21810	0.27745	97.03	0.1800	4.41181	129

Table 2 Thermodynamic Properties of Moist Air at Standard Atmospheric Pressure, 14.696 psi (29.921 in. Hg) (*Continued*)

Temp., °F	Humidity Ratio, lb_w/lb_{da}	Specific Volume, ft^3/lb_{da}			Specific Enthalpy, Btu/lb_{da}			Specific Entropy, $Btu/lb_{da}\cdot°F$			Condensed Water Specific Enthalpy, Btu/lb_w	Condensed Water Specific Entropy, $Btu/lb_w\cdot°F$	Condensed Water Vapor Pressure, in. Hg	Temp., °F
t	W_s	v_{da}	v_{as}	v_s	h_{da}	h_{as}	h_s	s_{da}	s_{as}	s_s	h_w	s_w	p_s	t
130	0.111738	14.865	2.655	17.520	31.249	124.828	156.076	0.05986	0.22470	0.28457	98.03	0.1817	4.53148	130
131	0.115322	14.891	2.745	17.635	31.489	128.880	160.370	0.06027	0.23162	0.29190	99.02	0.1834	4.65397	131
132	0.119023	14.916	2.837	17.753	31.730	133.066	164.796	0.06068	0.23876	0.29944	100.02	0.1851	4.77919	132
133	0.122855	14.941	2.934	17.875	31.971	137.403	169.374	0.06109	0.24615	0.30723	101.02	0.1868	4.90755	133
134	0.126804	14.966	3.033	17.999	32.212	141.873	174.084	0.06149	0.25375	0.31524	102.02	0.1885	5.03844	134
135	0.130895	14.992	3.136	18.127	32.452	146.504	178.957	0.06190	0.26161	0.32351	103.02	0.1902	5.17258	135
136	0.135124	15.017	3.242	18.259	32.693	151.294	183.987	0.06230	0.26973	0.33203	104.02	0.1919	5.30973	136
137	0.139494	15.042	3.352	18.394	32.934	156.245	189.179	0.06271	0.27811	0.34082	105.02	0.1935	5.44985	137
138	0.144019	15.067	3.467	18.534	33.175	161.374	194.548	0.06311	0.28707	0.35018	106.02	0.1952	5.59324	138
139	0.148696	15.093	3.585	18.678	33.415	166.677	200.092	0.06351	0.29602	0.35954	107.02	0.1969	5.73970	139
140	0.153538	15.118	3.708	18.825	33.656	172.168	205.824	0.06391	0.30498	0.36890	108.02	0.1985	5.88945	140
141	0.158643	15.143	3.835	18.978	33.897	177.857	211.754	0.06431	0.31456	0.37887	109.02	0.2002	6.04256	141
142	0.163748	15.168	3.967	19.135	34.138	183.754	217.892	0.06471	0.32446	0.38918	110.02	0.2019	6.19918	142
143	0.169122	15.194	4.103	19.297	34.379	189.855	224.233	0.06511	0.33470	0.39981	111.02	0.2035	6.35898	143
144	0.174694	15.219	4.245	19.464	34.620	196.183	230.802	0.06551	0.34530	0.41081	112.02	0.2052	6.52241	144
145	0.180467	15.244	4.392	19.637	34.860	202.740	237.600	0.06591	0.35626	0.42218	113.02	0.2068	6.68932	145
146	0.186460	15.269	4.545	19.815	35.101	209.550	244.651	0.06631	0.36764	0.43395	114.02	0.2085	6.86009	146
147	0.192668	15.295	4.704	19.999	35.342	216.607	251.949	0.06671	0.37941	0.44611	115.02	0.2101	7.03435	147
148	0.199110	15.320	4.869	20.189	35.583	223.932	259.514	0.06710	0.39160	0.45871	116.02	0.2118	7.21239	148
149	0.205792	15.345	5.040	20.385	35.824	231.533	267.356	0.06750	0.40424	0.47174	117.02	0.2134	7.39413	149
150	0.212730	15.370	5.218	20.589	36.064	239.426	275.490	0.06790	0.41735	0.48524	118.02	0.2151	7.57977	150
151	0.219945	15.396	5.404	20.799	36.305	247.638	283.943	0.06829	0.43096	0.49925	119.02	0.2167	7.76958	151
152	0.227429	15.421	5.596	21.017	36.546	256.158	292.705	0.06868	0.44507	0.51375	120.02	0.2184	7.96306	152
153	0.235218	15.446	5.797	21.243	36.787	265.028	301.816	0.06908	0.45973	0.52881	121.02	0.2200	8.16087	153
154	0.243309	15.471	6.005	21.477	37.028	274.245	311.273	0.06947	0.47494	0.54441	122.02	0.2216	8.36256	154
155	0.251738	15.497	6.223	21.720	37.269	283.849	321.118	0.06986	0.49077	0.56064	123.02	0.2233	8.56871	155
156	0.260512	15.522	6.450	21.972	37.510	293.849	331.359	0.07025	0.50723	0.57749	124.02	0.2249	8.77915	156
157	0.269644	15.547	6.686	22.233	37.751	304.261	342.012	0.07065	0.52434	0.59499	125.02	0.2265	8.99378	157
158	0.279166	15.572	6.933	22.505	37.992	315.120	353.112	0.07104	0.54217	0.61320	126.02	0.2281	9.21297	158
159	0.289101	15.598	7.190	22.788	38.233	326.452	364.685	0.07143	0.56074	0.63216	127.02	0.2297	9.43677	159
160	0.29945	15.623	7.459	23.082	38.474	338.263	376.737	0.07181	0.58007	0.65188	128.02	0.2314	9.6648	160
161	0.31027	15.648	7.740	23.388	38.715	350.610	389.325	0.07220	0.60025	0.67245	129.02	0.2330	9.8978	161
162	0.32156	15.673	8.034	23.707	38.956	363.501	402.457	0.07259	0.62128	0.69388	130.03	0.2346	10.1353	162
163	0.33336	15.699	8.341	24.040	39.197	376.979	416.175	0.07298	0.64325	0.71623	131.03	0.2362	10.3776	163
164	0.34572	15.724	8.664	24.388	39.438	391.095	430.533	0.07337	0.66622	0.73959	132.03	0.2378	10.6250	164
165	0.35865	15.749	9.001	24.750	39.679	405.865	445.544	0.07375	0.69022	0.76397	133.03	0.2394	10.8771	165
166	0.37220	15.774	9.355	25.129	39.920	421.352	461.271	0.07414	0.71535	0.78949	134.03	0.2410	11.1343	166
167	0.38639	15.800	9.726	25.526	40.161	437.578	477.739	0.07452	0.74165	0.81617	135.03	0.2426	11.3965	167
168	0.40131	15.825	10.117	25.942	40.402	454.630	495.032	0.07491	0.76925	0.84415	136.03	0.2442	11.6641	168
169	0.41698	15.850	10.527	26.377	40.643	472.554	513.197	0.07529	0.79821	0.87350	137.04	0.2458	11.9370	169
170	0.43343	15.875	10.959	26.834	40.884	491.372	532.256	0.07567	0.82858	0.90425	138.04	0.2474	12.2149	170
171	0.45079	15.901	11.414	27.315	41.125	511.231	552.356	0.07606	0.86058	0.93664	139.04	0.2490	12.4988	171
172	0.46905	15.926	11.894	27.820	41.366	532.138	573.504	0.07644	0.89423	0.97067	140.04	0.2506	12.7880	172
173	0.48829	15.951	12.400	28.352	41.607	554.160	595.767	0.07682	0.92962	1.00644	141.04	0.2521	13.0823	173
174	0.50867	15.976	12.937	28.913	41.848	577.489	619.337	0.07720	0.96707	1.04427	142.04	0.2537	13.3831	174
175	0.53019	16.002	13.504	29.505	42.089	602.139	644.229	0.07758	1.00657	1.08416	143.05	0.2553	13.6894	175
176	0.55294	16.027	14.103	30.130	42.331	628.197	670.528	0.07796	1.04828	1.12624	144.05	0.2569	14.0010	176
177	0.57710	16.052	14.741	30.793	42.572	655.876	698.448	0.07834	1.09253	1.17087	145.05	0.2585	14.3191	177
178	0.60274	16.078	15.418	31.496	42.813	685.260	728.073	0.07872	1.13943	1.21815	146.05	0.2600	14.6430	178
179	0.63002	16.103	16.139	32.242	43.054	716.524	759.579	0.07910	1.18927	1.26837	147.06	0.2616	14.9731	179
180	0.65911	16.128	16.909	33.037	43.295	749.871	793.166	0.07947	1.24236	1.32183	148.06	0.2632	15.3097	180
181	0.69012	16.153	17.730	33.883	43.536	785.426	828.962	0.07985	1.29888	1.37873	149.06	0.2647	15.6522	181
182	0.72331	16.178	18.609	34.787	43.778	823.487	867.265	0.08023	1.35932	1.43954	150.06	0.2663	16.0014	182
183	0.75885	16.204	19.551	35.755	44.019	864.259	908.278	0.08060	1.42396	1.50457	151.07	0.2679	16.3569	183
184	0.79703	16.229	20.564	36.793	44.260	908.061	952.321	0.08098	1.49332	1.57430	152.07	0.2694	16.7190	184
185	0.83817	16.254	21.656	37.910	44.501	955.261	999.763	0.08135	1.56797	1.64932	153.07	0.2710	17.0880	185
186	0.88251	16.280	22.834	39.113	44.742	1006.149	1050.892	0.08172	1.64834	1.73006	154.08	0.2725	17.4634	186
187	0.93057	16.305	24.111	40.416	44.984	1061.314	1106.298	0.08210	1.73534	1.81744	155.08	0.2741	17.8462	187
188	0.98272	16.330	25.498	41.828	45.225	1121.174	1166.399	0.08247	1.82963	1.91210	156.08	0.2756	18.2357	188
189	1.03951	16.355	27.010	43.365	45.466	1186.382	1231.848	0.08284	1.93221	2.01505	157.09	0.2772	18.6323	189
190	1.10154	16.381	28.661	45.042	45.707	1257.614	1303.321	0.08321	2.04412	2.12733	158.09	0.2787	19.0358	190
191	1.16965	16.406	30.476	46.882	45.949	1335.834	1381.783	0.08359	2.16684	2.25043	159.09	0.2803	19.4468	191
192	1.24471	16.431	32.477	48.908	46.190	1422.047	1468.238	0.08396	2.30193	2.38589	160.10	0.2818	19.8652	192
193	1.32788	16.456	34.695	51.151	46.431	1517.581	1564.013	0.08433	2.45144	2.53576	161.10	0.2834	20.2913	193
194	1.42029	16.481	37.161	53.642	46.673	1623.758	1670.430	0.08470	2.61738	2.70208	162.11	0.2849	20.7244	194
195	1.52396	16.507	39.928	56.435	46.914	1742.879	1789.793	0.08506	2.80332	2.88838	163.11	0.2864	21.1661	195
196	1.64070	16.532	43.046	59.578	47.155	1877.032	1924.188	0.08543	3.01244	3.09787	164.12	0.2880	21.6152	196
197	1.77299	16.557	46.580	63.137	47.397	2029.069	2076.466	0.08580	3.24914	3.33494	165.12	0.2895	22.0714	197
198	1.92472	16.583	50.636	67.218	47.638	2203.464	2251.102	0.08617	3.52030	3.60647	166.13	0.2910	22.5367	198
199	2.09975	16.608	55.316	71.923	47.879	2404.668	2452.547	0.08653	3.83275	3.91929	167.13	0.2926	23.0092	199
200	2.30454	16.633	60.793	77.426	48.121	2640.084	2688.205	0.08690	4.19787	4.28477	168.13	0.2941	23.4906	200

Table 3 Thermodynamic Properties of Water at Saturation

Temp., °F	Absolute Pressure		Specific Volume, ft³/lb$_w$			Specific Enthalpy, Btu/lb$_w$			Specific Entropy, Btu/lb$_w$·°F			Temp., °F
			Sat. Solid	Evap.	Sat. Vapor	Sat. Solid	Evap.	Sat. Vapor	Sat. Solid	Evap.	Sat. Vapor	
t	p, psi	p, in. Hg	v_i	v_{ig}	v_g	h_i	h_{ig}	h_g	s_i	s_{ig}	s_g	t
−80	0.000116	0.000236	0.01732	1953234	1953234	−193.50	1219.19	1025.69	−0.4067	3.2112	2.8045	−80
−79	0.000125	0.000254	0.01732	1814052	1814052	−193.11	1219.24	1026.13	−0.4056	3.2029	2.7972	−79
−78	0.000135	0.000275	0.01732	1685445	1685445	−192.71	1219.28	1026.57	−0.4046	3.1946	2.7900	−78
−77	0.000145	0.000296	0.01732	1566663	1566663	−192.31	1219.33	1027.02	−0.4036	3.1964	2.7828	−77
−76	0.000157	0.000319	0.01732	1456752	1456752	−191.92	1219.38	1027.46	−0.4025	3.1782	2.7757	−76
−75	0.000169	0.000344	0.01733	1355059	1355059	−191.52	1219.42	1027.90	−0.4015	3.1701	2.7685	−75
−74	0.000182	0.000371	0.01733	1260977	1260977	−191.12	1219.47	1028.34	−0.4005	3.1619	2.7615	−74
−73	0.000196	0.000399	0.01733	1173848	1173848	−190.72	1219.51	1028.79	−0.3994	3.1539	2.7544	−73
−72	0.000211	0.000430	0.01733	1093149	1093149	−190.32	1219.55	1029.23	−0.3984	3.1459	2.7475	−72
−71	0.000227	0.000463	0.01733	1018381	1018381	−189.92	1219.59	1029.67	−0.3974	3.1379	2.7405	−71
−70	0.000245	0.000498	0.01733	949067	949067	−189.52	1219.63	1030.11	−0.3963	3.1299	2.7336	−70
−69	0.000263	0.000536	0.01733	884803	884803	−189.11	1219.67	1030.55	−0.3953	3.1220	2.7267	−69
−68	0.000283	0.000576	0.01733	825187	825187	−188.71	1219.71	1031.00	−0.3943	3.1141	2.7199	−68
−67	0.000304	0.000619	0.01734	769864	769864	−188.30	1219.74	1031.44	−0.3932	3.1063	2.7131	−67
−66	0.000326	0.000664	0.01734	718508	718508	−187.90	1219.78	1031.88	−0.3922	3.0985	2.7063	−66
−65	0.000350	0.000714	0.01734	670800	670800	−187.49	1219.82	1032.32	−0.3912	3.0907	2.6996	−65
−64	0.000376	0.000766	0.01734	626503	626503	−187.08	1219.85	1032.77	−0.3901	3.0830	2.6929	−64
−63	0.000404	0.000822	0.01734	585316	585316	−186.67	1219.88	1033.21	−0.3891	3.0753	2.6862	−63
−62	0.000433	0.000882	0.01734	548041	547041	−186.26	1219.91	1033.65	−0.3881	3.0677	2.6730	−62
−61	0.000464	0.000945	0.01734	511446	511446	−185.85	1219.95	1034.09	−0.3870	3.0601	2.6730	−61
−60	0.000498	0.001013	0.01734	478317	478317	−185.44	1219.98	1034.54	−0.3860	3.0525	2.6665	−60
−59	0.000533	0.001086	0.01735	447495	447495	−185.03	1220.01	1034.98	−0.3850	3.0449	2.6600	−59
−58	0.000571	0.001163	0.01735	418803	418803	−184.61	1220.03	1035.42	−0.3839	3.0374	2.6535	−58
−57	0.000612	0.001246	0.01735	392068	392068	−184.20	1220.06	1035.86	−0.3829	3.0299	2.6470	−57
−56	0.000655	0.001333	0.01735	367172	367172	−183.78	1220.09	1036.30	−0.3819	3.0225	2.6406	−56
−55	0.000701	0.001427	0.01735	343970	343970	−183.37	1220.11	1036.75	−0.3808	3.0151	2.6342	−55
−54	0.000750	0.001526	0.01735	322336	322336	−182.95	1220.14	1037.19	−0.3798	3.0077	2.6279	−54
−53	0.000802	0.001632	0.01735	302157	302157	−182.53	1220.16	1037.63	−0.3788	3.0004	2.6216	−53
−52	0.000857	0.001745	0.01735	283335	283335	−182.11	1220.18	1038.07	−0.3778	2.9931	2.6153	−52
−51	0.000916	0.001865	0.01736	265773	265773	−181.69	1220.21	1038.52	−0.3767	2.9858	2.6091	−51
−50	0.000979	0.001992	0.01736	249381	249381	−181.27	1220.23	1038.96	−0.3757	2.9786	2.6029	−50
−49	0.001045	0.002128	0.01736	234067	234067	−180.85	1220.25	1039.40	−0.3747	2.9714	2.5967	−49
−48	0.001116	0.002272	0.01736	219766	219766	−180.42	1220.26	1039.84	−0.3736	2.9642	2.5906	−48
−47	0.001191	0.002425	0.01736	206398	206398	−180.00	1220.28	1040.28	−0.3726	2.9570	2.5844	−47
−46	0.001271	0.002587	0.01736	193909	193909	−179.57	1220.30	1040.73	−0.3716	2.9499	2.5784	−46
−45	0.001355	0.002760	0.01736	182231	182231	−179.14	1220.31	1041.17	−0.3705	2.9429	2.5723	−45
−44	0.001445	0.002943	0.01736	171304	171304	−178.72	1220.33	1041.61	−0.3695	2.9358	2.5663	−44
−43	0.001541	0.003137	0.01737	161084	161084	−178.29	1220.34	1042.05	−0.3685	2.9288	2.5603	−43
−42	0.001642	0.003343	0.01737	151518	151518	−177.86	1220.36	1042.50	−0.3675	2.9218	2.5544	−42
−41	0.001749	0.003562	0.01737	142566	142566	−177.43	1220.37	1042.94	−0.3664	2.9149	2.5485	−41
−40	0.001863	0.003793	0.01737	134176	134176	−177.00	1220.38	1043.38	−0.3654	2.9080	2.5426	−40
−39	0.001984	0.004039	0.01737	126322	126322	−176.57	1220.39	1043.82	−0.3644	2.9011	2.5367	−39
−38	0.002111	0.004299	0.01737	118959	118959	−176.13	1220.40	1044.27	−0.3633	2.8942	2.5309	−38
−37	0.002247	0.004574	0.01737	112058	112058	−175.70	1220.40	1044.71	−0.3623	2.8874	2.5251	−37
−36	0.002390	0.004866	0.01738	105592	105592	−175.26	1220.41	1045.15	−0.3613	2.8806	2.5193	−36
−35	0.002542	0.005175	0.01738	99522	99522	−174.83	1220.42	1045.59	−0.3603	2.8738	2.5136	−35
−34	0.002702	0.005502	0.01738	93828	93828	−174.39	1220.42	1046.03	−0.3592	2.8671	2.5078	−34
−33	0.002872	0.005848	0.01738	88489	88489	−173.95	1220.43	1046.48	−0.3582	2.8604	2.5022	−33
−32	0.003052	0.006213	0.01738	83474	83474	−173.51	1220.43	1046.92	−0.3572	2.8537	2.4965	−32
−31	0.003242	0.006600	0.01738	78763	78763	−173.07	1220.43	1047.36	−0.3561	2.8470	2.4909	−31
−30	0.003443	0.007009	0.01738	74341	74341	−172.63	1220.43	1047.80	−0.3551	2.8404	2.4853	−30
−29	0.003655	0.007441	0.01738	70187	70187	−172.19	1220.43	1048.25	−0.3541	2.8338	2.4797	−29
−28	0.003879	0.007898	0.01739	66282	66282	−171.74	1220.43	1048.69	−0.3531	2.8272	2.4742	−28
−27	0.004116	0.008380	0.01739	62613	62613	−171.30	1220.43	1049.13	−0.3520	2.8207	2.4687	−27
−26	0.004366	0.008890	0.01739	59161	59161	−170.86	1220.43	1049.57	−0.3510	2.8142	2.4632	−26
−25	0.004630	0.009428	0.01739	55915	55915	−170.41	1220.42	1050.01	−0.3500	2.8077	2.4577	−25
−24	0.004909	0.009995	0.01739	52861	52861	−169.96	1220.42	1050.46	−0.3489	2.8013	2.4523	−24
−23	0.005203	0.010594	0.01739	49986	49986	−169.51	1220.41	1050.90	−0.3479	2.7948	2.4469	−23
−22	0.005514	0.011226	0.01739	47281	47281	−169.07	1220.41	1051.34	−0.3469	2.7884	2.4415	−22
−21	0.005841	0.011892	0.01740	44733	44733	−168.62	1220.40	1051.78	−0.3459	2.7820	2.4362	−21
−20	0.006186	0.012595	0.01740	42333	42333	−168.16	1220.39	1052.22	−0.3448	2.7757	2.4309	−20
−19	0.006550	0.013336	0.01740	40073	40073	−167.71	1220.38	1052.67	−0.3438	2.7694	2.4256	−19
−18	0.006933	0.014117	0.01740	37943	37943	−167.26	1220.37	1053.11	−0.3428	2.7631	2.4203	−18
−17	0.007337	0.014939	0.01740	35934	35934	−166.81	1220.36	1053.55	−0.3418	2.7568	2.4151	−17
−16	0.007763	0.015806	0.01740	34041	34041	−166.35	1220.34	1053.99	−0.3407	2.7506	2.4098	−16
−15	0.008211	0.016718	0.01740	32256	32256	−165.90	1220.33	1054.43	−0.3397	2.7444	2.4046	−15
−14	0.008683	0.017678	0.01741	30572	30572	−165.44	1220.31	1054.87	−0.3387	2.7382	2.3995	−14

Table 3 Thermodynamic Properties of Water at Saturation (*Continued*)

Temp., °F	Absolute Pressure		Specific Volume, ft³/lb$_w$			Specific Enthalpy, Btu/lb$_w$			Specific Entropy, Btu/lb$_w$·°F			Temp., °F
			Sat. Solid/Liq.	Evap.	Sat. Vapor	Sat. Solid/Liq.	Evap.	Sat. Vapor	Sat. Solid/Liq.	Evap.	Sat. Vapor	
t	p, psi	p, in. Hg	v_i/v_f	v_{ig}	v_g	h_i/h_f	h_{ig}	h_g	s_i/s_f	s_{ig}	s_g	t
−13	0.009179	0.018689	0.01741	28983	28983	−164.98	1220.30	1055.32	−0.3377	2.7320	2.3943	−13
−12	0.009702	0.019753	0.01741	27483	27483	−164.52	1220.28	1055.76	−0.3366	2.7259	2.3892	−12
−11	0.010252	0.020873	0.01741	26067	26067	−164.06	1220.26	1056.20	−0.3356	2.7197	2.3841	−11
−10	0.010830	0.022050	0.01741	24730	24730	−163.60	1220.24	1056.64	−0.3346	2.7136	2.3791	−10
−9	0.011438	0.023288	0.01741	23467	23467	−163.14	1220.22	1057.08	−0.3335	2.7076	2.3740	−9
−8	0.012077	0.024590	0.01741	22274	22274	−162.68	1220.20	1057.53	−0.3325	2.7015	2.3690	−8
−7	0.012749	0.025958	0.01742	21147	21147	−162.21	1220.18	1057.97	−0.3315	2.6955	2.3640	−7
−6	0.013456	0.027396	0.01742	20081	20081	−161.75	1220.16	1058.41	−0.3305	2.6895	2.3591	−6
−5	0.014197	0.028906	0.01742	19074	19074	−161.28	1220.13	1058.85	−0.3294	2.6836	2.3541	−5
−4	0.014977	0.030493	0.01742	18121	18121	−160.82	1220.11	1059.29	−0.3284	2.6776	2.3492	−4
−3	0.015795	0.032159	0.01742	17220	17220	−160.35	1220.08	1059.73	−0.3274	2.6717	2.3443	−3
−2	0.016654	0.033908	0.01742	16367	16367	−159.88	1220.05	1060.17	−0.3264	2.6658	2.3394	−2
−1	0.017556	0.035744	0.01742	15561	15561	−159.41	1220.02	1060.62	−0.3253	2.6599	2.3346	−1
0	0.018502	0.037671	0.01743	14797	14797	−158.94	1220.00	1061.06	−0.3243	2.6541	2.3298	0
1	0.019495	0.039693	0.01743	14073	14073	−158.47	1219.96	1061.50	−0.3233	2.6482	2.3249	1
2	0.020537	0.041813	0.01743	13388	13388	−157.99	1219.93	1061.94	−0.3223	2.6424	2.3202	2
3	0.021629	0.044037	0.01743	12740	12740	−157.52	1219.90	1062.38	−0.3212	2.6367	2.3154	3
4	0.022774	0.046369	0.01743	12125	12125	−157.05	1219.87	1062.82	−0.3202	2.6309	2.3107	4
5	0.023975	0.048813	0.01743	11543	11543	−156.57	1219.83	1063.26	−0.3192	2.6252	2.3060	5
6	0.025233	0.051375	0.01743	10991	10991	−156.09	1219.80	1063.70	−0.3182	2.6194	2.3013	6
7	0.026552	0.054059	0.01744	10468	10468	−155.62	1219.76	1064.14	−0.3171	2.6138	2.2966	7
8	0.027933	0.056872	0.01744	9971	9971	−155.14	1219.72	1064.58	−0.3161	2.6081	2.2920	8
9	0.029379	0.059817	0.01744	9500	9500	−154.66	1219.68	1065.03	−0.3151	2.6024	2.2873	9
10	0.030894	0.062901	0.01744	9054	9054	−154.18	1219.64	1065.47	−0.3141	2.5968	2.2827	10
11	0.032480	0.066131	0.01744	8630	8630	−153.70	1219.60	1065.91	−0.3130	2.5912	2.2782	11
12	0.034140	0.069511	0.01744	8228	8228	−153.21	1219.56	1066.35	−0.3120	2.5856	2.2736	12
13	0.035878	0.073047	0.01745	7846	7846	−152.73	1219.52	1066.79	−0.3110	2.5801	2.2691	13
14	0.037696	0.076748	0.01745	7483	7483	−152.24	1219.47	1067.23	−0.3100	2.5745	2.2645	14
15	0.039597	0.080621	0.01745	7139	7139	−151.76	1219.43	1067.67	−0.3089	2.5690	2.2600	15
16	0.041586	0.084671	0.01745	6811	6811	−151.27	1219.38	1068.11	−0.3079	2.5635	2.2556	16
17	0.043666	0.088905	0.01745	6501	6501	−150.78	1219.33	1068.55	−0.3069	2.5580	2.2511	17
18	0.045841	0.093332	0.01745	6205	6205	−150.30	1219.28	1068.99	−0.3059	2.5526	2.2467	18
19	0.048113	0.097960	0.01745	5924	5924	−149.81	1219.23	1069.43	−0.3049	2.5471	2.2423	19
20	0.050489	0.102796	0.01746	5657	5657	−149.32	1219.18	1069.87	−0.3038	2.5417	2.2379	20
21	0.052970	0.107849	0.01746	5404	5404	−148.82	1219.13	1070.31	−0.3028	2.5363	2.2335	21
22	0.055563	0.113128	0.01746	5162	5162	−148.33	1219.08	1070.75	−0.3018	2.5309	2.2292	22
23	0.058271	0.118641	0.01746	4932	4932	−147.84	1219.02	1071.19	−0.3008	2.5256	2.2248	23
24	0.061099	0.124398	0.01746	4714	4714	−147.34	1218.97	1071.63	−0.2997	2.5203	2.2205	24
25	0.064051	0.130408	0.01746	4506	4506	−146.85	1218.91	1072.07	−0.2987	2.5149	2.2162	25
26	0.067133	0.136684	0.01747	4308	4308	−146.35	1218.85	1072.50	−0.2977	2.5096	2.2119	26
27	0.070349	0.143233	0.01747	4119	4119	−145.85	1218.80	1072.94	−0.2967	2.5044	2.2077	27
28	0.073706	0.150066	0.01747	3940	3940	−145.35	1218.74	1073.38	−0.2956	2.4991	2.2035	28
29	0.077207	0.157195	0.01747	3769	3769	−144.85	1218.68	1073.82	−0.2946	2.4939	2.1992	29
30	0.080860	0.164632	0.01747	3606	3606	−144.35	1218.61	1074.26	−0.2936	2.4886	2.1951	30
31	0.084669	0.172387	0.01747	3450	3450	−143.85	1218.55	1074.70	−0.2926	2.4834	2.1909	31
32	0.088640	0.180474	0.01747	3302	3302	−143.35	1218.49	1075.14	−0.2915	2.4783	2.1867	32
32*	0.08865	0.18049	0.01602	3302.07	3302.09	−0.02	1075.15	1075.14	0.0000	2.1867	2.1867	32
33	0.09229	0.18791	0.01602	3178.15	3178.16	0.99	1074.59	1075.58	0.0020	2.1811	2.1832	33
34	0.09607	0.19559	0.01602	3059.47	3059.49	2.00	1074.02	1076.01	0.0041	2.1756	2.1796	34
35	0.09998	0.20355	0.01602	2945.66	2945.68	3.00	1073.45	1076.45	0.0061	2.1700	2.1761	35
36	0.10403	0.21180	0.01602	2836.60	2836.61	4.01	1072.88	1076.89	0.0081	2.1645	2.1726	36
37	0.10822	0.22035	0.01602	2732.13	2732.15	5.02	1072.32	1077.33	0.0102	2.1590	2.1692	37
38	0.11257	0.22919	0.01602	2631.88	2631.89	6.02	1071.75	1077.77	0.0122	2.1535	2.1657	38
39	0.11707	0.23835	0.01602	2535.86	2535.88	7.03	1071.18	1078.21	0.0142	2.1481	2.1623	39
40	0.12172	0.24783	0.01602	2443.67	2443.69	8.03	1070.62	1078.65	0.0162	2.1426	2.1589	40
41	0.12654	0.25765	0.01602	2355.22	2355.24	9.04	1070.05	1079.09	0.0182	2.1372	2.1554	41
42	0.13153	0.26780	0.01602	2270.42	2270.43	10.04	1069.48	1079.52	0.0202	2.1318	2.1521	42
43	0.13669	0.27831	0.01602	2189.02	2189.04	11.04	1068.92	1079.96	0.0222	2.1265	2.1487	43
44	0.14203	0.28918	0.01602	2110.92	2110.94	12.05	1068.35	1080.40	0.0242	2.1211	2.1454	44
45	0.14755	0.30042	0.01602	2035.91	2035.92	13.05	1067.79	1080.84	0.0262	2.1158	2.1420	45
46	0.15326	0.31205	0.01602	1963.85	1963.87	14.05	1067.22	1081.28	0.0282	2.1105	2.1387	46
47	0.15917	0.32407	0.01602	1894.71	1894.73	15.06	1066.66	1081.71	0.0302	2.1052	2.1354	47
48	0.16527	0.33650	0.01602	1828.28	1828.30	16.06	1066.09	1082.15	0.0321	2.1000	2.1321	48
49	0.17158	0.34935	0.01602	1764.44	1764.46	17.06	1065.53	1082.59	0.0341	2.0947	2.1288	49
50	0.17811	0.36263	0.01602	1703.18	1703.20	18.06	1064.96	1083.03	0.0361	2.0895	2.1256	50
51	0.18484	0.37635	0.01602	1644.25	1644.26	19.06	1064.40	1083.46	0.0381	2.0843	2.1224	51
52	0.19181	0.39053	0.01603	1587.64	1587.65	20.07	1063.83	1083.90	0.0400	2.0791	2.1191	52

*Extrapolated to represent metastable equilibrium with undercooled liquid.

Table 3 Thermodynamic Properties of Water at Saturation (*Continued*)

Temp., °F	Absolute Pressure		Specific Volume, ft³/lb$_w$			Specific Enthalpy, Btu/lb$_w$			Specific Entropy, Btu/lb$_w$·°F			Temp., °F
			Sat. Liquid	Evap.	Sat. Vapor	Sat. Liquid	Evap.	Sat. Vapor	Sat. Liquid	Evap.	Sat. Vapor	
t	p, psi	p, in. Hg	v_f	v_{fg}	v_g	h_f	h_{fg}	h_g	s_f	s_{fg}	s_g	t
53	0.19900	0.40516	0.01603	1533.22	1533.24	21.07	1063.27	1084.34	0.0420	2.0740	2.1159	53
54	0.20643	0.42029	0.01603	1480.89	1480.91	22.07	1062.71	1084.77	0.0439	2.0689	2.1128	54
55	0.21410	0.43591	0.01603	1430.61	1430.62	23.07	1062.14	1085.21	0.0459	2.0637	2.1096	55
56	0.22202	0.45204	0.01603	1382.19	1382.21	24.07	1061.58	1085.65	0.0478	2.0586	2.1064	56
57	0.23020	0.46869	0.01603	1335.65	1335.67	25.07	1061.01	1086.08	0.0497	2.0536	2.1033	57
58	0.23864	0.48588	0.01603	1290.85	1290.87	26.07	1060.45	1086.52	0.0517	2.0485	2.1002	58
59	0.24735	0.50362	0.01603	1247.76	1247.78	27.07	1059.89	1086.96	0.0536	2.0435	2.0971	59
60	0.25635	0.52192	0.01604	1206.30	1206.32	28.07	1059.32	1087.39	0.0555	2.0385	2.0940	60
61	0.26562	0.54081	0.01604	1166.38	1166.40	29.07	1058.76	1087.83	0.0575	2.0334	2.0909	61
62	0.27519	0.56029	0.01604	1127.93	1127.95	30.07	1058.19	1088.27	0.0594	2.0285	2.0878	62
63	0.28506	0.58039	0.01604	1090.94	1090.96	31.07	1057.63	1088.70	0.0613	2.0235	2.0848	63
64	0.29524	0.60112	0.01604	1055.32	1055.33	32.07	1057.07	1089.14	0.0632	2.0186	2.0818	64
65	0.30574	0.62249	0.01604	1020.98	1021.00	33.07	1056.50	1089.57	0.0651	2.0136	2.0787	65
66	0.31656	0.64452	0.01604	987.95	987.97	34.07	1055.94	1090.01	0.0670	2.0087	2.0758	66
67	0.32772	0.66724	0.01605	956.11	956.12	35.07	1055.37	1090.44	0.0689	2.0039	2.0728	67
68	0.33921	0.69065	0.01605	925.44	925.45	36.07	1054.81	1090.88	0.0708	1.9990	2.0698	68
69	0.35107	0.71478	0.01605	895.86	895.87	37.07	1054.24	1091.31	0.0727	1.9941	2.0668	69
70	0.36328	0.73964	0.01605	867.34	867.36	38.07	1053.68	1091.75	0.0746	1.9893	2.0639	70
71	0.37586	0.76526	0.01605	839.87	839.88	39.07	1053.11	1092.18	0.0765	1.9845	2.0610	71
72	0.38882	0.79164	0.01606	813.37	813.39	40.07	1052.55	1092.61	0.0783	1.9797	2.0580	72
73	0.40217	0.81883	0.01606	787.85	787.87	41.07	1051.98	1093.05	0.0802	1.9749	2.0552	73
74	0.41592	0.84682	0.01606	763.19	763.21	42.06	1051.42	1093.48	0.0821	1.9702	2.0523	74
75	0.43008	0.87564	0.01606	739.42	739.44	43.06	1050.85	1093.92	0.0840	1.9654	2.0494	75
76	0.44465	0.90532	0.01606	716.51	726.53	44.06	1050.29	1094.35	0.0858	1.9607	2.0465	76
77	0.45966	0.93587	0.01607	694.38	699.80	45.06	1049.72	1094.78	0.0877	1.9560	2.0437	77
78	0.47510	0.96732	0.01607	673.05	673.06	46.06	1049.16	1095.22	0.0896	1.9513	2.0409	78
79	0.49100	0.99968	0.01607	652.44	652.46	47.06	1048.59	1095.65	0.0914	1.9466	2.0380	79
80	0.50736	1.03298	0.01607	632.54	632.56	48.06	1048.03	1096.08	0.0933	1.9420	2.0352	80
81	0.52419	1.06725	0.01608	613.35	613.37	49.06	1047.46	1096.51	0.0951	1.9373	2.0324	81
82	0.54150	1.10250	0.01608	594.82	594.84	50.05	1046.89	1096.95	0.0970	1.9327	2.0297	82
83	0.55931	1.13877	0.01608	576.90	576.92	51.05	1046.33	1097.38	0.0988	1.9281	2.0269	83
84	0.57763	1.17606	0.01608	559.63	559.65	52.05	1045.76	1097.81	0.1006	1.9235	2.0242	84
85	0.59647	1.21442	0.01609	542.93	542.94	53.05	1045.19	1098.24	0.1025	1.9189	2.0214	85
86	0.61584	1.25385	0.01609	526.80	526.81	54.05	1044.63	1098.67	0.1043	1.9144	2.0187	86
87	0.63575	1.29440	0.01609	511.21	511.22	55.05	1044.06	1099.11	0.1061	1.9098	2.0160	87
88	0.65622	1.33608	0.01609	496.14	496.15	56.05	1043.49	1099.54	0.1080	1.9053	2.0133	88
89	0.67726	1.37892	0.01610	481.60	481.61	57.04	1042.92	1099.97	0.1098	1.9008	2.0106	89
90	0.69889	1.42295	0.01610	467.52	467.53	58.04	1042.36	1100.40	0.1116	1.8963	2.0079	90
91	0.72111	1.46820	0.01610	453.91	453.93	59.04	1041.79	1100.83	0.1134	1.8918	2.0053	91
92	0.74394	1.51468	0.01611	440.76	440.78	60.04	1041.22	1101.26	0.1152	1.8874	2.0026	92
93	0.76740	1.56244	0.01611	428.04	428.06	61.04	1040.65	1101.69	0.1170	1.8829	2.0000	93
94	0.79150	1.61151	0.01611	415.74	415.76	62.04	1040.08	1102.12	0.1188	1.8785	1.9973	94
95	0.81625	1.66189	0.01612	403.84	403.86	63.03	1039.51	1102.55	0.1206	1.8741	1.9947	95
96	0.84166	1.71364	0.01612	392.33	392.34	64.03	1038.95	1102.98	0.1224	1.8697	1.9921	96
97	0.86776	1.76678	0.01612	381.20	381.21	65.03	1038.38	1103.41	0.1242	1.8653	1.9895	97
98	0.89456	1.82134	0.01612	370.42	370.44	66.03	1037.81	1103.84	0.1260	1.8610	1.9870	98
99	0.92207	1.87736	0.01613	359.99	360.01	67.03	1037.24	1104.26	0.1278	1.8566	1.9844	99
100	0.95031	1.93485	0.01613	349.91	349.92	68.03	1036.67	1104.69	0.1296	1.8523	1.9819	100
101	0.97930	1.99387	0.01613	340.14	340.15	69.03	1036.10	1105.12	0.1314	1.8479	1.9793	101
102	1.00904	2.05443	0.01614	330.69	330.71	70.02	1035.53	1105.55	0.1332	1.8436	1.9768	102
103	1.03956	2.11667	0.01614	321.53	321.55	71.02	1034.95	1105.98	0.1349	1.8393	1.9743	103
104	1.07088	2.18034	0.01614	312.67	312.69	72.02	1034.38	1106.40	0.1367	1.8351	1.9718	104
105	1.10301	2.24575	0.01615	304.08	304.10	73.02	1033.81	1106.83	0.1385	1.8308	1.9693	105
106	1.13597	2.31285	0.01615	295.76	295.77	74.02	1033.24	1107.26	0.1402	1.8266	1.9668	106
107	1.16977	2.38168	0.01616	287.71	287.73	75.01	1032.67	1107.68	0.1420	1.8223	1.9643	107
108	1.20444	2.45226	0.01616	279.91	279.92	76.01	1032.10	1108.11	0.1438	1.8181	1.9619	108
109	1.23999	2.52464	0.01616	272.34	272.36	77.01	1031.52	1108.54	0.1455	1.8139	1.9594	109
110	1.27644	2.59885	0.01617	265.02	265.03	78.01	1030.95	1108.96	0.1473	1.8097	1.9570	110
111	1.31381	2.67494	0.01617	257.91	257.93	79.01	1030.38	1109.39	0.1490	1.8055	1.9546	111
112	1.35212	2.75293	0.01617	251.02	251.04	80.01	1029.80	1109.81	0.1508	1.8014	1.9521	112
113	1.39138	2.83288	0.01618	244.36	244.38	81.01	1029.23	1110.24	0.1525	1.7972	1.9497	113
114	1.43162	2.91481	0.01618	237.89	237.90	82.00	1028.66	1110.66	0.1543	1.7931	1.9474	114
115	1.47286	2.99878	0.01619	231.62	231.63	83.00	1028.08	1111.09	0.1560	1.7890	1.9450	115
116	1.51512	3.08481	0.01619	225.53	225.55	84.00	1027.51	1111.51	0.1577	1.7849	1.9426	116
117	1.55842	3.17296	0.01619	219.63	219.65	85.00	1026.93	1111.93	0.1595	1.7808	1.9402	117
118	1.60277	3.26327	0.01620	213.91	213.93	86.00	1026.36	1112.36	0.1612	1.7767	1.9379	118
119	1.64820	3.35577	0.01620	208.36	208.37	87.00	1025.78	1112.78	0.1629	1.7726	1.9356	119
120	1.69474	3.45052	0.01620	202.98	202.99	88.00	1025.20	1113.20	0.1647	1.7686	1.9332	120

Table 3 Thermodynamic Properties of Water at Saturation (*Continued*)

Temp., °F	Absolute Pressure		Specific Volume, ft³/lb$_w$			Specific Enthalpy, Btu/lb$_w$			Specific Entropy, Btu/lb$_w$·°F			Temp., °F
			Sat. Liquid	Evap.	Sat. Vapor	Sat. Liquid	Evap.	Sat. Vapor	Sat. Liquid	Evap.	Sat. Vapor	
t	p, psi	p, in. Hg	v_f	v_{fg}	v_g	h_f	h_{fg}	h_g	s_f	s_{fg}	s_g	t
121	1.74240	3.54755	0.01621	197.76	197.76	89.00	1024.63	1113.62	0.1664	1.7645	1.9309	121
122	1.79117	3.64691	0.01621	192.69	192.69	90.00	1024.05	1114.05	0.1681	1.7605	1.9286	122
123	1.84117	3.74863	0.01622	187.78	187.78	90.99	1023.47	1114.47	0.1698	1.7565	1.9263	123
124	1.89233	3.85282	0.01622	182.98	182.99	91.99	1022.90	1114.89	0.1715	1.7525	1.9240	124
125	1.94470	3.95945	0.01623	178.34	178.36	92.99	1022.32	1115.31	0.1732	1.7485	1.9217	125
126	1.99831	4.06860	0.01623	173.85	173.86	93.99	1021.74	1115.73	0.1749	1.7445	1.9195	126
127	2.05318	4.18032	0.01623	169.47	169.49	94.99	1021.16	1116.15	0.1766	1.7406	1.9172	127
128	2.10934	4.29465	0.01624	165.23	165.25	95.99	1020.58	1116.57	0.1783	1.7366	1.9150	128
129	2.16680	4.41165	0.01624	161.11	161.12	96.99	1020.00	1116.99	0.1800	1.7327	1.9127	129
130	2.22560	4.53136	0.01625	157.11	157.12	97.99	1019.42	1117.41	0.1817	1.7288	1.9105	130
131	2.28576	4.65384	0.01625	153.22	153.23	98.99	1018.84	1117.83	0.1834	1.7249	1.9083	131
132	2.34730	4.77914	0.01626	149.44	149.46	99.99	1018.26	1118.25	0.1851	1.7210	1.9061	132
133	2.41025	4.90730	0.01626	145.77	145.78	100.99	1017.68	1118.67	0.1868	1.7171	1.9039	133
134	2.47463	5.03839	0.01627	142.21	142.23	101.99	1017.10	1119.08	0.1885	1.7132	1.9017	134
135	2.54048	5.17246	0.01627	138.74	138.76	102.99	1016.52	1119.50	0.1902	1.7093	1.8995	135
136	2.60782	5.30956	0.01627	135.37	135.39	103.98	1015.93	1119.92	0.1919	1.7055	1.8974	136
137	2.67667	5.44975	0.01628	132.10	132.12	104.98	1015.35	1120.34	0.1935	1.7017	1.8952	137
138	2.74707	5.59308	0.01628	128.92	128.94	105.98	1014.77	1120.75	0.1952	1.6978	1.8930	138
139	2.81903	5.73961	0.01629	125.83	125.85	106.98	1014.18	1121.17	0.1969	1.6940	1.8909	139
140	2.89260	5.88939	0.01629	122.82	122.84	107.98	1013.60	1121.58	0.1985	1.6902	1.8888	140
141	2.96780	6.04250	0.01630	119.90	119.92	108.98	1013.01	1122.00	0.2002	1.6864	1.8867	141
142	3.04465	6.19897	0.01630	117.05	117.07	109.98	1012.43	1122.41	0.2019	1.6827	1.8845	142
143	3.12320	6.35888	0.01631	114.29	114.31	110.98	1011.84	1122.83	0.2035	1.6789	1.8824	143
144	3.20345	6.52229	0.01631	111.60	111.62	111.98	1011.26	1123.24	0.2052	1.6752	1.8803	144
145	3.28546	6.68926	0.01632	108.99	109.00	112.98	1010.67	1123.66	0.2068	1.6714	1.8783	145
146	3.36924	6.85984	0.01632	106.44	106.45	113.98	1010.09	1124.07	0.2085	1.6677	1.8762	146
147	3.45483	7.03410	0.01633	103.96	103.98	114.98	1009.50	1124.48	0.2101	1.6640	1.8741	147
148	3.54226	7.21211	0.01633	101.55	101.57	115.98	1008.91	1124.89	0.2118	1.6603	1.8721	148
149	3.63156	7.39393	0.01634	99.21	99.22	116.98	1008.32	1125.31	0.2134	1.6566	1.8700	149
150	3.72277	7.57962	0.01634	96.93	96.94	117.98	1007.73	1125.72	0.2151	1.6529	1.8680	150
151	3.81591	7.76925	0.01635	94.70	94.72	118.99	1007.14	1126.13	0.2167	1.6492	1.8659	151
152	3.91101	7.96289	0.01635	92.54	92.56	119.99	1006.55	1126.54	0.2184	1.6455	1.8639	152
153	4.00812	8.16061	0.01636	90.44	90.46	120.99	1005.96	1126.95	0.2200	1.6419	1.8619	153
154	4.10727	8.36247	0.01636	88.39	88.41	121.99	1005.37	1127.36	0.2216	1.6383	1.8599	154
155	4.20848	8.56854	0.01637	86.40	86.41	122.99	1004.78	1127.77	0.2233	1.6346	1.8579	155
156	4.31180	8.77890	0.01637	84.45	84.47	123.99	1004.19	1128.18	0.2249	1.6310	1.8559	156
157	4.41725	8.99360	0.01638	82.56	82.58	124.99	1003.60	1128.59	0.2265	1.6274	1.8539	157
158	4.52488	9.21274	0.01638	80.72	80.73	125.99	1003.00	1128.99	0.2281	1.6238	1.8519	158
159	4.63472	9.43637	0.01639	78.92	78.94	126.99	1002.41	1129.40	0.2297	1.6202	1.8500	159
160	4.7468	9.6646	0.01639	77.175	77.192	127.99	1001.82	1129.81	0.2314	1.6167	1.8480	160
161	4.8612	9.8974	0.01640	75.471	75.488	128.99	1001.22	1130.22	0.2330	1.6131	1.8461	161
162	4.9778	10.1350	0.01640	73.812	73.829	130.00	1000.63	1130.62	0.2346	1.6095	1.8441	162
163	5.0969	10.3774	0.01641	72.196	72.213	131.00	1000.03	1131.03	0.2362	1.6060	1.8422	163
164	5.2183	10.6246	0.01642	70.619	70.636	132.00	999.43	1131.43	0.2378	1.6025	1.8403	164
165	5.3422	10.8768	0.01642	69.084	69.101	133.00	998.84	1131.84	0.2394	1.5989	1.8383	165
166	5.4685	11.1340	0.01643	67.587	67.604	134.00	998.24	1132.24	0.2410	1.5954	1.8364	166
167	5.5974	11.3963	0.01643	66.130	66.146	135.00	997.64	1132.64	0.2426	1.5919	1.8345	167
168	5.7287	11.6638	0.01644	64.707	64.723	136.01	997.04	1133.05	0.2442	1.5884	1.8326	168
169	5.8627	11.9366	0.01644	63.320	63.336	137.01	996.44	1133.45	0.2458	1.5850	1.8308	169
170	5.9993	12.2148	0.01645	61.969	61.986	138.01	995.84	1133.85	0.2474	1.5815	1.8289	170
171	6.1386	12.4983	0.01646	60.649	60.666	139.01	995.24	1134.25	0.2490	1.5780	1.8270	171
172	6.2806	12.7874	0.01646	59.363	59.380	140.01	994.64	1134.66	0.2506	1.5746	1.8251	172
173	6.4253	13.0821	0.01647	58.112	58.128	141.02	994.04	1135.06	0.2521	1.5711	1.8233	173
174	6.5729	13.3825	0.01647	56.887	56.904	142.02	993.44	1135.46	0.2537	1.5677	1.8214	174
175	6.7232	13.6886	0.01648	55.694	55.711	143.02	992.83	1135.86	0.2553	1.5643	1.8196	175
176	6.8765	14.0006	0.01648	54.532	54.549	144.02	992.23	1136.26	0.2569	1.5609	1.8178	176
177	7.0327	14.3186	0.01649	53.397	53.414	145.03	991.63	1136.65	0.2585	1.5575	1.8159	177
178	7.1918	14.6426	0.01650	52.290	52.307	146.03	991.02	1137.05	0.2600	1.5541	1.8141	178
179	7.3539	14.9727	0.01650	51.210	51.226	147.03	990.42	1137.45	0.2616	1.5507	1.8123	179
180	7.5191	15.3091	0.01651	50.155	50.171	148.04	989.81	1137.85	0.2632	1.5473	1.8105	180
181	7.6874	15.6518	0.01651	49.126	49.143	149.04	989.20	1138.24	0.2647	1.5440	1.8087	181
182	7.8589	16.0008	0.01652	48.122	48.138	150.04	988.60	1138.64	0.2663	1.5406	1.8069	182
183	8.0335	16.3564	0.01653	47.142	47.158	151.05	987.99	1139.03	0.2679	1.5373	1.8051	183
184	8.2114	16.7185	0.01653	46.185	46.202	152.05	987.38	1139.43	0.2694	1.5339	1.8034	184
185	8.3926	17.0874	0.01654	45.251	45.267	153.05	986.77	1139.82	0.2710	1.5306	1.8016	185
186	8.5770	17.4630	0.01654	44.339	44.356	154.06	986.16	1140.22	0.2725	1.5273	1.7998	186
187	8.7649	17.8455	0.01655	43.448	43.465	155.06	985.55	1140.61	0.2741	1.5240	1.7981	187
188	8.9562	18.2350	0.01656	42.579	42.595	156.07	984.94	1141.00	0.2756	1.5207	1.7963	188
189	9.1510	18.6316	0.01656	41.730	41.746	157.07	984.32	1141.39	0.2772	1.5174	1.7946	189

Table 3 Thermodynamic Properties of Water at Saturation (*Continued*)

Temp., °F	Absolute Pressure		Specific Volume, ft³/lb_w			Specific Enthalpy, Btu/lb_w			Specific Entropy, Btu/lb_w·°F			Temp., °F
			Sat. Liquid	Evap.	Sat. Vapor	Sat. Liquid	Evap.	Sat. Vapor	Sat. Liquid	Evap.	Sat. Vapor	
t	p, psi	p, in. Hg	v_f	v_{fg}	v_g	h_f	h_{fg}	h_g	s_f	s_{fg}	s_g	t
190	9.3493	19.0353	0.01657	40.901	40.918	158.07	983.71	1141.78	0.2787	1.5141	1.7929	190
191	9.5512	19.4464	0.01658	40.092	40.108	159.08	983.10	1142.18	0.2803	1.5109	1.7911	191
192	9.7567	19.8648	0.01658	39.301	39.317	160.08	982.48	1142.57	0.2818	1.5076	1.7894	192
193	9.9659	20.2907	0.01659	38.528	38.544	161.09	981.87	1142.95	0.2834	1.5043	1.7877	193
194	10.1788	20.7242	0.01659	37.774	37.790	162.09	981.25	1143.34	0.2849	1.5011	1.7860	194
195	10.3955	21.1653	0.01660	37.035	37.052	163.10	980.63	1143.73	0.2864	1.4979	1.7843	195
196	10.6160	21.6143	0.01661	36.314	36.331	164.10	980.02	1144.12	0.2880	1.4946	1.7826	196
197	10.8404	22.0712	0.01661	35.611	35.628	165.11	979.40	1144.51	0.2895	1.4914	1.7809	197
198	11.0687	22.5361	0.01662	34.923	34.940	166.11	978.78	1144.89	0.2910	1.4882	1.7792	198
199	11.3010	23.0091	0.01663	34.251	34.268	167.12	978.16	1145.28	0.2926	1.4850	1.7776	199
200	11.5374	23.4904	0.01663	33.594	33.610	168.13	977.54	1145.66	0.2941	1.4818	1.7759	200
201	11.7779	23.9800	0.01664	32.951	32.968	169.13	976.92	1146.05	0.2956	1.4786	1.7742	201
202	12.0225	24.4780	0.01665	32.324	32.340	170.14	976.29	1146.43	0.2971	1.4755	1.7726	202
203	12.2713	24.9847	0.01665	31.710	31.726	171.14	975.67	1146.81	0.2986	1.4723	1.7709	203
204	12.5244	25.5000	0.01666	31.110	31.127	172.15	975.05	1147.20	0.3002	1.4691	1.7693	204
205	12.7819	26.0241	0.01667	30.523	30.540	173.16	974.42	1147.58	0.3017	1.4660	1.7677	205
206	13.0436	26.5571	0.01667	29.949	29.965	174.16	973.80	1147.96	0.3032	1.4628	1.7660	206
207	13.3099	27.0991	0.01668	29.388	29.404	175.17	973.17	1148.34	0.3047	1.4597	1.7644	207
208	13.5806	27.6503	0.01669	28.839	28.856	176.18	972.54	1148.72	0.3062	1.4566	1.7628	208
209	13.8558	28.2108	0.01669	28.303	28.319	177.18	971.92	1149.10	0.3077	1.4535	1.7612	209
210	14.1357	28.7806	0.01670	27.778	27.795	178.19	971.29	1149.48	0.3092	1.4503	1.7596	210
212	14.7096	29.9489	0.01671	26.763	26.780	180.20	970.03	1150.23	0.3122	1.4442	1.7564	212
214	15.3025	31.1563	0.01673	25.790	25.807	182.22	968.76	1150.98	0.3152	1.4380	1.7532	214
216	15.9152	32.4036	0.01674	24.861	24.878	184.24	967.50	1151.73	0.3182	1.4319	1.7501	216
218	16.5479	33.6919	0.01676	23.970	23.987	186.25	966.23	1152.48	0.3212	1.4258	1.7469	218
220	17.2013	35.0218	0.01677	23.118	23.134	188.27	964.95	1153.22	0.3241	1.4197	1.7438	220
222	17.8759	36.3956	0.01679	22.299	22.316	190.29	963.67	1153.96	0.3271	1.4136	1.7407	222
224	18.5721	37.8131	0.01680	21.516	21.533	192.31	962.39	1154.70	0.3301	1.4076	1.7377	224
226	19.2905	39.2758	0.01682	20.765	20.782	194.33	961.11	1155.43	0.3330	1.4016	1.7347	226
228	20.0316	40.7848	0.01683	20.045	20.062	196.35	959.82	1156.16	0.3359	1.3957	1.7316	228
230	20.7961	42.3412	0.01684	19.355	19.372	198.37	958.52	1156.89	0.3389	1.3898	1.7287	230
232	21.5843	43.9461	0.01686	18.692	18.709	200.39	957.22	1157.62	0.3418	1.3839	1.7257	232
234	22.3970	45.6006	0.01688	18.056	18.073	202.41	955.92	1158.34	0.3447	1.3780	1.7227	234
236	23.2345	47.3060	0.01689	17.446	17.463	204.44	954.62	1159.06	0.3476	1.3722	1.7198	236
238	24.0977	49.0633	0.01691	16.860	16.877	206.46	953.31	1159.77	0.3505	1.3664	1.7169	238
240	24.9869	50.8738	0.01692	16.298	16.314	208.49	952.00	1160.48	0.3534	1.3606	1.7140	240
242	25.9028	52.7386	0.01694	15.757	15.774	210.51	950.68	1161.19	0.3563	1.3548	1.7111	242
244	26.8461	54.6591	0.01695	15.238	15.255	212.54	949.35	1161.90	0.3592	1.3491	1.7083	244
246	27.8172	56.6364	0.01697	14.739	14.756	214.57	948.03	1162.60	0.3621	1.3434	1.7055	246
248	28.8169	58.6717	0.01698	14.259	14.276	216.60	946.70	1163.29	0.3649	1.3377	1.7026	248
250	29.8457	60.7664	0.01700	13.798	13.815	218.63	945.36	1163.99	0.3678	1.3321	1.6998	250
252	30.9043	62.9218	0.01702	13.355	13.372	220.66	944.02	1164.68	0.3706	1.3264	1.6971	252
254	31.9934	65.1391	0.01703	12.928	12.945	222.69	942.68	1165.37	0.3735	1.3208	1.6943	254
256	33.1135	67.4197	0.01705	12.526	12.147	226.73	939.99	1166.72	0.3764	1.3153	1.6916	256
258	34.2653	69.7649	0.01707	12.123	12.140	226.76	939.97	1166.73	0.3792	1.3097	1.6889	258
260	35.4496	72.1760	0.01708	11.742	11.759	228.79	938.61	1167.40	0.3820	1.3042	1.6862	260
262	36.6669	74.6545	0.01710	11.376	11.393	230.83	937.25	1168.08	0.3848	1.2987	1.6835	262
264	37.9180	77.2017	0.01712	11.024	11.041	232.87	935.88	1168.74	0.3876	1.2932	1.6808	264
266	39.2035	79.8190	0.01714	10.684	10.701	234.90	934.50	1169.41	0.3904	1.2877	1.6781	266
268	40.5241	82.5078	0.01715	10.357	10.374	236.94	933.12	1170.07	0.3932	1.2823	1.6755	268
270	41.8806	85.2697	0.01717	10.042	10.059	238.98	931.74	1170.72	0.3960	1.2769	1.6729	270
272	43.2736	88.1059	0.01719	9.737	9.755	241.03	930.35	1171.38	0.3988	1.2715	1.6703	272
274	44.7040	91.0181	0.01721	9.445	9.462	243.07	928.95	1172.02	0.4016	1.2661	1.6677	274
276	46.1723	94.0076	0.01722	9.162	9.179	245.11	927.55	1172.67	0.4044	1.2608	1.6651	276
278	47.6794	97.0761	0.01724	8.890	8.907	247.16	926.15	1173.31	0.4071	1.2554	1.6626	278
280	49.2260	100.2250	0.01726	8.627	8.644	249.20	924.74	1173.94	0.4099	1.2501	1.6600	280
282	50.8128	103.4558	0.01728	8.373	8.390	251.25	923.32	1174.57	0.4127	1.2448	1.6575	282
284	52.4406	106.7701	0.01730	8.128	8.146	253.30	921.90	1175.20	0.4154	1.2396	1.6550	284
286	54.1103	110.1695	0.01731	7.892	7.910	255.35	920.47	1175.82	0.4182	1.2343	1.6525	286
288	55.8225	113.6556	0.01733	7.664	7.681	257.40	919.03	1176.44	0.4209	1.2291	1.6500	288
290	57.5780	117.2299	0.01735	7.444	7.461	259.45	917.59	1177.05	0.4236	1.2239	1.6476	290
292	59.3777	120.8941	0.01737	7.231	7.248	261.51	916.15	1177.66	0.4264	1.2187	1.6451	292
294	61.2224	124.6498	0.01739	7.026	7.043	263.56	914.69	1178.26	0.4291	1.2136	1.6427	294
296	63.1128	128.4987	0.01741	6.827	6.844	265.62	913.24	1178.86	0.4318	1.2084	1.6402	296
298	65.0498	132.4425	0.01743	6.635	6.652	267.68	911.77	1179.45	0.4345	1.2033	1.6378	298
300	67.0341	136.4827	0.01745	6.450	6.467	269.74	910.30	1180.04	0.4372	1.1982	1.6354	300

HUMIDITY PARAMETERS

Basic Parameters

Humidity ratio (alternatively, the moisture content or mixing ratio) ***W*** of a given moist air sample is defined as the ratio of the mass of water vapor to the mass of dry air contained in the sample:

$$W = M_w / M_{da} \tag{7}$$

The humidity ratio W is equal to the mole fraction ratio x_w/x_{da} multiplied by the ratio of molecular masses, namely, 18.01528/28.9645 = 0.62198:

$$W = 0.62198 x_w / x_{da} \tag{8}$$

Specific humidity γ is the ratio of the mass of water vapor to the total mass of the moist air sample:

$$\gamma = M_w / (M_w + M_{da}) \tag{9a}$$

In terms of the humidity ratio,

$$\gamma = W / (1 + W) \tag{9b}$$

Absolute humidity (alternatively, water vapor density) $\boldsymbol{d_v}$ is the ratio of the mass of water vapor to the total volume of the sample:

$$d_v = M_w / V \tag{10}$$

The **density** ρ of a moist air mixture is the ratio of the total mass to the total volume:

$$\rho = (M_{da} + M_w)/V = (1/v)(1 + W) \tag{11}$$

where v is the moist air specific volume, ft^3/lb_{da}, as defined by Equation (27).

Humidity Parameters Involving Saturation

The following definitions of humidity parameters involve the concept of moist air saturation:

Saturation humidity ratio $\boldsymbol{W_s(t, p)}$ is the humidity ratio of moist air saturated with respect to water (or ice) at the same temperature t and pressure p.

Degree of saturation μ is the ratio of the air humidity ratio W to the humidity ratio W_s of saturated moist air at the same temperature and pressure:

$$\mu = \left.\frac{W}{W_s}\right|_{t,p} \tag{12}$$

Relative humidity ϕ is the ratio of the mole fraction of water vapor x_w in a given moist air sample to the mole fraction x_{ws} in an air sample saturated at the same temperature and pressure:

$$\phi = \left.\frac{x_w}{x_{ws}}\right|_{t,p} \tag{13}$$

Combining Equations (8), (12), and (13),

$$\mu = \frac{\phi}{1 + (1 - \phi) W_s / 0.62198} \tag{14}$$

Dew-point temperature $\boldsymbol{t_d}$ is the temperature of moist air saturated at the same pressure p, with the same humidity ratio W as that of the given sample of moist air. It is defined as the solution $t_d(p, W)$ of the following equation:

$$W_s(p, t_d) = W \tag{15}$$

Thermodynamic wet-bulb temperature $\boldsymbol{t^*}$ is the temperature at which water (liquid or solid), by evaporating into moist air at a given dry-bulb temperature t and humidity ratio W, can bring air to saturation adiabatically at the same temperature t^* while the total pressure p is maintained constant. This parameter is considered separately in the section on Thermodynamic Wet-Bulb Temperature and Dew-Point Temperature.

PERFECT GAS RELATIONSHIPS FOR DRY AND MOIST AIR

When moist air is considered a mixture of independent perfect gases (i.e., dry air and water vapor), each is assumed to obey the perfect gas equation of state as follows:

$$\text{Dry air:} \quad p_{da} V = n_{da} R T \tag{16}$$

$$\text{Water vapor:} \quad p_w V = n_w R T \tag{17}$$

where

p_{da} = partial pressure of dry air
p_w = partial pressure of water vapor
V = total mixture volume
n_{da} = number of moles of dry air
n_w = number of moles of water vapor
R = universal gas constant, 1545.32 $ft \cdot lb_f/lb\ mol \cdot °R$
T = absolute temperature, °R

The mixture also obeys the perfect gas equation:

$$pV = nRT \tag{18}$$

or

$$(p_{da} + p_w)V = (n_{da} + n_w)RT \tag{19}$$

where $p = p_{da} + p_w$ is the total mixture pressure and $n = n_{da} + n_w$ is the total number of moles in the mixture. From Equations (16) through (19), the mole fractions of dry air and water vapor are, respectively,

$$x_{da} = p_{da} / (p_{da} + p_w) = p_{da} / p \tag{20}$$

and

$$x_w = p_w / (p_{da} + p_w) = p_w / p \tag{21}$$

From Equations (8), (20), and (21), the **humidity ratio** $\boldsymbol{W}$ is given by

$$W = 0.62198 \frac{p_w}{p - p_w} \tag{22}$$

The degree of saturation μ is, by definition, Equation (12):

$$\mu = \left.\frac{W}{W_s}\right|_{t,p}$$

where

$$W_s = 0.62198 \frac{p_{ws}}{p - p_{ws}} \tag{23}$$

The term p_{ws} represents the saturation pressure of water vapor in the absence of air at the given temperature t. This pressure p_{ws} is a function only of temperature and differs slightly from the vapor pressure of water in saturated moist air.

The **relative humidity** ϕ is, by definition, Equation (13):

$$\phi = \left.\frac{x_w}{x_{ws}}\right|_{t,p}$$

Substituting Equation (21) for x_w and x_{ws},

$$\phi = \left.\frac{p_w}{p_{ws}}\right|_{t,p} \tag{24}$$

Substituting Equation (21) for x_{ws} into Equation (14),

$$\phi = \frac{\mu}{1-(1-\mu)(p_{ws}/p)} \tag{25}$$

Both ϕ and μ are zero for dry air and unity for saturated moist air. At intermediate states their values differ, substantially so at higher temperatures.

The **specific volume** v of a moist air mixture is expressed in terms of a unit mass of dry air:

$$v = V/M_{da} = V/(28.9645n_{da}) \tag{26}$$

where V is the total volume of the mixture, M_{da} is the total mass of dry air, and n_{da} is the number of moles of dry air. By Equations (16) and (26), with the relation $p = p_{da} + p_w$,

$$v = \frac{RT}{28.9645(p-p_w)} = \frac{R_{da}T}{p-p_w} \tag{27}$$

Using Equation (22),

$$v = \frac{RT(1+1.6078W)}{28.964p} = \frac{R_{da}T(1+1.6078W)}{p} \tag{28}$$

In Equations (27) and (28), v is specific volume, T is absolute temperature, p is total pressure, p_w is the partial pressure of water vapor, and W is the humidity ratio.

In specific units, Equation (28) may be expressed as

$$v = 0.7543(t+459.67)(1+1.6078W)/p$$

where

v = specific volume, ft^3/lb_{da}
t = dry-bulb temperature, °F
W = humidity ratio, lb_w/lb_{da}
p = total pressure, in. Hg

The **enthalpy** of a mixture of perfect gases equals the sum of the individual partial enthalpies of the components. Therefore, the specific enthalpy of moist air can be written as follows:

$$h = h_{da} + Wh_g \tag{29}$$

where h_{da} is the specific enthalpy for dry air in Btu/lb_{da} and h_g is the specific enthalpy for saturated water vapor in Btu/lb_w at the temperature of the mixture. As an approximation,

$$h_{da} \approx 0.240t \tag{30}$$

$$h_g \approx 1061 + 0.444t \tag{31}$$

where t is the dry-bulb temperature in, °F. The moist air specific enthalpy in Btu/lb_{da} then becomes

$$h = 0.240t + W(1061 + 0.444t) \tag{32}$$

THERMODYNAMIC WET-BULB TEMPERATURE AND DEW-POINT TEMPERATURE

For any state of moist air, a temperature t^* exists at which liquid (or solid) water evaporates into the air to bring it to saturation at exactly this same temperature and total pressure (Harrison 1965). During the adiabatic saturation process, the saturated air is expelled at a temperature equal to that of the injected water. In this constant pressure process,

- Humidity ratio is increased from a given initial value W to the value W_s^* corresponding to saturation at the temperature t^*
- Enthalpy is increased from a given initial value h to the value h_s^* corresponding to saturation at the temperature t^*
- Mass of water added per unit mass of dry air is $(W_s^* - W)$, which adds energy to the moist air of amount $(W_s^* - W)h_w^*$, where h_w^* denotes the specific enthalpy in Btu/lb_w of the water added at the temperature t^*

Therefore, if the process is strictly adiabatic, conservation of enthalpy at constant total pressure requires that

$$h + (W_s^* - W)h_w^* = h_s^* \tag{33}$$

The properties W_s^*, h_w^*, and h_s^* are functions only of the temperature t^* for a fixed value of pressure. The value of t^*, which satisfies Equation (33) for given values of h, W, and p, is the **thermodynamic wet-bulb temperature**.

The **psychrometer** consists of two thermometers; one thermometer's bulb is covered by a wick that has been thoroughly wetted with water. When the wet bulb is placed in an airstream, water evaporates from the wick, eventually reaching an equilibrium temperature called the **wet-bulb temperature**. This process is not one of adiabatic saturation, which defines the thermodynamic wet-bulb temperature, but one of simultaneous heat and mass transfer from the wet bulb. The fundamental mechanism of this process is described by the Lewis relation [Equation (39) in Chapter 5]. Fortunately, only small corrections must be applied to wet-bulb thermometer readings to obtain the thermodynamic wet-bulb temperature.

As defined, thermodynamic wet-bulb temperature is a unique property of a given moist air sample independent of measurement techniques.

Equation (33) is exact since it defines the thermodynamic wet-bulb temperature t^*. Substituting the approximate perfect gas relation [Equation (32)] for h, the corresponding expression for h_s^*, and the approximate relation

$$h_w^* \approx t^* - 32 \tag{34}$$

into Equation (33), and solving for the humidity ratio,

$$W = \frac{(1093 - 0.556t^*)W_s^* - 0.240(t - t^*)}{1093 + 0.444t - t^*} \tag{35}$$

where t and t^* are in °F.

The **dew-point temperature** t_d of moist air with humidity ratio W and pressure p was defined earlier as the solution $t_d(p, w)$ of $W_s(p, t_d)$. For perfect gases, this reduces to

$$p_{ws}(t_d) = p_w = (pW)/(0.62198 + W) \tag{36}$$

where p_w is the water vapor partial pressure for the moist air sample and $p_{ws}(t_d)$ is the saturation vapor pressure at temperature t_d. The saturation vapor pressure is derived from Table 3 or from Equation

(5) or (6). Alternatively, the dew-point temperature can be calculated directly by one of the following equations (Peppers 1988):

For the dew-point temperature range of 32 to 200°F,

$$t_d = C_{14} + C_{15}\alpha + C_{16}\alpha^2 + C_{17}\alpha^3 + C_{18}(p_w)^{0.1984} \quad (37)$$

For temperatures below 32°F,

$$t_d = 90.12 + 26.142\alpha + 0.8927\alpha^2 \quad (38)$$

where

t_d = dew-point temperature, °F
α = ln p_w
p_w = water vapor partial pressure, psia
C_{14} = 100.45
C_{15} = 33.193
C_{16} = 2.319
C_{17} = 0.17074
C_{18} = 1.2063

NUMERICAL CALCULATION OF MOIST AIR PROPERTIES

The following are outlines, citing equations and tables already presented, for calculating moist air properties using perfect gas relations. These relations are sufficiently accurate for most engineering calculations in air-conditioning practice, and are readily adapted to either hand or computer calculating methods. For more details, refer to Tables 15 through 18 in Chapter 1 of Olivieri (1996). Graphical procedures are discussed in the section on Psychrometric Charts.

SITUATION 1.

Given: Dry-bulb temperature t, Wet-bulb temperature t^*, Pressure p

To Obtain	Use	Comments
$p_{ws}(t^*)$	Table 3 or Equation (5) or (6)	Sat. press. for temp. t^*
W_s^*	Equation (23)	Using $p_{ws}(t^*)$
W	Equation (35)	
$p_{ws}(t)$	Table 3 or Equation (5) or (6)	Sat. press. for temp. t
W_s	Equation (23)	Using $p_{ws}(t)$
μ	Equation (12)	Using W_s
ϕ	Equation (25)	Using $p_{ws}(t)$
v	Equation (28)	
h	Equation (32)	
p_w	Equation (36)	
t_d	Table 3 with Equation (36), (37), or (38)	

SITUATION 2.

Given: Dry-bulb temperature t, Dew-point temperature t_d, Pressure p

To Obtain	Use	Comments
$p_w = p_{ws}(t_d)$	Table 3 or Equation (5) or (6)	Sat. press. for temp. t_d
W	Equation (22)	
$p_{ws}(t)$	Table 3 or Equation (5) or (6)	Sat. press. for temp. t_d
W_s	Equation (23)	Using $p_{ws}(t)$
μ	Equation (12)	Using W_s
ϕ	Equation (25)	Using $p_{ws}(t)$
v	Equation (28)	
h	Equation (32)	
t^*	Equation (23) and (35) with Table 3 or with Equation (5) or (6)	Requires trial-and-error or numerical solution method

SITUATION 3.

Given: Dry-bulb temperature t, Relative humidity ϕ, Pressure p

To Obtain	Use	Comments
$p_{ws}(t)$	Table 3 or Equation (5) or (6)	Sat. press. for temp. t
p_w	Equation (24)	
W	Equation (22)	
W_s	Equation (23)	Using $p_{ws}(t)$
μ	Equation (12)	Using W_s
v	Equation (28)	
h	Equation (32)	
t_d	Table 3 with Equation (36), (37), or (38)	
t^*	Equation (23) and (35) with Table 3 or with Equation (5) or (6)	Requires trial-and-error or numerical solution method

Exact Relations for Computing W_s and ϕ

Corrections that account for (1) the effect of dissolved gases on properties of condensed phase; (2) the effect of pressure on properties of condensed phase; and (3) the effect of intermolecular force on properties of moisture itself, can be applied to Equations (23) and (25):

$$W_s = 0.62198 \frac{f p_{ws}}{p - f p_{ws}} \quad (23a)$$

$$\phi = \frac{\mu}{1 - (1 - \mu)(f p_{ws}/p)} \quad (25a)$$

Table 4 lists f values for a number of pressure and temperature combinations. Hyland and Wexler (1983a) give additional values.

Table 4 Values of f and Estimated Maximum Uncertainties (EMUs)

	14.50 psia		72.52 psia		145.04 psia	
T, °R	f	EMU E+04	f	EMU E+04	f	EMU E+04
311.67	1.0105	134	1.0540	66	1.1130	136
491.67	1.0039	2	1.0177	10	1.0353	19
671.67	1.0039	0.1	1.0180	4	1.0284	11

Moist Air Property Tables for Standard Pressure

Table 2 shows values of thermodynamic properties for standard atmospheric pressure at temperatures from −80 to 200°F. The properties of intermediate moist air states can be calculated using the degree of saturation μ:

Volume $$v = v_{da} + \mu v_{as} \quad (39)$$

Enthalpy $$h = h_{da} + \mu h_{as} \quad (40)$$

Entropy $$s = s_{da} + \mu s_{as} \quad (41)$$

These equations are accurate to about 160°F. At higher temperatures, the errors can be significant. Hyland and Wexler (1983a) include charts that can be used to estimate errors for v, h, and s for standard barometric pressure.

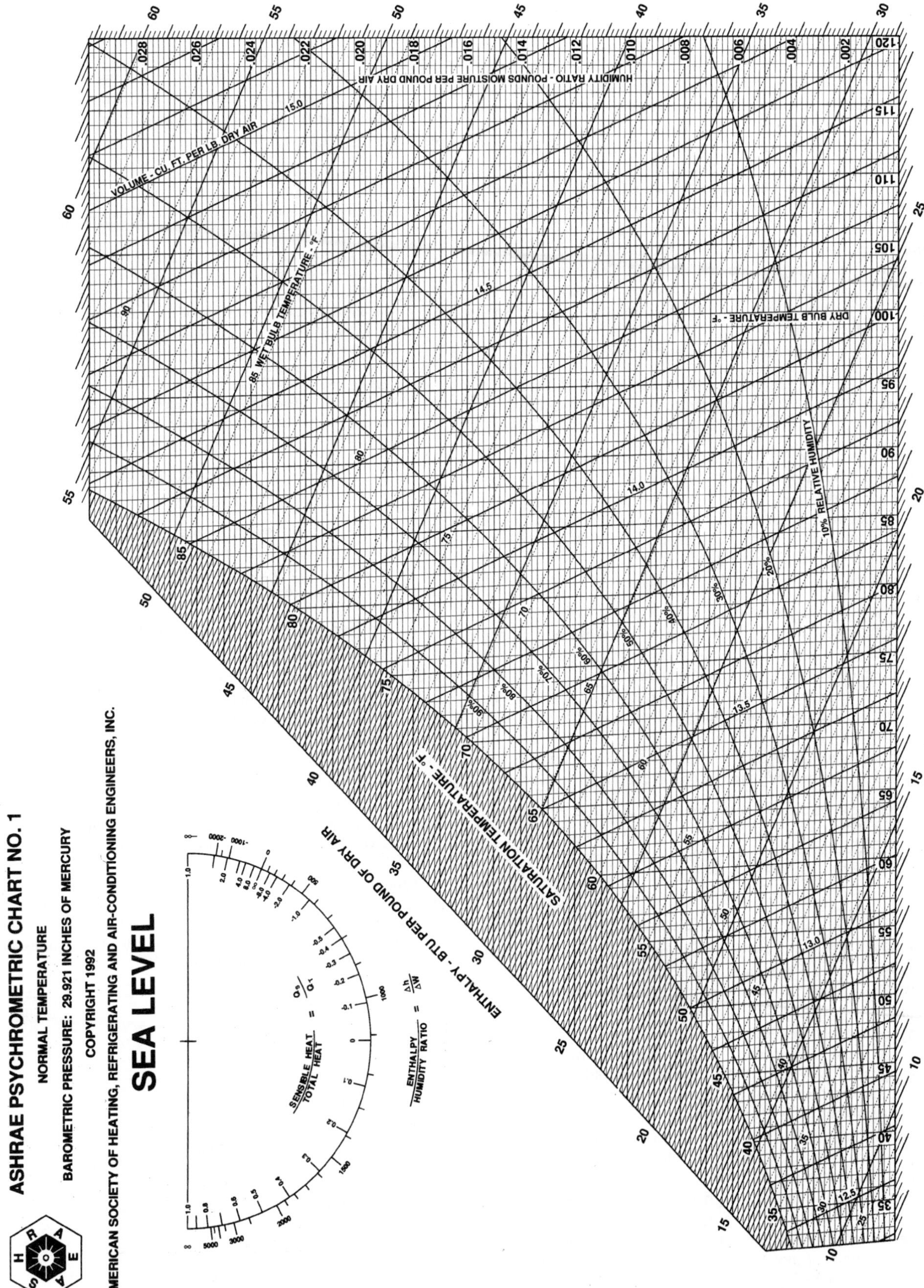

Fig. 1 ASHRAE Psychrometric Chart No. 1

PSYCHROMETRIC CHARTS

A psychrometric chart graphically represents the thermodynamic properties of moist air.

The choice of coordinates for a psychrometric chart is arbitrary. A chart with coordinates of enthalpy and humidity ratio provides convenient graphical solutions of many moist air problems with a minimum of thermodynamic approximations. ASHRAE developed five such psychrometric charts. Chart No. 1 is shown as Figure 1; the others may be obtained through ASHRAE.

Charts 1, 2, and 3 are for sea level pressure, Chart 4 is for 5000 ft altitude (24.89 in. Hg), and Chart 5 is for 7500 ft altitude (22.65 in. Hg). All charts use oblique-angle coordinates of enthalpy and humidity ratio, and are consistent with the data of Table 2 and the properties computation methods of Goff and Gratch (1945), and Goff (1949) as well as Hyland and Wexler (1983a). Palmatier (1963) describes the geometry of chart construction applying specifically to Charts 1 and 4.

The dry-bulb temperature ranges covered by the charts are

Charts 1, 4, 5	Normal temperature	32 to 120°F
Chart 2	Low temperature	−40 to 50°F
Chart 3	High temperature	60 to 250°F

Psychrometric properties or charts for other barometric pressures can be derived by interpolation. Sufficiently exact values for most purposes can be derived by methods described in the section on Perfect Gas Relationships for Dry and Moist Air. The construction of charts for altitude conditions has been treated by Haines (1961), Rohsenow (1946), and Karig (1946).

Comparison of Charts 1 and 4 by overlay reveals the following:

1. The dry-bulb lines coincide.
2. Wet-bulb lines for a given temperature originate at the intersections of the corresponding dry-bulb line and the two saturation curves, and they have the same slope.
3. Humidity ratio and enthalpy for a given dry- and wet-bulb temperature increase with altitude, but there is little change in relative humidity.
4. Volume changes rapidly; for a given dry-bulb and humidity ratio, it is practically inversely proportional to barometric pressure.

The following table compares properties at sea level (Chart 1) and 5000 ft (Chart 4):

Chart No.	db	wb	*h*	*W*	rh	*v*
1	100	81	44.6	0.0186	45	14.5
4	100	81	49.8	0.0234	46	17.6

Figure 1, which is ASHRAE Psychrometric Chart No. 1, shows humidity ratio lines (horizontal) for the range from 0 (dry air) to 0.03 lb_w/lb_{da}. Enthalpy lines are oblique lines drawn across the chart precisely parallel to each other.

Dry-bulb temperature lines are drawn straight, not precisely parallel to each other, and inclined slightly from the vertical position. Thermodynamic wet-bulb temperature lines are oblique lines that differ slightly in direction from that of enthalpy lines. They are straight but are not precisely parallel to each other.

Relative humidity lines are shown in intervals of 10%. The saturation curve is the line of 100% rh, while the horizontal line for $W = 0$ (dry air) is the line for 0% rh.

Specific volume lines are straight but are not precisely parallel to each other.

A narrow region above the saturation curve has been developed for fog conditions of moist air. This two-phase region represents a mechanical mixture of saturated moist air and liquid water, with the two components in thermal equilibrium. Isothermal lines in the fog region coincide with extensions of thermodynamic wet-bulb temperature lines. If required, the fog region can be further expanded by extension of humidity ratio, enthalpy, and thermodynamic wet-bulb temperature lines.

The protractor to the left of the chart shows two scales—one for sensible-total heat ratio, and one for the ratio of enthalpy difference to humidity ratio difference. The protractor is used to establish the direction of a condition line on the psychrometric chart.

Example 1 illustrates use of the ASHRAE Psychrometric Chart to determine moist air properties.

Example 1. Moist air exists at 100°F dry-bulb temperature, 65°F thermodynamic wet-bulb temperature, and 29.921 in. Hg pressure. Determine the humidity ratio, enthalpy, dew-point temperature, relative humidity, and specific volume.

Solution: Locate state point on Chart 1 (Figure 1) at the intersection of 100°F dry-bulb temperature and 65°F thermodynamic wet-bulb temperature lines. Read **humidity ratio** $W = 0.00523\ lb_w/lb_{da}$.

The **enthalpy** can be found by using two triangles to draw a line parallel to the nearest enthalpy line [30 Btu/lb_{da}] through the state point to the nearest edge scale. Read $h = 29.80\ Btu/lb_{da}$.

Dew-point temperature can be read at the intersection of $W = 0.00523\ lb_w/lb_{da}$ with the saturation curve. Thus, $t_d = 40°F$.

Relative humidity ϕ can be estimated directly. Thus, $\phi = 13\%$.

Specific volume can be found by linear interpolation between the volume lines for 14.0 and 14.5 ft^3/lb_{da}. Thus, $v = 14.22\ ft^3/lb_{da}$.

TYPICAL AIR-CONDITIONING PROCESSES

The ASHRAE psychrometric chart can be used to solve numerous process problems with moist air. Its use is best explained through illustrative examples. In each of the following examples, the process takes place at a constant total pressure of 29.921 in. Hg.

Moist Air Sensible Heating or Cooling

The process of adding heat alone to or removing heat alone from moist air is represented by a horizontal line on the ASHRAE chart, since the humidity ratio remains unchanged.

Figure 2 shows a device that adds heat to a stream of moist air. For steady flow conditions, the required rate of heat addition is

$$_1q_2 = \dot{m}_{da}(h_2 - h_1) \qquad (42)$$

Example 2. Moist air, saturated at 35°F, enters a heating coil at a rate of 20,000 cfm. Air leaves the coil at 100°F. Find the required rate of heat addition.

Solution: Figure 3 schematically shows the solution. State 1 is located on the saturation curve at 35°F. Thus, $h_1 = 13.01\ Btu/lb_{da}$, $W_1 = 0.00428\ lb_w/lb_{da}$, and $v_1 = 12.55\ ft^3/lb_{da}$. State 2 is located at the intersection of $t = 100°F$ and $W_2 = W_1 = 0.00428\ lb_w/lb_{da}$. Thus, $h_2 = 28.77\ Btu/lb_{da}$. The mass flow of dry air is

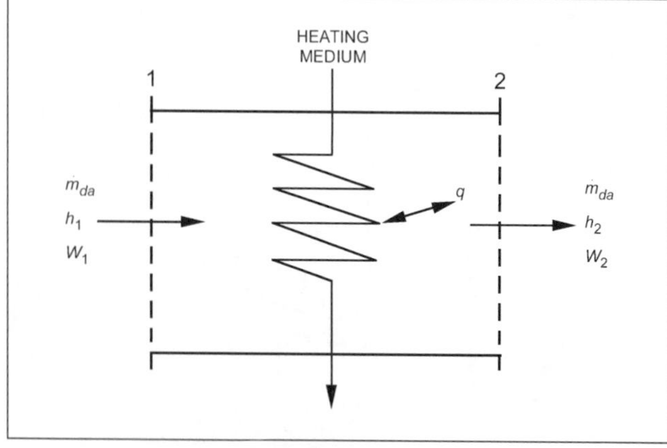

Fig. 2 Schematic of Device for Heating Moist Air

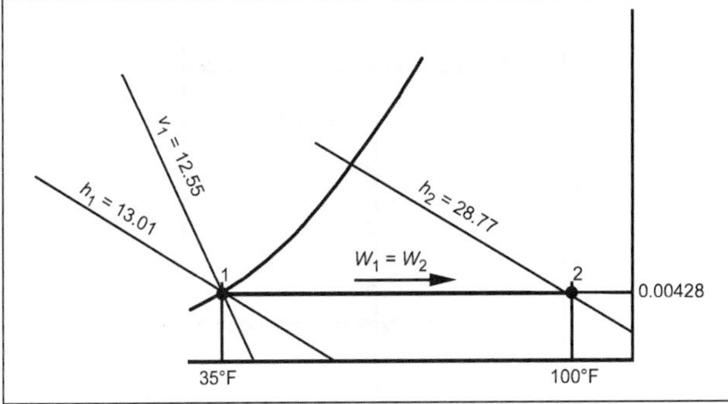

Fig. 3 Schematic Solution for Example 2

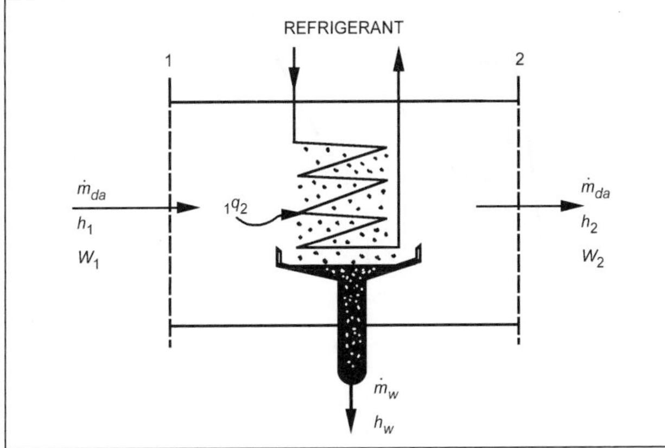

Fig. 4 Schematic of Device for Cooling Moist Air

$$^{\cdot}{}_{da} = (20{,}000 \times 60)/12.55 = 95{,}620 \text{ lb}_{da}/\text{h}$$

From Equation (42),

$$_1q_2 = (95{,}620)(28.77 - 13.01) = 1{,}507{,}000 \text{ Btu/h}$$

Moist Air Cooling and Dehumidification

Moisture condensation occurs when moist air is cooled to a temperature below its initial dew point. Figure 4 shows a schematic cooling coil where moist air is assumed to be uniformly processed. Although water can be removed at various temperatures ranging from the initial dew point to the final saturation temperature, it is assumed that condensed water is cooled to the final air temperature t_2 before it drains from the system.

For the system of Figure 4, the steady flow energy and material balance equations are

$$\dot{m}_{da}h_1 = \dot{m}_{da}h_2 + {}_1q_2 + \dot{m}_w h_{w2}$$

$$\dot{m}_{da}W_1 = \dot{m}_{da}W_2 + \dot{m}_w$$

Thus,

$$\dot{m}_w = \dot{m}_{da}(W_1 - W_2) \tag{43}$$

$$_1q_2 = \dot{m}_{da}[(h_1 - h_2) - (W_1 - W_2)h_{w2}] \tag{44}$$

Example 3. Moist air at 85°F dry-bulb temperature and 50% rh enters a cooling coil at 10,000 cfm and is processed to a final saturation condition at 50°F. Find the tons of refrigeration required.

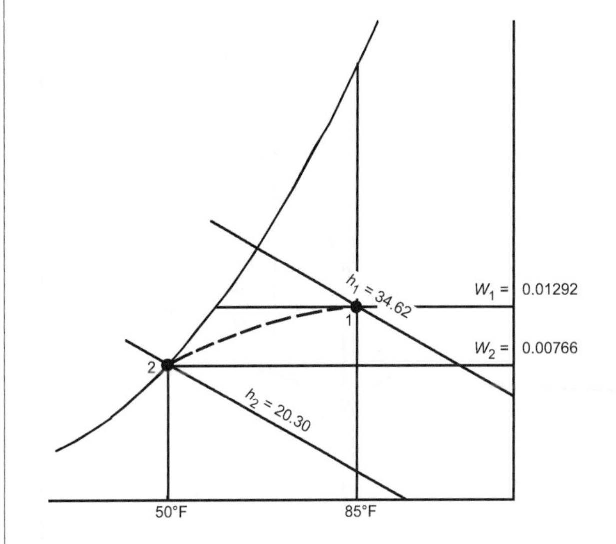

Fig. 5 Schematic Solution for Example 3

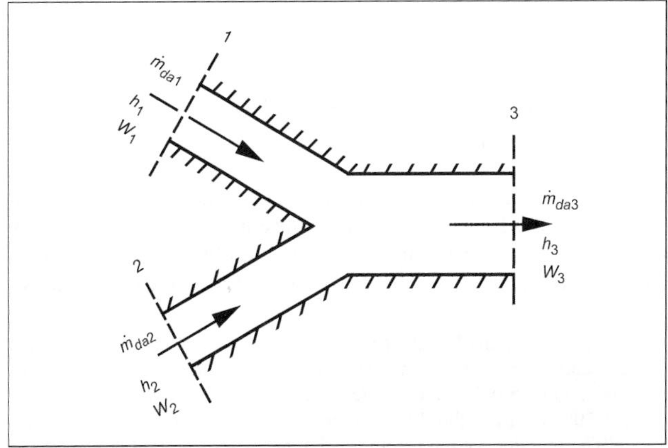

Fig. 6 Adiabatic Mixing of Two Moist Airstreams

Solution: Figure 5 shows the schematic solution. State 1 is located at the intersection of t = 85°F and ϕ = 50%. Thus, h_1 = 34.62 Btu/lb$_{da}$, W_1 = 0.01292 lb$_w$/lb$_{da}$, and v_1 = 14.01 ft^3/lb$_{da}$. State 2 is located on the saturation curve at 50°F. Thus, h_2 = 20.30 Btu/lb$_{da}$ and W_2 = 0.00766 lb$_w$/lb$_{da}$. From Table 2, h_{w2} = 18.11 Btu/lb$_w$. The mass flow of dry air is

$$\dot{m}_{da} = 10{,}000/14.01 = 713.8 \text{ lb}_{da}/\text{min}$$

From Equation (44),

$$_1q_2 = 713.8[(34.62 - 20.30) - (0.01292 - 0.00766)(18.11)]$$
$$= 10{,}150 \text{ Btu/min, or } 50.75 \text{ tons of refrigeration}$$

Adiabatic Mixing of Two Moist Airstreams

A common process in air-conditioning systems is the adiabatic mixing of two moist airstreams. Figure 6 schematically shows the problem. Adiabatic mixing is governed by three equations:

$$\dot{m}_{da1}h_1 + \dot{m}_{da2}h_2 = \dot{m}_{da3}h_3$$

$$\dot{m}_{da1} + \dot{m}_{da2} = \dot{m}_{da3}$$

$$\dot{m}_{da1}W_1 + \dot{m}_{da2}W_2 = \dot{m}_{da3}W_3$$

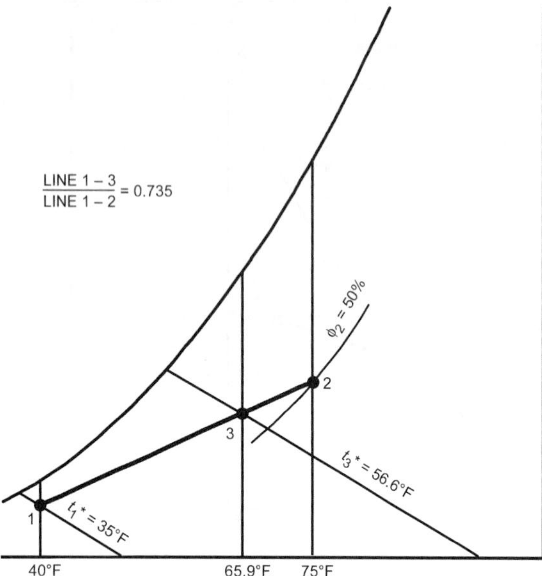

Fig. 7 Schematic Solution for Example 4

Eliminating $\dot{m}_{da3}$ gives

$$\frac{h_2 - h_3}{h_3 - h_1} = \frac{W_2 - W_3}{W_3 - W_1} = \frac{\dot{m}_{da1}}{\dot{m}_{da2}} \tag{45}$$

according to which, on the ASHRAE chart, the state point of the resulting mixture lies on the straight line connecting the state points of the two streams being mixed, and divides the line into two segments, in the same ratio as the masses of dry air in the two streams.

Example 4. A stream of 5000 cfm of outdoor air at 40°F dry-bulb temperature and 35°F thermodynamic wet-bulb temperature is adiabatically mixed with 15,000 cfm of recirculated air at 75°F dry-bulb temperature and 50% rh. Find the dry-bulb temperature and thermodynamic wet-bulb temperature of the resulting mixture.

Solution: Figure 7 shows the schematic solution. States 1 and 2 are located on the ASHRAE chart, revealing that v_1 = 12.65 ft³/lb$_{da}$, and v_2 = 13.68 ft³/lb$_{da}$. Therefore,

$$\dot{m}_{da1} = 5000/12.65 = 395 \text{ lb}_{da}/\text{min}$$

$$\dot{m}_{da2} = 15{,}000/13.68 = 1096 \text{ lb}_{da}/\text{min}$$

According to Equation (45),

$$\frac{\text{Line 3–2}}{\text{Line 1–3}} = \frac{\dot{m}_{da1}}{\dot{m}_{da2}} \text{ or } \frac{\text{Line 1–3}}{\text{Line 1–2}} = \frac{\dot{m}_{da2}}{\dot{m}_{da3}} = \frac{1096}{1491} = 0.735$$

Consequently, the length of line segment 1–3 is 0.735 times the length of entire line 1–2. Using a ruler, State 3 is located, and the values t_3 = 65.9°F and t_3^* = 56.6°F found.

Adiabatic Mixing of Water Injected into Moist Air

Steam or liquid water can be injected into a moist airstream to raise its humidity. Figure 8 represents a diagram of this common air-conditioning process. If the mixing is adiabatic, the following equations apply:

$$\dot{m}_{da}h_1 + \dot{m}_w h_w = \dot{m}_{da}h_2$$

$$\dot{m}_{da}W_1 + \dot{m}_w = \dot{m}_{da}W_2$$

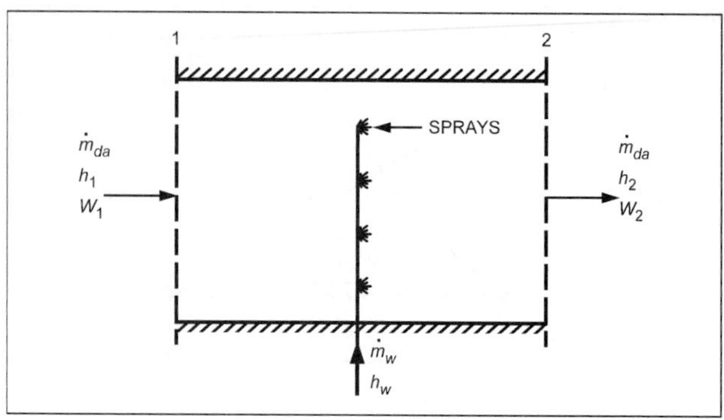

Fig. 8 Schematic Showing Injection of Water into Moist Air

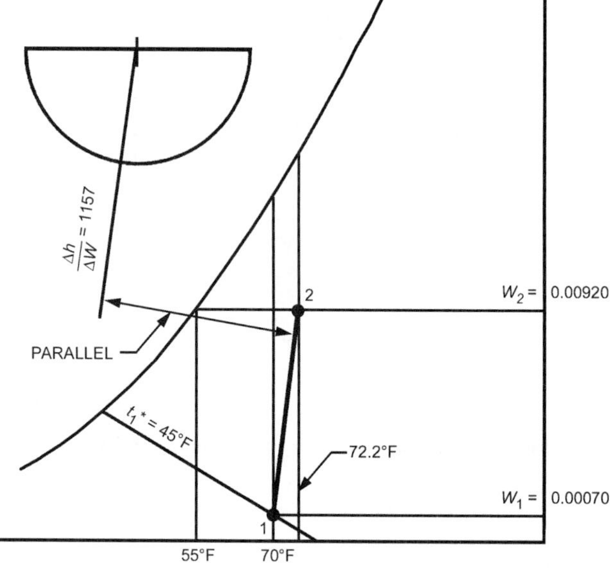

Fig. 9 Schematic Solution for Example 5

Therefore,

$$\frac{h_2 - h_1}{W_2 - W_1} = \frac{\Delta h}{\Delta W} = h_w \tag{46}$$

according to which, on the ASHRAE chart, the final state point of the moist air lies on a straight line whose direction is fixed by the specific enthalpy of the injected water, drawn through the initial state point of the moist air.

Example 5. Moist air at 70°F dry-bulb and 45°F thermodynamic wet-bulb temperature is to be processed to a final dew-point temperature of 55°F by adiabatic injection of saturated steam at 230°F. The rate of dry airflow $\dot{m}_{da}$ is 200 lb$_{da}$/min. Find the final dry-bulb temperature of the moist air and the rate of steam flow required.

Solution: Figure 9 shows the schematic solution. By Table 3, the enthalpy of the steam h_g = 1157 Btu/lb$_w$. Therefore, according to Equation (46), the condition line on the ASHRAE chart connecting States 1 and 2 must have a direction:

$$\Delta h/\Delta W = 1157 \text{ Btu/lb}_w$$

The condition line can be drawn with the $\Delta h/\Delta W$ protractor. First, establish the reference line on the protractor by connecting the origin with the value $\Delta h/\Delta W$ = 1157. Draw a second line parallel to the reference line and through the initial state point of the moist air. This second line is the condition line. State 2 is established at the intersection of the condition line with the horizontal line extended from the saturation curve at 55°F (t_{d2} = 55°F). Thus, t_2 = 72.2°F.

Values of W_2 and W_1 can be read from the chart. The required steam flow is

$$\dot{m}_w = \dot{m}_{da}(W_2 - W_1) = (200)(60)(0.00920 - 0.00070)$$
$$= 102 \text{ lb}_{steam}/\text{h}$$

Space Heat Absorption and Moist Air Moisture Gains

Air conditioning a space is usually determined by (1) the quantity of moist air to be supplied, and (2) the supply air condition necessary to remove given amounts of energy and water from the space at the exhaust condition specified.

Figure 10 schematically shows a space with incident rates of energy and moisture gains. The quantity q_s denotes the net sum of all rates of heat gain in the space, arising from transfers through boundaries and from sources within the space. This heat gain involves addition of energy alone and does not include energy contributions due to addition of water (or water vapor). It is usually called the **sensible heat gain**. The quantity $\Sigma\dot{m}_w$ denotes the net sum of all rates of moisture gain on the space arising from transfers through boundaries and from sources within the space. Each pound of water vapor added to the space adds an amount of energy equal to its specific enthalpy.

Assuming steady-state conditions, governing equations are

$$\dot{m}_{da}h_1 + q_s + \sum(\dot{m}_w h_w) = \dot{m}_{da}h_2$$
$$\dot{m}_{da}W_1 + \sum\dot{m}_w = \dot{m}_{da}W_2$$

or

$$q_s + \sum(\dot{m}_w h_w) = \dot{m}_{da}(h_2 - h_1) \quad (47)$$

$$\sum\dot{m}_w = \dot{m}_{da}(W_2 - W_1) \quad (48)$$

The left side of Equation (47) represents the total rate of energy addition to the space from all sources. By Equations (47) and (48),

$$\frac{h_2 - h_1}{W_2 - W_1} = \frac{\Delta h}{\Delta W} = \frac{q_s + \sum(\dot{m}_w h_w)}{\sum\dot{m}_w} \quad (49)$$

according to which, on the ASHRAE chart and for a given state of the withdrawn air, all possible states (conditions) for the supply air must lie on a straight line drawn through the state point of the withdrawn air, that has a direction specified by the numerical value of $[q_s + \Sigma(\dot{m}_w h_w)]/\Sigma\dot{m}_w$. This line is the condition line for the given problem.

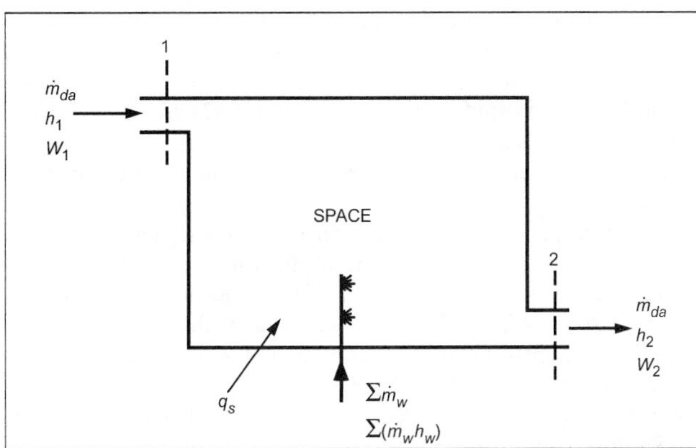

Fig. 10 Schematic of Air Conditioned Space

Example 6. Moist air is withdrawn from a room at 80°F dry-bulb temperature and 66°F thermodynamic wet-bulb temperature. The sensible rate of heat gain for the space is 30,000 Btu/h. A rate of moisture gain of 10 lb_w/h occurs from the space occupants. This moisture is assumed as saturated water vapor at 90°F. Moist air is introduced into the room at a dry-bulb temperature of 60°F. Find the required thermodynamic wet-bulb temperature and volume flow rate of the supply air.

Solution: Figure 11 shows the schematic solution. State 2 is located on the ASHRAE chart. From Table 3, the specific enthalpy of the added water vapor is h_g = 1100.40 Btu/lb_w. From Equation (49),

$$\frac{\Delta h}{\Delta W} = \frac{30{,}000 + (10)(1100.40)}{10} = 4100 \text{ Btu/lb}_w$$

With the $\Delta h/\Delta W$ protractor, establish a reference line of direction $\Delta h/\Delta W$ = 4100 Btu/lb_w. Parallel to this reference line, draw a straight line on the chart through State 2. The intersection of this line with the 60°F dry-bulb temperature line is State 1. Thus, t_1^* = 56.4°F.

An alternate (and approximately correct) procedure in establishing the condition line is to use the protractor's sensible-total heat ratio scale instead of the $\Delta h/\Delta W$ scale. The quantity $\Delta H_s/\Delta H_t$ is the ratio of the rate of sensible heat gain for the space to the rate of total energy gain for the space. Therefore,

$$\frac{\Delta H_s}{\Delta H_t} = \frac{q_s}{q_s + \Sigma(\dot{m}_w h_w)} = \frac{30{,}000}{30{,}000 + (10 \times 1100.40)} = 0.732$$

Note that $\Delta H_s/\Delta H_t$ = 0.732 on the protractor coincides closely with $\Delta h/\Delta W$ = 4100 Btu/lb_w.

The flow of dry air can be calculated from either Equation (47) or (48). From Equation (47),

$$\dot{m}_{da} = \frac{q_s + \Sigma(\dot{m}_w h_w)}{h_2 - h_1} = \frac{30{,}000 + (10 \times 1100.40)}{(60)(30.73 - 24.00)}$$
$$= 101.5 \text{ lb}_{da}/\text{min}$$

At State 1, v_1 = 13.29 $\text{ft}^3/\text{lb}_{da}$

Therefore, supply volume = $\dot{m}_{da}v_1$ = 101.5 × 13.29 = 1349 cfm

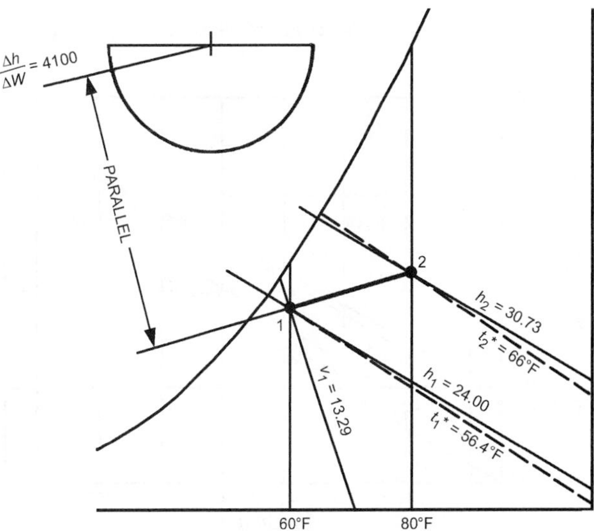

Fig. 11 Schematic Solution for Example 6

TRANSPORT PROPERTIES OF MOIST AIR

For certain scientific and experimental work, particularly in the heat transfer field, many other moist air properties are important.

Generally classified as transport properties, these include diffusion coefficient, viscosity, thermal conductivity, and thermal diffusion factor. Mason and Monchick (1965) derive these properties by calculation. Table 5 and Figures 12 and 13 summarize the authors' results on the first three properties listed. Note that, within the boundaries of ASHRAE Psychrometric Charts 1, 2, and 3, the viscosity varies little from that of dry air at normal atmospheric pressure, and the thermal conductivity is essentially independent of moisture content.

Table 5 Calculated Diffusion Coefficients for Water–Air at 29.921 in. Hg Barometric Pressure

Temp., °F	ft²/h	Temp., °F	ft²/h	Temp., °F	ft²/h
−100	0.504	40	0.884	140	1.205
−50	0.600	50	0.915	150	1.240
−40	0.655	60	0.942	200	1.414
−30	0.682	70	0.973	250	1.600
−20	0.709	80	1.008	300	1.794
−10	0.736	90	1.042	350	1.996
0	0.767	100	1.073	400	2.205
10	0.794	110	1.104	450	2.422
20	0.825	120	1.139	500	2.647
30	0.853	130	1.170		

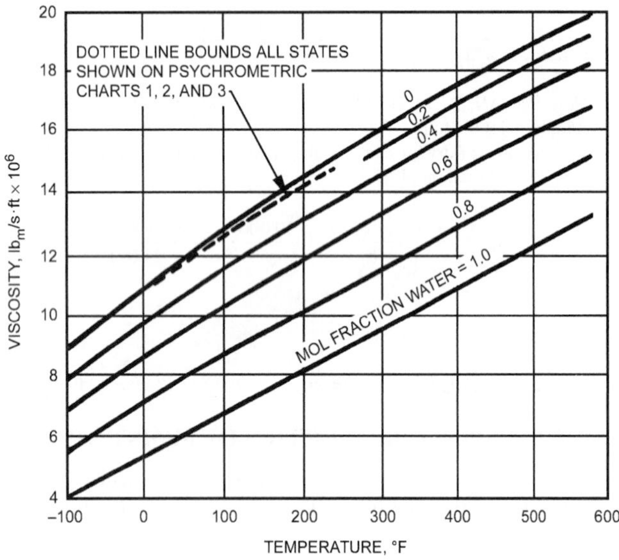

Fig. 12 Viscosity of Moist Air

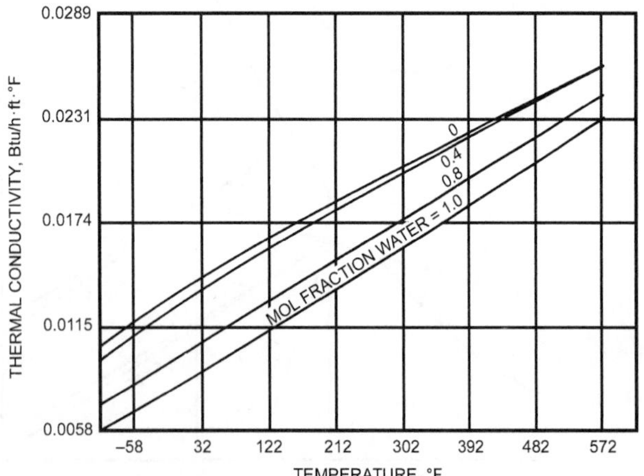

Fig. 13 Thermal Conductivity of Moist Air

REFERENCES FOR AIR, WATER, AND STEAM PROPERTIES

Coefficient f_w (over water) at pressures from 0.145 to 31.799 in. Hg for temperatures from −58 to 140°F (Smithsonian Institution).

Coefficient f_i (over ice) at pressures from 0.145 to 31.799 in. Hg for temperatures from 32 to 212°F (Smithsonian Institution).

Compressibility factor of dry air at pressures from 0.147 to 1470 psi and at temperatures from 90 to 5400°R (Hilsenrath et al. 1960).

Compressibility factor of moist air at pressures from 25 to 31 in. Hg at humidity ratios from 0.006 to 0.30 lb_w/lb_{da}, and for temperatures from 40 to 100°F. At pressures from 0 to 31.799 in. Hg, at values of degree of saturation from 0 to 100, and for temperatures from 32 to 140°F (Smithsonian Institution). [Note: At the time the Smithsonian Meteorological Tables were published, the value $\mu = W/W_s$ was known as relative humidity, in terms of a percentage. Since that time, there has been general agreement to designate the value μ as degree of saturation, usually expressed as a decimal and sometimes as a percentage. See Goff (1949) for more recent data and formulations.]

Compressibility factor for steam at pressures from 14.696 to 440 psi and at temperatures from 684 to 1530°R (Hilsenrath et al. 1960).

Density, enthalpy, entropy, Prandtl number, specific heat, specific heat ratio, and viscosity of dry air (Hilsenrath et al. 1960).

Density, enthalpy, entropy, specific heat, viscosity, thermal conductivity, and free energy of steam (Hilsenrath et al. 1960).

Dry air. Thermodynamic properties over a wide range of temperature (Keenan and Kaye 1945).

Enthalpy of saturated steam (Osborne et al. 1939).

Ideal-gas thermodynamic functions of dry air at temperatures from 18 to 5400°R (Hilsenrath et al. 1960).

Ideal-gas thermodynamic functions of steam at temperatures from 90 to 9000°R. Functions included are specific heat, enthalpy, free energy, and entropy (Hilsenrath et al. 1960).

Moist air properties from tabulated virial coefficients (Chaddock 1965).

Saturation humidity ratio over ice at pressures from 8.672 to 28.908 in. Hg and for temperatures from −128 to 32°F (Smithsonian Institution).

Saturation humidity ratio over water at pressures from 1.73 to 30.35 in. Hg and for temperatures from −58 to 138.2°F (Smithsonian Institution).

Saturation vapor pressure over water in in. Hg andfor temperatures from −60 to 212°F (Smithsonian Institution).

Speed of sound in dry air at pressures from 0.147 to 1470 psi for temperatures from 90 to 5400°R (Hilsenrath et al. 1960). At atmospheric pressure for temperatures from −130 to 140°F (Smithsonian Institution).

Speed of sound in moist air. Relations using the formulation of Goff and Gratch and studies by Hardy et al. (1942) give methods for calculating this speed (Smithsonian Institution).

Steam tables covering the range from −40 to 2400°F (Keenan et al. 1969).

Transport properties of moist air. Diffusion coefficient, viscosity, thermal conductivity, and thermal diffusion factor of moist air are listed (Mason and Monchick 1965). The authors' results are summarized in Table 5 and Figures 12 and 13.

Virial coefficients and other information for use with Goff and Gratch formulation (Goff 1949).

Volume of water in cubic feet for temperatures from 14 to 482°F (Smithsonian Institution 1954).

Water properties. Includes properties of ordinary water substance for the gaseous, liquid, and solid phases (Dorsey 1940).

SYMBOLS

C_1 to C_{18} = constants in Equations (5), (6), and (37)
d_v = absolute humidity of moist air, mass of water per unit volume of mixture
f = enhancement factor, used in Equations (23a) and (25a)
h = specific enthalpy of moist air
h_s^* = specific enthalpy of saturated moist air at thermodynamic wet-bulb temperature
h_w^* = specific enthalpy of condensed water (liquid or solid) at thermodynamic wet-bulb temperature and pressure of 29.921 in. Hg
H_s = rate of sensible heat gain for space
H_t = rate of total energy gain for space
$\dot{m}_{da}$ = mass flow of dry air, per unit time
$\dot{m}_w$ = mass flow of water (any phase), per unit time
M_{da} = mass of dry air in moist air sample
M_w = mass of water vapor in moist air sample
$n = n_{da} + n_w$, total number of moles in moist air sample
n_{da} = moles of dry air
n_w = moles of water vapor
p = total pressure of moist air
p_{da} = partial pressure of dry air
p_s = vapor pressure of water in moist air at saturation. Differs from saturation pressure of pure water because of presence of air.
p_w = partial pressure of water vapor in moist air
p_{ws} = pressure of saturated pure water
q_s = rate of addition (or withdrawal) of sensible heat
R = universal gas constant, 1545.32 $\text{ft}\cdot\text{lb}_f/\text{lb mole}\cdot°\text{R}$
R_{da} = gas constant for dry air
R_w = gas constant for water vapor
s = specific entropy
t = dry-bulb temperature of moist air
t_d = dew-point temperature of moist air
t^* = thermodynamic wet-bulb temperature of moist air
T = absolute temperature
v = specific volume
v_T = total gas volume
V = total volume of moist air sample
W = humidity ratio of moist air, mass of water per unit mass of dry air
W_s^* = humidity ratio of moist air at saturation at thermodynamic wet-bulb temperature
x_{da} = mole-fraction of dry air, moles of dry air per mole of mixture
x_w = mole-fraction of water, moles of water per mole of mixture
x_{ws} = mole-fraction of water vapor under saturated conditions, moles of vapor per mole of saturated mixture
Z = altitude
$\alpha = \ln(p_w)$, parameter used in Equations (37) and (38)
γ = specific humidity of moist air, mass of water per unit mass of mixture
μ = degree of saturation W/W_s, dimensionless
ρ = moist air density
ϕ = relative humidity

Subscripts

as = difference between saturated moist air and dry air
da = dry air
f = saturated liquid water
fg = difference between saturated liquid water and saturated water vapor
g = saturated water vapor
i = saturated ice
ig = difference between saturated ice and saturated water vapor
s = saturated moist air
t = total
w = water in any phase

REFERENCES

Chaddock, J.B. 1965. Moist air properties from tabulated virial coefficients. *Humidity and moisture measurement and control in science and industry* 3:273. A. Wexler and W.A. Wildhack, eds. Reinhold Publishing, New York.

Dorsey, N.E. 1940. *Properties of ordinary water substance*. Reinhold Publishing, New York.

Goff, J.A. 1949. Standardization of thermodynamic properties of moist air. *Heating, Piping, and Air Conditioning* 21(11):118.

Goff, J.A. and S. Gratch. 1945. Thermodynamic properties of moist air. ASHVE *Transactions* 51:125.

Goff, J.A., J.R. Anderson, and S. Gratch. 1943. Final values of the interaction constant for moist air. *ASHVE Transactions* 49:269.

Haines, R.W. 1961. How to construct high altitude psychrometric charts. *Heating, Piping, and Air Conditioning* 33(10):144.

Hardy, H.C., D. Telfair, and W.H. Pielemeier. 1942. The velocity of sound in air. *Journal of the Acoustical Society of America* 13:226.

Harrison, L.P. 1965. Fundamental concepts and definitions relating to humidity. *Humidity and moisture measurement and control in science and industry* 3:3. A. Wexler and W.A. Wildhack, eds. Reinhold Publishing, New York.

Hilsenrath, J. et al. 1960. Tables of thermodynamic and transport properties of air, argon, carbon dioxide, carbon monoxide, hydrogen, nitrogen, oxygen, and steam. National Bureau of Standards. *Circular* 564, Pergamon Press, New York.

Hyland, R.W. and A. Wexler. 1983a. Formulations for the thermodynamic properties of dry air from 173.15 K to 473.15 K, and of saturated moist air from 173.15 K to 372.15 K, at pressures to 5 MPa. *ASHRAE Transactions* 89(2A):520-35.

Hyland, R.W. and A. Wexler. 1983b. Formulations for the thermodynamic properties of the saturated phases of H_2O from 173.15 K to 473.15 K. *ASHRAE Transactions* 89(2A):500-519.

Karig, H.E. 1946. Psychrometric charts for high altitude calculations. *Refrigerating Engineering* 52(11):433.

Keenan, J.H. and J. Kaye. 1945. *Gas tables*. John Wiley and Sons, New York.

Keenan, J.H., F.G. Keyes, P.G. Hill, and J.G. Moore. 1969. *Steam tables*. John Wiley and Sons, New York.

Kuehn, T.H., J.W. Ramsey, and J.L. Threlkeld. 1998. *Thermal environmental engineering*, 3rd ed., p. 188. Prentice-Hall, Upper Saddle River, NJ.

Kusuda, T. 1970. Algorithms for psychrometric calculations. NBS *Publication* BSS21 (January) for sale by Superintendent of Documents, U.S. Government Printing Office, Washington, D.C.

Mason, E.A. and L. Monchick. 1965. Survey of the equation of state and transport properties of moist gases. *Humidity and moisture measurement and control in science and industry* 3:257. Reinhold Publishing, New York.

NASA. 1976. U.S. Standard atmosphere, 1976. National Oceanic and Atmospheric Administration, National Aeronautics and Space Administration, and the United States Air Force. Available from National Geophysical Data Center, Boulder, CO.

NIST. 1990. Guidelines for realizing the international temperature scale of 1990 (ITS-90). NIST *Technical Note* 1265. National Institute of Technology and Standards, Gaithersburg, MD.

Osborne, N.S. 1939. Stimson and Ginnings. Thermal properties of saturated steam. *Journal of Research*, National Bureau of Standards, 23(8):261.

Olivieri, J. 1996. *Psychrometrics—Theory and practice*. ASHRAE, Atlanta.

Palmatier, E.P. 1963. Construction of the normal temperature. ASHRAE psychrometric chart. *ASHRAE Journal* 5:55.

Peppers, V.W. 1988. Unpublished paper. Available from ASHRAE.

Preston-Thomas, H. 1990. The international temperature scale of 1990 (ITS-90). *Metrologia* 27(1):3-10.

Rohsenow, W.M. 1946. Psychrometric determination of absolute humidity at elevated pressures. *Refrigerating Engineering* 51(5):423.

Smithsonian Institution. 1954. *Smithsonian physical tables*, 9th rev. ed. Available from the Smithsonian Institution, Washington, D.C.

Smithsonian Institution. *Smithsonian meteorological tables*, 6th rev. ed. Out of print, but available in many libraries. Washington, D.C.

CHAPTER 7

SOUND AND VIBRATION

IF FUNDAMENTAL PRINCIPLES of sound and vibration control are applied in the design, installation, and use of HVAC and refrigeration systems, unacceptably high noise and vibration levels and the consequent complaints can be avoided. This chapter introduces these fundamental principles, including characteristics of sound, basic definitions and terminology, human response to sound, acoustical design goals, and vibration isolation fundamentals. Chapter 46 of the 1999 *ASHRAE Handbook—Applications* and the references listed at the end of this chapter contain technical discussions, tables, and design examples hellpful to HVAC designers.

ACOUSTICAL DESIGN OBJECTIVE

The primary objective for the acoustical design of HVAC systems and equipment is to ensure that the acoustical environment in a given space is not degraded. Sound and vibration are created by a **source**, are transmitted along one or more **paths**, and reach a **receiver**. Treatments and modifications can be applied to any or all of these elements to achieve an acceptable acoustical environment, although it is usually most effective and least expensive to reduce noise at the source.

CHARACTERISTICS OF SOUND

Sound is a propagating disturbance in a fluid (gas or liquid) or in a solid. In fluid media, the disturbance travels as a longitudinal compression wave. Sound in air is called **airborne sound** or simply sound. It is generated by a vibrating surface or a turbulent fluid stream. In solids, sound can travel as bending waves, compressional waves, torsional waves, shear waves and others. Sound in solids is generally called **structureborne sound**. In HVAC system design, **both** airborne and structureborne sound propagation are important.

Speed

The speed of a longitudinal wave in a fluid is a function of the fluid's density and bulk modulus of elasticity. In air, at room temperature, the speed of sound is about 1100 fps; in water, it is about 5000 fps. In solids, there are several different types of waves, each with a different speed and some that depend on frequency. In solids, the speed of sound is usually higher than that in air.

Sound Pressure and Sound Pressure Level

Sound waves in air are variations in pressure above and below atmospheric pressure. **Sound pressure** is measured in pascals (Pa) (SI units are used rather than I-P units because of international agreement). The human ear responds across a broad range of sound pressures; the threshold of hearing to the threshold of pain covers a range of approximately 10^{14}:1. Table 1 gives approximate values of sound pressure for various sources.

The preparation of this chapter is assigned to TC 2.6, Sound and Vibration Control.

Table 1 Typical Sound Pressures and Sound Pressure Levels

Source	Sound Pressure, Pa	Sound Pressure Level, dB re 20 μPa	Subjective Reaction
Military jet takeoff at 100 ft	200	140	Extreme danger
Artillery fire at 10 ft	63.2	130	
Passenger jet takeoff at 100 ft	20	120	Threshold of pain
Loud rock band	6.3	110	Threshold of discomfort
Platform of subway station (steel wheels)	2	100	
Unmuffled large diesel engine at 130 ft	0.6	90	Very loud
Computer printout room	0.2	80	
Freight train at 100 ft	0.06	70	
Conversational speech at 3 ft	0.02	60	
Window air conditioner at 3 ft	0.006	50	Moderate
Quiet residential area	0.002	40	
Whispered conversation at 6 ft	0.0006	30	
Buzzing insect at 3 ft	0.0002	20	Perceptible
Threshold of good hearing	0.00006	10	Faint
Threshold of excellent youthful hearing	0.00002	0	Threshold of hearing

The range of sound pressure in Table 1 is so large that it is more convenient to use a scale that is proportional to the logarithm of this quantity. The **decibel** (dB) scale is such a scale and is the preferred method of presenting quantities in acoustics. The term **level**, when used in relation to sound power, sound intensity, or sound pressure, indicates that dB notation is being used. Numerically, the decibel is ten times the base 10 logarithm of the ratio of two like quantities proportional to acoustical power or energy. Thus, the **sound pressure level** L_p (in dB) corresponding to a sound pressure is given by

$$L_p = 10\log(p/p_{ref})^2 \tag{1}$$

where p is the root mean square (rms) value of acoustic pressure in pascals. The ratio p/p_{ref} is squared to give quantities proportional to intensity or energy. The reference quantity p_{ref} is 20 μPa, which corresponds to the approximate threshold of hearing.

The decibel scale is used for many different descriptors relating to sound: source strength, sound level, and sound attenuation, among others; each has a different reference quantity. For this reason, it is important to be aware of the context in which the term decibel or level is used. For most acoustical quantities, there is an internationally accepted reference value. A reference quantity is always implied even if it does not appear.

Sound pressure level is relatively easy to measure and for this reason most noise codes and criteria use sound pressure level. (The human ear and microphones are pressure-sensitive devices.) Sound pressure levels for each source are also given in Table 1.

Frequency

Frequency is the number of oscillations (or cycles) completed per second by a vibrating object. The international unit for frequency is hertz (Hz) with dimension s^{-1}. When the motion of vibrating air particles is simple harmonic, the sound is said to be a **pure tone** and the sound pressure p as a function of time and frequency can be described by

$$p(\theta, f) = p_0 \sin 2\pi f \theta \quad (2)$$

where f is frequency in hertz and θ is time in seconds.

The **audible frequency range** for humans extends from about 20 Hz to 20 kHz. In some cases, infrasound (< 20 Hz) or ultrasound (> 20 kHz) are important, but methods and instrumentation for these frequency regions are specialized and are not considered here.

Wavelength

The wavelength of sound in a medium is the distance between successive maxima or minima of a simple harmonic disturbance propagating in that medium. Wavelength, speed, and frequency are related by

$$\lambda = c/f \quad (3)$$

where

λ = wavelength, ft
c = speed of sound, fps
f = frequency, Hz

Sound Power and Sound Power Level

The **sound power** of a source is its rate of emission of acoustical energy and is expressed in watts. The sound power of a source depends on the operating conditions. Approximate sound power outputs for common sources are shown in Table 2 together with the corresponding sound power levels. For **sound power level** L_w the power reference is 10^{-12} W. The definition of sound power level is therefore

$$L_w = 10 \log(w/10^{-12}) \quad (4)$$

where w is the acoustic power emitted by the source in watts. The acoustic power emitted by a source is not the same as the power consumed by the source. Only a small fraction of the consumed power is converted into sound. For example, a loudspeaker rated at 100 W may be only 1% to 5% efficient, generating only 1 to 5 W (acoustic).

Most mechanical equipment is rated according to sound power so equipment can be compared according to a common reference independent of distance and acoustical conditions in the room. AMCA *Publication* 303 provides guidelines for the application of sound power level ratings. In addition, AMCA *Standard* 301 provides standard methods for developing fan sound ratings from laboratory test data.

Sound Intensity and Sound Intensity Level

The **sound intensity** I at a point in a specified direction is the rate of flow of sound energy through unit area at that point. The unit area is perpendicular to the specified direction, and the units of intensity are W/m^2. (SI units are used rather than I-P units because of international agreement.) **Sound intensity level** L_I is expressed in dB with a reference quantity of 10^{-12} W/m^2, thus

$$L_I = \log(I/10^{-12})$$

Table 2 Typical Sound Power Outputs and Sound Power Levels

Source	Sound Power, W	Sound Power Level, dB re 10^{-12} W
Saturn rocket	10^8	200
Turbojet engine[a]	10^5	170
Jet aircraft at takeoff[b]	10^4	160
Turboprop at takeoff	1000	150
Prop aircraft at takeoff[c]	100	140
Large pipe organ	10	130
Small aircraft engine	1	120
Noisy HVAC fan	0.1	110
Automobile at highway speed	0.01	100
Voice, shouting	0.001	90
Garbage disposal unit	10^{-4}	80
Voice, conversation level	10^{-5}	70
Electronic equipment ventilation fan	10^{-6}	60
Office air diffuser	10^{-7}	50
Small electric clock	10^{-8}	40
Voice, soft whisper	10^{-9}	30
Rustling leaves	10^{-10}	20
Human breath	10^{-11}	10

[a] With afterburner [b] Four jet engines [c] Four propeller engines

Combining Sound Levels

Because the decibel is a logarithmic unit, two sound pressure levels cannot be added arithmetically. The levels must first be converted back to energy units, summed, and then converted to a level again. Thus, the combination of two levels, L_1 and L_2, produces a level L_{sum} given by

$$L_{sum} = 10 \log(10^{L_1/10} + 10^{L_2/10}) \quad (5)$$

This process may be extended to as many levels as needed. A simpler and slightly less accurate method is outlined in Table 3. This method, although not exact, results in errors of 1 dB or less. The process with a series of levels may be shortened by combining the largest with the next largest. Then this sum is combined with the third largest, then the fourth largest, and so on until the next level added has little or no influence. The process may then be stopped.

The procedures outlined in Table 3 and in Equation (5) are valid if the individual sound levels are not highly correlated. This is true for most (but not all) sounds encountered in HVAC work. One notable exception is the pure tone. If two or more noise signals contain pure tones at the same frequency, the combined sound level is a function of not only the level of each tone, but also the phase difference between the tones. The combined sound levels from two tones of equal amplitude can range from zero (if the two tones are 180° out of phase) up to 6 dB greater than the level of either tone (if the two tones are exactly in phase). When two tones of similar amplitude are very close in frequency, but not exactly the same, the combined sound level oscillates as the tones move in and out of phase. This effect creates an audible "beating" with a period equal to the inverse of the difference in frequency between the two tones.

Table 3 Combining Two Sound Levels

Difference between two levels to be combined, dB	0 to 1	2 to 4	5 to 9	10 and More
Number of decibels to be added to highest level to obtain combined level	3	2	1	0

MEASURING SOUND

Instrumentation

The basic instrument for measuring sound is a **sound level meter**, which includes a microphone, electronic circuitry, and a display device. Sound pressure at a point is converted to sound pressure level and displayed by analog or digital meters. These devices are usually light, battery-operated, hand-held units with outputs that vary in complexity depending on cost and the level of current technology.

Time-Averaging

No sounds are constant; the pressure fluctuates from moment to moment and the level can vary quickly or slowly. Sound level meters can show the time fluctuations of the sound pressure level using specified time constants (slow, fast, impulse), or can hold the maximum or minimum level recorded during some specified interval. All sound level meters perform some kind of **time-averaging**. Some **integrating sound level meters** take an average of the sound pressure level over a user-definable time, then hold and display the result. As a result, an integrating meter is easier to read and more repeatable (especially if the measurement period is long). The quantity measured by the integrating sound level meter is the **equivalent continuous sound pressure level** L_{eq}.

Spectra and Analysis Bandwidths

Real sounds are much more complex than simple pure tones. **Broadband sound** contains energy at many frequencies, usually covering most of the audible frequency range. All sounds, however, can be represented as a summation of pure tones with different amplitudes. This representation of a sound is called **frequency or spectral analysis** and is similar to spectral analysis in optics.

A **constant-bandwidth analysis** expresses the energy content of a sound as a spectrum where each data point represents the same spectral width, for example, 1 Hz. This kind of analysis is useful when an objectionable sound obviously contains strong tones and the frequencies need to be accurately identified before remedial action is taken. A constant-band spectrum usually contains too much information for typical noise control work.

Measurements for most HVAC noise control work are usually made with filters that extract the energy in either **octave bands** or **one-third octave bands**. An octave band is a frequency band with its upper frequency limit twice that of its lower frequency limit. Octave and 1/3 octave bands are identified by the mid-band frequency, which is the geometric mean of the upper and lower band limits (ANSI *Standards* S1.6, S1.11). Three 1/3 octave bands make up an octave band. Table 4 lists the preferred series of octave and one-third octave bands and the upper and lower band limit frequencies. For HVAC sound measurements, filters for the range of 16 Hz to 8000 Hz are usually adequate.

While analysis in octave bands is usually acceptable for rating acoustical environments in rooms, 1/3 octave band analysis is often useful in product development and for remedial investigations.

Some sound level meters have octave or 1/3 octave filters for determining the frequency content of the sound. Usually, standard broadband filters that simulate the response of the average human ear to sound are provided. The most commonly used filter is the **A-weighting** filter (Figure 1). This filter simulates the response of the human ear. It de-emphasizes the low-frequency portions of a sound spectrum, automatically compensating for the lower sensitivity of the human ear to low-frequency sound.

The **C-weighting** filter is sometimes used to estimate whether a particular sound has excessive low-frequency energy present when a spectrum analyzer is not available. If the difference between the C- and A-weighted levels for the sound exceeds about 30 dB, then the sound is likely to be annoying because of excessive low-frequency noise. Note (in Figure 1) that the C-weighting curve attenuates significantly at low and high frequencies; the instrument response is not flat when this filter is used.

Sound level meters are available in several accuracy grades (ANSI *Standard* S1.4). A Type 1 meter has a tolerance of about ±0.7 dB. The general-purpose meter, which is less expensive, is designated Type 2 with a tolerance of about ±1 dB, and is adequate for most HVAC sound measurements.

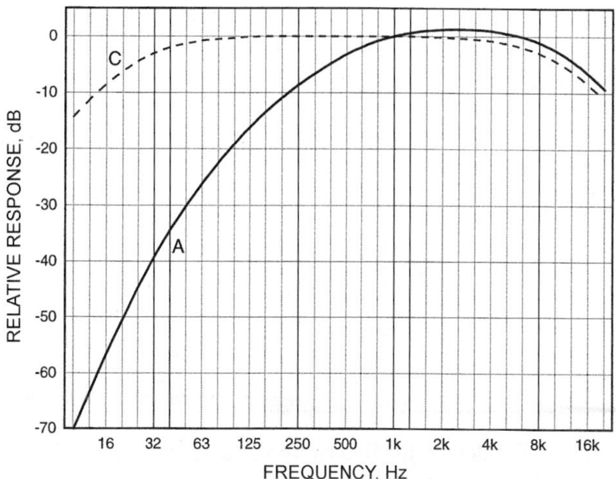

Fig. 1 Curves Showing A- and C-Weighting Responses for Sound Level Meters

Table 4 Mid-Band and Approximate Upper and Lower Cutoff Frequencies for Octave and 1/3 Octave Band Filters

Octave Bands, Hz			1/3 Octave Bands, Hz		
Lower	Mid-Band	Upper	Lower	Mid-Band	Upper
			11.2	12.5	14
11.2	16	22.4	14	16	18
			18	20	22.4
			22.4	25	28
22.4	31.5	45	28	31.5	35.5
			35.5	40	45
			45	50	56
45	63	90	56	63	71
			71	80	90
			90	100	112
90	125	180	112	125	140
			140	160	180
			180	200	224
180	250	355	224	250	280
			280	315	355
			355	400	450
355	500	710	450	500	560
			560	630	710
			710	800	900
710	1,000	1,400	900	1,000	1,120
			1,120	1,250	1,400
			1,400	1,600	1,800
1,400	2,000	2,800	1,800	2,000	2,240
			2,240	2,500	2,800
			2,800	3,150	3,550
2,800	4,000	5,600	3,550	4,000	4,500
			4,500	5,000	5,600
			5,600	6,300	7,100
5,600	8,000	11,200	7,100	8,000	9,000
			9,000	10,000	11,200
			11,200	12,500	14,000
11,200	16,000	22,400	14,000	16,000	18,000
			18,000	20,000	22,400

Table 5 Combining Decibels to Determine Overall Sound Pressure Level

Octave Band Frequency, Hz	Octave Band Level L_p, dB	$10^{L_p/10}$		
63	85	3.2×10^8	=	0.32×10^9
125	90	1.0×10^9	=	1.0×10^9
250	92	1.6×10^9	=	1.6×10^9
500	87	5.0×10^8	=	0.5×10^9
1000	82	1.6×10^8	=	0.16×10^9
2000	78	6.3×10^7	=	0.06×10^9
4000	65	3.2×10^6	=	0.003×10^9
8000	54	2.5×10^5	=	0.0002×10^9
				3.6432×10^9
				$10 \log (3.6 \times 10^9) = 96$ dB

Table 6 Guidelines for Determining Equipment Sound Levels in the Presence of Contaminating Background Sound

Measurement A minus Measurement B	Correction to Measurement A to Obtain Equipment Sound Level
10 dB or more	0 dB
6 to 9 dB	–1 dB
4 to 5 dB	–2 dB
Less than 4 dB	Equipment sound level is more than 2 dB below Measurement A

Measurement A = Tested equipment plus background sound
Measurement B = Background sound alone

Manually selecting filters sequentially to cover the frequency range from 16 Hz to 8000 Hz is time-consuming. An instrument that gives all filtered levels simultaneously is called a **real-time analyzer** (RTA). It speeds up measurement significantly and, on most models, the information can be saved to an internal or external digital storage device.

The process described in the previous section on Combining Sound Levels can be applied to a set of octave or 1/3 octave bands to calculate the overall broadband level (see Table 5 for an example).

Sound Measurement Basics

The sound pressure level in an occupied space can be measured directly with a sound level meter, or estimated from published sound power data after accounting for room volume, distance from the source, and other acoustical factors. Sound level meters measure the sound pressure at the microphone location. Estimation techniques calculate sound pressure at a specified point in an occupied space. Measured or estimated sound pressure levels in frequency bands can then be plotted, analyzed, and compared with established criteria for acceptance.

Measurements of HVAC sound must be done carefully to ensure repeatable and accurate results. The sound levels may not be steady, particularly at low frequencies (250 Hz and lower) and can vary significantly with time. In such cases, both the peak and average levels should be recorded.

Sophisticated sound measurements and their procedures should be carried out by experienced sound professionals. At present, only a few noise standards apply to measuring the interior noise from mechanical equipment (ASTM *Standard* E 1574, ASTM *Standard* E 1573). Most manuals for sound level meters include sections on how to measure sound.

Determining the sound spectrum in a room or investigating a noise complaint usually requires measuring the sound pressure levels in the octave bands from 16 Hz to 8000 Hz. In cases of tonal noise or rumble, narrow bands or 1/3 octave bands should be measured because their frequency resolution is higher. Whatever the measurement method, the sound pressure level is measured at a point. In a room, each measurement point often has a different sound pressure level, so the precise location of the measurement must be detailed in the report. One might survey the room and record the location and level of the loudest position. Also, one could establish a few representative locations where occupants are normally situated. In general, the most appropriate height is 4 to 6 ft above the floor. The exact geometric center of the room should be avoided, as should any location within 3 ft of a wall.

Wherever the location, it must be defined and recorded. If the meter has an integrating-averaging function, one can use a rotating boom to sample a large area or walk slowly around the room to measure the average sound pressure level over that path. However, care must be taken that no extraneous sounds are generated by microphone movement or by walking. Also, locations where the sound levels are notably higher than average should be recorded. The section on Measurement of Room Sound Pressure Level has more details.

The contribution of other sources (plumbing noise, business machines, nearby traffic, etc.) to the sound pressure levels must also be determined. Sound from sources other than the source to be measured is designated background sound.

Sometimes the sound from a particular piece of HVAC equipment must be measured in the presence of background sound from other sources that cannot be turned off, such as automobile traffic or certain business machines. Two sets of measurements are required to determine the sound level due to selected equipment: one set with both the HVAC equipment sound and the background sound and another set with only the background sound (the HVAC equipment is turned off). This situation might also occur, for example, when determining whether the noise exposure at the property line due to a cooling tower meets a local noise ordinance. The guidelines in Table 6 will help in determining the sound level of a particular machine in the presence of background sound.

The uncertainty associated with correcting for background sound depends on the uncertainty of the measuring instrument and the steadiness of the sounds being measured. In favorable circumstances, it might be possible to extend Table 6. In particularly unfavorable circumstances, even values obtained from the table could be substantially in error.

Measuring sound emissions from a particular piece of equipment or group of equipment requires a measurement plan specific to the situation. The section on Standards lists several sound level measurement procedures for various laboratory and field sound measurement situations.

Outdoor measurements are somewhat easier to make than indoor because typically there are few or no boundary surfaces to affect sound build-up or absorption. Nevertheless, important issues such as the effect of large, nearby sound-reflecting surfaces and weather conditions such as wind, temperature, and precipitation must be considered. (In cases where measurements are made close to extended surfaces, sound pressure levels can be significantly increased.) These effects can be estimated through guidelines in many sources such as Harris (1991).

Measurement of Room Sound Pressure Level

In the commissioning of HVAC systems in buildings, it often must be demonstrated that a specified room noise criterion has been met. Measurement procedures for obtaining the data to demonstrate compliance are often not specified. This can lead to confusion when different parties make measurements using different procedures, as the results often do not agree. The problem is that most rooms exhibit significant point-to-point variation in sound pressure level.

When a noise has no audible tonal components, the differences in measured sound pressure level at several locations in a room may be as high as 3 to 5 dB. However, when audible tonal components are present, especially at low frequencies, the variations due to standing waves may exceed 10 dB. These variations are generally noticeable to the average listener when moving through the room.

Although the commissioning procedures usually set precise limits for demonstrating compliance, the outcome can be controversial unless the measurement procedure has been specified in detail. At present, the industry has no general agreement regarding an acoustical measurement procedure for commissioning HVAC systems. However, ARI *Standard* 885 incorporates a "suggested procedure for field verification of NC/RC levels."

DETERMINING SOUND POWER

No instrument can measure the sound power of a source directly. Rather, sound power is calculated from several measurements of sound pressure or sound intensity created by a source in one of several test environments. The following four methods are commonly used.

1. Free-Field Method

A **free field** is a sound field where the effects of any boundaries are negligible over the frequency range of interest. In ideal conditions, there are no boundaries. Free-field conditions can be approximated in rooms having highly sound-absorbing walls, floor, and ceiling (**anechoic rooms**). In a free field the sound power of a sound source can be determined from a number of measurements of sound pressure level on an imaginary spherical surface centered on and surrounding the source. This method is based on the fact that, because absorption of sound in air can be practically neglected at small distances from the sound source, the sound power generated by a source must flow through an imagined sphere with the source at its center. The intensity of the sound is determined at each of the measuring points around the source and multiplied by the area of the imagined sphere associated with the measuring points. Total sound power is the sum of these products for each point.

ANSI *Standard* S12.35 describes various methods used to calculate the sound power level under free-field conditions. Measurement accuracy is limited at the lower frequencies because room surface treatments do not have high sound absorption coefficients at low frequencies. For example, a glass fiber wedge structure that gives significant absorption at 70 Hz must be at least 4 ft long.

Using values for the speed of sound for air at 68°F and 14.7 psi, the relationship between sound power level and sound pressure level for a nondirectional sound source is

$$L_w = L_p + 20\log r + 0.5 \tag{6}$$

where

L_w = sound power level, (dB re 10^{-12} W)
L_p = sound pressure level (dB re 20 μPa)
r = distance from nondirectional sound source, ft

Free-Field Over Reflecting Plane. In many cases, a completely free field is not available, and measurements can only be made in a free field over a reflecting plane. That is, the sound source is placed on a hard floor (in an otherwise sound-absorbing room) or on pavement outdoors. Since the sound is then radiated into a hemisphere rather than a full sphere, the relationship for L_w and L_p for a nondirectional sound source becomes

$$L_w = L_p + 20\log r - 2.5 \tag{7}$$

Source Directivity. A sound source may radiate different amounts of sound power in different directions because various areas of its surface do not vibrate at the same level or in phase. A directivity pattern can be established by measuring sound pressure under free-field conditions, either in an anechoic room or over a reflecting plane in a hemi-anechoic space at several points around the source. The **directivity factor** Q is defined as the ratio of sound pressure at a given angle from the sound source to the sound pressure that would be produced by the same source radiating uniformly in all directions. Q is a function of both frequency and direction. Chapter 46 of the 1999 *ASHRAE Handbook—Applications* provides more detailed information on sound source directivity.

2. Reverberation Room Method

Another method to determine sound power places the sound source in a reverberation room. Standardized methods for determining the sound power of HVAC equipment in reverberation rooms are given in ANSI *Standard* S12.31, when the sound source contains mostly broadband sound; ANSI *Standard* S12.32, when tonal sound is prominent; and AMCA *Standard* 300 for testing fans.

Sound sources that can be measured by these methods include room air conditioners, refrigeration compressors, components of central HVAC systems, and air terminal devices. AMCA *Standard* 300, ASHRAE *Standard* 130, and ARI *Standard* 880 establish special measuring procedures for some of these units. Large equipment that can operate on a large paved area, such as a parking lot, can also be measured under free-field conditions above a reflecting plane. Determining the sound power of large equipment outdoors is difficult; however, data may be available from some manufacturers.

A **diffuse field** is a sound field in which the sound intensity is the same in all directions and at every point. The **reverberant field** in a reverberation room, because of its sound reflecting walls, floor, and ceiling, provides an approximation to a diffuse field.

Direct Method. With the source in a reverberation room, the sound pressure level is measured at some minimum distance from the source and the surfaces of the room. The sound power level is calculated from the sound pressure level, if the rate of decay of sound (in dB/s) and the volume of the reverberation room are known.

Using the direct method, the relationship between sound power level and sound pressure level in a reverberation room is given by

$$L_w = L_p + 10\log V + 10\log D - 47.3 \quad \text{(direct method)} \tag{8}$$

where

L_p = sound pressure level averaged over room, dB re 20 μPa
V = volume of room, ft^3
D = decay rate, dB/s = 60/θ
θ = room reverberation time (time required for a 60 dB decay), s

Substitution Method. Most test standards use a calibrated **reference sound source** (RSS) to determine the sound power level of a device under test, such as a fan. The most common reference sound source is a small, direct-drive fan impeller that has no volute housing or scroll. The impeller is a forward-curved design, and a choke plate is installed on the inlet face of the impeller. The choke plate causes the fan to operate in a rotating-stall condition that is very noisy. The reference source is designed to have a stable sound power level output from 63 Hz to 8000 Hz and a relatively uniform frequency spectrum in each octave band.

To determine the sound power level of a given source, sound pressure level measurements are first made near the location of the given source with the reference sound source operating in the test room. Then the reference source is turned off and the measurements are repeated with the given source in operation. The differences in measured sound pressure levels between the reference source and unknown source represent differences in sound power level between the two. This procedure for sound power level determination is known as the **substitution method**.

Using this method, the relationship between sound power level and sound pressure level for the two sources is given by:

$$L_w = L_p + (L_w - L_p)_{ref} \quad \text{(substitution method)} \tag{9}$$

where

L_p = sound pressure level averaged over room, dB re 20 μPa
$(L_w - L_p)_{ref}$ = difference between sound power level and sound pressure level of reference sound source

3. Progressive Wave Method

By attaching a fan to one end of a duct, the sound energy is confined to a progressive wave field in the duct. Fan sound power can then be determined by measuring the sound pressure level inside the duct. The method is described in ASHRAE *Standard* 68 (AMCA *Standard* 330). This method is not commonly used because of difficulties in constructing the required duct termination.

4. Sound Intensity Method

Advances in acoustical instrumentation now permit the direct determination of sound intensity, defined as the sound power per unit area flowing through a small element of a surface surrounding a source. The average sound power radiated by the source can be determined by measuring the sound intensity over the sphere or hemisphere surrounding a sound source. One of the advantages of this method is that, with certain limitations, sound intensity (and, therefore, sound power) measurements can be made in the presence of steady background sound in ordinary rooms, thereby eliminating the need for a special testing environment. Another advantage is that by measuring sound intensity over restricted areas around a sound source, sound directivity can be determined. This procedure can be particularly useful in reducing noise of products during their development.

International and United States standards that prescribe methods for making sound power measurements with sound intensity probes have been issued (ISO *Standard* 9614-1, ISO *Standard* 9614-2, ANSI *Standard* S12.12). In some situations, the sound fields may be so complex that measurements become impractical. A particular concern is that small test rooms or those having somewhat flexible boundaries (sheet metal or thin drywall) can permit a reactive sound field to exist, one in which the room's acoustical characteristics cause it to affect the sound power output of the source.

Measurement Bandwidths for Sound Power

Sound power is normally determined in octave or 1/3 octave bands. Occasionally, a more detailed determination of the sound source spectrum is required. In these cases, narrow band analysis, using either constant fractional bandwidth (1/12 or 1/24 octave) or constant absolute bandwidth (e.g., 1 Hz) can be applied. The digital filter analyzer is most frequently used for constant percent bandwidth measurements, and the fast Fourier transform (FFT) analyzer is used for constant bandwidth measurements. Narrow band analysis results are used to determine the exact frequencies of pure tones and their harmonics in a sound spectrum.

CONVERTING FROM SOUND POWER TO SOUND PRESSURE

The designer is often required to use the sound power level information on a source to predict the sound pressure level at a given location. The sound pressure level at a given location in a room due to a source of known sound power level depends on: (1) room volume, (2) room furnishings and surface treatments, (3) magnitude of the sound source(s), and (4) distance from the sound source(s) to the point of observation.

The classic relationship between source sound power level and room sound pressure level at some frequency is

$$L_p = L_w + 10\log\left(\frac{Q}{4\pi r^2} + \frac{4}{R}\right) + 10.5 \qquad (10)$$

where

L_p = sound pressure level, dB re 20μPa
L_w = sound power level, dB re 10^{-12} W
Q = directivity of the sound source (dimensionless)
r = distance from the source, ft
R = room constant, $S\bar{\alpha}/(1-\bar{\alpha})$
S = sum of all surface areas, ft²
$\bar{\alpha}$ = average absorption coefficient of room surfaces at given frequency

If the source is outside, far from reflecting surfaces, this relationship simplifies to

$$L_p = L_w + 10\log(Q/4\pi r^2) + 10.5 \qquad (11)$$

This relationship does not account for atmospheric absorption, weather effects and barriers. Note that the r^2 term is present because the sound pressure in a free field decreases as $1/r^2$ (the **inverse-square law**). Each time the distance from the source is doubled, the sound pressure level decreases by 6 dB.

For a simple source centered in a large, flat, reflecting surface, Q may be taken as 2. At the junction of two large flat surfaces, Q is 4, and in a corner, Q is 8.

In most typical rooms, the presence of acoustically absorbent surfaces and sound-scattering elements, such as furniture, creates a relationship between sound power and sound pressure level that is almost independent of the absorptive properties of the space. For example, hospital rooms, which have only a small amount of absorption, and executive offices, which have substantial absorption, are similar when the comparison is based on the same room volume and distance between the source and point of observation.

Equation (12) can be used to estimate the sound pressure level at a chosen observation point in a normally furnished room. The estimate is accurate to ±2 dB (Schultz 1985).

$$L_p = L_w - 5\log V - 3\log f - 10\log r + 25 \qquad (12)$$

where

L_p = room sound pressure level at chosen reference point, dB re 20 μPa
L_w = source sound power level, dB re 10^{-12} W
V = room volume, ft³
f = octave band center frequency, Hz
r = distance from source to observation point, ft

Equation (12) applies to a single sound source in the room itself, not to sources above the ceiling. With more than one source, total sound pressure level at the observation point is obtained by adding the contribution from each source in energy or power-like units, not decibels, and then converting back to sound pressure level. Warnock (1998b), reporting on ASHRAE *Research Project* 755, indicated that sound sources above ceilings may not act as point sources, and Equation (12) may not apply. Chapter 46 of the 1999 *ASHRAE Handbook—Applications* provides more information on this topic.

SOUND TRANSMISSION PATHS

Sound from a source is transmitted along one or more paths to a receiver. Airborne and structureborne transmission paths are of principal concern for the HVAC system designer. Sound transmission between rooms occurs along both airborne and structureborne transmission paths. Chapter 46 of the 1999 *ASHRAE Handbook—Applications* has additional information on transmission paths.

Airborne Transmission

Atmospheric transmission. Sound transmits readily through air, both indoors and outdoors. Indoor sound transmission paths include the direct, line-of-sight path between the source and the receiver, as well as reflected paths introduced by the presence of a room's walls, floor, ceiling and furnishings, which cause multiple sound reflection paths.

Outdoors, the effects of the reflections are small, if the source is not located near large reflecting surfaces. However, sound outdoors can refract and change propagation direction, due to the presence of wind and temperature gradient effects. Sound propagation outdoors follows the inverse square law. Therefore, Equations (6) and (7) can generally be used to calculate the relationship between sound power level and sound pressure level for fully free-field and hemispherical free-field conditions, respectively.

Ductborne transmission. Ductwork can provide an effective sound transmission path because the sound is primarily contained within the boundaries of the ductwork. Sound can transmit both upstream and downstream from the source. A special case of ductborne transmission is **crosstalk**, where sound is transmitted from one room to another via the duct path.

Room-to-room transmission. Room-to-room sound transmission generally involves both airborne and structureborne sound paths. The sound power incident on a room surface element undergoes three processes: (1) some of the sound energy is reflected from the surface element back into the room; (2) a portion of the sound energy is lost due to energy transfer into the material comprising the element, and (3) the remainder of the sound energy is transmitted through the element to the other room. Airborne sound is radiated as the element vibrates and structureborne sound can be transmitted via the studs of a partition or the floor and ceiling surfaces.

Structureborne Transmission

Solid structures are efficient transmission paths for sound, which frequently originates as a vibration imposed on the transmitting structure. Typically, only a small amount of the input energy is radiated by the structure as airborne sound. A lightweight structure with little inherent damping radiates more sound than a massive structure with greater damping.

Flanking Transmission

Sound from the source room can bypass the primary separating element and get into the receiving room along other paths called **flanking paths**. Common sound flanking paths include return air plenums, doors, and windows. Less obvious paths are those along floor and adjoining wall structures. Such flanking paths can seriously reduce the sound isolation between rooms. Flanking can explain poor sound isolation between spaces when the partition between spaces is known to provide very good sound insulation. Flanking can also explain sounds being heard in one room at a great distance from another room. Determining whether flanking sound transmission is important and what paths are involved can be difficult. Experience with actual situations and the theoretical aspects of flanking transmission is helpful. Sound intensity methods may be useful in determining flanking paths.

TYPICAL SOURCES OF SOUND

Whenever mechanical power is generated or transmitted, a fraction of the power is converted into sound power and is radiated into the air. Therefore, virtually any major component of an HVAC system could be considered a sound source (e.g., fans, pumps, ductwork, piping, motors, etc.). The component's sound source characteristics depend upon its construction, its form of mechanical power and its integration with associated system components. The most important sound source characteristics include total sound power output, frequency distribution, and radiation directivity. All of these characteristics vary with frequency.

Sound sources in HVAC systems are so numerous that it is impractical to provide a complete listing here. Typical sources of sound and vibration in HVAC systems include

- Rotating and reciprocating equipment such as fans, motors, pumps, and chillers.
- Air and fluid sounds, such as those associated with flow through ductwork, piping systems, grilles, diffusers, terminal boxes, manifolds, and pressure-reducing stations.
- Excitation of surfaces—for example, friction; movement of mechanical linkages; turbulent flow impacts on ducts, plenum panels, and pipes; and impacts within equipment, such as cams and valve slap.
- Magnetostriction (transformer hum), which becomes significant in motor laminations, transformers, switchgear, lighting ballasts, and dimmers. A characteristic of magnetostrictive oscillations is that their fundamental frequency is twice the line frequency (120 Hz in a 60 Hz system.)

CONTROLLING SOUND

Terminology

The following noninterchangeable terms are used to describe the acoustical performance of many system components. ASTM *Standard* C 634 defines additional terms.

Sound attenuation is a general term describing the reduction of the level of sound as it travels from a source to a receiver.

Insertion loss (IL) of a silencer or other sound-attenuating element is expressed in dB and is defined as the decrease in sound pressure level or sound intensity level, measured at a fixed receiver location, when the sound-attenuating element is inserted into the path between the source and the receiver. For example, if a straight, unlined piece of ductwork were replaced with a duct silencer, the sound level difference at a fixed location would be considered the silencer's insertion loss. Measurements are typically made in either octave or 1/3 octave bands.

Sound transmission loss (TL) of a partition or other building element is equal to 10 times the logarithm (base 10) of the ratio of the airborne sound power incident on the partition to the sound power transmitted by the partition and radiated on the other side. The quantity so obtained is expressed in decibels. Measurements are typically made in octave or 1/3 octave bands. Chapter 46 of the 1999 ASHRAE *Handbook—Applications* defines the special case of breakout transmission loss through duct walls.

Noise reduction (NR) is the difference between the average sound pressure levels produced in two enclosed spaces or rooms—a receiving room and a source room—by one or more sound sources in the source room. An alternate, non-ASTM definition of NR is the difference in sound pressure levels measured upstream and downstream of a duct silencer or sound-attenuating element. Measurements are typically made in octave or 1/3 octave bands.

Random incidence sound absorption coefficient α is the fraction of the incident sound energy that is absorbed by a surface exposed to randomly incident sound. It is measured in a reverberation room using 1/3 octave bands of broadband sound (ASTM C 423). The sound absorption coefficient of a material in a specific 1/3 octave band depends on the material's thickness, airflow resistivity, stiffness, and method of attachment to the supporting structure.

Spherical spreading is the process by which sound level decreases with distance from a point source. It occurs when the sound source is located in free space or an anechoic room. Sound propagation in an anechoic space follows the *inverse square law*—6 dB level reduction per doubling of distance from a reference position.

Scattering is the change in direction of sound propagation due to an obstacle or inhomogeneity in the transmission medium. It results in the incident sound energy being dispersed in many directions.

Enclosures and Barriers

Enclosing a sound source is a common means of controlling airborne sound transmission. This may be done using single- or double-leaf partitions.

The term **single-leaf partition** refers to all types of solid homogeneous panels where both faces are rigidly connected. Examples are gypsum board, plywood, concrete block, brick, and poured concrete. The transmission loss of a single-leaf partition depends mainly on its surface mass (mass per unit area) because the heavier the partition, the less it vibrates in response to sound waves and the less sound it radiates on the side opposite the sound source. Increased surface mass can be achieved either by an increase in the partition's thickness or its density.

The **mass law** is a semi-empirical expression that may be used to predict transmission loss for randomly incident sound for thin, homogeneous single-leaf panels. It is written as

$$\text{TL} = 20\log(w_s f) - 33 \tag{13}$$

where

w_s = surface mass of panel, lb/ft^2
f = frequency, Hz

The mass law predicts that transmission loss increases by 6 dB for each doubling of surface mass or frequency. If sound is incident only perpendicularly on the panel, the TL is about 5 dB greater than that predicted by the mass law.

Transmission loss also depends on material properties, such as stiffness and internal damping. The transmission losses of three single-leaf walls are illustrated in Figure 2. For the 5/8 in. gypsum board, TL depends mainly on the surface mass of the wall at frequencies below about 1 kHz; agreement with the mass law is good. At higher frequencies, there is a dip in the TL curve, called the **coincidence dip** because it occurs at the frequency where the wavelength of flexural vibrations in the wall coincides with the wavelength of sound in the air. The frequency where the minimum value of TL occurs in the coincidence dip is called the **critical frequency**, which depends on material stiffness and thickness. The stiffer or thicker the layer of material, the lower the critical frequency. For example, the 6 in. concrete slab has a surface mass of about 75 lb/ft^2 and a coincidence frequency at 125 Hz. Thus, over most of the frequency range shown in Figure 2, the transmission loss for the 6 in. concrete slab is well below that predicted by mass law. The coincidence dip for the 25 gage steel sheet occurs at high frequencies not shown in the figure.

The **sound transmission class (STC) rating** of a partition or assembly is a single number rating often used in architecture to classify sound isolation for speech. (ASTM E 90, ASTM E 413) Because the STC rating system was developed to deal with sound sources in the speech frequency range (125 to 4000 Hz), the rating should not be used as an indicator of an assembly's ability to control sound of any source that is rich in low frequencies. Most fan sound spectra have dominant low-frequency sound; therefore, to control fan sound, walls and slabs should be selected only on the basis of 1/3 octave or octave band sound transmission loss values, particularly at low frequencies.

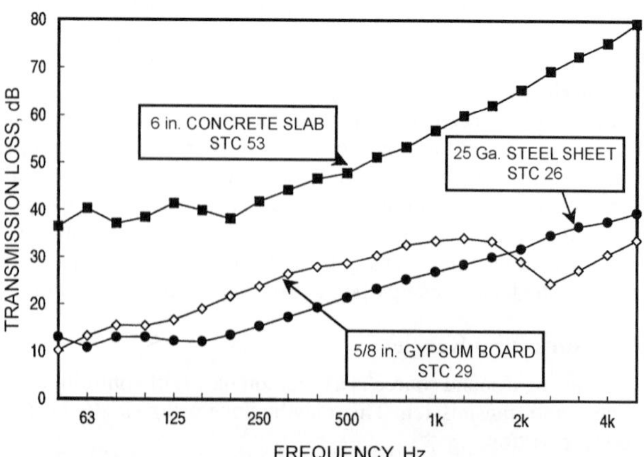

Fig. 2 Sound Transmission Loss Spectra for Single Layers of Some Common Materials

Sound transmission loss values for ceiling tile are also inappropriate for estimating the reduction of sound between a sound source located in a ceiling plenum and the room below. ARI *Standard* 885 has guidance on this topic.

Walls with identical STC ratings may not provide identical sound insulation at all frequencies. Because of the limited frequency range of most single number rating systems, designers should select partitions and floors on the basis of their 1/3 octave or octave band sound transmission loss values rather than single number ratings, especially when frequencies below 125 Hz are important.

For a given total mass in a wall or floor, much higher values of TL can be obtained by forming a double-leaf construction where each layer is independently or resiliently supported so vibration transmission between them is minimized. As well as mass, TL for such walls depends on cavity depth. Adding sound-absorbing material in the cavity significantly increases the TL relative to the unfilled cavity case. For further information on such walls, see Chapter 46 of the 1999 ASHRAE *Handbook—Applications.*

If the sound fields in the rooms on each side of a panel are diffuse and the panel is the only significant path for sound between the rooms, the noise reduction NR is a function of the panel area S_p and the total sound absorption A_r in the receiving space, according to

$$\text{NR} = \text{TL} - 10\log(S_p/A_r) \tag{14}$$

Because the total sound absorption in a room is expressed as the equivalent area of perfect sound absorption, both S_p and A_r are expressed in consistent units, usually square feet.

The sound reduction of an enclosure may be severely compromised by openings or leaks in the enclosure. Ducts that lead into or through a noisy space can carry sound to many areas of a building. Designers need to consider this factor when designing duct, piping, and electrical systems.

Attenuation of Sound in Ducts and Plenums

Most ductwork, even a sheet metal duct without acoustical lining or silencers, attenuates sound to some degree. The natural attenuation of unlined ductwork is minimal, but can, especially for long runs of rectangular ductwork, significantly reduce ductborne sound. Acoustic lining of ductwork can greatly attenuate the propagation of sound through ducts, particularly at mid to high frequencies. Chapter 46 of the 1999 ASHRAE *Handbook—Applications* has a detailed discussion of lined and unlined ductwork attenuation.

If analysis shows that lined ductwork will not reduce sound propagation adequately, commercially available sound attenuators (also known as sound traps or duct silencers) can be used. There are three types: dissipative, reactive, and active. The first two are commonly known as **passive attenuators**.

- **Dissipative silencers** use absorptive media such as glass or rock fiber as the principal sound-absorption mechanism. Thick, perforated sheet metal baffles filled with low-density fiber insulation restrict the air passage width within the attenuator housing. The fiber is sometimes protected from the airstream by cloths or films. This type of attenuator is most effective in reducing mid- and high-frequency sound energy.
- **Reactive silencers** do not use any absorptive media to dissipate sound. This attenuator is typically used in HVAC systems serving hospitals, laboratories, or other areas with strict air quality standards. They are constructed only of metal, both solid and perforated. Chambers of specially designed shapes and sizes behind the perforated metal are tuned as resonators or expansion chambers to react with and reduce the sound power at selected frequencies. When designed for a broad frequency range, they are usually not as effective as dissipative attenuators and so are longer and

have a greater pressure drop. However, they can be highly effective and compact if designed for a limited frequency range such as for a pure tone.

- **Active silencer** systems use microphones, loudspeakers, and appropriate electronics to reduce in-duct sound by generating inverse-phase sound waves that destructively interfere with the incident sound energy. Microphones sample the sound field in the duct and loudspeakers generate signals with the opposite phase to the noise. Controlled laboratory experiments have shown that active attenuators reduce both broadband and tonal sound, but they are typically only effective in the 31.5 Hz through 250 Hz octave bands. Insertion losses of as much as 30 dB have been achieved under controlled conditions. The microphones and loudspeakers create a negligible pressure drop because they are mounted flush with the duct wall. Because active attenuators are not effective in the presence of excessively turbulent airflow, their use is limited to relatively long, straight duct sections with an air velocity less than about 1500 fpm.

Silencers are available for fans, cooling towers, air-cooled condensers, compressors, gas turbines, and many other pieces of commercial and industrial equipment. Silencers are normally installed on the intake or the discharge side (or both) of a fan or air handling unit. Also, they may be used on the receiver side of other noise generators such as terminal boxes, valves, and dampers.

Self-noise can limit an attenuator's effective insertion loss for air velocities in excess of about 2000 fpm. Use extreme caution when reviewing manufacturers' performance data for attenuators and duct liner materials to be sure that the test conditions are comparable to the specific design conditions. Short sections (3 to 5 ft) of insulated flexible duct are often very effective as attenuators. (ARI *Standard* 885 and Chapter 46 of the 1999 *ASHRAE Handbook—Applications* have information on typical flexible duct attenuation factors).

End Reflections. End reflection losses due to abrupt area changes in duct cross-section are sometimes useful in controlling low frequencies. The end reflection effect can be maximized at the end of a duct run by designing the last few feet or so of duct with the characteristic dimension of less than 15 in. Low-frequency noise reduction is inversely proportional to the characteristic dimension of the duct. However, abrupt area changes can generate high frequency noise, especially at high flow rates.

Lined Plenums. Where space is available, a lined plenum can provide excellent attenuation across a broad frequency range. The combination of end reflection at the plenum's entrance and exit, a large distance between the entrance and exit, and sound-absorbing lining on the plenum walls can be as effective as a sound attenuator, but with less pressure drop.

Chapter 46 of the 1999 ASHRAE *Handbook—Applications* has additional information on the control of sound.

Standards for Testing Duct Silencers

Attenuators and duct liner materials are tested according to ASTM *Standard* E 477 in North America and ISO 7235 elsewhere. These define acoustical and aerodynamic performance in terms of insertion loss, self-generated noise (or self-noise), and airflow pressure drop. While many similarities exist, the ASTM and ISO standards produce differing results because of variations in loudspeaker location, orientation, duct termination conditions, and computation methods. Currently, no standard test methods are available to measure the attenuation of active silencers, although it is easy to measure in the field simply by turning the system on and off.

Insertion loss is measured in the presence of both forward and reverse flows. Forward flow occurs when the air and sound move in the same direction, as in a supply air or fan discharge system; reverse flow occurs when the air and sound travel in opposite directions, as in the case of a return air or fan intake system.

SYSTEM EFFECTS

The way the HVAC components are assembled into a **system** affects the sound level generated by the system. Many engineers believe that satisfactory noise levels in occupied spaces can be achieved solely by using a manufacturer's sound ratings as a design tool, without consideration of the system influence.

Sound data provided by most manufacturers is obtained under standard laboratory test conditions. If the equipment is installed in a manner that differs from the test configuration, different configurations of connected ductwork, and interactions with other components of the installation, often significantly increase the operating noise level. For example, aerodynamically clean fan inlet and outlet conditions are rarely found in typical field applications. Furthermore, components such as silencers are frequently installed too close to the fan to allow a uniform velocity profile to exist at the entrance to the silencer. This results in a significantly higher than anticipated pressure drop across that component. The combination of these two effects changes the operating point on the fan curve. As a result, airflow is reduced and must be compensated for by increasing the fan speed, which may increase noise.

HUMAN RESPONSE TO SOUND

Noise

Noise may be defined as any unwanted sound. Sound becomes noise when

- It is too loud—the sound is uncomfortable or makes speech difficult to understand
- It is unexpected (e.g., the sound of breaking glass)
- It is uncontrolled (e.g., a neighbor's lawn mower)
- it happens at the wrong time (e.g., a door slamming in the middle of the night)
- It contains pure tones (e.g., a whine, whistle, or hum)
- It contains unwanted information or is distracting (e.g., an adjacent telephone conversation or undesirable music)
- It is unpleasant (e.g., a dripping faucet)
- It connotes unpleasant experiences (e.g., a mosquito buzz or a siren wail)
- It is any combination of the above examples

To be noise, sound does not have to be loud, just unwanted. In addition to being annoying, loud noise can cause hearing loss, and, depending on other factors, it could affect stress level, sleep patterns and heart rate.

To increase privacy, broadband sound may be radiated into a room from a well-designed air-conditioning system to mask or hide low-level intrusive sounds from adjacent spaces. This controlled sound may be referred to as noise, but not in the context of unwanted sound; rather, it is a broadband, neutral sound that is frequently unobtrusive. Three types of broadband noise are frequently encountered in acoustics:

- **Random noise** is an oscillation, the instantaneous magnitude of which is not specified for any given instant. The instantaneous magnitudes of a random noise are specified only by probability distributions, giving the fraction of the total time that the magnitude, or some sequence of magnitudes, lies within a specified range (ANSI *Standard* S1.1).
- **White noise** is noise with a continuous frequency spectrum with equal energy per hertz over a specified frequency range. White noise is not necessarily random. Since octave bands double in width for each successive band, for white noise the energy also doubles in each successive octave band. Thus white noise displayed on a 1/3 octave or octave band chart increases by 3 dB per octave.
- **Pink noise** is noise with a continuous frequency spectrum but equal energy per constant-percentage bandwidth, such as per

octave or 1/3 octave band. Thus pink noise appears on a one-third octave or octave band chart as a horizontal line.

Predicting Human Response to Sound

Predicting the response of people to any given sound is, at best, only a statistical concept, and, at worst, very inaccurate. This is because response to sound is not only physiological but psychological and depends on the varying attitude of the listener. Hence, the effect of sound is often unpredictable. However, people's response is adverse if the sound is considered too loud for the situation or if it sounds "wrong." Therefore, most criteria are based on descriptors that account for level and spectrum shape.

Sound Quality

To determine the acoustic acceptability of a space to occupants, the sound pressure levels there must be known. This, however, is often not sufficient; the **sound quality** is important too. Factors influencing sound quality include (1) loudness, (2) tone perception, (3) frequency spectrum, (4) harshness, (5) time and frequency fluctuation, and (6) vibration.

People often perceive sounds with tones (like a whine or hum) as particularly annoying. A tone can cause a relatively low level sound to be perceived as noise. Studies have been done to characterize sounds with and without pure tones.

Loudness

The primary method used to determine a subjective estimate of loudness is to present sounds to a sample of human listeners under controlled conditions. To determine the loudness of a sound, listeners compare an unknown sound with a standard sound. (The accepted standard sound is a pure tone of 1000 Hz or a narrow band of random noise centered on 1000 Hz.) Loudness level is expressed in **phons**, and the loudness level of any sound in phons is equal to the sound pressure level in decibels of a standard sound deemed to be equally loud. Thus, a sound that is judged as loud as a 40 dB, 1000 Hz tone has a loudness level of 40 phons.

Average reactions of humans to tones are shown in Figure 3 (Robinson and Dadson 1956). The reaction changes when the sound is a band of random noise (Pollack 1952), rather than a pure tone (Figure 4). The figures indicate that people are most sensitive in the mid-frequency range. The contours in Figure 3 are closer together at low frequencies showing that at lower frequencies, although people are less sensitive to sound level, they are more sensitive to changes in level.

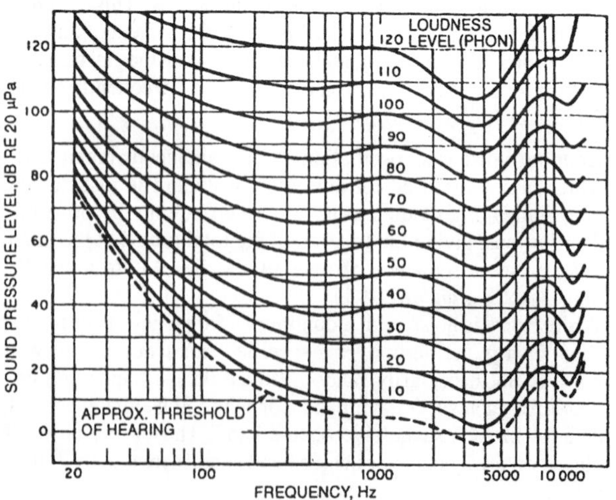

Fig. 3 Free-Field Equal Loudness Contours for Pure Tones

Table 7 Subjective Effect of Changes in Sound Pressure Level, Broadband Sounds (Frequency > 250 Hz)

Subjective Change	Objective Change in Sound Level (Aproximate)
Much louder	More than +10 dB
Twice as loud	+10 dB
Louder	+5 dB
Just perceptibly louder	+3 dB
Just perceptibly quieter	−3 dB
Quieter	−5 dB
Half as loud	−10 dB
Much quieter	Less than −10 dB

Under carefully controlled experimental conditions, humans can detect small changes in sound level. However, for humans to describe a sound as being half or twice as loud requires changes in overall sound pressure level of about 10 dB. For many people, a 3 dB change is the minimum perceptible difference. This means that halving the power output of the source causes a barely noticeable change in sound pressure level, and the power output must be reduced by a factor of 10 before humans determine that loudness has been halved. Table 7 summarizes the effect of changes in sound levels for simple sounds in the frequency range of 250 Hz and higher.

The phon scale covers the large dynamic range of the ear, but it does not fit a subjective linear loudness scale. Over most of the audible range, a doubling of loudness corresponds to a change of approximately 10 phons. To obtain a quantity proportional to the loudness sensation, a loudness scale is defined in which the unit of loudness is known as a **sone**. One sone equals the loudness level of 40 phons. A rating of two sones corresponds to 50 phons and so on.

The results of such work have led to the development of standard objective methods for calculating loudness. ANSI *Standard* S3.4 calculates loudness or loudness level by using octave-band sound pressure level data as a starting point. The loudness index for each octave band is obtained from a graph or by calculation. Total loudness is then calculated by combining the loudnesses for each band according to a formula given in the standard. A more complex calculation method using 1/3 octave band sound pressure levels by Zwicker (ISO *Standard* 532) or the German *Standard* DIN 45631 is

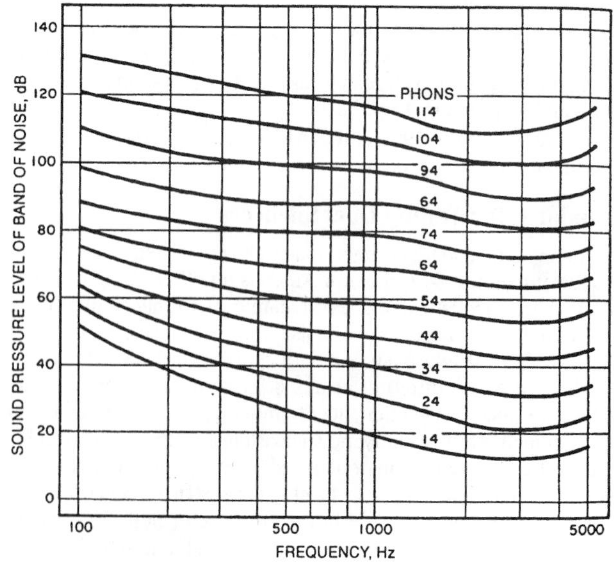

Fig. 4 Equal Loudness Contours for Relatively Narrow Bands of Random Noise

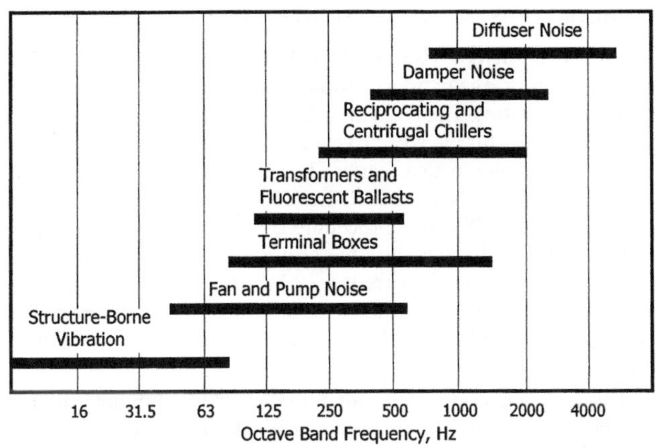

Fig. 5 Frequencies at Which Various Types of Mechanical and Electrical Equipment Generally Control Sound Spectra

tones. Due to its complexity, loudness has not been widely used in engineering practice in the past. However, with an increased awareness of sound quality and the availability of software for calculating loudness, this measure is now being used more frequently.

AMCA *Publication* 302 describes how the **sone method** is applied to rating the relative loudness of fans and ventilators. This calculation method is usually acceptable when the measured sound spectrum has no strong tonal components.

Acceptable Frequency Spectrum

The most acceptable frequency spectrum for HVAC sound is a balanced or neutral spectrum. This means that it is not too "hissy" (excessive high frequency content) or too "rumbly" (excessive low-frequency content). Unfortunately, achieving a balanced sound spectrum is not always easy—there may be a multiplicity of sound sources to consider. As a guide to the designer, Figure 5 shows the more common mechanical and electrical sound sources and frequency regions that control the indoor sound spectrum. Chapter 46 of the 1999 *ASHRAE Handbook—Applications* provides more detailed information on treating some of these sound sources.

SOUND RATING SYSTEMS AND ACOUSTICAL DESIGN GOALS

Several background sound rating methods are used to rate indoor sound. They include the A-weighted sound pressure level (dBA) and noise criteria (NC), the more recent room criteria (RC) and balanced noise criteria (NCB), and the new RC Mark II. Each sound rating method was developed from data for specific applications; not all methods are equally suitable for the rating of HVAC-related sound in the variety of applications encountered.

The degree of occupant satisfaction achieved with a given level of background sound is determined by many factors. For example, large conference rooms, auditoriums, and recording studios can tolerate only a low level of background sound. On the other hand, higher levels of background sound are acceptable and even desirable in certain situations, such as in open-plan offices where a certain amount of speech and activity masking is essential. Therefore, the system sound control goal varies depending on the required use of the space.

To be unobtrusive, background sound should have the following properties:

- A balanced distribution of sound energy over a broad frequency range
- No audible tonal or other characteristics such as whine, whistle, hum, or rumble

Table 8 Comparison of Sound Rating Methods

Method	Overview	Evaluates Sound Quality	Used For Rating of
dBA	• Can be determined using sound level meter • No quality assessment • Frequently used for outdoor noise ordinances	No	Cooling towers Water chillers Condensing units
NC	• Can rate components • No quality assessment • Does not evaluate low frequency rumble, frequencies <63 Hz	No	Air terminals Diffusers
NCB	• Can rate components • Some quality assessment	Yes	
RC	• Used to evaluate systems • Should not be used to evaluate components • Can be used to evaluate sound quality • Provides some diagnostic capability	Yes	
RC Mark II	• Evaluates sound quality • Provides improved diagnostics capability	Yes	

- No noticeable time-varying levels from beats or other system-induced aerodynamic instability
- No fluctuations in level such as a throbbing or pulsing

At present, no acceptable process easily characterizes the effects of audible tones and level fluctuations. The preferred sound rating methods generally comprise two distinct parts: a family of criteria curves (specifying sound levels by octave bands), and a companion procedure for rating the calculated or measured sound data relative to the criterion curves. A table of recommended design goals can be found in Chapter 46 of the 1999 ASHRAE *Handbook—Applications*.

Table 8 summarizes the essential differences, advantages and disadvantages of the rating methods that are used to characterize HVAC-related background sound. The text following the table gives more information on each rating. Note that all the ratings in the table consider speech interference effects and all are currently used for rating background noise.

A-Weighted Sound Level (dBA)

The A-weighted sound level L_A is widely used to state acoustical design goals as a single number, but its usefulness is limited because it gives no information on spectrum content. The rating is expressed as a number followed by dBA, for example 40 dBA.

A-weighted sound levels correlate well with human judgments of relative loudness, but give no information on spectral balance. Thus, they do not necessarily correlate well with the annoyance caused by the noise. Many different-sounding spectra can have the same numeric rating, but have quite different subjective qualities. A-weighted comparisons are best used with sounds that sound alike but differ in level. They should not be used to compare sounds with distinctly different spectral characteristics; that is, two sounds at the same sound level but with different spectral content are likely to be judged differently by the listener in terms of acceptability as a background sound. One of the sounds might be completely acceptable, while the other could be objectionable because its spectrum shape was rumbly, hissy, or tonal in character.

A-weighted sound levels are used extensively in outdoor environmental noise standards.

Noise Criteria (NC) Method

The NC method is a single-number rating that is somewhat sensitive to the relative loudness and speech interference properties of a given sound spectrum. The method consists of a family of criteria curves extending from 63 to 8000 Hz, and a **tangency rating procedure** (Beranek 1957). The criteria curves, shown in Figure 6, define the limits of octave band spectra that must not be exceeded to meet occupant acceptance in certain spaces. The rating is expressed as NC followed by a number. For example, the spectrum shown in Figure 6 is rated NC 45 because this is the lowest rating curve that falls entirely above the measured data. An NC 35 design goal is commonly used for private offices. The background sound level meets this goal if no portion of its spectrum lies above the designated NC 35 curve.

The NC method is sensitive to level but has the disadvantage that the tangency method used to determine the rating does not require that the sound spectrum approximate the shape of the NC curves. Thus, many different sounds can have the same numeric rating, but rank differently on the basis of subjective sound quality. In HVAC systems that do not produce excessive low frequency sound, the NC rating correlates relatively well with occupant satisfaction *if* sound quality is not a significant concern.

Two problems occur in using the NC procedure: (1) when the NC level is determined by a prominent peak in the spectrum, the actual level of resulting background sound may be quieter than that desired for masking unwanted speech and activity sounds, because the spectrum on either side of the tangent peak drops off too rapidly; and (2) when the measured spectrum closely matches the shape of the NC curve, the resulting sound is either rumbly or hissy or both.

The shape of the NC curve is not that of a well-balanced, neutral sound; thus, these curves should be used with caution in critical situations where the background sound of an air-conditioner is required to mask speech and activity sound.

NC contours are used to calculate ratings for some HVAC components such as terminal units and diffusers. NC ratings should not be used to characterize fans and air-handling units.

Balanced Noise Criteria (NCB) Method

The NCB method (Beranek 1989, ANSI *Standard* S12.2) is a specification or evaluation of room sound including noise due to occupant activities. The NCB criteria curves (Figure 7) are intended as replacements for the NC curves, and include both the addition of two low-frequency octave bands (16 and 31.5 Hz) and lower permissible sound levels in the high-frequency octave bands (4000 and 8000 Hz). The NCB rating procedure is based on the **speech interference level** (SIL), which is the arithmetic average of the sound pressure levels in the four frequency bands: 500, 1000, 2000, and 4000 Hz. Additional tests include rumble and hiss compliance. The rating is expressed as NCB followed by a number, for example, NCB 40.

The NCB method is better than the NC method in determining whether a sound spectrum has a shape sufficiently unbalanced to demand corrective action. Also, it addresses the issue of low-frequency sound. The rating procedure is somewhat more complicated than the tangency rating procedures.

Room Criteria (RC) Method

For some time, the RC method (Blazier 1981a,b; ANSI *Standard* S12.2) was recommended as the preferred method for rating HVAC-related sound. The RC curves were intended to establish HVAC system design goals. The revised RC Mark II method (discussed below) is now preferred.

The RC method consists of a family of criteria curves and a rating procedure. The shape of these curves differs from the NC curves to approximate a well-balanced neutral-sounding spectrum, and two additional octave bands (16 and 31.5 Hz) are added to deal with low-frequency sound. This rating procedure assesses background sound in spaces based on the effect of the sound on speech communication, and on estimates of subjective sound quality. The rating is

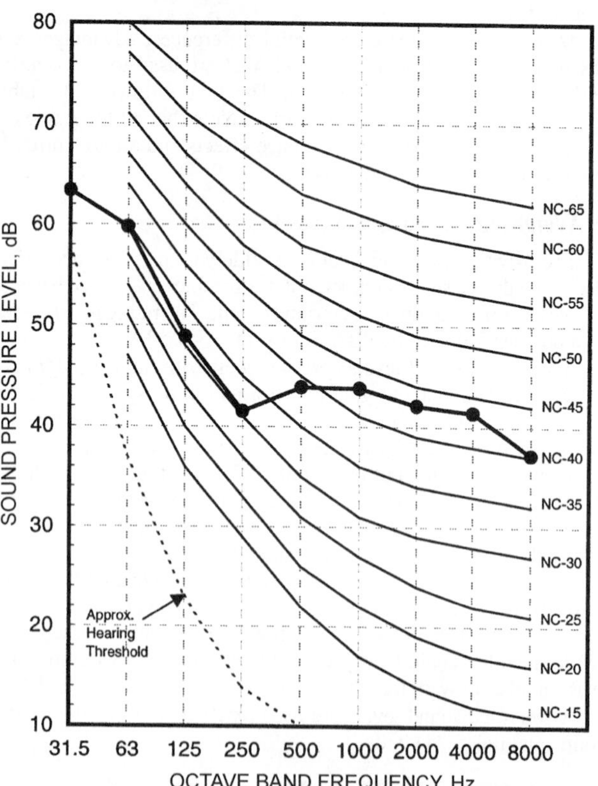

Fig. 6 NC (Noise Criteria) Curves and Sample Spectrum (Curve with Symbols)

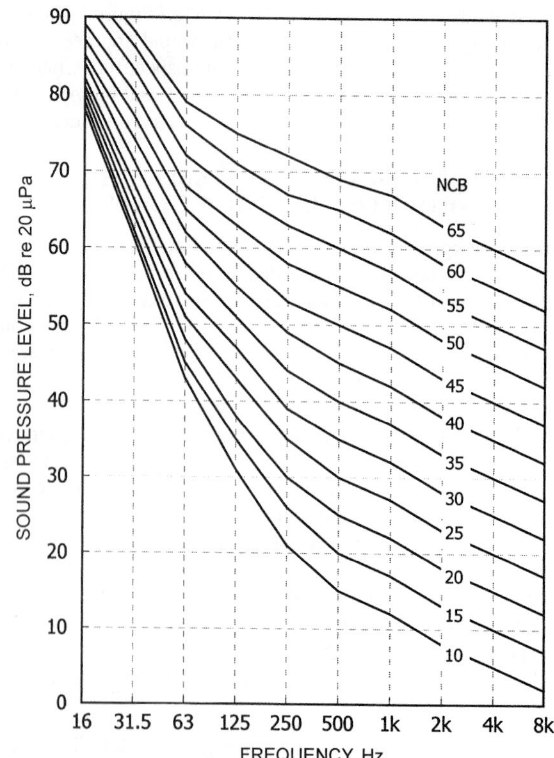

Fig. 7 NCB (Noise Criteria Balanced) Curves Drawn from ANSI *Standard* S12.2

expressed as RC followed by a number to show the level of the sound and a letter to indicate the quality, for example RC 35(N) where N denotes neutral.

RC Mark II Room Criteria Method

Based on experience and the findings from ASHRAE-sponsored research, the RC method was revised to the RC Mark II method (Blazier 1997). Like its predecessor, the RC Mark II method is intended for rating the sound performance of an HVAC system as a whole. The method can also be used as a diagnostic tool for analyzing sound problems in the field. The RC Mark II method is more complicated to use than the RC method, but spreadsheet macros are available to do the calculations and graphical analysis.

The RC Mark II method of rating HVAC system sound comprises three parts:

- Family of criteria curves (Figure 8)
- Procedure for determining the RC numerical rating and the sound spectral balance (quality)
- Procedure for estimating occupant satisfaction when the spectrum does not have the shape of an RC curve (Quality Assessment Index) (Blazier 1995)

The rating is expressed as RC followed by a number and a letter, for example, RC 45(N). The number is the arithmetic average rounded to the nearest integer of the sound pressure levels in the 500, 1000, and 2000 Hz octave bands (the principal speech frequency region). The letter is a qualitative descriptor that identifies the perceived character of the sound: (N) for neutral, (LF) for low-frequency rumble, (MF) for mid-frequency roar, and (HF) for high-frequency hiss. In addition, the low-frequency descriptor has two subcategories: (LF_B), denoting a moderate but perceptible degree of sound induced ceiling/wall vibration, and (LF_A), denoting a noticeable degree of sound induced vibration.

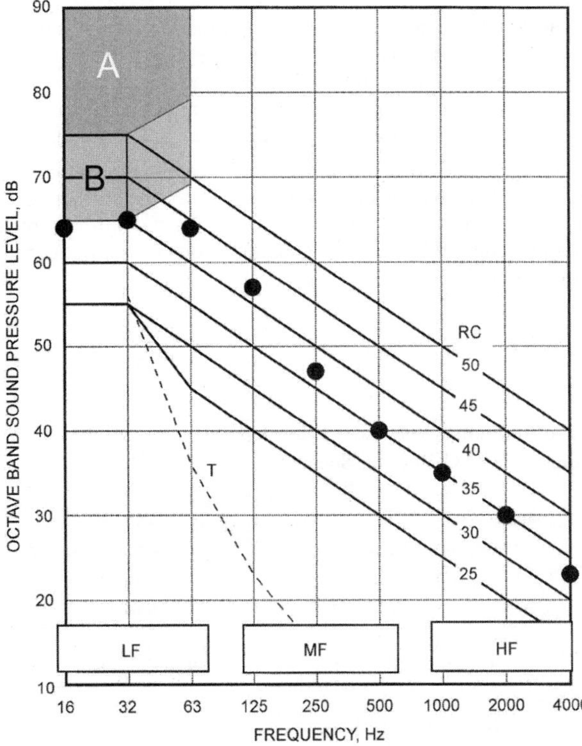

Sound levels in Region B may generate perceptible vibration in light wall and ceiling construction. Rattles in light fixtures, doors, windows, etc., are a slight possibility. Sound levels in Region A have a high probability of generating easily perceptible sound induced vibration in light wall and ceiling construction. Audible rattling in light fixtures, doors, windows etc. may be anticipated. The text explains Regions LF, MF, and HF. The solid dots are octave band sound pressure levels for the example in the text.

Fig. 8 Room Criteria Curves, Mark II

Each reference curve in Figure 8 identifies the shape of a neutral, bland-sounding spectrum, indexed to a curve number corresponding to the sound level in the 1000 Hz octave band. The shape of these curves is based on research by Blazier (1981a,b) and modified at 16 Hz following recommendations by Broner (1994). Regions A and B denote levels at which sound can induce vibration in light wall and ceiling constructions that can potentially cause rattles in light fixtures, furniture, etc. Curve T is the octave-band threshold of hearing as defined by ANSI *Standard* 12.2.

Procedure for Determining the RC Mark II Rating for a System

Step 1. Determine the appropriate RC reference curve. This is done by obtaining the arithmetic average of the sound levels in the principle speech frequency range represented by the levels in the 500, 1000, and 2000 Hz octave bands. The RC reference curve is chosen as that which has the same value at 1000 Hz as the calculated average value (rounded to the nearest integer). This curve is not to be confused with the speech-interference level (SIL), which is a four-band average obtained by including the 4000 Hz octave band.

Step 2. Assign a subjective quality by calculating the **Quality Assessment Index** (QAI) (Blazier 1995). This index is a measure of the degree the shape of the spectrum under evaluation deviates from the shape of the RC reference curve. The procedure requires calculation of the **energy-average spectral deviations** from the RC reference curve in each of three frequency groups: low frequency, LF (16–63 Hz), medium frequency, MF (125–500 Hz), and high frequency, HF (1000–4000 Hz). The procedure for the LF region is given by Equation (15) and is repeated in the MF and HF regions by substituting the corresponding values at each frequency. However, when evaluating typical HVAC-related sounds, a simple arithmetic average of these deviations is often adequate if the range of values does not exceed 3 dB.

$$\Delta LF = 10\log[10^{0.1\Delta L_{16}} + 10^{0.1\Delta L_{31.5}} + 10^{0.1\Delta L_{63}}/3] \qquad (15)$$

where the ΔL terms are the differences between the spectrum being evaluated and the RC reference curve in each frequency band. In this way, three spectral deviation factors (ΔLF, ΔMF, ΔHF), expressed in dB with either positive or negative values, are associated with the spectrum being rated. QAI is the **range** in dB between the highest and lowest values of the spectral deviation factors.

If QAI $\leq$ 5 dB, the spectrum is assigned a neutral (N) rating. If QAI exceeds 5 dB, the sound quality descriptor of the RC rating is the letter designation of the frequency region of the deviation factor having the highest positive value. As an example, the spectrum plotted in Figure 8 is processed in Table 9.

The arithmetic average of the sound levels in the 500, 1000, and 2000 Hz octave bands in Figure 8 is 35 dB, so the RC 35 curve is selected as the reference for spectrum quality evaluation. The spectral deviation factors in the LF, MF, and HF regions are 6.6, 4.0 and –0.6, respectively, giving a QAI of 7.2. The maximum positive deviation factor occurs in the LF region, and the QAI exceeds 5, resulting in a rating of RC 35(LF). An average room occupant should perceive this spectrum as marginally rumbly in character (see Table 10).

Estimating Occupant Satisfaction Using QAI

The quality assessment index (QAI) is useful in estimating the probable reaction of an occupant when the system does not produce optimum sound quality. The basis for the procedure outlined here for estimating occupant satisfaction is is as follows:

Table 9 Example Calculation of RC Mark II Rating for Sound Spectrum in Figure 8

	Frequency, Hz								
	16	31.5	63	125	250	500	1000	2000	4000
Sound pressure	64	65	64	57	47	40	35	30	23
Average sound pressure at 500-2000 Hz							35		
RC contour	60	60	55	50	45	40	35	30	25
Levels: RC contour	4	5	9	7	2	0	0	0	−2
	LF			MF			HF		
Spectral deviations	6.6			4.0			−0.6		
QAI	6.6 − (−0.6) = 7.2								
RC Mark II rating	**RC 35(LF)**								

- Changes in sound level of less than 5 dB do not cause subjects to change their ranking of sounds of similar spectral content. A QAI of 5 dB or less corresponds to a generally acceptable condition, *provided that the perceived level of the sound is in a range consistent with the given type of space occupancy.*
- A QAI that exceeds 5 dB but is less than or equal to 10 dB represents a marginal situation in which the acceptance by an occupant is questionable.
- A QAI greater than 10 dB will likely be objectionable to the average occupant.

Table 10 lists sound quality descriptors and QAI values and relates them to probable occupant reaction to the sound. An exception to this rule occurs when the sound pressure levels in the 16 Hz or 31.5 Hz octave-bands exceed 65 dB. In such cases, the potential for acoustically-induced vibration in typical lightweight office construction should be considered. If the levels in these bands exceed 75 dB, a significant problem with induced vibration is probable.

Undoubtedly situations will occur in the assessment of HVAC-related sound where the numerical part of the RC rating is less than the specified maximum for the space use, but the sound quality descriptor is other than the desirable (N). For example, a maximum of RC 40(N) is specified, but the actual sound environment turns out to be RC 35(MF). Knowledge in this area is insufficient to decide which spectrum is preferable.

Even at moderate levels, if the dominant portion of the background sound occurs in the very low-frequency region, some people experience a sense of oppressiveness or depression in the environment (Persson-Waye et al. 1997). In such situations, the basis for complaint may result from exposure to that environment for several hours, and thus may not be noticed during a short exposure period.

Criteria Selection Guidelines

In general, these basic guidelines are important:

- Sound levels below NCB 35 or RC 35 are not detrimental to good speech intelligibility; those at or above NCB 35 or RC 35 may interfere with or mask speech.
- Even if the occupancy sound will be significantly higher than the anticipated background sound level generated by mechanical equipment, the sound design goal should not necessarily be raised to levels approaching the occupancy sound. This avoids occupants having to raise their voices uncomfortably to be heard over the noise.

Table 11 gives recommended design criteria.

FUNDAMENTALS OF VIBRATION

A rigidly mounted machine transmits its internal vibratory forces directly to the supporting structure. However, by inserting resilient mountings, called **vibration isolators**, between the machine and supporting structure, the magnitude of transmitted vibration can be reduced to only a fraction of the original. Vibration isolators can also be used to protect sensitive equipment from disturbing vibrations that may be present in the floor of the building structure.

SINGLE-DEGREE-OF-FREEDOM MODEL

The simplest example of a vibration isolation system is the single-degree-of-freedom model illustrated in Figure 9. In this instance only motion along the vertical axis is considered and damping is disregarded. This is the model upon which most manufacturers of vibration isolation hardware base their catalog information.

The application of a single-degree-of-freedom model to the isolation of HVAC equipment is valid only when the stiffness of the supporting structure is large with respect to the stiffness of the vibration isolator. Under these conditions, the natural frequency f_n of the system is

$$f_n = \frac{1}{2\pi}\sqrt{\frac{k}{M}} \qquad (16)$$

Table 10 Definition of Sound Quality Descriptor and Quality Assessment Index (QAI) to Aid in Interpreting RC Mark II Ratings of HVAC-Related Sound

Sound Quality Descriptor	Description of Subjective Perception	Magnitude of QAI	Probable Occupant Evaluation, Assuming Level of Specified Criterion is Not Exceeded
(N) Neutral (Bland)	Balanced sound spectrum, no single frequency range dominant	$QAI \le 5$ dB, $L_{16}, L_{31.5} \le 65$ $QAI \le 5$ dB, $L_{16}, L_{31.5} > 65$	Acceptable Marginal
(LF) Rumble	Low-frequency range dominant (16 – 63 Hz)	5 dB < QAI ≤ 10 dB QAI > 10 dB	Marginal Objectionable
(LFV_B) Rumble, with moderately perceptible room surface vibration	Low-frequency range dominant (16 – 63 Hz)	$QAI \le 5$ dB, $65 < L_{16}, L_{31.5} < 75$ 5 dB < QAI ≤ 10 dB QAI > 10 dB	Marginal Marginal Objectionable
(LFV_A) Rumble, with clearly perceptible room surface vibration	Low-frequency range dominant (16 – 63 Hz)	$QAI \le 5$ dB, $L_{16}, L_{31.5} > 75$ 5 dB < QAI ≤ 10 dB QAI > 10 dB	Marginal Marginal Objectionable
(MF) Roar	Mid-frequency range dominant (125 – 500 Hz)	5 dB < QAI ≤ 10 dB QAI > 10 dB	Marginal Objectionable
(HF) Hiss	High-frequency range dominant (1000 – 4000 Hz)	5 dB < QAI ≤ 10 dB QAI > 10 dB	Marginal Objectionable

Table 11 Design Guidelines for HVAC-Related Background Sound in Rooms

Room Types	RC(N); QAI ≤ 5dB Criterion [a,b]
Residences, Apartments, Condominiums	25 – 35
Hotels/Motels	
Individual rooms or suites	25 – 35
Meeting/banquet rooms	25 – 35
Corridors, lobbies	35 – 45
Service/support areas	35 – 45
Office Buildings	
Executive and private offices	25 – 35
Conference rooms	25 – 35
Teleconference rooms	25 (max)
Open-plan offices	30 – 40
Corridors and lobbies	40 – 45
Hospitals and Clinics	
Private rooms	25 – 35
Wards	30 – 40
Operating rooms	25 – 35
Corridors and public areas	30 – 40
Performing Arts Spaces	
Drama theaters	25 (max)
Concert and recital halls [c]	
Music teaching studios	25 (max)
Music practice rooms	35 (max)
Laboratories (with fume hoods)	
Testing/research, minimal speech communication	45 – 55
Research, extensive telephone use, speech communication	40 – 50
Group teaching	35 – 45
Churches, Mosques, Synagogues	
General assembly	25 – 35
With critical music programs[c]	
Schools[d]	
Classrooms up to 750 ft^2	40 (max)
Classrooms over 750 ft^2	35 (max)
Large lecture rooms, without speech amplification	35 (max)
Libraries	30 – 40
Courtrooms	
Unamplified speech	25 – 35
Amplified speech	30 – 40
Indoor Stadiums, Gymnasiums	
Gymnasiums, natatoriums, and large seating-capacity spaces with speech amplification [e]	40 – 45

[a]The values and ranges are based on judgment and experience, not on quantitative evaluations of human reactions. They represent general limits of acceptability for typical building occupancies. Higher or lower values may be appropriate and should be based on a careful analysis of economics, space use, and user needs.

[b]When quality of sound in the space is important, specify criteria in terms of RC(N). If the quality of the sound in the space is of secondary concern, the criteria may be specified in terms of NC or NCB levels of similar magnitude.

[c]An experienced acoustical consultant should be retained for guidance on acoustically critical spaces (below RC 30) and for all performing arts spaces.

[d]HVAC-related sound criteria for schools, such as those listed in this table, may be too high and impede learning by children in primary grades whose vocabulary is limited, or whose first language is not the language of the class. Some educators and others believe that the HVAC-related background sound should not exceed RC 25(N).

[e]RC or NC criteria for these spaces need only be selected for the desired speech and hearing conditions.

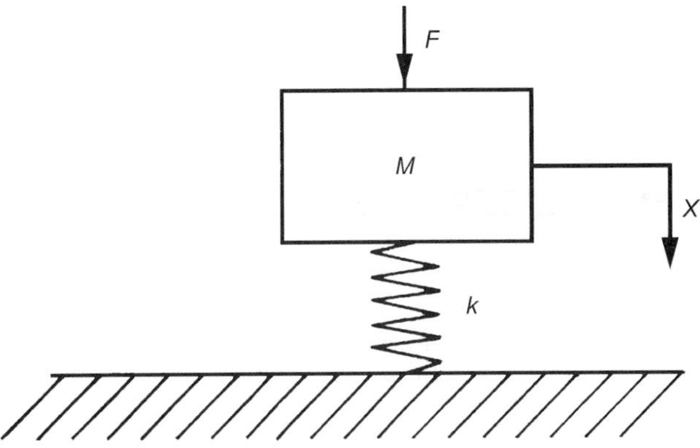

Fig. 9 Single-Degree-of-Freedom System

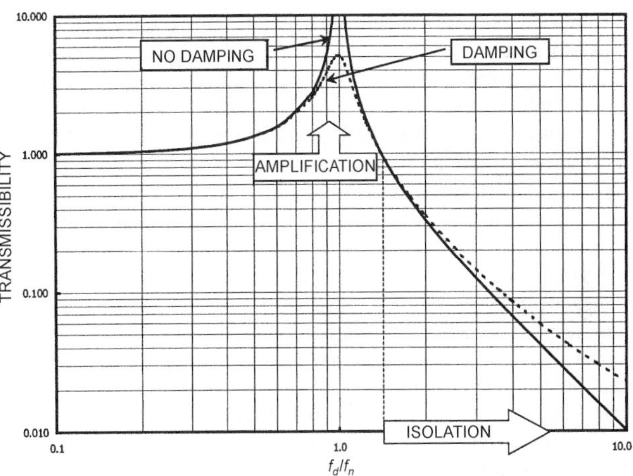

Fig. 10 Vibration Transmissibility T as a Function of f_d/f_n

where k is the stiffness of the vibration isolator (force per unit deflection) and M is the mass of the equipment supported by the isolator. This equation simplifies to

$$f_n = \frac{3.13}{\sqrt{\delta_{st}}} \tag{17}$$

where δ_{st} is the isolator static deflection in inches (the incremental distance the isolator spring compresses under the weight of the supported equipment, or $k/M = g/\delta_{st}$). Thus, to achieve the appropriate system natural frequency for a given application, the corresponding isolator static deflection and the load to be supported at each mounting point is specified.

The **transmissibility** is the ratio of the amplitudes of the force transmitted to the building structure to the exciting force produced by the vibrating equipment. Transmissibility T is inversely proportional to the square of the ratio of the disturbing frequency f_d to the system natural frequency f_n, or

$$T = \left| \frac{1}{1 - (f_d/f_n)^2} \right| \tag{18}$$

At $f_d = f_n$, **resonance** occurs (the denominator of Equation (18) equals zero), with theoretically infinite transmission of vibration. In practice, however, some limit on the transmission at resonance exists because inherent damping is always present to some degree. Thus, the magnitude of vibration amplification at resonance always has a finite value. Equation (18) is plotted in Figure 10.

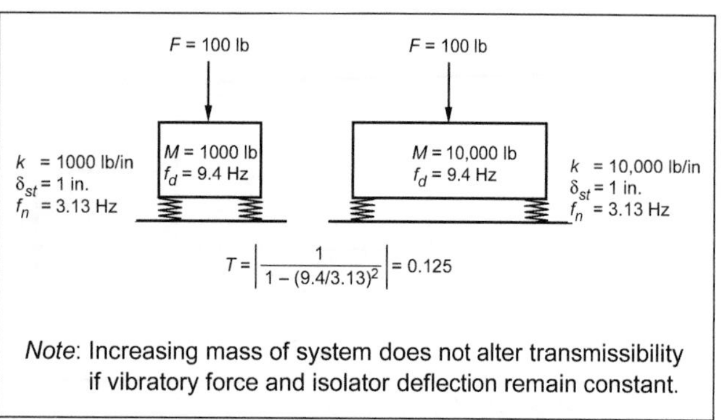

Fig. 11 Effect of Mass on Transmissibility

Vibration isolation does not begin to occur until $f_d/f_n > 1.4$. Above this ratio, the vibration transmissibility rapidly decreases. A frequency ratio of at least 3.5 is often specified, which corresponds to an isolation efficiency of about 90%, or 10% transmissibility. Higher ratios may be specified, but in practice this does not generally result in isolation efficiencies any greater than about 90%. The reason is that "wave-effects" and other nonlinear characteristics cause typical isolators to depart from the theoretical curve that limits performance.

If the mass of the equipment is increased, the resonance frequency decreases, thus increasing the isolation. In practice, the load-carrying capacity of isolators usually requires that their stiffness or their number be increased. Consequently, the static deflection and the transmissibility may remain unchanged. The use of stiffer springs leads, however, to smaller vibration amplitudes—less movement of the equipment. This is one of the main reasons for placing some high-power or highly eccentric equipment on inertia pads.

For example, as shown in Figure 11, a 1000 lb piece of equipment installed on isolators with stiffness k of 1000 lb_f/in. results in a 1 in. deflection and a system resonance frequency f_n of 3.13 Hz. If the equipment is operated at 564 rpm (9.4 Hz) and develops a force of 100 lb_f, a $100 \times 0.125 = 12.5$ lb_f force is transmitted to the structure. If the total mass is increased to 10,000 lb by placing the equipment on a concrete inertia base and the stiffness of the springs is increased to 10,000 lb_f/in., the deflection is still 1 in., the resonance frequency of the system is maintained at 3.13 Hz, and the force transmitted to the structure remains at 12.5 lb_f. The increased mass, however, reduces the equipment displacement.

TWO-DEGREE-OF-FREEDOM MODEL

The single-degree-of-freedom model is valid only when the stiffness of the supporting structure is large with respect to the stiffness of the vibration isolator. This condition is usually satisfied for mechanical equipment in on-grade or basement locations. However, when heavy mechanical equipment is installed on a structural floor, and in particular on the roof of a building, the relative stiffness of the supporting system can no longer be ignored. Significantly "softer" vibration isolators are usually required than in the on-grade or basement case. The appropriate model for the design of vibration isolation in upper-floor locations is the two-degree-of-freedom model illustrated in Figure 12.

The precise behavior of this system with respect to vibration isolation is difficult to determine. The objective is to minimize the motion of the supporting floor M_f in response to the exciting force F. This involves evaluating the interaction between two system natural frequencies and the frequency of the exciting force, which is mathematically complex. However, several engineering rules can simplify the calculations used to optimize the isolation system.

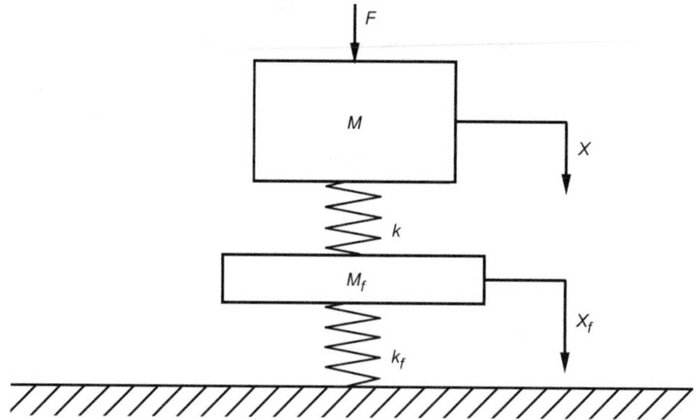

Note: M and M_f represent effective masses of vibrating equipment and supporting floor, respectively; k and k_f are corresponding stiffness of isolator and floor system.

Fig. 12 Two-Degree-of-Freedom System

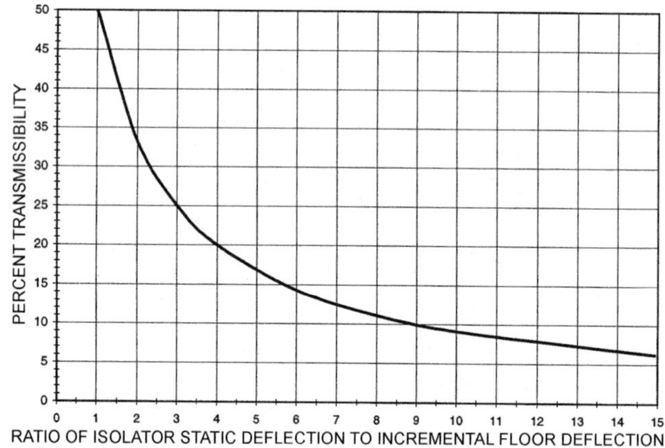

Fig. 13 Transmissibility of Two-Degree-of-Freedom System
Adapted from Plunkett (1958)

For example, the fraction of vibratory force transmitted across an isolator to the building structure (transmissibility) depends in part on the ratio between the isolator stiffness and that of the supporting floor at the point of loading. Because stiffness is inversely proportional to deflection under the applied load, this relationship can sometimes be expressed more conveniently as a ratio of deflections. To optimize isolation efficiency, the static deflection of the isolator, under the applied load, must be large with respect to the incremental static deflection of the floor that occurs due to the added equipment weight. Ideally, this ratio should be on the order of 10:1 to approach an isolation efficiency of about 90% (10% transmissibility). The relationship is illustrated in Figure 13.

Note that if the static deflection of the vibration isolator is similar to the incremental deflection of the supporting floor under the added weight of the equipment, 50% or more of the vibratory force will transmit directly to the building structure. This situation is a common problem in the field where excessive vibration is attributable to upper floor or rooftop mechanical installations. Frequently, the floor stiffness has been neglected and the static deflection on the installed vibration isolators is inadequate because the selection was made on the basis of the single-degree-of-freedom model.

Problems of this nature can usually be avoided by asking the structural engineer to estimate the incremental static deflection of the floor due to the added weight of the equipment at the point of loading, before selecting a vibration isolator. Then, choose an isolator that will provide a static deflection of 8 to 10 times that of the estimated incremental floor deflection.

VIBRATION MEASUREMENT BASICS

While the control of HVAC system sound and vibration are of equal importance, the measurement of vibration is not usually necessary for determining the sources or transmission paths of disturbing sound. Because the techniques and instrumentation used for vibration measurement and analysis are specialized, designers should consult other sources (e.g., Harris 1991) for thorough descriptions of vibration measurement and analysis methods.

The typical vibrations measured are periodic motions of a surface. This surface displacement oscillates with one or more frequencies produced by mechanical means (like gears), thermal means (like combustion), or fluid-dynamic means (like airflow through a duct or fan interactions with air). The displacement is generally inversely proportional to the frequency. In other words, if the displacements are high, the frequency is low. The frequencies of interest for most vibration measurements are between 5 Hz and 100 Hz.

A **transducer** can detect displacement, velocity, or acceleration of a surface and convert the motion to electrical signals. Displacement is the basic measure and good for low frequencies. Velocity is good for overall measurements, but requires large transducers. For most HVAC applications, the transducer of choice is an **accelerometer**, a device that detects acceleration. Readout may be as acceleration level in decibels, or acceleration with modifiers of peak, peak-to-peak, or rms.

The simplest measure is the overall signal as a function of time, be it acceleration, acceleration level, or another quantity. This is analogous to the unfiltered sound pressure level for sound. If a detailed frequency analysis is needed, there is a choice of filters similar to those available for sound measurements: octave band, 1/3 octave band or 1/12 octave band filters. In addition, narrow-band analyzers that use the fast Fourier transform (FFT) to analyze and filter a signal are available. While they are widely used, they should only be used by a specialist for accurate results.

The most important issues in vibration measurement include: (1) choosing a transducer with a frequency range appropriate to the measurement, (2) properly mounting the transducer to ensure that the frequency response claimed is achieved, and (3) not using hand-held probes for high frequencies where they are unreliable.

STANDARDS

AMCA. 1973. Application of sone ratings for non-ducted air moving devices. *Publication* 302-73. Air Movement and Control Association International, Arlington Heights, IL.

AMCA. 1979. Application of sound power level ratings for fans. *Publication* 303-79.

AMCA. 1986. Laboratory method of testing in-duct sound power measurement procedure for fans. *Standard* 330-86.

AMCA. 1990. Methods for calculating fan sound ratings from laboratory test data. *Standard* 301-90.

AMCA. 1996. Reverberant room method for sound testing of fans. *Standard* 300-96.

ANSI. 1980. Procedure for computation of loudness of noise. *Standard* S3.4-1980 (Reaffirmed 1997). American National Standards Institute, New York.

ANSI. 1983. Specifications for sound level meters. *Standard* S1.4-1983 (Amendment S1.4a-85) (R-1997).

ANSI. 1984. Preferred frequencies, frequency levels, and band numbers for acoustical measurements. *Standard* S1.6-1984 (R 1997).

ANSI. 1986. Specifications for octave-band and fractional octave-band analog and digital filters. *Standard* S1.11-1986 (R 1998).

ANSI. 1990. Precision methods for the determination of sound power levels of broadband noise sources in reverberation rooms. *Standard* S12.31-1990 (R 1996) (Supersedes ANSI S1.31-1980).

ANSI. 1990. Precision methods for the determination of sound power levels of discrete-frequency and narrow-band noise sources in reverberation rooms. *Standard* S12.32-1990 (R1996) (Supersedes ANSI S1.32-1980 and ASA 12.80).

ANSI. 1990. Precision methods for the determination of sound power levels of noise sources in anechoic and hemi-anechoic rooms. *Standard* S12.35-1990 (R 1996) (Supersedes ANSI S1.35-1979 and ASA 15-79).

ANSI. 1992. Engineering method for determination of sound power level of noise sources using sound intensity. *Standard* S12.12-1992 (R 1997).

ANSI. 1995. Criteria for evaluating room noise. *Standard* S12.2-1995.

ARI. 1998. Air terminals. *Standard* 880-1998. Air-Conditioning and Refrigeration Institute, Arlington, Virginia.

ARI. 1998. Procedure for estimating occupied space sound levels in the application of air terminals and air outlets. *Standard* 885-1998.

ASHRAE. 1995. Methods of testing for rating ducted air terminal units. *Standard* 130-1995.

ASTM. 1987. Classification for rating sound insulation. *Standard* E 413-87(1999). American Society for Testing and Materials, West Conshohocken, PA.

ASTM. 1993. Test method for evaluating masking sound in open offices using a-weighted and one-third octave band sound pressure levels. *Standard* E 1573-93(1998).

ASTM. 1998. Test method for measurement of sound in residential spaces. *Standard* E 1574-98.

ASTM. 1999. Standard test method for sound absorption and sound absorption coefficients by the reverberation room method. *Standard* C 423-99a.

ASTM. 1999. Standard test method for laboratory measurement of airborne sound transmission loss of building partitions and elements. *Standard* E 90-99.

ASTM. 1999. Standard test method for measuring acoustical and airflow performance of duct liner materials and prefabricated silencers. *Standard* E 477-99.

ASTM. 2000. Standard terminology relating to environmental acoustics. *Standard* C 634-00.

German *Standard* DIN 45631. Method for predicting loudness of sound spectra with tonal qualities.

ISO. 1975. Methods for calculating loudness level. *Standard* 532:1975. International Organization for Standardization, Geneva.

ISO. 1993. Determination of sound power levels of noise sources using sound intensity—Part 1: Measurements at discrete points. *Standard* 9614-1:1993.

ISO. 1996. Determination of sound power levels of noise sources using-sound intensity—Part 2: Measurements by scanning. *Standard* 9614-2:1993.

REFERENCES

ASHRAE. 1997. ASHRAE RP755. Sound transmission through ceilings from air terminal devices in the plenum.

Beranek, L.L. 1957. Revised criteria for noise in buildings. *Noise Control* 1:19.

Beranek, L.L. 1989. Balanced noise criterion (NCB) curves. *Journal of the Acoustic Society of America* (86):650-54.

Blazier, W.E., Jr. 1981a. Revised noise criteria for application in the acoustical design and rating of HVAC systems. *Noise Control Eng.* 16(2):64-73.

Blazier, W. E., Jr. 1981b. Revised noise criteria for design and rating of HVAC systems. *ASHRAE Transactions* 87(1).

Blazier, W.E., Jr. 1995. Sound quality considerations in rating noise from heating, ventilating and air-conditioning (HVAC) systems in buildings. *Noise Control Eng. J.* 43(3).

Blazier, W.E., Jr. 1997. RC Mark II: A refined procedure for rating the noise of heating, ventilating and air-conditioning (HVAC) systems in buildings. *Noise Control Eng. J.* 45(6), November/December.

Broner, N. 1994. Determination of the relationship between low-frequency HVAC noise and comfort in occupied spaces. *ASHRAE Research Project* 714 Objective.

Persson-Waye, K., et al. 1997. Effects on performance and work quality due to low-frequency ventilation noise. *Journal of Sound and Vibration* 205(4):467-474.

Plunkett, R. 1958. Interaction between a vibratory machine and its foundation. *Noise Control* 4(1).

Pollack, I. 1952. The loudness of bands of noise. *Journal of the Acoustical Society of America* 24(9):533.

Robinson, D.W. and R.S. Dadson. 1956. A redetermination of the equal loudness relations for pure tones. *British Journal of Applied Physics* 7(5):166.

Schultz, T.J. 1985. Relationship between sound power level and sound pressure level in dwellings and offices. *ASHRAE Transactions* 91(1):124-53.

Warnock, A.C.C. 1998a. Sound pressure level vs. distance from sources in rooms. *ASHRAE Transactions* 104(1A):643-649.

Warnock, A.C.C. 1998b. Transmission of sound from air terminal devices through ceiling systems. *ASHRAE Transactions* 104(1A):650-657.

BIBLIOGRAPHY

ASHRAE. 1998. Applications of manufacturers' sound data.

Beranek, L.L. 1988. *Noise and vibration control*. Institute of Noise Control Engineering, Washington, D.C.

Beranek, L.L. 1986. *Acoustics*. American Institute of Physics, Acoustical Society of America, New York.

Harris, C.M. 1991. *Handbook of acoustical measurements and noise control*. McGraw-Hill, New York.

Peterson, A.P.G. and E.E. Gross, Jr. 1974. *Handbook of noise measurement*. GenRad, Inc., Concord, MA.

Schaffer, M.E. 1991. *A practical guide to noise and vibration control for HVAC systems*. ASHRAE, Atlanta.

CHAPTER 8

THERMAL COMFORT

A PRINCIPAL purpose of heating, ventilating, and air-conditioning systems is to provide conditions for human thermal comfort. A widely accepted definition is, "Thermal Comfort is that condition of mind that expresses satisfaction with the thermal environment" (ASHRAE *Standard* 55). This definition leaves open what is meant by condition of mind or satisfaction, but it correctly emphasizes that the judgment of comfort is a cognitive process involving many inputs influenced by physical, physiological, psychological, and other processes.

The conscious mind appears to reach conclusions about thermal comfort and discomfort from direct temperature and moisture sensations from the skin, deep body temperatures, and the efforts necessary to regulate body temperatures (Hensel 1973, 1981; Hardy et al. 1971; Gagge 1937; Berglund 1995). In general, comfort occurs when body temperatures are held within narrow ranges, skin moisture is low, and the physiological effort of regulation is minimized.

Comfort also depends on behavioral actions that are initiated unconsciously or by the conscious mind and guided by thermal and moisture sensations to reduce discomfort. Some of the possible behavioral actions to reduce discomfort are altering clothing, altering activity, changing posture or location, changing the thermostat setting, opening a window, complaining, or leaving the space.

Surprisingly, although regional climate conditions, living conditions, and cultures differ widely throughout the world, the temperature that people choose for comfort under like conditions of clothing, activity, humidity, and air movement has been found to be very similar (Fanger 1972; de Dear et al. 1991; Busch 1992).

This chapter summarizes the fundamentals of human thermoregulation and comfort in terms useful to the engineer for operating systems and designing for the comfort and health of building occupants.

HUMAN THERMOREGULATION

The metabolic activities of the body result almost completely in heat that must be continuously dissipated and regulated to maintain normal body temperatures. Insufficient heat loss leads to overheating, also called **hyperthermia**, and excessive heat loss results in body cooling, also called **hypothermia**. Skin temperature greater than 113°F or less than 64.5°F causes pain (Hardy et al. 1952). Skin temperatures associated with comfort at sedentary activities are 91.5 to 93°F and decrease with increasing activity (Fanger 1968). In contrast, internal temperatures rise with activity. The temperature regulatory center in the brain is about 98.2°F at rest in comfort and increases to about 99.3°F when walking and 100.2°F when jogging. An internal temperature less than about 82°F can lead to serious cardiac arrhythmia and death, and a temperature greater than 115°F can cause irreversible brain damage. Therefore, the careful regulation of body temperature is critical to comfort and health.

The heat produced by a resting adult is about 350 Btu/h. Because most of this heat is transferred to the environment through the skin, it is often convenient to characterize metabolic activity in terms of heat production per unit area of skin. For the resting person, this is about 18.4 Btu/h·ft^2(50 kcal/h·m^2) and is called 1 met. This is based on the average male European, with a skin surface area of about 19.4 ft^2. For comparison, female Europeans have an average surface area of 17.2 ft^2. Systematic differences in this parameter may occur between ethnic and geographical groups. Higher metabolic rates are often described in terms of the resting rate. Thus, a person working at metabolic rate five times the resting rate would have a metabolic rate of 5 met.

The **hypothalamus**, located in the brain, is the central control organ for body temperature. It has hot and cold temperature sensors and is bathed by arterial blood. Since the recirculation rate of blood in the body is rapid and returning blood is mixed together in the heart before returning to the body, arterial blood is indicative of the average internal body temperature. The hypothalamus also receives thermal information from temperature sensors in the skin and perhaps other locations as well (spinal cord, gut), as summarized by Hensel (1981).

The hypothalamus controls various physiological processes of the body to regulate body temperature. Its control behavior is primarily proportional to deviations from set-point temperatures with some integral and derivative response aspects. The most important and often used of the physiological processes is regulating blood flow to the skin. When internal temperatures rise above a set point, an increasing proportion of the total blood is directed to the skin. This **vasodilation** of skin blood vessels can increase skin blood flow by 15 times (from 0.56 L/h·ft^2 at resting comfort to 8.4 L/h·ft^2 in extreme heat) to carry internal heat to the skin for transfer to the environment. When body temperatures fall below the set point, skin blood flow is reduced (vasoconstricted) to conserve body heat. The effect of maximum vasoconstriction is equivalent to the insulating effect of a heavy sweater. At temperatures less than the set point, muscle tension increases to generate additional heat; where muscle groups are opposed, this may increase to visible shivering. Shivering can double the resting rate of heat production.

At elevated internal temperatures, sweating occurs. This defense mechanism is a powerful way to cool the skin and increase heat loss from the core. The sweating function of the skin and its control is more advanced in humans than in other animals and is increasingly necessary for comfort at metabolic rates above resting level (Fanger 1968). Sweat glands pump perspiration onto the skin surface for evaporation. If conditions are good for evaporation, the skin can remain relatively dry even at high sweat rates with little perception of sweating. At skin conditions less favorable for evaporation, the sweat must spread out on the skin about the sweat gland until the sweat-covered area is sufficient to evaporate the sweat coming to the surface. The fraction of the skin that is covered with water to account for the observed total evaporation rate is termed **skin wettedness** (Gagge 1937).

The preparation of this chapter is assigned to TC 2.1, Physiology and Human Environment.

Humans are quite good at sensing skin moisture from perspiration (Berglund and Cunningham 1986; Berglund 1994), and skin moisture correlates well with warm discomfort and unpleasantness (Winslow et al. 1937). It is rare for a sedentary or slightly active person to be comfortable with a skin wettedness greater than 25%. In addition to the perception of skin moisture, skin wettedness increases the friction between skin and fabrics, making clothing feel less pleasant and fabrics feel more coarse (Gwosdow et al. 1987). This also occurs with architectural materials and surfaces, particularly smooth, nonhygroscopic surfaces.

With repeated intermittent heat exposure, the set point for the onset of sweating decreases and the proportional gain or temperature sensitivity of the sweating system increases (Gonzalez et al. 1978, Hensel 1981). However, under long-term exposure to hot conditions, the set point increases, perhaps to reduce the physiological effort of sweating. Perspiration as secreted has a lower salt concentration than interstitial body fluid or blood plasma. After prolonged heat exposure, sweat glands further reduce the salt concentration of sweat to conserve salt.

At the surface, the water in sweat evaporates while the dissolved salt and other constituents remain and accumulate. Because salt lowers the vapor pressure of water and thereby impedes its evaporation, the accumulating salt results in increased skin wettedness with time. Some of the relief and pleasure of washing after a warm day is related to the restoration of a hypotonic sweat film and decreased skin wettedness. Other adaptations to heat are increased blood flow and sweating in peripheral regions where heat transfer is better. Such adaptations are examples of **integral control**.

The role of thermoregulatory effort in comfort is highlighted by the experiments of Chatonnet and Cabanac (1965) and observations of Kuno (1995). Chatonnet's experiments compared the sensation of placing the subject's hand in relatively hot or cold water (86 to 100°F) for 30 s given the subject at different thermal states. When the person was overheated or hyperthermic, the cold water was pleasant and the hot water was very unpleasant, but when the subject was in a cold or hypothermic state, the hand felt pleasant in hot water and unpleasant in cold water. Kuno (1995) describes similar observations during transient whole body exposures to hot and cold environment. When a subject is in a state of thermal discomfort, any move away from the thermal stress of the uncomfortable environment is perceived as pleasant during the transition.

ENERGY BALANCE

Figure 1 shows the thermal interaction of the human body with its environment. The total metabolic rate of work M produced within the body is the metabolic rate required for the person's activity M_{act} plus the metabolic level required for shivering M_{shiv} (should shivering occur). A portion of the body's energy production may be expended as external work done by the muscles W; the net heat production $M - W$ is either stored (S), causing the body's temperature to rise, or dissipated to the environment through the skin surface (q_{sk}) and respiratory tract (q_{res}).

$$M - W = q_{sk} + q_{res} + S = (C + R + E_{sk}) + (C_{res} + E_{res}) + (S_{sk} + S_{cr}) \quad (1)$$

where

M = rate of metabolic heat production, Btu/h·ft²
W = rate of mechanical work accomplished, Btu/h·ft²
q_{sk} = total rate of heat loss from skin, Btu/h·ft²
q_{res} = total rate of heat loss through respiration, Btu/h·ft²
$C + R$ = sensible heat loss from skin, Btu/h·ft²
E_{sk} = total rate of evaporative heat loss from skin, Btu/h·ft²
C_{res} = rate of convective heat loss from respiration, Btu/h·ft²
E_{res} = rate of evaporative heat loss from respiration, Btu/h·ft²
S_{sk} = rate of heat storage in skin compartment, Btu/h·ft²
S_{cr} = rate of heat storage in core compartment, Btu/h·ft²

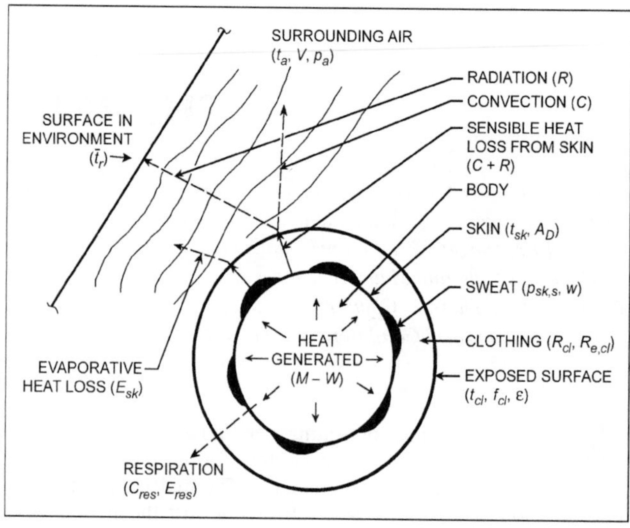

Fig. 1 Thermal Interaction of Human Body and Environment

Heat dissipation from the body to the immediate surroundings occurs by several modes of heat exchange: sensible heat flow $C + R$ from the skin; latent heat flow from the evaporation of sweat E_{rsw} and from evaporation of moisture diffused through the skin E_{dif}; sensible heat flow during respiration C_{res}; and latent heat flow due to evaporation of moisture during respiration E_{res}. Sensible heat flow from the skin may be a complex mixture of conduction, convection, and radiation for a clothed person; however, it is equal to the sum of the convection C and radiation R heat transfer at the outer clothing surface (or exposed skin).

Sensible and latent heat losses from the skin are typically expressed in terms of environmental factors, skin temperature t_{sk}, and skin wettedness w. The expressions also incorporate factors that account for the thermal insulation and moisture permeability of clothing. The independent environmental variables can be summarized as air temperature t_a, mean radiant temperature $\bar{t}_r$, relative air velocity V, and ambient water vapor pressure p_a. The independent personal variables that influence thermal comfort are activity and clothing.

The rate of heat storage in the body equals the rate of increase in internal energy. The body can be considered as two thermal compartments, the skin and the core (see the section on Two-Node Model under Prediction of Thermal Comfort). The rate of storage can be written separately for each compartment in terms of thermal capacity and time rate of change of temperature in each compartment:

$$S_{cr} = \frac{(1 - \alpha_{sk}) m c_{p,b}}{A_D} \frac{dt_{cr}}{d\theta} \quad (2)$$

$$S_{sk} = \frac{\alpha_{sk} m c_{p,b}}{A_D} \frac{dt_{sk}}{d\theta} \quad (3)$$

where

α_{sk} = fraction of body mass concentrated in skin compartment
m = body mass, lb
$c_{p,b}$ = specific heat capacity of body = 0.834 Btu/lb·°F
A_D = DuBois surface area, ft²
t_{cr} = temperature of core compartment, °F
t_{sk} = temperature of skin compartment, °F
θ = time, h

The fractional skin mass α_{sk} depends on the rate $\dot{m}_{bl}$ of blood flowing to the skin surface.

THERMAL EXCHANGES WITH THE ENVIRONMENT

Fanger (1967, 1970), Hardy (1949), Rapp and Gagge (1967), and Gagge and Hardy (1967) give quantitative information on calculating the heat exchange between people and the environment. A summary of the mathematical statements for various terms of heat exchange used in the heat balance equations (C, R, E_{sk}, C_{res}, E_{res}) follows. Terms describing the heat exchanges associated with the thermoregulatory control mechanisms ($q_{cr,sk}$, M_{shiv}, E_{rsw}), values for the coefficients, and appropriate equations for M_{act} and A_D are presented in later sections.

The mathematical description of the energy balance of the human body represents a combined rational/empirical approach to describing the thermal exchanges with the environment. Fundamental heat transfer theory is used to describe the various mechanisms of sensible and latent heat exchange, while empirical expressions are used to determine the values of the coefficients describing these rates of heat exchange. Empirical equations are also used to describe the thermophysiological control mechanisms as a function of skin and core temperatures in the body.

Body Surface Area

The terms in Equation (1) have units of power per unit area and refer to the surface area of the nude body. The most useful measure of nude body surface area, originally proposed by DuBois and DuBois (1916), is described by

$$A_D = 0.108 m^{0.425} l^{0.725} \tag{4}$$

where

A_D = DuBois surface area, ft²
m = mass, lb
l = height, in.

A correction factor $f_{cl} = A_{cl}/A_D$ must be applied to the heat transfer terms from the skin (C, R, and E_{sk}) to account for the actual surface area A_{cl} of the clothed body. This factor can be found in Table 7 for various clothing ensembles. For a 68 in. tall, 154 lb man, A_D = 19.6 ft². All terms in the basic heat balance equations are expressed per unit **DuBois surface area**.

Sensible Heat Loss from Skin

Sensible heat exchange from the skin surface must pass through clothing to the surrounding environment. These paths are treated in series and can be described in terms of heat transfer (1) from the skin surface, through the clothing insulation, to the outer clothing surface, and (2) from the outer clothing surface to the environment.

Both convective C and radiative R heat losses from the outer surface of a clothed body can be expressed in terms of a heat transfer coefficient and the difference between the mean temperature t_{cl} of the outer surface of the clothed body and the appropriate environmental temperature:

$$C = f_{cl} h_c (t_{cl} - t_a) \tag{5}$$

$$R = f_{cl} h_r (t_{cl} - \bar{t}_r) \tag{6}$$

where

h_c = convective heat transfer coefficient, Btu/h·ft²·°F
h_r = linear radiative heat transfer coefficient, Btu/h·ft²·°F
f_{cl} = clothing area factor A_{cl}/A_D, dimensionless

The coefficients h_c and h_r are both evaluated at the clothing surface. Equations (5) and (6) are commonly combined to describe the total sensible heat exchange by these two mechanisms in terms of an operative temperature t_o and a combined heat transfer coefficient h:

$$C + R = f_{cl} h (t_{cl} - t_o) \tag{7}$$

where

$$t_o = \frac{h_r \bar{t}_r + h_c t_a}{h_r + h_c} \tag{8}$$

$$h = h_r + h_c \tag{9}$$

Based on Equation (8), operative temperature t_o can be defined as the average of the mean radiant and ambient air temperatures, weighted by their respective heat transfer coefficients.

The actual transport of sensible heat through clothing involves conduction, convection, and radiation. It is usually most convenient to combine these into a single thermal resistance value R_{cl}, defined by

$$C + R = (t_{sk} - t_{cl})/R_{cl} \tag{10}$$

where R_{cl} is the thermal resistance of clothing in ft²·°F·h/Btu.

Since it is often inconvenient to include the clothing surface temperature in calculations, Equations (7) and (10) can be combined to eliminate t_{cl}:

$$C + R = \frac{t_{sk} - t_o}{R_{cl} + 1/(f_{cl} h)} \tag{11}$$

where t_o is defined in Equation (8).

Evaporative Heat Loss from Skin

Evaporative heat loss E_{sk} from skin depends on the amount of moisture on the skin and the difference between the water vapor pressure at the skin and in the ambient environment:

$$E_{sk} = \frac{w(p_{sk,s} - p_a)}{R_{e,cl} + 1/(f_{cl} h_e)} \tag{12}$$

where

w = skin wettedness, dimensionless
$p_{sk,s}$ = water vapor pressure at skin, normally assumed to be that of saturated water vapor at t_{sk}, psi
p_a = water vapor pressure in ambient air, psi
$R_{e,cl}$ = evaporative heat transfer resistance of clothing layer (analogous to R_{cl}), ft²·psi·h/Btu
h_e = evaporative heat transfer coefficient (analogous to h_c), Btu/h·ft²·psi

Procedures for calculating $R_{e,cl}$ and h_e are given in the section on Engineering Data and Measurements. The skin wettedness fraction is the ratio of the actual evaporative heat loss to the maximum possible evaporative heat loss E_{max} with the same conditions and a completely wet skin (w = 1). Skin wettedness is important in determining evaporative heat loss. Maximum evaporative potential E_{max} occurs when w = 1.

Evaporative heat loss from the skin is a combination of the evaporation of sweat secreted due to thermoregulatory control mechanisms E_{rsw} and the natural diffusion of water through the skin E_{dif}:

$$E_{sk} = E_{rsw} + E_{dif} \tag{13}$$

Evaporative heat loss by regulatory sweating is directly proportional to the rate of regulatory sweat generation:

$$E_{rsw} = \dot{m}_{rsw} h_{fg} \tag{14}$$

where

h_{fg} = heat of vaporization of water = 1045 Btu/lb at 86°F
$\dot{m}_{rsw}$ = rate at which regulatory sweat is generated, lb/h·ft²

The portion w_{rsw} of a body that must be wetted to evaporate the regulatory sweat is

$$w_{rsw} = E_{rsw}/E_{max} \quad (15)$$

With no regulatory sweating, skin wettedness due to diffusion is approximately 0.06 for normal conditions. For large values of E_{max} or long exposures to low humidities, the value may drop to as low as 0.02, since dehydration of the outer skin layers alters its diffusive characteristics. With regulatory sweating, the 0.06 value applies only to the portion of skin not covered with sweat $(1 - w_{rsw})$; the diffusion evaporative heat loss is

$$E_{dif} = (1 - w_{rsw})0.06E_{max} \quad (16)$$

These equations can be solved for w, given the maximum evaporative potential E_{max} and the regulatory sweat generation E_{rsw}:

$$w = w_{rsw} + 0.06(1 - w_{rsw}) = 0.06 + 0.94E_{rsw}/E_{max} \quad (17)$$

Once skin wettedness is determined, evaporative heat loss from the skin is calculated from Equation (12), or by

$$E_{sk} = wE_{max} \quad (18)$$

To summarize, the following calculations determine w and E_{sk}:

E_{max}	Equation (12), with $w = 1.0$
E_{rsw}	Equation (14)
w	Equation (17)
E_{sk}	Equation (18) or (12)

Although evaporation from the skin E_{sk} as described in Equation (12) depends on w, the body does not directly regulate skin wettedness but, rather, regulates sweat rate $\dot{m}_{rsw}$ [Equation (14)]. Skin wettedness is then an indirect result of the relative activity of the sweat glands and the evaporative potential of the environment. Skin wettedness of 1.0 is the upper theoretical limit. If the aforementioned calculations yield a wettedness of more than 1.0, then Equation (14) is no longer valid because not all the sweat is evaporated. In this case, $E_{sk} = E_{max}$.

Skin wettedness is strongly correlated with warm discomfort and is also a good measure of thermal stress. Theoretically, skin wettedness can approach 1.0 while the body still maintains thermoregulatory control. In most situations, it is difficult to exceed 0.8 (Berglund and Gonzalez 1978). Azer (1982) recommends 0.5 as a practical upper limit for sustained activity for a healthy acclimatized person.

Respiratory Losses

During respiration, the body loses both sensible and latent heat by convection and evaporation of heat and water vapor from the respiratory tract to the inhaled air. A significant amount of heat can be associated with respiration because the air is inspired at ambient conditions and expired nearly saturated at a temperature only slightly cooler than t_{cr}.

The total heat **and** moisture losses due to respiration are

$$q_{res} = C_{res} + E_{res} = \frac{\dot{m}_{res}(h_{ex} - h_a)}{A_D} \quad (19)$$

$$\dot{m}_{w,\,res} = \frac{\dot{m}_{res}(W_{ex} - W_a)}{A_D} \quad (20)$$

where

$\dot{m}_{res}$ = pulmonary ventilation rate, lb/h
h_{ex} = enthalpy of exhaled air, Btu/lb (dry air)
h_a = enthalpy of inspired (ambient) air, Btu/lb (dry air)
$\dot{m}_{w,\,res}$ = pulmonary water loss rate, lb/h
W_{ex} = humidity ratio of exhaled air, lb (water vapor)/lb (dry air)
W_a = humidity ratio of inspired (ambient) air, lb (water vapor)/lb (dry air)

Under normal circumstances, pulmonary ventilation rate is primarily a function of metabolic rate (Fanger 1970):

$$\dot{m}_{res} = K_{res}MA_D \quad (21)$$

where

M = metabolic rate, Btu/h·ft^2
K_{res} = proportionality constant (3.33 lb/Btu)

Respiratory air is nearly saturated and near the body temperature when exhaled. For typical indoor environments (McCutchan and Taylor 1951), the exhaled temperature and humidity ratio are given in terms of ambient conditions:

$$t_{ex} = 88.6 + 0.066t_a + 57.6W_a \quad (22)$$

$$W_{ex} = 0.0265 + 0.000036t_a + 0.2W_a \quad (23)$$

where ambient t_a and exhaled t_{ex} air temperatures are in °F. For extreme conditions, such as outdoor winter environments, different relationships may be required (Holmer 1984).

The humidity ratio of ambient air can be expressed in terms of total or barometric pressure p_t and ambient water vapor pressure p_a:

$$W_a = \frac{0.622p_a}{p_t - p_a} \quad (24)$$

Respiratory heat loss is often expressed in terms of sensible C_{res} and latent E_{res} heat losses. Two approximations are commonly used to simplify Equations (22) and (23) for that purpose. First, because the dry respiratory heat loss is relatively small compared to the other terms in the heat balance, an average value for t_{ex} is determined by evaluating Equation (22) at standard conditions of 68°F, 50% rh, sea level. Second, noting in Equation (23) that there is only a weak dependence on t_a, the second term in Equation (23) and the denominator in Equation (24) are evaluated at standard conditions. Using these approximations and substituting latent heat h_{fg} and specific heat of air $c_{p,a}$ at standard conditions, C_{res} and E_{res} can be determined by

$$C_{res} = 0.0084M(93.2 - t_a) \quad (25)$$

$$E_{res} = 1.28M(0.851 - p_a) \quad (26)$$

where p_a is expressed in psi and t_a is in °F.

Alternative Formulations

Equations (11) and (12) describe heat loss from skin for clothed people in terms of clothing parameters R_{cl}, $R_{e,cl}$, and f_{cl}; parameters h and h_e describe outer surface resistances. Other parameters and definitions are also used. Although these alternative parameters and definitions may be confusing, note that the information presented in one form can be converted to another form. Table 1 presents common parameters and their qualitative descriptions. Table 2 presents equations showing their relationship to each other. Generally, parameters related to dry or evaporative heat flows are not independent because they both rely, in part, on the same physical processes. The **Lewis relation** describes the relationship between convective heat transfer and mass transfer coefficients for a surface [see Equation (39) in Chapter 5]. The Lewis relation can be used to relate convective and evaporative heat transfer coefficients defined in Equations (5) and (12) according to

Table 1 Parameters Used to Describe Clothing

Sensible Heat Flow

R_{cl} = intrinsic clothing insulation, the thermal resistance of a uniform layer of insulation covering the entire body that has the same effect on sensible heat flow as the actual clothing.

R_t = total insulation, the total equivalent uniform thermal resistance between the body and the environment: clothing and boundary resistance.

R_{cle} = effective clothing insulation, the increased body insulation due to clothing as compared to the nude state.

R_a = boundary insulation, the thermal resistance at the skin boundary for a nude body.

$R_{a,cl}$ = outer boundary insulation, the thermal resistance at the outer boundary (skin or clothing).

h' = overall sensible heat transfer coefficient, overall equivalent uniform conductance between the body (including clothing) and the environment.

h'_{cl} = clothing conductance, the thermal conductance of a uniform layer of insulation covering the entire body that has the same effect on sensible heat flow as the actual clothing.

F_{cle} = effective clothing thermal efficiency, the ratio of the actual sensible heat loss to that of a nude body at the same conditions.

F_{cl} = intrinsic clothing thermal efficiency, the ratio of the actual sensible heat loss to that of a nude body at the same conditions including an adjustment for the increase in surface area due to the clothing.

Evaporative Heat Flow

$R_{e,cl}$ = evaporative heat transfer resistance of the clothing, the impedance to transport of water vapor of a uniform layer of insulation covering the entire body that has the same effect on evaporative heat flow as the actual clothing.

$R_{e,t}$ = total evaporative resistance, the total equivalent uniform impedance to the transport of water vapor from the skin to the environment.

F_{pcl} = permeation efficiency, the ratio of the actual evaporative heat loss to that of a nude body at the same conditions, including an adjustment for the increase in surface area due to the clothing.

Parameters Relating Sensible and Evaporative Heat Flows

i_{cl} = clothing vapor permeation efficiency, the ratio of the actual evaporative heat flow capability through the clothing to the sensible heat flow capability as compared to the Lewis ratio.

i_m = total vapor permeation efficiency, the ratio of the actual evaporative heat flow capability between the skin and the environment to the sensible heat flow capability as compared to the Lewis ratio.

i_a = air layer vapor permeation efficiency, the ratio of the actual evaporative heat flow capability through the outer air layer to the sensible heat flow capability as compared to the Lewis ratio.

Table 2 Relationships Between Clothing Parameters

Sensible Heat Flow

$$R_t = R_{cl} + 1/(hf_{cl}) = R_{cl} + R_a/f_{cl}$$
$$R_t = R_{cle} + 1/h = R_{cle} + R_a$$
$$h'_{cl} = 1/R_{cl}$$
$$h' = 1/R_t$$
$$h = 1/R_a$$
$$R_{a,cl} = R_a/f_{cl}$$
$$F_{cl} = h'/(hf_{cl}) = 1/(1 + f_{cl}hR_{cl})$$
$$F_{cle} = h'/h = f_{cl}/(1 + f_{cl}hR_{cl}) = f_{cl}F_{cl}$$

Evaporative Heat Flow

$$R_{e,t} = R_{e,cl} + 1/(h_e f_{cl}) = R_{e,cl} + R_{e,a}/f_{cl}$$
$$h_e = 1/R_{e,a}$$
$$h'_{e,cl} = 1/R_{e,cl}$$
$$h'_e = 1/R_{e,t} = f_{cl}F_{pcl}h_e$$
$$F_{pcl} = 1/(1 + f_{cl}h_e R_{e,cl})$$

Parameters Relating Sensible and Evaporative Heat Flows

$$i_{cl}\text{LR} = h'_{e,cl}/h'_{cl} = R_{cl}/R_{e,cl}$$
$$i_m\text{LR} = h'_e/h' = R_t/R_{e,t}$$
$$i_m = (R_{cl} + R_{a,cl})/[(R_{cl}/i_{cl}) + (R_{a,cl}/i_a)]$$
$$i_a\text{LR} = h_e/h$$
$$i_a = h_c/(h_c + h_r)$$

Table 3 Skin Heat Loss Equations

Sensible Heat Loss

$$C + R = (t_{sk} - t_o)/[R_{cl} + 1/(f_{cl}h)]$$
$$C + R = (t_{sk} - t_o)/R_t$$
$$C + R = F_{cle}h(t_{sk} - t_o)$$
$$C + R = F_{cl}f_{cl}h(t_{sk} - t_o)$$
$$C + R = h'(t_{sk} - t_o)$$

Evaporative Heat Loss

$$E_{sk} = w(p_{sk,s} - p_a)/[R_{e,cl} + 1/(f_{cl}h_e)]$$
$$E_{sk} = w(p_{sk,s} - p_a)/R_{e,t}$$
$$E_{sk} = wF_{pcl}f_{cl}h_e(p_{sk,s} - p_a)$$
$$E_{sk} = h'_e w(p_{sk,s} - p_a)$$
$$E_{sk} = h' w i_m \text{LR}(p_{sk,s} - p_a)$$

$$h_e/h_c = \text{LR} \tag{27}$$

where LR is the **Lewis ratio** and, at typical indoor conditions, equals approximately 205°F/psi. The Lewis relation applies to surface convection coefficients. Heat transfer coefficients that include the effects of insulation layers and/or radiation are still coupled, but the relationship may deviate significantly from that for a surface. The i terms in Tables 1 and 2 describe how the actual ratios of these parameters deviate from the ideal Lewis ratio (Woodcock 1962; Oohori et al. 1984).

Depending on the combination of parameters used, heat transfer from the skin can be calculated using several different formulations (see Tables 2 and 3). If the parameters are used correctly, the end result will be the same regardless of the formulation used.

Total Skin Heat Loss

Total skin heat loss—sensible heat plus evaporative heat—can be calculated from any combination of the equations presented in Table 3. Total skin heat loss is used as a measure of the thermal environment; two combinations of parameters that yield the same total heat loss for a given set of body conditions (t_{sk} and w) are considered to be approximately equivalent. The fully expanded skin heat loss equation, showing each parameter that must be known or specified, is as follows:

$$q_{sk} = \frac{t_{sk} - t_o}{R_{cl} + R_{a,cl}} + \frac{w(p_{sk,s} - p_a)}{R_{e,cl} + 1/(\text{LR}h_c f_{cl})} \tag{28}$$

where t_o is the operative temperature and represents the temperature of a uniform environment ($t_a = \bar{t}_r$) that will transfer dry heat at the same rate as in the actual environment [$t_o = (\bar{t}_r h_r + t_a h_c)/(h_c + h_r)$]. After rearranging, Equation (28) becomes

$$q_{sk} = F_{cl}f_{cl}h(t_{sk} - t_o) + w\text{LR}F_{pcl}h_c(p_{sk,s} - p_a) \tag{29}$$

This equation allows the trade-off between any two or more parameters to be evaluated under given conditions. If the trade-off between two specific variables is to be examined, then a simplified form of the equation suffices. The trade-off between operative temperature and humidity is often of interest. Equation (28) can be written in a simpler form for this purpose (Fobelets and Gagge 1988):

$$q_{sk} = h'[(t_{sk} + wi_m \mathrm{LR} p_{sk,s}) - (t_o + wi_m \mathrm{LR} p_a)] \quad (30)$$

Equation (30) can be used to define a combined temperature t_{com}, which reflects the combined effect of operative temperature and humidity for an actual environment:

$$t_{com} + wi_m \mathrm{LR} p_{t_{com}} = t_o + wi_m \mathrm{LR} p_a$$

or

$$t_{com} = t_o + wi_m \mathrm{LR} p_a - wi_m \mathrm{LR} p_{t_{com}} \quad (31)$$

where $p_{t_{com}}$ is a vapor pressure related in some fixed way to t_{com} and is analogous to $p_{wb,s}$ for t_{wb}. The term $wi_m \mathrm{LR} p_{t_{com}}$ is constant to the extent that i_m is constant, and any combination of t_o and p_a that gives the same t_{com} will result in the same total heat loss.

Environmental indices are discussed in the section on Environmental Indices. Two of these, the humid operative temperature t_{oh} and the effective temperature ET*, can be represented in terms of Equation (31). The humid operative temperature is that temperature which at 100% rh yields the same total heat loss as for the actual environment:

$$t_{oh} = t_o + wi_m \mathrm{LR}(p_a - p_{oh,s}) \quad (32)$$

where $p_{oh,s}$ is saturated vapor pressure, in psi, at t_{oh}.

The effective temperature is the temperature at 50% rh that yields the same total heat loss from the skin as for the actual environment:

$$\mathrm{ET}^* = t_o + wi_m \mathrm{LR}(p_a - 0.5 p_{\mathrm{ET}^*,s}) \quad (33)$$

where $p_{\mathrm{ET}^*,s}$ is saturated vapor pressure, in psi, at ET*.

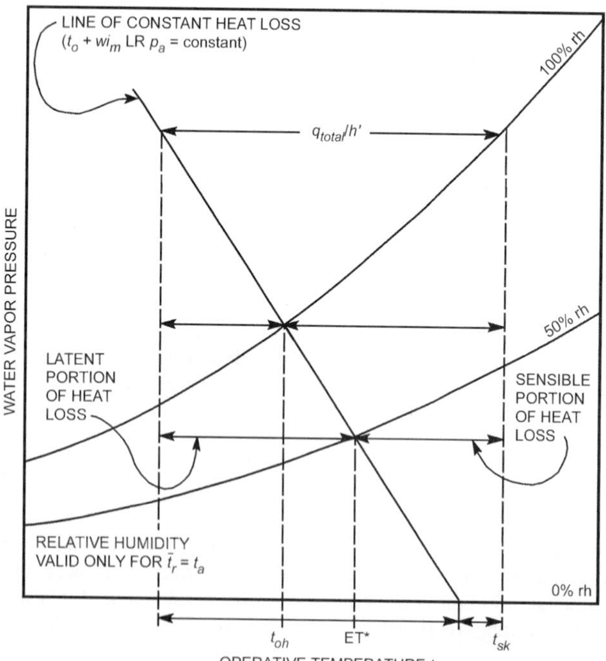

Fig. 2 Constant Skin Heat Loss Line and Its Relationship to t_{oh} and ET*

The psychrometric chart in Figure 2 shows a constant total heat loss line and the relationship between these indices. This line represents only one specific skin wettedness and permeation efficiency index. The relationship between indices depends on these two parameters (see the section on Environmental Indices).

ENGINEERING DATA AND MEASUREMENTS

Applying the preceding basic equations to practical problems of the thermal environment requires quantitative estimates of the body's surface area, metabolic requirements for a given activity and the mechanical efficiency for the work accomplished, evaluation of the heat transfer coefficients h_r and h_c, and the general nature of the clothing insulation used. This section provides the necessary data and describes methods used to measure the parameters of the heat balance equation.

Metabolic Rate and Mechanical Efficiency

Maximum Capacity. In choosing optimal conditions for comfort and health, the rate of work done during routine physical activities must be known, because metabolic power increases in proportion to exercise intensity. Metabolic rate varies over a wide range, depending on the activity, the person, and the conditions under which the activity is performed. Table 4 lists typical metabolic rates for an average adult (A_D = 19.6 ft^2) for activities performed continuously. The highest power a person can maintain for any continuous period is approximately 50% of the maximal capacity to use oxygen (maximum energy capacity).

A unit used to express the metabolic rate per unit DuBois area is the **met**, defined as the metabolic rate of a sedentary person (seated, quiet): 1 met = 18.4 Btu/h·ft^2 = 50 kcal/h·m^2. A normal, healthy man has a maximum capacity of approximately M_{act} = 12 met at age 20, which drops to 7 met at age 70. Maximum rates for women are about 30% lower. Long-distance runners and trained athletes have maximum rates as high as 20 met. An average 35 year old who does not exercise has a maximum rate of about 10 met, and activities with M_{act} > 5 met are likely to prove exhausting.

Intermittent Activity. The activity of many people consists of a mixture of activities or a combination of work-rest periods. A weighted average metabolic rate is generally satisfactory, provided that activities alternate frequently (several times per hour). For example, a person typing 50% of the time, filing while seated 25% of the time, and walking about 25% of the time would have an average metabolic rate of 0.50 × 20 + 0.25 × 22 + 0.25 × 31 = 23 Btu/h·ft^2 (see Table 4).

Accuracy. Estimating metabolic rates is difficult. The values given in Table 4 indicate metabolic rates only for the specific activities listed. Some entries give a range and some a single value, depending on the source of the data. The level of accuracy depends on the value of M_{act} and how well the activity can be defined. For well-defined activities with M_{act} < 1.5 met (e.g., reading), Table 4 is sufficiently accurate for most engineering purposes. For values of M_{act} > 3, where a task is poorly defined or where there are a variety of ways of performing a task (e.g., heavy machine work), the values may be in error by as much as ±50% for a given application. Engineering calculations should thus allow for potential variations.

Measurement. When metabolic rates must be determined more accurately than is possible with tabulated data, physiological measurements with human subjects may be necessary. The rate of metabolic heat produced by the body is most accurately measured by the rate of respiratory oxygen consumption and carbon dioxide production. An empirical equation for metabolic rate is given by Nishi (1981):

Table 4 Typical Metabolic Heat Generation for Various Activities

	Btu/h·ft²	met[a]
Resting		
Sleeping	13	0.7
Reclining	15	0.8
Seated, quiet	18	1.0
Standing, relaxed	22	1.2
Walking (on level surface)		
2.9 fps (2 mph)	37	2.0
4.4 fps (3 mph)	48	2.6
5.9 fps (4 mph)	70	3.8
Office Activities		
Reading, seated	18	1.0
Writing	18	1.0
Typing	20	1.1
Filing, seated	22	1.2
Filing, standing	26	1.4
Walking about	31	1.7
Lifting/packing	39	2.1
Driving/Flying		
Car	18 to 37	1.0 to 2.0
Aircraft, routine	22	1.2
Aircraft, instrument landing	33	1.8
Aircraft, combat	44	2.4
Heavy vehicle	59	3.2
Miscellaneous Occupational Activities		
Cooking	29 to 37	1.6 to 2.0
Housecleaning	37 to 63	2.0 to 3.4
Seated, heavy limb movement	41	2.2
Machine work		
sawing (table saw)	33	1.8
light (electrical industry)	37 to 44	2.0 to 2.4
heavy	74	4.0
Handling 110 lb bags	74	4.0
Pick and shovel work	74 to 88	4.0 to 4.8
Miscellaneous Leisure Activities		
Dancing, social	44 to 81	2.4 to 4.4
Calisthenics/exercise	55 to 74	3.0 to 4.0
Tennis, singles	66 to 74	3.6 to 4.0
Basketball	90 to 140	5.0 to 7.6
Wrestling, competitive	130 to 160	7.0 to 8.7

Sources: Compiled from various sources. For additional information, see Buskirk (1960), Passmore and Durnin (1967), and Webb (1964).
[a]1 met = 18.4 Btu/h·ft²

$$M = \frac{567(0.23\text{RQ} + 0.77)Q_{O_2}}{A_D} \tag{34}$$

where

M = metabolic rate, Btu/h·ft²
RQ = respiratory quotient; molar ratio of Q_{CO_2} exhaled to Q_{O_2} inhaled, dimensionless
Q_{O_2} = volumetric rate of oxygen consumption at conditions (STPD) of 32°F, 14.7 psi, ft³/h

The exact value of the respiratory quotient RQ used in Equation (34) depends on a person's activity, diet, and physical condition. It can be determined by measuring both carbon dioxide and oxygen in the respiratory airflows, or it can be estimated with reasonable accuracy. A good estimate for the average adult is RQ = 0.83 for light or sedentary activities (M < 1.5 met), increasing proportionately to RQ = 1.0 for extremely heavy exertion (M = 5.0 met). Estimation of RQ is generally sufficient for all except precision laboratory measurements since it does not strongly affect the value of the metabolic rate. A 10% error in estimating the respiratory quotient results in an error of less than 3% in the metabolic rate.

Table 5 Heart Rate and Oxygen Consumption at Different Activity Levels

Level of Exertion	Heart Rate, bpm	Oxygen Consumed, ft³/h
Light work	< 90	< 1
Moderate work	90 to 110	1 to 2
Heavy work	110 to 130	2 to 3
Very heavy work	130 to 150	3 to 4
Extremely heavy work	150 to 170	> 4

Source: Astrand and Rodahl (1977).

A second, much less accurate, method of estimating metabolic rate physiologically is to measure the heart rate. Table 5 shows the relationship between heart rate and oxygen consumption at different levels of physical exertion for a typical person. Once oxygen consumption is estimated from heart rate information, Equation (34) can be used to estimate the metabolic rate. A number of factors other than metabolic rate affect heart rate, such as physical condition, heat, emotional factors, muscles used, etc. Astrand and Rodahl (1977) show that heart rate is only a very approximate measure of metabolic rate and should not be the only source of information where accuracy is required.

Mechanical Efficiency. In the heat balance equation, the rate W of work accomplished must be in the same units as metabolism M and expressed in terms of A_D in Btu/h·ft². The mechanical work done by the muscles for a given task is often expressed in terms of the body's mechanical efficiency $\mu = W/M$. It is unusual for μ to be more than 0.05 to 0.10; for most activities, it is close to zero. The maximum value under optimal conditions (e.g., bicycle ergometer) is μ = 0.20 to 0.24 (Nishi 1981). It is common to assume that mechanical work is zero for several reasons: (1) the mechanical work produced is small compared to metabolic rate, especially for office activities; (2) estimates for metabolic rates can often be inaccurate; and (3) this assumption results in a more conservative estimate when designing air-conditioning equipment for upper comfort and health limits. More accurate calculation of heat generation may require estimation of the mechanical work produced for activities where it is significant (walking on a grade, climbing a ladder, bicycling, lifting, etc.). In some cases, it is possible to either estimate or measure the mechanical work. For example, a 200 lb person walking up a 5% grade at 4.4 fps (3 mph) would lift a 200 lb weight a height of 0.22 ft every second, for a work rate of 44 ft·lb$_f$/s, or 204 Btu/h. This rate of mechanical work would then be subtracted from M to determine the net heat generated.

Heat Transfer Coefficients

Values for the linearized radiative heat transfer coefficient, convective heat transfer coefficient, and evaporative heat transfer coefficient are required to solve the equations describing heat transfer from the body.

Radiative Heat Transfer Coefficient. The linearized radiative heat transfer coefficient can be calculated by

$$h_r = 4\varepsilon\sigma\frac{A_r}{A_D}\left[459.7 + \frac{t_{cl} + \bar{t}_r}{2}\right]^3 \tag{35}$$

where

h_r = radiative heat transfer coefficient, Btu/h·ft²·°F
ε = average emissivity of clothing or body surface, dimensionless
σ = Stefan-Boltzmann constant, 0.1712×10^{-8} Btu/h·ft²·°R⁴
A_r = effective radiation area of body, ft²

The ratio A_r/A_D is 0.70 for a sitting person and 0.73 for a standing person (Fanger 1967). Emissivity ε is close to unity (typically 0.95), unless special reflective materials are used or high-temperature sources are involved. It is not always possible to solve Equation (35) explicitly for h_r, since t_{cl} may also be an unknown. Some form of iteration may be necessary if a precise solution is required. Fortunately, h_r is nearly constant for typical indoor temperatures, and a value of 0.83 Btu/h·ft²·°F suffices for most calculations. If the emissivity is significantly less than unity, the value should be adjusted by

$$h_r = 0.83\varepsilon \tag{36}$$

where ε represents the area-weighted average emissivity for the clothing/body surface.

Convective Heat Transfer Coefficient. Heat transfer by convection is usually caused by air movement within the living space or by body movements. Equations for estimating h_c under various conditions are presented in Table 6. Where two conditions apply (e.g., walking in moving air), a reasonable estimate can be obtained by taking the larger of the two values for h_c. Limits have been given to all equations. If no limits were given in the source, reasonable limits have been estimated. Care should be exercised in using these values for seated and reclining persons. The heat transfer coefficients may be accurate, but the effective heat transfer area may be substantially reduced due to body contact with a padded chair or bed.

Quantitative values of h_c are important, not only in estimating convection loss, but in evaluating (1) operative temperature t_o, (2) clothing parameters I_t and i_m, and (3) rational effective temperatures t_{oh} and ET*. All heat transfer coefficients in Table 6 were evaluated at or near 14.7 psia. These coefficients should be corrected as follows for atmospheric pressure:

$$h_{cc} = h_c(p_t/14.7)^{0.55} \tag{37}$$

where

h_{cc} = corrected convective heat transfer coefficient, Btu/h·ft²·°F
p_t = local atmospheric pressure, psia

The combined coefficient h is the sum of h_r and h_c, described in Equation (35) and Table 6, respectively. The coefficient h governs exchange by radiation and convection from the exposed body surface to the surrounding environment.

Table 6 Equations for Convection Heat Transfer Coefficients

Equation	Limits	Condition	Remarks/Sources
$h_c = 0.061V^{0.6}$ $h_c = 0.55$	$40 < V < 800$ $0 < V < 40$	Seated with moving air	Mitchell (1974)
$h_c = 0.475 + 0.044V^{0.67}$ $h_c = 0.90$	$30 < V < 300$ $0 < V < 30$	Reclining with moving air	Colin and Houdas (1967)
$h_c = 0.092V^{0.53}$	$100 < V < 400$	Walking in still air	V is walking speed (Nishi and Gagge 1970)
$h_c = (M - 0.85)^{0.39}$	$1.1 < M < 3.0$	Active in still air	Gagge et al. (1976)
$h_c = 0.146V^{0.39}$	$100 < V < 400$	Walking on treadmill in still air	V is treadmill speed (Nishi and Gagge 1970)
$h_c = 0.068V^{0.69}$ $h_c = 0.70$	$30 < V < 300$ $0 < V < 30$	Standing person in moving air	Developed from data presented by Seppanen et al. (1972)

Note: h_c in Btu/h·ft²·°F, V in fpm, and M in met, where 1 met = 18.4 Btu/h·ft².

Evaporative Heat Transfer Coefficient. The evaporative heat transfer coefficient h_e for the outer air layer of a nude or clothed person can be estimated from the convective heat transfer coefficient using the Lewis relation given in Equation (27). If the atmospheric pressure is significantly different from the reference value (14.7 psia), the correction to the valueobtained from Equation (27) is

$$h_{ec} = h_e(14.7/p_t)^{0.45} \tag{38}$$

where h_{ec} is the corrected evaporative heat transfer coefficient in Btu/h·ft²·°F.

Clothing Insulation and Permeation Efficiency

Thermal Insulation. The most accurate methods for determining clothing insulation are (1) measurements on heated mannequins (McCullough and Jones 1984, Olesen and Nielsen 1983) and (2) measurements on active subjects (Nishi et al. 1975). For most routine engineering work, estimates based on tables and equations presented in this section are sufficient. Thermal mannequins can measure the sensible heat loss from the "skin" ($C + R$) in a given environment. Equation (11) can then be used to evaluate R_{cl} if the environmental conditions are well defined and f_{cl} is measured. Evaluation of clothing insulation on subjects requires measurement of t_{sk}, t_{cl}, and t_o. The clothing thermal efficiency is calculated by

$$F_{cl} = \frac{t_{cl} - t_o}{t_{sk} - t_o} \tag{39}$$

The intrinsic clothing insulation can then be calculated from mannequin measurements by the following relationship, provided f_{cl} is measured and conditions are sufficiently well defined to make an accurate determination of h:

$$R_{cl} = \frac{t_{sk} - t_o}{q} - \frac{1}{hf_{cl}} \tag{40}$$

where q is heat loss from the mannequin in Btu/h·ft².

Clothing insulation value may be expressed in clo units. In order to avoid confusion, the symbol I is used with the clo unit instead of the symbol R. The relationship between the two is

$$R = 0.88I \tag{41}$$

or 1.0 clo is equivalent to 0.88 ft²·°F·h/Btu.

Because clothing insulation cannot be measured for most routine engineering applications, tables of measured values for various clothing ensembles can be used to select an ensemble comparable to the one(s) in question. Table 7 gives values for typical indoor clothing ensembles. More detailed tables are presented by McCullough and Jones (1984) and Olesen and Nielsen (1983). Accuracies for I_{cl} on the order of ±20% are typical if good matches between ensembles are found.

Often it is not possible to find an already measured clothing ensemble that matches the one in question. In this case, the ensemble insulation can be estimated from the insulation of individual garments. Table 8 gives a list of individual garments commonly worn. The insulation of an ensemble is estimated from the individual values using a summation formula (McCullough and Jones 1984):

$$I_{cl} = 0.835\sum_i I_{clu,i} + 0.161 \tag{42}$$

where $I_{clu,i}$ is the effective insulation of garment i, and I_{cl}, as before, is the insulation for the entire ensemble. A simpler and nearly as accurate summation formula is (Olesen 1985)

Table 7 Typical Insulation and Permeation Efficiency Values for Clothing Ensembles

Ensemble Description[a]	I_{cl} (clo)	I_t[b] (clo)	f_{cl}	i_{cl}	i_m[b]
Walking shorts, short-sleeved shirt	0.36	1.02	1.10	0.34	0.42
Trousers, short-sleeved shirt	0.57	1.20	1.15	0.36	0.43
Trousers, long-sleeved shirt	0.61	1.21	1.20	0.41	0.45
Same as above, plus suit jacket	0.96	1.54	1.23		
Same as above, plus vest and T-shirt	1.14	1.69	1.32	0.32	0.37
Trousers, long-sleeved shirt, long-sleeved sweater, T-shirt	1.01	1.56	1.28		
Same as above, plus suit jacket and long underwear bottoms	1.30	1.83	1.33		
Sweat pants, sweat shirt	0.74	1.35	1.19	0.41	0.45
Long-sleeved pajama top, long pajama trousers, short 3/4 sleeved robe, slippers (no socks)	0.96	1.50	1.32	0.37	0.41
Knee-length skirt, short-sleeved shirt, panty hose, sandals	0.54	1.10	1.26		
Knee-length skirt, long-sleeved shirt, full slip, panty hose	0.67	1.22	1.29		
Knee-length skirt, long-sleeved shirt, half slip, panty hose, long-sleeved sweater	1.10	1.59	1.46		
Same as above, replace sweater with suit jacket	1.04	1.60	1.30	0.35	0.40
Ankle-length skirt, long-sleeved shirt, suit jacket, panty hose	1.10	1.59	1.46		
Long-sleeved coveralls, T-shirt	0.72	1.30	1.23		
Overalls, long-sleeved shirt, T-shirt	0.89	1.46	1.27	0.35	0.40
Insulated coveralls, long-sleeved thermal underwear, long underwear bottoms	1.37	1.94	1.26	0.35	0.39

Sources: McCullough and Jones (1984) and McCullough et al. (1989).

[a]All ensembles include shoes and briefs or panties. All ensembles except those with panty hose include socks unless otherwise noted.

[b]For $\bar{t}_r = t_a$ and air velocity less than 40 fpm (I_a = 0.72 clo and i_m = 0.48 when nude).

$$I_{cl} = \sum_i I_{clu,i} \tag{43}$$

Either Equation (42) or (43) gives acceptable accuracy for typical indoor clothing. The main source of inaccuracy is in determining the appropriate values for individual garments. Overall accuracies are on the order of ±25% if the tables are used carefully. If it is important to include a specific garment that is not included in Table 8, its insulation can be estimated by (McCullough and Jones 1984)

$$I_{clu,i} = (0.534 + 3.43x_f)(A_G/A_D) - 0.0549 \tag{44}$$

where

x_f = fabric thickness, in.
A_G = body surface area covered by garment, ft^2

Values in Table 7 may be adjusted by information in Table 8 and a summation formula. Using this method, values of $I_{clu,i}$ for the selected items in Table 8 are then added to or subtracted from the ensemble value of I_{cl} in Table 7.

When a person is sitting, the chair generally has the effect of increasing clothing insulation by up to 0.15 clo, depending on the contact area A_{ch} between the chair and body (McCullough et al. 1994). A string webbed or beach chair has little or no contact area, and the insulation actually decreases by about 0.1 clo due likely to compression of the clothing in the contact area. In contrast, a cushioned executive chair has a large contact area that can increase the intrinsic clothing insulation by 0.15 clo. For other chairs, the increase in intrinsic insulation (ΔI_{cl}) can be estimated from

$$\Delta I_{cl} = 6.95 \times 10^{-2} A_{ch} - 0.1 \tag{45}$$

where A_{ch} is in ft^2.

For example, a desk chair with a body contact area of 2.9 ft^2 has a ΔI_{cl} of 0.1 clo. This amount should be added to the intrinsic insulation of the standing clothing ensemble to obtain the insulation of the ensemble when sitting in the desk chair.

Although sitting has the effect of increasing clothing insulation, walking decreases it (McCullough and Hong 1994). The change in clothing insulation (ΔI_{cl}) can be estimated from the standing intrinsic insulation of the ensemble (I_{cl}) and the walking speed (Walkspeed) in steps per minute:

$$\Delta I_{cl} = -0.504 I_{cl} - 0.00281(\text{Walkspeed}) + 0.24 \tag{46}$$

For example, the clothing insulation of a person wearing a winter business suit with a standing intrinsic insulation of 1 clo would decrease by 0.52 clo when the person walks at 90 steps per minute (about 2.3 mph). Thus, when the person is walking, the intrinsic insulation of the ensemble would be 0.48 clo.

Permeation Efficiency. Permeation efficiency data for some clothing ensembles are presented in terms of i_{cl} and i_m in Table 7. The values of i_m can be used to calculate $R_{e,t}$ using the relationships in Table 2. Ensembles worn indoors generally fall in the range $0.3 < i_m < 0.5$, and assuming $i_m = 0.4$ is reasonably accurate (McCullough et al. 1989). This latter value may be used if a good match to ensembles in Table 7 cannot be made. The value of i_m or $R_{e,t}$ may be substituted directly into equations for body heat loss calculations (see Table 3). However, i_m for a given clothing ensemble is a function of the environment as well as the clothing properties. Unless i_m is evaluated at conditions very similar to the intended application, it is more rigorous to use i_{cl} to describe the permeation efficiency of the clothing. The value of i_{cl} is not as sensitive to environmental conditions; thus, given data are more accurate over a wider range of air velocity and radiant and air temperature combinations for i_{cl} than for i_m. The relationships in Table 2 can be used to determine $R_{e,cl}$ from i_{cl}, and i_{cl} or $R_{e,cl}$ can then be used for body heat loss calculations (see Table 3). McCullough et al. (1989) found an average value of i_{cl} = 0.34 for common indoor clothing; this value can be used when other data are not available.

Measurements of i_m or i_{cl} may be necessary if unusual clothing (e.g., impermeable or metallized) and/or extreme environments (e.g., high radiant temperatures or high air velocities) are to be addressed. There are three different methods for measuring the permeation efficiency of clothing: the first uses a wet mannequin to measure the effect of sweat evaporation on heat loss (McCullough 1986); the second uses permeation efficiency measurements on component fabrics as well as dry mannequin measurements (Umbach 1980); and the third uses measurements from sweating subjects (Nishi et al. 1975; Holmer 1984).

Clothing Surface Area. Many clothing heat transfer calculations require that clothing area factor f_{cl} be known. The most reliable approach is to measure it using photographic methods (Olesen et al. 1982). Other than actual measurements, the best method is to use previously tabulated data for similar clothing ensembles. Table 7 is adequate for most indoor clothing ensembles. No good method of estimating f_{cl} for a clothing ensemble from other information is available, although a rough estimate can be made by (McCullough and Jones 1984)

$$f_{cl} = 1.0 + 0.3 I_{cl} \tag{47}$$

Table 8 Garment Insulation Values

Garment Description[a]	$I_{clu,i}$, clo[b]	Garment Description[a]	$I_{clu,i}$, clo[b]	Garment Description[a]	$I_{clu,i}$, clo[b]
Underwear		Long-sleeved, flannel shirt	0.34	Long-sleeved (thin)	0.25
Men's briefs	0.04	Short-sleeved, knit sport shirt	0.17	Long-sleeved (thick)	0.36
Panties	0.03	Long-sleeved, sweat shirt	0.34	**Dresses and skirts[c]**	
Bra	0.01	**Trousers and Coveralls**	0.06	Skirt (thin)	0.14
T-shirt	0.08	Short shorts	0.08	Skirt (thick)	0.23
Full slip	0.16	Walking shorts	0.15	Long-sleeved shirtdress (thin)	0.33
Half slip	0.14	Straight trousers (thin)	0.24	Long-sleeved shirtdress (thick)	0.47
Long underwear top	0.20	Straight trousers (thick)	0.28	Short-sleeved shirtdress (thin)	0.29
Long underwear bottoms	0.15	Sweatpants	0.30	Sleeveless, scoop neck (thin)	0.23
Footwear		Overalls	0.49	Sleeveless, scoop neck (thick), i.e., jumper	0.27
Ankle-length athletic socks	0.02	Coveralls			
Calf-length socks	0.03	**Suit jackets and vests (lined)**		**Sleepwear and Robes**	
Knee socks (thick)	0.06	Single-breasted (thin)	0.36	Sleeveless, short gown (thin)	0.18
Panty hose	0.02	Single-breasted (thick)	0.44	Sleeveless, long gown (thin)	0.20
Sandals/thongs	0.02	Double-breasted (thin)	0.42	Short-sleeved hospital gown	0.31
Slippers (quilted, pile-lined)	0.03	Double-breasted (thick)	0.48	Long-sleeved, long gown (thick)	0.46
Boots	0.10	Sleeveless vest (thin)	0.10	Long-sleeved pajamas (thick)	0.57
Shirts and Blouses		Sleeveless vest (thick)	0.17	Short-sleeved pajamas (thin)	0.42
Sleeveless, scoop-neck blouse	0.12	**Sweaters**		Long-sleeved, long wrap robe (thick)	0.69
Short-sleeved, dress shirt	0.19	Sleeveless vest (thin)	0.13	Long-sleeved, short wrap robe (thick)	0.48
Long-sleeved, dress shirt	0.25	Sleeveless vest (thick)	0.22	Short-sleeved, short robe (thin)	0.34

[a]"Thin" garments are made of light, thin fabrics worn in summer; "thick" garments are made of heavy, thick fabrics worn in winter.

[b]1 clo = 0.88°F·ft²·h/Btu

[c]Knee-length

Total Evaporative Heat Loss

The total evaporative heat loss (latent heat) from the body due to both respiratory losses and skin losses, $E_{sk} + E_{res}$, can be measured directly from the body's rate of mass loss as observed by a sensitive scale:

$$E_{sk} + E_{res} = \frac{h_{fg}}{A_D}\frac{dm}{d\theta} \tag{48}$$

where

h_{fg} = latent heat of vaporization of water, Btu/lb
m = body mass, lb
θ = time, h

When using Equation (48), adjustments should be made for any materials consumed (e.g., food and drink), body effluents (e.g., wastes), and metabolic weight losses. Metabolism contributes slightly to weight loss primarily because the oxygen absorbed during respiration is converted to heavier CO_2 and exhaled. It can be calculated by

$$\frac{dm_{ge}}{d\theta} = 2.2Q_{O_2}(0.1225\text{RQ} - 0.0891) \tag{49}$$

where

$dm_{ge}/d\theta$ = rate of mass loss due to respiratory gas exchange, lb/h
Q_{O_2} = oxygen uptake at STPD, ft³/h
RQ = respiratory quotient
0.1225 = density of CO_2 at STPD, lb/ft³
0.0891 = density of O_2 at STPD, lb/ft³
STPD = standard temperature and pressure of dry air at 32°F and 14.7 psia

Environmental Parameters

The parameters describing the thermal environment that must be measured or otherwise quantified if accurate estimates of human thermal response are to be made are divided into two groups—those that can be measured directly and those that are calculated from other measurements.

Directly Measured. Seven of the parameters frequently used to describe the thermal environment are psychrometric and include (1) air temperature t_a; (2) wet-bulb temperature t_{wb}; (3) dew-point temperature t_{dp}; (4) water vapor pressure p_a; (5) total atmospheric pressure p_t; (6) relative humidity (rh); and (7) humidity ratio W_a. These parameters are discussed in detail in Chapter 6, and methods for measuring them are discussed in Chapter 14. Two other important parameters include air velocity V and mean radiant temperature $\bar{t}_r$. Air velocity measurements are also discussed in Chapter 14. The radiant temperature is the temperature of an exposed surface in the environment. The temperatures of individual surfaces are usually combined into a mean radiant temperature $\bar{t}_r$. Finally, globe temperature t_g, which can also be measured directly, is a good approximation of the operative temperature t_o and is also used with other measurements to calculate the mean radiant temperature.

Calculated Parameters. The **mean radiant temperature** $\bar{t}_r$ is a key variable in making thermal calculations for the human body. It is the uniform temperature of an imaginary enclosure in which radiant heat transfer from the human body equals the radiant heat transfer in the actual nonuniform enclosure. Measurements of the globe temperature, air temperature, and air velocity can be combined to estimate the mean radiant temperature (see Chapter 14). The accuracy of the mean radiant temperature determined this way varies considerably depending on the type of environment and the accuracy of the individual measurements. Since the mean radiant temperature is defined with respect to the human body, the shape of the sensor is also a factor. The spherical shape of the globe thermometer gives a reasonable approximation of a seated person; an ellipsoid-shaped sensor gives a better approximation of the shape of a human, both upright and seated.

The mean radiant temperature can also be calculated from measured values of the temperature of the surrounding walls and surfaces and their positions with respect to the person. As most building materials have a high emittance ε, all the surfaces in the room can be assumed to be black. The following equation is then used:

$$\bar{T}_r^4 = T_1^4 F_{p-1} + T_2^4 F_{p-2} + \dots + T_N^4 F_{p-N} \tag{50}$$

where

$\bar{T}_r$ = mean radiant temperature, °R
T_N = surface temperature of surface N, °R
F_{p-N} = angle factor between a person and surface N

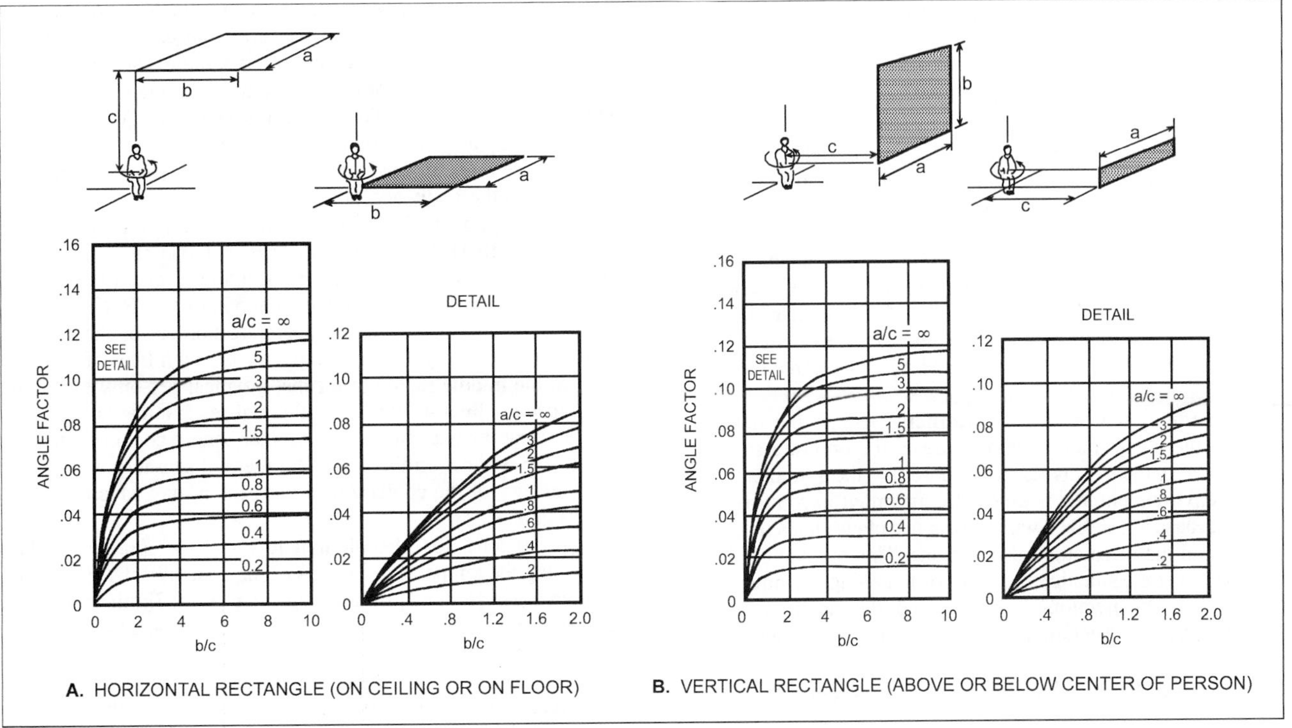

Fig. 3 Mean Value of Angle Factor Between Seated Person and Horizontal or Vertical Rectangle when Person is Rotated Around Vertical Axis

(Fanger 1982)

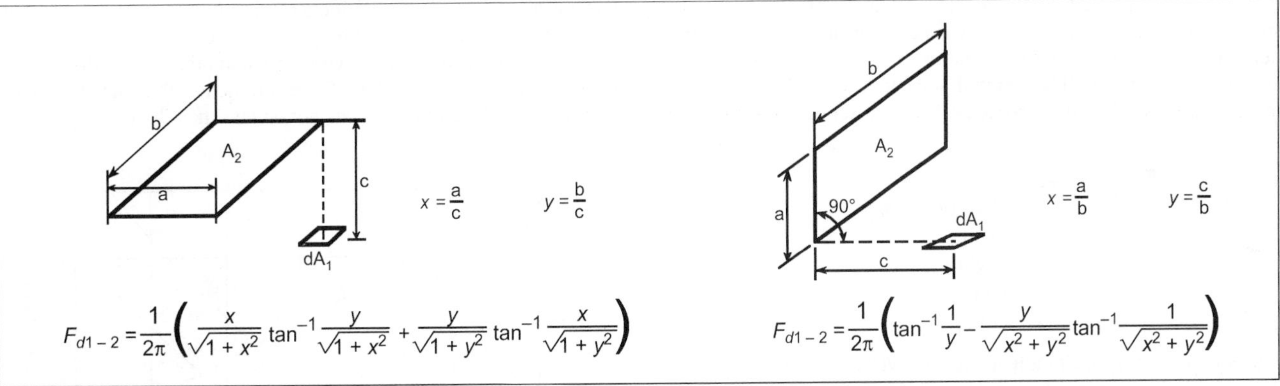

Fig. 4 Analytical Formulas for Calculating Angle Factor for Small Plane Element

Because the sum of the angle factors is unity, the fourth power of mean radiant temperature equals the mean value of the surrounding surface temperatures to the fourth power, weighted by the respective angle factors. In general, angle factors are difficult to determine, although Figures 3A and 3B may be used to estimate them for rectangular surfaces. The angle factor normally depends on the position and orientation of the person (Fanger 1982).

If relatively small temperature differences exist between the surfaces of the enclosure, Equation (50) can be simplified to a linear form:

$$\bar{t}_r = t_1 F_{p-1} + t_2 F_{p-2} + \ldots + t_N F_{p-N} \tag{51}$$

Equation (51) always gives a slightly lower mean radiant temperature than Equation (50), but in many cases the difference is small. If, for example, half the surroundings ($F_{p-N} = 0.5$) has a temperature 10°F higher than the other half, the difference between the calculated mean radiant temperatures—according to Equations (50) and (51)—is only 0.4°F. If, however, this difference is 200°F, the mean radiant temperature calculated by Equation (51) is 20°F too low.

The mean radiant temperature may also be calculated from the plane radiant temperature t_{pr} (defined below) in six directions (up, down, right, left, front, back) and for the projected area factors of a person in the same six directions. For a standing person, the mean radiant temperature may be estimated as

$$\begin{aligned}\bar{t}_r = \{&0.08[t_{pr}(\text{up}) + t_{pr}(\text{down})] + 0.23[t_{pr}(\text{right}) \\ &+ t_{pr}(\text{left})] + 0.35[t_{pr}(\text{front}) + t_{pr}(\text{back})]\} \\ &\div [2(0.08 + 0.23 + 0.35)]\end{aligned} \tag{52}$$

For a seated person, the mean radiant temperature may be estimated as

$$\bar{t}_r = \{0.18[t_{pr}(\text{up}) + t_{pr}(\text{down})] + 0.22[t_{pr}(\text{right}) + t_{pr}(\text{left})] + 0.30[t_{pr}(\text{front}) + t_{pr}(\text{back})]\} \div [2(0.18 + 0.22 + 0.30)] \quad (53)$$

The **plane radiant temperature** t_{pr}, first introduced by Korsgaard (1949), is the uniform temperature of an enclosure in which the incident radiant flux on one side of a small plane element is the same as that in the actual environment. The plane radiant temperature describes the thermal radiation in one direction, and its value thus depends on the direction. In comparison, the mean radiant temperature $\bar{t}_r$ describes the thermal radiation for the human body from all directions. The plane radiant temperature can be calculated using Equations (50) and (51) with the same limitations. Area factors are determined from Figure 4.

The **radiant temperature asymmetry** Δt_{pr} is the difference between the plane radiant temperature of the opposite sides of a small plane element. This parameter describes the asymmetry of the radiant environment and is especially important in comfort conditions. Because it is defined with respect to a plane element, its value depends on the orientation of that plane. This orientation may be specified in some situations (e.g., floor to ceiling asymmetry) and not in others. If direction is not specified, the radiant asymmetry should be for the orientation that gives the maximum value.

CONDITIONS FOR THERMAL COMFORT

In addition to the previously discussed independent environmental and personal variables influencing thermal response and comfort, other factors may also have some effect. These factors, such as nonuniformity of the environment, visual stimuli, age, and outdoor climate are generally considered secondary factors. Studies by Rohles and Nevins (1971) and Rohles (1973) on 1600 college-age students revealed correlations between comfort level, temperature, humidity, sex, and length of exposure. Many of these correlations are given in Table 9. The thermal sensation scale developed for these studies is called the **ASHRAE thermal sensation scale**:

+3 hot
+2 warm
+1 slightly warm
0 neutral
−1 slightly cool
−2 cool
−3 cold

The equations in Table 9 indicate that the women of this study were more sensitive to temperature and less sensitive to humidity than the men. But in general about a 5.4°F change in temperature or a 0.44 psi change in water vapor pressure is necessary to change a thermal sensation vote by one unit or temperature category.

Table 9 Equations for Predicting Thermal Sensation (Y) of Men, Women, and Men and Women Combined

Exposure Period, h	Subjects	Regression Equations[a, b] t = dry-bulb temperature, °F; p = vapor pressure, psi
1.0	Men	$Y = 0.122\,t + 1.61\,p - 9.584$
	Women	$Y = 0.151\,t + 1.71\,p - 12.080$
	Both	$Y = 0.136\,t + 1.71\,p - 10.880$
2.0	Men	$Y = 0.123\,t + 1.86\,p - 9.953$
	Women	$Y = 0.157\,t + 1.45\,p - 12.725$
	Both	$Y = 0.140\,t + 1.65\,p - 11.339$
3.0	Men	$Y = 0.118\,t + 2.02\,p - 9.718$
	Women	$Y = 0.153\,t + 1.76\,p - 13.511$
	Both	$Y = 0.135\,t + 1.92\,p - 11.122$

[a] Y values refer to the ASHRAE thermal sensation scale.
[b] For young adult subjects with sedentary activity and wearing clothing with a thermal resistance of approximately 0.5 clo, $\bar{t}_r \approx t_a$ and air velocities < 40 fpm.

Current and past studies are periodically reviewed to update *ASHRAE Standard* 55, Thermal Environmental Conditions for Human Occupancy. This standard specifies conditions or comfort zones where 80% of sedentary or slightly active persons find the environment thermally acceptable.

Because people typically change their clothing for the seasonal weather, ASHRAE *Standard* 55 specifies summer and winter comfort zones appropriate for clothing insulation levels of 0.5 and 0.9 clo, respectively (Figure 5) (Addendum 55a to ASHRAE *Standard* 55). The warmer and cooler temperature borders of the comfort zones are affected by humidity and coincide with lines of constant ET*. In the middle region of a zone, a typical person wearing the prescribed clothing would have a thermal sensation at or very near neutral. Near the boundary of the warmer zone, a person would feel about +0.5 warmer on the ASHRAE thermal sensation scale; near the boundary of the cooler zone, that person may have a thermal sensation of −0.5.

Comfort zones for other clothing levels can be approximated by decreasing the temperature borders of the zone by 1°F for each 0.1 clo increase in clothing insulation and vice versa. Similarly a zone's temperatures can be decreased by 2.5°F per met increase in activity above 1.2 met.

The upper and lower humidity levels of the comfort zones are less precise. Low humidity can lead to drying of the skin and mucous surfaces. Comfort complaints about dry nose, throat, eyes, and skin occur in low-humidity conditions, typically when the dew point is less than 32°F. Liviana et al. (1988) found eye discomfort increased with time in low-humidity environments (dew point < 36°F). Green (1982) quantified that respiratory illness and absenteeism increase in winter with decreasing humidity and found that any increase in humidity from very low levels decreased absenteeism in winter. In compliance with these and other discomfort observations, ASHRAE *Standard* 55 recommends that the dew-point temperature of occupied spaces not be less than 36°F.

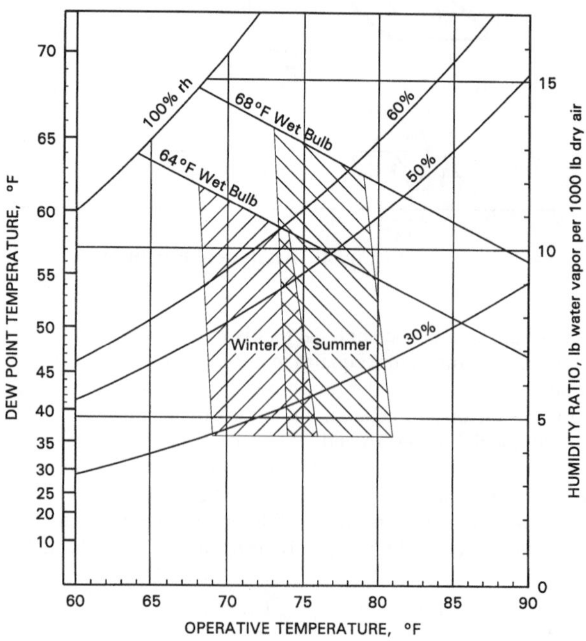

Fig. 5 ASHRAE Summer and Winter Comfort Zones
(Acceptable ranges of operative temperature and humidity for people in typical summer and winter clothing during primarily sedentary activity.)

At high humidity levels, too much skin moisture tends to increase discomfort (Gagge 1937, Berglund and Cunningham 1986), particularly skin moisture that is physiological in origin (water diffusion and perspiration). At high humidity levels, thermal sensation alone is not a reliable predictor of thermal comfort (Tanabe et al. 1987). The discomfort appears to be due to the feeling of the moisture itself, increased friction between skin and clothing with skin moisture (Gwosdow et al. 1986), and other factors. To prevent warm discomfort, Nevins et al. (1975) recommended that on the warm side of the comfort zone the relative humidity not exceed 60%.

The upper humidity limits of ASHRAE *Standard* 55 were developed theoretically from limited data. However, thermal acceptability data gathered at medium and high humidity levels at summer comfort temperatures with subjects wearing 0.55 clo corroborated the shape of the upper limit and found it corresponded to an 80% thermal acceptability level (Berglund 1995).

THERMAL NONUNIFORM CONDITIONS AND LOCAL DISCOMFORT

A person may feel thermally neutral as a whole but still feel uncomfortable if one or more parts of the body are too warm or too cold. Nonuniformities may be due to a cold window, a hot surface, a draft, or a temporal variation of these. Even small variations in heat flow cause the thermal regulatory system to compensate, thus increasing the physiological effort of maintaining body temperatures. The boundaries of the comfort zones (Figure 5) of ASHRAE *Standard* 55 provide a thermal acceptability level of 90% if the environment is thermally uniform. Because the standard's objective is to specify conditions for 80% acceptability, the standard permits nonuniformities to decrease acceptability by 10%. Fortunately for the designer and user, the effect of common thermal nonuniformities on comfort is quantifiable and predictable as discussed in the following sections. Furthermore, most humans are fairly insensitive to small nonuniformities.

Asymmetric Thermal Radiation

Asymmetric or nonuniform thermal radiation in a space may be caused by cold windows, uninsulated walls, cold products, cold or warm machinery, or improperly sized heating panels on the wall or ceiling. In residential buildings, offices, restaurants, etc., the most common reasons for discomfort due to asymmetric thermal radiation are large windows in the winter or improperly sized or installed ceiling heating panels. At industrial workplaces, the reasons include cold or warm products, cold or warm equipment, etc.

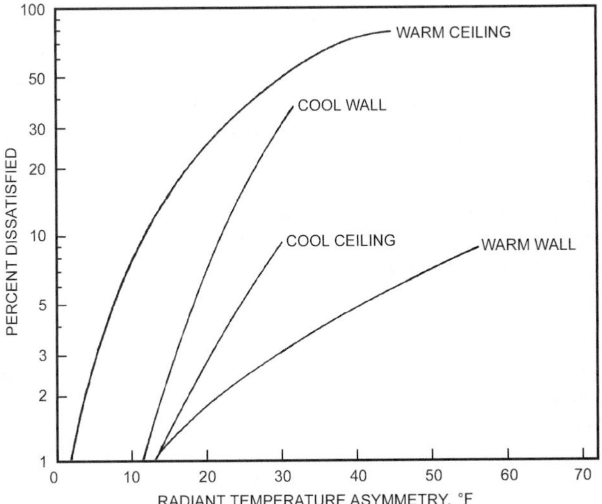

Fig. 6 Percentage of People Expressing Discomfort due to Asymmetric Radiation

The recommendations in ISO *Standard* 7730 and ASHRAE *Standard* 55 are based primarily on studies reported by Fanger et al. (1980). These standards include guidelines regarding the radiant temperature asymmetry from an overhead warm surface (heated ceiling) and a vertical cold surface (cold window). Among the studies conducted on the influence of asymmetric thermal radiation are those by McIntyre (1974, 1976), McIntyre and Griffiths (1975), Fanger and Langkilde (1975), McNall and Biddison (1970), and Olesen et al. (1972). These studies all used seated subjects. In these studies, the subjects were always in thermal neutrality and exposed only to the discomfort resulting from excessive asymmetry.

The subjects gave their reactions on their comfort sensation, and a relationship between the radiant temperature asymmetry and the number of subjects feeling dissatisfied was established (Figure 6). Radiant asymmetry, as defined in the section on Environmental Parameters, is the difference in radiant temperature of the environment on opposite sides of the person. More precisely, radiant asymmetry is the difference in radiant temperatures seen by a small flat element looking in opposite directions.

Figure 6 shows that people are more sensitive to asymmetry caused by an overhead warm surface than by a vertical cold surface. The influence of an overhead cold surface and a vertical warm surface is much less. These data are particularly important when applying radiant panels to provide comfort in spaces with large cold surfaces or cold windows.

Other studies of clothed persons in neutral environments found thermal acceptability unaffected by radiant temperature asymmetries of 18°F or less (Berglund and Fobelets 1987) and comfort unaffected by radiant temperature asymmetries of 36°F or less (McIntyre 1975).

Draft

Draft is an undesired local cooling of the human body caused by air movement. This is a serious problem, not only in many ventilated buildings but also in automobiles, trains, and aircraft. Draft has been identified as one of the most annoying factors in offices. When people sense draft, they often demand higher air temperatures in the room or that ventilation systems be stopped.

Fanger and Christensen (1986) aimed to establish the percentage of the population feeling draft when exposed to a given mean velocity. Figure 7 shows the percentage of subjects who felt draft on the head region (the dissatisfied) as a function of the mean air velocity at the neck. The head region comprises head, neck, shoulders, and back. The air temperature had a significant influence on the percentage of dissatisfied. There was no significant difference between responses of men and women to draft. The data in Figure 7 apply only to persons wearing normal indoor clothing and performing light, mainly sedentary work. Persons with higher activity levels are not as sensitive to draft (Jones et al. 1986).

A study of the effect of air velocity over the whole body found thermal acceptability unaffected in neutral environments by air speeds of 50 fpm or less (Berglund and Fobelets 1987). This study also found no interaction between air speed and radiant temperature asymmetry on subjective responses. This means that acceptability changes and the percent dissatisfied due to draft and radiant asymmetry are independent and additive.

Fanger et al. (1989) investigated the effect of turbulence intensity on sensation of draft. The turbulence intensity had a significant effect on the occurrence of draft sensation. The following model predicts the percentage of people dissatisfied because of draft intensity. The model can be used for quantifying draft risk in spaces and for developing air distribution systems with a low draft risk.

$$\text{PD} = 0.021(93.2 - t_a)(V - 9.8)^{0.62}(0.0019V\,\text{Tu} + 3.14) \quad (54)$$

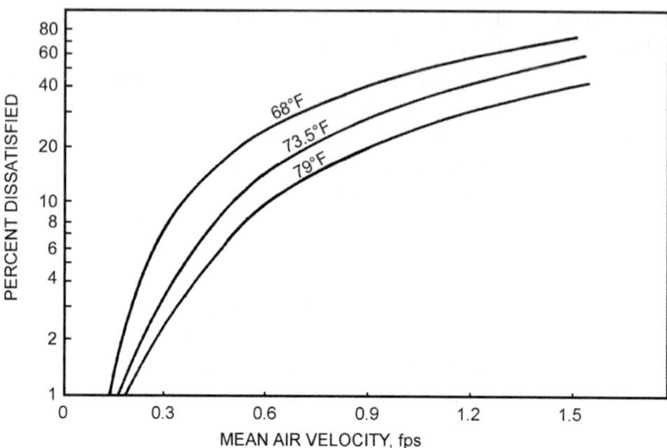

Fig. 7 Percentage of People Dissatisfied as Function of Mean Air Velocity

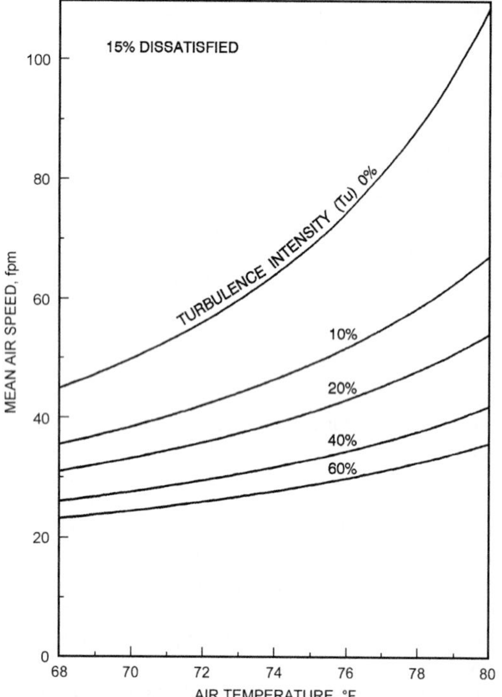

Fig. 8 Draft Conditions Dissatisfying 15% of Population

where Tu is the turbulence intensity in % defined by

$$Tu = 100\frac{V_{sd}}{V} \tag{55}$$

For $V < 9.8$ fpm, insert $V = 9.8$; and for PD > 100%, insert PD = 100%. V_{sd} is the standard deviation of the velocity measured with an omnidirectional anemometer having a 0.2 s time constant.

The model extends the Fanger and Christensen (1986) draft chart model to include turbulence intensity. In this study, Tu decreases when V increases. This means that the effect of V for the experimental data to which the model is fitted are: $68 < t_a < 79$°F, $10 < V < 100$ fpm, and $0 <$ Tu < 70%. Figure 8 gives more precisely the curves that result from intersections between planes of constant Tu and the surfaces of PD = 15%.

Vertical Air Temperature Difference

In most spaces in buildings, the air temperature normally increases with height above the floor. If the gradient is sufficiently large, local warm discomfort can occur at the head and/or cold discomfort can occur at the feet, although the body as a whole is thermally neutral. Among the few studies of vertical air temperature differences and the influence of thermal comfort reported are Olesen et al. (1979), McNair (1973), McNair and Fishman (1974), and Eriksson (1975). Subjects were seated in a climatic chamber so they were individually exposed to different air temperature differences between head and ankles (Olesen et al. 1979). During the tests, the subjects were in thermal neutrality because they were allowed to change the temperature level in the test room whenever they desired; the vertical temperature difference, however, was kept unchanged. The subjects gave subjective reactions to their thermal sensation; Figure 9 shows the percentage of dissatisfied as a function of the vertical air temperature difference between head (43 in. above the floor) and ankles (4 in. above the floor).

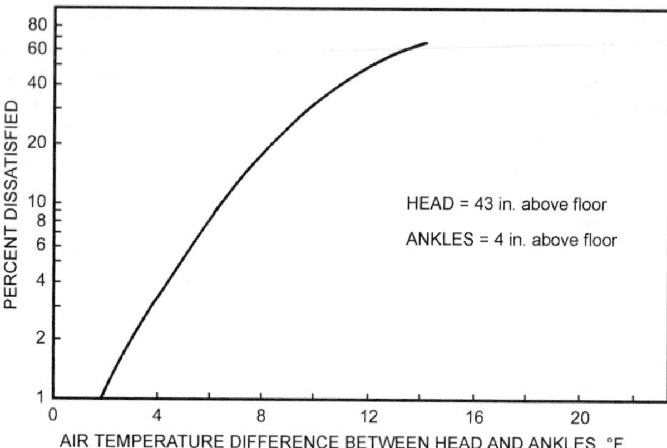

Fig. 9 Percentage of Seated People Dissatisfied as Function of Air Temperature Difference Between Head and Ankles

The case where the air temperature at head level is lower than that at ankle level will not be as critical for the occupants. Eriksson (1975) indicated that his subjects could tolerate much greater differences if the head were cooler. This observation is verified in the experiments with asymmetric thermal radiation from a cooled ceiling (Fanger et al. 1985).

Warm or Cold Floors

Due to the direct contact between the feet and the floor, local discomfort of the feet can often be caused by a too-high or too-low floor temperature. Also, the floor temperature has a significant influence on the mean radiant temperature in a room. The floor temperature is greatly influenced by the way a building is constructed (e.g., insulation of the floor, above a basement, directly on the ground, above another room, use of floor heating, floors in radiant heated areas). If a floor is too cold and the occupants feel cold discomfort in their feet, a common reaction is to increase the temperature level in the room; in the heating season, this also increases energy consumption. A radiant system, which radiates heat from the floor, can also prevent discomfort from cold floors.

The most extensive studies of the influence of floor temperature on feet comfort were performed by Olesen (1977a, 1977b), who, based on his own experiments and reanalysis of the data from Nevins and Flinner (1958), Nevins et al. (1964), and Nevins and Feyerherm (1967), recorded the following results. For floors occupied by people with bare feet (in swimming halls, gymnasiums, dressing rooms, bathrooms, and bedrooms), flooring material is important. Ranges for some typical floor materials are as follows

Textiles (rugs)	70 to 82°F
Pine floor	72.5 to 82°F
Oak floor	76 to 82°F

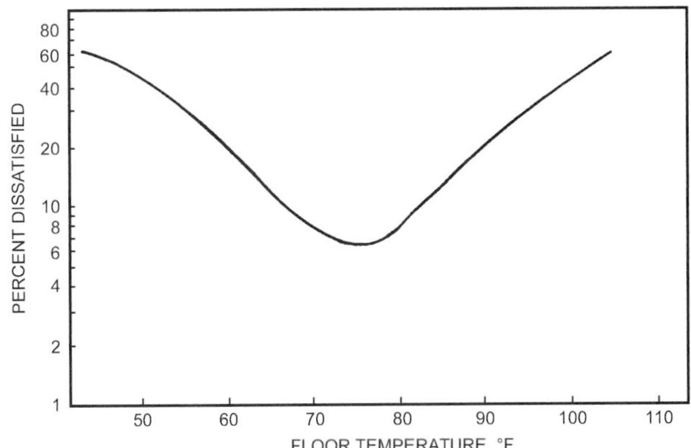

Fig. 10 Percentage of People Dissatisfied as Function of Floor Temperature

Hard linoleum	75 to 82°F
Concrete	79 to 83°F

To save energy, flooring materials with a low contact coefficient (cork, wood, carpets), radiant heated floors, or floor heating systems can be used to eliminate the desire for higher ambient temperatures caused by cold feet. These recommendations should also be followed in schools, where children often play directly on the floor.

For floors occupied by people with normal indoor footwear, flooring material is insignificant. Olesen (1977b) found an optimal temperature of 77°F for sedentary and 73.5°F for standing or walking persons. At the optimal temperature, 6% of the occupants felt warm or cold discomfort in the feet. Figure 10 shows the relationship between floor temperature and percentage of dissatisfied, combining data from experiments with seated and standing subjects. In all experiments, the subjects were in thermal neutrality; thus, the percentage of dissatisfied is only related to the discomfort due to cold or warm feet. No significant difference in floor temperature was preferred by females and males.

SECONDARY FACTORS AFFECTING COMFORT

Temperature, air speed, humidity, their variation, and personal parameters of metabolism and clothing insulation are primary factors that directly affect energy flow and thermal comfort. However, many secondary factors, some of which are discussed in this section, may more subtly influence comfort.

Day-to-Day Variations

Fanger (1973) conducted an experiment with a group of subjects, where the preferred ambient temperature for each subject under identical conditions was determined on four different days. Since the standard deviation was only 1.0°F, Fanger concluded that the comfort conditions for the individual can be reproduced and will vary only slightly from day to day.

Age

Because metabolism decreases slightly with age, many have stated that comfort conditions based on experiments with young and healthy subjects cannot be used for other age groups. Fanger (1982), Fanger and Langkilde (1975), Langkilde (1979), Nevins et al. (1966), and Rohles and Johnson (1972) conducted comfort studies in Denmark and the United States on different age groups (mean age 21 to 84). The studies revealed that the thermal environments preferred by older people do not differ from those preferred by younger people. The lower metabolism in older people is compensated for by a lower evaporative loss. Collins and Hoinville (1980) confirmed these results.

The fact that young and old people prefer the same thermal environment does not necessarily mean that they are equally sensitive to cold or heat. In practice, the ambient temperature level in the homes of older people is often higher than that for younger people. This may be explained by the lower activity level of elderly people, who are normally sedentary for a greater part of the day.

Adaptation

Many believe that people can acclimatize themselves by exposure to hot or cold surroundings, so that they prefer other thermal environments. Fanger (1982) conducted experiments involving subjects from the United States, Denmark, and tropical countries. The latter group was tested in Copenhagen immediately after their arrival by plane from the tropics where they had lived all their lives. Other experiments were conducted for two groups exposed to cold daily. One group comprised subjects who had been doing sedentary work in cold surroundings (in the meat-packing industry) for 8 h daily for at least 1 year. The other group consisted of winter swimmers who bathed in the sea daily.

Only slight differences in both the preferred ambient temperature and the physiological parameters in the comfort conditions were reported for the various groups. These results indicate that people cannot adapt to preferring warmer or colder environments. It is therefore likely that the same comfort conditions can be applied throughout the world. However, in determining the preferred ambient temperature from the comfort equations, a clo-value that corresponds to the local clothing habits should be used. A comparison of field comfort studies from different parts of the world shows significant differences in clothing habits depending on, among other things, the outdoor climate (Nicol and Humphreys 1972). According to these results, adaptation has little influence on the preferred ambient temperature. In uncomfortable warm or cold environments, however, adaptation will often have an influence. People used to working and living in warm climates can more easily accept and maintain a higher work performance in hot environments than people from colder climates.

Sex

Previously cited experiments by Fanger (1982), Fanger and Langkilde (1975), and Nevins et al. (1966) used equal numbers of male and female subjects, so comfort conditions for the two sexes can be compared. The experiments show that men and women prefer almost the same thermal environments. Women's skin temperature and evaporative loss are slightly lower than those for men, and this balances the somewhat lower metabolism of women. The reason that women often prefer higher ambient temperatures than men may be partly explained by the lighter clothing normally worn by women.

Seasonal and Circadian Rhythms

Since people cannot adapt to prefer warmer or colder environments, it follows that there is no difference between comfort conditions in winter and in summer. McNall et al. (1968) confirmed this in an investigation where results of winter and summer experiments showed no difference. On the other hand, it is reasonable to expect the comfort conditions to alter during the day because the internal body temperature has a daily rhythm—a maximum occurring late in the afternoon, and a minimum early in the morning.

In determining the preferred ambient temperature for each of 16 subjects both in the morning and in the evening, Fanger et al. (1974) and Ostberg and McNicholl (1973) observed no difference. Furthermore, Fanger et al. (1973) found only small fluctuations in the preferred ambient temperature during a simulated 8 h workday (sedentary work). There is a slight tendency to prefer somewhat warmer surroundings before lunch, but none of the fluctuations are significant.

PREDICTION OF THERMAL COMFORT

Thermal comfort and thermal sensation can be predicted several ways. One way is to use Figure 5 and Table 9 and adjust for clothing and activity levels that differ from those of the figure. More numerical and rigorous predictions are possible by using the PMV-PPD and two-node models described in this section.

Steady-State Energy Balance

Fanger (1982) related the comfort data to physiological variables. At a given level of metabolic activity M, and when the body is not far from thermal neutrality, the mean skin temperature t_{sk} and sweat rate E_{rsw} are the only physiological parameters influencing the heat balance. However, heat balance alone is not sufficient to establish thermal comfort. In the wide range of environmental conditions where heat balance can be obtained, only a narrow range provides thermal comfort. The following linear regression equations based on data from Rohles and Nevins (1971) indicate values of t_{sk} and E_{rsw} that provide thermal comfort.

$$t_{sk,req} = 96.3 - 0.156(M - W) \tag{56}$$

$$E_{rsw,req} = 0.42(M - W - 18.43) \tag{57}$$

At higher activity levels, sweat loss increases and the mean skin temperature decreases. Both reactions increase the heat loss from the body core to the environment. These two empirical relationships link the physiological and heat flow equations and thermal comfort perceptions. By substituting these values into Equation (11) for $C + R$, and into Equations (17) and (18) for E_{sk}, Equation (1), the energy balance equation, can be used to determine combinations of the six environmental and personal parameters that optimize comfort for steady-state conditions.

Fanger (1982) reduced these relationships to a single equation, which assumed all sweat generated is evaporated, eliminating clothing permeation efficiency i_{cl} as a factor in the equation. This assumption is valid for normal indoor clothing worn in typical indoor environments with low or moderate activity levels. At higher activity levels ($M_{act} > 3$ met), where a significant amount of sweating occurs even at optimum comfort conditions, this assumption may limit accuracy. The reduced equation is slightly different from the heat transfer equations developed here. The radiant heat exchange is expressed in terms of the Stefan-Boltzmann law (instead of using h_r), and diffusion of water vapor through the skin is expressed as a diffusivity coefficient and a linear approximation for saturated vapor pressure evaluated at t_{sk}. The combination of environmental and personal variables that produces a neutral sensation may be expressed as follows:

$$\begin{aligned} M - W = {} & 1.196 \times 10^{-9} f_{cl}[(t_{cl} + 460)^4 - (\bar{t}_r + 460)^4] \\ & + f_{cl}h_c(t_{cl} - t_a) \\ & + 0.97[5.73 - 0.022(M - W) - 6.9p_a] \\ & + 0.42[(M - W) - 18.43] \\ & + 0.0173M(5.87 - 6.9p_a) \\ & + 0.00077M(93.2 - t_a) \end{aligned} \tag{58}$$

where

$$\begin{aligned} t_{cl} = {} & 96.3 - 0.156(M - W) \\ & - R_{cl}\{(M - W) \\ & - 0.97[5.73 - 0.022(M - W) - 6.9p_a] \\ & - 0.42[(M - W) - 18.43] - 0.0173M(5.87 - 6.9p_a) \\ & - 0.00077M(93.2 - t_a)\} \end{aligned} \tag{59}$$

The values of h_c and f_{cl} can be estimated from tables and equations given in the section on Engineering Data and Measurements. Fanger used the following relationships:

$$h_c = \begin{cases} 0.361(t_{cl} - t_a)^{0.25} & 0.361(t_{cl} - t_a)^{0.25} > 0.151\sqrt{V} \\ 0.151\sqrt{V} & 0.361(t_{cl} - t_a)^{0.25} < 0.151\sqrt{V} \end{cases} \tag{60}$$

$$f_{cl} = \begin{cases} 1.0 + 0.2\,I_{cl} & I_{cl} < 0.5 \text{ clo} \\ 1.05 + 0.1\,I_{cl} & I_{cl} > 0.5 \text{ clo} \end{cases} \tag{61}$$

Figures 11and 12 show examples of how Equation (58) can be used.

Equation (58) is expanded to include a range of thermal sensations by using a **predicted mean vote (PMV) index**. The PMV index predicts the mean response of a large group of people according to the ASHRAE thermal sensation scale. Fanger (1970) related PMV to the imbalance between the actual heat flow from the body in a given environment and the heat flow required for optimum comfort at the specified activity by the following equation:

$$\text{PMV} = 3.155[0.303\exp(-0.114M) + 0.028]L \tag{62}$$

where L is the thermal load on the body, defined as the difference between internal heat production and heat loss to the actual environment for a person hypothetically kept at comfort values of t_{sk} and E_{rsw} at the actual activity level. Thermal load L is then the difference between the left and right sides of Equation (58) calculated for the actual values of the environmental conditions. As part of this calculation, the clothing temperature t_{cl} is found by iteration as

$$\begin{aligned} t_{cl} = {} & 96.3 - 0.156(M - W) \\ & - R_{cl}\{1.196 \times 10^{-9} f_{cl}[(t_{cl} + 460)^4 - (\bar{t}_r + 460)^4] \\ & + f_{cl}h_c(t_{cl} - t_a)\} \end{aligned} \tag{63}$$

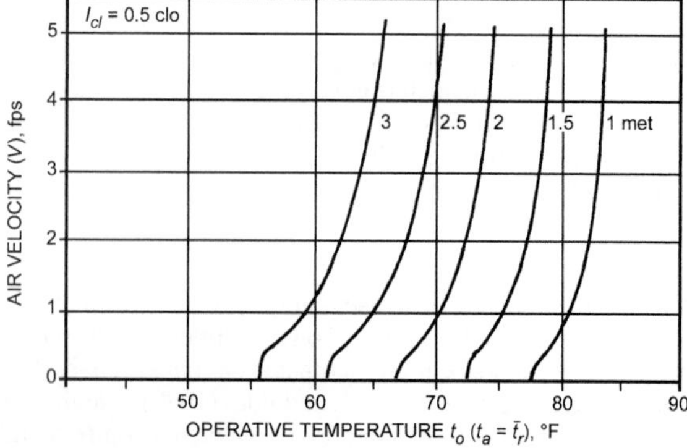

Fig. 11 Air Velocities and Operative Temperatures at 50% rh Necessary for Comfort (PMV = 0) of Persons in Summer Clothing at Various Levels of Activity

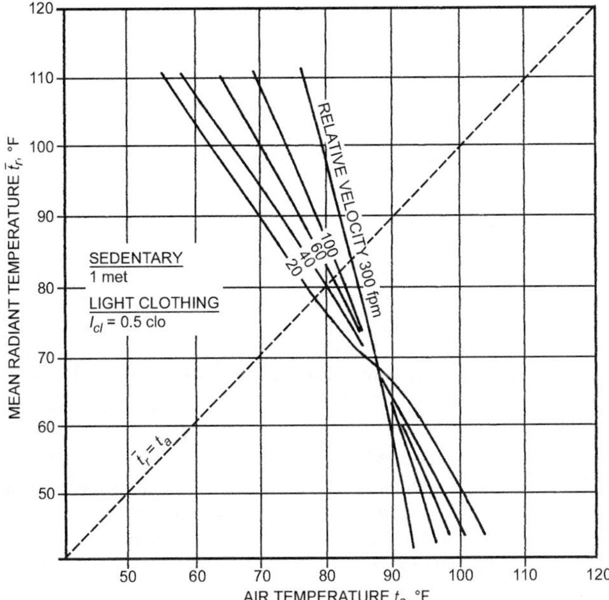

Fig. 12 Air Temperatures and Mean Radiant Temperatures Necessary for Comfort (PMV = 0) of Sedentary Persons in Summer Clothing at 50% rh

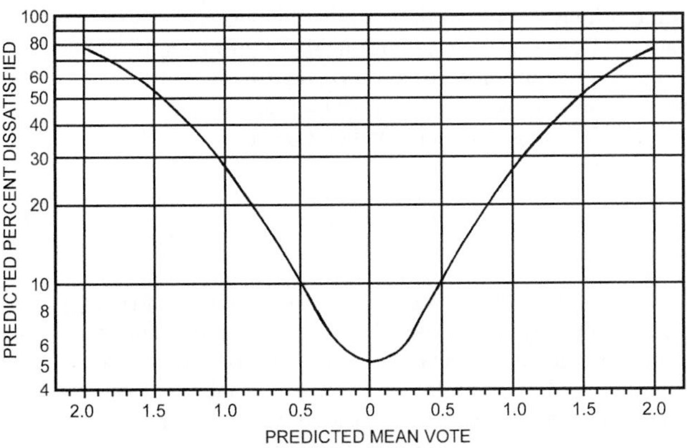

Fig. 13 Predicted Percentage of Dissatisfied (PPD) as Function of Predicted Mean Vote (PMV)

After estimating the PMV with Equation (62) or another method, the **predicted percent dissatisfied** (PPD) with a condition can also be estimated. Fanger (1982) related the PPD to the PMV as follows:

$$\text{PPD} = 100 - 95\exp[-(0.03353\text{PMV}^4 + 0.2179\text{PMV}^2)] \quad (64)$$

where dissatisfied is defined as anybody not voting −1, +1, or 0. This relationship is shown in Figure 13. A PPD of 10% corresponds to the PMV range of ±0.5, and even with PMV = 0, about 5% of the people are dissatisfied.

The **PMV-PPD model** is widely used and accepted for design and field assessment of comfort conditions. ISO *Standard* 7730 includes a short computer listing that facilitates computing PMV and PPD for a wide range of parameters.

Two-Node Model

The PMV model is useful only for predicting steady-state comfort responses. The two-node model can be used to predict physiological responses or responses to transient situations, at least for low and moderate activity levels in cool to very hot environments (Gagge et al. 1971, 1986). The two-node model is a simplification of more complex thermoregulatory models developed by Stolwijk and Hardy (1966). The simple, lumped parameter model considers a human as two concentric thermal compartments that represent the skin and the core of the body.

The **skin compartment** simulates the epidermis and dermis and is about 1/16 in. thick. Its mass, which is about 10% of the total body, depends on the amount of blood flowing through it for thermoregulation. The temperature in a compartment is assumed to be uniform so that the only temperature gradients are between compartments. In a cold environment, blood flow to the extremities may be reduced to conserve the heat of vital organs, resulting in axial temperature gradients in the arms, legs, hands, and feet. Heavy exercise with certain muscle groups or asymmetric environmental conditions may also cause nonuniform compartment temperatures and limit the accuracy of the model.

All the heat is assumed to be generated in the **core compartment**. In the cold, shivering and muscle tension may generate additional metabolic heat. This increase is related to skin and core temperature depressions from their set point values, or

$$M_{shiv} = 1.9(93.2 - t_{sk})(98.6 - t_{cr}) \quad (65)$$

where the temperature terms are set to zero if they become negative.

The core loses energy when the muscles do work on the surroundings. Heat is also lost from the core through respiration. The rate of respiratory heat loss is due to sensible and latent changes in the respired air and the ventilation rate as in Equations (19) and (20).

In addition, heat is conducted passively from the core to the skin. This is modeled as a massless thermal conductor (K = 0.93 Btu/h·ft²·°F). A controllable heat loss path from the core consists of pumping variable amounts of warm blood to the skin for cooling. This peripheral blood flow Q_{bl} in L/h·ft² depends on skin and core temperature deviations from their respective set points:

$$Q_{bl} = \frac{\text{BFN} + c_{dil}(t_{cr} - 98.6)}{1 + S_{tr}(93.2 - t_{sk})} \quad (66)$$

The temperature terms can only be > 0. If the deviation is negative, the term is set to zero. For average persons, the coefficients BFN, c_{dil}, and S_{tr} are 0.585, 9, and 0.28. Further, skin blood flow Q_{bl} is limited to a maximum of 8.4 L/h·ft².

Dry (sensible) heat loss q_{dry} from the skin flows through the clothing by conduction and then by parallel paths to the air and surrounding surfaces. Evaporative heat follows a similar path, flowing through the clothing and through the air boundary layer. Maximum evaporation E_{max} occurs if the skin is completely covered with sweat. The actual evaporation rate E_{sw} depends on the size w of the sweat film:

$$E_{sw} = wE_{max} \quad (67)$$

where w is E_{rsw}/E_{max}.

The rate of regulatory sweating E_{rsw} (rate at which water is brought to the surface of the skin in Btu/h·ft²) can be predicted by skin and core temperature deviations from their set points:

$$E_{rsw} = c_{sw}(t_b - t_{bset})\exp[-(t_{sk} - 93.2)/19.3] \quad (68)$$

where $t_b = (1 - \alpha_{sk})t_{cr} + \alpha_{sk}t_{sk}$ and is the mean body temperature, and c_{sw} = 30 Btu/h·ft²·°F. The temperature deviation terms are set to zero when negative. α_{sk} is the fraction of the total body mass that is considered to be thermally in the skin compartment.

$$\alpha_{sk} = 0.0418 + \frac{0.745}{10.8Q_{bl} - 0.585} \tag{69}$$

Regulatory sweating Q_{rsw} in the model is limited to 0.1 L/h·ft² or 200 Btu/h·ft². E_{rsw} evaporates from the skin, but if E_{rsw} is greater than E_{max}, the excess drips off.

An energy balance on the core yields

$$M + M_{shiv} = W + q_{res} + (K + \text{SKBF}c_{p,bl})(t_{cr} - t_{sk}) + m_{cr}c_{cr}\frac{dt_{cr}}{d\theta} \tag{70}$$

and for the skin,

$$(K + \text{SKBF}c_{p,bl})(t_{cr} - t_{sk}) = q_{dry} + q_{evap} + m_{sk}c_{sk}\frac{dt_{sk}}{d\theta} \tag{71}$$

where c_{cr}, c_{sk}, and $c_{p,bl}$ are specific heats of core, skin, and blood (0.83, 0.83, and 1.0 Btu/lb·°F, respectively), and SKBF is $\rho_{bl}Q_{bl}$, where ρ_{bl} is density of blood (28.4 lb/L).

Equations (70) and (71) can be rearranged in terms of $dt_{sk}/d\theta$ and $dt_{cr}/d\theta$ and numerically integrated with small time steps (10 to 60 s) from initial conditions or previous values to find t_{cr} and t_{sk} at any time.

After calculating values of t_{sk}, t_{cr}, and w, the model uses empirical expressions to predict thermal sensation (TSENS) and thermal discomfort (DISC). These indices are based on 11-point numerical scales, where positive values represent the warm side of neutral sensation or comfort, and negative values represent the cool side. TSENS is based on the same scale as PMV, but with extra terms for ±4 (very hot/cold) and ±5 (intolerably hot/cold). Recognizing the same positive/negative convention for warm/cold discomfort, DISC is defined as

5 intolerable
4 limited tolerance
3 very uncomfortable
2 uncomfortable and unpleasant
1 slightly uncomfortable but acceptable
0 comfortable

TSENS is defined in terms of deviations of mean body temperature t_b from cold and hot set points representing the lower and upper limits for the zone of evaporative regulation: $t_{b,c}$ and $t_{b,h}$, respectively. The values of these set points depend on the net rate of internal heat production and are calculated by

$$t_{b,c} = \frac{0.34}{58.15}(M - W) + 97.34 \tag{72}$$

$$t_{b,h} = \frac{0.608}{58.15}(M - W) + 98.0 \tag{73}$$

TSENS is then determined by

$$\text{TSENS} = \begin{cases} 0.26(t_b - t_{b,c}) & t_b < t_{b,c} \\ 4.7\eta_{ev}(t_b - t_{b,c})/(t_{b,h} - t_{b,c}) & t_{b,c} \le t_b \le t_{b,h} \\ 4.7\eta_{ev} + 0.26(t_b - t_{b,h}) & t_{b,h} < t_b \end{cases} \tag{74}$$

where η_{ev} is the evaporative efficiency (assumed to be 0.85).

DISC is numerically equal to TSENS when t_b is below its cold set point $t_{b,c}$ and it is related to skin wettedness when body temperature is regulated by sweating:

$$\text{DISC} = \begin{cases} 0.26(t_b - t_{b,c}) & t_b < t_{b,c} \\ \dfrac{4.7(E_{rsw} - E_{rsw,req})}{E_{max} - E_{rsw,req} - E_{dif}} & t_{b,c} \le t_b \end{cases} \tag{75}$$

where $E_{rsw,req}$ is calculated as in Fanger's model, using Equation (57).

Adaptive Models

Adaptive models do not actually predict comfort responses but rather the almost constant conditions under which people are likely to be comfortable in buildings. In general, people naturally adapt and may also make various adjustments to themselves and their surroundings to reduce discomfort and physiological strain. It has been observed that, through adaptive actions, an acceptable degree of comfort in residences and offices is possible over a range of air temperatures from about 63 to 88°F (Humphreys and Nicol 1998).

The adaptive adjustments are typically conscious behavioral actions such as altering clothing, posture, activity schedules, activity levels, rate of working, diet, ventilation, air movement, and local temperature. The adaptations may also include unconscious longer term changes to physiological set points and gains for the control of shivering, skin blood flow, and sweating, as well as adjustments to body fluid levels and salt loss. However, only limited documentation and information on such changes is available.

An important driving force behind the adaptive process is the pattern of outside weather conditions and the exposure to them. This is the principal input to the adaptive models that have evolved to date, and these models predict likely comfort temperatures t_c or ranges of t_c from monthly mean outdoor temperatures t_{out}. Such a model (Humphreys and Nicol 1998), based on data from a wide range of buildings, climates, and cultures is

$$t_c = 75.6 + 0.43(t_{out} - 71.6)\exp - \left(\frac{t_{out} - 71.6}{61.1}\right)^2 \tag{76}$$

The adaptive models are useful to guide design and energy decisions. They may also be useful to specify building temperatures set points throughout the year. A recent ASHRAE-sponsored study on adaptive models compiled an extensive database from past field studies to study, develop, and test adaptive models. For climates and buildings where cooling and central heating are not required, the study suggests the following model (de Dear and Brager 1998):

$$t_{oc} = 66 + 0.255(t_{out} - 32) \tag{77}$$

where t_{oc} is the operative comfort temperature.

In general, the value of using an adaptive model to specify set points or guide temperature control strategies is likely to increase with the freedom that occupants are given to adapt (e.g., by having flexible working hours, locations, or dress codes).

Zones of Comfort and Discomfort

The section on Two-Node Model shows that comfort and thermal sensation are not necessarily the same variable, especially for a person in the zone of evaporative thermal regulation. Figures 14 and 15 show this difference for the standard combination of met-clo-air movement used in the standard effective temperature. Figure 14 demonstrates that practically all basic physiological variables predicted by the two-node model are functions of ambient temperature and are relatively independent of vapor pressure. All exceptions occur at relative humidities above 80% and as the isotherms reach the ET* = 107°F line, where regulation by evaporation fails. Figure 15 shows that lines of constant ET* and wettedness are functions of both ambient temperature and vapor pressure. Thus, human thermal responses are divided into two classes—those in Figure 14, which respond only to heat stress from the environment, and those in Figure 15, which respond to both the heat stress from the environment and the resultant heat strain (Stolwijk et al. 1968).

For warm environments, any index with isotherms parallel to skin temperature is a reliable index of thermal sensation alone, and not of discomfort caused by increased humidity. Indices with isotherms parallel to ET* are reliable indicators of discomfort or dissatisfaction with thermal environments. For a fixed exposure time to cold, lines of constant t_{sk}, ET*, and t_o are essentially identical, and cold sensation is no different from cold discomfort. For a state of comfort with sedentary or light activity, lines of constant t_{sk} and ET* coincide. Thus comfort and thermal sensations coincide in this region as well. The upper and lower temperature limits for comfort at these levels can be specified either by thermal sensation (Fanger 1982) or by ET*, as is done in ASHRAE *Standard* 55, since lines of constant comfort and lines of constant thermal sensation should be identical.

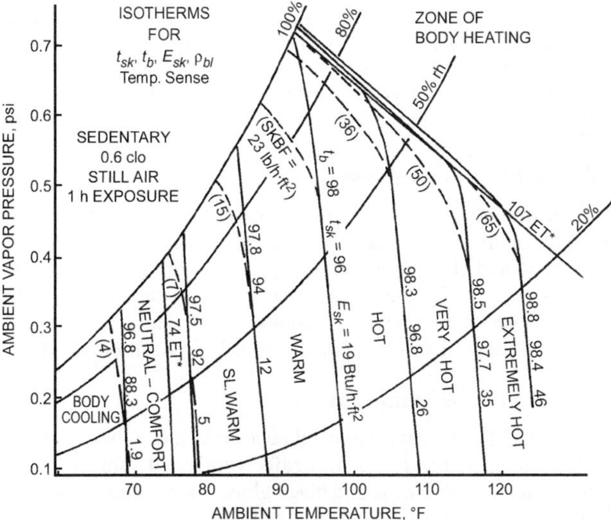

Fig. 14 Effect of Environmental Conditions on Physiological Variables

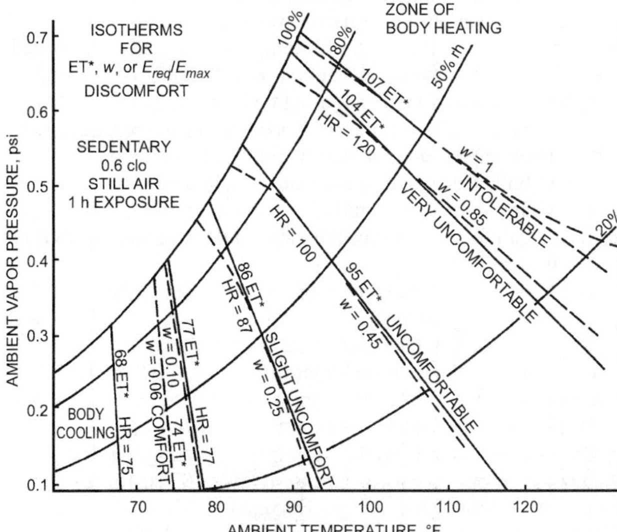

Fig. 15 Effect of Thermal Environment on Discomfort

ENVIRONMENTAL INDICES

An environmental index combines two or more parameters (e.g., air temperature, mean radiant temperature, humidity, or air velocity) into a single variable. Indices simplify the description of the thermal environment and the stress imposed by an environment. Environmental indices may be classified according to how they are developed. Rational indices are based on the theoretical concepts presented earlier. Empirical indices are based on measurements with subjects or on simplified relationships that do not necessarily follow theory. Indices may also be classified according to their application, generally either heat stress or cold stress.

Effective Temperature

The **effective temperature** ET* is probably the most common environmental index, and it has the widest range of application. It combines temperature and humidity into a single index, so two environments with the same ET* should evoke the same thermal response even though they have different temperatures and humidities; but they must have the same air velocities.

The original empirical effective temperature was developed by Houghten and Yaglou (1923). Gagge et al. (1971) defined a new effective temperature using a rational approach. Defined mathematically in Equation (33), this is the temperature of an environment at 50% rh that results in the same total heat loss E_{sk} from the skin as in the actual environment.

Because the index is defined in terms of operative temperature t_o, it combines the effects of three parameters ($\bar{t}_r$, t_a, and p_a) into a single index. Skin wettedness w and the permeability index i_m must be specified and are constant for a given ET* line for a particular situation. The two-node model is used to determine skin wettedness in the zone of evaporative regulation. At the upper limit of regulation, w approaches 1.0; at the lower limit, w approaches 0.06. Skin wettedness equals one of these values when the body is outside the zone of evaporative regulation. Since the slope of a constant ET* line depends on skin wettedness and clothing moisture permeability, effective temperature for a given temperature and humidity may depend on the clothing and activity of the person. This difference is shown in Figure 16. At low skin wettedness, the air humidity has little influence, and lines of constant ET* are nearly vertical. As skin wettedness increases due to activity and/or heat stress, the lines become more horizontal and the influence of humidity is much more pronounced. The ASHRAE comfort envelope shown in Figure 5 is described in terms of ET*.

Since ET* depends on clothing and activity, it is not possible to generate a universal ET* chart. Calculation of ET* can also be tedious, requiring the solution of multiple coupled equations to determine skin wettedness. A standard set of conditions representative of typical indoor applications is used to define a **standard effective temperature** SET*. The standard effective temperature is then defined as the equivalent air temperature of an isothermal environment at 50% rh in which a subject, while wearing clothing standardized for the activity concerned, has the same heat stress (skin temperature t_{sk}) and thermoregulatory strain (skin wettedness w) as in the actual environment.

Humid Operative Temperature

The **humid operative temperature** t_{oh} is the temperature of a uniform environment at 100% rh in which a person loses the same total amount of heat from the skin as in the actual environment. This index is defined mathematically in Equation (32). It is analogous to ET*, the only difference being that it is defined at 100% rh and 0% rh rather than at 50% rh. Figures 2 and 16 indicate that lines of constant ET* are also lines of constant t_{oh}. However, the values of these two indices differ for a given environment.

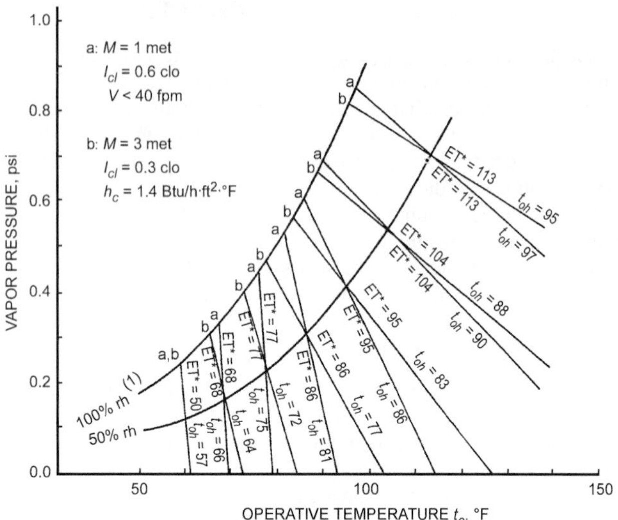

A. EFFECT OF CONDITIONS ON ET*, t_{oh}

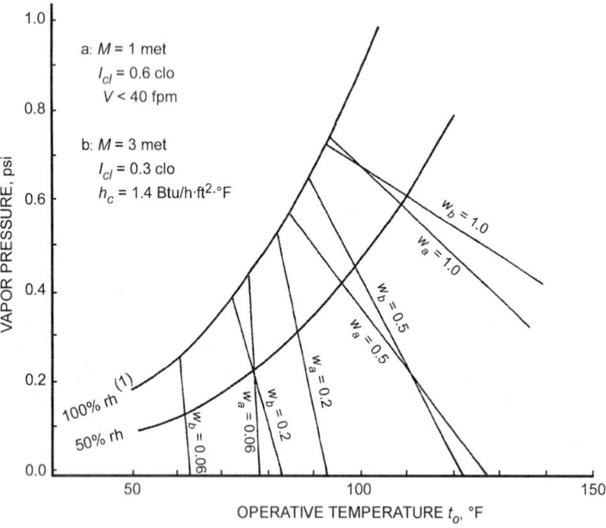

B. EFFECT OF CONDITIONS ON WETTEDNESS, w

Fig. 16 Effective Temperature ET* and Skin Wettedness w
[Adapted from Nishi et al. (1975) and Gonzalez et al. (1978)]

Heat Stress Index

Originally proposed by Belding and Hatch (1955), this rational index is the ratio of the total evaporative heat loss E_{sk} required for thermal equilibrium (the sum of metabolism plus dry heat load) to the maximum evaporative heat loss E_{max} possible for the environment, multiplied by 100, for steady-state conditions (S_{sk} and S_{cr} are zero) and with t_{sk} held constant at 95°F. The ratio E_{sk}/E_{max} equals skin wettedness w [Equation (18)]. When **heat stress index** (HSI) > 100, body heating occurs; when HSI < 0, body cooling occurs. Belding and Hatch (1955) limited E_{max} to 220 Btu/h·ft², which corresponds to a sweat rate of approximately 0.21 lb/h·ft². When t_{sk} is constant, loci of constant HSI coincide with lines of constant ET* on a psychrometric chart. Other indices based on wettedness have the same applications (Gonzalez et al. 1978, Belding 1970, ISO *Standard* 7933) but differ in their treatment of E_{max} and the effect of clothing. Table 10 describes physiological factors associated with HSI values.

Table 10 Evaluation of Heat Stress Index

Heat Stress Index	Physiological and Hygienic Implications of 8 h Exposures to Various Heat Stresses
0	No thermal strain.
10 20 30	Mild to moderate heat strain. If job involves higher intellectual functions, dexterity, or alertness, subtle to substantial decrements in performance may be expected. In performing heavy physical work, little decrement is expected, unless ability of individuals to perform such work under no thermal stress is marginal.
40 50 60	Severe heat strain involving a threat to health unless men are physically fit. Break-in period required for men not previously acclimatized. Some decrement in performance of physical work is to be expected. Medical selection of personnel desirable, because these conditions are unsuitable for those with cardiovascular or respiratory impairment or with chronic dermatitis. These working conditions are also unsuitable for activities requiring sustained mental effort.
70 80 90	Very severe heat strain. Only a small percentage of the population may be expected to qualify for this work. Personnel should be selected: (a) by medical examination, and (b) by trial on the job (after acclimatization). Special measures are needed to assure adequate water and salt intake. Amelioration of working conditions by any feasible means is highly desirable, and may be expected to decrease the health hazard while increasing job efficiency. Slight "indisposition," which in most jobs would be insufficient to affect performance, may render workers unfit for this exposure.
100	The maximum strain tolerated daily by fit, acclimatized young men.

Index of Skin Wettedness

Skin wettedness w is the ratio of observed skin sweating E_{sk} to the E_{max} of the environment as defined by t_{sk}, t_a, humidity, air movement, and clothing in Equation (12). Except for the factor of 100, it is essentially the same as HSI. Skin wettedness is more closely related to the sense of discomfort or unpleasantness than to temperature sensation (Gagge et al. 1969a,b; Gonzalez et al. 1978).

Wet-Bulb Globe Temperature

The WBGT is an environmental heat stress index that combines dry-bulb temperature t_{db}, a **naturally ventilated** (not aspirated) wet-bulb temperature t_{nwb}, and black globe temperature t_g, according to the relation (Dukes-Dobos and Henschel 1971, 1973)

$$\mathrm{WBGT} = 0.7t_{nwb} + 0.2t_g + 0.1t_a \qquad (78)$$

This form of the equation is usually used where solar radiation is present. The naturally ventilated wet-bulb thermometer is left exposed to the sunlight, but the air temperature t_a sensor is shaded. In enclosed environments, Equation (78) is simplified by dropping the t_a term and using a 0.3 weighting factor for t_g.

The black globe thermometer responds to air temperature, mean radiant temperature, and air movement, while the naturally ventilated wet-bulb thermometer responds to air humidity, air movement, radiant temperature, and air temperature. Thus, WBGT is a function of all four environmental factors affecting human environmental heat stress.

The WBGT is a better index of heat stress than the old ET; it shows almost as good a correlation with sweat rate as do the later corrected effective temperature (CET) and effective temperature with radiation (ETR) indices (Minard 1961); the CET and ETR both require direct measurement of wind velocity which, for accuracy, requires special instruments and trained technicians.

The WBGT index is widely used for estimating the heat stress potential of industrial environments (Davis 1976). In the United States, the National Institute of Occupational Safety and Health (NIOSH) developed criteria for a heat-stress-limiting standard (NIOSH 1986). ISO *Standard* 7243 also uses the WBGT. Figure 17 graphically summarizes the permissible heat exposure limits,

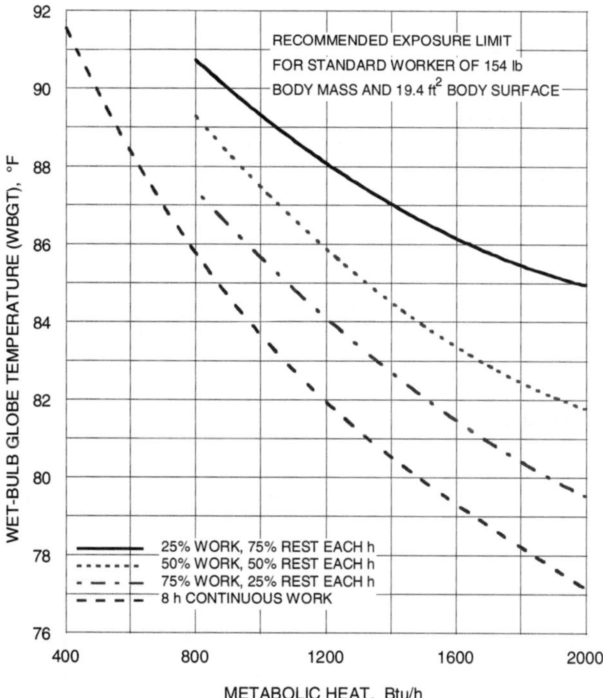

Fig. 17 Recommended Heat Stress Exposure Limits for Heat Acclimatized Workers
[Adapted from NIOSH (1986)]

expressed as working time per hour, for a fit individual, as specified for various WBGT levels. Values apply for normal permeable clothing (0.6 clo) and must be adjusted for heavy or partly vapor-permeable clothing. The USAF has recommended adjusting the measured WBGT upwards by 10°F for personnel wearing chemical protective clothing or body armor. This type of clothing increases the resistance to sweat evaporation about threefold (higher if it is totally impermeable), requiring an adjustment in WBGT level to compensate for reduced evaporative cooling at the skin.

Several mathematical models are available for predicting WBGT from the environmental factors: air temperature, psychrometric wet-bulb temperature, mean radiant temperature, and air motion (Azer and Hsu 1977; Sullivan and Gorton 1976). A simpler approach, involving plotting WBGT lines on a psychrometric chart, is recommended. Isotherms of WBGT are parallel and have negative slopes varying from 0.014 psi/°F for still air to 0.016 psi/°F for air motion greater than 200 fpm. By comparison, psychrometric wet-bulb lines have negative slopes of about 0.0056 psi/°F, or about 35% as steep.

Wet-Globe Temperature

The WGT, introduced by Botsford (1971), is a simpler approach to measuring environmental heat stress than the WBGT. The measurement is made with a wetted globe thermometer called a Botsball, which consists of a 2.5 in. black copper sphere covered with a fitted wet black mesh fabric, into which the sensor of a dial thermometer is inserted. A polished stem attached to the sphere supports the thermometer and contains a water reservoir for keeping the sphere covering wet. This instrument is suspended by the stem at the indoor (or outdoor) site to be measured.

Onkaram et al. (1980) have shown that WBGT can be predicted with reasonable accuracy from WGT for temperate to warm environments with medium to high humidities. With air temperatures between 68 and 95°F, dew points ranging from 45 to 77°F (relative humidities above 30%), and wind speeds of 15 mph or less, the experimental regression equation (r = 0.98) in °F for an outdoor environment is

$$\text{WBGT} = 1.044(\text{WGT}) - 1.745 \tag{79}$$

This equation should not be used outside the experimental range given because data from hot-dry desert environments show differences between WBGT and WGT that are too large (10°F and above) to be adjusted by Equation (79) (Matthew et al. 1986). At very low humidity and high wind, WGT approaches the psychrometric wet-bulb temperature, which is greatly depressed below t_a. However, in the WBGT, t_{nwb} accounts for only 70% of the index value, with the remaining 30% at or above t_a.

Ciriello and Snook (1977) handle the problem by providing a series of regression equations, the choice depending on the levels of wind speed, humidity, and radiant heat. They report an accuracy of conversion from WGT to WBGT within 0.7°F (90% confidence level), if good estimates of wind speed, humidity, and radiation level are available.

Wind Chill Index

The wind chill index (WCI) is an empirical index developed from cooling measurements obtained in Antarctica on a cylindrical flask partly filled with water (Siple and Passel 1945). The index describes the rate of heat loss from the cylinder by radiation and convection for a surface temperature of 91.4°F, as a function of ambient temperature and wind velocity. As originally proposed,

$$\text{WCI} = \frac{(10.45 - 0.447V + 6.686\sqrt{V})(91.4 - t_a)}{1.8} \tag{80}$$

where V and t_a are in mph and °F, respectively, and WCI units are kcal/(h·m^2). Multiply WCI by 0.368 to convert to Btu/h·ft^2. The 91.4°F surface temperature was chosen to be representative of the mean skin temperature of a resting human in comfortable surroundings.

A number of valid objections have been raised about this formulation. Cooling rate data from which it was derived were measured on a 2.24 in. diameter plastic cylinder, making it unlikely that WCI would be an accurate measure of heat loss from exposed flesh, which has different characteristics from the plastic (curvature, roughness, and radiation exchange properties) and is invariably below 91.4°F in a cold environment. Moreover, values given by the equation peak at 56 mph, then decrease with increasing velocity.

Nevertheless, for velocities below 50 mph, this index reliably expresses combined effects of temperature and wind on subjective discomfort. For example, if the calculated WCI is less than 1400 and actual air temperature is above 14°F, there is little risk of frostbite during brief exposures (1 h or less), even for bare skin. However, at a WCI of 2000 or more, the probability is high that exposed flesh will begin to freeze in 1 min or less unless measures are taken to shield the exposed skin (such as a fur ruff to break up the wind around the face).

Rather than using the WCI to express the severity of a cold environment, meteorologists use an index derived from the WCI called the **equivalent wind chill temperature**. This is the ambient temperature that would produce, in a calm wind (defined for this application as 4 mph), the same WCI as the actual combination of air temperature and wind velocity. Equivalent wind chill temperature $t_{eq,wc}$ can be calculated by

$$t_{eq,wc} = -0.0818(\text{WCI}) + 91.4 \tag{81}$$

Table 11 Equivalent Wind Chill Temperatures of Cold Environments

Wind Speed, mph	Actual Thermometer Reading, °F											
	50	40	30	20	10	0	−10	−20	−30	−40	−50	−60
	Equivalent Wind Chill Temperature, °F											
0	50	40	30	20	10	0	−10	−20	−30	−40	−50	−60
5	48	37	27	16	6	−5	−15	−26	−36	−47	−57	−68
10	40	28	16	3	−9	−21	−34	−46	−58	−71	−83	−95
15	36	22	9	−5	−18	−32	−45	−59	−72	−86	−99	−113
20	32	18	4	−11	−25	−39	−53	−68	−82	−96	−110	−125
25	30	15	0	−15	−30	−44	−59	−74	−89	−104	−119	−134
30	28	13	−3	−18	−33	−48	−64	−79	−94	−110	−125	−140
35	27	11	−4	−20	−36	−51	−67	−83	−98	−114	−129	−145
40	26	10	−6	−22	−38	−53	−69	−85	−101	−117	−133	−148
Little danger: In less than 5 h, with dry skin. Maximum danger from false sense of security. (WCI < 1400)					**Increasing danger:** Danger of freezing exposed flesh within 1 min. ($1400 \le$ WCI ≤ 2000)			**Great danger:** Flesh may freeze within 30 s. (WCI > 2000)				

Source: U.S. Army Research Institute of Environmental Medicine.
Notes: Cooling power of environment expressed as an equivalent temperature under calm conditions [Equation (81)]. Winds greater than 43 mph have little added chilling effect.

where $t_{eq,wc}$ is in °F (and frequently referred to as a **wind chill factor**), thus distinguishing it from WCI, which is given either as a cooling rate or as a plain number with no units. For velocities less than 4 mph, Equation (81) does not apply, and the wind chill temperature is equal to the air temperature.

Equation (81) does not imply cooling to below ambient temperature, but recognizes that, because of wind, the cooling rate is increased as though it were occurring at the lower equivalent wind chill temperature. Wind accelerates the rate of heat loss, so that the skin surface is cooling faster toward the ambient temperature. Table 11 shows a typical wind chill chart, expressed in equivalent wind chill temperature.

SPECIAL ENVIRONMENTS

Infrared Heating

Optical and thermal properties of skin must be considered in studies concerning the effects of infrared radiation in (1) producing changes in skin temperature and skin blood flow, and (2) evoking sensations of temperature and comfort (Hardy 1961). Although the body can be considered to have the properties of water, thermal sensation and heat transfer with the environment require a study of the skin and its interaction with visible and infrared radiation.

Figure 18 shows how skin reflectance and absorptance vary for a blackbody heat source at the temperature (in °R) indicated. These curves show that darkly pigmented skin is heated more by direct radiation from a high-intensity heater at 4500°R than is lightly pigmented skin. With low-temperature and low-intensity heating equipment used for total area heating, there is minimal, if any, difference. Also, in practice, clothing minimizes differences.

Changes in skin temperature caused by high-intensity infrared radiation depend on the thermal conductivity, density, and specific heat of the living skin (Lipkin and Hardy 1954). Modeling of skin heating with the heat transfer theory yields a parabolic relation between exposure time and skin temperature rise for nonpenetrating radiation:

$$t_{sf} - t_{si} = \Delta t = 2J\alpha\sqrt{\theta/(\pi k \rho c_p)} \qquad (82)$$

where

t_{sf} = final skin temperature, °F
t_{si} = initial skin temperature, °F
J = irradiance from source radiation temperatures, Btu/h·ft²
θ = time, h
k = specific thermal conductivity of tissue, Btu/h·ft·°R

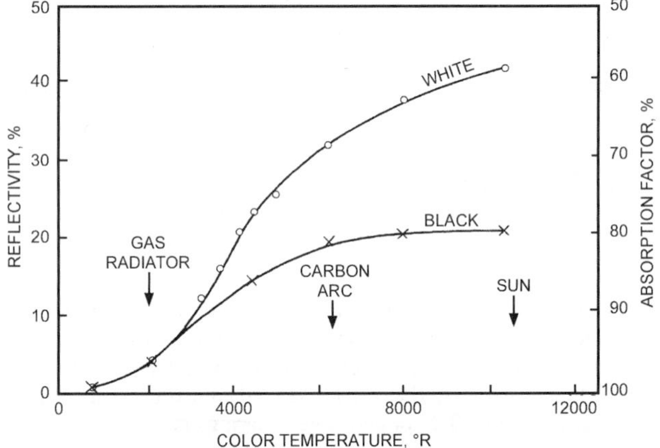

Fig. 18 Variation in Skin Reflection and Absorptivity for Blackbody Heat Sources

ρ = density, lb/ft³
c_p = specific heat, Btu/lb·°R
α = skin absorptance at radiation temperatures, dimensionless

Product $k\rho c_p$ is the physiologically important quantity that determines temperature elevation of skin or other tissue on exposure to nonpenetrating radiation. Fatty tissue, because of its relatively low specific heat, is heated more rapidly than moist skin or bone. Experimentally, $k\rho c_p$ values can be determined by plotting Δt^2 against $1.13J^2\theta$ (Figure 19). The relationship is linear, and the slopes are inversely proportional to the $k\rho c_p$ of the specimen. Comparing leather and water with body tissues suggests that thermal inertia values depend largely on tissue water content.

Living tissues do not conform strictly to this simple mathematical formula. Figure 20 compares excised skin with living skin with normal blood flow, and skin with blood flow occluded. For short exposure times, the $k\rho c_p$ of normal skin is the same as that in which blood flow has been stopped; excised skin heats more rapidly due to unavoidable dehydration that occurs postmortem. However, with longer exposure to thermal radiation, vasodilation increases blood flow, cooling the skin. For the first 20 s of irradiation, skin with normally constricted blood vessels has a $k\rho c_p$ value one-fourth that for skin with fully dilated vessels.

Skin temperature is the best single index of thermal comfort. The most rapid changes in skin temperature occur during the first 60 s of exposure to infrared radiation. During this initial period, thermal

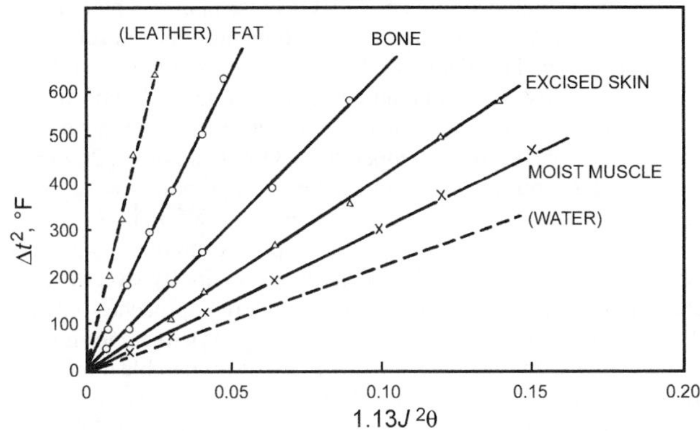

Fig. 19 Comparing Thermal Inertia of Fat, Bone, Moist Muscle, and Excised Skin to That of Leather and Water

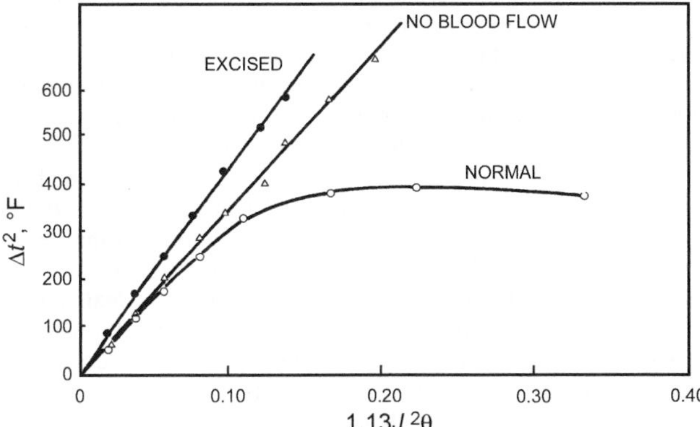

Fig. 20 Thermal Inertias of Excised, Bloodless, and Normal Living Skin

sensation and the heating rate of the skin vary with the quality of infrared radiation (color temperature in °R). Because radiant heat from a gas-fired heater is absorbed at the skin surface, the same unit level of absorbed radiation during the first 60 s of exposure can cause an even warmer initial sensation than penetrating solar radiation. Skin heating curves tend to level off after a 60 s exposure (Figure 20); this means that a relative balance is quickly created between heat absorbed, heat flow to the skin surface, and heat loss to the ambient environment. Therefore, the effects of radiant heating on thermal comfort should be examined for conditions approaching thermal equilibrium.

Stolwijk and Hardy (1966) described an unclothed subject's response for a 2 h exposure to temperatures of 41 to 95°F. Nevins et al. (1966) showed a relation between ambient temperatures and thermal comfort of clothed, resting subjects. For any given uniform environmental temperature, both initial physiological response and degree of comfort can be determined for a subject at rest.

Physiological implications for radiant heating can be defined by two environmental temperatures: (1) mean radiant temperature $\bar{t}_r$ and (2) ambient air temperature t_a. For this discussion on radiant heat, assume that (1) relative humidity is less than 50%, and (2) air movement is low and constant, with an equivalent convection coefficient of 0.51 Btu/h·ft²·°F.

The equilibrium equation, describing heat exchange between skin surface at mean temperature t_{sk} and the radiant environment, is given in Equation (28), and can be transformed to give (see Table 2)

$$M' - E_{sk} - F_{cle}[h_r(t_{sk} - \bar{t}_r) + h_c(t_{sk} - t_o)] = 0 \quad (83)$$

where M' is the net heat production $(M - W)$ less respiratory losses.

By algebraic transformation, Equation (83) can be rewritten as

$$M' + \text{ERF} \cdot F_{cle} = E_{sk} + (h_r + h_c)(t_{sk} - t_a)F_{cle} \quad (84)$$

where $\text{ERF} = h_r(\bar{t}_r - t_a)$ is the effective radiant field and represents the additional radiant exchange with the body when $\bar{t}_r \neq t_a$.

The last term in Equation (84) describes heat exchange with an environment uniformly heated to temperature t_a. The term h_r, evaluated in Equation (35), is also a function of posture, for which factor A_r/A_D can vary from 0.67 for crouching to 0.73 for standing. For preliminary analysis, a useful value for h_r is 0.83 Btu/h·ft²·°F, which corresponds to a normally clothed (at 75°F) sedentary subject. Ambient air movement affects h_c, which appears only in the right-hand term of Equation (84).

Although the linear radiation coefficient h_r is used in Equations (83) and (84), the same definition of ERF follows if the fourth power radiation law is used. By this law, assuming emissivity of the body surface is unity, the ERF term in Equation (84) is

$$\text{ERF} = \sigma(A_r/A_D)[(\bar{t}_r + 460)^4 - (t_a + 460)^4]F_{cle} \quad (85)$$

where σ is the Stefan-Boltzmann constant, 0.1712×10^{-8} Btu/h·ft²·°R⁴.

Because $\bar{t}_r$ equals the radiation of several surfaces at different temperatures $(T_1, T_2, \ldots, T_j)$,

$$\text{ERF} = (\text{ERF})_1 + (\text{ERF})_2 + \ldots + (\text{ERF})_j \quad (86)$$

where

$\text{ERF}_j = \sigma(A_r/A_D)\alpha_j F_{m-j}(T_j^4 - T_a^4)F_{cle}$

α_j = absorptance of skin or clothing surface for source radiating at temperature T_j

F_{m-j} = angle factor to subject m from source j

T_a = ambient air temperature, °R

ERF is the sum of the fields caused by each surface T_j [e.g., T_1 may be an infrared beam heater; T_2, a heated floor; T_3, a warm ceiling; T_4, a cold plate glass window $(T_4 < T_a)$; etc.]. Only surfaces with T_j differing from T_a contribute to the ERF.

Comfort Equations for Radiant Heating

The **comfort equation for radiant heat** (Gagge et al. 1967a,b) follows from definition of ERF and Equation (8):

$$t_o \text{ (for comfort)} = t_a + \text{ERF (for comfort)}/h \quad (87)$$

Thus, operative temperature for comfort is the temperature of the ambient air plus a temperature increment ERF/h, a ratio that measures the effectiveness of the incident radiant heating on occupants. Higher air movement (which increases the value of h or h_c) reduces the effectiveness of radiant heating systems. Clothing lowers t_o for comfort and for thermal neutrality.

Values for ERF and h must be determined to apply the comfort equation for radiant heating. Table 3 may be used to estimate h. One method of determining ERF is to calculate it directly from radiometric data that give (1) radiation emission spectrum of the source, (2) concentration of the beam, (3) radiation from the floor, ceiling, and windows, and (4) corresponding angle factors involved. This analytical approach is described in Chapter 52 of the 1999 *ASHRAE Handbook—Applications*.

For direct measurement, a skin-colored or black globe, 6 in. in diameter, can measure the radiant field ERF for comfort, by the following relation:

$$\text{ERF} = (A_r/A_D)[1.07 + 0.169\sqrt{V}](t_g - t_a) \quad (88)$$

where t_g is uncorrected globe temperature in °F and V is air movement in fpm. The average value of A_r/A_D is 0.7. For a skin-colored globe, no correction is needed for the quality of radiation. For a black globe, ERF must be multiplied by α for the exposed clothing/skin surface. For a subject with 0.6 to 1.0 clo, t_o for comfort should agree numerically with t_a for comfort in Figure 5. When t_o replaces t_a in Figure 5, humidity is measured in vapor pressure rather than relative humidity, which refers only to air temperature.

Other methods may be used to measure ERF. The most accurate is by physiological means. In Equation (84), when M, $t_{sk} - t_a$, and the associated transfer coefficients are experimentally held constant,

$$\Delta E = \Delta \text{ERF} \tag{89}$$

The variation in evaporative heat loss E (rate of weight loss) caused by changing the wattage of two T-3 infrared lamps is a measure in absolute terms of the radiant heat received by the body.

A third method uses a directional radiometer to measure ERF directly. For example, radiation absorbed at the body surface [in Btu/h·ft^2] is

$$\text{ERF} = \alpha(A_i/A_D)J \tag{90}$$

where irradiance J can be measured by a directional (Hardy-type) radiometer; α is the surface absorptance effective for the source used; and A_i is the projection area of the body normal to the directional irradiance. Equation (90) can be used to calculate ERF only for the simplest geometrical arrangements. For a human subject lying supine and irradiated uniformly from above, A_i/A_D is 0.3. Figure 18 shows variance of α for human skin with blackbody temperature (in °R) of the radiating source. When irradiance J is uneven and coming from many directions, as is usually the case, the previous physiological method can be used to obtain an effective A_i/A_D from the observed ΔE and $\Delta(\alpha J)$.

Hot and Humid Environments

Tolerance limits to high temperature vary with the ability to (1) sense temperature, (2) lose heat by regulatory sweating, and (3) move heat from the body core by blood flow to the skin surface, where cooling is the most effective. Many interrelating processes are involved in heat stress (Figure 21).

Skin surface temperatures of 113°F trigger pain receptors in the skin; direct contact with metal at this temperature is painful. However, since thermal insulation of the air layer around the skin is high, much higher dry air temperatures can be tolerated. For lightly clothed subjects at rest, tolerance times of nearly 50 min have been reported at 180°F dry-bulb temperature; 33 min at 200°F; 26 min at 220°F; and 24 min at 240°F. In each case, dew points were lower than 86°F. Many individuals are stimulated by brief periods of exposure to 185°F dry air in a sauna. Short exposures to these extremely hot environments are tolerable because of cooling by sweat evaporation. However, when ambient vapor pressure approaches 0.87 psi (97°F dew point, typically found on sweating skin), tolerance is drastically reduced. Temperatures of 122°F can be intolerable if the dew-point temperature is greater than 77°F, and both deep body temperature and heart rate rise within minutes (Gonzalez et al. 1978).

The rate at which and length of time a body can sweat are limited. The maximum rate of sweating for an average man is about 4 lb/h. If all this sweat evaporates from the skin surface under conditions of low humidity and air movement, maximum cooling is about 214 Btu/h·°F. However, this value does not normally occur because sweat rolls off the skin surface without evaporative cooling or is absorbed by or evaporated within clothing. A more typical cooling limit is 6 mets, 110 Btu/h·ft^2, representing approximately 2.2 lb/h of sweating for the average man.

Thermal equilibrium is maintained by dissipation of resting heat production (1 met) plus any radiant and convective load. If the environment does not limit heat loss from the body during heavy activity, decreasing skin temperature compensates for the core temperature rise. Therefore, mean body temperature is maintained, although the gradient from core to skin is increased. Blood flow through the skin is reduced, but muscle blood flow necessary for exercise is preserved. The upper limit of skin blood flow is about 200 lb/h (Burton and Bazett 1936).

Body heat storage of 318 Btu (or a rise in t_b of 2.5°F) for an average-sized man represents an average voluntary tolerance limit. Continuing work beyond this limit increases the risk of heat exhaustion. Collapse can occur at about 635 Btu of storage (5°F rise); few individuals can tolerate heat storage of 872 Btu (6.8°F above normal).

The cardiovascular system affects tolerance limits. In normal, healthy subjects exposed to extreme heat, heart rate and cardiac output increase in an attempt to maintain blood pressure and supply of

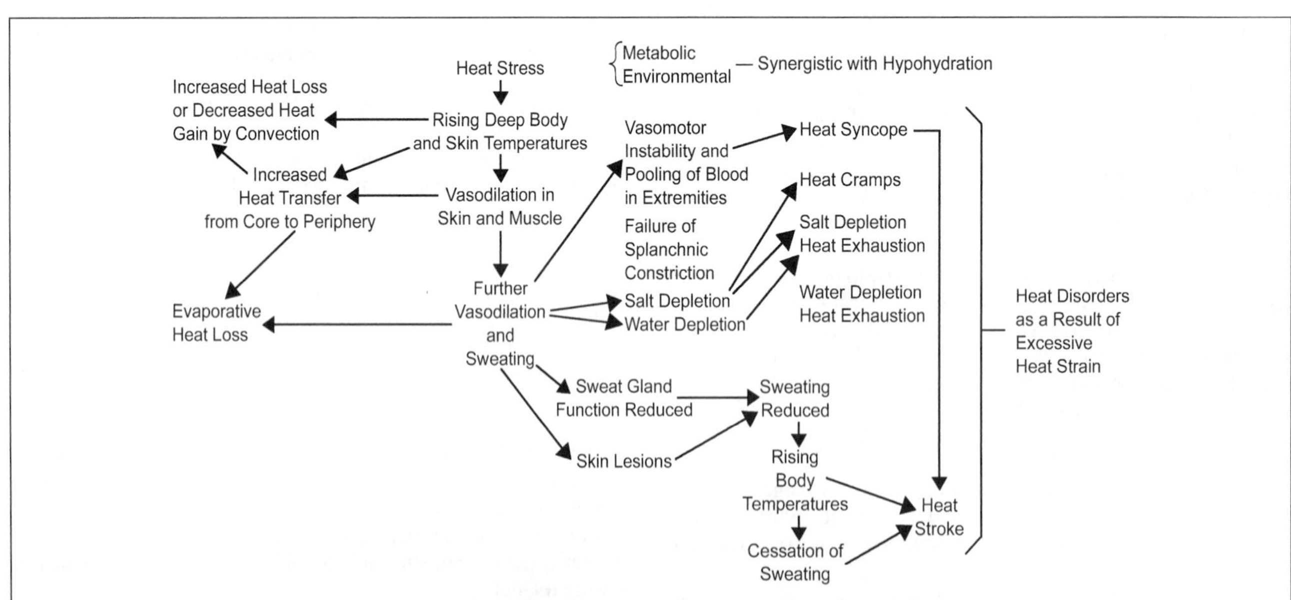

Fig. 21 Schematic Design of Heat Stress and Heat Disorders
[Modified by Buskirk from scale diagram by Belding (1967) and Leithead and Lind (1964)]

blood to the brain. At a heart rate of about 180 bpm, the short time between contractions prevents adequate blood supply to the heart chambers. As heart rate continues to increase, cardiac output drops, causing inadequate convective blood exchange with the skin and, perhaps more important, inadequate blood supply to the brain. Victims of this heat exhaustion faint or black out. Accelerated heart rate can also result from inadequate venous return to the heart caused by pooling of blood in the skin and lower extremities. In this case, cardiac output is limited because not enough blood is available to refill the heart between beats. This occurs most frequently when an overheated individual, having worked hard in the heat, suddenly stops working. The muscles no longer massage the blood back past the valves in the veins toward the heart. Dehydration compounds the problem, since fluid volume in the vascular system is reduced.

If core temperature t_{cr} increases above 106°F, critical hypothalamic proteins can be damaged, resulting in inappropriate vasoconstriction, cessation of sweating, increased heat production by shivering, or some combination of these. Heat stroke damage is frequently irreversible and carries a high risk of fatality.

A final problem, hyperventilation, occurs predominantly in hot-wet conditions, when too much CO_2 is washed from the blood. This can lead to tingling sensations, skin numbness, and vasoconstriction in the brain with occasional loss of consciousness.

Since a rise in heart rate or rectal temperature is essentially linear with ambient vapor pressure above a dew point of 77°F, these two changes can measure severe heat stress. Although individual heart rate and rectal temperature responses to mild heat stress vary, severe heat stress saturates physiological regulating systems, producing uniform increases in heart rate and rectal temperature. In contrast, sweat production measures stress under milder conditions but becomes less useful under more severe stress. The maximal sweat rate compatible with body cooling varies with (1) degree of heat acclimatization, (2) duration of sweating, and (3) whether the sweat evaporates or merely saturates the skin and drips off. Total sweat rates in excess of 4.4 lb/h can occur in short exposures, but about 2.2 lb/h is an average maximum sustainable level for an acclimatized man.

Figure 22 illustrates the decline in heart rate, rectal temperature, and skin temperature when exercising subjects are exposed to 104°F over a period of days. Acclimatization can be achieved by working in the heat for 100 min each day—30% improvement occurs after the first day, 50% after 3 days, and 95% after 6 or 7 days. Increased sweat secretion while working in the heat can be induced by rest. Although reducing salt intake during the first few days in the heat can conserve sodium, heat cramps may result. Working regularly in the heat improves cardiovascular efficiency, sweat secretion, and sodium conservation. Once induced, heat acclimatization can be maintained by as few as once-a-week workouts in the heat; otherwise, it diminishes slowly over a 2- to 3-week period and disappears.

Extreme Cold Environments

Human performance in extreme cold ultimately depends on maintaining thermal balance. Subjective discomfort is reported by a 154 lb man with 19.4 ft² of body surface area when a heat debt of about 100 Btu is incurred. A heat debt of about 600 Btu is acutely uncomfortable; this represents a drop of approximately 4.7°F (or about 7% of total heat content) in mean body temperature.

This loss can occur during 1 to 2 h of sedentary activity outdoors. A sleeping individual will awake after losing about 300 Btu, decreasing mean skin temperature by about 5.5°F and deep body temperature by about 1°F. A drop in deep body temperature (e.g., rectal temperature) below 95°F threatens a loss of body temperature regulation, while 82.4°F is considered critical for survival, despite recorded survival from a deep body temperature of 64.4°F. Temperature is more crucial than rate of temperature change; Witherspoon et al. (1971) observed a rate of fall in the core temperature of 5.4°F per hour in subjects immersed in 50°F water, without residual effect.

Activity level also affects human performance. Subjective sensations reported by sedentary subjects at a mean skin temperature of 92°F, are comfortable; at 88°F, uncomfortably cold; at 86°F, shivering cold; and at 84°F, extremely cold. The critical subjective tolerance limit (without numbing) for mean skin temperature appears to be about 77°F. However, during moderate to heavy activity, subjects reported the same skin temperatures as comfortable. Although mean skin temperature is significant, the temperature of the extremities is more frequently the critical factor for comfort in the cold. Consistent with this, one of the first responses to cold exposure is vasoconstriction, which reduces circulatory heat input to the hands and feet. A hand-skin temperature of 68°F causes a report of uncomfortably cold; 59°F, extremely cold; and 41°F, painful. Identical verbal responses for the foot surface occur at approximately 2.7 to 3.5°F warmer temperatures.

An ambient temperature of −30°F is the lower limit for useful outdoor activity, even with adequate insulative clothing. At −60°F, almost all outdoor effort becomes exceedingly difficult; even with appropriate protective equipment, only limited exposure is possible. Reported exposures of 30 min at −103°F have occurred in the Antarctic without injury.

In response to extreme heat loss, maximal heat production becomes very important. When the less efficient vasoconstriction cannot prevent body heat loss, shivering is an automatic, more efficient defense against cold. This can be triggered by low deep body temperature, low skin temperature, rapid change of skin temperature, or some combination of all three. Shivering is usually preceded

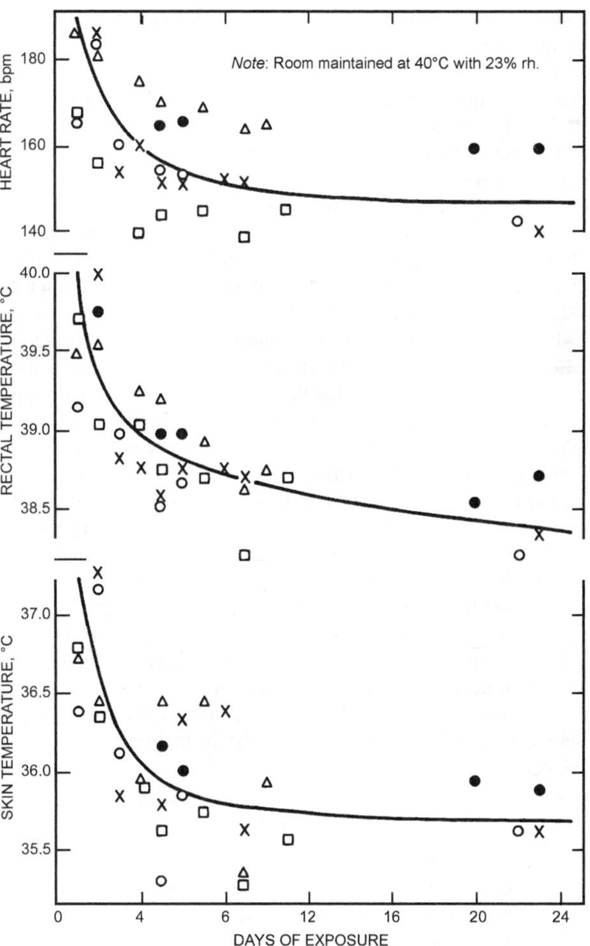

Fig. 22 Acclimatization to Heat Resulting from Daily Exposure of Five Subjects to Extremely Hot Room
(Robinson et al. 1943)

by an imperceptible increase in muscle tension and by noticeable gooseflesh produced by muscle contraction in the skin. It begins slowly in small muscle groups, initially increasing total heat production by 1.5 to 2 times resting levels. As body cooling increases, the reaction spreads to additional body segments. Ultimately violent, whole body shivering causes maximum heat production of about 6 times resting levels, rendering the individual totally ineffective.

Given sufficient cold exposure, the body undergoes changes that indicate cold acclimatization. These physiological changes include, (1) endocrine changes (e.g., sensitivity to norepinephrine), causing nonshivering heat production by metabolism of free fatty acids released from adipose tissue; (2) improved circulatory heat flow to skin, causing an overall sensation of greater comfort; and (3) improved circulatory heat flow to the extremities, reducing the risk of injury and permitting activities at what ordinarily would be severely uncomfortable temperatures in the extremities. Generally, these physiological changes are minor and are induced only by repeated extreme exposures. Nonphysiological factors, including training, experience, and selection of adequate protective clothing, are more useful and may be safer than dependence on physiological changes.

The caloric requirement for adequately clothed subjects in extreme cold is only slightly greater than that for subjects living and working in temperate climates. This greater requirement results from added work caused by (1) carrying the weight of heavy clothing (energy cost for heavy protective footwear may be six times that of an equivalent weight on the torso); and (2) the inefficiency of walking in snow, snowshoeing, or skiing, which can increase energy cost up to 300%.

To achieve proper protection in low temperatures, a person must either maintain high metabolic heat production by activity or reduce heat loss by controlling the body's microclimate with clothing. Other protective measures include spot radiant heating, showers of hot air for work at a fixed site, and warm-air-ventilated or electrically heated clothing. The extremities, such as fingers and toes, pose more of a problem than the torso because, as thin cylinders, they are particularly susceptible to heat loss and difficult to insulate without increasing the surface for heat loss. Vasoconstriction can reduce circulatory heat input to extremities by over 90%.

Although there is no ideal insulating material for protective clothing, radiation-reflective materials are promising. Insulation is primarily a function of clothing thickness; the thickness of trapped air, rather than fibers used, determines insulation effectiveness.

Protection for the respiratory tract seems unnecessary in healthy individuals, even at −50°F. However, asthmatics or individuals with mild cardiovascular problems may benefit from a face mask that warms inspired air. Masks are unnecessary for protecting the face because heat to facial skin is not reduced by local vasoconstriction, as it is for hands. If wind chill is great, there is always a risk of cold injury caused by freezing of exposed skin. Using properly designed torso clothing, such as a parka with a fur-lined hood to minimize wind penetration to the face, and 35 Btu/h of auxiliary heat to each hand and foot, inactive people can tolerate −67°F with a 10 mph wind for more than 6 h. As long as the skin temperature of fingers remains above 60°F, manual dexterity can be maintained and useful work performed without difficulty.

SYMBOLS

A = area, ft^2
- A_{ch} = contact area between chair and body
- A_{cl} = surface area of clothed body
- A_D = DuBois surface area of nude body
- A_G = body surface area covered by garment
- A_r = effective radiation area of body

BFN = neutral skin blood flow, L/h·ft^2

c = specific heat, Btu/lb·°F
- c_{cr} = body core
- c_{sk} = skin
- c_p = at constant pressure
- $c_{p,a}$ = of air
- $c_{p,b}$ = of body tissue
- $c_{p,bl}$ = of blood
- c_{dil} = constant for skin blood flow
- c_{sw} = proportionality constant for sweat control

C = convective heat loss, Btu/h·ft^2

C_{res} = sensible heat loss due to respiration, Btu/h·ft^2

$C + R$ = total sensible heat loss from skin, Btu/h·ft^2

E = evaporative heat loss, Btu/h·ft^2
- E_{dif} = due to moisture diffusion through skin
- E_{max} = maximum possible
- E_{rsw} = due to regulatory sweating
- E_{res} = due to respiration
- $E_{rsw,req}$ = required for comfort
- E_{sw} = actual evaporation rate
- E_{sk} = total from skin

ERF = effective radiant field, Btu/h·ft^2

ET = original effective temperature based on 100% rh, °F

ET* = effective temperature, based on 50% rh, °F

f_{cl} = clothing area factor, A_{cl}/A_D, dimensionless

F_{cl} = intrinsic clothing thermal efficiency, dimensionless

F_{cle} = effective clothing thermal efficiency, dimensionless

F_{pcl} = permeation efficiency, dimensionless

F_{m-j} = angle factor to person from source j, dimensionless

F_{p-N} = angle factor between person and source N, dimensionless

h = sensible heat transfer coefficient, Btu/h·ft^2·°F
- h = total at surface
- h' = overall including clothing

h_c = convection at surface
- h_{cc} = corrected convection at surface
- h_r = radiation

h_e = evaporative heat transfer coefficient, Btu/h·ft^2·psi
- h_e = at surface
- h_{ec} = at surface, corrected for atmospheric pressure
- $h'_{e,cl}$ = clothing evaporative conductance, the evaporative conductance of a uniform layer of insulation covering the entire body that has same effect on evaporative heat flow as the actual clothing.
- h'_e = overall including clothing

h_a = enthalpy of ambient air, Btu/lb (dry air)

h_{ex} = enthalpy of exhaled air, Btu/lb (dry air)

h_{fg} = heat of vaporization of water, Btu/lb

i = vapor permeation efficiency, dimensionless
- i_a = air layer
- i_{cl} = clothing
- i_m = total

I = thermal resistance in clo units, clo

All subscripts given for symbol R apply to symbol I
- $I^*_{clu,i}$ = effective insulation of garment i
- J = irradiance, Btu/h·ft^2
- k = thermal conductivity of body tissue, Btu/h·ft·°F
- K = effective conductance between core and skin, Btu/h·ft^2·°F
- l = height, ft or in.
- L = thermal load on body, Btu/h·ft^2
- LR = Lewis ratio, °F/psi
- m = body mass, lb
- m_{cr} = mass of body core
- m_{sk} = mass of skin

m_{ge} = mass to gas exchange, lb

$\dot{m}$ = mass flow, lb/h·ft^2

$\dot{m}_{res}$ = pulmonary ventilation flow rate, lb/h

$\dot{m}_{w,res}$ = pulmonary water loss rate, lb/h

M = metabolic heat production, Btu/h·ft^2
- M = total
- M' = net ($M - W$)
- M_{act} = due to activity
- M_{shiv} = due to shivering

p = water vapor pressure, psi
- p_a = in ambient air
- $p_{ET^*,s}$ = saturated at ET*
- $p_{oh,s}$ = saturated at t_{oh}

$p_{sk,s}$ = saturated at t_{sk}
p_t = atmospheric pressure, psi
$p_{t_{com}}$ = related to t_{com}
q = heat flow, Btu/h·ft^2
q_{dry} = sensible from skin
q_{evap} = latent from skin
$q_{cr,sk}$ = from core to skin
q_{res} = total due to respiration
q_{sk} = total from the skin
Q_{bl} = blood flow between core and skin, L/h·ft^2
Q_{CO_2} = volume rate of CO_2 produced, ft^3/h
Q_{O_2} = volume rate of O_2 consumed, ft^3/h
Q_{rsw} = rate of regulatory sweat generation, L/h·ft^2
R = radiative heat loss from skin, Btu/h·ft^2
R = thermal resistance, ft^2·°F·h/Btu
R_a = air layer on nude skin
$R_{a,cl}$ = air layer at outer surface
R_{cl} = clothing
R_{cle} = change due to clothing
R_e = evaporative resistance, ft^2·psi·h/Btu
$R_{e,a}$ = outer boundary layer air
$R_{e,cl}$ = clothing
$R_{e,t}$ = total
R_t = total
RQ = respiratory quotient, dimensionless
S = heat storage, Btu/h·ft^2
S_{cr} = in core compartment
S_{sk} = in skin compartment
S_{tr} = constriction constant for skin blood flow
SET* = standard effective temperature, °F
SKBF = skin blood flow lb/h·ft^2
t = temperature, °F
t_a = ambient air
t_b = average of body
$t_{b,c}$ = lower limit for evaporative regulation zone
$t_{b,h}$ = upper limit for evaporative regulation zone
t_c = comfort
t_{cl} = clothing surface
t_{com} = combined temperature
t_{cr} = core compartment
t_{db} = dry bulb
t_{dp} = dew point
$t_{eq,wc}$ = equivalent wind chill temperature
t_{ex} = of exhaled air
t_g = globe
t_N = of surface N
t_{nwb} = naturally ventilated wet bulb
t_o = operative
t_{oc} = operative comfort
t_{oh} = humid operation
t_{out} = monthly mean outside
t_{pr} = plane radiant
$\bar{t}_r$ = mean radiant
t_{sf} = final skin
t_{si} = initial skin
t_{sk} = skin compartment
$t_{sk,n}$ = at neutrality
$t_{sk,req}$ = required for comfort
t_{wb} = wet bulb
T = absolute temperature, °R

All subscripts that apply to symbol t may apply to symbol T.

Tu = turbulence intensity, %
V = air velocity, fpm
w = skin wettedness, dimensionless
w_{rsw} = required to evaporate regulatory sweat
W = external work accomplished, Btu/h·ft^2
W_a = humidity ratio of ambient air, lb (water vapor)/lb (dry air)
W_{ex} = humidity ratio of exhaled air, lb (water vapor)/lb (dry air)
WBGT = wet-bulb globe temperature, °F
WCI = wind chill index, kcal/(h·m^2)
WGT = wet-globe temperature, °F
α = skin absorptance, dimensionless
α_{sk} = fraction of total body mass concentrated in skin compartment, dimensionless
ε = emissivity, dimensionless
η_{ev} = evaporative efficiency, dimensionless
μ = body's mechanical efficiency = W/M, dimensionless
ρ = density, lb/ft^3
σ = Stefan-Boltzmann constant = 0.171×10^{-8} Btu/h·ft^2·°R^4
θ = time, h

CODES AND STANDARDS

ASHRAE. 1992. Thermal environmental conditions for human occupancy. ANSI/ASHRAE *Standard* 55-1992.

ASHRAE. 1994. Addendum 55a. ANSI/ASHRAE *Standard* 55-1992.

ISO. 1989. Hot environments—Analytical determination and interpretation of thermal stress using calculation of required sweat rates. *Standard* 7933. International Organization for Standardization, Geneva.

ISO. 1989. Hot environments—Estimation of the heat stress on working man, based on the WBGT-index (wet bulb globe temperature). *Standard* 7243.

ISO. 1994. Moderate thermal environments—Determination of the PMV and PPD indices and specification of the conditions for thermal comfort. *Standard* 7730.

REFERENCES

Astrand, P. and K. Rodahl. 1977. *Textbook of work physiology: Physiological bases of exercise*. McGraw-Hill, New York.

Azer, N.Z. 1982. Design guidelines for spot cooling systems: Part I—Assessing the acceptability of the environment. *ASHRAE Transactions* 88:1.

Azer, N.Z. and S. Hsu. 1977. OSHA heat stress standards and the WBGT index. *ASHRAE Transactions* 83(2):30.

Belding, H.D. 1967. Heat stress. In *Thermobiology*, ed. A.H. Rose. Academic Press, New York.

Belding, H.S. 1970. The search for a universal heat stress index. *Physiological and Behavioral Temperature Regulation*, eds. J.D. Hardy et al. Springfield, IL.

Belding, H.S. and T.F. Hatch. 1955. Index for evaluating heat stress in terms of resulting physiological strains. *Heating, Piping and Air Conditioning* 207:239.

Berglund, L.G. 1994. Common elements in the design and operation of thermal comfort and ventilation systems. *ASHRAE Transactions* 100(1).

Berglund, L.G. 1995. Comfort criteria: Humidity and standards. *Proceedings of Pan Pacific Symposium on Building and Urban Environmental Conditioning in Asia* 2:369-82. Architecture Department, University of Nagoya, Japan.

Berglund, L.G and D.J. Cunningham. 1986. Parameters of human discomfort in warm environments. *ASHRAE Transactions* 92(2):732-46.

Berglund, L.G. and A. Fobelets. 1987. A subjective human response to low level air currents and asymmetric radiation. *ASHRAE Transactions* 93(1):497-523.

Berglund, L.G. and R.R. Gonzalez. 1978. Evaporation of sweat from sedentary man in humid environments. *J. Appl. Physiol: Respirat, Environ, Exercise Physiol* 42(5): 767-72.

Botsford, J.H. 1971. A wet globe thermometer for environmental heat measurement. *American Industrial Hygiene Association Journal* 32:1-10.

Burton, A.C. and H.C. Bazett. 1936. A study of the average temperature of the tissues, the exchange of heat and vasomotor responses in man, by a bath calorimeter. *American Journal of Physiology* 117:36.

Busch, J.F. 1992. A tale of two populations: Thermal comfort in air-conditioned and naturally ventilated offices in Thailand. *Energy and Buildings* 18:235-49.

Buskirk, E.R. 1960. Problems related to the caloric cost of living. *Bulletin of the New York Academy of Medicine* 26:365.

Chatonnet, J. and M. Cabanac. 1965. The perception of thermal comfort. *International Journal of Biometeorology* 9:183-93.

Ciriello, V.M. and S.H. Snook. 1977. The prediction of WBGT from the Botsball. *American Industrial Hygiene Association Journal* 38:264.

Colin, J. and Y. Houdas. 1967. Experimental determination of coefficient of heat exchange by convection of the human body. *Journal of Applied Physiology* 22:31.

Collins, K.J. and E. Hoinville. 1980. Temperature requirements in old age. *Building Services Engineering Research and Technology* 1(4):165-72.

Davis, W.J. 1976. Typical WBGT indexes in various industrial environments. *ASHRAE Transactions* 82(2):303.

de Dear, R.J. and G.S. Brager. 1998. Developing an adaptive model of thermal comfort and preference. *ASHRAE Technical Data Bulletin* 14(1):27-49.

de Dear, R., K. Leow, and A. Ameen. 1991. Thermal comfort in the humid tropics— Part I. *ASHRAE Transactions* 97(1):874-79.

DuBois, D. and E.F. DuBois. 1916. A formula to estimate approximate surface area, if height and weight are known. *Archives of Internal Medicine* 17:863-71.

Dukes-Dobos, F. and A. Henschel. 1971. The modification of the WBGT index for establishing permissible heat exposure limits in occupational work. HEW, USPHE, NIOSH, TR-69.

Dukes-Dobos, F. and A. Henschel. 1973. Development of permissible heat exposure limits for occupational work. *ASHRAE Journal* 9:57.

Eriksson, H.A. 1975. Heating and ventilating of tractor cabs. Presented at the 1975 Winter Meeting, American Society of Agricultural Engineers, Chicago.

Fanger, P.O. 1967. Calculation of thermal comfort: Introduction of a basic comfort equation. *ASHRAE Transactions* 73(2):III.4.1.

Fanger, P.O. 1970. *Thermal comfort analysis and applications in environmental engineering*. McGraw-Hill, New York.

Fanger, P.O. 1972. *Thermal comfort*. McGraw-Hill, New York.

Fanger, P.O. 1973. The variability of man's preferred ambient temperature from day to day. *Archives des Sciences Physiologiques* 27(4):A403.

Fanger, P.O. 1982. *Thermal comfort*. Robert E. Krieger Publishing Company, Malabar, FL.

Fanger, P.O., L. Banhidi, B.W. Olesen, and G. Langkilde. 1980. Comfort limits for heated ceilings. *ASHRAE Transactions* 86.

Fanger, P.O. and N.K. Christensen. 1986. Perception of draught in ventilated spaces. *Ergonomics* 29(2):215-35.

Fanger, P.O., J. Hojbjerre, and J.O.B. Thomsen. 1974. Thermal comfort conditions in the morning and the evening. *International Journal of Biometeorology* 18(1):16.

Fanger, P.O., J. Hojbjerre, and J.O.B. Thomsen. 1973. Man's preferred ambient temperature during the day. *Archives des Sciences Physiologiques* 27(4): A395-A402.

Fanger, P.O., B.M. Ipsen, G. Langkilde, B.W. Olesen, N.K. Christensen, and S. Tanabe. 1985. Comfort limits for asymmetric thermal radiation. *Energy and Buildings*.

Fanger, P.O. and G. Langkilde. 1975. Interindividual differences in ambient temperature preferred by seated persons. *ASHRAE Transactions* 81(2): 140-47.

Fanger, P.O., A.K. Melikov, H. Hanzawa, and J. Ring, J. 1989. Turbulence and draft. *ASHRAE Journal* 31(4):18-25.

Fobelets, A.P.R. and A.P. Gagge. 1988. Rationalization of the ET* as a measure of the enthalpy of the human environment. *ASHRAE Transactions* 94:1.

Gagge, A.P. 1937. A new physiological variable associated with sensible and insensible perspiration. *American Journal of Physiology* 20(2):277-87.

Gagge, A.P., A.C. Burton, and H.D. Bazett. 1971. A practical system of units for the description of heat exchange of man with his environment. *Science* 94:428-30.

Gagge, A.P. and J.D. Hardy. 1967. Thermal radiation exchange of the human by partitional calorimetry. *Journal of Applied Physiology* 23:248.

Gagge, A.P., A.P. Fobelets, and L.G. Berglund. 1986. A standard predictive index of human response to the thermal environment. *ASHRAE Transactions* 92(2B).

Gagge, A.P., Y. Nishi, and R.G. Nevins. 1976. The role of clothing in meeting FEA energy conservation guidelines. *ASHRAE Transactions* 82(2): 234.

Gagge, A.P., G.M. Rapp, and J.D. Hardy. 1967a. The effective radiant field and operative temperature necessary for comfort with radiant heating. *ASHRAE Transactions* 73(1):I.2.1.

Gagge, A.P., G.M. Rapp, and J.D. Hardy. 1967b. The effective radiant field and operative temperature necessary for comfort with radiant heating. *ASHRAE Journal* 9(5):63.

Gagge, A.P., J. Stolwijk, and Y. Nishi. 1971. An effective temperature scale based on a simple model of human physiological regulatory response. *ASHRAE Transactions* 77(1):247-62.

Gagge, A.P., J.A.J. Stolwijk, and B. Saltin. 1969a. Comfort and thermal sensation and associated physiological responses during exercise at various ambient temperatures. *Environmental Research* 2:209.

Gagge, A.P., J.A.J. Stolwijk, and Y. Nishi. 1969b. The prediction of thermal comfort when thermal equilibrium is maintained by sweating. *ASHRAE Transactions* 75(2):108.

Gonzalez, R.R., L.G. Berglund, and A.P. Gagge. 1978. Indices of thermoregulatory strain for moderate exercise in the heat. *Journal of Applied Physiology* 44:889.

Green, G.H. 1982. Positive and negative effects of building humidification. *ASHRAE Transactions* 88(1):1049-61.

Gwosdow, A.R., J.C. Stevens, L. Berglund, and J.A.J. Stolwijk. 1986. Skin friction and fabric sensations in neutral and warm environments. *Textile Research Journal* 56:574-80.

Hardy, J.D. 1949. Heat transfer. In *Physiology of heat regulation and science of clothing*. eds. L.H. Newburgh and W.B. Saunders Ltd., London, 78.

Hardy, J.D. 1961. Physiological effects of high intensity infrared heating. *ASHRAE Journal* 4:11.

Hardy, J.D., J.A.J. Stolwijk, and A.P. Gagge. 1971. Man. In *Comparative physiology of thermoregulation*, Chapter 5. Charles C. Thomas, Springfield, IL.

Hardy, J.D., H.G. Wolf, and H. Goodell. 1952. *Pain sensations and reactions*. Williams and Wilkins, Baltimore.

Hensel, H. 1973. Temperature reception and thermal comfort. *Archives des Sciences Physiologiques* 27:A359-A370.

Hensel, H. 1981. *Thermoreception and temperature regulation*. Academic Press, London.

Holmer, I. 1984. Required clothing insulation (IREQ) as an analytical index of cold stress. *ASHRAE Transactions* 90(1B):1116-28.

Houghten, F.C. and C.P. Yaglou. 1923. ASHVE Research Report No. 673. *ASHVE Transactions* 29:361.

Humphreys, M. and J.F. Nicol. 1998. Understanding the adaptive approach to thermal comfort. *ASHRAE Technical Data Bulletin* 14(1):1-14.

Jones, B.W., K. Hsieh, and M. Hashinaga. 1986. The effect of air velocity on thermal comfort at moderate activity levels. *ASHRAE Transactions* 92:2.

Korsgaard, V. 1949. Necessity of using a directional mean radiant temperature to describe thermal conditions in rooms. *Heating, Piping and Air Conditioning* 21(6):117-20.

Kuno, S. 1995. Comfort and pleasantness. In *Proceedings of Pan Pacific Symposium on Building and Urban Environmental Conditioning in Asia*, 2:383-92. Architecture Department, University of Nagoya, Japan.

Langkilde, G. 1979. Thermal comfort for people of high age. In *Confort thermique: Aspects physiologiques et psychologiques*, INSERM, Paris 75:187-93.

Leithead, C.S. and A.R. Lind. 1964. *Heat stress and heat disorders*. Cassell & Co., London, 108. (F.A. Davis, Philadelphia, 1964).

Lipkin, M. and J.D. Hardy 1954. Measurement of some thermal properties of human tissues. *Journal of Applied Physiology* 7:212.

Liviana, J.E., F.H. Rohles, and O.D. Bullock. 1988. Humidity, comfort and contact lenses. *ASHRAE Transactions* 94(1):3-11.

Matthew, W.H. et al. 1986. Botsball (WGT) performance characteristics and their impact on the implementation of existing military hot weather doctrine. U.S. Army Reserves Institute of Environmental Medicine *Technical Report* T 9/86, April.

McCullough, E.A. 1986. An insulation data base for military clothing. Institute for Environmental Research *Report* 86-01, Kansas State University, Manhattan.

McCullough, E.A. and S. Hong. 1994. A data base for determining the decrease in clothing insulation due to body motion. *ASHRAE Transactions* 100(1):765.

McCullough, E.A. and B.W. Jones. 1984. A comprehensive data base for estimating clothing insulation. IER Technical *Report* 84-01, Institute for Environmental Research, Kansas State University, Manhattan. (Final report to *ASHRAE Research Project* 411-RP).

McCullough, E.A., B.W. Jones, and T. Tamura. 1989. A data base for determining the evaporative resistance of clothing. *ASHRAE Transactions* 95(2).

McCullough, E.A., B.W. Olesen, and S.W. Hong. 1994. Thermal insulation provided by chairs. *ASHRAE Transactions* 100(1):795-802.

McCutchan, J.W. and C.L. Taylor. 1951. Respiratory heat exchange with varying temperature and humidity of inspired air. *Journal of Applied Physiology* 4:121-35.

McIntyre, D.A. 1974. The thermal radiation field. *Building Science* 9:247-62.

McIntyre, D.A. 1976. Overhead radiation and comfort. *The Building Services Engineer* 44:226-32.

McIntyre, D.A. and I.D. Griffiths. 1975. The effects of uniform and asymmetric thermal radiation on comfort. *CLIMA 2000, 6th International Congress of Climatritics*, Milan.

McNall, P.E., Jr. and R.E. Biddison. 1970. Thermal and comfort sensations of sedentary persons exposed to asymmetric radiant fields. *ASHRAE Transactions* 76(1):123.

McNall, P.E., P.W. Ryan, and J. Jaax. 1968. Seasonal variation in comfort conditions for college-age persons in the Middle West. *ASHRAE Transactions* 74(1):IV.2.1-9.

McNair, H.P. 1973. A preliminary study of the subjective effects of vertical air temperature gradients. British Gas Corporation *Report* No. WH/T/R&D/73/94, London.

McNair, H.P. and D.S. Fishman. 1974. A further study of the subjective effects of vertical air temperature gradients. British Gas Corporation *Report* No. WH/T/R&D/73/94, London.

Minard, D. 1961. Prevention of heat casualties in Marine Corps recruits. *Military Medicine* 126:261.

Mitchell, D. 1974. Convective heat transfer in man and other animals. *Heat loss from animals and man*. Butterworth Publishing, London, 59.

Nevins, R., R.R. Gonzalez, Y. Nishi, and A.P. Gagge. 1975. Effect of changes in ambient temperature and level of humidity on comfort and thermal sensations. *ASHRAE Transactions* 81(2).

Nevins, R.G. and A.M. Feyerherm. 1967. Effect of floor surface temperature on comfort: Part IV, Cold floors. *ASHRAE Transactions* 73(2):III.2.1.

Nevins, R.G. and A.O. Flinner. 1958. Effect of heated-floor temperatures on comfort. *ASHRAE Transactions* 64:175.

Nevins, R.G., K.B. Michaels, and A.M. Feyerherm. 1964. The effect of floor surface temperature on comfort: Part 1, College age males; Part II, College age females. *ASHRAE Transactions* 70:29.

Nevins, R.G., F.H. Rohles, Jr., W.E. Springer, and A.M. Feyerherm. 1966. Temperature-humidity chart for thermal comfort of seated persons. *ASHRAE Transactions* 72(1):283.

Nicol, J.F. and M.A. Humphreys. 1972. Thermal comfort as part of a self-regulating system. Proceedings of CIB symposium on thermal comfort, Building Research Station, London.

NIOSH. 1986. Criteria for a recommended standard—Occupational exposure to hot environments, revised criteria. U.S. Dept. of Health and Human Services, USDHHS (NIOSH) *Publication* 86-113. Available from www.cdc.gov/NIOSH/86-113.html.

Nishi, Y. 1981. Measurement of thermal balance of man. In Bioengineering thermal physiology and comfort, K. Cena and J.A. Clark, Elsevier, eds. New York.

Nishi, Y. and A.P. Gagge. 1970. Direct evaluation of convective heat transfer coefficient by naphthalene sublimation. *Journal of Applied Physiology* 29:830.

Nishi, Y., R.R. Gonzalez, and A.P. Gagge. 1975. Direct measurement of clothing heat transfer properties during sensible and insensible heat exchange with thermal environment. *ASHRAE Transactions* 81(2):183.

Olesen, B.W. 1977a. Thermal comfort requirements for floors. In *Proceedings of Commissions B1, B2, E1 of the IIR*, Belgrade, 337-43.

Olesen, B.W. 1977b. Thermal comfort requirements for floors occupied by people with bare feet. *ASHRAE Transactions* 83(2).

Olesen, B.W. and R. Nielsen. 1983. Thermal insulation of clothing measured on a moveable manikin and on human subjects. Technical University of Denmark, Lyngby, Denmark.

Olesen, B.W., M. Scholer, and P.O. Fanger. 1979. Vertical air temperature differences and comfort. In *Indoor climate,* eds. P.O. Fanger and O. Valbjorn, 561-79. Danish Building Research Institute, Copenhagen.

Olesen, S., J.J. Bassing, and P.O. Fanger. 1972. Physiological comfort conditions at sixteen combinations of activity, clothing, air velocity and ambient temperature. *ASHRAE Transactions* 78(2):199.

Olesen, B.W., E. Sliwinska, T.L. Madsen, and P.O. Fanger. 1982. Effect of posture and activity on the thermal insulation of clothing. Measurement by a movable thermal manikin. *ASHRAE Transactions* 82(2):791-805.

Onkaram, B., L. Stroschein, and R.F. Goldman. 1980. Three instruments for assessment of WBGT and a comparison with WGT (Botsball). *American Industrial Hygiene Association* 41:634-41.

Oohori, T., L.G. Berglund, and A.P. Gagge. 1984. Comparison of current two-parameter indices of vapor permeation of clothing—As factors governing thermal equilibrium and human comfort. *ASHRAE Transactions* 90(2).

Ostberg, O. and A.G. McNicholl. 1973. The preferred thermal conditions for "morning" and "evening" types of subjects during day and night—Preliminary results. *Build International* 6(1):147-57.

Passmore, R. and J.V.G. Durnin. 1967. *Energy, work and leisure*. Heinemann Educational Books, London.

Rapp, G. and A.P. Gagge. 1967. Configuration factors and comfort design in radiant beam heating of man by high temperature infrared sources. *ASHRAE Transactions* 73(2):III.1.1.

Robinson, et al. 1943. *American Journal of Physiology* 140:168.

Rohles, F.H., Jr. 1973. The revised modal comfort envelope. *ASHRAE Transactions* 79(2):52.

Rohles, F.H., Jr. and M.A. Johnson. 1972. Thermal comfort in the elderly. *ASHRAE Transactions* 78(1):131.

Rohles, F.H., Jr. and R.G. Nevins. 1971. The nature of thermal comfort for sedentary man. *ASHRAE Transactions* 77(1):239.

Seppanen, O., P.E. McNall, D.M. Munson, and C.H. Sprague. 1972. Thermal insulating values for typical indoor clothing ensembles. *ASHRAE Transactions* 78(1):120-30.

Siple, P.A. and C.F. Passel. 1945. Measurements of dry atmospheric cooling in subfreezing temperatures. In *Proceedings of the American Philosophical Society* 89:177.

Stolwijk, J.A.J., A.P. Gagge, and B. Saltin. 1968. Physiological factors associated with sweating during exercise. *Journal of Aerospace Medicine* 39: 1101.

Stolwijk, J.A.J. and J.D. Hardy. 1966. Partitional calorimetric studies of response of man to thermal transients. *Journal of Applied Physiology* 21:967.

Sullivan, C.D. and R.L. Gorton. 1976. A method of calculating WBGT from environmental factors. *ASHRAE Transactions* 82(2):279.

Tanabe, S., K. Kimura, and T. Hara. 1987. Thermal comfort requirements during the summer season in Japan. *ASHRAE Transactions* 93(1):564-77.

Umbach, K.H. 1980. Measuring the physiological properties of textiles for clothing. *Melliand Textilberichte* (English edition) G1:543-48.

Webb, P. 1964. *Bioastronautics data base*, NASA.

Winslow, C.-E.A., L.P. Herrington, and A.P. Gagge. 1937. Relations between atmospheric conditions, physiological reactions and sensations of pleasantness. *American Journal of Hygiene* 26(1):103-15.

Witherspoon, J.M., R.F. Goldman, and J.R. Breckenridge. 1971. Heat transfer coefficients of humans in cold water. *Journal de Physiologie*, Paris, 63:459.

Woodcock, A.H. 1962. Moisture transfer in textile systems. *Textile Research Journal* 8:628-33.

BIBLIOGRAPHY

Fanger, P.O., A. Melikov, H. Hanzawa, and J. Ring. 1988. Air turbulence and sensation of draught. *Energy and Buildings* 12:21-39.

CHAPTER 9

INDOOR ENVIRONMENTAL HEALTH

INTRODUCTION

THIS CHAPTER INTRODUCES the field of environmental health as it pertains to the indoor environment of buildings, briefly covering problem identification and control and regulations. The role of HVAC designers in delivering clean, appropriately conditioned air and removing contaminants is vital, both in industrial and nonindustrial environments. Some knowledge of indoor environmental health will assist in that delivery. In many cases, architectural, structural, cleaning, maintenance, materials use, and other activities that affect the environment are outside the control of that individual. Nevertheless, whenever possible, the HVAC designer should encourage features and decisions that create a healthy building environment.

TERMINOLOGY

The most clearly defined area of indoor environmental health is occupational health, particularly as it pertains to workplace air contaminants. Poisoning incidents as well as human and animal laboratory studies have generated reasonable consensus on safe and unsafe workplace exposures to about one thousand chemicals and dusts. Consequently, many countries regulate exposures of workers. However, contaminant concentrations meeting occupational health criteria usually exceed levels found acceptable to occupants in nonindustrial spaces such as offices, schools, and residences, where exposures are often longer in duration and may involve mixtures of many contaminants and a less robust population (NAS 1981).

Operational definitions of health, disease, and discomfort are controversial (Cain et al. 1995). The World Health Organization (WHO) defines health as "a state of complete physical, mental, and social well-being and not merely the absence of disease or disability." Last (1983) defines health as "a state characterized by anatomic integrity, ability to perform personally valued family, work, and social roles; ability to deal with physical, biologic, and social stress; a feeling of well-being; and freedom from the risk of disease and untimely death." Higgins (1983) defines an adverse health effect as a biological change that reduces the level of well-being or functional capacity. These definitions indicate that good health is a function of freedom from active ill health or disease (i.e., short- and long-term disability or impairment, freedom from risk of disease in the future resulting from current exposures, and current subjective well-being).

The preparation of this chapter is assigned to the Environmental Health Committee.

Definitions of comfort are also varying. Traditionally comfort refers to immediate satisfaction. It encompasses perception of the environment (hot, cold, noisy, etc.) and a value rating of affective implications (too hot, too cold, etc.). Rohles et al. (1989) noted that acceptability may represent a more useful concept as it allows progression toward a concrete goal. This serves as the foundation of a number of standards including thermal comfort, acoustics, etc. Nevertheless, acceptability may change over time (secular drift) as expectations change.

SCOPE

This chapter covers indoor environmental health issues such as effects of air contaminants, thermal comfort, acoustics, exposure to radiation, ergonomics, and electrical safety. It presents information on the mechanisms by which the environment affects the human body, gives guidance on recognizing problems, and discusses standards and guidelines that provide protection. Since the number of air contaminants is very large, the introductory material on types, their characteristics, typical levels, and measurement methods is given in Chapter 12. Chapter 13 covers odors.

This chapter also covers indoor air quality (IAQ). The field of indoor air quality is distinct from that of occupational health in several ways, including locations of concern, which are mainly nonindustrial and include offices, schools, public and commercial buildings, and residences; existence of both comfort and health issues; presence of multiple contaminants at low concentrations rather than single ones at high concentration; and impacts of noncontaminant-related issues such as thermal comfort.

More is known about industrial exposures to air contaminants than is known about residential and light commercial exposures. Therefore, most of this chapter focusses on industrial environments.

DESCRIPTIONS OF SELECTED HEALTH SCIENCES

Normal interactions of the human body with its surrounding environment occur in predictable fashions. Light, heat, cold, and sound at extremes of the exposure range can result in such diseases as frostbite, burns, and noise-induced hearing loss. Some transitions between normal and disease states are more difficult to delineate. Pain from bright light, erythema from heat, and nausea from vibration represent reversible effects but are interpreted by health professionals as abnormal. Study of these effects includes a number of scientific disciplines.

EPIDEMIOLOGY AND BIOSTATISTICS

Epidemiology is the study of distributions and determinants of disease. It represents the application of quantitative methods to evaluate diseases or conditions of interest. The subjects may be humans, animals, or even buildings. Epidemiology is traditionally subdivided into observational and analytical components. It may be primarily descriptive, or it may attempt to identify causal associations. Some classical criteria for determining causal relationships in epidemiology are consistency, temporality, plausibility, specificity, strength of the association, and dose-response relationships.

Observational studies are generally performed by defining some group of interest because of a specific exposure or risk factor. A control group is selected on the basis of similar criteria, but without the factor of interest. Observations conducted at one point in time are considered cross-sectional studies. On the other hand, a group may be defined by some criteria at a specific time, for example, all employees who worked in a certain building for at least one month in 1982. They may then be followed over time, leading to an observational cohort study.

Analytical studies may be either experimental or case-control studies. In experimental studies, individuals are selectively exposed to specific agents or conditions. Such studies are generally performed with the consent of the participants, unless the conditions are part of their usual working conditions and known to be harmless. Sometimes exposures cannot be controlled on an individual basis, and the intervention must be applied to entire groups. Control groups must be observed in parallel. Case-control studies are conducted by identifying individuals with the condition of interest and comparing risk factors between these people and individuals without that condition.

All factors of interest must be measured in an unbiased fashion to avoid a subconscious influence of the investigator. The method of measurement should be repeatable and the technique meaningful. Statistical methods for data analysis follow standard procedures. Tests of hypotheses are performed at a specific probability level. They must have adequate power; that is, if a sample size or a measurement difference between the factors or groups is too small, a statistically significant association may not be found even if one is present.

Results obtained in a specific situation (i.e., in a sample of exposed individuals) may be generalized to others only if they share the same characteristics. For example, it may not be legitimate to assume that all individuals have the same tolerance of thermal conditions irrespective of their heritage. Therefore, the results of studies and groups must be evaluated as they apply to a specific situation.

TOXICOLOGY

Toxicology is the study of the influence of poisons on health. Most researchers hold that essentially all substances may function as poisons and that only smaller doses prevent them from becoming harmful. Of fundamental importance is defining which component of the chemical structure predicts the harmful effect. The second issue is defining the dose-response relationships. The definition of dose may refer to delivered dose (exposure that is presented to the lungs) or absorbed dose (the dose that is actually absorbed through the lungs into the body and available for metabolism). Measures of exposure may be quite distinct from measures of effect because of internal dose modifiers (e.g., the delayed metabolism of some poisons because of lack of enzymes to degrade them). In addition, the mathematical characteristics of a dose may vary, depending on whether a peak dose, a geometric or arithmetic mean dose, or an integral under the dose curve is used.

Because humans often cannot be exposed in experimental conditions, most toxicological literature is based on animal studies. Recent studies suggest that it is not easy to extrapolate between dose level effects from animals to man. Isolated animal systems, such as homogenized rat livers, purified enzyme systems, or other isolated living tissues, may be used to study the impact of chemicals.

MOLECULAR BIOLOGY

Molecular biology is the branch of science that studies the chemical and physical structure of biological macromolecules. (DNA, proteins, carbohydrates, etc.). It is interested in processes on a molecular level, identifying actual mechanisms of effect or toxicity, rather on the level of cells.

CELLULAR BIOLOGY

Cellular biology is the branch of science that studies cellular organelles, activities, and processes. Little indoor air quality related research has been done at this mechanistic level, but it offers the final evidence in postulated cause-and-effect relationships.

GENETICS

Genetics is a branch of science that examines heredity and variation among organisms at population, individual, and chromosomal levels. Newer studies in genetics appear to indicate that some individuals are more susceptible to or are at greater risk from certain exposures than the rest of the population. This susceptibility would explain why not all individuals react the same way to the same exposure or lifestyle.

INDUSTRIAL HYGIENE

Industrial hygiene is the science of anticipating, recognizing, evaluating, and controlling workplace conditions that may cause worker illness or injury. Important aspects of industrial hygiene included under these principles are (1) identification of toxic chemicals; (2) evaluation of the importance of the physical state and size of airborne particle to absorption by the lungs; (3) evaluation of the importance of airborne particle size to absorption by the lungs and the physical state of individuals; (4) evaluation of the importance of skin absorption and ingestion to exposure and absorption; (5) identification of chemicals to be collected and analyzed; (6) determination of methods for collection of air samples; (7) identification of analytic methods to be used or collaboration with a chemist to develop methods to be used; (8) evaluation of results of measurements; (9) identification of physical stressors, including noise, heat stress, ionizing radiation, nonionizing radiation, ergonomics, and illumination; and (10) development of control measures. In addition to examining the environment, interpreting collected data, and implementing control measures, the industrial hygienist has responsibilities related to creating regulatory standards for the work environment, preparing programs to comply with regulations, and collaborating in epidemiologic studies to document exposures and potential exposures to help determine occupation-related illness.

Hazard Recognition

Occupational hazards generally fall into one of four classes of environmental stressors: chemical, biological, physical, and ergonomic.

Chemical Hazards. Airborne chemical hazards exist as concentrations of mists, vapors, gases, fumes, or solids. Some are toxic through inhalation, some can irritate the skin on contact, some can be toxic by absorption through the skin or through ingestion, and some are corrosive to living tissue. The degree of risk from exposure to any given substance depends on the nature and potency of the toxic effects and the magnitude and duration of exposure. Air contaminants represent a very important subclass of chemical hazards due to their mobility and the ease of exposure through the lungs when they are inhaled. Air contaminants are commonly classified as either particulate or gaseous. Common particulate contam-

Table 1 Diseases Related to Buildings

Disease	History and Physical Examination	Laboratory Testing	Linkage	Causes
Rhinitis Sinusitis	Stuffy/ runny nose, post-nasal drip, pale or erythematous mucosa	Anterior and posterior rhino-manometry, acoustic rhino-metry, nasal lavage, biopsy, rhinoscopy, RAST or skin prick testing	Immunologic skin prick or RAST testing, bracketed physiology	Direct occupational exposures; molds in the workplace; specific occupational factors (laser toners, carbonless copy paper, cleaning agents), secondary occupational exposures; pet (e.g., cat) danders brought from home
Asthma	Coughing, wheezing, episodic dyspnea, wheezing on examination, chest tightness, temporal pattern at work	Spirometry before and after work on Monday, peak expiratory flow diary, methacholine challenge	Immunologic: skin prick or RAST testing Physiologic: related to work*	See rhinitis/sinusitis
Hypersensitivity pneumonitis	Cough, dyspnea, myalgia, weakness, rales, clubbing, feverishness	DLCO, FVC, TLC, CXR, lung biopsy	Immunologic: IgG ab to agents present, challenge testing Physiologic (in acute forms): spirometry, DLCO	Molds, moisture
Organic dust toxic syndrome	Cough, dyspnea, chest tightness, feverishness	DLCO, TLC	Temporal pattern related to work	Gram-negative bacteria
Contact dermatitis (allergic)	Dry skin, itching, scaling skin	Scaling rash, eczema biopsy	Patch testing	Molds, carbonless copy paper, laser toners
Contact urticaria	Hives	Inspection biopsy	Provocative testing	Office products (carbonless copy paper)
Eye irritation	Eye itching, irritation, dryness	Tear-film break-up time, conjunctival staining (fluorescein)	Temporal pattern	Low relative humidity, volatile organic compounds (allergic conjunctivitis), particulates
Nasal irritation	Stuffy, congested nose, rhinitis	Acoustic rhinometry, posterior and anterior rhinomanometry, nasal lavage, nasal biopsy	Temporal pattern	Low relative humidity, volatile organic compounds (allergic conjunctivitis), particulates
Central nervous system symptoms	Headache, fatigue, irritability, difficulty concentrating	Neuropsychological testing	Temporal pattern (epidemiology)	Volatile organic compounds, noise, lighting, work stress, carbon monoxide, cytokines from bioaerosol exposure
Legionnaires' disease	Pneumonic illness	History, *Legionella* culture from biopsy fluids	1) Organism isolated from patient and source, 2) immunologic watch	Aerosols from contaminated water sources, shower heads, water faucet aerators, humidifiers at home and at work, potable water sources (hot water heaters, etc.)

* (1) 10% decrement in FEV_1 across workday,
(2) peak flow changes suggestive of work relatedness, and
(3) methacholine reactivity resolving after six weeks away from exposure
RAST = radio allergen sorbent test
DLCO = single breath carbon monoxide diffusing capacity
FVC = forced vital capacity
TLC = total lung capacity
CXR = chest X-ray
IgG = class G immune globulins
FEV_1 = forced expiratory volume in the first second

inants include dusts, fumes, mists, aerosols, and fibers. Gaseous contaminants exist as vapors or gases. More background information on these is given in Chapter 12. Air contaminants are of concern in both industrial and nonindustrial environments. Table 1 summarizes diseases that have been associated with specific aspects of indoor environments, mostly in nonindustrial environments. For these diseases, diagnostic criteria may be used to distinguish between presence or absence of disease. These diseases come about because of the presence of an exposure, a susceptible host, and a vector of transmission.

Biological Hazards. These include bacteria, viruses, fungi, and other living or nonliving organisms that can cause acute and chronic infections by entering the body either directly or through breaks in the skin. In addition, some biological agents can cause allergic reactions or be toxic. Wastes, body parts, etc., of these organisms can also cause illness and allergic reactions.

Physical Hazards. These include excessive levels of ionizing and nonionizing electromagnetic radiation, noise, vibration, illumination, temperature, and force.

Ergonomic Hazards. These include tasks that involve repetitive motions, require excessive force, or must be carried out in awkward postures, all of which can result in damage to muscles, nerves, and joints.

Hazard Evaluation

Hazard evaluation determines the sources of potential problems. Safety and health professionals research, inspect, and analyze how the particular hazard affects worker health. Assessment of such exposures relies on qualitative, semiquantitative, or quantitative approaches. In many situations, air sampling will determine whether a hazardous condition exists. An appropriate sampling strategy must be used to ensure the validity of collected samples, determining worst-case (for compliance) or usual (average) exposures. Air sampling can be conducted to determine **time-weighted average** (TWA) exposures, which would cover an entire work shift; or **short-term exposures**, which would determine the magnitude of exposures to materials that are acutely hazardous. Samples may be collected for a single substance or a multicomponent mixture. Hazard evaluation also characterizes the workplace with respect to potential skin absorption or ingestion hazards. Analysis of bulk material samples and surface wipe samples could determine whether hazardous conditions exist. Physical agent characterization may require direct-reading sampling methods. After collection and analysis, the industrial hygienist must interpret the results and determine appropriate control strategies.

Hazard Control

The principles for controlling the occupational environment are substitution, isolation, ventilation, and air cleaning. Not all of these principles may be applied to all types of hazards, but all hazards can be controlled by using one of these principles. Engineering controls, work practices controls, administrative controls, and personal protective equipment are used to apply these principles. Source removal or substitution are customarily the most effective measure, but are not always feasible. Engineering controls such as ventilation and air cleaning may be effective for a range of hazards, but usually consume energy. Local exhaust ventilation is more effective for controlling point-source contaminants than is general ventilation. A building HVAC system is an example of a general ventilation system.

AIR CONTAMINANTS

Many of the same air contaminants cause problems in industrial and nonindustrial indoor environments. These include nonbiological particles (synthetic vitreous fibers, combustion nuclei, nuisance dust, and others); bioaerosols; and gases and vapors that may be generated due to industrial processes, by building materials, furnishings, and equipment, by occupants and their activities in a space, or brought in from the outdoors. In industrial environments, contaminants are usually known from the type of process, and exposures may be determined relatively easily by air sampling. Airborne contaminants in nonindustrial environments may (1) occur from emissions and/or shedding of building materials and systems, (2) originate in outside air, and/or (3) be from building operating and maintenance programs and procedures and conditions that may foster growth of biological organisms. In general, in nonindustrial environments, there are many more contaminants that may contribute to problems, they are more difficult to identify, and they are usually present in much smaller concentrations. More information on contaminant types, characteristics, typical levels, and measurement methods can be found in Chapter 12.

PARTICULATE MATTER

Particulate matter includes airborne solid or liquid particles. Typical examples of particulates include dust, smoke, fumes, and mists. Dusts range in size from 0.1 to 25 μm, smoke particulate is typically around 0.25 μm, and fumes are usually less than 0.1 μm in diameter (Zenz 1988). Bioaerosols and particles produced from abrasions are usually smaller than 1.0 μm.

Units of Measurement

The quantity of particulate matter in the air is frequently reported as the mass or the particle count in a given volume of air. Mass units are milligrams per cubic metre of air sampled (mg/m^3) or micrograms per cubic metre of air sampled (μg/m^3). 1 mg /m^3 = 1000 μg/m^3. Mass units are widely used in industrial environments as these units are used to express occupational exposure limits.

Particle counts are usually quoted for volumes of 1 cubic foot, 1 litre, or 1 cubic metre and are specified for a given range of particle diameter. Particle count measurements are useful in less contaminated environments such as office buildings and industrial clean rooms.

Particles in Industrial Environments

Particulates found in the work environment are generated as a result of work-related activities (i.e., adding batch ingredients for a manufacturing process, applying asphalt in a roofing operation, or drilling an ore deposit in preparation for blasting). The engineer must recognize sources of particulate generation in order to appropriately address exposure concerns.

The engineer or industrial hygienist determining worker exposure should assess particulate release by the activity, local air movement caused by makeup air and exhaust, and worker procedures for a complete evaluation (Burton 2000).

Health Effects of Exposure. The health effects of airborne particulates depend on several factors that include particle dimension, durability, dose, and toxicity of the materials in the particle. A particulate must first be inhalable to be potentially hazardous. Respirable particulates vary in size from < 1 to 10 μm (Alpaugh and Hogan 1988) depending on the source of the particulate. Methods for measuring particulates are discussed in Chapter 12. Durability, or how long the particle can live in the biological system before it dissolves, can determine relative toxicity. Lastly, the dose or the amount of exposure encountered by the worker must be considered. In some instances, very small exposures can cause adverse health effects (hazardous exposures) and in others, seemingly large exposures may not cause any adverse health effects (nuisance exposures).

Safety and health professionals are primarily concerned with particles smaller than 2 μm, since this is the range of sizes most likely to be retained in the lungs (Morrow 1964). Particles larger than 8 to 10 μm in aerodynamic diameter are primarily separated and retained by the upper respiratory tract. Intermediate sizes are deposited mainly in the conducting airways of the lungs, from which they are rapidly cleared and swallowed or coughed out. About 50% or less of the particles in inhaled air settle in the respiratory tract. Submicron particles penetrate deeper into the lungs, but many do not deposit and are exhaled.

Dusts

Dusts are coarse solid particles generated by handling, crushing, or grinding. Their size range is typically 1 to 100 μm. They may become airborne during the generation process or through handling. Any industrial process that produces dust fine enough to remain in the air long enough to be inhaled or ingested (size about 10 μm) should be regarded as hazardous until proven otherwise.

Fibers are solid particles whose length is several times greater than their diameter, such as asbestos, man-made mineral fibers, synthetic vitreous fibers, and refractory ceramic fibers.

Mechanisms of Health Effects. A disease associated with the inhalation of particulates in industrial settings is pneumoconiosis, a fibrous hardening of the lungs caused by the irritation created from the inhalation of dust. The most commonly known pneumoconioses are asbestosis, silicosis, and coal worker's pneumoconiosis.

Asbestosis results from the inhalation of asbestos fibers (chrysotile, crocidolite, amosite, actinolite, anthophyllite, and tremolite) found in the work environment. The onset of symptomatic asbestosis is uncommon under exposures encountered in the last 45 years before at least 20 to 30 years of exposure (Selikoff et al. 1965, Smith 1955). The asbestos fibers cause fibrosis (scarring) of the lung tissue, which clinically manifests itself as dyspnea (shortness of breath) and a nonproductive, irritating cough. Asbestos fiber is both dimensionally respirable as well as durable in the respiratory system.

Silicosis is probably the most common of all industrial occupational diseases of the lung. The hazard is created by inhalation of silica dust. The worker with silicosis usually is asymptomatic, and even the early stages of massive fibrosis are not associated with signs and symptoms (Leathart 1972). It is not considered a problem in nonindustrial indoor environments.

Coal worker's pneumoconiosis (CWP) results from the inhalation of dust generated in coal mining operations. The dust is composed of a combination of carbon and varying percentages of silica (usually <10%) (Alpaugh and Hogan 1988). Due to the confined underground work environment, exposures have the potential to be

Table 2 OSHA Permissible Exposure Limits (PELs) for Particulates (29 CFR 1910.1000, 29 CFR 1926.1101)

Substance	CAS #	PEL
Cadmium	7440-43-9	0.05 mg/m^3
Manganese fume	7439-96-5	1.0 mg/m^3
Plaster of Paris	Nuisance	10.0 mg/m^3
Emery	Nuisance	10.0 mg/m^3
Grain dust	Nuisance	10.0 mg/m^3
Crystalline silica (as quartz)	14808-60-7	0.1 mg/m^3
Asbestos	1332-21-4	0.1 fibers/cm^3
Total dust	Nuisance	15.0 mg/m^3
Respirable dust	Nuisance	5.0 mg/m^3

very high at times, thus creating very high doses. Data meanwhile show that workers may develop CWP at exposure below the current dust standard of 1 mg/m^3.

Exposure Control Strategies. Particulate or dust control strategies include source elimination or enclosure, local exhaust, general ventilation, wetting, filtration, and the use of personal protective devices such as respirators.

The most effective means of controlling exposures to a particulate is to totally eliminate it from the work environment. The best dust-control method is a total enclosure of the dust-producing process. A negative pressure is maintained inside the entire enclosure by exhaust ventilation (Alpaugh and Hogan 1988). This control strategy is typically found in manufacturing operations.

Local exhaust ventilation as an exposure control strategy is most frequently used where particulate is generated either at large volumes or with high velocities (i.e., lathe and grinding operations). In this situation, high-velocity air movement captures the particulate and removes it from the work environment. A number of recent studies show that push/pull methodology enhances the capture efficiency, but requires care in not "pushing" contaminants into the work environment. Counter airflow situations in source capture applications should be avoided.

General ventilation control of the work environment is defined as a dilution approach to reducing exposures. This type of ventilation is used when particulate sources are numerous and widely distributed over a large area. In this control strategy, the work environment is exhausted outside and resupplied with fresh air, thus diluting the work environment. Unfortunately this strategy is the least effective means of control and very costly because conditioned (warm or cold) air is exhausted and nonconditioned air is introduced.

Recirculation of indoor air through filters can be an effective method of reducing indoor particle concentrations. Filtration costs are often lower then costs of general ventilation.

The least desirable strategy used to control exposures is the use of personal protective equipment—a respirator. Respirators are appropriate as a primary control during intermittent maintenance or cleaning activities when fixed engineering (local or general ventilation) controls may not be feasible. Respirators can also be used as a supplement to good engineering and work practice controls to increase employee protection and comfort (Alpaugh and Hogan 1988).

Exposure Standards and Criteria. In the United States, the Occupational Safety and Health Administration (OSHA) has established Permissible Exposure Limits (PELs), which are published in the Code of Federal Regulations (CFR 1989a,b) under the authority of the Department of Labor. Table 2 lists PELs for several particulates commonly encountered in the workplace.

Synthetic Vitreous Fibers

Exposures and Exposure Sources. A fiber can be defined as a slender, elongated structure with substantially parallel sides. These parameters distinguish this form of particulate from a dust, which is more spherical. Synthetic vitreous fibers (SVFs) comprise a large number of important manufactured products, such as textile fibers; insulation and ceiling tile wool, including glass fibers, slag, and rock wool fibers; refractory ceramic fibers; and certain specialty glass fibers.

Exposures to SVFs primarily occur during manufacture, fabrication and installation, and demolition. Simultaneous exposures to other dusts (asbestos during manufacture, demolition products and bioaerosols during demolition) may be important as well. Facilities generally manufacture only one form. Generally, only spun glass and refractory ceramic fibers are in the respirable range. Manufacturing operations are most easily designed to assure a clean work environment, while product application operations are more difficult to control. Data on exposure likely to occur in buildings show that background levels are almost uniformly below 0.0001 fibers/cm^3.

Health Effects of Exposure. The possible effects of SVFs on health include the following:

Cancer. Respirable SVFs are considered to have the potential to cause carcinogenic and noncarcinogenic health effects. Although implantation studies have suggested the potential for carcinogenesis, this route of exposure is generally not pertinent for humans. Therefore, although SVFs are often classified as potential human carcinogens by regulatory and professional agencies and organizations, reviews of epidemiology studies generally fail to find convincing evidence that they are associated with excess rates of human cancer. Some mortality studies have identified mild excesses of respiratory cancer. These have been attributed to concurrent asbestos exposure and to smoking. Only refractory ceramic fibers are currently considered likely to represent true human carcinogens, although other very hard fibers are likely to have similar effects.

Nonmalignant respiratory disease. Cross-sectional surveys have suggested that few measurable adverse health effects are attributable to SVFs alone. The strongest evidence suggests that SVFs may exacerbate smoking-induced obstructive lung disease; some authors consider fiberglass, no different than any other dust, to cause excess rates of chest symptoms.

Dermatitis. SVFs may cause an irritant contact dermatitis through embedding in the skin or conjunctivae with local inflammation. Resin binders sometimes used to tie fibers together have, on rare occasions, been associated with allergic contact dermatitis.

Exposure Control Strategies. As with other particulates, SVF control strategies include source exclusion or enclosure, local exhaust, and the use of personal protective devices such as respirators. In indoor environments, SVFs may be identified in surface wipe samples. Appropriate intervention strategies focus on source control.

Exposure Standards and Criteria. At present, SVFs are regulated by OSHA as a "nuisance dust" with an 8-hour time-weighted average of 15 mg/m^3 for total dust and 5 mg/m^3 for respirable dust.

Combustion Nuclei

Exposures and Exposure Sources. Combustion nuclei can be defined as the particulate products of the combustion process. Combustion products from a material include water vapor, carbon dioxide, heat, oxides of carbon and nitrogen, and particulates known as combustion nuclei. In many situations combustion nuclei can be hazardous. They may contain potential carcinogens such as polycyclic aromatic hydrocarbons (PAHs).

Polycyclic aromatic compounds (PACs) are the nitrogen-, sulfur-, and oxygen-heterocyclic analogs of PAH and other related PAH derivatives. Depending on their relative molecular mass and vapor pressure, PACs are distributed between vapor and particulate phases. In general, combustion particulates are smaller than dusts generated by mechanical means.

Typical sources of combustion nuclei are tobacco smoke (cigarettes, pipes, and cigars), fossil-fuel-based heating devices such as unvented space heaters and gas ranges, and flue gas from improperly vented gas- or oil-fired furnaces and wood-burning fireplaces or stoves. Infiltration of outdoor combustion contaminants can also be a significant source of such contaminants in indoor air. Combustion nuclei are thus important in both industrial and nonindustrial settings.

Exposure Standards and Criteria. OSHA has established exposure limits for several of the carcinogens categorized as combustion nuclei (benzo(a)pyrene, cadmium, nickel, benzene, *n*-nitrosodimethylamine). These limits are established for industrial work environments and are not directly applicable to indoor air situations. Underlying atherosclerotic heart disease in individuals may be exacerbated by carbon monoxide (CO) exposures.

Exposure Control Strategies. Exposure control strategies for combustion nuclei are in many ways similar to those applied for other particles. For combustion nuclei derived from heating spaces, air contamination can be avoided by proper installation and ventilation of equipment to ensure that these contaminants cannot enter the work or personal environment. Proper equipment maintenance is also essential to minimize exposures to combustion nuclei. Changing makeup air availability, through the addition of enclosures, may be equally important.

Particles in Nonindustrial Environments

In the nonindustrial indoor environment, the indoor aerosol will be affected greatly by the outdoor particle environment. Indoor particulate sources may include cleaning, resuspension of particles from carpets and other surfaces, construction and renovation debris, paper dust, deteriorated insulation, office equipment, and combustion processes including cooking stoves and fires and environmental tobacco smoke. In general, source control is the preferred method. If a dust problem is identified, characterization of the nature of the dust will allow the development of an appropriate intervention strategy.

Although asbestos is encountered in insulation in many buildings, it generally does not represent a respiratory hazard except to individuals who actively disturb it in the course of maintenance and construction. School custodians, therefore, are recognized to be at risk for asbestos-related changes. Anderson et al. (1991) and Lilienfeld (1991) raise questions about risk to teachers.

The combustion nuclei of environmental tobacco smoke (ETS) consists of exhaled mainstream smoke from the smoker and sidestream smoke that is emitted from the smoldering tobacco. ETS consists of between 70 and 90% sidestream smoke and has a somewhat different chemical composition from mainstream smoke. More than 4700 compounds have been identified in laboratory-based studies, including known human toxic and carcinogenic compounds such as carbon monoxide, ammonia, formaldehyde, nicotine, tobacco-specific nitrosamines, benzo(a)pyrene, benzene, cadmium, nickel, and aromatic amines. Many of these toxic constituents are more concentrated in sidestream than in mainstream smoke (Glantz and Parmley 1991). In studies conducted in residences and office buildings with tobacco smoking, ETS was a substantial source of many gas and particulate PACs (Offermann et al. 1991).

Health Effects of Exposure. The health effects of exposure to combustion nuclei depend on many factors, including concentration, toxicity, and individual susceptibility or sensitivity to the particular substance. Polycyclic aromatic compounds generated by combustion processes include many PAHs and nitro-PAHs that have been shown to be carcinogenic in animals (NAS 1983). Other PACs are biologically active as tumor promoters and/or cocarcinogens. Mumford et al. (1987) reported high exposures to PAH and aza-arenes for a population in China with very high lung cancer rates.

ETS has been shown to be causally associated with lung cancer in adults (NRC 1986, DHHS 1986), and respiratory infections, asthma exacerbations, middle ear effusion (NRC 1986, DHHS 1986), and low birth weight in children (Martin and Bracken 1986). The U.S. Environmental Protection Agency classifies ETS as a known human carcinogen (EPA 1992). Health effects can also include headache and irritation. ETS is also a cause of sensory irritation and annoyance (odors and eye irritation).

Control of ETS is somewhat different in that it has been done primarily through regulatory mandates controlling the practice of tobacco smoking. Most states in the United States have passed laws to control tobacco smoking in public places such as restaurants and workplaces, and airlines have prohibited tobacco smoking on flights lasting 6 h or less. Where tobacco smoking is allowed, appropriate local and general ventilation can be used for control. OSHA has proposed that tobacco smoke in indoor environments be controlled through the use of separately ventilated and exhausted smoking lounges. These lounges are kept under negative pressure relative to all adjacent and communicating indoor spaces.

BIOAEROSOLS

Aerobiology is the study of airborne microorganisms or other biologically produced particles and the effects of these aerosols on other living organisms (people, animals, vegetation, etc.). Bioaerosols are airborne microbiological particulate matter derived from viruses, bacteria, fungi, protozoa, algae, mites, plants, insects, and their cellular or cell mass components. Bioaerosols are present in both indoor and outdoor environments. For the indoor environment, locations that provide appropriate temperature and humidity conditions for reproduction and a food source to support growth may become problematic.

Sources

Floors and floor coverings in hospitals can be reservoirs for organisms that are subsequently resuspended into the air. Carpet cleaning may even promote resuspension (Cox 1987). Some viruses may persist up to 8 weeks on nonporous surfaces (Mbithi et al. 1991).

Nonpotable water is a well-known source of infective agents, even by aerosolization. Baylor et al. (1977) demonstrated the sequestering of small particles by foam and their subsequent dispersal through a bubble burst phenomenon. Such dispersal may take place in surf, river sprays, or man-made sources such as whirlpools.

People are an important source of bacteria and viruses in indoor air. Contagious diseases occur when living organisms overcome the defense of the host and establish an infection in the host that may in turn infect another human. Infected humans are the primary sources for contagious disease and primary disseminators as well. Virulent agents can also be released from human skin when disease produces skin lesions, or dispersed from respiratory tract infection during coughing, sneezing, or talking. Other means for release directly from infected humans include sprays of saliva and other respiratory secretions during dental and respiratory therapy procedures. Blood sprays that occur during dental and surgical procedures are of potential concern for aerosol transmission of bloodborne diseases, including HIV and hepatitis viruses. Large droplets can transmit infectious particles to those close to the disseminator, while smaller particles can remain airborne for short or very long distances (Moser et al. 1979).

Both the physical and biological properties of the bioaerosols need to be understood. For a microorganism to cause a building-related illness, it must be transported in sufficient dose to the breathing zone of a susceptible occupant. Airborne infectious particles behave physically in the same way as any other aerosol-containing particles with similar size, density, electrostatic charge, etc. The major difference with bioaerosols is that they must remain viable to cause infection, although nonviable particles may promote an immunological response. An organism that does not remain virulent in the airborne state cannot cause infection, regardless of how many units of organisms are deposited in the human respiratory tract. Virulence depends on such factors as relative humidity, temperature, oxygen, pollutants, ozone, and ultraviolet light (Burge 1995). The effect of any one factor on survival and virulence can be different for different organisms.

Although microorganisms are normally present in indoor environments, the presence of abundant moisture and nutrients in interior niches amplifies the growth of some microbial agents. Thus certain types of humidifiers, water spray systems, and wet porous surfaces can be reservoirs and sites for growth of fungi, bacteria, protozoa, algae, or even nematodes (Strindehag et al. 1988, Arnow et al. 1978, Morey et al. 1986, Morey and Jenkins 1989). Excessive air moisture (Burge 1995) and floods (Hodgson et al. 1985) may cause the proliferation of microorganisms indoors. The turbulence associated with the start-up of air-handling unit plenums may also elevate concentrations of bacteria and fungi in occupied spaces (Yoshizawa et al. 1987).

Health Effects

Exposure to airborne fungal spores, hyphal fragments, or metabolites can cause a variety of respiratory diseases. These range from allergic diseases including allergic rhinitis, asthma, and hypersensitivity pneumonitis to infectious diseases such as histoplasmosis, blastomycosis, and aspergillosis. In addition, acute toxicosis and cancer have been ascribed to respiratory exposure to mycotoxins (Levetin 1995). A large body of literature supports an association between moisture indicators in the home and symptoms of coughing and wheezing (Spengler et al. 1992, Miller and Day 1997).

The presence of microorganisms in indoor environments may cause infective and/or allergic building-related illnesses (Morey and Feeley 1988, Burge 1989). Some microorganisms under certain conditions may produce volatile chemicals (Hyppel 1984) that are malodorous.

The diseases produced by the *Legionella* genus of bacteria are collectively called legionelloses. Presently more than 34 species of the *Legionella* family have been identified, of which over 20 have been isolated from both environmental and clinical sources. The diseases produced by *Legionella pneumophila* include the pneumonia form, Legionnaires' disease, and the flu-like form, Pontiac fever. *L. pneumophila* serogroup 1 is the most frequently isolated from nature and most frequently associated with disease. It has also been suggested that the host relationship affects the virulence of *Legionella spp*.

The fungal genus *Aspergillus* is widely distributed and is common in the soil and on decaying vegetation, dust, and other organic debris (Levitin 1995). The small spores are buoyant and remain airborne for long periods (Streifel et al. 1989). Most opportunistic fungal infections are caused by *Aspergillus fumigatus*. The literature on aspergillosis is extensive, particularly for hospitals, and in many cases the environmental source of the infection has been identified.

Histoplasmosis, an infective illness caused by the fungus *Histoplasma capsulatum*, has occurred (rarely) as a building-related illness among individuals involved in the removal of bat or bird droppings in abandoned buildings (Bartlett et al. 1982) and among chicken coop cleaners. Presumably asexual spores (conidia) from this fungus were inhaled by workers who removed the droppings without adequate respiratory protection.

Outbreaks of infective illness in the indoor air may be caused by other types of microorganisms, such as viruses. For example, most of the passengers in an airline cabin developed influenza following exposure to one acutely ill person (Moser et al. 1979). In this case, the plane had been parked on a runway for several hours with the ventilation system turned off.

Allergic respiratory illness may develop due to inhalation of particulates containing microorganisms or their components, such as spores, enzymes, mite excreta, and cell wall fragments. Numerous cases of allergic respiratory illness (humidifier fever, hypersensitivity pneumonitis) report affected people manifesting acute symptoms such as malaise, fever, chills, shortness of breath, and coughing (Edwards 1980, Morey 1988). In buildings, these illnesses may occur as a response to microbiological contaminants originating from HVAC system components, such as humidifiers and water spray systems, or other mechanical components that have been damaged by chronic water exposure (Hodgson et al. 1985, 1987). Affected individuals usually experience relief only after having left the building for an extended period in contrast to occupants with sick building syndrome, where relief is relatively rapid.

Crandall and Sieber (1996) showed that 47 of 104 problem buildings evaluated had water damage in occupied building areas. Other studies suggest that microorganisms in indoor air are important (Burge et al. 1987, Brundage et al. 1988, Burge 1995).

Guidelines

At present, numerical guidelines for bioaerosol exposure in indoor environments are not available for the following reasons (Morey 1990):

- Incomplete data on concentrations and types of microbial particulates indoors, especially as affected by geographical, seasonal, and type-of-building parameters
- Absence of data relating bioaerosol exposure to building-related illness
- Enormous variability in kinds of microbial particulates including viable cells, dead spores, toxins, antigens, and viruses
- Large variation in human susceptibility to microbial particulates, making estimates of health risk difficult

However, even in the absence of numerical guidelines, bioaerosol sampling data can be interpreted based on such factors as

- Rank order assessment of the kinds (genera/species) of microbial agents present in complainant and control locations (ACGIH 1989)
- Medical or laboratory evidence that a building-related illness is caused by a microorganism (ACGIH 1989)
- Indoor/outdoor concentration ratios for various microbial agents (Morey and Jenkins 1989, ACGIH 1989)

For a microorganism to cause a building-related illness, it must be transported in sufficient dose to the breathing zone of a susceptible occupant. Thus, the concepts of reservoir, amplifier, and disseminator need to be considered in interpreting data. Reservoirs allow microorganisms to survive, amplifiers allow microorganisms to proliferate, and disseminators effectively distribute bioaerosols. Some factors and systems may be all or only one of these. A cooling tower is all three for *Legionella*; that is, a cooling tower can harbor microorganisms in scale, allow them to proliferate, and generate an aerosol.

GASEOUS CONTAMINANTS

This category of indoor contaminants includes both true gases (which have boiling points less than room temperature) and vapors of liquids with boiling points above normal indoor temperatures. It also includes both volatile organic compounds (VOCs) and inorganic air contaminants.

Volatile organic compounds include 4- to 16-carbon alkanes, chlorinated hydrocarbons, alcohols, aldehydes, ketones, esters, terpenes, ethers, aromatic hydrocarbons (such as benzene and toluene), and heterocyclic hydrocarbons. Also included are chlorofluorocarbons (CFCs) and hydrochloroflurocarbons (HCFCs) used as refrigerants. More information on classifications, characteristics and measurement methods can be found in Chapter 12.

Inorganic gaseous air contaminants include ammonia, nitrogen oxides, ozone, sulfur dioxide, carbon monoxide, and carbon dioxide. Although the last two contain carbon, they are by tradition regarded as inorganic chemicals.

The most common units of measurement for gaseous contaminants are parts per million by volume (ppm) and milligrams per cubic metre (mg/m^3). For smaller quantities, parts per billion (ppb) and micrograms per cubic metre (μg/m^3) are used. The relationship between all of these is explained in Chapter 12.

GASEOUS CONTAMINANTS IN INDUSTRIAL ENVIRONMENTS

The Occupational Safety and Health Administration (OSHA) sets Permissible Exposure Limits (PELs), which are the only workplace regulatory standards in the United States. These are published yearly in the *Code of Federal Regulations* (29 CFR 1900, Part 1900.1000 ff) and intermittently in the *Federal Register*. Most of the levels were recommended by the American Conference of Governmental Industrial Hygienists (ACGIH) and the American National Standards Institute (ANSI). The health effects on which these standards were based can be found in their publications (ACGIH 1989). ACGIH reviews data on a regular basis and publishes Threshold Limit Values (TLVs) yearly.

The National Institute for Occupational Safety and Health (NIOSH) is charged with researching toxicity problems, and it influences the legally required levels. NIOSH publishes the *Registry of Toxic Effects and Chemical Substances* (RTECS) as well as numerous Criteria for Recommended Standard for Occupational Exposure to (compound). Some compounds not in the OSHA list are covered by NIOSH literature, and their Recommended Exposure Limits (RELs) are sometimes lower than the legal requirements set by OSHA. The NIOSH Pocket Guide to Chemical Hazards (NIOSH 1990) is a condensation of these references and is convenient for engineering purposes.

The harmful effects of gaseous pollutants on a person depend on both short-term peak concentrations and the time-integrated exposures received by the person. OSHA has defined three periods for concentration averaging and has assigned allowable levels that may exist in these categories in workplaces for over 490 compounds, mostly gaseous contaminants. The abbreviations for concentrations for the three averaging periods are

AMP = acceptable maximum peak for a short exposure

ACC = acceptable ceiling concentration, not to be exceeded during an 8-h shift, except for periods where AMP applies

TWA8 = time-weighted average, not to be exceeded in any 8-h shift of a 40-h week.

In non-OSHA literature, AMP is sometimes called STEL (short-term exposure limit), and TWA8 is sometimes called TLV (threshold limit value.) NIOSH (1990) also lists values for the toxic limit IDLH—immediately dangerous to life and health.

Table 3 lists values of the various exposure limits defined above for a selection of common gaseous industrial contaminants. Other countries have also issued standards for industrial exposure.

It is of interest to compare standards for industrial and nonindustrial environments. A Canadian National Task Force developed guideline criteria for residential indoor environments. Similarly, the World Health Organization has published indoor air quality guidelines for Europe. Table 4 compares these guidelines with occupational criteria for selected contaminants.

Table 3 Characteristics of Selected Gaseous Air Pollutants

Pollutant	Allowable Concentration, mg/m^3			
	IDLH[a]	AMP[a]	ACC[a]	TWA8[a]
Acetaldehyde	18,000			360
Acetone	4,800		3,200	2,400
Acetonitrile	7,000	105		70
Acrolein	13		0.75	0.25
Acrylonitrile	10			45
Allyl chloride	810		9	3
Ammonia	350		35	38
Benzene	10,000		25	5
Benzyl chloride	50			5
2-Butanone (MEK)	8,850			590
Carbon dioxide	90,000		54,000	9,000
Carbon monoxide	1,650		220	55
Carbon disulfide	1,500	300	90	60
Carbon tetrachloride	1,800	1,200	150	60
Chlorine	75		1.5	3
Chloroform	4,800		9.6	240
Chloroprene	1,440		3.6	90
p-Cresol	1,100			22
Dichlorodifluoromethane	250,000			4,950
Dioxane	720			360
Ethylene dibromide	3,110	271	233	155
Ethylene dichloride	4,100	818	410	205
Ethylene oxide	1,400		135	90
Formaldehyde	124	12	6	4
n-Heptane	17,000			2,000
Hydrogen chloride	140		7	7
Hydrogen cyanide	55			11
Hydrogen fluoride	13		5	2
Hydrogen sulfide	420	70	28	30
Mercury	28			0.1
Methane	ASPHY[b]			
Methanol	32,500			260
Methyl chloride	59,500		1,783	1,189
Methylene chloride	7,500		3,480	1,740
Nitric acid	250			5
Nitric oxide	120	45		30
Nitrogen dioxide	90		1.8	9
Ozone	20			2
Phenol	380		60	19
Phosgene	8		0.8	0.4
Propane	36,000			
Sulfur dioxide	260			13
Sulfuric acid	80			1
Tetrachloroethane	1,050			35
Tetrachloroethylene	3,430	2,060	1,372	686
o-Toluidene	440			22
Toluene	7,600	1,900	1,140	760
Toluene diisocyanate	70		0.14	0.14
1,1,1-Trichloroethane	2,250			45
Trichloroethylene	5,410	1,620	1,080	541
Vinyl chloride monomer			0.014	0.003
Xylene	43,500		870	435

[a]IDLH, AMP, ACC, and TWA8 are defined in the section on Gaseous Contaminants in Industrial Environments.

[b]ASPHY = Simple asphyxiant; causes breathing problems when concentration reaches about 1/3 atmospheric pressure.

GASEOUS CONTAMINANTS IN NONINDUSTRIAL ENVIRONMENTS

The gaseous contaminants that are of concern in nonindustrial environments are volatile organic compounds and inorganic gases.

Table 4 Comparison of Standards Pertinent to Indoor Environments

	Canadian	WHO/Europe	NAAQS/EPA[f]	NIOSH REL	OSHA	ACGIH	MAK[g]
Aldehydes							
Acrolein	0.02 ppm[a]			0.1 ppm 0.25 ppm (15 min)	0.1 ppm 0.3 ppm (15 min)	0.1 ppm 0.3 ppm (15 min)	0.1 ppm 0.2 ppm (15 min)
Acetaldehyde	5.0 ppm			ALARA[b]	100 ppm 150 ppm (15 min)	100 ppm 150 ppm (15 min)	50 ppm
Formaldehyde	0.1 ppm[c]	0.081 ppm		0.016 ppm 0.1 ppm (15 min)	0.75 ppm 2 ppm (15 min)	0.3 ppm	0.3 ppm
Carbon dioxide	3500 ppm			5000 ppm 30,000 ppm (15 min)	10,000 ppm 30,000 ppm (15 min)	5000 ppm 9000 ppm (15 min)	5000 ppm 9000 ppm (15 min)
Carbon monoxide	11 ppm (8 h) 25 ppm (1 h)	8.6 ppm (8 h) 25 ppm (1 h) 51 ppm (30 min) 86 ppm (15 min)	9 ppm (8 h) 35 ppm (1 h)	35 ppm (8 h) 200 ppm (15 min)	35 ppm (8 h) 200 ppm (15 min)	25 ppm (8 h)	30 ppm
Nitrogen dioxide	0.05 ppm 0.25 ppm (1 h)	0.08 ppm (24 h) 0.2 ppm (1 h)	0.053 ppm (1yr)		1 ppm (15 min)	3 ppm 5 ppm (15 min)	5 ppm
Ozone	0.12 ppm (1 h) no long-term level	0.08 ppm (8 h) 0.1 ppm (1 h)	0.12 ppm (1 h) 0.085 ppm (8 h)	0.1 ppm (15 min)	0.1 ppm (8 h) 0.3 ppm (15 min)	0.05 ppm (8 h) 0.2 ppm (15 min)	0.1 ppm
Particulate < 2.5 MMAD[d]	40 μg/m³ (8 h) 100 μg/m³ (1 h)		50 g/m³ (1 yr)		5 mg/m³ (8 h) (respirable dust)	3 mg/m³ (8 h) (no asbestos, <1% crystalline silica)	
Sulfur dioxide	0.019 ppm 0.38 ppm (5 min)			2 ppm (8 h) 5 ppm (15 min)	2 ppm (8 h) 5 ppm (15 min)	2 ppm (8 h) 5 ppm (15 min)	2 ppm
Radon	800 Bq/m³[e]						
Relative humidity	30-80% (summer) 30-55% (winter)						

( ) Numbers in parentheses represent averaging periods
[a]Parts per million (10^6)
[b]As low as reasonably achievable
[c]Target level of 0.05 ppm because of its carcinogenic effects
[d]Mass median aerodynamic diameter
[e]Mean in normal living areas
[f]U.S. EPA National Ambient Air Quality Standards
[g]German Maximale Arbeitsplatz Konzentrationen

The alternative Indoor Air Quality Procedure specified in ASHRAE *Standard* 62 sets limits for concentrations of several contaminants in nonindustrial environments and cautions that contaminants whose toxicity is well-known should be kept at or below one-tenth of the threshold limit value (TLV) specified by ACGIH (annual). When outdoor air quantities are reduced, the actual gaseous contaminant concentrations must be measured to ensure that *Standard* 62 is met.

Health Effects of Volatile Organic Compounds

Adverse health effects potentially caused by VOCs in nonindustrial indoor environments are not well understood but may include (1) irritant effects, including perception of unpleasant odors, mucous membrane irritation, and exacerbation of asthma; (2) systemic effects, such as fatigue and difficulty concentrating; and (3) toxic, chronic effects, such as carcinogenicity (Girman et al. 1989).

The chronic adverse health effects due to VOC exposure are of concern because some VOCs commonly found in indoor air are human (benzene) or animal (chloroform, trichloroethylene, carbon tetrachloride, *p*-dichlorobenzene) carcinogens. Some other VOCs are also genotoxic. Theoretical risk assessment studies suggest that chronic exposure risk due to VOCs in residential indoor air is greater than that associated with exposure to VOCs in the outdoor air or in drinking water (McCann et al. 1987, Tancrede et al. 1987).

Carbon tetrachloride (CCl_4) causes central nervous system (CNS) depression and significant liver and kidney damage. CCl_4 has also been shown to be an animal carcinogen and is classified as a potential human carcinogen.

A biological model for acute human response to low levels of VOCs indoors is based on three mechanisms: sensory perception of the environment, weak inflammatory reactions, and environmental stress reaction (Molhave 1991). A growing body of literature summarizes measurement techniques for the effects of VOCs on nasal (Koren 1990, Koren et al. 1992, Meggs 1994, Ohm et al. 1992, Molhave et al. 1993) and ocular (Kjaergard et al. 1991, Kjaergard 1992, Franck et al. 1993) mucosa. It is not well known how different sensory receptions to VOCs are combined into perceived comfort and the sensation of air quality. This perception is apparently interrelated to stimulation of the olfactory sense in the nasal cavity, to the gustatory sense on the tongue, and the common chemical sense (Molhave 1991, Cain 1989).

Cometto-Muñiz and Cain (1994a,b) addressed the independent contribution of the trigeminal and olfactory nerves to the detection of airborne chemicals. The sense of smell is experienced through receptors in the nose of the olfactory nerve. Nasal pungency, described as common chemical sensations including prickling, irritation, tingling, freshness, stinging, and burning among others, is experienced through the nonspecialized receptors of the trigeminal nerve in the face. Odor and pungency thresholds follow different patterns related to chemical concentration. Odor is often detected at much lower levels. A linear correlation between pungency thresholds of homologous series—of alcohols, acetates, ketones, and alkylbenzenes, relatively nonreactive agents—suggests that nasal pungency relies on a physicochemical interaction with a susceptible biophase within the cell membrane. It is postulated that through this nonspecific mechanism, low, subthreshold levels of a wide variety of VOCs—as found in many polluted indoor environments—can be additive in their sensory impact to produce noticeable sensory irritation.

Formaldehyde is a very reactive small molecule that requires different analytical techniques than those usually employed in VOC assessment. Primary sources include urea-formaldehyde resin-based particle and chipboard products used in indoor spaces. It is frequently encountered in indoor spaces in concentrations between 0.04 ppm, a frequently encountered lower limit of detection, and 0.1 ppm (Liu et al. 1991, Ritchie and Lehnen 1987). Many studies have demonstrated its ability to trigger mucous membrane irritation at levels below the ACGIH TLV, and even at levels below 0.1 ppm.

Standards for Volatile Organic Compounds

No standards for exposure to VOCs relevant to nonindustrial indoor environments are in place. NIOSH, OSHA, and the ACGIH have published regulatory standards or recommended limits for industrial occupational exposures (NIOSH 1992, ACGIH annual). With few exceptions, concentrations observed in nonindustrial indoor environments fall well below (100 to 1000 times lower) these published pollutant-specific occupational standards or recommended exposure limits. However, standards for the industrial workplace are higher than would be appropriate for the general population, which includes the elderly, children, and people who are more sensitive to VOCs than the average industrial worker.

Total VOC (TVOC) concentration has previously been used as an indicator of the potency of VOCs to cause health effects. This approach is no longer recommended since the toxicities of individual VOCs vary very widely, and concentrations differ depending on the measurement method used (Hodgson 1995). In controlled exposure experiments, odors are significant at 3 mg/m^3. At 5 mg/m^3, objective effects were seen in addition to the subjective irritation. Exposures for 50 min to 8 mg/m^3 of synthetic mixtures of 20 VOCs lead to significant irritation of mucous membranes in the eyes, nose, and throat.

Both OSHA and the ACGIH have set 8 h standards for formaldehyde as a ceiling level. California issued a residential air quality guideline of 0.1 ppm. In the setting of occupant complaints, the target guideline is 0.05 ppm.

Health Effects of Refrigerants

ASHRAE *Standard* 34 assigns **refrigerants** to one of two toxicity classes (A or B) based on allowable exposure. Fatalities have been reported following acute exposure to fluorocarbon refrigerants. Inhalation exposures to CFCs can cause cardiotoxicity at chronic, low-level exposures. Some are thought to be cardiac sensitizers to epinephrine and put occupants at risk for arrhythmias. Central nervous system (CNS) depression has been found at very high concentrations along with asphyxia. Proctor and Hughes (1991) found that volunteers exposed to 200,000 ppm of R-12 experienced significant eye irritation and CNS effects. Chronic exposure to 1000 ppm for 8 h per day for up to 17 days caused no subjective symptoms or changes in pulmonary function.

A significant hazard exists when chlorinated hydrocarbons (R-11, for example) are used in the vicinity of open flame or heated surfaces. Phosgene gas (carbonyl chloride), an extreme irritant to the lungs, and halogen acids may be generated when chlorinated or fluorinated solvents or gases decompose in the presence of heat.

CFC-containing systems may only be serviced by certified technicians. Controls for preventing exposures include selection and use of appropriate fittings and valves and insuring that compressed gas cylinders are secured when in use, in transport, and in storage. When repairs are made to leaking or defective components in HVAC equipment, adequate dilution ventilation should be provided to the work area. CFCs should never be used in the vicinity of open flame or heated materials due to the potential for the formation of phosgene gas. ASHRAE *Standard* 15 establishes specific requirements for designing, installing, operating, and servicing mechanical refrigeration equipment.

Health Effects of Inorganic Gases

Carbon monoxide is a chemical asphyxiant. Inhalation of CO causes a throbbing headache brought about because CO has a competitive preference for hemoglobin (about 240 times that of oxygen) and also a shift in the oxygen dissociation curve. Carbon monoxide inhibits oxygen transport in the blood through the formation of carboxyhemoglobin and inhibition of cytochrome oxidase at the cellular level. Deaths and adverse health effects from overexposures are attributed primarily to motor vehicles. Cobb and Etzel (1991) suggested that CO poisoning at home represented a major preventable disease. Moolenaar et al. (1995) subsequently identified similar data and suggested that motor vehicles and home furnaces were primary causes of mortality. Girman et al. (1996) identified both fatal outcomes and episodes. Respectively, 35.9% and 30.6% resulted from motor vehicles, 34.8% and 39.9% from appliance combustion, 4.5% and 5.2% from small appliances, 2.2% and 2.3% from camping equipment, 5.6% and 5.0% from fires, 13.4% and 13.3% from grills and hibachis, and the remainder were unknown.

Carbon dioxide can become dangerous not as a toxic agent but as a secondary asphyxiant. When concentrations exceed 35,000 ppm, central breathing receptors are triggered and cause the sensation of shortness of breath. At progressively higher concentrations, central nervous system dysfunction begins due to simple displacement of oxygen. Concentrations of CO_2 in the nonindustrial environment (office buildings and schools) are often measured in the range of 400 to 1500 depending on occupant density, ventilation distribution, and amount of outside air supplied to the occupied spaces.

Inhalation of **nitric oxide** (NO) causes the formation of methemoglobin, which adversely affects the body by interfering with oxygen transport at the cellular level. NO exposures of 3 ppm have been compared to carbon monoxide exposures of 10 to 15 ppm (Case et al. 1979 in EPA 1991).

Nitrogen dioxide (NO_2) is a corrosive gas with a pungent odor, the odor threshold of which is reported to be between 0.11 and 0.22 ppm (WHO 1987). NO_2 has a low water solubility and therefore can be inhaled into the deep lung where it causes a delayed inflammatory response. Increased airway resistance has been reported at 1.5 to 2 ppm (Bascom 1996). NO_2 is reported to be a potential carcinogen by way of free radical production (Burgess and Crutchfield 1995). At high concentrations, NO_2 causes lung damage directly by its oxidant properties and may cause health effects indirectly by increasing host susceptibility to respiratory infections. Health effects from exposures to ambient outdoor concentrations or in residential situations show inconsistency, especially studies relating to exposures from gas cooking stoves (Samet et al. 1987). Indoor concentrations of NO_2 often exceed ambient concentrations due to the presence of strong indoor sources and a trend toward more energy efficient (tighter) homes. Acute toxicity is seldom seen from NO_2 produced by unvented indoor combustion because of the insufficient quantities of NO_2 produced. Chronic pulmonary effects from exposure to combinations of low-level combustion pollutants are possible, however (Bascom et al. 1996).

Sulfur dioxide (SO_2) is a colorless gas with a pungent odor detected at about 0.5 ppm (EPA 1991). Because SO_2 is quite soluble in water, it can react with moisture in the upper respiratory tract to produce irritant effects on the upper respiratory mucosa. Concomitant exposure to fine particulates, an individual's depth and rate of breathing, and the presence of preexisting disease can influence the degree of SO_2 toxicity.

Ozone is a pulmonary irritant and causes changes in human pulmonary function at concentrations of approximately 0.12 ppm (Bates 1989). Exposure to ozone at 60 to 80 ppb causes inflammation, bronchoconstriction, and increased airway responsiveness.

Table 5 Inorganic Gas Comparative Criteria

Contaminant	OSHA/NIOSH TWA[a]	EPA NAAQS 1 Std.
Nitric oxide	1 h 2 ppm (5 mg/m^3) 24 h 25 ppm (30 mg/m^3)	None
Nitrogen dioxide	1 h 5 ppm (9 mg/m^3) 24 h 1 ppm (1.8 mg/m^3)	0.053 ppm (100 μg/m^3)
Sulfur dioxide	1 h 5 ppm (13 mg/m^3) 24 h 2 ppm (5 mg/m^3)	0.014 ppm (365 μg/m^3)
Ozone	1 h 0.1 ppm (0.2 mg/m^3) 24 h 0.1 ppm (0.2 mg/m^3)	0.12 ppm (235 μg/m^3)

[a]The values listed are the annual arithmetic mean unless otherwise listed. The first value listed is the 24 h average and the second value is the maximum 1 h average. (TWA = time weighted average)

Inhalation exposures to the gaseous oxides of nitrogen, sulfur (NO_x and SO_2) and ozone (O_3) can and do occur in residential and commercial buildings. These air pollutants are of considerable concern due to the potential for acute and chronic respiratory tract health effects in exposed individuals, particularly individuals with preexisting pulmonary disease.

Standards for Inorganic Gases

Currently, there are no specific United States government standards relative to nonindustrial occupational exposures to air contaminants. Occupational exposure criteria are health-based; that is, they consider healthy workers in an environment, and not necessarily individuals who may be unusually responsive to the effects of chemical exposures. The EPA National Ambient Air Quality Standards (NAAQS) (see Chapter 12) are also health-based standards designed to protect the general public health from the effects of hazardous airborne pollutants. However, there is a debate as to whether these standards truly represent health-based thresholds. Two of the six criteria, ozone and carbon monoxide, involve toxicologically based research for the development of the standards. The criteria (Table 5) are not meant to be health-based guidelines for the evaluation of exposures to inorganic gasses in the indoor environment. Table 5 is included for comparative use and consideration by investigators of the indoor environment with the understanding that these criteria may not be completely protective to all industrial workers.

PHYSICAL AGENTS

Physical factors in the indoor environment include thermal conditions (temperature, moisture, air velocity, and radiant energy), mechanical energy (noise and vibration), and electromagnetic radiation [ionizing (radon) and nonionizing (light, radio-frequency, and extremely low frequency magnetic and electric fields)]. Physical agents can act directly upon building occupants, interact with indoor air quality factors, or affect the way humans respond to the indoor environment. Physical agents, while not categorized as indoor air quality factors, often affect human perception of the quality of indoor air.

THERMAL ENVIRONMENT

The thermal environment affects human health in that it affects body temperatures, both internally and externally (of the skin). In the normal, healthy, resting adult, internal or core body temperatures are very stable, with variations seldom exceeding 1°F. The internal temperature of a resting adult, measured orally, averages about 98.6°F; measured rectally, it is about 1°F higher (Guyton 1972). The temperature of the core is carefully modulated by an elaborate physiological control system. In contrast, the temperature of the skin is basically unregulated and can vary from about 88 to 96°F in normal environments and activities, and also varies over different parts of the skin, with the greatest variation in the hands and feet.

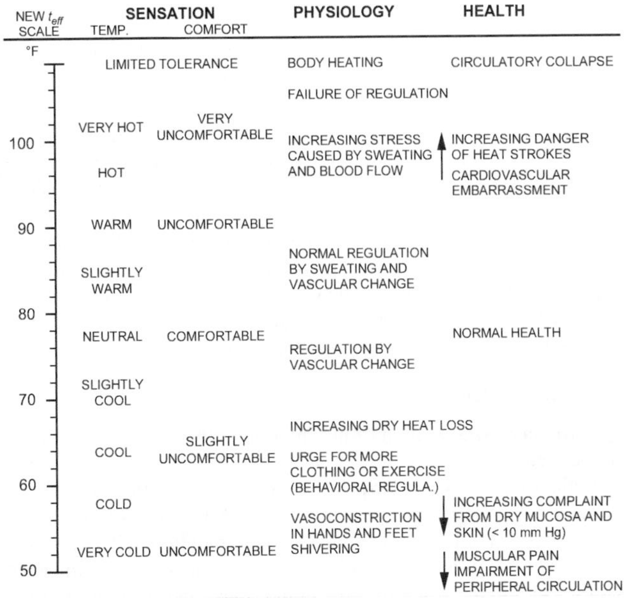

NEW t_{eff} SCALE °F	SENSATION TEMP.	SENSATION COMFORT	PHYSIOLOGY	HEALTH
	LIMITED TOLERANCE		BODY HEATING	CIRCULATORY COLLAPSE
			FAILURE OF REGULATION	
	VERY HOT	VERY UNCOMFORTABLE	INCREASING STRESS CAUSED BY SWEATING AND BLOOD FLOW ↑	INCREASING DANGER OF HEAT STROKES
100				
	HOT			CARDIOVASCULAR EMBARRASSMENT
90	WARM	UNCOMFORTABLE		
	SLIGHTLY WARM		NORMAL REGULATION BY SWEATING AND VASCULAR CHANGE	
80				
	NEUTRAL	COMFORTABLE	REGULATION BY VASCULAR CHANGE	NORMAL HEALTH
	SLIGHTLY COOL			
70				
			INCREASING DRY HEAT LOSS	
	COOL	SLIGHTLY UNCOMFORTABLE	URGE FOR MORE CLOTHING OR EXERCISE (BEHAVIORAL REGULA.)	
60				
	COLD		VASOCONSTRICTION IN HANDS AND FEET ↓	INCREASING COMPLAINT FROM DRY MUCOSA AND SKIN (< 10 mm Hg)
	VERY COLD	UNCOMFORTABLE	SHIVERING ↓	MUSCULAR PAIN IMPAIRMENT OF PERIPHERAL CIRCULATION
50				

Fig. 1 Related Human Sensory, Physiological, and Health Responses for Prolonged Exposure

Range of Healthy Living Conditions

The environmental conditions for thermal comfort are those that minimize the effort of the physiological control systems to maintain the internal temperature. The control system regulates the internal body temperature by varying the amount of blood flowing to different skin areas, thus increasing or decreasing heat lost to the environment, by secreting and evaporating sweat from the skin in warm or hot environments, and by increasing the metabolic heat production by shivering in the cold. For a resting person wearing trousers and a long-sleeved shirt, thermal comfort is experienced in a still air environment at 75°F. A zone of comfort extends about 3°F above and below this optimum level. An individual can minimize the need for physiological (involuntary) responses to the thermal environment that are generally perceived as uncomfortable, by a variety of behavioral responses. In a cool or cold environment, such responses include increased clothing, increased activity, or seeking or creating an environment that is warmer. In a warm or hot environment, the amount of clothing can be reduced, the level of physical activity can be reduced, or an environment that is more conducive to increased heat loss can be created. Some of the human responses to the thermal environment are shown in Figure 1.

Cardiovascular and other diseases and the inevitable processes of aging can reduce the capacity or ability of physiological processes to maintain internal body temperature through the balancing of heat gains and heat losses. Thus, some persons are less able to deal with thermal challenges and deviations from comfortable thermally neutral conditions. Metabolic heat production tends to decrease with age, as a result of decreasing basal metabolism together with decreased physical activity. Metabolic heat production at age 80 is about 20% less than that at 20 years old, for comparable size and weight. Persons in their eighties, therefore, prefer an environmental temperature about 3°F warmer than persons in their twenties. In any given environment near thermally neutral temperature, an older person is likely to have a lower core and skin temperature. Older people may have reduced capacity to secrete and evaporate sweat and to increase their skin blood flow and are therefore more likely to experience greater strain in warm and hot conditions as well as in cool and cold conditions.

Hyperthermia

Hyperthermia refers to the condition where body temperatures are above normal. A deep body temperature increase of 4°F above the normal range does not generally impair body function. For example, it is not unusual for runners to have rectal temperatures of 104°F after a long race. An elevated body temperature increases metabolism. Central nervous system function deteriorates at deep body temperatures above 106 to 108°F. Convulsions may occur above such temperatures and cells may be damaged. This condition is particularly dangerous for the brain, because lost neurons are not replaced. Thermoregulatory functions of sweating and peripheral vasodilation cease at about 110°F, after which body temperatures tend to rise rapidly if external cooling is not imposed.

Seasonal Patterns

Ordinary season changes in temperate climates are temporally associated with the prevalence of illness. Many acute and several chronic diseases vary in frequency or severity with time of year, and some are present only in certain seasons. Minor respiratory infections, such as colds and sore throats, occur mainly in fall and winter. More serious infections, such as pneumonia, have a somewhat shorter season in winter. Intestinal infections, such as dysentery and typhoid fever, are more prevalent in summer. Diseases transmitted by insects such as encephalitis and endemic typhus are limited to summer since insects are active in warm temperatures only.

Martinez et al. (1989), Hryhorczuk et al. (1992), and others describe a correlation between weather and seasonal illnesses; but such correlations do not necessarily establish a causal relationship. Daily or weekly mortality and heat stress in heat waves have a strong physiological basis directly linked to outdoor temperature. In indoor environments which are well controlled with respect to temperature and humidity, such temperature extremes and the possible adverse effects on health are strongly attenuated.

Increased Deaths in Heat Waves

The role of ambient temperature extremes produced by weather conditions in producing discomfort, incapacity, and death has been studied extensively (Katayama 1970). Military personnel, deep mine workers, and other workers occupationally exposed to extremes of high and low temperature have been studied, but the importance of thermal stress affecting both the sick and healthy is not sufficiently appreciated. Collins and Lehmann (1953) studied weekly deaths over many years in large cities in the United States and demonstrated the impact of heat waves in producing conspicuous periods of excess mortality. Excess mortality due to heat waves was of the same amplitude as that due to influenza epidemics, but tended to last one week instead of the 4 to 6 weeks duration of influenza epidemics.

Ellis (1972) reviewed heat wave related excess mortality in the United States. Mortality increases of 30% over background are commonly seen, especially in heat waves that occur early in the summer. Much of the increase occurs in the population over age 65, more of it in women than in men, and many deaths are due to cardiovascular and cerebrovascular causes. Oeschli and Buechley (1970) studied heat-related deaths in Los Angeles heat waves of 1939, 1955, and 1963. Kilbourne et al. (1982) suggested that the same risk factors (i.e., age, low income, and African-American derivation) persist in the heat death epidemics that continue to occur.

Hardy (1971) showed the relationship of health data to comfort on a psychrometric diagram (Figure 2). The diagram contains ASHRAE effective temperature ET* lines and lines of constant skin moisture level or skin wettedness w. Skin wettedness is defined as that fraction of the skin covered with water to account for the observed evaporation rate. The ET* lines are loci of constant physiological strain, and also correspond to constant levels of physiological discomfort—slightly uncomfortable, comfortable, and very

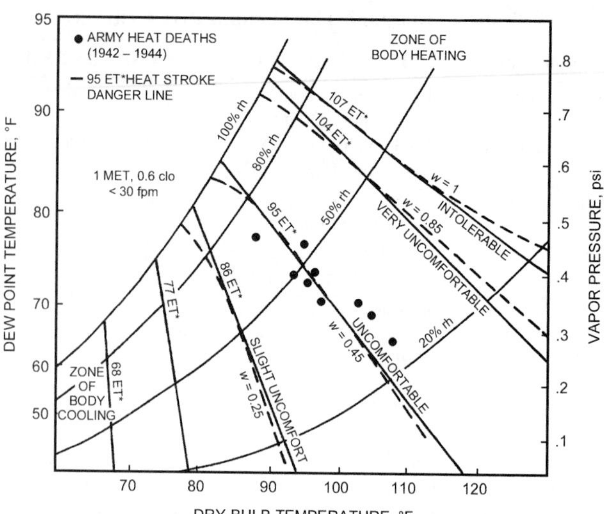

Fig. 2 Isotherms for Comfort, Discomfort, Physiological Strain, Effective Temperature (ET*), and Heat Stroke Danger Threshold

comfortable (Gonzalez et al. 1978). Skin wettedness, as an indicator of strain (Berglund and Gonzalez 1977, Berglund and Cunningham 1986) and the fraction of the skin wet with perspiration, is fairly constant along an ET* line. Numerically ET* is the equivalent temperature at 50% rh that produces the strain and discomfort of the actual condition. The summer comfort range is between an ET* of 73 and 79°F. In this region, skin wettedness is less than 0.2. Heat strokes occur generally when ET* exceeds 93°F (Bridger and Helfand 1968). Thus, the ET* line of 95°F is generally considered dangerous. At this point, skin wettedness will be 0.4 or higher.

The black dots in Figure 2 correspond to heat stroke deaths of healthy male U.S. soldiers assigned to sedentary duties in midwestern army camp offices (Shickele 1947). It is to be expected that older persons respond less well to thermal challenges than do healthy soldiers. This was apparently the case in the Illinois heat wave study mentioned earlier, where the first wave with a 33% increase in death rate and an ET* of 85°F affected mainly the over 65-year-old group. The studies suggest that the "danger line" represents a threshold of significant risk for young healthy people, and that the danger tends to move to lower values of ET* with increasing age.

Effects of Thermal Environment on Specific Diseases

Cardiovascular diseases are largely responsible for excess mortality during heat waves. For example, Burch and DePasquale (1962) found that the heart disease cases in whom decompensation is present are extremely sensitive to high temperatures and particularly to moist heat. However, both cold and hot temperature extremes are associated with increased coronary heart disease deaths and anginal symptoms (Teng and Heyer 1955).

Both acute and chronic respiratory diseases often increase in frequency and severity during extreme cold weather. No increase in these diseases has been noted in extreme heat. Additional studies of hospital admissions for acute respiratory illness show a negative correlation with temperature after removal of seasonal trends (Holland 1961). The symptoms of patients with chronic respiratory disease (bronchitis, emphysema) increase in cold weather. This is thought to be due to reflex constriction of the bronchi, adding to the obstruction already present. Greenberg (1964) revealed evidence of cold sensitivity in asthmatics; emergency room treatments for asthma increased abruptly in local hospitals with early and severe autumn cold spells. Later cold waves with even lower temperatures produced no such effects, and years without early extreme cold had no asthma epidemics of this type. Patients with cystic fibrosis are extremely sensitive to heat because their diminished sweat gland function greatly diminishes their ability to cope with increased temperature (Kessler and Anderson 1951).

Itching and chapping of the skin is influenced by (1) atmospheric factors, particularly cold and dry air, (2) frequent washing or wetting of skin, and (3) low indoor humidities. Although itching of the skin is usually a winter cold climate illness in the general population, it can be caused by excessive summer air conditioning (Susskind and Ishihara 1965, Gaul and Underwood 1952).

People suffering from chronic illness (heart disease) or serious acute illnesses that require hospitalization often manage to avoid serious thermal stress. Katayama et al. (1970) found that countries with the most carefully regulated indoor climates (such as the Scandinavian countries and the United States) have had only small seasonal fluctuations in mortality in recent decades, while countries with less space heating and cooling exhibit greater seasonal swings in seasonal mortality. For example, mandatory air conditioning in homes for the aged in the southwest United States has virtually eliminated previously observed mortality increases during heat waves.

Summer cooling reduces heat stress by removing both sensible and latent heat from the occupied space, but winter heating has a mixed effect. It reduces cold stress, but it usually does not increase the low water vapor pressure that occurs outdoors during the winter. This results in very low relative humidity in the heated space, which can contribute to dehydration and discomfort and cause injury to skin, eyes, nose, throat, and mucous membranes. These dry tissues may be less resistant to infection. Animal experiments also show that infection rates increase with low levels of either ventilation or relative humidity (Schulman and Kilbourne 1962).

In various tests conducted under identical conditions except humidity level, mechanical humidification raised the relative humidity in one space above that in the matched space; no humidified room was higher than 50% rh (Green 1979 and 1982, Gelperin 1973, Serati and Wuthrich 1969). In each investigation, the humidified rooms showed a reduction in absenteeism and upper respiratory infection—49% reduction in kindergarten children, 6% and 18% in office workers, and 8% and 18% in army recruits. Since occupants in each pair of spaces were subject to the same outdoor conditions and the same indoor air temperature, reductions were attributed to differences in humidity or a related factor (e.g., reduced dust levels and coughing). Therefore, while low humidity does not have a direct pathological effect, it is a factor contributing to disease. A more direct effect has been indicated among users of contact lenses on long airline flights in cabins at low humidity. Here, dehydration of the eyes has been blamed for causing irritation and corneal edema or even ulceration of the corneal epithelium (Laviana et al. 1988).

Injury from Hot and Cold Surfaces

The skin has cold, warm, and pain sensors to feed back thermal information about surface contacts. When the skin temperature rises above 113°F or falls below about 59°F, sensations from the skin's warm and cold receptors are replaced by those from the pain receptors to warn of thermal injury to the tissue (Guyton 1968). The temperature of the skin depends on the temperature of the contact surface, its conductivity, and the contact time. Table 6 gives approximate temperature limits to avoid pain and injury when contacting three classes of conductors for various contact times (CEN).

Table 6 Approximate Surface Temperature Limits to Avoid Pain and Injury

Material	Contact Time				
	1 s	10 s	1 min	10 min	8 h
Metal, water	149°F	133°F	124°F	118°F	109°F
Glass, concrete	176°F	151°F	129°F	118°F	109°F
Wood	248°F	190°F	140°F	118°F	109°F

ELECTRICAL HAZARDS

Electrical current can cause burns, neural disturbances, and cardiac fibrillation (Billings 1975). The threshold of perception is about 5 mA for direct current, with a feeling of warmth at the contact site. The threshold is 1 mA for alternating current, which causes a tingling sensation.

The resistance of the current pathway through the body is a combination of core and skin resistance. The core is basically a saline volume conductor with very little resistance; therefore, the skin resistance provides the largest component of the resistance. The skin resistance decreases with moisture. If the skin is moist, voltages as low as 2 V (ac) or 5 V (dc) are sufficient to be detected, and voltages as low as 20 V (ac) or 100 V (dc) can cause a 50% loss in muscular control.

The dangerous aspect of alternating electrical current is its ability to cause cardiac arrest by ventricular fibrillation. If a weak alternating current (100 mA for 2 s) passes through the heart (as it would in going from hand to foot), the current can force the heart muscle to fibrillate and lose the rhythmic contractions of the ventricles necessary to pump blood. Unconsciousness and death will soon follow if medical aid cannot rapidly restore normal rhythm.

MECHANICAL ENERGIES

Vibration

Vibration in a building originates from both outside and inside the building. Sources outside a building include blasting operations, road traffic, overhead aircraft, underground railways, earth movements, and weather conditions. Sources inside a building include doors closing, foot traffic, moving machinery, elevators, HVAC systems, and other building services. Vibration is an omnipresent, integral part of the built environment. The effects of the vibration on building occupants depend on whether it is perceived by those persons and on factors related to the building, the location of the building, the activities of the occupants in the building, and the perceived source and magnitude of the vibration. Factors influencing the acceptability of building vibration are presented in Figure 3.

The combination of hearing, seeing, or feeling vibration determines human response. Components concerned with hearing and seeing are part of the visual environment of a room and can be assessed as such. The perception of mechanical vibration by feeling is generally through the cutaneous and kinesthetic senses at high frequencies, and through the vestibular and visceral senses at low frequencies. Because of this and the nature of vibration sources and building responses, building vibration may be conveniently considered in two categories—low-frequency vibrations less than 1 Hz and high-frequency vibrations of 1 to 80 Hz.

Measurement and Assessment. Human response to vibration depends on the vibration of the body. The main vibrational characteristics are vibration level, frequency, axis (and area of the body), and exposure time. A root-mean-square (rms) averaging procedure (over the time of interest) is often used to represent vibration acceleration (ft/s^2 rms). Vibration frequency is measured in cycles per second (Hz), and the vibration axis is usually considered in three orthogonal, human-centered translational directions (up-and-down, side-to-side, and fore-and-aft). Although the coordinate system is centered inside the body, in practice, vibration is measured at the human surface and measurements are directly compared with relevant limit values or other data concerning human response.

Rotational motions of a building in roll, pitch, and yaw are usually about an axis of rotation some distance from the building

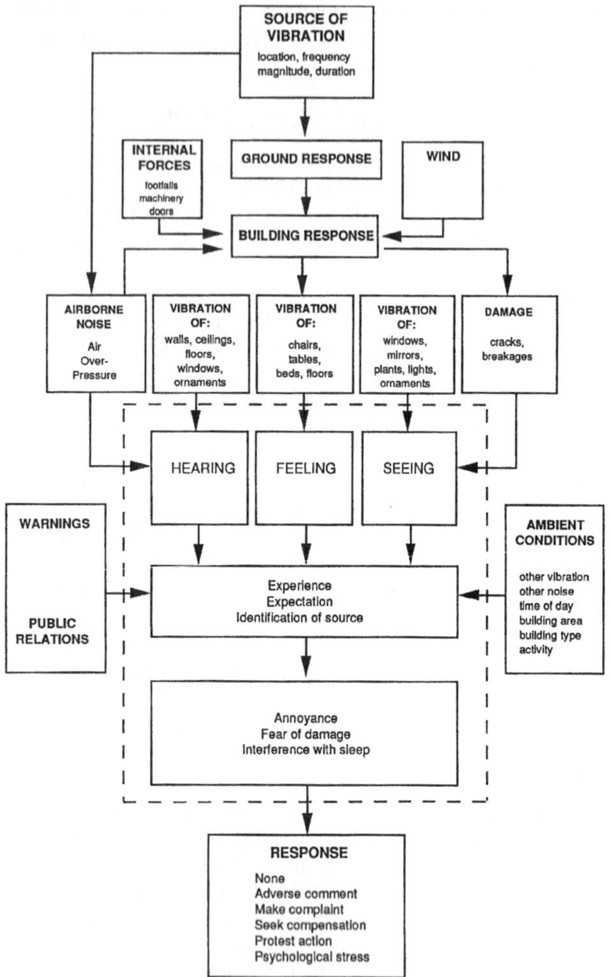

Fig. 3 Factors Affecting Acceptability of Building Vibration

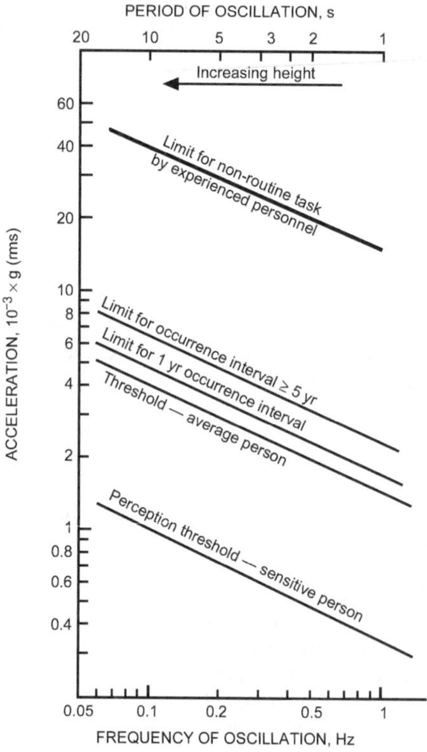

Fig. 4 Acceleration Perception Thresholds and Acceptability Limits for Horizontal Oscillations

occupants. For most purposes, these motions can be considered as the translational motions of the person. For example, a roll motion in a building about an axis of rotation some distance from a seated person will have a similar effect as side-to-side translational motions of that person, etc.

Most methods assess building vibrations with rms averaging and frequency analysis. However, human response is related to the time-varying characteristics of vibration as well. For example, many stimuli are transient, such as those caused by a train passing a building. The vibration event builds to a peak, followed by a decay in level over a total period of about 10 s. The nature of the time-varying event and the number of occasions it occurs during a day are important factors that might be overlooked if data are treated as steady-state and continuous.

Standard Limits

Low-Frequency Motion (1 Hz). The most commonly experienced form of slow vibration in buildings is building sway. This motion can be alarming to occupants if there is fear of building damage or injury. While occupants of two-story wood frame houses accept occasional creaks and motion from wind storms or a passing heavy vehicle, such events are not as accepted by occupants of high-rise buildings. Detected motion in tall buildings can cause discomfort and alarm. The perception thresholds of normal sensitive humans to low-frequency horizontal motion are given in Figure 4 (ISO 1984, Chen and Robertson 1972). The frequency range is from 0.06 to 1 Hz or, conversely, for oscillations with periods of 1 to 17 s. The natural frequency of sway of the Empire State Building in New York City, for example, has a period of 8.3 s (Davenport 1988). The thresholds are expressed in terms of relative acceleration, which is the actual acceleration divided by the standard acceleration of gravity (g = 32.2 ft/s^2). The perception threshold to sway in terms of building accelerations decreases with increasing frequency and ranges from 0.16 to 0.06 ft/s^2.

For tall buildings, the highest horizontal accelerations generally occur near the top at the building's natural frequency of oscillation. Other parts of the building may have high accelerations at multiples of the natural frequency. Tall buildings always oscillate at their natural frequency, but the deflection is small and the motion undetectable. In general, short buildings have a higher natural frequency of vibration than taller ones. However, strong wind forces energize the oscillation and increase the horizontal deflection, speed, and accelerations of the structure.

ISO (1984) states that building motions are not to produce alarm and adverse comment from more than 2% of the building's occupants. The level of alarm depends on the interval between events. If noticeable building sway occurs for at least 10 min at intervals of 5 years or more, the accept able acceleration limit is higher than if this sway occurs annually (Figure 4). For annual intervals, the acceptable limit is only slightly above the normal person's threshold of perception. Motion at the 5 year limit level is estimated to cause 12% to complain if it occurred annually. The recommended limits are for purely horizontal motion; rotational oscillations, wind noise, and/or visual cues of the building's motion exaggerate the sensation of motion, and, for such factors, the acceleration limit would be lower.

The upper line in Figure 4 is intended for offshore fixed structures such as oil drilling platforms. The line indicates the level of horizontal acceleration above which routine tasks by experienced personnel would be difficult to accomplish on the structure.

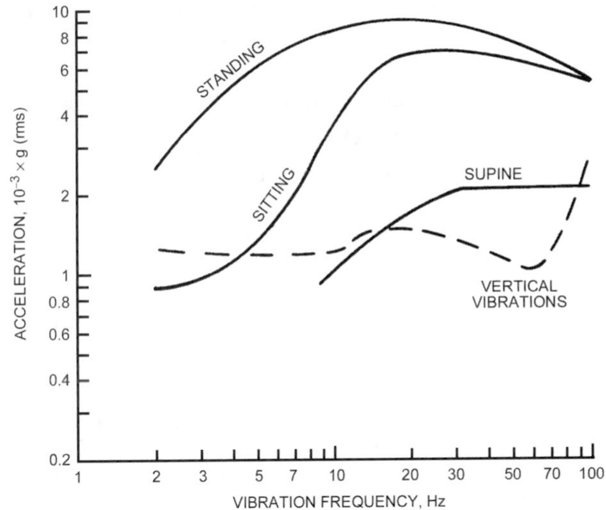

Fig. 5 Median Perception Thresholds to Horizontal (Solid Lines) and Vertical (Dashed Line) Vibrations

Table 7 Acceptable to Threshold Vibration Level Ratios

Place	Time	Continuous or Intermittent Vibration	Impulse or Transient Vibration Several Times per Day
Critical work areas	Day or night	1	1
Residential	Day/Night	2 to 4 / 1.4	30 to 90 / 1.4 to 20
Office	Day or night	4	60 to 128
Workshop	Day or night	8	90 to 128

Note: The ratios for continuous or intermittent vibration and repeated impulse shock are in the range of 0.7 to 1.0 for hospital operating theaters (room) and critical working areas. In other situations, impulse shock can generally be much higher than when the vibration is more continuous.

High-Frequency Motion (1 to 80 Hz). Higher frequency vibrations in buildings are caused by machinery, elevators, foot traffic, fans, pumps, and HVAC equipment. Further, the steel structures of modern buildings are good transmitters of high-frequency vibrations. The sensitivity to these higher frequency vibrations is indicated in Figure 5 (Parsons and Griffin 1988). Displayed are the median perception thresholds to vertical and horizontal vibrations in the 2 to 100 Hz frequency range. The average perception threshold for vibrations of this type is from 0.03 to 0.3 ft/s^2, depending on the frequency and on whether the person is standing, sitting, or lying down.

People detect horizontal vibrations at lower acceleration levels when lying down than when standing. However, a soft bed decouples and isolates a person fairly well from the vibrations of the structure. The threshold to vertical vibrations is nearly constant at approximately 0.04 ft/s^2 for both sitting and standing positions from 2 to 100 Hz. This agrees with earlier observations by Reiher and Meister (1931).

Many building spaces with critical work areas (surgery, precision laboratory work) are considered unacceptable if vibration is perceived by the occupants. In other situations and activities, perceived vibration may be acceptable. Parsons and Griffin (1988) found that accelerations twice the threshold level would be unacceptable to occupants in their homes. A method of assessing vibrational acceptability in buildings is to compare the vibration with perception threshold values (Table 7).

Sound and Noise

When the vibration of an object is transmitted to air particles, making them vibrate, a variation in normal atmospheric pressure is

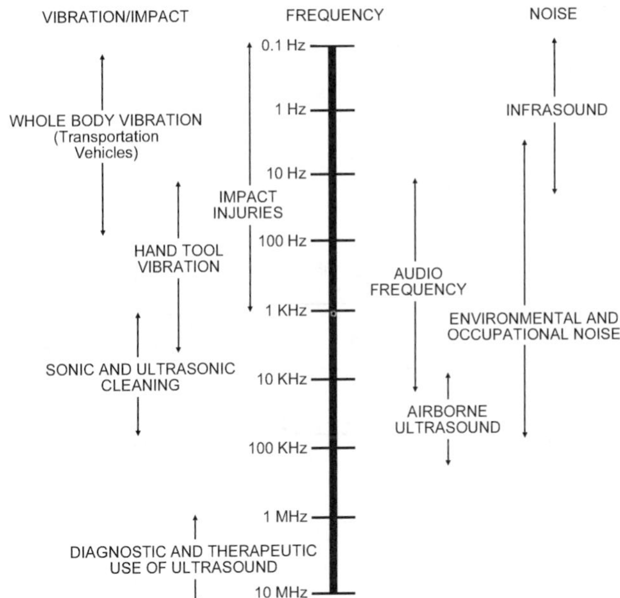

Fig. 6 Mechanical Energy Spectrum

created. When this disturbance spreads to the eardrum, it is vibrated, and this vibration is translated into the sensation called "sound." In general terms, sound in the physical sense is the vibration of particles in a gas, a liquid, or a solid. The entire mechanical energy spectrum includes include infrasound and ultrasound as well as audible sound (Figure 6).

Health Effects. Hearing loss is generally considered the most undesirable effect of noise exposure, although there are other effects. Tinnitus, a ringing in the ears, is really the hearing of sounds that do not exist. It often accompanies hearing loss. Paracusis is a disorder where a sound is heard incorrectly; that is, a tone is heard, but has an inappropriate pitch. Speech misperception occurs when an individual mistakenly hears one sound for another; for example, when the sound for "t" is heard as a "p."

Hearing loss can be categorized as conductive, sensory, or neural. Conductive hearing loss results from a general decrease in the amount of sound transmitted to the inner ear. Excessive ear wax, a ruptured ear drum, fluid in the middle ear, or missing elements of the bone structures in the middle ear are all associated with conductive hearing loss. These are generally not occupationally related and are generally reversible by medical or surgical means. Sensory hearing losses are associated with irreversible damage to the inner ear. Sensory hearing loss is further classified as (1) presbycusis, loss caused as the result of aging; (2) noise-induced hearing loss (industrial hearing loss and sociacusis, which is caused by noise in everyday life); and (3) nosoacusis, losses attributed to all other causes. Neural deficits are related to damage to higher centers of the auditory system.

Noise-induced hearing loss is believed to occur, in the most sensitive individuals, in those exposed for 8 h per day over a working lifetime at levels of 75 dBA and for most people similarly exposed to 85 dBA.

ELECTROMAGNETIC RADIATION

Radiation energy is emitted, transmitted, or absorbed in wave or particulate form. This energy consists of electric and magnetic forces which, when disturbed in some manner, produce electromagnetic radiation. Electromagnetic radiation is grouped into a spectrum arranged by frequency and/or wavelength. The product of frequency and wavelength is the speed of light (3×10^8 m/s). The

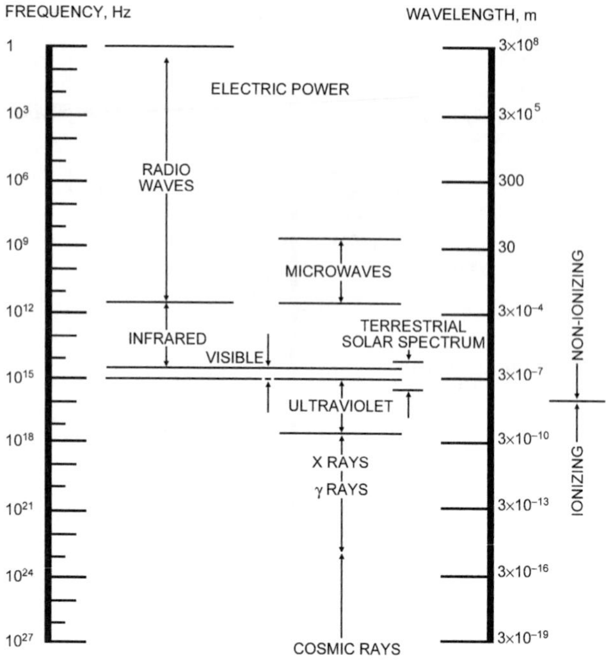

Fig. 7 Electromagnetic Spectrum

spectrum includes ionizing, ultraviolet, visible, infrared, microwave, radio-frequency, and extremely low frequency (Figure 7). Table 8 presents these electromagnetic radiations by their range of energies, frequencies, and wavelengths. The regions are not sharply delineated from each other and, in fact, often overlap. It is convenient to divide these regions as listed in Table 8, due to the nature of the physical and biological effects.

Ionizing Radiation

Ionizing radiation is that part of the electromagnetic spectrum that has very short wavelengths and high frequencies, and it has the ability to ionize matter. Such ionizations tend to be very damaging to living matter. Background radiation that occurs naturally in the environment is from cosmic rays and naturally occurring radionuclides. It has not been established whether exposure at the low dose rate of average background levels is harmful to humans.

The basic standards for permissible air concentrations of radioactive materials are those of the National Committee on Radiation Protection, published by the National Bureau of Standards as *Handbook* No. 69. Industries operating under licenses from the U.S. Nuclear Regulatory Commission or state licensing agencies must meet requirements of the *Code of Federal Regulations*, Title 10, Part 20. Some states have additional requirements.

An important naturally occurring radionuclide is radon (^{222}Rn), a decay product of uranium in the soil (^{238}U). Radon, denoted by the symbol Rn, is chemically inert. Details of units of measurement, typical radon levels, measurement methods and control strategies can be found in Chapter 12.

Health Effects of Radon. Studies of workers in uranium and other underground mines form the principal basis for knowledge about health risks due to radon. The radioactive decay of radon produces a series of radioactive isotopes of polonium, bismuth, and lead. Unlike their chemically inert radon parent, these progeny are chemically active. They can attach to airborne particles that subsequently deposit in the lung or deposit directly in the lung without prior attachment to particles. Some of these progeny, like radon, are alpha-particle emitters, and the passage of these alpha particles through lung cells can lead to cellular changes that may initiate lung cancer (Samet 1989). Thus, adverse health effects

Table 8 Energy, Wavelength, and Frequency Ranges for Electromagnetic Radiation

Radiation Type	Energy Range	Wavelength Range	Frequency Range
Ionizing	> 12.4 eV	< 100 nm	> 3.00 PHz
Ultraviolet (UV)	12.40 – 3.10 eV	100 – 400 nm	3.00 PHz – 0.75 PHz
Visible	3.10 – 1.63 eV	400 – 760 nm	750 THz – 395 THz
Infrared (IR)	1.63 – 1.24 meV	760 nm – 1 mm	395 THz – 0.30 THz
Microwave (MW)	1.24 meV – 1.24 eV	1 mm – 1 m	300 GHz – 300 MHz
Radio-frequency (RF)	1.24 eV – 1.24 peV	1 m – 1 Mm	300 MHz – 300 Hz
Extremely low frequency (ELF)	< 1.24 peV	> 1 Mm	< 300 Hz

Table 9 Action Levels for Radon Concentration Indoors

Country/Agency	Action Level Bq/m³	Action Level pCi/L
Australia	200	5.4
Austria	400	10.8
Belgium	400	10.8
CEC	400	10.8
Canada	800	21.6
Czech Republic	400	10.8
P.R. China	200	5.4
Finland	400	10.8
Germany	250	6.7
ICRP	200	5.4
Ireland	200	5.4
Italy	400	10.8
Norway	400	10.8
Sweden	400	10.8
United Kingdom	200	5.4
United States	148	4.0
World Health Organization	200	5.4

Source: DOE (1995).

associated with radon are due to exposures to radon decay products, and the amount of risk is assumed to be directly related to the total exposure. Even though it is the radon progeny that present the possibility of adverse health risks, radon itself is usually measured and used as a surrogate for progeny measurements because of the expense involved in accurate measurements of radon progeny.

Standards. Many countries besides the United States have established standards for exposure to radon. International action levels are listed in Table 9.

About 6% of homes in the United States (5.8 million homes) have annual average radon concentrations exceeding the action level of 148 Bq/m³ (4 pCi/L) set by the U.S. Environmental Protection Agency (Marcinowski et al. 1994).

Nonionizing Radiation

Ultraviolet radiation, visible light, and infrared radiation are components of sunlight and of all artificial light sources. Microwave radiation and radio-frequency radiation are essential in a wide range of communication technologies and are also in widespread use for heating as in microwave ovens and heat sealers, and for heat treatments of a variety of products. Power frequency fields are an essential and unavoidable consequence of the generation, transmission, distribution, and use of electrical power.

Optical Radiation. Ultraviolet (UV), visible, and infrared (IR) radiation compose the optical radiation region of the electromagnetic spectrum. The wavelengths range from 100 nm in the UV to 1 mm in the IR, with 100 nm generally considered to be the boundary between ionizing and nonionizing. The UV region wavelengths range from 100 to 400 nm, the visible region from 400 to 760 nm, and the IR from 760 nm to 1 mm.

Optical radiation can interact with a medium by reflection, absorption, or transmission. The skin and eyes are the organs at risk in humans. Optical radiation from any of the spectral regions can cause acute and/or chronic biologic effects given appropriate energy characteristics and exposure. These effects include tanning, burning (erythema), premature "aging," and cancer of the skin; and dryness, irritation, cataracts, and blindness in the eyes.

The region of the electromagnetic spectrum visible to humans is known as light. There can be biological, behavioral, psychological, and health effects from exposure to light. Assessment of these effects depends on the purpose and application of the illumination. Individual susceptibility varies, with other environmental factors (air quality, noise, chemical exposures, and diet) acting as modifiers. It is difficult, therefore, to generalize potential hazards. Light pollution results from the presence of unwanted light.

Light penetrating the retina not only allows the exterior world to be seen, but, like food and water, it is used in a variety of metabolic processes. Light stimulates the pineal gland to secrete melatonin, which regulates the human biological clock. This, in turn, influences reproductive cycles, sleeping, eating patterns, activity levels, and moods. The color of light affects the way the objects appear. The distortion of color rendition may result in disorientation, headache, dizziness, nausea, and fatigue.

As the daylight shortens, the human body may experience a gradual slowing down, loss of energy, and a need for more sleep. It becomes harder to get to work, and depression or even withdrawal may take place. This type of seasonal depression, brought on by changes in light duration and intensity, is called seasonal affective disorder (SAD). Sufferers of this syndrome also complain of anxiety, irritability, headache, weight gain, and lack of concentration and motivation. Treatment of this problem is through the manipulation of environmental lighting (exposure to full-spectrum lighting for extended periods, 12 h/day).

Radio-Frequency Radiation. Just as the body absorbs infrared and light energy, which can affect thermal balance, it can also absorb other longer wavelength electromagnetic radiation. For comparison, visible light has wavelengths in the range 0.4 to 0.7 μm and infrared from 0.7 to 10 μm, while the wavelength of K and X band radar is 12 and 28.6 mm. The wavelength of radiation in a typical microwave oven is 120 mm. Infrared is absorbed within 1 mm of the surface (Murray 1995).

The heat of the absorbed radiation raises the skin temperature and, if sufficient, is detected by the skin's thermoreceptors, warning the person of the possible thermal danger. With increasing wavelength, the radiation penetrates deeper into the body. The energy can thus be deposited well beneath the skin's thermoreceptors, making the person less able or slower to detect and be warned of the radiation (Justesen et al. 1982). Physiologically, these longer waves only heat the tissue and, because the heat may be deeper and less detectable, the maximum power density of such waves in occupied areas is regulated (ANSI 1991) (Figure 8). The maximum permitted power densities are less than half of sensory threshold values.

ERGONOMICS

Ergonomics may be defined as the scientific study of the relationship between man and his work environment to achieve optimum adjustment in terms of efficiency, health, and well-being. Ergonomic designs of tools, chairs, etc., help workers interact more comfortably and efficiently with their environment. In jobs that were ergonomically designed, productivity typically increased and the worker enjoyed a healthier working experience. More recently, researchers have distinguished intrinsic ergonomics from extrinsic, or traditional, ergonomics. Intrinsic ergonomics considers how the interface between an individual and the environment affects and relies on specific body parts (i.e., muscles, tendons, and bones) and work practices such as force of application, relaxation intervals, styles, and strength reserves that are not adequately considered in simple analyses of the physical environment.

The goal of ergonomic programs ranges from making work safe and humane, to increasing human efficiency, to creating human

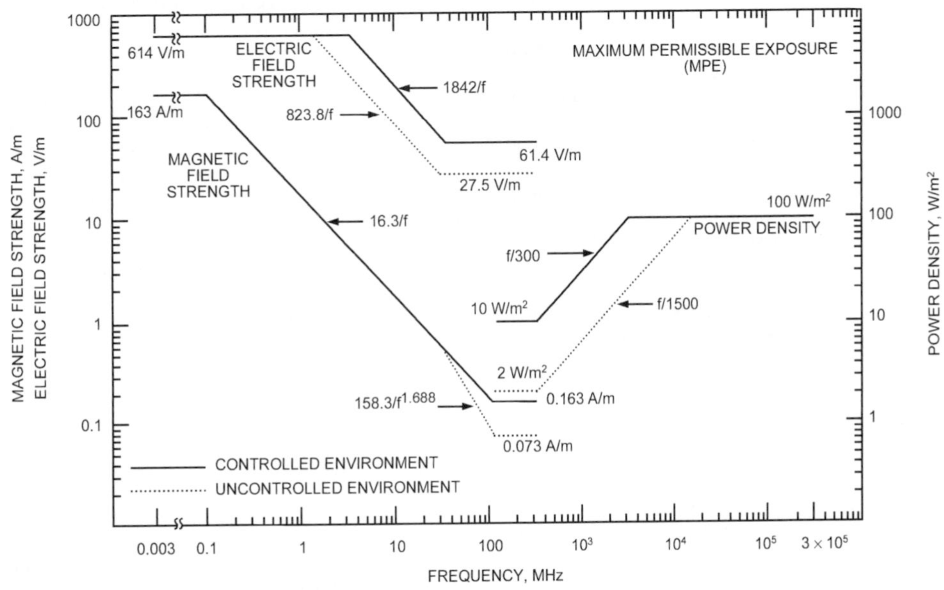

Fig. 8 Maximum Permissible Levels of Radio Frequency Radiation for Human Exposure

well-being. The successful application of ergonomic factors is measured by improved productivity, efficiency, safety, and acceptance of the resultant system design. The design engineer uses not only engineering skills, but also the sciences and principles of anatomy, orthopedics, physiology, medicine, psychology, and sociology to apply ergonomics to a design.

Implementing ergonomic principles in the workplace helps minimize on-the-job stress and strain, and prevents cumulative trauma disorders (CTDs). These disorders are subtle injuries that can affect the muscles, tendons, and nerves at body joints, especially the hands, wrists, elbows, shoulders, neck, back, and knees. Carpal tunnel syndrome is an example of a CTD. CTDs most frequently occur as a result of strain from performing the same task on a continuous or repetitive basis. This strain can slowly build over time, until the worker experiences pain and difficulty using the injured part of the body. Higher risks of developing CTDs are encountered when the work task requires repetitive motions, excessive force, or awkward postures. The ergonomics engineer addresses these risk factors by analyzing the task thoroughly and minimizing the repetitive motion, excessive force, and awkward posture.

REFERENCES

ACGIH. 1989. Guidelines for the assessment of bioaerosols in the indoor environment. American Conference of Government Industrial Hygienists, Cincinnati, OH.

ACGIH. Annual. Threshold Limit Values and Biological Exposure Indices for 1995-1996.

Alpaugh, E.L. and T.J. Hogan. 1988. *Fundamentals of industrial hygiene: Particulates*, 3rd Edition. National Safety Council.

Anderson, H.A., D. Higgins, L.P. Hanrahan, P. Sarow, and J. Schirmer. 1991. Mesothelioma among employees with likely contact with in-place asbestos-containing building materials. *Ann. N.Y. Acad. Sci.* 643:550-72.

ANSI. 1991. Safety level with respect to human exposure to radio frequency electromagnetic radiation of 3KH-300GH *Standard* C95.1-1991. American National Standards Institute, New York.

Arnow, P.M., J.N. Fink, D.P. Schlueter, J.J. Barboriak, G. Mallison, S.I. Said, S. Martin, G.F. Unger, G.T. Scanlon, and V.P. Kurup. 1978. *American Journal of Medicine* 64:237.

Bartlett, P.C., L.A. Vonbehren, R.P. Tewari, R.J. Martin, L. Eagleton, M.J. Isaac, and P.S. Kulkarni. 1982. *American Journal of Public Health* 72: 1369.

Bascom, R.A. 1996. Environmental factors and respiratory hypersensitivity: The Americas. *Toxicol. Lett.* 86:115-30.

Bascom, R., P. Bromberg, D.L. Costa, et al. 1996. Health effects of outdoor pollution, Parts I and II. *American Journal Respir. Crit. Care Med.* 153:3-50, 477-89.

Bates, D.V. 1989. Ozone—Myth and reality. *Environmental Research* 50: 230-37.

Baylor, E.R., V. Peters, and M.B. Baylor. 1977. Water-to-air transfer of virus. *Science*. 1977:763.

Berglund, L.G. and D. Cunningham. 1986. Parameters of human discomfort in warm environments. *ASHRAE Transactions* 92(2).

Berglund, L.G. and R.R. Gonzalez. 1977. Evaporation of sweat from sedentary man in humid environments. *Journal of Applied Physiology, Respiratory, Environmental and Exercise Physiology* 42(5):767-72.

Billings, C.E. 1975. Electrical shock. In *Textbook of medicine*. Saunders, Philadelphia, 72-73.

Bridger, C.A. and L.A. Helfand. 1968. Mortality from heat during July 1966 in Illinois. *International Journal of Biometeorology* 12:51.

Brundage, J.F., R.M. Scott, W.M. Lodnar, D.W. Smith, and R.N. Miller. 1988. *JAMA* 259:2108.

Burch, G.E. and N.P. DePasquale. 1962. *Hot climates, man and his heart.* Charles C. Thomas, Springfield, IL.

Burge, H.A. 1995. *Bioaerosols*. Lewis Publishers, Chelsea, MI.

Burge, H.A. 1989. *Occupational medicine: State of the art reviews.* J.E. Cone and M.J. Hodgson, eds. 4:713-21. Hanley and Belfus, Philadelphia.

Burge, S., A. Hedge, S. Wilson, J.H. Bass, and A. Robertson. 1987. *Annals of Occupational Hygiene* 31:493.

Burgess, J.L. and C.D. Crutchfield. 1995. Quantitative respirator fit tests of Tuscon fire fighters and measures of negative pressure excursions during exertion. *Appl. Occup. Env. Hyg.* 10(1):29-36.

Burton, D.J. 2000. Industrial ventilation, a self-directed learning workbook. Carr Printing.

Cain, W.S. 1989. Perceptual characteristics of nasal irritation. National Danish Institute on Occupational Health, Copenhagen.

Cain, W.S., J.M. Samet, and M.J. Hodgson. 1995. The quest for negligible health risk from indoor air. *ASHRAE Journal* 37(7):38.

CEN. Surface temperatures of touchable parts, a draft proposal. TC 114 N 122 D/E. European Standards Group.

CFR. 1989a. 29 CFR 1910.1000. U.S. Government Printing Office, Washington, DC.

CFR. 1989b. 29 CFR 1926.1101.

Chen, P.W. and L.E. Robertson. 1972. Human perception thresholds of horizontal motion. *ASCE Journal*, Structure Division, August.

Cobb, N. and R.A. Etzel. 1991. Unintentional carbon monoxide-related deaths in the United States, 1979 through 1988. *JAMA* 266:659-63.

Collins, S.D. and J. Lehmann. 1953. Excess deaths from influenza and pneumonia and from important chronic diseases during epidemic periods, 1918-51. Public Health Monograph 20.10, U.S. Public Health Service *Publication No.* 213.

Cometto-Muñiz, J.E. and W.S. Cain. 1994a. Sensory reactions of nasal pungency and odor to volatile organic compounds: The alkylbenzenes. *Am. Ind. Hyg. Assoc. J.* 55(9):811-17.

Cometto-Muñiz, J.E. and W.S. Cain. 1994b. Perception of odor and nasal pungency from homologous series of volatile organic compounds. *Indoor Air* 4:140-45.

Cox, C.S. 1987. *The aerobiological pathway of microorganisms.* John Wiley and Sons, New York.

Crandall, M.S. and W.K. Sieber. 1996. The NIOSH indoor environmental evaluation experience: Part one, building evaluations. *Applied Occupational and Environmental Hygiene.*

Davenport, A.G. 1988. The response of supertall buildings to wind. In *Second century of the skyscraper.* Van Nostrand Reinhold, New York, 705-26.

DHHS. 1986. The health consequences of involuntary smoking. A report of the surgeon general. *DHHS Publication No.* (PHS) 87-8398. U.S. Department of Health and Human Services, Public Health Services, Office of the Asst. Secretary for Health, Office of Smoking and Health.

DOE. 1995. *Symposium Series* CONF-77114, pp. 663-90. U.S. Dept. of Energy.

Edwards, J.H. 1980. Microbial and immunological investigations and remedial action after an outbreak of humidifier fever. *British Journal of Industrial Medicine* 37:55-62.

Ellis, F.P. 1972. Mortality from heat illness and heat aggravated illness in the United States. *Environmental Research* 5.

EPA. 1991. Introduction to indoor air quality, a reference manual. *Report* EPA 400/3-91/003

EPA. 1992. Respiratory health effects of passive smoking: Lung cancer and other disorders, review draft. EPA/6O0-6-90/006B. Office of Research and Development, Washington, DC.

Franck, C., P. Skov, and O. Bach. 1993. Prevalence of objective eye manifestations in people working in office buildings with different prevalences of the sick building syndrome compared with the general population. *Int. Arch. Occup. Environ. Health* 65:65-69.

Gaul, L.E. and G.B. Underwood. 1952. Relation of dew point and barometric pressure to chapping of normal skin. *Journal of Investigative Dermatology* 19:9.

Gelperin, A. 1973. Humidification and upper respiratory infection incidence. *Heating, Piping and Air Conditioning* 45:77.

Girman, J.R. 1989. Volatile organic compounds and building bake-out. In *Occupational medicine: State of the art reviews—Problem buildings.* J.E. Cone and M.J. Hodgson, eds. Hanley and Belfus, Philadelphia. 4(4):695-712.

Girman, J., Y.-L. Chang, S.B. Hayward, and K.S. Liu. 1996. Causes of unintentional deaths from carbon monoxide poisonings in California. Unpublished.

Glantz, S.A. and W.W. Parmley. 1991. Passive smoking and heart disease epidemiology, physiology, and biochemistry. *Circulation* 83:633-42.

Gonzalez, R.R., L.G. Berglund, and A.P. Gagge. 1978. Indices of thermoregulatory strain for moderate exercise. *Journal of Applied Physiology: Respiratory Environmental and Exercise Physiology* 44(6):889-99.

Green, G.H. 1979. The effect of indoor relative humidity on colds. *ASHRAE Transactions* 85(1).

Green, G.H. 1982. The positive and negative effects of building humidification. *ASHRAE Transactions* 88(1):1049.

Greenberg, L. 1964. Asthma and temperature change. *Archives of Environmental Health* 8:642.

Guyton, A.C. 1968. *Textbook of medical physiology*. Saunders, Philadelphia.

Hardy, J.D. 1971. Thermal comfort and health. *ASHRAE Journal* 13:43.

Higgins, I.T.T. 1983. What is an adverse health effect? *APCA Journal* 33:661-63.

Hodgson, A.T. 1995. A review and a limited comparison of methods for measuring total volatile organic compounds in indoor air. *Indoor Air* 5(4):247.

Hodgson, M.J., P.R. Morey, J.S. Simon, T.D. Waters, and J.N. Fink. 1987. *American Journal of Epidemiology* 125:631.

Hodgson, M.J., P.R. Morey, M. Attfield, W. Sorenson, J.N. Fink, W.W. Rhodes, and G.S. Visvesvara. 1985. *Archives of Environmental Health* 40:96.

Holland, W.W. 1961. Influence of the weather on respiratory and heart disease. *Lancet* 2:338.

Hryhorczuk, D.O., L.J. Frateschi, J.W. Lipscomb, and R. Zhang. 1992. Use of the scan statistic to detect temporal clustering of poisonings. *J. Tox-Clinical Toxicology* 30:459-65.

Hyppel, A. 1984. Proceedings of the 3rd International Conference on Indoor Air Quality and Climate, Stockholm, Sweden, 3:443-47.

ISO. 1984. Guideline to the evaluation of responses of occupants of fixed structures, especially buildings and off-shore structures, to low frequency horizontal motion, 0.063 to 1 Hz. ISO 6897. International Organization for Standardization, Geneva.

Justesen, D.R., E.R. Adair, J.C. Stevens and V. Bruce-Wolfe. 1982. A comparative study of human sensory thresholds: 2450 MHz microwaves vs. far-infrared radiation. *Bioelectromagnetics* 3:117-25.

Katayama, K. and M. Momiyana-Sakamoto. 1970. A biometeorological study of mortality from stroke and heart diseases. *Meteorological Geophysics* 21:127.

Kessler, W.R. and W.R. Anderson. 1951. Heat prostration in fibrocystic disease of pancreas and other conditions. *Pediatrics* 8:648.

Kilbourne, E.M., T.S. Jones, K. Choi, and S.B. Thacker. 1982. Risk factors for heatstroke: A case-control study. *JAMA* 247(24):3332-36.

Kjaergard, S., L. Molhave, and O.F. Pedersen. 1991. Human reactions to a mixture of indoor pollutants. *Atmos. Environ.* 25:1417-26.

Kjaergard, S. 1992. Assessment methods and causes of eye irritation in humans in indoor environments. H. Knoeppel and P. Wolkoff, eds. *Chemical, Microbiological, Health, and Comfort Aspects of Indoor Air Quality.* ECSC, EEC, EAEC, Brussels 115-27.

Koren, H. 1990. The inflammatory response of the human upper airways to volatile organic compounds. In: *Proceedings of Indoor Air* 90 1:325-330.

Koren, H., D.E. Graham, and R.B. Devlin. 1992. Exposure of humans to a volatile organic mixture III: Inflammatory response. *Arch. Environ. Health* 47:39-44.

Last, J.M. 1983. *Public health and human ecology.* Appleton & Lange, New York.

Laviana, J.E., F.H. Rohles, and P.E. Bullock. 1988. Humidity, comfort and contact lenses. *ASHRAE Transactions* 94(1).

Leathart, G.L. 1972. Clinical Aspects of Respiratory Disease Due to Mining. *Medicine in the Mining Industry.*

Levetin, E. 1995. Fungi. In: Burge, H.A. 1995. *Bioaerosols*. CRC Press. Lewis Publishers. Boca Raton, FL.

Lilienfeld, D.E. 1991. Asbestos-associated pleural mesothelioma in school teachers: A discussion of four cases. *Ann. N.Y. Acad. Sci.* 643:454-86.

Liu, K.S., J. Wesolowski, F.Y. Huang, K. Sexton, and S.B. Hayward. 1991. Irritant effects of formaldehyde exposure in mobile homes. *Environ. Health Perspect.* 94:91-94.

Marcinowski, F., R.M. Lucas, and W.M. Yeager 1994. National and regional distributions of airborne radon concentrations in U.S. homes. *Health Physics* 66(6):699-706.

Martin, T.R. and M.B. Braken. 1986. Association of low birth weight passive smoke and exposure in pregnancy. *American Journal of Epidemiology* 124:633-42.

Martinez, B.F., M.L. Kirk, J.L. Annest, K.J. Lui, E.M. Kilbourne, and S.M. Smith. 1989. Geographic distribution of heat-related deaths among elderly persons. Use of county-level dot maps for injury surveillance and epidemiologic research. *JAMA* 262:2246-50.

Mbithi, J.N., V.S. Springthorpe, and S.A. Sattar. 1991. Effect of relative humidity and air temperature on survival of hepatitis A virus on environmental surfaces. *Appl. Environ. Microbiol.* 57:1394.

McCann, J., L. Horn, J. Girman, and A.V. Nero. 1987. *Short-term bioassays in the analysis of complex mixtures*. Plenum Press, New York 5:325-54.

Meggs, W.J. 1994. RADS and RUDS—The toxic induction of asthma and rhinitis. *J. Toxicol. Clin. Toxicol.* 32:487-501.

Miller, J.D. and J. Day. 1997. Indoor mold exposure: epidemiology, consequences and immunotherapy. *Can. J. Allergy and Clinical Immunology* 2(1):25-32.

Molhave, L. 1991. Volatile organic compounds, indoor air quality and health. *Indoor Air* 1(4):357-76.

Molhave, L., Z. Liu, A.H. Jorgensen, O.F. Pederson, and S. Kjaergard. 1993. Sensory and physiologic effects on humans of combined exposures to air temperatures and volatile organic compounds. *Indoor Air* 3:155-69.

Moolenaar, R.L., R.A. Etzel, and R.G. Parrish. 1995. Unintentional deaths from carbon monoxide poisoning in New Mexico, 1980 to 1988. A comparison of medical examiner and national mortality data. *Western Journal of Medicine* 163(5):431-4.

Morey, P.R. 1988. Architectural design and indoor microbial pollution. Oxford University Press, New York, 40-80.

Morey, P.R. 1990. The practitioner's approach to indoor air quality investigations. *Proceedings* of the Indoor Air Quality International Symposium. American Industrial Hygiene Association, Akron, OH.

Morey, P.R. and J.C. Feeley. 1988. *ASTM Standardization News* 16:54.

Morey, P.R., M.J. Hodgson, W.G. Sorenson, G.J. Kullman, W.W. Rhodes, and G.S. Visvesvara. 1986 Environmental studies in moldy office buildings. *ASHRAE Transactions* 93(1B):399-419.

Morey, P.R. and B.A. Jenkins. 1989. What are typical concentrations of fungi, total volatile organic compounds, and nitrogen dioxide in an office environment. Proceedings of IAQ '89, The Human Equation: Health and Comfort. ASHRAE, Atlanta, 67-71.

Morrow, P.E. 1964. Evaluation of inhalation hazards based upon the respirable dust concept and the philosophy and application of selective sampling. AIHA *Journal* 25:213.

Moser, M.R., T. R. Bender, H.S. Margolis, et al. 1979. An outbreak of influenza aboard a commercial airliner. *Am. J. Epidemiol.* 110:1.

Mumford, J.L., X.Z. He, R.S. Chapman, S.R. Cao, D.B. Harris, K.M. Li, Y.L. Xian, W.Z. Jiang, C.W. Xu, J.C. Chang, W.E. Wilson, and M. Cooke. 1987. Lung cancer and indoor air pollution in Xuan Wei, China. *Science* 235:217-20.

Murray, W. 1995. Non-ionizing electromagnetic energies (Chapter 14). *Patty's Industrial Hygiene and Toxicology* 3B:623-727 (644).

NAS. 1981. *Indoor Pollutants*. National Research Council/National Academy of Sciences, Committee on Indoor Pollutants. National Academy of Sciences Press, Washington, D.C., Oct.

NAS. 1983. Polycyclic aromatic hydrocarbons: Evaluation of sources and effects. National Academy Press, Washington, DC.

NIOSH. 1990. NIOSH/OSHA pocket guide to chemical hazards. DHHS (NIOSH) *Publication* 90-117. U.S. Department of Labor, OSHA, Washington, DC. Users should read Cohen (1993) before using this reference. NIOSH. Annual registry of toxic effects of chemical substances. U.S. Department of Health and Human Services, National Institute for Occupational Safety and Health, Washington DC.

NIOSH. 1992. NIOSH recommendations for occupational safety and health compendium of policy documents and statements. DHHS (NIOSH) *Publication* 92-100. U.S. Dept. of Health and Human Services, Public Health Service, Centers for Disease Control, National Institute for Occupational Safety and Health, Atlanta.

NRC. 1986. Environmental tobacco smoke: Measuring exposures and assessing health effects. National Research Council. National Academy Press, Washington, DC.

Oeschli, F.W. and R.W. Beuchley. 1970. Excess mortality associated with three Los Angeles September hot spells. *Environmental Research* 3:277.

Ohm, M., J.E. Juto, and K. Andersson. 1992. Nasal hyperreactivity and sick building syndrome. *IAQ 92: Environments for People*. ASHRAE, Atlanta.

Offermann, F.J., S.A. Loiselle, A.T. Hodgson, L.A. Gundel, and J.M. Daisey. 1991. A pilot study to measure indoor concentrations and emission rates of polycyclic aromatic hydrocarbons. *Indoor Air* 4:497-512.

Parsons, K.C. and M.J. Griffin. 1988. Whole-body vibration perception thresholds. *Journal of Sound and Vibration* 121(2):237-58.

Proctor and Hughes' chemical hazards in the workplace, 3rd ed. 1991. G.J. Hathaway et al. eds. Van Nostrand Reinhold, New York.

Reiher, H. and F.J. Meister. 1931. Translation of Report Fts616RE 1946. *Forschung VDI* 2, 381-86, "The sensitivities of the human body to vibrations." Headquarters Air Material Command, Wright Field, Dayton, OH.

Ritchie, I.M. and R.E. Lehnen. 1987. Irritation from formaldehyde in mobile and regular homes. *Am J. Public Health* 77:323-28.

Rohles, F.H., J.A. Woods, and P.R. Morey. 1989. Indoor environmental acceptability: Development of a rating scale. *ASHRAE Transactions* 95(1):23-27.

Samet, J.M., M.C. Marbury, and J.D. Spengler. 1987. Health effects and sources of indoor air pollution. *Am. Rev. Respir. Disease* 136:1486-1508.

Samet, J.M. 1989. Radon and lung cancer. *J. Natl. Cancer Institute* 81:145.

Schulman, J.H. and E.M. Kilbourne. 1962. Airborne transmission of influenza virus infection in mice. *Nature* 195:1129.

Selikoff, I.J., J. Churg, and E.C. Hammond. 1965. The occurrence of asbestosis among insulation workers in the United States. *Ann. Rpt. NY Academy of Science*.

Serati, A. and M. Wuthrich. 1969. Luftfreughtigkeit und saison Krankenheit. *Schweizerische Medizinische Wochenscrift* 99:46.

Shickele, E. 1947. Environment and fatal heat stroke. *Military Surgeon* 100:235.

Smith, K.W. 1955. Pulmonary disability in asbestos workers. *Archives of Industrial Health.*

Spengler, J.D., H.A. Burge, and H.J. Su. 1992. Biological agents and the home environment. In: *Bugs, Mold and Rot (I): Proceedings of the Moisture Control Workshop*, E. Bales and W.B. Rose (eds). 11-18. Building Thermal Envelope Council, National Institute of Building Sciences, Washington, DC.

Strindehag, O., I. Josefsson, and E. Hennington. 1988. Healthy Buildings '88, 3:611-20. Stockholm, Sweden.

Streifel, A.J., D. Vesley, F.S. Rhame, and B. Murray. 1989. Control of airborne fungal spores in a university hospital. *Environ. Int.* 15:221.

Susskind, R.R. and M. Ishihara. 1965. The effects of wetting on cutaneous vulnerability. *Archives of Environmental Health* 11:529.

Tancrede, M., R. Wilson, L. Ziese, and E.A.C. Crouch. 1987. *Atmospheric Environment* 21:2187.

Teng, H.C. and H.E. Heyer, eds. 1955. The relationship between sudden changes in the weather and acute myocardial infarction. *American Heart Journal* 49:9.

WHO. 1987. Air Quality Guidelines for Europe. European Series No. 23. WHO, Copenhagen.

Yoshizawa, S., F. Surgawa, S. Ozawo, Y. Kohsaka, and A. Matsumae. 1987. *Proceedings* 4th International Conference on Indoor Air Quality and Climate, Berlin 1:627-31.

Zenz, C. 1988. Occupational safety in industry, occupational medicine, principles and practical applications.

CHAPTER 10

ENVIRONMENTAL CONTROL FOR ANIMALS AND PLANTS

THERMAL conditions, air quality, lighting, noise, ion concentration, and crowding are important in designing structures for animals and plants. Thermal environment influences heat dissipation by animals and chemical process rates in plants. Lighting influences photoperiodism in animals and plants, and photosynthesis and regulation in plants. Air quality, noise, ion concentrations, and crowding can affect the health and/or productivity of animals or plants. This chapter summarizes the published results from various research projects and provides a concept of the physiological factors involved in controlling the environment.

ANIMALS

Animal performance (growth, egg or milk production, wool growth, and reproduction) and their conversion of feed to useful products are closely tied to the thermal environment. For each homeothermic species, an optimum thermal environment permits necessary and desirable body functions with minimum energetic input (Figure 1A). The optimal thermal environment—in terms of an effective temperature that integrates the effects of dry-bulb temperature, humidity, air movement, and radiation—is less important to the designer than the range of conditions that provides acceptable animal performance, efficiency, well-being, and economic return for a given species. Figure 1A depicts this range as the **zone of nominal losses**, selected to limit losses in performance to a level acceptable to the livestock manager. Researchers have found that the zone of nominal losses corresponds to the welfare plateau (i.e., welfare is enhanced by maintaining environmental conditions within the zone of nominal losses). Milk and egg production by mature animals also shows an optional thermal environment zone, or zone of nominal losses (Figure 2).

Developed from actual measurements of swine growth, Figure 1B shows the relationships of energy, growth, and efficiency with air temperature. In the case of growing pigs in Figure 1B, the range of temperatures from 60 to 72°F, which includes both optimal productivity and efficiency levels, represents acceptable design conditions to achieve maximum performance and efficiency. Even beyond that temperature range, performance and efficiency do not markedly decline in the growing pig until near the lower critical temperature (LCT) or upper critical temperature (UCT), and potential performance losses within the temperature range from 50 to 76°F may be acceptable. Response relationships, as shown in Figure 1B, allow environmental selection and design criteria to be based on penalties to performance (i.e, economic costs) and animal well-being—particularly when used with climatological information to evaluate risks for a particular situation (Hahn et al. 1983). Choosing housing requires caution, because research indicates that factors such as group versus individual penning, feed intake, and floor type can affect the LCT by 10°F.

The limits of acceptable values of the LCT and UCT depend on such effects as the species, breed, genetic characteristics of an individual animal's age, weight, sex, level of feeding and type of feed, prior conditioning, parasites, disease, social factors such as space allocation, lactation or gestation, and physical features of the environment. The LCT and UCT vary among individuals; data reported are for group means. As a result, the limits become statistical values based on animal population and altered by time-dependent factors.

Acceptable conditions are most commonly established based on temperature because an animal's sensible heat dissipation is largely influenced by the temperature difference between the animal's surface and ambient air. Humidity and air movement are sometimes included as modifiers for an effective temperature. This has been a logical development. Air movement is a secondary but influential factor in sensible heat dissipation. The importance of air velocities is species and age dependent (e.g., swine under 8 weeks of age experience slower gains and increased disease susceptibility when air velocity is increased from 25 to 50 fpm).

With warm or hot ambient temperature, elevated humidity can restrict performance severely. Relative humidity has little effect on the animal's heat dissipation during cold temperatures, and it is usually only moderately important to thermal comfort during moderate temperatures. (Information is limited on such interactions, as well as on the effects of barometric pressure, air composition, and thermal radiation.)

Animals housed in a closed environment alter air composition by reducing oxygen content and increasing carbon dioxide and vapor content. Decomposing waste products add methane, hydrogen sulfide, and ammonia. Animal activities and air movement add microscopic particles of dust from feed, bedding, and fecal material. Generally, a ventilation rate sufficient to remove water vapor adequately controls gases. However, improper air movement patterns, certain waste-handling methods, and special circumstances (e.g., disease outbreak) may indicate that more ventilation is necessary. In many cases, ventilation is not as effective for dust control as for gas control. Alternative dust control strategies may be needed.

The preparation of this chapter is assigned to TC 2.2, Plant and Animal Environment, with cooperation of Committees SE-302 and SE-303 of the American Society of Agricultural Engineers.

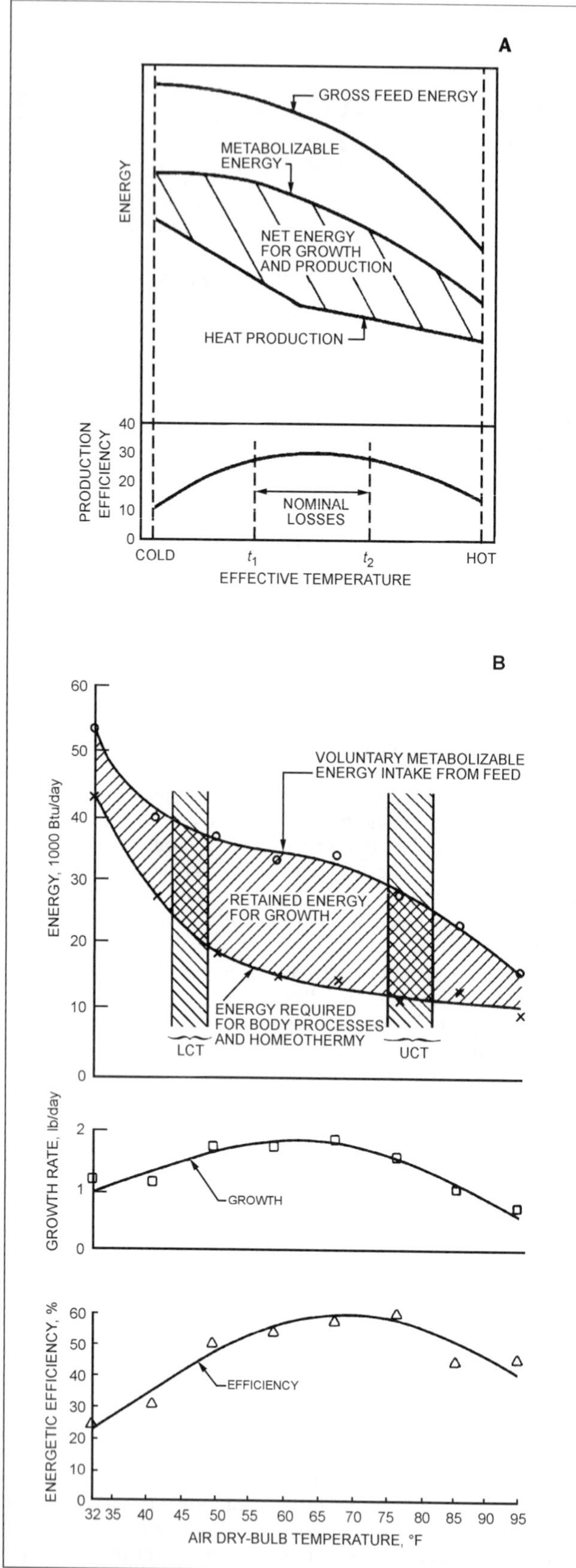

Fig. 1 Energetic and Performance Relationships Typical for Animals as Affected by Effective Environmental Temperatures
(Hahn et al. 1983)

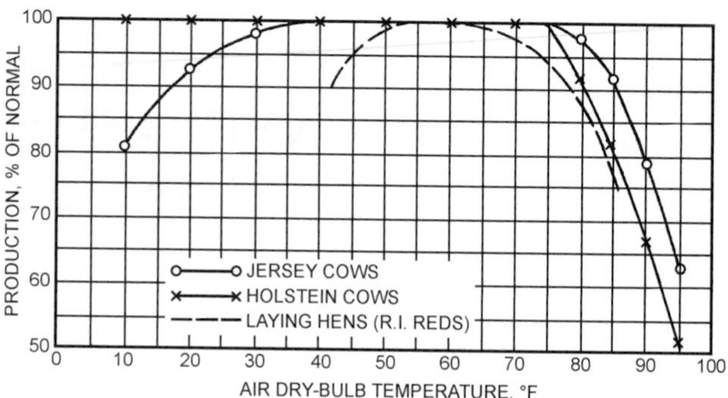

Fig. 2 Comparative Effect of Air Temperature in Mature Animal Production
(Hahn et al. 1983, Tienhoven et al. 1979)

ANIMAL CARE/WELFARE

Animal facilities that facilitate good animal care and welfare must be designed considering a wide range of environmental factors beyond thermal conditions. These include space requirements, flooring type, lighting, feed and water requirements, animal handling, and waste management. Facilities that meet animal care and welfare needs vary considerably by purpose, animal species, and geographic location. For additional information and guidelines for designing animal facilities to meet animal care and welfare needs, see the section on Bibliography.

PHYSIOLOGICAL CONTROL SYSTEMS

An animal has a phenomenally stable control system. Despite wide variations in environmental, nutritional, or pathological conditions, animals can control blood pressure and composition, body temperature, respiration, and cardiac output without conscious effort. Physiological control systems react to unfavorable conditions to ensure survival of the animal at the expense of production or reproduction.

One physiological control system important for air-conditioning design is the homeothermic system—the means by which an animal adapts to its thermal environment. The animal strives to control body temperature by adjusting both heat production in the tissues and transfer of heat to the environment from exposed surfaces. The homeothermic system in domestic animals is a closed-loop system and can be analyzed like any closed-loop control system.

Heat Production

Heat production data have been measured for many farm animals (Yeck and Stewart 1959, Longhouse et al. 1960, 1968, Bond et al. 1959). Much of the early data were obtained for a basal condition (i.e., all life processes at a minimum level). Heat production under conditions of normal metabolic activity is more useful to the engineer. Such data are now available for most farm animals (Figures 3 and 4). Where possible, the data show total or sensible and latent heat for animals fed in a typical housing situation. Thus, the sensible heat used to evaporate urine moisture (assuming no other moisture) appears as latent heat. For design purposes, such data are better reflections of total heat partitioning than metabolic heat production obtained by calorimetry. Scott et al. (1983) provide additional information on specific situations.

The rate of heat production is primarily a function of temperature, animal species, and animal size. Heat production varies diurnally, depending on animal activity and eating times, and may change dramatically under special circumstances (e.g., when animals are disease-stressed). Building type and waste-handling system can affect the conversion of sensible to latent heat by more than 50%. Heat and moisture production of animals changes with other

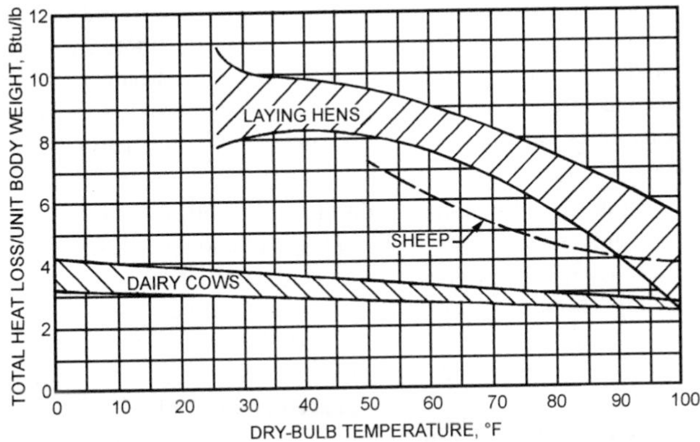

Fig. 3 Comparative Heat Loss of Mature, Producing Animals
(Tienhoven et al. 1979)

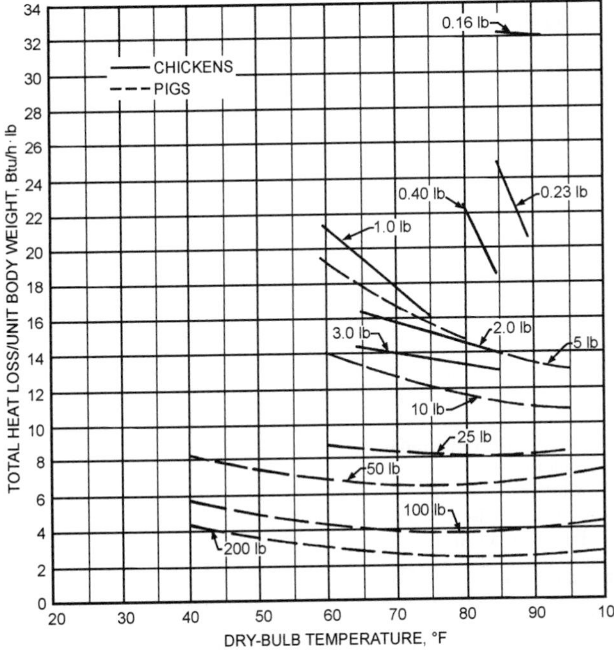

Fig. 4 Comparative Heat Loss of Growing Animals
(Bond et al. 1959)

variables such as their genotype and growth rate. For example, a broiler chicken grows to 9 lb in 7 weeks compared to 12 weeks 30 years ago. Faster growth rate is usually associated with higher heat production, so updated data is needed. Therefore, the ranges of ventilation require careful analysis.

Heat Transfer to Environment

An animal can control, to some extent, the amount of heat transferred from its body by chemical and physical regulations. In a cold environment, the animal's metabolism increases, which increases the amount of heat production, offsetting heat transfer to the environment. (The principal result of such chemical regulation is inefficient feed utilization.) Physical regulation is also used in a cold environment; blood circulation to subcutaneous capillaries decreases, hair or feathers are erected, and animals huddle together in an attempt to reduce sensible heat loss. In a hot environment, opposite physical and physiological responses generally occur to enable transfer to the environment of heat associated with necessary and productive life processes. In addition, as sensible heat dissipation becomes more difficult, evaporative heat loss increases and air moisture content becomes a factor in heat loss. Since the production of heat is a necessary by-product of growth and useful production, environmental limits to dissipating this heat cause a decrease in feed consumption and subsequent decrease in growth and production.

Most sensible heat of domestic animals is dissipated through the skin. Birds, sheep, and swine transfer most of their latent heat through the respiratory tract; cattle and horses transfer most of their latent heat through the skin. Blood transfers the heat produced by metabolism to the skin or to the lung surfaces, where it is dissipated by evaporation from the mucous layer coating the inside of the alveoli. Inspired air reaching the alveoli is heated almost to body temperature and may become nearly saturated with moisture. Expired air may not be saturated at body temperature—especially during periods of heat stress—because it is a combination of air, some of which has not reached deeply into the lungs. Minor amounts of heat are also transferred by ingestion of feed and water and through excretion.

Design factors affecting animal heat loss are (1) air temperature; (2) air vapor pressure; (3) air movement; (4) configuration, emissivity, absorptivity, and surface temperature of the surrounding shelter; and (5) temperature and conductivity of surfaces (e.g., floors with which the animal may be in contact).

Cyclic Conditions

The physiologic sensing elements respond both to environmental conditions and to changes in those conditions. Cycles of temperature, pressure, light, nutrients, parasites, magnetic fields, ionization, and other factors frequently occur with little engineering control.

Light is perhaps the earliest discovered and most important controlled environmental variable affecting reproductive processes (Farner 1961). However, the effects vary widely among animal species and age. Some animals grow and remain healthy with or without light (e.g., growing swine), while for others, lighting management is important. Studies have shown that (1) short photoperiods induce or accelerate estrus development in sheep; (2) day length affects semen production in sheep and horses (Farner 1961); (3) continuous white incandescent light during incubation caused White Leghorn eggs to hatch from 16 to 24 h earlier than eggs incubated in darkness (Shutze et al. 1962); and (4) red, yellow, or blue lights gave comparable results. Use of lights for only one of the three weeks also reduced the time required for hatching, but differences were not as marked. Percent hatchability was not affected by lighting treatments.

Light is used to delay sexual maturity in hens, which enhances subsequent production. This is done by gradually decreasing day length from hatching to 22 weeks of age or by abruptly decreasing day length to 9 h at 14 to 16 weeks of age. If pullets have previously been exposed to an increasing day length, the change should take place at 14 weeks of age; if exposed to a constant or decreasing schedule, a change at 16 weeks is adequate. Light intensities of 1 to 2 footcandles measured at bird height were adequate in all cases. Light can then be abruptly increased at 21 weeks of age to 14 or 16 h of light. The economic value of increasing day length beyond 14 h in a windowless poultry house, and 16 h in an open poultry house, has not been proved. Photoperiods such as 8 h of light (L), 10 h of darkness (D), 2 L and 4 D, and other cycles have improved feed utilization, egg production, and poultry growth (Buckland 1975, Riskowski et al. 1977).

Continuous light from hatching through 20 to 21 weeks of age markedly depressed subsequent egg production and caused a severe eye abnormality, but did not depress egg mass. (Light intensity was 1 to 3 footcandles, measured at bird height.)

Temperature cycles have been studied in cattle (Brody et al. 1955, Kibler and Brody 1956), swine (Bond et al. 1963, Nienaber et al. 1987), and poultry (Squibb 1959). The results differ somewhat among animals. Heat loss from cattle and swine can usually be calculated from average daily environmental temperatures with sufficient accuracy for design load calculation.

Productivity is only slightly different for averaged temperatures when cyclic conditions with a range less than 20°F are experienced (Squibb 1959). Above that range, productivity is depressed below that expected from the averaged temperature as determined in constant temperature tests, but under diurnally varying air conditions (68 to 77°F temperature and widely varying humidities), near-normal egg production is maintained.

Air Composition and Contaminants

The major contaminants in livestock housing are (1) respirable dusts from feed; manure; and animal skin, hair, and feathers; (2) microbes, both pathogenic and nonpathogenic, hosted in the respiratory tracts, animal wastes, or feed; and (3) several noxious gases of various concentrations. Respirable dust particles have diameters between approximately 0.5 and 5 μm. Gases are produced from the metabolic processes of animals, from the anaerobic microbial degradation of wastes, and from combustion processes. The gases of most concern are ammonia, hydrogen sulfide, carbon monoxide, and methane.

Dust levels in animal housing are high enough to create a nuisance in and near animal buildings, increase labor requirements for building and equipment maintenance, and interfere with the performance of heating and ventilating equipment. Dust has been implicated in poultry building fires. Dust is generated primarily by feed handling and increases with animal activity and air movement causing reentrainment of settled dust.

Contaminants are a concern because they predispose animals to disease and poorer performance and affect operator health. Animals experiencing stress (e.g., newborns, hot or cold, nutritionally limited, and sick animals) are more sensitive, and the presence of low levels of contaminants can have adverse effects on them.

Particulates (dust, endotoxins, live and dead microorganisms) and gases (ammonia, carbon dioxide, and hydrogen sulfide) in swine buildings (Zejda et al. 1993, Senthilselvan et al. 1997) have been implicated as contributors to the increased incidence of respiratory disorders among livestock producers compared to grain farmers and non-farm workers (Donham et al. 1989, Dosman et al. 1988). Young farmers may be at particular risk of developing chronic bronchitis, coughing, wheezing, toxic organic dust syndrome, and/or occupational chronic pulmonary disease (Zejda et al. 1994). For example, 33% of pork producers in Europe suffer chronic respiratory symptoms related to poor indoor air quality (Van't Klooster 1993). In the Netherlands, 10% of swine producers had to change jobs due to severe respiratory problems caused by poor air quality (Preller et al. 1992). In cold climates, the problems appear to be more serious because of (1) larger building size and longer working hours and (2) lower air exchange rates due to cold climates and energy conservation concerns.

Particle size is most commonly described using aerodynamic diameter, which is defined in terms of its aerodynamic behavior rather than its geometric properties. Aerodynamic diameter is a function of its equivalent geometric diameter, density, and a shape factor. The aerodynamic diameter of a particle is the diameter of a spherical water droplet having the same settling velocity as the particle of concern. Size distribution of dust particles influences the dust transportation and its effect on a human respiratory system. The human eye can only see a particle larger than 50 μm under normal light. For evaluating animal buildings, particle size is often divided into the following categories:

- Total dust = all sizes of particles suspended in the air
- Inhalable dust = all sizes of particles that are inhaled by a sampling instrument
- Respirable dust = particles smaller than 10 μm

In animal facilities, ultrafine particles (< 0.5 μm) are usually excluded from the evaluation because they are primarily background dust and not the concern of the indoor environment. If the dust concentration is measured in counts rather than mass, the ultrafine particles can dominate over the counts of the other particle sizes and cause large errors (Zhang et al. 1994).

Dust in animal buildings is different from other types of buildings in at least three aspects. First, animal building dust is biologically active in that it contains a variety of bacteria, microorganisms, and fungi (Martin et al. 1996, Wiegard and Hartung 1993). Second, its concentration is high, typically more than ten times higher than office buildings and residential buildings. Third, it is an odor carrier. Dust sources and flora are summarized in Table 1. Of those microbes, some are transmittable to people. For example, *Listeria monocytogenes* and *Streptococcus suis* are zoonotic agents that have caused fatal diseases in people (Fachlam and Carey 1985, Bortolussi et al. 1985).

Table 1 Dust Sources and Flora in Swine Buildings

Dust Sources
Feed particles:
Grain dust
Antibiotics
Growth promoters
Swine protein:
Urine
Dander
Serum
Other agents:
Swine feces
Mold
Pollen
Grain mites, insect parts
Mineral ash
Gram-negative bacteria
Endotoxin
Microbial proteases
Ammonia adsorbed to particles
Infectious agents
Microbial flora
Gram-positive cocci:
Staphylococcus species (coagulase-negative)
Staphylococcus haemolyticus
Staphylococcus hominis
Staphylococcus simulans
Staphylococcus sciuri
Staphylococcus warneri
Micrococcus species
Aerococcus species
Streptococcus equines
Streptococcus suis (presumptive)
Enterococcus durans
Gram-positive bacilli:
Corynebacterium species
Corynebacterium xerosis
Bacillus species
Gram-negative bacilli:
Acinetobacter calcoaceticus
Nonfermentative gram-negative bacillus:
Enterobacter calcoaceticus
Pasteurella species
Vibrio species
Fungi:
Alternaria species
Cladosporium species
Penicillium species

Source: Zhang (1995, 1996).

Threshold limit values for different types of exposure are available (Table 2). These limit values are likely additive. For example, someone exposed to 75% of maximum ammonia level may be exposed to no more than 25% of the maximum for any other contaminant or combination of contaminants. Due to its biological nature, a lower limit for animal building dust may be warranted. For example, Donham et al. (1989) proposed a more stringent threshold TWA value of 65 mg/ft^3 respirable dust for farm animal buildings.

Airborne dust and microbes have been linked to respiratory disease in cattle, poultry, swine, horses, and laboratory animals. Infectious diseases, such as viral pneumonia and diarrhea, in calf barns have caused mortality rates between 20 and 80%, with the highest mortality occurring in enclosed barns that were underventilated during cold weather. Many respiratory diseases of poultry have been found to be transmitted via pathogenic microbes that can travel over 30 miles and remain infectious for months (Siegmund 1979). Respiratory diseases are also costly to the swine industry—the incidence of enzootic pneumonia in pigs has ranged from 30 to 75%.

Gaseous ammonia is frequently a contaminant causing serious problems in swine and poultry housing. Ammonia levels of 50 ppm have been shown to reduce body mass gain rates of swine (Curtis 1983), and many researchers suspect that adverse health effects begin at levels below 25 ppm. Some physiologists and veterinary science researchers suggest that the level is between 5 and 10 ppm (Donham 1987). Ammonia at concentrations greater than 60 ppm has been implicated in reduced body weight and feed consumption, and in increased ocular and respiratory problems with broiler chickens (Carr and Nicholson 1980). Ammonia levels are influenced by animal diet, the ventilation rate, sanitation practices, and the waste-handling system. Ventilating for moisture control is usually adequate to control ammonia below 10 ppm in buildings sanitized monthly, where wastes are not allowed to accumulate for more than two weeks and where manure solids are covered by water.

Hydrogen sulfide is produced mainly from the wastes and is present at toxic levels that usually cause problems only when liquid wastes are agitated or pumped and emitted gases can filter into the animal occupied area. Hydrogen sulfide production can rise to lethal levels when manure in a pit below the animals is agitated, so extra precautions are necessary to ensure that ventilation is sufficient to remove the contamination.

Carbon monoxide is usually produced by malfunctioning fuel-burning heaters and internal combustion engines used for spray washing equipment. Many heaters are unventilated, which, when combined with the corrosive environment in livestock housing, makes this equipment susceptible to failure. Heaters should be vented to the outdoors or sensors should be installed to ensure complete combustion.

PRINCIPLES FOR AIR CONTAMINANT CONTROL

Reducing dust concentration by 85% or reducing dust inhalation by wearing a respirator improves human respiratory responses significantly (Senthilselvan et al. 1997, Zhang et al. 1998, Barber et al. 1999). Air contaminant control strategies include source suppression, air cleaning, and effective ventilation.

Source Suppression

Spreading (e.g., spraying, sprinkling, or fogging) a small quantity of oil, water, emulsifier or mixture of these can reduce dust concentration in the air (Takai et al. 1993, Zhang et al. 1995, 1996). Adding oil or fat to feed (Heber and Martin 1988) reduces emission of large feed dust particles. Other management practices, such as wet feeding, power washing, and maintaining sanitation in the facility also reduces the production and emission of air contaminants.

Table 2 Gas and Dust Threshold Values for People Exposure

Contaminant	Specific Density[a]	Odor	TWA[b]	STEV[b]	FEV[c]
Toxic gases (ppm in volume)					
Ammonia	0.6	sharp, pungent	25	35	300
Hydrogen sulfide	1.19	rotten eggs, nauseating	10	15	150
Methane	0.5	none	n/a	n/a	500,000
Carbon dioxide	1.53	none	5,000	30,000	50,000
Dust (mg/ft^3)					
Respirable			5	n/a	n/a

[a]Relative to the density of air.
[b]From American Conference of Governmental Industrial Hygienists (ACGIH 1998).
[c]From Canada Plan Service, M-8710 (1985), M-9707 (1992).
n/a = currently not available
TWA = time-weighted average, for 40 hours per week
STEV = short-term exposure value for not exceeding 30 minutes
FEV = fatal exposure values

Air Cleaning

Generally, air contaminants can be removed mechanically, electrically, chemically, or biologically. **Mechanical air cleaning** includes filtration, scrubbing, and aerodynamic separation (cyclones or dust separators). Existing mechanical air cleaning methods, including fiber filters, require large quantities of airflow through the filtration media. For many dusty environments such as found in animal buildings, this type of cleaner requires frequent maintenance.

Cyclones use centrifugal force to separate particles from the main air stream. But a high pressure drop (typically higher than 2 in. of water) across the cyclone is required to create the cyclone effect. This high pressure requires a large amount of power and creates high turbulence, which reduces the particle separation efficiency. Carpenter (1986) concluded that it is impractical to separate particles smaller than those that cause health concerns in animal buildings.

Electrical air cleaning includes ionization, electrostatic precipitation, and ozonation. The principle of the three technologies is to charge a dust particle with polarized electrical particles (ions or electrons) so that the charged dust particles can be attracted to a neutrally or opposite charged surface. Challenges of the ionization technology for livestock buildings include dust removal efficiency, initial and maintenance costs, and a potential problem of static electricity when used under cold and dry weather conditions.

The air cleaning efficiency of an electrostatic precipitator is proportional to the airflow rate through the precipitator. A large airflow rate in an airspace can reentrain a large amount of settled dust and thus create a high dust concentration in the air. Dust reduction efficiency, airflow rate, and power consumption of the air delivery through the precipitator must be balanced for optimum performance. Similar to air filters, an electrostatic precipitator requires frequent cleaning and maintenance. When dust particles accumulate on the precipitation surfaces of the device, the air cleaning efficiency rapidly decreases.

Ozone can effectively reduce airborne microorganisms in an animal building. Care must be taken to maintain an appropriate concentration because ozone is a potential irritant to humans and animals.

Ventilation Requirement

Ventilation is the primary method for controlling air quality in confinement buildings as it brings in fresh air to dilute the contaminants. The logic of ventilation is illustrated in Figure 5. Unlike residential or office buildings, the minimum ventilation for animal

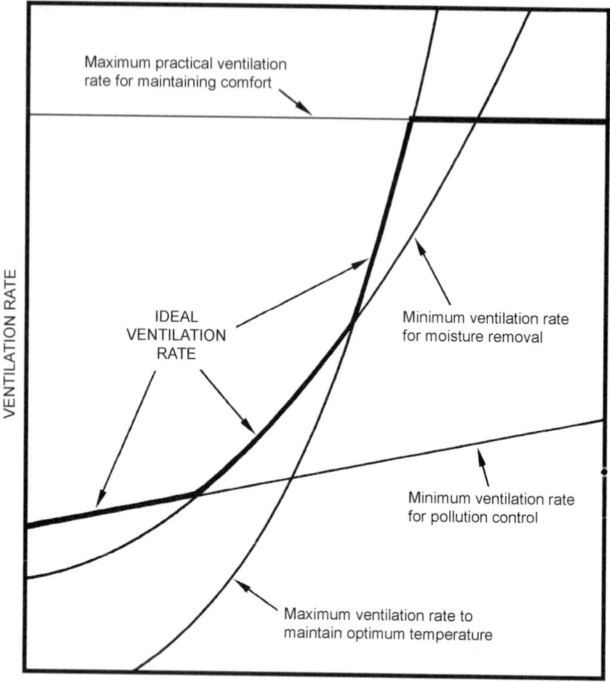

Fig. 5 Logic for Selecting the Appropriate Ventilation Rate in Livestock Buildings
(Christianson and Fehr 1983)

buildings is often determined by humidity or contaminant (such as ammonia) balances. Therefore, ventilation requirements must be calculated for sensible heat, humidity, and a given contaminant to determine the minimum ventilation for an animal building.

Volumetric ventilation requirement for sensible heat balance V_s, moisture balance V_w, and contaminant balance V_c (in cfm) can be calculated as follows:

$$V_s = \frac{q_n \text{v}}{(h_s - h_{so})} \tag{1}$$

$$V_w = \frac{W_n \text{v}}{(w - w_o)} \tag{2}$$

$$V_c = \frac{M_c \text{v}}{(c - c_o)} \tag{3}$$

where

v = specific volume of air, ft³/lb
q_n = net heat transfer rate through the building shelter, Btu/min
h_s = sensible heat content of room air, Btu/lb
h_{so} = sensible heat content of outside air or supply air, Btu/lb
w = moisture content of room air, lb/lb of air
w_o = moisture content of outside air, lb/lb of air
W_n = production rate of moisture within the room, lb/min
c = contaminant concentration of room air, lb/lb of air
c_o = contaminant concentration of outside or supply air, lb/lb of air
M_c = production rate of the contaminant in the room, lb/min

Ventilation Effectiveness

Two terms are often used to evaluate a ventilation system: ventilation efficiency and ventilation effectiveness. **Ventilation efficiency** is a criterion for energy and fan performance; it is not directly related to ventilation effectiveness and air quality control. Ventilation efficiency refers to the mass of air delivered per unit of power consumed by the ventilation system at a given pressure. The more air delivered per unit of power, the more efficient the ventilation system. It is also called the **ventilation efficiency ratio** (VER) or **energy efficiency ratio** (EER) (Pratt 1983).

Zhang et al. (2000) proposed a **ventilation effectiveness factor** (VEF) to quantitatively describe the ventilation effectiveness for controlling indoor air quality. The VEF can be defined as the ratio of two contaminant concentration differentials:

$$\text{VEF} = \frac{C_x - C_s}{C_m - C_s} \qquad (C_m > C_s \text{ and } C_x > C_s) \tag{4}$$

where C_x is the contaminant concentration under complete mixing conditions, which can be determined based on the mass balance of the contaminant of concern. For a ventilated airspace at steady state, the amount of contaminant removed by the exhaust air must equal the sum of the amount of contaminant brought in by the supply air plus the contaminant produced in the airspace, regardless of complete or incomplete mixing conditions. The contaminant concentration of the exhaust air C_e can be measured. Under complete mixing conditions, C_e is the same as the C_x; thus,

$$C_x = C_e \tag{5}$$

C_s is the contaminant concentration of the supply air, C_m is the mean contaminant concentration of the airspace at n measured locations.

$$C_m = \frac{1}{N}\sum_{i=1}^{N} C_i \tag{6}$$

Substituting Equations (5) and (6) into Equation (4), ventilation effectiveness factor becomes

$$\text{VEF} = \frac{N(C_e - C_s)}{\sum_{i=1}^{N} C_i - NC_s} \qquad (C_e > C_s) \tag{7}$$

The ventilation effectiveness factor is nondimensional and independent of the ventilation rate. The higher the value of the factor, the more effective the ventilation system for contaminant removal. When $C_e < C_s$, the airspace becomes a settling chamber or an air cleaner and the air cleaning efficiency $\xi = 1 - C_e/C_s$ should be used to evaluate the system. When the contaminant concentration of supply air is negligible, i.e., $C_s \approx 0$, the ventilation effectiveness factor becomes

$$\text{VEF} = \frac{C_e}{C_m} \tag{8}$$

One useful feature of VEF is that the system being evaluated can be its own control. This feature is particularly valuable for evaluating or troubleshooting in the field. From Equation (4) and Equation (8), the VEF compares the actual contaminant concentration with its own concentration under complete mixing conditions in the same airspace. The existing methods of evaluating ventilation effectiveness often require comparing two systems. In practice, it is usually difficult to find an identical system for comparison.

A ventilation effectiveness map (VEM) such as shown in Figure 6 is a useful tool for design, analysis, and modification of a ventilation system. It is a contour plot of the ventilation effectiveness factors calculated using Equation (8) for the entire airspace. Dust mass was measured concurrently at several locations throughout the building cross section and the VEF values were calculated from that

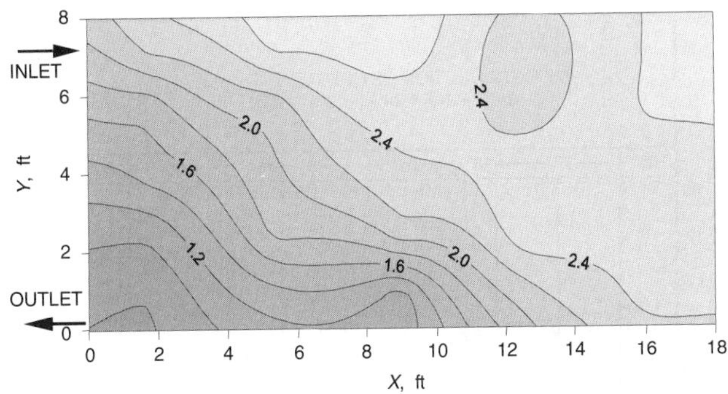

Fig. 6 Typical Ventilation Effectiveness Map for a Room
(Room is 18 ft long by 12 ft wide by 8 ft high with slot air inlet at top of one wall and air outlet near bottom of same wall.)

data. The higher the VEF values in the map, the more effective the ventilation system is in that area.

Improved ventilation effectiveness is an important strategy to reduce contaminant concentration in animal buildings. It is very expensive to clean large quantities of air and it is difficult to move the air through central cleaning equipment without creating drafts. Incomplete mixing ventilation such as **displacement** and **zonal ventilation** has great potential in creating a cleaner zone in a dusty airspace without changing the overall ventilation rates.

Interactive Stressors

Thermal (air temperature, air humidity, air velocity, and surrounding surface temperatures) and air quality (relative absence of dust, microbes, and contaminant gases) are the two stressors of most concern to ventilating a livestock structure. However, health status, animal age, stage of production, nutrition, and social conditions interact with the thermal and air quality conditions to determine animal health and performance.

Whether stressors are linearly additive or otherwise is an important research area. McFarlane (1987) found that ammonia, heat, acoustic noise, disease, and beak-trimming stressor effects on chickens were linearly additive in effect on the feed efficiencies and daily gain rates.

Comparing Individual Animal and Room Heat Production Data

Liquid wastes from urine, manure, waterer spillage, and cleaning affect the sensible and latent heat fraction in a room (ASAE 1991). More water on the floors requires more thermal energy from the animals or the building heating system to evaporate the moisture. This evaporated moisture must then be removed by ventilation to prevent condensation on building surfaces and equipment and minimize adverse health effects, which can occur when humidities exceed 80%.

Floor flushing or excessive water spillage by animals can increase latent heat production by one-third from animal moisture production data, with a corresponding reduction in sensible heat production on a per-animal basis (ASAE 1991). Partially slatted floors with underfloor storage of wastes reduce the room moisture production by approximately 35%, compared to solid floor systems. Rooms with slatted floors throughout the building may have moisture production rates as much as 50% lower than solid floors. Therefore, slatted floors reduce the ventilation rate required for moisture removal, while excessive water spillage increases the moisture removal ventilation rate.

Influence of Genetic Change and Breed on Heat Production

Researchers have noted significant genetic correlations with performance. For example, swine breeds with high fat content characteristics outgained low fat content breeds, but the low fat content lines used feed more efficiently (Bereskin et al. 1975). Barrows (male hogs castrated before sexual maturity) ate 6% more than gilts (young sows that have not farrowed) to sustain a 7% faster growth rate.

Genetic improvements within breeds suggest that performance data should be updated periodically. Between 1962 and 1977, for example, the average gain rate for swine increased 0.15 lb per day, and feed efficiency improved 15% for Missouri hogs (Thomeczek et al. 1977).

CATTLE

Growth

Figure 7 shows the general growth rate for both beef and dairy breeds. Efficiency of beef calf growth is of economic importance; the dairy calf is developed for adult productive and reproductive capacity. Fattening calves and yearlings could be fed increasing amounts of grain and hay, with an expected gain of about 2 lb per day. Figure 8 shows the effect of temperature on the growth rate of several breeds of beef calves.

Food and energy requirements for growth can be computed by taking the difference between the energy in the available ration consumed and the energy required for maintenance and growth. Energy requirements for maintenance at an air temperature of 68°F has been reported to be 8.65 Btu/h per pound of body weight (Blaxter and Wood 1951), 7.98 Btu/h (Bryant et al. 1967), and 8.34 Btu/h (Gebremedhin et al. 1983). Using 370 kcal/lb body tissue for conversion of feed energy to body tissue (Brody 1945), the growth rate of a calf at 70°F (5.4×10^{-4} lb/h gain per pound of body weight) can be nearly 30 times that of a calf being kept at 37°F (0.18×10^{-4} lb/h·lb). Alternatively, to get the same growth rate at 37°F as at 70°F, the same ration must be fed at the rate of 11.6% of body weight each day, a 16% increase (Gebremedhin et al. 1981).

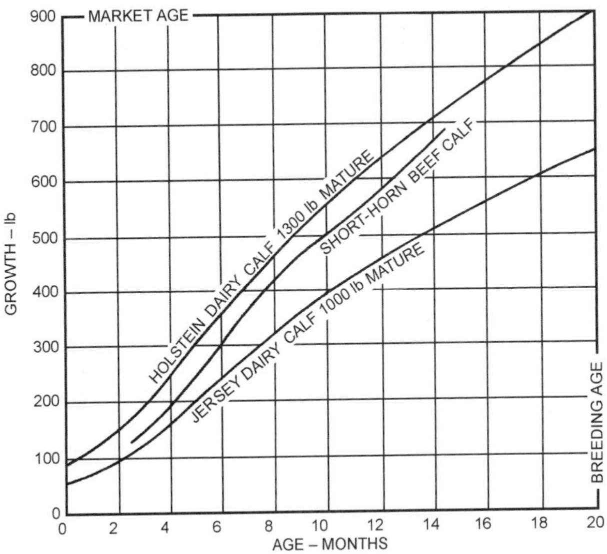

Fig. 7 General Growth Curves for Calves Fed Grain
(Johnson et al. 1958, Ragsdale 1960)

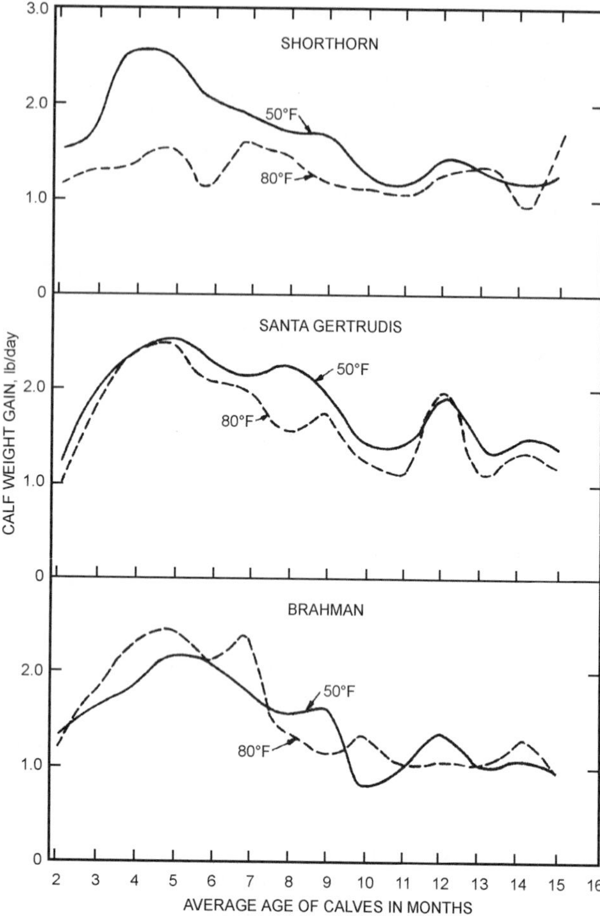

Fig. 8 Daily Gain of Beef Calves
(Longhouse et al. 1960)

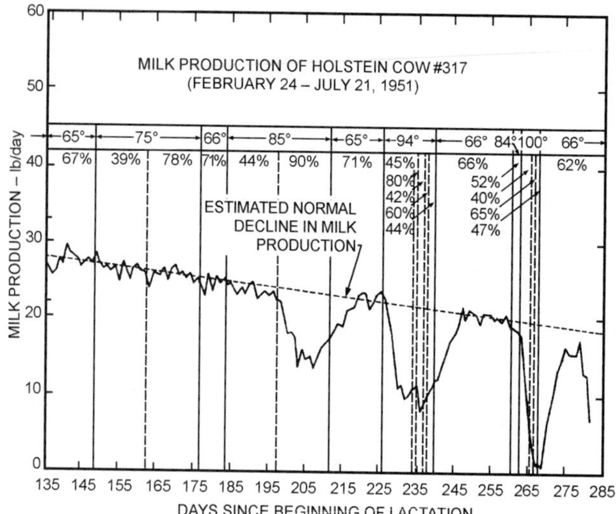

Fig. 9 Milk Production Decline of One Holstein During Temperature-Humidity Test
(Yeck and Stewart 1959)

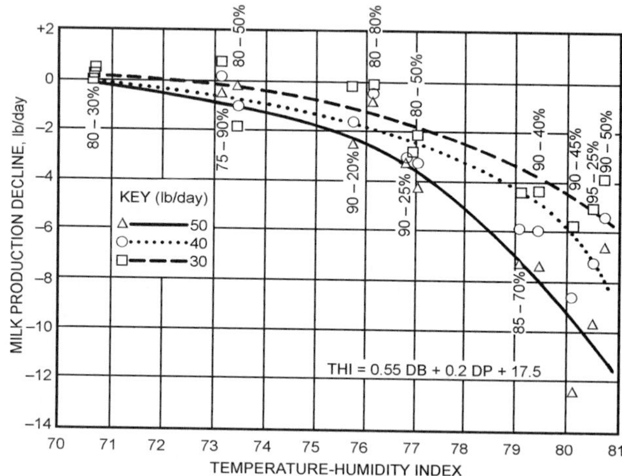

Fig. 10 Effect of Milk Production Level on Average Decline in Milk Production Versus Temperature-Humidity Index
(Johnson et al. 1962)

Lactation

Figure 9 illustrates the effect of high temperature on the milk production of one Holstein in a test designed to study the combined effects of temperature and humidity. The dotted line represents the normal decline in milk production (persistency) from an advancing stage of lactation.

The ideal environment for Holstein cattle should not exceed 75°F. Jerseys are somewhat more heat tolerant, and their limit can be 80°F (Yeck and Stewart 1959). At the lower end of the temperature scale, production decreases may be expected at 30°F for Jerseys and below 10°F for Holsteins and Brown Swiss.

A temperature-humidity index (THI) expresses the relationship of temperature and humidity to milk production. Temperature-humidity indexes have a greater effect on cows with a genetic potential for high milk production than for those with a lower potential (Figure 10).

Under high temperatures, a cooling system can help maintain milk production and reproductive function. In field studies conducted by Bucklin and Turner (1991) in both warm and hot, humid climates, a system providing evaporative cooling by forced air movement and direct sprinkling of lactating cows increased milk production by a range of 7 to 15% over cows not exposed to cooling.

Reproduction

Prolonged low temperatures, even those well below freezing, do not affect the reproductive performance of farm livestock, but breeding efficiency of both sexes decreases under summer conditions. Sustained temperatures above 85°F may decrease fertility, sperm production, and semen quality of males, and increase anestrus and embryonic death in females. In bulls, temperatures above 75°F decrease spermatogenesis, and long exposures at 85°F or above cause temporary sterility. The extent of the reaction depends on temperature rise and exposure duration.

In females, Guazdauskas (1985) observed that conception rates decreased from a range of 40 to 80% conception in thermoneutral environments (50 to 72°F) to a range of 10 to 51% conception in hot environments (>81.5°F). A cooling system involving a cooled shade and artificially induced air movement increased breeding efficiency by approximately 100% in a hot climate (Wiersma and Stott 1969).

Heat and Moisture Production

Sensible and latent heat production from individual animals (Figure 11) differ from the stable heat and moisture production of animals (Figures 12 and 13). The data in Figures 12 and 13 were obtained while ambient conditions were at constant temperatures and relative humidities were between 55 and 70%. The effects of evaporation from feces and urine are included in these data.

The rate of cutaneous water loss is very small at colder temperatures but rises sharply above 64°F. Cutaneous evaporation as a means of heat loss in calves becomes increasingly important as the air temperature rises above 75°F. As the air temperature rises, the proportion of nonevaporative cooling decreases. Above 86°F, about 80% of the heat transferred is by evaporative cooling. Gebremedhin et al. (1981) observed that cutaneous water loss from calves varied between 6×10^{-5} and 40×10^{-5} lb/h per pound of body weight for air temperatures between 32 and 64°F and increases steadily beyond 64°F (Figure 14). Water loss by respiration, the other avenue of loss, is shown in Figure 15.

SHEEP

Growth

In normal environments, an average daily gain of 0.5 lb per day can be expected of lambs marketed at 80 to 120 lb, depending on breed. Hampshires often gain more than 1 lb per day, and lambs can be fed to weigh over 200 lb while still under a year old.

Variations exist among breeds and strains of sheep in their ability to adjust to environmental changes. Temperature effects on sheep growth suggest a lower rate at elevated temperatures. A South African study shows that lack of shade in warm climates reduces growth rate. Australian studies indicate that Merino lambs survive for only about 2 h in an air temperature of 100°F.

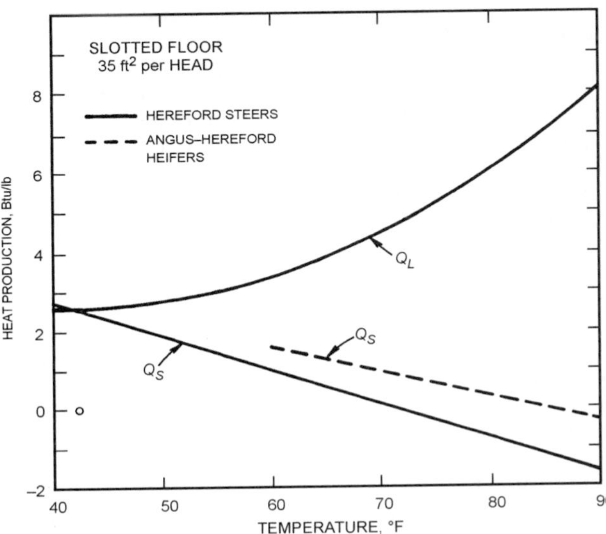

Fig. 11 Sensible and Latent Heat Production per Unit Livestock Mass (Weight)
(Hellickson et al. 1974)

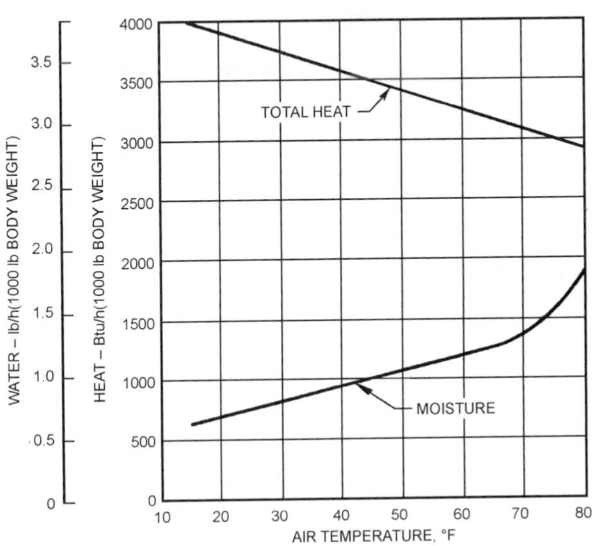

Fig. 12 Stable Heat and Moisture Dissipation Rates Dairy Cattle Stanchioned in Enclosed Stables
(Yeck and Stewart 1959)

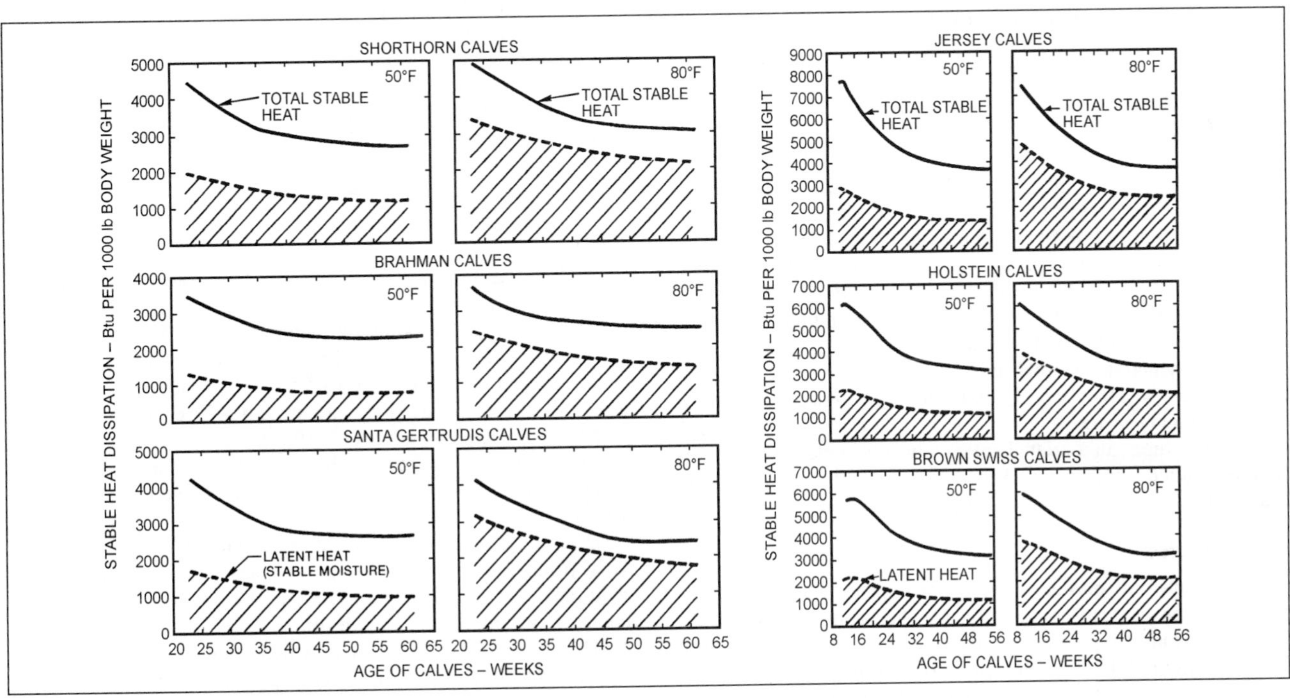

Fig. 13 Stable Heat and Moisture Loads for Beef and Dairy Calves in Pens
(Yeck 1957, Yeck and Stewart 1960)

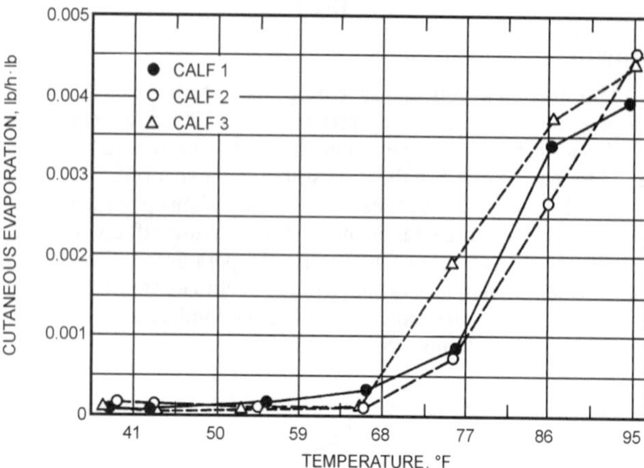

Fig. 14 Cutaneous Water Loss per Unit Body Weight of Holstein Calves
(Gebremedhin et al. 1981)

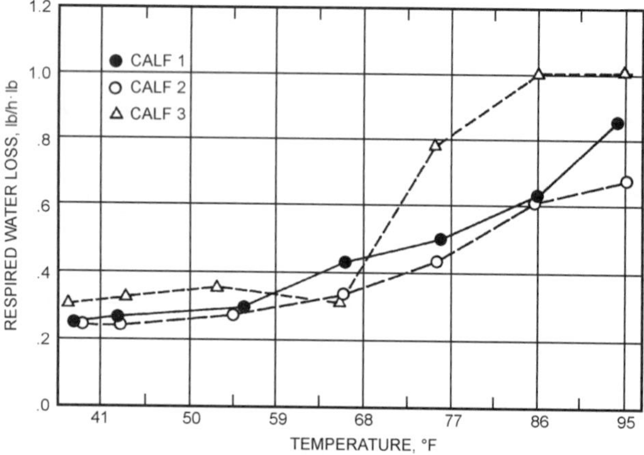

Fig. 15 Respiratory Water Loss per Unit Body Weight of Holstein Calves
(Gebremedhin et al. 1981)

Wool Production

The amount and quality of wool produced varies considerably among breeds, ranging from about 3 lb of poor quality wool from Dorset ewes (a breed developed primarily for mutton) to well over 10 lb of high quality fleece from dual-purpose breeds. Rambouillet, Merino, and Columbia-Southdale breeds supplied 7 lb per year, and about 4.5 lb was grown by a Hampshire, where monthly shearing increased wool production by about 1 lb per year.

Environmental factors such as photoperiod, nutritional level, and temperature also affect wool growth. Skin temperature is considered a dominant factor; high wool growth is associated with high skin temperatures. In a related study, low subcutaneous blood circulation limited wool growth. A thick fleece appears to limit radiant heat loads.

Reproduction

Sheep mate only during certain periods of the year, but they can sometimes be induced to mate outside the normal season if the natural environment is modified. Ewes may breed if exposed to an air temperature of 45°F before mating is attempted.

In studies on the effects of high temperatures on sheep reproduction, air temperature was reported to affect the spermatogenesis rate. Rams kept at 90°F environmental temperature showed a reduced rate of spermatogenesis, although a 3-week period at 70°F effected recovery to the normal rate. Later reports suggest that high air temperatures may cause lowered fertility for other reasons. Subjecting ewes to 100°F just before mating causes fertilization failure because of some form of ovum structure degeneration.

Under practical conditions, early embryonic death appears to be the greatest loss in potential offspring conceived in a high-temperature environment. Degree of susceptibility to high temperature is greatest near mating time but generally decreases as the length of time after mating increases. A temperature of 90°F and 65% rh at mating kills most embryos at an early age. The same conditions applied later in gestation do not cause death, but do cause the birth of small, weak lambs. The longer the high-temperature period, especially during the last third of gestation, the greater the number of weak lambs born. All reproductive processes appear to be more adversely affected by high air temperature when accompanied by high relative humidity.

Table 3 Heat Production of Sheep

Fleece Length, in.	At 46°F Total[a]	At 46°F Latent[b]	At 68°F Total[a]	At 68°F Latent[b]	At 90°F Total[a]	At 90°F Latent[b]
Mature, maintenance-fed						
Shorn	4.0	8%	2.7	12%	2.0	38%
1.2	2.2	29%	2.0	28%	2.0	65%
2.4	2.0	23%	1.9	43%	1.9	76%
Lambs, 1 to 14 days						
Normal	10.5	—	8.0	—	—	—

Source: Scott et al. (1983). [a]Btu/h·lb of body mass [b]Percent of total heat

Heat Production

Only a few heat-production tests have been conducted, mostly for correlation with other physiological data. A lamb's ability to generate heat is important because it is normally born during the most severe climatic season. One study indicated newborn lambs were limited in heat production to 30 Btu/h·lb—five times the basal level. Test conditions were from 14 to 32°F with a 12 mph wind, but lambs are capable of withstanding temperatures as low as –40°F. Table 3 provides heat production data for sheep; while the total heat production data are reliable, the latent heat proportions do not reflect the portions of sensible heat used for water and urine evaporation from flooring and bedding.

SWINE

Growth

Ambient air temperature affects the feed conversion and daily weight of growing swine. As shown in Figure 1B, a temperature range of 60 to 72°F produces maximum rates of gain and feed use for 155 to 220 lb hogs, while a broader range of 50 to 76°F reduces performance only slightly. For 45 to 130 lb animals, the optimal and nominal loss ranges are about 62 to 73°F and 55 to 76°F, respectively (Kibler and Brody 1956). Younger animals require temperatures of 73 to 82°F for best performance, and piglets from 3 days to 2 weeks of age should have 86 to 90°F conditions (Hahn 1983).

Daily air temperature cycles of more than 10°F on either side of 70°F result in reduced daily gain by pigs and an increased feed requirement per unit of gain (Nienaber et al. 1987). Reasonably constant air temperatures are desirable.

The level of air temperature in which swine grow affects deposition and retention of protein (carcass quality). Lean meat formation is reportedly highest in pigs raised between 60 and 70°F (Mount 1963). However, the ratio of protein to fat decreases at air temperatures above 59°F.

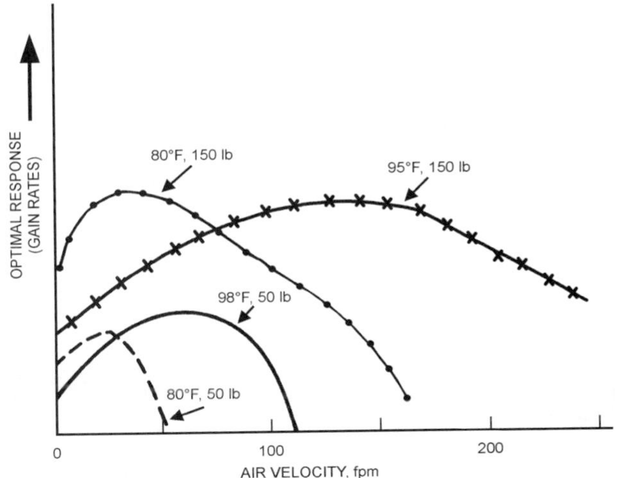

Fig. 16 Swine Response to Air Velocity
(Riskowski and Bundy 1986, Bond et al. 1965)

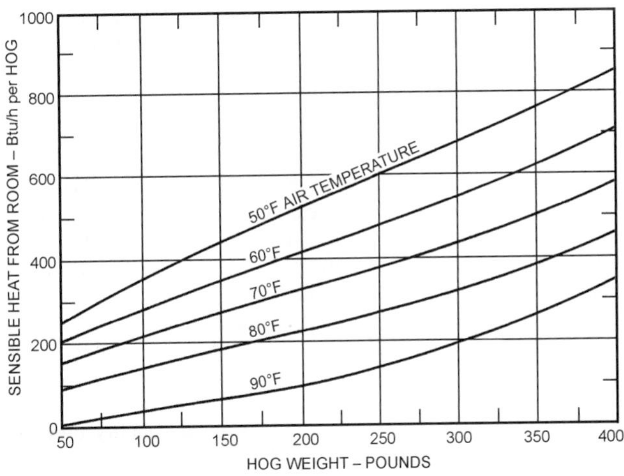

Fig. 17 Room Sensible Heat in Hog House
(Bond et al. 1959)

Table 4 Heat Production of Grouped Nursery Pigs

Weight Range, lb	Temperature, °F	Total Heat Production, Btu/h · lb	Latent Heat, % of Total
9 to 13	85	5.1	33
13 to 24	75	7.0	31
24 to 37	65	7.8	30

Source: Ota et al. (1982).

Swine gains and feed conversion rates are highly sensitive to air velocities and, in many cases, are affected adversely at velocities as low as 50 fpm. Swine less than 8 weeks of age should not be exposed to velocities greater than 50 fpm; a velocity lower than 25 fpm is preferred when temperatures are in the recommended range. Even in hot conditions, pigs are affected adversely by air velocities greater than 200 fpm (Bond et al. 1965, Gunnarson et al. 1967, Riskowski and Bundy 1991). Feed utilization and gain are much better at low air velocities (35 fpm) than at high (300 fpm) air velocities when temperatures are optimal (Figure 16).

Reproduction

Merkle and Hazen (1967) and Heard et al. (1986) have shown that sows benefit from some type of cooling in hot weather. In these studies, cool, dry air was directed at the sow to relieve heat stress by increasing evaporative and convective heat dissipation. Field studies of breeding problems and resultant small litters have established that both sexes suffer losses from high temperature. Sprinkling the sow during hot weather at breeding time and shortly after resulted in more live births than with unsprinkled sows. Sprinkling boars before mating also increased the number of live births per litter. Conception rate varied from about 100% of normal at 70°F to only 70% at 90°F. Breeding difficulties and a decrease in live embryos were observed in tests with controlled temperatures and relative humidities (Roller and Teague 1966). In some species, spontaneous abortion under severe heat stress may save the mother's life; however, sows appear to die of heat prostration, due to extra metabolic heat generation in late pregnancy, before spontaneous abortion occurs.

Little information is available on the value of temperature control for sows during cold weather. Forcing sows to produce the necessary heat to maintain body temperature may require rigid nutritional management to avoid the animals from becoming overweight—a condition resulting in poorer conception and smaller litters.

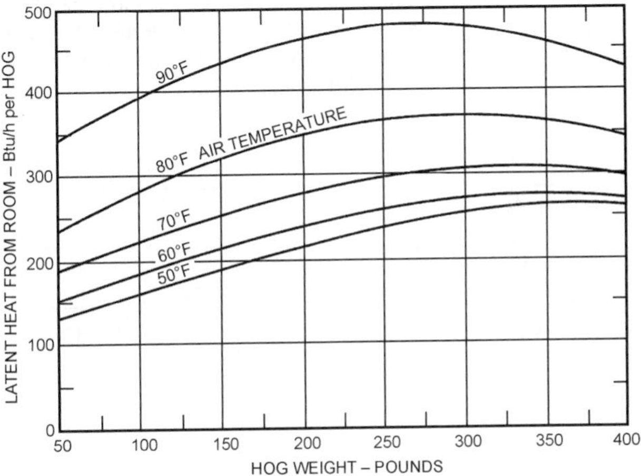

Fig. 18 Room Latent Heat in Hog House
(Bond et al. 1959)

Heat and Moisture Production

Table 4 (Butchbaker and Shanklin 1964, Ota et al. 1982) shows heat production for piglets. Figure 17 and Figure 18 (Bond et al. 1959) show heat and moisture that must be accounted for in ventilating or air conditioning older swine housing for swine between 50 and 400 lb, housed at temperatures from 50 to 90°F. These data are the sensible and latent heat levels measured in a room containing hogs. For design purposes, theroom heat data reflect swine housing conditions rather than metabolic heat production from an individual animal.

The same data can be applied even where building temperatures are cycling, if the average air temperature during the cycle is used for design. The total animal heat load and the latent load that the ventilation system must remove are greater than the values in Figure 17 and Figure 18, if the air velocity around the animals is more than about 50 fpm.

The lower critical temperature (the ambient temperature below which heat production increases) is about 77 to 86°F for a group of newborn pigs; for a single piglet, the critical temperature is about 93 to 95°F. As the pigs grow, the critical temperature falls—for 4.5 to 9 lb pigs, it is between 86 and 95°F; for 9 to 18 lb pigs, between 77 and 86°F. The floor bedding materials for growing piglets influence these critical temperatures. For a group of nine pigs averaging 88 lb, the critical temperature is 53 to 55°F on straw, 57 to 59°F on asphalt, and 66 to 68°F on concrete slabs (Butchbaker and Shanklin 1964).

CHICKENS

Growth

The tremendous worldwide production of broiler chickens is a result of improvements in genetics, nutrition, and housing. Four-pound broilers are produced in less than 6 weeks. Intermittent light and darkness (photoperiod) improve growth and conserve feed and energy (McDaniel and Brewer 1975).

Supplemental heat is generally needed for newly hatched chicks. Broiler strains respond well to brooding temperatures of 92°F for the first few days. Thereafter, air temperatures may be decreased at a rate of 7°F per week until 70°F is reached. Relative humidities of 65 to 70% promote good feathering and market-quality broilers. Effects of air velocity on heat dissipation of broilers is widely acknowledged, although not completely quantified, and has led to the widespread adoption of tunnel ventilation in new housing. Some benefit from increased velocity (up to 500 fpm) is obtained at air temperatures from 75 to 95°F (Drury 1966). High-velocity air, at a temperature above the feather temperatures, causes more, not less, heat stress. Observations showed that 28-day-old broilers rested in areas with airflow temperature combinations of 55 fpm at 60°F, 100 fpm at 70°F, and 150 fpm at 75°F.

Reproduction

Adverse effects of high thermal environments (above 85°F) on egg production include fewer eggs, reduced egg weight, and thinner shells. Hens over 1 year of age are more adversely affected by high thermal environments than younger hens. Larger hens are more adversely affected than smaller hens. Hatchability also declines as temperatures increase. Although the fertility rate is good at 70°F, it declines at 86°F. Feed requirements increase markedly below 45°F; activity and productivity decline below 32°F. The suggested ideal environment is between 55 and 75°F.

Increasing relative humidity above 80%, at 85°F dry-bulb temperature, produces an increased respiration rate and drooping wings. After 7 to 10 days, acclimatization may reduce the respiration rate, but production will still be less than optimal. Generally, relative humidity can vary from around 40% to 75% with little impact on bird performance. Lower relative humidities may be associated with dusty conditions and signal excessive minimum ventilation; higher relative humidities are conducive to mold growth and building deterioration from excess moisture.

Heat and Moisture Production

Figure 19 shows the total sensible and latent heat produced by laying hens during day and night at various temperatures. Figure 20

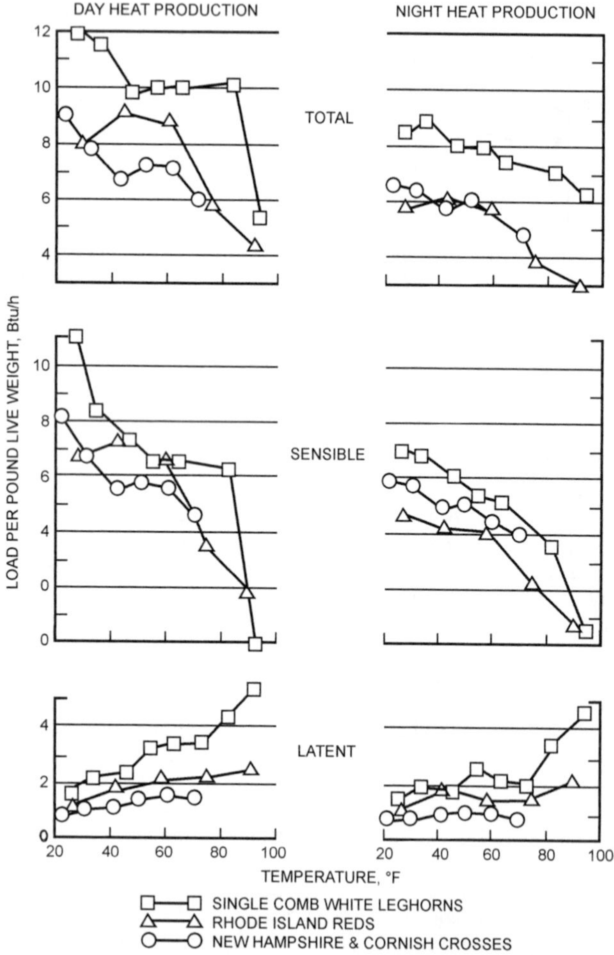

Fig. 19 Heat and Moisture Loads for Caged Laying Hens at Various Air Temperatures
(Ota and McNally 1961)

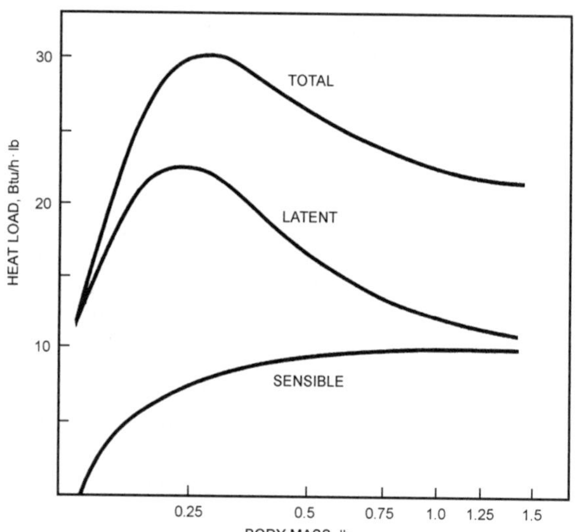

Fig. 20 Sensible, Latent, and Total Heat for Chicks Brooded on Litter
(Reece and Lott 1982)

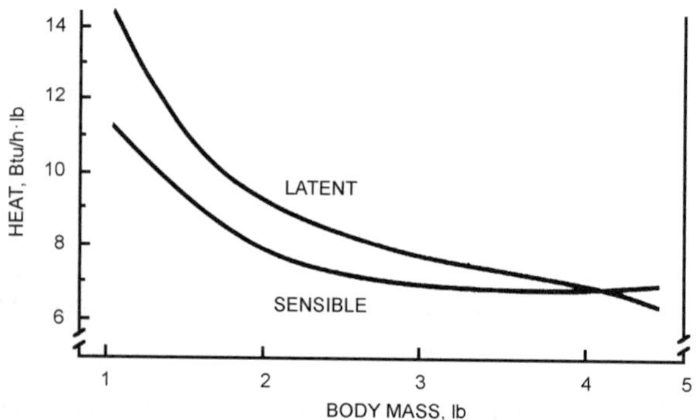

Fig. 21 Sensible and Latent Heat for Broilers Raised at 60°F on Litter
(Reece and Lott 1982)

shows the same data for chicks. Figure 21 and Figure 22 show sensible and latent heat production for broilers grown on litter at typical stocking rates at two temperatures. The effects of heat and moisture absorption or release by the litter are included in these figures. (*Note*: modern bird growth rates have increased approximately threefold since the work reported in Figures 20, 21, and 22 was performed.)

The continued improvement in nutrition, genetics, and housing for poultry means that currently available data for heat and moisture production is outdated. Research by Gates et al. (1996) suggests that total specific heat production (i.e., expressed per pound of body mass) may be reasonably similar, but that partitioning between sensible and latent components has changed. In general, latent loads have increased during brooding and decreased during growout, presumably because of new water delivery systems with fewer leaks.

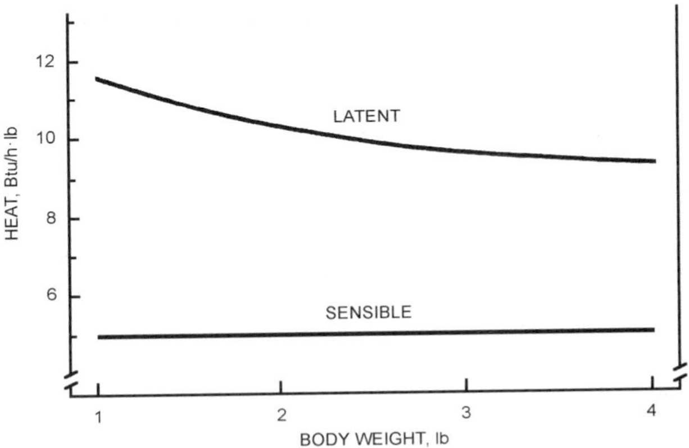

Fig. 22 Sensible and Latent Heat for Broilers Raised at 80°F on Litter
(Reece and Lott 1982)

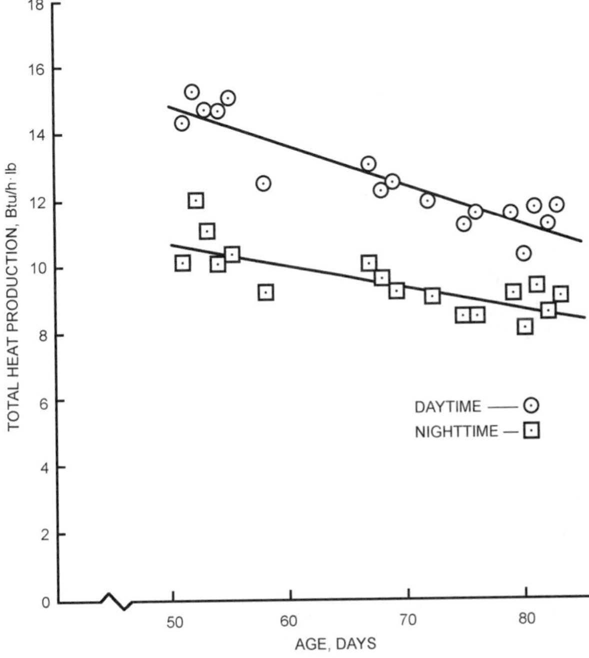

Fig. 23 Total Heat Production of Medium Breed Turkeys Versus Age During Daytime and Nighttime
(Buffington et al. 1974)

TURKEYS

Growth

The turkey poult has the best rate and efficiency of gain at a brooder room temperature of 70 to 75°F for the first 2 weeks, and 65°F thereafter, with about 70% rh. Initial brooder temperature should be 100°F, reduced 5°F per week until room temperature is reached. Proper control of photoperiod stimulates the growth rate in turkeys. Table 5 shows weekly cumulative average live weight for both sexes (Sell 1990). Figure 23 provides heat production data for medium breed turkeys.

Reproduction

Mature turkeys tolerate temperature and humidity over a range of at least 20 to 90°F and 35 to 85% rh. Since mature birds are kept only for fertile egg production, lighting for off-season egg production becomes very important. The young stock are raised to 20 weeks on natural light. At 20 weeks, females are placed in a

Table 5 Weekly Average Live Mass of Male and Female Turkeys

Age, Weeks	Male, lb	Female, lb
0	0.07	0.07
1	0.33	0.31
2	0.66	0.64
3	1.15	1.10
4	1.85	1.76
5	2.80	2.58
6	3.92	3.55
7	5.27	4.65
8	6.75	5.89
9	8.33	7.19
10	10.10	8.56
11	11.95	9.97
12	14.00	11.40
13	16.07	12.81
14	18.26	14.18
15	20.46	15.52
16	22.67	16.80
17	24.87	17.99
18	27.06	19.14
19	29.24	20.20
20	31.40	21.17
21	33.52	
22	35.54	
23	37.51	
24	39.42	

Source: Sell (1990).

Table 6 Heat Loss of Growing Turkeys at Various Air Temperatures and Relative Humidities

Age, Days	Weight of Poultry, lb	Dry Bulb, °F	Relative Humidity, %	Heat Loss, Btu/h · lb Sensible	Heat Loss, Btu/h · lb Latent	Total Live Weight, lb
15	0.487	100	23	2.3	16.5	18.8
6	0.233	95	26	7.7	18.2	25.9
19	0.802	95	26	3.3	10.1	13.4
14	0.518	90	31	9.2	10.9	20.1
29	1.629	90	31	5.6	6.1	11.7
7	0.244	85	36	12.2	14.1	26.3
21	0.923	85	36	7.7	8.0	15.7
36	2.000	85	36	6.8	4.3	11.1
28	1.385	80	42	9.0	4.1	13.1
23	0.963	75	50	10.8	6.2	17.0
27	1.250	75	50	11.7	3.1	14.8
35	2.120	75	50	9.5	2.6	12.1

Source: DeShazer et al. (1974).

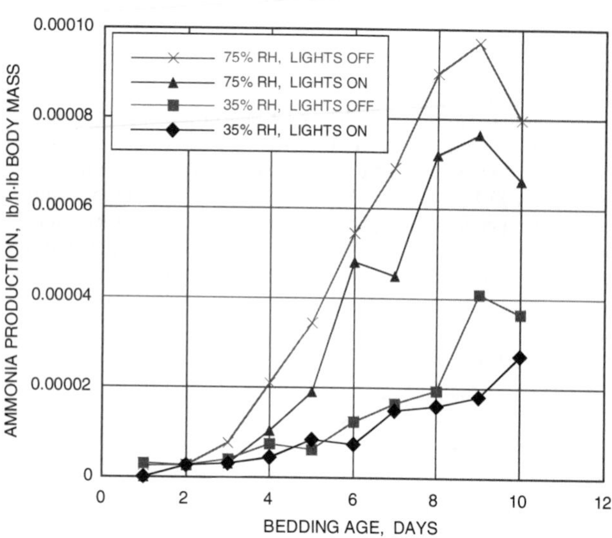

Fig. 24 Ammonia Generated by Laboratory Mice

totally darkened pen and given 8 h per day illumination at an intensity of 2 to 5 footcandles at bird's-eye level. At 30 weeks, the sexes are mixed, and the lighting period is increased to 13 to 15 h. The breeding season then continues for 12 to 26 weeks.

Heat and Moisture Production

Table 6 gives limited heat loss data on large breed growing turkeys (DeShazer et al. 1974); however, an estimate from data on heavy chickens, applied to turkeys on a live-weight basis, should be satisfactory for design purposes. The reduction in heat loss during the dark period of the day is 25% for large breeds (Buffington et al. 1974) and between 5 and 40% for small breeds.

The latent heat production of turkeys may be significantly higher than the 1970s data show if the growth rate or feed consumption is significantly higher in the 1990s than it was in the 1970s (see the section on Chickens, Heat and Moisture Production).

LABORATORY ANIMALS

Significant environmental conditions for facilities that house laboratory animals include temperature, humidity, air motion, illumination, noise, and gaseous and viable particulate contaminants (Moreland 1975). Design conditions vary widely, depending on whether the animals are experiencing disease-induced stress (which alters environmental needs), subjected to test environments, or simply housed (Besch 1975, Murakami 1971, Nienaber and Hahn 1983). This fact reflects differing housing and ventilation guidelines for animals used in research (NIH 1985), compared to recommendations by the ASAE (1991) and Midwest Plan Service (MWPS 1983) for production agriculture. Little is known about the influence of disease on environmental requirements, animal performance, and well-being. Because significantly different conditions may exist between animal cages and the animal room (macroenvironment), control of cage environments is essential to ensure the animal's physiological well-being. Temperature and velocity gradient controls require low supply air-to-room air temperature differential, overhead high induction diffusion, uniform horizontal and vertical air distribution, and low return outlets.

Heat and Moisture Production

Memarazadeh (1998) reported CO_2, H_2O and NH_3 generation rates, O_2 consumption rates, and heat generation rates of laboratory mice. The rates were measured at two room relative humidities (35% and 75%), and during light and dark periods. The production rate of CO_2 and H_2O and the consumption of O_2 were constant over the 10 day period once the mice became acclimatized to cage conditions (Table 7). However, the generation rate of ammonia depended on the cage relative humidity, light phase, and day in the experiment (Figure 24).

Table 7 Average Production/Consumption Rates for Laboratory Mice

Production/Consumption Rates per Pound of Body Mass	Light Period	Dark Period	Light/Dark Average
Total heat, Btu/h	36	42	—
Water generated, lb/h	—	—	0.08
CO_2 generated, lb/h	0.0068	0.0090	—
O_2 consumed, lb/h	0.0059	0.0066	—

Notes:
1. Mice were white laboratory females in shoebox cages with chipped hardwood bedding.
2. Heat production rates were calculated from O_2 consumption data. A mixed respiratory quotient (RQ) was assumed (RQ = 0.82), and heat production was based on a rate of 540 Btu/ft^3 of O_2 consumed.

Table 8 Heat Generated by Laboratory Animals

		Heat Generation, Btu/h per Animal			
			Normally Active[b,c]		
Animal	Weight, lb	Basal[a]	Sensible	Latent	Total
Mouse	0.046	0.66	1.11	0.54	1.65
Hamster	0.260	2.40	4.02	1.98	6.00
Rat	0.62	4.64	7.77	3.83	11.60
Guinea pig	0.90	6.10	10.22	5.03	15.25
Rabbit	5.41	23.41	39.22	19.31	58.53
Cat	6.61	27.21	45.57	22.45	68.02
Primate	12	42.55	71.27	35.10	106.38
Dog	22.7	68.64	104.8	56.4	161.2
Dog	50.0	124.10	230.7	124.2	354.9
Goat	79.3	175.39	293.78	144.70	438.48
Sheep	99	207.14	346.96	170.89	517.85
Pig	150	282.09	371	292	663
Chicken	4.0	18.67	12.9	21.9	34.7

[a]Based on standard metabolic rate $M = 6.6W^{0.75}$ Btu/h per animal (Kleiber 1961) or appropriate reference (W = animal mass, lb).
[b]Referenced according to availability of heat generation data. Otherwise, heat generations is calculated on basis of ATHG = 2.5M (Gordon et al. 1976). Latent heat is assumed to be 33% of total heat and sensible heat is 67% of total heat (Besch 1973, Woods et al. 1972).
[c]Data taken from Runkle (1964), Kleiber (1961), Besch (1973), Woods and Besch (1974), Woods et al. (1972), Bond et al. (1959), and Ota and McNally (1961).

Table 8 lists the approximate amount of heat released by laboratory animals at rest and during normal activity. For load calculation purposes, heat gain from all laboratory animal species can be estimated (Wood et al. 1972, Gordon et al. 1976) with an acceptable level of error from

$$\text{ATHG} = 2.5M \quad \text{where } M = 6.6W^{0.75} \tag{9}$$

and where

ATHG = average total heat gain, Btu/h per animal
M = basal metabolic rate of animal, Btu/h per animal
W = mass of animal, lb

Conditions in animal rooms must be maintained continuously. This requires year-round availability of refrigeration and, in some cases, dual air-conditioning facilities and emergency power for motor drives and control instruments. Chapter 21 of the 1999 *ASHRAE Handbook—Applications* has additional information on laboratory animal facilities.

PLANTS: GREENHOUSES, GROWTH CHAMBERS, AND OTHER FACILITIES

Most agronomically important plant crops are produced outdoors in favorable climates and seasons. Greenhouses and other indoor facilities are used for the out-of-season production of horticultural crops for both commercial sales and research purposes, and for producing food, floricultural, and other crops in conditions that permit the highest quality by buffering the crops from the vagaries of weather. The industry that produces crops in greenhouses may be termed controlled environment agriculture (CEA).

Historically, many cold-climate commercial greenhouses were operated only from late winter into early summer, and during autumn. Greenhouses were too warm during midsummer; during winter in some cold-climate locations, light levels were too low and the day length inadequate for many crops. Mechanical ventilation, evaporative cooling, centralized heating systems, movable insulations, carbon dioxide enrichment, and supplemental lighting have extended the use of greenhouses to year-round cropping on a relatively large scale.

Growth chambers, growth rooms, and propagation units are environmentally controlled spaces used for either research or commercial crop production. Environmentally controlled chambers may include highly sophisticated facilities used for micropropagation (e.g., tissue culture), or may be simple boxes in which air temperature and lights are controlled. Indoor facilities having controlled temperature and humidity environments may be used as warehouses to hold plants and plant products prior to commercial sale. Often these are simple refrigerated storage rooms or chambers.

Primary atmospheric requirements for plant production include: (1) favorable temperatures, (2) adequate light intensity and suitable radiation spectrum, and (3) favorable air composition and circulation. Engineering design to meet these requirements typically is based on steady-state assumptions. The thermal and ventilation time constants of most greenhouses are sufficiently short that transient conditions are seldom considered.

TEMPERATURE

Plant Requirements

Leaf and root temperatures are dominant environmental factors for plant growth and flowering. Factors in the energy balance of a plant canopy include air temperature, relative humidity, air movement, thermal radiation exchange, and convective exchange coefficients for sensible and latent heat. Therefore, leaf temperature is affected by such environmental factors as the type of heating and ventilating systems, supplemental lighting, light transmittance characteristics of the greenhouse cover, misting or evaporative cooling, location of the leaf on the plant, and the geometry of the surrounding leaf canopy.

Most information on plant responses to temperature is based on air temperature rather than plant temperature. Leaf temperature is difficult to measure, and one or several leaves represent neither the average nor the extreme temperatures of the plant. Since plants cannot actively regulate their cell and tissue temperatures in response to changing ambient conditions (passive regulation by opening and closing leaf stomata, which controls evapotranspiration, provides a small degree of control), their leaves and stems are usually within a few degrees of the surrounding air temperature (above during times of solar insolation, below at other times due to thermal reradiation and evapotranspiration).

All plants have minimum, optimum, and maximum temperatures for growth. Optimum temperature depends on the physiological process desired. Thermoperiodic species have different optimum day and night temperatures for each stage of growth, and each stage of plant growth may have its own unique optimum temperature influenced by radiant flux density, the ambient carbon dioxide level, and water and nutrient availability.

Historically, plants have been grown with night temperatures lower than day temperatures. In practice, many greenhouse crops are grown at standard (blueprint) night temperatures. Day temperatures are increased from 10 to 20°F (depending on solar intensity) above night temperatures. Table 9 presents recommended ranges of night temperatures for a selection of greenhouse crops.

Night temperature (NT) and day temperature (DT) are manipulated to provide nonchemical means to control plant height and development in some greenhouse crops. The effects of both DT and NT and their difference (DIF = DT − NT) are being investigated. Stem elongation, internode spacing and elongation, rate of elongation, photosynthesis rate, mature stem length, and fruit yield can be affected with DIF control (Jacobson et al. 1998, Neily et al. 1997, Erwin et al. 1991, Moe and Heins 1990). Typically, a negative DIF results in shorter crops. For example, Gent and Ma (1998) increased the yield of greenhouse tomatoes with a larger positive DIF (25°F) as compared to a lower positive DIF (9°F).

Heating Greenhouses

Heat loss from greenhouses is caused primarily by conduction through the structural cover and infiltration of outdoor air. The heating system is designed to meet the sum of the two. Perimeter heat loss is generally only a few percent of the total and is often neglected in design. When movable night insulation is used to conserve energy, heating systems are still designed to match conduction and infiltration losses without insulation because movable insulation may be opened early in the morning when outdoor air temperature is near its minimum, and excess heating capacity may be useful to melt snow in climatic regions where this occurs. Greenhouses are not designed to carry heavy snow loads.

Energy Balance

Radiation energy exchange. Solar gain can be estimated using procedures presented in Chapter 30. Not all insolation appears as sensible heat, however. As a general rule, from two-thirds to three-quarters of ambient insolation is available inside a typical commercial greenhouse. (Highly detailed models for calculating solar transmittance may be found in the literature.) If a greenhouse is filled with mature plants, approximately one-half of the available insolation (transmitted) may be converted to latent heat, one-quarter to one-third released as sensible heat, and the rest either reflected back outdoors or converted through photosynthesis (perhaps 3%).

Supplemental lighting can add significantly to the thermal load in a greenhouse. If movable night insulation is used, venting may be required during times of lighting, even during cold weather. The components of heat addition from supplemental lighting are divided between sensible and latent loads, with approximately one-quarter to one-third of the total heat load in latent form.

Reradiation heat loss from greenhouses comprises complex processes and may involve both reradiation from the structural cover and reradiation from inside the greenhouse if the cover is not thermally opaque. Glass is nearly thermally opaque, but many plastics are not. Newer plastic films may contain IR-inhibiting substances and can save a significant amount of heating energy, while adding only slightly to summer ventilation needs. Condensation on plastic films also reduces transmittance of thermal radiation, while diffusing but not seriously impairing transmittance of solar radiation. Generally, heat loss coefficients used in greenhouse design include the effects of thermal radiation exchange by the structural cover.

Structural heat loss. Conduction q_c plus infiltration q_i determine total heating requirements q_t. While infiltration heat loss is most accurately calculated using enthalpy differences, in practice only air temperature changes are considered, but with the apparent specific heat of air adjusted upward to account for the latent heat component.

Table 9 Recommended Night Temperatures for Greenhouse Crops

Crop Species	Night Temperatures, °F	Remarks
Aster *Callistephus chinensis*	50-55	Long days during early stages of growth
Azalea *Rhododendron* spp.	61-64	Vegetative growth and forcing specific temperatures required for flower initiation and development
Calceolaria *C. herbeohydrida*	61 50	Vegetative growth Flower initiation and development; initiation also occurs with long days and high temperatures if photon flux density is high
Calendula *C. officinallis*	39-45	
Calla *Zantedeschia* spp.	55-61	Decrease to 55°F as plants bloom
Carnation *Dianthus caryophyllus*	50-52 winter 55 spring 55-61 summer	Night temperatures adjusted seasonally in relation to photon energy flux density
Chrysanthemum *C. morifolium*	61 cut flowers 63-64 pot plants	Temperatures during flower initiation especially critical; uniform initiation very important for pot mums; cultivars classified on basis of temperature for development
Cineraria *Senecio cruentus*	61 48-52	Vegetative growth Flower initiation and development; plant quality best at low temperatures
Crossandra *C. infundibuliforms*	75-81 64	Germination Growth and flowering
Cyclamen *C. indicum*	61-64 55 50-52	Germination Seedlings Growth and flowering
Foliage plants	64-70	Species differ in their temperature and radiant energy requirements
Fuchsia *F. hydrida*	52-61	Long days for flower initiation
Geranium *Pelargonium hortorum*	55-61	61 to 64°F for fast crops at high radiant energy flux
Gardenia *G. grandiflora* *G. jasminoides*	 61-63 61-63	Lower temperatures result in iron chlorosis; higher temperatures increase bud abscission
Gloxinia *Sinningia speciosa*	64-70	Lower temperatures increase bud brittleness
Hydrangea *H. macrophylla*	55-61 61-63 (forcing)	Specific temperature for flower initiation and development
Iris *I. tingitana* (*Wedgewood*)	45-61 (forcing)	Forcing temperature 55 to 57°F for 10/11 bulbs; 50 to 54°F for 9/10 bulbs
Kalanchoe (*K. Blossfeldiana*)	61	Temperatures influence rate of flower development and incidence of powdery mildew
Lily *Lilium longiflorum*	61	Temperatures manipulated to alter rate of flower development; specific temperatures for flower initiation
Orchida *Cattleya* spp.	61	Temperature requirement of hybrids related in parental species
Orchids *Phalaenopsis* spp. *Cymbidium* spp. *Cypripedium*	 64 50 50-55	
Poinsettia *Euphorbia pulcherrima*	64 61-63	Vegetative growth Photoperiod requirement changes with temperature; bract development influenced by temperature
Roses *Rosa* spp.	61-63	
Saintpaulia *S. ionantha*	64-70	Below 61°F, growth is slow, hard and brittle
Snapdragon *Antirrhinum maias*	48-50 55-61	Winter Spring and Fall seedlings benefit from 61 to 64°F temperatures
Stock *Matthiola incana*	45-50	Buds fail to set if temperatures are above 64°F for 6 h or more per day. Grown mainly as field crop
Tomato	61-66	Dry temperatures from 70 to 81°F on sunny days
Lettuce	55	63 to 64°F on cloudy days 70 to 79°F on sunny days
Cucumber	64	75°F on cloudy days 81°F on sunny days

Table 10 Suggested Heat Transmission Coefficients

		U, Btu/h · ft² · °F
Glass		
	Single-glazing	1.13
	Double-glazing	0.70
Plastic film		Manufacturer's Data
	Single film[a]	1.20
	Double film, inflated	0.70
	Single film over glass	0.85
	Double film over glass	0.60
Corrugated glass fiber		
	Reinforced panels	1.20
Plastic structured sheet[b]		
	16 mm thick	0.58
	8 mm thick	0.65
	6 mm thick	0.72

[a]Infrared barrier polyethylene films reduce heat loss; however, use this coefficient when designing heating systems because the structure could occasionally be covered with non-IR materials.

[b]Plastic structured sheets are double-walled, rigid plastic panels.

Table 11 Typical Infiltration Rates for Various Construction Types (ach)

Metal frame and glazing system, 16 to 24 in. spacing	1.08
Metal frame and glazing system, 48 in. spacing	1.05
Fiberglass on metal frame	1.03
Film plastic on metal frame	1.02
Film or fiberglass on wood	1.00

$$q_t = q_c + q_i \quad (10)$$

$$q_c = \sum UA\Delta t \quad (11)$$

$$q_i = c_p VN\Delta t \quad (12)$$

where

U = heat loss coefficient, Btu/h·ft²·°F (Table 10)
A = exposed surface area, ft²
Δt = inside minus outside air temperature, °F
c_p = volumetric specific heat of air (adjusted upward to account for latent heat component), 0.03 Btu/ft³·°F
V = greenhouse internal volume, ft³
N = number of air exchanges per hour (Table 11)
Σ = summation of all exposed surfaces of the greenhouse and perimeter heat losses

When design conditions are assumed for indoor and outdoor air temperatures and air exchange rate, the resulting heat loss may be assumed equal to the peak heating requirement for the greenhouse.

No universally accepted method exists to determine season-long heating needs for greenhouses. The heating degree-day method may be applied, but heating degree-day data must be adjusted to a base lower than 65°F because of the significant passive solar heating effect in greenhouses. The proper base must be determined locally to reflect the expected solar climate of the region and the expected greenhouse operating temperature. These difficulties often lead designers to obtain season-long heating data from comparable, existing greenhouses in the region, and apply them to new designs.

LIGHT AND RADIATION

Plant Requirements

Light (400 to 700 nm) is essential for plant vegetative growth and reproduction (Figure 25 and Figure 26). Intensity integrated over time provides the energy for growth and development, while

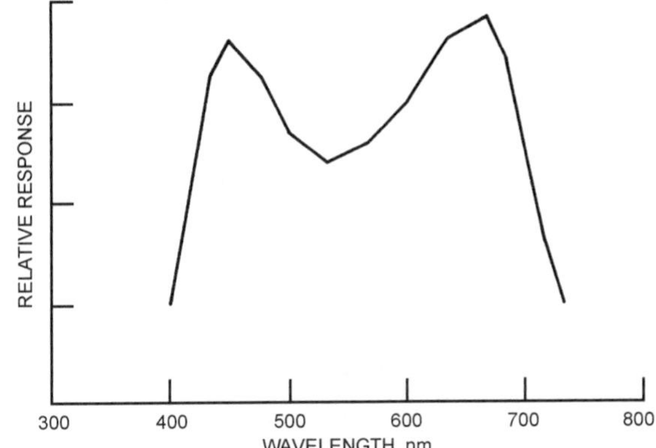

Fig. 25 Traditional Photosynthesis Action Spectra Based on Chlorophyll Absorption

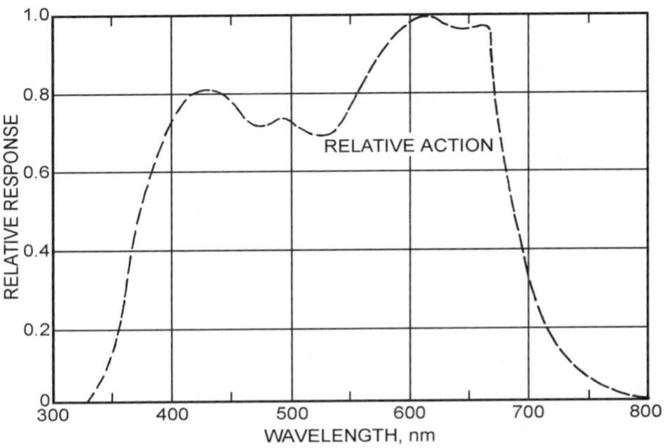

Fig. 26 Relative Photosynthetic Response

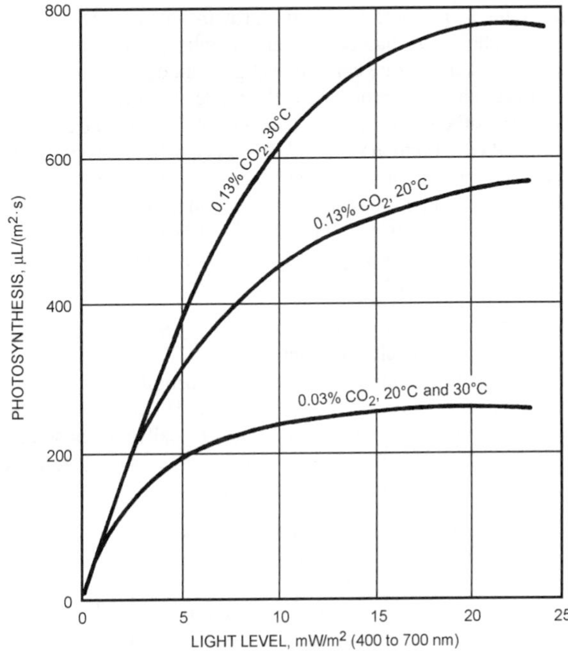

Fig. 27 Photosynthesis of Cucumber Leaf at Limiting and Saturating Carbon Dioxide Concentrations under Incandescent Light

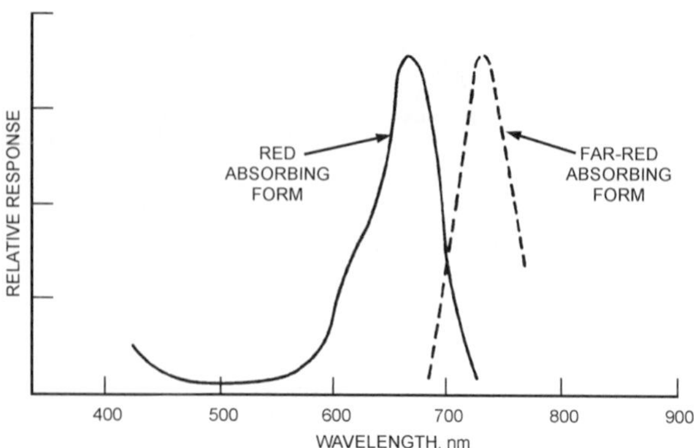

Fig. 28 Phytochrome Action Spectra

duration (either long or short, depending on species) may be essential for certain physiological processes such as flowering. High light intensity may exceed the ability of individual leaves to photosynthesize. However, if there is a dense canopy, excess light may be beneficial to lower leaves even when upper leaves are light saturated. The intensity at which light saturates a leaf depends on various environmental factors, such as the concentration of carbon dioxide in the ambient air, as well as biological factors (Figure 27).

Spectral distribution of light can affect plant development, but sunlight's spectral distribution need not be duplicated by artificial lighting to have suitable growth and development. Certain reproductive changes are initiated by red (660 nm) and far red (730 nm) light (Figure 28), and excessive ultraviolet light (290 to 390 nm) may be detrimental to growth.

Plants that respond to the durations of light and dark periods are termed photoperiodic (photoperiodic effects generally relate to flowering). Some plant species are long-day obligates, some are short-day obligates, some are day length-intermediate, and others are day-neutral. Such responses are usually (relatively) independent of light intensity. Photoperiodic effects can be initiated by very low light levels (less than 0.09 W/ft^2), such as that provided to chrysanthemums by incandescent lights for a short period during the middle of the night to promote vegetative growth and inhibit flowering (until a suitable size has been attained) during the winter. Some plant species can tolerate continuous light, but others require some period of darkness for proper growth and development.

Sunlight is the most common source of photosynthetically active radiation (PAR, 400 to 700 nm). Although specially designed lamp sources may provide light similar to sunlight, no single source or combination of sources has spectral radiation exactly like the emission of the sun from 300 to 2700 nm. Table 12 summarizes the spectral distribution of various light sources. Three systems of measuring radiation are as follows:

1. Radiometric units (irradiance) in watts per square foot (W/ft^2), with specified wavelength intervals.
2. Quantum units as photon flux density in μmol/(s·m^2) (at 400 to 700 nm unless otherwise specified). A mole of photons delivered in one second over a square meter may be called an einstein.
3. Photometric units (illuminance) as one lumen per square meter, or lux (lx). One lumen per square foot is one footcandle (fc).

Plant scientists use photosynthetic photon flux density (PPFD) in μmol/(s·m^2) (400 to 700 nm). Engineering organizations and manufacturers of light sources use photometric and radiometric units. Because of the variation in spectral power distribution, conversion from one system of units to another must be made individually for each light source for the wavelength interval included (Table 13). To obtain comparable plant growth from different light sources, the same radiation levels (PAR) and red/far-red ratios must be maintained.

Table 12 Radiation Power Distribution of Light Sources

Light Sources	UV 300-400 nm	PAR + FAR 400-850 nm	IR 850-2700 nm	Thermal 2700 + nm	Total Radiation
FCW	2	36	1	61	100
HG/DX	3	19	18	60	100
MH	4	41	8	47	100
HPS	0.4	50	12	38	100
LPS	0.1	56	3	41	100
INC	0.2	17	74	9	100
SUN	6	59	33	2	100

Values are in watts per 100 W of radiation.

Table 13 Light Conversion Factors

Light Source	Multiply W/ft^2 (400-850 nm) to Obtain: μmol/(s·ft^2) (400-700 nm)	Divide lux by Constant to Obtain: μmol/(s·ft^2) (400-700 nm)	Divide lux by Constant to Obtain: μmol/(s·ft^2) (400-850 nm)
Sun and sky, daylight	4.57	581	388
Blue sky only	4.24	560	441
High-pressure sodium	4.98	883	581
Metal halide	4.59	764	657
Mercury deluxe	4.52	904	829
Warm white fluorescent	4.67	818	797
Cool white fluorescent	4.59	797	775
Plant growth fluorescent A	4.80	355	334
Plant growth fluorescent B	4.69	581	506
Incandescent	5.00	538	215
Low-pressure sodium	4.92	1141	958

Adapted from Thimijan and Heins (1983).

Radiation Levels for Plant Growth

Display 0.03 W/ft^2. For display purposes, plants can exist at an irradiance of 0.03 W/ft^2. The preferred lamp has changed with technological advances in efficiency and distribution. The emphasis, however, has always been on color rendering and the type of atmosphere created in the display space. Low-wattage incandescent and fluorescent lamps are preferred. At this irradiance, plants can be displayed (seen), but little or no significant positive effect on plants can be expected. Extended holding in such low light conditions will have a negative effect on many plant species. Timing (light-dark durations) and temperature interaction are not a concern.

Photoperiod Response 0.08 W/ft^2 (4 to 6 h). For a photoperiod response, plant growth can be regulated at an irradiance of 0.08 W/ft^2 for as little time as 1 h. This irradiance is called a low light intensity system. The range of plant responses (promote or delay flowering, promote growth) that can be regulated is extensive, and this lighting is widely used by commercial growers.

Survival 0.3 W/ft^2 (8 h). Plants can survive at an irradiance of 0.3 W/ft^2 for 8 or more hours daily. This level enables many green plants to maintain their color. However, stem lengthening (etieliolation) and reduction in new leaf size and thickness occur under this irradiance level. In time, the overall development of the plants falls behind that of other plants grown under higher radiation levels. Photoperiod responses do not function well at this irradiance. However, strong interactions occur between this irradiance and temperature, watering frequency, and nutrition. Cooler temperatures (less than 63°F) help conserve previously stored material, while frequent watering and fertilization aggravate stem lengthening and senescence of older foliage.

Growth Maintenance 0.8 W/ft^2 (12 h). Plants maintain growth over many months when exposed to an irradiance of 0.8 W/ft^2 of 12-h duration daily. This is the intensity at which many indoor gardeners (professional or hobbyists) grow their plants when starting them from seeds, cuttings, or meristems. Interactions with the environment (temperature, airflow, relative humidity, and pollutants) can vary among installations. Simple facilities with good air exchange and limited lamp concentration can grow a wide range of plant species. The rate of development, particularly as the plants grow in size, can be slow compared to plants grown at higher irradiances.

Propagation 1.7 W/ft^2 (6 to 8 h). Plants propagate rapidly when exposed to an irradiance of 1.7 W/ft^2 for a minimum of 6 to 8 h daily, but they prefer 12 h. Above this level, many propagators attempt to shade their greenhouses with one or several layers of neutral filters (painted films on glazing, or movable or semipermanent plastic or other fabric shade cloth materials) to restrict light (and heat) in the propagation area.

Cuttings rooted at this intensity maintain a growth rate much like that of similar tissue on a stock plant. Stem length, branching, and leaf color, however, can be regulated by manipulating temperature, moisture, stress, and nutrients. Most plants grown for their flowers and fruits can be brought to maturity by increasing the day length to 16 to 18 h for flower initiation (or rapid growth) and then reducing the day length to 8 to 12 h for development. The growth rate, however, is relatively slow. For quickest development (leaf number, number of branches, and early flower initiation), the plants must be transferred to a higher lighting regime—2.2 to 4.6 W/ft^2.

Greenhouse Supplemental Light 2.2 W/ft^2 (8 to 16 h). When natural light is inadequate, it may be supplemented up to approximately 2.2 W/ft^2 for 8 to 16 h daily. When coupled with the ambient sunlight (shaded by clouds, greenhouse structures, and lamp fixtures), this irradiance simulates many of the growth responses and rates associated with growth chamber studies. Plants grown in greenhouses without supplemental lighting grow slower and flower later than lighted ones in cloudy regions or in northern areas during winter. Duration (in hours) and timing (day-night) of lighting is critical.

Supplemental lighting for 8 h, particularly during the day (0800 to 1600) may not be as cost-effective as lighting at night (2000 to 0400) if off-peak electric rates are available. Neither of these lighting regimes, however, is as effective as lighting for 16 h from morning to midnight.

Lighting short-day plants, such as chrysanthemums and poinsettias, is relatively inefficient because they can be lighted only during the 8- to 12-h day, followed by an obligatory 12- to 16-h daily dark period.

Growth Chambers 4.6 W/ft^2 (8 to 24 h). Plants grow in growth chambers or growth rooms if the light irradiance is a minimum of 4.6 W/ft^2 for 8 to 24 h daily. This irradiance is approximately one-fourth that of outdoor sunlight. Cool, white fluorescent lamps, combined with incandescent lamps, are widely used. More recently, HID lamps have been substituted for fluorescent lamps. For consistent results, all require a barrier of glass or other material between the lamp and the plants, and a separate ventilating system to remove the heat from such enclosed spaces.

Since filters cannot remove infrared completely, chambers are difficult to standardize. This often leads to confusing information on plant growth and flowering of plants grown in greenhouses and outdoors. When the total irradiance is 4.6 W/ft^2 and 10 to 20% of the total radiation is provided by incandescent lamps, most kinds of plants can be grown. In typical plant forms, flowering and fruiting responses occur when the plants are subjected to the following parameters:

- Day length, 8 to 24 h
- Temperature, 50 to 95°F
- Carbon dioxide, 300 to 2000 ppm (540 to 3600 mg/m^3)
- Relative humidity, 20 to 80%

Photoperiod

Day length affects the performance of some plants. There are four basic day length plant groups:

1. *Short-day plants* flower only when the length of the daily light period is less than the critical number of hours. Daily light periods longer than the critical length inhibit flowering.
2. *Long-day plants* flower only when the daily light period is longer than the critical number of hours. They become dormant or remain vegetative when the daily light period is shorter than the critical length.
3. *Day length-intermediate plants* flower only within a narrow range of day length, usually between about 10 and 14 h. If the day length is shorter than the optimum day length for flowering, the plants stop growing.
4. *Day-neutral plants* continue in vegetative growth or flower regardless of the day length.

Continuous light inhibits flowering and promotes vegetative growth of short-day plants, but encourages continued vegetative growth and early flowering of long-day plants, blocks the flowering of day length-intermediate plants, and in many instances, increases the stem length of day-neutral plants. Plants vary in their responsiveness to light source, duration, and intensity. The technology that has evolved to control the photoperiod of plants is based primarily on the incandescent-filament lamp. Of all the light sources available, this lamp creates the regulating mechanism most similar to that of sunlight. This is because the red/far-red wavelength ratio of light from an incandescent lamp is similar to the ratio of sunlight.

The effectiveness for photoperiod response in plants peaks at wavelengths of 660 nm (red) and 730 nm (far-red). The relative order of activity in regulating photoperiod responses by lamp type is as follows: Incandescent (INC) > High-pressure sodium (HPS) >> Metal halide (MN) = Cool white fluorescent (F) = Low-pressure sodium (LPS) >> Clear mercury (Hg). Photoperiod lighting is always used in combination with daylight or another main light source. Short days (less than normal day length) are created in the greenhouse with opaque materials that surround the plants.

RELATIVE HUMIDITY

Relative humidity affects the rate at which plants take water up, the rate of latent heat transfer, and certain diseases. Normal plant growth generally occurs at relative humidities between 20 and 80% if the plants have a well-developed root system, although relative humidities above 40% are preferred to avoid water stress conditions.

Transpiration, the movement of water vapor and gases from the plant to its surroundings, is controlled by the plant's stomatal openings. It is a function of air velocity and the vapor pressure difference between water at saturation at the leaf temperature and the actual water vapor partial pressure in the air. Generally, as relative humidity decreases, transpiration increases. Very low relative humidities (less than 20%) can cause wilting, since evaporation losses may be higher than the plant can replace, especially when light intensity is high.

High humidity provides a good environment for pathogenic organisms. Many pathogenic spores do not germinate unless relative humidity is 96% or more and many require a film of water on the leaves. Somewhat lower relative humidities may support other pathogen growth stages.

Still air surrounding a plant may be much wetter than the general atmosphere because evapotranspiration from the leaves raises the relative humidity in interfoliage air. The lower leaves, which stay moist longer, are more susceptible to disease. The upper leaves are dried by radiation and air currents.

AIR COMPOSITION

Carbon dioxide, which comprises about 0.035% of ambient air, is essential for plant growth. There are basically three ways to obtain carbon dioxide for greenhouse enrichment: pure in solid, liquid, or gaseous form; from burning fuels such as propane, natural gas, or kerosene; and by the aerobic breakdown of organic matter. The three ways are listed in order of purity and reliability. Carbon dioxide enters plants through stomata and is converted to carbohydrates through photosynthesis. The carbon dioxide concentration in air surrounding a plant, as well as light level, affects the rate of photosynthesis. The concentration for maximum growth depends on many factors, including the stage of growth, leaf area, light intensity, temperature, and air velocity past the stomatal openings.

An important relationship exists between light level and carbon dioxide uptake (Figure 27). As light level increases, carbon dioxide concentration must increase concurrently to take maximum advantage of the greater photosynthetic potential. In plastic greenhouses, and in glass greenhouses sealed against infiltration, the carbon dioxide level can drop below 200 ppm when the weather is cold, light levels are moderate, and the greenhouse is not ventilated. Carbon dioxide enrichment just to maintain normal levels can then be beneficial. During times of high light levels, carbon dioxide enrichment gives maximum benefit from the available light and may even be economically desirable when greenhouse ventilation is modest. However, carbon dioxide concentrations above 1500 ppm are seldom recommended; levels between 800 and 1200 ppm are typically used.

The effects of enrichment are not always positive. Without proper crop management, the yield, quality, or both may decrease, and timing of crop maturity may change.

Pollutants

Plants are sensitive to atmospheric pollutants such as ethylene, ammonia, gaseous fuels, ozone, fluorides, photochemical smog, and oxidants (from nitrogen and sulfur). Pollution damage can range from small spots on leaves, to yellowing of leaves, to severe foliage burn, and, ultimately, to plant death in extreme but not rare situations. The effect occurs both outdoors and in greenhouses; however, this is more common in greenhouses, because of their closed nature. Pollutants indoors can be removed by activated charcoal filters in the ventilation system; however, these are seldom used in commercial greenhouses. Economically, the more feasible approach is to limit pollutant production within, or introduction into, the greenhouse air space.

Ethylene is produced naturally by plants and leads to flower and whole plant senescence. It is also produced by combustion of gaseous and liquid fuels and can rapidly cause plant damage. Concentrations above 0.2 ppm can have a detrimental effect on plant growth. Unvented heaters, air currents that bring vented combustion products back into the greenhouse, and burners for carbon dioxide production are common sources of ethylene injury. Liquefied carbon dioxide may be used to supplement natural levels rather than combustion, specifically to avoid introducing ethylene into the greenhouse air, but even liquefied carbon dioxide should be carefully selected to avoid residual amounts of ethylene that may be contained within it.

Nitrogen oxides, common components of air pollution, can cause serious plant damage. Greenhouse locations near highways, nearby industrial complexes, and even a truck left running for an extended time near a greenhouse air intake vent may lead to leaf damage from NO and NO_2.

Sulfur dioxide, produced by the burning of sulfur containing fuels, causes injury to many plants within a short time. Sources of sulfur dioxide may be nearby, such as an industrial area, or may be within the greenhouse complex, such as the vented combustion products from a central heating facility, combustion products from carbon dioxide burners (using kerosene as a fuel, for example), and sulfur burned for mildew control.

Ozone is widely recognized as a serious pollutant affecting the production of many agronomic crops. Although few research results exist to quantify the effect of ozone on greenhouse crops, damage is likely to occur when greenhouses are located near ozone sources.

Phenolics and certain other organic vapors are phytotoxic. Phenolics, as volatiles from certain wood preservatives (creosote and pentachlorophenol), can cause leaf and petal damage. Vapors from some paints can also be damaging. Misuse of herbicides and pesticides can lead to plant injury, either through spray drift or volatilization.

Covering and sealing greenhouses for energy conservation can increase concentrations of ethylene and other air pollutants if their sources are within the air space, since infiltration and ventilation are decreased. Sealing to reduce infiltration can also lead to rapid carbon dioxide depletion and inhibited plant growth during cold temperatures when, even with bright light, ventilation is not required.

AIR MOVEMENT

Air movement influences transpiration, evaporation, and the availability of carbon dioxide. Air speed affects the thickness of the boundary layer at the leaf surface, which in turn influences the transport resistance between the ambient air and the leaf stomatal cavities. Air speed of 100 to 150 fpm is commonly accepted as suitable for plant growth under CEA conditions. Air speeds across the leaf of 6 to 20 fpm are needed to facilitate carbon dioxide uptake. Air speeds above 200 fpm can induce excessive transpiration, cause the stomatal guard cells to close, reduce carbon dioxide uptake, and inhibit plant growth. Air speeds above 1000 fpm may cause physical damage to plants. Generally, if plants within a greenhouse move noticeably due to ventilation, air speed is excessive.

Air circulation within greenhouses may be created to reduce thermal stratification and maintain suitable levels of carbon dioxide within the leaf canopy. Horizontal air flow, produced by small propeller fans that move air around the greenhouse in a racetrack pattern, has been found to be effective. Such fans are approximately 14 in. in diameter, with approximately one-sixth horsepower motors, spaced at approximately 50 ft intervals. Total fan capacity in cubic feet per minute (cfm) should equal about 25% of the greenhouse volume in cubic feet.

REFERENCES

ASAE. 1991. Design of ventilation systems for poultry and livestock shelters. *ASAE Standard* D270.5. American Society of Agricultural Engineers, St. Joseph, MI.

Barber, E.M., A. Senthilselvan, S. Kirychuk, S. Lemay, J.A. Dosman, Y. Zhang, L. Holfeld, D. Bono, and Y. Comier. 1999. Evaluation of two strategies for reducing exposure of swine barn workers to dust. *Proceedings* International Symposium on Dust Control for Animal Production Facilities. Aarhus, Denmark.

Bereskin, B., R.J. Davey, W.H. Peters, and H.O. Hetzer. 1975. Genetic and environmental effects and interactions in swine growth and feed utilization. *Journal of Animal Science* 40(1):53.

Besch, E.L. 1973. Final report to Animal Resources Branch, Division of Research Resources, National Institutes of Health. NIH Contract 71-2511.

Besch, E.L. 1975. Animal cage room dry-bulb and dew point temperature differentials. *ASHRAE Transactions* 81(2):549.

Blaxter, K.L.and W.A. Wood. 1951. The nutrition of the young Ayrshire calf, IV some factors affecting the biological value of protein determined by nitrogen-balance methods. *Brit. J. Nutrition* 5:55.

Bond, T.E., H. Heitman, Jr., and C.F. Kelly. 1965. Effects of increased air velocities on heat and moisture loss and growth of swine. *Transactions of ASAE* 8:167.

Bond, T.E., C.F. Kelly, and H. Heitman, Jr. 1959. Hog house air conditioning and ventilation data. *Transactions of ASAE* 2:1.

Bond, T.E., C.F. Kelly, and H. Heitman, Jr. 1963. Effect of diurnal temperature upon swine heat loss and well-being. *Transactions of ASAE* 6:132.

Bortolussi, R., W.F. Schlech, and W.L Albritton, eds. 1985. Listeria. In: *Manual of clinical microbiology*. Amer. Soc. Microbiology, Washington D.C.

Brody, S. 1945. *Bioenergetics and growth*. Reinhold Publishing Co., New York.

Brody, S., A.C. Ragsdale, R.G. Yeck, and D. Worstell. 1955. Milk production, feed and water consumption, and body weight of Jersey and Holstein cows in relation to several diurnal temperature rhythms. University of Missouri Agricultural Experiment Station *Research Bulletin* No. 578.

Bryant, J.M., C.F. Foreman, N.L. Jacobson, and A.D. McGilliard. 1967. Protein and energy requirementsof the young calf. *Journal of Dairy Science* 50(10):1645-1653.

Buckland, R.B. 1975. The effect of intermittent lighting programmes on the production of market chickens and turkeys. *World Poultry Science Journal* 31(4):262.

Bucklin, R.A. and L.W. Turner. 1991. Methods to relieve heat stress in hot, humid climates. *Applied Engineering in Agriculture* 7. ASAE.

Buffington, D.E., K.A. Jordan, W.A. Junnila, and L.L. Boyd. 1974. Heat production of active, growing turkeys. *Transactions of ASAE* 17:542.

Bundy, D.S. 1984. Rate of dust decay as affected by relative humidity, ionization and air movement. *Transactions of ASAE* 27(3):865-70.

Bundy, D.S. 1986. Sound preventive measures to follow when working in confinement buildings. Presented at the American Pork Congress, St. Louis, MO.

Butchbaker, A.F. and M.D. Shanklin. 1964. Partitional heat losses of newborn pigs as affected by air temperature, absolute humidity, age and body weight. *Transactions of ASAE* 7(4):380.

Carpenter, G.A. 1986. Dust in livestock buildings–Review of some aspects. *J. Agr. Engng. Res.* 33:227-41.

Carr, L.E. and J.L. Nicholson. 1980. Broiler response to three ventilation ranges. *Transactions of ASAE* 22(2):414-18.

Christianson, L.L. and R.L. Fehr. 1983. Ventilation energy and economics. In *Ventilation of agricultural structures*, M.A. Hellickson and J.N. Walker, eds. *ASAE Nomograph* No. 6, 336.

Curtis, S.E. 1983. *Environmental management in animal agriculture*. Iowa State University Press, Ames, IA.

DeShazer, J.A., L.L. Olson, and F.B. Mather. 1974. Heat losses of large white turkeys—6 to 36 days of age. *Poultry Science* 53(6):2047.

Donham, K.J. 1987. Human health and safety for workers in livestock housing. CIGR Proceedings. *ASAE Special Public.* 6-87.

Donham, K.J., P. Haglind, Y. Peterson, R. Rylander, and L. Belin. 1989. Environmental and health studies of workers in Swedish swine confinement buildings. *British J. Ind. Med.* 40:31-37.

Donham, J.K., P. Haglind, Y. Peterson, R. Rylander, and L. Belin. 1989. Environmental and health studies of workers in Swedish swine confinement buildings. *British Journal of Industrial Medicine* 40:31-37.

Dosman, J.A., B.L. Graham, D. Hall, P. Pahwa, H.H. Mcduffie, and M. Lucewicz. 1988. Respiratory symptoms and alterations in pulmonary function test in swine producers in Sakstchewan: Results of a survey of farmers. *J. Occup. Med.* 30:715-20.

Drury, L.N. 1966. The effect of air velocity of broiler growth in a diurnally cycling hot humid environment. *Transactions of ASAE* 9:329.

Erwin, J.E., R.D. Heins, and R. Moe. 1991. Temperature and photoperiod effects of Fuschia X hybrida morphology. *J. Am. Soc. Hortic. Sci* 116(6): 955-60.

Fachlam, R.R. and R.B. Carey. 1985. Streptococci and aerococci. In: *Manual of clinical microbiology*. American Society of Microbiology.

Farner, D.S. 1961. Comparative physiology: Photoperiodicity. *Annual Review of Physiology* 23:71.

Gates, R.S., D.G. Overhults, and S.H. Zhang. 1996 Minimum ventilation for modern broiler facilities. *Transactions of ASAE* 38(5):1479-86.

Gebremedhin, K.G., C.O. Cramer, and W.P. Porter. 1981. Predictions and measurements of heat production and food and water requirements of Holstein calves in different environments. *Transactions of ASAE* 24(3): 715-20.

Gebremedhin, K G., W.P. Porter, and C.O. Cramer. 1983. Quantitative analysis of the heat exchange through the fur layer of Holstein calves. *Transactions of ASAE* 26(1):188-93.

Gent, M.P.N. and Y.Z. Ma. 1998. Diurnal temperature variation of the root and shoot affects yield of greenhouse tomato. *HortScience* 33(1):47-51.

Gordon, R.L., J.E. Woods, and E.L. Besch. 1976. System load characteristics and estimation of animal heat loads for laboratory animal facilities. *ASHRAE Transactions* 82(1):107.

Guazdauskas, F.C. 1985. Effects of climate on reproduction in cattle. *Journal of Dairy Science* 68:1568-78.

Gunnarson, H.J. et al. 1967. Effect of air velocity, air temperatures and mean radiant temperature on performance of growing-finishing swine. *Transactions of ASAE* 10:715.

Hahn, G.L. 1983. Management and housing of farm animals environments. *Stress physiology in livestock*. CRC Press, Boca Raton, FL.

Hahn, G.L., A. Nygaard, and E. Simensen. 1983. Toward establishing rational criteria for selection and design of livestock environments. *ASAE Paper* 83-4517.

Heard, L., D. Froehlich, L. Christianson, R. Woerman, and W. Witmer. 1986. Snout cooling effects on sows and litters. *Transactions of ASAE* 29(4):1097.

Heber, A.J., M. Stroik, J.L. Nelssen, and D.A. Nichols. 1988. Influence of environmental factors on concentrations and inorganic content of aerial dust in swine finishing buildings. *Transactions of ASAE* 31(3):875-881.

Heber, A.J. and C.R. Martin. 1988. Effect of additives on aerodynamic segregation of dust from swine feed. *Transactions of ASAE* 31(2):558-563.

Hellickson, M.A., H.G. Young, and W.B. Witmer. 1974. Ventilation design for closed beef buildings. Proceedings of the International Livestock Environment Symposium, SP-0174, 123.

Hillman, P.E., K.G. Gebremedhin, and R.G. Warner. 1992. Ventilation system to minimize airborne bacteria, dust, humidity, and ammonia in calf nurseries. *Journal of Dairy Science* 75:1305-12.

Jacobson, B.M., D.H. Willits, and P.V. Nelson. 1998. Dynamic internode elongation in chrysanthemum. *Appl. Engr. Agric.* 14(3):311-16.

Janni, K.A., P.T. Redig, J. Newmen, and J. Mulhausen. 1984. Respirable aerosol concentration in turkey grower building. *ASAE Paper* No. 84-4522.

Johnson, H.D., A.C. Ragsdale, and R.G. Yeck. 1958. Effects of constant environmental temperature of 50°F and 80°F on the feed and water consumption of Brahman, Santa Gertrudis and Shorthorn calves during growth. University of Missouri Agricultural Experiment Station *Research Bulletin* 683.

Johnson, H.D., A.C. Ragsdale, I.L. Berry, and M.D. Shanklin. 1962. Effect of various temperature-humidity combinations on milk production of Holstein cattle. University of Missouri Agricultural Experiment Station *Research Bulletin* 791.

Kibler, H.H. and S. Brody. 1956. Influence of diurnal temperature cycles on heat production and cardiorespirativities in Holstein and Jersey cows. University of Missouri Agricultural Experiment Station *Research Bulletin* 601.

Kleiber, M. 1961. *The fire of life: An introduction to animal energetics*. John Wiley and Sons, New York.

Longhouse, A.D. et al. 1968. Heat and moisture design data for broiler houses. *Transactions of ASAE* 41(5):694.

Longhouse, A.D., H. Ota, and W. Ashby. 1960. Heat and moisture design data for poultry housing. *Agricultural Engineering* 41(9):567.

Martin, W.T., Y. Zhang, P.J. Willson, T.P. Archer, C. Kinahan, and E.M. Barber. 1996. Bacterial and Fungal flora of dust deposits in a swine building. *British J. Occupational and Environmental Medicine* 53:484-487.

McDaniel, G.R. and R.N. Brewer. 1975. Intermittent light speeds broiler growth and improves efficiency. *Highlights of Agricultural Research* 4:9. Auburn University, Auburn, AL.

McFarlane, J. 1987. Linear additivity of multiple concurrent environmental stressors effects on chick performance, physiology, histopathology and behavior. *Ph.D. thesis*, Animal Sciences Dept., Univ. of Illinois, Urbana.

Memarzadeh, F. 1998. *Design handbook on animal research facilities using static microisolators*, Volumes I and II. National Institute of Health. Bethesda, MD.

Merkle, J.A. and T.E. Hazen. 1967. Zone cooling for lactating sows.*Transactions of ASAE* 10:444.

Moe, R. and R. Heins. 1990. Control of plant morphogenesis and flowering by light quality and temperature. *Acta Hortic.* (July): 81-89.

Moreland, A.F. 1975. Characteristics of the research animal bioenvironment. *ASHRAE Transactions* 81(2):542.

Mount, L.E. 1963. Food, meat, and heat conservation. Pig Industry Development Authority Conference Circulat. Buxton, Derbyshire, England.

Murakami, H. 1971. Differences between internal and external environments of the mouse cage. *Laboratory Animal Science* 21:680.

MWPS. 1983. *Structures and environment handbook*. Midwest Plan Service, Ames, IA.

Neily, W.G., P.R. Hickleton, and D.N. Kristie. 1997. Temperature and developmental state influence diurnal rhythms of stem elongation in snapdragon and zinnia. *J. Am. Soc. Hortic. Sci.* 122(6):778-83.

Nienaber, J.A. and G.L. Hahn. 1983. Temperature distribution within controlled-environment animal rooms. *Transactions of ASAE* 26:895.

Nienaber, J.A., G.L. Hahn, H.G. Klencke, B.A. Becker, and F. Blecha. 1987. Cyclic temperature effects on growing-finishing swine. CIGR Proceedings. *ASAE Special Public.* 687:312.

NIH. 1985. Guide for the care and use of laboratory animals. National Institutes of Health Publication 85-23. Bethesda, MD.

OSHA. 1985. OSHA Safety and Health Standard. U.S. Department of Labor Code 1910.1000, 653-59.

Ota, H.J. and E.H. McNally. 1961. Poultry respiration calorimetric studies of laying hens. ARS-USDA 42-13, June.

Ota, H., J.A. Whitehead, and R.J. Davey. 1982. Heat production of male and female piglets. *Proceedings of Second International Livestock Environment Symposium*, SP-03-82. ASAE.

Pratt, G.L. 1983. Ventilation equipment and controls. In: *Ventilation of agricultural structures*, eds. M.A. Hellickson and J.N. Walker. ASAE, pp. 54.

Preller, L., J. Boley, D. Heederik, S. Gielen, and M. Tielen. 1992. *Proceedings* 3rd Inter. Symp: Issues in Health, Safety and Agriculture. Center for Agr. Medicine, University of Saskatchewan, Saskatoon, pp. 23-24.

Reece, F.N. and B.D. Lott. 1982. Heat and moisture production of broiler chickens. *Proceedings of the Second International Livestock Environment Symposium*, SP-03-82. ASAE.

Riskowski, G.L., J.A. DeShazer, and F.B. Mather. 1977. Heat losses of White Leghorn hens as affected by intermittent lighting schedules. *Transactions of ASAE* 20(4):727-731.

Riskowski, G.L. and D.S. Bundy. 1991. Response of young pigs to various air temperatures and velocities. *ASHRAE Transactions* 97(2):543-549.

Roller, W.L. and H.S. Teague. 1966. Effect of controlled thermal environment on reproductive performance of swine. *Proceedings of the Fourth International Biometeorological Congress.* Rutgers University, New Brunswick, NJ.

Runkle, R.S. 1964. Laboratory animal housing, Part II. *AIA Journal* 4:73.

Scott, N.R., J.A. DeShazer, and W.L. Roller. 1983. Effects of the thermal and gaseous environment on livestock. *ASAE Monograph No.* 6. ASAE.

Sell, J.L. 1990. Faster growing, more efficient turkeys in 1989. 66(1):12.

Senthilselvan, A., Y. Zhang, J.A. Dosman, E.M. Barber, L.E. Holfeld, S.P. Kirychuk, Y. Comier, T.S. Hurst, and C.S. Rhodes. 1997. Positive human health effects of dust suppression with canola oil in swine barns. *American Journal for Respiratory and Critical Care Medicine* 156:410-17.

Shutze, J.V., J.K. Lauber, M. Kato, and W.O. Wilson. 1962. Influence of incandescent and colored light on chicken embryos during incubation. *Nature* 196(4854):594.

Siegmund, H., ed. 1979. *The Merck veterinary manual: A handbook of diagnosis and therapy for the veterinarian.* Merck & Co., Inc., Rahway, NJ.

Squibb, R.L. 1959. Relation of diurnal temperature and humidity ranges to egg production and feed efficiency of New Hampshire hens. *Journal of Agricultural Science* 52(2):217.

Sutton, A.L., S.R. Nichols, D.D. Jones, D.T. Kelley, and A.B. Scheidt. 1987. Survey of seasonal atmospheric changes in confinement farrowing houses. In *Latest developments in livestock housing*. International Commission of Agricultural Engineering (CIGR).

Takai, H., F. Moller, M. Iversen, S.E. Jorsal, and V. Bille-Hansen. 1993. Dust control in swine buildings by spraying of rapeseed oil. *Proceedings of the Fourth Livestock Environment Symposium.* ASAE, pp. 726-33.

Thomeczek, F.J., M.R. Ellersieck, R.K. Leavitt, and J.F. Lasley. 1977. Trends in economic traits of production tested boars in the Missouri Evaluation Station. University of Missouri *Research Bulletin* No. 1021.

Tienhoven, A.W., N.R. Scott, and P.E. Hillman. 1979. The hypothalamus and thermoregulation: A review. *Poultry Science* 52(6):1633.

Van't Klooster, C.E. and J.A.M. Voermans. 1993. European perspectives—How are they solving their problems? *Symposium* "Meeting the Environmental Challenge." National Pork Producers Council.

Wiegard, B. and J. Hartung. 1993. Bacterial contaminant of air and floor surfaces in an animal house of a cattle clinic. Proceedings of 4th International Livestock Environment Symposium ASAE, St. Joseph, MI, pp. 643-49.

Wiersma, F. and G.H. Stott. 1969. New concepts in the physiology of heat stress in dairy cattle of interest of engineers. *Transactions of ASAE* 12(1): 130-32.

Woods, J.E. and E.L. Besch. 1974. Influence of group size on heat dissipation from dogs in a controlled environment. *Laboratory Animal Science* 24:72.

Woods, J.E., E.L. Besch, and R.G. Nevins. 1972. A direct calorimetric analysis of heat and moisture dissipated from dogs. *ASHRAE Transactions* 78(2):170-83.

Yeck, R.G. 1957. Stable heat and moisture dissipation with beef calves at temperatures of 50° and 80°F. University of Missouri Agricultural Experiment Station *Research Bulletin* 645.

Yeck, R.G. and R.E. Stewart. 1959. A ten-year summary of the psychroenergetic laboratory dairy cattle research at the University of Missouri. *Transactions of ASAE* 2(1):71.

Yeck, R.G. and R.E. Stewart. 1960. Stable heat and moisture dissipation with dairy calves at temperatures of 50° and 80°F. University of Missouri *Research Bulletin* No. 759.

Zejda, J.E, T.S. Hurst, C.S. Rhodes, E.M. Barber, H.H. Mcduffie, and J.A. Dosman. 1993. Respiratory health of swine producers, focus on young workers. CHEST 103:702-09.

Zejda, J.E., E.M. Barber, J.A. Dosman, S.A. Olenchock, H.H. McDuffie, C.S. Rhodes, and T.S. Hurst. 1994. Respiratory health status in swine producers relates to endotoxin exposure in the presence of low dust levels. *J. Occup. Med.* 36(1):49-56.

Zhang, Y. 1994. *Swine building ventilation,* pp.14-15, Prairie Swine Centre, Saskatoon, SK, Canada.

Zhang, Y., E.M. Barber, J.J.R. Feddes, and J.F. Patience. 1995. Identification of oils to be sprinkled in swine barns to reduce dust. *ASHRAE Transactions* 101(2):1179-91.

Zhang, Y., A. Tanaka, E.M. Barber, and J.J.R. Feddes. 1996. Effect of frequency and quantity of sprinkling canola oil on dust reduction in swine buildings. *Transactions of ASAE* 39(3):1077-81.

Zhang, Y., A. Tanaka, J.A. Dosman, A. Senthilselvan, E.M. Barber, S. Kirychuk, and T. Hurst. 1998. Acute respiratory responses of human subjects to air quality in a swine building. *J. Agr. Engng. Res.* 70:367-73.

Zhang, Y., X. Wang, G.L. Riskowski, and L.L. Christianson. 2000. A method to quantify ventilation effectiveness for air quality control. *Proceedings* 2nd International Conference on Air Pollution in Agricultural Operations, Des Moines, IA.

Zulovich, J.M., M.B. Manbeck, and W.B. Roush. 1987. Whole-house heat and moisture production of young floor brood layer pullets. *Transactions of ASAE* 30(2)455-58.

BIBLIOGRAPHY

ASAE. 1991. Design of ventilation systems for poultry and livestock shelters. ASAE *Standard* D270.5. American Society of Agricultural Engineers, St. Joseph, MI.

Consortium for Developing a Guide for the Care and Use of Agricultural Animals in Agricultural Research and Teaching. 1988. *Guide for the care and use of agricultural animals in agricultural research and teaching.* Agricultural Animal Care Guide Division of Agriculture, NASULGC, Washington, D.C. 20036-1191.

Esmay, M.E. and J.E. Dixon. 1986. *Environmental control for agricultural buildings.* AVI Publications, Westport, CT.

Hellickson, M.A. and J.N. Walker, eds. 1983. Ventilation of agricultural structures. *ASAE Monograph* No. 6. ASAE.

MWPS. 2000. The Midwest Plan Service handbooks and plans series. Midwest Plan Service, Ames, IA.

CHAPTER 11

PHYSIOLOGICAL FACTORS IN DRYING AND STORING FARM CROPS

THIS CHAPTER focuses on the drying and storage of grains, oil-seeds, hay, cotton, and tobacco. However, the primary focus is on grains and oilseeds, collectively referred to as **grain**. Major causes of postharvest losses in these products are fungi, insects, and rodents. Substantial deterioration of grain can occur in storage. However, where the principles of good grain storage are applied, losses are minimal.

Preharvest invasion of grains by storage insects is usually not a problem in the midwestern United States. Field infestations can occur in grains when they are dried in the field at warm temperatures during harvest. Preharvest invasion by storage fungi is possible and does occur if appropriate weather conditions prevail when the grain is ripening. For example, preharvest invasion of corn by *Aspergillus flavus* occurs when hot weather is prevalent during grain ripening; it is, therefore, more common in the southeastern United States (McMillan et al. 1985). Invasion of wheat, soybeans, and corn by other fungi can occur when high ambient relative humidities prevail during grain ripening (Christensen and Meronuck 1986). However, the great majority of damage occurs during storage, due to improper conditions that permit storage fungi or insects to develop.

Deterioration from fungi during storage is prevented or minimized by (1) reduction of grain moisture content to below limits for growth of fungi, (2) maintenance of low grain temperatures throughout the storage period, (3) chemical treatment to prevent the development of fungi or to reduce the rate of fungal growth while the grain moisture content is being lowered to a safe level, and (4) airtight storage in which initial microbial and seed respiration reduces the oxygen level so that further activity by potentially harmful aerobic fungi is reduced.

Reduction of moisture by artificial drying is the most commonly used technique. The longer grain is stored, the lower its storage moisture should be. Some of the basic principles of grain drying and a summary of methods for predicting grain drying rate are included in the section on Drying Theory.

Reduction of grain temperature by aeration is practical in temperate climates and for grains that are harvested during cooler seasons. Fans are operated when ambient temperature is lower than grain temperature. Basic information on aeration is summarized in the section on Drying Theory. Use of refrigeration systems to reduce temperature is not generally cost-effective for feed grains but may have application for higher value food grains.

Chemical treatment of grain is becoming more common and is briefly described in the section on Prevention of Deterioration.

When grain is placed in airtight silos, the oxygen level is rapidly reduced, and carbon dioxide increases. Although many fungi will not grow under ideal hermetic conditions, some will grow initially in imperfectly sealed bins, and this growth can reduce the feeding value of the grain for some animals. Partially emptied bins may support harmful mold, yeast, and bacterial growth, which makes the grain unsuitable for human consumption. Airtight storage is briefly addressed in the section on Oxygen and Carbon Dioxide under Factors Determining Safe Storage.

Deterioration from insects can also be prevented by a combination of reducing moisture and lowering temperature. Lowering of temperatures is best achieved by aeration with cool ambient air during cool nights and periods of cool weather. Both the use of clean storage structures and the segregation of new crop grain from carryover grain or grain contaminated with insects are important. If insect infestation has already occurred, fumigation is often used to kill the insects. Aeration with cold air may retard the development of the insect population. Prevention and control of insect infestations are addressed in the section on Prevention of Deterioration.

For information on rodent problems, see the section on Prevention of Deterioration.

Moisture content is the most important factor determining successful storage. Although some grains are harvested at safe storage moistures, other grains (notably corn, rice, and most oilseeds) must usually be artificially dried prior to storage. During some harvest seasons, wheat and soybeans are harvested at moistures above those safe for storage and, therefore, also require drying.

Sauer (1992), Brooker et al. (1992), Hall (1980), Christensen and Meronuck (1986), and Gunasekaran (1986) summarize the basic aspects of grain storage and grain drying. Chapter 22 of the 1999 *ASHRAE Handbook—Applications* covers crop-drying equipment and aeration systems.

FACTORS DETERMINING SAFE STORAGE

Moisture Content

Grain is bought and sold on the basis of characteristics of representative samples. Probes or samplers, such as diverters, are used to obtain representative subsamples. Often representative subsamples must be taken from a large quantity (several tons) of grain. Manis (1992) summarizes sampling procedures and equipment. For safe storage, it is necessary to know the range in moisture content within a given bulk and whether any of the grain in the bulk has a moisture content high enough to permit damaging fungal growth. This range can be determined by taking probe samples from different portions of the bulk. Commonly, in large quantities of bulk-stored grain, some portions have moisture contents 2 to 3% higher than the average (Brusewitz 1987). If the moisture content anywhere in the bulk is too high, fungi will grow, regardless of the average. Therefore, the moisture content of a single representative sample is not a reliable measure of storage risk or spoilage hazard. Measurement of moisture content and the precision of various moisture-measuring methods are covered in the section on Moisture Measurement.

Table 1 summarizes recommended safe storage moistures for several common grains. Note that for long-term storage, lower

The preparation of this chapter is assigned to TC 2.2, Plant and Animal Environment.

moistures are recommended. Most storage fungi will not grow in environments where the relative humidity of the air between kernels is lower than 60%. The relationship between grain moisture and the relative humidity of air between kernels is addressed in the section on Equilibrium Moisture. Table 2 summarizes the relative humidities and temperatures that permit the growth of common storage fungi. Table 3 summarizes the relative humidities and temperatures that permit growth of common storage insects.

Moisture Transfer

If temperatures vary within bulk-stored grain, moisture migrates from warmer to cooler portions. The rate of movement depends on the gradients in moisture content and temperature. Sellam and Christensen (1976) studied moisture transfer in a sample of one cubic foot of shelled corn initially at 15.5% moisture. They used heat lamps to produce a temperature differential of 18°F along the length of a sealed plastic container 1.22 ft long. After 2 days, this gradient (approximately 15°F/ft) caused the moisture content at the cool end to increase by 1.2% and the moisture content at the warm end to decrease by 1.1%.

Thorpe (1982) developed an equation to describe moisture transfer caused by a temperature gradient. The equation was solved numerically, and laboratory experiments of moisture transfer in wheat were successfully modeled initially at 12% moisture content. In the experiments, an 18°F temperature gradient was developed across a column of wheat 0.66 ft thick (equivalent to a gradient of 27°F/ft). After one month, the moisture content of the warmest grain dropped to 10.6%, while the moisture content of the coolest grain increased to 14%.

Smith and Sokhansanj (1990a) provided a method of approximate analysis of the energy and velocity equations of the natural convection in grain bins. They showed that for small cereals such as wheat, the influence of convection on temperature gradients may not be significant, whereas for larger cereals such as corn, the effect of convection is more noticeable. Smith and Sokhansanj (1990b) also showed that convection flows in a grain bin are significant if the radius of the storage bin is approximately equal to the height of the bin.

Christensen and Meronuck (1986) cite an example of heating that developed in wheat initially at 13.2% in a nonaerated bin. Specially prepared samples were placed at various positions in the bin at the time the bin was filled. After 3 months, the grain began to heat from fungal activity. Moisture content in some of the samples had increased to 18%, while in others it had decreased to 10%.

These examples illustrate the importance of aeration in long-term storage. Aeration is generally required for storage structures with capacities exceeding 2000 bu or 50 tons. Moisture migration can initiate fungal and insect growth, and the heat of respiration generated by these organisms accelerates their growth and leads to spoilage. Studies suggest that temperature gradients could promote spoilage of grain loaded into a ship or barge—even if the grain is initially at a uniform moisture. Most shipments do not spoil because they remain in the ship or barge for a short time and because large temperature gradients do not develop. Christensen and Meronuck (1986) report studies of grain quality in barges and ships.

Table 1 Safe Storage Moisture for Aerated Good-Quality Grain

Grain	Maximum Safe Moisture Content, % wet basis
Shelled corn and sorghum	
To be sold as #2 grain or equivalent by spring	15
To be stored up to 1 year	14
To be stored more than 1 year	13
Soybeans	
To be sold by spring	14
To be stored up to 1 year	12
Wheat	13
Small grain (oats, barley, etc.)	13
Sunflower	
To be stored up to 6 months	10
To be stored up to 1 year	8

Source: McKenzie (1980).

Table 2 Approximate Temperature and Relative Humidity Requirements for Spore Germination and Growth of Fungi Common on Corn Kernels

Fungus	Minimum Relative Humidity for Spore Germination,[b] %	Grain Moisture,[a] % w.b.	Growth Temperature, °F		
			Lower Limit	Optimum	Upper Limit
Alternaria	91	19	25	68	97 to 104
Aspergillus glaucus	70 to 72	13.5 to 14	46	75	100
Aspergillus flavus	82	16 to 17	42 to 43	97 to 100	111 to 115
Aspergillus fumigatus	82	16 to 17	54	104 to 108	131
Cephalosporium acremonium[c]	97	22	46	75	104
Cladosporium	88	18	23	75 to 77	86 to 90
Epicoccum	91	19	25	75	82
Fusarium moniliforme	91	19	39	82	97
Fusarium graminearum, Fusarium roseum (Gibberella zeae)	94	20 to 21	39	75	90
Mucor	91	19	25	82	97
Nigrospora oryzae[c]	91	19	39	82	90
Penicillium funiculosum[c] *(field)*	91	19	46	86	97
Penicillium oxalicum[c] *(field)*	86	17 to 18	46	86	97
Penicillium brevicompactum (storage)	81[b]	16	28	73	86
Penicillium cyclopium (storage)	81[b]	16	28	73	86
Penicillium viridicatum (storage)	81[b]	16	28	73	97

Source: Stroshine et al. (1984).

[a] Approximate corn moisture content at 77°F, which gives an interseed relative humidity equal to the minimum at which fungus can germinate. It is probably *below* the moisture content at which the fungus would be able to compete with other fungi on grain, except for *Aspergillus glaucus*. The latter has no real competitor at 72% rh, except occasionally *Aspergillus restrictus*.

[b] Approximately 5% or more of the population can germinate at this relative humidity.

[c] Rarely found growing in stored grain, regardless of moisture and temperature.

Temperature

Most processes that cause spoilage in stored grains are accompanied by a temperature rise. Therefore, temperatures should be monitored throughout the bulk. Temperature monitoring is commonly done by attaching thermocouples to cables that extend through the bulk from the top to the bottom, with thermocouples about 3 to 6 ft apart on each cable. Single cables are used in the center of circular bins up to 25 ft in diameter. In large bins or flat storage structures, cables are spaced 20 to 25 ft apart. Relatively dry grain is a good insulator, so a **hot spot** can develop without being detected immediately (Foster and Mayes 1962). However, when these thermocouple spacings are used, extensive spoilage can usually be detected by a temperature rise at a nearby thermocouple. A temperature rise of even a few degrees is evidence that grain has spoiled or is spoiling. Forced aeration maintains a uniform and preferably low temperature throughout the bulk.

Table 2 summarizes minimum, optimum, and maximum temperatures for the growth of some common storage fungi. Storage molds grow slowly at 32 to 40°F. However, at higher moisture contents, some species of *Penicillium* will grow when the temperature is slightly below freezing. Grains with a moisture content high enough for invasion by *Aspergillus glaucus* will deteriorate rapidly at temperatures of 75 to 85°F but can be kept for months without damage at 40 to 50°F. Most grain-infesting insects become inactive below about 50°F. Mites remain active but cannot develop rapidly below about 40°F. Control of fungi and insects is described further in the section on Prevention of Deterioration.

Oxygen and Carbon Dioxide

Only a few fungi that cause stored grain deterioration can grow in an atmosphere containing only 0.1 to 0.2% oxygen or more than 60% carbon dioxide. Some yeasts can grow in grain stored in airtight storage at moisture contents above 18 to 19% and temperatures above 40°F, producing flavors that make the grain unsuitable as food. However, the grain remains suitable feed for cattle and swine (Bell and Armitage 1992), and its nutritional value may be enhanced (Beeson and Perry 1958).

Dry grain is stored in airtight underground or earth-sheltered structures in many parts of the world (Dunkel 1985, Bell and Armitage 1992). Janardhana et al. (1998) reported that while storing shelled corn at 15 to 20% moisture in a warm, high-humidity environment, visible molding and the loss of food reserves can be postponed up to 45 days by using high carbon dioxide in the storage. Alagusundarum et al. (1996) studied the diffusion of carbon dioxide introduced into bulk stored grain. Bell and Armitage (1992) and Shejbal (1980) cover controlled atmosphere storage in more detail.

Insects present when dry grain is put into storage usually die when oxygen has been depleted, and do not usually reproduce if grain is sufficiently dry and in good condition. Insects can also be controlled in conventional storage structures by forcing carbon dioxide or other gases such as nitrogen through the grain (Jay 1980, Ripp 1984, White and Jayas 1991). Generally, carbon dioxide environments are effective for insect control only when concentrations are greater than 40% for long periods. Paster et al. (1990, 1991) investigated biogenerated modified atmospheres for insect control. Athie et al. (1998) reported that using the toxic chemical phosphine with a high carbon dioxide environment increased the effectiveness of the phosphine, particularly in resistant populations. However, the costs of controlled-atmosphere storage may be uneconomic unless the structure can be inexpensively sealed or the gases can be inexpensively generated or purchased.

Grain Condition

Grain that has been stored for several months may already be invaded by storage fungi and partly deteriorated, whether or not this is evident to the naked eye. Molding occurs more rapidly in partially deteriorated grain than in sound grain when the grain is exposed to conditions favorable to mold growth. Microscopic examination and plating techniques can often reveal the fungal infection of grain in its early stages (Sauer et al. 1992, Christensen and Meronuck 1986, Stroshine et al. 1984). Accelerated storage tests, in which samples of grain are stored at a moisture content in equilibrium with air at 80% rh and 85°F and examined periodically, are useful in evaluating storability. These tests enable a manager to estimate the risk of spoilage during storage and to take appropriate action.

Equilibrium Moisture

If air remains in contact with a product for sufficient time, the partial pressure of the water vapor in the air reaches equilibrium

Table 3 Estimates of Optimum and Minimum Temperatures and Relative Humidity Conditions for Population Increase of Grain-Infesting Insects

Insect Type: In Regard to Temperature	Insect Type: In Regard to Relative Humidity	Species	Temperature, °F: Minimum	Temperature, °F: Optimum	Minimum Relative Humidity, %
		Species Needing High Temperatures			
Cold hardy	Tolerant of low	*Trogoderma granarium*	75	91 to 99	1
		Cryptolestes ferrugineus	73	90 to 95	10
		Oryzaephilus surinamensis	70	88 to 93	10
	Need moderate	*Plodia interpunctella*	64	82 to 90	40
	Need high	*Cryptolestes turcicus*	70	86 to 91	50
Moderately cold hardy	Tolerant of low	*Tribolium confusum*	70	86 to 91	1
	Need moderate	*Rhyzopertha dominica*	73	90 to 95	30
		Lasioderma serricorne	72	90 to 95	30
Cold susceptible	Tolerant of low	*Tribolium castaneum*	72	90 to 95	1
		Oryzaephilus mercator	70	88 to 93	10
	Need high	*Cryptolestes pusillus*	72	82 to 91	60
		Species Thriving at Moderate Temperatures			
Cold hardy	Need moderate	*Sitotroga cerealella*	61	79 to 86	30
	Need high	*Sitophilus granarius*	59	79 to 86	50
		Stegobium paniceum	63	77 to 82	60
		Acarus siro	45	70 to 81	65
Moderately cold hardy	Need high	*Sitophilus oryzae*	63	81 to 88	60

Source: Pederson (1992). Reprinted with permission.

with the partial pressure of the water vapor in the material. The relative humidity of the air at equilibrium with a material of a given moisture is the **equilibrium relative humidity**. The moisture content of a hygroscopic material in equilibrium with air of a given relative humidity is the **equilibrium moisture content** M_e.

Several theoretical, semitheoretical, and empirical models have been proposed for calculating the M_e of grains. Morey et al. (1978) report that the modified Henderson equation is among the best equations available:

$$M_e = \frac{1}{100}\left[\frac{\ln(1.0-\phi)}{-K(t+C)}\right]^{1/N} \quad (1)$$

where

M_e = equilibrium moisture content, decimal, dry basis
t = temperature, °C = (°F − 32)/1.8
ϕ = relative humidity, decimal equivalent
K, N, C = empirical constants

Table 4 lists values for K, N, and C for various crops. ASAE *Standard* D245.5 also gives the Chung-Pfost equation, another equation used to predict M_e. Figure 1, based on the Chung-Pfost equation, shows equilibrium moisture content curves for shelled corn, wheat, soybeans, and rice. Note that equilibrium moisture depends strongly on temperature. ASAE *Standard* D245.5 gives additional curves drawn from the Chung-Pfost equation and tabulated experimental data. Pfost et al. (1976) summarize variations in reported values of M_e for several grains. Locklair et al. (1957) give data for tobacco.

The modified Henderson and the Chung-Pfost equations give only approximate values of M_e and are for desorption. When grain is rewetted after it has been dried to a low moisture, the value of M_e is generally lower for a given relative humidity. Sun and Woods (1993) reviewed the equilibrium relationship between wheat moisture content and air relative humidity. They reported that the hysteresis effect was greatest at 20 to 40% rh. They also observed that the modified Henderson equation was least effective for wheat.

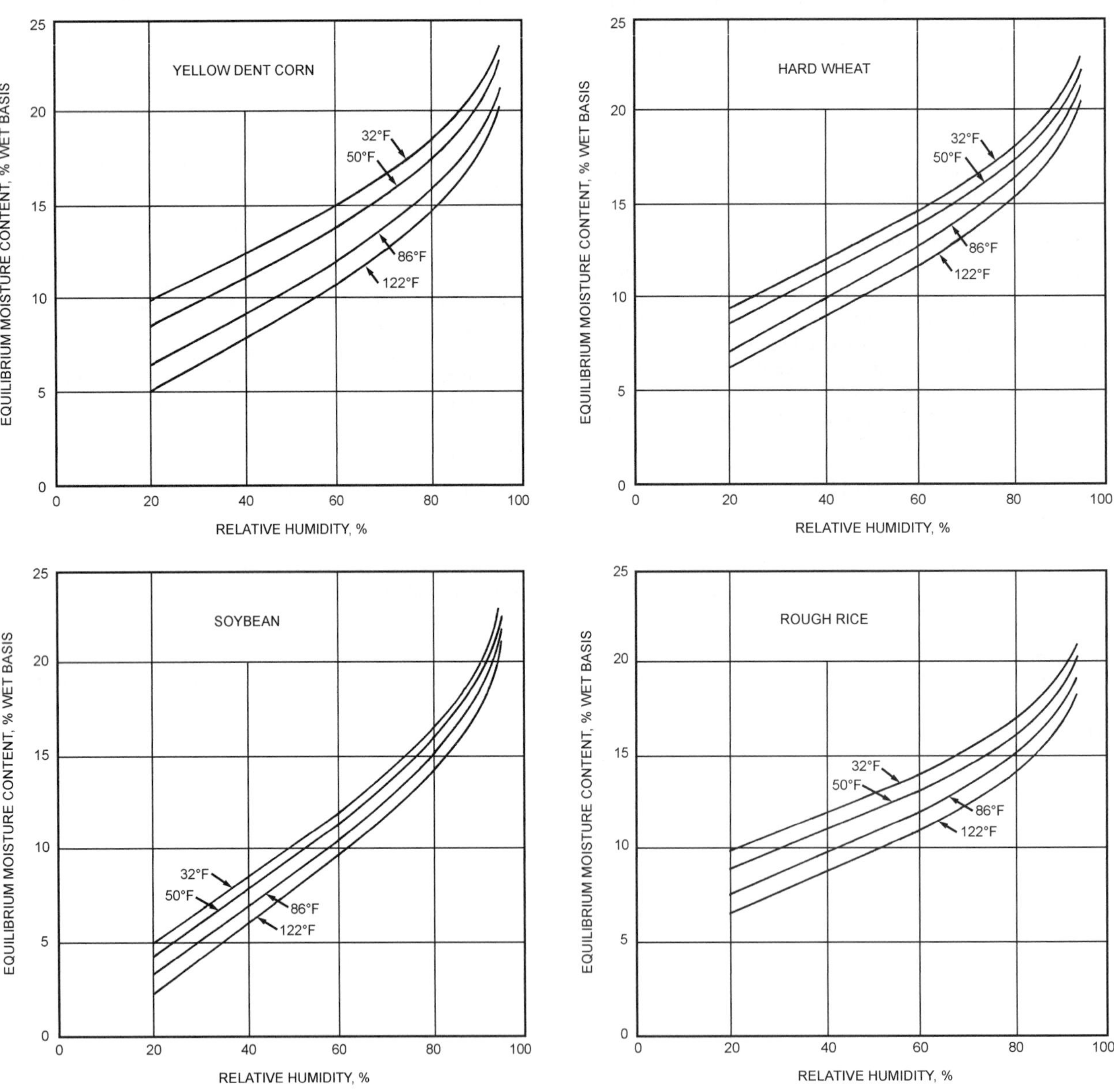

Fig. 1 Equilibrium Moisture Relationships for Certain Crops
(ASAE *Standard* D245.5)

Table 4 Desorption Equilibrium Moisture Constants for Modified Henderson Equation [Equation (1)] for Various Crops

Product	K	N	C
Barley	0.000022919	2.0123	195.267
Beans, edible	0.000020899	1.8812	254.230
Canola (rapeseed)	0.000505600	1.5702	40.1205
Corn, yellow dent	0.000086541	1.8634	49.810
Peanut, kernel	0.000650413	1.4984	50.561
Peanut, pod	0.000066587	2.5362	23.318
Rice, rough	0.000019187	2.4451	51.161
Sorghum	0.000085320	2.4757	113.725
Soybean	0.000305327	1.2164	134.136
Wheat, durum	0.000025738	2.2110	70.318
Wheat, hard	0.000023007	2.2857	55.815
Wheat, soft	0.000012299	2.5558	64.346

Source: ASAE *Standard* D245.5.

Variations of as much as 0.5 to 1.0% can result from differences in variety; maturity; and relative starch, protein, and oil content. Shivhare et al. (1992) reported that, under a microwave drying regime, M_e increased with drying air velocity and decreased with microwave power.

High-temperature drying can decrease the M_e of shelled corn by 0.5 to 1.0% for a given relative humidity (Tuite and Foster 1963). Sun and Woods (1994) reported that the equilibrium moisture content for wheat dried at low temperatures was generally higher than the data reported in the literature. Chung and Verma (1991) studied the M_e of rice during drying and storage.

MOISTURE MEASUREMENT

Rapid and accurate measurement of moisture of grains, seeds, and other farm crops determines whether they can be safely stored. Allowable upper limits for moisture are set by the market, and discounts and/or drying charges are usually imposed for higher moistures. Drying the grain to moistures below the accepted market limit or the limit for safe storage moisture results in additional drying expense and may actually decrease the value of the grain.

If shelled corn is dried to 12% moisture, it becomes brittle and breaks more easily during handling. The moisture removal also reduces the total weight of grain. After drying 1 bu (56.0 lb) of shelled corn at 18% moisture to 15%, only 0.965 bu (54.0 lb) remains. However, at high moistures, seed respiration and fungal growth can cause greater loss in value.

Moisture content can be expressed on a wet or dry basis. The wet basis is used by farmers and the grain trade, while dry-basis moistures are often used by engineers and scientists to describe drying rates. Unless otherwise noted, moisture contents in this chapter are on a **wet basis** and are calculated by dividing the weight of water in the material by the total weight. The **dry basis** is calculated by dividing the weight of water by the weight of dry matter.

$$\text{\% Moisture (wet basis): } M_w = \frac{100W_w}{W_w + W_d} \tag{2}$$

$$\text{\% Moisture (dry basis): } M_d = \frac{100W_w}{W_d} \tag{3}$$

where

W_w = weight of water
W_d = weight of dry matter

Percent moisture on a wet basis M_w can be converted to percent moisture on a dry basis M_d and vice versa by the following formulas:

$$M_d = \frac{100M_w}{100 - M_w} \tag{4}$$

$$M_w = \frac{100M_d}{100 + M_d} \tag{5}$$

The weight change resulting from a change in moisture can be determined by assuming that the weight of dry matter is constant. The dry matter is calculated by multiplying the weight of grain by the quantity $(1 - M_w/100)$. For example, 1000 lb of wheat at 15% moisture has 1000 (1 – 15/100) = 850 lb of dry matter and 150 lb water. As grain dries, the dry matter remains constant and the mass of water is reduced. If the dry matter is constant, the mass or moisture content of the dried grain is

$$W_w\left(1 - \frac{M_w}{100}\right) = W_d\left(1 - \frac{M_d}{100}\right) \tag{6}$$

The weight W_d of the dried grain after it reaches a moisture content of M_d is

$$W_d = W_w\left(\frac{100 - M_w}{100 - M_d}\right) \tag{7}$$

For example, if 1000 lb of grain is dried from 15% to 13% moisture, the dried weight is

$$W_d = 1000\left(\frac{100 - 15}{100 - 13}\right) = 977 \text{ lb}$$

Similarly, to find the moisture content M_d of grain after it reaches a dried weight of W_d

$$M_d = 100 - \frac{W_w}{W_d}(100 - M_w) \tag{8}$$

For example, 100 lb of grain at 15% is dried to a weight of 950 lb. The dried moisture content is

$$M_d = 100 - \frac{1000}{950}(100 - 15) = 10.5\%$$

If two quantities of grain at differing moistures are mixed, the final moisture of the mixture can be determined by calculating the weight of water in each, adding these together, and dividing by the total weight. The weight of water is the product of the decimal equivalent of M_w and the total weight. For example, if 500 lb of grain at 16% moisture is mixed with 1000 lb of grain at 14% moisture, the weight of water in each sample is

$$(500 \text{ lb})(0.16) = 80 \text{ lb}$$

$$(1000 \text{ lb})(0.14) = 140 \text{ lb}$$

The moisture content after mixing will be

$$\left[\frac{80 + 140}{500 + 1000}\right] \times 100 = 14.7\%$$

Either a **direct method** or an **indirect method** is used to determine moisture content. Direct methods involving the use of an oven determine moisture content based on the loss in product weight caused by evaporation of the water. The Karl Fischer method, a basic reference method involving a chemical reaction of water and a reagent, is classified as a direct method.

Indirect methods such as moisture meters measure the properties of the material that are functions of moisture content. Indirect moisture measurement methods are used in commercial practice. Direct methods are used in research and to calibrate indirect methods. Christensen et al. (1992) summarize approved methods used in Europe and the United States.

Direct Methods

Christensen et al. (1992) describe the fundamental or basic methods of moisture determination as (1) drying in a vacuum with a desiccant and (2) titration with a Karl Fischer reagent. It is assumed that these methods measure the true water content and can be used to verify measurements obtained with routine reference methods, including oven drying and the Brown-Duval distillation method. The Brown-Duval method, not commonly used, involves heating the grain in a special apparatus and condensing and collecting the vaporized water.

Oven techniques use either forced-convection air ovens or vacuum ovens and either ground or whole kernels. Drying times and temperatures vary considerably, and the different techniques can give significantly different results. Oven techniques are used to calibrate moisture meters (see the section on Indirect Methods). As a result, during the export of grain, the meter moisture measurements can vary between arrival and destination if the importing country uses a different standard oven technique than the exporting country.

ASAE *Standard* S352.2 is a widely used standard that recommends heating temperatures and times for various grains. The temperatures may be either 217°F (shelled corn, soybeans, sunflower) or 266°F (wheat, barley, onion). Heating times vary between 50 min (onion seeds) and 72 h (soybeans, shelled corn).

Grinding samples and using a vacuum oven reduce heating time. When initial moistures are high, a two-stage method may be used (USDA 1971). A weighed sample of whole grain is partially dried to a moisture content of 13% or below, weighed, and then ground and completely dried as in the one-stage method. The mass lost in both stages is used to calculate moisture content.

Indirect Methods

Electronic moisture meters are simple to operate and give readings within minutes. Direct methods of moisture measurement are used to calibrate the meters for each type of grain. Meters are sensitive to grain temperature, and calibration must include a temperature correction factor. The newer automatic meters or **moisture computers** sense and correct for sample temperature and print or display the corrected moisture.

Near infrared reflectance (NIR) instruments have been developed that measure moisture, protein, starch, and oil content of ground samples (Butler 1983, Cooper 1983, Watson 1977). **Near infrared transmittance (NIT) instruments** measure the properties of whole seeds.

Conductance meters measure resistance, which varies with grain moisture. The practical range of moisture content measurable by conductance meters is approximately 7 to 23%. For up to 72 h after moisture addition or removal, the moisture at the surface of the kernels differs from the moisture in the interior. Therefore, recently dried grain reads low and recently wetted grain reads high. Mixing wet and dry grain and mixing good grain with partially deteriorated grain also result in erroneous readings. Martin et al. (1986) measured the signal from the conductance Tag-Heppenstall meter and related this to individual kernel moisture variations in mixtures of wet and dry corn.

The dielectric properties of products depend largely on moisture content. The **capacitance meter** uses this relationship by using grain as the dielectric in a capacitor in a high-frequency electrical circuit. Although the capacitive reactance is the primary portion of the overall impedance measured, the resistive component is also significant in many capacitance meters. At higher frequencies and in instruments with insulated electrodes, the relative effect of the resistance is reduced, which is important in reducing errors introduced by unusual product surface conditions. Capacitance meters are affected less than conductance meters by uneven moisture distribution within kernels. Sokhansanj and Nelson (1988a) showed that the capacitance meters give low and high readings, respectively, on recently dried or rewetted grain. The range of measurable moisture content is slightly wider than that for conductance meters.

Moisture measurement by capacitance meters is sensitive to temperature, product weight, and product density (Sokhansanj and Nelson 1988b). To reduce these sources of error, a weighed sample is introduced into the measuring cell by reproducible mechanical means. Calibration, including temperature correction, is required. At least one commercially available unit measures bulk density and corrects for this factor as well as temperature. Tests of moisture meter accuracy have been reported by Hurburgh et al. (1985, 1986). Accuracy of moisture readings can be improved by taking multiple samples from a grain lot and averaging the meter measurements. Equipment for continuous measurement of moisture in flowing grain is available commercially but is not widely used in the grain trade.

Equilibrium relative humidity (described in the section on Equilibrium Moisture) can be used to indicate moisture content. It also indicates storability independent of the actual moisture content because the equilibrium relative humidity of the air surrounding the grain, to a large extent, determines whether mold growth can occur (see Table 2). Measurement of equilibrium relative humidity at specific points within a grain mass requires specialized sampling equipment and has been used primarily for research.

Hay moisture content does not receive the consideration devoted to grains. Oven methods (ASAE *Standard* S358.2) are used extensively, as well as some commercial hay and forage moisture meters. Several conductance moisture meters are available for both hay and forages, but the extreme variability of the moisture and density of the material tested lead to great variability in the readings obtained. A reasonable indication of the average moisture content of a mass of hay can be obtained if many (25 or more) measurements are taken and averaged.

Microwave radiation can be used to sense properties of grains. Kraszewski and Nelson (1992) reviewed the use of resonant microwave cavities to determine simultaneously both mass and moisture content in grain kernels.

PREVENTION OF DETERIORATION

Fungal Growth and Mycotoxins

Fungal growth is the most important limitation to successful drying and storage of grain. Sauer et al. (1992) provide a good review of microflora in grains. Early detection of mold growth would provide the storage manager with a management tool. Magan (1993) reviews early detection methods, including enzyme and biochemical tests, fungal volatiles, and respiration activity.

Mycotoxins (fungal metabolites) may affect the marketability and use of moldy grains. Wicklow (1988) reported that climate and other natural processes influenced the distribution of aflatoxigenic strains of microflora. Choudhary and Sinha (1993) reported that aflatoxigenic *Aspergillus* species were positively correlated with grain moisture content in the field and afternoon relative humidity.

In the United States, corn is one of the major crops that must be harvested above safe storage moistures. Shelled corn can be held at these higher moistures for a limited time before it must be dried. Mold growth produces carbon dioxide (CO_2). Allowable storage time at moistures above those for safe storage can be estimated by measuring CO_2 production of samples. By assuming that a simple sugar is being oxidized by microbial respiration, CO_2 production can be expressed in terms of dry matter loss in percent by weight.

Saul and Steele (1966) and Steele et al. (1969) studied the production of CO_2 in shelled corn, mostly on samples above 18%. Based on changes in the official grade of shelled corn, Saul and Steele (1966) established a criterion for acceptable deterioration of quality as 0.5% dry matter loss. This is equivalent to the production of 0.00735 lb of CO_2 per pound of dry matter. Thompson (1972) expressed Saul's data on dry matter loss per pound of dry matter as a function of moisture, time, and temperature using the following mathematical expression:

$$\text{DML} = 1.3\left[\exp\left(\frac{0.006\theta}{K_m K_t}\right) - 1.0\right] + \frac{0.015\theta}{K_m K_t} \tag{9}$$

with

$$K_m = 0.103\left[\exp\left(\frac{455}{M_d^{1.53}}\right) - 0.00845 M_d + 1.558\right] \tag{10}$$

$$K_t = A\ \exp B(0.03t + 0.533) + C\ \exp[0.0183t - 0.285)] \tag{11}$$

where

DML = dry matter loss per pound of dry matter, lb/lb
θ = time in storage, h
t = grain temperature, °F
M_d = moisture content, % dry basis

Table 5 lists values for A, B, and C. According to Steele (1967), the damage level effect can be determined for dry matter losses of 0.1, 0.5, and 1.0% by multiplying θ from Equation (9) by K_d, where K_d is calculated as follows:

$$0.1\%\ \text{DML: } K_d = 1.82\exp(-0.0143d) \tag{12a}$$

$$0.5\%\ \text{DML: } K_d = 2.08\exp(-0.0239d) \tag{12b}$$

$$1.0\%\ \text{DML: } K_d = 2.17\exp(-0.0254d) \tag{12c}$$

where d = mechanical damage, % by weight.

Seitz et al. (1982a, 1982b) found unacceptable levels of aflatoxin production prior to the time when 0.5% dry matter loss occurred. Nevertheless, Equations (9) through (12) give approximate predictions of mold activity, and they have been used in several computer simulation studies (Thompson 1972, Pierce and Thompson 1979, Brooker and Duggal 1982). Based on a simulation, Thompson (1972) concluded that for airflow rates between 0.5 and 2.0 cfm/bu, grain deterioration in the top layer during low-temperature drying is doubled when the airflow rate is halved. Thompson also concluded that weather variations during harvest and storage seasons can cause up to a twofold difference in deterioration. Pierce and Thompson (1979) recommend airflow rates for several common low-temperature drying systems and for various locations in the midwestern United States.

Table 5 Constants for Dry Matter Loss of Shelled Corn [Equation (11)]

Temperature Range, °F	Moisture Range, % Wet Bulb	A	B	C
$t < 60$	All moistures	128.76	−4.68	0
$t \geq 60$	$M_w \leq 19$	32.3	−3.48	0
$t \geq 60$	$19 < M_w \leq 28$	32.3	−3.48	$(M_w - 19)/100$
$t \geq 60$	$M_w > 28$	32.3	−3.48	0.09

Acceptable dry matter losses for wheat and barley are much lower than those for shelled corn—0.085% and 0.10%, respectively (Brook 1987). Hamer et al. (1992) reported visible mold growth at 0.15% dry matter loss. Brook reported reasonable agreement with published experimental data for the following equation (Frazer and Muir 1981) for allowable storage time as a function of percent wet basis moisture and temperature based on the development of visible mold:

$$\log\theta_D = A + BM_w + Ct + G \tag{13}$$

where

θ_D = allowable storage time, days
t = temperature, °F
A,B,C,G= empirical constants, defined as follows:

Moisture Range, % w.b.	A	B	C	G
$12.0 < M_w \leq 19.0$	6.234	−0.2118	−0.0293	0.937
$19.0 < M_w < 24.0$	4.129	−0.0997	−0.0315	1.008

Brook (1987) also reported that an adaptation of Equation (9) by Morey et al. (1981) gave reasonable results for storage time of wheat. Morey's method predicts dry matter loss by adjusting M_d for differences between corn and wheat equilibrium relative humidities.

Table 2 can be used to gain insight into the deterioration of stored grain. *Aspergillus* and *Penicillium* sp. are primarily responsible for deterioration because some of their species can grow at storage moistures and temperatures frequently encountered in commercial storage. In temperate climates, shelled corn is often harvested at relatively high moistures; during the harvest and storage season, ambient temperatures can be relatively low. Aeration of the grain during cold weather and cool nights can reduce the temperature of the grain to 40 to 60°F. This is below the optimum temperature for growth of *Aspergillus* sp. (Table 2). However, *Penicillium* sp. can still grow if grain moisture is above 16 to 17%; therefore, its growth is a persistent problem in temperate climates. If hot weather prevails prior to harvest, *Aspergillus flavus*, which competes effectively at warmer temperatures and higher moistures, can begin to grow in the field and continue to grow in stored shelled corn. In growing seasons when shelled corn must be harvested at moistures above 22%, *Fusarium*, *Alternaria*, *Epicoccum* and *Mucor* can compete with *Penicillium* sp.

Chemical Treatment. Application of chemicals slows deterioration until grain can be either dried or fed to animals. Preservatives include propionic acid, acetic acid, isobutyric acid, butyric acid (Sauer and Burroughs 1974), a combination of sorbic acid and carbon dioxide (Danziger et al. 1973), ammonia (Peplinski et al. 1978), and sulfur dioxide (Eckhoff et al. 1984). Propionic acid (Hall et al. 1974) or propionic-acetic acid mixtures, although not extensively used, are perhaps the most popular in the United States with high-moisture corn. Acetic acid and formic acid are popular in Europe. Grain treated with propionic acid can be used only as animal feed.

Hertung and Drury (1974) summarize fungicidal levels needed to preserve grain at various moistures. Both ammonia (Nofsinger et al. 1979, Nofsinger 1982) and sulfur dioxide (Eckhoff et al. 1984, Tuite et al. 1986) treatments require considerable management. Attention must be given to uniform application of the chemicals to the entire quantity of stored grain.

Antimicrobial properties occurring naturally in plants have been studied (Beuchar and Golden 1989, Shelef 1984). Some of these also inhibit mycotoxin formation (Bullerman et al. 1984, Rusal and Marth 1988). The essential oils of oregano and thyme were tested as fumigants against *Aspergillus* species and natural microflora of wheat (Paster et al. 1995). Oregano oil provided complete control at 2.0 mL/m^3; thyme oil was not completely effective at 4.0 mL/m^3 and affected seed germination at 5.0 mL/m^3.

Insect Infestation

Insects cause major losses of stored grain. Grain containing live insects or insect fragments in sufficient numbers is unsuitable for human food. When grain is stored for long periods (a year or more), insects can infest the grain and cause significant amounts of deterioration. Traps and chemical attractants have been developed that monitor insects in storage facilities (Barak and Harein 1982, Barak and Burkholder 1985, Burkholder and Ma 1985). Detection in samples of grain taken for grading and inspection is often difficult. Many of the insects are relatively small and can be seen easily only with a magnifying lens. Many of the insect larvae develop within the kernels and cannot be detected without staining techniques or grinding of the grain sample. Infested grain mixed with good grain in marketing channels compounds the infestation problem.

Sanitation is one of the most effective methods of insect control. Cleaning of bins after removal of old-crop grain and prior to filling with new-crop grain is essential. In bins containing perforated floors, fine material that collects beneath the floors can harbor insects, which infest new-crop grain when it is added. Control by aeration is feasible in temperate climates because insect activity is reduced greatly at temperatures below 50°F. The effectiveness of temperature control has been documented by Bloome and Cuperus (1984) and Epperly et al. (1987). Chemicals have frequently been used to control live insects in grain, and methods are described by Harein and Davis (1982). Thermal treatments have also been investigated (Lapp et al. 1986). Pederson (1992) summarizes the types of grain insects, the ecology of insect growth, and the methods of detecting insects in samples of grain. Athie et al. (1998) reviewed the status and future of chemical grain protectants. Armitage et al. (1994) proposed an integrated pest management strategy combining surface insecticide treatment and aeration. Control of insects in farm-stored grain is detailed by Storey et al. (1979), Quinlan (1982), and Harein and Davis (1992).

Rodents

The shift from ear corn harvesting and storage to field shelling and the introduction of metal bins have helped to reduce rodent problems. However, significant problems can arise when rodents consume grain and contaminate it with their hair and droppings. Storage structures should be made rodent-proof whenever possible. Rats can reach 13 in. up a wall, so storage structures should have concrete foundations and metal sides that resist gnawing.

In some countries, smaller on-farm storage structures are often elevated 18 in. to give protection from rodents. Double-wall construction and false ceilings should be avoided, and vents and holes should be covered with wire grates. Proper sanitation can help prevent rodent problems by eliminating areas where rodents can nest and hide. Rodents need water to survive, so elimination of available water is also effective. Techniques for killing rodents include trapping, poisoning with bait, and fumigation. Harris and Bauer (1992) address rodent problems and control in more detail.

DRYING THEORY

In ordinary applications, drying is a heat and mass transfer process that vaporizes liquid water, mixes the vapor with the drying air, and removes the vapor by carrying away the mixture mechanically. In forced-convection drying, sufficient heat for vaporization of product moisture (about 1100 Btu/lb of water) comes from the sensible heat in the drying air.

The most common mode of drying uses the sensible heat content of the air. The method can be diagrammed on the psychrometric chart by locating the state points for the air as it is heated from ambient temperature to plenum temperature and then exhausted from the grain. The process is assumed to be adiabatic (i.e., all the sensible heat lost by the air is used for moisture vaporization and converted to latent heat of the water vapor in the drying air). Therefore, the state point of the air can be considered to move along adiabatic saturation lines on the psychrometric chart. In the simplified psychrometric chart in Figure 2, the ambient air at dry-bulb temperature t_a and dew-point temperature t_{dp} is heated to drying air temperature t_d, where it has a relative humidity ϕ_1. As the air passes through the grain, its sensible heat provides the latent heat of vaporization of the water. When the air exits from the grain, its temperature has dropped to t_e, and its relative humidity has increased to ϕ_2. The moisture gained by each pound of drying air is the difference $W_2 - W_1$ in humidity ratio. If the air has sufficient contact time with the grain, the value for ϕ_2 will be the equilibrium relative humidity of the grain at that moisture and temperature t_e.

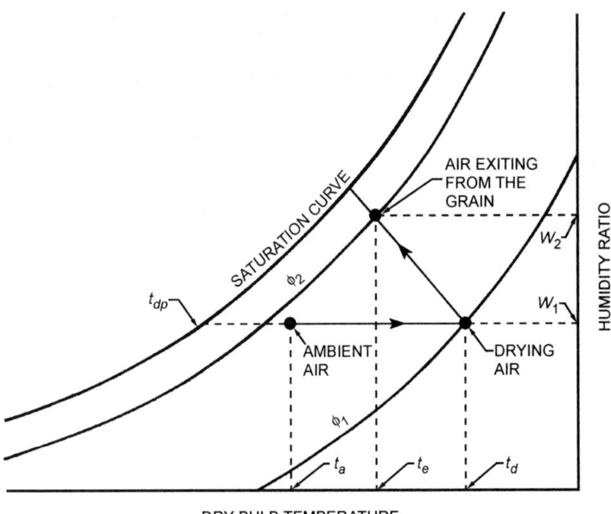

Fig. 2 Drying Process Diagrammed on Psychrometric Chart Showing Adiabatic Evaporation of Moisture from Grain

Example 1. Shelled corn at 20% moisture content is dried with air heated to 160°F. The air has an ambient temperature of 68°F with a dew point of 50°F. The air is observed to exhaust from the shelled corn at 86°F. Find the amount of energy needed to heat the air and the amount of water removed per pound of dry air.

Solution: Estimate the psychrometric air conditions, using information contained in Chapter 6 and assuming a standard atmospheric pressure of 14.696 psi.

At 68°F, the enthalpy of the dry air is h_a = 16.337 Btu/lb (Table 2 in Chapter 6) and the saturation vapor pressure p_{ws} = 0.33921 psi (Table 3 in Chapter 6). At 50°F dew point, the vapor pressure p_w = 0.17811 psi, and the enthalpy of the water vapor h_g = 1083.03 Btu/lb (Table 3 in Chapter 6). The relative humidity is then ϕ = 53% [Equation (24) in Chapter 6]; humidity ratio W = 0.0076 lb/lb [Equation (22) in Chapter 6]; and enthalpy h = 24.6 Btu/lb [Equation (29) in Chapter 6]. As the air is heated, the humidity ratio is assumed to remain constant. At 160°F, the enthalpy of the dry air is h_a = 38.474 Btu/lb (Table 2 in Chapter 6) and the saturation vapor pressure p_{ws} = 4.7468 psi (Table 3 in Chapter 6). The relative humidity has been reduced to ϕ = 4% [Equation (24) in Chapter 6]; enthalpy increased to h = 46.8 Btu/lb [Equation (29) in Chapter 6]; and the wet-bulb temperature of the drying air is t^* = 84°F [iterative solution to Equation (35) in Chapter 6]. The amount of energy needed to heat each pound of dry air is then 46.8 – 24.6 = 22.2 Btu.

As the heated air passes through the grain, it increases in moisture and decreases in temperature until it comes into equilibrium with the corn at the point of air exhaust (initially 20%). The exhaust air relative humidity can be estimated by reading the equilibrium relative humidity from the curve for shelled corn shown in Figure 1. Enter the curve for shelled corn at 20% equilibrium moisture content and a temperature of 86°F. The equilibrium relative humidity is approximately 92%. At 86°F, the saturation vapor pressure of the air p_{ws} = 0.61584 psi (Table 3 in Chapter 6); the vapor pressure p_w = 0.5666 psi [Equation (24) in Chapter 6]; and the humidity ratio W = 0.0249 lb/lb [Equation (22) in

Chapter 6]. Each pound of dry air carries with it 0.0249 – 0.0076 = 0.0173 lb of water from the grain.

After the grain at the air exhaust has dried to 15%, the equilibrium moisture content curve from Figure 1 can be used to estimate the exhaust air relative humidity. If the temperature of the air were 86°F, then the equilibrium relative humidity would be approximately 76%; if the temperature of the air were 122°F, then the equilibrium relative humidity would be approximately 81%. From Equation (35) in Chapter 6, the wet-bulb temperatures associated with these two points are 80°F and 115°F, respectively. A linear interpolation between these two points results in an air temperature of 90°F and an equilibrium relative humidity of 77%. At 90°F, the saturation vapor pressure of the air p_{ws} = 0.69889 psi (Table 3 in Chapter 6); the vapor pressure p_w = 0.5381 psi [Equation (24) in Chapter 6]; and the humidity ratio W = 0.0236 lb/lb [Equation (22) in Chapter 6]. Each pound of dry air carries with it 0.0236 – 0.0076 = 0.016 lb of water from the grain.

Thin Layer Drying

A *thin layer* of grain is a layer of grain no more than several kernels deep. The ratio of grain to air is such that there is only a small change in temperature and relative humidity of the drying air when it exits the grain. The maximum rate ($dM/d\theta$) at which a thin layer of a granular hygroscopic material (such as grain) transfers moisture to or from air can be approximated by the following equation (Hukill 1947):

$$\frac{dM}{d\theta} = -C(p_g - p_a) \qquad (14)$$

where

C = constant representing vapor conductivity of kernel and surrounding air film
p_g = partial pressure of water vapor in grain
p_a = partial pressure of water vapor in drying air

If $p_g > p_a$, drying takes place. If $p_g = p_a$, moisture equilibrium exists and no moisture transfer occurs. If $p_g < p_a$, wetting occurs. The assumption of a linear relationship between (1) water vapor pressure and equilibrium relative humidity and (2) equilibrium relative humidity and moisture content over the range in which drying occurs lead to the following equation:

$$\frac{dM}{d\theta} = -k(M - M_e) \qquad (15)$$

where

M = moisture content (dry basis) of material at time θ
M_e = equilibrium moisture content (dry basis) of material in reference to drying air
k = constant dependent on material

The solution to this differential equation is

$$\frac{M - M_e}{M_o - M_e} = \exp(-k\theta) \qquad (16)$$

where M_o = moisture content, dry basis, when $\theta = 0$.

In later work (Hukill and Schmidt 1960, Troeger and Hukill 1971), Hukill recognized that Equation (16) did not describe the drying rate of grain adequately. Misra and Brooker (1980) identified the following model as more promising for shelled corn:

$$\frac{M - M_e}{M_o - M_e} = \exp(-K\theta^N) \qquad (17)$$

They give an equation for K, which is a function of drying air temperature and velocity, and another equation for N as a function of drying air relative humidity and initial grain moisture. Their equations are valid for drying air temperatures of 36 to 160°F, drying air relative humidities of 3 to 83%, drying air velocities of 5 to 459 fpm, and initial moistures of 18 to 60% (dry basis).

Li and Morey (1984) also fit their data to Equation (17) and found that within the limits of drying airflow rates and air relative humidities used, K and N can be expressed as functions of air temperature and initial grain moisture only. Their equations for K and N apply to air temperatures ranging from 80 to 240°F, initial grain moistures of 23 to 36% dry basis, airflows of 20 to 100 cfm/bu, and air relative humidities of 5 to 40%.

Other forms of the thin layer drying equation have also been proposed. Thompson et al. (1968) fitted data for shelled corn to the following equation, which is applicable in the range of 140 to 300°F:

$$\theta = A \ln MR + B(\ln MR)^2 \qquad (18)$$

where

A = −1.86178 + 0.00488t
B = 427.3740 exp (−0.03301t)
MR = $(M - M_e)/(M_o - M_e)$
θ = time, h
t = temperature, °F

Martins and Stroshine (1987) describe the effects of hybrid and damage on the thin layer drying rate and give values for constants A and B in Equation (18) for several hybrids and damage levels.

Results of thin layer drying tests for other grains have also been reported. Data are available for the following grains:

- Wheat (Watson and Bhargava 1974, Sokhansanj et al. 1984, Bruce and Sykes 1983)
- Soybeans (Hukill and Schmidt 1960, Overhults et al. 1973, Sabbah et al. 1976)
- Barley (O'Callaghan et al. 1971, Sokhansanj et al. 1984, Bruce 1985)
- Sorghum (Hukill and Schmidt 1960, Paulsen and Thompson 1973)
- Rice (Agrawal and Singh 1977, Noomhorm and Verma 1986, Banaszek and Siebenmorgen 1990)
- Sunflower (Syarief et al. 1984, Li et al. 1987)
- Canola (Sokhansanj et al. 1984, Pathak et al. 1991)
- Oats (Hukill and Schmidt 1960)
- Lentil seeds (Tang et al. 1989)

Sokhansanj and Bruce (1987) developed more rigorous thin layer drying equations based on simultaneous heat and mass transfer through a single kernel and demonstrated that such a model accurately predicts the temperature and moisture content of the grain throughout the drying process. Jayas et al. (1991) reviewed thin-layer drying models, and Parti (1993) presented comparisons of models under different conditions.

Equations (16) through (18) do not describe the usual drying process, where grain is in a deep bed and where drying air changes condition but does not necessarily reach moisture equilibrium with the grain. Those models, which are formulated using thin layer drying equations such as these, are summarized in the section on Deep Bed Drying.

Airflow Resistance

Data on resistance of grain to airflow are used for a variety of design calculations such as selecting fans, determining optimum depths for drying bins, predicting airflow paths in bins with aeration ducts, and determining the practical limitations on airflow caused by fan power requirements. For a given fan and dryer or bin, airflow resistance can change with the type of grain being dried, the depth of grain, and the amount of fine material in the grain. In many grain-drying applications, such as when air is forced through a grain bin that has a uniform grain depth and a full perforated floor, airflow is one-dimensional and the pressure drop per unit depth of grain can be assumed to be constant. Shedd (1953) determined pressure drop per unit depth versus airflow for a number of grains and seeds and summarized by plotting them on logarithmic axes. These curves (commonly called **Shedd's curves**) are included in ASAE *Standard* D272.3. They can also be calculated from the following equation (ASAE 1996, Sokhansanj and Yand 1996):

$$\frac{\Delta p}{L} = \frac{aQ^2}{\ln(1 + bQ)} \tag{19}$$

where

Δp = pressure, in. of water
L = bed depth, ft
Q = airflow rate, cfm/ft²
a, b = empirical constants

Table 6 summarizes the constants for Equation (19) for some of the more common grains. Constants for grass seeds and some vegetables are included in ASAE *Standard* D272.3. Jayas and Mann (1994) reviewed the presentation of airflow resistance data for 22 different seeds, including grains. They reported that the mean relative percent error for each grain could be significantly reduced if the airflow range were divided into two subranges: 0.8 to 10 cfm/ft² and 10 to 70 cfm/ft².

Equation (19) gives the airflow resistance for clean, dry grain when the bin is loaded by allowing the grain to flow into the bin through a chute from a relatively low height. Kumar and Muir (1986) reported on the effect of filling method on the airflow resistance of wheat and barley. Jayas et al. (1987) showed that the resistance of canola to airflow in a horizontal direction was 0.5 to 0.7 times the resistance to airflow for the vertical direction.

The presence of fine material in grain can significantly alter the airflow resistance. Fine material is generally defined as broken kernels and other matter that can pass through a round hole sieve with a hole size slightly less than the kernel size. For shelled corn, a 0.1875 in. (3/16 in.) round hole sieve is used to measure fines.

The pressure drop per foot is routinely increased by multiplying the value from Equation (19) by a packing factor. A factor of 1.5 is used for shelled corn; 1.2 for other grains. Haque et al. (1978) developed a correction factor for airflow resistance in shelled corn with fine material fractions from 0 to 20%.

Table 6 Constants for Airflow Resistance [Equation (19)]

Material	Value of *a*, in. of water·min²/ft²	Value of *b*, ft²/cfm	Range of *Q*, cfm/ft²
Barley	6.76×10^{-4}	6.71×10^{-2}	1.1 to 40
Canola (rapeseed)	1.65×10^{-3}	3.69×10^{-2}	4.77 to 52
Ear corn	3.29×10^{-4}	1.65	10 to 69
Lentils	1.72×10^{-3}	1.87×10^{-1}	0.55 to 116
Oats	7.62×10^{-4}	7.06×10^{-2}	1.1 to 40
Peanuts	1.20×10^{-4}	5.64×10^{-1}	6 to 60
Popcorn, white	6.92×10^{-4}	5.99×10^{-2}	1.1 to 40
Popcorn, yellow	5.63×10^{-4}	8.94×10^{-2}	1.1 to 40
Rice, rough	8.12×10^{-4}	6.71×10^{-2}	1.1 to 30
Rice, long brown	6.48×10^{-4}	3.93×10^{-2}	1.1 to 32
Rice, long milled	6.89×10^{-4}	4.24×10^{-2}	1.1 to 32
Rice, medium brown	1.10×10^{-3}	5.53×10^{-2}	1.1 to 32
Rice, medium milled	9.16×10^{-4}	5.38×10^{-2}	1.1 to 32
Shelled corn	6.54×10^{-4}	1.54×10^{-1}	1.1 to 60
Shelled corn, low airflow	3.09×10^{-4}	4.34×10^{-2}	0.05 to 4
Sorghum	6.70×10^{-4}	4.09×10^{-2}	1.1 to 40
Soybeans	3.22×10^{-4}	8.13×10^{-2}	1.1 to 60
Sunflower, confectionery	3.48×10^{-4}	9.19×10^{-2}	1.1 to 35
Sunflower, oil	7.87×10^{-4}	1.20×10^{-1}	5 to 112
Wheat	8.53×10^{-4}	4.45×10^{-2}	1.1 to 40
Wheat, low airflow	2.66×10^{-4}	1.38×10^{-2}	0.05 to 4

Source: ASAE *Standard* D272.3.

Grama et al. (1984) reported the effect of fine material particle size distribution on resistance in shelled corn. They also reported the effect of the increased resistance from fines on fan power requirements. Kumar and Muir (1986) reported the effects of fines in wheat. Bern and Hurburgh (1992) reviewed the characteristics of fines in shelled corn, including their composition, size distribution, density, airflow resistance, and nutritive and economic value. They concluded that fines can increase airflow resistance up to 200%.

Bulk density can have a significant effect on airflow resistance. For moderate heights of 14 to 24 ft, drop height does not affect bulk density in bins filled with a spout (Chang et al. 1986). Bern et al. (1982) reported that auger stirring can decrease the bulk density of bins filled with a grain spreader but has no effect on or increases bulk density in bins filled by gravity. Magnitudes of the increase in bulk density caused by grain spreaders have been reported by Stephens and Foster (1976b, 1978) and Chang et al. (1983).

Moisture content also affects airflow resistance. Its effect may, in part, be caused by its influence on bulk density. Shedd's curves include a footnote recommending that for loose fill of clean grain, airflow resistance should be multiplied by 0.80 if the grain is in equilibrium with air at relative humidities greater than 85% (ASAE *Standard* D272.2). At 70°F, this corresponds to a moisture of 18% or more for shelled corn (Figure 1). Haque et al. (1982) give equations that correct for the effects of moisture content of shelled corn, sorghum, and wheat.

Li and Sokhansanj (1994) argued that a generalized equation for airflow resistance, a modification of Leva's equation (Leva 1959), could account for airflow resistance differences due to variations in grain density, moisture content, and fines. Supporting constants for nine seeds are presented. Giner and Denisienia (1996) proposed a modified Ergun equation for the quadratic moisture effects and linear fines effects in wheat.

When the flow lines are parallel and airflow is linear (as is the case in a drying bin with a full perforated floor), calculation of the airflow is a straightforward application of Equation (16). For a given fan attached to a particular bin filled to a uniform depth with grain, the operating point of the fan can be determined as follows. A curve is plotted showing the total static pressure in the bin plenum versus airflow to the bin. Airflow rate is calculated by dividing the total air volume supplied to the plenum by the cross-sectional area of the bin. Using Equation (16), the pressure drop per unit depth can be calculated and multiplied by the total depth of grain in the bin to give total static pressure in the plenum. The fan curve showing air delivery volume versus static pressure can be plotted on the same axes. The intersection of the curves is the operating point for the fan. These calculations can also be done on a computer, and the point of intersection of the curves can be determined using appropriate numerical methods. McKenzie et al. (1980) and Hellevang (1983) summarize airflow resistances for various bin and fan combinations in tabular and graphical form. Sokhansanj and Woodward (1991) developed a design procedure for use on personal computers to select fans for near-ambient drying of grain.

In cases where airflow is nonlinear, as in conical piles or systems with air ducts, computation is complex (Miketinac and Sokhansanj 1985). Numerical methods for predicting airflow patterns have been developed and applied to bins aerated with ducts (Brooker 1969, Segerlind 1982, Khompos et al. 1984), conical-shaped piles (Jindal and Thompson 1972), and bins in which porosity varies within the bed (Lai 1980). Lai's study applies to bins in which filling methods have created differences in bulk density within the bin or where fine material is unevenly distributed. Alagusundaram et al. (1994) studied airflow patterns through wheat, barley, and canola in bins with different patterns of partially perforated floors.

Analysis of Deep Bed Drying

The ability to predict the rate at which grain dries in a given type of dryer operating in specific weather conditions with a specified airflow and air temperature can assist designers in developing dryers for maximum efficiency. It can also guide operators in finding the optimum way to operate their particular dryers for given weather conditions. Computer simulations have helped researchers understand the mechanisms and processes involved in drying.

Two relatively simple prediction equations can be solved on a hand calculator. Hukill (1947) developed a widely known and used method that predicts the moisture distribution in a bed of grain during drying. A graphical presentation of one of the equations, which further simplifies calculations, is available. Hukill's method is summarized by Brooker et al. (1992), who give an example calculation for shelled corn drying. Barre et al. (1971) made further adaptations of Hukill's method, and Foster (1986) gives a historical perspective on the development and utility of the method. Brooker et al. (1992) also present a technique called the *heat balance equation*, which equates the heat available in the air for drying with the amount of heat needed to evaporate the desired amount of water from the grain. Both of the above methods take into account airflow, drying air temperature and relative humidity, exit air conditions, grain moisture, and the amount of grain to be dried.

Thompson et al. (1968) considered a deep bed of grain as a series of thin layers of grain stacked one on top of another. Algebraic heat and mass balances were applied to each layer, with the exit air conditions of one layer becoming the input conditions of the next layer. Thompson et al. (1969) used the model to predict concurrent-flow, crossflow, and counterflow drying of shelled corn. Paulsen and Thompson (1973) used it to evaluate crossflow drying of sorghum. Stephens and Thompson (1976) and Pierce and Thompson (1981) used the model to make recommendations about optimum design of high-temperature grain dryers.

Bakker-Arkema et al. (1978) used simultaneous heat and mass transfer equations in a series of coupled partial differential equations to describe deep bed drying. The equations, solved using a finite difference technique, predict grain temperature, grain moisture content, and air temperature and humidity ratio. Bakker-Arkema et al. (1979, 1984) give solutions for in-bin, batch, continuous crossflow, and continuous concurrent-flow dryers. Morey et al. (1976) used the model to evaluate energy requirements for drying. Morey and Li (1984) and Bakker-Arkema et al. (1983) demonstrated the effect of thin layer drying rate on the model predictions. Bridges et al. (1980) used the Thompson model for simulation of batch-in-bin drying. Morey et al. (1978) and Parry (1985) review many of the mathematical models used for high-temperature grain drying.

Computer simulations have also been developed for low-temperature and solar drying. Some of these models have been referenced in the section on Fungal Growth and Mycotoxins under Prevention of Deterioration. Thompson (1972) developed a model that was later used by Pierce and Thompson (1979) to make recommendations on airflow in solar grain drying and by Pierce (1986) to evaluate natural air drying. Sabbah et al. (1979) used the logarithmic model of Barre et al. (1971) for simulation of solar grain drying. Bridges et al. (1984) used a model to evaluate the economics of stirring devices in in-bin drying systems. Morey et al. (1979), Frazer and Muir (1981), Bowden et al. (1983), and Smith and Bailey (1983) have also modeled low-temperature drying. Sharp (1982) reviewed low-temperature drying simulation models.

Aeration of Grain

Aeration involves forcing small amounts of air through the stored grain to maintain a uniform temperature. Prior to the development of this concept, grain was turned by moving it from one storage bin to another. Foster (1986) credits Hukill (1953) with developing the concept of aeration. As mentioned in the sections on Fungal Growth and Mycotoxins and Insect Infestation, lowering of the grain temperature during winter in temperate climates can reduce the rate of deterioration from molds and insects. Aeration can also prevent temperature gradients from developing within the grain mass. Such gradients can cause moisture migration, which results in unacceptably high moistures in certain portions of the bin.

Aeration is used to cool stored grain in the fall. A typical practice is to aerate the grain when the difference between grain temperature and the average daily outside temperature exceeds 10°F. In the United States, grain is usually not warmed in the spring unless it is to be stored past early June. Foster and McKenzie (1979) and McKenzie (1980) give practical recommendations for aeration of grain. Airflow rates of 0.025 to 0.5 cfm per bushel are normally used. Air is usually distributed through the bottom of the bin using ducts. Duct spacing and fan selection are related to bin size and shape and to the airflow rate. Foster and Tuite (1992) give an overview of the topic and include information and charts used for design of such systems. Peterson (1982) gives recommendations for duct spacing in flat storages.

Several computer simulations have been developed to study the effects of heat buildup from microbial activity with and without aeration (Thompson 1972, Brooker and Duggal 1982, Metzger and Muir 1983, Lissik 1986). Aldis and Foster (1977) and Schultz et al. (1984) studied the effect of aeration on grain moisture changes.

DRYING SPECIFIC CROPS

Hay

Forage crops can be either harvested, dried, and stored as hay or harvested and stored under anaerobic conditions as silage. Hay quality can be judged by its color, leafiness, and appearance. Laboratory tests and feeding trials give a more detailed picture of hay quality. The traditional method of making hay is to mow the forage and allow it to field cure or dry in the swath and windrow. Harvesting at higher moistures with subsequent artificial drying may be economically feasible, depending on the local weather conditions.

Basic principles of hay drying and storage are covered by Hall (1980), FEC (1985), and Schuler et al. (1986). Forage must be harvested in the proper stage of maturity to attain maximum feeding value. Leaf loss from alfalfa is high when it is handled at moistures below 39%. Therefore, if it is baled at 40% moisture and dried artificially to the recommended storage moisture of 20% (Schuler et al. (1986), a significantly higher feeding value can be achieved. Both Schuler et al. (1986) and Hall (1980) give sketches for batch and in-storage hay dryers. They recommend airflows of 15 to 20 cfm per square foot of mow floor area or 200 to 500 cfm per ton.

Dehydrated alfalfa meal supplies provitamin A (carotene), vitamin E, xanthophylls (poultry pigmenting factors), vitamin K, vitamin C, and B vitamins. Figure 3 shows losses from field drying of hay found in tests conducted by Shepherd (1954). The rapid loss of carotene immediately after the forage is cut indicates the need for

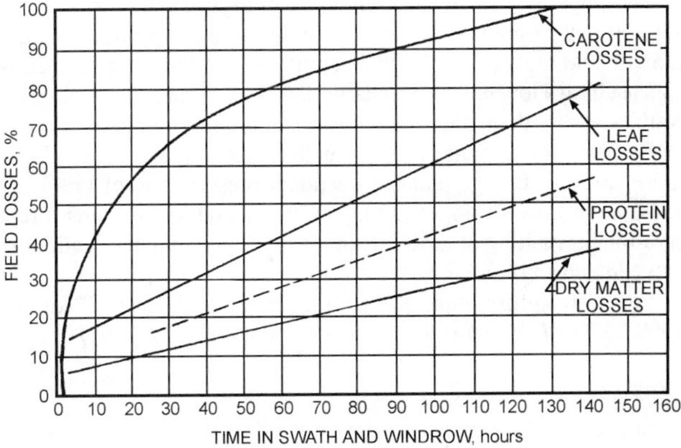

Fig. 3 Time in Swath and Windrow Versus Field Losses of Leaves, Dry Matter, Protein, and Carotene for Hay Drying

Table 7 Effect of Heating Chopped Alfalfa on Carotene Loss During Subsequent Storage of Meal

Hours in Oven at 212°F	Initial Carotene, ppm	Carotene Retained 7 Days at 149°F, %
0.75	229	37
1	228	37
2	197	37
3	176	28
4	149	21
5.5	112	18
7.5	86	15

Source: Thompson et al. (1960).
Note: Alfalfa is fresh frozen from Ryer Island, CA.

rapid transport to the dehydrator when alfalfa meal with high vitamin content is desired.

Several factors influence retention of vitamins during storage, including the starting plant material, dehydration conditions, addition of stabilizers, and storage conditions. Lowering the temperature reduces the loss rate. Inert gas atmosphere in storage also reduces losses (Hoffman et al. 1945). According to Shepherd (1954), blanching of fresh alfalfa before drying does not alter the storage stability of carotene. Table 7 shows Shepherd's results on the effect of prolonged heating at 212°F. The alfalfa was dried after 45 min; heating beyond this time represented excessive exposure to this temperature. Carotene retention in the intact meal at 149°F storage temperature was considered a measure of storage stability. Normal storage moisture is 8 to 9%. Thompson et al. (1960) summarize the effects of over- and underdrying on carotene stability.

Drying and handling of large round bales has been researched. These bales may weigh from 850 to 1500 lb and are handled individually with forklifts. Verma and Nelson (1983) studied storage of large round bales and found that dry matter loss was the primary component of the total storage losses. Bales stored so that they were protected from the weather had lower losses of dry matter than bales exposed to the weather. They were also higher in total protein. Jones et al. (1985) found significant dry matter loss in large round bales of mature fescue hay. Harrison (1985) found that addition of sulfur dioxide at the rate of 1% of dry matter had little effect on dry matter loss and nutrient contents for a mixture of alfalfa and bromegrass. However, bales protected with plastic bags did have significantly lower dry matter loss. Jones et al. (1985) found that bales of mature fescue hay stored inside and bales treated with ammonia had less dry matter loss and higher in vitro dry matter digestibility. Henry et al. (1977) and Frisby et al. (1985) developed and tested solar dryers for large round bales.

Grain

The physiological factors involved in drying and storing grain are different from those of forages. Grain is the end product of plant growth, and most physiological activity within the grain or seed is approaching a low level when harvested. With forage, the biological activity within the plant is at or near its peak at the time of harvest.

Both the deterioration of grain harvested at moistures above those safe for storage and the chemical preservation of grain are addressed in the section on Fungal Growth and Mycotoxins. Preservation by ensiling or airtight storage is addressed in the section on Oxygen and Carbon Dioxide.

For more information on grain drying, see Chapter 22 of the 1999 *ASHRAE Handbook—Applications* and Brook (1992).

Corn

Shelled field corn is used primarily as livestock feed, but some is used by milling or processing industries for manufacturing starch, corn oil, and other products. Little information is available on the relationship between the drying method and the feed value of corn. Market grade, as established by the Agricultural Marketing Service of the United States Department of Agriculture (USDA), is the primary criterion for determining corn value. Tests by Cabell et al. (1958) indicated that shelled corn with a moisture content of 29 to 32% can be dried without loss of protein nutritive value by air with temperatures as high as 240°F, provided the airflow rate is approximately 110 cfm/bu.

Breakage Susceptibility. The market grade of dry corn is affected more by the amount of fine material than by other grading factors. Fine material is defined as the broken grain and other material that passes through a 12/64 in. round-hole sieve. The physical damage done to wet corn or the brittleness imparted to the corn during drying causes it to break each time it is handled. The propensity of corn to break during subsequent handling, called **breakage susceptibility**, can be measured with a multiple-impact device called the *Stein breakage tester*. Stephens and Foster (1976a) demonstrated that corn breakage in the tester was correlated with damage during handling. Watson et al. (1986) give a standardized procedure for using the Stein breakage tester, and Watson and Herum (1986) describe and compare other devices developed for measurement of breakage susceptibility. They concluded that a device developed by Singh and Finner (1983) offers great potential for testing of grain for breakage susceptibility in commercial situations.

Paulsen et al. (1983) found significant variations in breakage susceptibility among hybrids. Corn dried with air at high temperatures (140°F) was two to six times more susceptible to breakage than corn dried at near-ambient temperatures. Gustafson and Morey (1979) found that delayed cooling (maintaining the corn at or near its temperature at the end of drying for 6 to 12 h) reduced breakage susceptibility and improved the test weight.

In a study of combination drying, Gustafson et al. (1978) found that combination drying (high-temperature drying to 18% followed by low-temperature drying to 16.6% moisture or below) significantly reduced the increase in breakage susceptibility normally caused by high-temperature drying.

Quality. Both drying temperature and corn hybrid can affect the quality of shelled corn for specific end uses. Brekke et al. (1973) found that drying at temperatures above 140°F reduced the quality of the corn for dry milling. Peplinski et al. (1982) found that optimum dry milling quality could be achieved by harvesting corn at moistures below 25%, minimizing machinery-induced damage to the kernels, and drying at air temperatures below 180°F. Paulsen and Hill (1985) found that the yield of flaking grits from dry milling of corn was significantly greater for corn that had a high test weight and relatively low breakage susceptibility. Weller et al. (1987) found that corn variety affected wet milling quality. At drying temperatures between 120 and 160°F, protein conformational changes occurred and decreased the ethanol soluble protein. Hybrids differ in resistance to storage mold (Tuite and Foster 1979), thin layer drying rate, and dry milling quality (Stroshine et al. 1986). Watson (1987) gives an extensive summary of measurement and maintenance of quality of corn, and Foster (1975) summarizes approaches to reducing damage during harvesting, handling, and drying.

Cotton

The lint moisture content for best results in ginning cotton appears to be 5 to 7%, with an optimum moisture content of 6% (Franks and Shaw 1962). Cotton, like grain, is hygroscopic and should be dried just prior to ginning. The wide variation in incoming moisture content usually requires different amounts of drying for each load. Rapid changes in the amount of drying required can best be handled by using a multipath drying tower in which the cotton is exposed for various lengths of time (2 to 10 s) at temperatures not exceeding 350°F. The air-to-cotton ratio can range from 40 to 100 cfm/lb of cotton (Franks and Shaw 1962). Laird and Baker (1983) found that substantial amounts of heat could be reclaimed and used for drying in commercial cotton gin plants. Equilibrium moisture content data for newly harvested cotton fibers are given by Griffin (1974). Anthony (1982) studied moisture gain of cotton bales during storage.

Cottonseed removed from the fibers is also dried. The germination of cottonseed is unimpaired by drying if the internal cottonseed temperatures do not exceed 140°F (Shaw and Franks 1962). This temperature is not exceeded in the tower dryer described previously. However, the moisture content of the seed can be above the recommended level of 12% following the multipath tower drying. Drying seed in a triple-pass drum at 250 to 300°F with an exposure time of 4 min, followed by cooling, reduces moisture content, inhibits the formation of free fatty acids, and improves germination compared to undried seed. Anthony (1983) dried cottonseed in a vacuum microwave dryer. The cottonseed would not germinate, but its oil properties were not harmed as long as lower temperatures were used. The drying rate was increased by reducing pressure below atmospheric. Rayburn et al. (1978) studied preservation of high-moisture cottonseed with propionic acid.

Peanuts

Peanuts in the shell normally have a moisture content of about 50% at the time of digging. Allowing peanuts to dry on the vines in the windrows for a few days removes much of this water. However, peanuts normally contain 20 to 30% moisture when removed from the vines, and some artificial drying in the shell is necessary. Drying should begin within 6 h after harvesting in order to prevent peanuts from self-heating. The maximum temperature and rate of drying must be controlled to maintain quality. High temperatures result in off-flavor or bitterness. Overly rapid drying without high temperatures results in blandness or inability to develop flavor when roasted (Bailey et al. 1954). High temperatures and rapid or excessive drying also cause the skin to slip easily and the kernels to become brittle. These conditions result in damage in the shelling operation and can be avoided if the moisture removal rate does not exceed 0.5% per hour. Because of these limitations, continuous-flow drying is not usually recommended.

Young (1984) found energy savings up to 26% when comparing recirculating dryers with conventional peanut dryers. Smith and Davidson (1982) and Smith et al. (1985) address the aeration of peanuts during warehouse storage.

Rice

Of all grains, rice is possibly the most difficult to process without quality loss. Rice containing more than 12.5% moisture cannot be stored safely for long periods, yet the recommended harvest moisture content for best milling and germination ranges from 20 to 26% (Kramer 1951). If the rice is harvested at this moisture content, drying must begin promptly to prevent heat-related damage, which can result in "stack-burn," a yellowing of the kernel. To prevent excessive internal fissuring, which results in broken kernels during milling, multiple-pass drying is usually necessary (Calderwood and Webb 1971). Kunze and Calderwood (1980) summarize rice-drying techniques.

Because the market demands polished whole kernels of rice, it is necessary to prevent damage in the form of fissures. Rapid moisture removal or addition can create moisture gradients within kernels. According to Kunze (1984), gradients can develop in the field on a humid night before harvest, in a hopper containing a mixture of rice kernels at varying moistures, and in certain types of dryers. Banaszek and Siebenmorgen (1990) quantified the rate at which moisture absorption reduces head rice yields. Velupillai and Verma (1986) report that drying at 200°F followed by tempering in a sealed container for 24 h gave good kernel strength and head rice yields. They also found that storing the rice after drying for 3 weeks gave optimum grain quality. Bakker-Arkema et al. (1984) achieved good rice quality with concurrent-flow drying of rice.

Soybean, Sunflower, and Edible Beans

Prolonged periods of extremely wet weather during the harvest season can make artificial drying of soybeans necessary. Like peanuts and other oilseeds, soybeans cannot be dried satisfactorily with the high-temperature, high-speed methods used for cereal grains. Because of the different seed structure, rapid drying splits the seed coat and reduces quality and storage life. Overhults et al. (1975) reported a significant decrease in the quality of oil extracted from soybeans dried at temperatures above 160°F. Soybeans have one of the slowest thin layer drying rates of commonly grown cereals and oilseeds (Bakker-Arkema et al. 1983). Therefore, they dry more slowly and require more energy when dried in continuous-flow dryers.

Sunflower is a major crop in some areas of the United States. Hellevang (1987) recommends maximum drying temperatures of 200°F for continuous-flow drying of oil sunflower and 180°F for nonoil sunflower to prevent scorching of the seed meat. Schuler (1974) gives data on equilibrium moisture, airflow resistance, and specific heat of sunflower seeds. Because sunflower is about half the density of shelled corn, moisture can be removed more rapidly, and there is a tendency to overdry. This factor, along with accumulation of foreign material when drying, causes an increased fire hazard (Hellevang 1982). Schmidt and Backer (1980) attribute most of the problems encountered with storage of sunflower seed to improper drying and/or aeration.

Edible beans, a major crop in several states, should be dried with air at relative humidities above 40% to prevent stress cracking. Natural air or low-temperature drying is best (Hellevang 1987). If dried at high rates, seed coats may crack, and beans may split during subsequent handling (Otten et al. 1984, Radajewski et al. 1992). Broken beans can develop a bitter or undesirable flavor and spoil more easily during storage (Uebersax and Bedford 1980).

Wheat and Barley

In northern regions of the United States, wheat and barley may be harvested above safe storage moistures to prevent excessive field losses. Bruce (1992) modeled the effect of heated-air drying on the bread baking quality of wheat. The quality indicator was loaf volume. Moilanen et al. (1973) recommended that hard red spring wheat be dried at temperatures below 160, 140, and 120°F, respectively, for harvest moistures of 16, 20, and 24% wet basis. These data assumed airflow of 100 to 150 cfm/ft^2. For airflow of 50 cfm/ft^2, the authors recommended that the drying air temperature be reduced by 10 to 15°F. In the case of barley used for malting, the seed must be able to germinate. Therefore, the maximum recommended drying air temperature is 110 to 120°F (Hellevang 1987, Jilak 1993). Watson et al. (1962) studied the effects of harvest moisture and drying temperature on barley malting quality and recommended harvesting below 20% moisture. If wheat or barley is used for seed, the maximum recommended drying air temperature is 110°F.

In regions where soft wheat is grown, it may be economical to harvest at 20 to 24% moisture to allow double cropping with soybeans; this allows wheat harvest to begin 5 to 7 days earlier than normal and increases the yield of the soybeans (Swearingen 1979). In areas where double cropping is feasible, soft wheat can be dried using low-temperature solar drying or ambient drying with intermittent fan operation (Barrett et al. 1981). High-speed and continuous-flow systems with reduced drying air temperatures can also be used (Parsons et al. 1979). Kirleis et al. (1982) harvested soft red winter wheat at moistures of 25% or below and dried with air temperatures of 150°F or below without adverse effects on milling or cookie baking quality.

In high-temperature continuous-flow dryers, wheat and barley reduce airflow because they have a high airflow resistance. Bakker-Arkema et al. (1983) report that thin layer drying rates for barley and wheat are much faster than for corn. Barley dries more slowly than wheat, presumably because the kernels are larger. In their computer simulations of a concurrent-flow dryer, wet bushel capacity for wheat was about 80% of the capacity for shelled corn when moisture content was reduced by 4.7%. The drying capacity difference was probably caused by a decrease in airflow.

Tobacco (Curing)

Tobacco leaves normally have a moisture content of about 85% at harvest. The major methods of tobacco drying are air curing and flue curing (Johnson et al. 1960).

For **air curing**, whole plants are cut and allowed to wilt in the field until the leaves reach about 70% moisture (Walton et al. 1994). The plants are then hung in open barns, where temperatures range from 60 to 90°F and humidities from 65 to 70%. The curing period is 28 to 56 days (Jefries 1940). The desired end product for air curing is a tan leaf. Overdrying at low temperatures results in green color and low sugar content; overdrying at high temperatures results in yellow color (Walton and Henson 1971). Both conditions are undesirable because the normal chemical changes are arrested prematurely. Subsequent drying at optimum rates can reverse some damage. Underdrying at all temperatures results in undesirable dark color and damage from mold and bacterial growth (Walton et al. 1973).

Flue curing uses artificial heat. The leaves are harvested and hung in closed barns where temperatures are increased gradually during the curing period. Normally, 3 days of drying at temperatures of 90 to 120°F brings about yellowing. For the next 2 days, temperatures of 120 to 140°F are used for leaf drying; then, stems are dried at 170°F for 1 to 2 days. A bright yellow to orange color is desirable in flue-cured or bright-leaf tobacco.

REFERENCES

Agrawal, Y.C. and R.P. Singh. 1977. Thin-layer drying studies on short grain rice. *Paper* 77-3531. American Society of Agricultural Engineers, St. Joseph, MI.

Alagusundaram, K., D.S. Jayas, O.H. Friesen, and N.D.G. White. 1994. Airflow patterns through wheat, barley and canola in bins with partially perforated floors: an experimental investigation. *Applied Engineering in Agriculture* 10(6):791-96.

Alagusundaram, K., D.S. Jayas, W.E. Muir, and N.D.G. White. 1996. Convective-diffusive transport of carbon dioxide through stored-grain bulks. *Transactions of ASAE* 39(4):1505-10.

Aldis, D.F. and G.H. Foster. 1977. Moisture changes in grain from exposure to ambient air. *Paper* 77-3524. ASAE.

Anthony, W.S. 1982. Moisture gain and resilient forces of cotton bales during equilibration. *Transactions of ASAE* 25(4):1066-70.

Anthony, W.S. 1983. Vacuum microwave drying of cotton: Effect on cottonseed. *Transactions of ASAE* 26(1):275-78.

Armitage, D.M., P.M. Cogan, and D.R. Wilkin. 1994. Integrated pest management in stored grain: Combining surface insecticide treatments with aeration. *Journal of Stored Products Research* 30(4):303-19.

Arthur, F.H. 1996. Grain protectants: Current status and prospects for the future. *Journal of Stored Products Research* 32(4):293-302.

ASAE. 1995. Moisture relationships of grains. *Standard* D245.5. American Society of Agricultural Engineers, St. Joseph, MI.

ASAE. 1996. Resistance to airflow of grains, seeds, other agricultural products, and perforated metal sheets. *Standard* D272.3.

ASAE. 1997. Moisture measurement—Unground grain and seeds. *Standard* S352.2.

ASAE. 1998. Moisture measurement—Forages. *Standard* S358.2.

Athie, I., R.A.R. Gomes, S. Bolonhezi, S.R.T. Valentini, and M.F.P.M. de Castro. 1998. Effects of carbon dioxide and phosphine mixtures on resistant populations of stored-grain insects. *Journal of Stored Products Research* 34(1):27-32.

Bailey, W.K., T.A. Pickett, and J.G. Futral. 1954. Rapid curing adversely affects quality of peanuts. *Peanut Journal and Nut World* 33(8):37-39.

Bakker-Arkema, F.W., R.C. Brook, and L.E. Lerew. 1978. Cereal grain drying. In *Advances in cereal science and technology*, ed. Y. Pomeranz, pp. 1-90. American Association of Cereal Chemists, St. Paul, MN.

Bakker-Arkema, F.W., S. Fosdick, and J. Naylor. 1979. Testing of commercial crossflow dryers. *Paper* 79-3521. ASAE.

Bakker-Arkema, F.W., C. Fontana, R.C. Brook, and C.W. Westlake. 1984. Concurrent flow rice drying. *Drying Technology* 1(2):171-91.

Bakker-Arkema, F.W., C. Fontana, G.L. Fedewa, and I.P. Schisler. 1983. A comparison of drying rates of different grains. *Paper* 83-3009. ASAE.

Banaszek, M.M. and T.J. Siebenmorgen. 1990. Head rice yield reduction rate caused by moisture absorption. *Transactions of ASAE* 33(4):1263-69.

Barak, A.V. and W.E. Burkholder. 1985. A versatile and effective trap for detecting and monitoring stored-product coleoptera. *Agricultural Ecosystems and Environment* 12:207-18.

Barak, A.V. and P.K. Harein. 1982. Trap detection of stored-grain insects in farm-stored shelled corn. *Journal of Economic Entomology* 75(1):108-11.

Barre, H.J., G.R. Baughman, and M.Y. Hamdy. 1971. Application of the logarithmic model to cross-flow deep-bed grain drying. *Transactions of ASAE* 14(6):1061-64.

Barrett, Jr., J.R., M.R. Okos, and J.B. Stevens. 1981. Simulation of low temperature wheat drying. *Transactions of ASAE* 24(4):1042-46.

Beeson, W.M. and T.W. Perry. 1958. The comparative feeding value of high moisture corn and low moisture corn with different feed additives for fattening beef cattle. *Journal of Animal Science* 17(2):368-73.

Bell, C.H. and D.M. Armitage. 1992. Alternative storage practices. In *Storage of cereal grains and their products*, ed. D.B. Sauer, pp. 249-312.

Bern, C.J. and L.F. Charity. 1975. Airflow resistance characteristics of corn as influenced by bulk density. *Paper* 75-3510. ASAE.

Bern, C.J. and C.R. Hurburgh, Jr. 1992. Characteristics of fines in corn: Review and analysis. *Transactions of ASAE* 35(6):1859-67.

Bern, C.J., M.E. Anderson, W.F. Wilcke, and C.R. Hurburgh. 1982. Auger-stirring wet and dry corn—Airflow resistance and bulk density effects. *Transactions of ASAE* 25(1):217-20.

Beuchar, L.R. and D.A. Golden. 1989. Antimicrobials occurring naturally in foods. *Food Technology* 43:135-42.

Bloome, P.D. and G.W. Cuperus. 1984. Aeration for management of stored grain insects in wheat. *Paper* 84-3517. ASAE.

Bowden, P.J., W.J. Lamond, and E.A. Smith. 1983. Simulation of near-ambient grain drying: I, Comparison of simulations with experimental results. *Journal of Agricultural Engineering Research* 28:279-300.

Brekke, O.L., E.L. Griffin, Jr., and G.C. Shove. 1973. Dry milling of corn artificially dried at various temperatures. *Transactions of ASAE* 16(4):761-65.

Bridges, T.C., D.G. Colliver, G.M. White, and O.J. Loewer. 1984. A computer aid for evaluation of on-farm stir drying systems. *Transactions of ASAE* 27(5):1549-55.

Bridges, T.C., I.J. Ross, G.M. White, and O.J. Loewer. 1980. Determination of optimum drying depth for batch-in bin corn drying systems. *Transactions of ASAE* 23(1):228-33.

Brook, R.C. 1987. Modelling grain spoilage during near-ambient grain drying. *Divisional Note* DN 1388, AFRC Institute of Engineering Research, Wrest Park, Silsoe, Bedford, MK45 4HS, England, 20 p.

Brook, R.C. 1992. Drying cereal grains. In *Storage of cereal grains and their products*, ed. D.B. Sauer, pp. 183-218.

Brooker, D.B. 1969. Computing air pressure and velocity distribution when air flows through a porous medium and nonlinear velocity-pressure relationships exist. *Transactions of ASAE* 12(1):118-20.

Brooker, D.B. and A.K. Duggal. 1982. Allowable storage time of corn as affected by heat buildup, natural convection and aeration. *Transactions of ASAE* 25(3):806-10.

Brooker, D.B., F.W. Bakker-Arkema, and C.W. Hall. 1992. *Drying and storage of grains and oilseeds*. Van Nostrand Reinhold, New York.

Bruce, D.M. 1985. Exposed-layer barley drying: Three models fitted to new data up to 150°C. *Journal of Agricultural Engineering Research* 32:337-47.

Bruce, D.M. 1992. A model of the effect of heated-air drying on the bread baking quality of wheat. *Journal of Agricultural Engineering Research* 52(1):53-76.

Bruce, D.M. and R.A. Sykes. 1983. Apparatus for determining mass transfer coefficients at high temperatures for exposed particulate crops, with initial results for wheat and hops. *Journal of Agricultural Engineering Research* 28:385-400.

Brusewitz, G.H. 1987. Corn moisture variability during drying, mixing and storage. *Journal of Agricultural Engineering Research* 38:281-88.

Bullerman, L.B., L.L. Schroeder, and K. Park. 1984. Formation and control of mycotoxins in food. *Journal of Food Protection* 47:637-46.

Burkholder, W.E. and M. Ma. 1985. Pheromones for monitoring and control of stored-product insects. *Annual Review of Entomology* 30:257-72.

Butler, L.A. 1983. The history and background of NIR. *Cereal Foods World* 28(4):238-40.

Cabell, C.A., R.E. Davis, and R.A. Saul. 1958. Relation of drying air temperature, time and air flow rate to the nutritive value of field-shelled corn. *Technical Progress Report* 1957-58 ARS 44-41. USDA, Washington, D.C.

Calderwood, D.L. and B.D. Webb. 1971. Effect of the method of dryer operation on performance and on the milling and cooking characteristics of rice. *Transactions of ASAE* 14(1):142-46.

Chang, C.S., H.H. Converse, and F.S. Lai. 1986. Technical Notes: Distribution of fines and bulk density of corn as affected by choke-flow, spout-flow, and drop-height. *Transactions of ASAE* 29(2):618-20.

Chang, C.S., H.H. Converse, and C.R. Martin. 1983. Bulk properties of grain as affected by self-propelled rotational type grain spreaders. *Transactions of ASAE* 26(5):1543-50.

Choudhary, A.K. and K.K. Sinha. 1993. Competition between a toxigenic *Aspergillus flavus* strain and other fungi on stored maize kernels. *Journal of Stored Products Research* 29(1):75-80.

Christensen, C.M. and R.A. Meronuck. 1986. *Quality maintenance in stored grains and seeds*. University of Minnesota Press, Minneapolis.

Christensen, C.M., B.S. Miller, and J.A. Johnston. 1992. Moisture and its measurement. In *Storage of cereal grains and their products*, ed. D.B. Sauer, pp. 39-54.

Chung, J.H. and L.R. Verma. 1991. Dynamic and quasi-static rice moisture models using humidity sensors. *Transactions of ASAE* 34(6):2477-83.

Colliver, D.G., R.M. Peart, R.C. Brook, and J.R. Barrett, Jr. 1983. Energy usage for low temperature grain drying with optimized management. *Transactions of ASAE* 26(2):594-600.

Cooper, P.J. 1983. NIR analysis for process control. *Cereal Foods World* 28(4):241-45.

Danziger, M.T., M.P. Steinberg, and A.I. Nelson. 1973. Effect of CO_2, moisture content, and sorbate on safe storage of wet corn. *Transactions of ASAE* 16(4):679-82.

Dunkel, F.V. 1985. Underground and earth sheltered food storage: Historical, geographic, and economic considerations. *Underground Space* 9:310-15.

Eckhoff, S.R., J. Tuite, G.H. Foster, R.A. Anderson, and M.R. Okos. 1984. Inhibition of microbial growth during ambient air corn drying using sulfur dioxide. *Transactions of ASAE* 27(3):907-14.

Epperly, D.R., R.T. Noyes, G.W. Cuperus, and B.L. Clary. 1987. Control stored grain insects by grain temperature management. *Paper* 87-6035. ASAE.

FEC. 1985. Hay drying: A guide to the practical design of installations. Farm Electric Center, Kenilworth, Warwickshire, England.

Foster, G.H. 1975. Causes and cures of physical damage to corn. In *Corn quality in world markets*, L.D. Hill, ed. Interstate Printers and Publishers, Danville, IL.

Foster, G.H. 1986. William V. Hukill, a pioneer in crop drying and storage. *Drying Technology* 4(3):461-71.

Foster, G.H. and H.F. Mayes. 1962. Temperature effects of an artificial hotspot embedded in stored grain. AMS-479. U.S. Department of Agriculture, Washington, D.C.

Foster, G.H. and B.A. McKenzie. 1979. Managing grain for year-round storage. AE-90. Cooperative Extension Service, Purdue University, West Lafayette, IN.

Foster, G.H. and J. Tuite. 1992. Aeration and stored grain management. In *Storage of cereal grains and their products*, ed. D.B. Sauer, pp. 219-48.

Franks, G.N. and C.S. Shaw. 1962. Multipath drying for controlling moisture in cotton. ARS 42-69. USDA, Washington, D.C.

Frazer, B.M. and W.E. Muir. 1981. Airflow requirements for drying grain with ambient and solar-heated air in Canada. *Transactions of ASAE* 24(1):208-10.

Frisby, J.C., J.T. Everett, and R.M. George. 1985. A solar dryer for large, round alfalfa bales. *Applied Engineering in Agriculture* 1(2):50-52.

Giner, S.A. and E. Denisienia. 1996. Pressure drop through wheat as affected by air velocity, moisture content and fines. *Journal of Agricultural Engineering Research* 63(1):73-85.

Grama, S.N., C.J. Bern, and C.R. Hurburgh, Jr. 1984. Airflow resistance of moistures of shelled corn and fines. *Transactions of ASAE* 27(1):268-72.

Griffin, A.C., Jr. 1974. The equilibrium moisture content of newly harvested cotton fibers. *Transactions of ASAE* 17(2):327-28.

Gunasekaran, S. 1986. Optimal energy management in grain drying. *Critical Reviews in Food Science and Nutrition* 25(1):1-48.

Gustafson, R.J. and R.V. Morey. 1979. Study of factors affecting quality changes during high-temperature drying. *Transactions of ASAE* 22(4): 926-32.

Gustafson, R.J., R.V. Morey, C.M. Christensen, and R.A. Meronuck. 1978. Quality changes during high-low temperature drying. *Transactions of ASAE* 21(1):162-69.

Hall, C.W. 1980. *Drying and storage of agricultural crops*. AVI Publishing Company, Westport, CT.

Hall, G.E., L.D. Hill, E.E. Hatfield, and A.H. Jenson. 1974. Propionic-acetic acid for high-moisture preservation. *Transactions of ASAE* 17(2):379-82, 387.

Hamer, A., J. Lacey, and N. Magan. 1992. Respiration and fungal deterioration of stored grains. eds. D.S. Jayas et al., pp. 65-66.

Haque, E., Y.N. Ahmed, and C.W. Deyoe. 1982. Static pressure drop in a fixed bed of grain as affected by grain moisture content. *Transactions of ASAE* 25(4):1095-98.

Haque, E., G.H. Foster, D.S. Chung, and F.S. Lai. 1978. Static pressure drop across a bed of corn mixed with fines. *Transactions of ASAE* 21(5):997-1000.

Harein, P.K. and R. Davis. 1992. Control of stored grain insects. In *Storage of cereal grains and their products*, ed. D.B. Sauer, pp. 491-534.

Harris, K.L. and F.J. Bauer. 1992. Rodents. In *Storage of cereal grains and their products*, ed. D.B. Sauer, pp. 393-434.

Harrison, H.P. 1985. Preservation of large round bales at high moisture. *Transactions of ASAE* 28(3):675-79, 686.

Hellevang, K.J. 1982. Crop dryer fires while drying sunflower. *Paper* 82-3563. ASAE.

Hellevang, K.J. 1983. Natural air/low temperature crop drying. *Bulletin* 35. Cooperative Extension Service, North Dakota State University, Fargo, ND.

Hellevang, K.J. 1987. Grain drying. *Publication* AE-701. Cooperative Extension Service, North Dakota State University, Fargo, ND.

Henry, Z.A., B.L. Bledsoe, and D.D. Eller. 1977. Drying of large hay packages with solar heated air. *Paper* 77-3001. ASAE.

Hertung, D.C. and E.E. Drury. 1974. Antifungal activity of volatile fatty acids on grains. *Cereal Chemistry* 51(1):74-83.

Hoffman, E.J., G.F. Lum, and A.L. Pitman. 1945. Retention of carotene in alfalfa stored in atmospheres of low oxygen content. *Journal of Agricultural Research* 71:361-73.

Hukill, W.V. 1947. Basic principles in drying corn and grain sorghum. *Agricultural Engineering* 28(8):335-38, 340.

Hukill, W.V. 1953. Grain cooling by air. *Agricultural Engineering* 34(7): 456-58.

Hukill, W.V. and J.L. Schmidt. 1960. Drying rate of fully exposed grain kernels. *Transactions of ASAE* 3(2): 71-77, 80.

Hurburgh, C.R., T.E. Hazen, and C.J. Bern. 1985. Corn moisture measurement accuracy. *Transactions of ASAE* 28(2):634-40.

Hurburgh, C.R., L.N. Paynter, S.G. Schmitt, and C.J. Bern. 1986. Performance of farm-type moisture meters. *Transactions of ASAE* 29(4): 1118-23.

Janardhana, G.R., K.A. Raveesha, and H.S. Shetty. 1998. Modified atmosphere storage to prevent mould-induced nutritional moss in maize. *J. Sci. Food Agric.* 76(4):573-78.

Jay, E. 1980. Methods of applying carbon dioxide for insect control in stored grain. Science and Education Administration, Advances in Agricultural Technology, Southern Series, AAT-S-13, Agricultural Research (Southern Region), SEA, USDA, P.O. Box 53326, New Orleans, LA 70153.

Jayas, D.S. and D.D. Mann. 1994. Presentation of airflow resistance data of seed bulks. *Appl. Eng. Agric.* 10(1):79-83.

Jayas, D.S. and S. Sokhansanj. 1989. Design data on the airflow resistance to canola (rapeseed). *Transactions of ASAE* 32(1):295-96.

Jayas, D.S., S. Sokhansanj, E.B. Moysey, and E.M. Barber. 1987. The effect of airflow direction on the resistance of canola (rapeseed) to airflow. *Canadian Agricultural Engineering* 29(2):189-92.

Jayas, D.S., S. Cenkowski, S. Pabis, and W.E. Muir. 1991. Review of thin-layer drying and wetting equations. *Drying Tech.* 9(3):551-88.

Jayas, D.S., N.D.G. White, W.E. Muir, and R.N. Sinha, eds. 1992. *International Symposium on Stored Grain Ecosystems.* University of Manitoba, Winnipeg.

Jefries, R.N. 1940. The effect of temperature and relative humidity during curing upon the quality of white burley tobacco. *Bulletin* No. 407. Kentucky Agricultural Experiment Station, Lexington, KY.

Jilek, J. 1993. Critical temperature for barley drying. *Drying Tech.* 11(1):183-193.

Jindal, V.K. and T.L. Thompson. 1972. Air pressure patterns and flow paths in two-dimensional triangular-shaped piles of sorghum using forced convection. *Transactions of ASAE* 15(4):737-44.

Johnson, W.H., W.H. Henson, Jr., F.J. Hassler, and R.W. Watkins. 1960. Bulk curing of bright-leaf tobacco. *Agricultural Engineering* 41(8):511-15, 517.

Jones, A.L., R.E. Morrow, W.G. Hires, G.B. Garner, and J.E. Williams. 1985. Quality evaluation of large round bales treated with sodium diacetate or anhydrous ammonia. *Transactions of ASAE* 28(4):1043-45.

Khompos, V., L.J. Segerlind, and R.C. Brook. 1984. Pressure patterns in cylindrical grain storages. *Paper* 84-3011. ASAE.

Kirleis, A.W., T.L. Housley, A.M. Emam, F.L. Patterson, and M.R. Okos. 1982. Effect of preripe harvest and artificial drying on the milling and baking quality of soft red winter wheat. *Crop Science* 22:871-76.

Kramer, H.A. 1951. Engineering aspects of rice drying. *Agricultural Engineering* 32(1):44-45, 50.

Kraszewski, A.W. and S.O. Nelson. 1992. Resonant microwave cavities for sensing properties of agricultural products. *Transactions of ASAE* 35(4):1315-1321.

Kumar, A. and W.E. Muir. 1986. Airflow resistance of wheat and barley affected by airflow direction, filling method and dockage. *Transactions of ASAE* 29(5):1423-26.

Kunze, O.R. 1984. Physical properties of rice related to drying the grain. *Drying Technology* 2(3):369-87.

Kunze, O.R. and D.L. Calderwood. 1980. Systems for drying of rice. In *Drying and storage of agriculture crops*, C.W. Hall. AVI Publishing Company, Westport, CT.

Lai, F.S. 1980. Three dimensional flow of air through nonuniform grain beds. *Transactions of ASAE* 23(3):729-34.

Laird, W. and R.V. Baker. 1983. Heat recapture for cotton gin drying systems. *Transactions of ASAE* 26(3):912-17.

Lapp, H.M., F.J. Madrid, and L.B. Smith. 1986. A continuous thermal treatment to eradicate insects from stored wheat. *Paper* 86-3008. ASAE.

Leva, M. 1959. *Fluidization*. McGraw-Hill, Inc. New York.

Li, H. and R.V. Morey. 1984. Thin-layer drying of yellow dent corn. *Transactions of ASAE* 27(2):581-85.

Li, W. and S. Sokhansanj. 1994. Generalized equation for airflow resistance of bulk grains with variable density, moisture content and fines. *Drying Tech.* 12(3):649-667.

Li, Y., R.V. Morey, and M. Afinrud. 1987. Thin-layer drying rates of oilseed sunflower. *Transactions of ASAE* 30(4):1172-75, 1180.

Lissik, E.A. 1986. A model for the removal of heat in respiring grains. *Paper* 86-6509. ASAE.

Locklair, E.E., L.G. Veasey, and M. Samfield. 1957. Equilibrium desorption of water vapor on tobacco. *Journal of Agricultural and Food Chemistry* 5:294-98.

Magan, N. 1993. Early detection of fungi in stored grain. *Int. J. Biodeterioration and Biodegredation* 32 (1/3):145-160.

Manis, J.M. 1992. Sampling, inspection and grading. In *Storage of cereal grains and their products*, ed. D.B. Sauer, pp. 563-88

Martin, C.R., Z. Czuchajowska, and Y. Pomeranz. 1986. Aquagram standard deviations of moisture in mixtures of wet and dry corn. *Cereal Chemistry* 63(5):442-45.

Martins, J. and R.L. Stroshine. 1987. Difference in drying efficiencies among corn hybrids dried in a high-temperature column-batch dryer. *Paper* 87-6559. ASAE.

McKenzie, B.A. 1980. Managing dry grain in storage. AED-20. Midwest Plan Service, Iowa State University, Ames, IA.

McKenzie, B.A., G.H. Foster, and S.S. DeForest. 1980. Fan sizing and application for bin drying/cooling of grain. AE-106. Cooperative Extension Service, Purdue University, West Lafayette, IN.

McMillan, W.W., D.M. Wilson, and N.W. Widstrom. 1985. Aflatoxin contamination of preharvest corn in Georgia—A six-year study of insect damage and visible *Aspergillus flavus*. *Journal of Environmental Quality* 14:200-02.

Metzger, J.F. and W.E. Muir. 1983. Computer model of two-dimensional conduction and forced convection in stored grain. *Canadian Agricultural Engineering* 25:119-25.

Miketinac, M.J. and S. Sokhansanj. 1985. Velocity-pressure distribution in grain bins—Brooker model. *International Journal of Applied Numerical Analysis in Engineering* 21:1067-75.

Misra, M.K. and D.B. Brooker. 1980. Thin-layer drying and rewetting equations for shelled yellow corn. *Transactions of ASAE* 23(5):1254-60.

Moilanen, C.W., R.T. Schuler, and E.R. Miller. 1973. Effect on wheat quality of air flow and temperatures in mechanical dryers. *North Dakota Farm Research* 30(6):15-19.

Morey, R.V. and H.A. Cloud. 1980. Combination high-speed, natural-air corn drying. M-163. Agricultural Extension Service, University of Minnesota, St. Paul.

Morey, R.V. and H. Li. 1984. Thin-layer equation effects on deep-bed drying prediction. *Transactions of ASAE* 27(6):1924-28.

Morey, R.V., H.A. Cloud, and D.J. Hansen. 1981. Ambient air wheat drying. *Transactions of ASAE* 24(5):1312-16.

Morey, R.V., H.A. Cloud, and W.E. Lueschen. 1976. Practices for the efficient utilization of energy from drying corn. *Transactions of ASAE* 19(1): 151-55.

Morey, R.V., H.A. Cloud, R.J. Gustafson, and D.W. Peterson. 1979. Management of ambient air drying systems. *Transactions of ASAE* 22(6): 1418-25.

Morey, R.V., H.M. Keener, T.L. Thompson, G.M. White, and F.W. Bakker-Arkema. 1978. The present status of grain drying simulation. *Paper* 78-3009. ASAE.

Nofsinger, G.W. 1982. The trickle ammonia process—An update. Grain Conditioning Conference Proceedings, Agricultural Engineering Department, University of Illinois, Champaign-Urbana.

Nofsinger, G.W., R.J. Bothast, and R.A. Anderson. 1979. Field trials using extenders for ambient-conditioning high-moisture corn. *Transactions of ASAE* 22(5):1208-13.

Noomhorm, A. and L.R. Verma. 1986. Generalized single-layer rice drying models. *Transactions of ASAE* 29(2):587-91.

O'Callaghan, J.R., D.J. Menzies, and P.H. Bailey. 1971. Digital simulation of agricultural dryer performance. *Journal of Agricultural Engineering Research* 16:223-44.

Otten, L., R. Brown, and W.S. Reid. 1984. Drying of white beans—Effects of temperature and relative humidity on seed coat damage. *Canadian Agricultural Engineering* 26(2):101-04.

Overhults, D.G., G.M. White, M.E. Hamilton, and I.J. Ross. 1973. Drying soybeans with heated air. *Transactions of ASAE* 16(1):112-13.

Overhults, D.G., G.M. White, M.E. Hamilton, I.J. Ross, and J.D. Fox. 1975. Effect of heated air drying on soybean oil quality. *Transactions of ASAE* 16(1):112-13.

Parry, J.L. 1985. Mathematical modelling and computer simulation of heat and mass transfer in agricultural grain drying: A review. *Journal of Agricultural Engineering Research* 32:1-29.

Parsons, S.D., B.A. McKenzie, and J.R. Barrett, Jr. 1979. Harvesting and drying high-moisture wheat. In *Double cropping winter wheat and soybeans in Indiana*. ID 96, Cooperative Extension Service, Purdue University, West Lafayette, IN.

Parti, M. 1993. Selection of mathematical models for drying grain in thin-layers. *Journal of Agricultural Engineering Research* 54(4):339-52.

Paster, N., M. Calderon, M. Menasherov, and M. Mora. 1990. Biogeneration of modified atmospheres in small storage containers using plant wastes. *Crop Protection* 9:235-238.

Paster, N., M. Calderon, M. Menasherov, V. Barak, and M. Mora. 1991. Application of biogenerated modified atmospheres for insect control in small grain bins. *Tropical Sci.* 31(4):355-358.

Paster, N., M. Menasherov, U. Ravid, and B. Juven. 1995. Antifungal activity of oregano and thyme essential oils applied as fumigants against fungi attaching stored grain. *J. Food Protect.* 58(1):81-85.

Pathak, P.K., Y.C. Agrawal, and B.P.N. Singh. 1991. Thin-layer drying model for rapeseed. *Trans ASAE*. 34(6):2505-2508.

Paulsen, M.R. and L.D. Hill. 1985. Corn quality factors affecting dry milling performance. *Journal of Agricultural Engineering Research* 31:255-63.

Paulsen, M.R. and T.L. Thompson. 1973. Drying analysis of grain sorghum. *Transactions of ASAE* 16(3):537-40.

Paulsen, M.R., L.D. Hill, D.G. White, and G.F. Sprague. 1983. Breakage susceptibility of corn-belt genotypes. *Transactions of ASAE* 26(6):1830-36.

Pederson, J.R. 1992. Insects: Identification, damage, and detection. In *Storage of cereal grains and their products*, ed. D.B. Sauer, pp. 435-90.

Peplinski, A.J., R.A. Anderson, and O.L. Brekke. 1982. Corn dry milling as influenced by harvest and drying conditions. *Transactions of ASAE* 25(4):1114-17.

Peplinski, A.J., O.L. Brekke, R.J. Bothast, and L.T. Black. 1978. High moisture corn—An extended preservation trial with ammonia. *Transactions of ASAE* 21(4): 773-76, 781.

Peterson, W.H. 1982. Design principles for grain aeration in flat storages. Illinois Farm Electrification Council *Fact Sheet* No. 9. University of Illinois, Agricultural Engineering Department, Urbana.

Pfost, H.B., S.G. Mauer, D.S. Chung, and G.A. Milliken. 1976. Summarizing and reporting equilibrium moisture data for grains. *Paper* 76-3520. ASAE.

Pierce, R.O. 1986. Economic consideration for natural air corn drying in Nebraska. *Transactions of ASAE* 29(4):1131-35.

Pierce, R.O. and T.L. Thompson. 1979. Solar grain drying in the North Central Region—Simulation results. *Transactions of ASAE* 15(1):178-87.

Pierce, R.O. and T.L. Thompson. 1981. Energy use and performance related to crossflow dryer design. *Transactions of ASAE* 24 (1):216-20.

Quinlan, J.K. 1982. Grain protectants for insect control. *Marketing Bulletin* 72. Agricultural Research Service, U.S. Department of Agriculture, Washington, D.C.

Radajewski, W., T. Jensen, G.Y. Abawi, and E.J. McGahan. 1992. Drying rate and damage to navy beans. *Transactions of ASAE* 35(2):583-90.

Rayburn, S.T., A.C. Griffin, Jr., and M.E. Whitten. 1978. Storing cottonseed with propionic acid. *Transactions of ASAE* 21(5):990-92.

Ripp, B.E., ed. 1984. Controlled atmosphere and fumigation in grain storages. Proceedings of an International Symposium, Practical Aspects of Controlled Atmosphere and Fumigation in Grain Storages, in Perth, Western Australia. Elsevier Science Publishing Company, New York.

Rusal, G. and E.H. Marth. 1988. Food additives and plant components control growth and aflatoxin production in toxigenic *Aspergilli*: A review. *Mycopathologia* 101:13-23.

Sabbah, M.A., H.M. Keener, and G.E. Meyer. 1979. Simulation of solar drying of shelled corn using the logarithmic model. *Transactions of ASAE* 22(3):637-43.

Sabbah, M.A., G.E. Meyer, H.M. Keener, and W.L. Roller. 1976. Reversed-air drying for fixed bed of soybean seed. *Paper* 76-3023. ASAE.

Sauer, D.B., ed. 1992. *Storage of cereal grains and their products*. American Association of Cereal Chemists, St. Paul, MN.

Sauer, D.B. and R. Burroughs. 1974. Efficacy of various chemicals as grain mold inhibitors. *Transactions of ASAE* 17(3):557-59.

Sauer, D.B., R.A. Meronuck, and C.M. Christensen. 1992. Microflora. In *Storage of cereal grains and their products*, ed. D.B. Sauer, pp. 313-40.

Saul, R.A. and J.L. Steele. 1966. Why damaged shelled corn costs more to dry. *Agricultural Engineering* 47:326-29, 337.

Schmidt, B.J. and L.F. Backer. 1980. Results of a sunflower storage monitoring program in North Dakota. *Paper* 80-6033, ASAE.

Schuler, R.T. 1974. Drying related properties of sunflower seeds. *Paper* 74-3534. ASAE.

Schuler, R.T., B.J. Holmes, R.J. Straub, and D.A. Rohweder. 1986. Hay drying. *Publication* A3380. Cooperative Extension Service, University of Wisconsin, Madison.

Schultz, L.J., M.L. Stone, and P.D. Bloome. 1984. A comparison of simulation techniques for wheat aeration. *Paper* 84-3012. ASAE.

Segerlind, L.J. 1982. Solving the nonlinear airflow equation. *Paper* 82-3017. ASAE.

Seitz, L.M., D.B. Sauer, and H.E. Mohr. 1982a. Storage of high-moisture corn: Fungal growth and dry matter loss. *Cereal Chemistry* 59(2):100-105.

Seitz, L.M., D.B. Sauer, H.E. Mohr, and D.F. Aldis. 1982b. Fungal growth and dry matter loss during bin storage of high-moisture corn. *Cereal Chemistry* 59(1):9-14.

Sellam, M.A. and C.M. Christensen. 1976. Temperature differences, moisture transfer and spoilage in stored corn. *Feedstuffs* 48(36):28, 33.

Sharp, J.R. 1982. A review of low-temperature drying simulation models. *Journal of Agricultural Engineering Research* 27(3):169-90.

Shaw, C.S. and G.N. Franks. 1962. Cottonseed drying and storage at cotton gins. *Technical Bulletin* 1262. USDA, ARS, Washington, D.C.

Shedd, C.K. 1953. Resistance of grains and seeds to air flow. *Agricultural Engineering* 34(9):616-18.

Shejbal, J., ed. 1980. *Controlled atmosphere storage of grains*. Elsevier Science Publishing Company, New York.

Shelef, L.A. 1984. Antimicrobial affects of spices. *Journal of Food Safety* 6:29-44.

Shepherd, J.B. 1954. Experiments in harvesting and preserving alfalfa for dairy cattle feed. *Technical Bulletin* 1079. USDA, Washington, D.C.

Shivhare, U.S., G.S.V. Raghaven, and R.G. Bosisio. 1992. Microwave drying of corn. I. Equilibrium moisture content. *Transactions of ASAE* 35(3):947-50.

Singh, S.S. and M.F. Finner. 1983. A centrifugal impacter for damage susceptibility evaluation of shelled corn. *Transactions of ASAE* 26(6):1858-63.

Smith, E.A. and P.H. Bailey. 1983. Simulation of near-ambient grain drying. II, Control strategies for drying barley in Northern Britain. *Journal of Agricultural Engineering Research* 28:301-17.

Smith, E.A. and S. Sokhansanj. 1990a. Moisture transport due to natural convection in grain stores. *Journal of Agricultural Engineering Research* 47:23-34.

Smith, E.A. and S. Sokhansanj. 1990b. Natural convection and temperature of stored products—A theoretical analysis. *Canadian Agricultural Engineering* 32(1):91-97.

Smith, J.S., Jr. and J.I. Davidson, Jr. 1982. Psychrometrics and kernel moisture content as related to peanut storage. *Transactions of ASAE* 25(1): 231-36.

Smith, J.S., Jr., J.I. Davidson, Jr., T.H. Sanders, and R.J. Cole. 1985. Storage environment in a mechanically ventilated peanut warehouse. *Transactions of ASAE* 28(4):1248-52.

Sokhansanj, S. and D.M. Bruce. 1987. A conduction model to predict grain drying simulation. *Transactions of ASAE* 30(4):1181-84.

Sokhansanj, S. and S.O. Nelson. 1988a. Dependence of dielectric properties of whole-grain wheat on bulk density. *Journal of Agricultural Engineering Research* 39:173-79.

Sokhansanj, S. and S.O. Nelson. 1988b. Transient dielectric properties of wheat associated with non-equilibrium kernel moisture conditions. *Transactions of ASAE* 31(4):1251-54.

Sokhansanj, S. and G.E. Woodward. 1991. Computer assisted fan selection for natural grain drying—A teaching and extension tool. *Applied Engineering in Agriculture* 6(6):782-84.

Sokhansanj, S., D. Singh, and J.D. Wasserman. 1984. Drying characteristics of wheat, barley and canola subjected to repetitive wetting and drying cycles. *Transactions of ASAE* 27(3):903-906, 914.

Sokhansanj, S., W. Zhijie, D.S. Jayas, and T. Kameoka. 1986. Equilibrium relative humidity moisture content of rapeseed from 5 to 25°C. *Transactions of ASAE* 29(3):837-39.

Sokhansanj, S., A.A. Falacinski, F.W. Sosulski, D.S. Jayas, and J. Tang. 1990. Resistance of bulk lentils to airflow. *Transactions of ASAE* 33(4):1281-85.

Steele, J.L. 1967. Deterioration of damaged shelled corn as measured by carbon dioxide production. Unpublished Ph.D. diss., Department of Agricultural Engineering, Iowa State University, Ames, IA.

Steele, J.L., R.A. Saul, and W.V. Hukill. 1969. Deterioration of shelled corn as measured by carbon dioxide production. *Transactions of ASAE* 12(5):685-89.

Stephens, G.R. and T.L. Thompson. 1976. Improving crossflow grain dryer design using simulation. *Transactions of ASAE* 19(4):778-81.

Stephens, L.E. and G.H. Foster. 1976a. Breakage tester predicts handling damage in corn. ARS-NC-49. ARS, USDA, Washington, D.C.

Stephens, L.E. and G.H. Foster. 1976b. Grain bulk properties as affected by mechanical grain spreaders. *Transactions of ASAE* 19(2):354-58.

Stephens, L.E. and G.H. Foster. 1978. Bulk properties of wheat and grain sorghum as affected by a mechanical grain spreader. *Transactions of ASAE* 21(6):1217-18.

Storey, C.L., R.D. Speirs, and L.S. Henderson. 1979. Insect control in farm-stored grain. *Farmers Bulletin* 2269. USDA-SEA (Available for sale from Superintendent of Documents, U.S. Government Printing Office, Washington, D.C. 20402).

Stroshine, R.L., A.W. Kirleis, J.F. Tuite, L.F. Bauman, and A. Emam. 1986. Differences in grain quality among selected corn hybrids. *Cereal Foods World* 31(4):311-16.

Stroshine, R.L., J. Tuite, G.H. Foster, and K. Baker. 1984. Self-study guide for grain drying and storage. Department of Agricultural Engineering, Purdue University, West Lafayette, IN.

Sun, D.W. and J.L. Woods. 1993. The moisture content/relative humidity equilibrium relationship of wheat: A review. *Drying Tech.* 11(7):1523-51.

Sun, D.W. and J.L. Woods. 1994. Low temperature moisture transfer characteristics of wheat in thin layers. *Transactions of ASAE* 37(6):1919-26.

Swearingen, M.L. 1979. A practical guide to no-till double cropping. In *Double cropping winter wheat and soybeans in Indiana.* ID 96. Cooperative Extension Service, Purdue University, West Lafayette, IN.

Syarief, A.M., R.V. Morey, and R.J. Gustafson. 1984. Thin-layer drying rates of sunflower seed. *Transactions of ASAE* 27(1):195-200.

Tang, J., S. Sokhansanj, and F.W. Sosulski. 1989. Thin-layer drying of lentil. *Paper* 89-6607. ASAE.

Thompson, C.R., E.M. Bickoff, G.R. VanAtta, G.O. Kohler, J. Guggolz, and A.L. Livingston. 1960. Carotene stability in alfalfa as affected by laboratory- and industrial-scale processing. *Technical Bulletin* 1232. ARS, USDA, Washington, D.C.

Thompson, T.L. 1972. Temporary storage of high-moisture shelled corn using continuous aeration. *Transactions of ASAE* 15(2):333-37.

Thompson, T.L., G.H. Foster, and R.M. Peart. 1969. Comparison of concurrent-flow, crossflow and counterflow grain drying methods. *Marketing Research Report* 841. USDA-ARS, Washington, D.C.

Thompson, T.L., R.M. Peart, and G.H. Foster. 1968. Mathematical simulation of corn drying—A new model. *Transactions of ASAE* 11(4):582-86.

Thorpe, G.R. 1982. Moisture diffusion through bulk grain subjected to a temperature gradient. *Journal of Stored Products Research* 18:9-12.

Troeger, J.M. and W.V. Hukill. 1971. Mathematical description of the drying rate of fully exposed corn. *Transactions of ASAE* 14(6):1153-56, 1162.

Tuite, J. and G.H. Foster. 1963. Effect of artificial drying on the hygroscopic properties of corn. *Cereal Chemistry* 40:630-37.

Tuite, J. and G.H. Foster. 1979. Control of storage diseases of grain. *Annual Review of Phytopathology* 17:343.

Tuite, J., G.H. Foster, S.R. Eckhoff, and O.L. Shotwell. 1986. Sulfur dioxide treatment to extend corn drying time (note). *Cereal Chemistry* 63(5): 462-64.

Uebersax, M.A. and C.L. Bedford. 1980. Navy bean processing: Effect of storage and soaking methods on quality of canned beans. *Research Report* 410. Agricultural Experiment Station, Michigan State University, East Lansing.

USDA. 1971. Oven methods for determining moisture content of grain and related agricultural commodities, Chapter 12. *Equipment manual*, GR Instruction 916-6. U.S. Department of Agriculture, Consumer and Marketing Service, Grain Division, Hyattsville, MD.

Velupillai, L. and L.R. Verma. 1986. Drying and tempering effects on parboiled rice quality. *Transactions of ASAE* 29(1):312-19.

Verma, L.R. and B.D. Nelson. 1983. Changes in round bales during storage. *Transactions of ASAE* 26(2):328-32.

Walton, L.R. and W.H. Henson, Jr. 1971. Effect of environment during curing on the quality of burley tobacco: Effect of low humidity curing on support price. *Tobacco Science* 15:54-57.

Walton, L.R., W.H. Henson, Jr., and J.M. Bunn. 1973. Effect of environment during curing on the quality of burley tobacco: Effect of high humidity curing on support price. *Tobacco Science* 17:25-27.

Walton, L.R., L.D. Swetnam, and J.H. Casada. 1994. Curing burley tobacco in a field curing structure. *Applied Engineering in Agric.* 10(3):385-89.

Watson, C.A. 1977. Near infrared reflectance spectro-photometric analysis of agricultural products. *Analytical Chemistry* 49(9):835A-40A.

Watson, C.A., O.J. Banasick, and G.L. Pratt. 1962. Effect of drying temperature on barley malting quality. *Brewers Digest* 37(7):44-48.

Watson, E.L. and V.K. Bhargava. 1974. Thin-layer drying studies on wheat. *Canadian Agricultural Engineering* 16(1):18-22.

Watson, S.A. 1987. Measurement and maintenance of quality. In *Corn: Chemistry and technology,* eds. S.A. Watson and P.E. Ramstad, pp. 125-83. American Association of Cereal Chemists, St. Paul, MN.

Watson, S.A. and F.L. Herum. 1986. Comparison of eight devices for measuring breakage susceptibility of shelled corn. *Cereal Chemistry* 63(2):139-42.

Watson, S.A. and P.E. Ramstad. 1987. *Corn: Chemistry and technology.* American Association of Cereal Chemists, St. Paul, MN.

Watson, S.A., L.L. Darrah, and F.L. Herum. 1986. Measurement of corn breakage susceptibility with the Stein breakage tester: A collaborative study. *Cereal Foods World* 31(5):366-72.

Weller, C.L., M.R. Paulsen, and M.P. Steinberg. 1987. Varietal, harvest moisture and drying air temperature effects on quality factors affecting corn wet milling. *Paper* 87-6046. ASAE.

White, N.D.G. and D.S. Jayas. 1991. Control of insects and mites with carbon dioxide in wheat stored at cool temperatures in nonairtight bins. *J. Econ. Entomology* 84(6):1933-42.

Wicklow, D.T. 1988. Patterns of fungal association within maize kernels harvested in North Carolina. *Plant Disease* 72:113-15.

Young, J.H. 1984. Energy conservation by partial recirculation of peanut drying air. *Transactions of ASAE* 27(3):928-34.

CHAPTER 12

AIR CONTAMINANTS

AIR IS COMPOSED mainly of gases. The major gaseous components of clean, dry air near sea level are approximately 21% oxygen, 78% nitrogen, 1% argon, and 0.04% carbon dioxide.

Normal outdoor air contains varying amounts of foreign materials (permanent atmospheric impurities). These materials can arise from natural processes such as wind erosion, sea spray evaporation, volcanic eruption, and metabolism or decay of organic matter. The natural contaminant concentrations in the air that we breathe vary but are usually lower than those caused by human activity.

Man-made outdoor contaminants are many and varied, originating from numerous types of human activity. Electric power-generating plants, various modes of transportation, industrial processes, mining and smelting, construction, and agriculture generate large amounts of contaminants.

Contaminants that present particular problems in the indoor environment include, among others, tobacco smoke, radon, and formaldehyde.

Air composition may be changed accidentally or deliberately. In sewers, sewage treatment plants, tunnels, and mines, the oxygen content of air can become so low that people cannot remain conscious or survive. Concentrations of people in confined spaces (theaters, survival shelters, submarines) require that carbon dioxide given off by normal respiratory functions be removed and replaced with oxygen. Pilots of high-altitude aircraft, breathing at greatly reduced pressure, require systems that increase oxygen concentration. Conversely, for divers working at extreme depths, it is common to increase the percentage of helium in the atmosphere and reduce nitrogen and sometimes oxygen concentrations.

At atmospheric pressure, oxygen concentrations less than 12% or carbon dioxide concentrations greater than 5% are dangerous, even for short periods. Lesser deviations from normal composition can be hazardous under prolonged exposures. Chapter 9 further details environmental health issues.

CLASSES OF AIR CONTAMINANTS

The major classes of air contaminants are particulate and gaseous. The **particulate** class covers a vast range of particle sizes from dust large enough to be visible to the eye to submicroscopic particles that elude most filters. Particulates may be solid or liquid. The following traditional contaminant classifications are subclasses of particulates:

- **Dusts**, **fumes**, and **smokes**, which are mostly solid particulate matter, although smoke often contains liquid particles
- **Mists**, **fogs**, and **smogs**, which are mostly suspended liquid particles smaller than those in dusts, fumes, and smokes
- **Bioaerosols**, including viruses, bacteria, fungal spores, and pollen, whose primary impact is related to their biological origin
- Particle size definitions such as **coarse** or **fine**, **visible** or **invisible**, and **macroscopic**, **microscopic**, or **submicroscopic**
- Definitions that relate to particle interaction with the human respiratory system, such as **inhalable** and **respirable**

These classes, their characteristics, units of measurement, and measurement methods are discussed in more detail in this chapter.

The **gaseous** class covers chemical contaminants that can exist as free molecules or atoms in air. Molecules and atoms are smaller than particles and may behave differently as a result. This class covers two important subclasses:

- **Gases**, which are naturally gaseous under ambient indoor or outdoor conditions
- **Vapors**, which are normally solid or liquid under ambient indoor or outdoor conditions, but which evaporate readily

Through evaporation, liquids change into vapors and mix with the surrounding atmosphere. Like gases, they are formless fluids that expand to occupy the space or enclosure in which they are confined. Typical gaseous contaminants, their characteristics, units of measurement, and measurement methods are discussed in detail later in this chapter.

Air contaminants can also be classified according to their sources; their properties; or the health, safety, and engineering issues faced by people exposed to them. Any of these can form a convenient classification system because they allow grouping of applicable standards, guidelines, and control strategies. Most of the following classes include both particulate and gaseous contaminants. This chapter covers the background information for these classes, while Chapter 9 deals with the applicable indoor health and comfort regulations. The classes are

- Industrial air contaminants
- Nonindustrial indoor air contaminants (including indoor air quality)
- Flammable gases and vapors
- Combustible dusts
- Radioactive contaminants
- Soil gases

The preparation of this chapter is assigned to TC 2.3, Gaseous Air Contaminants and Gas Contaminant Removal Equipment, in conjunction with TC 2.4, Particulate Air Contaminants and Particulate Contaminant Removal Equipment.

PARTICULATE CONTAMINANTS

PARTICULATE MATTER

Airborne particulate matter is not a single substance but a complex mixture of many different components, usually from several different sources. Particles can be anthropogenic or natural in origin. **Anthropogenic** particles are those produced by human activities, including fossil fuel combustion, industrial processes, and road dust. Particles from **natural** sources do not involve human activity. They include wind-blown dust and smoke from forest fires.

Particles can be generated by primary or secondary processes. Particles from **primary processes** are those emitted directly into the air. Particles from **secondary processes** are those formed from condensable vapors and chemical reactions.

Details on the different particle types are given below.

Types of Solid Particles

Dusts are solid particles projected into the air by natural forces such as wind, volcanic eruption, or earthquakes, or by mechanical processes including crushing, grinding, demolition, blasting, drilling, shoveling, screening, and sweeping. Some of these forces produce dusts by reducing larger masses, while others disperse materials that have already been reduced. Particles are not considered to be dust unless they are smaller than about 100 µm. Dusts can be mineral, such as rock, metal, or clay; vegetable, such as grain, flour, wood, cotton, or pollen; or animal, including wool, hair, silk, feathers, and leather.

Fumes are solid particles formed by condensation of vapors of solid materials. Metallic fumes are generated from molten metals and usually occur as oxides because of the highly reactive nature of finely divided matter. Fumes can also be formed by sublimation, distillation, or chemical reaction. Such processes create airborne particles smaller than 1 µm. Fumes permitted to age may agglomerate into larger clusters.

Bioaerosols include airborne viruses, bacteria, pollen, and fungus spores. **Viruses** range in size from 0.003 to 0.06 µm, although they usually occur in colonies or attached to other particles. Most **bacteria** range between 0.4 and 5 µm and are usually associated with large particles. **Fungus** spores are usually 2 to 10 µm, while **pollen** grains are 10 to 100 µm, with many common varieties in the 20 to 40 µm range.

Types of Liquid Particles

Mists are small airborne droplets of materials that are ordinarily liquid at normal temperatures and pressure. They can be formed by atomizing, spraying, mixing, violent chemical reactions, evolution of gas from liquid, or escape as a dissolved gas when pressure is released. Small droplets expelled or atomized by sneezing constitute mists.

Fogs are fine airborne droplets, usually formed by condensation of vapor, which remain airborne longer than mists. Fog nozzles are named for their ability to produce extra fine droplets, as compared with mists from ordinary spray devices. Many droplets in fogs or clouds are microscopic and submicroscopic and serve as a transition stage between larger mists and vapors.

The volatile nature of most liquids reduces the size of their airborne droplets from the mist to the fog range and eventually to the vapor phase, until the air becomes saturated with that liquid. If solid material is suspended or dissolved in the liquid droplet, it remains in the air as particulate contamination. For example, sea spray evaporates fairly rapidly, generating a large number of fine salt particles that remain suspended in the atmosphere.

Complex Particles

Smokes are small solid and/or liquid particles produced by incomplete combustion of organic substances such as tobacco, wood, coal, oil, and other carbonaceous materials. The term smoke is applied to a mixture of solid, liquid, and gaseous products, although technical literature distinguishes between such components as soot or carbon particles, fly ash, cinders, tarry matter, unburned gases, and gaseous combustion products. Smoke particles vary in size, the smallest being much less than 1 µm in diameter. The average is often in the range of 0.1 to 0.3 µm.

Environmental tobacco smoke consists of a suspension of 0.01 to 1.0 µm (mass median diameter of 0.3 µm) liquid particles that form as the superheated vapors leaving the burning tobacco condense. Also produced are numerous gaseous contaminants including carbon monoxide.

Smog commonly refers to air pollution; it implies an air mixture of smoke particles, mists, and fog droplets of such concentration and composition as to impair visibility, in addition to being irritating or harmful. The composition varies among different locations and at different times. The term is often applied to haze caused by a sunlight-induced photochemical reaction involving the materials in automobile exhausts. Smog is often associated with temperature inversions in the atmosphere that prevent normal dispersion of contaminants.

Sizes of Airborne Particles

Particle size can be defined in several different ways. These depend, for example, on the source or method of generation, the visibility, or the effects.

Particles can be classified as coarse or fine. Coarse particles are larger, and are generally formed by mechanical breaking up of solids. They generally have a minimum size of 1 to 3 µm (EPA 1996). Fine particles are generally formed from chemical reactions or condensing gases. These particles have a maximum size of about 1 to 3 µm. Fine particles are usually more chemically complex, anthropogenic, and secondary in origin, while the coarse particles are predominantly primary, natural, and chemically inert. Coarse particles also include bioaerosols such as mold spores, pollen, animal dander, and dust mite particles that can affect the immune system.

The differences between fine and coarse particles lead to a bimodal distribution of particles in most environments with concentration peaks at about 0.25 µm and 5 µm. Figure 1 shows a typical urban distribution including the chemical species present in each mode.

The size of a particle determines where in the human respiratory system the particle is deposited. The **inhalable mass** is made up of

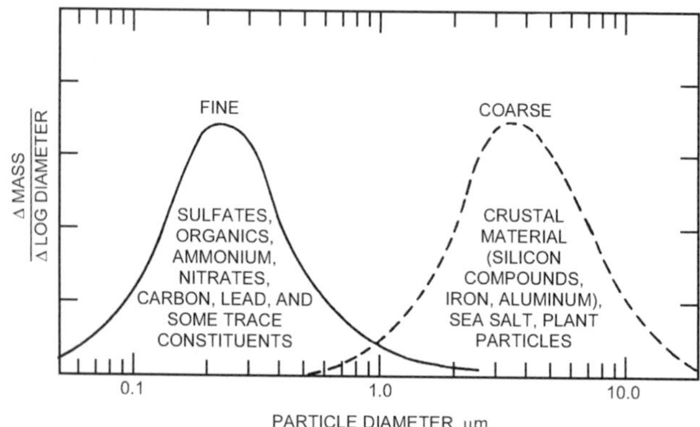

Fig. 1 Typical Urban Aerosol Composition by Particle Size Fraction
(EPA 1982, Willeke and Baron 1993)

particles that may deposit anywhere in the respiratory system and is represented by a sample with a median cut point of 100 µm. The **thoracic particle mass** is that fraction which can penetrate to the lung airways and is represented by a sample with a median cut point of 10 µm. The **respirable particulate mass** is that fraction that can penetrate to the gas-exchange region of the lungs and is represented by a sample with a median cut point of 4 µm (ACGIH 1998). Figure 2 illustrates the relative deposition efficiencies of various sizes of particles. The characteristics described above were used by the EPA in development of their outdoor PM_{10} and $PM_{2.5}$ standards.

Most particles are irregular in shape, and it is useful to characterize their size in terms of some standard particle. For example, the **aerodynamic (equivalent) diameter** of a particle is defined as the diameter of a unit-density sphere having the same gravitational settling velocity as the particle in question (Willeke and Baron 1993).

The tendency of particles to settle on surfaces is a property of interest. Figure 3 shows the sizes of typical indoor airborne solid and liquid particles. Particles smaller than 0.1 µm behave like gas molecules, exhibiting Brownian motion due to collisions with air molecules and having no measurable settling velocity. Particles in

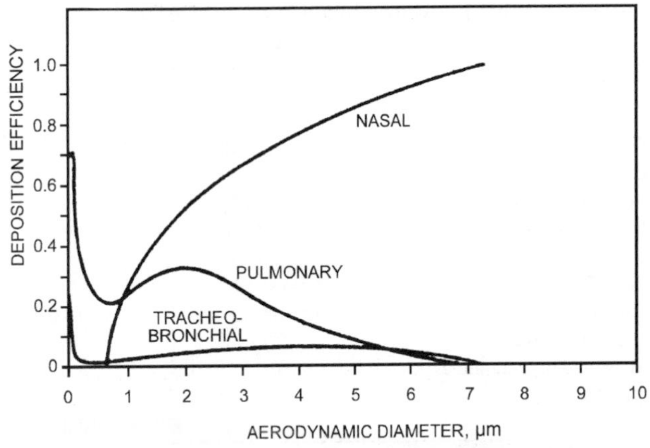

Fig. 2 Relative Deposition Efficiencies of Different Sized Particles in the Three Main Regions of the Human Respiratory System, Calculated for Moderate Activity Level
(Task Group on Lung Dynamics 1966)

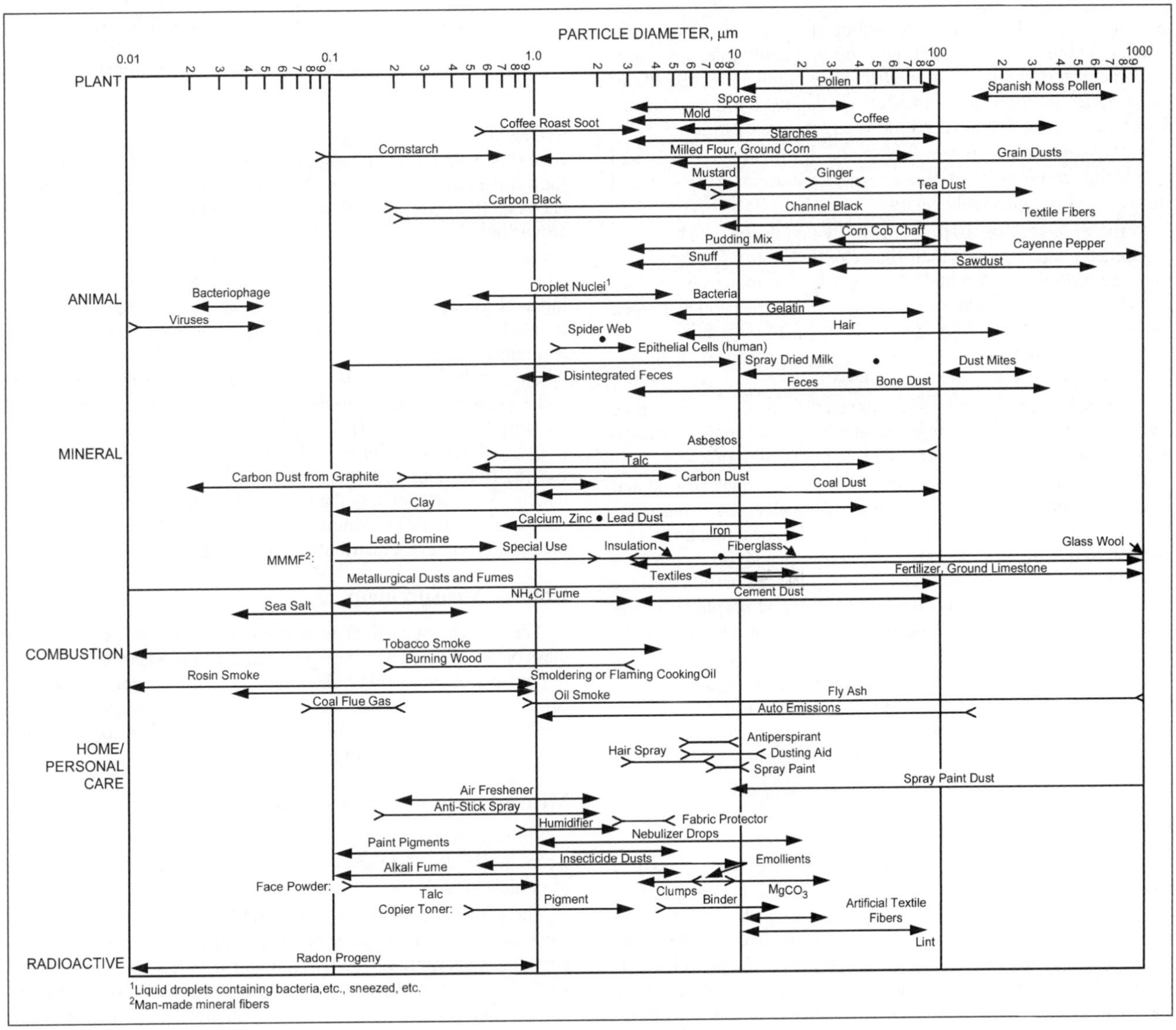

Fig. 3 Sizes of Indoor Particles
(Owen et al. 1992)

Table 1 Approximate Particle Sizes and Their Times to Settle One Metre

Type of Particle	Diameter, µm	Settling Time
Human hair	100 to 150	5 s
Skin flakes	20 to 40	
Observable dust in air	>10	
Common pollens	15 to 25	
Mite allergens	10 to 20	5 min
Common spores	2 to 10	
Bacteria	1 to 5	
Cat dander	1± 0.5	10 h
Tobacco smoke	0.1 to 1	
Metal and organic fumes	<0.1 to 1	
Cell debris	0.01 to 1	
Viruses	<0.1	10 days

Source: J.D. Spengler, Harvard School of Public Health.

Table 2 Relation of Screen Mesh to Particle Size

U.S. Standard sieve mesh	400	325	200	140	100	60	35	18
Nominal sieve opening, µm	37	44	74	105	149	250	500	1000

the range from 0.1 to 1 µm have settling velocities that can be calculated but are so low that settling is usually negligible, since normal air currents counteract any settling. By number, over 99.9% of the particles in a typical atmosphere are below 1 µm (this means fewer than 1 particle in every 1000 is larger than 1 µm.).

Particles in the 1 to 10 µm range settle in still air at constant and appreciable velocity. However, normal air currents keep them in suspension for appreciable periods.

Particles larger than 10 µm settle fairly rapidly and can be found suspended in air only near their source or under strong wind conditions. Exceptions are lint and other light fibrous materials, such as portions of certain weed seeds, which remain suspended longer. Table 1 shows settling times for various types of particles.

Most individual particles 10 µm or larger are visible to the naked eye under favorable conditions of lighting and contrast. Smaller particles are visible only in high concentrations. Cigarette smoke (with an average particle size less than 0.5 µm) and clouds are common examples.

Direct fallout in the vicinity of the dispersing stack or flue and other nuisance problems of air pollution involve larger particles. Smaller particles, as well as mists, fogs, and fumes, remain in suspension longer. In this size range, meteorology and topography are more important than physical characteristics of the particles. Since settling velocities are small, the ability of the atmosphere to disperse these small particles depends largely on local weather conditions.

Comparison is often made to screen sizes used for grading useful industrial dusts and granular materials. Table 2 illustrates the relationship of U.S. standard sieve mesh to particle size in micrometres. Particles above 40 µm are the screen sizes, and those below are the subscreen or microscopic sizes.

Particle Size Distribution

The particle size distribution in any sample can be expressed as the percentage of the number of particles smaller than a specified size, area, or mass. The upper curve of Figure 4 shows these data plotted for typical atmospheric contamination. The middle curve shows the percentage of the total projected area of the particles contributed by particles less than a specified size. The lower curve shows the percentage of the total particle mass contributed by those particles less than a given size. Thus, for particles 10 µm and larger, 4% by mass, a 90% arrestance filter essentially removes 90% of 4% for an efficiency by mass of 3.6%.

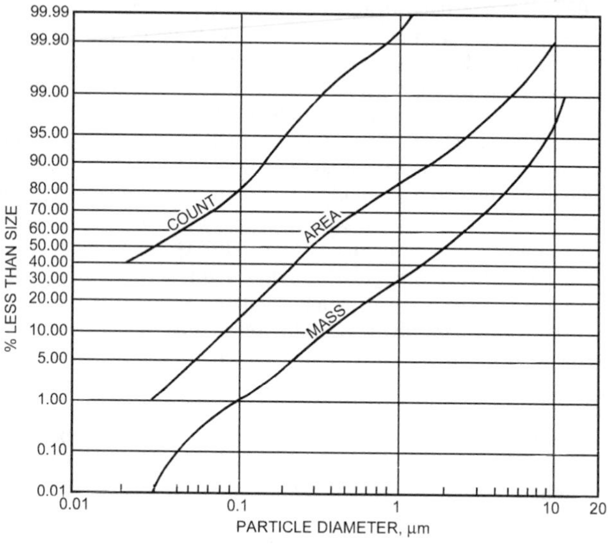

Count curve: Based on measurements by electron microscope
Area curve: Calculated
Mass curve: Solid section based on measurements by sedimentation

Fig. 4 Particle Size Distribution of Atmospheric Dust
(Whitby et al. 1955, 1957)

The differences among values presented by the three curves should be noted. For example, particles 0.1 µm or less in diameter (but still above electron microscope minimum detection size of about 0.005 µm) make up 80% of the number of particles in the atmosphere but contribute only 1% of the mass. Also, the 0.1% of particles by number larger than 1 µm carry 70% of the total mass, which is the direct result of the mass of a spherical particle increasing as the cube of its diameter. Although most of the mass is contributed by intermediate and larger particles, over 80% of the area (staining) contamination is supplied by particles less than 1 µm in diameter, which is in the center of the respirable particle size range and is the size most likely to remain in the lungs. (See Chapter 9.) Of possible concern to the HVAC industry is the fact that most of the staining effect on ceilings, walls, windows, and light fixtures, as well as fouling of heat transfer devices and rotating equipment, results from particles less than 1 µm in diameter. Suspended particles in urban air are predominantly smaller than 1 µm (aerodynamic diameter) and have a distribution that is approximately log-normal.

Units of Measurement

The quantity of particulate matter in the air can be determined as mass or particle count in a given volume of air. Mass units are milligrams per cubic metre of air sampled (mg/m^3) or micrograms per cubic metre of air sampled (µg/m^3). 1 mg/m^3 = 1000 µg/m^3. Particle counts are usually quoted for volumes of 0.1 ft^3, 1 ft^3, or 1 litre and are specified for a given range of particle diameter.

Measurement of Airborne Particles

Suitable methods for determining the quantity of particulate matter in the air vary depending on the amount present and on the size of particles involved.

The particle mass (total or respirable) is easily determined by measuring the mass of a filter before and after drawing a known volume of dusty air through it. This method is widely used in industrial workplaces, where there may be significant numbers of large particles, but is not sensitive enough for evaluating office environments.

Optical particle counters are widely used and likely to become more so since ASHRAE adopted *Standard* 52.2. This standard defines a laboratory method for assessing the performance of media filters using an optical particle counter to measure particle counts upstream and downstream of the filter in 12 size ranges between 0.3 and 10 μm. Filters are then given a **minimum efficiency reporting value** (MERV) rating based on the count data.

Counters are also used to test cleanrooms for compliance with Federal *Standard* 209E and ISO *Standard* 14644-1. Cleanrooms are defined in terms of the number of particles in certain size ranges that they contain. Chapter 15 of the 1999 *ASHRAE Handbook—Applications* has further information on cleanrooms.

Optical particle counters use laser light-scattering to continuously count and size airborne particles and can detect particles down to 0.1 μm (ASTM *Standard* F 50). A **condensation nucleus counter** can count particles to below 0.01 μm. These particles, present in great numbers in the atmosphere, serve as nuclei for condensation of water vapor (Scala 1963).

Another indirect method measures the **optical density** of the collected dust based on the projected area of the particles. Dust particles can be sized with graduated scales or optical comparisons using a standard **microscope**. The lower limit for sizing with the light field microscope is approximately 0.9 μm, depending on the vision of the observer, the dust color, and the contrast available. This size can be reduced to about 0.4 μm by using oil immersion objective techniques. Dark field microscopic techniques reveal particles smaller than these, to a limit of approximately 0.1 μm. Smaller submicroscopic dusts can be sized and compared with the aid of an electron microscope. Other sizing techniques may take into account velocity of samplings in calibrated devices and actual settlement measurements in laboratory equipment. The electron microscope and various sampling instruments such as the **cascade impactor** have been successful in sizing particulates, including fogs and mists.

Each of the various methods of measuring particle size distribution gives a different value for the same size particle, since different properties are actually measured. For example, a microscopic technique may measure longest dimension, while impactor results are based on aerodynamic behavior (ACGIH 2001).

Typical Particle Levels

Particle counters, which detect particles larger than about 0.1 μm, have indicated that the number of suspended particles is enormous. A room with heavy cigarette smoke has a particle concentration of 3×10^7 particles per cubic foot. Even clean air typically contains over 10^6 particles/ft^3. If smaller particles detectable by other means, such as an electron microscope or condensation nucleus counter, are also included, the total particle concentration would be greater than the above concentrations by a factor of 10 to 100.

Extensive measurements have been made of outdoor pollution, but limited data have been gathered on indoor pollution not associated with specific industrial processes. Indoor levels are influenced by the number of people and their activities, building materials and construction, outside conditions, ventilation rate, and the air-conditioning and filtration system. For further information, see the section on Nonindustrial Indoor Air Contaminants, Spengler et al. (1982), and NRC (1981).

BIOAEROSOLS

Bioaerosols are airborne microbiological particulate matter derived from fungi, bacteria, viruses, protozoa, algae, pollen, mites, and their cellular or cell mass components. Bioaerosols are universally present in both indoor and outdoor environments. Problems of concern to engineers occur when microorganisms grow and reproduce indoors.

Microorganisms break down complex molecules found in dead organic materials to simple substances such as carbon dioxide, water, and nitrates. These components are then used by photosynthetic organisms such as plants and algae. Thus, the presence of bacteria and fungi in soil, water, and atmospheric habitats is normal. Spores of *Cladosporium*, a fungus commonly found on leaves and dead vegetation, are almost always found in outdoor air samples. They are found in variable numbers in indoor air, depending on the amount of outdoor air that infiltrates into interior spaces or is brought in by the HVAC system. Outdoor microorganisms can also enter on shoes and clothing and be transferred to other surfaces in buildings.

Public interest has focused on airborne microorganisms responsible for diseases and infections, primarily bacteria and viruses. This is discussed in more detail in Chapter 9. A variety of airborne microorganisms are of economic significance and can cause product contamination or loss. In the food processing industry, yeast and mold can reduce the shelf life of some products. Refined syrups can be damaged by mold scums. Wild yeast can destroy a batch of beer. Antibiotic yields can be reduced by foreign organisms in the culture mix.

Fungi

Much attention has been given to fungi, which include yeasts, molds, and mildews, as well as large mushrooms, puffballs, and bracket fungi. All fungi depend on external sources of organic material for both energy requirements and carbon skeletons. Thus, they cannot increase in number unless supplied with a suitable food source such as small quantities of dust, paper, or wood. Reproduction also requires appropriate temperatures (70 to 100°F) and the presence of water, high air humidity (typically greater than 60%), and/or high moisture content in the food source.

Bacteria

Cooling towers, evaporative condensers, and domestic water service systems all provide water and nutrients for amplification of microorganisms such as *Legionella pneumophila*. Growth of microbial populations to excessive concentrations is generally associated with inadequate preventive maintenance of these systems, at least in cooling towers. A body of literature has identified characteristics of indoor plumbing and heating systems associated with frequent isolation of *Legionella* species, including blind ends, scale, upright electric water heaters, and lower water temperatures. The survival of *Legionella* is enhanced by a variety of parameters including but not limited to warm temperatures, particular algal and protozoan associations, and symbiotic relationships with certain aquatic plants (Fliermans 1985). Evidence has indicated that amoebae and other protozoa act as natural hosts and amplifiers for *Legionella* in the environment (Barbaree et al. 1986). ASHRAE (1989) has further information on the topic.

Bacteria from the soil are likely to be spore-formers and are capable of surviving in hostile environments. Other airborne bacteria, especially within closed occupied spaces, may originate from droplet nuclei caused by actions such as sneezing or be carried on human or animal skin scales.

Pollen

Pollen grains discharged by weeds, grasses, and trees (Hewson et al. 1967, Jacobson and Morris 1977, Solomon and Mathews 1978) and capable of causing hay fever have properties of special interest to air-cleaning equipment designers (see Chapter 24 of the 2000 *ASHRAE Handbook—Systems and Equipment*). Whole grains and fragments transported by air range between 10 and 50 μm; however, some measure as small as 5 μm, and others measure over 100 μm in diameter. Ragweed pollen grains are fairly uniform in size, ranging from 15 to 25 μm.

Most pollen grains are hygroscopic and, therefore, vary in mass with the humidity. Illustrations and data on pollen grains are available in botanical literature. Geographical distribution of plants that produce hay fever is also recorded.

The quantity of pollen in the air is generally estimated by exposing an adhesive-coated glass plate outdoors for 24 h, then counting calibrated areas under the microscope. Methods are available for determining the number of pollen grains in a measured volume of air. However, despite their greater accuracy, these methods have not replaced the simpler gravity slide method used for most pollen counts. Counting techniques vary, but daily pollen counts reported in local newspapers during hay fever season usually represent the number of grains found on 0.28 in^2 of a 24 h gravity slide.

Hay fever sufferers may experience the first symptoms when the pollen count is 10 to 25; in some localities, maximum figures for the seasonal peak may approach 1000 or more for a 24 h period, depending on the sampling and reporting methods used by the laboratory. Translation of gravity counts by special formulas to a volumetric basis (i.e., number of grains per unit volume of air) is unreliable, due to the complexity of the modifying factors. When such information is important, it should be obtained directly by a volumetric instrument. The number of pollen grains per cubic yard of air varies from 2 to 20 times the number found on 0.16 in^2 of a 24 h gravity slide, depending on grain diameter, shape, specific gravity, wind velocity, humidity, and physical placement of the collecting plate.

Pollen grains can be removed from the air more readily than the dust particles prevalent in outdoor air or those produced by dusty processes, since a larger fraction of them are in the size range easily removed by building HVAC filters.

Whole-grain pollens are easily removed from the outside air entering a ventilation system with medium-efficiency [35 to 45% (MERV 9 to 11)] filters selected to remove 99% of particles 10 µm and greater. Once they have entered a building, whole-grain pollens settle rapidly, reducing concentrations without the need for air cleaning. On resuspension from occupant activities, the whole-grain pollens may disintegrate into fragments, which may possibly be controlled effectively with a high-efficiency [70 to 95% (MERV 12 to 14)] filter capable of removing a high percentage of particles in the 0.3 to 3.0 µm range.

Sampling for Bioaerosols

Sampling for biological agents such as fungi and bacteria can include visual observation of colonies, collection of bulk or surface samples, or air sampling. The principles of sampling and analysis for microorganisms are reviewed by Chatigny (1983). ACGIH (1989) and AIHA (1996) have developed assessment guidelines for the collection of microbiological particulates. An introduction is given below.

Preassessment. Sampling for microorganisms should be undertaken when medical evidence indicates the occurrence of diseases such as humidifier fever, hypersensitivity pneumonitis, allergic asthma, and allergic rhinitis. A walk-through examination of the indoor environment for visual detection of possible microbial reservoirs and amplification sites should be performed before sampling. Note that a visual examination will miss reservoirs that are behind walls. If a reservoir or amplifier is visually identified, it is useful to obtain bulk or source samples from it. Also, removal of clearly identified reservoirs and amplifiers is preferable to complicated and costly air-sampling procedures.

Air Sampling. The same principles that affect the collection of an inert particulate aerosol also govern air sampling for microorganisms. Air sampling is not likely to yield useful data and information unless the sample collected is representative of exposure. The most representative samples are those that are collected in breathing zones over the range of aerosol concentrations. Presently, no personal sampling method has been proposed that is sensitive enough for any bioaerosol (Burge 1995). Thus, ambient sampling designs that obtain reasonable estimates of exposures of given populations over representative periods are necessary. Because *Legionella* requires special nutrients for growth and does not produce resistant spores, this bacterium is difficult to recover from air.

Concentrations of microorganisms in the atmosphere vary from a few to several hundred per cubic foot, depending on many factors. The sampling method for microorganisms has an effect on the measured count. Collection on **dry filter paper** can cause count degradation because of the dehydration loss of some organisms. **Glass impingers** may give high counts because agitation can cause clusters to break up into smaller individual organisms. **Slit samples** may give a more accurate colony count.

The viability and/or antigenicity of the microbial particulates must be protected during sampling. In general, **culture plate impactors**, including multiple- and single-stage devices as well as **slit-to-agar samplers**, are most useful in office environments where low concentrations of bacteria and fungi are expected. Because not all microorganisms will grow on the same media, liquid impingement subculturing may be more suitable. **Filter cassette samplers** are useful for microorganisms or components of microorganisms (i.e., endotoxins), although binding to glass and plastic has been reported (Milton et al. 1990). Filter cassettes can also be used for spore counts. Area sampling is often used. Some investigators attempt to replicate exposure conditions through disturbance of the environment (semiaggressive sampling) such as occurs through walking on carpets, slamming doors, and opening books or file cabinets.

Viruses, many bacteria, algae, and protozoa are more difficult to culture than fungi, and air-sampling methodology for these agents is less well known and defined (ASTM 1990).

Data Interpretation. Rank order assessment is a method used to interpret air-sampling data for microorganisms (ACGIH 1989). Individual organisms are listed in descending order of abundance for a complainant indoor site and for one or more control locations. The predominance of one or more microbes in the complainant site, but not in the control sites or outdoors, suggests the presence of an amplifier for that organism. In the example in Table 3, *Tritirachium* and *Aspergillus* were the predominant fungi represented in complainant locations in an office building, where *Cladosporium* and *Fusarium* dominated outdoor collections. In this case, *Tritirachium* and *Aspergillus* were being amplified in the building. In addition to comparing individual organisms, indoor-outdoor ratios of overall quantities of culturable microorganisms are useful.

Control of Bioaerosols

When maximum removal of airborne microorganisms is either necessary or desirable, high-efficiency particulate air (HEPA) or ultralow penetration air (ULPA) filters are used. These filters create essentially sterile atmospheres and are more frequently employed than chemical scrubbers and ultraviolet radiation for control of airborne microorganisms. They have been used to prevent cross-infection in hospitals, to protect clean rooms from contamination, and to assemble and launch space probes under sterile conditions.

Table 3 Example Case of Airborne Fungi in Building and in Outdoor Air

Location	cfu/m^3	Rank Order Taxa
Outdoors	210	*Cladosporium* > *Fusarium* > *Epicoccum* > *Aspergillus*
Complainant Office #1	2500	*Tritirachium* > *Aspergillus* > *Cladosporium*
Complainant Office #2	3000	*Tritirachium* > *Aspergillus* > *Cladosporium*

Notes:
cfu/m^3 = Colony-forming units per cubic metre of air.
Culture media was malt extract agar (ACGIH 1989).

In many situations, total control of airborne microorganisms is not required. For these applications, there are various other types of high-efficiency dry media extended-surface filters that will provide the necessary efficiency. These filters have lower pressure differentials than HEPA filters operating at the same face velocity and, when properly selected, will remove the contaminants of concern.

GASEOUS CONTAMINANTS

The terms gas and vapor are both used to describe the gaseous state of a substance. **Gas** is the correct term for describing any pure substance or mixture that naturally exists in the gaseous state at normal atmospheric conditions. Examples are oxygen, helium, ammonia, and nitrogen. **Vapor** is used to describe a substance in the gaseous state whose natural state is a liquid or solid at normal atmospheric conditions. Examples include benzene, carbon tetrachloride, and water. There are differences between the two classes that reflect their preferred states:

- The concentration of gases in air is limited only by the strength of the source, so that gases can completely fill a space, driving out the oxygen necessary for survival.
- Vapors can never exceed their saturated vapor pressure in air. The most familiar example of a vapor is water, with relative humidity expressing the air concentration as a percentage of the saturated vapor pressure.
- Vapors, because their natural state is liquid or solid, have a tendency to condense on surfaces and be adsorbed.

Gaseous contaminants can also be divided into organic and inorganic types. **Organic** compounds include all chemicals based on a skeleton of carbon atoms. Since carbon atoms easily combine to form chain, branched, and ring structures, there is a bewildering variety of organic compounds. They include gases such as methane, but the majority are vapors. All other gaseous contaminants are classified as **inorganic**. Most of the inorganic air contaminants of interest to engineers are gases.

Major chemical families of gaseous pollutants and examples of specific compounds are shown in Table 4. The *Merck Index* (Budavi 1996), the *Toxic Substances Control Act Chemical Substance Inventory* (EPA 1979), and *Dangerous Properties of Industrial Materials* (Sax and Lewis 1988) are all useful in identifying contaminants,

Table 4 Major Chemical Families of Gaseous Air Pollutants (with Examples)

Inorganic Pollutants

1. Single-element atoms and molecules
   - chlorine
   - radon
   - mercury
2. Oxidants
   - ozone
   - nitrogen dioxide
3. Reducing agents
   - carbon monoxide
4. Acid gases
   - carbon dioxide
   - sulfur dioxide
   - sulfuric acid
   - hydrochloric acid
   - hydrogen sulfide
   - nitric acid
5. Nitrogen compounds
   - ammonia
   - hydrazine
   - nitrous oxide
6. Miscellaneous
   - arsine

Organic Pollutants

7. *n*-Alkanes
   - methane
   - *n*-butane
   - *n*-hexane
   - *n*-octane
   - *n*-hexadecane
8. Branched alkanes
   - 2-methyl pentane
   - 2-methyl hexane
9. Alkenes and cyclic hydrocarbons
   - butadiene
   - 1-octene
   - cyclohexane
   - 4-phenyl cyclohexene (4-PC)
10. Chlorofluorocarbons
    - R-11 (trichlorofluoromethane)
    - R-114 (dichlorotetrafluoroethane)
11. Chlorinated hydrocarbons
    - carbon tetrachloride
    - chloroform
    - 1,1,1-trichloroethane
    - tetrachloroethylene
    - dichlorobenzene
12. Halide compounds
    - methyl bromide
    - methyl iodide
13. Alcohols
    - methanol
    - ethanol
    - 2-propanol (isopropanol)
    - phenol
    - ethylene glycol
14. Ethers
    - ethyl ether
    - methoxyvinyl ether
    - *n*-butoxyethanol
15. Aldehydes
    - formaldehyde
    - acetaldehyde
    - acrolein
    - benzaldehyde
16. Ketones
    - 2-propanone (acetone)
    - 2-butanone (MEK)
    - methyl isobutyl ketone (MIBK)
    - chloroacetophenone
17. Esters
    - ethyl acetate
    - vinyl acetate
    - methyl formate
    - dioctyl phthalate (DOP)
18. Nitrogen compounds other than amines
    - nitromethane
    - acetonitrile
    - acrylonitrile
    - urea
    - uric acid
    - skatole
    - putrescine
    - hydrogen cyanide
    - peroxyacetal nitrite (PAN)
19. Aromatic hydrocarbons
    - benzene
    - toluene
    - *p*-xylene
    - naphthalene
    - benz-α-pyrene
20. Terpenes
    - 2-pinene
    - limonene
21. Heterocyclics
    - ethylene oxide
    - tetrahydrofuran
    - 1, 4-dioxane
    - pyrrole
    - pyridine
    - nicotine
    - caffeine
22. Organophosphates
    - malathion
    - tabun
    - sarin
    - soman
23. Amines
    - trimethylamine
    - ethanolamine
    - cyclohexylamine
    - morpholine
24. Monomers
    - vinyl chloride
    - methyl formate
    - ethylene
    - methyl methacrylate
25. Mercaptans and other sulfur compounds
    - bis-2-chloroethyl sulfide (mustard gas)
    - ethyl mercaptan
    - carbon disulfide
    - carbonyl sulfide
26. Organic acids
    - formic acid
    - acetic acid
    - butyric acid
27. Miscellaneous
    - phosgene

including some known by trade names only. Chemical and physical properties can be found in reference books such as the *Handbook of Chemistry and Physics* (Lide 1996). Note that a single chemical compound, especially if it is an organic compound, may have several scientific names. To reduce confusion, the Chemical Abstracts Service (CAS) has assigned each chemical a unique identifier number containing five to nine digits.

Harmful Effects of Gaseous Contaminants

Harmful effects may be divided into four categories: toxicity, odor, irritation, and material damage.

Toxicity. The harmful effects of gaseous pollutants on a person depend on both short-term peak concentrations and the time-integrated exposure received by the person. Toxic effects are generally considered to be proportional to the exposure dose, although individual response variation can obscure the relationship. The allowable concentration for short exposures is higher than that for long exposures. Safe exposure limits have been set for a number of common gaseous contaminants in industrial settings. This topic is covered in more detail in the section on Industrial Air Contaminants and in Chapter 9.

Irritation. Although gaseous pollutants may have no discernible continuing health effects, exposure may cause physical irritation to building occupants. This phenomenon has been studied principally in laboratories and nonindustrial work environments and is discussed in more detail in the section on Nonindustrial Indoor Air Contaminants and in Chapter 9.

Odors. Gaseous contaminant problems often appear as complaints about odors, and these usually are the result of concentrations considerably below industrial exposure limits. Odors are discussed in more detail in Chapter 13.

Damage to Materials. Material damage from gaseous pollutants may take such forms as corrosion, embrittlement, or discoloration. Because such effects usually involve chemical reactions that need water, material damage from air pollutants is less severe in the relatively dry indoor environment than outdoors, even at similar gaseous contaminant concentrations. To maintain this advantage, indoor condensation should be avoided. However, some dry materials can be significantly damaged. These effects are most serious in museums, as any loss of color or texture changes the essence of the object. Libraries and archives are also vulnerable, as are pipe organs and textiles. Ventilation is often a poor method of protecting collections of rare objects because these facilities are usually located in the centers of cities that have relatively polluted ambient outdoor air. Gaseous compounds known to be harmful include ozone, nitrous oxides, sulfur dioxide, hydrochloric acid, many volatile organic compounds, and hydrogen sulfide. Various concerns about material damage by indoor air pollutants are discussed by Lull (1995), Walsh et al. (1977), Chiarenzelli and Joba (1966), NTIS (1982, 1984), Braun and Wilson (1970), Jaffe (1967), Grosjean et al. (1987), American Guild of Organists (1966), Mathey et al. (1983), Haynie (1978), Graminski et al. (1978), and Thomson (1986).

Unlike the human body, conservation items have no repair mechanism, yet they are expected to have a longer life than people. This is one reason the acceptable levels of contaminants in museums, archives, and other conservation applications are very low—1 to 2 orders of magnitude below U.S. EPA standards (Mathey et al. 1983; Thomson 1986). Achieving these levels and measuring performance is difficult. Lull (1995) discusses ventilation and air cleaning specifically for conservation environments.

Units of Measurement

Concentrations of gaseous contaminants are usually expressed in the following units:

ppm = parts of contaminant by volume per million parts of air by volume

ppb = parts of contaminant by volume per billion parts of air by volume

1000 ppb = 1 ppm

mg/m^3 = milligrams of contaminant per cubic metre of air

$\mu g/m^3$ = micrograms of contaminant per cubic metre of air

The conversions between ppm and mg/m^3 are

$$\text{ppm} = [0.6699(459.7 + t)/Mp]\ (\text{mg/m}^3) \tag{1}$$

$$\text{mg/m}^3 = [1.493(Mp)/(459.7 + t)]\ (\text{ppm}) \tag{2}$$

where

M = relative molecular mass of contaminant
p = mixture pressure, psia
t = mixture temperature, °F

Concentration data are often reduced to standard temperature and pressure (i.e., 25°C and 101.325 kPa), in which case

$$\text{ppm} = (24.46/M)\ (\text{mg/m}^3) \tag{3}$$

Measurement of Gaseous Contaminants

The concentration of contaminants in the air needs to be measured to determine whether the indoor air quality conforms to occupational health standards (in industrial environments) and is acceptable (in nonindustrial environments).

Measurement methods for airborne chemicals that are important industrially have been published by several organizations including OSHA (1995) and NIOSH (1994). Methods typically involve sampling air with pumps for several hours to capture the contaminant on a filter or in an adsorbent tube, followed by laboratory analysis for detection and determination of contaminant concentration. Concentrations measured in this way can usefully be compared to 8 h industrial exposure limits.

The measurement of gaseous contaminants at the lower levels acceptable for indoor air is not always as straightforward. Relatively costly analytical equipment may be needed, and it must be calibrated and operated by experienced personnel.

Currently available sampling techniques are listed in Table 5 with details of their advantages and disadvantages. Analytical techniques are shown in Table 6 with information on the types of contaminants to which they apply.

The first two techniques in Table 5 combine sampling and analysis in one piece of equipment and give immediate results on-site. The other sampling methods require laboratory analysis following the field work. Equipment employing the first technique in Table 5 can be coupled with a data logger to perform continuous monitoring and to obtain average concentrations over a time period. Most of the sampling techniques can capture several contaminants. Several allow pollutants to be accumulated or concentrated over time so that very low concentrations can be measured.

Some analytical techniques are specific for a single pollutant, while others are capable of providing concentrations for many contaminants simultaneously. Note that formaldehyde requires different measurement methods from other volatile organic compounds.

Measurement instruments used in industrial situations should be able to detect contaminants of interest at about one-tenth of threshold limit value (TLV) levels. If odors are of concern, detection sensitivity must be at odor threshold levels. Procedures for evaluating odor levels are given in Chapter 13.

When sampling and analytical procedures appropriate to the application have been selected, a pattern of sampling locations and times must be carefully planned. The building and air-handling system layout and the space occupancy and use patterns must be considered so that representative concentrations will be measured.

Table 5 Gaseous Contaminant Sampling Techniques

Technique[a]	Advantages	Disadvantages
1. Direct flow to detectors	Real-time readout, continuous monitoring possible Several pollutants possible with one sample (when coupled with chromatograph, spectroscope, or multiple detectors)	Average concentration must be determined by integration No preconcentration possible before detector; sensitivity may be inadequate On-site equipment complicated, expensive, intrusive
2. Colorimetric detector tubes	Very simple, relatively inexpensive equipment and materials Immediate readout Integration over time	Rather long sampling period normally required One pollutant per sample Relatively high detection limit Poor precision
3. Capture by pumped flow through solid absorbent; subsequent desorption for concentration measurement	On-site sampling equipment relatively simple and inexpensive Preconcentration and integration over time inherent in method Several pollutants possible with one sample	Sampling media and desorption techniques are compound-specific Interaction between captured compounds and between compounds and sampling media; bias may result Gives only average over sampling period, no peaks
4. Collection in evacuated containers	Very simple on-site equipment No pump (silent) Several pollutants possible with one sample	Gives only average over sampling period, no peaks
5. Collection in nonrigid containers (plastic bags) held in an evacuated box	Simple, inexpensive on-site equipment (pumps required) Several pollutants possible with one sample	Cannot hold same pollutants
6. Cryogenic condensation	Wide variety of organic pollutants can be captured Minimal problems with interferences and media interaction Several pollutants possible with one sample	Water vapor interference
7. Passive diffusional samplers	Simple, unobtrusive, inexpensive No pumps; mobile; may be worn by occupants to determine average exposure	Gives only average over sampling period, no peaks
8. Liquid impingers (bubblers)	Integration over time Several pollutants possible with one sample	May be noisy

Sources: NIOSH (1977, 1994), Lodge (1988), Taylor et al. (1977), and ATC (1990).

[a]All techniques except 1 and 2 require laboratory work after completion of field sampling. Only technique 1 is adaptable to continuous monitoring and able to detect short-term excursions.

Traynor (1987) and Nagda and Rector (1983) offer guidance in planning such surveys.

VOLATILE ORGANIC COMPOUNDS

The entire range of organic indoor pollutants has been categorized as indicated in Table 7 (WHO 1989). No sharp limits exist between the categories, which are defined by boiling-point ranges. **Volatile organic compounds** (VOCs) have attracted considerable attention in nonindustrial environments. They have boiling points in the range of approximately 120 to 480°F and vapor pressures greater than about 4×10^{-5} to 4×10^{-6} in. Hg.

Sources of VOCs include solvents, reagents, and degreasers in industrial environments; and furniture, furnishings, wall and floor finishes, cleaning and maintenance products, and office and hobby activities in nonindustrial environments.

Berglund et al. (1988) found that the sources of VOCs in nonindustrial indoor environments are confounded by the variable nature of emissions from potential sources. Emissions of VOCs from indoor sources can be classified by their presence and rate patterns. For example, emissions are continuous and regular from building materials and furnishings (e.g., carpet and composite-wood furniture); whereas emissions from other sources can be continuous but irregular (e.g., paints used in renovation work), intermittent and regular (e.g., VOCs in combustion products from gas stoves or cleaning products), or intermittent and irregular (e.g., VOCs from carpet shampoos) (Morey and Singh 1991).

Many "wet" emission sources (paints and adhesives) have very high emission rates immediately after application, and rates drop quite steeply with time until the product has cured or dried. New "dry" materials (carpets, wall coverings, and furnishings) also emit chemicals at higher rates until aged. Decay of these elevated VOC concentrations to what can be considered normal constant-source levels can take weeks to months, depending on emission rates, surface areas of materials, and ventilation protocols. Renovation activities can cause similar increases of somewhat lower magnitude. The total VOC concentration in new office buildings at the time of initial occupancy can be 50 to 100 times that present in outdoor air (Sheldon et al. 1988a, 1988b). In new office buildings with adequate outdoor air ventilation, these ratios often fall to less than 5:1 after 4 or 5 months of aging. In older buildings with continuous, regular, and irregular emission sources, indoor/outdoor ratios of total VOCs may vary from nearly 1:1 when maximum amounts of outdoor air are being used in HVAC systems to greater than 10:1 during winter and summer months when minimum amounts of outdoor air are being used (Morey and Jenkins 1989, Morey and Singh 1991).

While direct VOC emissions are usually the primary source of VOCs in a space, some materials act as sinks for emissions and then become secondary sources as they reemit adsorbed chemicals (Berglund et al. 1988). Adsorption may lower the peak concentrations achieved, but the subsequent desorption prolongs the presence of indoor air pollutants. Sink materials include carpet, fabric partitions, and other fleecy materials, as well as ceiling tiles and wallboard. The type of material and compound affects the rate of adsorption and desorption (Colombo et al. 1991). Indoor air quality models using empirically derived adsorption and desorption rates have been developed to predict the behavior of sinks. Experiments conducted in an IAQ test house confirmed the importance of sinks when trying to control the level of indoor VOCs (Tichenor et al. 1991). Longer periods of increased ventilation lessen sink and reemission effects.

Because of the large number of VOCs usually found indoors and the impossibility of identifying all of them in samples, the concept of total VOC or TVOC was developed. Some researchers have used it to represent the sum of all detected VOCs. TVOC concentrations can also be determined by using methods such as photoionization

Table 6 Gaseous Contaminant Concentration Measurement Methods

Method	Description	Typical Application
1. Gas chromatography (Using the following detectors)	Separation of gas mixtures by time of passage down an absorption column	
Flame ionization	Change in flame electrical resistance due to ions of pollutant	Volatile, nonpolar organics
Flame photometry	Measures light produced when pollutant is ionized by a flame	Sulfur, phosphorous compounds
Photoionization	Measures ion current for ions created by ultraviolet light	Most organics (except methane)
Electronic capture	Radioactively generated electrons attach to pollutant atoms; current measured	Halogenated organics Nitrogenated organics
Mass spectroscopy	Pollutant molecules are charged, passed through electrostatic magnetic fields in a vacuum; path curvature depends on mass of molecule, allowing separation and counting of each type	Volatile organics
2. Infrared spectroscopy and Fourier transform infrared (FTIR) spectroscopy	Absorption of infrared light by pollutant gas in a transmission cell; a range of wavelengths is used, allowing identification and measurement of individual pollutants	Acid gases Many organics
3. High-performance liquid chromatography (HPLC)	Pollutant is captured in a liquid, which is then passed through a liquid chromatograph (analogous to a gas chromatograph)	Aldehydes, ketones Phosgene Nitrosamines Cresol, phenol
4. Colorimetry	Chemical reaction with pollutant in solution yields a colored product whose light absorption is measured	Ozone Oxides of nitrogen Formaldehyde
5. Fluorescence and pulsed fluorescence	Pollutant atoms are stimulated by a monochromatic light beam, often ultraviolet; they emit light at characteristic fluorescent wavelengths, whose intensity is measured	Sulfur dioxide Carbon monoxide
6. Chemiluminescence	Reaction (usually with a specific injected gas) results in photon emission proportional to concentration	Ozone Nitrogen compounds Several organics
7. Electrochemical	Pollutant is bubbled through reagent/water solution, changing its conductivity or generating a voltage	Ozone Hydrogen sulfide Acid gases
8. Titration	Pollutant is absorbed into water	Acid gases
9. Ultraviolet absorption	Absorption of UV light by a cell through which the polluted air passes	Ozone Aromatics Sulfur dioxide Oxides of nitrogen Carbon monoxide
10. Atomic absorption	Contaminant is burned in a hydrogen flame; a light beam with a spectral line specific to the pollutant is passed through the flame; optical absorption of the beam is measured	Mercury vapor

Sources: NIOSH (1977,1994), Lodge (1988), Taylor et al (1977), and ATC (1990).

and flame ionization that do not separate chemicals. All methods for TVOC determination are intrinsically of low to moderate accuracy due to variations in detector response to different classes of VOCs. Both theoretical and practical limitations of the TVOC approach have been discussed (Hodgson 1995, Otson and Fellin 1993).

Wallace et al. (1991) showed that individual VOC concentrations in homes and buildings are 2 to 5 times those of outdoors, and personal TVOC exposures were estimated to be 2 to 3 times greater than general indoor air concentrations. Normal daily activities carried out by individuals are the cause of these higher personal exposures. Personal activities frequently bring individuals close to air contaminant sources. The degree of exposure also depends on how air flows around the body due to convective forces, air turbulence, and obstructions nearby (Rodes et al. 1991).

Individual organic compounds seldom exceed 0.05 mg/m^3 (50 μg/m^3) in air. An upper extreme average concentration of TVOCs in normally occupied houses is approximately 20 mg/m^3. The Large Buildings Study by the U.S. EPA (Brightman et al. 1996) developed the VOC sample target list shown in Table 8 to identify common VOCs that should be measured. Lists of common indoor VOCs prepared by other organizations are similar.

A VOC enrichment factor (VEF) has been described to quantify VOC concentrations due to building sources beyond those expected as bioeffluents (Batterman and Peng 1995). A VEF > 1 describes the overabundance, while a VEF < 1 denotes the depletion of VOCs compared to VOCs expected as bioeffluents. Bioeffluents alone should produce a VEF of 1. The VEF ranged from 0.6 to 17.1 for the 20 office building studies reviewed.

Because chlorofluorocarbons (CFCs) are hydrocarbons with some hydrogen atoms replaced by chlorine, bromine, and fluorine atoms, they are classed as organic chemicals. They have been widely used as heat transfer gases in refrigeration applications, blowing agents, and propellants in aerosol products (including medications and consumer products) and as expanders in plastic foams. CFCs are discussed in Chapter 19 and in ASHRAE *Standards* 15 and 34. During the 1970s and 1980s, strong evidence was found linking use of CFCs to depletion of the earth's stratospheric ozone layer. CFCs add chlorine and bromine atoms to the atmosphere, which accelerate the natural ozone destruction rate. Ozone deple-

Table 7 Classification of Indoor Organic Pollutants

Description	Abbreviation	Boiling Point Range, °F
Very volatile (gaseous) organic compounds	VVOC	<32 to 120–212
Volatile organic compounds	VOC	120–212 to 460–500
Semivolatile organics (pesticides, polynuclear aromatic compounds, plasticizers)	SVOC	460–500 to 720–750

Sources: WHO (1989).
Note: Polar compounds and higher molecular weight VOCs with appear at the higher end of each boiling-point range.

tion is linked to increases in the amount of ultraviolet radiation (UV-B) reaching the earth's surface. Increasing UV-B can adversely affect human health, most notably by causing malignant melanoma and cataracts, harm the environment, and decrease crop production. As a result, CFCs are now being phased out. New refrigerants with lower ozone reduction potential are being used, including hydrochlorofluorocarbons (HCFCs). In addition, there has been a return to refrigerants such as ammonia and propellants such as butane and hexane.

Exposure to CFCs and HCFCs occurs mainly through inhalation and can occur from leaks in refrigeration equipment or during servicing of HVAC systems.

Controlling Exposures to Volatile Organic Compounds

Much can be done to reduce building occupants' exposures to emissions of VOCs from building materials and products and to prevent outdoor VOCs from being brought into buildings. The control principles of substitution, isolation, and ventilation apply. Control measures include careful planning; specifications; and selection, modification, and treatment of products, as well as special installation procedures and proper ventilation system operation. Chapter 44 of the 1999 *ASHRAE Handbook —Applications* provides full details.

Levin (1989, 1991) has written extensively about designing new buildings for good indoor air quality. Reducing VOC emissions by careful selection and installation of building materials and furnishings is a most effective strategy for controlling IAQ. Advances in product formulation and emission testing are leading to products claimed to be low-polluting, nontoxic, and environmentally safe. Requiring the submission of emission testing data by manufacturers for building products, whether for a new building, for a building renovation or remodeling, or for substitution of a consumable product (housekeeping supplies), is becoming accepted practice. Eliminating the sources of VOCs prevents them from becoming a problem in the first place.

Gas-phase air filtration has been applied to control industrial gaseous contaminants for many years. The application of this technology to nonindustrial building HVAC is of interest for improving IAQ, whether it is to provide ventilation without the need to use more outdoor air or to help clean poor-quality outdoor air. Activated carbon and potassium permanganate-impregnated alumina are effective adsorbents that can be used, based on the contaminant mixture present (Liu and Huza 1995, Muller and England 1995, VanOsdell and Sparks 1995). Portable air cleaners with sorbent sections are only marginally effective (Shaughnessy et al. 1994). Photocatalytic reactors capable of destroying VOCs are being studied (Peral et al. 1997). These reactors use ultraviolet light and a catalytic surface, such as titanium dioxide, to convert organic pollutants to CO_2 and water.

Ventilation has traditionally been considered the primary means for controlling indoor VOC contaminants. Dilution ventilation is an effective way to control normal constant-emission sources present in buildings, assuming no unusually strong sources. Compliance

Table 8 Example Sample Contaminant Target List

benzene	styrene
m-,*p*-xylene	*p*-dichlorobenzene
1,2,4-trimethylbenzene	*n*-undecane
n-octane	*n*-nonane
n-decane	ethylacetate
n-dodecane	dichloromethane
butylacetate	1,1,1-trichloroethane
chloroform	tetrachloroethylene
trichloroethylene	carbon disulfide
trichlorofluoromethane	acetone
dimethyl disulfide	2-butanone
4-methyl-2-pentanone	methyl tertiary butyl ether
limonene	naphthalene
α-,β-pinene	4-phenylcyclohexene
propane	butane
butyl cellosolve	ethanol
isopropanol	phenol
formaldehyde	siloxanes
toluene	

with ASHRAE *Standard* 62 should satisfy indoor dilution ventilation requirements. Local exhaust ventilation is effective for controlling known, unavoidable point emissions sources. It is prudent to isolate office machines, such as photocopiers and laser printers, food service equipment, such as microwave ovens and coffee makers, and work areas, such as graphics and photographic labs, using dedicated local exhaust systems that vent to the outside and away from outdoor air intakes.

A good ventilation protocol during renovation or remodeling includes using a single-pass (100% outdoor air) system during and at the finish of these activities, continuing until enough time has passed to lower emitted concentrations to near background. This practice minimizes sink effects and secondary emissions.

Prudent practice and administrative control should be used to minimize the generation of VOCs in indoor air during occupied hours whenever possible. Scheduling the use of volatile organic products, housekeeping activities, and pesticide application when occupant density is lowest should be considered. VOC-containing supplies should be stored in well-ventilated areas other than HVAC mechanical rooms or plenums.

INORGANIC GASES

Several inorganic gases are of concern because of their effects on human health and comfort and on materials. These include carbon dioxide, carbon monoxide, oxides of nitrogen, sulfur dioxide, ozone, and ammonia. Most have both outdoor and indoor sources.

Carbon dioxide (CO_2) or carbonic acid gas is produced by human respiration. It is not normally considered to be a toxic air contaminant, but it can be a simple asphyxiant (by oxygen displacement) in confined spaces such as submarines. CO_2 is found in the ambient environment at 330 to 370 ppm. Levels in the urban environment may be higher due to emissions from gasoline and, more often, diesel engines. Measurement of CO_2 in occupied spaces has been widely used to evaluate the amount of outdoor air supplied to indoor spaces. In ASHRAE *Standard* 62, a level of 1000 ppm (or 650 ppm above outdoor air) has been suggested as being representative of delivery rates of 15 cfm per person of outside air when CO_2 is measured at equilibrium concentrations and at occupant densities of 10 people per 1000 ft^2 of floor space. Measuring CO_2 level before it has reached steady-state conditions can lead to inaccurate conclusions regarding the amount of outside air used in the building.

Carbon monoxide (CO) is an odorless, colorless, and tasteless gas produced by the incomplete combustion of hydrocarbons. It is a common ambient air pollutant and is very toxic. Common indoor sources of CO include gas stoves, kerosene lanterns and heaters, mainstream and sidestream tobacco smoke, woodstoves, and unvented or improperly vented combustion sources. Building makeup air intakes located at street level or near parking garages can entrain CO from automobiles and carry it to the indoor environment.

The major predictors of indoor CO concentrations are indoor fossil fuel sources, such as gas furnaces, hot water heaters, and other combustion appliances; attached garages; and weather inversions. Levels in homes only rarely exceed 5 ppm. In one sample of randomly selected homes, 10% failed a backdrafting test (Conibear et al. 1996). Under backdrafting conditions, indoor CO sources may contribute to much higher, dangerous levels of CO.

Oxides of nitrogen (NO_x) indoors result mainly from cooking appliances, pilot lights, and unvented heaters. Sources generating carbon monoxide (CO) often produce nitric oxide (NO) as well. Underground or attached parking garages can also contribute to indoor concentrations of NO_x. An unvented gas cookstove contributes approximately 0.025 ppm of nitrogen dioxide (NO_2) to a home. During cooking, 0.2 ppm to 0.4 ppm peak levels may be reached (Samet et al. 1987). Ambient air pollution from vehicle exhausts in urban locations can contribute NO_x to the indoor environment in makeup air. Oxides of nitrogen also are present in mainstream and sidestream tobacco smoke. Nitric oxide and nitrogen dioxide are of most concern.

Sulfur dioxide (SO_2) can result from the emissions of kerosene space heaters; the combustion of fossil fuels such as coal, heating oil, and gasoline; or burning any material containing sulfur. As a result, sulfur dioxide is a common ambient air pollutant in many urban areas.

Ozone (O_3) is a photochemical oxidant that forms at ground level when hydrocarbons and oxides of nitrogen react with ultraviolet radiation in sunlight to produce photochemical smog. Ozone can be emitted by the electrical or coronal discharges from office equipment including laser printers and photocopiers. It can also be generated when ozone-generating devices—often marketed as portable air cleaners and ionizers—are used in the indoor environment (Esswein and Boeniger 1994).

Controlling Exposures to Inorganic Gases

Three methods of control for inorganic gaseous contaminants should be considered: source control, ventilation control, and removal by filters. **Source control** involves limiting (or removing) the source of the problem; for example, gas cookstoves should not be used for space heating (often a problem in low-income urban residences). Another example is limiting automobile parking around building makeup air intakes. Source control should always be the primary consideration. But source control is not always feasible when there are many diverse contaminant sources as in new buildings where the building itself or building furnishings may be the prime contributors to the problem.

Ventilation control involves bringing clean dilution air into the occupied space or directly exhausting air contaminants at the point of generation. ASHRAE *Standard* 62 provides guidance in applying the Ventilation Rate Procedure and the Indoor Air Quality Procedure for ventilation control.

Where neither source control nor ventilation control appear likely to control gaseous air contaminants, **air filtration** should be investigated. Gas-phase air filtration involves dry scrubbing to remove contaminants by adsorption onto several sorbents, including granular activated carbon (GAC), potassium permanganate-impregnated alumina (PIA), and alkaline-impregnated carbon filters. Muller and England (1995), Liu and Huza (1995), VanOsdell (1995), and Coutant et al. (1994) review various filtration procedures.

No one media is effective for the broad range of gaseous contaminants found indoors. Granular activated charcoal is generally an agent of choice for nonpolar compounds and is a suitable choice for O_3 and NO_2 but not for SO_x and NO. Permanganate-impregnated alumina is more appropriate for SO_x and NO.

Ozone can be best controlled by local exhaust ventilation for demonstrated sources of ozone, such as photocopiers and equipment creating coronal discharges. Routine cleaning of attractor plates in an electrostatic precipitator and ensuring adequate prefilters can reduce ozone generation and limit arcing in this type of particle removal equipment. The use of ozone-generating devices as a means of air cleaning or purification has not been documented as a prudent means of air contaminant control considering the potential health effects of the use of ozone indoors (Esswein and Boeniger 1994). The Food and Drug Administration (FDA 1990) specifically limits the use of ozone in concentrations greater than 50 ppb in areas intended for continuous occupancy, such as residences, offices, schools, and hospitals.

Carbon monoxide exposure control strategies primarily involve identification and control of CO emissions directly at their source. Local exhaust ventilation is an appropriate and effective control in most cases. For example, automobile repair garages commonly use a tailpipe exhaust extension to control the CO exposure of the mechanics working in the repair bays. Relocating building makeup air intakes or limiting vehicle access are reasonable means to prevent entrainment of automobile exhausts into building HVAC systems.

Carbon monoxide, however, is a common pollutant of ambient air. As a result, direct control by dilution may not be feasible if ambient air is heavily contaminated with CO. Diesel or natural gas may be substituted for gasoline engines to reduce CO where specific sources from engine exhaust are identified or are a concern. Adequate venting of any combustion sources is critical to prevent the buildup of CO indoors. CO may be monitored by a properly calibrated, direct-reading CO monitor, colorimetric indicator tubes, or passive diffusion sampling badges.

Exposure controls for carbon dioxide are generally limited to situations where exposure concentrations are expected to exceed 3 to 5%. CO_2 is not encountered at levels harmful to humans in the ambient environment. It is normally present at 350 to 375 ppm, and slightly higher in congested cities. With the exception of an intentional or accidental CO_2 "dump" from a fire suppression system or in a dry ice manufacturing facility, CO_2 is not encountered in significant concentrations that require specific engineering controls. However, CO_2 is denser than air and can persist for some time in low areas such as trenches, depressions, and pits. This characteristic creates a simple asphyxiation hazard because CO_2 displaces oxygen.

SPECIAL TYPES OF AIR CONTAMINANTS

OUTDOOR AIR CONTAMINANTS

The total amount of suspended particulate matter in the atmosphere can influence the loading rate of air filters and their selection. The amount of soot that falls in U.S. cities ranges from 7 to 70 Mg/km^2 per month. Soot fall data indicate effectiveness of smoke abatement and proper combustion methods and serve as comparative indices of such control programs. However, the data are of limited value to the ventilating and air-conditioning engineer, since they do not accurately represent airborne soot concentrations.

Concentrations of outdoor pollutants are important, as they may determine indoor concentrations in the absence of indoor sources. Table 9 presents typical urban outdoor concentrations of some com-

Table 9 Typical Outdoor Concentrations of Selected Gaseous Air Pollutants

Pollutant	Typical Concentration, µg/m³	Pollutant	Typical Concentration, µg/m³
Acetaldehyde	20	Methylene chloride	2.4
Acetone	3	Nitric acid	6
Ammonia	1.2	Nitric oxide	10
Benzene	8	Nitrogen dioxide	51
2-Butanone (MEK)	0.3	Ozone	40
Carbon dioxide	612,000 [a]	Phenol	20
Carbon monoxide	3,000	Propane	18
Carbon disulfide	310	Sulfur dioxide	240
Carbon tetrachloride	2	Sulfuric acid	6
Chloroform	1	Tetrachloroethylene	2.5
Ethylene dichloride	10	Toluene	20
Formaldehyde	20	1,1,1-Trichloroethane	4
n-Heptane	29	Trichloroethylene	15
Mercury (vapor)	0.005	Vinyl chloride monomer	0.8
Methane	1,100		
Methyl chloride	9	Xylene	10

[a]Normal concentration of carbon dioxide in air. The concentration in occupied spaces should be maintained at no greater than three times this level (1000 ppm).
Sources: Braman and Shelley (1980), Casserly and O'Hara (1987), Chan et al. (1990), Cohen et al. (1989), Coy (1987), Fung and Wright (1990), Hakov et al. (1987), Hartwell et al. (1985), Hollowell et al. (1982), Lonnemann et al. (1974), McGrath and Stele (1987), Nelson et al. (1987), Sandalls and Penkett (1977), Shah and Singh (1988), Singh et al. (1981), Wallace et al. (1983), and Weschler and Shields (1989).

Table 10 Primary Ambient Air Quality Standards for the United States

	Long Term		Short Term	
Contaminant	Concentration, µg/m³	Averaging Period	Concentration, µg/m³	Averaging Period, h
Sulfur dioxide	80	1 year	365	24
Carbon monoxide			10,000	1
			40,000	8
Nitrogen dioxide	100	1 year		
Ozone[a]			235	1
Hydrocarbons				
Total particulate (PM_{10})[b]	75	1 year	260	24
Lead particulate	1.5	3 months		

[a]Standard is met when the number of days per year with maximum hour-period concentration above 235 µg/m³ is less than one.
[b]PM_{10} = particulates below 10 µm diameter.

mon gaseous pollutants. Higher levels might be found if the building under consideration were located near a major source of contamination such as a power plant, a refinery, or a sewage treatment plant. Note that levels of sulfur dioxide and nitrogen dioxide, which are often attached to particles, may be reduced by about half by building filtration systems. Also, ozone is a reactive gas that can be significantly reduced by contact with ventilation system components (Weschler et al. 1989).

The U.S. Environmental Protection Agency has identified several important outdoor contaminants as **criteria pollutants**. The list includes suspended particulate matter, lead particulate matter, ozone, nitrogen dioxide, sulfur dioxide, carbon monoxide, and total hydrocarbons. Standards have been set for these contaminants, as shown in Table 10, and levels measured at a large number of locations in the United States are published by the EPA each year (40 CFR 50).

Daily concentrations of VOCs in outdoor air can vary drastically (Ekberg 1994). These variations are due to vehicle traffic density, wind direction, industrial emissions, and photochemical reactions.

Table 11 Characteristics of Selected Gaseous Air Pollutants

Pollutant	Chemical and Physical Properties: Family[a]	BP[b], °F	M[c]
Acetaldehyde	15	70	44
Acetone	16	133	58
Acetonitrile	18	180	41
Acrolein	15	126	56
Acrylonitrile	18	171	53
Allyl chloride	12	111	77
Ammonia	5	–27	17
Benzene	19	176	78
Benzyl chloride	12	354	127
2-Butanone (MEK)	16	174	72
Carbon dioxide	4	–108	44
Carbon monoxide	3	–314	28
Carbon disulfide	25	115	76
Carbon tetrachloride	11	171	154
Chlorine	1	–29	71
Chloroform	11	255	119
Chloroprene	12	248	89
p-Cresol	13	581	108
Dichlorodifluoromethane	10	–22	121
Dioxane	21	212	68
Ethylene dibromide	12	268	188
Ethylene dichloride	12	183	99
Ethylene oxide	21	50	44
Formaldehyde	15	207	30
n-Heptane	7	208	100
Hydrogen chloride	4	–186	37
Hydrogen cyanide	18	79	27
Hydrogen fluoride	4	66	20
Hydrogen sulfide	4	–76	34
Mercury	1	675	201
Methane	7	–263	16
Methanol	13	147	32
Methyl chloride	12	165	133
Methylene chloride	12	104	85
Nitric acid	4	183	63
Nitric oxide	2	–242	30
Nitrogen dioxide	2	70	46
Ozone	2	–170	48
Phenol	13	360	94
Phosgene	27	46	90
Propane	7	–44	44
Sulfur dioxide	4	14	64
Sulfuric acid	4	518	98
Tetrachloroethane	11	295	108
Tetrachloroethylene	11	250	166
o-Toluidene	23	390	107
Toluene	19	232	92
Toluene diisocyanate	18	454	174
1,1,1-Trichloroethane	11	235	133
Trichloroethylene	11	189	131
Vinyl chloride monomer	24	7	63
Xylene	19	279	106

Source: See Table 2, Chapter 44, 1999 *ASHRAE Handbook*.
[a]Chemical family numbers are as given in Table 4.
[b]BP = boiling point at 14.7psia (1 atm) pressure
[c]M = molecular mass

INDUSTRIAL AIR CONTAMINANTS

Many industrial processes produce significant quantities of air contaminants in the form of dusts, fumes, smokes, mists, vapors, and gases. Table 11 lists chemical and physical properties of some common gaseous industrial contaminants.

Particulate and gaseous contaminants are best controlled at the source so that they are neither dispersed through the factory nor allowed to increase to toxic concentration levels. Dilution ventilation is much less effective than local exhaust for reducing contamination from point source emissions and is used for control only when sources are distributed and not amenable to capture by an exhaust hood. For sources generating high levels of contaminants, it may also be necessary to provide equipment that reduces the amount of material discharged to the atmosphere (e.g., a dust collector for particulate contaminants and/or a high dwell time gas-phase media bed for solvent vapors). Control methods are covered in Chapters 24 and 25 of the 2000 *ASHRAE Handbook—Systems and Equipment* and Chapters 29 and 44 of the 1999 *ASHRAE Handbook—Applications*.

Zero concentration of all contaminants is not economically feasible. Absolute control of all contaminants cannot be maintained, and workers can assimilate small quantities of various toxic materials without injury. Industrial hygiene science is based on the fact that most air contaminants become toxic only if their concentration exceeds a maximum allowable limit for a specified period. Allowable limits in industrial environments are covered in Chapter 9.

Although the immediately dangerous to life and health (IDLH) toxicity limit is rarely a factor in HVAC design, HVAC engineers should consider it when deciding how much recirculation is safe in a given system. Ventilation airflow must never be so low that the concentration of any gaseous contaminant could rise to the IDLH level. Another toxic effect that may influence design is the loss of sensory acuity due to gaseous contaminant exposure. For example, high concentrations of hydrogen sulfide, which has a very unpleasant odor, effectively eliminate a person's ability to smell the gas. Carbon monoxide, which has no odor to alert people to its presence, affects psychomotor responses and could be a problem in areas such as air traffic control towers. Clearly, waste anesthetic gases should not be allowed to reach levels in operating suites such that the alertness of any of the personnel is affected. NIOSH recommendations are frequently based on such subtle effects.

NONINDUSTRIAL INDOOR AIR CONTAMINANTS

Indoor air quality in residences, offices, and other indoor, nonindustrial environments has become a widespread concern (Spengler et al. 1982, NRC 1981). Exposure to indoor pollutants can be as important as exposure to outdoor pollutants because a large portion of the population spends up to 90% of their time indoors and because indoor pollutant concentrations are frequently higher than corresponding outdoor contaminant levels.

Symptoms of exposure include coughing; sneezing; eye, throat, and skin irritation; nausea; breathlessness; drowsiness; headaches; and depression. Rask (1988) suggests that when 20% of a single building's occupants suffer such irritations, the structure is suffering from **sick building syndrome** (SBS). Case studies of such occurrences have consisted of analyses of questionnaires submitted to building occupants, measurements of contaminant levels, or both of these. Some attempts to relate irritations to gaseous contaminant concentrations are reported (Lamm 1986, Cain et al. 1986, Berglund et al. 1986, Molhave et al. 1982). The correlation of reported complaints with gaseous pollutant concentrations is not strong; many factors affect these less serious responses to pollution. In general, physical irritation does not occur at odor threshold concentrations.

Characterization of indoor air quality has been the subject of numerous recent studies. ASHRAE *Indoor Air Quality (IAQ) Conference Proceedings* discuss indoor air quality problems and some practical controls. ASHRAE *Standard* 62 addresses many indoor air quality concerns. Table 12 illustrates the sources, levels, and indoor-to-outdoor concentration ratios of several contaminants found in indoor environments. Chapter 9 has further information on indoor health issues.

Table 12 Sources, Possible Concentrations, and Indoor-to-Outdoor Concentration Ratios of Some Indoor Pollutants

Pollutant	Sources of Indoor Pollution	Possible Indoor Concentration	I/O Concentration Ratio	Location
Carbon monoxide	Combustion equipment, engines, faulty heating systems	100 mg/kg	>>1	Skating rinks, offices, homes, cars, shops
Respirable particles	Stoves, fireplaces, cigarettes, condensation of volatiles, aerosol sprays, resuspension, cooking	100 to 500 μg/m³	>>1	Homes, offices, cars, public facilities, bars, restaurants
Organic vapors	Combustion, solvents, resin products, pesticides, aerosol sprays	NA	>1	Homes, restaurants, public facilities, offices, hospitals
Nitrogen dioxide	Combustion, gas stoves, water heaters, driers, cigarettes, engines	200 to 1000 μg/m³	>>1	Homes, skating rinks
Sulfur dioxide	Heating system	20 μg/m³	<1	Removal inside
Total suspended particles without smoking	Combustion, resuspension, heating system	100 μg/m³	1	Homes, offices, transportation, restaurants
Sulfate	Matches, gas stoves	5 μg/m³	<1	Removal inside
Formaldehyde	Insulation, product binders, particleboard	0.05 to 1.0 mg/kg	>>1	Homes, offices
Radon and progeny	Building materials, groundwater, soil	0.1 to 100 nCi/m³	>>1	Homes, buildings
Asbestos	Fireproofing	<10⁶ fiber/m³	1	Homes, schools, offices
Mineral and synthetic fibers	Products, cloth, rugs, wallboard	NA	—	Homes, schools, offices
Carbon dioxide	Combustion, humans, pets	3000 mg/kg	>>1	Homes, schools, offices
Viable organisms	Humans, pets, rodents, insects, plants, fungi, humidifiers, air conditioners	NA	>1	Homes, hospitals, schools, offices, public facilities
Ozone	Electric arcing Ultraviolet light sources	20 μg/kg 200 μg/kg	<1 >1	Airplanes Offices

Source: NRC (1981).

[a]Concentrations listed are only those reported indoors. Both higher and lower concentrations have been measured. No averaging times are given. NA indicates that it is not appropriate to list a concentration.

A knowledge of sources frequently present in different types of buildings can be useful during investigations of the causes of SBS. Common nonindustrial indoor sources are discussed in some detail below. Technical advances have allowed generation rates to be measured for a number of these sources. These rates are necessary inputs for design of control equipment, and full details are given in Chapter 44 of the 1999 *ASHRAE Handbook—Applications*.

Building materials and **furnishing** sources have been well studied. Particleboard, which is usually made from wood chips bonded with a phenol-formaldehyde or other resin, is widely used in current construction, especially for mobile homes, carpet underlay, and case goods. These materials, along with ceiling tiles, carpeting, wall coverings, office partitions, adhesives, and paint finishes emit formaldehyde and other VOCs. Latex paints containing mercury have been shown to emit mercury vapor. While the emission rates for these materials decline steadily with age, the half-life of emissions is surprisingly long. Black and Bayer (1986), Nelms et al. (1986), and Molhave et al. (1982) report on these sources.

Ventilation systems may be a source of VOCs (Molhave and Thorsen 1990). The interior of the HVAC system can have large areas of porous material used as acoustical liner that can adsorb odorous compounds. This material can also hold nutrients and, with moisture, can become a reservoir for microorganisms. Microbial contaminants produce characteristic VOCs, called microbial VOCs (MVOCs), associated with their metabolism. Other HVAC components, such as condensate drain pans, fouled cooling coils, and some filter media, may support microbiological life. Deodorants, sealants, and encapsulants are also sources of VOCs in HVAC systems.

Equipment sources in commercial and residential spaces have generation rates that are usually substantially lower than in the industrial environment. As these sources are rarely hooded, emissions go directly to the occupants. In commercial spaces, the chief sources of gaseous contaminants are office equipment, including dry-process copiers (ozone), liquid-process copiers (VOCs), diazo printers (ammonia and related compounds), carbonless copy paper (formaldehyde), correction fluids, inks, and adhesives (various VOCs), spray cans, cosmetics, and so forth (Miksch et al. 1982). Medical and dental activities generate pollutants from the escape of anesthetic gases (nitrous oxide and isoflurene) and from sterilizers (ethylene oxide). The potential for asphyxiation is always a concern when compressed gases are present, even if that gas is nitrogen.

In residences, the main sources of equipment-derived pollutants are gas ranges, wood stoves, and kerosene heaters. Venting is helpful, but some pollutants escape into the occupied area. The pollutant contribution by gas ranges is somewhat mitigated by the fact that they operate for shorter periods than heaters. The same is true of showers, which can contribute to radon and halocarbon concentrations indoors.

Cleaning agents and **other consumer products** can act as contaminant sources. Commonly used liquid detergents, waxes, polishes, spot removers, and cosmetics contain organic solvents that volatilize slowly or quickly. Mothballs and other pest control agents emit organic vapors. Knoeppel and Schauenburg (1989), Black and Bayer (1986), and Tichenor (1989) report data on the release of these volatile organic compounds (VOCs). Field studies have shown that such products contribute significantly to indoor pollution; however, a large variety of compounds is in use, and few studies have been made that allow calculation of typical emission rates. Pesticides, both those applied indoors and those applied outdoors to control termites, also pollute building interiors.

Tobacco smoke is a prevalent and potent source of indoor air pollutants. Almost all tobacco smoke arises from cigarette smoking. Environmental tobacco smoke (ETS), sometimes called secondhand smoke, is the aged and diluted combination of sidestream smoke (the smoke from the lit end of a cigarette and the smoke that escapes from the filter between puffs) and mainstream smoke (the smoke exhaled by a smoker). Emission factors for ETS components, the ratio of ETS components to marker compounds, and apportionment of ETS components in indoor air are reported in the literature by Heavner et al. (1996), Hodgson et al. (1996), Martin et al. (1996), and Nelson et al. (1994).

Occupants, both humans and animals, emit a wide array of pollutants by breath, sweat, and flatus. Some of these emissions are conversions from solids or liquids within the body. Many volatile organics emitted are, however, reemissions of pollutants inhaled earlier, with the tracheobronchial system acting like a physical adsorber.

Floor dust, which is different from the dust in the air, has been found to be a sink (adsorption medium) and secondary emission source for VOCs. Floor dust is a mixture of organic and inorganic particles, hair and skin scales, and textile fibers. The fiber portion of floor dust has been shown to contain 169 ppm TVOC, and the particle portion 148 ppm (Gyntelberg et al. 1994). These VOCs were correlated to the prevalence of irritative (sore throat) and cognitive (concentration problems) symptoms among building occupants. One hundred eighty-eight compounds were identified from thermal desorption of office dust at 250°F (Wilkins et al. 1993). Household dust was found to be similar in composition (Wolkoff and Wilkins 1994).

Contaminants from other sources include chloroform from water; tetrachloroethylene and 1,1,1-trichloroethane from cleaning solvents; methylene chloride from paint strippers, fresheners, cleaners, and polishers; *a*-pinene and limonene from floor waxes; and 1-methoxy-2-propanol from spray carpet cleaners. Formaldehyde, a major VOC, has many sources, but pressed wood products appear to be the most significant.

FLAMMABLE GASES AND VAPORS

The use of flammable materials is widespread. Flammable gases and vapors (NFPA *Standard* 325M) can be found in sewage treatment plants, sewage and utility tunnels, dry-cleaning plants, automobile garages, and industrial finishing process plants.

A flammable liquid's vapor pressure and volatility or rate of evaporation determine its ability to form an explosive mixture. These properties can be expressed by the **flash point**, which is the temperature to which a combustible liquid must be heated to produce a flash when a small flame is passed across the surface of the liquid. Depending on the test methods, either the open cup or closed cup flash point may be listed. The higher the flash point, the more safely the liquid can be handled. Liquids with flash points under 70°F should be regarded as highly flammable.

In addition to having a low flash point, the air-vapor or air-gas mixture must have a concentration in the explosive range before it can be ignited. The **explosive range** is the range between the upper and lower explosive limits, expressed as percent by volume in air. Concentrations of material above the higher range or below the lower range will not explode. The range for many chemicals is found in *National Fire Code* bulletins published by the National Fire Protection Association (NFPA). Some representative limits of flammability are listed in Table 13.

In designing ventilation systems to control flammable gases and vapors, the engineer must consider the following:

Most safety authorities and fire underwriters prefer to limit concentrations to 20 to 25% of the lower explosive limit of a material. The resulting safety factor of 4 or 5 allows latitude for imperfections in air distribution and variations of temperature or mixture and guards against unpredictable or unrecognized sources of ignition. Operation at concentrations above the upper explosive limit should be resorted to only in rare instances. To reach the upper explosive limit, the flammable gas or vapor must pass through the active explosive range, in which any source of ignition can cause an explosion. In addition, a drop in gas concentration due to unforeseen dilu-

Table 13 Flammable Limits of Some Gases and Vapors

Gas or Vapor	Flash Point, °F	Flammable Limits, % by Volume	
		Lower	Upper
Acetone	0	2.6	12.8
Ammonia	Gas	16	25
Benzene (benzol)	12	1.3	7.5
n-Butane	Gas	1.9	8.5
Carbon disulfide	–22	1.3	44
Carbon monoxide	Gas	12.5	74
1,2-Dichloroethylene	43	9.7	12
Diethylether	–49	1.9	48
Ethyl alcohol	55	4.3	19
Ethylene	Gas	3.1	32
Gasoline	–45	1.4	7.6
Hydrogen	Gas	4.0	75
Hydrogen sulfide	Gas	4.3	45
Isopropyl alcohol	53	2.0	12
Methyl alcohol	52	7.3	36
Methyl ethyl ketone	21	1.8	10
Natural gas (variable)	Gas	3.8	17
Naphtha (benzine)	50	0.9	6.7
Propane	Gas	2.2	9.5
Toluene (toluol)	39	1.2	7.1
o-Xylene	90	1.0	6.0

tion or reduced evaporation rate may place a system in the dangerous explosive range.

In occupied places where ventilation is applied for proper health control, the danger of an explosion is minimized. In most instances, flammable gases and vapors are also toxic, and the maximum allowable concentrations are far below the lower explosive limit of the material. For example, proper ventilation for acetone vapors keeps the concentration below 1000 mg/m^3. This is equivalent to 0.1% by mass. The lower explosive limit for acetone is 2.5% by volume.

Proper location of exhaust and supply ventilation equipment depends primarily on how a contaminant is given off and on other problems of the process, and secondarily on the relative density of flammable vapor.

If the specific density of the explosive mixture is the same as that of air, cross drafts, equipment movement, and temperature differentials may cause sufficient mixing to produce explosive concentrations and disperse these throughout the atmosphere. In reasonably still air, heavier-than-air vapors may pool at floor level. Therefore, the engineer must either provide proper exhaust and supply air patterns to control the hazardous material, preferably at its source, or offset the effects of drafts, equipment movement, and convective forces by providing good distribution of exhaust and supply air for general dilution and exhaust. The intake duct should be positioned so that it does not bring in exhaust gases or emissions from ambient sources.

Adequate ventilation minimizes the risk of or prevents fires and explosions and is necessary, regardless of other precautions, such as elimination of the ignition sources, safe building construction, and the use of automatic alarm and extinguisher systems.

Chapter 29 of the 1999 *ASHRAE Handbook—Applications* gives more details about equipment for control of combustible materials.

COMBUSTIBLE DUSTS

Many organic and some mineral dusts can produce dust explosions (Hartmann 1958). Often, a primary explosion results from a small amount of dust in suspension that has been exposed to a source of ignition; the pressure and vibration created can dislodge large accumulations of dust on horizontal surfaces, creating a larger secondary explosion.

For ignition, dust clouds require high temperatures and sufficient dust concentration. These temperatures and concentrations and the minimum spark energy can be found in Avallone and Baumeister (1987).

Explosive dusts are potential hazards whenever uncontrolled dust escapes, disperses in the atmosphere or settles on horizontal surfaces such as beams and ledges. Proper exhaust ventilation design involves the principles covered in Chapter 29 of the 1999 *ASHRAE Handbook—Applications*. The ventilation systems and equipment chosen must prevent the pocketing of dust inside the equipment. When local exhaust ventilation is used, separation equipment should be installed as close to the dust source as possible to prevent transport of dust in the exhaust system.

RADIOACTIVE AIR CONTAMINANTS

Radioactive contaminants (Jacobson and Morris 1977) can be particulate or gaseous and are similar to ordinary industrial contaminants. Many radioactive materials would be chemically toxic if present in high concentrations; however, in most cases, the radioactivity necessitates limiting their concentration in air.

Most radioactive air contaminants affect the body when they are absorbed and retained. This is known as the **internal radiation hazard**. Radioactive particulates may settle to the ground, where they contaminate plants and eventually enter the food chain and the human body. Deposited material on the ground increases **external radiation exposure**. However, except for fallout from nuclear weapons or a serious reactor accident, such exposure is insignificant.

Radioactive air contaminants can emit alpha, beta, or gamma rays. The alpha rays penetrate poorly and present no hazard, except when the material is deposited inside or on the body. Beta rays are somewhat more penetrating and can be both an internal and an external hazard. The penetration of gamma rays depends on their energy, which varies from one type of radioactive element or isotope to another. A distinction should be made between the radioactive material itself and the radiation it gives off. Radioactive particles can be removed from air by devices such as HEPA and ULPA filters, and radioactive gases by impregnated carbon or alumina (radioactive iodine) and absorption traps, but the gamma radiation from such material is capable of penetrating solid materials. This distinction is frequently overlooked. The amount of radioactive material in air is measured in becquerels per cubic metre (1 becquerel equals 2.702702×10^{-11} curies), while the dose of radiation from deposited material is measured in rads.

Radioactive materials present problems that make them distinctive. High concentrations of radioactivity can generate enough heat to damage filtration equipment or ignite the material spontaneously. The concentrations at which most radioactive materials are hazardous are much lower than those of ordinary materials; as a result, special electronic instruments that respond to radioactivity must be used to detect these hazardous levels.

The ventilation engineer faces difficulty in dealing with radioactive air contamination because of the extremely low permissible concentrations for radioactive materials. For certain sensitive industrial plants, such as those in the photographic industry, contaminants must be kept from entering the plant. If radioactive materials are handled inside the plant, the problem is to collect the contaminated air as close to the source as possible, and then remove the contaminant from the air with a high degree of efficiency, before releasing it to the outdoors. Filters are generally used for particulate materials, but venturi scrubbers, wet washers, and other devices can be used as prefilters to meet special needs.

The design of equipment and systems for control of radioactive particulates and gases in nuclear laboratories, power plants, and fuel-processing facilities is a highly specialized technology. Careful attention must be given to the reliability, as well as the contaminant-removal ability, of the equipment under the special environmental stresses involved. Various publications of the U.S. Department of Energy can provide guidance in this field.

Radon

A major source of airborne radioactive exposure to the population comes from radon. Radon (Rn) is a naturally occurring, chemically inert, colorless, odorless, tasteless radioactive gas. It is produced from the radioactive decay of radium, which is formed through several intermediate steps of the decay of uranium and thorium. Radon is widely found in the natural environment, as uranium salt precursors are widespread. Radon-222 is the most common isotope of radon. Before it decays, radon can move limited distances through very small spaces, such as those between particles of soil and rock, and enter indoor environments (Nazaroff et al. 1988, Tanner 1980). Additional but secondary sources of indoor radon include groundwater (radon is quite soluble in water) and radium-containing building materials.

Radon gas enters a house or building primarily through leakage paths in the foundation and is transported by pressure-driven flow. Entry occurs through cracks, joints, and other holes in concrete foundations; directly through porous concrete blocks; through the joints and openings in crawl space ceilings; and through leakage points in HVAC ductwork that is embedded in slab floors or located in crawl spaces. Pressure-driven flow is the dominant radon entry mechanism in houses with elevated radon concentrations (Nazaroff et al. 1987). Pressure differences are caused by several factors, including the thermal stack effect, wind, and operation of HVAC equipment. In addition to pressure-driven radon entry, Rn can also diffuse directly through substructural materials (e.g., concrete). The diffusive Rn entry rate is often a significant portion of the total entry rate in houses with low Rn concentrations.

Measurement. Indoor concentrations of radon can vary hourly, daily, and seasonally, in some cases by as much as a factor of 10 to 20 on a daily basis (Turk et al. 1990). Thus, long-term measurements (3 months to 1 year) made during normal home activities generally provide more reliable estimates of the average indoor concentration than do short-term measurements. Two techniques widely used for homeowner measurements are the short-term charcoal canister (up to 7 days), and the long-term alpha-track methods (90 days to 1 year). Generally, short-term measurements should only be used as a screening technique to determine whether a long-term measurement is necessary. The great uncertainties in measurement accuracy with these devices, up to 50% at the radon levels typically found in homes, as well as the natural variability of radon concentrations should be considered in interpreting the results.

Ideally, long-term measurements should be the basis for decisions on installation of radon mitigation systems, and short-term measurements should only be used as a screening method to identify buildings with Rn concentrations that are very high, justifying immediate remedial action. In practice, short-term measurements at the time a building is sold are the basis for most decisions about remedial action.

Typical Levels. The outdoor radon concentration is about 15 Bq/m^3 (0.4 pCi/L). The annual average concentration of radon in homes in the United States is about 46 Bq/m^3 (1.25 pCi/L) (EPA 1989). While several sources of radon may contribute to the annual indoor average, pressure-driven flow of soil gas constitutes the principal source for elevated concentrations; nonmunicipal water supplies can be a source of elevated indoor radon, but only in isolated instances.

Control. Exposure to indoor Rn may be reduced by (1) inhibiting Rn entry into the building or (2) removing or diluting Rn decay products in indoor air. The most effective and energy-efficient control measures are generally those that reduce Rn entry rates (Henschel 1993). One of the most common effective techniques is active subslab depressurization, in which a fan and piping system draw soil gas from beneath the slab and exhaust the gas outside. This technique reduces or reverses the pressure gradient that normally draws soil gas and Rn into the building and often reduces indoor Rn concentrations by a large factor (e.g., 5 to 10). Passive control methods such as Rn-resistant construction techniques and/or passive stack subslab depressurization systems are also used; however, the performance of these control methods is not well characterized, and average reductions in Rn concentrations may be 50% or less. Sealing cracks and joints in slab floors improve the performance of subslab depressurization systems. Sealing by itself is often not very effective in reducing indoor Rn.

In houses with crawl spaces, active (fan-forced) or passive crawl space ventilation is often effective in maintaining low indoor Rn concentrations, although other techniques are also used (Henschel 1988, 1993).

SOIL GASES

The radioactive gas radon (Rn) is the best-known soil gas. However, gaseous contaminants other than radon may enter buildings from surrounding soil. Methane from landfills has reached explosive levels in some buildings. Potentially toxic or carcinogenic VOCs in the soil as a consequence of spills, improper disposal, leaks from storage tanks, and disposal in landfills can also be transported into buildings (Wood and Porter 1987, Hodgson et al. 1992, Garbesi and Sextro 1989, Kullman and Hill 1990). Pesticides applied to the soil beneath or adjacent to houses have also been detected in indoor air (Livingston and Jones 1981, Wright and Leidy 1982), and pressure-driven flow is a suspected entry mechanism. The broad significance of the health effects due to exposure to these soil contaminants is not well understood. Techniques that reduce Rn entry from the soil should also be effective in reducing the entry of these contaminants into buildings. Another approach may be to increase ventilation in the building, such as by slightly opening a window. That change in home-use behavior can also help reduce the negative pressure in the house, with respect to the soil gas pressure, created by the stack effect.

REFERENCES

ACGIH. 1985. *Particle size-selective sampling in the workplace*. American Conference of Governmental Industrial Hygienists, Cincinnati, OH.

ACGIH. 1989. Guidelines for the assessment of bioaerosols in the indoor environment.

ACGIH. 1998. TLVs and BEIs: Threshold limit values for chemical substances and physical agents.

ACGIH. 2001. *Air sampling instruments*, 9th ed.

AIHA. 1996. *Field guide for the determination of biological contaminants in environmental samples*. American Industrial Hygiene Association, Fairfax, VA.

American Guild of Organists. 1966. Air pollution and organ leathers: Panel discussion. *AGO Quarterly* II(2):62-73.

ASHRAE. 1989. Legionellosis position paper.

ASHRAE. 1994. Safety code for mechanical refrigeration. ANSI/ASHRAE *Standard* 15-1994.

ASHRAE. 1997. Designation and safety classification of refrigerants. ANSI/ASHRAE *Standard* 34-1997.

ASHRAE. 1999. Method of testing general ventilation air-cleaning devices for removal efficiency by particle size. ANSI/ASHRAE *Standard* 52.2-1999.

ASHRAE. 1999. Ventilation for acceptable indoor air quality. ANSI/ASHRAE *Standard* 62-1999.

ASTM. 1983. Practice for continuous sizing and counting of airborne particles in dust-controlled areas using instruments based upon light-scattering principles. ASTM *Standard* F 50-83. American Society for Testing and Materials, West Conshohocken, PA.

ASTM. 1990. Biological contaminants in indoor environments. STP 1071.

ATC. 1990. Technical assistance document for sampling and analysis of toxic organic compounds in ambient air. Environmental Protection Agency, Research Triangle Park, NC.

Avallone, E.A. and T. Baumeister. 1987. *Marks' standard handbook for mechanical engineers*. McGraw-Hill Book Company, New York.

Barbaree, J.M., B.S. Fields, J.C. Feeley, G.W. Gorman, and W.T. Martin. 1986. Isolation of protozoa from water associated with a Legionellosis outbreak and demonstration of intracellular multiplication of Legionella pneumophila. *Applied Environmental Microbiology* 51:422-24.

Batterman, S. and C. Peng. 1995. TVOC and CO_2 as indicators in indoor air quality studies. *American Industrial Hygiene Association Journal* 56(1):55-65.

Berglund, B., U. Berglund, and T. Lindvall. 1986. Assessment of discomfort and irritation from the indoor air. In IAQ '86: *Managing Indoor Air for Health and Energy Conservation*, 138-49. ASHRAE, Atlanta.

Berglund, B., I. Johansson, and T. Lindvall. 1988. *Healthy Buildings* 88, 3:299-309. Stockholm, Sweden.

Black, M.S. and C.W. Bayer. 1986. Formaldehyde and other VOC exposures from consumer products. In IAQ '86: *Managing Indoor Air for Health and Energy Conservation*. ASHRAE, Atlanta.

Braman, R.S. and T.J. Shelley. 1980. Gaseous and particulate and ammonia and nitric acid concentrations. Columbus, Ohio area—Summer 1980. PB 81-125007. National Technical Information Service, Springfield, VA.

Brightman, H.S., S.E. Womble, E.L. Ronca, and J.R. Girman. 1996. Baseline information on indoor air quality in large buildings (BASE '95). In *Proceedings of Indoor Air '96*, Vol. 3:1033-38.

Braun, R.C. and M.J.G. Wilson. 1970. The removal of atmospheric sulphur by building stones. *Atmospheric Environment* 4:371-78.

Budavi, S., ed. 1996. *The Merck index*, 12th ed. Merck and Company, White Station, NJ.

Burge, H.A. 1995. *Bioaerosols*. Lewis Publishers, Chelsea, MA.

Cain, W.S., L.C. See, and T. Tosun. 1986. Irritation and odor from formaldehyde chamber studies, 1986. In IAQ '86: *Managing Indoor Air for Health and Energy Conservation*. ASHRAE, Atlanta.

Casserly, D.M. and K.K. O'Hara. 1987. Ambient exposures to benzene and toluene in southwest Louisiana. *Paper* 87-98.1. Air and Waste Management Association, Pittsburgh, PA.

Chan, C.C., L. Vanier, J.W. Martin, and D.T. William. 1990. Determination of organic contaminants in residential indoor air using an adsorption-thermal desorption technique. *Journal of the Air and Waste Management Association* 40(1):62-67.

Chatigny, M. 1983. *Air sampling instruments*, 6th ed. American Conference of Governmental Industrial Hygienists. Cincinnati, OH.

Chiarenzelli, R.V. and E.L. Joba. 1966. The effects of air pollution on electrical contact materials: A field study. *Journal of the Air Pollution Control Association* 16(3):123-27.

Code of Federal Regulations. *National primary and secondary ambient air quality standards*. 40 CFR 50. U.S. Government Printing Office, Washington, D.C. Revised annually.

Code of Federal Regulations. 29 CFR 1900. U.S. Government Printing Office, Washington, D.C. Revised annually.

Cohen, M.A., P.B. Ryan, Y. Yanigisawa, J.D. Spengler, H. Ozkaynak, and P.S. Epstein. 1989. Indoor/outdoor measurements of volatile organic compounds in the Kanawha Valley of West Virginia. *Journal of the Air Pollution Control Association* 39(8):1986-93.

Colombo, A., M. DeBortoli, H. Knöppel, H. Schauenburg, and H. Vissers. 1991. Small chamber tests and headspace analysis of volatile organic compounds emitted from household products. *Indoor Air* 1:13-21.

Conibear, S., S. Geneser, and B.W. Carnow. 1996. Carbon monoxide levels and sources found in a random sample of households in Chicago during the 1994-1995 heating season. *Proceedings* of IAQ 95. ASHRAE, 111-18.

Coutant, R.W., G.F. Ward, C.W. Spicer, et al. 1994. Control of NO_2 and nitrogen acids using alkaline-impregnated carbon filters. *Proceedings* of IAQ 94. ASHRAE, 139-45.

Coy, C.A. 1987. Regulation and control of air contaminants during hazardous waste site redemption. *Paper* 87-18.1. Air and Waste Management Association, Pittsburgh, PA.

Ekberg, L.E. 1994. Outdoor air contaminants and indoor air quality under transient conditions. *Indoor Air* 4:189-196.

EPA. 1979. Toxic substances control act chemical substance inventory, Volumes I-IV. Environmental Protection Agency, Office of Toxic Substances, Washington, D.C.

EPA. 1982. Air quality criteria for particulate matter and sulfur. EPA-600/8-82-029b.

EPA. 1989. Radon and radon reduction technology. EPA-600/9-89/006a, 1:4-15.

EPA. 1996. Air quality criteria for particulate matter. Office of Research and Development.

Esswein, E.J. and M.F. Boeniger. 1994. Effect of an ozone generating air purifying device on reducing concentrations of formaldehyde in air. *Applied Occupational Environmental Hygiene* 9(2).

FDA. 1990. Maximum acceptable level of ozone. 21 CFR Ch. 1. Sec. 801.415: 24-25. U.S. Food and Drug Administration.

Federal *Standard* 209E. 1992. Airborne particulate cleanliness classes in cleanrooms and clean zones.

Fliermans, C.B. 1985. Ecological niche of *Legionella pneumophila*. In: *Critical Reviews of Microbiology*, 75-116.

Fung, K. and B. Wright. 1990. Measurement of formaldehyde and acetaldehyde using 2-4 dinitrophenylhydrazine-impregnated cartridges. *Aerosol Science and Technology* 12(1):44-48.

Garbesi, K. and R.G. Sextro. 1989. Modeling and field evidence of pressure-driven entry of soil gas into a home through permeable below-grade walls. *Environment Science and Technology* 23:1481-87.

Graminski, E.L., E.J. Parks, and E.E. Toth. 1978. The effects of temperature and moisture on the accelerated aging of paper. NBSIR 78-1443. National Institute of Standards and Technology, Gaithersburg, MD.

Grosjean, D., P.M. Whitmore, C.P. de Moor, G.R. Cass, and J.R. Druzik. 1987. Fading of alizarinad-related artist's pigments by atmospheric ozone. *Environmental Science and Technology* 21:635-43.

Gyntelberg, F., P. Suadicami, J.W. Nielsen, P. Skov, O. Valbjorn, T. Nielsen, T.O. Schneider, O. Jorgenson, P. Wolkoff, C. Wilkins, S. Gravesen, and S. Nom. 1994. Dust and the sick-building syndrome. *Indoor Air* 4: 223-38.

Hakov, R.J., J. Kemlis, and C. Ruggeri. 1987. Volatile organic compounds in the air near a regional sewage treatment plant in New Jersey. *Paper* 87-95.1. Air and Waste Management Association, Pittsburgh, PA.

Hartmann, I. 1958. Explosion and fire hazards of combustible dusts. *Industrial Hygiene and Toxicology* 1:2 Interscience Publishers, New York.

Hartwell, T.D., J.H. Crowder, L.S. Sheldon, and E.D. Pellizzari. 1985. Levels of volatile organics in indoor air. *Paper* 85-30B.3. Air and Waste Management Association, Pittsburgh, PA.

Haynie, F.H. 1978. Theoretical air pollution and climate effects on materials confirmed by zinc corrosion data. In Durability of building materials and components (ASTM STP 691). American Society for Testing and Materials, West Conshohocken, PA.

Heavner, D.L., W.T. Morgan, and M.W. Ogden. 1996. Determination of volatile organic compounds and respirable particulate matter in New Jersey and Pennsylvania homes and workplaces. *Environment International* 22:159-183.

Henschel, D.B. 1988. Radon reduction techniques for detached houses—Technical guidance, 2nd ed. EPA/625/5-87/019.

Henschel, D.B. 1993. Radon reduction techniques for existing detached houses—Technical guidance, 3rd ed. EPA/625/R-93/011.

Hewson, E.W. et al. 1967. Air pollution by ragweed pollen. *Journal of the Air Pollution Control Association* 17(10):651.

Hodgson, A.T., K. Garbesi, R.G. Sextro, and J.M. Daisey. 1988. Transport of volatile organic compounds from soil into a residential basement. Paper 88-95B.1, Proceedings of the 81st Annual Meeting of the Air Pollution Control Association. Also LBL *Report* No. 25267.

Hodgson, A.T., K. Garbesi, R.G. Sextro, and J.M. Daisey 1992. Soil gas contamination and entry of volatile organic compounds into a house near a landfill. *Journal of the Air and Waste Management Association* 42: 277-83.

Hodgson, A.T. 1995. A review and a limited comparison of methods for measuring total volatile organic compounds in indoor air. *Indoor Air* 5(4):247.

Hodgson, A.T., J.M. Daisey, K.R.R. Mahanama, J.T. Brinke, and L.E. Alevantis. 1996. Use of volatile tracers to determine the contribution of environmental tobacco smoke to concentrations of volatile organic compounds in smoking environments. *Environment International* 22:295-307.

Hollowell, C.D., R.A. Young, J.V. Berk, and S.R. Brown. 1982. Energy-conserving retrofits and indoor air quality in residential housing. *ASHRAE Transactions* 88(1):875-93.

ISO. 1999. Cleanrooms and associated controlled environments—Part 1: Classification of air cleanliness. *Standard* 14644-1. International Organization for Standardization, Geneva.

Jacobson, A.R. and S.C. Morris. 1977. The primary pollutants, viable particulates, their occurrence, sources and effects. In *Air pollution*, 3rd ed., Academic Press, New York, 169.

Jaffe, L.S. 1967. The effects of photochemical oxidants on materials. *Journal of the Air Pollution Control Association* 17(6):375-78.

Knoeppel, H. and H. Schauenburg. 1989. Screening of household products for the emission of volatile organic compounds. *Environment International* 15:413-18.

Kullman, G.J. and R.A. Hill 1990. Indoor air quality affected by abandoned gasoline tanks. *Applied Occupational Environmental Hygiene* 5:36-37.

Lamm, S.H. 1986. Irritancy levels and formaldehyde exposures in U.S. mobile homes. In *Indoor Air Quality in Cold Climates*, 137-47. Air and Waste Management Association, Pittsburgh, PA.

Levin, H. 1989. Building materials and indoor air quality. In: *Occupational Medicine: State of the Art Reviews—Problem Buildings,* J.E. Cone and M.J. Hodgson, eds. Hanley & Belfus, Philadelphia. 4(4):667-94.

Levin, H. 1991. Critical building design factors for indoor air quality and climate: current status and predicted trends. *Indoor Air* 1(1):79-92.

Lide, D.R., ed. 1996. *Handbook of chemistry and physics*, 77th ed. Chemical Rubber Company Press, Boca Raton, FL.

Liu, R. and M.A. Huza. 1995. Filtration and indoor air quality: A practical approach. *ASHRAE Journal* 37(2):18-23.

Livingston, J.M. and C.R. Jones. 1981. Living area contamination by chlordane used for termite treatment. *Bulletin of Environmental Contaminant Toxicology* 27:406-11.

Lodge, J.E., ed. 1988. *Methods of air sampling and analysis*, 3rd ed. Lewis Publishers, Chelsea, MD.

Lonnemann, W.A., S.L. Kopczynski, P.E. Darley, and P.D. Sutterfield. 1974. Hydrocarbon composition of urban air pollution. *Environmental Science and Technology* 8(3):229-35.

Lull, W.P. 1995. Conservation environment guidelines for libraries and archives. Canadian Council of Archives.

Martin, P., D.L. Heavner, P.R. Nelson, K.C. Maiolo, C.H. Risner, P.S. Simmons, W.T. Morgan, and M.W. Ogden. 1996. *Environment International* 22.

Mathey, R.G., T.K. Faison, and S. Silberstein. 1983. Air quality criteria for storage of paper-based archival records. NBSIR 83-2795. National Institute of Standards and Technology, Gaithersburg, MD.

McGrath, T.R. and D.B. Stele. 1987. Characterization of phonemic odors in a residential neighborhood. *Paper* 87-75A.5. Air and Waste Management Association, Pittsburgh, PA.

Miksch, R.R., C.D. Hollowell, and H.E. Schmidt. 1982. Trace organic chemical contaminants in office spaces. *Atmospheric Environment* 8:129-37.

Milton, D.K., R.J. Gere, H.A. Feldman, and I.A. Greaves. 1990. Endotoxin measurment: Aerosol sampling and application of a new limulus method. *American Industrial Hygiene Association Journal* 51:331.

Molhave, L. and M. Thorsen. 1990. A model for investigations of ventilation systems as sources for volatile organic compounds in indoor climate. *Atmospheric Environment* 25A:241-49.

Molhave, L., L. Anderson, G.R. Lundquist, and O. Nielson. 1982. Gas emission from building materials. *Report* 137. Danish Building Research Institute, Copenhagen.

Morey, P.R. and B.A. Jenkins. 1989. What are typical concentrations of fungi, total volatile organic compounds, and nitrogen dioxide in an office environment. Proceedings of IAQ '89, *The Human Equation: Health and Comfort*. ASHRAE, Atlanta, 67-71.

Morey, P.R. and J. Singh. 1991. Indoor air quality in non-industrial occupational environments. In: *Patty's industrial hygiene and toxicology*, 4th ed. 1(1).

Muller, C.O. and W.G. England. 1995. Achieving your indoor air quality goals: Which filtration system works best? *ASHRAE Journal* 27(2):24-32.

Nagda, N.L. and H.E. Rector. 1983. Guidelines for monitoring indoor-air quality. EPA 600/4-83-046. Environmental Protection Agency, Research Triangle Park, NC.

Nazaroff, W.W., S.R. Lewis, S.M. Doyle, B.A. Moed, and A.V. Nero. 1987. Experiments on pollutant transport from soil into residential basements by pressure-driven air flow. *Environment Science and Technology* 21:459.

Nazaroff, W.W., B.A. Moed, and R.G. Sextro. 1988. Soil as a source of indoor radon: Generation, migration and entry. In: *Radon and its decay products in indoor air.* Wiley, New York, 57-112.

Nelms, L.H., M.A. Mason, and B.A. Tichenor. 1986. The effects of ventilation rates and product loading on organic emission rates from particleboard. In IAQ '86: *Managing Indoor Air for Health and Energy Conservation*, 469-85. ASHRAE.

Nelson, E.D.P., D. Shikiya and C.S. Liu. 1987. Multiple air toxins exposure and risk assessment in the south coast air basin. *Paper* 87-97.4. Air and Waste Management Association, Pittsburgh, PA.

Nelson, P.R., P. Martin, M.W. Ogden, D.L. Heavner, C.H. Risner, K.C. Maiolo, P.S. Simmons, and W.T. Morgan. 1994. Environmental tobacco smoke characteristics of different commercially available cigarettes. *Proceedings of the Fourth International Aerosol Conference*, Vol. 1: 454-55.

NFPA. 1984. Fire hazard properties of flammable liquids, gases and volatile solids. NFPA *Standard* 325M. National Fire Protection Association, Quincy, MA.

NIOSH. 1977. *NIOSH manual of sampling data sheets*. U.S. Department of Health and Human Services, National Institute for Occupational Safety and Health, Washington, D.C.

NIOSH. 1994. *NIOSH manual of analytical methods*, 4th ed. Cassellini, M.E. and P.F. O'Connor, eds. DHHS (NIOSH) *Publication* 94-113.

NRC. 1981. Indoor pollutants. National Research Council, National Academy Press, Washington, D.C.

NTIS. 1982. Air pollution effects on materials. *Published Search* PB 82-809427. National Technical Information Service, Springfield, VA.

NTIS. 1984. Air pollution effects on materials. Published Search PB 84-800267.

OSHA. 1995. OSHA computerized information system chemical sampling information. U.S. Government Printing Office, Washington, D.C.

Otson, R. and P. Fellin. 1993. TVOC measurements: Relevance and limitations. *Proceedings of Indoor Air* '93, Vol. 2:281-85.

Owen et al. 1992. Airborne particle sizes and sources found in indoor air. *Atmospheric Environment* 26A(12):2149-62.

Peral, J., X. Domenech, and D.F. Ollis. 1997. Heterogeneous photocatalysis for purification, decontamination and deodorization of air. *J. Chem. Technol. Biotechnol.* 70:117-40.

Rask, D. 1988. Indoor air quality and the bottom line. *Heating, Piping and Air Conditioning* 60(10).

Rodes, C.E., R.M. Kamens, and R.W. Wiener. 1991. The significance and characteristics of the personal activity cloud on exposure assessment methods for indoor contaminants. *Indoor Air* 2:123-145.

Samet, J.M., M.C. Marbury, and J.D. Spengler. 1987. Health effects and sources of indoor air pollution. *Am. Rev. Respir. Disease* 136:1486-1508.

Sandalls, E.J. and S.A. Penkett. 1977. Measurements of carbonyl sulfide and carbon disulphide in the atmosphere. *Atmospheric Environment* 11:197-99.

Sax, N.I. and R.J. Lewis, Sr. 1988. *Dangerous properties of industrial materials*, 6th ed, 3 vols. Van Nostrand Reinhold, New York.

Scala, G.F. 1963. A new instrument for the continuous measurement of condensation nuclei. *Analytical Chemistry* 35(5):702.

Shah, J.J. and H.B. Singh. 1988. Distribution of volatile organic chemicals in outdoor and indoor air. *Environmental Science and Technology* 22(12):1391-88.

Shaughnessy, R.J., E. Levetin, J. Glocker, and K.L. Sublette. 1994. Effectiveness of portable indoor air cleaners: Sensory testing results. *Indoor Air* 4:179-88.

Sheldon, L., R.W. Handy, T. Hartwell, R.W. Whitmore, H. Zelon, and E.D. Pellizzari. 1988a. *Indoor air quality in public buildings*, Vol. I. EPA/600/S6-88/009a. Environmental Protection Agency, Washington, D.C.

Sheldon, L., H. Zelon, J. Sickles, C. Easton, T. Hartwell, and L. Wallace. 1988b. *Indoor air quality in public buildings*, Vol. II. EPA/600/S6-88/009b. Environmental Protection Agency, Research Triangle Park, NC.

Singh. H.B., L.J. Salas, A. Smith, R. Stiles, and H. Shigeishi. 1981. Atmospheric measurements of selected hazardous organic chemicals. PB 81200628. National Technical Information Service, Springfield, VA.

Solomon, W.R. and K.P. Mathews. 1978. Aerobiology and inhalant allergens. Allergy, principles and practices, St. Louis, MO.

Spengler, J., C. Hallowell, D. Moschandreas, and O. Fanger. 1982. Environment international. *Indoor air pollution.* Pergamon Press, Oxford, England.

Tanner, A.B. 1980. Radon migration in the ground: A supplementary review. In: *Natural radiation environment III.* U.S. Department of Commerce, NTIS, Springfield, VA.

Task Group on Lung Dyamics. 1966. Deposition and retention models for internal dosimetry of the human respiratory tract. *Health Physics* 12:173-207.

Taylor, D.G., R.E. Kupel, and J.M. Bryant. 1977. Documentation of the NIOSH validation tests.

Thomson, G. 1986. *The museum environment*, 2nd ed. Butterworths Publishers, Boston.

Tichenor, B.A. 1989. Measurement of organic compound emissions using small test chambers. *Environment International* 15:389-96.

Tichenor, B.A., G. Guo, J.E. Dunn, L.E. Sparks, and M.A. Mason. 1991. The interaction of vapour phase organic compounds with indoor sinks. *Indoor Air* 1:23-35.

Traynor, G.W. 1987. Field monitoring design considerations for assessing indoor exposures to combustion pollutants. *Atmospheric Environment* 21(2):377-83.

Turk, B.H., R.J. Prill, D.T. Grimsrud, B.A. Moed, and R.G. Sextro. 1990. Characterizing the occurrence, sources and variability of radon in Pacific Northwest homes. *Journal of the Air and Waste Management Association* 40:498-506.

VanOsdell, D.W. and L.E. Sparks. 1995. Carbon absorption for indoor air cleaning. *ASHRAE Journal* 27(2):34-40.

Wallace, L.A., E.D. Pellizari, T.D. Hartwell, C. Sparacino, and H. Zelon. 1983. Personal exposure to volatile organic and other compounds indoors and outdoors—the team study. APCA *Meeting Paper*, and later *Report*.

Wallace, L.A., E. Pellizzari, and C. Wendel. 1991. Total volatile organic concentrations in 2700 personal, indoor, and outdoor air samples collected in the US EPA Team Studies. *Indoor Air* 4:465-77.

Walsh, M., A. Black, A. Morgan, and G.H. Crawshaw. 1977. Sorption of SO_2 by typical indoor surfaces including wool carpets, wallpaper and paint. *Atmospheric Environment* 11:1107-11.

Weschler, C.J. and H.C. Shields. 1989. The effects of ventilation, filtration, and outdoor air on the composition of indoor air at a telephone office building. *Environment International* 15:593-604.

Weschler, C.J., H.C. Shields, and D.V. Naik. 1989. Indoor ozone exposures. *Journal of the Air Pollution Control Association* 39:1562-68.

Whitby, K.T., A.B. Algren, and R.C. Jordan. 1955. Size distribution and concentration of airborne dust. *Heating, Piping and Air Conditioning* 27:121.

Whitby, K.T., A.B. Algren, and R.C. Jordan. 1957. The ASHRAE airborne dust survey. *Heating, Piping and Air Conditioning* 29(11):185.

WHO. 1989. Indoor air quality: Organic pollutants. *Report* on a WHO-meeting, Euro Report and Studies III. WHO Regional Office for Europe, Copenhagen.

Wilkins, C.K., P. Wolkoff, F. Gyntelberg, P. Skov, and O. Valbjørn. 1993. Characterization of office dust by VOCs and TVOC release—Identification of potential irritant VOCs by partial least squares analysis. *Indoor Air* 3:283-90.

Willeke, K. and P.A. Baron, eds. 1993. *Aerosol measurement—Principles, techniques and applications.* Van Nostrand Reinhold, New York.

Wolkoff, P. and C.K. Wilkins. 1994. Indoor VOCs from household floor dust: Comparison of headspace with desorbed VOCs; method for VOC release determination. *Indoor Air* 4:248-54.

Wood, J.A. and M.L. Porter. 1987. Hazardous pollutants in Class II landfills. *Journal of the Air Pollution Control Association* 37:609-15.

Wright, C.G. and R.B. Leidy. 1982. Chlordane and heptachlor in the ambient air of houses treated for termites. *Bulletin of Environmental Contaminant Toxicology* 28:617-23.

CHAPTER 13

ODORS

VARIOUS factors make odor control a primary consideration in ventilation engineering: (1) contemporary construction methods result in buildings that permit less air infiltration through the building envelope; (2) indoor sources of odors associated with modern building materials, furnishings, and office equipment have increased; (3) outdoor air is often polluted; and (4) energy costs have encouraged ventilation rate reductions at a time when requirements for a relatively odor-free environment are greater than ever.

Since Yaglou's classic studies (1936), the philosophy behind the ventilation of nonindustrial buildings has mainly been to provide indoor air that is acceptable to occupants. Air is evaluated by the olfactory sense, although the general chemical sense, which is sensitive to irritants in the air, also plays a role.

This chapter reviews how odoriferous substances are perceived. Chapter 44 of the 1999 *ASHRAE Handbook—Applications* covers control methods. Chapter 9 of this volume has more information on indoor environmental health.

ODOR SOURCES

Outdoor sources of odors include automotive and diesel exhausts, hazardous waste sites, sewage treatment plants, compost piles, refuse facilities, printing plants, refineries, chemical plants, and many other stationary and mobile sources. These sources produce both inorganic compounds (e.g., ammonia and hydrogen sulfide) and volatile organic compounds (VOCs), including some which evaporate from solid or liquid particulate matter. Odors emitted by outdoor sources eventually enter the indoor environment.

Indoor sources also emit odors. Sources include tobacco products; bathrooms and toilets; building materials (adhesives, paints, caulks, processed wood, carpets, plastic sheeting, and insulation board); consumer products (e.g., food, toiletries, cleaning materials, and polishes); hobby materials; fabrics; and foam cushions. In offices, offset printing processes, copiers, and computer printers may produce odors. If electrostatic processes are involved, ozone may be emitted. Humans emit a wide range of odorants, including acetaldehyde, ammonia, ethanol, hydrogen sulfide, and mercaptans.

Mildew and other processes of decay often produce odors in occupied spaces (home and office), damp basements, and ventilation systems (e.g., from wetted air-conditioning coils and spray dehumidifiers).

Chapter 44 of the 1999 *ASHRAE Handbook—Applications* gives further information on contaminant sources and generation rates.

The preparation of this chapter is assigned to TC 2.3, Gaseous Air Contaminants and Gas Contaminant Removal Equipment.

SENSE OF SMELL

Olfactory Stimuli

Among organic substances, those with molecular weights greater than 300 are generally odorless. Some substances with molecular weights less than 300 are such potent olfactory stimuli that they can be perceived at concentrations too low to be detected with direct-reading instruments. Trimethylamine, for example, can be recognized as a fishy odor by a human observer at a concentration of about 10^{-4} ppm.

Table 1 shows **threshold concentrations** for selected compounds. These threshold values are not precise numbers and may vary by several orders of magnitude. The **threshold limit value** (TLV) is the concentration of a compound that should have no adverse health consequences if a worker is regularly exposed for 8 h periods (ACGIH, revised annually). Table 1 also includes the ratio

Table 1 Odor Thresholds, ACGIH TLVs, and TLV:Threshold Ratios of Selected Gaseous Air Pollutants

Compound	Odor Threshold,[a] ppmv	TLV, ppmv	Ratio
Acetaldehyde	0.067	40	597
Acetone	62	500	8.1
Acetonitrile	1600	40	0.025
Acrolein	1.8	0.1	0.06
Ammonia	17	25	1.5
Benzene	61	0.5	0.01
Benzyl chloride	0.041	1	24
Carbon tetrachloride	250	5	0.02
Chlorine	0.08	0.5	6
Chloroform	192	10	0.05
Dioxane	12	25	2
Ethylene dichloride	26	10	0.4
Hydrogen sulfide	0.0094	10	1064
Methanol	160	200	1.25
Methylene chloride	160	50	0.3
Methyl ethyl ketone	16	200	12.5
Phenol	0.06	5	83
Sulfur dioxide	2.7	2	0.74
Tetrachloroethane	7.3	1	0.14
Tetrachloroethylene	47	25	0.5
Toluene	1.6	50	31
1,1,1-Trichloroethane	390	350	0.9
Trichloroethylene	82	50	0.6
Xylene (isomers)	20	100	5

Sources: AIHA (1989), ACGIH (1998).
[a]All thresholds are detection thresholds (ED_{50}).

of the TLV to the odor threshold for each compound. For ratios greater than 1, most occupants can detect the odor and leave the area long before the compound becomes a health risk. As the ratio increases, the safety factor provided by the odor also increases. Table 1 is not a comprehensive list of the chemicals found in indoor air. AIHA (1989) and EPA (1992) have published lists of odor thresholds for selected chemicals.

Olfactory sensitivity often makes it possible to detect potentially harmful substances at concentrations below dangerous levels so that they can be eliminated. Foul-smelling air is often assumed to be unhealthy. In reality, however, there is little correlation between odor perception and toxicity, and there is considerable individual variation in the perception of pleasantness/unpleasantness of odors. When symptoms such as nausea, headache, and loss of appetite are caused by an unpleasant odor, it may not matter whether the air is toxic but whether the odor is perceived to be unpleasant, associated with an unpleasant experience, or simply felt to be out of appropriate context. The magnitude of the symptoms is related to the magnitude of the odor. Even a room with a low but recognizable odor can arouse uneasiness among occupants. Several papers review sensory irritation and its relation to indoor air pollution (Cain and Cometto-Muñiz 1995; Cometto-Muñiz and Cain 1992; Cometto-Muñiz et al. 1997; Shams Esfandabad 1993).

Anatomy and Physiology

The **olfactory receptors** lie in the **olfactory cleft**, which is high in the nasal cavity. About five million olfactory **neurons** (a small cluster of nerve cells inside the nasal cavity above the bridge of the nose) each send an **axon** (an extension of the neuron) into the olfactory bulb of the brain. Information received from the receptors is passed to various central structures of the brain (e.g., the olfactory cortex, the hippocampus, and the amygdala). One sniff of an odorant can often evoke a complex, emotion-laden memory, such as a scene from childhood.

The surrounding nasal tissue contains other diffusely distributed nerve endings of the trigeminal nerve that also respond to airborne vapors. These receptors mediate the chemosensory responses such as tickling, burning, cooling, and, occasionally, painful sensations that accompany olfactory sensations. Most odorous substances at sufficient concentration will also stimulate these nerve endings.

Olfactory Acuity

The olfactory acuity of the population is normally distributed. Most people have an average ability to smell substances or to respond to odoriferous stimuli, a few people are very sensitive or hypersensitive, and a few others are insensitive, including some who are totally unable to smell (**anosmic**). The olfactory acuity of an individual varies with the odorant.

Hormonal factors, which often influence emotional states, can modulate olfactory sensitivity. Although the evidence is not uniformly compelling, research has found that (1) the sensitivity of females varies during the menstrual cycle, reaching a peak just before and during ovulation (Schneider 1974); (2) females are generally more sensitive than males, but this difference only emerges around the time of sexual maturity (Koelega and Koster 1974); (3) sensitivity is altered by certain diseases (Schneider 1974); and (4) various hormones and drugs (e.g., estrogen and alcohol) alter sensitivity (Schneider 1974; Engen et al. 1975).

Other factors that may affect olfactory perception include the olfactory acuity of an individual, the magnitude of the flow rate toward the olfactory receptors, the temperature, and the relative humidity. Olfactory acuity can also vary with age (Stevens et al. 1989; Wysocki and Gilbert 1989), genetics (Wysocki and Beauchamp 1984), exposure history (Dalton and Wysocki 1996; Wysocki et al. 1997), and disease or injury (Cowart et al. 1993, 1997). Humans are able to perceive a large number of odors, yet untrained individuals are able to name only a few (Ruth 1986).

Individuals exhibiting a total inability to detect odors are relatively rare (Cowart et al. 1997). A more common occurrence is an inability to detect one or a very limited number of odors, a condition known as **specific anosmia**. Although the huge number of possible chemicals makes for an untestable hypothesis, it has been posited that most, if not all, individuals have a specific anosmia to one or more compounds (Wysocki and Beauchamp 1984). The fact that individuals with specific anosmias have normal olfactory acuity for all other odors suggests that such anosmias may be due to genetic differences.

Olfactory science uses the term **adaptation** to refer to decreased sensitivity or responsiveness to an odor as a function of prolonged exposures. Such exposure can selectively impair the perception of the exposure odorant, but there are also examples of cross-adaptation, where exposure to one odorant can result in adaptation to other odors as well. Adaptation can occur in the short term, where the perception of a room's odor begins to fade within seconds of entering the room (Cometto-Muñiz and Cain 1995; Pierce et al. 1996). With long-term adaptation, an individual who habitually returns to the same environment does not smell odors that are quite discernible to a naive observer. This effect appears to shift both the threshold and the **suprathreshold** (stimuli above the threshold level) response to the odor (Dalton and Wysocki 1996). This is an important phenomenon for indoor air quality (IAQ) personnel to be aware of because it is often one of the biggest reasons for the variation in detectability or response in real-world environments and makes the choice of test population or panelists for air quality evaluations a critical one.

FACTORS AFFECTING ODOR PERCEPTION

Humidity and Temperature

Temperature and humidity can both affect the perception of odors. Cain et al. (1983) reported that a combination of high temperature (78°F) and high humidity exacerbates odor problems. Berglund and Cain (1989) found that air was generally perceived to be fresher and less stuffy with decreasing temperature and humidity. Fang et al. (1998a,b) and Toftum (1998) found little or no increase in odor intensity with increasing enthalpy (temperature and humidity) but reported a very significant decrease in odor acceptability with increasing enthalpy.

Not all researchers have supported these findings. Kerka and Humphreys (1956) reported a decrease in odor intensity with increasing humidity. Berg-Munch and Fanger (1982) found no increase in odor intensity with increasing temperature (73.5 to 89.5°F). Clausen et al. (1985) found no significant change in odor intensity with increasing relative humidity (30% to 80%).

While the findings are not homogeneous, they do show that temperature and humidity can act together to affect one's perception of odors. Air that is cooler and drier is generally perceived to be fresher and more acceptable even if the odor intensity is not affected.

Sorption and Release of Odors

Due to sorption of odors by furnishings and interior surfaces during occupancy, with later desorption, spaces frequently retain normal occupancy odor levels long after occupancy has ceased. This phenomenon is observed when furnaces or radiators, after a long shutdown, are heated at winter start-up and when evaporator coils warm up. The rate of desorption can be decreased by decreasing temperature and relative humidity, and increased (as for cleaning) by the reverse.

Environmental tobacco smoke may desorb from surfaces long after smoking has taken place. This phenomenon has caused many hotels to establish nonsmoking rooms.

Where the odor source is intrinsic to the materials (as in linoleum, paint, rubber, and upholstery), reducing the relative humidity decreases the rate of odor release. Quantitative values should not be used without considering the sorption-desorption phenomenon.

Emotional Responses to Odors

There can be considerable variation between individuals regarding the perceived pleasantness or unpleasantness of a given odor. Responses to odors may be determined by prior experiences and can include strong emotional reactions. This is because one of the brain structures involved in the sense of smell is the **amygdala**, a regulator of emotional behaviors (Frey 1995). Some IAQ complaints can involve emotional responses completely out of proportion to the concentration of the odorant or the intensity of the odor it produces.

Two theories describe physiological reasons for these strong responses. One of these is **kindling**, in which repeated, intermittent stimuli produce an amplification of nerve responses. The other is **response facilitation**, in which an initial stimulus perceived as strong is facilitated (becomes greater) rather than adapted to (Frey 1995).

Because of this emotional aspect, IAQ complaints involving odors can be very difficult to solve, especially if they are coming from a few sensitized individuals. It is important to respond quickly to complaints in order to minimize the risk of kindling or response facilitation.

ODOR SENSATION ATTRIBUTES

Odor sensation has four components or attributes: detectability, intensity, character, and hedonic tone.

Detectability (or **threshold**) is the minimum concentration of an odorant that provokes detection by some predetermined segment of the population. Two types of thresholds exist: the detection threshold and the recognition threshold.

The **detection threshold** is the lowest level that elicits response by a segment of the population. If that segment is 50%, the detection threshold is denoted by ED_{50}. **Recognition threshold** is the lowest level at which a segment of the population can recognize a given odor. Thresholds can be attributed to 100%, which includes all olfactory sensitivities, or to 10%, which includes only the most sensitive segment of the population. Threshold values are not physical constants; rather they are statistical measurements of best estimates.

Intensity is a quantitative aspect of a descriptive analysis, stating the degree or magnitude of the perception elicited. Intensity of the perceived odor is, therefore, the strength of the odoriferous sensation. Detection threshold values and, most often, odor intensity determine the need for indoor odor controls.

Character defines the odor as similar to some familiar smell (e.g., fishy, sour, flowery, and the like). **Hedonics**, or the hedonic tone of an odor, is the degree to which an odor is perceived as pleasant or unpleasant. Hedonic judgments include both a *category* judgment (pleasant, neutral, unpleasant) and a *magnitude* judgment (very unpleasant, slightly pleasant).

Important questions are

1. What is the minimum concentration of odorant that can be detected?
2. How does perceived odor magnitude grow with concentration above the threshold?

No universal method has been accepted to measure either the threshold or the perceived magnitude of the odor above threshold. However, certain guidelines and conventions simplify the choice of methods.

Detectability

The perception of weak odoriferous signals is probabilistic: at one moment odor may be perceptible, and at the next moment it may not. Factors affecting this phenomenon include the moment-to-moment variability in the number of molecules striking the olfactory receptors, the variability in which of the receptors are stimulated, the concentration of the odor, the individual's style of breathing, and the individual's previous experience with the odor. The combined effect of these factors may prevent an individual from perceiving an odor during the entire time of the stimulus. During odor evaluation, dilution to detection or recognition threshold values allows determination of the largest number of dilutions that still permits half of the panelists to detect or recognize the odor.

Determination of Odor Thresholds. Odor threshold testing is over a century old. The process is complex, and several different methods are used. Partly due to variations in measurement techniques, reported threshold values can vary by several orders of magnitude for a given substance. In order to minimize variation due to experimental techniques, a standard set of criteria has been developed. These criteria include the panel, the presentation apparatus, and the presentation method (AIHA 1989; EPA 1992).

The **panel** should

- Include at least six members per group.
- Be selected based on odor sensitivity. Factors to be considered include anosmia, pregnancy, drug use, and smoking.
- The panel should be calibrated to document individual and group variability.

Considerations concerning the **presentation apparatus** include

- Vapor modality—choice of a gas-air mixture, water vapor, or other substance.
- Diluent—choice of diluent (e.g., air, nitrogen), how it is treated, and what its source is.
- Presentation mode—delivery systems can be nose ports, vents into which the head is inserted, flasks, whole rooms.
- Analytic measurement of odorant concentration.
- System calibration—flow rate and face velocity. Flow rate should be approximately 0.10 cfm. Face velocity should be low enough to be barely perceptible to the panelists.

Criteria for the **presentation method** include

- Threshold type—detection or recognition.
- Concentration presentation—this must take adaptation into account. Presenting ascending concentrations or allowing longer periods between concentrations helps avoid adaptation.
- Number of trials—test-retest reliability for thresholds is low. Increasing the number of trials helps correct for this.
- Forced-choice procedure—panelists must choose between the stimuli and one or two blanks. This helps eliminate false positive responses.
- Concentration steps—odorant should be presented successively at concentrations no more than three times the preceding one.

For more details regarding psychophysical procedures, ways to sample odoriferous air, handling samples, means of stimulus presentation, and statistical procedures, consult ASTM STP 434.

Intensity

Psychophysical Power Law. The relation between **perceived odor magnitude** S and **concentration** C conforms to a power function:

$$S = kC^n \tag{1}$$

where

S = perceived intensity (magnitude) of sensation
k = characteristic constant
C = odorant concentration
n = exponent of psychophysical function (slope on a log-log scale)

This exemplifies the psychophysical power law, also called **Stevens' law** (Stevens 1957). In the olfactory realm, $n < 1.0$. Accordingly, a given percentage change in odorant concentration causes a smaller percentage change in perceived odor magnitude.

Scaling Methods. Of the various ways to scale perceived magnitude, a **category scale**, which can be either number- or word-categorized, is commonly used. Numerical values on this scale do not reflect ratio relations among odor magnitudes (e.g., a value of 2 does not represent a perceived magnitude twice as great as a value of 1). Table 2 gives four examples of category scales.

Although category scaling procedures can be advantageous in field situations, **ratio scaling** techniques are used frequently in the laboratory (Cain and Moskowitz 1974). Ratio scaling procedures require observers to assign numbers proportional to perceived magnitude. For example, if the observer is instructed to assign the number 10 to one concentration and a subsequently presented concentration seems three times as strong, the observer calls it 30; if another seems one-half as strong, the observer assigns it 5. This ratio scaling procedure, called **magnitude estimation**, was used to derive the power function for butanol (Figure 1). Ratio scaling techniques allow for such relationships because they require subjects to produce numbers to match perceived sensations in which the numbers emitted reflect the ratio relations among the sensations.

More recently, a hybrid of category and ratio scales known as the **labeled magnitude scale** has been developed (Green et al. 1996). This scale is intended to yield ratio-level data with a true zero and an orderly relationship among the scale values, such that any stimulus can be expressed as being proportionately more or less intense than another. Because it allows subjects to use natural language descriptors to scale perceived experience, it often requires less training than ratio scales and produces absolute intensity estimates of perceived sensation (Figure 2).

Table 2 Examples of Category Scales

Number Category		Word Category	
Scale I	Scale II	Scale I	Scale II
0	0	None	None at all
1	1	Threshold	Just detectable
2	2.5	Very slight	Very mild
3	5	Slight	Mild
4	7.5	Slight-moderate	Mild-distinct
5	10	Moderate	Distinct
6	12.5	Moderate-strong	Distinct-strong
7	15	Strong	Strong

Source: Meilgaard et al. (1987).

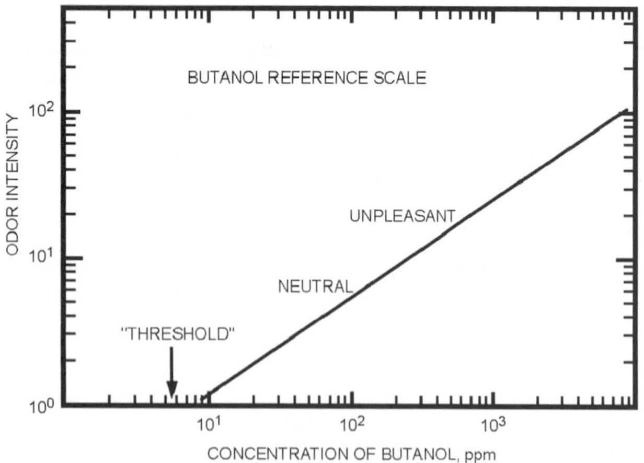

Fig. 1 Standardized Function Relating Perceived Magnitude to Concentration of 1-Butanol
(Moskowitz et al. 1974)

A fourth way to measure suprathreshold odor intensity is to **match the intensity of odorants**. An observer can be given a concentration series of a matching odorant (e.g., 1-butanol) to choose the member that matches most closely the intensity of an *unknown* odorant. The matching odorant can be generated by a relatively inexpensive olfactometer such as that shown in Figure 3. Figure 4 shows, in logarithmic coordinates, functions for various odorants obtained by the matching method (Dravnieks and Laffort 1972). The left-hand ordinate expresses intensity in terms of concentration of butanol, and the right-hand ordinate expresses intensity in terms of perceived magnitude. The two ordinates are related by the function in Figure 1, the standardized function for butanol. The matching method illustrated here has been incorporated into ASTM

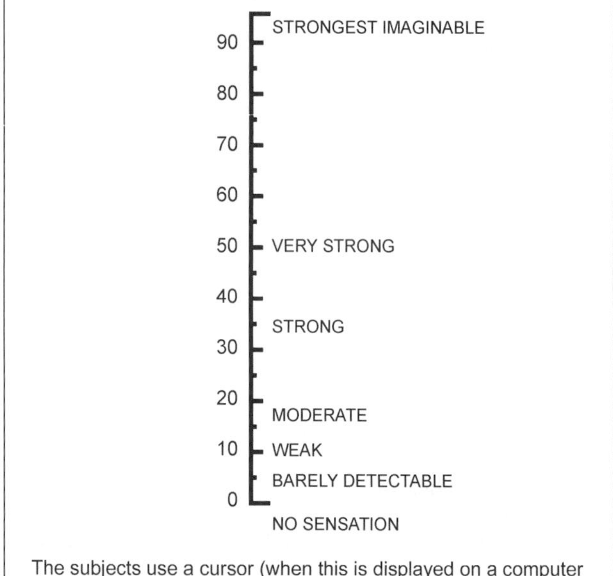

The subjects use a cursor (when this is displayed on a computer screen) or a pencil to mark the location on the scale that represents their judgment of intensity. They do not see numbers, only labels, and can place the mark anywhere on the scale.

Fig. 2 Labeled Magnitude Scale

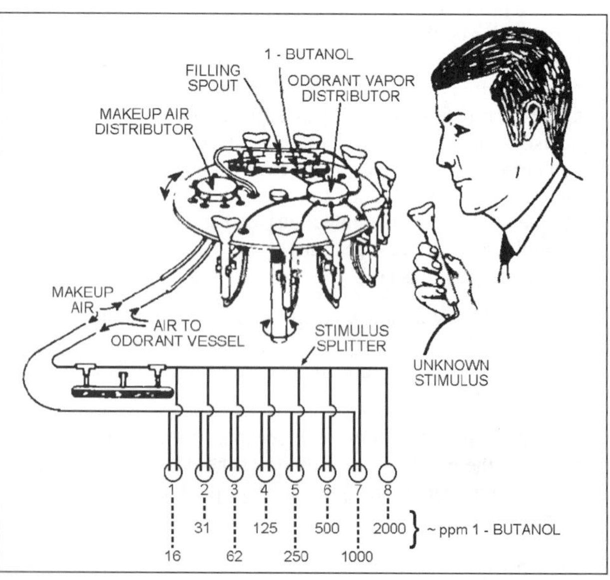

Fig. 3 Panelist Using Dravnieks Binary Dilution Olfactometer
(Dravnieks 1975)

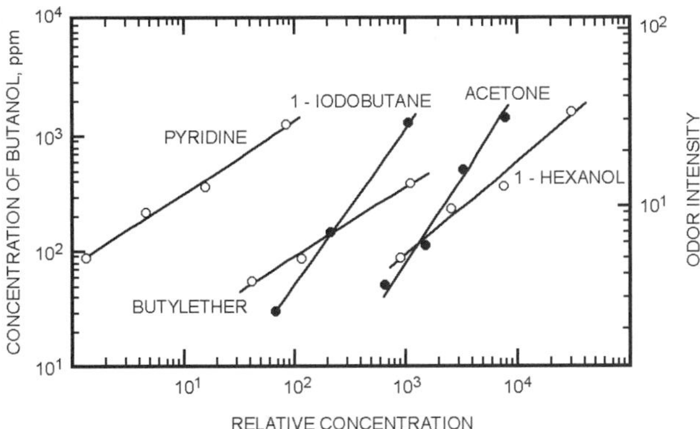

Fig. 4 Matching Functions Obtained with Dravnieks Olfactometer
(Dravnieks and Laffort 1972; Cain 1978)

Standard E 544, Standard Practices for Referencing Suprathreshold Odor Intensity.

Character

The quality or character of an odor is difficult to assess quantitatively. A primary difficulty is that odors can vary along many dimensions. One way to assess quality is to ask panelists to judge the similarity between a test sample and various reference samples, using a five-point category scale. For certain applications, reference odorants can be chosen to represent only that portion of the qualitative range relevant to the odor problem under investigation (e.g., animal odors). Another procedure is to ask panelists to assess the degree of association between a test sample's quality and certain verbal descriptors (e.g., sweaty, woody, chalky, sour).

The number of odorant descriptors and the descriptors to be used have been subjects of disagreement (Harper et al. 1968). The number of descriptors varies from a minimum of seven (Amoore 1962) to as many as 830 used by an ASTM subcommittee. An atlas of odor characters, containing 146 descriptors, was compiled for 180 chemicals by ASTM (1985).

An odor can be characterized either by an open-ended word description or by multidimensional scaling. **Multidimensional scaling** is based on similarity and dissimilarity judgments in comparison to a set of standard odors or to various descriptors.

In some instances, the interest may be merely whether an odor's quality has changed as a result of some treatment (e.g., use of a bacteriostat). Under these circumstances, samples of air taken before and after treatment can be compared directly (using a simple scale of similarity) or indirectly (with appropriate verbal descriptors).

Hedonics

The acceptability or pleasantness of an odor can be measured psychophysically in the same way as odor intensity. Both ratio and category scaling procedures can be adapted to odor acceptability.

Odors do not always cause adverse reactions. Products are manufactured to elicit favorable responses. Acceptance tests may involve product comparison (frequently used in the perfume industry) or a hedonic scale. The premise of acceptance tests is that the larger the segment of subjects accepting the odor, the better the odor. A hedonic scale that allows for negative as well as positive responses is likely to better answer the question of how acceptable the odor is.

All persons exposed to a given odor are not likely to agree on its acceptability. Acceptability of a given odor to a person is based on a complex combination of associations and is not simply a characteristic of the odor itself (Beck and Day 1991). Responses to odors are determined by both **bottom-up factors** (attributes or properties of the odorant) and **top-down factors** (expectations, attitudes, and associations from prior experience stored in memory, and appropriateness of the odor in its present context). Both of these factors are activated when an individual detects an odor, and the individual's ultimate response (e.g., perception of intensity, hedonics, irritation, or symptoms) is a joint function of both (Dalton 1996; Dalton et al. 1997). In some cases, the interpretation provided by the top-down process appears to override the outcome from the bottom-up process, resulting in complaints, symptoms, and reports of illness.

DILUTION OF ODORS BY VENTILATION

The size of the exponent n in Stevens' law [Equation (1)] varies from one odorant to another, ranging from less than 0.2 to about 0.7 (Cain and Moskowitz 1974). This determines the slope or dose response of the odor intensity/odorant concentration function and has important consequences for malodor control. A low slope value indicates an odor that requires greater relative dilution for the odor to dissipate; a high slope value indicates an odor that can be more quickly reduced by ventilation. For example, an exponent of 0.7 implies that in order to reduce perceived intensity by a factor of 5, the concentration must be reduced by a factor of 10; an exponent of 0.2 implies that a reduction in perceived magnitude by a factor of 5 would require a reduction in concentration by a factor of more than 3000. Examples of compounds with low slope values include hydrogen sulfide, butyl acetate and amines. Compounds with high slope values include ammonia and aldehydes.

The ability of ventilation to control odors also depends on the strength of the source generating the odorant(s) and the nature of the odor. An odorant with a stronger source requires proportionately more ventilation to achieve the same reduction in concentration. Odors that are perceived as unpleasant may require substantially greater reduction before being perceived as acceptable. In addition, some sources, such as painted walls and flooring materials, may show increased emission rates in response to increased ventilation rates, which further complicates this picture (Gunnarsen 1997).

ODOR CONCENTRATION

Analytical Measurement

Performance data on the control of specific odorants can be obtained using suitable analytical methods. Detectors permit detection of substances in amounts as little as 1 ng. Air contains many minor components, so gas chromatographic separation of the components must precede detection. Because odor thresholds for some compounds are low, preconcentration of the minor components is necessary. **Preconcentration** consists of adsorption or absorption by a stable, sufficiently nonvolatile material, followed by thermal desorption or extraction. The state of the art for sampling and analysis of VOCs in indoor air is reviewed in NIOSH (1993).

Mass spectrometry can be used with **gas chromatography** to identify constituents of complex mixtures. The chromatograph resolves a mixture into its constituents, and the spectrometer provides identification and concentration of selected constituents.

Several other detectors are sufficiently sensitive and specific to detect resolved components. **Hydrogen flame ionization detectors** respond adequately and nearly mass-proportionally to almost all organic compounds. **Flame photometric detectors** can pinpoint with equal sensitivity compounds that contain sulfur; many sulfur compounds are strongly odorous and are of interest in odor work. A **Coulson conductometric detector** is specifically and adequately sensitive to ammonia and organic nitrogen compounds. **Thermal conductivity detectors** are generally not sensitive enough for analytical work on odors.

Frequently, a **sniffing port** (Dravnieks and O'Donnell 1971; Dravnieks and Krotoszynski 1969) is installed in parallel with the detector(s). Part of the resolved effluent exhausts through the port and allows the components that are particularly odorous or carry some relevant odor quality to be annotated. Usually, only a fraction of all components studied exhibits odors.

Airborne VOCs cause odors, but the correlation between indoor VOC concentrations and odor complaints in indoor environments is poor. Considerable work is being done on **artificial noses**, which may be the future of objective determination of odorants (Bartlett et al. 1997; Freund and Lewis 1995; Moy et al. 1994; Taubes 1996). However, because the physicochemical correlates of olfaction are poorly understood, no simple analytical means to predict the perceived quality and intensity of an odorant exists. Moreover, since acceptability of an odorant depends strongly on context, it is unlikely that analytical instruments will supplant human evaluation.

Odor Units

Odor concentration can be expressed as the number of unit volumes that a unit volume of odorous sample occupies when diluted to the odor threshold with nonodorous air. If a sample of odorous air can be reduced to threshold by a tenfold dilution with pure air, the concentration of the original sample is said to be 10 odor units. Hence, odor units are equivalent to multiples of threshold concentrations. Odor units are not units of perceived magnitude.

Odor units are widely used to express legal limits for emission of odoriferous materials. For example, the law may state that a factory operation may not cause the ambient odor level to exceed 15 odor units. For every odorant (chemical), odor units and parts per million (ppm) are proportional. The proportionality constant varies from one odorant to another, depending on the number of ppm needed to evoke a threshold response. Perceived odor magnitude (intensity), however, does not grow proportionally with concentration expressed in ppm. Therefore, it cannot grow proportionally with concentration expressed in odor units. For example, a sample of 20 odor units is always perceived as less than twice as strong as a sample of 10 odor units. Moreover, because the psychophysical function (slope) varies from one odorant to another, samples of two odorants, each at 20 odor units, may have unequal perceived intensities.

Although odor units are not equivalent to units of perceived magnitude, they can be useful. Most indoor and outdoor contaminants are complex mixtures, so that the actual concentration of the odoriferous portion of a sample cannot be expressed with certainty. Thus, the odor unit is a useful measure of concentration of the mixture when evaluating, for example, the efficiency of a filter or ventilation system to remove or dilute the odor.

OLF UNITS

Sometimes IAQ scientists cannot successfully resolve complaints about the air in offices, schools, and other nonindustrial environments. Customarily, complaints are attributed to elevated pollutant concentrations; frequently, however, such high concentrations are not found, yet complaints persist.

On the assumption that the inability to find a difference between air pollutant levels in buildings with registered complaints and those without complaints is due to inadequacies of prevailing measurement techniques, Fanger and others changed the focus from chemical analysis to sensory analysis (Fanger 1987, 1988; Fanger et al. 1988). Fanger quantified air pollution sources by comparing them with a well-known source: a sedentary person in thermal comfort. A new unit, the olf, was introduced. An **olf** is defined as the emission rate of air pollutants (bioeffluents) from a standard person. A **decipol** is one olf ventilated at a rate of 20 cfm of unpolluted air.

To use these units, Fanger generated a curve that relates the percentage of persons dissatisfied with air polluted by human bioeffluents as a function of the outdoor air ventilation rate and obtained the following expression:

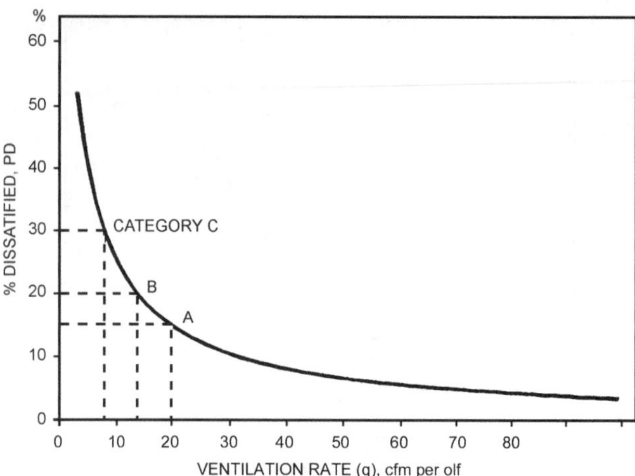

Fig. 5 Percentage of Dissatisfied Persons as a Function of Ventilation Rate per Standard Person (i.e., per Olf)
(CEN 1998)

Table 3 Sensory Pollution Load from Different Pollution Sources

Source	Sensory Load
Sedentary person (1 to 1.5 met)	1 olf
Person exercising	
Low level (3 met)	4 olf
Medium level (6 met)	10 olf
Children, kindergarten (3 to 6 yrs)	1.2 olf
Children, school (4 to 16 yrs)	1.3 olf
Low-polluting building	0.01 olf/ft^3
Non-low-polluting building	0.02 olf/ft^2

Source: CEN (1998).

$$D = 395\exp(-3.66q^{0.36}) \quad \text{for} \quad q \geq 0.332$$
$$D = 100 \quad \text{for} \quad q < 0.332 \qquad (2)$$

where

D = percentage of persons dissatisfied
q = ventilation-emission ratio, cfm

This curve, which is shown in Figure 5, is based on experiments involving more than 1000 European subjects (Fanger and Berg-Munch 1983). Experiments with American (Cain et al. 1983) and Japanese (Iwashita et al. 1990) subjects show very similar results.

The idea behind the olf concept is that sensory sources other than humans also be expressed in equivalent standard persons (i.e., in olfs). A room should therefore be ventilated to handle the total sensory load from persons and building. The olf concept is used in European publications for ventilation (CEN 1998; ECA 1992) to determine the required ventilation and in several national standards (DIN 1994; Norwegian Building Code 1996). Table 3 shows the sensory loads from different pollution sources used in CEN (1998).

Example. Office, low-polluting building, occupancy 0.007 persons/ft^2

Occupants	0.007 olf/ft^2
Building	0.01 olf/ft^2
Total sensory load	0.017 olf/ft^2

30% dissatisfied requires 8 cfm per olf ventilation rate (Figure 5)
Required ventilation: $8 \times 0.017 = 0.14$ cfm/ft^2

The sensory load on the air in a space can be determined from Figure 5 by measuring the outdoor ventilation rate and determining the percent dissatisfied, using an untrained panel with a minimum of 20 impartial persons (ASHRAE *Standard* 62; Gunnarsen and Fanger 1992). The panel judges the acceptability of the air just after entering the space. The required ventilation rate depends on the desired percentage of occupant satisfaction. In ASHRAE *Standard* 62, 80% acceptability (20% dissatisfied) is the goal; European guidelines offer three quality levels: 15%, 20%, and 30% dissatisfied.

Although this system has much to offer from a theoretical standpoint, its use is controversial in some areas. Problems have been found in cultural differences among panel members and access to outdoor air for dilution (Aizlewood et al. 1996). The trend is now to use untrained panels as described in the previous paragraph. Knudsen et al. (1998) have shown that the curve giving the relation between percent dissatisfied and ventilation rate for some building materials is less steep than that in Figure 5, while other materials show a steeper curve. The sensory load in this case depends on the ventilation rate. The constant sensory loads given in Table 3 should therefore be seen as a first approximation.

CODES AND STANDARDS

ASHRAE. 1999. Ventilation for acceptable indoor air quality. ANSI/ASHRAE *Standard* 62-1989.

ASTM. 1997. Standard practices for referencing suprathreshold odor intensity. ASTM *Standard* E 544-75 (R 1993). American Society for Testing and Materials, West Conshohocken, PA.

DIN. 1994. Raumlufttechnik; Gesundheitstechnische Anforderungen, VDI-Lüftungsregeln. DIN *Standard* 1946-2. Deutsches Institut für Normung, Berlin.

Norwegian Building Code. 1996. Oslo, Statens Hygningstekniske Eur.

REFERENCES

ACGIH. 1998 (revised annually). *Threshold limit values for chemical substances and physical agents and biological exposure indices.* American Conference of Governmental Industrial Hygienists, Cincinnati, OH.

AIHA. 1989. *Odor thresholds for chemicals with established occupational health standards.* American Industrial Hygiene Association, Akron, OH.

Aizlewood, C.E., G.J Raw, and N.A. Oseland. 1996. Decipols: Should we use them? *Indoor Built Environment* 5:263-69.

Amoore, J.E. 1962. The stereochemical theory of olfaction 1, Identification of seven primary odors. *Proc. Sci. Sect. Toilet Goods Assoc.* 37:1-12.

ASTM. 1985. Atlas of odor character profiles. *Data Series* 61. American Society for Testing and Materials, West Conshohocken, PA.

ASTM. 1996. Manual on sensory testing methods. *Special Technical Publication* STP 434-96.

Bartlett, P., J. Elliott, and J. Gardner. 1997. Electronic noses and their applications in the food industry. *Food Technology* 51(12):44-48.

Beck, L. and V. Day. 1991. New Jersey's approach to odor problems. From *Transactions: Recent developments and current practices and odor regulations control and technology*, D.R. Derenzo and A. Gnyp, eds. Air and Waste Management Association, Pittsburgh, PA.

Berglund, L. and W.S. Cain. 1989. Perceived air quality and the thermal environment. In *Proceedings of IAQ '89: The Human Equation: Health and Comfort,* San Diego, pp. 93-99.

Berg-Munch, B. and P.O. Fanger. 1982. The influence of air temperature on the perception of body odour. *Environment International* 8:333-35.

Cain, W.S. 1978. The odoriferous environment and the application of olfactory research. In *Handbook of perception*, Vol. 6, Tasting, smelling, feeling, and hurting, eds. Carterette and Friedman. Academic Press, New York.

Cain, W.S. and J.E. Cometto-Muñiz. 1995. Irritation and odors as indicators of indoor pollution. *Occupational Medicine* 10(1):133-35.

Cain, W.S. and H.R. Moskowitz. 1974. Psychophysical scaling of odor. In *Human responses to environmental odors*, eds. Turk, Johnson, and Moulton. Academic Press, New York.

Cain, W.S., B.P. Leaderer, R. Isseroff, L.G. Berglund, R.I. Huey, E.D. Lipsitt, and D. Perlman. 1983. Ventilation requirements in buildings—I: Control of occupancy odor and tobacco smoke odor. *Atmospheric Environment* 17:1183-97.

CEN. 1998. Ventilation for buildings: Design criteria for the indoor environment. *Technical Report* CR1752. European Committee for Standardization, Brussels.

Clausen, G., P.O Fanger, W.S. Cain, and B.P. Leaderer. 1985. The influence of aging particle filtration and humidity on tobacco smoke odour. In *Proceedings of CLIMA 2000* 4:345-49. Copenhagen.

Cometto-Muñiz, J.E. and W.S. Cain. 1992. Sensory irritation, relation to indoor air pollution in sources of indoor air contaminants—characterizing emissions and health effects. *Annals of the New York Academy of Sciences*, Vol. 641.

Cometto-Muñiz, J.E. and W.S. Cain. 1995. Olfactory adaptation. In *Handbook of Olfaction and Gustation*, ed. R.L. Doty. Marcel Dekker, New York.

Cometto-Muñiz, J.E., W.S. Cain, and H.K. Hudnell. 1997. Agonistic sensory effects of airborne chemicals in mixtures: Odor, nasal pungency and eye irritation. *Perception and Psychophysics* 59(5):665-74.

Cowart, B.J., K. Flynn-Rodden, S.J. McGeady, and L.D. Lowry. 1993. Hyposmia in allergic rhinitis. *Journal of Allergy and Clinical Immunology* 91:747-51.

Cowart, B.J., I.M. Young, R.S. Feldman, and L.D. Lowry. 1997. Clinical disorders of smell and taste. *Occupational Medicine* 12:465-83.

Dalton, P. 1996. Odor perception and beliefs about risk. *Chemical Senses* 21:447-58.

Dalton, P. and C.J. Wysocki. 1996. The nature and duration of adaptation following long-term exposure to odors. *Perception & Psychophysics* 58(5):781-92.

Dalton, P., C.J. Wysocki, M.J. Brody, and H.J. Lawley. 1997. The influence of cognitive bias on the perceived odor, irritation and health symptoms from chemical exposure. *International Archives of Occupational and Environmental Health* 69:407-17.

Dravnieks, A. 1975. Evaluation of human body odors, methods and interpretations. *Journal of the Society of Cosmetic Chemists* 26:551.

Dravnieks, A. and B. Krotoszynski. 1969. Analysis and systematization of data for odorous compounds in air. *ASHRAE Symposium Bulletin*, Odor and odorants: The engineering view.

Dravnieks, A. and P. Laffort. 1972. Physicochemical basis of quantitative and qualitative odor discrimination in humans. *Olfaction and Taste* IV:142, ed. Schneider. Wissenschaftliche Verlagsgesellschaft mbH, Stuttgart.

Dravnieks, A. and A. O'Donnell. 1971. Principles and some techniques of high resolution headspace analysis. *Journal of Agricultural and Food Chemistry* 19:1049.

ECA. 1992. Guidelines for ventilation requirements in buildings. European Collaborative Action *Indoor Air Quality and its Impact on Man. Report* No. 11. EUR 14449 EN. Office for Official Publications of the European Committees, Luxembourg.

Engen, T., R.A. Kilduff, and N.J. Rummo. 1975. The influence of alcohol on odor detection. *Chemical Senses and Flavor* 1:323.

EPA. 1992. *Reference guide to odor thresholds for hazardous air pollutants listed in the* Clean Air Act Amendments of 1990. EPA/600/R-92/047. Office of Research and Development, U.S. Environmental Protection Agency, Washington, D.C.

Fang, L., G. Clausen, and P.O. Fanger. 1998a. Impact of temperature and humidity on the perception of indoor air quality. *Indoor Air* 8(2):80-90.

Fang, L., G. Clausen, and P.O. Fanger. 1998b. Impact of temperature and humidity on perception of indoor air quality during immediate and longer whole-body exposures. *Indoor Air* 8(4):276-84.

Fanger, P.O. 1987. A solution to the sick building mystery. In *Indoor Air '87, Proceedings of the International Conference on Indoor Air and Climate.* Institute of Water, Soil and Air Hygiene, Berlin.

Fanger, P.O. 1988. Introduction of the olf and decipol units to quantify air pollution perceived by humans indoors and outdoors. *Energy and Buildings* 12:1-6.

Fanger, P.O. and B. Berg-Munch. 1983. Ventilation and body odor. In *Proceedings of Engineering Foundation Conference on Management of Atmospheres in Tightly Enclosed Spaces,* pp. 45-50. ASHRAE, Atlanta.

Fanger, P.O., J. Lauridsen, P. Bluyssen, and G. Clausen. 1988. Air pollution sources in offices and assembly halls quantified by the olf unit. *Energy and Buildings* 12:7-19.

Freund, M.S. and N.S. Lewis. 1995. A chemically diverse conduction polymer-based "electronic nose". *Proceedings of the National Academy of Sciences* 92:2652-56.

Frey, A.F. 1995. A review of the nature of odour perception and human response. *Indoor Environment* 4:302-05.

Green, B.G., P. Dalton, B. Cowart, G. Shaffer, K.R. Rankin, and J. Higgins. 1996. Evaluating the "labeled magnitude scale" for measuring sensations of taste and smell. *Chemical Senses* 21(3):323-34.

Gunnarsen, L. 1997. The influence of area-specific rate on the emissions from construction products. *Indoor Air* 7:116-20.

Gunnarsen, L. and P.O. Fanger. 1992. Adaptation to indoor air pollution. *Energy and Buildings* 18:43-54.

Harper, R., E.C. Bate Smith, and D.G. Land. 1968. *Odour description and odour classification.* American Elsevier Publishing, New York. Distributors for Churchill Livingston Publishing, Edinburgh, Scotland.

Iwashita, G., K. Kimura, S. Tanabe, S. Yoshizawa, and K. Ikeda. 1990. Indoor air quality assessment based on human olfactory sensation. *Journal of Architecture, Planning and Environmental Engineering* 410:9-19.

Kerka, W.F. and C.M. Humphreys. 1956. Temperature and humidity effect on odour perception. *ASHRAE Transactions* 62:531-52.

Knudsen, H.N. et al. 1998. Determination of exposure-response relationships for emissions from building products. *Indoor Air* 8(4):264-75.

Koelega, H.S. and E.P. Koster. 1974. Some experiments on sex differences in odor perception. *Annals of the New York Academy of Sciences* 237:234.

Meilgaard, M., G.V. Civille, and B.T. Carr. 1987. *Sensory evaluation techniques.* CRC Press, Boca Raton, FL.

Moskowitz, H.R., A. Dravnieks, W.S. Cain, and A. Turk. 1974. Standardized procedure for expressing odor intensity. *Chemical Senses and Flavor* 1:235.

Moy, L., T. Tan, and J.W. Gardner. 1994. Monitoring the stability of perfume and body odors with an "electronic nose". *Perfumer and Flavorist* 19:11-16.

NIOSH. 1993. Case studies—Indoor environmental quality "from the ground up." *Applied Occupational and Environmental Hygiene* 8:677-80.

Pierce, J.J.D., C.J. Wysocki, E.V. Aronov, J.B. Webb, and R.M. Boden. 1996. The role of perceptual and structural similarity in cross adaptation. *Chemical Senses* 21:223-27.

Ruth, J.H. 1986. Odor thresholds and irritation levels of several chemical substances: A review. *American Industrial Hygiene Association Journal* 47:142-51.

Schneider, R.A. 1974. Newer insights into the role and modifications of olfaction in man through clinical studies. *Annals of the New York Academy of Sciences* 237:217.

Shams Esfandabad, H. 1993. Perceptual analysis of odorous irritants in indoor air. Ph.D. diss., Department of Psychology, Stockholm University.

Stevens, J.C., W.S. Cain, F.T. Schiet, and M.W. Oatley. 1989. Olfactory adaptation and recovery in old age. *Perception* 18(22):265-76.

Stevens, S.S. 1957. On the psychophysical law. *Psychological Review* 64:153.

Taubes, G. 1996. The electronic nose. *Discover* (September):40-50.

Toftum, J. et al. 1998. Upper limits for air humidity for preventing warm respiratory discomfort. *Energy and Buildings* 28(3):15-23.

Wysocki, C.J. and G.K. Beauchamp. 1984. Ability to smell androstenone is genetically determined. In *Proceeding of the National Academy of Sciences of the United States of America* 81:4899-4902.

Wysocki, C.J. and A.N. Gilbert. 1989. National geographic smell survey: Effects of age are heterogeneous. In *Nutrition and the Chemical Senses in Aging: Recent Advances and Current Research*, eds. C.L. Murphy, W.S. Cain, and D.M. Hegsted, pp. 12-28. New York Academy of Sciences, New York.

Wysocki, C.J., P. Dalton, P., M.J. Brody, and H.J. Lawley. 1997. Acetone odor and irritation thresholds obtained from acetone-exposed factory workers and from control (occupationally non-exposed) subjects. *American Industrial Hygiene Association Journal* 58:704-12.

Yaglou, C.P., E.C. Riley, and D.I. Coggins. 1936. Ventilation requirements. *ASHRAE Transactions* 42:133-62.

BIBLIOGRAPHY

ACGIH. 1988. *Advances in air sampling*. American Conference of Government Industrial Hygienists, Cincinnati, OH.

Clemens, J.B. and R.G. Lewis. 1988. Sampling for organic compounds. In *Principles of environmental sampling* 20:147-57. American Chemical Society, Washington, D.C.

Moschandreas, D.J. and S.M. Gordon. 1991. Volatile organic compounds in the indoor environment: Review of characterization methods and indoor air quality studies. In *Organic chemistry of the atmosphere*. CRC Press, Boca Raton, FL.

CHAPTER 14

MEASUREMENT AND INSTRUMENTS

HEATING, refrigerating, and air-conditioning engineers and technicians require instruments for both laboratory work and fieldwork. Precision is more essential in the laboratory, where research and development are undertaken, than in the field, where acceptance and adjustment tests are conducted. This chapter describes the characteristics and uses of some of these instruments.

TERMINOLOGY

The following definitions are generally accepted.

Accuracy. The capability of an instrument to indicate the true value of measured quantity. This is often confused with inaccuracy, which is the departure from the true value to which all causes of error (e.g., hysteresis, nonlinearity, drift, temperature effect, and other sources) contribute.

Amplitude. The magnitude of variation from its zero value in an alternating quantity.

Average. The sum of a number of values divided by the number of values.

Bandwidth. The range of frequencies over which a given device is designed to operate within specified limits.

Bias. The tendency of an estimate to deviate in one direction from a true value (a systematic error).

Calibration. (1) The process of comparing a set of discrete magnitudes or the characteristic curve of a continuously varying magnitude with another set or curve previously established as a standard. Deviation between indicated values and their corresponding standard values constitutes the correction (or calibration curve) for inferring true magnitude from indicated magnitude thereafter; (2) the process of adjusting an instrument to fix, reduce, or eliminate the deviation defined in (1).

Calibration curve. (1) The path or locus of a point that moves so that its coordinates on a graph correspond to values of input signals and output deflections; (2) the plot of error versus input (or output).

Confidence. The degree to which a statement (measurement) is believed to be true.

Dead band. The range of values of the measured variable to which an instrument will not effectively respond.

Deviate. Any item of a statistical distribution that differs from the selected measure of control tendency (average, median, mode).

Deviation, standard. The square root of the average of the squares of the deviations from the mean (root mean square deviation). A measure of dispersion of a population.

The preparation of this chapter is assigned to TC 1.2, Instruments and Measurements.

Distortion. An unwanted change in wave form. Principal forms of distortion are inherent nonlinearity of the device, nonuniform response at different frequencies, and lack of constant proportionality between phase-shift and frequency. (A wanted or intentional change might be identical, but it is called **modulation.**)

Drift. A gradual, undesired change in output over a period of time that is unrelated to input, environment, or load. Drift is gradual; if variation is rapid, the fluctuation is referred to as **cycling**.

Dynamic error band. The spread or band of output-amplitude deviation incurred by a constant amplitude sine wave as its frequency is varied over a specified portion of the frequency spectrum (see *Static error band*).

Error. The difference between the true or actual value to be measured (input signal) and the indicated value (output) from the measuring system. Errors can be systematic or random.

Error, accuracy. See *Error, systematic*.

Error, fixed. See *Error, systematic*.

Error, instrument. The error of an instrument's measured value that includes random or systematic errors.

Error, precision. See *Error, random*.

Error, probable. An error with a 50% or higher chance of occurrence. A statement of probable error is of little value.

Error, random. A statistical error caused by chance and not recurring. This term is a general category for errors that can take values on either side of an average value. To describe a random error, its distribution must be known.

Error, root mean square, or **RMS.** An accuracy statement of a system comprising several items. For example, a laboratory potentiometer, volt box, null detector, and reference voltage source have individual accuracy statements assigned to them. These errors are generally independent of one another, so a system of these units displays an accuracy given by the square root of the sum of the squares of the individual limits of error. For example, four individual errors of 0.1% yield a calibrated accuracy of 0.4% but an RMS error of only 0.2%.

Error, systematic. A persistent error not due to chance; systematic errors are causal. A systematic error is likely to have the same magnitude and sign for every instrument constructed with the same components and procedures. Errors in calibrating equipment cause systematic errors because all instruments calibrated are biased in the direction of the calibrating equipment error. Voltage and resistance drifts over time are generally in one direction and are classed as systematic errors.

Frequency response (flat). The portion of the frequency spectrum over which the measuring system has a constant value of amplitude response and a constant value of time lag. Input signals that have frequency components within this range are indicated by the measuring system (without distortion).

Hysteresis. The summation of all effects, under constant environmental conditions, that cause the output of an instrument to assume different values at a given stimulus point when that point is approached with increasing stimulus and with decreasing stimulus. Hysteresis includes backlash. It is usually measured as a percent of full scale when input varies over the full increasing and decreasing range. In instrumentation, hysteresis and dead band are similar quantities.

Linearity. The straight-lineness of the transfer curve between an input and an output; that condition prevailing when output is directly proportional to input (see *Nonlinearity*).

Loading error. A loss of output signal from a device caused by a current drawn from its output. It increases the voltage drop across the internal impedance, where no voltage drop is desired.

Mean. See *Average*.

Median. The middle value in a distribution, above and below which lie an equal number of values.

Noise. Any unwanted disturbance or spurious signal that modifies the transmission, measurement, or recording of desired data.

Nonlinearity. The prevailing condition (and the extent of its measurement) under which the input-output relationship (known as the input-output curve, transfer characteristic, calibration curve, or response curve) fails to be a straight line. Nonlinearity is measured and reported in several ways, and the way, along with the magnitude, must be stated in any specification.

Minimum-deviation-based nonlinearity. The maximum departure between the calibration curve and a straight line drawn to give the greatest accuracy; expressed as a percent of full-scale deflection.

Slope-based nonlinearity. The ratio of maximum slope error anywhere on the calibration curve to the slope of the nominal sensitivity line; usually expressed as a percent of nominal slope.

Most variations beyond these two definitions result from the many ways in which the straight line can be arbitrarily drawn. All are valid as long as construction of the straight line is explicit.

Population. A group of individual persons, objects, or items from which samples may be taken for statistical measurement.

Precision. The repeatability of measurements of the same quantity under the same conditions; not a measure of absolute accuracy. The precision of a measurement is used here to describe the relative tightness of the distribution of measurements of a quantity about their mean value. Therefore, precision of a measurement is associated more with its repeatability than its accuracy. It combines uncertainty caused by random differences in a number of identical measurements and the smallest readable increment of the scale or chart. Precision is given in terms of deviation from a mean value.

Primary calibration. A calibration procedure in which the instrument output is observed and recorded while the input stimulus is applied under precise conditions—usually from a primary external standard traceable directly to the National Institute of Standards and Technology (NIST).

Range. A statement of the upper and lower limits between which an instrument's input can be received and for which the instrument is calibrated.

Reliability. The probability that an instrument's precision and accuracy will continue to fall within specified limits.

Repeatability. See *Precision*.

Reproducibility. In instrumentation, the closeness of agreement among repeated measurements of the output for the same value of input made under the same operating conditions over a period of time, approaching from both directions; it is usually measured as a nonreproducibility and expressed as reproducibility in percent of span for a specified time period. Normally, this implies a long period of time, but under certain conditions, the period may be a short time so that drift is not included. Reproducibility includes hysteresis, dead band, drift, and repeatability. Between repeated measurements, the input may vary over the range, and operating conditions may vary within normal limits.

Resolution. The smallest change in input that produces a detectable change in instrument output. Resolution differs from precision in that it is a psychophysical term referring to the smallest increment of humanly perceptible output (rated in terms of the corresponding increment of input). The precision, the resolution, or both may be better than the accuracy. An ordinary six-digit (or dial) instrument has a resolution of one part per million (ppm) of full scale; however, it is possible that the accuracy is no better than 25 ppm (0.0025%). Note that the practical resolution of an instrument cannot be any better than the resolution of the indicator or detector, whether internal or external.

Scale factor. (1) The amount by which a measured quantity must change to produce unity output; (2) the ratio of real to analog values.

Sensitivity. The property of an instrument that determines scale factor. The word is often short for maximum sensitivity or the minimum scale factor with which an instrument can respond. The minimum input signal strength required to produce a desired value of output signal (e.g., full scale or unit output or the ratio of output to input values).

Sensitivity inaccuracy. The maximum error in sensitivity displayed as a result of the summation of the following: frequency response; attenuator inaccuracy; hysteresis or dead band; amplitude distortion (sensitivity nonlinearity); phase distortion (change in phase relationship between input signal and output deflection); and gain instability. Only by taking into account all these factors can nominal sensitivity, as indicated by the numeral on the attenuator readout, be discounted for accurate interpretation.

Stability. (1) Independence or freedom from changes in one quantity as the result of a change in another; (2) the absence of drift.

Static error band. (1) The spread of error present if the indicator (pen, needle) stopped at some value (e.g., at one-half of full scale). It is normally reported as a percent of full scale; (2) a specification or rating of maximum departure from the point where the indicator must be when an on-scale signal is stopped and held at a given signal level. This definition stipulates that the stopped position can be approached from either direction in following any random waveform. Therefore, it is a quantity that includes hysteresis and nonlinearity but excludes items such as chart paper accuracy or electrical drift (see *Dynamic error band*).

Step-function response. The characteristic curve or output plotted against time resulting from the input application of a step function (a function that is zero for all values of time before a certain instant, and a constant for all values of time thereafter).

Threshold. The smallest stimulus or signal that results in a detectable output.

Time constant. The time required for an exponential quantity to change by an amount equal to 0.632 times the total change required to reach steady state for first-order systems.

Transducer. A device for translating the changing magnitude of one kind of quantity into corresponding changes of another kind of quantity. The second quantity often has dimensions different from the first and serves as the source of a useful signal. The first quantity may be considered an input and the second an output. Significant energy may or may not transfer from the transducer's input to output.

Uncertainty. An estimated value for the error (i.e., what an error might be if it were measured by calibration). Although uncertainty may be the result of both systematic and precision errors, only precision error can be treated by statistical methods.

Zero shift. Drift in the zero indication of an instrument without any change in the measured variable.

UNCERTAINTY ANALYSIS

Uncertainty Sources

Measurement generally consists of a sequence of operations or steps. Virtually every step introduces a conceivable source of uncertainty, the effect of which must be assessed. The following list is representative of the most common, but not all, sources of uncertainty.

- Inaccuracy in the mathematical model that describes the physical quantity
- Inherent stochastic variability of the measurement process
- Uncertainties in measurement standards and calibrated instrumentation
- Time-dependent instabilities due to gradual changes in standards and instrumentation
- Effects of environmental factors such as temperature, humidity, and pressure
- Values of constants and other parameters obtained from outside sources
- Uncertainties arising from interferences, impurities, inhomogeneity, inadequate resolution, and incomplete discrimination
- Computational uncertainties and data analysis
- Incorrect specifications and procedural errors
- Laboratory practice, including handling techniques, cleanliness, and operator techniques, etc.
- Uncertainty in corrections made for known effects, such as installation effect corrections

Uncertainty of a Measured Variable

For a measured variable X_i, the total error is caused by both **precision (random)** and **systematic (bias) errors**. This relationship is shown in Figure 1. The possible measurement values of the variable are scattered in a distribution around the parent population mean μ_i (Figure 1A). The curve is the normal or Gaussian distribution and is the theoretical distribution function for the infinite population of measurements that generated X_i. The parent population mean differs from $(X_i)_{true}$ by an amount called the systematic (or bias) error β_i (Figure 1B). The quantity β_i is the total fixed error that remains after all calibration corrections have been made. In general, there are several sources of bias error, such as calibration standard errors, data acquisition errors, data reduction errors, and test technique errors. There is usually no direct way to measure these errors. These errors are unknown and are assumed to be zero; otherwise, an additional correction would be applied to reduce them to as close to zero as possible.

The **precision uncertainty** for a variable, which is an estimate of the possible error associated with the repeatability of a particular measurement, is determined from the sample standard deviation, or the estimate of the error associated with the repeatability of a particular measurement. Unlike the systematic error, the precision error varies from reading to reading. As the number of readings of a particular variable tends to infinity, the distribution of these possible errors becomes Gaussian.

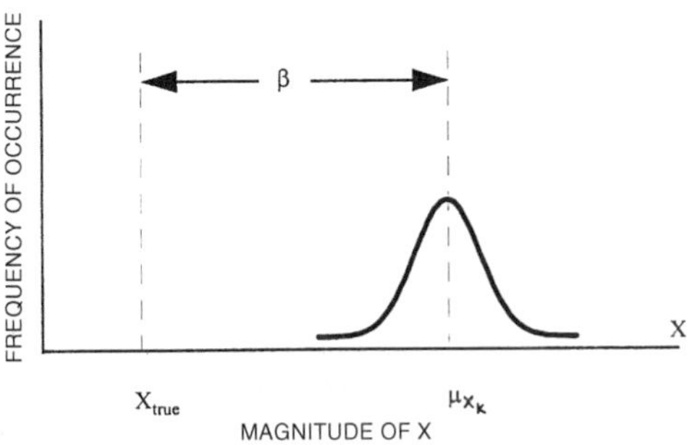

A. INFINITE NUMBER OF READINGS

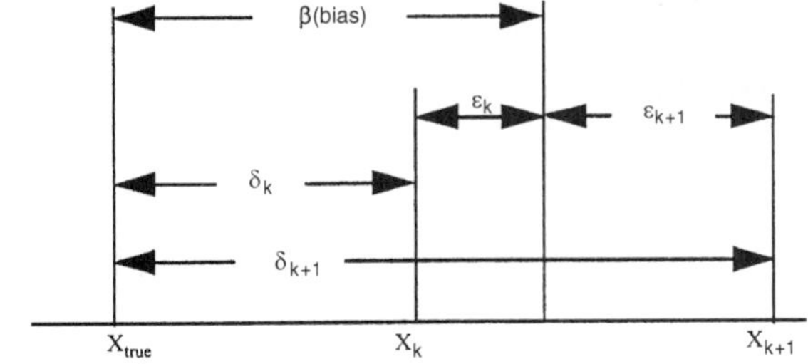

B. TWO READINGS

Fig. 1 Errors in the Measurement of a Variable *X*

For each bias error source, the experimenter must estimate a **systematic uncertainty**. Systematic uncertainties are usually estimated from previous experience, calibration data, analytical models, and engineering judgment. For a discussion on estimating systematic uncertainties (bias limits), see Coleman and Steele (1989).

For further information on measurement uncertainty, see ASME *Standards* MFC-2M and PTC 19.1 and Coleman and Steele (1995).

TEMPERATURE MEASUREMENT

Instruments for measuring temperature are listed in Table 1. Temperature sensor output must be related to an accepted temperature scale. This is achieved by manufacturing the instrument according to certain specifications or by calibrating it against a temperature standard. To help users conform to standard temperatures and temperature measurements, the International Committee of Weights and Measures (CIPM) has adopted the International Temperature Scale of 1990 (ITS-90).

The unit of temperature of the ITS-90 is the kelvin (K) and has a size equal to the fraction 1/273.16 of the thermodynamic temperature of the triple point of water.

The ITS-90 is maintained in the United States by the National Institute of Standards and Technology (NIST), and any laboratory may obtain calibrations from NIST based on this scale.

Benedict (1984), Considine (1985), Quinn (1990), Schooley (1986, 1992), and DeWitt and Nutter (1988) cover temperature measurement in more detail.

Sampling and Averaging

Although temperature is usually measured within, and is associated with, a relatively small volume (depending on the size of the thermometer), it can also be associated with an area (e.g., on a surface or in a flowing stream). To determine average stream temperature, the cross section must be divided into smaller areas and the temperature of each area measured. The temperatures measured are then combined into a weighted mass flow average by either (1) using equal areas and multiplying each temperature by the fraction of total mass flow in its area or (2) using areas of size inversely proportional to mass flow and taking a simple arithmetic average of the temperatures in each. A means of mixing or selective sampling may be preferable to these cumbersome procedures. While mixing can occur from turbulence alone, **transposition** is much more effective. In transposition, the stream is divided into parts determined by the type of stratification, and alternate parts pass through one another.

Static Temperature Versus Total Temperature

When a fluid stream impinges on a temperature-sensing element such as a thermometer or thermocouple, the element is at a temperature greater than the true stream temperature. The difference is a fraction of the temperature equivalent of the stream velocity t_e.

$$t_e = \frac{3600V^2}{2g_cJc_p} \tag{1}$$

where

t_e = temperature equivalent of stream velocity, °F
V = velocity of stream, fpm
g_c = gravitational constant = 32.174 $lb_m \cdot ft/lb_f \cdot s^2$
J = mechanical equivalent of heat, 778.3 $ft \cdot lb_f/Btu$
c_p = specific heat of stream at constant pressure, $Btu/lb_m \cdot °F$

Table 1 Temperature Measurement

Measurement Means	Application	Approximate Range, °F	Uncertainty, °F	Limitations
Liquid-in-glass thermometers				
Mercury-in-glass	Temperature of gases and liquids by contact	−36/1000	0.05 to 3.6	In gases, accuracy affected by radiation
Organic	Temperature of gases and liquids by contact	−330/400	0.05 to 3.6	In gases, accuracy affected by radiation
Gas thermometer	Primary standard	−456/1200	Less than 0.02	Requires considerable skill to use
Resistance thermometers				
Platinum	Precision; remote readings; temperature of fluids or solids by contact	−430/1800	Less than 0.0002 to 0.2	High cost; accuracy affected by radiation in gases
Rhodium-iron	Transfer standard for cryogenic applications	−460/−400	0.0002 to 0.2	High cost
Nickel	Remote readings; temperature by contact	−420/400	0.02 to 2	Accuracy affected by radiation in gases
Germanium	Remote readings; temperature by contact	−460/−400	0.0002 to 0.2	
Thermistors	Remote readings; temperature by contact	Up to 400	0.0002 to 0.2	
Thermocouples				
Pt-Rh/Pt (type S)	Standard for thermocouples on IPTS-68, not on ITS-90	32/2650	0.2 to 5	High cost
Au/Pt	Highly accurate reference thermometer for laboratory applications	−60/1800	0.1 to 2	High cost
Types K and N	General testing of high temperature; remote rapid readings by direct contact	Up to 2300	0.2 to 18	Less accurate than listed above thermocouples
Iron/Constantan (type J)	Same as above	Up to 1400	0.2 to 10	Subject to oxidation
Copper/Constantan (type T)	Same as above, especially suited for low temperature	Up to 660	0.2 to 5	
Ni-Cr/Constantan (type E)	Same as above, especially suited for low temperature	Up to 1650	0.2 to 13	
Beckman thermometers (metastatic)	For differential temperature in same applications as in glass-stem thermometer	10°F scale, used 32 to 212°F	0.01	Must be set for temperature to be measured
Bimetallic thermometers	For approximate temperature	−4/1200	2, usually much more	Time lag; unsuitable for remote use
Pressure-bulb thermometers				
Gas-filled bulb	Remote testing	−100/1200	4	Caution must be exercised so that installation is correct
Vapor-filled bulb	Remote testing	−25/500	4	Caution must be exercised so that installation is correct
Liquid-filled bulb	Remote testing	−60/2100	4	Caution must be exercised so that installation is correct
Optical pyrometers	For intensity of narrow spectral band of high-temperature radiation (remote)	1500 and up	30	
Radiation pyrometers	For intensity of total high-temperature radiation (remote)	Any range		
Seger cones (fusion pyrometers)	Approximate temperature (within temperature source)	1200/3600	90	
Triple points, freezing/melting points, and boiling points of materials	Standards	All except extremely high temperatures	Extremely precise	For laboratory use only

This fraction of the temperature equivalent of the velocity is the **recovery factor**, and it varies from 0.6 to 0.8°F for bare thermometers to 1.0°F for aerodynamically shielded thermocouples. For precise temperature measurement, each temperature sensor must be calibrated to determine its recovery factor. However, for most applications where air velocities are below 2000 fpm, the recovery factor can be omitted.

Various temperature sensors are available for temperature measurement in fluid streams. The principal sensors are the **static temperature thermometer**, which indicates true stream temperature but is cumbersome, and the **thermistor**, used for accurate temperature measurement within a limited range.

LIQUID-IN-GLASS THERMOMETERS

Any device that changes monotonically with temperature is a thermometer; however, the term usually signifies an ordinary liquid-in-glass temperature-indicating device. Mercury-filled thermometers have a useful range from −37.8°F, the freezing point of mercury, to about 1000°F, near which the glass usually softens. Lower temperatures can be measured with organic-liquid-filled thermometers (e.g., alcohol-filled), with ranges of −330 to 400°F. During manufacture, thermometers are roughly calibrated for at least two temperatures, often the freezing and boiling points of water; space between the calibration points is divided into desired scale divisions. Thermometers that are intended for precise measurement applications have scales etched into the glass that forms their stems. The probable error for as-manufactured, etched-stem thermometers is ±1 scale division. The highest quality mercury thermometers may have uncertainties of ±0.06 to ±4°F if they have been calibrated by comparison against primary reference standards.

Liquid-in-glass thermometers are used for many applications within the HVAC industry. Some of these uses include local temperature indication of process fluids related to HVAC systems, such as cooling and heating fluids and air.

The use of mercury-in-glass thermometers as temperature measurement standards is fairly common because of their relatively high accuracy and low cost. Such thermometers used as references must be calibrated on the ITS-90 by comparison in a uniform bath with a standard platinum resistance thermometer that has been calibrated either by the appropriate standards agency or by a laboratory that has direct traceability to the standards agency and the ITS-90. Such a calibration is necessary in order to determine the proper corrections to be applied to the scale readings. For application and calibration of liquid-in-glass thermometers, refer to NIST (1976, 1986).

Liquid-in-glass thermometers are calibrated by the manufacturer for total or partial stem immersion. If a thermometer calibrated for total immersion is used at partial immersion (i.e., with a portion of the liquid column at a temperature different from that of the bath), an emergent stem correction must be made. This correction can be calculated as follows:

$$\text{Stem correction} = Kn(t_b - t_s) \tag{2}$$

where

K = differential expansion coefficient of mercury or other liquid in glass. K is 0.00009 for Fahrenheit mercurial thermometers. For K values for other liquids and specific glasses, refer to Schooley (1992).
n = number of degrees that liquid column emerges from bath
t_b = temperature of bath, °F
t_s = average temperature of emergent liquid column of n degrees, °F

Sources of Thermometer Errors

A thermometer measuring gas temperatures can be affected by radiation from surrounding surfaces. If the gas temperature is approximately the same as that of the surrounding surfaces, radiation effects can be ignored. If the temperature differs considerably from that of the surroundings, radiation effects should be minimized by shielding or aspiration (ASME *Standard* PTC 193). **Shielding** may be provided by highly reflective surfaces placed between the thermometer bulb and the surrounding surfaces such that air movement around the bulb is not appreciably restricted (Parmelee and Huebscher 1946). Improper shielding can increase errors. **Aspiration** results from passing a high-velocity stream of air or gas over the thermometer bulb.

When a **thermometer well** within a container or pipe under pressure is required, the thermometer should fit snugly and be surrounded with a high thermal conductivity material (oil, water, or mercury, if suitable). Liquid in a long, thin-walled well is advantageous for rapid response to temperature changes. The surface of the pipe or container around the well should be insulated to eliminate heat transfer to or from the well.

Industrial thermometers are available for permanent installation in pipes or ducts. These instruments are fitted with metal guards to prevent breakage and are useful for many other purposes. The considerable heat capacity and conductance of the guards or shields can cause errors, however.

Allowing ample time for the thermometer to attain temperature equilibrium with the surrounding fluid prevents excessive errors in temperature measurements. When reading a liquid-in-glass thermometer, the eye should be kept at the same level as the top of the liquid column to avoid parallax.

RESISTANCE THERMOMETERS

Resistance thermometers depend on a change of the electrical resistance of a sensing element (usually metal) with a change in temperature; resistance increases with increasing temperature. The use of resistance thermometers largely parallels that of thermocouples, although readings are usually unstable above about 1000°F. Two-lead temperature elements are not recommended because they do not permit correction for lead resistance. Three leads to each resistor are necessary to obtain consistent readings, and four leads are preferred. Wheatstone bridge circuits or 6-1/2-digit multimeters can be used for measurements.

A typical circuit used by several manufacturers is shown in Figure 2. In this design, a differential galvanometer is used in which coils L and H exert opposing forces on the indicating needle. Coil L is in series with the thermometer resistance AB, and coil H is in series with the constant resistance R. As the temperature falls, the resistance of AB decreases, allowing more current to flow through coil L than through coil H. This causes an increase in the force exerted by coil L, pulling the needle down to a lower reading. Likewise, as the temperature rises, the resistance of AB increases, causing less current to flow through coil L than through coil H. This forces the indicating needle to a higher reading. Rheostat S must be adjusted occasionally to maintain a constant current.

The resistance thermometer is more costly to make and likely to have considerably longer response times than thermocouples. A resistance thermometer gives best results when used to measure steady or slowly changing temperature.

Resistance Temperature Devices

Resistance temperature devices (RTDs) are typically constructed from platinum, rhodium-iron, nickel, nickel-iron, tungsten, or copper. These devices are further characterized by their simple circuit designs, high degree of linearity, good sensitivity, and excellent stability. The choice of materials for an RTD usually depends on the intended application; temperature range, corrosion protection, mechanical stability, and cost are some of the selection criteria.

Platinum RTDs. Presently, for HVAC applications, RTDs constructed of platinum are the most widely used. Platinum is extremely stable and corrosion-resistant. Platinum RTDs are highly malleable and can thus be drawn into fine wires; they can also be manufactured at low cost as thin films. They have a high melting point and can be refined to a high degree of purity, thus attaining highly reproducible results. Due to these properties, platinum RTDs are used to define the ITS-90 for the range of 13.8033 K (triple point of equilibrium hydrogen) to 1234.93 K (freezing point of silver).

Platinum resistance temperature devices can measure the widest range of temperatures and are the most accurate and stable temperature sensors. Their resistance-temperature relationship is one of the most linear. The higher the purity of the platinum, the more stable and accurate the sensor. With high-purity platinum, primary grade platinum RTDs are capable of achieving reproducibility of ±0.00002°F, whereas the minimum uncertainty of a recently calibrated thermocouple is ±0.4°F.

Platinum RTD Design. The most widely used RTD is designed with a resistance of 100 Ω at 32°F (R_0 = 100 Ω). Other RTDs are available that use lower resistances at temperatures above 1100°F. The lower the resistance value, the faster the response time for sensors of the same size.

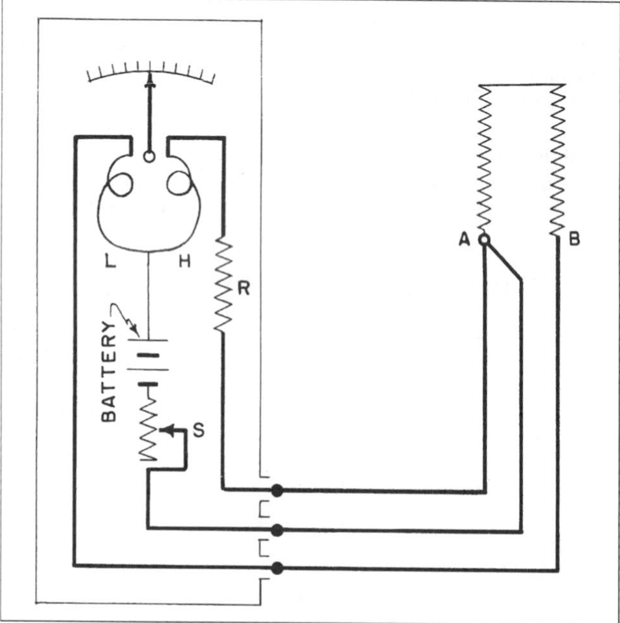

Fig. 2 Typical Resistance Thermometer Circuit

Thin-Film Platinum RTDs. Thin-film 1000 Ω platinum RTDs are readily available. They have the excellent linear properties of lower resistance platinum RTDs and are more cost-effective because they are mass produced and have lower platinum purity. However, the problem with many platinum RTDs with R_0 values of greater than 100 Ω is the difficulty in obtaining transmitters or electronic interface boards from sources other than the RTD manufacturer. In addition to a nonstandard interface, higher R_0 value platinum RTDs may have higher self-heating losses if the excitation current is not controlled properly.

Thin-film RTDs have the advantages of lower cost and smaller sensor size. They are specifically adapted to surface mounting. Thin-film sensors tend to have an accuracy limitation of ±0.1% or ±0.2°F. This may prove to be adequate for most HVAC applications; only in tightly controlled facilities may users wish to install the standard wire-wound platinum RTDs with accuracies of 0.01% or ±0.02°F (these are available upon special request for certain temperature ranges).

Assembly and Construction. Regardless of the R_0 resistance value of RTDs, their assembly and construction are relatively simple. The electrical connections come in three basic types, depending on the number of wires to be connected to the resistance measurement circuitry. Two, three, or four wires are used for electrical connection using a Wheatstone bridge or a variation of it (Figure 3).

In the basic two-wire configuration, the resistance of the RTD is measured through the two connecting wires. Because the connecting wires extend from the site of the temperature measurement, any additional changes in resistivity due to a change in temperature may affect the measured resistance. Three- and four-wire assemblies are built to compensate for the connecting lead resistance values. The original three-wire circuit improved the resistance measurement by adding a compensating wire to the voltage side of the circuit. This helps reduce part of the connecting wire resistance. When more accurate measurements (better than ±0.2°F) are required, the four-wire bridge is recommended. The four-wire bridge eliminates all connecting wire resistance errors.

All the bridges discussed here are direct current (dc) circuits and were used extensively until the advent of precision alternating current (ac) circuits using microprocessor-controlled ratio transformers, dedicated analog-to-digital converters, and other solid-state

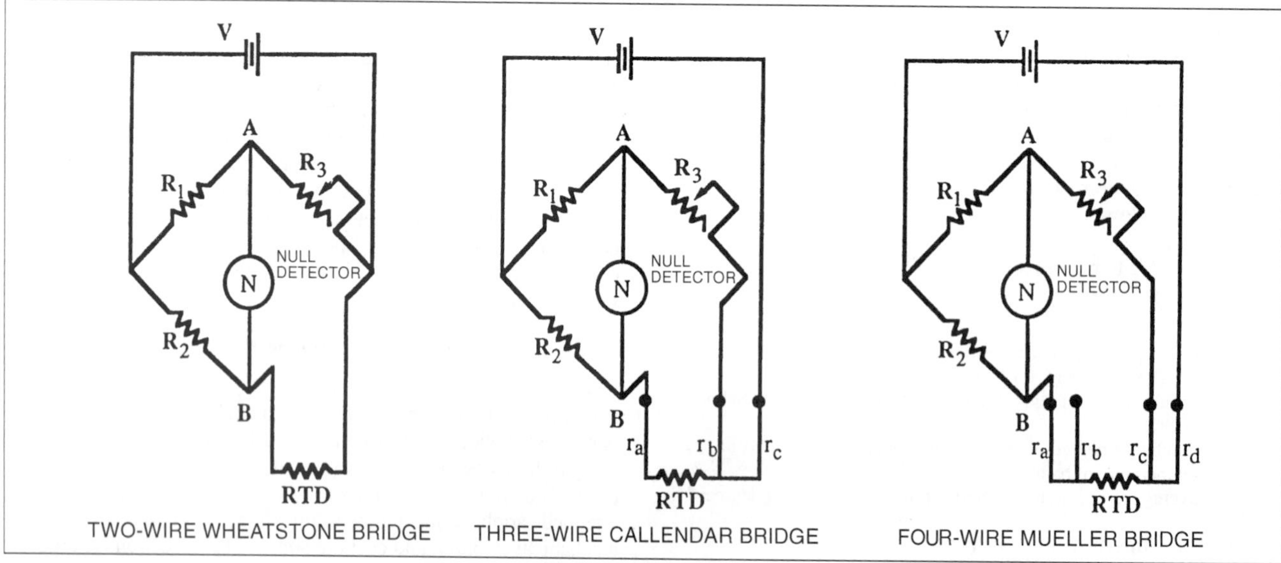

Fig. 3 Typical Resistance Temperature Device Bridge Circuits

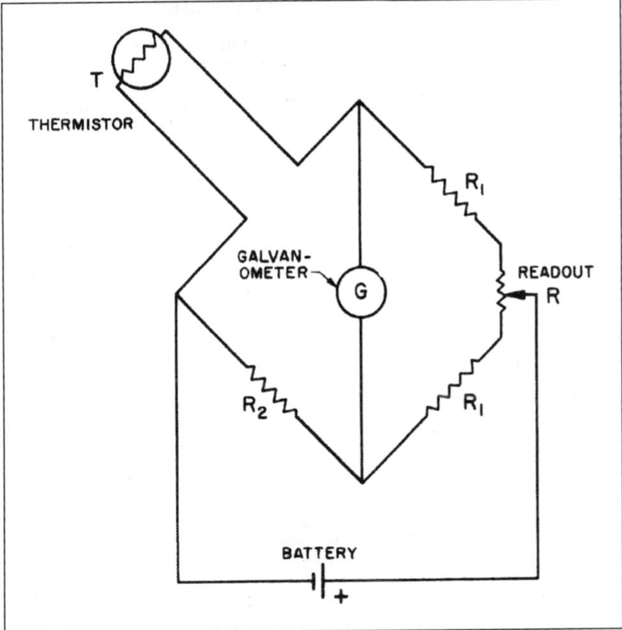

Fig. 4 Basic Thermistor Circuit

devices that measure resistance with uncertainties of less than 1 ppm. Resistance measurement technology now allows more portable thermometers, lower cost, ease of use, and high-precision temperature measurement in industrial uses.

Thermistors

Certain semiconductor compounds (usually sintered metallic oxides) exhibit large changes in resistance with temperature, usually decreasing as the temperature increases. For use, the thermistor element may be connected by lead wires into a galvanometer bridge circuit and calibrated. Alternatively, a 6-1/2-digit multimeter and a constant-current source with a means for reversing the current to eliminate thermal electromotive force (emf) effects may also be used. This method of measurement is easier and faster, and it may be more precise and accurate. Thermistors are usually applied to electronic temperature compensation circuits, such as thermocouple reference junction compensation, or to other applications where high resolution and limited operating temperature ranges exist. Figure 4 illustrates a typical thermistor circuit.

Semiconductor Devices

In addition to the positive resistance coefficient RTDs and the negative resistance coefficient thermistor, there are two other types of devices that vary resistance or impedance with temperature. Although the principle of their operation has long been known, their reliability was questioned due to imprecise manufacturing techniques. Improvements in silicon microelectronics manufacturing techniques have brought semiconductors to the point where low-cost, precise temperature sensors are commercially available.

Elemental Semiconductors. Due to controlled doping of impurities into elemental germanium, a germanium semiconductor is a reliable temperature sensor for cryogenic temperature measurement in the range of 1.8 to 150°R.

Junction Semiconductors. The first simple junction semiconductor device consisted of a single diode or transistor, in which the forward-connected base emitter voltage is very sensitive to temperature. Today the more common form is a pair of diode-connected transistors, which make the device suitable for ambient temperature measurement. Applications include thermocouple reference junction compensation.

The primary advantages of silicon transistor temperature sensors are their extreme linearity and exact R_0 value. Another advantage is the incorporation of signal conditioning circuitry into the same device as the sensor element. As with thermocouples, these semiconductors require highly precise manufacturing techniques, extremely precise voltage measurements, multiple-point calibration, and temperature compensation to achieve an accuracy as high as ±0.02°F, but with a much higher cost. Lower cost devices achieve accuracies of ±0.2°F using mass manufacturing techniques and single-point calibration. A mass-produced silicon temperature sensor can be interchanged easily. If one device fails, only the sensor element need be changed. Electronic circuitry can be used to recalibrate the new device.

Winding Temperature. The winding temperature of electrical operating equipment is usually determined from the resistance change of these windings in operation. With copper windings, the relation between these parameters is

$$\frac{R_1}{R_2} = \frac{148 + t_1}{148 + t_2} \tag{3}$$

where

R_1 = winding resistance at temperature t_1, Ω
R_2 = winding resistance at temperature t_2, Ω
t_1, t_2 = winding temperatures, °F

The classical method of determining winding temperature is to measure the equipment when it is inoperative and temperature-stabilized at room temperature. After the equipment has operated sufficiently to cause temperature stabilization under load conditions, the winding resistance should be measured again. The latter value is obtained by taking resistance measurements at known short time intervals after shutdown. These values may be extrapolated to zero time to indicate the winding resistance at the time of shutdown. The obvious disadvantage of this method is that the device must be shut down to determine winding temperature. A circuit described by Seely (1955), however, makes it possible to measure resistances while the device is operating.

THERMOCOUPLES

When two wires of dissimilar metals are joined by soldering, welding, or twisting, they form a thermocouple junction or **thermojunction**. An emf that depends on the wire materials and the junction temperature exists between the wires. This is known as the **Seebeck voltage**.

Thermocouples for temperature measurement yield less precise results than platinum resistance thermometers, but except for glass thermometers, thermocouples are the most common instruments of temperature measurement for the range of 32 to 1800°F. Due to their low cost, moderate reliability, and ease of use, thermocouples continue to maintain widespread acceptance.

The most commonly used thermocouples in industrial applications are assigned letter designations. The tolerances of such commercially available thermocouples are given in Table 2.

Because the measured emf is a function of the difference in temperature and the type of dissimilar metals used, a known temperature at one junction is required, whereas the remaining junction temperature may be calculated. It is common practice to call the one with a known temperature the (cold) **reference** junction and the one with the unknown temperature the (hot) **measured** junction. The reference junction is typically kept at a reproducible temperature, such as the ice point of water.

Various systems are used to maintain the reference junction temperature—a mixture of ice and water contained in an insulated flask or commercially available thermoelectric coolers to maintain the ice point temperature automatically within a reference chamber. When

Table 2 Thermocouple Tolerances on Initial Values of Electromotive Force Versus Temperature

Thermocouple Type	Material Identification	Temperature Range, °F	Reference Junction Tolerance at 32°F[a]	
			Standard Tolerance (whichever is greater)	Special Tolerance (whichever is greater)
T	Copper versus Constantan	32 to 700	±1.8°F or ±0.75%	±0.9°F or ±0.4%
J	Iron versus Constantan	32 to 1400	±4°F or ±0.75%	±2°F or ±0.4%
E	Nickel-10% Chromium versus Constantan	32 to 1600	±3.1°F or ±0.5%	±1.8°F or ±0.4%
K	Nickel-10% Chromium versus 5% Aluminum, Silicon	32 to 2300	±4°F or ±0.75%	±2°F or ±0.4%
N	Nickel-14% Chromium, 1.5% Silicon versus Nickel-4.5% Silicon, 0.1% Magnesium	32 to 2300	±4°F or ±0.75%	±2°F or ±0.4%
R	Platinum-13% Rhodium versus Platinum	32 to 2700	±2.7°F or ±0.25%	±1.1°F or ±0.1%
S	Platinum-10% Rhodium versus Platinum	32 to 2700	±2.7°F or ±0.25%	±1.1°F or ±0.1%
B	Platinum-30% Rhodium versus Platinum-6% Rhodium	1600 to 3100	±0.5%	±0.25%
T[b]	Copper versus Constantan	−328 to 32	±1.8°F or ±1.5%	c
E[b]	Nickel-10% Chromium versus Constantan	−328 to 32	±3.1°F or ±1%	c
K[b]	Nickel-10% Chromium versus 5% Aluminum, Silicon	−328 to 32	±4°F or ±2%	c

Source: ASTM *Standard* E 230, Temperature-Electromotive Force (EMF) Tables for Standardized Thermocouples.

[a]Tolerances in this table apply to new thermocouple wire, normally in the size range of 0.01 to 0.1 in. diameter and used at temperatures not exceeding the recommended limits. Thermocouple wire is available in two grades: standard and special.

[b]Thermocouples and thermocouple materials are normally supplied to meet the tolerance specified in the table for temperatures above 32°F. The same materials, however, may not fall within the tolerances given in the second section of the table when operated below freezing (32°F). If materials are required to meet tolerances at subfreezing temperatures, the purchase order must state so.

[c]Little information is available to justify establishing special tolerances for below-freezing temperatures. Limited experience suggests the following special tolerances for types E and T thermocouples:

Type E −328 to 32°F; ±2°F or ±0.5% (whichever is greater)

Type T −328 to 32°F; ±1°F or ±0.8% (whichever is greater)

These tolerances are given only as a guide for discussion between purchaser and supplier.

these systems cannot be used in an application, measuring instruments with automatic reference junction temperature compensation may be used.

As previously described, the principle for measuring temperature with a thermocouple is based on the accurate measurement of the Seebeck voltage. The acceptable dc voltage measurement methods are (1) millivoltmeter, (2) millivolt potentiometer, and (3) a high-input impedance digital voltmeter. Many digital voltmeters include built-in software routines for the direct calculation and display of temperature. Regardless of the method selected, many options to simplify the measurement process are available.

Solid-state digital readout devices in combination with a millivolt- or microvoltmeter, as well as packaged thermocouple readouts with built-in cold junction and linearization circuits, are available. The latter requires a proper thermocouple to provide direct meter reading of temperature. Accuracy approaching or surpassing that of potentiometers can be attained, depending on the instrument quality. This method is popular because it eliminates the null balancing requirement and reads temperature directly in a digital readout.

Wire Diameter and Composition

Thermocouple wire is selected by considering the temperature to be measured, the corrosion protection afforded to the thermocouple, and the precision and service life required. Type T thermocouples are suitable for temperatures up to 700°F; type J, up to 1400°F; and types K and N, up to 2300°F. Higher temperatures require noble metal thermocouples (type S, R, or B), which have a higher initial cost and do not develop as high an emf as the base metal thermocouples. Thermocouple wires of the same type have small compositional variation from lot to lot from the same manufacturer and especially among different manufacturers. Consequently, calibrating samples from each wire spool is essential for precision. Calibration data on wire may be obtained from the manufacturer.

Reference functions are available for relating temperature and emf of letter-designated thermocouple types. Such functions are easy to use with computers. The functions depend on thermocouple type and on the range of temperature; they are used to generate reference tables of emf as a function of temperature but are not well suited for calculating temperatures directly from values of emf. Approximate inverse functions are available, however, for calculating temperature and are of the form

$$t = \sum_{i=0}^{n} a_i E^i \tag{4}$$

where t = temperature, a = thermocouple constant, and E = voltage. Burns et al. (1992) give the reference functions and approximate inverses for all letter-designated thermocouples.

The emf of a thermocouple, as measured with a high-input impedance device, is independent of the diameters of its constituent wires. Thermocouples with small-diameter wires respond faster to temperature changes and are less affected by radiation than larger ones. Large-diameter wire thermocouples, however, are necessary for high-temperature work when wire corrosion is a problem. For use in heated air or gases, thermocouples are often shielded and sometimes aspirated. An arrangement for avoiding error due to radiation involves using several thermocouples of different wire sizes and estimating the true temperature by extrapolating readings to zero diameter.

With thermocouples, temperatures can be indicated or recorded remotely on conveniently located instruments. Because thermocouples can be made of small-diameter wire, they can be used to measure temperatures within thin materials, within narrow spaces, or in otherwise inaccessible locations.

Multiple Thermocouples

Thermocouples in series, with alternate junctions maintained at a common temperature, produce an emf that, when divided by the number of thermocouples, gives the average emf corresponding to the temperature difference between two sets of junctions. This series arrangement of thermocouples, often called a **thermopile**, is used to increase sensitivity and is often used for measuring small temperature changes and differences.

Connecting a number of thermocouples of the same type in parallel with a common reference junction is useful for obtaining an average temperature of an object or volume. In such measurements, however, it is important that the electrical resistances of the individual thermocouples be the same. The use of thermocouples in series and parallel arrangements is discussed in ASTM *Manual* 12.

Surface Temperature Measurement

The thermocouple is useful in determining surface temperature. It can be attached to a metal surface in several ways. For permanent installations, soldering, brazing, or peening is suggested. For peening, a small hole is drilled and the thermocouple measuring junction is driven into it. For temporary arrangements, thermocouples can be attached by tape, adhesive, or putty-like material. For boiler or furnace surfaces, furnace cement should be used. To minimize the possibility of error caused by heat conduction along wires, a surface thermocouple should be made of fine wires placed in close contact with the surface being measured for about an inch from the junction to ensure good thermal contact. The wires must be insulated electrically from each other and from the metal surface (except at the junction).

Thermocouple Construction

The thermocouple wires are typically insulated with fibrous glass, fluorocarbon resin, or ceramic insulators. In another form of thermocouple, the thermocouple wires are insulated with compacted ceramic insulation inside a metal sheath. This form of thermocouple provides both mechanical protection and protection from stray electromagnetic fields. The measuring junction may be exposed or enclosed within the metal sheath. An enclosed junction may be either grounded or ungrounded to the metal sheath.

For the exposed junction type, the measuring junction is in direct contact with the process stream; it is therefore subject to corrosion or contamination but provides a fast temperature response. The grounded enclosed junction type, in which the thermocouple wires are welded to the metal sheath, provides electrical grounding, as well as mechanical and corrosion protection. This type, however, has a slower response time than the exposed junction type. With the ungrounded enclosed junction construction, the response time is even slower, but the thermocouple wires are isolated electrically and are less susceptible to some forms of mechanical strain than those with grounded construction.

INFRARED RADIOMETERS

Infrared radiation thermometers, also known as remote temperature sensors (Hudson 1969), permit noncontact measurement of surface temperature over a wide range. In these instruments, radiant flux from the observed object is focused by an optical system onto an infrared detector that generates an output signal proportional to the incident radiation that can be read from a meter or display unit. Point and scanning radiometers are available; the latter are able to display the temperature variation existing within the field of view.

Radiometers are usually classified according to the detector used—either thermal or photon. In thermal detectors, a change in electrical property is caused by the heating effect of the incident radiation. Examples of thermal detectors are the thermocouple, the thermopile, and metallic and semiconductor bolometers. In photon detectors, a change in electrical property is caused by the surface absorption of incident photons. Because these detectors do not require an increase in temperature for activation, their response time is much shorter than that of thermal detectors. Scanning radiometers usually use photon detectors.

A radiometer only measures the power level of the radiation incident on the detector; this incident radiation is a combination of the thermal radiation emitted by the object and the surrounding background radiation reflected from the surface of the object. An accurate measurement of the temperature, therefore, requires knowledge of the long-wavelength emissivity of the object as well as the effective temperature of the thermal radiation field surrounding the object. Calibration against an internal or external source of known temperature and emissivity is required in order to obtain true surface temperature from the radiation measurements.

The temperature resolution of a radiometer decreases as the object temperature decreases. For example, a radiometer that can resolve a temperature difference of 0.5°F on an object near 70°F may only resolve a difference of 2°F on an object at 32°F.

INFRARED THERMOGRAPHY

Infrared thermography is the discipline concerned with the acquisition and analysis of thermal information in the form of images from an infrared imaging system. An infrared imaging system consists of (1) an infrared television camera and (2) a display unit. The infrared television camera scans a surface and senses the self-emitted and reflected radiation viewed from the surface. The display unit contains either a cathode-ray tube (CRT) that displays a gray-tone or color-coded thermal image of the surface or a color liquid crystal display (LCD) screen. A photograph of the image on the CRT is called a **thermogram**. An introductory treatise on infrared thermography is given by Paljak and Pettersson (1972).

Thermography has been used successfully to detect missing insulation and air infiltration paths in building envelopes (Burch and Hunt 1978). Standard practices for conducting thermographic inspections of buildings are given in ASTM *Standard* C 1060. A technique for quantitatively mapping the heat loss in building envelopes is given by Mack (1986).

Aerial infrared thermography of buildings is effective in identifying regions of an individual built-up roof that have wet insulation (Tobiasson and Korhonen 1985), but it is ineffective in ranking a group of roofs according to their thermal resistance (Goldstein 1978, Burch 1980). In this latter application, the emittances of the separate roofs and outdoor climate (i.e., temperature and wind speed) throughout the microclimate often produce changes in the thermal image that may be incorrectly attributed to differences in thermal resistance.

Industrial applications include locating defective or missing pipe insulation in buried heat distribution systems, surveys of manufacturing plants to quantify energy loss from equipment, and locating defects in coatings (Bentz and Martin 1987).

HUMIDITY MEASUREMENT

Any instrument capable of measuring the humidity or psychrometric state of air is a hygrometer, and many are available. The indication sensors used on the instruments respond to different moisture property contents. These responses are related to factors such as wet-bulb temperature, relative humidity, humidity (mixing) ratio, dew point, and frost point.

Table 3 lists instruments for measuring humidity. Each is capable of accurate measurement under certain conditions and within specific limitations. The following sections describe various instruments used to measure humidity.

PSYCHROMETERS

A typical industrial psychrometer consists of a pair of matched electrical or mechanical temperature sensors, one of which is kept wet with a moistened wick. A blower aspirates the sensor, which lowers the temperature at the moistened temperature sensor. The lowest temperature depression occurs when the evaporation rate required to saturate the moist air adjacent to the wick is constant. This is a steady-state, open-loop, nonequilibrium process, which depends on the purity of the water, the cleanliness of the wick, the ventilation rate, radiation effects, the size and accuracy of the temperature sensors, and the transport properties of the gas.

Table 3 Humidity Sensor Properties

Type of Sensor	Sensor Category	Method of Operation	Approximate Range	Some Uses	Approximate Accuracy
Psychrometer	Evaporative cooling	Temperature measurement of wet bulb	32 to 180°F	Measurement, standard	±3 to ±7% rh
Adiabatic saturation psychrometer	Evaporative cooling	Temperature measurement of thermodynamic wet bulb	40 to 85°F	Measurement, standard	±0.2 to ±2% rh
Chilled mirror	Dew point	Optical determination of moisture formation	−110 to 200°F dp	Measurement, control, meteorology	±0.4 to ±4°F
Heated saturated salt solution	Water vapor pressure	Vapor pressure depression in salt solution	−20 to 160°F dp	Measurement, control, meteorology	±3°F
Hair	Mechanical	Dimensional change	5 to 100% rh	Measurement, control	±5% rh
Nylon	Mechanical	Dimensional change	5 to 100% rh	Measurement, control	±5% rh
Dacron thread	Mechanical	Dimensional change	5 to 100% rh	Measurement	±7% rh
Goldbeater's skin	Mechanical	Dimensional change	5 to 100% rh	Measurement	±7% rh
Cellulosic materials	Mechanical	Dimensional change	5 to 100% rh	Measurement, control	±5% rh
Carbon	Mechanical	Dimensional change	5 to 100% rh	Measurement	±5% rh
Dunmore type	Electrical	Impedance	7 to 98% rh at 40 to 140°F	Measurement, control	±1.5% rh
Ion exchange resin	Electrical	Impedance or capacitance	10 to 100% rh at −40 to 190°F	Measurement, control	±5% rh
Porous ceramic	Electrical	Impedance or capacitance	Up to 400°F	Measurement, control	±1 to ±1.5% rh
Aluminum oxide	Electrical	Capacitance	5 to 100% rh	Measurement, control	±3% rh
Aluminum oxide	Electrical	Capacitance	−110 to 140°F dp	Trace moisture measurement, control	±2°F dp
Electrolytic hygrometer	Electrical	Capacitance			
Coulometric	Electrolytic cell	Electrolyzes due to adsorbed moisture	1 to 1000 ppm	Measurement	
Infrared laser diode	Electrical	Optical diodes	0.1 to 100 ppm	Trace moisture measurement	±0.1 ppm
Surface acoustic wave	Electrical	SAW attenuation	85 to 98% rh	Measurement, control	±1% rh
Piezoelectric	Mass sensitive	Mass changes due to adsorbed moisture	−100 to 0°F	Trace moisture measurement, control	±2 to ±10°F dp
Radiation absorption	Moisture absorption	Moisture absorption of UV or IR radiation	0 to 180°F dp	Measurement, control, meteorology	±4°F dp, ±5% rh
Gravimetric	Direct measurement of mixing ratio	Comparison of sample gas with dry airstream	120 to 20,000 ppm mixing ratio	Primary standard, research and laboratory	±0.13% of reading
Color change	Physical	Color changes	10 to 80% rh	Warning device	±10% rh

Notes:
1. This table does not include all available technology for humidity measurement.
2. The approximate range for the device types listed is based on surveys of device manufacturers.
3. The approximate accuracy is based on manufacturers' data.
4. Presently, NIST only certifies instruments whose operating range is within −103 to 212°F dew point.

ASHRAE *Standard* 41.6 recommends an airflow over both the wet and dry bulbs of 600 to 1000 fpm for transverse ventilation and 300 to 500 fpm for axial ventilation.

The **sling psychrometer** consists of two thermometers mounted side by side in a frame fitted with a handle for whirling the device through the air. The thermometers are spun until their readings become steady. In the **ventilated** or **aspirated psychrometer**, the thermometers remain stationary, and a small fan, blower, or syringe moves the air across the thermometer bulbs. Various designs are used in the laboratory, and commercial models are available.

Other temperature sensors, such as thermocouples and thermistors, are also used and can be adapted for recording temperatures or for use where a small instrument is required. Small-diameter wet-bulb sensors operate with low ventilation rates.

Charts and tables showing the relationship between the temperatures and humidity are available. Data are usually based on a barometric pressure equal to one standard atmosphere. To meet special needs, charts can be produced that apply to nonstandard pressure (e.g., the ASHRAE 7500 ft psychrometric chart). Alternatively, mathematical calculations can be made (Kusuda 1965). Uncertainties of 3 to 7% rh are typical for psychrometer-based derivation. The degree of uncertainty is a function of the accuracy of the temperature measurements, wet and dry bulb, knowledge of the barometric pressure, and conformance to accepted operational procedures such as those outlined in ASHRAE *Standard* 41.6.

In air temperatures below 32°F, the water on the wick may either freeze or supercool. Because the wet-bulb temperature is different for ice and water, the state must be known and the proper chart or table used. Some operators remove the wick from the wet-bulb for freezing conditions and dip the bulb in water a few times; this allows water to freeze on the bulb between dips, forming a film of ice. Because the wet-bulb depression is slight at low temperatures, precise temperature readings are essential. A psychrometer can be used at high temperatures, but if the wet-bulb depression is large, the wick must remain wet and water supplied to the wick must be cooled so as not to influence the wet-bulb temperature by carrying sensible heat to it (Richardson 1965, Worrall 1965).

Greenspan and Wexler (1968) and Wentzel (1961) developed devices to measure adiabatic saturation temperature.

DEW-POINT HYGROMETERS

Condensation Dew-Point Hygrometers

The condensation (chilled mirror) dew-point hygrometer is an accurate and reliable instrument with a wide humidity range. However, these features are obtained through an increase in complexity and cost compared to the psychrometer. In the condensation hygrometer, a surface is cooled (thermoelectrically, mechanically, or chemically) until dew or frost begins to condense out. The condensate surface is maintained electronically in vapor pressure equilibrium with the surrounding gas, while surface condensation is detected by optical, electrical, or nuclear techniques. The measured surface temperature is then the dew-point temperature.

The largest source of error in a condensation hygrometer stems from the difficulty in measuring condensate surface temperature accurately. Typical industrial versions of the instrument are accurate to ±1.0°F over wide temperature spans. With proper attention to the condensate surface temperature measuring system, errors can be reduced to about ±0.4°F. Condensation hygrometers can be made surprisingly compact using solid-state optics and thermoelectric cooling.

Wide span and minimal errors are two of the main features of this instrument. A properly designed condensation hygrometer can measure dew points from 200°F down to frost points of −100°F. Typical condensation hygrometers can cool to 150°F below the ambient temperature, establishing lower limits of the instrument to dew points corresponding to approximately 0.5% rh. Accuracies for measurements above −40°F can be ±2°F or better, deteriorating to ±4°F at lower temperatures.

The response time of a condensation dew-point hygrometer is usually specified in terms of its cooling/heating rate, typically 4°F/s for thermoelectric cooled mirrors. This makes it somewhat faster than a heated salt hygrometer. Perhaps the most significant feature of the condensation hygrometer is its fundamental measuring technique, which essentially renders the instrument self-calibrating. For calibration, it is necessary only to manually override the surface cooling control loop, causing the surface to heat, and witness that the instrument recools to the same dew point when the loop is closed. Assuming that the surface temperature measuring system is correct, this is a reasonable check on the instrument's performance.

Although condensation hygrometers can become contaminated, they can easily be cleaned and returned to service with no impairment to performance.

Salt-Phase Heated Hygrometers

Another instrument in which the temperature varies with ambient dew-point temperature is variously designated as a self-heating salt-phase transition hygrometer or a heated electrical hygrometer. This device usually consists of a tubular substrate covered by glass fiber fabric, with a spiral bifilar winding for electrodes. The surface is covered with a salt solution, usually lithium chloride. The sensor is connected in series with a ballast and a 24 V (ac) supply. When the instrument is in operation, electrical current flowing through the salt film heats the sensor. The electrical resistance characteristics of the salt are such that a balance is reached with the salt at a critical moisture content corresponding to a saturated solution. The sensor temperature adjusts automatically so that the water vapor pressures of the salt film and ambient atmosphere are equal.

With lithium chloride, this sensor cannot be used to measure relative humidity below approximately 12% (the equilibrium relative humidity of this salt), and it has an upper dew-point limit of about 160°F. The regions of highest precision are between −10 and 93°F, and above 105°F dew point. Another problem is that the lithium chloride solution can be washed off when exposed to water. In addition, this type of sensor is subject to contamination problems, which limits its accuracy. Its response time is also very slow; it takes approximately 2 min for a 67% step change.

MECHANICAL HYGROMETERS

Many organic materials change in dimension with changes in humidity; this action is used in a number of simple and effective humidity indicators, recorders, and controllers (see Chapter 15). They are coupled to pneumatic leak ports, mechanical linkages, or electrical transduction elements to form hygrometers.

Commonly used organic materials are human hair, nylon, Dacron, animal membrane, animal horn, wood, and paper. Their inherent nonlinearity and hysteresis must be compensated for within the hygrometer. These devices are generally unreliable below 32°F. The response is generally inadequate for monitoring a changing process. Responses can be affected significantly by exposure to extremes of humidity. Mechanical hygrometers require initial calibration and frequent recalibration; however, they are useful because they can be arranged to read relative humidity directly, and they are simpler and less expensive than most other types.

ELECTRICAL IMPEDANCE AND CAPACITANCE HYGROMETERS

Many substances adsorb or lose moisture with changing relative humidity and exhibit corresponding changes in electrical impedance or capacitance.

Dunmore Hygrometers

This sensor consists of dual electrodes on a tubular or flat substrate; it is coated with a film containing salt, such as lithium chloride, in a binder to form an electrical connection between windings. The relation of sensor resistance to humidity is usually represented by graphs. Because the sensor is highly sensitive, the graphs are a series of curves, each for a given temperature, with intermediate values found by interpolation. Several resistance elements, called Dunmore elements, cover a standard range. Systematic calibration is essential because the resistance grid varies with time and contamination as well as with exposure to temperature and humidity extremes.

Polymer Film Electronic Hygrometers

These devices consist of a hygroscopic organic polymer deposited by means of thin or thick film processing technology on a water-permeable substrate. Both capacitance and impedance sensors are available. The impedance devices may be either ionic or electronic conduction types. These hygrometers typically have integrated circuits that provide temperature correction and signal conditioning. The primary advantages of this sensor technology are small size; low cost; fast response times (on the order of 1 to 120 s for 64% change in relative humidity); and good accuracy over the full range, including the low end (1 to 15% h), where most other devices are less accurate.

Ion Exchange Resin Electric Hygrometers

A conventional ion exchange resin consists of a polymer having a high relative molecular mass and polar groups of positive or negative charge in cross-link structure. Associated with these polar groups are ions of opposite charge that are held by electrostatic forces to the fixed polar groups. In the presence of water or water vapor, the electrostatically held ions become mobile; thus, when a voltage is impressed across the resin, the ions are capable of electrolytic conduction. The **Pope cell** is one example of an ion exchange element. It is a wide-range sensor, typically covering 15 to 95% rh; therefore, one sensor can be used where several Dunmore elements would be required. The Pope cell, however, has a nonlinear characteristic from approximately 1000 Ω at 100% rh to several megohms at 10% rh.

Impedance-Based Porous Ceramic Electronic Hygrometers

Using the adsorption characteristics of oxides, humidity-sensitive ceramic oxide devices employ either ionic or electronic measurement techniques to relate adsorbed water to relative humidity. Ionic conduction is produced by dissociation of water molecules forming surface hydroxyls. The dissociation causes migration of protons such that the impedance of the device decreases with increasing water content. The ceramic oxide is sandwiched between porous metal electrodes that connect the device to an impedance-measuring circuit for linearizing and signal conditioning. These sensors have excellent sensitivity, are resistant to contamination and high temperature (up to 400°F), and may get fully wet without sensor degradation. These sensors are accurate to about ±1.5% rh, and ±1% rh when temperature compensated. These sensors have a moderate cost.

Aluminum Oxide Capacitive Sensor

This sensor consists of an aluminum strip that is anodized by a process that forms a porous oxide layer. A very thin coating of cracked chromium or gold is then evaporated over this structure. The aluminum base and the cracked chromium or gold layer form the two electrodes of what is essentially an aluminum oxide capacitor.

Water vapor is rapidly transported through the cracked chromium or gold layer and equilibrates on the walls of the oxide pores in a manner functionally related to the vapor pressure of water in the atmosphere surrounding the sensor. The number of water molecules adsorbed on the oxide structure determines the capacitance between the two electrodes.

ELECTROLYTIC HYGROMETERS

In electrolytic hygrometers, air is passed through a tube, where moisture is adsorbed by a highly effective desiccant (usually phosphorous pentoxide) and electrolyzed. The airflow is regulated to 0.0035 cfm at a standard temperature and pressure. As the incoming water vapor is absorbed by the desiccant and electrolyzed into hydrogen and oxygen, the current of electrolysis determines the mass of water vapor entering the sensor. The flow rate of the entering gas is controlled precisely to maintain a standard sample mass flow rate into the sensor. The instrument is usually designed for use with moisture-air ratios in the range of less than 1 ppm to 1000 ppm but can be used with higher humidities.

PIEZOELECTRIC SORPTION

This hygrometer compares the changes in frequency of two hygroscopically coated quartz crystal oscillators. As the mass of the crystal changes due to the absorption of water vapor, the frequency changes. The amount of water sorbed on the sensor is a function of relative humidity (i.e., partial pressure of water as well as ambient temperature).

A commercial version uses a hygroscopic polymer coating on the crystal. The humidity is measured by monitoring the change in the vibration frequency of the quartz crystal when the crystal is alternately exposed to wet and dry gas.

SPECTROSCOPIC (RADIATION ABSORPTION) HYGROMETERS

Radiation absorption devices operate on the principle that selective absorption of radiation is a function of frequency for different media. Water vapor absorbs **infrared** radiation at 2 to 3 μm wavelengths and **ultraviolet** radiation centered about the Lyman-alpha line at 0.122 μm. The amount of absorbed radiation is directly related to the absolute humidity or water vapor content in the gas mixture according to Beer's law. The basic unit consists of an energy source and optical system for isolating wavelengths in the spectral region of interest and a measurement system for determining the attenuation of radiant energy caused by the water vapor in the optical path. The absorbed radiation is measured extremely quickly and independent of the degree of saturation of the gas mixture. Response times of 0.1 to 1 s for 90% change in moisture content are common. Spectroscopic hygrometers are primarily used where a noncontact application is required; this may include atmospheric studies, industrial drying ovens, and harsh environments. The primary disadvantages of this device are its high cost and relatively large size.

GRAVIMETRIC HYGROMETERS

Humidity levels can be measured by extracting and finding the mass of water vapor in a known quantity or atmosphere. For precise laboratory work, powerful desiccants, such as phosphorous pentoxide and magnesium perchlorate, are used for the extraction process; for other purposes, calcium chloride or silica gel is satisfactory.

When the highest level of accuracy is required, the gravimetric hygrometer, developed and maintained by NIST, is the ultimate in the measurement hierarchy. The gravimetric hygrometer gives the absolute water vapor content, where the mass of the absorbed water and the precise measurement of the gas volume associated with the water vapor determine the mixing ratio or absolute humidity of the sample. This system has been chosen as the primary standard because the required measurements of mass, temperature, pressure, and volume can be made with extreme precision. However, its complexity and required attention to detail limit the usefulness of the gravimetric hygrometer.

CALIBRATION

For many hygrometers, the need for recalibration depends on the accuracy required, the stability of the sensor, and the conditions to which the sensor is being subjected. Many hygrometers should be calibrated regularly by exposure to an atmosphere maintained at a known humidity and temperature, or by comparison with a transfer standard hygrometer. Complete calibration usually requires observation of a series of temperatures and humidities. Methods for producing known humidities include saturated salt solutions (Greenspan 1977, Huang and Whetstone 1985); sulfuric acid solutions, and mechanical systems, such as the divided flow, two-pressure (Amdur 1965); two-temperature (Till and Handegord 1960); and NIST two-pressure humidity generator (Hasegawa 1976). All these systems rely on precise methods of temperature and pressure control within a controlled environment to produce a known humidity, usually with accuracies of 0.5 to 1.0%. The operating range for the precision generator is typically 5 to 95% rh.

PRESSURE MEASUREMENT

Pressure is the force exerted per unit area by a medium, generally a liquid or gas. Pressure so defined is sometimes called **absolute pressure**. Thermodynamic and material properties are expressed in terms of absolute pressures; thus, the properties of a refrigerant will be given in terms of absolute pressures. **Vacuum** refers to pressures below atmospheric.

Differential pressure is the difference between two absolute pressures. In many cases, the differential pressure can be very small compared to either of the absolute pressures (these are often referred to as low-range, high-line differential pressures). A common example of differential pressure is the pressure drop, or difference between inlet and outlet pressures, across a filter or flow element.

Gage pressure is a special case of differential pressure where one of the pressures (the reference pressure) is atmospheric pressure. Many pressure gages, including most refrigeration test sets, are designed to make gage pressure measurements, and there are probably more gage pressure measurements made than any other. Gage pressure measurements are often used as surrogates for absolute pressures. However, because of variations in atmospheric pressure due to elevation (atmospheric pressure in Denver, Colorado, is about 81% of sea-level pressure) and weather changes, the measurement of gage pressures to determine absolute pressures can significantly restrict the accuracy of the measured pressure, unless corrections are made for the local atmospheric pressure at the time of the measurement.

Pressures can be further classified as static or dynamic. **Static pressures** have a small or undetectable change with time; **dynamic pressures** include a significant pulsed, oscillatory, or other time-dependent component. Static pressure measurements are the most common, but equipment such as blowers and compressors can generate significant oscillatory pressures at discrete frequencies. Flow in pipes and ducts can generate resonant pressure changes, as well as turbulent "noise" that can span a wide range of frequencies.

Units

A plethora of pressure units, many of them poorly defined, are in common use. The international (SI) unit is the newton per square metre, called the pascal (Pa). While the bar and the standard atmosphere are used, they should not be introduced where they are not used at present. Although not internationally recognized, the pound per square inch (psi) is a widely used unit. Also widely used, but not as rigorously defined and, therefore, a potential source of error, are units based on the length of liquid columns, including inches of mercury (in. Hg), mm of mercury (mm Hg), and inches of water (in. of water). The latter is often used for low-range differential pressure measurements. In the case of pounds per square inch, the type of pressure measurement is often indicated by a modification of the unit (i.e., both psi and psia are used to indicate absolute pressure measurements, psid indicates a differential measurement, and psig indicates a gage measurement). No such standard convention exists for other units, and unless explicitly stated, reported values are assumed to be absolute pressures. Conversion factors for different pressure units can be found in Chapter 37.

The difference between the conversion factors for inches of mercury and inches of water at the different temperatures is indicative of the errors that can arise from uncertainties about the definitions of these units.

Types of Pressure-Measuring Instruments

Broadly speaking, pressure instruments can be divided into three different categories—standards, mechanical gages, and electromechanical transducers. Standards instruments are used for the most accurate calibrations. The liquid-column manometer, which is the most common and potentially the most accurate standard, is used for a variety of applications, including field applications. Mechanical pressure gages are generally the least expensive and the most common pressure instruments. However, electromechanical transducers have become much less expensive and are easier to use, so they are being used more often.

PRESSURE STANDARDS

Liquid-column manometers measure pressure by determining the vertical displacement of a liquid of known density in a known gravitational field. Typically they are constructed as a U-tube of transparent material (glass or plastic). The pressure to be measured is applied to one side of the U-tube. If the other (reference) side is evacuated (zero pressure), the manometer measures absolute pressure; if the reference side is open to the atmosphere, it measures gage pressure; if the reference side is connected to some other pressure, the manometer measures the differential between the two pressures. Manometers filled with water and different oils are often used to measure low-range differential pressures. In some low-range instruments, one tube of the manometer is inclined in order to enhance the readability. Mercury-filled manometers are used for higher range differential and absolute pressure measurements. In the latter case, the reference side is evacuated, generally with a mechanical vacuum pump. Typical full-scale ranges for manometers vary from 10 in. of water to 3 atm.

For pressures above the range of manometers, standards are generally of the piston-gage, pressure-balance, or deadweight-tester type. These instruments apply pressure to the bottom of a vertical piston, which is surrounded by a close-fitting cylinder (typical clearances are millionths of an inch). The pressure generates a force approximately equal to the pressure times the area of the piston. This force is balanced by weights stacked on the top of the piston. If the mass of the weights, the local acceleration of gravity, and the area of the piston (or more properly, the "effective area" of the piston and cylinder assembly) are known, the applied pressure can be calculated. Piston gages generally generate gage pressures with respect to the atmospheric pressure above the piston. They can be used to measure absolute pressures either indirectly by separately measuring the atmospheric pressure and adding it to the gage pressure determined by the piston gage, or directly by surrounding the top of the piston and weights with an evacuated bell jar. Piston gage full-scale ranges vary from 5 to 200,000 psi.

At the other extreme, very low absolute pressures (below about 0.4 in. of water), a number of different types of standards are used. These tend to be specialized and expensive instruments found only in major standards laboratories. However, one low-pressure standard, the **McLeod gage**, has been used for field applications. Unfortunately, although the theory of the McLeod gage is simple and straightforward, it is difficult to make accurate measurements with this instrument, and major errors can occur when it is used to measure gases that condense or are adsorbed (e.g., water). In general, gages other than the McLeod gage should be used for most low-pressure or vacuum applications.

MECHANICAL PRESSURE GAGES

Mechanical pressure gages couple a pressure sensor to a mechanical readout, typically a pointer and dial. The most common type employs a **Bourdon tube** sensor, which is essentially a coiled metal tube of circular or elliptical cross section. Increasing pressure applied to the inside of the tube causes it to uncoil. A mechanical linkage translates the motion of the end of the tube to the rotation of a pointer. In most cases, the Bourdon tube is surrounded by atmospheric pressure, so that the gages measure gage pressure. A few instruments surround the Bourdon tube with a sealed enclosure that can be evacuated for absolute measurements or connected to another pressure for differential measurements. Available instruments vary widely in cost, size, pressure range, and accuracy. Full-scale ranges can vary from 5 to 100,000 psi. Accuracy of properly calibrated and used instruments can vary from 0.1 to 10% of full scale. Generally there is a strong correlation between size, accuracy, and price; larger instruments are more accurate and expensive.

To achieve better sensitivity, some low-range mechanical gages, sometimes called **aneroid gages**, employ corrugated diaphragms or capsules as sensors. The capsule is basically a short bellows sealed with end caps. These sensors are more compliant than a Bourdon tube, and a given applied pressure will cause a larger deflection of the sensor. The inside of a capsule can be evacuated and sealed in order to measure absolute pressures or connected to an external fitting to allow differential pressures to be measured. Typically, these gages are used for low-range measurements of 1 atm or less. In instruments of better quality, accuracies of 0.1% of reading or better can be achieved.

ELECTROMECHANICAL TRANSDUCERS

Mechanical pressure gages are generally limited by inelastic behavior of the sensing element, friction in the readout mechanism, and limited resolution of the pointer and dial. These effects can be eliminated or reduced by using electronic techniques to sense the distortion or stress of a mechanical sensing element and electronically convert that stress or distortion to a pressure reading. A wide variety of sensors is used, including Bourdon tubes, capsules, diaphragms, and different resonant structures whose vibration frequency varies with the applied pressure. Capacitive, inductive, and optical lever sensors are used to measure the displacement of the sensor element. In some cases, feedback techniques may be used to constrain the sensor in a null position, minimizing distortion and hysteresis of the sensing element. Temperature control or compensation is often included. Readout may be in the form of a digital display, analog voltage or current, or a digital code. Size varies, but in the case of transducers employing a diaphragm fabricated as part of a silicon chip, the sensor and signal-conditioning electronics can be contained in a small transistor package, and the largest part of the device is the pressure fitting. The best of these instruments achieve long-term instabilities of 0.01% or less of full scale, and corresponding accuracies when properly calibrated. Performance of the less expensive instruments can be more on the order of several percent.

While the dynamic response of most mechanical gages is limited by the sensor and readout, the response of some electromechanical transducers can be much faster, allowing measurements of dynamic pressures at frequencies up to 1 kHz and beyond in the case of transducers specifically designed for dynamic measurements. Manufacturers' literature should be consulted as a guide to the dynamic response of specific instruments.

As the measured pressure is reduced below about 1.5 psia, it becomes increasingly difficult to sense mechanically. A variety of gages have been developed that measure some other property of the gas that is related to the pressure. In particular, thermal conductivity gages, known as thermocouple, thermistor, Pirani, and convection gages, are used for pressures down to about 0.0004 in. of water. These gages have a sensor tube with a small heated element and a temperature sensor; the temperature of the heated element is determined by the thermal conductivity of the gas, and the output of the temperature sensor is displayed on an analog or digital electrical meter contained in an attached electronics unit. The accuracy of thermal conductivity gages is limited by their nonlinearity, dependence on gas species, and tendency to read high when contaminated. Oil contamination is a particular problem. However, these gages are small, reasonably rugged, and relatively inexpensive; in the hands of a typical user, they will give far more reliable results than a McLeod gage. They can be used to check the base pressure in a system that is being evacuated prior to being filled with refrigerant. They should be checked periodically for contamination by comparing the reading with that from a new, clean sensor tube.

GENERAL CONSIDERATIONS

Accurate values of atmospheric or barometric pressure are required for weather prediction and aircraft altimetry. In the United States, a network of calibrated instruments, generally accurate to within 0.1% of reading and located at airports, is maintained by the National Weather Service, the Federal Aviation Administration, and local airport operating authorities. These agencies are generally cooperative in providing current values of atmospheric pressure that can be used to check the calibration of absolute pressure gages or to correct gage pressure readings to absolute pressures. However, the pressure readings generally reported for weather and altimetry purposes are not the true atmospheric pressure, but rather a value adjusted to an equivalent sea level pressure. Therefore, unless the location is near sea level, it is important to ask for the station or true atmospheric pressure rather than using the adjusted values broadcast by radio stations. Further, the atmospheric pressure decreases with increasing elevation at a rate (near sea level) of about 0.001 in. Hg/ft, and corresponding corrections should be made to account for the difference in elevation between the instruments being compared.

As noted before, gage-pressure instruments are sometimes used to measure absolute pressures, and the accuracy of these measurements can be compromised by uncertainties in the atmospheric pressure. This error can be particularly serious when gage-pressure instruments are used to measure a vacuum (negative gage pressures). For all but the most crude measurements, absolute-pressure gages should be used for vacuum measurements; for pressures below about 0.4 in. of water, a thermal conductivity gage should be used.

All pressure gages are susceptible to temperature errors. Several techniques are used to minimize these errors—sensor materials are generally chosen to minimize temperature effects, mechanical readouts can include temperature compensation elements, electromechanical transducers may include a temperature sensor and compensation circuit, and some transducers are operated at a controlled temperature. Clearly, temperature effects are of greater concern for field applications, and it is prudent to check the manufacturers' literature for the temperature range over which the specified accuracy can be maintained. Abrupt temperature changes can also cause large transient errors that may take some time to decay.

The readings of some electromechanical transducers with a resonant or vibrating sensor can depend on the gas species. Although some of these units can achieve calibrated accuracies of the order of 0.01% of reading, they are typically calibrated with dry air or nitrogen, and the readings for other gases can be in error by several percent, quite possibly much more for refrigerants and other high-density gases. High-accuracy readings can be maintained by calibrating these devices with the gas to be measured. Manufacturer's literature should be consulted.

The measurement of dynamic pressures is limited not just by the frequency response of the pressure gage, but also by the hydraulic or pneumatic time constant of the connection between the gage and the system to be monitored. As a general rule, the longer the connecting lines and the smaller their diameter, the lower the frequency response of the system. Further, even if only the static component of the pressure is of interest, and a gage with a low-frequency response is used, a significant pulsating or oscillating pressure component can cause significant errors in pressure gage readings and, in some cases, can damage the gage, particularly gages with a mechanical readout mechanism. In these cases, a filter or snubber should be used to reduce the higher frequency components.

VELOCITY MEASUREMENT

HVAC engineers measure the flow of air more often than any other gas, and the air is usually measured at or near atmospheric pressure. Under this condition, the air can be treated as an incompressible fluid, and simple formulas give sufficient precision to solve many problems. Instruments that measure fluid velocity and their application range and precision are listed in Table 4.

Table 4 Velocity Measurement

Measurement Means	Application	Range, fpm	Precision	Limitations
Smoke puff or airborne solid tracer	Low air velocities in rooms; highly directional	5 to 50	10 to 20%	Awkward to use but valuable in tracing air movement
Deflecting vane anemometer	Air velocities in rooms, at outlets, etc.; directional	30 to 24,000	5%	Needs periodic check calibration
Revolving vane anemometer	Moderate air velocities in ducts and rooms; somewhat directional	100 to 3000	2 to 5%	Extremely subject to error with variations in velocities with space or time; easily damaged; needs periodic calibration
Hot-wire anemometer	a. Low air velocities; directional and nondirectional available	1 to 2000	2 to 5%	Requires accurate calibration at frequent intervals. Some are relatively costly.
	b. High air velocities	Up to 60,000	0.2 to 5%	
	c. Transient velocity and turbulence			
Pitot tube	Standard instrument for measuring duct velocities	180 to 10,000 with micromanometer; 600 to 10,000 with draft gages; 10,000 up with manometer	1 to 5%	Accuracy falls off at low end of range
Impact tube and sidewall or other static tap	High velocities, small tubes and where air direction may be variable	120 to 10,000 with micromanometer; 600 to 10,000 with draft gages; 10,000 up with manometer	1 to 5%	Accuracy depends on constancy of static pressure across stream section
Cup anemometer	Meteorological	Up to 12,000	2 to 5%	Poor accuracy at low air velocity (<500 fpm)
Laser Doppler velocimeter	Calibration of air velocity instruments	1 to 6000	1 to 3%	High cost and complexity limit LDVs to laboratory applications

AIRBORNE TRACER TECHNIQUES

Tracer techniques are suitable for measuring velocity in an open space. Typical tracers include smoke, feathers, pieces of lint, and radioactive or nonradioactive gases. Measurements are made by timing the rate of movement of solid tracers or by monitoring the change in concentration level of gas tracers.

Smoke is a useful qualitative tool in studying air movements. Smoke can be obtained from titanium tetrachloride (irritating to nasal membranes) or by mixing potassium chlorate and powdered sugar (a nonirritating smoke) and firing the mixture with a match. The latter process produces considerable heat and should be confined to a pan away from flammable materials. Titanium tetrachloride smoke works well for spot tests, particularly for leakage through casings and ducts, because it can be handled easily in a small, pistol-like ejector.

The fumes of ammonia water and sulfuric acid, if permitted to mix, form a white precipitate. Two bottles, one containing ammonia water and the other containing acid, are connected to a common nozzle by rubber tubing. A syringe forces air over the liquid surfaces in the bottles; the two streams mix at the nozzle and form a white cloud.

A satisfactory test smoke also can be made by bubbling an airstream through ammonium hydroxide and then hydrochloric acid (Nottage et al. 1952). Smoke tubes, smoke candles, and smoke bombs are available for studying airflow patterns.

ANEMOMETERS

Deflecting Vane Anemometers

The deflecting vane anemometer consists of a pivoted vane enclosed in a case. Air exerts pressure on the vane as it passes through the instrument from an upstream to a downstream opening. A hair spring and a damping magnet resist vane movement. The instrument gives instantaneous readings of directional velocities on an indicating scale. With fluctuating velocities, it is necessary to average the needle swings visually to obtain average velocities. This instrument is useful for studying air motion in a room; locating objectionable drafts; measuring air velocities at supply and return diffusers and grilles; and measuring laboratory hood face velocities.

Propeller or Revolving Vane Anemometers

The propeller anemometer consists of a light, revolving wind-driven wheel connected through a gear train to a set of recording dials that read linear feet of air passing in a measured length of time. It is made in various sizes—3, 4, and 6 in. are the most common. Each instrument requires individual calibration. At low velocities, the friction drag of the mechanism is considerable. To compensate for this, a gear train that overspeeds is commonly used. For this reason, the correction is often additive at the lower range and subtractive at the upper range, with the least correction in the middle range of velocities. The best of these instruments have starting speeds of 50 fpm or higher; therefore, they cannot be used below that air speed. Electronic revolving vane anemometers, with optical or magnetic pickups to sense the rotation of the vane, are available. Sizes for the vanes range as small as 1/2 in. in diameter for the electronic versions.

Cup Anemometers

The cup anemometer is primarily used to measure outdoor, meteorological wind speeds. It consists of three or four hemispherical cups mounted radially from a vertical shaft. Wind from any direction with a vector component in the plane of cup rotation causes the cups and shaft to rotate. Because the primary use of this anemometer is to make meteorological wind speed measurements, the instrument is usually constructed so that wind speeds can be recorded or indicated electrically at a remote point.

Thermal Anemometers

The thermal or hot-wire anemometer consists of a heated RTD, thermocouple junction, or thermistor sensor constructed at the end of a probe; it is designed to provide a direct, simple method of determining air velocity at a point in the flow field. The probe is placed into an airstream, and the movement of air past the electrically heated velocity sensor tends to cool the sensor in proportion to the speed of the airflow. The electronics and sensor are commonly combined into a portable, hand-held device that interprets the sensor signal and provides a direct reading of air velocity in either analog or digital display format. Often the sensor probe also incorporates an ambient temperature-sensing RTD or thermistor, in which case the indicated air velocity is "temperature compensated" to "standard" air density conditions (typically 0.0748 lb/ft^3).

Hot-wire anemometers have long been used in the fluid flow research field. Research anemometer sensors have been constructed using very fine wires in configurations that allow the researcher to characterize fluid flows in one, two, and three dimensions with sensor/electronics response rates up to several hundred kilohertz. This technology has been incorporated into more ruggedized sensors suitable for measurements in the HVAC field, primarily for unidirectional airflow measurement. Omnidirectional sensing instruments suitable for thermal comfort studies are also available.

The principal advantages of thermal anemometers are their wide dynamic range and their ability to sense extremely low velocities. Typical accuracy (including repeatability) of 2 to 5% of reading over the entire velocity range is often achieved in commercially available portable instruments.

Among the limitations of thermal anemometers are the following: (1) the unidirectional sensor must be carefully aligned in the airstream (typically to within ±20° rotation) to achieve accurate results; (2) the velocity sensor must be kept clean because contaminant buildup will cause the calibration to change; and (3) due to the inherent high speed of response of thermal anemometers, measurements in turbulent flows can yield fluctuating velocity measurements. Electronically controlled time-integrated functions are now available in many digital air velocity meters to help smooth these turbulent flow measurements.

In the HVAC field, thermal anemometers are suitable for use in a variety of applications. They are particularly well-suited to the low velocities associated with laboratory fume hood face velocity measurements (typically in the 60 to 200 fpm range). Thermal anemometers can also be used for taking multipoint traverse measurements in ventilation ductwork.

Laser Doppler Velocimeters (or Anemometers)

The laser Doppler velocimeter (LDV) or laser Doppler anemometer (LDA) is an extremely complex system that collects scattered light produced by a particle passing through the intersection volume of two intersecting laser beams of the same light frequency (Mease et al. 1992). The scattered light consists of bursts containing a regularly spaced fringe pattern whose frequency is linearly proportional to the speed of the particle. Due to the cost and complexity of these systems, they are usually not suitable for in situ field measurements. Rather, the primary application of LDV systems in the HVAC industry is the calibration of systems used to calibrate other air velocity instruments.

The greatest advantage of an LDV is its performance at low air speeds. It is capable of reading air speeds as low as 15 fpm with uncertainty levels of 1% or less (Mease et al. 1992). In addition, it is nonintrusive in the flow—only optical access is required. It can be used to measure fluctuating components as well as mean speeds and is available in one-, two-, and even three-dimensional configurations. Its biggest disadvantages are its high cost and extreme technological complexity, which requires highly skilled operators. Modern fiber optic systems require less-skilled operators but at a considerable increase in cost.

PITOT-STATIC TUBES

The pitot-static tube, in conjunction with a suitable manometer or differential pressure transducer, provides a simple method of determining air velocity at a point in a flow field. Figure 5 shows the construction of a standard pitot tube (ASHRAE *Standard* 51) and the method of connecting it with inclined manometers to display both static pressure and velocity pressure. The equation for determining air velocity from measured velocity pressure is

$$V = C\sqrt{\frac{2p_w g_c}{\rho}} \tag{5}$$

where

V = velocity, fpm
p_w = velocity pressure (pitot-tube manometer reading), in. of water
ρ = density of air, lb$_m$/ft^3
g_c = gravitational constant = 32.174 lb$_m$·ft/lb$_f$·s^2
C = unit conversion factor = 136.8

The type of manometer or differential pressure transducer used with a pitot-static tube depends on the magnitude of velocity pressure being measured and on the desired accuracy. At velocities greater than 1500 fpm, a draft gage of appropriate range is usually satisfactory. If the pitot-static tube is used to measure air velocities lower than 1500 fpm, a precision manometer or comparable pressure differential transducer is essential.

Other pitot-static tubes have been used and calibrated. To meet special conditions, various sizes of pitot-static tubes geometrically similar to the standard tube can be used. For relatively high velocities in ducts of small cross-sectional area, total pressure readings can be obtained with an impact (pitot) tube. Where static pressure across the stream is relatively constant, as in turbulent flow in a straight duct, a sidewall tap to obtain static pressure can be used with the impact tube to obtain the velocity pressure head. One form of impact tube is a small streamlined tube with a fine hole in its upstream end and with its axis placed parallel to the stream.

If the Mach number of the flow is greater than about 0.3, the effects of compressibility should be included in the computation of the air speed from pitot-static and impact (stagnation or pitot) tube measurements (Mease et al. 1992).

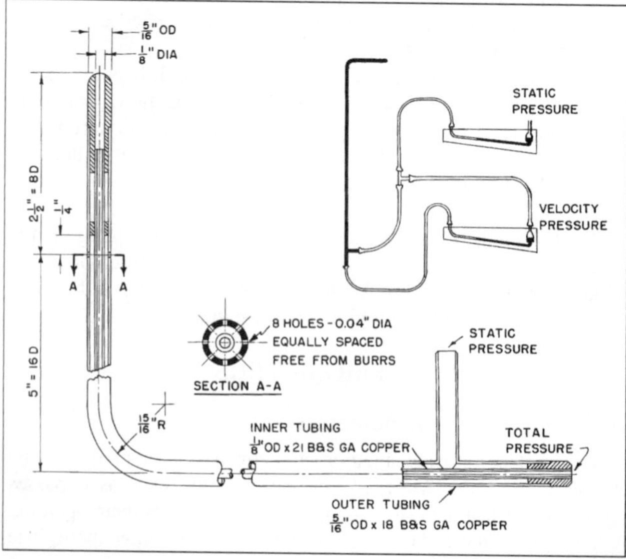

Fig. 5 Standard Pitot Tube

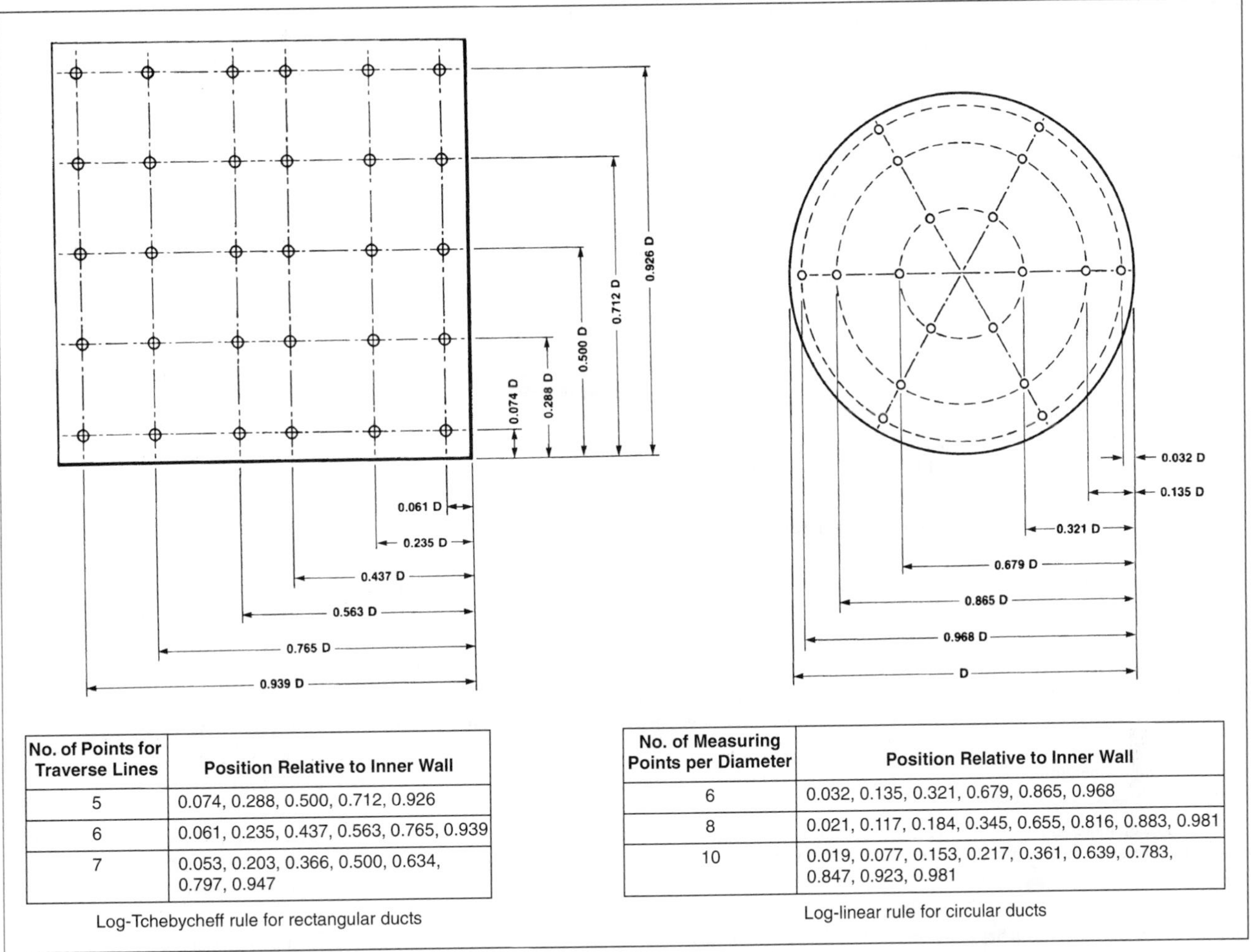

No. of Points for Traverse Lines	Position Relative to Inner Wall
5	0.074, 0.288, 0.500, 0.712, 0.926
6	0.061, 0.235, 0.437, 0.563, 0.765, 0.939
7	0.053, 0.203, 0.366, 0.500, 0.634, 0.797, 0.947

Log-Tchebycheff rule for rectangular ducts

No. of Measuring Points per Diameter	Position Relative to Inner Wall
6	0.032, 0.135, 0.321, 0.679, 0.865, 0.968
8	0.021, 0.117, 0.184, 0.345, 0.655, 0.816, 0.883, 0.981
10	0.019, 0.077, 0.153, 0.217, 0.361, 0.639, 0.783, 0.847, 0.923, 0.981

Log-linear rule for circular ducts

Fig. 6 Measuring Points for Rectangular and Round Duct Traverse

MEASURING FLOW IN DUCTS

Because velocity in a duct is seldom uniform across any section, and a pitot tube reading or thermal anemometer indicates velocity at only one location, a traverse is usually made to determine average velocity. Generally, velocity is lowest near the edges or corners and greatest at or near the center.

To determine the velocity in a traverse plane, a straight average of individual point velocities will give satisfactory results when point velocities are determined by the **log-Tchebycheff rule** (ISO *Standard* 3966) or, if care is taken, by the **equal area method**. Figure 6 shows suggested sensor locations for traversing round and rectangular ducts. The log-Tchebycheff rule provides the greatest accuracy because its location of traverse points accounts for the effect of wall friction and the fall-off of velocity near wall ducts. This method is now recommended for rectangular ducts, although for circular ducts the log-Tchebycheff and log-linear traverse methods are similar. Log-Tchebycheff minimizes the positive error (measured greater than actual) caused by the failure to account for losses at the duct wall. This error can occur when using the older method of equal subareas to traverse rectangular ducts.

For a rectangular duct traverse, a minimum of 25 points should be measured. For a duct size less than 18 in., the points should be located at the center of equal areas not more than 6 in. apart, and a minimum of 2 points per side should be used. For a duct side greater than 56 in., the maximum distance between points is 8 in. For a circular duct traverse, the log-linear rule and three symmetrically disposed diameters may be used (Figure 6). Points on two perpendicular diameters may be used where access is limited.

If possible, measuring points should be located at least 7.5 diameters downstream and 3 diameters upstream from a disturbance (e.g., caused by a turn). Compromised traverses as close as 2 diameters downstream and 1 diameter upstream can be performed with an increase in measurement error. Because field-measured airflows are rarely steady and uniform, particularly near disturbances, accuracy can be improved by increasing the number of measuring points. Straightening vanes (ASHRAE *Standard* 51) located 1.5 duct diameters ahead of the traverse plane improve measurement precision.

When velocities at a traverse plane fluctuate, the readings should be averaged on a time-weighted basis. Two traverse readings in short succession also help to average out velocity variations that occur with time. If negative velocity pressure readings are encountered, they are considered a measurement value of zero and calculated in the average velocity pressure. ASHRAE *Standard* 111 has further information on measuring flow in ducts.

FLOW RATE MEASUREMENT

Various means of measuring fluid flow rate are listed in Table 5. The values for volumetric or mass flow rate measurement (ASME *Standard* PTC 19.5, Benedict 1984) are often determined

Table 5 Volumetric or Mass Flow Rate Measurement

Measurement Means	Application	Range	Precision	Limitations
Orifice and differential pressure measurement system	Flow through pipes, ducts, and plenums for all fluids	Above Reynolds number of 5000	1 to 5%	Discharge coefficient and accuracy influenced by installation conditions
Nozzle and differential pressure measurement system	Flow through pipes, ducts, and plenums for all fluids	Above Reynolds number of 5000	0.5 to 2.0%	Discharge coefficient and accuracy influenced by installation conditions
Venturi tube and differential pressure measurement system	Flow through pipes, ducts, and plenums for all fluids	Above Reynolds number of 5000	0.5 to 2.0%	Discharge coefficient and accuracy influenced by installation conditions
Timing given mass or volumetric flow	Liquids or gases; used to calibrate other flowmeters	Any	0.1 to 0.5%	System is bulky and slow
Rotameters	Liquids or gases	Any	0.5 to 5.0%	Should be calibrated for fluid being metered
Displacement meter	Relatively small volumetric flow with high pressure loss	As high as 1000 cfm depending on type	0.1 to 2.0% depending on type	Most types require calibration with fluid being metered
Gasometer or volume displacement	Short-duration tests; used to calibrate other flowmeters	Total flow limited by available volume of containers	0.5 to 1.0%	—
Thomas meter (temperature rise of stream due to electrical heating)	Elaborate setup justified by need for good accuracy	Any	1%	Uniform velocity; usually used with gases
Element of resistance to flow and differential pressure measurement system	Used for check where system has calibrated resistance element	Lower limit set by readable pressure drop	1 to 5%	Secondary reading depends on accuracy of calibration
Turbine flowmeters	Liquids or gases	Any	0.25 to 2.0%	Uses electronic readout
Instrument for measuring velocity at point in flow	Primarily for installed systems with no special provision for flow measurement	Lower limit set by accuracy of velocity measurement	2 to 4%	Accuracy depends on uniformity of flow and completeness of traverse
Heat input and temperature changes with steam and water coil	Check value in heater or cooler tests	Any	1 to 3%	—
Laminar flow element and differential pressure measurement system	Measure liquid or gas volumetric flow rate; nearly linear relationship with pressure drop; simple and easy to use	0.0001 to 2000 cfm	1%	Fluid must be free of dirt, oil, and other impurities that could plug meter or affect its calibration
Magnetohydrodynamic flowmeter (electromagnetic)	Measures electrically conductive fluids, slurries; meter does not obstruct flow; no moving parts	0.1 to 10,000 gpm	1%	At present state of the art, conductivity of fluid must be greater than 5 μmho/cm
Swirl flowmeter and vortex shedding meter	Measure liquid or gas flow in pipe; no moving parts	Above Reynolds number of 10^4	1%	—

by measuring pressure difference across an orifice, nozzle, or venturi tube. The various meters have different advantages and disadvantages. For example, the orifice plate is more easily changed than the complete nozzle or venturi tube assembly. However, the nozzle is often preferred to the orifice because its discharge coefficient is more precise. The venturi tube is a nozzle followed by an expanding recovery section to reduce net pressure loss. Differential pressure-type flow measurement has benefited through workshops addressing fundamental issues, textbooks, research, and improved standards (Miller 1983, DeCarlo 1984, Mattingly 1984, ASME *Standard* B40.1, ASME *Standard* MFC-9M, ASME *Standard* MFC-10M, ASME *Standard* MFC-1M).

Fluid meters use a wide variety of physical techniques to make flow measurements (ASME *Standard* PTC 19.5, Miller 1983, DeCarlo 1984); those more prevalently used are described in this section. To assure and validate the accuracy of flow rate measurement instruments, appropriate calibration procedures should include documentation of traceability to the calibration facility. The calibration facility should, in turn, provide documentation of traceability to national standards.

Flow Measurement Methods

Direct. Both gas and liquid flow can be measured quite accurately by timing a collected amount of fluid that is measured gravimetrically or volumetrically. While this method is commonly used for calibrating other metering devices, it is particularly useful where the flow rate is low or intermittent and where a high degree of accuracy is required. These systems are generally large and slow, but in their simplicity, they can be considered primary devices.

The variable area meter or rotameter is a convenient direct-reading flowmeter for liquids and gases. This is a vertical, tapered tube in which the flow rate is indicated by the position of a float suspended in the upward flow. The position of the float is determined by its buoyancy and the upwardly directed fluid drag.

Displacement meters measure total liquid or gas flow over time. The two major types of displacement meters used for gases are the conventional gas meter, which uses a set of bellows, and the wet test meter, which uses a water displacement principle.

Indirect. The **Thomas meter** is used in laboratories to measure high gas flow rates with low pressure losses. The gas is heated by electric heaters, and the temperature rise is measured by two resistance thermometer grids. When the heat input and temperature rise are known, the mass flow of gas is calculated as the quantity of gas that will remove the equivalent heat at the same temperature rise.

A velocity traverse (made using a pitot tube or other velocity-measuring instrument) measures airflow rates in the field or calibrates large nozzles. This method can be imprecise at low velocities and impracticable where many test runs are in progress.

Another field-estimating method measures the pressure drop across elements with known pressure drop characteristics, such as heating and cooling coils or fans. If the pressure drop/flow rate relationship has been calibrated, the results can be precise. If the method depends on rating data, it should be used for check purposes only.

VENTURI, NOZZLE, AND ORIFICE FLOWMETERS

Flow in a pipeline can be measured by a venturi meter (Figure 7), flow nozzle (Figure 8), or orifice plate (Figure 9). American Society of Mechanical Engineers (ASME) *Standard* MFC-3M describes measurement of fluid flow in pipes using the orifice, nozzle, and venturi; ASME *Standard* PTC 19.5 specifies their construction.

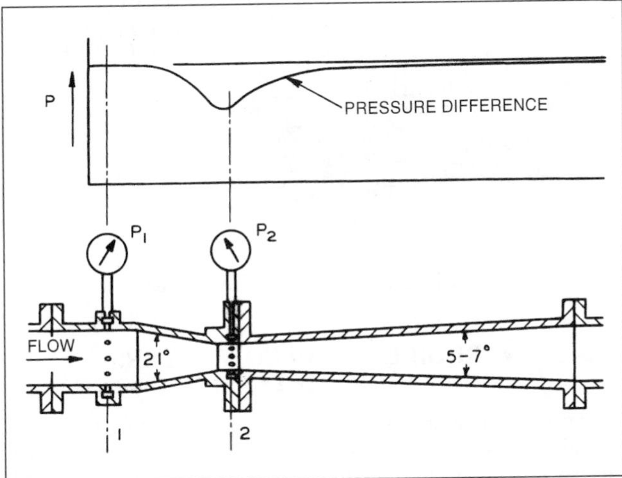

Fig. 7 Typical Herschel Type Venturi Meter

Assuming an incompressible fluid (liquid or slow-moving gas), uniform velocity profile, frictionless flow, and no gravitational effects, the principle of conservation of mass and energy can be applied to the venturi and nozzle geometries to give

$$w = \rho V_1 A_1 = \rho V_2 A_2 = A_2 \sqrt{\frac{2g_c \rho (p_1 - p_2)}{1 - \beta^4}} \tag{6}$$

where

- w = mass flow rate, lb_m/s
- V = velocity of stream, fps
- A = flow area, ft^2
- g_c = gravitational constant = 32.174 $lb_m \cdot ft/lb_f \cdot s^2$
- ρ = density of fluid, lb_m/ft^3
- p = absolute pressure, lb_f/ft^2
- β = (D_2/D_1) for venturi and sharp edge orifice and d/D for flow nozzle

Note: Subscript 1 refers to the entering conditions; subscript 2 refers to the throat conditions.

Because the flow through the meter is not frictionless, a correction factor C is defined to account for friction losses. If the fluid is at a high temperature, an additional correction factor F_a should be included to account for thermal expansion of the primary element. Because this amounts to less than 1% at 500°F, it can usually be omitted. Equation (6) then becomes

$$w = CA_2 \sqrt{\frac{2g_c \rho (p_1 - p_2)}{1 - \beta^4}} \tag{7}$$

where C is the friction loss correction factor.

The factor C is a function of geometry and Reynolds number. Values of C are given in ASME *Standard* PTC 19.5. The jet passing through an orifice plate contracts to a minimum area at the vena contracta located a short distance downstream from the orifice

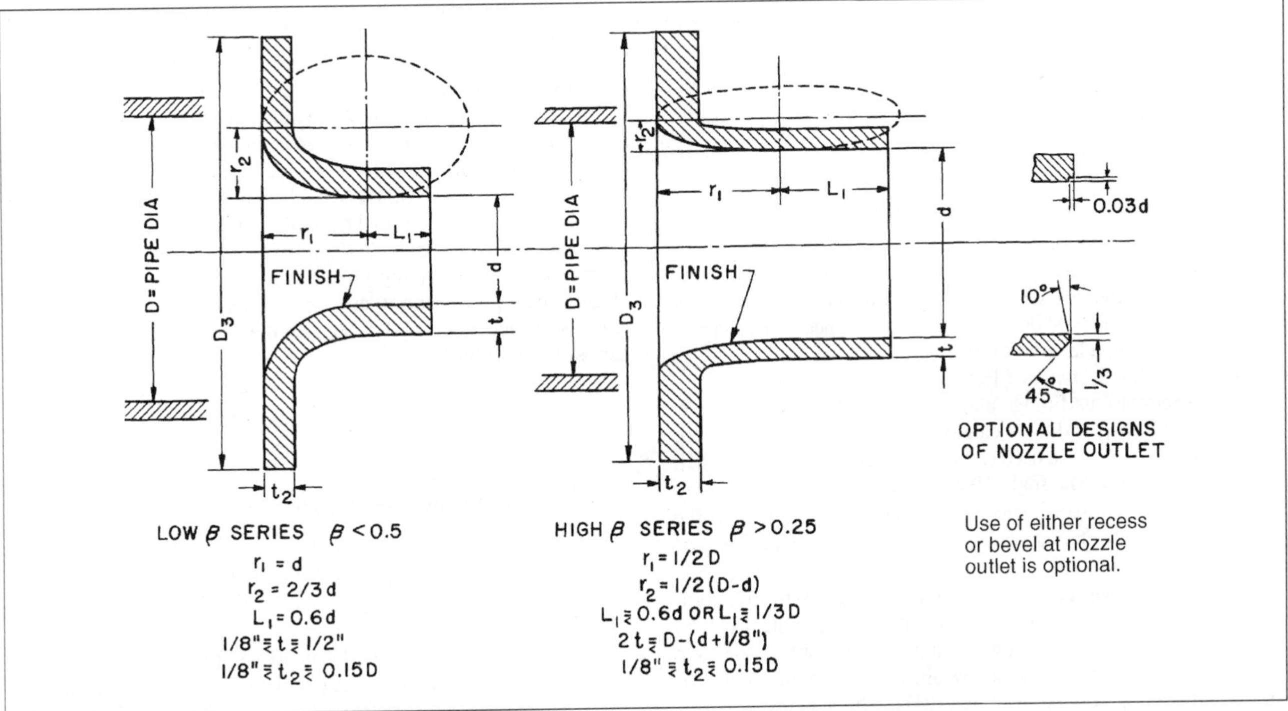

Fig. 8 Dimensions of ASME Long-Radius Flow Nozzles
From ASME PTC 19.5. Reprinted with permission of ASME.

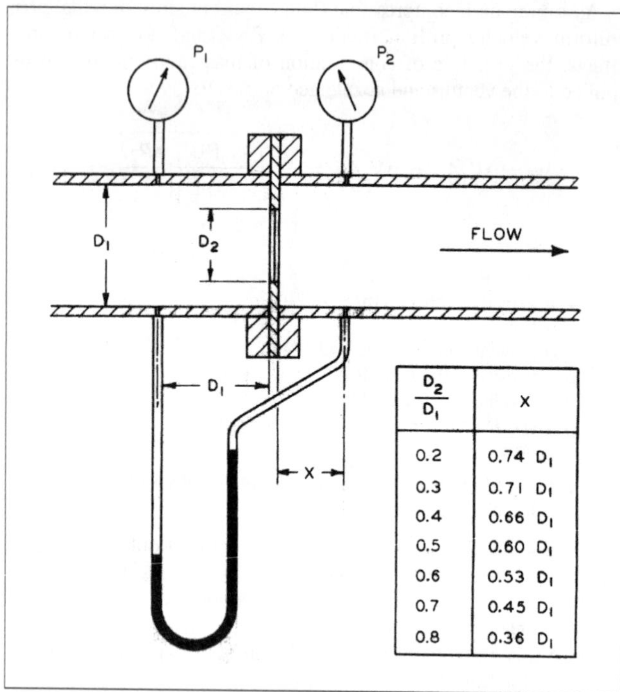

D_2/D_1	X
0.2	0.74 D_1
0.3	0.71 D_1
0.4	0.66 D_1
0.5	0.60 D_1
0.6	0.53 D_1
0.7	0.45 D_1
0.8	0.36 D_1

Fig. 9 Sharp Edge Orifice with Pressure Tap Locations
From ASME PTC 19.5. Reprinted with permission of ASME.

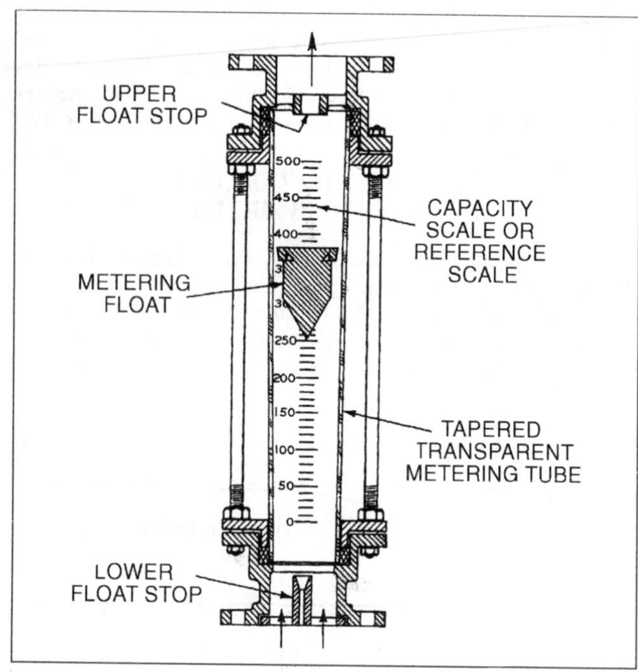

Fig. 10 Variable Area Flowmeter

plate. The contraction coefficient, friction loss coefficient C, and approach factor $1/(1 - \beta^4)^{0.5}$ can be combined into a single constant K, which is a function of geometry and Reynolds number. The orifice flow rate equations then become

$$Q = KA_2\sqrt{\frac{2g_c(p_1 - p_2)}{\rho}} \tag{8}$$

where

Q = discharge flow rate, cfs
A_2 = orifice area, ft^2
$p_1 - p_2$ = pressure drop as obtained by pressure taps, lb$_f$/ft^2

Values of K are shown in ASME *Standard* PTC 19.5.

Valves, bends, and fittings upstream from the flowmeter can cause errors. Long, straight pipes should be installed upstream and downstream from the flow devices to assure fully developed flow for proper measurement. ASHRAE *Standard* 41.8 specifies upstream and downstream pipe lengths for measuring flow of liquids with an orifice plate. ASME *Standard* PTC 19.5 gives the piping requirements between various fittings and valves and the venturi, nozzle, and orifice. If these conditions cannot be met, flow conditioners or straightening vanes can be used (ASME *Standard* PTC 19.5, ASME *Standard* MFC-10M, Mattingly 1984, Miller 1983).

Compressibility effects must be considered for gas flow if the pressure drop across the measuring device is more than a few percent of the initial pressure.

Nozzles are sometimes arranged in parallel pipes from a common manifold; thus, the capacity of the testing equipment can be changed by shutting off the flow through one or more nozzles. An apparatus designed for testing airflow and capacity of air-conditioning equipment is described by Wile (1947), who also presents pertinent information on nozzle discharge coefficients, Reynolds numbers, and resistance of perforated plates. Some laboratories refer to this apparatus as a code tester.

VARIABLE AREA FLOWMETERS (ROTAMETERS)

In permanent installations where high precision, ruggedness, and operational ease are important, the variable area flowmeter is satisfactory. It is frequently used to measure liquids or gases in small-diameter pipes. For ducts or pipes over 6 in. in diameter, the expense of this meter may not be warranted. In larger systems, however, the meter can be placed in a bypass line and used with an orifice.

The variable area meter (Figure 10) commonly consists of a float that is free to move vertically in a transparent tapered tube. The fluid to be metered enters at the narrow bottom end of the tube and moves upward, passing at some point through the annulus formed between the float and the inside wall of the tube. At any particular flow rate, the float assumes a definite position in the tube; a calibrated scale on the tube shows the float's location and the fluid flow rate.

The position of the float is established by a balance between the fluid pressure forces across the annulus and gravity on the float. The buoyant force supporting the float, $v_f(\rho_f - \rho)g/g_c$, is balanced by the pressure difference acting on the cross-sectional area of the float $A_f\Delta p$, where ρ_f, A_f, and v_f are, respectively, the float density, float cross-sectional area, and float volume. The pressure difference across the annulus is

$$\Delta p = \frac{v_f(\rho_f - \rho)g}{A_f g_c} \tag{9}$$

The mass flow follows from Equation (8) as

$$w = KA_2\sqrt{\frac{2v_f(\rho_f - \rho)g\rho}{A_f}} \tag{10}$$

The flow for any selected fluid is nearly proportional to the area, so that calibration of the tube is convenient. To use the meter for different fluids, the flow coefficient variation for any float must be known. Float design can reduce variation of the flow coefficient with Reynolds number; float materials can reduce the dependence of mass flow calibration on fluid density.

POSITIVE DISPLACEMENT METERS

Many positive displacement meters are available for measuring total liquid or gas volumetric flow rates. The fluid measured in these meters flows progressively into compartments of definite size. As the compartments are filled, they are rotated so that the fluid discharges from the meter. The flow rate through the meter is equal to the product of the compartment volume, the number of compartments, and the rotation rate of the rotor. Most of these meters have a mechanical register calibrated to show total flow.

TURBINE FLOWMETERS

Turbine flowmeters are volumetric flow rate sensing meters with a magnetic stainless steel turbine rotor suspended in the flow stream of a nonmagnetic meter body. The fluid stream exerts a force on the blades of the turbine rotor, setting it in motion and converting the fluid's linear velocity to an angular velocity. Design motivation for turbine meters is to have the rotational speed of the turbine proportional to the average fluid velocity and thus to the volume rate of fluid flow (Miller 1983, DeCarlo 1984, Mattingly 1992).

The rotational speed of the rotor is monitored by an externally mounted pickoff assembly. Magnetic and radio frequency are the most commonly used pickoffs. The magnetic pickoff contains a permanent magnet and coil. As the turbine rotor blades pass through the field produced by the permanent magnet, a shunting action induces ac voltage in the winding of the coil wrapped around the magnet. A sine wave with a frequency proportional to the flow rate develops. With the radio frequency pickoff, an oscillator applies a high-frequency carrier signal to a coil in the pickoff assembly. The rotor blades pass through the field generated by the coil and modulate the carrier signal by shunting action on the field shape. The carrier signal is modulated at a rate corresponding to the rotor speed, which is proportional to the flow rate. With both pickoffs, frequency of the pulses generated becomes a measure of flow rate, and the total number of pulses measures total volume (Woodring 1969, Shafer 1961, Mattingly 1992).

Because output frequency of the turbine flowmeter is proportional to flow rate, every pulse from the turbine meter is equivalent to a known volume of fluid that has passed through the meter; the sum of these pulses yields total volumetric flow. Summation is accomplished by electronic counters designed for use with turbine flowmeters; they combine a mechanical or electronic register with the basic electronic counter.

Turbine flowmeters should be installed with straight lengths of pipe upstream and downstream from the meter. The length of the inlet and outlet pipes should be according to manufacturers' recommendations or pertinent standards. Where recommendations of standards cannot be accommodated, the meter installation should be calibrated. Some turbine flowmeters can be used in bidirectional flow applications. A fluid strainer, used with liquids of poor or marginal lubricity, minimizes bearing wear.

The lubricity of the process fluid and the type and quality of rotor bearings determine whether the meter is satisfactory for the particular application. When choosing turbine flowmeters for use with fluorocarbon refrigerants, attention must be paid to the type of bearings used in the meter and to the oil content of the refrigerant. For these applications, sleeve-type rather than standard ball bearings are recommended. The amount of oil in the refrigerant can severely affect calibration and bearing life.

In metering liquid fluorocarbon refrigerants, the liquid must not flash to a vapor (cavitate). This would cause a tremendous increase in flow volume. Flashing results in erroneous measurements and rotor speeds that can damage the bearings or cause a failure. Flashing can be avoided by maintaining an adequate back pressure on the downstream side of the meter (Liptak 1972).

AIRFLOW-MEASURING HOODS

Flow-measuring hoods are portable instruments designed to measure supply or exhaust airflow through diffusers and grilles in HVAC systems. A flow-measuring hood assembly typically consists of a fabric hood section, a plastic or metal base, an airflow-measuring manifold, a meter, and handles for carrying and holding the hood in place.

For volumetric airflow measurements, the flow-measuring hood is placed over a diffuser or grille. The fabric hood captures and directs airflow from the outlet or inlet across the flow-sensing manifold in the base of the instrument. The manifold consists of a number of tubes containing upstream and downstream holes in a grid pattern designed to simultaneously sense and average multiple velocity points across the base of the flow-measuring hood. Air from the upstream holes flows through the tubes past a sensor and then exits through the downstream holes. Sensors employed by different manufacturers include swinging vane anemometers, electronic micromanometers, and thermal anemometers. In the case of the electronic micromanometer sensor, air does not actually flow through the manifold, but the airtight sensor senses the pressure differential from the upstream to downstream series of holes. The meter on the base of the flow-measuring hood interprets the signal from the sensor and provides a direct reading of volumetric flow in either an analog or digital display format.

As a performance check in the field, the indicated flow of a measuring hood can be compared to a duct traverse flow measurement (using a pitot-tube or a thermal anemometer). All flow-measuring hoods induce some back pressure on the air-handling system because the hood restricts the flow coming out of the diffuser. This added resistance alters the true amount of air coming out of the diffuser. In most cases, this error is negligible and is less than the accuracy of the instrument. For proportional balancing, this error need not be taken into account because all similar diffusers will have about the same amount of back pressure. To determine whether back pressure is significant, a velocity traverse can be made in the duct ahead of the diffuser with and without the flow-measuring hood in place. The difference in the average velocity of the traverse indicates the degree of back-pressure compensation required on similar diffusers in the system. For example, if the average velocity is 800 fpm with the hood in place and 820 fpm without the hood, the indicated flow reading can be multiplied by 1.025 on similar diffusers in the system (820/800 = 1.025). As an alternative, the designer of the air-handling system can predict the reduction in airflow due to the additional pressure of the hood by using a curve supplied by the flow-measuring hood manufacturer. This curve indicates the pressure drop through the hood for different flow rates.

AIR INFILTRATION, AIRTIGHTNESS, AND OUTDOOR AIR VENTILATION RATE MEASUREMENT

Two major characteristics describe air infiltration in buildings—air exchange rate and envelope air leakage. The measurement approaches used to determine these factors are described in Chapter 26. The air exchange rate of a building refers to the rate at which outdoor air enters the building under normal weather and ventilation system operation. In general, the air change rate includes both outdoor air taken in through the air handlers and air leakage through the building envelope (infiltration). The outdoor air intake rate is determined by the design, installation, and operation of the mechanical ventilation system. Infiltration is determined by the extent and distribution of leaks over the building envelope and the pressure differences across these leaks. These pressure differences are induced by wind, inside-outside temperature differences, and the operation of building mechanical equipment. To fully characterize the air exchange performance of a building, the air exchange rate must be measured over a range of weather and equipment operation.

The outdoor air ventilation rate is an indicator of the rate of dilution of occupant- and building-generated contaminants. Building air exchange rates can be measured by injecting a tracer gas into a building and monitoring and analyzing the tracer gas concentration response. The equipment required for tracer testing includes (1) a means of injecting the tracer gas and (2) a tracer gas monitor. A variety of tracer gas techniques are used. They are distinguished by their injection strategy and analysis approach. These techniques include the constant concentration (equilibrium tracer), tracer decay (ASTM *Standard* E 741), and outside air fraction (air ratio) methods. Tracer decay is the simplest and most accurate of these techniques (as per Chapter 26), but the other methods may be satisfactory if care is taken. Carbon dioxide is often used as a tracer gas because CO_2 gas monitors are relatively inexpensive and easy to use, and occupant-generated CO_2 can be used for most tracer gas techniques. Bottled CO_2 or CO_2 fire extinguishers are also readily available for tracer gas injection.

The airtightness of a building envelope can be measured relatively quickly using building pressurization techniques, which are described in Chapter 26. In the pressurization technique, a large fan or blower mounted in a door or window induces a large and roughly uniform pressure difference across the building shell. The airflow required to maintain this pressure difference is then measured. The more leakage in the building, the more airflow is required to induce a specific indoor-outdoor pressure difference. The building airtightness is characterized by the airflow rate at a reference pressure, normalized by the building volume or surface area. Under proper test conditions, the results of a pressurization test are independent of weather conditions. The instrumentation requirements for pressurization testing include air-moving equipment, a device to measure airflow, and a differential pressure gage.

CARBON DIOXIDE MEASUREMENT

Carbon dioxide has become an important measurement parameter for air-conditioning, heating, and refrigerating engineers, particularly for use in indoor air quality (IAQ) applications. Although CO_2 is generally not of concern as a specific toxin in indoor air, it is used as a surrogate indicator of odor related to human occupancy. ASHRAE *Standard* 62 states that maintaining CO_2 concentrations below 1000 ppm (based on a differential of 700 ppm between indoor and outdoor CO_2 concentrations) usually results in conditions conducive to comfort and reduced odor from human-generated pollutants. *Standard* 62 also recommends specific minimum outdoor air ventilation rates to ensure adequate indoor air quality. Carbon dioxide is often used as a tracer gas when quantifying outdoor air ventilation rates. Carbon dioxide sensors are also used in building control strategies to optimize ventilation as a function of occupancy.

NONDISPERSIVE INFRARED CO_2 DETECTORS

The technology in most widespread use for IAQ applications is the nondispersive infrared (NDIR) sensor (Figure 11). This device makes use of the strong absorption band that CO_2 produces at 4.2 μm when excited by an infrared light source. Indoor air quality-specific NDIR instruments, when calibrated between 0 and 5000 ppm, are typically accurate within 150 ppm, but the accuracy of some sensors can be improved to within 50 ppm if the instrument is calibrated for a narrower range. Portable NDIR meters are available with direct-reading digital displays; however, response time varies significantly among different instruments. While most NDIR cell designs facilitate very rapid CO_2 sample diffusion, some of the instruments now in widespread use for indoor air quality measurement exhibit slower sensor response, resulting in stabilization times greater than 5 min (up to 15 min), which may complicate walk-through inspections.

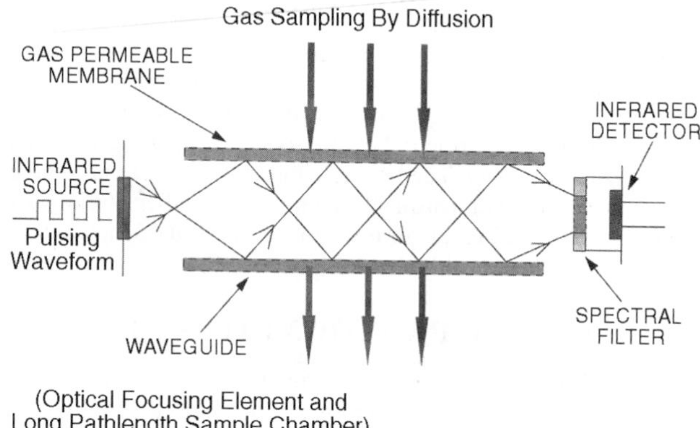

Fig. 11 Nondispersive Infrared Carbon Dioxide Sensor

Calibration

In a clean, stable environment, NDIR sensors can hold calibration for months, but condensation, dust, dirt, and mechanical shock may offset calibration. As with all other CO_2 sensor technologies, NDIR sensor readings are proportional to pressure due to the change in density of the gas molecules that results from a change in the sample pressure. This leads to errors in CO_2 readings when the barometric pressure changes from the calibration pressure. Weather-induced errors will be small, but all CO_2 instruments should be recalibrated if used at an altitude that is significantly different from the calibration altitude. Some NDIR sensors are sensitive to cooling effects when placed in an airstream. This is an important consideration when locating a fixed sensor or when using a portable system to evaluate air-handling system performance because airflow in supply and return ducts may significantly shift readings.

Applications

Nondispersive infrared sensors are well suited for equilibrium tracer and tracer decay ventilation studies, and faster response models are ideal for a quick, basic evaluation of human-generated pollution and ventilation adequacy. When properly located, these sensors are also appropriate for continuous monitoring and for control strategies using equilibrium tracer and air fraction tracer calculations.

AMPEROMETRIC ELECTROCHEMICAL CO_2 DETECTORS

Amperometric electrochemical CO_2 sensors (Figure 12) use a measured current driven between two electrodes by the reduction of CO_2 that diffuses across a porous membrane. Unlike NDIR sensors, which normally last the lifetime of the instrument, electrochemical CO_2 sensors may change in electrolyte chemistry over time (typically 12 to 18 months); so sensors should be replaced periodically. These sensors typically hold their calibration for several weeks, but they may drift more if exposed to low humidity; this drift makes them less suitable for continuous monitoring applications. At low humidity (below 30% rh), the sensors must be kept moist to maintain specified accuracy.

Amperometric electrochemical sensors have a lower power requirement than NDIR sensors, usually operating continuously for weeks where NDIR instruments typically operate for 6 hours (older models) to 150 hours (newer models). The longer battery life can be advantageous for spot checks and walk-throughs and for measuring CO_2 distribution throughout a building and within a zone. Unlike most NDIR sensors, amperometric electrochemical sensors are not affected by high humidity, although readings may be affected if condensate is allowed to form on the sensor.

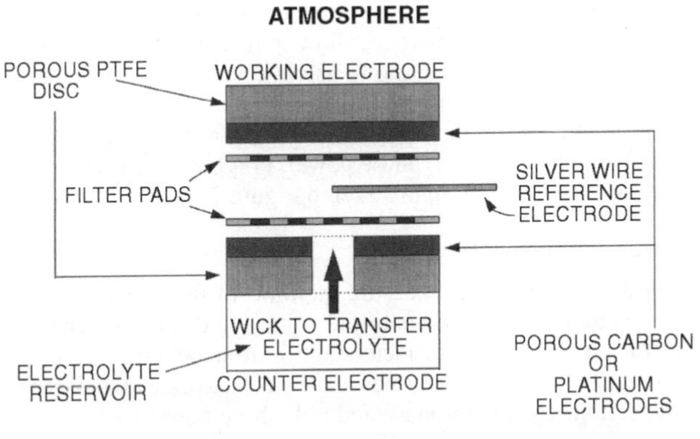

Fig. 12 Amperometric Carbon Dioxide Sensor

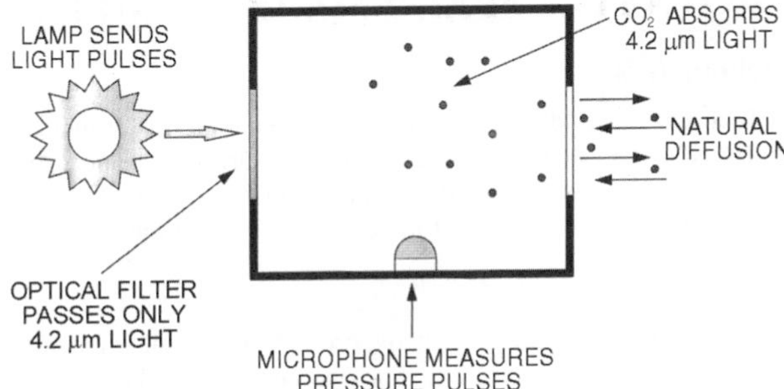

Fig. 13 Open-Cell Photoacoustic Carbon Dioxide Sensor

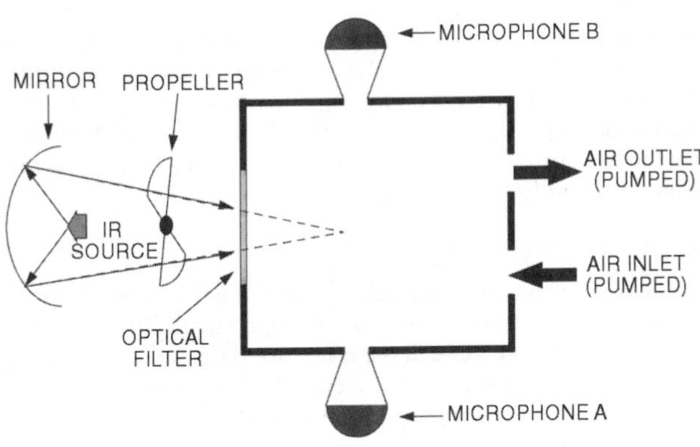

Fig. 14 Closed-Cell Photoacoustic Carbon Dioxide Sensor

PHOTOACOUSTIC CO_2 DETECTORS

Both open- and closed-cell photoacoustic sensors are available.

Open-Cell Sensors

Open-cell photoacoustic CO_2 sensors (Figure 13) operate as air diffuses through a permeable membrane into a chamber that is pulsed with filtered light at the characteristic CO_2 absorption frequency of 4.2 μm. The light energy absorbed by the CO_2 heats the sample chamber, causing a pressure pulse, which is sensed by a piezoresistor. Open-cell photoacoustic CO_2 sensors are presently unavailable in portable instruments, in part because any vibration that might occur during transportation would affect calibration and might affect the signal obtained for a given concentration of CO_2. Ambient acoustical noise may also influence readings. For continuous monitoring, vibration is a concern, as are temperature and airflow cooling effects. However, if a sensor is located properly and if the optical filter is kept relatively clean, photoacoustic CO_2 sensors may be very stable. Commercially available open-cell photoacoustic transmitters do not allow recalibration to adjust for pressure differences, so an offset should be incorporated in any control system using these sensors at an altitude or duct pressure other than calibration conditions.

Closed-Cell Sensors

Closed-cell photoacoustic sensors (Figure 14) operate under the same principle as the open-cell version, except that samples are pumped into a sample chamber that is sealed and environmentally stabilized. Two acoustic sensors are sometimes used in the chamber to minimize vibration effects. Closed-cell units, available as portable or fixed monitors, come with particle filters that are easily replaced (typically at 3- to 6-month intervals) if dirt or dust accumulates on them. Closed-cell photoacoustic monitors permit recalibration to correct for drift, pressure effects, or other environmental factors that might influence accuracy.

POTENTIOMETRIC ELECTROCHEMICAL CO_2 DETECTORS

Potentiometric electrochemical CO_2 sensors use a porous fluorocarbon membrane that is permeable to CO_2, which diffuses into a carbonic acid electrolyte, changing the electrolyte pH. This pH change is monitored by a pH electrode inside the cell. The pH electrode isopotential drift prohibits long-term monitoring to the accuracy and resolution required for continuous measurement or control or for detailed IAQ evaluations, although accuracy within 100 ppm, achievable short-term over the 2000 ppm range, may be adequate for basic ventilation and odor evaluations. In addition, this type of sensor exhibits slow response, which increases the operator time necessary for field applications or for performing a walk-through of a building.

COLORIMETRIC DETECTOR TUBES

Colorimetric detector tubes contain a chemical compound that discolors in the presence of CO_2 gas, with the amount of discoloration related to the CO_2 concentration. These detector tubes are often used to spot check CO_2 levels; when used properly, they are accurate to within 25%. If numerous samples are taken (i.e., six or more), uncertainty may be reduced; this may render the tubes, if used in the late afternoon, adequate as a very basic determination of odor and discomfort related to human occupation. However, CO_2 detector tubes are generally not appropriate for specific ventilation assessment because of their inaccuracy and inability to record concentration changes over time.

LABORATORY MEASUREMENTS

Laboratory techniques for measuring CO_2 concentration include mass spectroscopy, thermal conductivity, infrared spectroscopy, and gas chromatography. These techniques typically require taking on-site **grab samples** for laboratory analysis. Capital costs for each piece of equipment are high, and significant training is required. A considerable drawback to grab sampling is that CO_2 levels change significantly during the day and over the course of a week, making it sensible to place sensors on site with an instrument capable of recording or data logging measurements continuously over the course of a workweek. An automated grab sampling system capturing many samples of data would be quite cumbersome and expensive if designed to provide CO_2 trend information over time. However, an advantage to laboratory techniques is that they can be highly accurate. A mass spectrometer, for example, can measure CO_2 concentration to within 5 ppm from 0 to 2000 ppm. All laboratory measurement techniques are subject to errors resulting from interfering agents. A gas chromatograph is typically used in conjunction with the mass spectrometer to eliminate interference from nitrous oxide (N_2O), which has an equivalent mass, if samples are collected in a hospital or in another location where N_2O might be present.

ELECTRIC MEASUREMENT

Ammeters

Ammeters are low-resistance instruments for measuring current. They should be connected in series with the circuit being measured (Figure 15). Ideally, they have the appearance of a short circuit, but in practice, all ammeters have a nonzero input impedance that influences the measurement to some extent.

Ammeters often have several ranges, and it is good practice when measuring unknown currents to start with the highest range and then reduce the range to the appropriate value to obtain the most sensitive reading. Ammeters with range switches maintain circuit continuity during switching. On some older instruments, it may be necessary to short-circuit the ammeter terminals when changing the range.

Current transformers are often used to increase the operating range of ammeters. They may also provide isolation and thus protection from a high-voltage line. Current transformers have at least two separate windings on a magnetic core (Figure 16). The primary winding is connected in series with the circuit in which the current is measured. In the case of a clamp-on probe, the transformer core is actually opened and then connected around a single conductor carrying the current to be measured. That conductor serves as the primary winding. The secondary winding carries a scaled-down version of the primary current, which is connected to an ammeter. Depending on the type of instrument, the ammeter reading may have to be multiplied by the ratio of the transformer.

When using an auxiliary current transformer, the secondary circuit must not be open when current is flowing in the primary winding; dangerous high voltage may exist across the secondary terminals. A short-circuiting blade between the secondary terminals should be closed before the secondary circuit is opened at any point.

Transformer accuracy can be impaired by the residual magnetism in the core when the primary circuit is opened at an instant when the flux is large. The transformer core may be left magnetized, resulting in ratio and phase angle errors. The primary and secondary windings should be short-circuited before making changes.

Voltmeters

Voltmeters are high-resistance instruments that should be connected across the load (in parallel), as shown in Figure 17. Ideally they have the appearance of an open circuit, but in practice all voltmeters have some finite impedance that influences the measurement to some extent.

Voltage transformers are often used to increase the operating range of a voltmeter (Figure 18). They also provide isolation from high voltages and prevent injury to the operator. Like current transformers, voltage transformers consist of two or more windings on a magnetic core. The primary winding is generally connected across the high voltage to be measured, and the secondary winding is connected to the voltmeter. It is important not to short-circuit the secondary winding of a voltage transformer.

Wattmeters

Wattmeters are instruments that measure the active power of an ac circuit, which equals the voltage multiplied by that part of the current in phase with the voltage. There are generally two sets of terminals—one to connect the load voltage and the other to connect in series with the load current. Current and voltage transformers can be used to extend the range of a wattmeter or to isolate it from high voltage. Figure 19 and Figure 20 show connections for single-phase wattmeters, and Figure 21 shows use of current and voltage transformers with a single-phase wattmeter.

Wattmeters with multiple current and voltage elements are available to measure polyphase power. Polyphase wattmeter connections are shown in Figure 22 and Figure 23.

Power-Factor Meters

Power-factor meters measure the ratio of the active power to the apparent power (product of the voltage and current). The connections for power-factor meters and wattmeters are similar, and current and voltage transformers can be used to extend their range. Connections for single-phase and polyphase power-factor meters are shown in Figure 24 and Figure 25, respectively.

ROTATIVE SPEED MEASUREMENT

Tachometers

Tachometers, or direct-measuring rpm counters, vary from hand-held mechanical or electric meters to shaft-driven and electronic pulse counters. They are used in general laboratory and shop work to check the rotative speeds of motors, engines, and turbines.

Stroboscopes

Optical rpm counters work by producing a controlled high-speed electronic flashing light. The operator directs the light on a rotating member and increases the rate of flashes until the optical effect of stopping rotation of the member is achieved. At this point, the rpm measured is equal to the flashes per minute emitted by the strobe unit. Care must be taken to start at the bottom of the instrument scale and work up because multiples of the rpm produce almost the same optical effect as true synchronism. Multiples can be indicated by positioning suitable marks on the shaft, such as a bar on one side and a circle on the opposite side. If, for example, the two are seen superimposed, then the strobe light is flashing at an even multiple of the true rpm.

AC Tachometer-Generators

A tachometer-generator consists of a rotor and a stator. The rotor is a permanent magnet driven by the equipment. The stator is a winding with a hole through the center for the rotor. Concentricity is not critical; bearings are not required between rotor and stator. The output can be a single-cycle-per-revolution signal whose voltage is a linear function of rotor speed. The polypole configuration that generates 10 cycles per revolution permits measurement of speeds as low as 20 rpm without causing the indicating needle to flutter. The output of the ac tachometer-generator is rectified and connected to a dc voltmeter.

SOUND AND VIBRATION MEASUREMENT

Measurement systems for determining sound pressure, sound intensity, and mechanical vibration generally involve the use of transducers to convert mechanical signals into electrical signals, which are then processed electronically in order to characterize the measured mechanical signals. These measurement systems contain one or more of the following elements, which may or may not be contained in a single instrument:

1. A transducer, or an assembly of transducers, to convert sound pressure, sound intensity, or mechanical vibration (time-varying strain, displacement, velocity, acceleration, or force) into an electrical signal that is quantitatively related to the mechanical quantity being measured.

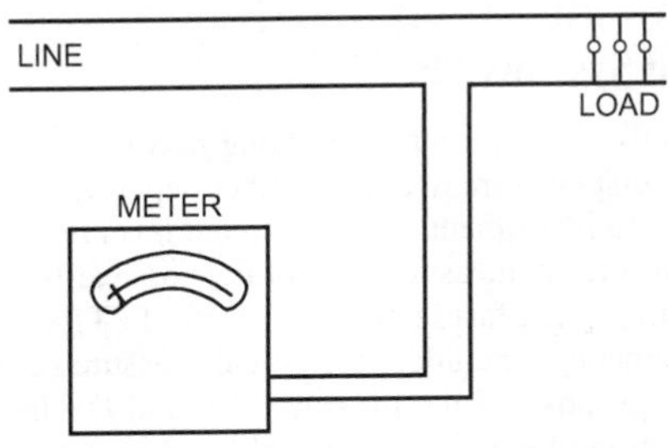

Fig. 15 Ammeter Connected in Power Circuit

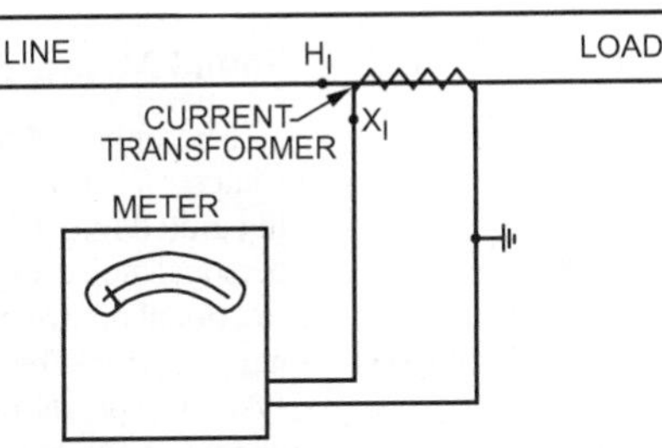

Fig. 16 Ammeter with Current Transformer

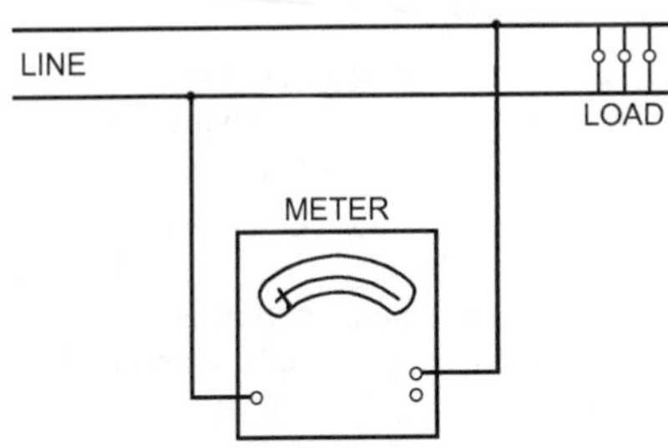

Fig. 17 Voltmeter Connected Across Load

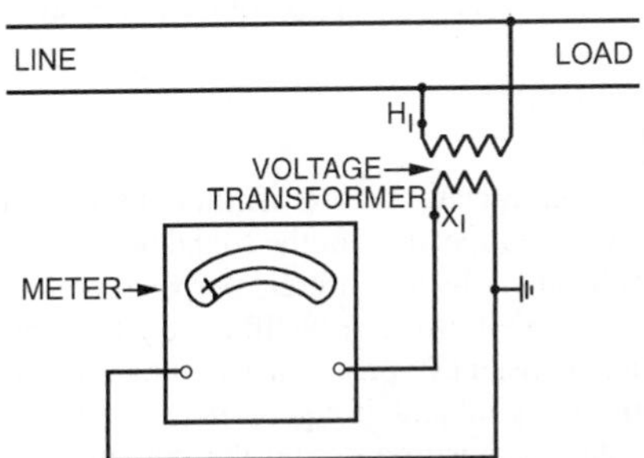

Fig. 18 Voltmeter with Potential Transformer

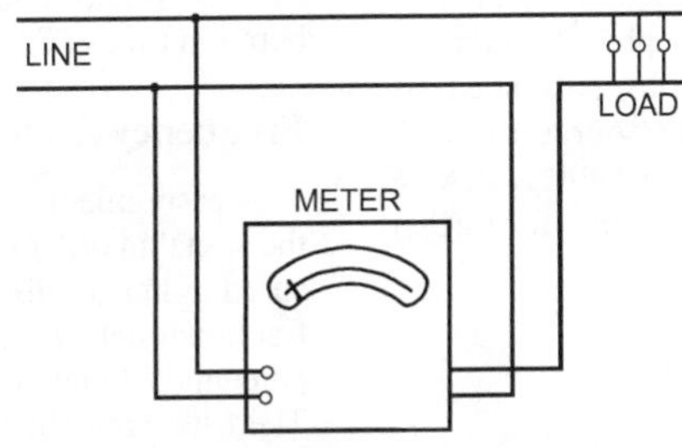

Fig. 19 Wattmeter in Single-Phase Circuit Measuring Power Load plus Loss in Current-Coil Circuit

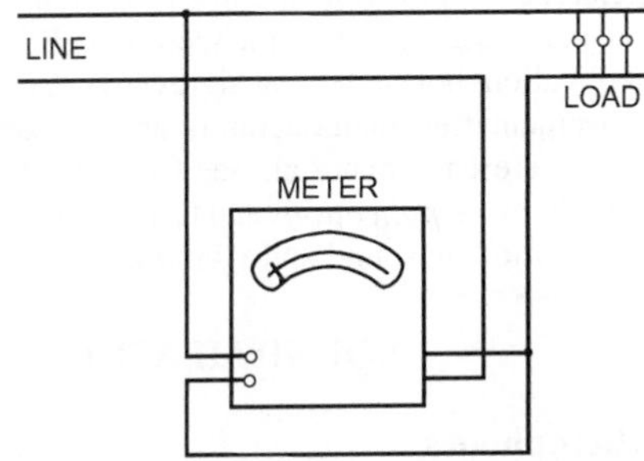

Fig. 20 Wattmeter in Single-Phase Circuit Measuring Power Load plus Loss in Potential-Coil Circuit

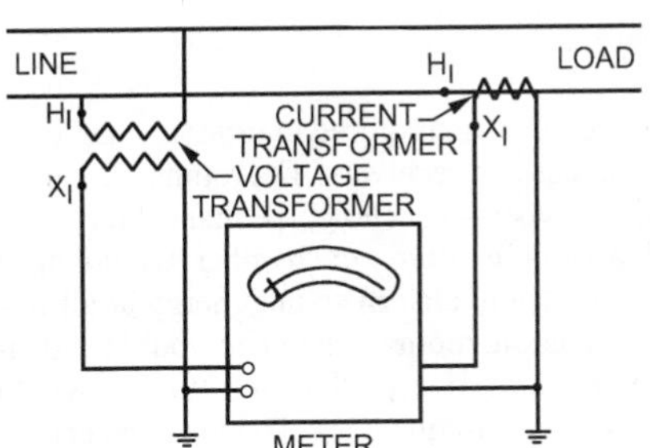

Fig. 21 Wattmeter with Current and Potential Transformer

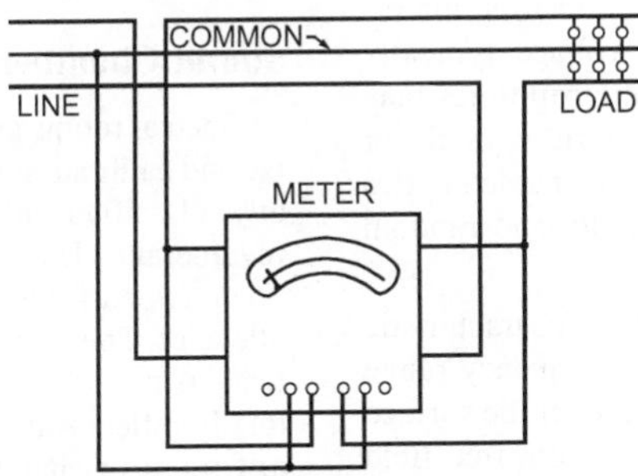

Fig. 22 Polyphase Wattmeter in Two-Phase, Three-Wire Circuit with Balanced or Unbalanced Voltage or Load

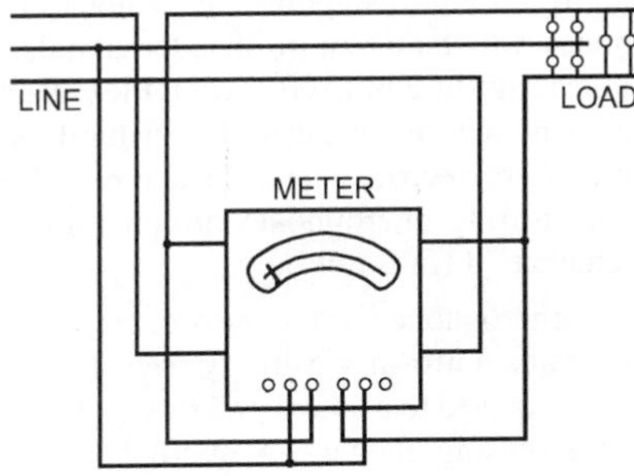

Fig. 23 Polyphase Wattmeter in Three-Phase, Three-Wire Circuit

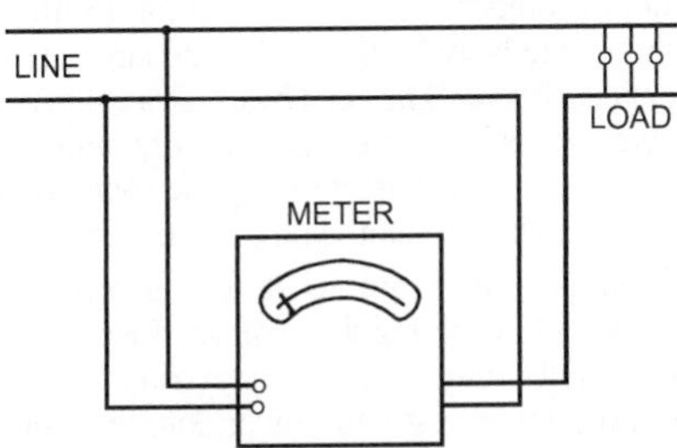

Fig. 24 Single-Phase Power-Factor Meter

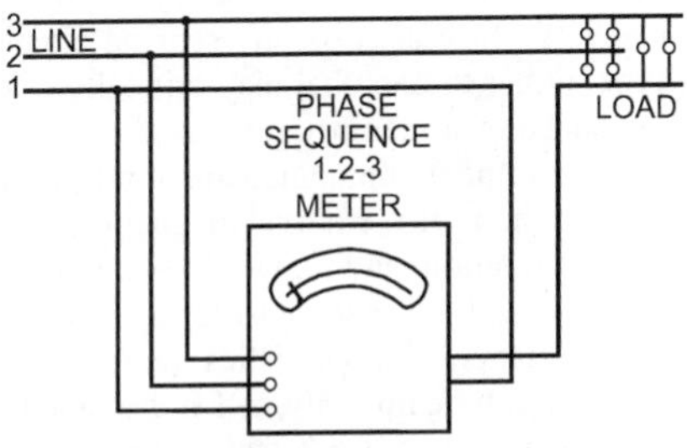

Fig. 25 Three-Wire, Three-Phase Power-Factor Meter

2. Amplifiers and networks to provide such functions as electrical impedance matching, signal conditioning, integration, differentiation, frequency weighting, and gain.
3. Signal-processing equipment to detect and quantify those aspects of the signal that are being measured (peak value, rms value, time-weighted average level, power spectral density, or magnitude or phase of a complex linear spectrum or transfer function).
4. A device such as a meter, oscilloscope, digital display, or level recorder to display the signal or the aspects of it that are being quantified.

The relevant range of sound and vibration signals can vary over more than 12 orders of magnitude in amplitude and more than 8 orders of magnitude in frequency, depending on the application. References on instrumentation, measurement procedures, and signal analysis are given in the section on Bibliography. Product and application notes, technical reviews, and books published by instrumentation manufacturers are an excellent source of additional reference material. See Chapter 46 of the 1999 *ASHRAE Handbook—Applications* and Chapter 7 of this volume for further information on sound and vibration.

SOUND MEASUREMENT

Microphones

A microphone is an electroacoustical transducer that transforms an acoustical signal into an electrical signal. The two predominant transduction principles used in the measurement of sound (as opposed to broadcasting or recording) are the electrostatic and the piezoelectric. **Electrostatic (capacitor) microphones** are available either as electric microphones, which do not require an external polarizing voltage, or as condenser microphones, which do require an external polarizing voltage, typically in the range of 28 to 200 V (dc). **Piezoelectric microphones** may be manufactured using either natural piezoelectric crystals or poled ferroelectric crystals. The types of response characteristics of measuring microphones are pressure, free field, and random incidence (diffuse field).

A microphone with a uniform pressure response characteristic maintains uniform sensitivity over its operating frequency range when exposed to a sound pressure that is uniform over the surface of the sensing element. A microphone with a uniform free-field response characteristic maintains uniform sensitivity over its operating frequency range when exposed to a plane progressive sound wave at a specified angle of incidence to the surface of the sensing element. A microphone with a uniform random-incidence response characteristic maintains uniform sensitivity over its operating frequency range when exposed to a diffuse sound field.

The sensitivity and the frequency range over which the microphone has uniform sensitivity (flat frequency response) vary with both the diameter (surface area) of the sensing element and the microphone type. Other factors that may critically affect the performance or response of a measuring microphone and preamplifier in a given measurement application are atmospheric pressure, temperature, relative humidity, external magnetic and electrostatic fields, mechanical vibration, and radiation. A microphone should be selected based on its long- and short-term stability; the match between its performance characteristics (e.g., sensitivity, frequency response, amplitude linearity, self-noise) and the expected amplitude of sound pressure, frequency, range of analysis, and expected environmental conditions of measurement; and any other pertinent considerations, such as size and directional characteristics.

Sound Measurement Systems

Microphone preamplifiers, amplifiers, weighting networks (see Chapter 7), filters, and displays are available either separately or integrated into a measuring instrument such as a sound level meter, personal noise exposure meter, measuring amplifier, or real-time fractional octave or Fourier [e.g., fast Fourier transform (FFT)] signal analyzer. The instrument(s) included in a sound measurement system depends on the purpose of the measurement and the frequency range and resolution of signal analysis. In the case of community and industrial noise measurements for regulatory purposes, the instrument, signal processing, and quantity to be measured are usually dictated by the pertinent regulation. The optimal set of instruments generally varies for measurement of different characteristics such as sound power in HVAC ducts, sound power emitted by machinery, noise criteria (NC) numbers, sound absorption coefficients, sound transmission loss of building partitions, and reverberation times (T_{60}).

Frequency Analysis

Measurement criteria often dictate the use of filters to analyze the signal in order to indicate the spectrum of the sound being measured. Filters of different bandwidths for different purposes include fractional octave band (one, one-third, one-twelfth, etc.), constant percentage bandwidth, and constant (typically narrow) bandwidth. The filters may be analog or digital and, if digital, may or may not be capable of real-time data acquisition during the measurement period, depending on the bandwidth of frequency analysis. FFT signal analyzers are generally used in situations that require very narrow band signal analysis when the amplitudes of the sound spectra vary significantly with respect to frequency. This may occur in regions of resonance or when it is necessary to identify narrow-band or discrete sine-wave signal components of a spectrum in the presence of other such components or of broadband noise.

Sound Chambers

Special rooms and procedures are required in order to characterize and calibrate sound sources and receivers. The rooms are generally classified into three types—anechoic, semianechoic, and reverberant. The ideal **anechoic** room or chamber would have boundary surfaces that completely absorb sound energy at all frequencies. The ideal **semianechoic** room or chamber would be identical to the ideal anechoic room, except that one surface would totally reflect sound energy at all frequencies. The ideal **reverberant** room or chamber would have boundary surfaces that totally reflect sound energy at all frequencies.

Anechoic chambers are used to perform measurements under conditions approximating those of a free sound field. They can be used in calibrating and characterizing individual microphones, microphone arrays, acoustic intensity probes, reference sound power sources, loudspeakers, sirens, and other individual or complex sources of sound.

Semianechoic chambers are built with a hard reflecting floor in order to accommodate heavy machinery or to simulate large factory floor or outdoor conditions. They can be used in calibrating and characterizing reference sound power sources, obtaining sound power levels of noise sources, and characterizing the sound output of emergency vehicle sirens when mounted on an emergency motor vehicle.

Reverberation chambers are used to perform measurements under conditions approximating those of a diffuse sound field. They can be used in calibrating and characterizing random-incidence microphones and reference sound power sources, obtaining sound power ratings of equipment and sound power levels of noise sources, measuring sound absorption coefficients of building materials and panels, and measuring the transmission loss through building partitions and components such as doors and windows.

Calibration

A measurement system should be calibrated as a system from microphone or probe to indicating device before it is used to perform absolute measurements of sound. Acoustic calibrators and pistonphones of fixed or variable frequency and amplitude are available for this purpose. These calibrators should be used at a frequency low enough that the pressure, free-field, and random-incidence response characteristics of the measuring microphone(s) are, for practical purposes, equivalent, or at least related in a known quantitative manner for that specific measurement system. In general, the sound pressure produced by these calibrators may vary, depending on the microphone type, whether the microphone has a protective grid, atmospheric pressure, temperature, and relative humidity. Correction factors and coefficients are required for conditions of use that differ from those existing during the calibration of the acoustic calibrator or pistonphone. For demanding applications, precision sound sources and measuring microphones should periodically be sent to the manufacturer, a private testing laboratory, or a national standards laboratory for calibration.

VIBRATION MEASUREMENT

With the exception of seismic instruments that record or indicate vibration directly via a mechanical or optomechanical device connected to the test surface, vibration measurements involve the use of an electromechanical or interferometric vibration transducer. Here, the term vibration transducer refers to a generic mechanical vibration transducer. Electromechanical and interferometric vibration transducers belong to a large and varied group of transducers that detect mechanical motion and furnish an electrical signal that is quantitatively related to a particular physical characteristic of the motion. Depending on the design of the transducer, the electrical signal may be related to mechanical strain, displacement, velocity, acceleration, or force. The operating principles of vibration transducers may involve optical interference; electrodynamic coupling; piezoelectric (including poled ferroelectric) or piezoresistive crystals; or variable capacitance, inductance, reluctance, or resistance. A considerable variety of vibration transducers with a wide range of sensitivities and bandwidths is commercially available. Vibration transducers may be contacting (e.g., seismic transducers) or noncontacting (e.g., interferometric or capacitive).

Transducers

Seismic transducers use a spring mass resonator within the transducer. At frequencies much greater than the fundamental natural frequency of the mechanical resonator, the relative displacement between the base and the seismic mass of the transducer is nearly proportional to the displacement of the transducer base. At frequencies much lower than the fundamental resonant frequency, the relative displacement between the base and the seismic mass of the transducer is nearly proportional to the acceleration of the transducer base. Therefore, seismic displacement transducers and seismic electrodynamic velocity transducers tend to have a relatively compliant suspension with a low resonant frequency; piezoelectric accelerometers and force transducers have a relatively stiff suspension with a high resonant frequency.

Strain transducers include the metallic resistance gage and the piezoresistive strain gage. For dynamic strain measurements, these are usually of the bonded type, where the gages are bonded directly to the test surface. The accuracy with which a bonded strain gage replicates strain occurring in the test structure is largely a function of how well the strain gage was oriented and bonded to the test surface.

Displacement transducers include the capacitance gage, fringe-counting interferometer, seismic displacement transducer, and linear variable differential transformer (LVDT). Velocity transducers include the reluctance (magnetic) gage, laser Doppler interferometer, and seismic electrodynamic velocity transducer. Accelerometers and force transducers include the piezoelectric, piezoresistive, and force-balance servo.

Vibration Measurement Systems

The sensitivity, frequency limitations, bandwidth, and amplitude linearity of vibration transducers vary greatly with the transduction mechanism and the manner in which the transducer is applied in a given measurement apparatus. The performance of contacting transducers can be significantly affected by the mechanical mounting methods and points of attachment of the transducer and connecting cable and by the mechanical impedance of the structure loading the transducer. Amplitude linearity varies significantly over the operating range of the transducer, with some transducer types or configurations being inherently more linear than others. Other factors that may critically affect the performance or response of a vibration transducer in a given measurement application are temperature; relative humidity; external acoustic, magnetic, and electrostatic fields; transverse vibration; base strain; chemicals; and radiation. A vibration transducer should be selected based on its long- and short-term stability; the match between its performance characteristics (e.g., sensitivity, frequency response, amplitude linearity, self-noise) and the expected amplitude of vibration, frequency range of analysis, and expected environmental conditions of measurement; and any other pertinent considerations (e.g., size, mass, and resonant frequency).

Vibration exciters, or **shakers**, are used in structural analysis, vibration analysis of machinery, fatigue testing, mechanical impedance measurements, and vibration calibration systems. Vibration exciters have a table or moving element with a drive mechanism that may be mechanical, electrodynamic, piezoelectric, or hydraulic. They range from relatively small, low-power units for calibrating transducers such as accelerometers to relatively large, high-power units for structural and fatigue testing.

Conditioning amplifiers, power supplies, preamplifiers, charge amplifiers, voltage amplifiers, power amplifiers, filters, controllers, and displays are available either separately or integrated into a measuring instrument or system, such as a structural analysis system, vibration analyzer, vibration monitoring system, vibration meter, measuring amplifier, multichannel data-acquisition and modal analysis system, or real-time fractional-octave or FFT signal analyzer. The choice of instrument(s) to include in a vibration measurement system depends on the mechanical quantity to be determined, the purpose of the measurement, and the frequency range and resolution of signal analysis. In the case of vibration measurements, the signal analysis is relatively narrow in bandwidth and may be relatively low in frequency in order to accurately characterize structural resonances. Accelerometers with internal integrated circuitry are available to provide impedance matching or servo control for measuring very low frequency acceleration (servo accelerometers). Analog integration and differentiation of vibration signals is available through integrating and differentiating networks and amplifiers, and digital is available through FFT analyzers. Vibration measurements made for different purposes (e.g., machinery diagnostics and health monitoring, balancing rotating machinery, analysis of torsional vibration, analysis of machine-tool vibration, modal analysis, analysis of vibration isolation, stress monitoring, industrial control) will generally each dictate different mechanical measurement requirements and a different optimal set of instrumentation.

Calibration

Because of their inherent long- and short-term stability, amplitude linearity, wide bandwidth, wide dynamic range, low noise, and wide range of sensitivities, seismic accelerometers have traditionally been used as a reference standard for dynamic mechanical measurements. A measurement system should be calibrated as a system from transducer to indicating device before it is used to perform absolute dynamic measurements of mechanical quantities. Calibrated reference vibration exciters, standard reference accelerometers, precision conditioning amplifiers, and precision calibration exciters are available for this purpose. These exciters and standard reference accelerometers can be used to transfer a calibration to another transducer. For demanding applications, either a calibrated exciter or a standard reference accelerometer with connecting cable and conditioning amplifier should periodically be sent to the manufacturer, a private testing laboratory, or a national standards laboratory for calibration.

LIGHTING MEASUREMENT

Light level, or illuminance, is usually measured with a photocell made from a semiconductor such as silicon or selenium. Such photocells produce an output current proportional to incident luminous flux; when linked with a microammeter, color- and cosine-corrected filters, and multirange switches, they are used in inexpensive hand-held light meters and more precise instruments. Different cell heads allow multirange use in precision meters.

Cadmium sulfide photocells, in which the resistance varies with illumination, are also used in light meters. Both gas-filled and vacuum photoelectric cells are in use.

Small survey-type meters are not as accurate as laboratory meters; their readings should be considered approximate, although consistent, for a given condition. Their range is usually from 5 to 5000 footcandles. Precision low-level meters have cell heads with ranges down to 0 to 2 footcandles.

A photometer installed in a revolving head is called a goniophotometer and is used to measure the distribution of light sources or luminaires. To measure total luminous flux, the luminaire is placed in the center of a sphere painted inside with a high-reflectance white with a near perfect diffusing matte surface. Total light output is measured through a small baffled window in the sphere wall.

To measure irradiation from germicidal lamps, a filter of fused quartz with fluorescent phosphor is placed over the light meter cell.

If meters are used to measure the number of lumens per unit area diffusely leaving a surface, luminance (cd/in^2) instead of illumination (footcandles) is read. Light meters can be used to measure luminance; or electronic lux meters containing a phototube, an amplifier, and a microammeter can read luminance directly. In the case of a perfectly diffuse reflecting surface, which has a constant luminance regardless of viewing angle, the unit of footlamberts in lumens/ft^2 is sometimes applied.

THERMAL COMFORT MEASUREMENT

Thermal comfort depends on the combined influence of clothing, activity, air temperature, air velocity, mean radiant temperature, and air humidity. Thermal comfort is influenced by heating or cooling of particular body parts. This is due to radiant temperature asymmetry (plane radiant temperature), draft (air temperature, air velocity, turbulence), vertical air temperature differences, and floor temperature (surface temperature).

A general description of thermal comfort is given in Chapter 8, and guidelines for an acceptable thermal environment are given in ASHRAE *Standard* 55 and ISO *Standard* 7730. ASHRAE *Standard* 55 also includes required measuring accuracy. In addition to specified accuracy, ISO *Standard* 7726 includes recommended measuring locations and a detailed description of instruments and methods.

Clothing and Activity Level

These values are estimated from tables (Chapter 8, ISO *Standard* 9920, ISO *Standard* 8996). The thermal insulation of clothing (clo-value) can be measured on a thermal mannequin (McCullough et al. 1985, Olesen 1985). The activity (met-value) can be estimated from measuring CO_2 and O_2 in a person's expired air.

Air Temperature

Various types of thermometers may be used to measure air temperature. Placed in a room, the sensor registers a temperature between air temperature and mean radiant temperature. One way of reducing the radiant error is to make the sensor as small as possible because the convective heat transfer coefficient increases as the size decreases while the radiant heat transfer coefficient is constant. A smaller sensor also provides a favorably low time constant. The radiant error can also be reduced by using a shield (an open, polished aluminum cylinder) around the sensor, by using a sensor with a low-emittance surface, or by artificially increasing the air velocity around the sensor (aspirating air through a tube in which the sensor is placed).

Air Velocity

In occupied zones, air velocities are usually small (0 to 100 fpm) but have an effect on human thermal sensation. Because the velocity fluctuates, the mean value should be measured over a suitable period, typically 3 min. Velocity fluctuations with frequencies up to 1 Hz significantly increase human discomfort due to draft, which is a function of air temperature, mean air velocity, and turbulence (see Chapter 8). The fluctuations can be given as the standard deviation of the air velocity over the measuring period (3 min) or as the turbulence intensity (standard deviation divided by mean air velocity). Velocity direction may change and is difficult to identify at low air velocities. An omnidirectional sensor with a short response time should be used. A thermal anemometer is suitable. If a hot-wire anemometer is used, the direction of the flow being measured must be perpendicular to the hot wire. Smoke puffs can be used to identify the direction.

Plane Radiant Temperature

This refers to the uniform temperature of an enclosure in which the radiant flux on one side of a small plane element is the same as in the actual nonuniform environment. It describes the radiation in one direction. The plane radiant temperature can be calculated from the surface temperatures of the environment (half-room) and the angle factors between the surfaces and a plane element (ASHRAE *Standard* 55). The plane radiant temperature may also be measured by a net-radiometer or a radiometer with a sensor consisting of a reflective disk (polished) and an absorbent disk (painted black) (Olesen et al. 1989).

Mean Radiant Temperature

This is the uniform temperature of an imaginary black enclosure in which an occupant would exchange the same amount of radiant heat as in the actual nonuniform enclosure. The mean radiant temperature can be calculated from measured surface temperatures and the corresponding angle factors between the person and the surfaces (Chapter 8).

The mean radiant temperature can also be determined from the plane radiant temperature in six opposite directions weighted according to the projected area factors for a person (Chapter 8).

Because of its simplicity, the instrument most commonly used to determine the mean radiant temperature is a **black globe thermometer** (Vernon 1932, Bedford and Warmer 1935). This thermometer consists of a hollow sphere usually 6 in. in diameter coated in flat black paint with a thermocouple or thermometer bulb at its center.

The temperature assumed by the globe at equilibrium results from a balance between heat gained and lost by radiation and convection.

The mean radiant temperatures are calculated from

$$\bar{t}_r = \left[(t_g + 460)^4 + \frac{4.74 \times 10^7 V_a^{0.6}}{\varepsilon D^{0.4}}(t_g - t_a)\right]^{1/4} - 460 \quad (11)$$

where

$\bar{t}_r$ = mean radiant temperature, °F
t_g = globe temperature, °F
V_a = air velocity, fpm
t_a = air temperature, °F
D = globe diameter, ft
ε = emissivity (0.95 for black globe)

According to Equation (11), air temperature and air velocity around the globe must also be determined. The globe thermometer is spherical, while mean radiant temperature is defined in relation to the human body. For sedentary people, the globe represents a good approximation. For people who are standing, the globe, in a radiant nonuniform environment, overestimates the radiation from floor or ceiling. An ellipsoid-shaped sensor gives a closer approximation to the human shape. A black globe will also overestimate the influence of short-wave radiation (e.g., sunshine). A flat gray color better represents the radiant characteristic of normal clothing (Olesen et al. 1989). The hollow sphere is usually made of copper, which results in an undesirable high time constant. This can be overcome by using lighter materials (e.g., a thin plastic bubble).

Air Humidity

The water vapor pressure (absolute humidity) is usually uniform in the occupied zone of a space; therefore, it is sufficient to measure absolute humidity at one location. Many of the instruments listed in Table 3 are applicable. At ambient temperatures that provide comfort or slight discomfort, the thermal effect of humidity is only moderate, and highly accurate humidity measurements are unnecessary.

CALCULATING THERMAL COMFORT

When the thermal parameters have been measured, their combined effect can be calculated by the thermal indices in Chapter 8. For example, the effective temperature (Gagge et al. 1971) can be determined from air temperature and humidity. Based on the four environmental parameters and an estimation of clothing and activity, the **predicted mean vote** (PMV) can be determined with the aid of tables (Fanger 1982, ISO *Standard* 7730, Chapter 8). The PMV is an index predicting the average thermal sensation that a group of occupants may experience in a given space.

For certain types of normal activity and clothing, the environmental parameters measured can be compared directly with those described in ASHRAE *Standard* 55 or ISO *Standard* 7730.

INTEGRATING INSTRUMENTS

Several instruments have been developed to evaluate the combined effect of two or more thermal parameters on human comfort. Madsen (1976) developed an instrument that gives information on the occupants' expected thermal sensation by direct measurement of the PMV value. The comfort meter has a heated ellipsoid-shaped sensor that simulates the body (Figure 26). The estimated clothing (insulation value), activity in the actual space, and humidity are set on the instrument. The sensor then integrates the thermal effect of the air temperature, mean radiant temperature, and air velocity in approximately the same way the body does. The electronic instrument gives the measured operative and equivalent temperature, calculated PMV, and predicted percentage of dissatisfied (PPD).

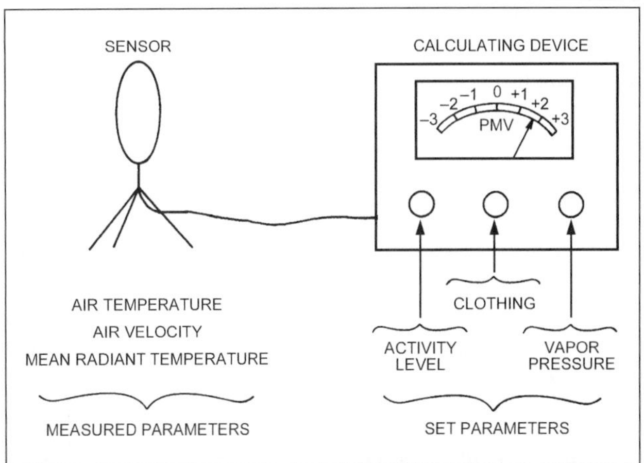

Fig. 26 Madsen's Comfort Meter
(Madsen 1976)

MOISTURE CONTENT AND TRANSFER MEASUREMENT

Little off-the-shelf instrumentation exists to measure the moisture content of porous materials or the moisture transfer through those materials. However, many measurements can be set up with a small investment of time and money. Three moisture properties are most commonly sought—(1) the sorption isotherm, a measure of the amount of water vapor a hygroscopic material will adsorb from humid air; (2) vapor permeability, a measure of the rate at which water vapor will pass through a given material; and (3) liquid diffusivity, a measure of the rate at which liquid water will pass through a porous material.

Sorption Isotherm

A sorption isotherm relates the **equilibrium moisture content** (EMC) of a hygroscopic material to the ambient relative humidity under conditions of constant temperature. Moisture content is the ratio of the total mass of water in a sample to the dry mass of the sample. Determining a sorption isotherm involves exposing a sample of material to a known relative humidity at a known temperature and then measuring the sample's moisture content after a sufficient period of time has elapsed for the sample to reach equilibrium with its surroundings. Hysteresis in the sorption behavior of most hygroscopic materials requires that measurements be made both for increasing relative humidity (the adsorption isotherm) and for decreasing relative humidity (the desorption isotherm).

The ambient relative humidity can be controlled using saturated salt solutions or mechanical refrigeration equipment (Tveit 1966, Cunningham and Sprott 1984, Carotenuto et al. 1991). Precise measurements of the relative humidity produced by various salt solutions have been reported by Greenspan (1977). ASTM *Standard* E 104 describes the use of saturated salt solutions. The EMC of a sample is usually determined gravimetrically using a precision balance. The sample dry mass, necessary to calculate moisture content, can be found by oven drying or desiccant drying. Oven dry mass may be lower than desiccant dry mass because of the loss of volatiles other than water in the oven (Richards et al. 1992).

A major difficulty in the measurement of the sorption isotherms of engineering materials is the long time required for many materials to reach equilibrium—often as long as weeks or months. The rate-limiting mechanism for these measurements is usually the slow process of vapor diffusion into the pores of the material. The use of smaller samples can help reduce the diffusion time.

Vapor Permeability

The diffusive transfer of water vapor through porous materials is often described by a modified form of Fick's law:

$$w''_v = -\mu\frac{dp}{dx} \tag{12}$$

where

w''_v = mass of vapor diffusing through unit area in unit time, gr/h·ft²
dp/dx = vapor pressure gradient, in. Hg/in.
μ = vapor permeability, gr·in/h·ft²·in. Hg

In engineering practice, permeance may be used instead of permeability. **Permeance** is simply permeability divided by the material thickness in the direction of the flow of vapor; thus, while permeability is a material property, permeance depends on thickness.

Measurement of permeability is made with wet-cup, dry-cup, or modified cup tests. Wet- and dry-cup tests are described in Chapter 23.

For many engineering materials, vapor permeability is a strong function of mean relative humidity. Wet and dry cups cannot adequately characterize this dependence on relative humidity. Instead, a modified cup method can be used (McLean et al. 1990, Burch et al. 1992). In the modified cup method, the pure water or desiccant within a cup is replaced with a saturated salt solution. A second saturated salt solution is used to condition the environment external to the cup. With such an arrangement, the relative humidities on both sides of the sample material can be varied from 0 to 100%. Several cups with a range of mean relative humidities are used to map out the dependence of vapor permeability on relative humidity.

In measuring materials of high permeability, the finite rate of vapor diffusion through air in the cup may become a factor. The air-film resistance could then be a significant fraction of the resistance to vapor flow presented by the sample material. An accurate measurement of high-permeability materials may require an accounting of diffusive rates across all air gaps (Fanney et al. 1991).

Liquid Diffusivity

The transfer of liquid water through porous materials may be characterized as a diffusion-like process:

$$w_l'' = -\rho D_l\frac{d\gamma}{dx} \tag{13}$$

where

w''_l = mass of liquid transferred through unit area per unit time, lb/h·ft²
ρ = liquid density, lb/ft³
D_l = liquid diffusivity, ft²/h
$d\gamma/dx$ = moisture content gradient, ft⁻¹

D_l typically shows a strong dependence on moisture content.

Transient measurement methods deduce the functional form of $D_l\gamma$ by observing the evolution of a one-dimensional moisture content profile over time. An initially dry specimen is brought into contact with liquid water. The free water will migrate into the specimen, drawn in by surface tension. The resulting moisture content profile, which changes with time, must be differentiated to find the liquid diffusivity of the material (Bruce and Klute 1956).

Determining the transient moisture content profile typically involves the use of a noninvasive and nondestructive method of measuring local moisture content. Gamma ray absorption (Freitas et al. 1991, Kumaran and Bomberg 1985, Quenard and Sallee 1989), X-ray radiography (Ambrose et al. 1990), neutron radiography (Prazak et al. 1990), and nuclear magnetic resonance (NMR) (Gummerson et al. 1979) have all been employed.

Uncertainty in the resulting measurement of the liquid diffusivity is often large because of the necessity to differentiate noisy experimental data.

HEAT TRANSFER THROUGH BUILDING MATERIALS

Thermal Conductivity

The thermal conductivity of a heat insulator, as defined in Chapter 23, is a unit heat transfer factor. Two methods of determining the thermal conductivity of flat insulation are the **guarded hot plate** and the **heat flow meter apparatus**, according to ASTM *Standards* C 177 and C 518, respectively. Both methods use parallel isothermal plates to induce a steady temperature gradient across the thickness of the specimen(s). The guarded hot plate is considered an absolute method for determining thermal conductivity. The heat flow meter apparatus requires calibration with a specimen having a known thermal conductivity, usually determined in the guarded hot plate. The heat flow meter apparatus is calibrated by determining the voltage output of its heat flux transducer(s) as a function of the heat flux through the transducer(s).

The basic design of the guarded hot plate consists of an electrically heated plate and two liquid-cooled plates. Two similar specimens of a material are required for a test; one is mounted on each side of the hot plate. A cold plate is then pressed against the outside of each specimen by a clamp screw. The heated plate consists of two sections separated by a small gap. During tests, the central (metering) section and the outer (guard) section are maintained at the same temperature to minimize errors caused by edge effects. The electric energy required to heat the metering section is measured carefully and converted to heat flow. The thermal conductivity of the material can be calculated under steady-state conditions using this heat flow quantity, the area of the metering section, the temperature gradient, and the specimen thickness. The thermal conductivity of cylindrical or pipe insulation (Chapter 23) is determined in a similar manner, but an equivalent thickness must be calculated to account for the cylindrical shape (ASTM *Standard* C 335). Transient methods have been developed by D'Eustachio and Schreiner (1952), Hooper and Lepper (1950), and Hooper and Chang (1953) using a line heat source within a slender probe. These instruments are available commercially and have the advantages of rapidity and a small test specimen requirement. The probe is a useful research and development tool, but it has not been as accepted as the guarded hot plate, heat flow meter apparatus, or pipe insulation apparatus.

Thermal Conductance and Resistance

Thermal conductances (C-factors) and resistances (R-values) of many building assemblies can be calculated from the conductivities and dimensions of their components, as described in Chapter 25. Test values can also be determined experimentally using large specimens representative of the building assemblies tested in the hot box apparatus described in ASTM *Standards* C 236 and C 976. This laboratory apparatus allows measurement of heat transfer through a specimen under controlled air temperature, air velocity, and radiation conditions. It is specially suited for large nonhomogeneous specimens.

For in situ measurements, heat flux and temperature transducers are useful in measuring the dynamic or steady-state behavior of opaque building components (ASTM *Standard* C 1046). A heat flux transducer is simply a differential thermopile within a core or substrate material. Two types of construction are used: (1) multiple thermocouple junctions wrapped around a core material, or (2) printed circuits with a uniform array of thermocouple junctions. The transducer is calibrated by determining its voltage output as a function of the heat flux through the transducer. For in situ measurements, the transducer is installed in either the wall or roof, or mounted on an exterior surface with tape or adhesive. The data obtained can be used to compute the thermal conductance or resistance of the building component (ASTM *Standard* C 1155).

AIR CONTAMINANT MEASUREMENT

Three measures of particulate air contamination include the number, projected area, and mass of particles per unit volume of air (ASTM Volume 11.03). Each requires an appropriate sampling technique.

Particles are counted by capturing them in impingers, impactors, membrane filters, or thermal or electrostatic precipitators. Counting may be done by microscope, using stage counts if the sample covers a broad range of sizes.

Electronic particle counters can give rapid data on particle size distribution and concentration. However, their accuracy depends on careful calibration, appropriate maintenance, and proper application. Particle counters have been used in indoor office environments as well as in clean rooms.

Projected area determinations are usually made by sampling onto a filter paper and comparing the light transmitted or scattered by this filter to a standard filter. The staining ability of dusts depends on the projected area and refractive index per unit volume. For sampling, filters must collect the minimum sized particle of interest. In this respect, membrane or glass fiber filters are recommended.

To determine particle mass, a measured quantity of air is drawn through filters, preferably of membrane or glass fiber, and the filter mass is compared to the mass before sampling. Electrostatic or thermal precipitators and various impactors have also been used. For further information, see ACGIH (1983), Lundgren et al. (1979), and Lodge (1989).

Chapter 44 of the 1999 *ASHRAE Handbook—Applications* presents information on measuring and monitoring gaseous contaminants. Relatively costly analytical equipment, which must be calibrated and operated carefully by experienced personnel, is needed. Numerous methods of sampling the contaminants, as well as the laboratory analysis techniques used after sampling, are specified. Some of the analytical methods are specific to a single pollutant; others are capable of presenting a concentration spectrum for many compounds simultaneously.

COMBUSTION ANALYSIS

Two approaches are used to measure the thermal output or capacity of a boiler, furnace, or other fuel-burning device. The direct or **calorimetric test** measures change in enthalpy or heat content of the fluid, air, or water heated by the device, and multiplies this by the flow rate to arrive at the unit's capacity. The indirect test or **flue gas analysis** method determines the heat losses in the flue gases and the jacket and deducts them from the heat content (higher heating value) of the measured fuel input to the appliance. A **heat balance** simultaneously applies both tests to the same device. The indirect test usually indicates the greater capacity, and the difference is credited to radiation from the casing or jacket and unaccounted-for losses.

With small equipment, the expense of the direct test is usually not justified, and the indirect test is used with an arbitrary radiation and unaccounted-for loss factor.

FLUE GAS ANALYSIS

The flue gases from burning fossil fuels generally contain carbon dioxide (CO_2), water, and hydrogen (H_2) with some small amounts of carbon monoxide (CO), nitrogen oxides (NO_x), sulfur oxides (SO_x), and unburned hydrocarbons. However, generally only CO_2 (or O_2) and CO are measured to determine completeness of combustion and efficiency.

In the laboratory, the instruments most commonly used to measure CO and CO_2 are nondispersive infrared (NDIR) analyzers. The NDIR instruments have several advantages: (1) they are not very sensitive to flow rate, (2) no wet chemicals are required, (3) they have a relatively fast response, (4) measurements can be made over a wide range of concentrations, and (5) they are not sensitive to the presence of contaminants in the ambient air.

In the laboratory, oxygen is generally measured with an instrument that makes use of the paramagnetic properties of oxygen. The paramagnetic instruments are generally used because of their excellent accuracy and because they can be made specific to the measurement of oxygen.

For field testing and burner adjustment, portable combustion testing equipment is available. These instruments generally measure O_2 and CO with electrochemical cells. The CO_2 is then calculated by an on-board microprocessor and, together with temperature, is used to calculate thermal efficiency. If a less expensive approach is required, a portable Orsat apparatus can be used to measure CO_2, and a length-of-stain tube to measure CO.

DATA ACQUISITION AND RECORDING

Almost every type of transducer and sensor is available with the necessary interface system to make it computer-compatible. The transducer itself begins to lose its identity when integrated into a system that incorporates such features as linearization, offset correction, self-calibration, and so forth. This has eliminated the concern regarding the details of signal conditioning and amplification of basic transducer outputs. The personal computer is integrated into every aspect of data recording, including sophisticated graphics, acquisition and control, and analysis. Modems connected to the Internet or an internal network allow easy access to remote personal computer-based data-recording systems from virtually any locale.

Other means of recording, such as chart recorders, which can be either multipurpose or specifically designed for a given sensor, are available. Chart recorders provide a visual indication and a hard copy record of the data. Rarely is the output of a chart recorder used to process data. Simple indicators and readouts are used mostly to monitor the output of a sensor visually. In most situations, analog indicators such as d'Arsonval movement meters have been rendered obsolete by modern digital indicators. Industrial environments commonly employ signal transmitters for control or computer data-handling systems to convert the signal output of the primary sensor into a compatible common signal span (e.g., the standard 4-20 mA current loop). All signal conditioning (ranging, zero suppression, reference-junction compensation) is provided at the transmitter. Thus, all recorders and controllers in the system can have an identical electrical span, with variations only in charts and scales offering the advantages of interchangeability and economy in equipment cost. Long signal transmission lines can be used, and receiving devices can be added to the loop without degrading performance.

The vast selection of available hardware, an often confusing set of terminology, and the challenge of optimizing the performance/cost ratio for a specific application make the task of configuring a data acquisition system difficult. A system specifically configured to meet a particular measurement need can quickly become obsolete if it has inadequate flexibility. Memory size, recording speed, and signal processing capability are major considerations in determining the correct recording system. Thermal, mechanical, electromagnetic interference, portability, and meteorological factors also influence the selection.

Digital Recording

A digital data acquisition system must contain an interface, which is a system involving one or several analog-to-digital converters, and in the case of multichannel inputs, circuitry for multiplexing. The interface may also provide excitation for transducers, calibration, and conversion of units. The digital data are arranged into one or several standard digital bus formats. Many data acquisition systems are designed to acquire data rapidly and store large records of data for later recording and analysis. Once the input signals have been digitized, the digital data are essentially immune to noise and can be transmitted over great distances.

Information is transferred to a computer/recorder from the interface as a pulse train, which can be transmitted as 4-, 8-, 12-, 16-, or 32-bit words. An 8-bit word is a byte; many communications methods are rated according to their bytes per second transfer rate. Digital data are transferred in either serial or parallel mode. Serial transmission means that the data are sent as a series of pulses, one bit at a time. Although slower than parallel systems, serial interfaces require only two wires, which lowers their cabling cost. The speed of serial transmissions is rated according to the symbols per second rate, or baud rate. In parallel transmission, the entire data word is transmitted at one time. To do this, each bit of a data word has to have its own transmission line; other lines are needed for clocking and control. Parallel mode is used for short distances or when high data transmission rates are required. Serial mode must be used for long-distance communications where wiring costs are prohibitive.

The two most popular interface bus standards currently used for data transmission are the IEEE 488, or general-purpose interface bus (GPIB), and the RS232 serial interface. The IEEE 488 bus system feeds data down eight parallel wires, one data byte at a time. This parallel operation allows it to transfer data rapidly at up to 1 million characters per second. However, the IEEE 488 bus is limited to a cable length of 65 ft and requires an interface connection on every meter for proper termination. The RS232 system feeds data serially down two wires, one bit at a time. An RS232 line may be over 1000 ft long. For longer distances, it may feed a modem to send data over standard telephone lines. A local area network (LAN) may be available in a facility for transmitting information. With appropriate interfacing, transducer data are available to any computer connected to the network.

Bus measurements can greatly simplify three basic applications—data gathering, automated limit testing, and computer-controlled processes. In data gathering applications, readings are collected over time. The most common applications include aging tests in quality control, temperature tests in quality assurance, and testing for intermittents in service. A controller can monitor any output indefinitely and then display the data directly on its screen or record it on magnetic tape or disks for future use.

In automated limit testing, the computer simply compares each measurement with programmed limits. The controller converts the readings to a good/bad readout. Automatic limit testing becomes highly cost-effective when working with large number of parameters of a particular unit under test.

In computer-controlled processes, the IEEE 488 bus system becomes a permanent part of a larger, completely automated system. For example, a large industrial process may require many electrical sensors that feed a central computer controlling many parts of the manufacturing process. An IEEE 488 bus controller collects readings from several sensors and saves the data until asked to dump an entire batch of readings to a larger central computer at one time. Used in this manner, the IEEE 488 bus controller serves as a slave of the central computer.

Dynamic range and accuracy are two important parameters that must be considered in a digital recording system. **Dynamic range** refers to the ratio of the maximum input signal for which the system is useful to the noise floor of the system. The **accuracy** figure for a system is impacted by the signal noise level, nonlinearity, temperature, time, crosstalk, and so forth. In selecting an 8-, 12-, or 16-bit analog-to-digital converter, the designer cannot assume that the system accuracy will necessarily be determined by the resolution of the encoders (i.e., 0.4%, 0.025%, and 0.0016%, respectively. If the sensor preceding the converter is limited to 1% full-scale accuracy, for example, no significant benefits are gained by using a 12-bit system over an 8-bit system and suppressing the least significant bit. However, a greater number of bits may be required to cover a larger dynamic range.

Data Logging Devices

Data loggers digitally store electrical signals (analog or digital) to an internal memory storage component. The signal from connected sensors is typically stored to memory at timed intervals ranging from MHz to hourly sampling. Some data loggers store data based on an event (e.g., button push, contact closure). Many data loggers can perform linearization, scaling, or other signal conditioning and permit logged readings to be either instantaneous or averaged values. Most data loggers have built-in clocks that record the time and date together with transducer signal information. Data loggers range from single-channel input to 256 or more channels. Some are general-purpose devices that will accept a multitude of analog and/or digital inputs, while others are more specialized to a specific measurement (e.g., a portable anemometer with built-in data logging capability) or for a specific application (e.g., a temperature, relative humidity, CO_2, and CO monitor with data logging for IAQ applications). Stored data are generally downloaded using a serial interface with a temporary direct connection to a personal computer. Remote data loggers may also download via modem through telephone lines. Some data loggers are designed to allow downloading directly to a printer.

With the reduction in size of personal computers (laptops, notebooks, hand-held PCs, and palmtops), the computer itself is now being used as the data logger. These "mobile" computers may be left in the field storing measurements from sensors directly interfaced into the computer.

Chart Recorders

Chart recorders convert electrical signals (analog or digital) to records of the data versus time on a hard copy, usually paper. Mechanical styluses use ink, hot wire, pressure, or electrically sensitive paper to provide a continuous trace. They are useful up to a few hundred hertz. Thermal and ink recorders are confined to chart speeds of a few inches per second for recording relatively slow processes. Newer advances in portable recorders provide multichannel inputs and up to 25 kHz real-time frequency response without using a pen or pen motor. Both *x-y* recorders and plotters allow two variables to be recorded with respect to one another. Their response times are generally limited to that of thermal and ink recorders. Oscillographic recorders have largely been made obsolete by digital oscilloscopes.

STANDARDS

ASA. 1980. Techniques of machinery vibration measurement. ANSI *Standard* S2.17-80 (R 1986). Acoustical Society of America, New York.

ASA. 1984. Mechanical vibration of rotating and reciprocating machinery—Requirements for instruments for measuring vibration severity. ANSI *Standard* S2.40-84 (R 1990).

ASA. 1984. Specification for acoustical calibrators. ANSI *Standard* S1.40-84 (R 1994).

ASA. 1985. Statistical methods for determining and verifying stated noise emission values of machinery and equipment. ANSI *Standard* S12.3-85 (R 1990).

ASA. 1987. Methods for determination of insertion loss of outdoor noise barriers. ANSI *Standard* S12.8-87.

ASA. 1989. Guide to the mechanical mounting of accelerometers. ANSI *Standard* S2.61-89 (R 1991).

ASA. 1989. Method for the designation of sound power emitted by machinery and equipment. ANSI *Standard* S12.23-89.

ASA. 1989. Reference quantities for acoustical levels. ANSI *Standard* S1.8-89.

ASA. 1990. Survey methods for the determination of sound power levels of noise sources. ANSI *Standard* S12.36-90.

ASA. 1990. Vibrations of buildings—Guidelines for the measurements of vibrations and evaluation of their effects on buildings. ANSI *Standard* S2.47-90.

ASA. 1995. Measurement of sound pressure levels in air. ANSI *Standard* S1.13-95.

ASHRAE. 1984. Standard method for measurement of flow of gas. ANSI/ASHRAE *Standard* 41.7-1984 (RA 91).

ASHRAE. 1985. Laboratory methods of testing fans for rating. ANSI/ASHRAE *Standard* 51-1985, also ANSI/AMCA *Standard* 210-85.

ASHRAE. 1986. Engineering analysis of experimental data. *Guideline* 2-1986 (RA 96).

ASHRAE. 1986. Laboratory method of testing in-duct sound power measurement procedure for fans. ANSI/ASHRAE *Standard* 68-1986, also ANSI/AMCA *Standard* 330-86.

ASHRAE. 1986. Standard method for temperature measurement. ANSI/ASHRAE *Standard* 41.1-1986 (RA 91).

ASHRAE. 1987. Standard methods for laboratory air flow measurement. ANSI/ASHRAE *Standard* 41.2-1987 (RA 92).

ASHRAE. 1988. Practices for measurement, testing, adjusting, and balancing of building heating, ventilation, air-conditioning and refrigeration systems. ANSI/ASHRAE *Standard* 111-1988.

ASHRAE. 1988. A standard calorimeter test method for flow measurement of a volatile refrigerant. ANSI/ASHRAE *Standard* 41.9-1988.

ASHRAE. 1989. Standard method for pressure measurement. ANSI/ASHRAE *Standard* 41.3-1989.

ASHRAE. 1989. Standard methods of measurement of flow of liquids in pipes using orifice flowmeters. ANSI/ASHRAE *Standard* 41.8-1989.

ASHRAE. 1992. Thermal environmental conditions for human occupancy. ANSI/ASHRAE *Standard* 55-1992.

ASHRAE. 1994. Standard method for measurement of moist air properties. ANSI/ASHRAE *Standard* 41.6-1994.

ASHRAE. 1996. Method for measurement of proportion of lubricant in liquid refrigerant. *Standard* 41.4-1996.

ASME. 1972. Application of fluid meters, Part II, 6th ed. (Interim Supplement 19.5 on Instruments and Apparatus) *Standard* PTC 19.5-72. American Society of Mechanical Engineers, New York.

ASME. 1974. Temperature measurement instruments and apparatus. ANSI/ASME *Standard* PTC 19.3-74 (R 1986).

ASME. 1983. Measurement uncertainty for fluid flow in closed conduits. ANSI/ASME *Standard* MFC-2M-83.

ASME. 1985. Part I: Measurement uncertainty instruments and apparatus. ANSI/ASME *Standard* PTC 19.1-85.

ASME. 1988. Measurement of liquid flow in closed conduits by weighing methods. ANSI/ASME *Standard* MFC-9M-88.

ASME. 1989. Measurement of fluid flow in pipes using orifice, nozzle, and venturi. *Standard* MFC-3M-85.

ASME. 1991. Gauges—Pressure indicating dial type—Elastic elements. ANSI/ASME *Standard* B40.1-91.

ASME. 1991. Glossary of terms used in the measurement of fluid flow in pipes. ANSI/ASME *Standard* MFC-1M-91.

ASME. 1994. Method for establishing installation effects on flowmeters. ANSI/ASME *Standard* MFC-10M-94.

ASTM. 1985. Standard practice for maintaining constant relative humidity by means of aqueous solutions. *Standard* E 104-85 (R 1996). American Society for Testing and Materials, West Conshohocken, PA.

ASTM. 1989. Standard test method for steady-state thermal performance of building assemblies by means of a guarded hot box. *Standard* C 236-89 (1993)e1.

ASTM. 1990. Standard practice for thermographic inspection of insulation installations in envelope cavities of frame buildings. *Standard* C 1060-90 (1997)e1.

ASTM. 1990. Standard test method for thermal performance of building assemblies by means of a calibrated hot box. *Standard* C 976-90 (1996)e1.

ASTM. 1993. Temperature-electromotive force (EMF) tables for standardized thermocouples. *Standard* E 230-93.

ASTM. 1995. Standard practice for in-situ measurement of heat flux and temperature on building envelope components. *Standard* C 1046-95.

ASTM 1995. Standard test method for determining air change in a single zone by means of gas dilution. *Standard* E 741-95.

ASTM. 1995. Standard test method for steady-state heat transfer properties of horizontal pipe insulation. *Standard* C 335-95.

ASTM. 1995. Standard practice for determining thermal resistance of building envelope components from the in-situ data. *Standard* C 1155-95.

ASTM. 1995. Standard test methods for water vapor transmission of materials. *Standard* E 96-95.

ASTM. 1997. Standard test method for steady-state heat flux measurements and thermal transmission properties by means of the guarded-hot-plate apparatus. *Standard* C 177-97.

ASTM. 1998. Standard test method for steady-state thermal transmission properties by means of the heat flow meter apparatus. *Standard* C 518-98.

ASTM. 2000. *Atmospheric analysis; occupational health and safety; protective clothing*. Vol. 11.03. (182 standards).

ISO. 1977. Measurement of fluid flow in closed conduits—Velocity area method using Pitot static tubes. *Standard* 3966. International Organization for Standardization, Geneva.

ISO. 1985. Thermal environments—Instruments and methods for measuring physical quantities. *Standard* 7726.

ISO. 1990. Ergonomics—Determination of metabolic heat production. *Standard* 8996.

ISO. 1994. Moderate thermal environments—Determination of the PMV and PPD indices and specification of the conditions for thermal comfort. *Standard* 7730.

ISO. 1995. Ergonomics of the thermal environment—Estimation of the thermal insulation and evaporative resistance of a clothing ensemble. *Standard* 9920.

REFERENCES

ACGIH. 1983. *Air sampling instruments for evaluation of atmospheric contaminants*, 6th ed. American Conference of Governmental Industrial Hygienists, Cincinnati, OH.

ASTM. 1993. *Manual on the use of thermocouples in temperature measurement. Manual* 12.

Ambrose, J.H., L.C. Chow, and J.E. Beam. 1990. Capillary flow properties of mesh wicks. AIAA *Journal of Thermophysics* 4:318-24.

Amdur, E.J. 1965. Two-pressure relative humidity standards. In *Humidity and Moisture* 3:445. Reinhold Publishing Corporation, New York.

Bedford, T. and C.G. Warmer. 1935. The globe thermometer in studies of heating and ventilating. *Journal of the Institution of Heating and Ventilating Engineers* 2:544.

Benedict, R.P. 1984. *Fundamentals of temperature, pressure and flow measurements*, 3rd ed. John Wiley and Sons, New York.

Bentz, D.P. and J.W. Martin. 1987. Using the computer to analyze coating defects. *Journal of Protective Coatings and Linings* 4(5).

Bruce, R.R. and A. Klute. 1956. The measurement of soil moisture diffusivity. Proceedings of the Soil Science Society of America 20:458-62.

Burch, D.M. 1980. Infrared audits of roof heat loss. *ASHRAE Transactions* 86(2).

Burch, D.M. and C.M. Hunt. 1978. Retrofitting an existing residence for energy conservation—An experimental study. *Building Science Series* 105. National Institute of Standards and Technology, Gaithersburg, MD.

Burch, D.M., W.C. Thomas, and A.H. Fanney. 1992. Water vapor permeability measurements of common building materials. *ASHRAE Transactions* 98(1).

Burns, G.W., M.G. Scroger, G.F. Strouse, M.C. Croarkin, and W.F. Guthrie. 1992. Temperature-electromotive force reference functions and tables for the letter-designated thermocouple types based on the ITS-90. NIST *Monograph* 175. U.S. Government Printing Office, Washington, D.C.

Carotenuto, A., F. Fucci, and G. LaFianzi. 1991. Adsorption phenomena in porous media in the presence of moist air. *International Journal of Heat and Mass Transfer* 18:71-81.

Cohen, E.R. 1990. The expression of uncertainty in physical measurements. 1990 Measurement Science Conference Proceedings, Anaheim, CA.

Coleman, H.W. and W.G. Steele. 1989. *Experimentation and uncertainty analysis for engineers*. John Wiley and Sons, New York.

Coleman, H.W. and W.G. Steele. 1995. Engineering application of experimental uncertainty analysis. *AIAA Journal* 33(10).

Considine, D.M. 1985. *Process instruments and controls handbook*, 3rd ed. McGraw-Hill, New York.

Cunningham, M.J. and T.J. Sprott. 1984. Sorption properties of New Zealand building materials. Building Research Association of New Zealand *Research Report* No. 45, Judgeford.

D'Eustachio, D. and R.E. Schreiner. 1952. A study of transient heat method for measuring thermal conductivity. *ASHVE Transactions* 58:331.

DeCarlo, J.P. 1984. *Fundamentals of flow measurement.* Instrumentation Society of America, Research Triangle Park, NC.

DeWitt, D.P. and G.D. Nutter. 1988. *Theory and practice of radiation thermometry.* John Wiley and Sons, New York.

Fanger, P.O. 1982. *Thermal comfort.* Robert E. Krieger, Malabar, FL.

Fanney, A.H., W.C. Thomas, D.M. Burch, and L.R. Mathena. 1991. Measurements of moisture diffusion in building materials. *ASHRAE Transactions* 97:99-113.

Freitas, V., P. Crausse, and V. Abrantes. 1991. Moisture diffusion in thermal insulating materials. In *Insulation materials: Testing and applications,* Vol. 2. ASTM *Special Technical Publication* STP 1116. American Society for Testing and Materials, West Conshohocken, PA.

Gagge, A.P., J.A.J. Stolwijk, and Y. Nishi. 1971. An effective temperature scale based on a simple model of human physiological regulatory response. *ASHRAE Transactions* 77(1).

Goldstein, R.J. 1978. Application of aerial infrared thermography. *ASHRAE Transactions* 84(1).

Greenspan, L. 1977. Humidity fixed points of binary saturated aqueous solutions. *Journal of Research of the National Bureau of Standards* 81A:89-95.

Greenspan, L. and A. Wexler. 1968. An adiabatic saturation psychrometer. *Journal of Research of the National Bureau of Standards* 72C(1):33.

Gummerson, R.J., C. Hall, W.D. Hoff, R. Hawkes, G.N. Holland, and W.S. Moore. 1979. Unsaturated water flow within porous materials observed by NMR imaging. *Nature* 281:56-57.

Hasegawa, S. 1976. The NBS two-pressure humidity generator, Mark 2. *Journal of Research of the National Bureau of Standards* 81A:81.

Hooper, F.C. and F.C. Lepper. 1950. Transient heat flow apparatus for the determination of thermal conductivity. *ASHVE Transactions* 56:309.

Hooper, F.C. and S.C. Chang. 1953. Development of thermal conductivity probe. *ASHVE Transactions* 59:463.

Hudson, R.D., Jr. 1969. *Infrared system engineering.* John Wiley & Sons, NY.

Kumaran, M.K. and M. Bomberg. 1985. A gamma-spectrometer for determination of density distribution and moisture distribution in building materials. Proceedings of the International Symposium on Moisture and Humidity, Washington, D.C., pp. 485-90.

Kusuda, T. 1965. Calculation of the temperature of a flat-plate wet surface under adiabatic conditions with respect to the Lewis relation. *Humidity and Moisture* I:16. Reinhold Publishing Corporation, New York.

Liptak, B.G., ed. 1972. *Instrument engineers handbook,* Vol. 1. Chilton Book Company, Philadelphia, PA.

Lodge, J.P., ed. 1989. *Methods of air sampling and analysis,* 3rd ed. Lewis Publishers, MI.

Lundgren, D.A. et al., eds. 1979. *Aerosol measurement.* University Presses of Florida, Gainesville.

Mack, R.T. 1986. Energy loss profiles: Foundation for future profit in thermal imager, sales, and service. Proceedings of the 5th Infrared Information Exchange, Book 1, AGEMA Infrared Systems, Secaucus, NJ.

Madsen, T.L. 1976. Thermal comfort measurements. *ASHRAE Transactions* 82(1).

Mattingly, G.E. 1984. Workshop on fundamental research issues in orifice metering. GRI *Report* 84/0190. Gas Research Institute, Chicago, IL.

Mattingly, G.E. 1992. The characterization of a piston displacement-type flowmeter calibration facility and the calibration and use of pulsed output type flowmeters. *Journal of Research of the National Institute of Standards and Technology* 97(5):509.

McCullough, E.A., B.W. Jones, and J. Huck. 1985. A comprehensive data base for estimating clothing insulation. *ASHRAE Transactions* 92:29-47.

McLean, R.C., G.H. Galbraith, and C.H. Sanders. 1990. Moisture transmission testing of building materials and the presentation of vapour permeability values. *Building Research and Practice.* 82-103.

Mease, N.E., W.G. Cleveland, Jr., G.E. Mattingly, J.M. Hall. 1992. Air speed calibrations at the National Institute of Standards and Technology. Proceedings of the 1992 Measurement Science Conference, Anaheim, CA.

Miller, R.W. 1983. Measurement engineering handbook. McGraw-Hill, NY.

NIST. 1976. Liquid-in-glass thermometry. NIST *Monograph* 150. National Institute of Standards and Technology, Gaithersburg, MD.

NIST. 1986. Thermometer calibrations. NIST *Monograph* 174.

Nottage, H.B., J.G. Slaby, and W.P. Gojsza. 1952. A smoke-filament technique for experimental research in room air distribution. *ASHVE Transactions* 58:399.

Olesen, B.W. 1985. A new and simpler method for estimating the thermal insulation of a clothing ensemble. *ASHRAE Transactions* 92:478-92.

Olesen, B.W., J. Rosendahl, L.N. Kalisperis, L.H. Summers, and M. Steinman. 1989. Methods for measuring and evaluating the thermal radiation in a room. *ASHRAE Transactions* 95(1).

Paljak, I. and B. Pettersson. 1972. *Thermography of buildings.* National Swedish Institute for Materials Testing, Stockholm.

Parmelee, G.V. and R.G. Huebscher. 1946. The shielding of thermocouples from the effects of radiation. *ASHVE Transactions* 52:183.

Prazak, J., J. Tywoniak, F. Peterka, and T. Slonc. 1990. Description of transport of liquid in porous media—A study based on neutron radiography data. *International Journal of Heat and Mass Transfer* 33:1105-20.

Quenard, D. and H. Sallee. 1989. A gamma-ray spectrometer for measurement of the water diffusivity of cementitious materials. Proceedings of the Materials Research Society Symposium, Vol. 137.

Quinn, T.J. 1990. *Temperature,* 2nd ed. Academic Press, New York.

Richards, R.F., D.M. Burch, W.C. Thomas. 1992. Water vapor sorption measurements of common building materials. *ASHRAE Transactions* 98(1).

Richardson, L. 1965. A thermocouple recording psychrometer for measurement of relative humidity in hot, arid atmosphere. In *Humidity and Moisture* I:101. Reinhold Publishing Corporation, New York.

Schooley, J.F. 1986. *Thermometry.* CRC Press, Boca Raton, FL.

Schooley, J.F., ed. 1992. *Temperature: Its measurement and control in science and in industry,* Vol. 6. American Institute of Physics, New York.

Seely, R.E. 1955. A circuit for measuring the resistance of energized A-C windings. *AIEE Transactions,* 214.

Shafer, M.R. 1961. Performance characteristics of turbine flowmeters. Proceedings of the Winter Annual Meeting, *Paper* No. 61-WA-25. American Society of Mechanical Engineers, New York.

Till, C.E. and G.E. Handegord. 1960. Proposed humidity standard. *ASHRAE Transactions* 66:288.

Tobiasson, W. and C. Korhonen. 1985. Roofing moisture surveys: Yesterday, today, and tomorrow. Proceedings of the Second International Symposium on Roofing Technology, Gaithersburg, MD.

Tveit, A. 1966. Measurement of moisture sorption and moisture permeability of porous materials. *Report* No. 45. Norwegian Building Research Institute, Oslo.

Vernon, H.M. 1932. The globe thermometer. *Proceedings of the Institution of Heating and Ventilating Engineers* 39:100.

Wentzel, J.D. 1961. An instrument for measurement of the humidity of air. *ASHRAE Journal* 11:67.

Wile, D.D. 1947. Air flow measurement in the laboratory. *Refrigerating Engineering* 6:515.

Woodring, E.D. 1969. Magnetic turbine flowmeters. *Instruments and Control Systems* 6:133.

Worrall, R.W. 1965. Psychrometric determination of relative humidities in air with dry-bulb temperatures exceeding 212°F. In *Humidity and Moisture* I:105. Reinhold Publishing Corporation, New York.

BIBLIOGRAPHY

Beranek, L.L. 1988. *Acoustical measurements.* Published for the Acoustical Society of America by the American Institute of Physics, New York.

Beranek, L.L. 1989. *Noise and vibration control.* Institute of Noise Control Engineering, Poughkeepsie, NY.

EPA. 1991. Introduction to indoor air quality: A self-paced learning module. EPA/400/3-91/002, U.S. Environmental Protection Agency, Washington.

Harris, C.M. 1987. *Shock and vibration handbook,* 3rd ed. McGraw-Hill, New York.

IEEE. 1987. Standard digital interface for programmable instrumentation. ANSI/IEEE *Standard* 488.1-87 (R 1994).

IESNA. 1999. *Lighting handbook,* 9th ed. Illuminating Engineering Society of North America, New York.

Lord, H.W. et al. 1987. *Noise control for engineers.* Krieger Publishing, Melbourne, FL.

Morrison, R. 1986. *Grounding and shielding techniques in instrumentation,* 3rd ed. John Wiley and Sons, New York.

Spitzer, D.W., ed. 1991. *Flow measurement.* Instrumentation Society of America, Research Triangle Park, NC.

Steele, W.G., R.A. Ferguson, R.P. Taylor, and H.W. Coleman. 1994. Comparison of ANSI/ASME and ISO models for calculation of uncertainty. *ISA Transactions* 33:339-52.

Tilford, C.R. 1992. Pressure and vacuum measurements. In *Physical methods of chemistry,* 2nd ed., Vol. 6, pp. 106-73. John Wiley and Sons, New York.

CHAPTER 15

FUNDAMENTALS OF CONTROL

AUTOMATIC HVAC control systems are designed to maintain temperature, humidity, pressure, flow, power, lighting levels, and safe levels of indoor contaminants. Automatic control primarily modulates, stages, or sequences mechanical and electrical equipment to meet load requirements and provide safe operation of the equipment. It can use digital, pneumatic, mechanical, electrical, and electric control devices; human intervention is limited to starting and stopping equipment and adjusting control set points.

This chapter focuses on the fundamental concepts and devices normally used by a control system designer. It covers (1) control fundamentals, including terminology; (2) the types of control components; (3) the methods of connecting these components to form various individual control loops or subsystems; and (4) commissioning, operation, and maintenance. Chapter 45 of the 1999 *ASHRAE Handbook—Applications* discusses the design of controls for specific HVAC applications.

TERMINOLOGY

A **closed loop** or **feedback** control measures actual changes in the controlled variable and actuates the controlled device to bring about a change. The corrective action continues until the variable is brought to a desired value within the design limitations of the controller. This arrangement of having the controller sense the value of the controlled variable is known as **feedback**. Figure 1 shows a feedback control.

An **open loop** control does not have a direct link between the value of the controlled variable and the controller. An open loop control anticipates the effect of an external variable on the system and adjusts the set point to avoid excessive offset. An example is an outdoor thermostat arranged to control heat to a building in proportion to the calculated load caused by changes in outdoor temperature. In essence, the designer presumes a fixed relationship between outside air temperature and the heat requirement of the building and specifies control action based on the outdoor air temperature. The actual space temperature has no effect on this controller. Because there is no feedback on the controlled variable (space temperature), the control is an open loop.

Figure 1 illustrates the components of the typical **control loop**. The **sensor** measures the controlled variable and transmits to the controller a signal (pneumatic, electric, or electronic) having a pressure, voltage, or current value proportional to the value of the variable being measured. The **controller** compares this value with the set point and signals to the controlled device for corrective action. A controller can be hardware or software. A **hardware controller** is an analog device that continuously receives and acts on data. Thermostats, humidistats, and pressure controls are examples of hardware controllers. A **software controller** is a digital device that receives and acts on data on a sample-rate basis. Digital algorithms are examples of software controllers.

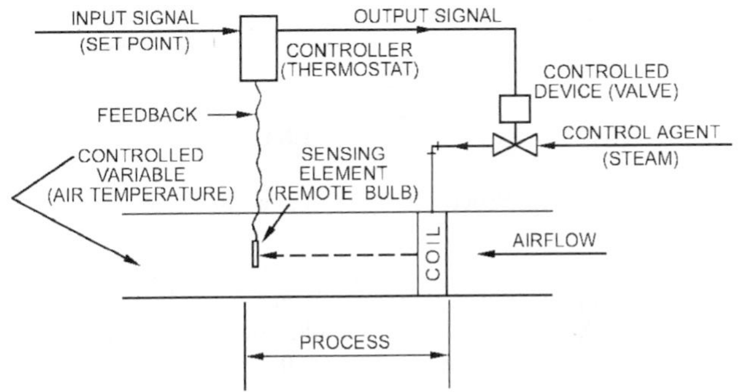

Fig. 1 Discharge Air Temperature Control
(Example of Feedback Control)

The **set point** is the desired value of the controlled variable. The controller seeks to maintain this set point. The **controlled device** reacts to signals received from the controller to vary the control agent. The controlled device may be a valve, a damper, a heating element, or a motor driving a pump or a fan.

The **control agent** is the medium manipulated by the controlled device. It may be air or gas flowing through a damper; gas, steam, or water flowing through a valve; or an electric current.

The **process** is the air-conditioning apparatus being controlled. It reacts to the output of the control agent and effects the change in the controlled variable. It may be a coil, fan, or humidifier.

The **controlled variable** is the temperature, humidity, pressure, etc., being controlled.

A control loop can be represented in the form of a **block diagram**, in which each component is modeled and represented in its own block. Figure 2 is a block diagram of the control loop shown in Figure 1. The flow of information from one component to the next is shown by lines between the blocks. The figure shows the set point being compared to the controlled variable. The difference is the **offset error**, also known as offset drift, deviation, droop, or steady-state error. The offset error is fed into the controller, which sends a control signal to the controlled device. In this case, the controlled device is a valve that can change the amount of steam flow through the coil of Figure 1. The amount of steam flow is the input to the next block, which represents the process. From the process block comes the controlled variable, which is temperature. The controlled variable is sensed by the sensing element and fed to the controller as feedback, completing the loop.

The preparation of this chapter is assigned to TC 1.4, Control Theory and Application.

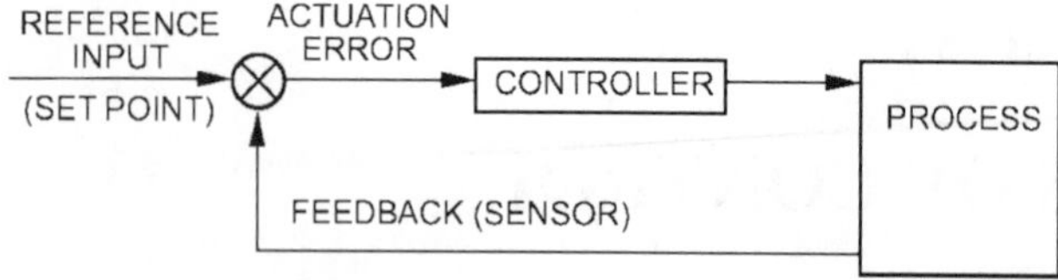

Fig. 2 Block Diagram of Discharge Air Temperature Control

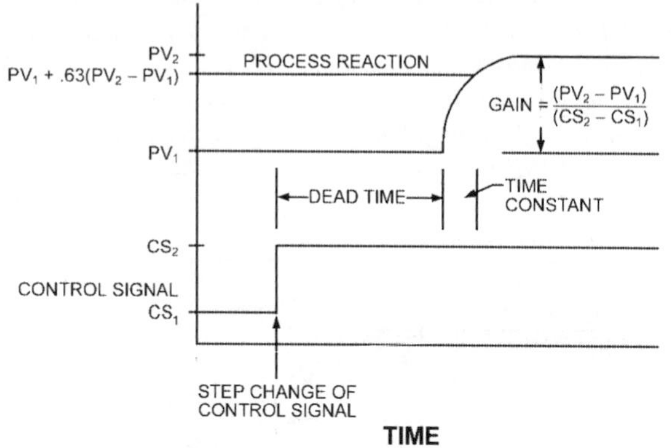

Fig. 3 Process Subjected to a Step Input

Each component of Figure 2 can be represented by a **transfer function**, which is an idealized mathematical representation of the relationship between the input and the output variables of the component. The transfer function must be sufficiently detailed to cover both the dynamic and static characteristics of the device. The dynamics of the component are represented in the time domain by a differential equation. In environmental control, the transfer function of many of the components can be adequately described by a first-order differential equation, implying that the dynamic behavior is dominated by a single capacitance factor. For a solution, the differential equation is converted to its Laplace transform or z-transform.

The **time constant** is defined as the time it takes for the output to reach 63.2% of its final value when a step change in the input is effected. Components with small time constants alter their output rapidly to reflect changes in the input; conversely, components with a larger time constant are sluggish in responding to changes in the input.

Dead time is a phase shift that can cause control and modeling problems. Dead time (or **time lag**) is the time between a change in the process input and when the change affects the output of the process. Dead time can occur in the control loop of Figure 1 due to the transportation time of the air from the coil to the space. After a coil temperature is changed, there is dead time while the affected supply air travels the distribution system and finally reaches the sensor in the space. The mass of air within the space further delays detection of the coil temperature change. Dead time can also be caused by a slow sensor, or a time lag in the signal from the controller. If the dead time is small, it may be ignored in the model of the control; if it is significant, it must be considered.

The **gain** of a transfer function is the amount the output of the component changes for a given change of input under steady-state conditions. If the element is linear, its gain remains constant. However, many control components are nonlinear, and have gains that depend on the operating conditions. Figure 3 shows the response of the first-order-plus-dead-time process to a step change of the input signal. Notice that the process shows no reaction during the dead time, followed by a response that resembles a first order exponential.

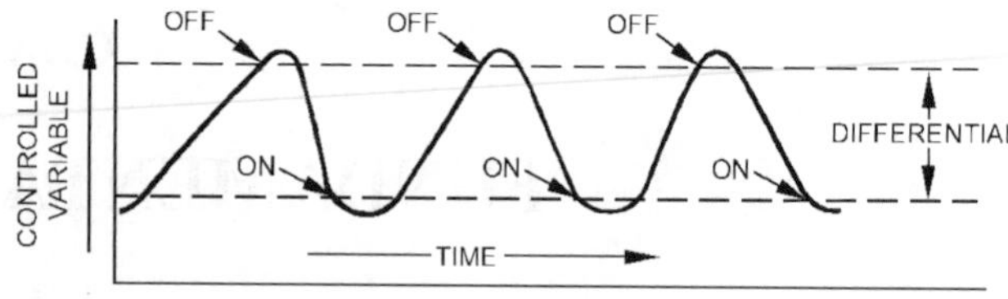

Fig. 4 Two-Position Control

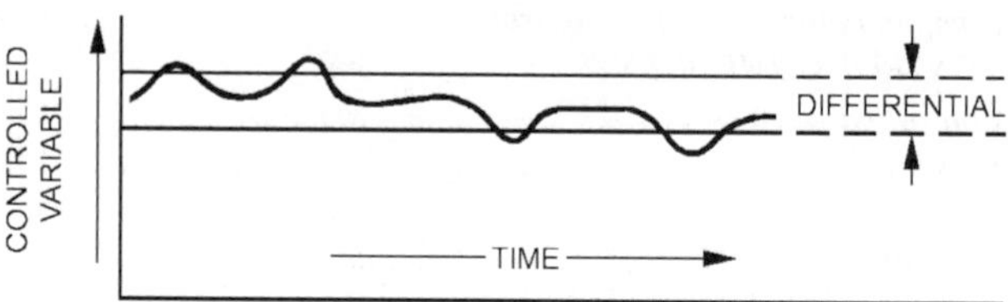

Fig. 5 Floating Control Showing Variations in Controlled Variable as Load Changes

TYPES OF CONTROL ACTION

Closed loop controls are commonly classified by the type of corrective action the controller is programmed to take when it senses a deviation of the controlled variable from the set point. Both hardware and software controllers can be classified according to the following most common types of control action.

Two-Position Action

The control device shown in Figure 4 can be positioned only to a maximum or minimum state (e.g., on or off). Because two-position control is simple and inexpensive, it is used extensively for both industrial and commercial control. A typical home thermostat that starts and stops a furnace is an example of two-position action.

Controller differential, as it applies to two-position control action, is the difference between a setting at which the controller operates to one position and a setting at which it operates to the other. Thermostat ratings usually refer to the differential (in degrees) that becomes apparent by raising and lowering the dial setting. This differential is known as the **manual differential** of the thermostat. When the same thermostat is applied to an operating system, the total change in temperature that occurs between a "turn-on" state and a "turn-off" state is usually different from the mechanical differential. The **operating differential** may be greater due to thermostat lag or hysteresis or to heating or cooling anticipators built into the thermostat.

Anticipation Applied to Two-Position Action. This common variation of strictly two-position action is often used on room thermostats to reduce the operating differential. In heating thermostats, a heater element in the thermostat is energized during "on" periods, thus shortening the on-time because the heater warms the thermostat. This is known as **heat anticipation**. The same anticipation action can be obtained in cooling thermostats by energizing a heater thermostat at "off" periods. In both cases, the percentage of on-time is varied in proportion to the load, while the total cycle time remains relatively constant.

Timed Two-Position Action. This action occurs when a heating or cooling element is turned on for a time interval proportional to the deviation from set point. For example, an element may be turned on for two minutes and off for one minute when the deviation from set point is 3°F. This is similar to incremental action applied to floating control, except the time interval is usually shorter for incremental action.

Floating Action

In floating action, the controller can perform only two operations—moving the controlled device toward either its open or closed position, usually at a constant rate (Figure 5). Generally, a neutral zone between the two positions allows the controlled device to stop at any position when the controlled variable is within the differential of the controller. When the controlled variable falls outside the differential of the controller, the controller moves the controlled device in the proper direction. In order to function properly, the sensing element must react faster than the actuator drive time. If not, the control functions the same as a two-position control.

Incremental Action. This action is a variation of floating control. Incremental action varies the pulse action to open or close an actuator depending on how close the controlled variable is to the set point. As the controlled variable comes close to the set point, the pulses become shorter. This action allows closer control using floating motor actuators.

Modulating Control

With modulating control, the output of the controller can vary infinitely over the range of the controller. The following terms are used to describe this type of control:

- **Throttling range** is the amount of change in the controlled variable required to cause the controller to move the controlled device from one extreme to the other. It can be adjusted to meet job requirements. Throttling range is inversely proportional to proportional gain.
- **Control point** is the actual value of the controlled variable at which the instrument is controlling. It varies within the throttling range of the controller and changes with changing load on the system and other variables.
- **Offset**, or error signal, is the difference between the set point and the actual control point under stable conditions. This is sometimes called drift, deviation, droop, or steady-state error.

The following are the three typical modulating control modes:

Proportional Action. In proportional action, the controlled device is positioned proportionally in response to changes in the controlled variable (Figure 6). A proportional controller can be described mathematically by

$$V_p = K_p e + V_o \tag{1}$$

where

V_p = output of controller
K_p = proportional gain (inversely proportional to throttling range)
e = error signal or offset
V_o = offset adjustment parameter

The output of the controller is proportional to the difference between the sensed value, the controlled variable, and its set point. The controlled device is normally adjusted to be in the middle of its control range at set point by using an offset adjustment. This control is similar to that shown in Figure 1.

Proportional plus Integral (PI) Action. This type of control improves on simple proportional control by adding another component to the control action that eliminates the offset typical of proportional control (Figure 7). **Reset action** may be described by

$$V_p = K_p e + K_i \int e d\theta + V_o \tag{2}$$

where

K_i = integral gain
θ = time

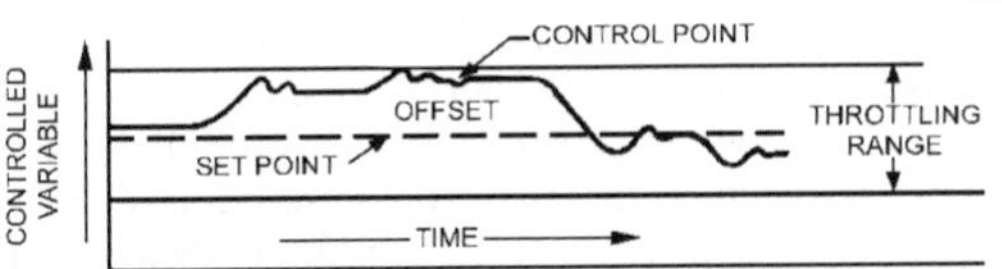

Fig. 6 Proportional Control Showing Variations in Controlled Variable as Load Changes

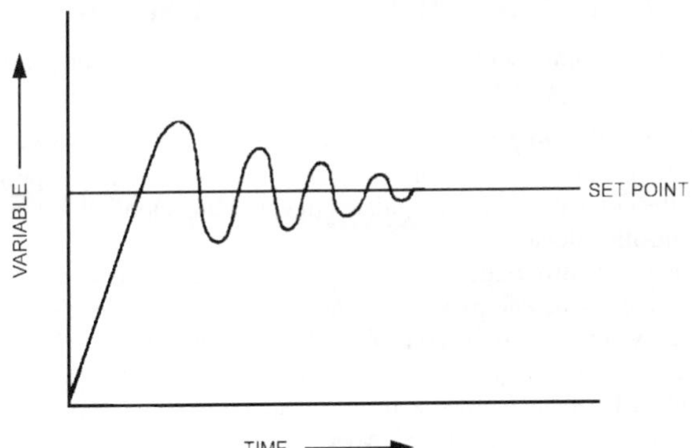

Fig. 7 Proportional plus Integral (PI) Control

The second term in Equation (2) implies that the longer error e exists, the more the controller output will change in attempting to eliminate the error. Proper selection of proportional and integral gain constants increases stability, and eliminates offset, giving greater control accuracy. PI control can also improve energy efficiency in applications such as VAV fan control, chiller control, and hot and cold deck control of the air handler.

Proportional-Integral-Derivative (PID) Action. This type of control is PI control with a derivative term added to the controller. It varies with the value of the derivative of the error. The equation for PID control is

$$V_p = K_p e + K_i \int e d\theta + K_a \frac{de}{d\theta} + V_o \tag{3}$$

where

K_a = derivative gain of controller
$de/d\theta$ = time derivative of error

Adding the derivative term gives some anticipatory action to the controller, which results in a faster response and greater stability. However, the derivative term also makes the controller more sensitive to noisy signals and harder to tune than a PI controller. Most HVAC control loops perform satisfactorily with PI control alone. **Adaptive control**, or self tuning, is a form of digital PID control, where the gain factors (K_p, K_i, and K_a) are continuously or periodically modified automatically to compensate for the control loop offset.

Fuzzy Logic

Fuzzy logic control offers an alternative to traditional control algorithms. A fuzzy logic controller uses a series of "if-then" rules that emulates the way a human operator might control the process. Examples of fuzzy logic might include

1. IF the room temperature is high AND the rate of change is decreasing, THEN increase cooling a little.
2. IF the room temperature is high AND the rate of change is increasing, THEN increase cooling a lot.

The designer of a fuzzy logic controller must first define the rules and then define such terms as *high, increasing, decreasing, a lot,* and *a little*. The room temperature, for instance, might be mapped into a series of functions that include *very low, low, OK, high,* and *very high*. The "fuzzy" element is introduced when the functions overlap and the room temperature is, for example, 70% high and 30% OK. In this case, multiple rules are combined to determine the appropriate control action.

CLASSIFICATION BY ENERGY SOURCE

Control components may be classified according to the primary source of energy as follows:

- **Pneumatic components** use compressed air, usually at a pressure of 15 to 20 psig, as an energy source. The air is generally supplied to the controller, which regulates the pressure supplied to the controlled device.
- **Electric components** use electrical energy, either low- or line-voltage, as the energy source. The controller regulates electrical energy supplied to the controlled device. Controlled devices in this category include relays; electromechanical, electromagnetic, and hydraulic actuators; and solid-state regulating devices. Components that include signal conditioning, modulation, and amplification in their operation are classified as **electronic**.

  A **digital controller** receives electronic signals from the sensors, converts the electronic signals to numbers, and performs mathematical operations on these numbers inside a microprocessor. The output from the digital controller takes the form of a number, which is then converted to an electronic signal to operate the actuator. The digital controller must sample its data because the microprocessor requires time for other operations than reading data. If the sampling interval for the digital controller is properly chosen to avoid second- and third-order harmonics, there will be no significant degradation in control performance due to sampling.
- **Self-powered components** apply the power of the measured system to induce the necessary corrective action. The measuring system derives its energy from the process under control, without any auxiliary source of energy. Temperature changes at the sensor result in pressure or volume changes of the enclosed media that are transmitted directly to the operating device of the valve or damper. A component using a thermopile in a pilot flame to generate electrical energy is also self-powered.

This method of classification can be extended to individual control loops and to complete control systems. For example, the room temperature control for a particular room that includes a pneumatic room thermostat and a pneumatically actuated reheat coil would be referred to as a pneumatic control loop. Many control systems use a combination of controls and are called **hybrid** systems.

COMPUTERS FOR AUTOMATIC CONTROL

Computers can perform the control described in this chapter. Chapter 38 of the 1999 *ASHRAE Handbook—Applications* covers computer components and some of the ways computers are being used in the HVAC control industry.

CONTROL COMPONENTS

This section groups components by their function in a complete control system. The section on Controlled Devices considers the controlled device or final control element, examples of which are relays, valves, and dampers. Actuators, which are used to drive the valve or damper assembly, are also covered.

The section on Sensors considers the sensing element that measures changes in the controlled variable. Some of the sensor types included are temperature, humidity, flow, and pressure. While many other sensors are available, these represent the majority of those found in HVAC control systems.

The section on Controllers reviews various controllers. Controllers are classified according to energy source; pneumatic, electric/electronic, and digital. They are also classified according to the control action they cause to maintain the desired condition (set point)—two-position, floating, proportional, proportional plus integral (PI), or proportional plus integral plus derivative (PID). Thermostats (devices that combine a temperature sensor and controller in a single unit) are also described as well.

Fundamental control systems can be constructed using only the components described in the first three subsections. In practice, however, a fourth group is sometimes necessary. The section on Auxiliary Control Devices covers transducers, switches, power supplies, and air compressors.

CONTROLLED DEVICES

A controlled device regulates the flow of steam, water, electricity, or air in an HVAC system. Water and steam flow regulators are known as **valves**, and airflow control devices are called **dampers**; both devices perform essentially the same function and must be properly sized and selected for the particular application. The control link to the valve or damper is called an **actuator** or **operator**. This device uses electricity, compressed air, or hydraulic fluid to power the motion of the valve stem or damper linkage through its operating range.

Valves

An automatic valve is designed to control the flow of steam, water, gas, or other fluids. It can be considered a variable orifice positioned by an electric or pneumatic actuator in response to impulses or signals from the controller. It may be equipped with a throttling plug or V-port specially designed to provided a desired flow characteristic.

Renewable composition disks are common. They are made of materials best suited to the media handled by the valve, the operating temperature, and the pressure. For high pressure or for superheated steam, metal disks are often used. Internal parts, such as the seat ring, throttling plug, or V-port skirt, disk holder, and stem, are sometimes made of stainless steel or other hard and corrosion-resistant metal for use in severe service.

Various types of automatic valves include the following:

A **single-seated valve** (Figure 8A) is designed for tight shutoff. Appropriate disk materials for various pressures and media are used.

A **double-seated** or **balanced valve** (Figure 8B) is designed so that the media pressure acting against the valve disk is essentially balanced, reducing the actuator force required. It is widely used where fluid pressure is too high to permit a single-seated valve to close. It cannot be used where a tight shutoff is required.

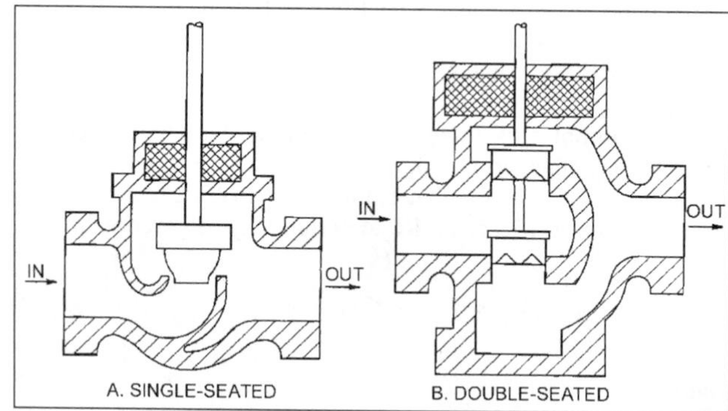

Fig. 8 Typical Single- and Double-Seated Two-Way Valves

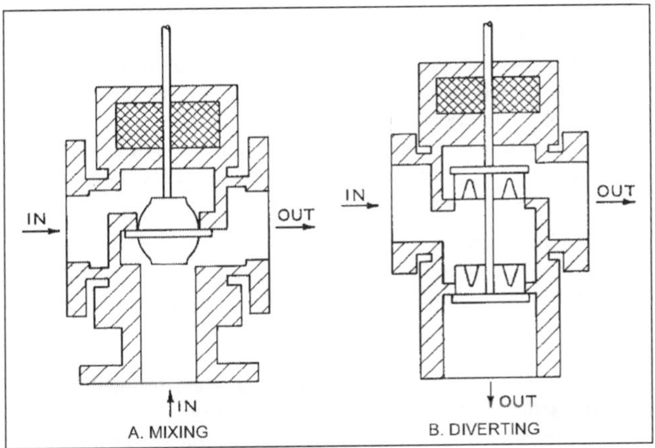

Fig. 9 Typical Three-Way Mixing and Diverting Valves

A **three-way mixing valve** (Figure 9A) has two inlet connections and one outlet connection and a double-faced disk operating between two seats. It is used to mix two fluids entering through the inlet connections and leaving through the common outlet, according to the position of the valve stem and disk.

A **three-way diverting valve** (Figure 9B) has one inlet connection and two outlet connections and two separate disks and seats. It is used to divert the flow to either of the outlets or to proportion the flow to both outlets.

A **butterfly valve** consists of a heavy ring enclosing a disk that rotates on an axis at or near its center and is similar to a round single-blade damper. In principle, the disk seats against a ring machined within the body or a resilient liner in the body. Two butterfly valves can be used together to act like a three-way valve for mixing or diverting.

Flow Characteristics. The performance of a valve is expressed in terms of its flow characteristics as it operates through its stroke, based on a constant pressure drop. Three common characteristics are shown in Figure 10 and are defined as follows:

- **Quick opening.** Maximum flow is approached rapidly as the device begins to open.
- **Linear.** Opening and flow are related in direct proportion.
- **Equal percentage.** Each equal increment of opening increases the flow by an equal percentage over the previous value.

Because the pressure drop across a valve seldom remains constant as its opening changes, actual performance usually deviates from the published characteristic curve. The magnitude of the deviation is determined by the overall design. For example, in a system arranged so that control valves or dampers can shut off all flow, the pressure drop across a controlled device increases from a minimum at design conditions to the total pressure drop at no flow. Figure 11 shows the extent of the resulting deviations for a valve or damper designed with a linear characteristic, when selection is based on various percentages of total system pressure drop. To allow for adequate control by valve or damper, the design pressure drop should be a reasonably large percentage of the total system pressure drop, or the system should be designed and controlled so that the pressure drop remains relatively constant.

Selection and Sizing. Higher pressure drops for controlled devices are obtained by using smaller sizes with a possible increase in size of other equipment in the system. Because sizing techniques are different for steam, water, and air, each is discussed separately.

Steam Valves. Steam-to-water and steam-to-air heat exchangers are typically controlled through regulation of steam flow using a two-way throttling valve. One-pipe steam systems require a line-size, two-position valve for proper condensate drainage and steam flow, while two-pipe steam systems can be controlled by two-position or modulating (throttling) valves.

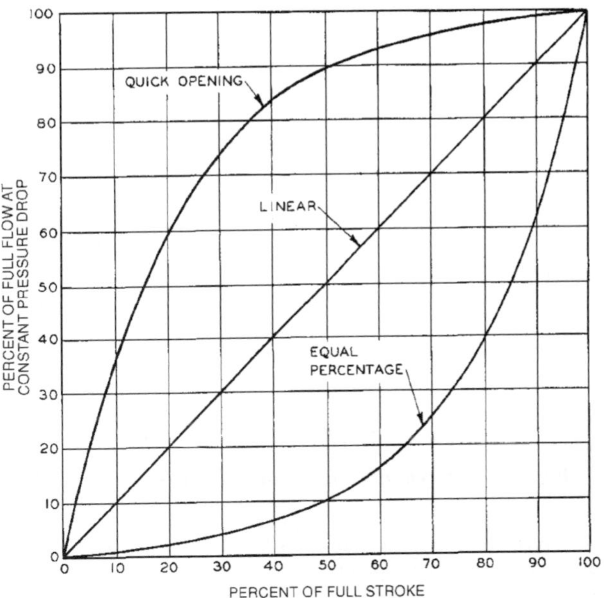

Fig. 10 Typical Flow Characteristics of Valves

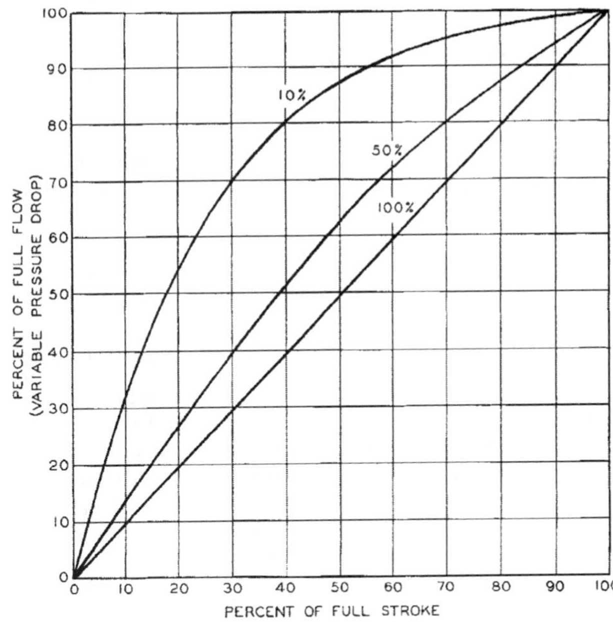

Fig. 11 Typical Performance Curves for Linear Devices at Various Percentages of Total System Pressure Drop

Water Valves. Valves for water service may be two- or three-way and two-position or proportional. Proportional valves are used most often, but two-position valves are not unusual and are sometimes essential. While it is possible to design a water system in which the pressure differential from supply to return is kept constant, it is seldom done. It is safer to assume that the pressure drop across the valve increases as it modulates from fully open to fully closed. Figure 12 shows the effect in a simple system with one pump, one two-way control valve, and a heat exchanger. The system curve represents the pressure loss in the piping and heat

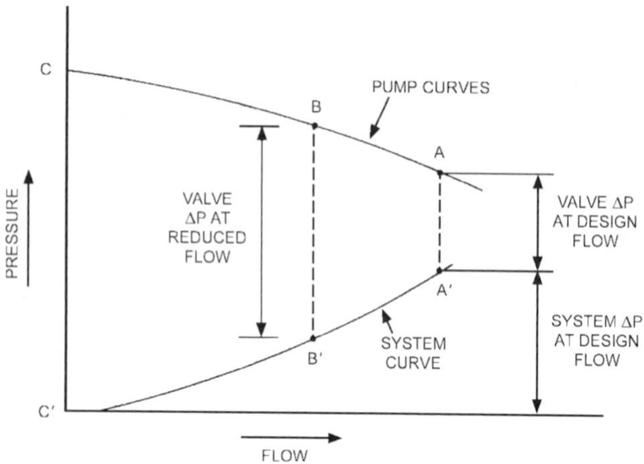

Fig. 12 Pump and System Curves with Valve Control

exchanger at various flow rates. The pump curve is the typical curve for a centrifugal pump. At design flow rates, the valve is selected for a specific pressure drop A – A′. At part load, the valve must partially close to provide a higher pressure drop B –B′. The ratio between the design pressure drop A – A′ and the zero Flow pressure drop C – C′ influences the control capability of the valve.

Equal percentage valves provide better control at part load, particularly in hot water coils where the heat output of the coil is not linearly related to flow. As flow is reduced, a greater amount of heat is transferred from each unit of water, counteracting the reduction in flow. Equal percentage valves are used in an attempt to linearize the heat transfer from the coil with respect to the control signal.

Two-way control valves should be sized to provide from 20 to 60% of the total system pressure drop. The valve operator should be sized to close the valve against the full pump head pressure to ensure complete shut off during no-flow condition.

For additional information on control valve sizing and selection, see Chapters 12 and 42 of the 2000 *ASHRAE Handbook—Systems and Equipment*.

Actuators. Valve actuators include the following general types:

- A **pneumatic valve actuator** consists of a spring-opposed, flexible diaphragm or bellows attached to the valve stem. An increase in air pressure above the minimum point of the spring range compresses the spring and simultaneously moves the valve stem. Springs of various pressure ranges can sequence the operation of two or more devices, if properly selected or adjusted. For example, a chilled water valve actuator may modulate the valve from fully closed to fully open over a spring range of 3 to 8 psig, while a sequenced steam valve may actuate from 8 to 13 psig.

  Two-position pneumatic control is accomplished using a two-position pneumatic relay to apply either full air pressure or no pressure to the valve actuator. Pneumatic valves and valves with spring-return electric actuators can be classified as normally open or normally closed.

  A **normally open valve** assumes an open position, providing full flow, when all actuating force is removed.

  A **normally closed valve** assumes a closed position, stopping flow, when all actuating force is removed.
- **Springless pneumatic valve actuators**, which use two opposed diaphragms or two sides of a single diaphragm, are generally limited to special applications involving large valves or high fluid pressure.
- An **electric-hydraulic valve actuator** is similar to a pneumatic one, except that it uses an incompressible fluid circulated by an internal electric pump.
- A **solenoid** consists of a magnetic coil operating a movable plunger. Most are for two-position operation, but modulating solenoid valves are available with a pressure equalization bellows or piston to achieve modulation. Solenoid valves are generally limited to relatively small sizes (up to 4 in.).
- An **electric motor** actuates the valve stem through a gear train and linkage. Electric motor actuators are classified in the following three types:

  Unidirectional—for two-position operation. The valve opens during one half-revolution of the output shaft and closes during the other half-revolution. Once started, it continues until the half-revolution is completed, regardless of subsequent action by the controller. Limit switches in the actuator stop the motor at the end of each stroke. If the controller has been satisfied during this interval, the actuator continues to the other position.

  Spring-return—for two-position operation. Electric energy drives the valve to one position and a spring returns the valve to its normal position.

  Reversible—for floating and proportional operation. The motor can run in either direction and can stop in any position. It is sometimes equipped with a return spring. In proportional control applications, a feedback potentiometer for rebalancing the control circuit is also driven by the motor.

Dampers

Types and Characteristics. Automatic dampers are used in air-conditioning and ventilation to control airflow. They may be used (1) for modulating control to maintain a controlled variable such as mixed air temperature or supply air duct static pressure; or (2) for two-position control to initiate operation such as opening minimum outside air dampers when a fan is started.

Multiblade dampers are available in two arrangements—parallel blade and opposed blade (Figure 13). They are used to control flow through large openings typical of those in air handlers. **Parallel-blade dampers** are adequate for two-position control and can be used for modulating control when the pressure drop remains relatively constant (i.e., outdoor air and return air dampers on air-handling unit mixing boxes). However, **opposed-blade dampers** are

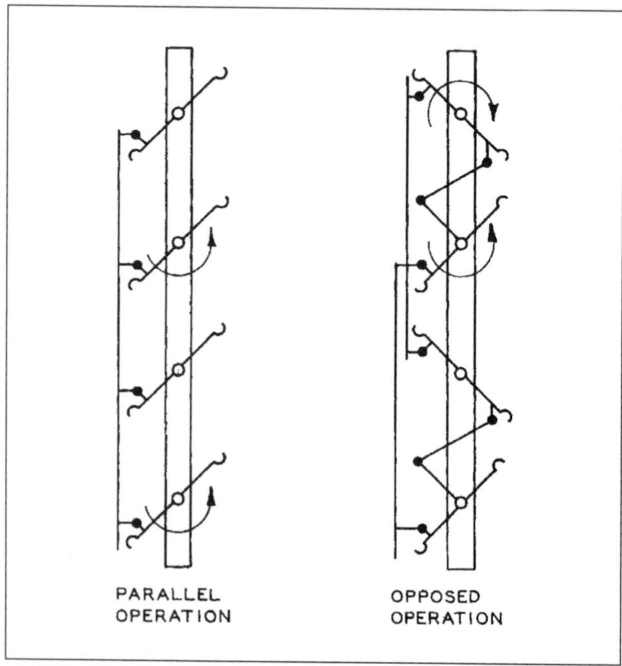

Fig. 13 Typical Multiblade Dampers

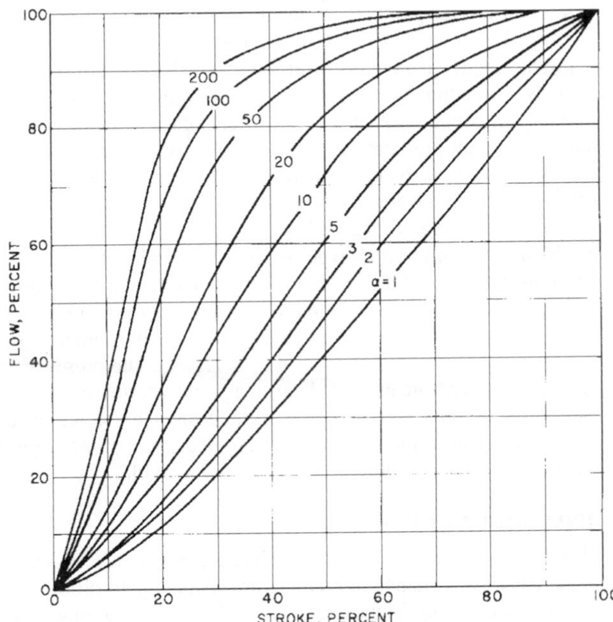

Fig. 14 Characteristic Curves of Installed Parallel-Blade Dampers

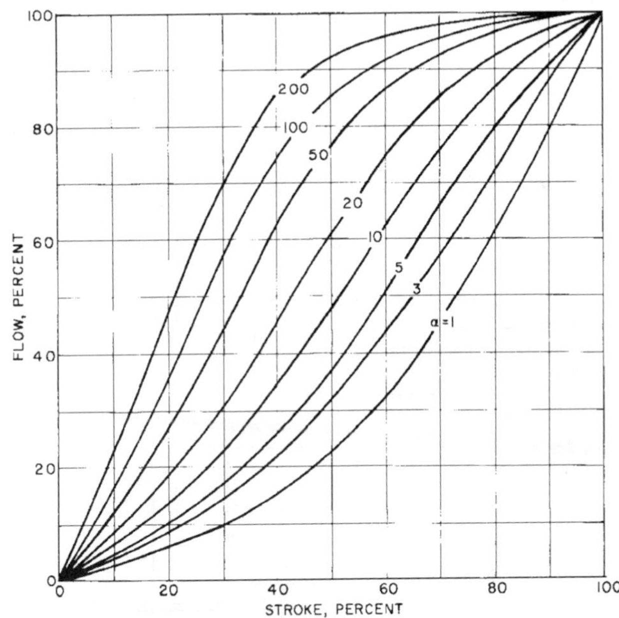

Fig. 15 Characteristic Curves of Installed Opposed-Blade Dampers

preferable for throttling control because they normally provide better control when the ratio of the pressure drop between closed and fully open is large (Figures 14 and 15). In these figures, α is the ratio of the system pressure drop to the drop across the damper at maximum (fully open) flow. **Single-blade dampers** are typically used for flow control at the zone.

Damper leakage is a concern, particularly where tight shutoff is necessary to reduce energy consumption significantly. Also, outdoor air dampers in cold climates must close tightly to prevent coils and pipes from freezing. Low-leakage dampers cost more and require larger actuators because of the friction of the seals in the closed position; therefore, they should be used only as necessary.

Actuators. Either electricity or compressed air is used to actuate dampers.

- **Pneumatic damper actuators** are similar to pneumatic valve actuators, except that they have a longer stroke or the stroke is increased by a multiplying lever. Increasing the air pressure produces a linear motion of the shaft, which, through a linkage, moves the crank arm to open or close the dampers.
- **Electric damper actuators** can be unidirectional, spring-return, or reversible. A **reversible** actuator, which has two sets of motor windings, is frequently used for accurate control in modulating damper applications. Energizing one set of windings turns the actuator output shaft clockwise; energizing the other turns the shaft counter-clockwise.

  When neither set of windings is energized, the shaft remains in its last position. The simplest form of control for this actuator is a floating point controller, which causes a contact closure to drive the motor clockwise or counter-clockwise. This type of actuator is available with a wide range of options for rotational shaft travel (expressed in degrees of rotation) and timing (expressed in the number of seconds to move through the rotational range). In addition, a variety of standard electronics signals from electronic controllers, such as 4–20 mA (dc) or 0–10 V (dc), can be used to control this type of modulating actuator.

  A two-position **spring-return** actuator moves in one direction when power is applied to its internal windings; when no power is present, the actuator returns (via spring force) to its normal position. Depending on how the actuator is connected to the dampers, this action opens or closes the dampers. A modulating actuator may also have spring-return action.

Actuator Mounting. Damper actuators are mounted in different ways, depending on the size, and accessibility of the damper, and the power required to move the damper. The most common method of mounting electric actuators is directly over the damper shaft with no external linkage. Actuators can also be mounted in the airflow on the damper frame and be linked directly to a damper blade; or they can be mounted outside the duct and connected to a crank arm attached to a shaft extension of one of the blades. On large dampers, two or more actuators may be needed. In this case, they are usually mounted at separate points on the damper. An alternative is to install the damper in two or more sections, each section being controlled by a single damper actuator; however, proper flow control is easier with a single modulating damper. Positive positioners may be required for proper sequencing. A small damper with a two-position spring-return actuator may be used for minimum outside flow, with a large damper being independently controlled for economy cycle cooling.

Positive Positioners

A pneumatic actuator may not respond quickly or accurately enough to small changes in control pressure due to friction in the actuator or load, or to changing load conditions such as wind acting on a damper blade. Where accurate positioning of a modulating damper or valve in response to load is required, positive positioners should be used. A positive positioner provides up to full supply control air pressure to the actuator for any change in position required by the controller (Figure 16). An increase in branch pressure from the controller (A) moves the relay lever (B), which opens the supply valve (C). This allows supply air to flow into the relay chamber and the actuator cylinder, moving the pistons. A linkage and spring (D) transmit the piston movement to the other end of lever (B), and when the force due to movement balances the control force, the supply valve closes, leaving the actuator in the new position. A decrease in control pressure allows the exhaust valve (E) to open until a new balance is obtained.

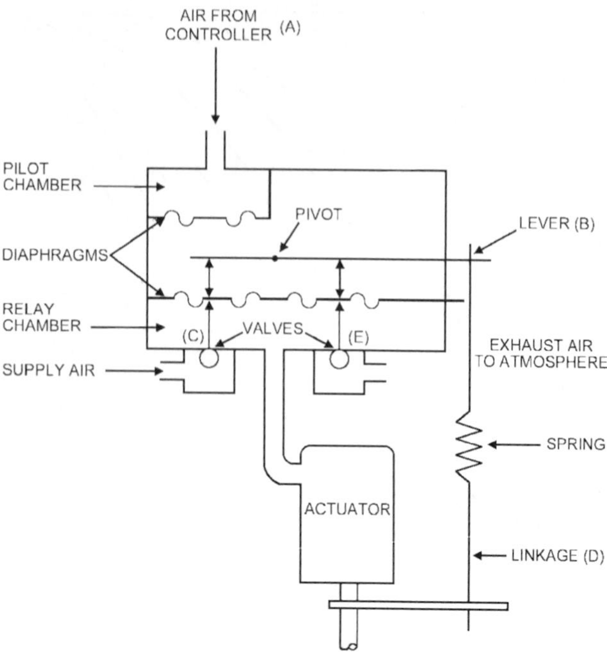

Fig. 16 Pilot Positioner (Positive Positioner)

A positive positioner provides finite and repeatable positioning change and permits adjustment of the control range (spring range of the actuator) to provide a proper sequencing control of two or more controlled devices.

SENSORS

A sensor is a device that responds to a change in the controlled variable. The response, which is a change in some physical or electrical property of the primary sensing element, is available for translation or amplification by mechanical or electrical signal. This signal is sent to the controller.

Chapter 45 of the 1999 *ASHRAE Handbook—Applications* and manufacturer's catalogs and tutorials include information on specific applications. In selecting a sensor for a specific application, the following should be considered:

- **Operating range of controlled variable.** The sensor must be capable of providing an adequate change in its output signal over the expected input range.
- **Compatibility of controller input.** Electronic and digital controllers accept various ranges and types of electronic signals. The specific controller to be used must be considered in the selection of an electronic sensor; if this is not known, an industry standard signal, such as 4–20 mA (dc) or 0–10 V (dc), should be used.
- **Accuracy and repeatability.** For some control applications, the controlled variable must be maintained within a narrow band around a desired set point. Both the accuracy and the sensitivity of the sensor selected must reflect this requirement. However, even an accurate sensor can not maintain the set point if (1) the controller is unable to resolve the input signal, (2) the controlled device can not be positioned accurately, (3) the controlled device exhibits excessive hysteresis, or (4) disturbances drive its system faster than the controls can regulate it.
- **System response time (or process dynamics).** Associated with a sensor/transducer arrangement is a response curve, which describes the response of the sensor output to change in the controlled variable. If the time constant of the process being controlled is short, and stable accurate control is important, the sensor selected must have a fast response time.
- **Control agent properties and characteristics.** The control agent is the medium to which the sensor is exposed, or with which it comes in contact, for measuring a controlled variable such as temperature or pressure. If the agent corrodes the sensor or otherwise degrades its performance, a different sensor should be selected, or the sensor must be isolated or protected from direct contact with the control agent.
- **Ambient environment characteristics.** Even when the sensor's components are isolated from direct contact with the control agent, the ambient environment must be considered. The temperature and humidity range of the ambient environment must not reduce the accuracy of the sensor. Likewise, the presence of certain gases, chemicals, and electromagnetic interference (EMI) can cause component degradation. In such cases, a special sensor or transducer housing can be used to protect the element, while ensuring a true indication of the controlled variable.

Temperature Sensors

Temperature-sensing elements fall into three general categories: (1) those that use a change in relative dimension due to differences in thermal expansion, (2) those that use a change in state of a vapor or liquid, and (3) those that use a change in some electrical property. Within each category, there are a variety of sensing elements to measure room, duct, water, and surface temperatures. Temperature-sensing technologies commonly used in HVAC applications are as follows:

- A **bimetal element** is composed of two thin strips of dissimilar metals fused together. Because the two metals have different coefficients of thermal expansion, the element bends and changes position as the temperature varies. Depending on the space available and the movement required, it may be straight, U-shaped, or wound into a spiral. This element is commonly used in room, insertion, and immersion thermostats.
- A **rod-and-tube element** consists of a high-expansion metal tube containing a low-expansion rod. One end of the rod is attached to the rear of the tube. The tube changes length with changes in temperature, causing the free end of the rod to move. This element is commonly used in certain insertion and immersion thermostats.
- A **sealed bellows element** is either vapor-, gas-, or liquid-filled. Temperature changes vary the pressure and volume of the gas or liquid, resulting in a change in force or a movement.
- A **remote bulb element** is a bulb or capsule that is connected to a sealed bellows or diaphragm by a capillary tube; the entire system is filled with vapor, gas, or liquid. Temperature changes at the bulb cause volume or pressure changes that are conveyed to the bellows or diaphragm through the capillary tube. The remote bulb element is useful where the temperature measuring point is remote from the desired thermostat location.
- A **thermistor** is a semiconductor that changes electrical resistance with temperature. It has a negative temperature coefficient (i.e., the resistance decreases as the temperature increases). Its characteristic curve of temperature versus resistance is nonlinear over a wide range. Several techniques are used to convert its response to a linear change over a particular temperature range. With digital control, one technique is to store a computer "look-up table" that maps the temperature corresponding to the measured resistance. The table breaks the curve into small segments, and each segment is assumed to be linear over its range. Thermistors are used because of their relatively low cost and the large charge in resistance possible for a small change in temperature.
- A **resistance temperature device (RTD)** is another sensor that changes resistance with temperature. Most metallic materials increase in resistance with increasing temperature. Over limited ranges, this variation is linear for certain metals such as platinum,

copper, tungsten, and nickel/iron alloys. Platinum, for example, is linear within ±0.3% from 0 to 300°F. The RTD sensing element is available in several forms for surface or immersion mounting. Flat grid windings are used for measurements of surface temperatures. For direct measurement of fluid temperatures, the windings are encased in a stainless steel bulb to protect them from corrosion.

Humidity Sensors

Humidity sensors, or **hygrometers**, are used to measure relative humidity, dew point, or absolute humidity of ambient or moving air. Two types that detect relative humidity are mechanical hygrometers and electronic hygrometers.

A **mechanical hygrometer** operates on the principle that a hygroscopic material, usually a moisture-sensitive nylon or bulk polymer material, retains moisture and expands when exposed to water vapor. The change in size or form is detected by a mechanical linkage and converted to a pneumatic or electronic signal. Mechanical sensors using hair, wood, paper, or cotton do not match the performance of moisture-sensitive nylon or bulk polymer sensors and are not widely used.

Electronic hygrometers can use either resistance or capacitance sensing elements. The resistance element is a conductive grid coated with a hygroscopic (water-absorbent) substance. The conductivity of the grid varies with the water retained; thus, the resistance varies according to the relative humidity. The conductive element is arranged in an alternating current excited Wheatstone bridge and responds rapidly to humidity changes.

The capacitance element is a stretched membrane of nonconductive film. It is coated on both sides with metal electrodes and mounted in a perforated plastic capsule. The response of the sensor's capacity to rising relative humidity is nonlinear. The signal is linearized and temperature is compensated in the amplifier circuit to provide an output signal as the relative humidity changes from 0 to 100%.

The **chilled mirror humidity sensor** determines dew point rather than relative humidity. Air flows across a small mirror in the sensor. A thermoelectric cooler lowers the surface temperature of the mirror. The mirror surface condenses until it reaches the dew point of the air. The condensation from the surface reduces the amount of light reflected from the mirror compared to a reference light level.

Dispersive infrared (DIR) technology can be used to sense absolute humidity or dew point. It is similar to technology used to sense carbon dioxide or other gases. **Infrared water vapor sensors** are optical sensors that detect the amount of water vapor in air based on the infrared light absorption characteristics of water molecules. Light absorption is proportional to the number of molecules present. The output of an **infrared hygrometer** is typically a value of absolute humidity or dew point. They can operate in diffusion or flow-through sample mode. This type of humidity sensor is unique in that the sensing element (a light detector and an infrared filter) is behind a transparent window that is never exposed directly to the sample environment. As a result, this sensor has excellent long-term stability and life and fast response time, is not subject to saturation, and operates equally well in very high or low humidity. Previously used solely for high-end applications, infrared hygrometers are now commonly used in HVAC applications because they cost about the same as mid-range accuracy (1 to 3%) humidity sensors.

Pressure Transmitters and Transducers

A pneumatic pressure transmitter converts a change in absolute, gage, or differential pressure to a mechanical motion using a bellows, diaphragm, or Bourdon tube mechanism. When corrected through appropriate links, this mechanical motion produces a change in the air pressure to a controller. In some instances, the sensing and control functions are combined in a single component, a **pressure controller**.

An electronic pressure transducer may use the mechanical actuation of a diaphragm or Bourdon tube to operate a potentiometer or differential transformer. Another type of transducer uses a strain gage bonded to a diaphragm. The strain gage detects the displacement resulting from the force applied to the diaphragm. Electronic circuits provide temperature compensation and amplification to produce a standard output signal.

Flow Rate Sensors

Orifice plate, pitot tube, venturi, turbine, magnetic flow, vortex shedding, and Doppler effect meters are used to sense fluid flow. In general, the pressure differential devices (orifice plates, venturi and pitot tubes) are less expensive and simpler to use but have limited range; thus, their accuracy depends on how they are applied and where in a system they are located.

More sophisticated flow devices, such as turbine, magnetic, and vortex shedding meters, usually have better range and are more accurate over a wide range. If an existing piping system is being considered for retrofit with a flow device, the expense of shutting down the system and cutting into a pipe must be considered. In this case, a noninvasive meter, such as a Doppler effect meter, can be cost-effective.

Indoor Air Quality Sensors

Indoor air quality control can be divided into two categories—ventilation control and contamination protection. Ventilation control measures levels of carbon dioxide or other contaminants in a space and controls the amount of outdoor air introduced into the occupied space. Demand control of ventilation helps maintain proper ventilation rates at all levels of occupancy. Typical control set point levels for carbon dioxide are 800 to 1000 ppm. ASHRAE *Standard* 62 provides further information on ventilation for acceptable indoor air quality.

Contamination protection sensors monitor levels of hazardous or toxic substances and issue warning signals and/or initiate corrective actions through the building automation system (BAS). Sensors are available for many different gases. The carbon monoxide (CO) sensor is one of the most common. The CO sensor is used to control and alarm CO levels in parking garages. Oxygen depletion sensors are used to measure, alarm, and initiate ventilation purging in enclosed spaces that house refrigeration equipment to prevent suffocation of occupants upon a refrigeration leak. The application of these sensors determines the type selected, the substances monitored, and the action taken.

Lighting Level Sensors

Analog lighting level transmitters packaged in various configurations allow control of ambient lighting levels using building automation strategies for energy conservation. Some examples include ceiling-mounted indoor light sensors used to measure room lighting levels; outdoor ambient lighting sensors used to control parking, general exterior, security, and sign lighting; and interior skylight sensors used to monitor and control light levels in skylight wells and other atrium spaces.

Power Sensing and Transmission

Passive electronic devices that sense the magnetic field around a conductor carrying current allow low-cost instrumentation of power circuits. A wire in the sensor forms an inductive coupling that powers the internal function and senses the level of the power signal. These devices can provide an analog output signal to monitor current flow or operate a switch at a user-set level to turn on an alarm or other device.

CONTROLLERS

A controller compares the sensor's signal with a desired set point and regulates an output signal to a controlled device. Digital controllers perform the control function using a microprocessor and control algorithm. The sensor and controller can be combined in a single instrument, such as a room thermostat, or they may be two separate devices.

Pneumatic Receiver-Controllers

Pneumatic receiver-controllers are normally combined with pneumatic elements that use a force or position reaction to the sensed variable to obtain a variable output air pressure. The control mode is usually proportional, but other modes such as proportional-plus-integral can be used. These controllers are generally classified as nonrelay, relay direct, or reverse-acting.

The nonrelay pneumatic controller uses low-volume output. A relay-type pneumatic controller actuates a relay device that amplifies the air volume available for control. The relay provides quicker response to a variable change.

Direct-acting controllers increase the output signal as the controlled variable increases. Reverse-acting controllers increase the output signal as the controlled variable decreases. A reverse-acting thermostat increases output pressure when the temperature drops.

Electric/Electronic Controllers

For two-position control, the controller output may be a simple electrical contact that starts a burner or pump, or one that actuates a spring-return valve or damper actuator. Single-pole, double-throw (SPDT) switching circuits are used to control a three-wire unidirectional motor actuator. SPDT circuits are also used for heating and cooling applications. Both single-pole, single-throw (SPST) and SPDT circuits can be modified for timed two-position action.

Output for floating control is a SPDT switching circuit with a neutral zone where neither contact is made. This control is used with reversible motors; it has slow response and a wide throttling range.

Pulse modulation control is an improvement over floating control. It provides closer control by varying the duration of the contact closure. As the actual condition moves closer to the set point, the pulse duration shortens for closer control. As the actual condition moves farther from the set point, the pulse duration lengthens.

Proportional control gives continuous or incremental changes in output signal to position an electrical actuator or controlled device.

Digital Controllers

A microprocessor in the digital controller executes control algorithms on one or multiple control loops. This controller is fundamentally different from pneumatic or electronic controllers in that the control algorithm is stored as a set of program instructions in memory (software or firmware). The controller itself calculates the proper control signals digitally rather than using an analog circuit or mechanical change.

A digital controller can be either single-loop or multiloop. Interface hardware allows the digital computer to process signals from various input devices such as electronic temperature, humidity, or pressure sensors described in the section on Sensors. Based on digitized equivalents of the input voltage or current signals, the control software calculates the required state of the output devices, such as valve and damper actuators and fan starters. The output devices are then positioned to the calculated state via interface hardware that converts the digital signal from the computer to an analog voltage or current required to position the actuator or energize a relay. It is common in both new and retrofit applications to use an additional interface device to convert the analog voltage or current into a pneumatic signal, which operates a pneumatic actuator. Such devices are called **electronic-to-pneumatic (E/P) transducers**.

The operator enters parameters such as set points, proportional or integral gains, minimum on and off times, or high and low limits, but the control algorithms stored in the computer's memory make the control decisions. The computer scans the input devices, executes the control algorithms, and then positions the output device(s), in a stepwise scheme. Digital controllers can be classified with regard to the way control algorithms are stored in memory (such as in firmware and software) and their ability to communicate to higher level devices such as terminals and computers.

Firmware and Software. Preprogrammed control routines, known as firmware, are typically stored in permanent memory such as electrically erasable programmable read-only memory (EEPROM). The operator can modify parameters such as set points, limits, and minimum off times within the control routines, but the program logic cannot be changed without replacing the memory chips.

User-programmable controllers allow the algorithms to be changed by the user. The programming language provided with the controller can vary from (1) a derivation of a standard language (such as Pascal or BASIC) to (2) a custom language developed by the controller's manufacturer to (3) graphically based programming. Preprogrammed routines for proportional, proportional plus integral, Boolean logic, timers, and so forth, are typically included in the language. Standard energy management routines may also be preprogrammed and may interact with other control loops where appropriate.

Digital controllers can be furnished with both preprogrammed firmware and user-programmed routines. These routines can automatically modify the parameters of the firmware according to user-defined conditions to accomplish the sequence of control designed by the control engineer.

Operator Interface. Some digital controllers are designed for dedicated purposes and are adjustable only through manual switches and potentiometers mounted on the controller. This type of controller cannot be networked with other controllers. An example is the programmable room thermostat. A **direct digital controller** can have manual adjustable features, but it more typically is adjusted through a built-in LED or LCD display, a hand-held device, or a terminal or computer. The DDC controller can communicate digitally, which allows remote connection to other controllers and to higher level computing devices and host operating stations.

A **terminal** allows the user to communicate with the controller and, where applicable, modify the program in the controller. Terminals can range from hand-held units with an LCD display and several buttons to a full-size console with a video monitor and keyboard. The terminal can be limited in function to allow only the display of sensor and parameter values or powerful enough to allow changing or reprogramming the control strategies. In some instances, a terminal can communicate remotely with one or more controllers, thus allowing central displays, alarms, and commands. Usually, hand-held terminals are used by technicians for troubleshooting, and full consoles at a fixed location are used to monitor the entire digital control system.

Thermostats

Thermostats combine sensing and control functions in a single device. Microprocessor-based thermostats have many of the features described in the following paragraphs.

- The **occupied-unoccupied** or **dual-temperature room thermostat** controls at a reduced temperature at night. It may be indexed (changed from occupied to unoccupied) individually or in groups by a manual switch or time switch from a remote point. Some electric units have an individual clock and switch built into the thermostat.
- The **pneumatic day-night thermostat** uses a two-pressure air supply system—the two pressures often being 13 and 17 psig, or 15 and 20 psig. Changing the pressure at a central point from one

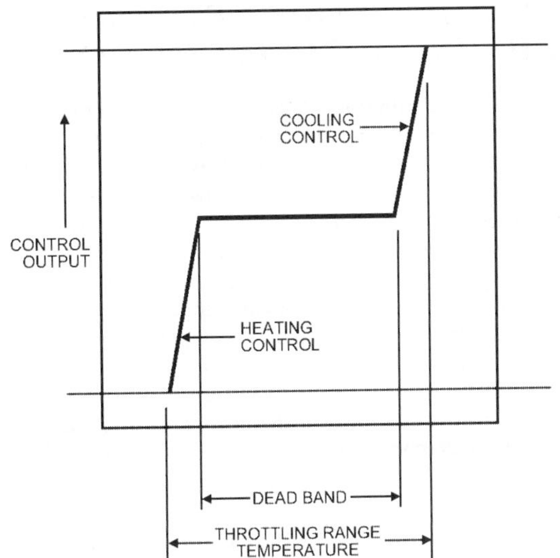

Fig. 17 Dead Band Thermostat

value to the other actuates switching devices in the thermostat that index it from occupied to unoccupied or vice versa.

- The **heating-cooling** or **summer-winter thermostat** can have its action reversed and its set point changed by indexing. It is used to actuate controlled devices, such as valves or dampers, that regulate a heating source at one time and a cooling source at another.
- **Multistage thermostats** are arranged to operate two or more successive steps in sequence.
- A **submaster thermostat** has its set point raised or lowered over a predetermined range in accordance with variations in output from a master controller. The master controller can be a thermostat, manual switch, pressure controller, or similar device.
- A **dead band thermostat** has a wide differential over which the thermostat remains neutral, requiring neither heating nor cooling. This differential may be adjustable up to 10°F. The thermostat then controls to maximum or minimum output over a small differential at the end of each dead band (Figure 17).

AUXILIARY CONTROL DEVICES

Auxiliary control devices for electric systems include

- **Transformers** to provide current at the required voltage.
- **Occupancy sensors** to automatically adjust controlled variables based on occupancy such as lighting, ventilation rate, and temperature.
- **Signal transducers** to change one standard signal into another. The popularity of digital control and other electric-based control systems has generated a variety of transducers. The variables usually transformed include voltage [0–10, 0–5, 2–10 V (dc)], current [4–20 mA], resistance [0–135 Ω], pressure (3–15, 0–20 psig), phase cut voltage [0–20 V (dc)], pulse-width modulation, and time duration pulse. Signal transducers allow the use of an existing control device in a retrofit application.
- **Electric relays** to control electric heaters or to start and stop burners, compressors, fans, pumps, or other apparatus for which the electrical load is too large to be handled directly by the controller. Other uses include time-delay and circuit-interlocking safety applications.
- **Potentiometers** for manual positioning of proportional control devices, for remote set point adjustment of electronic controllers, and for feedback.
- **Manual switches**, either two-position or multiple-position with single or multiple poles.
- **Auxiliary switches** on valve and damper actuators for selecting a sequence of operation.

Auxiliary control devices for pneumatic systems include

- **Air compressors** and accessories, including dryers and filters, to provide a source of clean, dry air at the required pressure.
- **Electropneumatic relays**, electrically actuated air valves for operating pneumatic equipment according to variations in electrical input to the relay.
- **Pneumatic-electric switches**, which are actuated by the pressure from a controller to permit a controller actuating a proportional device to also actuate one or more two-position devices.
- **Pneumatic transducers**, which are used to reverse the action of a proportional controller, select the higher or lower of two or more pressures, average two or more pressures, respond to the difference between two pressures, add or subtract pressures, and amplify or retard pressure changes.
- **Positive positioning relays** to ensure accurate positioning of a valve or damper actuator in response to changes in pressure from a controller.
- **Switching relays**, which are pneumatically operated air valves used to divert air from one circuit to another or to open and close air circuits.
- **Pneumatic switches**, which are manually operated devices used to divert air from one circuit to another or to open and close air circuits. They can be two-position or multiple- position.
- **Gradual switches**, which are proportional devices used to manually vary air pressure in a circuit.

Auxiliary control devices common to both electric and pneumatic systems include the following:

- **Step controllers** to operate several switches in sequence by means of a proportional electric or pneumatic actuator. They are commonly used to control several steps of refrigeration capacity. They may be arranged to prevent simultaneous starting of compressors and to alternate the sequence to equalize wear. These controllers may also be used for sequenced operation of electric heating elements and other equipment.
- **Power controllers** to control electric power input to resistance heating elements. The final controlled device may be a variable autotransformer, a saturable-core reactor, or a solid-state power controller. They are available with various ratings for single- or three-phase heater loads and are usually arranged to regulate power input to the heater in response to the demands of the proportional electronic or pneumatic controllers. However, solid-state controllers may also be used in two-position control modes.
- **Clocks** or **timers** to turn apparatus on and off at predetermined times, to switch control systems from day to night operation, and to regulate other time sequence functions.
- **Transducers**, which consist of combinations of electric or pneumatic control devices, may be required. For these applications, transducers are used to convert electric signals to pneumatic output or vice versa. Transducers may convert proportional input to either proportional or two-position output.

The electronic-to-pneumatic (E/P) transducer is used in many applications. It converts a proportional electronic output signal into a proportional pneumatic signal (as illustrated in Figure 18) and can be used to combine electronic and pneumatic control components to form a control loop, as illustrated in Figure 19. Electronic components are used for sensing and signal conditioning, while pneumatic components are used for actuation. The electronic controller can be either analog or digital.

The E/P transducer presents a special option for retrofit applications. An existing HVAC system with pneumatic controls can be retrofitted with electronic sensors and controllers while retaining the existing pneumatic actuators (Figure 20).

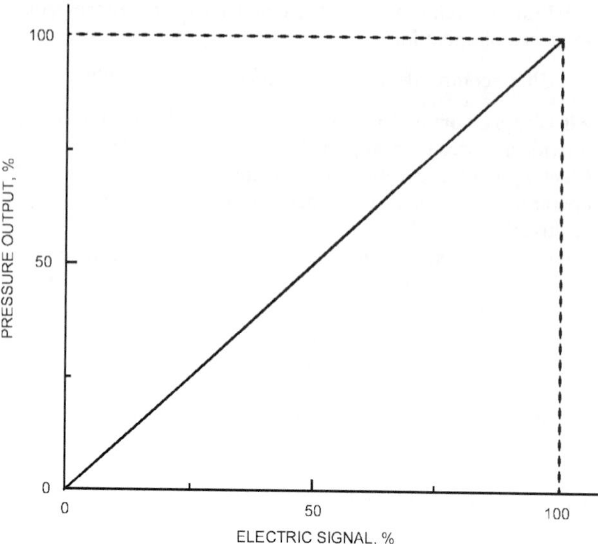

Fig. 18 Response of Electronic-to-Pneumatic (E/P) Transducer

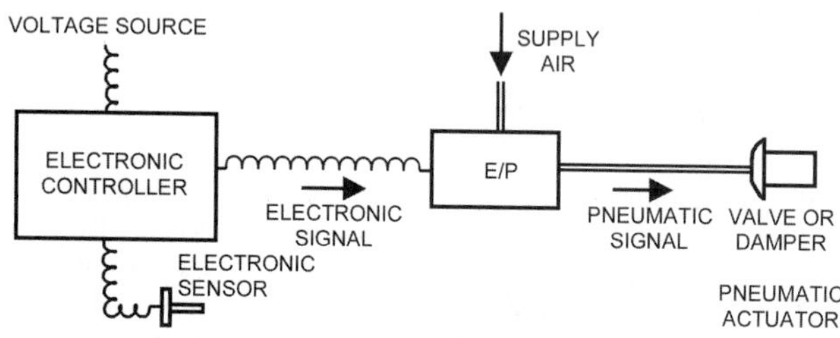

Fig. 19 Electronic and Pneumatic Control Components Combined with Electronic-to-Pneumatic (E/P) Transducer

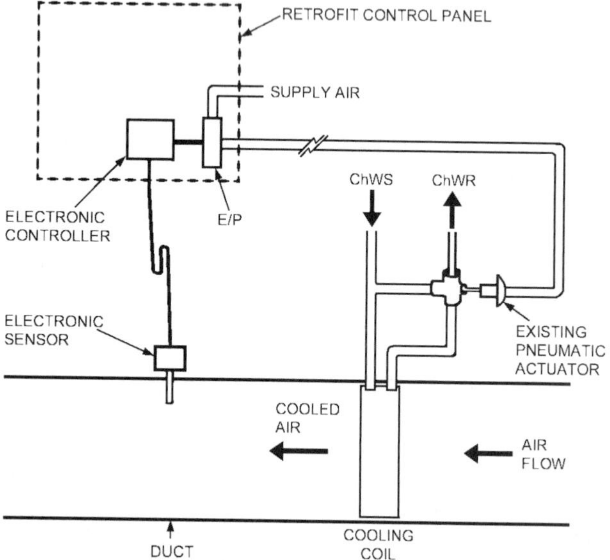

Fig. 20 Retrofit of Existing Pneumatic Control with Electronic Sensors and Controllers

COMMUNICATION NETWORKS FOR BUILDING AUTOMATION SYSTEMS

A **building automation system (BAS)** is a centralized control and/or monitoring system for many or all building systems (e.g., HVAC, electrical, life safety, security). A BAS may link together information from control systems actuated by different technologies.

One important characteristic of direct digital control (DDC) is the ability to share information. Information is transferred (1) between controllers to coordinate their action, (2) between controllers and building operator interfaces to monitor and command systems, and (3) between controllers and other computers for off-line calculation. This information is typically shared over communication networks. DDC systems nearly always involve at least one network and commonly involve more than one. A **network** is a set of connections between controllers, routers, bridges, and computers that enables them to exchange digital information.

COMMUNICATION PROTOCOLS

A **protocol** is a set of rules that define the communication behavior of each element in a communication network. The word may describe the communication at one layer of the network [e.g., **Internet protocol (IP)**, which defines the network layer in the Internet suite of protocols], or it may refer to the entire communication process. In discussions of BASs, most communication needs are described in terms of the entire process, but it is sometimes necessary to discuss the protocols at a particular layer.

There is great interest in open protocols for BASs to facilitate communication among devices from different suppliers. Although there is no commonly accepted definition of "openness," the Institute of Electrical and Electronics Engineers defines three classes of protocols (IEEE *Standard* 802):

- **Standard protocol.** Published and controlled by a standards body. Examples include BACnet by ASHRAE, LonTalk by Electronic Industries Alliance, and TCP/IP by Internet Engineering Task Force.
- **Public protocol.** Published but controlled by a private organization.
- **Private protocol.** Unpublished; use and specification controlled by a private organization. Examples include the proprietary communications used by many building automation devices.

Multivendor communication is possible with any of these three classes, but the challenges vary. Specifying a common protocol does not ensure that the end user's requirement for interoperability is met. The engineer may select an open protocol and specify the interaction between devices. This limits the bidders, but assures the engineer of certain communication characteristics. On the other hand, the engineer can specify the required interoperation and put the burden on the suppliers to select combinations of products that meet the need. This is likely to result in a wider range of options, but they may be more difficult to compare.

THE OSI NETWORK MODEL

ISO *Standard* 7498-1 presents a seven-layer model of information exchange called the **Open Systems Interconnection (OSI) Reference Model** (Figure 21). Most descriptions of computer networks, especially open networks, are based on this reference model. The layers can be thought of as steps in the translation of a message from something with meaning at the application layer, to something measurable at the physical layer, and back to meaningful information at the application layer.

The layered approach to network design is valuable because it allows developers to take advantage of existing standards such as IP at the network layer or Electronic Industries Association (EIA)

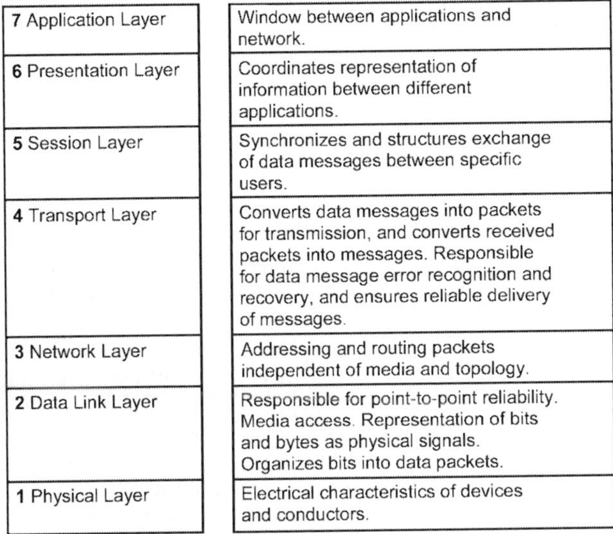

7 Application Layer	Window between applications and network.
6 Presentation Layer	Coordinates representation of information between different applications.
5 Session Layer	Synchronizes and structures exchange of data messages between specific users.
4 Transport Layer	Converts data messages into packets for transmission, and converts received packets into messages. Responsible for data message error recognition and recovery, and ensures reliable delivery of messages.
3 Network Layer	Addressing and routing packets independent of media and topology.
2 Data Link Layer	Responsible for point-to-point reliability. Media access. Representation of bits and bytes as physical signals. Organizes bits into data packets.
1 Physical Layer	Electrical characteristics of devices and conductors.

Fig. 21 OSI Reference Model

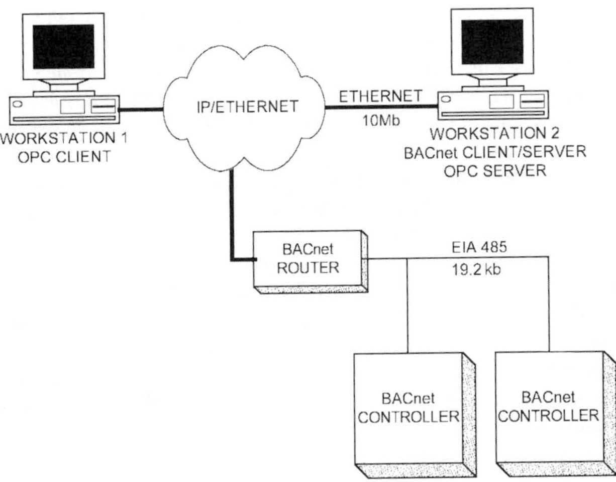

Fig. 22 Portion of a BAS Network

Standard 485, a signaling standard, at the physical layer without becoming tied to one technology.

The full seven-layer model does not apply to every network, but it is still used to describe the aspects that do fit. When describing DDC networks that use the same technology throughout the system, the seven-layer model is relatively unimportant. For systems that employ various technologies at different points in the network, the model helps describe where and how the pieces are bound together.

The portion of network shown in Figure 22 illustrates how the OSI Reference Model describes communications. The system includes a number of controllers on a BACnet network and several operator workstations connected by an ISO 8802-3 backbone or local area network (LAN), in this case an Ethernet. A BACnet router links the BACnet controllers to the Ethernet.

Figure 23 shows the network layers in each device that make communication possible. Workstation 2 and the BACnet controller both show communication stacks with BACnet at the application layer; however, these devices do not communicate directly because they use different protocols at the lower layers. If the stacks are different at any layer, a device that bridges the gap is required. The BACnet router shows two stacks. One matches the physical, data

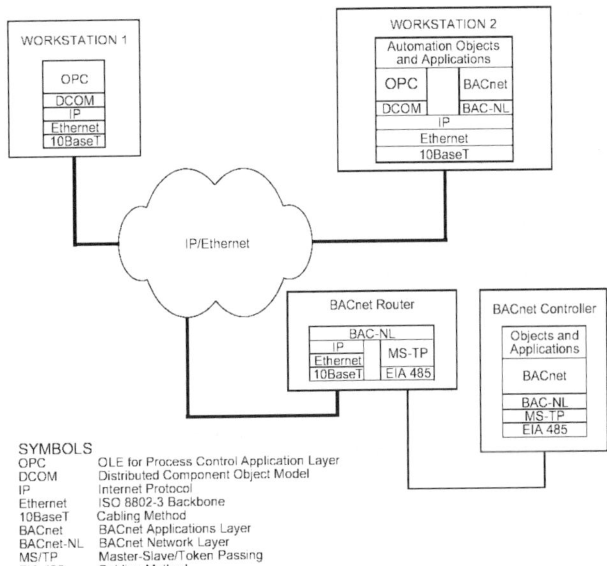

Fig. 23 Network Layers in BAS Devices

link, and network layers of the BACnet controllers; the other matches the bottom three layers of the BACnet stack on Workstation 2. The BACnet network layer is the common layer that delivers BACnet messages from one end of the router to the other. When a message reaches the router, it travels up through the stack. The router reformats it and sends the same message out through the other stack to the other network. This makes it possible for application data to pass between the controller and BACnet devices on the Ethernet. Bushby (1998) described this process more fully.

Both workstations contain an OPC stack. **OPC** (OLE for process control) is a standard for communicating automation data between computers or processes; it is based on distributed component object model (DCOM) technology. DCOM is a software development standard that facilitates interchangeability of software components.

The workstations communicate with each other over the Ethernet with OPC messages. Workstation 2 and the BACnet router communicate with each other over the Ethernet using BACnet messages. Although they are on the same network, Workstation 1 and the BACnet router do not communicate directly with each other because their protocol stacks do not match at the upper layers. If BACnet data is required at Workstation 1, it must pass through the OPC server on Workstation 2.

NETWORK STRUCTURE

Often, a single DDC system applies several different network technologies at different points in the system. For example, a relatively low-speed, inexpensive network with relatively primitive functions may link a group of room controllers to a larger equipment controller. A faster, more sophisticated network links the large controller to its peers and to an operator's workstation. Several workstations communicate over a high-speed, relatively expensive, general-purpose office automation network. Figure 24 illustrates this sort of high-speed **hierarchical** network. Structures like it have been popular in DDC for years. Frequently, the network hierarchy corresponds roughly to a hierarchy related to the control function of the devices. Variations on this hierarchical structure will continue to emerge. The opposite extreme is a completely **flat** network architecture. A flat architecture links all the devices through the same network without altering any other hierarchy that exists among the devices. A flat architecture is more viable in small systems than in large ones due to the considerations discussed in the following paragraph.

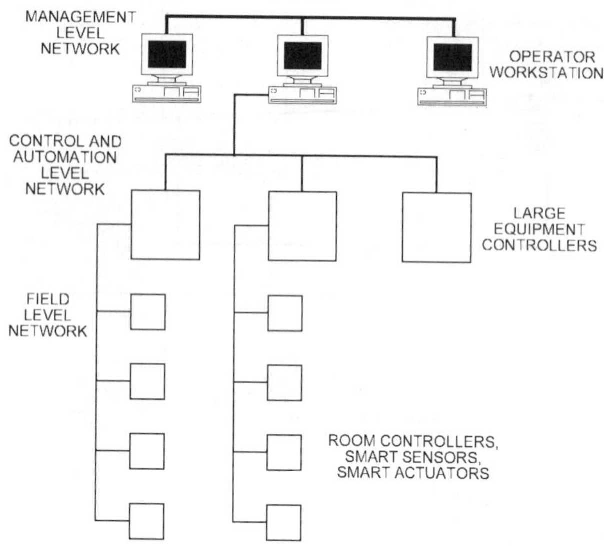

Fig. 24 Hierarchical Network

Network structure sometimes affects the cost, operation, and opportunities for expansion of a BAS. The structure can affect the reliability and failure modes of the system. It may be appropriate to separate sections of a network in order to isolate failures. Structure can affect the way devices load the information carrying capacity of the network. It can isolate one busy branch from the rest of the system. It can isolate branches from the high-speed backbone. Structure affects the cost of the system because it determines the mix of low-speed and high-speed devices. Network structure can influence system data security and access control. The relative merits of one structure versus another depend on the communication functions required, the hardware and software available for the task, and cost. For a given job, there is probably more than one suitable structure. Product capabilities change quickly. Engineers who choose to specify network structure must be aware of new technologies to take advantage of the most cost-effective solutions.

Connections Between Networks and Network Segments

Some BAS networks use other networks to connect segments of the BAS. This occurs

- Within a building, using the information technology network
- Between buildings, using telephone lines
- Between buildings, using the Internet

In each case, the link between BAS segments must be considered part of the BAS network when evaluating function, security, and performance. The link also raises new issues. The connecting segment is likely to be outside the control of the owner of the BAS, which could affect availability of service. It may mean that traffic and bandwidth issues have to be addressed outside the facilities department.

The connection may be **switched** (dial-up) or **dedicated**. In the case of a switched connection, the function of the network depends on which segment may dial the other and the circumstances that trigger the call. Switched connections are most commonly used to handle remote buildings or to serve a remote operator (i.e., one who is on call over the weekend).

Transmission Media

The transmission medium is the foundation of the network. It is usually, but not always, cable. In cases where physical cable connection is not possible or practical, devices may transfer information using wireless technologies, such as radio waves or infrared light. However, this section covers only physical cabling media.

Twisted-Pair Copper Cable. A twisted-pair cable consists of multiple twisted pairs (typically 24 AWG) of wire covered by an overall sheath or jacket. Varying the number of twists for each pair relative to the other pairs in the cable can greatly reduce crosstalk (interference between signals on different pairs).

In shielded twisted pair (STP) cable, each wire pair, as well as the combined grouping of all pairs, is covered with a layer of shielding to minimize interference-related problems. STP cable performs better than **unshielded** twisted pair (UTP) cable in environments where a high level of immunity and/or a low level of emissions is critical. It also allows less crosstalk than UTP. However, STP requires a more labor-intensive installation, and any break or improper grounding of the shield reduces its overall effectiveness.

Category 5 (defined by TIA/EIA *Standard* 568-A) UTP cable is currently the most common medium. RJ-45 jacks and plugs are specified and have standard pinouts. This cable is rated at up to 1 gigabit per second for 330 ft and can be used over much longer distances for lower speed applications (e.g., EIA *Standard* 485).

Fiber Optic Cable. Fiber optic cable uses glass or plastic fibers to transfer data in the form of light pulses, which are typically generated by either a laser or an LED. Fiber optic cable systems are classified as either single-mode fiber or multimode fiber systems. Table 1 compares their characteristics.

Table 1 Comparison of Fiber Optic Technology

	Multimode Fiber	Single-Mode Fiber
Light source	LED	Laser
Cable designation (core/cladding diameter)	62.6/125	8.3/125
Transmission distance	660 ft	980 ft
Data rate	>10 gigabit/s and increasing	Even higher
Relative cost	Less per connection, more per data rate	More per connection, less per data rate

Light in a fiber optic system experiences less energy loss than electrical signals traveling through copper and no capacitance. This translates into greater transmission distances and dramatically higher data transfer rates. With the rapid advances in this technology, the data transfer rate of a fiber cable imposes no limits on a BAS. Fiber optics also have exceptional noise immunity. However, the necessary conversions between light-based signaling and electricity-based computing make fiber optics more expensive per device, which sometimes offsets the other advantages.

Structured Cabling. TIA/EIA *Standard* 568-A, Commercial Building Telecommunications Cabling Standard, permits cable planning and installation to begin before the network engineering is finalized. It supports both voice and data. The standard was written for the telecommunications industry, but cabling is gaining recognition as building infrastructure, and the standard is being applied to BAS networks as well.

TIA/EIA *Standard* 568-A specifies **star topology** (each device individually cabled to a hub) because connectivity is more robust and management is simpler than for busses and rings. If the wires in a leg are shorted, only that leg fails, making fault isolation easier; with a bus, all drops would fail.

The basic structure specified is a **backbone**, which typically runs from floor to floor within a building and possibly between buildings. The **horizontal cabling** runs between the distribution frames on each floor and the information outlets in the work areas. The maximum length of horizontal cabling recommended is 330 ft.

SPECIFYING BAS NETWORKS

Specifying a DDC system includes specifying a network. The many network technologies available deliver many performance levels at many different prices. A rational selection requires an assessment of the requirements (i.e., what information will pass between devices and at what rates). In some cases, the new equipment is required to interface with existing devices, which may limit networking options.

Specification Method

As with other aspects of an HVAC system, an engineer must choose a method of specification. The Construction Specification Institute lists four methods. The following list relates those methods to BAS networks.

- **Descriptive.** Calls out the exact properties of the products. Properties could include communication protocols and data transfer rates.
- **Performance.** Tells what result is required and the criteria by which performance will be verified. Allows bidders to propose products to meet the need.
- **Reference standard.** Requires products to conform to an established standard. Does not oblige contractor to meet end user's needs not addressed in the standard.
- **Proprietary.** Calls out brand names. May be necessary in expansion of existing systems.

To write a descriptive network specification, the designer must know the details of network technology. To succeed with any specification, the designer must articulate the end user's needs. Typically, a performance-based specification results in the best value for the customer (Ehrlich and Pittel 1999).

Communication Tasks

Determining network performance requirements means identifying and quantifying the communication functions required. Ehrlich and Pittel (1999) identified five basic communication tasks. To establish network requirements, the specifier must elaborate on each basic task. They are listed here along with some of the questions an engineer can use to identify the client's needs.

Data Exchange. What data passes between which devices? What control and optimization data passes between controllers? What update rates are required? What data does an operator need to reach? How much delay is acceptable in retrieving values? What update rates are required on "live" data displays? Within one system, the answers may vary according to the use of the data. Which set points and control parameters do operators need to adjust over the network?

Alarms and Events. Where do alarms originate? Where are they logged and displayed? How much delay is acceptable? Where are they acknowledged? What information must be delivered along with the alarm? (Depending on the design of the system, alarm messages may be passed over the network along with the alarms.) Where are alarm summary reports required? How and where do operators need to adjust alarm limits, etc?

Schedules. For the HVAC equipment that runs on schedules, where can the schedules be read? Where can they be modified?

Trends. Where does trend data originate? Where is it stored? How much will be transmitted? Where is it displayed and processed? Which user interfaces can set and modify trend collection parameters?

Network Management. What network diagnostic and maintenance functions are required at which user interfaces? Data access and security functions may be handled as network management functions.

Bushby et al. (1999) refer to the same five communication tasks as **interoperability areas** and list many more specific considerations in each area.

APPROACHES TO INTEROPERABILITY

In the surge toward interoperability, many approaches have been proposed and applied, each with varying degrees of success under various circumstances. The field changes quickly as product lines emerge and standards develop and gain acceptance. The building automation world continues to evaluate the options project by project.

Typically, an interoperable system uses one of two approaches: standard protocols or special-purpose gateways. With a standard, the supplier is responsible for compliance with the standard; the system specifier or integrator is responsible for interoperation. With a gateway, the supplier takes responsibility for interoperation. The majority of integrated building automation systems currently depend on gateways, especially where the job requires interoperation with existing equipment. Bushby (1998) addressed this issue and some of the limitations associated with gateways. To date, interoperability by any method requires solid field engineering and capable system integration; the issues extend well beyond the selection of a communication protocol.

Standard Protocols

Several standard protocols have been applied successfully in building automation systems. Their different characteristics make some more suited to particular tasks than others. The European Committee for Standardization (CEN) discusses characteristics of protocols appropriate for different building functions (CEN 1999). Table 2 lists some of the applicable standard protocols..

Table 2 Some Standard Communication Protocols Applicable to BAS

Protocol	Definition
BACnet	ASHRAE *Standard* 135
LonTalk	EIA *Standard* 709.1
PROFIBUS FMS	EN 50170:1996 Volume 2
EIB	ENV 13154-2 Annex C
EIBnet	ENV 18321-2

Gateways and Interfaces

Rather than conforming to a published standard, a supplier can design a specific device to exchange data with another specific device. This typically requires cooperation between two manufacturers. It can be simpler and more cost-effective than for both manufacturers to conform to an agreed-upon standard. Sometimes the device is developed for one particular installation; other times it is an off-the-shelf product. In either case, the communication tasks must be carefully specified to ensure that the gateway performs as needed.

Choosing a system that supports a variety of gateways may be a way to maintain a flexible position as products and standards continue to develop.

COMMISSIONING

A successful control system requires a proper start-up and testing, not merely the adjustment of a few parameters (set points and throttling ranges) and a few quick checks. With the services of an experienced control professional, the typical DDC system can be used effectively in the commissioning process to test and document the peformance of the HVAC system. In general, the increased use of VAV systems and digital controls has increased the importance of and need for commissioning.

Design and construction specifications should include specific commissioning procedures. In addition, commissioning should be coordinated with testing, adjusting, and balancing (TAB) because each affects the other. The TAB procedure begins by checking each control device to ensure that it is installed and connected according to approved drawings. Each electrical and pneumatic connection is verified, and all interlocks to fan and pump motors and primary heating and cooling equipment are checked. *ASHRAE Guideline* 1 explains how commissioning starts with project conception and continues for the life of the building.

TUNING

The systematic tuning of controllers improves the performance of all controls and is particularly important for digital control. First, the controlled process should be controlled manually between various set points to evaluate the following questions:

- Is the process noisy (rapid fluctuations in controlled variable)?
- Is there appreciable hysteresis (backlash) in the actuator?
- How easy (or difficult) is it to maintain and change set point?
- In which operating region is the process most sensitive (highest gain)?

If the process cannot be controlled manually, the reason should be identified and corrected before the controller is tuned.

Tuning selects control parameters that determine the steady-state and transient characteristics of the control system. HVAC processes are nonlinear, and characteristics change on a seasonal basis. Controllers tuned under one operating condition may become unstable as conditions change. A well-tuned controller (1) minimizes the steady-state error for set point, (2) responds quickly to disturbances, and (3) remains stable under all operating conditions. Tuning proportional controllers is a compromise between minimizing steady-state error and maintaining margins of stability. Proportional plus integral (PI) control minimizes this compromise because the integral action reduces steady-state error, while the proportional term determines the controller's response to disturbances.

Tuning Proportional, PI, and PID Controllers

Popular methods of determining proportional, PI, and PID controller tuning parameters include closed- and open-loop process identification methods and trial-and-error methods. Two of the most widely used techniques for tuning these controllers are ultimate oscillation and first order plus dead time. There are many optimization calculations for these two techniques, but the most widely used is the Ziegler-Nichols, which is given here.

Ultimate Oscillation (Closed-Loop) Method. The closed-loop method increases the gain of the controller in proportional-only mode until the equipment continuously cycles after a set point change (Figure 25, where $K_p = 40$). Proportional and integral terms are then computed from the cycle's period of oscillation and the K_p value that caused cycling. The ultimate oscillation method is as follows:

1. Adjust control parameters so that all are essentially off. This corresponds to a proportion band (gain) at its maximum (minimum), the reset (repeats per minute) to maximum (minimum), and derivative to its minimum.
2. Adjust the manual output of the controller to give a measurement as close to midscale as possible.
3. Put the controller in automatic.
4. Slowly and gradually increase the proportional constant effect (this corresponds to reducing the proportional band or increasing the proportional gain) until the observed oscillations neither grow nor diminish in amplitude. If the response saturates at either extreme, start over at Step 2 to obtain a stable response. If no oscillations are observed, change the set point and try again.
5. Record the proportional band as PB_u and the period of the oscillations as T_u.
6. Use the recorded proportional band and oscillation period to calculate controller settings as follows:

Proportional only:

$$PB = 1.8(PB_u) \quad \text{percent} \tag{4}$$

Proportional plus integral (PI):

$$PB = 2.22(PB_u) \quad \text{percent} \tag{5}$$

$$T_i = 0.83T_u \quad \text{minute per repeat} \tag{6}$$

Proportional plus integral plus derivative (PID):

$$PB = 1.67(PB_u) \quad \text{percent} \tag{7}$$

$$T_i = 0.50T_u \quad \text{minute per repeat} \tag{8}$$

$$T_d = 0.125T_u \quad \text{minute} \tag{9}$$

First-Order-plus-Dead-Time (Open-Loop) Method. The open-loop method introduces a step change in input into the opened control loop. A graphical technique is used to estimate the process transfer function parameters. Proportional and integral terms are calculated from the estimated process parameters using a series of equations.

The value of the process variable must be recorded over time, and the dead time and time constant must be determined from it. This can be accomplished graphically as seen in Figure 26. The first-order-plus-dead-time method is as follows:

1. Adjust the controller manual output to give a midscale measurement.
2. Arrange for the recording of the process variable over time.
3. Move the manual output of the controller by 10% as rapidly as possible to approximate a step change.
4. Record the value of the process variable over time until it reaches a new steady state value.
5. Determine the dead time and time constant.
6. Use the dead time (TD) and time constant (TC) values to calculate PID values as follows:

$$\text{Gain} = \frac{\%\text{ change in controlled variable}}{\%\text{ change in control signal}} \tag{10}$$

Proportional only:

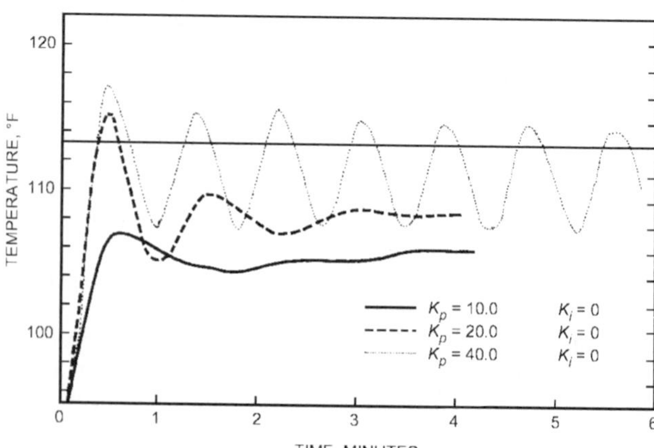

Fig. 25 Response of Discharge Air Temperature to Step Change in Set Points at Various Proportional Constants with No Integral Action

$$PB = Gain/(TC/TD) \quad (11)$$

Proportional plus integral (PI):

$$PB = 0.9(Gain)/(TC/TD) \quad (12)$$

$$T_i = 3.33(TD) \quad (13)$$

Proportional-integral-derivative (PID):

$$PB = 1.2(Gain)/(TC/TD) \quad (14)$$

$$T_i = 2(TD) \quad (15)$$

$$T_d = 0.5(TD) \quad (16)$$

Trial and Error Method. This method involves adjusting the gain of the proportion-only controller until the desired response to a set point is observed. Conservative tuning dictates that this response should have a small initial overshoot and quickly damp to steady-state conditions. Set point changes should be made in the range where controller saturation, or output limit, is avoided. The integral term is then increased until changes in set point produce the same dynamic response as the controller under proportional control, but with the response now centered about the set point (Figure 27).

Tuning Digital Controllers

In tuning digital controllers, additional parameters may need to be specified. The digital controller sampling interval is critical because it can introduce harmonic distortion if not selected properly. This sampling interval is usually set at the factory and may not be adjustable. A controller sampling interval of about one-half the time constant of the controlled process usually provides adequate control. Many digital control algorithms include an error dead band to eliminate unnecessary control actions when the process is near set point. Hysteresis compensation is possible with digital controllers, but it must be carefully applied because overcompensation can cause continuous cycling of the control loop.

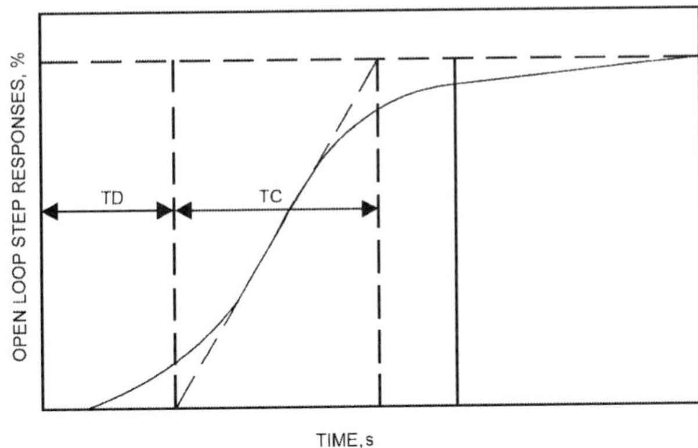

Fig. 26 Open Loop Step Response Versus Time

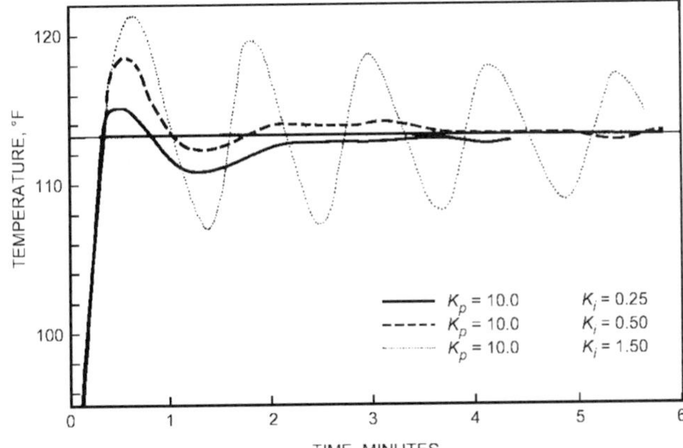

Fig. 27 Response of Discharge Air Temperature to Step Change in Set Points at Various Integral Constants with Fixed Proportional Constant

CODES AND STANDARDS

ASHRAE. 1995. BACnet—A data communication protocol for building automation and control networks. ANSI/ASHRAE *Standard* 135-1995.

EIA. 1995. Commercial building telecommunications cabling standard. TIA/EIA *Standard* 568-A-95. Electronic Industries Alliance, Arlington, VA.

EIA. 1998. Electrical characteristics of generators and receivers for use in balanced digital multipoint systems. TIA/EIA *Standard* 485-98.

EIA. 1999. Control network protocol specification. ANSI/EIA *Standard* 709.1-99

IEEE. 1990. Local and metropolitan area networks: Overview and architecture. IEEE *Standard* 802-1990. Institute of Electrical and Electronic Engineers, Piscataway, NJ.

ISO. 1994. Information processing systems—Open systems interconnection—Basic reference model: The basic model. ISO/IEC 7498-1:1994. International Organization for Standardization, Geneva, Switzerland.

REFERENCES

ASHRAE. 1996. The HVAC commissioning process. *Guideline* 1-1996.

Avery, G. 1989. Updating the VAV outside air economizer controls. *ASHRAE Journal* (April).

Avery, G. 1992. The instability of VAV systems. *Heating, Piping and Air Conditioning* (February).

BICSI. 1999. *LAN and internetworking design manual*, 3rd ed. BICSI: A Telecommunications Association, Tampa, FL.

Bushby, S.T. 1998. Friend or foe? Communication gateways. *ASHRAE Journal* 40(4):50-53.

Bushby, S.T., H.M. Newman, and M.A. Applebaum. 1999. GSA guide to specifying interoperable building automation and control systems using ANSI/ASHRAE *Standard* 135-1995, BACnet. NISTIR 6392. National Institute of Standards and Technology, Gaithersburg, MD. Available from National Technical Information Service, Springfield, VA.

CEN. 1999. Building control systems, Part 1: Overview and definitions. prEN ISO16484-1.

Ehrlich, P. and O. Pittel. 1999. Specifying interoperability. *ASHRAE Journal* 41(4):25-29.

Haines, R.W. and D.C. Hittle. 1993. *Control systems for heating, ventilating and air conditioning*, 5th ed. Kluwer Academic Publishing.

Hartman, T.B. 1993. *Direct digital controls for HVAC systems*. McGraw-Hill, New York.

Kettler, J.P. 1998. Controlling minimum ventilation volume in VAV systems. *ASHRAE Journal* (May).

Levenhagen, J.I. and D.H. Spethmann. 1993. *HVAC controls and systems*. McGraw-Hill, New York.

LonMark. 1998. LonMark application layer interoperability guidelines. LonMark Interoperability Association, Sunnyvale, CA.

LonMark. 1999. LonMark functional profile: Space comfort controller. 8500-10.

Newman, H.M. 1994. *Direct digital control for building systems, theory and Practice*. John Wiley and Sons, New York.

OPC Foundation. 1998. OPC overview. OPC Foundation, Boca Raton, FL.

Rose, M.T. 1990. *The open book: A practical perspective on OSI*. Prentice-Hall, Englewood Cliffs, NJ.

Seem, J.E., J.M. House, and R.H. Monroe. 1999. On-line monitoring and fault detection, *ASHRAE Journal* (July).

Starr, R. 1999. Pneumatic controls in a digital age. *Heating, Piping and Air Conditioning* (November).

Tillack, L. and J.B. Rishel. 1998. Proper control of HVAC variable speed pumps. *ASHRAE Journal* (November).

CHAPTER 16

AIRFLOW AROUND BUILDINGS

AIRFLOW around buildings affects worker safety, process and building equipment operation, weather and pollution protection at inlets, and the ability to control environmental factors of temperature, humidity, air motion, and contaminants. Wind causes variable surface pressures on buildings that change intake and exhaust system flow rates, natural ventilation, infiltration and exfiltration, and interior pressures. The mean flow patterns and turbulence of wind passing over a building can cause recirculation of exhaust gases to air intakes. This chapter contains information for evaluating flow patterns, estimating wind pressures, and identifying problems caused by the effects of wind on intakes, exhausts, and equipment. Related information can be found in Chapters 12, 14, 26, and 27 of this volume; in Chapters 28, 29, 43, and 51 of the 1999 *ASHRAE Handbook—Applications*; and in Chapters 25, 30, and 36 of the 2000 *ASHRAE Handbook—Systems and Equipment*.

FLOW PATTERNS

Buildings having even moderately complex shapes, such as L- or U-shaped structures formed by two or three rectangular blocks, can generate flow patterns too complex to generalize for design. To determine flow conditions influenced by surrounding buildings or topography, wind tunnel or water channel tests of scale models or tests of existing buildings are required. However, if a building is oriented perpendicular to the wind, it can be considered as consisting of several independent rectangular blocks. Only isolated rectangular block buildings are discussed here. Saunders and Melbourne (1979), Hosker (1984, 1985), Walker et al. (1996), and English and Fricke (1997) review the effects of nearby buildings.

The **mean speed of wind** U_H approaching a building increases with height H above the ground (Figure 1). Both the upwind velocity profile shape and its turbulence intensity strongly influence flow patterns and surface pressures (Melbourne 1979). A **stagnation zone** exists on the upwind wall. The flow separates at the sharp edges to generate **recirculating flow zones** that cover the downwind surfaces of the building (roof, sides, and leeward walls) and extend for some distance into the **wake**. If the building has sufficient length L in the windward direction, the flow will reattach to the building (Figure 2) and may generate two distinct regions of separated recirculating flow—on the building and in its wake.

Surface flow patterns on the upwind wall are largely influenced by **approach wind characteristics**. Higher wind speed at roof level causes a larger **stagnation pressure** on the upper part of the wall than near the ground, which leads to downwash on the lower one-half to two-thirds of the building (Figure 1). On the upper one-quarter to one-third of the building, the surface flow is directed upward over the roof. For a building whose height H is three or more times the width W of the upwind face, an **intermediate zone** can exist between the **upwash** and **downwash** regions, where the **surface streamlines** pass horizontally around the building. The downwash on the lower surface of the upwind face separates from the building before it reaches ground level and moves upwind to form a **vortex** that can generate high velocities close to the ground. This **ground level upwind vortex** is carried around the sides of the building in a U shape (Figure 1) and is responsible for the suspension of dust and debris that can contaminate air intakes close to ground level.

For wind perpendicular to a building wall, the height H and width W of the upwind building face determine the flow patterns shown in Figure 2. According to Wilson (1979), the **scaling length** R is

$$R = B_s^{0.67} B_L^{0.33} \tag{1}$$

where

B_S = smaller of upwind building face dimensions H and W
B_L = larger of upwind building face dimensions H and W

When B_L is larger than $8B_S$, use $B_L = 8B_S$ in Equation (1). For buildings with varying roof levels or with wings separated by at least a distance B_S, only the height and width of the building face below the portion of the roof in question should be used to calculate R.

Streamline patterns are independent of wind speed and depend mainly on building shape and upwind conditions. Because of the three-dimensional flow around a building, the shape and size of the **recirculation airflow** is not constant over the surface. The airflow reattaches closer to the upwind building face along the edges of the building than it does near the middle of the roof and sidewalls (Figure 2). The height H_c of the **recirculation region** (Figures 1 and 3) also decreases near roof edges.

The wind above the roof recirculation region is affected by the presence of the building. The flow accelerates as the **streamlines** curve upward over the roof and decelerates as they curve downward over the wake on the downwind side of the building. The distance above roof level where a building influences the flow is approximately 1.5R (Figure 1). The roof pitch begins to affect flow when it exceeds about 15° (1:4). When roof pitch reaches 20° (1:3), the flow remains attached to the upwind pitched roof and produces a recirculation region downwind of the roof ridge that is larger than that for a flat roof.

The downwind wall of a building exhibits a region of low average velocity and high turbulence. Velocities near the downwind wall are typically one-quarter of those at the corresponding upwind wall location. Figures 1 through 3 show that an upward flow exists over most of the downwind walls. A flow recirculation region extends for an approximate distance $L_r = 1.0R$ downwind.

If the angle of the approach wind is not perpendicular to the upwind face, complex flow patterns result. Strong vortices develop from the upwind edges of a roof, causing a strong downwash into the building wake above the roof. High speeds in these vortices cause large negative pressures near roof corners that can be a hazard to roof-mounted equipment during high winds. When the angle

The preparation of this chapter is assigned to TC 2.5, Air Flow Around Buildings.

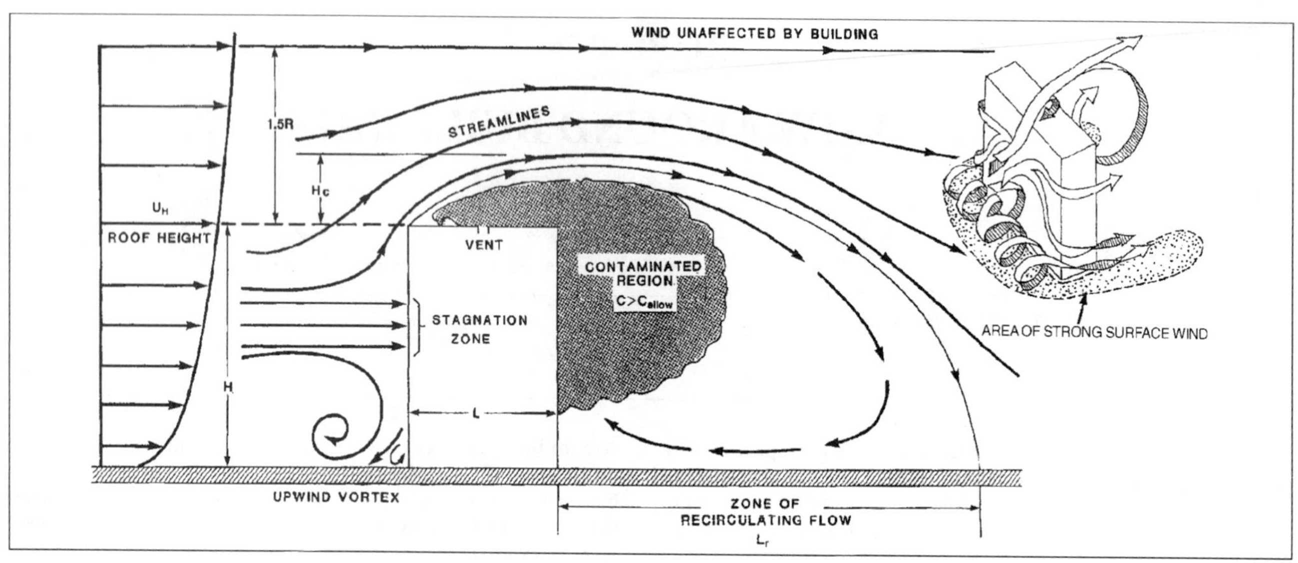

Fig. 1 Flow Patterns Around Rectangular Building

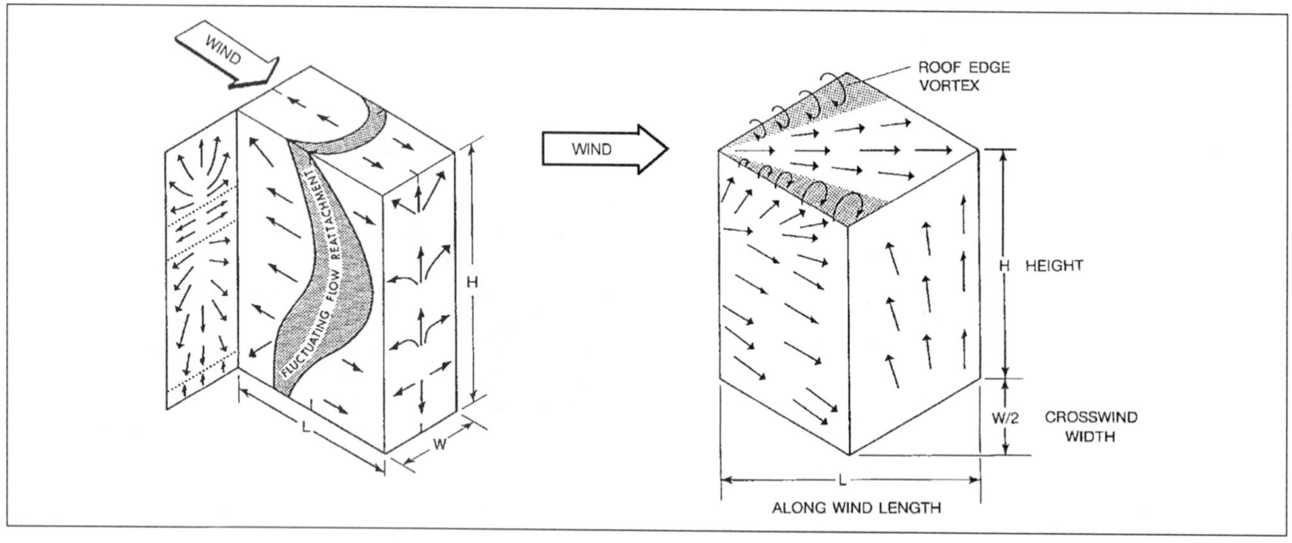

Fig. 2 Surface Flow Patterns and Building Dimensions

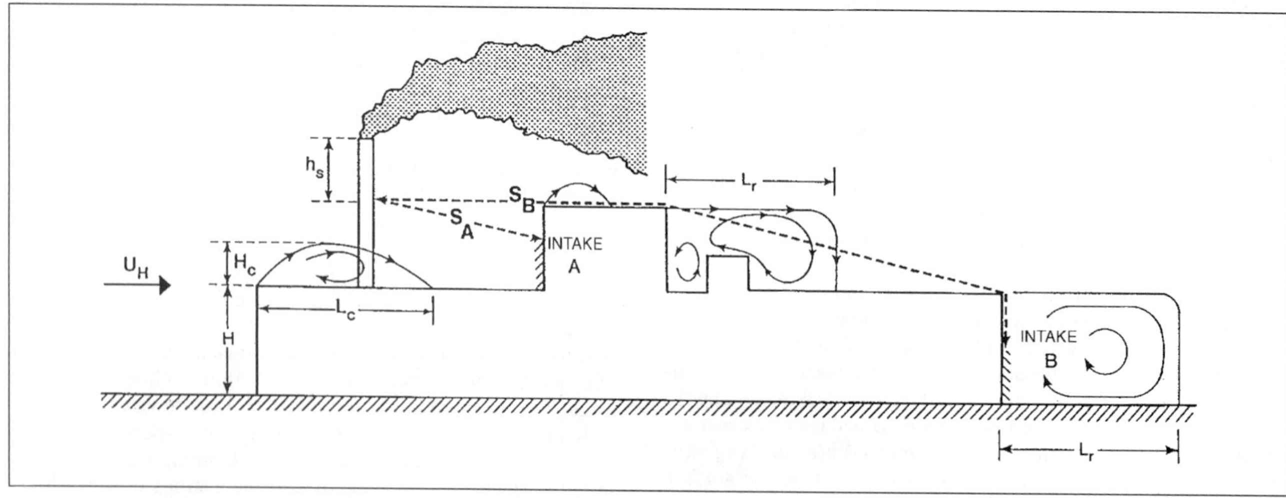

Fig. 3 Flow Recirculation Regions and Exhaust-to-Intake Stretched-String Distances
(Wilson 1982)

between the wind direction and the upwind face of the building is less than about 70°, the downwash-upwash patterns on the upwind face of the building are less pronounced, as is the ground level vortex shown in Figure 1. For an approach flow angle of 45°, streamlines remain close to the horizontal in their passage around the sides of the building (Figure 2), except near roof level where the flow is sucked upward into the roof edge vortices (Cochran 1992).

WIND PRESSURE ON BUILDINGS

In addition to the flow patterns described previously, the **turbulence** or **gustiness** of the approaching wind and the unsteady character of the separated flows cause surface pressures to fluctuate. The pressures discussed here are **time-averaged values**, with an averaging period of about 600 s. **Instantaneous pressures** may vary significantly above and below these averages, and **peak pressures** two or three times the mean values are possible. Although peak pressures are important with regard to structural loads, mean values are more appropriate for computing infiltration and ventilation rates. The time-averaged surface pressures are proportional to the wind velocity pressure p_v given by **Bernoulli's equation**:

$$p_v = \frac{\rho_a U_H^2}{2g_c} \tag{2}$$

where

U_H = approach wind speed at upwind wall height H
ρ_a = ambient (outdoor) air density
g_c = gravitational proportionality constant

The difference p_s between the pressure on the building surface and the local **outdoor atmospheric pressure** at the same level in an undisturbed wind approaching the building is

$$p_s = C_p p_v \tag{3}$$

where C_p is the local wind pressure coefficient for the building surface.

The local wind speed U_H at the top of the wall that is required for Equation (2) is estimated by applying terrain and height corrections to the hourly wind speed U_{met} from a nearby meteorological station.

Table 1 Atmospheric Boundary Layer Parameters

Terrain Category	Description	Exponent a	Layer Thickness δ, ft
1	Large city centers, in which at least 50% of buildings are higher than 70 ft, over a distance of at least 0.5 mi or 10 times the height of the structure upwind, whichever is greater	0.33	1500
2	Urban and suburban areas, wooded areas, or other terrain with numerous closely spaced obstructions having the size of single-family dwellings or larger, over a distance of at least 0.5 mi or 10 times the height of the structure upwind, whichever is greater	0.22	1200
3	Open terrain with scattered obstructions having heights generally less than 30 ft, including flat open country typical of meteorological station surroundings	0.14	900
4	Flat, unobstructed areas exposed to wind flowing over water for at least 1 mi, over a distance of 1500 ft or 10 times the height of the structure inland, whichever is greater	0.10	700

U_{met} is generally measured in flat, open terrain. The anemometer that records U_{met} is located at a height H_{met}, usually 33 ft above ground level. The hourly average wind speed U_H at wall height H in the undisturbed wind approaching a building in its local terrain (Figures 1 and 3) can be calculated from U_{met} as follows:

$$U_H = U_{met}\left(\frac{\delta_{met}}{H_{met}}\right)^{a_{met}}\left(\frac{H}{\delta}\right)^a \tag{4}$$

The wind boundary layer thickness δ and exponent a for the local building terrain and a_{met} and δ_{met} for the meteorological station are determined from Table 1. Typical values for meteorological stations located in flat, open terrain (Category 3 in Table 1) are a_{met} = 0.14 and δ_{met} = 900 ft. The values and terrain categories in Table 1 are consistent with those adopted in other engineering applications, for example ASCE *Standard* 7. Equation (4) gives the wind speed at height H above the average height of local obstacles, such as buildings and vegetation, weighted by the plan-area. At heights at or below this average obstacle height (e.g., at roof height in densely built-up suburbs), the speed depends on the geometrical arrangement of the buildings, and Equation (4) is less reliable.

An alternative mathematical description of the **atmospheric boundary layer**, which uses a logarithmic function, is given by Deaves and Harris (1978). While this model is more complicated than the power law used in Equation (4), it more closely models the real physics of the atmosphere and has been adopted by several foreign codes (e.g., SAA *Standard* AS-1170 from Australia).

Local Wind Pressure Coefficients

Values of the mean local wind pressure coefficient C_p used in Equation (3) depend on building shape, wind direction, and the influence of nearby buildings, vegetation, and terrain features. Accurate determination of C_p can be obtained only from wind tunnel model tests of the specific site and building. Ventilation rate calculations for single, unshielded rectangular buildings can be reasonably estimated using existing wind tunnel data. Many wind load codes (e.g., ASCE 7-98, AS-1170) give mean pressure coefficients for common building shapes.

Figure 4 shows pressure coefficients for walls of a tall rectangular cross section building (high-rise) sited in urban terrain (Davenport and Hui 1982). Figure 5 shows pressure coefficients for walls of a low-rise building (Holmes 1986). Generally, high-rise buildings are those where the height H is more than three times the crosswind width W. For $H > 3W$, use Figure 4, and for $H < 3W$, use Figure 5. At a wind angle $\theta = 0°$ (e.g., wind perpendicular to the face in question), the pressure coefficients are positive, and their magnitudes decrease near the sides and the top as the flow velocities increase.

As can be seen in Figure 4, C_p generally increases with height, which reflects the increasing velocity pressure in the approach flow as wind speed increases with height. As the wind direction moves off normal ($\theta = 0°$), the region of maximum pressure occurs closer to the upwind edge (B in Figure 4) of the building. At a wind angle of $\theta = 45°$, the pressures become negative at the downwind edge (A in Figure 4) of the front face. At some angle θ between 60° and 75°, the pressures become negative over the whole front face. For θ = 90°, maximum suction (negative) pressure occurs near the upwind edge (B in Figure 4) of the building side and then recovers towards $C_p = 0$ towards the downwind edge (A in Figure 4). The degree of this recovery depends on the length of the side in relation to the width W of the structure. For wind angles larger than $\theta = 100°$, the side is completely within the separated flow of the wake and the spatial variations in pressure over the face are not as great. The average pressure on a face is positive for wind angles from $\theta = 0°$ to almost 60° and negative (suction) for $\theta = 60°$ to 180°.

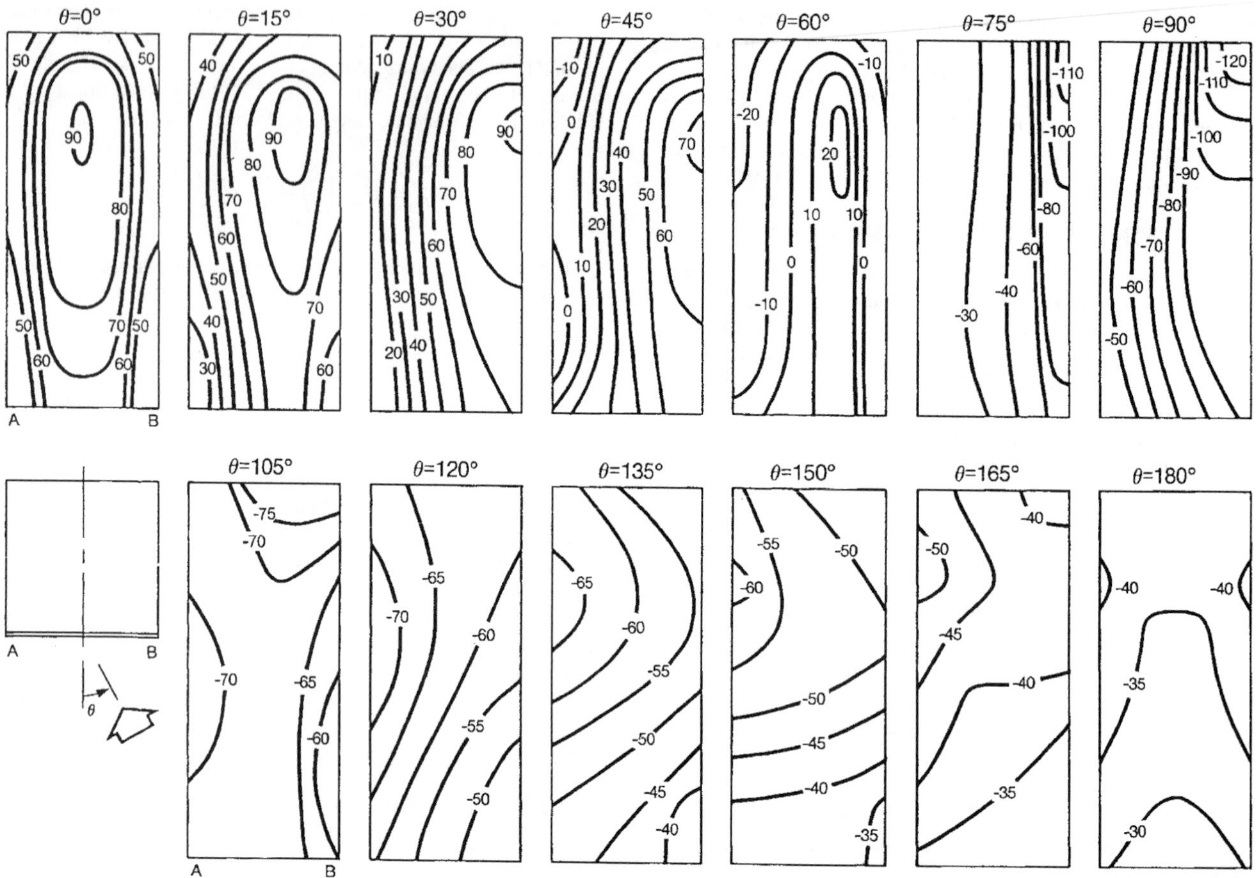

Fig. 4 Local Pressure Coefficients ($C_p \times 100$) for Tall Building with Varying Wind Direction
(Davenport and Hui 1982)

A similar pattern of behavior in the wall pressure coefficients for a low-rise building is shown in Figure 5. Here, the recovery from the strong suction with distance from the upwind edge is more rapid.

Surface Averaged Wall Pressures

Surface averaged pressure coefficients may be used in determining ventilation and/or infiltration rates, as discussed in Chapter 26. Figure 6 shows the surface pressure coefficient C_s averaged over a complete wall of a low-rise building (Swami and Chandra 1987). The figure also includes the values calculated from the pressure distributions shown in Figure 5. Similar results for a tall building are shown in Figure 7 (Akins et al. 1979).

The wind-induced indoor-outdoor pressure difference is found using the coefficient $C_{p\,(in\text{-}out)}$, which is defined as

$$C_{p(in\text{-}out)} = C_p - C_{in} \tag{5}$$

where C_{in} is the internal wind-induced pressure coefficient. For uniformly distributed air leakage sites in all the walls, C_{in} is about −0.2.

Roof Pressures

Surface pressures on the roof of a low-rise building depend strongly on roof slope. Figure 8 shows typical distributions for a wind direction normal to a side of the building. For very low slopes, the pressures are negative over the whole roof surface. The magnitude is greatest within the separated flow zone near the leading edge and recovers toward the free stream pressure downwind of the edge. For steeper slopes, the pressures are weakly positive on the windward slope and negative within the separated flow over the leeward slope. With a wind angle of about 45°, the vortices originating at the leading corner of a roof with a low slope can induce very large localized negative pressures (Figure 2). A similar vortex forms on the downwind side of a leading ridge end on a steep roof. A discussion of roof corner vortices and how to disrupt their influence may be found in Cochran and Cermak (1992) and Cochran and English (1997), respectively. Figure 9 shows the average pressure coefficient over the roof of a tall building (Akins et al. 1979).

The section on Combining Driving Forces in Chapter 26 discusses the effect of stack pressure and mechanical systems on infiltration and ventilation in a building.

Interference and Shielding Effects on Pressures

Nearby structures strongly influence surface pressures on both high- and low-rise buildings. These effects are very strong for spacing-to-height ratios less than five, where the distributions of pressure shown in Figures 4 through 9 do not apply. Although the effect of shielding is for low-rise buildings still significant at larger spacing, it is largely accounted for by the reduction in p_v with increased terrain roughness. Saunders and Melbourne (1979), Sherman and Grimsrud (1980), Bailey and Kwok (1985), and Walker et al. (1996) discuss interference. English and Fricke (1997) discuss shielding through use of an interference index, while Walker et al. (1996) present a wind shadow model for predicting shelter factors. Chapter 26 gives shielding classes for air infiltration and ventilation applications.

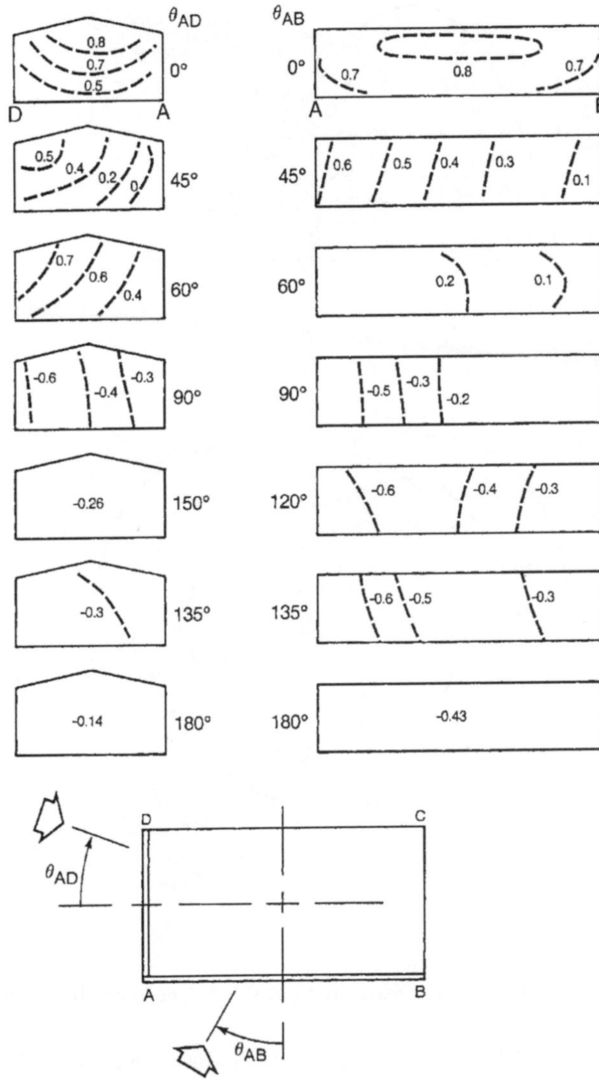

Fig. 5 Local Pressure Coefficients for Walls of Low-Rise Building with Varying Wind Direction
(Holmes 1986)

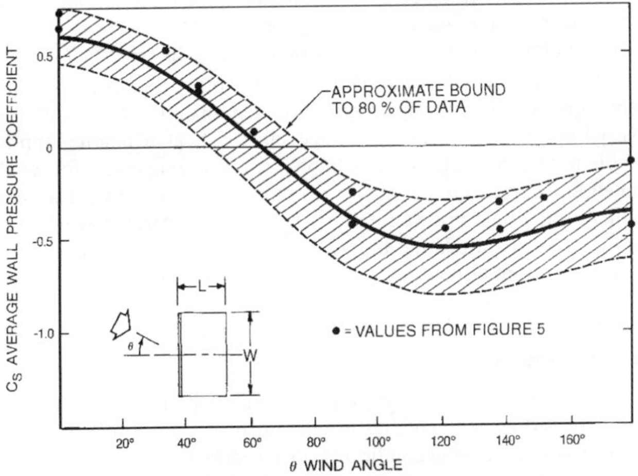

Fig. 6 Variation of Surface Averaged Wall Pressure Coefficients for Low-Rise Buildings
(Swami and Chandra 1987)

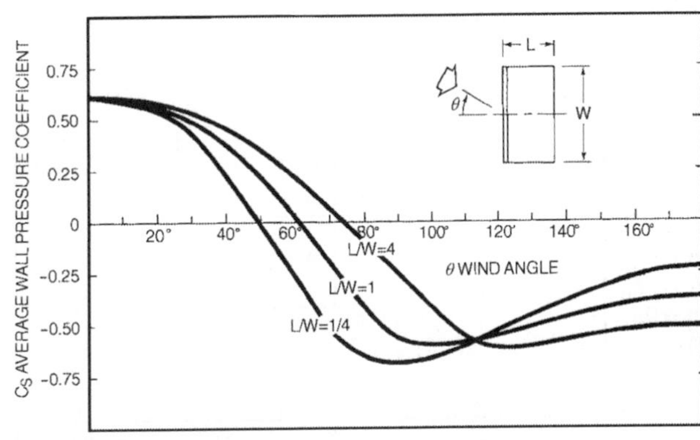

Fig. 7 Surface Averaged Wall Pressure Coefficients for Tall Buildings
(Akins et al. 1979)

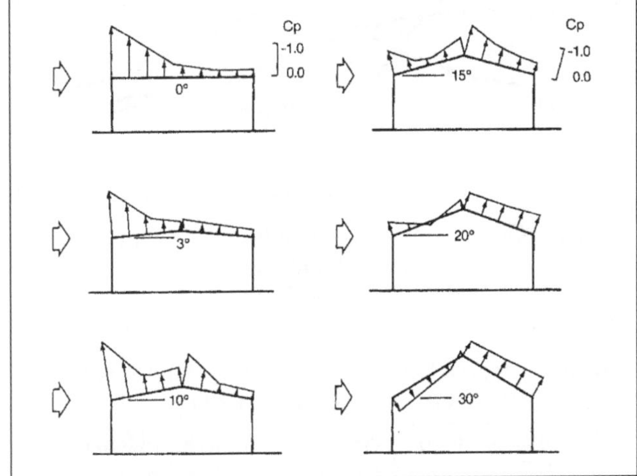

Fig. 8 Local Roof Pressure Coefficients for Roof of Low-Rise Buildings
(Holmes 1986)

Sources of Wind Data

In order to design for the effects of airflow around buildings, wind speed and direction frequency data should be obtained. The simplest forms of wind data are tables or charts of climatic normals, which give hourly average wind speeds, prevailing wind directions, and peak gust wind speeds for each month of the year. This information can be found in sources such as The *Weather Almanac* (Bair 1992) and the *Climatic Atlas of the United States* (DOC 1968). Climatic design information, including wind speed at various frequencies of occurrence, is included in Chapter 27. A current source, which contains information on wind speed and direction frequencies, is the *International Station Meteorological Climatic Summary* available in CD-ROM format from the National Climatic Data Center (NCDC) in Asheville, North Carolina. Where more detailed information is required, digital records of hourly winds and other meteorological parameters are available (on magnetic tape or CD-ROM) from the NCDC for stations throughout the world. Most countries also have weather services that provide data. For example, in Canada, the Atmospheric Environment Service in Downsview, Ontario, provides hourly meteorological data and summaries.

When an hourly wind speed U_{met} at a specified probability level (e.g., the wind speed that is exceeded 1% of the time) is desired, but only the average annual wind speed U_{annual} is available for a given meteorological station, U_{met} may be estimated from Table 2. The

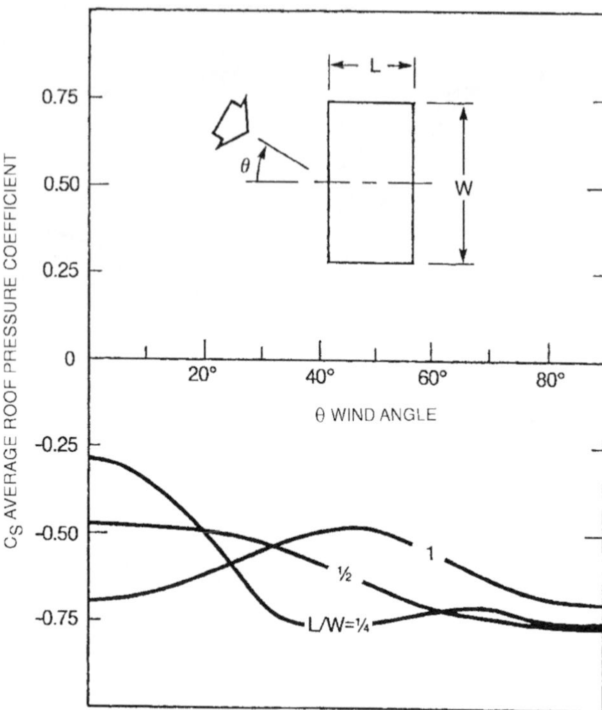

Fig. 9 Surface Averaged Roof Pressure Coefficients for Tall Buildings
(Akins et al. 1979)

Table 2 Typical Relationship of Hourly Wind Speed U_{met} to Annual Average Wind Speed U_{annual}

Percentage of Hourly Values That Exceed U_{met}	Wind Speed Ratio U_{met}/U_{annual}
90%	0.2 ± 0.1
75%	0.5 ± 0.1
50%	0.8 ± 0.1
25%	1.2 ± 0.15
10%	1.6 ± 0.2
5%	1.9 ± 0.3
1%	2.5 ± 0.4

ratios U_{met}/U_{annual} are based on long-term data from 24 weather stations widely distributed over North America. At these stations, U_{met} ranges from 7 to 14 mph. The uncertainty ranges listed in Table 2 are one standard deviation of the wind speed ratios. The following example demonstrates the use of Table 2.

Example 1. The wind speed U_{met} that is exceeded 1% of the time (88 hours per year) is needed for a building pressure or exhaust dilution calculation. If U_{annual} = 9 mph, find U_{met}.

Solution: From Table 2, the wind speed U_{met} exceeded 1% of the time is 2.5 ± 0.4 times U_{annual}. For U_{annual} = 9 mph, U_{met} is 23 mph with an uncertainty range of 19 to 26 mph at one standard deviation.

Using a single prevailing wind direction for design can cause serious errors. For any set of wind direction frequencies, one direction always has a somewhat higher frequency of occurrence. Thus, it is often called the prevailing wind, even though winds from other directions may be almost as frequent.

When using long-term meteorological records, check the anemometer location history as the instrument may have been relocated and its height varied. This can affect its directional exposure and the recorded wind speeds. Equation (4) can be used to correct wind data

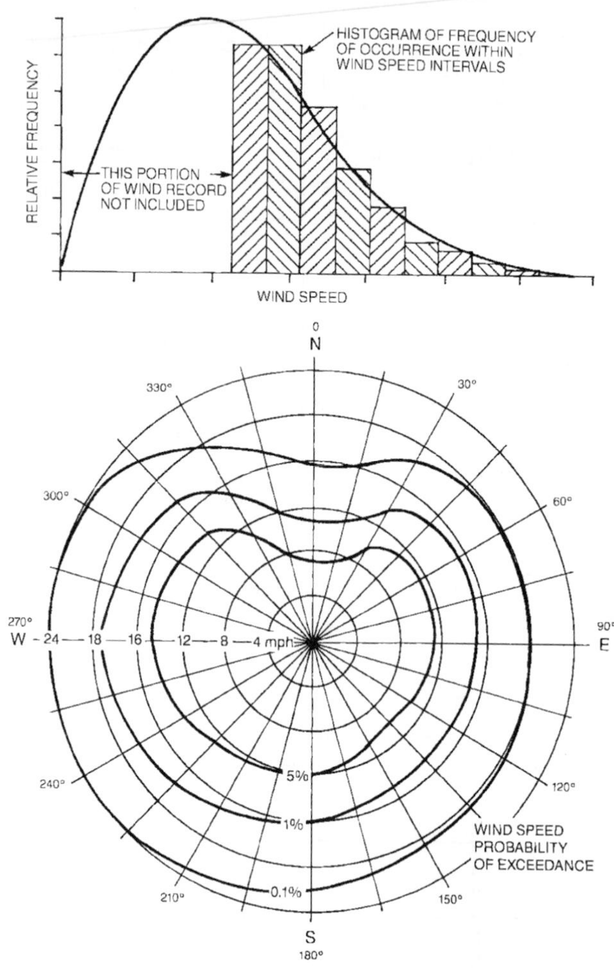

Fig. 10 Frequency Distribution of Wind Speed and Direction

collected at different mounting heights. Poor anemometer exposure due to obstructions or mounting on top of a building cannot be easily corrected, and the records for that period should be deleted.

If an estimate of the probability of an extreme wind speed outside the range of the recorded values at a site is required, the observations may be fit to an appropriate probability distribution (e.g., a Weibull distribution) and the particular probabilities calculated from the resulting function (see Figure 10). This process is usually repeated for each of 16 wind directions (e.g., 22.5° intervals).

Where estimates at extremely low probability (high wind speed) are required, curve fitting at the tail of the probability distribution is very important and may require special statistical techniques applicable to extreme values (see Chapter 27). Building codes for wind loading on structures contain information on estimating extreme wind conditions. For ventilation applications, extreme winds are usually not required, and the 99 percentile limit can be accurately estimated from airport data averaged over less than 10 years.

Estimating Wind at Sites Remote from Recording Stations

Many building sites are located far from the nearest long-term wind recording site, which is usually an airport meteorological station. To estimate wind conditions at such sites, the terrain surrounding both the anemometer site and the building site should be checked. In the simplest case of flat or slightly undulating terrain with few obstructions extending for large distances around and between the anemometer site and building site, recorded wind data can be assumed to be representative of that at the building site. Wind direction occurrence frequency at a building site should be inferred from airport data only if the two locations are on the same terrain, with no terrain features that could alter wind direction between them.

In cases where the only significant difference between the anemometer site terrain and the building site terrain is surface roughness, the mean wind speed can be adjusted using Equation (4) and Table 1, to yield approximate wind velocities at the building site. Wind direction frequencies at the site are assumed to be the same as at the recording station.

In using Equation (4), cases may be encountered where, for a given wind direction, the terrain upwind of either the building site or the recording site does not fall into just one of the categories in Table 1. The terrain immediately upwind of the site may fall into one category, while that somewhat further upwind falls into a different category. For example, at a downtown airport the terrain may be flat and open (Category 3) immediately around the recording instrument, but urban or suburban (Category 2) a relatively short distance away. This difference in terrains also occurs when a building site or recording site is in an urban area near open water or at the edge of town. In these cases, the suggested approach is to use the terrain category that is most representative of the average condition within approximately 1 mile upwind of the site (Deaves 1981). If the average condition is somewhere between two categories described in Table 1, the values of a and δ can be interpolated from those given in the table.

A rough guideline is that only wind speeds U_H of 9 mph or greater at the building site can be estimated reliably using Equation (4) and Table 2 for building and meteorological stations in different terrain categories.

In addition to changes in **surface roughness**, several other factors are important in causing the wind speed and direction at a building site to differ from values recorded at a nearby meteorological station. Wind speeds for buildings on hillcrests or in valleys where the wind is accelerated or channeled can be 1.5 times higher than meteorological station data. Wind speeds for buildings sheltered in the lee of hills and escarpments can be reduced to 0.5 times the values at nearby flat meteorological station terrain.

Solar heating of valley slopes can cause light winds of 2 to 9 mph to occur as warm air flows upslope. At night, radiant cooling of the ground can produce similar speeds as cold air drains downslope. In general, rolling terrain experiences a smaller fraction of low speeds than nearly flat terrain.

When the wind is calm or light in the rural area surrounding a city, urban air tends to rise in a buoyant plume over the city center. This rising air, heated by man-made sources and higher solar absorption in the city, is replaced by air pushed toward the city center from the edges. In this way, the **urban heat island** can produce light wind speeds and direction frequencies significantly different than those at a rural meteorological station.

In more **complex terrain**, both wind speed and direction may be significantly different from those at the distant recording site. In these cases, building site wind conditions should not be estimated from airport data. Options are either to establish an on-site wind recording station or to commission a detailed wind tunnel correlation study between the building site and long-term meteorological station wind observations.

WIND EFFECTS ON SYSTEM OPERATION

With few exceptions, building intakes and exhausts cannot be located or oriented such that a prevailing wind ensures ventilation and air-conditioning system operation. Wind can assist or hinder inlet and exhaust fans, depending on their positions on the building, but even in locations with a predominant wind direction, the ventilating system must perform adequately for all other directions. To

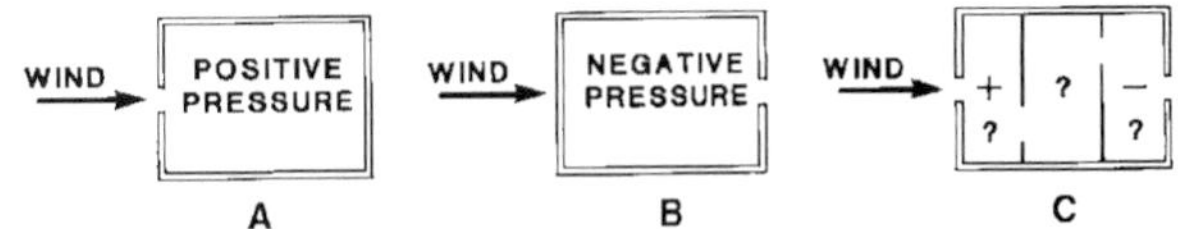

Pressures in Building Resulting from Wind:
A. With upstream opening only, pressure is positive.
B. With downstream opening only, pressure is negative.
C. Pressures are as shown if openings are equal in shape and area. With unequal openings, pressures can be either negative or positive in each space, depending on relative areas of openings.

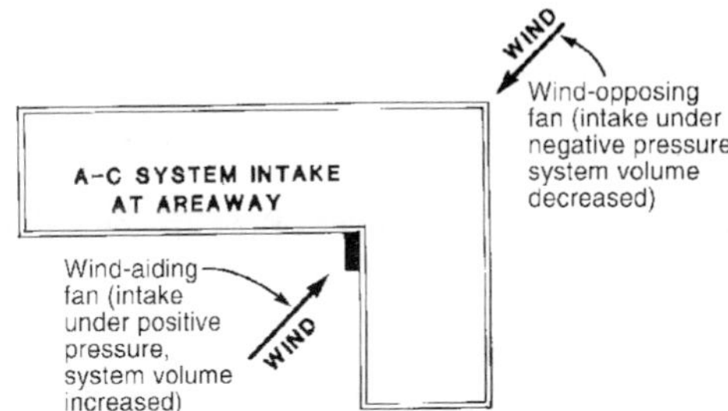

Fig. 11 Sensitivity of System Volume to Locations of Building Openings, Intakes, and Exhausts

avoid variability in system flow rates, use Figures 4, 5, and 8 as a guide to placing inlets and exhausts in locations where the surface pressure coefficients do not vary greatly with the wind direction.

Cooling towers and similar equipment should be oriented to take advantage of prevailing wind directions, based on careful study of the meteorological data and flow patterns on the building for the area and time of year involved.

A building with only upwind openings is under a positive pressure (Figure 11). Building pressures are negative when there are only downwind openings. A building with internal partitions and openings is under various pressures depending on the relative sizes of the openings and the wind direction. With larger openings on the windward face, the building interior tends to remain under positive pressure; the reverse is also true (see Figures 4 through 9, and Chapter 26).

Airflow through a wall opening results from differential pressures, which may exceed 0.5 in. of water during high winds. Supply and exhaust systems, openings, dampers, louvers, doors, and windows make the building flow conditions too complex for direct calculation. Iterative calculations are required because of the nonlinear dependence of volume flow rate on the differential pressure across an opening. Several **multizone airflow models** are available for these iterative calculations (Walton 1997, Feustel and Dieris 1992). The opening and closing of doors and windows by building occupants add further complications. In determining $C_{p(in\text{-}out)}$ from Equation (5), the wind direction is more important than the position of an opening on a wall, as shown in Figures 4 and 5. Please see Chapter 26 for more detailed information regarding wind effects on building ventilation, including natural and mechanical systems.

Natural and Mechanical Ventilation

With **natural ventilation**, wind may augment, impede, or sometimes reverse the airflow through a building. For large roof areas (Figure 2), the wind can reattach to the roof downwind of the leading edge. Thus, any natural ventilation openings could see either a positive or negative pressure, dependent on wind speed and wind direction. Positive pressure existing where negative pressures were expected could lead to a reversal of expected natural ventilation. These reversals can be avoided by using stacks, continuous roof ventilators, or other exhaust devices in which the flow is augmented by the wind.

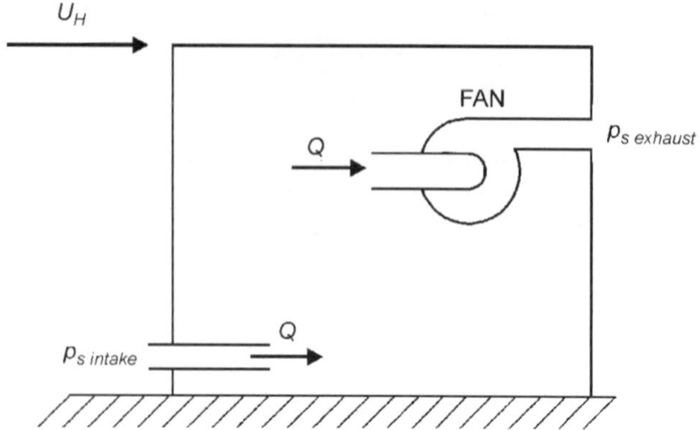

Fig. 12 Intake and Exhaust Pressures on Exhaust Fan in Single Zone Building

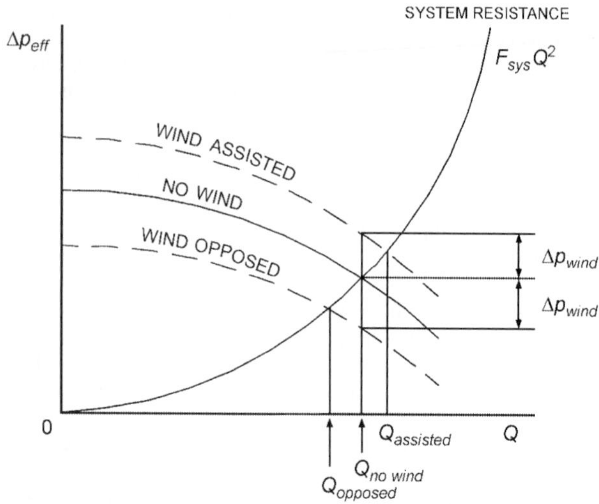

Fig. 13 Effect of Wind-Assisted and Wind-Opposed Flow

Mechanical ventilation is also affected by wind conditions. A low-pressure wall exhaust fan (0.05 to 0.1 in. of water) can suffer drastic reduction in capacity. Flow can be reversed by wind pressures on windward walls, or its rate can be increased substantially when subjected to negative pressures on the lee and other sides. Clarke (1967), when measuring medium-pressure air-conditioning systems (1 to 1.5 in. of water), found flow rate changes of 25% for wind blowing into intakes on an L-shaped building compared to wind blowing away from intakes. Such changes in flow rate can cause noise at the supply outlets and drafts in the space served.

For mechanical systems, the wind can be thought of as an additional pressure source in series with a system fan, either assisting or opposing it (Houlihan 1965). Where system stability is essential, the supply and exhaust systems must be designed for high pressures (about 3 to 4 in. of water) or must use devices to actively minimize unacceptable variations in flow rate. To conserve energy, the system pressure selected should be consistent with system needs.

Quantitative estimates of the effect of wind on a mechanical ventilation system can be made by using the pressure coefficients in Figures 4 through 9 to calculate the wind pressure on air intakes and exhausts. A simple worst-case estimate is to assume a system with 100% makeup air supplied by a single intake and exhausted from a single outlet. The building is treated as a single zone, with an exhaust-only fan as shown in Figure 12. This will overestimate the effect of wind on system volume flow.

Combining Equations (2) and (3), the wind pressures at the air intake and exhaust locations are

$$p_{s\ intake} = C_{p\ intake} \frac{\rho_a U_h^2}{2g_c} \tag{6}$$

$$p_{s\ exhaust} = C_{p\ exhaust} \frac{\rho_a U_h^2}{2g_c} \tag{7}$$

For the single zone building shown in Figure 12, a worst-case estimate of wind effect neglects any flow resistance in the intake grill and duct, making the interior building pressure $p_{interior}$ equal to the outdoor wind pressure on the intake, $p_{interior} = p_{s\ intake}$. Then, with all the system flow resistance assigned to the exhaust duct in Figure 12, and a pressure rise Δp_{fan} across the fan, the pressure drop from outdoor intake to outdoor exhaust yields

$$(p_{s\ intake} - p_{s\ exhaust}) + \Delta p_{fan} = F_{sys} \frac{\rho Q^2}{A_L^2 g_c} \tag{8}$$

where F_{sys} is the system flow resistance, A_L is the flow leakage area, and Q is the system volume flow rate. This result shows that for the worst case estimate, the wind induced pressure difference simply adds to or subtracts from the fan pressure rise. With the inlet and exhaust pressures from Equations (6) and (7), the effective fan pressure rise $\Delta p_{fan\ eff}$ is

$$\Delta p_{fan\ eff} = \Delta p_{fan} + \Delta p_{wind} \tag{9}$$

where

$$\Delta p_{wind} = (C_{p\ intake} - C_{p\ exhaust}) \frac{\rho_a U_h^2}{2g_c} \tag{10}$$

The fan will be wind-assisted when $C_{p\ intake} > C_{p\ exhaust}$ and wind-opposed when the wind direction changes causing $C_{p\ intake} < C_{p\ exhaust}$. The effect of wind-assisted and wind-opposed pressure differences is illustrated in Figure 13.

Example 2. Make a worst-case estimate for the effect of wind on the supply fan for a low-rise building with a height H = 50 ft located in a city suburb. Use the hourly average wind speed that will be exceeded only 1% of the time and assume an annual hourly average speed of U_{annual} = 8 mph measured on a meteorological tower at height H_{met} = 33 ft at a nearby airport. Outdoor air density is ρ_a = 0.075 lb$_m$/ft^3.

Solution: From Table 2 the wind speed that is exceeded only 1% of the hours each year will be a factor of 2.5 ± 0.4 higher than the annual average of 8 mph, so the 1% maximum speed at the airport meteorological station is

$$U_{met} = 2.5 \times 8 = 20 \text{ mph}$$

From Table 1, the airport meteorological station is in terrain category 3, with a boundary layer thickness δ_{met} = 900 ft and a velocity profile exponent a = 0.14. The suburban location of the building places it in terrain category 2 in Table 2, with δ = 1200 ft and a = 0.22. Using Equation (4) to determine the wind speed U_H at roof level H = 50 ft in the flow approaching the building,

$$U_H = 20\left(\frac{900}{33}\right)^{0.14}\left(\frac{50}{1200}\right)^{0.22} = 15.8 \text{ mph} = 23.2 \text{ ft/s}$$

A worst-case estimate of wind effect must assume intake and exhaust locations on the building that produce the largest difference ($C_{p\ intake} - C_{p\ exhaust}$) in Equations (9) and (10). From Figure 5, the largest difference occurs for the intake on the upwind wall AB and the exhaust on the downwind wall CD, with a wind angle θ_{AB} = 0°. For this worst case, $C_{p\ intake}$ = +0.8 on the upwind wall and $C_{p\ exhaust}$ = −0.43 on the downwind wall. Using these coefficients in Equations (9) and (10) to evaluate the effective fan pressure $\Delta p_{fan\ eff}$,

$$\Delta p_{fan\ eff} = \Delta p_{fan} + [0.8 - (-0.43)]\frac{0.075(23.2)^2}{2(32.2)}$$

$$= \Delta p_{fan} + 0.77 \mathrm{lb_f/ft^2}$$

Expressing this as in. of water pressure for water density of 62.4 $\mathrm{lb_m/ft^3}$ and gravitational acceleration of 32.2 $\mathrm{ft/s^2}$,

$$\Delta p_{wind} = \frac{0.77 \times 32.2}{62.4 \times 32.2} = 0.012 \text{ ft} = 0.15 \text{ in. of water}$$

This wind-assisted hourly averaged pressure is exceeded only 1% of the time (88 hours per year). When the wind direction reverses, the outlet will be on the upwind wall and the inlet on the downwind wall, producing wind-opposed flow, changing the sign from +0.15 to –0.15 in of water. The importance of these pressures depends on their size relative to the fan pressure rise Δp_{fan}, as shown in Figure 13.

Building Pressure Balance

Proper **building pressure balance** avoids flow conditions that make doors hard to open, cause drafts, and prevent the confinement of contaminants to specific areas. Although the supply and exhaust systems in an area may be in nominal balance, wind can upset this balance, not only because of the changes in fan capacity but also by superimposing infiltrated or exfiltrated air or both on the area. These effects can make it impossible to control environmental conditions. Where building balance and minimum infiltration are important, consider the following:

- Design HVAC system with pressure adequate to minimize wind effects
- Include controls to regulate flow rate or pressure or both
- Separate supply and exhaust systems to serve each building area requiring control or balance
- Effect of doors (possibly self-closing) or double-door air locks to noncontrolled adjacent areas, particularly outside doors
- Sealing of windows and other leakage sources and closing natural ventilation openings.

System volume and pressure control is described in Chapter 45 of the 1999 *ASHRAE Handbook—Applications*. This control is not possible without adequate system pressure for both the supply and exhaust systems to overcome wind effects. Such a control system may require fan inlet or discharge dampers, fan speed or pitch control, or both.

Fume Hood Operation

Wind effects can interfere with safe fume hood operation. Supply volume variations can cause both disturbances at hood faces and a lack of adequate fume hood makeup air. Volume surges, due to fluctuating wind pressures acting on the exhaust system, can cause momentary inadequate hood exhaust. If highly toxic contaminants are involved, surging is unacceptable. The system should be designed to eliminate this condition. On low-pressure exhaust systems, it is impossible to test the hoods under wind-induced, surging conditions. These systems should be tested during calm conditions for safe flow into the hood faces; they should be rechecked by smoke tests during high wind conditions. For more information, see Chapter 13 of the 1999 *ASHRAE Handbook—Applications*.

Minimizing Wind Effect on System Volume

Wind effect can be reduced by careful selection of inlet and exhaust locations. Because wall surfaces are subject to a wide variety of positive and negative pressures, wall openings should be avoided when possible. When they are required, wall openings should be away from corners formed by building wings (Figure 11). Mechanical ventilation systems should operate at a pressure high enough to minimize wind effect. **Low-pressure systems** and **propeller exhaust fans** should not be used with wall openings unless their ventilation rates are small or they are used in noncritical services such as storage areas.

Although roof air intakes in flow recirculation zones best minimize wind effect on system flow rates, current and future air quality in these zones must be considered. These locations should be avoided if a source of contamination exists or may be added in the future. The best area is near the middle of the roof because the negative pressure there is small and least affected by changes in wind direction (Figure 8). Avoid the edges of the roof and walls, where large pressure fluctuations occur. Either vertical or horizontal (mushroom) openings can be used. On roofs having large areas, where the intake may be outside the roof recirculation zone, mushroom or 180° gooseneck designs minimize impact pressure from wind flow. The 135° gooseneck that is frequently used or vertical louvered openings are undesirable for this purpose or for rain protection.

Heated air or contaminants should be exhausted vertically through stacks, above the roof recirculation zone. Horizontal, louvered (45° down), and 135° gooseneck discharges are undesirable, even for heat removal systems, because of their sensitivity to wind effects. A 180° gooseneck for systems handling hot air may be undesirable because of air impingement on tar and felt roofs. Vertically discharging stacks located in a recirculation region (except near a wall) have the advantage of being subjected only to negative pressure created by wind flow over the tip of the stack. See Chapter 43 of the 1999 *ASHRAE Handbook—Applications* for information regarding stack design.

BUILDING INTERNAL PRESSURE AND FLOW CONTROL

In air-conditioning and ventilation systems for a building containing airborne contaminants, the correct internal airflow is toward the contaminated areas. Airflow direction is maintained by controlling pressure differentials between spaces. In a laboratory building, for example, peripheral rooms such as offices and conference rooms are maintained at a positive pressure, and laboratories at a negative pressure, both with reference to corridor pressure. Pressure differentials between spaces are normally obtained by balancing the air-conditioning and ventilation supply system airflows in the spaces in conjunction with the exhaust systems in the laboratories, with differential pressure instrumentation to control the airflow. Chapter 45 of the 1999 *ASHRAE Handbook—Applications* has further information on controls.

Airflow in corridors is sometimes controlled by an outdoor reference probe that senses static pressure at doorways and air intakes. The differential pressure measured between the corridor and the outside may then signal a controller to increase or decrease airflow to the corridor. Unfortunately, it is difficult to locate an external probe where it will sense the proper external static pressure. High wind velocity and resulting pressure changes around entrances can cause great variations in pressure. Care must be taken to ensure that the probe is unaffected by wind pressure.

The pressure differential for a room adjacent to a corridor can be controlled using the corridor pressure as the reference. Outdoor pressure cannot control pressure differentials within rooms, even during periods of relatively constant wind velocity (wind-induced pressure). A single pressure sensor can measure the outside pressure at one point only and may not be representative of pressures elsewhere.

SCALE MODEL SIMULATION AND TESTING

For many routine design applications, the flow patterns and wind pressures can be estimated using the data and equations presented in the previous sections. Exhaust dilution for simple building geometries located in homogeneous terrain environments (e.g., no larger buildings or terrain features nearby) can be estimated using the data and equations presented in the previous sections and in Chapter 43 of the 1999 *ASHRAE Handbook—Applications*. However, in critical applications, such as where health and safety are of concern, **physical modeling** or **full-scale field evaluations** may be required to obtain more accurate estimates. Measurements on small-scale models in wind tunnels or water channels can provide information for design prior to construction. These measurements can also be used as an economical method of performance evaluation for existing facilities. Full-scale testing is not generally useful in the initial design phase because of the time and expense required to obtain meaningful information. On the other hand, full-scale testing is useful for verifying data derived from physical modeling and for planning remedial changes to improve existing facilities.

Detailed accounts of physical modeling, field measurements and applications, and engineering problems resulting from atmospheric flow around buildings are available in the proceedings of conferences on wind engineering (see the section on Bibliography).

The wind tunnel is the main tool used to assess and understand the airflow around buildings. Water channels or tanks can also be used. However, the water methods are more difficult to implement and give only qualitative results for some cases. Models of buildings, complexes, and the local surrounding topography are constructed and tested in a simulated turbulent atmospheric boundary layer. The airflow, wind pressures, snow loads, structural response, or pollutant concentrations can then be measured directly by properly scaling the wind, building geometry, and exhaust flow characteristics. Weil et al. (1981), Petersen (1987a), and Dagliesh (1975) found generally good agreement between the results of wind tunnel simulations and corresponding full-scale data. Cochran (1992) and Cochran and Cermak (1992) have found good agreement between the model- and full-scale measurements of low-rise architectural aerodynamics and cladding pressures, respectively.

Similarity Requirements

Physical modeling is most appropriate for applications involving small-scale atmospheric motions, such as recirculation of exhaust downwind of a laboratory, wind loads on structures, wind speeds around building clusters, snow loads on roofs, and airflow over hills or other terrain features. Winds associated with tornadoes, thunderstorms, and large-scale atmospheric motion cannot currently be simulated accurately.

Snyder (1981) gives guidelines for fluid modeling of atmospheric diffusion. This report contains explicit directions and should be used whenever designing wind tunnel studies to assess concentration levels due to air pollutants. ASCE *Standard* 7 and ASCE *Manual of Practice* 67 (ASCE 1999) also provide guidance when wind tunnels are used for evaluating wind effects on structures.

A complete and exact simulation of the airflow over buildings and the resulting concentration or pressure distributions cannot be achieved in a physical model. However, this is not a serious limitation. Cermak (1971, 1975, 1976a,b), Snyder (1981), and Petersen (1987a,b) found that an accurate simulation of the transport and dispersion of laboratory exhaust can be achieved if the following criteria are met in the model and full scale:

1. Match exhaust velocity to wind speed ratios, V_e/U_H.
2. Match exhaust to ambient air density ratios, ρ_e/ρ_a.
3. Match exhaust Froude numbers. $Fr^2 = \rho_a V_e^2/[(\rho_e - \rho_a)gd]$, where d is the effective exhaust stack diameter.
4. Ensure fully turbulent stack gas flow by ensuring stack flow Reynolds number ($Re_s = V_e d/\nu$) is greater than 2000 (where ν is the kinematic viscosity of ambient (outdoor) air), or by placing an obstruction inside the stack to enhance turbulence.
5. Ensure fully turbulent wind flow.
6. Scale all dimensions and roughness by a common factor.
7. Match atmospheric stability by the bulk Richardson number (Cermak 1975). For most applications related to airflow around buildings, neutral stratification is assumed, and no Richardson number matching is required.
8. Match mean velocity and turbulence distributions in the wind.
9. Ensure building wind Reynolds number ($Re_b = U_H R/\nu$) is greater than 11,000 for sharp-edged structures, or greater than 90,000 for round-edged structures.
10. Ensure less than 5% blockage of wind tunnel cross section.

For wind speeds, flow patterns, or pressure distributions around buildings, only Conditions 5 through 10 are necessary. Usually, each wind tunnel study requires a detailed assessment to determine the appropriate parameters to match in the model and full scale.

In wind tunnel simulations of exhaust gas recirculation, the buoyancy of the exhaust gas (Condition 3) is often not modeled. This allows using a high wind tunnel speed or a smaller model to achieve high enough Reynolds numbers (Conditions 4, 5, and 9). Neglecting buoyancy is justified if the density of building exhaust air is within 10% of the ambient (outdoor) air. Also, critical minimum dilution D_{crit} occurs at wind speeds high enough to produce a well-mixed, neutrally stable atmosphere, allowing stability matching (Condition 7) to be neglected (see Chapter 43 of the 1999 *ASHRAE Handbook—Applications* for discussion of D_{crit}). Omission of Conditions 3 and 7 simplifies the test procedure considerably, reducing both testing time and cost.

Buoyancy should be properly simulated for high-temperature exhausts such as boilers and diesel generators. Equality of model and prototype Froude numbers (Condition 3) requires tunnel speeds of less than 100 fpm for testing. However, greater tunnel speeds may be needed to meet the minimum building Reynolds number requirement (Condition 4).

Wind Simulation Facilities

Boundary layer wind tunnels are required for conducting most wind studies. The wind tunnel test section should be long enough so that a deep boundary layer that slowly changes with downwind distance can be established upwind of the model building.

Other important wind tunnel characteristics include the width and height of the test section, range of wind speeds, roof adjustability, and temperature control. Larger models can be used in tunnels that are wider and taller, which, in turn, give better measurement resolution. Model blockage effects can be minimized by an adjustable roof height. Temperature control of the tunnel surface and airflow is required when atmospheric conditions other than neutral stability are to be simulated. Boundary layer characteristics appropriate for the site are established by using roughness elements on the tunnel floor that produce mean velocity and turbulence intensity profiles characteristic of the full scale.

Water can also be used for the modeling fluid if an appropriate flow facility is available. Flow facilities may be in the form of a tunnel, tank, or open channel. Water tanks with a free surface ranging in size up to that of a wind tunnel test section have been used by towing a model (upside down) through the nonflowing fluid. Stable stratification can be obtained by adding a salt solution. This technique (towed model in a tank) does not permit development of a boundary layer and therefore yields only approximate, qualitative information on flow around buildings. Water channels can be designed to develop thick turbulent boundary layers similar to those developed in the wind tunnel. One advantage of such a flow system is ease of flow visualization, but this is offset by a greater difficulty in developing the correct turbulence structure and the measurement of flow variables and concentrations.

Designing Model Test Programs

The first step in planning a test program is selection of the model length scale. Choice of this scale depends on cross-sectional dimensions of the test section, dimensions of the buildings to be included in the model, and/or topographic features and thickness of the simulated atmospheric boundary layer. Typical geometric scales range from about 120:1 to 1000:1.

Because a large model size is desirable to meet minimum Reynolds number and Froude number requirements, a wide test section is advantageous. In general, the model at any section should be small compared to the test section area so that blockage is less than 5% (Melbourne 1982).

The test program must include specifications of the meteorological variables to be considered. These include wind direction, wind speed, and thermal stability. Data taken at the nearest meteorological station should be reviewed to obtain a realistic assessment of wind climate for a particular site. Ordinarily, local winds around a building, pressures, and/or concentrations are measured for 16 wind directions (e.g., 22.5° intervals). This is easily accomplished by mounting the building model and its nearby surroundings on a turntable. More than 16 wind directions are required for highly toxic exhausts or for finding peak fluctuating pressures on a building. If only local wind information and pressures are of interest, testing at one wind speed with neutral stability is sufficient.

SYMBOLS

A_L = flow leakage area, Equation (8), ft^2
a = exponent in power law wind speed profile for local building terrain, Equation (4) and Table 1, dimensionless
a_{met} = exponent a for the meteorological station, Equation (4) and Table 1, dimensionless
B_L = larger of the two upwind building face dimensions H and W, Equation (1), ft
B_s = smaller of the two upwind building face dimensions H and W, Equation (1), ft
$C_{p\,in}$ = internal wind-induced pressure coefficient, Equation (5), dimensionless
C_p = local wind pressure coefficient for building surface, Equation (3), dimensionless
$C_{p(in\text{-}out)}$ = difference between outdoor and indoor pressure coefficients, Equation (5), dimensionless
C_s = surface-averaged pressure coefficient, Figure 6, dimensionless
d = effective stack diameter, ft
D_{crit} = critical dilution factor at roof level for uncapped vertical exhaust at critical wind speed (see Chapter 43 of the 1999 *ASHRAE Handbook—Applications*), dimensionless
Fr = Froude number, dimensionless
F_{sys} = system flow resistance, Equation (8), dimensionless
g = acceleration of gravity, 32.2 ft/s^2
g_c = gravitational proportionality constant, 32.2 $ft \cdot lb_m/lb_f \cdot s^2$, Equations (2), (6), (7), (10)
H = wall height above ground on upwind building face, Equation (4) and Figure 1, ft
H_c = maximum height above roof level of upwind roof edge flow recirculation zone, Figures 1 and 3, ft
H_{met} = height of anemometer at meteorological station, Equation (4), ft
L = length of building in wind direction, Figures 1 and 2, ft
L_c = length of upwind roof edge recirculation zone, Figure 3, ft
L_r = length of flow recirculation zone behind rooftop obstacle or building, Figures 1 and 3, ft
p_s = wind pressure difference between exterior building surface and local ambient (outdoor) atmospheric pressure at the same elevation in an undisturbed approach wind, Equation (3), lb_f/ft^2
p_v = wind velocity pressure at roof level, Equation (2), lb_f/ft^2
Q = volumetric flow rate, Equation (8), cfm
R = scaling length for roof flow patterns, Equation (1), ft
Re_b = building Reynolds number, dimensionless
Re_s = stack flow Reynolds number, dimensionless
U_{annual} = annual average of hourly wind speeds U_{met}, Table 2, mph
U_H = mean wind speed at height H of upwind wall in undisturbed flow approaching building, Equation (2) and Figures 1, 2 and 3, mph
U_{met} = meteorological station hourly wind speed, measured at height H_{met} above ground in smooth terrain, Equation (4) and Table 2, mph
V_e = exhaust face velocity, ft/s
W = width of upwind building face, Figure 2, ft
δ = fully developed atmospheric boundary layer thickness, Equation (4) and Table 1, ft
δ_{met} = atmospheric boundary layer thickness at meteorological station, Equation (4) and Table 1, ft
Δp_{fan} = pressure rise across fan, Equation (8), psi
$\Delta p_{fan\,eff}$ = effective pressure rise across fan, Equation (9), psi
Δp_{wind} = wind-induced pressure, Equations (9) and (10), psi
ν = kinematic viscosity of ambient (outdoor) air, ft^2/s
ρ_a = ambient (outdoor) air density, Equation (2), lb_m/ft^3
ρ_e = density of exhaust gas mixture, lb_m/ft^3
θ = angle between perpendicular line from upwind building face and wind direction, Figures 4 through 7, degrees

REFERENCES

AIHA. 1992. Laboratory ventilation. ANSI/AIHA *Standard* Z9.5-1992. American Industrial Hygiene Association, Fairfax, VA.

Akins, R.E., J.A. Peterka, and J.E. Cermak. 1979. Averaged pressure coefficients for rectangular buildings. *Wind Engineering*. Proceedings of the Fifth International Conference 7:369-80, Fort Collins, CO. Pergamon Press, NY.

ASCE. 1998. Minimum design loads for buildings and other structures. *Standard* 7-1998. American Society of Civil Engineers, New York.

ASCE. 1999. Wind tunnel model studies of buildings and structures. *Manual of Practice* 67.

Bailey, P.A. and K.C.S. Kwok. 1985. Interference excitation of twin fall buildings. *Wind Engineering and Industrial Aerodynamics* 21:323-338.

Bair, F.E. 1992. *The weather almanac*, 6th ed. Gale Research Inc., Detroit, MI.

Cermak, J.E. 1971. Laboratory simulation of the atmospheric boundary layer. *AIAA Journal* 9(9):1746.

Cermak, J.E. 1975. Applications of fluid mechanics to wind engineering. Journal of Fluid Engineering, *Transactions of ASME* 97:9.

Cermak, J.E. 1976a. Nature of airflow around buildings. *ASHRAE Transactions* 82(1):1044-60.

Cermak, J.E. 1976b. Aerodynamics of buildings. *Annual Review of Fluid Mechanics* 8:75.

Clarke, J.H. 1967. Airflow around buildings. *Heating Piping and Air Conditioning* 39(5):145.

Cochran, L.S. 1992. Low-rise architectural aerodynamics: The Texas Tech University experimental building. *Architectural Science Review* 35(4):131-36.

Cochran, L.S. and J.E. Cermak. 1992. Full and model scale cladding pressures on the Texas Tech University experimental building. *Journal of Wind Engineering and Industrial Aerodynamics* 41-44:1589-1600.

Cochran, L.S. and E.C. English. 1997. Reduction of wind loads by architectural features. *Architectural Science Review* 40(3):79-87.

Dagliesh, W.A. 1975. Comparison of model/full-scale wind pressures on a high-rise building. *J. Industrial Aerodynamics* 1:55-66.

Davenport, A.G. and H.Y.L. Hui. 1982. External and internal wind pressures on cladding of buildings. Boundary Layer Wind Tunnel Laboratory, University of Western Ontario, London, Ontario, Canada. BLWT-820133.

Deaves, D.M. 1981. Computations of wind flow over changes in surface roughness. *J. Wind Engineering and Industrial Aerodynamics* 7:65-94.

Deaves, D.M. and R.I. Harris. 1978. A mathematical model of the structure of strong winds. *Report* 76. Construction Industry Research and Information Association (UK).

DOC. 1968. *Climatic atlas of the United States*. U.S. Department of Commerce, Washington, D.C.

English, E.C. and F.R. Fricke. 1997. The interference index and its prediction using a neural network analysis of wind tunnel data. *Fourth Asia-Pacific Symposium on Wind Engineering*. APSOWE IV University of Queensland 363-366.

Feustel, H.E. and J. Dieris. 1992. A survey of airflow models for multizone buildings. *Energy and Buildings* 18:79-100.

Holmes, J.D. 1986. Wind loads on low-rise buildings: The structural and environmental effects of wind on buildings and structures, Chapter 12. Faculty of Engineering, Monash University, Melbourne, Australia.

Hosker, R.P. 1984. Flow and diffusion near obstacles. *Atmospheric science and power production*. U.S. Department of Energy DOE/TIC-27601 (DE 84005177).

Hosker, R.P. 1985. Flow around isolated structures and building clusters: A review. *ASHRAE Transactions* 91(2b):1671-92.

Houlihan, T.F. 1965. Effects of relative wind on supply air systems. *ASHRAE Journal* 7(7):28.

Melbourne, W.H. 1979. Turbulence effects on maximum surface pressures; A mechanism and possibility of reduction. *Proceedings* of Fifth International Conference on Wind Engineering. J.E. Cermak, ed. Fort Collins, Colorado. 541-551.

Melbourne, W.H. 1982. Wind tunnel blockage effects and corrections. *Proceedings* of the International Workshop on Wind Tunnel Modeling Criteria and Techniques in Civil Engineering Applications. T.A. Reinhold, ed. Maryland, USA. 197-216.

Meroney, R.N., and B. Bienkiewicz. 1997. *Computational wind engineering* 2. Elsevier, Amsterdam.

NCDC. Updated periodically. International station meteorological climatic summary (CD-ROM). National Climatic Data Center, Asheville, NC. Published jointly with U.S. Air Force and U.S. Navy.

Petersen, R.L. 1987a. Wind tunnel investigation of the effect of platform-type structures on dispersion of effluents from short stacks. *Journal of Air Pollution Control Association* 36:1347-52.

Petersen, R.L. 1987b. Designing building exhausts to achieve acceptable concentrations of toxic effluent. *ASHRAE Transactions* 93(2):2165-85.

SAA. 1989. Minimum design loads on structures – Part 2: Wind loads. *Standard* AS-1170 (known as the SAA Loading Code). Standards Association of Australia, North Sydney, NSW.

Saunders, J.W. and W.H. Melbourne. 1979. Buffeting effect of upwind buildings. Fifth International Conference on Wind Engineering. Pergammon Press.

Sherman, M.H. and D.T. Grimsrud. 1980. The measurement of infiltration using fan pressurization and weather data. *Report* # LBL-10852. Lawrence Berkeley Laboratory, University of California, Berkeley.

Snyder, W.H. 1981. Guideline for fluid modeling of atmospheric diffusion. Environmental Protection Agency *Report* EPA-600/881-009.

Swami, H.V. and S. Chandra. 1987. Procedures for calculating natural ventilation airflow rates in buildings. *Final Report* FSEC-CR-163-86. Florida Solar Energy Center, Cape Canaveral.

Walker, I.S., D.J. Wilson, and T.W. Forest. 1996. Wind shadow model for air infiltration sheltering by upwind obstacles. *Int. J. of HVAC&R Research* 2(4):265-283.

Walton, G.N. 1997. CONTAM96 user manual. National Institute of Standards and Technology NISTIR 6056. Gaithersburg, Maryland.

Wilson, D.J. 1979. Flow patterns over flat roofed buildings and application to exhaust stack design. *ASHRAE Transactions* 85:284-95.

BIBLIOGRAPHY

ASCE. 1987. Wind tunnel model studies of building and structures. ASCE Manuals and Reports on *Engineering Practice* No. 67. American Society of Civil Engineers, New York.

Cermak, J.E. 1977. Wind-tunnel testing of structures. *Journal of the Engineering Mechanics Division*, ASCE 103, EM6:1125.

Cermak, J.E., ed. 1979. *Wind engineering*. Proceedings of Fifth International Conference, Fort Collins, CO. Pergamon Press, New York.

Clarke, J.H. 1965. The design and location of building inlets and outlets to minimize wind effect and building reentry of exhaust fumes. *Journal of American Industrial Hygiene Association* 26:242.

Cochran, L.S. and Cermak, J.E. 1992. Full and model-scale cladding pressures on the Texas Tech University experimental building. *Journal of Wind Engineering and Industrial Aerodynamics* 41-44, 1589-1600.

Defant, F. 1951. Local winds. *Compendium of Meteorology*, pp. 655-72. American Meteorology Society, Boston.

Elliot, W.P. 1958. The growth of the atmospheric internal boundary layer. *Transactions of the American Geophysical Union* 39:1048-54.

ESDU. 1990. Strong winds in the atmospheric boundary layer. Part 1: Mean hourly wind speeds, pp. 15-17. *Engineering Science Data Unit*, Item 82-26, London, UK.

Geiger, R. 1966. *The climate near the ground*. Harvard University Press, Cambridge, MA.

Houghton, E.L. and N.B. Carruthers. 1976. *Wind forces on buildings and structures: An introduction*. Edward Arnold, London.

Landsberg, H. 1981. *The urban climate*. Academic Press, New York.

Panofsky, H.A. and J.A. Dutton. 1984. *Atmospheric turbulence: Models and methods for engineering applications*. John Wiley and Sons, New York.

Proceedings of the Fifth National Conference on Wind Engineering. 1985. Texas Tech University, Lubbock, TX.

Simiu, V. and R. Scanlan. 1986. *Wind effects on structures: An introduction to wind engineering*, 2nd ed. Wiley Interscience, New York.

CHAPTER 17

ENERGY RESOURCES

BUILDINGS and facilities of various types may be heated, ventilated, air conditioned, and refrigerated—using systems and equipment designed for that purpose and using the site energy forms commonly available—without concern for the original **energy resources** from whence those energy forms came. Since the energy used in buildings and facilities comprises a significant amount of the total energy used for all purposes, and since the use of this energy has an impact on energy resources, ASHRAE recognizes the "effect of its technology on the environment and natural resources to protect the welfare of posterity" (ASHRAE 1990).

Many governmental agencies regulate energy conservation legislation for obtaining building permits (Conover 1984). The application of specific values to building energy use situations has a considerable effect on the selection of HVAC&R systems and equipment and how they are applied.

CHARACTERISTICS OF ENERGY AND ENERGY RESOURCE FORMS

The HVAC&R industry deals with energy forms as they occur on or arrive at a building site. Generally, these forms are fossil fuels (natural gas, oil, and coal) and electricity. Solar energy and wind energy are also available at most sites, as is low-level geothermal energy (**energy source** for heat pumps). Direct-use (high-temperature) geothermal energy is available at some. These are the prime forms of energy used to power or heat the improvements on a site.

Forms of On-Site Energy

Fossil fuels and electricity are commodities that are usually metered or measured for payment by the facility owner or operator. On the other hand, solar or wind, each of which might be considered a dispersed energy form in its natural state (i.e., requiring neither central processing nor a distribution network) costs nothing for the commodity itself, but does incur cost for the means to make use of it. High-temperature geothermal energy, which is not universally available, may or may not be a sold commodity, depending on the particular locale and local regulations. Chapter 31 of the 1999 *ASHRAE Handbook—Applications* has more information on geothermal energy.

Some prime on-site energy forms require further processing or conversion into other forms more directly suited for the particular systems and equipment needed in a building or facility. For instance, natural gas or oil is burned in a boiler to produce steam or hot water, a form of thermal energy which is then distributed to various use points (such as heating coils in air-handling systems, unit heaters, convectors, fin-tube elements, steam-powered cooling units, humidifiers, and kitchen equipment) throughout the building. Although electricity is not converted in form on-site, it is nevertheless used in a variety of ways, including lighting, running motors for fans and pumps, powering electronic equipment and office machinery, and space heating. While the methods and efficiencies with which these processes take place fall within the scope of the HVAC&R designer, the process by which a prime energy source arrives at a given facility site is not under direct control of the professional. On-site energy choices, if available, may be controlled by the designer based in part on the present and future availability of the associated resource commodities.

The basic energy source for heating may be natural gas, oil, coal, or electricity. Cooling may be produced by electricity, thermal energy, or natural gas. If electricity is generated on-site, the generator may be turned by an engine using natural gas or oil, or by a turbine using steam or gas directly.

The term **energy source** refers to on-site energy in the form in which it arrives at or occurs on a site (e.g., electricity, gas, oil, or coal). **Energy resource** refers to the raw energy, which (1) is extracted from the earth (wellhead or mine-mouth), (2) is used in the generation of the energy source delivered to a building site (coal used to generate electricity), or (3) occurs naturally and is available at a site (solar, wind, or geothermal energy).

Nonrenewable and Renewable Energy Resources

From the standpoint of energy conservation, energy resources may be classified in two broad categories: (1) nonrenewable (or discontinuous) resources, which have definite, although sometimes unknown, limitations; and (2) renewable (or continuous) resources, which can generally be freely used without depletion or have the potential to renew in a reasonable period. Resources used most in industrialized countries, both now and in the past, are nonrenewable (Gleeson 1951).

Nonrenewable resources of energy include

- Coal
- Crude oil
- Natural gas
- Uranium-235 (atomic energy)

Renewable resources of energy include

- Hydropower
- Solar
- Wind
- Earth heat (geothermal)
- Biomass (wood, wood wastes, and municipal solid waste)
- Tidal power

The preparation of this chapter is assigned to TC 1.10, Energy Resources.

- Ocean thermal
- Atmosphere or large body of water (as used by the heat pump)
- Crops (for alcohol production)

Characteristics of Fossil Fuels and Electricity

Most on-site energy for buildings in developed countries involves electricity and fossil fuels as the prime on-site energy sources. Both fossil fuels and electricity can be described in terms of their energy content (Btu). This implies that the two energy forms are comparable and that an equivalence can be established. In reality, however, fossil fuels and electricity are only comparable in energy terms when they are used to generate heat. Fossil fuels, for example, cannot directly drive motors or energize light bulbs. Conversely, electricity gives off heat as a byproduct regardless of whether it is used for running a motor or lighting a light bulb, and regardless of whether that heat is needed. Thus, electricity and fossil fuels have different characteristics, uses, and capabilities aside from any differences relating to their derivation.

Beyond the building site, further differences between these energy forms may be observed, such as methods of extraction, transformation, transportation, and delivery, and the characteristics of the resource itself. Natural gas arrives at the site in virtually the same form in which it was extracted from the earth. Oil is processed (distilled) before arriving at the site; having been extracted as crude oil, it arrives at a given site as, for example, No. 2 oil or diesel fuel. Electricity is created (converted) from a different energy form, often a fossil fuel, which itself may first be converted to a thermal form. The total electricity conversion, or generation, process includes energy losses governed largely by the laws of thermodynamics.

Fuel cells, which are used only on a small scale, convert a fossil fuel to electricity by chemical means.

Fossil fuels undergo a conversion process by combustion (oxidation) and heat transfer to thermal energy in the form of steam or hot water. The conversion equipment used is a boiler or a furnace in lieu of a generator, and conversion usually occurs on a project site rather than off-site. (District heating is an exception.) Inefficiencies of the fossil fuel conversion occur on-site, while the inefficiencies of most electricity generation occur off-site, before the electricity arrives at the building site. (Cogeneration is an exception.)

SUSTAINABILITY

As the world has increased in population and developed technologically, the consequences of uncontrolled growth are being recognized: pollution, toxic waste creation, waste disposal, global climate change, ozone depletion, deforestation, and resource depletion. Continuation of current trends without implementation of mitigation strategies will adversely impact the ability of the earth's ecosystem to regenerate and remain viable for future generations.

The built environment contributes significantly to these effects, accounting for one-sixth of the world's fresh water use, one-quarter of its wood harvest, and two-fifths of its material and energy flows. Other impacts include air quality, transportation patterns, and watersheds. The resources required to serve this sector are considerable—and many of them are diminishing (Gottfried 1996).

The building industry's recognition of the impacts of its activities is changing the way it approaches the design, construction, operation, maintenance, reuse, and demolition of what it creates—namely, toward addressing the environmental and long-term economic consequences of its actions. While this **sustainable design** ethic—or **sustainability**—covers things beyond the purview of the heating, ventilating, air-conditioning (HVAC) industry alone, design for the efficient use of energy resources would certainly be a key element of any sustainable design, and it is certainly very much under the control of the HVAC designer.

The following section of this chapter provides guidance in achieving a sustainable design.

DESIGNING FOR EFFECTIVE ENERGY RESOURCE UTILIZATION

The preponderance of energy used in buildings is from nonrenewable energy resources, the cost of which historically has not included considerations of replenishment or environmental impact. As a result, energy use resulting from many building system designs has been based primarily upon economic considerations, which unfortunately are biased to encourage more rather than less use.

As these resources become less readily available and practitioners look toward the use of more exotic and replenishable sources, the need to operate buildings effectively using less energy becomes paramount. Extensive study since the mid-1970s has revealed that significant reductions in building energy use can be achieved by the application of some fundamental principles.

THE ENERGY ETHIC: RESOURCE CONSERVATION DESIGN PRINCIPLES

The basic approach to achieving an energy-efficient design is reducing loads (power), improving transport systems, and providing efficient components and "intelligent" controls. Design concepts applicable to successful energy-efficient design include understanding the relationship between *energy* and *power*, maintaining *simplicity*, using *self-imposed budgets*, and applying *energy-smart design practices*.

Energy and Power

From the standpoint of economics, more energy-efficient systems need not be more expensive than less efficient systems; quite the opposite is true. This observation results from the simple relationship between energy and power in which power is simply the time rate of energy use (or, conversely, energy is power times time). Power, in turn, describes the size of something. For example, power terms such as horsepower, ton of refrigeration, Btu per hour, or kilowatt might be used in expressing the size of a motor, chiller, boiler, or transformer, respectively. Of course, the smaller something is, the less it costs; then, other things being equal, as the unit of a smaller size operates over time, it consumes less energy. Thus, in designing for energy efficiency, the first objective is always to reduce the power required to the bare minimum necessary to provide the desired performance—starting with the building's heating and cooling loads (a power term, in Btu/h) and working through the various systems and subsystems.

Simplicity

Complex designs to save energy will seldom function in the manner intended unless the systems are continually managed and operated by technically skilled individuals. Experience in designing energy-efficient systems has shown that achieving energy-efficient performance over a long period of time with a complex system is seldom achievable; further, when these complex systems are operated by minimally skilled individuals, not only are energy efficiencies not achieved but performance suffers as well. The majority of the techniques discussed subsequently can be implemented with a high degree of simplicity.

Self-Imposed Budgets

Just as an engineer must work to a cost budget with most designs, self-imposed power budgets can be similarly helpful in achieving an energy-efficient design. Examples of some budgets that designers have set for themselves for office buildings in a typical midwestern or northeastern temperate climate are

Installed lighting (overall)	1.3 W/ft^2
Space sensible cooling	20 Btu/h·ft^2
Space heating load	15 Btu/h·ft^2

Fan system pressure	4 in. of water
Air circulation rates	1 cfm/ft^2
Electric power (overall)	4.5 W/ft^2
Thermal power (overall)	30 Btu/h·ft^2
Hydronic system head	70 ft of water
Water chiller (water cooled)	0.6 kW/ton
Chilled water system auxiliaries	0.15 kW/ton
Unitary air-conditioning systems	1.0 kW/ton
Annual electric energy	25 kWh/ft^2·yr
Annual thermal energy	10 Btu/ft^2·yr·°F·day

Then, as the building and its systems are designed, all decisions become interactive as the result of each subsystem's power or energy peformance being continually compared to the "budget."

THE ENERGY-EFFICIENT DESIGN PROCESS

Energy efficiency should be considered at the beginning of the building design process because energy-efficient features are most easily and effectively incorporated at that time. Active participation of all members of the design team—including owner, architect, engineer, and often the contractor—should be sought early in the design process. Consider building attributes such as building function, form, orientation, window/wall ratio, and HVAC system types early as each has major energy implications.

Address a building's energy requirements in the following sequence:

1. **Minimize the impact of the building's functional requirements** by analyzing how the building relates to its external environment. Advocate changes in building form, aspect ratio, and other attributes that reduce, redistribute, or delay (shift) loads. The load calculation should be interactive so that the impact of those factors can be seen immediately.
2. **Minimize loads** by analyzing the external and internal loads imposed on the building energy-using subsystems, both for peak-load and part-load conditions.
3. **Maximize subsystem efficiency** by analyzing the diversified energy and power requirements of each energy-using subsystem serving the building's functional requirements. Consider static and dynamic efficiencies of energy conversion and energy transport subsystems, and consider opportunities to reclaim, redistribute, and store energy for later use.
4. **Study alternate ways to integrate subsystems** into the building by considering both power and time components of energy use. Identify, evaluate, and design each of these components to control overall design energy consumption. The following should be considered when integrating major building subsystems:
    - Address more than one problem at time when developing design solutions, and make maximum use of building components incorporated for nonenergy reasons (e.g., windows, structural mass).
    - Examine design solutions that consider time (i.e., *when* the energy use takes place) since sufficient energy may already be present from the environment (e.g., solar heat, night cooling) or from internal equipment (e.g., lights, computers) but available at times different from when needed. Thus, active (e.g., heat pumps with water tanks) and passive (e.g., building mass) storage techniques may be considered.
    - Examine design solutions that consider the anticipated use of space. For example, in large but relatively unoccupied spaces, task or zone lighting may be considered. Transporting excess energy (light and heat) from locations of production and availability to locations of need may be considered instead of purchasing additional energy.
    - Never reject waste energy at temperatures usable for space conditioning or other practical purposes without calculating the economic benefit of energy recovery or treatment for reuse.
    - Consider or advocate design solutions that provide more comfortable surface temperatures or increase the availability of controlled daylight in buildings where human occupancy is a primary function.
    - Use easily understood design solutions as they have a greater probability of use by building operators and occupants.
    - Where the functional requirements of a building are known to be likely to change over time, design the installed environmental system to adapt to meet those changes that can be anticipated as well as to provide flexibility in meeting future changes in use, occupancy, or other functions.

BUILDING ENERGY USE ELEMENTS

Envelope

- Control thermal conductivity through the use of insulation (including movable insulation), thermal mass, and/or phase change thermal storage at levels that minimize net heating and cooling loads on a time-integrated (annual) basis.
- Minimize unintentional or uncontrolled thermal bridges, and consider them in energy-related calculations because they can radically alter the conductivity of a building envelope. Examples include wall studs, balconies, ledges, and extensions of building slabs.
- Minimize infiltration so that it approaches zero. (An exception is when infiltration provides the sole means of ventilation such as in small residential units.) This effort will minimize fan energy consumption in pressurized buildings during occupied periods and minimize heat loss (or unwanted heat gain in warm climates) during unoccupied periods. An additional benefit of a tight envelope in warm humid climates is that it improves indoor air quality. Infiltration should be reduced through design details that enhance the fit and integrity of building envelope joints in ways that may be readily achieved during construction. This includes infiltration control by caulking, weatherstripping, vestibule doors, and/or revolving doors—with construction meeting accepted specifications.
- Consider operable windows to allow occupant-controlled ventilation. If this is done, the design of the building's mechanical system must be carefully executed to minimize unnecessary HVAC energy consumption, and building operators and occupants should be cautioned about how operable windows may be improperly used.
- Strive to maintain occupant radiant comfort regardless of whether the building envelope is designed to be a static or dynamic membrane. Opaque surfaces should be designed so that the average inside surface temperatures remain within 5°F of room temperature in the coldest anticipated weather (i.e., winter design conditions) and so that the coldest inside surface will remain within 25°F of room temperature (but always above the indoor dew point). In a building with time-varying internal heat generation, consider thermal mass for controlling radiant comfort. In the perimeter zone, thermal mass is more effective when it is positioned inside the envelope's insulation.
- Effective control of solar radiation is critical to the design of energy-efficient buildings due to the high level of internal heat production already present in most commercial buildings. In some climates, lighting energy consumption savings due to daylighting techniques can be greater than the heating and cooling energy penalties that result from additional glazed surface area required, provided that the building envelope is properly designed for daylighting and that lighting controls are installed and used. (In other climates, such net savings may not be realized.) Daylighting designs are most effective if direct solar beam radiation is not allowed to cause glare in building spaces.
- Design the transparent portions of the building envelope to prevent solar radiant gain above that necessary for effective daylighting

and solar heating. On south-facing facades, the use of low shading coefficients is generally not as effective as external physical shading devices in achieving this balance. Low-emissivity, high-visible-transmittance glazings may be considered for effective control of radiant heat gains and losses. For shading control, designers may consider the judicious use of vegetation to block excess gain year-round or seasonally, depending on the plant species chosen.

Lighting

Lighting is both a major energy end use in commercial buildings (especially office buildings) and a major contributor to internal loads by increasing cooling loads and decreasing heating loads. Therefore, it is important to produce (or advocate) a design that meets the lighting functional criteria of the space as well as one that minimizes energy use. The *IESNA Lighting Handbook* (IESNA 2000) recommends illuminance levels for visual tasks and surrounding lighted areas. Principles of energy-conserving design within that context are described as follows:

- Energy use is determined by the lighting load (demand power) and its duration of use (time). Minimize the actual demand load rather than just the apparent connected load. Control the load rather than just area switching, if switching may adversely affect the quality of the luminous environment.
- Consider daylighting along with the proper use of controls to reduce costs of electric lighting. Design should be sensitive to window glare, sudden changes in luminances, and general user acceptance of daylighting controls. Window treatment (blinds, drapes, and shades) and glazing should be carefully selected to control direct solar penetration and luminance extremes while still maintaining the view and daylight penetration.
- Design the lighting system so that illumination required for tasks is primarily limited to the location of the task and comes from a direction that minimizes direct glare and veiling reflections on the task. When the design concept is based on nonuniform illuminance, walls should be a light to medium color or otherwise illuminated to provide visual comfort. In densely occupied work spaces, uniform distribution of general lighting may be most appropriate. Where necessary, provide supplementary task illumination. General ambient illumination should not be lower than a third of the luminance required for the task; this will help maintain visually comfortable luminance ratios.
- Use local task lighting to accommodate the need for higher lighting levels due to task visual difficulty, glare, intermittently changing requirements, or individual visual differences (poor or aging eyesight).
- Group similar activities so that high illuminance or special lighting for particular tasks can be localized in certain rooms or areas, and so that less efficient fixtures required for critical glare control do not have to be installed uniformly when they are only required sparsely.
- Use lighting controls throughout so lighting is available when and where it is needed, but not wasted during those times when tasks being performed are less critical or spaces are not fully occupied. Also consider user acceptance of control strategies to maximize energy saving.
- Limit use of lower efficiency lamps (such as incandescent) to those applications where their color, lumens, or distribution characteristics cannot be duplicated by other sources. Due to their lower efficiency, limit the use of extended service incandescent lamps to those applications where fixtures are difficult to reach and/or maintenance costs for replacing lamps would be excessive.
- Continue carrying through the lighting design process as the building's interior design is occurring. Reduced light absorption may be achieved by using lighter finishes, particularly on ceilings, walls, and partitions.

Other Loads

- Minimize the thermal impact of equipment and appliances on HVAC systems by the use of hoods, radiation shields, or other confining techniques, and by using controls to ensure that such equipment is turned off when not needed. In addition, locate, where practical, major heat-generating equipment where it can balance other heat losses. Computer centers or kitchen areas usually have separate, dedicated HVAC equipment. In addition, consider heat recovery for this equipment.
- Use storage techniques to level or distribute loads that vary on a time or spatial basis to allow operation of a device at maximum (often full-load) efficiency.

HVAC System Design

- Consider separate HVAC systems to serve areas expected to operate on widely differing operating schedules or design conditions. For instance, systems serving office areas should generally be separate from those serving retail areas.
- Arrange systems so that spaces with relatively constant and weather-independent loads are served with systems separate from those serving perimeter spaces. Areas with special temperature or humidity requirements, such as computer rooms, should be served by systems separate from those serving areas that require comfort heating and cooling only. Alternatively, these areas may be served by supplementary or auxiliary systems.
- Sequence the supply of zone cooling and heating to prevent the simultaneous operation of heating and cooling systems for the same space to the extent possible. Where this is not possible due to ventilation, humidity control, or air circulation requirements, reduce air quantities as much as possible before incorporating reheating, recooling, or mixing hot and cold airstreams. As an example, if reheat is necessary to provide dehumidification and prevent overcooling, *only* the ventilation air need be so treated rather than the entire recirculated air quantity. Finally, reset the supply air temperature up to the extent possible to reduce reheating, recooling, or mixing losses.
- Provide controls to allow operation in an occupied mode and an unoccupied mode. In the occupied mode, controls may provide for a gradually changing control point as system demands change from cooling to heating. In the unoccupied mode, ventilation and exhaust systems should be shut off if possible, and comfort heating and cooling systems should be shut off except to maintain space conditions ready for the next occupancy cycle.
- In geographical areas where diurnal temperature swings and humidity levels permit, consider the judicious coupling of air distribution and building structural mass to allow nighttime cooling to reduce the requirement for daytime mechanical cooling.
- High ventilation rates, where required for special applications, can impose enormous heating and cooling loads on HVAC equipment. In these cases, consider recirculation of filtered and cleaned air to the extent possible rather than 100% outside air. Also, consider preheating outside air with reclaimed heat from other sources.

HVAC Equipment Selection

- To allow for HVAC equipment operation at the highest efficiencies, match conversion devices to load increments, and sequence the operation of modules. Oversized or large-scale systems should never serve small seasonal loads (e.g., a large heating boiler serving a summer service water-heated load). Include specific low-load units and auxiliaries where prolonged use at minimal capacities is expected.
- Select the most efficient (or highest COP) equipment practical at both design and reduced capacity (part-load) operating conditions.

- In the selection of large power devices such as chillers (including their auxiliary energy burdens), seriously consider life-cycle purchasing techniques.
- Keep fluid temperatures for heating equipment devices as low as practical and for cooling equipment as high as practical, while still meeting loads and minimizing flow quantities.

Energy Transport Systems

Energy should be transported by the most energy-efficient means possible. The following options are listed in order of efficiency from the lowest energy transport burden (most efficient) to the highest (least efficient):

1. Electric wire or fuel pipe
2. Two-phase fluid pipe (steam or refrigerant)
3. Single-phase liquid/fluid pipe (water, glycol, etc.)
4. Air duct

Select a distribution system that complements other parameters such as control strategies, storage capabilities, conversion efficiency, and utilization efficiency.

The following are specific design techniques that may be applied to thermal energy transport systems:

Steam Systems

- Include provisions for seasonal or non-use shutdown.
- Minimize the venting of steam and ingestion of air with the design directed toward full-vapor performance.
- Avoid subcooling, if practical.
- Return condensate to boilers or source devices at the highest possible temperature.

Hydronic Systems

- Minimize flow quantity by designing for the maximum practical temperature range (Δt).
- Vary flow quantity with load where possible.
- Design for the lowest practical pressure rise (or drop).
- Provide OPERATING and IDLE control modes.
- When locating equipment, identify the critical pressure path and size runs for the minimum reasonable pressure drop.

Air Systems

- Minimize airflow by careful load analysis and an effective distribution system. If the application allows, the supply air quantity should vary with the sensible load (i.e., VAV systems). Hold the fan pressure requirement to the lowest practical value and avoid using fan pressure as a source for control power.
- Provide NORMAL and IDLE control modes for the fan systems as well as the psychrometric systems.
- Keep duct runs as short as possible, and keep runs on the critical pressure path sized for minimum practical pressure drop.

Power Distribution

- Size transformers and generating units as close as possible to the actual anticipated load (i.e., avoid oversizing to minimize fixed thermal losses).
- Consider distribution of electric power at the highest practical voltage and load selection at the maximum power factor consistent with safety.
- Consider tenant submetering in commercial and multifamily buildings as a cost-effective energy conservation measure. (A large portion of energy use in tenant facilities occurs simply because there is no economic incentive to conserve.)

Domestic Hot Water Systems

- Choose shower heads that provide and maintain user comfort and energy savings. They should not have removable flow-restricting inserts to meet flow limitation requirements.
- Consider point-of-use water heaters where their use will reduce energy consumption and annual energy cost.
- Consider using storage facilitate heat recovery when the heat to be recovered is out of phase with the demand for hot water or when energy use for water heating can be shifted to take advantage of off-peak rates.

Controls

- A well-designed digital control system provides information to managers and operators as well as to the data processor that serves as the intelligent controller. Include the energy saving concepts discussed previously throughout the operating sequences and control logic. However, energy conservation should not be sought at the expense of inadequate performance; in a well-designed system, these two parameters are compatible.

ON-SITE ENERGY/ENERGY RESOURCE RELATIONSHIPS

An HVAC&R designer sooner or later must consider the use of one or more forms of prime energy. Most likely, these would be fossil fuels and electricity, although installations are sometimes designed using a single energy source (e.g., only a fossil fuel or only electricity).

Solar energy normally impinges on the site (and on the facilities to be put there), so it has an impact on the energy consumption of the facility. The designer must account for this impact and may have to decide whether to make active use of solar energy. Other naturally occurring and distributed renewable forms such as wind power and earth heat (if available) might also be considered.

The designer should be aware of the relationship between on-site energy sources and raw energy resources—including how these resources are used and what they are used for. The relationship between energy sources and energy resources involves two parts: (1) quantifying the energy resource units expended and (2) considering the societal impact of the depletion of one energy resource (caused by on-site energy use) with respect to others. The following two sections describe those parts in more specific terms.

Quantifiable Relationships

As on-site energy sources are consumed, a corresponding amount of resources are consumed to produce that on-site energy. For instance, for every 1000 gal of No. 2 oil consumed by a boiler at a building site, some greater volume of crude oil is extracted from the earth. On leaving the well, the crude oil is transported and processed into its final form, perhaps stored, and then transported to the site where it will be used.

Even though gas requires no significant processing, it is transported, often over long distances, to reach its final destination, which causes some energy loss. Electricity may have as its raw energy resource a fossil fuel, uranium, or an elevated body of water (hydroelectric generating plant).

Data to assist in determining the amount of resource use per delivered on-site energy source unit are available. In the United States, data are available from entities within the U.S. Department of Energy and from the agencies and associations listed at the end of this chapter.

A **resource utilization factor** (RUF) is the ratio of resources consumed to energy delivered (for each form of energy) to a building site. Specific RUFs may be determined for various energy sources normally consumed on-site, including nonrenewable sources such as coal, gas, oil, and electricity, and renewable sources such as solar, geothermal, waste, and wood energy. With electricity, which may derive from several resources depending on the particular fuel mix of the generating stations in the region served, the overall RUF is the weighted combination of individual factors applicable to electricity and a particular energy resource. Grumman (1984) gives specific formulas for calculating RUFs.

While a designer is usually not required to determine the amount of energy resources attributable to a given building or building site for its design or operation, this information may be helpful when assessing the long-range availability of energy for a building or the building's impact on energy resources. Factors or fuel-quantity-to-energy resource ratios or factors are often used, which suggests that energy resources are of concern to the HVAC&R industry.

Intangible Relationships

Energy resources should not simply be converted into common energy units [e.g., quadrillion (10^{15}) Btu or quad] because the commonality gives a misleading picture of the equivalence of these resources. Other differences and limitations of each of the resources defy easy quantification but are nonetheless real. For instance, electricity that arrives and is used on a site can be generated from coal, oil, natural gas, uranium, or hydropower. The end result is the same: electricity at x kV, y Hz. However, is a kilowatt-hour of electricity generated by hydropower equal in societal impact to that same kilowatt-hour generated by coal, by uranium, by domestic oil, or by imported oil?

Intangible factors such as safety, environmental acceptability, availability, and national interest also are affected in different ways by the consumption of each resource. Heiman (1984) proposes a procedure for weighting the following intangible factors:

National/Global Considerations

- Balance of trade
- Environmental impacts
- International policy
- Employment
- Minority employment
- Availability of supply
- Alternative uses
- National defense
- Domestic policy
- Effect on capital markets

Local Considerations

- Exterior environmental impact—air
- Exterior environmental impact—solid waste
- Exterior environmental impact—water resources
- Local employment
- Local balance of trade
- Use of distribution infrastructure
- Local energy independence
- Land use
- Exterior safety

Site Considerations

- Reliability of supply
- Indoor air quality
- Aesthetics
- Interior safety
- Anticipated changes in energy resource prices

SUMMARY

In designing HVAC&R systems, the need to address immediate issues such as economics, performance, and space constraints often prevents designers from fully considering the energy resources affected. Today's energy resources are less certain because of issues such as availability, safety, national interest, environmental concerns, and the world political situation. As a result, the reliability, economics, and continuity of many common energy resources over the potential life of a building being designed are unclear. For this reason, the designer of building energy systems must consider the energy resources on which the long-term operation of the building will depend. If the continued viability of those resources is reason for concern, the design should provide for, account for, or address such an eventuality.

OVERVIEW OF GLOBAL ENERGY RESOURCES

WORLD ENERGY RESOURCES

Production

Energy production trends for the world, by leading producers and world areas, from 1989 to 1998, are shown in Figure 1. World primary energy production, following essentially no increase in the

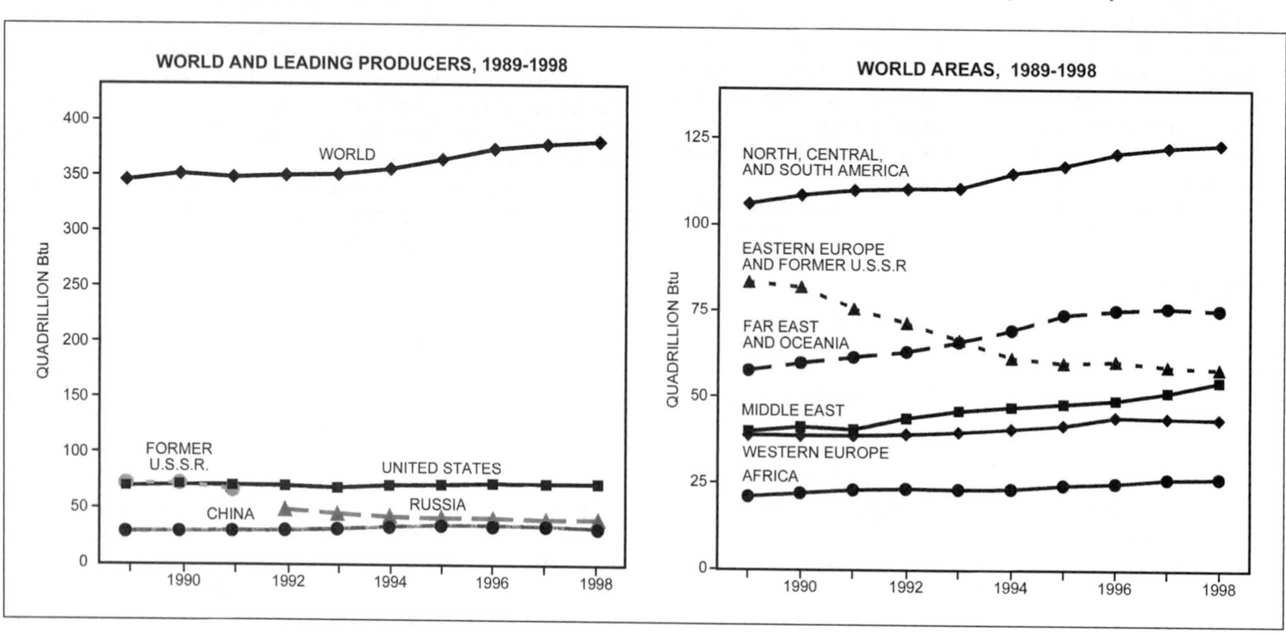

Fig. 1 World Primary Energy Production
(Basis: Energy Information Administration/*Annual Energy Review 1999*)

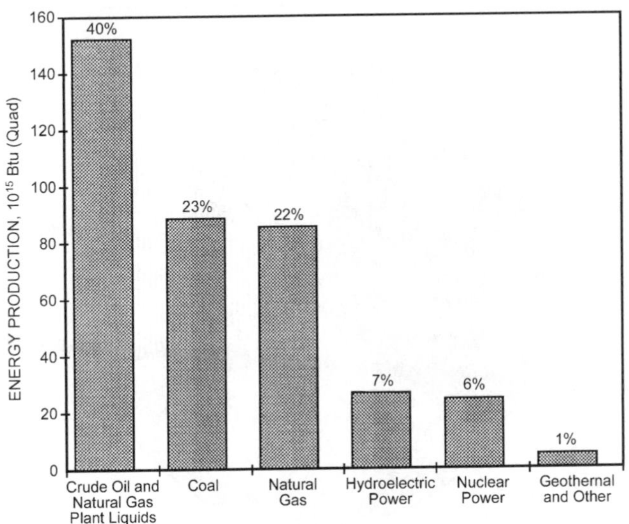

Fig. 2 World Primary Energy Production by Resource: 1998

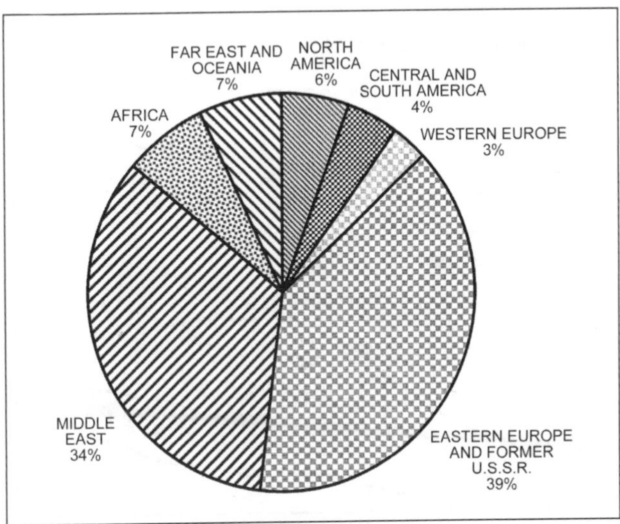

Fig. 4 World Natural Gas Reserves: January 1, 1999
(Basis: *Oil and Gas Journal*)

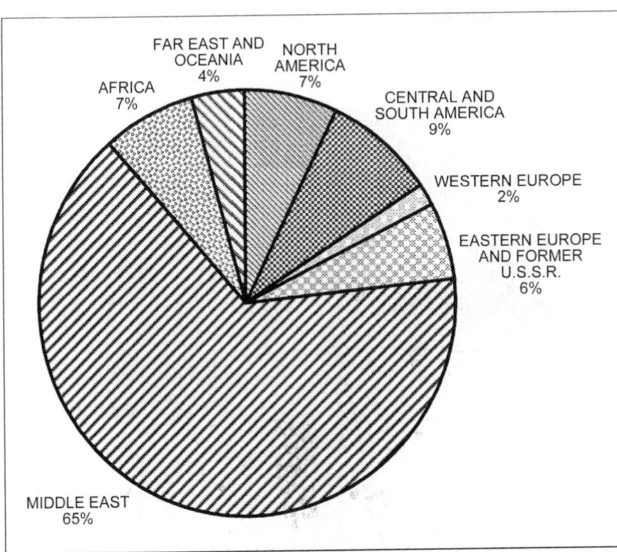

Fig. 3 World Crude Oil Reserves: January 1, 1999
(Basis: *Oil and Gas Journal*)

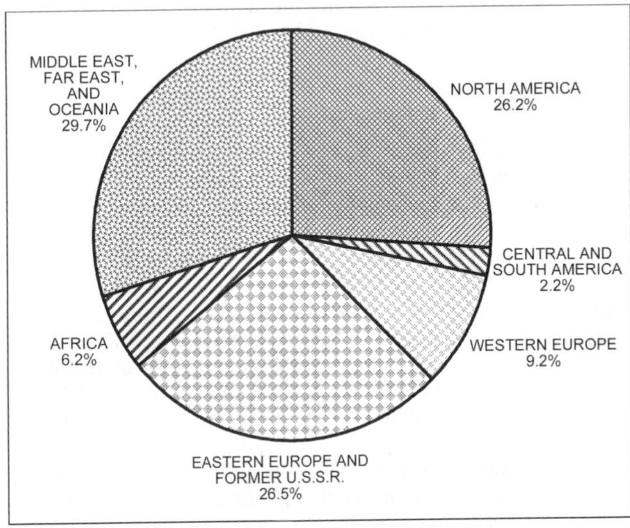

Fig. 5 World Recoverable Coal Reserves: December 31, 1996 and January 1, 1999 (U.S. Only)
(Basis: Energy Information Administration)

early 1990s, has risen subsequently at about 1.7% per year. The largest total energy producers in 1998 were the United States (20%), Russia (11%), and China (9%). Saudi Arabia (5%+) was the world's fourth largest producer. Together these producers accounted for almost half of the world's energy. Total world energy production by resource type is shown in Figure 2.

Crude Oil. World crude oil production was 66 million barrels per day in 1999—up 18% since 1973. The biggest crude oil producers in 1999 were the eight nations comprising the Organization of Petroleum Exporting Countries (OPEC) at 42% (including Saudi Arabia at 12%+), Russia (9%), and the United States (9%). Oil production declined most noticeably since 1992 in Russia (20%) and in the United States (17%). The other primary non-OPEC producers were China, Mexico, Norway, the United Kingdom, and Canada.

Natural Gas. World production reached 83×10^{12} ft^3 in 1998—up 15% from the 1989 level. The biggest producers in 1998 were Russia (25%) and the United States (23%).

Coal. At 5.0×10^9 short tons in 1998, coal production was lower by 0.3% than in 1989 and comprised 23% of the world's energy production. The leading producers of coal were China (27%), the United States (22%), India (7%), Russia (5%+), Germany (5%), and Poland (4%).

Reserves

On January 1, 1999, the estimated world reserves of crude oil and gas were distributed by world region as shown in Figures 3 and 4. Saudi Arabia was estimated to have 39% of the Middle Eastern crude oil reserves. Iraq, the United Arab Emirates, Kuwait, and Iran were each estimated to have more crude oil reserves than any world region outside the Middle East. Outside the Middle East, the biggest reserves were estimated to be in Venezuela, Russia, and Mexico. The country with the largest gas reserves by far was Russia.

World coal reserves as of January 1, 1996, are shown by world area in Figure 5 (note that the U.S. data are as of January 1, 1999). The countries with the most plentiful reserves, as a percent of total, were Russia (16%), the United States (25%), China (12%), and Australia (9%).

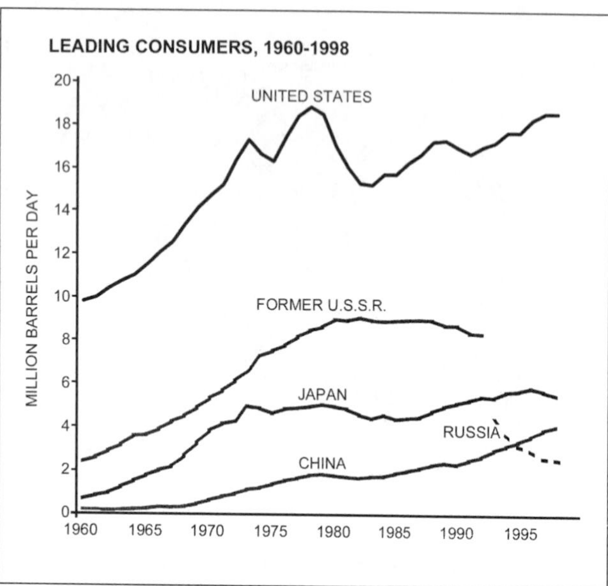

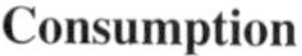

Fig. 6 World Petroleum Consumption: 1960-1998
(Basis: Energy Information Administration/*Annual Energy Review 1999*)

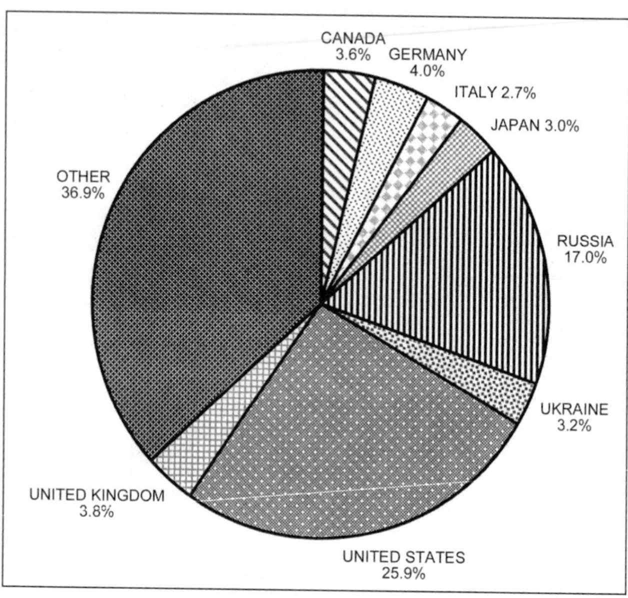

Fig. 7 World Natural Gas Consumption: 1998
(Basis: Energy Information Administration)

Consumption

Data on world energy consumption are available only by type of resource rather than by total energy consumed.

Petroleum. The consumption trends of the leading consumers from 1960 to 1998 are depicted in Figure 6. In 1998, the United States consumed far more than any other country—26% of world consumption and 40% of the consumption of Organization for Economic Cooperation and Development (OECD) countries. By contrast, Japan, the second largest consumer among OECD countries, consumed just 7.5% of the world total and 12% of that of the OECD countries. Of the non-OECD countries, Russia and China were the biggest consumers (3% and 6%, respectively, of world consumption).

Natural Gas. Transport pipelines notwithstanding, this energy resource tends to be consumed close to the site of production; and indeed, in 1998 the two biggest natural gas producers were also the two biggest consumers. Figure 7 depicts natural gas consumption by the leading consumer countries as a percentage of world consumption. Of the major consumers, the United States consumed more than it produced (114%), and the former U.S.S.R. consumed less (67%). Germany and the United Kingdom, the third and fourth largest consumers, produced very little. Canadian consumption was 49% of its production. World consumption of natural gas increased 55% between 1980 and 1998, with the three major gas-consuming states of the former U.S.S.R. (Russia, Ukraine, and Uzbekistan) up 35% and the United States up 7%. After the United States and Russia, no single country consumed more than 5% of the world total.

Coal. Here, the three largest producers were also the three largest consumers. Figure 8 depicts the percentage of world consumption by the leading consumers during 1998. Since 1980, world coal consumption has increased 21%, with most of that increase in the early 1980s. Over the same period, consumption by China increased 93%, the United States 48%, India 185%, and Japan 43%. Significant drops occurred in Germany, Poland, and Russia.

Electricity. Figure 9 shows the world's electricity generation by energy resource in 1998. Figure 10 shows installed capacity for the same resources at the beginning of 1998. Both net generation and installed capacity were dominated by the United States (26%+ and 25%, respectively). Comparable figures for the next largest are China (8% for both), Japan (7% and 7+%), and Russia (6% and 7%). China, which has grown almost four times in net electric generating capacity since 1980, has surpassed Japan in the last few years.

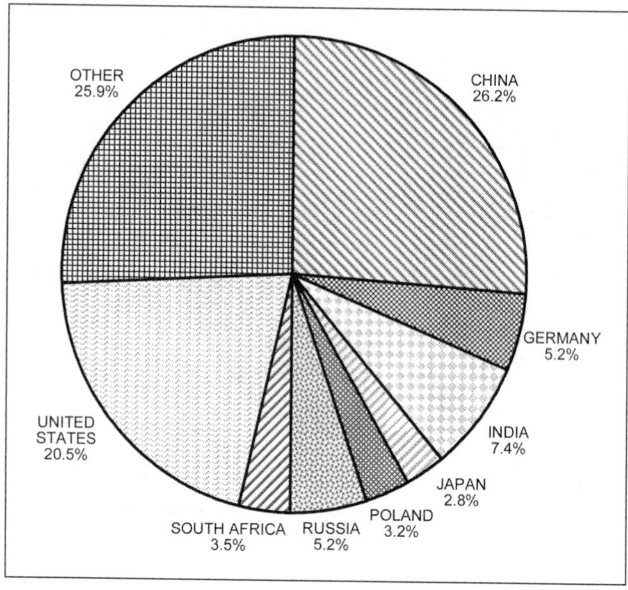

Fig. 8 World Coal Consumption: 1998
(Basis: Energy Information Administration)

Electricity generated by hydroelectric means increased in the world by 47% between 1980 and 1998, with the largest increase occurring in Central and South America. The top countries for hydroelectric generation in 1998 were the United States, Canada, and Brazil—collectively accounting for 37% of the world total quantity of electricity generated by that means.

Total world electricity generation from nuclear resources increased 239% between 1980 and 1998, with higher-than-average increases occurring in the Far East/Oceania, Western Europe, and Africa. The top-generating countries in 1998 were the United States (27% of world total), China (8%), and Japan (7%).

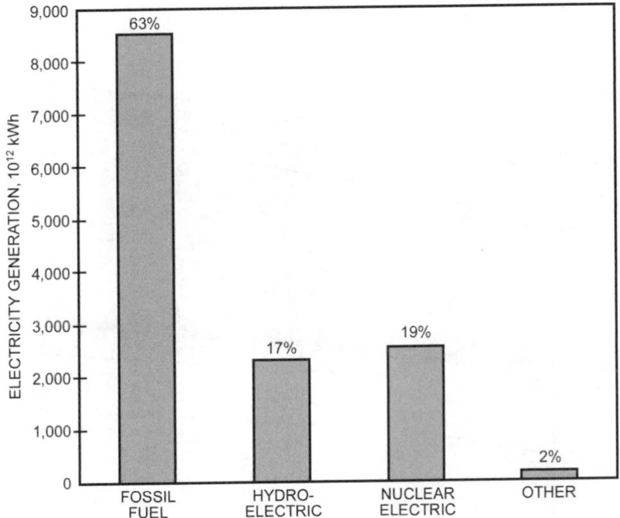

Fig. 9 World Electricity Generation by Resource: 1998

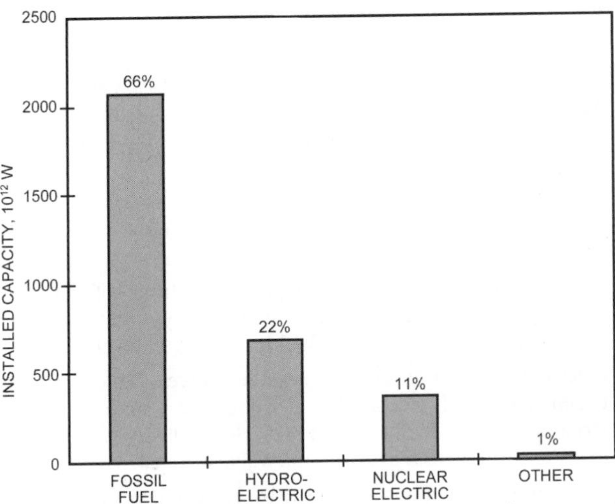

Fig. 10 World Installed Electricity Generation Capacity by Resource: January 1, 1998

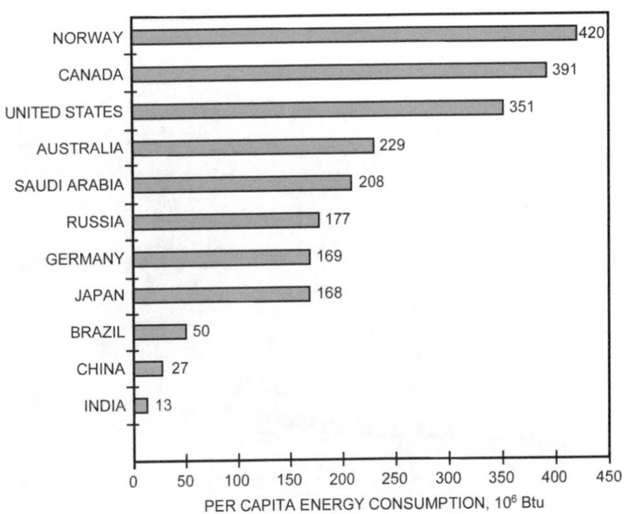

Fig. 11 Per Capita Energy Consumption by Selected Developed Countries in 1998

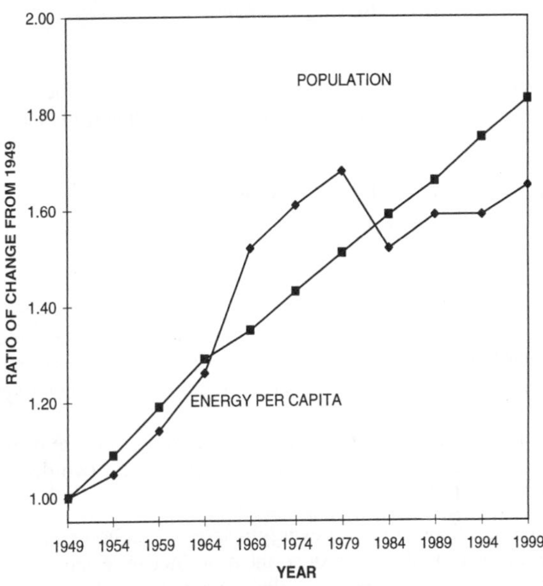

Fig. 12 Per Capita End-Use Energy Consumption Trends in the United States

Per Capita. Figure 11 compares the per capita commercial energy consumption of selected developed countries for 1998. As is apparent, the per capita energy consumption in cold-climate countries tends to be highest; also, the level in more developed countries is vastly different from that in less developed countries and differs considerably even among the more developed countries.

UNITED STATES ENERGY USE

Per Capita Energy Consumption

Figure 12, which is based on EIA (1999), presents a capsule overview of past U.S. energy use intensity by relating the growth in per capita energy use since 1949 to population growth. The growth in per capita energy use has varied significantly from the population growth rate. The 1960s experienced a sharp increase in the per capita energy use growth rate, which leveled off during the 1970s due to higher energy prices and the emphasis on energy conservation. [No change in the intensity of energy use would result in a flat (horizontal) energy-per-capita curve]. In the early 1980s, however, a significant drop in the per capita energy use growth rate occurred as industrial output decreased, efficiency of use improved, and global economic pressures mounted. The period since the mid-1980s shows a resumption of per capita energy use growth at a rate at first paralleling that of the population.

The *Annual Energy Outlook* is the basic source of data for projecting the use of energy in the United States (EIA 2000). Figure 13 and Figure 14 are summaries of data from this source.

EIA (2000) presents forecasts of energy trends that are based on macroeconomic growth scenarios prepared by several outside sources. EIA has also added a low and a high world oil price scenario to its mid-growth economic scenario. Thus, the National Energy Modeling System (NEMS), which EIA uses for these forecasts, has yielded three scenarios. Figure 13 and Figure 14 present the baseline or reference case. It assumes average annual growth of the real gross domestic product (GDP) at 2.2%, of the labor force at 0.9%, and of productivity at 1.3%. The forecast, in order to be

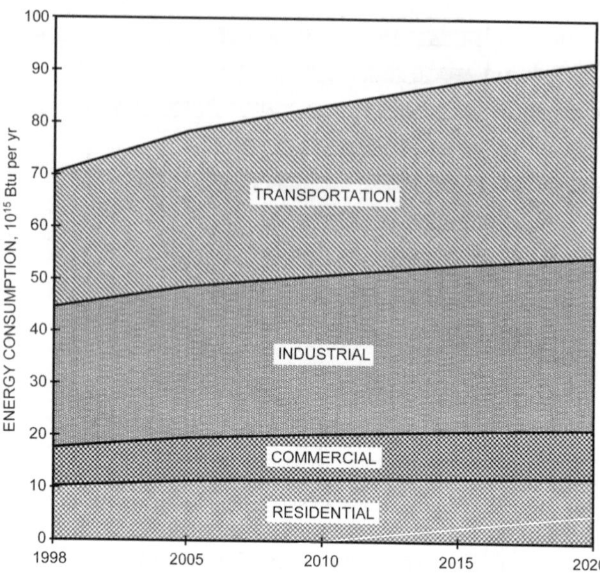

Fig. 13 Projected Total U.S. Energy Consumption by End-Use Sector

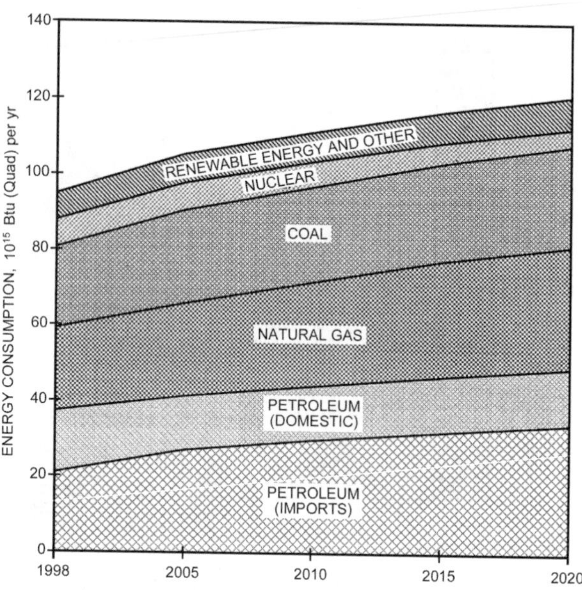

Fig. 14 Projected Total U.S. Energy Consumption by Resource

policy-neutral, also assumes that all federal, state, and local laws and regulations in effect as of July 1, 1999, remain unchanged through 2020.

Projected Overall Energy Consumption

Figure 13 shows energy use by end-use sector, with the major end-use sectors being residential, commercial, industrial, and transportation. HVAC&R engineers are primarily concerned with the first three sectors. Figure 14 shows energy consumption by type of resource. Figure 13 shows less total energy consumption than Figure 14 primarily because it excludes the thermodynamic losses of electricity generation and the processing and delivery burdens of various energy forms.

The following observations apply to the overall picture of projected energy use in the United States over the next two decades (Figures 13 and 14):

- Although a major issue in energy markets is carbon emissions, impacts of programs to reduce them are not reflected in these forecasts because no specific policies for carbon reductions have been enacted.
- Given the above premise, carbon emissions from energy use will increase by an average of 1.3% per year through 2020 due to rising energy demand, declining nuclear power generation, improvements in efficiency, and slow growth in the use of renewable sources. The 2020 level of carbon emissions will thus be almost 33% higher than the 1998 levels.
- Though average crude oil prices are expected to remain low over the next several years, unanticipated changes in supply or political events could change that scenario.
- The wellhead price of natural gas is expected to increase at an average annual rate of 1.7% up to 2020, reflecting exploration and production improvements balanced by greater demand.
- The price of coal, on the other hand, is expected to decline by 28% over the same period as a result of better productivity, more low-cost western coal production, and competitive labor pressures.
- Ongoing changes in the financial structure of the electricity industry and resulting cost reductions due to increased competition are reflected. Transitions to retail competitive pricing have been assumed in areas where legislation has been enacted, and stranded cost recovery will be phased out by 2008.
- Electricity prices will decline an average of 19% per kilowatt-hour between 1997 and 2020 due to lower operating, maintenance, and administrative costs resulting from industry restructuring.
- More than 40% of the nuclear electricity generating capacity existing in 1998 will be retired by 2020 as operating licenses expire with no new plants having been built.
- Offsetting this decline, electricity generation from both coal and natural gas is projected to increase significantly, with the share of natural gas in 2020 more than doubling the 1998 level.
- Electricity generation using renewable sources (which includes cogenerators) will increase by only 0.5% per year, held back in large measure by the fact that the electric industry restructuring tends to favor natural gas over renewables or coal.
- Petroleum consumption will grow by 1.3% annually, led by the transportation sector, which is where most of it (70%) is used.
- The share of petroleum consumption met by net imports will reach 64% in 2020. (This is up from 52% in 1998.) Over this period, U.S. crude oil production will decline at an annual rate of 0.8%, with declining resources overshadowing improvements in exploration and processing technologies.
- Natural gas consumption will increase by 1.8% per year, with demand increases in all sectors and led by its use for electricity generation.
- Coal consumption will increase at an average annual rate of 0.9%. Most of it (90%) will be used for electricity generation, but its share for that purpose will decline by 2020.
- Consumption of renewable energy (including using ethanol in gasoline) will increase by only 0.8% per year, however, with about 60% going toward electricity generation.
- Electricity demand will grow by 1.4% annually, with efficiency gains offset by growth in the use of new electricity-using equipment.
- Total energy demand in the commercial and residential sectors will grow at 0.9% and 0.8% per year, respectively. This results from greater use of computers, telecommunications, and other office appliances, but it is offset by somewhat improved building and building equipment efficiencies.

- Energy use by the transportation sector will grow at an average of 1.7% per year, with variations from this average very dependent on prevailing fuel prices.
- Per capita energy use will remain stable through 2020 as overall efficiency gains offset higher per capita demand.
- Total energy use per dollar of gross domestic product (energy intensity), however, will continue to fall at an average rate of about 1.1% per year through 2020.

Outlook Summary

In general, the following key issues will dominate energy matters in the next two decades:

- Rising dependency of the United States on imported oil
- Expected retirement of more than half of the existing nuclear power plants
- Role of technology developments, including energy conservation and energy efficiency as alternatives to energy production
- Deregulation of the utility industry and the growth of independent power producers
- Dealing with the increasing rate of carbon emissions, especially in light of international agreements with, and moral pressure from, others of the global community
- Growth of population, coupled with the shift of "baby boomers" into retirement

ENERGY RESOURCE PLANNING

The energy supplier (or suppliers) in a particular jurisdiction must plan for the future energy needs of that jurisdiction. For competitive energy markets where these decisions do not have high societal costs, these plans are made by the energy suppliers as part of their business. These decisions are not revealed to governmental authorities or the public more than is absolutely necessary, because significant competitive advantages could be gained by knowledge of a competitor's future plans. For electricity (and to a lesser extent natural gas) significant societal issues are involved in energy resource planning decisions, and these decisions cannot be made by energy suppliers without approval by many different groups and entities. These issues include:

- Reliability, which is affected by the diversity of supply sources available. This would include the number of geographic supply sources and pipelines in the case of gas and would include the percentage of generation from various fuel sources in the case of electricity. Due consideration should be given to the projected future supply and reliability of energy resources, including the possibility of supply disruption due to natural or political events, and to the likelihood of future supply shortages, which could reduce reliability.
- Reserve margins, or the ratio of total supply sources to expected peak supply source needs. Reserve levels that are too high result in the waste of resources, higher environmental costs, and possibly poor financial health of the energy suppliers. Reserves that are too low result in volatile and very high peak energy prices and reduced reliability.
- The production and transmission of energy often requires governmental cooperation to condemn private property for use in energy production and transmission facilities. The construction and maintenance of this property are also regulated due to wetlands, prevention of toxic waste releases, and other environmental issues.

Integrated Resource Planning

Integrated resource planning (IRP) is a process commonly used for the planning of significant new energy facilities. The steps in this planning process include: (1) forecasting the amount of new resources needed and (2) determining the type of resource to be provided and who will provide this resource. Traditionally, the local utility provider forecasts the future needs of a given energy resource, then either builds the necessary facility with the approval of regulators or uses a standard offer bid to determine what non-utility provider (or the utility itself) would provide the new energy resource. However, with increasing deregulation of energy providers, a single energy provider may no longer be responsible for maintaining an adequate supply. This forecasting task may be provided by a regulatory body, or it could be by a consensus decision of the various competing energy suppliers, with oversight and approval of regulators.

Supplying new energy resources through either a standard bid process by a supplier or traditional utility regulation will usually result in the selection of the lowest cost supply option, without regard for environmental costs or other societal needs. IRP allows for the choice of a greater variety of resource options and allows environmental costs and other indirect costs to society to be given greater consideration.

IRP addresses a wider population of stakeholders then do most other planning processes. Many regulatory agencies involve the public in the formulation and review of integrated resource plans. Customers, environmentalists, and other public interest groups are often prominent in these proceedings.

Prior to the deregulation of energy suppliers, **demand-side management** (DSM) was a common option for providing new energy resources, especially for electricity. These are actions taken to reduce the demand for energy, rather than increase the supply of energy. DSM is desirable because the environmental costs of DSM are almost always lower than the environmental costs of building new energy facilities. However, several factors have caused a decline in the number of DSM programs:

- Lower cost supply options, especially for natural gas-fired generation, make it more difficult to implement DSM options that do not result in higher total costs.
- Building and equipment codes and standards are a highly efficient form of DSM, providing reduced energy use with much lower administrative costs than programs that reward the installation of more efficient equipment at a single site. However, they are more subtle than traditional DSM programs and may not always be recognized as a form of DSM.
- The opening of markets to competing suppliers makes it more difficult to administer and implement DSM programs. However, they are still possible if the regulators wish to continue them, and set appropriate rules and regulations for the energy market to allow the implementation of DSM programs.

Many participants in IRP processes may be interested in only one aspect of the process. For example, the energy industry's main interest may be cost minimization, while environmentalists may want to minimize pollutant emissions and prevent environmental damage from the construction of energy facilities. Participation by all affected interest groups helps provide the best overall solution for society, including the indirect costs and benefits from these energy resource decisions.

TRADABLE EMISSION CREDITS

Increasingly, quotas and limits apply to the emissions of various pollutants. These currently include sulphur dioxide (SO_2) and nitrogen oxides (NO_x) but may someday include carbon dioxide (CO_2). Often a market-based system of **tradable credits** is used with these quotas. A company is given the right to produce a given level of emissions and it earns a credit, which can be sold to others, if it produces fewer emissions than that level. If one company can reduce its emissions at a lower cost than another, it can do so and sell the emissions credit to the second company and earn a profit from its pollution control efforts. To date, this type of activity has largely involved large industrial plants, but it can also involve commercial buildings with on-site emissions, such as generation equipment or gas engine-driven cooling.

Designers must be aware of any regulations concerning pollutant emissions; failure to comply with these regulations may result in civil or criminal penalties for designers or their clients. However, designers should understand the options available under these regulations. The purchase or sale of emissions credits may allow reduced construction or building operations costs if the equipment can "overcomply" at a lower cost than the cost of another source of emissions to comply, or vice versa. In some cases, documentation of energy savings beyond what codes and regulations require can result in receiving emissions credits that may be sold later.

AGENCIES AND ASSOCIATIONS IN THE UNITED STATES

American Gas Association (AGA), Arlington, VA
American Petroleum Institute (API), Washington, D.C.
Bureau of Mines, Department of Interior, Washington, D.C.
Council on Environmental Quality (CEQ), Washington, D.C.
Edison Electric Institute (EEI), Washington, D.C.
Electric Power Research Institute (EPRI), Palo Alto, CA
Energy Information Administration (EIA), Washington, D.C.
Gas Research Institute (GRI), Chicago, IL
National Coal Association (NCA), Washington, D.C.
North American Electric Reliability Council (NAERC), Princeton, NJ
Organization of Petroleum Exporting Countries (OPEC), Vienna, Austria
United States Green Building Council (USGBC), San Francisco, CA

REFERENCES

ASHRAE. 1990. Energy position statement.

Conover, D.R. 1984. Accounting for energy resource use in building regulations. *ASHRAE Transactions* 90(1B):547-63.

EIA. 1999. *Annual energy review.* DOE/EIA—0384(99). Energy Information Administration, Office of Energy Markets and End Use, U.S. Department of Energy, Washington, D.C.

EIA. 2000. *Annual energy outlook 2000.* DOE/EIA—0383(2000). Energy Information Administration, U.S. Department of Energy, Washington, D.C.

Gleeson, G.W. 1951. Energy—Choose it wisely today for safety tomorrow. *ASHVE Transactions* 57:523-40.

Gottfried, D.A. 1996. Sustainable building technical manual. U.S. Green Building Council, San Francisco, CA.

Grumman, D.L. 1984. Energy resource accounting: ASHRAE *Standard* 90C-1977R. *ASHRAE Transactions* 90(1B):531-46.

Heiman, J.L. 1984. Proposal for a simple method for determining resource impact factors. *ASHRAE Transactions* 90(1B):564-70.

IESNA. 2000. *The IESNA lighting handbook.* IESNA, New York.

BIBLIOGRAPHY

DOE. 1979. *Impact assessment of a mandatory source-energy approach to energy conservation in new construction.* U.S. Department of Energy, Washington, D.C.

Pacific Northwest Laboratory. 1987. *Development of whole-building energy design targets for commercial buildings phase 1 planning.* PNL-5854, Vol. 2. U.S. Department of Energy, Washington, D.C.

USGBC. 1999. LEED™ reference guide. U.S. Green Building Council, San Francisco.

CHAPTER 18

COMBUSTION AND FUELS

PRINCIPLES OF COMBUSTION

COMBUSTION is a chemical reaction in which an oxidant reacts rapidly with a fuel to liberate stored energy as thermal energy, generally in the form of high-temperature gases. Small amounts of electromagnetic energy (light), electric energy (free ions and electrons), and mechanical energy (noise) are also produced during combustion. Except in special applications, the oxidant for combustion is oxygen in the air.

Conventional hydrocarbon fuels contain primarily hydrogen and carbon, in elemental form or in various compounds. Their complete combustion produces mainly carbon dioxide (CO_2) and water (H_2O); however, small quantities of carbon monoxide (CO) and partially reacted flue gas constituents (gases and liquid or solid aerosols) may form. Most conventional fuels also contain small amounts of sulfur, which is oxidized to sulfur dioxide (SO_2) or sulfur trioxide (SO_3) during combustion, and noncombustible substances such as mineral matter (ash), water, and inert gases. Flue gas is the product of complete or incomplete combustion and includes excess air (if present), but not dilution air.

Fuel combustion rate depends on (1) the rate of the chemical reaction of the combustible fuel constituents with oxygen, (2) the rate at which oxygen is supplied to the fuel (the mixing of air and fuel), and (3) the temperature in the combustion region. The reaction rate is fixed by fuel selection. Increasing the mixing rate or temperature increases the combustion rate.

With **complete combustion** of hydrocarbon fuels, all hydrogen and carbon in the fuel are oxidized to H_2O and CO_2. Generally, for complete combustion, excess oxygen or excess air must be supplied beyond the amount theoretically required to oxidize the fuel. Excess air is usually expressed as a percentage of the air required to completely oxidize the fuel.

In **stoichiometric combustion** of a hydrocarbon fuel, fuel is reacted with the exact amount of oxygen required to oxidize all carbon, hydrogen, and sulfur in the fuel to CO_2, H_2O, and SO_2. Therefore, exhaust gas from stoichiometric combustion theoretically contains no incompletely oxidized fuel constituents and no unreacted oxygen (i.e., no carbon monoxide and no excess air or oxygen). The percentage of CO_2 contained in products of stoichiometric combustion is the maximum attainable and is referred to as the **stoichiometric CO_2**, **ultimate CO_2**, or **maximum theoretical percentage of CO_2**.

Stoichiometric combustion is seldom realized in practice because of imperfect mixing and finite reaction rates. For economy and safety, most combustion equipment should operate with some excess air. This ensures that fuel is not wasted and that combustion is complete despite variations in fuel properties and in the supply rates of fuel and air. The amount of excess air to be supplied to any combustion equipment depends on (1) expected variations in fuel properties and in fuel and air supply rates, (2) equipment application, (3) degree of operator supervision required or available, and (4) control requirements. For maximum efficiency, combustion at low excess air is desirable.

Incomplete combustion occurs when a fuel element is not completely oxidized during combustion. For example, a hydrocarbon may not completely oxidize to carbon dioxide and water, but may form partially oxidized compounds, such as carbon monoxide, aldehydes, and ketones. Conditions that promote incomplete combustion include (1) insufficient air and fuel mixing (causing local fuel-rich and fuel-lean zones), (2) insufficient air supply to the flame (providing less than the required quantity of oxygen), (3) insufficient reactant residence time in the flame (preventing completion of combustion reactions), (4) flame impingement on a cold surface (quenching combustion reactions), or (5) flame temperature that is too low (slowing combustion reactions).

Incomplete combustion uses fuel inefficiently, can be hazardous because of carbon monoxide production, and contributes to air pollution.

Combustion Reactions

The reaction of oxygen with the combustible elements and compounds in fuels occurs according to fixed chemical principles, including

- Chemical reaction equations
- Law of matter conservation: the mass of each element in the reaction products must equal the mass of that element in the reactants
- Law of combining masses: chemical compounds are formed by elements combining in fixed mass relationships
- Chemical reaction rates

Oxygen for combustion is normally obtained from air, which is a physical mixture of nitrogen, oxygen, small amounts of water vapor, carbon dioxide, and inert gases. For practical combustion calculations, dry air consists of 20.95% oxygen and 79.05% inert gases (nitrogen, argon, and so forth) by volume, or 23.15% oxygen and 76.85% inert gases by mass. For calculation purposes, nitrogen is assumed to pass through the combustion process unchanged (although small quantities of nitrogen oxides form). Table 1 lists oxygen and air requirements for stoichiometric combustion and the products of stoichiometric combustion of some pure combustible materials (or constituents) found in common fuels.

Flammability Limits

Fuel burns in a self-sustained reaction only when the volume percentages of fuel and air in a mixture at standard temperature and pressure are within the upper and lower flammability limits or explosive limits (UEL and LEL). See Table 2. Both temperature and pressure affect these limits. As the temperature of the mixture increases, the upper limit increases and the lower limit decreases. As the pressure of the mixture decreases below atmospheric pressure, the upper limit decreases and the lower limit increases.

The preparation of this chapter is assigned to TC 6.10, Fuels and Combustion.

Table 1 Combustion Reactions of Common Fuel Constituents

Constituent	Molecular Formula	Combustion Reactions	Stoichiometric Oxygen and Air Requirements: lb/lb Fuel[a] O_2	lb/lb Fuel[a] Air	ft³/ft³ Fuel O_2	ft³/ft³ Fuel Air	Flue Gas from Stoichiometric Combustion with Air: Ultimate CO_2, %	Dew Point,[c] °F	ft³/ft³ Fuel CO_2	ft³/ft³ Fuel H_2O	lb/lb Fuel CO_2	lb/lb Fuel H_2O
Carbon (to CO)	C	$C + 0.5O_2 \rightarrow CO$	1.33	5.75	b	b	—	—	—	—	—	—
Carbon (to CO_2)	C	$C + O_2 \rightarrow CO_2$	2.66	11.51	b	b	29.30	—	—	—	3.664	—
Carbon monoxide	CO	$CO + 0.5O_2 \rightarrow CO$	0.57	2.47	0.50	2.39	34.70	—	1.0	—	1.571	—
Hydrogen	H_2	$H_2 + 0.5O_2 \rightarrow H_2O$	7.94	34.28	0.50	2.39	—	162	—	1.0	—	8.937
Methane	CH_4	$CH_4 + 2O_2 \rightarrow CO_2 + 2H_2O$	3.99	17.24	2.00	9.57	11.73	139	1.0	2.0	2.744	2.246
Ethane	C_2H_6	$C_2H_6 + 3.5O_2 \rightarrow 2CO_2 + 3H_2O$	3.72	16.09	3.50	16.75	13.18	134	2.0	3.0	2.927	1.798
Propane	C_3H_8	$C_3H_8 + 5O_2 \rightarrow 3CO_2 + 4H_2O$	3.63	15.68	5.00	23.95	13.75	131	3.0	4.0	2.994	1.634
Butane	C_4H_{10}	$C_4H_{10} + 6.5O_2 \rightarrow 4CO_2 + 5H_2O$	3.58	15.47	6.50	31.14	14.05	129	4.0	5.0	3.029	1.550
Alkanes	C_nH_{2n+2}	$C_nH_{2n+2} + (1.5n + 0.5)O_2 \rightarrow nCO_2 + (n+1)H_2O$	—	—	$1.5n + 0.5$	$7.18n + 2.39$	—	128 to 127	n	$n + 1$	$\frac{44.01n}{14.026n + 2.016}$	$\frac{18.01(n+1)}{14.026n + 2.016}$
Ethylene	C_2H_4	$C_2H_4 + 3O_2 \rightarrow 2CO_2 + 2H_2O$	3.42	14.78	3.00	14.38	15.05	125	2.0	2.0	3.138	1.285
Propylene	C_3H_6	$C_3H_6 + 4.5O_2 \rightarrow 3CO_2 + 3H_2O$	3.42	14.78	4.50	21.53	15.05	125	3.0	3.0	3.138	1.285
Alkenes	C_nH_{2n}	$C_nH_{2n} + 1.5nO_2 \rightarrow nCO_2 + nH_2O$	3.42	14.78	$1.50n$	$7.18n$	15.05	125	n	n	3.138	1.285
Acetylene	C_2H_2	$C_2H_2 + 2.5O_2 \rightarrow 2CO_2 + H_2O$	3.07	13.27	2.50	11.96	17.53	103	2.0	1.0	3.834	0.692
Alkynes	C_nH_{2m}	$C_nH_{2m} + (n + 0.5m)O_2 \rightarrow nCO_2 + mH_2O$	—	—	$n + 0.5m$	$4.78n + 2.39m$	—	—	n	m	$\frac{22.005n}{6.005n + 1.008m}$	$\frac{9.008m}{6.005n + 1.008m}$
									SO_x	H_2O	SO_x	H_2O
Sulfur (to SO_2)	S	$S + O_2 \rightarrow SO_2$	1.00	4.31	b	b	—	—	1.0 SO_2	—	1.998 (SO_2)	—
Sulfur (to SO_3)	S	$S + 1.5O_2 \rightarrow SO_3$	1.50	6.47	b	b	—	—	1.0 SO_3	—	2.497 (SO_3)	—
Hydrogen sulfide	H_2S	$H_2S + 1.5O_2 \rightarrow SO_2 + H_2O$	1.41	6.08	1.50	7.18	—	125	1.0 SO_2	1.0	1.880 (SO_2)	0.528

Adapted, in part, from *Gas Engineers Handbook* (1965)

[a]Atomic masses: H = 1.008, C = 12.01, O = 16.00, S = 32.06.

[b]Volume ratios are not given for fuels that do not exist in vapor form at reasonable temperatures or pressure.

[c] Dew point is determined from Figure 2.

Table 2 Flammability Limits and Ignition Temperatures of Common Fuels in Fuel-Air Mixtures

Substance	Molecular Formula	Lower Flammability Limit, %	Upper Flammability Limit, %	Ignition Temperature, °F	References
Carbon (activated coke)	C	—	—	1220	Hartman (1958)
Carbon monoxide	CO	12.5	74	1128	Scott et al. (1948)
Hydrogen	H_2	4.0	75.0	968	Zabetakis (1956)
Methane	CH_4	5.0	15.0	1301	*Gas Engineers Handbook* (1965)
Ethane	C_2H_6	3.0	12.5	968 to 1166	Trinks (1947)
Propane	C_3H_8	2.1	10.1	871	NFPA (1962)
n-Butane	C_4H_{10}	1.86	8.41	761	NFPA (1962)
Ethylene	C_2H_4	2.75	28.6	914	Scott et al. (1948)
Propylene	C_3H_6	2.00	11.1	856	Scott et al. (1948)
Acetylene	C_2H_2	2.50	81	763 to 824	Trinks (1947)
Sulfur	S	—	—	374	Hartman (1958)
Hydrogen sulfide	H_2S	4.3	45.50	558	Scott et al. (1948)

Flammability limits adapted from Coward and Jones (1952). All values corrected to 60°F, 30 in. Hg, dry.

However, as pressure increases above atmospheric pressure, the upper limit increases and the lower limit is relatively constant.

Ignition Temperature

Ignition temperature is the lowest temperature at which heat is generated by combustion faster than heat is lost to the surroundings and combustion becomes self-propagating. See Table 2. The fuel-air mixture will not burn freely and continuously below the ignition temperature unless heat is supplied, but chemical reaction between the fuel and air may occur. Ignition temperature is affected by a large number of factors.

The ignition temperature and flammability limits of a fuel-air mixture, together, are a measure of the potential for ignition (*Gas Engineers Handbook* 1965).

Combustion Modes

Combustion reactions occur in either continuous or pulse flame modes. **Continuous combustion** burns fuel in a sustained manner as long as fuel and air are continuously fed to the combustion zone and the fuel-air mixture is within the flammability limits. Continuous combustion is more common than pulse combustion and is used in most fuel-burning equipment.

Pulse combustion is an acoustically resonant process that burns various fuels in small, discrete fuel-air mixture volumes in a very rapid series of combustions.

The introduction of fuel and air into the pulse combustor is controlled by mechanical or aerodynamic valves. Typical combustors consist of one or more valves, a combustion chamber, an exit pipe, and a control system (ignition means, fuel-metering devices, etc.). Typically, combustors for warm air furnace, hot water boiler, and commercial cooking equipment use mechanical valves. Aerodynamic valves are usually used in higher pressure applications, such as thrust engines. Separate valves for air and fuel, a single valve for premixed air and fuel, or multiple valves of either type can be used. Premix valve systems may require a flame trap at the combustion chamber entrance to prevent flashback.

In a mechanically valved pulse combustor, air and fuel are forced into the combustion chamber through the valves under pressures less than 0.5 psi. An ignition source, such as a spark, ignites the fuel-air mixture, causing a positive pressure buildup in the combustion chamber. The positive pressure causes the valves to close, leaving only the exit pipe of the combustion chamber as a pressure relief opening. The combustion chamber and exit pipe geometry determine the resonant frequency of the combustor.

The pressure wave from the initial combustion travels down the exit pipe at sonic velocity. As this wave exits the combustion chamber, most of the flue gases present in the chamber are carried with it into the exit pipe. Flue gases remaining in the combustion chamber begin to cool immediately. The contraction of the cooling gases and the momentum of gases in the exit pipe create a vacuum inside the chamber that opens the valves and allows more fuel and air into the chamber. While the fresh charge of fuel-air enters the chamber, the pressure wave reaches the end of the exit pipe and is partially reflected from the open end of the pipe. The fresh fuel-air charge is ignited by residual combustion and/or heat. The resulting combustion starts another cycle.

Typical pulse combustors operate at 30 to 100 cycles per second and emit resonant sound, which must be considered in their application. The pulses produce high convective heat transfer rates.

Heating Value

Combustion produces thermal energy or heat. The quantity of heat generated by complete combustion of a unit of specific fuel is constant and is termed the **heating value**, **heat of combustion,** or **caloric value** of that fuel. The heating value of a fuel can be determined by measuring the heat evolved during combustion of a known quantity of the fuel in a calorimeter, or it can be estimated from chemical analysis of the fuel and the heating values of the various chemical elements in the fuel. For information on calculating heating values, see the sections on Characteristics of Fuel Oils and Characteristics of Coals.

Higher heating value, **gross heating value**, or **total heating value** includes the latent heat of vaporization and is determined when water vapor in the fuel combustion products is condensed. Conversely, **lower heating value** or **net heating value** is obtained when the latent heat of vaporization is *not* included. When the heating value of a fuel is specified without designating higher or lower, it generally means the higher heating value in the United States. (Lower heating value is mainly used for internal combustion engine fuels.)

Heating values are usually expressed in Btu/ft^3 for gaseous fuels, Btu/gal for liquid fuels, and Btu/lb for solid fuels. Heating values are always given in relation to a certain reference temperature and pressure, usually 60, 68, or 77°F and 14.696 psia, depending on the particular industry practice. Heating values of several substances in common fuels are listed in Table 3.

With incomplete combustion, not all fuel is completely oxidized, and the heat produced is less than the heating value of the fuel. Therefore, the quantity of heat produced per unit of fuel consumed decreases, implying lower combustion efficiency.

Not all heat produced during combustion can be used effectively. The greatest heat loss is the thermal energy of the increased temperature of hot exhaust gases above the temperature of incoming air and fuel. Other heat losses include radiation and convection heat transfer from the outer walls of combustion equipment to the environment.

Table 3 Heating Values of Substances Occurring in Common Fuels

Substance	Molecular Formula	Higher Heating Values,[a] Btu/lb	Lower Heating Values,[a] Btu/lb	Specific Volume,[b] ft^3/lb
Carbon (to CO)	C	3,950	3,950	—
Carbon (to CO_2)	C	14,093	14,093	—
Carbon monoxide	CO	4,347	4,347	13.5
Hydrogen	H_2	61,095	51,623	188.0
Methane	CH_4	23,875	21,495	23.6
Ethane	C_2H_6	22,323	20,418	12.5
Propane	C_3H_8	21,669	19,937	8.36
Butane	C_4H_{10}	21,321	19,678	6.32
Ethylene	C_2H_4	21,636	20,275	—
Propylene	C_3H_6	21,048	19,687	9.01
Acetylene	C_2H_2	21,502	20,769	14.3
Sulfur (to SO_2)	S	3,980	3,980	—
Sulfur (to SO_3)	S	5,940	5,940	—
Hydrogen sulfide	H_2S	7,097	6,537	11.0

Adapted from *Gas Engineers Handbook* (1965).
[a]All values corrected to 60°F, 30 in. Hg, dry. For gases saturated with water vapor at 60°F, deduct 1.74% of the value to adjust for gas volume displaced by water vapor.
[b]At 32°F and 29.92 in. Hg.

Altitude Compensation

Air at altitudes above sea level is less dense and has less oxygen per unit volume. Therefore, combustion at altitudes above sea level has less available oxygen to burn with the fuel unless compensation is made for the altitude. Combustion occurs, but the amount of excess air is reduced. If excess air is reduced enough by an increase in altitude, combustion is incomplete or ceases.

Altitude compensation is achieved by matching the fuel and air supply rates to attain complete combustion without too much excess air or too much fuel. Fuel and air supply rates can be matched by increasing the air supply rate to the combustion zone or by decreasing the fuel supply rate to the combustion zone. The air supply rate can be increased with a combustion air blower, and the fuel supply rate can be reduced by decreasing the fuel input (derating).

Power burners use combustion air blowers and can increase the air supply rate to compensate for altitude. The combustion zone can be pressurized to attain the same air density in the combustion chamber as that attained at sea level.

Derating can be used as an alternative to power combustion. In the United States, the fuel gas codes generally do not require derating of nonpower burners at altitudes up to 2000 ft. At altitudes above 2000 ft, burners should be derated 4% for each 1000 ft above sea level (NFPA/IAS *National Fuel Gas Code*). Chimney or vent operation also must be considered at high altitudes (see Chapter 30 of the 2000 *ASHRAE Handbook—Systems and Equipment*).

FUEL CLASSIFICATION

Generally, hydrocarbon fuels are classified according to physical state (gas, liquid, or solid). Different types of combustion equipment are usually needed to burn fuels in the different physical states. Gaseous fuels can be burned in premix or diffusion burners. Liquid fuel burners must include a means for atomizing or vaporizing fuel into small droplets or a vapor and must provide adequate mixing of fuel and air. Solid fuel combustion equipment must (1) heat fuel to vaporize sufficient volatiles to initiate and sustain combustion, (2) provide residence time to complete combustion, and (3) provide space for ash containment.

Principal fuel applications include space heating and cooling of residential, commercial, industrial, and institutional buildings; service water heating; steam generation; and refrigeration. Major fuels for these applications are natural and liquefied petroleum gases, fuel oils, diesel and gas turbine fuels (for on-site energy applications), and coal.

Fuels of limited use, such as manufactured gases, kerosene, briquettes, wood, and coke, are not discussed here. Fuel choice is based on one or more of the following:

Fuel factors

- Availability, including dependability of supply
- Convenience of use and storage
- Economy
- Cleanliness

Combustion equipment factors

- Operating requirements
- Cost
- Service requirements
- Ease of control

GASEOUS FUELS

Although various gaseous fuels have been used as energy sources in the past, heating and cooling applications are presently limited to natural gas and liquefied petroleum gases.

Types and Properties

Natural gas is a nearly odorless and colorless gas that accumulates in the upper parts of oil and gas wells. Raw natural gas is a mixture of methane (55 to 98%), higher hydrocarbons (primarily ethane), and noncombustible gases. Some constituents, principally water vapor, hydrogen sulfide, helium, and gases for liquefied petroleum gases and gasoline are removed prior to distribution.

Natural gas used as fuel typically contains methane, CH_4 (70 to 96%); ethane, C_2H_6 (1 to 14%); propane, C_3H_8 (0 to 4%); butane, C_4H_{10} (0 to 2%); pentane, C_5H_{12} (0 to 0.5%); hexane, C_6H_{14} (0 to 2%); carbon dioxide, CO_2 (0 to 2%); oxygen, O_2 (0 to 1.2%); and nitrogen, N_2 (0.4 to 17%).

The composition of natural gas depends on its geographical source. Because the gas is drawn from various sources, the composition of gas distributed in a given location can vary slightly, but a fairly constant heating value is usually maintained for control and safety. Local gas utilities are the best sources of current gas composition data for a particular area.

Heating values of natural gases vary from 900 to 1200 Btu/ft^3; the usual range is 1000 to 1050 Btu/ft^3 at sea level. The heating value for a particular gas can be calculated from the composition data and values in Table 3.

For safety purposes, odorants (such as mercaptans) are added to natural gas and LPG to give them noticeable odors.

Liquefied petroleum gases (LPG) consist primarily of propane and butane, and are usually obtained as a byproduct of oil refinery operations or by stripping natural gas. Propane and butane are gaseous under usual atmospheric conditions, but can be liquefied under moderate pressures at normal temperatures.

Commercial propane consists primarily of propane but generally contains about 5 to 10% propylene. It has a heating value of about 21,560 Btu/lb, about 2500 Btu/ft^3 of gas, or about 91,000 Btu/gal of liquid propane. At atmospheric pressure, commercial propane has a boiling point of about −40°F. The low boiling point of propane allows it to be used during winter in the northern United States and in Canada. Tank heaters and vaporizers permit its use in colder climates and where high fuel flow rates are required. Propane is available in cylinders, bottles, tank trucks, or tank cars.

Propane-air mixtures are used in place of natural gas in small communities and by natural gas companies to supplement normal supplies at peak loads. Table 4 lists heating values and specific gravities for various fuel-air ratios.

Table 4 Propane-Air and Butane-Air Gas Mixtures

Heating Value, Btu/ft^3	Propane-Air[a]			Butane-Air[b]		
	% Gas	% Air	Sp Gr	% Gas	% Air	Sp Gr
500	19.8	80.2	1.103	15.3	84.7	1.155
600	23.8	76.2	1.124	18.4	81.6	1.186
700	27.8	72.2	1.144	21.5	78.5	1.216
800	31.7	68.3	1.165	24.5	75.5	1.248
900	35.7	64.3	1.185	27.6	72.4	1.278
1000	39.7	60.3	1.206	30.7	69.3	1.310
1100	43.6	56.4	1.227	33.7	66.3	1.341
1200	47.5	52.5	1.248	36.8	63.2	1.372
1300	51.5	48.5	1.268	39.8	60.2	1.402
1400	55.5	44.5	1.288	42.9	57.1	1.433
1500	59.4	40.6	1.309	46.0	54.0	1.464
1600	63.4	36.6	1.330	49.0	51.0	1.495
1700	67.4	32.6	1.350	52.1	47.9	1.526
1800	71.3	28.7	1.371	55.2	44.8	1.557

Adapted from *Gas Engineers Handbook* (1965).
[a]Values used for calculation: 2522 Btu/ft^3; 1.52 specific gravity.
[b]Values used for calculation: 3261 Btu/ft^3; 2.01 specific gravity.

Commercial butane consists primarily of butane but may contain up to 5% butylene. It has a heating value of about 21,180 Btu/lb, about 3200 Btu/ft^3 of gas, or about 102,000 Btu/gal of liquid butane. At atmospheric pressure, commercial butane has a relatively high boiling point of about 32°F. Therefore, butane cannot be used in cold weather unless the gas temperature is maintained above 32°F or the partial pressure is decreased by dilution with a gas having a lower boiling point. Butane is usually available in bottles, tank trucks, or tank cars, but not in cylinders.

Butane-air mixtures are used in place of natural gas in small communities and by natural gas companies to supplement normal supplies at peak loads. Table 4 lists heating values and specific gravities for various fuel-air ratios.

Commercial propane-butane mixtures with various ratios of propane and butane are available. Their properties generally fall between those of the unmixed fuels.

Manufactured gases are combustible gases produced from coal, coke, oil, liquefied petroleum gases, or natural gas. For more detailed information, see *Gas Engineers Handbook* (1965). These fuels are used primarily for industrial in-plant operations or as specialty fuels (e.g., acetylene for welding).

LIQUID FUELS

Significant liquid fuels include various fuel oils for firing combustion equipment and engine fuels for on-site energy systems. Liquid fuels, with few exceptions, are mixtures of hydrocarbons derived by refining crude petroleum. In addition to hydrocarbons, crude petroleum usually contains small quantities of sulfur, oxygen, nitrogen, vanadium, other trace metals, and impurities such as water and sediment. Refining produces a variety of fuels and other products. Nearly all lighter hydrocarbons are refined into fuels (e.g., liquefied petroleum gases, gasoline, kerosene, jet fuels, diesel fuels, and light heating oils). Heavy hydrocarbons are refined into residual fuel oils and other products (e.g., lubricating oils, waxes, petroleum coke, and asphalt).

Crude petroleums from different oil fields vary in hydrocarbon molecular structure. Crude is paraffin-base (principally chain-structured paraffin hydrocarbons), naphthene- or asphaltic-base (containing relatively large quantities of saturated ring-structural naphthenes), aromatic-base (containing relatively large quantities of unsaturated, ring-structural aromatics), or mixed- or intermediate-base (between paraffin- and naphthene-base crudes). Except for heavy fuel oils, the crude type has little significant effect on resultant products and combustion applications.

Types of Fuel Oils

Fuel oils for heating are broadly classified as **distillate fuel oils** (lighter oils) or **residual fuel oils** (heavier oils). ASTM *Standard* D 396 has specifications for fuel oil properties that subdivide the oils into various grades. Grades No. 1 and 2 are distillate fuel oils. Grades 4, 5 (Light), 5 (Heavy), and 6 are residual fuel oils. Specifications for the grades are based on required characteristics of fuel oils for use in different types of burners.

Grade No. 1 is a light distillate intended for vaporizing-type burners. High volatility is essential to continued evaporation of the fuel oil with minimum residue.

Grade No. 2 is a heavier distillate than No. 1. It is used primarily with pressure-atomizing (gun) burners that spray the oil into a combustion chamber. The atomized oil vapor mixes with air and burns. This grade is used in most domestic burners and many medium-capacity commercial-industrial burners. A dewaxed No. 2 oil with a pour point of −58°F is supplied only to areas where regular No. 2 oil would jell.

Grade No. 4 is an intermediate fuel that is considered either a heavy distillate or a light residual. Intended for burners that atomize oils of higher viscosity than domestic burners can handle, its permissible viscosity range allows it to be pumped and atomized at relatively low storage temperatures.

Grade No. 5 (Light) is a residual fuel of intermediate viscosity for burners that handle fuel more viscous than No. 4 without preheating. Preheating may be necessary in some equipment for burning and, in colder climates, for handling.

Grade No. 5 (Heavy) is a residual fuel more viscous than No. 5 (Light), but intended for similar purposes. Preheating is usually necessary for burning and, in colder climates, for handling.

Grade No. 6, sometimes referred to as Bunker C, is a high-viscosity oil used mostly in commercial and industrial heating. It requires preheating in the storage tank to permit pumping, and additional preheating at the burner to permit atomizing.

Low-sulfur residual oils are marketed in many areas to permit users to meet sulfur dioxide emission regulations. These fuel oils are produced (1) by refinery processes that remove sulfur from the oil (hydrodesulfurization), (2) by blending high-sulfur residual oils with low-sulfur distillate oils, or (3) by a combination of these methods. These oils have significantly different characteristics from regular residual oils. For example, the viscosity-temperature relationship can be such that low-sulfur fuel oils have viscosities of No. 6 fuel oils when cold, and of No. 4 when heated. Therefore, normal guidelines for fuel handling and burning can be altered when using these fuels.

Fuel oil grade selection for a particular application is usually based on availability and economic factors, including fuel cost, clean air requirements, preheating and handling costs, and equipment cost. Installations with low firing rates and low annual fuel consumption cannot justify the cost of preheating and other methods that use residual fuel oils. Large installations with high annual fuel consumption cannot justify the premium cost of distillate fuel oils.

Characteristics of Fuel Oils

Characteristics that determine grade classification and suitability for given applications are (1) viscosity, (2) flash point, (3) pour point, (4) water and sediment content, (5) carbon residue, (6) ash, (7) distillation qualities or distillation temperature ranges, (8) specific gravity, (9) sulfur content, (10) heating value, and (11) carbon-hydrogen content. Not all of these are included in ASTM *Standard* D 396.

Viscosity is an oil's resistance to flow. It is significant because it indicates the ease at which oil flows or can be pumped and the ease of atomization. Differences in fuel oil viscosities are caused by variations in the concentrations of fuel oil constituents and different refining methods. Approximate viscosities of fuel oils are shown in Figure 1.

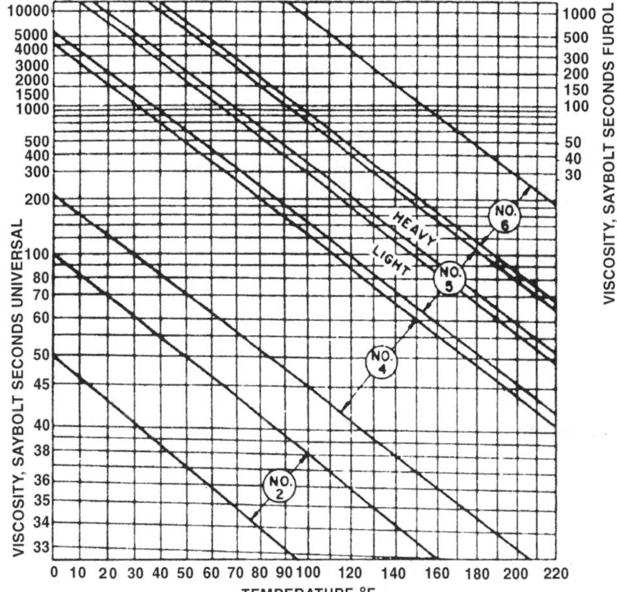

Fig. 1 Approximate Viscosity of Fuel Oils

Flash point is the lowest temperature to which an oil must be heated for its vapors to ignite in a flame. Minimum permissible flash point is usually prescribed by state and municipal laws.

Pour point is the lowest temperature at which a fuel can be stored and handled. Fuels with higher pour points can be used when heated storage and piping facilities are provided.

Water and **sediment content** should be low to prevent fouling the facilities. Sediment accumulates on filter screens and burner parts. Water in distillate fuels can cause tanks to corrode and emulsions to form in residual oil.

Carbon residue is obtained by a test in which the oil sample is destructively distilled in the absence of air. When commercial fuels are used in proper burners, this residue has almost no relationship to soot deposits, except indirectly when deposits are formed by vaporizing burners.

Ash is the noncombustible material in an oil. An excessive amount indicates the presence of materials that cause high wear on burner pumps.

The **distillation** test shows the volatility and ease of vaporization of a fuel.

Specific gravity is the ratio of the density of a fuel oil to the density of water at a specific temperature. **Specific gravities** cover a range in each grade, with some overlap between distillate and residual grades. **API gravity** (developed by the American Petroleum Institute) is a parameter widely used in place of specific gravity. It is obtained by the following formula:

$$\text{Degrees API} = \frac{141.5}{\text{Sp Gr at } 60/60°\text{F}} - 131.5 \quad (1)$$

where Sp Gr at 60/60°F is the ratio of the mass of a given volume of oil at 60°F to the mass of the same volume of water at 60°F.

The API gravity of water at 60°F is 10.0.

Air pollution considerations are important in determining the allowable **sulfur content** of fuel oils. Sulfur content is frequently limited by legislation aimed at reducing sulfur oxide emissions from combustion equipment. These laws require sulfur content to be below a certain level, usually 1.0, 0.5, or 0.3%. Table 5 lists sulfur levels of some marketed fuel oils.

Table 5 Sulfur Content of Marketed Fuel Oils

Grade of Oil	No. 1	No. 2	No. 4	No. 5 (Light)	No. 5 (Heavy)	No. 6
Total fuel samples	31	61	13	15	16	96
Sulfur content,						
% mass minimum	0.001	0.03	0.46	0.90	0.57	0.32
maximum	0.120	0.50	1.44	3.50	2.92	4.00
average	0.023	0.20	0.83	1.46	1.46	1.41
No. samples with S						
over 0.3%	0	17	13	15	16	96
over 0.5%	0	2	11	15	16	93
over 1.0%	0	0	3	9	11	60
over 3.0%	0	0	0	2	0	8

Data for No. 1 and No. 2 oil derived from Dickson and Sturm (1994).
Data for No. 4, 5, and 6 oil derived from Shelton (1974).

Table 6 Typical API Gravity, Density, and Higher Heating Value of Standard Grades of Fuel Oil

Grade No.	API Gravity	Density, lb/gal	Higher Heating Value, Btu/gal
1	38 to 45	6.950 to 6.675	137,000 to 132,900
2	30 to 38	7.296 to 6.960	141,800 to 137,000
4	20 to 28	7.787 to 7.396	148,100 to 143,100
5L	17 to 22	7.940 to 7.686	150,000 to 146,800
5H	14 to 18	8.080 to 7.890	152,000 to 149,400
6	8 to 15	8.448 to 8.053	155,900 to 151,300

Sulfur in fuel oils is also undesirable because of the corrosiveness of sulfur compounds in the flue gas. Although low-temperature corrosion can be minimized by maintaining the stack at temperatures above the dew point of the flue gas, this limits the overall thermal efficiency of combustion equipment.

For certain industrial applications, the sulfur content of a fuel must be limited because of adverse effects on product quality. The applications include direct-fired metallurgy where work is performed in the combustion zone.

Heating value is an important property, although ASTM *Standard* D 396 does not list it as one of the criteria for fuel oil classification. Heating value can generally be correlated with the API gravity. Table 6 shows the relationship between heating value, API gravity, and density for several oil grades. In the absence of more specific data, heating values can be calculated as shown in the *North American Combustion Handbook* (1978):

$$\text{Higher heating value, Btu/lb} = 22{,}320 - 3{,}780(\text{Specific gravity})^2 \quad (2)$$

Distillate fuel oils (Grades 1 and 2) have a carbon-hydrogen content of 84 to 86% carbon, with the remainder predominantly hydrogen. The heavier residual fuel oils (Grades 4, 5, and 6) may contain up to 88% carbon and as little as 11% hydrogen. An approximate relationship for determining the hydrogen content of fuel oils is

$$\text{Hydrogen, \%} = 26 - (15 \times \text{Specific gravity}) \quad (3)$$

ASTM *Standard* D 396 is more a classification than a specification, distinguishing between six generally nonoverlapping grades, one of which characterizes any commercial fuel oil. Quality is not defined, as a refiner might control it; for example, the standard lists the distillation temperature 90% point for Grade No. 2 as having a maximum of 640°F, whereas commercial practice rarely exceeds 600°F.

Types and Properties of Liquid Fuels for Engines

The primary stationary engine fuels are diesel and gas turbine oils, natural gases, and liquefied petroleum gases. Other fuels include sewage gas, manufactured gas, and gas mixtures. Gasoline and the JP series of gas turbine fuels are rarely used for stationary engines.

Only properties of diesel and gas turbine fuel oils are covered here; properties of natural and liquefied petroleum gases are found in the section on Gaseous Fuels. For properties of gasolines and JP turbine fuel, consult texts on internal combustion engines and gas turbines. Properties of currently marketed gasolines can be found in the latest volumes of *Mineral Industry Surveys*, *Motor Gasolines*, issued semiannually by the U.S. Bureau of Mines.

Properties of the three grades of diesel fuel oils (1-D, 2-D, and 4-D) are listed in ASTM *Standard* D 975.

Grade No. 1-D includes the class of volatile fuel oils from kerosene to intermediate distillates. These fuels are used in high-speed engines with frequent and relatively wide variations in loads and speeds and where abnormally low fuel temperatures are encountered.

Grade No. 2-D includes the class of lower volatility distillate gas oils. These fuels are used in high-speed engines with relatively high loads and uniform speeds, or in engines not requiring fuels with the higher volatility or other properties specified for Grade No. 1-D.

Grade No. 4-D covers the class of more viscous distillates and blends of these distillates with residual fuel oils. These fuels are used in low- and medium-speed engines involving sustained loads at essentially constant speed.

Property specifications and test methods for Grade No. 1-D, 2-D, and 4-D diesel fuel oils are essentially identical to specifications of Grade No. 1, 2, and 4 fuel oils, respectively. However, diesel fuel oils have an additional specification for **cetane number**, which measures ignition quality and influences combustion roughness. Cetane number requirements depend on engine design, size, speed and load variations, and starting and atmospheric conditions. An increase in cetane number over values actually required does not improve engine performance. Thus, the cetane number should be as low as possible to assure maximum fuel availability. ASTM *Standard* D 975 provides several methods for estimating cetane number from other fuel oil properties.

ASTM *Standard* D 2880 for gas turbine fuel oils relates gas turbine fuel oil grades to fuel and diesel fuel oil grades. Test methods for determining properties of gas turbine fuel oils are essentially identical to those for fuel oils. However, gas turbine specifications contain quantity limits on some trace elements that may be present. These limits are intended to prevent excessive corrosion in gas turbine engines. For a detailed discussion of fuels for gas turbines and combustion in gas turbines, see Chapters 5 and 9, respectively, in Hazard (1971).

SOLID FUELS

Solid fuels include coal, coke, wood, and waste products of industrial and agricultural operations. Of these, only coal is widely used for heating and cooling applications.

The complex composition of coal makes classification difficult. Chemically, coal consists of carbon, hydrogen, oxygen, nitrogen, sulfur, and a mineral residue, ash. Chemical analysis provides some indication of coal quality, but does not define its burning characteristics sufficiently. The coal user is principally interested in the available heat per unit mass of coal and the amount of ash and dust produced, but is also interested in burning characteristics and handling and storing properties. A description of coal qualities and their characteristics can be obtained from the U.S. Bureau of Mines.

Table 7 Classification of Coals by Rank[a]

	Class		Group	Limits of Fixed Carbon or Energy Content, Mineral-Matter-Free Basis	Requisite Physical Properties
I	Anthracite	1.	Metaanthracite	Dry FC, 98% or more (Dry VM, 2% or less)	Nonagglomerating
		2.	Anthracite	Dry FC, 92% or more, and less than 98% (Dry VM, 8% or less, and more than 2%)	
		3.	Semianthracite	Dry FC, 86% or more, and less than 92% (Dry VM, 14% or less, and more than 8%)	
II	Bituminous[d]	1.	Low-volatile bituminous coal	Dry FC, 78% or more, and less than 86% (Dry VM, 22% or less, and more than 14%)	Either agglomerating[b] or nonweathering[f]
		2.	Medium-volatile bituminous coal	Dry FC, 69% or more, and less than 78% (Dry VM, 31% or less, and more than 22%)	
		3.	High-volatile *A* bituminous coal	Dry FC, less than 69% (Dry VM, more than 31%), and moist[c], about 14,000 Btu/lb[e] or more	
		4.	High-volatile *B* bituminous coal	Moist[c], about 13,000 Btu/lb or more, and less than 14,000 Btu/lb[e]	
		5.	High-volatile *C* bituminous coal	Moist[c], about 11,000 Btu/lb or more, and less than 13,000 Btu/lb[e]	
III	Subbituminous	1.	Subbituminous *A* coal	Moist[c], about 11,000 Btu/lb or more, and less than 13,000 Btu/lb[e]	Both weathering and nonagglomerating[b]
		2.	Subbituminous *B* coal	Moist[c], about 9,500 Btu/lb or more, and less than 11,000 Btu/lb[e]	
		3.	Subbituminous *C* coal	Moist[c], about 8,300 Btu/lb or more, and less than 9,500 Btu/lb[e]	
IV	Lignitic	1.	Lignite	Moist[c], less than 8,300 Btu/lb	Consolidated
		2.	Brown coal	Moist[c], less than 8,300 Btu/lb	Unconsolidated

Source: Adapted from ASTM *Standard* D 388, Standard Classification of Coals by Rank.

FC = Fixed Carbon; VM = Volatile Matter.

[a]This classification does not include a few coals of unusual physical and chemical properties which come within the limits of fixed carbon or Btu of high-volatile bituminous and subbituminous ranks. All these coals either contain less than 48% dry, mineral-matter-free fixed carbon, or have more than about 15,500 Btu/lb, which is moist, mineral-matter-free.

[b]If agglomerating, classify in low-volatile group of the bituminous class.

[c]Moist refers to coal containing its natural bed moisture but not including visible water on the coal surface.

[d]There may be noncaking varieties in each group of the bituminous class.

[e]Coals having 69% or more fixed carbon on the dry, mineral-matter-free basis shall be classified according to fixed carbon, regardless of energy content.

[f]There are three varieties of coal in the high-volatile *C* bituminous coal group: Variety 1, agglomerating and nonweathering; Variety 2, agglomerating and weathering; and Variety 3, nonagglomerating and nonweathering.

Types of Coals

Commonly accepted definitions for classifying coals are listed in Table 7. This classification is arbitrary because there are no distinct demarcation lines between coal types.

Anthracite is a clean, dense, hard coal that creates little dust in handling. It is comparatively hard to ignite, but burns freely once started. It is noncaking and burns uniformly and smokelessly with a short flame.

Semianthracite has a higher volatile content than anthracite. It is not as hard and ignites more easily. Otherwise, its properties are similar to those of anthracite.

Bituminous coal includes many types of coal with distinctly different compositions, properties, and burning characteristics. Coals range from high-grade bituminous, such as those found in the eastern United States, to low-rank coals, such as those found in the western United States. Caking properties range from coals that melt or become fully plastic, to those from which volatiles and tars are distilled without changing form (classed as noncaking or free-burning). Most bituminous coals are strong and nonfriable enough to permit screened sizes to be delivered free of fines. Generally, they ignite easily and burn freely. Flame length is long and varies with different coals. If improperly fired, much smoke and soot are possible, especially at low burning rates.

Semibituminous coal is soft and friable, and handling creates fines and dust. It ignites slowly and burns with a medium-length flame. Its caking properties increase as volatile matter increases, but the coke formed is weak. With only half the volatile matter content of bituminous coals, burning produces less smoke; hence, it is sometimes called smokeless coal.

Subbituminous coal, such as that found in the western United States, is high in moisture when mined and tends to break up as it dries or is exposed to the weather; it is likely to ignite spontaneously when piled or stored. It ignites easily and quickly, has a medium-length flame, and is noncaking and free-burning. The lumps tend to break into small pieces if poked. Very little smoke and soot are formed.

Lignite is woody in structure, very high in moisture when mined, of low heating value, and clean to handle. It has a greater tendency than subbituminous coals to disintegrate as it dries and is also more likely to ignite spontaneously. Because of its high moisture, freshly mined lignite ignites slowly and is noncaking. The char left after moisture and volatile matter are driven off burns very easily, like charcoal. The lumps tend to break up in the fuel bed and pieces of char that fall into the ash pit continue to burn. Very little smoke or soot forms.

Characteristics of Coal

The characteristics of coals that determine classification and suitability for given applications are the proportions of (1) volatile matter, (2) fixed carbon, (3) moisture, (4) sulfur, and (5) ash. Each of these is reported in the proximate analysis. Coal analyses can be reported on several bases: as-received, moisture-free (or dry), and mineral-matter-free (or ash-free). As-received is applicable for combustion calculations; moisture-free and mineral-matter-free, for classification purposes.

Volatile matter is driven off as gas or vapor when the coal is heated according to a standard temperature test. It consists of a variety of organic gases, generally resulting from distillation and decomposition. Volatile products given off by coals when heated differ materially in the ratios (by mass) of the gases to oils and tars. No heavy oils or tars are given off by anthracite, and very small quantities are given off by semianthracite. As volatile matter in the coal increases to as much as 40% of the coal (dry and ash-free basis), increasing amounts of oils and tars are released. However, for coals of higher volatile content, the quantity of oils and tars decreases and is relatively low in the subbituminous coals and in lignite.

Table 8 Typical Ultimate Analyses for Coals

Rank	As Received, Btu/lb	Constituents, Percent by Mass					
		Oxygen	Hydrogen	Carbon	Nitrogen	Sulfur	Ash
Anthracite	12,700	5.0	2.9	80.0	0.9	0.7	10.5
Semianthracite	13,600	5.0	3.9	80.4	1.1	1.1	8.5
Low-volatile bituminous	14,350	5.0	4.7	81.7	1.4	1.2	6.0
Medium-volatile bituminous	14,000	5.0	5.0	81.4	1.4	1.5	6.0
High-volatile bituminous *A*	13,800	9.3	5.3	75.9	1.5	1.5	6.5
High-volatile bituminous *B*	12,500	13.8	5.5	67.8	1.4	3.0	8.5
High-volatile bituminous *C*	11,000	20.6	5.8	59.6	1.1	3.5	9.4
Subbituminous *B*	9,000	29.5	6.2	52.5	1.0	1.0	9.8
Subbituminous *C*	8,500	35.7	6.5	46.4	0.8	1.0	9.6
Lignite	6,900	44.0	6.9	40.1	0.7	1.0	7.3

Fixed carbon is the combustible residue left after the volatile matter is driven off. It is not all carbon. Its form and hardness are an indication of fuel coking properties and, therefore, guide the choice of combustion equipment. Generally, fixed carbon represents that portion of fuel that must be burned in the solid state.

Moisture is difficult to determine accurately because a sample can lose moisture on exposure to the atmosphere, particularly when reducing the sample size for analysis. To correct for this loss, total moisture content of a sample is customarily determined by adding the moisture loss obtained when air-drying the sample to the measured moisture content of the dried sample. Moisture does not represent all of the water present in coal; water of decomposition (combined water) and of hydration are not given off under standardized test conditions.

Ash is the noncombustible residue remaining after complete coal combustion. Generally, the mass of ash is slightly less than that of mineral matter before burning.

Sulfur is an undesirable constituent in coal, because the sulfur oxides formed when it burns contribute to air pollution and cause combustion system corrosion. Table 8 lists the sulfur content of typical coals. Legislation has limited the sulfur content of coals burned in certain locations.

Heating value may be reported on an as-received, dry, dry and mineral-matter-free, or moist and mineral-matter-free basis. Higher heating values of coals are frequently reported with their proximate analysis. When more specific data are lacking, the higher heating value of higher quality coals can be calculated by the Dulong formula:

$$\text{Higher heating value, Btu/lb} = 14{,}544C + 62{,}028[H - (O/8)] + 4{,}050S \tag{4}$$

where C, H, O, and S are the mass fractions of carbon, hydrogen, oxygen, and sulfur in the coal.

Other important parameters in judging coal suitability include

1. **Ultimate analysis**, which is another method of reporting coal composition. Percentages of C, H, O, N, S, and ash in the coal sample are reported. Ultimate analysis is used for detailed fuel studies and for computing a heat balance when required in heating device testing. Typical ultimate analyses of various coals are shown in Table 8.
2. **Ash-fusion temperature**, which indicates the fluidity of the ash at elevated temperatures. It is helpful in selecting coal to be burned in a particular furnace and in estimating the possibility of ash handling and slagging problems.
3. The **grindability index**, which indicates the ease with which a coal can be pulverized and is helpful in estimating ball mill capacity with various coals. There are two common methods for determining the index—Hardgrove and ball mill.
4. The **free-swelling index**, which denotes the extent of coal swelling on combustion on a fuel bed and indicates the coking characteristics of coal.

COMBUSTION CALCULATIONS

Calculations of the quantity of air required for combustion and the quantity of flue gas products generated during combustion are frequently needed for sizing system components and as input to efficiency calculations. Other calculations, such as values for excess air and theoretical CO_2, are useful in estimating combustion system performance.

Frequently, combustion calculations can be simplified by using molecular mass. The molecular mass of a compound equals the sum of the atomic masses of the elements in the compound. Molecular mass can be expressed in any mass units. The pound molecular weight or pound mole is the molecular weight of the compound expressed in pounds. The molecular mass of any substance contains the same number of molecules as the molecular mass of any other substance.

Corresponding to measurement standards common to the industries, calculations involving gaseous fuels are generally based on volume, and calculations involving liquid and solid fuels are generally based on mass.

Some calculations described here require data on concentrations of carbon dioxide, carbon monoxide, and oxygen in the flue gas. Gas analyses for CO_2, CO, and O_2 can be obtained by volumetric chemical analysis and other analytical techniques, including electromechanical cells used in portable electronic flue gas analyzers.

Air Required for Combustion

Stoichiometric or theoretical air is the exact quantity of air required to provide oxygen for complete combustion.

The three most prevalent components in hydrocarbon fuels (C, H_2, and S) are completely combusted as in the following reactions:

$$C + O_2 \rightarrow CO_2$$

$$H_2 + 0.5O_2 \rightarrow H_2O$$

$$S + O_2 \rightarrow SO_2$$

In the reactions, C, H_2, and S can be taken to represent 1 lb mole of carbon, hydrogen, and sulfur, respectively. Using approximate atomic masses (C = 12, H = 1, S = 32, and O = 16), 12 lb of C are oxidized by 32 lb of O_2 to form 44 lb of CO_2, 2 lb of H_2 are oxidized by 16 lb of O_2 to form 18 lb of H_2O, and 32 lb of S are oxidized by 32 lb of O_2 to form 64 lb of SO_2. These relationships can be extended to include hydrocarbons.

The mass of dry air required to supply a given quantity of oxygen is 4.32 times the mass of the oxygen. The mass of air required to oxidize the fuel constituents listed in Table 1 was calculated on this basis. Oxygen contained in the fuel, except that contained in ash, should be deducted from the amount of oxygen required, because this oxygen is already combined with fuel components. In addition, when calculating the mass of air to be supplied for combustion, allowance should be made for water vapor, which is always present in atmospheric air.

As stated previously, combustion calculations for gaseous fuels are based on volume. Avogadro's law states that, for any gas, one mole occupies the same volume at a given temperature and pressure. Therefore, in reactions involving gaseous compounds, the gases react in volume ratios identical to the pound mole ratios. That is, for the oxidation of hydrogen in the above reaction, one volume (or one lb mole) of hydrogen reacts with one-half volume (or one-half lb mole) of oxygen to form one volume (or one lb mole) of water vapor.

The volume of air required to supply a given volume of oxygen is 4.78 times the volume of oxygen. The volumes of dry air required to oxidize the fuel constituents listed in Table 1 were calculated on this basis. Volume ratios are not given for fuels that do not exist in vapor form at reasonable temperatures or pressures. Again, oxygen contained in the fuel should be deducted from the quantity of oxygen required, because this oxygen is already combined with fuel components. Allowance should be made for water vapor, which increases the volume of dry air by 1 to 3%.

From the relationships just described, the theoretical mass m_a of dry air required for stoichiometric combustion of a unit mass of any hydrocarbon fuel is

$$m_a = 0.0144(8C + 24H + 3S - 3O) \tag{5}$$

where C, H, S, and O are the mass percentages of carbon, hydrogen, sulfur, and oxygen in the fuel.

Analyses of gaseous fuels are generally based on hydrocarbon components rather than elemental content.

If the fuel analysis is based on mass, the theoretical mass m_a of dry air required for stoichiometric combustion of a unit mass of gaseous fuel is

$$\begin{aligned} m_a = {} & 2.47CO + 34.28H_2 + 17.24CH_4 + 16.09C_2H_6 \\ & + 15.68C_3H_8 + 15.47C_4H_{10} + 13.27C_2H_2 \\ & + 14.78C_2H_4 + 6.08H_2S - 4.32O_2 \end{aligned} \tag{6}$$

If the fuel analysis is reported on a volumetric or molecular basis, it is simplest to calculate air requirements based on volume and, if necessary, convert to mass. The theoretical volume V_a of air required for stoichiometric combustion of a unit volume of gaseous fuels is

$$\begin{aligned} V_a = {} & 2.39CO + 2.39H_2 + 9.57CH_4 + 16.75C_2H_6 \\ & + 23.95C_3H_8 + 31.14C_4H_{10} + 11.96C_2H_2 \\ & + 14.38C_2H_4 + 7.18H_2S - 4.78O_2 \\ & + 30.47 \text{ illuminants} \end{aligned} \tag{7}$$

where CO, H_2, and so forth are the volumetric fractions of each constituent in the fuel gas.

Illuminants include a variety of compounds not separated by usual gas analysis. In addition to ethylene (C_2H_4) and acetylene (C_2H_2), the principal illuminants included in Equation (7), and the dry air required for combustion, per unit volume of each gas, are: propylene (C_3H_6), 21.44; butylene (C_4H_8), 28.58; pentene (C_5H_{10}), 35.73; benzene (C_6H_6); 35.73, toluene (C_7H_8), 42.88; and xylene (C_8H_{10}), 50.02. Because toluene and xylene are normally scrubbed from the gas before distribution, they can be disregarded in computing air required for combustion of gaseous fuels. The percentage of illuminants present in gaseous fuels is small, so the values can be lumped together, and an approximate value of 30 unit volumes of dry air per unit volume of gas can be used. If ethylene and acetylene are included as illuminants, a value of 20 unit volumes of dry air per unit volume of gaseous illuminants can be used.

For many combustion calculations, only approximate values of air requirements are necessary. If approximate values for theoretical air are sufficient, or if complete information on the fuel is not available, the values in Tables 9 and 10 can be used. Another value used for estimating air requirements is 0.9 ft^3 of air for 100 Btu of fuel.

Table 9 Approximate Air Requirements for Stoichiometric Combustion of Fuels

Type of Fuel	Air Required		Approx. Precision,	
	lb/lb Fuel	ft^3/Unit Fuel[a]	%	Exceptions
Solid	Btu/lb × 0.00073	Btu/lb × 0.0097	3	Fuels containing more than 30% water
Liquid	Btu/lb × 0.00071	Btu/lb × 0.0094	3	Results low for gasoline and kerosene
Gas	Btu/lb × 0.00067	Btu/ft^3 × 0.0089	5	300 Btu/ft^3 or less

Source: Data based on Shnidman (1954).
[a]Unit fuel for solid and liquid fuels in lb, for gas in ft^3.

Table 10 Approximate Air Requirements for Stoichiometric Combustion of Various Fuels

Type of Fuel	Theoretical Air Required for Combustion
Solid fuels	lb/lb fuel
Anthracite	9.6
Semibituminous	11.2
Bituminous	10.3
Lignite	6.2
Coke	11.2
Liquid fuels	lb/gal fuel
No. 1 fuel oil	103
No. 2 fuel oil	106
No. 5 fuel oil	112
No. 6 fuel oil	114
Gaseous fuels	ft^3/ft^3 fuel
Natural gas	9.6
Butane	31.1
Propane	24.0

In addition to the amount theoretically required for combustion, **excess air** must be supplied to most practical combustion systems to ensure complete combustion.

$$\text{Excess air, \%} = \frac{\text{Air supplied} - \text{Theoretical air}}{\text{Theoretical air}} \tag{8}$$

The excess air level at which a combustion process operates significantly affects its overall efficiency. Too much excess air dilutes flue gas excessively, lowering its heat transfer temperature and increasing sensible flue gas loss. Conversely, if the level of excess air is too low, incomplete combustion and loss of unburned combustible gases from the equipment can result. The highest combustion efficiency is usually obtained when just enough excess air is supplied and properly mixed with combustible gases to ensure complete combustion. The general practice is to supply from 5 to 50% excess air, the exact amount depending on the type of fuel burned, combustion equipment, and other factors.

The amount of dry air supplied per unit mass of fuel burned can be obtained from the following equation, which is reasonably precise for most solid and liquid fuels.

$$\text{Dry air supplied} = \frac{C\,(3.04N_2)}{CO_2 + CO} \tag{9}$$

where

Dry air supplied = unit mass per unit mass of fuel
C = unit mass of carbon burned per unit mass of fuel, corrected for carbon in the ash
CO_2, CO, N_2 = percentages by volume from the flue gas analysis

These values of dry air supplied and theoretical air can be used in Equation (8) to determine excess air.

Excess air can also be calculated from unit volumes of stoichiometric combustion products and air, and from volumetric analysis of the flue gas:

$$\text{Excess air, \%} = 100\left(\frac{P}{A}\right)\left(\frac{U - CO_2}{CO_2}\right) \qquad (10)$$

where

U = ultimate carbon dioxide of flue gases resulting from stoichiometric combustion, %
CO_2 = carbon dioxide content of flue gases, %
P = dry products from stoichiometric combustion, unit volume per unit volume of gas burned
A = air required for stoichiometric combustion, unit volume per unit volume of gas burned

As the ratio P/A is approximately 0.9 for most natural gases, a value of 90 can be substituted for 100 (P/A) in Equation (10) for rough calculation.

Because excess air calculations are almost invariably made from flue gas analysis results and theoretical air requirements are not always known, another convenient method of expressing the relation of Equation (8) is

$$\text{Excess air, \%} = \frac{100[O_2 - (CO/2)]}{0.264N_2 - [O_2 - (CO/2)]} \qquad (11)$$

where O_2, CO, and N_2 are percentages by volume from the flue gas analysis.

Theoretical CO_2

The theoretical CO_2, ultimate CO_2, stoichiometric CO_2, or maximum CO_2 concentration attainable in the products from the combustion of a hydrocarbon fuel with air is obtained when the fuel is completely burned with the theoretical quantity of air and zero excess air. Theoretical CO_2 varies with the carbon-hydrogen ratio of the fuel. For combustion with excess air present, theoretical CO_2 values can be calculated from the flue gas analysis:

$$\text{Theoretical } CO_2\text{, \%} = U = \frac{CO_2}{1 - (O_2/20.95)} \qquad (12)$$

where CO_2 and O_2 are percentages by volume from the flue gas analysis.

Table 11 gives approximate theoretical CO_2 values for stoichiometric combustion of several common types of fuel, as well as CO_2 values attained with different amounts of excess air. In practice, desirable CO_2 values depend on the excess air, fuel, firing method, and other considerations.

Quantity of Flue Gas Produced

The mass of dry flue gas produced per mass of fuel burned is required in heat loss and efficiency calculations. This mass is equal to the sum of the mass of (1) fuel (minus ash retained in the furnace), (2) air theoretically required for combustion, and (3) excess air. For solid fuels, this mass, determined from the flue gas analysis, is

$$\text{Dry flue gas} = \frac{11CO_2 + 8O_2 + 7(CO + N_2)}{3(CO_2 + CO)}C \qquad (13)$$

where

Dry flue gas = lb/lb of fuel
C = lb of carbon burned per lb of fuel, corrected for carbon in the ash
CO_2, O_2, CO, N_2 = percentages by volume from flue gas analysis

Table 11 Approximate Maximum Theoretical (Stoichiometric) CO_2 Values, and CO_2 Values of Various Fuels with Different Percentages of Excess Air

Type of Fuel	Theoretical or Maximum CO_2, %	Percent CO_2 at Given Excess Air Values		
		20%	40%	60%
Gaseous Fuels				
Natural gas	12.1	9.9	8.4	7.3
Propane gas (commercial)	13.9	11.4	9.6	8.4
Butane gas (commercial)	14.1	11.6	9.8	8.5
Mixed gas (natural and carbureted water gas)	11.2	12.5	10.5	9.1
Carbureted water gas	17.2	14.2	12.1	10.6
Coke oven gas	11.2	9.2	7.8	6.8
Liquid Fuels				
No. 1 and 2 fuel oil	15.0	12.3	10.5	9.1
No. 6 fuel oil	16.5	13.6	11.6	10.1
Solid Fuels				
Bituminous coal	18.2	15.1	12.9	11.3
Anthracite	20.2	16.8	14.4	12.6
Coke	21.0	17.5	15.0	13.0

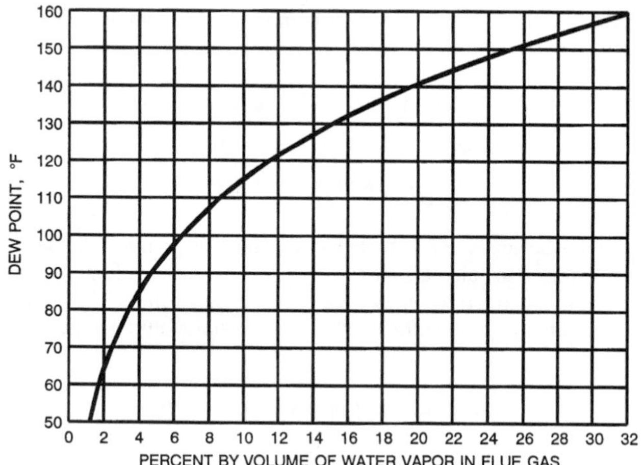

Fig. 2 Water Vapor and Dew Point of Flue Gas
Adapted from *Gas Engineers Handbook* (1965). Printed with permission of Industrial Press and American Gas Association.

The total dry gas volume of flue gases from combustion of one unit volume of gaseous fuels for various percentages of CO_2 is

$$\text{Dry flue gas} = \left(\frac{\text{Volume of } CO_2 \text{ produced}}{\text{Unit vol. of gas burned}}\right)\left(\frac{100}{CO_2}\right) \qquad (14)$$

where

Dry flue gas = unit volume per unit volume of gaseous fuel
CO_2 = percentage by volume from the flue gas analysis

Excess air quantity can be estimated by subtracting the quantity of dry flue gases resulting from stoichiometric combustion from the total volume of flue gas.

Water Vapor and Dew Point of Flue Gas

Water vapor in flue gas is the total of (1) the water contained in the fuel; (2) the water contained in the stoichiometric and excess air; and (3) the water produced from the combustion of hydrogen or hydrocarbons in the fuel. The amount of water vapor in the stoichiometric combustion products may be calculated from the fuel burned by using the water data in Table 1.

The dew point is the temperature at which condensation begins and can be determined using Figure 2. The volume fraction of water vapor in the flue gas can be determined as follows:

$$P_{wv} = \frac{V_w}{(100\,V_c/P_c) + V_w} \tag{15}$$

where

V_w = total water vapor volume (from fuel; from stoichiometric, excess, and dilution air; and from combustion)
V_c = unit volume of CO_2 produced per unit volume of gaseous fuel
P_c = percent CO_2 in flue gas

Using Figure 3, the dew points of solid, liquid, or gaseous fuels may be estimated. For example, to find the dew point of flue gas resulting from the combustion of a solid fuel with a weight ratio (hydrogen to carbon-plus-sulfur) of 0.088 and sufficient excess air to produce 11.4% oxygen in the flue gas, start with the weight ratio of 0.088. Proceed vertically to the intersection of the solid fuels curve and then to the theoretical dew point of 115°F on the dew-point scale (see dotted lines in Figure 3). Follow the curve fixed by this point (down and to the right) to 11.4% oxygen in the flue gas (on the abscissa). The actual dew point is 93°F and is found on the dew-point scale.

An estimation can be made of the dew point of the flue gas from natural gas having a higher heating value (HHV) of 1020 Btu/ft^3 with 6.3% oxygen or 31.5% air. Start with 1020 Btu/ft^3 and proceed vertically to the intersection of the gaseous fuels curve and then to the theoretical dew point of 139°F on the dew-point scale. Follow the curve fixed by this point to 6.3% oxygen or 31.5% air in the flue gas. The actual dew point is 127°F.

The presence of sulfur dioxide, and particularly sulfur trioxide, influences the vapor pressure of condensate in flue gas, and the dew point can be raised by as much as 25 to 75°F, as shown in Figure 4. To illustrate the use of Figure 4, for a manufactured gas with a HHV of 550 Btu/ft^3 containing 15 grains of sulfur per 100 ft^3 being burned with 40% excess air, the proper curve in Figure 4 is determined as follows:

$$\frac{\text{Grains of S per 100 ft}^3\text{ of fuel}}{\text{Btu per ft}^3\text{ of fuel}} \times 100 = \frac{15}{550} \times 100 = 2.73 \tag{16}$$

This curve lies between the 0 and 3 curves and is close to the 3 curve. The dew point for any percentage of excess air from zero to 100% can be determined on this curve. For this flue gas with 40% excess air, the dew point is about 160°F, instead of 127°F for zero sulfur at 40% excess air.

Sample Combustion Calculations

Example 1. Analysis of flue gases from the burning of a natural gas shows 10.0% CO_2, 3.1% O_2, and 86.9% N_2 by volume. Analysis of the fuel is 90% CH_4, 5% N_2, and 5% C_2H_6 by volume. Find U (maximum theoretical percent CO_2), and the percentage of excess air.

Solution: From Equation (12),

$$U = \frac{10.0}{1-(3.1/20.95)} = 11.74\%\ CO_2$$

From Equation (10), using 100 (P/A) = 90,

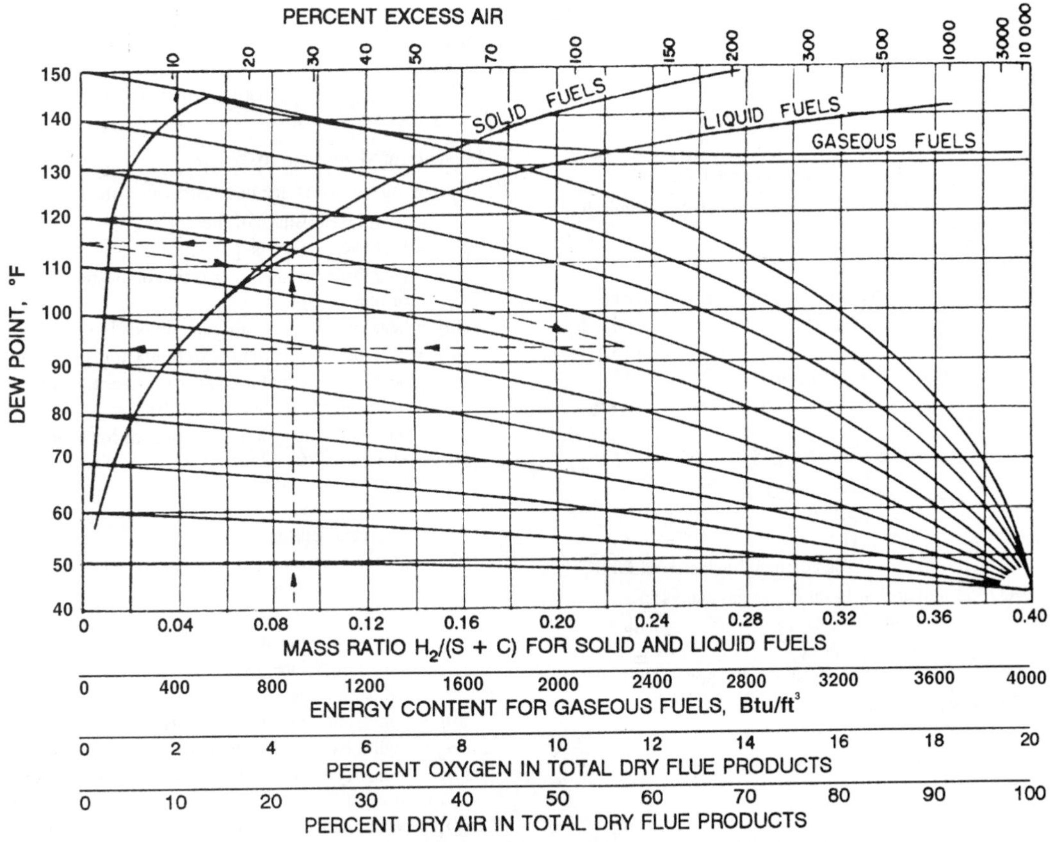

Fig. 3 Theoretical Dew Points of Combustion Products of Industrial Fuels
Adapted from *Gas Engineers Handbook* (1965). Printed with permission of Industrial Press and American Gas Association.

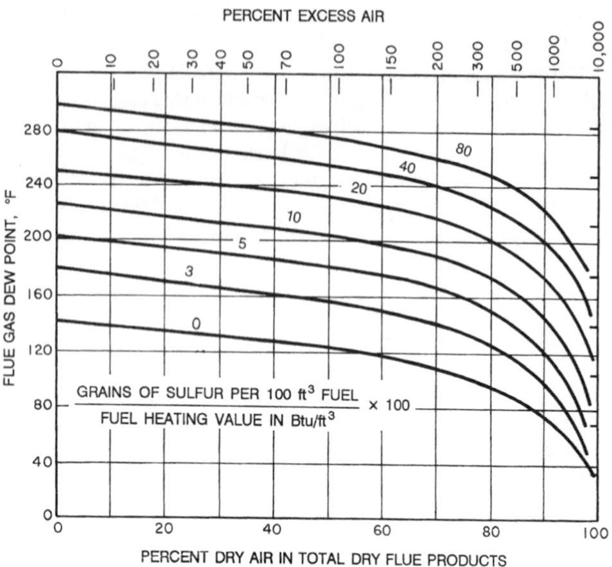

Fig. 4 Influence of Sulfur Oxides on Flue Gas Dew Point
(Stone 1969)

$$\text{Excess air} = \frac{(11.74 - 10.0)90}{10} = 15.7\%$$

Example 2. For the same analysis as in Example 1, find, per cubic foot of fuel gas, the volume of dry air required for combustion, the volume of each constituent in the flue gases, and the total volume of dry and wet flue gases.

Solution: From Equation (7), the volume of dry air required for combustion is:

$$9.57CH_4 + 16.75C_2H_6 = (9.57 \times 0.90) + (16.75 \times 0.05)$$
$$= 9.45 \text{ ft}^3 \text{ per ft}^3 \text{ of fuel gas}$$

(The volume of dry air may also be calculated using Table 10.)

From Table 1, the cubic feet of flue gas constituents per cubic foot of fuel gas are as follows:

Nitrogen, N_2			
From methane	$(0.9CH_4)(9.57 - 2.0)$	=	6.81
From ethane	$(0.05C_2H_6)(16.75 - 3.5)$	=	0.66
Nitrogen in fuel		=	0.05
Nitrogen in excess air	$0.791 \times 0.157 \times 9.45$	=	1.17
	Total nitrogen	=	8.69 ft³
Oxygen, O_2			
In excess air	$0.209 \times 0.157 \times 9.45$	=	0.31 ft³
Carbon dioxide, CO_2			
From methane	$(0.9CH_4)(1.0)$	=	0.90
From ethane	$(0.05C_2H_6)(2.0)$	=	0.10
	Total carbon dioxide	=	1.00 ft³
Water vapor, H_2O (does not appear in some flue gas analyses)			
From methane	$(0.9CH_4)(2.0)$	=	1.8
From ethane	$(0.05C_2H_6)(3.0)$	=	0.15
	Total water vapor	=	1.95 ft³

Total volume of dry gas per cubic foot of fuel gas

$$8.69 + 0.31 + 1.00 = 10.0 \text{ ft}^3$$

Total volume of wet gases per cubic foot of fuel gas (neglecting water vapor in combustion air)

$$10.0 + 1.95 = 11.95 \text{ ft}^3$$

The cubic feet of dry flue gas per cubic foot of fuel gas can also be computed from Equation (14):

$$(1.00)(100)/10.0 = 10.0 \text{ ft}^3$$

EFFICIENCY CALCULATIONS

In analyzing heating appliance efficiency, an energy balance is made that accounts (as much as possible) for disposition of all thermal energy released by combustion of the fuel quantity consumed. The various components of this balance are generally expressed in terms of Btu/lb of fuel burned or as a percentage of its higher heating value. The following are major components of an energy balance and their calculation methods:

1. Useful heat, or heat transferred to the heated medium; for convection heating equipment, this value q_1 is computed as the product of the mass rate of flow and enthalpy change.
2. Heat loss as sensible heat in the dry flue gases

$$q_2 = m_g c_{pg}(t_g - t_a) \tag{17}$$

where m_g is calculated as in Equation (13).

3. Heat loss in water vapor in products formed by combustion of hydrogen

$$q_3 = (9H_2/100)[(h)_{tg} - (h_f)_{ta}] \tag{18}$$

4. Heat loss in water vapor in the combustion air

$$q_4 = Mm_a[(h)_{tg} - (h_g)_{ta}] \tag{19}$$

where m_a is calculated as in Equations (5) and (6).

5. Heat loss from incomplete combustion of carbon

$$q_5 = 10{,}143\,C\left(\frac{CO}{CO_2 + CO}\right) \tag{20}$$

6. Heat loss from unburned carbon in the ash or refuse

$$q_6 = 14{,}600[(C_u/100) - C] \tag{21}$$

7. Unaccounted-for heat losses, q_7

The following symbols are used in Equations (17) through (21):

q_1 = useful heat, Btu/lb of fuel
q_2 = heat loss in dry flue gases, Btu/lb of fuel
q_3 = heat loss in water vapor from combustion of hydrogen, Btu/lb of fuel
q_4 = heat loss in water vapor in combustion air, Btu/lb of fuel
q_5 = heat loss from incomplete combustion of carbon, Btu/lb of fuel
q_6 = heat loss from unburned carbon in ash, Btu/lb of fuel
q_7 = unaccounted-for heat losses, Btu/lb of fuel
c_{pg} = mean specific heat of flue gases at constant pressure (c_{pg} ranges from 0.242 to 0.254 Btu/lb·°F for flue gas temperatures from 300 to 1000°F), Btu/lb·°F
$(h)_{tg}$ = enthalpy of superheated steam at flue gas temperature and 14.7 psia, Btu/lb
$(h_f)_{ta}$ = enthalpy of saturated water vapor at air temperature, Btu/lb
$(h_g)_{ta}$ = enthalpy of saturated steam at combustion air temperature, Btu/lb
m_a = mass of combustion air per mass of fuel used, lb/lb of fuel
m_g = mass of dry flue gas per mass of fuel, lb/lb of fuel
t_a = temperature of combustion air, °F
t_g = temperature of flue gases at exit of heating device, °F
H_2 = hydrogen in fuel, % by mass (from ultimate analysis of fuel)
M = humidity ratio of combustion air, mass of water vapor per mass of dry air
CO, CO_2 = carbon monoxide and carbon dioxide in flue gases, % by volume
C = mass of carbon burned per unit of mass of fuel, corrected for carbon in ash, lb/lb of fuel

$$C = \frac{WC_u - W_aC_a}{100W} \qquad (22)$$

where

C_u = percentage of carbon in fuel by mass from ultimate analysis
W_a = mass of ash and refuse
C_a = percent of combustible in ash by mass (combustible in ash is usually considered to be carbon)
W = mass of fuel used

Useful heat (item 1) is generally measured for a particular piece of combustion equipment.

Flue gas loss is the sum of items 2 through 6. However, for clean-burning gas- and oil-fired equipment, items 5 and 6 are usually negligible and flue gas loss is the sum of items 2, 3, and 4.

Flue gas losses (the sum of items 2, 3, and 4) can be determined with sufficient precision for most purposes from the curves in Figure 5, if O_2 content and flue gas temperature are known. Values of the losses were computed from typical ultimate analyses, assuming 1% water vapor (by mass) in the combustion air. Curves for medium-volatile bituminous coal can be used for high-volatile bituminous coal with no appreciable error.

Generally, item 5 is negligible for modern combustion equipment in good operating condition.

Item 6 is generally negligible for gas and oil firing, but should be determined for coal-firing applications.

Item 7 consists primarily of radiation and convection losses from combustion equipment surfaces and losses caused by incomplete combustion not included in items 5 and 6. Heat loss from incomplete combustion is determined by subtracting the sum of items 1 through 6 from the fuel heating value.

Radiation and convection losses are not usually determined by direct measurement. But if the heating appliance is located within the heated space, radiation and convection losses can be considered useful heat rather than lost heat and can be omitted from heat loss calculations or added to item 1.

If CO is present in flue gases, small amounts of unburned hydrogen and hydrocarbons may also be present. The small losses caused by incomplete combustion of these gases would be included in item 7, if item 7 was determined by subtracting items 1 through 6 from the fuel heating value.

The overall thermal efficiency of combustion equipment is defined as

$$\text{Thermal efficiency, \%} = 100\,\frac{\text{Useful heat}}{\text{Heating value of fuel}} \qquad (23)$$

The following equation can be used to estimate efficiency for equipment where item 7 is small or radiation and convection are useful heat:

Thermal efficiency, % =

$$100\,\frac{\text{Heating value of fuel} - (q_2 + q_3 + q_4 + q_5 + q_6)}{\text{Heating value of fuel}} \qquad (24)$$

Using heating values based on gas volume, the thermal efficiency of a gas appliance can be computed with sufficient precision by the following equation:

$$\eta = \frac{100(Q_h - Q_{fl})}{Q_h} \qquad (25)$$

where

η = thermal efficiency, %
Q_h = higher heating value of fuel gas per unit volume
Q_{fl} = flue gas losses per unit volume of fuel gas

To produce heat efficiently by burning any common fuel, flue gas losses must be minimized by (1) providing adequate heat-absorbing surface in the appliance, (2) maintaining clean heat transfer surfaces on both fire and water or air sides, and (3) reducing excess air to the minimum level consistent with complete combustion and discharge of combustion products.

Seasonal Efficiency

The method just presented is useful for calculating the steady-state efficiency of a heating system or device. Unfortunately, the seasonal efficiency of a combustion heating system can be significantly different from the steady-state efficiency. The primary factor affecting the seasonal efficiency is flue loss during the burner-off period. The warm stack that exists at the end of the firing period can cause airflow in the stack while the burner is off. This airflow can remove heat from furnace and heat exchanger components, from the structure itself, and from pilot flames. Also, if combustion air is drawn from the heated space within the structure, the heated air lost must be at least partly replaced with cold infiltrated air. For further discussion of seasonal efficiency, see Chapter 9 of the 2000 *ASHRAE Handbook—Systems and Equipment.*

COMBUSTION CONSIDERATIONS

Air Pollution

Combustion processes constitute the largest single source of air pollution. Pollutants can be grouped into four categories:

1. Products of incomplete fuel combustion
   - Combustible aerosols (solid and liquid), including smoke, soot, and organics, but excluding ash
   - Carbon monoxide, CO
   - Gaseous hydrocarbons
2. Oxides of nitrogen (generally grouped and referred to as NO_x)
   - Nitric oxide, NO
   - Nitrogen dioxide, NO_2
3. Emissions resulting from fuel contaminants
   - Sulfur oxides, primarily sulfur dioxide, SO_2, and small quantities of sulfur trioxide, SO_3
   - Ash
   - Trace metals
4. Emissions resulting from additives
   - Combustion-controlling additives
   - Other additives

Emission levels of nitrogen oxides and products of incomplete combustion are directly related to the combustion process and can be controlled, to some extent, by process modification. Emissions due to fuel contaminants are related to fuel selection and are slightly affected by the combustion process. Emissions due to additives must be considered in the overall evaluation of the merits of using additives.

Nitrogen oxides are produced during the combustion process, either (1) by thermal fixation (reaction of nitrogen and oxygen at high combustion temperatures), or (2) from fuel nitrogen (oxidation of organic nitrogen in fuel molecules). Unfortunately, high excess air and high flame temperature techniques, which ensure complete fuel combustion, tend to promote NO_x formation.

Table 12 lists NO_x emission factors for uncontrolled fuel-burning equipment (i.e., equipment that does not have exhaust gas recirculation, low-NO_x burners, or other emission controls). Differences in the NO_x emissions of fuels are caused by the flame temperature and different levels of fuel nitrogen. The data in Table 12 are adapted from EPA (1993), Compilation of Air Pollutant Emission Factors, which lists emission factors of a wide variety of equipment, as well as emission reduction options. Carbon monoxide emissions are less dependent on fuel type and typically range

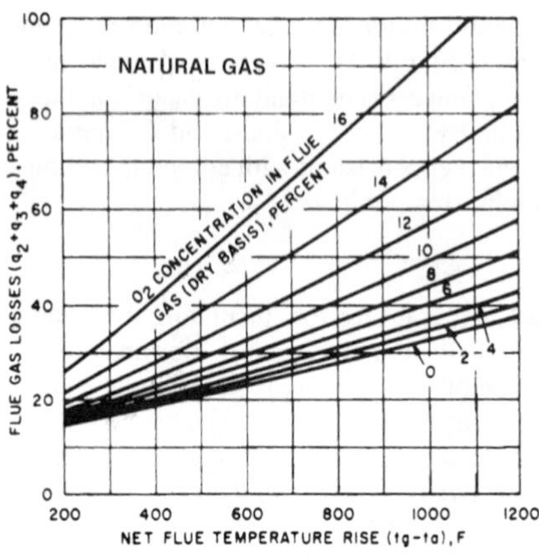

FUEL: NATURAL GAS
HEATING VALUE: 22,345 BTU/LB COMBUSTIBLES
H/C RATIO: 3.8

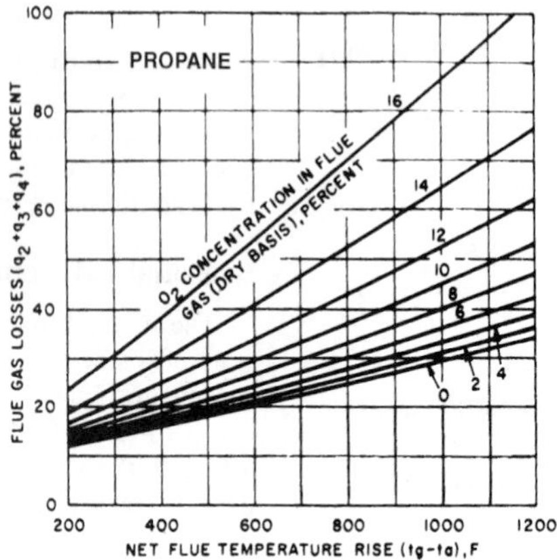

FUEL: PROPANE
HEATING VALUE: 21,646 BTU/LB COMBUSTIBLES
H/C RATIO: 2.667

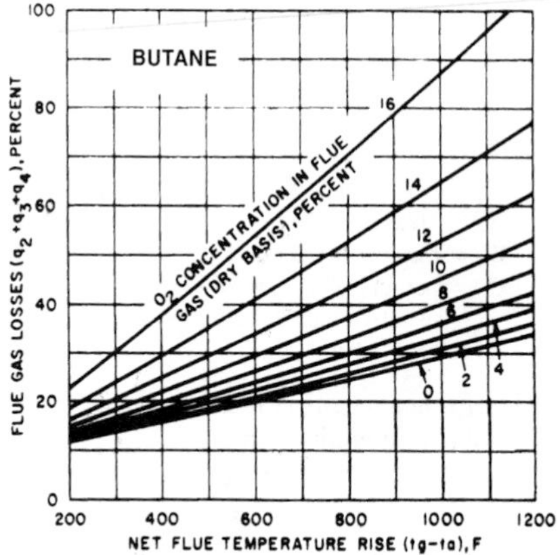

FUEL: BUTANE
HEATING VALUE: 21,180 BTU/LB COMBUSTIBLES
H/C RATIO: 2.5

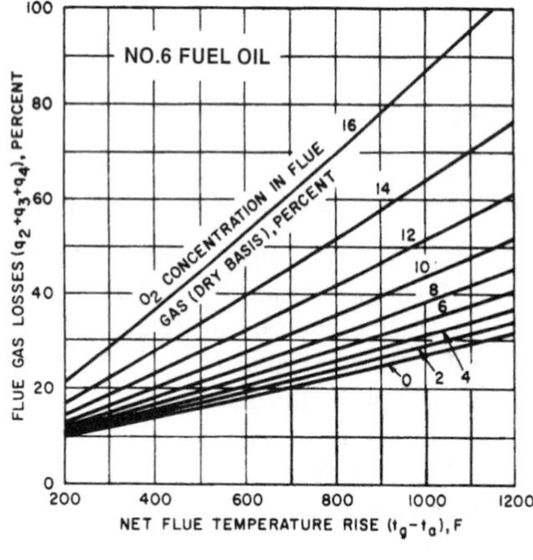

FUEL: NO. 6 FUEL OIL
HEATING VALUE: 19,000 BTU/LB COMBUSTIBLES
H/C RATIO: 1.5

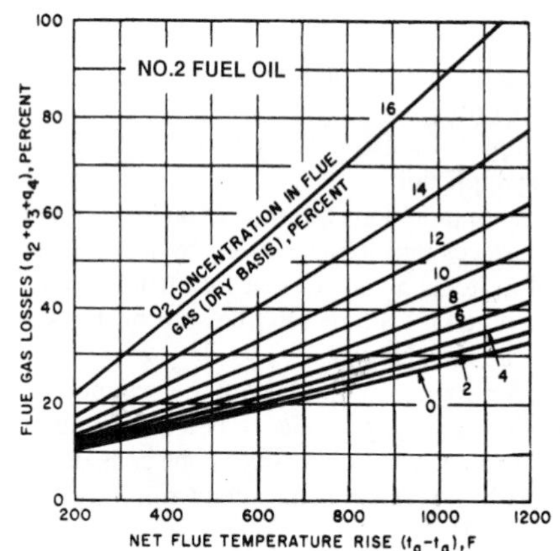

FUEL: NO. 2 FUEL OIL
HEATING VALUE: 19,500 BTU/LB COMBUSTIBLES
H/C RATIO: 1.8

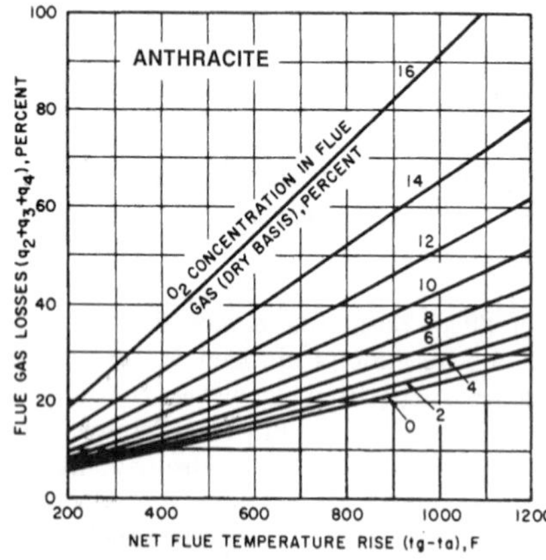

FUEL: ANTHRACITE
HEATING VALUE: 14,980 BTU/LB COMBUSTIBLES
H/C RATIO: 0.28

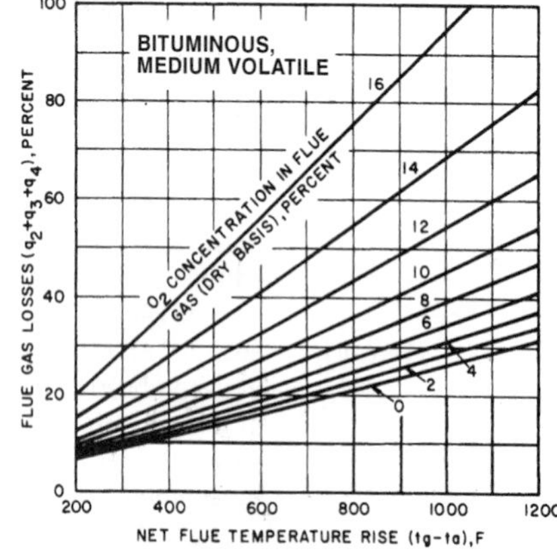

FUEL: BITUMINOUS, MEDIUM VOLATILE
HEATING VALUE: 15,500 BTU/LB COMBUSTIBLES
H/C RATIO: 0.68

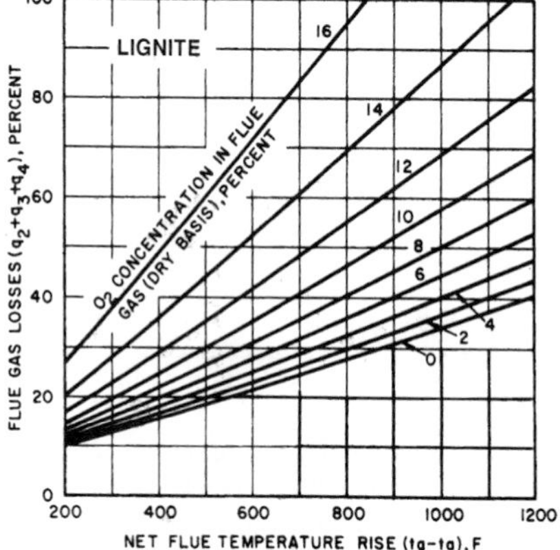

FUEL: LIGNITE
HEATING VALUE: 12,560 BTU/LB COMBUSTIBLES
H/C RATIO: 0.79

Fig. 5 Flue Gas Losses with Various Fuels
(Flue gas temperature rise shown. Loss is based on 65°F room temperature.)

Table 12 NO_x Emission Factors for Combustion Sources Without Emission Controls

Source	NO_x Emission Factor, lb/10^6 Btu of Heat Input
Gas-Fired Equipment	
Small industrial boilers	0.14
Commercial boilers	0.10
Residential furnaces	0.09
Distillate-oil-fired small industrial boilers, commercial boilers, and residential furnaces	0.14
Residual oil-fired small industrial boilers and commercial boilers	0.37

from 0.03 to 0.04 lb/10^6 Btu of heat input. For gas-fired commercial and industrial boilers, particulate emissions range from 0.005 to 0.006 lb/10^6 Btu. For distillate-oil-fired commercial and industrial boilers, particulates are typically 0.014 lb/10^6 Btu. For residential oil-fired equipment, particulate emission factors are 0.003 lb/10^6 Btu. For residual-oil-fired equipment, particulate emissions depend on the sulfur content. For a sulfur content of 1%, the particulate emission rate is typically 0.083 lb/10^6 Btu.

Emission levels of products of incomplete fuel combustion can be reduced by reducing burner cycling, ensuring adequate excess air, improving the mixing of air and fuel (by increasing turbulence, improving distribution, and improving liquid fuel atomization), increasing residence time in the hot combustion zone (possibly by decreasing the firing rate), increasing combustion zone temperatures (to speed reactions), and avoiding quenching the flame before reactions are completed.

The relative contribution of each of these mechanisms to the total NO_x emissions depends on the amount of organic nitrogen in the fuel. Natural gas contains very little nitrogen. Virtually all NO_x emissions with gas firing are due to the thermal mechanism. The nitrogen content of distillate oil varies, but an average of 20 ppm of fuel NO_x is produced (about 20 to 30% of the total NO_x). The fuel nitrogen in residual oil can be significantly higher, with fuel NO_x contributing heavily to the total emissions.

Thermal fixation is strongly dependent on flame maximum temperature. For example, increasing the flame temperature from 2600 to 2800°F increases thermal NO_x tenfold. Therefore, methods to control thermal NO_x are based on methods to reduce the maximum flame temperature. Flue gas recirculation is perhaps the most effective method of reducing thermal NO_x in commercial and industrial boilers. In gas-fired boilers, NO_x reductions of 70% can be realized with 15-20% recirculation of flue gas into the flame. The NO_x reduction decreases with increasing fuel nitrogen content. With distillate-oil firing, reductions of 60-70% can be achieved. In residual-oil-fired boilers, flue gas recirculation can reduce NO_x emissions by 15 to 30%. The maximum rate of flue gas recirculation is limited by combustion instability and CO production.

Two-stage firing is the only technique that reduces NO_x produced both by thermal fixation and fuel nitrogen in industrial and utility applications. The fuel-rich or air-deficient primary combustion zone retards NO_x formation early in the combustion process (when NO_x forms most readily from fuel nitrogen), and avoids peak temperatures, reducing thermal NO_x. Retrofit low-NO_x burners that control air distribution and fuel air mixing in the flame zone can be used to achieve staged combustion. With oil firing, NO_x reductions of 20 to 50% can be obtained with low-NO_x burners. The application of flue gas recirculation and other control methods to residential, oil-fired warm air furnaces was reviewed by Butcher et al. (1994).

The following are some methods of reducing NO_x emissions from gas-fired appliances (Murphy and Putnam 1985):

- Burner adjustment
- Flame inserts (radiation screens or rods)
- Staged combustion and delayed mixing
- Secondary air baffling
- Catalytic and radiant burners
- Total premix
- Pulse

Radiation screens or rods (flame inserts) surrounding or inserted into the flame absorb radiation to reduce flame temperature and retard NO_x formation. Proprietary appliance burners with no flame inserts have been developed and produced to comply with the very strict NO_x emission limitations of California's Air Quality Management Districts.

The U.S. EPA sets limits on air pollutant emissions (Source Performance Standards) from boilers larger than 10 million Btu/h of heat input. In addition, states set emission regulations that are at least as strict at the federal limits and may apply to smaller equipment.

The EPA's automobile emission standard is 1.0 g of NO_2 per mile, which is equivalent to 750 ng/J of NO_x emission. California's maximum is 0.4 g/mile, equivalent to 300 ng/J. California's Air Quality Management Districts for the South Coast (Los Angeles) and the San Francisco Bay Area limit NO_x emission to 40 ng/J of useful heat for some natural gas-fired residential heating appliances.

For further discussion of air pollution aspects of fuel combustion, see EPA (1971a and 1971b).

Condensation and Corrosion

Fuel-burning systems that cycle on and off to meet demand cool down during the off-cycle. When the appliance starts again, condensate forms briefly on surfaces until they are heated above the dew-point temperature. Low-temperature corrosion occurs in system components (heat exchangers, flues, vents, chimneys) when their surfaces remain below the dew-point temperature of flue gas constituents (water vapor, sulfides, chlorides, fluorides, etc.) long enough to cause condensation. Corrosion increases as condensate dwell time increases.

Acids in the flue gas condensate are the principle substances responsible for low-temperature corrosion in fuel-fired systems. Sulfuric, hydrochloric, and other acids are formed when acidic compounds in fuel and air combustion products combine with condensed moisture in appliance heat exchangers, flues, or vents. Corrosion can be avoided by maintaining these surfaces above the flue gas dew point.

In high-efficiency, condensing-type appliances and economizers, flue gas temperatures are intentionally reduced below the flue gas dew-point temperatures to achieve efficiencies approaching 100%. In these systems, the surfaces subjected to condensate must be made of corrosion-resistant materials. The most corrosive conditions exist at the leading edge of the condensing region, especially those areas that experience evaporation during each cycle (Strickland et al. 1987). Drainage of condensate retards the concentration of acids on system surfaces. Regions from which condensate partially or completely drains away before evaporation are less severely attacked than regions from which condensate does not drain before evaporation.

The metals most resistant to condensate corrosion are stainless-steel alloys with high chromium and molybdenum content, and nickel-chromium alloys with high molybdenum content (Stickford et al. 1988). Aluminum experiences general corrosion rather than pitting when exposed to flue gas condensate. If applied in sufficiently thick cross section to allow for metal loss, aluminum can be used in condensing regions. Most ceramic and high-temperature polymer materials resist the corrosive effects of flue gas condensate. These materials may have application in the condensing regions, if they can meet the structural and temperature requirements of a particular application.

In coal-fired power plants, the rate of corrosion for carbon steel condensing surfaces by the mixed acids (primarily sulfuric and hydrochloric) is reported to be maximum at about 122 ± 18°F (Davis 1987). Mitigation techniques include (1) acid neutralization with a base such as NH_3 or $Ca(OH)_2$; (2) use of protective linings of glass-filled polyester or coal-tar epoxy; and (3) replacement of steel with molybdenum-bearing stainless steels, nickel alloys, polymers, or other corrosion-resistant materials. Other elements in residual fuel oils and coals that contribute to high-temperature corrosion include sodium, potassium, and vanadium. Each fuel-burning system component should be evaluated during installation, or when modified, to determine the potential for corrosion and the means to retard corrosion (Paul et al. 1988).

Soot

Soot deposits on flue surfaces of a boiler or heater act as an insulating layer over the surface, reducing heat transfer to the water or air. Soot can also clog flues, reduce draft and available air, and prevent proper combustion. Proper burner adjustment can minimize soot accumulation. The use of off-specification fuel can contribute to the generation of soot.

REFERENCES

ASTM. 1995. Standard specification for fuel oils. ANSI/ASTM *Standard* D 396-95. American Society for Testing and Materials, West Conshohocken, PA.

ASTM. 1996. Standard specification for diesel fuel oils. ANSI/ASTM *Standard* D 975-96.

ASTM. 1996. Standard specification for gas turbine fuel oils. ANSI/ASTM *Standard* D 2880-96.

ASTM. 1995. Standard classification of coals by rank. ASTM *Standard* D 388-95.

Butcher, T.A., L. Fisher, B. Kamath, T. Kirchstetter, and J. Batey. 1994. Nitrogen oxides (NO_x) and oil burners. Proceedings of the 1994 Oil Heat Technology Conference and Workshops. BNL *Report* No. 52430. Brookhaven National Laboratory, Upton, NY.

Coward, H.F. and G.W. Jones. 1952. Limits of flammability of gases and vapors. *Bulletin* 503. U.S. Bureau of Mines, Washington, D.C.

Davis, J.R., ed. 1987. *Metals handbook*, 9th ed., Vol. 13. ASM International, Metals Park, OH.

Dickson, C.L. and G.P. Sturm, Jr. 1994. *Heating oils*. National Institute for Petroleum and Energy Research, Bartlesville, Oklahoma.

EPA. 1971a. Standards of performance for new stationary sources, Group I, Federal Register 36, August 17. U.S. Environmental Protection Agency, Washington, D.C.

EPA. 1971b. Standards of performance for new stationary sources, Group I, Part II, Federal Register 36, December 23. U.S. Environmental Protection Agency, Washington, D.C.

EPA. 1993. Compilation of air pollutant emission factors. *Report* AP-42. U.S. Environmental Protection Agency, Washington, D.C.

Gas engineers handbook. 1965. The Industrial Press, New York.

Hartman, I. 1958. "Dust explosions." In *Mechanical engineers' handbook*, 6th ed., Section 7, pp. 41-48. McGraw-Hill, New York.

Hazard, H.R. 1971. "Gas turbine fuels." In *Gas turbine handbook*, Gas Turbine Publications, Stamford, CT.

Murphy, M.J. and A.A. Putnam. 1985. Burner technology bulletin: Control of NO_x emissions from residential gas appliances. *Report* GRI-85/0132. Battelle Columbus Division for Gas Research Institute.

NFPA. 1962. Fire-hazard properties of flammable liquids, gases and volatile solids, Tables 6-126, pp. 6-131 ff. In *Fire protection handbook*, 12th ed., National Fire Protection Association, Quincy, MA.

NFPA/IAS. 1999. ANSI/NFPA *Standard* 54-1992. National Fuel Gas Code, Section 8.1.2. National Fire Protection Association, Quincy, MA. ANSI/IAS *Standard* Z223.1-1999. American Gas Association, Arlington, VA.

North American combustion handbook. 1978. The North American Manufacturing Co., Cleveland, OH.

Paul, D.D., A.L. Rutz, S.G. Talbert, J.J. Crisafolli, G.R. Whitacre, and R.D. Fischer. 1988. User's manual for Vent-II Ver. 3.0—A dynamic microcomputer program for analyzing gas venting systems. *Report* GRI-88/0304. Battelle Columbus Division for Gas Research Institute.

Scott, G.S., et al. 1948. Determination of ignition temperatures of combustible liquids and gases. *Analytical Chemistry* 20:238-41.

Shelton, E.M. 1974. *Burner oil fuels*. Petroleum Products Survey 86. U.S. Bureau of Mines, Washington, D.C.

Shnidman, L. 1954. *Gaseous fuels*. American Gas Association, Arlington, VA.

Stickford, G.H., S.G. Talbert, B. Hindin, and D.W. Locklin. 1988. Research on corrosion-resistant materials for condensing heat exchangers. Proceedings of the 39th Annual International Appliance Technical Conference.

Stone, R.L. 1969. Fireplace operation depends upon good chimney design. *ASHRAE Journal* 11(February):63-69.

Trinks, W. 1947. Simplified calculation of radiation from non-luminous furnace gases. *Industrial Heating* 14:40-46.

U.S. Bureau of Mines. Semiannually. *Mineral Industry Surveys, Motor Gasolines*. Washington, D.C.

Zabetakis, M.G. 1956. Research on the combustion and explosion hazards of hydrogen-water vapor-air mixtures. Division of Explosives Technology, *Progress Report* 1. U.S. Bureau of Mines, Washington, D.C.

BIBLIOGRAPHY

Bonne, U. and A. Patani. 1982. Combustion system performance analysis and simulation study. *Report* GRI-81/0093 (PB 83-161 406). Honeywell SSPL, Bloomington, MN.

Gas Appliance Technology Center, Gas Research Institute, Manufacturer update on status of GATC research on heat-exchanger corrosion, May 1984. Battelle Columbus Laboratories and American Gas Association Laboratories.

Lewis, B. and G. von Elbe. 1987. *Combustion, flames, and explosion of gases*, 3rd ed. Academic Press, New York.

Stickford, G.H., S.G. Talbert, and D.W. Locklin. 1987. Condensate corrosivity in residential condensing appliances. Proceedings of the International Symposium on Condensing Heat Exchangers, Paper 3, BNL *Report* No. 52068, 1 and 2. Brookhaven National Laboratory, Upton, NY.

CHAPTER 19

REFRIGERANTS

REFRIGERANTS are the working fluids in refrigeration, air-conditioning, and heat pumping systems. They absorb heat from one area, such as an air-conditioned space, and reject it into another, such as outdoors, usually through evaporation and condensation, respectively. These phase changes occur both in absorption and mechanical vapor compression systems, but they do not occur in systems operating on a gas cycle using a fluid such as air. (See Chapter 1 for more information on refrigeration cycles.) The design of the refrigeration equipment depends strongly on the properties of the selected refrigerant. Table 1 lists ASHRAE standard refrigerant designations from ASHRAE *Standard* 34.

Refrigerant selection involves compromises between conflicting desirable thermodynamic properties. A refrigerant must satisfy many requirements, some of which do not directly relate to its ability to transfer heat. Chemical stability under conditions of use is the most important characteristic. Safety codes may require a nonflammable refrigerant of low toxicity for some applications. Cost, availability, efficiency, and compatibility with compressor lubricants and materials with which the equipment is constructed are other concerns.

The environmental consequences of a refrigerant that leaks from a system must also be considered. Because of their great stability, fully halogenated compounds, such as **chlorofluorocarbons** (CFCs), persist in the atmosphere for many years and eventually diffuse into the stratosphere. The molecules of CFCs, such as R-11 and R-12, contain only carbon and the halogens chlorine and fluorine. Once in the upper atmosphere, CFC molecules break down and release chlorine, which destroys ozone (**ozone depletion**). In the lower atmosphere, these molecules absorb infrared radiation, which may contribute to the warming of the earth. Substitution of a hydrogen atom for one or more of the halogens in a CFC molecule greatly reduces its atmospheric lifetime and lessens its environmental impact. These compounds are called **hydrochlorofluorocarbons** (HCFCs). A similar class of compounds used as fire extinguishing agents and called halons also cause ozone depletion. **Halons** are compounds containing bromine, fluorine, and carbon. Like CFCs, halons break down, but release bromine, which is even more destructive to stratospheric ozone than chlorine.

Latent heat of vaporization is another important property. On a molar basis, fluids with similar boiling points have almost the same latent heat. Since the compressor operates on volumes of gas, refrigerants with similar boiling points produce similar capacities in a given compressor. On a mass basis, latent heat varies widely among fluids. The maximum efficiency of a theoretical vapor compression cycle is achieved by fluids with low vapor heat capacity. This property is associated with fluids having a simple molecular structure and low molecular weight.

The preparation of this chapter is assigned to TC 3.1, Refrigerants and Secondary Coolants.

Transport properties of thermal conductivity and viscosity affect the performance of heat exchangers and piping. High thermal conductivity and low viscosity are desirable.

No single fluid satisfies all the attributes desired of a refrigerant; as a result, a variety of refrigerants is used. This chapter describes the basic characteristics of various refrigerants, and Chapter 20 lists thermophysical properties.

PHASEOUT OF REFRIGERANTS

The Montreal Protocol is an international treaty that controls the production of ozone-depleting substances, including refrigerants containing chlorine and/or bromine (U.N. 1994, 1996). The original Protocol was signed September 16, 1987, by the European Economic Community (currently the European Union) and 24 nations, including the United States. It entered into force on January 1, 1989, and limits the 1998 production of specified CFCs to 50% of their 1986 levels. Starting in 1992, the production of specified halons (including R-13B1) was frozen at 1986 levels. Developing countries were granted additional time to meet these deadlines.

The original Protocol contained provisions for periodic revision. Four such revisions, referred to as the London, Copenhagen, Montreal, and Beijing Amendments, were agreed to in 1990, 1992, 1997 and 1999, respectively. As of February, 2000, the Montreal Protocol had been ratified by 172 parties, the London Amendment by 138 parties, and the Copenhagen Amendment by 104 parties; the Beijing amendment has yet to be ratified.

The Copenhagen Amendment entered into force on June 14, 1994. It called for a complete cessation of the production of CFCs by January 1, 1996, and of halons by January 1, 1994. Continued use from existing (reclaimed or recycled) stock is permitted. Allowance is also provided for continued production for very limited essential uses. In addition, HCFCs (such as R-22 and R-123) are to be phased out relative to a 1989 reference level for developed countries. Production was frozen at the reference level on January 1, 1996. Production will be limited to 65% of the reference level by January 1, 2004; to 35% by January 1, 2010; to 10% by January 1, 2015; and to 0.5% of the reference level by January 1, 2020. Complete cessation of the production of HCFCs is called for by January 1, 2030. In addition to the international agreement, individual countries may have domestic regulations for ozone-depleting compounds.

The Beijing Amendment will regulate the production of HCFCs in developed countries. A production cap will begin in 2004 and will be equal to the original HCFC use cap plus an additional 15% allowance to meet developing country needs. At this time, there is no provision for reductions to this production cap.

The production and use of hydrofluorocarbon (HFC) refrigerants (such as R-32, R-125, R-134a, R-143a, and their mixtures, including R-404A, R-407C, and R-410A) are not regulated by the Montreal Protocol.

Table 1 Standard Designation of Refrigerants (ASHRAE *Standard* 34)

Refrigerant Number	Chemical Name or Composition (% by mass)	Chemical Formula
Methane Series		
10	tetrachloromethane (carbon tetrachloride)	CCl_4
11	trichlorofluoromethane	CCl_3F
12	dichlorodifluoromethane	CCl_2F_2
12B1	bromochlorodifluoromethane	$CBrClF_2$
12B2	dibromodifluoromethane	CBr_2F_2
13	chlorotrifluoromethane	$CClF_3$
13B1	bromotrifluoromethane	$CBrF_3$
14	tetrafluoromethane (carbon tetrafluoride)	CF_4
20	trichloromethane (chloroform)	$CHCl_3$
21	dichlorofluoromethane	$CHCl_2F$
22	chlorodifluoromethane	$CHClF_2$
22B1	bromodifluoromethane	$CHBrF_2$
23	trifluoromethane	CHF_3
30	dichloromethane (methylene chloride)	CH_2Cl_2
31	chlorofluoromethane	CH_2ClF
32	difluoromethane (methylene fluoride)	CH_2F_2
40	chloromethane (methyl chloride)	CH_3Cl
41	fluoromethane (methyl fluoride)	CH_3F
50	methane	CH_4
Ethane Series		
110	hexachloroethane	CCl_3CCl_3
111	pentachlorofluoroethane	CCl_3CCl_2F
112	1,1,2,2-tetrachloro-1,2-difluoroethane	CCl_2FCCl_2F
112a	1,1,1,2-tetrachloro-2,2-difluoroethane	CCl_3CClF_2
113	1,1,2-trichloro-1,2,2-trifluoroethane	CCl_2FCClF_2
113a	1,1,1-trichloro-2,2,2-trifluoroethane	CCl_3CF_3
114	1,2-dichloro-1,1,2,2-tetrafluoroethane	$CClF_2CClF_2$
114a	1,1-dichloro-1,2,2,2-tetrafluoroethane	CCl_2FCF_3
114B2	1,2-dibromo-1,1,2,2-tetrafluoroethane	$CBrF_2CBrF_2$
115	chloropentafluoroethane	$CClF_2CF_3$
116	hexafluoroethane	CF_3CF_3
120	pentachloroethane	$CHCl_2CCl_3$
123	2,2-dichloro-1,1,1-trifluoroethane	$CHCl_2CF_3$
123a	1,2-dichloro-1,1,2-trifluoroethane	$CHClFCClF_2$
124	2-chloro-1,1,1,2-tetrafluoroethane	$CHClFCF_3$
124a	1-chloro-1,1,2,2-tetrafluoroethane	CHF_2CClF_2
125	pentafluoroethane	CHF_2CF_3
133a	2-chloro-1,1,1-trifluoroethane	CH_2ClCF_3
134a	1,1,1,2-tetrafluoroethane	CH_2FCF_3
140a	1,1,1-trichloroethane (methyl chloroform)	CH_3CCl_3
141b	1,1-dichloro-1-fluoroethane	CCl_2FCH_3
142b	1-chloro-1,1-difluoroethane	$CClF_2CH_3$
143a	1,1,1-trifluoroethane	CF_3CH_3
150a	1,1-dichloroethane	$CHCl_2CH_3$
152a	1,1-difluoroethane	CHF_2CH_3
160	chloroethane (ethyl chloride)	CH_3CH_2Cl
170	ethane	CH_3CH_3
Propane Series		
216ca	1,3-dichloro-1,1,2,2,3,3-hexafluoropropane	$CClF_2CF_2CClF_2$
218	octafluoropropane	$CF_3CF_2CF_3$
245cb	1,1,1,2,2-pentafluoropropane	$CF_3CF_2CH_3$
290	propane	$CH_3CH_2CH_3$
Cyclic Organic Compounds		
C316	1,2-dichloro-1,2,3,3,4,4-hexafluorocyclobutane	$C_4Cl_2F_6$
C317	chloroheptafluorocyclobutane	C_4ClF_7
C318	octafluorocyclobutane	C_4F_8
Zeotropic Blends (% by mass)		
400	R-12/114 (must be specified)	
401A	R-22/152a/124 (53/13/34)	
401B	R-22/152a/124 (61/11/28)	
401C	R-22/152a/124 (33/15/52)	
402A	R-125/290/22 (60/2/38)	
402B	R-125/290/22 (38/2/60)	
403A	R-290/22/218 (5/75/20)	
403B	R-290/22/218 (5/56/39)	
404A	R-125/143a/134a (44/52/4)	
405A	R-22/152a/142b/C318 (45/7/5.5/42.5)	
406A	R-22/600a/142b (55/4/41)	
407A	R-32/125/134a (20/40/40)	
407B	R-32/125/134a (10/70/20)	
407C	R-32/125/134a (23/25/52)	
407D	R-32/125/134a (15/15/70)	
408A	R-125/143a/22 (7/46/47)	
409A	R-22/124/142b (60/25/15)	
409B	R-22/124/142b (65/25/10)	
410A	R-32/125 (50/50)	
410B	R-32/125 (45/55)	
411A	R-1270/22/152a (1.5/87.5/11.0)	
411B	R-1270/22/152a (3/94/3)	
412A	R-22/218/142b (70/5/25)	
413A	R-218/134a/600a (9/88/3)	
Azeotropic Blends (% by mass)		
500	R-12/152a (73.8/26.2)	
501	R-22/12 (75.0/25.0)*	
502	R-22/115 (48.8/51.2)	
503	R-23/13 (40.1/59.9)	
504	R-32/115 (48.2/51.8)	
505	R-12/31 (78.0/22.0)*	
506	R-31/114 (55.1/44.9)	
507A	R-125/143a (50/50)	
508A	R-23/116 (39/61)	
508B	R-23/116 (46/54)	
509A	R-22/218 (44/56)	
Miscellaneous Organic Compounds		
Hydrocarbons		
600	butane	$CH_3CH_2CH_2CH_3$
600a	2-methyl propane (isobutane)	$CH(CH_3)_3$
Oxygen Compounds		
610	ethyl ether	$C_2H_5OC_2H_5$
611	methyl formate	$HCOOCH_3$
Sulfur Compounds		
620	(Reserved for future assignment)	
Nitrogen Compounds		
630	methyl amine	CH_3NH_2
631	ethyl amine	$C_2H_5NH_2$
Inorganic Compounds		
702	hydrogen	H_2
704	helium	He
717	ammonia	NH_3
718	water	H_2O
720	neon	Ne
728	nitrogen	N_2
732	oxygen	O_2
740	argon	Ar
744	carbon dioxide	CO_2
744A	nitrous oxide	N_2O
764	sulfur dioxide	SO_2
Unsaturated Organic Compounds		
1112a	1,1-dichloro-2,2-difluoroethene	$CCl_2{=}CF_2$
1113	1-chloro-1,2,2-trifluoroethene	$CClF{=}CF_2$
1114	tetrafluoroethene	$CF_2{=}CF_2$
1120	trichloroethene	$CHCl{=}CCl_2$
1130	1,2-dichloroethene (trans)	$CHCl{=}CHCl$
1132a	1,1 difluoroethene (vinylidene fluoride)	$CF_2{=}CH_2$
1140	1-chloroethene (vinyl chloride)	$CHCl{=}CH_2$
1141	1-fluoroethene (vinyl fluoride)	$CHF{=}CH_2$
1150	ethene (ethylene)	$CH_2{=}CH_2$
1270	propene (propylene)	$CH_3CH{=}CH_2$

*The exact composition of this azeotrope is in question.

Table 2 Physical Properties of Selected Refrigerants[a]

Refrigerant									
No.	Chemical Name or Composition (% by mass)	Chemical Formula	Molecular Mass	Boiling Pt. (NBP) at 14.696 psia, °F	Freezing Point, °F	Critical Temperature, °F	Critical Pressure, psia	Critical Volume, ft^3/lb	Refractive Index of Liquid[b,c]
704	Helium	He	4.0026	–452.1	None	–450.3	33.21	0.2311	1.021 (NBP) 5461 Å
702p	Hydrogen, para	H_2	2.0159	–423.2	–434.8	–400.3	187.5	0.5097	1.09 (NBP)[f]
702n	Hydrogen, normal	H_2	2.0159	–423.0	–434.5	–399.9	190.8	0.5320	1.097 (NBP) 5791 Å
720	Neon	Ne	20.183	–410.9	–415.5	–379.7	493.1	0.03316	—
728	Nitrogen	N_2	28.013	–320.4	–346.0	–232.4	492.9	0.05092	1.205 (83 K) 5893 Å
729	Air	—	28.97	–317.8	—	–220.95	548.9	0.0530	—
						–221.1	546.3	0.05007	—
740	Argon	Ar	39.948	–302.55	–308.7	–188.48	704.9	0.0301	1.233 (84 K) 5893 Å
732	Oxygen	O_2	31.9988	–297.332	–361.8	–181.424	731.4	0.03673	1.221 (92 K) 5893 Å
50	Methane	CH_4	16.04	–258.7	–296	–116.5	673.1	0.099	—
14	Tetrafluoromethane	CF_4	88.01	–198.3	–299	–50.2	543	0.0256	—
1150	Ethylene	C_2H_4	28.05	–154.7	–272	48.8	742.2	0.070	1.363(–148)[1]
744A [2]	Nitrous oxide	N_2O	44.02	–129.1	–152	97.7	1048	0.0355	—
170	Ethane	C_2H_6	30.07	–127.85	–297	90.0	709.8	0.0830	—
503	R-23/13 (40.1/59.9)	—	87.5	–127.6	—	67.1	607	0.0326	—
508A [9]	R-23/116 (39/61)	—	100.1	–125.34	—	51.82	536.78	0.0279	—
508B [9]	R-23/116 (46/54)	—	95.39	–125.28	—	53.71	556.07	0.0280	—
23	Trifluoromethane	CHF_3	70.02	–115.7	–247	78.1	701.4	0.0311	—
13	Chlorotrifluoromethane	$CClF_3$	104.47	–114.6	–294	83.9	561	0.0277	1.146 (77)[4]
744	Carbon dioxide	CO_2	44.01	–109.2[d]	–69.9[e]	87.9	1070.0	0.0342	1.195 (59)
13B1	Bromotrifluoromethane	$CBrF_3$	148.93	–71.95	–270	152.6	575	0.0215	1.239 (77)[4]
504	R-32/115 (48.2/51.8)	—	79.2	–71.0	—	151.5	690.5	0.0324	—
32	Difluoromethane	CH_2F_2	52.02	–61.1	–213	173.14	845.6	0.03726	—
410A [9]	R-32/125 (50/50)	—	72.6	–60.83	—	158.4	694.87	0.0293	—
125	Pentafluoroethane	C_2HF_5	120.03	–55.43	–153.67	151.34	526.57	—	—
1270	Propylene	C_3H_6	42.09	–53.86	–301	197.2	670.3	0.0720	1.3640 (–58)[1]
143a [9]	Trifluoroethane	CH_3CF_3	84	–53.039	–169.26	162.87	545.49	0.0372	—
507A [9]	R-125/143a (50/50)	—	98.9	–52.80	—	159.34	538.97	0.0325	—
404A [9]	R-125/143a/134a (44/52/4)	—	97.6	–51.66	—	162.5	597.5	0.0279	—
502 [5]	R-22/115 (48.8/51.2)	—	111.63	–49.8	—	179.9	591.0	0.0286	—
407C [9]	R-32/125/134a (23/25/52)	—	86.2	–46.22	—	186.9	672.2	0.0317	—
290	Propane	C_3H_8	44.10	–43.76	–305.8	206.1	616.1	0.0726	1.3397 (–43)
22	Chlorodifluoromethane	$CHClF_2$	86.48	–41.36	–256	204.8	721.9	0.0305	1.234 (77)[4]
115	Chloropentafluoroethane	$CClF_2CF_3$	154.48	–38.4	–159	175.9	457.6	0.0261	1.221 (77)[4]
500	R-12/152a (73.8/26.2)	—	99.31	–28.3	–254	221.9	641.9	0.0323	—
717	Ammonia	NH_3	17.03	–28.0	–107.9	271.4	1657	0.068[d]	1.325 (61.7)
12	Dichlorodifluoromethane	CCl_2F_2	120.93	–21.62	–252	233.6	596.9	0.0287	1.288 (77)[4]
134a	Tetrafluoroethane	CF_3CH_2F	102.03	–15.08	–141.9	214.0	589.8	0.029	—
152a	Difluoroethane	CHF_2CH_3	66.05	–13.0	–178.6	236.3	652	0.0439	—
40 [2]	Methyl chloride	CH_3Cl	50.49	–11.6	–144	289.6	968.7	0.0454	—
124	Chlorotetrafluoroethane	$CHClFCF_3$	136.47	8.26	–326.47	252.5	530.84	—	—
600a	Isobutane	C_4H_{10}	58.13	10.89	–255.5	275.0	529.1	0.0725	1.3514 (–13)[1]
764 [6]	Sulfur dioxide	SO_2	64.07	14.0	–103.9	315.5	1143	0.0306	—
142b	Chlorodifluoroethane	$CClF_2CH_3$	100.5	14.4	–204	278.8	598	0.0368	—
630 [6]	Methyl amine	CH_3NH_2	31.06	19.9	–134.5	314.4	1082		1.432 (63.5)
C318	Octafluorocyclobutane	C_4F_8	200.04	21.5	–42.5	239.6	403.6	0.0258	—
600	Butane	C_4H_{10}	58.13	31.1	–217.3	305.6	550.7	0.0702	1.3562 (5)[1]
114	Dichlorotetrafluoroethane	$CClF_2CClF_2$	170.94	38.8	–137	294.3	473	0.0275	1.294 (77)
21 [7]	Dichlorofluoromethane	$CHCl_2F$	102.92	47.8	–211	353.3	750	0.0307	1.332 (77)[4]
160 [2]	Ethyl chloride	C_2H_5Cl	64.52	54.32	–216.9	369.0	764.4	0.0485	—
631 [6]	Ethyl amine	$C_2H_5NH_2$	45.08	61.88	–113	361.4	815.6	—	—
11	Trichlorofluoromethane	CCl_3F	137.38	74.87	–168	388.4	639.5	0.0289	1.362 (77)[4]
123	Dichlorotrifluoroethane	$CHCl_2CF_3$	152.93	82.17	–160.87	362.82	532.87	—	—
611 [6]	Methyl formate	$C_2H_4O_2$	60.05	89.2	–146	417.2	870	0.0459	—
141b	Dichlorofluoroethane	CCl_2FCH_3	116.95	89.6	—	399.6	616.4	—	—
610 [6]	Ethyl ether	$C_4H_{10}O$	74.12	94.3	–177.3	381.2	523	0.0607	1.3526 (68)
216ca	Dichlorohexafluoropropane	$C_3Cl_2F_6$	220.93	96.24	–193.7	356.0	399.5	0.0279	—
30 [6]	Methylene chloride	CH_2Cl_2	84.93	104.4	–142	458.6	882	—	1.4244 (68)[3]
113	Trichlorotrifluoroethane	CCl_2FCClF_2	187.39	117.63	–31	417.4	498.9	0.0278	1.357 (77)[4]
1130 [8]	Dichloroethylene	$CHCl{=}CHCl$	96.95	118	–58	470	795	—	—
1120 [6]	Trichloroethylene	$CHCl{=}CCl_2$	131.39	189.0	–99	520	728	—	1.4782(68)[3]
718 [6]	Water	H_2O	18.02	212	32	705.18	3200	0.0498	—

Notes for Table 2

[a] Data from ASHRAE *Thermodynamic Properties of Refrigerants* (Stewart et al. 1986) or from McLinden (1990), unless otherwise noted.
[b] Temperature of measurement (°F, unless kelvin is noted) shown in parentheses. Data from CRC *Handbook of Chemistry and Physics* (CRC 1987), unless otherwise noted.
[c] For the sodium D line.
[d] Sublimes.
[e] At 76.4 psia.
[f] Dielectric constant data.

References

[1] Kirk and Othmer (1956).
[2] *Matheson Gas Data Book* (1966).
[3] Electrochemicals Department, E.I. duPont de Nemours & Co.
[4] *Bulletin* B-32A (duPont).
[5] *Bulletin* T-502 (duPont 1980).
[6] *Handbook of Chemistry* (1967).
[7] *Bulletin* G-1 (duPont).
[8] CRC *Handbook of Chemistry and Physics* (CRC 1987).
[9] NIST *Standard Reference Database* 23, Version 6.01.

REFRIGERANT PROPERTIES

Physical Properties

Table 2 lists some physical properties of commonly used refrigerants, a few very low-boiling cryogenic fluids, some newer refrigerants, and some older refrigerants of historical interest. These refrigerants are arranged in increasing order of atmospheric boiling point, from helium at −452.1°F to water at 212°F.

Table 2 also includes the freezing point, critical properties, and refractive index. Of these properties, the boiling point is most important because it is a direct indicator of the temperature level at which a refrigerant can be used. The freezing point must be lower than any contemplated usage. The critical properties describe a material at the point where the distinction between liquid and gas is lost. At higher temperatures, no separate liquid phase is possible. In refrigeration cycles involving condensation, a refrigerant must be chosen that allows this change of state to occur at a temperature somewhat below the critical. Cycles that reject heat at supercritical temperatures (such as cycles using carbon dioxide) are also possible.

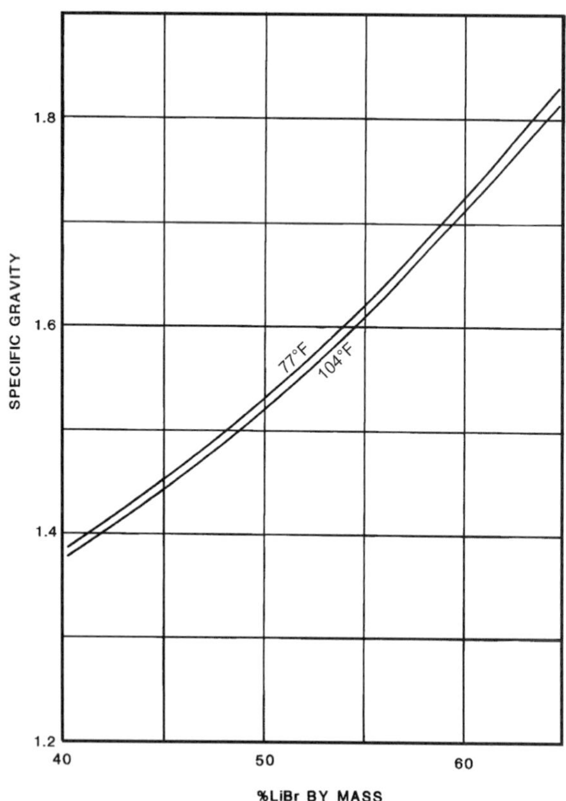

Fig. 1 Specific Gravity of Aqueous Solutions of Lithium Bromide

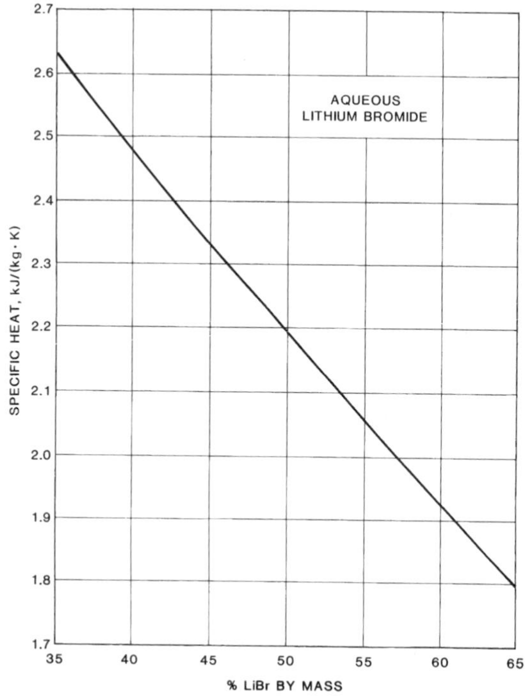

Fig. 2 Specific Heat of Aqueous Lithium Bromide Solutions

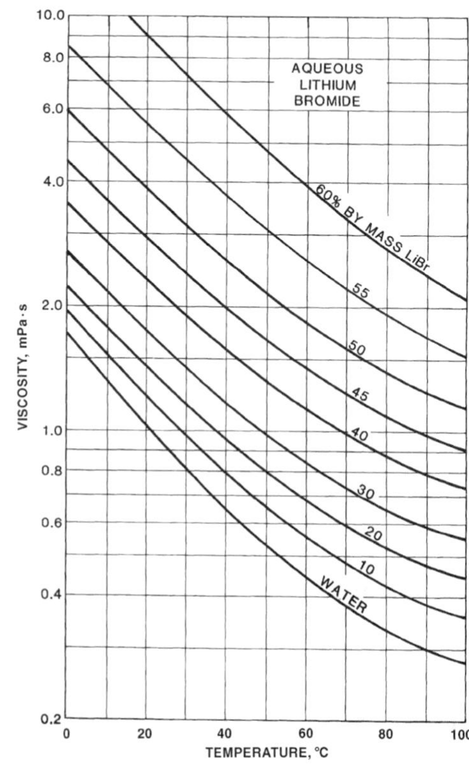

Fig. 3 Viscosity of Aqueous Solutions of Lithium Bromide

Lithium Bromide-Water and Ammonia-Water Solutions. These are the most commonly used working fluids in absorption refrigeration systems. Figure 1 shows specific gravity, Figure 2 shows specific heat, and Figure 3 shows viscosity of lithium bromide-water solutions. Chapter 20 has an enthalpy-concentration diagram and a vapor pressure diagram for lithium bromide-water solutions. Chapter 20 also has equilibrium properties of water-ammonia solutions.

Electrical Properties

Table 3 and Table 4 list the electrical characteristics of refrigerants that are especially important in hermetic systems.

Table 3 Electrical Properties of Liquid Refrigerants

No.	Chemical Name or Composition (% by mass)	Temp., °F	Dielectric Constant	Volume Resistivity, MΩ·m	Ref.
11	Trichlorofluoromethane	84	2.28		1
		a	1.92	63680	2
		77	2.5	90	3
12	Dichlorodifluoromethane	84	2.13		1
		a	1.74	53900	2
		77	2.1	> 120	3
		77	2.100		4
13	Chlorotrifluoromethane	–22	2.3	120	4
		68	1.64		
22	Chlorodifluoromethane	75	6.11		1
		a	6.12	0.83	2
		77	6.6	75	3
23	Trifluoromethane	–22	6.3		3
		68	5.51		4
32	Difluoromethane	a	14.27	-	6
113	Trichlorotrifluoroethane	86	2.44		1
		a	1.68	45490	2
		77	2.6	> 120	3
114	Dichlorotetrafluoroethane	88	2.17		1
		a	1.83	66470	2
		77	2.2	> 70	3
123	2,2-dichloro-1,1,1-trifluoroethane	a	4.50	14700	7
124a	Chlorotetrafluoroethane	77	4.0	50	3
125	Pentafluoroethane	68	4.94	-	8
134a	1,1,1,2-tetrafluoroethane	a	9.51	17700	7
290	Propane	a	1.27	73840	2
404A	R-125/143a/134a (44/52/4)	a	7.58	8450	9
407C	R-32/125/134a (23/25/52)	a	8.74	7420	9
410A	R-32/125 (50/50)	a	7.78	3920	9
500	R-12/152a (73.8/26.2)	a	1.80	55750	2
507A	R-125/143a (50/50)	a	6.97	5570	9
508A	R-23/116 (39/61)	–22	6.60	-	1
		32	5.02		1
508B	R-23/116 (46/54)	–22	7.24	-	1
		32	5.48		1
717	Ammonia	69	15.5		5
744	Carbon dioxide	32	1.59		5

a = ambient temperature

References:

1 Data from E.I. duPont de Nemours & Co., Inc. Used by permission.
2 Beacham and Divers (1955)
3 Eiseman (1955)
4 Makita et al. (1976)
5 CRC *Handbook of Chemistry and Physics* (CRC 1987).
6 Bararo et al. (1997)
7 Fellows et al. (1991)
8 Pereira et al. (1999)
9 Meurer et al. (2000)

Table 4 Electrical Properties of Refrigerant Vapors

No.	Chemical Name or Composition (% by mass)	Pressure, atm.	Temp., °F	Dielectric Constant	Relative Dielectric Strength, Nitrogen = 1	Volume Resistivity, GΩ·m	Ref.
11	Trichlorofluoromethane	0.5	79	1.0019			3
		a	b	1.009		74.35	2
		1.0	73		3.1		4
12	Dichlorodifluoromethane	0.5	84	1.0016			3
		a	b	1.012	452[c]	72.77	2
		1.0	73		2.4		4
		4.9	68	1.019			5
13	Chlorotrifluoromethane	0.5	84	1.0013			3
		1.0	73		1.4		4
		4.9	68	1.013			5
		19.5	90	1.055			6
14	Tetrafluoromethane	0.5	76	1.0006			3
		1.0	73		1.0		4
22	Chlorodifluoromethane	0.5	78	1.0035			3
		a	b	1.004	460[c]	2113	2
		1.0	73		1.3		4
		4.9	68	1.033			5
23	Trifluoromethane	4.9	68	1.042			5
113	Trichlorotrifluoroethane	a	b	1.010	440[c]	94.18	2
		0.4	73		2.6		4
114	Dichlorotetrafluoroethane	0.5	80	1.0021			3
		a	b	1.002	295[c]	148.3	2
		1.0	73		2.8		4
116	Hexafluoroethane	0.94	73	1.002			3
133a	Chlorotrifluoroethane	0.94	80	1.010			3
142b	Chlorodifluoroethane	0.93	81	1.013			3
143a	Trifluoroethane	0.85	77	1.013			3
170	Ethane	1.0	32	1.0015			1
290	Propane	a	b	1.009	440[c]	105.3	2
500	R-12/152a (73.8/26.2)	a	b	1.024	470[c]	76.45	2
508A	R-23/116 (39/61)	a	–22	1.12			7
		a	32	1.31			7
508B	R-23/116 (46/54)	a	–22	1.13			7
		a	32	1.34			7
717	Ammonia	1.0	32	1.0072			1
		a	32		0.82		4
729	Air	1.0	32	1.00059			1
744	Carbon dioxide	1.0	32	1.00099			1
		1.0	b		0.88		4
1150	Ethylene	1.0	32	1.00144			1
		1.0	73		1.21		4

Notes:
a = saturation vapor pressure
b = ambient temperature
c = measured breakdown voltage, volts/mil

References:

1 CRC *Handbook of Chemistry and Physics* (CRC 1987)
2 Beacham and Divers (1955)
3 Fuoss (1938)
4 Charlton and Cooper (1937)
5 Makita et al. (1976)
6 Hess et al. (1962)
7 Data from E.I. duPont de Nemours & Co., Inc. Used by permission.

Sound Velocity

Table 5 gives examples of the velocity of sound in the vapor phase of various fluorinated refrigerants. Chapter 20 has sound velocity data for many refrigerants. The velocity increases when the temperature is increased and decreases when the pressure is increased. The velocity of sound can be calculated from the equation

$$V_a = \left(g_c \frac{dp}{d\rho}\right)_S^{0.5} = \left[\gamma g_c \left(\frac{dp}{d\rho}\right)_T\right]^{0.5} \quad (1)$$

where

V_a = sound velocity, ft/s
g_c = gravitational constant = 32.1740 $lb_m \cdot ft/lb_f \cdot s^2$
p = absolute pressure, lb_f/ft^2
ρ = density, lb_m/ft^2
$\gamma = c_p/c_v$ = ratio of specific heats
S = entropy, Btu/lb·°R
T = temperature, °R

The sound velocity can be estimated from the tables of thermodynamic properties. The change in pressure with a change in density ($dp/d\rho$) can be estimated either at constant entropy or at constant temperature. It is simpler to estimate at constant temperature but then the ratio of specific heats must also be known. The practical velocity of a gas in piping or through openings is limited by the velocity of sound in the gas.

Latent Heat of Vaporization

An empirical rule of chemistry (Trouton's rule) states that the latent heat of vaporization at the boiling point on a molar basis, divided by the temperature in absolute units, is a constant for most materials. This rule is applied to refrigerants in Table 6. It applies fairly well to these refrigerants, although the result is not entirely constant. The rule helps in comparing different refrigerants and in understanding the operation of refrigeration systems.

REFRIGERANT PERFORMANCE

Chapter 1 describes several methods of calculating refrigerant performance, and Chapter 20 includes tables of thermodynamic properties of the various refrigerants.

Table 7 shows the theoretical calculated performance of a number of refrigerants for the U.S. standard cycle of 5°F evaporation and 86°F condensation. Calculated data for other conditions are given in Table 8. The tables can be used to compare the properties of different refrigerants, but actual operating conditions are somewhat different from the calculated data. In most cases, the suction vapor is assumed to be saturated, and the compression is assumed adiabatic or at constant entropy. For R-113 and R-114, these assumptions would cause some liquid in the discharge vapor. In these cases, it is assumed that the discharge vapor is saturated and that the suction vapor is slightly superheated. In Section F of Table 8, the temperature of the suction gas is assumed to be 65°F (−10°F saturated evaporating plus 75°F superheat). Comparison with Section E illustrates the effect of suction gas superheating on refrigerant performance.

SAFETY

Table 9 summarizes the toxicity and flammability characteristics of many refrigerants. In ASHRAE *Standard* 34, refrigerants are classified according to the hazard involved in their use. The toxicity and flammability classifications yield six safety groups (A1, A2, A3, B1, B2, and B3) for refrigerants. Group A1 refrigerants are the least hazardous, Group B3 the most hazardous.

The safety classification in ASHRAE *Standard* 34 consists of a capital letter and a numeral. The capital letter designates the toxicity of the refrigerant at concentrations below 400 ppm by volume:

Table 5 Velocity of Sound in Refrigerant Vapors

		Temperature, °F		
	Pressure,	50	100	150
Refrigerant	psia	Velocity of Sound, ft/s		
11	10	b	469	490
12	10	480	503	525
	100	b	457	490
	200	b	b	442
22	10	583	610	635
	100	b	574	607
	200	b	523	572
23	10	657	685	712
	100	631	666	699
	200	600	644	682
32	10	775	809	840
	100	726	774	815
	200	b	730	784
113	10	b	435	456
114	10	391	411	430
123	10	b	435	456
	100	b	b	b
	200	b	b	b
124	10	443	465	486
	100	b	b	439
	200	b	b	b
125	10	477	500	521
	100	b	466	497
	200	b	420	467
134a	10	517	543	566
	100	b	490	528
	200	b	b	476
143a	10	576	603	629
	100	513	558	595
	200	b	495	552
290	10	799	835	870
	100	b	771	820
	200	b	b	754
404A	10	532	557	581
	100	473	515	549
	200	b	456	509
407C	10	573	600	625
	100	b	558	594
	200	b	500	555
410A	10	635	664	691
	100	587	629	665
	200	b	585	634
502	10	501	525	547
	100	450	488	519
	200	b	435	483
507A	10	529	554	577
	100	471	512	546
	200	b	454	507
508A	10	n.a.	n.a.	n.a.
	100	538	538	538
	200	549	549	549
508B	10	n.a.	572	595
	100	n.a.	553	581
	200	489	531	565
600	10	678	712	743
	100	b	b	652
	200	b	b	b
600a	10	680	713	744
	100	b	b	666
	200	b	b	b
717	10	1388	1453	1513
	100	b	1403	1477
	200	b	1336	1432
744	10	862	899	935
	100	843	885	924
	200	820	869	912

Source: NIST *Standard Reference Database* 23, Version 6.01 (NIST 1996).
b = Below saturation temperature. n.a. = Not available

Table 6 Latent Heat of Vaporization Versus Boiling Point

Refrigerant		Normal Boiling Point, °F	Latent Heat λ at NBP, Btu/lb·mol	Trouton Constant, λ/°R [b]	Ref.
No.	Chemical Name or Composition (% by mass)				
717	Ammonia	–28.0	10,036	23.256	1
630	Methyl amine [a]	23.0	11,141	23.086	4
764	Sulfur dioxide	13.6	10,705	22.626	2
631	Ethyl amine	68.0	11,645	22.076	4
611	Methyl formate [a]	100.0	12,094	21.616	4
134a	Tetrafluoroethane	–15.07	9,531	21.44	5
504	R-32/115 (48.2/51.8)	–71.0	8,282	21.316	1
23	Trifluoromethane	–115.7	7,325	21.29	1
124	Chlorotetrafluoroethane	8.26	9,742	20.82	5
C318	Octafluorocyclobutane	21.5	10,017	20.81	1
21	Dichlorofluoromethane	47.8	10,557	20.80	3
22	Chlorodifluoromethane	–41.4	8,687	20.76	1
40	Methyl chloride	–10.8	9,305	20.73	3
123	Dichlorotrifluoroethane	82.17	11,215	20.70	5
506	R-31/114 (55.1/44.9)	9.9	9,644	20.54	3
125	Pentafluoroethane	–55.43	8,295	20.52	5
113	Trichlorotrifluoroethane	117.6	11,828	20.49	1
152a	Difluoroethane	–13.0	9,045	20.25	1
502	R-22/115 (48.8/51.2)	–49.9	8,280	20.21	3
114	Dichlorotetrafluoroethane	38.8	10,005	20.07	1
216ca	Dichlorohexafluoropropane	96.2	11,154	20.07	1
505	R-12/31 (78.0/22.0) [c]	–21.8	8,735	19.95	3
11	Trichlorofluoromethane	74.9	10,648	19.92	1
500	R-12/152a (73.8/26.2)	–28.3	8,588	19.91	1
14	Tetrafluoromethane	–198.3	5,146	19.69	1
30	Methylene chloride [a]	120.0	11,398	19.66	4
600	Butane	31.1	9,641	19.64	1
13B1	Bromotrifluoromethane	–72.0	7,607	19.62	1
12	Dichlorodifluoromethane	–21.6	8,591	19.61	1
142b	Chlorodifluoroethane	14.4	9,297	19.61	1
115	Chloropentafluoroethane	–38.4	8,245	19.57	1
1270	Propylene	–53.9	7,931	19.55	1
503	R-23/13 (40.1/59.9)	–126.1	6,483	19.43	1
600a	Isobutane	10.9	9,103	19.34	1
13	Chlorotrifluoromethane	–114.6	6,670	19.33	1
290	Propane	–43.7	8,026	19.29	1
1150	Ethylene	–154.7	5,793	19.00	1
170	Ethane	–127.9	6,296	18.98	1
50	Methane	–258.7	3,521	17.52	1

Notes:
[a] Not at normal atmospheric pressure
[b] Normal boiling temperatures
[c] The exact composition of this azeotrope is in question.
References:
1 ASHRAE *Thermodynamic Properties of Refrigerants* (Stewart et al. 1986)
2 CRC *Handbook of Chemistry and Physics* (CRC 1987)
3 ASHRAE (1977)
4 *Chemical Engineer's Handbook* (1973)
5 NIST *Standard Reference Database* 23 (NIST 1996)

- Class A Toxicity not identified
- Class B Evidence of toxicity identified

The numeral denotes the flammability of the refrigerant:

- Class 1 No flame propagation in air at 65°F and 14.7 psia
- Class 2 Lower flammability limit (LFL) greater than 0.00625 lb/ft^3 at 70°F and 14.7 psia *and* heat of combustion less than 8174 Btu/lb
- Class 3 Highly flammable as defined by LFL less than or equal to 0.00625 lb/ft^3 at 70°F and 14.7 psia *or* heat of combustion greater than or equal to 8174 Btu/lb

LEAK DETECTION

Leak detection in refrigeration equipment is a major problem for manufacturers and service engineers. The following sections describe several leak detection methods.

Electronic Detection

The electronic detector is widely used in the manufacture and assembly of refrigeration equipment. Instrument operation depends on the variation in current flow caused by ionization of decomposed refrigerant between two oppositely charged platinum electrodes. This instrument can detect any of the halogenated refrigerants except R-14; however, *it is not recommended for use in atmospheres that contain explosive or flammable vapors.* Other vapors, such as alcohol and carbon monoxide, may interfere with the test.

The electronic detector is the most sensitive of the various leak detection methods, reportedly capable of sensing a leak of 1/100 oz of R-12 per year. A portable model is available for field testing. Other models are available with automatic balancing systems that correct for refrigerant vapors that might be present in the atmosphere around the test area.

Halide Torch

The halide torch is a fast and reliable method of detecting leaks of chlorinated refrigerants. Air is drawn over a copper element heated by a methyl alcohol or hydrocarbon flame. If halogenated vapors are present, they decompose, and the color of the flame changes to bluish-green. Although not as sensitive as the electronic detector, this method is suitable for most purposes.

Bubble Method

The object to be tested is pressurized with air or nitrogen. A pressure corresponding to operating conditions is generally used. The object is immersed in water, and any leaks are detected by observing bubbles in the liquid. Adding a detergent to the water decreases the surface tension, prevents escaping gas from clinging to the side of the object, and promotes the formation of a regular stream of small bubbles. Kerosene or other organic liquids are sometimes used for the same reason. A solution of soap or detergent can be brushed or poured onto joints or other spots where leakage is suspected. Leaking gas forms soap bubbles that can be readily detected.

Leaks can also be determined by pressurizing or evacuating and observing the change in pressure or vacuum over a period of time. This is effective in checking the tightness of the system but does not locate the point of leakage.

Ammonia and Sulfur Dioxide Leaks

Ammonia can be detected by burning a sulfur candle in the vicinity of the suspected leak or by bringing a solution of hydrochloric acid near the object. If ammonia vapor is present, a white cloud or smoke of ammonium sulfite or ammonium chloride forms. Ammonia can also be detected with indicator paper that changes color in the presence of a base.

Sulfur dioxide can be detected by the appearance of white smoke when aqueous ammonia is brought near the leak.

Table 7 Comparative Refrigerant Performance per Ton of Refrigeration

Refrigerant No.	Chemical Name or Composition (% by mass)	Evaporator Pressure, psia	Condenser Pressure, psia	Compression Ratio	Net Refrigerating Effect, Btu/lb_m	Refrigerant Circulated, lb_m/min	Liquid Circulated, in^3/min	Specific Volume of Suction Gas, ft^3/lb_m	Compressor Displacement, cfm	Power Consumption, hp	Coefficient of Performance	Comp. Discharge Temp., °F
170	Ethane	236.41	674.71	2.85	69.27	2.887	289.13	0.534	1.543	1.73	2.72	123
744	Carbon dioxide	332.38	1045.36	3.15	57.75	3.463	158.53	0.264	0.914	1.68	2.81	156
13B1	Bromotrifluoromethane	77.82	264.13	3.39	28.45	7.029	129.78	0.380	2.669	1.13	4.16	104
1270	Propylene	52.70	189.44	3.59	123.15	1.624	90.71	2.049	3.327	1.04	4.56	108
290	Propane	42.37	156.82	3.70	120.30	1.663	95.04	2.459	4.088	1.03	4.57	98
502	R-22/115 (48.8/51.2)	50.56	191.29	3.78	44.91	4.453	103.35	0.802	3.569	1.07	4.42	98
507A	R125/R-143a (50/50)	55.3	212.4	3.84	47.28	4.230	114.03	0.810	3.427	1.13	4.18	95
125	Pentafluoroethane	58.87	228.11	3.87	37.69	5.306	126.82	0.628	3.333	1.28	3.67	108
404A	R125/143a/134a (44/52/4)	53.3	206.8	3.88	48.98	4.083	110.79	0.856	3.494	1.12	4.21	96
410A	R-32/125 (50/50)	69.7	272.6	3.91	72.09	2.775	74.33	0.868	2.409	1.07	4.41	124
22	Chlorodifluoromethane	42.94	172.63	4.02	70.46	2.838	66.88	1.258	3.573	1.02	4.65	129
12	Dichlorodifluoromethane	26.51	107.99	4.07	50.25	3.980	85.23	1.465	5.830	0.99	4.75	100
500	R-12/152a (73.8/26.2)	31.06	127.50	4.10	60.64	3.298	80.19	1.502	4.955	1.01	4.69	105
407C	R-32/125/134a (23/25/52)	42.0	183.4	4.37	69.77	2.867	70.34	1.280	3.656	1.05	4.51	117
600a	Isobutane	12.92	59.29	4.59	113.00	1.770	90.01	6.419	11.361	1.07	4.41	80
134a	Tetrafluoroethane	23.79	111.62	4.69	64.51	3.100	72.37	1.959	6.076	1.02	4.60	96
717	Ammonia	34.17	168.80	4.94	474.20	0.422	19.61	8.179	3.450	0.99	4.77	210
124	Chlorotetrafluoroethane	12.96	64.59	4.98	50.93	3.927	81.16	2.714	10.658	1.05	4.47	90
600	Butane	8.18	41.19	5.04	125.55	1.593	77.78	10.206	16.258	0.95	4.95	88
114	Dichlorotetrafluoroethane [a]	6.75	36.49	5.41	43.02	4.649	89.56	4.340	20.176	1.02	4.65	86
11	Trichlorofluoromethane	2.94	18.32	6.24	67.21	2.976	56.26	12.240	36.425	0.94	5.02	110
123	Dichlorotrifluoroethane	2.26	15.93	7.06	61.42	3.256	62.19	14.337	46.684	0.97	4.86	90
113	Trichlorotrifluoroethane [a]	1.01	7.88	7.83	52.08	3.840	68.60	26.285	100.945	1.11	4.27	86

Notes: Data based on 5°F evaporation, 86°F condensation, 0°F subcool, and 0°F superheat.

[a] Saturated suction except R-113 and R-114. Enough superheat was added to give saturated discharge.

Table 8 Comparative Refrigerant Performance per Ton at Various Evaporating and Condensing Temperatures

Refrigerant No.	Chemical Name or Composition (% by mass)	Suction Temp., °F	Evaporator Pressure, psia	Condenser Pressure, psia	Compression Ratio	Net Refrigerating Effect, Btu/lb_m	Refrigerant Circulated, lb_m/min	Specific Volume of Suction Gas, ft^3/lb_m	Compressor Displacement, cfm	Power Consumption, hp
A. –130°F Saturated Evaporating, 0°F Suction Superheat, –40°F Saturated Condensing										
1150	Ethylene	–130	30.89	210.67	6.82	142.01	1.408	3.853	5.43	1.756
170	Ethane	–130	13.62	112.79	8.28	156.58	1.277	8.357	10.68	1.633
13	Chlorotrifluoromethane	–130	9.06	88.04	9.72	45.82	4.365	3.625	15.82	1.685
23	Trifluoromethane	–130	9.06	103.03	11.37	79.38	2.520	5.458	13.75	1.753
508A	R-23/116 (39/61)	–130	12.6	122.2	9.70	44.12	4.533	2.68	12.15	1.738
508B	R-23/116 (46/54)	–130	12.4	122.8	9.90	47.50	4.210	2.86	12.05	1.734
B. –100°F Saturated Evaporating, 0°F Suction Superheat, –30°F Saturated Condensing										
170	Ethane	–100	31.27	134.73	4.31	157.76	1.268	3.867	4.90	1.118
23	Trifluoromethane	–100	23.74	125.99	5.31	79.37	2.520	2.219	5.59	1.178
13	Chlorotrifluoromethane	–100	22.28	106.29	4.77	46.23	4.326	1.563	6.76	1.153
125	Pentafluoroethane	–100	3.78	27.76	7.34	56.43	3.544	8.390	29.73	1.101
22	Chlorodifluoromethane	–100	2.38	19.63	8.25	90.75	2.204	18.558	40.90	1.074
508A	R-23/116 (39/61)	–100	31.0	147.5	4.76	44.28	4.517	1.15	5.18	1.180
508B	R-23/116 (46/54)	–100	30.9	148.4	4.80	47.53	4.208	1.21	5.10	1.173
C. –76°F Saturated Evaporating, 0°F Suction Superheat, 5°F Saturated Condensing										
1150	Ethylene	–76	109.37	416.24	3.81	116.95	1.710	1.167	1.99	1.478
170	Ethane	–76	54.63	235.44	4.31	322.65	0.620	2.291	1.42	0.566
23	Trifluoromethane	–76	45.41	237.18	5.22	69.60	2.874	1.203	3.46	1.394
13	Chlorotrifluoromethane	–76	40.87	192.14	4.70	39.42	5.074	0.880	4.47	1.382
125	Pentafluoroethane	–76	8.21	58.87	7.17	50.62	3.951	4.072	16.09	1.277
290	Propane	–76	6.15	42.37	6.89	147.39	1.357	14.856	20.16	1.196
22	Chlorodifluoromethane	–76	5.44	42.96	7.90	84.24	2.374	8.593	20.40	1.195
717	Ammonia	–76	3.18	34.26	10.79	540.63	0.37	75.784	28.04	1.247
12	Dichlorodifluoromethane	–76	3.28	26.50	8.09	58.61	3.412	10.245	34.96	1.191
134a	Tetrafluoroethane	–76	2.3	23.77	10.32	78.1	2.561	17.304	44.32	1.182
410A	R-32/125 (50/50)	–76	9.4	69.7	7.41	92.86	2.153	5.84	12.57	1.204
407C	R-32/125/134a (23/25/52)	–76	4.9	43.5	8.88	86.88	2.302	9.74	22.43	1.200
404A	R125/143a/134a (44/52/4)	–76	7.2	53.4	7.42	65.25	3.065	5.73	17.55	1.219

Table 8 Comparative Refrigerant Performance per Ton at Various Evaporating and Condensing Temperatures *(Continued)*

Refrigerant No.	Chemical Name or Composition (% by mass)	Suction Temp., °F	Evaporator Pressure, psia	Condenser Pressure, psia	Compression Ratio	Net Refrigerating Effect, Btu/lb_m	Refrigerant Circulated, lb_m/min	Specific Volume of Suction Gas, ft^3/lb_m	Compressor Displacement, cfm	Power Consumption, hp
C. –76°F Saturated Evaporating, 0°F Suction Superheat, 5°F Saturated Condensing *(Concluded)*										
507A	R125/143a (50/50)	–76	7.6	55.3	7.28	63.43	3.153	5.34	16.84	1.219
508A	R-23/116 (39/61)	–76	56.8	267.8	4.71	35.23	5.677	0.642	3.64	1.468
508B	R-23/116 (46/54)	–76	56.8	270.0	4.75	38.22	5.233	0.675	3.53	1.447
D. –40°F Saturated Evaporating, 0°F Suction Superheat, 68°F Saturated Condensing										
744	Carbon dioxide	–40	145.77	830.5	5.70	77.22	2.590	0.613	1.59	2.208
23	Trifluoromethane	–40	103.03	597.9	5.80	45.67	4.379	0.545	2.39	2.442
125	Pentafluoroethane	–40	21.84	175.1	8.02	37.44	5.342	1.625	8.68	1.962
290	Propane	–40	16.01	121.6	7.55	119.33	1.676	6.083	10.20	1.670
22	Chlorodifluoromethane	–40	15.27	132.0	8.65	70.65	2.831	3.280	9.29	1.606
717	Ammonia	–40	10.4	124.3	11.95	486.55	0.411	25.144	10.33	1.576
12	Dichlorodifluoromethane	–40	9.30	82.3	8.84	49.44	4.046	3.887	15.73	1.596
134a	Tetrafluoroethane	–40	7.42	83.0	11.19	63.17	3.166	5.790	18.33	1.597
410A	R-32/125 (50/50)	–40	25.5	208.9	8.19	74.42	2.69	2.28	6.12	1.643
407C	R-32/125/134a (23/25/52)	–40	14.1	138.7	9.84	70.34	2.84	3.60	10.23	1.624
404A	R125/143a/134a (44/52/4)	–40	19.5	158.9	8.15	49.36	4.05	2.24	9.07	1.735
507A	R125/143a (50/50)	–40	20.4	163.4	8.01	47.68	4.19	2.10	8.82	1.747
E. –10°F Saturated Evaporating, 0°F Suction Superheat, 100°F Saturated Condensing										
124	Chlorotetrafluoroethane	–10	8.95	80.9	9.04	44.99	4.445	3.841	17.08	1.649
134a	Tetrafluoroethane	–10	16.62	139.0	8.36	56.57	3.535	2.711	9.59	1.589
12	Dichlorodifluoromethane	–10	19.20	131.7	6.86	44.89	4.456	1.980	8.82	1.606
717	Ammonia	–10	23.73	212.0	8.93	461.25	0.434	11.677	5.07	1.494
22	Chlorodifluoromethane	–10	31.23	210.7	6.75	64.07	3.122	1.676	5.23	1.602
502	R-22/115 (48.8/51.2)	–10	37.26	230.9	6.20	39.05	5.122	1.073	5.49	1.904
125	Pentafluoroethane	–10	43.32	277.0	6.39	31.09	6.433	0.846	5.44	2.172
410A	R-32/125 (50/50)	–10	51.1	331.6	6.49	64.70	3.09	1.17	3.63	1.672
407C	R-32/125/134a (23/25/52)	–10	29.7	225.0	7.58	62.45	3.20	1.78	5.69	1.627
404A	R125/143a/134a (44/52/4)	–10	39.0	251.0	6.44	41.62	4.80	1.16	5.55	1.814
507A	R125/143a (50/50)	–10	40.7	257.6	6.33	39.96	5.00	1.09	5.46	1.834
F. –10°F Saturated Evaporating, 75°F Suction Superheat (Included in Refrigeration Effect), 100°F Saturated Condensing										
123	Dichlorotrifluoroethane	65	1.48	20.8	14.07	67.3	2.972	24.797	73.70	1.387
11	Trichlorofluoromethane	65	1.92	23.4	12.2	71.88	2.783	21.276	59.21	1.403
124	Chlorotetrafluoroethane	65	8.95	80.9	9.04	57.33	3.489	4.531	15.81	1.506
134a	Tetrafluoroethane	65	16.62	139.0	8.36	71.25	2.807	3.236	9.08	1.513
12	Dichlorodifluoromethane	65	19.20	131.7	6.86	55.83	3.583	2.360	8.45	1.539
717	Ammonia	65	23.73	212.0	8.93	498.44	0.401	13.751	5.51	1.612
22	Chlorodifluoromethane	65	31.23	210.7	6.75	75.95	2.633	2.012	5.30	1.623
502	R-22/115 (48.8/51.2)	65	37.26	230.9	6.20	51.23	3.904	1.302	5.08	1.761
125	Pentafluoroethane	65	43.32	277.0	6.39	45.13	4.432	1.028	4.56	1.773
410A	R-32/125 (50/50)	65	51.1	331.6	6.49	80.50	2.48	1.44	3.58	1.657
407C	R-32/125/134a (23/25/52)	65	29.7	225.0	7.58	77.48	2.58	2.14	5.51	1.585
404A	R125/143a/134a (44/52/4)	65	39.0	251.0	6.44	57.29	3.49	1.41	4.91	1.636
507A	R125/143a (50/50)	65	40.7	257.6	6.33	55.56	3.60	1.33	4.79	1.644
G. 40°F Saturated Evaporating, 0°F Suction Superheat, 100°F Saturated Condensing										
125	Pentafluoroethane	40	111.7	275.7	2.47	35.85	5.58	0.331	1.84	0.788
290	Propane	40	78.6	188.6	2.40	119.47	1.67	1.356	2.27	0.692
22	Chlorodifluoromethane	40	83.25	210.7	2.53	68.71	2.911	0.656	1.91	0.696
717	Ammonia	40	73.3	212.0	2.89	480.33	0.416	4.084	1.70	0.653
500	R-12/152a (73.8/26.2)	40	60.72	155.8	2.57	60.54	3.303	0.792	2.62	0.692
12	Dichlorodifluoromethane	40	51.71	131.7	2.55	50.50	3.960	0.778	3.08	0.689
134a	Tetrafluoroethane	40	49.77	139.0	2.79	63.72	3.139	0.952	2.99	0.679
124	Chlorotetrafluoroethane	40	27.89	80.9	2.90	52.06	3.842	1.318	5.06	0.698
600a	Isobutane	40	26.75	73.4	2.74	115.83	1.727	3.256	5.62	0.693
600	Butane	40	17.68	51.7	2.92	129.22	1.548	4.975	7.70	0.669
11	Trichlorofluoromethane	40	6.99	23.4	3.34	68.04	2.939	5.455	16.03	0.624
123	Dichlorotrifluoroethane	40	5.79	20.8	3.59	62.82	3.184	5.921	18.85	0.635
113	Trichlorotrifluoroethane	40	2.7	10.5	3.89	54.58	3.67	10.5	38.44	0.638
410A	R-32/125 (50/50)	40	132.9	331.6	2.50	69.08	2.89	0.456	1.32	0.721
407C	R-32/125/134a (23/25/52)	40	84.2	225.0	2.67	68.60	2.92	0.648	1.89	0.699

Table 9 Comparison of Safety Group Classifications in ASHRAE *Standard* 34-1989 and ASHRAE *Standard* 34-1997

Refrigerant Number	Chemical Formula	Safety Group Old	Safety Group New
10	CCl_4	2	B1
11	CCl_3F	1	A1
12	CCl_2F_2	1	A1
13	$CClF_3$	1	A1
13B1	$CBrF_3$	1	A1
14	CF_4	1	A1
21	$CHCl_2F$	2	B1
22	$CHClF_2$	1	A1
23	CHF_3		A1
30	CH_2Cl_2	2	B2
32	CH_2F_2		A2
40	CH_3Cl	2	B2
50	CH_4	3a	A3
113	CCl_2FCClF_2	1	A1
114	$CClF_2CClF_2$	1	A1
115	$CClF_2CF_3$	1	A1
116	CF_3CF_3		A1
123	$CHCl_2CF_3$		B1
124	$CHClFCF_3$		A1
125	CHF_2CF_3		A1
134a	CF_3CH_2F		A1
142b	$CClF_2CH_3$	3b	A2
143a	CF_3CH_3		A2
152a	CHF_2CH_3	3b	A2
170	CH_3CH_3	3a	A3
218	$CF_3CF_2CF_3$		A1
290	$CH_3CH_2CH_3$	3a	A3
C318	C_4F_8	1	A1
400	R-12/114 (must be specified)	1	A1/A1
500	R-12/152a (73.8/26.2)	1	A1
501	R-22/12 (75.0/25.0)*	1	A1
502	R-22/115 (48.8/51.2)	1	A1
507A	R-125/143a (50/50)		A1
508A	R-23/116 (39/61)		A1
508B	R-23/116 (46/54)		A1/A1
509A	R-22/218 (44/56)		A1
600	$CH_3CH_2CH_2CH_3$	3a	A3
600a	$CH(CH_3)_3$	3a	A3
611	$HCOOCH_3$	2	B2
702	H_2		A3
704	He		A1
717	NH_3	2	B2
718	H_2O		A1
720	Ne		A1
728	N_2		A1
740	Ar		A1
744	CO_2	1	A1
764	SO_2	2	B1
1140	$CHCl{=}CH_2$		B3
1150	$CH_2{=}CH_2$	3a	A3
1270	$CH_3CH{=}CH_2$	3a	A3

*The exact composition of this azeotrope is in question.

Table 10 Swelling of Elastomers in Liquid Refrigerants at Room Temperature

	Linear Swell, %							
Refrig. No.	Buna N	Buna S (GR-S)	Butyl (GR-1)	Natural Rubber	Neoprene GN	Thiokol FA	Viton B	Silicone
11	6	21	41	23	17	2	6	38
12	2	3	6	6	0	1	9	—
13	1	1	0	1	0	0	4	—
13B1	1	1	2	1	2	—	7	—
21	48	49	24	34	28	28	22	—
22	26	4	1	6	2	4	20	20
30	52	26	23	34	37	59	—	—
40	35	20	16	26	22	11	—	—
113	1	9	21	17	3	1	7	34
114	0	2	2	2	0	0	9	—
502	7	3	—	4	1	—	—	—
600	1	8	20	16	3	0	—	—

Adapted from Eiseman (1949).

Table 11 Diffusion of Water and R-22 Through Elastomers

	Diffusion Rate	
Elastomer	Water[a]	R-22[b]
Neoprene	0.717	1.31
Buna N	0.109	19.7
Hypalon 40	0.457	0.52
Butyl	0.043	0.30
Viton	—	3.61
Polyethylene	0.123	—
Natural	1.428	—

Adapted from Eiseman (1966).

[a] 0.003 in. film, 100% rh at 100°F. Water diffusion rate is in pounds per hour per 1000 ft^2 of elastomer.

[b] Film thickness = 0.001 in.; temperature = 77°F. Gas at 1 atm. and 32°F. Diffusion rate per day in ft^3 of gas per ft^2 of elastomer.

Table 12 Swelling of Plastics in Liquid Refrigerants at Room Temperature

	Linear Swell, %						
	Refrigerant						
Plastic	11	12	21	30	113	114a	22
Phenol formaldehyde resin	0	0	0	0	−0.2	−0.2	n.a.
Cellulose acetate	0.4	0	b	b	0	−0.1	n.a.
Cellulose nitrate	0.6	0	b	b	0	−0.1	n.a.
Nylon	0	0	0	0	0	−0.2	1
Methyl methacrylate resin	0	−0.1	b	b	−0.2	−0.2	a
Polyethylene	6.7	0.4	4.5	4.6	2.3	0.6	2
Polystyrene	b	−0.1	b	b	−0.2	−0.2	n.a.
Polyvinyl alcohol	0.3	−0.7	12.9	9.1	−0.1	0.2	n.a.
Polyvinyl chloride	0	0	15.1	b	0	0.1	n.a.
Polyvinylidene chloride	−0.2	0	1.0	2.4	−0.1	0	4
Polytetrafluoroethylene	0	−0.7	0.1	0	0	−0.3	1

Adapted from Brown (1960)
n.a. = data not available
a = sample completely disintegrated

EFFECT ON CONSTRUCTION MATERIALS

Metals

Halogenated refrigerants can be used satisfactorily under normal conditions with most common metals, such as steel, cast iron, brass, copper, tin, lead, and aluminum. Under more severe conditions, various metals affect such properties as hydrolysis and thermal decomposition in varying degrees. The tendency of metals to promote thermal decomposition of halogenated compounds is in the following order:

(least decomposition) Inconel < 18-8 stainless steel < nickel < copper < 1340 steel < aluminum < bronze < brass < zinc < silver (most decomposition)

This order is only approximate, and exceptions may be found for individual compounds or for special use conditions. The effect of metals on hydrolysis is probably similar.

Magnesium, zinc, and aluminum alloys containing more than 2% magnesium are not recommended for use with halogenated compounds where even trace amounts of water may be present.

Warning: Never use methyl chloride with aluminum in any form. A highly flammable gas is formed, and the explosion hazard is great.

Ammonia should never be used with copper, brass, or other alloys containing copper. When water is present in sulfur dioxide systems, sulfurous acid is formed and can attack iron or steel rapidly and other metals at a slower rate.

Further discussion of the compatibility of refrigerants and lubricants with construction materials may be found in Chapter 5 of the 1998 *ASHRAE Handbook—Refrigeration*.

Elastomers

The linear swelling of some elastomers in the liquid phase of various refrigerants is shown in Table 10. Swelling data can be used to a limited extent in comparing the effect of refrigerants on elastomers. However, other factors, such as the amount of extraction, tensile strength, and degree of hardness of the exposed elastomer must be considered. When other fluids are present in addition to the refrigerant, the combined effect on elastomers should be determined. In some instances, somewhat higher swelling of elastomers is found in mixtures of R-22 and lubricating oil than in either fluid alone. Table 11 shows the diffusion rate of water and R-22 through elastomers.

Plastics

The effect of a refrigerant on a plastic material should be thoroughly examined under the conditions of intended use. Plastics are often mixtures of two or more basic types, and it is difficult to predict the effect of the refrigerant. The linear swelling of some plastic materials in refrigerants is shown in Table 12. Swelling data can be used as a guide but, as with elastomers, the effect on the properties of the plastic should also be examined. Comparable data for R-22 is limited, but the effect on plastics is generally more severe than that of R-12, but not as severe as that of R-21. The effect of R-114 is very similar to that of R-114a.

REFERENCES

ASHRAE. 1977. *ASHRAE Handbook—Fundamentals*, Chapter 16.

ASHRAE. 1997. Number designation and safety classification of refrigerants. ANSI/ASHRAE *Standard* 34-1997.

Bararo, M.T., U.V. Mardolcar, and C.A. Nieto de Castro. 1997. Molecular properties of alternative refrigerants derived from dielectric-constant measurements. *Journal of Thermophysics* 18(2):419-438.

Beacham, E.A. and R.T. Divers. 1955. Some aspects of the dielectric properties of refrigerants. *Refrigerating Engineering* 7:33.

Brown, J.A. 1960. Effect of propellants on plastic valve components. *Soap and Chemical Specialties* 3:87.

Charlton, E.E. and F.S. Cooper. 1937. Dielectric strengths of insulating fluids. *General Electric Review* 865(9):438.

Chemical engineer's handbook, 5th ed. 1973. McGraw-Hill, New York.

CRC Handbook of chemistry and physics, 68th ed. 1987. CRC Press, Boca Raton, FL.

duPont. *Bulletin* B-32A. Freon Products Division. E.I. duPont de Nemours & Co., Inc., Wilmington, DE.

duPont. *Bulletin* G-1. Freon Products Division. E.I. duPont de Nemours & Co., Inc., Wilmington, DE.

duPont. 1980. *Bulletin* T-502. Freon Products Division. E.I. duPont de Nemours & Co., Inc., Wilmington, DE.

Eiseman, B.J., Jr. 1949. Effect on elastomers of Freon compounds and other halohydrocarbons. *Refrigerating Engineering* 12:1171.

Eiseman, B.J., Jr. 1955. How electrical properties of Freon compounds affect hermetic system's insulation. *Refrigerating Engineering* 4:61.

Fellows, B.R., R.G. Richard, and I.R. Shankland. 1991. Electrical characterization of alternate refrigerants. *Actes Congr. Int. Froid*, 18th, Vol. 2. International Institute of Refrigeration, Paris.

Fuoss, R.M. 1938. Dielectric constants of some fluorine compounds. *Journal of the American Chemical Society*, 1633.

Handbook of chemistry, 10th ed. 1967. McGraw-Hill, New York.

Handbook of chemistry and physics, 41st ed. 1959-60. The Chemical Rubber Publishing Co., Cleveland, OH.

Kirk and Othmer. 1956. *The encyclopedia of chemical technology*. The Interscience Encyclopedia, Inc., New York.

Matheson gas data book. 1966. The Matheson Company, Inc., East Rutherford, NJ.

McLinden, M.O. 1990. *International Journal of Refrigeration* 13:149-62.

Meurer C., G. Pietsch, and M. Haacke M. 2000. Electrical properties of CFC- and HCFC-substitutes. *Int. Journal of Refrigeration*.

NIST. 1996. *Standard Reference Database* 23, Version 6.01. National Institute of Standards and Technology, Gaithersburg, MD.

Pereira L.F., F.E. Brito, A.N. Gurova, U.V. Mardolcar, and C.A. Nieta de Castro. 1999. Dipole moment, expansivity and compressibility coefficients of HFC 125 derived from dielectric constant measurements. 1st Int. Workshop on Thermochemical, Thermodynamic and Transport Properties of Halogenated Hydrocarbons and Mixtures, Pisa, Italy.

Stewart, R.B., R.T. Jacobsen, and S.G. Penoncello. 1986. *ASHRAE Thermodynamic properties of refrigerants*. ASHRAE.

U.N. 1994. *1994 Report of the refrigeration, air conditioning, and heat pumps technical options committee*. United Nations Environment Programme, Nairobi, Kenya. ISBN 92-807-1455-4.

U.N. 1996. OzonAction (The Newsletter of the United Nations Environment Programme Industry and Environment OzonAction Programme). October (No. 20):10.

CHAPTER 20

THERMOPHYSICAL PROPERTIES OF REFRIGERANTS

THIS CHAPTER presents tabular data for the thermodynamic and transport properties of refrigerants arranged for the occasional user. Most of the refrigerants have a thermodynamic property chart on pressure-enthalpy coordinates with an abbreviated set of tabular data for the saturated liquid and vapor on the facing page. In addition, tabular data in the superheated vapor region are given for R-134a to assist students working on compression cycle examples.

For each of the cryogenic fluids, a second table of properties is provided for the vapor at a pressure of one standard atmosphere; these tables provide data needed when these gases are used in heat transfer or purge gas applications. For the zeotropic blends, including R-729 (air), tables are incremented in pressure with properties given for the liquid on the bubble line and vapor on the dew line. This arrangement is chosen because pressure is more commonly measured in the field while servicing equipment; it also highlights the difference between the bubble and dew point temperatures—the so-called temperature glide experienced with blends.

For a few fluids there are gaps in some properties at low temperatures due to limitations in the data and/or the models used.

New for the 2001 *ASHRAE Handbook* are R-143a and R-245fa. Most of the CFC refrigerants have been deleted. Tables for R-11, R-13, R-113, R-114, R-141b, R-142b, R-500, R-502, R-503, and R-720 (neon) may be found in the 1997 *ASHRAE Handbook*. R-12 has been retained to assist in making comparisons. Revised formulations have been used for many of the HFC refrigerants; these conform to international standards, where applicable. The formulations used are detailed in the References section.

The reference states used for most of the refrigerants correspond to the American convention of 0 Btu/lb for enthalpy and 0 Btu/lb·°F for entropy for the saturated liquid at –40°F. The exceptions are water and fluids that have very low critical temperatures, such as ethylene and the cryogens.

These data are intended to help engineers make preliminary comparisons among unfamiliar fluids. For greater detail and a wider range of data, consult the sources listed in the References section.

The preparation of this chapter is assigned to TC 3.1, Refrigerants and Secondary Coolants.

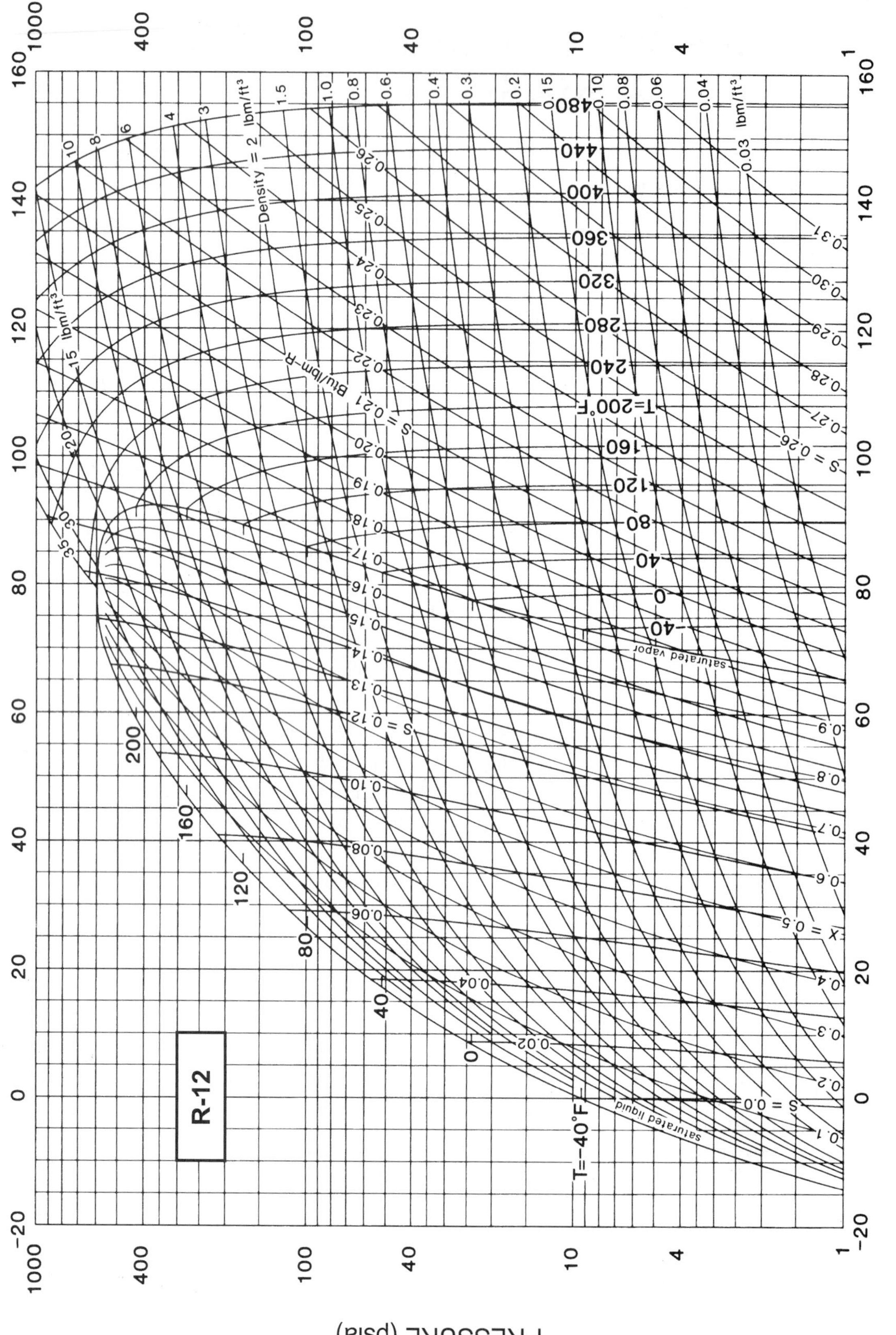

Fig. 1 Pressure-Enthalpy Diagram for Refrigerant 12

Refrigerant 12 (Dichlorodifluoromethane) Properties of Saturated Liquid and Saturated Vapor

Temp.,* °F	Pres-sure, psia	Denisty, lb/ft³ Liquid	Volume, ft³/lb Vapor	Enthalpy, Btu/lb Liquid	Enthalpy, Btu/lb Vapor	Entropy, Btu/lb·°F Liquid	Entropy, Btu/lb·°F Vapor	Specific Heat c_p, Btu/lb·°F Liquid	Specific Heat c_p, Btu/lb·°F Vapor	c_p/c_v Vapor	Vel. of Sound, ft/s Liquid	Vel. of Sound, ft/s Vapor	Viscosity, lb_m/ft·h Liquid	Viscosity, lb_m/ft·h Vapor	Thermal Cond., Btu/h·ft·°F Liquid	Thermal Cond., Btu/h·ft·°F Vapor	Surface Tension, dyne/cm	Temp., °F
–150.00	0.155	105.01	176.84	–22.133	60.979	–0.06101	0.20738	0.1955	0.1069	1.1831	3412.	387.7	2.493	0.0163	0.0678	0.00244	26.65	–150.00
–140.00	0.258	104.10	109.67	–20.175	62.044	–0.05479	0.20241	0.1962	0.1090	1.1795	3329.	393.0	2.212	0.0168	0.0663	0.00257	25.77	–140.00
–130.00	0.415	103.18	70.398	–18.209	63.125	–0.04873	0.19798	0.1970	0.1111	1.1764	3247.	398.3	1.981	0.0174	0.0647	0.00270	24.90	–130.00
–120.00	0.644	102.26	46.615	–16.234	64.219	–0.04284	0.19402	0.1979	0.1133	1.1736	3166.	403.3	1.789	0.0179	0.0633	0.00283	24.03	–120.00
–110.00	0.973	101.34	31.744	–14.250	65.326	–0.03708	0.19050	0.1989	0.1154	1.1711	3085.	408.3	1.626	0.0184	0.0618	0.00296	23.17	–110.00
–100.00	1.430	100.41	22.173	–12.255	66.444	–0.03146	0.18735	0.2000	0.1175	1.1691	3005.	413.0	1.487	0.0190	0.0604	0.00310	22.32	–100.00
–90.00	2.052	99.48	15.847	–10.248	67.571	–0.02596	0.18455	0.2013	0.1197	1.1675	2926.	417.5	1.366	0.0195	0.0590	0.00323	21.47	–90.00
–80.00	2.880	98.54	11.565	–8.228	68.705	–0.02057	0.18206	0.2026	0.1218	1.1663	2848.	421.8	1.261	0.0200	0.0577	0.00337	20.63	–80.00
–75.00	3.387	98.06	9.9506	–7.213	69.274	–0.01791	0.18092	0.2033	0.1229	1.1659	2809.	423.8	1.213	0.0203	0.0570	0.00344	20.22	–75.00
–70.00	3.963	97.59	8.6006	–6.194	69.844	–0.01529	0.17985	0.2040	0.1240	1.1655	2770.	425.8	1.167	0.0206	0.0563	0.00352	19.80	–70.00
–65.00	4.616	97.11	7.4658	–5.171	70.415	–0.01268	0.17884	0.2047	0.1252	1.1653	2732.	427.8	1.125	0.0208	0.0557	0.00359	19.39	–65.00
–60.00	5.353	96.63	96.63	–4.145	70.986	–0.01010	0.17788	0.2055	0.1263	1.1653	2693.	429.6	1.084	0.0211	0.0550	0.00366	18.98	–60.00
–55.00	6.181	96.14	5.6943	–3.115	71.558	–0.00754	0.17699	0.2062	0.1274	1.1653	2655.	431.4	1.046	0.0214	0.0544	0.00373	18.57	–55.00
–50.00	7.108	95.66	5.0014	–2.081	72.130	–0.00501	0.17614	0.2070	0.1286	1.1654	2617.	433.1	1.010	0.0216	0.0537	0.00381	18.16	–50.00
–45.00	8.144	8.144	4.4085	–1.043	72.702	–0.00249	0.17535	0.2078	0.1297	1.1657	2579.	434.7	0.976	0.0219	0.0531	0.00388	17.75	–45.00
–40.00	9.295	94.68	3.8992	0.000	73.273	0.00000	0.17460	0.2087	0.1309	1.1662	2541.	436.3	0.943	0.0222	0.0525	0.00396	17.35	–40.00
–35.00	10.571	94.18	3.4599	1.047	73.844	0.00247	0.17389	0.2095	0.1321	1.1667	2503.	437.8	0.912	0.0224	0.0518	0.00403	16.95	–35.00
–30.00	11.982	93.68	3.0797	2.098	74.414	0.00493	0.17323	0.2104	0.1333	1.1674	2466.	439.2	0.882	0.0227	0.0512	0.00411	16.55	–30.00
–25.00	13.536	93.18	2.7494	3.154	74.982	0.00736	0.17261	0.2113	0.1346	1.1683	2428.	440.5	0.854	0.0229	0.0506	0.00418	16.15	–25.00
–21.55b	14.696	92.83	2.5469	3.884	75.373	0.00903	0.17221	0.2119	0.1354	1.1690	2403.	441.3	0.835	0.0231	0.0502	0.00424	15.88	–21.55
–20.00	15.244	92.67	2.4615	4.214	75.549	0.00978	0.17203	0.2122	0.1358	1.1693	2391.	441.7	0.827	0.0232	0.0500	0.00426	15.76	–20.00
–15.00	17.115	92.16	2.2098	5.280	76.115	0.01218	0.17148	0.2131	0.1371	1.1705	2354.	442.8	0.801	0.0235	0.0494	0.00434	15.36	–15.00
–10.00	19.159	91.65	1.9889	6.350	76.678	0.01457	0.17097	0.2141	0.1384	1.1718	2317.	443.9	0.776	0.0237	0.0488	0.00442	14.97	–10.00
–5.00	21.388	91.13	1.7945	7.425	77.239	0.01693	0.17048	0.2151	0.1397	1.1733	2279.	444.8	0.752	0.0240	0.0482	0.00450	14.58	–5.00
0.00	23.812	90.61	1.6230	8.505	77.797	0.01929	0.17003	0.2161	0.1411	1.1750	2243.	445.6	0.729	0.0243	0.0476	0.00458	14.19	0.00
5.00	26.442	90.08	1.4711	9.591	78.352	0.02162	0.16960	0.2171	0.1424	1.1769	2206.	446.4	0.707	0.0245	0.0470	0.00466	13.81	5.00
10.00	29.290	89.55	1.3364	10.682	78.904	0.02395	0.16920	0.2182	0.1439	1.1790	2169.	447.0	0.686	0.0248	0.0464	0.00474	13.43	10.00
15.00	32.365	89.02	1.2165	11.778	79.453	0.02625	0.16883	0.2193	0.1453	1.1813	2132.	447.5	0.666	0.0251	0.0458	0.00482	13.05	15.00
20.00	35.682	88.48	1.1095	12.880	79.998	0.02855	0.16847	0.2204	0.1468	1.1838	2095.	447.9	0.646	0.0253	0.0452	0.00491	12.67	20.00
25.00	39.250	87.93	1.0138	13.988	80.539	0.03083	0.16814	0.2216	0.1483	1.1866	2059.	448.2	0.627	0.0256	0.0447	0.00499	12.29	25.00
30.00	43.083	87.38	0.9281	15.102	81.075	0.03310	0.16783	0.2228	0.1499	1.1896	2022.	448.4	0.609	0.0259	0.0441	0.00508	11.92	30.00
35.00	47.192	86.82	0.8510	16.222	81.606	0.03536	0.16754	0.2240	0.1515	1.1929	1986.	448.5	0.591	0.0261	0.0435	0.00516	11.55	35.00
40.00	51.590	86.25	0.7816	17.348	82.133	0.03761	0.16726	0.2253	0.1532	1.1965	1949.	448.4	0.574	0.0264	0.0429	0.00525	11.18	40.00
45.00	56.289	85.68	0.7190	18.481	82.654	0.03984	0.16700	0.2266	0.1549	1.2004	1913.	448.3	0.557	0.0267	0.0424	0.00534	10.82	45.00
50.00	61.303	85.10	0.6624	19.621	83.169	0.04207	0.16675	0.2279	0.1567	1.2047	1876.	448.0	0.541	0.0270	0.0418	0.00543	10.45	50.00
55.00	66.643	84.52	0.6110	20.767	83.678	0.04428	0.16652	0.2294	0.1585	1.2093	1840.	447.5	0.526	0.0272	0.0413	0.00552	10.09	55.00
60.00	72.323	83.92	0.5645	21.921	84.180	0.04649	0.16630	0.2308	0.1604	1.2143	1803.	447.0	0.511	0.0275	0.0407	0.00562	9.74	60.00
65.00	78.357	83.32	0.5221	23.082	84.675	0.04869	0.16608	0.2323	0.1624	1.2197	1766.	446.3	0.496	0.0278	0.0401	0.00571	9.38	65.00
70.00	84.757	82.71	0.4835	24.251	85.163	0.05088	0.16588	0.2339	0.1645	1.2256	1730.	445.4	0.482	0.0281	0.0396	0.00581	9.03	70.00
75.00	91.536	82.10	0.4482	25.427	85.642	0.05306	0.16568	0.2356	0.1666	1.2319	1693.	444.5	0.468	0.0284	0.0390	0.00591	8.68	75.00
80.00	98.710	81.47	0.4159	26.612	86.113	0.05524	0.16549	0.2373	0.1689	1.2389	1656.	443.3	0.454	0.0287	0.0385	0.00601	8.33	80.00
85.00	106.29	80.83	0.3864	27.805	86.576	0.05740	0.16531	0.2391	0.1712	1.2464	1619.	442.0	0.441	0.0290	0.0379	0.00612	7.99	85.00
90.00	114.29	80.18	0.3593	29.007	87.028	0.05957	0.16512	0.2410	0.1737	1.2546	1582.	440.6	0.428	0.0293	0.0374	0.00623	7.65	90.00
95.00	122.73	79.52	0.3344	30.218	87.470	0.06173	0.16494	0.2430	0.1763	1.2635	1545.	439.0	0.415	0.0296	0.0369	0.00634	7.31	95.00
100.00	131.62	78.85	0.3114	31.439	87.902	0.06388	0.16477	0.2452	0.1791	1.2733	1508.	437.3	0.403	0.0299	0.0363	0.00645	6.98	100.00
105.00	140.97	78.16	0.2903	32.669	88.322	0.06603	0.16459	0.2474	0.1820	1.2839	1470.	435.4	0.391	0.0302	0.0358	0.00657	6.65	105.00
110.00	150.81	77.46	0.2707	33.911	88.729	0.06818	0.16440	0.2498	0.1851	1.2956	1433.	433.3	0.379	0.0306	0.0352	0.00670	6.32	110.00
115.00	161.13	76.75	0.2527	35.163	89.123	0.07032	0.16422	0.2523	0.1884	1.3085	1395.	431.0	0.367	0.0309	0.0347	0.00682	6.00	115.00
120.00	171.97	76.02	0.2359	36.427	89.503	0.07247	0.16403	0.2551	0.1920	1.3226	1357.	428.5	0.356	0.0313	0.0341	0.00696	5.68	120.00
125.00	183.33	75.28	0.2204	37.703	89.868	0.07461	0.16383	0.2580	0.1958	1.3383	1318.	425.9	0.345	0.0316	0.0336	0.00710	5.36	125.00
130.00	195.24	74.51	0.2060	38.992	90.216	0.07676	0.16362	0.2611	0.2000	1.3556	1279.	423.0	0.334	0.0320	0.0331	0.00724	5.05	130.00
135.00	207.70	73.73	0.1926	40.295	90.546	0.07890	0.16341	0.2645	0.2045	1.3750	1240.	420.0	0.323	0.0324	0.0325	0.00740	4.74	135.00
140.00	220.74	72.93	0.1801	41.613	90.857	0.08106	0.16317	0.2683	0.2094	1.3966	1201.	416.7	0.312	0.0328	0.0320	0.00756	4.44	140.00
145.00	234.37	72.10	0.1684	42.947	91.148	0.08321	0.16293	0.2724	0.2149	1.4210	1161.	413.3	0.302	0.0333	0.0315	0.00773	4.14	145.00
150.00	248.61	71.24	0.1575	44.298	91.415	0.08538	0.16266	0.2769	0.2209	1.4486	1120.	409.6	0.291	0.0337	0.0309	0.00792	3.85	150.00
155.00	263.47	70.36	0.1472	45.667	91.657	0.08755	0.16237	0.2819	0.2277	1.4800	1079.	405.6	0.281	0.0342	0.0304	0.00811	3.56	155.00
160.00	278.99	69.45	0.1377	47.056	91.872	0.08973	0.16206	0.2875	0.2353	1.5161	1037.	401.4	0.271	0.0347	0.0298	0.00833	3.27	160.00
165.00	295.17	68.50	0.1286	48.467	92.056	0.09193	0.16171	0.2938	0.2440	1.5578	995.	397.0	0.261	0.0352	0.0293	0.00855	2.99	165.00
170.00	312.04	67.51	0.1202	49.903	92.206	0.09415	0.16133	0.3011	0.2540	1.6065	952.	392.3	0.251	0.0358	0.0288	0.00880	2.72	170.00
175.00	329.62	66.47	0.1122	51.366	92.318	0.09638	0.16091	0.3094	0.2657	1.6642	908.	387.3	0.240	0.0364	0.0283	0.00908	2.45	175.00
180.00	347.93	65.39	0.1046	52.859	92.385	0.09865	0.16044	0.3192	0.2795	1.7333	864.	382.0	0.230	0.0371	0.0278	0.00938	2.19	180.00
185.00	367.00	64.24	0.0974	54.386	92.403	0.10094	0.15991	0.3309	0.2962	1.8175	818.	376.4	0.220	0.0379	0.0273	0.00973	1.93	185.00
190.00	386.85	63.03	0.0906	55.954	92.361	0.10327	0.15931	0.3450	0.3168	1.9220	772.	370.5	0.210	0.0387	0.0268	0.01012	1.68	190.00
195.00	407.52	61.73	0.0840	57.568	92.250	0.10565	0.15863	0.3625	0.3427	2.0549	724.	364.2	0.200	0.0396	0.0264	0.01057	1.44	195.00
200.00	429.04	60.34	0.0778	59.237	92.054	0.10810	0.15784	0.3848	0.3766	2.2291	676.	357.7	0.189	0.0407	0.0261	0.01109	1.21	200.00
210.00	474.80	57.15	0.0658	62.795	91.317	0.11323	0.15582	0.4555	0.4887	2.8082	575.	343.4	0.167	0.0434	0.0262	0.01250	0.77	210.00
220.00	524.53	53.07	0.0540	66.829	89.806	0.11896	0.15277	0.6261	0.7725	4.2679	468.	327.4	0.143	0.0475	0.0291	0.01489	0.38	220.00
230.00	579.05	46.36	0.0404	72.258	86.082	0.12659	0.14664	1.866	2.734	14.143	345.	308.3	0.112	0.0568	0.0657	0.02282	0.07	230.00
233.55c	599.89	35.27	0.0284	79.118	79.118	0.13637	0.13637	∞	∞	∞	0.	0.0	–	–	∞	∞	0.00	233.55

*Temperatures are on the ITS-90 scale   b = normal boiling point   c = critical point

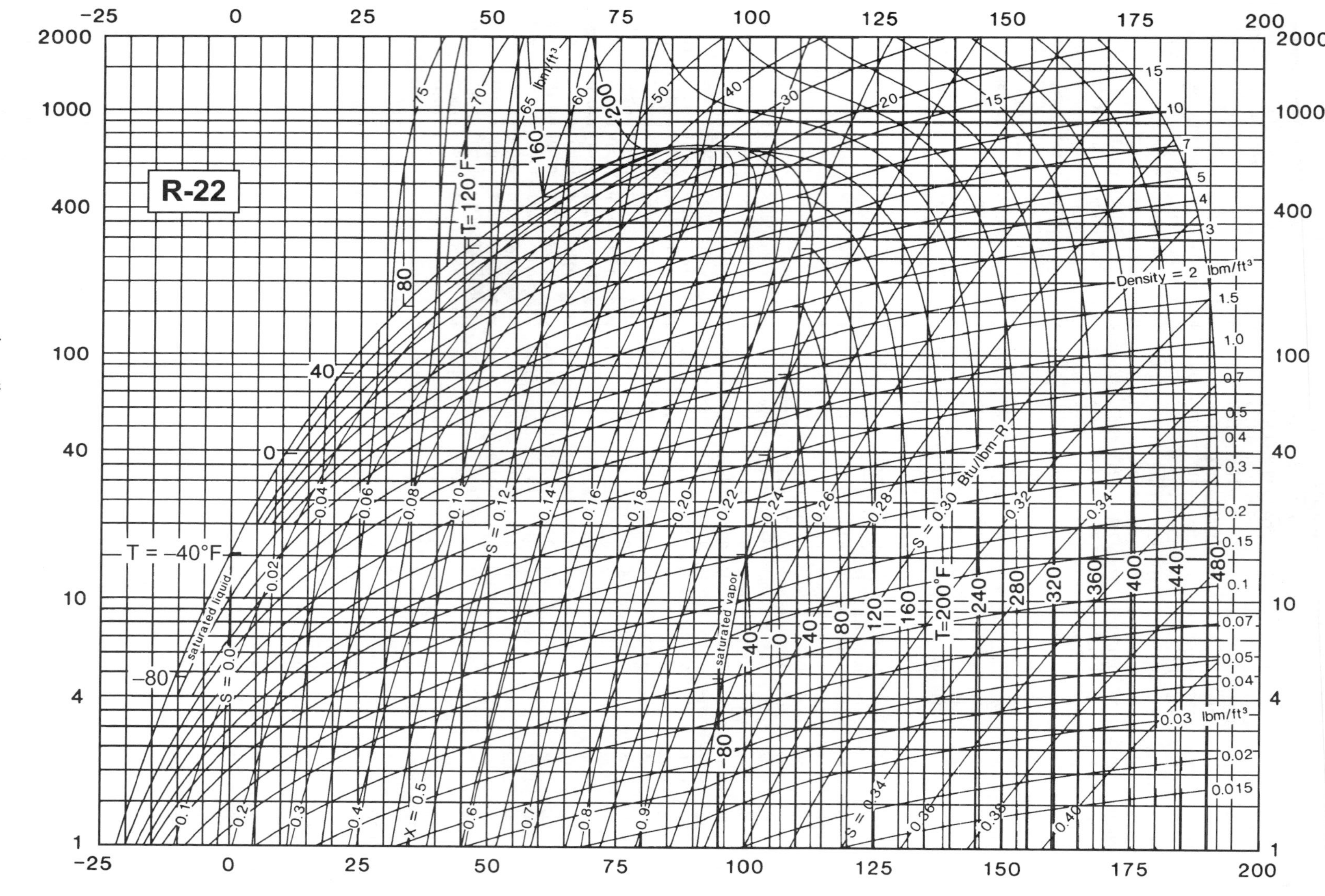

Fig. 2 Pressure-Enthalpy Diagram for Refrigerant 22

Refrigerant 22 (Chlorodifluoromethane) Properties of Saturated Liquid and Saturated Vapor

Temp.,* °F	Pressure, psia	Density, lb/ft³ Liquid	Volume, ft³/lb Vapor	Enthalpy, Btu/lb Liquid	Enthalpy, Btu/lb Vapor	Entropy, Btu/lb·°F Liquid	Entropy, Btu/lb·°F Vapor	Specific Heat c_p, Btu/lb·°F Liquid	Specific Heat c_p, Btu/lb·°F Vapor	c_p/c_v Vapor	Vel. of Sound, ft/s Liquid	Vel. of Sound, ft/s Vapor	Viscosity, lb_m/ft·h Liquid	Viscosity, lb_m/ft·h Vapor	Thermal Cond., Btu/h·ft·°F Liquid	Thermal Cond., Btu/h·ft·°F Vapor	Surface Tension, dyne/cm	Temp., °F
–150.00	0.263	98.28	146.06	–28.119	87.566	–0.07757	0.29600	0.2536	0.1185	1.2437	3716.	469.7	2.093	0.0174	0.0831	0.00255	28.31	–150.00
–140.00	0.436	97.36	90.759	–25.583	88.729	–0.06951	0.28808	0.2536	0.1204	1.2404	3630.	476.2	1.874	0.0180	0.0814	0.00267	27.34	–140.00
–130.00	0.698	96.44	58.384	–23.046	89.899	–0.06170	0.28090	0.2536	0.1223	1.2375	3544.	482.4	1.692	0.0186	0.0797	0.00280	26.36	–130.00
–120.00	1.082	95.52	38.745	–20.509	91.074	–0.05412	0.27439	0.2537	0.1244	1.2350	3458.	488.5	1.537	0.0191	0.0780	0.00293	25.40	–120.00
–110.00	1.629	94.59	26.444	–17.970	92.252	–0.04675	0.26846	0.2540	0.1265	1.2330	3373.	494.2	1.405	0.0197	0.0763	0.00306	24.44	–110.00
–100.00	2.388	93.66	18.511	–15.427	93.430	–0.03959	0.26307	0.2543	0.1288	1.2315	3287.	499.7	1.290	0.0203	0.0747	0.00320	23.49	–100.00
–95.00	2.865	93.19	15.623	–14.154	94.018	–0.03608	0.26055	0.2546	0.1300	1.2310	3245.	502.4	1.238	0.0206	0.0739	0.00327	23.02	–95.00
–90.00	3.417	92.71	13.258	–12.880	94.605	–0.03261	0.25815	0.2549	0.1312	1.2307	3202.	504.9	1.189	0.0208	0.0731	0.00334	22.55	–90.00
–85.00	4.053	92.24	11.309	–11.604	95.191	–0.02918	0.25585	0.2552	0.1324	1.2305	3160.	507.4	1.144	0.0211	0.0723	0.00341	22.08	–85.00
–80.00	4.782	91.76	9.6939	–10.326	95.775	–0.02580	0.25366	0.2556	0.1337	1.2304	3118.	509.8	1.101	0.0214	0.0715	0.00348	21.61	–80.00
–75.00	5.615	91.28	8.3487	–9.046	96.357	–0.02245	0.25155	0.2561	0.1350	1.2305	3075.	512.2	1.060	0.0217	0.0708	0.00355	21.15	–75.00
–70.00	6.561	90.79	7.2222	–7.763	96.937	–0.01915	0.24954	0.2566	0.1363	1.2308	3033.	514.4	1.021	0.0220	0.0700	0.00363	20.68	–70.00
–65.00	7.631	90.31	6.2744	–6.477	97.514	–0.01587	0.24761	0.2571	0.1377	1.2313	2990.	516.5	0.985	0.0223	0.0692	0.00370	20.22	–65.00
–60.00	8.836	89.82	5.4730	–5.189	98.087	–0.01264	0.24577	0.2577	0.1392	1.2320	2948.	518.6	0.951	0.0225	0.0684	0.00378	19.76	–60.00
–55.00	10.190	89.33	4.7924	–3.897	98.657	–0.00943	0.24400	0.2583	0.1406	1.2328	2906.	520.5	0.918	0.0228	0.0677	0.00386	19.30	–55.00
–50.00	11.703	88.83	4.2119	–2.602	99.224	–0.00626	0.24230	0.2591	0.1422	1.2339	2863.	522.4	0.887	0.0231	0.0669	0.00394	18.85	–50.00
–45.00	13.390	88.33	3.7147	–1.303	99.786	–0.00311	0.24067	0.2598	0.1438	1.2352	2821.	524.1	0.857	0.0234	0.0661	0.00402	18.40	–45.00
–41.46b	14.696	87.97	3.4054	–0.381	100.181	–0.00091	0.23955	0.2604	0.1449	1.2362	2791.	525.3	0.837	0.0236	0.0656	0.00407	18.08	–41.46
–40.00	15.262	87.82	3.2872	0.000	100.343	0.00000	0.23910	0.2606	0.1454	1.2367	2778.	525.8	0.829	0.0237	0.0654	0.00410	17.94	–40.00
–35.00	17.336	87.32	2.9181	1.308	100.896	0.00309	0.23759	0.2615	0.1471	1.2384	2736.	527.3	0.802	0.0240	0.0646	0.00418	17.49	–35.00
–30.00	19.624	86.80	2.5984	2.620	101.443	0.00615	0.23615	0.2625	0.1488	1.2404	2694.	528.7	0.776	0.0242	0.0639	0.00426	17.05	–30.00
–25.00	22.142	86.29	2.3204	3.937	101.984	0.00918	0.23475	0.2635	0.1506	1.2426	2651.	530.0	0.751	0.0245	0.0631	0.00435	16.60	–25.00
–20.00	24.906	85.76	2.0778	5.260	102.519	0.01220	0.23341	0.2645	0.1525	1.2451	2609.	531.2	0.728	0.0248	0.0624	0.00444	16.16	–20.00
–15.00	27.929	85.24	1.8656	6.588	103.048	0.01519	0.23211	0.2656	0.1544	1.2479	2566.	532.3	0.705	0.0251	0.0617	0.00452	15.72	–15.00
–10.00	31.230	84.71	1.6792	7.923	103.570	0.01815	0.23086	0.2668	0.1564	1.2510	2524.	533.2	0.683	0.0254	0.0609	0.00461	15.28	–10.00
–5.00	34.824	84.17	1.5150	9.263	104.085	0.02110	0.22965	0.2681	0.1585	1.2544	2481.	534.0	0.662	0.0257	0.0602	0.00471	14.85	–5.00
0.00	38.728	83.63	1.3701	10.610	104.591	0.02403	0.22848	0.2694	0.1607	1.2581	2438.	534.7	0.642	0.0260	0.0595	0.00480	14.41	0.00
5.00	42.960	83.08	1.2417	11.964	105.090	0.02694	0.22735	0.2708	0.1629	1.2622	2396.	535.3	0.622	0.0262	0.0587	0.00489	13.98	5.00
10.00	47.536	82.52	1.1276	13.325	105.580	0.02983	0.22625	0.2722	0.1652	1.2666	2353.	535.7	0.603	0.0265	0.0580	0.00499	13.55	10.00
15.00	52.475	81.96	1.0261	14.694	106.061	0.03270	0.22519	0.2737	0.1676	1.2714	2310.	536.0	0.585	0.0268	0.0573	0.00509	13.13	15.00
20.00	57.795	81.39	0.9354	16.070	106.532	0.03556	0.22415	0.2753	0.1702	1.2767	2268.	536.1	0.568	0.0271	0.0566	0.00519	12.70	20.00
25.00	63.514	80.82	0.8543	17.455	106.994	0.03841	0.22315	0.2770	0.1728	1.2824	2225.	536.1	0.551	0.0274	0.0558	0.00530	12.28	25.00
30.00	69.651	80.24	0.7815	18.848	107.445	0.04124	0.22217	0.2787	0.1755	1.2886	2182.	535.9	0.534	0.0277	0.0551	0.00540	11.86	30.00
35.00	76.225	79.65	0.7161	20.250	107.884	0.04406	0.22121	0.2806	0.1783	1.2953	2139.	535.6	0.518	0.0280	0.0544	0.00551	11.45	35.00
40.00	83.255	79.05	0.6572	21.662	108.313	0.04686	0.22028	0.2825	0.1813	1.3026	2096.	535.1	0.503	0.0283	0.0537	0.00562	11.04	40.00
45.00	90.761	78.44	0.6040	23.083	108.729	0.04966	0.21936	0.2845	0.1844	1.3105	2053.	534.4	0.488	0.0286	0.0530	0.00574	10.63	45.00
50.00	98.763	77.83	0.5558	24.514	109.132	0.05244	0.21847	0.2866	0.1877	1.3191	2010.	533.6	0.473	0.0289	0.0522	0.00586	10.22	50.00
55.00	107.28	77.20	0.5122	25.956	109.521	0.05522	0.21758	0.2889	0.1911	1.3284	1967.	532.6	0.459	0.0292	0.0515	0.00598	9.82	55.00
60.00	116.33	76.57	0.4725	27.409	109.897	0.05798	0.21672	0.2913	0.1947	1.3385	1924.	531.5	0.445	0.0296	0.0508	0.00611	9.41	60.00
65.00	125.94	75.92	0.4364	28.874	110.257	0.06074	0.21586	0.2938	0.1985	1.3495	1880.	530.1	0.432	0.0299	0.0501	0.00625	9.02	65.00
70.00	136.13	75.27	0.4035	30.350	110.602	0.06350	0.21501	0.2964	0.2025	1.3615	1836.	528.6	0.419	0.0302	0.0494	0.00638	8.62	70.00
75.00	146.92	74.60	0.3734	31.839	110.929	0.06625	0.21417	0.2992	0.2067	1.3746	1793.	526.9	0.406	0.0305	0.0487	0.00653	8.23	75.00
80.00	158.33	73.92	0.3459	33.342	111.239	0.06899	0.21333	0.3022	0.2112	1.3889	1749.	525.0	0.394	0.0309	0.0479	0.00668	7.84	80.00
85.00	170.38	73.23	0.3207	34.859	111.530	0.07173	0.21250	0.3054	0.2160	1.4046	1705.	522.9	0.381	0.0312	0.0472	0.00684	7.46	85.00
90.00	183.09	72.52	0.2975	36.391	111.801	0.07447	0.21166	0.3089	0.2212	1.4218	1660.	520.6	0.369	0.0316	0.0465	0.00701	7.08	90.00
95.00	196.50	71.80	0.2762	37.938	112.050	0.07721	0.21083	0.3126	0.2267	1.4407	1615.	518.1	0.358	0.0320	0.0458	0.00718	6.70	95.00
100.00	210.61	71.06	0.2566	39.502	112.276	0.07996	0.20998	0.3166	0.2327	1.4616	1570.	515.4	0.346	0.0324	0.0450	0.00737	6.33	100.00
105.00	225.46	70.30	0.2385	41.084	112.478	0.08270	0.20913	0.3209	0.2391	1.4849	1525.	512.4	0.335	0.0328	0.0443	0.00757	5.96	105.00
110.00	241.06	69.52	0.2217	42.686	112.653	0.08545	0.20827	0.3257	0.2461	1.5107	1479.	509.2	0.324	0.0332	0.0436	0.00778	5.60	110.00
115.00	257.45	68.72	0.2062	44.308	112.799	0.08821	0.20739	0.3309	0.2538	1.5396	1433.	505.8	0.313	0.0336	0.0428	0.00801	5.24	115.00
120.00	274.65	67.90	0.1918	45.952	112.914	0.09098	0.20649	0.3367	0.2623	1.5722	1387.	502.1	0.302	0.0341	0.0421	0.00825	4.88	120.00
125.00	292.69	67.05	0.1785	47.621	112.996	0.09376	0.20557	0.3431	0.2717	1.6090	1340.	498.1	0.292	0.0346	0.0413	0.00851	4.53	125.00
130.00	311.58	66.18	0.1660	49.316	113.040	0.09656	0.20462	0.3504	0.2822	1.6509	1292.	493.9	0.281	0.0351	0.0406	0.00880	4.19	130.00
135.00	331.37	65.27	0.1544	51.041	113.043	0.09937	0.20364	0.3585	0.2941	1.6990	1244.	489.4	0.271	0.0356	0.0399	0.00911	3.85	135.00
140.00	352.08	64.32	0.1435	52.798	113.000	0.10222	0.20261	0.3679	0.3076	1.7548	1195.	484.6	0.260	0.0362	0.0391	0.00946	3.51	140.00
145.00	373.74	63.34	0.1334	54.591	112.907	0.10509	0.20153	0.3787	0.3233	1.8201	1146.	479.5	0.250	0.0369	0.0383	0.00984	3.18	145.00
150.00	396.38	62.31	0.1238	56.425	112.756	0.10800	0.20040	0.3913	0.3416	1.8976	1095.	474.1	0.240	0.0375	0.0376	0.01027	2.86	150.00
155.00	420.04	61.22	0.1149	58.305	112.539	0.11096	0.19919	0.4063	0.3633	1.9907	1044.	468.4	0.230	0.0383	0.0368	0.01076	2.54	155.00
160.00	444.75	60.07	0.1064	60.240	112.247	0.11397	0.19790	0.4243	0.3897	2.1047	992.	462.3	0.219	0.0391	0.0361	0.01131	2.24	160.00
165.00	470.56	58.84	0.0984	62.237	111.866	0.11705	0.19650	0.4467	0.4225	2.2474	939.	455.8	0.209	0.0400	0.0353	0.01195	1.93	165.00
170.00	497.50	57.53	0.0907	64.309	111.378	0.12022	0.19497	0.4750	0.4643	2.4310	884.	449.0	0.198	0.0410	0.0346	0.01270	1.64	170.00
175.00	525.62	56.10	0.0834	66.474	110.760	0.12350	0.19328	0.5124	0.5198	2.6759	828.	441.7	0.188	0.0422	0.0340	0.01360	1.36	175.00
180.00	554.98	54.52	0.0764	68.757	109.976	0.12693	0.19136	0.5641	0.5972	3.0184	769.	433.9	0.176	0.0436	0.0335	0.01470	1.09	180.00
185.00	585.63	52.74	0.0695	71.196	108.972	0.13056	0.18916	0.6410	0.7132	3.5317	706.	425.6	0.165	0.0452	0.0332	0.01609	0.83	185.00
190.00	617.64	50.67	0.0626	73.859	107.654	0.13450	0.18651	0.7681	0.9067	4.3857	639.	416.6	0.152	0.0474	0.0334	0.01793	0.58	190.00
195.00	651.12	48.14	0.0556	76.875	105.835	0.13893	0.18316	1.020	1.295	6.090	565.	406.9	0.138	0.0502	0.0347	0.02061	0.35	195.00
200.00	686.20	44.68	0.0479	80.593	103.010	0.14437	0.17835	1.778	2.472	11.190	480.	395.8	0.121	0.0547	0.0395	0.02574	0.15	200.00
205.06c	723.74	32.70	0.0306	91.208	91.208	0.16012	0.16012	∞	∞	∞	0.	0.0	–	–	∞	∞	0.00	205.06

*Temperatures are on the ITS-90 scale b = normal boiling point c = critical point

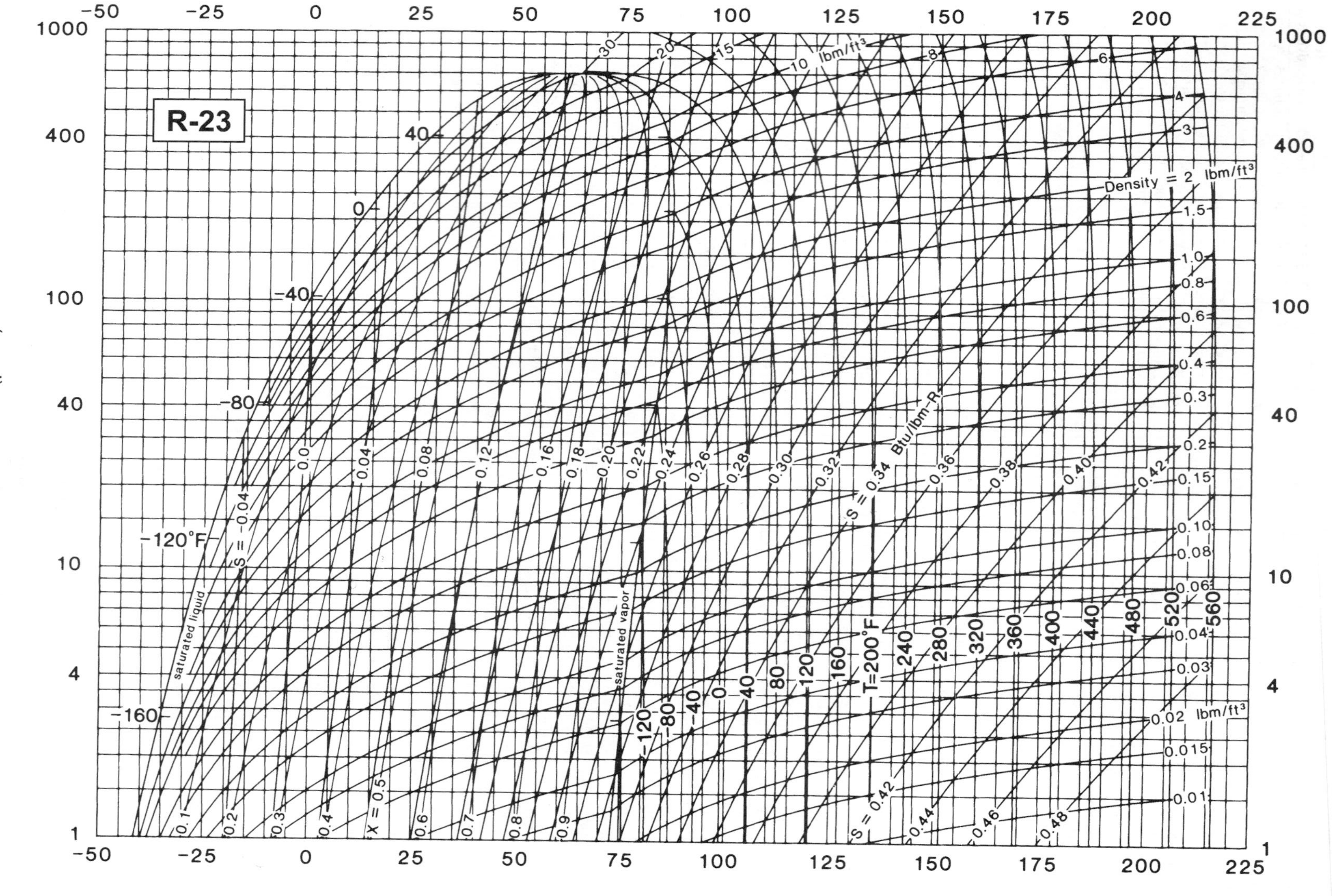

Fig. 3 Pressure-Enthalpy Diagram for Refrigerant 23

Refrigerant 23 (Trifluoromethane) Properties of Saturated Liquid and Saturated Vapor

Temp.,* °F	Pressure, psia	Denisty, lb/ft³ Liquid	Volume, ft³/lb Vapor	Enthalpy, Btu/lb Liquid	Enthalpy, Btu/lb Vapor	Entropy, Btu/lb·°F Liquid	Entropy, Btu/lb·°F Vapor	Specific Heat c_p, Btu/lb·°F Liquid	Specific Heat c_p, Btu/lb·°F Vapor	c_p/c_v Vapor	Vel. of Sound, ft/s Liquid	Vel. of Sound, ft/s Vapor	Viscosity, lb_m/ft·h Liquid	Viscosity, lb_m/ft·h Vapor	Thermal Cond., Btu/h·ft·°F Liquid	Thermal Cond., Btu/h·ft·°F Vapor	Surface Tension, dyne/cm	Temp., °F
–247.23a	0.0085	105.02	3815.	–61.278	64.550	–0.19959	0.39274	0.2974	0.1187	1.3153	3290.	445.3	4.2331	0.01293	0.1489	0.00219	32.34	–247.23
–240.00	0.0169	104.26	1989.	–59.141	65.403	–0.18970	0.37727	0.2934	0.1196	1.3128	3317.	452.3	3.6342	0.01350	0.1412	0.00232	31.51	–240.00
–230.00	0.0402	103.20	875.7	–56.229	66.587	–0.17673	0.35802	0.2893	0.1211	1.3093	3337.	461.7	2.9904	0.01428	0.1318	0.00249	30.36	–230.00
–220.00	0.0878	102.11	418.0	–53.350	67.774	–0.16446	0.34091	0.2866	0.1228	1.3059	3339.	470.8	2.5014	0.01505	0.1237	0.00267	29.20	–220.00
–210.00	0.1785	101.00	213.9	–50.492	68.959	–0.15278	0.32566	0.2850	0.1249	1.3026	3325.	479.4	2.1227	0.01583	0.1166	0.00284	28.05	–210.00
–200.00	0.3407	99.88	116.4	–47.646	70.141	–0.14161	0.31200	0.2843	0.1273	1.2997	3296.	487.7	1.8243	0.01660	0.1103	0.00301	26.90	–200.00
–190.00	0.6155	98.73	66.779	–44.804	71.314	–0.13087	0.29973	0.2842	0.1302	1.2973	3256.	495.6	1.5856	0.01736	0.1048	0.00319	25.76	–190.00
–180.00	1.0587	97.57	40.159	–41.959	72.475	–0.12051	0.28866	0.2847	0.1335	1.2956	3204.	503.0	1.3917	0.01812	0.0998	0.00336	24.61	–180.00
–170.00	1.7438	96.39	25.170	–39.108	73.617	–0.11050	0.27865	0.2855	0.1374	1.2946	3144.	510.0	1.2322	0.01888	0.0953	0.00354	23.47	–170.00
–160.00	2.7636	95.20	16.362	–36.246	74.737	–0.10079	0.26956	0.2867	0.1419	1.2947	3076.	516.5	1.0995	0.01963	0.0913	0.00372	22.34	–160.00
–150.00	4.2314	93.98	10.986	–33.371	75.830	–0.09136	0.26127	0.2881	0.1469	1.2959	3002.	522.5	0.9877	0.02037	0.0876	0.00391	21.21	–150.00
–140.00	6.2822	92.75	7.5919	–30.478	76.891	–0.08219	0.25369	0.2899	0.1526	1.2985	2922.	527.9	0.8925	0.02112	0.0842	0.00410	20.08	–140.00
–130.00	9.0723	91.49	5.3819	–27.567	77.917	–0.07324	0.24673	0.2918	0.1589	1.3025	2838.	532.8	0.8108	0.02185	0.0810	0.00429	18.96	–130.00
–120.00	12.779	90.21	3.9029	–24.634	78.902	–0.06449	0.24032	0.2940	0.1657	1.3081	2749.	537.0	0.7399	0.02259	0.0781	0.00449	17.85	–120.00
–115.72b	14.696	89.66	3.4227	–23.370	79.312	–0.06081	0.23773	0.2951	0.1689	1.3110	2711.	538.6	0.7124	0.02290	0.0769	0.00458	17.38	–115.72
–110.00	17.601	88.91	2.8883	–21.677	79.844	–0.05594	0.23439	0.2965	0.1732	1.3155	2658.	540.6	0.6779	0.02332	0.0754	0.00470	16.75	–110.00
–106.00	19.890	88.38	2.5739	–20.486	80.208	–0.05257	0.23214	0.2976	0.1763	1.3190	2620.	541.9	0.6553	0.02361	0.0743	0.00478	16.31	–106.00
–102.00	22.406	87.85	2.3001	–19.291	80.564	–0.04923	0.22996	0.2987	0.1795	1.3228	2582.	543.0	0.6337	0.02390	0.0733	0.00487	15.87	–102.00
–98.00	25.165	87.31	2.0608	–18.092	80.912	–0.04591	0.22783	0.2999	0.1828	1.3270	2544.	544.0	0.6131	0.02420	0.0723	0.00496	15.43	–98.00
–94.00	28.182	86.77	1.8510	–16.887	81.251	–0.04261	0.22577	0.3011	0.1862	1.3315	2506.	544.9	0.5935	0.02449	0.0714	0.00505	15.00	–94.00
–90.00	31.473	86.22	1.6666	–15.677	81.582	–0.03934	0.22376	0.3024	0.1897	1.3364	2466.	545.7	0.5747	0.02478	0.0704	0.00514	14.57	–90.00
–88.00	33.227	85.95	1.5828	–15.070	81.744	–0.03771	0.22277	0.3030	0.1915	1.3390	2447.	546.0	0.5656	0.02493	0.0699	0.00519	14.35	–88.00
–86.00	35.055	85.67	1.5041	–14.462	81.904	–0.03609	0.22180	0.3037	0.1933	1.3417	2427.	546.3	0.5567	0.02508	0.0695	0.00523	14.14	–86.00
–84.00	36.960	85.39	1.4300	–13.852	82.062	–0.03448	0.22084	0.3044	0.1952	1.3445	2407.	546.6	0.5480	0.02522	0.0690	0.00528	13.92	–84.00
–82.00	38.945	85.11	1.3603	–13.241	82.217	–0.03286	0.21989	0.3051	0.1970	1.3475	2387.	546.8	0.5395	0.02537	0.0685	0.00533	13.71	–82.00
–80.00	41.010	84.83	1.2947	–12.629	82.370	–0.03126	0.21895	0.3059	0.1989	1.3505	2367.	547.1	0.5312	0.02552	0.0681	0.00538	13.49	–80.00
–78.00	43.158	84.55	1.2329	–12.014	82.520	–0.02966	0.21803	0.3066	0.2008	1.3536	2347.	547.2	0.5230	0.02566	0.0676	0.00543	13.28	–78.00
–76.00	45.392	84.26	1.1746	–11.399	82.668	–0.02806	0.21711	0.3074	0.2027	1.3569	2327.	547.4	0.5150	0.02581	0.0672	0.00548	13.07	–76.00
–74.00	47.714	83.98	1.1196	–10.781	82.813	–0.02647	0.21621	0.3082	0.2047	1.3602	2307.	547.5	0.5072	0.02596	0.0668	0.00553	12.86	–74.00
–72.00	50.126	83.69	1.0677	–10.162	82.956	–0.02488	0.21532	0.3090	0.2066	1.3637	2287.	547.6	0.4995	0.02611	0.0663	0.00558	12.64	–72.00
–70.00	52.630	83.40	1.0186	–9.542	83.096	–0.02330	0.21444	0.3099	0.2086	1.3674	2266.	547.6	0.4920	0.02626	0.0659	0.00563	12.43	–70.00
–68.00	55.228	83.11	0.9723	–8.920	83.234	–0.02172	0.21356	0.3107	0.2107	1.3711	2246.	547.6	0.4846	0.02640	0.0655	0.00569	12.22	–68.00
–66.00	57.923	82.82	0.9285	–8.295	83.369	–0.02015	0.21270	0.3116	0.2127	1.3750	2226.	547.6	0.4773	0.02655	0.0650	0.00574	12.01	–66.00
–64.00	60.718	82.52	0.8870	–7.670	83.501	–0.01858	0.21184	0.3125	0.2148	1.3790	2205.	547.6	0.4702	0.02670	0.0646	0.00579	11.80	–64.00
–62.00	63.613	82.23	0.8477	–7.042	83.630	–0.01701	0.21100	0.3135	0.2170	1.3832	2184.	547.5	0.4632	0.02685	0.0642	0.00585	11.59	–62.00
–60.00	66.613	81.93	0.8105	–6.412	83.757	–0.01545	0.21016	0.3144	0.2191	1.3875	2164.	547.3	0.4563	0.02701	0.0638	0.00590	11.39	–60.00
–58.00	69.719	81.63	0.7753	–5.781	83.880	–0.01389	0.20933	0.3154	0.2213	1.3920	2143.	547.1	0.4495	0.02716	0.0634	0.00596	11.18	–58.00
–56.00	72.933	81.32	0.7418	–5.147	84.001	–0.01234	0.20851	0.3164	0.2235	1.3967	2122.	546.9	0.4429	0.02731	0.0630	0.00602	10.97	–56.00
–54.00	76.259	81.02	0.7101	–4.511	84.119	–0.01078	0.20769	0.3175	0.2258	1.4015	2101.	546.7	0.4364	0.02746	0.0625	0.00608	10.77	–54.00
–52.00	79.697	80.71	0.6800	–3.874	84.233	–0.00923	0.20689	0.3186	0.2281	1.4065	2080.	546.4	0.4300	0.02761	0.0621	0.00613	10.56	–52.00
–50.00	83.252	80.40	0.6514	–3.234	84.344	–0.00769	0.20609	0.3197	0.2304	1.4117	2059.	546.1	0.4237	0.02777	0.0617	0.00619	10.36	–50.00
–48.00	86.926	80.09	0.6242	–2.592	84.452	–0.00615	0.20530	0.3208	0.2328	1.4171	2038.	545.7	0.4175	0.02792	0.0613	0.00626	10.15	–48.00
–46.00	90.720	79.78	0.5983	–1.947	84.557	–0.00460	0.20451	0.3220	0.2352	1.4227	2017.	545.3	0.4114	0.02808	0.0609	0.00632	9.95	–46.00
–44.00	94.638	79.46	0.5737	–1.301	84.658	–0.00307	0.20373	0.3233	0.2377	1.4285	1996.	544.9	0.4053	0.02824	0.0605	0.00638	9.74	–44.00
–42.00	98.681	79.14	0.5503	–0.652	84.756	–0.00153	0.20295	0.3245	0.2402	1.4345	1975.	544.4	0.3994	0.02839	0.0601	0.00644	9.54	–42.00
–40.00	102.85	78.82	0.5280	0.000	84.850	0.00000	0.20218	0.3259	0.2427	1.4407	1954.	543.9	0.3936	0.02855	0.0597	0.00651	9.34	–40.00
–36.00	111.59	78.17	0.4865	1.311	85.027	0.00306	0.20066	0.3286	0.2480	1.4539	1911.	542.7	0.3822	0.02887	0.0589	0.00664	8.94	–36.00
–32.00	120.88	77.51	0.4488	2.633	85.189	0.00611	0.19915	0.3316	0.2535	1.4682	1868.	541.3	0.3711	0.02920	0.0581	0.00678	8.54	–32.00
–28.00	130.73	76.83	0.4144	3.966	85.335	0.00916	0.19766	0.3348	0.2593	1.4837	1825.	539.8	0.3603	0.02953	0.0574	0.00692	8.15	–28.00
–24.00	141.17	76.15	0.3831	5.311	85.463	0.01221	0.19618	0.3382	0.2653	1.5005	1781.	538.1	0.3498	0.02987	0.0566	0.00707	7.76	–24.00
–20.00	152.23	75.45	0.3544	6.670	85.573	0.01525	0.19471	0.3419	0.2717	1.5187	1738.	536.2	0.3395	0.03022	0.0558	0.00722	7.37	–20.00
–16.00	163.91	74.73	0.3281	8.042	85.664	0.01829	0.19324	0.3459	0.2784	1.5386	1694.	534.1	0.3295	0.03058	0.0550	0.00738	6.99	–16.00
–12.00	176.25	74.00	0.3040	9.429	85.735	0.02133	0.19178	0.3502	0.2856	1.5604	1649.	531.8	0.3197	0.03094	0.0542	0.00755	6.61	–12.00
–8.00	189.27	73.25	0.2818	10.832	85.783	0.02438	0.19032	0.3549	0.2932	1.5842	1605.	529.3	0.3101	0.03132	0.0535	0.00772	6.23	–8.00
–4.00	202.99	72.48	0.2614	12.253	85.808	0.02743	0.18886	0.3601	0.3014	1.6105	1560.	526.6	0.3007	0.03171	0.0527	0.00790	5.86	–4.00
0.00	217.44	71.69	0.2426	13.691	85.807	0.03050	0.18738	0.3658	0.3102	1.6394	1515.	523.7	0.2915	0.03211	0.0519	0.00809	5.50	0.00
4.00	232.64	70.87	0.2252	15.150	85.780	0.03357	0.18590	0.3720	0.3197	1.6716	1470.	520.6	0.2824	0.03253	0.0511	0.00829	5.14	4.00
8.00	248.62	70.04	0.2092	16.630	85.723	0.03666	0.18440	0.3789	0.3301	1.7074	1425.	517.2	0.2734	0.03297	0.0503	0.00849	4.78	8.00
12.00	265.41	69.17	0.1943	18.134	85.634	0.03977	0.18287	0.3865	0.3415	1.7475	1379.	513.6	0.2646	0.03343	0.0495	0.00871	4.43	12.00
16.00	283.02	68.28	0.1804	19.663	85.510	0.04289	0.18132	0.3950	0.3541	1.7928	1333.	509.7	0.2558	0.03391	0.0486	0.00894	4.09	16.00
20.00	301.49	67.35	0.1676	21.220	85.349	0.04605	0.17974	0.4046	0.3681	1.8442	1287.	505.6	0.2472	0.03442	0.0478	0.00919	3.75	20.00
24.00	320.85	66.39	0.1556	22.809	85.145	0.04923	0.17812	0.4154	0.3838	1.9029	1240.	501.3	0.2386	0.03497	0.0469	0.00945	3.42	24.00
28.00	341.13	65.39	0.1444	24.431	84.895	0.05246	0.17644	0.4277	0.4017	1.9708	1193.	496.6	0.2300	0.03555	0.0461	0.00973	3.09	28.00
32.00	362.35	64.35	0.1339	26.091	84.593	0.05572	0.17471	0.4420	0.4221	2.0498	1146.	491.7	0.2215	0.03618	0.0452	0.01002	2.78	32.00
36.00	384.56	63.25	0.1241	27.794	84.233	0.05904	0.17291	0.4586	0.4459	2.1432	1098.	486.5	0.2131	0.03687	0.0442	0.01034	2.47	36.00
40.00	407.78	62.09	0.1149	29.546	83.806	0.06242	0.17102	0.4783	0.4739	2.2549	1050.	481.0	0.2045	0.03762	0.0433	0.01068	2.17	40.00
50.00	470.53	58.87	0.0941	34.196	82.374	0.07126	0.16579	0.5508	0.5737	2.6626	919.	465.7	0.1829	0.03987	0.0407	0.01169	1.46	50.00
60.00	540.50	54.86	0.0756	39.448	80.179	0.08102	0.15940	0.7070	0.7659	3.4728	760.	448.2	0.1596	0.04305	0.0376	0.01301	0.83	60.00
70.00	618.48	48.85	0.0581	46.073	76.466	0.09311	0.15049	–	–	–	–	–	–	–	–	–	0.30	70.00
79.03c	700.14	32.83	0.0305	60.818	60.818	0.11999	0.11999	∞	∞	∞	0.	0.0	–	–	∞	∞	0.00	79.03

*Temperatures are on the ITS-90 scale

b = normal boiling point

c = critical point

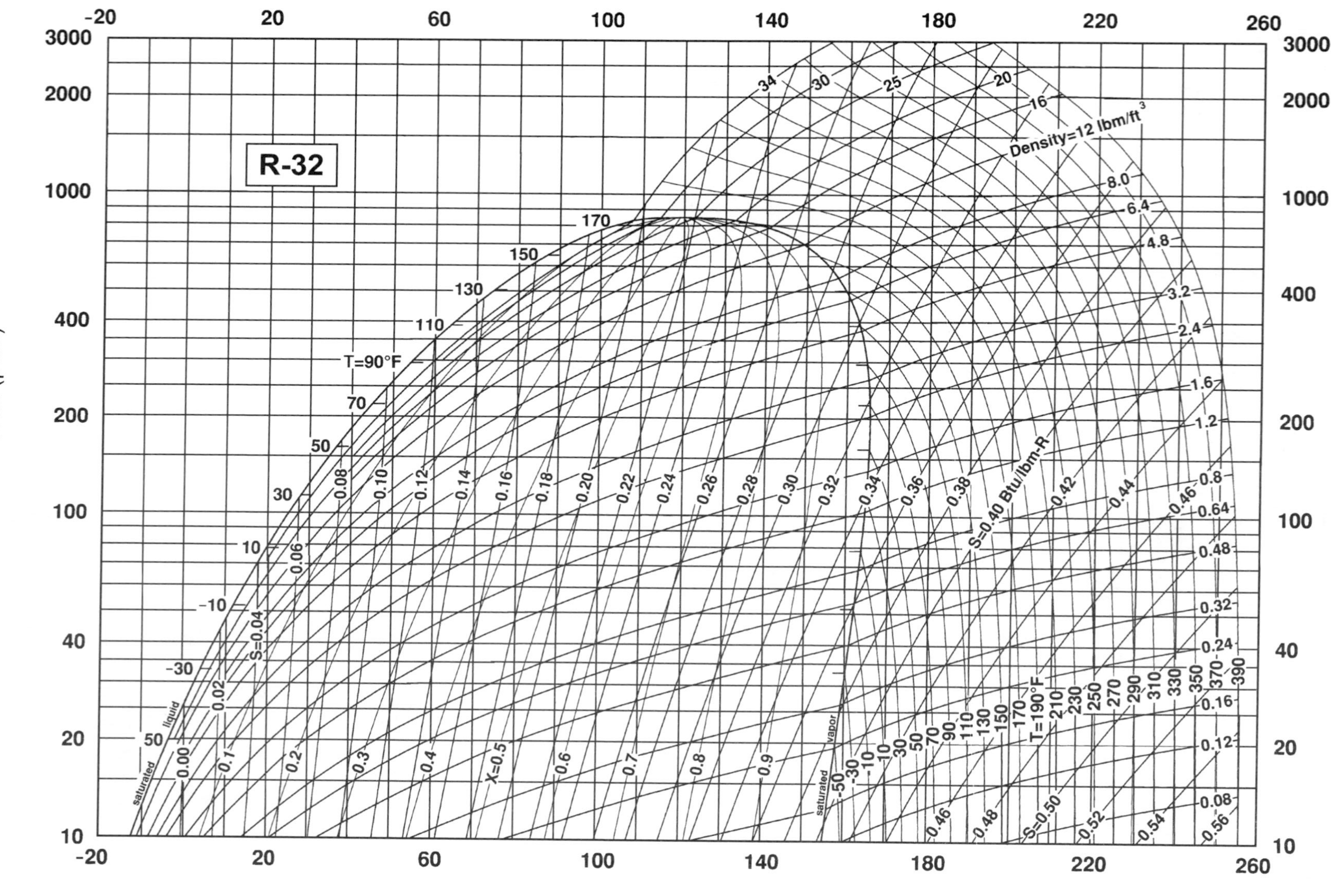

Fig. 4 Pressure-Enthalpy Diagram for Refrigerant 32

Refrigerant 32 (Difluoromethane) Properties of Saturated Liquid and Saturated Vapor

Temp.,* °F	Pres- sure, psia	Denisty, lb/ft³ Liquid	Volume, ft³/lb Vapor	Enthalpy, Btu/lb Liquid	Enthalpy, Btu/lb Vapor	Entropy, Btu/lb·°F Liquid	Entropy, Btu/lb·°F Vapor	Specific Heat c_p, Btu/lb·°F Liquid	Specific Heat c_p, Btu/lb·°F Vapor	c_p/c_v Vapor	Vel. of Sound, ft/s Liquid	Vel. of Sound, ft/s Vapor	Viscosity, lb$_m$/ft·h Liquid	Viscosity, lb$_m$/ft·h Vapor	Thermal Cond., Btu/h·ft·°F Liquid	Thermal Cond., Btu/h·ft·°F Vapor	Surface Tension, dyne/cm	Temp., °F
−214.26a	0.007	89.23	7270.1	−65.520	133.830	−0.20152	0.61079	0.3806	0.1577	1.3206	4641.	556.4	–	–	–	–	39.01	−214.26
−210.00	0.010	88.87	5150.0	−63.901	134.497	−0.19498	0.59966	0.3798	0.1581	1.3199	4599.	561.0	–	–	–	–	38.47	−210.00
−200.00	0.022	88.03	2413.3	−60.111	136.065	−0.18010	0.57539	0.3781	0.1591	1.3180	4503.	571.6	–	–	–	–	37.22	−200.00
−190.00	0.046	87.18	1207.6	−56.338	137.633	−0.16584	0.55345	0.3766	0.1602	1.3163	4407.	581.9	–	–	–	–	35.97	−190.00
−180.00	0.090	86.33	640.30	−52.578	139.199	−0.15215	0.53358	0.3754	0.1616	1.3146	4311.	591.9	–	–	–	–	34.73	−180.00
−170.00	0.167	85.48	357.44	−48.829	140.760	−0.13898	0.51552	0.3743	0.1633	1.3130	4216.	601.6	–	–	–	–	33.51	−170.00
−160.00	0.295	84.63	208.90	−45.090	142.311	−0.12629	0.49907	0.3735	0.1652	1.3117	4121.	611.0	–	–	–	–	32.29	−160.00
−150.00	0.500	83.77	127.18	−41.358	143.849	−0.11404	0.48404	0.3729	0.1675	1.3106	4026.	620.0	1.387	0.0173	0.1293	0.00350	31.07	−150.00
−140.00	0.816	82.90	80.317	−37.630	145.370	−0.10219	0.47027	0.3726	0.1702	1.3100	3931.	628.6	1.259	0.0179	0.1271	0.00363	29.87	−140.00
−130.00	1.286	82.02	52.414	−33.904	146.869	−0.09072	0.45763	0.3725	0.1734	1.3097	3836.	636.8	1.150	0.0185	0.1249	0.00376	28.68	−130.00
−120.00	1.966	81.14	35.229	−30.177	148.341	−0.07959	0.44598	0.3727	0.1770	1.3100	3742.	644.6	1.055	0.0190	0.1228	0.00390	27.50	−120.00
−110.00	2.923	80.25	24.316	−26.447	149.783	−0.06877	0.43522	0.3731	0.1811	1.3109	3648.	651.9	0.972	0.0196	0.1206	0.00405	26.33	−110.00
−100.00	4.235	79.35	17.190	−22.711	151.189	−0.05824	0.42525	0.3738	0.1858	1.3125	3553.	658.7	0.899	0.0202	0.1184	0.00419	25.17	−100.00
−95.00	5.053	78.90	14.573	−20.840	151.877	−0.05308	0.42054	0.3743	0.1884	1.3136	3506.	662.0	0.865	0.0205	0.1173	0.00427	24.59	−95.00
−90.00	5.996	78.44	12.417	−18.966	152.556	−0.04798	0.41600	0.3748	0.1910	1.3149	3459.	665.1	0.834	0.0207	0.1162	0.00434	24.02	−90.00
−85.00	7.078	77.98	10.633	−17.089	153.223	−0.04295	0.41162	0.3754	0.1938	1.3165	3412.	668.0	0.804	0.0210	0.1151	0.00442	23.45	−85.00
−80.00	8.312	77.52	9.1470	−15.209	153.879	−0.03797	0.40738	0.3760	0.1968	1.3182	3365.	670.9	0.775	0.0213	0.1141	0.00449	22.88	−80.00
−75.00	9.716	77.05	7.9036	−13.325	154.523	−0.03305	0.40329	0.3768	0.1998	1.3203	3318.	673.6	0.748	0.0216	0.1130	0.00457	22.31	−75.00
−70.00	11.304	76.58	6.8579	−11.437	155.155	−0.02818	0.39934	0.3776	0.2030	1.3226	3270.	676.1	0.723	0.0219	0.1119	0.00465	21.75	−70.00
−65.00	13.095	76.11	5.9744	−9.544	155.774	−0.02337	0.39551	0.3785	0.2063	1.3252	3223.	678.5	0.698	0.0222	0.1108	0.00473	21.19	−65.00
−60.97b	14.696	75.72	5.3611	−8.016	156.263	−0.01953	0.39251	0.3793	0.2091	1.3276	3185.	680.3	0.679	0.0224	0.1099	0.00480	20.74	−60.97
−60.00	15.105	75.63	5.2246	−7.647	156.379	−0.01860	0.39180	0.3795	0.2097	1.3282	3176.	680.7	0.675	0.0224	0.1097	0.00482	20.63	−60.00
−55.00	17.354	75.15	4.5854	−5.744	156.971	−0.01389	0.38821	0.3805	0.2133	1.3314	3129.	682.8	0.653	0.0227	0.1086	0.00490	20.08	−55.00
−50.00	19.861	74.66	4.0383	−3.836	157.549	−0.00922	0.38472	0.3817	0.2170	1.3350	3081.	684.8	0.632	0.0230	0.1075	0.00499	19.53	−50.00
−45.00	22.646	74.17	3.5681	−1.921	158.111	−0.00459	0.38134	0.3829	0.2208	1.3390	3034.	686.5	0.611	0.0233	0.1064	0.00508	18.98	−45.00
−40.00	25.731	73.67	3.1625	0.000	158.658	0.00000	0.37805	0.3842	0.2247	1.3433	2986.	688.1	0.592	0.0236	0.1053	0.00516	18.44	−40.00
−35.00	29.137	73.17	2.8114	1.928	159.189	0.00455	0.37486	0.3857	0.2287	1.3481	2939.	689.6	0.573	0.0239	0.1042	0.00526	17.89	−35.00
−30.00	32.887	72.67	2.5062	3.864	159.703	0.00906	0.37175	0.3872	0.2329	1.3532	2891.	690.8	0.555	0.0242	0.1031	0.00535	17.36	−30.00
−25.00	37.003	72.16	2.2402	5.808	160.200	0.01353	0.36872	0.3888	0.2372	1.3589	2843.	691.9	0.538	0.0245	0.1020	0.00545	16.82	−25.00
−20.00	41.511	71.64	2.0076	7.761	160.679	0.01797	0.36577	0.3906	0.2416	1.3651	2795.	692.8	0.521	0.0248	0.1009	0.00555	16.29	−20.00
−15.00	46.433	71.12	1.8034	9.723	161.139	0.02238	0.36289	0.3924	0.2462	1.3718	2747.	693.6	0.505	0.0251	0.0998	0.00565	15.76	−15.00
−10.00	51.796	70.60	1.6237	11.694	161.579	0.02676	0.36008	0.3944	0.2510	1.3790	2699.	694.1	0.490	0.0254	0.0987	0.00575	15.24	−10.00
−5.00	57.625	70.06	1.4652	13.676	161.999	0.03111	0.35733	0.3965	0.2559	1.3869	2651.	694.4	0.475	0.0257	0.0976	0.00586	14.72	−5.00
0.00	63.947	69.52	1.3248	15.669	162.397	0.03543	0.35463	0.3987	0.2610	1.3955	2602.	694.6	0.461	0.0260	0.0964	0.00597	14.20	0.00
5.00	70.789	68.97	1.2002	17.673	162.774	0.03973	0.35199	0.4011	0.2662	1.4047	2553.	694.5	0.447	0.0263	0.0953	0.00609	13.69	5.00
10.00	78.179	68.42	1.0894	19.690	163.127	0.04400	0.34940	0.4037	0.2717	1.4148	2504.	694.3	0.434	0.0266	0.0942	0.00621	13.18	10.00
15.00	86.144	67.86	0.9905	21.719	163.455	0.04825	0.34685	0.4064	0.2774	1.4256	2455.	693.8	0.421	0.0269	0.0930	0.00634	12.67	15.00
20.00	94.715	67.28	0.9021	23.762	163.758	0.05248	0.34434	0.4093	0.2834	1.4375	2406.	693.1	0.408	0.0273	0.0919	0.00647	12.17	20.00
25.00	103.92	66.71	0.8228	25.820	164.034	0.05670	0.34187	0.4124	0.2897	1.4503	2356.	692.2	0.396	0.0276	0.0907	0.00661	11.67	25.00
30.00	113.79	66.12	0.7516	27.893	164.281	0.06090	0.33943	0.4157	0.2963	1.4642	2306.	691.1	0.384	0.0279	0.0895	0.00676	11.18	30.00
35.00	124.35	65.52	0.6874	29.983	164.499	0.06508	0.33701	0.4192	0.3032	1.4794	2256.	689.7	0.372	0.0283	0.0884	0.00691	10.69	35.00
40.00	135.65	64.91	0.6296	32.090	164.686	0.06926	0.33462	0.4230	0.3106	1.4959	2206.	688.1	0.361	0.0286	0.0872	0.00707	10.21	40.00
45.00	147.70	64.29	0.5773	34.215	164.839	0.07342	0.33225	0.4271	0.3184	1.5140	2155.	686.3	0.350	0.0290	0.0860	0.00724	9.73	45.00
50.00	160.54	63.65	0.5298	36.360	164.957	0.07758	0.32989	0.4316	0.3266	1.5337	2104.	684.2	0.339	0.0294	0.0848	0.00743	9.25	50.00
55.00	174.21	63.01	0.4868	38.526	165.039	0.08173	0.32754	0.4363	0.3355	1.5553	2052.	681.9	0.329	0.0297	0.0836	0.00764	8.78	55.00
60.00	188.74	62.35	0.4477	40.714	165.081	0.08588	0.32519	0.4415	0.3450	1.5790	2000.	679.3	0.319	0.0301	0.0823	0.00786	8.32	60.00
65.00	204.17	61.68	0.4120	42.927	165.081	0.09002	0.32284	0.4471	0.3552	1.6052	1947.	676.4	0.309	0.0305	0.0811	0.00811	7.86	65.00
70.00	220.52	60.99	0.3794	45.165	165.036	0.09418	0.32049	0.4533	0.3663	1.6341	1894.	673.2	0.299	0.0310	0.0799	0.00838	7.40	70.00
75.00	237.85	60.28	0.3496	47.431	164.944	0.09833	0.31812	0.4600	0.3783	1.6662	1840.	669.8	0.289	0.0314	0.0786	0.00868	6.96	75.00
80.00	256.18	59.56	0.3223	49.727	164.800	0.10250	0.31573	0.4674	0.3915	1.7019	1785.	666.1	0.280	0.0318	0.0773	0.00901	6.51	80.00
85.00	275.56	58.81	0.2972	52.055	164.602	0.10669	0.31332	0.4756	0.4060	1.7419	1730.	662.1	0.271	0.0323	0.0760	0.00938	6.07	85.00
90.00	296.02	58.04	0.2741	54.419	164.343	0.11089	0.31087	0.4847	0.4221	1.7869	1674.	657.7	0.262	0.0328	0.0747	0.00980	5.64	90.00
95.00	317.61	57.25	0.2529	56.821	164.020	0.11511	0.30838	0.4950	0.4400	1.8378	1617.	653.0	0.253	0.0333	0.0733	0.01028	5.22	95.00
100.00	340.37	56.43	0.2333	59.266	163.627	0.11937	0.30584	0.5066	0.4602	1.8958	1560.	648.0	0.244	0.0338	0.0719	0.01082	4.80	100.00
105.00	364.34	55.58	0.2151	61.758	163.155	0.12366	0.30323	0.5197	0.4831	1.9624	1501.	642.6	0.235	0.0344	0.0705	0.01145	4.39	105.00
110.00	389.58	54.69	0.1983	64.303	162.598	0.12800	0.30054	0.5349	0.5094	2.0397	1441.	636.8	0.227	0.0350	0.0691	0.01218	3.99	110.00
115.00	416.13	53.77	0.1826	66.907	161.945	0.13239	0.29777	0.5525	0.5399	2.1302	1379.	630.5	0.218	0.0356	0.0676	0.01303	3.59	115.00
120.00	444.04	52.80	0.1681	69.577	161.184	0.13685	0.29488	0.5734	0.5759	2.2378	1317.	623.9	0.209	0.0363	0.0661	0.01403	3.21	120.00
125.00	473.37	51.77	0.1545	72.325	160.301	0.14139	0.29186	0.5984	0.6189	2.3675	1253.	616.7	0.201	0.0371	0.0646	0.01523	2.83	125.00
130.00	504.17	50.69	0.1418	75.163	159.275	0.14603	0.28867	0.6290	0.6715	2.5268	1187.	609.1	0.192	0.0379	0.0630	0.01666	2.46	130.00
135.00	536.51	49.53	0.1298	78.108	158.082	0.15080	0.28529	0.6675	0.7373	2.7273	1119.	600.9	0.183	0.0388	0.0613	0.01841	2.11	135.00
140.00	570.47	48.28	0.1185	81.181	156.689	0.15573	0.28165	0.7172	0.8225	2.9873	1049.	592.0	0.175	0.0398	0.0596	0.02060	1.76	140.00
145.00	606.10	46.91	0.1077	84.415	155.048	0.16088	0.27769	0.7842	0.9372	3.3381	977.	582.3	0.166	0.0410	0.0577	0.02340	1.43	145.00
150.00	643.51	45.40	0.0974	87.858	153.089	0.16630	0.27329	0.880	1.101	3.838	902.	571.8	0.156	0.0424	0.0558	0.02720	1.11	150.00
160.00	724.09	41.65	0.0775	95.724	147.670	0.17853	0.26236	1.288	1.802	5.966	740.	547.0	0.135	0.0464	0.0515	0.04271	0.54	160.00
170.00	813.58	35.24	0.0549	107.269	136.620	0.19632	0.24293	4.865	7.673	23.309	545.	509.3	0.106	0.0557	0.0477	0.22608	0.07	170.00
172.59c	838.61	26.47	0.0378	120.855	120.855	0.21760	0.21760	∞	∞	∞	0.	0.0	–	–	∞	∞	0.00	172.59

*Temperatures are on the ITS-90 scale a = triple point b = normal boiling point c = critical point

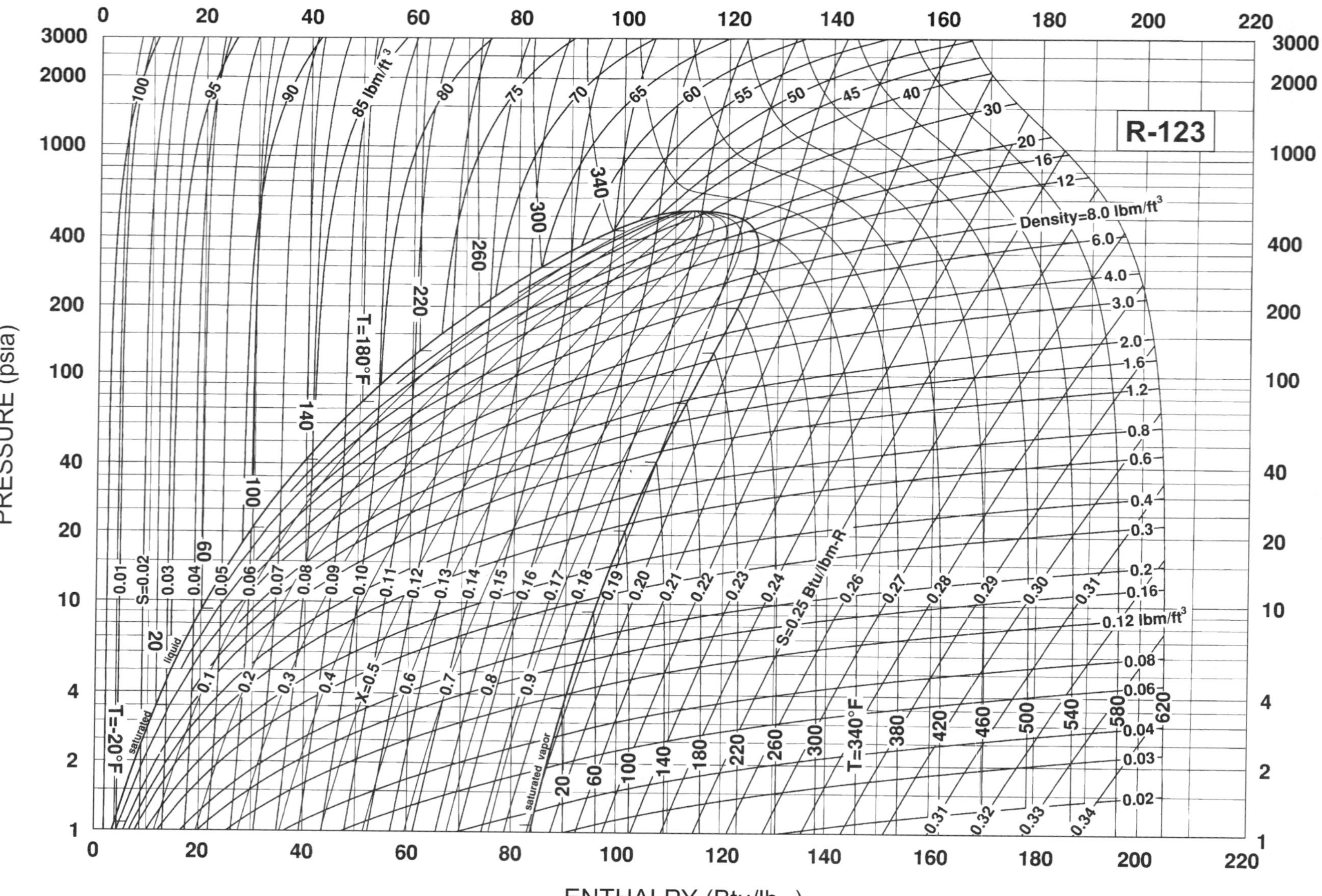

Fig. 5 Pressure-Enthalpy Diagram for Refrigerant 123

Refrigerant 123 (2,2-Dichloro-1,1,1-trifluoroethane) Properties of Saturated Liquid and Saturated Vapor

Temp.,* °F	Pressure, psia	Denisty, lb/ft³ Liquid	Volume, ft³/lb Vapor	Enthalpy, Btu/lb Liquid	Enthalpy, Btu/lb Vapor	Entropy, Btu/lb·°F Liquid	Entropy, Btu/lb·°F Vapor	Specific Heat c_p, Btu/lb·°F Liquid	Specific Heat c_p, Btu/lb·°F Vapor	c_p/c_v Vapor	Vel. of Sound, ft/s Liquid	Vel. of Sound, ft/s Vapor	Viscosity, $lb_m/ft·h$ Liquid	Viscosity, $lb_m/ft·h$ Vapor	Thermal Cond., Btu/h·ft·°F Liquid	Thermal Cond., Btu/h·ft·°F Vapor	Surface Tension, dyne/cm	Temp., °F
−140.00	0.003	108.90	7431.6	−22.241	71.783	−0.06050	0.23363	0.2210	0.1181	1.1237	3928.	341.7	7.731	0.0146	0.0645	0.00135	30.50	−140.00
−130.00	0.006	108.12	3871.0	−20.033	72.974	−0.05370	0.22843	0.2207	0.1203	1.1212	3854.	346.6	6.547	0.0151	0.0636	0.00153	29.75	−130.00
−120.00	0.011	107.35	2111.6	−17.826	74.187	−0.04710	0.22379	0.2206	0.1226	1.1187	3778.	351.4	5.645	0.0157	0.0628	0.00171	29.01	−120.00
−110.00	0.020	106.57	1201.0	−15.619	75.421	−0.04070	0.21966	0.2208	0.1248	1.1165	3702.	356.2	4.934	0.0163	0.0619	0.00190	28.27	−110.00
−100.00	0.036	105.80	709.46	−13.410	76.676	−0.03447	0.21600	0.2211	0.1270	1.1144	3626.	360.8	4.359	0.0168	0.0611	0.00208	27.53	−100.00
−90.00	0.060	105.03	433.83	−11.195	77.950	−0.02840	0.21275	0.2217	0.1291	1.1124	3549.	365.4	3.885	0.0174	0.0602	0.00226	26.80	−90.00
−80.00	0.097	104.26	273.77	−8.975	79.244	−0.02247	0.20989	0.2224	0.1313	1.1106	3472.	369.9	3.488	0.0179	0.0593	0.00244	26.07	−80.00
−70.00	0.154	103.48	177.81	−6.746	80.556	−0.01668	0.20737	0.2233	0.1334	1.1090	3394.	374.3	3.150	0.0185	0.0584	0.00263	25.34	−70.00
−60.00	0.236	102.70	118.57	−4.509	81.885	−0.01101	0.20516	0.2243	0.1356	1.1075	3317.	378.5	2.860	0.0190	0.0575	0.00281	24.62	−60.00
−50.00	0.354	101.92	80.999	−2.260	83.231	−0.00545	0.20323	0.2254	0.1377	1.1061	3240.	382.7	2.607	0.0196	0.0565	0.00299	23.91	−50.00
−40.00	0.519	101.13	56.576	0.000	84.592	0.00000	0.20157	0.2266	0.1398	1.1050	3164.	386.8	2.386	0.0201	0.0555	0.00317	23.19	−40.00
−30.00	0.744	100.34	40.333	2.272	85.967	0.00535	0.20014	0.2279	0.1420	1.1040	3087.	390.7	2.191	0.0206	0.0546	0.00335	22.48	−30.00
−20.00	1.046	99.54	29.299	4.558	87.355	0.01061	0.19892	0.2292	0.1441	1.1032	3012.	394.5	2.018	0.0211	0.0536	0.00353	21.78	−20.00
−10.00	1.445	98.73	21.655	6.857	88.754	0.01578	0.19790	0.2306	0.1463	1.1026	2936.	398.2	1.864	0.0217	0.0526	0.00371	21.08	−10.00
0.00	1.963	97.92	16.264	9.170	90.163	0.02086	0.19706	0.2320	0.1484	1.1022	2862.	401.7	1.725	0.0222	0.0515	0.00390	20.38	0.00
5.00	2.274	97.51	14.174	10.332	90.871	0.02337	0.19670	0.2327	0.1495	1.1021	2825.	403.3	1.661	0.0224	0.0510	0.00399	20.03	5.00
10.00	2.625	97.10	12.396	11.498	91.582	0.02587	0.19638	0.2334	0.1506	1.1020	2788.	405.0	1.601	0.0227	0.0505	0.00408	19.69	10.00
15.00	3.019	96.69	10.878	12.667	92.294	0.02834	0.19609	0.2341	0.1517	1.1020	2751.	406.6	1.543	0.0229	0.0501	0.00417	19.34	15.00
20.00	3.460	96.28	9.5779	13.840	93.008	0.03080	0.19585	0.2349	0.1528	1.1020	2714.	408.1	1.488	0.0232	0.0496	0.00426	19.00	20.00
25.00	3.952	95.86	8.4595	15.017	93.723	0.03324	0.19563	0.2356	0.1540	1.1021	2678.	409.6	1.435	0.0235	0.0491	0.00435	18.66	25.00
30.00	4.499	95.44	7.4943	16.198	94.440	0.03566	0.19544	0.2364	0.1551	1.1023	2641.	411.0	1.385	0.0237	0.0486	0.00444	18.32	30.00
35.00	5.106	95.02	6.6586	17.382	95.158	0.03806	0.19529	0.2371	0.1562	1.1025	2605.	412.4	1.337	0.0239	0.0481	0.00453	17.98	35.00
40.00	5.778	94.60	5.9327	18.570	95.877	0.04045	0.19517	0.2379	0.1574	1.1028	2569.	413.7	1.292	0.0242	0.0476	0.00462	17.64	40.00
45.00	6.519	94.17	5.3002	19.762	96.597	0.04282	0.19507	0.2387	0.1585	1.1031	2533.	414.9	1.248	0.0244	0.0471	0.00471	17.30	45.00
50.00	7.334	93.74	4.7474	20.958	97.317	0.04518	0.19500	0.2394	0.1597	1.1035	2498.	416.1	1.207	0.0247	0.0467	0.00481	16.97	50.00
55.00	8.229	93.31	4.2629	22.158	98.038	0.04752	0.19495	0.2402	0.1609	1.1040	2462.	417.3	1.167	0.0249	0.0462	0.00490	16.63	55.00
60.00	9.208	92.88	3.8371	23.362	98.760	0.04984	0.19493	0.2410	0.1621	1.1046	2427.	418.3	1.129	0.0252	0.0457	0.00499	16.30	60.00
65.00	10.278	92.44	3.4617	24.570	99.481	0.05215	0.19493	0.2418	0.1633	1.1052	2392.	419.4	1.092	0.0254	0.0453	0.00508	15.97	65.00
70.00	11.445	92.01	3.1301	25.782	100.203	0.05444	0.19495	0.2426	0.1645	1.1059	2357.	420.3	1.057	0.0256	0.0448	0.00518	15.64	70.00
75.00	12.713	91.56	2.8362	26.998	100.924	0.05673	0.19499	0.2434	0.1657	1.1067	2322.	421.2	1.023	0.0259	0.0444	0.00527	15.31	75.00
80.00	14.090	91.12	2.5753	28.218	101.645	0.05899	0.19505	0.2442	0.1669	1.1075	2287.	422.0	0.991	0.0261	0.0439	0.00537	14.98	80.00
82.08b	14.696	90.94	2.4753	28.728	101.945	0.05993	0.19508	0.2445	0.1675	1.1079	2273.	422.3	0.978	0.0262	0.0437	0.00540	14.84	82.08
85.00	15.580	90.67	2.3429	29.443	102.365	0.06124	0.19513	0.2450	0.1682	1.1085	2252.	422.7	0.960	0.0264	0.0435	0.00546	14.65	85.00
90.00	17.192	90.22	2.1356	30.671	103.085	0.06348	0.19522	0.2458	0.1695	1.1095	2218.	423.3	0.930	0.0266	0.0430	0.00556	14.33	90.00
95.00	18.931	89.77	1.9503	31.904	103.804	0.06571	0.19534	0.2467	0.1707	1.1106	2184.	423.9	0.901	0.0268	0.0426	0.00565	14.00	95.00
100.00	20.804	89.31	1.7841	33.141	104.521	0.06792	0.19546	0.2475	0.1720	1.1119	2149.	424.4	0.874	0.0271	0.0422	0.00575	13.68	100.00
105.00	22.819	88.85	1.6349	34.383	105.238	0.07012	0.19560	0.2484	0.1734	1.1132	2115.	424.8	0.847	0.0273	0.0418	0.00585	13.36	105.00
110.00	24.980	88.39	1.5006	35.628	105.953	0.07231	0.19576	0.2492	0.1747	1.1146	2081.	425.2	0.822	0.0275	0.0413	0.00595	13.04	110.00
115.00	27.297	87.92	1.3795	36.879	106.666	0.07449	0.19593	0.2501	0.1761	1.1162	2047.	425.4	0.797	0.0277	0.0409	0.00604	12.72	115.00
120.00	29.776	87.45	1.2701	38.134	107.377	0.07665	0.19611	0.2510	0.1775	1.1178	2014.	425.6	0.773	0.0280	0.0405	0.00614	12.41	120.00
125.00	32.425	86.98	1.1710	39.393	108.086	0.07881	0.19630	0.2520	0.1789	1.1196	1980.	425.6	0.751	0.0282	0.0401	0.00625	12.09	125.00
130.00	35.251	86.50	1.0812	40.657	108.792	0.08095	0.19650	0.2529	0.1803	1.1215	1946.	425.6	0.728	0.0284	0.0397	0.00635	11.78	130.00
135.00	38.261	86.01	0.9996	41.926	109.497	0.08308	0.19671	0.2539	0.1818	1.1236	1913.	425.5	0.707	0.0287	0.0393	0.00645	11.47	135.00
140.00	41.464	85.52	0.9253	43.200	110.198	0.08520	0.19693	0.2548	0.1833	1.1258	1879.	425.3	0.687	0.0289	0.0389	0.00656	11.16	140.00
145.00	44.868	85.03	0.8577	44.479	110.896	0.08732	0.19716	0.2559	0.1848	1.1281	1846.	425.0	0.667	0.0291	0.0385	0.00666	10.85	145.00
150.00	48.479	84.53	0.7959	45.763	111.591	0.08942	0.19739	0.2569	0.1863	1.1306	1813.	424.6	0.648	0.0293	0.0381	0.00677	10.54	150.00
160.00	56.360	83.52	0.6876	48.347	112.970	0.09359	0.19788	0.2591	0.1896	1.1362	1746.	423.5	0.611	0.0298	0.0374	0.00699	9.93	160.00
170.00	65.173	82.49	0.5965	50.953	114.333	0.09773	0.19839	0.2614	0.1929	1.1426	1680.	422.0	0.577	0.0303	0.0366	0.00722	9.33	170.00
180.00	74.986	81.43	0.5195	53.583	115.678	0.10184	0.19892	0.2638	0.1965	1.1499	1614.	420.1	0.545	0.0307	0.0359	0.00745	8.74	180.00
190.00	85.868	80.34	0.4539	56.237	117.001	0.10592	0.19945	0.2665	0.2004	1.1583	1548.	417.7	0.515	0.0312	0.0352	0.00769	8.15	190.00
200.00	97.892	79.23	0.3979	58.918	118.300	0.10997	0.19999	0.2694	0.2045	1.1681	1482.	414.7	0.487	0.0317	0.0345	0.00795	7.57	200.00
210.00	111.13	78.08	0.3497	61.627	119.572	0.11400	0.20053	0.2726	0.2089	1.1793	1416.	411.3	0.460	0.0322	0.0338	0.00821	7.00	210.00
220.00	125.66	76.89	0.3080	64.367	120.813	0.11801	0.20106	0.2761	0.2138	1.1925	1349.	407.3	0.435	0.0328	0.0331	0.00849	6.44	220.00
230.00	141.56	75.66	0.2719	67.141	122.019	0.12201	0.20158	0.2800	0.2191	1.2079	1283.	402.8	0.411	0.0334	0.0324	0.00877	5.88	230.00
240.00	158.91	74.38	0.2404	69.952	123.184	0.12599	0.20207	0.2845	0.2251	1.2262	1216.	397.6	0.388	0.0340	0.0317	0.00908	5.34	240.00
250.00	177.80	73.04	0.2128	72.805	124.303	0.12997	0.20254	0.2896	0.2319	1.2482	1148.	391.7	0.367	0.0347	0.0310	0.00940	4.81	250.00
260.00	198.31	71.64	0.1885	75.704	125.367	0.13396	0.20296	0.2956	0.2398	1.2749	1080.	385.1	0.346	0.0355	0.0303	0.00974	4.29	260.00
270.00	220.53	70.16	0.1670	78.655	126.368	0.13795	0.20334	0.3026	0.2490	1.3079	1012.	377.7	0.326	0.0364	0.0296	0.01010	3.78	270.00
280.00	244.58	68.60	0.1479	81.666	127.294	0.14196	0.20365	0.3110	0.2603	1.3496	942.	369.4	0.307	0.0374	0.0289	0.01050	3.28	280.00
290.00	270.54	66.92	0.1309	84.749	128.128	0.14600	0.20387	0.3215	0.2742	1.4035	871.	360.2	0.288	0.0386	0.0282	0.01092	2.80	290.00
300.00	298.53	65.11	0.1155	87.916	128.851	0.15010	0.20398	0.3349	0.2922	1.4755	800.	349.9	0.270	0.0400	0.0275	0.01139	2.33	300.00
310.00	328.69	63.12	0.1016	91.188	129.431	0.15426	0.20395	0.3529	0.3166	1.5762	726.	338.5	0.252	0.0418	0.0267	0.01191	1.88	310.00
320.00	361.16	60.91	0.0889	94.594	129.822	0.15853	0.20372	0.3785	0.3520	1.7258	650.	325.7	0.234	0.0440	0.0259	0.01251	1.45	320.00
330.00	396.11	58.37	0.0770	98.186	129.950	0.16297	0.20320	0.4186	0.4084	1.9693	571.	311.5	0.216	0.0469	0.0251	0.01321	1.04	330.00
340.00	433.76	55.33	0.0658	102.059	129.670	0.16769	0.20222	0.4925	0.5138	2.4318	488.	295.4	0.196	0.0510	0.0243	0.01411	0.66	340.00
350.00	474.41	51.32	0.0544	106.459	128.628	0.17298	0.20036	0.6830	0.7861	3.6383	397.	277.0	0.173	0.0575	0.0234	0.01539	0.32	350.00
360.00	518.66	43.97	0.0403	112.667	125.064	0.18039	0.19551	2.507	3.263	14.633	290.	255.2	0.138	0.0730	0.0227	0.01819	0.05	360.00
362.63c	531.10	34.34	0.0291	118.800	118.800	0.18779	0.18779	∞	∞	∞	0.	0.0	–	–	∞	∞	0.00	362.63

*Temperatures are on the ITS-90 scale

b = normal boiling point

c = critical point

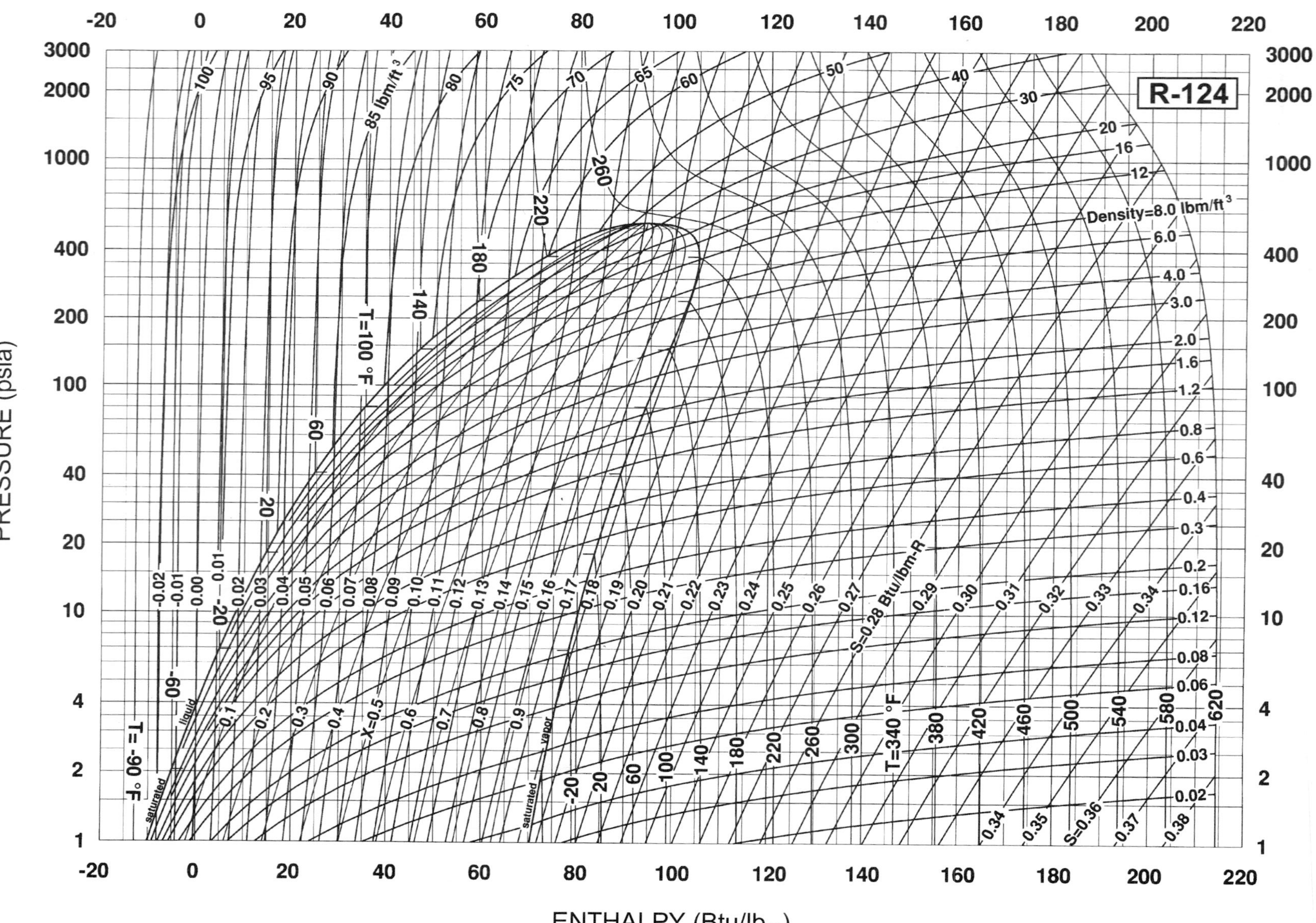

Fig. 6 Pressure-Enthalpy Diagram for Refrigerant 124

Refrigerant 124 (2-Chloro-1,1,1,2-tetrafluoroethane) Properties of Saturated Liquid and Saturated Vapor

Temp.,* °F	Pres-sure, psia	Denisty, lb/ft³	Volume, ft³/lb	Enthalpy, Btu/lb		Entropy, Btu/lb·°F		Specific Heat c_p, Btu/lb·°F		c_p/c_v	Vel. of Sound, ft/s		Viscosity, lb_m/ft·h		Thermal Cond., Btu/h·ft·°F		Surface Tension, dyne/cm	Temp., °F
		Liquid	Vapor	Liquid	Vapor	Liquid	Vapor	Liquid	Vapor	Vapor	Liquid	Vapor	Liquid	Vapor	Liquid	Vapor		
−150.00	0.030	107.22	800.14	−25.872	61.843	−0.07135	0.21191	0.2276	0.1271	1.1298	3468.	356.9	–	–	–	–	26.26	−150.00
−140.00	0.056	106.32	450.73	−23.590	63.119	−0.06409	0.20715	0.2288	0.1294	1.1274	3386.	362.1	–	–	–	–	25.47	−140.00
−130.00	0.098	105.41	264.79	−21.295	64.415	−0.05703	0.20296	0.2300	0.1317	1.1253	3304.	367.2	–	–	–	–	24.69	−130.00
−120.00	0.165	104.51	161.57	−18.988	65.729	−0.05013	0.19928	0.2313	0.1340	1.1233	3223.	372.2	–	–	–	–	23.92	−120.00
−110.00	0.269	103.59	102.02	−16.668	67.061	−0.04340	0.19605	0.2327	0.1364	1.1215	3142.	377.1	2.921	0.0181	0.0607	0.00337	23.15	−110.00
−100.00	0.424	102.68	66.458	−14.334	68.410	−0.03682	0.19323	0.2341	0.1387	1.1200	3062.	381.8	2.541	0.0186	0.0594	0.00352	22.38	−100.00
−90.00	0.650	101.76	44.529	−11.985	69.773	−0.03038	0.19078	0.2356	0.1411	1.1187	2982.	386.4	2.238	0.0192	0.0581	0.00368	21.62	−90.00
−80.00	0.969	100.83	30.612	−9.621	71.149	−0.02407	0.18866	0.2372	0.1435	1.1176	2902.	390.8	1.992	0.0197	0.0568	0.00384	20.86	−80.00
−70.00	1.411	99.89	21.542	−7.241	72.537	−0.01789	0.18685	0.2387	0.1460	1.1168	2823.	395.0	1.788	0.0202	0.0555	0.00400	20.11	−70.00
−60.00	2.008	98.95	15.485	−4.844	73.935	−0.01182	0.18529	0.2404	0.1485	1.1163	2745.	398.9	1.616	0.0207	0.0543	0.00416	19.36	−60.00
−50.00	2.801	98.00	11.350	−2.431	75.342	−0.00586	0.18398	0.2421	0.1512	1.1162	2667.	402.7	1.469	0.0213	0.0531	0.00433	18.61	−50.00
−45.00	3.284	97.52	9.7829	−1.218	76.047	−0.00292	0.18341	0.2430	0.1525	1.1162	2629.	404.5	1.403	0.0215	0.0525	0.00441	18.24	−45.00
−40.00	3.832	97.03	8.4679	0.000	76.754	0.00000	0.18289	0.2439	0.1539	1.1164	2590.	406.2	1.342	0.0218	0.0519	0.00450	17.88	−40.00
−35.00	4.454	96.55	7.3594	1.222	77.462	0.00289	0.18242	0.2448	0.1553	1.1166	2552.	407.8	1.285	0.0220	0.0513	0.00459	17.51	−35.00
−30.00	5.154	96.06	6.4208	2.449	78.171	0.00576	0.18199	0.2457	0.1567	1.1169	2514.	409.4	1.232	0.0223	0.0507	0.00467	17.14	−30.00
−25.00	5.941	95.57	5.6227	3.681	78.881	0.00861	0.18161	0.2466	0.1581	1.1173	2476.	410.9	1.181	0.0226	0.0502	0.00476	16.78	−25.00
−20.00	6.821	95.08	4.9412	4.918	79.590	0.01143	0.18127	0.2476	0.1596	1.1179	2438.	412.3	1.134	0.0228	0.0496	0.00485	16.41	−20.00
−15.00	7.803	94.58	4.3571	6.159	80.300	0.01424	0.18097	0.2486	0.1611	1.1185	2400.	413.6	1.090	0.0231	0.0490	0.00494	16.05	−15.00
−10.00	8.896	94.08	3.8545	7.406	81.010	0.01702	0.18071	0.2496	0.1626	1.1193	2362.	414.9	1.048	0.0233	0.0485	0.00503	15.69	−10.00
−5.00	10.106	93.57	3.4204	8.657	81.720	0.01978	0.18048	0.2506	0.1642	1.1201	2324.	416.1	1.008	0.0236	0.0479	0.00512	15.33	−5.00
0.00	11.444	93.06	3.0442	9.914	82.429	0.02253	0.18028	0.2516	0.1658	1.1211	2287.	417.1	0.971	0.0238	0.0474	0.00521	14.97	0.00
5.00	12.918	92.55	2.7171	11.176	83.137	0.02525	0.18011	0.2527	0.1674	1.1222	2250.	418.1	0.935	0.0241	0.0468	0.00530	14.62	5.00
10.00	14.537	92.04	2.4317	12.444	83.843	0.02796	0.17998	0.2538	0.1690	1.1235	2212.	419.0	0.902	0.0244	0.0463	0.00539	14.26	10.00
10.47b	14.696	91.99	2.4070	12.563	83.909	0.02821	0.17997	0.2539	0.1692	1.1236	2209.	419.1	0.899	0.0244	0.0462	0.00540	14.23	10.47
15.00	16.312	91.51	2.1820	13.717	84.549	0.03065	0.17987	0.2549	0.1708	1.1249	2175.	419.8	0.870	0.0246	0.0458	0.00549	13.91	15.00
20.00	18.252	90.99	1.9627	14.996	85.253	0.03332	0.17979	0.2560	0.1725	1.1264	2138.	420.5	0.839	0.0249	0.0452	0.00558	13.56	20.00
25.00	20.368	90.46	1.7697	16.281	85.955	0.03597	0.17973	0.2572	0.1743	1.1281	2101.	421.1	0.810	0.0251	0.0447	0.00568	13.21	25.00
30.00	22.670	89.92	1.5993	17.572	86.655	0.03861	0.17969	0.2583	0.1761	1.1300	2064.	421.6	0.782	0.0254	0.0442	0.00577	12.86	30.00
35.00	25.169	89.38	1.4484	18.868	87.352	0.04124	0.17968	0.2595	0.1780	1.1320	2028.	422.0	0.756	0.0256	0.0437	0.00587	12.51	35.00
40.00	27.876	88.84	1.3145	20.171	88.047	0.04385	0.17969	0.2608	0.1799	1.1342	1991.	422.3	0.730	0.0259	0.0431	0.00597	12.16	40.00
45.00	30.802	88.29	1.1953	21.481	88.739	0.04644	0.17971	0.2621	0.1819	1.1365	1955.	422.4	0.706	0.0261	0.0426	0.00607	11.82	45.00
50.00	33.959	87.73	1.0890	22.797	89.427	0.04902	0.17976	0.2634	0.1839	1.1391	1918.	422.4	0.682	0.0264	0.0421	0.00617	11.48	50.00
55.00	37.358	87.17	0.9939	24.119	90.112	0.05159	0.17982	0.2647	0.1859	1.1419	1882.	422.4	0.660	0.0267	0.0416	0.00627	11.14	55.00
60.00	41.011	86.60	0.9087	25.449	90.793	0.05415	0.17989	0.2661	0.1881	1.1450	1845.	422.1	0.638	0.0269	0.0411	0.00637	10.80	60.00
65.00	44.931	86.02	0.8322	26.786	91.470	0.05669	0.17998	0.2675	0.1903	1.1482	1809.	421.8	0.618	0.0272	0.0406	0.00648	10.46	65.00
70.00	49.129	85.44	0.7633	28.129	92.143	0.05922	0.18008	0.2690	0.1925	1.1517	1773.	421.3	0.598	0.0275	0.0401	0.00659	10.12	70.00
75.00	53.619	84.85	0.7011	29.481	92.810	0.06174	0.18019	0.2706	0.1948	1.1556	1737.	420.7	0.578	0.0277	0.0396	0.00669	9.79	75.00
80.00	58.413	84.25	0.6449	30.840	93.472	0.06426	0.18031	0.2721	0.1972	1.1597	1700.	420.0	0.560	0.0280	0.0391	0.00680	9.46	80.00
85.00	63.524	83.65	0.5940	32.207	94.129	0.06676	0.18044	0.2738	0.1997	1.1641	1664.	419.1	0.542	0.0283	0.0386	0.00691	9.13	85.00
90.00	68.966	83.03	0.5478	33.582	94.779	0.06925	0.18058	0.2755	0.2023	1.1689	1628.	418.1	0.524	0.0285	0.0381	0.00703	8.80	90.00
95.00	74.750	82.41	0.5058	34.966	95.423	0.07173	0.18073	0.2773	0.2049	1.1741	1592.	416.9	0.507	0.0288	0.0376	0.00714	8.48	95.00
100.00	80.893	81.77	0.4676	36.358	96.060	0.07420	0.18088	0.2791	0.2077	1.1797	1555.	415.6	0.491	0.0291	0.0372	0.00726	8.15	100.00
105.00	87.406	81.13	0.4327	37.760	96.689	0.07667	0.18103	0.2810	0.2106	1.1858	1519.	414.1	0.475	0.0294	0.0367	0.00738	7.83	105.00
110.00	94.304	80.47	0.4008	39.171	97.311	0.07913	0.18119	0.2831	0.2136	1.1925	1483.	412.4	0.460	0.0297	0.0362	0.00750	7.51	110.00
115.00	101.60	79.81	0.3716	40.592	97.923	0.08158	0.18135	0.2852	0.2167	1.1997	1446.	410.6	0.445	0.0300	0.0357	0.00763	7.20	115.00
120.00	109.31	79.13	0.3448	42.023	98.526	0.08403	0.18151	0.2875	0.2200	1.2075	1410.	408.6	0.431	0.0303	0.0352	0.00776	6.88	120.00
125.00	117.45	78.44	0.3202	43.465	99.120	0.08648	0.18167	0.2898	0.2235	1.2161	1373.	406.4	0.417	0.0306	0.0347	0.00789	6.57	125.00
130.00	126.04	77.73	0.2976	44.918	99.702	0.08892	0.18182	0.2924	0.2271	1.2255	1336.	404.0	0.403	0.0309	0.0343	0.00803	6.26	130.00
135.00	135.08	77.01	0.2767	46.383	100.272	0.09135	0.18197	0.2950	0.2310	1.2358	1299.	401.4	0.389	0.0312	0.0338	0.00817	5.95	135.00
140.00	144.60	76.28	0.2575	47.860	100.830	0.09379	0.18212	0.2979	0.2352	1.2472	1262.	398.7	0.376	0.0316	0.0333	0.00832	5.65	140.00
145.00	154.60	75.52	0.2397	49.351	101.374	0.09622	0.18226	0.3009	0.2396	1.2598	1224.	395.7	0.364	0.0319	0.0328	0.00847	5.35	145.00
150.00	165.12	74.75	0.2232	50.855	101.903	0.09866	0.18239	0.3042	0.2444	1.2737	1186.	392.5	0.351	0.0323	0.0324	0.00862	5.05	150.00
155.00	176.16	73.96	0.2080	52.374	102.416	0.10110	0.18251	0.3078	0.2496	1.2892	1148.	389.1	0.339	0.0327	0.0319	0.00879	4.75	155.00
160.00	187.74	73.14	0.1938	53.909	102.911	0.10354	0.18261	0.3117	0.2552	1.3066	1110.	385.4	0.327	0.0331	0.0314	0.00896	4.46	160.00
165.00	199.87	72.30	0.1806	55.460	103.387	0.10598	0.18270	0.3159	0.2614	1.3262	1071.	381.5	0.315	0.0335	0.0309	0.00915	4.17	165.00
170.00	212.59	71.44	0.1683	57.030	103.841	0.10843	0.18277	0.3206	0.2682	1.3484	1031.	377.4	0.303	0.0339	0.0305	0.00934	3.88	170.00
175.00	225.90	70.54	0.1568	58.619	104.272	0.11089	0.18282	0.3258	0.2757	1.3737	992.	373.0	0.292	0.0344	0.0300	0.00955	3.60	175.00
180.00	239.83	69.61	0.1461	60.229	104.676	0.11336	0.18284	0.3316	0.2842	1.4028	951.	368.3	0.281	0.0349	0.0295	0.00977	3.32	180.00
185.00	254.39	68.65	0.1361	61.863	105.051	0.11584	0.18283	0.3382	0.2939	1.4365	911.	363.3	0.270	0.0354	0.0290	0.01001	3.05	185.00
190.00	269.61	67.64	0.1266	63.522	105.394	0.11834	0.18279	0.3456	0.3049	1.4759	869.	358.0	0.259	0.0360	0.0286	0.01027	2.78	190.00
195.00	285.52	66.59	0.1178	65.210	105.698	0.12086	0.18271	0.3543	0.3177	1.5227	827.	352.4	0.248	0.0366	0.0281	0.01056	2.51	195.00
200.00	302.12	65.48	0.1094	66.931	105.960	0.12341	0.18257	0.3644	0.3329	1.5788	785.	346.4	0.237	0.0373	0.0276	0.01088	2.25	200.00
210.00	337.55	63.06	0.0940	70.487	106.326	0.12861	0.18212	0.3910	0.3736	1.7330	698.	333.3	0.215	0.0389	0.0267	0.01164	1.74	210.00
220.00	376.14	60.29	0.0801	74.247	106.409	0.13401	0.18133	0.4329	0.4389	1.9870	607.	318.4	0.193	0.0409	0.0258	0.01266	1.25	220.00
230.00	418.15	56.94	0.0671	78.313	106.051	0.13975	0.17997	0.5103	0.5632	2.4776	512.	301.5	0.169	0.0436	0.0252	0.01414	0.80	230.00
240.00	463.99	52.51	0.0543	82.943	104.866	0.14619	0.17753	0.7123	0.8961	3.8005	406.	282.0	0.143	0.0479	0.0258	0.01668	0.39	240.00
250.00	514.35	44.21	0.0388	89.526	100.836	0.15526	0.17119	3.110	4.631	18.477	277.	257.1	0.106	0.0590	0.0349	0.02673	0.05	250.00
252.10c	525.66	34.96	0.0286	95.009	95.009	0.16290	0.16290	∞	∞	∞	0.	0.0	–	–	∞	∞	0.00	252.10

*Temperatures are on the ITS-90 scale b = normal boiling point c = critical point

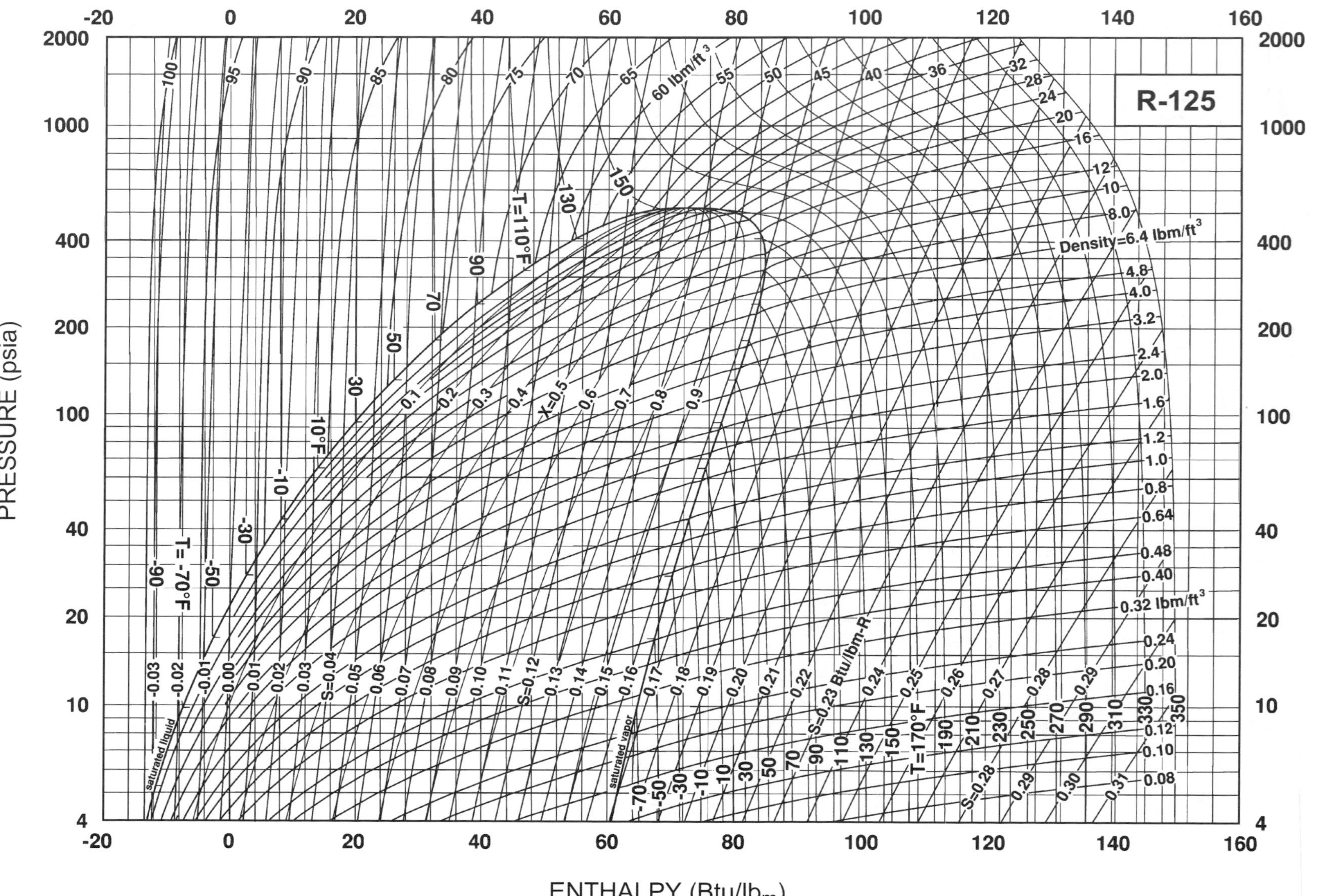

Fig. 7 Pressure-Enthalpy Diagram for Refrigerant 125

Refrigerant 125 (Pentafluoroethane) Properties of Saturated Liquid and Saturated Vapor

Temp.,* °F	Pres-sure, psia	Denisty, lb/ft³ Liquid	Volume, ft³/lb Vapor	Enthalpy, Btu/lb Liquid	Enthalpy, Btu/lb Vapor	Entropy, Btu/lb·°F Liquid	Entropy, Btu/lb·°F Vapor	Specific Heat c_p, Btu/lb·°F Liquid	Specific Heat c_p, Btu/lb·°F Vapor	c_p/c_v Vapor	Vel. of Sound, ft/s Liquid	Vel. of Sound, ft/s Vapor	Viscosity, lb_m/ft·h Liquid	Viscosity, lb_m/ft·h Vapor	Thermal Cond., Btu/h·ft·°F Liquid	Thermal Cond., Btu/h·ft·°F Vapor	Surface Tension, dyne/cm	Temp., °F
−149.13a	0.427	105.56	64.848	−28.057	53.649	−0.07717	0.18595	0.2494	0.1360	1.1416	3132.	382.0	2.945	0.0180	0.0696	0.00331	21.79	−149.13
−145.00	0.526	105.09	53.294	−27.026	54.199	−0.07387	0.18426	0.2495	0.1372	1.1407	3086.	384.2	2.746	0.0182	0.0688	0.00338	21.42	−145.00
−140.00	0.671	104.52	42.385	−25.778	54.867	−0.06993	0.18234	0.2497	0.1387	1.1398	3031.	386.8	2.536	0.0185	0.0679	0.00347	20.97	−140.00
−135.00	0.849	103.95	34.002	−24.529	55.540	−0.06606	0.18056	0.2500	0.1402	1.1389	2978.	389.3	2.352	0.0188	0.0669	0.00356	20.53	−135.00
−130.00	1.065	103.38	27.501	−23.277	56.215	−0.06223	0.17889	0.2504	0.1417	1.1382	2926.	391.7	2.190	0.0191	0.0660	0.00365	20.08	−130.00
−125.00	1.325	102.81	22.416	−22.024	56.894	−0.05846	0.17735	0.2510	0.1433	1.1376	2875.	394.1	2.046	0.0194	0.0651	0.00374	19.64	−125.00
−120.00	1.636	102.24	18.405	−20.767	57.575	−0.05474	0.17591	0.2516	0.1449	1.1371	2824.	396.4	1.918	0.0197	0.0642	0.00383	19.20	−120.00
−115.00	2.005	101.67	15.217	−19.507	58.259	−0.05106	0.17457	0.2522	0.1466	1.1368	2775.	398.7	1.802	0.0200	0.0633	0.00392	18.76	−115.00
−110.00	2.440	101.10	12.664	−18.244	58.944	−0.04742	0.17333	0.2530	0.1483	1.1366	2726.	400.9	1.698	0.0203	0.0624	0.00402	18.32	−110.00
−105.00	2.950	100.52	10.605	−16.976	59.631	−0.04382	0.17217	0.2538	0.1500	1.1365	2678.	402.9	1.603	0.0206	0.0615	0.00411	17.89	−105.00
−100.00	3.544	99.94	8.9339	−15.705	60.318	−0.04026	0.17110	0.2547	0.1518	1.1367	2631.	405.0	1.517	0.0209	0.0607	0.00421	17.46	−100.00
−95.00	4.232	99.35	7.5682	−14.428	61.007	−0.03674	0.17012	0.2556	0.1537	1.1369	2584.	406.9	1.438	0.0211	0.0599	0.00430	17.03	−95.00
−90.00	5.026	98.77	6.4454	−13.147	61.696	−0.03326	0.16920	0.2566	0.1556	1.1374	2538.	408.7	1.365	0.0214	0.0590	0.00440	16.60	−90.00
−85.00	5.935	98.18	5.5170	−11.860	62.384	−0.02981	0.16836	0.2577	0.1576	1.1380	2492.	410.4	1.298	0.0217	0.0582	0.00450	16.17	−85.00
−80.00	6.973	97.58	4.7449	−10.568	63.073	−0.02638	0.16758	0.2588	0.1596	1.1388	2447.	412.0	1.236	0.0220	0.0574	0.00460	15.75	−80.00
−75.00	8.152	96.98	4.0995	−9.270	63.760	−0.02299	0.16686	0.2599	0.1617	1.1398	2402.	413.6	1.178	0.0223	0.0566	0.00470	15.33	−75.00
−70.00	9.485	96.38	3.5572	−7.966	64.446	−0.01963	0.16620	0.2611	0.1639	1.1410	2357.	415.0	1.124	0.0226	0.0558	0.00481	14.91	−70.00
−65.00	10.986	95.77	3.0993	−6.656	65.131	−0.01630	0.16559	0.2624	0.1661	1.1424	2313.	416.3	1.073	0.0229	0.0550	0.00491	14.49	−65.00
−60.00	12.670	95.16	2.7109	−5.339	65.813	−0.01299	0.16504	0.2637	0.1684	1.1440	2269.	417.4	1.026	0.0232	0.0542	0.00502	14.07	−60.00
−55.00	14.551	94.54	2.3799	−4.015	66.494	−0.00971	0.16453	0.2651	0.1707	1.1458	2226.	418.5	0.982	0.0235	0.0535	0.00512	13.66	−55.00
−54.64b	14.696	94.50	2.3579	−3.919	66.543	−0.00947	0.16449	0.2652	0.1709	1.1459	2223.	418.6	0.979	0.0235	0.0534	0.00513	13.63	−54.64
−50.00	16.646	93.92	2.0967	−2.684	67.171	−0.00645	0.16407	0.2665	0.1731	1.1478	2182.	419.4	0.940	0.0237	0.0527	0.00523	13.25	−50.00
−45.00	18.971	93.28	1.8534	−1.346	67.846	−0.00321	0.16365	0.2680	0.1756	1.1501	2139.	420.2	0.901	0.0240	0.0519	0.00534	12.84	−45.00
−40.00	21.542	92.65	1.6435	0.000	68.517	0.00000	0.16326	0.2695	0.1781	1.1526	2096.	420.9	0.864	0.0243	0.0512	0.00545	12.44	−40.00
−35.00	24.377	92.00	1.4618	1.354	69.184	0.00319	0.16292	0.2711	0.1807	1.1554	2053.	421.4	0.829	0.0246	0.0504	0.00556	12.03	−35.00
−30.00	27.493	91.35	1.3039	2.716	69.847	0.00637	0.16261	0.2728	0.1834	1.1584	2011.	421.7	0.795	0.0249	0.0497	0.00568	11.63	−30.00
−25.00	30.909	90.69	1.1661	4.087	70.506	0.00952	0.16233	0.2745	0.1861	1.1617	1968.	422.0	0.764	0.0252	0.0490	0.00579	11.24	−25.00
−20.00	34.643	90.02	1.0457	5.466	71.159	0.01266	0.16208	0.2762	0.1889	1.1654	1926.	422.0	0.734	0.0255	0.0482	0.00591	10.84	−20.00
−15.00	38.715	89.34	0.9399	6.854	71.807	0.01578	0.16185	0.2781	0.1918	1.1694	1883.	421.9	0.705	0.0258	0.0475	0.00603	10.45	−15.00
−10.00	43.144	88.65	0.8468	8.252	72.449	0.01889	0.16165	0.2800	0.1947	1.1737	1841.	421.7	0.677	0.0261	0.0468	0.00615	10.06	−10.00
−5.00	47.949	87.95	0.7646	9.660	73.084	0.02198	0.16147	0.2819	0.1978	1.1785	1798.	421.2	0.651	0.0264	0.0460	0.00628	9.67	−5.00
0.00	53.152	87.24	0.6917	11.077	73.712	0.02505	0.16132	0.2840	0.2009	1.1836	1756.	420.6	0.626	0.0267	0.0453	0.00641	9.29	0.00
5.00	58.773	86.52	0.6271	12.505	74.333	0.02812	0.16118	0.2861	0.2042	1.1893	1713.	419.8	0.602	0.0271	0.0446	0.00653	8.91	5.00
10.00	64.834	85.79	0.5695	13.944	74.945	0.03117	0.16105	0.2884	0.2075	1.1954	1671.	418.9	0.579	0.0274	0.0439	0.00667	8.53	10.00
15.00	71.355	85.05	0.5180	15.394	75.549	0.03421	0.16094	0.2907	0.2110	1.2021	1628.	417.7	0.557	0.0277	0.0432	0.00680	8.16	15.00
20.00	78.359	84.29	0.4720	16.856	76.142	0.03724	0.16084	0.2932	0.2146	1.2095	1585.	416.3	0.535	0.0281	0.0425	0.00694	7.79	20.00
25.00	85.868	83.51	0.4307	18.329	76.725	0.04026	0.16075	0.2958	0.2184	1.2176	1542.	414.7	0.515	0.0284	0.0418	0.00709	7.42	25.00
30.00	93.906	82.73	0.3936	19.816	77.297	0.04328	0.16067	0.2985	0.2224	1.2265	1499.	412.9	0.495	0.0262	0.0411	0.00722	7.06	30.00
35.00	102.49	81.92	0.3601	21.316	77.856	0.04629	0.16059	0.3014	0.2266	1.2363	1456.	410.9	0.476	0.0271	0.0404	0.00736	6.70	35.00
40.00	111.66	81.10	0.3298	22.829	78.402	0.04929	0.16051	0.3044	0.2311	1.2471	1413.	408.7	0.457	0.0277	0.0397	0.00750	6.34	40.00
45.00	121.42	80.25	0.3024	24.357	78.933	0.05229	0.16043	0.3077	0.2358	1.2592	1369.	406.2	0.439	0.0283	0.0390	0.00765	5.99	45.00
50.00	131.81	79.39	0.2775	25.901	79.448	0.05528	0.16035	0.3112	0.2409	1.2726	1325.	403.5	0.422	0.0288	0.0383	0.00781	5.64	50.00
55.00	142.85	78.50	0.2549	27.460	79.946	0.05828	0.16026	0.3150	0.2463	1.2876	1281.	400.5	0.405	0.0293	0.0376	0.00798	5.29	55.00
60.00	154.56	77.59	0.2343	29.037	80.424	0.06127	0.16016	0.3190	0.2522	1.3044	1236.	397.2	0.388	0.0298	0.0369	0.00816	4.95	60.00
65.00	166.97	76.65	0.2155	30.633	80.882	0.06427	0.16004	0.3235	0.2587	1.3235	1191.	393.7	0.372	0.0304	0.0362	0.00834	4.62	65.00
70.00	180.12	75.69	0.1982	32.248	81.316	0.06727	0.15991	0.3284	0.2658	1.3451	1145.	389.8	0.357	0.0309	0.0355	0.00854	4.29	70.00
75.00	194.02	74.68	0.1824	33.885	81.724	0.07029	0.15976	0.3337	0.2736	1.3699	1099.	385.7	0.341	0.0315	0.0348	0.00875	3.96	75.00
80.00	208.71	73.65	0.1679	35.544	82.103	0.07331	0.15958	0.3398	0.2825	1.3985	1053.	381.2	0.326	0.0320	0.0341	0.00898	3.64	80.00
85.00	224.22	72.57	0.1545	37.229	82.450	0.07634	0.15937	0.3466	0.2925	1.4318	1006.	376.4	0.312	0.0327	0.0334	0.00922	3.32	85.00
90.00	240.58	71.44	0.1421	38.942	82.761	0.07940	0.15911	0.3543	0.3040	1.4710	958.	371.3	0.297	0.0333	0.0327	0.00949	3.01	90.00
95.00	257.83	70.26	0.1307	40.686	83.030	0.08247	0.15881	0.3632	0.3174	1.5176	910.	365.7	0.283	0.0340	0.0320	0.00979	2.71	95.00
100.00	276.00	69.02	0.1201	42.466	83.252	0.08558	0.15846	0.3737	0.3332	1.5739	861.	359.8	0.269	0.0348	0.0313	0.01012	2.41	100.00
105.00	295.13	67.71	0.1101	44.285	83.419	0.08872	0.15803	0.3862	0.3523	1.6431	810.	353.5	0.255	0.0356	0.0306	0.01050	2.12	105.00
110.00	315.25	66.32	0.1009	46.150	83.521	0.09191	0.15752	0.4016	0.3758	1.7299	759.	346.7	0.241	0.0365	0.0299	0.01093	1.84	110.00
115.00	336.42	64.82	0.0922	48.070	83.545	0.09517	0.15690	0.4208	0.4058	1.8417	707.	339.5	0.227	0.0376	0.0292	0.01143	1.56	115.00
120.00	358.69	63.20	0.0840	50.056	83.474	0.09849	0.15614	0.4458	0.4452	1.9905	653.	331.7	0.213	0.0387	0.0285	0.01203	1.30	120.00
125.00	382.10	61.42	0.0762	52.125	83.282	0.10193	0.15522	0.4800	0.4997	2.1979	598.	323.4	0.199	0.0401	0.0279	0.01276	1.04	125.00
130.00	406.73	59.42	0.0687	54.303	82.932	0.10551	0.15406	0.5298	0.5800	2.5057	541.	314.4	0.184	0.0418	0.0274	0.01367	0.80	130.00
135.00	432.64	57.12	0.0614	56.632	82.357	0.10930	0.15256	0.6102	0.7109	3.0086	481.	304.6	0.169	0.0439	0.0271	0.01488	0.57	135.00
140.00	459.93	54.33	0.0541	59.198	81.434	0.11345	0.15053	0.7649	0.9626	3.9759	417.	293.9	0.152	0.0467	0.0274	0.01659	0.35	140.00
145.00	488.74	50.60	0.0463	62.221	79.843	0.11830	0.14744	1.197	1.651	6.606	348.	281.5	0.132	0.0512	0.0294	0.01953	0.16	145.00
150.00	519.36	43.07	0.0352	67.112	75.717	0.12615	0.14027	8.301	11.291	42.778	269.	263.8	0.101	0.0627	0.0409	0.03336	0.01	150.00
150.83c	526.34	35.46	0.0282	71.345	71.345	0.13305	0.13305	∞	∞	∞	0.	0.0	–	–	∞	∞	0.00	150.83

*Temperatures are on the ITS-90 scale   a = triple point   b = normal boiling point   c = critical point

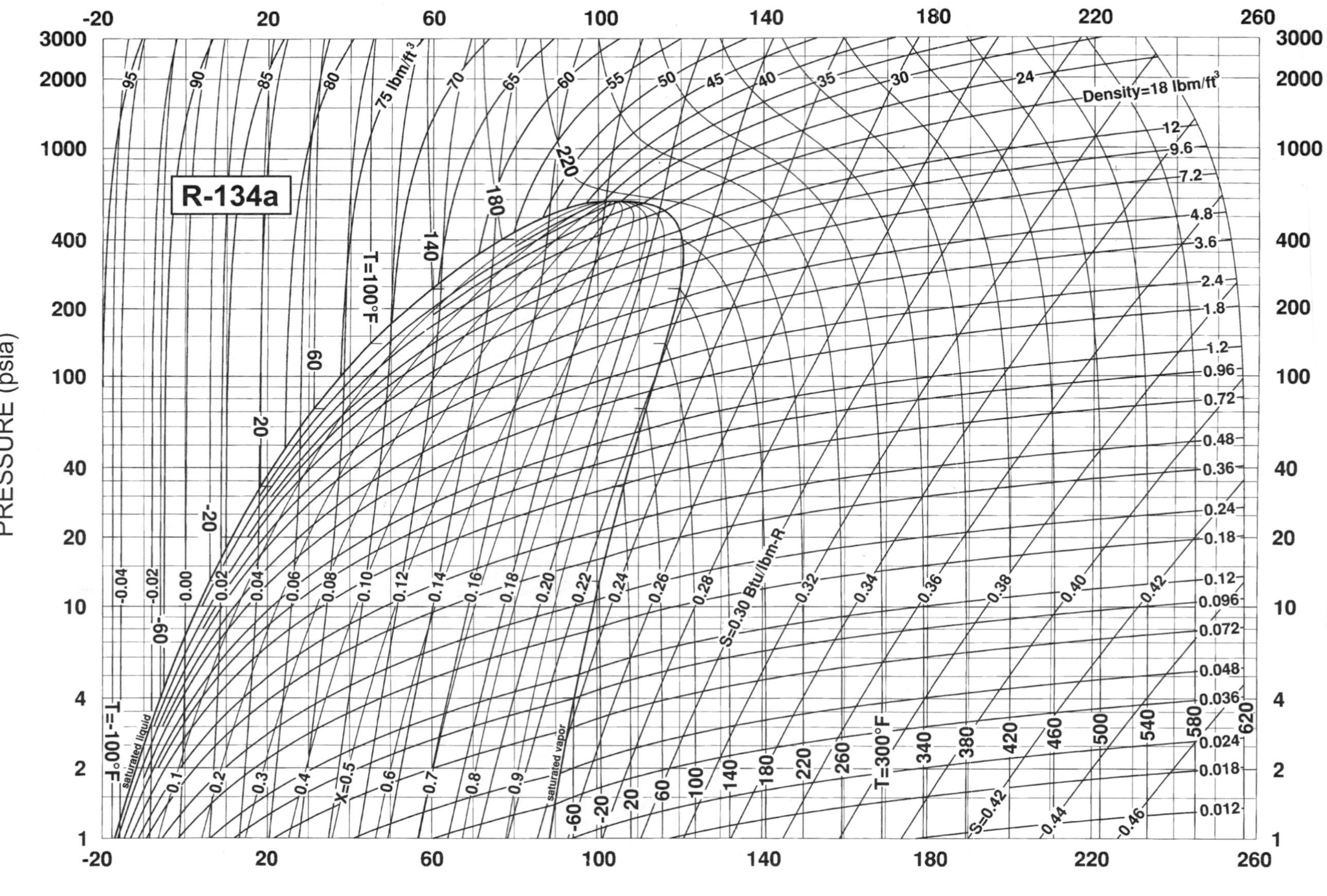

Fig. 8 Pressure-Enthalpy Diagram for Refrigerant 134a

Refrigerant 134a (1,1,1,2-Tetrafluoroethane) Properties of Saturated Liquid and Saturated Vapor

Temp.,* °F	Pressure, psia	Density, lb/ft³ Liquid	Volume, ft³/lb Vapor	Enthalpy, Btu/lb Liquid	Enthalpy, Btu/lb Vapor	Entropy, Btu/lb·°F Liquid	Entropy, Btu/lb·°F Vapor	Specific Heat c_p, Btu/lb·°F Liquid	Specific Heat c_p, Btu/lb·°F Vapor	c_p/c_v Vapor	Vel. of Sound, ft/s Liquid	Vel. of Sound, ft/s Vapor	Viscosity, lb_m/ft·h Liquid	Viscosity, lb_m/ft·h Vapor	Thermal Cond., Btu/h·ft·°F Liquid	Thermal Cond., Btu/h·ft·°F Vapor	Surface Tension, dyne/cm	Temp., °F
–153.94a	0.057	99.33	568.59	–32.992	80.362	–0.09154	0.27923	0.2829	0.1399	1.1637	3674.	416.0	5.262	0.0156	0.0840	0.00178	28.07	–153.94
–150.00	0.072	98.97	452.12	–31.878	80.907	–0.08791	0.27629	0.2830	0.1411	1.1623	3638.	418.3	4.790	0.0159	0.0832	0.00188	27.69	–150.00
–140.00	0.129	98.05	260.63	–29.046	82.304	–0.07891	0.26941	0.2834	0.1443	1.1589	3545.	424.2	3.880	0.0164	0.0813	0.00214	26.74	–140.00
–130.00	0.221	97.13	156.50	–26.208	83.725	–0.07017	0.26329	0.2842	0.1475	1.1559	3452.	429.9	3.238	0.0170	0.0794	0.00240	25.79	–130.00
–120.00	0.365	96.20	97.481	–23.360	85.168	–0.06166	0.25784	0.2853	0.1508	1.1532	3360.	435.5	2.762	0.0176	0.0775	0.00265	24.85	–120.00
–110.00	0.583	95.27	62.763	–20.500	86.629	–0.05337	0.25300	0.2866	0.1540	1.1509	3269.	440.8	2.396	0.0182	0.0757	0.00291	23.92	–110.00
–100.00	0.903	94.33	41.637	–17.626	88.107	–0.04527	0.24871	0.2881	0.1573	1.1490	3178.	446.0	2.105	0.0187	0.0739	0.00317	22.99	–100.00
–90.00	1.359	93.38	28.381	–14.736	89.599	–0.03734	0.24490	0.2898	0.1607	1.1475	3087.	450.9	1.869	0.0193	0.0722	0.00343	22.07	–90.00
–80.00	1.993	92.42	19.825	–11.829	91.103	–0.02959	0.24152	0.2916	0.1641	1.1465	2998.	455.6	1.673	0.0199	0.0705	0.00369	21.16	–80.00
–75.00	2.392	91.94	16.711	–10.368	91.858	–0.02577	0.23998	0.2925	0.1658	1.1462	2954.	457.8	1.587	0.0201	0.0696	0.00382	20.71	–75.00
–70.00	2.854	91.46	14.161	–8.903	92.614	–0.02198	0.23854	0.2935	0.1676	1.1460	2909.	460.0	1.509	0.0204	0.0688	0.00395	20.26	–70.00
–65.00	3.389	90.97	12.060	–7.432	93.372	–0.01824	0.23718	0.2945	0.1694	1.1459	2866.	462.1	1.436	0.0207	0.0680	0.00408	19.81	–65.00
–60.00	4.002	90.49	10.321	–5.957	94.131	–0.01452	0.23590	0.2955	0.1713	1.1460	2822.	464.1	1.369	0.0210	0.0671	0.00420	19.36	–60.00
–55.00	4.703	90.00	8.8733	–4.476	94.890	–0.01085	0.23470	0.2965	0.1731	1.1462	2778.	466.0	1.306	0.0212	0.0663	0.00433	18.92	–55.00
–50.00	5.501	89.50	7.6621	–2.989	95.650	–0.00720	0.23358	0.2976	0.1751	1.1466	2735.	467.8	1.248	0.0215	0.0655	0.00446	18.47	–50.00
–45.00	6.406	89.00	6.6438	–1.498	96.409	–0.00358	0.23252	0.2987	0.1770	1.1471	2691.	469.6	1.193	0.0218	0.0647	0.00460	18.03	–45.00
–40.00	7.427	88.50	5.7839	0.000	97.167	0.00000	0.23153	0.2999	0.1790	1.1478	2648.	471.2	1.142	0.0221	0.0639	0.00473	17.60	–40.00
–35.00	8.576	88.00	5.0544	1.503	97.924	0.00356	0.23060	0.3010	0.1811	1.1486	2605.	472.8	1.095	0.0223	0.0632	0.00486	17.16	–35.00
–30.00	9.862	87.49	4.4330	3.013	98.679	0.00708	0.22973	0.3022	0.1832	1.1496	2563.	474.2	1.050	0.0226	0.0624	0.00499	16.73	–30.00
–25.00	11.299	86.98	3.9014	4.529	99.433	0.01058	0.22892	0.3035	0.1853	1.1508	2520.	475.6	1.007	0.0229	0.0616	0.00512	16.30	–25.00
–20.00	12.898	86.47	3.4449	6.051	100.184	0.01406	0.22816	0.3047	0.1875	1.1521	2477.	476.8	0.968	0.0231	0.0608	0.00525	15.87	–20.00
–15.00	14.671	85.95	3.0514	7.580	100.932	0.01751	0.22744	0.3060	0.1898	1.1537	2435.	477.9	0.930	0.0234	0.0601	0.00538	15.44	–15.00
–14.93b	14.696	85.94	3.0465	7.600	100.942	0.01755	0.22743	0.3061	0.1898	1.1537	2434.	477.9	0.929	0.0234	0.0601	0.00538	15.44	–14.93
–10.00	16.632	85.43	2.7109	9.115	101.677	0.02093	0.22678	0.3074	0.1921	1.1554	2393.	478.9	0.894	0.0237	0.0593	0.00552	15.02	–10.00
–5.00	18.794	84.90	2.4154	10.657	102.419	0.02433	0.22615	0.3088	0.1945	1.1573	2350.	479.8	0.860	0.0240	0.0586	0.00565	14.60	–5.00
0.00	21.171	84.37	2.1579	12.207	103.156	0.02771	0.22557	0.3102	0.1969	1.1595	2308.	480.5	0.828	0.0242	0.0578	0.00578	14.18	0.00
5.00	23.777	83.83	1.9330	13.764	103.889	0.03107	0.22502	0.3117	0.1995	1.1619	2266.	481.1	0.798	0.0245	0.0571	0.00592	13.76	5.00
10.00	26.628	83.29	1.7357	15.328	104.617	0.03440	0.22451	0.3132	0.2021	1.1645	2224.	481.6	0.769	0.0248	0.0564	0.00605	13.35	10.00
15.00	29.739	82.74	1.5623	16.901	105.339	0.03772	0.22403	0.3147	0.2047	1.1674	2182.	482.0	0.741	0.0250	0.0556	0.00619	12.94	15.00
20.00	33.124	82.19	1.4094	18.481	106.056	0.04101	0.22359	0.3164	0.2075	1.1705	2140.	482.2	0.715	0.0253	0.0549	0.00632	12.53	20.00
25.00	36.800	81.63	1.2742	20.070	106.767	0.04429	0.22317	0.3181	0.2103	1.1740	2098.	482.2	0.689	0.0256	0.0542	0.00646	12.12	25.00
30.00	40.784	81.06	1.1543	21.667	107.471	0.04755	0.22278	0.3198	0.2132	1.1777	2056.	482.2	0.665	0.0258	0.0535	0.00660	11.72	30.00
35.00	45.092	80.49	1.0478	23.274	108.167	0.05079	0.22241	0.3216	0.2163	1.1818	2014.	481.9	0.642	0.0261	0.0528	0.00674	11.32	35.00
40.00	49.741	79.90	0.9528	24.890	108.856	0.05402	0.22207	0.3235	0.2194	1.1862	1973.	481.5	0.620	0.0264	0.0521	0.00688	10.92	40.00
45.00	54.749	79.32	0.8680	26.515	109.537	0.05724	0.22174	0.3255	0.2226	1.1910	1931.	481.0	0.598	0.0267	0.0514	0.00703	10.53	45.00
50.00	60.134	78.72	0.7920	28.150	110.209	0.06044	0.22144	0.3275	0.2260	1.1961	1889.	480.3	0.578	0.0270	0.0507	0.00717	10.14	50.00
55.00	65.913	78.11	0.7238	29.796	110.871	0.06362	0.22115	0.3297	0.2294	1.2018	1847.	479.4	0.558	0.0273	0.0500	0.00732	9.75	55.00
60.00	72.105	77.50	0.6625	31.452	111.524	0.06680	0.22088	0.3319	0.2331	1.2079	1805.	478.3	0.539	0.0275	0.0493	0.00747	9.36	60.00
65.00	78.729	76.87	0.6072	33.120	112.165	0.06996	0.22062	0.3343	0.2368	1.2145	1763.	477.0	0.520	0.0278	0.0486	0.00762	8.98	65.00
70.00	85.805	76.24	0.5572	34.799	112.796	0.07311	0.22037	0.3368	0.2408	1.2217	1721.	475.6	0.503	0.0281	0.0479	0.00777	8.60	70.00
75.00	93.351	75.59	0.5120	36.491	113.414	0.07626	0.22013	0.3394	0.2449	1.2296	1679.	474.0	0.485	0.0284	0.0472	0.00793	8.23	75.00
80.00	101.39	74.94	0.4710	38.195	114.019	0.07939	0.21989	0.3422	0.2492	1.2382	1636.	472.2	0.469	0.0287	0.0465	0.00809	7.86	80.00
85.00	109.93	74.27	0.4338	39.913	114.610	0.08252	0.21966	0.3451	0.2537	1.2475	1594.	470.1	0.453	0.0291	0.0458	0.00825	7.49	85.00
90.00	119.01	73.58	0.3999	41.645	115.186	0.08565	0.21944	0.3482	0.2585	1.2578	1551.	467.9	0.437	0.0294	0.0451	0.00842	7.13	90.00
95.00	128.65	72.88	0.3690	43.392	115.746	0.08877	0.21921	0.3515	0.2636	1.2690	1509.	465.4	0.422	0.0297	0.0444	0.00860	6.77	95.00
100.00	138.85	72.17	0.3407	45.155	116.289	0.09188	0.21898	0.3551	0.2690	1.2813	1466.	462.7	0.407	0.0301	0.0437	0.00878	6.41	100.00
105.00	149.65	71.44	0.3148	46.934	116.813	0.09500	0.21875	0.3589	0.2747	1.2950	1423.	459.8	0.393	0.0304	0.0431	0.00897	6.06	105.00
110.00	161.07	70.69	0.2911	48.731	117.317	0.09811	0.21851	0.3630	0.2809	1.3101	1380.	456.7	0.378	0.0308	0.0424	0.00916	5.71	110.00
115.00	173.14	69.93	0.2693	50.546	117.799	0.10123	0.21826	0.3675	0.2875	1.3268	1337.	453.2	0.365	0.0312	0.0417	0.00936	5.36	115.00
120.00	185.86	69.14	0.2493	52.382	118.258	0.10435	0.21800	0.3723	0.2948	1.3456	1294.	449.6	0.351	0.0316	0.0410	0.00958	5.03	120.00
125.00	199.28	68.32	0.2308	54.239	118.690	0.10748	0.21772	0.3775	0.3026	1.3666	1250.	445.6	0.338	0.0320	0.0403	0.00981	4.69	125.00
130.00	213.41	67.49	0.2137	56.119	119.095	0.11062	0.21742	0.3833	0.3112	1.3903	1206.	441.4	0.325	0.0324	0.0396	0.01005	4.36	130.00
135.00	228.28	66.62	0.1980	58.023	119.468	0.11376	0.21709	0.3897	0.3208	1.4173	1162.	436.8	0.313	0.0329	0.0389	0.01031	4.04	135.00
140.00	243.92	65.73	0.1833	59.954	119.807	0.11692	0.21673	0.3968	0.3315	1.4481	1117.	432.0	0.301	0.0334	0.0382	0.01058	3.72	140.00
145.00	260.36	64.80	0.1697	61.915	120.108	0.12010	0.21634	0.4048	0.3435	1.4837	1072.	426.8	0.288	0.0339	0.0375	0.01089	3.40	145.00
150.00	277.61	63.83	0.1571	63.908	120.366	0.12330	0.21591	0.4138	0.3571	1.5250	1027.	421.2	0.276	0.0344	0.0368	0.01122	3.09	150.00
155.00	295.73	62.82	0.1453	65.936	120.576	0.12653	0.21542	0.4242	0.3729	1.5738	980.	415.3	0.264	0.0350	0.0361	0.01158	2.79	155.00
160.00	314.73	61.76	0.1343	68.005	120.731	0.12979	0.21488	0.4362	0.3914	1.6318	934.	409.1	0.253	0.0357	0.0354	0.01199	2.50	160.00
165.00	334.65	60.65	0.1239	70.118	120.823	0.13309	0.21426	0.4504	0.4133	1.7022	886.	402.4	0.241	0.0364	0.0346	0.01245	2.21	165.00
170.00	355.53	59.47	0.1142	72.283	120.842	0.13644	0.21356	0.4675	0.4400	1.7889	837.	395.3	0.229	0.0372	0.0339	0.01297	1.93	170.00
175.00	377.41	58.21	0.1051	74.509	120.773	0.13985	0.21274	0.4887	0.4733	1.8984	786.	387.7	0.218	0.0381	0.0332	0.01358	1.66	175.00
180.00	400.34	56.86	0.0964	76.807	120.598	0.14334	0.21180	0.5156	0.5159	2.0405	734.	379.6	0.206	0.0391	0.0325	0.01430	1.39	180.00
185.00	424.36	55.38	0.0881	79.193	120.294	0.14693	0.21069	0.5512	0.5729	2.2321	680.	371.0	0.194	0.0403	0.0318	0.01516	1.14	185.00
190.00	449.52	53.76	0.0801	81.692	119.822	0.15066	0.20935	0.6012	0.6532	2.5041	624.	361.8	0.182	0.0417	0.0311	0.01623	0.90	190.00
195.00	475.91	51.91	0.0724	84.343	119.123	0.15459	0.20771	0.6768	0.7751	2.9192	565.	352.0	0.169	0.0435	0.0304	0.01760	0.67	195.00
200.00	503.59	49.76	0.0647	87.214	118.097	0.15880	0.20562	0.8062	0.9835	3.6309	502.	341.3	0.155	0.0457	0.0300	0.01949	0.45	200.00
205.00	532.68	47.08	0.0567	90.454	116.526	0.16353	0.20275	1.083	1.425	5.136	436.	329.4	0.140	0.0489	0.0300	0.02240	0.26	205.00
210.00	563.35	43.20	0.0477	94.530	113.746	0.16945	0.19814	2.113	3.008	10.512	363.	315.5	0.120	0.0543	0.0316	0.02848	0.09	210.00
213.91c	588.75	31.96	0.0313	103.894	103.894	0.18320	0.18320	∞	∞	∞	0.	0.0	–	–	∞	∞	0.00	213.91

*Temperatures are on the ITS-90 scale a = triple point b = normal boiling point c = critical point

Refrigerant 134a Properties of Superheated Vapor

Pressure = 14.696 psia
Saturation temperature = −14.92°F

Temp.,* °F	Density, lb/ft³	Enthalpy, Btu/lb	Entropy, Btu/lb·°F	Vel. Sound, ft/s
Saturated				
Liquid	85.7972	7.53	0.01739	2451.2
Vapor	0.3283	100.81	0.22713	478.0
0.00	0.3158	103.62	0.23335	487.2
20.00	0.3008	107.45	0.24149	499.0
40.00	0.2874	111.34	0.24944	510.2
60.00	0.2753	115.31	0.25723	521.0
80.00	0.2642	119.35	0.26486	531.5
100.00	0.2541	123.47	0.27236	541.6
120.00	0.2448	127.68	0.27974	551.4
140.00	0.2362	131.96	0.28700	561.0
160.00	0.2282	136.32	0.29416	570.4
180.00	0.2208	140.77	0.30122	579.5
200.00	0.2139	145.30	0.30819	588.5
220.00	0.2074	149.90	0.31507	597.3
240.00	0.2013	154.59	0.32187	606.0
260.00	0.1955	159.36	0.32858	614.5
280.00	0.1901	164.20	0.33522	622.8
300.00	0.1850	169.12	0.34178	631.1

Pressure = 25.00 psia
Saturation temperature = 7.22°F

Temp.,* °F	Density, lb/ft³	Enthalpy, Btu/lb	Entropy, Btu/lb·°F	Vel. Sound, ft/s
Saturated				
Liquid	83.4823	14.32	0.03224	2263.9
Vapor	0.5426	104.07	0.22446	481.5
20.00	0.5245	106.60	0.22982	489.9
40.00	0.4991	110.61	0.23800	502.4
60.00	0.4765	114.66	0.24596	514.1
80.00	0.4563	118.78	0.25373	525.4
100.00	0.4379	122.96	0.26135	536.2
120.00	0.4212	127.22	0.26881	546.6
140.00	0.4058	131.55	0.27615	556.7
160.00	0.3916	135.95	0.28337	566.5
180.00	0.3786	140.43	0.29048	576.0
200.00	0.3663	144.98	0.29750	585.3
220.00	0.3549	149.61	0.30441	594.4
240.00	0.3443	154.32	0.31124	603.3
260.00	0.3343	159.10	0.31798	612.0
280.00	0.3248	163.96	0.32464	620.6
300.00	0.3160	168.90	0.33122	629.0

Pressure = 50.00 psia
Saturation temperature = 40.29°F

Temp.,* °F	Density, lb/ft³	Enthalpy, Btu/lb	Entropy, Btu/lb·°F	Vel. Sound, ft/s
Saturated				
Liquid	79.8125	24.79	0.05377	1982.3
Vapor	1.0545	108.74	0.22170	481.7
60.00	0.9982	113.00	0.23005	496.2
80.00	0.9489	117.32	0.23822	509.8
100.00	0.9055	121.68	0.24614	522.5
120.00	0.8670	126.07	0.25385	534.5
140.00	0.8322	130.51	0.26139	545.8
160.00	0.8008	135.01	0.26877	556.7
180.00	0.7718	139.57	0.27601	567.2
200.00	0.7454	144.20	0.28313	577.4
220.00	0.7208	148.89	0.29014	587.2
240.00	0.6980	153.65	0.29704	596.8
260.00	0.6768	158.48	0.30385	606.1
280.00	0.6569	163.38	0.31056	615.2
300.00	0.6383	168.35	0.31719	624.1

Pressure = 75.00 psia
Saturation temperature = 62.24°F

Temp.,* °F	Density, lb/ft³	Enthalpy, Btu/lb	Entropy, Btu/lb·°F	Vel. Sound, ft/s
Saturated				
Liquid	77.1862	31.98	0.06775	1793.6
Vapor	1.5686	111.67	0.22042	478.1
80.00	1.4873	115.74	0.22809	492.7
100.00	1.4092	120.30	0.23639	507.7
120.00	1.3416	124.85	0.24439	521.6
140.00	1.2822	129.43	0.25215	534.5
160.00	1.2294	134.04	0.25971	546.6
180.00	1.1817	138.69	0.26710	558.2
200.00	1.1383	143.39	0.27434	569.2
220.00	1.0984	148.15	0.28145	579.8
240.00	1.0620	152.97	0.28843	590.1
260.00	1.0280	157.85	0.29531	600.0
280.00	0.9966	162.79	0.30208	609.7
300.00	0.9671	167.80	0.30876	619.0
320.00	0.9398	172.87	0.31535	628.2
340.00	0.9138	178.01	0.32186	637.1
360.00	0.8895	183.21	0.32828	645.9
380.00	0.8665	188.48	0.33463	654.5
400.00	0.8448	193.82	0.34091	662.9

Pressure = 100.00 psia
Saturation temperature = 79.17°F

Temp.,* °F	Density, lb/ft³	Enthalpy, Btu/lb	Entropy, Btu/lb·°F	Vel. Sound, ft/s
Saturated				
Liquid	75.0245	37.69	0.07840	1646.8
Vapor	2.0917	113.78	0.21960	472.8
80.00	2.0858	113.98	0.21998	473.6
100.00	1.9576	118.80	0.22874	491.6
120.00	1.8509	123.55	0.23709	507.8
140.00	1.7597	128.29	0.24512	522.4
160.00	1.6800	133.02	0.25288	536.0
180.00	1.6094	137.78	0.26044	548.8
200.00	1.5463	142.57	0.26781	560.8
220.00	1.4891	147.40	0.27502	572.3
240.00	1.4368	152.27	0.28210	583.3
260.00	1.3886	157.21	0.28905	593.9
280.00	1.3444	162.19	0.29588	604.1
300.00	1.3031	167.24	0.30261	614.0
320.00	1.2647	172.35	0.30925	623.6
340.00	1.2287	177.52	0.31579	632.9
360.00	1.1950	182.75	0.32225	642.1
380.00	1.1633	188.04	0.32863	651.0
400.00	1.1334	193.39	0.33494	659.8

Pressure = 125.00 psia
Saturation temperature = 93.15°F

Temp.,* °F	Density, lb/ft³	Enthalpy, Btu/lb	Entropy, Btu/lb·°F	Vel. Sound, ft/s
Saturated				
Liquid	73.1279	42.53	0.08715	1524.7
Vapor	2.6279	115.41	0.21898	466.7
100.00	2.5638	117.16	0.22212	473.9
120.00	2.4025	122.16	0.23090	492.9
140.00	2.2694	127.08	0.23924	509.7
160.00	2.1561	131.96	0.24725	525.0
180.00	2.0577	136.83	0.25498	539.0
200.00	1.9710	141.71	0.26250	552.2
220.00	1.8935	146.62	0.26983	564.6
240.00	1.8233	151.56	0.27700	576.4
260.00	1.7592	156.55	0.28402	587.6
280.00	1.7006	161.59	0.29093	598.4
300.00	1.6463	166.67	0.29771	608.8
320.00	1.5959	171.82	0.30440	618.9
340.00	1.5492	177.02	0.31098	628.7
360.00	1.5055	182.27	0.31747	638.2
380.00	1.4644	187.59	0.32388	647.5
400.00	1.4258	192.97	0.33021	656.6

Pressure = 150.00 psia
Saturation temperature = 105.17°F

Temp.,* °F	Density, lb/ft³	Enthalpy, Btu/lb	Entropy, Btu/lb·°F	Vel. Sound, ft/s
Saturated				
Liquid	71.4013	46.78	0.09464	1419.1
Vapor	3.1801	116.71	0.21844	460.0
120.00	3.0077	120.64	0.22530	476.6
140.00	2.8181	125.78	0.23403	496.0
160.00	2.6620	130.83	0.24231	513.3
180.00	2.5295	135.83	0.25026	528.9
200.00	2.4146	140.82	0.25794	543.3
220.00	2.3132	145.82	0.26539	556.7
240.00	2.2223	150.83	0.27267	569.3
260.00	2.1401	155.88	0.27978	581.3
280.00	2.0658	160.97	0.28675	592.7
300.00	1.9971	166.10	0.29360	603.7
320.00	1.9338	171.28	0.30033	614.2
340.00	1.8751	176.51	0.30696	624.5
360.00	1.8208	181.80	0.31349	634.4
380.00	1.7695	187.14	0.31993	644.0
400.00	1.7216	192.54	0.32628	653.4
420.00	1.6766	198.00	0.33256	662.6
440.00	1.6341	203.51	0.33876	671.6
460.00	1.5940	209.08	0.34488	680.4
480.00	1.5558	214.71	0.35094	689.0
500.00	1.5197	220.40	0.35692	697.4

Pressure = 175.00 psia
Saturation temperature = 115.76°F

Temp.,* °F	Density, lb/ft³	Enthalpy, Btu/lb	Entropy, Btu/lb·°F	Vel. Sound, ft/s
Saturated				
Liquid	69.7902	50.62	0.10126	1325.3
Vapor	3.7511	117.76	0.21794	453.0
120.00	3.6836	118.95	0.21999	458.4
140.00	3.4148	124.38	0.22921	481.3
160.00	3.2025	129.64	0.23783	500.9
180.00	3.0271	134.79	0.24602	518.3
200.00	2.8785	139.90	0.25388	534.0
220.00	2.7494	144.99	0.26148	548.5
240.00	2.6349	150.08	0.26887	562.1
260.00	2.5328	155.19	0.27607	574.8
280.00	2.4403	160.34	0.28312	586.9
300.00	2.3558	165.51	0.29003	598.5
320.00	2.2785	170.73	0.29681	609.5
340.00	2.2071	176.00	0.30348	620.2
360.00	2.1411	181.32	0.31004	630.5
380.00	2.0795	186.69	0.31651	640.5
400.00	2.0216	192.11	0.32290	650.3
420.00	1.9675	197.59	0.32920	659.7
440.00	1.9164	203.12	0.33542	669.0
460.00	1.8683	208.71	0.34156	678.0
480.00	1.8228	214.36	0.34763	686.9
500.00	1.7797	220.05	0.35363	695.6

Pressure = 200.00 psia
Saturation temperature = 125.27°F

Temp.,* °F	Density, lb/ft³	Enthalpy, Btu/lb	Entropy, Btu/lb·°F	Vel. Sound, ft/s
Saturated				
Liquid	68.2602	54.14	0.10721	1240.5
Vapor	4.3437	118.61	0.21743	445.6
140.00	4.0726	122.86	0.22460	465.2
160.00	3.7850	128.36	0.23363	487.8
180.00	3.5561	133.70	0.24210	507.2
200.00	3.3656	138.94	0.25018	524.5
220.00	3.2036	144.14	0.25793	540.2
240.00	3.0623	149.31	0.26544	554.7
260.00	2.9371	154.50	0.27274	568.3
280.00	2.8247	159.69	0.27987	581.1
300.00	2.7234	164.92	0.28684	593.2
320.00	2.6305	170.18	0.29368	604.8
340.00	2.5455	175.49	0.30039	615.9
360.00	2.4668	180.83	0.30700	626.7
380.00	2.3934	186.23	0.31350	637.0
400.00	2.3254	191.68	0.31991	647.1
420.00	2.2614	197.18	0.32624	656.9
440.00	2.2017	202.73	0.33248	666.4
460.00	2.1453	208.34	0.33864	675.7
480.00	2.0920	214.00	0.34473	684.8
500.00	2.0417	219.71	0.35075	693.7

*Temperatures are on the ITS–90 scale

Refrigerant 134a Properties of Superheated Vapor (*Concluded*)

Pressure = 225.00 psia
Saturation temperature = 133.93°F

Temp.,* °F	Density, lb/ft^3	Enthalpy, Btu/lb	Entropy, Btu/lb·°F	Vel. Sound, ft/s
Saturated				
Liquid	66.7870	57.42	0.11266	1162.8
Vapor	4.9609	119.30	0.21690	438.1
140.00	4.8123	121.16	0.22002	447.3
160.00	4.4191	126.99	0.22959	473.6
180.00	4.1206	132.54	0.23840	495.5
200.00	3.8796	137.94	0.24671	514.6
220.00	3.6784	143.25	0.25465	531.6
240.00	3.5058	148.52	0.26229	547.2
260.00	3.3542	153.78	0.26970	561.6
280.00	3.2202	159.04	0.27691	575.1
300.00	3.0995	164.32	0.28395	587.9
320.00	2.9899	169.62	0.29084	600.0
340.00	2.8897	174.97	0.29761	611.7
360.00	2.7978	180.35	0.30425	622.8
380.00	2.7122	185.77	0.31079	633.6
400.00	2.6330	191.24	0.31723	644.0
420.00	2.5592	196.77	0.32358	654.0
440.00	2.4900	202.34	0.32984	663.9
460.00	2.4249	207.96	0.33603	673.4
480.00	2.3636	213.64	0.34213	682.8
500.00	2.3057	219.36	0.34816	691.9

Pressure = 250.00 psia
Saturation temperature = 141.89°F

Temp.,* °F	Density, lb/ft^3	Enthalpy, Btu/lb	Entropy, Btu/lb·°F	Vel. Sound, ft/s
Saturated				
Liquid	65.3526	60.50	0.11770	1090.7
Vapor	5.6060	119.84	0.21634	430.3
160.00	5.1189	125.49	0.22560	458.2
180.00	4.7275	131.31	0.23484	483.1
200.00	4.4239	136.89	0.24343	504.2
220.00	4.1756	142.34	0.25156	522.8
240.00	3.9664	147.71	0.25935	539.5
260.00	3.7854	153.05	0.26688	554.9
280.00	3.6265	158.37	0.27418	569.2
300.00	3.4847	163.71	0.28129	582.6
320.00	3.3571	169.06	0.28824	595.3
340.00	3.2408	174.44	0.29506	607.4
360.00	3.1342	179.85	0.30175	618.9
380.00	3.0359	185.31	0.30832	630.1
400.00	2.9451	190.81	0.31479	640.8
420.00	2.8604	196.35	0.32117	651.2
440.00	2.7813	201.94	0.32745	661.3
460.00	2.7072	207.58	0.33365	671.1
480.00	2.6374	213.27	0.33977	680.7
500.00	2.5717	219.02	0.34582	690.1

Pressure = 275.00 psia
Saturation temperature = 149.27°F

Temp.,* °F	Density, lb/ft^3	Enthalpy, Btu/lb	Entropy, Btu/lb·°F	Vel. Sound, ft/s
Saturated				
Liquid	63.9423	63.43	0.12241	1023.4
Vapor	6.2831	120.25	0.21572	422.3
160.00	5.9060	123.82	0.22155	441.2
180.00	5.3869	129.98	0.23133	469.9
200.00	5.0031	135.78	0.24026	493.4
220.00	4.6978	141.38	0.24862	513.7
240.00	4.4465	146.87	0.25658	531.7
260.00	4.2314	152.30	0.26423	548.0
280.00	4.0446	157.69	0.27162	563.1
300.00	3.8803	163.08	0.27881	577.2
320.00	3.7317	168.49	0.28583	590.5
340.00	3.5987	173.91	0.29270	603.1
360.00	3.4764	179.36	0.29943	615.1
380.00	3.3646	184.84	0.30604	626.6
400.00	3.2612	190.37	0.31254	637.7
420.00	3.1653	195.93	0.31894	648.4
440.00	3.0758	201.55	0.32525	658.8
460.00	2.9922	207.20	0.33147	668.9
480.00	2.9136	212.91	0.33761	678.7
500.00	2.8397	218.67	0.34368	688.3

Pressure = 300.00 psia
Saturation temperature = 156.16°F

Temp.,* °F	Density, lb/ft^3	Enthalpy, Btu/lb	Entropy, Btu/lb·°F	Vel. Sound, ft/s
Saturated				
Liquid	62.5436	66.23	0.12686	959.8
Vapor	6.9967	120.54	0.21505	414.2
160.00	6.8168	121.92	0.21730	422.0
180.00	6.1118	128.55	0.22782	455.8
200.00	5.6239	134.61	0.23715	482.1
220.00	5.2494	140.39	0.24578	504.3
240.00	4.9472	146.00	0.25393	523.7
260.00	4.6939	151.52	0.26171	541.1
280.00	4.4758	157.00	0.26921	557.0
300.00	4.2852	162.45	0.27649	571.8
320.00	4.1160	167.90	0.28357	585.7
340.00	3.9631	173.37	0.29049	598.8
360.00	3.8247	178.85	0.29727	611.2
380.00	3.6981	184.37	0.30392	623.1
400.00	3.5816	189.92	0.31045	634.6
420.00	3.4737	195.51	0.31688	645.6
440.00	3.3735	201.15	0.32321	656.3
460.00	3.2799	206.83	0.32945	666.6
480.00	3.1922	212.55	0.33561	676.7
500.00	3.1098	218.32	0.34169	686.5

Pressure = 325.00 psia
Saturation temperature = 162.62°F

Temp.,* °F	Density, lb/ft^3	Enthalpy, Btu/lb	Entropy, Btu/lb·°F	Vel. Sound, ft/s
Saturated				
Liquid	61.1446	68.92	0.13110	899.5
Vapor	7.7526	120.71	0.21431	405.9
180.00	6.9220	126.96	0.22423	440.3
200.00	6.2928	133.36	0.23408	470.2
220.00	5.8341	139.34	0.24301	494.5
240.00	5.4723	145.10	0.25136	515.5
260.00	5.1741	150.73	0.25930	534.0
280.00	4.9208	156.29	0.26692	550.9
300.00	4.7017	161.81	0.27428	566.4
320.00	4.5082	167.31	0.28144	580.9
340.00	4.3352	172.82	0.28841	594.5
360.00	4.1790	178.35	0.29524	607.4
380.00	4.0368	183.90	0.30193	619.7
400.00	3.9063	189.48	0.30849	631.5
420.00	3.7859	195.09	0.31495	642.8
440.00	3.6743	200.75	0.32131	653.8
460.00	3.5703	206.44	0.32757	664.4
480.00	3.4731	212.19	0.33375	674.7
500.00	3.3819	217.97	0.33984	684.7

Pressure = 350.00 psia
Saturation temperature = 168.71°F

Temp.,* °F	Density, lb/ft^3	Enthalpy, Btu/lb	Entropy, Btu/lb·°F	Vel. Sound, ft/s
Saturated				
Liquid	59.7334	71.54	0.13516	841.7
Vapor	8.5577	120.76	0.21349	397.5
180.00	7.8491	125.18	0.22046	423.2
200.00	7.0242	132.01	0.23098	457.6
220.00	6.4561	138.24	0.24029	484.4
240.00	6.0219	144.17	0.24888	507.1
260.00	5.6728	149.91	0.25698	526.9
280.00	5.3805	155.56	0.26472	544.7
300.00	5.1295	161.15	0.27218	561.0
320.00	4.9098	166.72	0.27941	576.1
340.00	4.7148	172.27	0.28644	590.2
360.00	4.5396	177.84	0.29332	603.6
380.00	4.3807	183.42	0.30005	616.3
400.00	4.2355	189.03	0.30665	628.4
420.00	4.1019	194.67	0.31313	640.1
440.00	3.9784	200.35	0.31952	651.4
460.00	3.8636	206.06	0.32580	662.2
480.00	3.7564	211.82	0.33199	672.8
500.00	3.6561	217.62	0.33811	683.0

Pressure = 375.00 psia
Saturation temperature= 174.46°F

Temp.,* °F	Density, lb/ft^3	Enthalpy, Btu/lb	Entropy, Btu/lb·°F	Vel. Sound, ft/s
Saturated				
Liquid	58.2974	74.09	0.13908	785.9
Vapor	9.4209	120.69	0.21256	389.0
180.00	8.9498	123.10	0.21634	403.8
200.00	7.8311	130.54	0.22781	444.1
220.00	7.1211	137.08	0.23758	474.0
240.00	6.6028	143.19	0.24644	498.5
260.00	6.1926	149.07	0.25473	519.6
280.00	5.8555	154.82	0.26260	538.4
300.00	5.5694	160.49	0.27016	555.5
320.00	5.3212	166.11	0.27747	571.3
340.00	5.1022	171.72	0.28457	586.0
360.00	4.9066	177.32	0.29149	599.8
380.00	4.7300	182.94	0.29826	612.9
400.00	4.5692	188.58	0.30490	625.4
420.00	4.4217	194.24	0.31141	637.4
440.00	4.2857	199.94	0.31782	648.9
460.00	4.1596	205.68	0.32413	660.1
480.00	4.0421	211.46	0.33034	670.8
500.00	3.9323	217.28	0.33647	681.3

Pressure = 400.00 psia
Saturation temperature = 197.93°F

Temp.,* °F	Density, lb/ft^3	Enthalpy, Btu/lb	Entropy, Btu/lb·°F	Vel. Sound, ft/s
Saturated				
Liquid	56.8213	76.60	0.14289	731.8
Vapor	10.3541	120.50	0.21152	380.4
180.00	10.3454	120.53	0.21158	380.6
200.00	8.7370	128.93	0.22451	429.5
220.00	7.8399	135.85	0.23484	463.0
240.00	7.2145	142.18	0.24403	489.7
260.00	6.7351	148.21	0.25252	512.3
280.00	6.3472	154.06	0.26055	532.1
300.00	6.0221	159.81	0.26821	550.0
320.00	5.7425	165.49	0.27560	566.5
340.00	5.4977	171.15	0.28277	581.7
360.00	5.2802	176.80	0.28975	596.0
380.00	5.0848	182.45	0.29656	609.5
400.00	4.9075	188.12	0.30323	622.4
420.00	4.7454	193.82	0.30978	634.7
440.00	4.5964	199.54	0.31621	646.5
460.00	4.4584	205.30	0.32254	657.9
480.00	4.3303	211.09	0.32878	669.0
500.00	4.2107	216.93	0.33492	679.6

Pressure = 600.00 psia
Saturation temperature = n/a (supercritical)

Temp.,* °F	Density, lb/ft^3	Enthalpy, Btu/lb	Entropy, Btu/lb·°F	Vel. Sound, ft/s
Saturated				
Liquid				
Vapor				
220.00	19.6784	118.27	0.20421	340.3
240.00	14.2159	131.50	0.22343	409.1
260.00	12.2674	139.92	0.23530	449.7
280.00	11.0672	147.15	0.24522	480.8
300.00	10.2049	153.83	0.25413	506.7
320.00	9.5351	160.21	0.26241	529.2
340.00	8.9895	166.39	0.27024	549.2
360.00	8.5305	172.45	0.27774	567.5
380.00	8.1351	178.45	0.28496	584.4
400.00	7.7885	184.40	0.29197	600.1
420.00	7.4804	190.34	0.29879	615.0
440.00	7.2035	196.26	0.30546	629.0
460.00	6.9523	202.20	0.31198	642.4
480.00	6.7229	208.15	0.31838	655.3
500.00	6.5118	214.12	0.32467	667.6

*temperatures are on the ITS-90 scale

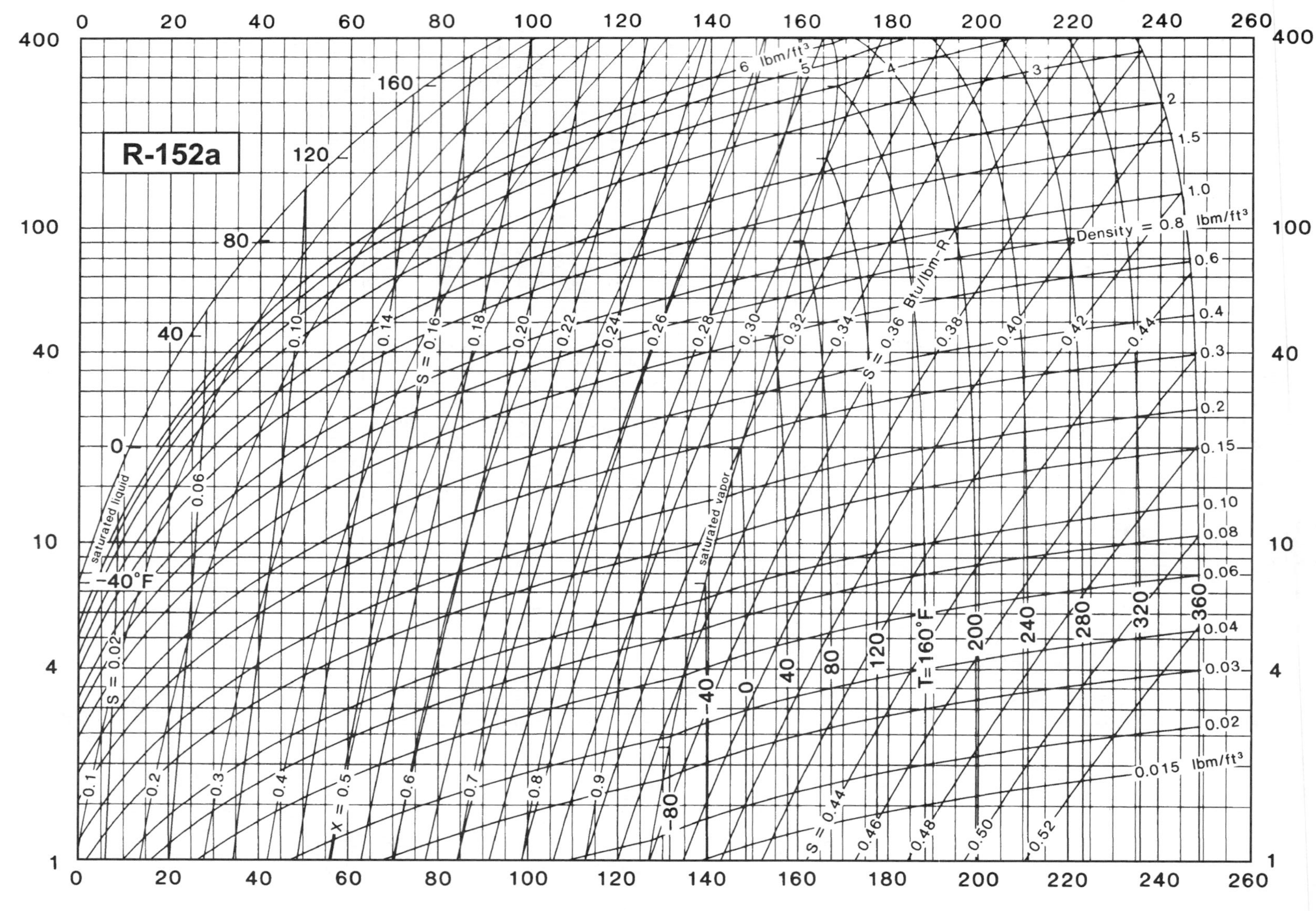

Fig. 9 Pressure-Enthalpy Diagram for Refrigerant 152a

Refrigerant 152a (1,1-Difluoroethane) Properties of Saturated Liquid and Saturated Vapor

Temp.,* °F	Pres-sure, psia	Denisty, lb/ft³ Liquid	Volume, ft³/lb Vapor	Enthalpy, Btu/lb Liquid	Enthalpy, Btu/lb Vapor	Entropy, Btu/lb·°F Liquid	Entropy, Btu/lb·°F Vapor	Specific Heat c_p, Btu/lb·°F Liquid	Specific Heat c_p, Btu/lb·°F Vapor	c_p/c_v Vapor	Vel. of Sound, ft/s Liquid	Vel. of Sound, ft/s Vapor	Viscosity, lb_m/ft·h Liquid	Viscosity, lb_m/ft·h Vapor	Thermal Cond., Btu/h·ft·°F Liquid	Thermal Cond., Btu/h·ft·°F Vapor	Surface Tension, dyne/cm	Temp., °F
−181.46a	0.009	74.47	4858.2	−51.885	122.577	−0.15045	0.47664	0.3531	0.1670	1.2201	4596.	505.4	–	–	0.1019	0.00006	31.65	−181.46
−180.00	0.010	74.38	4379.8	−51.368	122.821	−0.14859	0.47424	0.3540	0.1674	1.2194	4575.	506.6	–	–	0.1016	0.00010	31.52	−180.00
−170.00	0.021	73.74	2227.6	−47.803	124.502	−0.13607	0.45876	0.3586	0.1703	1.2152	4438.	514.6	–	–	0.0993	0.00042	30.60	−170.00
−160.00	0.041	73.10	1195.0	−44.203	126.208	−0.12385	0.44481	0.3611	0.1732	1.2112	4316.	522.4	–	–	0.0971	0.00073	29.68	−160.00
−150.00	0.075	72.46	672.43	−40.584	127.938	−0.11197	0.43222	0.3626	0.1762	1.2074	4204.	530.0	4.334	0.0140	0.0949	0.00104	28.76	−150.00
−140.00	0.131	71.82	394.96	−36.953	129.689	−0.10044	0.42086	0.3635	0.1793	1.2039	4099.	537.5	3.527	0.0144	0.0928	0.00136	27.86	−140.00
−130.00	0.222	71.17	241.12	−33.315	131.460	−0.08923	0.41059	0.3642	0.1825	1.2006	3999.	544.7	2.955	0.0149	0.0907	0.00167	26.96	−130.00
−120.00	0.361	70.52	152.41	−29.669	133.249	−0.07834	0.40130	0.3649	0.1857	1.1977	3902.	551.7	2.530	0.0153	0.0887	0.00198	26.06	−120.00
−110.00	0.569	69.87	99.413	−26.015	135.052	−0.06774	0.39289	0.3658	0.1892	1.1952	3808.	558.5	2.202	0.0157	0.0867	0.00230	25.17	−110.00
−100.00	0.871	69.22	66.714	−22.351	136.868	−0.05741	0.38527	0.3670	0.1927	1.1931	3715.	565.0	1.941	0.0162	0.0848	0.00261	24.28	−100.00
−90.00	1.297	68.56	45.938	−18.674	138.692	−0.04732	0.37837	0.3684	0.1965	1.1913	3623.	571.2	1.728	0.0166	0.0829	0.00292	23.40	−90.00
−80.00	1.886	67.89	32.380	−14.981	140.522	−0.03747	0.37210	0.3700	0.2004	1.1900	3532.	577.2	1.552	0.0171	0.0811	0.00324	22.53	−80.00
−70.00	2.680	67.22	23.313	−11.270	142.353	−0.02783	0.36641	0.3720	0.2045	1.1893	3442.	582.8	1.404	0.0176	0.0793	0.00355	21.66	−70.00
−60.00	3.731	66.54	17.113	−7.538	144.184	−0.01838	0.36124	0.3742	0.2088	1.1890	3352.	588.0	1.277	0.0180	0.0776	0.00387	20.80	−60.00
−50.00	5.099	65.86	12.784	−3.782	146.010	−0.00911	0.35653	0.3766	0.2134	1.1892	3263.	592.9	1.168	0.0185	0.0758	0.00419	19.94	−50.00
−45.00	5.921	65.51	11.117	−1.894	146.920	−0.00454	0.35434	0.3779	0.2157	1.1896	3218.	595.2	1.119	0.0187	0.0750	0.00435	19.52	−45.00
−40.00	6.847	65.16	9.7045	0.000	147.828	0.00000	0.35225	0.3793	0.2181	1.1901	3174.	597.4	1.072	0.0189	0.0742	0.00451	19.09	−40.00
−35.00	7.887	64.82	8.5025	1.902	148.732	0.00450	0.35025	0.3807	0.2206	1.1908	3129.	599.4	1.029	0.0192	0.0733	0.00467	18.67	−35.00
−30.00	9.051	64.46	7.4754	3.811	149.634	0.00896	0.34834	0.3822	0.2232	1.1916	3085.	601.4	0.989	0.0194	0.0725	0.00483	18.25	−30.00
−25.00	10.348	64.11	6.5943	5.728	150.531	0.01339	0.34652	0.3838	0.2258	1.1926	3040.	603.2	0.854	0.0197	0.0717	0.00499	17.83	−25.00
−20.00	11.789	63.75	5.8357	7.653	151.424	0.01778	0.34478	0.3854	0.2285	1.1937	2996.	604.9	0.802	0.0200	0.0709	0.00516	17.42	−20.00
−15.00	13.386	63.40	5.1800	9.586	152.312	0.02214	0.34311	0.3870	0.2313	1.1951	2952.	606.5	0.758	0.0202	0.0701	0.00532	17.00	−15.00
−11.24b	14.696	63.12	4.7450	11.046	152.977	0.02540	0.34191	0.3883	0.2334	1.1962	2918.	607.6	0.729	0.0204	0.0695	0.00544	16.69	−11.24
−10.00	15.149	63.04	4.6114	11.528	153.196	0.02647	0.34152	0.3887	0.2341	1.1966	2907.	608.0	0.720	0.0205	0.0693	0.00548	16.59	−10.00
−5.00	17.092	62.67	4.1167	13.479	154.073	0.03077	0.34000	0.3905	0.2370	1.1983	2863.	609.3	0.686	0.0207	0.0685	0.00565	16.18	−5.00
0.00	19.226	62.30	3.6848	15.439	154.945	0.03505	0.33854	0.3924	0.2400	1.2002	2818.	610.5	0.657	0.0210	0.0678	0.00582	15.77	0.00
5.00	21.564	61.93	3.3066	17.408	155.810	0.03929	0.33714	0.3943	0.2431	1.2024	2774.	611.6	0.630	0.0212	0.0670	0.00598	15.36	5.00
10.00	24.119	61.56	2.9745	19.388	156.667	0.04352	0.33580	0.3962	0.2463	1.2047	2729.	612.5	0.606	0.0215	0.0662	0.00615	14.96	10.00
15.00	26.904	61.18	2.6819	21.378	157.518	0.04771	0.33452	0.3982	0.2495	1.2073	2685.	613.3	0.584	0.0218	0.0655	0.00632	14.55	15.00
20.00	29.934	60.80	2.4234	23.378	158.360	0.05188	0.33329	0.4003	0.2529	1.2101	2640.	613.9	0.563	0.0220	0.0648	0.00649	14.15	20.00
25.00	33.223	60.42	2.1945	25.389	159.194	0.05603	0.33211	0.4025	0.2564	1.2132	2596.	614.3	0.544	0.0223	0.0640	0.00666	13.75	25.00
30.00	36.784	60.03	1.9912	27.411	160.019	0.06016	0.33097	0.4047	0.2599	1.2166	2551.	614.6	0.526	0.0225	0.0633	0.00684	13.36	30.00
35.00	40.634	59.64	1.8103	29.444	160.834	0.06427	0.32988	0.4070	0.2636	1.2202	2506.	614.8	0.509	0.0228	0.0626	0.00701	12.96	35.00
40.00	44.787	59.24	1.6488	31.490	161.639	0.06836	0.32883	0.4094	0.2673	1.2242	2461.	614.8	0.493	0.0230	0.0619	0.00719	12.57	40.00
45.00	49.259	58.84	1.5043	33.547	162.433	0.07243	0.32781	0.4119	0.2712	1.2285	2416.	614.6	0.478	0.0233	0.0612	0.00737	12.18	45.00
50.00	54.065	58.44	1.3748	35.618	163.216	0.07648	0.32683	0.4145	0.2753	1.2331	2371.	614.2	0.463	0.0235	0.0604	0.00755	11.79	50.00
55.00	59.222	58.03	1.2585	37.701	163.987	0.08051	0.32589	0.4171	0.2794	1.2381	2326.	613.7	0.449	0.0238	0.0597	0.00774	11.40	55.00
60.00	64.746	57.61	1.1537	39.798	164.745	0.08454	0.32497	0.4199	0.2837	1.2436	2281.	612.9	0.436	0.0240	0.0590	0.00792	11.02	60.00
65.00	70.654	57.19	1.0591	41.909	165.490	0.08854	0.32408	0.4228	0.2882	1.2494	2235.	612.0	0.423	0.0243	0.0584	0.00811	10.64	65.00
70.00	76.963	56.76	0.9737	44.034	166.221	0.09253	0.32322	0.4258	0.2928	1.2557	2190.	610.9	0.411	0.0245	0.0577	0.00831	10.26	70.00
75.00	83.691	56.33	0.8963	46.175	166.936	0.09651	0.32238	0.4289	0.2976	1.2626	2144.	609.7	0.399	0.0248	0.0570	0.00851	9.88	75.00
80.00	90.854	55.89	0.8260	48.331	167.636	0.10048	0.32156	0.4322	0.3026	1.2700	2098.	608.2	0.387	0.0251	0.0563	0.00871	9.51	80.00
85.00	98.470	55.44	0.7621	50.503	168.319	0.10444	0.32075	0.4357	0.3079	1.2780	2052.	606.5	0.375	0.0253	0.0556	0.00892	9.14	85.00
90.00	106.56	54.98	0.7039	52.692	168.985	0.10840	0.31996	0.4393	0.3133	1.2867	2006.	604.6	0.364	0.0256	0.0549	0.00913	8.77	90.00
95.00	115.14	54.52	0.6508	54.899	169.631	0.11234	0.31919	0.4431	0.3191	1.2961	1960.	602.5	0.354	0.0259	0.0543	0.00935	8.40	95.00
100.00	124.23	54.05	0.6023	57.124	170.258	0.11628	0.31842	0.4471	0.3251	1.3063	1913.	600.1	0.343	0.0261	0.0536	0.00957	8.04	100.00
105.00	133.84	53.57	0.5578	59.369	170.864	0.12021	0.31766	0.4513	0.3314	1.3175	1866.	597.6	0.333	0.0264	0.0529	0.00980	7.68	105.00
110.00	144.01	53.08	0.5171	61.633	171.447	0.12414	0.31691	0.4558	0.3381	1.3296	1819.	594.8	0.323	0.0267	0.0523	0.01004	7.33	110.00
115.00	154.74	52.58	0.4797	63.919	172.006	0.12807	0.31616	0.4606	0.3451	1.3429	1772.	591.7	0.313	0.0270	0.0516	0.01029	6.97	115.00
120.00	166.07	52.07	0.4453	66.227	172.539	0.13200	0.31540	0.4657	0.3527	1.3574	1725.	588.5	0.304	0.0273	0.0510	0.01055	6.62	120.00
125.00	178.00	51.55	0.4135	68.558	173.044	0.13593	0.31464	0.4711	0.3607	1.3734	1677.	584.9	0.294	0.0276	0.0503	0.01082	6.27	125.00
130.00	190.56	51.01	0.3843	70.915	173.520	0.13987	0.31387	0.4770	0.3693	1.3911	1629.	581.1	0.285	0.0279	0.0497	0.01110	5.93	130.00
135.00	203.77	50.47	0.3572	73.297	173.964	0.14381	0.31309	0.4833	0.3785	1.4106	1581.	577.1	0.276	0.0283	0.0490	0.01140	5.59	135.00
140.00	217.66	49.90	0.3322	75.708	174.374	0.14776	0.31229	0.4902	0.3886	1.4323	1532.	572.7	0.267	0.0286	0.0484	0.01171	5.25	140.00
145.00	232.25	49.32	0.3090	78.149	174.746	0.15172	0.31148	0.4977	0.3995	1.4566	1483.	568.1	0.258	0.0290	0.0477	0.01204	4.92	145.00
150.00	247.55	48.73	0.2874	80.621	175.078	0.15570	0.31063	0.5059	0.4114	1.4838	1434.	563.1	0.249	0.0294	0.0471	0.01239	4.59	150.00
155.00	263.60	48.11	0.2674	83.128	175.365	0.15970	0.30975	0.5149	0.4245	1.5145	1384.	557.9	0.241	0.0298	0.0464	0.01277	4.27	155.00
160.00	280.42	47.48	0.2487	85.673	175.603	0.16371	0.30884	0.5249	0.4392	1.5495	1334.	552.3	0.232	0.0302	0.0458	0.01317	3.95	160.00
165.00	298.04	46.82	0.2313	88.258	175.787	0.16776	0.30788	0.5361	0.4556	1.5896	1283.	546.4	0.224	0.0307	0.0451	0.01361	3.63	165.00
170.00	316.48	46.13	0.2150	90.887	175.911	0.17183	0.30686	0.5488	0.4742	1.6359	1232.	540.1	0.216	0.0312	0.0445	0.01408	3.32	170.00
175.00	335.77	45.42	0.1997	93.566	175.968	0.17595	0.30578	0.5634	0.4954	1.6900	1180.	533.4	0.207	0.0318	0.0438	0.01460	3.02	175.00
180.00	355.95	44.67	0.1854	96.300	175.948	0.18011	0.30462	0.5801	0.5202	1.7540	1127.	526.4	0.199	0.0324	0.0432	0.01517	2.72	180.00
185.00	377.04	43.88	0.1719	99.095	175.841	0.18432	0.30337	0.5999	0.5493	1.8307	1074.	518.9	0.191	0.0330	0.0425	0.01580	2.42	185.00
190.00	399.08	43.06	0.1592	101.961	175.633	0.18861	0.30200	0.6235	0.5844	1.9244	1020.	510.9	0.183	0.0338	0.0419	0.01651	2.14	190.00
195.00	422.10	42.17	0.1471	104.910	175.306	0.19297	0.30050	0.6524	0.6276	2.0411	965.	502.5	0.175	0.0346	0.0413	0.01731	1.85	195.00
200.00	446.15	41.23	0.1356	107.955	174.835	0.19744	0.29883	0.6889	0.6821	2.1906	908.	493.5	0.166	0.0355	0.0407	0.01823	1.58	200.00
210.00	497.53	39.10	0.1141	114.433	173.320	0.20683	0.29477	0.8017	0.8519	2.6633	791.	473.8	0.149	0.0379	0.0395	0.02057	1.06	210.00
220.00	553.69	36.41	0.0935	121.740	170.564	0.21725	0.28909	1.054	1.231	3.733	663.	451.1	0.131	0.0415	0.0388	0.02418	0.58	220.00
230.00	615.42	32.30	0.0714	131.138	164.733	0.23049	0.27920	2.214	2.919	8.532	518.	423.7	0.108	0.0482	0.0402	0.03208	0.17	230.00
235.87c	655.10	22.97	0.0435	147.629	147.629	0.25390	0.25390	∞	∞	∞	0.	0.0	–	–	∞	∞	0.00	235.87

*Temperatures are on the ITS-90 scale a = triple point b = normal boiling point c = critical point

Refrigerant 143a (1,1,1-Trifluoroethane) Properties of Saturated Liquid and Saturated Vapor

Temp.,* °F	Pres-sure, psia	Denisty, lb/ft³ Liquid	Volume, ft³/lb Vapor	Enthalpy, Btu/lb Liquid	Enthalpy, Btu/lb Vapor	Entropy, Btu/lb·°F Liquid	Entropy, Btu/lb·°F Vapor	Specific Heat c_p, Btu/lb·°F Liquid	Specific Heat c_p, Btu/lb·°F Vapor	c_p/c_v Vapor	Vel. of Sound, ft/s Liquid	Vel. of Sound, ft/s Vapor	Viscosity, lb$_m$/ft·h Liquid	Viscosity, lb$_m$/ft·h Vapor	Thermal Cond., Btu/h·ft·°F Liquid	Thermal Cond., Btu/h·ft·°F Vapor	Surface Tension, dyne/cm	Temp., °F
–169.26a	0.156	83.06	237.18	–39.089	75.807	–0.11084	0.28480	0.2895	0.1506	1.1924	3472.	451.4	2.206	0.0143	0.0792	0.00283	13.72	–169.26
–165.00	0.199	82.71	188.87	–37.856	76.423	–0.10662	0.28120	0.2897	0.1522	1.1908	3434.	454.1	2.069	0.0145	0.0783	0.00291	13.75	–165.00
–160.00	0.261	82.29	145.98	–36.406	77.150	–0.10174	0.27719	0.2901	0.1541	1.1890	3391.	457.4	1.926	0.0148	0.0773	0.00299	13.78	–160.00
–155.00	0.340	81.87	113.97	–34.954	77.881	–0.09694	0.27341	0.2906	0.1560	1.1873	3347.	460.5	1.799	0.0150	0.0762	0.00308	13.80	–155.00
–150.00	0.438	81.44	89.843	–33.499	78.618	–0.09220	0.26985	0.2913	0.1580	1.1858	3304.	463.6	1.687	0.0152	0.0752	0.00316	13.80	–150.00
–145.00	0.559	81.02	71.465	–32.041	79.358	–0.08753	0.26649	0.2920	0.1599	1.1843	3260.	466.7	1.586	0.0155	0.0743	0.00325	13.80	–145.00
–140.00	0.707	80.59	57.335	–30.579	80.102	–0.08292	0.26331	0.2929	0.1619	1.1830	3217.	469.6	1.496	0.0157	0.0733	0.00334	13.78	–140.00
–135.00	0.887	80.17	46.374	–29.112	80.850	–0.07837	0.26032	0.2938	0.1639	1.1818	3174.	472.5	1.413	0.0160	0.0723	0.00343	13.75	–135.00
–130.00	1.103	79.74	37.797	–27.640	81.602	–0.07387	0.25749	0.2948	0.1660	1.1807	3130.	475.4	1.339	0.0162	0.0714	0.00353	13.71	–130.00
–125.00	1.362	79.30	31.032	–26.163	82.357	–0.06943	0.25483	0.2958	0.1680	1.1797	3087.	478.1	1.271	0.0165	0.0705	0.00362	13.66	–125.00
–120.00	1.670	78.87	25.654	–24.681	83.114	–0.06504	0.25232	0.2969	0.1701	1.1789	3044.	480.8	1.208	0.0167	0.0696	0.00372	13.60	–120.00
–115.00	2.033	78.43	21.347	–23.193	83.874	–0.06069	0.24995	0.2981	0.1722	1.1783	3001.	483.4	1.150	0.0170	0.0687	0.00381	13.53	–115.00
–110.00	2.459	78.00	17.874	–21.699	84.636	–0.05639	0.24771	0.2994	0.1743	1.1777	2957.	485.9	1.097	0.0172	0.0678	0.00391	13.44	–110.00
–105.00	2.955	77.56	15.054	–20.198	85.400	–0.05213	0.24561	0.3006	0.1765	1.1773	2914.	488.3	1.048	0.0175	0.0669	0.00402	13.35	–105.00
–100.00	3.530	77.11	12.751	–18.691	86.165	–0.04791	0.24362	0.3020	0.1787	1.1771	2870.	490.7	1.002	0.0177	0.0660	0.00412	13.24	–100.00
–95.00	4.194	76.67	10.857	–17.176	86.931	–0.04374	0.24175	0.3034	0.1809	1.1770	2827.	492.9	0.959	0.0180	0.0652	0.00422	13.13	–95.00
–90.00	4.955	76.22	9.2915	–15.655	87.697	–0.03960	0.23998	0.3048	0.1832	1.1771	2784.	495.0	0.919	0.0182	0.0644	0.00433	13.01	–90.00
–85.00	5.825	75.77	7.9899	–14.126	88.464	–0.03550	0.23832	0.3062	0.1855	1.1774	2740.	497.0	0.881	0.0185	0.0635	0.00444	12.87	–85.00
–80.00	6.813	75.32	6.9019	–12.590	89.230	–0.03143	0.23675	0.3077	0.1879	1.1779	2697.	498.9	0.846	0.0187	0.0627	0.00455	12.73	–80.00
–75.00	7.931	74.86	5.9879	–11.046	89.995	–0.02740	0.23527	0.3093	0.1904	1.1786	2654.	500.7	0.813	0.0190	0.0619	0.00466	12.58	–75.00
–70.00	9.191	74.40	5.2164	–9.494	90.759	–0.02340	0.23388	0.3109	0.1928	1.1794	2610.	502.4	0.782	0.0192	0.0611	0.00477	12.42	–70.00
–65.00	10.606	73.94	4.5621	–7.934	91.521	–0.01943	0.23257	0.3125	0.1954	1.1805	2567.	503.9	0.752	0.0195	0.0604	0.00488	12.25	–65.00
–60.00	12.187	73.47	4.0046	–6.365	92.280	–0.01549	0.23133	0.3142	0.1980	1.1818	2524.	505.3	0.724	0.0197	0.0596	0.00500	12.07	–60.00
–55.00	13.950	73.00	3.5277	–4.787	93.037	–0.01157	0.23016	0.3159	0.2007	1.1833	2480.	506.6	0.698	0.0200	0.0588	0.00512	11.88	–55.00
–53.03b	14.696	72.82	3.3592	–4.164	93.334	–0.01004	0.22972	0.3166	0.2018	1.1840	2463.	507.1	0.688	0.0201	0.0585	0.00517	11.81	–53.03
–50.00	15.907	72.53	3.1181	–3.201	93.790	–0.00769	0.22906	0.3177	0.2035	1.1851	2437.	507.7	0.672	0.0202	0.0581	0.00524	11.69	–50.00
–45.00	18.074	72.05	2.7648	–1.605	94.539	–0.00383	0.22802	0.3195	0.2064	1.1871	2393.	508.7	0.648	0.0204	0.0573	0.00536	11.48	–45.00
–40.00	20.464	71.57	2.4589	0.000	95.284	0.00000	0.22705	0.3214	0.2093	1.1894	2350.	509.6	0.625	0.0207	0.0566	0.00549	11.27	–40.00
–35.00	23.094	71.08	2.1932	1.615	96.024	0.00381	0.22612	0.3234	0.2123	1.1920	2307.	510.2	0.604	0.0209	0.0558	0.00561	11.05	–35.00
–30.00	25.979	70.59	1.9615	3.240	96.759	0.00759	0.22525	0.3254	0.2155	1.1949	2263.	510.8	0.583	0.0212	0.0551	0.00574	10.82	–30.00
–25.00	29.135	70.09	1.7589	4.875	97.487	0.01136	0.22442	0.3274	0.2187	1.1981	2220.	511.1	0.563	0.0214	0.0544	0.00588	10.59	–25.00
–20.00	32.579	69.59	1.5811	6.520	98.209	0.01510	0.22364	0.3295	0.2220	1.2017	2176.	511.3	0.544	0.0221	0.0537	0.00602	10.35	–20.00
–15.00	36.329	69.08	1.4246	8.177	98.923	0.01883	0.22290	0.3317	0.2255	1.2056	2133.	511.3	0.525	0.0224	0.0530	0.00616	10.10	–15.00
–10.00	40.401	68.57	1.2864	9.845	99.629	0.02253	0.22220	0.3340	0.2290	1.2098	2089.	511.2	0.508	0.0226	0.0523	0.00630	9.84	–10.00
–5.00	44.813	68.04	1.1640	11.524	100.327	0.02622	0.22153	0.3364	0.2327	1.2146	2046.	510.9	0.491	0.0229	0.0516	0.00644	9.58	–5.00
0.00	49.584	67.52	1.0554	13.216	101.016	0.02989	0.22090	0.3388	0.2365	1.2197	2002.	510.3	0.474	0.0231	0.0509	0.00659	9.31	0.00
5.00	54.733	66.98	0.9587	14.920	101.695	0.03355	0.22029	0.3414	0.2405	1.2253	1958.	509.6	0.458	0.0234	0.0502	0.00673	9.04	5.00
10.00	60.277	66.44	0.8724	16.637	102.363	0.03719	0.21971	0.3440	0.2446	1.2315	1914.	508.7	0.443	0.0237	0.0495	0.00689	8.76	10.00
15.00	66.237	65.89	0.7951	18.367	103.020	0.04082	0.21916	0.3468	0.2489	1.2382	1870.	507.6	0.429	0.0239	0.0488	0.00705	8.47	15.00
20.00	72.632	65.33	0.7259	20.112	103.664	0.04444	0.21862	0.3497	0.2534	1.2456	1826.	506.3	0.414	0.0242	0.0482	0.00721	8.19	20.00
25.00	79.482	64.76	0.6636	21.870	104.295	0.04804	0.21811	0.3528	0.2580	1.2536	1782.	504.7	0.401	0.0245	0.0475	0.00737	7.89	25.00
30.00	86.808	64.18	0.6075	23.644	104.912	0.05164	0.21761	0.3560	0.2629	1.2624	1738.	502.9	0.387	0.0248	0.0468	0.00755	7.59	30.00
35.00	94.630	63.59	0.5569	25.434	105.514	0.05523	0.21712	0.3594	0.2681	1.2721	1693.	501.0	0.374	0.0251	0.0462	0.00772	7.29	35.00
40.00	102.97	62.99	0.5110	27.241	106.099	0.05882	0.21664	0.3630	0.2735	1.2827	1649.	498.7	0.362	0.0254	0.0455	0.00791	6.98	40.00
45.00	111.85	62.37	0.4695	29.064	106.667	0.06240	0.21617	0.3668	0.2793	1.2944	1604.	496.3	0.349	0.0257	0.0448	0.00811	6.68	45.00
50.00	121.29	61.75	0.4317	30.906	107.215	0.06597	0.21569	0.3709	0.2854	1.3072	1559.	493.5	0.337	0.0260	0.0442	0.00831	6.36	50.00
55.00	131.32	61.10	0.3973	32.768	107.743	0.06955	0.21522	0.3752	0.2919	1.3215	1514.	490.6	0.326	0.0264	0.0435	0.00853	6.05	55.00
60.00	141.95	60.45	0.3659	34.649	108.248	0.07312	0.21475	0.3799	0.2989	1.3373	1468.	487.3	0.314	0.0267	0.0429	0.00875	5.73	60.00
65.00	153.22	59.77	0.3373	36.553	108.730	0.07670	0.21427	0.3849	0.3065	1.3549	1422.	483.8	0.303	0.0271	0.0422	0.00900	5.41	65.00
70.00	165.14	59.08	0.3110	38.479	109.184	0.08029	0.21378	0.3904	0.3146	1.3746	1376.	480.0	0.293	0.0275	0.0416	0.00925	5.09	70.00
75.00	177.74	58.36	0.2869	40.430	109.611	0.08388	0.21327	0.3963	0.3235	1.3968	1329.	475.8	0.282	0.0279	0.0409	0.00953	4.77	75.00
80.00	191.05	57.63	0.2648	42.407	110.005	0.08748	0.21274	0.4028	0.3333	1.4218	1282.	471.4	0.272	0.0283	0.0403	0.00983	4.45	80.00
85.00	205.09	56.87	0.2444	44.413	110.365	0.09109	0.21218	0.4100	0.3441	1.4504	1235.	466.7	0.261	0.0287	0.0396	0.01015	4.13	85.00
90.00	219.89	56.08	0.2256	46.449	110.686	0.09473	0.21159	0.4180	0.3562	1.4831	1187.	461.6	0.251	0.0292	0.0390	0.01050	3.81	90.00
95.00	235.48	55.26	0.2082	48.519	110.965	0.09838	0.21096	0.4269	0.3698	1.5209	1138.	456.1	0.241	0.0297	0.0383	0.01089	3.50	95.00
100.00	251.89	54.41	0.1920	50.626	111.196	0.10206	0.21029	0.4371	0.3853	1.5650	1089.	450.3	0.232	0.0302	0.0376	0.01131	3.18	100.00
105.00	269.14	53.52	0.1770	52.773	111.372	0.10578	0.20955	0.4487	0.4033	1.6171	1039.	444.0	0.222	0.0308	0.0370	0.01179	2.87	105.00
110.00	287.29	52.59	0.1631	54.966	111.486	0.10953	0.20875	0.4622	0.4244	1.6795	988.	437.4	0.212	0.0314	0.0363	0.01233	2.56	110.00
115.00	306.35	51.61	0.1500	57.210	111.530	0.11334	0.20786	0.4782	0.4496	1.7553	937.	430.3	0.203	0.0320	0.0356	0.01295	2.26	115.00
120.00	326.37	50.56	0.1378	59.512	111.489	0.11720	0.20687	0.4974	0.4804	1.8494	884.	422.7	0.193	0.0328	0.0349	0.01365	1.96	120.00
125.00	347.38	49.45	0.1263	61.884	111.350	0.12114	0.20575	0.5211	0.5190	1.9689	829.	414.7	0.184	0.0336	0.0343	0.01448	1.68	125.00
130.00	369.44	48.25	0.1155	64.336	111.090	0.12517	0.20446	0.5514	0.5690	2.1254	773.	406.1	0.174	0.0346	0.0336	0.01546	1.40	130.00
135.00	392.60	46.95	0.1051	66.889	110.680	0.12933	0.20297	0.5918	0.6366	2.3388	715.	396.8	0.164	0.0357	0.0329	0.01665	1.13	135.00
140.00	416.90	45.50	0.0952	69.570	110.074	0.13366	0.20120	0.6487	0.7334	2.6465	655.	386.9	0.154	0.0370	0.0322	0.01814	0.87	140.00
145.00	442.43	43.86	0.0856	72.424	109.201	0.13822	0.19904	0.7365	0.8842	3.1275	591.	376.2	0.143	0.0386	0.0315	0.02009	0.63	145.00
150.00	469.27	41.93	0.0760	75.535	107.928	0.14315	0.19629	0.892	1.152	3.984	523.	364.5	0.132	0.0407	0.0309	0.02286	0.41	150.00
155.00	497.54	39.48	0.0660	79.091	105.965	0.14875	0.19247	1.251	1.765	5.933	450.	351.3	0.119	0.0438	0.0305	0.02750	0.21	155.00
160.00	527.46	35.70	0.0541	83.797	102.284	0.15614	0.18597	3.007	4.581	14.798	367.	334.6	0.101	0.0496	0.0319	0.03771	0.06	160.00
162.87c	545.49	26.91	0.0372	92.722	92.722	0.17033	0.17033	∞	∞	∞	0.	0.0	–	–	∞	∞	0.00	162.87

*Temperatures are on the ITS-90 scale a = triple point b = normal boiling point c = critical point

Refrigerant 245fa (1,1,1,3,3-Pentafluoropropane) Properties of Saturated Liquid and Saturated Vapor

Temp.,* °F	Pres-sure, psia	Denisty, lb/ft³ Liquid	Volume, ft³/lb Vapor	Enthalpy, Btu/lb Liquid	Enthalpy, Btu/lb Vapor	Entropy, Btu/lb·°F Liquid	Entropy, Btu/lb·°F Vapor	Specific Heat c_p, Btu/lb·°F Liquid	Specific Heat c_p, Btu/lb·°F Vapor	c_p/c_v Vapor	Vel. of Sound, ft/s Liquid	Vel. of Sound, ft/s Vapor	Viscosity, lb_m/ft·h Liquid	Viscosity, lb_m/ft·h Vapor	Thermal Cond., Btu/h·ft·°F Liquid	Thermal Cond., Btu/h·ft·°F Vapor	Surface Tension, dyne/cm	Temp., °F
−50.00	0.574	94.47	56.869	−2.904	95.409	−0.00700	0.23298	0.2888	0.1739	1.0960	3355.	406.4	3.679	0.0190	0.0609	0.00448	22.94	−50.00
−45.00	0.698	94.07	47.310	−1.456	96.260	−0.00349	0.23216	0.2904	0.1754	1.0955	3295.	408.5	3.396	0.0193	0.0603	0.00457	22.60	−45.00
−40.00	0.844	93.67	39.573	0.000	97.115	0.00000	0.23141	0.2919	0.1769	1.0951	3237.	410.6	3.148	0.0195	0.0597	0.00466	22.26	−40.00
−35.00	1.015	93.27	33.275	1.463	97.974	0.00346	0.23073	0.2932	0.1784	1.0947	3182.	412.6	2.929	0.0197	0.0591	0.00475	21.92	−35.00
−30.00	1.214	92.86	28.121	2.933	98.838	0.00690	0.23011	0.2945	0.1799	1.0944	3129.	414.6	2.736	0.0200	0.0585	0.00485	21.58	−30.00
−25.00	1.445	92.46	23.880	4.409	99.705	0.01032	0.22956	0.2957	0.1814	1.0942	3078.	416.5	2.563	0.0202	0.0579	0.00494	21.24	−25.00
−20.00	1.712	92.05	20.373	5.891	100.576	0.01371	0.22906	0.2969	0.1830	1.0940	3028.	418.4	2.407	0.0204	0.0573	0.00503	20.89	−20.00
−15.00	2.018	91.64	17.458	7.378	101.450	0.01707	0.22862	0.2980	0.1845	1.0938	2980.	420.3	2.267	0.0207	0.0567	0.00513	20.55	−15.00
−10.00	2.368	91.22	15.024	8.871	102.327	0.02041	0.22824	0.2991	0.1861	1.0937	2932.	422.1	2.140	0.0209	0.0561	0.00523	20.20	−10.00
−5.00	2.767	90.81	12.982	10.370	103.208	0.02372	0.22791	0.3002	0.1877	1.0937	2886.	423.8	2.024	0.0212	0.0555	0.00532	19.85	−5.00
0.00	3.220	90.39	11.262	11.874	104.091	0.02701	0.22762	0.3013	0.1893	1.0937	2841.	425.5	1.918	0.0214	0.0549	0.00542	19.50	0.00
5.00	3.732	89.97	9.8066	13.384	104.977	0.03027	0.22739	0.3024	0.1909	1.0938	2797.	427.1	1.820	0.0216	0.0544	0.00552	19.15	5.00
10.00	4.309	89.55	8.5702	14.899	105.865	0.03351	0.22719	0.3034	0.1926	1.0940	2753.	428.6	1.731	0.0219	0.0538	0.00562	18.80	10.00
15.00	4.956	89.12	7.5157	16.420	106.755	0.03673	0.22704	0.3045	0.1942	1.0942	2710.	430.1	1.648	0.0221	0.0532	0.00572	18.45	15.00
20.00	5.680	88.69	6.6129	17.946	107.648	0.03993	0.22693	0.3057	0.1959	1.0945	2668.	431.5	1.571	0.0223	0.0527	0.00582	18.10	20.00
25.00	6.488	88.26	5.8371	19.478	108.542	0.04310	0.22686	0.3068	0.1976	1.0948	2626.	432.9	1.499	0.0226	0.0521	0.00592	17.75	25.00
30.00	7.386	87.83	5.1682	21.016	109.437	0.04625	0.22683	0.3079	0.1994	1.0953	2584.	434.1	1.432	0.0228	0.0516	0.00602	17.39	30.00
35.00	8.381	87.39	4.5894	22.560	110.334	0.04939	0.22683	0.3091	0.2011	1.0958	2543.	435.3	1.370	0.0230	0.0510	0.00613	17.04	35.00
40.00	9.482	86.95	4.0870	24.109	111.232	0.05250	0.22686	0.3103	0.2029	1.0964	2503.	436.4	1.312	0.0233	0.0505	0.00623	16.68	40.00
45.00	10.694	86.50	3.6494	25.665	112.130	0.05559	0.22692	0.3115	0.2047	1.0970	2462.	437.5	1.257	0.0235	0.0500	0.00634	16.33	45.00
50.00	12.028	86.06	3.2672	27.227	113.030	0.05867	0.22701	0.3128	0.2065	1.0978	2422.	438.4	1.206	0.0237	0.0494	0.00644	15.97	50.00
55.00	13.490	85.61	2.9324	28.795	113.930	0.06172	0.22714	0.3140	0.2084	1.0986	2382.	439.3	1.157	0.0240	0.0489	0.00655	15.62	55.00
58.81b	14.696	85.26	2.7047	29.995	114.616	0.06404	0.22725	0.3150	0.2098	1.0993	2352.	439.9	1.122	0.0242	0.0485	0.00663	15.34	58.81
60.00	15.089	85.15	2.6382	30.370	114.829	0.06476	0.22728	0.3153	0.2103	1.0995	2342.	440.0	1.112	0.0242	0.0484	0.00666	15.26	60.00
65.00	16.834	84.69	2.3789	31.952	115.729	0.06778	0.22746	0.3167	0.2122	1.1005	2303.	440.7	1.069	0.0244	0.0479	0.00677	14.90	65.00
70.00	18.735	84.23	2.1499	33.540	116.629	0.07079	0.22766	0.3180	0.2142	1.1017	2263.	441.3	1.028	0.0247	0.0474	0.00688	14.55	70.00
75.00	20.800	83.76	1.9471	35.136	117.528	0.07378	0.22788	0.3194	0.2161	1.1029	2224.	441.8	0.989	0.0249	0.0469	0.00699	14.19	75.00
80.00	23.038	83.29	1.7670	36.738	118.427	0.07675	0.22812	0.3208	0.2182	1.1042	2185.	442.1	0.952	0.0252	0.0464	0.00710	13.83	80.00
85.00	25.461	82.82	1.6066	38.348	119.324	0.07971	0.22838	0.3223	0.2202	1.1056	2146.	442.4	0.917	0.0254	0.0459	0.00722	13.48	85.00
90.00	28.077	82.34	1.4636	39.965	120.220	0.08265	0.22866	0.3238	0.2223	1.1072	2107.	442.6	0.884	0.0256	0.0454	0.00733	13.12	90.00
95.00	30.897	81.85	1.3357	41.590	121.115	0.08559	0.22896	0.3253	0.2245	1.1089	2068.	442.6	0.852	0.0259	0.0449	0.00745	12.77	95.00
100.00	33.931	81.36	1.2210	43.223	122.008	0.08850	0.22927	0.3269	0.2266	1.1107	2030.	442.6	0.822	0.0261	0.0444	0.00756	12.41	100.00
105.00	37.189	80.87	1.1181	44.864	122.899	0.09141	0.22960	0.3285	0.2289	1.1127	1991.	442.4	0.793	0.0263	0.0439	0.00768	12.06	105.00
110.00	40.684	80.37	1.0254	46.513	123.787	0.09430	0.22995	0.3302	0.2311	1.1148	1952.	442.1	0.765	0.0266	0.0434	0.00780	11.70	110.00
115.00	44.425	79.86	0.9418	48.170	124.673	0.09718	0.23031	0.3319	0.2335	1.1171	1914.	441.7	0.739	0.0268	0.0429	0.00792	11.35	115.00
120.00	48.424	79.35	0.8663	49.837	125.556	0.10006	0.23068	0.3337	0.2358	1.1195	1875.	441.1	0.713	0.0271	0.0425	0.00804	11.00	120.00
125.00	52.692	78.83	0.7979	51.512	126.436	0.10292	0.23106	0.3355	0.2383	1.1222	1837.	440.4	0.689	0.0273	0.0420	0.00817	10.64	125.00
130.00	57.241	78.30	0.7359	53.196	127.312	0.10577	0.23146	0.3374	0.2408	1.1250	1799.	439.6	0.665	0.0276	0.0415	0.00829	10.29	130.00
135.00	62.083	77.77	0.6795	54.890	128.184	0.10861	0.23186	0.3393	0.2434	1.1281	1760.	438.7	0.643	0.0278	0.0411	0.00842	9.94	135.00
140.00	67.230	77.22	0.6282	56.593	129.051	0.11144	0.23227	0.3413	0.2460	1.1314	1722.	437.6	0.621	0.0281	0.0406	0.00855	9.59	140.00
145.00	72.693	76.67	0.5814	58.306	129.914	0.11426	0.23269	0.3434	0.2488	1.1350	1684.	436.3	0.600	0.0283	0.0401	0.00868	9.25	145.00
150.00	78.485	76.12	0.5387	60.030	130.771	0.11708	0.23311	0.3456	0.2516	1.1389	1645.	434.9	0.580	0.0286	0.0397	0.00881	8.90	150.00
155.00	84.620	75.55	0.4996	61.764	131.623	0.11989	0.23354	0.3478	0.2545	1.1430	1607.	433.4	0.560	0.0289	0.0392	0.00895	8.56	155.00
160.00	91.109	74.97	0.4638	63.510	132.468	0.12269	0.23397	0.3501	0.2575	1.1476	1568.	431.6	0.541	0.0292	0.0387	0.00908	8.21	160.00
165.00	97.965	74.39	0.4309	65.266	133.306	0.12548	0.23440	0.3526	0.2607	1.1525	1530.	429.7	0.523	0.0294	0.0383	0.00922	7.87	165.00
170.00	105.20	73.79	0.4007	67.035	134.136	0.12827	0.23484	0.3551	0.2640	1.1578	1492.	427.7	0.505	0.0297	0.0378	0.00937	7.53	170.00
175.00	112.83	73.18	0.3729	68.815	134.959	0.13106	0.23528	0.3578	0.2674	1.1637	1453.	425.4	0.488	0.0300	0.0374	0.00951	7.20	175.00
180.00	120.87	72.56	0.3473	70.609	135.772	0.13384	0.23571	0.3606	0.2710	1.1700	1415.	423.0	0.471	0.0303	0.0369	0.00966	6.86	180.00
185.00	129.34	71.93	0.3236	72.415	136.576	0.13662	0.23615	0.3636	0.2748	1.1770	1376.	420.4	0.455	0.0306	0.0365	0.00982	6.53	185.00
190.00	138.23	71.28	0.3017	74.236	137.369	0.13940	0.23658	0.3667	0.2788	1.1846	1338.	417.6	0.439	0.0309	0.0360	0.00998	6.20	190.00
195.00	147.58	70.62	0.2814	76.071	138.150	0.14218	0.23700	0.3701	0.2830	1.1930	1299.	414.5	0.424	0.0313	0.0356	0.01014	5.87	195.00
200.00	157.40	69.94	0.2627	77.921	138.918	0.14495	0.23742	0.3736	0.2875	1.2023	1260.	411.3	0.409	0.0316	0.0351	0.01031	5.55	200.00
205.00	167.69	69.25	0.2452	79.787	139.673	0.14773	0.23783	0.3774	0.2923	1.2126	1221.	407.9	0.394	0.0320	0.0347	0.01049	5.23	205.00
210.00	178.49	68.53	0.2290	81.670	140.412	0.15051	0.23823	0.3814	0.2975	1.2241	1182.	404.2	0.380	0.0324	0.0342	0.01067	4.91	210.00
215.00	189.79	67.80	0.2139	83.570	141.134	0.15329	0.23861	0.3858	0.3030	1.2370	1143.	400.2	0.366	0.0327	0.0337	0.01086	4.60	215.00
220.00	201.63	67.05	0.1998	85.490	141.837	0.15608	0.23898	0.3905	0.3091	1.2514	1103.	396.0	0.353	0.0332	0.0333	0.01106	4.29	220.00
225.00	214.01	66.27	0.1867	87.430	142.520	0.15887	0.23933	0.3957	0.3158	1.2678	1064.	391.6	0.339	0.0336	0.0328	0.01128	3.98	225.00
230.00	226.96	65.47	0.1744	89.391	143.179	0.16167	0.23966	0.4013	0.3231	1.2864	1024.	386.9	0.326	0.0340	0.0324	0.01150	3.68	230.00
235.00	240.49	64.64	0.1628	91.376	143.812	0.16448	0.23997	0.4076	0.3313	1.3078	984.	381.8	0.313	0.0345	0.0319	0.01174	3.39	235.00
240.00	254.63	63.78	0.1520	93.385	144.417	0.16731	0.24024	0.4146	0.3404	1.3325	943.	376.5	0.301	0.0351	0.0315	0.01199	3.10	240.00
245.00	269.38	62.88	0.1419	95.423	144.988	0.17015	0.24049	0.4225	0.3508	1.3613	902.	370.9	0.288	0.0356	0.0310	0.01227	2.81	245.00
250.00	284.79	61.95	0.1323	97.491	145.522	0.17301	0.24069	0.4315	0.3628	1.3954	861.	364.9	0.276	0.0362	0.0306	0.01257	2.53	250.00
260.00	317.62	59.94	0.1147	101.732	146.455	0.17880	0.24095	0.4541	0.3934	1.4854	776.	351.8	0.251	0.0376	0.0297	0.01325	1.99	260.00
270.00	353.32	57.68	0.0989	106.150	147.152	0.18474	0.24094	0.4867	0.4383	1.6232	689.	337.0	0.227	0.0392	0.0288	0.01411	1.48	270.00
280.00	392.15	55.08	0.0843	110.813	147.506	0.19092	0.24052	0.5392	0.5120	1.8571	597.	320.2	0.202	0.0413	0.0280	0.01523	1.01	280.00
290.00	434.39	51.91	0.0706	115.861	147.305	0.19750	0.23944	0.6414	0.6576	2.3306	500.	301.0	0.176	0.0443	0.0274	0.01685	0.59	290.00
300.00	480.51	47.56	0.0568	121.683	145.986	0.20498	0.23697	0.942	1.085	3.743	393.	278.9	0.147	0.0492	0.0280	0.01970	0.23	300.00
309.29c	527.94	32.28	0.0310	134.804	134.804	0.22184	0.22184	∞	∞	∞	0.	0.0	–	–	∞	∞	0.00	309.29

*Temperatures are on the ITS-90 scale b = normal boiling point c = critical point

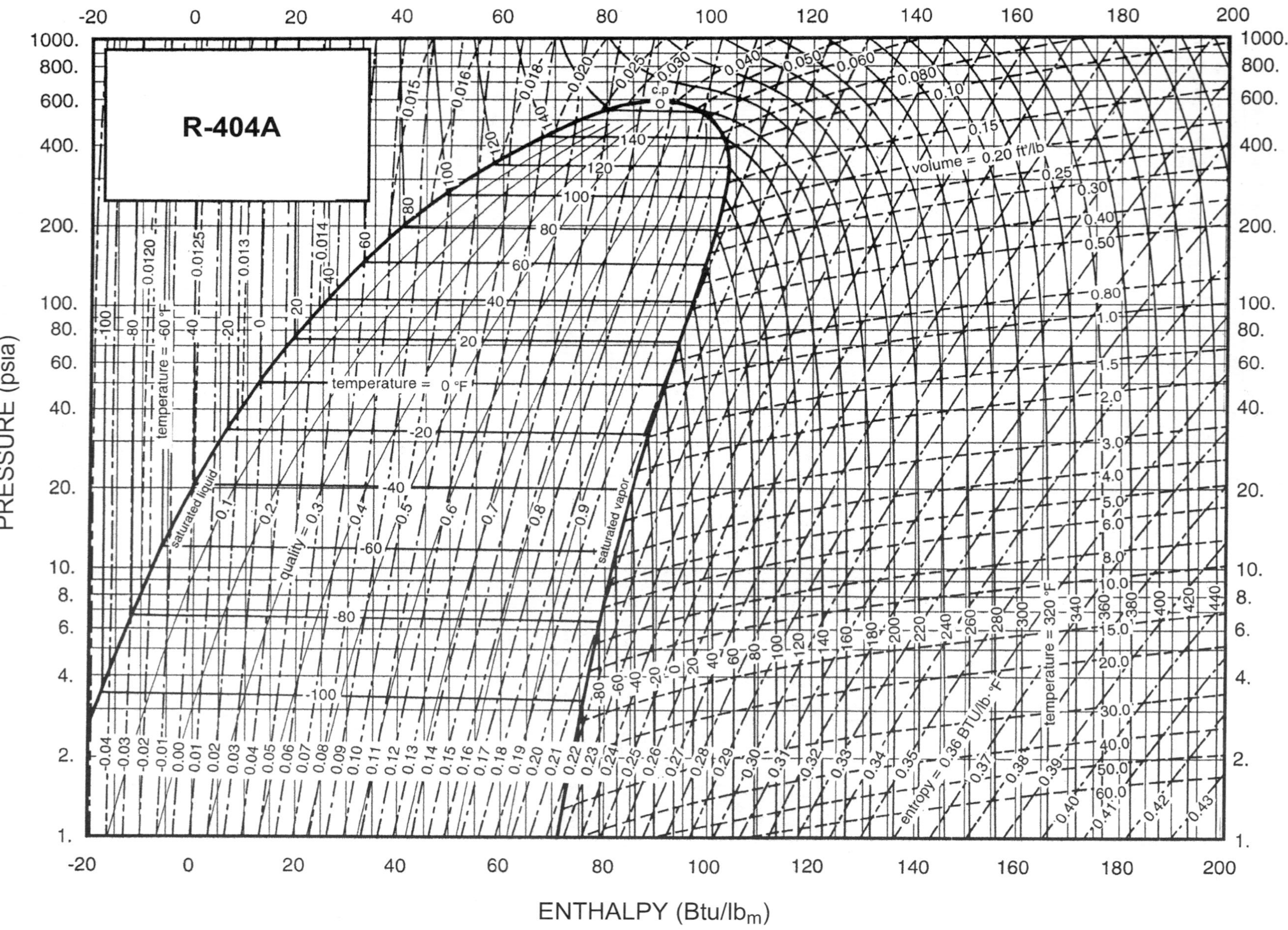

Fig. 10 Pressure-Enthalpy Diagram for Refrigerant 404A
Reprinted with permission from E.I. duPont de Nemours

Refrigerant 404A [R-125/143a/134a (44/52/4)] Properties of Liquid on the Bubble Line and Vapor on the Dew Line

Pressure, psia	Temp.*, °F Bubble	Temp.*, °F Dew	Denisty, lb/ft³ Liquid	Volume, ft³/lb Vapor	Enthalpy, Btu/lb Liquid	Enthalpy, Btu/lb Vapor	Entropy, Btu/lb·°F Liquid	Entropy, Btu/lb·°F Vapor	Specific Heat c_p, Btu/lb·°F Liquid	Specific Heat c_p, Btu/lb·°F Vapor	c_p/c_v Vapor	Vel. of Sound, ft/s Liquid	Vel. of Sound, ft/s Vapor	Viscosity, lb_m/ft·h Liquid	Viscosity, lb_m/ft·h Vapor	Thermal Cond., Btu/h·ft·°F Liquid	Thermal Cond., Btu/h·ft·°F Vapor	Surface Tension, dyne/cm	Pressure, psia
1.00	−130.39	−128.34	89.49	36.1508	−25.77	71.14	−0.06894	0.22484	0.2751	0.1549	1.161	3077.	439.4	1.6838	0.0178	0.0701	0.00360	17.40	1.00
1.50	−120.83	−118.88	88.56	24.7288	−23.13	72.51	−0.06105	0.22074	0.2766	0.1584	1.160	2987.	444.2	1.5074	0.0183	0.0684	0.00378	16.90	1.50
2.00	−113.63	−111.76	87.86	18.8932	−21.14	73.55	−0.05522	0.21796	0.2779	0.1610	1.159	2921.	447.7	1.3938	0.0187	0.0671	0.00391	16.51	2.00
2.50	−107.80	−105.98	87.29	15.3350	−19.51	74.39	−0.05057	0.21588	0.2790	0.1633	1.159	2867.	450.4	1.3116	0.0191	0.0661	0.00403	16.19	2.50
3.00	−102.86	−101.09	86.81	12.9318	−18.13	75.11	−0.04667	0.21424	0.2801	0.1652	1.158	2822.	452.6	1.2478	0.0193	0.0652	0.00412	15.92	3.00
4.00	−94.72	−93.02	86.00	9.8828	−15.84	76.30	−0.04033	0.21175	0.2819	0.1685	1.158	2749.	456.1	1.1529	0.0198	0.0638	0.00429	15.47	4.00
5.00	−88.09	−86.45	85.34	8.0221	−13.97	77.26	−0.03525	0.20991	0.2835	0.1712	1.159	2689.	458.7	1.0837	0.0202	0.0627	0.00442	15.09	5.00
6.00	−82.47	−80.87	84.78	6.7646	−12.37	78.08	−0.03099	0.20847	0.2849	0.1737	1.159	2639.	460.8	1.0299	0.0205	0.0618	0.00451	14.77	6.00
7.00	−77.56	−75.99	84.28	5.8562	−10.97	78.80	−0.02730	0.20729	0.2862	0.1758	1.160	2596.	462.5	0.9860	0.0207	0.0610	0.00462	14.48	7.00
8.00	−73.18	−71.65	83.84	5.1681	−9.71	79.43	−0.02403	0.20630	0.2874	0.1778	1.161	2557.	463.9	0.9493	0.0210	0.0603	0.00471	14.22	8.00
10.00	−65.59	−64.12	83.06	4.1930	−7.52	80.53	−0.01843	0.20472	0.2895	0.1813	1.162	2491.	466.2	0.8902	0.0214	0.0591	0.00488	13.77	10.00
12.00	−59.13	−57.70	82.39	3.5336	−5.64	81.47	−0.01371	0.20349	0.2914	0.1844	1.164	2434.	467.9	0.8440	0.0217	0.0581	0.00503	13.38	12.00
14.00	−53.48	−52.08	81.79	3.0571	−3.99	82.28	−0.00962	0.20250	0.2931	0.1872	1.166	2385.	469.3	0.8063	0.0220	0.0572	0.00516	13.04	14.00
14.70b	−51.66	−50.27	81.60	2.9207	−3.45	82.55	−0.00831	0.20220	0.2937	0.1881	1.166	2369.	469.7	0.7947	0.0221	0.0569	0.00520	12.93	14.70
16.00	−48.43	−47.06	81.26	2.6959	−2.50	83.01	−0.00599	0.20168	0.2947	0.1898	1.168	2341.	470.3	0.7745	0.0223	0.0564	0.00527	12.73	16.00
18.00	−43.85	−42.51	80.77	2.4125	−1.14	83.66	−0.00273	0.20098	0.2962	0.1922	1.169	2302.	471.2	0.7473	0.0225	0.0557	0.00538	12.45	18.00
20.00	−39.65	−38.34	80.32	2.1840	0.10	84.26	0.00025	0.20038	0.2977	0.1945	1.171	2265.	471.8	0.7234	0.0227	0.0551	0.00548	12.18	20.00
22.00	−35.77	−34.48	79.90	1.9956	1.26	84.81	0.00298	0.19985	0.2990	0.1966	1.173	2232.	472.3	0.7022	0.0230	0.0545	0.00558	11.94	22.00
24.00	−32.16	−30.89	79.50	1.8376	2.35	85.32	0.00552	0.19938	0.3003	0.1986	1.175	2201.	472.7	0.6831	0.0231	0.0540	0.00566	11.71	24.00
26.00	−28.77	−27.52	79.13	1.7031	3.37	85.79	0.00789	0.19896	0.3015	0.2006	1.177	2172.	473.1	0.6659	0.0233	0.0535	0.00575	11.50	26.00
28.00	−25.58	−24.35	78.78	1.5872	4.33	86.24	0.01011	0.19858	0.3027	0.2024	1.179	2144.	473.3	0.6502	0.0235	0.0530	0.00583	11.29	28.00
30.00	−22.56	−21.35	78.44	1.4862	5.25	86.66	0.01221	0.19824	0.3038	0.2042	1.181	2118.	473.4	0.6357	0.0237	0.0526	0.00590	11.10	30.00
32.00	−19.69	−18.49	78.12	1.3974	6.13	87.05	0.01419	0.19793	0.3049	0.2059	1.183	2093.	473.5	0.6224	0.0238	0.0521	0.00598	10.92	32.00
34.00	−16.96	−15.77	77.81	1.3186	6.96	87.43	0.01607	0.19764	0.3060	0.2076	1.185	2070.	473.6	0.6100	0.0240	0.0517	0.00605	10.74	34.00
36.00	−14.35	−13.18	77.51	1.2484	7.77	87.78	0.01787	0.19738	0.3071	0.2093	1.187	2048.	473.5	0.5984	0.0241	0.0514	0.00611	10.57	36.00
38.00	−11.85	−10.69	77.22	1.1852	8.54	88.12	0.01959	0.19713	0.3081	0.2108	1.189	2026.	473.5	0.5876	0.0243	0.0510	0.00617	10.41	38.00
40.00	−9.44	−8.30	76.94	1.1282	9.28	88.45	0.02123	0.19690	0.3091	0.2124	1.191	2005.	473.4	0.5774	0.0244	0.0507	0.00624	10.25	40.00
42.00	−7.13	−5.99	76.68	1.0764	10.00	88.76	0.02281	0.19669	0.3101	0.2139	1.193	1985.	473.3	0.5677	0.0244	0.0503	0.00638	10.10	42.00
44.00	−4.90	−3.77	76.41	1.0291	10.69	89.06	0.02433	0.19649	0.3110	0.2154	1.195	1966.	473.1	0.5586	0.0245	0.0500	0.00644	9.96	44.00
46.00	−2.74	−1.63	76.16	0.9858	11.36	89.35	0.02579	0.19630	0.3120	0.2168	1.197	1948.	472.9	0.5500	0.0247	0.0497	0.00650	9.82	46.00
48.00	−0.66	0.44	75.91	0.9460	12.02	89.62	0.02721	0.19612	0.3129	0.2183	1.199	1930.	472.7	0.5418	0.0248	0.0494	0.00656	9.68	48.00
50.00	1.36	2.45	75.67	0.9092	12.65	89.89	0.02858	0.19595	0.3138	0.2197	1.202	1912.	472.5	0.5340	0.0249	0.0491	0.00661	9.55	50.00
55.00	6.15	7.22	75.10	0.8286	14.16	90.52	0.03181	0.19557	0.3160	0.2231	1.207	1871.	471.8	0.5159	0.0252	0.0485	0.00675	9.23	55.00
60.00	10.62	11.67	74.55	0.7610	15.58	91.09	0.03482	0.19523	0.3182	0.2263	1.212	1833.	471.0	0.4996	0.0255	0.0478	0.00688	8.93	60.00
65.00	14.82	15.85	74.04	0.7035	16.92	91.63	0.03763	0.19493	0.3203	0.2295	1.217	1796.	470.2	0.4849	0.0257	0.0473	0.00700	8.66	65.00
70.00	18.78	19.79	73.54	0.6538	18.20	92.12	0.04028	0.19466	0.3223	0.2326	1.223	1762.	469.2	0.4714	0.0260	0.0467	0.00712	8.39	70.00
75.00	22.53	23.52	73.06	0.6106	19.41	92.59	0.04278	0.19441	0.3244	0.2357	1.228	1730.	468.2	0.4589	0.0262	0.0462	0.00723	8.14	75.00
80.00	26.09	27.07	72.61	0.5725	20.58	93.02	0.04516	0.19417	0.3264	0.2387	1.234	1699.	467.1	0.4473	0.0264	0.0457	0.00735	7.91	80.00
85.00	29.50	30.46	72.16	0.5388	21.69	93.43	0.04742	0.19396	0.3283	0.2416	1.239	1669.	466.0	0.4366	0.0267	0.0452	0.00746	7.68	85.00
90.00	32.75	33.70	71.73	0.5087	22.76	93.82	0.04958	0.19376	0.3303	0.2445	1.245	1641.	464.8	0.4265	0.0269	0.0448	0.00756	7.46	90.00
95.00	35.87	36.81	71.32	0.4816	23.80	94.18	0.05165	0.19357	0.3322	0.2474	1.251	1613.	463.6	0.4170	0.0271	0.0444	0.00767	7.25	95.00
100.00	38.87	39.79	70.91	0.4572	24.80	94.53	0.05364	0.19339	0.3341	0.2503	1.257	1587.	462.3	0.4081	0.0273	0.0440	0.00777	7.05	100.00
110.00	44.55	45.45	70.13	0.4147	26.71	95.16	0.05739	0.19305	0.3380	0.2559	1.269	1537.	459.7	0.3917	0.0276	0.0432	0.00798	6.67	110.00
120.00	49.86	50.73	69.39	0.3790	28.51	95.74	0.06089	0.19274	0.3418	0.2616	1.282	1490.	457.0	0.3768	0.0280	0.0425	0.00818	6.32	120.00
130.00	54.84	55.69	68.67	0.3487	30.22	96.26	0.06418	0.19245	0.3457	0.2673	1.296	1446.	454.2	0.3632	0.0284	0.0418	0.00838	5.99	130.00
140.00	59.55	60.37	67.97	0.3225	31.85	96.74	0.06728	0.19217	0.3496	0.2730	1.310	1403.	451.3	0.3508	0.0287	0.0412	0.00858	5.68	140.00
150.00	64.01	64.81	67.30	0.2996	33.41	97.17	0.07022	0.19189	0.3535	0.2789	1.324	1363.	448.3	0.3392	0.0291	0.0406	0.00877	5.38	150.00
160.00	68.25	69.04	66.64	0.2795	34.91	97.57	0.07302	0.19163	0.3576	0.2848	1.340	1325.	445.3	0.3285	0.0294	0.0400	0.00897	5.10	160.00
170.00	72.30	73.07	66.00	0.2616	36.36	97.93	0.07570	0.19136	0.3617	0.2909	1.356	1288.	442.2	0.3184	0.0297	0.0395	0.00917	4.84	170.00
180.00	76.18	76.93	65.37	0.2457	37.75	98.25	0.07826	0.19110	0.3659	0.2972	1.373	1252.	439.1	0.3090	0.0301	0.0390	0.00937	4.58	180.00
190.00	79.89	80.63	64.75	0.2313	39.11	98.55	0.08073	0.19083	0.3703	0.3036	1.391	1218.	436.0	0.3000	0.0304	0.0385	0.00958	4.34	190.00
200.00	83.47	84.18	64.14	0.2183	40.42	98.82	0.08311	0.19057	0.3748	0.3103	1.410	1185.	432.7	0.2916	0.0307	0.0380	0.00979	4.11	200.00
220.00	90.23	90.91	62.95	0.1957	42.95	99.29	0.08762	0.19002	0.3844	0.3246	1.452	1121.	426.2	0.2759	0.0314	0.0371	0.01022	3.68	220.00
240.00	96.54	97.20	61.78	0.1767	45.36	99.66	0.09187	0.18944	0.3948	0.3402	1.499	1060.	419.5	0.2615	0.0320	0.0362	0.01067	3.28	240.00
260.00	102.47	103.10	60.62	0.1604	47.67	99.94	0.09590	0.18883	0.4063	0.3577	1.553	1002.	412.7	0.2482	0.0327	0.0354	0.01116	2.91	260.00
280.00	108.07	108.67	59.46	0.1463	49.90	100.14	0.09974	0.18818	0.4192	0.3773	1.616	947.	405.8	0.2358	0.0334	0.0347	0.01168	2.57	280.00
300.00	113.37	113.94	58.30	0.1338	52.07	100.26	0.10342	0.18748	0.4338	0.3999	1.689	893.	398.7	0.2242	0.0342	0.0340	0.01225	2.26	300.00
320.00	118.40	118.94	57.13	0.1228	54.18	100.29	0.10698	0.18672	0.4506	0.4262	1.776	841.	391.6	0.2132	0.0349	0.0333	0.01287	1.97	320.00
340.00	123.20	123.71	55.94	0.1130	56.25	100.24	0.11043	0.18588	0.4703	0.4574	1.880	790.	384.3	0.2026	0.0358	0.0326	0.01356	1.70	340.00
360.00	127.78	128.27	54.72	0.1040	58.28	100.11	0.11379	0.18497	0.4939	0.4953	2.008	740.	376.9	0.1925	0.0367	0.0320	0.01433	1.44	360.00
380.00	132.17	132.62	53.46	0.0959	60.30	99.88	0.11709	0.18395	0.5229	0.5423	2.169	691.	369.3	0.1826	0.0376	0.0314	0.01519	1.21	380.00
400.00	136.37	136.80	52.15	0.0884	62.31	99.55	0.12036	0.18282	0.5597	0.6026	2.375	642.	361.7	0.1730	0.0387	0.0308	0.01618	0.99	400.00
450.00	146.19	146.54	48.49	0.0715	67.41	98.13	0.12855	0.17924	0.7205	0.8698	3.298	521.	341.7	0.1491	0.0421	0.0296	0.01951	0.53	450.00
500.00	155.11	155.36	43.59	0.0557	73.13	95.22	0.13758	0.17351	1.3043	1.8188	6.562	396.	319.5	0.1227	0.0478	0.0297	0.02605	0.17	500.00
547.45c	162.5	162.5	35.84	0.0279	80.5	80.5	0.1491	0.1491	–	–	–	–	–	–	–	–	–	0.00	547.45

*temperatures are on the ITS-90 scale b = bubble and dew points at one standard atmosphere c = critical point

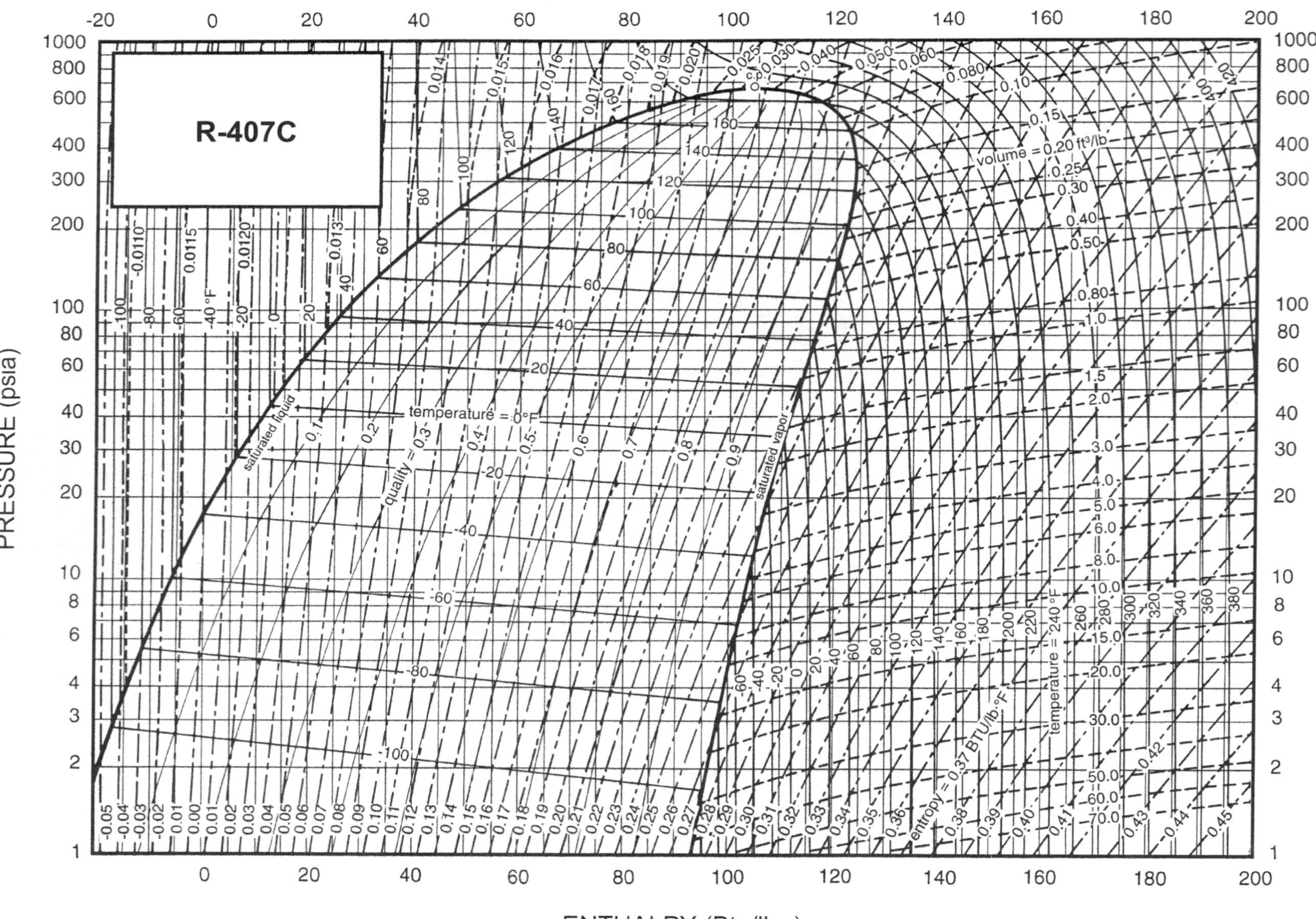

Fig. 11 Pressure-Enthalpy Diagram for Refrigerant 407C

Reprinted with permission from E.I. duPont de Nemours

Refrigerant 407C [R-32/125/134a (23/25/52)] Properties of Liquid on the Bubble Line and Vapor on the Dew Line

Pressure, psia	Temp.*, °F Bubble	Temp.*, °F Dew	Denisty, lb/ft³ Liquid	Volume, ft³/lb Vapor	Enthalpy, Btu/lb Liquid	Enthalpy, Btu/lb Vapor	Entropy, Btu/lb·°F Liquid	Entropy, Btu/lb·°F Vapor	Specific Heat c_p, Btu/lb·°F Liquid	Specific Heat c_p, Btu/lb·°F Vapor	c_p/c_v Vapor	Vel. of Sound, ft/s Liquid	Vel. of Sound, ft/s Vapor	Viscosity, lb_m/ft·h Liquid	Viscosity, lb_m/ft·h Vapor	Thermal Cond., Btu/h·ft·°F Liquid	Thermal Cond., Btu/h·ft·°F Vapor	Surface Tension, dyne/cm	Pressure, psia
1.00	−125.91	−111.58	94.30	43.0639	−26.04	93.36	−0.06927	0.28125	0.2970	0.1565	1.182	3366.	484.2	1.9459	0.0194	0.0873	0.00354	25.71	1.00
1.50	−116.25	−102.11	93.35	29.4288	−23.17	94.74	−0.06079	0.27589	0.2979	0.1596	1.181	3273.	489.4	1.7307	0.0199	0.0854	0.00373	24.72	1.50
2.00	−108.99	−95.00	92.63	22.4684	−21.00	95.79	−0.05455	0.27221	0.2987	0.1620	1.181	3204.	493.2	1.5942	0.0203	0.0841	0.00388	23.98	2.00
2.50	−103.11	−89.24	92.05	18.2270	−19.24	96.63	−0.04958	0.26941	0.2994	0.1641	1.180	3148.	496.1	1.4962	0.0207	0.0829	0.00400	23.39	2.50
3.00	−98.14	−84.36	91.56	15.3639	−17.75	97.34	−0.04543	0.26717	0.3000	0.1658	1.180	3101.	498.5	1.4209	0.0209	0.0820	0.00410	22.88	3.00
4.00	−89.95	−76.34	90.74	11.7335	−15.29	98.51	−0.03870	0.26373	0.3012	0.1688	1.180	3025.	502.3	1.3095	0.0214	0.0805	0.00427	22.06	4.00
5.00	−83.30	−69.82	90.06	9.5196	−13.28	99.46	−0.03332	0.26113	0.3022	0.1714	1.181	2964.	505.2	1.2290	0.0217	0.0793	0.00441	21.40	5.00
6.00	−77.65	−64.29	89.49	8.0243	−11.57	100.26	−0.02882	0.25905	0.3032	0.1737	1.182	2912.	507.5	1.1666	0.0221	0.0782	0.00453	20.84	6.00
7.00	−72.73	−59.47	88.99	6.9445	−10.07	100.96	−0.02493	0.25733	0.3041	0.1757	1.182	2867.	509.4	1.1161	0.0223	0.0773	0.00463	20.36	7.00
8.00	−68.34	−55.17	88.54	6.1270	−8.74	101.58	−0.02150	0.25587	0.3049	0.1776	1.183	2828.	511.1	1.0739	0.0226	0.0766	0.00473	19.93	8.00
10.00	−60.75	−47.73	87.75	4.9691	−6.41	102.64	−0.01563	0.25348	0.3064	0.1809	1.185	2759.	513.7	1.0063	0.0230	0.0752	0.00489	19.19	10.00
12.00	−54.29	−41.40	87.07	4.1866	−4.43	103.54	−0.01070	0.25158	0.3078	0.1839	1.187	2702.	515.7	0.9537	0.0233	0.0741	0.00503	18.56	12.00
14.00	−48.64	−35.86	86.47	3.6213	−2.68	104.32	−0.00644	0.25000	0.3091	0.1866	1.189	2651.	517.3	0.9110	0.0236	0.0731	0.00515	18.02	14.00
14.70b	−46.82	−34.08	86.28	3.4596	−2.12	104.56	−0.00508	0.24951	0.3095	0.1875	1.189	2635.	517.8	0.8978	0.0237	0.0728	0.00519	17.84	14.70
16.00	−43.59	−30.92	85.94	3.1932	−1.12	105.00	−0.00267	0.24867	0.3103	0.1890	1.191	2606.	518.6	0.8751	0.0239	0.0722	0.00526	17.53	16.00
18.00	−39.03	−26.45	85.45	2.8573	0.30	105.62	0.00072	0.24751	0.3114	0.1914	1.193	2566.	519.7	0.8443	0.0242	0.0714	0.00537	17.10	18.00
20.00	−34.85	−22.36	85.00	2.5866	1.61	106.19	0.00380	0.24649	0.3125	0.1935	1.194	2529.	520.6	0.8174	0.0244	0.0707	0.00546	16.70	20.00
22.00	−30.98	−18.57	84.57	2.3635	2.82	106.70	0.00663	0.24558	0.3135	0.1956	1.196	2495.	521.3	0.7936	0.0246	0.0700	0.00555	16.34	22.00
24.00	−27.38	−15.05	84.18	2.1764	3.95	107.18	0.00925	0.24476	0.3145	0.1975	1.198	2463.	521.9	0.7723	0.0248	0.0694	0.00563	16.00	24.00
26.00	−24.01	−11.74	83.81	2.0172	5.02	107.62	0.01169	0.24402	0.3155	0.1994	1.200	2434.	522.4	0.7530	0.0250	0.0688	0.00571	15.69	26.00
28.00	−20.83	−8.64	83.46	1.8800	6.02	108.04	0.01398	0.24333	0.3164	0.2012	1.202	2406.	522.9	0.7355	0.0252	0.0683	0.00578	15.39	28.00
30.00	−17.83	−5.70	83.12	1.7605	6.98	108.43	0.01613	0.24270	0.3173	0.2029	1.204	2379.	523.2	0.7193	0.0254	0.0678	0.00585	15.11	30.00
32.00	−14.98	−2.91	82.80	1.6555	7.88	108.79	0.01817	0.24212	0.3182	0.2045	1.206	2354.	523.5	0.7044	0.0255	0.0673	0.00592	14.85	32.00
34.00	−12.26	−0.25	82.49	1.5624	8.75	109.14	0.02011	0.24158	0.3190	0.2062	1.208	2331.	523.7	0.6906	0.0257	0.0669	0.00598	14.60	34.00
36.00	−9.67	2.29	82.20	1.4793	9.58	109.47	0.02195	0.24107	0.3198	0.2077	1.210	2308.	523.9	0.6778	0.0258	0.0664	0.00604	14.36	36.00
38.00	−7.18	4.72	81.92	1.4047	10.38	109.78	0.02371	0.24059	0.3207	0.2092	1.212	2286.	524.0	0.6657	0.0260	0.0660	0.00610	14.13	38.00
40.00	−4.79	7.06	81.64	1.3372	11.15	110.08	0.02539	0.24014	0.3215	0.2107	1.214	2265.	524.1	0.6544	0.0261	0.0656	0.00616	13.92	40.00
42.00	−2.50	9.30	81.38	1.2760	11.89	110.36	0.02700	0.23972	0.3223	0.2122	1.216	2245.	524.1	0.6438	0.0262	0.0652	0.00622	13.71	42.00
44.00	−0.28	11.47	81.12	1.2202	12.60	110.64	0.02856	0.23932	0.3230	0.2136	1.218	2226.	524.1	0.6337	0.0264	0.0648	0.00627	13.51	44.00
46.00	1.86	13.56	80.87	1.1690	13.30	110.90	0.03005	0.23893	0.3238	0.2150	1.220	2207.	524.1	0.6241	0.0265	0.0645	0.00632	13.31	46.00
48.00	3.92	15.58	80.63	1.1220	13.97	111.15	0.03149	0.23857	0.3245	0.2163	1.222	2189.	524.1	0.6151	0.0266	0.0641	0.00637	13.13	48.00
50.00	5.93	17.53	80.39	1.0786	14.62	111.39	0.03289	0.23822	0.3253	0.2177	1.224	2171.	524.0	0.6064	0.0267	0.0638	0.00642	12.95	50.00
55.00	10.68	22.18	79.83	0.9835	16.17	111.96	0.03618	0.23741	0.3271	0.2209	1.229	2130.	523.7	0.5865	0.0270	0.0630	0.00654	12.52	55.00
60.00	15.11	26.50	79.30	0.9037	17.63	112.48	0.03924	0.23668	0.3289	0.2240	1.234	2091.	523.3	0.5686	0.0273	0.0623	0.00666	12.13	60.00
65.00	19.27	30.56	78.79	0.8358	19.01	112.95	0.04210	0.23601	0.3306	0.2271	1.239	2054.	522.8	0.5524	0.0275	0.0616	0.00677	11.76	65.00
70.00	23.19	34.39	78.31	0.7772	20.31	113.40	0.04478	0.23540	0.3323	0.2300	1.244	2020.	522.3	0.5376	0.0277	0.0610	0.00687	11.42	70.00
75.00	26.90	38.01	77.85	0.7263	21.55	113.81	0.04731	0.23483	0.3340	0.2329	1.249	1987.	521.6	0.5239	0.0280	0.0603	0.00697	11.10	75.00
80.00	30.43	41.46	77.41	0.6814	22.73	114.19	0.04971	0.23429	0.3356	0.2357	1.255	1956.	520.9	0.5113	0.0282	0.0598	0.00707	10.79	80.00
85.00	33.80	44.74	76.98	0.6417	23.87	114.55	0.05200	0.23379	0.3372	0.2384	1.260	1927.	520.2	0.4996	0.0284	0.0592	0.00716	10.50	85.00
90.00	37.02	47.88	76.57	0.6062	24.96	114.89	0.05418	0.23332	0.3388	0.2411	1.265	1898.	519.3	0.4886	0.0286	0.0587	0.00725	10.22	90.00
95.00	40.11	50.89	76.17	0.5744	26.01	115.21	0.05626	0.23287	0.3404	0.2438	1.270	1871.	518.5	0.4783	0.0288	0.0582	0.00734	9.96	95.00
100.00	43.08	53.78	75.78	0.5456	27.02	115.51	0.05826	0.23245	0.3420	0.2464	1.276	1845.	517.6	0.4686	0.0290	0.0577	0.00743	9.71	100.00
110.00	48.70	59.24	75.03	0.4956	28.96	116.07	0.06203	0.23165	0.3452	0.2516	1.287	1795.	515.7	0.4508	0.0294	0.0568	0.00760	9.23	110.00
120.00	53.95	64.35	74.32	0.4537	30.78	116.56	0.06555	0.23092	0.3483	0.2567	1.298	1748.	513.8	0.4347	0.0297	0.0559	0.00776	8.80	120.00
130.00	58.87	69.13	73.64	0.4180	32.50	117.01	0.06884	0.23024	0.3515	0.2618	1.309	1704.	511.7	0.4201	0.0300	0.0551	0.00792	8.39	130.00
140.00	63.53	73.65	72.99	0.3872	34.14	117.41	0.07194	0.22960	0.3546	0.2669	1.321	1662.	509.5	0.4066	0.0304	0.0543	0.00808	8.01	140.00
150.00	67.94	77.93	72.36	0.3604	35.71	117.78	0.07488	0.22899	0.3578	0.2720	1.333	1623.	507.3	0.3943	0.0307	0.0536	0.00824	7.65	150.00
160.00	72.13	81.99	71.75	0.3368	37.21	118.11	0.07767	0.22841	0.3610	0.2771	1.346	1585.	505.0	0.3828	0.0310	0.0529	0.00840	7.32	160.00
170.00	76.14	85.87	71.16	0.3159	38.66	118.41	0.08034	0.22785	0.3643	0.2823	1.359	1548.	502.7	0.3720	0.0313	0.0522	0.00855	7.00	170.00
180.00	79.97	89.58	70.58	0.2972	40.06	118.68	0.08289	0.22731	0.3676	0.2875	1.373	1513.	500.4	0.3619	0.0316	0.0516	0.00871	6.69	180.00
190.00	83.65	93.13	70.01	0.2804	41.41	118.93	0.08533	0.22679	0.3709	0.2928	1.387	1479.	498.0	0.3525	0.0319	0.0510	0.00886	6.41	190.00
200.00	87.18	96.55	69.46	0.2653	42.72	119.15	0.08769	0.22628	0.3744	0.2982	1.401	1447.	495.5	0.3435	0.0322	0.0504	0.00902	6.13	200.00
220.00	93.88	103.00	68.38	0.2389	45.23	119.53	0.09215	0.22529	0.3815	0.3094	1.432	1385.	490.5	0.3269	0.0328	0.0493	0.00934	5.62	220.00
240.00	100.14	109.02	67.33	0.2167	47.61	119.83	0.09633	0.22433	0.3890	0.3211	1.465	1326.	485.5	0.3119	0.0334	0.0482	0.00968	5.15	240.00
260.00	106.02	114.67	66.31	0.1978	49.89	120.06	0.10028	0.22338	0.3969	0.3335	1.502	1270.	480.3	0.2981	0.0339	0.0472	0.01002	4.72	260.00
280.00	111.58	119.99	65.31	0.1815	52.08	120.22	0.10403	0.22245	0.4054	0.3468	1.542	1217.	475.0	0.2853	0.0345	0.0462	0.01039	4.31	280.00
300.00	116.85	125.03	64.32	0.1672	54.19	120.32	0.10761	0.22151	0.4145	0.3612	1.585	1165.	469.7	0.2734	0.0351	0.0453	0.01078	3.94	300.00
320.00	121.86	129.81	63.34	0.1547	56.23	120.36	0.11104	0.22057	0.4244	0.3768	1.634	1116.	464.3	0.2623	0.0357	0.0444	0.01119	3.59	320.00
340.00	126.65	134.36	62.37	0.1434	58.22	120.35	0.11434	0.21962	0.4352	0.3939	1.688	1068.	458.9	0.2517	0.0363	0.0436	0.01163	3.26	340.00
360.00	131.23	138.70	61.39	0.1334	60.16	120.28	0.11753	0.21864	0.4471	0.4128	1.749	1022.	453.3	0.2417	0.0369	0.0428	0.01211	2.95	360.00
380.00	135.63	142.86	60.42	0.1243	62.06	120.15	0.12063	0.21764	0.4604	0.4339	1.817	976.	447.8	0.2322	0.0376	0.0420	0.01263	2.66	380.00
400.00	139.85	146.84	59.43	0.1161	63.92	119.97	0.12365	0.21660	0.4753	0.4577	1.895	932.	442.1	0.2231	0.0382	0.0412	0.01319	2.39	400.00
450.00	149.77	156.11	56.90	0.0983	68.48	119.26	0.13093	0.21382	0.5226	0.5333	2.145	824.	427.8	0.2016	0.0401	0.0394	0.01484	1.78	450.00
500.00	158.90	164.54	54.21	0.0834	72.98	118.14	0.13799	0.21065	0.5938	0.6471	2.527	718.	413.0	0.1815	0.0423	0.0377	0.01699	1.25	500.00
550.00	167.37	172.23	51.20	0.0706	77.57	116.46	0.14507	0.20686	0.7163	0.8408	3.180	614.	397.6	0.1617	0.0451	0.0363	0.01997	0.79	550.00
600.00	175.31	179.23	47.59	0.0589	82.49	113.92	0.15257	0.20192	0.9852	1.2509	4.558	509.	381.2	0.1411	0.0490	0.0354	0.02465	0.41	600.00
672.20c	186.9	186.9	31.6	0.0317	98.5	98.5	0.1771	0.1771	–	–	–	–	–	–	–	–	–	0.00	672.20

*Temperatures are on the ITS-90 scale b = bubble and dew points at one standard atmosphere c = critical point

Refrigerant 410A [R-32/125 (50/50)] Properties of Liquid on the Bubble Line and Vapor on the Dew Line

Pressure, psia	Temp.*, °F Bubble	Temp.*, °F Dew	Denisty, lb/ft³ Liquid	Volume, ft³/lb Vapor	Enthalpy, Btu/lb Liquid	Enthalpy, Btu/lb Vapor	Entropy, Btu/lb·°F Liquid	Entropy, Btu/lb·°F Vapor	Specific Heat c_p, Btu/lb·°F Liquid	Specific Heat c_p, Btu/lb·°F Vapor	c_p/c_v Vapor	Vel. of Sound, ft/s Liquid	Vel. of Sound, ft/s Vapor	Viscosity, lb_m/ft·h Liquid	Viscosity, lb_m/ft·h Vapor	Thermal Cond., Btu/h·ft·°F Liquid	Thermal Cond., Btu/h·ft·°F Vapor	Surface Tension, dyne/cm	Pressure, psia
1.00	−135.75	−135.69	92.11	47.5564	−30.50	99.85	−0.08227	0.32010	0.3135	0.1564	1.228	3466.	518.2	1.4851	0.0193	0.0993	0.00369	24.83	1.00
1.50	−126.58	−126.52	91.19	32.5257	−27.62	101.14	−0.07352	0.31303	0.3137	0.1596	1.227	3373.	524.1	1.3443	0.0199	0.0973	0.00383	23.86	1.50
2.00	−119.70	−119.64	90.50	24.8468	−25.46	102.11	−0.06710	0.30811	0.3140	0.1622	1.227	3305.	528.3	1.2532	0.0203	0.0959	0.00393	23.15	2.00
2.50	−114.12	−114.06	89.93	20.1648	−23.71	102.89	−0.06199	0.30435	0.3144	0.1644	1.227	3250.	531.6	1.1869	0.0206	0.0948	0.00402	22.57	2.50
3.00	−109.40	−109.34	89.45	17.0030	−22.22	103.54	−0.05773	0.30131	0.3147	0.1664	1.227	3204.	534.3	1.1352	0.0209	0.0938	0.00410	22.08	3.00
4.00	−101.65	−101.58	88.66	12.9918	−19.78	104.61	−0.05083	0.29658	0.3155	0.1699	1.228	3130.	538.5	1.0580	0.0214	0.0923	0.00422	21.28	4.00
5.00	−95.34	−95.28	88.00	10.5444	−17.79	105.47	−0.04532	0.29297	0.3162	0.1730	1.229	3070.	541.7	1.0013	0.0218	0.0910	0.00433	20.64	5.00
6.00	−90.00	−89.93	87.45	8.8907	−16.09	106.19	−0.04071	0.29006	0.3169	0.1757	1.231	3019.	544.3	0.9569	0.0221	0.0899	0.00441	20.10	6.00
7.00	−85.33	−85.26	86.96	7.6962	−14.61	106.82	−0.03674	0.28762	0.3175	0.1782	1.232	2975.	546.5	0.9206	0.0224	0.0890	0.00449	19.63	7.00
8.00	−81.18	−81.11	86.52	6.7915	−13.29	107.37	−0.03323	0.28554	0.3182	0.1804	1.233	2937.	548.3	0.8901	0.0226	0.0882	0.00456	19.21	8.00
10.00	−74.00	−73.92	85.76	5.5099	−11.00	108.31	−0.02724	0.28208	0.3194	0.1845	1.236	2870.	551.3	0.8408	0.0231	0.0868	0.00469	18.49	10.00
12.00	−67.89	−67.81	85.10	4.6435	−9.04	109.10	−0.02222	0.27930	0.3206	0.1882	1.239	2814.	553.6	0.8019	0.0234	0.0857	0.00480	17.88	12.00
14.00	−62.54	−62.46	84.52	4.0174	−7.32	109.78	−0.01788	0.27698	0.3217	0.1916	1.241	2764.	555.4	0.7701	0.0238	0.0846	0.00489	17.36	14.00
14.70b	−60.83	−60.74	84.33	3.8382	−6.77	110.00	−0.01649	0.27625	0.3221	0.1926	1.242	2749.	556.0	0.7602	0.0239	0.0843	0.00492	17.19	14.70
16.00	−57.78	−57.69	84.00	3.5431	−5.79	110.38	−0.01404	0.27498	0.3228	0.1946	1.244	2721.	556.9	0.7432	0.0241	0.0837	0.00498	16.89	16.00
18.00	−53.46	−53.37	83.53	3.1709	−4.39	110.92	−0.01059	0.27324	0.3238	0.1975	1.247	2681.	558.2	0.7199	0.0243	0.0829	0.00506	16.47	18.00
20.00	−49.51	−49.42	83.09	2.8708	−3.11	111.40	−0.00746	0.27169	0.3248	0.2002	1.249	2645.	559.3	0.6995	0.0246	0.0822	0.00513	16.09	20.00
22.00	−45.86	−45.76	82.68	2.6236	−1.92	111.84	−0.00458	0.27030	0.3257	0.2027	1.252	2612.	560.1	0.6814	0.0248	0.0815	0.00520	15.73	22.00
24.00	−42.46	−42.36	82.30	2.4162	−0.81	112.25	−0.00192	0.26903	0.3267	0.2051	1.255	2581.	560.9	0.6650	0.0250	0.0808	0.00527	15.41	24.00
26.00	−39.28	−39.18	81.94	2.2397	0.24	112.63	0.00056	0.26788	0.3276	0.2074	1.257	2552.	561.5	0.6502	0.0252	0.0802	0.00533	15.10	26.00
28.00	−36.28	−36.18	81.60	2.0876	1.22	112.98	0.00288	0.26681	0.3284	0.2096	1.260	2525.	562.1	0.6366	0.0254	0.0797	0.00539	14.81	28.00
30.00	−33.45	−33.34	81.27	1.9551	2.15	113.31	0.00507	0.26583	0.3293	0.2117	1.262	2499.	562.5	0.6241	0.0255	0.0791	0.00544	14.55	30.00
32.00	−30.76	−30.65	80.96	1.8387	3.04	113.61	0.00714	0.26490	0.3302	0.2137	1.265	2475.	562.9	0.6125	0.0257	0.0786	0.00550	14.29	32.00
34.00	−28.20	−28.09	80.67	1.7354	3.89	113.91	0.00910	0.26404	0.3310	0.2157	1.267	2451.	563.2	0.6017	0.0259	0.0781	0.00555	14.05	34.00
36.00	−25.75	−25.64	80.38	1.6433	4.70	114.18	0.01097	0.26323	0.3318	0.2176	1.270	2429.	563.5	0.5916	0.0260	0.0777	0.00560	13.82	36.00
38.00	−23.41	−23.29	80.11	1.5605	5.48	114.44	0.01275	0.26247	0.3326	0.2194	1.272	2408.	563.7	0.5822	0.0262	0.0772	0.00565	13.60	38.00
40.00	−21.16	−21.04	79.84	1.4857	6.23	114.69	0.01445	0.26174	0.3334	0.2212	1.275	2387.	563.9	0.5733	0.0263	0.0768	0.00569	13.39	40.00
42.00	−19.00	−18.88	79.59	1.4178	6.96	114.92	0.01609	0.26106	0.3342	0.2230	1.278	2367.	564.0	0.5648	0.0264	0.0764	0.00574	13.19	42.00
44.00	−16.91	−16.79	79.34	1.3558	7.66	115.15	0.01766	0.26040	0.3349	0.2247	1.280	2348.	564.1	0.5569	0.0266	0.0760	0.00578	12.99	44.00
46.00	−14.90	−14.77	79.10	1.2991	8.33	115.36	0.01918	0.25978	0.3357	0.2263	1.283	2330.	564.2	0.5493	0.0267	0.0757	0.00583	12.81	46.00
48.00	−12.95	−12.82	78.87	1.2469	8.99	115.57	0.02064	0.25919	0.3365	0.2280	1.285	2312.	564.2	0.5421	0.0268	0.0753	0.00587	12.63	48.00
50.00	−11.07	−10.94	78.64	1.1988	9.63	115.77	0.02205	0.25861	0.3372	0.2296	1.288	2295.	564.2	0.5352	0.0269	0.0749	0.00591	12.45	50.00
55.00	−6.59	−6.46	78.10	1.0932	11.14	116.22	0.02538	0.25728	0.3390	0.2334	1.294	2254.	564.1	0.5193	0.0272	0.0741	0.00601	12.04	55.00
60.00	−2.42	−2.29	77.58	1.0047	12.56	116.64	0.02848	0.25607	0.3408	0.2372	1.300	2216.	563.9	0.5049	0.0275	0.0733	0.00610	11.66	60.00
65.00	1.49	1.63	77.10	0.9293	13.90	117.02	0.03137	0.25496	0.3426	0.2407	1.306	2180.	563.6	0.4919	0.0274	0.0726	0.00625	11.31	65.00
70.00	5.17	5.32	76.63	0.8644	15.17	117.38	0.03408	0.25392	0.3443	0.2442	1.313	2146.	563.1	0.4799	0.0277	0.0719	0.00635	10.97	70.00
75.00	8.66	8.81	76.19	0.8078	16.38	117.70	0.03665	0.25296	0.3460	0.2476	1.319	2114.	562.6	0.4688	0.0279	0.0712	0.00644	10.66	75.00
80.00	11.98	12.13	75.77	0.7580	17.53	118.00	0.03907	0.25206	0.3477	0.2509	1.325	2083.	562.0	0.4585	0.0281	0.0706	0.00652	10.37	80.00
85.00	15.14	15.30	75.36	0.7139	18.64	118.28	0.04138	0.25121	0.3494	0.2541	1.332	2054.	561.4	0.4489	0.0284	0.0700	0.00661	10.09	85.00
90.00	18.17	18.32	74.96	0.6746	19.70	118.54	0.04358	0.25040	0.3511	0.2573	1.338	2026.	560.7	0.4399	0.0286	0.0694	0.00669	9.82	90.00
95.00	21.06	21.22	74.58	0.6392	20.72	118.78	0.04569	0.24964	0.3527	0.2604	1.345	1999.	560.0	0.4315	0.0288	0.0689	0.00678	9.57	95.00
100.00	23.85	24.01	74.21	0.6073	21.70	119.01	0.04771	0.24891	0.3544	0.2634	1.351	1973.	559.2	0.4235	0.0290	0.0683	0.00686	9.32	100.00
110.00	29.12	29.28	73.49	0.5518	23.58	119.42	0.05152	0.24755	0.3577	0.2695	1.364	1923.	557.6	0.4087	0.0294	0.0673	0.00702	8.87	110.00
120.00	34.03	34.20	72.82	0.5052	25.35	119.78	0.05506	0.24630	0.3609	0.2755	1.378	1876.	555.8	0.3954	0.0297	0.0664	0.00718	8.45	120.00
130.00	38.65	38.82	72.17	0.4656	27.02	120.10	0.05838	0.24513	0.3642	0.2814	1.392	1832.	554.0	0.3832	0.0301	0.0655	0.00734	8.05	130.00
140.00	43.00	43.18	71.55	0.4314	28.61	120.38	0.06151	0.24403	0.3675	0.2873	1.406	1790.	552.0	0.3720	0.0305	0.0646	0.00750	7.69	140.00
150.00	47.13	47.31	70.95	0.4017	30.13	120.62	0.06447	0.24299	0.3708	0.2932	1.421	1751.	550.0	0.3617	0.0308	0.0638	0.00766	7.34	150.00
160.00	51.05	51.23	70.37	0.3755	31.59	120.84	0.06729	0.24201	0.3742	0.2992	1.436	1712.	548.0	0.3520	0.0311	0.0631	0.00782	7.02	160.00
170.00	54.79	54.98	69.80	0.3523	32.99	121.03	0.06997	0.24106	0.3776	0.3052	1.452	1675.	545.9	0.3430	0.0315	0.0623	0.00798	6.71	170.00
180.00	58.37	58.56	69.25	0.3315	34.35	121.19	0.07254	0.24016	0.3811	0.3112	1.468	1640.	543.7	0.3345	0.0318	0.0616	0.00815	6.42	180.00
190.00	61.80	61.99	68.72	0.3129	35.65	121.34	0.07500	0.23929	0.3846	0.3174	1.484	1606.	541.5	0.3265	0.0321	0.0609	0.00832	6.15	190.00
200.00	65.10	65.29	68.19	0.2961	36.92	121.46	0.07737	0.23844	0.3882	0.3237	1.502	1573.	539.3	0.3189	0.0324	0.0603	0.00850	5.88	200.00
220.00	71.34	71.54	67.18	0.2668	39.34	121.64	0.08186	0.23682	0.3956	0.3366	1.538	1509.	534.7	0.3049	0.0330	0.0590	0.00886	5.39	220.00
240.00	77.16	77.36	66.20	0.2423	41.65	121.76	0.08607	0.23528	0.4034	0.3503	1.577	1449.	530.0	0.2921	0.0336	0.0578	0.00925	4.94	240.00
260.00	82.63	82.83	65.24	0.2213	43.84	121.81	0.09004	0.23379	0.4117	0.3647	1.619	1391.	525.3	0.2803	0.0341	0.0567	0.00967	4.53	260.00
280.00	87.79	87.99	64.30	0.2032	45.95	121.81	0.09380	0.23234	0.4206	0.3801	1.665	1336.	520.4	0.2694	0.0347	0.0556	0.01011	4.14	280.00
300.00	92.68	92.88	63.38	0.1874	47.98	121.75	0.09739	0.23092	0.4301	0.3967	1.716	1283.	515.5	0.2592	0.0352	0.0545	0.01060	3.78	300.00
320.00	97.32	97.53	62.48	0.1735	49.95	121.64	0.10083	0.22952	0.4403	0.4147	1.771	1231.	510.5	0.2497	0.0358	0.0535	0.01112	3.45	320.00
340.00	101.75	101.95	61.58	0.1611	51.86	121.48	0.10413	0.22813	0.4515	0.4344	1.832	1181.	505.5	0.2406	0.0364	0.0526	0.01169	3.13	340.00
360.00	105.99	106.19	60.68	0.1500	53.72	121.27	0.10733	0.22674	0.4637	0.4560	1.900	1132.	500.4	0.2321	0.0369	0.0516	0.01231	2.84	360.00
380.00	110.05	110.24	59.78	0.1400	55.54	121.02	0.11042	0.22534	0.4773	0.4800	1.975	1085.	495.2	0.2239	0.0375	0.0507	0.01299	2.56	380.00
400.00	113.94	114.13	58.88	0.1309	57.32	120.71	0.11343	0.22393	0.4923	0.5069	2.061	1038.	490.0	0.2161	0.0381	0.0498	0.01374	2.30	400.00
450.00	123.06	123.24	56.60	0.1113	61.67	119.73	0.12068	0.22030	0.5390	0.5911	2.331	925.	476.7	0.1977	0.0398	0.0477	0.01593	1.72	450.00
500.00	131.41	131.58	54.21	0.0952	65.93	118.39	0.12765	0.21640	0.6062	0.7144	2.727	816.	462.9	0.1806	0.0416	0.0456	0.01877	1.21	500.00
550.00	139.12	139.27	51.61	0.0814	70.21	116.61	0.13455	0.21203	0.7141	0.9152	3.373	709.	448.3	0.1640	0.0439	0.0442	0.02256	0.78	550.00
600.00	146.28	146.40	48.65	0.0690	74.66	114.16	0.14164	0.20682	0.9238	1.3087	4.629	601.	432.5	0.1472	0.0468	0.0447	0.02804	0.42	600.00
694.87c	158.4	158.4	34.2	0.0293	90.2	90.2	0.1663	0.1663	–	–	–	–	–	–	–	–	–	0.00	694.87

*Temperatures are on the ITS-90 scale

b = bubble and dew points at one standard atmosphere

c = critical point

Refrigerant 507A [R-125/143a (50/50)] Properties of Saturated Liquid and Saturated Vapor

Temp.,* °F	Pres-sure**, psia	Denisty, lb/ft³ Liquid	Volume, ft³/lb Vapor	Enthalpy, Btu/lb Liquid	Enthalpy, Btu/lb Vapor	Entropy, Btu/lb·°F Liquid	Entropy, Btu/lb·°F Vapor	Specific Heat c_p, Btu/lb·°F Liquid	Specific Heat c_p, Btu/lb·°F Vapor	c_p/c_v Vapor	Vel. of Sound, ft/s Liquid	Vel. of Sound, ft/s Vapor	Viscosity, lb_m/ft·h Liquid	Viscosity, lb_m/ft·h Vapor	Thermal Cond., Btu/h·ft·°F Liquid	Thermal Cond., Btu/h·ft·°F Vapor	Surface Tension, dyne/cm	Temp., °F
–150.00	0.410	92.02	81.746	–30.900	66.427	–0.08508	0.22923	0.2709	0.1468	1.1649	3246.	424.1	2.131	0.0169	0.0730	0.00321	18.41	–150.00
–145.00	0.526	91.54	64.705	–29.545	67.130	–0.08074	0.22650	0.2713	0.1485	1.1636	3196.	427.0	1.985	0.0172	0.0720	0.00330	18.16	–145.00
–140.00	0.668	91.06	51.673	–28.187	67.837	–0.07646	0.22394	0.2717	0.1502	1.1625	3147.	429.8	1.856	0.0174	0.0710	0.00339	17.90	–140.00
–135.00	0.842	90.57	41.614	–26.827	68.548	–0.07224	0.22153	0.2723	0.1520	1.1615	3098.	432.5	1.742	0.0177	0.0701	0.00348	17.64	–135.00
–130.00	1.051	90.08	33.780	–25.464	69.262	–0.06807	0.21927	0.2730	0.1538	1.1606	3050.	435.2	1.639	0.0180	0.0692	0.00357	17.37	–130.00
–125.00	1.303	89.59	27.627	–24.097	69.979	–0.06396	0.21715	0.2737	0.1556	1.1598	3002.	437.8	1.546	0.0182	0.0682	0.00367	17.10	–125.00
–120.00	1.603	89.10	22.757	–22.726	70.699	–0.05989	0.21516	0.2745	0.1575	1.1591	2955.	440.3	1.461	0.0185	0.0673	0.00376	16.83	–120.00
–115.00	1.959	88.61	18.872	–21.351	71.421	–0.05588	0.21329	0.2754	0.1593	1.1586	2908.	442.7	1.384	0.0188	0.0664	0.00386	16.56	–115.00
–110.00	2.377	88.12	15.751	–19.971	72.145	–0.05191	0.21153	0.2763	0.1613	1.1582	2862.	445.1	1.314	0.0190	0.0656	0.00395	16.28	–110.00
–105.00	2.866	87.62	13.226	–18.586	72.871	–0.04798	0.20989	0.2774	0.1632	1.1580	2816.	447.4	1.249	0.0193	0.0647	0.00405	16.00	–105.00
–100.00	3.434	87.12	11.170	–17.196	73.599	–0.04409	0.20835	0.2784	0.1652	1.1579	2771.	449.6	1.190	0.0196	0.0638	0.00415	15.72	–100.00
–95.00	4.090	86.62	9.4858	–15.800	74.327	–0.04024	0.20691	0.2796	0.1673	1.1580	2725.	451.6	1.134	0.0198	0.0630	0.00425	15.43	–95.00
–90.00	4.846	86.11	8.0970	–14.398	75.055	–0.03642	0.20556	0.2807	0.1694	1.1583	2680.	453.6	1.083	0.0201	0.0622	0.00435	15.14	–90.00
–85.00	5.710	85.61	6.9457	–12.990	75.784	–0.03265	0.20429	0.2820	0.1715	1.1587	2636.	455.5	1.035	0.0204	0.0614	0.00446	14.85	–85.00
–80.00	6.695	85.10	5.9860	–11.576	76.512	–0.02890	0.20311	0.2833	0.1737	1.1593	2591.	457.3	0.991	0.0206	0.0605	0.00456	14.55	–80.00
–75.00	7.812	84.58	5.1819	–10.155	77.239	–0.02519	0.20200	0.2846	0.1760	1.1601	2547.	459.0	0.949	0.0209	0.0597	0.00467	14.25	–75.00
–70.00	9.072	84.06	4.5048	–8.727	77.965	–0.02151	0.20097	0.2860	0.1783	1.1611	2503.	460.6	0.910	0.0211	0.0590	0.00478	13.95	–70.00
–65.00	10.490	83.54	3.9319	–7.292	78.690	–0.01786	0.20000	0.2874	0.1807	1.1623	2459.	462.0	0.873	0.0215	0.0582	0.00489	13.65	–65.00
–60.00	12.078	83.01	3.4449	–5.850	79.412	–0.01423	0.19910	0.2889	0.1831	1.1637	2415.	463.3	0.838	0.0218	0.0574	0.00500	13.34	–60.00
–55.00	13.851	82.48	3.0292	–4.399	80.132	–0.01064	0.19825	0.2904	0.1856	1.1653	2372.	464.5	0.805	0.0220	0.0566	0.00511	13.03	–55.00
–52.80b	14.695	82.25	2.8653	–3.757	80.449	–0.00906	0.19790	0.2911	0.1868	1.1661	2353.	465.0	0.791	0.0221	0.0563	0.00516	12.89	–52.80
–50.00	15.822	81.95	2.6728	–2.941	80.849	–0.00707	0.19747	0.2920	0.1882	1.1671	2328.	465.6	0.774	0.0223	0.0559	0.00523	12.72	–50.00
–45.00	18.008	81.41	2.3661	–1.475	81.563	–0.00352	0.19673	0.2937	0.1908	1.1692	2285.	466.5	0.745	0.0225	0.0551	0.00535	12.41	–45.00
–40.00	20.423	80.86	2.1010	0.000	82.273	0.00000	0.19604	0.2953	0.1936	1.1716	2242.	467.3	0.717	0.0228	0.0544	0.00546	12.09	–40.00
–35.00	23.083	80.31	1.8712	1.484	82.979	0.00350	0.19540	0.2971	0.1963	1.1742	2199.	467.9	0.690	0.0230	0.0537	0.00558	11.77	–35.00
–30.00	26.005	79.75	1.6712	2.976	83.679	0.00698	0.19481	0.2989	0.1992	1.1771	2156.	468.4	0.665	0.0233	0.0529	0.00571	11.45	–30.00
–25.00	29.205	79.19	1.4965	4.478	84.375	0.01044	0.19425	0.3008	0.2022	1.1802	2113.	468.7	0.640	0.0235	0.0522	0.00583	11.13	–25.00
–20.00	32.701	78.62	1.3434	5.990	85.065	0.01387	0.19373	0.3027	0.2052	1.1837	2070.	468.9	0.617	0.0238	0.0515	0.00596	10.81	–20.00
–15.00	36.512	78.04	1.2089	7.511	85.749	0.01730	0.19324	0.3047	0.2083	1.1876	2027.	468.9	0.595	0.0241	0.0508	0.00609	10.48	–15.00
–10.00	40.653	77.46	1.0904	9.043	86.425	0.02070	0.19279	0.3067	0.2115	1.1918	1984.	468.7	0.574	0.0244	0.0501	0.00622	10.15	–10.00
–5.00	45.146	76.87	0.9855	10.585	87.095	0.02409	0.19237	0.3089	0.2149	1.1964	1941.	468.4	0.554	0.0244	0.0494	0.00642	9.82	–5.00
0.00	50.007	76.26	0.8926	12.138	87.756	0.02746	0.19197	0.3111	0.2183	1.2015	1898.	467.8	0.534	0.0247	0.0487	0.00656	9.49	0.00
5.00	55.257	75.66	0.8099	13.703	88.409	0.03081	0.19159	0.3135	0.2219	1.2070	1855.	467.1	0.515	0.0250	0.0480	0.00670	9.16	5.00
10.00	60.916	75.04	0.7362	15.279	89.053	0.03416	0.19124	0.3159	0.2255	1.2130	1811.	466.2	0.497	0.0253	0.0473	0.00685	8.83	10.00
15.00	67.002	74.41	0.6704	16.868	89.687	0.03749	0.19091	0.3184	0.2294	1.2195	1768.	465.1	0.480	0.0257	0.0466	0.00699	8.49	15.00
20.00	73.538	73.77	0.6114	18.469	90.310	0.04081	0.19059	0.3211	0.2334	1.2267	1725.	463.8	0.463	0.0260	0.0459	0.00714	8.16	20.00
25.00	80.544	73.12	0.5585	20.083	90.921	0.04412	0.19029	0.3238	0.2375	1.2345	1682.	462.3	0.447	0.0263	0.0452	0.00730	7.83	25.00
30.00	88.040	72.46	0.5108	21.711	91.519	0.04743	0.19000	0.3268	0.2419	1.2431	1638.	460.6	0.431	0.0266	0.0445	0.00746	7.49	30.00
35.00	96.049	71.78	0.4678	23.354	92.105	0.05072	0.18971	0.3299	0.2465	1.2526	1594.	458.7	0.416	0.0270	0.0439	0.00763	7.16	35.00
40.00	104.59	71.09	0.4289	25.011	92.675	0.05401	0.18944	0.3331	0.2514	1.2629	1550.	456.5	0.401	0.0273	0.0432	0.00781	6.82	40.00
45.00	113.69	70.39	0.3937	26.684	93.230	0.05730	0.18916	0.3366	0.2565	1.2744	1506.	454.1	0.386	0.0277	0.0425	0.00799	6.49	45.00
50.00	123.38	69.67	0.3617	28.374	93.768	0.06058	0.18889	0.3403	0.2620	1.2871	1462.	451.4	0.372	0.0281	0.0419	0.00818	6.15	50.00
55.00	133.66	68.93	0.3326	30.082	94.287	0.06386	0.18862	0.3443	0.2678	1.3011	1417.	448.5	0.359	0.0284	0.0412	0.00838	5.82	55.00
60.00	144.57	68.17	0.3061	31.808	94.786	0.06714	0.18834	0.3485	0.2741	1.3168	1372.	445.4	0.346	0.0288	0.0405	0.00859	5.49	60.00
65.00	156.14	67.39	0.2819	33.553	95.263	0.07042	0.18804	0.3532	0.2809	1.3343	1327.	441.9	0.333	0.0292	0.0399	0.00882	5.16	65.00
70.00	168.38	66.60	0.2597	35.319	95.716	0.07371	0.18774	0.3582	0.2882	1.3540	1282.	438.2	0.320	0.0297	0.0392	0.00906	4.83	70.00
75.00	181.33	65.77	0.2393	37.108	96.143	0.07700	0.18742	0.3636	0.2963	1.3763	1236.	434.2	0.308	0.0301	0.0385	0.00932	4.50	75.00
80.00	195.01	64.92	0.2207	38.921	96.540	0.08030	0.18708	0.3697	0.3052	1.4017	1189.	429.8	0.296	0.0306	0.0379	0.00959	4.18	80.00
85.00	209.45	64.04	0.2034	40.760	96.906	0.08362	0.18671	0.3764	0.3152	1.4307	1142.	425.2	0.284	0.0311	0.0372	0.00990	3.86	85.00
90.00	224.68	63.13	0.1876	42.627	97.235	0.08695	0.18630	0.3839	0.3264	1.4643	1095.	420.2	0.272	0.0316	0.0365	0.01023	3.54	90.00
95.00	240.72	62.18	0.1729	44.526	97.525	0.09030	0.18586	0.3923	0.3391	1.5034	1046.	414.8	0.261	0.0321	0.0358	0.01059	3.23	95.00
100.00	257.62	61.18	0.1593	46.459	97.768	0.09368	0.18536	0.4020	0.3538	1.5495	997.	409.1	0.249	0.0327	0.0352	0.01100	2.92	100.00
105.00	275.40	60.14	0.1466	48.430	97.961	0.09709	0.18481	0.4133	0.3710	1.6045	948.	403.0	0.238	0.0334	0.0345	0.01146	2.61	105.00
110.00	294.10	59.05	0.1348	50.444	98.093	0.10054	0.18419	0.4265	0.3915	1.6713	897.	396.4	0.227	0.0341	0.0338	0.01197	2.31	110.00
115.00	313.75	57.89	0.1238	52.508	98.157	0.10404	0.18348	0.4424	0.4163	1.7537	846.	389.4	0.216	0.0348	0.0331	0.01257	2.02	115.00
120.00	334.41	56.65	0.1135	54.630	98.138	0.10760	0.18266	0.4620	0.4473	1.8578	793.	382.0	0.205	0.0357	0.0324	0.01325	1.74	120.00
125.00	356.10	55.32	0.1037	56.819	98.020	0.11123	0.18171	0.4868	0.4870	1.9931	739.	374.0	0.194	0.0367	0.0318	0.01406	1.46	125.00
130.00	378.89	53.87	0.0945	59.093	97.779	0.11497	0.18059	0.5196	0.5402	2.1758	683.	365.5	0.183	0.0378	0.0311	0.01503	1.19	130.00
135.00	402.83	52.27	0.0856	61.472	97.381	0.11885	0.17924	0.5651	0.6151	2.4353	625.	356.3	0.171	0.0391	0.0304	0.01622	0.94	135.00
140.00	427.99	50.46	0.0771	63.992	96.769	0.12292	0.17758	0.6336	0.7290	2.8316	564.	346.4	0.159	0.0407	0.0298	0.01775	0.70	140.00
145.00	454.44	48.36	0.0687	66.715	95.848	0.12728	0.17546	0.7501	0.9238	3.5105	500.	335.6	0.146	0.0428	0.0293	0.01981	0.47	145.00
150.00	482.31	45.74	0.0600	69.777	94.416	0.13214	0.17256	0.998	1.335	4.939	432.	323.6	0.132	0.0457	0.0292	0.02292	0.27	150.00
159.34c	538.97	30.74	0.0325	82.615	82.615	0.15256	0.15256	∞	∞	∞	0.	0.0	–	–	∞	∞	0.00	159.34

*Temperatures are on the ITS-90 scale

**Small deviations from azeotropic behavior occur at some conditions; tabulated pressures are the average of the bubble and dew point pressures

b = normal boiling point

c = critical point

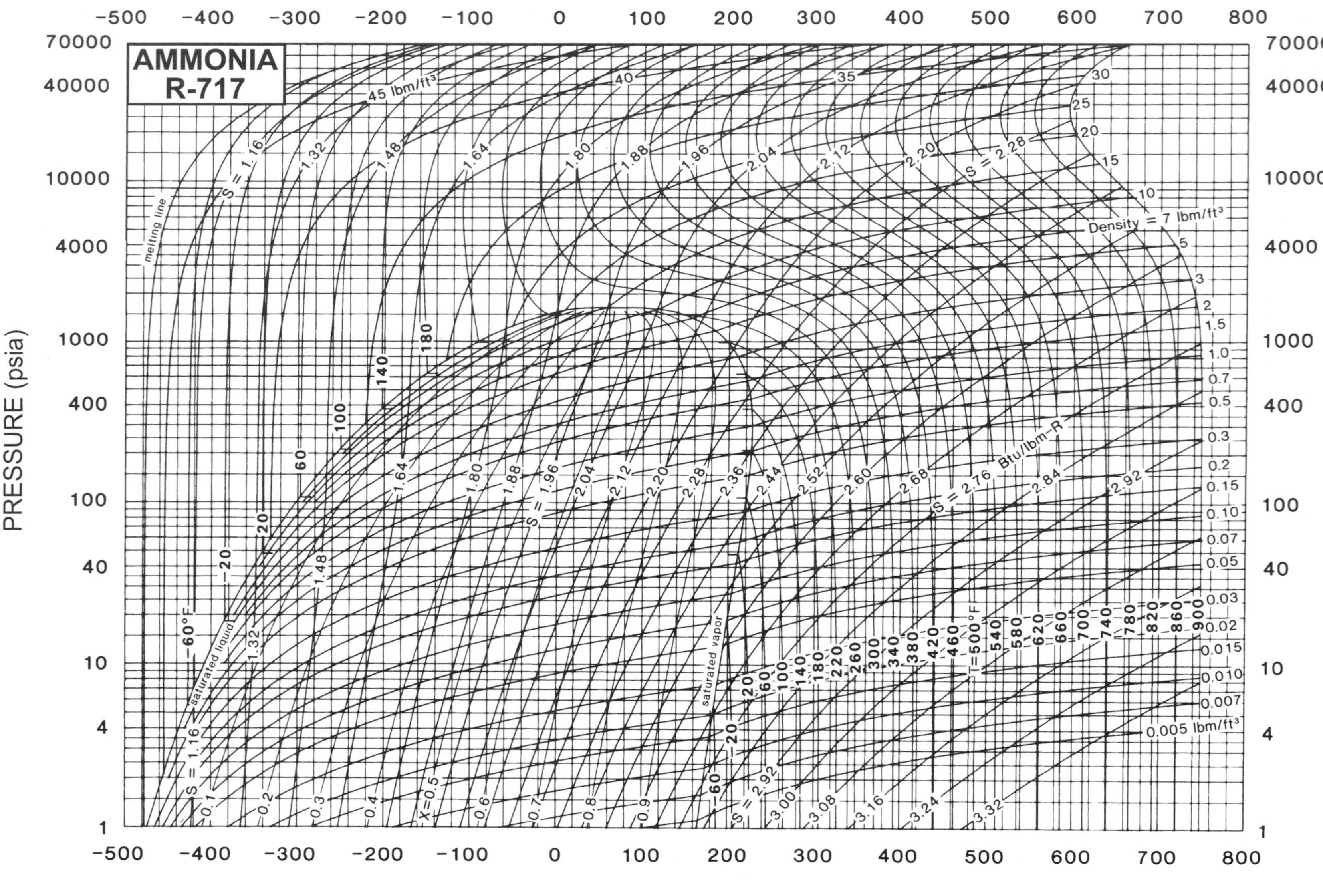

Fig. 12 Pressure-Enthalpy Diagram for Refrigerant 717 (Ammonia)

Note: The reference states for enthalpy and entropy differ from those in the table.

Refrigerant 717 (Ammonia) Properties of Saturated Liquid and Saturated Vapor

Temp.,* °F	Pres-sure, psia	Denisty, lb/ft³ Liquid	Volume, ft³/lb Vapor	Enthalpy, Btu/lb Liquid	Enthalpy, Btu/lb Vapor	Entropy, Btu/lb·°F Liquid	Entropy, Btu/lb·°F Vapor	Specific Heat c_p, Btu/lb·°F Liquid	Specific Heat c_p, Btu/lb·°F Vapor	c_p/c_v Vapor	Vel. of Sound, ft/s Liquid	Vel. of Sound, ft/s Vapor	Viscosity, lb_m/ft·h Liquid	Viscosity, lb_m/ft·h Vapor	Thermal Cond., Btu/h·ft·°F Liquid	Thermal Cond., Btu/h·ft·°F Vapor	Surface Tension, dyne/cm	Temp., °F
−107.78a	0.883	45.75	249.92	−69.830	568.765	−0.18124	1.63351	1.0044	0.4930	1.3252	6969.	1161.8	1.354	0.0165	0.4735	0.01135	62.26	−107.78
−100.00	1.237	45.47	182.19	−61.994	572.260	−0.15922	1.60421	1.0100	0.4959	1.3262	6830.	1173.8	1.232	0.0168	0.4647	0.01138	60.47	−100.00
−90.00	1.864	45.09	124.12	−51.854	576.688	−0.13142	1.56886	1.0176	0.5003	1.3278	6666.	1188.6	1.099	0.0171	0.4534	0.01143	58.19	−90.00
−80.00	2.739	44.71	86.546	−41.637	581.035	−0.10416	1.53587	1.0254	0.5056	1.3296	6513.	1202.9	0.986	0.0175	0.4422	0.01149	55.94	−80.00
−70.00	3.937	44.31	61.647	−31.341	585.288	−0.07741	1.50503	1.0331	0.5118	1.3319	6367.	1216.7	0.891	0.0179	0.4310	0.01158	53.73	−70.00
−60.00	5.544	43.91	44.774	−20.969	589.439	−0.05114	1.47614	1.0406	0.5190	1.3346	6228.	1229.7	0.810	0.0182	0.4198	0.01168	51.54	−60.00
−50.00	7.659	43.50	33.105	−10.521	593.476	−0.02534	1.44900	1.0478	0.5271	1.3379	6092.	1242.2	0.741	0.0186	0.4088	0.01180	49.39	−50.00
−40.00	10.398	43.08	24.881	0.000	597.387	0.00000	1.42347	1.0549	0.5364	1.3419	5959.	1253.9	0.680	0.0190	0.3978	0.01193	47.26	−40.00
−30.00	13.890	42.66	18.983	10.592	601.162	0.02491	1.39938	1.0617	0.5467	1.3465	5827.	1264.9	0.628	0.0194	0.3870	0.01209	45.17	−30.00
−27.99b	14.696	42.57	18.007	12.732	601.904	0.02987	1.39470	1.0631	0.5490	1.3475	5801.	1267.1	0.618	0.0195	0.3849	0.01212	44.75	−27.99
−25.00	15.962	42.45	16.668	15.914	602.995	0.03720	1.38784	1.0651	0.5524	1.3491	5762.	1270.2	0.604	0.0196	0.3817	0.01217	44.14	−25.00
−20.00	18.279	42.23	14.684	21.253	604.789	0.04939	1.37660	1.0684	0.5583	1.3520	5697.	1275.2	0.582	0.0198	0.3764	0.01226	43.11	−20.00
−15.00	20.858	42.01	12.976	26.609	606.544	0.06148	1.36567	1.0716	0.5646	1.3550	5632.	1280.0	0.561	0.0200	0.3711	0.01236	42.09	−15.00
−10.00	23.723	41.79	11.502	31.982	608.257	0.07347	1.35502	1.0749	0.5711	1.3584	5567.	1284.7	0.541	0.0202	0.3658	0.01246	41.08	−10.00
−5.00	26.895	41.57	10.226	37.372	609.928	0.08536	1.34463	1.0782	0.5781	1.3619	5503.	1289.1	0.522	0.0204	0.3606	0.01256	40.08	−5.00
0.00	30.397	41.34	9.1159	42.779	611.554	0.09715	1.33450	1.0814	0.5853	1.3657	5438.	1293.3	0.505	0.0206	0.3555	0.01267	39.08	0.00
5.00	34.253	41.12	8.1483	48.203	613.135	0.10885	1.32462	1.0847	0.5929	1.3698	5373.	1297.3	0.488	0.0208	0.3503	0.01279	38.10	5.00
10.00	38.487	40.89	7.3020	53.644	614.669	0.12045	1.31496	1.0880	0.6009	1.3742	5308.	1301.1	0.472	0.0210	0.3453	0.01291	37.12	10.00
15.00	43.126	40.66	6.5597	59.103	616.154	0.13197	1.30552	1.0914	0.6092	1.3789	5243.	1304.7	0.457	0.0212	0.3402	0.01304	36.15	15.00
20.00	48.194	40.43	5.9067	64.579	617.590	0.14340	1.29629	1.0948	0.6179	1.3840	5178.	1308.0	0.443	0.0214	0.3352	0.01317	35.19	20.00
25.00	53.720	40.20	5.3307	70.072	618.974	0.15474	1.28726	1.0983	0.6271	1.3894	5113.	1311.1	0.429	0.0216	0.3302	0.01331	34.23	25.00
30.00	59.730	39.96	4.8213	75.585	620.305	0.16599	1.27842	1.1019	0.6366	1.3951	5048.	1314.0	0.416	0.0218	0.3253	0.01345	33.29	30.00
35.00	66.255	39.72	4.3695	81.116	621.582	0.17717	1.26975	1.1056	0.6465	1.4012	4982.	1316.6	0.404	0.0220	0.3204	0.01360	32.35	35.00
40.00	73.322	39.48	3.9680	86.666	622.803	0.18827	1.26125	1.1094	0.6569	1.4078	4916.	1319.0	0.392	0.0222	0.3155	0.01376	31.42	40.00
45.00	80.962	39.24	3.6102	92.237	623.967	0.19929	1.25291	1.1134	0.6678	1.4147	4850.	1321.1	0.381	0.0224	0.3107	0.01392	30.50	45.00
50.00	89.205	38.99	3.2906	97.828	625.072	0.21024	1.24472	1.1175	0.6791	1.4222	4784.	1323.0	0.370	0.0227	0.3059	0.01409	29.59	50.00
55.00	98.083	38.75	3.0045	103.441	626.115	0.22111	1.23667	1.1218	0.6909	1.4301	4717.	1324.6	0.360	0.0229	0.3012	0.01426	28.69	55.00
60.00	107.63	38.50	2.7479	109.076	627.097	0.23192	1.22875	1.126	0.703	1.438	4650.	1325.9	0.350	0.0231	0.2965	0.01445	27.79	60.00
65.00	117.87	38.25	2.5172	114.734	628.013	0.24266	1.22095	1.131	0.716	1.447	4583.	1327.0	0.340	0.0233	0.2918	0.01464	26.90	65.00
70.00	128.85	37.99	2.3094	120.417	628.864	0.25334	1.21327	1.136	0.730	1.457	4515.	1327.8	0.331	0.0235	0.2872	0.01483	26.03	70.00
75.00	140.59	37.73	2.1217	126.126	629.647	0.26396	1.20570	1.141	0.744	1.467	4447.	1328.3	0.322	0.0237	0.2825	0.01504	25.16	75.00
80.00	153.13	37.47	1.9521	131.861	630.359	0.27452	1.19823	1.147	0.758	1.478	4378.	1328.6	0.313	0.0239	0.2780	0.01525	24.30	80.00
85.00	166.51	37.21	1.7983	137.624	630.999	0.28503	1.19085	1.153	0.774	1.490	4309.	1328.5	0.305	0.0241	0.2734	0.01548	23.44	85.00
90.00	180.76	36.94	1.6588	143.417	631.564	0.29549	1.18356	1.159	0.790	1.502	4240.	1328.2	0.297	0.0244	0.2689	0.01571	22.60	90.00
95.00	195.91	36.67	1.5319	149.241	632.052	0.30590	1.17634	1.166	0.807	1.515	4170.	1327.5	0.289	0.0246	0.2644	0.01595	21.77	95.00
100.00	212.01	36.40	1.4163	155.098	632.460	0.31626	1.16920	1.173	0.824	1.529	4099.	1326.6	0.282	0.0248	0.2600	0.01620	20.94	100.00
105.00	229.09	36.12	1.3108	160.990	632.785	0.32659	1.16211	1.180	0.843	1.544	4028.	1325.3	0.274	0.0250	0.2556	0.01646	20.13	105.00
110.00	247.19	35.83	1.2144	166.919	633.025	0.33688	1.15508	1.188	0.862	1.561	3956.	1323.7	0.267	0.0253	0.2512	0.01673	19.32	110.00
115.00	266.34	35.55	1.1262	172.887	633.175	0.34713	1.14809	1.197	0.883	1.578	3884.	1321.8	0.260	0.0255	0.2468	0.01702	18.53	115.00
120.00	286.60	35.26	1.0452	178.896	633.232	0.35736	1.14115	1.206	0.905	1.597	3811.	1319.5	0.254	0.0257	0.2424	0.01732	17.74	120.00
125.00	307.98	34.96	0.9710	184.949	633.193	0.36757	1.13423	1.216	0.928	1.617	3737.	1316.9	0.247	0.0260	0.2381	0.01763	16.96	125.00
130.00	330.54	34.66	0.9026	191.049	633.053	0.37775	1.12733	1.227	0.952	1.638	3662.	1313.9	0.241	0.0262	0.2338	0.01795	16.19	130.00
135.00	354.32	34.35	0.8397	197.199	632.807	0.38792	1.12044	1.239	0.978	1.662	3587.	1310.6	0.235	0.0265	0.2295	0.01829	15.44	135.00
140.00	379.36	34.04	0.7817	203.403	632.451	0.39808	1.11356	1.251	1.006	1.687	3511.	1306.9	0.229	0.0267	0.2253	0.01865	14.69	140.00
145.00	405.70	33.72	0.7280	209.663	631.978	0.40824	1.10666	1.265	1.035	1.715	3434.	1302.8	0.223	0.0270	0.2210	0.01903	13.95	145.00
150.00	433.38	33.39	0.6785	215.984	631.383	0.41840	1.09975	1.280	1.067	1.745	3356.	1298.3	0.217	0.0273	0.2168	0.01943	13.22	150.00
155.00	462.45	33.06	0.6325	222.370	630.659	0.42857	1.09281	1.296	1.101	1.778	3277.	1293.4	0.211	0.0276	0.2125	0.01986	12.51	155.00
160.00	492.95	32.72	0.5899	228.827	629.798	0.43875	1.08582	1.313	1.138	1.813	3198.	1288.1	0.206	0.0279	0.2083	0.02031	11.80	160.00
165.00	524.94	32.37	0.5504	235.359	628.791	0.44896	1.07878	1.333	1.178	1.853	3117.	1282.4	0.200	0.0282	0.2041	0.02079	11.10	165.00
170.00	558.45	32.01	0.5136	241.973	627.630	0.45919	1.07167	1.354	1.222	1.896	3035.	1276.2	0.195	0.0285	0.1999	0.02130	10.42	170.00
175.00	593.53	31.64	0.4793	248.675	626.302	0.46947	1.06447	1.377	1.270	1.944	2952.	1269.6	0.190	0.0288	0.1957	0.02185	9.75	175.00
180.00	630.24	31.26	0.4473	255.472	624.797	0.47980	1.05717	1.403	1.322	1.998	2868.	1262.4	0.185	0.0292	0.1916	0.02245	9.09	180.00
185.00	668.63	30.87	0.4174	262.374	623.100	0.49019	1.04974	1.432	1.381	2.058	2783.	1254.8	0.179	0.0296	0.1874	0.02310	8.44	185.00
190.00	708.74	30.47	0.3895	269.390	621.195	0.50066	1.04217	1.465	1.446	2.126	2696.	1246.7	0.174	0.0300	0.1832	0.02381	7.80	190.00
195.00	750.64	30.05	0.3633	276.530	619.064	0.51121	1.03443	1.502	1.519	2.203	2608.	1238.0	0.169	0.0304	0.1790	0.02458	7.18	195.00
200.00	794.38	29.62	0.3387	283.809	616.686	0.52188	1.02649	1.543	1.602	2.290	2519.	1228.7	0.165	0.0309	0.1748	0.02545	6.56	200.00
205.00	840.03	29.17	0.3156	291.240	614.035	0.53267	1.01831	1.591	1.697	2.392	2428.	1218.9	0.160	0.0314	0.1706	0.02641	5.97	205.00
210.00	887.64	28.70	0.2938	298.842	611.081	0.54360	1.00986	1.646	1.806	2.509	2336.	1208.4	0.155	0.0320	0.1663	0.02749	5.38	210.00
215.00	937.28	28.21	0.2733	306.637	607.788	0.55472	1.00109	1.711	1.935	2.648	2243.	1197.2	0.150	0.0326	0.1621	0.02872	4.81	215.00
220.00	989.03	27.69	0.2538	314.651	604.112	0.56605	0.99193	1.788	2.088	2.814	2147.	1185.4	0.145	0.0333	0.1578	0.03013	4.26	220.00
225.00	1042.96	27.15	0.2354	322.918	599.996	0.57763	0.98232	1.882	2.272	3.015	2050.	1172.7	0.140	0.0340	0.1536	0.03178	3.72	225.00
230.00	1099.14	26.57	0.2178	331.483	595.371	0.58953	0.97216	1.999	2.501	3.265	1950.	1159.1	0.136	0.0349	0.1492	0.03372	3.20	230.00
235.00	1157.69	25.95	0.2010	340.404	590.142	0.60182	0.96133	2.148	2.790	3.582	1848.	1144.5	0.131	0.0359	0.1449	0.03607	2.70	235.00
240.00	1218.68	25.28	0.1849	349.766	584.183	0.61462	0.94966	2.346	3.171	4.000	1743.	1128.8	0.126	0.0370	0.1406	0.03895	2.22	240.00
245.00	1282.24	24.55	0.1693	359.695	577.309	0.62809	0.93690	2.624	3.693	4.575	1634.	1111.6	0.120	0.0383	0.1363	0.04261	1.76	245.00
250.00	1348.49	23.72	0.1540	370.391	569.240	0.64249	0.92269	3.047	4.460	5.420	1520.	1092.6	0.115	0.0399	0.1320	0.04744	1.33	250.00
260.00	1489.71	21.60	0.1233	395.943	547.139	0.67662	0.88671	5.273	8.106	9.439	1271.	1045.9	0.102	0.0446	0.1250	0.06473	0.55	260.00
270.05c	1643.71	14.05	0.0712	473.253	473.253	0.78093	0.78093	∞	∞	∞	0.	0.0	–	–	∞	∞	0.00	270.05

*Temperatures are on the ITS-90 scale a = triple point b = normal boiling point c = critical point

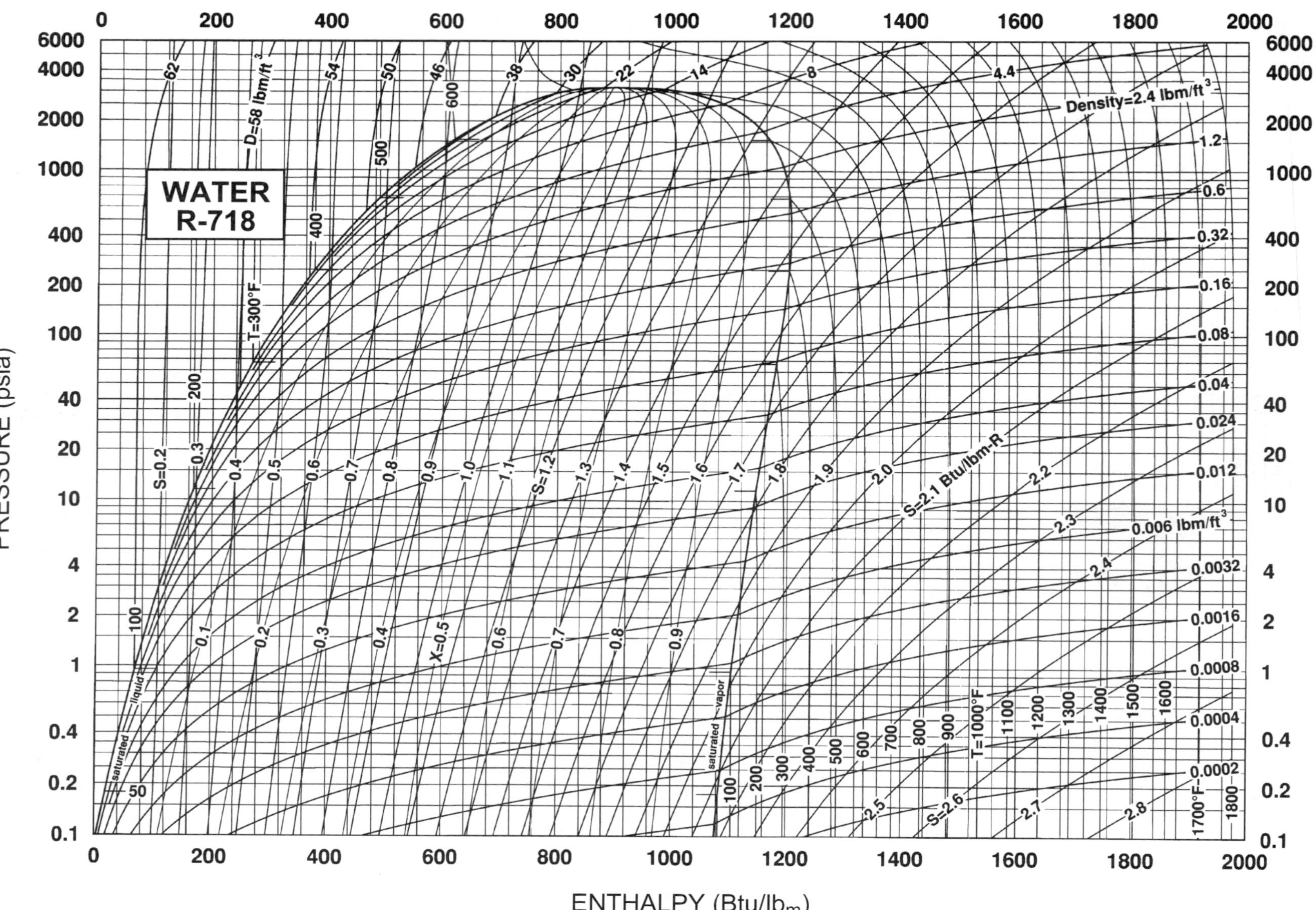

Fig. 13 Pressure-Enthalpy Diagram for Refrigerant 718 (Water/Steam)

Refrigerant 718 (Water/Steam) Properties of Saturated Liquid and Saturated Vapor

Temp.,* °F	Pres-sure, psia	Denisty, lb/ft³ Liquid	Volume, ft³/lb Vapor	Enthalpy, Btu/lb Liquid	Enthalpy, Btu/lb Vapor	Entropy, Btu/lb·°F Liquid	Entropy, Btu/lb·°F Vapor	Specific Heat c_p, Btu/lb·°F Liquid	Specific Heat c_p, Btu/lb·°F Vapor	c_p/c_v Vapor	Vel. of Sound, ft/s Liquid	Vel. of Sound, ft/s Vapor	Viscosity, lb_m/ft·h Liquid	Viscosity, lb_m/ft·h Vapor	Thermal Cond., Btu/h·ft·°F Liquid	Thermal Cond., Btu/h·ft·°F Vapor	Surface Tension, dyne/cm	Temp., °F
32.02a	0.089	62.42	3299.7	0.00	1075.92	0.0000	2.1882	1.0086	0.4504	1.3285	4601.	1342.	4.333	0.0223	0.3244	0.00987	75.65	32.02
40.00	0.122	62.42	2443.3	8.04	1079.42	0.0162	2.1604	1.0055	0.4514	1.3282	4670.	1352.	3.738	0.0226	0.3293	0.01001	75.02	40.00
50.00	0.178	62.41	1702.8	18.08	1083.79	0.0361	2.1271	1.0028	0.4528	1.3278	4748.	1365.	3.159	0.0229	0.3353	0.01019	74.22	50.00
60.00	0.256	62.36	1206.0	28.10	1088.15	0.0556	2.0954	1.0010	0.4543	1.3275	4815.	1378.	2.712	0.0232	0.3413	0.01038	73.40	60.00
70.00	0.363	62.30	867.11	38.10	1092.50	0.0746	2.0653	0.9999	0.4558	1.3273	4874.	1391.	2.359	0.0236	0.3471	0.01058	72.57	70.00
80.00	0.507	62.21	632.38	48.10	1096.83	0.0933	2.0366	0.9993	0.4574	1.3272	4924.	1403.	2.074	0.0240	0.3527	0.01079	71.71	80.00
90.00	0.699	62.11	467.40	58.09	1101.15	0.1117	2.0093	0.9990	0.4591	1.3271	4967.	1416.	1.841	0.0244	0.3579	0.01101	70.84	90.00
100.00	0.951	61.99	349.84	68.08	1105.44	0.1297	1.9832	0.9989	0.4609	1.3272	5003.	1428.	1.648	0.0248	0.3628	0.01124	69.96	100.00
110.00	1.277	61.86	264.97	78.07	1109.71	0.1474	1.9583	0.9991	0.4628	1.3273	5033.	1440.	1.486	0.0252	0.3672	0.01148	69.05	110.00
120.00	1.695	61.71	202.95	88.06	1113.95	0.1648	1.9346	0.9993	0.4648	1.3276	5056.	1452.	1.348	0.0256	0.3713	0.01172	68.13	120.00
130.00	2.226	61.55	157.09	98.06	1118.17	0.1819	1.9118	0.9997	0.4671	1.3280	5075.	1463.	1.230	0.0260	0.3750	0.01198	67.19	130.00
140.00	2.893	61.38	122.82	108.06	1122.35	0.1987	1.8901	1.0003	0.4696	1.3285	5088.	1475.	1.128	0.0265	0.3783	0.01225	66.24	140.00
150.00	3.723	61.19	96.930	118.07	1126.49	0.2152	1.8693	1.0009	0.4723	1.3291	5097.	1486.	1.040	0.0269	0.3813	0.01253	65.27	150.00
160.00	4.747	61.00	77.184	128.08	1130.59	0.2315	1.8493	1.0016	0.4753	1.3299	5101.	1497.	0.962	0.0273	0.3839	0.01282	64.28	160.00
170.00	6.000	60.79	61.980	138.11	1134.65	0.2476	1.8302	1.0025	0.4787	1.3309	5101.	1508.	0.894	0.0278	0.3862	0.01312	63.28	170.00
180.00	7.520	60.58	50.169	148.14	1138.65	0.2634	1.8118	1.0035	0.4824	1.3320	5098.	1518.	0.834	0.0282	0.3881	0.01343	62.26	180.00
190.00	9.350	60.35	40.916	158.19	1142.60	0.2789	1.7942	1.0046	0.4865	1.3333	5090.	1528.	0.780	0.0287	0.3898	0.01375	61.23	190.00
200.00	11.538	60.12	33.609	168.24	1146.48	0.2943	1.7772	1.0059	0.4911	1.3348	5080.	1538.	0.732	0.0291	0.3912	0.01409	60.19	200.00
210.00	14.136	59.88	27.794	178.31	1150.30	0.3094	1.7609	1.0073	0.4961	1.3366	5066.	1547.	0.690	0.0296	0.3924	0.01444	59.13	210.00
211.95b	14.696	59.83	26.802	180.28	1151.04	0.3124	1.7578	1.0076	0.4971	1.3369	5063.	1549.	0.682	0.0297	0.3926	0.01451	58.92	211.95
220.00	17.201	59.63	23.133	188.40	1154.05	0.3244	1.7451	1.0088	0.5016	1.3386	5049.	1557.	0.651	0.0300	0.3934	0.01480	58.05	220.00
230.00	20.795	59.37	19.371	198.51	1157.72	0.3391	1.7299	1.0106	0.5077	1.3408	5029.	1565.	0.616	0.0305	0.3941	0.01517	56.96	230.00
240.00	24.986	59.10	16.314	208.63	1161.31	0.3537	1.7153	1.0125	0.5145	1.3434	5007.	1574.	0.585	0.0310	0.3947	0.01556	55.86	240.00
250.00	29.844	58.82	13.815	218.78	1164.81	0.3680	1.7011	1.0147	0.5218	1.3464	4981.	1582.	0.556	0.0314	0.3951	0.01596	54.74	250.00
260.00	35.447	58.53	11.759	228.95	1168.21	0.3823	1.6874	1.0170	0.5299	1.3496	4953.	1590.	0.530	0.0319	0.3953	0.01638	53.62	260.00
270.00	41.878	58.24	10.058	239.14	1171.52	0.3963	1.6741	1.0196	0.5387	1.3533	4923.	1597.	0.506	0.0324	0.3953	0.01680	52.47	270.00
280.00	49.222	57.94	8.6431	249.37	1174.71	0.4102	1.6612	1.0224	0.5483	1.3574	4890.	1604.	0.484	0.0328	0.3952	0.01725	51.32	280.00
290.00	57.574	57.63	7.4600	259.62	1177.79	0.4239	1.6487	1.0254	0.5586	1.3620	4855.	1611.	0.464	0.0333	0.3949	0.01770	50.16	290.00
300.00	67.029	57.31	6.4658	269.91	1180.75	0.4375	1.6365	1.0287	0.5698	1.3671	4817.	1617.	0.445	0.0338	0.3944	0.01817	48.98	300.00
310.00	77.691	56.99	5.6263	280.23	1183.58	0.4510	1.6246	1.0323	0.5818	1.3727	4777.	1623.	0.428	0.0342	0.3939	0.01866	47.79	310.00
320.00	89.667	56.65	4.9142	290.60	1186.28	0.4643	1.6131	1.0362	0.5947	1.3790	4735.	1628.	0.412	0.0347	0.3931	0.01916	46.59	320.00
330.00	103.07	56.31	4.3075	301.00	1188.84	0.4775	1.6018	1.0404	0.6085	1.3858	4691.	1633.	0.397	0.0351	0.3923	0.01967	45.38	330.00
340.00	118.02	55.95	3.7884	311.45	1191.25	0.4906	1.5908	1.0449	0.6231	1.3934	4644.	1638.	0.383	0.0356	0.3912	0.02020	44.16	340.00
350.00	134.63	55.59	3.3425	321.95	1193.51	0.5036	1.5800	1.0497	0.6386	1.4018	4596.	1642.	0.370	0.0361	0.3901	0.02074	42.93	350.00
360.00	153.03	55.22	2.9580	332.50	1195.61	0.5164	1.5694	1.0550	0.6551	1.4109	4546.	1645.	0.358	0.0365	0.3888	0.02130	41.69	360.00
370.00	173.36	54.85	2.6252	343.11	1197.54	0.5292	1.5591	1.0606	0.6725	1.4210	4493.	1648.	0.347	0.0370	0.3873	0.02187	40.45	370.00
380.00	195.74	54.46	2.3361	353.77	1199.29	0.5419	1.5489	1.0666	0.6910	1.4320	4438.	1651.	0.337	0.0375	0.3857	0.02246	39.19	380.00
390.00	220.33	54.06	2.0841	364.50	1200.86	0.5545	1.5388	1.0732	0.7105	1.4441	4382.	1653.	0.327	0.0379	0.3839	0.02307	37.93	390.00
400.00	247.26	53.65	1.8638	375.30	1202.24	0.5670	1.5290	1.0802	0.7311	1.4573	4323.	1655.	0.318	0.0384	0.3819	0.02369	36.66	400.00
410.00	276.68	53.23	1.6706	386.17	1203.42	0.5795	1.5192	1.0878	0.7529	1.4718	4263.	1656.	0.309	0.0389	0.3798	0.02433	35.38	410.00
420.00	308.76	52.80	1.5006	397.12	1204.38	0.5919	1.5096	1.0959	0.7761	1.4876	4200.	1656.	0.300	0.0393	0.3776	0.02499	34.10	420.00
430.00	343.64	52.36	1.3505	408.15	1205.13	0.6042	1.5000	1.1047	0.8007	1.5050	4136.	1656.	0.292	0.0398	0.3751	0.02568	32.81	430.00
440.00	381.48	51.91	1.2177	419.27	1205.64	0.6165	1.4906	1.1143	0.8268	1.5240	4069.	1655.	0.285	0.0403	0.3725	0.02638	31.51	440.00
450.00	422.46	51.45	1.0999	430.49	1205.91	0.6287	1.4811	1.1246	0.8547	1.5448	4000.	1654.	0.278	0.0407	0.3696	0.02711	30.22	450.00
460.00	466.75	50.97	0.9951	441.81	1205.93	0.6409	1.4718	1.1358	0.8846	1.5678	3930.	1652.	0.271	0.0412	0.3666	0.02786	28.92	460.00
470.00	514.52	50.48	0.9015	453.24	1205.68	0.6531	1.4625	1.1479	0.9167	1.5930	3857.	1650.	0.264	0.0417	0.3633	0.02865	27.61	470.00
480.00	565.95	49.98	0.8179	464.78	1205.13	0.6652	1.4531	1.1612	0.9513	1.6208	3782.	1647.	0.258	0.0422	0.3599	0.02947	26.30	480.00
490.00	621.23	49.46	0.7429	476.46	1204.29	0.6774	1.4438	1.1757	0.9888	1.6515	3706.	1643.	0.252	0.0427	0.3562	0.03033	25.00	490.00
500.00	680.55	48.92	0.6756	488.27	1203.13	0.6895	1.4344	1.1916	1.0295	1.6856	3626.	1638.	0.246	0.0432	0.3522	0.03124	23.69	500.00
510.00	744.11	48.37	0.6149	500.23	1201.62	0.7017	1.4250	1.2091	1.0741	1.7235	3545.	1632.	0.240	0.0438	0.3481	0.03220	22.38	510.00
520.00	812.10	47.80	0.5601	512.35	1199.75	0.7138	1.4155	1.2285	1.1231	1.7658	3461.	1626.	0.235	0.0443	0.3436	0.03323	21.08	520.00
530.00	884.74	47.20	0.5105	524.65	1197.49	0.7260	1.4059	1.2500	1.1772	1.8133	3375.	1619.	0.229	0.0449	0.3389	0.03434	19.77	530.00
540.00	962.24	46.59	0.4655	537.14	1194.80	0.7383	1.3962	1.2740	1.2374	1.8667	3286.	1611.	0.224	0.0455	0.3340	0.03554	18.47	540.00
550.00	1044.8	45.95	0.4247	549.84	1191.66	0.7506	1.3863	1.3011	1.3048	1.9274	3195.	1601.	0.219	0.0461	0.3288	0.03685	17.18	550.00
560.00	1132.7	45.29	0.3874	562.77	1188.02	0.7630	1.3762	1.3317	1.3810	1.9966	3101.	1591.	0.214	0.0467	0.3233	0.03830	15.89	560.00
570.00	1226.2	44.60	0.3534	575.97	1183.83	0.7755	1.3658	1.3668	1.4677	2.0763	3003.	1580.	0.209	0.0474	0.3177	0.03992	14.61	570.00
580.00	1325.5	43.88	0.3223	589.44	1179.04	0.7881	1.3552	1.4072	1.5675	2.1687	2903.	1567.	0.204	0.0481	0.3118	0.04175	13.35	580.00
590.00	1430.8	43.12	0.2937	603.25	1173.59	0.8009	1.3443	1.4543	1.6838	2.2772	2799.	1553.	0.199	0.0489	0.3057	0.04385	12.09	590.00
600.00	1542.5	42.32	0.2674	617.42	1167.39	0.8139	1.3329	1.5100	1.8210	2.4061	2691.	1537.	0.194	0.0497	0.2995	0.04627	10.85	600.00
610.00	1660.9	41.47	0.2431	632.02	1160.34	0.8271	1.3210	1.5769	1.9855	2.5615	2580.	1520.	0.189	0.0506	0.2931	0.04912	9.62	610.00
620.00	1786.2	40.57	0.2206	647.11	1152.31	0.8406	1.3085	1.6588	2.1869	2.7524	2463.	1501.	0.183	0.0516	0.2867	0.05252	8.42	620.00
630.00	1918.9	39.61	0.1997	662.79	1143.14	0.8545	1.2953	1.7618	2.4392	2.9922	2341.	1480.	0.178	0.0527	0.2801	0.05664	7.23	630.00
640.00	2059.3	38.57	0.1802	679.19	1132.60	0.8689	1.2812	1.8958	2.7654	3.3023	2213.	1456.	0.173	0.0540	0.2735	0.06174	6.08	640.00
650.00	2207.8	37.42	0.1618	696.48	1120.40	0.8839	1.2659	2.0791	3.2045	3.7194	2076.	1430.	0.167	0.0555	0.2668	0.06819	4.96	650.00
660.00	2364.9	36.15	0.1444	714.96	1106.08	0.8997	1.2491	2.3480	3.8305	4.3113	1928.	1399.	0.161	0.0572	0.2600	0.07660	3.88	660.00
670.00	2531.2	34.69	0.1277	735.12	1088.91	0.9169	1.2301	2.7832	4.8020	5.2224	1762.	1363.	0.154	0.0594	0.2531	0.08808	2.84	670.00
680.00	2707.3	32.94	0.1113	757.89	1067.56	0.9361	1.2078	3.5861	6.5383	6.8275	1574.	1320.	0.146	0.0622	0.2461	0.10495	1.88	680.00
690.00	2894.0	30.69	0.0946	785.02	1039.02	0.9589	1.1798	5.3920	10.639	10.516	1365.	1263.	0.136	0.0663	0.2404	0.13393	1.00	690.00
700.00	3093.0	27.28	0.0748	823.00	991.66	0.9907	1.1361	15.579	32.942	29.223	1119.	1162.	0.122	0.0740	0.2547	0.21590	0.26	700.00
705.10c	3200.1	20.10	0.0497	896.67	896.67	1.0533	1.0533	∞	∞	∞	0.	0.	–	–	∞	∞	0.00	705.10

*Temperatures are on the ITS-90 scale a = triple point b = normal boiling point c = critical point

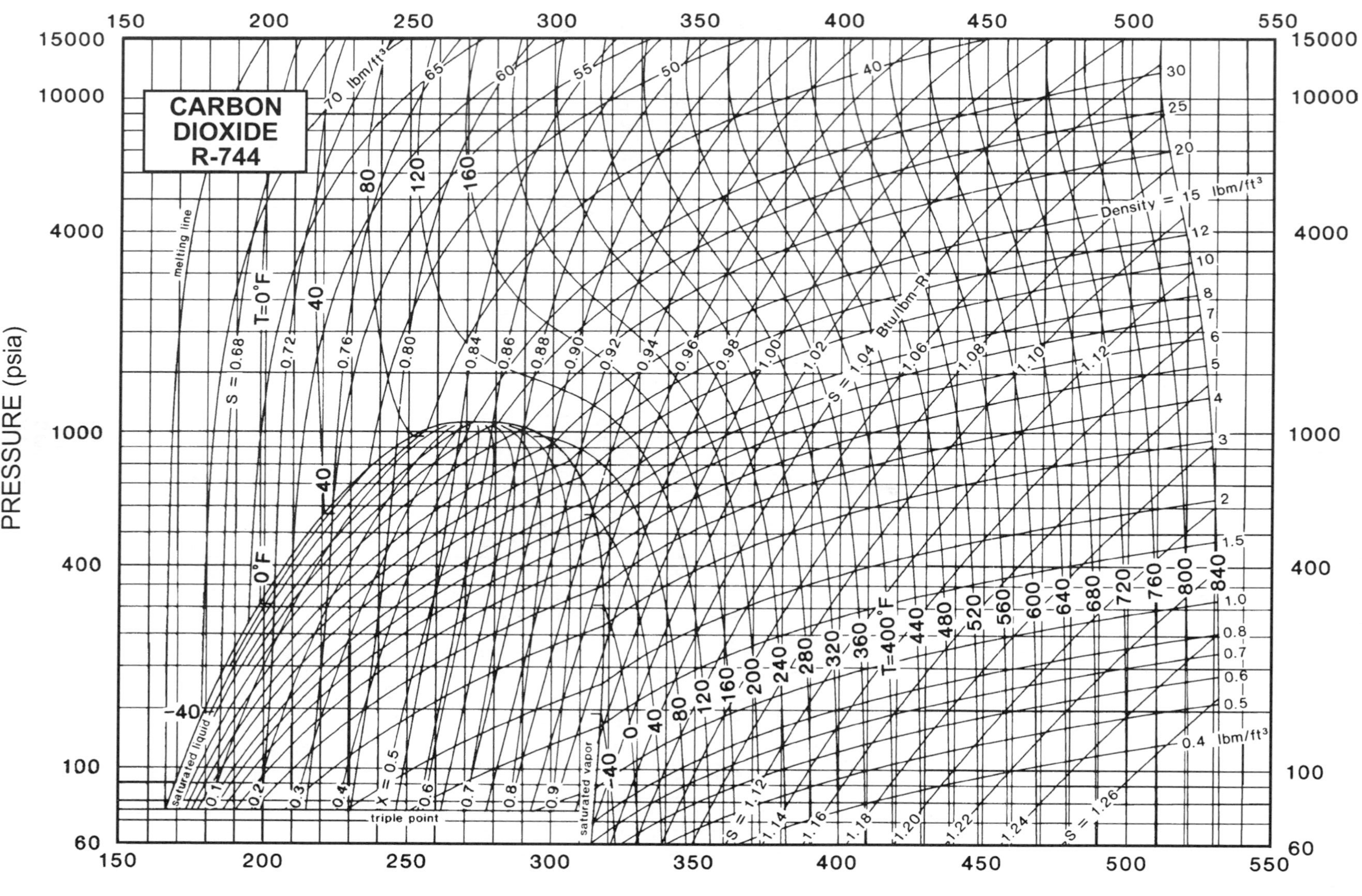

Fig. 14 Pressure-Enthalpy Diagram for Refrigerant 744 (Carbon Dioxide)

Note: The reference states for enthalpy and entropy differ from those in the table.

Refrigerant 744 (Carbon Dioxide) Properties of Saturated Liquid and Saturated Vapor

Temp.,* °F	Pres-sure, psia	Denisty, lb/ft³ Liquid	Volume, ft³/lb Vapor	Enthalpy, Btu/lb Liquid	Enthalpy, Btu/lb Vapor	Entropy, Btu/lb·°F Liquid	Entropy, Btu/lb·°F Vapor	Specific Heat c_p, Btu/lb·°F Liquid	Specific Heat c_p, Btu/lb·°F Vapor	c_p/c_v Vapor	Vel. of Sound, ft/s Liquid	Vel. of Sound, ft/s Vapor	Viscosity, lb_m/ft·h Liquid	Viscosity, lb_m/ft·h Vapor	Thermal Cond., Btu/h·ft·°F Liquid	Thermal Cond., Btu/h·ft·°F Vapor	Surface Tension, dyne/cm	Temp., °F
–69.80a	75.124	73.57	1.1641	–14.140	136.598	–0.03449	0.35215	0.4668	0.2172	1.4442	3202.	730.9	0.621	0.0265	0.1044	0.00637	17.16	–69.80
–65.00	84.234	72.97	1.0434	–11.886	137.013	–0.02881	0.34847	0.4684	0.2212	1.4534	3138.	731.9	0.593	0.0268	0.1024	0.00650	16.49	–65.00
–60.00	94.573	72.33	0.9336	–9.532	137.417	–0.02294	0.34473	0.4703	0.2257	1.4638	3073.	732.7	0.565	0.0272	0.1003	0.00664	15.81	–60.00
–55.00	105.84	71.69	0.8375	–7.167	137.790	–0.01714	0.34107	0.4724	0.2304	1.4754	3007.	733.2	0.539	0.0276	0.0982	0.00678	15.12	–55.00
–50.00	118.08	71.04	0.7532	–4.791	138.130	–0.01138	0.33749	0.4749	0.2355	1.4882	2941.	733.5	0.514	0.0279	0.0962	0.00693	14.45	–50.00
–48.00	123.26	70.77	0.7224	–3.837	138.257	–0.00909	0.33608	0.4760	0.2377	1.4937	2915.	733.5	0.505	0.0281	0.0954	0.00699	14.18	–48.00
–46.00	128.61	70.51	0.6930	–2.881	138.379	–0.00681	0.33467	0.4771	0.2399	1.4994	2889.	733.5	0.496	0.0283	0.0945	0.00706	13.91	–46.00
–44.00	134.13	70.24	0.6651	–1.923	138.494	–0.00453	0.33328	0.4783	0.2422	1.5054	2862.	733.5	0.486	0.0284	0.0937	0.00712	13.65	–44.00
–42.00	139.82	69.97	0.6386	–0.963	138.604	–0.00226	0.33189	0.4795	0.2445	1.5116	2836.	733.4	0.477	0.0286	0.0929	0.00718	13.38	–42.00
–40.00	145.69	69.70	0.6132	0.000	138.708	0.00000	0.33052	0.4808	0.2470	1.5180	2809.	733.3	0.469	0.0287	0.0921	0.00725	13.12	–40.00
–38.00	151.74	69.42	0.5891	0.965	138.806	0.00226	0.32915	0.4821	0.2495	1.5247	2783.	733.1	0.460	0.0289	0.0913	0.00732	12.86	–38.00
–36.00	157.98	69.15	0.5661	1.933	138.898	0.00451	0.32779	0.4836	0.2520	1.5317	2756.	732.9	0.452	0.0290	0.0905	0.00739	12.60	–36.00
–34.00	164.40	68.87	0.5442	2.904	138.983	0.00675	0.32643	0.4850	0.2547	1.5390	2730.	732.6	0.443	0.0292	0.0897	0.00746	12.34	–34.00
–32.00	171.02	68.59	0.5233	3.877	139.062	0.00899	0.32509	0.4866	0.2574	1.5466	2703.	732.3	0.435	0.0293	0.0889	0.00753	12.08	–32.00
–30.00	177.83	68.31	0.5033	4.854	139.134	0.01123	0.32375	0.4882	0.2603	1.5545	2677.	732.0	0.427	0.0295	0.0881	0.00760	11.82	–30.00
–28.00	184.83	68.02	0.4842	5.833	139.199	0.01346	0.32241	0.4899	0.2632	1.5628	2650.	731.6	0.420	0.0297	0.0873	0.00768	11.56	–28.00
–26.00	192.04	67.74	0.4659	6.816	139.258	0.01568	0.32108	0.4917	0.2662	1.5714	2623.	731.1	0.412	0.0298	0.0865	0.00775	11.31	–26.00
–24.00	199.46	67.45	0.4485	7.802	139.309	0.01790	0.31975	0.4935	0.2694	1.5804	2596.	730.6	0.405	0.0300	0.0857	0.00783	11.06	–24.00
–22.00	207.08	67.16	0.4318	8.791	139.353	0.02012	0.31843	0.4955	0.2726	1.5898	2569.	730.1	0.397	0.0302	0.0849	0.00791	10.80	–22.00
–20.00	214.91	66.86	0.4158	9.784	139.389	0.02234	0.31711	0.4975	0.2760	1.5996	2542.	729.5	0.390	0.0303	0.0841	0.00799	10.55	–20.00
–18.00	222.97	66.56	0.4005	10.781	139.418	0.02455	0.31580	0.4996	0.2795	1.6099	2515.	728.9	0.383	0.0305	0.0833	0.00807	10.30	–18.00
–16.00	231.24	66.27	0.3859	11.781	139.438	0.02675	0.31448	0.5018	0.2831	1.6206	2488.	728.2	0.376	0.0307	0.0825	0.00816	10.05	–16.00
–14.00	239.73	65.96	0.3718	12.786	139.451	0.02896	0.31317	0.5042	0.2869	1.6318	2461.	727.5	0.369	0.0308	0.0818	0.00825	9.81	–14.00
–12.00	248.45	65.66	0.3584	13.794	139.455	0.03116	0.31186	0.5066	0.2908	1.6435	2433.	726.7	0.363	0.0310	0.0810	0.00834	9.56	–12.00
–10.00	257.40	65.35	0.3455	14.807	139.450	0.03336	0.31055	0.5091	0.2949	1.6557	2405.	725.9	0.356	0.0312	0.0802	0.00843	9.32	–10.00
–8.00	266.58	65.04	0.3331	15.824	139.437	0.03556	0.30924	0.5118	0.2991	1.6685	2378.	725.0	0.350	0.0314	0.0794	0.00853	9.07	–8.00
–6.00	276.01	64.72	0.3212	16.846	139.415	0.03776	0.30793	0.5146	0.3035	1.6820	2350.	724.1	0.343	0.0315	0.0786	0.00863	8.83	–6.00
–4.00	285.67	64.40	0.3098	17.873	139.383	0.03996	0.30662	0.5175	0.3082	1.6960	2321.	723.1	0.337	0.0317	0.0778	0.00873	8.59	–4.00
–2.00	295.58	64.08	0.2989	18.905	139.342	0.04216	0.30531	0.5206	0.3130	1.7108	2293.	722.1	0.331	0.0319	0.0771	0.00883	8.35	–2.00
0.00	305.74	63.76	0.2884	19.942	139.291	0.04435	0.30399	0.5238	0.3180	1.7262	2264.	721.0	0.325	0.0321	0.0763	0.00894	8.11	0.00
2.00	316.15	63.43	0.2782	20.985	139.230	0.04655	0.30267	0.5272	0.3233	1.7425	2235.	719.8	0.319	0.0323	0.0755	0.00905	7.88	2.00
4.00	326.82	63.09	0.2685	22.033	139.158	0.04875	0.30135	0.5307	0.3288	1.7596	2206.	718.6	0.313	0.0325	0.0747	0.00916	7.64	4.00
6.00	337.75	62.76	0.2591	23.088	139.075	0.05095	0.30003	0.5345	0.3346	1.7776	2176.	717.4	0.307	0.0327	0.0740	0.00928	7.41	6.00
8.00	348.94	62.42	0.2501	24.148	138.981	0.05315	0.29869	0.5384	0.3406	1.7965	2146.	716.1	0.302	0.0329	0.0732	0.00941	7.18	8.00
10.00	360.41	62.07	0.2414	25.215	138.876	0.05535	0.29736	0.5425	0.3470	1.8166	2116.	714.7	0.296	0.0331	0.0724	0.00953	6.95	10.00
12.00	372.14	61.72	0.2331	26.289	138.758	0.05756	0.29601	0.5469	0.3537	1.8377	2085.	713.2	0.291	0.0333	0.0716	0.00967	6.72	12.00
14.00	384.16	61.36	0.2250	27.369	138.628	0.05977	0.29466	0.5514	0.3607	1.8601	2054.	711.8	0.286	0.0335	0.0709	0.00981	6.50	14.00
16.00	396.45	61.00	0.2173	28.457	138.485	0.06198	0.29329	0.5563	0.3681	1.8837	2023.	710.2	0.280	0.0338	0.0701	0.00995	6.27	16.00
18.00	409.03	60.63	0.2098	29.552	138.328	0.06420	0.29192	0.5614	0.3759	1.9089	1991.	708.6	0.275	0.0340	0.0693	0.01010	6.05	18.00
20.00	421.91	60.26	0.2025	30.656	138.158	0.06642	0.29054	0.5669	0.3841	1.9356	1959.	706.9	0.270	0.0342	0.0685	0.01026	5.83	20.00
22.00	435.07	59.89	0.1956	31.768	137.973	0.06865	0.28915	0.5726	0.3928	1.9640	1926.	705.2	0.265	0.0345	0.0677	0.01042	5.61	22.00
24.00	448.54	59.50	0.1888	32.889	137.772	0.07089	0.28774	0.5787	0.4021	1.9942	1894.	703.4	0.260	0.0347	0.0670	0.01059	5.39	24.00
26.00	462.30	59.11	0.1823	34.019	137.556	0.07313	0.28632	0.5853	0.4120	2.0266	1861.	701.6	0.255	0.0350	0.0662	0.01077	5.17	26.00
28.00	476.38	58.71	0.1760	35.159	137.323	0.07538	0.28488	0.5922	0.4225	2.0611	1827.	699.7	0.250	0.0352	0.0654	0.01096	4.96	28.00
30.00	490.77	58.31	0.1699	36.309	137.072	0.07764	0.28342	0.5997	0.4337	2.0982	1794.	697.7	0.245	0.0355	0.0646	0.01116	4.75	30.00
32.00	505.48	57.90	0.1640	37.470	136.803	0.07991	0.28195	0.6076	0.4457	2.1380	1760.	695.7	0.240	0.0358	0.0638	0.01137	4.54	32.00
34.00	520.51	57.48	0.1583	38.643	136.514	0.08220	0.28045	0.6162	0.4586	2.1808	1726.	693.6	0.236	0.0361	0.0631	0.01160	4.33	34.00
36.00	535.86	57.05	0.1528	39.828	136.206	0.08449	0.27893	0.6254	0.4725	2.2271	1692.	691.4	0.231	0.0364	0.0623	0.01183	4.13	36.00
38.00	551.55	56.61	0.1475	41.025	135.875	0.08680	0.27739	0.6353	0.4875	2.2771	1657.	689.1	0.227	0.0367	0.0615	0.01208	3.92	38.00
40.00	567.58	56.16	0.1423	42.237	135.522	0.08912	0.27582	0.6460	0.5038	2.3314	1623.	686.8	0.222	0.0370	0.0607	0.01235	3.72	40.00
42.00	583.95	55.71	0.1373	43.464	135.145	0.09147	0.27422	0.6577	0.5215	2.3905	1588.	684.4	0.217	0.0373	0.0599	0.01263	3.53	42.00
44.00	600.67	55.24	0.1324	44.706	134.741	0.09383	0.27259	0.6704	0.5408	2.4551	1553.	681.9	0.213	0.0377	0.0591	0.01294	3.33	44.00
46.00	617.75	54.76	0.1276	45.965	134.310	0.09621	0.27092	0.6843	0.5620	2.5260	1518.	679.3	0.209	0.0381	0.0583	0.01326	3.14	46.00
48.00	635.18	54.27	0.1230	47.242	133.850	0.09861	0.26921	0.6996	0.5854	2.6040	1482.	676.7	0.204	0.0384	0.0575	0.01362	2.94	48.00
50.00	652.99	53.76	0.1185	48.539	133.357	0.10104	0.26746	0.7164	0.6113	2.6903	1447.	673.9	0.200	0.0388	0.0567	0.01400	2.76	50.00
52.00	671.16	53.24	0.1141	49.858	132.830	0.10350	0.26566	0.7352	0.6402	2.7863	1411.	671.0	0.195	0.0393	0.0559	0.01441	2.57	52.00
54.00	689.72	52.70	0.1099	51.200	132.266	0.10599	0.26381	0.7562	0.6725	2.8937	1375.	668.1	0.191	0.0397	0.0551	0.01485	2.39	54.00
56.00	708.67	52.14	0.1057	52.568	131.661	0.10852	0.26190	0.7798	0.7091	3.0147	1338.	665.0	0.187	0.0402	0.0543	0.01534	2.21	56.00
58.00	728.01	51.56	0.1017	53.964	131.012	0.11109	0.25992	0.8065	0.7507	3.1519	1302.	661.8	0.182	0.0407	0.0535	0.01588	2.03	58.00
60.00	747.75	50.96	0.0977	55.392	130.313	0.11370	0.25787	0.8370	0.7984	3.3088	1264.	658.4	0.178	0.0413	0.0527	0.01647	1.86	60.00
62.00	767.91	50.34	0.0938	56.855	129.560	0.11637	0.25574	0.8722	0.8538	3.4899	1227.	654.9	0.173	0.0419	0.0519	0.01713	1.69	62.00
64.00	788.48	49.69	0.0900	58.358	128.745	0.11910	0.25351	0.9131	0.9188	3.7014	1188.	651.3	0.169	0.0425	0.0511	0.01786	1.52	64.00
66.00	809.48	49.00	0.0862	59.906	127.860	0.12190	0.25117	0.9613	0.9962	3.9514	1148.	647.4	0.165	0.0432	0.0503	0.01869	1.36	66.00
68.00	830.93	48.28	0.0825	61.505	126.896	0.12478	0.24871	1.019	1.090	4.252	1108.	643.4	0.160	0.0440	0.0495	0.01963	1.20	68.00
70.00	852.82	47.52	0.0788	63.165	125.840	0.12776	0.24609	1.089	1.205	4.618	1066.	639.0	0.155	0.0448	0.0488	0.02070	1.05	70.00
75.00	909.62	45.36	0.0697	67.656	122.671	0.13578	0.23867	1.363	1.659	6.027	951.	626.5	0.143	0.0474	0.0472	0.02430	0.69	75.00
80.00	969.57	42.62	0.0603	72.945	118.309	0.14515	0.22921	2.005	2.726	9.198	816.	609.5	0.129	0.0512	0.0466	0.03046	0.36	80.00
85.00	1033.07	38.41	0.0493	80.262	111.006	0.15811	0.21455	5.226	8.106	23.712	636.	576.6	0.111	0.0582	0.0510	0.04701	0.10	85.00
87.76c	1069.99	29.19	0.0343	94.364	94.364	0.18355	0.18355	∞	∞	∞	0.	0.0	–	–	∞	∞	0.00	87.76

*Temperatures are on the ITS-90 scale

a = triple point

c = critical point

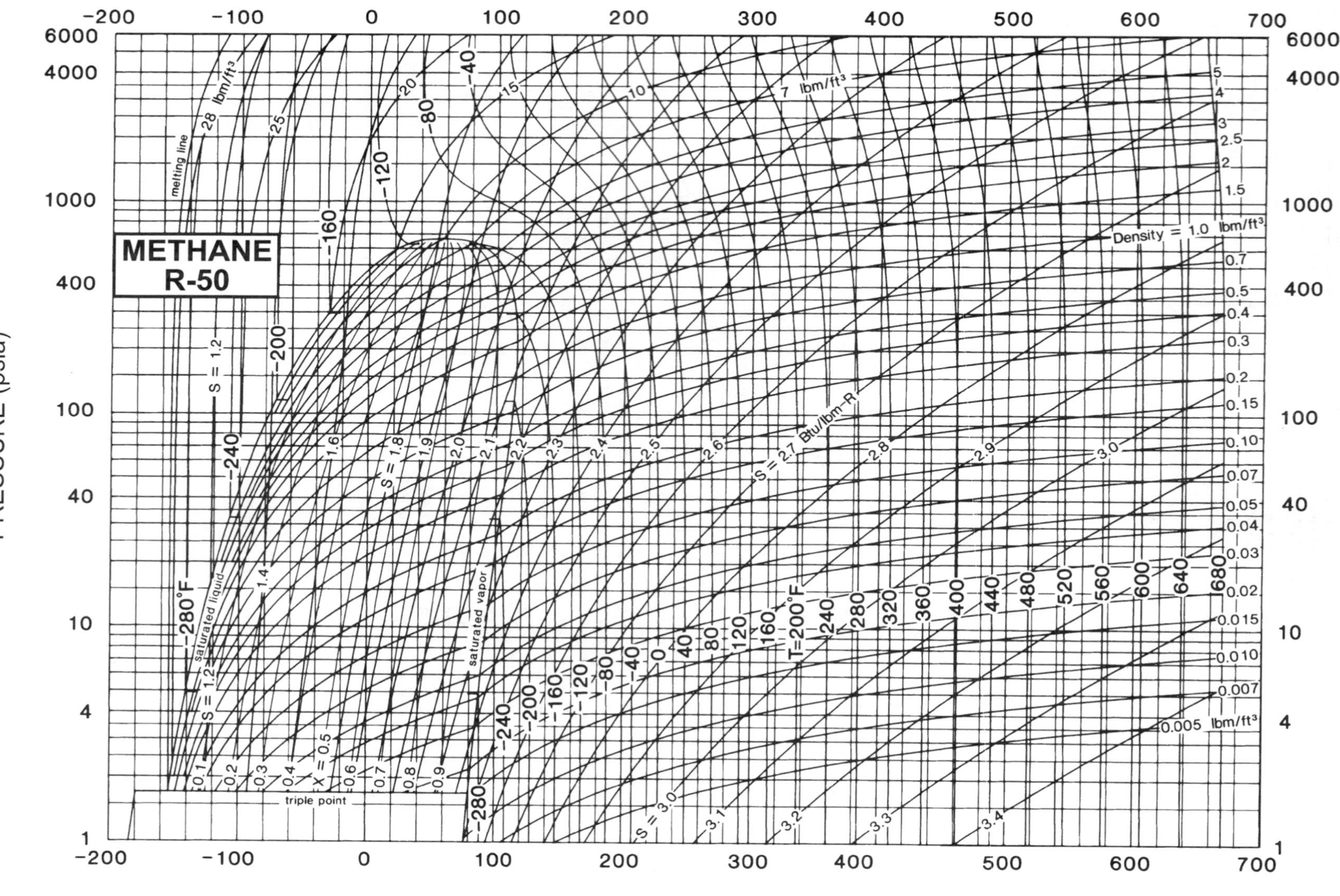

Fig. 15 Pressure-Enthalpy Diagram for Refrigerant 50 (Methane)

Note: The reference states for enthalpy and entropy differ from those in the table.

Refrigerant 50 (Methane) Properties of Saturated Liquid and Saturated Vapor

Temp.,* °F	Pressure, psia	Density, lb/ft³ Liquid	Volume, ft³/lb Vapor	Enthalpy, Btu/lb Liquid	Enthalpy, Btu/lb Vapor	Entropy, Btu/lb·°F Liquid	Entropy, Btu/lb·°F Vapor	Specific Heat c_p, Btu/lb·°F Liquid	Specific Heat c_p, Btu/lb·°F Vapor	c_p/c_v Vapor	Vel. of Sound, ft/s Liquid	Vel. of Sound, ft/s Vapor	Viscosity, lb_m/ft·h Liquid	Viscosity, lb_m/ft·h Vapor	Thermal Cond., Btu/h·ft·°F Liquid	Thermal Cond., Btu/h·ft·°F Vapor	Surface Tension, dyne/cm	Temp., °F
−296.42a	1.696	28.18	63.884	−30.898	203.250	−0.16968	1.26461	0.8049	0.5043	1.3410	5048.	817.3	0.709	0.0085	0.1584	0.00590	18.76	−296.42
−290.00	2.635	27.89	42.619	−25.716	206.228	−0.13858	1.22845	0.8081	0.5071	1.3440	4941.	831.7	0.443	0.0091	0.1197	0.00537	17.78	−290.00
−280.00	4.891	27.41	24.167	−17.593	210.749	−0.09216	1.17874	0.8144	0.5127	1.3505	4770.	852.7	0.379	0.0097	0.1156	0.00578	16.30	−280.00
−270.00	8.466	26.93	14.619	−9.396	215.095	−0.04789	1.13569	0.8219	0.5198	1.3594	4594.	871.6	0.328	0.0102	0.1113	0.00620	14.87	−270.00
−260.00	13.822	26.43	9.3292	−1.111	219.231	−0.00551	1.09801	0.8307	0.5287	1.3712	4414.	888.5	0.287	0.0107	0.1069	0.00664	13.48	−260.00
−250.00	21.476	25.92	6.2231	7.279	223.120	0.03522	1.06465	0.8409	0.5398	1.3868	4231.	903.2	0.254	0.0113	0.1024	0.00710	12.14	−250.00
−258.67b	14.696	26.37	8.8187	0.000	219.763	0.00000	1.09335	0.8320	0.5300	1.3731	4390.	890.6	0.282	0.0108	0.1063	0.00670	13.30	−258.67
−245.00	26.341	25.66	5.1555	11.517	224.960	0.05503	1.04932	0.8467	0.5462	1.3963	4137.	909.7	0.239	0.0116	0.1001	0.00734	11.49	−245.00
−240.00	31.996	25.39	4.3071	15.788	226.724	0.07451	1.03475	0.8529	0.5534	1.4070	4043.	915.5	0.226	0.0119	0.0979	0.00759	10.85	−240.00
−235.00	38.517	25.12	3.6261	20.095	228.407	0.09368	1.02087	0.8598	0.5613	1.4191	3948.	920.8	0.214	0.0121	0.0956	0.00785	10.23	−235.00
−230.00	45.984	24.84	3.0742	24.439	230.002	0.11256	1.00760	0.8672	0.5701	1.4329	3851.	925.4	0.202	0.0124	0.0933	0.00811	9.62	−230.00
−225.00	54.475	24.56	2.6230	28.824	231.505	0.13117	0.99486	0.8754	0.5800	1.4484	3753.	929.4	0.192	0.0127	0.0910	0.00839	9.02	−225.00
−220.00	64.073	24.27	2.2512	33.255	232.908	0.14955	0.98258	0.8845	0.5909	1.4660	3655.	932.7	0.182	0.0131	0.0887	0.00868	8.43	−220.00
−215.00	74.859	23.97	1.9423	37.734	234.206	0.16770	0.97071	0.8945	0.6032	1.4860	3554.	935.4	0.173	0.0134	0.0864	0.00898	7.86	−215.00
−210.00	86.918	23.67	1.6839	42.267	235.391	0.18566	0.95918	0.9057	0.6170	1.5087	3452.	937.4	0.164	0.0137	0.0841	0.00929	7.31	−210.00
−205.00	100.33	23.36	1.4662	46.858	236.456	0.20345	0.94793	0.9181	0.6325	1.5346	3349.	938.7	0.156	0.0140	0.0818	0.00962	6.76	−205.00
−200.00	115.19	23.04	1.2817	51.514	237.390	0.22109	0.93691	0.9322	0.6502	1.5643	3244.	939.3	0.149	0.0144	0.0795	0.00997	6.23	−200.00
−195.00	131.58	22.71	1.1243	56.241	238.185	0.23862	0.92605	0.9480	0.6703	1.5984	3137.	939.2	0.141	0.0147	0.0773	0.01033	5.72	−195.00
−190.00	149.59	22.37	0.9893	61.047	238.828	0.25605	0.91531	0.9661	0.6935	1.6378	3028.	938.3	0.135	0.0151	0.0750	0.01072	5.22	−190.00
−185.00	169.30	22.01	0.8728	65.940	239.308	0.27343	0.90461	0.9868	0.7202	1.6837	2917.	936.7	0.128	0.0155	0.0727	0.01113	4.74	−185.00
−180.00	190.81	21.64	0.7718	70.932	239.607	0.29078	0.89390	1.011	0.7515	1.7375	2804.	934.3	0.122	0.0159	0.0704	0.01157	4.27	−180.00
−175.00	214.22	21.26	0.6836	76.035	239.709	0.30815	0.88311	1.039	0.7886	1.8014	2688.	931.1	0.116	0.0163	0.0681	0.01204	3.82	−175.00
−170.00	239.62	20.86	0.6064	81.263	239.590	0.32557	0.87215	1.072	0.8329	1.8780	2570.	927.1	0.110	0.0168	0.0658	0.01256	3.38	−170.00
−165.00	267.11	20.44	0.5383	86.636	239.224	0.34312	0.86095	1.112	0.8868	1.9714	2448.	922.2	0.104	0.0173	0.0635	0.01314	2.96	−165.00
−160.00	296.80	19.99	0.4781	92.177	238.576	0.36085	0.84938	1.160	0.9538	2.0871	2322.	916.4	0.099	0.0178	0.0612	0.01378	2.56	−160.00
−155.00	328.79	19.52	0.4243	97.917	237.600	0.37885	0.83733	1.221	1.039	2.2340	2193.	909.7	0.094	0.0183	0.0589	0.01452	2.17	−155.00
−150.00	363.20	19.01	0.3761	103.898	236.236	0.39725	0.82460	1.299	1.151	2.4258	2058.	901.9	0.088	0.0190	0.0565	0.01538	1.80	−150.00
−140.00	439.83	17.84	0.2929	116.835	231.967	0.43590	0.79606	1.553	1.523	3.0583	1768.	882.8	0.078	0.0205	0.0517	0.01776	1.13	−140.00
−130.00	527.92	16.34	0.2216	131.950	224.363	0.47950	0.75982	2.175	2.478	4.6351	1435.	857.4	0.067	0.0227	0.0469	0.02236	0.55	−130.00
−120.00	629.39	13.83	0.1511	153.364	207.529	0.53971	0.69917	6.640	9.528	15.490	995.	814.0	0.052	0.0272	0.0456	0.04087	0.10	−120.00
−116.65c	667.06	10.15	0.0985	178.791	178.791	0.61244	0.61244	∞	∞	∞	0.	0.0	–	–	∞	∞	0.00	−116.65

*Temperatures are on the ITS-90 scale   a = triple point   b = normal boiling point   c = critical point

Refrigerant 50 (Methane) Properties of Gas at 14.696 psia (one standard atmosphere)

Temp., °F	Density, lb/ft³	Enthalpy, Btu/lb	Entropy, Btu/lb·°F	Specific Heat c_p, Btu/lb·°F	c_p/c_v	Vel. of Sound, ft/s	Viscosity, lb/ft·h	Thermal Cond., Btu/h·ft·°F
−258.7a	0.1134	219.76	1.0933	0.5300	1.373	890.6	0.0108	0.00670
−250.0	0.1082	224.33	1.1156	0.5235	1.368	912.4	0.0112	0.00700
−240.0	0.1028	229.54	1.1398	0.5183	1.363	936.5	0.0118	0.00737
−230.0	0.0980	234.70	1.1628	0.5145	1.359	959.7	0.0123	0.00774
−220.0	0.0936	239.83	1.1847	0.5116	1.356	982.2	0.0128	0.00811
−200.0	0.0860	250.02	1.2255	0.5077	1.351	1025.2	0.0138	0.00887
−180.0	0.0796	260.14	1.2631	0.5052	1.347	1066.1	0.0149	0.00963
−160.0	0.0741	270.23	1.2979	0.5038	1.345	1105.2	0.0159	0.01039
−140.0	0.0694	280.30	1.3305	0.5030	1.342	1142.6	0.0169	0.01116
−120.0	0.0652	290.36	1.3610	0.5029	1.340	1178.6	0.0179	0.01193
−100.0	0.0615	300.42	1.3898	0.5034	1.338	1213.2	0.0189	0.01267
−80.0	0.0582	310.50	1.4170	0.5044	1.335	1246.6	0.0199	0.01343
−60.0	0.0552	320.60	1.4430	0.5060	1.333	1278.8	0.0208	0.01419
−40.0	0.0526	330.74	1.4677	0.5082	1.330	1309.8	0.0218	0.01497
−20.0	0.0501	340.93	1.4914	0.5110	1.326	1339.8	0.0227	0.01576
0.0	0.0479	351.18	1.5142	0.5145	1.323	1368.7	0.0236	0.01656
20.0	0.0459	361.51	1.5362	0.5185	1.319	1396.6	0.0245	0.01738
40.0	0.0441	371.93	1.5575	0.5232	1.315	1423.6	0.0254	0.01822
60.0	0.0424	382.45	1.5781	0.5285	1.310	1449.8	0.0263	0.01908
80.0	0.0408	393.07	1.5982	0.5343	1.305	1475.1	0.0272	0.01996
100.0	0.0393	403.82	1.6178	0.5406	1.300	1499.7	0.0280	0.02087
120.0	0.0379	414.70	1.6369	0.5474	1.295	1523.5	0.0289	0.02180
140.0	0.0367	425.72	1.6556	0.5547	1.290	1546.7	0.0297	0.02275
160.0	0.0355	436.89	1.6739	0.5624	1.285	1569.3	0.0306	0.02373
180.0	0.0344	448.22	1.6919	0.5704	1.280	1591.3	0.0314	0.02473
200.0	0.0333	459.71	1.7096	0.5788	1.274	1612.8	0.0322	0.02576
220.0	0.0323	471.37	1.7270	0.5875	1.269	1633.9	0.0330	0.02680
240.0	0.0314	483.21	1.7441	0.5964	1.264	1654.5	0.0337	0.02787
260.0	0.0305	495.23	1.7611	0.6055	1.259	1674.7	0.0345	0.02896
280.0	0.0297	507.43	1.7778	0.6148	1.254	1694.6	0.0353	0.03008
300.0	0.0289	519.82	1.7943	0.6243	1.249	1714.1	0.0360	0.03121
320.0	0.0282	532.40	1.8107	0.6339	1.244	1733.3	0.0368	0.03236
340.0	0.0275	545.18	1.8269	0.6436	1.239	1752.2	0.0375	0.03353
360.0	0.0268	558.15	1.8429	0.6534	1.235	1770.9	0.0382	0.03472
380.0	0.0262	571.32	1.8587	0.6633	1.231	1789.3	0.0389	0.03592
400.0	0.0256	584.68	1.8745	0.6732	1.226	1807.4	0.0397	0.03714
420.0	0.0250	598.24	1.8901	0.6831	1.222	1825.4	0.0404	0.03837
440.0	0.0244	612.00	1.9055	0.6931	1.218	1843.1	0.0411	0.03962
460.0	0.0239	625.97	1.9209	0.7031	1.215	1860.6	0.0417	0.04088
480.0	0.0234	640.13	1.9361	0.7130	1.211	1877.9	0.0424	0.04216
500.0	0.0229	654.49	1.9512	0.7230	1.207	1895.1	0.0431	0.04345

a = saturated vapor at the normal boiling point

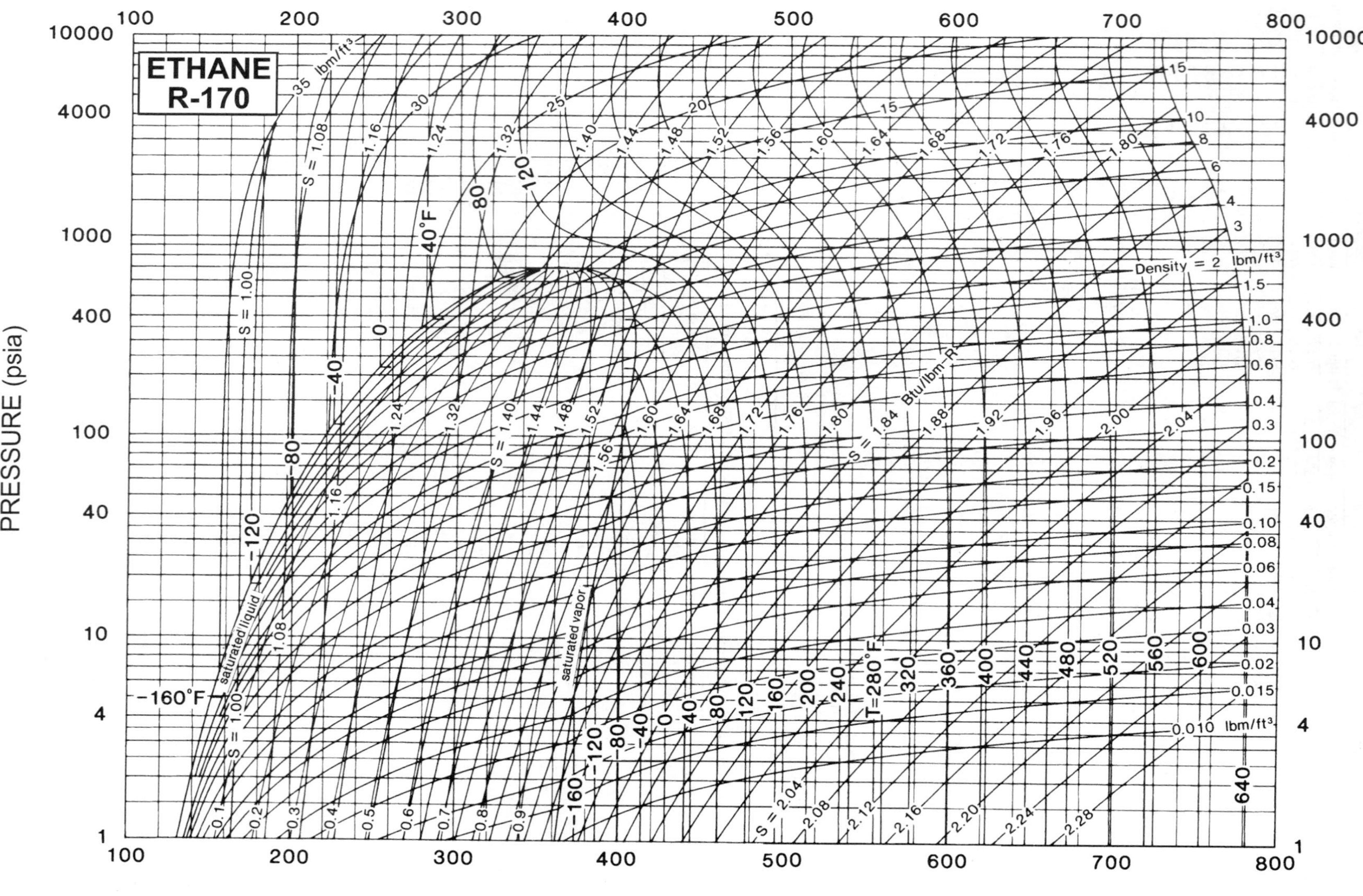

Fig. 16 Pressure-Enthalpy Diagram for Refrigerant 170 (Ethane)

Note: The reference states for enthalpy and entropy differ from those in the table.

Refrigerant 170 (Ethane) Properties of Saturated Liquid and Saturated Vapor

Temp.,* °F	Pressure, psia	Denisty, lb/ft³ Liquid	Volume, ft³/lb Vapor	Enthalpy, Btu/lb Liquid	Enthalpy, Btu/lb Vapor	Entropy, Btu/lb·°F Liquid	Entropy, Btu/lb·°F Vapor	Specific Heat c_p, Btu/lb·°F Liquid	Specific Heat c_p, Btu/lb·°F Vapor	c_p/c_v Vapor	Vel. of Sound, ft/s Liquid	Vel. of Sound, ft/s Vapor	Viscosity, lb_m/ft·h Liquid	Viscosity, lb_m/ft·h Vapor	Thermal Cond., Btu/h·ft·°F Liquid	Thermal Cond., Btu/h·ft·°F Vapor	Surface Tension, dyne/cm	Temp., °F
−250.00	0.031	38.90	2432.6	−68.960	174.898	−0.25844	0.90461	0.5617	0.2845	1.2964	5975.	670.6	1.290	0.0092	0.1351	0.00256	27.27	−250.00
−245.00	0.046	38.71	1655.0	−66.153	176.373	−0.24521	0.88455	0.5608	0.2840	1.2955	5908.	678.4	1.202	0.0094	0.1336	0.00266	26.80	−245.00
−240.00	0.068	38.52	1148.9	−63.352	177.855	−0.23231	0.86573	0.5599	0.2833	1.2947	5842.	686.0	1.125	0.0096	0.1320	0.00276	26.34	−240.00
−235.00	0.099	38.33	812.62	−60.555	179.344	−0.21972	0.84806	0.5590	0.2824	1.2940	5775.	693.6	1.055	0.0098	0.1305	0.00287	25.87	−235.00
−230.00	0.140	38.13	584.87	−57.762	180.839	−0.20743	0.83145	0.5582	0.2814	1.2935	5707.	701.1	0.992	0.0100	0.1290	0.00297	25.41	−230.00
−225.00	0.196	37.94	427.81	−54.972	182.339	−0.19542	0.81584	0.5575	0.2804	1.2930	5640.	708.5	0.936	0.0102	0.1274	0.00307	24.94	−225.00
−220.00	0.269	37.75	317.69	−52.186	183.842	−0.18367	0.80114	0.5570	0.2794	1.2925	5573.	715.8	0.884	0.0104	0.1258	0.00318	24.48	−220.00
−215.00	0.365	37.55	239.25	−49.401	185.348	−0.17217	0.78728	0.5567	0.2786	1.2921	5505.	722.9	0.838	0.0106	0.1243	0.00329	24.01	−215.00
−210.00	0.488	37.36	182.56	−46.618	186.854	−0.16091	0.77421	0.5565	0.2779	1.2918	5438.	729.9	0.795	0.0108	0.1227	0.00339	23.55	−210.00
−205.00	0.643	37.16	141.02	−43.835	188.360	−0.14988	0.76187	0.5565	0.2775	1.2914	5370.	736.7	0.756	0.0110	0.1211	0.00350	23.09	−205.00
−200.00	0.839	36.96	110.19	−41.051	189.863	−0.13906	0.75020	0.5568	0.2774	1.2911	5302.	743.4	0.720	0.0112	0.1195	0.00362	22.63	−200.00
−195.00	1.082	36.76	87.029	−38.265	191.361	−0.12844	0.73916	0.5572	0.2776	1.2908	5234.	749.8	0.687	0.0114	0.1179	0.00373	22.16	−195.00
−190.00	1.381	36.56	69.431	−35.477	192.854	−0.11801	0.72870	0.5578	0.2783	1.2905	5166.	756.1	0.656	0.0116	0.1163	0.00384	21.70	−190.00
−185.00	1.744	36.36	55.916	−32.685	194.340	−0.10775	0.71878	0.5586	0.2793	1.2902	5098.	762.2	0.627	0.0118	0.1147	0.00396	21.25	−185.00
−180.00	2.183	36.16	45.433	−29.888	195.818	−0.09767	0.70937	0.5596	0.2808	1.2900	5030.	768.0	0.601	0.0120	0.1131	0.00408	20.79	−180.00
−175.00	2.708	35.96	37.222	−27.085	197.286	−0.08775	0.70043	0.5607	0.2826	1.2898	4962.	773.6	0.576	0.0122	0.1115	0.00420	20.33	−175.00
−170.00	3.331	35.75	30.735	−24.276	198.743	−0.07798	0.69193	0.5620	0.2849	1.2897	4893.	779.0	0.553	0.0124	0.1099	0.00433	19.87	−170.00
−165.00	4.066	35.55	25.565	−21.460	200.189	−0.06835	0.68384	0.5635	0.2877	1.2898	4825.	784.1	0.531	0.0126	0.1084	0.00445	19.42	−165.00
−160.00	4.927	35.34	21.413	−18.635	201.621	−0.05886	0.67614	0.5652	0.2908	1.2899	4757.	789.1	0.511	0.0129	0.1068	0.00458	18.96	−160.00
−155.00	5.928	35.13	18.052	−15.801	203.041	−0.04950	0.66879	0.5670	0.2943	1.2902	4688.	793.8	0.492	0.0131	0.1052	0.00471	18.51	−155.00
−150.00	7.084	34.92	15.312	−12.958	204.445	−0.04026	0.66179	0.5690	0.2982	1.2907	4619.	798.2	0.474	0.0133	0.1036	0.00484	18.06	−150.00
−145.00	8.414	34.71	13.063	−10.103	205.835	−0.03114	0.65510	0.5712	0.3024	1.2914	4550.	802.5	0.456	0.0135	0.1021	0.00498	17.61	−145.00
−140.00	9.934	34.50	11.206	−7.236	207.208	−0.02213	0.64871	0.5735	0.3069	1.2923	4481.	806.5	0.440	0.0137	0.1005	0.00512	17.16	−140.00
−135.00	11.661	34.29	9.6620	−4.357	208.565	−0.01322	0.64259	0.5759	0.3118	1.2935	4412.	810.2	0.425	0.0139	0.0990	0.00526	16.71	−135.00
−130.00	13.617	34.07	8.3715	−1.465	209.904	−0.00441	0.63674	0.5785	0.3169	1.2950	4343.	813.8	0.410	0.0141	0.0974	0.00541	16.26	−130.00
−127.48b	14.696	33.96	7.8007	0.000	210.573	0.00000	0.63389	0.5799	0.3196	1.2959	4308.	815.5	0.403	0.0142	0.0966	0.00548	16.04	−127.48
−125.00	15.819	33.85	7.2866	1.441	211.225	0.00430	0.63114	0.5813	0.3223	1.2968	4274.	817.1	0.396	0.0143	0.0959	0.00555	15.82	−125.00
−120.00	18.288	33.63	6.3699	4.363	212.527	0.01293	0.62577	0.5843	0.3280	1.2989	4204.	820.1	0.383	0.0146	0.0943	0.00571	15.37	−120.00
−115.00	21.046	33.41	5.5913	7.300	213.809	0.02147	0.62062	0.5874	0.3339	1.3013	4135.	822.9	0.370	0.0148	0.0928	0.00586	14.93	−115.00
−110.00	24.113	33.18	4.9269	10.254	215.070	0.02993	0.61567	0.5906	0.3401	1.3042	4065.	825.5	0.358	0.0150	0.0913	0.00602	14.49	−110.00
−105.00	27.513	32.95	4.3574	13.225	216.311	0.03831	0.61092	0.5941	0.3465	1.3074	3995.	827.8	0.346	0.0152	0.0898	0.00618	14.05	−105.00
−100.00	31.267	32.73	3.8670	16.215	217.529	0.04662	0.60634	0.5977	0.3531	1.3110	3925.	829.8	0.335	0.0154	0.0883	0.00635	13.62	−100.00
−95.00	35.400	32.49	3.4431	19.224	218.724	0.05487	0.60194	0.6015	0.3599	1.3151	3854.	831.6	0.324	0.0157	0.0868	0.00652	13.18	−95.00
−90.00	39.933	32.26	3.0750	22.254	219.895	0.06305	0.59769	0.6056	0.3670	1.3197	3784.	833.1	0.314	0.0159	0.0853	0.00669	12.75	−90.00
−85.00	44.893	32.02	2.7544	25.305	221.042	0.07117	0.59359	0.6098	0.3743	1.3247	3713.	834.3	0.304	0.0161	0.0839	0.00687	12.32	−85.00
−80.00	50.302	31.78	2.4740	28.378	222.162	0.07923	0.58963	0.6142	0.3819	1.3303	3642.	835.3	0.295	0.0164	0.0824	0.00706	11.89	−80.00
−75.00	56.186	31.54	2.2279	31.475	223.255	0.08725	0.58580	0.6189	0.3897	1.3364	3571.	836.0	0.286	0.0166	0.0810	0.00724	11.46	−75.00
−70.00	62.570	31.29	2.0112	34.595	224.320	0.09521	0.58209	0.6238	0.3978	1.3432	3500.	836.4	0.277	0.0168	0.0795	0.00744	11.04	−70.00
−65.00	69.479	31.04	1.8197	37.742	225.355	0.10313	0.57850	0.6290	0.4062	1.3506	3428.	836.4	0.268	0.0171	0.0781	0.00764	10.62	−65.00
−60.00	76.940	30.78	1.6501	40.915	226.359	0.11101	0.57500	0.6345	0.4149	1.3588	3356.	836.2	0.260	0.0173	0.0767	0.00784	10.20	−60.00
−55.00	84.978	30.53	1.4993	44.117	227.331	0.11885	0.57159	0.6402	0.4239	1.3677	3283.	835.7	0.252	0.0176	0.0753	0.00806	9.78	−55.00
−50.00	93.620	30.26	1.3650	47.348	228.268	0.12665	0.56828	0.6464	0.4334	1.3775	3211.	834.8	0.244	0.0178	0.0739	0.00828	9.37	−50.00
−45.00	102.89	30.00	1.2449	50.610	229.169	0.13443	0.56503	0.6528	0.4433	1.3883	3138.	833.6	0.236	0.0181	0.0725	0.00850	8.96	−45.00
−40.00	112.82	29.72	1.1372	53.905	230.032	0.14218	0.56186	0.6597	0.4536	1.4001	3064.	832.1	0.229	0.0184	0.0712	0.00874	8.55	−40.00
−35.00	123.44	29.45	1.0405	57.234	230.854	0.14991	0.55874	0.6671	0.4645	1.4131	2990.	830.2	0.222	0.0186	0.0698	0.00898	8.15	−35.00
−30.00	134.77	29.17	0.9534	60.599	231.634	0.15762	0.55568	0.6749	0.4760	1.4274	2916.	828.0	0.215	0.0189	0.0685	0.00923	7.75	−30.00
−25.00	146.84	28.88	0.8747	64.002	232.368	0.16531	0.55265	0.6833	0.4882	1.4431	2841.	825.4	0.208	0.0192	0.0671	0.00950	7.35	−25.00
−20.00	159.68	28.59	0.8035	67.446	233.053	0.17300	0.54966	0.6923	0.5012	1.4606	2766.	822.5	0.202	0.0195	0.0658	0.00977	6.96	−20.00
−15.00	173.31	28.29	0.7389	70.933	233.687	0.18069	0.54670	0.7020	0.5151	1.4800	2690.	819.1	0.196	0.0198	0.0645	0.01006	6.57	−15.00
−10.00	187.78	27.98	0.6802	74.465	234.264	0.18837	0.54374	0.7125	0.5301	1.5015	2613.	815.4	0.189	0.0202	0.0632	0.01035	6.18	−10.00
−5.00	203.10	27.66	0.6267	78.047	234.782	0.19607	0.54079	0.7239	0.5463	1.5256	2536.	811.3	0.183	0.0205	0.0619	0.01067	5.80	−5.00
0.00	219.31	27.34	0.5778	81.680	235.234	0.20378	0.53783	0.7364	0.5639	1.5527	2458.	806.7	0.177	0.0209	0.0606	0.01100	5.42	0.00
5.00	236.45	27.01	0.5330	85.370	235.615	0.21151	0.53485	0.7502	0.5832	1.5832	2379.	801.7	0.171	0.0212	0.0593	0.01135	5.05	5.00
10.00	254.53	26.67	0.4919	89.121	235.920	0.21927	0.53183	0.7654	0.6046	1.6179	2299.	796.3	0.166	0.0216	0.0581	0.01173	4.68	10.00
15.00	273.60	26.31	0.4542	92.938	236.139	0.22707	0.52876	0.7825	0.6284	1.6575	2218.	790.4	0.160	0.0220	0.0568	0.01212	4.32	15.00
20.00	293.69	25.95	0.4194	96.827	236.265	0.23492	0.52562	0.8016	0.6552	1.7032	2135.	784.0	0.154	0.0225	0.0556	0.01255	3.96	20.00
25.00	314.83	25.57	0.3872	100.797	236.287	0.24284	0.52239	0.8234	0.6858	1.7563	2052.	777.1	0.149	0.0229	0.0543	0.01301	3.61	25.00
30.00	337.07	25.17	0.3574	104.855	236.192	0.25084	0.51905	0.8486	0.7210	1.8188	1966.	769.6	0.144	0.0234	0.0530	0.01352	3.27	30.00
35.00	360.43	24.76	0.3298	109.014	235.965	0.25894	0.51557	0.8780	0.7622	1.8931	1879.	761.7	0.138	0.0240	0.0518	0.01407	2.93	35.00
40.00	384.97	24.32	0.3040	113.287	235.586	0.26716	0.51192	0.9129	0.8112	1.9829	1789.	753.1	0.133	0.0246	0.0505	0.01468	2.60	40.00
45.00	410.72	23.86	0.2800	117.692	235.033	0.27554	0.50805	0.9553	0.8708	2.0933	1697.	744.0	0.127	0.0252	0.0493	0.01537	2.28	45.00
50.00	437.74	23.36	0.2574	122.252	234.273	0.28411	0.50390	1.008	0.945	2.232	1601.	734.2	0.122	0.0259	0.0480	0.01616	1.97	50.00
55.00	466.06	22.83	0.2361	126.999	233.267	0.29293	0.49941	1.075	1.040	2.412	1501.	723.8	0.116	0.0267	0.0468	0.01706	1.66	55.00
60.00	495.74	22.25	0.2159	131.975	231.956	0.30208	0.49448	1.164	1.166	2.652	1397.	712.6	0.111	0.0276	0.0455	0.01814	1.37	60.00
65.00	526.84	21.61	0.1967	137.242	230.260	0.31167	0.48896	1.287	1.343	2.990	1287.	700.7	0.105	0.0287	0.0442	0.01946	1.09	65.00
70.00	559.43	20.88	0.1780	142.887	228.051	0.32184	0.48262	1.468	1.609	3.499	1172.	687.9	0.099	0.0300	0.0430	0.02114	0.82	70.00
75.00	593.59	20.03	0.1597	149.055	225.115	0.33284	0.47510	1.761	2.055	4.347	1049.	674.0	0.092	0.0316	0.0419	0.02341	0.57	75.00
80.00	629.43	19.01	0.1412	156.012	221.020	0.34516	0.46562	2.329	2.953	6.041	917.	658.8	0.085	0.0338	0.0413	0.02684	0.34	80.00
85.00	667.14	17.63	0.1208	164.489	214.599	0.36009	0.45209	4.028	5.689	11.115	773.	640.7	0.076	0.0371	0.0422	0.03344	0.14	85.00
89.92c	706.59	12.90	0.0775	188.545	188.545	0.40315	0.40315	∞	∞	∞	0.	0.0	–	–	∞	∞	0.00	89.92

*Temperatures are on the ITS-90 scale

b = normal boiling point

c = critical point

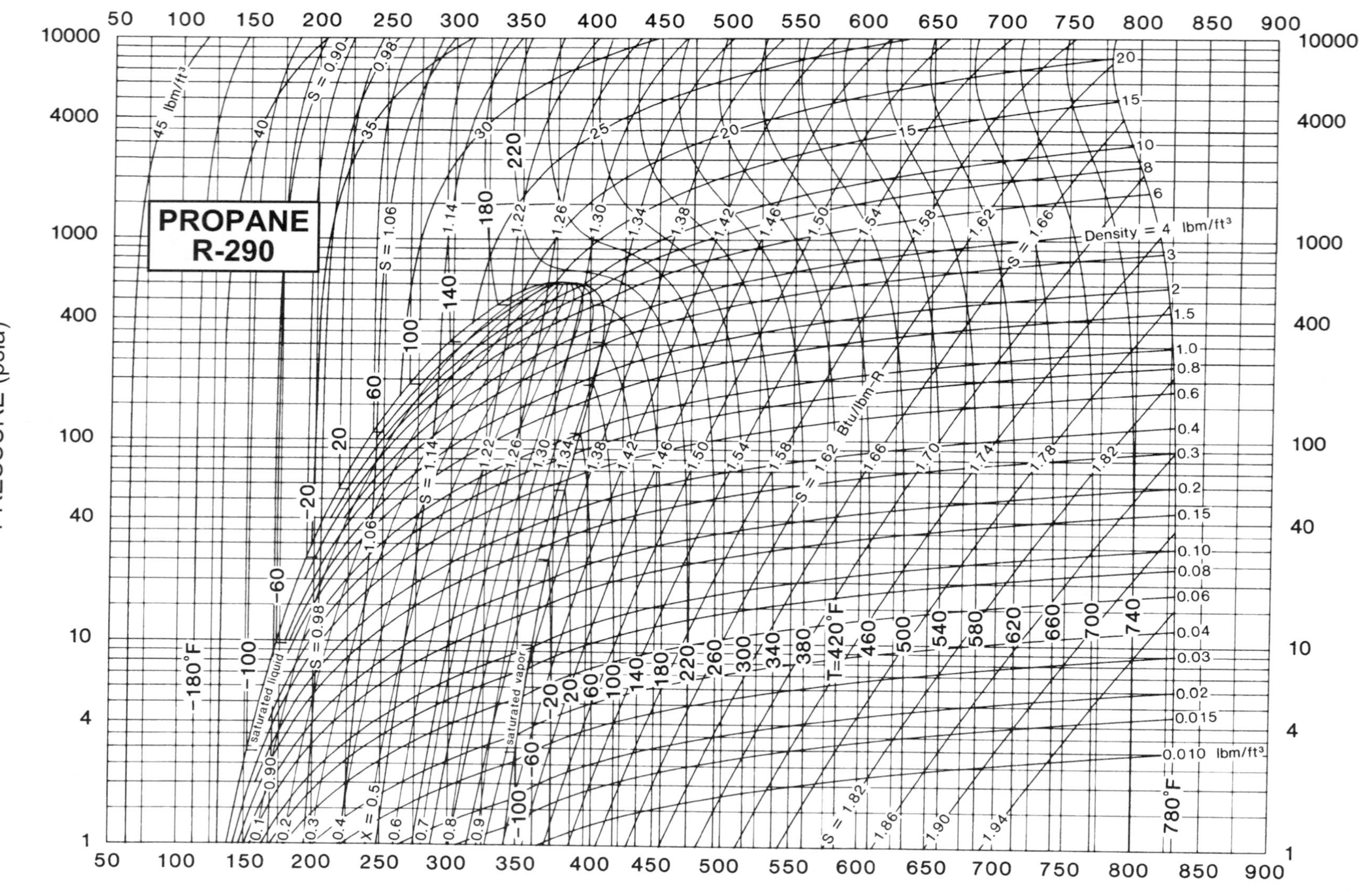

Fig. 17 Pressure-Enthalpy Diagram for Refrigerant 290 (Propane)

Note: The reference states for enthalpy and entropy differ from those in the table.

Refrigerant 290 (Propane) Properties of Saturated Liquid and Saturated Vapor

Temp.,* °F	Pres-sure, psia	Denisty, lb/ft³ Liquid	Volume, ft³/lb Vapor	Enthalpy, Btu/lb Liquid	Enthalpy, Btu/lb Vapor	Entropy, Btu/lb·°F Liquid	Entropy, Btu/lb·°F Vapor	Specific Heat c_p, Btu/lb·°F Liquid	Specific Heat c_p, Btu/lb·°F Vapor	c_p/c_v Vapor	Vel. of Sound, ft/s Liquid	Vel. of Sound, ft/s Vapor	Viscosity, lb_m/ft·h Liquid	Viscosity, lb_m/ft·h Vapor	Thermal Cond., Btu/h·ft·°F Liquid	Thermal Cond., Btu/h·ft·°F Vapor	Surface Tension, dyne/cm	Temp., °F
−200.00	0.020	42.05	3140.2	−80.254	137.547	−0.23938	0.59946	0.4748	0.2597	1.2104	5655.	595.1	1.806	0.0099	0.1051	0.00287	28.59	−200.00
−190.00	0.040	41.70	1639.9	−75.496	140.152	−0.22140	0.57835	0.4770	0.2639	1.2066	5532.	605.4	1.595	0.0102	0.1033	0.00308	27.74	−190.00
−180.00	0.075	41.35	903.80	−70.714	142.794	−0.20399	0.55951	0.4793	0.2682	1.2031	5411.	615.4	1.422	0.0106	0.1014	0.00330	26.90	−180.00
−170.00	0.135	40.99	522.49	−65.908	145.468	−0.18710	0.54268	0.4819	0.2726	1.1998	5290.	625.2	1.278	0.0109	0.0995	0.00351	26.06	−170.00
−160.00	0.231	40.64	315.21	−61.074	148.175	−0.17069	0.52762	0.4847	0.2771	1.1967	5170.	634.6	1.157	0.0113	0.0975	0.00374	25.23	−160.00
−150.00	0.380	40.28	197.57	−56.212	150.910	−0.15474	0.51416	0.4877	0.2819	1.1938	5051.	643.8	1.053	0.0116	0.0956	0.00397	24.39	−150.00
−145.00	0.482	40.10	158.46	−53.769	152.287	−0.14691	0.50797	0.4893	0.2843	1.1925	4991.	648.3	1.007	0.0118	0.0946	0.00408	23.98	−145.00
−140.00	0.605	39.92	128.13	−51.318	153.671	−0.13918	0.50211	0.4910	0.2868	1.1912	4932.	652.7	0.963	0.0119	0.0936	0.00420	23.56	−140.00
−135.00	0.753	39.74	104.41	−48.858	155.061	−0.13155	0.49657	0.4927	0.2893	1.1900	4872.	657.0	0.923	0.0121	0.0926	0.00432	23.15	−135.00
−130.00	0.931	39.56	85.699	−46.389	156.457	−0.12401	0.49133	0.4945	0.2919	1.1889	4813.	661.2	0.885	0.0123	0.0916	0.00444	22.74	−130.00
−125.00	1.143	39.37	70.830	−43.911	157.857	−0.11655	0.48638	0.4964	0.2946	1.1879	4754.	665.3	0.850	0.0125	0.0906	0.00456	22.33	−125.00
−120.00	1.393	39.19	58.927	−41.424	159.262	−0.10917	0.48169	0.4983	0.2973	1.1869	4695.	669.3	0.816	0.0126	0.0896	0.00468	21.92	−120.00
−115.00	1.686	39.01	49.331	−38.927	160.672	−0.10188	0.47726	0.5003	0.3001	1.1860	4636.	673.2	0.785	0.0128	0.0886	0.00480	21.51	−115.00
−110.00	2.029	38.82	41.543	−36.419	162.086	−0.09466	0.47307	0.5024	0.3030	1.1853	4577.	677.1	0.755	0.0130	0.0877	0.00493	21.10	−110.00
−105.00	2.427	38.64	35.182	−33.900	163.503	−0.08751	0.46910	0.5045	0.3060	1.1846	4518.	680.8	0.727	0.0132	0.0867	0.00505	20.70	−105.00
−100.00	2.886	38.45	29.955	−31.370	164.924	−0.08043	0.46536	0.5067	0.3090	1.1840	4459.	684.4	0.701	0.0133	0.0857	0.00518	20.29	−100.00
−95.00	3.414	38.27	25.636	−28.829	166.348	−0.07342	0.46182	0.5090	0.3121	1.1834	4401.	687.9	0.676	0.0135	0.0847	0.00531	19.89	−95.00
−90.00	4.017	38.08	22.046	−26.276	167.774	−0.06648	0.45848	0.5114	0.3153	1.1830	4342.	691.2	0.652	0.0137	0.0837	0.00544	19.49	−90.00
−85.00	4.704	37.89	19.046	−23.710	169.202	−0.05959	0.45532	0.5138	0.3185	1.1827	4283.	694.5	0.629	0.0138	0.0827	0.00557	19.08	−85.00
−80.00	5.482	37.70	16.528	−21.132	170.631	−0.05277	0.45234	0.5164	0.3219	1.1825	4225.	697.6	0.608	0.0140	0.0817	0.00570	18.68	−80.00
−75.00	6.360	37.51	14.403	−18.541	172.062	−0.04600	0.44952	0.5190	0.3253	1.1825	4166.	700.6	0.587	0.0142	0.0808	0.00583	18.29	−75.00
−70.00	7.347	37.32	12.601	−15.937	173.493	−0.03928	0.44687	0.5216	0.3288	1.1825	4108.	703.5	0.568	0.0144	0.0798	0.00596	17.89	−70.00
−65.00	8.452	37.12	11.067	−13.318	174.925	−0.03262	0.44436	0.5244	0.3324	1.1826	4049.	706.2	0.549	0.0145	0.0788	0.00610	17.49	−65.00
−60.00	9.684	36.93	9.7549	−10.685	176.356	−0.02601	0.44200	0.5273	0.3361	1.1829	3990.	708.8	0.531	0.0147	0.0779	0.00624	17.10	−60.00
−55.00	11.053	36.73	8.6283	−8.037	177.788	−0.01944	0.43978	0.5302	0.3399	1.1833	3932.	711.2	0.514	0.0149	0.0769	0.00637	16.70	−55.00
−50.00	12.571	36.53	7.6570	−5.374	179.218	−0.01292	0.43769	0.5332	0.3437	1.1838	3873.	713.5	0.498	0.0150	0.0760	0.00651	16.31	−50.00
−45.00	14.247	36.33	6.8166	−2.695	180.646	−0.00644	0.43572	0.5364	0.3477	1.1845	3815.	715.6	0.482	0.0152	0.0750	0.00665	15.92	−45.00
−43.75b	14.691	36.28	6.6248	−2.024	181.002	−0.00483	0.43524	0.5372	0.3487	1.1847	3800.	716.1	0.479	0.0153	0.0748	0.00669	15.82	−43.75
−40.00	16.092	36.13	6.0867	0.000	182.073	0.00000	0.43386	0.5396	0.3518	1.1853	3756.	717.5	0.467	0.0154	0.0741	0.00680	15.53	−40.00
−35.00	18.118	35.93	5.4506	2.712	183.497	0.00640	0.43212	0.5429	0.3559	1.1863	3697.	719.3	0.453	0.0156	0.0732	0.00694	15.15	−35.00
−30.00	20.336	35.73	4.8943	5.442	184.918	0.01276	0.43048	0.5464	0.3602	1.1874	3639.	721.0	0.439	0.0157	0.0723	0.00709	14.76	−30.00
−25.00	22.758	35.52	4.4062	8.189	186.337	0.01909	0.42895	0.5499	0.3646	1.1887	3580.	722.4	0.426	0.0159	0.0713	0.00724	14.38	−25.00
−20.00	25.397	35.32	3.9767	10.955	187.751	0.02539	0.42751	0.5535	0.3690	1.1902	3521.	723.7	0.413	0.0161	0.0704	0.00739	13.99	−20.00
−15.00	28.264	35.11	3.5976	13.739	189.160	0.03165	0.42616	0.5573	0.3736	1.1918	3462.	724.8	0.401	0.0163	0.0695	0.00754	13.61	−15.00
−10.00	31.371	34.90	3.2621	16.543	190.565	0.03788	0.42489	0.5612	0.3783	1.1936	3403.	725.7	0.389	0.0164	0.0686	0.00769	13.23	−10.00
−5.00	34.733	34.68	2.9642	19.368	191.964	0.04409	0.42370	0.5651	0.3832	1.1957	3344.	726.4	0.378	0.0166	0.0677	0.00785	12.86	−5.00
0.00	38.361	34.47	2.6990	22.212	193.357	0.05027	0.42260	0.5693	0.3881	1.1979	3285.	726.9	0.367	0.0168	0.0668	0.00801	12.48	0.00
5.00	42.269	34.25	2.4625	25.078	194.742	0.05643	0.42156	0.5735	0.3932	1.2004	3226.	727.2	0.356	0.0170	0.0660	0.00817	12.11	5.00
10.00	46.471	34.03	2.2508	27.966	196.121	0.06256	0.42059	0.5779	0.3984	1.2031	3167.	727.3	0.346	0.0172	0.0651	0.00834	11.74	10.00
15.00	50.980	33.80	2.0610	30.877	197.491	0.06867	0.41968	0.5825	0.4038	1.2061	3108.	727.2	0.336	0.0174	0.0642	0.00850	11.37	15.00
20.00	55.809	33.58	1.8904	33.811	198.851	0.07476	0.41884	0.5872	0.4094	1.2094	3049.	726.9	0.326	0.0175	0.0634	0.00868	11.00	20.00
25.00	60.972	33.35	1.7368	36.768	200.202	0.08084	0.41805	0.5920	0.4151	1.2130	2989.	726.3	0.317	0.0177	0.0625	0.00885	10.64	25.00
30.00	66.485	33.12	1.5980	39.751	201.542	0.08690	0.41731	0.5971	0.4210	1.2169	2930.	725.5	0.308	0.0179	0.0617	0.00903	10.27	30.00
35.00	72.361	32.88	1.4725	42.759	202.870	0.09294	0.41662	0.6023	0.4271	1.2212	2870.	724.5	0.299	0.0181	0.0609	0.00921	9.91	35.00
40.00	78.614	32.64	1.3586	45.794	204.185	0.09898	0.41597	0.6077	0.4334	1.2258	2810.	723.2	0.291	0.0183	0.0600	0.00940	9.55	40.00
45.00	85.259	32.40	1.2552	48.856	205.487	0.10500	0.41536	0.6133	0.4399	1.2309	2750.	721.7	0.283	0.0185	0.0592	0.00959	9.20	45.00
50.00	92.311	32.15	1.1611	51.946	206.773	0.11101	0.41479	0.6192	0.4466	1.2364	2690.	719.9	0.275	0.0188	0.0584	0.00979	8.84	50.00
55.00	99.785	31.90	1.0753	55.065	208.044	0.11702	0.41425	0.6253	0.4537	1.2425	2630.	717.9	0.267	0.0190	0.0576	0.00999	8.49	55.00
60.00	107.70	31.65	0.9968	58.215	209.296	0.12302	0.41374	0.6316	0.4610	1.2490	2570.	715.5	0.259	0.0192	0.0568	0.01020	8.15	60.00
65.00	116.06	31.39	0.9250	61.396	210.530	0.12902	0.41325	0.6383	0.4687	1.2563	2510.	712.9	0.252	0.0194	0.0561	0.01042	7.80	65.00
70.00	124.89	31.12	0.8592	64.609	211.742	0.13501	0.41279	0.6452	0.4767	1.2641	2449.	710.0	0.245	0.0197	0.0553	0.01064	7.46	70.00
75.00	134.21	30.85	0.7988	67.856	212.932	0.14101	0.41234	0.6525	0.4851	1.2728	2388.	706.9	0.238	0.0199	0.0545	0.01087	7.12	75.00
80.00	144.02	30.58	0.7432	71.138	214.098	0.14701	0.41190	0.6602	0.4940	1.2823	2327.	703.4	0.231	0.0202	0.0538	0.01111	6.78	80.00
85.00	154.35	30.30	0.6919	74.456	215.237	0.15301	0.41148	0.6683	0.5034	1.2928	2266.	699.5	0.224	0.0204	0.0530	0.01136	6.45	85.00
90.00	165.21	30.01	0.6446	77.813	216.347	0.15902	0.41105	0.6768	0.5133	1.3043	2205.	695.4	0.218	0.0207	0.0523	0.01161	6.12	90.00
95.00	176.63	29.72	0.6009	81.209	217.426	0.16505	0.41062	0.6858	0.5239	1.3171	2144.	690.9	0.211	0.0210	0.0515	0.01188	5.79	95.00
100.00	188.61	29.42	0.5604	84.647	218.470	0.17108	0.41019	0.6954	0.5353	1.3313	2082.	686.1	0.205	0.0213	0.0508	0.01216	5.46	100.00
110.00	214.33	28.79	0.4880	91.656	220.442	0.18320	0.40927	0.7166	0.5606	1.3648	1958.	675.4	0.193	0.0219	0.0494	0.01276	4.83	110.00
120.00	242.54	28.13	0.4253	98.858	222.234	0.19542	0.40825	0.7411	0.5905	1.4072	1832.	663.2	0.181	0.0226	0.0480	0.01343	4.21	120.00
130.00	273.39	27.43	0.3706	106.279	223.804	0.20776	0.40706	0.7703	0.6269	1.4618	1705.	649.3	0.169	0.0234	0.0466	0.01418	3.60	130.00
140.00	307.04	26.67	0.3227	113.950	225.102	0.22027	0.40562	0.8059	0.6726	1.5343	1576.	633.8	0.158	0.0243	0.0452	0.01504	3.02	140.00
150.00	343.66	25.85	0.2803	121.915	226.055	0.23301	0.40382	0.8514	0.7326	1.6342	1444.	616.4	0.147	0.0253	0.0439	0.01604	2.46	150.00
160.00	383.47	24.96	0.2425	130.243	226.559	0.24609	0.40151	0.9129	0.8159	1.7787	1307.	597.0	0.136	0.0266	0.0426	0.01725	1.92	160.00
170.00	426.68	23.94	0.2084	139.042	226.450	0.25965	0.39846	1.004	0.9416	2.0046	1165.	575.3	0.125	0.0281	0.0413	0.01878	1.41	170.00
180.00	473.57	22.76	0.1769	148.512	225.435	0.27398	0.39423	1.158	1.1574	2.4031	1013.	551.0	0.113	0.0301	0.0400	0.02085	0.93	180.00
190.00	524.49	21.29	0.1469	159.110	222.880	0.28975	0.38791	1.495	1.6294	3.2933	846.	523.3	0.100	0.0330	0.0389	0.02408	0.51	190.00
200.00	579.96	19.05	0.1147	172.492	216.579	0.30941	0.37624	2.986	3.6492	7.1557	651.	490.4	0.083	0.0381	0.0395	0.03135	0.15	200.00
206.06c	616.07	13.76	0.0727	195.434	195.434	0.34339	0.34339	∞	∞	∞	0.	0.0	–	–	∞	∞	0.00	206.06

*Temperatures have been converted from the IPTS-68 scale of the original formulation to the ITS-90 scale

b = normal boiling point

c = critical point

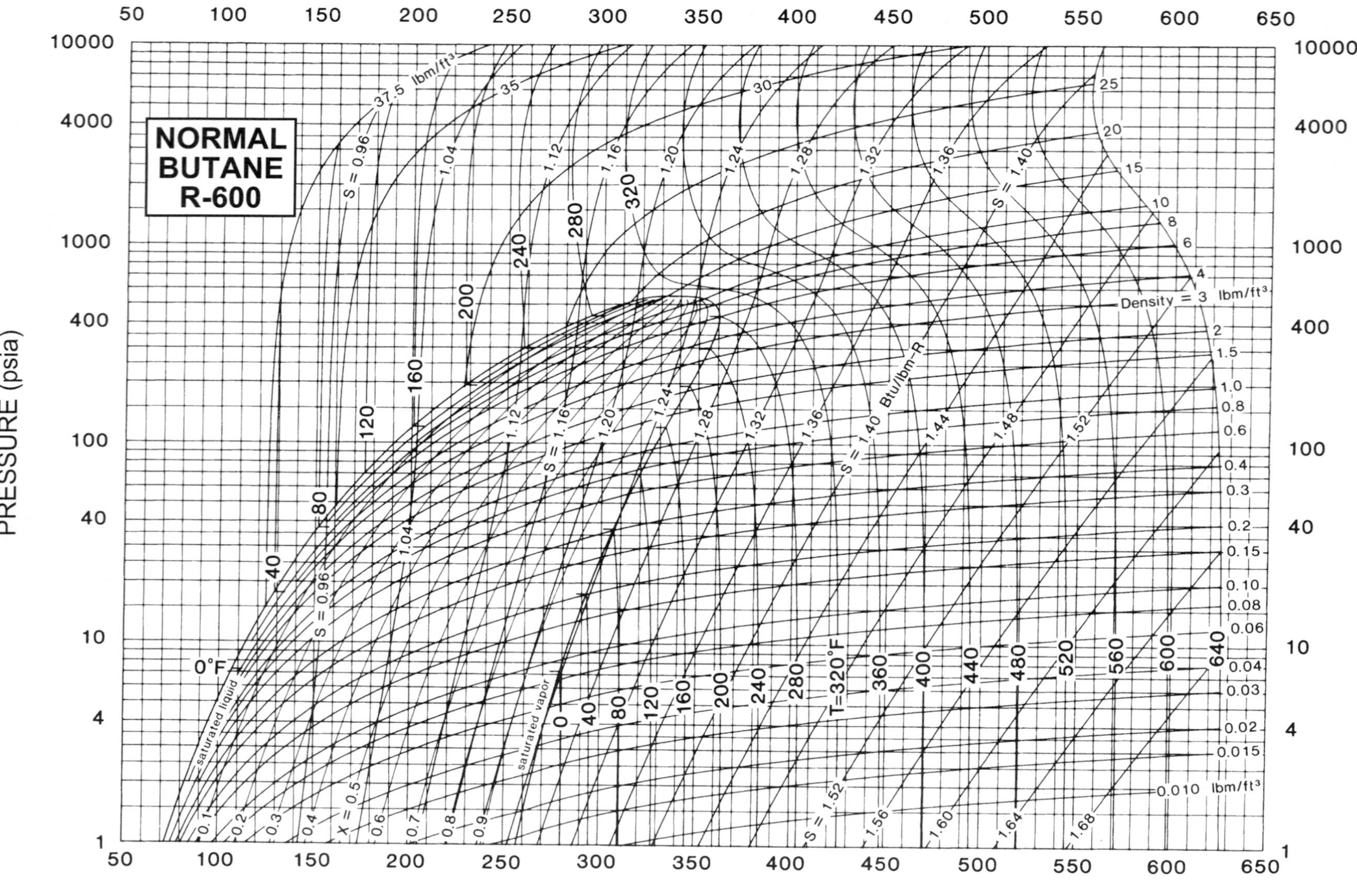

Fig. 18 Pressure-Enthalpy Diagram for Refrigerant 600 (*n*-Butane)

Note: The reference states for enthalpy and entropy differ from those in the table.

Refrigerant 600 (*n*-Butane) Properties of Saturated Liquid and Saturated Vapor

Temp.,* °F	Pres-sure, psia	Denisty, lb/ft^3 Liquid	Volume, ft^3/lb Vapor	Enthalpy, Btu/lb Liquid	Enthalpy, Btu/lb Vapor	Entropy, Btu/lb·°F Liquid	Entropy, Btu/lb·°F Vapor	Specific Heat c_p, Btu/lb·°F Liquid	Specific Heat c_p, Btu/lb·°F Vapor	c_p/c_v Vapor	Vel. of Sound, ft/s Liquid	Vel. of Sound, ft/s Vapor	Viscosity, lb$_m$/ft·h Liquid	Viscosity, lb$_m$/ft·h Vapor	Thermal Cond., Btu/h·ft·°F Liquid	Thermal Cond., Btu/h·ft·°F Vapor	Surface Tension, dyne/cm	Temp., °F
–150.00	0.021	43.72	2684.3	–54.039	145.704	–0.14907	0.49600	0.4804	0.2930	1.1323	4976.	547.5	1.945	0.0109	0.0991	0.00362	28.19	–150.00
–140.00	0.038	43.39	1539.1	–49.233	148.646	–0.13379	0.48526	0.4807	0.2971	1.1304	4906.	555.7	1.738	0.0112	0.0967	0.00381	27.41	–140.00
–130.00	0.066	43.06	918.21	–44.422	151.625	–0.11897	0.47574	0.4815	0.3012	1.1286	4833.	563.7	1.564	0.0115	0.0944	0.00399	26.64	–130.00
–120.00	0.110	42.74	567.75	–39.601	154.639	–0.10456	0.46732	0.4829	0.3055	1.1268	4757.	571.6	1.417	0.0118	0.0922	0.00419	25.87	–120.00
–110.00	0.178	42.41	362.61	–34.763	157.690	–0.09052	0.45989	0.4848	0.3099	1.1251	4678.	579.2	1.291	0.0121	0.0901	0.00439	25.11	–110.00
–100.00	0.278	42.08	238.50	–29.901	160.775	–0.07682	0.45335	0.4873	0.3145	1.1236	4597.	586.6	1.182	0.0124	0.0881	0.00459	24.35	–100.00
–90.00	0.422	41.75	161.12	–25.013	163.894	–0.06341	0.44763	0.4903	0.3192	1.1221	4514.	593.8	1.087	0.0127	0.0862	0.00480	23.59	–90.00
–80.00	0.626	41.41	111.54	–20.092	167.046	–0.05028	0.44264	0.4938	0.3241	1.1208	4428.	600.8	1.004	0.0130	0.0843	0.00502	22.85	–80.00
–70.00	0.906	41.08	78.954	–15.134	170.229	–0.03739	0.43832	0.4976	0.3292	1.1196	4341.	607.5	0.930	0.0133	0.0826	0.00524	22.10	–70.00
–60.00	1.284	40.74	57.038	–10.135	173.442	–0.02473	0.43461	0.5019	0.3344	1.1185	4252.	614.0	0.865	0.0136	0.0809	0.00547	21.36	–60.00
–50.00	1.785	40.40	41.981	–5.091	176.684	–0.01227	0.43146	0.5065	0.3399	1.1177	4162.	620.1	0.806	0.0139	0.0793	0.00571	20.63	–50.00
–45.00	2.090	40.23	36.251	–2.552	178.314	–0.00611	0.43007	0.5089	0.3427	1.1173	4116.	623.1	0.779	0.0141	0.0785	0.00583	20.26	–45.00
–40.00	2.437	40.06	31.432	0.000	179.951	0.00000	0.42881	0.5114	0.3456	1.1170	4070.	625.9	0.754	0.0143	0.0777	0.00595	19.90	–40.00
–35.00	2.829	39.89	27.360	2.565	181.594	0.00607	0.42766	0.5139	0.3485	1.1168	4024.	628.7	0.729	0.0144	0.0769	0.00608	19.54	–35.00
–30.00	3.271	39.71	23.905	5.142	183.244	0.01210	0.42662	0.5165	0.3515	1.1166	3977.	631.4	0.706	0.0146	0.0762	0.00621	19.18	–30.00
–25.00	3.768	39.54	20.961	7.734	184.898	0.01809	0.42569	0.5192	0.3545	1.1164	3930.	634.1	0.684	0.0148	0.0754	0.00634	18.82	–25.00
–20.00	4.325	39.36	18.444	10.339	186.559	0.02404	0.42485	0.5219	0.3576	1.1163	3883.	636.6	0.663	0.0149	0.0747	0.00647	18.46	–20.00
–15.00	4.945	39.19	16.283	12.957	188.224	0.02996	0.42412	0.5247	0.3608	1.1163	3836.	639.0	0.643	0.0151	0.0740	0.00660	18.10	–15.00
–10.00	5.636	39.01	14.421	15.590	189.894	0.03584	0.42347	0.5275	0.3640	1.1163	3789.	641.3	0.623	0.0153	0.0733	0.00674	17.75	–10.00
–5.00	6.402	38.83	12.810	18.238	191.569	0.04169	0.42292	0.5304	0.3673	1.1164	3741.	643.6	0.605	0.0154	0.0726	0.00687	17.40	–5.00
0.00	7.249	38.66	11.413	20.901	193.248	0.04750	0.42245	0.5334	0.3706	1.1166	3693.	645.7	0.587	0.0156	0.0719	0.00701	17.04	0.00
5.00	8.182	38.48	10.197	23.578	194.930	0.05329	0.42205	0.5364	0.3740	1.1168	3645.	647.7	0.570	0.0158	0.0712	0.00715	16.69	5.00
10.00	9.209	38.30	9.1351	26.271	196.617	0.05904	0.42174	0.5395	0.3774	1.1171	3597.	649.6	0.554	0.0159	0.0705	0.00730	16.34	10.00
15.00	10.335	38.12	8.2048	28.980	198.307	0.06477	0.42150	0.5426	0.3810	1.1175	3549.	651.4	0.539	0.0161	0.0698	0.00745	16.00	15.00
20.00	11.567	37.93	7.3877	31.704	200.000	0.07046	0.42132	0.5457	0.3845	1.1180	3500.	653.1	0.524	0.0163	0.0691	0.00760	15.65	20.00
25.00	12.912	37.75	6.6678	34.444	201.696	0.07613	0.42122	0.5489	0.3882	1.1185	3452.	654.6	0.509	0.0165	0.0685	0.00775	15.31	25.00
30.00	14.376	37.56	6.0318	37.202	203.394	0.08178	0.42118	0.5522	0.3919	1.1191	3403.	656.1	0.495	0.0166	0.0678	0.00790	14.96	30.00
31.04b	14.696	37.52	5.9091	37.776	203.747	0.08295	0.42118	0.5529	0.3927	1.1193	3393.	656.4	0.492	0.0167	0.0677	0.00793	14.89	31.04
35.00	15.967	37.38	5.4684	39.975	205.095	0.08740	0.42120	0.5555	0.3957	1.1198	3354.	657.4	0.482	0.0168	0.0672	0.00806	14.62	35.00
40.00	17.692	37.19	4.9681	42.766	206.797	0.09299	0.42127	0.5588	0.3995	1.1206	3305.	658.6	0.469	0.0170	0.0665	0.00822	14.28	40.00
45.00	19.558	37.00	4.5228	45.574	208.501	0.09857	0.42140	0.5623	0.4035	1.1215	3256.	659.6	0.456	0.0172	0.0659	0.00838	13.94	45.00
50.00	21.573	36.81	4.1254	48.399	210.206	0.10412	0.42159	0.5657	0.4075	1.1225	3207.	660.5	0.444	0.0174	0.0652	0.00854	13.61	50.00
55.00	23.745	36.61	3.7699	51.242	211.912	0.10965	0.42182	0.5692	0.4115	1.1236	3158.	661.3	0.432	0.0176	0.0646	0.00871	13.27	55.00
60.00	26.082	36.42	3.4512	54.103	213.618	0.11516	0.42211	0.5728	0.4157	1.1248	3108.	662.0	0.421	0.0178	0.0639	0.00888	12.94	60.00
65.00	28.591	36.22	3.1649	56.983	215.325	0.12065	0.42244	0.5765	0.4199	1.1261	3059.	662.4	0.410	0.0179	0.0633	0.00905	12.61	65.00
70.00	31.281	36.03	2.9071	59.881	217.031	0.12612	0.42281	0.5802	0.4242	1.1276	3009.	662.8	0.400	0.0181	0.0627	0.00923	12.28	70.00
75.00	34.160	35.83	2.6746	62.798	218.736	0.13157	0.42322	0.5839	0.4286	1.1291	2960.	663.0	0.389	0.0183	0.0621	0.00941	11.95	75.00
80.00	37.236	35.63	2.4644	65.734	220.440	0.13701	0.42367	0.5878	0.4331	1.1308	2910.	663.0	0.379	0.0185	0.0614	0.00959	11.63	80.00
85.00	40.518	35.42	2.2741	68.689	222.142	0.14243	0.42416	0.5917	0.4377	1.1326	2860.	662.9	0.369	0.0187	0.0608	0.00978	11.30	85.00
90.00	44.015	35.22	2.1013	71.665	223.843	0.14783	0.42468	0.5956	0.4424	1.1346	2810.	662.6	0.360	0.0189	0.0602	0.00996	10.98	90.00
95.00	47.735	35.01	1.9443	74.660	225.541	0.15322	0.42523	0.5997	0.4472	1.1367	2760.	662.1	0.351	0.0191	0.0596	0.01016	10.66	95.00
100.00	51.688	34.80	1.8013	77.677	227.236	0.15860	0.42581	0.6038	0.4521	1.1390	2710.	661.5	0.342	0.0194	0.0590	0.01035	10.34	100.00
105.00	55.882	34.59	1.6708	80.714	228.927	0.16396	0.42643	0.6081	0.4571	1.1415	2660.	660.7	0.333	0.0196	0.0584	0.01055	10.03	105.00
110.00	60.326	34.37	1.5517	83.773	230.614	0.16931	0.42707	0.6124	0.4622	1.1442	2610.	659.7	0.325	0.0198	0.0578	0.01075	9.71	110.00
115.00	65.030	34.15	1.4426	86.853	232.296	0.17465	0.42773	0.6168	0.4675	1.1470	2559.	658.6	0.316	0.0200	0.0572	0.01096	9.40	115.00
120.00	70.002	33.93	1.3426	89.956	233.974	0.17998	0.42842	0.6213	0.4729	1.1501	2509.	657.2	0.308	0.0202	0.0566	0.01117	9.09	120.00
125.00	75.253	33.71	1.2508	93.081	235.645	0.18530	0.42912	0.6260	0.4785	1.1534	2458.	655.7	0.300	0.0204	0.0560	0.01138	8.78	125.00
130.00	80.792	33.48	1.1664	96.229	237.309	0.19061	0.42985	0.6307	0.4842	1.1570	2408.	653.9	0.293	0.0207	0.0554	0.01160	8.48	130.00
135.00	86.628	33.25	1.0887	99.401	238.965	0.19591	0.43059	0.6356	0.4900	1.1609	2357.	651.9	0.285	0.0209	0.0548	0.01182	8.17	135.00
140.00	92.771	33.02	1.0170	102.597	240.614	0.20120	0.43134	0.6407	0.4961	1.1650	2306.	649.8	0.278	0.0211	0.0542	0.01205	7.87	140.00
145.00	99.230	32.78	0.9508	105.818	242.253	0.20649	0.43211	0.6459	0.5024	1.1695	2256.	647.4	0.271	0.0214	0.0536	0.01228	7.58	145.00
150.00	106.02	32.54	0.8896	109.064	243.881	0.21177	0.43289	0.6512	0.5089	1.1744	2205.	644.7	0.264	0.0216	0.0530	0.01251	7.28	150.00
160.00	120.61	32.05	0.7804	115.635	247.104	0.22233	0.43447	0.6625	0.5226	1.1853	2103.	638.8	0.250	0.0221	0.0518	0.01300	6.69	160.00
170.00	136.64	31.54	0.6863	122.317	250.273	0.23287	0.43607	0.6747	0.5375	1.1982	2000.	631.8	0.237	0.0227	0.0507	0.01350	6.12	170.00
180.00	154.18	31.01	0.6047	129.118	253.376	0.24342	0.43766	0.6880	0.5538	1.2135	1897.	623.8	0.224	0.0232	0.0496	0.01403	5.56	180.00
190.00	173.32	30.45	0.5337	136.044	256.399	0.25399	0.43923	0.7027	0.5719	1.2318	1794.	614.6	0.212	0.0238	0.0484	0.01458	5.01	190.00
200.00	194.17	29.87	0.4715	143.108	259.328	0.26458	0.44075	0.7191	0.5924	1.2541	1690.	604.2	0.200	0.0244	0.0473	0.01516	4.47	200.00
210.00	216.80	29.26	0.4168	150.320	262.140	0.27522	0.44218	0.7378	0.6158	1.2815	1585.	592.5	0.188	0.0251	0.0462	0.01577	3.94	210.00
220.00	241.34	28.62	0.3685	157.700	264.809	0.28592	0.44350	0.7595	0.6434	1.3159	1479.	579.3	0.177	0.0258	0.0451	0.01641	3.43	220.00
230.00	267.88	27.93	0.3255	165.267	267.303	0.29672	0.44466	0.7855	0.6767	1.3601	1372.	564.5	0.165	0.0266	0.0440	0.01710	2.93	230.00
240.00	296.55	27.19	0.2869	173.052	269.574	0.30765	0.44559	0.8175	0.7184	1.4186	1263.	547.9	0.154	0.0274	0.0429	0.01784	2.45	240.00
250.00	327.49	26.39	0.2522	181.097	271.558	0.31876	0.44622	0.8589	0.7729	1.4991	1152.	529.3	0.143	0.0284	0.0419	0.01867	1.99	250.00
260.00	360.86	25.50	0.2206	189.465	273.160	0.33014	0.44642	0.9159	0.8487	1.6163	1037.	508.5	0.132	0.0295	0.0407	0.01962	1.55	260.00
270.00	396.83	24.50	0.1914	198.261	274.225	0.34190	0.44600	1.002	0.9639	1.8009	918.	485.1	0.121	0.0308	0.0396	0.02082	1.13	270.00
280.00	435.64	23.31	0.1639	207.679	274.476	0.35431	0.44461	1.150	1.1643	2.1324	791.	458.6	0.109	0.0325	0.0387	0.02257	0.75	280.00
290.00	477.60	21.79	0.1368	218.172	273.293	0.36794	0.44146	1.489	1.6165	2.8969	654.	428.2	0.096	0.0348	0.0385	0.02573	0.40	290.00
300.00	523.22	19.39	0.1067	231.451	268.460	0.38498	0.43370	3.153	3.7350	6.5277	497.	392.9	0.078	0.0393	0.0411	0.03394	0.11	300.00
305.62c	550.56	14.22	0.0703	251.565	251.565	0.41095	0.41095	∞	∞	∞	0.	0.0	–	–	∞	∞	0.00	305.62

*Temperatures have been converted from the IPTS-68 scale of the original formulation to the ITS-90 scale b = normal boiling point c = critical point

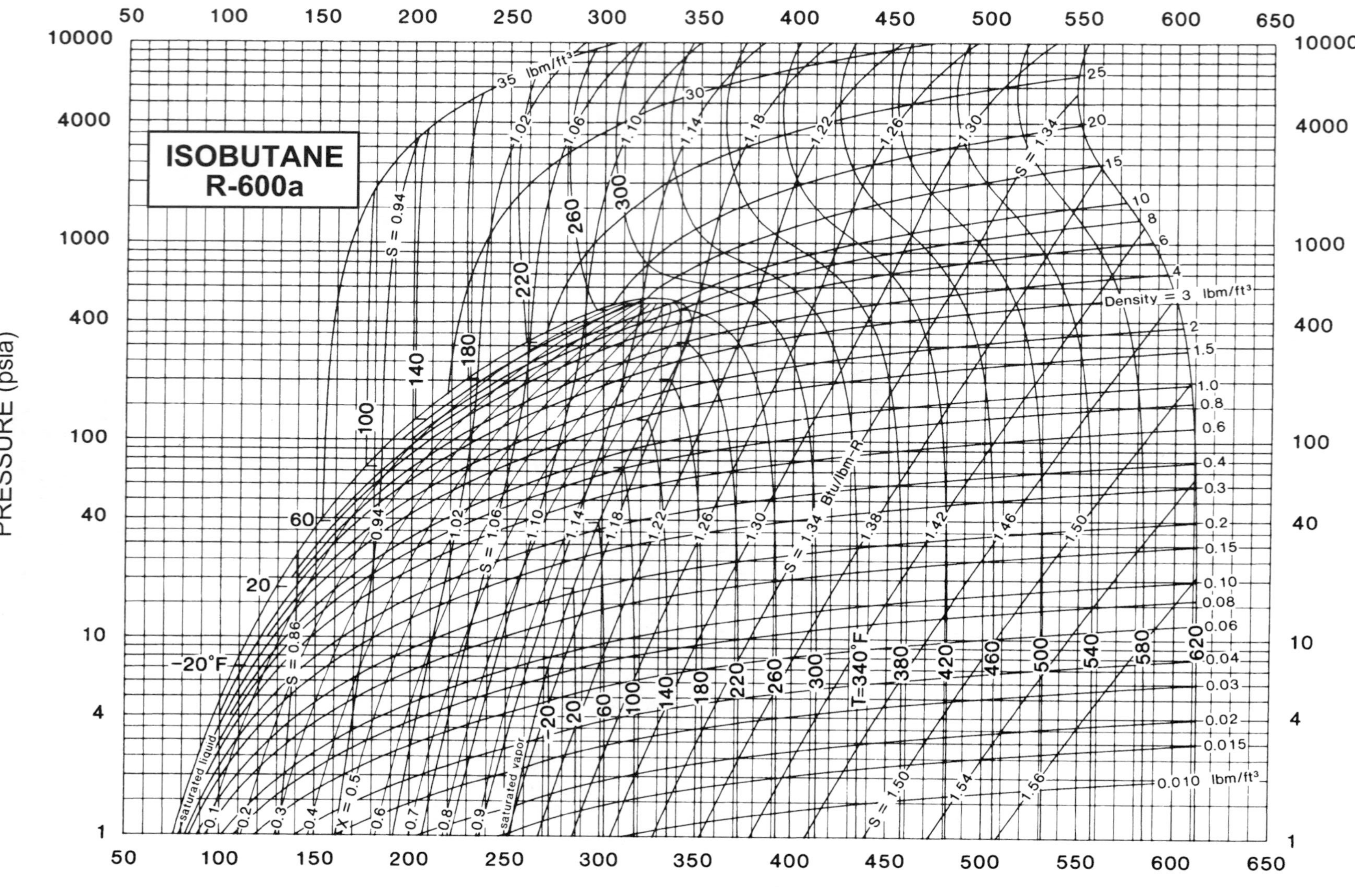

Fig. 19 Pressure-Enthalpy Diagram for Refrigerant 600a (Isobutane)

Note: The reference states for enthalpy and entropy differ from those in the table.

Refrigerant 600a (Isobutane) Properties of Saturated Liquid and Saturated Vapor

Temp.,* °F	Pres-sure, psia	Denisty, lb/ft³ Liquid	Volume, ft³/lb Vapor	Enthalpy, Btu/lb Liquid	Enthalpy, Btu/lb Vapor	Entropy, Btu/lb·°F Liquid	Entropy, Btu/lb·°F Vapor	Specific Heat c_p, Btu/lb·°F Liquid	Specific Heat c_p, Btu/lb·°F Vapor	c_p/c_v Vapor	Vel. of Sound, ft/s Liquid	Vel. of Sound, ft/s Vapor	Viscosity, lb_m/ft·h Liquid	Viscosity, lb_m/ft·h Vapor	Thermal Cond., Btu/h·ft·°F Liquid	Thermal Cond., Btu/h·ft·°F Vapor	Surface Tension, dyne/cm	Temp., °F
–150.00	0.047	42.76	1221.5	–52.324	136.454	–0.14411	0.46554	0.4486	0.2693	1.1460	4760.	550.6	2.313	0.0108	0.0799	0.00345	25.67	–150.00
–140.00	0.082	42.42	723.13	–47.814	139.156	–0.12978	0.45514	0.4532	0.2747	1.1431	4675.	558.5	2.041	0.0111	0.0800	0.00363	24.95	–140.00
–130.00	0.137	42.08	444.42	–43.259	141.903	–0.11575	0.44595	0.4579	0.2801	1.1404	4589.	566.3	1.814	0.0114	0.0798	0.00382	24.24	–130.00
–120.00	0.222	41.74	282.54	–38.656	144.694	–0.10200	0.43783	0.4626	0.2856	1.1379	4503.	573.8	1.623	0.0117	0.0794	0.00402	23.53	–120.00
–110.00	0.348	41.40	185.21	–34.005	147.527	–0.08850	0.43068	0.4675	0.2911	1.1356	4417.	581.1	1.462	0.0120	0.0789	0.00422	22.82	–110.00
–100.00	0.530	41.06	124.83	–29.304	150.399	–0.07525	0.42441	0.4724	0.2968	1.1336	4329.	588.2	1.323	0.0123	0.0783	0.00442	22.11	–100.00
–90.00	0.787	40.72	86.295	–24.554	153.309	–0.06222	0.41894	0.4775	0.3026	1.1317	4242.	595.1	1.204	0.0126	0.0775	0.00464	21.40	–90.00
–80.00	1.140	40.37	61.046	–19.751	156.255	–0.04941	0.41419	0.4827	0.3086	1.1301	4153.	601.6	1.099	0.0130	0.0766	0.00486	20.70	–80.00
–70.00	1.617	40.02	44.103	–14.896	159.234	–0.03679	0.41009	0.4880	0.3148	1.1288	4064.	607.9	1.008	0.0133	0.0755	0.00509	20.00	–70.00
–60.00	2.246	39.67	32.481	–9.987	162.245	–0.02436	0.40659	0.4934	0.3212	1.1277	3974.	613.9	0.928	0.0136	0.0744	0.00533	19.30	–60.00
–50.00	3.063	39.32	24.346	–5.022	165.285	–0.01210	0.40363	0.4990	0.3277	1.1269	3883.	619.5	0.856	0.0139	0.0732	0.00557	18.60	–50.00
–45.00	3.554	39.14	21.207	–2.518	166.815	–0.00603	0.40234	0.5018	0.3311	1.1266	3837.	622.2	0.823	0.0141	0.0726	0.00570	18.26	–45.00
–40.00	4.106	38.96	18.544	0.000	168.352	0.00000	0.40117	0.5047	0.3345	1.1263	3791.	624.8	0.792	0.0143	0.0719	0.00583	17.91	–40.00
–35.00	4.726	38.78	16.275	2.533	169.895	0.00599	0.40010	0.5076	0.3380	1.1262	3745.	627.3	0.763	0.0144	0.0713	0.00596	17.57	–35.00
–30.00	5.418	38.60	14.334	5.081	171.444	0.01195	0.39915	0.5106	0.3416	1.1261	3698.	629.7	0.735	0.0146	0.0706	0.00609	17.22	–30.00
–25.00	6.189	38.41	12.667	7.644	172.998	0.01787	0.39830	0.5136	0.3452	1.1260	3651.	632.0	0.709	0.0148	0.0699	0.00622	16.88	–25.00
–20.00	7.046	38.23	11.231	10.222	174.559	0.02376	0.39754	0.5166	0.3489	1.1261	3605.	634.1	0.684	0.0150	0.0693	0.00636	16.54	–20.00
–15.00	7.994	38.04	9.9881	12.816	176.124	0.02962	0.39688	0.5197	0.3526	1.1262	3557.	636.2	0.660	0.0151	0.0686	0.00650	16.19	–15.00
–10.00	9.039	37.86	8.9093	15.426	177.695	0.03544	0.39631	0.5228	0.3564	1.1264	3510.	638.2	0.637	0.0153	0.0678	0.00664	15.85	–10.00
–5.00	10.190	37.67	7.9695	18.052	179.270	0.04124	0.39583	0.5260	0.3602	1.1267	3463.	640.0	0.616	0.0155	0.0671	0.00679	15.51	–5.00
0.00	11.453	37.48	7.1485	20.694	180.849	0.04700	0.39542	0.5292	0.3642	1.1271	3415.	641.7	0.595	0.0157	0.0664	0.00693	15.17	0.00
5.00	12.835	37.29	6.4288	23.352	182.433	0.05274	0.39510	0.5325	0.3682	1.1276	3367.	643.3	0.575	0.0158	0.0657	0.00708	14.84	5.00
10.00	14.344	37.10	5.7960	26.027	184.020	0.05845	0.39485	0.5358	0.3722	1.1281	3319.	644.8	0.556	0.0160	0.0650	0.00724	14.50	10.00
11.10b	14.694	37.06	5.6670	26.619	184.371	0.05971	0.39480	0.5366	0.3731	1.1283	3308.	645.1	0.552	0.0161	0.0648	0.00727	14.43	11.10
15.00	15.987	36.91	5.2381	28.720	185.612	0.06414	0.39467	0.5392	0.3764	1.1288	3271.	646.1	0.538	0.0162	0.0642	0.00739	14.16	15.00
20.00	17.773	36.72	4.7448	31.430	187.206	0.06979	0.39455	0.5427	0.3806	1.1295	3222.	647.3	0.521	0.0164	0.0635	0.00755	13.83	20.00
25.00	19.708	36.52	4.3075	34.157	188.803	0.07543	0.39451	0.5462	0.3849	1.1304	3174.	648.3	0.504	0.0166	0.0628	0.00771	13.50	25.00
30.00	21.802	36.32	3.9187	36.902	190.403	0.08104	0.39452	0.5497	0.3892	1.1313	3125.	649.3	0.488	0.0168	0.0620	0.00787	13.16	30.00
35.00	24.063	36.12	3.5722	39.666	192.005	0.08664	0.39460	0.5534	0.3936	1.1324	3076.	650.0	0.473	0.0170	0.0613	0.00803	12.83	35.00
40.00	26.498	35.92	3.2627	42.448	193.609	0.09221	0.39473	0.5571	0.3982	1.1336	3027.	650.6	0.458	0.0172	0.0606	0.00820	12.50	40.00
45.00	29.117	35.72	2.9855	45.249	195.214	0.09776	0.39491	0.5608	0.4028	1.1349	2978.	651.1	0.444	0.0174	0.0598	0.00837	12.17	45.00
50.00	31.928	35.52	2.7368	48.070	196.821	0.10329	0.39515	0.5647	0.4075	1.1363	2928.	651.4	0.431	0.0176	0.0591	0.00855	11.85	50.00
55.00	34.940	35.31	2.5130	50.910	198.429	0.10881	0.39543	0.5686	0.4122	1.1379	2879.	651.5	0.418	0.0178	0.0584	0.00872	11.52	55.00
60.00	38.163	35.10	2.3112	53.770	200.036	0.11430	0.39576	0.5726	0.4171	1.1396	2829.	651.5	0.405	0.0180	0.0577	0.00890	11.20	60.00
65.00	41.604	34.89	2.1290	56.650	201.644	0.11978	0.39613	0.5767	0.4221	1.1415	2779.	651.3	0.393	0.0182	0.0570	0.00908	10.87	65.00
70.00	45.273	34.68	1.9641	59.551	203.251	0.12525	0.39655	0.5808	0.4272	1.1435	2729.	650.9	0.381	0.0184	0.0562	0.00927	10.55	70.00
75.00	49.179	34.46	1.8145	62.474	204.857	0.13070	0.39700	0.5851	0.4324	1.1457	2679.	650.3	0.370	0.0186	0.0555	0.00946	10.23	75.00
80.00	53.332	34.25	1.6785	65.417	206.461	0.13614	0.39749	0.5894	0.4377	1.1481	2629.	649.5	0.359	0.0189	0.0548	0.00965	9.92	80.00
85.00	57.741	34.03	1.5548	68.383	208.062	0.14157	0.39801	0.5939	0.4431	1.1507	2578.	648.6	0.348	0.0191	0.0541	0.00984	9.60	85.00
90.00	62.417	33.80	1.4419	71.371	209.661	0.14698	0.39856	0.5985	0.4487	1.1536	2528.	647.4	0.338	0.0193	0.0535	0.01004	9.28	90.00
95.00	67.367	33.58	1.3387	74.383	211.256	0.15239	0.39914	0.6032	0.4544	1.1566	2477.	646.0	0.328	0.0195	0.0528	0.01024	8.97	95.00
100.00	72.603	33.35	1.2443	77.418	212.847	0.15778	0.39975	0.6080	0.4602	1.1600	2426.	644.4	0.319	0.0198	0.0521	0.01044	8.66	100.00
105.00	78.135	33.11	1.1578	80.477	214.432	0.16317	0.40039	0.6130	0.4663	1.1636	2375.	642.6	0.309	0.0200	0.0514	0.01065	8.35	105.00
110.00	83.971	32.88	1.0783	83.561	216.011	0.16855	0.40104	0.6181	0.4725	1.1675	2324.	640.5	0.300	0.0203	0.0508	0.01086	8.04	110.00
115.00	90.123	32.64	1.0053	86.670	217.583	0.17392	0.40172	0.6234	0.4789	1.1718	2273.	638.3	0.291	0.0205	0.0501	0.01108	7.74	115.00
120.00	96.600	32.39	0.9379	89.806	219.147	0.17929	0.40241	0.6288	0.4856	1.1764	2222.	635.7	0.283	0.0208	0.0495	0.01129	7.44	120.00
125.00	103.41	32.15	0.8758	92.967	220.701	0.18465	0.40311	0.6345	0.4924	1.1815	2171.	632.9	0.275	0.0211	0.0488	0.01152	7.13	125.00
130.00	110.57	31.90	0.8184	96.157	222.245	0.19001	0.40383	0.6403	0.4996	1.1870	2119.	629.9	0.266	0.0213	0.0482	0.01174	6.84	130.00
135.00	118.09	31.64	0.7653	99.375	223.777	0.19537	0.40456	0.6464	0.5070	1.1930	2067.	626.6	0.259	0.0216	0.0476	0.01197	6.54	135.00
140.00	125.98	31.38	0.7161	102.621	225.296	0.20073	0.40529	0.6527	0.5148	1.1996	2016.	623.0	0.251	0.0219	0.0470	0.01221	6.25	140.00
145.00	134.25	31.11	0.6704	105.898	226.799	0.20609	0.40603	0.6593	0.5229	1.2068	1964.	619.1	0.243	0.0222	0.0464	0.01245	5.96	145.00
150.00	142.90	30.84	0.6280	109.206	228.286	0.21145	0.40676	0.6662	0.5315	1.2148	1911.	615.0	0.236	0.0225	0.0458	0.01269	5.67	150.00
160.00	161.44	30.28	0.5517	115.920	231.201	0.22219	0.40822	0.6811	0.5500	1.2333	1807.	605.7	0.222	0.0232	0.0446	0.01319	5.10	160.00
170.00	181.69	29.69	0.4852	122.775	234.023	0.23297	0.40963	0.6978	0.5711	1.2561	1701.	595.1	0.209	0.0239	0.0434	0.01371	4.54	170.00
180.00	203.75	29.08	0.4270	129.783	236.728	0.24379	0.41097	0.7167	0.5954	1.2846	1594.	583.1	0.196	0.0246	0.0423	0.01426	4.00	180.00
190.00	227.73	28.42	0.3757	136.962	239.287	0.25468	0.41218	0.7386	0.6241	1.3209	1486.	569.6	0.183	0.0254	0.0411	0.01484	3.47	190.00
200.00	253.76	27.73	0.3302	144.332	241.663	0.26568	0.41321	0.7648	0.6593	1.3684	1376.	554.4	0.171	0.0263	0.0400	0.01546	2.95	200.00
210.00	281.96	26.97	0.2897	151.924	243.804	0.27681	0.41400	0.7972	0.7038	1.4324	1264.	537.6	0.159	0.0273	0.0389	0.01614	2.46	210.00
220.00	312.48	26.16	0.2533	159.779	245.641	0.28814	0.41446	0.8390	0.7631	1.5221	1149.	518.8	0.147	0.0285	0.0378	0.01691	1.98	220.00
230.00	345.48	25.25	0.2202	167.962	247.072	0.29974	0.41444	0.8968	0.8470	1.6550	1030.	498.1	0.136	0.0299	0.0367	0.01782	1.53	230.00
240.00	381.15	24.21	0.1899	176.578	247.936	0.31176	0.41374	0.9844	0.9770	1.8685	905.	475.1	0.124	0.0315	0.0356	0.01903	1.10	240.00
250.00	419.71	22.98	0.1614	185.827	247.944	0.32445	0.41197	1.140	1.2092	2.2606	772.	449.4	0.112	0.0337	0.0347	0.02085	0.71	250.00
260.00	461.44	21.39	0.1336	196.186	246.451	0.33846	0.40830	1.511	1.7541	3.2004	627.	420.3	0.098	0.0368	0.0345	0.02406	0.36	260.00
270.00	506.72	18.73	0.1024	209.683	240.972	0.35650	0.39938	3.808	4.7861	8.4943	460.	385.1	0.080	0.0427	0.0369	0.03216	0.08	270.00
274.46c	527.94	14.01	0.0714	226.832	226.832	0.37959	0.37959	∞	∞	∞	0.	0.0	–	–	∞	∞	0.00	274.46

*Temperatures have been converted from the IPTS-68 scale of the original formulation to the ITS-90 scale b = normal boiling point c = critical point

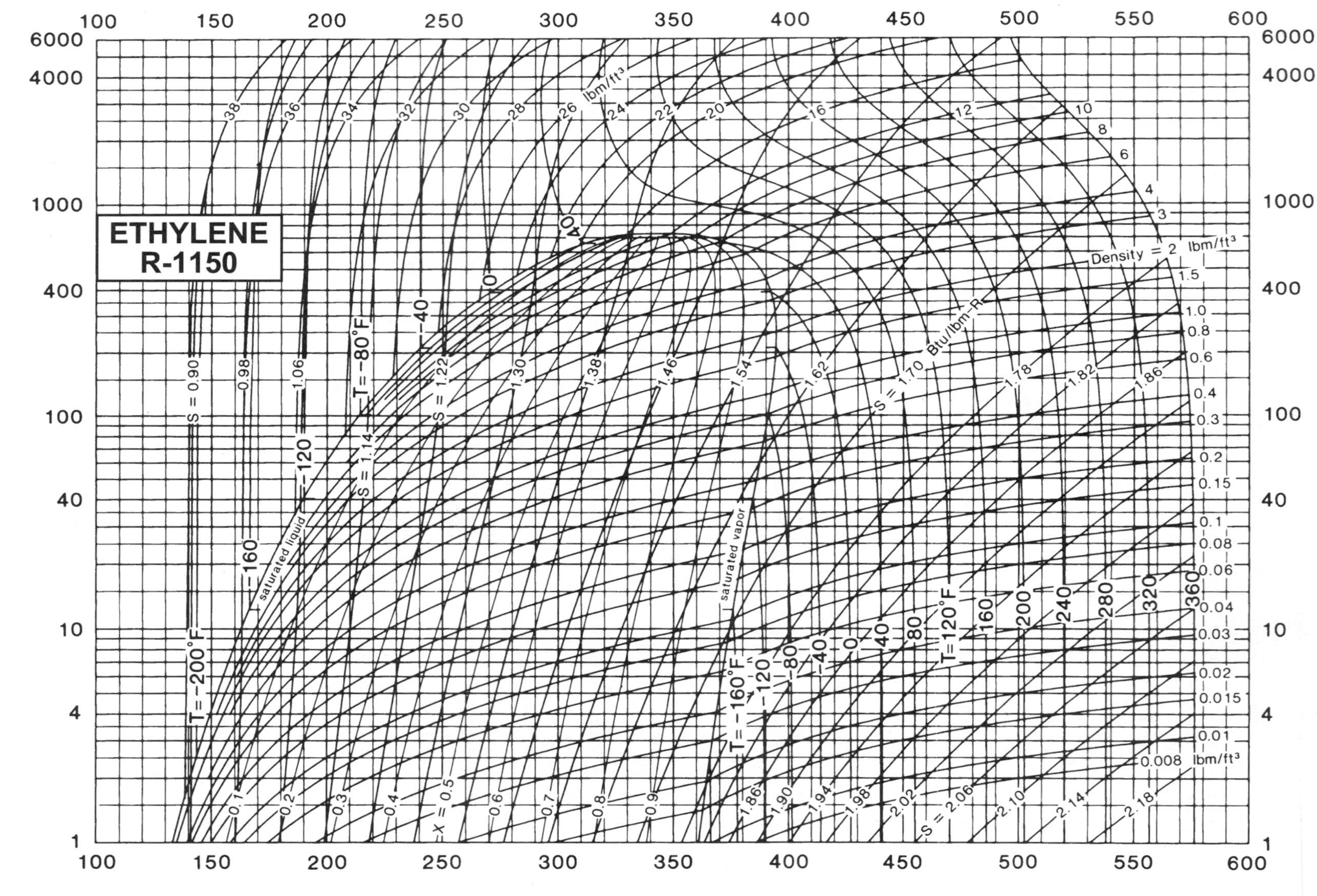

Fig. 20 Pressure-Enthalpy Diagram for Refrigerant 1150 (Ethylene)
Note: The reference states for enthalpy and entropy differ from those in the table.

Refrigerant 1150 (Ethylene) Properties of Saturated Liquid and Saturated Vapor

Temp.,* °F	Pres-sure, psia	Denisty, lb/ft³ Liquid	Volume, ft³/lb Vapor	Enthalpy, Btu/lb Liquid	Enthalpy, Btu/lb Vapor	Entropy, Btu/lb·°F Liquid	Entropy, Btu/lb·°F Vapor	Specific Heat c_p, Btu/lb·°F Liquid	Specific Heat c_p, Btu/lb·°F Vapor	c_p/c_v Vapor	Vel. of Sound, ft/s Liquid	Vel. of Sound, ft/s Vapor	Viscosity, lb_m/ft·h Liquid	Viscosity, lb_m/ft·h Vapor	Thermal Cond., Btu/h·ft·°F Liquid	Thermal Cond., Btu/h·ft·°F Vapor	Surface Tension, dyne/cm	Temp., °F
–272.50a	0.018	40.87	4047.0	–68.014	176.139	–0.28177	1.02264	0.5807	0.2837	1.3333	5796.	664.9	3.013	0.0085	0.1561	0.00345	28.14	–272.50
–270.00	0.023	40.76	3219.5	–66.565	176.844	–0.27408	1.00925	0.5810	0.2837	1.3333	5767.	669.3	1.584	0.0026	0.1554	0.00389	27.86	–270.00
–265.00	0.036	40.54	2077.4	–63.659	178.256	–0.25896	0.98374	0.5813	0.2838	1.3334	5708.	678.0	1.447	0.0040	0.1532	0.00383	27.31	–265.00
–260.00	0.056	40.32	1374.6	–60.752	179.667	–0.24421	0.95987	0.5814	0.2840	1.3334	5648.	686.6	1.328	0.0051	0.1510	0.00381	26.77	–260.00
–255.00	0.084	40.09	930.98	–57.846	181.075	–0.22984	0.93751	0.5811	0.2842	1.3335	5587.	695.0	1.225	0.0062	0.1488	0.00382	26.22	–255.00
–250.00	0.124	39.87	644.22	–54.941	182.481	–0.21582	0.91654	0.5808	0.2845	1.3336	5526.	703.3	1.134	0.0071	0.1466	0.00385	25.68	–250.00
–245.00	0.180	39.65	454.73	–52.038	183.883	–0.20214	0.89686	0.5802	0.2848	1.3338	5464.	711.5	1.054	0.0079	0.1444	0.00390	25.14	–245.00
–240.00	0.257	39.43	326.95	–49.138	185.281	–0.18878	0.87836	0.5796	0.2852	1.3340	5402.	719.5	0.983	0.0086	0.1422	0.00397	24.60	–240.00
–235.00	0.359	39.20	239.12	–46.241	186.674	–0.17575	0.86095	0.5790	0.2857	1.3343	5339.	727.4	0.920	0.0092	0.1401	0.00405	24.06	–235.00
–230.00	0.493	38.98	177.68	–43.348	188.061	–0.16301	0.84456	0.5783	0.2862	1.3347	5276.	735.1	0.864	0.0098	0.1379	0.00413	23.53	–230.00
–225.00	0.668	38.75	133.99	–40.457	189.442	–0.15056	0.82911	0.5776	0.2868	1.3351	5212.	742.6	0.813	0.0103	0.1358	0.00423	23.00	–225.00
–220.00	0.891	38.52	102.44	–37.570	190.816	–0.13840	0.81452	0.5769	0.2875	1.3357	5148.	750.0	0.767	0.0107	0.1337	0.00432	22.47	–220.00
–215.00	1.174	38.29	79.326	–34.686	192.181	–0.12649	0.80074	0.5763	0.2884	1.3364	5084.	757.3	0.726	0.0112	0.1316	0.00442	21.94	–215.00
–210.00	1.527	38.06	62.164	–31.805	193.536	–0.11484	0.78771	0.5758	0.2893	1.3372	5020.	764.3	0.688	0.0115	0.1295	0.00452	21.42	–210.00
–205.00	1.963	37.83	49.260	–28.926	194.881	–0.10343	0.77538	0.5753	0.2904	1.3381	4955.	771.2	0.654	0.0119	0.1274	0.00462	20.90	–205.00
–200.00	2.495	37.60	39.441	–26.048	196.214	–0.09225	0.76369	0.5749	0.2916	1.3393	4890.	777.9	0.622	0.0122	0.1254	0.00471	20.38	–200.00
–195.00	3.141	37.37	31.888	–23.172	197.534	–0.08129	0.75260	0.5747	0.2929	1.3406	4825.	784.3	0.593	0.0125	0.1234	0.00481	19.87	–195.00
–190.00	3.915	37.14	26.016	–20.297	198.840	–0.07055	0.74206	0.5745	0.2944	1.3421	4760.	790.6	0.567	0.0128	0.1214	0.00491	19.36	–190.00
–185.00	4.835	36.90	21.407	–17.421	200.131	–0.06000	0.73205	0.5745	0.2960	1.3438	4694.	796.6	0.542	0.0131	0.1194	0.00501	18.85	–185.00
–180.00	5.922	36.66	17.754	–14.545	201.405	–0.04964	0.72252	0.5747	0.2978	1.3457	4628.	802.5	0.519	0.0134	0.1174	0.00511	18.34	–180.00
–175.00	7.195	36.42	14.835	–11.667	202.662	–0.03946	0.71344	0.5749	0.2997	1.3479	4561.	808.1	0.498	0.0136	0.1154	0.00520	17.83	–175.00
–170.00	8.675	36.18	12.483	–8.786	203.900	–0.02946	0.70478	0.5754	0.3018	1.3504	4494.	813.4	0.478	0.0139	0.1135	0.00530	17.33	–170.00
–165.00	10.386	35.94	10.572	–5.903	205.118	–0.01962	0.69651	0.5760	0.3041	1.3531	4427.	818.5	0.459	0.0141	0.1116	0.00540	16.84	–165.00
–160.00	12.351	35.69	9.0094	–3.015	206.315	–0.00993	0.68860	0.5768	0.3066	1.3562	4360.	823.4	0.442	0.0143	0.1097	0.00550	16.34	–160.00
–155.00	14.594	35.45	7.7219	–0.122	207.490	–0.00040	0.68103	0.5778	0.3093	1.3596	4292.	828.0	0.425	0.0146	0.1079	0.00559	15.85	–155.00
–154.79b	14.696	35.44	7.6726	0.000	207.539	0.00000	0.68072	0.5778	0.3095	1.3598	4289.	828.1	0.425	0.0146	0.1078	0.00560	15.83	–154.79
–150.00	17.143	35.20	6.6543	2.777	208.641	0.00900	0.67378	0.5789	0.3123	1.3634	4224.	832.3	0.410	0.0148	0.1061	0.00569	15.36	–150.00
–145.00	20.022	34.95	5.7635	5.684	209.766	0.01826	0.66682	0.5803	0.3154	1.3676	4155.	836.3	0.395	0.0150	0.1042	0.00580	14.87	–145.00
–140.00	23.259	34.69	5.0159	8.598	210.866	0.02739	0.66013	0.5819	0.3188	1.3722	4086.	840.1	0.382	0.0153	0.1025	0.00590	14.39	–140.00
–135.00	26.883	34.44	4.3848	11.522	211.938	0.03641	0.65370	0.5837	0.3224	1.3773	4016.	843.6	0.369	0.0155	0.1007	0.00600	13.91	–135.00
–130.00	30.922	34.18	3.8493	14.457	212.981	0.04531	0.64750	0.5858	0.3263	1.3829	3946.	846.8	0.356	0.0158	0.0990	0.00611	13.44	–130.00
–125.00	35.406	33.92	3.3926	17.404	213.994	0.05411	0.64153	0.5881	0.3304	1.3890	3876.	849.6	0.344	0.0160	0.0973	0.00622	12.96	–125.00
–120.00	40.365	33.65	3.0012	20.363	214.975	0.06281	0.63575	0.5907	0.3349	1.3957	3805.	852.2	0.333	0.0162	0.0956	0.00634	12.49	–120.00
–115.00	45.829	33.38	2.6643	23.337	215.923	0.07141	0.63016	0.5936	0.3397	1.4031	3733.	854.5	0.322	0.0165	0.0939	0.00646	12.03	–115.00
–110.00	51.829	33.11	2.3730	26.327	216.836	0.07993	0.62475	0.5968	0.3448	1.4112	3661.	856.4	0.312	0.0167	0.0922	0.00658	11.57	–110.00
–105.00	58.398	32.84	2.1200	29.334	217.712	0.08836	0.61950	0.6003	0.3503	1.4200	3588.	858.0	0.302	0.0170	0.0906	0.00671	11.11	–105.00
–100.00	65.567	32.56	1.8995	32.360	218.551	0.09672	0.61439	0.6041	0.3562	1.4297	3515.	859.3	0.292	0.0172	0.0890	0.00685	10.66	–100.00
–95.00	73.369	32.27	1.7065	35.407	219.350	0.10501	0.60942	0.6083	0.3625	1.4403	3441.	860.3	0.283	0.0175	0.0874	0.00699	10.21	–95.00
–90.00	81.836	31.99	1.5370	38.476	220.106	0.11324	0.60457	0.6129	0.3694	1.4519	3367.	860.9	0.274	0.0178	0.0858	0.00714	9.76	–90.00
–85.00	91.001	31.69	1.3876	41.569	220.819	0.12140	0.59982	0.6180	0.3767	1.4647	3292.	861.1	0.265	0.0181	0.0842	0.00729	9.32	–85.00
–80.00	100.90	31.40	1.2555	44.688	221.485	0.12952	0.59518	0.6235	0.3846	1.4787	3216.	861.0	0.257	0.0184	0.0827	0.00746	8.88	–80.00
–75.00	111.56	31.09	1.1383	47.835	222.103	0.13759	0.59062	0.6295	0.3932	1.4942	3140.	860.5	0.249	0.0186	0.0812	0.00763	8.45	–75.00
–70.00	123.03	30.78	1.0340	51.012	222.670	0.14562	0.58614	0.6361	0.4025	1.5113	3063.	859.7	0.241	0.0190	0.0796	0.00781	8.02	–70.00
–65.00	135.33	30.47	0.9409	54.222	223.182	0.15361	0.58172	0.6433	0.4126	1.5301	2985.	858.5	0.233	0.0193	0.0781	0.00801	7.60	–65.00
–60.00	148.50	30.14	0.8576	57.468	223.636	0.16158	0.57735	0.6513	0.4236	1.5511	2906.	856.8	0.226	0.0196	0.0766	0.00821	7.18	–60.00
–55.00	162.58	29.81	0.7827	60.752	224.029	0.16953	0.57301	0.6600	0.4357	1.5744	2827.	854.8	0.219	0.0200	0.0752	0.00843	6.76	–55.00
–50.00	177.60	29.48	0.7154	64.077	224.357	0.17747	0.56871	0.6696	0.4489	1.6003	2747.	852.4	0.212	0.0203	0.0737	0.00866	6.35	–50.00
–45.00	193.59	29.13	0.6546	67.447	224.616	0.18540	0.56442	0.6802	0.4635	1.6295	2666.	849.5	0.205	0.0207	0.0722	0.00891	5.95	–45.00
–40.00	210.61	28.77	0.5996	70.867	224.799	0.19333	0.56013	0.6920	0.4797	1.6622	2584.	846.2	0.198	0.0211	0.0707	0.00917	5.55	–40.00
–35.00	228.68	28.41	0.5496	74.339	224.901	0.20128	0.55582	0.7051	0.4978	1.6993	2500.	842.5	0.191	0.0215	0.0693	0.00946	5.16	–35.00
–30.00	247.84	28.03	0.5042	77.870	224.915	0.20925	0.55148	0.7198	0.5181	1.7416	2416.	838.3	0.184	0.0219	0.0678	0.00976	4.78	–30.00
–25.00	268.14	27.64	0.4628	81.465	224.833	0.21726	0.54709	0.7364	0.5411	1.7901	2331.	833.7	0.178	0.0224	0.0663	0.01009	4.40	–25.00
–20.00	289.61	27.23	0.4249	85.131	224.646	0.22531	0.54263	0.7553	0.5674	1.8461	2244.	828.5	0.171	0.0229	0.0648	0.01045	4.02	–20.00
–15.00	312.30	26.81	0.3902	88.875	224.343	0.23343	0.53808	0.7769	0.5978	1.9115	2156.	822.9	0.165	0.0234	0.0634	0.01084	3.66	–15.00
–10.00	336.25	26.38	0.3582	92.708	223.909	0.24163	0.53340	0.8020	0.6334	1.9885	2067.	816.7	0.159	0.0240	0.0619	0.01128	3.30	–10.00
–5.00	361.50	25.92	0.3288	96.639	223.329	0.24993	0.52857	0.8314	0.6756	2.0805	1975.	809.9	0.152	0.0246	0.0604	0.01177	2.95	–5.00
0.00	388.12	25.43	0.3015	100.684	222.582	0.25835	0.52354	0.8665	0.7264	2.1920	1882.	802.6	0.146	0.0253	0.0588	0.01232	2.60	0.00
5.00	416.15	24.92	0.2762	104.860	221.642	0.26694	0.51827	0.9091	0.7889	2.3297	1785.	794.6	0.140	0.0260	0.0573	0.01295	2.27	5.00
10.00	445.64	24.38	0.2526	109.190	220.476	0.27574	0.51268	0.9622	0.8678	2.5037	1686.	786.0	0.133	0.0268	0.0557	0.01369	1.95	10.00
15.00	476.65	23.79	0.2304	113.706	219.037	0.28480	0.50670	1.030	0.9705	2.730	1583.	776.6	0.127	0.0277	0.0541	0.01458	1.64	15.00
20.00	509.26	23.15	0.2096	118.451	217.263	0.29420	0.50020	1.121	1.110	3.037	1474.	766.3	0.120	0.0287	0.0524	0.01568	1.33	20.00
25.00	543.54	22.45	0.1897	123.488	215.057	0.30407	0.49300	1.249	1.309	3.473	1360.	755.1	0.113	0.0299	0.0507	0.01714	1.05	25.00
30.00	579.58	21.65	0.1706	128.921	212.271	0.31460	0.48482	1.444	1.620	4.145	1239.	742.5	0.106	0.0313	0.0489	0.01918	0.78	30.00
35.00	617.49	20.72	0.1517	134.928	208.638	0.32613	0.47514	1.782	2.170	5.309	1107.	728.0	0.098	0.0332	0.0473	0.02237	0.52	35.00
40.00	657.42	19.56	0.1324	141.901	203.590	0.33942	0.46288	2.522	3.397	7.823	961.	709.7	0.089	0.0357	0.0461	0.02838	0.29	40.00
45.00	699.58	17.86	0.1106	151.091	195.355	0.35688	0.44459	5.514	8.383	17.35	782.	679.7	0.078	0.0399	0.0496	0.04548	0.10	45.00
48.56c	731.25	13.37	0.0748	171.840	171.840	0.39709	0.39709	∞	∞	∞	0.	0.0	–	–	∞	∞	0.00	48.56

*Temperatures are on the ITS-90 scale a = triple point b = normal boiling point c = critical point

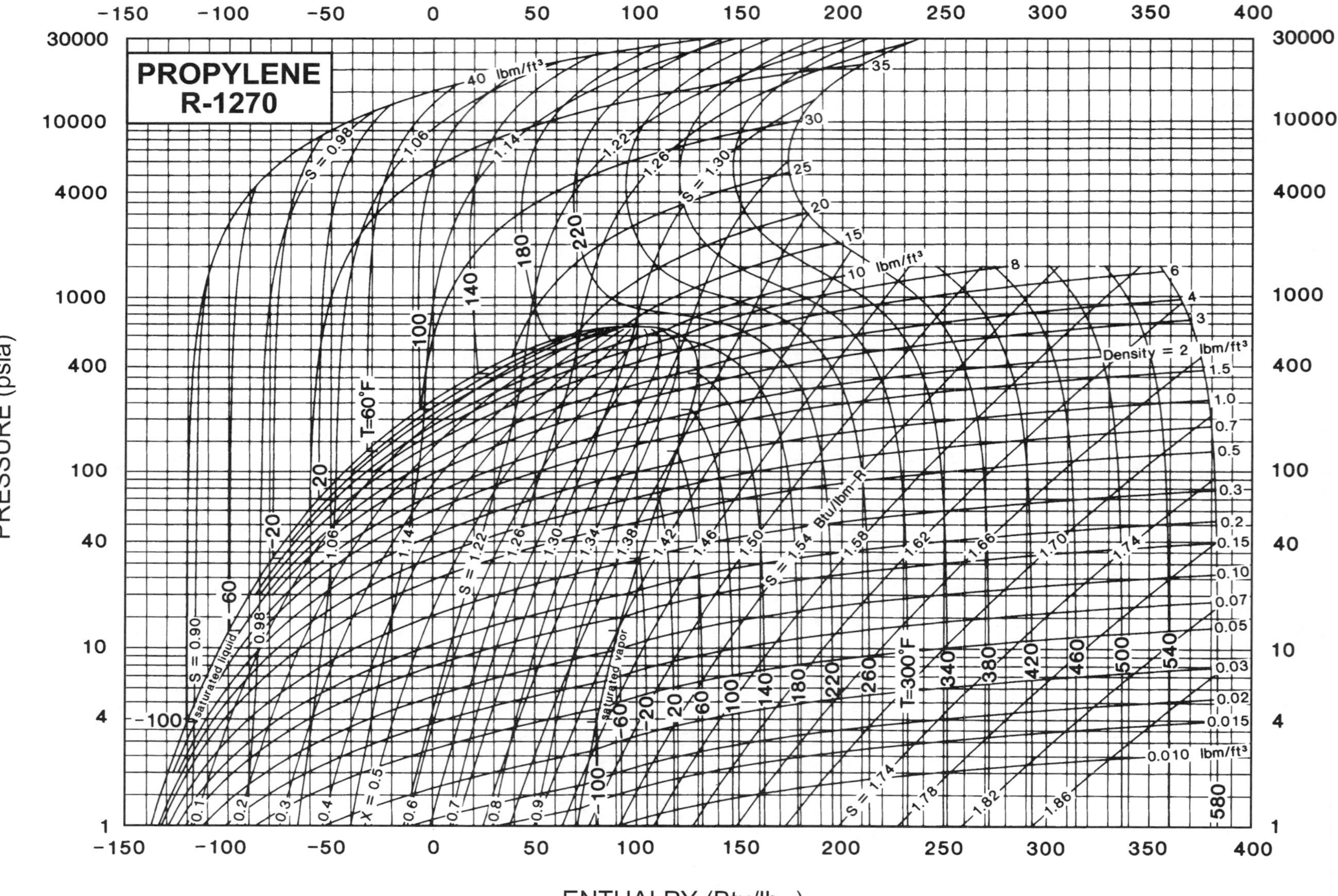

Fig. 21 Pressure-Enthalpy Diagram for Refrigerant 1270 (Propylene)

Note: The reference states for enthalpy and entropy differ from those in the table.

Refrigerant 1270 (Propylene) Properties of Saturated Liquid and Saturated Vapor

Temp.,* °F	Pressure, psia	Denisty, lb/ft³ Liquid	Volume, ft³/lb Vapor	Enthalpy, Btu/lb Liquid	Enthalpy, Btu/lb Vapor	Entropy, Btu/lb·°F Liquid	Entropy, Btu/lb·°F Vapor	Specific Heat c_p, Btu/lb·°F Liquid	Specific Heat c_p, Btu/lb·°F Vapor	c_p/c_v Vapor	Vel. of Sound, ft/s Liquid	Vel. of Sound, ft/s Vapor	Viscosity, lb_m/ft·h Liquid	Viscosity, lb_m/ft·h Vapor	Thermal Cond., Btu/h·ft·°F Liquid	Thermal Cond., Btu/h·ft·°F Vapor	Surface Tension, dyne/cm	Temp., °F
−200.00	0.030	43.93	2190.7	−79.208	143.636	−0.23626	0.62201	0.4646	0.2489	1.2345	5497.	615.2	3.072	0.0099	0.1174	0.00346	29.86	−200.00
−190.00	0.059	43.54	1162.7	−74.538	146.132	−0.21861	0.59976	0.4693	0.2522	1.2310	5375.	625.9	2.666	0.0103	0.1146	0.00363	28.92	−190.00
−180.00	0.110	43.14	650.36	−69.823	148.654	−0.20144	0.57983	0.4735	0.2555	1.2277	5258.	636.4	2.325	0.0107	0.1118	0.00381	27.99	−180.00
−170.00	0.194	42.75	381.16	−65.070	151.201	−0.18474	0.56193	0.4771	0.2590	1.2245	5144.	646.5	2.037	0.0110	0.1092	0.00399	27.06	−170.00
−160.00	0.327	42.35	232.88	−60.280	153.772	−0.16849	0.54586	0.4805	0.2626	1.2215	5031.	656.3	1.793	0.0114	0.1065	0.00418	26.13	−160.00
−150.00	0.533	41.96	147.70	−55.459	156.364	−0.15266	0.53142	0.4837	0.2664	1.2188	4919.	665.8	1.585	0.0118	0.1040	0.00437	25.22	−150.00
−145.00	0.671	41.76	119.13	−53.036	157.666	−0.14490	0.52474	0.4852	0.2683	1.2175	4864.	670.4	1.493	0.0120	0.1027	0.00447	24.76	−145.00
−140.00	0.838	41.56	96.853	−50.605	158.973	−0.13724	0.51841	0.4868	0.2703	1.2163	4808.	675.0	1.407	0.0122	0.1014	0.00457	24.31	−140.00
−135.00	1.039	41.36	79.334	−48.166	160.284	−0.12967	0.51241	0.4884	0.2724	1.2151	4753.	679.4	1.328	0.0124	0.1002	0.00467	23.86	−135.00
−130.00	1.277	41.16	65.449	−45.719	161.597	−0.12220	0.50671	0.4900	0.2745	1.2141	4697.	683.7	1.255	0.0126	0.0989	0.00477	23.41	−130.00
−125.00	1.560	40.96	54.359	−43.264	162.914	−0.11481	0.50130	0.4916	0.2767	1.2131	4642.	687.9	1.186	0.0128	0.0977	0.00487	22.96	−125.00
−120.00	1.892	40.76	45.439	−40.801	164.232	−0.10751	0.49616	0.4933	0.2789	1.2122	4586.	692.1	1.123	0.0130	0.0965	0.00497	22.51	−120.00
−115.00	2.280	40.56	38.214	−38.330	165.553	−0.10029	0.49128	0.4950	0.2812	1.2114	4530.	696.1	1.064	0.0132	0.0953	0.00507	22.07	−115.00
−110.00	2.730	40.36	32.324	−35.849	166.875	−0.09315	0.48665	0.4967	0.2836	1.2107	4474.	700.0	1.010	0.0134	0.0941	0.00518	21.63	−110.00
−105.00	3.252	40.16	27.492	−33.359	168.198	−0.08608	0.48224	0.4985	0.2860	1.2101	4418.	703.8	0.959	0.0136	0.0929	0.00529	21.19	−105.00
−100.00	3.851	39.95	23.504	−30.860	169.522	−0.07909	0.47806	0.5004	0.2886	1.2096	4362.	707.4	0.911	0.0138	0.0918	0.00539	20.75	−100.00
−95.00	4.536	39.75	20.196	−28.351	170.844	−0.07217	0.47409	0.5023	0.2911	1.2093	4306.	711.0	0.867	0.0140	0.0906	0.00550	20.31	−95.00
−90.00	5.316	39.54	17.435	−25.832	172.167	−0.06532	0.47031	0.5043	0.2938	1.2090	4249.	714.4	0.825	0.0142	0.0895	0.00561	19.87	−90.00
−85.00	6.201	39.34	15.119	−23.302	173.488	−0.05854	0.46672	0.5064	0.2966	1.2089	4192.	717.6	0.786	0.0144	0.0883	0.00572	19.44	−85.00
−80.00	7.199	39.13	13.167	−20.762	174.807	−0.05181	0.46331	0.5085	0.2994	1.2089	4136.	720.8	0.750	0.0146	0.0872	0.00584	19.01	−80.00
−75.00	8.321	38.92	11.513	−18.211	176.123	−0.04515	0.46007	0.5107	0.3023	1.2091	4078.	723.8	0.716	0.0148	0.0861	0.00595	18.58	−75.00
−70.00	9.578	38.71	10.107	−15.648	177.437	−0.03855	0.45698	0.5130	0.3053	1.2094	4021.	726.6	0.684	0.0150	0.0850	0.00607	18.15	−70.00
−65.00	10.980	38.50	8.9046	−13.072	178.747	−0.03200	0.45405	0.5154	0.3084	1.2099	3964.	729.3	0.654	0.0152	0.0839	0.00619	17.73	−65.00
−60.00	12.538	38.29	7.8733	−10.485	180.052	−0.02550	0.45125	0.5179	0.3116	1.2105	3906.	731.9	0.626	0.0154	0.0828	0.00631	17.31	−60.00
−55.00	14.265	38.07	6.9847	−7.884	181.353	−0.01906	0.44860	0.5204	0.3150	1.2113	3848.	734.3	0.600	0.0157	0.0817	0.00643	16.88	−55.00
−53.84b	14.690	38.02	6.7968	−7.280	181.654	−0.01757	0.44800	0.5210	0.3157	1.2115	3835.	734.8	0.594	0.0157	0.0814	0.00646	16.79	−53.84
−50.00	16.172	37.86	6.2161	−5.270	182.649	−0.01266	0.44607	0.5231	0.3184	1.2123	3791.	736.5	0.575	0.0159	0.0806	0.00656	16.47	−50.00
−45.00	18.272	37.64	5.5490	−2.642	183.938	−0.00631	0.44366	0.5258	0.3219	1.2134	3732.	738.5	0.552	0.0161	0.0795	0.00668	16.05	−45.00
−40.00	20.577	37.42	4.9679	0.000	185.221	0.00000	0.44136	0.5287	0.3255	1.2148	3674.	740.4	0.530	0.0163	0.0785	0.00681	15.63	−40.00
−35.00	23.101	37.20	4.4599	2.657	186.496	0.00627	0.43918	0.5316	0.3292	1.2163	3616.	742.2	0.509	0.0165	0.0774	0.00695	15.22	−35.00
−30.00	25.856	36.97	4.0144	5.330	187.764	0.01249	0.43709	0.5347	0.3331	1.2181	3557.	743.7	0.489	0.0167	0.0764	0.00708	14.81	−30.00
−25.00	28.857	36.75	3.6224	8.019	189.023	0.01868	0.43510	0.5379	0.3371	1.2201	3498.	745.0	0.471	0.0170	0.0754	0.00722	14.41	−25.00
−20.00	32.117	36.52	3.2765	10.725	190.272	0.02483	0.43321	0.5411	0.3412	1.2223	3440.	746.2	0.453	0.0172	0.0744	0.00736	14.00	−20.00
−15.00	35.651	36.29	2.9705	13.447	191.512	0.03095	0.43140	0.5445	0.3454	1.2248	3381.	747.2	0.437	0.0174	0.0733	0.00751	13.60	−15.00
−10.00	39.472	36.06	2.6989	16.187	192.741	0.03703	0.42967	0.5481	0.3498	1.2275	3322.	747.9	0.421	0.0177	0.0723	0.00766	13.20	−10.00
−5.00	43.596	35.83	2.4571	18.946	193.959	0.04308	0.42801	0.5517	0.3544	1.2306	3262.	748.5	0.406	0.0179	0.0713	0.00781	12.80	−5.00
0.00	48.036	35.59	2.2414	21.723	195.164	0.04911	0.42643	0.5555	0.3591	1.2339	3203.	748.9	0.392	0.0181	0.0704	0.00797	12.40	0.00
5.00	52.808	35.35	2.0486	24.520	196.356	0.05511	0.42491	0.5594	0.3640	1.2375	3144.	749.0	0.378	0.0184	0.0694	0.00813	12.01	5.00
10.00	57.928	35.11	1.8756	27.337	197.535	0.06108	0.42346	0.5635	0.3690	1.2415	3084.	749.0	0.365	0.0186	0.0684	0.00830	11.62	10.00
15.00	63.411	34.86	1.7201	30.175	198.698	0.06703	0.42206	0.5677	0.3743	1.2459	3024.	748.7	0.353	0.0189	0.0674	0.00847	11.23	15.00
20.00	69.271	34.61	1.5801	33.034	199.847	0.07295	0.42072	0.5721	0.3797	1.2506	2965.	748.1	0.341	0.0191	0.0665	0.00864	10.85	20.00
25.00	75.525	34.36	1.4537	35.916	200.978	0.07886	0.41943	0.5766	0.3854	1.2558	2905.	747.4	0.330	0.0194	0.0655	0.00882	10.47	25.00
30.00	82.190	34.10	1.3393	38.821	202.092	0.08475	0.41818	0.5814	0.3913	1.2614	2845.	746.4	0.319	0.0196	0.0646	0.00900	10.09	30.00
35.00	89.280	33.84	1.2356	41.749	203.187	0.09062	0.41698	0.5863	0.3975	1.2675	2785.	745.2	0.309	0.0199	0.0636	0.00919	9.71	35.00
40.00	96.812	33.58	1.1414	44.702	204.263	0.09648	0.41581	0.5915	0.4039	1.2742	2725.	743.7	0.299	0.0202	0.0627	0.00939	9.34	40.00
45.00	104.80	33.31	1.0556	47.681	205.317	0.10232	0.41467	0.5968	0.4107	1.2815	2665.	741.9	0.290	0.0204	0.0618	0.00959	8.97	45.00
50.00	113.27	33.04	0.9774	50.687	206.349	0.10815	0.41357	0.6025	0.4177	1.2894	2605.	739.9	0.280	0.0207	0.0608	0.00980	8.60	50.00
55.00	122.23	32.77	0.9060	53.720	207.357	0.11398	0.41249	0.6083	0.4251	1.2980	2544.	737.6	0.272	0.0210	0.0599	0.01001	8.24	55.00
60.00	131.69	32.49	0.8406	56.782	208.340	0.11979	0.41144	0.6145	0.4330	1.3074	2484.	735.0	0.263	0.0213	0.0590	0.01023	7.88	60.00
65.00	141.69	32.20	0.7806	59.874	209.296	0.12561	0.41040	0.6210	0.4412	1.3177	2423.	732.2	0.255	0.0216	0.0581	0.01046	7.52	65.00
70.00	152.22	31.91	0.7255	62.996	210.223	0.13141	0.40937	0.6279	0.4499	1.3289	2362.	729.0	0.247	0.0219	0.0572	0.01069	7.17	70.00
75.00	163.32	31.61	0.6749	66.151	211.119	0.13722	0.40835	0.6352	0.4591	1.3413	2301.	725.6	0.239	0.0222	0.0563	0.01094	6.82	75.00
80.00	175.00	31.31	0.6282	69.340	211.983	0.14303	0.40734	0.6429	0.4690	1.3548	2240.	721.8	0.232	0.0225	0.0554	0.01119	6.47	80.00
85.00	187.27	31.00	0.5851	72.564	212.810	0.14884	0.40632	0.6510	0.4795	1.3698	2179.	717.7	0.225	0.0229	0.0545	0.01146	6.13	85.00
90.00	200.16	30.68	0.5452	75.826	213.600	0.15466	0.40530	0.6598	0.4907	1.3863	2118.	713.3	0.218	0.0232	0.0536	0.01173	5.79	90.00
95.00	213.69	30.36	0.5083	79.128	214.348	0.16049	0.40427	0.6691	0.5029	1.4045	2056.	708.5	0.211	0.0236	0.0527	0.01202	5.45	95.00
100.00	227.87	30.03	0.4741	82.471	215.053	0.16634	0.40322	0.6792	0.5160	1.4249	1994.	703.4	0.204	0.0239	0.0519	0.01232	5.12	100.00
110.00	258.27	29.34	0.4127	89.292	216.312	0.17808	0.40104	0.7021	0.5458	1.4731	1870.	692.1	0.192	0.0247	0.0501	0.01296	4.48	110.00
120.00	291.55	28.60	0.3594	96.314	217.343	0.18993	0.39871	0.7295	0.5819	1.5345	1743.	679.3	0.179	0.0256	0.0484	0.01369	3.85	120.00
130.00	327.86	27.82	0.3127	103.571	218.096	0.20193	0.39614	0.7637	0.6270	1.6148	1615.	664.9	0.167	0.0265	0.0466	0.01452	3.24	130.00
140.00	367.42	26.98	0.2716	111.108	218.505	0.21416	0.39324	0.8079	0.6855	1.7233	1483.	648.7	0.156	0.0276	0.0449	0.01550	2.65	140.00
150.00	410.43	26.05	0.2350	118.997	218.474	0.22670	0.38986	0.8686	0.7656	1.8774	1348.	630.6	0.144	0.0289	0.0432	0.01670	2.09	150.00
160.00	457.14	25.01	0.2022	127.349	217.855	0.23974	0.38579	0.9591	0.8835	2.1120	1207.	610.4	0.133	0.0305	0.0414	0.01823	1.56	160.00
170.00	507.83	23.79	0.1720	136.371	216.389	0.25356	0.38064	1.112	1.0787	2.5111	1058.	587.8	0.121	0.0325	0.0396	0.02036	1.07	170.00
180.00	562.86	22.28	0.1436	146.509	213.541	0.26884	0.37362	1.438	1.4775	3.3437	898.	562.4	0.108	0.0352	0.0378	0.02379	0.62	180.00
190.00	622.70	20.08	0.1144	159.185	207.737	0.28768	0.36241	2.638	2.8289	6.2076	719.	533.8	0.091	0.0399	0.0363	0.03168	0.23	190.00
198.36c	676.54	13.95	0.0717	186.439	186.439	0.32843	0.32843	∞	∞	∞	0.	0.0	–	–	∞	∞	0.00	198.36

*Temperatures have been converted from the IPTS-68 scale of the original formulation to the ITS-90 scale. b = normal boiling point. c = critical point

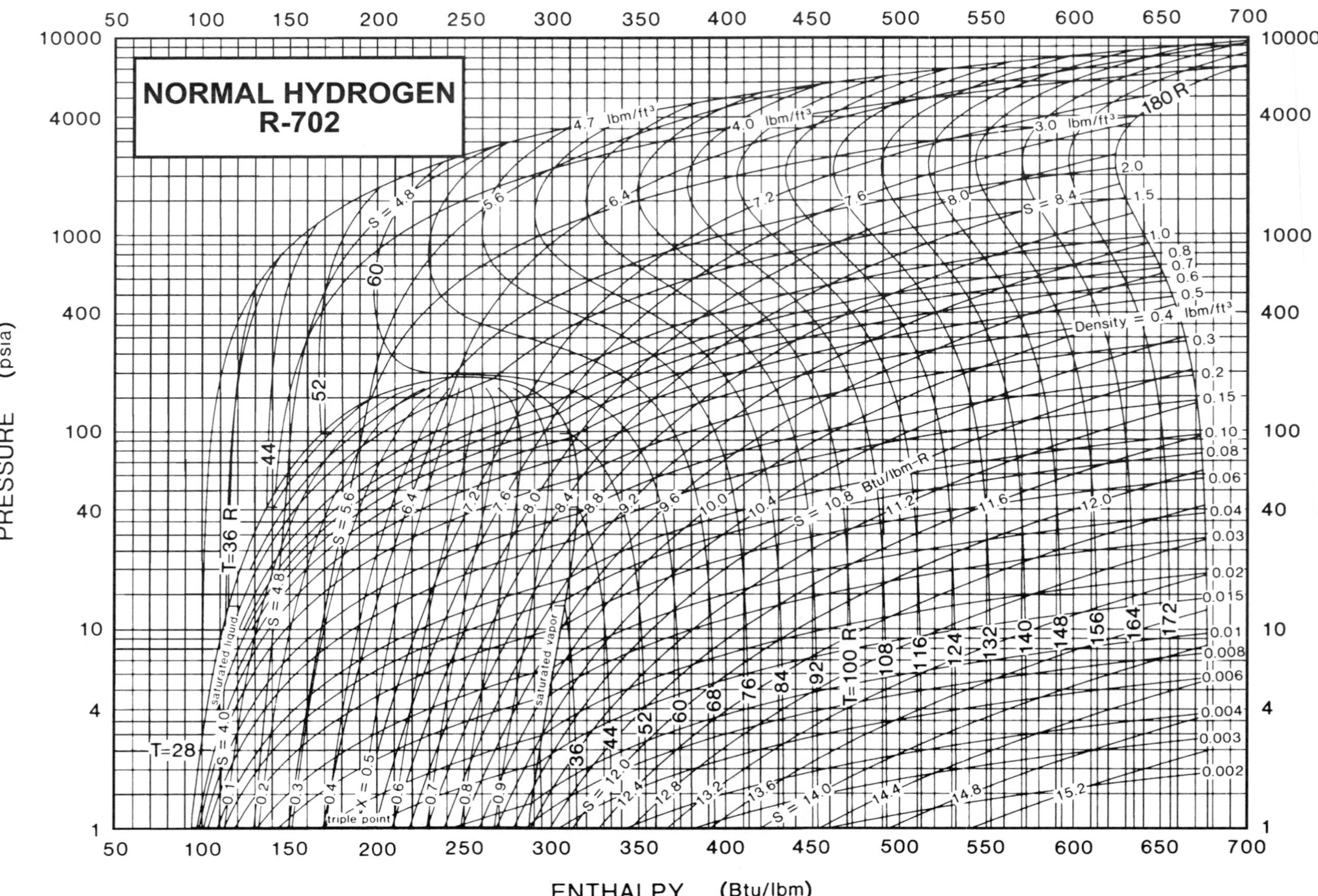

Fig. 22 Pressure-Enthalpy Diagram for Refrigerant 702 (Normal Hydrogen)

Refrigerant 702 (Normal Hydrogen) Properties of Saturated Liquid and Saturated Vapor

Temp.,* °R	Pressure, psia	Density, lb/ft³ Liquid	Volume, ft³/lb Vapor	Enthalpy, Btu/lb Liquid	Enthalpy, Btu/lb Vapor	Entropy, Btu/lb·°R Liquid	Entropy, Btu/lb·°R Vapor	Specific Heat c_p, Btu/lb·°R Liquid	Specific Heat c_p, Btu/lb·°R Vapor	c_p/c_v Vapor	Velocity of Sound, ft/s Liquid	Velocity of Sound, ft/s Vapor	Viscosity, lb_m/ft·h Liquid	Viscosity, lb_m/ft·h Vapor	Thermal Cond., Btu/h·ft·°R Liquid	Thermal Cond., Btu/h·ft·°R Vapor	Surface Tension, dyne/cm	Temp.,* °R
25.11a	1.125	4.801	116.73	93.84	287.11	3.3656	11.0477	1.858	2.605	1.669	4467.	998.9	0.0617	0.00160	0.0440	0.00599	3.181	25.11
26.00	1.464	4.776	92.561	95.60	289.08	3.4270	10.8701	1.751	2.590	1.688	4405.	1018.5	0.0577	0.00170	0.0463	0.00635	3.081	26.00
28.00	2.513	4.715	57.548	99.07	293.31	3.5540	10.5077	1.716	2.611	1.711	4240.	1054.2	0.0503	0.00189	0.0506	0.00706	2.859	28.00
30.00	4.060	4.651	37.736	102.54	297.29	3.6755	10.1889	1.807	2.658	1.732	4072.	1084.8	0.0446	0.00206	0.0539	0.00771	2.640	30.00
31.00	5.062	4.618	31.072	104.38	299.18	3.7357	10.0431	1.872	2.686	1.744	3992.	1098.9	0.0422	0.00215	0.0553	0.00803	2.532	31.00
32.00	6.239	4.584	25.834	106.31	301.00	3.7960	9.9051	1.944	2.717	1.758	3917.	1112.3	0.0400	0.00223	0.0564	0.00835	2.424	32.00
33.00	7.611	4.549	21.668	108.34	302.74	3.8565	9.7740	2.020	2.752	1.773	3845.	1125.0	0.0380	0.00231	0.0574	0.00867	2.318	33.00
34.00	9.195	4.513	18.319	110.47	304.39	3.9173	9.6491	2.100	2.790	1.791	3776.	1137.0	0.0361	0.00239	0.0583	0.00899	2.212	34.00
35.00	11.013	4.477	15.600	112.68	305.96	3.9785	9.5297	2.182	2.832	1.811	3709.	1148.3	0.0345	0.00247	0.0590	0.00932	2.107	35.00
36.00	13.082	4.439	13.372	114.98	307.42	4.0401	9.4152	2.267	2.879	1.833	3644.	1159.0	0.0329	0.00256	0.0595	0.00966	2.003	36.00
36.70b	14.696	4.412	12.045	116.66	308.39	4.0836	9.3376	2.328	2.915	1.851	3598.	1166.1	0.0319	0.00261	0.0598	0.00990	1.931	36.70
37.00	15.425	4.400	11.531	117.38	308.79	4.1023	9.3051	2.355	2.931	1.858	3579.	1169.0	0.0315	0.00264	0.0600	0.01000	1.900	37.00
38.00	18.061	4.359	9.9976	119.86	310.04	4.1650	9.1989	2.445	2.989	1.887	3514.	1178.2	0.0301	0.00272	0.0603	0.01036	1.798	38.00
39.00	21.010	4.317	8.7112	122.43	311.18	4.2284	9.0961	2.540	3.054	1.919	3448.	1186.8	0.0289	0.00281	0.0605	0.01073	1.698	39.00
40.00	24.295	4.273	7.6245	125.10	312.19	4.2924	8.9963	2.640	3.127	1.955	3380.	1194.8	0.0277	0.00289	0.0606	0.01111	1.598	40.00
41.00	27.935	4.227	6.7006	127.87	313.06	4.3573	8.8991	2.746	3.209	1.995	3311.	1202.0	0.0266	0.00298	0.0607	0.01150	1.499	41.00
42.00	31.952	4.179	5.9104	130.75	313.80	4.4229	8.8041	2.860	3.302	2.042	3239.	1208.7	0.0255	0.00307	0.0606	0.01191	1.402	42.00
43.00	36.367	4.129	5.2308	133.75	314.38	4.4894	8.7109	2.982	3.408	2.094	3166.	1214.7	0.0245	0.00316	0.0605	0.01233	1.306	43.00
44.00	41.202	4.077	4.6430	136.88	314.80	4.5567	8.6192	3.116	3.528	2.154	3089.	1220.0	0.0236	0.00326	0.0603	0.01278	1.211	44.00
45.00	46.477	4.022	4.1320	140.14	315.04	4.6250	8.5285	3.262	3.666	2.224	3011.	1224.7	0.0227	0.00335	0.0600	0.01324	1.117	45.00
46.00	52.214	3.965	3.6857	143.56	315.10	4.6944	8.4386	3.425	3.827	2.304	2929.	1228.8	0.0218	0.00345	0.0596	0.01373	1.025	46.00
47.00	58.434	3.905	3.2939	147.13	314.94	4.7649	8.3490	3.609	4.014	2.398	2845.	1232.3	0.0209	0.00356	0.0591	0.01425	0.935	47.00
48.00	65.158	3.842	2.9484	150.87	314.56	4.8368	8.2592	3.817	4.235	2.509	2757.	1235.2	0.0201	0.00367	0.0586	0.01480	0.846	48.00
49.00	72.407	3.776	2.6423	154.81	313.92	4.9103	8.1690	4.059	4.499	2.642	2666.	1237.4	0.0193	0.00378	0.0580	0.01540	0.758	49.00
50.00	80.202	3.705	2.3698	158.95	313.01	4.9858	8.0776	4.344	4.820	2.803	2572.	1239.1	0.0185	0.00390	0.0573	0.01603	0.673	50.00
52.00	97.512	3.547	1.9068	167.99	310.19	5.1449	7.8888	5.123	5.718	3.252	2367.	1240.8	0.0170	0.00417	0.0556	0.01750	0.508	52.00
54.00	117.25	3.356	1.5282	178.43	305.67	5.3220	7.6857	6.484	7.246	4.013	2133.	1240.5	0.0154	0.00451	0.0535	0.01933	0.352	54.00
56.00	139.59	3.102	1.2092	191.38	298.64	5.5351	7.4556	9.791	10.325	5.537	1851.	1238.9	0.0138	0.00500	0.0507	0.02187	0.208	56.00
58.00	164.68	2.702	0.9281	210.22	287.39	5.8358	7.1718	–	–	–	–	–	–	–	–	–	0.082	58.00
59.74c	190.75	1.880	0.5320	248.30	248.30	6.4441	6.4441	∞	∞	∞	0.	0.0	–	–	∞	∞	0.000	59.74

*Temperatures are on the IPTS-68 scale a = triple point b = normal boiling point c = critical point

Refrigerant 702 (Normal Hydrogen) Properties of Gas at 14.696 psia (one standard atmosphere)

Temp., °F	Density, lb/ft³	Enthalpy, Btu/lb	Entropy, Btu/lb·°F	c_p, Btu/lb·°F	c_p/c_v	Vel. Sound, ft/s	Viscosity, lb/ft·h	Thermal Cond., Btu/h·ft·°F
−423.0a	0.08302	308.39	9.3376	2.915	1.851	1166.1	0.0026	0.0099
−420.0	0.07529	316.85	9.5592	2.792	1.812	1226.5	0.0028	0.0106
−410.0	0.05801	343.71	10.1639	2.617	1.748	1401.0	0.0035	0.0132
−400.0	0.04750	369.53	10.6376	2.555	1.720	1549.3	0.0042	0.0156
−390.0	0.04032	394.93	11.0312	2.527	1.704	1681.7	0.0048	0.0179
−380.0	0.03507	420.11	11.3690	2.512	1.694	1802.7	0.0054	0.0201
−370.0	0.03105	445.19	11.6655	2.505	1.686	1914.8	0.0060	0.0221
−360.0	0.02787	470.22	11.9302	2.503	1.679	2019.3	0.0065	0.0241
−340.0	0.02314	520.39	12.3889	2.518	1.663	2208.2	0.0075	0.0281
−320.0	0.01980	571.09	12.7806	2.556	1.641	2373.7	0.0084	0.0318
−300.0	0.01730	622.76	13.1262	2.613	1.615	2520.1	0.0093	0.0355
−280.0	0.01537	675.70	13.4385	2.683	1.588	2651.9	0.0101	0.0394
−260.0	0.01382	730.11	13.7256	2.759	1.561	2773.0	0.0109	0.0437
−240.0	0.01256	786.05	13.9925	2.835	1.537	2886.5	0.0117	0.0479
−220.0	0.01151	843.50	14.2428	2.910	1.515	2994.3	0.0124	0.0521
−200.0	0.01063	902.40	14.4788	2.979	1.497	3098.0	0.0131	0.0562
−180.0	0.00986	962.63	14.7022	3.043	1.481	3198.3	0.0138	0.0604
−160.0	0.00921	1024.07	14.9143	3.100	1.468	3295.9	0.0145	0.0645
−140.0	0.00863	1086.59	15.1163	3.151	1.457	3391.1	0.0151	0.0686
−120.0	0.00812	1150.06	15.3089	3.196	1.447	3484.2	0.0158	0.0726
−100.0	0.00767	1214.38	15.4929	3.235	1.439	3575.5	0.0164	0.0765
−80.0	0.00727	1279.43	15.6689	3.269	1.432	3664.9	0.0170	0.0802
−60.0	0.00690	1345.13	15.8375	3.299	1.427	3752.7	0.0176	0.0839
−40.0	0.00657	1411.38	15.9992	3.325	1.422	3838.9	0.0182	0.0875
−20.0	0.00627	1478.11	16.1546	3.348	1.418	3923.7	0.0188	0.0910
0.0	0.00600	1545.27	16.3039	3.367	1.414	4006.9	0.0194	0.0944
10.0	0.00587	1578.98	16.3765	3.376	1.413	4048.1	0.0197	0.0961
20.0	0.00575	1612.78	16.4477	3.384	1.411	4088.9	0.0200	0.0978
30.0	0.00563	1646.66	16.5176	3.391	1.410	4129.4	0.0203	0.0994
40.0	0.00552	1680.61	16.5862	3.398	1.409	4169.5	0.0205	0.1010
50.0	0.00541	1714.62	16.6536	3.405	1.408	4209.4	0.0208	0.1026
60.0	0.00531	1748.69	16.7198	3.410	1.407	4248.9	0.0211	0.1042
70.0	0.00521	1782.83	16.7849	3.416	1.406	4288.2	0.0214	0.1057
80.0	0.00511	1817.01	16.8488	3.421	1.405	4327.1	0.0216	0.1073
90.0	0.00502	1851.24	16.9117	3.425	1.404	4365.8	0.0219	0.1088
100.0	0.00493	1885.51	16.9735	3.429	1.404	4404.2	0.0222	0.1104
120.0	0.00476	1954.17	17.0940	3.437	1.402	4480.2	0.0227	0.1134
140.0	0.00460	2022.97	17.2107	3.443	1.401	4555.2	0.0233	0.1166
160.0	0.00445	2091.87	17.3237	3.447	1.401	4629.1	0.0238	0.1199
180.0	0.00431	2160.85	17.4333	3.451	1.400	4702.1	0.0243	0.1231
200.0	0.00418	2229.91	17.5396	3.454	1.399	4774.1	0.0248	0.1261
220.0	0.00406	2299.02	17.6428	3.457	1.399	4845.2	0.0253	0.1292
240.0	0.00394	2368.17	17.7431	3.459	1.399	4915.3	0.0258	0.1322
260.0	0.00383	2437.36	17.8406	3.460	1.398	4984.6	0.0263	0.1352
280.0	0.00373	2506.57	17.9354	3.461	1.398	5053.0	–	–
300.0	0.00363	2575.80	18.0278	3.462	1.398	5120.4	–	–
320.0	0.00354	2645.06	18.1178	3.463	1.398	5187.1	–	–
340.0	0.00345	2714.33	18.2055	3.464	1.398	5252.8	–	–
360.0	0.00337	2783.62	18.2911	3.465	1.398	5317.8	–	–
380.0	0.00329	2852.93	18.3746	3.466	1.397	5381.9	–	–
400.0	0.00321	2922.25	18.4562	3.467	1.397	5445.2	–	–

a = saturated vapor at normal boiling point

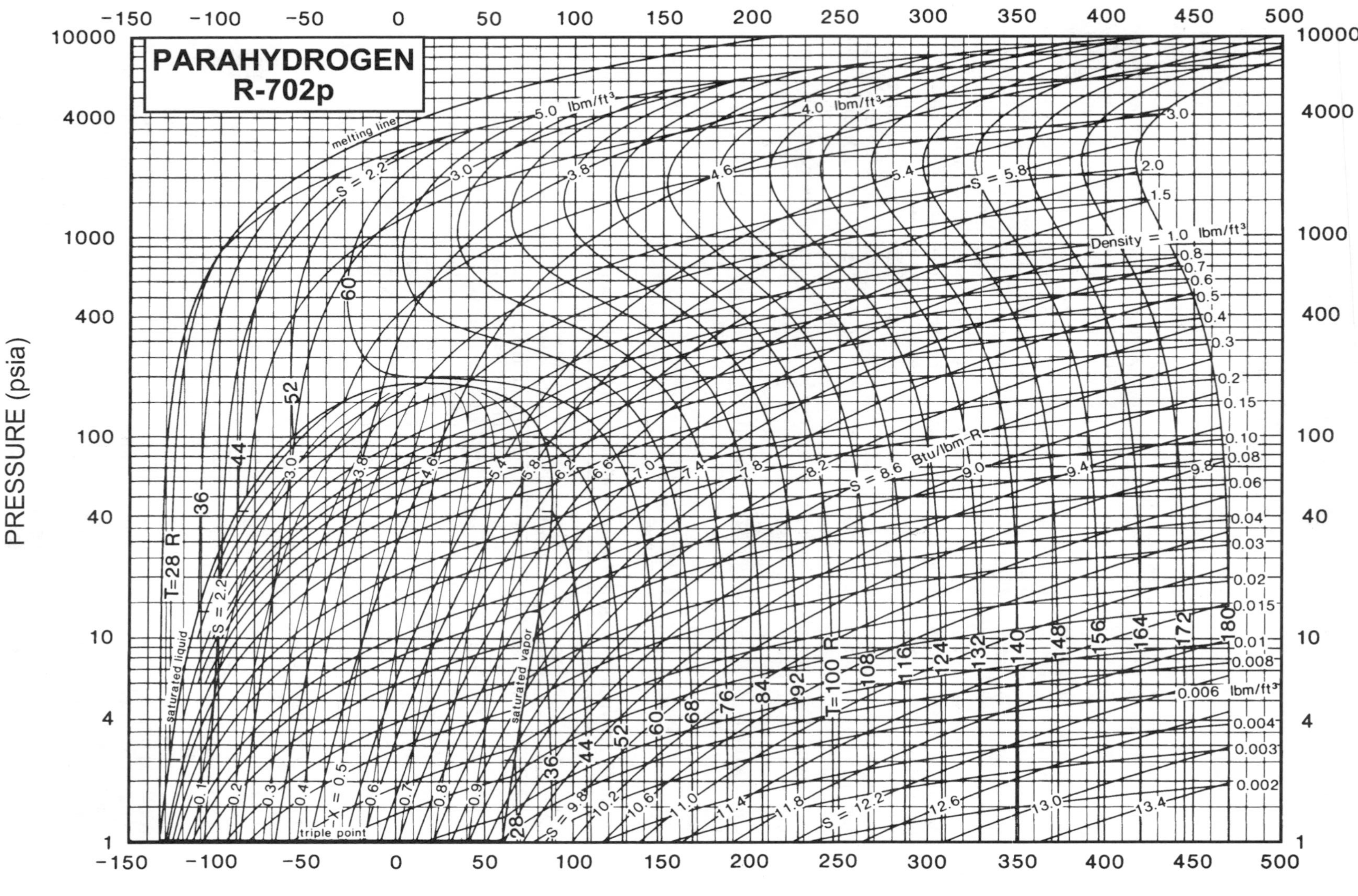

Fig. 23 Pressure-Enthalpy Diagram for Refrigerant 702p (Parahydrogen)

Refrigerant 702p (Parahydrogen) Properties of Saturated Liquid and Saturated Vapor

Temp.,* °R	Pressure, psia	Density, lb/ft³ Liquid	Volume, ft³/lb Vapor	Enthalpy, Btu/lb Liquid	Enthalpy, Btu/lb Vapor	Entropy, Btu/lb·°R Liquid	Entropy, Btu/lb·°R Vapor	Specific Heat c_p, Btu/lb·°R Liquid	Specific Heat c_p, Btu/lb·°R Vapor	c_p/c_v Vapor	Velocity of Sound, ft/s Liquid	Velocity of Sound, ft/s Vapor	Viscosity, lb_m/ft·h Liquid	Viscosity, lb_m/ft·h Vapor	Thermal Cond., Btu/h·ft·°R Liquid	Thermal Cond., Btu/h·ft·°R Vapor	Surface Tension, dyne/cm	Temp.,* °R
24.84a	1.022	4.809	125.6440	−132.18	60.28	1.1902	8.9456	1.845	2.552	1.690	4503.	1000.6	0.0630	0.00157	0.0435	0.00604	3.124	24.84
25.00	1.075	4.805	120.3520	−131.88	60.63	1.2018	8.9123	1.823	2.554	1.690	4490.	1003.5	0.0622	0.00159	0.0439	0.00609	3.107	25.00
26.00	1.462	4.775	92.4333	−130.06	62.81	1.2710	8.7083	1.733	2.571	1.696	4406.	1021.2	0.0577	0.00170	0.0464	0.00640	3.001	26.00
27.00	1.948	4.745	71.9555	−128.33	64.94	1.3356	8.5153	1.705	2.590	1.704	4319.	1038.1	0.0538	0.00180	0.0486	0.00672	2.894	27.00
28.00	2.548	4.715	56.8038	−126.61	67.00	1.3976	8.3337	1.716	2.612	1.712	4231.	1054.2	0.0503	0.00189	0.0506	0.00703	2.789	28.00
29.00	3.279	4.683	45.4567	−124.86	69.01	1.4583	8.1635	1.754	2.636	1.722	4146.	1069.6	0.0473	0.00198	0.0523	0.00735	2.684	29.00
30.00	4.158	4.651	36.8429	−123.07	70.94	1.5185	8.0038	1.807	2.663	1.733	4063.	1084.2	0.0446	0.00206	0.0539	0.00766	2.579	30.00
31.00	5.201	4.618	30.2129	−121.21	72.81	1.5786	7.8539	1.872	2.693	1.746	3985.	1098.1	0.0422	0.00215	0.0552	0.00798	2.475	31.00
32.00	6.427	4.585	25.0403	−119.26	74.59	1.6388	7.7125	1.943	2.726	1.761	3910.	1111.3	0.0400	0.00223	0.0564	0.00831	2.371	32.00
33.00	7.854	4.550	20.9531	−117.24	76.30	1.6993	7.5789	2.019	2.762	1.777	3839.	1123.7	0.0380	0.00231	0.0574	0.00863	2.268	33.00
34.00	9.500	4.514	17.6851	−115.13	77.92	1.7602	7.4522	2.099	2.803	1.796	3770.	1135.5	0.0361	0.00239	0.0582	0.00897	2.166	34.00
35.00	11.384	4.477	15.0434	−112.93	79.44	1.8215	7.3315	2.181	2.848	1.817	3703.	1146.6	0.0345	0.00247	0.0589	0.00931	2.064	35.00
36.00	13.526	4.439	12.8866	−110.64	80.87	1.8833	7.2162	2.266	2.897	1.840	3637.	1157.0	0.0329	0.00256	0.0595	0.00965	1.963	36.00
36.50b	14.696	4.420	11.9588	−109.47	81.54	1.9143	7.1606	2.310	2.924	1.853	3604.	1161.9	0.0322	0.00260	0.0597	0.00983	1.913	36.50
37.00	15.945	4.400	11.1093	−108.26	82.19	1.9456	7.1057	2.354	2.952	1.867	3572.	1166.7	0.0315	0.00264	0.0600	0.01001	1.863	37.00
38.00	18.660	4.359	9.6324	−105.79	83.40	2.0084	6.9993	2.445	3.014	1.897	3506.	1175.7	0.0301	0.00272	0.0603	0.01037	1.763	38.00
39.00	21.691	4.317	8.3954	−103.21	84.49	2.0718	6.8966	2.540	3.083	1.930	3440.	1184.0	0.0289	0.00281	0.0605	0.01074	1.664	39.00
40.00	25.058	4.273	7.3516	−100.54	85.46	2.1357	6.7970	2.639	3.160	1.968	3373.	1191.8	0.0277	0.00289	0.0607	0.01112	1.565	40.00
41.00	28.781	4.227	6.4648	−97.76	86.29	2.2003	6.7002	2.744	3.246	2.011	3305.	1198.8	0.0266	0.00298	0.0607	0.01152	1.468	41.00
42.00	32.880	4.180	5.7066	−94.87	86.98	2.2656	6.6057	2.856	3.344	2.059	3235.	1205.2	0.0255	0.00307	0.0606	0.01193	1.371	42.00
43.00	37.375	4.130	5.0544	−91.87	87.52	2.3317	6.5131	2.977	3.455	2.115	3163.	1211.0	0.0245	0.00316	0.0605	0.01235	1.275	43.00
44.00	42.287	4.079	4.4901	−88.74	87.89	2.3986	6.4219	3.108	3.582	2.178	3088.	1216.2	0.0236	0.00326	0.0603	0.01280	1.180	44.00
45.00	47.637	4.025	3.9992	−85.48	88.08	2.4665	6.3319	3.252	3.728	2.251	3011.	1220.7	0.0227	0.00335	0.0600	0.01326	1.086	45.00
46.00	53.446	3.968	3.5699	−82.07	88.09	2.5354	6.2424	3.412	3.897	2.336	2931.	1224.6	0.0218	0.00345	0.0596	0.01375	0.993	46.00
47.00	59.734	3.909	3.1926	−78.51	87.87	2.6056	6.1532	3.593	4.096	2.435	2847.	1227.9	0.0209	0.00356	0.0592	0.01427	0.901	47.00
48.00	66.525	3.846	2.8593	−74.78	87.43	2.6773	6.0638	3.798	4.332	2.553	2760.	1230.6	0.0201	0.00367	0.0586	0.01482	0.811	48.00
49.00	73.840	3.779	2.5634	−70.87	86.73	2.7507	5.9735	4.037	4.615	2.695	2670.	1232.6	0.0193	0.00378	0.0580	0.01541	0.721	49.00
50.00	81.703	3.708	2.2994	−66.74	85.73	2.8262	5.8818	4.320	4.962	2.868	2575.	1234.0	0.0185	0.00390	0.0573	0.01604	0.633	50.00
52.00	99.171	3.550	1.8483	−57.74	82.68	2.9853	5.6909	5.089	5.952	3.361	2370.	1235.0	0.0170	0.00417	0.0556	0.01750	0.462	52.00
54.00	119.139	3.362	1.4749	−47.40	77.70	3.1604	5.4812	6.388	7.721	4.238	2142.	1233.4	0.0154	0.00451	0.0535	0.01933	0.299	54.00
56.00	141.856	3.124	1.1525	−34.92	69.57	3.3637	5.2331	9.186	11.753	6.220	1882.	1229.1	0.0138	0.00500	0.0507	0.02186	0.146	56.00
58.00	167.654	2.772	0.8435	−17.82	54.28	3.6351	4.8811	20.040	29.226	14.714	1576.	1223.1	0.0118	0.00593	0.0466	0.02617	0.015	58.00
59.29c	186.195	1.958	0.5108	17.32	17.32	4.2101	4.2101	∞	∞	∞	0.	0.0	–	–	∞	∞	0.000	59.29

*Temperatures are on the IPTS-68 scale a = triple point b = normal boiling point c = critical point

Refrigerant 702p (Parahydrogen) Properties of Gas at 14.696 psia (one standard atmosphere)

Temp., °F	Density, lb/ft³	Enthalpy, Btu/lb	Entropy, Btu/lb·°F	c_p, Btu/lb·°F	c_p/c_v	Vel. Sound, ft/s	Viscosity, lb/ft·h	Thermal Cond., Btu/h·ft·°F
−423.2a	0.08362	81.54	7.1606	2.924	1.853	1161.9	0.0026	0.0098
−420.0	0.07527	90.59	7.3986	2.791	1.811	1226.7	0.0028	0.0107
−410.0	0.05800	117.45	8.0031	2.616	1.748	1401.1	0.0035	0.0132
−400.0	0.04750	143.26	8.4768	2.555	1.720	1549.3	0.0042	0.0156
−390.0	0.04032	168.65	8.8703	2.526	1.704	1681.8	0.0048	0.0179
−380.0	0.03507	193.86	9.2084	2.517	1.691	1801.4	0.0054	0.0201
−370.0	0.03105	219.03	9.5060	2.519	1.680	1911.1	0.0060	0.0222
−360.0	0.02787	244.28	9.7729	2.534	1.665	2011.1	0.0065	0.0244
−340.0	0.02314	295.67	10.2426	2.618	1.622	2180.8	0.0075	0.0291
−320.0	0.01980	349.47	10.6580	2.773	1.563	2316.2	0.0084	0.0343
−300.0	0.01730	406.94	11.0422	2.979	1.502	2430.0	0.0093	0.0400
−280.0	0.01537	468.76	11.4068	3.205	1.449	2533.3	0.0101	0.0464
−260.0	0.01382	535.09	11.7566	3.424	1.408	2633.2	0.0109	0.0532
−240.0	0.01256	605.50	12.0925	3.610	1.378	2733.3	0.0117	0.0597
−220.0	0.01151	679.18	12.4135	3.749	1.359	2835.2	0.0124	0.0657
−200.0	0.01063	755.20	12.7180	3.844	1.346	2938.1	0.0131	0.0712
−180.0	0.00986	832.67	13.0054	3.896	1.340	3042.0	0.0138	0.0760
−160.0	0.00921	910.82	13.2753	3.913	1.338	3146.5	0.0145	0.0802
−140.0	0.00863	989.03	13.5280	3.905	1.339	3250.9	0.0151	0.0839
−120.0	0.00812	1066.89	13.7642	3.879	1.341	3354.7	0.0158	0.0870
−100.0	0.00767	1144.11	13.9851	3.842	1.346	3457.5	0.0164	0.0899
−80.0	0.00727	1220.56	14.1920	3.801	1.351	3558.8	0.0170	0.0924
−60.0	0.00690	1296.16	14.3861	3.759	1.356	3658.5	0.0176	0.0948
−40.0	0.00657	1370.93	14.5686	3.718	1.361	3756.1	0.0182	0.0972
−20.0	0.00627	1444.91	14.7408	3.680	1.366	3851.6	0.0188	0.0995
0.0	0.00600	1518.17	14.9038	3.646	1.371	3944.9	0.0194	0.1017
10.0	0.00587	1554.55	14.9821	3.630	1.373	3990.7	0.0197	0.1029
20.0	0.00575	1590.78	15.0584	3.616	1.375	4036.0	0.0200	0.1040
30.0	0.00563	1626.87	15.1329	3.602	1.377	4080.6	0.0203	0.1052
40.0	0.00552	1662.83	15.2056	3.590	1.379	4124.8	0.0205	0.1063
50.0	0.00541	1698.67	15.2766	3.578	1.381	4168.3	0.0208	0.1075
60.0	0.00531	1734.40	15.3460	3.567	1.382	4211.4	0.0211	0.1087
70.0	0.00521	1770.02	15.4139	3.558	1.383	4253.9	0.0214	0.1099
80.0	0.00511	1805.55	15.4804	3.549	1.385	4295.9	0.0216	0.1110
90.0	0.00502	1841.00	15.5455	3.541	1.386	4337.4	0.0219	0.1123
100.0	0.00493	1876.37	15.6092	3.533	1.387	4378.3	0.0222	0.1135
110.0	0.00484	1911.66	15.6717	3.526	1.388	4418.9	0.0225	0.1148
120.0	0.00476	1946.90	15.7330	3.520	1.389	4458.9	0.0227	0.1160
130.0	0.00468	1982.07	15.7932	3.515	1.390	4498.5	0.0230	0.1173
140.0	0.00460	2017.20	15.8523	3.510	1.391	4537.7	0.0233	0.1186
150.0	0.00453	2052.27	15.9103	3.506	1.391	4576.5	0.0235	0.1199
160.0	0.00445	2087.31	15.9673	3.501	1.392	4614.9	0.0238	0.1212
170.0	0.00438	2122.30	16.0233	3.498	1.393	4652.8	0.0241	0.1226
180.0	0.00431	2157.27	16.0784	3.495	1.393	4690.4	0.0243	0.1239
190.0	0.00425	2192.20	16.1326	3.492	1.393	4727.7	0.0246	0.1252
200.0	0.00418	2227.10	16.1859	3.489	1.394	4764.6	0.0248	0.1265
210.0	0.00412	2261.98	16.2384	3.487	1.394	4801.2	0.0251	0.1278
220.0	0.00406	2296.84	16.2900	3.485	1.395	4837.5	0.0253	0.1292
230.0	0.00400	2331.67	16.3409	3.483	1.395	4873.4	0.0256	0.1306
240.0	0.00394	2366.49	16.3910	3.481	1.395	4909.1	0.0258	0.1319
250.0	0.00389	2401.29	16.4404	3.479	1.395	4944.5	0.0261	0.1333

a = saturated vapor at normal boiling point

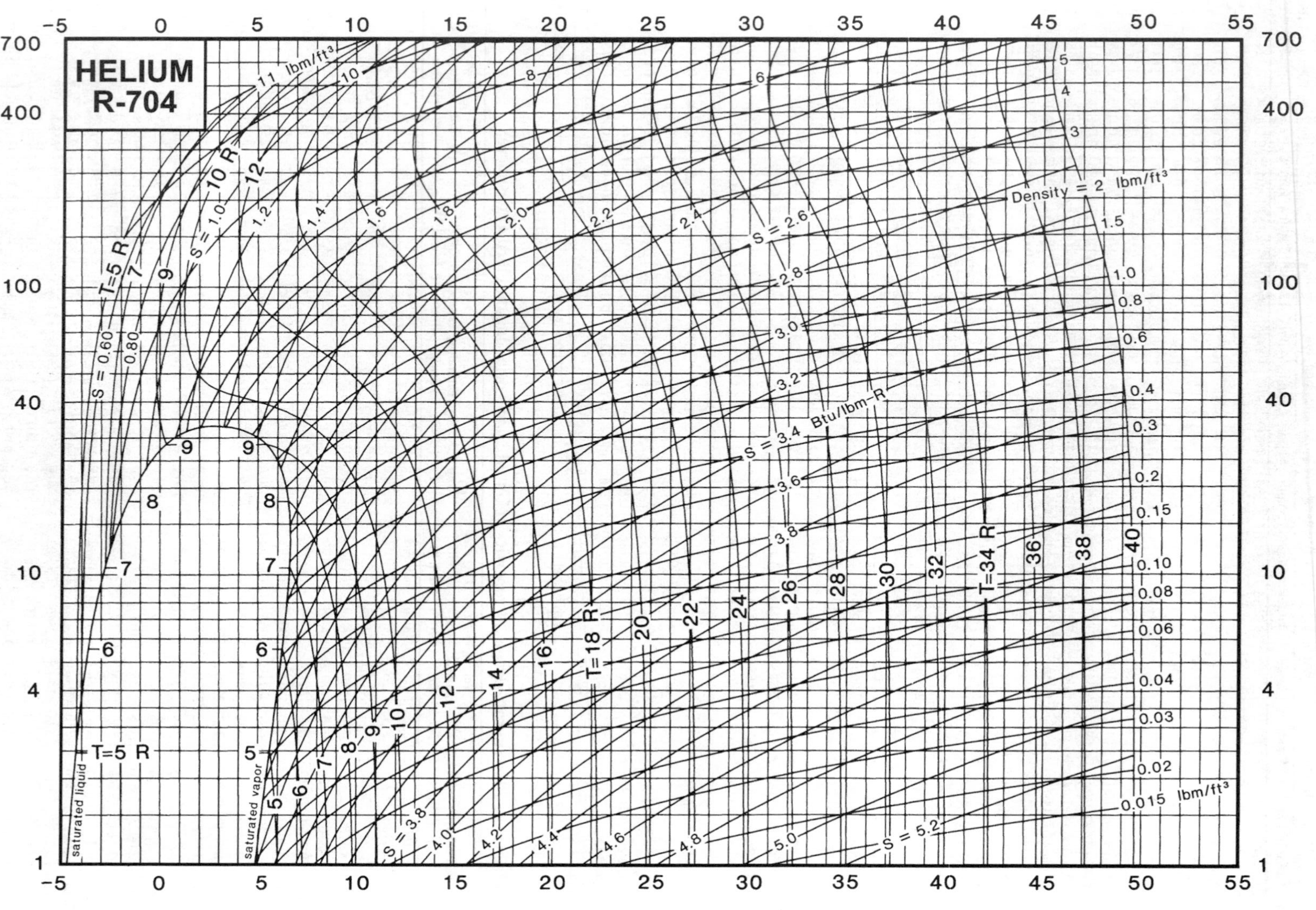

Fig. 24 Pressure-Enthalpy Diagram for Refrigerant 704 (Helium)

Note: The reference states for enthalpy and entropy differ from those in the table.

Refrigerant 704 (Helium) Properties of Saturated Liquid and Saturated Vapor

Temp.,* °R	Pressure, psia	Density, lb/ft³ Liquid	Volume, ft³/lb Vapor	Enthalpy, Btu/lb Liquid	Enthalpy, Btu/lb Vapor	Entropy, Btu/lb·°R Liquid	Entropy, Btu/lb·°R Vapor	Specific Heat c_p, Btu/lb·°R Liquid	Specific Heat c_p, Btu/lb·°R Vapor	c_p/c_v Vapor	Velocity of Sound, ft/s Liquid	Velocity of Sound, ft/s Vapor	Viscosity, lb_m/ft·h Liquid	Viscosity, lb_m/ft·h Vapor	Thermal Cond., Btu/h·ft·°R Liquid	Thermal Cond., Btu/h·ft·°R Vapor	Surface Tension, dyne/cm	Temp.,* °R
3.92a	0.704	9.130	13.9837	1.005	10.998	0.33557	2.88589	1.5100	1.4485	1.747	711.	273.0	–	–	–	–	0.388	3.92
4.00	0.789	9.122	12.7008	1.121	11.076	0.36428	2.85306	1.2809	1.4556	1.752	710.	275.2	–	–	–	–	0.382	4.00
4.20	1.021	9.099	10.1926	1.340	11.263	0.41671	2.77927	0.9051	1.4720	1.768	708.	280.3	–	–	–	–	0.366	4.20
4.40	1.296	9.068	8.3274	1.504	11.445	0.45355	2.71288	0.7039	1.4879	1.786	708.	285.1	–	–	–	–	0.350	4.40
4.60	1.617	9.030	6.9024	1.640	11.621	0.48225	2.65223	0.6064	1.5037	1.805	710.	289.7	–	–	–	–	0.334	4.60
4.80	1.988	8.987	5.7905	1.763	11.792	0.50693	2.59623	0.5700	1.5199	1.827	711.	294.0	–	–	–	–	0.319	4.80
5.00	2.413	8.938	4.9078	1.884	11.955	0.52983	2.54407	0.5694	1.5371	1.851	711.	298.0	–	–	–	–	0.303	5.00
5.20	2.896	8.883	4.1970	2.008	12.111	0.55216	2.49514	0.5893	1.5554	1.879	708.	301.7	–	–	–	–	0.287	5.20
5.40	3.442	8.824	3.6172	2.138	12.260	0.57454	2.44894	0.6209	1.5755	1.910	703.	305.2	–	–	–	–	0.272	5.40
5.60	4.053	8.760	3.1392	2.276	12.400	0.59727	2.40509	0.6590	1.5978	1.944	696.	308.5	–	–	–	–	0.256	5.60
5.80	4.734	8.690	2.7410	2.423	12.531	0.62047	2.36323	0.7009	1.6229	1.984	688.	311.5	–	–	–	–	0.241	5.80
6.00	5.488	8.616	2.4063	2.579	12.652	0.64419	2.32304	0.7453	1.6513	2.029	679.	314.3	–	–	–	–	0.226	6.00
6.20	6.319	8.536	2.1226	2.744	12.763	0.66842	2.28427	0.7918	1.6839	2.080	669.	316.8	–	–	–	–	0.211	6.20
6.40	7.230	8.451	1.8803	2.920	12.862	0.69316	2.24664	0.8407	1.7217	2.139	659.	319.2	0.00844	0.00238	0.0103	0.00416	0.196	6.40
6.60	8.224	8.360	1.6717	3.106	12.950	0.71841	2.20990	0.8927	1.7657	2.207	647.	321.3	0.00832	0.00248	0.0104	0.00432	0.181	6.60
6.80	9.306	8.262	1.4910	3.303	13.025	0.74419	2.17383	0.9489	1.8177	2.286	635.	323.2	0.00819	0.00258	0.0105	0.00448	0.166	6.80
7.00	10.480	8.158	1.3333	3.511	13.085	0.77053	2.13816	1.0109	1.8798	2.379	623.	324.9	0.00807	0.00268	0.0106	0.00466	0.152	7.00
7.20	11.748	8.047	1.1948	3.732	13.129	0.79748	2.10264	1.0810	1.9547	2.491	609.	326.3	0.00793	0.00279	0.0107	0.00483	0.138	7.20
7.40	13.114	7.927	1.0724	3.965	13.155	0.82515	2.06700	1.1621	2.0466	2.626	595.	327.6	0.00780	0.00290	0.0107	0.00502	0.123	7.40
7.60	14.584	7.797	0.9634	4.214	13.161	0.85365	2.03091	1.2586	2.1613	2.792	580.	328.7	0.00766	0.00301	0.0108	0.00522	0.109	7.60
7.61b	14.696	7.787	0.9559	4.233	13.160	0.85578	2.02824	1.2664	2.1708	2.806	579.	328.8	0.00765	0.00302	0.0108	0.00524	0.108	7.61
7.80	16.161	7.656	0.8657	4.479	13.143	0.88319	1.99398	1.3768	2.3077	3.003	564.	329.6	0.00752	0.00313	0.0108	0.00544	0.095	7.80
8.00	17.850	7.501	0.7774	4.764	13.098	0.91401	1.95573	1.5272	2.5001	3.278	547.	330.3	0.00737	0.00326	0.0108	0.00568	0.082	8.00
8.20	19.657	7.329	0.6969	5.072	13.018	0.94651	1.91551	1.7271	2.7626	3.650	529.	330.8	0.00721	0.00340	0.0109	0.00596	0.068	8.20
8.40	21.587	7.135	0.6226	5.410	12.895	0.98128	1.87237	2.0092	3.1400	4.181	509.	331.2	0.00704	0.00354	0.0109	0.00629	0.055	8.40
8.60	23.650	6.911	0.5529	5.788	12.716	1.01927	1.82486	2.4415	3.7248	5.003	487.	331.6	0.00686	0.00371	0.0110	0.00670	0.042	8.60
8.80	25.854	6.641	0.4861	6.222	12.453	1.06226	1.77037	3.1940	4.7457	6.434	463.	332.2	0.00664	0.00389	0.0110	0.00724	0.030	8.80
9.00	28.215	6.294	0.4192	6.751	12.055	1.11412	1.70345	4.8375	6.9536	9.531	436.	333.4	0.00638	0.00411	0.0112	0.00801	0.018	9.00
9.20	30.755	5.771	0.3451	7.493	11.364	1.18708	1.60775	11.0204	14.9798	20.845	404.	336.9	0.00603	0.00440	0.0115	0.00932	0.007	9.20
9.35c	32.990	4.348	0.2300	9.339	9.339	1.37760	1.37760	∞	∞	∞	0.	0.0	–	–	∞	∞	0.000	9.35

*Temperatures are on the EPT-76 scale a = lower lambda point b = normal boiling point c = critical point

Refrigerant 704 (Helium) Properties of Gas at 14.696 psia (one standard atmosphere)

Temp., °F	Density, lb/ft³	Enthalpy, Btu/lb	Entropy, Btu/lb·°F	c_p, Btu/lb·°F	c_p/c_v	Vel. Sound, ft/s	Viscosity, lb/ft·h	Thermal Cond., Btu/h·ft·°F
−452.1a	1.04613	13.16	2.0282	2.1708	2.806	328.8	0.0030	0.00523
−450.0	0.66840	16.73	2.4459	1.5277	2.037	415.4	0.0035	0.00624
−440.0	0.28462	30.24	3.4138	1.2880	1.729	636.4	0.0058	0.01035
−430.0	0.18546	42.96	3.9370	1.2616	1.691	786.5	0.0076	0.01343
−420.0	0.13809	55.52	4.3020	1.2525	1.679	910.2	0.0092	0.01608
−400.0	0.09163	80.49	4.8118	1.2459	1.671	1115.8	0.0119	0.02074
−380.0	0.06862	105.39	5.1716	1.2437	1.669	1288.4	0.0143	0.02493
−360.0	0.05486	130.25	5.4500	1.2427	1.668	1440.3	0.0164	0.02882
−340.0	0.04571	155.10	5.6772	1.2421	1.667	1577.5	0.0183	0.03248
−320.0	0.03917	179.94	5.8691	1.2418	1.667	1703.7	0.0202	0.03596
−300.0	0.03427	204.77	6.0353	1.2416	1.667	1821.2	0.0219	0.03931
−280.0	0.03046	229.60	6.1818	1.2415	1.667	1931.5	0.0236	0.04254
−260.0	0.02741	254.43	6.3129	1.2414	1.667	2035.8	0.0248	0.04566
−240.0	0.02492	279.26	6.4314	1.2413	1.667	2135.1	0.0264	0.04870
−220.0	0.02285	304.08	6.5395	1.2413	1.667	2229.9	0.0280	0.05166
−200.0	0.02109	328.91	6.6390	1.2413	1.666	2320.9	0.0295	0.05454
−180.0	0.01958	353.73	6.7311	1.2412	1.666	2408.4	0.0310	0.05737
−160.0	0.01827	378.56	6.8168	1.2412	1.666	2492.9	0.0324	0.06013
−140.0	0.01713	403.38	6.8970	1.2412	1.666	2574.6	0.0338	0.06284
−120.0	0.01613	428.21	6.9724	1.2412	1.666	2653.7	0.0352	0.06551
−100.0	0.01523	453.03	7.0434	1.2412	1.666	2730.6	0.0366	0.06813
−80.0	0.01443	477.85	7.1105	1.2412	1.667	2805.4	0.0380	0.07070
−60.0	0.01371	502.68	7.1743	1.2412	1.667	2878.2	0.0393	0.07324
−40.0	0.01305	527.50	7.2349	1.2412	1.667	2949.3	0.0406	0.07574
−20.0	0.01246	552.32	7.2927	1.2412	1.667	3018.7	0.0419	0.07821
0.0	0.01192	577.15	7.3479	1.2412	1.667	3086.5	0.0432	0.08064
20.0	0.01142	601.97	7.4007	1.2412	1.667	3152.8	0.0445	0.08304
40.0	0.01096	626.79	7.4514	1.2412	1.667	3217.8	0.0457	0.08542
60.0	0.01054	651.62	7.5001	1.2412	1.667	3281.5	0.0470	0.08776
80.0	0.01015	676.44	7.5470	1.2411	1.667	3344.0	0.0482	0.09008
100.0	0.00979	701.26	7.5922	1.2411	1.667	3405.3	0.0494	0.09238
120.0	0.00945	726.09	7.6357	1.2411	1.667	3465.6	0.0506	0.09465
140.0	0.00914	750.91	7.6778	1.2411	1.667	3524.8	0.0518	0.09690
160.0	0.00884	775.73	7.7186	1.2411	1.667	3583.1	0.0530	0.09912
180.0	0.00857	800.55	7.7580	1.2411	1.667	3640.4	0.0542	0.10133
200.0	0.00831	825.38	7.7962	1.2411	1.667	3696.8	0.0553	0.10351
240.0	0.00783	875.02	7.8693	1.2411	1.667	3807.1	0.0576	0.10783
280.0	0.00741	924.67	7.9383	1.2411	1.667	3914.4	0.0599	0.11207
320.0	0.00703	974.32	8.0036	1.2411	1.667	4018.8	0.0621	0.11624
360.0	0.00669	1023.96	8.0657	1.2411	1.667	4120.5	0.0643	0.12036
400.0	0.00637	1073.61	8.1249	1.2411	1.667	4219.8	0.0665	0.12441
440.0	0.00609	1123.25	8.1813	1.2411	1.667	4316.8	0.0686	0.12841
480.0	0.00583	1172.90	8.2353	1.2411	1.667	4411.6	0.0707	0.13236
520.0	0.00559	1222.55	8.2871	1.2411	1.667	4504.5	0.0728	0.13626
560.0	0.00537	1272.19	8.3367	1.2411	1.667	4595.5	0.0749	0.14011
600.0	0.00517	1321.84	8.3845	1.2411	1.667	4684.7	0.0769	0.14392
640.0	0.00498	1371.48	8.4305	1.2412	1.667	4772.3	0.0789	0.14768
680.0	0.00481	1421.13	8.4748	1.2412	1.667	4858.3	0.0809	0.15141
720.0	0.00465	1470.78	8.5176	1.2412	1.667	4942.8	0.0829	0.15509
760.0	0.00449	1520.42	8.5590	1.2412	1.667	5025.8	0.0849	0.15874
800.0	0.00435	1570.07	8.5991	1.2412	1.667	5107.6	0.0868	0.16236

a = saturated vapor at normal boiling point

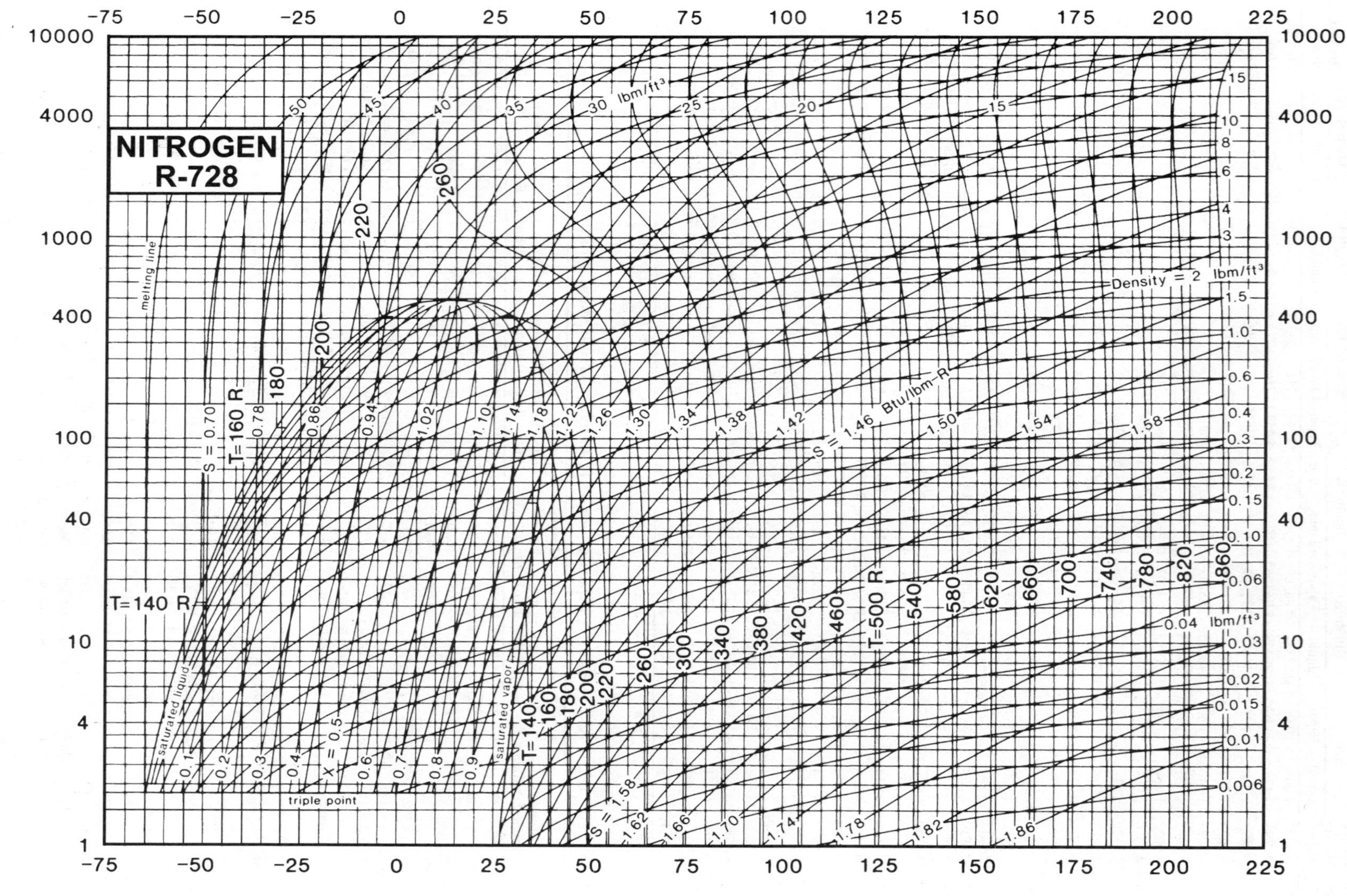

Fig. 25 Pressure-Enthalpy Diagram for Refrigerant 728 (Nitrogen)

Refrigerant 728 (Nitrogen) Properties of Saturated Liquid and Saturated Vapor

Temp.,* °R	Pressure, psia	Density, lb/ft³	Volume, ft³/lb	Enthalpy, Btu/lb		Entropy, Btu/lb·°R		Specific Heat c_p, Btu/lb·°R		c_p/c_v	Velocity of Sound, ft/s		Viscosity, lb_m/ft·h		Thermal Cond., Btu/h·ft·°R		Surface Tension, dyne/cm	Temp.,* °R
		Liquid	Vapor	Liquid	Vapor	Liquid	Vapor	Liquid	Vapor	Vapor	Liquid	Vapor	Liquid	Vapor	Liquid	Vapor		
113.67a	1.82	54.29	23.738	−64.91	27.76	0.57825	1.39348	0.4826	0.2838	1.390	3352.	523.2	0.6916	0.0101	0.1009	0.00311	12.24	113.67
115.00	2.08	54.09	20.970	−64.26	28.06	0.58387	1.38665	0.4821	0.2857	1.390	3318.	525.7	0.6639	0.0102	0.1000	0.00315	12.06	115.00
120.00	3.34	53.34	13.542	−61.85	29.16	0.60435	1.36278	0.4811	0.2929	1.394	3197.	535.0	0.5749	0.0107	0.0963	0.00333	11.39	120.00
125.00	5.15	52.58	9.0919	−59.44	30.22	0.62397	1.34131	0.4815	0.3002	1.400	3084.	543.7	0.5044	0.0111	0.0928	0.00350	10.72	125.00
130.00	7.66	51.81	6.3130	−57.03	31.25	0.64286	1.32189	0.4830	0.3075	1.408	2976.	551.7	0.4472	0.0116	0.0893	0.00369	10.06	130.00
135.00	11.04	51.03	4.5130	−54.60	32.22	0.66109	1.30423	0.4854	0.3146	1.419	2873.	559.2	0.3999	0.0121	0.0859	0.00388	9.42	135.00
139.23b	14.696	50.36	3.4662	−52.53	33.01	0.67606	1.29047	0.4881	0.3205	1.430	2787.	565.0	0.3659	0.0125	0.0831	0.00404	8.88	139.23
140.00	15.45	50.23	3.3089	−52.16	33.15	0.67875	1.28806	0.4887	0.3215	1.432	2772.	566.0	0.3602	0.0126	0.0826	0.00407	8.78	140.00
145.00	21.11	49.42	2.4801	−49.69	34.01	0.69590	1.27315	0.4928	0.3285	1.449	2673.	572.1	0.3262	0.0131	0.0793	0.00428	8.15	145.00
150.00	28.20	48.58	1.8949	−47.20	34.81	0.71259	1.25931	0.4977	0.3355	1.470	2575.	577.5	0.2968	0.0137	0.0761	0.00449	7.53	150.00
155.00	36.93	47.73	1.4722	−44.69	35.53	0.72888	1.24637	0.5035	0.3430	1.495	2478.	582.1	0.2710	0.0142	0.0729	0.00472	6.92	155.00
160.00	47.53	46.85	1.1605	−42.13	36.16	0.74482	1.23417	0.5104	0.3513	1.526	2380.	585.9	0.2482	0.0148	0.0698	0.00496	6.32	160.00
165.00	60.20	45.94	0.9264	−39.54	36.71	0.76044	1.22256	0.5186	0.3610	1.563	2281.	588.9	0.2277	0.0153	0.0667	0.00521	5.74	165.00
170.00	75.19	44.99	0.7476	−36.91	37.14	0.77580	1.21141	0.5284	0.3726	1.608	2180.	591.1	0.2093	0.0159	0.0636	0.00548	5.16	170.00
175.00	92.71	44.02	0.6089	−34.22	37.46	0.79095	1.20059	0.5404	0.3870	1.663	2078.	592.3	0.1925	0.0166	0.0605	0.00577	4.60	175.00
180.00	113.01	42.99	0.4997	−31.48	37.65	0.80595	1.18999	0.5553	0.4051	1.731	1972.	592.6	0.1770	0.0172	0.0575	0.00609	4.05	180.00
185.00	136.32	41.92	0.4127	−28.65	37.69	0.82086	1.17946	0.5741	0.4284	1.817	1862.	592.0	0.1627	0.0179	0.0545	0.00644	3.52	185.00
190.00	162.89	40.78	0.3424	−25.74	37.55	0.83577	1.16887	0.5984	0.4591	1.927	1748.	590.3	0.1493	0.0187	0.0515	0.00683	3.00	190.00
195.00	192.99	39.56	0.2850	−22.71	37.21	0.85077	1.15805	0.6309	0.5007	2.074	1627.	587.5	0.1366	0.0195	0.0484	0.00727	2.50	195.00
200.00	226.88	38.23	0.2374	−19.54	36.62	0.86603	1.14680	0.6762	0.5590	2.275	1499.	583.5	0.1245	0.0204	0.0453	0.00778	2.02	200.00
205.00	264.86	36.76	0.1975	−16.16	35.71	0.88175	1.13481	0.7430	0.6455	2.570	1361.	578.2	0.1127	0.0215	0.0421	0.00841	1.56	205.00
210.00	307.25	35.07	0.1634	−12.52	34.39	0.89828	1.12165	0.8496	0.7847	3.035	1212.	571.5	0.1009	0.0228	0.0388	0.00925	1.13	210.00
215.00	354.43	33.06	0.1336	−8.45	32.47	0.91622	1.10651	1.0416	1.0417	3.877	1049.	563.0	0.0889	0.0244	0.0354	0.01052	0.73	215.00
220.00	406.84	30.49	0.1065	−3.66	29.52	0.93683	1.08763	1.4809	1.6619	5.856	867.	552.1	0.0761	0.0266	0.0326	0.01310	0.38	220.00
225.00	465.28	26.51	0.0784	2.91	23.94	0.96462	1.05807	3.8790	5.1786	16.586	654.	535.3	0.0606	0.0309	0.0358	0.02362	0.08	225.00
227.15c	492.81	19.55	0.0512	12.83	12.83	1.00750	1.00750	∞	∞	∞	0.	0.0	–	–	∞	∞	0.00	227.15

*Temperatures are on the IPTS-68 scale   a = triple point   b = normal boiling point   c = critical point

Refrigerant 728 (Nitrogen) Properties of Gas at 14.696 psia (one standard atmosphere)

Temp., °F	Density, lb/ft³	Enthalpy, Btu/lb	Entropy, Btu/lb·°F	c_p, Btu/lb·°F	c_p/c_v	Vel. Sound, ft/s	Viscosity, lb/ft·h	Thermal Cond., Btu/h·ft·°F
−320.4a	0.2885	33.01	1.2905	0.3205	1.430	565.0	0.0125	0.00404
−300.0	0.2471	38.64	1.3283	0.2595	1.436	619.8	0.0144	0.00470
−275.0	0.2116	45.05	1.3656	0.2546	1.425	670.7	0.0167	0.00550
−250.0	0.1853	51.39	1.3978	0.2527	1.418	717.1	0.0189	0.00631
−225.0	0.1650	57.69	1.4262	0.2516	1.413	760.2	0.0211	0.00711
−200.0	0.1487	63.97	1.4516	0.2508	1.410	800.8	0.0232	0.00783
−180.0	0.1379	68.98	1.4702	0.2504	1.408	831.7	0.0248	0.00841
−160.0	0.1286	73.99	1.4875	0.2500	1.407	861.4	0.0264	0.00898
−140.0	0.1204	78.99	1.5036	0.2498	1.406	890.1	0.0279	0.00954
−120.0	0.1132	83.98	1.5188	0.2496	1.405	917.9	0.0294	0.01008
−100.0	0.1069	88.97	1.5331	0.2494	1.404	944.8	0.0309	0.01061
−80.0	0.1012	93.96	1.5466	0.2493	1.404	970.9	0.0323	0.01113
−60.0	0.0961	98.94	1.5594	0.2492	1.403	996.3	0.0337	0.01163
−40.0	0.0915	103.92	1.5715	0.2491	1.403	1021.1	0.0351	0.01213
−20.0	0.0873	108.90	1.5831	0.2490	1.403	1045.2	0.0364	0.01261
0.0	0.0835	113.88	1.5942	0.2490	1.402	1068.8	0.0378	0.01309
10.0	0.0817	116.37	1.5996	0.2489	1.402	1080.4	0.0384	0.01332
20.0	0.0800	118.86	1.6048	0.2489	1.402	1091.9	0.0390	0.01355
30.0	0.0784	121.35	1.6099	0.2489	1.402	1103.3	0.0397	0.01378
40.0	0.0768	123.84	1.6150	0.2489	1.402	1114.5	0.0403	0.01401
50.0	0.0753	126.33	1.6199	0.2489	1.402	1125.6	0.0409	0.01424
60.0	0.0738	128.82	1.6247	0.2489	1.402	1136.6	0.0416	0.01446
70.0	0.0724	131.31	1.6295	0.2489	1.401	1147.5	0.0422	0.01469
80.0	0.0711	133.80	1.6341	0.2489	1.401	1158.3	0.0428	0.01491
90.0	0.0698	136.28	1.6387	0.2489	1.401	1169.0	0.0434	0.01513
100.0	0.0686	138.77	1.6432	0.2489	1.401	1179.6	0.0440	0.01534
120.0	0.0662	143.75	1.6519	0.2489	1.401	1200.4	0.0452	0.01578
140.0	0.0640	148.73	1.6604	0.2490	1.400	1220.9	0.0463	0.01620
160.0	0.0619	153.71	1.6685	0.2491	1.400	1241.0	0.0475	0.01662
180.0	0.0600	158.69	1.6765	0.2492	1.400	1260.8	0.0486	0.01704

Temp., °F	Density, lb/ft³	Enthalpy, Btu/lb	Entropy, Btu/lb·°F	c_p, Btu/lb·°F	c_p/c_v	Vel. Sound, ft/s	Viscosity, lb/ft·h	Thermal Cond., Btu/h·ft·°F
200.0	0.0581	163.68	1.6841	0.2493	1.399	1280.2	0.0497	0.01745
220.0	0.0564	168.66	1.6916	0.2494	1.399	1299.4	0.0508	0.01786
240.0	0.0548	173.65	1.6988	0.2496	1.399	1318.2	0.0519	0.01826
260.0	0.0533	178.65	1.7059	0.2497	1.398	1336.7	0.0530	0.01866
280.0	0.0518	183.64	1.7127	0.2499	1.398	1354.9	0.0540	0.01906
300.0	0.0505	188.65	1.7194	0.2502	1.397	1372.8	0.0551	0.01945
320.0	0.0492	193.65	1.7259	0.2504	1.396	1390.5	0.0561	0.01985
340.0	0.0480	198.66	1.7322	0.2507	1.396	1407.9	0.0571	0.02024
360.0	0.0468	203.68	1.7384	0.2510	1.395	1425.0	0.0581	0.02062
380.0	0.0457	208.70	1.7445	0.2514	1.394	1441.9	0.0591	0.02101
400.0	0.0446	213.73	1.7504	0.2517	1.393	1458.6	0.0601	0.02140
420.0	0.0436	218.77	1.7562	0.2521	1.392	1475.0	0.0611	0.02178
440.0	0.0426	223.82	1.7619	0.2525	1.391	1491.2	0.0620	0.02216
460.0	0.0417	228.87	1.7674	0.2529	1.391	1507.2	0.0630	0.02254
480.0	0.0408	233.93	1.7729	0.2534	1.390	1522.9	0.0639	0.02292
500.0	0.0400	239.01	1.7782	0.2538	1.389	1538.5	0.0648	0.02330
520.0	0.0391	244.09	1.7834	0.2543	1.387	1553.9	0.0658	0.02368
540.0	0.0384	249.18	1.7886	0.2548	1.386	1569.0	0.0667	0.02406
560.0	0.0376	254.28	1.7936	0.2553	1.385	1584.0	0.0676	0.02444
580.0	0.0369	259.39	1.7986	0.2559	1.384	1598.8	0.0685	0.02482
600.0	0.0362	264.51	1.8035	0.2564	1.383	1613.5	0.0693	0.02519
620.0	0.0355	269.65	1.8083	0.2570	1.382	1627.9	0.0702	0.02557
640.0	0.0349	274.79	1.8130	0.2575	1.381	1642.2	0.0711	0.02594
660.0	0.0342	279.95	1.8176	0.2581	1.379	1656.3	0.0720	0.02632
680.0	0.0336	285.12	1.8222	0.2587	1.378	1670.3	0.0728	0.02670
700.0	0.0331	290.30	1.8267	0.2593	1.377	1684.2	0.0737	0.02707
720.0	0.0325	295.49	1.8312	0.2599	1.376	1697.9	0.0745	0.02745
740.0	0.0320	300.69	1.8355	0.2606	1.374	1711.4	0.0754	0.02782
760.0	0.0314	305.91	1.8399	0.2612	1.373	1724.8	0.0762	0.02820
780.0	0.0309	311.14	1.8441	0.2618	1.372	1738.1	0.0770	0.02857
800.0	0.0304	316.38	1.8483	0.2624	1.371	1751.3	0.0778	0.02895

a = saturated vapor at normal boiling point

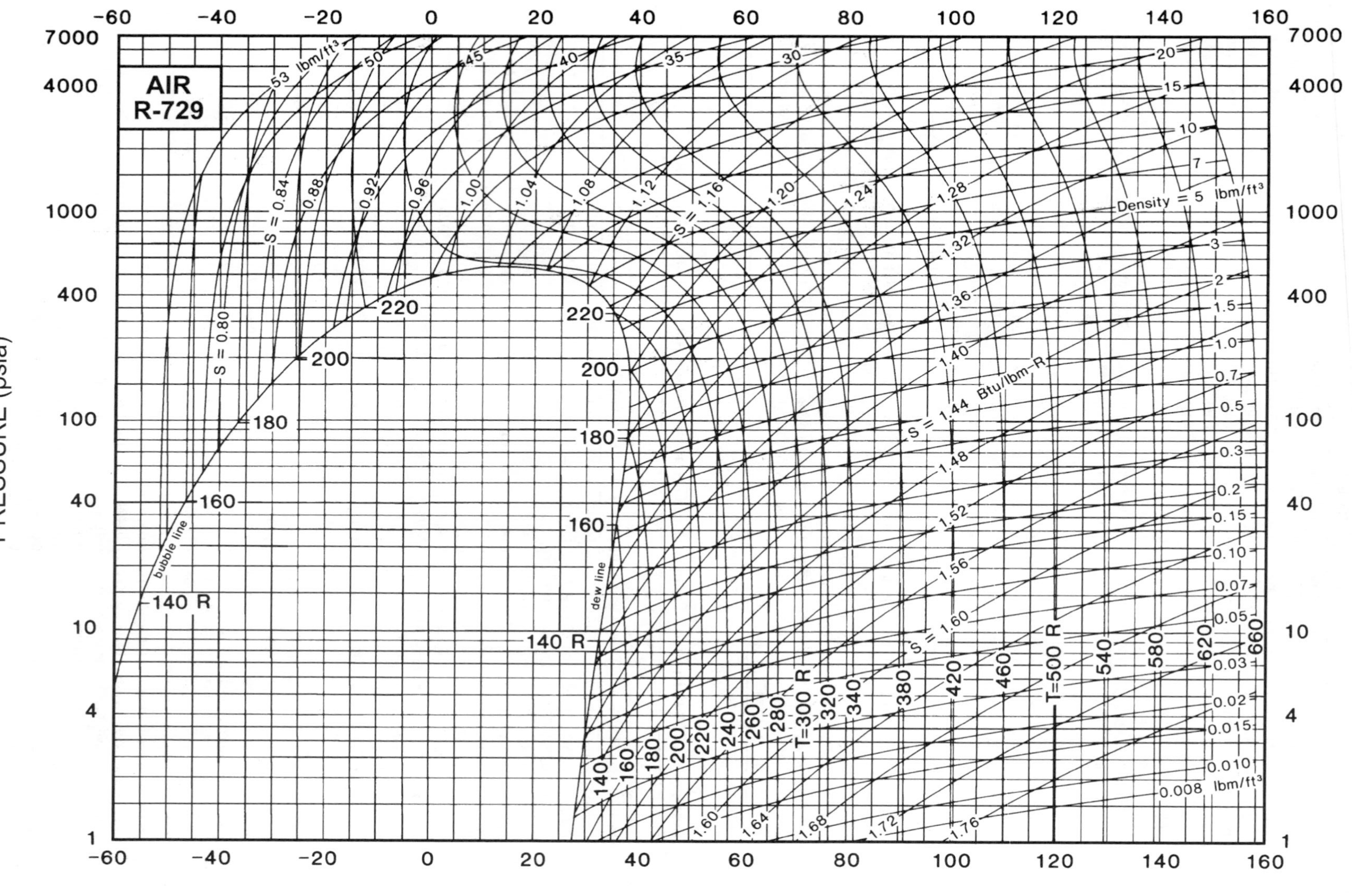

Fig. 26 Pressure-Enthalpy Diagram for Refrigerant 729 (Air)

Refrigerant 729 (Air) Properties of Liquid on the Bubble Line and Vapor on the Dew Line

Pressure, psia	Temp.*, °R Bubble	Temp.*, °R Dew	Density, lb/ft³ Liquid	Volume, ft³/lb Vapor	Enthalpy, Btu/lb Liquid	Enthalpy, Btu/lb Vapor	Entropy, Btu/lb·°R Liquid	Entropy, Btu/lb·°R Vapor	Specific Heat c_p, Btu/lb·°R Liquid	Specific Heat c_p, Btu/lb·°R Vapor	c_p/c_v Vapor	Vel. of Sound, ft/s Liquid	Vel. of Sound, ft/s Vapor	Viscosity, lb_m/ft·h Liquid	Viscosity, lb_m/ft·h Vapor	Thermal Cond., Btu/h·ft·°R Liquid	Thermal Cond., Btu/h·ft·°R Vapor	Surface Tension, dyne/cm
0.91a	107.55	115.19	59.79	46.8083	−70.04	27.29	0.58484	1.46086	0.4591	0.2447	1.406	3322.	524.	1.0650	0.0106	0.1009	0.00325	14.12
1.00	108.49	116.03	59.65	42.7531	−69.61	27.48	0.58883	1.45585	0.4588	0.2450	1.406	3312.	526.	1.0248	0.0107	0.1003	0.00328	14.00
2.00	115.58	122.47	58.62	22.4796	−66.36	28.92	0.61781	1.42069	0.4573	0.2475	1.410	3225.	539.	0.7885	0.0114	0.0961	0.00350	13.04
4.00	123.53	129.89	57.44	11.8472	−62.72	30.51	0.64820	1.38636	0.4573	0.2511	1.417	3115.	553.	0.6173	0.0121	0.0914	0.00375	11.98
6.00	128.66	134.75	56.66	8.1498	−60.37	31.51	0.66682	1.36662	0.4581	0.2539	1.423	3039.	562.	0.5382	0.0126	0.0885	0.00393	11.31
8.00	132.56	138.46	56.07	6.2499	−58.57	32.24	0.68050	1.35274	0.4591	0.2563	1.429	2979.	568.	0.4892	0.0129	0.0863	0.00406	10.80
10.00	135.75	141.49	55.58	5.0865	−57.10	32.83	0.69142	1.34203	0.4602	0.2585	1.435	2928.	572.	0.4546	0.0133	0.0845	0.00417	10.39
12.00	138.47	144.09	55.15	4.2980	−55.84	33.31	0.70054	1.33332	0.4613	0.2605	1.440	2885.	576.	0.4283	0.0135	0.0830	0.00427	10.04
14.696b	141.63	147.10	54.65	3.5634	−54.38	33.86	0.71094	1.32367	0.4627	0.2630	1.448	2833.	580.	0.4009	0.0138	0.0812	0.00439	9.64
20.00	146.73	151.96	53.84	2.6779	−52.00	34.69	0.72728	1.30905	0.4657	0.2675	1.461	2749.	586.	0.3624	0.0144	0.0784	0.00458	9.00
40.00	159.68	164.35	51.68	1.4015	−45.87	36.53	0.76687	1.27616	0.4765	0.2819	1.511	2524.	599.	0.2875	0.0157	0.0715	0.00510	7.42
60.00	168.40	172.73	50.13	0.9542	−41.64	37.50	0.79222	1.25669	0.4875	0.2952	1.561	2364.	605.	0.2492	0.0167	0.0669	0.00549	6.39
80.00	175.19	179.27	48.87	0.7234	−38.27	38.08	0.81141	1.24256	0.4988	0.3085	1.612	2235.	608.	0.2239	0.0175	0.0633	0.00582	5.61
100.00	180.84	184.71	47.77	0.5815	−35.40	38.43	0.82710	1.23128	0.5107	0.3222	1.666	2124.	609.	0.2051	0.0182	0.0604	0.00612	4.98
150.00	192.04	195.53	45.40	0.3865	−29.49	38.66	0.85774	1.20957	0.5440	0.3609	1.819	1893.	609.	0.1724	0.0197	0.0544	0.00680	3.77
200.00	200.80	203.97	43.34	0.2851	−24.59	38.32	0.88164	1.19252	0.5846	0.4089	2.009	1701.	606.	0.1497	0.0210	0.0497	0.00745	2.88
250.00	208.12	210.98	41.41	0.2222	−20.23	37.56	0.90188	1.17767	0.6368	0.4719	2.258	1530.	601.	0.1322	0.0223	0.0457	0.00815	2.18
300.00	214.47	217.03	39.51	0.1787	−16.17	36.45	0.91999	1.16383	0.7075	0.5593	2.602	1372.	596.	0.1175	0.0236	0.0420	0.00895	1.60
350.00	220.11	222.38	37.57	0.1464	−12.26	34.96	0.93692	1.15024	0.8101	0.6893	3.109	1223.	590.	0.1046	0.0250	0.0387	0.00998	1.12
400.00	225.21	227.19	35.51	0.1210	−8.33	33.03	0.95342	1.13618	0.9738	0.9040	3.934	1079.	584.	0.0927	0.0266	0.0357	0.01149	0.72
450.00	229.89	231.56	33.18	0.0998	−4.19	30.48	0.97040	1.12061	1.2881	1.3302	5.534	936.	578.	0.0813	0.0287	0.0335	0.01417	0.39
500.00	234.23	235.51	30.19	0.0802	0.65	26.68	0.99001	1.10078	2.2460	2.6522	10.317	773.	571.	0.0694	0.0315	0.0336	0.02069	0.12
548.97c	238.71	238.71	18.89	0.0529	16.38	16.38	1.05489	1.05489	–	–	–	–	–	–	–	–	–	0.00

*Temperatures are on the IPTS-68 scale a = triple point b = bubble and dew points at 14.696 psia c = critical point

Refrigerant 729 (Air) Properties of Gas at 14.696 psia (one standard atmosphere)

Temp., °F	Density, lb/ft³	Enthalpy, Btu/lb	Entropy, Btu/lb·°F	c_p, Btu/lb·°F	c_p/c_v	Vel. Sound, ft/s	Viscosity, lb/ft·h	Thermal Cond., Btu/ft·h·°F
−312.6a	0.2806	33.86	1.3237	0.2630	1.448	580.3	0.0138	0.00439
−300.0	0.2561	37.11	1.3449	0.2547	1.438	608.6	0.0150	0.00480
−280.0	0.2255	42.13	1.3745	0.2483	1.428	649.6	0.0169	0.00545
−260.0	0.2017	47.06	1.4005	0.2454	1.422	687.3	0.0188	0.00611
−240.0	0.1827	51.95	1.4239	0.2437	1.417	722.5	0.0206	0.00677
−220.0	0.1670	56.81	1.4451	0.2427	1.414	755.9	0.0224	0.00764
−200.0	0.1538	61.66	1.4645	0.2420	1.412	787.7	0.0241	0.00799
−180.0	0.1426	66.50	1.4824	0.2415	1.410	818.2	0.0258	0.00856
−160.0	0.1329	71.32	1.4991	0.2412	1.409	847.5	0.0275	0.00913
−140.0	0.1245	76.14	1.5147	0.2409	1.408	875.7	0.0291	0.00968
−120.0	0.1171	80.96	1.5293	0.2407	1.407	903.0	0.0307	0.01023
−100.0	0.1105	85.77	1.5431	0.2405	1.406	929.5	0.0322	0.01076
−80.0	0.1047	90.58	1.5561	0.2404	1.405	955.2	0.0337	0.01128
−60.0	0.0994	95.39	1.5684	0.2403	1.405	980.2	0.0351	0.01179
−40.0	0.0946	100.19	1.5801	0.2403	1.404	1004.6	0.0366	0.01229
−20.0	0.0903	105.00	1.5913	0.2402	1.404	1028.3	0.0380	0.01278
0.0	0.0863	109.81	1.6020	0.2402	1.403	1051.5	0.0393	0.01326
20.0	0.0827	114.61	1.6122	0.2402	1.403	1074.2	0.0407	0.01372
40.0	0.0794	119.42	1.6221	0.2403	1.403	1096.3	0.0420	0.01419
60.0	0.0763	124.22	1.6315	0.2403	1.402	1118.0	0.0433	0.01465
80.0	0.0735	129.03	1.6406	0.2404	1.402	1139.3	0.0446	0.01510
100.0	0.0709	133.84	1.6493	0.2405	1.401	1160.1	0.0458	0.01554
120.0	0.0684	138.65	1.6578	0.2407	1.401	1180.6	0.0470	0.01599
140.0	0.0661	143.47	1.6659	0.2408	1.400	1200.6	0.0483	0.01642
160.0	0.0640	148.28	1.6738	0.2410	1.400	1220.3	0.0495	0.01685
180.0	0.0620	153.11	1.6815	0.2412	1.399	1239.6	0.0506	0.01728
200.0	0.0601	157.93	1.6889	0.2414	1.398	1258.6	0.0518	0.01771
220.0	0.0583	162.76	1.6961	0.2417	1.398	1277.3	0.0529	0.01813
240.0	0.0567	167.60	1.7032	0.2420	1.397	1295.6	0.0541	0.01854
260.0	0.0551	172.44	1.7100	0.2423	1.396	1313.7	0.0552	0.01896
280.0	0.0536	177.29	1.7166	0.2426	1.395	1331.5	0.0563	0.01937
300.0	0.0522	182.15	1.7231	0.2430	1.394	1348.9	0.0574	0.01978
320.0	0.0508	187.01	1.7294	0.2433	1.394	1366.2	0.0584	0.02019
340.0	0.0496	191.88	1.7356	0.2437	1.393	1383.1	0.0595	0.02059
360.0	0.0484	196.76	1.7416	0.2442	1.392	1399.8	0.0605	0.02099
380.0	0.0472	201.65	1.7475	0.2446	1.391	1416.3	0.0616	0.02140
400.0	0.0461	206.54	1.7533	0.2451	1.389	1432.5	0.0626	0.02180
420.0	0.0451	211.45	1.7589	0.2455	1.388	1448.5	0.0636	0.02220
440.0	0.0441	216.36	1.7644	0.2460	1.387	1464.3	0.0646	0.02260
460.0	0.0431	221.29	1.7698	0.2465	1.386	1479.9	0.0656	0.02299
480.0	0.0422	226.22	1.7752	0.2471	1.385	1495.3	0.0666	0.02339
500.0	0.0413	231.17	1.7804	0.2476	1.384	1510.5	0.0675	0.02378
520.0	0.0405	236.13	1.7855	0.2482	1.382	1525.5	0.0685	0.02418
540.0	0.0397	241.10	1.7905	0.2487	1.381	1540.2	0.0694	0.02457
560.0	0.0389	246.08	1.7954	0.2493	1.380	1554.9	0.0704	0.02496
580.0	0.0381	251.07	1.8003	0.2499	1.379	1569.3	0.0713	0.02536
600.0	0.0374	256.08	1.8051	0.2505	1.377	1583.6	0.0722	0.02575
620.0	0.0367	261.09	1.8097	0.2511	1.376	1597.8	0.0732	0.02614
640.0	0.0360	266.12	1.8144	0.2517	1.375	1611.8	0.0741	0.02653
660.0	0.0354	271.16	1.8189	0.2524	1.374	1625.6	0.0750	0.02692
680.0	0.0348	276.22	1.8234	0.2530	1.372	1639.3	0.0759	0.02731
700.0	0.0342	281.28	1.8278	0.2536	1.371	1652.8	0.0768	0.02770
720.0	0.0336	286.36	1.8321	0.2543	1.370	1666.2	0.0776	0.02808
740.0	0.0330	291.45	1.8364	0.2549	1.368	1679.5	0.0785	0.02847
760.0	0.0325	296.56	1.8406	0.2555	1.367	1692.7	0.0794	0.02885
780.0	0.0320	301.68	1.8448	0.2562	1.366	1705.7	0.0802	0.02924
800.0	0.0315	306.81	1.8489	0.2568	1.365	1718.6	0.0811	0.02962
820.0	0.0310	311.95	1.8529	0.2574	1.363	1731.4	0.0820	0.03001
840.0	0.0305	317.10	1.8569	0.2581	1.362	1744.1	0.0828	0.03039
860.0	0.0300	322.27	1.8609	0.2587	1.361	1756.7	0.0836	0.03078
880.0	0.0296	327.45	1.8648	0.2594	1.360	1769.1	0.0845	0.03116
900.0	0.0292	332.64	1.8686	0.2600	1.359	1781.5	0.0853	0.03154

a = dew point temperature

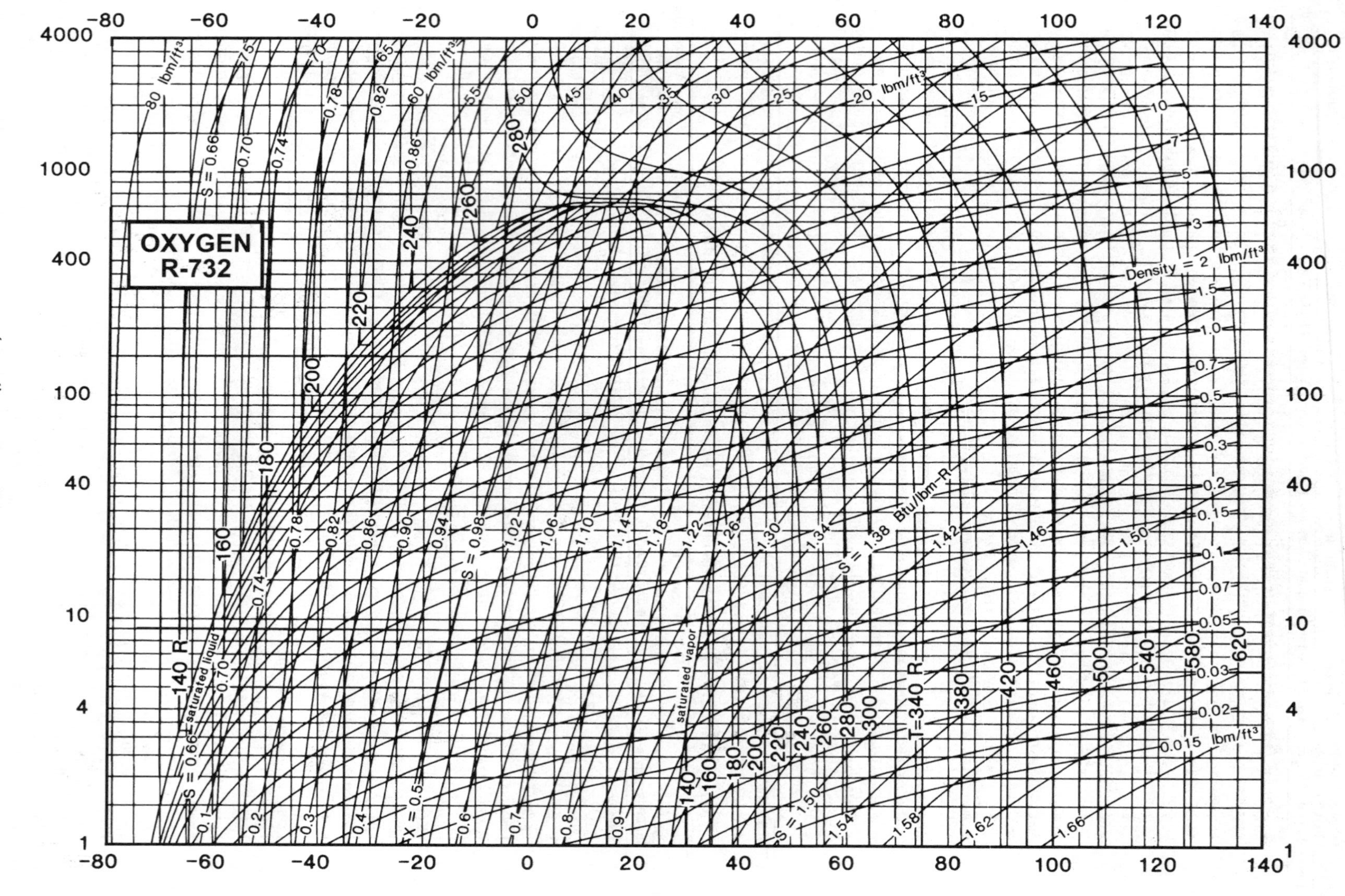

Fig. 27 Pressure-Enthalpy Diagram for Refrigerant 732 (Oxygen)

Refrigerant 732 (Oxygen) Properties of Saturated Liquid and Saturated Vapor

Temp.,* °R	Pressure, psia	Density, lb/ft³ Liquid	Volume, ft³/lb Vapor	Enthalpy, Btu/lb Liquid	Enthalpy, Btu/lb Vapor	Entropy, Btu/lb·°R Liquid	Entropy, Btu/lb·°R Vapor	Specific Heat c_p, Btu/lb·°R Liquid	Specific Heat c_p, Btu/lb·°R Vapor	c_p/c_v Vapor	Velocity of Sound, ft/s Liquid	Velocity of Sound, ft/s Vapor	Viscosity, lb_m/ft·h Liquid	Viscosity, lb_m/ft·h Vapor	Thermal Cond., Btu/h·ft·°R Liquid	Thermal Cond., Btu/h·ft·°R Vapor	Surface Tension, dyne/cm	Temp.,* °R
97.85a	0.02	81.54	1546.5	−83.30	21.13	0.49922	1.56638	0.3999	0.2213	1.395	3686.	460.4	2.3755	0.0097	0.1173	0.00291	22.68	97.85
100.00	0.03	81.23	1092.6	−82.44	21.59	0.50790	1.54816	0.3993	0.2223	1.394	3704.	465.2	2.1422	0.0099	0.1164	0.00297	22.35	100.00
110.00	0.14	79.73	265.18	−78.44	23.74	0.54599	1.47490	0.4002	0.2276	1.389	3685.	486.6	1.4145	0.0109	0.1119	0.00326	20.82	110.00
120.00	0.48	78.17	83.783	−74.43	25.87	0.58085	1.41673	0.4010	0.2322	1.386	3578.	506.8	1.0216	0.0119	0.1074	0.00356	19.31	120.00
130.00	1.34	76.58	32.306	−70.42	27.98	0.61295	1.36988	0.4011	0.2341	1.388	3443.	526.3	0.7868	0.0129	0.1028	0.00386	17.82	130.00
140.00	3.20	74.97	14.5114	−66.40	30.04	0.64267	1.33157	0.4015	0.2335	1.397	3298.	545.1	0.6339	0.0139	0.0982	0.00417	16.36	140.00
150.00	6.73	73.33	7.3371	−62.38	32.02	0.67039	1.29969	0.4028	0.2320	1.411	3151.	562.8	0.5270	0.0149	0.0937	0.00449	14.92	150.00
155.00	9.39	72.49	5.4088	−60.36	32.96	0.68360	1.28566	0.4039	0.2316	1.421	3077.	571.0	0.4848	0.0154	0.0914	0.00465	14.20	155.00
160.00	12.80	71.64	4.0699	−58.33	33.87	0.69643	1.27267	0.4054	0.2317	1.431	3002.	578.8	0.4480	0.0160	0.0891	0.00482	13.50	160.00
162.34b	14.696	71.24	3.5859	−57.38	34.28	0.70230	1.26692	0.4062	0.2320	1.437	2967.	582.2	0.4324	0.0162	0.0881	0.00490	13.17	162.34
165.00	17.11	70.78	3.1187	−56.29	34.74	0.70890	1.26060	0.4072	0.2325	1.444	2927.	586.0	0.4156	0.0165	0.0869	0.00499	12.80	165.00
170.00	22.45	69.90	2.4290	−54.24	35.56	0.72105	1.24932	0.4094	0.2342	1.457	2851.	592.6	0.3868	0.0170	0.0846	0.00517	12.11	170.00
175.00	28.97	69.01	1.9195	−52.18	36.34	0.73291	1.23873	0.4121	0.2369	1.473	2775.	598.5	0.3609	0.0176	0.0823	0.00535	11.43	175.00
180.00	36.84	68.10	1.5366	−50.10	37.06	0.74451	1.22875	0.4153	0.2405	1.491	2697.	603.9	0.3376	0.0181	0.0801	0.00554	10.75	180.00
185.00	46.21	67.17	1.2444	−48.00	37.73	0.75587	1.21928	0.4190	0.2453	1.510	2619.	608.5	0.3163	0.0187	0.0778	0.00573	10.09	185.00
190.00	57.25	66.22	1.0182	−45.88	38.34	0.76701	1.21026	0.4234	0.2512	1.533	2540.	612.4	0.2968	0.0193	0.0755	0.00593	9.43	190.00
195.00	70.13	65.24	0.8409	−43.74	38.88	0.77796	1.20163	0.4284	0.2583	1.559	2460.	615.7	0.2787	0.0198	0.0732	0.00615	8.77	195.00
200.00	85.02	64.24	0.7002	−41.56	39.35	0.78874	1.19333	0.4343	0.2668	1.589	2378.	618.2	0.2620	0.0204	0.0709	0.00637	8.13	200.00
205.00	102.10	63.20	0.5873	−39.36	39.75	0.79938	1.18529	0.4411	0.2768	1.623	2294.	620.0	0.2464	0.0210	0.0686	0.00661	7.50	205.00
210.00	121.55	62.13	0.4958	−37.12	40.07	0.80990	1.17746	0.4491	0.2884	1.663	2209.	621.1	0.2318	0.0216	0.0663	0.00686	6.87	210.00
215.00	143.55	61.02	0.4208	−34.84	40.30	0.82033	1.16980	0.4585	0.3020	1.711	2122.	621.4	0.2180	0.0223	0.0640	0.00714	6.26	215.00
220.00	168.28	59.87	0.3590	−32.51	40.43	0.83069	1.16224	0.4696	0.3181	1.767	2033.	621.1	0.2049	0.0229	0.0616	0.00743	5.66	220.00
225.00	195.93	58.66	0.3074	−30.12	40.46	0.84101	1.15474	0.4830	0.3372	1.835	1942.	619.9	0.1924	0.0236	0.0593	0.00776	5.07	225.00
230.00	226.69	57.40	0.2641	−27.68	40.38	0.85134	1.14722	0.4991	0.3603	1.919	1848.	618.1	0.1805	0.0243	0.0569	0.00813	4.49	230.00
235.00	260.77	56.07	0.2274	−25.16	40.15	0.86170	1.13962	0.5190	0.3886	2.023	1750.	615.4	0.1690	0.0251	0.0545	0.00855	3.93	235.00
240.00	298.37	54.65	0.1961	−22.55	39.78	0.87216	1.13185	0.5442	0.4244	2.155	1649.	611.9	0.1578	0.0259	0.0521	0.00903	3.38	240.00
245.00	339.72	53.13	0.1692	−19.83	39.22	0.88279	1.12381	0.5768	0.4710	2.328	1544.	607.5	0.1470	0.0268	0.0496	0.00962	2.84	245.00
250.00	385.04	51.47	0.1457	−16.97	38.45	0.89367	1.11535	0.6208	0.5342	2.563	1434.	602.2	0.1363	0.0278	0.0471	0.01034	2.33	250.00
260.00	488.67	47.59	0.1068	−10.68	35.97	0.91683	1.09626	0.7794	0.7669	3.421	1192.	588.2	0.1148	0.0303	0.0423	0.01257	1.36	260.00
270.00	611.86	42.17	0.0745	−2.87	31.22	0.94436	1.07061	1.3059	1.5835	6.314	898.	567.0	0.0914	0.0342	0.0383	0.01751	0.51	270.00
278.25c	731.42	27.23	0.0367	13.95	13.95	1.00320	1.00320	∞	∞	∞	0.	0.0	—	—	∞	∞	0.00	278.25

*Temperatures are on the IPTS-68 scale   a = triple point   b = normal boiling point   c = critical point

Refrigerant 732 (Oxygen) Properties of Gas at 14.696 psia (one standard atmosphere)

Temp., °F	Density, lb/ft³	Enthalpy, Btu/lb	Entropy, Btu/lb·°F	c_p, Btu/lb·°F	c_p/c_v	Vel. Sound, ft/s	Viscosity, lb/ft·h	Thermal Cond., Btu/h·ft·°F
−297.3a	0.2789	34.28	1.2669	0.2320	1.437	582.2	0.0162	0.00490
−280.0	0.2499	38.20	1.2898	0.2236	1.433	617.3	0.0180	0.00542
−260.0	0.2234	42.66	1.3134	0.2226	1.424	652.8	0.0200	0.00603
−240.0	0.2022	47.10	1.3346	0.2217	1.418	686.2	0.0221	0.00664
−220.0	0.1848	51.53	1.3539	0.2208	1.414	718.0	0.0241	0.00724
−200.0	0.1702	55.94	1.3716	0.2201	1.411	748.3	0.0261	0.00783
−180.0	0.1577	60.33	1.3879	0.2196	1.409	777.3	0.0280	0.00843
−160.0	0.1470	64.72	1.4030	0.2192	1.407	805.2	0.0300	0.00899
−140.0	0.1377	69.10	1.4172	0.2189	1.406	832.1	0.0318	0.00954
−120.0	0.1295	73.48	1.4304	0.2187	1.405	858.1	0.0337	0.01009
−100.0	0.1222	77.85	1.4430	0.2186	1.404	883.2	0.0355	0.01062
−80.0	0.1157	82.23	1.4548	0.2185	1.404	907.7	0.0372	0.01115
−60.0	0.1099	86.60	1.4660	0.2185	1.403	931.4	0.0389	0.01167
−40.0	0.1046	90.97	1.4767	0.2185	1.402	954.5	0.0406	0.01217
−20.0	0.0998	95.34	1.4869	0.2186	1.401	977.0	0.0423	0.01267
0.0	0.0955	99.71	1.4966	0.2188	1.401	998.9	0.0439	0.01316
20.0	0.0915	104.09	1.5059	0.2190	1.400	1020.3	0.0455	0.01365
40.0	0.0878	108.47	1.5149	0.2192	1.399	1041.1	0.0470	0.01413
60.0	0.0844	112.86	1.5235	0.2195	1.398	1061.5	0.0485	0.01460
80.0	0.0812	117.25	1.5318	0.2199	1.396	1081.4	0.0500	0.01507
100.0	0.0783	121.65	1.5398	0.2203	1.395	1100.9	0.0515	0.01553
120.0	0.0756	126.06	1.5475	0.2207	1.394	1120.0	0.0530	0.01600
140.0	0.0731	130.48	1.5550	0.2212	1.392	1138.7	0.0544	0.01646
160.0	0.0707	134.91	1.5623	0.2218	1.391	1157.0	0.0558	0.01692
180.0	0.0685	139.35	1.5693	0.2223	1.389	1174.9	0.0572	0.01738
200.0	0.0664	143.80	1.5762	0.2230	1.388	1192.5	0.0586	0.01784
220.0	0.0645	148.27	1.5828	0.2236	1.386	1209.7	0.0599	0.01829
240.0	0.0626	152.75	1.5893	0.2243	1.384	1226.7	0.0613	0.01874
260.0	0.0609	157.24	1.5957	0.2251	1.382	1243.3	0.0626	0.01920
280.0	0.0592	161.75	1.6018	0.2258	1.381	1259.7	0.0639	0.01964
300.0	0.0577	166.28	1.6079	0.2266	1.379	1275.8	0.0652	0.02009
320.0	0.0562	170.81	1.6138	0.2274	1.377	1291.6	0.0665	0.02054
340.0	0.0548	175.37	1.6196	0.2282	1.375	1307.2	0.0677	0.02099
360.0	0.0535	179.94	1.6252	0.2290	1.373	1322.6	0.0690	0.02143
380.0	0.0522	184.53	1.6307	0.2298	1.371	1337.7	0.0702	0.02188
400.0	0.0510	189.13	1.6361	0.2307	1.369	1352.6	0.0714	0.02232
420.0	0.0498	193.75	1.6415	0.2315	1.367	1367.3	0.0726	0.02276
440.0	0.0487	198.39	1.6467	0.2323	1.366	1381.9	0.0738	0.02320
460.0	0.0476	203.05	1.6518	0.2332	1.364	1396.2	0.0750	0.02364
480.0	0.0466	207.72	1.6568	0.2340	1.362	1410.4	0.0761	0.02407
500.0	0.0457	212.41	1.6618	0.2349	1.360	1424.4	0.0773	0.02451
520.0	0.0447	217.12	1.6666	0.2357	1.358	1438.3	0.0784	0.02494
540.0	0.0438	221.84	1.6714	0.2365	1.357	1452.0	0.0795	0.02537
560.0	0.0430	226.58	1.6761	0.2374	1.355	1465.5	0.0807	0.02580
580.0	0.0421	231.33	1.6807	0.2382	1.353	1478.9	0.0818	0.02622
600.0	0.0413	236.10	1.6852	0.2390	1.352	1492.2	0.0829	0.02665
650.0	0.0395	248.10	1.6963	0.2409	1.348	1524.8	0.0856	0.02770
700.0	0.0378	260.20	1.7070	0.2428	1.344	1556.6	0.0882	0.02873
750.0	0.0362	272.38	1.7172	0.2447	1.340	1587.7	0.0908	0.02975
800.0	0.0348	284.66	1.7272	0.2464	1.337	1618.3	0.0934	0.03076

a = saturated vapor at normal boiling point

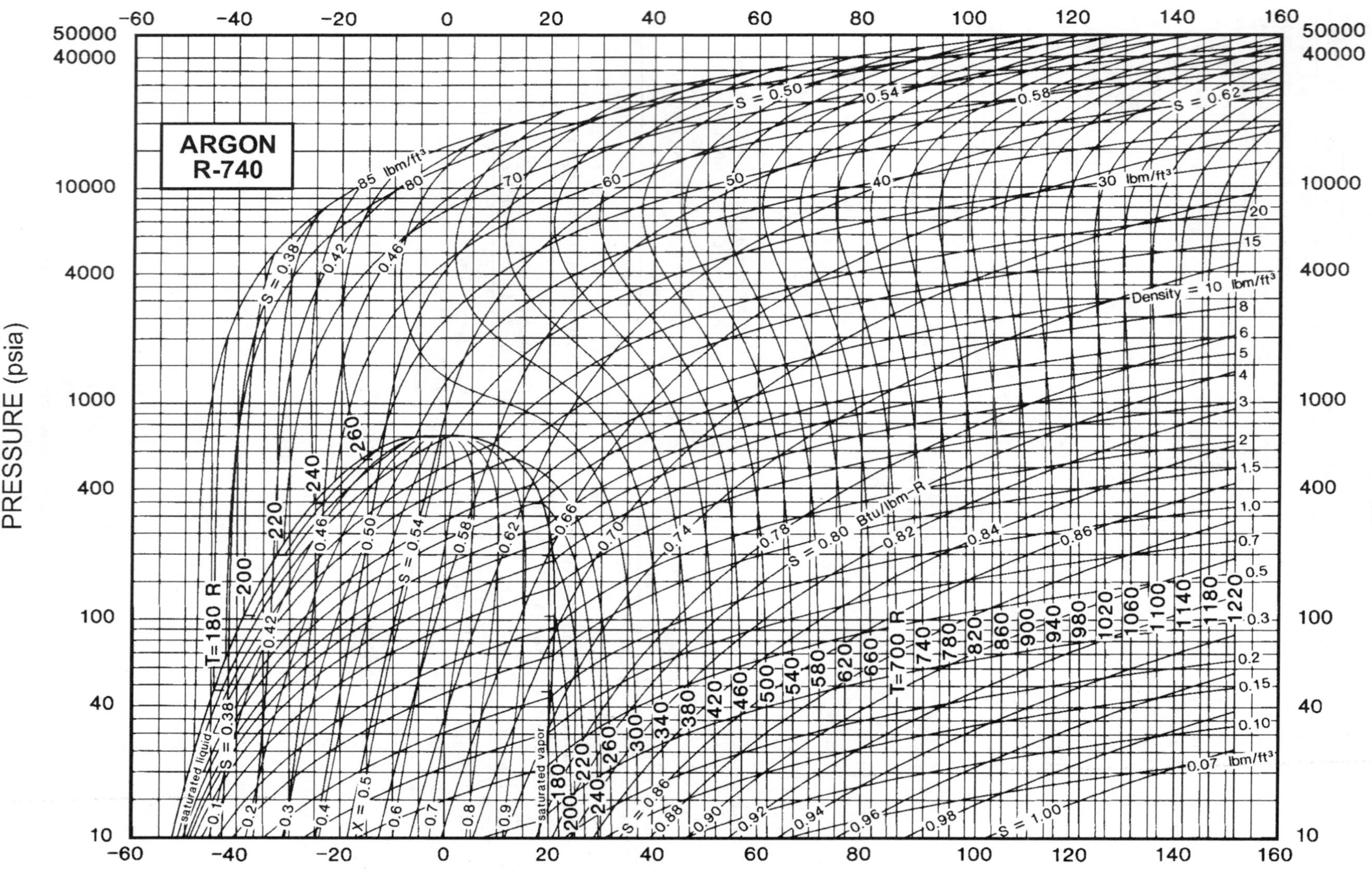

Fig. 28 Pressure-Enthalpy Diagram for Refrigerant 740 (Argon)

Refrigerant 740 (Argon) Properties of Saturated Liquid and Saturated Vapor

Temp.,* °R	Pressure, psia	Density, lb/ft³ Liquid	Volume, ft³/lb Vapor	Enthalpy, Btu/lb Liquid	Enthalpy, Btu/lb Vapor	Entropy, Btu/lb·°R Liquid	Entropy, Btu/lb·°R Vapor	Specific Heat c_p, Btu/lb·°R Liquid	Specific Heat c_p, Btu/lb·°R Vapor	c_p/c_v Vapor	Velocity of Sound, ft/s Liquid	Velocity of Sound, ft/s Vapor	Viscosity, lb_m/ft·h Liquid	Viscosity, lb_m/ft·h Vapor	Thermal Cond., Btu/h·ft·°R Liquid	Thermal Cond., Btu/h·ft·°R Vapor	Surface Tension, dyne/cm	Temp.,* °R
150.85a	10.00	88.47	3.9491	−52.08	18.32	0.31822	0.78492	0.2549	—	—	2799.	685.0	0.6840	0.0162	0.0784	0.00304	13.42	150.85
155.00	12.95	87.61	3.1154	−51.01	18.64	0.32515	0.77453	0.2568	—	—	2749.	644.0	0.6382	0.0166	0.0764	0.00313	12.83	155.00
157.13b	14.696	87.17	2.7731	−50.46	18.80	0.32865	0.76944	0.2577	—	—	2723.	630.8	0.6166	0.0169	0.0754	0.00318	12.53	157.13
160.00	17.34	86.56	2.3822	−49.72	19.00	0.33332	0.76282	0.2592	—	—	2688.	618.0	0.5889	0.0172	0.0740	0.00325	12.13	160.00
165.00	22.78	85.49	1.8525	−48.41	19.34	0.34130	0.75192	0.2619	—	—	2626.	604.2	0.5449	0.0178	0.0716	0.00337	11.44	165.00
170.00	29.43	84.38	1.4624	−47.08	19.66	0.34913	0.74173	0.2649	—	—	2563.	596.8	0.5053	0.0184	0.0693	0.00350	10.75	170.00
175.00	37.45	83.25	1.1701	−45.74	19.95	0.35681	0.73217	0.2683	0.1379	1.866	2500.	593.1	0.4696	0.0190	0.0670	0.00364	10.08	175.00
180.00	46.99	82.09	0.9474	−44.38	20.21	0.36436	0.72316	0.2721	0.1474	1.854	2436.	591.6	0.4371	0.0196	0.0647	0.00378	9.41	180.00
185.00	58.23	80.90	0.7753	−43.00	20.43	0.37179	0.71463	0.2763	0.1563	1.859	2370.	591.4	0.4074	0.0203	0.0624	0.00392	8.76	185.00
190.00	71.32	79.68	0.6405	−41.59	20.61	0.37913	0.70650	0.2810	0.1647	1.875	2304.	592.1	0.3802	0.0210	0.0601	0.00408	8.11	190.00
195.00	86.44	78.42	0.5335	−40.16	20.74	0.38638	0.69870	0.2864	0.1731	1.904	2235.	593.2	0.3551	0.0217	0.0579	0.00425	7.48	195.00
200.00	103.76	77.12	0.4477	−38.70	20.82	0.39356	0.69119	0.2924	0.1818	1.944	2165.	594.5	0.3319	0.0224	0.0557	0.00442	6.86	200.00
205.00	123.47	75.77	0.3782	−37.21	20.85	0.40069	0.68389	0.2992	0.1912	1.995	2093.	595.8	0.3103	0.0231	0.0535	0.00462	6.25	205.00
210.00	145.72	74.38	0.3212	−35.68	20.80	0.40778	0.67676	0.3072	0.2016	2.060	2018.	597.0	0.2901	0.0239	0.0514	0.00482	5.65	210.00
215.00	170.72	72.93	0.2742	−34.12	20.69	0.41485	0.66975	0.3166	0.2136	2.140	1941.	598.0	0.2711	0.0247	0.0492	0.00504	5.07	215.00
220.00	198.64	71.41	0.2349	−32.51	20.48	0.42193	0.66279	0.3278	0.2279	2.239	1860.	598.6	0.2532	0.0256	0.0471	0.00529	4.50	220.00
225.00	229.66	69.82	0.2019	−30.84	20.19	0.42905	0.65583	0.3415	0.2453	2.363	1775.	598.9	0.2362	0.0266	0.0450	0.00556	3.95	225.00
230.00	264.00	68.13	0.1739	−29.11	19.78	0.43624	0.64881	0.3586	0.2671	2.519	1686.	598.6	0.2200	0.0276	0.0428	0.00587	3.41	230.00
235.00	301.85	66.34	0.1499	−27.31	19.25	0.44355	0.64165	0.3805	0.2952	2.721	1591.	597.8	0.2043	0.0287	0.0407	0.00623	2.89	235.00
240.00	343.44	64.40	0.1292	−25.41	18.56	0.45104	0.63424	0.4097	0.3328	2.989	1490.	596.4	0.1892	0.0300	0.0385	0.00665	2.39	240.00
245.00	388.99	62.27	0.1111	−23.39	17.68	0.45881	0.62646	0.4505	0.3853	3.360	1380.	594.2	0.1742	0.0314	0.0363	0.00717	1.91	245.00
250.00	438.76	59.88	0.0952	−21.21	16.56	0.46701	0.61811	0.5110	0.4636	3.904	1260.	591.2	0.1593	0.0331	0.0341	0.00785	1.46	250.00
255.00	493.05	57.12	0.0809	−18.81	15.12	0.47586	0.60889	0.6095	0.5914	4.778	1129.	587.3	0.1440	0.0352	0.0319	0.00882	1.03	255.00
260.00	552.18	53.78	0.0678	−16.05	13.19	0.48581	0.59826	0.7963	0.8335	6.393	984.	582.1	0.1278	0.0379	0.0299	0.01043	0.64	260.00
265.00	616.56	49.38	0.0553	−12.65	10.43	0.49785	0.58497	1.2601	1.4494	10.381	824.	574.9	0.1099	0.0418	0.0293	0.01386	0.30	265.00
271.19c	704.88	33.14	0.0302	−1.53	−1.53	0.53776	0.53776	∞	∞	∞	0.	0.0	–	–	∞	∞	0.00	271.19

*Temperatures are on the IPTS-68 scale a = triple point b = normal boiling point c = critical point

Refrigerant 740 (Argon) Properties of Gas at 14.696 psia (one standard atmosphere)

Temp., °F	Density, lb/ft³	Enthalpy, Btu/lb	Entropy, Btu/lb·°F	c_p Btu/lb·°F	c_p/c_v	Vel. Sound, ft/s	Viscosity, lb/ft·h	Thermal Cond., Btu/h·ft·°F
−302.5a	0.3606	18.80	0.7694	–	–	630.8	0.0169	0.00318
−300.0	0.3545	19.06	0.7711	–	–	597.4	0.0171	0.00323
−275.0	0.3028	22.35	0.7902	0.1335	1.689	609.4	0.0199	0.00373
−250.0	0.2648	25.62	0.8068	0.1291	1.689	654.0	0.0226	0.00425
−225.0	0.2356	28.82	0.8213	0.1273	1.686	694.2	0.0254	0.00478
−200.0	0.2124	31.99	0.8341	0.1264	1.682	731.6	0.0281	0.00534
−180.0	0.1969	34.52	0.8435	0.1260	1.680	759.9	0.0303	0.00578
−160.0	0.1835	37.03	0.8521	0.1257	1.678	787.0	0.0324	0.00609
−140.0	0.1719	39.54	0.8603	0.1255	1.676	813.2	0.0345	0.00647
−120.0	0.1616	42.05	0.8679	0.1253	1.675	838.6	0.0366	0.00684
−100.0	0.1525	44.56	0.8750	0.1252	1.674	863.1	0.0386	0.00721
−80.0	0.1444	47.06	0.8818	0.1251	1.673	887.0	0.0406	0.00759
−60.0	0.1372	49.56	0.8882	0.1250	1.672	910.2	0.0426	0.00795
−40.0	0.1306	52.06	0.8943	0.1249	1.672	932.8	0.0445	0.00831
−20.0	0.1246	54.56	0.9001	0.1248	1.671	954.9	0.0464	0.00866
0.0	0.1192	57.05	0.9057	0.1248	1.671	976.5	0.0482	0.00900
10.0	0.1166	58.30	0.9084	0.1248	1.671	987.1	0.0491	0.00917
20.0	0.1142	59.55	0.9110	0.1248	1.670	997.5	0.0500	0.00934
30.0	0.1118	60.79	0.9136	0.1247	1.670	1007.9	0.0509	0.00950
40.0	0.1096	62.04	0.9161	0.1247	1.670	1018.2	0.0518	0.00967
50.0	0.1074	63.29	0.9186	0.1247	1.670	1028.4	0.0527	0.00983
60.0	0.1053	64.54	0.9210	0.1247	1.670	1038.4	0.0535	0.00999
70.0	0.1034	65.78	0.9234	0.1247	1.670	1048.4	0.0544	0.01016
80.0	0.1014	67.03	0.9257	0.1247	1.669	1058.3	0.0553	0.01031
90.0	0.0996	68.28	0.9280	0.1246	1.669	1068.0	0.0561	0.01047
100.0	0.0978	69.52	0.9302	0.1246	1.669	1077.7	0.0569	0.01063
120.0	0.0944	72.01	0.9346	0.1246	1.669	1096.8	0.0586	0.01094
140.0	0.0913	74.51	0.9388	0.1246	1.669	1115.6	0.0602	0.01125
160.0	0.0883	77.00	0.9429	0.1246	1.669	1134.1	0.0619	0.01155
180.0	0.0855	79.49	0.9469	0.1246	1.668	1152.3	0.0635	0.01184
200.0	0.0829	81.98	0.9507	0.1245	1.668	1170.2	0.0650	0.01214
220.0	0.0805	84.47	0.9544	0.1245	1.668	1187.8	0.0666	0.01243
240.0	0.0782	86.96	0.9580	0.1245	1.668	1205.2	0.0681	0.01271
260.0	0.0760	89.45	0.9616	0.1245	1.668	1222.3	0.0696	0.01299
280.0	0.0740	91.94	0.9650	0.1245	1.668	1239.2	0.0711	0.01327
300.0	0.0720	94.43	0.9683	0.1245	1.668	1255.8	0.0726	0.01354
320.0	0.0702	96.92	0.9715	0.1245	1.668	1272.2	0.0740	0.01382
340.0	0.0684	99.41	0.9747	0.1245	1.668	1288.4	0.0755	0.01408
360.0	0.0667	101.90	0.9778	0.1245	1.668	1304.5	0.0769	0.01435
380.0	0.0651	104.39	0.9808	0.1245	1.668	1320.3	0.0783	0.01461
400.0	0.0636	106.88	0.9837	0.1245	1.668	1335.9	0.0797	0.01487
420.0	0.0622	109.37	0.9865	0.1244	1.668	1351.3	0.0811	0.01513
440.0	0.0608	111.86	0.9893	0.1244	1.667	1366.6	0.0824	0.01538
460.0	0.0595	114.34	0.9921	0.1244	1.667	1381.7	0.0838	0.01563
480.0	0.0582	116.83	0.9948	0.1244	1.667	1396.7	0.0851	0.01588
500.0	0.0570	119.32	0.9974	0.1244	1.667	1411.5	0.0864	0.01613
520.0	0.0558	121.81	0.9999	0.1244	1.667	1426.1	0.0878	0.01637
540.0	0.0547	124.30	1.0025	0.1244	1.667	1440.6	0.0891	0.01662
560.0	0.0536	126.79	1.0049	0.1244	1.667	1454.9	0.0903	0.01685
580.0	0.0526	129.28	1.0073	0.1244	1.667	1469.1	0.0916	0.01709
600.0	0.0516	131.76	1.0097	0.1244	1.667	1483.2	0.0929	0.01733
620.0	0.0507	134.25	1.0120	0.1244	1.667	1497.1	0.0941	0.01756
640.0	0.0497	136.74	1.0143	0.1244	1.667	1510.9	0.0954	0.01779
660.0	0.0488	139.23	1.0166	0.1244	1.667	1524.5	0.0966	0.01802
680.0	0.0480	141.72	1.0188	0.1244	1.667	1538.1	0.0978	0.01825
700.0	0.0472	144.21	1.0209	0.1244	1.667	1551.5	0.0990	0.01847
720.0	0.0464	146.69	1.0231	0.1244	1.667	1564.8	0.1002	0.01870
740.0	0.0456	149.18	1.0252	0.1244	1.667	1578.0	0.1014	0.01892
760.0	0.0448	151.67	1.0272	0.1244	1.667	1591.2	0.1026	0.01914
780.0	0.0441	154.16	1.0292	0.1244	1.667	1604.1	0.1037	0.01936
800.0	0.0434	156.65	1.0312	0.1244	1.667	1617.0	0.1049	0.01957

a = saturated vapor at normal boiling point

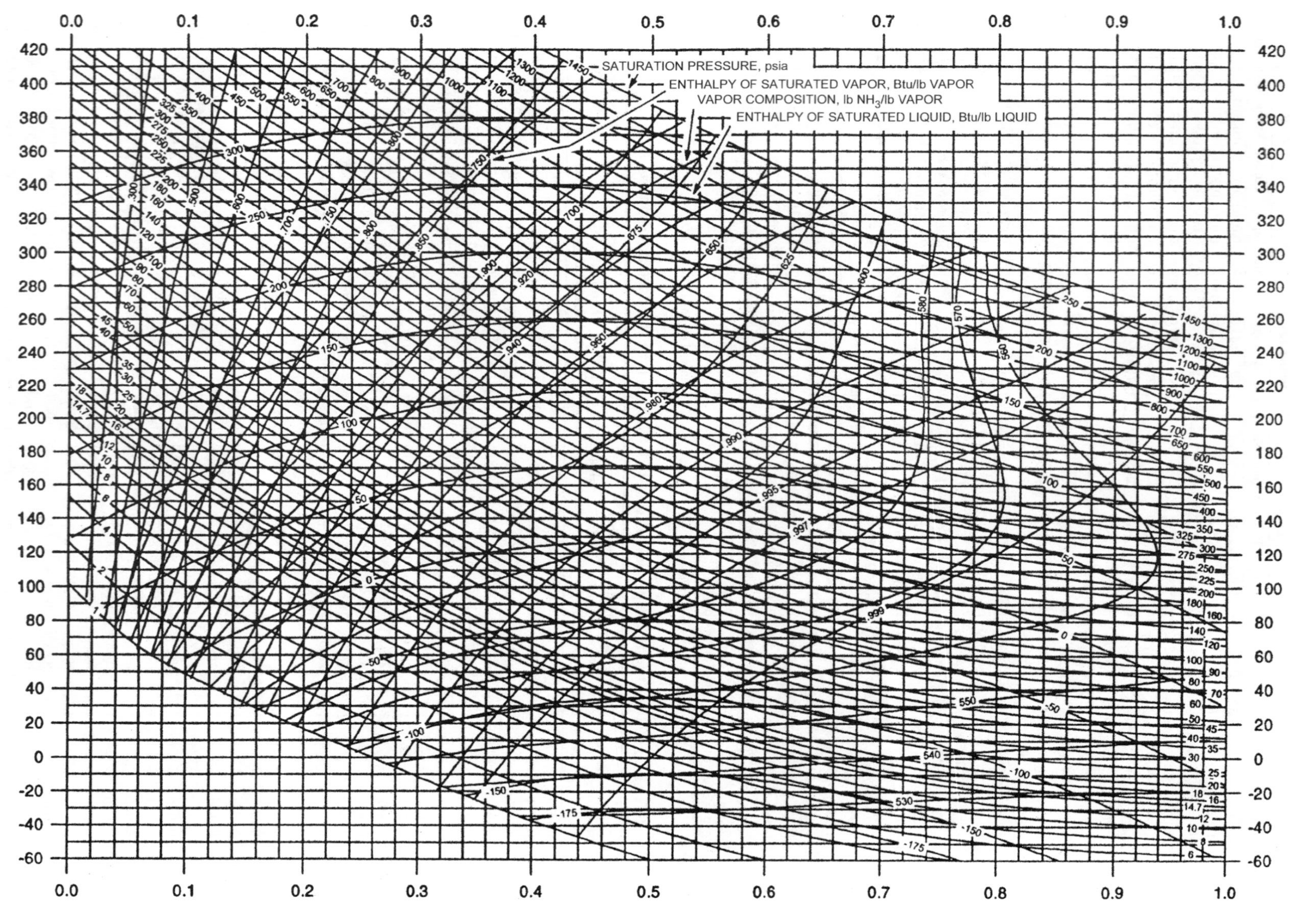

Fig. 29 Enthalpy Diagram for Ammonia-Water Solutions

Prepared by Kwang Kim and Keith Herold, Center for Environmental Energy Engineering, University of Maryland at College Park

Specific Volume of Saturated Ammonia-Water Solutions, ft³/lb

Temp., °F	Concentration, Ammonia (Mass Basis)											Temp., °F
	0	10	20	30	40	50	60	70	80	90	100	
20	0.0160	0.0165	0.0170	0.0176	0.0182	0.0190	0.0197	0.0207	0.0217	0.0230	0.0245	20
40	0.0160	0.0165	0.0171	0.0177	0.0184	0.0191	0.0200	0.0209	0.0221	0.0236	0.0253	40
60	0.0160	0.0166	0.0172	0.0178	0.0186	0.0193	0.0202	0.0212	0.0225	0.0241	0.0260	60
80	0.0161	0.0167	0.0173	0.0180	0.0188	0.0196	0.0205	0.0216	0.0230	0.0247	0.0267	80
100	0.0161	0.0168	0.0174	0.0182	0.0190	0.0198	0.0208	0.0220	0.0235	0.0254	0.0275	100
120	0.0162	0.0169	0.0176	0.0184	0.0192	0.0201	0.0211	0.0224	0.0241	0.0261	0.0284	120
140	0.0163	0.0170	0.0177	0.0185	0.0194	0.0203	0.0215	0.0229	0.0247	0.0268	0.0294	140
160	0.0164	0.0172	0.0179	0.0187	0.0196	0.0206	0.0219	0.0235	0.0254	0.0277	0.0306	160
180	0.0165	0.0173	0.0181	0.0190	0.0199	0.0210	0.0223	0.0241	0.0262	0.0286	0.0320	180
200	0.0166	0.0175	0.0183	0.0192	0.0202	0.0213	0.0228	0.0247	0.0270	0.0298	0.0338	200
220	0.0168	0.0176	0.0185	0.0194	0.0205	0.0217	0.0234	0.0255	0.0279	0.0312	0.0361	220

Prepared under Research Project No. 271-RP, sponsored by TC 8.3.
Data reference: B.H. Jennings, Ammonia water properties (paper presented at ASHRAE meeting, January 1965).

Refrigerant Temperature (t' = °F) and Enthalpy (h = Btu/lb) of Lithium Bromide Solutions

Temp., (t = °F)		Percent LiBr										
		0	10	20	30	40	45	50	55	60	65	70
80	t'	80.0	78.2	75.6	70.5	60.9	53.5	42.1	28.6	13.8	–0.2#	–11.6#
	h	48.0	39.2	31.8	25.6	21.6	21.2	23.0	28.7	38.9	52.7#	67.1#
100	t'	100.0	98.1	95.3	89.9	79.6	71.8	60.0	46.1	30.9	16.2#	3.8#
	h	68.0	56.6	47.0	38.7	33.2	32.1	33.2	38.2	47.8	61.1#	75.1#
120	t'	120.0	117.9	114.9	109.2	98.3	90.1	77.9	63.6	48.1	32.7	19.1#
	h	87.9	73.6	61.7	51.7	44.7	43.0	43.6	48.0	56.9	69.4	83.0#
140	t'	140.0	137.8	134.6	128.5	117.1	108.5	95.8	81.2	65.2	49.1	34.4#
	h	107.9	91.0	77.0	65.1	56.5	54.1	54.1	57.9	66.1	78.0	91.1#
160	t'	160.0	157.7	154.3	147.9	135.8	126.8	113.8	98.7	82.3	65.6	49.7#
	h	127.9	108.2	92.0	78.2	68.1	65.1	64.7	67.9	75.4	86.6	99.2#
180	t'	180.0	177.5	173.9	167.2	154.5	145.1	131.7	116.2	99.5	82.0	65.1#
	h	147.9	125.4	107.9	91.9	80.4	76.6	75.3	77.7	84.6	95.1	107.2#
200	t'	200.0	197.4	193.6	186.5	173.3	163.5	149.6	133.7	116.6	98.5	80.4#
	h	168.0	143.4	123.3	105.3	92.1	87.4	85.9	87.8	94.1	104.0	115.6#
220	t'	220.0	217.2	213.3	205.8	192.0	181.8	167.5	151.3	133.7	114.9	95.7
	h	188.1	160.7	138.2	119.0	104.1	99.0	96.5	97.8	103.3	112.5	123.6
240	t'	240.0*	237.1*	232.9	225.2	210.7	200.2	185.4	168.8	150.9	131.4	111.0
	h	208.3*	178.4*	154.0	132.6	116.0	110.3	107.1	107.7	112.5	121.1	131.6
260	t'	260.0*	256.9*	252.6*	244.5*	229.4	218.5	203.3	186.3	168.0	147.9	126.4
	h	228.6*	195.7*	169.1*	146.2*	128.1	121.6	117.6	117.6	121.6	129.5	139.5
280	t'	280.0*	276.8*	272.3*	263.8*	248.2*	236.8*	221.2	203.9	185.1	164.3	141.7
	h	249.1*	213.8*	185.1*	159.7*	140.0*	132.8*	128.1	127.5	130.6	137.9	147.6
300	t'	300.0*	296.7*	291.9*	283.1*	266.9*	255.2*	239.2*	221.4	202.3	180.8	157.0
	h	269.6*	231.6*	200.7*	173.5*	152.1*	144.1*	138.9*	137.3	139.8	146.5	155.5
320	t'	320.0*	316.5*	311.6*	302.5*	285.6*	273.5*	257.1*	238.9*	219.4	197.2	172.4
	h	290.3*	249.7*	216.3*	187.2*	164.2*	155.3*	149.5*	147.1*	148.8	154.9	163.4
340	t'	340.0*	336.4*	331.3*	321.8*	304.4*	291.9*	275.0*	256.4*	236.5*	213.7	187.7
	h	311.1*	267.9*	232.1*	201.0*	176.1*	166.6*	160.1*	157.0*	158.0*	163.5	171.0
360	t'	360.0*	356.2*	350.9*	341.1*	323.1*	310.2*	292.9*	274.0*	253.7*	230.1	203.0
	h	332.2*	286.1*	248.0*	214.9*	188.2*	178.0*	170.6*	166.8*	167.0*	171.9	178.3

*Extensions of data above 235°F are well above the original data and should be used with care.
#Supersaturated solution.

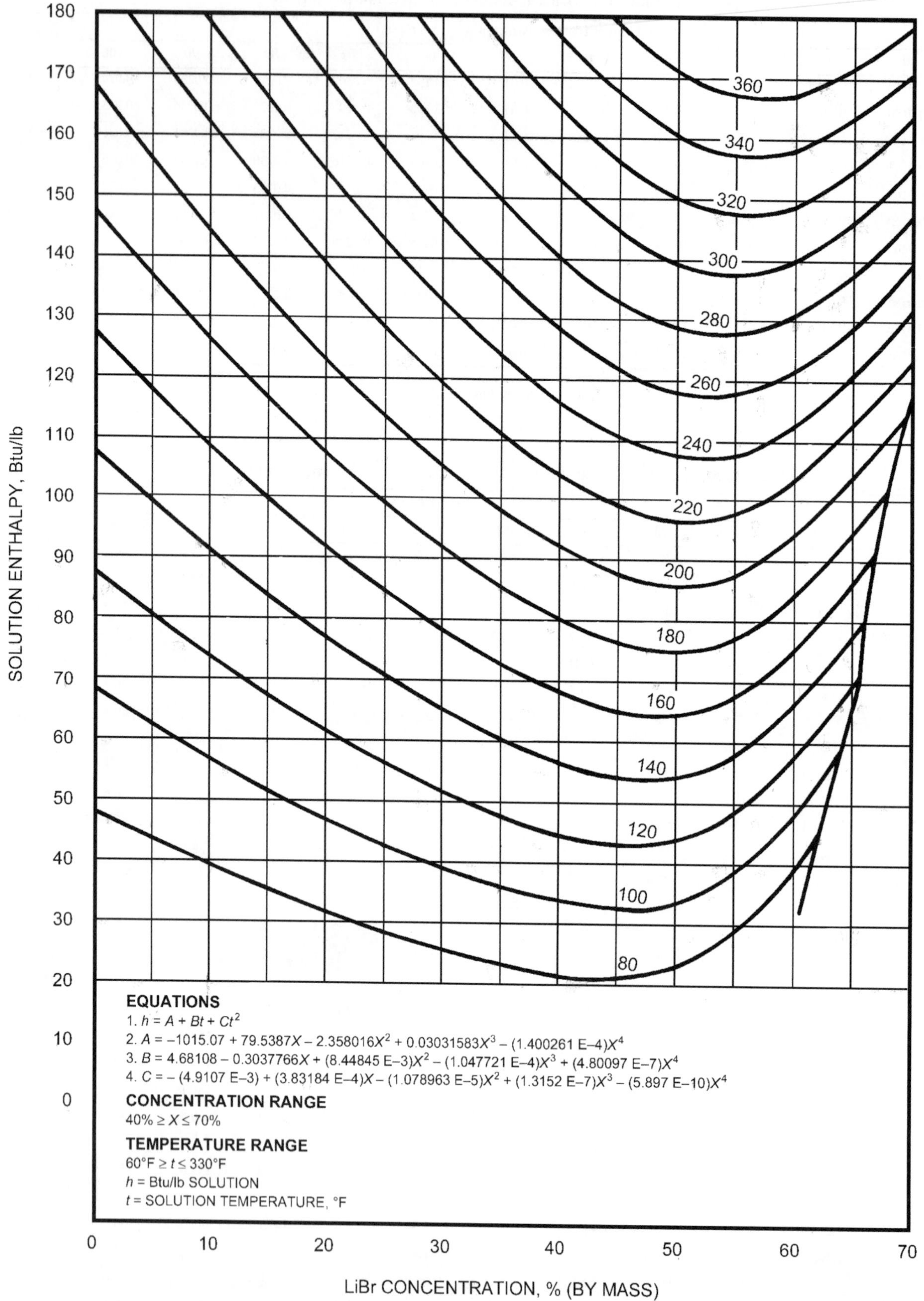

Fig. 30 Enthalpy-Concentration Diagram for Water-Lithium Bromide Solutions

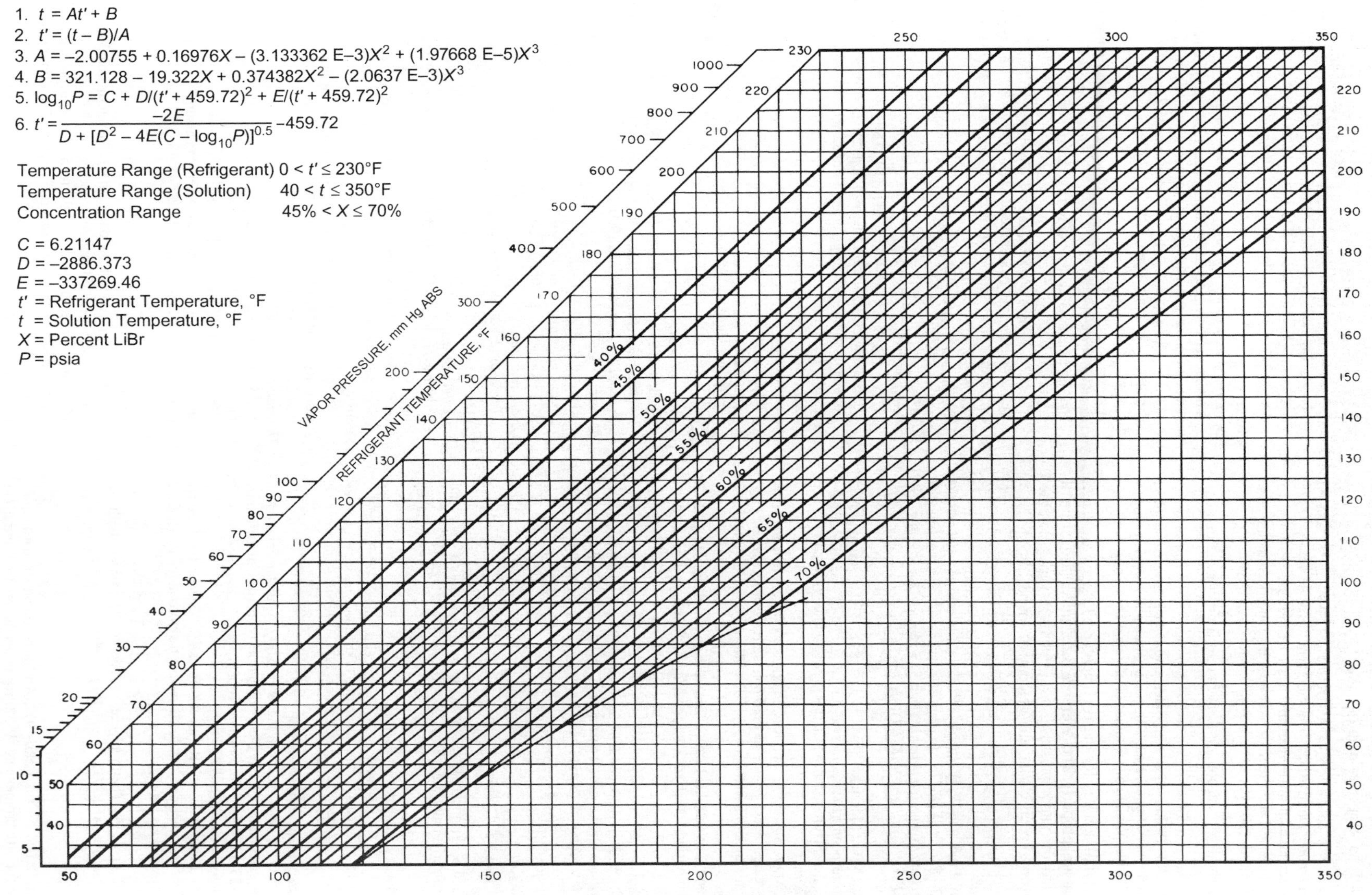

Fig. 31 Equilibrium Chart for Aqueous Lithium Bromide Solutions

Reprinted by permission of Carrier Corp.

REFERENCES

Except where noted below, tables for the halocarbon refrigerants and their blends have been calculated using:

McLinden, M.O., S.A. Klein, E.W. Lemmon, and A.P. Peskin. 2000a. *NIST Standard Reference Database* 23: Thermodynamic and transport properties of refrigerants and refrigerant mixtures—REFPROP, Version 6.10. Standard Reference Data Program, National Institute of Standards and Technology, Gaithersburg, MD.

Tables for the hydrocarbons and inorganic fluids have been calculated using:

Lemmon, E.W., A.P. Peskin, M.O. McLinden, and D.G. Friend. 2000. *NIST Standard Reference Database* 12: Thermodynamic and transport properties of pure fluids, Version 5.0. Standard Reference Data Program.

The underlying sources for these computer packages are listed below by fluid and property. The reference listed under "Equation of state" was used for vapor pressure, liquid density, vapor volume, enthalpy, entropy, specific heat, and velocity of sound.

R-12

Equation of state

Marx, V., A. Pruß, and W. Wagner. 1992. *Neue Zustandsgleichungen für R-12, R-22, R-11 und R-113. Beschreibung des thermodynamishchen Zustandsverhaltens bei Temperaturen bis 525 K und Drücken bis 200 MPa. VDI-Fortschritt-Ber* 19(57). VDI Verlag, Düsseldorf.

Viscosity

Klein, S.A., M.O. McLinden, and A. Laesecke. 1997. An improved extended corresponding states method for estimation of viscosity of pure refrigerants and mixtures. *Int. Journal of Refrigeration* 20:208-17.

Thermal conductivity

McLinden, M.O., S.A. Klein, and R.A. Perkins. 2000b. An extended corresponding states model for the thermal conductivity of refrigerants and refrigerant mixtures. *International Journal of Refrigeration* 23:43-63.

Surface tension

Okada, M. and K. Watanabe. 1988. Surface tension correlations for several fluorocarbon refrigerants, *Heat Transfer—Japanese Research* 17:35-52.

R-22

Equation of state

Kamei, A., S.W. Beyerlein, and R.T. Jacobsen. 1995. Application of nonlinear regression in the development of a wide range formulation for HCFC-22. *International Journal of Thermophysics* 16(5):1155-64.

Viscosity

Klein et al. 1997. *op. cit.*

Thermal conductivity

McLinden et al. 2000b. *op. cit.*

Surface tension

Okada, M. and K. Watanabe. 1988. Surface tension correlations for several fluorocarbon refrigerants. *Heat Transfer—Japanese Research* 17:35-52.

R-23

Data calculated using computer progams based on

Lemmon, E.W., R.T. Jacobsen., S.G. Penoncello, and S.W. Beyerlein. 1994. Computer programs for the calculation of thermodynamic properties of cryogens and other fluids. *Advances in Cryogenic Engineering* 39:1891-97.

Equation of state

Penoncello, S.G., Z. Shan, and R.T. Jacobsen. 2000. A fundamental equation for the calculation of the thermodynamic properties of trifluoromethane (R-23). *ASHRAE Transactions* 106(1).

Viscosity and thermal conductivity

Shan, Z., S.G. Penoncello, and R.T. Jacobsen. 2000. A generalized model for viscosity and thermal conductivity of trifluoromethane (R-23). *ASHRAE Transactions* 106(1).

Surface tension

Penoncello, S.G. 1999. Thermophysical properties of trifluoromethane (R-23). Final report *ASHRAE Research Project* 997.

R-32

Equation of state

Tillner-Roth, R. and A. Yokozeki. 1997. An international standard equation of state for difluoromethane (R-32) for temperatures from the triple point at 136.34 K to 435 K and pressures up to 70 MPa. *Journal of Physical and Chemical Reference Data* 26:1273-1328.

Viscosity

Klein et al. 1997. *op. cit.*

Thermal conductivity

McLinden et al. 2000b. *op. cit.*

Surface tension

Okada, M. and Y. Higashi. 1995. Experimental surface tensions for HFC-32, HCFC-124, HFC-125, HCFC-141b, HCFC-142b, and HFC-152a. *International Journal of Thermophysics* 16(3):791-800.

R-123

Equation of state

Younglove, B.A. and M.O. McLinden. 1994. An international standard equation-of-state formulation of the thermodynamic properties of refrigerant 123 (2,2-dichloro-1,1,1-trifluoroethane). *Journal of Physical and Chemical Reference Data* 23(5):731-79.

Viscosity

Tanaka, Y. and T. Sotani. 1995. Chapter 2: Transport properties (thermal conductivity and viscosity), R-123. *Thermodynamic and Physical Properties*. International Institute of Refrigeration, Paris.

Thermal conductivity

Laesecke, A., R.A. Perkins, and J.B. Howley. 1996. An improved correlation for the thermal conductivity of HCFC-123 (2,2-dichloro-1,1,1-trifluoroethane). *International Journal of Refrigeration* 19:231-38.

Surface tension

Okada and Higashi. 1995. *op. cit.*

R-124

Equation of state

de Vries, B., R. Tillner-Roth, and H.D. Baehr. 1995. Thermodynamic properties of HCFC-124. 19th International Congress of Refrigeration, International Institute of Refrigeration, IVa:582-89.

Viscosity

Klein et al. 1997. *op. cit.*

Thermal conductivity

McLinden et al. 2000b. *op. cit.*

Surface tension

Okada and Higashi. 1995. *op. cit.*

R-125

Equation of state

Sunaga, H., R. Tillner-Roth, H. Sato, and K. Watanabe. 1998. A thermodynamic equation of state for pentafluoroethane (R-125). *International Journal of Thermophysics* 19:1623-35.

Viscosity

Klein et al. 1997. *op. cit.*

Thermal conductivity

McLinden et al. 2000b. *op. cit.*

Surface tension

Okada and Higashi. 1995. *op. cit.*

R-134a

Equation of state

Tillner-Roth, R. and H.D. Baehr. 1994. An international standard formulation of the thermodynamic properties of 1,1,1,2-tetrafluoroethane (HFC-134a) covering temperatures from 170 K to 455 K at pressures up to 70 MPa. *Journal of Physical and Chemical Reference Data* 23:657-729.

Viscosity

Laesecke, A. 2000. Data reassessment and full surface correlation of the viscosity of HFC-134a (1,1,1,2-tetrafluoroethane). *Journal of Physical and Chemical Reference Data* (submitted).

Thermal conductivity

McLinden et al. 2000b. *op. cit.*

Surface tension

Okada, M. and Y. Higashi. 1994. Surface tension correlation of HFC-134a and HCFC-123. CFCs, The Day After. Proceedings of Joint Meeting of IIR Commissions B1, B2, E1, and E2, 541-48.

R-143a

Equation of state

Lemmon, E.W. and R.T. Jacobsen. 2001. An international standard formulation for the thermodynamic properties of 1,1,1-trifluoroethane (HFC-143a) for temperatures from 161 to 500 K and pressures to 60 MPa. *Journal of Physical and Chemical Reference Data* (in press).

Viscosity

Klein et al. 1997. *op. cit.*

Thermal conductivity

McLinden et al. 2000b. *op. cit.*

Surface tension

Schmidt, J.W., E. Carrillo-Nava, and M.R. Moldover. 1996. Partially halogenated hydrocarbons CHFCl-CF_3, CF_3-CH_3, CF_3-CHF-CHF_2, CF_3-CH_2-CF_3, CHF_2-CF_2-CH_2F, CF_3-CH_2-CHF_2, CF_3-O-CHF_2: Critical temperature, refractive indices, surface tension and estimates of liquid, vapor and critical densities. *Fluid Phase Equilibria* 122:187-206.

R-152a

Equation of state

Outcalt, S.L. and M.O. McLinden. 1996. A modified Benedict-Webb-Rubin equation of state for the thermodynamic properties of R152a (1,1-difluoroethane). *Journal of Physical and Chemical Reference Data* 25(2):605-36.

Viscosity and thermal conductivity

Krauss, R., V.C. Weiss, T.A. Edison, J.V. Sengers, and K. Stephan. 1996. Transport properties of 1,1-Difluoroethane (R-152a). *International Journal of Thermophysics* 17:731-57.

Surface tension

Okada and Higashi. 1995. *op. cit.*

R-245fa

Equation of state

Extended corresponding states model of:

Huber, M.L. and J.F. Ely. 1994. A predictive extended corresponding states model for pure and mixed refrigerants including an equation of state for R-134a. *International Journal of Refrigeration* 17:18-31.

Fitted to the data of:

Defibaugh, D.R. and M.R. Moldover. 1997. Compressed and saturated liquid densities for 18 halogenated organic compounds. *Journal of Chemical and Engineering Data* 42:160-68.

Viscosity

Klein et al. 1997. *op. cit.*

Thermal conductivity

McLinden et al. 2000b. *op. cit.*

Surface tension

Schmidt et al. 1996. *op. cit.*

R-404A

Equation of state

Lemmon, E.W. and R.T. Jacobsen. 1999. Thermodynamic properties of mixtures of R-32, R-125, R-134a, and R-152a. *International Journal of Thermophysics* 20:1629-38.

Viscosity

Klein et al. 1997. *op. cit.*

Thermal conductivity

McLinden et al. 2000b. *op. cit.*

Surface tension

Moldover, M.R. and J.C. Rainwater. 1988. Interfacial tension and vapor-liquid equilibria in the critical region of mixtures. *J. Chem. Phys.* 88: 7772-80.

R-407C

Equation of state

Lemmon and Jacobsen. 1999. *op. cit.*

Viscosity

Klein et al. 1997. *op. cit.*

Thermal conductivity

McLinden et al. 2000b. *op. cit.*

Surface tension

Moldover and Rainwater. 1988. *op. cit.*

R-410A

Equation of state

Lemmon and Jacobsen. 1999. *op. cit.*

Viscosity

Klein et al. 1997. *op. cit.*

Thermal conductivity

McLinden et al. 2000b. *op. cit.*

Surface tension

Moldover and Rainwater. 1988. *op. cit.*

R-507A

Equation of state

Lemmon and Jacobsen. 1999. *op. cit.*

Viscosity

Klein et al. 1997. *op. cit.*

Thermal conductivity

McLinden et al. 2000b. *op. cit.*

Surface tension

Moldover and Rainwater. 1988. *op. cit.*

R-717 (Ammonia)

Equation of state

Tillner-Roth, R., F. Harms-Watzenberg, and H.D. Baehr. 1993. Eine neue Fundamentalgleichung für Ammoniak. DKV-*Tagungsbericht* 20(II): 167-81.

Viscosity

Fenghour, A., W.A. Wakeham, V. Vesovic, J.T.R. Watson, J. Millat, and E. Vogel. 1995. The viscosity of ammonia. *Journal of Physical and Chemical Reference Data* 24:1649-67.

Thermal conductivity

Tufeu, R., D.Y. Ivanov, Y. Garrabos, and B. Le Neindre. 1984. Thermal conductivity of ammonia in a large temperature and pressure range including the critical region. *Ber. Bunsen-Ges. Phys. Chem.* 88:422-27.

Surface tension

Stairs, R.A. and M.J. Sienko. 1956. Surface tension of ammonia and of solutions of alkalai halides in ammonia. *J. Am. Chem. Soc.* 78: 920-923.

R-718 (Water/Steam)

Data computed using

Harvey, A.H., S.A. Klein, and A.P. Peskin. 1999. NIST *Standard Reference Database* 10. NIST/ASME Steam properties database, Version 2.2. Standard Reference Data Program.

Equation of state

Wagner, W. and A. Pruß. 1999. New international formulation for the thermodynamic properties of ordinary water substance for general and scientific use. *Journal of Physical and Chemical Reference Data* (to be submitted).

Viscosity and thermal conductivity

Kestin, J., J.V. Sengers, B. Kamgar-Parsi, and J.M.H. Levelt Sengers. 1984. Thermophysical properties of fluid H_2O. *Journal of Physical and Chemical Reference Data* 13:175.

Surface tension

IAPWS. 1995. Physical chemistry of aqueous systems: Meeting the needs of industry. Proceedings 12th International Conference on the Properties of Water and Steam, Orlando. Begell House, Inc., A139-A142. International Association for the Properties of Steam.

R-744 (Carbon Dioxide)

Equation of state

Span, R. and W. Wagner. 1996. A new equation of state for carbon dioxide covering the fluid region from the triple-point temperature to 1100 K at pressures up to 800 MPa. *Journal of Physical and Chemical Reference Data* 26:1509-96.

Viscosity

Fenghour et al. 1995. *op. cit.*

Thermal conductivity

Vesovic, V., W.A. Wakeham, G.A. Olchowy, J.V. Sengers, J.T.R. Watson, and J. Millat. 1990. The transport properties of carbon dioxide. *Journal of Physical and Chemical Reference Data* 19:763-808.

Surface tension

Rathjen, W. and J. Straub. 1977. Chapter 18, Temperature dependence of surface tension, coexistence curve, and vapor pressure of CO_2, $CClF_3$, $CBrF_3$, and SF_6. *Heat transfer in boiling*. Academic Press, New York.

R-50 (Methane)

Equation of state

Setzmann, U. and W. Wagner. 1991. A new equation of state and tables of thermodynamic properties for methane covering the range from the melting line to 625 K at pressures to 1000 MPa. *Journal of Physical and Chemical Reference Data* 20:1061-1151.

Viscosity

Younglove, B.A. and J.F. Ely. 1987. Thermophysical properties of fluids. II. Methane, ethane, propane, isobutane and normal butane. *Journal of Physical and Chemical Reference Data* 16:577-798.

Thermal conductivity

Friend, D.G., J.F. Ely, and H. Ingham. 1989. Thermophysical properties of methane. *Journal of Physical and Chemical Reference Data* 18(2):583-638.

Surface tension

Somayajulu, G.R. 1988. A generalized equation for surface tension from the triple point to the critical point. *Int. J. of Thermophysics* 9:559-66.

R-170 (Ethane)

Equation of state, viscosity, and thermal conductivity

Friend, D.G., H. Ingham, and J.F. Ely. 1991. Thermophysical properties of ethane. *Journal of Physical and Chemical Reference Data* 20(2):275-347.

Surface tension

Soares, V.A.M., B.d.J.V.S. Almeida, I.A. McLure, and R.A. Higgins. 1986. Surface tension of pure and mixed simple substances at low temperature. *Fluid Phase Equilibria* 32:9-16.

R-290 (Propane)

Equation of state and thermal conductivity

Younglove and Ely. 1987. *op. cit.*

Viscosity

Vogel, E., C. Kuchenmeister, E. Bich, and A. Laesecke. 1998. Reference correlation of the viscosity of propane. *Journal of Physical and Chemical Reference Data* 27:947-70.

Thermal conductivity

Marsh, K., R. Perkins, and M.L.V. Ramires. 2001. Measurement and correlation of the thermal conductivity of propane from 86 to 600 K at pressures to 70 MPa. *Journal of Chemical and Engineering Data* (in press).

Surface tension:

Baidakov, V.G. and I.I. Sulla. 1985. Surface tension of propane and isobutane at near-critical temperatures. *Russian Journal of Physical Chemistry* 59:551-54.

R-600 (*n*-Butane)

Equation of state, viscosity, and thermal conductivity

Younglove and Ely. 1987. *op. cit.*

Surface tension

Calado, J.C.G., I.A. McLure, and V.A.M. Soares. 1978. Surface tension for octafluorocyclobutane, n-butane and their mixtures from 233 K to 254 K, and vapour pressure, excess gibbs function and excess volume for the mixture at 233 K. *Fluid Phase Equilibria* 2:199-213.

Coffin, C.C. and O. Maass. 1928. The preparation and physical properties of a-, b- and g-butylene and normal and isobutane. *Journal of the American Chemical Society* 50:1427-37.

R-600a (Isobutane)

Equation of state, viscosity, and thermal conductivity

Younglove and Ely. 1987. *op. cit.*

Surface tension

Baidakov and Sulla. 1985. *op. cit.*

R-1150 (Ethylene)

Equation of state

Smukala, J., R. Span, and W. Wagner. 1999. A new fundamental equation for the fluid state of ethylene for temperatures from the melting line to 450 K and pressure to 300 MPa. *Fortschritt-Berichte VDI*, Duesseldorf, Series 3, Number 616.

Viscosity and thermal conductivity

Holland, P.M., Eaton, B.E. and Hanley, H.J.M. 1983. A correlation of the viscosity and thermal conductivity data of gaseous and liquid ethylene. *Journal of Physical and Chemical Reference Data* 12:917-32.

Surface tension

Soares, V.A.M., B.d.J.V.S. Almeida, I.A. McLure, and R.A. Higgins. 1986. Surface tension of pure and mixed simple substances at low temperature. *Fluid Phase Equilibria* 32:9-16.

R-1270 (Propylene)

Equation of state

Angus, S., B. Armstrong, and K.M. de Reuck. 1980. International thermodynamic tables of the fluid state—7: Propylene. Pergammon Press, Oxford, England.

Surface tension

Maass, O. and C.H. Wright. 1921. Some physical properties of hydrocarbons containing two and three carbon atoms. *Journal of the American Chemical Society* 43:1098-1111.

†R-702 (Hydrogen)

Thermodynamic data computed using the ALLPROPS database, Version 4.0:

Lemmon, E.W., R.T. Jacobsen, S.G. Penoncello, and S.W. Beyerlein. 1994. Computer programs for the calculation of thermodynamic properties of cryogens and other fluids. *Advances in Cryogenic Engineering* 39:1891-97.

Transport data computed using the NIST 12 database, Version 3.0:

Friend, D.G., R.D. McCarty, and V. Arp. 1992. NIST Thermophysical properties of pure fluids database, Version 3.0. Standard Reference Data Program.

†Indicates data reprinted from the 1997 *ASHRAE Handbook*.

Equation of state, viscosity, and thermal conductivity

McCarty, R.D. 1975. Hydrogen: Technology survey—Thermophysical properties. NASA SP-3089.

Surface tension

Liley, P.E. and P.D. Desai. 1993. *ASHRAE Thermophysical properties of refrigerants.*

†R-702p (Parahydrogen)

Thermodynamic data computed using the ALLPROPS database, Version 4.0:

Lemmon et al. 1994. *op. cit.*

Transport data computed using the NIST12 database, Version 3.0:

Friend et al. 1992. *op. cit.*

Equation of state, viscosity, and thermal conductivity

Younglove, B.A. 1982. Thermophysical properties of fluids. I. Argon, ethylene, parahydrogen, nitrogen, nitrogen trifluoride, and oxygen. *Journal of Physical and Chemical Reference Data* 11(Supplement No. 1).

Surface tension

Liley and Desai. 1993. *op. cit.*

†R-704 (Helium)

Thermodynamic data computed using the ALLPROPS database, Version 4.0:

Lemmon et al. 1994. *op. cit.*

Transport data computed using the NIST12 database, Version 3.0:

Friend et al. 1992. *op. cit.*

Equation of state, viscosity, and thermal conductivity

Arp, V.D., R.D. McCarty, and D.G. Friend. 1995. Thermophysical properties of helium-4 from 0.8 to 1500 K with pressures to 2000 MPa. NIST *Technical Note* 1334 (revised).

Surface tension

Liley and Desai. 1993. *op. cit.*

†R-728 (Nitrogen)

Thermodynamic data computed using the ALLPROPS database, Version 4.0:

Lemmon et al. 1994. *op. cit.*

Equation of state

Jacobsen, R.T., R.B. Stewart, and M. Jahangiri. 1986. Thermodynamic properties of nitrogen from the freezing line to 2000 K at pressures to 1000 MPa. *Journal of Physical and Chemical Reference Data* 15(2):735-909.

Viscosity and thermal conductivity

Ely, J.F. 1997. Correlation contained in the AIRPROPS database, Version 1.0.

Surface tension

Lemmon, E.W. and S.G. Penoncello. 1994. The surface tension of air and air component mixtures. *Advances in Cryogenic Engineering* 39:1927-34.

†R-729 (Air)

Data computed using the AIRPROPS database, Version 1.0.

Lemmon, E.W. 1997. NIST thermophysical properties of air and air component mixtures, Version 1.0. Standard Reference Data Program.

Equation of state, viscosity, and thermal conductivity

Jacobsen, R.T., S.G. Penoncello, S.W. Beyerlein, D.G. Friend, J.F. Ely, J.C. Rainwater, and W.M. Haynes. 1995. Thermophysical properties of air. National Institute of Standards and Technology, Supplement to NASP *Technical Memorandum* 1005, NASA Langley Research Center.

Surface tension

Lemmon and Penoncello. 1994. *op.cit.*

†R-732 (Oxygen)

Thermodynamic data computed using the ALLPROPS database, Version 4.0:

Lemmon et al. 1994. *op. cit.*

Equation of state

Schmidt, R. and W. Wagner. 1985. A new form of the equation of state for pure substances and its application to oxygen. *Fluid Phase Equilibria* 19:175-200.

Viscosity and thermal conductivity

Ely. 1997. *op. cit.*

Surface tension

Lemmon and Penoncello. 1994. *op.cit.*

†R-740 (Argon)

Thermodynamic data computed using the ALLPROPS database, Version 4.0:

Lemmon et al. 1994. *op. cit.*

Equation of state

Stewart, R.B. and R.T. Jacobsen. 1989. Thermodynamic properties of argon from the triple point to 1200 K with pressures to 1000 MPa. *Journal of Physical and Chemical Reference Data* 18(2):639-798.

Viscosity and thermal conductivity

Ely. 1997. *op. cit.*

Surface tension

Lemmon and Penoncello. 1994. *op.cit.*

CHAPTER 21

PHYSICAL PROPERTIES OF SECONDARY COOLANTS (BRINES)

IN MANY refrigeration applications, heat is transferred to a **secondary coolant**, which can be any liquid cooled by the refrigerant and used to transfer heat without changing state. These liquids are also known as **heat transfer fluids**, **brines**, or **secondary refrigerants**.

Other ASHRAE Handbooks describe various applications for secondary coolants. In the 1998 *ASHRAE Handbook—Refrigeration*, refrigeration systems are discussed in Chapter 4, their uses in food processing are found in Chapters 14 through 28, and ice rinks are discussed in Chapter 34. In the 1999 *ASHRAE Handbook—Applications*, solar energy use is discussed in Chapter 32, thermal storage in Chapter 33, and snow melting in Chapter 49.

This chapter describes the physical properties of several secondary coolants and provides information on their use. The chapter also includes information on corrosion protection. Additional information on corrosion inhibition can be found in Chapter 47 of the 1999 *ASHRAE Handbook—Applications* and Chapter 4 of the 1998 *ASHRAE Handbook—Refrigeration.*

BRINES

Physical Properties

Water solutions of calcium chloride and sodium chloride are the most common refrigeration brines. Tables 1 and 2 list the properties of pure calcium chloride brine and sodium chloride brine. For commercial grades, use the formulas in the footnotes to these tables. Figures 1 and 5 give the specific heats for calcium chloride and sodium chloride brines and are used for computation of heat loads with ordinary brine (Carrier 1959). Figures 2 and 6 show the ratio of the mass of the solution to that of water, which is commonly used as the measure of salt concentration. Viscosities are given in Figures 3 and 7. Figures 4 and 8 show thermal conductivity of calcium and sodium brines at varying temperatures and concentrations.

Brine applications in refrigeration are mainly in the industrial machinery field and in skating rinks. Corrosion is the principal problem for calcium chloride brines, especially in ice-making tanks where galvanized iron cans are immersed.

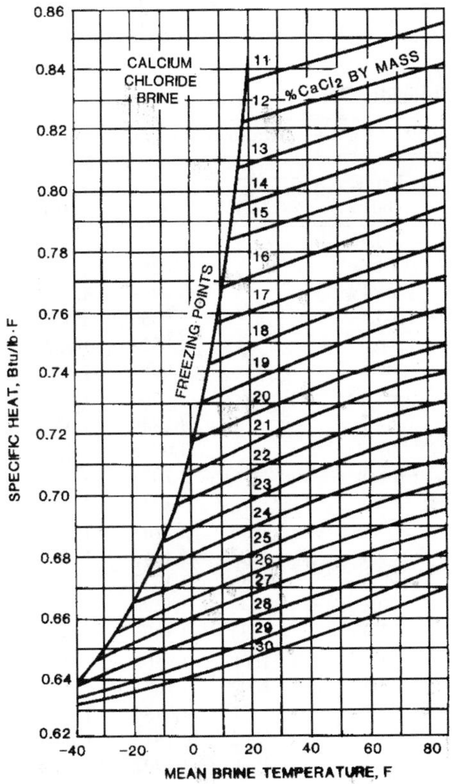

Fig. 1 Specific Heat of Calcium Chloride Brines

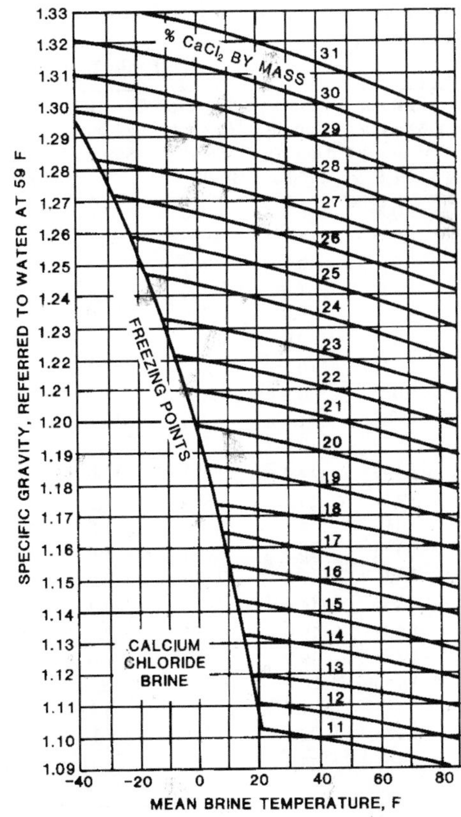

Fig. 2 Specific Gravity of Calcium Chloride Brines

The preparation of this chapter is assigned to TC 3.1, Refrigerants and Brines.

Table 1 Properties of Pure Calcium Chloride[a] Brines

Pure $CaCl_2$, % by Mass	Ratio of Mass to Water at 60°F	Baume Density at 60°F	Specific Heat at 60°F, Btu/lb·°F	Crystallization Starts, °F	Mass per Unit Volume[b] at 60°F				Ratio of Mass at Various Temperatures to Water at 60°F			
					$CaCl_2$, lb/gal	Brine, lb/gal	$CaCl_2$, lb/ft³	Brine, lb/ft³	–4°F	14°F	32°F	50°F
0	1.000	0.0	1.000	32.0	0.000	8.34	0.00	62.40				
5	1.044	6.1	0.924	27.7	0.436	8.717	3.26	65.15			1.043	1.042
6	1.050	7.0	0.914	26.8	0.526	8.760	3.93	65.52			1.052	1.051
7	1.060	8.2	0.898	25.9	0.620	8.851	4.63	66.14			1.061	1.060
8	1.069	9.3	0.884	24.6	0.714	8.926	5.34	66.70			1.071	1.069
9	1.078	10.4	0.869	23.5	0.810	9.001	6.05	67.27			1.080	1.078
10	1.087	11.6	0.855	22.3	0.908	9.076	6.78	67.83			1.089	1.087
11	1.096	12.6	0.842	20.8	1.006	9.143	7.52	68.33			1.098	1.096
12	1.105	13.8	0.828	19.3	1.107	9.227	8.27	68.95			1.108	1.105
13	1.114	14.8	0.816	17.6	1.209	9.302	9.04	69.51			1.117	1.115
14	1.124	15.9	0.804	15.5	1.313	9.377	9.81	70.08			1.127	1.124
15	1.133	16.9	0.793	13.5	1.418	9.452	10.60	70.64		1.139	1.137	1.134
16	1.143	18.0	0.779	11.2	1.526	9.536	11.40	71.26		1.149	1.146	1.143
17	1.152	19.1	0.767	8.6	1.635	9.619	12.22	71.89		1.159	1.156	1.153
18	1.162	20.2	0.756	5.9	1.747	9.703	13.05	72.51		1.169	1.166	1.163
19	1.172	21.3	0.746	2.8	1.859	9.786	13.90	73.13		1.180	1.176	1.173
20	1.182	22.1	0.737	–0.4	1.970	9.853	14.73	73.63		1.190	1.186	1.183
21	1.192	23.0	0.729	–3.9	2.085	9.928	15.58	74.19				
22	1.202	24.4	0.716	–7.8	2.208	10.037	16.50	75.00	1.215	1.211	1.207	1.203
23	1.212	25.5	0.707	–11.9	2.328	10.120	17.40	75.63				
24	1.223	26.4	0.697	–16.2	2.451	10.212	18.32	76.32	1.236	1.232	1.228	1.224
25	1.233	27.4	0.689	–21.0	2.574	10.295	19.24	76.94				
26	1.244	28.3	0.682	–25.8	2.699	10.379	20.17	77.56				
27	1.254	29.3	0.673	–31.2	2.827	10.471	21.13	78.25				
28	1.265	30.4	0.665	–37.8	2.958	10.563	22.10	78.94				
29	1.276	31.4	0.658	–49.4	3.090	10.655	23.09	79.62				
29.87	1.290	32.6	0.655	–67.0	3.16	10.75	23.65	80.45				
30	1.295	33.0	0.653	–50.8	3.22	10.80	24.06	80.76				
32	1.317	34.9	0.640	–19.5	3.49	10.98	26.10	82.14				
34	1.340	36.8	0.630	4.3	3.77	11.17	28.22	83.57				

[a]Mass of Type 1 (77% min.) $CaCl_2$ = (mass of pure $CaCl_2$)/(0.77). Mass of Type 2 (94% min.) $CaCl_2$ = (mass of pure $CaCl_2$)/(0.94).
[b]Mass of water per unit volume = Brine mass minus $CaCl_2$ mass.

Table 2 Properties of Pure Sodium Chloride[a] Brines

Pure NaCl, % by Mass	Ratio of Mass to Water at 59°F	Baume Density at 60°F	Specific Heat at 59°F, Btu/lb·°F	Crystallization Starts, °F	Mass per Unit Volume[b] at 60°F				Ratio of Mass at Various Temperatures to Water at 60°F			
					NaCl, lb/gal	Brine, lb/gal	NaCl, lb/ft³	Brine, lb/ft³	14°F	32°F	50°F	68°F
0	1.000	0.0	1.000	32.0	0.000	8.34	0.000	62.4				
5	1.035	5.1	0.938	26.7	0.432	8.65	3.230	64.6		1.0382	1.0366	1.0341
6	1.043	6.1	0.927	25.5	0.523	8.71	3.906	65.1		1.0459	1.0440	1.0413
7	1.050	7.0	0.917	24.3	0.613	8.76	4.585	65.5		1.0536	1.0515	1.0486
8	1.057	8.0	0.907	23.0	0.706	8.82	5.280	66.0		1.0613	1.0590	1.0559
9	1.065	9.0	0.897	21.6	0.800	8.89	5.985	66.5		1.0691	1.0665	1.0633
10	1.072	10.1	0.888	20.2	0.895	8.95	6.690	66.9		1.0769	1.0741	1.0707
11	1.080	10.8	0.879	18.8	0.992	9.02	7.414	67.4		1.0849	1.0817	1.0782
12	1.087	11.8	0.870	17.3	1.090	9.08	8.136	67.8		1.0925	1.0897	1.0857
13	1.095	12.7	0.862	15.7	1.188	9.14	8.879	68.3		1.1004	1.0933	1.0971
14	1.103	13.6	0.854	14.0	1.291	9.22	9.632	68.8		1.1083	1.1048	1.1009
15	1.111	14.5	0.847	12.3	1.392	9.28	10.395	69.3	1.1195	1.1163	1.1126	1.1086
16	1.118	15.4	0.840	10.5	1.493	9.33	11.168	69.8	1.1277	1.1243	1.1205	1.1163
17	1.126	16.3	0.833	8.6	1.598	9.40	11.951	70.3	1.1359	1.1323	1.1284	1.1241
18	1.134	17.2	0.826	6.6	1.705	9.47	12.744	70.8	1.1442	1.1404	1.1363	1.1319
19	1.142	18.1	0.819	4.5	1.813	9.54	13.547	71.3	1.1535	1.1486	1.1444	1.1398
20	1.150	19.0	0.813	2.3	1.920	9.60	14.360	71.8	1.1608	1.1568	1.1542	1.1478
21	1.158	19.9	0.807	0.0	2.031	9.67	15.183	72.3	1.1692	1.1651	1.1606	1.1559
22	1.166	20.8	0.802	–2.3	2.143	9.74	16.016	72.8	1.1777	1.1734	1.1688	1.1640
23	1.175	21.7	0.796	–5.1	2.256	9.81	16.854	73.3	1.1862	1.1818	1.1771	1.1721
24	1.183	22.5	0.791	3.8	2.371	9.88	17.712	73.8	1.1948	1.1902	1.1854	1.1804
25	1.191	23.4	0.786	16.1	2.488	9.95	18.575	74.3				
25.2	1.200			32.0								

[a]Mass of commercial NaCl required = (mass of pure NaCl required)/(% purity).
[b]Mass of water per unit volume = Brine mass minus NaCl mass.

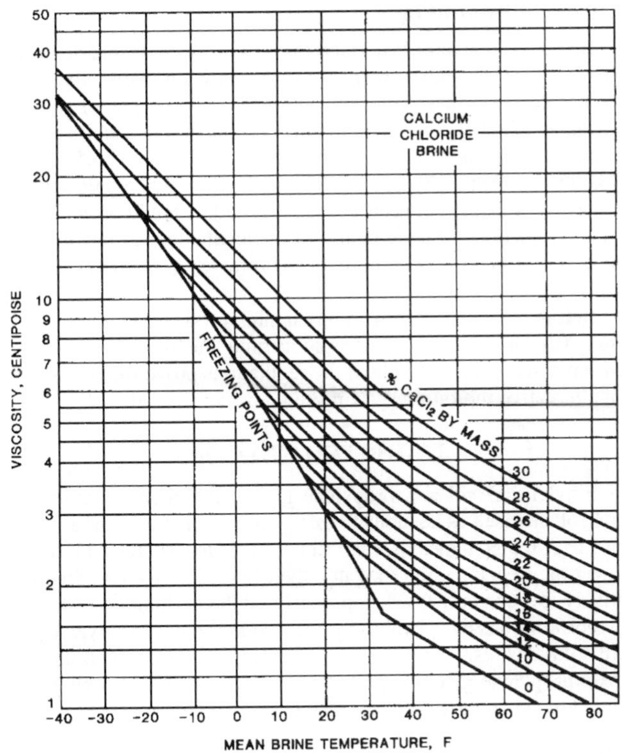

Fig. 3 Viscosity of Calcium Chloride Brines

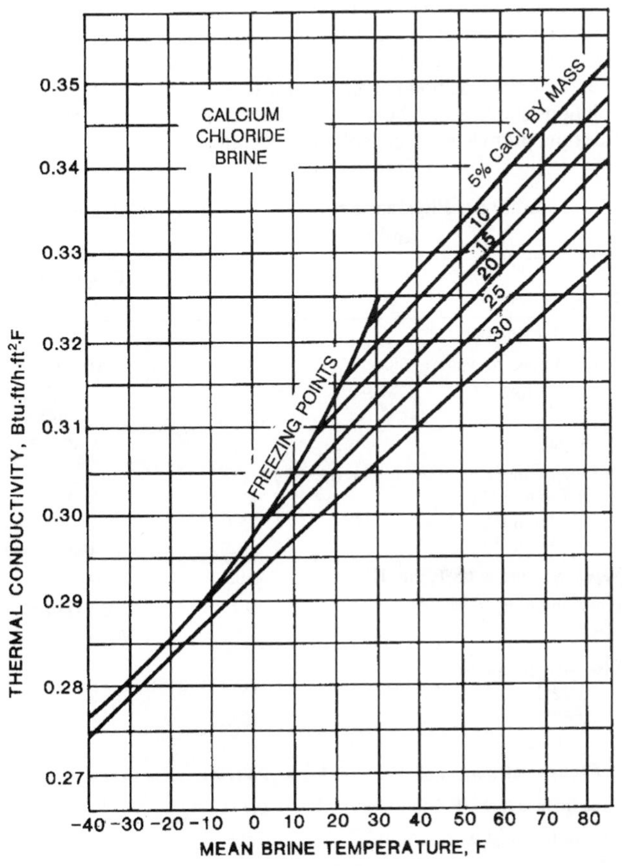

Fig. 4 Thermal Conductivity of Calcium Chloride Brines

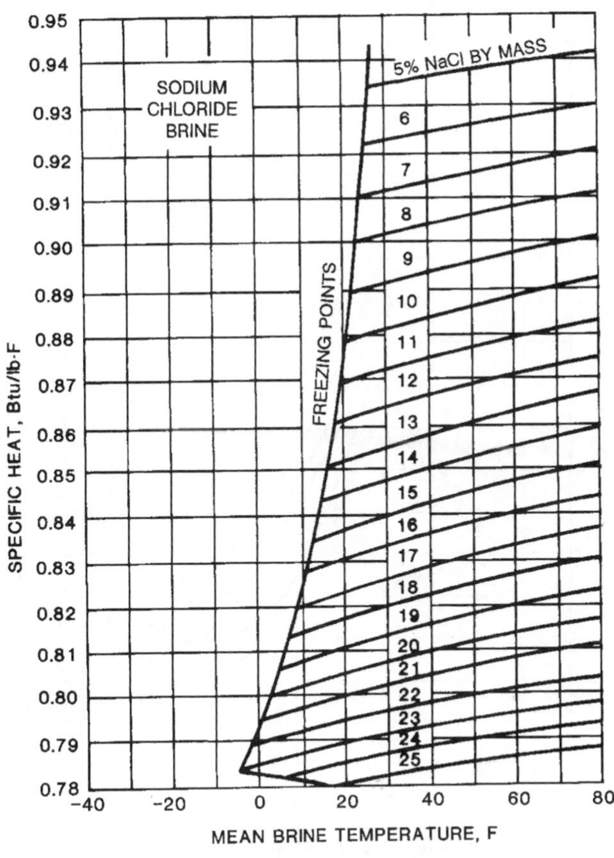

Fig. 5 Specific Heat of Sodium Chloride Brines

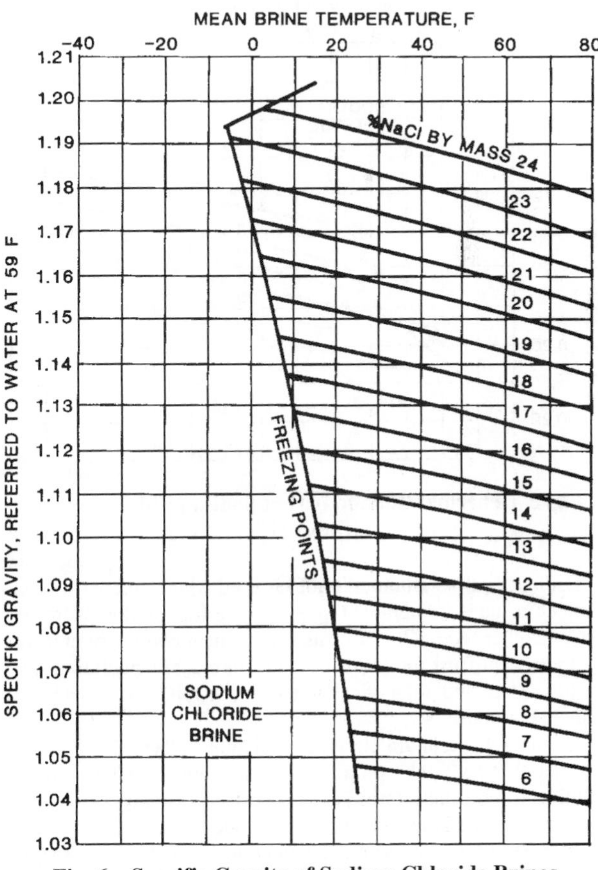

Fig. 6 Specific Gravity of Sodium Chloride Brines

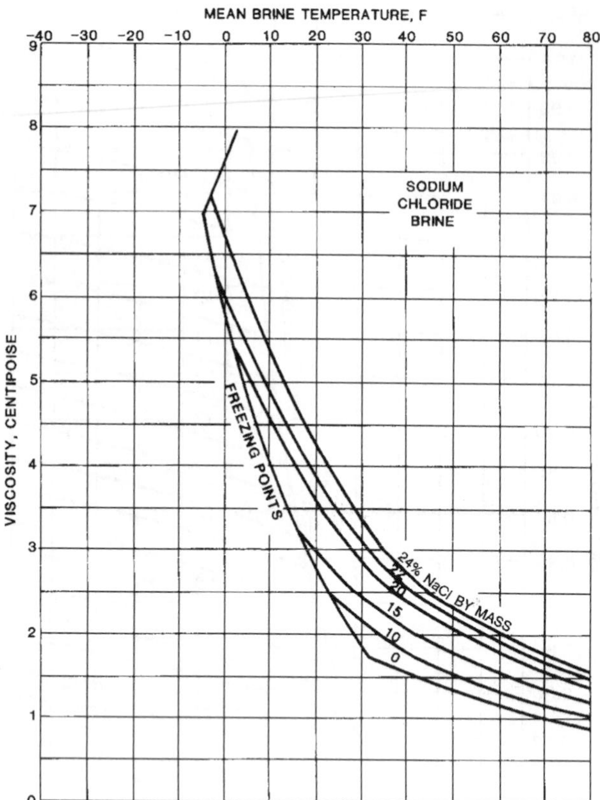

Fig. 7 Viscosity of Sodium Chloride Brines

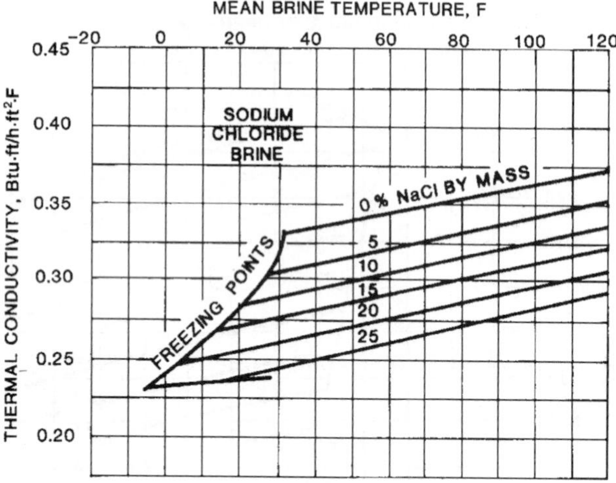

Fig. 8 Thermal Conductivity of Sodium Chloride Brines
(Carrier 1959)

Ordinary salt (sodium chloride) is used where contact with calcium chloride is intolerable (e.g., the brine fog method of freezing fish and other foods). It is used as a spray in air cooling of unit coolers to prevent frost formation on coils. In most refrigerating work, the lower freezing point of calcium chloride solution makes it more convenient to use.

Commercial calcium chloride, available as Type 1 (77% minimum) and Type 2 (94% minimum), is marketed in flake, solid, and solution forms; flake form is used most extensively. Commercial sodium chloride is available both in crude (rock salt) and refined grades. Because magnesium salts tend to form sludge, their presence in sodium or calcium chloride is undesirable.

Corrosion Inhibition

Brine systems must be treated to control corrosion and deposits. The standard chromate treatment program is the most effective. Calcium chloride brines require a minimum of 1800 ppmof sodium chromate with pH 6.5 to 8.5. Sodium chloride brines require a minimum of 3600 ppmof sodium chromate and a pH of 6.5 to 8.5. Sodium nitrite at 3000 ppmin calcium brines or 4000 ppmin sodium brines controls pH between 7.0 and 8.5, and it should provide adequate protection. Organic inhibitors are available that may provide adequate protection where neither chromates nor nitrites can be used.

Before using any chromate-based inhibitor package, review federal, state, and local regulations concerning the use and disposal of chromate-containing fluids. If the regulations prove too restrictive, an alternative inhibition system should be considered.

INHIBITED GLYCOLS

Ethylene glycol and propylene glycol, inhibited for corrosion control, are used as aqueous freezing point depressants (antifreeze) and heat transfer media. Their chief attributes are their ability to lower the freezing point of water, their low volatility, and their relatively low corrosivity when properly inhibited. Inhibited ethylene glycol solutions have better physical properties than propylene glycol solutions, especially at lower temperatures. However, the less toxic propylene glycol is preferred for applications involving possible human contact or where mandated by regulations.

Physical Properties

Ethylene glycol and propylene glycol are colorless, practically odorless liquids that are miscible with water and many organic compounds. Table 3 shows properties of the pure materials.

The freezing and boiling points of aqueous solutions of ethylene glycol and propylene glycol are given in Tables 4 and 5. Note that increasing the concentration of ethylene glycol above 60% by mass causes the freezing point of the solution to increase. Propylene glycol solutions above 60% by mass do not have freezing points. Instead of freezing, propylene glycol solutions become a glass (glass being an

Table 3 Physical Properties of Ethylene Glycol and Propylene Glycol

Property	Ethylene Glycol	Propylene Glycol
Molecular weight	62.07	76.10
Ratio of mass to water at 68/68°F	1.1155	1.0381
Density at 68°F		
lb/ft^3	69.50	64.68
lb/gal	9.29	8.65
Boiling point, °F		
at 760 mm Hg	388	369
at 50 mm Hg	253	241
at 10 mm Hg	192	185
Vapor pressure at 68°F, mm Hg	0.05	0.07
Freezing point, °F	9.1	Sets to glass below −60°F
Viscosity, centipoise		
at 32°F	57.4	243
at 68°F	20.9	60.5
at 104°F	9.5	18.0
Refractive index n_D at 68°F	1.4319	1.4329
Specific heat at 68°F, Btu/lb·°F	0.561	0.593
Heat of fusion at 9.1°F, Btu/lb	80.5	—
Heat of vaporization at 1 atm, Btu/lb	364	296
Heat of combustion at 68°F, Btu/lb	8,280	10,312

Table 4 Freezing and Boiling Points of Aqueous Solutions of Ethylene Glycol

Percent Ethylene Glycol		Freezing Point, °F	Boiling Point, °F at 14.6 psia
By Mass	By Volume		
0.0	0.0	32.0	212
5.0	4.4	29.4	213
10.0	8.9	26.2	214
15.0	13.6	22.2	215
20.0	18.1	17.9	216
21.0	19.2	16.8	216
22.0	20.1	15.9	216
23.0	21.0	14.9	217
24.0	22.0	13.7	217
25.0	22.9	12.7	218
26.0	23.9	11.4	218
27.0	24.8	10.4	218
28.0	25.8	9.2	219
29.0	26.7	8.0	219
30.0	27.7	6.7	220
31.0	28.7	5.4	220
32.0	29.6	4.2	220
33.0	30.6	2.9	220
34.0	31.6	1.4	220
35.0	32.6	−0.2	221
36.0	33.5	−1.5	221
37.0	34.5	−3.0	221
38.0	35.5	−4.5	221
39.0	36.5	−6.4	221
40.0	37.5	−8.1	222
41.0	38.5	−9.8	222
42.0	39.5	−11.7	222
43.0	40.5	−13.5	223
44.0	41.5	−15.5	223
45.0	42.5	−17.5	224
46.0	43.5	−19.8	224
47.0	44.5	−21.6	224
48.0	45.5	−23.9	224
49.0	46.6	−26.7	224
50.0	47.6	−28.9	225
51.0	48.6	−31.2	225
52.0	49.6	−33.6	225
53.0	50.6	−36.2	226
54.0	51.6	−38.8	226
55.0	52.7	−42.0	227
56.0	53.7	−44.7	227
57.0	54.7	−47.5	228
58.0	55.7	−50.0	228
59.0	56.8	−52.7	229
60.0	57.8	−54.9	230
65.0	62.8	a	235
70.0	68.3	a	242
75.0	73.6	a	248
80.0	78.9	−52.2	255
85.0	84.3	−34.5	273
90.0	89.7	−21.6	285
95.0	95.0	−3.0	317

[a]Freezing points are below −60°F.

Table 5 Freezing and Boiling Points of Aqueous Solutions of Propylene Glycol

Percent Propylene Glycol		Freezing Point, °F	Boiling Point, °F at 14.6 psia
By Mass	By Volume		
0.0	0.0	32.0	212
5.0	4.8	29.1	212
10.0	9.6	26.1	212
15.0	14.5	22.9	212
20.0	19.4	19.2	213
21.0	20.4	18.3	213
22.0	21.4	17.6	213
23.0	22.4	16.6	213
24.0	23.4	15.6	213
25.0	24.4	14.7	214
26.0	25.3	13.7	214
27.0	26.4	12.6	214
28.0	27.4	11.5	215
29.0	28.4	10.4	215
30.0	29.4	9.2	216
31.0	30.4	7.9	216
32.0	31.4	6.6	216
33.0	32.4	5.3	216
34.0	33.5	3.9	216
35.0	34.4	2.4	217
36.0	35.5	0.8	217
37.0	36.5	−0.8	217
38.0	37.5	−2.4	218
39.0	38.5	−4.2	218
40.0	39.6	−6.0	219
41.0	40.6	−7.8	219
42.0	41.6	−9.8	219
43.0	42.6	−11.8	219
44.0	43.7	−13.9	219
45.0	44.7	−16.1	220
46.0	45.7	−18.3	220
47.0	46.8	−20.7	220
48.0	47.8	−23.1	221
49.0	48.9	−25.7	221
50.0	49.9	−28.3	222
51.0	50.9	−31.0	222
52.0	51.9	−33.8	222
53.0	53.0	−36.7	223
54.0	54.0	−39.7	223
55.0	55.0	−42.8	223
56.0	56.0	−46.0	223
57.0	57.0	−49.3	224
58.0	58.0	−52.7	224
59.0	59.0	−56.2	224
60.0	60.0	−59.9	225
65.0	65.0	a	227
70.0	70.0	a	230
75.0	75.0	a	237
80.0	80.0	a	245
85.0	85.0	a	257
90.0	90.0	a	270
95.0	95.0	a	310

[a]Above 60% by mass, solutions do not freeze but become a glass.

amorphous, undercooled liquid of extremely high viscosities that has all the appearances of a solid). On the dilute side of the eutectic, ice forms on freezing; on the concentrated side, solid glycol separates from solution on freezing. The freezing velocity of such solutions is often quite slow; but, in time, they set to a hard, solid mass.

Physical properties (i.e., density, specific heat, thermal conductivity, and viscosity) for aqueous solutions of ethylene glycol can be found in Tables 6 through 9 and Figures 9 through 12; similar data for aqueous solutions of propylene glycol can be found in Tables 10 through 13 and Figures 13 through 16. Densities are for aqueous solutions of industrially inhibited glycols. These densities are somewhat higher than those for pure glycol and water alone. Typical corrosion inhibitor packages do not significantly affect the other physical properties. The physical properties for the two fluids are similar, with the exception of viscosity. At the same concentration, aqueous solutions of propylene glycol are more viscous than solutions of ethylene glycol. This higher viscosity accounts for the majority of the performance difference between the two fluids.

The choice of glycol concentration depends on the type of protection required by the application. If the fluid is being used to prevent equipment damage during idle periods in cold weather, such as winterizing coils in an HVAC system, 30% ethylene glycol or 35% propylene glycol is sufficient. These concentrations will allow the fluid to freeze. As the fluid freezes, it forms a slush that expands and flows into any available space. Therefore, expansion volume must be included with this type of protection. If the application requires that the fluid remain entirely liquid, a concentration with a freezing point 5°F below the lowest expected temperature should be chosen. Avoid excessive glycol concentration because it increases initial cost and adversely affects the physical properties of the fluid.

Table 6 Density of Aqueous Solutions of Ethylene Glycol

	Concentrations in Volume Percent Ethylene Glycol								
Temperature, °F	10%	20%	30%	40%	50%	60%	70%	80%	90%
−30					68.12	69.03	69.90	70.75	
−20					68.05	68.96	69.82	70.65	71.45
−10				67.04	67.98	68.87	69.72	70.54	71.33
0				66.97	67.90	68.78	69.62	70.43	71.20
10			65.93	66.89	67.80	68.67	69.50	70.30	71.06
20		64.83	65.85	66.80	67.70	68.56	69.38	70.16	70.92
30	63.69	64.75	65.76	66.70	67.59	68.44	69.25	70.02	70.76
40	63.61	64.66	65.66	66.59	67.47	68.31	69.10	69.86	70.59
50	63.52	64.56	65.55	66.47	67.34	68.17	68.95	69.70	70.42
60	63.42	64.45	65.43	66.34	67.20	68.02	68.79	69.53	70.23
70	63.31	64.33	65.30	66.20	67.05	67.86	68.62	69.35	70.04
80	63.19	64.21	65.17	66.05	66.90	67.69	68.44	69.15	69.83
90	63.07	64.07	65.02	65.90	66.73	67.51	68.25	68.95	69.62
100	62.93	63.93	64.86	65.73	66.55	67.32	68.05	68.74	69.40
110	62.79	63.77	64.70	65.56	66.37	67.13	67.84	68.52	69.17
120	62.63	63.61	64.52	65.37	66.17	66.92	67.63	68.29	68.92
130	62.47	63.43	64.34	65.18	65.97	66.71	67.40	68.05	68.67
140	62.30	63.25	64.15	64.98	65.75	66.48	67.16	67.81	68.41
150	62.11	63.06	63.95	64.76	65.53	66.25	66.92	67.55	68.14
160	61.92	62.86	63.73	64.54	65.30	66.00	66.66	67.28	67.86
170	61.72	62.64	63.51	64.31	65.05	65.75	66.40	67.01	67.58
180	61.51	62.42	63.28	64.07	64.80	65.49	66.12	66.72	67.28
190	61.29	62.19	63.04	63.82	64.54	65.21	65.84	66.42	66.97
200	61.06	61.95	62.79	63.56	64.27	64.93	65.55	66.12	66.65
210	60.82	61.71	62.53	63.29	63.99	64.64	65.24	65.81	66.33
220	60.57	61.45	62.27	63.01	63.70	64.34	64.93	65.48	65.99
230	60.31	61.18	61.99	62.72	63.40	64.03	64.61	65.15	65.65
240	60.05	60.90	61.70	62.43	63.10	63.71	64.28	64.81	65.29
250	59.77	60.62	61.40	62.12	62.78	63.39	63.94	64.46	64.93

Note: Density in lb/ft^3.

Table 7 Specific Heat of Aqueous Solutions of Ethylene Glycol

	Concentrations in Volume Percent Ethylene Glycol								
Temperature, °F	10%	20%	30%	40%	50%	60%	70%	80%	90%
−30					0.734	0.680	0.625	0.567	
−20					0.739	0.686	0.631	0.574	0.515
−10				0.794	0.744	0.692	0.638	0.581	0.523
0				0.799	0.749	0.698	0.644	0.588	0.530
10			0.849	0.803	0.754	0.703	0.651	0.595	0.538
20		0.897	0.853	0.808	0.759	0.709	0.657	0.603	0.546
30	0.940	0.900	0.857	0.812	0.765	0.715	0.664	0.610	0.553
40	0.943	0.903	0.861	0.816	0.770	0.721	0.670	0.617	0.561
50	0.945	0.906	0.864	0.821	0.775	0.727	0.676	0.624	0.569
60	0.947	0.909	0.868	0.825	0.780	0.732	0.683	0.631	0.576
70	0.950	0.912	0.872	0.830	0.785	0.738	0.689	0.638	0.584
80	0.952	0.915	0.876	0.834	0.790	0.744	0.696	0.645	0.592
90	0.954	0.918	0.880	0.839	0.795	0.750	0.702	0.652	0.600
100	0.957	0.922	0.883	0.843	0.800	0.756	0.709	0.659	0.607
110	0.959	0.925	0.887	0.848	0.806	0.761	0.715	0.666	0.615
120	0.961	0.928	0.891	0.852	0.811	0.767	0.721	0.673	0.623
130	0.964	0.931	0.895	0.857	0.816	0.773	0.728	0.680	0.630
140	0.966	0.934	0.898	0.861	0.821	0.779	0.734	0.687	0.638
150	0.968	0.937	0.902	0.865	0.826	0.785	0.741	0.694	0.646
160	0.971	0.940	0.906	0.870	0.831	0.790	0.747	0.702	0.654
170	0.973	0.943	0.910	0.874	0.836	0.796	0.754	0.709	0.661
180	0.975	0.946	0.913	0.879	0.842	0.802	0.760	0.716	0.669
190	0.978	0.949	0.917	0.883	0.847	0.808	0.766	0.723	0.677
200	0.980	0.952	0.921	0.888	0.852	0.813	0.773	0.730	0.684
210	0.982	0.955	0.925	0.892	0.857	0.819	0.779	0.737	0.692
220	0.985	0.958	0.929	0.897	0.862	0.825	0.786	0.744	0.700
230	0.987	0.961	0.932	0.901	0.867	0.831	0.792	0.751	0.708
240	0.989	0.964	0.936	0.905	0.872	0.837	0.799	0.758	0.715
250	0.992	0.967	0.940	0.910	0.877	0.842	0.805	0.765	0.723

Note: Specific heat in Btu/lb·°F.

Table 8 Thermal Conductivity of Aqueous Solutions of Ethylene Glycol

	Concentrations in Volume Percent Ethylene Glycol								
Temperature, °F	10%	20%	30%	40%	50%	60%	70%	80%	90%
−30					0.190	0.178	0.167	0.158	
−20					0.193	0.181	0.170	0.160	0.151
−10				0.212	0.197	0.184	0.172	0.161	0.152
0				0.216	0.200	0.186	0.174	0.163	0.153
10			0.238	0.220	0.204	0.189	0.176	0.164	0.154
20		0.264	0.243	0.224	0.207	0.191	0.178	0.166	0.155
30	0.294	0.269	0.247	0.227	0.210	0.194	0.180	0.167	0.156
40	0.300	0.274	0.251	0.231	0.212	0.196	0.182	0.169	0.157
50	0.305	0.279	0.255	0.234	0.215	0.198	0.183	0.170	0.158
60	0.311	0.284	0.259	0.237	0.218	0.200	0.185	0.171	0.159
70	0.316	0.288	0.263	0.240	0.220	0.202	0.186	0.172	0.160
80	0.320	0.292	0.266	0.243	0.223	0.204	0.188	0.173	0.161
90	0.325	0.296	0.269	0.246	0.225	0.206	0.189	0.174	0.161
100	0.329	0.299	0.272	0.248	0.227	0.208	0.190	0.175	0.162
110	0.333	0.302	0.275	0.251	0.229	0.209	0.192	0.176	0.163
120	0.336	0.305	0.277	0.253	0.230	0.210	0.193	0.177	0.163
130	0.339	0.308	0.280	0.255	0.232	0.212	0.194	0.178	0.164
140	0.342	0.311	0.282	0.256	0.233	0.213	0.195	0.179	0.165
150	0.345	0.313	0.284	0.258	0.235	0.214	0.196	0.180	0.165
160	0.347	0.315	0.285	0.259	0.236	0.215	0.197	0.180	0.166
170	0.349	0.316	0.287	0.261	0.237	0.216	0.197	0.181	0.166
180	0.351	0.318	0.288	0.262	0.238	0.217	0.198	0.181	0.167
190	0.352	0.319	0.289	0.263	0.239	0.218	0.199	0.182	0.167
200	0.353	0.320	0.290	0.263	0.240	0.218	0.199	0.182	0.168
210	0.354	0.321	0.291	0.264	0.240	0.219	0.200	0.183	0.168
220	0.355	0.321	0.291	0.265	0.240	0.219	0.200	0.183	0.168
230	0.355	0.322	0.291	0.265	0.241	0.219	0.200	0.183	0.169
240	0.355	0.322	0.291	0.265	0.241	0.219	0.200	0.184	0.169
250	0.354	0.321	0.291	0.265	0.241	0.220	0.201	0.184	0.169

Note: Thermal conductivity in Btu·ft/h·ft^2·°F.

Table 9 Viscosity of Aqueous Solutions of Ethylene Glycol

	Concentrations in Volume Percent Ethylene Glycol								
Temperature, °F	10%	20%	30%	40%	50%	60%	70%	80%	90%
−30					63.69	89.67	128.79	185.22	
−20					40.38	60.46	89.93	131.32	284.48
−10				19.58	27.27	42.05	63.50	91.88	169.83
0				13.76	19.34	30.08	45.58	65.04	107.77
10			6.83	10.13	14.26	22.06	33.31	46.89	71.87
20		3.90	5.38	7.74	10.85	16.56	24.79	34.48	49.94
30	2.16	3.14	4.33	6.09	8.48	12.68	18.77	25.84	35.91
40	1.82	2.59	3.54	4.91	6.77	9.90	14.45	19.71	26.59
50	1.56	2.18	2.95	4.04	5.50	7.85	11.31	15.29	20.18
60	1.35	1.86	2.49	3.38	4.55	6.33	8.97	12.05	15.65
70	1.18	1.61	2.13	2.87	3.81	5.17	7.22	9.62	12.37
80	1.04	1.41	1.84	2.46	3.23	4.28	5.88	7.79	9.93
90	0.93	1.24	1.60	2.13	2.76	3.58	4.85	6.38	8.10
100	0.83	1.11	1.41	1.87	2.39	3.03	4.04	5.28	6.68
110	0.75	0.99	1.25	1.64	2.08	2.58	3.40	4.41	5.58
120	0.68	0.90	1.11	1.46	1.82	2.23	2.88	3.73	4.71
130	0.62	0.81	1.00	1.30	1.61	1.93	2.47	3.17	4.01
140	0.57	0.74	0.90	1.17	1.43	1.69	2.13	2.72	3.45
150	0.53	0.68	0.82	1.05	1.28	1.49	1.86	2.35	2.98
160	0.49	0.63	0.75	0.95	1.15	1.32	1.63	2.05	2.60
170	0.46	0.58	0.68	0.87	1.04	1.18	1.43	1.80	2.28
180	0.43	0.54	0.63	0.79	0.94	1.06	1.27	1.58	2.01
190	0.40	0.50	0.58	0.73	0.85	0.95	1.14	1.40	1.79
200	0.37	0.47	0.54	0.67	0.78	0.86	1.02	1.25	1.60
210	0.35	0.43	0.50	0.61	0.71	0.78	0.92	1.12	1.43
220	0.33	0.41	0.46	0.57	0.66	0.72	0.83	1.01	1.29
230	0.32	0.38	0.43	0.53	0.60	0.66	0.76	0.91	1.16
240	0.30	0.36	0.40	0.49	0.56	0.61	0.69	0.83	1.06
250	0.29	0.34	0.38	0.45	0.52	0.56	0.63	0.75	0.96

Note: Viscosity in centipoise.

Table 10 Density of Aqueous Solutions of an Industrially Inhibited Propylene Glycol

	Concentrations in Volume Percent Propylene Glycol								
Temperature, °F	10%	20%	30%	40%	50%	60%	70%	80%	90%
−30						67.05	67.47	68.38	68.25
−20					66.46	66.93	67.34	68.13	68.00
−10					66.35	66.81	67.20	67.87	67.75
0				65.71	66.23	66.68	67.05	67.62	67.49
10			65.00	65.60	66.11	66.54	66.89	67.36	67.23
20		64.23	64.90	65.48	65.97	66.38	66.72	67.10	66.97
30	63.38	64.14	64.79	65.35	65.82	66.22	66.54	66.83	66.71
40	63.30	64.03	64.67	65.21	65.67	66.05	66.35	66.57	66.44
50	63.20	63.92	64.53	65.06	65.50	65.87	66.16	66.30	66.18
60	63.10	63.79	64.39	64.90	65.33	65.68	65.95	66.04	65.91
70	62.98	63.66	64.24	64.73	65.14	65.47	65.73	65.77	65.64
80	62.86	63.52	64.08	64.55	64.95	65.26	65.51	65.49	65.37
90	62.73	63.37	63.91	64.36	64.74	65.04	65.27	65.22	65.09
100	62.59	63.20	63.73	64.16	64.53	64.81	65.03	64.95	64.82
110	62.44	63.03	63.54	63.95	64.30	64.57	64.77	64.67	64.54
120	62.28	62.85	63.33	63.74	64.06	64.32	64.51	64.39	64.26
130	62.11	62.66	63.12	63.51	63.82	64.06	64.23	64.11	63.98
140	61.93	62.46	62.90	63.27	63.57	63.79	63.95	63.83	63.70
150	61.74	62.25	62.67	63.02	63.30	63.51	63.66	63.55	63.42
160	61.54	62.03	62.43	62.76	63.03	63.22	63.35	63.26	63.13
170	61.33	61.80	62.18	62.49	62.74	62.92	63.04	62.97	62.85
180	61.11	61.56	61.92	62.22	62.45	62.61	62.72	62.68	62.56
190	60.89	61.31	61.65	61.93	62.14	62.29	62.39	62.39	62.27
200	60.65	61.05	61.37	61.63	61.83	61.97	62.05	62.10	61.97
210	60.41	60.78	61.08	61.32	61.50	61.63	61.69	61.81	61.68
220	60.15	60.50	60.78	61.00	61.17	61.28	61.33	61.51	61.38
230	59.89	60.21	60.47	60.68	60.83	60.92	60.96	61.21	61.08
240	59.61	59.91	60.15	60.34	60.47	60.55	60.58	60.91	60.78
250	59.33	59.60	59.82	59.99	60.11	60.18	60.19	60.61	60.48

Note: Density in lb/ft^3.

Table 11 Specific Heat of Aqueous Solutions of Propylene Glycol

	Concentrations in Volume Percent Propylene Glycol								
Temperature, °F	10%	20%	30%	40%	50%	60%	70%	80%	90%
−30						0.741	0.680	0.615	0.542
−20					0.799	0.746	0.687	0.623	0.550
−10					0.804	0.752	0.693	0.630	0.558
0				0.855	0.809	0.758	0.700	0.637	0.566
10			0.898	0.859	0.814	0.764	0.707	0.645	0.574
20		0.936	0.902	0.864	0.820	0.770	0.713	0.652	0.583
30	0.966	0.938	0.906	0.868	0.825	0.776	0.720	0.660	0.591
40	0.968	0.941	0.909	0.872	0.830	0.782	0.726	0.667	0.599
50	0.970	0.944	0.913	0.877	0.835	0.787	0.733	0.674	0.607
60	0.972	0.947	0.917	0.881	0.840	0.793	0.740	0.682	0.615
70	0.974	0.950	0.920	0.886	0.845	0.799	0.746	0.689	0.623
80	0.976	0.953	0.924	0.890	0.850	0.805	0.753	0.696	0.631
90	0.979	0.956	0.928	0.894	0.855	0.811	0.760	0.704	0.639
100	0.981	0.959	0.931	0.899	0.861	0.817	0.766	0.711	0.647
110	0.983	0.962	0.935	0.903	0.866	0.823	0.773	0.718	0.656
120	0.985	0.965	0.939	0.908	0.871	0.828	0.779	0.726	0.664
130	0.987	0.967	0.942	0.912	0.876	0.834	0.786	0.733	0.672
140	0.989	0.970	0.946	0.916	0.881	0.840	0.793	0.740	0.680
150	0.991	0.973	0.950	0.921	0.886	0.846	0.799	0.748	0.688
160	0.993	0.976	0.953	0.925	0.891	0.852	0.806	0.755	0.696
170	0.996	0.979	0.957	0.929	0.896	0.858	0.812	0.762	0.704
180	0.998	0.982	0.961	0.934	0.902	0.864	0.819	0.770	0.712
190	1.000	0.985	0.964	0.938	0.907	0.869	0.826	0.777	0.720
200	1.002	0.988	0.968	0.943	0.912	0.875	0.832	0.784	0.729
210	1.004	0.991	0.971	0.947	0.917	0.881	0.839	0.792	0.737
220	1.006	0.994	0.975	0.951	0.922	0.887	0.845	0.799	0.745
230	1.008	0.996	0.979	0.956	0.927	0.893	0.852	0.806	0.753
240	1.011	0.999	0.982	0.960	0.932	0.899	0.859	0.814	0.761
250	1.013	1.002	0.986	0.965	0.937	0.905	0.865	0.821	0.769

Note: Specific heat in Btu/lb·°F.

Table 12 Thermal Conductivity of Aqueous Solutions of Propylene Glycol

	Concentrations in Volume Percent Propylene Glycol								
Temperature, °F	10%	20%	30%	40%	50%	60%	70%	80%	90%
−30						0.171	0.159	0.147	0.137
−20					0.188	0.174	0.160	0.148	0.137
−10					0.191	0.176	0.161	0.148	0.136
0				0.211	0.194	0.178	0.162	0.149	0.136
10			0.235	0.215	0.196	0.179	0.163	0.149	0.136
20		0.262	0.239	0.218	0.199	0.181	0.164	0.150	0.136
30	0.293	0.267	0.243	0.222	0.201	0.183	0.165	0.150	0.135
40	0.299	0.272	0.247	0.225	0.204	0.184	0.166	0.150	0.135
50	0.304	0.277	0.251	0.227	0.206	0.186	0.167	0.150	0.135
60	0.310	0.281	0.254	0.230	0.208	0.187	0.168	0.150	0.134
70	0.315	0.285	0.258	0.233	0.210	0.188	0.168	0.151	0.134
80	0.319	0.289	0.261	0.235	0.211	0.189	0.169	0.151	0.134
90	0.323	0.292	0.263	0.237	0.213	0.190	0.169	0.151	0.133
100	0.327	0.295	0.266	0.239	0.214	0.191	0.170	0.151	0.133
110	0.331	0.298	0.268	0.241	0.215	0.192	0.170	0.151	0.132
120	0.334	0.301	0.270	0.243	0.217	0.193	0.170	0.150	0.132
130	0.338	0.304	0.272	0.244	0.218	0.193	0.170	0.150	0.131
140	0.340	0.306	0.274	0.245	0.218	0.194	0.171	0.150	0.131
150	0.343	0.308	0.276	0.246	0.219	0.194	0.171	0.150	0.130
160	0.345	0.309	0.277	0.247	0.220	0.194	0.171	0.150	0.130
170	0.347	0.311	0.278	0.248	0.220	0.195	0.171	0.149	0.129
180	0.348	0.312	0.279	0.249	0.221	0.195	0.170	0.149	0.129
190	0.350	0.313	0.280	0.249	0.221	0.195	0.170	0.148	0.128
200	0.351	0.314	0.280	0.249	0.221	0.194	0.170	0.148	0.127
210	0.351	0.314	0.280	0.249	0.221	0.194	0.169	0.147	0.127
220	0.352	0.314	0.280	0.249	0.220	0.194	0.169	0.147	0.126
230	0.352	0.314	0.280	0.249	0.220	0.193	0.168	0.146	0.125
240	0.351	0.314	0.280	0.249	0.220	0.193	0.168	0.146	0.125
250	0.351	0.314	0.279	0.248	0.219	0.192	0.167	0.145	0.124

Note: Thermal conductivity in Btu·ft/h·ft^2·°F.

Table 13 Viscosity of Aqueous Solutions of Propylene Glycol

	Concentrations in Volume Percent Propylene Glycol								
Temperature, °F	10%	20%	30%	40%	50%	60%	70%	80%	90%
−30						497.57	864.87	1363.75	3555.22
−20					156.08	298.75	493.93	820.58	1819.72
−10					95.97	182.96	291.28	495.68	983.05
0				40.99	61.32	114.90	177.73	303.94	558.32
10			13.44	27.17	40.62	74.19	112.20	190.41	332.02
20		5.36	9.91	18.64	27.83	49.29	73.22	122.30	205.91
30	2.80	4.23	7.47	13.20	19.66	33.68	49.32	80.66	132.67
40	2.28	3.41	5.75	9.63	14.28	23.65	34.22	54.64	88.51
50	1.89	2.79	4.52	7.22	10.65	17.05	24.41	37.99	60.93
60	1.60	2.32	3.61	5.55	8.13	12.59	17.86	27.10	43.16
70	1.38	1.95	2.94	4.36	6.34	9.51	13.38	19.79	31.37
80	1.20	1.66	2.43	3.50	5.04	7.34	10.25	14.79	23.35
90	1.05	1.43	2.04	2.86	4.08	5.77	8.00	11.29	17.75
100	0.93	1.25	1.73	2.37	3.35	4.62	6.37	8.79	13.76
110	0.83	1.10	1.49	2.00	2.79	3.76	5.15	6.97	10.86
120	0.75	0.97	1.30	1.71	2.36	3.11	4.23	5.62	8.71
130	0.68	0.87	1.14	1.49	2.02	2.61	3.53	4.60	7.09
140	0.62	0.78	1.01	1.30	1.75	2.22	2.98	3.82	5.85
150	0.57	0.71	0.90	1.16	1.53	1.91	2.54	3.22	4.89
160	0.52	0.64	0.82	1.03	1.35	1.66	2.19	2.75	4.13
170	0.48	0.59	0.74	0.93	1.20	1.45	1.91	2.37	3.52
180	0.44	0.54	0.68	0.85	1.08	1.29	1.69	2.07	3.04
190	0.41	0.50	0.62	0.78	0.97	1.15	1.50	1.82	2.64
200	0.38	0.46	0.58	0.72	0.88	1.04	1.34	1.61	2.31
210	0.36	0.43	0.54	0.67	0.81	0.94	1.21	1.45	2.04
220	0.34	0.40	0.50	0.62	0.74	0.86	1.10	1.31	1.82
230	0.32	0.38	0.47	0.59	0.69	0.79	1.00	1.19	1.63
240	0.30	0.36	0.45	0.55	0.64	0.73	0.92	1.09	1.47
250	0.28	0.34	0.42	0.52	0.59	0.68	0.85	1.00	1.33

Note: Viscosity in centipoise.

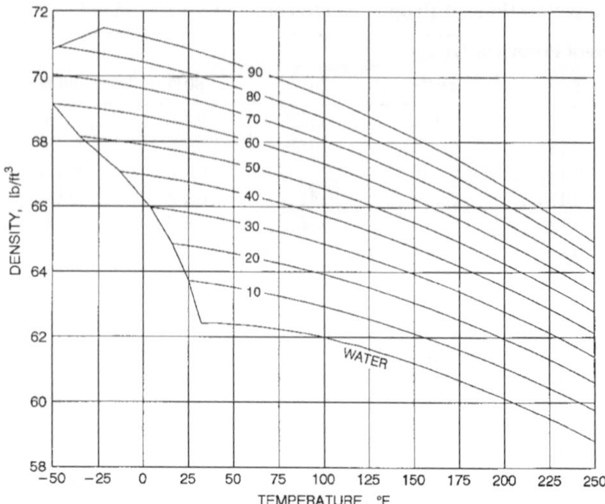

Fig. 9 Density of Aqueous Solutions of Industrially Inhibited Ethylene Glycol (vol. %)

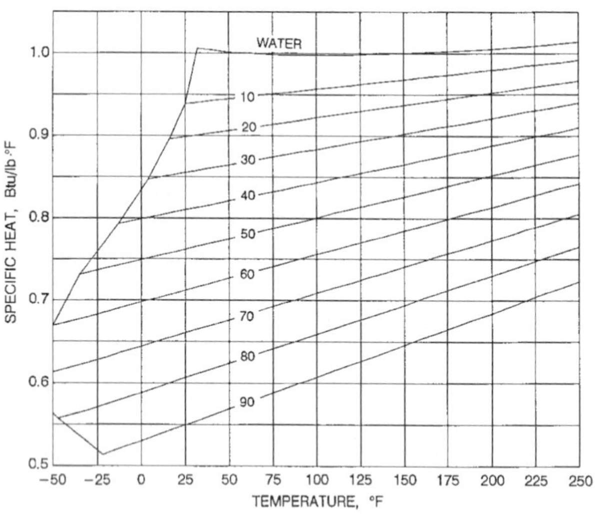

Fig. 10 Specific Heat of Aqueous Solutions of Industrially Inhibited Ethylene Glycol (vol. %)

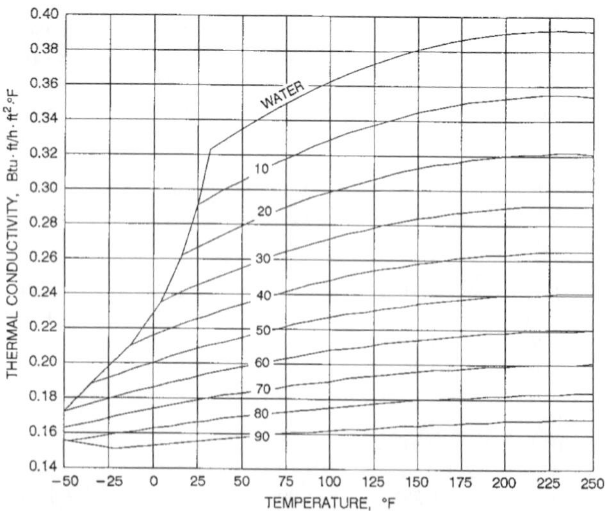

Fig. 11 Thermal Conductivity of Aqueous Solutions of Industrially Inhibited Ethylene Glycol (vol. %)

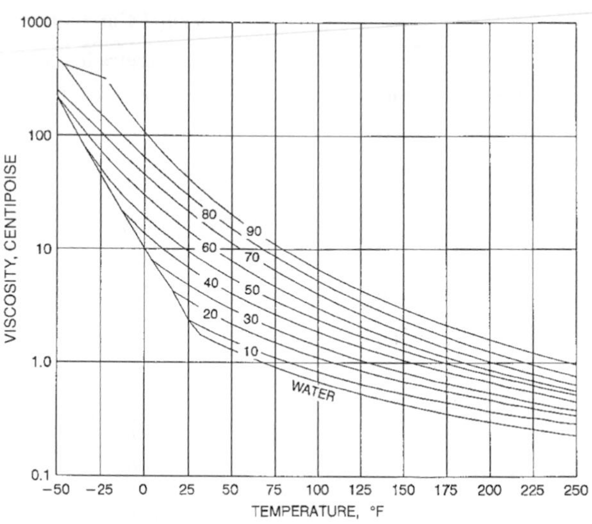

Fig. 12 Viscosity of Aqueous Solutions of Industrially Inhibited Ethylene Glycol (vol. %)

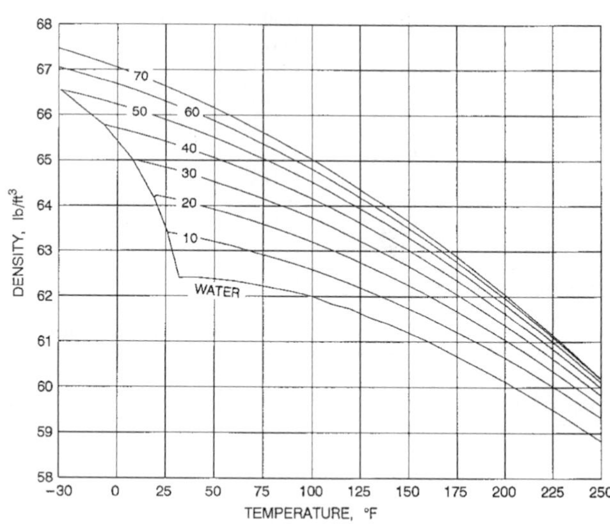

Fig. 13 Density of Aqueous Solutions of Industrially Inhibited Propylene Glycol (vol. %)

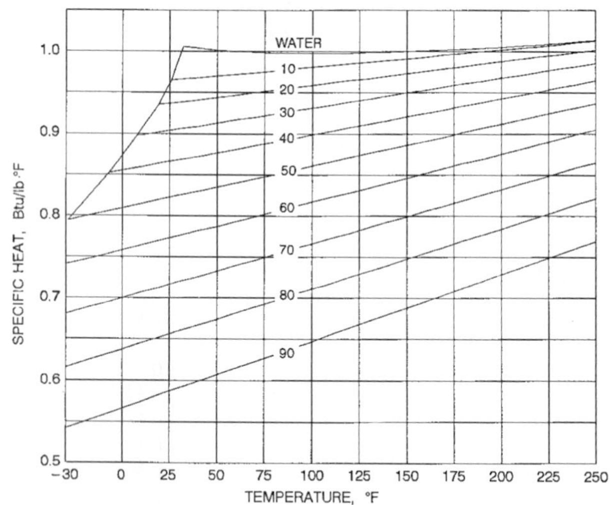

Fig. 14 Specific Heat of Aqueous Solutions of Industrially Inhibited Propylene Glycol (vol. %)

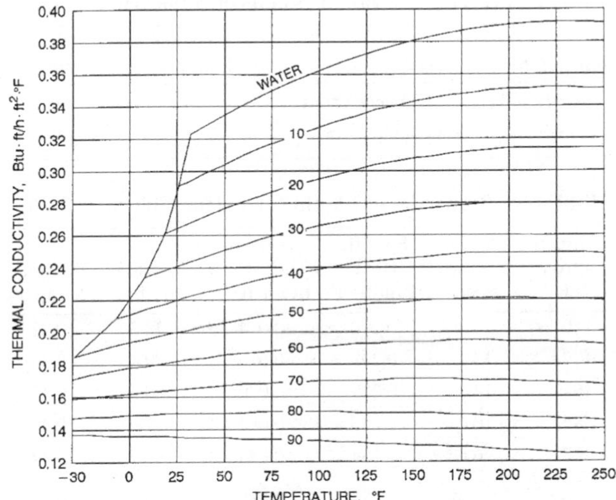

Fig. 15 Thermal Conductivity of Aqueous Solutions of Industrially Inhibited Propylene Glycol (vol. %)

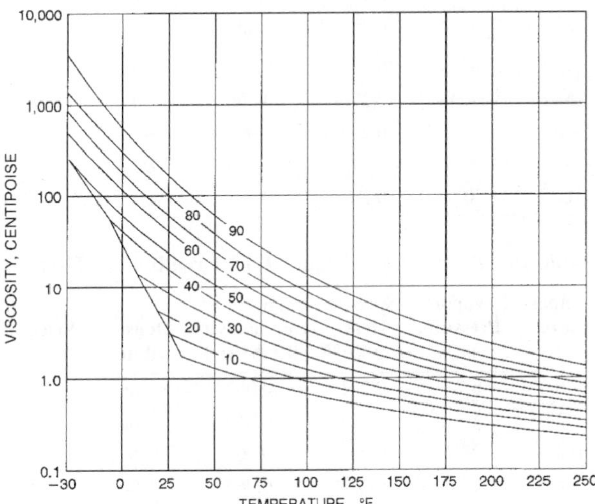

Fig. 16 Viscosity of Aqueous Solutions of Industrially Inhibited Propylene Glycol (vol. %)

Additional physical property data is available from suppliers of industrially inhibited ethylene and propylene glycol.

Corrosion Inhibition

Commercial ethylene glycol or propylene glycol, when pure, is generally less corrosive than water to common metals used in construction. However, aqueous solutions of these glycols assume the corrosivity of the water from which they are prepared and can become increasingly corrosive with use if they are not properly inhibited. Without inhibitors, glycols oxidize into acidic end products. The amount of oxidation is influenced by temperature, degree of aeration, and, to some extent, the particular combination of metal components to which the glycol solution is exposed.

Corrosion inhibition can be described by classifying additives as either (1) corrosion inhibitors, or (2) environmental stabilizers and adjusters. **Corrosion inhibitors** form a surface barrier that protects the metal from attack. These barriers are usually formed by adsorption of the inhibitor by the metal, by reaction of the inhibitor with the metal, or by the incipient reaction product. In most cases, metal surfaces are covered by films of their oxides that inhibitors reinforce.

Environmental stabilizers or adjusters, while not corrosion inhibitors in the strict sense, decrease corrosion by stabilizing or favorably altering the overall environment. An alkaline buffer such as borax is an example of an environmental stabilizer, since its prime purpose is to maintain an alkaline condition (pH above 7). Some chelating agents function as stabilizers by removing from the solution certain deleterious ions that accelerate the corrosion process or mechanism; however, exercise caution in their use because improper combinations of pH and concentration may lead to excessive corrosion.

Certain oxidants, such as sodium chromate, should not be used with glycol solutions, because the glycol can oxidize prematurely. Generally, combinations of the two types of additives, inhibitors, and stabilizers offer the best corrosion resistance in a given system. Commercial inhibited glycols are available from several suppliers.

Service Considerations

Design Considerations. Inhibited glycols can be used at temperatures as high as 350°F. However, maximum-use temperatures vary from fluid to fluid. Therefore, the manufacturer's suggested temperature-use ranges should be followed. In systems with a high degree of aeration, the bulk fluid temperature should not exceed 180°F; however, temperatures up to 350°F are permissible in a pressurized system if air intake is eliminated. Maximum film temperatures should not exceed 50°F above the bulk temperature. Nitrogen blanketing minimizes oxidation when the system operates at elevated temperatures for extended periods.

Minimum operating temperatures are typically −10°F for ethylene glycol solutions and 0°F for propylene glycol solutions. Operation below these temperatures is generally impractical, because the viscosity of the fluids builds dramatically, thus increasing pumping horsepower requirements and reducing heat transfer film coefficients.

Standard materials can be used with most inhibited glycol solutions except galvanized steel, because the galvanizing material, zinc, reacts with a portion of the inhibitor package found in most formulated glycols.

Because the removal of sludge and other contaminants is critical, install suitable filters. If inhibitors are rapidly and completely adsorbed by such contamination, the fluid is ineffective for corrosion inhibition. Consider such adsorption when selecting filters.

Storage and Handling. Inhibited glycol concentrates are stable, relatively noncorrosive materials with high flash points. These fluids can be stored in mild steel, stainless steel, or aluminum vessels. However, aluminum should be used only when the fluid temperature is below 150°F. Corrosion in the vapor space of vessels may be a problem, because the fluid's inhibitor package cannot reach these surfaces to protect them. To prevent this problem, a coating may be used. Suitable coatings include novolac-based vinyl ester resins, high-bake phenolic resins, polypropylene, and polyvinylidene fluoride. To ensure the coating is suitable for a particular application and temperature, the manufacturer should be consulted. Since the chemical properties of an inhibited glycol concentrate differ from those of its dilutions, the effect of the concentrate on different containers should be known when selecting storage.

Choose transfer pumps only after considering temperature-viscosity data. Centrifugal pumps with electric motor drives are often used. Materials compatible with ethylene or propylene glycol should be used for pump packing material. Mechanical seals are also satisfactory. Welded mild steel transfer piping with a minimum diameter is normally used in conjunction with the piping, although flanged and gasketed joints are also satisfactory.

Preparation Before Application. Before an inhibited glycol is charged into a system, remove residual contaminants such as sludge, rust, brine deposits, and oil so the contained inhibitor functions properly. Avoid strong acid cleaners; if they are required, consider inhibited acids. Completely remove the cleaning agent before charging with inhibited glycol.

Dilution Water. Use distilled, deionized, or condensate water, because water from some sources contains elements that reduce the effectiveness of the inhibited formulation. If water of this quality is unavailable, water containing less than 25 ppm chloride, less than 25 ppm sulfate, and less than 100 ppm of total hardness may be used.

Fluid Maintenance. Glycol concentrations can be determined by refractive index, gas chromatography, or Karl Fischer analysis for water (assuming that the concentration of other fluid components, such as inhibitor, is known). Using density to determine glycol concentration is unsatisfactory because (1) density measurements are temperature sensitive, (2) inhibitor concentrations can change density, (3) values for propylene glycol are close to those of water, and (4) propylene glycol values are maximum at 70 to 75% concentration.

A rigorous inhibitor monitoring and maintenance schedule is essential to maintain a glycol solution in relatively noncorrosive condition for a long period. However, a specific schedule is not always easy to establish, because inhibitor depletion rate depends on the particular conditions of use. Analysis of samples immediately after installation, after two to three months, and after six months should establish the pattern for the schedule. Visually inspecting the solution and filter residue can detect active corrosion.

Many manufacturers of inhibited glycol-based heat transfer fluids provide analytical service to ensure that their heat transfer fluid remains in good condition. This analysis may include some or all of the following: percent of ethylene and/or propylene glycol, freezing point, pH, reserve alkalinity, corrosion inhibitor evaluation, contaminants, total hardness, metal content, and degradation products. If maintenance on the fluid is required, recommendations may be given along with the analysis results.

Properly inhibited and maintained glycol solutions provide better corrosion protection than brine solutions in most systems. A long, though not indefinite, service life can be expected. Avoid indiscriminate mixing of inhibited formulations. Exercise caution in replacing brine systems with inhibited glycols because brine components are incompatible with glycol formulations.

HALOCARBONS

Many common refrigerants are used as secondary coolants as well as primary refrigerating media. Their favorable properties as heat transfer fluids include low freezing points, low viscosities, nonflammability, and good stability. Chapters 19 and 20 present physical and thermodynamic properties for common refrigerants. Table 14 lists two halocarbon compounds that are commonly used as secondary coolants. Table 15 gives vapor pressure, specific heat, thermal conductivity, density, and viscosity values for methylene chloride (R-30). Table 16 gives the same properties for trichloroethylene (R-1120).

Table 9 in Chapter 19 summarizes comparative safety characteristics for halocarbons. *Threshold Limit Values and Biological Exposure Indices* (ACGIH) has more information on halocarbon toxicity.

Construction materials and stability factors in halocarbon use are discussed in Chapter 19 of this volume and Chapter 5 of the 1998 *ASHRAE Handbook—Refrigeration*. Note particularly that methylene chloride and trichloroethylene should not be used in contact with aluminum components.

NONHALOCARBON, NONAQUEOUS FLUIDS

In addition to the aforementioned fluids, numerous other secondary refrigerants are available. These fluids have been used primarily by the chemical processing and pharmaceutical industries. They

Table 14 Freezing and Boiling Points of Halocarbon Coolants

Refrigerant	Name	Freezing Point, °F	Boiling Point, °F
30	Methylene chloride	−142	103.6
1120	Trichloroethylene	−123	189

Table 15 Properties of Liquid Methylene Chloride (R-30)

Temperature, °F	Vapor Pressure, psia	Specific Heat, Btu/lb·°F	Thermal Conductivity, Btu/h·ft·°F	Density, lb/ft³	Viscosity, Centipoise
140	25.4	0.296	0.074	78.3	0.32
122	19.9	0.293	0.076	79.4	0.34
104	14.5	0.289	0.079	80.5	0.37
86	10.2	0.286	0.081	81.6	0.40
68	6.82	0.284	0.083	82.7	0.44
50	4.39	0.282	0.085	83.8	0.48
32	2.73	0.280	0.087	84.9	0.53
14	1.64	0.278	0.089	86.0	0.59
−4	0.97	0.277	0.091	87.1	0.66
−22	0.55	0.275	0.093	88.2	0.76
−40	0.32	0.274	0.094	89.3	0.88
−58	0.18	0.273	0.096	90.4	1.05
−76	0.10	0.273	0.098	91.5	1.29
−94	0.06	0.273	0.099	92.6	1.68
−112	0.03	0.272	0.101	93.7	2.50

Table 16 Properties of Liquid Trichloroethylene (R-1120)

Temperature, °F	Vapor Pressure, psia	Specific Heat, Btu/lb·°F	Thermal Conductivity, Btu/h·ft·°F	Density, lb/ft³	Viscosity, Centipoise
140	5.73	0.231	0.062	86.8	0.40
122	4.21	0.228	0.063	88.0	0.44
104	2.87	0.225	0.065	89.0	0.48
86	1.86	0.223	0.066	90.1	0.52
68	1.13	0.220	0.068	91.3	0.57
50	0.667	0.218	0.069	92.4	0.63
32	0.370	0.216	0.071	93.5	0.70
14	0.199	0.213	0.073	94.6	0.78
−4	0.102	0.211	0.074	95.6	0.87
−22	0.052	0.209	0.076	96.6	0.99
−40	0.024	0.207	0.077	97.7	1.14
−58	0.011	0.206	0.079	98.7	1.33
−76	0.005	0.204	0.080	99.7	1.60
−94	0.002	0.202	0.082	100.6	1.93
−112	0.001	0.201	0.084	101.6	2.45

Table 17 Summary of Physical Properties of Polydimethylsiloxane Mixture and d-Limonene

	Polydimethylsiloxane Mixture	d-Limonene
Flash point, °F, closed cup	116	115
Boiling point, °F	347	310
Freezing point, °F	−168	−142
Operational temperature range, °F	−100 to 500	None published

Table 18 Properties of a Polydimethylsiloxane Heat Transfer Fluid

Temperature, °F	Vapor Pressure, psia	Viscosity, Centipoise	Density, lb/ft^3	Heat Capacity, Btu/lb·°F	Thermal Conductivity, Btu/h·ft·°F
−100	0.00	12.5	57.8	0.337	0.0748
−90	0.00	10.5	57.5	0.340	0.0742
−80	0.00	8.82	57.2	0.344	0.0736
−70	0.00	7.50	56.9	0.347	0.0730
−60	0.00	6.43	56.6	0.350	0.0724
−50	0.00	5.55	56.3	0.354	0.0717
−40	0.00	4.83	56.0	0.357	0.0711
−30	0.00	4.22	55.7	0.361	0.0705
−20	0.00	3.72	55.4	0.364	0.0699
−10	0.00	3.29	55.1	0.367	0.0692
0	0.00	2.93	54.8	0.371	0.0686
10	0.00	2.62	54.5	0.374	0.0679
20	0.00	2.36	54.2	0.378	0.0673
30	0.00	2.13	53.9	0.381	0.0666
40	0.01	1.93	53.6	0.384	0.0659
50	0.01	1.76	53.3	0.388	0.0652
60	0.02	1.60	53.0	0.391	0.0646
70	0.03	1.47	52.7	0.395	0.0639
80	0.04	1.35	52.4	0.398	0.0632
90	0.05	1.25	52.1	0.402	0.0625
100	0.08	1.15	51.8	0.405	0.0618
110	0.11	1.07	51.5	0.408	0.0610
120	0.15	0.993	51.1	0.412	0.0603
130	0.20	0.926	50.8	0.415	0.0596
140	0.27	0.865	50.5	0.419	0.0589
150	0.35	0.810	50.2	0.422	0.0581
160	0.46	0.760	49.8	0.425	0.0574
170	0.60	0.715	49.5	0.429	0.0567
180	0.76	0.673	49.2	0.432	0.0559
190	0.96	0.635	48.8	0.436	0.0551
200	1.20	0.601	48.5	0.439	0.0544
210	1.49	0.569	48.1	0.442	0.0536
220	1.84	0.540	47.8	0.446	0.0528
230	2.24	0.513	47.4	0.449	0.0521
240	2.72	0.488	47.0	0.453	0.0513
250	3.27	0.465	46.7	0.456	0.0505
260	3.91	0.443	46.3	0.459	0.0497
270	4.65	0.424	45.9	0.463	0.0489
280	5.50	0.405	45.5	0.466	0.0481
290	6.46	0.388	45.1	0.470	0.0473
300	7.55	0.372	44.7	0.473	0.0465
310	8.78	0.357	44.3	0.476	0.0457
320	10.16	0.343	43.9	0.480	0.0449
330	11.71	0.330	43.5	0.483	0.0441
340	13.43	0.317	43.1	0.487	0.0432
350	15.33	0.306	42.6	0.490	0.0424
360	17.45	0.295	42.2	0.494	0.0416
370	19.77	0.285	41.7	0.497	0.0407
380	22.32	0.275	41.3	0.500	0.0399
390	25.12	0.266	40.8	0.504	0.0390
400	28.17	0.257	40.4	0.507	0.0382
410	31.49	0.249	39.9	0.511	0.0373
420	35.10	0.242	39.4	0.514	0.0365
430	39.00	0.234	38.9	0.517	0.0356
440	43.21	0.227	38.4	0.521	0.0348
450	47.75	0.221	37.9	0.524	0.0339
460	52.63	0.214	37.4	0.528	0.0330
470	57.86	0.209	36.8	0.531	0.0321
480	63.46	0.203	36.3	0.534	0.0313
490	69.44	0.197	35.8	0.538	0.0304
500	75.81	0.192	35.2	0.541	0.0295

Table 19 Physical Properties of d-Limonene

Temperature, °F	Specific Heat, Btu/lb·°F	Viscosity, Centipoise	Density, lb/ft^3	Thermal Conductivity, Btu/h·ft·°F
−100	0.3	3.8	57.1	0.0794
−50	0.34	2.8	55.8	0.0764
0	0.37	2.1	54.5	0.0734
50	0.41	1.6	53.2	0.0704
100	0.44	1.2	51.8	0.0674
150	0.48	0.9	50.4	0.0644
200	0.51	0.7	49	0.0614
250	0.54	0.6	47.6	0.0584
300	0.58	0.4	46	0.0554

Note: Properties are estimated or based on incomplete data.

have been used rarely in the HVAC and allied industries due to their cost and relative novelty. Before choosing these types of fluids, consider electrical classifications, disposal, potential worker exposure, process containment, and other relevant issues.

Tables 17 through 19 contain physical property information on a mixture of dimethylsiloxane polymers of various relative molecular masses (Dow Corning 1989) and d-limonene. Information on d-limonene is limited; it is based on measurements made over small data temperature ranges or simply on standard physical property estimation techniques. The compound is an optically active terpene (molecular formula $C_{10}H_{16}$) derived as an extract from orange and lemon oils. The "d" indicates that the material is dextrorotatory, which is a physical property of the material that does not affect the transport properties of the material significantly.

The mixture of dimethylsiloxane polymers can be used with most standard construction materials; d-limonene, however, can be quite corrosive, easily autooxidizing at ambient temperatures. This fact should be understood and considered before using d-limonene in a system.

REFERENCES

ACGIH. 1998. Threshold limit values and biological exposure indices. Published annually by the American Conference of Governmental Industrial Hygienists, Cincinnati, OH.

Carrier Air Conditioning Company. 1959. Basic data, Section 17M. Syracuse, NY.

Dow Corning USA. 1989. Syltherm heat transfer liquids. Midland, MI.

BIBLIOGRAPHY

Born, D.W. 1989. Inhibited glycols for corrosion and freeze protection in water-based heating and cooling systems. Midland, MI.

CCI. Calcium chloride for refrigeration brine. Manual RM-1. Calcium Chloride Institute.

Dow Chemical USA. 1994. Engineering manual for DOWFROST and DOWFROST HD heat transfer fluids. Midland, MI.

Dow Chemical USA. 1996. Engineering manual for Dowtherm SR-1 and Dowtherm 4000 heat transfer fluids. Midland, MI.

Fontana, M.G. 1986. *Corrosion engineering*. McGraw-Hill, New York.

NACE. 1973. *Corrosion inhibitors.* National Association of Corrosion Engineers, Houston, TX.

NACE. 1991. *NACE corrosion engineer's reference book.*

NACE. 2000. *Corrosion: Understanding the basics.*

Union Carbide Corporation. 1994. Ucartherm heat transfer fluids. South Charleston, WV.

CHAPTER 22

SORBENTS AND DESICCANTS

SORPTION refers to the binding of one substance to another. **Sorbents** are materials that have an ability to attract and hold other gases or liquids. They can be used to attract gases or liquids other than water vapor, a characteristic that makes them very useful in chemical separation processes. **Desiccants** are a subset of sorbents; they have a particular affinity for water.

Virtually all materials are desiccants; that is, they attract and hold water vapor. Wood, natural fibers, clays, and many synthetic materials attract and release moisture as commercial desiccants do, but they lack the holding capacity. For example, woolen carpet fibers attract up to 23% of their dry weight in water vapor, and nylon can take up almost 6% of its weight in water. In contrast, a commercial desiccant takes up between 10 and 1100% of its dry weight in water vapor, depending on its type and on the moisture available in the environment. Furthermore, commercial desiccants continue to attract moisture even when the surrounding air is quite dry, a characteristic that other materials do not share.

All desiccants behave in a similar way—they attract moisture until they reach equilibrium with the surrounding air. Moisture is usually removed from the desiccant by heating it to temperatures between 120 and 500°F and exposing it to a scavenger airstream. After the desiccant dries, it must be cooled so that it can attract moisture once again. Sorption always generates sensible heat equal to the latent heat of the water vapor taken up by the desiccant plus an additional heat of sorption that varies between 5 and 25% of the latent heat of the water vapor. This heat is transferred to the desiccant and to the surrounding air.

The process of attracting and holding moisture is described as either adsorption or absorption, depending on whether the desiccant undergoes a chemical change as it takes on moisture. **Adsorption** does not change the desiccant, except by the addition of the weight of water vapor; it is similar in some ways to a sponge soaking up water. **Absorption**, on the other hand, changes the desiccant. An example of an absorbent is table salt, which changes from a solid to a liquid as it absorbs moisture.

DESICCANT APPLICATIONS

Desiccants can dry either liquids or gases, including ambient air, and are used in many air-conditioning applications, particularly when

- The latent load is large in comparison to the sensible load.
- The cost of energy to regenerate the desiccant is low compared to the cost of energy to dehumidify the air by chilling it below its dew point.
- The moisture control level for the space would require chilling the air to subfreezing dew points if compression refrigeration alone were used to dehumidify the air.
- The temperature control level for the space or process requires continuous delivery of air at subfreezing temperatures.

In any of these situations, the cost of running a vapor compression cooling system can be very high. A desiccant process may offer considerable advantages in energy, initial cost of equipment, and maintenance.

Because desiccants are able to attract and hold more than simply water vapor, they can remove contaminants from airstreams to improve indoor air quality. Desiccants have been used to remove organic vapors and, in special circumstances, to control microbiological contaminants (Batelle 1971, Buffalo Testing Laboratory 1974). Hines et al. (1991) have also confirmed the usefulness of desiccants in removing vapors that can degrade indoor air quality. Desiccant materials are capable of adsorbing hydrocarbon vapors while they are collecting moisture from air. These desiccant cosorption phenomena show promise of improving indoor air quality in typical building HVAC systems.

Desiccants are also used in drying compressed air to low dew points. In this application, moisture can be removed from the desiccant without heat. Desorption is accomplished using differences in vapor pressures compared to the total pressures of the compressed and ambient pressure airstreams.

Finally, desiccants are used to dry the refrigerant circulating in air-conditioning and refrigeration systems. This reduces corrosion in refrigerant piping and prevents valves and capillaries from becoming clogged with ice crystals. In this application, the desiccant is not regenerated; it is discarded when it has adsorbed its limit of water vapor.

This chapter discusses the water sorption characteristics of desiccant materials and explains some of the implications of those characteristics in ambient pressure air-conditioning applications. Information on other applications for desiccants can be found in Chapters 14 and 29 of this volume, Chapters 6, 25, 34, 41, and 46 of the 1998 *ASHRAE Handbook—Refrigeration*, Chapters 1, 2, 5, 8, 15, 17, 20, 27, and 44 of the 1999 *ASHRAE Handbook—Applications,* and Chapters 22 and 44 of the 2000 *ASHRAE Handbook—Systems and Equipment.*

DESICCANT CYCLE

Practically speaking, all desiccants function by the same mechanism—transferring moisture because of a difference between the water vapor pressure at their surface and that of the surrounding air. When the vapor pressure at the desiccant surface is lower than that of the air, the desiccant attracts moisture. When the surface vapor pressure is higher than that of the surrounding air, the desiccant releases moisture.

Figure 1 shows the relationship between the moisture content of the desiccant and its surface vapor pressure. As the moisture content of the desiccant rises, so does the water vapor pressure at its surface. At some point, the vapor pressure at the desiccant surface is the

The preparation of this chapter is assigned to TC 3.5, Desiccant and Sorption Technology.

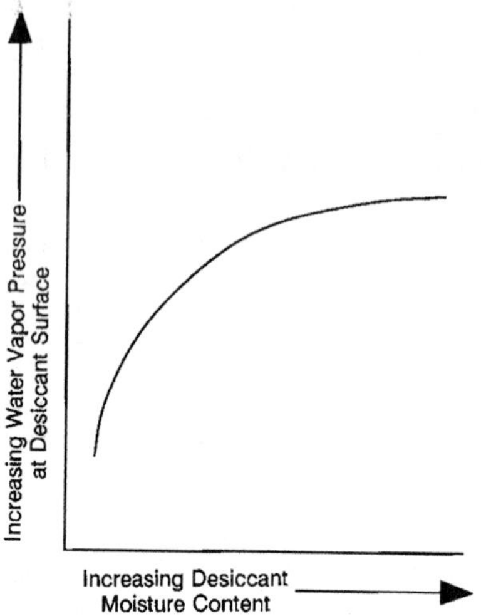

Fig. 1 Desiccant Water Vapor Pressure as Function of Moisture Content
(Harriman 1990)

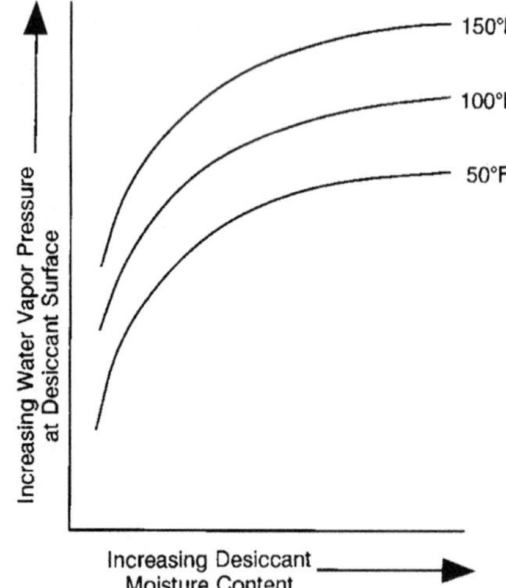

Fig. 2 Desiccant Water Vapor Pressure as Function of Desiccant Moisture Content and Temperature
(Harriman 1990)

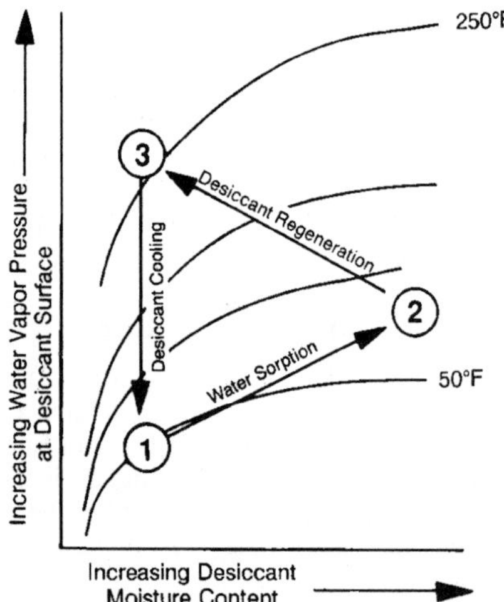

Fig. 3 Desiccant Cycle
(Harriman 1990)

same as that of the air—the two are in equilibrium. Then, moisture cannot move in either direction until some external force changes the vapor pressure at the desiccant or in the air.

Figure 2 shows the effect of temperature on the vapor pressure at the desiccant surface. Both higher temperature and increased moisture content increase the vapor pressure at the surface. When the surface vapor pressure exceeds that of the surrounding air, moisture leaves the desiccant—a process called **reactivation** or **regeneration**. After the desiccant is dried (reactivated) by the heat, its vapor pressure remains high, so that it has very little ability to absorb moisture. **Cooling** the desiccant reduces its surface vapor pressure so that it can absorb moisture once again. The complete cycle is illustrated in Figure 3.

The economics of desiccant operation depend on the energy cost of moving a given material through this cycle. The dehumidification of air (loading the desiccant with water vapor) generally proceeds without energy input other than fan and pump costs. The major portion of energy is invested in regenerating the desiccant (moving from point 2 to point 3) and cooling the desiccant (point 3 to point 1).

Regeneration energy is equal to the sum of three variables:

1. The heat necessary to raise the desiccant to a temperature high enough to make its surface vapor pressure higher than that of the surrounding air
2. The heat necessary to vaporize the moisture it contains (about 1060 Btu/lb)
3. The small amount of heat from desorption of the water from the desiccant

The **cooling energy** is proportional to (1) the mass of the desiccant and (2) the difference between its temperature after regeneration and the lower temperature that allows the desiccant to remove water from the airstream once again.

The cycle is similar when desiccants are regenerated using pressure differences in a compressed air application. The desiccant is saturated in a high-pressure chamber (i.e., that of the compressed air). Then valves open, isolating the compressed air from the material, and the desiccant is exposed to air at ambient pressure. The vapor pressure of the saturated desiccant is much higher than ambient air at normal pressures; thus the moisture leaves the desiccant for the surrounding air. An alternate desorption strategy uses a small portion of the dried air, returning it to the moist desiccant bed to reabsorb the moisture, then venting the air to the atmosphere at ambient pressures.

Table 1 shows the range of vapor pressures over which the desiccant must operate in space-conditioning applications. It converts the relative humidity at 70°F to dew point and the corresponding vapor pressure. The greater the difference between the air and desiccant surface vapor pressures, the greater the ability of the material to absorb moisture from the air at that moisture content.

Table 1 Vapor Pressures of Different Relative Humidities at 70°F

Relative Humidity at 70°F, %	Dew Point, °F	Vapor Pressure, in. Hg
10	12	0.07
20	28	0.15
30	37	0.22
40	45	0.30
50	51	0.37
60	55	0.44
70	60	0.52
80	64	0.59
90	67	0.67
100	70	0.74

The ideal desiccant for a particular application depends on the range of water vapor pressures likely to occur in the air, the temperature of the regeneration heat source, and the moisture sorption and desorption characteristics of the desiccant within those constraints. In commercial practice, however, most desiccants can be made to perform well in a wide variety of operating situations through careful engineering of the mechanical aspects of the dehumidification system. Some of these hardware issues are discussed in Chapter 22 of the 2000 *ASHRAE Handbook—Systems and Equipment*.

TYPES OF DESICCANTS

Desiccants can be liquids or solids and can hold moisture through absorption or adsorption, as described earlier. Most absorbents are liquids, and most adsorbents are solids.

Liquid Absorbents

Liquid absorption dehumidification can best be illustrated by comparison to the operation of an air washer. When air passes through an air washer, its dew point approaches the temperature of the water supplied to the machine. Air that is more humid is dehumidified and air that is less humid is humidified. In a similar manner, a liquid absorption dehumidifier brings air into contact with a liquid desiccant solution. The liquid has a vapor pressure lower than water at the same temperature, and the air passing over the solution approaches this reduced vapor pressure; it is dehumidified.

The vapor pressure of a liquid absorption solution is directly proportional to its temperature and inversely proportional to its concentration. Figure 4 illustrates the effect of increasing desiccant concentration on the water vapor pressure at its surface. The figure shows the vapor pressure of various solutions of water and triethylene glycol, a common commercial desiccant. As the glycol content of the mixture increases, the vapor pressure of the mixture decreases. This pressure difference allows the glycol solution to absorb moisture from the air whenever the vapor pressure of the air is greater than that of the solution.

From a slightly different perspective, the vapor pressure of a given concentration of absorbent solution approximates the vapor pressure values of a fixed relative humidity line on a psychrometric chart. Higher solution concentrations give lower equilibrium relative humidities, which allow the absorbent to dry air to lower levels.

Figure 5 illustrates the effect of temperature on the vapor pressure of various solutions of water and lithium chloride (LiCl), another liquid desiccant in common use. A solution that is 25% lithium chloride has a vapor pressure of 0.37 in. Hg at a temperature of 70°F. If the same 25% solution is heated to 100°F, its vapor pressure more than doubles to 0.99 in. Hg. Expressed another way, the 70°F, 25% solution is in equilibrium with air at a 51°F dew point. The same 25% solution at 100°F is at equilibrium with an airstream at a 79°F dew point. The warmer the desiccant, the less moisture it can attract from the air.

In standard practice, the behavior of a liquid desiccant is controlled by adjusting its temperature, its concentration, or both. Desiccant temperature is controlled by simple heaters and coolers. Concentration is controlled by heating the desiccant to drive moisture out into a waste airstream or directly to the ambient.

Commercially available liquid desiccants have an especially high water-holding capacity. Each molecule of LiCl, for example, can hold two water molecules, even in the dry state. Above two water molecules per molecule of LiCl, the desiccant becomes a liquid and continues to absorb water. If the solution is in equilibrium with air at 90% rh, approximately 26 water molecules are attached to each molecule of LiCl. This represents a water absorption of more than 1000% on a dry weight basis.

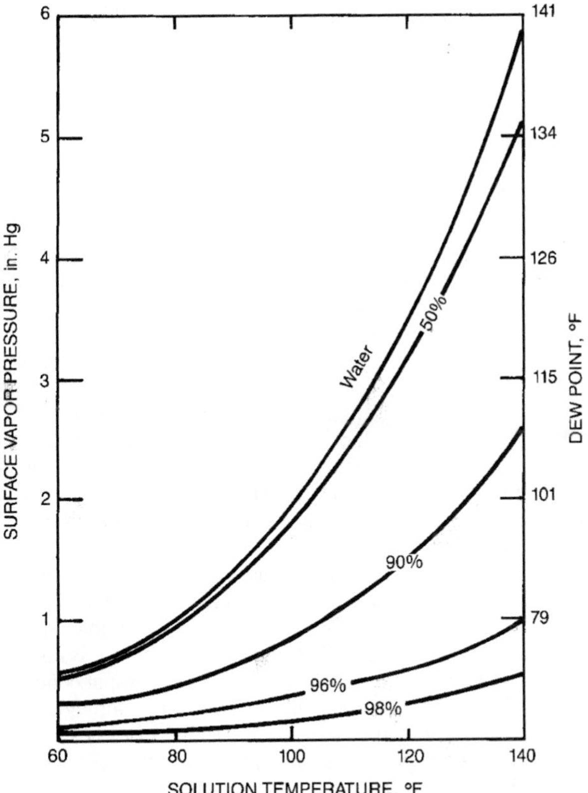

Fig. 4 Surface Vapor Pressure of Water-Triethylene Glycol Solutions
(Dow 1981)

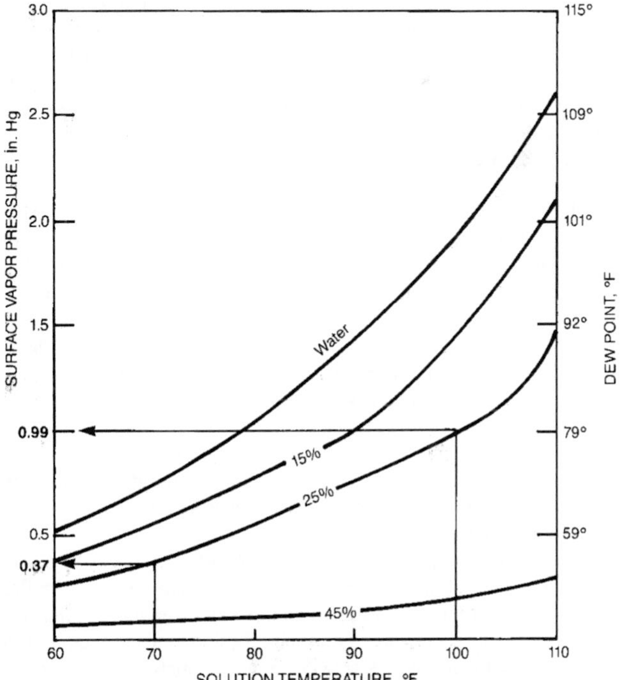

Fig. 5 Surface Vapor Pressure of Water-Lithium Chloride Solutions
(Foote Mineral 1988)

As a practical matter, however, the absorption process is limited by the exposed surface area of the desiccant and by the contact time allowed for the reaction. More surface area and more contact time allow the desiccant to approach its theoretical capacity. Commercial desiccant systems stretch these limits by flowing the liquid desiccant onto an extended surface much like in a cooling tower.

Solid Adsorbents

Adsorbents are solid materials with a tremendous internal surface area per unit of mass; a single gram can have more than 50,000 ft^2 of surface area. Structurally, adsorbents resemble a rigid sponge, and the surface of the sponge in turn resembles the ocean coastline of a fjord. This analogy indicates the scale of the different surfaces in an adsorbent. The fjords can be compared to the **capillaries** in the adsorbent. The spaces between the grains of sand on the fjord beaches can be compared to the spaces between the individual molecules of the adsorbent, all of which have the capacity to hold water molecules. The bulk of the adsorbed water is contained by condensation into the capillaries, and the majority of the surface area that attracts individual water molecules is in the crystalline structure of the material itself.

Adsorbents attract moisture because of the electrical field at the desiccant surface. The field is not uniform in either force or charge, so specific sites on the desiccant surface attract water molecules that have a net opposite charge. When the complete surface is covered, the adsorbent can hold still more moisture because vapor condenses into the first water layer and fills the capillaries throughout the material. As with liquid absorbents, the ability of an adsorbent to attract moisture depends on the difference in vapor pressure between its surface and the air.

The capacity of solid adsorbents is generally less than the capacity of liquid absorbents. For example, a typical molecular sieve adsorbent can hold 17% of its dry weight in water when the air is at 70°F and 20% rh. In contrast, LiCl can hold 130% of its mass at the same temperature and relative humidity. But solid adsorbents have several other favorable characteristics.

For example, molecular sieves continue to adsorb moisture even when they are quite hot, allowing dehumidification of very warm airstreams. Also, several solid adsorbents can be manufactured to precise tolerances, with pore diameters that can be closely controlled. This means they can be tailored to adsorb molecules of a specific diameter. Water, for example, has an effective molecular diameter of 3.2 nm. A molecular sieve adsorbent with an average pore diameter of 4.0 nm adsorbs water but has almost no capacity for larger molecules, such as organic solvents. This selective adsorption characteristic is useful in many applications. For example, several desiccants with different pore sizes can be combined in series to remove first water and then other specific contaminants from an airstream.

Adsorption Behavior. The adsorption behavior of solid adsorbents depends on (1) total surface area, (2) total volume of capillaries, and (3) range of capillary diameters. A large surface area gives the adsorbent a larger capacity at low relative humidities. Large capillaries provide a high capacity for condensed water, which gives the adsorbent a higher capacity at high relative humidities. A narrow range of capillary diameters makes an adsorbent more selective in the vapor molecules it can hold.

In designing a desiccant, some trade-offs are necessary. For example, materials with large capillaries necessarily have a smaller surface area per unit of volume than those with smaller capillaries. As a result, adsorbents are sometimes combined to provide a high adsorption capacity across a wide range of operating conditions. Figure 6 illustrates this point using three noncommercial silica gel adsorbents prepared for use in laboratory research. Each has a different internal structure, but since they are all silicas, they have similar surface adsorption characteristics. Gel 1 has large capillaries, making its total volume large but its total surface area small. It has a large adsorption capacity at high relative humidities but adsorbs a small amount at low relative humidities.

In contrast, Gel 8 has a capillary volume one-seventh the size of Gel 1, but a total surface area almost twice as large. This gives it a higher capacity at low relative humidities but a lower capacity to hold the moisture that condenses at high relative humidities.

Silica gels and most other adsorbents can be manufactured to provide optimum performance in a specific application, balancing capacity against strength, weight, and other favorable characteristics (Bry-Air 1986).

Types of Solid Adsorbents. General classes of solid adsorbents include

- Silica gels
- Zeolites
- Synthetic zeolites (molecular sieves)
- Activated aluminas
- Carbons
- Synthetic polymers

Silica gels are amorphous solid structures formed by condensing soluble silicates from solutions of water or other solvents. They have the advantages of relatively low cost and relative simplicity of structural customizing. They are available as large as spherical beads about 3/16 in. in diameter or as small as grains of a fine powder.

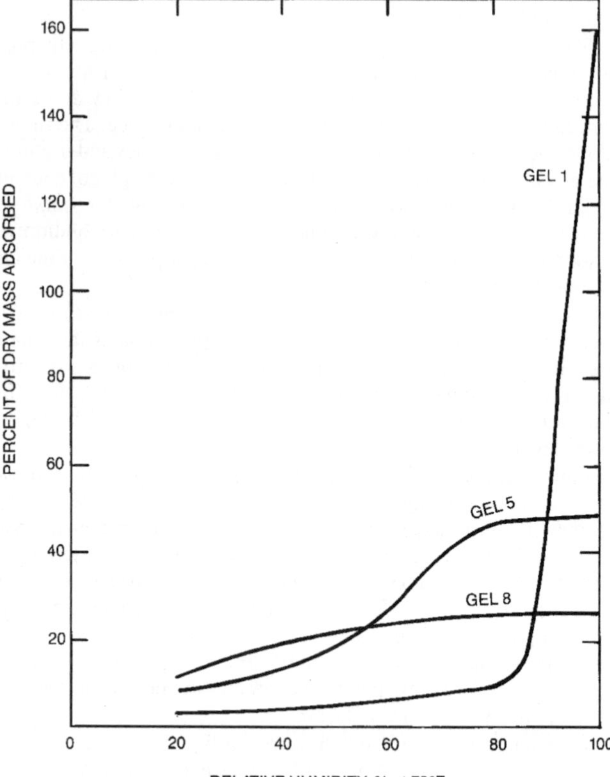

Gel Number	Total Surface Area, m^2/g	Average Capillary Diameter, nm	Total Volume of Capillaries, mm^3/g
1	315	21	1700
5	575	3.8	490
8	540	2.2	250

Fig. 6 Adsorption and Structural Characteristics of Some Experimental Silica Gels
(Oscic and Cooper 1982)

Zeolites are aluminosilicate minerals. They occur in nature and are mined rather than synthesized. Zeolites have a very open crystalline lattice that allows molecules like water vapor to be held inside the crystal itself like an object in a cage. Particular atoms of an aluminosilicate determine the size of the openings between the "bars" of the cage, which in turn governs the maximum size of the molecule that can be adsorbed into the structure.

Synthetic zeolites, also called **molecular sieves**, are crystalline aluminosilicates manufactured in a thermal process. Controlling the temperature of the process and the composition of the ingredient materials allows close control of the structure and surface characteristics of the adsorbent. At a somewhat higher cost, this provides a much more uniform product than naturally occurring zeolites.

Activated aluminas are oxides and hydrides of aluminum that are manufactured in thermal processes. Their structural characteristics can be controlled by the gases used to produce them and by the temperature and duration of the thermal process.

Carbons are most frequently used for adsorption of gases other than water vapor because they have a greater affinity for the nonpolar molecules typical of organic solvents. Like other adsorbents, carbons have a large internal surface and especially large capillaries. This capillary volume gives them a high capacity to adsorb water vapor at relative humidities of 45 to 100%.

Synthetic polymers have potential for use as desiccants as well. Long molecules, like those found in polystyrenesulfonic acid sodium salt (PSSASS), are twisted together like the strands of string. Each of the many sodium ions in the long PSSASS molecules has the potential to bind several water molecules, and the spaces between the packed strings can also contain condensed water, giving the polymer a capacity exceeding that of many other solid adsorbents.

DESICCANT ISOTHERMS

Figure 7 shows a rough comparison of the sorption characteristics of different desiccants. Large variations from these isotherms occur because manufacturers use different methods to optimize the materials for different applications. The suitability of a given desiccant to a particular application is generally governed as much by the engineering of the mechanical system that presents the material to the airstreams as by the characteristics of the material itself.

Several sources give details of desiccant equipment design and information about desiccant isotherm characteristics. Brunauer (1945) considers five basic isotherm shape types. Each isotherm shape is determined by the dominant sorption mechanisms of the desiccant, which give rise to its specific capacity characteristics at different vapor pressures. Isotherm shape can be important in designing the optimum desiccant for applications where a narrow range of operating conditions can be expected. Collier (1986, 1988) illustrates how an optimum isotherm shape can be used to ensure a maximum coefficient of performance in one particular air-conditioning desiccant application.

DESICCANT LIFE

The useful life of desiccant materials depends largely on the quantity and type of contamination in the airstreams they dry. In commercial equipment, desiccants last from 10,000 to 100,000 h and longer before they need replacement. Normally, two mechanisms cause the loss of desiccant capacity: (1) change in desiccant sorption characteristics through **chemical reactions** with contaminants and (2) loss of effective surface area through clogging or **hydrothermal degradation**.

Liquid absorbents are more susceptible to chemical reaction with airstream contaminants other than water vapor than are solid adsorbents. For example, certain sulfur compounds can react with LiCl to form lithium sulfate, which is not a desiccant. If the concentration of sulfur compounds in the airstream were below 10 ppm and the desiccant were in use 24 h a day, the capacity reduction would be approximately 10 to 20% after three years of operation. If the concentration were 30 ppm, this reduction would occur after one year.

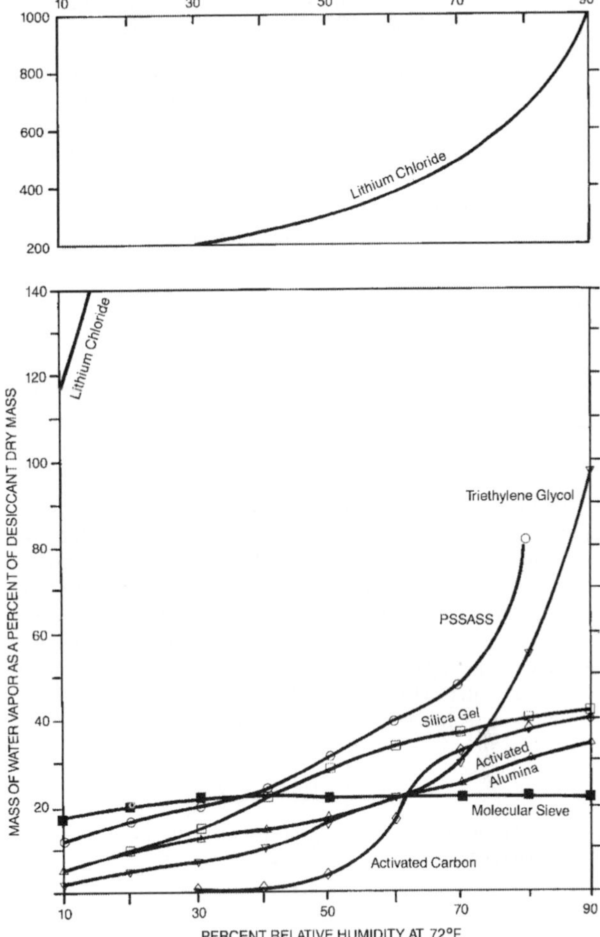

The sources for isotherms presented in the figure include
PSSASS: Czanderna (1988)
Lithium chloride: Munters Corporation—Cargocaire Division and Kathabar, Inc.
Triethylene glycol: Dow Chemical Corporation
Silica gel: Davison Chemical Division of W.R. Grace Co.
Activated carbon: Calgon Corporation
Activated alumina: LaRoche Industries Inc.
Molecular sieve: Davison Chemical Division of W.R. Grace Co.

Fig. 7 Sorption Isotherms of Various Desiccants

Solid adsorbents tend to be less chemically reactive and more sensitive to clogging, a function of the type and quantity of particulate material in the airstream. In some situations, certain adsorbents are sensitive to hydrothermal stress due to the thermal expansion and contraction of the desiccant material on rapid changes in desiccant moisture content. For example, silica gel that must move between an airstream above 95% rh at low temperatures and a reactivating airstream at high temperatures six times per hour, 24 h a day can partly fracture; this may mean a 10% reduction in capacity over the course of one year. For applications where such capacity reduction is undesirable, thermally stabilized desiccants are used in place of more sensitive materials.

In air-conditioning applications, desiccant equipment is designed to minimize the need for desiccant replacement in much the same way that vapor compression cooling systems are designed to avoid the need for compressor replacement. Unlike filters, desiccants are seldom intended to be frequently replaced during normal service in an air-drying application.

COSORPTION OF WATER VAPOR AND INDOOR AIR CONTAMINANTS

Hines et al. (1991) have confirmed that many desiccant materials can collect common indoor pollutants at the same time they collect water vapor from ambient air. This characteristic promises to become useful in future air-conditioning systems where the quality of indoor air is especially important.

The behavior of different desiccant and vapor mixtures is complex, but in general, pollutant sorption reactions can be classified into five categories:

- Humidity-neutral sorption
- Humidity-reduced sorption
- Humidity-enhanced sorption
- Humidity-pollutant displacement
- Desiccant-catalyzed pollutant conversion

Humidity-reduced sorption is illustrated by the behavior of water vapor and chloroform on activated carbon. Sorption is humidity-neutral until relative humidity exceeds 45%, when the uptake of chloroform is reduced. The adsorbed water blocks sites that would otherwise attract and hold chloroform. In contrast, water and carbonyl chloride mixtures on activated carbon demonstrate humidity-enhanced sorption. Here, sorption of the pollutant increases at high relative humidities. Hines et al. (1991) attribute this phenomenon to the high water solubility of carbonyl chloride.

REFERENCES

Batelle. 1971. Project No. N-0914-5200-1971. Batelle Memorial Institute, Columbus, OH.

Brunauer, S. 1945. *The adsorption of gases and vapors*, Vol. I. Princeton University Press, Princeton, NJ. This information is quoted and expanded in *The physical chemistry of surfaces*, by Arthur W. Adamson. John Wiley and Sons, New York, 1982.

Bry-Air. 1986. MVB Series engineering data. Bry-Air Inc., Sunbury, OH.

Buffalo Testing Laboratory. 1974. *Report* No. 65711-1974.

Collier, R.K. Advanced desiccant materials assessment. *Research Report* 5084-243-1089. Phase I-1986, Phase II-1988. Gas Research Institute, Chicago.

Czanderna, A.W. 1988. Polymers as advanced materials for desiccant applications. *Research Report* NREL/PR-255-3308. National Renewable Energy Laboratory, Golden, CO.

Dow. 1981. Guide to glycols. Dow Chemical Corporation, Organic Chemicals Division, Midland, MI.

Foote Mineral. 1988. Lithium chloride technical data. *Bulletin* 151. Foote Mineral Corporation, Exton, PA.

Harriman, L.G., III. 1990. *The dehumidification handbook*, 2nd ed. Munters Cargocaire, Amesbury, MA.

Hines, A.J., T.K. Ghosh, S.K. Loyalka, and R.C. Warder, Jr. 1991. Investigation of co-sorption of gases and vapors as a means to enhance indoor air quality. ASHRAE *Research Project* 475-RP and Gas Research Institute *Project* GRI-90/0194. Gas Research Institute, Chicago.

Oscic, J. and I.L. Cooper. 1982. *Adsorption*. John Wiley and Sons, New York.

BIBLIOGRAPHY

Adamson, A.W. 1982. *The physical chemistry of surfaces*. John Wiley and Sons, New York.

Falcone, J.S., Jr., ed. 1982. Soluble silicates. *Symposium Series* 194. American Chemical Society, Washington, D.C.

Ruthven, D.M. 1984. *Principles of adsorption and adsorption processes*. John Wiley and Sons, New York.

SUNY Buffalo School of Medicine. Effects of glycol solution on microbiological growth. Niagrara Blower *Report* No. 03188.

Valenzuela, D. and A. Myers. 1989. *Adsorption equilibrium data handbook*. Simon and Schuster/Prentice-Hall, Englewood Cliffs, NJ.

CHAPTER 23

THERMAL AND MOISTURE CONTROL IN INSULATED ASSEMBLIES—FUNDAMENTALS

PROPER DESIGN of space heating, air-conditioning, refrigeration, and other industrial systems requires knowledge of thermal insulations and thermal-moisture behavior of building structures. This chapter deals with heat and moisture transfer definitions, fundamentals and properties of thermal insulation materials, heat flow calculations, economic thickness of insulation, and the fundamentals of moisture as it relates to building components and systems.

TERMINOLOGY AND SYMBOLS

The following heat and moisture transfer definitions and symbols are commonly used in the building industry. The following I-P units are used in the definitions: British thermal units (Btu), hour (h), temperature (°F), absolute temperature (°R), temperature difference (°F), feet (ft), inches (in.), pounds (lb), grains (gr), and inches of mercury (in. Hg).

Thermal transmission, heat transfer, or rate of heat flow. The flow of heat energy induced by a temperature difference. Heat may be transferred by conduction, convection, radiation, and mass transfer. These can occur separately or in combinations, depending on specific circumstances.

Thermal conductivity, *k*. The time rate of heat flow through a unit area (1 ft^2) of 1 in. or 1 ft thick homogeneous material in a direction perpendicular to isothermal planes, induced by a unit temperature gradient. (ASTM *Standard* C 168 defines homogeneity.) Units are Btu·in/h·ft^2·°F or Btu/h·ft·°F. Thermal conductivity must be evaluated for a specific mean temperature and moisture content, because in most materials it varies with temperature and moisture content.

For porous materials, heat flows by a combination of modes and may depend on orientation, direction, or both. The measured property of such materials may be called **effective** or **apparent thermal conductivity**. The specific test conditions (i.e., sample thickness, orientation, environment, environmental pressure, surface temperature, mean temperature, temperature difference, and moisture content) should be reported with the values. With thermal conductivity, the symbol k_{app} is used to denote the lack of pure conduction or to indicate that all values reported are apparent.

The preparation of this chapter is assigned to TC 4.4, Building Materials and Building Envelope Performance.

Thermal resistivity, R_u. The reciprocal of thermal conductivity. Units are °F·ft^2·h/Btu·in.

Thermal conductance, C-factor, *C*. The time rate of heat flow through a unit area of a body induced by a unit temperature difference between the body surfaces. Units are Btu/h·ft^2·°F.

When the two defined surfaces of mass-type (i.e., nonreflective) thermal insulation have unequal areas, as in the case of radial heat flow through a curved block or through a pipe covering (see Table 2 in Chapter 3), or through materials of nonuniform thickness, an appropriate mean area and mean thickness must be given. Heat flow formulas involving materials that are not uniform slabs must contain *shape factors* to account for the area variation involved.

When heat flow occurs by conduction alone, the thermal conductance of a material may be obtained by dividing the thermal conductivity of the material by its thickness. When several modes of heat transfer are involved, the **apparent** or **effective thermal conductance** may be obtained by dividing the apparent thermal conductivity by the thickness.

Where air circulates within or passes through insulation, as it may with low-density fibrous materials, the effective thermal conductance is affected.

Thermal conductances and resistances of the more common building materials and industrial insulations are tabulated in Table 4 in Chapter 25.

Heat transfer film coefficient (or surface coefficient of heat transfer or surface film conductance), ***h*** or ***f***. Heat transferred between a surface and a fluid per unit time per unit area driven by a unit temperature difference between the surface and the fluid in contact with it, in Btu/h·ft^2·°F.

Surface film resistance. The reciprocal of the heat transfer film coefficient, in °F·ft^2·h/Btu. Subscripts *i* and *o* often denote inside and outside surface resistances and conductances, respectively.

The surrounding space must be air or other fluids for convection to take place. If the space is evacuated, the heat flow only occurs by radiation.

Thermal resistance R-value, *R*. Under steady-state conditions, the mean temperature difference between two defined surfaces of material or construction that induces unit heat flow through a unit area, in °F·ft^2·h/Btu.

Thermal transmittance, U-factor, *U*. The rate of heat flow per unit area under steady-state conditions from the fluid on the warm side of a barrier to the fluid on the cold side, per unit temperature difference between the two fluids. It is determined by first evaluating the R-value, including the surface film resistances, and then computing its reciprocal, U, in Btu/h·ft^2·°F. The U-factor is sometimes called the **overall coefficient of heat transfer**. In building practice, the heat transfer fluid is air. The temperature of the fluid is obtained by averaging its temperature over a finite region near the surface involved.

Thermal emittance, ε. The ratio of the radiant flux emitted by a body to that emitted by a blackbody at the same temperature and under the same conditions.

Effective emittance of an air space, E. The combined effect of emittances from the boundary surfaces of an air space, where the boundaries are parallel and of a dimension much larger than the distance between them. Table 2 in Chapter 25 lists values of E for various air spaces.

Surface reflectance, ρ. The fraction of the radiant flux falling on a surface that is reflected by it.

Water vapor permeance, M. The rate of water vapor transmission by diffusion per unit area of a body between two specified parallel surfaces, induced by unit vapor pressure difference between the two surfaces, gr/(h·ft^2·in. Hg) or perm.

Water vapor permeability, μ. The rate of water vapor transmission by diffusion per unit area of flat material of unit thickness induced by unit vapor pressure difference between two surfaces, under specified temperature and humidity conditions. When permeability varies with psychrometric conditions, the spot or specific permeability defines the property at a specific condition in gr·in/(h·ft^2·in. Hg) or perm·in., where the vapor pressure difference is in in. Hg and 1 grain = 1/7000 lb.

Water vapor resistance, Z. The reciprocal of permeance—signifies a resistance to moisture flow, in. Hg·ft^2·h/gr. A **rep** is a unit of resistance to water vapor flow, where rep = 1/perm.

THERMAL INSULATION

Thermal insulations are materials or combinations of materials that, when properly applied, retard the rate of heat by conductive, convective, and/or radiative transfer modes. Thermal insulations can be fibrous, particulate, film or sheet, block or monolithic, open-cell or closed-cell, or composites of these materials that can be chemically or mechanically bound or supported.

By retarding heat flow, thermal insulations can serve one or more of the following functions:

- Conserve energy by reducing heat loss or gain of piping, ducts, vessels, equipment, and structures
- Control surface temperatures of equipment and structures for personnel protection and comfort
- Help control the temperature of a chemical process, a piece of equipment, or a structure
- Prevent vapor condensation on surfaces. However, thermal insulation may promote moisture condensation and subsequent damage in a building envelope, if the insulation application is improperly installed or poorly designed.
- Reduce temperature fluctuations within an enclosure when heating or cooling is not needed or available
- Reduce temperature variations within a conditioned space for increased personal comfort
- Provide fire protection

Thermal insulation can serve additional functions, although such secondary functions should be consistent with its capabilities and primary purpose. Under certain conditions, insulations may

- Add structural strength to a wall, ceiling, or floor section
- Provide support for a surface finish
- Impede water vapor transmission and air infiltration
- Prevent or reduce damage to equipment and structures from exposure to fire and freezing conditions
- Reduce noise and vibration
- Reduce growth of mold and mildew

Thermal insulation is used to control heat flow at all temperatures, the limiting value being its survival temperature.

BASIC MATERIALS

Thermal insulations normally consist of the following basic materials and composites:

- Inorganic, fibrous, or cellular materials such as glass, rock, or slag wool; and calcium silicate, bonded perlite, vermiculite, and ceramic products. In the past, asbestos insulations were applied, but, because asbestos has been shown to be a carcinogen, extreme caution must be used if it is encountered.
- Organic fibrous materials such as cellulose, cotton, animal hair, wood, pulp, or cane; synthetic fibers; and organic cellular materials such as cork, foamed rubber, polystyrene, polyurethane, and other polymers.
- Metallic or metallized organic reflective membranes. These surfaces must face an air-filled, gas-filled, or evacuated space to be effective.

PHYSICAL STRUCTURE AND FORM

Mass-type insulation can be cellular, granular, or fibrous solid material. Reflective insulation consists of smooth-surfaced sheets of metal foil or foil-surfaced material separated by air spaces.

The physical forms of industrial and building insulations include the following:

Loose-fill insulations consist of fibers, powders, granules, or nodules. They are usually poured or blown into walls or other spaces.

Insulating cement is a loose material that is mixed with water or a suitable binder to obtain plasticity and adhesion. It is troweled or blown wet on a surface and dried in place. Both loose-fill and insulating cement are suited for covering irregular spaces.

Flexible and semirigid insulations consist of organic and inorganic materials with and without binders and with varying degrees of compressibility and flexibility. These insulations are generally available as blanket, batt, or felt insulation, and in either sheets or rolls. Coverings and facings may be fastened to one or both sides and serve as reinforcing, air or vapor retarders or both, reflective surfaces, or surface finishes. These coverings include combinations of laminated foil, glass, cloth or plastics and paper, wire mesh, or metal lath. Although standard sizes are generally used, thickness and shape of insulation can be any dimension handled conveniently.

Rigid materials are available in rectangular blocks, boards, or sheets, which are preformed during manufacture to standard lengths, widths, and thicknesses. Insulation for pipes and curved surfaces is supplied in sections or segments, with radii of curvature available to suit all standard sizes of pipe and tubing.

Reflective materials are available in sheets and rolls of single layer or multilayer construction and in preformed shapes with integral air spaces.

Formed-in-place insulations are available as liquid components or expandable pellets that can be poured, frothed, or sprayed in place to form rigid or semirigid foam insulation. Fibrous materials mixed with liquid binders can also be sprayed in place, and in some products, the binder is also a foam.

Accessory materials for thermal insulation include mechanical and adhesive fasteners, exterior and interior finishes, vapor and air retarder coatings, jackets and weather coatings, sealants, lagging adhesives, membranes, and flashing compounds. ASTM *Standard* C 168 defines terms related to thermal insulating materials.

PROPERTIES

In addition to low thermal conductivity, the selection of thermal insulation may involve secondary criteria. Characteristics such as resiliency or rigidity, acoustical energy absorption, water vapor permeability, airflow resistance, fire hazard and fire resistance, ease of application, applied cost, health and safety aspects, or other parameters may influence the choice among materials that have almost equal thermal performance values.

Thermal Properties

Thermal resistance is a measure of the effectiveness of thermal insulation to retard heat flow. A material with a high thermal resistance (low thermal conductance) is an effective insulator.

Heat flows through most thermal insulations by a combination of gas and solid conduction, radiation, and convection. Heat transfer through materials or systems is controlled by factors such as length of heat flow paths, temperature, temperature difference characteristics of the system, and environmental conditions.

Although heat transmission characteristics are usually determined by measuring thermal conductivity, this property does not strictly apply to thermal insulation. A particular sample of a material has a unique value of thermal conductivity for a particular set of conditions. This value may not be representative of the material at other conditions and should be called **apparent thermal conductivity**. For details, refer to ASTM *Standards* C 168, C 177, C 236, C 335, C 518, C 976, and C 1045.

Reflective insulations impede radiant heat transfer because the surfaces have high reflectance and low emittance values. (Table 1 and Table 2 in Chapter 25 give typical design values.) To be effective, the reflective face of both single and multiple layer reflective insulations must face an air or evacuated space. Multiple layers of reflective materials and smooth and parallel sealed air spaces increase overall thermal resistance. Air exchange and movement must be inhibited.

Mass-type insulation can be combined with reflective surfaces and air spaces to obtain a higher thermal resistance. However, each design must be evaluated because maximum thermal performance depends on such factors as condition of the insulation, shape and form of the construction, the means to avoid air leakage and movement, and the condition of the installed reflective surfaces and their aging characteristics.

Design values of effective or apparent thermal conductivity, thermal conductance, and thermal resistance for the most common insulations are listed in Table 4 in Chapter 25. These values have been selected as typical and useful for engineering calculations. The manufacturer or test results of the insulation under appropriate conditions can give values for a particular insulation.

Other thermal properties that can be important are specific heat, heat capacity, thermal diffusivity, the coefficient of thermal expansion, and the maximum temperature limit. **Heat capacity** is the product of specific heat and mass. Because the rate of temperature change within an insulation is proportional to its thermal diffusivity for a given thickness, **thermal diffusivity** becomes important for applications where temperature varies with time. Chapter 3 covers symbols, definitions, and methods of calculation in steady-state heat transfer.

Mechanical Properties

Some insulations have sufficient structural strength for load bearing. They are used occasionally to support load-bearing roofs and floors, form self-supporting partitions, or stiffen structural panels. For such applications, one or more of the following properties of an insulation may be important: strength in compression, tension, shear, impact, flexure, and resistance to vibration. These mechanical properties vary with basic composition, density, cell size, fiber diameter and orientation, type and amount of binder (if any), and temperature and environmental conditioning.

Health and Safety

Most thermal insulations have good resistance to fire, vermin, rot, objectionable odors, and vapors; some are a potential risk to health and safety. These risks can result from direct exposure to these materials and accessories while they are being stored or transported, during or after installation, or as a result of intervening or indirect actions or events, such as aging, fire, or physical disturbance. The potential health and safety effects of thermal insulation can be considered in two categories: (1) those related to storage, handling, and installation operations and (2) those that occur after installation (such as aging). Potential hazards during manufacture are not discussed here. Correct handling, installation, and precautionary measures can reduce or eliminate these problems.

Potential health effects range from temporary irritation to serious changes in body functions. The principal concerns are with insulation containing asbestos. Questions have also been raised about man-made fibers. To date, research is inconclusive as to their potential hazard; however, they can be very annoying in installation, and the use of proper safeguards is desirable. Potential traumatic injury can occur from direct contact with materials that are sharp, rough, have protrusions or abrasive surfaces, permit overheating, or transmit electrical energy.

Combustion of insulation materials and accessories may release heat, hazardous gases, fibers, and particulates. Manufacturers' recommendations and applicable government codes and standards (ASTM *Standard* C 930) have more details.

Acoustics

Some thermal insulations are used as acoustical control materials, whether or not thermal performance is a design requirement. Acoustical efficiency depends on the physical structure of the material. Materials with open, porous surfaces have sound absorption capability. Those with high density and resilient characteristics can act as vibration insulators; either alone or in combination with other materials, some are effective barriers to sound transmission. Insulations for sound conditioning include flexible and semirigid, formed-in-place fibrous materials, and rigid fibrous insulation.

Sound absorption insulations are normally installed on interior surfaces or used as interior surfacing materials. Rigid insulations are fabricated into tile or blocks, edge-treated to facilitate mechanical or adhesive application, and prefinished during manufacture. Some insulation units have a natural porous surface, and others include mechanical perforations to facilitate the entry of sound waves. Still others have a diaphragm or decorative film surfacing attached only to the edges of the units, which allows the sound waves to reach the fibrous backing by diaphragm action.

Flexible, semirigid, and formed-in-place fibrous materials used for sound absorption are available in a variety of thicknesses and densities that determine their sound absorption characteristics.

When density is increased by reducing the thickness of the material, sound absorption is generally reduced; however, as thickness increases, the influence of density decreases.

Thermal insulations improve sound transmission loss when installed in **discontinuous construction**. A wall of staggered-stud construction that uses resilient clips or channels on one side of the stud or resilient insulation boards of special manufacture to prevent acoustical coupling mechanically between the surfaces reduces sound transmission. A sound absorption thermal insulation blanket in a wall cavity reduces sound transmission, depending on the type of construction.

In floor construction, resilient channels or separate floor and ceiling joists form a discontinuous construction. Sound-absorbing thermal insulation placed within this construction further reduces sound transmission. Sound-deadening boards underlying finish flooring absorb impact sounds and improve the airtightness of the construction, thus reducing airborne sound transmission.

Thermal insulation boards can be placed under mechanical equipment to isolate vibration. The imposed loading and natural resonant frequency of materials are critical for proper design. Because material must deflect properly under load to provide isolation, the system should be neither overloaded nor underloaded.

For further information on sound and vibration control, refer to Chapter 46 of the 1999 *ASHRAE Handbook—Applications*.

Other Properties

Other properties of insulating materials that can be important, depending on the application, include density, resilience, resistance to settling, permanence, reuse or salvage value, ease of handling, dimensional uniformity and stability, resistance to chemical action and chemical change, resistance to moisture penetration, ease in fabrication, application of finishes, and sizes and thicknesses obtainable.

HEAT FLOW

FACTORS AFFECTING THERMAL PERFORMANCE

A wide variety of physical, environmental, application, and, in some cases, aging factors affect the thermal performance of insulations.

Thermal conductivity k is a property of a homogeneous material. Building materials, such as lumber, brick, and stone, are usually considered homogeneous. Most thermal insulation and many other building materials are porous and consist of combinations of solid matter with small voids.

For many materials, conduction is not the only mode of heat transfer. Consequently, the term **apparent thermal conductivity** describes the heat flow properties of materials. Some materials with low thermal conductivities are almost purely conductive (silica opacified aerogel, corkboard, etc.).

The apparent thermal conductivity of insulation varies with form and physical structure, environment, and application conditions. Form and physical structure vary with the basic material and manufacturing process. Typical variations include density, cell size, diameter and arrangement of fibers or particles, degree and extent of bonding materials, transparency to thermal radiation, and the type and pressure of gas within the insulation.

Environment and application conditions include mean temperature, temperature gradient, moisture content, air infiltration, orientation, and direction of heat flow. Thermal performance values for insulation materials and systems are usually obtained by standard methods listed in Volume 04.06 of the *Annual Book of ASTM Standards*. The methods apply mainly to laboratory measurements on dried or conditioned samples at specific mean temperatures and temperature gradient conditions. Although the fundamental heat transmission characteristics of a material or system can be determined accurately, actual performance in a structure may vary from that indicated in the laboratory. The design of the envelope, its construction, and the materials used may all affect the test procedure. These factors are detailed in ASTM STP 544, STP 660, STP 718, STP 789, STP 879, STP 885, STP 922, STP 1030, and STP 1116.

The effective thermal conductivity of some thermal insulation materials varies with density. Figure 1 illustrates this variation at one mean temperature for a number of materials currently used to insulate building envelopes. For most mass-type insulations, there is a minimum in the respective apparent thermal conductivity versus density. This minimum depends on the type and form of material,

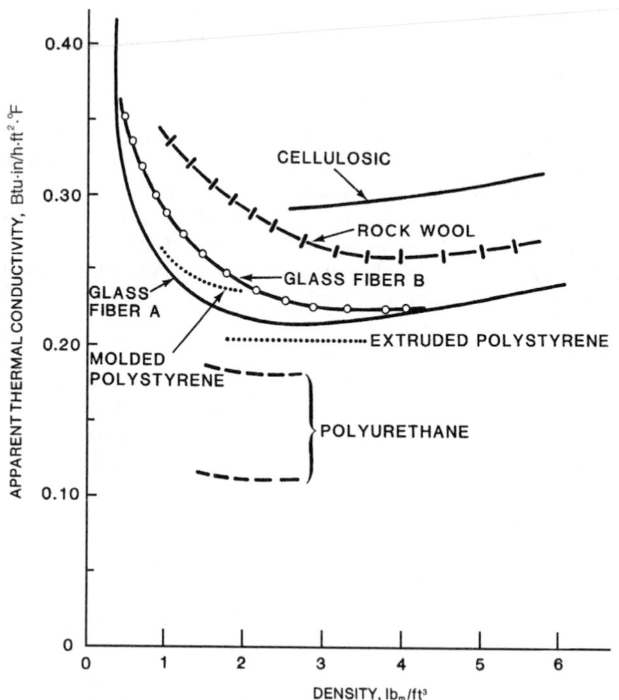

Fig. 1 Apparent Thermal Conductivity Versus Density of Several Thermal Insulations Used as Building Insulations

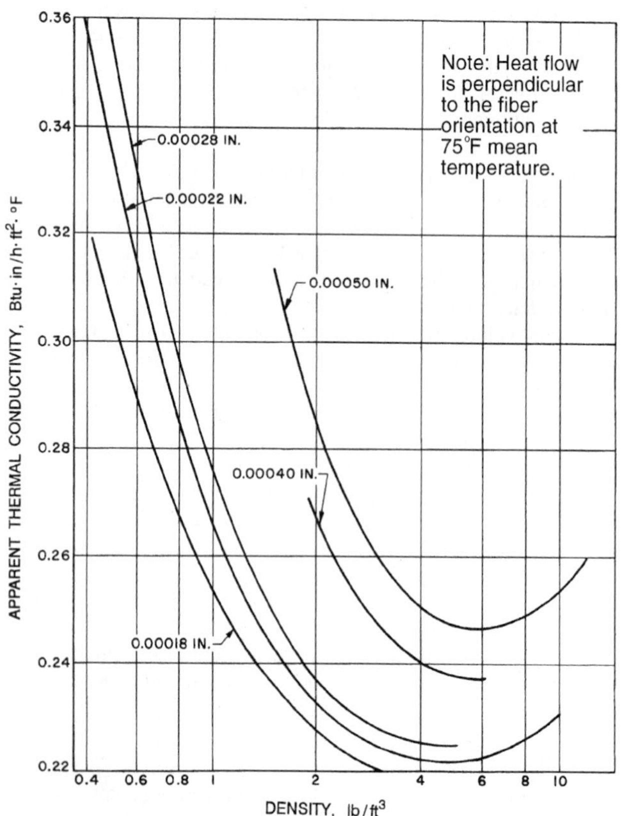

Fig. 2 Typical Variation of Apparent Thermal Conductivity with Fiber Diameter and Density

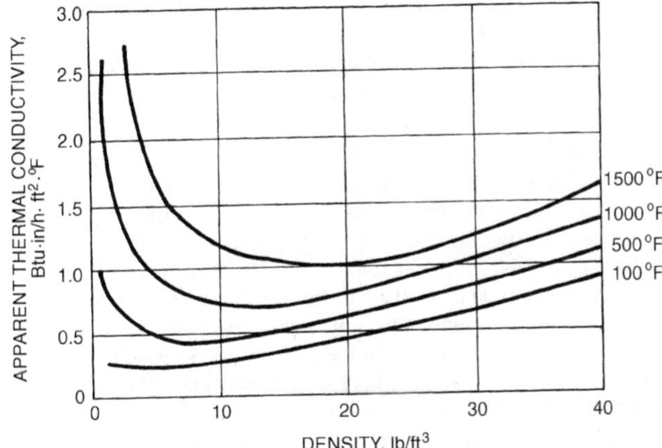

Fig. 3 Typical Variation of Apparent Thermal Conductivity with Mean Temperature and Density for Fibrous Insulations

temperature, and direction of heat flow. For fibrous materials, the values of density at which the minimum value occurs increase as both the fiber diameter or cell size and the mean temperature increase. These effects are shown in Figure 2 (Lotz 1969) and Figure 3, respectively.

Other factors that affect thermal performance include compaction and settling of insulation, air permeability, type and amount of binder used, additives that may influence the bond or contact between fibers or particles, and the type and form of radiation transfer inhibitor, if used. In cellular materials, the factors that influence thermal performance and strength properties are the same as those that control the thermal conductivity of the basic structured material: size and shape of the cells, thickness of the cell walls, gas contained in the cells, orientation of the cells, and radiation characteristics of the cell surfaces.

A change in density caused by the degree of compaction of insulation powders affects their effective thermal conductivity. Insulating concretes made from lightweight aggregates can be produced in a wide range of densities, with corresponding thermal conductances.

Fibrous insulations reach a minimum conductivity when fibers are uniformly spaced and perpendicular to the direction of heat flow. Generally, a decrease in the diameter of the fiber lowers the conductivity for the same density (Figure 2). For cellular insulation, a specific combination of cell size, density, and gas composition produces optimum thermal conductivity.

At temperatures below 400 to 570°F, a large portion of heat transfer across most insulations occurs by conduction through the air or other gas in the insulations (Rowley et al. 1952, Lander 1955, Simons 1955, Verschoor and Greebler 1952). The overall heat transfer can be closely approximated by supposition of gas conduction with other mechanisms of heat transfer, each determined separately. If the gas in the insulation is replaced by another gas with a different thermal conductivity, the apparent thermal conductivity of the insulation is changed by an amount approximately equal to the difference in conductivity of the two gases. For example, replacing air with a fluorinated hydrocarbon gas can lower the apparent thermal conductivity of the insulation by as much as 50%.

Fluorocarbon-expanded cellular plastic foams with a high proportion (greater than 90%) of closed cells retain the fluorocarbon within the cells for extended periods. As these products are initially produced, they have apparent thermal conductivities of approximately 0.111 Btu·in/h·ft^2·°F at 75°F when the gas contained has a mean-free-path greater than the dimensions within the cells. However, this value increases with time as atmospheric gases diffuse into the cells and, over a long period of time, the fluorocarbon gas dissolves in the polymer or diffuses out.

The rates of diffusion and increase in apparent thermal conductivity depend on several factors, including permeance of the cell walls to the gases involved, age of the foam, temperature, geometry of the insulation (thickness), and integrity of the surface protection provided. The significance of the surface protection is apparent when foams are encased in gas-impermeable membranes or some water vapor retarders (Brandreth 1986, Tye 1987). For estimating the long-term change in thermal resistance of unfaced rigid closed-cell plastic foams, a test method (ASTM *Standard* C 1303) has been developed that involves slicing and scaling under controlled laboratory conditions. Christian et al. (1995) provides an example of an application of this accelerated aging test method.

Brandreth (1986) and Tye (1988) show that the aging process of polyurethane and polyisocyanurate materials is reasonably well understood analytically and confirmed experimentally. The dominant parameters for minimum aging are as follows:

- Closed-cell content > 90%, preferably > 95%
- Small uniform cell diameter << 0.04 in., with a larger proportion of polymer in windows
- Small anisotropy in cell structure
- High density
- Increased thickness
- High initial pressure of fluorocarbon blowing agent in cell
- Polymer highly resistant to gas diffusion and solubility
- Polymer distributed evenly in struts and windows of cells
- Low aging temperature

Aging may be further reduced, particularly for laminated and spray-applied products, with higher density polymer skins, or by well-adhered facings and coverings with low gas and moisture permeance characteristics. An oxygen diffusion rate of less than 0.02 in^3/1000 ft^2·day for 0.001 in. thickness of barrier is one criterion used by some industry organizations manufacturing laminated products. The adhesion of any facing must be continuous, and every effort must be made in the manufacturing process to eliminate or minimize the shear plane layer at the foam/substrate interface (Ostrogorsky and Glicksman 1986).

Closed-cell phenolic-type materials and products, which are blown with similar gases, age differently and much more slowly. The reasons for this are believed to be higher material density, smaller more uniform cell size with a larger proportion of polymer in windows, and a basic polymer more resistant to gas diffusion.

The average distance, or **mean free path**, that an enclosed gas molecule travels before striking another gas molecule increases as pressure within an insulation decreases. When the mean free path equals the average distance a gas molecule travels before striking a solid part of the insulation, the apparent thermal conductivity of the insulation decreases with decreasing pressure. Correspondingly, for materials such as silica gel and fine carbon black, which have an average pore size smaller than the mean free path of air at atmospheric pressure, it is possible to attain thermal conductivity values lower than those for still air (Verschoor and Greebler 1952).

For homogeneous, dense materials, the primary mode of heat transfer is conduction. However, as the temperature increases, the heat transmission by thermal radiation and possible convection becomes a greater part of the total heat transferred. The magnitude of radiation and convection transfer depends on temperature difference, direction of heat flow, the nature of materials involved, and geometric considerations.

Because heat is transferred partly by radiation in low-density insulation, measured apparent thermal conductivity depends on test thickness. The thickness effect increases the apparent thermal conductivity measured at installed thickness over that commonly determined at 1 in. (Pelanne 1979). From a thermal resistance standpoint, the effect is small, typically less than 10% even for thin (e.g., 1 in.)

low-density (e.g., 0.35 lb/ft³) insulation. The effect becomes negligible for typical building applications (e.g., 6 in. insulation with a density of 0.7 lb/ft³).

Environmental and Application Conditions

The apparent conductivity of insulating materials generally increases with increasing temperature. Figure 4 shows typical variations with mean temperature. However, some materials, such as fluorocarbon-expanded, closed-cell urethanes, have an inflection in the curve over the temperature range where there is a change of phase of the fluorocarbon from gas to liquid (see Table 10 in Chapter 25).

The apparent thermal conductivity of a sample at one mean temperature (average of the two surface temperatures) only applies to the material at the particular thickness tested. Further testing is required to obtain values suitable for all thicknesses. The rate of change of apparent thermal conductivity with temperature and environmental conditions varies with the type and density of material.

Insulating materials that permit a large percentage of heat transfer by radiation, such as low-density fibrous and cellular products, show the greatest change in apparent thermal conductivity with changes of temperature and surrounding surface emittance. The ASTM Standard Test Methods recognize the importance of radiation in heat transmission and require that the surfaces of all test apparatus plates be painted or otherwise treated to have a total emittance greater than 0.8 at operating temperature.

The effect of temperature alone on structural integrity is not ordinarily important for most materials in low-temperature insulation applications. Decomposition, excessive linear shrinkage, softening, or some other effects of temperature alone limit the maximum temperature for which a material is suited. At extreme temperatures, selecting materials for a specific service becomes more critical and must be based on experience and actual performance data (see Table 4 and Table 10 in Chapter 25).

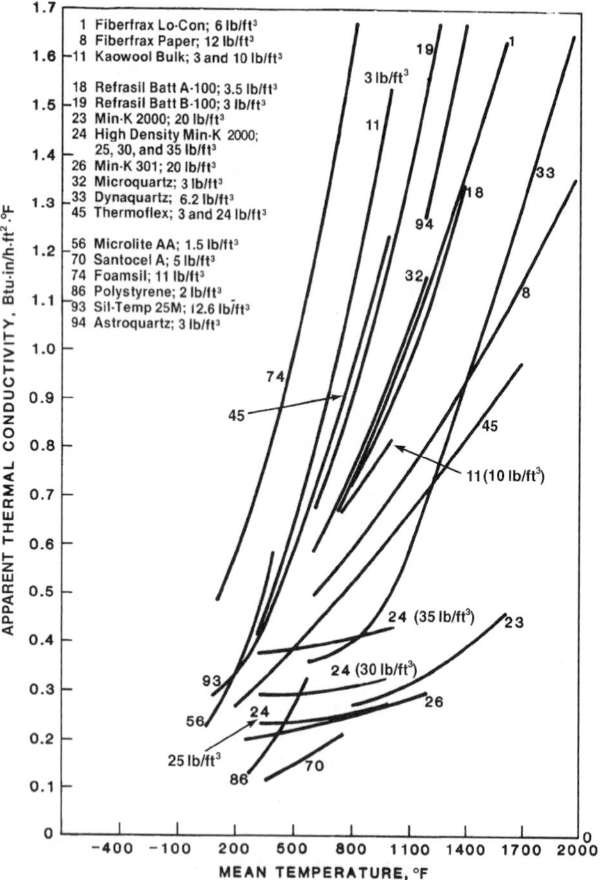

Fig. 4 Apparent Thermal Conductivity Versus Mean Temperature for Various Materials (in Air at Atmospheric Pressure)
(Glaser et al. 1967, Pelanne 1977)

Convection and air infiltration in or through some insulation systems may increase heat transfer across them. Low-density, loose-fill, large open-cell and fibrous insulations, and poorly designed or installed reflective systems are most susceptible to increased heat transfer by air filtration and convection. The temperature difference across an insulation, as well as the height, thickness, or width of the insulated space, influences the amount of convection. In some cases, natural convection may be inherent in the systems (Wilkes and Rucker 1983, Wilkes and Childs 1992). However, in many cases, the effect of convection in or through insulation can be minimized by careful design of an insulated structure (Donnelly et al. 1976).

Gaps between both board- and batt-type insulation can lower the effectiveness. Board-type insulation may not be perfectly square, may be installed improperly, and may be applied to uneven surfaces. For example, a 4% void area around batt insulation can produce a 50% loss in effective thermal resistance of an R-19 ceiling (Verschoor 1977). Similar results have been obtained with different test conditions and for wall configurations (Lewis 1979, Hedlin 1985, Tye et al. 1981). Preformed joints in board-type insulation allow it to fit together without air gaps. Boards and batts can be installed in two layers, with joints between layers offset and staggered to eliminate gaps. Chapter 24 has further details on these effects.

The presence of moisture may decrease the thermal performance of the installed insulation. The apparent thermal conductivity of construction materials increases with moisture content. If moisture condenses in the insulation, it may further reduce thermal resistance, and, perhaps physically damage the system. The reduction in thermal resistance depends on the material, the moisture content, and its distribution.

More information on the effects of moisture is provided in the section on Effect of Moisture on Heat Flow. Section A3 of the CIBSE *Guide* (1986) and Chapter 24 cover the thermal properties of building structures affected by moisture.

THERMAL TRANSMITTANCE

The method of calculating an overall coefficient of heat transmission requires knowledge of (1) the apparent thermal conductivity and thickness of homogeneous components, (2) thermal conductance of nonhomogeneous components, (3) surface conductances of both sides of the construction, and (4) conductance of air spaces in the construction. Procedures for calculating thermal conductance and resistance, and definitions of heat transfer terms and symbols are included in this chapter and in Chapter 3. In some construction, multidimensional heat flow effects are significant. Parallel heat flow paths of different resistances occur in wood frame house construction, for example. In such cases, the hot box method (ASTM *Standard* C 1363) or a multi-dimensional computer model may be used to determine overall thermal transmittance.

Surface Conductance

Surface conductance is the heat transfer to or from the surface by the combined effects of radiation, convection, and conduction. Each of these transport modes can vary independently. Heat transfer by radiation between two surfaces is controlled by the character of the surfaces (emittance and reflectivity), the temperature difference between them, and the solid angle through which they see each other. Heat transfer by convection and conduction is controlled by surface roughness, air movement, and temperature difference between the air and surface. Table 1 illustrates the importance of the

Table 1 Variation in Surface Heat Flux for Vertical Surfaces at 80°F with Different Temperatures of Surrounding Surface (70°F Ambient Still Air; 0.83 Emittance)

Surrounding Surface Temperature, °F	Surface Heat Flux, Btu/h·ft² 75	70	65	60	50
Convection	6.6	6.6	6.6	6.6	6.6
Radiation	4.4	8.6	12.8	17.0	24.9
Total	11.0	15.2	19.4	23.6	31.5

effect of temperature of surrounding surfaces on surface heat flux caused by radiation.

In many cases, because the thermal resistance (reciprocal of conductance) of the internal parts of the wall is high compared with the surface resistance, the surface factors are of minor importance. However, surface resistances on a window with a single pane of glass constitute almost the entire resistance and are very important. Raber and Hutchinson (1945) analyzed the factors affecting surface conductance and the difference between surface and air temperatures.

The convective part of the surface conductance is markedly affected by the nature of the air movement on the surface, illustrated by Figure 1 in Chapter 25. On smooth surfaces, surface length (Parmelee and Huebscher 1947) also affects the convection part of conductance; the average value decreases as the surface length increases. Moreover, Parmelee and Aubele (1950) observed that only under certain conditions can outdoors be treated as a blackbody radiating at an effective air temperature. Therefore, selection of surface conductance coefficients for a building becomes a matter of judgment. Surface conductances in Table 1 in Chapter 25 are applicable to ordinary building materials. In special cases, where surface conductances become important factors in the overall rates of heat transfer, more accurate coefficients may be required. Principles and data given in Chapter 3 can be applied in such cases.

FACTORS AFFECTING HEAT TRANSFER ACROSS AIR SPACES

Heat transfer across an air space is affected by the nature of the boundary surfaces, the intervening air space, the orientation of the air space, the distance between boundary surfaces, and the direction of heat flow. Air space conductance coefficients represent the total conductance from one surface bounding the air space to the other. The total conductance is the sum of a radiation component and a convection and conduction component. In all cases, the spaces are considered airtight with no through air leakage.

The radiation portion of the coefficient is affected by the temperature of the two boundary surfaces and by their respective surface properties. For surfaces that can be considered ideal graybodies, the surface properties can be characterized by emissivity. The combined effect of the emittances of the boundary surfaces of an air space is expressed by the effective emittance E of the air space. The radiation component is practically not affected by the thickness of the space, its orientation, the direction of heat flow, or the order of emittance (hot or cold surface). In contrast the heat transfer by convection and conduction combined is affected markedly by orientation of the air space and the direction of heat flow, by the temperature difference across the space, and, in some cases, by the thickness of the space. It is also slightly affected by the mean temperature of its surfaces. For air spaces in building construction, the radiation and convection-conduction components can vary independently of each other.

Table 3 in Chapter 25 lists the thermal resistance values of sealed air spaces of uniform thickness and moderately smooth, plane, parallel surfaces. These data are based on experimental measurements (Robinson et al. 1954). Resistance values for systems with air spaces can be estimated from these results if emissivity values are corrected for field conditions. However, the resistance value of some common composite building insulation systems involving mass-type insulation with a reflective surface in conjunction with an air space may be appreciably lower than the estimated value (Palfey 1980). This is true particularly if the air space is not sealed or of uniform thickness. The thermal resistance values for plane air spaces in Table 3 in Chapter 25 represent typical values; for critical applications, the effectiveness of a particular design should be confirmed by actual test data undertaken by using ASTM hot box methods (ASTM *Standards* C 236 and C 976). This test is especially necessary for constructions combining reflective and nonreflective thermal insulation.

For narrow air spaces, defined as those for which the product of the temperature difference (in degrees Fahrenheit) and the cube of the space thickness (in inches) is less than 3 for heat flow horizontally or downward, or less than 1 for heat flow upward, convection is practically suppressed. The conductance for these spaces is the sum of the radiative heat transfer coefficient and that for heat conduction alone through air. The radiation and conduction component can be computed by the method shown in the footnote to Table 3 in Chapter 25. Effects of different mean temperatures, temperature differences, and effective emittances may be found in this table.

To obtain high thermal resistance with reflective insulation, a series of multiple air layers bounded by reflective surfaces is needed. The total resistance equals the sum of the resistance values across each air space. All air layers or spaces must be sealed because air moving between the layers can increase the heat flow. Depending on the type of reflective insulation, one or both sides may have highly reflective surfaces. Except for thick horizontal air spaces with heat flow down, little is gained thermally by the addition of a second highly reflective surface to the same air space. If an air space has only one reflective surface, and conditions are completely dry, the side on which the reflective surface is placed makes no appreciable difference in the rate of heat transfer. In typical building situations moisture is present and the reflective facing should always be placed on the warm side of the air space, else trace condensation will at times increase the emittance and reduce the insulation value (see Chapter 25, Table 2). A reflective surface placed on the warm side of an air space may also act as a water vapor retarder if the material and its joints have sufficiently low permeance (see the section on Vapor Retarder Functions and Properties).

The emittance of a surface is the measure of its ability to emit radiant energy and, for the same temperature and wavelength, is equal to its absorptance (ratio of the radiant energy absorbed by a surface to the total radiant energy falling on it). The ratio of the energy reflected by the surface to that falling on it is the reflectance. For an opaque surface, reflectance is equal to one minus the emittance. This emittance varies with surface type and condition and radiation wavelength.

For reflective insulation used with heating, air-conditioning, and refrigeration applications, the emittance value for long-wavelength (infrared) radiation is important, not the value for the shorter wavelengths of the visible spectrum. Visible brightness is not a true measure of the reflectance for thermal radiation because the reflectance for light and for long-wavelength radiation is unrelated. Table 2 in Chapter 25 lists typical emittance values for reflective surfaces and building materials, and the corresponding emittance factors for air spaces.

Chemical action, dust or oil accumulation, or the presence of condensation or frost can change a reflective surface enough to reduce its reflectance and increase its emittance. Chemical changes include oxidation, corrosion, or tarnishing caused by air, moisture, wet plaster, or the chemical treatment of wood spacing strips or other adjoining structural members. Surface emittance values should be obtained by tests.

CALCULATING OVERALL THERMAL RESISTANCE

Using the principles of heat flow presented in Chapter 3, calculating heat flow by the overall thermal resistance method is preferred.

The total resistance to heat flow through a flat building construction composed of parallel layers such as a flat ceiling, floor, or wall (or curved surface if the curvature is small) is the numerical sum of the resistances (R-values) of all layers of the construction in series:

$$R = R_1 + R_2 + R_3 + R_4 + \ldots + R_n \quad (1)$$

where

$R_1, R_2, \ldots, R_n$ = individual resistances of the layers
R = resistance of construction from inside surface to outside surface

However, in buildings, to obtain the overall resistance R_T, the film resistances R_i and R_o from Table 1 in Chapter 25 must be added to R:

$$R_T = R_i + R + R_o \quad (2)$$

The U-factor (thermal transmittance) is the reciprocal of R_T, or

$$U = \frac{1}{R_T} \quad (3)$$

Thus, for a wall with air space construction, consisting of two homogeneous materials of conductivities k_1 and k_2 and thickness x_1 and x_2, respectively, and separated by an air space of conductance C, the overall resistance is

$$R_T = \frac{1}{h_i} + \frac{x_1}{k_1} + \frac{1}{C} + \frac{x_2}{k_2} + \frac{1}{h_o} \quad (4)$$

where h_i and h_o are the heat transfer film coefficients.

Series and Parallel Heat Flow Paths

In many installations, components are arranged so that heat flows in parallel paths of different conductances. If no heat flows between lateral paths, heat flow in each path may be calculated using Equations (1) and (2). The average transmittance is then

$$U_{av} = aU_a + bU_b + \ldots + nU_n \quad (5)$$

where $a, b, \ldots, n$ are respective fractions of a typical basic area composed of several different paths with transmittances $U_a, U_b, \ldots, U_n$.

If heat can flow laterally with little resistance in any continuous layer so that transverse isothermal planes result, total average resistance $R_{T(av)}$ is the sum of the resistance of the layers between such planes. Each layer is calculated by the appropriate Equation (1) or modification of Equation (4), using the resistance values. This is a series combination of layers, of which one (or more) provides parallel paths.

The calculated heat flow, assuming parallel heat flow only, is usually considerably lower than that calculated with the assumption of combined series-parallel heat flow. The actual heat flow is some value between the two calculated values. In the absence of test values for the combination, an intermediate value should be used. Examination of the construction usually reveals whether a value closer to the higher or lower calculated value should be used. Generally, if the construction contains any highly conductive layer in which lateral conduction is very high compared to heat flow through the wall, a value closer to the series-parallel calculation should be used. If, however, no layer has a high lateral conductance, a value closer to the parallel heat flow calculation should be used. In the case where the presence or absence of high lateral conductance is unclear, use the arithmetic mean of the isothermal-planes and parallel-path methods. A more precise value of the thermal transmittance may be obtained by using an appropriate computer program for two- or three-dimensional heat flow.

CALCULATING INTERFACE TEMPERATURES

The temperature at any interface can be calculated, since the temperature drop through any component of the wall is proportional to its resistance. Thus, the temperature drop Δt_n through R_n in Equation (1) is

$$\Delta t_n = \frac{R_n(t_i - t_o)}{R_T} \quad (6)$$

where t_i and t_o are the indoor and outdoor temperatures, respectively. Hence, the temperature at the interface between R_n and R_{n+1} is

$$t_{n-n+1} = t_i - \sum_{m=1}^{n} \Delta t_m \quad (7)$$

For building materials having nonuniform or irregular sections such as hollow clay tile or concrete blocks, the R-value of the unit as manufactured should be used.

If the resistances of materials in a wall are highly dependent on temperature, the mean temperature must be known to assign the correct value. In such cases, it is perhaps most convenient to use a trial-and-error procedure for the calculation of the total resistance R_T. First, the mean operating temperature for each layer is estimated, and R-values for the particular materials are selected. The total resistance R_T is then calculated as in Equation (4), and the temperature at each interface is calculated using Equations (6) and (7). The mean temperature of each component (arithmetic mean of its surface temperatures) can then be used to obtain second generation R-values. This procedure can then be repeated until the R-values have been correctly selected for the resulting mean temperatures. Generally, this can be done in two or three trial calculations.

HEAT FLOW CALCULATIONS

Equation (8) is used to calculate heat flow through flat surfaces; Equation (9) is used for cylindrical surfaces (Figure 5).

$$q_s = \frac{t_{is} - t_{os}}{R} \quad (8)$$

$$q_s = \frac{t_{is} - t_{os}}{r_s \sum_{m=1}^{N} \ln(r_{o,n}/r_{i,n})/k_n} \quad (9)$$

where

q_s = rate of heat transfer per unit area of outer surface of insulation
R = surface-to-surface thermal resistance from Equation (1)
k_n = thermal conductivity of insulation layer n at calculated mean temperature
t_{is} = temperature of inner surface
t_{os} = temperature of outer surface
$r_{i,n}$ = inner radius of insulation layer n
$r_{o,n}$ = outer radius of insulation layer n
r_s = outer radius of outer insulation layer
N = total number of insulation layers
ln = natural or Naperian logarithm

Heat flow per unit length of pipe is the preferred unit. The terms that appear in the denominator of Equation (9) represent the resistances to heat flow based on logarithmic mean thickness. The thermal resistances for flat surfaces should not be confused with the

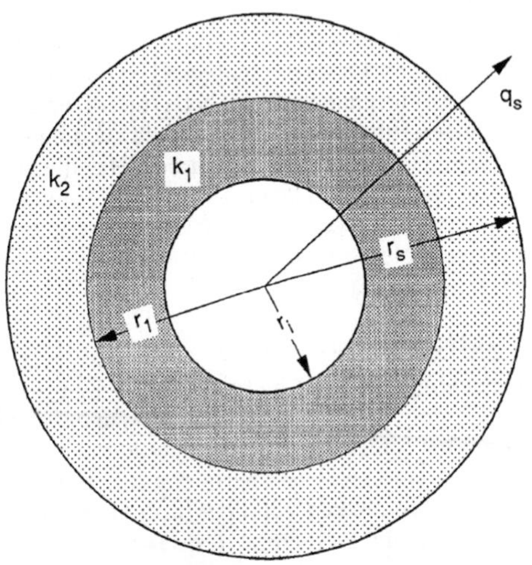

Fig. 5 Heat Flow Through Cylindrical Surfaces

thermal resistance of a cylindrical insulation where the surface areas involved are never equal.

INSULATION THICKNESS

ECONOMIC THICKNESS

Economics can be used (1) to select the optimum insulation thickness for a specific insulation, or (2) to evaluate two or more insulation materials for least cost for a given thermal performance. In either case, economics determine the most cost-effective solution for insulating over a specific period (FEA 1976). This solution can be reached by different techniques, but only one solution exists for a given set of economic variables.

Greater than optimum insulation thickness may also require a greater capital investment for the structure and piping. However, other concerns such as limited energy availability, sustainability, or comfort may alter the results of an economic analysis alone.

Life-cycle costing spreads the initial cost of the insulation over the number of years the insulation is expected to be in service. The life cycle selected affects the economic thickness of the insulation. Thinner insulation pays back on a short life cycle, and thicker insulation pays back over a longer cycle. An insulation system designed to pay for itself with energy savings in a short time, and that stays in service longer, does not necessarily produce the lowest total cost over the service period.

The annualized cost of the installed insulation must also be adjusted for the cost of money, which can be based on a desired rate of return on the insulation investment. Insulation system maintenance costs should also be included in the annual cost.

Because fuel cost is likely to change during the depreciation period (the life of the facility or payback period), the average cost should be estimated and this value used rather than current cost.

The total annual cost is the sum of the annual cost of insulation and the annual cost of lost energy. For each incremental increase in insulation thickness, the corresponding change in the total cost is $m_t = m_c + m_s$ where m_t is the incremental change in total annual costs, m_c is the incremental change in cost of insulation, and m_s is the incremental change in cost of lost energy plus any change in the cost of the mechanical equipment needed.

Initially, as insulation is applied, the total cost decreases because the incremental energy savings is greater than the incremental cost of insulation. Additional insulation reduces this cost up to a thickness where the change in the total cost is equal to zero. At this point, $m_t = m_c + m_s = 0$, and no further reduction can be obtained. Beyond this thickness, the incremental insulation costs become greater than the additional energy savings derived by adding another increment of insulation. However, if total insulation costs then exceed total insulation benefits, the optimal level of insulation for a given component from a purely economics viewpoint is no insulation at all. This may occur if the startup costs are extremely high, as for installation of insulation in a sealed wall cavity.

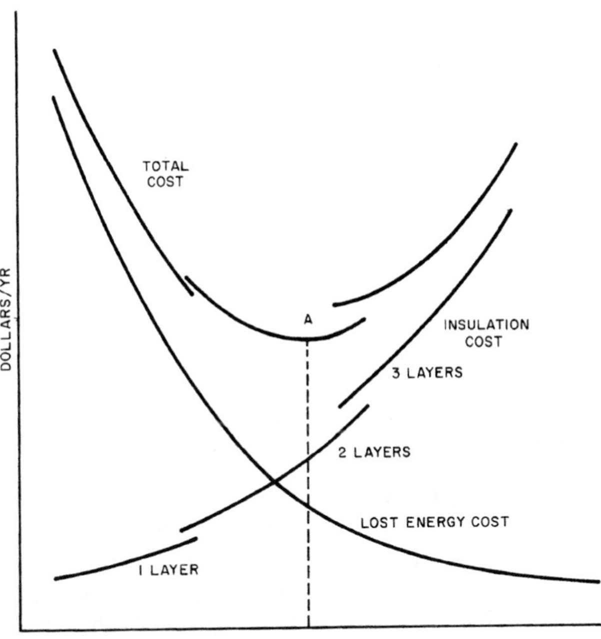

Fig. 6 Determination of Economic Thickness of Installed Insulation of Mechanical Equipment

ECONOMIC THICKNESS: MECHANICAL SYSTEMS

Determining the most profitable thickness of insulation for mechanical systems is difficult. The economics of each plant (including cost of producing energy, cost of insulating, discount rate or cost of capital, and potential for energy loss) indicate the preferred amount of insulation. Various types of equipment and piping also require different economic thicknesses. This analysis is further complicated because future energy cost and the life of the facility and insulation must also be considered. For every plant, these factors dictate different solutions to the economic analysis. Furthermore, additional insulation may lower heat loss or gain, possibly allowing for lower capacity equipment and thus for lower first costs.

Economic Analysis

The cost of installed insulation increases with thickness. This incremental cost is for both labor and material. Insulation is often applied in multiple layers (1) because materials are not manufactured in single layers of sufficient thickness and, (2) in many cases, to compensate for expansion and contraction. Figure 6 shows installed costs for a multilayer application. The average slope of the curves increases with the number of layers because labor and material costs increase at a more rapid rate as thickness increases.

Because the optimal economic thickness is the lowest total cost of lost energy and installed insulation over the life of the insulation, these two costs must be compared in similar terms. Either the annual cost of the insulation must be compared to the average annual cost of lost energy, or the cost of the energy lost each year must be expressed in present dollars and compared with the total cost of the insulation investment. The former method, annualizing the insulation cost and comparing this with the average annual cost of lost energy, is easier to compute.

Insulation reduces the size and capital costs of the heating and cooling equipment required for an installation because it lowers energy demand. This capital cost may be annualized by considering the plant depreciation period, cost of money, annual energy output for the plant, and operational expenses.

Figure 6 shows curves of total annual costs of operation, insulation costs, and lost energy costs. Point A on the total cost curve corresponds to the economic insulation thickness, which, in this example, is in the double-layer range. Viewing the calculated economic thickness as a minimum thickness provides a hedge against unforeseen fuel price increases and conserves energy.

ECONOMIC THICKNESS: BUILDING ENVELOPES

In buildings such as residences and warehouses, the internal energy gains are insignificant compared with the heat losses and gains through the envelope. For these buildings, the heating and cooling requirements are roughly proportional to the difference between the indoor and outdoor temperature. For commercial, industrial, and institutional buildings, internal heat loads can be significant, and the heating and cooling requirements are not as directly related to the indoor/outdoor temperature difference. In both types of buildings, solar heat can be an important factor and should be evaluated.

Dominant Heat Loss and Gain Through Envelope

Thermal insulation is generally installed in building envelope components (e.g., ceilings, walls, and floors) to reduce space heating and cooling costs on a long-term basis. Additional benefits may include increased occupant comfort, reduced heating and cooling system capacity, and elimination of condensation on wall surfaces in cold climates. When possible these benefits should be considered. The economically optimal insulation thickness (best measured in terms of thermal resistance) in an envelope component minimizes total life-cycle space heating and cooling costs attributable to it. Total life-cycle costs are the sum of present-value heating and cooling costs over the useful lifetime of the insulation plus the installed cost of the insulation and the installed cost of the heating and cooling equipment.

If the R-value of the insulation used is continuously variable (e.g., loose-fill insulation in attics), uniformly small increments of insulation can be used to determine appropriate optimal thicknesses; or calculus can be used to determine an exact optimum. If the insulation materials used are available only in discrete levels of thermal resistance (R-12, R-18, R-27), the increment used in determining optimal thickness should be based on differences between those levels (R-12 over R-0, R-18 over R-12, R-27 over R-18). Where discrete increments of resistance are used, determining the resistance level for which incremental savings equal incremental costs may not be feasible. In such cases, the selection should be left to the judgment of the analyst based on the level of conservation desired. In addition, an increase in insulation may make it possible to reduce the size of the heating and cooling equipment, which becomes a discrete reduction in equipment cost.

If a building envelope component requires structural modifications to accommodate increased insulation thickness, this cost must be included in the installed cost of the additional insulation. Generally, such modifications should only be considered when they are less costly than the use of more efficient (i.e., lower thermal conductivity) but more expensive insulation materials than those ordinarily used.

Typically, the incremental energy savings and insulation costs differ for each building envelope component; therefore, the optimal insulation level differs for each component in the same building. Less efficient heating plants and higher costs of heating energy necessitate higher optimal insulation levels in each building envelope component. Conversely, more efficient heating equipment reduces the optimal insulation level. The effects of climate, cooling energy costs, and cooling equipment efficiency on optimal insulation levels are less clear and differ widely, depending on overall building design and operational profile.

Dominant Internal Heat Loads

In buildings with dominant internal loads, the energy requirements vary so widely that no generalizations can be made regarding insulation. This contrasts with envelope-dominated structures, in which more insulation reduces energy consumption (Hart 1981).

In internal-load-dominated buildings with both annual heating and cooling loads, higher thermal resistance increases cooling energy consumption while reducing heating energy consumption. Therefore, the calculation of economically optimum resistance becomes quite complex and involves multiple measure or hourly methods described in Chapter 31. Spielvogel (1974), Burch and Hunt (1978), and Rudoy (1975) give more details.

Figure 7 shows the results of these calculations for a building in Columbus, Ohio, with 8.2 Btu/h·ft^2 of internal heat gains that operates 24 h per day (Spielvogel 1974). This solution is not the only one possible, but illustrates problems faced by the designer. In this case, thermal resistance increases, the U-factor decreases, annual heating energy decreases, and annual cooling energy increases. The energy optimum exists at Point Y in Figure 7, where the total heating and cooling energy is at a minimum. Because the cost of cooling energy differs from the cost of heating energy, the economic optimum will not be the same as the energy optimum.

These results occur in some localities where there are far more hours per year with outdoor temperatures between 50 and 75°F than between 75 and 100°F. At temperatures between 75 and 100°F, low U-factors result in less energy consumption for cooling. However, for temperatures between 50 and 75°F, low

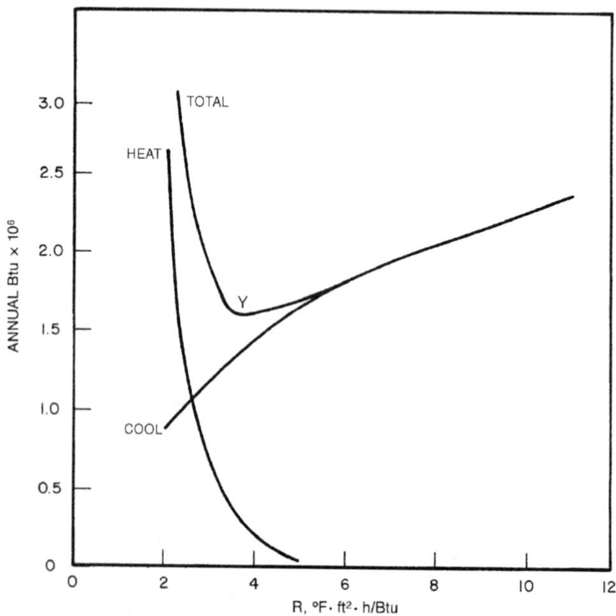

Fig. 7 Example of Optimal Thermal Resistance for Building with Internal Heat Gains
(Adapted from Spielvogel 1974)

U-factors inhibit the flow of internal heat from the building, thereby creating higher cooling loads and higher energy requirements than those in buildings with higher U-factors. What might be saved at outdoor temperatures over 75°F can be more than spent in additional cooling energy at temperatures below 75°F. Economizer cycles could offset these excess internal gains with ventilation air, however.

Where the hours of use or the quantity of internal heat gains vary from room to room, the optimum thermal resistance also varies. For example, in a cold climate, a hotel kitchen requires little or no insulation, because the internal heat is sufficient to heat the space almost all year. In a meeting room adjacent to the kitchen, substantially more insulation is justified. Thus, the economic thermal resistance of any envelope element, such as a roof, will not be the same throughout the entire building.

This type of analysis must include the level and duration of internal gains and the nature of the energy consumption of the heating and cooling systems. Most buildings need evaluation of walls, roofs, and floors on a room-by-room basis. Computer programs make these more complex analyses possible. Due to the wide diversity of building types, internal gains, system types, and operating conditions, no simple rules can establish U-factors for minimum energy consumption.

Effectiveness of Added Insulation

The effectiveness of added insulation varies with many factors, including climate, original insulation level, preparation costs, and predicted life, based on payback calculations. Building codes generally balance life-cycle costs between construction, financing, and energy expenditures. Figure 8 shows typical relationships between life-cycle costs and energy consumption. Individual points on the curve represent different combinations of ceiling, wall, and floor insulation in R-values and glazing types (single, double, or triple). Because life-cycle costs vary not only with construction and energy costs but also with climatic factors, the profiles of this curve vary according to locality. In Figure 8, the optimal condition for this example is attained with R-30 attic insulation, R-19 wall insulation, R-11 floor insulation, and double glazing. However, the effectiveness of added insulation can be determined only by analyzing actual conditions.

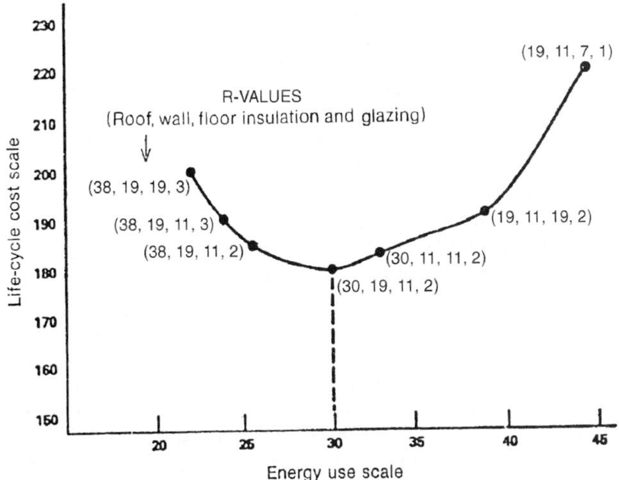

Fig. 8 Typical Relationship of Life-Cycle Cost to Energy Use

MOISTURE IN BUILDINGS

MOISTURE PROBLEMS IN BUILDINGS

Moisture control is necessary to avoid moisture-related problems with building energy performance, building maintenance and durability, and human comfort and health. Moisture degradation is the largest factor limiting the useful life of a building and can be visible or invisible. Invisible degradation includes degradation of the thermal resistance of building materials and decrease in the strength and stiffness of some materials. Visible moisture degradation may be in the form of (1) mold and mildew, (2) the decay of wood-based materials, (3) spalling of masonry and concrete caused by freeze-thaw cycles, (4) hydration of plastic materials, (5) corrosion of metals, (6) damage due to expansion of materials (e.g., buckling of wood floors), and (7) a decline in visual appearance (e.g., buckling of wood siding or efflorescence of masonry materials, which is the formation of a salt crust from the leaching of free alkalies). In addition, high moisture levels can lead to odors and mold spores in indoor air, which can seriously affect occupant health and comfort. Short summaries of such moisture conditions and related performance and health issues follow.

Mold, Mildew, Dust Mites, and Human Health

Mold and mildew in buildings are offensive, and the spores can cause respiratory problems and other allergic reactions in humans. Mold and mildew will grow on most surfaces if the relative humidity at the surface is above a critical value and the surface temperature is conducive to growth. The longer the relative humidity remains above the critical value, the more likely is visible mold growth; and the higher the humidity or temperature, the shorter is the time needed for germination. The surface relative humidity is a complex function of material moisture content, material properties, and local temperature and humidity conditions. In addition, mold growth depends on the type of surface. Fully recognizing the complexity of the issue, the International Energy Agency Annex 14 (1990) nevertheless established a surface humidity criterion for design purposes: The monthly average surface relative humidity should remain below 80%. Others have proposed more stringent criteria, the most stringent requiring that surface relative humidity remain below 70% *at all times*. Although there still is no agreement on which criterion is most appropriate, mold and mildew can usually be avoided by limiting surface moisture conditions over 80% to short time periods. These criteria should only be relaxed for nonporous surfaces that are regularly cleaned. Hukka and Viitanen (1999) developed a mathematical model for the prediction of a mold growth index. This model was successfully implemented and linked to a hygrothermal model by Karagiozis and Salonvaara (1998). Most molds grow at temperatures above approximately 40°F. Moisture accumulation below 40°F may not cause mold and mildew if the material is allowed to dry below the critical moisture content before the temperature rises above 40°F.

Dust mites can trigger allergies and asthma (Burge et al. 1994). Dust mites thrive at high relative humidities (over 70%) at room temperature, but will not survive sustained relative humidities below 50% (Burge et al. 1994). These relative humidities relate to local conditions in typical places that mites tend to inhabit such as mattresses, carpets, soft furniture, etc.

Paint Failure and Other Appearance Problems

Moisture trapped behind paint films may cause failure of the paint. Water or condensation may also cause streaking or staining. Excessive swings in the moisture content of wood-based panels or boards may cause buckling or warping. Excessive moisture in masonry and concrete may cause efflorescence, a white powdery area or lines, or, when combined with low temperatures, may cause freeze-thaw damage and spalling (chipping).

Structural Failures

Structural failures due to decay of wood are rare but have occurred (e.g., Merrill and TenWolde 1989). Decay generally requires wood moisture content at fiber saturation (usually about 30%) or higher and temperatures between 50 and 100°F. Wood moisture contents above fiber saturation are only possible in green lumber or by absorption of liquid water from condensation, leaks, ground water, or other saturated materials in contact with the wood. To maintain a safety margin, 20% moisture content is sometimes used during field inspections as the maximum allowable moisture level. Because wood moisture content can vary widely with sample location, a local moisture content of 20% or higher may indicate fiber saturation elsewhere. Once established, decay fungi produce water that enables them to maintain moisture conditions conducive to their own growth.

Rusting or corrosion of nails, nail plates, or other metal building components is also a potential cause of structural failure. Corrosion may occur at high relative humidities near the metal surface or as a result of liquid water from elsewhere. Wood moisture content over 20% encourages corrosion of steel fasteners in wood, especially if the wood is treated with preservatives. In buildings, metal fasteners are often the coldest surfaces, encouraging condensation on and corrosion of the fasteners.

Effect of Moisture on Heat Flow

Moisture in the building envelope can significantly degrade the thermal performance of most insulation materials. Bomberg and Shirtliffe (1978), Epstein and Putnam (1977), Hedlin (1977, 1983, 1987, 1988a, 1988b), Jespersen (1960), Joy (1957), Knab et al. (1980), Kumaran (1989), Kyle and Desjarlais (1994), Langlais et al. (1983), Paljack (1973), Pedersen (1990), Pedersen et al. (1991), Shapiro and Motakef (1990), Thomas et al. (1983), Tobiasson and Richard (1979), Tobiasson (1987, 1991), and Tye (1987) investigated the effect of moisture content on heat flow and showed that the effect depends on the type of insulation material, the moisture content, the temperature of the insulation material, the insulation material's thermal history, and the building envelope's interior and exterior environments. The reported relationships between the thermal performance of the insulation material and heat flow can vary significantly; the variations are more pronounced in open-cell and fibrous insulations. Rapid vapor transfer through vapor-permeable insulations during testing accounts for these variations. Variations can also be due to the location of water in the insulation layer. Kyle and Desjarlais (1994) estimated that water distribution can account for a difference of up to 25% in heat flow in certain cases.

Moisture can contribute to both sensible and latent heat flow, as well as through mass transfer by diffusion or convection. Evaporation on the warm side of the insulation and condensation or adsorption on the cold side add a latent heat component to the heat flow.

Under steady-state conditions, the effect of moisture on thermal resistance may be small. Verschoor (1985) showed that an insulated residential-type stud wall panel with a poor vapor retarder on the warm side accumulated 0.3 lb of moisture per square foot when exposed to conditions continuously below freezing (steady-state) for 31 days. All of the accumulated moisture was located in the 0.5 in. layer of mineral fiber insulation immediately adjacent to the cold-side sheathing and at the interface between the insulation and the sheathing. In a continuation of the test program, to a total exposure period of 60 days, the rate of moisture gain remained constant during the entire period. Subsequently, the same test wall was subjected to diurnal outside temperature cycles through the freezing point. With this exposure, most of the accumulated moisture was found in the sheathing and the bottom of the test wall rather than in the coldest insulation layer. Except for the sheathing, there was negligible change in the overall thermal performance of the wall.

The behavior may be different for closed-cell plastic foam insulations. Field observations of low-temperature insulated tanks have shown that when no air space exists between the cold surface and the insulation, the expected accumulation of condensed moisture did not occur.

Degradation of thermal resistance is more pronounced when daily reversals in temperature across the insulation drive moisture back and forth through the insulation layer and is exacerbated when the insulation material has a high water vapor permeance. A wetted compact low-slope roof is a good example of these phenomena. During the nighttime, moisture migrates upward through the insulation layer and can condense in the upper part of the insulation layer on the underside of the membrane. The following day, the increase in ambient temperature coupled with solar radiation heats the membrane and reverses the vapor pressure difference, evaporating some or all of the condensed water and driving it downward into the roofing system, transporting heat in the process. If sufficient water vapor can be driven downward, it may condense somewhere in the lower portion of the insulation layer, releasing its heat of condensation.

Hedlin (1988b) and Shuman (1980) experimentally showed that, for building envelopes containing permeable fibrous insulations that underwent temperature reversals, the rate of energy transfer increased sharply as the moisture content (MC) increased to approximately 1% by volume. The rate of increase in energy transfer diminished rapidly with further increases in moisture content. Energy transfer for permeable insulation with 1% MC by volume was roughly double that of dry insulation. Pedersen et al. (1991) analytically reproduced Hedlin's results; he demonstrated the high mobility of moisture in a permeable insulation and showed that latent effects are appreciable for a wide variety of North American climates. The latent effects typically add to the energy load and can increase peak energy demand. The extra load is added to the building load in the warm afternoon, and nearly the same amount of heat is removed in the cool evening.

Certain organic insulations such as wood fiberboard and perlite are hygroscopic and can contain 1% MC by volume if installed at equilibrium conditions; water leakage into the building envelope component is *not* a prerequisite for significant increases in heat flow.

In studies of hourly temperature and moisture content variations in wood frame wall cavities, Duff (1971) showed measurable levels of daily moisture migration across the cavity. Pedersen et al. (1991) demonstrated the same phenomenon in compact low-slope roofs.

The building envelope is exposed to ever-changing exterior conditions; fluctuations in the outdoor air temperature and the amount of solar radiation affect the temperature profile through the building envelope. These temperature changes affect the magnitude and the direction of the vapor pressure gradients. Water vapor transfer rates and direction are constantly changing, adjusting not only to the thermal changes but to changes in moisture concentration. However, conditions may exist where the *average* vapor pressure drive is in one direction, for example, upward in a low-slope roof during winter. In this situation, moisture accumulates in the insulation just below the impermeable roof membrane, and, assuming these conditions can be maintained for a reasonable length of time, water accumulates or frost forms unless the top layer of the insulation has adequate moisture absorption capacity (e.g., perlite or wood fiberboard). Under similar conditions in the summer, moisture is driven down to the vapor retarder or deck (if the deck is less permeable than the insulation layer). If there are no layers in the system that are less permeable than the insulation, the water vapor simply diffuses into the building interior.

Under conditions where the vapor pressure changes slowly or where the insulation layer has an extremely low water vapor permeance, little water vapor is transported. Moisture still affects the sensible heat transfer in the building envelope component. Epstein and Putnam (1977) and Larsson et al. (1977) showed a nearly linear increase in sensible energy transfer of approximately 3 to 5% for each volume percent increase in moisture content in cellular plastic insulations. For example, an insulation material with a 5% MC by volume has 15 to 25% greater energy transfer than the dry insulation material. Other field studies by Dechow and Epstein (1978) and Ovstaas et al. (1983) have shown similar results for insulations installed in below-grade applications such as foundation walls.

Moisture Effect on Heat Storage

Moisture also affects the thermal storage capacity of certain hygroscopic building materials. At 10% MC, nearly 30% of the heat storage capacity of wood is in the water held in the cell walls. Since the specific heat of wood is a function of its temperature and moisture content (but almost independent of density and species), heat storage calculations must include an estimate of the equilibrium in-service moisture content of wood building components.

PROPERTIES OF WATER VAPOR IN AIR

Chapter 6 describes the properties of moist air and defines the various terms associated with water vapor in air, such as dew-point temperature, dry-bulb and wet-bulb temperatures, and relative humidity. Chapter 6 also explains the use of the psychrometric chart, as well as the physics of heating and cooling moist air.

MOISTURE IN BUILDING MATERIALS

Many common building materials are porous. The pores provide a large internal surface area, which generally has an affinity for water. In some materials, such as wood, moisture may also be adsorbed in the cell wall itself. The amount of water in these **hygroscopic** (water-attracting) materials is related to the relative humidity (rh) of the surrounding atmosphere. When relative humidity of the surrounding air rises, hygroscopic materials gain moisture (**adsorption**), and when the relative humidity drops these materials lose moisture (**desorption**). The relationship between relative humidity and the moisture content (MC) at a particular temperature can be represented in a graph called a sorption isotherm. Often isotherms obtained by adsorption are not identical to isotherms obtained by desorption, because the material tends to retain moisture when it is drying. This difference between desorption and adsorption isotherms is called **hysteresis**. At high relative humidity, small pores become entirely filled with water. The maximum moisture content is reached when all pores are filled with water. But experimentally this can only be achieved in a vacuum, by boiling the material or by keeping the material in contact with water for an extremely long time (i.e., many years). In practice, the maximum moisture content of porous materials is lower. This lower maximum moisture content is sometimes referred to as the **capillary saturation moisture content**. Figure 9 shows typical sorption curves, giving the **equilibrium moisture content** (EMC) as a function of relative humidity. The EMC increases with relative humidity, especially above 80% rh. The EMC decreases slightly with increasing temperature.

Chapter 22 describes hygroscopic substances and their use as dehumidifying agents. Table 2 in Chapter 11 of the 1999 *ASHRAE Handbook—Applications* has data on the moisture content of various common materials in equilibrium with the atmosphere at various relative humidities.

Wood and many other hygroscopic materials change dimensions with variations in moisture content. The moisture content at which cell walls of wood are saturated but no free water exists in cell cavities is the **fiber saturation point**. It represents the upper limit for moisture gain from the air as water vapor. It is also the upper limit of swelling; although more water can be absorbed in cell cavities, additional swelling does not occur. Wood moisture content is expressed as a percentage of its oven-dry weight. The average fiber saturation point for most species is about 30%. The average EMC at 68°F and 45% rh (heating season indoor conditions) is about 8.5%. At 75°F and 70% rh (summer conditions), the EMC is about 13%.

The resulting dimensional changes in wood are proportional to the change in moisture content, but vary with species and direction of grain. For white oak, which is representative of interior trim material, 4.5% MC variation causes 2.5% volumetric shrinkage or swelling. Longitudinal dimensional change in straight-grained wood is insignificant, but it increases with crossgrain and other irregularities.

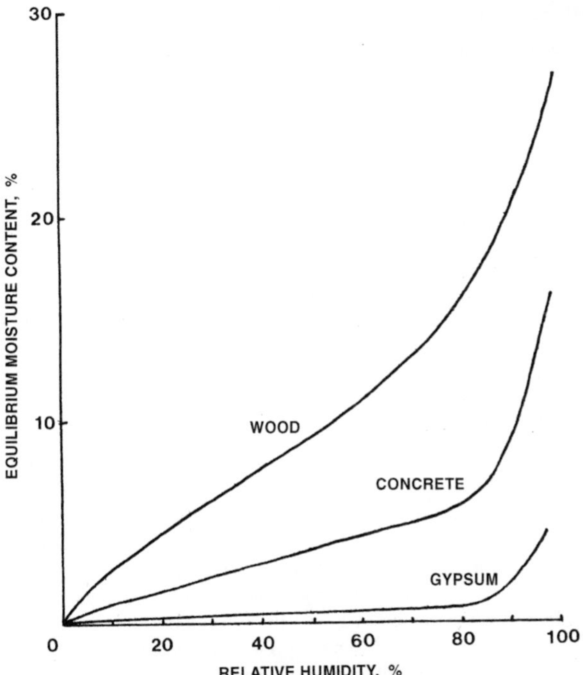

Fig. 9 Typical Sorption Isotherms for Wood, Concrete, and Gypsum (Hysteresis Is Ignored)

Table 2 Linear and Volumetric Shrinkage Values of Wood, from Green to Oven Dry Moisture Content

Wood Species	Radial %	Tangential %	Volumetric %
Hardwoods			
Birch			
Yellow	7.3	9.5	16.8
Oak			
Northern red	4.0	8.6	13.7
Southern red	4.7	11.3	16.1
White	5.6	10.5	16.3
Softwoods			
Cedar			
Northern white	2.2	4.9	7.2
Western red cedar	2.4	5.0	6.8
Douglas fir			
Coast	4.8	7.6	12.4
Fir			
White	3.3	7.0	9.8
Hemlock			
Western	4.2	7.8	12.4
Pine			
Eastern white	2.1	6.1	8.2
Longleaf	5.1	7.5	12.2
Ponderosa	3.9	6.2	9.7
Redwood			
Young growth	2.2	4.9	7.0
Imported wood			
Lauan	3.8	8.0	—

Note: Values expressed as a percentage of the green dimension (FPL 1999).

Accordingly, residential wood trusses with top and bottom chords exposed to different temperature and moisture regimes can show measurable seasonal vertical movements (Plewes 1976).

Thermal expansion of wood is usually outweighed by shrinkage or swelling due to moisture content changes. The linear thermal expansion coefficient for wood across the grain ranges from about 7×10^{-6} per °F for light wood species to about 25×10^{-6} per °F for dense wood species, which is small compared to that of many other materials. The thermal expansion coefficient parallel to the grain is between 1.7×10^{-6} and 2.5×10^{-6} per °F.

Most plant and animal fibers undergo dimensional changes similar to those noted in wood. Typical values for wood can be found in Table 2; the *Wood Handbook* (FPL 1999) gives more detailed information. Related changes in other materials are not as well documented. However, different expansion rates caused by temperature and moisture changes in different materials used in composite constructions should be considered (Baker 1964, BRS 1974).

Porous materials also absorb water in liquid form when in contact with it. Liquid water may be present because of leaks, rain penetration, flooding, or surface condensation. Wetting may be so complete that the material reaches the capillary saturation moisture content.

MOISTURE MIGRATION

Liquid water and water vapor migrate by a variety of moisture transport mechanisms. The following are some of the most important mechanisms:

- Liquid flow by gravity or air pressure differences
- Capillary suction of liquid water in porous building materials
- Movement of water vapor by air movement
- Water vapor diffusion by vapor pressure differences

Although in the past many moisture control strategies focused on control of vapor diffusion through the installation of vapor (diffusion) retarders, the other mechanisms, when present, can move far greater amounts of moisture. Therefore, liquid flow, capillary suction, and air movement should be controlled first.

Liquid flow by gravity and air pressure difference is not discussed here, but a short description of the other mechanisms follows. A more comprehensive treatment of moisture transport and storage may be found in Kumaran et al. (1994) and Kuenzel (1998).

Capillary Suction

Within small pores of diameter less than approximately 0.1 μm, molecular attraction between the surface of the capillary and the water molecules causes a capillary suction defined by

$$s = \frac{2\sigma\cos\theta}{r} \tag{10}$$

where

s = capillary suction
σ = surface tension of water
r = radius of the capillary
θ = contact wetting angle

The **contact wetting angle** is the angle between the water meniscus and the capillary surface. The smaller this angle is, the larger the capillary suction. In **hydrophilic** (water-attracting) materials, the contact wetting angle is less than 90°, and in **hydrophobic** (water-repelling) materials, the angle is between 90° and 180°. Capillary suction is greater in smaller capillaries, so capillary suction moves moisture from larger to smaller capillaries. Although surface tension is a function of temperature (the higher the temperature, the lower the surface tension), this effect is very small.

Isothermal and nonisothermal movement may occur in the liquid phase, or in the vapor phase if the capillaries are not completely filled. The transfer in the vapor phase is by vapor diffusion and is caused by a difference in vapor saturation pressure in the capillaries. **Thompson's law** states that the saturation vapor pressure in equilibrium with the water in a capillary is given by

$$p'' = p'\exp\left(\frac{s}{\rho RT}\right) = p'\exp\left(\frac{2\sigma\cos\theta}{r\rho RT}\right) \tag{11}$$

where

p'' = saturation vapor pressure in capillary
p' = saturation vapor pressure in ambient air at same temperature as p''
ρ = density of water
R = gas constant
T = absolute temperature

Equation (11) shows that the saturation vapor pressure is lower in smaller capillaries than in larger capillaries, which causes vapor diffusion from larger water-filled capillaries to smaller capillaries. Saturation vapor pressure in the capillaries is lower at lower temperatures, which causes diffusion from higher to lower temperatures.

If the vapor pressure in the ambient air is in equilibrium with the saturation vapor pressure in the capillaries, Equation (11) can be rewritten as

$$s = \rho RT \ln\phi \tag{12}$$

where ϕ is the relative humidity of the ambient air.

Capillary flow of water can be expressed as a function of suction pressure gradients follows:

$$w_m = -k_m \frac{ds}{dx} \tag{13}$$

where

w_m = water flux
k_m = water permeability coefficient

Air Movement

Water vapor movement by air can be represented by

$$w = W\rho v \tag{14}$$

where

w = water vapor flux (flow per unit area)
W = humidity ratio of source air
ρ = density of air
v = airflow velocity

Even small airflows can carry large amounts of water vapor when compared to vapor diffusion. Airflow retarders are designed to inhibit the flow of air and thereby the transport of water vapor into the construction.

Water Vapor Diffusion

Water vapor also migrates by diffusion through air and building materials. Although moisture diffusion through air is relatively rapid, distribution of vapor in air is dominated by convection. This results in minimal vapor pressure differentials between connected spaces and rapid flow to condensing surfaces, such as cold glass or dehumidifier coils. Although movement by air convection usually dominates, when present, vapor diffusion can be an important mode of transportation in industrial applications such as cold storage facilities or built-in refrigerators. The control of diffusion also becomes more important with increasingly airtight construction.

The equation used to calculate water vapor diffusion flux through materials is based on a form of Fick's law:

$$w = -\mu \frac{dp}{dx} \tag{15}$$

where

w = water vapor flux (flow rate per unit area)
p = water vapor pressure
x = distance along flow path
μ = water vapor permeability

According to Equation (15), water vapor flow by diffusion is proportional to the water vapor pressure gradient and closely parallels Fourier's equation for heat flow. The actual diffusion of vapor through a material is complex. The apparent water vapor permeability is a function of relative humidity and temperature and may vary spatially due to variations in material properties.

ASTM *Standard* E 96 describes test methods for the measurement of water vapor permeability. The rate of moisture transfer through a material is determined gravimetrically while maintaining a steady temperature and vapor pressure differential across the specimen. The standard identifies two tests: a dry cup (0% to 50% rh) and a wet cup (50% to 100% rh). In the wet-cup tests, liquid water movement may occur and the resulting water vapor transport coefficients should be treated with care (Kuenzel 1995).

Permeance is usually expressed in perms (grains/h·ft^2·in. Hg), and permeability in perm·in. Whereas permeability refers to water vapor flux per unit thickness, the term permeance is used in reference to a material of a specific thickness. For example, a material that is 2 in. thick generally is assumed to have half the permeance of 1 in. thick material, even though permeance of many materials often is not strictly proportional to thickness. In many cases, this assumption ignores the effect of moisture content variations within the material, and the effect of cracks or holes in the surface. It is inappropriate to use the term permeability in reference to inhomogeneous or composite materials, such as plywood or gypsum board with paper facings. Whenever a building product is made of two dissimilar materials, only water vapor permeance is meaningful for characterizing vapor transfer through that product.

Methods have been developed that allow measurement of water vapor transport with temperature gradients across the specimen (Douglas et al. 1992, Krus 1996, Galbraith et al. 1998). These test methods promise more accurate data on isothermal vapor transfer through materials and will eventually allow better distinction between various transport modes.

Combined Liquid and Vapor Flow

It is nearly impossible to experimentally distinguish between liquid flows and vapor flows in hygroscopic materials. To distinguish the moisture flow in each phase, nonisothermal tests may be performed. In most practical applications in moisture analysis both vapor and liquid flow may be treated as parallel processes. Therefore, moisture flow is often expressed as the sum of two transport equations—one using vapor pressure as a driving potential for vapor flow and the other using either capillary suction or relative humidity as driving potentials for liquid moisture flow. These conservation equations may be written as follows:

$$\frac{\partial m}{\partial t} = -\nabla(w_v + w_w) + S_w \tag{16}$$

where

m = moisture content of the building material
w_v = vapor diffusion flux
w_w = liquid transport flux
S_w = moisture source or sink

The symbol ∇ represents the nabla operator, which denotes divergence, or the gradients along the three spatial coordinates. The vapor and liquid transport flux densities may be given by

$$w_v = -\mu \nabla p \text{ or } w_v = -\mu \nabla(\phi p_{sat}) \tag{17}$$

$$w_w = -D_p \nabla s \text{ or } w_w = -D_\phi \nabla \phi \tag{18}$$

where

μ = water vapor permeability of building materials
s = capillary suction pressure
p_{sat} = water vapor saturation pressure
D_p = liquid transport coefficient related to capillary suction
D_ϕ = liquid transport coefficient related to relative humidity
ϕ = relative humidity

Previous methods of calculation used moisture content and temperature as driving potentials for combined vapor and liquid movement. This approach has the disadvantage that moisture content is discontinuous at interfaces between different materials and, therefore, cannot be used as a moisture flow potential at those interfaces. Equation (17) and Equation (18) provide the advantage of continuity at material interfaces and maintain physically meaningful material parameters.

WATER VAPOR RETARDERS AND AIRFLOW RETARDERS

Water vapor retarders and airflow retarders combine to control the movement of moisture and air. Although their functions are different, a single component may serve both functions. The designer assesses the needs for moisture and air movement control in the building envelope and provides a system that combines the required vapor retarder and airflow retarder properties.

Airflow Retarder Functions and Properties

In addition to a vapor retarder, control of moisture requires an effective airflow retarder (also referred to as an air barrier or air infiltration barrier). Without effective control of airflow, vapor retarders are completely ineffective. However, airflow may accelerate the drying of a wet building component, as described by Karagiozis and Salonvaara (1999a,b).

A vapor retarder may also be an airflow retarder, that is, an air/vapor retarder. In the past, many designs were based on this, and measures were taken to ensure that the vapor retarder was continuous in order to control airflow through it. This remains a valid approach. Some recent designs treat airflow retarders and vapor retarders as separate entities, but an airflow retarder should not be placed in a location where it can cause moisture to condense if it also has vapor retarder properties. For example, an airflow retarder placed on the cold side of a building envelope may cause condensation, particularly if the vapor retarder is ineffective and the airflow retarder is impermeable to moisture.

However, a cold-side air/vapor retarder that also has sufficient insulation may result in a lower potential for condensation by raising the temperature of the surface of the air/vapor retarder, but this requires careful installation and sealing of joints.

Air leakage characteristics of airflow retarders can be determined with ASTM *Standard* E 283 or ASTM *Standard* E 1424. Di Lenardo et al. (1995) discuss specific air leakage criteria for airflow retarders in cold climates. These specifications call for maximum permissible air leakage rates between 0.01 and 0.04 cfm/ft^2 (as measured with an air pressure difference of 0.3 in. of water), depending on the water vapor permeance of the outermost layer of the building envelope. The highest permissible air leakage rate of the airflow retarder applies if the permeance of the outermost layer is greater than 10 perms, and the lowest rate applies if the permeance is less than 1 perm. Intermediate values are provided as well. The recommendations apply only to heating climates.

The effectiveness of an airflow retarder can be greatly reduced if openings, even small ones, exist in it. Such openings can be caused by poor design, poor workmanship during application, poorly sealed joints and edges, insufficient coating thickness, improper caulking and flashing, uncompensated thermal expansion, mechanical forces, aging, and other forms of degradation. Common faults or leaks occur at electrical boxes, plumbing penetrations, telephone and television wiring, and other unsealed openings in the structure.

A ceiling airflow retarder needs to be continuous at chases for plumbing, ducts, flues, and electrical wiring. In flat roofing, mechanical fasteners are sometimes used to adhere the system to the deck. These often penetrate the airflow retarder, and the resulting hole may allow air and accompanying moisture leakage into the roof.

Because it resists airflow, an airflow retarder must withstand pressures exerted by chimney (stack) effects, wind effects, or both, during construction and over the life of the building. The magnitude of the pressure varies, depending on the type of building and the sequence of construction. At one extreme, single-family dwellings may be built with the exterior cladding partly or entirely installed and insulation in place before the airflow retarder is added. Chimney effects in such buildings are small even in cold weather, so stresses on the airflow retarder during construction are small.

At the other extreme, in tall buildings, wind and chimney-effect forces are much greater than they are in single-family or other low-rise buildings. A fragile, unprotected sheet material should not be used as an airflow retarder (or vapor retarder) in a tall building because it will probably be torn by the wind before construction is completed.

In summary, an airflow retarder must

- Meet air permeability requirements
- Be continuous, i.e.,
  - Tight joints in the airflow retarder must be constructible
  - Effective bonds in the airflow retarder must be made at intersections (e.g., wall/roof)
  - Dimensional changes due to temperature changes or shrinkage must be accommodated without damage to joints or the retarder material
- Be strong enough to support the stresses applied to it, i.e.,
  - It must not be ruptured or excessively deformed by air pressures due to wind and stack effects
  - Where an adhesive is used to complete a joint, it must be designed to withstand forces that might gradually peel it away

A small penetration across an airflow retarder may seriously affect its performance. A penetration concentrates the air/vapor flow in such a way that large local deposits of water and ice are possible. In this situation calculations of moisture flow and accumulation using permeance values are useless when airflow is involved.

In addition, the following properties of an airflow retarder may be important, depending on the application:

- Elasticity
- Thermal stability
- Fire and flammability resistance
- Inertness to deteriorating elements
- Ease of installation

More information on air leakage in buildings may be found in Chapter 26.

Vapor Retarder Functions and Properties

A vapor retarder retards water vapor diffusion but does not totally prevent its transmission. The requirements for buildings are entirely different than those for pipes and equipment. Conditions on the inside and outside of buildings vary continually, and air movement and ventilation can provide wetting as well as drying at various times. Moisture entering one side of a wall cavity can be stored and released at a later time, or transmitted immediately out of the cavity through the other side. Requirements for vapor retarders in building components are therefore not extremely stringent.

In contrast, moisture that enters the insulation of cold storage facilities, cold pipes, or equipment is unlikely to escape, except during periods when the facilities, pipes, or equipment are not in use and are allowed to warm up. Vapor retarders for cold pipe or equipment applications must therefore have an extremely low permeance (e.g., less that than 0.05 perm, or lower for severe conditions). However, Korsgaard (1993) demonstrated that water condensing on an insulated cold pipe can be removed continuously by wicking action with a specially designed and installed wick system, as long as pipe temperatures are above freezing.

In HVAC applications, vapor retarders are applied to thermal insulation on tanks, cold pipes, ducts, refrigerated enclosures, and buildings. If conditions are conducive to condensation, water vapor retarders help (1) keep the insulation dry, thereby reducing the cooling load; (2) prevent structural damage by rot, corrosion, or the expansion of freezing water; and (3) reduce paint problems on exterior wall construction (ASTM *Standard* C 755).

In addition to vapor permeance, the following properties of vapor retarders are important, depending on the application:

- Mechanical strength in tension, shear, impact, and flexure
- Adhesion
- Elasticity
- Thermal stability
- Fire and flammability resistance
- Inertness to other deteriorating elements
- Ease of fabrication, application, and joint sealing

Any vapor retarder's effectiveness depends on its vapor permeance, installation, and location within the insulated section. The vapor retarder is usually located at or near the surface exposed to the higher water vapor pressure. For residences in heating climates, this is usually the winter-warm side of the insulation.

Under conditions of reversible water vapor flow that can occur during temperature cycling of industrial insulations or of special-purpose buildings, the selection and location of water vapor retarders require special study and treatment (Stachelek 1955).

Vapor retarder material is usually a thin sheet or coating. However, a construction of several materials, some perhaps of substantial thickness, could also constitute a vapor retarder system.

Water vapor permeances and permeabilities of some vapor retarders and other building materials are given in Table 9 in Chapter 25.

Vapor retarders that allow substantial summer drying while functioning as effective vapor retarders during the heating season are sometimes called "smart" vapor retarders. One type of smart vapor retarder has a low permeance to vapor but is permeable to liquid water, allowing the drying of condensed moisture. Korsgaard and Pedersen (1989, 1992) describe such a vapor retarder composed of a synthetic fabric sandwiched between staggered strips of plastic film. The fabric wicks free water from the building envelope, while the plastic film retards vapor flow into it.

Another smart vapor retarder provides low vapor permeance at low relative humidities but much higher permeance at high relative humidity. During the heating season, indoor humidity usually is below 50%, and the permeance of the smart vapor retarder is low. In the summer and even during winter days with high solar heat gains, when the temperature gradient is reversed, moisture moving from the exterior of the wall or roof raises the relative humidity at the vapor retarder. This leads to a higher vapor permeance of the vapor retarder and the potential for the wall or roof to dry out. One such vapor retarder, a nylon film, is described by Kuenzel (1998). Below 50% rh, the permeance of the film is less than 1 perm, but it becomes more permeable at above 60% rh, reaching 36 perms at 90% rh.

Vapor retarder material for pipes is usually sheet metal with soldered seams; heavy foil with wide, sealed overlaps; plastic pipe; or other very low permeance systems.

Classification of Vapor Retarders

Historically, a material or system with a permeance of 1 perm or less qualifies as a vapor retarder. More recently, further classification of vapor retarders has been proposed. For example, the Canadian General Standards Board (CGSB) has specified Type I vapor retarders as retarders with a permeance of 0.25 perm or less, and Type II as retarders with a permeance of 0.75 perm or less before aging and 1 perm or less after aging.

Water vapor retarders are classified as rigid, flexible, or coating materials. Rigid retarders include reinforced plastics, aluminum, and stainless steel. These retarders usually are mechanically fastened in place and are vapor-sealed at the joints.

Flexible retarders include metal foils, laminated foil and treated papers, coated felts and papers, and plastic films or sheets. Such retarders are supplied in roll form or as an integral part of a building material (e.g., insulation). Accessory materials are required for sealing joints.

Coating retarders may be semifluid or mastic; paint (arbitrarily called surface coatings); or hot melt, including thermofusible sheet materials. Their basic composition may be asphaltic, resinous, or polymeric, with or without pigments and solvents, as required to meet design conditions. They can be applied by spray, brush, trowel, roller, dip or mop, or in sheet form, depending on the type of coating and surface to which it is applied. Potentially, each of these materials is an airflow retarder; however, to meet airflow retarder specifications, it must satisfy the requirements for strength, continuity, and air permeance.

Designers have many options. For example, the conditions for control of airflow and moisture movement might be achieved using an interior finish, such as drywall, to provide strength and stiffness, along with a low-permeability coating, such as a vapor retarder paint, to provide the required low level of permeance. In this case, edge sealing is needed to establish continuity with adjacent airflow retarder/vapor retarder components.

Other designs may use more than one component. However, (1) any component that qualifies as a vapor retarder usually also impedes airflow, and is thus subject to pressure differences that it must resist, and (2) any component that impedes airflow often also retards vapor movement and may promote condensation or frost formation.

Additional information regarding the control of moisture and airflow through the use of vapor retarders and airflow retarders may be found in Chapters 24 and 26 and in Construction Specifications Canada (1990).

Several studies found a significant increase in the apparent permeance of vapor retarders as a result of small holes in the vapor retarder. For example, Seiffert (1970) reports a 100-fold increase in the permeance of aluminum foil when it is 0.014% perforated, and a 4000-fold increase when 0.22% of the surface is perforated. In general, penetrations particularly degrade a vapor retarder's effectiveness if the vapor retarder has a very low permeance (e.g., polyethylene or aluminum foil). In addition, perforations may lead to additional air leakage, which further erodes the effectiveness of the vapor retarder.

STEADY-STATE DESIGN TOOLS

Traditional methods for moisture design of the exterior building envelope all have severe limitations, and the results are sometimes difficult to interpret. However, these methods are used by design professionals and form the basis for current codes dealing with moisture control and vapor retarders.

The three best-known manual steady-state design tools for evaluating the probability of condensation within exterior envelopes (exterior walls, roofs, or ceilings) are (1) the dew-point method, (2) the Glaser diagram, and (3) the Kieper diagram. All three methods compare the vapor pressures within the envelope, as calculated by simple vapor diffusion equations, with the saturation pressures, which are based on the calculated temperatures within the envelope. If the calculated vapor pressure is above the saturation pressure at any point within the envelope, condensation is indicated.

The **dew-point method**, used in North America, and the **Glaser diagram**, commonly used in Europe and elsewhere, are almost identical. They differ slightly in the formulation of the vapor diffusion equation for flow through a building material and in definition of terms; the main difference lies in the graphical procedures. These methods are often misused, especially when condensation is present.

The **Kieper diagram**, a variant of the previous two methods, was introduced by Kieper et al. (1976) and described in greater detail by Trethowen (1979) and TenWolde (1983, 1994). As with the dew-point method and the Glaser diagram, the Kieper diagram is based entirely on vapor diffusion theory. The advantages of this method are that (1) the same diagram can be used for different wall configurations, as long as indoor and outdoor conditions are not changed; and (2) the calculation does not need to be repeated if condensation is indicated.

Both the dew-point method and the Kieper diagram allow simple estimation of the effect of wall or roof cavity ventilation by representing the effect of ventilation on thermal and vapor transport through the addition of parallel thermal and vapor diffusion resistances (Trethowen 1979, TenWolde and Carll 1992). The parallel resistances account for the heat and vapor bypassing the exterior material layers with outside ventilation air. The magnitude of the parallel resistances may be determined from the following equations:

$$R_{t,\,par} = \frac{S}{\underline{Q}\rho c_p} \tag{19}$$

$$R_{v,\,par} = \frac{S}{\underline{Q}\rho c} \tag{20}$$

where

$R_{t,par}$ = parallel thermal resistance, °F·ft²·h/Btu
$R_{v,par}$ = parallel vapor flow resistance, rep
S = surface area of the wall or ceiling, ft²
$\underline{Q}$ = cavity ventilation airflow rate, ft³/h
ρ = density of air, lb/ft³
c = ratio of humidity ratio and vapor pressure, approximately 145 gr/lb·in. Hg
c_p = specific heat, Btu/lb·°F

Although this method of including ventilation only approximates the actual effects of ventilation, it can be a useful tool.

Many people advocate abandoning steady-state design tools because of their severe limitations. Perhaps the greatest limitation is that their focus is restricted to prevention of sustained interstitial condensation. Many building failures (such as mold and mildew, buckling of siding, and paint failure) are not necessarily related to surface condensation.

Conversely, limited condensation can often be tolerated, depending on the materials involved, the temperature conditions, and the speed at which the material dries out. Wetting and drying cycles cannot be accurately analyzed with steady-state tools because these tools neglect moisture storage in the building materials. Another weakness is that these methods exclude all moisture transfer mechanisms other than vapor diffusion. Results obtained with any of these methods should therefore be considered as approximations and be used with extreme care.

The validity and usefulness of any of these methods depend on the judicious selection of boundary conditions and material properties. Specifically, they should only be used to estimate seasonal mean conditions, rather than daily or even weekly mean conditions. Furthermore, permeances may vary with moisture content and the effect of rain splash, flashing imperfections, leaky or poorly formed joints, weather exposure, and sunshine can have overriding effects.

For those who want to use these simple tools, despite their shortcomings, a short description of the dew-point method is presented, with an example of its use. The Glaser diagram closely parallels the dew-point method. A comprehensive description of these methods can be found in TenWolde (1994).

Dew-Point Method

The dew-point method is based on a slight modification of diffusion Equation (15):

$$w = -\mu \frac{\Delta p}{d} \qquad (21)$$

where

w = water vapor flux through a layer of material, grains/h·ft²
μ = water vapor permeability of material, perm·in.
Δp = vapor pressure difference across the layer, in. Hg
d = thickness of the layer, in.

The term μ/d represents the permeance of the material. Water vapor resistance Z is defined as the inverse of the permeance and is expressed in reps (where rep = 1/perm). Then Equation (21) can be written as

$$w = -\frac{\Delta p}{Z} \qquad (22)$$

Example 1. The dew-point method is explained and demonstrated for the frame wall construction and materials described in Table 3. Assume 70°F, 40% rh indoors, and 20°F, 50% rh outdoors.

Solution:

Step 1. Calculate the temperature drop across each material. The temperature drop is proportional to the R-value as follows:

$$\frac{\Delta T_{material}}{\Delta T_{wall}} = \frac{R_{material}}{R_{wall}}$$

Table 4 lists the resulting temperature drops and resulting temperatures at each surface.

Step 2. Find the saturation vapor pressures corresponding to the surface temperatures (Table 4). These values can be found in Table 2 in Chapter 6, or in other psychrometric tables or charts.

Step 3. Calculate the vapor pressure drops across each material. These are calculated in much the same way as the temperature drops in step 1.

$$\frac{\Delta p_{material}}{\Delta p_{wall}} = \frac{Z_{material}}{Z_{wall}}$$

where

p = vapor pressure, in. of Hg
Z = vapor diffusion resistance, rep

From Table 3, the total resistance of the wall with the vapor retarder is

$$Z_{wall} = 1/160 + 1/5 + 1/30 + 1/0.5 + 1/35 + 1/1000 = 2.27 \text{ reps}$$

The vapor pressure drop across the wall is calculated from indoor and outdoor relative humidities and the indoor and outdoor saturation vapor pressures (see Table 4).

$$\Delta p_{wall} = p_{indoor} - p_{outdoor}$$
$$= (40/100)\,0.7392 - (50/100)\,0.1096 = 0.2409 \text{ in. Hg}$$

As with temperatures, the vapor pressures at the surfaces of each material can easily be determined from the vapor pressure drops. Table 5 lists the results for the example wall in the initial calculation column.

Table 3 Approximate Thermal and Vapor Diffusion Properties of Wall in Example 1

Air Film or Material	Thermal Resistance R, °F·ft²·h/Btu	Permeance M, perm	Diffusion Resistance Z, rep
Air film (still)	0.68	160[a]	0.0063[a]
Gypsum board, painted	0.45	5	0.2
Insulation	11	30	0.033
Plywood sheathing	0.62	0.5	2
Wood siding	1[b]	35[b]	0.029[b]
Air film (wind)	0.17	1000[a]	0.001[a]
Total	13.92	Not applicable	2.27

[a]Approximate values; permeances of surface air films are very large compared to those of other materials and do not affect results of calculations.
[b]Approximate values; permeance reflects limited ventilation of back of siding.

Table 4 Calculated Temperature Drops, Surface Temperatures, and Saturation Vapor Pressures in Example 1

Air Film or Material or Surface	Temperature Drop, °F	Surface Temperature, °F	Saturation Vapor Pressure at Surface, in. Hg
Indoor air	—	70	0.7392
Surface air film	2.4	—	—
Interior wall surface	—	67.6	0.6807
Gypsum board	1.7	—	—
Gypsum board/Insulation	—	65.9	0.6419
Insulation	39.5	—	—
Insulation/Sheathing	—	26.4	0.1434
Plywood sheathing	2.2	—	—
Sheathing/Siding	—	24.2	0.1309
Wood siding	3.6	—	—
Exterior wall surface	—	20.6	0.1124
Surface air film	0.6	—	—
Outdoor air	—	20	0.1096

Table 5 Initial and Final Calculations of Vapor Pressure Drops and Surface Vapor Pressures in Example 1

Air Film or Material or Surface	Saturation Vapor Pressure, in. Hg	Initial Calculation Vapor Pressure, in. Hg: Drop	Initial: At Surface	Final Calculation Vapor Pressure, in. Hg: Drop	Final: At Surface
Indoor air (40% rh)	0.7392	—	0.2957	—	0.2957
Surface air film	—	0.0007	—	0.0040	—
Interior wall surface	0.6807	—	0.2950	—	0.2917
Gypsum board	—	0.0212	—	0.1271	—
Gyp. board/Insulation	0.6419	—	0.2738	—	0.1646
Insulation	—	0.0036	—	0.0212	—
Insulation/Sheathing	0.1434	—	0.2702	—	0.1434
Plywood sheathing	—	0.2123	—	0.0874	—
Sheathing/Siding	0.1309	—	0.0579	—	0.0560
Wood siding	—	0.0030	—	0.0012	—
Exterior wall surface	0.1124	—	0.0549	—	0.0548
Surface air film	—	0.0001	—	0.00004	—
Outdoor air (50% rh)	0.1096	—	0.0548	—	0.0548

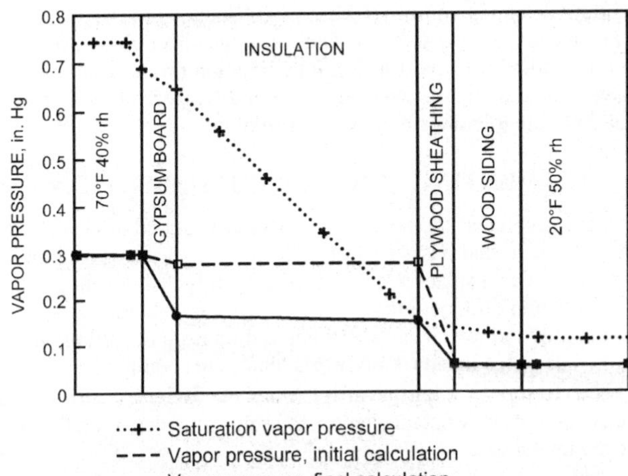

Fig. 10 Dew-Point Calculation in Example 1

Step 4. Figure 10 shows the saturation and calculated vapor pressures. Comparison with saturation pressures reveals that the calculated vapor pressure on the interior surface of the sheathing (0.2702 in. Hg) is well above the saturation pressure at that location (0.1434 in. Hg). This indicates condensation, probably on the surface of the sheathing. *It does not indicate condensation within the insulation.*

If the location of the condensation or the condensation rate is of interest, additional calculations (steps 5 and 6) are necessary.

Step 5. Figure 10 shows that the calculated vapor pressure exceeds the saturation vapor pressure by the greatest amount at the interior surface of the plywood sheathing. Therefore, this is the most likely location for condensation to occur. With condensation at that surface, the vapor pressure should equal the saturation vapor pressure at that location (see the final calculation column in Table 5).

Step 6. The change of vapor pressure on the plywood sheathing alters all other vapor pressures as well as the vapor flow through the wall. The calculation of vapor pressures is similar to the calculation in step 3, but the wall is now divided into two parts: one part on the interior of the condensation plane (i.e., gypsum board and insulation) and the other part on the exterior (plywood sheathing and wood siding). The vapor pressure drop over the first part of the wall is

$$\Delta p_1 = 0.2957 - 0.1434 = 0.1523 \text{ in. Hg}$$

and that over the second part is

$$\Delta p_2 = 0.1434 - 0.0548 = 0.0886 \text{ in. Hg}$$

The vapor diffusion resistances of both parts of the wall are

$$Z_1 = 1/160 + 1/5 + 1/30 = 0.24 \text{ rep}$$

$$Z_2 = 1/0.5 + 1/35 + 1/1000 = 2.03 \text{ reps}$$

The vapor pressure drops across each material can now be calculated from

$$\frac{\Delta p_{material}}{\Delta p_i} = \frac{Z_{material}}{Z_i} \quad \text{for } i = 1, 2$$

Final calculations of vapor pressure are shown in Table 5. The vapor pressure no longer exceeds the saturation vapor pressure, which means that the condensation plane was chosen correctly. Figure 10 shows the vapor pressure profile (labeled as Vapor pressure, final calculation).

Vapor flow is no longer the same throughout the wall—vapor flow into the wall from the indoor air increases as a result of the lower vapor pressure at the plywood surface, while flow from the wall to the outside decreases. The difference between the two flows is the rate of moisture accumulation:

$$w_c = \Delta p_1/Z_1 - \Delta p_2/Z_2 = 0.1523/0.24 - 0.0886/2.03 = 0.59 \text{ gr/h}\cdot\text{ft}^2$$

This amounts to about 0.002 lb/day·ft^2. In our example, the plywood surface is below freezing, and this moisture would probably accumulate as frost. The moisture content of the plywood would be increased by 1% after about a week of condensation at this rate.

The dew-point method can be summarized as follows:

1. Calculate temperature drops and surface temperatures.
2. Find corresponding saturation vapor pressures.
3. Calculate vapor pressure drops and vapor pressures.
4. Check whether the saturation pressure is above the vapor pressure at all surfaces; if so, no condensation is indicated. Vapor flow through the wall may be determined if desired.

If condensation is indicated, the following steps may be followed:

1. Select the condensation surface. In most cases it is the surface where the difference between the calculated vapor pressure and the saturation vapor pressure is the highest. Vapor pressure at this surface should be set equal to the saturation vapor pressure.
2. Recalculate the vapor pressures; if any vapor pressures are above saturation, steps 5 and 6 should be repeated with a different choice for the condensation surface.
3. If needed, calculate the rate of condensation.

MATHEMATICAL MODELS

The rapid advance in computer technology has made it possible to develop computer models capable of analyzing and predicting the thermal and moisture behavior of building components. A small number are extensions of the steady-state analysis methods described before, but with the **transient models** it is possible to predict hourly drying and wetting of building components in a variety of conditions, climates, and design configurations. The information in this section is largely based on a review of the state of the art of heat and moisture transport modeling for buildings, which identified 37 different models of various complexity (Hens 1996). Some of those models do not include moisture transport and are therefore not included here. Most of the models are research tools that are not readily available, but some are available either commercially, free of charge, or through a consultant. Karagiozis (2001) describes the more sophisticated design tools, indicating both the capabilities and the limitations. ASTM *Manual* 40 covers available design tools and approaches to investigate the performance of building envelope systems in terms of heat and moisture transport.

For many applications, the actual behavior of an assembly under transient climatic conditions must be simulated in order to account for short-term processes like driving rain absorption, summer condensations, and phase changes. Understanding the application limits of each model is an important part of applying mathematical models to develop design guidelines.

The features of a complete moisture analysis model include:

1. Transient heat, air, and moisture transport formulation, incorporating physics of
   - Vapor transport
   - Liquid transport
   - Airflow
   - Heat and moisture storage/capacity
   - Condensation and evaporation
   - Freezing and thawing
2. One- or two-dimensional spatial formulation
3. Material properties as functions of moisture content, relative humidity, or temperature
   - Thermal properties (density, heat capacity, thermal conductivity)
   - Moisture properties (porosity, sorption, water retention, vapor permeability, liquid diffusivity)
   - Airflow properties (air permeability)

4. Boundary conditions (generally on an hourly basis)
   - Incident solar radiation and sky radiation (depending on inclination and orientation)
   - Wind-driven rain at exterior surfaces (depending on location, and aerodynamics)
   - Wind speed, orientation, and pressure
   - Interior and exterior temperature and relative humidity
   - Interior moisture sources and stratification
5. Surface conditions
   - Convective heat transfer coefficients
   - Mass transfer coefficients
   - Short wave absorptivity
   - Long wave emissivity
   - Precipitation absorptivity
6. Building systems and subsystem effects
   - Water penetration rates through subsystems (joints, cracks, etc)
   - Air leakage (cracks, joints, e.g., around a window)
   - Additional sources of moisture

While not all these features are required for every analysis, additional features may be required in some applications, such as moisture flow through unintentional cracks and intentional openings. To accurately model these cracks and openings, supplementary experiments may be used to define their performance under various loads. In Feature 6, additional laboratory-controlled tests provide performance data specific to each subsystem (e.g., water entry through a crack) (Straube and Burnett 1997). In most cases, field measurement of system and subsystem effects is preferred.

The validation, verification, and benchmarking of hygrothermal models is a formidable task. Little internationally accepted experimental data exist to benchmark hygrothermal models. The main difficulty lies in the fact that it is difficult to measure moisture flows and moisture transport potentials even in laboratory-controlled conditions. Even a validated model should be verified for each new application. Two-dimensional and three-dimensional hygrothermal models are extremely complex and mostly require the direct expertise of the authors.

In most hygrothermal models, moisture content, temperature, relative humidity, and related fluxes are common outputs of simulations. Results have to be checked for consistency, accuracy, grid independency, and sensitivity to parameter changes. The results may be used to determine the moisture tolerance of an envelope subjected to various interior and exterior loads. The heat fluxes may be used to determine the thermal performance under the influence of moisture and airflow. Furthermore, the transient output data may be used to assess durability and indoor air quality. Post-processing tools concerning durability (e.g., corrosion, mold growth, freezing and thawing, hygrothermal dilation, and indoor air humidity) are currently being developed. For instance, Karagiozis and Kuenzel (1999) developed a model to estimate the rate of mold growth and corrosion. It is expected that hygrothermal models will be incorporated in whole-building simulation tools (Karagiozis and Salonvaara 1999b).

Transient models enable an hour-by-hour analysis of moisture conditions in building components and a much more realistic calculation than steady-state diffusion models. However, many are not easy to use and require judgment and expertise on the part of the user. Existing moisture transport models are one- or two-dimensional, which requires the user to devise a realistic representation of a three-dimensional building component. Users need to be aware of which transport phenomena and types of boundary conditions are included and which are not. For instance, some models are not able to handle rain wetting of the exterior.

Results from an analysis tend to be very sensitive to the choice of indoor and outdoor conditions, but exact conditions are seldom known. Indoor and outdoor conditions to be used for the purpose of building design have not yet been established, although standards are being developed. Accurate data for all the materials in a component is difficult to find. Finally, interpretation of the results is not easy; accurate data on what moisture and temperature conditions materials can tolerate are not often available.

PREVENTING SURFACE CONDENSATION

Surface condensation occurs when water vapor comes in contact with a surface that has a temperature lower than the dew point of that vapor. The insulation should be sufficiently thick to ensure that the insulation surface temperature always exceeds the dew-point temperature to prevent surface condensation from occurring on the warm side of insulated rooms, pipes, ducts, and equipment.

The concept of a **temperature index coefficient** τ, also called the condensation resistance factor, is useful for calculating the surface temperature:

$$\tau = \frac{t_s - t_c}{t_a - t_c} \tag{23}$$

where

t_s = surface temperature on warm side of insulation
t_c = temperature on cold side of insulation
t_a = ambient temperature on warm side of insulation

The minimum temperature index coefficient to avoid surface condensation is

$$\tau_{min} = \frac{t_d - t_c}{t_a - t_c} \tag{24}$$

where t_d is the dew point of ambient air.

For flat surfaces and insulation, the temperature index coefficient can be stated as

$$\tau = \frac{L/k}{L/k + 1/h_i} \tag{25}$$

where

L = thickness of the insulation
k = thermal conductivity of the insulation
h_i = surface heat transfer coefficient

The minimum insulation thickness to avoid surface condensation on a flat surface can be calculated from

$$L_{min} = \frac{\tau_{min}}{1 - \tau_{min}} \frac{k}{h_i} \tag{26}$$

The minimum insulation thickness for pipe insulation is given by

$$r_{s,min} \ln\left(\frac{r_{s,min}}{r_i}\right) = \frac{\tau_{min}}{1 - \tau_{min}} \frac{k}{h_i} \tag{27}$$

where

$r_{s,min}$ = minimum outer radius of the insulation
r_i = inner radius of the insulation

The term to the left of the equal sign in Equation (27) is called equivalent thickness. **Equivalent thickness** is the thickness of insulation on a flat surface required to give the same rate of heat transmission per unit area of outer surface of insulation as on a cylinder or pipe. Figure 11 provides a convenient way to convert equivalent thickness to actual thickness.

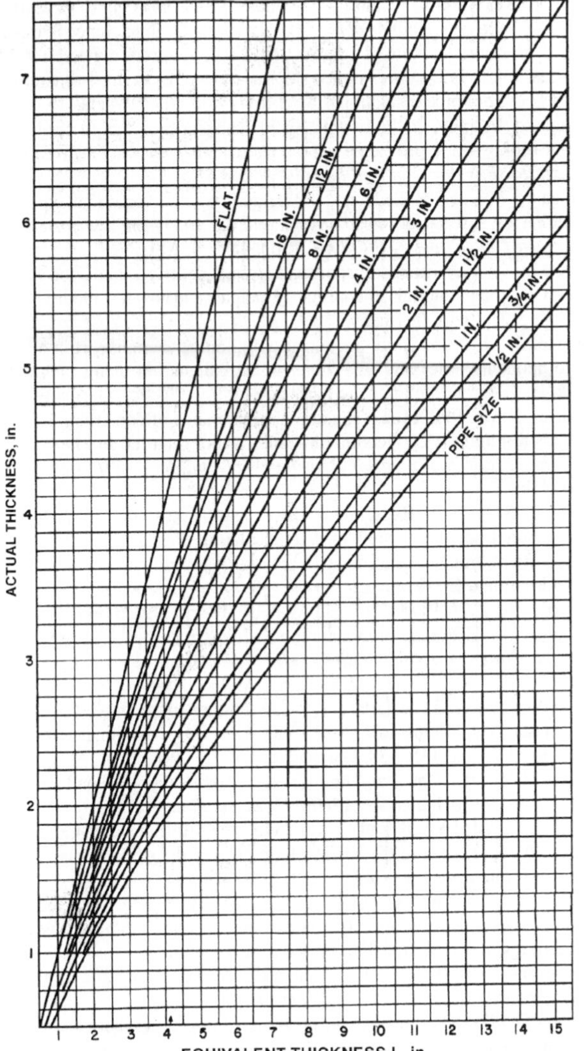

Fig. 11 Conversion of Equivalent Thickness to Actual Thickness for Pipe Insulation

STANDARDS

ASTM. 1997. Standard terminology relating to thermal insulating materials. *Standard* C 168-97.

ASTM. 1997. Standard test method for steady-state heat flux measurements and thermal transmission properties by means of the guarded-hot-plate apparatus. *Standard* C 177-97.

ASTM. 1989. Standard test method for steady-state thermal performance of building assemblies by means of guarded hot box. *Standard* C 236-89(1993).

ASTM. 1995. Standard test method for steady-state heat transfer properties of horizontal pipe insulation. *Standard* C 335-95.

ASTM. 1998. Standard test method for steady-state thermal transmission properties by means of the heat flow meter apparatus. *Standard* C 518-98.

ASTM. 1997. Standard practice for selection of vapor retarders for thermal insulation. *Standard* C 755-97.

ASTM. 1999. Standard classification of potential health and safety concerns associated with thermal insulation materials and accessories. *Standard* C 930-99.

ASTM. 1990. Standard test method for thermal performance of building assemblies by means of a calibrated hot box. *Standard* C 976-90(1996).

ASTM 1997. Standard practice for calculating thermal transmission properties from steady-state heat flux measurements. *Standard* C 1045-97.

ASTM. 2000. Standard test method for estimating the long-term change in the thermal resistance of unfaced rigid closed cell plastic foams by slicing and scaling under controlled laboratory conditions. *Standard* C 1303-00.

ASTM 1997. Standard test method for the thermal performance of building assemblies by means of a hot box apparatus. Standard C 1363-97.

ASTM. 1995. Standard test methods for water vapor transmission of materials. *Standard* E 96-95.

ASTM. 1991. Standard test method for determining the rate of air leakage through exterior windows, curtain walls, and doors under specified pressure differences across the specimen. *Standard* E 283-91.

ASTM. 1991. Standard test method for determining the rate of air leakage through exterior windows, curtain walls, and doors under specified pressure and temperature differences across the specimen. *Standard* E 1424-91(2000).

ASTM. Annual. *Annual book of ASTM standards.* Volume 4.06, *Thermal insulation and environmental acoustics.*

REFERENCES

ASHRAE. 1989. *Thermal Performance of Exterior Envelopes of Buildings IV.*

ASHRAE. 1992. *Thermal Performance of Exterior Envelopes of Buildings V.*

ASTM. 1974. Heat transmission measurements in thermal insulations. *Special Technical Publication* STP 544. American Society for Testing and Materials, West Conshohocken, PA.

ASTM. 1979. Thermal transmission measurements of insulation. *Special Technical Publication* STP 660.

ASTM. 1980. Thermal insulation performance. *Special Technical Publication* STP 718.

ASTM. 1983. Thermal insulation materials and systems for energy conservation in the 80s. *Special Technical Publication* STP 789.

ASTM. 1985. Guarded hot plate and heat flow meter methodology. *Special Technical Publication* STP 879.

ASTM. 1985. Building applications of heat flux transducers. *Special Technical Publication* STP 885.

ASTM. 1988. Thermal insulation—Materials and systems. *Special Technical Publication* STP 922.

ASTM. 1990. Insulation materials, testing, and applications. *Special Technical Publication* STP 1030.

ASTM. 1991. Insulation materials, testing, and applications, Second volume. *Special Technical Publication* STP 1116.

Baker, M.C. 1964. Thermal and moisture deformation in building materials. *Canadian Building Digest* 56. National Research Council of Canada, Ottawa.

Bomberg, M. and C. Shirtliffe. 1978. Influence of moisture and moisture gradients on heat transfer through porous building materials. ASTM *Special Technical Publication* STP 660:211-33.

Brandreth, D.A., ed. 1986. Advances in foam aging—A topic in energy conservation series. Caissa Editions, Yorklyn, DE.

BRS. 1974. Estimation of thermal and moisture movements and stresses, Parts 1, 2, and 3. *Building Research Station Digest* 227, 228, and 229, Watford, England.

Burch, D.M. and C.M. Hunt. 1978. Retrofitting an existing wood frame residence to reduce its heating and cooling energy requirements. *ASHRAE Transactions* 84:176.

Burge, H.A., H.J. Su, and J.D. Spengler. 1994. Moisture, organisms, and health effects. In *Moisture control in buildings.* ASTM Manual *Series*: MNL 18, Chapter 6.

Christian, J.E., A. Desjarlais, R. Graves, and T.L. Smith. 1995. Five-year field study confirms the PIMA standard for estimating polyisocyanurate insulation long-term thermal performance. 11th Conference of Roofing Technology, National Roofing Contractors Association, Rosemont, IL.

CIBSE. 1986. Thermal properties of building structures. In *CIBSE Guide: Reference Data.* The Chartered Institution of Building Services Engineers, London, England.

Construction Specifications Canada. 1990. Tek-AID on Air Barrier Systems. 1 St. Clair Street W., Toronto, M4W 1K6, Canada.

Dechow, F.J. and K.A. Epstein. 1978. Laboratory and field investigations of moisture absorption and its effect on thermal performance of various insulations. ASTM *Special Technical Publication* STP 660:234-60.

Di Lenardo, B., W.C. Brown, W.A. Dalgleish, K. Kumaran, and G.F. Poirier. 1995. Technical guide for air barrier systems for exterior walls of low-rise buildings. Canadian Construction Materials Centre, National Research Council Canada, Ottawa, Ontario.

Donnelly, R.G., V.J. Tennery, D.L. McElroy, T.G. Godfrey, and J.O. Kolb. 1976. Industrial thermal insulation, An assessment. Oak Ridge National Laboratory *Report* TM-5283 and TM-5515, TID-27120.

Douglas, J.S., T.H. Kuehn, and J.W. Ramsey. 1992. A new moisture permeability measurement method and representative test data. *ASHRAE Transactions* 98(2):513-19.

Duff, J.E. 1971. The effect of air conditioning on the moisture conditions in wood walls. USDA Forest Service Research *Paper* SE-78. Madison, WI.

Epstein, K.A. and L.E. Putnam. 1977. Performance criteria for the protected membrane roof system. Proceedings of the Symposium on Roofing Technology. National Institute of Standards and Technology, Gaithersburg, MD, and National Roofing Contractors Association, Rosemont, IL.

FEA. 1976. Economic thickness for industrial insulation. GPO No. 41-018-00115-8. U.S. Government Printing Office, Washington, D.C.

FPL. 1999. *Wood handbook*. FPL-GTR-113, Forest Products Laboratory, Forest Service, U.S. Department of Agriculture.

Galbraith, G.H., R.C. McLean, and J.S. Guo. 1998. Moisture permeability data presented as a mathematical relationship. *Building Research and Information* 20 (6):364-72.

Glaser, P.E., I.A. Black, R.S. Lindstrom, F.E. Ruccia, and A.E. Wechsler. 1967. Thermal insulation systems—A survey. NASA SP5027.

Hart, G.H. 1981. Heating the perimeter zone of an office building: An analytical study using the proposed ASHRAE comfort standard (55-74R). *ASHRAE Transactions* 87(2):529.

Hedlin, C.P. 1977. Moisture gains by foam plastic roof insulations under controlled temperature gradients. *Journal of Cellular Plastics* 13(5): 313-19.

Hedlin, C.P. 1983. Effect of moisture on thermal resistances of some insulations in a flat roof system under field-type conditions. ASTM *Special Technical Publication* STP 789:602-25.

Hedlin, C.P. 1985. Effect of insulation joints on heat loss through flat roofs. *ASHRAE Transactions* 91(2B):608-22.

Hedlin, C.P. 1987. Seasonal variations in the modes of heat transfer in a moist porous thermal insulation in a flat roof. *Journal of Thermal Insulation* 11:54-66.

Hedlin, C.P. 1988a. Heat flow through a roof insulation having moisture contents between 0 and 1% by volume, in summer. *ASHRAE Transactions* 94(2):1579-94.

Hedlin, C.P. 1988b. Heat transfer in a wet porous thermal insulation in a flat roof. *Journal of Thermal Insulation* 11:165-88.

Hens, H. 1996. Heat, air and moisture transfer in highly insulated envelope parts, task 1: Modelling. *Final Report*, Volume 1, International Energy Agency, Annex 24. Catholic University-Leuven, Laboratorium for Building Physics, Leuven, Belgium.

Hukka, A. and H.A. Viitanen. 1999. A mathematical model of mold growth on wooden materials. *Wood Science and Technology* 300, 475-85.

International Energy Agency. 1990. *Guidelines and practice*, Volume 2. International Energy Agency, Annex XIV, Leuven, Belgium.

Jespersen, H.B. 1960. The effect of moisture on insulating materials made of plastic foam, impregnated mineral wool, vermiculite, and concrete. Denmark Technological Institute.

Joy, F.A. 1957. Thermal effect of moisture on insulation containing moisture. In *Thermal conductivity measurements and applications of thermal insulations*. ASTM *Special Technical Publication* STP 217.

Karagiozis, A.N. 2001. Advanced hygrothermal modeling for hygrothermal research. In *Moisture Analysis Manual*, Chapter 10. ASTM *Manual* 40, American Society for Testing and Materials.

Karagiozis, A.N. and H.M. Kuenzel. 1999. A state-of-the-art moisture engineering model. *ASHRAE Transactions* 105(1).

Karagiozis, A.N. and Kuenzel H.M. 2001. Moisture engineering practice. *ASHRAE Transactions* 107(1).

Karagiozis, A.N. and H.M. Salonvaara. 1998. EIFS hygrothermal performance due to initial construction moisture as a function of air leakage, interior cavity insulation, and climate conditions. *Thermal Performance of the Exterior Envelopes of Buildings VII*, pp. 179-88. ASHRAE.

Karagiozis, A.N. and H.M. Salonvaara. 1999a. Hygrothermal performance of EIFS-clad walls: Effect of vapor diffusion and air leakage on the drying of construction moisture. ASTM STP 1352, pp 32-51.

Karagiozis, A.N. and H.M. Salonvaara. 1999b. Whole building hygrothermal performance. *Building Physics in the Nordic Countries* 2:745-53.

Kieper, G., W. Caemmerer, and A. Wagner. 1976. A new diagram to evaluate the performance of building constructions with a view to water vapor diffusion. Presented at CIB Working Group W40 Meeting, Washington, DC.

Knab, L.I., D.R. Jenkins, and R.G. Mathey. 1980. *The effect of moisture on the thermal conductance of roofing systems*. Building Science Series No. 123. National Institute of Standards and Technology, Gaithersburg, MD.

Korsgaard, V. 1993. Innovative concept to prevent moisture formation and icing of cold pipe insulation. *ASHRAE Transactions* 99(1):270-73.

Korsgaard, V. and C.R. Pedersen. 1989. Transient moisture distribution in flat roofs with hygro diode vapor retarder. In *Thermal Performance of Exterior Envelopes of Buildings IV*. ASHRAE.

Korsgaard, V. and C.R. Pedersen.1992. Laboratory and practical experience with a novel water-permeable vapor retarder. In *Thermal Performance of Exterior Envelopes of Buildings V*, pp. 480-90. ASHRAE.

Krus, M. 1996. Moisture transport and storage coefficients of porous mineral building materials, theoretical principals and new test methods, IRB-Verlag, Stuttgart, Germany.

Kuenzel, H.M. 1995. Simultaneous heat and moisture transport in building components, one- and two-dimensional calculation using parameters. Ph.D. thesis. Fraunhofer Institute for Building Physics, Germany.

Kuenzel, H.M. 1998. More moisture load tolerance of construction assemblies through the application of a smart vapor retarder. In *Thermal Performance of the Exterior Envelopes of Buildings VII*, pp. 129-32. ASHRAE.

Kumaran, M.K. 1989. Experimental investigation on simultaneous heat and moisture transport through thermal insulation. *Proceedings* of the (Conseil International du Batiment-International Building Council) CIB 11th International Conference 2:275-84.

Kumaran, M.K., G.P. Mitalas, and M.T. Bomberg. 1994. Fundamentals of transport and storage of moisture in building materials and components. In *Moisture control in buildings*. ASTM Manual *Series*: MNL 18, Chapter 1.

Kyle, D.M. and A.O. Desjarlais. 1994. Assessment of technologies for constructing self-drying low-slope roofs. Oak Ridge National Laboratory *Report* ORNL/CON-380. Oak Ridge, TN.

Lander, R.M.1955. Gas is an important factor in the thermal conductivity of most insulating materials. *ASHRAE Transactions* 61:151.

Langlais, C., M. Hyrien, and S. Klarsfeld. 1983. Influence of moisture on heat transfer through fibrous insulating materials. ASTM *Special Technical Publication* STP 789:563-81.

Larsson, L.E., J. Ondrus, and B.A. Petersson. 1977. The protected membrane roof (PMR)—A study combining field and laboratory tests. Proceedings of the Symposium on Roofing Technology. National Institute of Standards and Technology, Gaithersburg, MD, and National Roofing Contractors Association, Rosemont, IL.

Lewis, J.E. 1979. Thermal evaluation of the effects of gaps between adjacent roof insulation panels. *Journal of Thermal Insulation* (October): 80-103.

Lotz, W.A. 1969. Facts about thermal insulation. *ASHRAE Journal* (June): 83-84.

Merrill, J.L. and A. TenWolde. 1989. Overview of moisture-related damage in one group of Wisconsin manufactured homes. *ASHRAE Transactions* 95(1):405-14.

Ojanen, T., R. Kohonen, and M.K Kumaran. 1994. Modeling heat, air, and moisture transport through building materials and components. In *Moisture control in buildings*. ASTM Manual *Series*: MNL 18, Chapter 2.

Ostrogorsky, A.G. and L.R. Glicksman. 1986. Laboratory tests of effectiveness of diffusion barriers. *Journal of Cellular Plastics* 22:303.

Ovstaas, G., S. Smith, W. Strzepek, and G. Titley. 1983. Thermal performance of various insulations in below-earth-grade perimeter application. ASTM *Special Technical Publication* STP 789:435-54.

Palfey, A.J. 1980. Thermal performance of low emittance building sheathing. *Journal of Thermal Insulation* 3.

Paljak, I. 1973. Condensation in slabs of cellular plastics. *Materiaux et Constructions* 6(31):53.

Parmelee, G.V. and W.W. Aubele. 1950. ASHVE Research *Report* No.1399. Heat flow through unshaded glass: Design data for load calculations. *ASHVE Transactions* 56:371.

Parmelee, G.V. and R.G. Huebscher. 1947. Forced convection heat transfer from flat surfaces. *ASHVE Transactions* 53:245.

Pedersen-Rode, C. 1990. Combined heat and moisture transfer in building construction. *Report* No.214. Thermal Insulation Laboratory, Technical University of Denmark, Lyngby, Denmark.

Pedersen-Rode, C., T.W. Petrie, G.E. Courville, P.W. Childs, and K.E. Wilkes. 1991. Moisture migration and drying rates for low slope roofs—Preliminary results. Proceedings of the 3rd International Symposium on Roofing Technology. National Roofing Contractors Association, Rosemont, IL.

Pelanne, C.M. 1977. Heat flow principles in thermal insulation. *Journal of Thermal Insulation* 1:48.

Pelanne, C.M. 1979. Thermal insulation heat flow measurements: Requirements for implementation. *ASHRAE Journal* 21(3):51.

Plewes, W.G. 1976. Upward deflection of wood trusses in winter. Building Research *Note* 107, National Research Council of Canada, Ottawa.

Raber, B.F. and F.W. Hutchinson. 1945. Radiation corrections for basic constants used in the design of all types of heating systems. *ASHVE Transactions* 51:213.

Robinson, H.E., F.J. Powlitch, and R.S. Dill. 1954. The thermal insulating value of airspaces. Housing and Home Finance Agency, Housing Research *Paper* No. 32, U.S. Government Printing Office, Washington, D.C.

Rowley, F.B., R.C. Jordan, C.E. Lund, and R.M. Lander. 1952. Gas is an important factor in the thermal conductivity of most insulating materials. *ASHVE Transactions* 58:15.

Rudoy, W. 1975. Effect of building envelope parameters on annual heating/cooling load. *ASHRAE Journal* 17(7):19.

Seiffert, K. 1970. *Damp diffusion and buildings.* Elsevier, Amsterdam, Netherlands.

Shapiro, A.P. and S. Motakef. 1990. Unsteady heat and mass transfer with phase change in a porous slab: Analytical solutions and experimental results. *International Journal of Heat and Mass Transfer* 33(1):163-73.

Shuman, E.C. 1980. Field measurement of heat flux through a roof with saturated thermal insulation and covered with black and white granules. ASTM *Special Technical Publication* STP 718:519-39.

Simons, E. 1955. In-place studies of insulated structures. *Refrigerating Engineering* 63:40 and 63:128.

Spielvogel, L.G. 1974. More insulation can increase energy consumption. *ASHRAE Journal* 16(1):61.

Stachelek, S.J. 1955. Vapor barrier requirements for cold storage structures. *Industrial Refrigeration* 34.

Straube, J.F. and E.F.P. Burnett. 1997. Rain control and screened wall systems. 7th Conference on Building Science and Technology, Durability of Buildings-Design, Maintenance, Codes and Practices, pp. 17-37. University of Waterloo, Ontario, Canada.

TenWolde, A. 1983. The Kieper and MOISTWALL moisture analysis methods for walls. In *Thermal Performance of Exterior Envelopes of Buildings II*, pp. 1033-51. ASHRAE.

TenWolde, A. 1994. Design tools. In *Moisture control in buildings.* ASTM *Manual* 18, Chapter 11.

TenWolde A. and C. Carll. 1992. Effect of cavity ventilation on moisture in walls and roofs. In *Thermal Performance of the Exterior Envelopes of Buildings V*, pp. 555-62. ASHRAE.

Thomas, W.C., G.P. Bal, and R.J. Onega. 1983. Heat and moisture transfer in glass fiber roof-insulating material. ASTM *Special Technical Publication* STP 789:582-601.

Tobiasson, W. and J. Richard. 1979. Moisture gain and its thermal consequences for common roof insulations. Proceedings of the 5th Conference on Roofing Technology, National Institute of Standards and Technology, Gaithersburg, MD, and National Roofing Contractors Association, Rosemont, IL.

Tobiasson, W., A. Greatorex, and D. Van Pelt. 1987. Wetting of polystyrene and urethane roof insulations in the laboratory and on a protected membrane roof. In *Thermal Insulation, Materials and Systems.* ASTM *Special Technical Publication* STP 922:421-30.

Tobiasson, W., A. Greatorex, and D. Van Pelt. 1991. New wetting curves for common insulations. In *International Symposium on Roofing Technology.* National Institute of Standards and Technology, Gaithersburg, MD, and National Roofing Contractors Association, Rosemont, IL.

Trechsel, H.R. and M. Bomberg, eds. 1989. *Water vapor transmission through materials and systems.* ASTM *Special Technical Publication* STP 1039.

Trethowen, H.A. 1979. The Kieper method for building moisture design. BRANZ Reprint 12, Building Research Association of New Zealand, New Zealand.

Tye, R.P. 1987. Assessment of foam in-place urethane foam insulation used in buildings. ORNL/Sub-86/56525/1. Oak Ridge National Laboratory, Oak Ridge, TN.

Tye, R.P. 1988. Aging of cellular plastics: A comprehensive bibliography. *Journal of Thermal Insulation* 11:196-222.

Tye, R.P. and C.F. Baker. 1987. Development of experimental data on cellular plastic insulation under simulated winter exposure conditions. In *Thermal Insulation, Materials and Systems*, ASTM *Special Technical Publication* STP 922:518-37.

Tye, R.P. and A.O. Desjarlais. 1981. Performance characteristics of foam-in-place urea formaldehyde insulation. ORNL/Sub-78/86993/1. Oak Ridge National Laboratory, Oak Ridge, TN.

Verschoor, J.D. 1977. Effectiveness of building insulation applications. USN/CEL *Report* No. CR78.006—NTIS No. AD-AO53 452/9ST.

Verschoor J.D. 1985. Measurement of water vapor migration and storage in composite building construction. *ASHRAE Transactions* 91(2):390-403.

Verschoor, J.D. and P. Greebler. 1952. Heat transfer by gas conductivity and radiation in fibrous insulations. *ASME Transactions* 74(6):961-68.

Wilkes, K.E. and P.W. Childs. 1992. Thermal performance of fiberglass and cellulose attic insulations. In *Thermal Performance of the Exterior Envelopes of Buildings V*, pp. 357-67. ASHRAE.

Wilkes, K.E. and J.L. Rucker. 1983. Thermal performance of residential attic insulation E. *Energy and Buildings* 5:263-77.

BIBLIOGRAPHY

ASHRAE. 1979. *Thermal Performance of Exterior Envelopes of Buildings.*

ASHRAE. 1982. *Thermal Performance of Exterior Envelopes of Buildings II.*

ASHRAE. 1985. *Thermal Performance of Exterior Envelopes of Buildings III.*

ASTM. 1989. Standard practice for determination of heat gain or loss and the surface temperature of insulated pipe and equipment systems by the use of a computer program. *Standard* C 680-89(1995). American Society for Testing and Materials, West Conshohocken, PA.

ASTM. 1994. Standard test method for water vapor transmission rate of flexible barrier materials using an infrared detection technique. *Standard* F 372-94.

Bomberg, M., ed. 1991. Moisture research in North America: 25 Years of building science. Lund University, Lund, Sweden.

ECON-I. 1973. How to determine economic thickness of thermal insulation. Thermal Insulation Manufacturers Association, Mt. Kisco, NY.

FEA. 1974. Retrofitting existing housing for energy conservation: An economic analysis. Federal Energy Administration, Washington, D.C.

Hite, S.C. and J.L. Dray. 1948. Research in home humidity control. Research Series No. 106. Purdue University, Engineering Experiment Station, West Lafayette, IN.

Rose, W.B. and A. TenWolde, eds. 1993. *Bugs, mold & rot II: Workshop proceedings.* National Institute of Building Sciences, Washington, D.C.

Trechsel, H.R., ed. 1994. *Moisture control in buildings.* ASTM Manual *Series*: MNL 18.

CHAPTER 24

THERMAL AND MOISTURE CONTROL IN INSULATED ASSEMBLIES—APPLICATIONS

IN THE ORIGINAL planning phase of buildings, the thermal and moisture design and long-term performance must be considered. Installation of adequate insulation and moisture control assemblies during construction can be much more economical than installation later. Proper selection of thermal insulation and moisture control assemblies must be based on

- Thermal and moisture properties of the materials
- Other properties required by the location of the materials
- Space availability
- Compatibility of the materials with adjacent materials
- Interior and exterior climate
- Performance expectations

Types of thermal insulation, their properties, economic thickness, and principles of moisture control and moisture transport are discussed in Chapters 23 and 25. Insulation in various assemblies that can be used interchangeably for a given construction, as well as specific moisture control options for various climatic regions, are discussed in this chapter. For specific industrial applications of insulated assemblies see the appropriate chapter in other ASHRAE Handbooks. In the 1998 *ASHRAE Handbook—Refrigeration,* for refrigerators and freezers, see Chapters 47, 48, and 49; for insulation systems for refrigerant piping, see Chapter 32 and this chapter; for refrigerated facility design, see Chapters 13 and 39; for trucks, trailers, and containers, see Chapter 29; for marine refrigeration, see Chapter 30; and for environmental test facilities, see Chapter 37.

GENERAL BUILDING INSULATION PRACTICE

WOOD FRAME CONSTRUCTION

Wood framing members and structural panels such as plywood, particleboard, and fiberboard only provide limited resistance to heat flow; therefore, wood frame construction is well suited to application of both cavity insulation and surface-applied insulation. The most common materials for cavity insulation are glass fiber, mineral fiber, cellulose, and spray-applied foams. For surface applications, a wide variety of sheathing insulations exists.

Roof decks of wood, metal, or preformed units may be insulated on top of or below the deck.

Attic construction with conventional rafters and ceiling joists or roof trusses can be insulated between framing members with batt, blanket, or loose-fill insulation. In warm climates, radiant barriers, low emissivity surfaces, and reflective insulations further reduce cooling loads. The Radiant Barrier Attic Fact Sheet (DOE 1991) provides information on climatic areas best suited for radiant barrier applications. This document also provides comparative information on the relative performance of these products versus conventional fibrous insulations.

The cavities of **cathedral ceiling construction** (in which the ceiling insulation and interior finish are parallel to the roof plane) can be insulated using glass fiber, cellulose, rigid foam, or spray-applied insulation. The surface above cathedral ceiling framing may be insulated with insulating panels or structural insulated panels (SIPs). The placement of insulation directly beneath a sloped roof deck in standard flat-ceiling construction, with or without ventilation, has been called "cathedralized" construction.

The **wall cavities** of wood frame construction can be insulated with batt, blanket, and loose-fill or spray-applied insulation. When using insulation materials in wall applications, extra care must be taken during the installation to eliminate voids within the wall cavity. When installing loose-fill insulation during retrofit of existing construction, all cavities should be checked prior to installation for obstructions such as fire stop headers and wiring that could prevent complete filling of the cavity. In addition, the material must be installed at the manufacturer's recommended density to ensure the desired thermal performance.

In addition to being properly insulated, the **exterior envelope** of a building should be constructed to minimize airflow into or through the building envelope. Airflow may degrade the thermal performance of insulation and cause excessive moisture accumulation in the building envelope. The use and function of airflow retarders are discussed in both in Chapter 23 and in this chapter.

COLD-FORMED STEEL FRAME CONSTRUCTION

Conventional light frame construction with cold-formed steel framing has many characteristics in common with light frame wood construction. The greatest differences are the increased thermal conductivity and the dimension and shape characteristics of steel framing members. Barbour et al. (1994) and Tuluca and Gorthala (1999) found that the conductivity and framing member spacing

The preparation of this chapter is assigned to TC 4.4, Building Materials and Building Envelope Performance.

have the most significant roles in determining overall R-value (U-factor). They determined that the thickness of cold-formed steel does not play a significant role in the R-value (U-factor). The parallel path method overestimates the R-values for building assemblies containing cold-formed steel framing, and the isothermal planes calculation methods underestimates the R-value. For methods to calculate heat flow in framing assemblies using cold-formed steel, see Chapter 25.

The most common cavity insulation materials are glass fiber, mineral fiber, cellulose, and spray-applied foams. Cavity insulation should be full width to cover between the steel framing members; it should not be a nominal width such as used for wood stud construction. Surface applications of continuous insulation sheathing, such as rigid foam board (i.e., extruded or expanded polyisocyanurate or polystyrene), may be applied to the exterior or interior of the assembly framing to provide a thermal break. Rigid board insulation applied as a sheathing does not provide structural lateral bracing.

Thermal bridging may be more severe with cold-formed steel-framed trusses than with conventional joist and rafter framing. Roof trusses have chords and web members that extend through the insulation into an unconditioned attic space and act as a thermal bridge to the bottom chord. A continuous insulating sheathing on the exterior or interior of the roof framing provides a thermal break.

HEAVY STEEL FRAME CONSTRUCTION

Buildings and structures with heavy, steel framing supports and exterior metal cladding are usually insulated between the frame and cladding with faced blanket or spray insulation. In heating climates, the facing may serve as a combination vapor retarder and interior finish, which must be protected against physical damage. Other types of insulation can be used by adding framing or furring to the inside of frame or exterior siding. Insulation securements that compress batt and blanket insulations reduce thermal resistance. This reduction, plus the thermal bridging caused by the screw or bolt penetrating the insulation, may cause condensation during cold weather.

The exterior cladding of structural steel-framed buildings may be (1) custom walls for specific projects, where most construction components are developed on a one-time basis; (2) commercial walls, where standard components are adapted to a particular building design; and (3) industrial-type walls, where standard metal sheets are fluted or ribbed to form field-assembled sandwiches to meet job conditions.

The cladding may consist of prefabricated panels or sections that may be classified as one of the following general types:

1. To withstand the elements, **single-thickness facings** are usually inserted in subframing with the windows to form a veneer over separate backup walls. Thermal insulation is integral with, or added to, the exterior of the backup wall and covers the edge of the floor slab or spandrel beam. In heating climates, this method reduces heat loss and keeps floors warm at their exterior edge. Insulating the exterior of the framing reduces thermal movement of the building structure.
2. **Sandwich construction** or adhesive-bonded panels are generally three-ply: exterior skin, core materials, and interior skin. This type of panel, when manufactured with concrete faces and an insulation core, comes in large sizes and can be attached directly to the exterior of the building frame. When using metal facings, the width of the panels is usually restricted to match that of the standard formed sheet. These panels are installed on the exterior of the building framing. The thermal performance required for a particular installation (including the interfaces between walls, floors, ceiling, doors, and windows) should be checked carefully because adding insulation later is difficult.
3. **Mechanically fastened panels** generally have a hollow box shape, with the exterior facing nesting over the interior facing. The cavity can be filled with flexible or semirigid insulation. The edges should vent to the exterior of the building to allow the panel to serve as a rain screen. The panels are normally installed in a subframing system.
4. **Industrial metal panels** have an exterior facing of standard ribbed, corrugated V-beam materials and an interior facing of proprietary metal pans or standard corrugated or ribbed sheeting. Insulation varies from semirigid to rigid, depending on the design of the inner surfacing materials. It provides good thermal and moisture control when installed with tight inner surface construction.

All curtain wall panels with insulation in the cavity, or as a core, should be sufficiently tight at their edges to prevent the entrance of free water or moisture. However, because some moisture may enter the wall from the inside or outside, the wall should be capable of drying out. Panel edges should not be hermetically sealed. For cold climates, the subframing members or wall mullions should be of noncontinuous construction. The exterior to interior path should have a thermal break or insulated mullion cover on the interior or exterior to reduce the hazard of condensation.

MASONRY AND CONCRETE CONSTRUCTION

Because concrete masonry unit walls and masonry cavity walls have hollow vertical cavities, insulating materials can be placed within the wall itself. Loose-fill insulation, such as water-repellent perlite and vermiculite, as well as foam inserts and foamed-in-place insulations, are used. This insulation method is more effective with low-density masonry units. Rigid insulation can be placed between the wythes in a cavity wall or veneered construction, and furring on the inside of the wall is still used in many areas.

The thickness and density of concrete masonry wall construction coupled with its interior air cavities moderate the heat transmission. The thermal mass or inertia of such heavy construction causes a time lag in heat migration, which may lower the peak gains or losses. Insulation placed on the outside of a concrete masonry wall enhances this time lag effect and helps protect the structure from expansion and contraction caused by temperature extremes.

Precast or poured-in-place solid concrete walls are insulated similarly to solid masonry construction. The design and method of fabricating the panels dictates the type of insulation selected.

Insulating concrete forms are manufactured from foam plastic. Generally they are made in units that resemble concrete masonry units, although they are typically somewhat larger. The individual units, which have interlocking edges, are stacked to the desired height to create a form for the concrete for a wall. Where openings for doors and windows are desired, units are omitted. The top, bottom, and sides of the opening are formed and braced with framing lumber to retain the concrete in the same way that an opening would be formed in a conventionally formed concrete wall.

Steel reinforcing and anchor bolts, if required, are placed, the forms are plumbed and braced, and concrete is placed in the form. To this point, construction of the concrete wall proceeds in much the same way as it would for a conventionally formed concrete wall. However, the foam plastic forming material is left in place to serve as building insulation and to reduce sound transmission through the wall after the concrete has hardened.

Although the concrete contributes to the R-value of the wall, the primary insulation value is obtained from the foam plastic forms. The R-value for the walls depends on the thickness and type of foam plastic, but the typical R-value for an approximately 8 in. thick wall ranges from 20 to 30. The method used to tie the foam plastic panels together to create the form also affects the R-value. Ties to connect the outside panels are usually made of plastic or metal, but some manufactured forms use continuous foam plastic across the width of the wall.

FOUNDATION AND FLOOR SYSTEMS

Perimeter foundation insulation may be applied on the outside or inside of foundation walls. Rigid mineral fiber or cellular plastic insulation is commonly used as perimeter exterior insulation. Exterior insulation may be applied below grade vertically or may extend as skirting down and outward from the building. These two profiles have comparable heat loss. Insulation applied on the exterior, especially skirting, should resist compressive forces from the soil and from the backfilling process. The thermal performance of vertically applied insulation may be degraded if it remains wet, so drainage at the base of the foundation may be required by manufacturers of some perimeter insulations. Some insulation products are designed to facilitate vertical drainage.

Perimeter insulation usually needs protection from physical damage and ultraviolet radiation if it extends above grade level. The exposed section of a foundation that extends from grade up to the top of the foundation can be a source of considerable heat loss. However, in buildings subject to termite infestation, the exposed foundation at grade should remain uncovered to facilitate inspection.

Exterior insulation that extends as skirting from a **shallow foundation** can prevent frost heave of the slab without a perimeter footing down to frost depth. The insulation thickness is selected so that even during the coldest weather, the soil beneath the insulation remains above freezing.

A **slab foundation** may be insulated by vertical and horizontal insulation, usually of rigid foam. Insulation at the perimeter of the building provides more resistance to heat loss than beneath the slab inside the building. Perimeter insulation usually includes vertical (installed either inside or outside the stem wall) and horizontal panels. The horizontal panels may extend from the foundation wall inward for a distance of 2 ft or more, or outward as skirting.

In **crawl space construction**, insulation may be applied either to the perimeter walls or to the floor framing. Rigid foam panels are often installed at the inside of the crawl space walls. The cavities at the band joist may be insulated with batts or custom fitted panels of rigid insulation. In vented crawl space construction, the floor framing may be insulated—commonly with batt or blanket insulation. The facing is stapled to the underside of the floor framing and the exposed batt faces down. Unless additional measures are taken to keep the insulation in place, such as wire strapping, rigid insulation panels, or a reinforced plastic membrane (like the "belly paper" used in manufactured housing), it may fall down.

Another floor insulation method uses double-sided, perforated aluminum foil draped and stapled over floor joists (BRANZ 1983). For a sag depth of about 4 in., laboratory R-values had a mean of 6.8 and ranged from 5.7 to 14.2 $ft^2 \cdot °F \cdot h/Btu$, depending on the humidity above the floor space. Field measurements have shown R-values up to 14 $ft^2 \cdot °F \cdot h/Btu$ for carpeted floors over good installations with a sag of 4 in.

Basement insulation may be applied at the interior as well as the exterior. Insulation should not be placed on the interior of a basement unless extra measures have been taken to ensure proper drainage. Insulating concrete forms (see above) are often used for basement wall construction. Control of pests around below-grade insulation, especially termite and insects, is a continuing concern.

Details on foundation insulation may be found in the *Building Foundation Design Handbook* (Labs et al. 1988).

LOW-SLOPE ROOF DECK CONSTRUCTION

Almost all low-slope structural roof deck construction requires thermal insulation to economically maintain the design indoor environment. For low-slope roofs, insulation should be placed on top of the deck, on the outside of the structure. This location moderates the deck temperature, which reduces thermal movement of the deck and the potential for underdeck condensation. Traditionally, the insulation is placed under the roof membrane so that it functions as a base for the built-up roof (BUR) or single-ply membrane. For a stable base, a good bond must be established between the insulation and the roof deck, and between the insulation and the BUR. Ineffective bonds that allow the BUR to move in reaction to stresses from temperature changes often cause the roof to fail. Single-ply membrane roofs are frequently installed with ballasted, mechanically fastened or fully adhered systems.

Insulation can also be placed above the membrane in a protected or inverted roof system. With this approach, the roof membrane is installed on the deck, where it functions as a waterproofing membrane and vapor retarder. Insulation can be placed above and below membranes to function both as the base and protector for the membrane. Some insulation can be wetted and then successfully dried; however, while wet, it has a greatly reduced insulation value.

INSULATION FIELD PERFORMANCE CHARACTERISTICS

Convection and air infiltration in some insulation systems may increase the heat transfer across them. Low-density loose-fill, large open-cell, and fibrous insulations, and poorly designed or installed reflective systems are most susceptible to increased heat transfer caused by natural and forced convection (air infiltration). A temperature differential across the insulation, as well as the height, thickness, or width of the insulated space, influences the amount of convection. When a membrane with low air permeance is applied to one surface or when the cavity is filled with insulation, natural convection is reduced significantly and apparent thermal conductivities measured by standard test methods apply. The heat loss due to air convection may not be significant for many types of insulation products, such as batts and higher density loose-fill insulations. Convective heat loss potentials should be obtained from insulation manufacturers.

The effectiveness of thermal insulation is seriously impaired when it is installed incorrectly. For example, a 4% void area in R-11 wall insulation increases heat loss by 15%. A 4% void area in R-19 ceiling insulation causes a 50% increase in heat loss. Verschoor (1977) found that air interchange around thin wall insulation installed vertically with air spaces on both sides increases heat loss by 60%. Lecompte (1990) found significant losses (up to 300%) as a function of the size and distribution of openings around insulation materials. Other factors, including vibration, temperature cycling, and other mechanical forces, can affect thermal performance by causing settling or other dimensional changes. Chapter 23 gives more information on the effect of moisture on thermal properties of building structures.

MOISTURE CONTROL IN BUILDINGS

Not all moisture problems can be avoided at all times. Proper design can help reduce the risk and make a building more tolerant to moisture. The recommendations in this chapter are intended to provide guidance.

Strategies to control moisture accumulation fall into two general categories: (1) minimizing moisture entry into the building envelope and (2) removing moisture from the building envelope. Once basic moisture transport mechanisms and specific moisture control practices are understood, roof, wall, and foundation constructions for various climates can be reviewed to determine whether each significant moisture transport mechanism is controlled. Because it is not possible to prevent moisture migration completely, construction should include drainage, ventilation, removal by capillary suction, or other provisions to carry away unwanted water.

CONTROL OF LIQUID WATER ENTRY

Moisture problems in buildings are frequently caused by liquid water entering through leaking roofs or the foundation, or through the walls due to wind-driven rain or rain splashing. Poor flashing details are often a major cause of water entry into walls and roofs. Rainwater should be carried away from the foundation through gutters, downspouts, and positive grading. A rain screen can minimize penetration of walls due to raindrop momentum, capillarity, gravity, and air pressure difference. The rain screen wall is designed so that the air pressure difference across the exterior rain screen is nearly zero at all times. A rain screen wall contains three components: an airflow retarder system, a pressure-equalization chamber, and a rain screen. The airflow retarder must be able to resist pressures from wind, the stack effect, and mechanical ventilation. The pressure-equalization chamber separates the rain screen and the air flow retarder system. It may be an air cavity or may be filled with a self-draining material to prevent water that penetrates the rain screen from reaching the airflow retarder system. The chamber should consist of separate compartments to avoid lateral airflow, especially around corners of the building. Each chamber compartment is vented to the outside through the rain screen to provide pressure equalization and must be flashed to the outside to drain water that has penetrated the rain screen. The rain screen must contain sufficient vents to provide pressure equalization; that is, the airflow resistance of the rain screen must be much lower than that of the airflow retarder.

CONTROL OF WATER VAPOR MIGRATION

Water vapor entry into the building envelope can be limited by airflow retarders and water vapor retarders. As described in Chapter 23, airflow retarders are intended to restrict airflow, and thereby water vapor flow, whereas water vapor retarders are designed to restrict vapor flow by diffusion.

Air Leakage Control

Past research demonstrated that air movement is more effective than water vapor diffusion for transporting water vapor within the building envelope. In order to minimize moisture penetration by air leakage, the building envelope should be as airtight as possible. The airflow retarder must also be sufficiently strong and well supported to resist wind loads.

In the past, air leakage in residential buildings provided sufficient ventilation, and the air leakage paths rarely led to interstitial condensation. However, in airtight buildings mechanical ventilation must be provided to ensure acceptable air quality and prevent moisture and health problems caused by excessive indoor humidity. Ventilation or drainage must go to the outside of the airtight layer of construction or it will increase air leakage of the building. To avoid condensation on the airtight layer, either the temperature of the layer must be kept above the dew point by locating it on the warm side of the insulation, or the permeance of the layer must be adequate to permit vapor transmission.

As described in the section on Leakage Distribution in Residential Buildings in Chapter 26, air leakage through the building envelope is not confined to doors and windows. Although 6 to 22% of the air leakage occurs at windows and doors, 18 to 50% typically takes place through walls, and 3 to 30% through the ceiling. Leakage often occurs between the sill plate and the foundation, through interior walls, electrical outlets, plumbing penetrations, and cracks at the top and bottom of the exterior walls. More detailed information can be found in Chapter 26.

Not all cracks and openings can be sealed in existing buildings, nor can absolutely tight construction be achieved in new buildings. However, an effort should be made to provide as tight an enclosure as possible to reduce leakage and minimize potential condensation within the envelope. Such measures also reduce energy loss.

Moisture accumulation in the building envelope can also be minimized by controlling the dominant direction of airflow. This can be accomplished by operating the building at a small negative or positive air pressure, depending on climate. In cooling climates, the pressure should be positive to prevent the entry of humid outside air into the envelope. In heating climates, the building pressure should be neither strongly negative, which could risk drawing soil gas or combustion products to the indoors, nor strongly positive, which could risk driving moisture into building envelope cavities.

MOISTURE CONTROL OPTIONS

Options for moisture control under heating conditions often differ from those under cooling conditions, even though the physical principles of moisture movement are the same. Therefore, the selection of moisture control options depends on whether the local climate is predominantly a heating or cooling climate. The *Moisture Control Handbook* (Lstiburek and Carmody 1991) recommends a three-step procedure for designing energy-efficient roofs, walls, and foundations with inherent moisture control capabilities:

1. Identify the climate: heating, cooling, or mixed
2. Determine the potential moisture transport mechanisms in each part of the exterior envelope: liquid flow, capillary suction, air movement, and vapor diffusion
3. Select the moisture control strategies: control moisture entry, control liquid moisture accumulation (condensation), or remove moisture by draining, venting, or diffusion

The definitions of climate zones are somewhat arbitrary. Lstiburek and Carmody (1991) recommend that heating climates be defined as climates with 4000 heating degree-days (base 65°F) or more. Cooling climates are defined as warm, humid climates where one or both of the following conditions occur: (1) a 67°F or higher wet-bulb temperature for 3000 or more hours during the warmest six consecutive months of the year; (2) a 73°F or higher wet-bulb temperature for 1750 or more hours during the warmest six consecutive months of the year.

Mixed climates are all climates that do not fall under the definitions of heating or cooling. Regions with heating climates in North America generally include the northern half of the United States, Alaska, and all of Canada. The climate in southeastern coastal regions of the United States generally can be characterized as cooling. However, the local climate should be evaluated to determine whether to design for heating, cooling, or mixed-climate conditions.

MOISTURE CONTROL OPTIONS FOR HEATING CLIMATES

Surface Condensation

Heating climates are defined as climates with 4000 heating degree-days (base 65°F) or more. In such climates, occasional window condensation is common in buildings during winter and fall.

Lowering indoor humidity to minimize surface condensation on windows is one approach, but increasing the interior surface temperature of the window using multiple glazing, low-emittance glazing, low-conductivity spacers, appropriate selection of window frame, or gas-filled glazing may be more effective. Higher thermal resistance in windows has the added advantages of saving energy, improving occupant comfort, and reducing the possibility of condensate damage to the interior adjacent to the window (e.g., staining of the wall, rotting of the window sill, and mold growth).

Windows should remain clear most of the time. Some condensation may appear around the window perimeter, but should disappear with a warming trend. This criterion should be used to decide which glazing should be installed to maintain the desired humidity, or whether to reduce the humidity to avoid condensation during the coldest periods.

Local condensation and mildew growth on walls and ceilings is often the result of low inside surface temperatures due to insufficient or faulty insulation. Increasing the thermal insulation or eliminating the voids in the insulation is the obvious remedy. If the problem is due to infiltration of cold air, an attempt should be made to eliminate the air leakage. However, in existing buildings these measures are often difficult or too expensive. In these cases, the only alternatives are lowering the indoor humidity, raising the indoor temperature, or increasing the air circulation near the surface.

Indoor Humidity Control

A common cause of moisture problems during the heating season is excessive indoor humidity. This is caused by an improper balance between moisture generation and moisture removal. This balance can be changed by reducing the sources of moisture or by increasing the removal rate, usually by ventilation or dehumidification. However, it is important to avoid lowering the relative humidity too far below the lower comfort limit, which is generally about 25 to 30% rh.

Because water vapor is introduced into the building from various sources, the moisture content of the air in an occupied building without dehumidification is always higher than that of the outdoor air. Christian (1994) provides a detailed discussion of various individual moisture sources, primarily in residences. The section on Internal Moisture Gains in Chapter 20 of the 2000 *ASHRAE Handbook—Systems and Equipment* states that a family of four produces an average of 0.7 lb/h of moisture. TenWolde (1988, 1994) reports production rates between 0.3 and 0.73 lb/h for one to two adults, with an average of 0.53 lb/h. European sources report rates between 0.6 and 1.2 lb/h for families without children, and rates between 0.46 and 2.1 lb/h for families with one to three children (Christian 1994). These numbers demonstrate that moisture production rates in residences can vary widely. Moisture production rates for various kinds of livestock and plants can be found in Chapter 10.

A residential crawl space or basement can contribute significant amounts of additional water vapor. Trethowen (1994) reported an average moisture release of 0.082 lb/ft^2·day from moist or wet crawl spaces. Moisture released from building materials that are drying (construction moisture) in a new building can add large amounts of water vapor. Exposed soil surfaces in crawl spaces or cellars should be covered with vapor retarder membranes (see the section on Crawl Spaces).

If indoor humidity is excessive, and source reduction is impossible, increasing the ventilation rate should be considered. Ventilation may be natural or mechanical, and mechanical ventilation may be exhaust, inlet, or balanced. An air-to-air heat recovery device can be included to reduce the heating energy penalty from an exhaust fan. Other approaches include the use of mechanical dehumidifiers and insulated vent stacks that extend from the living space through the roof.

Short-term ventilation procedures such as occasionally opening a window or door may lower humidity momentarily, but it will rise to its original level soon after the window or door has been closed. This is due to the evaporation of stored moisture into the indoor air. When water vapor is released from showers or cooking, much of it is adsorbed by hygroscopic materials (paper, wood, fabrics, etc.) in the building. Some temporary storage in the form of surface condensation may also occur. This moisture is released more slowly at a later time. Moisture storage effectively dampens the effect of short-term (hourly or daily) changes in moisture release or weather conditions on indoor humidity. Stored moisture also slows the effect of ventilation and dehumidification because this moisture needs to be released and removed before the indoor humidity can be lowered permanently. The evaporation of moisture stored during the summer's periods of high relative humidities is the cause of high relative humidity and window condensation in early fall.

In cold or cool winter climates, house ventilation can be an effective method for moisture removal. In these climates, a ventilation level of 0.35 air changes per hour (ACH) (as recommended in ASHRAE *Standard* 62) is generally sufficient to prevent excessive indoor humidity and most window condensation (TenWolde 1994). Ventilation is primarily required to ensure acceptable indoor air quality. If the recommended minimum ventilation levels are achieved, additional ventilation is probably unnecessary and ineffective for humidity control. In mild, humid climates, ventilation rates greater than 0.35 ACH may be needed for humidity control, but in such climates other means of moisture removal, such as dehumidifiers, should be considered.

Some suggest that high indoor humidity is caused by vapor retarders that lock in moisture. However, only a small fraction of the total moisture generated can be removed by vapor diffusion through the building envelope. Most high indoor humidity is due to inadequate ventilation, inadequate dehumidification and air conditioning, or an unusually large moisture source in the building.

Vapor Retarders and Airflow Retarders

Vapor retarders are recommended and often mandated in heating climates, and should be placed on the interior (warm) side of the insulation. Airflow retarders are also necessary, but their placement in the wall or ceiling assembly is probably less critical and still subject to debate. Vapor retarder and airflow retarder functions may be combined in one material. Airflow retarder placement on the exterior prevents cold air from penetrating the insulation (wind washing) and therefore improves thermal performance of the building envelope. However, in heating climates, airflow retarders on the exterior should have a high water vapor permeance. Special airflow retarder materials for exterior use with sufficiently high water vapor permeance are commonly available. Exterior airflow retarders do not prevent penetration and circulation of warm indoor air inside the wall or ceiling/roof cavity. Conversely, interior airflow retarders do not prevent wind washing. Interior airflow retarders do not need to have a high water vapor permeance. For additional general guidance on the placement and properties of airflow retarders, see Chapter 23.

The use of vapor retarders in compact low-slope roofing systems has been a long-standing issue for the roofing industry. Unlike other portions of the building envelope, water intrusion into a low-slope roof due to membrane failure is inevitable. Wet insulation performs below thermal performance levels specified during design. A survey by Kyle and Desjarlais (1994) has indicated that the average energy efficiency of the entire roofing inventory in the United States is reduced by approximately 40% due to moisture contamination.

Powell and Robinson (1971) studied these problems and stated that the "most practical and economical solution to the problem of moisture in insulated flat-roof constructions (is) to provide a design that would have in-service self-drying characteristics." A self-drying roof uses the local meteorological conditions to create a vapor drive into the building interior. Desjarlais (1995) demonstrated that climates with up to 9000 heating degree-days create annually averaged downward vapor drives. In a self-drying roof, any leakage into the roofing system is passively driven into the building interior; if the leak is repaired, the roof system will dry.

A vapor retarder prevents including the self-drying characteristics in the roofing design by placing an impermeable layer between the roof insulation and the building interior. A vapor retarder should only be placed in a roofing system when the amount of wintertime water uptake that the roofing system experiences exceeds the moisture limit of the insulation material in the roof. Desjarlais (1995) offers guidelines on how to determine these limits.

Penetrations through the airflow retarder (such as electrical outlets, light fixtures, or plumbing stacks) should be minimized. Any penetration should be sealed carefully. Special airtight electrical boxes are available. Limited or minor penetrations of the vapor retarder are not of great concern, if an effective airflow retarder is placed elsewhere in the wall or ceiling.

Attics and Cathedral Ceilings

Attics and cathedral ceilings are protected from interior moisture in heating climates first by limiting the entry of moisture into roof cavities, and second by ventilating the cavities to minimize accumulation of moisture. Entry of interior moisture is limited by designing and constructing effective airflow and vapor retarders between the interior space and roof cavities.

Attics. Ventilation is required in all United States and Canadian Model Building Codes. Studies on roof cavity and attic ventilation aimed at reducing paint peeling were conducted by Rowley et al. (1939), Britton (1948), and Jordan et al. (1948). These studies concluded with recommendations for venting of attics. Britton found that air movement from a wet foundation through chases and wall space could reach the attic and cause moisture damage to roof sheathing. Jordan et al. and later reports demonstrated the use of air barriers as the primary requirement for keeping moisture out of attics. Colder attics were found to require more ventilation to prevent frost. A review of the studies (Rose 1992) concluded that support for attic ventilation was at times contradictory and the specific requirement for 1 ft^2 of vent area for 300 ft^2 of floor space not resolved by the findings.

The four commonly cited reasons for attic ventilation are (1) preventing moisture damage, (2) enhancing the service life of temperature-sensitive roofing materials, (3) preventing ice dams, and (4) reducing the cooling load (TenWolde and Rose 1999). In some cases, venting may be inconsistent with the moisture control design approach. If the attic is vented, care should be taken to prevent entry of snow and to prevent airflow that might degrade the thermal performance of insulating materials (Hens and Janssens 1999).

Moisture. Vents have been shown to provide effective lowering of moisture levels in roof sheathing for attics constructed with a single unconditioned space, sloped roof, and a tight ceiling plane (Jordan 1948). It is relatively easy and inexpensive to install vents in such an attic without compromising the effectiveness of the ceiling insulation. In heating climates, attic ventilation usually provides a measure of protection from excessive moisture accumulation in the roof sheathing. If indoor humidity is high and humid indoor air leaks into the attic, attic vents by themselves may not prevent moisture accumulation.

Moisture control in attics in heating climates depends primarily on (1) maintaining lower indoor humidity levels during cold weather, (2) assuring maintainable airtightness and vapor resistance in the ceiling, and (3) attic ventilation (NRC 1963).

Temperature. A ventilated attic is cooler in the summer than an unventilated attic, and ventilation can reduce the temperature of shingles during daylight hours. Asphalt shingle manufacturers encourage ventilation as a prescriptive practice (ARMA 1997). In one study, the temperature difference due to power or turbine ventilation over soffit ventilation led to significant differences in maximum attic air temperatures, but was not shown be an effective energy conservation method in moderately or heavily insulated ceilings (Burch and Treado 1978). It is not clear that attic air temperature reduction is a significant factor in extending the service life of shingles (TenWolde and Rose 1999), since the long term studies on the temperature effects on shingle service life are incomplete.

Ice Dams. Ventilation of roofs, coupled with additional insulation and reductions in air exfiltration, reduces ice dam damage during winter in cold regions (Buska et al 1998). Where heat sources are located in the unconditioned attic space, large amounts of ventilation may be needed to prevent ice dams, necessitating mechanical attic ventilation (Tobiasson et al. 1994). Such heat sources may include furnaces, air handlers, or ductwork with conductive or convective heat losses.

Reducing heat loss into the attic by effective insulation, air leakage control, and avoidance of heat sources such as uninsulated or leaky heating ducts in the attic, possibly coupled with ventilation, is a positive way of reducing ice dams and moisture damage (Fugler 1999). Damage due to ice damming in roof valleys and eaves can also be prevented by installing a waterproof underlayment of sufficient width beneath the shingles.

Other Considerations. Roofs with absorbent claddings, such as wood shingles or cement or clay tiles are subject to solar-driven moisture penetration (Cunningham et al. 1990). Moisture is driven into the roof when it is wetted by rain or dew and subsequently exposed to sunshine. When the moisture source is from the exterior, an impermeable membrane under the shingles or tiles can greatly reduce moisture transfer into the roof; but measures should be taken to prevent water accumulation on the underside of this membrane.

Leaks cause another moisture load on roofs. Roof leaks are properly addressed by repair rather than by ventilation.

Cathedral Ceilings. Cathedral ceiling construction is inherently prone to a wider range of conditions than attic construction because this type of construction has isolated air cavities in each rafter bay. Vented attics perform better than vented cathedral ceilings for the same framing type (Goldberg 1999). Although providing effective ventilation to attics with simple geometries is relatively easy and inexpensive, providing soffit and ridge ventilation to each individual cavity in a cathedral ceiling may be difficult or impractical. Improperly installed insulation can obstruct the area designed or intended to provide ventilation. (Tobiasson 1994). An airtight ceiling plane, a vapor retarder, and foam air chutes between the sheathing and the top of the insulation effectively control moisture in cathedral ceilings with fiberglass insulation (Rose 1995). Hens and Jannsens (1999) pointed out that moisture control is assured only if airtightness is effective and can be maintained. They showed that the consequences of air entry and wind washing in insulated cathedral ceilings are detrimental, leading to degraded thermal performance, moisture response and overall durability. TenWolde and Carll (1992) showed that ventilation of roof cavities may cause increased air leakage, and that the net moisture effect depends on whether the principal source of makeup air is from indoors or outdoors.

Goldberg et al. (1999) noted that unvented attics and cathedral ceilings show a better retention of thermal resistance of the fibrous insulation than similar vented assemblies, though this benefit is smaller for attics than for cathedral ceilings. With careful attention to design for air- and vapor-tightness, unvented cathedral ceilings can be expected to perform satisfactorily in cold heating climates.

Operating Practices

Details of indoor humidity control are discussed in the section on Indoor Humidity Control. Buildings in heating climates should not be operated at substantial positive indoor air pressures, which drive moist air into the building envelope and increase the potential for moisture accumulation. Large negative pressures should also be avoided if any unsealed combustion equipment is operated in the building. Negative pressure in the basement or in slab-on-grade buildings should also be avoided when there is potential radon leakage from the soil into the building, unless a subslab depressurization system has been installed.

Other Considerations

In heating climates, it is important to design for excessive indoor humidity. If the anticipated indoor humidity will be high, then extra care must be taken in design and construction by using air barriers in conjunction with building pressure regulation.

In general, mechanical equipment should be kept within the conditioned space of the building. This approach reduces the number of openings through the building envelope and reduces the energy losses associated with exterior equipment and ductwork. Several design options permit installation of insulation below the roof plane, as in cathedralized construction (Rose 1995).

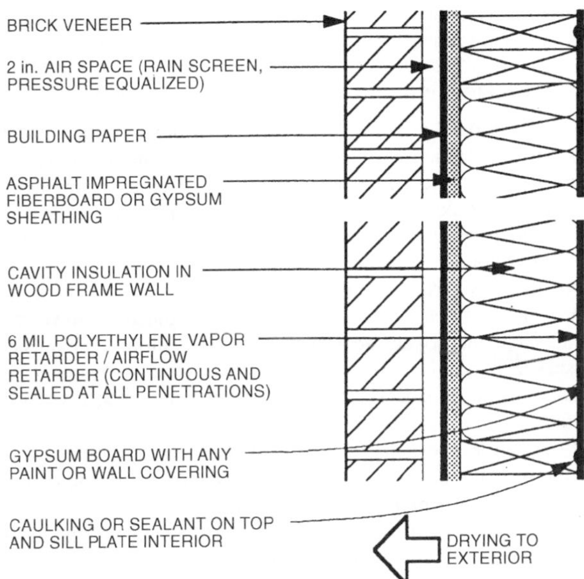

Fig. 1 Example of Residential Wall Construction for Heating Climates

Source: Lstiburek and Carmody (1991). Reprinted with permission.

Example of Residential Wall Construction for Heating Climates

Figure 1 shows the cross section of a residential wall for heating climates. Moisture control is handled in the following ways:

- **Rain.** The brick veneer, an air space, and the building paper form an effective rain screen. The air space behind the brick veneer provides a capillary break for any rainwater absorbed by the brick and mortar. Mortar should not breach the air space and touch the building paper, as this would allow rainwater to bypass the capillary break. The building paper protects the fiberboard or gypsum from any water penetrating the rain screen
- **Air movement.** The sheathing and building paper serve as an airflow retarder. Sufficient airtightness can be obtained by airtight installation of the sheathing (i.e., installed vertically with joints over the studs, with sealant or caulk used at the joints).
- **Vapor diffusion.** Vapor diffusion from the inside is inhibited by the polyethylene vapor retarder.

MOISTURE CONTROL OPTIONS FOR MIXED CLIMATES

Mixed or temperate climates fall neither under the definition of a heating climate, nor under the definition of a hot, humid climate. Mixed climates may be heating- or cooling-dominated. This zone includes areas with hot and dry climates (e.g., Arizona). Buildings in mixed climates may encounter high interior levels of humidity during winter and high exterior levels of humidity during summer.

Summer cooling or winter heating for comfort in mixed climates does not usually create serious vapor problems in exterior walls and ceilings. The summer outdoor dew point, especially during peak values, may exceed the design dew-point temperature in common use, but it seldom exceeds 75°F for a prolonged period. Condensation within exterior walls exposed to an indoor temperature of 75°F is seldom as serious as winter condensation.

In a study of a wood-sided house in Athens, Georgia, Duff (1956) showed that under summer cooling conditions, temperatures were lower outside than inside long enough to prevent moisture buildup from damaging the structure. This was true regardless of whether or not a low-permeance material was placed near the inner surface. However, masonry or brick-veneered structures with a low-permeance vapor retarder (e.g., vinyl wallpaper or polyethylene) near inner surfaces do have a moisture buildup under summer cooling conditions.

Vapor Retarders and Airflow Retarders

Airtight construction is recommended in all climates. Airflow retarders provide protection from excessive moisture accumulation in the building envelope during cooling and heating, and may reduce energy consumption. In mixed climates, the need for low-permeance vapor retarders in most types of buildings is less pronounced than in heating climates or in warm, humid climates.

However, if a vapor retarder is deemed necessary in a mixed climate zone, its placement presents somewhat of a dilemma. Under cooling conditions, a vapor retarder would normally be located on the outside of the insulation. But under winter conditions, it would be located on the inner side. Use of vapor retarders at both locations is undesirable because it restricts moisture movement into the insulation as well as the escape of any moisture. In dwelling construction, the vapor retarder should be placed to protect against the more serious condensation (winter or summer). However, if indoor humidity is kept below 35% (at 70°F) during winter, a vapor retarder on the inside of the insulation is probably not necessary in mixed climates.

The choice and placement of a vapor retarder, airflow retarder, and other materials minimize the potential for condensation while allowing for some drying. For example, if a vapor retarder is installed on the interior, an exterior airflow retarder and/or sheathing should have sufficient permeance to allow drying. The corresponding situation in cold storage buildings, in which a more serious reversal of vapor flow conditions from winter to summer may occur, requires individual analysis.

Attics and Cathedral Ceilings

Venting of attics and cathedral ceilings during winter in a mixed climate has similar benefits and drawbacks as in a heating climate. Venting may provide benefits for moisture control in attics where effective vents can be installed relatively easily and cheaply, and where the ceiling is tightened against air leakage. Unvented cathedral ceilings can provide satisfactory moisture performance in mixed climates when the system (1) is designed to control moisture migration, and (2) contains an airflow retarder that is maintained. More detailed discussion of ventilation of attics and cathedral ceilings can be found in the sections on Attics and Cathedral Ceilings under Moisture Control Options for Heating Climates.

Both vented and unvented construction should be designed and constructed to exclude interior moisture from cathedral ceiling cavities. As in heating climates, vents in cathedral ceilings may be less effective and beneficial than vents in attics; therefore, vents should be considered a design option.

Ductwork should be kept in the conditioned space of the building in order to improve energy efficiency. In hot, dry climates, energy losses through ductwork located in unconditioned attics is greater than energy losses in attics using cathedral construction, in which the insulated envelope is located at the roof (Rudd 1998).

Example of Residential Wall Construction for Mixed Climates

Figure 2 shows an example of a residential wall detail for mixed climates, with rigid insulating sheathing serving as a vapor retarder and air retarder. Moisture control is handled in the following ways:

- **Rain.** The combination of siding and airtight foam sheathing serves as a screen system and controls rain penetration. The air cavities behind the siding should be sufficient to act as a capillary break. If the air space is insufficient, the siding may be installed

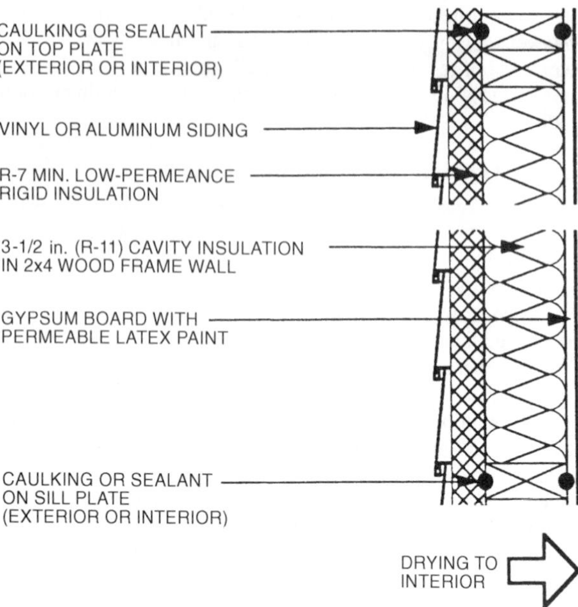

Fig. 2 Example of Residential Wall Construction for Mixed Climates

Source: Lstiburek and Carmody (1991). Reprinted with permission.

on furring strips to provide the air space. With vinyl or aluminum siding, liquid absorption and capillary moisture transfer are not a concern.

- **Air movement.** The rigid insulating sheathing can be caulked at the top and bottom plates and at the joints to provide an exterior air flow retarder. Alternatively, caulking of the gypsum board can provide an interior airflow retarder.
- **Vapor diffusion.** The impermeable rigid insulation acts as a vapor retarder. During cooling periods, vapor diffusion from the outside is impeded at the exterior sheathing surface. During heating periods, vapor diffusion from the inside is inhibited at the interior surface of the foam sheathing. To keep moisture condensation to a minimum, this first condensing surface temperature should be elevated through the use of foam sheathing with a high R-value. For mixed climates, the thermal resistance of the insulating sheathing in this example should be R-7 or greater, with R-11 thermal resistance in the cavity.

MOISTURE CONTROL OPTIONS FOR WARM, HUMID CLIMATES

Warm, humid cooling climates are defined as climates where one or both of the following conditions occur: (1) a 67°F or higher wet-bulb temperature for 3000 or more hours during the warmest six consecutive months of the year; (2) a 73°F or higher wet-bulb temperature for 1500 or more hours during the warmest six consecutive months of the year. Depending on local experience with moisture problems, humid climate design criteria may also be desirable in locations that do not quite meet the foregoing conditions.

In warm, humid climates dehumidification by air conditioning or other means is the most practical approach to moisture removal from the conditioned space. The overall latent-cooling load is composed of diffusion, ventilation, infiltration, and internally generated latent cooling loads. Because the latent-cooling load on an air conditioner in high-humidity climates frequently exceeds the sensible load, a system should be capable of handling the latent load without overcooling. In residential buildings, oversized air conditioners may not alleviate the problem of high humidity due to short cycling. Approaches to solving this problem include proper sizing of the system, the use of reheat, or design for variable flow rates.

Airflow Retarders and Water Vapor Retarders

Construction should be airtight, as in all other climates. Many moisture and condensation problems in cooling climates have been found to be caused by excessive leakage of outside air into the building envelope. Airflow retarders in cooling climates are best placed on the exterior. Negative pressures of the indoor space should be avoided. Under high-humidity conditions, ambient water vapor diffuses through building materials from the outside into air-conditioned spaces. Exterior surfaces should have lower permeance than interior surfaces in high-humidity climates. Paints and finishes can provide the necessary permeance, with lower permeance at the outside surface and higher permeance toward the inside.

Low-permeance paints, vinyl wall paper, or any other similar low-permeance material should not be used on the inside of walls and ceilings in warm, humid cooling climates.

Vapor retarders, if used, should be on the outside of the insulation. Then, any water vapor that enters the construction can flow to the inside of the building, where it can be removed by the air conditioner instead of accumulating in the wall or roof. Note that this recommendation is the reverse of the recommended practice for cold climates.

Attics and Cathedral Ceilings

The commonly stated rules for attic and cathedral ceiling construction—ventilation and vapor retarder toward the inside—pertain to cold climates and not to warm, humid climates with indoor air conditioning. Common sense suggests that venting with relatively humid outdoor air means higher levels of moisture in the attic or cathedral ceiling. Higher moisture levels in vented attics in hot, humid climates do not lead to moisture damage in sheathing or framing. However, higher moisture levels in attic cavities may affect chilled surfaces of the ceiling and cold surfaces of mechanical equipment. When cooling ducts are located in the attic space, attic ventilation with humid outdoor air may increase the chance of condensation on the ducts.

As in all climates, airtight construction is desirable. In warm, humid climates, airtight construction usually reduces the latent load. Insulation and interior finishes should be selected and installed with an understanding that vapor diffusion is primarily inward.

As with other climates, a ventilated attic in a warm, humid climate is noticeably cooler in the summer than an unventilated attic. Beal and Chandra (1995) found that venting can greatly affect the temperature difference across the ceiling.

Other Considerations

To encourage drying, shaded exterior surfaces that do not benefit from the evaporative effects of sun and wind (such as inside corners) should be avoided or minimized. Building components that are prone to thermal bridging (such as exterior cantilevers, columns, foundations, or window and door frames) are of special concern. Although these solutions may not totally eliminate mold and mildew growth, they substantially reduce the potential.

Serious wetting within walls can occur in summer under certain conditions. The National Research Council of Canada tested the walls of huts of brick masonry finished inside with furring, insulation, a vapor retarder, and plasterboard. The walls were opened during a sunny period following rain. Extensive wetting was observed in the insulation, particularly on the back of the vapor retarder. The absorptive brick wall had accumulated substantial quantities of water during the rainfall. Subsequent heating by the sun had driven the moisture as vapor into the wall, where it condensed and caused serious wetting. The construction had no protection in the form of parging or paper on the inside of the brick.

The study showed that walls with exterior coverings capable of absorbing and storing considerable quantities of water during a rain, and providing little resistance to vapor flow into the insulation from outdoors, may experience serious interior wetting by condensation. No wetting occurred in a similar construction when a saturated sheathing paper was placed between the insulation and the brick. Thus, a moderate vapor flow resistance, such as that provided by parging or a good sheathing paper on the outside of the insulation, can effectively stop vapor flow in such cases.

Operation and Maintenance

Because the potential for damage to a building and its contents is substantial in an air-conditioned building in humid climates, it is more important to properly operate and maintain the building and its air conditioning system in humid climates than it is in others. The chilled water supply temperature and flow should be reliable, and multiple chillers and pumps should be considered to ensure continuous uninterrupted dehumidification.

Raising the chilled water supply temperature to conserve energy should not be attempted under these conditions, as this would impair the dehumidification capacity of the air-conditioning system.

Lowering the cooling thermostat setting generally increases the chance for mold and condensation in exterior walls, especially in locations where the cooled air is blown directly towards the wall.

Example of Residential Wall Construction for Warm, Humid Climates

Figure 3 shows an example of a residential wall detail for warm, humid climates, with rigid insulation serving as a vapor retarder and airflow retarder. Moisture control is handled in the following ways:

- **Rain.** The combination of airtight foam sheathing and siding serves as a rain screen system and controls rain penetration. The air cavities behind the siding should be sufficient to act as a capillary break. If the air space is insufficient, the siding may be installed on furring strips to provide the air space. With vinyl or aluminum siding, liquid absorption and capillary moisture transfer are not a concern. Wood siding may be backprimed to prevent moisture absorption through the back, and installation of wedges and clips on wood lapped siding should be considered to minimize capillary transport between the boards.
- **Air movement.** The exterior sheathing is the best location for an air seal, using either an adhesive or caulk to fasten the sheathing to the framing members.
- **Vapor diffusion.** In warm, humid climates, the dominant source of moisture is the outside air, and moisture is typically driven toward the interior. This means that the best location for the vapor retarder is at or near the exterior wall surface. Vapor-permeable paint should be used on the interior gypsum wallboard.

MEMBRANE ROOF SYSTEMS

Because most membrane roof systems in commercial and institutional construction are highly resistant to vapor leakage, condensation must be prevented when insulation is placed between the heated interior and the roof membrane. Wet insulation in low-slope roof construction is difficult to dry. Drainage is likely to be so slow as to be ineffective. Ventilation to the outside is not effective for drying roof insulation, because forces acting to remove the moisture are small. The vents themselves may present a hazard to the insulation by admitting moisture and drifting snow. Also, water leaks can occur where the vents penetrate the roof unless they are properly installed. Finally, vents may allow chimney action to carry air upward through openings in the deck and ceiling. Then as air flows to the outside, further moisture is drawn into the roof with the replacement air and may condense.

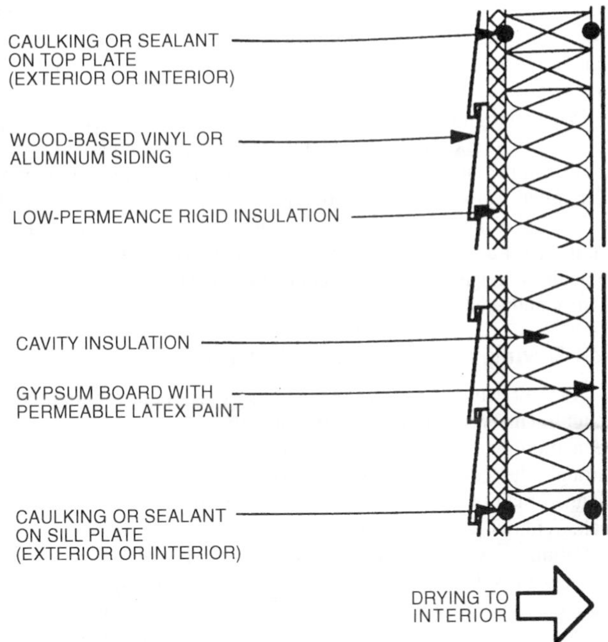

Fig. 3 Example of Residential Wall Construction for Warm, Humid Climates

Source: Lstiburek and Carmody (1991). Reprinted with permission.

A vapor retarder in a conventional flat roof can trap moisture in the roof cavity. The decision whether to use a vapor retarder depends on interior humidity and climate. The absence of a vapor retarder allows vapor to enter a roof during the heating season, but it also facilitates the removal of moisture in warm weather. This may not be acceptable in buildings with high indoor humidity or in extremely cold climates, when a large accumulation of frost or liquid condensation results in dripping. Where humidities are lower, or the climate less severe, the roof system may successfully store moisture through the heating season without problems (Baker 1980). The success of this strategy, however, also depends on the airtightness of the roof assembly. More information on this can be found in the section on Self-Drying, Low-Slope Roof Systems.

Regular inspection of the membrane and flashings helps prevent water leakage into the roof. Infrared scanners or capacitance meters can help detect wet insulation, which can be removed or possibly dried out.

Inverted Roof Systems

The top layers in protected membrane or inverted roof systems are not waterproof; therefore, insulation is exposed to rainwater. To remain effective, it must be able to resist moisture penetration. Extruded polystyrene board has been used extensively. Insulation moisture content commonly ranges up to 4 or 5% by volume.

Some insulations are damaged by freezing and thawing, which fracture cell walls and allow water into an otherwise low-permeance material. When free moisture is available, the rate of entry increases rapidly as the temperature gradient increases (Hedlin 1977). Even when insulation is immersed in ponded water, moisture absorption through the edges is less than through the upper and lower surfaces, because the temperature gradient is normal to the roof surface.

Protective measures can reduce moisture gains. Roof slope performs much the same function for protected membrane roofs as it does for conventional ones. Covering the bottom surface of the insulation with a low-permeance layer inhibits moisture entry there. The upper surface should be open to the atmosphere so that water can evaporate freely. If it is trapped against the upper surface (e.g., by paving stones), solar heating may drive the water into the insulation.

Where a high thermal resistance is required, roofs may combine conventional and protected membrane systems when they are applied in two separate lifts. The protected membrane system may be applied to existing conventional roofs to increase the thermal resistance, if the roof structure can support the added weight. This addition keeps the roof membrane warmer, so that the chance of moisture condensation on the underside of the roof membrane is significantly reduced.

Self-Drying, Low-Slope Roof Systems

A major cause of roof replacement is excessive accumulation of water in the roofing system. Historically, this accumulation has been minimized by delaying its ingress into the roofing system through the use of improved roofing membranes and periodic planned maintenance. Of course, most roofing systems eventually leak. Without periodic inspection, small leaks in a roofing system containing a vapor retarder or some other low-permeance layer (such as an asphalt mopping) can go undetected for long periods and lead to a major roof system failure. The self-drying roof facilitates the controlled out flow of water vapor into the building interior, preventing any long-term accumulation of water in the roof. Although they have not been optimized, the roofing industry has constructed self-drying roofs for many years. A roof installed without a vapor retarder or a low-permeance layer is effectively a self-drying roof.

A self-drying roof should be considered whenever the average yearly vapor drive is into the building interior. Tobiasson and Harrington (1985) have produced vapor drive maps for the continental United States. Desjarlais (1995) reported that this condition (vapor drive to the interior) is satisfied for climatic regions having less than 8800 heating degree-days (65°F base).

The self-drying roof system must be carefully designed and include special features. The deck system must be reasonably permeable to water vapor so that downward drying can be maximized. The water vapor permeance of the insulation materials must be selected so that the anticipated wintertime wetting is maintained at a level that the insulation materials can tolerate. Water vapor absorptive layer(s) should be included in the roof system so that a major leak into the roof can be controlled without leakage into the building interior. The self-drying roof must not contain a vapor retarder or any layers that are relatively impermeable. A suggested roof design procedure for self-drying roofs has been proposed by Kyle and Desjarlais (1994). The drying rate at the bottom depends on the airflow and the drying potential beneath the roof.

MOISTURE CONTROL IN FOUNDATIONS

Grading

Many of the problems related to moisture in foundations are due to the failure to discharge rainwater away from the building foundation.

Good construction practice generally requires exposed foundation between soil grade and the top of the foundation, as an inspection site for pest control. Traditionally the height of exposed foundation has been 8 in., although recent codes have reduced this requirement to 6 in. Because of the likelihood of heat loss through this exposed foundation, covering with an insulating material is desirable wherever pest inspections are not necessary.

The soil should slope away from the building at a 5% grade for the first 10 ft around the building perimeter. That is, the fall from grade to a point 10 ft away from the building should be a minimum of 6 in. The soil should be covered with a cap of relatively impermeable soil, in order to maximize surface flow of water away from the foundation. The soil backfill around the foundation is likely to settle in the first years after construction, requiring correction to achieve the proper grade. If the general slope of the soil on the site is toward the building, then swales or drains should be used to divert surface flow around the building.

Rainwater discharge from the downspouts should be managed to ensure that it does not contribute to saturation of any soil that is in contact with the foundation. The discharge may consist of extenders, splash blocks, or designed drainage systems to carry the roof runoff away from the building. On sloped lots, downspouts should discharge on the downhill side where possible.

Footing drains are traditionally installed to ensure against a rising water table; nevertheless, they may assist in draining water that may accumulate directly from surface rains. The footing drains should discharge water to an appropriate discharge site such as a storm sewer, a sump pump, or, if the site permits, daylight. Any footings for basements or crawl spaces, where the interior grade is below the exterior, should be provided with footing drains.

Gutters, downspouts, and below-grade drainage systems require maintenance. Below-grade drainage systems should be designed with cleanouts.

Floor Slab

Summer surface condensation may form on concrete floors on grade, especially during the first few years after construction. Carpeting tends to lower the interior slab surface temperature, increasing the condensation potential. Dehumidification and ventilation may be sufficient to avoid odors or floor cover bonding problems caused by moisture, which are generally more objectionable than actual damage to the floor covering.

Entry of ground moisture can be further reduced by isolating the slab with the placement of a low-permeability membrane over the soil beneath the slab and by using coarse gravel to break the capillary moisture rise. Application of a membrane is difficult, however, because it can be easily damaged during construction.

Sealing of floor slabs and basements against the entry of radon should also be considered. Although soil cover sheets are commonly referred to as vapor retarders, they can also act as a waterproofing membrane when exposed to liquid water.

Control of the slab surface temperature is important in order to minimize the need for mechanical dehumidification, particularly with solar-oriented designs that emphasize the effects of thermal mass. In localities with severe summer surface condensation problems, lightweight concrete should be considered for floor slabs to increase their insulating value, or insulation should be added under the slab.

Crawl Spaces

Moisture problems generally occur when improper drainage or grading around the house leads to wet soil or even standing water in the crawl space. Evaporation of moisture then causes high humidity in the crawl space and often in the rest of the building. Sometimes the wet soil leads to high moisture content in wood framing members in the floor and in the band joist (header joist). Any source of subfloor warmth (heating ducts, furnaces) is likely to seriously increase the evaporation from wet subfloor soil (Trethowen and Middlemass 1988, Trethowen 1988). Providing proper drainage of water away from the foundation is critical for moisture control (ASHRAE 1994).

Dewatering techniques, including sump pumps, drain tiles, etc., should be used to keep the soil in the crawl space as dry as possible. Ground covers that restrict evaporation of water from the soil into the crawl space provide an effective way to prevent moisture and humidity problems. It is important to seal any ducts in the crawl space, to avoid venting clothes dryers into the crawl space, and to repair any leaking water pipes. A minimum clearance of 18 in.

between the crawl space soil and the underside of any wood framing members is recommended and often required. Good access into and around the crawl space is very important.

Whether or not to ventilate a crawl space has been a controversial issue. Most building codes require installation of vents to provide ventilation with outside air, but a symposium on crawl space design concluded that there is no compelling technical basis for crawl space ventilation requirements (ASHRAE 1994). A distinction must be made between conditioned and unconditioned crawl spaces. Conditioned crawl spaces have insulated perimeter walls and may contain plumbing and heating runs. Conditioned crawl spaces should not be ventilated with outdoor air. If air circulation is desired, indoor air should be used. One way to accomplish this is by exhausting indoor air through the insulated crawl space, which may be done in conjunction with an air-to-air heat exchanger for energy efficiency (Samuelson 1994). Unconditioned crawl spaces have an insulated floor above the crawl space. Ventilation with outside air is permitted but not always necessary. Unvented crawl spaces must have a ground cover, which should cover 100% of the crawl space soil. Ground cover treatments for conditioned and unconditioned crawl spaces are similar.

The ground cover material should have a permeance of no more than 1.0 perms and must be rugged enough to withstand foot and knee traffic. The most commonly used material is 6 mil (0.006 in.) polyethylene. The membrane ground cover may be covered with a thin slab of concrete to prevent entry of rodents. Before laying the membrane, all debris must be removed and the soil leveled. Edges need only be lapped 4 to 6 in., and no sealing is required. The membrane need not be carried up the face of the wall.

If control of entry of radon or other soil gases is desired, a minimum 6 mil polyethylene ground cover is recommended. Some have recommended that the ground cover should be weighted down and sealed at the perimeter and overlapped to retard radon entry, but others argue that weighting and sealing may lead to water ponding on top of the ground cover. If radon control is not of primary importance, the ground cover may be cut in several low spots to provide drainage should ponding occur. The primary function of the ground cover (i.e., moisture control or radon control) should govern the decision.

Example of Residential Foundation Construction Details

Figure 4 shows a cross section detail of a typical residential basement for mixed climates. Moisture control has been handled in the following ways:

- **Rain and groundwater.** Rain is carried away by gutters, downspouts, grading away from the building, and a cap of low-permeance backfill material. Subgrade drainage prevents water from reaching the foundation wall by use of a drain screen (gravel and footer drain connected to daylight, sump, or storm sewer).
- **Liquid moisture transport.** A dampproof coating is installed on the exterior of the foundation wall and over the top of the footing to control water entry. Capillary moisture movement into the slab is inhibited by a gravel pad 4 in. thick.
- **Air movement.** All air leakage openings (i.e., floor slab/wall intersection, rim joist area) are caulked and sealed.
- **Vapor diffusion.** Dampproofing on the wall and polyethylene under the slab inhibit vapor diffusion into the slab and foundation walls. During heating periods, vapor may diffuse from the interior into the rim joist framing, where it may accumulate. To reduce moisture accumulation, the temperature of the rim joist is raised through the installation of exterior insulation.

Figure 5 shows an example of a moisture-controlling residential slab-on-grade foundation detail for warm, humid climates, with insulation laid horizontally beneath the perimeter of the floor. Rigid insulation is also placed in the vertical joint between the wall and the slab. Because the rigid insulation can act as a conduit for insects

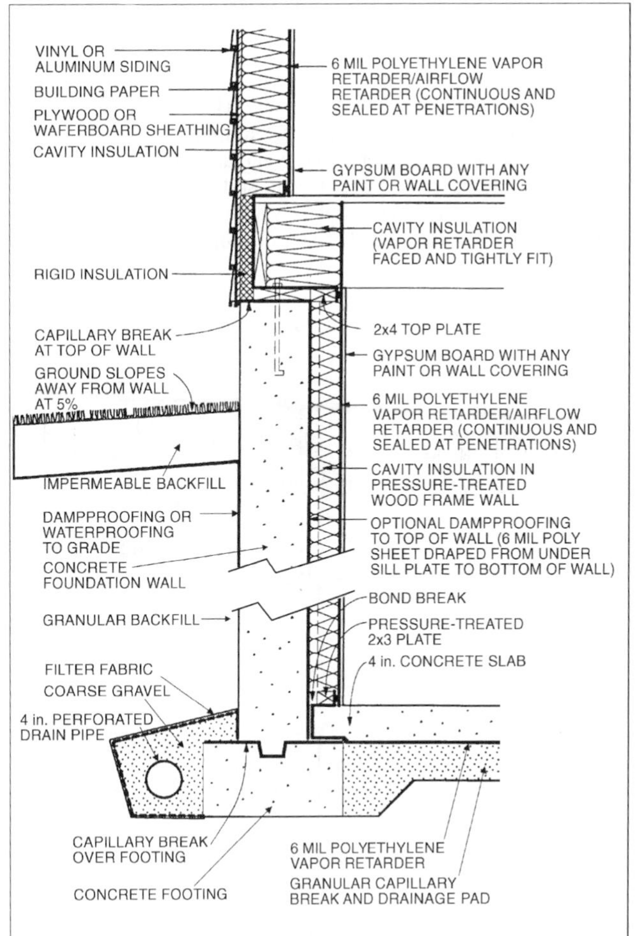

Fig. 4 Example of Residential Basement Construction for Mixed Climates
Source: Lstiburek and Carmody (1991). Reprinted with permission.

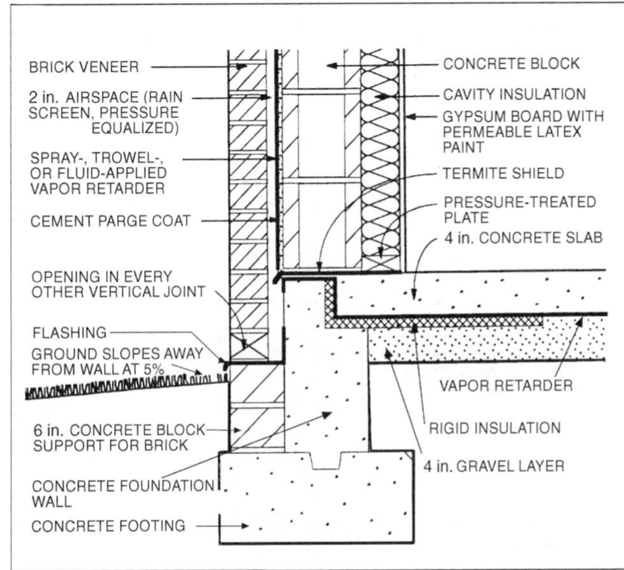

Fig. 5 Example of Residential Slab-on-Grade Construction in Warm, Humid Climates
Source: Lstiburek and Carmody (1991). Reprinted with permission.

into the building, additional protection such as metal flashings or other treatments may be necessary.

- **Rain and groundwater.** The bottom of the gravel layer is the grade level adjacent to the perimeter. The ground should be graded to direct water away from the building.
- **Liquid moisture transport.** The granular layer under the slab provides a capillary break between the soil and the slab. This pad can also be integrated into a subslab ventilation system to provide radon mitigation, if needed. Extension of the vapor diffusion retarder over the top of the foundation wall and appropriate flashing for the brick facing serve as a capillary break protecting the above-grade wall from ground moisture.
- **Air movement and vapor diffusion.** The vapor retarder placed under the slab restricts both moist soil gas entry and vapor diffusion through the slab. Ductwork located in slabs increases the risk of ground source moisture entering the conditioned space if groundwater and soil gas are permitted to seep into the ducts.

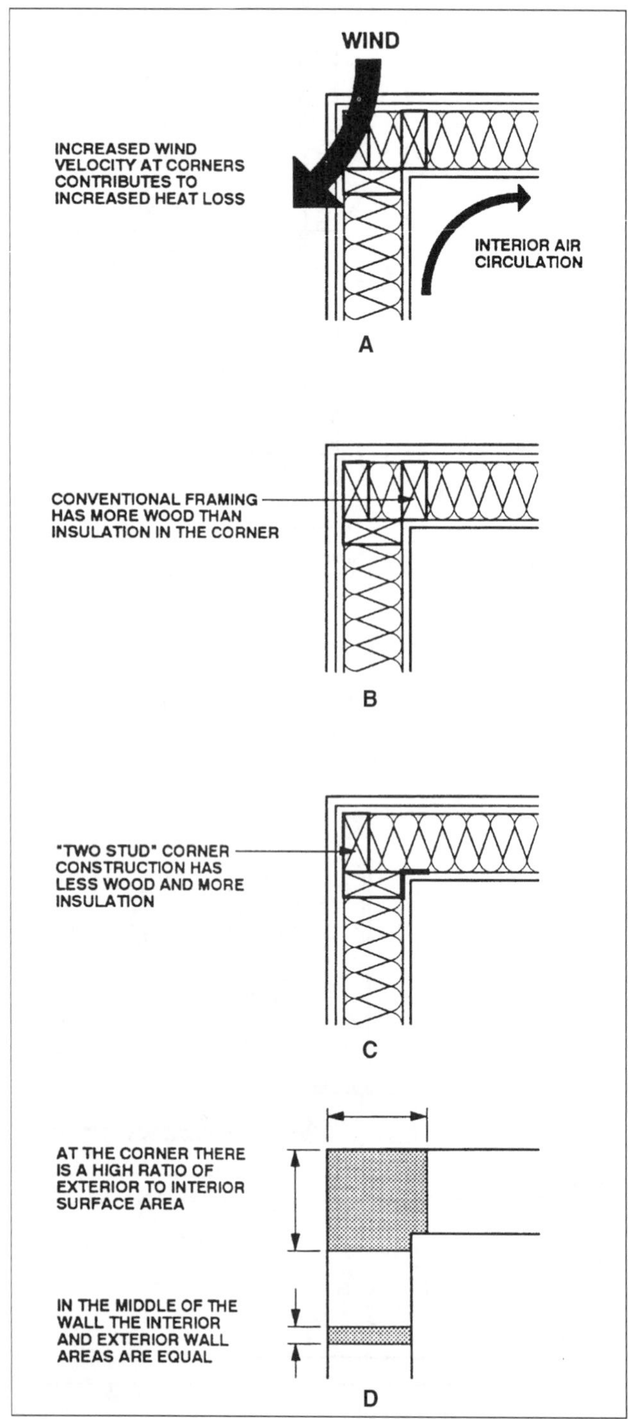

Fig. 6 Heat Loss at Building Corners
Source: Lstiburek and Carmody (1991). Reprinted with permission.

ENVELOPE COMPONENT INTERSECTIONS

A moisture control strategy must consider not only envelope components, but also how these components come together. Component intersections are especially prone to air leakage and thermal bridging and therefore require special care.

Exterior Wall Corners. Mold and mildew often grow in exterior corners during heating periods due to cold surfaces caused by (1) thermal bridges, where structural members penetrate the insulation and provide a low-resistance heat flow path; (2) wind washing; (3) increased heat loss due to the fin effect; and (4) poor circulation of indoor air. Figure 6 shows heat loss effects at building corners. Insulating sheathings and two-stud corners help prevent cold interior surfaces and corner moisture problems.

Wall/Window Intersections. Restricting air-transported moisture at all potential openings makes a major contribution to the overall building tightness. The airflow retarder must be continuous. Figure 7 shows several details that help form a continuous airflow retarder at a window jamb.

Wall/Roof Intersections. Exterior wall/ceiling intersections are other common cool spots during the heating season caused by reduced attic insulation at the eaves and wind washing. High-heel trusses that allow installation of more insulation, wind baffles, and rigid insulation exterior sheathing all help control moisture at these locations. Figure 8 shows the heat loss mechanisms at attic/wall intersections and how to minimize the risk of moisture problems.

Wall/Floor Intersections. Air leakage at rim joist assemblies is avoided by making sure the airflow retarder is continuous. Caulking and sealing are necessary at all polyethylene seams, as shown in Figure 4. Floor structural members penetrating the insulation can cause thermal bridging, but the use of insulating sheathing helps minimize this (see Figure 4).

Wall/Foundation Intersections. Concrete footings are frequently poured directly in contact with the ground, which occasionally becomes damp or wet. Concrete used for most residential foundations has the right degree of porosity to provide capillary suction, which draws water into the footings and then into the foundation wall. This water usually evaporates into the inside space undetected. However, the moisture occasionally manifests itself as a ring of dampness visible at the bottom interior surface of a basement or crawl space wall. Gravel and capillary breaks installed between the footing and the foundation wall are effective moisture control strategies in these below-grade envelope intersections. Several techniques to control capillary moisture below grade are shown in Figure 4 and Figure 5.

MOISTURE CONTROL IN COMMERCIAL AND INSTITUTIONAL BUILDINGS

Moisture control in commercial and institutional buildings often requires approaches different from those in residential buildings. Indoor humidity conditions can vary greatly from one building to the next, depending on the use and requirements. The building envelope should be designed to perform well with these indoor conditions. Special thought should be given to the moisture control features of the HVAC equipment and the building envelope for certain buildings with special indoor humidity conditions or requirements, such as swimming pools, hospitals, and museums.

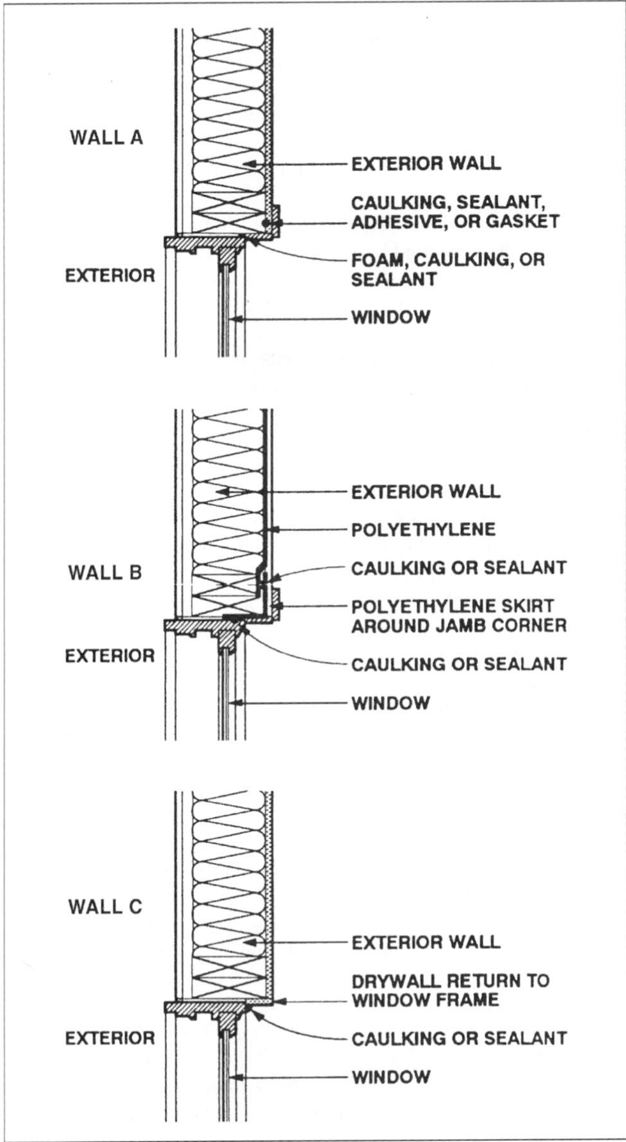

Fig. 7 Interior Airflow Retarder Details at Window Jamb
Source: Lstiburek and Carmody (1991). Reprinted with permission.

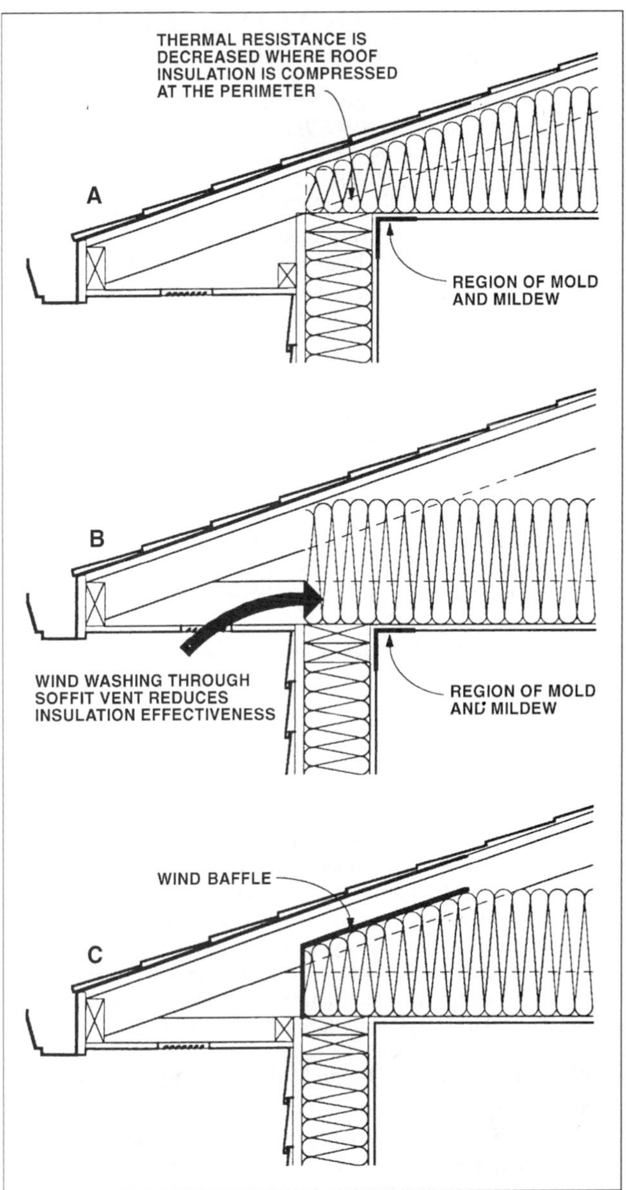

Fig. 8 Heat Loss Effect at Ceiling Edge
Source: Lstiburek and Carmody (1991). Reprinted with permission.

Materials commonly used in commercial and institutional construction tend to be more moisture-tolerant and decay-resistant than those used in residential construction. Airflow retarder systems are often poorly designed and executed. As a result, air leakage through the building envelope is common. Air leakage through hollow concrete masonry is often greater than in other types of construction. Upward airflow in the cavities is not always adequately blocked, and parallel random leakage paths are found between gypsum wallboard or other finishes and the block face.

Air can leak through exterior walls where the structural system or services penetrate the air barrier or at joints between dissimilar materials or components. For example, masonry cannot form a tight seal with structural steel columns and beams. To reduce this problem, the structural frame should be inside and separate from the exterior wall. The resulting curtain wall can then incorporate a more continuous air barrier and be protected from the fluctuating weather by insulation applied to the outside. Exterior cladding to control rain penetration is best applied following the weather-tightening system. The interior wythe (masonry course) and air barrier should be accessible for maintenance of the air seal and joints.

The deterioration of exterior structural elements of a building and damage to the interior through condensation from air leakage has an important bearing on the operation and maintenance costs of the building. Improving the airtightness of internal floors and partitions, particularly in high-rise buildings, redistributes the pressure differences caused by the stack effect and reduces the pressure difference across the exterior wall on each floor. This approach also improves ventilation and air distribution and reduces the air circulation between occupancies on different floors. It also helps control smoke movement in the case of fire and may enable a more equitable apportioning of energy charges for space heating between individual units in apartment buildings.

Leakage in actual buildings often occurs through holes cut accidentally or deliberately in a reasonably tight membrane or component, such as the penetration of services through specified air or vapor retarders or solid components. Other leakage openings in the exterior envelope result from dimensional changes in improperly placed materials or from inadequate sealants or membranes applied to bridge joints or cracks that will eventually open.

INDUSTRIAL AND COMMERCIAL INSULATION PRACTICE

For pipe materials, selection, application, and installation, see Chapter 41 of the 2000 *ASHRAE Handbook—Systems and Equipment.* For insulation systems for refrigerant piping, see also Chapter 32 of the 1998 *ASHRAE Handbook—Refrigeration.*

If the equipment is subject to wide changes in temperature, the insulation installation should be designed to accommodate the associated dimensional change. If the equipment is operated at below ambient temperature, the application should be designed to resist the accumulation of moisture in the insulation.

PIPES

Small pipes are insulated with cylindrical half-sections of insulation, with factory-applied jackets that form a hinge-and-lap, or with flexible, closed-cell material. Large pipes can be insulated with flexible material or with curved, flat segmented, or cylindrical half-, third-, or quarter-sections of rigid insulation, particularly where removal for frequent servicing of the pipe is necessary. Fittings (valves, tees, crosses, and elbows) are insulated with preformed fitting insulation, fabricated fitting insulation, individual pieces cut from sectional straight pipe insulation, or insulating cements. Fitting insulations should always be equal in thermal performance to the pipe insulation.

Securing Methods

The method of securement varies with the size of pipe, form and weight of the insulation, and the type of jacketing (i.e., separate or factory-applied). Insulation with certain factory-applied jacketing can be secured on small piping by cementing the overlapping jacket. Large piping may require supplemental wiring or banding. Insulation on large piping requiring separate jacketing is wired or banded in place, and the jacket is cemented, wired, or banded, depending on the type. Insulation with factory-applied metal or PVC jacketing is secured by specific design of the jacket and its joint closure. The flexible closed-cell materials require no jacket for most applications and are applied using specially formulated contact adhesives.

Insulation Finish for Above-Ambient Temperatures

Pipe insulation finishes for indoor use are usually governed by location. The finishes are factory-applied jackets designed to meet fire safety requirements. For maximum fire safety, unusual exposure conditions, or appearance, factory or separately applied metal or PVC jackets may be used. An outdoor finish protects the insulation from the weather. Chemical exposure, mechanical abuse, and appearance are additional considerations.

Pipes that operate at temperatures above 500°F may require two layers of insulation in order to accommodate the large dimension change of pipe and insulation materials.

Insulation Finish for Below-Ambient Temperatures

Piping at temperatures below ambient is insulated to control heat gain and prevent condensation of moisture from the ambient air. Because piping is an absolute barrier to the passage of water vapor, the outer surface of the insulation must be covered by an impervious membrane or cover, which also helps protect the pipe against corrosion.

Retarder treatment should be recommended by the insulation manufacturer as established by performance testing. The insulation should be as dry as possible, and therefore should be protected from undue weather exposure.

Vapor seals for straight pipe insulation are generally designed to meet operating temperature, fire safety, and appearance requirements. Jacketing commonly consists of various combinations of laminates of paper, aluminum foil, plastic film, and glass fiber reinforcing. An important feature of such jacketing is very low permeance in a relatively thin layer, which provides flexibility for ease of cementing and sealing laps and end-joint strips. This type of jacketing is commonly used indoors without additional treatment. In some cases of operating temperatures below 0°F, multilayer insulation and jacketing may be used. Flexible closed-cell materials must be carefully cemented to avoid openings in the insulations.

Insulation fittings are usually vapor sealed by applying suitable materials in the field, and may vary with the type of insulation and operating temperature. For temperatures above 10°F, the vapor seal can be a lapped spiral wrap of plastic film adhesive tape or a relatively thin coat of vapor-seal mastic. For temperatures below 10°F, common practice is to apply two coats of vapor-seal mastic reinforced with open weave glass or other fabric. The thickness of the mastic increases with decreasing temperature. With long lines of piping, the insulation should be sealed off every 15 or 20 ft to limit water penetration if vapor seal damage occurs.

For dual-temperature service, where piping is alternately cold and hot, the vapor-seal finish, including mastics, must withstand pipe movement and exposure temperatures without deterioration. When flexible closed-cell insulation is used, it should be applied slightly compressed to prevent it from being strained when the piping expands.

Outdoor pipe insulation may be vapor-sealed in the same manner as indoor piping, by applying added weather protection jacketing without damage to the retarder and sealing it to keep out water. In some instances, heavy-duty weather and vapor-seal finish may be used.

Because cold piping frequently operates year-round, a constant vapor drive exists under humid conditions. Even with vapor retarder insulation, jackets, and vapor sealing of joints and fittings, moisture inevitably accumulates in permeable insulations. This moisture not only reduces the thermal resistance of the insulation, it also accelerates condensation on the jacket surface. Since periodic insulation replacement is the only known solution, the piping installation should be accessible for such replacement and should have a means for draining water. As an alternative, very low-permeance insulating materials (e.g., materials not exceeding 0.1 perm by the wet cup method) have been used to extend the life of the system and reduce replacement frequency. The lower the permeance of the insulation material, the longer its life, provided good workmanship is practiced during installation.

Surface Temperature

In elevated-temperature applications, the surface temperature of the insulation system should be below that at which personnel coming into contact with the surface could be harmed. In below-ambient temperature applications, the surface temperature of the system should be above the dew point to prevent condensation.

Compared to a jacket with a nonreflective surface, a jacket with a reflective surface has a higher surface temperature for hot applications and a lower surface temperature for cold-temperature applications, because the lower emissivity reduces the rate of heat exchange. Therefore, adding a reflective jacket could produce a surface temperature capable of burning personnel on hot applications and causing condensation on cold applications. The jacketing material used also contributes to the relative safety at a given surface temperature. For example, at 175°F, a stainless steel jacket blisters skin more severely than a canvas jacket does.

Insulating Pipes to Prevent Freezing

If the surrounding air temperature remains sufficiently low for an extended period, insulation cannot prevent freezing of still water or of water flowing at a rate insufficient for the available heat content to offset the heat loss. Insulation can only prolong the time required

Table 1 Estimated Requirements to Prevent Freezing of Water in Pipes

Steel Pipe Nominal Diameter, in.	Insulation Thickness, in.					
	2	3	4	2	3	4
	Time to Cool Water to Freezing, h			Water Mass Flow Rate per Unit Length of Exposed Pipe to Prevent Freezing, lb/h·ft		
0.5	0.27	0.32	0.36	0.54	0.45	0.39
1	0.61	0.75	0.85	0.70	0.55	0.48
1.5	1.16	1.46	1.69	0.89	0.68	0.58
2	1.67	2.13	2.49	1.05	0.79	0.66
3	2.83	3.71	4.42	1.44	1.03	0.84
4	4.07	5.43	6.54	1.83	1.26	1.00
5	5.45	7.36	8.96	2.30	1.52	1.19
6	6.86	9.37	11.5	2.83	1.80	1.37
8	9.59	13.3	16.5	4.08	2.39	1.76
10	12.6	17.6	22.1	5.94	3.13	2.21
12	15.4	21.7	27.4	8.63	3.99	2.70

Design Conditions: Surrounding air temperature $t_a = -18$°F, initial water temperature $t_i = 42$°F, and insulation thermal conductivity $k_I = 0.025$ Btu/h·ft·°F. Thermal resistances of pipe and air film at surface of insulation are ignored.
Calculations are for 40ST steel pipe. See Table 2 in Chapter 41 of the 2000 *ASHRAE Handbook—Systems and Equipment* for actual pipe dimensions.

for water to freeze or prevent freezing if water flow is maintained at a sufficient rate. The first section of Table 1 can be used to estimate the thickness of insulation necessary to prevent freezing of still water in pipes. The second section of Table 1 gives the minimum flow of water at an initial temperature of 42°F to prevent the temperature of the pipe from reaching 32°F at the end of the exposed length.

To calculate time θ (in hours) required for water to cool to 32°F, the following equation can be used:

$$\theta = \rho c_p \pi \left(\frac{D_i}{2}\right)^2 R_T \ln\left[\frac{t_i - t_a}{t_f - t_a}\right] \qquad (1)$$

where

θ = time for water to cool to freezing, h
ρ = density of water = 62.4 lb/ft^3
c_p = specific heat of water = 1.0 Btu/lb·°F
D_i = inside diameter of pipe, ft
D_p = outer diameter of pipe or inner diameter of insulation, ft
D_I = outer diameter of insulation, ft
$R_T = R_p + R_I + R_a$ = combined thermal resistance of pipe wall, insulation, and exterior air film per foot of pipe, ft·°F·h/Btu
$R_a = 1/(h_a \pi D_I)$ = resistance between ambient air and outer surface of insulation per foot of pipe, ft·°F·h/Btu
h_a = air heat transfer coefficient (see Chapter 3 for values)
$R_I = \ln(D_I/D_p)/(2\pi k_I)$ = resistance of thermal insulation per foot of pipe, ft·°F·h/Btu
$R_p = \ln(D_p/D_i)/(2\pi k_p)$ = resistance of pipe per foot of pipe, ft·°F·h/Btu ($R_p \approx 0$ for metal pipe)
k_I = thermal conductivity of insulation, Btu/h·ft·°F
k_p = thermal conductivity of pipe material, Btu/h·ft·°F (see Table 7 in Chapter 41 of the 2000 *ASHRAE Handbook—Systems and Equipment* for thermal conductivity of various plastic pipes)
t_a = ambient air temperature, °F
t_i = initial water temperature, °F
t_f = freezing temperature, °F

When unusual conditions make it impractical to maintain protection with insulation alone, a hot trace pipe or, preferably, electric resistance heating cable is required along the bottom or top of the water pipe. The heating system then supplies the heat lost through the insulation. The insulation thickness is determined by the cost of the heating system, the insulation, and the heat loss.

Pipe bursting is not an immediate consequence of water pipes reaching freezing temperatures. Clean water and pipes usually supercool several degrees below freezing before any ice is formed. Then, upon nucleation, dendritic ice forms in the water and the temperature rises to freezing. Ice can be formed from water only by the release of the latent heat of fusion (144 Btu/lb) through the pipe insulation. With well-insulated pipes, this release of latent heat may be greatly retarded. Pipe bursting in water pipes has been shown (Gordon 1996) to be a consequence not of ice crystal growth in the pipe, but of elevated fluid pressure within a confined pipe section occluded by a growing ice blockage.

Underground Pipe Insulation

Both heated and cooled underground piping systems are insulated. Protecting underground insulated piping is more difficult than protecting aboveground piping. Groundwater conditions, including chemical or electrolytic contributions by the soil and the existence of water pressure, require a special design to protect insulated pipes from corrosion. Walk-through tunnels, conduits, or integral protective coverings are generally provided to protect the pipe and insulation from water. Examples and general design features of conduits and a description of tunnels can be found in Chapter 11 of the 2000 *ASHRAE Handbook—Systems and Equipment*.

Temperatures Above Ambient. Piping for heated systems in walk-through tunnels is usually covered with sectional insulation and finished with effective mechanical protection such as metal or waterproofing jackets. The use of walk-through tunnels is declining because of cost.

Conduit systems are generally used for underground insulated piping systems. The most successful application is sectional insulation with the conduit sized for drainage and adequate drying of insulation on heated piping in the event of accidental flooding. BRAB (1975) gives detailed design criteria for conduit systems. The criteria require that (1) all systems provide for draining and insulation drying, (2) the insulation withstand boiling and drying without physical damage and loss of insulating value, and (3) the conduit casing is watertight in the field. The insulation should be a nonconductor of electricity, verminproof, and chemically and dimensionally stable at the operating temperature of the pipe.

BRAB (1964) describes evaluative tests and field investigations, which have shown that calcium silicate is resistant to severe boiling action. Fibrous glass (density of 4 to 7 lb/ft^3) will not withstand boiling when a conduit becomes flooded, and wet poured-in-place insulations are likely to remain partially wet for their installed life.

The thickness of insulation for underground piping is not determined on the same basis as above ground piping. Chapter 11 of the 2000 *ASHRAE Handbook—Systems and Equipment* provides details for determining thickness.

Temperatures Below Ambient. Integrally protected, insulated piping buried directly in the ground is commonly used for chilled water. However, since no heat is available to drive out moisture, an absolute protective covering against water and insulation with low permeance and water absorption is extremely important. Cellular glass with proper protection has been widely used for this type of application. The acceptance of plastic foams is increasing, but their long-term performance has not yet been established.

Conduit for chilled water piping requires a different approach than for hot piping. Insulation must have low conductivity, and conduit design must use this low conductivity and maintain continuing performance. More recent designs use low-conductivity plastic foam insulation with plastic pipe as the internal water-carrying piping and as the external conduit.

Where the temperature difference between the pipe at 40°F and the soil at 55 to 60°F is small, pipe size, length, and flow rate may economically justify direct burial without insulation. However, good piping protection may be required.

DUCTS

The need for duct insulation is influenced by (1) duct location, whether indoors or outdoors; (2) the effect of heat loss or gain on equipment size and operating cost; (3) the need to prevent condensation on low-temperature ducts; (4) the need to control temperature change in long duct lengths; and (5) the need to control noise with interior duct lining.

All ducts exposed to outdoor conditions, as well as those passing through unconditioned spaces, should be insulated. While analyses of temperature change, heat loss or gain, and other factors affecting the economics of thermal insulation are seldom made for residential installations, they are essential for large commercial and industrial projects.

The U-factor for uninsulated sheet metal ducts is affected by air velocity, the emittance of the metal, and the shape of the duct. An approximate value of 1.0 Btu/h·ft^2·°F may be used. For insulated ducts, U-factors of 0.24 and 0.13 Btu/h·ft^2·°F represent 1 and 2 in. thick rigid insulation with a thermal conductivity of 0.27 Btu·in/h·ft^2·°F at 75°F mean temperature. A method for determining heat loss or gain for ducts is given in Chapter 34.

Materials for Ducts, Insulations, and Liners

Ducts within buildings can be of insulated sheet metal, fibrous glass, or insulated flexible ducts, all of which provide combined air barrier, thermal insulation, and sound absorption. Ducts embedded in or below floor slabs may be of compressed fiber, ceramic tile, or other rigid materials.

Duct insulations include semirigid boards and flexible blanket types, composed of organic and inorganic materials in fibrous, cellular, or bonded particle forms. Insulations for exterior surfaces may have attached vapor barriers or facings, or vapor barriers may be applied separately. When applied to the duct interior as a liner, insulation both insulates thermally and absorbs sound. Duct liner insulations have sound-permeable coatings or other treatment on the side facing the airstream to withstand air velocities or duct cleaning without deterioration.

Per UL *Standard* 181, fibrous glass air ducts are tested to 12,500 fpm and are rated at 5000 fpm and at a pressure of at least 2 in. of water. Primary use is for low-pressure systems tested at 1.5 times the recommended static pressure. Maximum recommended air temperature is 250°F. A complete system provides greater decibel attenuation than is usually provided by standard duct liners, with greater reduction in airborne equipment noise and crosstalk. Higher design velocities are also possible.

To satisfy most building codes, duct insulation and fibrous glass duct materials must meet the fire hazard requirements of (1) NFPA *Standard* 90A, to restrict spread of smoke, heat, and fire through duct systems, and to minimize ignition sources; and (2) NFPA *Standard* 90B, on supply ducts, controls, clearances, heating panels, return ducts, air filters, and heat pumps. Local code authorities should also be consulted.

Where thermally insulated air-conditioning ducts pass through unconditioned spaces, such as attics, the maximum allowable heat flux should be no greater than that required by NFPA *Standard* 90A.

Securing Methods

Exterior rigid duct insulation can be attached with adhesive, with supplemental preattached pins and clips, or with wiring or banding. Liners can be attached with adhesive and supplemental pins and clips. Flexible duct wraps do not require attachment except on bottom duct panels greater than 24 in. wide. For larger ducts a pin no more than 24 in. on center is sufficient.

Manufacturers provide information on the construction of fibrous glass duct systems. Preformed round duct for straight runs is combined with fittings fabricated from straight duct. Rectangular ducts and fittings are fabricated by grooving, folding, and taping. Metal accessories such as turning vanes, splitters, and dampers are incorporated into the system. When rectangular ducts exceed predetermined dimensions for particular static pressures, ductwork must be reinforced. The Sheet Metal and Air Conditioning Contractors National Association's (SMACNA) *Fibrous Glass Duct Construction Standards* (1992) have further information.

Heating Ducts

The effect of duct insulation on residential heating system equipment size can be marginal. However, insulation can reduce operating costs significantly, depending on unit costs for heating and the extent of duct exposed to outside conditions. In addition, duct insulation maintains the supply air temperature, which may keep the air entering the conditioned space within a more comfortable range.

Vapor retarders are not required on exterior insulation of ducts used only for heating, but they must be provided for ducts used for alternate heating and cooling.

Cooling Ducts

Insulation can significantly reduce operating costs and cooling equipment size. The advantage of adequate insulation is especially significant in areas with high dry-bulb and dew-point temperatures. Ducts for summer air conditioning are insulated with the same materials as heating ducts. Ducts in any unconditioned space should be insulated and have vapor retarders to prevent condensation. Joints and laps in the vapor retarder must be sealed. Flexible closed-cell insulation does not always need a supplemental vapor retarder, but care must be taken to form vapor-tight seams at joints.

REFERENCES

ARMA 1997. Ventilation and moisture control for residential roofing. *Technical Bulletin* 209-RR-86. Asphalt Roofing Manufacturers Association, Calverton, MD.

ASHRAE. 1994. Recommended practices for controlling moisture in crawl spaces. *Technical Data Bulletin* 10(3).

ASHRAE. 1999. Ventilation for acceptable indoor air quality. ANSI/ASHRAE *Standard* 62-1999.

Baker, M.C. 1980. *Roofs*, Chapters 6 and 8. Multiscience Publications, Montreal, Canada.

Barbour, E.J. Goodrow, J. Kosny, and J.E. Christian. 1994 thermal performance of steel-framed walls. Prepared for American Iron and Steel Institute by NAHB Research Center.

Beal, D. and S. Chandra. 1995. The measured summer performance of tile roof systems and attic ventilation strategies in hot, humid climates. In: *Thermal Performance of the Exterior Envelopes of Buildings VI*, pp. 753-760. ASHRAE.

BRAB. 1964. Field evaluation of underground heat distribution systems. *Technical Report* 47. Building Research Advisory Board, Federal Construction Council, National Academy of Sciences, National Research Council, Washington, D.C.

BRAB. 1975. Criteria for underground distribution. *Technical Report* 66. Standing Committee on Mechanical Engineering of the FCC, BRAB, and NRC.

BRANZ. 1983. Insulating suspended timber frame floors. *Bulletins* 177 and 233. Building Research Association, New Zealand.

Britton, R.R. 1948. Condensation in walls and roofs. *Technical Papers No.* 1, 2, 3, and 8. Housing and Home Finance Agency, Washington, D.C.

Burch, D.M. and S.J. Treado. 1978. Ventilating residences and their attics for energy conservation—An experimental study. In: Summer attic and whole-house ventilation. *Special Publication* 548. National Institute of Standards and Technology, Gaithersburg, MD.

Buska, J., W. Tobiasson, and A. Greatorex. 1998. Roof ventilation to prevent problematic icings at eaves. *ASHRAE Transactions* 104(2).

Christian, J.E. 1994. Moisture sources. In: *Moisture Control in Buildings*, pp. 176-82. ASTM *Manual* 18. American Society for Testing and Materials, West Conshohocken, PA.

Cunningham, M.J., G.A. Tsongas, and D. McQuade. 1990. Solar-driven moisture transfer through absorbent roofing materials. *ASHRAE Transactions* 96(2):465-71.

Desjarlais, A.O. 1995. Self-drying roofs: What! No dripping! In: Proceedings, *Thermal Performance of the Exterior Envelopes of Buildings VI*, pp. 763-73. ASHRAE.

DOE. 1991. Radiant barrier attic fact sheet. DOE/CE-0335P. U.S. Dept. of Energy.

Duff, J.E. 1956. The effect of air conditioning of water vapor permeability on temperature and humidity. *ASHRAE Transactions* 62:437.

Fugler, D. 1999. Conclusions from ten years of Canadian attic research. *ASHRAE Transactions* 105(1).

Goldberg, L., P. Huelman, and B. Bridges. 1999. A preliminary experimental assessment of the comparative thermal performance of attics and cathedral ceilings in a cold climate. *ASHRAE Transactions* 105(1).

Gordon, J. 1996. An investigation into freezing and bursting water pipes in residential construction. *Research Report* 96-1. University of Illinois Building Research Council.

Hedlin, C.P. 1977. Moisture gains by foam plastic roof insulations under controlled temperature gradients. *Journal of Cellular Plastics* 13(5). NRCC 16317.

Hens, H, and A. Janssens. 1999. Heat and moisture response of vented and compact cathedral ceilings: A test house evaluation. *ASHRAE Transactions* 105(1).

Jordan, C.A., E.C. Peck, F.A. Strange, and L.V. Teesdale. 1948. Attic condensation in tightly built houses. Housing and Home Finance Agency *Technical Bulletin* No 6.

Korsgaard, V. and C.R. Pedersen. 1992. Laboratory and practical experience with a novel water-permeable vapor retarder. In: *Thermal Performance of the Exterior Envelopes of Buildings V*, pp. 480-90. ASHRAE.

Kyle, D.M. and A.O. Desjarlais. 1994. Assessment of technologies for constructing self-drying low-slope roofs. *Report* ORNL/CON-380. Oak Ridge National Laboratory, Oak Ridge, TN.

Labs, K., J. Carmody, R. Sterling, L. Shen, Y.J. Huang, and D. Parker. 1988. *Building foundation design handbook. Report* ORNL/sub/86-72143/1. Oak Ridge National Laboratory.

Lecompte, J. 1990. Influence of natural convection in an insulated cavity on the thermal performance of a wall. *Insulation Materials, Testing and Applications*, pp. 397-420. ASTM STP 1030. American Society for Testing and Materials, West Conshohocken, PA.

Lstiburek, J. and J. Carmody. 1991. *The moisture control handbook—New low-rise, residential construction.* ORNL/Sub/89-SD350/1. Martin Marietta Energy Systems, Oak Ridge National Laboratory, TN.

NFPA. 1993. Standard for the installation of air conditioning and ventilating systems. *Standard* 90A-93, National Fire Protection Association, Quincy, MA.

NFPA. 1993. Standard for the installation of warm air heating and air conditioning systems. *Standard* 90B-93. National Fire Protection Association, Quincy, MA.

NRC 1963. Humidified buildings. *Canadian Building Digest* 42.

Powell, F.J. and H.E. Robinson. 1971. The effect of moisture on the heat transfer performance of insulated flat-roof constructions. Building Science Series 37. National Institute of Standards and Technology, Gaithersburg, MD.

Rose, W. 1992. Measured values of temperature and sheathing content in residential attic assemblies. In: *Thermal Performance of the Exterior Envelopes of Buildings V*, pp. 379-90. ASHRAE.

Rose, W.B. 1995. Attic construction with sheathing-applied insulation. *ASHRAE Transactions* 101(2):789-98.

Rowley, F.B., A.B. Algren, and C.E. Lund. 1939. Condensation of moisture and its relation to building construction and operation. *ASHVE Transactions* 45:231-52.

Rudd, A.F. and J.W. Lstiburek. 1998. Vented and sealed attics in hot climates. *ASHRAE Transactions* 104(2):1199-1210.

Samuelson, I. 1994. Moisture control in crawl spaces. In: Recommended practices for controlling moisture in crawl spaces. *ASHRAE Technical Data Bulletin* 10(3):pp. 58-64.

Sanders, C. 1996. Final report: Volume 2, Task 2: Environmental conditions. International Energy Agency *Annex* 24: Heat, Air and Moisture Transfer Through New and Retrofitted Insulated Envelope Parts. K.U. Leuven, Belgium.

SMACNA. 1992. *Fibrous glass duct construction standards*, 6th ed. Sheet Metal and Air Conditioning Contractor's National Association, Chantilly, VA.

TenWolde, A. 1988. A mathematical model for indoor humidity in homes during winter. In: *Proceedings of symposium on air infiltration, ventilation, and moisture transfer*, pp. 3-32. BTECC-National Institute for Building Science, Washington D.C.

TenWolde, A. and C. Carll. 1992. Effect of cavity ventilation on moisture in walls and roofs. In: *Thermal Performance of the Exterior Envelopes of Buildings V*, pp. 555-62. ASHRAE.

TenWolde, A. 1994. Ventilation, humidity, and condensation in manufactured houses during winter. *ASHRAE Transactions* 100(1):103-15.

TenWolde, A. and W. Rose. 1999. Issues related to venting of attics and cathedral ceilings. *ASHRAE Transactions* 105(1).

Tobiasson, W. and M. Harrington. 1985. Vapor drive maps of the U.S. *Thermal Performance of the Exterior Envelopes of Buildings III*, pp. 663-72.

Tobiasson, W., J. Buska, and A. Greatorex. 1994. Ventilating attics to minimize icing at eaves. *Energy and Buildings* 20:229-34.

Trethowen, H.A. 1988. A survey of subfloor ground evaporation rates. *Study Report* 13. Building Research Association of New Zealand.

Trethowen, H.A. 1994. Three surveys of subfloor moisture in New Zealand. *ASHRAE Transactions* 100(1):1427-38.

Trethowen, H.A. and G. Middlemass. 1988. A survey of moisture damage in southern New Zealand buildings. *Study Report* 7. Building Research Association of New Zealand.

Tuluca, A. and R. Gorthala. 1999. Thermal performance of cold-formed steel ceiling/roof framing assemblies. *ASHRAE Research Project* 981.

Verschoor, J.D. 1977. Effectiveness of building insulation applications. USN/CEL Report No. CR78.006-NTIS No. AD-A053 452/9ST.

BIBLIOGRAPHY

IEA. 1991. Vol. 2, Guidelines and practice. Annex XIV, Condensation and Energy. International Energy Agency, Leuven, Belgium.

Trechsel, H.R., ed. 1994. Moisture control in buildings. *Manual* 18. American Society for Testing and Materials, West Conshohocken, PA.

CHAPTER 25

THERMAL AND WATER VAPOR TRANSMISSION DATA

THIS CHAPTER presents thermal and water vapor transmission data based on steady-state or equilibrium conditions. Chapter 3 covers heat transfer under transient or changing temperature conditions. Chapter 23 discusses selection of insulation materials and procedures for determining overall thermal resistances by simplified methods.

BUILDING ENVELOPES

Thermal Transmission Data for Building Components

The steady-state thermal resistances (R-values) of building components (walls, floors, windows, roof systems, etc.) can be calculated from the thermal properties of the materials in the component; or the heat flow through the assembled component can be measured directly with laboratory equipment such as the guarded hot box (ASTM *Standard* C 236) or the calibrated hot box (ASTM *Standard* C 976).

Tables 1 through 6 list thermal values, which may be used to calculate thermal resistances of building walls, floors, and ceilings. The values shown in these tables were developed under ideal conditions. In practice, overall thermal performance can be reduced significantly by such factors as improper installation and shrinkage, settling, or compression of the insulation (Tye and Desjarlais 1983; Tye 1985, 1986).

Most values in these tables were obtained by accepted ASTM test methods described in ASTM *Standards* C 177 and C 518 for materials and ASTM *Standards* C 236 and C 976 for building envelope components. Because commercially available materials vary, not all values apply to specific products.

The most accurate method of determining the overall thermal resistance for a combination of building materials assembled as a building envelope component is to test a representative sample by a hot box method. However, all combinations may not be conveniently or economically tested in this manner. For many simple constructions, calculated R-values agree reasonably well with values determined by hot box measurement.

The performance of materials fabricated in the field is especially subject to the quality of workmanship during construction and installation. Good workmanship becomes increasingly important as the insulation requirement becomes greater. Therefore, some engineers include additional insulation or other safety factors based on experience in their design.

Figure 1 shows how convection affects surface conductance of several materials. Other tests on smooth surfaces show that the average value of the convection part of the surface conductance decreases as the length of the surface increases.

Vapor retarders, which are discussed in Chapters 23 and 24, require special attention. Moisture from condensation or other sources may reduce the thermal resistance of insulation, but the effect of moisture must be determined for each material. For example, some materials with large air spaces are not affected significantly if the moisture content is less than 10% by weight, while the effect of moisture on other materials is approximately linear.

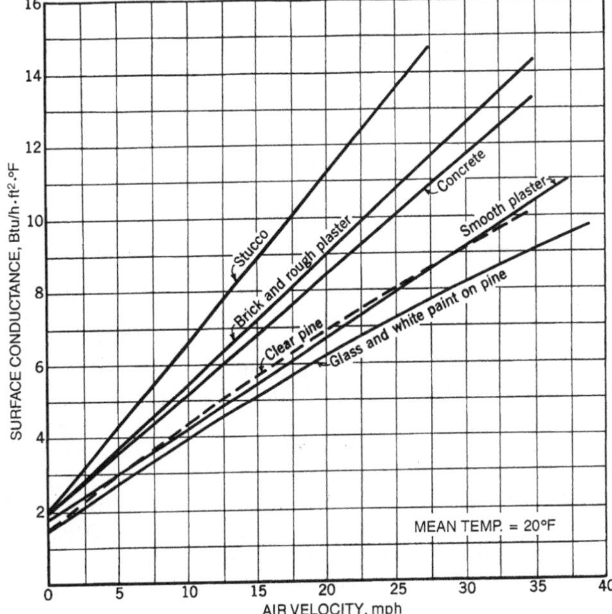

Fig. 1 Surface Conductance for Different Surfaces as Affected by Air Movement

Ideal conditions of components and installations are assumed in calculating overall R-values (i.e., insulating materials are of uniform nominal thickness and thermal resistance, air spaces are of uniform thickness and surface temperature, moisture effects are not involved, and installation details are in accordance with design). The National Institute of Standards and Technology Building Materials and Structures Report BMS 151 shows that measured values differ from calculated values for certain insulated constructions. For this reason, some engineers decrease the calculated R-values a moderate amount to account for departures of constructions from requirements and practices.

Tables 3 and 2 give values for well-sealed systems constructed with care. Field applications can differ substantially from laboratory test conditions. Air gaps in these insulation systems can seriously degrade thermal performance as a result of air movement due to both natural and forced convection. Sabine et al. (1975) found that the tabular values are not necessarily additive for multiple-layer, low-emittance air spaces, and tests on actual constructions should be conducted to accurately determine thermal resistance values.

The preparation of this chapter is assigned to TC 4.4, Thermal Insulation and Moisture Retarders.

Table 1 Surface Conductances and Resistances for Air

Position of Surface	Direction of Heat Flow	Surface Emittance, ε: Non-reflective ε = 0.90 h_i	R	Reflective ε = 0.20 h_i	R	Reflective ε = 0.05 h_i	R
STILL AIR							
Horizontal	Upward	1.63	0.61	0.91	1.10	0.76	1.32
Sloping—45°	Upward	1.60	0.62	0.88	1.14	0.73	1.37
Vertical	Horizontal	1.46	0.68	0.74	1.35	0.59	1.70
Sloping—45°	Downward	1.32	0.76	0.60	1.67	0.45	2.22
Horizontal	Downward	1.08	0.92	0.37	2.70	0.22	4.55
MOVING AIR (Any position)		h_o	R				
15 mph wind (for winter)	Any	6.00	0.17	—	—	—	—
7.5 mph wind (for summer)	Any	4.00	0.25	—	—	—	—

Notes:
1. Surface conductance h_i and h_o measured in Btu/h·ft²·°F; resistance R in ft²·°F·h/Btu.
2. No surface has both an air space resistance value and a surface resistance value.
3. For ventilated attics or spaces above ceilings under summer conditions (heat flow down), see Table 5.
4. Conductances are for surfaces of the stated emittance facing virtual blackbody surroundings at the same temperature as the ambient air. Values are based on a surface-air temperature difference of 10°F and for surface temperatures of 70°F.
5. See Chapter 3 for more detailed information, especially Tables 5 and 6, and see Figure 1 for additional data.
6. Condensate can have a significant impact on surface emittance (see Table 2).

Values for foil insulation products supplied by manufacturers must also be used with caution because they apply only to systems that are identical to the configuration in which the product was tested. In addition, surface oxidation, dust accumulation, condensation, and other factors that change the condition of the low-emittance surface can reduce the thermal effectiveness of these insulation systems (Hooper and Moroz 1952). Deterioration results from contact with several types of solutions, either acidic or basic (e.g., wet cement mortar or the preservatives found in decay-resistant lumber). Polluted environments may cause rapid and severe material degradation. However, site inspections show a predominance of well-preserved installations and only a small number of cases in which rapid and severe deterioration has occurred. An extensive review of the reflective building insulation system performance literature is provided by Goss and Miller (1989).

CALCULATING OVERALL THERMAL RESISTANCES

Relatively small, highly conductive elements in an insulating layer called thermal bridges can substantially reduce the average thermal resistance of a component. Examples include wood and metal studs in frame walls, concrete webs in concrete masonry walls, and metal ties or other elements in insulated wall panels. The following examples illustrate the calculation of R-values and U-factors for components containing thermal bridges.

The following conditions are assumed in calculating the design R-values:

- Equilibrium or steady-state heat transfer, disregarding effects of thermal storage
- Surrounding surfaces at ambient air temperature
- Exterior wind velocity of 15 mph for winter (surface with $R = 0.17$ ft²·°F·h/Btu) and 7.5 mph for summer (surface with $R = 0.25$ ft²·°F·h/Btu)
- Surface emittance of ordinary building materials is 0.90

Table 2 Emittance Values of Various Surfaces and Effective Emittances of Air Spaces[a]

Surface	Average Emittance ε	Effective Emittance ε_{eff} of Air Space: One Surface Emittance ε; Other, 0.9	Both Surfaces Emittance ε
Aluminum foil, bright	0.05	0.05	0.03
Aluminum foil, with condensate just visible (> 0.7 gr/ft²)	0.30[b]	0.29	—
Aluminum foil, with condensate clearly visible (> 2.9 gr/ft²)	0.70[b]	0.65	—
Aluminum sheet	0.12	0.12	0.06
Aluminum coated paper, polished	0.20	0.20	0.11
Steel, galvanized, bright	0.25	0.24	0.15
Aluminum paint	0.50	0.47	0.35
Building materials: wood, paper, masonry, nonmetallic paints	0.90	0.82	0.82
Regular glass	0.84	0.77	0.72

[a]These values apply in the 4 to 40 μm range of the electromagnetic spectrum.
[b]Values are based on data presented by Bassett and Trethowen (1984).

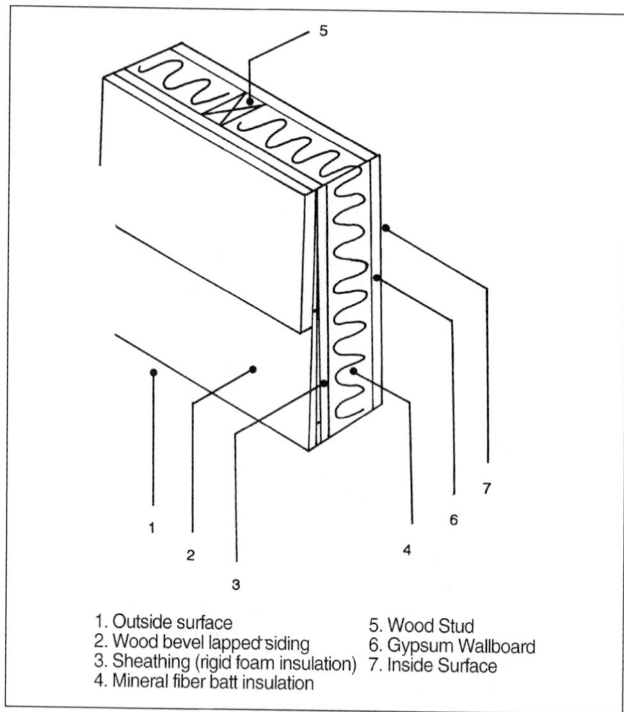

Fig. 2 Insulated Wood Frame Wall (Example 1)

Wood Frame Walls

The average overall R-values and U-factors of wood frame walls can be calculated by assuming either parallel heat flow paths through areas with different thermal resistances or by assuming isothermal planes. Equations (1) through (5) from Chapter 23 are used.

The framing factor or fraction of the building component that is framing depends on the specific type of construction, and it may vary based on local construction practices—even for the same type of construction. For stud walls 16 in. on center (OC), the fraction of insulated cavity may be as low as 0.75, where the fraction of studs, plates, and sills is 0.21 and the fraction of headers is 0.04. For studs 24 in. OC, the respective values are 0.78, 0.18, and 0.04. These fractions contain an allowance for multiple studs, plates, sills, extra framing around windows, headers, and band joists. These assumed framing fractions are used in the following example, to illustrate the importance of including the effect of framing in determining the overall thermal conductance of a building. The actual framing fraction should be calculated for each specific construction.

Example 1. Calculate the U-factor of the 2 by 4 stud wall shown in Figure 2. The studs are at 16 in. OC. There is 3.5 in. mineral fiber batt insulation (R-13) in the stud space. The inside finish is 0.5 in. gypsum wallboard; the outside is finished with rigid foam insulating sheathing (R-4) and 0.5 in. by 8 in. wood bevel lapped siding. The insulated cavity occupies approximately 75% of the transmission area; the studs, plates, and sills occupy 21%; and the headers occupy 4%.

Solution: Obtain the R-values of the various building elements from Tables 1 and 4. Assume $R = 1.25$ per inch for the wood framing. Also, assume the headers are solid wood, in this case, and group them with the studs, plates, and sills.

Because the U-factor is the reciprocal of R-value, $U_1 = 0.052$ and $U_2 = 0.095$ Btu/h·ft²·°F.

If the wood framing (thermal bridging) is not included, Equation (3) from Chapter 23 may be used to calculate the U-factor of the wall as follows

$$U_{av} = U_1 = \frac{1}{R_1} = 0.052 \text{ Btu/h} \cdot \text{ft}^2 \cdot {}^\circ\text{F}$$

Element	R (Insulated Cavity)	R (Studs, Plates, and Headers)
1. Outside surface, 15 mph wind	0.17	0.17
2. Wood bevel lapped siding	0.81	0.81
3. Rigid foam insulating sheathing	4.0	4.0
4. Mineral fiber batt insulation, 3.5 in.	13.0	—
5. Wood stud, nominal 2 × 4	—	4.38
6. Gypsum wallboard, 0.5 in.	0.45	0.45
7. Inside surface, still air	0.68	0.68
	$R_1 = 19.11$	$R_2 = 10.49$

If the wood framing is accounted for using the parallel-path flow method, the U-factor of the wall is determined using Equation (5) from Chapter 23 as follows:

$$U_{av} = (0.75 \times 0.052) + (0.25 \times 0.095) = 0.063 \text{ Btu/h} \cdot \text{ft}^2 \cdot {}^\circ\text{F}$$

If the wood framing is included using the isothermal planes method, the U-factor of the wall is determined using Equations (2) and (3) from Chapter 23 as follows:

$$R_{T(av)} = 4.98 + 1/[(0.75/13.0) + (0.25/4.38)] + 1.13$$

$$= 14.82 \text{ ft}^2 \cdot {}^\circ\text{F} \cdot \text{h/Btu}$$

$$U_{av} = 0.067 \text{ Btu/h} \cdot \text{ft}^2 \cdot {}^\circ\text{F}$$

For a frame wall with a 24 in. OC stud space, the average overall R-value is 15.18 ft²·°F·h/Btu. Similar calculation procedures may be used to evaluate other wall designs, except those with thermal bridges.

Masonry Walls

The average overall R-values of masonry walls can be estimated by assuming a combination of layers in series, one or more of which provides parallel paths. This method is used because heat flows laterally through block face shells so that transverse isothermal planes result. Average total resistance $R_{T(av)}$ is the sum of the resistances of the layers between such planes, each layer calculated as shown in Example 2.

Example 2. Calculate the overall thermal resistance and average U-factor of the 7-5/8 in. thick insulated concrete block wall shown in Figure 3. The two-core block has an average web thickness of 1 in. and a face shell thickness of 1-1/4 in. Overall block dimensions are 7-5/8 by 7-5/8 by 15-5/8 in. Measured thermal resistances of 112 lb/ft³ concrete and 7 lb/ft³ expanded perlite insulation are 0.10 and 2.90 ft²·°F·h/Btu per inch, respectively.

Solution: The equation used to determine the overall thermal resistance of the insulated concrete block wall is derived from Equations (2) and (5) from Chapter 23 and is given below:

$$R_{T(av)} = R_i + R_f + \left(\frac{a_w}{R_w} + \frac{a_c}{R_c}\right)^{-1} + R_o$$

where

$R_{T(av)}$ = overall thermal resistance based on assumption of isothermal planes
R_i = thermal resistance of inside air surface film (still air)
R_o = thermal resistance of outside air surface film (15 mph wind)
R_f = total thermal resistance of face shells
R_c = thermal resistance of cores between face shells
R_w = thermal resistance of webs between face shells
a_w = fraction of total area transverse to heat flow represented by webs of blocks
a_c = fraction of total area transverse to heat flow represented by cores of blocks

From the information given and the data in Table 1, determine the values needed to compute the overall thermal resistance.

$R_i = 0.68$
$R_o = 0.17$
$R_f = (2)(1.25)(0.10) = 0.25$
$R_c = (5.125)(2.90) = 14.86$
$R_w = (5.125)(0.10) = 0.51$
$a_w = 3/15.625 = 0.192$
$a_c = 12.625/15.625 = 0.808$

Using the equation given, the overall thermal resistance and average U-factor are calculated as follows:

$$R_{T(av)} = 0.68 + 0.25 + \frac{0.51 \times 14.86}{(0.808 \times 0.51) + (0.192 \times 14.86)} + 0.17$$

$$= 3.43 \text{ ft}^2 \cdot {}^\circ\text{F} \cdot \text{h/Btu}$$

$$U_{av} = 1/3.43 = 0.29 \text{ Btu/h} \cdot \text{ft}^2 \cdot {}^\circ\text{F}$$

Based on guarded hot box tests of this wall without mortar joints, Tye and Spinney (1980) measured the average R-value for this insulated concrete block wall as 3.13 ft²·°F·h/Btu.

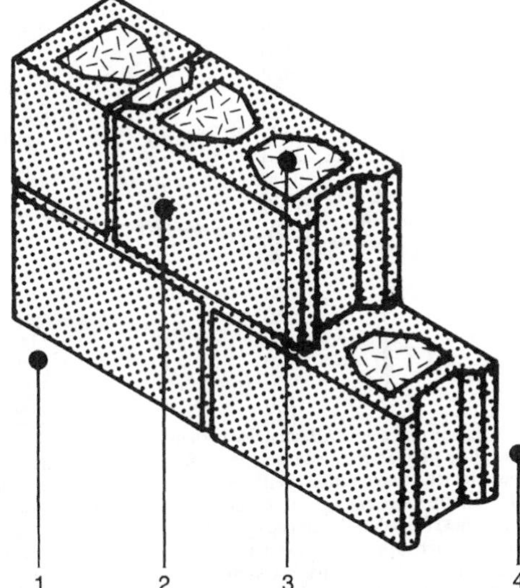

Fig. 3 Insulated Concrete Block Wall (Example 2)

Table 3 Thermal Resistances of Plane Air Spaces[a,b,c], ft²·°F·h/Btu

Position of Air Space	Direction of Heat Flow	Air Space Mean Temp.[d], °F	Air Space Temp. Diff.[d], °F	0.5-in. Air Space[c] Effective Emittance ε_{eff}[d,e] 0.03	0.05	0.2	0.5	0.82	0.75-in. Air Space[c] Effective Emittance ε_{eff}[d,e] 0.03	0.05	0.2	0.5	0.82
		90	10	2.13	2.03	1.51	0.99	0.73	2.34	2.22	1.61	1.04	0.75
		50	30	1.62	1.57	1.29	0.96	0.75	1.71	1.66	1.35	0.99	0.77
		50	10	2.13	2.05	1.60	1.11	0.84	2.30	2.21	1.70	1.16	0.87
Horiz.	Up ↑	0	20	1.73	1.70	1.45	1.12	0.91	1.83	1.79	1.52	1.16	0.93
		0	10	2.10	2.04	1.70	1.27	1.00	2.23	2.16	1.78	1.31	1.02
		−50	20	1.69	1.66	1.49	1.23	1.04	1.77	1.74	1.55	1.27	1.07
		−50	10	2.04	2.00	1.75	1.40	1.16	2.16	2.11	1.84	1.46	1.20
		90	10	2.44	2.31	1.65	1.06	0.76	2.96	2.78	1.88	1.15	0.81
		50	30	2.06	1.98	1.56	1.10	0.83	1.99	1.92	1.52	1.08	0.82
		50	10	2.55	2.44	1.83	1.22	0.90	2.90	2.75	2.00	1.29	0.94
45° Slope	Up ↗	0	20	2.20	2.14	1.76	1.30	1.02	2.13	2.07	1.72	1.28	1.00
		0	10	2.63	2.54	2.03	1.44	1.10	2.72	2.62	2.08	1.47	1.12
		−50	20	2.08	2.04	1.78	1.42	1.17	2.05	2.01	1.76	1.41	1.16
		−50	10	2.62	2.56	2.17	1.66	1.33	2.53	2.47	2.10	1.62	1.30
		90	10	2.47	2.34	1.67	1.06	0.77	3.50	3.24	2.08	1.22	0.84
		50	30	2.57	2.46	1.84	1.23	0.90	2.91	2.77	2.01	1.30	0.94
		50	10	2.66	2.54	1.88	1.24	0.91	3.70	3.46	2.35	1.43	1.01
Vertical	Horiz. →	0	20	2.82	2.72	2.14	1.50	1.13	3.14	3.02	2.32	1.58	1.18
		0	10	2.93	2.82	2.20	1.53	1.15	3.77	3.59	2.64	1.73	1.26
		−50	20	2.90	2.82	2.35	1.76	1.39	2.90	2.83	2.36	1.77	1.39
		−50	10	3.20	3.10	2.54	1.87	1.46	3.72	3.60	2.87	2.04	1.56
		90	10	2.48	2.34	1.67	1.06	0.77	3.53	3.27	2.10	1.22	0.84
		50	30	2.64	2.52	1.87	1.24	0.91	3.43	3.23	2.24	1.39	0.99
		50	10	2.67	2.55	1.89	1.25	0.92	3.81	3.57	2.40	1.45	1.02
45° Slope	Down ↘	0	20	2.91	2.80	2.19	1.52	1.15	3.75	3.57	2.63	1.72	1.26
		0	10	2.94	2.83	2.21	1.53	1.15	4.12	3.91	2.81	1.80	1.30
		−50	20	3.16	3.07	2.52	1.86	1.45	3.78	3.65	2.90	2.05	1.57
		−50	10	3.26	3.16	2.58	1.89	1.47	4.35	4.18	3.22	2.21	1.66
		90	10	2.48	2.34	1.67	1.06	0.77	3.55	3.29	2.10	1.22	0.85
		50	30	2.66	2.54	1.88	1.24	0.91	3.77	3.52	2.38	1.44	1.02
		50	10	2.67	2.55	1.89	1.25	0.92	3.84	3.59	2.41	1.45	1.02
Horiz.	Down ↓	0	20	2.94	2.83	2.20	1.53	1.15	4.18	3.96	2.83	1.81	1.30
		0	10	2.96	2.85	2.22	1.53	1.16	4.25	4.02	2.87	1.82	1.31
		−50	20	3.25	3.15	2.58	1.89	1.47	4.60	4.41	3.36	2.28	1.69
		−50	10	3.28	3.18	2.60	1.90	1.47	4.71	4.51	3.42	2.30	1.71

Position of Air Space	Direction of Heat Flow	Air Space Mean Temp.[d], °F	Air Space Temp. Diff.[d], °F	1.5 in. Air Space[c] 0.03	0.05	0.2	0.5	0.82	3.5 in. Air Space[c] 0.03	0.05	0.2	0.5	0.82
		90	10	2.55	2.41	1.71	1.08	0.77	2.84	2.66	1.83	1.13	0.80
		50	30	1.87	1.81	1.45	1.04	0.80	2.09	2.01	1.58	1.10	0.84
		50	10	2.50	2.40	1.81	1.21	0.89	2.80	2.66	1.95	1.28	0.93
Horiz.	Up ↑	0	20	2.01	1.95	1.63	1.23	0.97	2.25	2.18	1.79	1.32	1.03
		0	10	2.43	2.35	1.90	1.38	1.06	2.71	2.62	2.07	1.47	1.12
		−50	20	1.94	1.91	1.68	1.36	1.13	2.19	2.14	1.86	1.47	1.20
		−50	10	2.37	2.31	1.99	1.55	1.26	2.65	2.58	2.18	1.67	1.33
		90	10	2.92	2.73	1.86	1.14	0.80	3.18	2.96	1.97	1.18	0.82
		50	30	2.14	2.06	1.61	1.12	0.84	2.26	2.17	1.67	1.15	0.86
		50	10	2.88	2.74	1.99	1.29	0.94	3.12	2.95	2.10	1.34	0.96
45° Slope	Up ↗	0	20	2.30	2.23	1.82	1.34	1.04	2.42	2.35	1.90	1.38	1.06
		0	10	2.79	2.69	2.12	1.49	1.13	2.98	2.87	2.23	1.54	1.16
		−50	20	2.22	2.17	1.88	1.49	1.21	2.34	2.29	1.97	1.54	1.25
		−50	10	2.71	2.64	2.23	1.69	1.35	2.87	2.79	2.33	1.75	1.39
		90	10	3.99	3.66	2.25	1.27	0.87	3.69	3.40	2.15	1.24	0.85
		50	30	2.58	2.46	1.84	1.23	0.90	2.67	2.55	1.89	1.25	0.91
		50	10	3.79	3.55	2.39	1.45	1.02	3.63	3.40	2.32	1.42	1.01
Vertical	Horiz. →	0	20	2.76	2.66	2.10	1.48	1.12	2.88	2.78	2.17	1.51	1.14
		0	10	3.51	3.35	2.51	1.67	1.23	3.49	3.33	2.50	1.67	1.23
		−50	20	2.64	2.58	2.18	1.66	1.33	2.82	2.75	2.30	1.73	1.37
		−50	10	3.31	3.21	2.62	1.91	1.48	3.40	3.30	2.67	1.94	1.50
		90	10	5.07	4.55	2.56	1.36	0.91	4.81	4.33	2.49	1.34	0.90
		50	30	3.58	3.36	2.31	1.42	1.00	3.51	3.30	2.28	1.40	1.00
		50	10	5.10	4.66	2.85	1.60	1.09	4.74	4.36	2.73	1.57	1.08
45° Slope	Down ↘	0	20	3.85	3.66	2.68	1.74	1.27	3.81	3.63	2.66	1.74	1.27
		0	10	4.92	4.62	3.16	1.94	1.37	4.59	4.32	3.02	1.88	1.34
		−50	20	3.62	3.50	2.80	2.01	1.54	3.77	3.64	2.90	2.05	1.57
		−50	10	4.67	4.47	3.40	2.29	1.70	4.50	4.32	3.31	2.25	1.68
		90	10	6.09	5.35	2.79	1.43	0.94	10.07	8.19	3.41	1.57	1.00
		50	30	6.27	5.63	3.18	1.70	1.14	9.60	8.17	3.86	1.88	1.22
		50	10	6.61	5.90	3.27	1.73	1.15	11.15	9.27	4.09	1.93	1.24
Horiz.	Down ↓	0	20	7.03	6.43	3.91	2.19	1.49	10.90	9.52	4.87	2.47	1.62
		0	10	7.31	6.66	4.00	2.22	1.51	11.97	10.32	5.08	2.52	1.64
		−50	20	7.73	7.20	4.77	2.85	1.99	11.64	10.49	6.02	3.25	2.18
		−50	10	8.09	7.52	4.91	2.89	2.01	12.98	11.56	6.36	3.34	2.22

[a] See Chapter 23, section on Factors Affecting Heat Transfer Across Air Spaces. Thermal resistance values were determined from the relation, $R = 1/C$, where $C = h_c + \varepsilon_{eff} h_r$, h_c is the conduction-convection coefficient, $\varepsilon_{eff} h_r$ is the radiation coefficient $\approx 0.0068\varepsilon_{eff}[(t_m + 460)/100]^3$, and t_m is the mean temperature of the air space. Values for h_c were determined from data developed by Robinson et al. (1954). Equations (5) through (7) in Yarbrough (1983) show the data in this table in analytic form. For extrapolation from this table to air spaces less than 0.5 in. (as in insulating window glass), assume $h_c = 0.159(1 + 0.0016\,t_m)/l$ where l is the air space thickness in inches, and h_c is heat transfer through the air space only.

[b] Values are based on data presented by Robinson et al. (1954). (Also see Chapter 3, Tables 3 and 4, and Chapter 38). Values apply for ideal conditions (i.e., air spaces of uniform thickness bounded by plane, smooth, parallel surfaces with no air leakage to or from the space). When accurate values are required, use overall U-factors determined through calibrated hot box (ASTM C 976) or guarded hot box (ASTM C 236) testing. Thermal resistance values for multiple air spaces must be based on careful estimates of mean temperature differences for each air space.

[c] A single resistance value cannot account for multiple air spaces; each air space requires a separate resistance calculation that applies only for the established boundary conditions. Resistances of horizontal spaces with heat flow downward are substantially independent of temperature difference.

[d] Interpolation is permissible for other values of mean temperature, temperature difference, and effective emittance ε_{eff}. Interpolation and moderate extrapolation for air spaces greater than 3.5 in. are also permissible.

[e] Effective emittance ε_{eff} of the air space is given by $1/\varepsilon_{eff} = 1/\varepsilon_1 + 1/\varepsilon_2 - 1$, where ε_1 and ε_2 are the emittances of the surfaces of the air space (see Table 2).

Table 4 Typical Thermal Properties of Common Building and Insulating Materials—Design Values[a]

Description	Density, lb/ft³	Conductivity[b] (*k*), Btu·in/h·ft²·°F	Conductance (*C*), Btu/h·ft²·°F	Resistance[c] (*R*) Per Inch Thickness (1/*k*), ft²·°F·h/Btu·in	Resistance[c] (*R*) For Thickness Listed (1/*C*), ft²·°F·h/Btu	Specific Heat, Btu/lb·°F
BUILDING BOARD						
Asbestos-cement board	120	4.0	—	0.25	—	0.24
Asbestos-cement board ... 0.125 in.	120	—	33.00	—	0.03	
Asbestos-cement board ... 0.25 in.	120	—	16.50	—	0.06	
Gypsum or plaster board ... 0.375 in.	50	—	3.10	—	0.32	0.26
Gypsum or plaster board ... 0.5 in.	50	—	2.22	—	0.45	
Gypsum or plaster board ... 0.625 in.	50	—	1.78	—	0.56	
Plywood (Douglas fir)[d]	34	0.80	—	1.25	—	0.29
Plywood (Douglas fir) ... 0.25 in.	34	—	3.20	—	0.31	
Plywood (Douglas fir) ... 0.375 in.	34	—	2.13	—	0.47	
Plywood (Douglas fir) ... 0.5 in.	34	—	1.60	—	0.62	
Plywood (Douglas fir) ... 0.625 in.	34	—	1.29	—	0.77	
Plywood or wood panels ... 0.75 in.	34	—	1.07	—	0.93	0.29
Vegetable fiber board						
Sheathing, regular density[e] ... 0.5 in.	18	—	0.76	—	1.32	0.31
... 0.78125 in.	18	—	0.49	—	2.06	
Sheathing intermediate density[e] ... 0.5 in.	22	—	0.92	—	1.09	0.31
Nail-base sheathing[e] ... 0.5 in.	25	—	0.94	—	1.06	0.31
Shingle backer ... 0.375 in.	18	—	1.06	—	0.94	0.31
Shingle backer ... 0.3125 in.	18	—	1.28	—	0.78	
Sound deadening board ... 0.5 in.	15	—	0.74	—	1.35	0.30
Tile and lay-in panels, plain or acoustic	18	0.40	—	2.50	—	0.14
... 0.5 in.	18	—	0.80	—	1.25	
... 0.75 in.	18	—	0.53	—	1.89	
Laminated paperboard	30	0.50	—	2.00	—	0.33
Homogeneous board from repulped paper	30	0.50	—	2.00	—	0.28
Hardboard[e]						
Medium density	50	0.73	—	1.37	—	0.31
High density, service-tempered grade and service grade	55	0.82	—	1.22	—	0.32
High density, standard-tempered grade	63	1.00	—	1.00	—	0.32
Particleboard[e]						
Low density	37	0.71	—	1.41	—	0.31
Medium density	50	0.94	—	1.06	—	0.31
High density	62	.5	1.18	—	0.85	—
Underlayment ... 0.625 in.	40	—	1.22	—	0.82	0.29
Waferboard	37	0.63	—	1.59	—	—
Wood subfloor ... 0.75 in.	—	—	1.06	—	0.94	0.33
BUILDING MEMBRANE						
Vapor—permeable felt	—	—	16.70	—	0.06	
Vapor—seal, 2 layers of mopped 15 lb felt	—	—	8.35	—	0.12	
Vapor—seal, plastic film	—	—	—	—	Negl.	
FINISH FLOORING MATERIALS						
Carpet and fibrous pad	—	—	0.48	—	2.08	0.34
Carpet and rubber pad	—	—	0.81	—	1.23	0.33
Cork tile ... 0.125 in.	—	—	3.60	—	0.28	0.48
Terrazzo ... 1 in.	—	—	12.50	—	0.08	0.19
Tile—asphalt, linoleum, vinyl, rubber	—	—	20.00	—	0.05	0.30
vinyl asbestos						0.24
ceramic						0.19
Wood, hardwood finish ... 0.75 in.	—	—	1.47	—	0.68	
INSULATING MATERIALS						
Blanket and Batt[f,g]						
Mineral fiber, fibrous form processed from rock, slag, or glass						
approx. 3-4 in.	0.4-2.0	—	0.091	—	11	
approx. 3.5 in.	0.4-2.0	—	0.077	—	13	
approx. 3.5 in.	1.2-1.6	—	0.067	—	15	
approx. 5.5-6.5 in.	0.4-2.0	—	0.053	—	19	
approx. 5.5 in.	0.6-1.0	—	0.048	—	21	
approx. 6-7.5 in.	0.4-2.0	—	0.045	—	22	
approx. 8.25-10 in.	0.4-2.0	—	0.033	—	30	
approx. 10-13 in.	0.4-2.0	—	0.026	—	38	
Board and Slabs						
Cellular glass	8.0	0.33	—	3.03	—	0.18
Glass fiber, organic bonded	4.0-9.0	0.25	—	4.00	—	0.23
Expanded perlite, organic bonded	1.0	0.36	—	2.78	—	0.30
Expanded rubber (rigid)	4.5	0.22	—	4.55	—	0.40
Expanded polystyrene, extruded (smooth skin surface) (CFC-12 exp.)	1.8-3.5	0.20	—	5.00	—	0.29

Table 4 Typical Thermal Properties of Common Building and Insulating Materials—Design Values[a] *(Continued)*

Description	Density, lb/ft³	Conductivity[b] (k), Btu·in/h·ft²·°F	Conductance (C), Btu/h·ft²·°F	Resistance[c] (R) Per Inch Thickness (1/k), ft²·°F·h/Btu·in	Resistance[c] (R) For Thickness Listed (1/C), ft²·°F·h/Btu	Specific Heat, Btu/lb·°F
Expanded polystyrene, extruded (smooth skin surface) (HCFC-142b exp.)[h]	1.8-3.5	0.20	—	5.00	—	0.29
Expanded polystyrene, molded beads	1.0	0.26	—	3.85	—	—
	1.25	0.25	—	4.00	—	—
	1.5	0.24	—	4.17	—	—
	1.75	0.24	—	4.17	—	—
	2.0	0.23	—	4.35	—	—
Cellular polyurethane/polyisocyanurate[i] (CFC-11 exp.) (unfaced)	1.5	0.16-0.18	—	6.25-5.56	—	0.38
Cellular polyisocyanurate[i] (CFC-11 exp.) (gas-permeable facers)	1.5-2.5	0.16-0.18	—	6.25-5.56	—	0.22
Cellular polyisocyanurate[j] (CFC-11 exp.) (gas-impermeable facers)	2.0	0.14	—	7.04	—	0.22
Cellular phenolic (closed cell) (CFC-11, CFC-113 exp.)[k]	3.0	0.12	—	8.20	—	—
Cellular phenolic (open cell)	1.8-2.2	0.23	—	4.40	—	—
Mineral fiber with resin binder	15.0	0.29	—	3.45	—	0.17
Mineral fiberboard, wet felted						
Core or roof insulation	16-17	0.34	—	2.94	—	—
Acoustical tile	18.0	0.35	—	2.86	—	0.19
Acoustical tile	21.0	0.37	—	2.70	—	—
Mineral fiberboard, wet molded						
Acoustical tile[l]	23.0	0.42	—	2.38	—	0.14
Wood or cane fiberboard						
Acoustical tile[l] 0.5 in.	—	—	0.80	—	1.25	0.31
Acoustical tile[l] 0.75 in.	—	—	0.53	—	1.89	—
Interior finish (plank, tile)	15.0	0.35	—	2.86	—	0.32
Cement fiber slabs (shredded wood with Portland cement binder)	25-27.0	0.50-0.53		2.0-1.89	—	
Cement fiber slabs (shredded wood with magnesia oxysulfide binder)	22.0	0.57	—	1.75	—	0.31
Loose Fill						
Cellulosic insulation (milled paper or wood pulp)	2.3-3.2	0.27-0.32	—	3.70-3.13	—	0.33
Perlite, expanded	2.0-4.1	0.27-0.31	—	3.7-3.3	—	0.26
	4.1-7.4	0.31-0.36	—	3.3-2.8	—	—
	7.4-11.0	0.36-0.42	—	2.8-2.4	—	—
Mineral fiber (rock, slag, or glass)[g]						
approx. 3.75-5 in.	0.6-2.0	—	—	—	11.0	0.17
approx. 6.5-8.75 in.	0.6-2.0	—	—	—	19.0	—
approx. 7.5-10 in.	0.6-2.0	—	—	—	22.0	—
approx. 10.25-13.75 in.	0.6-2.0	—	—	—	30.0	—
Mineral fiber (rock, slag, or glass)[g]						
approx. 3.5 in. (closed sidewall application)	2.0-3.5	—	—	—	12.0-14.0	—
Vermiculite, exfoliated	7.0-8.2	0.47	—	2.13	—	0.32
	4.0-6.0	0.44	—	2.27	—	—
Spray Applied						
Polyurethane foam	1.5-2.5	0.16-0.18	—	6.25-5.56	—	—
Ureaformaldehyde foam	0.7-1.6	0.22-0.28	—	4.55-3.57	—	—
Cellulosic fiber	3.5-6.0	0.29-0.34	—	3.45-2.94	—	—
Glass fiber	3.5-4.5	0.26-0.27	—	3.85-3.70	—	—
Reflective Insulation						
Reflective material ($\varepsilon < 0.5$) in center of 3/4 in. cavity forms two 3/8 in. vertical air spaces[m]	—	—	0.31	—	3.2	—
METALS						
(See Chapter 38, Table 3)						
ROOFING						
Asbestos-cement shingles	120	—	4.76	—	0.21	0.24
Asphalt roll roofing	70	—	6.50	—	0.15	0.36
Asphalt shingles	70	—	2.27	—	0.44	0.30
Built-up roofing 0.375 in.	70	—	3.00	—	0.33	0.35
Slate 0.5 in.	—	—	20.00	—	0.05	0.30
Wood shingles, plain and plastic film faced	—	—	1.06	—	0.94	0.31
PLASTERING MATERIALS						
Cement plaster, sand aggregate	116	5.0	—	0.20	—	0.20
Sand aggregate 0.375 in.	—	—	13.3	—	0.08	0.20
Sand aggregate 0.75 in.	—	—	6.66	—	0.15	0.20

Table 4 Typical Thermal Properties of Common Building and Insulating Materials—Design Values[a] (*Continued*)

Description	Density, lb/ft³	Conductivity[b] (*k*), Btu·in/h·ft²·°F	Conductance (*C*), Btu/h·ft²·°F	Resistance[c] (*R*) Per Inch Thickness (1/*k*), ft²·°F·h/Btu·in	Resistance[c] (*R*) For Thickness Listed (1/*C*), ft²·°F·h/Btu	Specific Heat, Btu/lb·°F
Gypsum plaster:						
Lightweight aggregate ....0.5 in.	45	—	3.12	—	0.32	—
Lightweight aggregate ....0.625 in.	45	—	2.67	—	0.39	—
Lightweight aggregate on metal lath....0.75 in.	—	—	2.13	—	0.47	—
Perlite aggregate....	45	1.5	—	0.67	—	0.32
Sand aggregate....	105	5.6	—	0.18	—	0.20
Sand aggregate....0.5 in.	105	—	11.10	—	0.09	—
Sand aggregate....0.625 in.	105	—	9.10	—	0.11	—
Sand aggregate on metal lath....0.75 in.	—	—	7.70	—	0.13	—
Vermiculite aggregate	45	1.7	—	0.59	—	—
MASONRY MATERIALS						
Masonry Units						
Brick, fired clay	150	8.4-10.2	—	0.12-0.10	—	—
	140	7.4-9.0	—	0.14-0.11	—	—
	130	6.4-7.8	—	0.16-0.12	—	—
	120	5.6-6.8	—	0.18-0.15	—	0.19
	110	4.9-5.9	—	0.20-0.17	—	—
	100	4.2-5.1	—	0.24-0.20	—	—
	90	3.6-4.3	—	0.28-0.24	—	—
	80	3.0-3.7	—	0.33-0.27	—	—
	70	2.5-3.1	—	0.40-0.33	—	—
Clay tile, hollow						
1 cell deep ....3 in.	—	—	1.25	—	0.80	0.21
1 cell deep ....4 in.	—	—	0.90	—	1.11	—
2 cells deep....6 in.	—	—	0.66	—	1.52	—
2 cells deep....8 in.	—	—	0.54	—	1.85	—
2 cells deep....10 in.	—	—	0.45	—	2.22	—
3 cells deep....12 in.	—	—	0.40	—	2.50	—
Concrete blocks[n, o]						
Limestone aggregate						
8 in., 36 lb, 138 lb/ft³ concrete, 2 cores	—	—	—	—	—	—
Same with perlite filled cores	—	—	0.48	—	2.1	—
12 in., 55 lb, 138 lb/ft³ concrete, 2 cores	—	—	—	—	—	—
Same with perlite filled cores	—	—	0.27	—	3.7	—
Normal weight aggregate (sand and gravel)						
8 in., 33-36 lb, 126-136 lb/ft³ concrete, 2 or 3 cores	—	—	0.90-1.03	—	1.11-0.97	0.22
Same with perlite filled cores	—	—	0.50	—	2.0	—
Same with vermiculite filled cores	—	—	0.52-0.73	—	1.92-1.37	—
12 in., 50 lb, 125 lb/ft³ concrete, 2 cores	—	—	0.81	—	1.23	0.22
Medium weight aggregate (combinations of normal weight and lightweight aggregate)						
8 in., 26-29 lb, 97-112 lb/ft³ concrete, 2 or 3 cores..	—	—	0.58-0.78	—	1.71-1.28	—
Same with perlite filled cores	—	—	0.27-0.44	—	3.7-2.3	—
Same with vermiculite filled cores	—	—	0.30	—	3.3	—
Same with molded EPS (beads) filled cores	—	—	0.32	—	3.2	—
Same with molded EPS inserts in cores....	—	—	0.37	—	2.7	—
Lightweight aggregate (expanded shale, clay, slate or slag, pumice)						
6 in., 16-17 lb 85-87 lb/ft³ concrete, 2 or 3 cores	—	—	0.52-0.61	—	1.93-1.65	—
Same with perlite filled cores	—	—	0.24	—	4.2	—
Same with vermiculite filled cores	—	—	0.33	—	3.0	—
8 in., 19-22 lb, 72-86 lb/ft³ concrete	—	—	0.32-0.54	—	3.2-1.90	0.21
Same with perlite filled cores	—	—	0.15-0.23	—	6.8-4.4	—
Same with vermiculite filled cores	—	—	0.19-0.26	—	5.3-3.9	—
Same with molded EPS (beads) filled cores	—	—	0.21	—	4.8	—
Same with UF foam filled cores	—	—	0.22	—	4.5	—
Same with molded EPS inserts in cores....	—	—	0.29	—	3.5	—
12 in., 32-36 lb, 80-90 lb/ft³ concrete, 2 or 3 cores...	—	—	0.38-0.44	—	2.6-2.3	—
Same with perlite filled cores	—	—	0.11-0.16	—	9.2-6.3	—
Same with vermiculite filled cores	—	—	0.17	—	5.8	—
Stone, lime, or sand....	180	72	—	0.01	—	—
Quartzitic and sandstone	160	43	—	0.02	—	—
	140	24	—	0.04	—	—
	120	13	—	0.08	—	0.19
Calcitic, dolomitic, limestone, marble, and granite	180	30	—	0.03	—	—
	160	22	—	0.05	—	—
	140	16	—	0.06	—	—
	120	11	—	0.09	—	0.19
	100	8	—	0.13	—	—

Table 4 Typical Thermal Properties of Common Building and Insulating Materials—Design Values[a] (*Continued*)

Description	Density, lb/ft³	Conductivity[b] (*k*), Btu·in / h·ft²·°F	Conductance (*C*), Btu / h·ft²·°F	Resistance[c] (*R*): Per Inch Thickness (1/*k*), ft²·°F·h / Btu·in	Resistance[c] (*R*): For Thickness Listed (1/*C*), ft²·°F·h / Btu	Specific Heat, Btu / lb·°F
Gypsum partition tile						
3 by 12 by 30 in., solid	—	—	0.79	—	1.26	0.19
3 by 12 by 30 in., 4 cells	—	—	0.74	—	1.35	—
4 by 12 by 30 in., 3 cells	—	—	0.60	—	1.67	—
Concretes[o]						
Sand and gravel or stone aggregate concretes (concretes with more than 50% quartz or quartzite sand have conductivities in the higher end of the range)	150	10.0-20.0	—	0.10-0.05	—	—
	140	9.0-18.0	—	0.11-0.06	—	0.19-0.24
	130	7.0-13.0	—	0.14-0.08	—	—
Limestone concretes	140	11.1	—	0.09	—	—
	120	7.9	—	0.13	—	—
	100	5.5	—	0.18	—	—
Gypsum-fiber concrete (87.5% gypsum, 12.5% wood chips)	51	1.66	—	0.60	—	0.21
Cement/lime, mortar, and stucco	120	9.7	—	0.10	—	—
	100	6.7	—	0.15	—	—
	80	4.5	—	0.22	—	—
Lightweight aggregate concretes						
Expanded shale, clay, or slate; expanded slags; cinders; pumice (with density up to 100 lb/ft³); and scoria (sanded concretes have conductivities in the higher end of the range)	120	6.4-9.1	—	0.16-0.11	—	—
	100	4.7-6.2	—	0.21-0.16	—	0.20
	80	3.3-4.1	—	0.30-0.24	—	0.20
	60	2.1-2.5	—	0.48-0.40	—	—
	40	1.3	—	0.78	—	—
Perlite, vermiculite, and polystyrene beads	50	1.8-1.9	—	0.55-0.53	—	—
	40	1.4-1.5	—	0.71-0.67	—	0.15-0.23
	30	1.1	—	0.91	—	—
	20	0.8	—	1.25	—	—
Foam concretes	120	5.4	—	0.19	—	—
	100	4.1	—	0.24	—	—
	80	3.0	—	0.33	—	—
	70	2.5	—	0.40	—	—
Foam concretes and cellular concretes	60	2.1	—	0.48	—	—
	40	1.4	—	0.71	—	—
	20	0.8	—	1.25	—	—
SIDING MATERIALS (on flat surface)						
Shingles						
Asbestos-cement	120	—	4.75	—	0.21	—
Wood, 16 in., 7.5 exposure	—	—	1.15	—	0.87	0.31
Wood, double, 16 in., 12 in. exposure	—	—	0.84	—	1.19	0.28
Wood, plus ins. backer board, 0.312 in.	—	—	0.71	—	1.40	0.31
Siding						
Asbestos-cement, 0.25 in., lapped	—	—	4.76	—	0.21	0.24
Asphalt roll siding	—	—	6.50	—	0.15	0.35
Asphalt insulating siding (0.5 in. bed.)	—	—	0.69	—	1.46	0.35
Hardboard siding, 0.4375 in.	—	—	1.49	—	0.67	0.28
Wood, drop, 1 by 8 in.	—	—	1.27	—	0.79	0.28
Wood, bevel, 0.5 by 8 in., lapped	—	—	1.23	—	0.81	0.28
Wood, bevel, 0.75 by 10 in., lapped	—	—	0.95	—	1.05	0.28
Wood, plywood, 0.375 in., lapped	—	—	1.69	—	0.59	0.29
Aluminum, steel, or vinyl[p, q], over sheathing						
Hollow-backed	—	—	1.64	—	0.61	0.29[q]
Insulating-board backed nominal 0.375 in.	—	—	0.55	—	1.82	0.32
Insulating-board backed nominal 0.375 in., foil backed	—	—	0.34	—	2.96	—
Architectural (soda-lime float) glass	158	6.9	—	—	—	0.21
WOODS (12% moisture content)[e,r]						
Hardwoods						0.39[s]
Oak	41.2-46.8	1.12-1.25	—	0.89-0.80	—	
Birch	42.6-45.4	1.16-1.22	—	0.87-0.82	—	
Maple	39.8-44.0	1.09-1.19	—	0.92-0.84	—	
Ash	38.4-41.9	1.06-1.14	—	0.94-0.88	—	
Softwoods						0.39[s]
Southern pine	35.6-41.2	1.00-1.12	—	1.00-0.89	—	
Douglas fir-Larch	33.5-36.3	0.95-1.01	—	1.06-0.99	—	
Southern cypress	31.4-32.1	0.90-0.92	—	1.11-1.09	—	
Hem-Fir, Spruce-Pine-Fir	24.5-31.4	0.74-0.90	—	1.35-1.11	—	
West coast woods, Cedars	21.7-31.4	0.68-0.90	—	1.48-1.11	—	
California redwood	24.5-28.0	0.74-0.82	—	1.35-1.22	—	

Notes for Table 4

[a]Values are for a mean temperature of 75°F. Representative values for dry materials are intended as design (not specification) values for materials in normal use. Thermal values of insulating materials may differ from design values depending on their in-situ properties (e.g., density and moisture content, orientation, etc.) and variability experienced during manufacture. For properties of a particular product, use the value supplied by the manufacturer or by unbiased tests.

[b]To obtain thermal conductivities in Btu/h·ft·°F, divide the *k*-factor by 12 in/ft.

[c]Resistance values are the reciprocals of *C* before rounding off *C* to two decimal places.

[d]Lewis (1967).

[e]U.S. Department of Agriculture (1974).

[f]Does not include paper backing and facing, if any. Where insulation forms a boundary (reflective or otherwise) of an airspace, see Tables 2 and 3 for the insulating value of an airspace with the appropriate effective emittance and temperature conditions of the space.

[g]Conductivity varies with fiber diameter. (See Chapter 23, Factors Affecting Thermal Performance.) Batt, blanket, and loose-fill mineral fiber insulations are manufactured to achieve specified R-values, the most common of which are listed in the table. Due to differences in manufacturing processes and materials, the product thicknesses, densities, and thermal conductivities vary over considerable ranges for a specified R-value.

[h]This material is relatively new and data are based on limited testing.

[i]For additional information, see Society of Plastics Engineers (SPI) *Bulletin* U108. Values are for aged, unfaced board stock. For change in conductivity with age of expanded polyurethane/polyisocyanurate, see Chapter 23, Factors Affecting Thermal Performance.

[j]Values are for aged products with gas-impermeable facers on the two major surfaces. An aluminum foil facer of 0.001 in. thickness or greater is generally considered impermeable to gases. For change in conductivity with age of expanded polyisocyanurate, see Chapter 23, Factors Affecting Thermal Performance, and SPI *Bulletin* U108.

[k]Cellular phenolic insulation may no longer be manufactured. The thermal conductivity and resistance values do not represent aged insulation, which may have a higher thermal conductivity and lower thermal resistance.

[l]Insulating values of acoustical tile vary, depending on density of the board and on type, size, and depth of perforations.

[m]Cavity is framed with 0.75 in. wood furring strips. Caution should be used in applying this value for other framing materials. The reported value was derived from tests and applies to the reflective path only. The effect of studs or furring strips must be included in determining the overall performance of the wall.

[n]Values for fully grouted block may be approximated using values for concrete with a similar unit weight.

[o]Values for concrete block and concrete are at moisture contents representative of normal use.

[p]Values for metal or vinyl siding applied over flat surfaces vary widely, depending on amount of ventilation of airspace beneath the siding; whether airspace is reflective or nonreflective; and on thickness, type, and application of insulating backing used. Values are averages for use as design guides, and were obtained from several guarded hot box tests (ASTM C 236) or calibrated hot box (ASTM C 976) on hollow-backed types and types made using backing-boards of wood fiber, foamed plastic, and glass fiber. Departures of ±50% or more from these values may occur.

[q]Vinyl specific heat = 0.25 Btu/lb·°F

[r]See Adams (1971), MacLean (1941), and Wilkes (1979). The conductivity values listed are for heat transfer across the grain. The thermal conductivity of wood varies linearly with the density, and the density ranges listed are those normally found for the wood species given. If the density of the wood species is not known, use the mean conductivity value. For extrapolation to other moisture contents, the following empirical equation developed by Wilkes (1979) may be used:

$$k = 0.1791 + \frac{(1.874 \times 10^{-2} + 5.753 \times 10^{-4} M)\rho}{1 + 0.01M}$$

where ρ is density of the moist wood in lb/ft^3, and M is the moisture content in percent.

[s]From Wilkes (1979), an empirical equation for the specific heat of moist wood at 75°F is as follows:

$$c_p = \frac{(0.299 + 0.01M)}{(1 + 0.01M)} + \Delta c_p$$

where Δc_p accounts for the heat of sorption and is denoted by

$$\Delta c_p = M(1.921 \times 10^{-3} - 3.168 \times 10^{-5} M)$$

where M is the moisture content in percent by mass.

Assuming parallel heat flow only, the calculated resistance is higher than that calculated on the assumption of isothermal planes. The actual resistance generally is some value between the two calculated values. In the absence of test values, examination of the construction usually reveals whether a value closer to the higher or lower calculated R-value should be used. Generally, if the construction contains a layer in which lateral conduction is high compared with transmittance through the construction, the calculation with isothermal planes should be used. If the construction has no layer of high lateral conductance, the parallel heat flow calculation should be used.

Hot box tests of insulated and uninsulated masonry walls constructed with block of conventional configuration show that thermal resistances calculated using the isothermal planes heat flow method agree well with measured values (Van Geem 1985, Valore 1980, Shu et al. 1979). Neglecting horizontal mortar joints in conventional block can result in thermal transmittance values up to 16% lower than actual, depending on the density and thermal properties of the masonry, and 1 to 6% lower, depending on the core insulation material (Van Geem 1985, McIntyre 1984). For aerated concrete block walls, other solid masonry, and multicore block walls with full mortar joints, neglecting mortar joints can cause errors in R-values up to 40% (Valore 1988). Horizontal mortar joints usually found in concrete block wall construction are neglected in Example 2.

Constructions Containing Metal

Curtain and metal stud-wall constructions often include metallic and other thermal bridges, which can significantly reduce the thermal resistance. However, the capacity of the adjacent facing materials to transmit heat transversely to the metal is limited, and some contact resistance between all materials in contact limits the reduction. Contact resistances in building structures are only 0.06 to 0.6 ft^2·°F·h/Btu—too small to be of concern in many cases. However, the contact resistances of steel framing members may be important. Also, in many cases (as illustrated in Example 3), the area of metal in contact with the facing greatly exceeds the thickness of the metal, which mitigates the contact reistance effects.

Thermal characteristics for panels of sandwich construction can be computed by combining the thermal resistances of the various layers. R-values for the assembled sections should be determined on a representative sample by using a hot box method. If the sample is a wall section with air cavities on both sides of fibrous insulation, the sample must be of representative height since convective airflow can contribute significantly to heat flow through the test section. Computer modeling can also be useful, but all heat transfer mechanisms must be considered.

In Example 3, the metal member is only 0.020 in. thick, but it is in contact with adjacent facings over a 1.25 in.-wide area. The steel member is 3.50 in. deep, has a thermal resistance of approximately 0.011 ft^2·°F·h/Btu, and is virtually isothermal. The calculation involves careful selection of the appropriate thickness for the steel member. If the member is assumed to be 0.020 in. thick, the fact that the flange transmits heat to the adjacent facing is ignored, and the heat flow through the steel is underestimated. If the member is assumed to be 1.25 in. thick, the heat flow through the steel is overestimated. In Example 3, the steel member behaves in much the same way as a rectangular member 1.25 in. thick and 3.50 in. deep with a thermal resistance of $(1.25/0.020) \times 0.011 = 0.69$ ft²·°F·h/Btu does. The Building Research Association of New Zealand (BRANZ) commonly uses this approximation.

Example 3. Calculate the C-factor of the insulated steel frame wall shown in Figure 4. Assume that the steel member has an R-value of 0.69 ft²·°F·h/Btu and that the framing behaves as though it occupies approximately 8% of the transmission area.

Solution: Obtain the R-values of the various building elements from Table 4.

Element	*R* (Insul.)	*R* (Framing)
1. 0.5-in. gypsum wallboard	0.45	0.45
2. 3.5-in. mineral fiber batt insulation	11	—
3. Steel framing member	—	0.69
4. 0.5-in. gypsum wallboard	0.45	0.45
	$R_1 = 11.90$	$R_2 = 1.59$

Therefore, $C_1 = 0.084$; $C_2 = 0.629$ Btu/h·ft²·°F.

If the steel framing (thermal bridging) is not considered, the C-factor of the wall is calculated using Equation (3) from Chapter 23 as follows:

$$C_{av} = C_1 = 1/R_1 = 0.084 \text{ Btu/h}\cdot\text{ft}^2\cdot°\text{F}$$

If the steel framing is accounted for using the parallel flow method, the C-factor of the wall is determined using Equation (5) from Chapter 23 as follows:

$$C_{av} = (0.92 \times 0.084) + (0.08 \times 0.629)$$

$$= 0.128 \text{ Btu/h}\cdot\text{ft}^2\cdot°\text{F}$$

$$R_{T(av)} = 7.81 \text{ ft}^2\cdot°\text{F}\cdot\text{h/Btu}$$

If the steel framing is included using the isothermal planes method, the C-factor of the wall is determined using Equations (2) and (3) from Chapter 23 as follows:

$$R_{T(av)} = 0.45 + 1/[(0.92/11.00) + (0.08/0.69)] + 0.45$$

$$= 5.91 \text{ ft}^2\cdot°\text{F}\cdot\text{h/Btu}$$

$$C_{av} = 0.169 \text{ Btu/h}\cdot\text{ft}^2\cdot°\text{F}$$

For this insulated steel frame wall, Farouk and Larson (1983) measured an average R-value of 6.61 ft²·°F·h/Btu.

In ASHRAE/IESNA *Standard* 90.1-1989, one method given for determining the thermal resistance of wall assemblies containing metal framing involves using a parallel path correction factor F_c, which is listed in Table 8C-2 of the standard. For 2 by 4 steel framing, 16 in. OC, $F_c = 0.50$. Using the correction factor method, an R-value of 6.40 ft²·°F·h/Btu [0.45 + 11(0.50) + 0.45] is obtained for the wall described in Example 3.

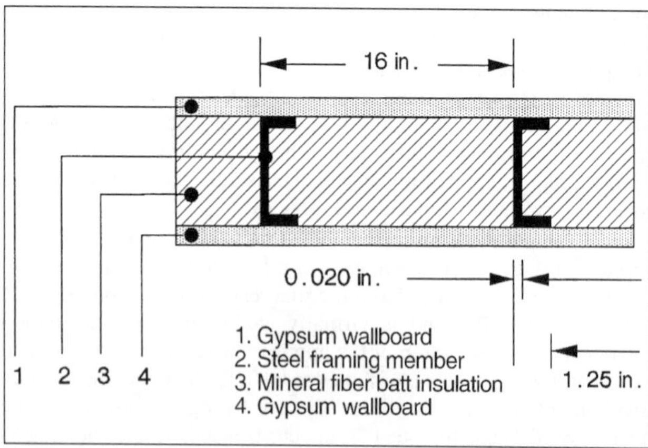

Fig. 4 Insulated Steel Frame Wall (Example 3)

Zone Method of Calculation

For structures with widely spaced metal members of substantial cross-sectional area, calculation by the isothermal planes method can result in thermal resistance values that are too low. For these constructions, the **zone method** can be used. This method involves two separate computations—one for a chosen limited portion, Zone A, containing the highly conductive element; the other for the remaining portion of simpler construction, Zone B. The two computations are then combined using the parallel flow method, and the average transmittance per unit overall area is calculated. The basic laws of heat transfer are applied by adding the area conductances *CA* of elements in parallel, and adding area resistances *R/A* of elements in series.

The surface shape of Zone A is determined by the metal element. For a metal beam (see Figure 5), the Zone A surface is a strip of width *W* that is centered on the beam. For a rod perpendicular to panel surfaces, it is a circle of diameter *W*. The value of *W* is calculated from Equation (1), which is empirical. The value of *d* should not be less than 0.5 in. for still air.

$$W = m + 2d \tag{1}$$

where

m = width or diameter of metal heat path terminal, in.
d = distance from panel surface to metal, in.

Generally, the value of *W* should be calculated using Equation (1) for each end of the metal heat path; the larger value, within the limits of the basic area, should be used as illustrated in Example 4.

Example 4. Calculate transmittance of the roof deck shown in Figure 5. Tee-bars at 24 in. OC support glass fiber form boards, gypsum concrete, and built-up roofing. Conductivities of components are: steel, 314.4 Btu·in/h·ft²·°F; gypsum concrete, 1.66 Btu·in/h·ft²·°F; and glass fiber form board, 0.25 Btu·in/h·ft²·°F. Conductance of built-up roofing is 3.00 Btu/h·ft²·°F.

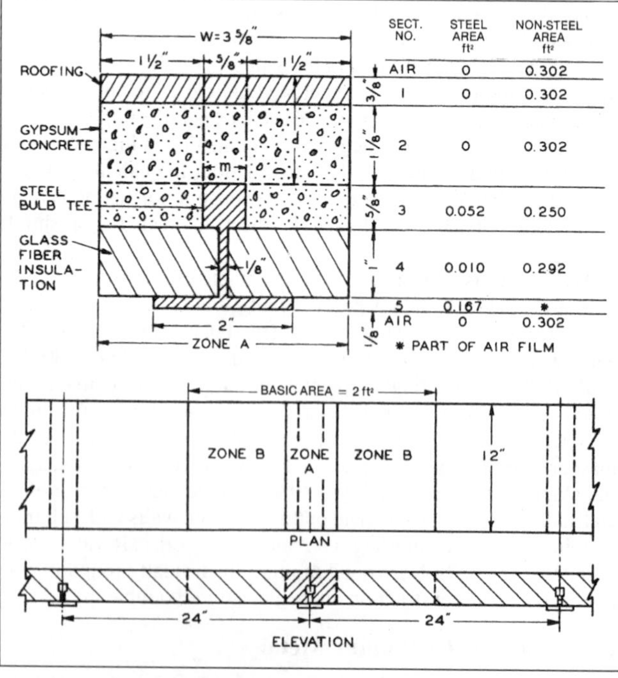

Fig. 5 Gypsum Roof Deck on Bulb Tees (Example 4)

Solution: The basic area is 2 ft² (24 in. by 12 in.) with a tee-bar (12 in. long) across the middle. This area is divided into Zones A and B.

Zone A is determined from Equation (1) as follows:

$$\text{Top side } W = m + 2d = 0.625 + (2 \times 1.5) = 3.625 \text{ in.}$$
$$\text{Bottom side } W = m + 2d = 2.0 + (2 \times 0.5) = 3.0 \text{ in.}$$

Using the larger value of W, the area of Zone A is $(12 \times 3.625)/144 = 0.302$ ft². The area of Zone B is $2.0 - 0.302 = 1.698$ ft².

To determine area transmittance for Zone A, divide the structure within the zone into five sections parallel to the top and bottom surfaces (Figure 5). The area conductance CA of each section is calculated by adding the area conductances of its metal and nonmetal paths. Area conductances of the sections are converted to area resistances R/A and added to obtain the total resistance of Zone A.

Section	Area × Conductance =	CA	$\frac{1}{CA} = \frac{R}{A}$
Air (outside, 15 mph)	0.302 × 6.00	1.81	0.55
No. 1, Roofing	0.302 × 3.00	0.906	1.10
No. 2, Gypsum concrete	0.302 × 1.66/1.125	0.446	2.24
No. 3, Steel	0.052 × 314.4/0.625	26.2	0.04
No. 3, Gypsum concrete	0.250 × 1.66/0.625	0.664	
No. 4, Steel	0.010 × 314.4/1.00	3.14	0.31
No. 4, Glass fiberboard	0.292 × 0.25/1.00	0.073	
No. 5, Steel	0.167 × 314.4/0.125	420.0	0.002
Air (inside)	0.302 × 1.63	0.492	2.03
		Total R/A = 6.27	

Area transmittance of Zone A = $1/(R/A) = 1/6.27 = 0.159$.

For Zone B, the unit resistances are added and then converted to area transmittance, as shown in the following table.

Section	Resistance, R
Air (outside, 15 mph)	1/6.00 = 0.17
Roofing	1/3.00 = 0.33
Gypsum concrete	1.75/1.66 = 1.05
Glass fiberboard	1.00/0.25 = 4.00
Air (inside)	1/1.63 = 0.61
Total resistance	= 6.16

Because unit transmittance = $1/R = 0.162$, the total area transmittance UA is calculated as follows:

$$\text{Zone B} = 1.698 \times 0.162 = 0.275$$
$$\text{Zone A} = 0.159$$
$$\text{Total area transmittance of basic area} = 0.434$$
$$\text{Transmittance per ft}^2 = 0.434/2.0 = 0.217$$
$$\text{Resistance per ft}^2 = 4.61$$

Overall R-values of 4.57 and 4.85 ft²·°F·h/Btu have been measured in two guarded hot box tests of a similar construction.

When the steel member represents a relatively large proportion of the total heat flow path, as in Example 4, detailed calculations of resistance in sections 3, 4, and 5 of Zone A are unnecessary; if only the steel member is considered, the final result of Example 4 is the same. However, if the heat flow path represented by the steel member is small, as for a tie rod, detailed calculations for sections 3, 4, and 5 are necessary. A panel with an internal metallic structure and bonded on one or both sides to a metal skin or covering presents special problems of lateral heat flow not covered in the zone method.

Modified Zone Method for Metal Stud Walls with Insulated Cavities

The modified zone method is similar to the parallel path method and the zone method. All three methods are based on parallel-path calculations. Figure 6 shows the width w of the zone of thermal anomalies around a metal stud. This zone can be assumed to equal the length of the stud flange L (parallel path method), or can be calculated as a sum of the length of stud flange and a distance double that from wall surface to metal Σd_i (zone method). In the modified zone method the width of the zone depends on the following three parameters:

- Ratio between thermal resistivity of sheathing material and cavity insulation
- Size (depth) of stud
- Thickness of sheathing material

The modified zone method is explained in Figure 6 (which can be copied and used as a calculation form). The wall cross section

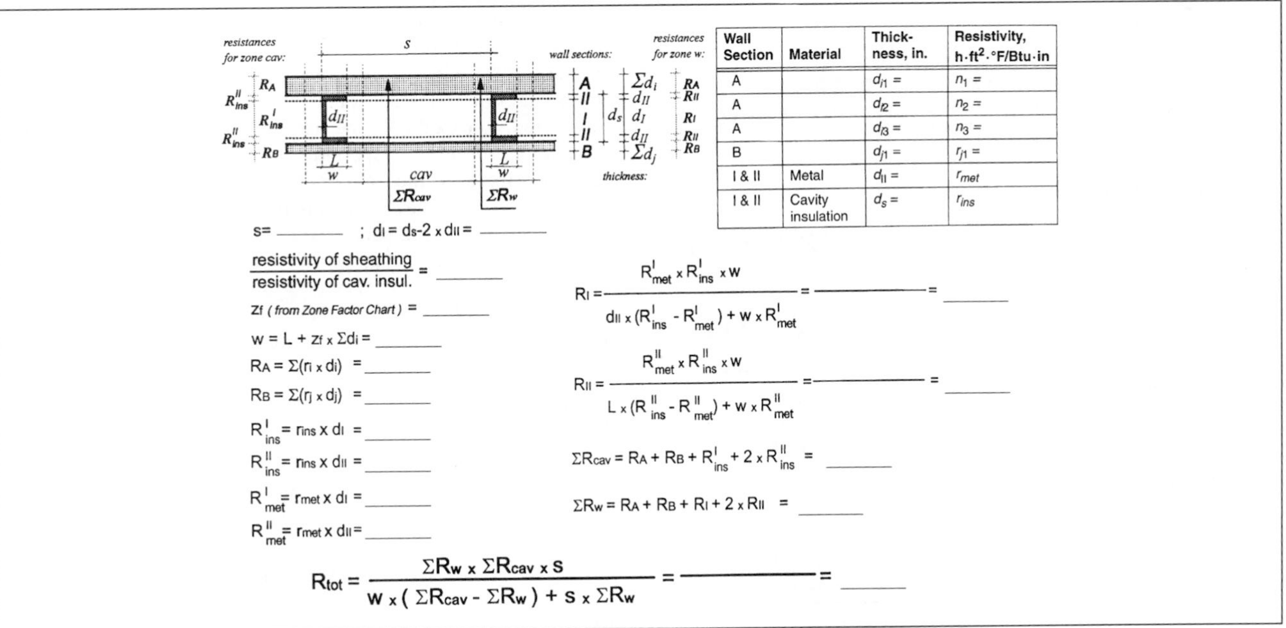

Fig. 6 Modified Zone Method R-Value Calculation Form for Metal Stud Walls

shown in Figure 6, is divided into two zones: the zone of thermal anomalies around metal stud w and the cavity zone cav. Wall material layers are grouped into an exterior and interior surface sections—A (sheathing, siding) and B (wallboard)—and interstitial sections I and II (cavity insulation, metal stud flange).

Assuming that the layers or layer of wall materials in wall section A are thicker than those in wall section B, as show by the cross section in Figure 6, they can be described as follows:

$$\sum_{i=1}^{n} d_i \geq \sum_{j=1}^{m} d_j \qquad (2)$$

where

n = number of material layer (of thickness d_i) between metal stud flange and wall surface for section A

m = number of material layer (of thickness d_j) for section B

Then, the width of the zone of thermal anomalies around the metal stud w can be estimated by

$$w = L + z_f \sum_{i=1}^{n} d_i \qquad (3)$$

where

L = stud flange size

d_i = thickness of material layer in section A

z_f = zone factor, which is shown in Figure 7 (z_f = 2 for zone method)

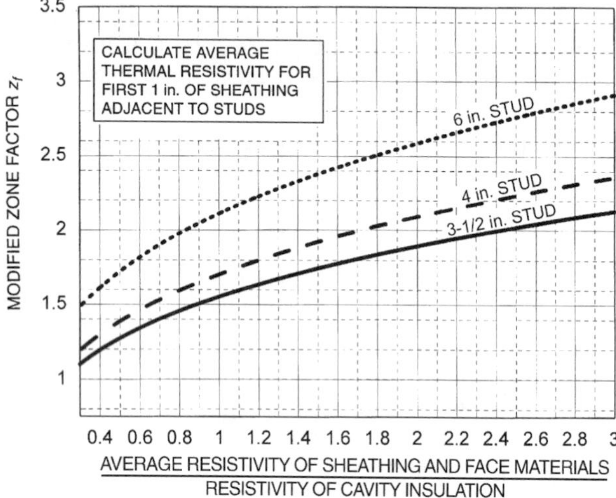

Use $z_f = -0.5$ for walls when total thickness of layer of materials attached to one side of metal frame ≤ **5/8 in. and** thermal resistivity of sheathing ≤ **1.5 h·ft²·°F/Btu·in**.

Use $z_f = +0.5$ for walls when total thickness of layer of materials attached to one side of metal frame ≤ **5/8 in. and** thermal resistivity of sheathing > **1.5 h·ft²·°F/Btu·in**

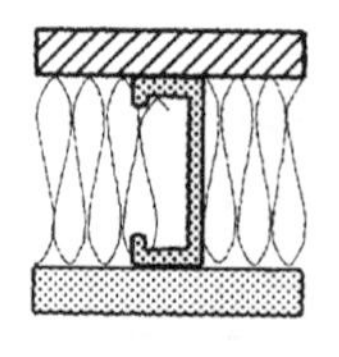

Find z_f in chart above for walls when total thickness of layer of materials attached to one side of metal frame > **5/8 in.**

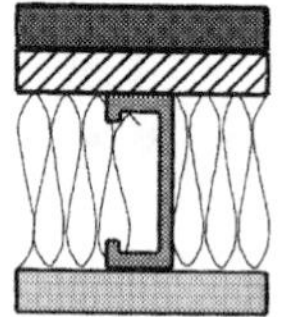

Fig. 7 Modified Zone Factor for Calculating R-Value of Metal Stud Walls with Cavity Insulation

Kosny and Christian (1995) verified the accuracy of the modified zone method for over 200 simulated cases of metal frame walls with insulated cavities. For all configurations considered the discrepancy between results were within ±2%. Hot box measured R-values for 15 metal stud walls tested by Barbour et al. (1994) were compared with results obtained by Kosny and Christian (1995) and McGowan and Desjarlais (1997). The modified zone method was found to be the most accurate simple method for estimating the clear wall R-value of light-gage steel stud walls with insulated cavities. However, this analysis does not apply to construction with metal sheathing. Also, *ASHRAE Standard* 90.1 may require a different method of analysis.

Ceilings and Roofs

The overall R-value for ceilings of wood frame flat roofs can be calculated using Equations (1) through (5) from Chapter 23. Properties of the materials are found in Tables 1, 3, 2, and 4. The fraction of framing is assumed to be 0.10 for joists at 16 in. OC and 0.07 for joists at 24 in. OC. The calculation procedure is similar to that shown in Example 1. Note that if the ceiling contains plane air spaces (see Table 3), the resistance depends on the direction of heat flow, i.e., whether the calculation is for a winter (heat flow up) or summer (heat flow down) condition.

For ceilings of pitched roofs under winter conditions, calculate the R-value of the ceiling using the procedure for flat roofs. Table 5 can be used to determine the effective resistance of the attic space under summer conditions for varying conditions of ventilation air temperature, airflow direction and rates, ceiling resistance, roof or sol-air temperatures, and surface emittances (Joy 1958).

The R-value is the total resistance obtained by adding the ceiling and effective attic resistances. The applicable temperature difference is that difference between room air and sol-air temperatures or between room air and roof temperatures (see Table 5, footnote f). Table 5 can be used for pitched and flat residential roofs over attic spaces. When an attic has a floor, the ceiling resistance should account for the complete ceiling-floor construction.

Windows and Doors

Table 4 of Chapter 30 lists U-factors for various fenestration products. Table 6 lists U-factors for exterior wood and steel doors. All U-factors are approximate, because a significant portion of the resistance of a window or door is contained in the air film resistances, and some parameters that may have important effects are not considered. For example, the listed U-factors assume the surface temperatures of surrounding bodies are equal to the ambient air temperature. However, the indoor surface of a window or door in an actual installation may be exposed to nearby radiating surfaces, such as radiant heating panels, or opposite walls with much higher or lower temperatures than the indoor air. Air movement across the indoor surface of a window or door, such as that caused by nearby heating and cooling outlet grilles, increases the U-factor; and air movement (wind) across the outdoor surface of a window or door also increases the U-factor.

U_o Concept

U_o is the combined thermal transmittance of the respective areas of gross exterior wall, roof or ceiling or both, and floor assemblies. The U_o equation for a wall is as follows:

$$U_o = (U_{wall}A_{wall} + U_{window}A_{window} + U_{door}A_{door})/A_o \qquad (4)$$

where

U_o = average thermal transmittance of gross wall area

A_o = gross area of exterior walls

Table 5 Effective Thermal Resistance of Ventilated Attics[a] (Summer Condition)

Ventilation Air Temperature, °F	Sol-Air[f] Temperature, °F	No Ventilation[b]		Natural Ventilation		Power Ventilation[c]					
		Ventilation Rate, cfm/ft²									
		0		0.1[d]		0.5		1.0		1.5	
		Ceiling Resistance R^e, ft²·°F·h/Btu									
		10	20	10	20	10	20	10	20	10	20
NONREFLECTIVE SURFACES											
80	120	1.9	1.9	2.8	3.4	6.3	9.3	9.6	16	11	20
	140	1.9	1.9	2.8	3.5	6.5	10	9.8	17	12	21
	160	1.9	1.9	2.8	3.6	6.7	11	10	18	13	22
90	120	1.9	1.9	2.5	2.8	4.6	6.7	6.1	10	6.9	13
	140	1.9	1.9	2.6	3.1	5.2	7.9	7.6	12	8.6	15
	160	1.9	1.9	2.7	3.4	5.8	9.0	8.5	14	10	17
100	120	1.9	1.9	2.2	2.3	3.3	4.4	4.0	6.0	4.1	6.9
	140	1.9	1.9	2.4	2.7	4.2	6.1	5.8	8.7	6.5	10
	160	1.9	1.9	2.6	3.2	5.0	7.6	7.2	11	8.3	13
REFLECTIVE SURFACES[g]											
80	120	6.5	6.5	8.1	8.8	13	17	17	25	19	30
	140	6.5	6.5	8.2	9.0	14	18	18	26	20	31
	160	6.5	6.5	8.3	9.2	15	18	19	27	21	32
90	120	6.5	6.5	7.5	8.0	10	13	12	17	13	19
	140	6.5	6.5	7.7	8.3	12	15	14	20	16	22
	160	6.5	6.5	7.9	8.6	13	16	16	22	18	25
100	120	6.5	6.5	7.0	7.4	8.0	10	8.5	12	8.8	12
	140	6.5	6.5	7.3	7.8	10	12	11	15	12	16
	160	6.5	6.5	7.6	8.2	11	14	13	18	15	20

[a] Although the term effective resistance is commonly used when there is attic ventilation, this table includes values for situations with no ventilation. The effective resistance of the attic added to the resistance ($1/U$) of the ceiling yields the effective resistance of this combination based on sol-air (see Chapter 29) and room temperatures. These values apply to wood frame construction with a roof deck and roofing that has a conductance of 1.0 Btu/h·ft²·°F.

[b] This condition cannot be achieved in the field unless extreme measures are taken to tightly seal the attic.

[c] Based on air discharging outward from attic.

[d] When attic ventilation meets the requirements stated in Chapter 26, 0.1 cfm/ft² is assumed as the natural summer ventilation rate.

[e] When determining ceiling resistance, do not add the effect of a reflective surface facing the attic, as it is accounted for in the Reflective Surfaces part of the table.

[f] Roof surface temperature rather than sol-air temperature (see Chapter 29) can be used if 0.25 is subtracted from the attic resistance shown.

[g] Surfaces with effective emittance $\varepsilon_{eff} = 0.05$ between ceiling joists facing attic space.

Table 6 Transmission Coefficients U for Wood and Steel Doors, Btu/h·ft²·°F

Nominal Door Thickness, in.	Description	No Storm Door	Wood Storm Door[c]	Metal Storm Door[d]
Wood Doors[a,b]				
1-3/8	Panel door with 7/16 in. panels[e]	0.57	0.33	0.37
1-3/8	Hollow core flush door	0.47	0.30	0.32
1-3/8	Solid core flush door	0.39	0.26	0.28
1-3/4	Panel door with 7/16 in. panels[e]	0.54	0.32	0.36
1-3/4	Hollow core flush door	0.46	0.29	0.32
1-3/4	Panel door with 1-1/8 in. panels[e]	0.39	0.26	0.28
1-3/4	Solid core flush door	0.40	—	0.26
2-1/4	Solid core flush door	0.27	0.20	0.21
Steel Doors[b]				
1-3/4	Fiberglass or mineral wool core with steel stiffeners, no thermal break[f]	0.60	—	—
1-3/4	Paper honeycomb core without thermal break[f]	0.56	—	—
1-3/4	Solid urethane foam core without thermal break[a]	0.40	—	—
1-3/4	Solid fire rated mineral fiberboard core without thermal break[f]	0.38	—	—
1-3/4	Polystyrene core without thermal break (18 gage commercial steel)[f]	0.35	—	—
1-3/4	Polyurethane core without thermal break (18 gage commercial steel)[f]	0.29	—	—
1-3/4	Polyurethane core without thermal break (24 gage residential steel)[f]	0.29	—	—
1-3/4	Polyurethane core with thermal break and wood perimeter (24 gage residential steel)[f]	0.20	—	—
1-3/4	Solid urethane foam core with thermal break[a]	0.20	—	0.16

Note: All U-factors for exterior doors in this table are for doors with no glazing, except for the storm doors which are in addition to the main exterior door. Any glazing area in exterior doors should be included with the appropriate glass type and analyzed as a window (see Chapter 30). Interpolation and moderate extrapolation are permitted for door thicknesses other than those specified.

[a] Values are based on a nominal 32 in. by 80 in. door size with no glazing.

[b] Outside air conditions: 15 mph wind speed, 0°F air temperature; inside air conditions: natural convection, 70°F air temperature.

[c] Values for wood storm door are for approximately 50% glass area.

[d] Values for metal storm door are for any percent glass area.

[e] 55% panel area.

[f] ASTM C 236 hot box data on a nominal 3 ft by 7 ft door size with no glazing.

U_{wall} = thermal transmittance of all elements of opaque wall area
A_{wall} = opaque wall area
U_{window} = thermal transmittance of window area (including frame)
A_{window} = window area (including frame)
U_{door} = thermal transmittance of door area
A_{door} = door area (including frame)

Where more than one type of wall, window, or door is used, the *UA* term for that exposure should be expanded into its subelements, as shown in Equation (3).

$$\begin{aligned} U_o A_o = {} & U_{wall\ 1} A_{wall\ 1} + U_{wall\ 2} A_{wall\ 2} + \ldots + U_{wall\ m} A_{wall\ m} \\ & + U_{window\ 1} A_{window\ 1} + U_{window\ 2} A_{window\ 2} + \ldots \\ & + U_{window\ n} A_{window\ n} + U_{door\ 1} A_{door\ 1} \\ & + U_{door\ 2} A_{door\ 2} + \ldots + U_{door\ o} A_{door\ o} \end{aligned} \tag{5}$$

Example 5. Calculate U_o for a wall 30 ft by 8 ft, constructed as in Example 1. The wall contains two double-glazed (0.5 in. airspace) fixed windows with wood/vinyl frames. (From Table 4 in Chapter 30, U = 0.52 Btu/h·ft²·°F.) One window is 60 in. by 34 in. and the second 36 in. by 30 in. The wall also contains a 1.75 in. solid core flush door with a metal storm door 34 in. by 80 in. (U = 0.26 Btu/h·ft²·°F from Table 6).

Solution: The U-factor for the wall was obtained in Example 1. The areas of the different components are

$$A_{window} = [(60 \times 34) + (36 \times 30)]/144 = 21.7 \text{ ft}^2$$

$$A_{door} = (34 \times 80)/144 = 18.9 \text{ ft}^2$$

$$A_{wall} = (30 \times 8) - (21.7 + 18.9) = 199.4 \text{ ft}^2$$

Therefore, the combined thermal transmittance for the wall is

$$U_o = \frac{(0.063 \times 199.4) + (0.52 \times 21.7) + (0.26 \times 18.9)}{(30 \times 8)}$$

$$= 0.119 \text{ Btu/h} \cdot \text{ft}^2 \cdot {}^\circ\text{F}$$

Slab-on-Grade and Below-Grade Construction

Heat transfer through basement walls and floors to the ground depends on the following factors: (1) the difference between the air temperature within the room and that of the ground and outside air, (2) the material of the walls or floor, and (3) the thermal conductivity of the surrounding earth. The latter varies with local conditions and is usually unknown. Because of the great thermal inertia of the surrounding soil, ground temperature varies with depth, and there is a substantial time lag between changes in outdoor air temperatures and corresponding changes in ground temperatures. As a result, ground-coupled heat transfer is less amenable to steady-state representation than above-grade building elements. However, several simplified procedures for estimating ground-coupled heat transfer have been developed. These fall into two principal categories: (1) those that reduce the ground heat transfer problem to a closed form solution, and (2) those that use simple regression equations developed from statistically reduced multidimensional transient analyses.

Closed form solutions, including the ASHRAE arc-length procedure discussed in Chapter 28 by Latta and Boileau (1969), generally reduce the problem to one-dimensional, steady-state heat transfer. These procedures use simple, "effective" U-factors or ground temperatures or both. Methods differ in the various parameters averaged or manipulated to obtain these effective values. Closed form solutions provide acceptable results in climates that have a single dominant season, because the dominant season persists long enough to permit a reasonable approximation of steady-state conditions at shallow depths. The large errors (percentage) that are likely during transition seasons should not seriously affect building design decisions, since these heat flows are relatively insignificant when compared with those of the principal season.

The ASHRAE arc-length procedure is a reliable method for wall heat losses in cold winter climates. Chapter 28 discusses a slab-on-grade floor model developed by one study. Although both procedures give results comparable to transient computer solutions for cold climates, their results for warmer U.S. climates differ substantially.

Research conducted by Hougten et al. (1942) and Dill et al. (1945) indicates a heat flow of approximately 2.0 Btu/h·ft² through an uninsulated concrete basement floor with a temperature difference of 20°F between the basement floor and the air 6 in. above it. A U-factor of 0.10 Btu/h·ft²·°F is sometimes used for concrete basement floors on the ground. For basement walls below grade, the temperature difference for winter design conditions is greater than for the floor. Test results indicate that at the midheight of the below-grade portion of the basement wall, the unit area heat loss is approximately twice that of the floor.

For concrete slab floors in contact with the ground at grade level, tests indicate that for small floor areas (equal to that of a 25 ft by 25 ft house) the heat loss can be calculated as proportional to the length of exposed edge rather than total area. This amounts to 0.81 Btu/h per linear foot of exposed edge per degree Fahrenheit difference between the indoor air temperature and the average outdoor air temperature. This value can be reduced appreciably by installing insulation under the ground slab and along the edge between the floor and abutting walls. In most calculations, if the perimeter loss is calculated accurately, no other floor losses need to be considered. Chapter 28 contains data for load calculations and heat loss values for below-grade walls and floors at different depths.

The second category of simplified procedures uses transient two-dimensional computer models to generate the ground heat transfer data that are then reduced to compact form by regression analysis (Mitalas 1982, 1983; Shipp 1983). These are the most accurate procedures available, but the database is very expensive to generate. In addition, these methods are limited to the range of climates and constructions specifically examined. Extrapolating beyond the outer bounds of the regression surfaces can produce significant errors.

Apparent Thermal Conductivity of Soil

Effective or apparent soil thermal conductivity is difficult to estimate precisely and may change substantially in the same soil at different times due to changed moisture conditions and the presence of freezing temperatures in the soil. Figure 8 shows the typical apparent soil thermal conductivity as a function of moisture content for different general types of soil. The figure is based on data presented in Salomone and Marlowe (1989) using envelopes of thermal behavior coupled with field moisture content ranges for different soil types. In Figure 8, the term well-graded applies to granular soils with good representation of all particle sizes from largest to smallest. The term poorly graded refers to granular soils with either a uniform gradation, in which most particles are about the same size, or a skip (or gap) gradation, in which particles of one or more intermediate sizes are not present.

Although thermal conductivity varies greatly over the complete range of possible moisture contents for a soil, this range can be narrowed if it is assumed that the moisture contents of most field soils lie between the "wilting point" of the soil (i.e., the moisture content of a soil below which a plant cannot alleviate its wilting symptoms) and the "field capacity" of the soil (i.e., the moisture content of a soil that has been thoroughly wetted and then drained until the drainage rate has become negligibly small). After a prolonged dry spell, the moisture will be near the wilting point, and after a rainy period, the soil will have a moisture content near its field capacity. The moisture contents at these limits have been studied by many agricultural researchers, and data for different types of soil are given by Salomone and Marlowe (1989) and Kersten (1949). The shaded

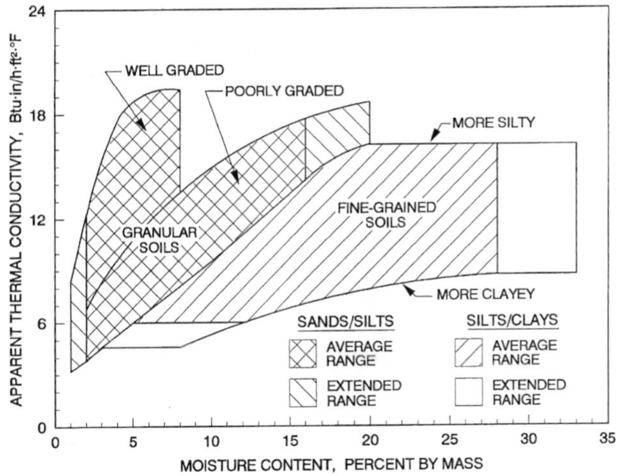

Fig. 8 Trends of Apparent Thermal Conductivity of Moist Soils

Table 7 Typical Apparent Thermal Conductivity Values for Soils, Btu·in/h·ft²·°F

		Recommended Values for Design[a]	
	Normal Range	Low[b]	High[c]
Sands	4.2 to 17.4	5.4	15.6
Silts	6 to 17.4	11.4	15.6
Clays	6 to 11.4	7.8	10.8
Loams	6 to 17.4	6.6	15.6

[a]Reasonable values for use when no site- or soil-specific data are available.
[b]Moderately conservative values for minimum heat loss through soil (e.g., use in soil heat exchanger or earth-contact cooling calculations). Values are from Salomone and Marlowe (1989).
[c]Moderately conservative values for maximum heat loss through soil (e.g., use in peak winter heat loss calculations). Values are from Salomone and Marlowe (1989).

Table 8 Typical Apparent Thermal Conductivity Values for Rocks, Btu·in/h·ft²·°F

	Normal Range
Pumice, tuff, obsidian	3.6 to 15.6
Basalt	3.6 to 18.0
Shale	6 to 27.6
Granite	12 to 30
Limestone, dolomite, marble	8.4 to 30
Quartzose sandstone	9.6 to 54

areas on Figure 8 approximate (1) the full range of moisture contents for different soil types and (2) a range between average values of each limit.

Table 7 gives a summary of design values for thermal conductivities of the basic soil classes. Table 8 gives ranges of thermal conductivity for some basic classes of rock. The value chosen depends on whether heat transfer is being calculated for minimum heat loss through the soil, as in a ground heat exchange system, or a maximum value, as in peak winter heat loss calculations for a basement. Hence, a high and a low value are given for each soil class.

As heat flows through the soil, the moisture tends to move away from the source of heat. This moisture migration provides initial mass transport of heat, but it also dries the soil adjacent to the heat source, hence lowering the apparent thermal conductivity in that zone of soil.

Trends typical in a soil when other factors are held constant are:

- k increases with moisture content
- k increases with increasing dry density of a soil
- k decreases with increasing organic content of a soil
- k tends to decrease for soils with uniform gradations and rounded soil grains (because the grain-to-grain contacts are reduced)
- k of a frozen soil may be higher or lower than that of the same unfrozen soil (because the conductivity of ice is higher than that of water but lower than that of the typical soil grains). Differences in k below moisture contents of 7 to 8% are quite small. At approximately 15% moisture content, differences in k-factors may vary up to 30% from unfrozen values.

When calculating annual energy use, values that represent typical site conditions as they vary during the year should be chosen. In climates where ground freezing is significant, accurate heat transfer simulations should include the effect of the latent heat of fusion of water. The energy released during this phase change significantly retards the progress of the frost front in moist soils.

Water Vapor Transmission Data for Building Components

Table 9 gives typical water vapor permeance and permeability values for common building materials. These values can be used to calculate water vapor flow through building components and assemblies using equations in Chapter 23.

MECHANICAL AND INDUSTRIAL SYSTEMS

Thermal Transmission Data

Table 10 lists the thermal conductivities of various materials used as industrial insulations. These values are functions of the arithmetic mean of the temperatures of the inner and outer surfaces for each insulation.

Heat Loss from Pipes and Flat Surfaces

Tables 11A, 11B, and 12 give heat losses from bare steel pipes and flat surfaces and bare copper tubes. These tables were calculated using ASTM *Standard* C 680. User inputs for the programs described in the standard include operating temperature, ambient temperature, pipe size, insulation type, number of insulation layers, and thickness for each layer. A program option allows the user to input a surface coefficient or surface emittance, surface orientation, and wind speed. The computer uses this information to calculate the heat flow and the surface temperature. The programs calculate the surface coefficients if the user has not already supplied them.

The equations used in ASTM C 680 are

$$h_{cv} = C\left(\frac{1}{d}\right)^{0.2}\left(\frac{1}{T_{avg}}\right)^{0.181}(\Delta T^{0.266})\sqrt{1 + 1.277(\text{Wind})} \quad (6)$$

where

h_{cv} = convection surface coefficient, Btu/h·ft²·°F
d = diameter for cylinder, in. For flat surfaces and large cylinders ($d > 24$ in.), use $d = 24$ in.
T_{avg} = average temperature of air film = $(T_a + T_s)/2$, °R
T_a = temperature of ambient air, °R
T_s = temperature of surface, °R
ΔT = surface to air temperature difference, °R
Wind = air speed, mph
C = constant depending on shape and heat flow condition
= 1.016 for horizontal cylinders
= 1.235 for longer vertical cylinders
= 1.394 for vertical plates
= 1.79 for horizontal plates, warmer than air, facing upward
= 0.89 for horizontal plates, warmer than air, facing downward
= 0.89 for horizontal plates, cooler than air, facing upward
= 1.79 for horizontal plates, cooler than air, facing downward

Table 9 Typical Water Vapor Permeance and Permeability Values for Common Building Materials[a]

Material	Thickness, in.	Permeance, Perm	Resistance[h], Rep	Permeability, Perm-in.	Resistance/in.[h], Rep/in.
Construction Materials					
Concrete (1:2:4 mix)				3.2	0.31
Brick masonry	4	0.8[f]	1.3		
Concrete block (cored, limestone aggregate)	8	2.4[f]	0.4		
Tile masonry, glazed	4	0.12[f]	8.3		
Asbestos cement board	0.12	4-8[d]	0.1-0.2		
With oil-base finishes		0.3-0.5[d]	2-3		
Plaster on metal lath	0.75	15[f]	0.067		
Plaster on wood lath		11[e]	0.091		
Plaster on plain gypsum lath (with studs)		20[f]	0.050		
Gypsum wall board (plain)	0.375	50[f]	0.020		
Gypsum sheathing (asphalt impregnated)	0.5			20[d]	0.050
Structural insulating board (sheathing quality)				20-50[f]	0.050-0.020
Structural insulating board (interior, uncoated)	0.5	50-90[f]	0.020-0.011		
Hardboard (standard)	0.125	11[f]	0.091		
Hardboard (tempered)	0.125	5[f]	0.2		
Built-up roofing (hot mopped)		0			
Wood, sugar pine				0.4-5.4[b]	2.5-0.19
Plywood (douglas fir, exterior glue)	0.25	0.7[f]	1.4		
Plywood (douglas fir, interior glue)	0.25	1.9[f]	0.53		
Acrylic, glass fiber reinforced sheet	0.056	0.12[d]	8.3		
Polyester, glass fiber reinforced sheet	0.048	0.05[d]	20		
Thermal Insulations					
Air (still)				120[f]	0.0083
Cellular glass				0[d]	∞
Corkboard				2.1-2.6[d]	0.48-0.38
				9.5[e]	0.11
Mineral wool (unprotected)				116[e]	0.0086
Expanded polyurethane (R-11 blown) board stock				0.4-1.6[d]	2.5-0.62
Expanded polystyrene—extruded				1.2[d]	0.83
Expanded polystyrene—bead				2.0-5.8[d]	0.50-0.17
Phenolic foam (covering removed)				26	0.038
Unicellular synthetic flexible rubber foam				0.02-0.15[d]	50-6.7
Plastic and Metal Foils and Films[c]					
Aluminum foil	0.001	0.0[d]	∞		
Aluminum foil	0.00035	0.05[d]	20		
Polyethylene	0.002	0.16[d]	6.3		3100
Polyethylene	0.004	0.08[d]	12.5		3100
Polyethylene	0.006	0.06[d]	17		3100
Polyethylene	0.008	0.04[d]	25		3100
Polyethylene	0.010	0.03[d]	33		3100
Polyvinylchloride, unplasticized	0.002	0.68[d]	1.5		
Polyvinylchloride, plasticized	0.004	0.8-1.4[d]	1.3-0.72		
Polyester	0.001	0.73[d]	1.4		
Polyester	0.0032	0.23[d]	4.3		
Polyester	0.0076	0.08[d]	12.5		
Cellulose acetate	0.01	4.6[d]	0.2		
Cellulose acetate	0.125	0.32[d]	3.1		

$$h_{rad} = \frac{\varepsilon\sigma(T_a^4 - T_s^4)}{T_a - T_s} \qquad (7)$$

where

h_{rad} = radiation surface coefficient, Btu/h·ft²·°F
ε = surface emittance
σ = Stefan-Boltzmann constant = 0.1712×10^{-8} Btu/h·ft²·°R⁴

Example 6. Compute the total annual heat loss from 165 ft of nominal 2 in. bare steel pipe in service 4000 h per year. The pipe is carrying steam at 10 psig and is exposed to an average air temperature of 80°F.

Solution: The pipe temperature is taken as the steam temperature, which is 239.4°F, obtained by interpolation from Steam Tables. By interpolation in Table 11A between 180°F and 280°F, heat loss from a nominal 2 in. pipe is 285 Btu/h·ft. Total annual heat loss from the entire line is 285 Btu/h·ft × 165 ft × 4000 h = 188×10^6 Btu.

In calculating heat flow, Equations (8) and (9) from Chapter 23 generally are used. For dimensions of standard pipe and fitting sizes, refer to the *Piping Handbook*. For insulation product dimensions, refer to ASTM *Standard* C 585, or to the insulation manufacturers' literature.

Table 9 Typical Water Vapor Permeance and Permeability Values for Common Building Materials (*Concluded*)[a]

Material	Weight, lb/100 ft²	Permeance, Perms			Resistance[h] Rep		
		Dry-Cup	Wet-Cup	Other	Dry-Cup	Wet-Cup	Other
Building Paper, Felts, Roofing Papers[g]							
Duplex sheet, asphalt laminated, aluminum foil one side	8.6	0.002	0.176		500	5.8	
Saturated and coated roll roofing	65	0.05	0.24		20	4.2	
Kraft paper and asphalt laminated, reinforced 30-120-30	6.8	0.3	1.8		3.3	0.55	
Blanket thermal insulation backup paper, asphalt coated	6.2	0.4	0.6-4.2		2.5	1.7-0.24	
Asphalt-saturated and coated vapor retarder paper	8.6	0.2-0.3	0.6		5.0-3.3	1.7	
Asphalt-saturated, but not coated, sheathing paper	4.4	3.3	20.2		0.3	0.05	
15-lb asphalt felt	14	1.0	5.6		1.0	0.18	
15-lb tar felt	14	4.0	18.2		0.25	0.055	
Single-kraft, double	3.2	31	42		0.032	0.024	
Liquid-Applied Coating Materials	**Thickness, in.**						
Commercial latex paints (dry film thickness)[i]							
Vapor retarder paint	0.0031			0.45			2.22
Primer-sealer	0.0012			6.28			0.16
Vinyl acetate/acrylic primer	0.002			7.42			0.13
Vinyl-acrylic primer	0.0016			8.62			0.12
Semi-gloss vinyl-acrylic enamel	0.0024			6.61			0.15
Exterior acrylic house and trim	0.0017			5.47			0.18
Paint—2 coats							
Asphalt paint on plywood			0.4			2.5	
Aluminum varnish on wood		0.3-0.5			3.3-2.0		
Enamels on smooth plaster				0.5-1.5			2.0-0.66
Primers and sealers on interior insulation board				0.9-2.1			1.1-0.48
Various primers plus 1 coat flat oil paint on plaster				1.6-3.0			0.63-0.33
Flat paint on interior insulation board				4			0.25
Water emulsion on interior insulation board				30-85			0.03-0.012
	Weight, oz/ft²						
Paint-3 coats							
Exterior paint, white lead and oil on wood siding		0.3-1.0			3.3-1.0		
Exterior paint, white lead-zinc oxide and oil on wood		0.9			1.1		
Styrene-butadiene latex coating	2	11			0.09		
Polyvinyl acetate latex coating	4	5.5			0.18		
Chlorosulfonated polyethylene mastic	3.5	1.7			0.59		
	7.0	0.06			16		
Asphalt cutback mastic, 1/16 in., dry		0.14			7.2		
3/16 in., dry		0.0			—		
Hot melt asphalt	2	0.5			2		
	3.5	0.1			10		

[a]This table permits comparisons of materials; but in the selection of vapor retarder materials, exact values for permeance or permeability should be obtained from the manufacturer or from laboratory tests. The values shown indicate variations among mean values for materials that are similar but of different density, orientation, lot, or source. The values should not be used as design or specification data. Values from dry-cup and wet-cup methods were usually obtained from investigations using ASTM E 96 and C 355; values shown under others were obtained by two-temperature, special cell, and air velocity methods. Permeance, resistance, permeability, and resistance per unit thickness values are given in the following units:

Permeance	Perm	= gr/h·ft²·in. Hg
Resistance	Rep	= in. Hg·ft²·h/gr
Permeability	Perm-in.	= gr/h·ft²·(in. Hg/in.)
Resistance/unit thickness	Rep/in.	= (in. Hg·ft²·h/gr)/in.

[b]Depending on construction and direction of vapor flow.

[c]Usually installed as vapor retarders, although sometimes used as an exterior finish and elsewhere near the cold side, where special considerations are then required for warm side barrier effectiveness.

[d]Dry-cup method.

[e]Wet-cup method.

[f]Other than dry- or wet-cup method.

[g]Low permeance sheets used as vapor retarders. High permeance used elsewhere in construction.

[h]Resistance and resistance/in. values have been calculated as the reciprocal of the permeance and permeability values.

[i]Cast at 10 mils (0.01 in.) wet film thickness.

Table 10 Typical Thermal Conductivity for Industrial Insulations at Various Mean Temperatures—Design Values[a]

Material	Max. Temp.,[b] °F	Typical Density, lb/ft³	Typical Conductivity k in Btu·in/h·ft²·°F at Mean Temperature, °F													
			−100	−75	−50	−25	0	25	50	75	100	200	300	500	700	900
BLANKETS AND FELTS																
ALUMINOSILICATE FIBER																
7 to 10 μm diameter fiber	1800	4								0.24		0.32		0.54	0.99	1.03
	2000	6-8								0.25		0.30		0.48	0.78	0.95
3 μm diameter fiber	2200	4								0.22		0.29		0.45	0.59	0.74
MINERAL FIBER (Rock, slag, or glass)																
Blanket, metal reinforced	1200	6-12									0.26	0.32	0.39	0.54		
	1000	2.5-6									0.24	0.31	0.40	0.61		
Blanket, flexible, fine-fiber organic bonded	350	0.75				0.25	0.26	0.28	0.30	0.33	0.36	0.53				
		0.75				0.24	0.25	0.27	0.29	0.32	0.34	0.48				
		1.0				0.23	0.24	0.25	0.27	0.29	0.32	0.43				
		1.5				0.21	0.22	0.23	0.25	0.27	0.28	0.37				
		2.0				0.20	0.21	0.22	0.23	0.25	0.26	0.33				
		3.0				0.19	0.20	0.21	0.22	0.23	0.24	0.31				
Blanket, flexible, textile fiber, organic bonded	350	0.65				0.27	0.28	0.29	0.30	0.31	0.32	0.50	0.68			
		0.75				0.26	0.27	0.28	0.29	0.31	0.32	0.48	0.66			
		1.0				0.24	0.25	0.26	0.27	0.29	0.31	0.45	0.60			
		1.5				0.22	0.23	0.24	0.25	0.27	0.29	0.39	0.51			
		3.0				0.20	0.21	0.22	0.23	0.24	0.25	0.32	0.41			
Felt, semirigid organic bonded	400	3-8						0.24	0.25	0.26	0.27	0.35	0.44			
Laminated and felted without binder	850	3	0.16	0.17	0.18	0.19	0.20	0.21	0.22	0.23	0.24	0.35	0.55			
	1200	7.5											0.35	0.45	0.60	
BLOCKS, BOARDS, AND PIPE INSULATION																
MAGNESIA	600	11-12									0.35	0.38	0.42			
85% CALCIUM SILICATE	1200	11-15									0.38	0.41	0.44	0.52	0.62	0.72
	1800	12-15												0.63	0.74	0.95
CELLULAR GLASS	900	7.8-8.2	0.24	0.25	0.26	0.28	0.29	0.30	0.32	0.33	0.34	0.41	0.49	0.70	1.01	
DIATOMACEOUS SILICA	1600	21-22												0.64	0.68	0.72
	1900	23-25												0.70	0.75	0.80
MINERAL FIBER (Glass)																
Organic bonded, block and boards	400	3-10	0.16	0.17	0.18	0.19	0.20	0.22	0.24	0.25	0.26	0.33	0.40			
Nonpunking binder	1000	3-10									0.26	0.31	0.38	0.52		
Pipe insulation, slag, or glass	350	3-4					0.20	0.21	0.22	0.23	0.24	0.29				
	500	3-10					0.20	0.22	0.24	0.25	0.26	0.33	0.40			
Inorganic bonded block	1000	10-15									0.33	0.38	0.45	0.55		
	1800	15-24									0.32	0.37	0.42	0.52	0.62	0.74
Pipe insulation, slag, or glass	1000	10-15									0.33	0.38	0.45	0.55		
Resin binder		15	0.23	0.24	0.25	0.26	0.28	0.29								
RIGID POLYSTYRENE																
Extruded (CFC-12 exp.)(smooth skin surface)	165	1.8-3.5	0.16	0.16	0.17	0.16	0.17	0.18	0.19	0.20						
Molded beads	165	1	0.17	0.19	0.20	0.21	0.22	0.24	0.25	0.26	0.28					
		1.25	0.17	0.18	0.19	0.20	0.22	0.23	0.24	0.25	0.27					
		1.5	0.16	0.17	0.19	0.20	0.21	0.22	0.23	0.24	0.26					
		1.75	0.16	0.17	0.18	0.19	0.20	0.22	0.23	0.24	0.25					
		2.0	0.15	0.16	0.18	0.19	0.20	0.21	0.22	0.23	0.24					
RIGID POLYURETHANE/POLYISOCYANURATE[c,d]																
Unfaced (CFC-11 exp.)	210	1.5-2.5	0.16	0.17	0.18	0.18	0.18	0.17	0.16	0.16	0.17					
RIGID POLYISOCYANURATE																
Gas-impermeable facers (CFC-11 exp.)	250	2.0						0.12	0.13	0.14	0.15					
RIGID PHENOLIC																
Closed cell (CFC-11, CFC-113 exp.)		3.0						0.11	0.115	0.12	0.125					
RUBBER, Rigid foamed	150	4.5						0.20	0.21	0.22	0.23					
VEGETABLE AND ANIMAL FIBER																
Wool felt (pipe insulation)	180	20						0.28	0.30	0.31	0.33					
INSULATING CEMENTS																
MINERAL FIBER (Rock, slag, or glass)																
With colloidal clay binder	1800	24-30									0.49	0.55	0.61	0.73	0.85	
With hydraulic setting binder	1200	30-40									0.75	0.80	0.85	0.95		
LOOSE FILL																
Cellulose insulation (milled pulverized paper or wood pulp)		2.5-3							0.26	0.27	0.29					
Mineral fiber, slag, rock, or glass		2-5			0.19	0.21	0.23	0.25	0.26	0.28	0.31					
Perlite (expanded)		3-5	0.22	0.24	0.25	0.27	0.28	0.30	0.31	0.33	0.35					
Silica aerogel		7.6			0.13	0.14	0.15	0.15	0.16	0.17	0.18					
Vermiculite (expanded)		7-8.2			0.39	0.40	0.42	0.44	0.45	0.47	0.49					
		4-6			0.34	0.35	0.38	0.40	0.42	0.44	0.46					

[a]Representative values for dry materials, which are intended as design (not specification) values for materials in normal use. Insulation materials in actual service may have thermal values that vary from design values depending on their in-situ properties (e.g., density and moisture content). For properties of a particular product, use the value supplied by the manufacturer or by unbiased tests.

[b]These temperatures are generally accepted as maximum. When operating temperature approaches these limits, follow the manufacturers' recommendations.

[c]Some polyurethane foams are formed by means that produce a stable product (with respect to k), but most are blown with refrigerant and will change with time.

[d]See Table 4, footnote i.

[e]See Table 4, footnote j.

Examples 7 and 8 illustrate how Equations (8) and (9) from Chapter 23 can be used to determine heat loss from both flat and cylindrical surfaces. Figure 9 shows surface resistance as a function of heat transmission for both flat and cylindrical surfaces. The surface emittance is assumed to be 0.85 to 0.90 in still air at 80°F.

Example 7. Compute the heat loss from a boiler wall if the interior insulation surface temperature is 1100°F and ambient still air temperature is 80°F. The wall is insulated with 4.5 in. of mineral fiber block and 0.5 in. of mineral fiber insulating and finishing cement.

Solution: Assume that the mean temperature of the mineral fiber block is 700°F, the mean temperature of the insulating cement is 200°F, and the surface resistance R_s is 0.60 ft²·°F·h/Btu.

From Table 10, $k_1 = 0.62$ and $k_2 = 0.80$. Using Equation (8) from Chapter 23:

$$q_s = \frac{1100 - 80}{(4.5/0.62) + (0.5/0.80) + 0.60} = 120.2 \text{ Btu/h} \cdot \text{ft}^2$$

As a check, from Figure 9, at 120.2 Btu/h·ft², $R_s = 0.56$. The mean temperature of the mineral fiber block is

$$4.5/0.62 = 7.26;\ 7.26/2 = 3.63$$

$$1100 - \frac{3.63}{8.48}(1020) = 663°\text{F}$$

and the mean temperature of the insulating cement is

$$0.5/0.80 = 0.63;\ 0.63/2 = 0.31;\ 7.26 + 0.31 = 7.57$$

$$1100 - \frac{7.57}{8.48}(1020) = 189°\text{F}$$

From Table 10, at 663°F, $k_1 = 0.60$; at 189°F, $k_2 = 0.79$.

Using these adjusted values to recalculate q_s

$$q_s = \frac{1020}{(4.5/0.60) + (0.5/0.79) + 0.56} = \frac{1020}{8.69}$$

$$= 117.4 \text{ Btu/h} \cdot \text{ft}^2$$

From Figure 9, at 117.4 Btu/h·ft², $R_s = 0.56$. The mean temperature of the mineral fiber block is

$$4.5/0.6 = 7.50;\ 7.50/2 = 3.75$$

$$1100 - \frac{3.75}{8.69}(1020) = 660°\text{F}$$

and the mean temperature of the insulating cement is

$$0.5/0.79 = 0.63;\ 0.63/2 = 0.31;\ 7.50 + 0.31 = 7.81$$

$$1100 - \frac{7.81}{8.69}(1020) = 183°\text{F}$$

From Table 10, at 660°F, $k_1 = 0.60$; at 183°F, $k_2 = 0.79$.

Because R_s, k_1, and k_2 do not change at these values, $q_s = 117.4$ Btu/h·ft.

Example 8. Compute heat loss per square foot of outer surface of insulation if pipe temperature is 1200°F and ambient still air temperature is 80°F. The pipe is nominal 6 in. steel pipe, insulated with a nominal 3 in. thick diatomaceous silica as the inner layer and a nominal 2-in. thick calcium silicate as the outer layer.

Solution: From Chapter 41 of the 2000 *ASHRAE Handbook—Equipment*, $r_o = 3.31$ in. A nominal 3 in. thick diatomaceous silica insulation to fit a nominal 6 in. steel pipe is 3.02 in. thick. A nominal 2 in. thick calcium silicate insulation to fit over the 3.02 in. diatomaceous silica is 2.08 in. thick. Therefore, $r_i = 6.33$ in. and $r_s = 8.41$ in.

Assume that the mean temperature of the diatomaceous silica is 600°F, the mean temperature of the calcium silicate is 250°F and the surface resistance R_s is 0.50. From Table 10, $k_1 = 0.66$; $k_2 = 0.42$. By Equation (9) from Chapter 23,

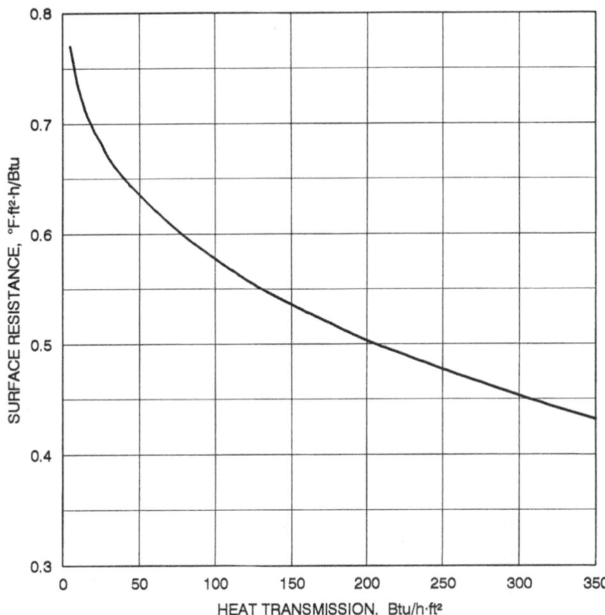

Fig. 9 Surface Resistance as Function of Heat Transmission for Flat Surfaces and Cylindrical Surfaces Greater than 24 in. in Diameter

$$q_s = \frac{1200 - 80}{[8.41 \ln(6.33/3.31)/0.66] + [8.41 \ln(8.41/3.31)/0.40] + 0.50}$$

$$= \frac{1120}{(5.45/0.66) + (2.39/0.40) + 0.50} = 76.0 \text{ Btu/h} \cdot \text{ft}^2$$

From Figure 9, at 76.0 Btu/h·ft², $R_s = 0.60$. The mean temperature of the diatomaceous silica is

$$5.45/0.66 = 8.26;\ 8.26/2 = 4.13$$

$$1200 - \frac{4.13}{14.83}(1120) = 888°\text{F}$$

and the mean temperature of the calcium silicate is

$$2.39/0.40 = 5.98;\ 5.98/2 = 2.99;\ 8.26 + 2.99 = 11.25$$

$$1200 - \frac{11.25}{14.83}(1120) = 350°\text{F}$$

From Table 10, $k_1 = 0.72$; $k_2 = 0.46$. Recalculating,

$$q_s = \frac{1120}{(5.45/0.72) + (2.39/0.46) + 0.60} = 83.8 \text{ Btu/h} \cdot \text{ft}^2$$

From Figure 9 at 83.8 Btu/h·ft², $R_s = 0.59$. The mean temperature of the diatomaceous silica is

$$5.45/0.72 = 7.57;\ 7.57/2 = 3.78$$

$$1200 - \frac{3.78}{13.36}(1120) = 883°\text{F}$$

and the mean temperature of the calcium silicate is

$$2.39/0.40 = 5.98;\ 5.98/2 = 2.99;\ 8.26 + 2.99 = 11.25$$

$$1200 - \frac{11.25}{14.83}(1120) = 350°\text{F}$$

From Table 10, $k_1 = 0.72$; $k_2 = 0.46$. Recalculating,

$$2.39/0.46 = 5.20;\ 5.20/2 = 2.60;\ 7.57 + 2.60 = 10.17$$

$$1200 - \frac{10.17}{13.36}(1120) = 347°\text{F}$$

Table 11A Heat Loss from Bare Steel Pipe to Still Air at 80°F[a], Btu/h·ft

Nominal Pipe Size[b], in.	Pipe Inside Temperature, °F									
	180	280	380	480	580	680	780	880	980	1080
0.50	59.3	147.2	263.2	412.3	600.9	836.8	1128.6	1485.6	1918.0	2436.8
0.75	72.5	180.1	322.6	506.2	739.2	1031.2	1392.9	1836.0	2373.5	3018.8
1.00	88.8	220.8	396.1	622.7	910.9	1272.6	1721.2	2271.5	2939.4	3741.6
1.25	109.7	272.8	490.4	772.3	1131.7	1583.8	2145.6	2835.4	3673.4	4680.9
1.50	123.9	308.5	555.1	875.1	1283.8	1798.3	2438.2	3224.6	4180.5	5330.0
2.00	151.8	378.1	681.4	1076.3	1581.5	2218.9	3012.6	3989.2	5177.2	6606.8
2.50	180.5	450.0	811.9	1284.0	1888.8	2652.6	3604.3	4775.3	6199.5	7912.5
3.00	215.9	538.8	973.5	1541.8	2271.4	3194.0	4344.9	5762.2	7486.9	9562.3
3.50	243.9	609.0	1101.4	1746.1	2574.7	3623.6	4933.0	6546.4	8510.4	10874.3
4.00	271.6	678.6	1228.2	1948.7	2875.9	4050.5	5517.5	7326.0	9528.1	12178.9
4.50	299.2	747.7	1354.4	2150.9	3176.8	4477.7	6103.8	8109.5	10553.2	13496.2
5.00	329.8	824.7	1494.8	2375.4	3510.6	4950.7	6751.3	8972.5	11678.4	14936.3
6.00	387.1	968.7	1757.8	2796.8	4138.0	5841.4	7972.7	10603.1	13808.2	17667.6
7.00	440.5	1102.8	2003.0	3189.9	4723.9	6673.5	9114.2	12127.4	15799.4	20220.8
8.00	493.3	1235.7	2246.1	3580.0	5305.5	7500.0	10248.4	13642.2	17778.2	22758.0
9.00	545.9	1368.1	2488.8	3970.2	5888.7	8331.0	11392.1	15174.5	19787.1	25343.6
10.00	604.3	1514.8	2757.2	4400.7	6530.1	9241.1	12638.6	16835.1	21949.2	28104.9
11.00	656.0	1644.8	2995.5	4783.8	7102.1	10054.9	13756.2	18328.4	23900.3	30606.1
12.00	704.0	1762.3	3203.8	5104.9	7557.3	10661.8	14524.9	19256.7	24967.6	31766.8
14.00	771.0	1934.2	3525.9	5636.0	8373.9	11862.4	16235.5	21635.6	28212.3	36120.3
16.00	872.2	2189.0	3993.2	6387.4	9495.9	13458.0	18424.8	24556.6	32021.1	40990.7
18.00	972.5	2441.7	4456.7	7132.9	10609.4	15041.3	20596.7	27453.2	35795.6	45813.1
20.00	1072.1	2692.4	4916.8	7873.2	11715.1	16613.4	22752.5	30326.8	39537.6	50590.0
24.00	1269.3	3188.9	5828.3	9339.9	13905.5	19726.9	27019.7	36010.1	46930.3	60014.7

Table 11B Heat Loss from Flat Surfaces to Still Air at 80°F, Btu/h·ft²

	Surface Inside Temperature, °F									
	180	280	380	480	580	680	780	880	980	1080
Vertical surface	212.2	533.1	973.3	1558.6	2321.2	3298.0	4530.1	6062.8	7945.5	10231.5
Horizontal surface										
Facing up	234.7	586.4	1061.1	1683.5	2484.9	3501.9	4775.4	6350.4	8276.3	10606.1
Facing down	183.6	465.3	861.4	1399.6	2112.8	3038.4	4217.8	5696.7	7524.5	9754.7

[a]Calculations from ASTM C 680; steel: k = 314.4 Btu·in/h·ft²·°F; $\varepsilon = 0.94$.

[b]Losses per square foot of pipe for pipes larger than 24 in. can be considered the same as losses per square foot for 24 in. pipe.

Because R_s, k_1, and k_2 do not change at 83.8 Btu/h·ft², this is q_s. The heat flow per ft² of the inner surface of the insulation is

$$q_o = q_s(r_s/r_o) = 83.8(8.41/3.31) = 213 \text{ Btu/h} \cdot \text{ft}^2$$

Because trial-and-error techniques are tedious, the computer programs previously described should be used to estimate heat flows per unit area of flat surfaces or per unit length of piping, and interface temperatures including surface temperatures.

Several methods can be used to determine the most effective thickness of insulation for piping and equipment. Table 13 shows the recommended insulation thicknesses for three different pipe and equipment insulations. Installed cost data can be developed using procedures described by the Federal Energy Administration (1976). Computer programs capable of calculating thickness information are available from several sources. Also, manufacturers of insulations offer computerized analysis programs for designers and owners to evaluate insulation requirements. For more information on determining economic insulation thickness, see Chapter 23.

Chapters 3 and 23 give guidance concerning process control, personnel protection, condensation control, and economics. For specific information on sizes of commercially available pipe insulation, see ASTM *Standard* C 585 and consult with the North American Insulation Manufacturers Association (NAIMA) and its member companies.

CALCULATING HEAT FLOW FOR BURIED PIPELINES

In calculating heat flow to or from buried pipelines, the thermal properties of the soil must be assumed. Table 7 gives the apparent thermal conductivity values of various soil types, and Figure 8 shows the typical trends of apparent soil thermal conductivity with moisture content for various soil types. Table 8 provides ranges of apparent thermal conductivity for various types of rock. Kernsten (1949) also discusses thermal properties of soils. Carslaw and Jaeger (1959) give methods for calculating the heat flow taking place between one or more buried cylinders and the surroundings.

Table 12 Heat Loss from Bare Copper Tube to Still Air at 80°F[a], Btu/h·ft

Nominal Tube Size, in.	Tube Inside Temperature, °F								
	120	150	180	210	240	270	300	330	
0.250	7.1	14.1	21.9	30.6	39.9	49.9	60.6	71.9	
0.375	9.1	18.0	28.1	39.1	51.1	63.9	77.6	92.2	
0.500	11.0	21.8	34.0	47.4	61.9	77.5	94.1	111.8	
0.750	14.7	29.1	45.4	63.3	82.7	103.6	126.0	149.8	
1.000	18.3	36.2	56.4	78.7	102.8	128.9	156.7	186.5	
1.250	21.8	43.1	67.2	93.6	122.4	153.4	186.7	222.2	
1.500	25.2	49.8	77.6	108.3	141.5	177.4	216.0	257.1	
2.000	31.8	62.9	98.0	136.7	178.8	224.3	273.1	325.4	
2.500	38.3	75.6	117.9	164.4	215.1	269.8	328.7	391.8	Dull ε = 0.44
3.000	44.6	88.1	137.2	191.5	250.5	314.4	383.2	456.9	
3.500	50.8	100.3	156.3	218.0	285.4	358.2	436.7	520.8	
4.000	57.0	112.3	175.0	244.2	319.7	401.4	489.4	583.9	
5.000	69.0	135.9	211.7	295.5	386.9	486.0	592.8	707.6	
6.000	80.7	159.0	247.7	345.7	452.8	568.9	694.2	829.0	
8.000	103.7	204.1	317.8	443.7	581.3	730.7	892.1	1066.0	
10.000	126.1	247.9	386.1	539.1	706.5	888.4	1085.2	1297.4	
12.000	148.0	290.9	453.0	632.5	829.2	1043.1	1274.6	1524.4	
0.250	5.4	10.8	16.9	23.5	30.5	37.9	45.5	53.5	
0.375	6.8	13.7	21.4	29.7	38.6	47.9	57.6	67.6	
0.500	8.2	16.4	25.7	35.7	46.3	57.4	69.1	81.2	
0.750	10.7	21.6	33.8	46.9	60.9	75.6	90.9	106.8	
1.000	13.2	26.5	41.4	57.6	74.7	92.8	111.6	131.2	
1.250	15.5	31.3	48.8	67.8	88.0	109.3	131.6	154.7	
1.500	17.8	35.8	56.0	77.8	100.9	125.3	150.8	177.4	
2.000	22.2	44.6	69.7	96.8	125.7	156.1	187.9	221.1	Bright ε = 0.08
2.500	26.4	53.0	82.8	115.1	149.5	185.6	223.5	263.0	
3.000	30.5	61.2	95.6	132.8	172.4	214.2	257.9	303.5	
3.500	34.4	69.1	107.9	150.0	194.8	242.0	291.4	342.9	
4.000	38.3	76.8	120.0	166.8	216.6	269.1	324.1	381.4	
5.000	45.7	91.8	143.4	199.3	258.8	321.6	387.4	456.1	
6.000	53.0	106.3	166.0	230.7	299.7	372.5	448.7	528.3	
8.000	66.8	134.1	209.4	291.1	378.2	470.1	566.5	667.2	
10.000	80.2	160.8	251.0	349.0	453.4	563.7	679.5	800.4	
12.000	93.0	186.5	291.3	404.9	526.1	654.2	788.7	929.3	

[a]Calculations from ASTM C 680; for copper: $k = 2784$ Btu·in/h·ft^2·°F.

CODES AND STANDARDS

ASTM. 1990. Standard practice for inner and outer diameters of rigid thermal insulation for nominal sizes of pipe and tubing. *Standard* C585-90. American Society for Testing and Materials, West Conshohocken, PA.

ASTM. 1991. Standard test method for steady-state heat flux measurements and thermal transmission properties by means of the heat flow meter apparatus. *Standard* C 518-91.

ASTM. 1993. Standard test method for steady-state heat flux measurements and thermal transmission properties by means of the guarded-hot-plate apparatus. *Standard* C 177-85 (Revised 1993).

ASTM. 1993. Standard test method for steady-state thermal performance of building assemblies by means of a guarded hot box. *Standard* C 236-89 (Revised 1993).

ASTM. 1995. Standard practice for determination of heat gain or loss and the surface temperatures of insulated pipe and equipment systems by the use of a computer program. *Standard* C 680-89 (Revised 1995).

ASTM. 1996. Standard test method for thermal performance of building assemblies by means of a calibrated hot box. *Standard* C 976-90 (Revised 1996).

REFERENCES

Adams, L. 1971. Supporting cryogenic equipment with wood. *Chemical Engineering* (May):156-58.

Barbour, E., J. Goodrow, J. Kosny, and J.E. Christian. 1994. Thermal performance of steel-framed walls. Prepared for American Iron and Steel Institute by NAHB Research Center.

Bassett, M.R. and H.A. Trethowen. 1984. Effect of condensation on emittance of reflective insulation. *Journal of Thermal Insulation* 8 (October):127.

Carslaw, H.S. and J.C. Jaeger. 1959. Conduction of heat in solids. Oxford University Press, Amen House, London, England, 449.

Dill, R.S., W.C. Robinson, and H.E. Robinson. 1945. Measurements of heat losses from slab floors. National Bureau of Standards. Building Materials and Structures Report, BMS 103.

Economic thickness for industrial insulation. 1976. GPO No. 41-018-001 15-8, Federal Energy Administration, Washington, D.C.

Farouk, B. and D.C. Larson. 1983. Thermal performance of insulated wall systems with metal studs. Proceedings of the 18th Intersociety Energy Conversion Engineering Conference, Orlando, FL.

HEAT LOSS IN BTU/(HR-FT)

Table 13 Recommended Thicknesses for Pipe and Equipment Insulation

Nom Dia., in.		MINERAL FIBER (Fiberglass and Rock Wool)										CALCIUM		
		Process Temperature, °F										Process Temp., °F		
		150	250	350	450	550	650	750	850	950	1050	150	250	350
	Thickness	1	1½	2	2½	3	3½	4	4	4½	5½	1	1½	2
½	Heat loss	8	16	24	33	43	54	66	84	100	114	13	24	34
	Surface temp.	72	75	76	78	79	81	82	86	87	87	75	78	80
	Thickness	1	1½	2	2½	3½	4	4	4½	5	5½	1	2	2½
1	Heat loss	11	21	30	41	49	61	79	96	114	135	16	26	38
	Surface temp.	73	76	78	80	79	81	84	86	88	89	76	76	79
	Thickness	1	2	2½	3	4	4	4	5½	5½	6	1½	2½	3
1½	Heat loss	14	22	33	45	54	73	94	103	128	152	17	29	42
	Surface temp.	73	74	77	79	79	82	86	84	88	90	73	75	78
	Thickness	1½	2	3	3½	4	4	4	5½	6	6	1½	2½	3
2	Heat loss	13	25	34	47	61	81	105	114	137	168	19	32	47
	Surface temp.	71	75	75	77	79	83	87	85	87	91	74	76	79
	Thickness	1½	2½	3½	4	4	4½	4½	6	6½	7	2	3	3½
3	Heat loss	16	28	39	54	75	94	122	133	154	184	21	37	54
	Surface temp.	72	74	75	77	81	83	87	86	87	90	73	75	78
	Thickness	1½	3	4	4	4	5	5½	6	7	7½	2	3	4
4	Heat loss	19	29	42	63	88	102	126	152	174	206	25	43	58
	Surface temp.	72	73	74	78	82	86	85	87	88	90	70	76	77
	Thickness	2	3	4	4	4½	5	5½	6½	7½	8	2	3½	4
6	Heat loss	21	38	54	81	104	130	159	181	208	246	33	51	75
	Surface temp.	71	74	75	79	82	84	87	88	89	91	74	75	79
	Thickness	2	3½	4	4	5	5	5½	7	8	8½	2½	3½	4
8	Heat loss	26	42	65	97	116	155	189	204	234	277	35	62	90
	Surface temp.	71	73	76	80	81	86	89	88	89	92	73	76	79
	Thickness	2	3½	4	4	5	5½	5½	7½	8½	9	2½	4	4
10	Heat loss	32	50	77	115	136	170	220	226	259	307	41	66	106
	Surface temp.	72	74	77	81	82	85	90	87	89	91	73	75	80
	Thickness	2	3½	4	4	5	5½	5½	7½	8½	9½	2½	4	4
12	Heat loss	36	57	87	131	154	192	249	253	290	331	47	75	121
	Surface temp.	72	74	77	82	82	86	91	88	89	91	73	76	81
	Thickness	2	3½	4	4	5	5½	6½	7½	9	9½	2½	4	4
14	Heat loss	40	61	94	141	165	206	236	271	297	352	51	81	130
	Surface temp.	72	74	77	82	83	86	87	89	89	91	73	76	81
	Thickness	2½	3½	4	4	5½	5½	7	8	9	10	3	4	4
16	Heat loss	37	68	105	157	171	228	247	284	326	372	50	90	144
	Surface temp.	71	74	78	83	82	87	86	88	89	91	72	76	82
	Thickness	2½	3½	4	4	5½	5½	7	8	9	10	3	4	4
18	Heat loss	41	75	115	173	187	250	270	310	354	404	55	99	159
	Surface temp.	71	74	78	83	83	87	87	88	90	91	73	76	82
	Thickness	2½	3½	4	4	5½	5½	7	8	9	10	3	4	4
20	Heat loss	45	82	126	189	204	272	292	335	383	436	60	108	174
	Surface temp.	71	75	78	83	83	87	87	89	90	92	73	77	82
	Thickness	2½	4	4	4	5½	6	7½	8	9	10	3	4	4
24	Heat loss	53	86	147	221	237	295	320	386	439	498	71	127	203
	Surface temp.	71	74	78	83	83	86	86	89	91	93	73	77	82
	Thickness	2½	4	4	4	5½	6½	7½	8½	10	10	3	4	4
30	Heat loss	65	105	179	268	286	332	383	439	481	591	86	154	247
	Surface temp.	71	74	79	84	84	85	87	89	89	94	73	77	83
	Thickness	2½	4	4	4	5½	7	8	9	10	10	2½	4	4
36	Heat loss	77	123	211	316	335	364	422	486	556	683	119	181	291
	Surface temp.	71	74	79	84	84	84	86	88	90	94	74	77	83
	Thickness	2	3½	4	4½	5½	8½	9½	10	10	10	2½	3½	4
Flat	Heat loss	10	14	20	27	31	27	31	38	47	58	12	20	28
	Surface temp.	72	74	77	80	82	80	82	85	89	93	73	77	81

Consult manufacturer's literature for product temperature limitations. Table is based on typical operating conditions, e.g., 65°F ambient temperature and 7.5 mph wind speed, and may not represent actual conditions of use. Units for thickness, heat loss, and surface temperature are in inches, Btu/h·ft (Btu/h·ft^2 for flat surfaces), and °F, respectively.

Table 13 Recommended Thicknesses for Pipe and Equipment Insulation (*Concluded*)

Nom. Dia., in.		SILICATE							CELLULAR GLASS						
		Process Temperature, °F							Process Temperature, °F						
		450	550	650	750	850	950	1050	150	250	350	450	550	650	750
½	Thickness	2½	3	3½	4	4	4	4	1½	1½	2	2½	3	3½	4
	Heat loss	42	53	63	75	90	108	128	9	23	34	48	62	78	92
	Surface temp.	81	82	83	84	87	91	94	70	76	78	82	83	85	84
1	Thickness	3	3½	4	4	4	4	4	1½	2	2½	3	3½	4	4
	Heat loss	49	60	72	89	109	130	154	12	25	38	52	68	86	112
	Surface temp.	80	82	83	86	90	94	98	71	75	77	79	81	83	88
1½	Thickness	3½	4	4	4	4	5	5	1½	2½	3	4	4	4	4
	Heat loss	54	68	86	106	128	139	164	15	28	44	56	79	105	137
	Surface temp.	80	81	85	88	92	91	94	72	75	77	78	82	87	92
2	Thickness	3½	4	4½	5	5½	6	6	1½	2½	3	4	4	4	4½
	Heat loss	61	75	90	106	123	142	167	17	31	47	61	84	113	140
	Surface temp.	81	82	84	85	87	88	91	72	74	77	78	82	86	89
3	Thickness	4	4½	5	5½	6	6	6	1½	3	3½	4	4	4½	5
	Heat loss	71	87	105	123	143	71	202	22	35	54	75	105	132	161
	Surface temp.	80	82	84	85	87	90	94	73	74	77	79	84	86	89
4	Thickness	4	4½	5	5½	6	6½	7	2	3	4	4	4	4½	5
	Heat loss	82	101	121	142	164	187	213	22	41	59	87	122	150	185
	Surface temp.	81	83	85	87	89	90	92	71	74	76	80	85	87	90
6	Thickness	4	4½	5	5½	6	7	8	2	3½	4	4	4½	5½	6
	Heat loss	105	129	153	178	205	224	245	30	48	74	111	144	171	212
	Surface temp.	83	85	87	89	91	91	91	72	74	77	82	85	86	89
8	Thickness	4½	5	5	6	7	8	8½	2½	3½	4	4	5	5½	6½
	Heat loss	117	144	183	200	220	243	277	30	58	90	134	161	203	238
	Surface temp.	82	85	89	89	89	90	92	71	74	78	83	84	87	89
10	Thickness	4	5	5½	6	7½	8½	9	2½	4	4	4	5½	5½	7
	Heat loss	149	168	200	233	243	269	306	37	63	106	159	178	238	264
	Surface temp.	85	86	88	90	89	89	91	71	74	79	84	84	87	88
12	Thickness	4	5	5½	7	8	8½	9½	2½	4	4	4	5½	5½	7½
	Heat loss	170	191	266	236	262	300	330	42	71	121	181	201	269	284
	Surface temp.	86	86	89	88	88	90	91	71	74	79	85	84	90	88
14	Thickness	4	5	5½	7	8	9	9½	2½	4	4	4	5½	5½	8
	Heat loss	183	205	242	252	262	308	352	47	79	134	199	219	293	293
	Surface temp.	86	87	89	88	88	89	91	72	74	80	85	85	91	87
16	Thickness	4	5½	6½	7½	8	9	10	2½	4	4	4	5½	5½	8
	Heat loss	204	211	237	265	307	338	372	53	88	149	222	242	325	322
	Surface temp.	87	85	86	87	89	90	91	72	75	80	86	86	91	88
18	Thickness	4	5½	6½	7½	8½	9	10	2½	4	4	4	5½	5½	8
	Heat loss	225	232	259	289	320	367	403	59	96	164	245	266	356	351
	Surface temp.	87	86	87	87	88	90	91	72	75	80	86	86	92	88
20	Thickness	4	5½	6½	7½	8½	9½	10	2½	4	4	4½	5½	5½	8
	Heat loss	245	252	281	312	346	381	435	64	105	179	243	289	387	379
	Surface temp.	87	86	87	88	89	90	92	72	75	81	84	86	92	88
24	Thickness	4	5½	6½	7½	8½	9½	10	2½	4	4	5	5½	5½	8
	Heat loss	287	293	325	360	397	437	497	76	123	209	260	336	449	436
	Surface temp.	88	87	88	88	89	90	93	72	75	81	83	87	93	89
30	Thickness	4	5½	7	8	9	10	10	2½	4	4	5½	5½	5½	8
	Heat loss	349	353	368	409	452	498	589	93	150	254	290	405	542	521
	Surface temp.	88	87	87	88	89	90	94	72	75	81	82	87	93	90
36	Thickness	4	6½	7½	8	9	10	10	2½	4	4	5½	5½	5½	8
	Heat loss	410	359	406	475	524	576	681	110	176	229	340	474	635	606
	Surface temp.	89	84	86	88	89	91	94	73	76	81	82	88	94	90
Flat	Thickness	5½	6½	7½	8½	9½	10	10	2½	4	4	5½	5½	7½	8½
	Heat loss	29	33	36	39	43	49	58	11	17	29	31	44	43	50
	Surface temp.	81	83	84	85	87	89	93	73	76	83	84	90	90	93

Farouki, O.T. 1981. Thermal properties of soil. CRREL *Monograph* 81-1, United States Army Corps of Engineers Cold Regions Research and Engineering Laboratory, December.

Fishenden, M. 1962. Tables of emissivity of surfaces. *International Journal of Heat and Mass Transfer* 5:67-76.

Goss, W.P. and R.G. Miller. 1989. Literature review of measurement and prediction of reflective building insulation system performance: 1900-1989. *ASHRAE Transactions* 95(2).

Hooper, F.C. and W.J. Moroz. 1952. The impact of aging factors on the emissivity of reflective insulations. ASTM *Bulletin* (May):92-95.

Hougten, F.C., S.I. Taimuty, C. Gutberlet, and C.J. Brown. 1942. Heat loss through basement walls and floors. *ASHVE Transactions* 48:369.

Joy, F.A. 1958. Improving attic space insulating values. *ASHAE Transactions* 64:251.

Kersten, M.S. 1949. Thermal properties of soils. University of Minnesota, Engineering Experiment Station Bulletin 28, June.

Kosny, J. and J.E. Christian. 1995. Reducing the uncertainties associated with using the ASHRAE zone method for R-value calculations of metal frame walls. *ASHRAE Transactions* 101(2).

Latta, J.K. and G.G. Boileau. 1969. Heat losses from house basements. *Canadian Building* 19(10).

Lewis, W.C. 1967. Thermal conductivity of wood-base fiber and particle panel materials. Forest Products Laboratory, Research Paper FPL 77, June.

MacLean, J.D. 1941. Thermal conductivity of wood. *ASHVE Transactions* 47:323.

McElroy, D.L., D.W. Yarbrough, and R.S. Graves. 1987. Thickness and density of loose-fill insulations after installation in residential attics. *Thermal insulation: Materials and systems.* F.J. Powell and S.L. Matthews, eds. ASTM STP 922:423-505.

McGowan, A. and A.O. Desjarlais. 1997. An investigation of common thermal bridges in walls. *ASHRAE Transactions* 103(2).

McIntyre, D.A. 1984. The increase in U-value of a wall caused by mortar joints, ECRC/M1843. The Electricity Council Research Centre, Copenhurst, England, June.

Mitalas, G.P. 1982. Basement heat loss studies at DBR/NRC, NRCC 20416. Division of Building Research, National Research Council of Canada, September.

Mitalas, G.P. 1983. Calculation of basement heat loss. *ASHRAE Transactions* 89(1B):420.

Prangnell, R.D. 1971. The water vapor resistivity of building materials—A literature survey. *Materiaux et Constructions* 4:24 (November).

Robinson, H.E., F.J. Powell, and L.A. Cosgrove. 1957. Thermal resistance of airspaces and fibrous insulations bounded by reflective surfaces. National Bureau of Standards, Building Materials and Structures Report BMS 151.

Robinson, H.E., F.J. Powlitch, and R.S. Dill. 1954. The thermal insulation value of airspaces. Housing and Home Finance Agency, Housing Research Paper No. 32.

Sabine, H.J., M.B. Lacher, D.R. Flynn, and T.L. Quindry. 1975. Acoustical and thermal performance of exterior residential walls, doors and windows. *NBS Building Science Series* 77. National Institute of Standards and Technology, Gaithersburg, MD.

Salomone, L.A. and J.I. Marlowe. 1989. Soil and rock classification according to thermal conductivity: Design of ground-coupled heat pump systems. EPRI CU-6482, Electric Power Research Institute, August.

Shipp, P.H. 1983. Basement, crawlspace and slab-on-grade thermal performance. Proceedings of the ASHRAE/DOE Conference, Thermal Performance of the Exterior Envelopes of Buildings II, ASHRAE SP 38:160-79.

Shu, L.S., A.E. Fiorato, and J.W. Howanski. 1979. Heat transmission coefficients of concrete block walls with core insulation. Proceedings of the ASHRAE/DOE-ORNL Conference, Thermal Performance of the Exterior Envelopes of Buildings, ASHRAE SP 28:421-35.

Tye, R.P. 1985. Upgrading thermal insulation performance of industrial processes. *Chemical Engineering Progress* (February):30-34.

Tye, R.P. 1986. Effects of product variability on thermal performance of thermal insulation. Proceedings of the First Asian Thermal Properties Conference, Beijing, People's Republic of China.

Tye, R.P. and A.O. Desjarlais. 1983. Factors influencing the thermal performance of thermal insulations for industrial applications. *Thermal insulation, materials, and systems for energy conservation in the '80s.* F.A. Govan, D.M. Greason, and J.D. McAllister, eds. ASTM STP 789:733-48.

Tye, R.P. and S.C. Spinney. 1980. A study of various factors affecting the thermal performance of perlite insulated masonry construction. Dynatech Report No. PII-2. Holometrix, Inc. (formerly Dynatech R/D Company), Cambridge, MA.

USDA. 1974. *Wood handbook.* Wood as an engineering material. Forest Products Laboratory, U.S. Department of Agriculture Handbook No. 72, Tables 3-7 and 4-2, and Figures 3-4 and 3-5.

Valore, R.C. 1980. Calculation of U-values of hollow concrete masonry. American Concrete Institute, *Concrete International* 2(2):40-62.

Valore, R.C. 1988. Thermophysical properties of masonry and its constituents, Parts I and II. International Masonry Institute, Washington, D.C.

Valore, R., A. Tuluca, and A. Caputo. 1988. Assessment of the thermal and physical properties of masonry block products (ORNL/Sub/86-22020/1), September.

Van Geem, M.G. 1985. Thermal transmittance of concrete block walls with core insulation. *ASHRAE Transactions* 91(2).

Wilkes, K.E. 1979. Thermophysical properties data base activities at Owens-Corning Fiberglas. Proceedings of the ASHRAE/DOE-ORNL Conference, Thermal Performance of the Exterior Envelopes of Buildings, ASHRAE SP 28:662-77.

Yarbrough, E.W. 1983. Assessment of reflective insulations for residential and commercial applications (ORNL/TM-8891), October.

Yellott, J.I. 1965. Thermal and mechanical effects of solar radiation on steel doors. *ASHRAE Transactions* 71(2):42.

CHAPTER 26

VENTILATION AND INFILTRATION

PROVIDING a comfortable and healthy indoor environment for building occupants is the primary concern of HVAC engineers. Comfort and indoor air quality (IAQ) depend on many factors, including thermal regulation, control of internal and external sources of pollutants, supply of acceptable air, removal of unacceptable air, occupants' activities and preferences, and proper operation and maintenance of building systems. Ventilation and infiltration are only part of the acceptable indoor air quality and thermal comfort problem. HVAC designers, occupants, and building owners must be aware of and address other factors as well. Choosing appropriate ventilation and infiltration rates to solve thermal comfort problems and to reduce energy consumption can affect indoor air quality and may be against code, so such procedures should be approached with care and be under the direction of a registered professional engineer with expertise in HVAC analysis and design.

HVAC design engineers and others concerned with building ventilation and indoor air quality should obtain a copy of ASHRAE *Standard* 62. This standard is reviewed regularly and contains ventilation design and evaluation requirements for commercial and residential buildings. In the design of a new building or the analysis of an existing building, the version of *Standard* 62 that has been adopted by the local code authority must be determined. An existing building may be required to meet current code, or it may be grandfathered under an older code. If a project involves infiltration in residences, then ASHRAE *Standards* 119 and 136 should be consulted. The last chapter of each year's *ASHRAE Handbook* (Chapter 39 of this volume) has a list of current standards.

This chapter focuses on commercial and institutional buildings, where ventilation concerns usually dominate, and on single- and multifamily residences, where infiltration is important. The basic concepts and terminology for both are presented before more advanced analytical and design techniques are given. Ventilation of industrial buildings is covered in Chapter 28 of the 1999 *ASHRAE Handbook—Applications*. However, many of the fundamental ideas and terminology covered in this chapter can also be applied to industrial buildings.

BASIC CONCEPTS AND TERMINOLOGY

Outdoor air that flows through a building is often used to dilute and remove indoor air contaminants. However, the energy required to condition this outdoor air can be a significant portion of the total space-conditioning load. The magnitude of the outdoor airflow into the building must be known for proper sizing of the HVAC equipment and evaluation of energy consumption. For buildings without mechanical cooling and dehumidification, proper ventilation and infiltration airflows are important for providing comfort for occupants. ASHRAE *Standard* 55 specifies conditions under which 80% or more of the occupants in a space will find it thermally acceptable. Chapter 8 of this volume also addresses thermal comfort. Additionally, the flow of air into buildings and between zones will affect fires and the movement of smoke. Smoke management is addressed in Chapter 51 of the 1999 *ASHRAE Handbook—Applications*.

Ventilation and Infiltration

Air exchange of outdoor air with the air already in a building can be divided into two broad classifications: ventilation and infiltration. **Ventilation air** is air used to provide acceptable indoor air quality. It may be composed of forced or natural ventilation, infiltration, suitably treated recirculated air, transfer air, or an appropriate combination (ASHRAE *Standard* 62). **Ventilation** includes the intentional introduction of air from the outside into a building; it is further subdivided into natural ventilation and forced ventilation. **Natural ventilation** is the flow of air through open windows, doors, grilles, and other planned building envelope penetrations, and it is driven by natural and/or artificially produced pressure differentials. **Forced ventilation**, shown in Figure 1, is the intentional movement of air into and out of a building using fans and intake and exhaust vents; it is also called **mechanical ventilation**.

Infiltration is the flow of outdoor air into a building through cracks and other unintentional openings and through the normal use of exterior doors for entrance and egress. Infiltration is also known as **air leakage** into a building. **Exfiltration**, depicted in Figure 1, is the leakage of indoor air out of a building through similar types of openings. Like natural ventilation, infiltration and exfiltration are driven by natural and/or artificial pressure differences. These forces are discussed in detail in the section on Driving Mechanisms for Ventilation and Infiltration. **Transfer air** is air that moves from one interior space to another, either intentionally or not.

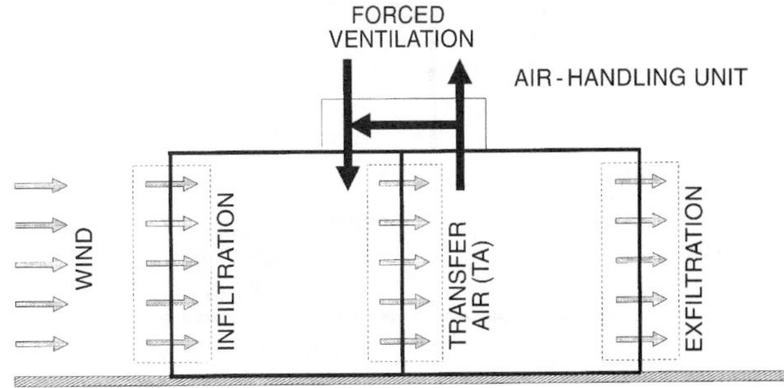

Fig. 1 Two-Space Building with Forced Ventilation, Infiltration, and Exfiltration

The preparation of this chapter is assigned to TC 4.3, Ventilation Requirements and Infiltration.

These modes of air exchange differ significantly in how they affect energy consumption, air quality, and thermal comfort, and they can each vary with weather conditions, building operation, and use. Although one mode may be expected to dominate in a particular building, all must be considered for the proper design and operation of an HVAC system.

Modern commercial and institutional buildings are normally required to have forced ventilation and are usually pressurized somewhat to reduce or eliminate infiltration. Forced ventilation has the greatest potential for control of air exchange when the system is properly designed, installed, and operated; it should provide acceptable indoor air quality and thermal comfort when ASHRAE *Standard* 62 and *Standard* 55 requirements are followed. Forced ventilation equipment and systems are described in Chapters 1, 2, and 9 of the 2000 *ASHRAE Handbook—Systems and Equipment.*

In commercial and institutional buildings, natural ventilation, such as through operable windows, may not be desirable from the point of view of energy conservation and comfort. In commercial and institutional buildings with mechanical cooling and forced ventilation, an air- or water-side economizer cycle may be preferable to operable windows for taking advantage of cool outdoor conditions when interior cooling is required. Infiltration may be significant in commercial and institutional buildings, especially in tall, leaky, or partially pressurized buildings and in lobby areas.

In most of the United States, residential buildings typically rely on infiltration and natural ventilation to meet their ventilation needs. Neither is reliable for ventilation purposes because they depend on weather conditions, building construction, and maintenance. However, natural ventilation, usually through operable windows, is more likely to allow occupants to control airborne contaminants and interior air temperature, but it can have a substantial energy cost if used while the residence's heating or cooling equipment is operating.

In place of operable windows, small exhaust fans may be provided for localized venting in residential spaces, such as kitchens and bathrooms. Not all local building codes require that the exhaust be vented to the outside. Instead, the code may allow the air to be treated and returned to the space or to be discharged to an attic space. Poor maintenance of these treatment devices can make nonducted vents ineffective for ventilation purposes. Condensation in attics should be avoided. In northern Europe and in Canada, some building codes require general forced ventilation in residences, and heat recovery heat exchangers are popular for reducing the energy impact. Residential buildings with low rates of infiltration and natural ventilation require forced ventilation at rates given in ASHRAE *Standard* 62.

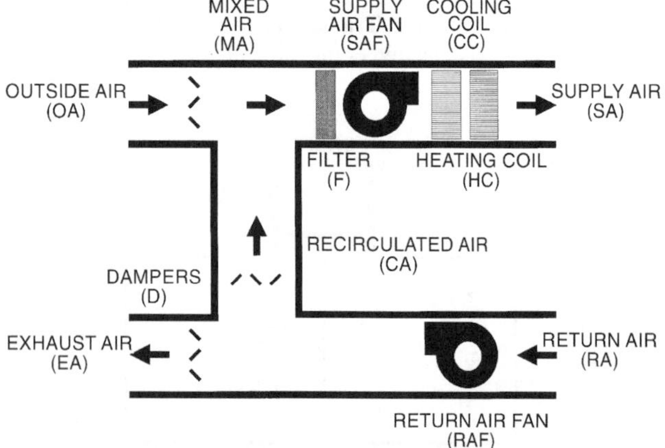

Fig. 2 Simple All-Air Air-Handling Unit with Associated Airflows

Forced-Air Distribution Systems

Figure 2 shows a simple **air-handling unit** (AHU) or **air-handler** that conditions air for a building. Air brought back to the air handler from the conditioned space is **return air** (ra). The portion of the return air that is discharged to the environment is **exhaust air** (ea), and the part of the return air that is reused is **recirculated air** (ca). Air brought in intentionally from the environment is **outdoor** or **outside air** (oa). Because outside air may need treatment to be acceptable for use in a building, it should not be called "fresh air." The outside air and the recirculated air are combined to form **mixed air** (ma), which is then conditioned and delivered to the thermal zone as **supply air** (sa). Any portion of the mixed air that intentionally or unintentionally circumvents conditioning is **bypass air** (ba). Due to the wide variety of air-handling systems, the airflows shown in Figure 2 may not all be present in a particular system as defined here. Also, more complex systems may have additional airflows.

Outside Air Fraction

The outside airflow being introduced to a building or zone by an air-handling unit can also be described by the **outside air fraction** X_{oa}, which is the ratio of the volumetric flow rate of outside air brought in by the air handler to the total supply airflow rate:

$$X_{oa} = \frac{Q_{oa}}{Q_{sa}} = \frac{Q_{oa}}{Q_{ma}} = \frac{Q_{oa}}{Q_{oa} + Q_{ca}} \tag{1}$$

When expressed as a percentage, the outside air fraction is called the **percent outside air**. The design outside airflow rate for a building's or zone's ventilation system is found through evaluating the requirements of ASHRAE *Standard* 62. The supply airflow rate is that required to meet the thermal load. The outside air fraction and percent outside air then describe the degree of recirculation, where a low value indicates a high rate of recirculation, and a high value shows little recirculation. Conventional all-air air-handling systems for commercial and institutional buildings have approximately 10 to 40% outside air.

100% outside air means no recirculation of return air through the air-handling system. Instead, all the supply air is treated outside air, also known as **makeup air** (ka), and all return air is discharged directly to the outside as **relief air** (la), or exhausted by separate exhaust fans. An air-handling unit that provides 100% outside air to offset air that is exhausted is typically called a **makeup air unit** (MAU).

When outside air via forced ventilation is used to provide ventilation air (as is common in commercial and institutional buildings), this outside air is delivered to spaces as all or part of the supply air. With a variable air volume (VAV) system, the outside air fraction of the supply air may need to be increased when the flow rate of the supply air is reduced to meet a particular thermal load.

Room Air Movement

Air movement within spaces affects the diffusion of ventilation air and, therefore, indoor air quality and comfort. Two distinct flow patterns are commonly used to characterize air movement in rooms: displacement flow and entrainment flow. **Displacement flow**, shown in Figure 3, is the movement of air within a space in a piston- or plug-type motion. No mixing of the room air occurs in ideal displacement flow, which is desirable for removing pollutants generated within a space. A laminar flow air distribution system that sweeps air across a space may produce displacement flow.

Entrainment flow, shown in Figure 4, is also known as **conventional mixing**. Systems with ceiling-based supply air diffusers and return air grilles are common examples of air distribution systems that produce entrainment flow. Entrainment flow with very poor mixing within the room has been called short-circuiting flow

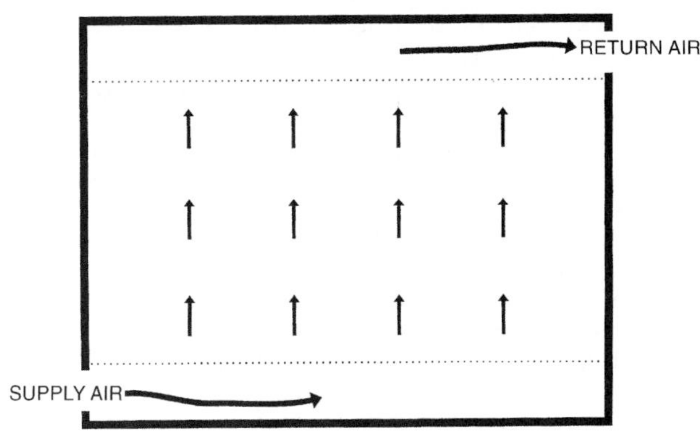

Fig. 3 Displacement Flow Within a Space

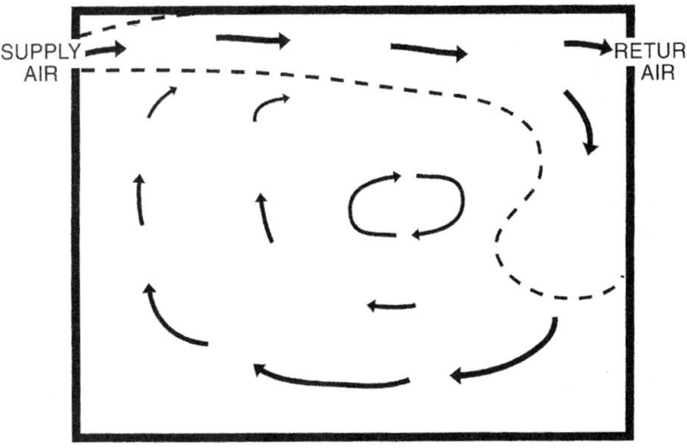

Fig. 4 Entrainment Flow Within a Space

because much of the supply air leaves the room without mixing with the room air. There is little evidence that properly designed, installed, and operated air distribution systems exhibit short-circuiting behavior. There is some evidence that poorly designed, installed, or operated systems can exhibit short-circuiting behavior, especially ceiling-based systems in the heating mode (Offermann and Int-Hout 1989).

Perfect mixing occurs when supply air is instantly and evenly distributed throughout a space. Perfect mixing is also known as **complete** or **uniform mixing**; the air may be called **well stirred** or **well mixed**. This theoretical performance is approached by entrainment flow systems that have good mixing and by displacement flow systems that allow too much mixing (Rock et al. 1995). The outdoor air requirements given in Table 2 of ASHRAE *Standard* 62 assume delivery of ventilation air with perfect mixing within spaces. For more detailed information on space air diffusion, see Chapter 32.

The supply air that enters a space through a diffuser is also known as **primary air**. A **jet** is formed as this primary air leaves the diffuser. **Secondary air** is the room air entrained into the jet. **Total air** is the combination of primary air and secondary air at a specific point in a jet. The term primary air is also used to describe the supply air provided to fan-powered mixing boxes by a central air-handling unit.

Air Exchange Rate

The **air exchange rate** I compares airflow to volume and is

$$I = \frac{Q}{V} \tag{2}$$

where

Q = volumetric flow rate of air into space, cfm
V = interior volume of space, ft^3

The air exchange rate has units of 1/time. When the time unit is hours, the air exchange rate is also called **air changes per hour (ACH)**. The air exchange rate may be defined for several different situations. For example, the air exchange rate for an entire building or zone served by an air-handling unit compares the amount of outside air brought into the building or zone to the total interior volume. This **nominal air exchange rate** I_N is

$$I_N = \frac{Q_{oa}}{V} \tag{3}$$

where Q_{oa} is the outdoor airflow rate including ventilation and infiltration. The nominal air exchange rate describes the outside air ventilation rate entering a building or zone. It does not describe recirculation or the distribution of the ventilation air to each space within a building or zone.

For a particular space, the **space air exchange rate** I_S compares the supply airflow rate Q_{sa} to the volume of that space:

$$I_S = \frac{Q_{sa}}{V} \tag{4}$$

The space air exchange rate for a particular space or zone includes recirculated as well as outside air in the supply air, and it is used frequently in the evaluation of supply air diffuser performance and space air mixing.

Time Constants

Time constants τ, which have units of time (usually in hours or seconds), are also used to describe ventilation and infiltration. One time constant is the time required for one air change in a building, zone, or space if ideal displacement flow existed. It is the inverse of the air exchange rate:

$$\tau = \frac{1}{I} = \frac{V}{Q} \tag{5}$$

The **nominal time constant** compares the interior volume of a building or zone to the volumetric outdoor airflow rate:

$$\tau_N = \frac{V}{Q_{oa}} \tag{6}$$

Like the nominal air exchange rate, the nominal time constant does not describe recirculation of air within a building or zone. It also does not characterize the distribution of the outside air to individual spaces within a building or zone.

The **space time constant** compares the interior volume of a particular space to the total supply airflow rate to that space. The space time constant is the inverse of the space air exchange rate:

$$\tau_S = \frac{V}{Q_{sa}} \tag{7}$$

The space time constant includes the effect of recirculated air, if present, as well as that of outside air introduced to the space through the supply air. If infiltration is significant in a space, then the infiltration flow rate should be included when determining both the space air exchange rate and the space time constant.

Averaging Time-Varying Ventilation

When assessing time-varying ventilation in terms of controlling indoor air quality, the quantity of interest is often the temporal average rather than the peak. The concept of **effective ventilation** (Sherman and Wilson 1986, Yuill 1986, and Yuill 1991) describes the proper ventilation rate averaging process. In this concept, the average (effective) rate is the steady-state rate that would yield the same average contaminant concentration over the period of interest in the occupied space as does the actual sequence of time-varying discrete ventilation rates over the same period and in the same space. This effective rate is only equal to the simple arithmetic average rate when the discrete ventilation rates are constant over the period of interest and the contaminant concentration has reached its steady-state value. Simple arithmetic averaging of instantaneous ventilation rates or concentrations cannot generally be used to determine these averages due to the nonlinear response of indoor concentrations to the ventilation rate variations.

An important constraint in the effective ventilation concept is that the contaminant source strength F must be constant over the period of interest or must be uncorrelated with the ventilation rate. These conditions are satisfied in many residential and commercial buildings because the emission rates of many contaminants that are controlled by whole-building ventilation vary slowly. Sherman and Wilson (1986) describe how to deal with pollutants that have stepwise constant emission rates. Pollutants such as carbon monoxide, radon, and formaldehyde, whose emission rates can be affected by ventilation, cannot be analyzed with this concept and require more complex analyses.

For constant source-strength pollutants, the relationship between effective air exchange rate, effective ventilation rate, volumetric flow, source strength, average concentration, and time-averaged effective turnover time is given by

$$I_m = \frac{\bar{Q}}{V} = \frac{F}{V\bar{C}} = \frac{1}{\bar{\tau}_e} \tag{8}$$

The time-averaged effective turnover time $\bar{\tau}_e$ in Equation (8) represents the characteristic time for the concentration in the occupied space to approach steady state over the period of interest. It can be determined from a sequence of discrete instantaneous ventilation air change rates I_i using the following (Sherman and Wilson 1986):

$$\bar{\tau}_e = \frac{1}{N}\sum_{i=1}^{N} \tau_{e,i} \tag{9}$$

$$\text{for } I_i > 0, \tau_{e,i} = \frac{1 - \exp(-I_i \Delta t)}{I_i} + \tau_{e,i-1}\exp(-I_i \Delta t) \tag{10}$$

$$\text{for } I_i = 0, \tau_{e,i} = \Delta t + \tau_{e,i-1} \tag{11}$$

where

Δt = length of each discrete time period
$\bar{\tau}_e$ = time-averaged effective turnover time
$\tau_{e,i}$ = instantaneous turnover time in period i
$\tau_{e,i-1}$ = instantaneous turnover time in previous period

ASHRAE *Standard* 136 provides a set of factors to assist in calculating the annual effective air exchange rate for houses. These factors were determined using Equations (9) through (11).

Age of Air

The **age of air** θ_{age} (Sandberg 1981) is the length of time that some quantity of outside air has been in a building, zone, or space. The "youngest" air is at the point where outside air enters the building by forced or natural ventilation or through infiltration (Grieve 1989). The "oldest" air may be at some location in the building or in the exhaust air. When the characteristics of the air distribution system are varied, a longer age of air suggests poorer outside air delivery compared to a shortage of air for the same location. The age of air has units of time, usually in seconds or minutes, so it is not a true "efficiency" or "effectiveness" measure. The age of air concept, however, has gained wide acceptance in Europe and is used increasingly in North America.

The age of air can be evaluated for existing buildings using tracer gas methods. Using either the decay (step-down) or the growth (step-up) tracer gas method, the zone average or **nominal age of air** $\theta_{age,N}$ can be determined by taking concentration measurements in the exhaust air. The **local age of air** $\theta_{age,L}$ is evaluated through tracer gas measurements at any desired point in a space, such as at a worker's desk. Once time-dependent data of tracer gas concentration are available, the age of air can be calculated from

$$\theta_{age} = \int_{\theta=0}^{\infty} \frac{C_{in} - C}{C_{in} - C_o}\, d\theta \tag{12}$$

where C_{in} = concentration of tracer gas being injected.

Because evaluation of the age of air requires integration to infinite time, an exponential tail is usually added to the known concentration data (Farrington et al. 1990).

Air Change Effectiveness

Ventilation effectiveness is a description of an air distribution system's ability to remove internally generated pollutants from a building, zone, or space. **Air change effectiveness** is a description of an air distribution system's ability to deliver ventilation air to a building, zone, or space. The HVAC design engineer does not have knowledge or control of actual pollutant sources within buildings, so Table 2 of ASHRAE *Standard* 62 defines outdoor air requirements for typical, expected building uses. For most projects, therefore, the air change effectiveness is of more relevance to HVAC system design than the ventilation effectiveness. Various definitions for air change effectiveness have been proposed. The specific measure that meets the local code requirements must be determined, if any is needed at all.

Air change effectiveness measures ε_I are nondimensional gages of ventilation air delivery. One common definition of air change effectiveness is the ratio of a time constant to an age of air:

$$\varepsilon_I = \frac{\tau}{\theta_{age}} \tag{13}$$

The **nominal air change effectiveness** $\varepsilon_{I,N}$ shows the effectiveness of outside air delivery to the entire building, zone, or space:

$$\varepsilon_{I,N} = \frac{\tau_N}{\theta_{age,N}} \tag{14}$$

where the nominal time constant τ_N is usually calculated from measured airflow rates.

The **local air change effectiveness** $\varepsilon_{I,L}$ shows the effectiveness of outside air delivery to one specific point in a space:

$$\varepsilon_{I,L} = \frac{\tau_N}{\theta_{age,L}} \tag{15}$$

where τ_N is found either through airflow measurements or from tracer gas concentration data. An $\varepsilon_{I,L}$ value of 1.0 indicates that the air distribution system delivers air equivalent to that of a system with perfectly mixed air in the spaces. A value less than 1.0 shows less than perfect mixing with some degree of stagnation. A value of $\varepsilon_{I,L}$ greater than 1.0 suggests that a degree of plug or displacement flow is present at that point (Rock 1992).

Currently, the HVAC design engineer must assume that a properly designed, installed, operated, and maintained air distribution system provides an air change effectiveness of about 1. Therefore, the Table 2 values of ASHRAE *Standard* 62 are appropriate for the design of commercial, institutional, and residential buildings when the Ventilation Rate Procedure is used. If the Indoor Air Quality Procedure of *Standard* 62 is used, then actual pollutant sources and the air change effectiveness must be known for the successful design of HVAC systems that have fixed ventilation airflow rates.

ASHRAE *Standard* 129 describes a method for measuring air change effectiveness of mechanically vented spaces and buildings with limited air infiltration, exfiltration, and air leakage with surrounding indoor spaces.

DRIVING MECHANISMS FOR VENTILATION AND INFILTRATION

Natural ventilation and infiltration are driven by pressure differences across the building envelope caused by wind and air density differences due to temperature differences between indoor and outdoor air (buoyancy, or the stack effect). Mechanical air-moving systems also induce pressure differences across the envelope due to the operation of appliances, such as combustion devices, leaky forced-air thermal distribution systems, and mechanical ventilation systems. The indoor-outdoor pressure difference at a location depends on the magnitude of these driving mechanisms as well as on the characteristics of the openings in the building envelope (i.e., their locations and the relationship between pressure difference and airflow for each opening).

Stack Pressure

Stack pressure is the hydrostatic pressure caused by the weight of a column of air located inside or outside a building. It can also occur within a flow element such as a duct or chimney that has vertical separation between its inlet and outlet. The hydrostatic pressure in the air depends on density and the height of interest above a reference point.

Air density is a function of local barometric pressure, temperature, and humidity ratio, as described by Chapter 6. As a result, standard conditions should not be used to calculate the density. For example, a building site at 5000 ft has an air density that is about 20% less than if the building were at sea level. An air temperature increase from −20 to 70°F causes a similar air density difference. Combined, these elevation and temperature effects reduce the air density about 45%. Moisture effects on density are generally negligible, so the dry air density can be used instead, except in hot, humid climates when the air is hot and close to saturation. For example, saturated air at 105°F has a density about 5% greater than that of dry air.

Assuming temperature and barometric pressure are constant over the height of interest, the stack pressure decreases linearly as the separation above the reference point increases. For a single column of air, the stack pressure can be calculated as

$$p_s = p_r - C_1 \rho g H \tag{16}$$

where

p_s = stack pressure, in. of water
p_r = stack pressure at reference height, in. of water
g = gravitational constant, 32.2 ft/s^2
ρ = indoor or outdoor air density, lb$_m$/ft^3
H = height above reference plane, ft
C_1 = unit conversion factor = 0.00598 (in. of water)·ft·s^2/lb$_m$

For tall buildings or when significant temperature stratification occurs indoors, Equation (16) should be modified to include the density gradient over the height of the building.

Temperature differences between indoors and outdoors cause stack pressure differences that drive airflows across the building envelope. Sherman (1991) showed that any single-zone building can be treated as an equivalent box from the point of view of stack effect, if its leaks follow the power law. The building is then characterized by an effective stack height and neutral pressure level (NPL) or leakage distribution (see the section on Neutral Pressure Level). Once calculated, these parameters can be used in physical, single-zone models to estimate infiltration.

Neglecting vertical density gradients, the stack pressure difference for a horizontal leak at any vertical location is given by

$$\Delta p_s = C_1(\rho_o - \rho_i)g(H_{NPL} - H) = C_1\rho_o\left(\frac{T_o - T_i}{T_i}\right)g(H_{NPL} - H) \tag{17}$$

where

T_o = outdoor temperature, °R
T_i = indoor temperature, °R
ρ_o = outdoor air density, lb/ft^3
ρ_i = indoor air density, lb/ft^3
H_{NPL} = height of neutral pressure level above reference plane without any other driving forces, ft

Chastain and Colliver (1989) showed that when there is stratification, the average of the vertical distribution of temperature differences is more appropriate to use in Equation (17) than the localized temperature difference near the opening of interest.

By convention, stack pressure differences are positive when the building is pressurized relative to outdoors, which causes flow out of the building. Therefore, in the absence of other driving forces and assuming no stack effect is within the flow elements themselves, when the indoor air is warmer than outdoors, the base of the building is depressurized and the top is pressurized relative to outdoors; when the indoor air is cooler than outdoors, the reverse is true.

In the absence of other driving forces, the location of the NPL is influenced by leakage distribution over the building exterior and by interior compartmentation. As a result, the NPL is not necessarily located at the mid-height of the building nor is it necessarily unique. NPL location and leakage distribution are described later in the section on Combining Driving Forces.

For a penetration through the building envelope for which (1) there is a vertical separation between its inlet and outlet and (2) the air inside the flow element is not at the indoor or outdoor temperature, such as in a chimney, more complex analyses than Equation (17) are required to determine the stack effect at any location on the building envelope.

Wind Pressure

When wind impinges on a building, it creates a distribution of static pressures on the building's exterior surface that depends on the wind direction, wind speed, air density, surface orientation, and surrounding conditions. Wind pressures are generally positive with respect to the static pressure in the undisturbed airstream on the windward side of a building and negative on the leeward sides. However, pressures on these sides can be negative or positive, depending on wind angle and building shape. Static pressures over building surfaces are almost proportional to the velocity head of the undisturbed airstream. The wind pressure or velocity head is given by the Bernoulli equation, assuming no height change or pressure losses:

$$p_w = C_2 C_p \rho \frac{U^2}{2} \tag{18}$$

where

p_w = wind surface pressure relative to outdoor static pressure in undisturbed flow, in. of water
ρ = outside air density, lb$_m$/ft^3 (about 0.075)
U = wind speed, mph
C_p = wind surface pressure coefficient, dimensionless
C_2 = unit conversion factor = 0.0129 (in. of water)·ft^3/lb$_m$·mph^2

C_p is a function of location on the building envelope and wind direction. Chapter 16 provides additional information on the values of C_p.

Most pressure coefficient data are for winds normal to building surfaces. Unfortunately, for a real building, this fixed wind direction rarely occurs, and when the wind is not normal to the upwind wall, these pressure coefficients do not apply. A harmonic trigonometric function was developed by Walker and Wilson (1994) to interpolate between the surface average pressure coefficients on a wall that were measured with the wind normal to each of the four building surfaces. This function was developed for low-rise buildings three stories or less in height. For each wall of the building, C_p is given by

$$C_p(\phi) = \frac{1}{2}\{ [C_p(1) + C_p(2)](\cos^2\phi)^{1/4} + [C_p(1) - C_p(2)](\cos\phi)^{3/4} + [C_p(3) + C_p(4)](\sin^2\phi)^2 + [C_p(3) - C_p(4)]\sin\phi \} \quad (19)$$

where

$C_p(1)$ = pressure coefficient when wind is at 0°
$C_p(2)$ = pressure coefficient when wind is at 180°
$C_p(3)$ = pressure coefficient when wind is at 90°
$C_p(4)$ = pressure coefficient when wind is at 270°
ϕ = wind angle measured clockwise from the normal to Wall 1

The measured data used to develop the harmonic function from Akins et al. (1979) and Wiren (1985) show that typical values for the pressure coefficients are $C_p(1) = 0.6$, $C_p(2) = -0.3$, $C_p(3) = C_p(4) = -0.65$. Because of the geometry effects on flow around a building, the application of this interpolation function is limited to low-rise buildings that are of rectangular plan form (i.e., not L-shaped) with the longest wall less than three times the length of the shortest wall. For less regular buildings, simple correlations are inadequate and building-specific pressure coefficients are required. Chapter 16 discusses wind pressures for complex building shapes and for high-rise buildings in more detail.

The wind speed most commonly available for infiltration calculations is the wind speed measured at the local weather station, typically the nearest airport. This wind speed needs to be corrected for reductions due to the difference between the height where the wind speed is measured and the height of the building and reductions due to shelter effects.

The reference wind speed used to determine pressure coefficients is usually the wind speed at the eaves height for low-rise buildings and the building height for high-rise buildings. However, meteorological wind speed measurements are made at a different height (typically 33 ft) and at a different location. The difference in terrain between the measurement station and the building under study must also be accounted for. Chapter 16 shows how to calculate the effective wind speed U_H from the reference wind speed U_{met} using boundary layer theory and estimates of terrain effects.

In addition to the reduction in wind pressures due to the reduction in wind speed, the effects of local shelter also act to reduce wind pressures. The shielding effects of trees, shrubbery, and other buildings within several building heights of a particular building produce large-scale turbulence eddies that not only reduce effective wind speed but also alter wind direction. Thus, meteorological wind speed data must be reduced carefully when applied to low buildings.

Ventilation rates measured by Wilson and Walker (1991) for a row of houses showed reductions in ventilation rates of up to a factor of three when the wind changed direction from perpendicular to parallel to the row. They recommended estimating wind shelter for winds perpendicular to each side of the building and then using the interpolation function in Equation (20) to find the wind shelter for intermediate wind angles:

$$s = \frac{1}{2}\left\{ \begin{matrix} [s(1) + s(2)]\cos^2\phi + [s(1) - s(2)]\cos\phi \\ + [s(3) + s(4)]\sin^2\phi + [s(3) - s(4)]\sin\phi \end{matrix} \right\} \quad (20)$$

where

s = shelter factor for the particular wind direction ϕ
$s(i)$ = shelter factor when wind is normal to Wall i
(i = 1 to 4, for four sides of a building)

Although the above method gives a realistic variation of wind shelter effects with wind direction, estimates for the numerical values of wind shelter factor s for each of the four cardinal directions must be provided. Table 11 in the section on Residential Calculation Examples lists typical shelter factors. The wind speed used in Equation (18) is then given by

$$U = sU_H \quad (21)$$

The magnitude of the pressure differences found on the surfaces of buildings varies rapidly with time because of turbulent fluctuations in the wind (Grimsrud et al. 1979, Etheridge and Nolan 1979). However, the use of average wind pressures to calculate pressure differences is usually sufficient to calculate average infiltration values.

Mechanical Systems

The operation of mechanical equipment, such as supply or exhaust systems and vented combustion devices, affects pressure differences across the building envelope. The interior static pressure adjusts such that the sum of all airflows through the openings in the building envelope plus equipment-induced airflows balance to zero. To predict these changes in pressure differences and airflow rates caused by mechanical equipment, the location of each opening in the envelope and the relationship between pressure difference and airflow rate for each opening must be known. The interaction between mechanical ventilation system operation and envelope airtightness has been discussed for low-rise buildings (Nylund 1980) and for office buildings (Tamura and Wilson 1966, 1967b; Persily and Grot 1985a).

Air exhausted from a building by a whole-building exhaust system must be balanced by increasing the airflow into the building through other openings. As a result, the airflow at some locations changes from outflow to inflow. For supply fans, the situation is reversed and envelope inflows become outflows. Thus, the effects a mechanical system has on a building must be considered. Depressurization caused by an improperly designed exhaust system can increase the rate of radon entry into a building and interfere with the proper operation of combustion device venting or other exhaust systems. Depressurization can also force moist outdoor air through the building envelope; for example, during the cooling season in hot humid climates, moisture may condense within the building envelope. A similar phenomenon, but in reverse, can occur during the heating season in cold climates if the building is depressurized.

The interaction between mechanical systems and the building envelope also pertains to systems serving zones of buildings. The performance of zone-specific exhaust or pressurization systems is affected by the leakage in zone partitions as well as in exterior walls.

Mechanical systems can also create infiltration-driving forces in single-zone buildings. Specifically, some single-family houses with central forced-air duct systems have multiple supply registers, yet only a central return register. When internal doors are closed in these houses, large positive indoor-outdoor pressure differentials are created for rooms with only supply registers, whereas the room with the return duct tends to depressurize relative to outside. This is caused by the resistance of internal door undercuts to flow from the supply register to the return (Modera et al. 1991). The magnitudes of the indoor-outdoor pressure differentials created have been measured to average 0.012 to 0.024 in. of water (Modera et al. 1991).

Building envelope airtightness and interzonal airflow resistance can also affect the performance of mechanical systems. The actual airflow rate delivered by these systems, particularly ventilation systems, depends on the pressure they work against. This effect is the same as the interaction of a fan with its associated ductwork, which is discussed in Chapter 34 of this volume and Chapter 18 of the 2000 *ASHRAE Handbook—Systems and Equipment*. The building envelope and its leakage must be considered part of the ductwork in determining the pressure drop of the system.

Duct leakage can cause similar problems. Supply leaks to the outside will tend to depressurize the building; return leaks to the outside will tend to pressurize it.

Combining Driving Forces

The pressure differences due to wind pressure, stack pressure, and mechanical systems are considered in combination by adding them together and then determining the airflow rate through each opening due to this total pressure difference. The air flows must be determined in this manner, as opposed to adding the airflow rates due to the separate driving forces, because the airflow rate through each opening is not linearly related to pressure difference.

For uniform indoor air temperatures, the total pressure difference across each leak can be written in terms of a reference wind parameter P_U and stack effect parameter P_T common to all leaks:

$$P_U = \rho_o \frac{U_H^2}{2} \quad (22)$$

$$P_T = g\rho_o\left(\frac{T_o - T_i}{T_i}\right) \quad (23)$$

where T = air temperature, °R.

The pressure difference across each leak (with positive pressures for flow into the building) is then given by

$$\Delta p = C_2 s^2 C_p P_U + HP_T + \Delta p_I \quad (24)$$

where Δp_I = pressure that acts to balance inflows and outflows (including mechanical system flows). Equation (24) can then be applied to every leak for the building with the appropriate values of C_p, s, and H. Thus, each leak is defined by its pressure coefficient, shelter, and height. Where indoor pressures are not uniform, more complex analyses are required.

Neutral Pressure Level

The neutral pressure level (NPL) is that location or locations in the building envelope where there is no pressure difference. Internal partitions, stairwells, elevator shafts, utility ducts, chimneys, vents, operable windows, and mechanical supply and exhaust systems complicate the analysis of NPL location. An opening with a large area relative to the total building leakage causes the NPL to shift toward the location of the opening. In particular, chimneys and openings at or above roof height raise the NPL in small buildings. Exhaust systems increase the height of the NPL; outdoor air supply systems lower it.

Figure 5 qualitatively shows the addition of driving forces for a building with uniform openings above and below mid-height and without significant internal resistance to airflow. The slopes of the pressure lines are a function of the densities of the indoor and outdoor air. In Figure 5A, with inside air warmer than outside and pressure differences caused solely by thermal forces, the NPL is at mid-height, with inflow through lower openings and outflow through higher openings. Direction of flow is always from the higher to the lower pressure region.

Figure 5B presents qualitative uniform pressure differences caused by wind alone, with opposing effects on the windward and leeward sides. When the temperature difference and wind effects both exist, the pressures due to each are added together to determine the total pressure difference across the building envelope. In Figure 5B, there is no NPL because no locations on the building envelope have zero pressure difference. Figure 5C shows the combination, where the wind force of Figure 5B has just balanced the thermal force of Figure 5A, causing no pressure difference at the top windward or bottom leeward side.

The relative importance of the wind and stack pressures in a building depends on building height, internal resistance to vertical airflow, location and flow resistance characteristics of envelope

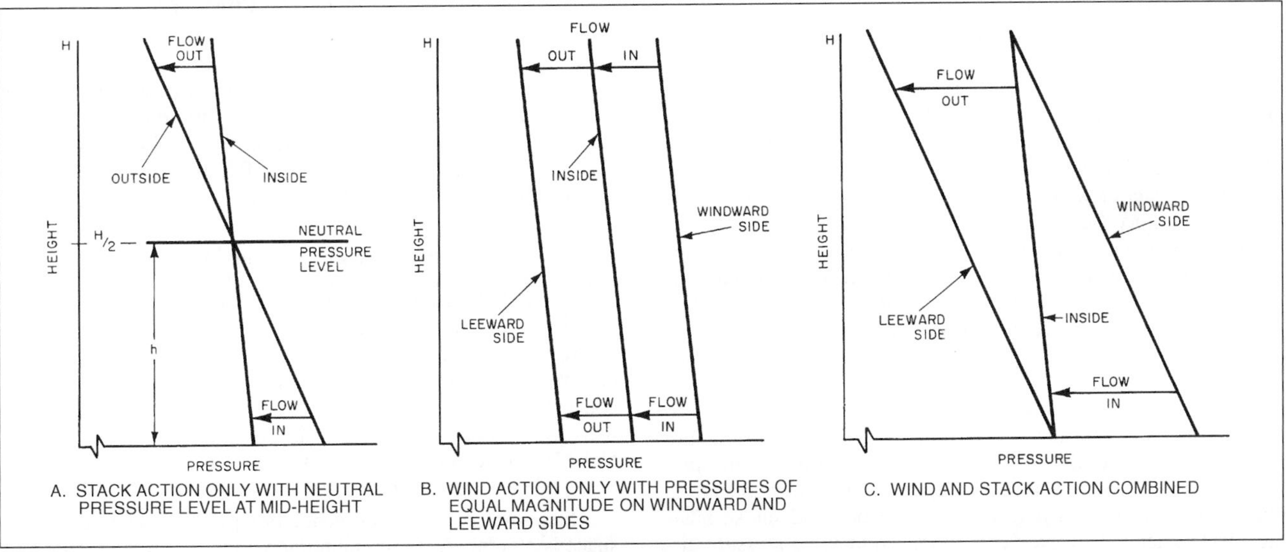

Fig. 5 Distribution of Inside and Outside Pressures over Height of Building

openings, local terrain, and the immediate shielding of the building. The taller the building is and the smaller its internal resistance to airflow, the stronger the stack effect will be. The more exposed a building is, the more susceptible it will be to wind. For any building, there will be ranges of wind speed and temperature difference for which the building's infiltration is dominated by the stack effect, the wind, or the driving pressures of both (Sinden 1978a). These building and terrain factors determine, for specific values of temperature difference and wind speed, in which regime the building's infiltration lies.

The effect of mechanical ventilation on envelope pressure differences is more complex and depends on both the direction of the ventilation flow (exhaust or supply) and the differences in these ventilation flows among the zones of the building. If mechanically supplied outdoor air is provided uniformly to each story, the change in the exterior wall pressure difference pattern is uniform. With a nonuniform supply of outdoor air (for example, to one story only), the extent of pressurization varies from story to story and depends on the internal airflow resistance. Pressurizing all levels uniformly has little effect on the pressure differences across floors and vertical shaft enclosures, but pressurizing individual stories increases the pressure drop across these internal separations. Pressurization of the ground level is often used in tall buildings to reduce the stack pressures across entries.

Available data on the NPL in various kinds of buildings are limited. The NPL in tall buildings varies from 0.3 to 0.7 of total building height (Tamura and Wilson 1966, 1967a). For houses, especially houses with chimneys, the NPL is usually above mid-height. Operating a combustion heat source with a flue raises the NPL further, sometimes above the ceiling (Shaw and Brown 1982).

Thermal Draft Coefficient

Compartmentation of a building also affects the NPL location. Equation (17) provides a maximum stack pressure difference, given no internal airflow resistance. The sum of the pressure differences across the exterior wall at the bottom and at the top of the building, as calculated by these equations, equals the total theoretical draft for the building. The sum of the actual top and bottom pressure differences, divided by the total theoretical draft pressure difference, equals the **thermal draft coefficient**. The value of the thermal draft coefficient depends on the airflow resistance of the exterior walls relative to the airflow resistance between floors. For a building without internal partitions, the total theoretical draft is achieved across the exterior walls (Figure 6A), and the thermal draft coefficient equals 1. In a building with airtight separations at each floor, each story acts independently, its own stack effect being unaffected by that of any other floor (Figure 6B). The theoretical draft is minimized in this case, and each story has an NPL.

Real multistory buildings are neither open inside (Figure 6A), nor airtight between stories (Figure 6B). Vertical air passages, stairwells, elevators, and other service shafts allow airflow between floors. Figure 6C represents a heated building with uniform openings in the exterior wall, through each floor, and into the vertical shaft at each story. Between floors, the slope of the line representing the inside pressure is the same as that shown in Figure 6A, and the discontinuity at each floor (Figure 6B) represents the pressure difference across it. Some of the pressure difference maintains flow through openings in the floors and vertical shafts. As a result, the pressure difference across the exterior wall at any level is less than it would be with no internal flow resistance.

Maintaining airtightness between floors and from floors to vertical shafts is a means of controlling indoor-outdoor pressure differences due to the stack effect and therefore infiltration (Lovaett and Wilson 1994). Good separation is also conducive to the proper operation of mechanical ventilation and smoke management systems. However, care is needed to avoid pressure differences that could prevent door opening in an emergency. Tamura and Wilson (1967b) showed that when vertical shaft

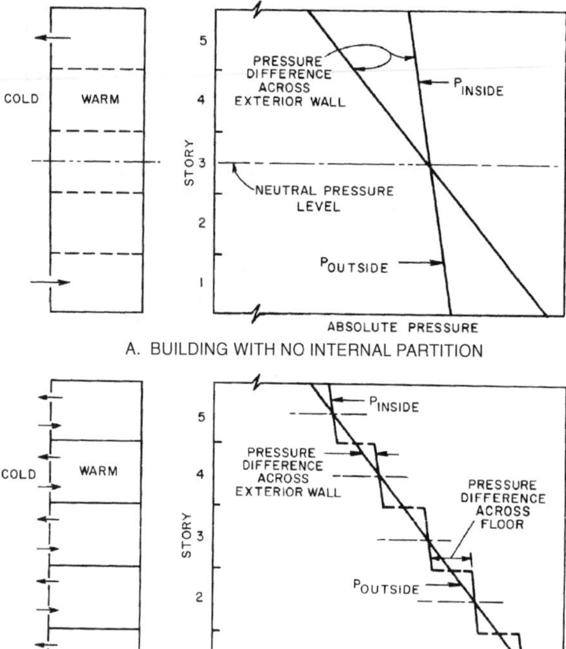

A. BUILDING WITH NO INTERNAL PARTITION

B. BUILDING WITH AIRTIGHT SEPARATION OF EACH STORY

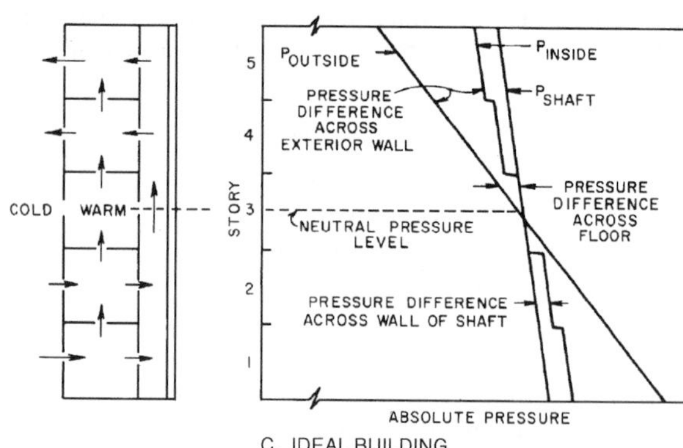

C. IDEAL BUILDING

Fig. 6 Compartmentation Effect in Buildings

leakage is at least two times the envelope leakage, the thermal draft coefficient is almost one and the effect of compartmentation is negligible. Measurements of pressure differences in three tall office buildings by Tamura and Wilson (1967a) indicated that the thermal draft coefficient ranged from 0.8 to 0.9 with the ventilation systems off.

ASHRAE *Standard* 62

ASHRAE *Standard* 62 provides guidance on ventilation and indoor air quality and includes two alternative procedures for determining design ventilation rates. In the **Ventilation Rate Procedure**, indoor air quality is assumed to be acceptable if (1) the concentrations of six pollutants in the incoming outdoor air meet the U.S. EPA national ambient air quality standards (EPA 1986), and (2) the outdoor air ventilation rates meet or exceed values (which depend on the space type) provided in a table.

The minimum outside air ventilation per person for any type of space is 15 cfm. This minimum rate will maintain the indoor CO_2 concentration within 0.07% (700 parts per million) of the outdoor concentration, assuming a typical CO_2 generation rate per occupant

(Janssen 1989). This minimum outside air ventilation rate was based, in part, on research by Berg-Munch et al. (1986), Yaglou et al. (1936), Iwashita et al. (1989), and Cain et al. (1993) that indicated that 15 cfm was required to satisfy the odor perceptions of 80% or more of visitors.

The other design approach in ASHRAE *Standard* 62 is the **Indoor Air Quality Procedure**. In this procedure, acceptable indoor air quality is achieved through the control of indoor contaminant concentrations. Such control can be realized through source control, air cleaning, and ventilation. It allows for either or both improved indoor air quality and reduced energy consumption. Chapter 24 of the 2000 *ASHRAE Handbook—Systems and Equipment* has information on air cleaning.

A combination of source control and local exhaust, as opposed to dilution with ventilation air, is the method of choice in industrial environments. Industrial ventilation is discussed in Chapters 28 and 29 of the 1999 *ASHRAE Handbook—Applications* and in *Industrial Ventilation: A Manual of Recommended Practice* (ACGIH 1998).

INDOOR AIR QUALITY

Outdoor air requirements for acceptable indoor air quality (IAQ) have long been debated, and different rationales have produced radically different ventilation standards (Grimsrud and Teichman 1989, Janssen 1989, Klauss et al. 1970, Yaglou et al. 1936, Yaglou and Witheridge 1937). Historically, the major considerations have included the amount of outdoor air required to control moisture, carbon dioxide (CO_2), odors, and tobacco smoke generated by occupants. These considerations have led to prescriptions of a minimum rate of outdoor air supply per occupant. More recently, the maintenance of acceptable indoor concentrations of a variety of additional pollutants that are not generated primarily by occupants has been a major concern. Information on contaminants can be found in Chapter 12, and odors are covered in Chapter 13.

Indoor pollutant concentrations depend on the strength of pollutant sources and the total rate of pollutant removal. Pollutant sources include the outdoor air; indoor sources such as occupants, furnishings, and appliances; and the soil adjacent to the building. Pollutant removal processes include dilution with outside air, local exhaust ventilation, deposition on surfaces, chemical reactions, and air-cleaning processes. If (1) general building ventilation is the only significant pollutant removal process, (2) the indoor air is thoroughly mixed, and (3) the pollutant source strength and ventilation rate have been stable for a sufficient period; then the steady-state indoor pollutant concentration is given by

$$C_i = C_o + S/Q_{oa} \tag{25}$$

where

C_i = steady-state indoor concentration, µg/ft³
C_o = outdoor concentration, µg/ft³
S = total pollutant source strength, µg/min
Q_{oa} = ventilation rate, cfm

Variation in pollutant source strengths (rather than variation in ventilation rate) is considered the largest cause of building-to-building variation in the concentrations of pollutants that are not generated by occupants. Turk et al. (1989) found that a lack of correlation between average indoor respirable particle concentrations and whole-building outdoor ventilation rate indicated that source strength, high outdoor concentrations, building volume, and removal processes are important. Because pollutant source strengths are highly variable, maintenance of minimum ventilation rates does not ensure acceptable indoor air quality in all situations. The lack of health-based concentration standards for many indoor air pollutants, primarily due to the lack of health data, makes the specification of minimum ventilation rates even more difficult.

In cases of high contaminant source strengths, impractically high rates of ventilation are required to control contaminant levels, and other methods of control are more effective. Removal or reduction of contaminant sources is a very effective means of control. Controlling a localized source by means of local exhaust, such as range hoods or bathroom exhaust fans, can also be effective.

Particles can be removed with various types of air filters. Gaseous contaminants with higher molecular weight can be controlled with activated carbon or alumina pellets impregnated with a substance such as potassium permanganate. Chapter 24 of the 2000 *ASHRAE Handbook—Systems and Equipment* has information on air cleaning.

THERMAL LOADS

Outdoor air introduced into a building constitutes a large part of the total space-conditioning (heating, cooling, humidification, and dehumidification) load, which is one reason to limit air exchange rates in buildings to the minimum required. Air exchange typically represents 20 to 50% of a building's thermal load. Chapter 28 and Chapter 29 cover thermal loads in more detail.

Air exchange increases a building's thermal load in three ways. First, the incoming air must be heated or cooled from the outdoor air temperature to the indoor or supply air temperature. The rate of energy consumption due to this sensible heating or cooling is given by

$$q_s = 60Q\rho c_p\Delta t = 1.1Q\Delta t \tag{26}$$

where

q_s = sensible heat load, Btu/h
Q = airflow rate, cfm
ρ = air density, lb/ft³ (about 0.075)
c_p = specific heat of air, Btu/lb·°F (about 0.24)
Δt = temperature difference between indoors and outdoors, °F

Second, air exchange modifies the moisture content of the air in a building. The rate of energy consumption associated with these latent loads (neglecting the energy associated with any condensate) is given by

$$q_l = Q\Delta W[3{,}011{,}000(2173\Delta t)] \tag{27}$$

where

q_l = latent heat load, Btu/h
ΔW = humidity ratio difference between indoors and outdoors, lb_m water/lb_m dry air

Finally, air exchange can change a building's thermal load by altering the performance of the envelope insulation system. Airflow through the insulation can decrease the thermal load due to heat exchange between infiltrating or exfiltrating air and the thermal insulation. Conversely, air moving in and out of the insulation from outside can increase the thermal load. Experimental and numerical studies have demonstrated that significant thermal coupling can occur between air leakage and insulation layers, thereby modifying the heat transmission in building envelopes. In particular, a number of researchers (Wolf 1966; Berlad et al. 1978; Bankvall 1987; Lecompte 1987) have shown that convective airflow through air-permeable insulation in an envelope assembly may degrade its effective thermal resistance. This R-value degradation occurs when outside air moves through and/or around the insulation within the wall cavity and returns to the outdoors without reaching the conditioned space. A literature review by Powell et al. (1989) summarized the findings about air movement effects on the effective thermal resistance of porous insulation under various conditions. The effect of such airflow on insulation system performance is difficult to quantify but should be considered. Airflow within the insulation system can also decrease the system's performance due to moisture condensation in and on the insulation.

Even if air flows only through cracks instead of through the insulation, the actual heating/cooling load due to the combined effect of conduction and airflow heat transfer can be lower than the heating/cooling load calculated by Equation (26). This reduction in total heating/cooling load is a consequence of the thermal coupling between conduction and convection heat transfer and is called **infiltration heat recovery** (IHR). This effect appears to be significant in building envelopes, based on preliminary laboratory and numerical work under controllable conditions by several investigators.

Using a computer simulation, Kohonen et al. (1987) found that the conduction/infiltration thermal interaction reduced total heating load by 15%. Several experimental studies by Claridge et al. (1988), Claridge and Bhattacharyya (1990), Liu and Claridge (1992a, 1992b, 1992c, 1995), Timusk et al. (1992), and others using a test cell under both steady-state and dynamic conditions found that the actual energy attributed to air infiltration can be 20 to 80% of the values given by Equation (27). Judkoff et al. (1997) measured heat recovery using a mobile home under steady-state conditions. They found that up to 40% heat recovery occurs during exfiltration through the envelope of the mobile home. Buchanan and Sherman (2000) carried out two- and three-dimensional computational fluid dynamics simulations to study the fundamental physics of the IHR process and developed a simple macro-scale mathematical model based on the steady-state one-dimensional convection-diffusion equation to predict a heat recovery factor. Their results show that the traditional method may overpredict the infiltration energy load.

Infiltration Degree-Days

Heating and cooling degree-days are a simple way to characterize the severity of a particular climate. Heating and cooling degree-day values are based on sensible temperature data, but infiltration loads are both sensible and latent. **Infiltration degree days** (IDDs) more fully describe a climate and can be used to estimate heat loss or gain due to infiltration in residences (Sherman 1986). Total infiltration degree-days is the sum of the heating and cooling infiltration degree-days and is calculated from hour-by-hour weather data and base conditions using weather weighted by infiltration rate. The selection of base conditions is an important part of the calculation of the IDDs. ASHRAE *Standard* 119 lists IDDs for many locations with a particular set of base conditions.

NATURAL VENTILATION

Natural ventilation is the flow of outdoor air due to wind and thermal pressures through intentional openings in the building's shell. Under some circumstances, it can effectively control both temperature and contaminants in mild climates, but it is not considered practical in hot and humid climates or in cold climates. Temperature control by natural ventilation is often the only means of providing cooling when mechanical air conditioning is not available. The arrangement, location, and control of ventilation openings should combine the driving forces of wind and temperature to achieve a desired ventilation rate and good distribution of ventilation air through the building. However, intentional openings cannot always guarantee adequate temperature and humidity control or indoor air quality because of the dependence on natural (wind and stack) effects to drive the flow (Wilson and Walker 1992).

Natural Ventilation Openings

Natural ventilation openings include (1) windows, doors, dormer (monitor) openings, and skylights; (2) roof ventilators; (3) stacks connecting to registers; and (4) specially designed inlet or outlet openings.

Windows transmit light and provide ventilation when open. They may open by sliding vertically or horizontally; by tilting on horizontal pivots at or near the center; or by swinging on pivots at the top, bottom, or side. The type of pivoting used is important for weather protection and affects airflow rate.

Roof ventilators provide a weather-resistant air outlet. Capacity is determined by the ventilator's location on the roof; the resistance to airflow of the ventilator and its ductwork; the ventilator's ability to use kinetic wind energy to induce flow by centrifugal or ejector action; and the height of the draft.

Natural-draft or gravity roof ventilators can be stationary, pivoting, oscillating, or rotating. Selection criteria include ruggedness, corrosion resistance, stormproofing features, dampers and operating mechanisms, noise, cost, and maintenance. Natural ventilators can be supplemented with power-driven supply fans; the motors need only be energized when the natural exhaust capacity is too low. Gravity ventilators can have manual dampers or dampers controlled by thermostat or wind velocity.

A natural-draft roof ventilator should be positioned so that it receives the full, unrestricted wind. Turbulence created by surrounding obstructions, including higher adjacent buildings, impairs a ventilator's ejector action. The ventilator inlet should be conical or bell mounted to give a high flow coefficient. The opening area at the inlet should be increased if screens, grilles, or other structural members cause flow resistance. Building air inlets at lower levels should be larger than the combined throat areas of all roof ventilators.

Stacks or vertical flues should be located where wind can act on them from any direction. Without wind, stack effect alone removes air from the room with the inlets.

Required Flow for Indoor Temperature Control

The ventilation airflow rate required to remove a given amount of heat from a building can be calculated from Equations (26) and (27) if the quantity of heat to be removed and the indoor-outdoor temperature difference are known.

Airflow Through Large Intentional Openings

The relationship describing the airflow through a large intentional opening is based on the Bernoulli equation with steady, incompressible flow. The general form that includes stack, wind, and mechanical ventilation pressures across the opening is

$$Q = C_3 C_D A \sqrt{2\Delta p / \rho} \tag{28}$$

where

Q = airflow rate, cfm
C_D = discharge coefficient for opening, dimensionless
A = cross-sectional area of opening, ft^2
ρ = air density, lb_m/ft^3
Δp = pressure difference across opening, in. of water
C_3 = unit conversion factor = 776

The discharge coefficient C_D is a dimensionless number that depends on the geometry of the opening and the Reynolds number of the flow.

Flow Caused by Wind Only

Factors due to wind forces that affect the ventilation rate include average speed, prevailing direction, seasonal and daily variation in speed and direction, and local obstructions such as nearby buildings, hills, trees, and shrubbery. Liddament (1988) reviewed the relevance of wind pressure as a driving mechanism. A multiflow path simulation model was developed and used to illustrate the effects of wind on air exchange rate.

Wind speeds may be lower in summer than in winter; directional frequency is also a function of season. Natural ventilation systems are often designed for wind speeds of one-half the seasonal average. The following equation shows the rate of air forced through ventilation inlet openings by wind or determines the proper size of openings to produce given airflow rates:

$$Q = C_4 C_v A U \quad (29)$$

where

Q = airflow rate, cfm
C_v = effectiveness of openings (C_v is assumed to be 0.5 to 0.6 for perpendicular winds and 0.25 to 0.35 for diagonal winds)
A = free area of inlet openings, ft^2
U = wind speed, mph
C_4 = unit conversion factor = 88.0

Inlets should face directly into the prevailing wind. If they are not advantageously placed, flow will be less than that predicted by Equation (29); if the inlets are unusually well placed, flow will be slightly more. Desirable outlet locations are (1) on the leeward side of the building directly opposite the inlet, (2) on the roof, in the low-pressure area caused by a flow discontinuity of the wind, (3) on the side adjacent to the windward face where low-pressure areas occur, (4) in a dormer on the leeward side, (5) in roof ventilators, or (6) by stacks. Chapter 16 gives a general description of the wind pressure distribution on a building. The inlets should be placed in the exterior high-pressure regions; the outlets should be placed in the exterior low-pressure regions.

Flow Caused by Thermal Forces Only

If building internal resistance is not significant, the flow caused by stack effect can be expressed by

$$Q = 60 C_D A \sqrt{2g\Delta H_{NPL}(T_i - T_o)/T_i} \quad (30)$$

where

Q = airflow rate, cfm
C_D = discharge coefficient for opening
ΔH_{NPL} = height from midpoint of lower opening to NPL, ft
T_i = indoor temperature, °R
T_o = outdoor temperature, °R

Equation (30) applies when $T_i > T_o$. If $T_i < T_o$, replace T_i in the denominator with T_o, and replace $(T_i - T_o)$ in the numerator with $(T_o - T_i)$. An average temperature should be used for T_i if there is thermal stratification. If the building has more than one opening, the outlet and inlet areas are considered equal. The discharge coefficient C_D accounts for all viscous effects such as surface drag and interfacial mixing.

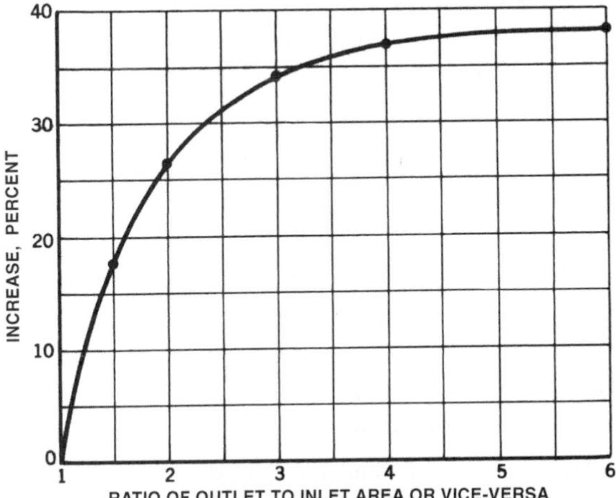

Fig. 7 Increase in Flow Caused by Excess Area of One Opening over the Other

Estimation of ΔH_{NPL} is difficult for naturally ventilated buildings. If one window or door represents a large fraction (approximately 90%) of the total opening area in the envelope, then the NPL is at the mid-height of that aperture, and ΔH_{NPL} equals one-half the height of the aperture. For this condition, flow through the opening is bidirectional (i.e., air from the warmer side flows through the top of the opening, and air from the colder side flows through the bottom). Interfacial mixing occurs across the counterflow interface, and the orifice coefficient can be calculated according to the following equation (Kiel and Wilson 1986):

$$C_D = 0.40 + 0.0025|T_i - T_o| \quad (31)$$

If enough other openings are available, the airflow through the opening will be unidirectional, and mixing cannot occur. A discharge coefficient of C_D = 0.65 should then be used. Additional information on stack-driven airflows for natural ventilation can be found in Foster and Down (1987).

Greatest flow per unit area of openings is obtained when inlet and outlet areas are equal; Equations (30) and (31) are based on this equality. Increasing the outlet area over inlet area (or vice versa) increases airflow but not in proportion to the added area. When openings are unequal, use the smaller area in Equation (30) and add the increase as determined from Figure 7.

Natural Ventilation Guidelines

Several general guidelines should be observed in designing for natural ventilation. Some of these may conflict with other climate-responsive strategies (such as using orientation and shading devices to minimize solar gain) or other design considerations.

1. In hot, humid climates, use mechanical cooling. If mechanical cooling is not available, air velocities should be maximized in the occupied zones. In hot, arid climates, consider evaporative cooling. Airflow throughout the building should be maximized for structural cooling, particularly at night when the temperature is low.
2. Topography, landscaping, and surrounding buildings should be used to redirect airflow and give maximum exposure to breezes. Vegetation can funnel breezes and avoid wind dams, which reduce the driving pressure differential around the building. Site objects should not obstruct inlet openings.
3. The building should be shaped to expose maximum shell openings to breezes.
4. Architectural elements such as wing walls, parapets, and overhangs should be used to promote airflow into the building interior.
5. The long facade of the building and the majority of the door and window openings should be oriented with respect to the prevailing summer breezes. If there is no prevailing direction, openings should be sufficient to provide ventilation regardless of wind direction.
6. Windows should be located in opposing pressure zones. Two openings on opposite sides of a space increase the ventilation flow. Openings on adjacent sides force air to change direction, providing ventilation to a greater area. The benefits of the window arrangement depend on the outlet location relative to the direction of the inlet airstream.
7. If a room has only one external wall, better airflow is achieved with two widely spaced windows.
8. If the openings are at the same level and near the ceiling, much of the flow may bypass the occupied level and be ineffective in diluting contaminants there.
9. Vertical distance between openings is required to take advantage of the stack effect; the greater the vertical distance, the greater the ventilation.

10. Openings in the vicinity of the NPL are least effective for thermally induced ventilation. If the building has only one large opening, the NPL tends to move to that level, which reduces the pressure across the opening.
11. Greatest flow per unit area of total opening is obtained by inlet and outlet openings of nearly equal areas. An inlet window smaller than the outlet creates higher inlet velocities. An outlet smaller than the inlet creates lower but more uniform airspeed through the room.
12. Openings with areas much larger than calculated are sometimes desirable when anticipating increased occupancy or very hot weather.
13. Horizontal windows are generally better than square or vertical windows. They produce more airflow over a wider range of wind directions and are most beneficial in locations where prevailing wind patterns shift.
14. Window openings should be accessible to and operable by occupants.
15. Inlet openings should not be obstructed by indoor partitions. Partitions can be placed to split and redirect airflow but should not restrict flow between the building's inlets and outlets.
16. Vertical airshafts or open staircases can be used to increase and take advantage of stack effects. However, enclosed staircases intended for evacuation during a fire should not be used for ventilation.

RESIDENTIAL AIR LEAKAGE

Most infiltration in residential buildings in the U.S. is dominated by envelope leakage. However, trends in new construction are towards tighter envelopes such that envelope leakage is reduced in newer housing.

Envelope Leakage Measurement

Envelope leakage of a building can be measured with **pressurization testing** (commonly called a **blower-door test**). Fan pressurization is relatively quick and inexpensive, and it characterizes building envelope airtightness independent of weather conditions. In this procedure, a large fan or blower is mounted in a door or window and induces a large and roughly uniform pressure difference across the building shell (ASTM *Standard* E 779, ASTM *Standard* E 1827, ISO *Standard* 9972, and CGSB *Standard* 149.10). The airflow required to maintain this pressure difference is then measured. The leakier the building is, the more airflow is necessary to induce a specific indoor-outdoor pressure difference. The airflow rate is generally measured at a series of pressure differences ranging from about 0.04 to 0.30 in. of water.

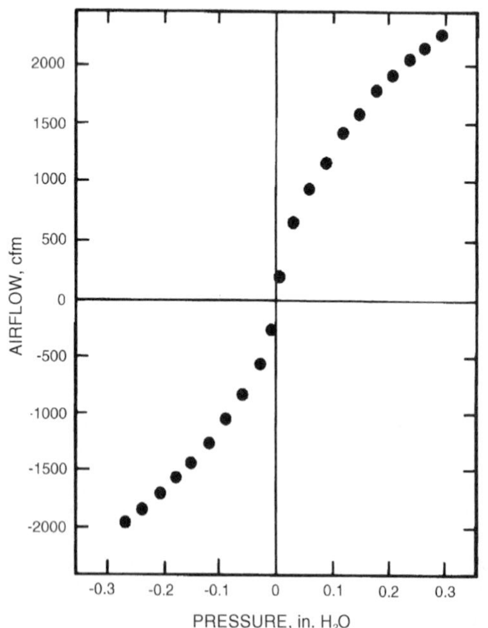

Fig. 8 Airflow Rate Versus Pressure Difference Data from Whole-House Pressurization Test

The results of a pressurization test, therefore, consist of several combinations of pressure difference and airflow rate data. An example of typical data is shown in Figure 8. These data points characterize the air leakage of a building and are generally converted to a single value that serves as a measure of the building's airtightness. There are several different measures of airtightness, most of which involve fitting the data to a curve describing the relationship between the airflow Q through an opening in the building envelope and the pressure difference Δp across it. This relationship is called the **leakage function** of the opening. The form of the leakage function depends on the geometry of the opening. Background theoretical material relevant to leakage functions may be found in Hopkins and Hansford (1974), Etheridge (1977), Kronvall (1980), Chastain et al. (1987), and Walker et al. (1997).

The openings in a building envelope are not uniform in geometry and, generally, the flow never becomes fully developed. Each opening in the building envelope can be described by Equation (32), commonly called the **power law equation**:

$$Q = c(\Delta p)^n \tag{32}$$

where

c = flow coefficient, cfm/(in. of water)^n
n = pressure exponent, dimensionless

Sherman (1992b) showed how the power law can be developed analytically by looking at developing laminar flow in short pipes. Equation (32) only approximates the relationship between Q and Δp. Measurements of single cracks (Honma 1975, Krieth et al. 1957) have shown that n can vary if Δp changes over a wide range. Additional investigation of pressure/flow data for simple cracks by Chastain et al. (1987) further indicated the importance of adequately characterizing the three-dimensional geometry of openings and the entrance and exit effects. Walker et al. (1997) showed that for the arrays of cracks in a building envelope over the range of pressures acting during infiltration, n is constant. A typical value for n is about 0.65. c and n can be determined for a building using fan pressurization techniques.

Airtightness Ratings

In some cases, the predicted airflow rate is converted to an equivalent or effective air leakage area as follows:

$$A_L = C_5 Q_r \frac{\sqrt{\rho / 2\Delta p_r}}{C_D} \tag{33}$$

where

A_L = equivalent or effective air leakage area, in^2
Q_r = predicted airflow rate at Δp_r (from curve fit to pressurization test data), cfm
ρ = air density, lb$_m$/ft^3
Δp_r = reference pressure difference, in. of water
C_D = discharge coefficient
C_5 = unit conversion factor = 0.186

All the openings in the building shell are combined into an overall opening area and discharge coefficient for the building when the equivalent or effective air leakage area is calculated. Some users of the leakage area approach set $C_D = 1$. Others set $C_D \approx 0.6$ (i.e., the discharge coefficient for a sharp-edged orifice). The air leakage area of a building is, therefore, the area of an orifice (with an assumed value of C_D) that would produce the same amount of leakage as the building envelope at the reference pressure.

An airtightness rating, whether based on an air leakage area or a predicted airflow rate, is generally normalized by some factor to account for building size. Normalization factors include floor area, exterior envelope area, and building volume.

With the wide variety of possible approaches to normalization and reference pressure difference, and the use of the air leakage area concept, many different airtightness ratings are being used. Reference pressure differences include 0.016, 0.04, 0.10, 0.20, and 0.30 in. of water. Reference pressure differences of 0.016 and 0.04 in. of water are advocated because they are closer to the pressure differences that actually induce air exchange and, therefore, better model the flow characteristics of the openings. While this may be true, they are outside the range of measured values in the test; therefore, the predicted airflow rates at 0.016 and 0.04 in. of water are subject to significant uncertainty. The uncertainty in these predicted airflow rates and the implications for quantifying airtightness are discussed in Persily and Grot (1985b), Chastain (1987), and Modera and Wilson (1990). Round robin tests by Murphy et al. (1991) to determine the repeatability and reproducibility of fan pressurization devices found that subtle errors in fan calibration or operator technique are greatly exaggerated when extrapolating the pressure versus flow curve out to 0.016 in. of water, with errors as great as ±40%, mainly due to the fan calibration errors at low flow.

Some common airtightness ratings include the effective air leakage area at 0.016 in. of water assuming $C_D = 1.0$ (Sherman and Grimsrud 1980); the equivalent air leakage area at 0.04 in. of water assuming $C_D = 0.611$ (CGSB *Standard* 149.10); and the airflow rate at 0.20 in. of water, divided by the building volume to give units of air changes per hour (Blomsterberg and Harrje 1979).

Conversion Between Ratings

Air leakage areas at one reference pressure difference can be converted to air leakage areas at an other reference pressure difference according to:

$$A_{r,2} = A_{r,1}\left(\frac{C_{D,1}}{C_{D,2}}\right)\left(\frac{\Delta p_{r,2}}{\Delta p_{r,1}}\right)^{n-0.5} \tag{34}$$

where

$A_{r,1}$ = air leakage area at reference pressure difference $\Delta p_{r,1}$, in²
$A_{r,2}$ = air leakage area at reference pressure difference $\Delta p_{r,2}$, in²
$C_{D,1}$ = discharge coefficient used to calculate $A_{r,1}$
$C_{D,2}$ = discharge coefficient used to calculate $A_{r,2}$
n = pressure exponent from Equation (32)

An air leakage area at one reference pressure difference can be converted to an airflow rate at some other reference pressure difference according to

$$Q_{r,2} = C_6 C_{D,1} A_{r,1}\sqrt{\frac{2}{\rho}}(\Delta p_{r,1})^{0.5-n}(\Delta p_{r,2})^n \tag{35}$$

where

$Q_{r,2}$ = airflow rate at reference pressure difference $\Delta p_{r,2}$, cfm
C_6 = unit conversion factor = 5.39

The flow coefficient c in Equation (32) may be converted to an air leakage area according to

$$A_L = \frac{c}{C_6 C_D}\sqrt{\frac{\rho}{2}}\Delta p_r^{(n-0.5)} \tag{36}$$

Finally, an air leakage area may be converted to the flow coefficient c in Equation (32) according to

$$c = C_6 C_D A_L\sqrt{\frac{2}{\rho}}(\Delta p_r)^{0.5-n} \tag{37}$$

Equations (34) through (37) require the assumption of a value of n, unless it is reported with the measurement results. When whole-building pressurization test data are fitted to Equation (32), the value of n generally lies between 0.6 and 0.7. Therefore, using a value of n in this range is reasonable.

Building Air Leakage Data

Fan pressurization measures a building property that ideally varies little with time and weather conditions. In reality, unless the wind and temperature differences during the measurement period are sufficiently mild, the pressure differences they induce during the test will interfere with the test pressures and cause measurement errors. Persily (1982) and Modera and Wilson (1990) studied the effects of wind speed on pressurization test results. Several experimental studies have also shown variations on the order of 20 to 40% over a year in the measured airtightness in homes (Persily 1982, Kim and Shaw 1986, Warren and Webb 1986).

Figure 9 summarizes envelope leakage measured North American housing (Sherman and Dickerhoff 1998) and from several European and Canadian sources (AIVC 1994). This figure shows the large range of measured envelope tightness but can still be used to illustrate typical and extreme values in the housing stock.

ASHRAE *Standard* 119 establishes air leakage performance levels for residential buildings. These levels are in terms of the normalized leakage area A_n:

$$A_n = 6.944\left(\frac{A_L}{A_f}\right)\left(\frac{H}{H_o}\right)^{0.3} \tag{38}$$

where

A_n = normalized leakage area, dimensionless
A_L = effective leakage area at 0.016 in. of water ($C_D = 1.0$), in²
A_f = gross floor area (within exterior walls), ft²
H = building height, ft
H_o = reference height of one-story building = 8 ft

Air Leakage of Building Components

The fan pressurization procedure discussed in the section on Envelope Leakage Measurement enables the measurement of whole-building air leakage. The location and size of individual openings in building envelopes are extremely important because they influence the air infiltration rate of a building as well as the heat and moisture transfer characteristics of the envelope. Additional test procedures exist for pressure-testing individual building components such as windows, walls, and doors; they are discussed in ASTM *Standards* E 283 and E 783 for laboratory and field tests, respectively.

Leakage Distribution

Dickerhoff et al. (1982) and Harrje and Born (1982) studied the air leakage of individual building components and systems. The following points summarize the percentages of whole-building air leakage area associated with various components and systems. The values in parentheses include the range determined for each component and the mean of the range.

Walls (18 to 50%; 35%). Both interior and exterior walls contribute to the leakage of the structure. Leakage between the sill plate and the foundation, cracks below the bottom of the gypsum wallboard, electrical outlets, plumbing penetrations, and leaks into the attic at the top plates of walls all occur.

Ceiling details (3 to 30%; 18%). Leakage across the top ceiling of the heated space is particularly insidious because it reduces the effectiveness of insulation on the attic floor and contributes to infiltration heat loss. Ceiling leakage also reduces the effectiveness of

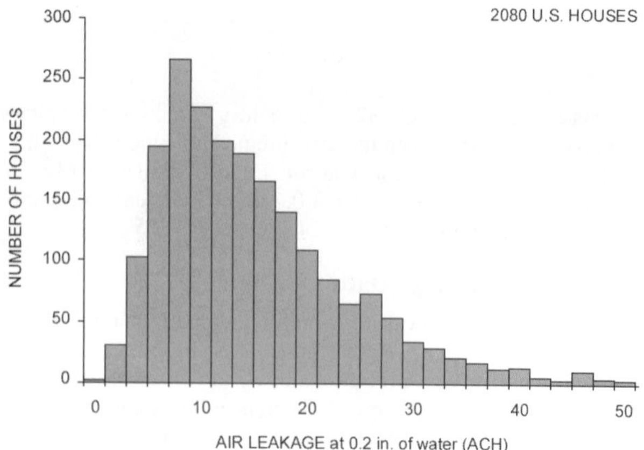

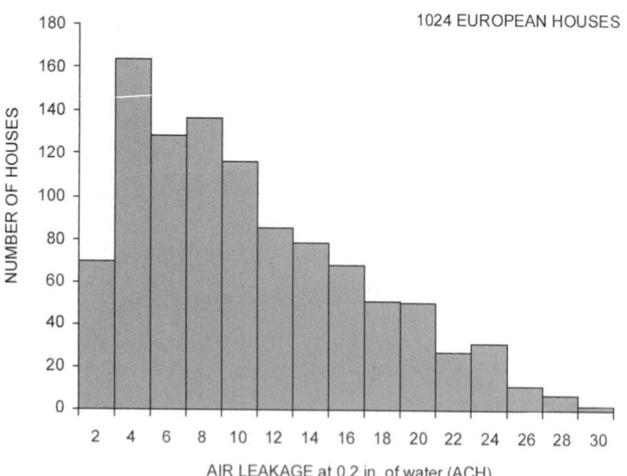

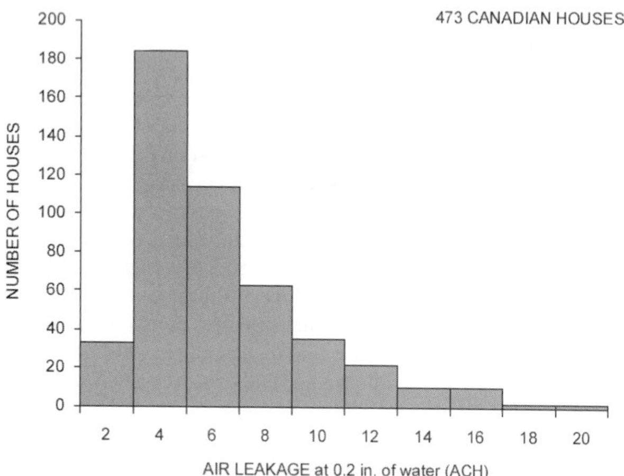

Fig. 9 Envelope Leakage Measurements

ceiling insulation in buildings without attics. Recessed lighting, plumbing, and electrical penetrations leading to the attic are some particular areas of concern.

Forced-air heating and/or cooling systems (3 to 28%; 18%). The location of the heating or cooling equipment, air handler, or ductwork in conditioned or unconditioned spaces; the venting arrangement of a fuel-burning device; and the existence and location of a combustion air supply all affect leakage. Modera et al. (1991) and Robison and Lambert (1989), among others, have shown that the variability of leakage in ducts passing through unconditioned spaces is high, the coefficient of variation being on the order of 50%. Field studies have also shown that in-situ repairs can eliminate one-quarter to two-thirds of the observed leakage (Cummings and Tooley 1989, Cummings et al. 1990, Robison and Lambert 1989, Jump et al. 1996). The 18% contribution of ducts to total leakage significantly underestimates their impact because during system operation, the pressure differentials across the duct leaks are approximately ten times higher than typical pressure differences across the envelope leaks (Modera 1989, Modera et al. 1991) and result in large (factors of two to three) changes in ventilation rate (Walker 1999, Walker et al. 1999, Cummings et al. 1990).

Windows and doors (6 to 22%; 15%). More variation in window leakage is seen among window types (e.g., casement versus double-hung) than among new windows of the same type from different manufacturers (Weidt et al. 1979). Windows that seal by compressing the weather strip (casements, awnings) show significantly lower leakage than windows with sliding seals.

Fireplaces (0 to 30%; 12%). When a fireplace is not in use, poorly fitting dampers allow air to escape. Glass doors reduce excess air while a fire is burning but rarely seal the fireplace structure more tightly than a closed damper does. Chimney caps or fireplace plugs (with signs that warn they are in place) effectively reduce leakage through a cold fireplace.

Vents in conditioned spaces (2 to 12%; 5%). Exhaust vents in conditioned spaces frequently have either no dampers or dampers that do not close properly.

Diffusion through walls (<1%). Diffusion, in comparison to infiltration through holes and other openings in the structure, is not an important flow mechanism. At 0.02 in. of water, the permeability of building materials produces an air exchange rate of less than 0.01 ACH by wall diffusion in a typical house.

Component leakage areas. Table 1 shows effective air leakage areas for a variety of residential building components at 0.016 in. of water with C_D assumed equal to 1 (Colliver et al. 1992). The values in the table present results in terms of air leakage area per unit component. Per unit component means per component, per unit surface area, or per unit length of crack or sash, whichever is appropriate. These air leakage areas may be converted to air leakage areas at other reference pressures, airflow rates, or flow coefficients using Equations (34) through (37). Table 1 can be used to estimate the air leakage area of the building if test data are not available. To obtain the building's total air leakage area, multiply the overall dimensions or number of occurrences of each building component by the appropriate table entry. The sum of the resulting products is the total building air leakage area. Table 2 gives the result of an example calculation of the effective air leakage area of a residence. Each leakage component is identified in the first column and described in the second. The length, area, or number of the component is in the third column. The fourth column contains the air leakage area per unit component, from Table 1, and the fifth contains the total air leakage area associated with that component. The sum of the terms in the last column is the total air leakage area of the building, in this case 131 in^2. Klote and Milke (1992) describe a method for estimating airflows through gaps such as those found around doors.

Multifamily Building Leakage

Leakage distribution is particularly important in multifamily apartment buildings. These buildings often cannot be treated as single zones due to the internal resistance between apartments. Moreover, the leakage between apartments varies widely, tending to be small in modern construction, and ranging as high as 60% of the total apartment leakage in turn-of-the-century brick walk-up apartment buildings (Modera et al. 1991, Diamond et al. 1986). Little information on interzonal leakage has been reported because of the difficulty and expense of these measurements.

Table 1 Effective Air Leakage Areas (Low-Rise Residential Applications Only)

	Units (see note)	Best Estimate	Minimum	Maximum
Ceiling				
General	in^2/ft^2	0.026	0.011	0.04
Drop	in^2/ft^2	0.0027	0.00066	0.003
Ceiling penetrations				
Whole-house fans	in^2 ea	3.1	0.25	3.3
Recessed lights	in^2 ea	1.6	0.23	3.3
Ceiling/Flue vent	in^2 ea	4.8	4.3	4.8
Surface-mounted lights	in^2 ea	0.13		
Chimney	in^2 ea	4.5	3.3	5.6
Crawl space				
General (area for exposed wall)	in^2/ft^2	0.144	0.1	0.24
8 in. by 16 in. vents	in^2 ea	20		
Door frame				
General	in^2 ea	1.9	0.37	3.9
Masonry, not caulked	in^2/ft^2	0.07	0.024	0.07
Masonry, caulked	in^2/ft^2	0.014	0.004	0.014
Wood, not caulked	in^2/ft^2	0.024	0.009	0.024
Wood, caulked	in^2/ft^2	0.004	0.001	0.004
Trim	in^2/lftc	0.05		
Jamb	in^2/lftc	0.4	0.3	0.5
Threshold	in^2/lftc	0.1	0.06	1.1
Doors				
Attic/crawl space, not weatherstripped	in^2 ea	4.6	1.6	5.7
Attic/crawl space, weatherstripped	in^2 ea	2.8	1.2	2.9
Attic fold down, not weatherstripped	in^2 ea	6.8	3.6	13
Attic fold down, weatherstripped	in^2 ea	3.4	2.2	6.7
Attic fold down, with insulated box	in^2 ea	0.6		
Attic from unconditioned garage	in^2 ea	0	0	0
Double, not weatherstripped	in^2/ft^2	0.16	0.1	0.32
Double, weatherstripped	in^2/ft^2	0.12	0.04	0.33
Elevator (passenger)	in^2 ea	0.04	0.022	0.054
General, average	in^2/lftc	0.015	0.011	0.021
Interior (pocket, on top floor)	in^2 ea	2.2		
Interior (stairs)	in^2/lftc	0.04	0.012	0.070
Mail slot	in^2/lftc	0.2		
Sliding exterior glass patio	in^2 ea	3.4	0.46	9.3
Sliding exterior glass patio	in^2/ft^2	0.079	0.009	0.22
Storm (difference between with and without)	in^2 ea	0.9	0.46	0.96
Single, not weatherstripped	in^2 ea	3.3	1.9	8.2
Single, weatherstripped	in^2 ea	1.9	0.6	4.2
Vestibule (subtract per each location)	in^2 ea	1.6		
Electrical outlets/Switches				
No gaskets	in^2 ea	0.38	0.08	0.96
With gaskets	in^2 ea	0.023	0.012	0.54
Furnace				
Sealed (or no) combustion	in^2 ea	0	0	0
Retention head or stack damper	in^2 ea	4.6	3.1	4.6
Retention head and stack damper	in^2 ea	3.7	2.8	4.6
Floors over crawl spaces				
General	in^2/ft^2	0.032	0.006	0.071
Without ductwork in crawl space	in^2/ft^2	0.0285		
With ductwork in crawl space	in^2/ft^2	0.0324		
Fireplace				
With damper closed	in^2/ft^2	0.62	0.14	1.3
With damper open	in^2/ft^2	5.04	2.09	5.47
With glass doors	in^2/ft^2	0.58	0.06	0.58
With insert and damper closed	in^2/ft^2	0.52	0.37	0.66
With insert and damper open	in^2/ft^2	0.94	0.58	1.3
Gas water heater	in^2 ea	3.1	2.3	3.9
Joints				
Ceiling-wall	in^2/lftc	0.070	0.0075	0.12
Sole plate, floor/wall, uncaulked	in^2/lftc	0.2	0.018	0.26
Sole plate, floor/wall, caulked	in^2/lftc	0.04	0.0035	0.056
Top plate, band joist	in^2/lftc	0.005	0.0035	0.018
Piping/Plumbing/Wiring penetrations				
Uncaulked	in^2 ea	0.9	0.31	3.7
Caulked	in^2 ea	0.3	0.16	0.3
Vents				
Bathroom with damper closed	in^2 ea	1.6	0.39	3.1
Bathroom with damper open	in^2 ea	3.1	0.95	3.4
Dryer with damper	in^2 ea	0.46	0.45	1.1
Dryer without damper	in^2 ea	2.3	1.9	5.3
Kitchen with damper open	in^2 ea	6.2	2.2	11
Kitchen with damper closed	in^2 ea	0.8	0.16	1.1
Kitchen with tight gasket	in^2 ea	0.16		
Walls (exterior)				
Cast-in-place concrete	in^2/ft^2	0.007	0.0007	0.026
Clay brick cavity wall, finished	in^2/ft^2	0.0098	0.0007	0.033
Precast concrete panel	in^2/ft^2	0.017	0.0004	0.024
Lightweight concrete block, unfinished	in^2/ft^2	0.05	0.019	0.058
Lightweight concrete block, painted or stucco	in^2/ft^2	0.016	0.0075	0.016
Heavyweight concrete block, unfinished	in^2/ft^2	0.0036		
Continuous air infiltration barrier	in^2/ft^2	0.0022	0.0008	0.003
Rigid sheathing	in^2/ft^2	0.005	0.0042	0.006
Window framing				
Masonry, uncaulked	in^2/ft^2	0.094	0.082	0.148
Masonry, caulked	in^2/ft^2	0.019	0.016	0.03
Wood, uncaulked	in^2/ft^2	0.025	0.022	0.039
Wood, caulked	in^2/ft^2	0.004	0.004	0.007
Windows				
Awning, not weatherstripped	in^2/ft^2	0.023	0.011	0.035
Awning, weatherstripped	in^2/ft^2	0.012	0.006	0.017
Casement, weatherstripped	in^2/lftc	0.011	0.005	0.14
Casement, not weatherstripped	in^2/lftc	0.013		
Double horizontal slider, not weatherstripped	in^2/lftc	0.052	0.0009	0.16
Double horizontal slider, wood, weatherstripped	in^2/lftc	0.026	0.0070	0.081
Double horizontal slider, aluminum, weatherstripped	in^2/lftc	0.034	0.027	0.038
Double-hung, not weatherstripped	in^2/lftc	0.12	0.040	0.29
Double-hung, weatherstripped	in^2/lftc	0.031	0.009	0.089
Double-hung with storm, not weatherstripped	in^2/lftc	0.046	0.023	0.080
Double-hung with storm, weatherstripped	in^2/lftc	0.037	0.021	0.05
Double-hung with pressurized track, weatherstripped	in^2/lftc	0.023	0.018	0.026
Jalousie	in^2/louver	0.524		
Lumped	in^2/lfts	0.022	0.00042	0.097
Single horizontal slider, weatherstripped	in^2/lfts	0.031	0.009	0.097
Single horizontal slider, aluminum	in^2/lfts	0.04	0.013	0.097
Single horizontal slider, wood	in^2/lfts	0.021	0.013	0.047
Single horizontal slider, wood clad	in^2/lfts	0.030	0.025	0.038
Single-hung, weatherstripped	in^2/lfts	0.041	0.029	0.058
Sill	in^2/lftc	0.0099	0.0065	0.010
Storm inside, heat shrink	in^2/lfts	0.00085	0.00042	0.00085
Storm inside, rigid sheet with magnetic seal	in^2/lfts	0.0056	0.00085	0.011
Storm inside, flexible sheet with mechanical seal	in^2/lfts	0.0072	0.00085	0.039
Storm inside, rigid sheet with mechanical seal	in^2/lfts	0.019	0.0021	0.039
Storm outside, pressurized track	in^2/lftc	0.025		
Storm outside, 2-track	in^2/lftc	0.058		
Storm outside, 3-track	in^2/lftc	0.116		

Note: Air leakage areas are based on values found in the literature. The effective air leakage area (in square inches) is based on a pressure difference of 0.016 in. of water and $C_D = 1$.

Abbreviations: ft^2 = gross area in square feet; ea = each; lftc = linear foot of crack; lfts = linear foot of sash

Table 2 Example of Calculation of Building Effective Air Leakage Area Based on Component Leakage Areas

Component	Description	Size or Number ×	A_L per unit	= A_L, in²
Sills	Uncaulked	142 ft	0.19 in²/ft	27.0
Electrical outlets		20	0.08 in² ea	1.6
Windows	Sliding	141 ft²	0.057 in²/ft²	11.4
Framing		141 ft²	0.024 in²/ft²	
Exterior doors	Single	62 ft²	0.11 in²/ft²	8.3
Framing		62 ft²	0.024 in²/ft²	
Fireplace	Without damper	1	54.0 in² ea	54.0
Penetrations	Pipes	7	0.93 in² ea	6.5
Heating ducts	Ducts untaped, in basement	1	22.0 in² ea	22.0
	Calculated total building air leakage area A_c =			130.8 in²

Controlling Air Leakage

New Buildings. It is much easier to build a tight building than to tighten an existing building. Elmroth and Levin (1983), Eyre and Jennings (1983), Marbek Resource Consultants (1984), and Nelson et al. (1985) provide information and construction details on airtight building design for houses.

A continuous air infiltration retarder is one of the most effective means of reducing air leakage through walls, around window and door frames, and at joints between major building elements. Particular care must be taken to ensure its continuity at all wall, floor, and ceiling joints; at window and door frames; and at all penetrations of the retarder, such as electrical outlets and switches, plumbing connections, and utility service penetrations. Joints in the **air-vapor retarder** must be lapped and sealed. Plastic vapor retarders installed in the ceiling should be tightly sealed with the vapor retarder in the outside walls and should be continuous over the partition walls. A seal at the top of the partition walls prevents leakage into the attic; a plate on top of the studs generally gives a poor seal. The air infiltration retarder can be installed either on the inside of the wall framing, in which case it usually functions as a vapor retarder as well, or on the outside of the wall framing, in which case it should have a permeance rating high enough to permit diffusion of water vapor from the wall. For a discussion of moisture transfer in building envelopes, see Chapter 23 and Chapter 24.

A continuous air infiltration retarder installed on the outside of wall framing can cover many difficult construction details associated with the installation of continuous air-vapor retarders. Interior air-vapor retarders must be lapped and sealed at electrical outlets and switches, at joints between walls and floors and between walls and ceilings, and at plumbing connections penetrating the wall's interior finish. The exterior air infiltration retarder can cover these problem areas continuously. Joints in the air infiltration retarder should be lapped and sealed or taped. Exterior air infiltration retarders are generally made of a material stronger than plastic film and are more likely to withstand damage during construction. Sealing the wall against air leakage at the exterior of the insulation also cuts down on convection currents within the wall cavity, allowing insulation to retain more of its effectiveness.

Existing Buildings. The air leakage sites must first be located in order to tighten the envelope of an existing building. As discussed earlier, air leakage in buildings is due not only to windows and doors, but to a wide range of unexpected and unobvious construction defects. Many important leakage sites can be very difficult to find. A variety of techniques developed to locate leakage sites are described in ASTM *Standard* E 1186 and Charlesworth (1988).

Once leakage sites are located, they can be repaired with materials and techniques appropriate to the size and location of the leak. Harrje et al. (1979), Diamond et al. (1982), and Energy Resource Center (1982) include information on airtightening in existing residential buildings. With these procedures, the air leakage of residential buildings can be reduced dramatically. Depending on the extent of the tightening effort and the experience of those doing the work, residential buildings can be tightened anywhere from 5% to more than 50% (Blomsterberg and Harrje 1979, Harrje and Mills 1980, Jacobson et al. 1986, Verschoor and Collins 1986, Giesbrecht and Proskiw 1986). Much less information is available for airtightening large, commercial buildings, but the same general principles apply (Parekh et al. 1991, Persily 1991).

RESIDENTIAL VENTILATION

Typical infiltration values in housing in North America vary by a factor of about ten, from tightly constructed housing with seasonal average air exchange rates of about 0.2 air changes per hour (ACH) to loosely constructed housing with air exchange rates as great as 2.0 ACH. Figure 10 and Figure 11 show histograms of infiltration rates measured in two different samples of North American housing (Grimsrud et al. 1982, Grot and Clark 1979). Figure 10 shows the average seasonal infiltration of 312 houses located in different areas in North America. The median infiltration value of this sample is 0.5 ACH. Figure 11 represents measurements in 266 houses located in 16 cities in the United States. The median value of this sample is 0.9 ACH. The group of houses contained in the Figure 10 sample is biased toward new, energy-efficient houses, while the group in Figure 11 represents older, low-income housing in the United States.

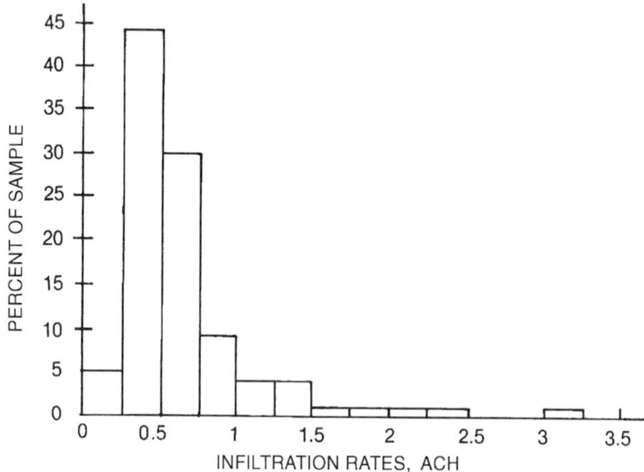

Fig. 10 Histogram of Infiltration Values—New Construction

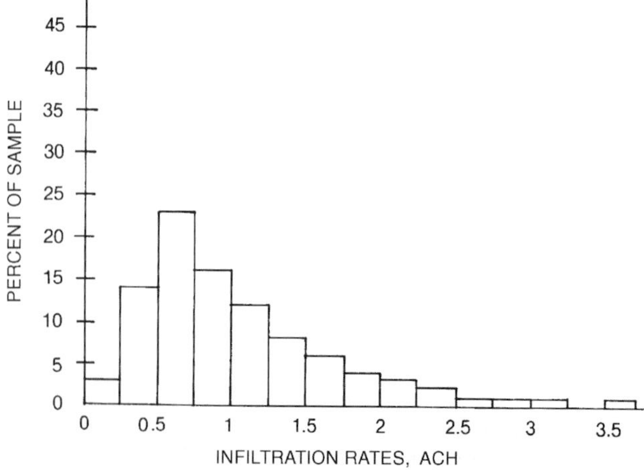

Fig. 11 Histogram of Infiltration Values—Low-Income Housing

Additional studies have found average values for houses in regional areas. Palmiter and Brown (1989) and Parker et al. (1990) found a heating season average of 0.40 ACH (range: 0.13 to 1.11 ACH) for 134 houses in the Pacific Northwest. In a comparison of 292 houses incorporating energy-efficient features (including measures to reduce air infiltration and provide ventilation heat recovery) with 331 control houses, Parker et al. (1990) found an average of about 0.25 ACH (range: 0.02 to 1.63 ACH) for the energy-efficient houses versus 0.49 (range: 0.05 to 1.63 ACH) for the control. Ek et al. (1990) found an average of 0.5 ACH (range: 0.26 to 1.09) for 93 double-wide manufactured homes also in the Pacific Northwest. Canadian housing stock has been characterized by Yuill and Comeau (1989) and Riley (1990). While these studies do not represent random samples of North American housing, they indicate the distribution of infiltration rates expected in a group of buildings.

Occupancy influences have not been measured directly and vary widely. Desrochers and Scott (1985) estimated that they add an average of 0.10 to 0.15 ACH to unoccupied values. Kvisgaard and Collet (1990) found that in 16 Danish dwellings, the users on average provided 63% of the total air exchange rate.

Ventilation air requirements for houses in the U.S. have traditionally been met on the assumption that the building envelope is leaky enough that infiltration will suffice. Possible difficulties with this approach include low infiltration when natural forces (temperature difference and wind) are weak, unnecessary energy consumption when such forces are strong, drafts in cold climates, lack of control of ventilation rates to meet changing needs, poor humidity control, potential for interstitial condensation from exfiltration in cold climates or infiltration in hot humid climates, and lack of opportunity to recover the energy used to condition the ventilation air. The solution to these concerns is to have a reasonably tight building envelope and a properly designed and operated mechanical ventilation system.

ASHRAE *Standard* 119 and the National Building Code of Canada (NRCC 1995) encourage the transition to tighter envelope construction. Hamlin (1991) shows a 30% increase in airtightness of tract-built Canadian houses between 1982 and 1989. Also, 82% of the newer houses had natural air exchange rates below 0.3 ACH in March. Yuill (1991) derived a procedure to show the extent to which infiltration contributes toward meeting ventilation air requirements. As a result, the National Building Code of Canada has requirements for mechanical ventilation capability in all new dwelling units.

ASHRAE *Standard* 62 gives ventilation air requirements for houses, essentially 0.35 ACH with at least 15 cfm per occupant. Canadian Standards Association (CSA) *Standard* F326 expands the requirements for residential mechanical ventilation systems to cover air distribution within the house, thermal comfort, minimum temperatures for equipment and ductwork, system controls, pressurization and depressurization of the dwelling, installation requirements, and verification of compliance. Verification can be by design or by test, but the total rate of outside air delivery must be measured.

Mechanical ventilation is being used in houses, especially in energy-efficient housing demonstration programs (Riley 1990, Palmiter et al. 1991). Possible systems can be characterized as local or central; exhaust, supply, or balanced; with forced-air or radiant/ hydronic heating/cooling systems; with or without heat recovery; and with continuous operation or controlled by occupants, demand (i.e., by pollutant sensing), timers or humidity. Note that not all combinations are viable. Various options are described by Fisk et al. (1984), Hekmat et al. (1986), Sibbitt and Hamlin (1991), Palmiter et al. (1991), Yuill et al. (1991), Holton et al. (1997), Sherman and Matson (1997), Reardon and Shaw (1997), and Lubliner et al. (1997).

The simplest systems use bathroom and kitchen fans to augment infiltration. Noise, installed capacity, durability under continuous operation, distribution to all rooms (especially bedrooms), envelope moisture, combustion safety, and energy efficiency issues need to be addressed. Many present bath and kitchen fans are ineffective ventilators because of poor installation and design. However, properly specified and installed exhaust fans can form part of good whole-house ventilation systems and are so specified in some Canadian building codes.

Some central supply systems use a central air-handling unit blower to induce air from the outdoors and distribute it. However, the blower operates intermittently if thermostatically controlled and provides little ventilation in mild weather. Continuous blower operation increases energy consumption. If the blower operates continuously when the heat source is off, the combination of lower mixed air temperature and high air speed can cause cold air drafts. To offset these problems, some systems use electronically commutated blower motors, which allow efficient continuous operation at lower speeds. Some others use a timer to cycle the blower when thermostatic demands are inadequate to cause the blower to operate when needed for ventilation (Rudd 1998).

Central exhaust systems use leakage sites and, in some cases, intentional and controllable openings in the building envelope as the supply. Such systems are suitable for retrofit in existing houses. Energy can be recovered from the exhaust airstream with a heat pump to supplement domestic hot water and/or space heating.

For new houses with tightly constructed envelopes, balanced ventilation with passive heat recovery (air-to-air heat exchangers or heat recovery ventilators) can be appropriate in some climates. Fan-induced supply and exhaust air flows at nearly equal rates over a heat exchanger, where heat and sometimes moisture is transferred between the airstreams. This reduces the energy required to condition the ventilation air by typically 60 to 80% (Cutter 1987). It also reduces the thermal comfort problem that occurs when untempered air is introduced directly into the house. Airflow balance, leakage between streams, biological contamination of wet surfaces, frosting, and first cost are concerns associated with these systems.

The type of ventilation system can be selected based on house leakage class as defined in ASHRAE *Standard* 119. Balanced air-to-air systems with heat recovery are optimal for tight houses (leakage classes A–C). The leakier the house is, the larger is the contribution from infiltration and the less effective is heat recovery ventilation. Tightening the envelope beyond the level of ASHRAE *Standard* 119 may be warranted in extreme climates to better use the heat recovery effect (Sherman and Matson 1997). In mild climates, these systems can also effectively be used in leakage classes D–F. Central exhaust systems should not be used for leakage classes A–C unless special provisions are made for air inlets; otherwise their operation may depressurize the house enough to cause backdrafting through fossil-fueled appliances. Unbalanced systems (either supply or exhaust) are optimal for leakage classes D–F. Ventilation systems are normally not needed for leakage classes G–J, but for those cases in which they are, an unbalanced system is usually the best choice.

Residential Ventilation Zones

For guidance in the selection of residential ventilation systems, Sherman (1995) developed four climatic zones for the United States. These zones are shown in Figure 12 for the continental United States. Alaska is in Zone 1, and Hawaii is in Zone 4.

Zone 1 includes the severe climates of the northern tier of states. A Zone 1 residence that meets airtightness and energy conservation standards probably cannot meet its ventilation needs through infiltration and will require forced (mechanical) ventilation. Zone 2 includes the moderate climates where careful design and construction may allow buildings to simultaneously meet energy standards and ventilation needs through infiltration and mechanical exhaust. The mild climates in Zone 3 allow residences to meet both ASHRAE *Standards* 119 and 62 over a substantial range of airtightness. Zone 4 residences have relatively small energy penalties associated

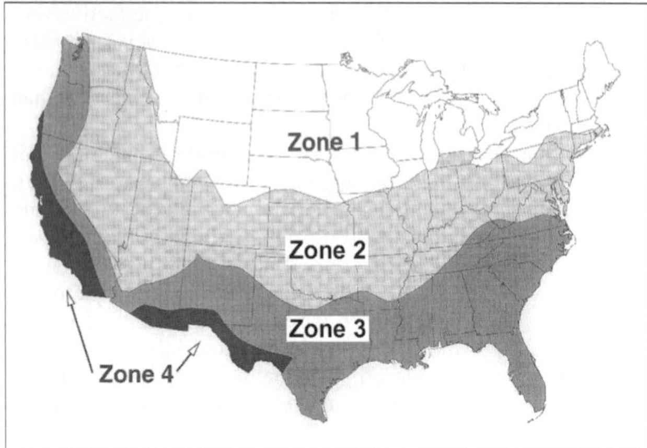

Fig. 12 Airtightness Zones for Residences in the United States
(Sherman 1995)

with infiltration or ventilation. In this zone, natural ventilation is usually preferred to forced ventilation as a technique to supplement infiltration.

RESIDENTIAL VENTILATION REQUIREMENTS

Traditionally, residential ventilation has been provided by natural ventilation and infiltration. Sherman and Matson (1997) showed that most of the older building stock is sufficiently leaky that infiltration alone can meet the minimum requirements of ASHRAE *Standard* 62. Houses built or retrofitted to new standards have substantially tighter envelopes and insufficient infiltration to meet ventilation standards. In most circumstances, concerns over safety, noise, comfort, air quality, and energy minimize occupant use of operable windows. As a result, these houses require supplemental mechanical ventilation to satisfy these standards.

Simply meeting the minimum residential ventilation rates is not always sufficient to adequately dilute all contaminants. For some buildings, such ventilation may not meet the requirements of individuals with allergies or chemical sensitivities or when there are unusual sources. In these cases, source control or extra ventilation is required to manage the contaminant levels. Therefore, especially in single-family dwellings, occupants must be responsible for introducing, monitoring, and controlling the sources in the indoor environment, as well as for operating the dwelling unit to meet their individual needs.

Source Control

When considering how much whole-house ventilation should be supplied, typical and unusual significant sources of indoor pollution need to be controlled. This can be done either by mitigating the source itself or by using local exhaust to extract the contaminants before they can mix into the indoor environment. Typical sources that should be considered include the following:

Clothes Dryers. Clothes dryer exhaust is heavily laden with moisture and laundry by-products. Many moisture problems have been traced to clothes dryers vented indoors. Exhaust from clothes dryers, which is typically about 150 cfm, should be vented directly to the outdoors.

Combustion. Water and carbon dioxide are always emitted during combustion. Other more dangerous compounds can be emitted as well. All these by products should be vented directly outdoors. Venting of combustion appliances should meet all applicable codes, but for buildings with naturally aspirated combustion appliances within the pressure boundary, excessive depressurization due to exhaust systems should be avoided. In addition, a depressurization safety test should be considered, such as described in ASTM *Standard* E 1998 or CGSB *Standard* 51.71.

Carbon monoxide is one of the most pervasive indoor contaminants. It can come from virtually any source of combustion, including automobiles. Because even combustion appliances that meet manufacturers specifications can interact with the building and emit carbon monoxide, at least one carbon monoxide alarm meeting UL *Standard* 2034 should be installed in each dwelling that has combustion appliances (including fireplaces) within the pressure boundary or attached garages.

Garages. Garages contain many sources of contaminants. Doors between garages and occupied space should be well sealed (with gaskets or weatherstripping) and possibly be self-closing. Depressurized sections of HVAC systems, such as air handlers or return ducts, should not be located in garages. If such sections must pass through garages, they should be well sealed.

Particulates. The ventilation system should be designed such that return and outdoor air is filtered before passing through the thermal conditioning components. Pressure drops associated with this filtration should be considered in the design of the air-handling system. Particulate filters or air cleaners should have a minimum efficiency of 60% for 0.125 in. particles, which is equivalent to a MERV 6 designated filter according to ASHRAE *Standard* 52.2.

Outdoor Air. Outdoor air may contain unacceptably high levels of pollutants, including ozone, pollen, carbon monoxide, particulate matter, odors, etc. In such cases, it may be impossible to provide acceptable indoor air quality; increased ventilation rates can actually decrease indoor air quality. In areas in which this problem may be anticipated, controls should be provided to allow the occupants to temporarily reduce the ventilation rate. Air cleaning should be considered for sensitive individuals.

Local Exhaust Ventilation

The single most important source control mechanism in dwellings apart from source elimination is local exhaust. All wet rooms and other spaces designed to allow specific contaminant release should be provided with local exhaust. These spaces include kitchens, utility rooms, bathrooms, and toilets. Workshops, recreation rooms, smoking areas, art studios, and hobby rooms may also require local ventilation and/or air cleaning to remove contaminants generated by the activities involved. If unvented combustion appliances must be used, then rooms with these appliances should meet the same general ventilation requirements for kitchens because such appliances generate significant amounts of moisture even when burning properly.

Mechanical ventilation is the preferred method of providing local ventilation. Normally, it is designed to operate intermittently to exhaust the contaminated air outside when the contaminant is being produced. However, in many circumstances, a continuous and lower-flow-rate exhaust can work as well.

Continuous Local Mechanical Ventilation. A continuously-operating mechanical exhaust is intended to operate without occupant intervention. This exhaust may be part of a balanced mechanical ventilation system. The system should be designed to operate during all hours in which the dwelling is occupied. Override control should be provided if needed. The minimum delivered ventilation should be at least that given in Table 3.

Table 3 Continuous Exhaust Airflow Rates

Application	Continuous Flow	Notes
Kitchen	5 ACH	Based on kitchen volume
Utility room	20 cfm	Not less than 2 room air changes per hour
Bathroom	20 cfm	
Toilet	20 cfm	

Table 4 Intermittent Exhaust Airflow Rates

Application	Continuous Flow	Notes
Kitchen	100 cfm	Vented range hood required if less than 5 kitchen air changes per hour
Utility room	50 cfm	Not less than 2 room air changes per hour
Bathroom	50 cfm	
Toilet	50 cfm	

Table 5 Total Ventilation Air Requirements

Area Based	Occupancy Based
2 cfm/100 ft^2 of floor space	15 cfm per person, based on normal occupancy

Intermittent Local Mechanical Ventilation. An intermittently operating local mechanical exhaust is intended to be operated as needed by the occupant and should be designed with this intent. Shut-off timers, occupancy controls, multiple speed fans, and switching integral with room lighting are helpful, provided they do not impede occupant control. The minimum airflow rating should be at least that given in Table 4, or as mandated by local codes.

Alternative Local Ventilation. Cleaning of recirculated air can sometimes be substituted for local ventilation, if it can be shown to be effective in removing contaminants of concern. Natural ventilation is not generally a suitable method for local ventilation in most climates and spaces. Use of natural ventilation can cause re-entrainment problems when air flows into rather than out of the space, and contaminated exfiltrating air reenters the building. In milder climates, it may be acceptable to use natural ventilation, when the contaminant of concern is related to odor rather than health or safety. Purpose-designed passive exhaust systems have shown acceptable ventilation in European settings and may be considered in lieu of mechanical systems.

Whole-House Ventilation

While control of significant sources of pollution in a dwelling is important, whole-house ventilation may still be needed. Each dwelling should be provided with outdoor air according to Table 5. The rate is the **sum** of the Area-Based and Occupancy-Based columns. Design occupancy can be based on the number of bedrooms as follows: first bedroom, two persons; each additional bedroom, one person.

Natural whole-house ventilation that relies on occupant operation should not be used to make up any part of the minimum total whole-house ventilation. However, because occupancy and sources vary significantly, the capacity to ventilate above minimum rates can be provided by operable openings such as doors and windows.

Continuous whole-house mechanical ventilation systems are intended to operate without occupant intervention. The ventilation effects of multiple local systems may be combined to meet whole-house requirements.

Intermittent whole-house mechanical ventilation systems are intended to be operated automatically and regularly, but not necessarily continuously. The system may consist of supply, exhaust, or balanced mechanical systems. It should be designed so that it can operate automatically based on a timer. A supplementary control mechanism such as a humidistat or indoor air quality sensor may be used.

The effective ventilation rate of an intermittent system is the product of its delivered capacity, its fractional on-time, and the temporal ventilation effectiveness from Table 6. Systems cycling at least once every 3 h may be assumed to have perfect ventilation effectiveness. Multiple systems may be combined, but the combination should only be supply or exhaust values and not both.

Table 6 Ventilation Effectiveness for Intermittent Fans

Fractional On-Time, f	Temporal Ventilation Effectiveness
$f \leq 35\%$	0.33
$35\% < f \leq 60\%$	0.50
$60\% < f \leq 80\%$	0.75
$80\% < f$	1.0

Air Distribution

Ventilation air should be provided to each habitable room through mechanical and natural air distribution. If a room does not have a balanced air supply (or inlet) and return (or exhaust), pathways for transfer air should be provided. These pathways may be door undercuts, transfer ducts with grilles, or simply grilles where ducts are not necessary.

In houses without central air handlers, special provisions to distribute outdoor air may be required. Rooms in which occupants spend many continuous hours, such as bedrooms, may require special consideration.

Selection Principles for Ventilation Systems

Determining the right ventilation approach is part of optimizing the entire building, and it can rarely be done without considering the building envelope, the climate, the needs of the users, and costs. For example, mechanical exhaust systems should not be used in hot, humid climates (because of possible condensation problems) when the building has or is likely to have air conditioning installed. Similarly, mechanical supply ventilation systems should not be used in very cold climates.

Occupant comfort, energy efficiency, ease of use, service life, first and life-cycle cost, value-added features, and indoor environmental quality should be considered when selecting a strategy and system. HVAC (and related) systems can be a potential cause of poor indoor air quality. For example, occupants may not use the ventilation systems as intended if operation results in discomfort (e.g., drafts) or excessive energy use. The resulting lack of ventilation might produce poor indoor air quality. Therefore, careful design, operation, and maintenance is necessary to provide optimum effectiveness.

All exhaust, supply, or air-handler fans have the potential to change the pressure of the living space relative to the outside. High-volume fans, such as the air handler and some cooking exhaust fans, can cause high levels of depressurization, particularly in tightly constructed homes. Considering these effects is essential in design. Depressurization of the living space relative to outside may cause backdrafting of combustion appliances and the migration of contaminants (such as radon or other soil gases, car exhaust, insulation particles, etc.) into the living space. Depressurization can also result in moisture intrusion into building cavities in warm, moist climates, which may cause structural damage and fungal growth. Pressurization of the living space can cause condensation in building cavities in cold climates, resulting in damage to the structural integrity of the home. Excess pressure can best be prevented by balanced ventilation systems and nonleaky duct systems. In addition, adequate pathways must be available for all return air to the air-handling devices.

Occupant activities and operation of fans that exhaust air from the home (including leaky ducts on air conditioners, furnaces, or heat pumps) may produce depressurization of the structure. Several options to address backdrafting concerns include

- Using combustion appliances with isolated (or sealed) combustion systems
- Locating combustion appliances in a ventilated room isolated from depressurized zones by well-sealed partitions
- Installing supply fans to balance or partially balance the exhaust from the zone
- Testing to ensure that depressurization will not be excessive

The system must be designed, built, operated, and maintained in a way that discourages the growth of biological contaminants. Typical precautions include sloping condensate drain pans toward the drain, keeping condensate drains free of obstructions, maintaining cooling coils free of dirt and other obstructions, and checking and eliminating any cause of moisture inside ducts.

Selecting a Whole-House Ventilation System

Whole-house ventilation can be provided through mechanical or natural means or by a combination thereof. Regardless of the approach chosen, it is necessary to consider where the outdoor air comes from, how it enters the house, how it is distributed, and how it leaves the house. Systems that are uncomfortable, expensive to operate, unsafe, noisy, or in other ways unacceptable to the occupants are not likely to be used.

Natural Ventilation. The use of operable windows and other natural ventilation openings is rarely an acceptable means of providing base levels of whole-house ventilation (Wilson and Walker 1992), but when it is, it can be a cost-effective alternative. Operable windows allow the user to control the ventilation by the size of the opening. Windows are present for light and egress and can serve as a ventilation system; most detached dwellings have sufficient operable windows for natural ventilation. Environmental factors may inhibit effective occupant use of the windows. During cold periods, open windows can cause thermal discomfort. Noise or security issues may also reduce the desirability of opening windows. Finally, the uncontrolled nature of the natural ventilation may increase energy use in many climates, but it can be energy-efficient in some mild climates.

Mechanical Ventilation. Whole-house systems may run continuously or intermittently. Intermittent systems must supply more ventilation air and, thus, may cost more to temper the outside air and to run the fans. The system can consist of supply, exhaust, or a balanced combination of the two. Fans that are noisy are likely to be unacceptable to many occupants and will be disabled; noise should be reduced by using quiet fans or by remote mounting. Continuous ventilation requires less power if the system is designed with low-resistance ducting. The electrical energy required for operation should be calculated together with the energy required to temper the outside air introduced through ventilation.

Supply Ventilation. A supply ventilation system usually has a single intake for the ventilation air. This air is distributed through the house either by a dedicated duct system or by the thermal distribution system. The outdoor air may be filtered in a supply ventilation system to remove pollen and dust.

Envelope leakage, exhaust stacks, and flues provide pathways for exhaust air. Supply ventilation can result in indoor pressurization. This is generally unacceptable in very cold climates because the exfiltrating air can cause condensation in the building envelope. However, continuous supply ventilation systems can be designed in conjunction with specific building envelope features to account for this condensation.

Supply ventilation can mitigate radon entry or backdrafting problems and may reduce interstitial moisture problems in hot, humid climates. In temperate or severe climates, the supply air, if delivered directly to rooms without tempering, can cause thermal discomfort or draftiness. A dedicated duct system increases the cost of a supply ventilation system.

Exhaust Ventilation. An exhaust ventilation system usually has a central mechanical exhaust. Air enters the house though envelope leakage, open windows, or designed inlets. Because the air intake is dispersed, thermal discomfort is less than that with supply ventilation. An exhaust air heat pump can recover energy, but it may not be economical in many climates. The exhaust fan depressurizes the house, which can aggravate radon or backdrafting problems. Because of potential interstitial space moisture problems, exhaust ventilation may be unacceptable in hot, moist climates when the indoor air is mechanically cooled.

Although air intake through building leaks reduces the particulate concentration somewhat, filtration of outdoor air is not generally possible with exhaust ventilation. In unusually tight houses, envelope leakage may be insufficient to provide enough supply air, so designed intakes may be needed.

Balanced Ventilation. In a balanced mechanical ventilation system, there is usually a mechanical exhaust either centrally located or ducted from locations likely to have high contaminant levels. There is a single outside air intake for the ventilation air, which is then dispersed through the house. The systems are designed to produce equal supply and exhaust flows. With equal flows, the system creates neither pressurization nor depressurization, so problems associated with house pressures are reduced. It is also possible to use slight positive pressurization to reduce infiltration. Filtration and tempering of the incoming air can be accomplished at the central unit.

Most balanced systems feature either sensible heat recovery or total heat recovery. Sensible heat recovery systems provide tempering of the incoming air using the exiting air as a source for heat (or cooling). Total heat recovery systems provide both sensible and latent tempering of the incoming air.

Because of its energy recovery properties, a balanced system becomes more attractive as total space-conditioning costs increase, such as in severe climates. The balanced system is initially the most costly of the three systems, but total operating costs may be less, particularly if high-efficiency fan/motor assemblies are used.

Selecting a Local Ventilation System

The two most common methods for providing local ventilation to damp rooms are natural ventilation through an operable window and mechanical exhaust fans. These methods have the same design issues as whole-house systems. Balanced or supply-only systems should not be used for local ventilation to reduce spreading of moisture and contaminants through the house.

SIMPLIFIED MODELS OF RESIDENTIAL VENTILATION AND INFILTRATION

This section describes several calculation procedures, ranging from simple estimation techniques to more physical models. Orme (1999) provides a more thorough review of simplified models. The air exchange rate of a building cannot be reliably deduced from the building's construction or age or from a simple visual inspection. Some measurement is necessary, such as a pressurization test of envelope airtightness or a detailed quantification of the leakage sites and their magnitude. The air exchange rate of a building may be calculated given (1) the location and leakage function for every opening in the building envelope and between major building zones, (2) the wind pressure coefficients over the building envelope, and (3) any mechanical ventilation airflow rates. These inputs are generally unavailable for all except very simple structures or extremely well studied buildings. Therefore, assumptions as to their values must be made. The appropriateness of these assumptions determines the accuracy of predictions of air exchange rates.

Empirical Models

These models of residential infiltration are based on statistical fits of infiltration rate data for specific houses. They use pressurization test results to account for house airtightness and take the form of simple relations between infiltration rate, an airtightness rating, and, in most cases, weather conditions. Empirical models account for envelope infiltration only and do not deal with intentional ventilation. In one approach, the calculated air exchange rate at 0.20 in. of water based on a pressurization test is simply divided by a constant approximately equal to 20 (Sherman 1987). This technique does not account for the effect of infiltration driving mechanisms on air exchange. Empirical models that do account for weather effects have been developed by Reeves et al. (1979), Kronvall (1980), and Shaw (1981).

The latter two models account for building air leakage using the values of c and n from Equation (32). The only other inputs required are wind speed and temperature difference. Such empirical models predict long-term (one-week) infiltration rates very well in the houses from which they were developed; they do not, however, work as well in other houses due to the building-specific nature of leakage distribution, wind pressure, and internal partitioning. Persily and Linteris (1983) and Persily (1986) show comparisons between measured and predicted house infiltration rates for these and other models. The average long-term differences between measurements and predictions are generally on the order of 40%, although individual predictions can be off by 100% or more.

Multizone Models

Multicell models of air exchange treat buildings as a series of interconnected zones and assume that the air within each zone is well mixed. Several such models have been developed by Allard and Herrlin (1989), Etheridge and Alexander (1980), Liddament and Allen (1983), Walton (1984, 1989), Herrlin (1985), and Feustel and Raynor-Hoosen (1990). They are all based on a mass balance for each zone of the building. These mass balances are used to solve for interior static pressures within the building by requiring that the inflows and outflows for each zone balance to zero. The models require the user to input a location and leakage function for every opening in the building envelope and in relevant interior partitions, a value for the wind pressure coefficient C_p at the location of each building envelope leakage site, temperatures for each zone, and any mechanical ventilation airflow rates. Such detailed information is difficult to obtain for a building. Wind pressure coefficient data in the literature, air leakage measurement results from the building or its components, and air leakage data from the literature can be used. These models not only solve for whole-building and individual zone air exchange rates, but also determine airflow rates and pressure differences between zones. These interzone airflow rates are useful for predicting pollutant transport within buildings with well mixed zones. Caution should be used when applying them to the prediction of smoke movement patterns in the event of a fire. Multizone models have the advantage of being able to model very complex representations of buildings by using personal computers; however, determining the correct inputs to these models is difficult. As a result of these uncertainties, multizone models are best used to bound a solution rather than determining an absolute solution. Monte Carlo simulation coupled with the multizone modeling is a useful technique to determine these bounds, if the probability distribution for the uncertain parameters can be defined.

Single-Zone Models

Several procedures have been developed to calculate building air exchange rates that are based on physical models of the building interior as a single zone. These single-zone models are only appropriate to buildings with no internal resistance to airflow and are therefore inappropriate to large, multizone buildings. Models of this type have been developed by the Institute of Gas Technology (IGT) (Cole et al. 1980), the Building Research Establishment (Warren and Webb 1980), the Lawrence Berkeley National Laboratory (LBNL) (Sherman and Grimsrud 1980), and the University of Alberta (AIM-2) (Walker and Wilson 1998). These models have exhibited average errors on the order of 40% for many measurements on groups of houses and can be off by 100% in individual cases (Persily 1986, Walker and Wilson 1998). The following section on Residential Calculation Examples uses a basic model and an enhanced model (Walker and Wilson 1998, Palmiter and Bond 1994, Hamlin and Pushka 1994, CHBA 1994, Bradley 1993).

The **basic model** uses the effective air leakage area A_L at 0.016 in. of water, which can be obtained from a whole-building pressurization test. If a test value is not available, the data in Table 1 can be used to estimate the air leakage area of the building. To obtain the building's total air leakage area, multiply the overall dimensions or number of occurrences of each building component by the appropriate table entry. The sum of the resulting products is the total building air leakage area. See Table 2 for an example.

The **enhanced model** uses pressurization test results to characterize house air leakage through the leakage coefficient c and the pressure exponent n. The enhanced model improves on the basic model by using a power law to represent the envelope leakage, including a flue as a separate leakage site, and having separate wind effects for houses with crawl spaces or slab/basement foundations.

For both models, the user must input wind speed, temperature difference, information on distribution of leakage over the building envelope, a wind shelter (or local shielding for the basic model) parameter, and a terrain coefficient. The predictive accuracy of the enhanced model can be very good, typically ±10% when the parameters are well known for the building in question (Walker and Wilson 1998, Palmiter and Bond 1994, Sherman and Modera 1986). All these single-zone models are sensitive to the values of the inputs, which are quite difficult to determine.

Superposition of Wind and Stack Effects

Simplified physical models of infiltration solve the problem of two natural driving forces (wind and stack) separately and then combine them in a process called **superposition**. Superposition is necessary because each physical process can affect the internal and external pressures on the structure, which can cause interactions between physical processes that are otherwise independent. An exact solution is impossible because detailed properties of all the building leaks are unknown and because leakage is a nonlinear process. For this reason, most modelers have developed a simplified superposition process to combine stack and wind effects. Sherman (1992a) compares various superposition procedures and derives a generalized superposition equation that involves some simple leakage distribution parameters. He shows that the result is always subadditive. Typically only 35% of the infiltration due to the smaller effect can be added to the larger effect. Depending on the details, that percentage could go as high as 85% or as low as zero. Walker and Wilson (1993) compared several superposition techniques to measured data. Sherman, as well as Walker and Wilson, found quadrature [shown in Equation (39)] to be a robust superposition technique:

$$Q = \sqrt{Q_s^2 + Q_w^2} \tag{39}$$

The following sections discuss how this superposition is combined with the calculation of the wind and stack flows to determine the total flow.

Residential Calculation Examples

Basic Model. The following calculations are based on the LBNL model (Sherman and Grimsrud 1980), which uses the effective air leakage area at 0.016 in. of water. This leakage area can be obtained from a whole-building pressurization test. Using the effective air leakage area, the airflow rate due to infiltration is calculated according to

$$Q = A_L\sqrt{C_s \Delta t + C_w U^2} \tag{40}$$

where

Q = airflow rate, cfm
A_L = effective air leakage area, in²
C_s = stack coefficient, cfm²/(in⁴·°F)

Δt = average indoor-outdoor temperature difference for time interval of calculation, °F
C_w = wind coefficient, cfm²/(in⁴·mph²)
U = average wind speed measured at local weather station for time interval of calculation, mph

Table 7 presents values of C_s for one-, two-, and three-story houses. The value of the wind coefficient C_w depends on the local shelter class of the building (described in Table 8) and the building height. Table 9 presents values of C_w for one-, two-, and three-story houses in shelter classes 1 through 5. In calculating the values in Table 7 and Table 9, the following assumptions were made regarding input to the basic model:

- Terrain used for converting meteorological to local wind speeds is that of a rural area with scattered obstacles
- $R = 0.5$ (half of the building leakage in the walls)
- $X = 0$ (equal amounts of leakage in the floor and ceiling)
- Heights of one-, two-, and three-story buildings = 8, 16, and 24 ft, respectively

Example 1. Estimate the infiltration at design conditions for a two-story house in Lincoln, Nebraska. The house has an effective air leakage area of 77 in² and a volume of 12,000 ft³, and the predominant wind is perpendicular to the street (shelter class 3). The indoor air temperature is 68 °F.

Solution: The 99% design temperature for Lincoln is −2°F. Assume a design wind speed of 15 mph. From Equation (40), with C_s = 0.0299 from Table 7 and C_w = 0.0086 from Table 9, the airflow rate due to infiltration is

$$Q = 77\sqrt{(0.0299)(70) + (0.0086)(15^2)}$$

$$= 155 \text{ cfm} = 9300 \text{ ft}^3/\text{h}$$

From Equation (2), the air exchange rate I is equal to Q divided by the building volume:

$$I = (9300 \text{ ft}^3/\text{h})/12{,}000 \text{ ft}^3$$

$$= 0.78 \text{ h}^{-1} = 0.78 \text{ ACH}$$

Example 2. Calculate the average infiltration during a one-week period in January for a one-story house in Portland, Oregon. During this period, the average indoor-outdoor temperature difference is 30°F, and the average wind speed is 6 mph. The house has a volume of 9000 ft³ and an effective air leakage area of 107 in², and it is located in an area with buildings and trees within 30 ft in most directions (shelter class 4).

Solution: From Equation (40), the airflow rate due to infiltration is

$$Q = 107\sqrt{(0.0150)(30) + (0.0039)(6^2)}$$

$$= 82.2 \text{ cfm} = 4930 \text{ ft}^3/\text{h}$$

The air exchange rate is therefore

$$I = 4930/9000 = 0.55 \text{ h}^{-1} = 0.55 \text{ ACH}$$

Example 3. Estimate the average infiltration over the heating season in a two-story house with a volume of 11,000 ft³ and the air leakage area calculated in Table 2 (131 in²). The house is located on a lot with several large trees but no other close buildings (shelter class 3). The average wind speed during the heating season is 7 mph, while the average indoor-outdoor temperature difference is 36°F.

Solution: From Equation (40), the airflow rate due to infiltration is

$$Q = 131\sqrt{(0.0299)(36) + (0.0086)(7^2)}$$

$$= 160 \text{ cfm} = 9620 \text{ ft}^3/\text{h}$$

The average air exchange rate is therefore

$$I = 9620/11{,}000 = 0.87 \text{ h}^{-1} = 0.87 \text{ ACH}$$

Enhanced Model. This section presents a simple, single-zone approach to calculating air infiltration rates in houses based on the AIM-2 model (Walker and Wilson 1998). The airflow rate due to infiltration is calculated using

$$Q_s = cC_s\Delta t^n \tag{41}$$

$$Q_w = cC_w(sU)^{2n} \tag{42}$$

where

Q_s = stack airflow rate, cfm
Q_w = wind airflow rate, cfm
c = flow coefficient, cfm/(in. of water)^n
C_s = stack coefficient, (in. of water)/°F)^n
C_w = wind coefficient, (in. of water/mph²)^n
s = shelter factor

In calculating the tabulated values of C_s, C_w, and s, the following assumptions were made:

- Each story is 8 ft high.
- The flue is 6 in. in diameter and reaches 6 ft above the upper ceiling.
- The flue is unsheltered.
- Half of the envelope leakage (not including the flue) is in the walls and one-quarter each is at the floor and ceiling, respectively.
- $n = 0.67$

Using typical values for terrain factors, house height, and wind speed measurement height, the windspeed multiplier G (given in Table 10) uses a relationship based on equations found in Chapter 16.

Example 4. Estimate the infiltration at design conditions for a two-story slab-on-grade house with a flue in Lincoln, Nebraska. The house has a flow coefficient of c = 2.6 cfm/(in. of water)^n and a pressure expo-

Table 7 Basic Model Stack Coefficient C_s

	House Height (Stories)		
	One	Two	Three
Stack coefficient	0.0150	0.0299	0.0449

Table 8 Local Shelter Classes

Shelter Class	Description
1	No obstructions or local shielding
2	Typical shelter for an isolated rural house
3	Typical shelter caused by other buildings across the street from the building under study
4	Typical shelter for urban buildings on larger lots where sheltering obstacles are more than one building height away
5	Typical shelter produced by buildings or other structures that are immediately adjacent (closer than one house height): e.g., neighboring houses on the same side of the street, trees, bushes, etc.

Table 9 Basic Model Wind Coefficient C_w

Shelter Class	House Height (Stories)		
	One	Two	Three
1	0.0119	0.0157	0.0184
2	0.0092	0.0121	0.0143
3	0.0065	0.0086	0.0101
4	0.0039	0.0051	0.0060
5	0.0012	0.0016	0.0018

Table 10 Enhanced Model Wind Speed Multiplier G

	House Height (Stories)		
	One	Two	Three
Wind speed multiplier G	0.48	0.59	0.67

Table 11 Enhanced Model Shelter Factor s

Shelter Class	No Flue	One Story with Flue	Two Story with Flue	Three Story with Flue
1	1.00	1.10	1.07	1.06
2	0.90	1.02	0.98	0.97
3	0.70	0.86	0.81	0.79
4	0.50	0.70	0.64	0.61
5	0.30	0.54	0.47	0.43

Table 12 Enhanced Model Stack and Wind Coefficients

	One Story		Two Story		Three Story	
	No Flue	With Flue	No Flue	With Flue	No Flue	With Flue
C_s	1.46	1.87	2.13	2.41	2.68	2.92
C_w for basement/ slab	2.14	1.95	2.34	2.14	2.34	2.29
C_w for crawl space	1.75	1.75	1.95	1.95	2.07	2.11

nent of $n = 0.67$ (this corresponds to effective leakage area of 77 in^2 at 0.016 in. of water). The building volume is 12,000 ft^3. The 97.5% design temperature is –2°F, and the design wind speed is 15 mph.

Solution: For a slab-on-grade two-story house with a flue, Table 12 gives C_s = 2.41 [(in. of water)/°F]^n and C_w = 2.14 [(in. of water)/mph^2]^n. The house is maintained at 68°F indoors. The building wind speed is determined by taking the design wind speed and multiplying by the wind speed multiplier G from Table 10:

$$U = GU_{met} = 0.59(15) = 8.9 \text{ mph}$$

From Table 8, the shelter class for a typical urban house is 4. Table 11 gives the shelter factor for a two-story house with a flue and shelter class 4 as $s = 0.64$. The stack flow is calculated using Equation (41):

$$Q_s = (2.6)(2.41)[68 - (-2)]^{0.67} = 108 \text{ cfm}$$

The wind flow is calculated using Equation (42):

$$Q_w = (2.6)(2.14)(0.64 \times 8.9)^{1.34} = 57 \text{ cfm}$$

Substituting Q_s and Q_w into Equation (39) gives Q = 122 cfm = 7300 ft^3/h. From Equation (2), the air exchange rate I is equal to Q divided by the building volume:

$$I = (7300 \text{ ft}^3/\text{h})/12{,}000 \text{ ft}^3$$
$$= 0.61 \text{ h}^{-1} = 0.61 \text{ ACH}$$

Example 5. Estimate the average infiltration over a one-week period for a single-story crawl space house in Redmond, Washington. The house has a flow coefficient of c = 4.1 cfm/(in. of water)^n and a pressure exponent of $n = 0.6$ (this corresponds to effective leakage area of 107 in^2 at 0.016 in. of water). The building volume is 9000 ft^3. During this period, the average indoor-outdoor temperature difference is 29°F, and the wind speed is 6 mph. The house is electrically heated and has no flue.

Solution: For a single-story house with no flue, C_s = 1.46 [(in. of water)/F]^n. For a crawl space, C_w = 1.75 [(in. of water)/mph^2]^n. From Table 10, for a one-story house, $G = 0.48$.

$$U = GU_{met} = 0.48(6) = 2.9 \text{ mph}$$

Table 11 gives shelter factor $s = 0.50$ for a house with no flue and shelter class 4. The stack flow is calculated using Equation (41):

$$Q_s = (4.1)(1.46)(29)^{0.6} = 45 \text{ cfm}$$

The wind flow is calculated using Equation (42):

$$Q_w = (4.1)(1.75)(0.50 \times 2.9)^{1.2} = 11.2 \text{ cfm}$$

Substituting Q_s and Q_w into Equation (39) gives Q = 46 cfm = 2800 ft^3/h. From Equation (2), the air exchange rate I is equal to Q divided by the building volume:

$$I = (2800 \text{ ft}^3/\text{h})/9000 \text{ ft}^3$$
$$= 0.31 \text{ h}^{-1} = 0.31 \text{ ACH}$$

Example 6. Estimate the infiltration for a three-story house in San Francisco, California. The house has a flow coefficient of c = 5.2 cfm/(in. of water)^n and a pressure exponent of $n = 0.67$ (this corresponds to effective leakage area of 155 in^2 at 0.016 in. of water). The building volume is 14,200 ft^3. The indoor-outdoor temperature difference is 9°F and the wind speed is 10 mph. The house has a flue and a crawl space foundation.

Solution: For a three-story house with a flue, C_s = 2.92 [(in. of water)/°F]^n. For a crawl space, C_w = 2.11 [(in. of water)/mph^2]^n. From Table 10, for a three-story house, $G = 0.67$.

$$U = GU_{met} = 0.67(10) = 6.7 \text{ mph}$$

The prevailing wind blows along the row of houses parallel to the street, so the house has a shelter class of 5. Table 11 gives the shelter factor for a three-story house with a flue and shelter class 5 as $s = 0.43$.

$$Q_s = (5.2)(2.92)(9)^{0.67} = 66 \text{ cfm}$$

$$Q_w = (5.2)(2.11)(0.43 \times 6.7)^{1.34} = 45 \text{ cfm}$$

Substituting Q_s and Q_w in Equation (39) gives Q = 80 cfm = 4800 ft^3/h.

$$I = (4800 \text{ ft}^3/\text{h})/(14{,}200)\text{ft}^3$$
$$= 0.34 \text{ h}^{-1} = 0.34 \text{ ACH}$$

Combining Residential Infiltration and Mechanical Ventilation

Significant infiltration and mechanical ventilation often occur simultaneously in residences. The pressure difference from Equation (24) can be used for each building leak, and the flow network (including mechanical ventilation) for the building can be solved to find the flow through all the leaks while accounting for the effect of the mechanical ventilation. However, for simplified models, the natural infiltration and mechanical ventilation are usually determined separately and require a superposition method to combine the flow rates.

Sherman (1992) compares various superposition procedures and derives a generalized superposition equation that involves some simple leakage distribution parameters. He shows that the result is always subadditive. For small unbalanced fans, typically only half the flow contributes to the total, but this fraction can be anywhere between 0% and 100% depending on the leakage distribution. When the fan flow becomes large, infiltration may be ignored.

In special cases when the leakage distribution is known and highly skewed, it may be necessary to work through the superposition method in more detail. For example, in a wind-dominated situation, a supply fan will have a much bigger effect than an exhaust fan on changing the total ventilation rate; the same situation holds for houses having high neutral levels in cold climates. For the general case, when the details are not known or can be assumed to be broad and typical, the following superposition gives good results:

$$Q_{comb} = Q_{bal} + \sqrt{Q_{unbal}^2 + Q_{infiltration}^2} \qquad (43)$$

NONRESIDENTIAL AIR LEAKAGE

Commercial Building Envelope Leakage

There is currently only one industry standard for the measurement of envelope leakage in tall buildings: CGSB *Standard* 149.15. It utilizes the building's own air handlers. ASTM *Standard* E779 and CGSB *Standard* 149.10 are intended for small detached buildings such as houses and provide no guidelines for dealing with problems arising in tall buildings, such as stack and wind effects. Tall buildings require refinement and extensions of established procedures because of obstacles to accurate measurement not present in small buildings, including large envelope leakage area, interfloor leakage, vertical shafts, and large wind and stack pressures. Recent work by Bahnfleth et al. (1999) also shows how to deal with some of these issues.

The building envelopes of large commercial buildings are often thought to be quite airtight. The National Association of Architectural Metal Manufacturers specifies a maximum leakage per unit of exterior wall area of 0.060 cfm/ft^2 at a pressure difference of 0.30 in. of water exclusive of leakage through operable windows. Tamura and Shaw (1976a) found that, assuming a flow exponent n of 0.65 in Equation (32), air leakage measurements in eight Canadian office buildings with sealed windows ranged from 0.120 to 0.480 cfm/ft^2. Persily and Grot (1986) ran whole-building pressurization tests in large office buildings that showed that pressurization airflow rate divided by building volume is relatively low compared to that of houses. However, if these airflow rates are normalized by building envelope area instead of by volume, the results indicate envelope airtightness levels similar to those in typical American houses. In a study of eight U.S. office buildings, Persily and Grot (1986) found air leakage ranging from 0.213 to 1.028 cfm/ft^2 at 0.30 in. of water. This means that office building envelopes are leakier than expected. Typical air leakage values per unit wall area at 0.30 in. of water are 0.10, 0.30, and 0.60 cfm/ft^2 for tight, average, and leaky walls, respectively (Tamura and Shaw 1976a).

Grot and Persily (1986) also found that eight recently constructed office buildings had infiltration rates ranging from 0.1 to 0.6 ACH with no outdoor air intake. The infiltration rates of these buildings exhibited varying degrees of weather dependence, generally much lower than that measured in houses.

Air Leakage Through Internal Partitions

In large buildings, the air leakage associated with internal partitions becomes very important. Elevator, stair, and service shaft walls; floors; and other interior partitions are the major separations of concern in these buildings. Their leakage characteristics are needed to determine infiltration through exterior walls and airflow patterns within a building. These internal resistances are also important in the event of a fire to predict smoke movement patterns and evaluate smoke management systems.

Table 13 gives air leakage areas (calculated at 0.30 in. of water with C_D = 0.65) for different internal partitions of commercial buildings (Klote and Milke 1992). Figure 13 presents examples of measured air leakage rates of elevator shaft walls (Tamura and Shaw 1976b), the type of data used to derive the values in Table 13.

Table 13 Air Leakage Areas for Internal Partitions in Commercial Buildings (at 0.30 in. of water and C_D = 0.65)

Construction Element	Wall Tightness	Area Ratio
		A_L/A_w
Stairwell walls	Tight	0.14×10^{-4}
	Average	0.11×10^{-3}
	Loose	0.35×10^{-3}
Elevator shaft walls	Tight	0.18×10^{-3}
	Average	0.84×10^{-3}
	Loose	0.18×10^{-2}
		A_L/A_f
Floors	Average	0.52×10^{-4}

A_L = air leakage area A_w = wall area A_f = floor area

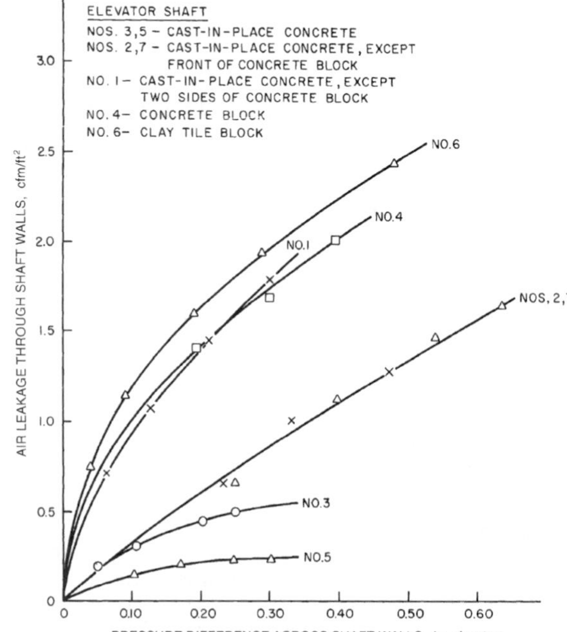

Fig. 13 Air Leakage Rates of Elevator Shaft Walls

Chapter 51 of the 1999 *ASHRAE Handbook—Applications* should be consulted for performance models and applications of smoke management systems.

Leakage openings at the top of elevator shafts are equivalent to orifice areas of 620 to 1550 in^2. Air leakage rates through stair shaft and elevator doors are shown in Figure 14 as a function of average crack width around the door. The air leakage areas associated with other openings within commercial buildings are also important for air movement calculations. These include interior doors and partitions, suspended ceilings in buildings where the space above the ceiling is used in the air distribution system, and other components of the air distribution system.

Air Leakage Through Exterior Doors

Door infiltration depends on the type of door, room, and building. In residences and small buildings where doors are used infrequently, the air exchange associated with a door can be estimated based on air leakage through cracks between the door and the frame. A frequently opened single door, as in a small retail store, has a much larger amount of airflow than a closed door.

Air Leakage Through Automatic Doors

Automatic doors are a major source of air leakage in buildings. They are normally installed in locations where large numbers of

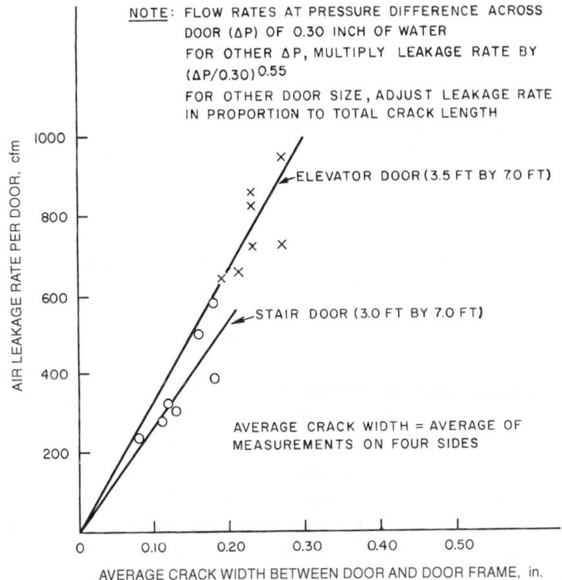

Fig. 14 Air Leakage Rate of Door Versus Average Crack Width

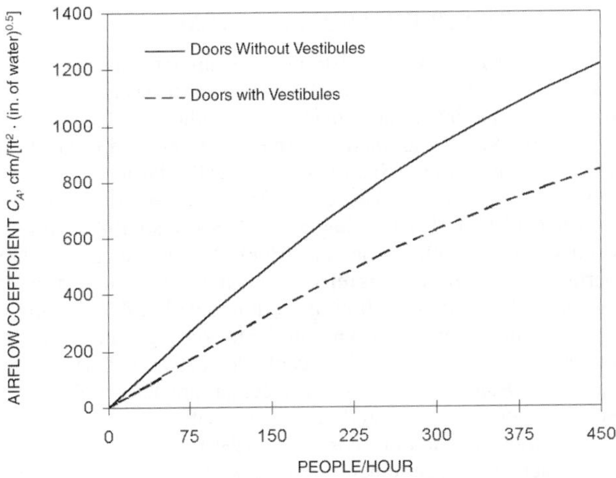

Fig. 15 Airflow Coefficient for Automatic Doors

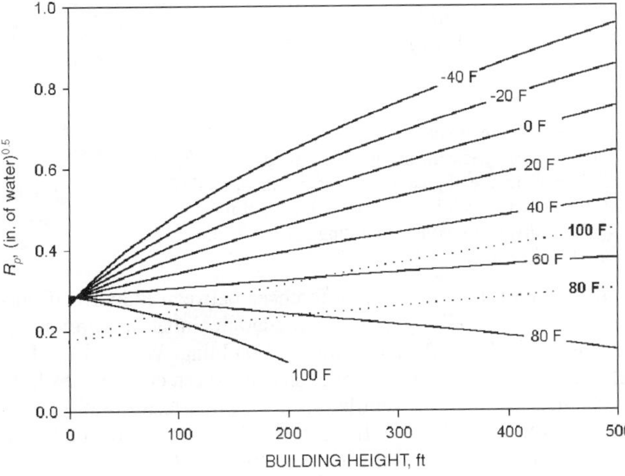

Fig. 16 Pressure Factor for Automatic Doors

people use the doors. They stay open longer with each use than manual doors. The air leakage through automatic doors can be reduced by the installation of a vestibule. However, pairs of automatic doors on the inside and outside of a vestibule normally have overlapping open periods, even when used by only one person at a time. Therefore, it is important that designers take into account the airflow through automatic doors when calculating the heating and cooling loads in the spaces next to them.

To calculate the average airflow rate through an automatic door, the designer must take into account the area of the door, the pressure difference across it, the discharge coefficient of the door when it is open, and the fraction of time that it is open. Obtaining the discharge coefficient is complicated by the fact that it changes as the door opens and closes.

To simplify this calculation, Figure 15 has been developed (Yuill 1996) to combine the discharge coefficients of doors as they open and close with the fraction of time that doors are open at a particular level of use. This figure presents an overall airflow coefficient as a function of the people using a door per hour. To obtain the average infiltration rate through an automatic door, the user must multiply this coefficient by the opening area of the door and by the square root of the pressure difference between the outdoor air and the indoor air at the location of the door. The pressure difference across a door in a building depends on the wind pressure on the building, the stack effect due to the indoor-outdoor temperature difference, and the effect of the operation of the air-handling system. It also depends on the leakage characteristics of the exterior walls of the building and of internal partitions.

Two simple methods are presented here. The first method uses simplifying assumptions to determine design values for R_p, the square root of the pressure difference across the automatic door, given in Figure 16. The second method requires explicit calculation of envelope pressures.

In Figure 16, the airflows shown for ambient temperatures of 80 and 100°F, represented by dotted lines, are outward flows. They intercept the vertical axis at a lower point than the other lines because the wind pressure coefficients on the downwind face of the building (where the greatest outward flows will occur) are lower than on the upward face. In many buildings, the pressure in the building is controlled by varying the flow rate through the return fan(s) or by controlling the relief air dampers. These systems are usually set to maintain a pressure above ambient in the lobby. Subtracting the interior pressure maintained in the lobby from the wind pressure gives the net pressure for estimating airflow through the door.

Method 1. For the first method, the infiltration rate through the automatic door is given by

$$Q = C_A A R_p \tag{44}$$

where

Q = airflow rate, cfm
C_A = airflow coefficient from Figure 15, cfm/[ft²(in. of water)$^{0.5}$]
A = area of the door opening, ft²
R_p = pressure factor from Figure 16, (in. of water)$^{0.5}$

Method 2. The airflow Q is given by

$$Q = C_A A \sqrt{\Delta p} \tag{45}$$

where

Q = airflow rate, cfm
C_A = airflow coefficient from Figure 15, cfm/[ft²(in. of water)$^{0.5}$]
A = area of the door opening, ft²
Δp = pressure difference across the door, in. of water

To find Δp, it is necessary to find the pressure differential due to wind and that due to the stack effect. In order to give the largest possible pressure difference across the door, there are no interactions between the two natural pressures:

$$\Delta p = p_w - \Delta p_s \quad (46)$$

where

p_w = wind-induced surface pressure relative to static pressure, in. of water
Δp_s = pressure difference due to stack effect, in. of water

Example Calculations

It is desired to find the maximum possible infiltration through an automatic door located on the ground floor of a 20-story building. The area of the door is 36 in. × 84 in. = 3024 in^2 = 21 ft^2. Each floor is 13 ft high. Approximately 300 people per hour pass through the door. The design wind conditions are 15 mph, the indoor temperature is 70°F, and the outdoor temperature is 20°F. The airflow coefficient from Figure 15 (using the line for doors without vestibules) is approximately 920 cfm/[ft^2(in. of water)$^{0.5}$].

Method 1:

The pressure factor from Figure 16 is 0.5 (in. of water)$^{0.5}$. Equation (44) gives the door flow as

$$Q = 920(21)0.5 = 9660 \text{ cfm}$$

Method 2:

The worst possible case for the wind surface pressure coefficient C_p at any point and in any position on the ground floor of the building is inferred from figures in Chapter 16 to be about 0.75. Using this in Equation (18), together with the specified wind speed, results in p_w = 0.082 in. of water. Assume that H is one-half the height of the door (42 in.). In order to have the maximum pressure across the door, assume the neutral pressure plane is located halfway up the building such that

$$H_{NPL} = \frac{1}{2}(20 \text{ stories})\frac{13 \text{ ft}}{\text{story}} = 130 \text{ ft}$$

Substituting these values into Equation (17) gives Δp_s = −0.19 in. of water. This is the maximum stack pressure difference given no internal resistance to airflow. To find the actual stack pressure difference, it is necessary to multiply this by a draft coefficient. We will assume that this coefficient is 0.9, which is the highest value that has been found for tall buildings. Therefore, Δp_s = 0.9(−0.19 in. of water) = −0.17 in. of water. The total pressure is then Δp = 0.082 − (−0.17) = 0.252 in. of water. Substituting into Equation (45),

$$Q = 920(21)\sqrt{0.252} = 9700 \text{ cfm}$$

If the building had a vestibule, the airflow coefficient would be read from Figure 15 using the line for doors with vestibules, and it would be approximately 626 cfm/[ft^2(in. of water)$^{0.5}$], reducing the airflow to 6600 cfm into the building.

NONRESIDENTIAL VENTILATION

Commercial and institutional building ventilation systems are typically designed to provide a slight pressurization to minimize infiltration. This pressurization is achieved by having the outside or makeup airflow rate higher than the exhaust or relief airflow rate. In these buildings, infiltration is usually neglected except in areas such as lobbies, where infiltration can be important due to doors. As discussed in the section on Driving Mechanisms for Ventilation and Infiltration, wind and the stack effect can also cause significant infiltration and exfiltration. Ventilation airflow rates for commercial and institutional buildings are typically determined using procedures in ASHRAE *Standard* 62, Ventilation for Acceptable Indoor Air Quality. In these procedures for designing forced ventilation systems, no credit is given for infiltration. However, weather-driven pressure differentials may be significant and need to be considered when designing the ventilation system.

ASHRAE *Standard* 62 includes two procedures for obtaining acceptable indoor air quality: the Ventilation Rate Procedure and the Indoor Air Quality Procedure. The Ventilation Rate Procedure is by far the more commonly used.

Ventilation Rate Procedure

In the Ventilation Rate Procedure, the design ventilation rate is determined based on a table of minimum ventilation requirements for different space types. These requirements are expressed as an outdoor airflow rate per occupant or per unit floor area, depending on the space type. These ventilation rates are based on air pollutants generated by people, activities, and building materials and furnishings.

The HVAC designer faces several challenges in designing an air distribution system to deliver outdoor air to the occupants of a building. The first is to determine whether the outdoor air is acceptable for use and to design a system for cleaning the air if it is not acceptable. A second goal is to design an air intake and distribution system that will deliver the required level of outdoor air to the occupied portions of the building. This outdoor air must be delivered not only at the design conditions, but throughout the year. The task is complicated by weather-related variations in indoor-outdoor pressure difference. Other complications include pressure variations due to building components such as exhaust fans or dirty filters, and probably most significantly by supply flow variations associated with the operation of variable air volume (VAV) systems (Mumma and Wong 1990, Janu et al. 1995). After the design level of outdoor air is brought into the building, it must then be delivered to the occupants. This issue is related to the discussion in the section on Air Change Effectiveness presented earlier in this chapter.

TRACER GAS MEASUREMENTS

The only reliable way to determine the air exchange rate of an existing building is to measure it. Several tracer gas measurement procedures exist (including a standard test method: ASTM *Standard* E 741), all involving an inert or nonreactive gas used to label the indoor air (Hunt 1980; Sherman et al. 1980; Harrje et al. 1981; Lagus and Persily 1985; Dietz et al. 1986; Charlesworth 1988; Persily 1988; Fisk et al. 1989; Lagus 1989; Sherman 1989a, 1989b; Fortmann et al. 1990; Harrje et al. 1990; Persily and Axley 1990; Sherman 1990). The tracer is released into the building in a specified manner, and the concentration of the tracer within the building is monitored and related to the building's air exchange rate. A variety of tracer gases and associated concentration detection devices have been used. Desirable qualities of a tracer gas are detectability, nonreactivity, nontoxicity, neutral buoyancy, relatively low concentration in ambient air, and low cost (Hunt 1980).

All tracer gas measurement techniques are based on a mass balance of the tracer gas within the building. Assuming the outdoor concentration is zero and the indoor air is well mixed, this total balance takes the following form:

$$V\left(\frac{dC}{d\theta}\right) = F(\theta) - Q(\theta)C(\theta) \quad (47)$$

where

V = volume of space being tested, ft^3
$C(\theta)$ = tracer gas concentration at time θ
$dC/d\theta$ = time rate of change of concentration, min^{-1}
$F(\theta)$ = tracer gas injection rate at time θ, cfm
$Q(\theta)$ = airflow rate out of building at time θ, cfm
θ = time, min

In Equation (47), density differences between indoor and outdoor air are generally ignored for moderate climates; therefore, Q also refers to the airflow rate into the building. While Q is often referred to as the infiltration rate, any measurement includes both mechanical and natural ventilation in addition to infiltration. The ratio of Q to the volume V being tested has units of 1/time (often converted to ACH) and is the air exchange rate I.

Equation (47) is based on the assumptions that (1) no unknown tracer gas sources exist; (2) the airflow out of the building is the dominant means of removing the tracer gas from the space (i.e., the tracer gas does not react chemically within the space and/or is not adsorbed onto or by interior surfaces), and (3) the tracer gas concentration within the building can be represented by a single value (i.e., the tracer gas is uniformly mixed within the space).

Three different tracer gas procedures are used to measure air exchange rates: (1) decay or growth, (2) constant concentration, and (3) constant injection.

Decay or Growth

Decay. The simplest tracer gas measurement technique is the decay method (also known as the step-down method). A small amount of tracer gas is injected into the space and is allowed to mix with the interior air. After the injection, $F = 0$ and then the solution to Equation (47) is

$$C(\theta) = C_o e^{-I\theta} \tag{48}$$

where C_o is the concentration of the tracer in the space at $\theta = 0$.

Equation (48) is generally used to solve for I by measuring the tracer gas concentration periodically during the decay and fitting the data to the logarithmic form of Equation (48):

$$\ln C(\theta) = \ln C_o - I\theta \tag{49}$$

Like all tracer gas techniques, the decay method has advantages and disadvantages. One advantage is that because logarithms of concentration are taken, only relative concentrations are needed, which can simplify the calibration of the concentration-measuring equipment. Also, the tracer gas injection rate need not be measured, although it must be controlled so that the tracer gas concentrations are within the range of the concentration-measuring device. The concentration-measuring equipment can be located on site, or building samples can be collected in suitable containers and analyzed elsewhere.

The most serious problem with the decay technique is imperfect mixing of the tracer gas with the interior air, both at initial injection and during the decay. Equations (47) and (48) employ the assumption that the tracer gas concentration within the building is uniform. If the tracer is not well mixed, this assumption is not appropriate and the determination of I will be subject to errors. It is difficult to estimate the magnitude of the errors due to poor mixing, and little analysis of this problem has been performed.

Growth. The growth or step-up method is similar to the decay method except that the initial tracer gas concentration is low and is increased during the test.

Constant Concentration

In the constant concentration technique, the tracer gas injection rate is adjusted to maintain a constant concentration within the building. If the concentration is truly constant, then Equation (47) reduces to

$$Q(\theta) = F(\theta)/C \tag{50}$$

There is less experience with this technique than with the decay procedure, but an increasing number of applications exist (Kumar et al. 1979, Collet 1981, Bohac et al. 1985, Fortmann et al. 1990, Walker and Wilson 1998, Wilson and Walker 1993, Walker and Forest 1995).

Because the tracer gas injection is continuous, no initial mixing period is required. Another advantage is that the tracer gas injection into each zone of the building can be separately controlled; thus, the amount of outdoor air flowing into each zone can be determined. This procedure is best suited for longer term continuous monitoring of fluctuating infiltration rates. This procedure has the disadvantage of requiring the measurement of absolute tracer concentrations and injection rates. Also, imperfect mixing of the tracer and the interior air causes a delay in the response of the concentration to changes in the injection rate.

Constant Injection

In the constant injection procedure, the tracer is injected at a constant rate, and the solution to Equation (47) becomes

$$C(\theta) = (F/Q)(1 - e^{-I\theta}) \tag{51}$$

After sufficient time, the transient term reduces to zero, the concentration attains equilibrium, and Equation (51) reduces to

$$Q = F/C \tag{52}$$

Equation (52) is valid only when air exchange rate I and airflow rate Q are constant; thus, this technique is only appropriate for systems at or near equilibrium. It is particularly useful in spaces with mechanical ventilation or with high air exchange rates. Constant injection requires the measurement of absolute concentrations and injection rates.

Dietz et al. (1986) used a special case of the constant injection technique. This technique uses permeation tubes as a tracer gas source. The tubes release the tracer at an ideally constant rate into the building being tested. A sampling tube packed with an adsorbent collects the tracer from the interior air at a constant rate by diffusion. After a sampling period of one week or more, the sampler is removed and analyzed to determine the average tracer gas concentration within the building during the sampling period.

Solving Equation (47) for C and taking the time average gives

$$\langle C \rangle = \langle F/Q \rangle = F\langle 1/Q \rangle \tag{53}$$

where $< \ldots >$ denotes time average. (Note that the time average of $dC/d\theta$ is assumed to equal zero.)

Equation (53) shows that the average tracer concentration $\langle C \rangle$ and the injection rate F can be used to calculate the average of the inverse airflow rate. The average of the inverse is less than the inverse of the actual average, with the magnitude of this difference depending on the distribution of airflow rates during the measurement period. Sherman and Wilson (1986) calculated these differences to be about 20% for one-month averaging periods. Differences greater than 30% have been measured when there were large changes in air exchange rate due to occupant airing of houses; errors from 5 to 30% were measured when the variation was due to weather effects (Bohac et al. 1987). Longer averaging periods and large changes in air exchange rates during the measurement periods generally lead to larger differences between the average inverse exchange rate and the inverse of the actual average rate.

Multizone Air Exchange Measurement

Equation (47) is based on the assumption of a single, well-mixed enclosure, and the techniques described are for single-zone measurements. Airflow between internal zones and between the exterior and individual internal zones has led to the development of multizone measurement techniques (Harrje et al. 1985, Sherman and Dickerhoff 1989, Fortmann et al. 1990, Harrje et al. 1990). These techniques are important when considering the transport of pollutants from one room of a building to another. For a theoretical development, see Sinden (1978b). Multizone measurements typically use either multiple tracer gases for the different zones or the constant concentration technique. A proper error analysis is essential in all multizone flow determination (Charlesworth 1988, D'Ottavio et al. 1988).

SYMBOLS

A = area, ft^2 or in^2
c = flow coefficient, cfm/(in. of water)^n
c_p = specific heat, Btu/lb·°F
C = concentration
$\overline{C}$ = time averaged concentration
C_A = airflow coefficient for automatic doors, cfm/[ft^2 (in. of water)$^{0.5}$]
C_D = discharge coefficient
C_p = pressure coefficient
C_s = stack flow coefficient, cfm^2/(in^4·°F) or (in. of water/°F)^n
C_v = effectiveness of openings
C_w = wind flow coefficient, cfm^2/(in^4·mph^2) or (in. of water/mph^2)^n
F = tracer gas injection rate, cfm
$\overline{F}$ = time-averaged contaminant source strength, cfm
f = fractional on-time
g = wind speed reduction factor
h = specific enthalpy, Btu/lb
H = height, ft
i = hour of year
I = air exchange rate, 1/time
I_i = instantaneous air exchange rate, 1/time
I_m = effective air exchange rate, 1/time
IDD = infiltration degree-days, °F
n = pressure exponent
N = number of discrete time periods in period of interest
p = pressure, in. of water
P = parameter
q = heat rate, Btu/h
Q = volumetric flow rate, cfm
$\overline{Q}$ = effective volumetric flow rate, cfm
s = shelter factor
S = source strength, μg/min
t = relative temperature, °F
T = absolute temperature, °R
U = wind speed, mph
V = volume, ft^3
W = humidity ratio, lb$_m$ water/lb$_m$ dry air
ε_I = air change effectiveness
ϕ = wind angle, degrees
ρ = air density, lb$_m$/ft^3
θ = time
θ_{age} = age of air
τ = time constant

Subscripts

b = base
ba = bypass air
c = calculated
ca = recirculated air
e = effective
ea = exhaust air
f = floor
i = indoor or time counter for summation (instantaneous)
H = building height, eaves or roof
ka = makeup air
l = latent
la = relief air
L = leakage or local
ma = mixed air
met = meteorological station location
n = normalized
N = nominal
NPL = neutral pressure level
o = outdoor, initial condition, or reference
oa = outside air
p = pressure
r = reference
s = sensible or stack
sa = supply air
S = space or source
w = wind

REFERENCES

ACGIH. 1998. *Industrial ventilation: A manual of recommended practice*, 23rd ed. American Conference of Governmental Industrial Hygienists, Lansing, MI.

AIVC. 1994. *An analysis and data summary of the AIVC's numerical database*. AIVC *Technical Note* 44.

AIVC. 1999. *AIRBASE bibliographic database*. International Energy Agency Air Infiltration and Ventilation Centre, Coventry, UK.

Akins, R.E., Peterka, J.A., and Cermak, J.E. 1979. Averaged pressure coefficients for rectangular buildings. Vol.1, Proceeding of the *Fifth International Wind Engineering Conference*, Fort Collins, U.S., July, pp. 369-80.

Allard, F. and M. Herrlin. 1989. Wind-induced ventilation. *ASHRAE Transactions* 95(2):722-28.

ASHRAE. 1988. Air leakage performance for detached single-family residential buildings. ANSI/ASHRAE *Standard* 119-1988 (RA 94).

ASHRAE. 1992. Thermal environmental conditions for human occupancy. ANSI/ASHRAE *Standard* 55-92.

ASHRAE. 1997. Measuring air change effectiveness. ANSI/ASHRAE *Standard* 129-97

ASHRAE. 1999. Method of testing general ventilation air cleaning devices for removal efficiency by particle size. *Standard* 52.2-1999.

ASHRAE. 1999. Ventilation for acceptable indoor air quality. ANSI/ASHRAE *Standard* 62-1999.

ASHRAE. 2000. A method of determining air change rates in detached dwellings. ANSI/ASHRAE *Standard* 136-2000.

ASTM. 1987. Practices for air leakage site detection in building envelopes. *Standard* E 1186-87(R 1992). American Society for Testing and Materials, West Conshohocken, PA.

ASTM. 1991. Test method for determining the rate of air leakage through exterior windows, curtain walls, and doors under specified pressure differences across the specimen. *Standard* E 283-91.

ASTM. 1993. Test method for field measurement of air leakage through installed exterior windows and doors. *Standard* E 783-93.

ASTM. 1993. Test methods for determining air change in a single zone by means of a tracer gas dilution. *Standard* E 741-93.

ASTM. 1996. Standard test methods for determining airtightness of buildings using an orifice blower door. *Standard* E 1827-96.

ASTM. 1999. Measuring air leakage by the fan pressurization method. *Standard* E 779-99.

ASTM. 1999. Standard guide for assessing depressurization-induced backdrafting and spillage from vented combustion appliances. ASTM *Standard* E 1998-99.

Bahnfleth, W.P., G.K. Yuill, and B.W. Lee. 1999. Protocol for field testing of tall buildings to determine envelope air leakage rates. *ASHRAE Transactions* 105(2).

Bankvall, C.G. 1987. Air movements and thermal performance of the building envelope. *Thermal Insulation: Materials and Systems*. F.J. Powell and S.L. Mathews, eds. ASTM, Philadelphia, pp. 124-31.

Berg-Munch, B., G. Clausen, and P.O. Fanger. 1986. Ventilation requirements for the control of body odor in spaces occupied by women. *Environmental International* 12(1-4):195.

Berlad, A.L., N. Tutu, Y. Yeh, R. Jaung, R. Krajewski, R. Hoppe and F. Salzano. 1978. Air intrusion effects on the performance of permeable insulation systems. In *Thermal insulation performance*, D. McElroy and R. Tye, eds. ASTM *STP* 718, pp. 181-94 (October).

Blomsterberg, A.K. and D.T. Harrje. 1979. Approaches to evaluation of air infiltration energy losses in buildings. *ASHRAE Transactions* 85(1):797.

Bohac, D., D.T. Harrje, and G.S. Horner. 1987. Field study of constant concentration and PFT infiltration measurements. Proceedings of the *8th IEA Conference of the Air Infiltration and Ventilation Centre*, Uberlingen, Germany.

Bohac, D., D.T. Harrje, and L.K. Norford. 1985. Constant concentration infiltration measurement technique: An analysis of its accuracy and field measurements, 176. Proceedings of the *ASHRAE-DOE-BTECC Conference on the Thermal Performance of the Exterior Envelopes of Buildings III,* Clearwater Beach, FL.

Bradley, B. 1993. Implementation of the AIM-2 infiltration model in HOT-2000. Report for Department of Natural Resources Canada.

Buchanan, C.R. and M. H. Sherman. 2000. A mathematical model for infiltration heat recovery. Proceedings of the *21st AVIC Annual Conference, Innovations in Ventilation Technology,* 26-29 September, Air Infiltration and Ventilation Centre, Coventry, UK. Lawrence Berkeley National Laboratory *Report* LBNL-44294.

Cain, W.S. et al. 1983. Ventilation requirements in buildings—I. Control of occupancy odor and tobacco smoke odor. *Atmos. Environ.* 17(6): 1183-97.

CGSB. 1986. Determination of the airtightness of building envelopes by the fan depressurization method. CGSB *Standard* 149.10-M86. Canadian General Standards Board, Ottawa.

CGSB. 1995. The spillage test: Method to determine the potential for pressure-induced spillage from vented, fuel-fired, space heating appliances, water heaters, and fireplaces. CAN/CGSB *Standard* 51.71-95.

CGSB. 1996. Determination of the overall envelope airtightness of buildings by the fan pressurization method using the building's air handling systems. CGSB *Standard* 149.15-96.

CHBA. 1994. *HOT2000 technical manual*. Canadian Home Builders Association, Ottawa, Canada.

Charlesworth, P.S. 1988. Measurement of air exchange rates. Chapter 2 in *Air exchange rate and airtightness measurement techniques—An applications guide.* Air Infiltration and Ventilation Centre, Coventry, Great Britain.

Chastain, J.P. 1987. Pressure gradients and the location of the neutral pressure axis for low-rise structures under pure stack conditions. Unpublished M.S. thesis. University of Kentucky, Lexington.

Chastain, J.P., D.G. Colliver, and P.W. Winner, Jr. 1987. Computation of discharge coefficients for laminar flow in rectangular and circular openings. *ASHRAE Transactions* 93(2):2259-83.

Chastain, J.P. and D.G. Colliver. 1989. Influence of temperature stratification on pressure differences resulting from the infiltration stack effect. *ASHRAE Transactions* 95(1):256-68.

Claridge, D.E., M. Krarti, and S. Bhattacharyya. 1988. Preliminary measurements of the energy impact of infiltration in a test cell. Proceedings of *Fifth Annual Symposium on Improving Building Energy Efficiency in Hot and Humid Climates*, pp. 308-17.

Claridge, D.E., and S. Bhattacharyya. 1990. The measured impact of infiltration in a test cell. *ASME Journal of Solar Energy Engineering* 112: 123-26.

Cole, J.T., T.S. Zawacki, R.H. Elkins, J.W. Zimmer, and R.A. Macriss. 1980. Application of a generalized model of air infiltration to existing homes. *ASHRAE Transactions* 86(2):765.

Collet, P.F. 1981. Continuous measurements of air infiltration in occupied dwellings. Proceedings of the *2nd IEA Conference of the Air Infiltration Centre*, p.147. Stockholm, Sweden.

Colliver, D.G., W.E. Murphy, and W. Sun. 1992. Evaluation of the techniques for the measurement of air leakage of building components. *Final Report* of ASHRAE Research Project RP-438. University of Kentucky, Lexington.

CSA. 1991. Residential mechanical ventilation systems. CAN/CSA-F326-M91. Canadian Standards Association, Toronto.

Cummings, J.B. and J.J. Tooley, Jr. 1989. Infiltration and pressure differences induced by forced air systems in Florida residences. *ASHRAE Transactions* 96(20):551-60.

Cummings, J.B., J.J. Tooley, Jr., and R. Dunsmore. 1990. Impacts of duct leakage on infiltration rates, space conditioning energy use and peak electrical demand in Florida homes. Proceedings of *ACEEE Summer Study*, Pacific Grove, CA. American Council for an Energy-Efficient Economy, Washington, D.C.

Cutter. 1987. Air-to-air heat exchangers. In *Energy design update.* Cutter Information Corporation, Arlington, MA.

Desrochers, D. and A.G. Scott. 1985. Residential ventilation rates and indoor radon daughter levels. Transactions of the *APCA Specialty Conference, Indoor Air Quality in Cold Climates: Hazards and Abatement Measures,* p. 362. Ottawa, Canada.

Diamond, R.C., M.P. Modera, and H.E. Feustel. 1986. Ventilation and occupant behavior in two apartment buildings. Lawrence Berkeley National Laboratory *Report* LBL-21862. Presented at the 7th AIC Conference, Stratford-upon-Avon, UK.

Diamond, R.C., J.B. Dickinson, R.D. Lipschutz, B. O'Regan, and B. Shohl. 1982. The house doctor's manual. *Report* PUB 3017. Lawrence Berkeley National Laboratory, Berkeley, CA.

Dickerhoff, D.J., D.T. Grimsrud, and R.D. Lipschutz. 1982. Component leakage testing in residential buildings. Proceedings of the *American Council for an Energy-Efficient Economy, 1982 Summer Study*, Santa Cruz, CA. *Report* LBL 14735. Lawrence Berkeley National Laboratory, Berkeley, CA.

Dietz, R.N., R.W. Goodrich, E.A. Cote, and R.F. Wieser. 1986. Detailed description and performance of a passive perfluorocarbon tracer system for building ventilation and air exchange measurement. In *Measured air leakage of buildings*, p. 203. ASTM STP 904. H.R. Trechsel and P.L. Lagus, eds.

D'Ottavio, T.W., G.I. Senum, and R.N. Dietz. 1988. Error analysis techniques for perfluorocarbon tracer derived multizone ventilation rates. *Building and Environment* 23(40).

Ek, C.W., S.A. Anisko, and G.O. Gregg. 1990. Air leakage tests of manufactured housing in the Northwest United States. In *Air change rate and airtightness in buildings*, pp. 152-64. ASTM STP 1067. M.H. Sherman, ed.

Elmroth, A. and P. Levin. 1983. Air infiltration control in housing, a guide to international practice. *Report* D2:1983. Air Infiltration Centre, Swedish Council for Building Research, Stockholm.

Energy Resource Center. 1982. How to house doctor. University of Illinois, Chicago.

Etheridge, D.W. 1977. Crack flow equations and scale effect. *Building and Environment* 12:181.

Etheridge, D.W. and D.K. Alexander. 1980. The British gas multi-cell model for calculating ventilation. *ASHRAE Transactions* 86(2):808.

Etheridge, D.W. and J.A. Nolan. 1979. Ventilation measurements at model scale in a turbulent flow. *Building and Environment* 14(1):53.

Eto, J. 1990. The HVAC costs of increased fresh air ventilation rates in office buildings. Proceedings of *Indoor Air '90 4:53-58.* International Conference on Indoor Air Quality and Climate, Ottawa, Canada.

Eto, J. and C. Meyer. 1988. The HVAC costs of increased fresh air ventilation rates in office buildings. *ASHRAE Transactions* 94(2):331-45.

Eyre, D. and D. Jennings. 1983. Air-vapour barriers—A general perspective and guidelines for installation. Energy, Mines, and Resources Canada, Ottawa.

Farrington, R., D. Martin, and R. Anderson. 1990. A comparison of displacement efficiency, decay time constant, and age of air for isothermal flow in an imperfectly mixed enclosure. Proceedings of the *ACEEE 1990 Summer Study on Energy Efficiency in Buildings*, pp. 4.35-4.43. American Council for an Energy-Efficient Economy, Washington, D.C.

Feustel, H.E. and A. Raynor-Hoosen, eds. 1990. Fundamentals of the multizone air flow model—COMIS. *Technical Note* 29. Air Infiltration and Ventilation Centre, Coventry, Great Britain.

Fisk, W.J., R.J. Prill, and O. Steppanen. 1989. A multi-tracer technique for studying rates of ventilation, air distribution patterns and air exchange efficiencies. Proceedings of *Conference on Building Systems—Room Air and Air Contaminant Distribution,* pp. 237-40. ASHRAE, Atlanta.

Fisk, W.J., R.K. Spencer, D.T. Grimsrud, F.J. Offermann, B. Pedersen, and R. Sextro. 1984. Indoor air quality control techniques: A critical review. *Report* LBL 16493. Lawrence Berkeley National Laboratory, Berkeley, CA.

Fortmann, R.C., N.L. Nagda, and H.E. Rector. 1990. Comparison of methods for the measurement of air change rates and interzonal airflows to two test residences. In *Air change rate and airtightness in buildings*, pp. 104-18. ASTM STP 1067. M.H. Sherman, ed.

Foster, M.P. and M.J. Down. 1987. Ventilation of livestock buildings by natural convection. *Journal of Agricultural Engineering Research* 37:1.

Giesbrecht, P. and G. Proskiw. 1986. An evaluation of the effectiveness of air leakage sealing. In *Measured air leakage of buildings*, p. 312. ASTM STP 904. H.R. Trechsel and P.L. Lagus, eds.

Grieve, P.W. 1989. *Measuring ventilation using tracer-gases*. Brüel and Kjær, Denmark.

Grimsrud, D.T. and K.Y. Teichman. 1989. The scientific basis of *Standard* 62-1989. *ASHRAE Journal* 31(10):51-54.

Grimsrud, D.T., M.H. Sherman, and R.C. Sonderegger. 1982. Calculating infiltration: Implications for a construction quality standard. Proceedings of the *ASHRAE-DOE Conference on the Thermal Performance of the Exterior Envelope of Buildings II*, p. 422. Las Vegas, NV.

Grimsrud, D.T., M.H. Sherman, R.C. Diamond, P.E. Condon, and A.H. Rosenfeld. 1979. Infiltration-pressurization correlations: Detailed measurements in a California house. *ASHRAE Transactions* 85(1):851.

Grot, R.A. and R.E. Clark. 1979. Air leakage characteristics and weatherization techniques for low-income housing. Proceedings of the *ASHRAE-DOE Conference on the Thermal Performance of the Exterior Envelopes of Buildings*, p. 178. Orlando, FL.

Grot, R.A. and A.K. Persily. 1986. Measured air infiltration and ventilation rates in eight large office buildings. In *Measured air leakage of buildings*, p. 151. ASTM STP 904. H.R. Trechsel and P.L. Lagus, eds.

Hamlin, T.L. 1991. Ventilation and airtightness in new, detached Canadian housing. *ASHRAE Transactions* 97(2):904-10.

Hamlin, T. and W. Pushka. 1994. Predicted and measured air change rates in houses with predictions of occupant IAQ comfort. Proceedings of the *15th AIVC Conference*, Buxton, UK, pp. 771-75.

Harrje, D.T. and G.J. Born. 1982. Cataloguing air leakage components in houses. Proceedings of the *ACEEE 1982 Summer Study*, Santa Cruz, CA. American Council for an Energy-Efficient Economy, Washington, D.C.

Harrje, D.T. and T.A. Mills, Jr. 1980. Air infiltration reduction through retrofitting, 89. *Building air change rate and infiltration measurements*. ASTM STP 719. C.M. Hunt, J.C. King, and H.R. Trechsel, eds.

Harrje, D.T., G.S. Dutt, and J. Beyea. 1979. Locating and eliminating obscure but major energy losses in residential housing. *ASHRAE Transactions* 85(2):521.

Harrje, D.T., R.A. Grot, and D.T. Grimsrud. 1981. Air infiltration site measurement techniques. Proceedings of the *2nd IEA Conference of the Air Infiltration Centre*, p. 113. Stockholm, Sweden.

Harrje, D.T., G.S. Dutt, D.L. Bohac, and K.J. Gadsby. 1985. Documenting air movements and infiltration in multicell buildings using various tracer techniques. *ASHRAE Transactions* 91(2):2012-27.

Harrje, D.T., R.N. Dietz, M. Sherman, D.L. Bohac, T.W. D'Ottavio, and D.J. Dickerhoff. 1990. Tracer gas measurement systems compared in a multifamily building. In *Air change rate and airtightness in buildings*, p. 5-12. ASTM STP 1067. M.H. Sherman, ed.

Hekmat, D., H.E. Feustel, and M.P. Modera. 1986. Impacts of ventilation strategies on energy consumption and indoor air quality in single-family residences. *Energy and Buildings* 9(3):239.

Herrlin, M.K. 1985. MOVECOMP: A static-multicell-airflow-model. *ASHRAE Transactions* 91(2B):1989.

Holton, J.K., M.J. Kokayko, and T.R. Beggs. 1997. Comparative ventilation system evaluations. *ASHRAE Transactions* 103(2):675-92.

Honma, H. 1975. *Ventilation of dwellings and its disturbances*. Faibo Grafiska, Stockholm.

Hopkins, L.P. and B. Hansford. 1974. Air flow through cracks. *Building Service Engineer* 42(September):123.

Hunt, C.M. 1980. Air infiltration: A review of some existing measurement techniques and data. *Building air change rate and infiltration measurements*, p. 3. ASTM STP 719. C.M. Hunt, J.C. King, and H.R. Trechsel, eds.

ISO. 1995. Thermal insulation—Determination of building air tightness—Fan pressurization method. ISO *Standard* 9972. International Organization for Standardization, Geneva.

Iwashita, G., K. Kimura., et al. 1989. Pilot study on addition of old units for perceived air pollution sources. Proceedings of the *SHASE Annual Meeting*, pp. 3221-324. Society of Heating, Air-Conditioning and Sanitary Engineers of Japan, Tokyo.

Jacobson, D.I., G.S. Dutt, and R.H. Socolow. 1986. Pressurization testing, infiltration reduction, and energy savings. In *Measured air leakage of buildings*, p. 265. ASTM STP 904. H.R. Trechsel and P.L. Lagus, eds.

Janssen, J.E. 1989. Ventilation for acceptable indoor air quality. *ASHRAE Journal* 31(10):40-48.

Janu, G.J., J.D. Wegner, C.G. Nesler. 1995. Outdoor air flow control for VAV systems. *ASHRAE Journal* 37(4):62-68.

Judkoff, R., J.D. Balcomb, C.E. Handcock, G. Barker, and K. Subbarao. 1997. Side-by-side thermal tests of modular offices: A validation study of the STEM method. *Report*. NREL, Golden, CO.

Jump, D.A., I.S. Walker, and M.P. Modera. 1996. Field measurements of efficiency and duct retrofit effectiveness in residential forced air distribution systems. Proceedings of the *1996 ACEEE Summer Study*, pp.1.147-1.156. ACEEE, Washington, D.C.

Kiel, D.E. and D.J. Wilson. 1986. Gravity driven airflows through open doors, 15.1. Proceedings of the *7th IEA Conference of the Air Infiltration Centre*, Stratford-upon-Avon, United Kingdom.

Kim, A.K. and C.Y. Shaw. 1986. Seasonal variation in airtightness of two detached houses. In *Measured air leakage of buildings*, p. 17. ASTM STP 904. H.R. Trechsel and P.L. Lagus, eds.

Klauss, A.K., R.H. Tull, L.M. Rootsd, and J.R. Pfafflino. 1970. History of the changing concepts in ventilation requirements. *ASHRAE Journal* 12(6):51-55.

Kohonen, R., T. Ojanen, and M. Virtanen. 1987. Thermal coupling of leakage flows and heating load of buildings. *8th AIVC Conference*, Berlin.

Kohonen, R., and T. Ojanen. 1987. Non-steady state coupled diffusion and convection heat and mass transfer in porous media. *5th International Conference on Numerical Methods in Thermal Problems*, Montreal.

Kreith, F., and R. Eisenstadt. 1957. Pressure drop and flow characteristics of short capillary tubes at low Reynolds numbers. *ASME Transactions*, pp. 1070-78.

Kronvall, J. 1980. Correlating pressurization and infiltration rate data—Tests of an heuristic model. Lund Institute of Technology, Division of Building Technology, Lund, Sweden.

Kumar, R., A.D. Ireson, and H.W. Orr. 1979. An automated air infiltration measuring system using SF6 tracer gas in constant concentration and decay methods. *ASHRAE Transactions* 85(2):385.

Kvisgaard, B. and P.F. Collet. 1990. The user's influence on air change. In *Air change rate and airtightness in buildings*, pp. 67-76. ASTM STP 1067. M.H. Sherman, ed.

Lagus, P.L. 1989. Tracer measurement instrumentation suitable for infiltration, air leakage, and air flow pattern characterization. Proceedings of the *Conference on Building Systems—Room Air and Air Contaminant Distribution*, pp. 97-102. ASHRAE, Atlanta.

Lagus, P. and A.K. Persily. 1985. A review of tracer-gas techniques for measuring airflows in buildings. *ASHRAE Transactions* 91(2B):1075.

Lamming, S., and Salmon, J. 1998. Wind data for design of smoke control systems. ASHRAE *Transactions* 104(1).

Lecompte, J.G.N. 1987. The influence of natural convection in an insulated cavity on thermal performance of a wall. *Insulation Materials Testing and Applications*. ASTM, Florida.

Liddament, M.W. 1988. The calculation of wind effect on ventilation. *ASHRAE Transactions* 94(2):1645-60.

Liddament, M. and C. Allen. 1983. The validation and comparison of mathematical models of air infiltration. *Technical Note* 11. Air Infiltration Centre, Bracknell, Great Britain.

Liu, M. and D.E. Claridge. 1992a. The measured energy impact of infiltration under dynamic conditions. Proceedings of the *8th Symposium on Improving Building Systems in Hot and Humid Climates*, Dallas, TX.

Liu, M. and D.E. Claridge. 1992b. The measured energy impact of infiltration in a test cell. Proceedings of the *8th Symposium on ImprovingBuilding Systems in Hot and Humid Climates*, Dallas, TX.

Liu, M. and D.E. Claridge. 1992c. The energy impact of combined solar radiation/infiltration/conduction effects in walls and attics. Proceedings of *Thermal Performance of Exterior Envelopes of Buildings, 5th ASHRAE/DOE/BTECC Conference*, Clearwater Beach, FL.

Liu, M. and D.E. Claridge. 1995. Experimental methods for identifying infiltration heat recovery in building. Proceedings of the *Thermal Performance of Exterior Envelopes of Buildings, 6th ASHRAE/DOE/BTECC Conference*, Clearwater Beach, FL.

Lubliner, M., D.T. Stevens, and B. Davis. 1997. Mechanical ventilation in HVD-code manufactured housing in the Pacific Northwest. *ASHRAE Transactions* 103(1):693-705.

Marbek Resource Consultants. 1984. Air sealing homes for energy conservation. Energy, Mines and Resources Canada, Buildings Energy Technology Transfer Program, Ottawa.

Modera, M.P. 1989. Residential duct system leakage: Magnitude, impacts, and potential for reduction. *ASHRAE Transactions*. 96(2):561-69.

Modera, M.P. and D.J. Wilson. 1990. The effects of wind on residential building leakage measurements. *Air change rate and airtightness in buildings*, pp. 132-45. ASTM STP 1067. M.H. Sherman, ed.; Lawrence Berkeley National Laboratory *Report* LBL-24195.

Modera, M.P., D. Dickerhoff, R. Jansky, and B. Smith. 1991. Improving the energy efficiency of residential air distribution systems in California. *Report* LBL-30866. Lawrence Berkeley National Laboratory, Berkeley, CA.

Mumma, S.A., and R.J. Bolin. 1994. Real-time, on-line optimization of VAV system control to minimize the energy consumption rate and to satisfy ASHRAE *Standard* 62-1989 for all occupied zones. *ASHRAE Transactions* 94(1):168-79.

Mumma, S.A., and Y.M. Wong. 1990. Analytical evaluation of outdoor airflow rate variation vs. supply airflow rate variation in VAV systems when the outside air damper position is fixed. *ASHRAE Transactions* 90(1):1197-1208.

Murphy, W.E., D.G. Colliver, and L.R. Piercy. 1991. Repeatability and reproducibility of fan pressurization devices in measuring building air leakage. *ASHRAE Transactions* 97(2):885-95.

Nelson, B.D., D.A. Robinson, and G.D. Nelson. 1985. Designing the envelope—Guidelines for buildings. Thermal Performance of the Exterior Envelopes of Buildings III. Proceedings of the *ASHRAE/DOE/BTECC Conference*, Florida. ASHRAE SP 49, pp. 1117-22.

NRCC. 1995. *National Building Code of Canada.* National Research Council of Canada, Ottawa.

Nylund, P.O. 1980. Infiltration and ventilation. *Report* D22:1980. Swedish Council for Building Research, Stockholm.

Offermann, F., and D. Int-Hout. 1989. Ventilation effectiveness measurements of three supply/return air configurations. *Environment International* 15(1-6):585-92.

Orme, M. 1999. Applicable models for air infiltration and ventilation calculations. AIVC *Technical Note* 51. AIVC, Coventry. UK.

Palmiter, L. and I. Brown. 1989. The Northwest Residential Infiltration Survey, description and summary of results. Proceedings of the *ASHRAE/DOE/BTECC/CIBSE Conference—Thermal Performance of the Exterior Envelopes of Buildings IV*, Florida, pp. 445-57.

Palmiter, L., I.A. Brown, and T.C. Bond. 1991. Measured infiltration and ventilation in 472 all-electric homes. *ASHRAE Transactions* 97(2):979-87.

Palmiter, L. and T. Bond. 1994. Modeled and measured infiltration II—A detailed case study of three homes. EPRI *Report* TR 102511. Electric Power Research Institute, Palo Alto, CA.

Parekh, A., K. Ruest, and M. Jacobs. 1991. Comparison of airtightness, indoor air quality and power consumption before and after air-sealing of high-rise residential buildings. Proceedings of the *12th AIVC Conference: Air Movement and Ventilation Control within Buildings, Air Infiltration and Ventilation Centre*, Coventry, Great Britain.

Parker, G.B., M. McSorley, and J. Harris. 1990. The Northwest Residential Infiltration Survey: A field study of ventilation in new houses in the Pacific Northwest. In *Air change rate and airtightness in buildings*, pp. 93-103. ASTM STP 1067. M.H. Sherman, ed.

Persily, A. 1982. Repeatability and accuracy of pressurization testing. Proceedings of the *ASHRAE-DOE Conference, Thermal Performance of the Exterior Envelopes of Buildings II*, Las Vegas, NV.

Persily, A.K. 1986. Measurements of air infiltration and airtightness in passive solar homes. In *Measured air leakage of buildings*, p. 46. ASTM STP 904. H.R. Trechsel and P.L. Lagus, eds.

Persily, A.K. 1988. Tracer gas techniques for studying building air exchange. *Report* NBSIR 88-3708. National Institute of Standards and Technology, Gaithersburg, MD.

Persily, A.K. 1991. Design guidelines for thermal envelope integrity in office buildings. Proceedings of the *12th AIVC Conference: Air Movement and Ventilation Control within Buildings, Air Infiltration and Ventilation Centre*, Coventry, Great Britain.

Persily, A.K. and J. Axley. 1990. Measuring airflow rates with pulse tracer techniques. In *Air change rate and airtightness in buildings*, pp. 31-51. ASTM STP 1067. M.H. Sherman, ed.

Persily, A.K. and R.A. Grot. 1985a. The airtightness of office building envelopes. Proceedings of the *ASHRAE-DOE-BTECC Conference on the Thermal Performance of the Exterior Envelopes of Buildings III,* Clearwater Beach, FL. p. 125.

Persily, A.K. and R.A. Grot. 1985b. Accuracy in pressurization data analysis. *ASHRAE Transactions* 91(2B):105.

Persily, A.K. and R.A. Grot. 1986. Pressurization testing of federal buildings. In *Measured air leakage of buildings*, p. 184. ASTM STP 904. H.R. Trechsel and P.L. Lagus, eds.

Persily, A.K. and G.T. Linteris. 1983. A comparison of measured and predicted infiltration rates. *ASHRAE Transactions* 89(2):183.

Powell, F., M. Krarti, and A. Tuluca. 1989. Air movement influence on the effective thermal resistance of porous insulations: A literature survey. *Journal of Thermal Insulation* 12:239-251.

Reardon, J.T. and C.-Y. Shaw. 1997. Evaluation of five simple ventilation strategies suitable for houses without forced-air heating. *ASHRAE Transactions* 103(1):731-744.

Reeves, G., M.F. McBride, and C.F. Sepsy. 1979. Air infiltration model for residences. *ASHRAE Transactions* 85(1):667.

Riley, M. 1990. Indoor air quality and energy conservation: The R-2000 home program experience. Proceedings of *Indoor Air '90 5:143*. International Conference on Indoor Air Quality and Climate, Ottawa, Canada.

Robison, P.E. and L.A. Lambert. 1989. Field investigation of residential infiltration and heating duct leakage. *ASHRAE Transactions* 95(2):542-50.

Rock, B.A. 1992. Characterization of transient pollutant transport, dilution, and removal for the study of indoor air quality. Ph.D. diss., University of Colorado at Boulder. University Microfilms International.

Rock, B.A., M.J. Brandemuehl, and R. Anderson. 1995. Toward a simplified design method for determining the air change effectiveness. *ASHRAE Transactions* 101(1):217-27.

Rudd, A.F. 1998. Design/sizing methodology and economic evaluation of central-fan-integrated supply ventilation systems. *ACEEE 1998 Summer Study on Energy Efficiency in Buildings*. 23-28 August, Pacific Grove, California. American Council for an Energy Efficient Economy, Washington, D.C.

Sandberg, M.H. 1981. What is ventilation efficiency? *Building and Environment* 16:123-35.

Shaw, C.Y. 1981. A correlation between air infiltration and air tightness for a house in a developed residential area. *ASHRAE Transactions* 87(2):333.

Shaw, C.Y. and W.C. Brown. 1982. Effect of a gas furnace chimney on the air leakage characteristic of a two-story detached house, 12.1. Proceedings of the *3rd IEA Conference of the Air Infiltration Centre*, London.

Sherman, M.H. 1986. Infiltration degree-days: A statistic for quantifying infiltration-related climate. *ASHRAE Transactions* 92(2):161-81.

Sherman, M.H. 1987. Estimation of infiltration from leakage and climate indications. *Energy and Buildings* 10(1):81.

Sherman, M.H. 1989a. Uncertainty in airflow calculations using tracer gas measurements. *Building and Environment* 24(4):347-54.

Sherman, M.H. 1989b. On the estimation of multizone ventilation rates from tracer gas measurements. *Building and Environment* 24(4):355-62.

Sherman, M.H. 1990. Tracer gas techniques for measuring ventilation in a single zone. *Building and Environment* 25(4):365-74.

Sherman, M.H. 1991. Single-zone stack-dominated infiltration modeling. Proceedings of the *12th Air Infiltration and Ventilation Conference,* Ottawa, Canada.

Sherman, M.H. 1992a. Superposition in infiltration modeling. *Indoor Air* 2:101-14.

Sherman, M.H. 1992b. A power law formulation of laminar flow in short pipes. *Journal of Fluids Engineering* 114:601-05. LBL *Report* 29414, Lawrence Berkeley Laboratory, University of California.

Sherman, M.H. 1995. The use of blower door data. *Indoor Air* 5:215-24.

Sherman, M.H. and Dickerhoff, D.J. 1998. Airtightness of U.S. dwellings. *ASHRAE Transactions* 104(2).

Sherman, M.H. and D. Dickerhoff. 1989. Description of the LBL multitracer measurement system. Proceedings of the *ASHRAE/DOE/BTECC/CIBSE Conference—Thermal Performance of the Exterior Envelopes of Buildings IV*, pp. 417-32.

Sherman, M.H. and D.T. Grimsrud. 1980. Infiltration-pressurization correlation: Simplified physical modeling. *ASHRAE Transactions* 86(2):778.

Sherman, M.H., D.T. Grimsrud, P.E. Condon, and B.V. Smith. 1980. Air infiltration measurement techniques, 9. Proceedings of the *1st IEA Symposium of the Air Infiltration Centre*, London. *Report* LBL 10705. Lawrence Berkeley National Laboratory, Berkeley, CA.

Sherman, M.H. and N. Matson. 1997. Residential ventilation and energy characteristics. *ASHRAE Transactions* 103(1):717-30.

Sherman, M.H. and M.P. Modera. 1986. Comparison of measured and predicted infiltration using the LBL infiltration model. In *Measured air leakage of buildings*, p. 325. ASTM STP 904. H.R. Trechsel and P.L. Lagus, eds.

Sherman, M.H. and D.J. Wilson. 1986. Relating actual and effective ventilation in determining indoor air quality. *Building and Environment* 21(3/4):135.

Sibbitt, B.E., and T. Hamlin. 1991. *Meeting Canadian residential ventilation standard requirements with low-cost systems.* Canada Mortgage and Housing Corporation, Ottawa.

Sinden, F.W. 1978a. Wind, temperature and natural ventilation—Theoretical considerations. *Energy and Buildings* 1(3):275.

Sinden, F.W. 1978b. Multi-chamber theory of air infiltration. *Building and Environment* 13:21-28.

Subbarao, K., J.D. Burch, C.E. Handcock, A. Lekov, and J.D. Balcomb. 1990. Measuring the energy performance of buildings through short-term tests. *ACEEE 1990 Summer Study on Energy Efficiency in Buildings* 10: 245-252.

Swedish Building Code. 1980. Thermal insulation and air tightness. SBN 1980.

Tamura, G.T. and C.Y. Shaw. 1976a. Studies on exterior wall airtightness and air infiltration of tall buildings. *ASHRAE Transactions* 82(1):122.

Tamura, G.T. and C.Y. Shaw. 1976b. Air leakage data for the design of elevator and stair shaft pressurization system. *ASHRAE Transactions* 82(2):179.

Tamura, G.T. and A.G. Wilson. 1966. Pressure differences for a nine-story building as a result of chimney effect and ventilation system operation. *ASHRAE Transactions* 72(1):180.

Tamura, G.T. and A.G. Wilson. 1967a. Pressure differences caused by chimney effect in three high buildings. *ASHRAE Transactions* 73(2):II.1.1.

Tamura, G.T. and A.G. Wilson. 1967b. Building pressures caused by chimney action and mechanical ventilation. *ASHRAE Transactions* 73(2): II.2.1.

Timusk, J., A.L. Seskus, and K. Linger. 1992. A systems approach to extend the limit of envelope performance. Proceedings of the *Thermal Performance of Exterior Envelopes of Buildings, 6th ASHRAE/DOE/BTECC Conference*, Clearwater Beach, FL.

Turk, B.T., D.T. Grimsrud, J.T. Brown, K.L. Geisling-Sobotka, J. Harrison, and R.J. Prill. 1989. Commercial building ventilation rates and particle concentrations. *ASHRAE Transactions* 95(1):422-33.

U.S. Environmental Protection Agency. *Code of Federal regulations.* Title 40. Part 50 (current national ambient air quality standards) July 1, 1986.

Verschoor, J.D. and J.O. Collins. 1986. Demonstration of air leakage reduction program in navy family housing. In *Measured air leakage of buildings*, p. 294. ASTM STP 904. H.R. Trechsel and P.L. Lagus, eds.

Walker, I.S. 1999. Distribution system leakage impacts on apartment building ventilation rates. *ASHRAE Transactions*105(1):943-50.

Walker, I.S. and T.W. Forest. 1995. Field measurements of ventilation rates in attics. *Building and Environment* 30(3):333-47, Elsevier Science Ltd., Pergamon Press, U.K.

Walker, I.S. and D.J. Wilson. 1993. Evaluating models for superposition of wind and stack effects in air infiltration. *Building and Environment* 28(2): 201-10, Pergamon Press.

Walker, I.S. and D.J. Wilson. 1994. Simple methods for improving estimates of natural ventilation rates. Proceedings of the *15th AIVC Conference*, Buxton, UK. September.

Walker, I.S. and D.J. Wilson. 1998. Field validation of equations for stack and wind driven air infiltration calculations. *International Journal of HVAC&R Research* 4(2).

Walker, I.S., D.J. Wilson., and M.H. Sherman. 1997. A comparison of the power law to quadratic formulations for air infiltration calculations. *Energy and Buildings* 27(3).

Walker, I., M. Sherman, J. Siegel, D. Wang, C. Buchanan, and M. Modera. 1999. Leakage diagnostics, sealant longevity, sizing and technology transfer in residential thermal distribution systems: Part II. LBNL *Report* 42691.

Walton, G.N. 1984. A computer algorithm for predicting infiltration and interroom airflows. *ASHRAE Transactions* 90(1B):601.

Walton, G.N. 1989. Airflow network models for element-based building airflow modeling. *ASHRAE Transactions* 95(2):611-20.

Warren, P.R. and B.C. Webb. 1980. The relationship between tracer gas and pressurization techniques in dwellings. Proceedings of the *1st IEA Symposium of the Air Infiltration Centre*, London.

Warren, P.R. and B.C. Webb. 1986. Ventilation measurements in housing. CIBSE Symposium, Natural Ventilation by Design. Chartered Institution of Building Services Engineers, London.

Weidt, J.L., J. Weidt, and S. Selkowitz. 1979. Field air leakage of newly installed residential windows. Proceedings of the *ASHRAE-DOE Conference, Thermal Performance of the Exterior Envelopes of Buildings*, p. 149. Orlando, FL.

Wilson, D.J. and I.S. Walker. 1991. Wind shelter effects on a row of houses. Proceedings of the *12th AIVC Conference*, Ottawa, Canada.

Wilson, D.J. and I.S. Walker. 1992. Feasibility of passive ventilation by constant area vents to maintain indoor air quality in houses. Proceedings of *Indoor Air Quality '92, ASHRAE/ACGIH/AIHA Conference*, San Francisco. October.

Wilson, D.J. and I.S. Walker. 1993. Infiltration data from the Alberta Home Heating Research Facility. Air Infiltration and Ventilation Centre *Technical Note* 41. AIVC, Coventry, England.

Wiren, B.G. 1984. Wind pressure distributions and ventilation losses for a single-family house as influenced by surrounding buildings—A wind tunnel study. Proceedings of the *Air Infiltration Centre Wind Pressure Workshop*, pp. 75-101. Brussels, Belgium.

Wolf, S. 1996. A theory of the effects of convective air flow through fibrous thermal insulation. *ASHRAE Transactions* 72(1):III 2.1-III 2.9.

Yaglou, C.P. and W.N. Witheridge. 1937. Ventilation requirements. *ASHVE Transactions* 43:423.

Yaglou, C.P., E.C. Riley, and D.I. Coggins. 1936. Ventilation requirements. *ASHVE Transactions* 42:133.

Yuill, G.K. 1986. The variation of the effective natural ventilation rate with weather conditions. *Renewable Energy Conference '86*. Solar Energy Society of Canada Inc., pp. 70-75.

Yuill, G.K. 1991. The development of a method of determining air change rates in detached dwellings for assessing indoor air quality. *ASHRAE Transactions* 97(2).

Yuill, G.K. 1996. Impact of high use automatic doors on infiltration. *Report* to ASHRAE (763-RP), July.

Yuill, G.K. and G.M. Comeau. 1989. Investigation of the indoor air quality, air tightness and air infiltration rates of a random sample of 78 houses in Winnipeg, pp. 122-27. Proceedings of *IAQ '89, The Human Equation—Health and Comfort*. ASHRAE, Atlanta.

Yuill, G.K., M.R. Jeanson, and C.P. Wray. 1991. Simulated performance of demand-controlled ventilation systems using carbon dioxide as an occupancy indicator. *ASHRAE Transactions* 97(2):963-68.

BIBLIOGRAPHIC DATABASE

AIRBASE, a database of bibliographic references that contains abstracts in English of technical papers covering air infiltration in buildings, has been developed by the Air Infiltration and Ventilation Centre (AIVC 1992). Most of the articles are concerned with the prediction, measurement, and reduction of air infiltration and leakage rates. Abstracts are also included for selected papers relating to indoor air quality, occupant behavior, thermal comfort, ventilation efficiency, natural and mechanical ventilation, wind pressure and its influence on infiltration, energy-saving measures, and moisture and condensation.

CHAPTER 27

CLIMATIC DESIGN INFORMATION

THIS CHAPTER provides tables of climatic conditions for 1459 locations in the United States, Canada, and around the world. These summaries include values of dry-bulb, wet-bulb and dew-point temperature and wind speed with direction at various frequencies of occurrence. This information is commonly used for design, sizing, distribution, installation, and marketing of heating, ventilating, air-conditioning, and dehumidification equipment; as well as for other energy-related processes in residential, agricultural, commercial, and industrial applications. Sources of other information such as degree-days and typical weather years for energy calculations are also described.

The design information in this chapter was developed largely as part of a research project (ASHRAE 1997c, Colliver et al. 2000). The information includes design values of dry-bulb with mean coincident wet-bulb temperature, design wet-bulb with mean coincident dry-bulb temperature, and design dew-point with mean coincident dry-bulb temperature and corresponding humidity ratio. These data allow the designer to consider various operational peak conditions. Design values of wind speed (ASHRAE 1994, Lamming and Salmon 1998) allow for the design of smoke management systems in buildings.

The design conditions in this chapter are provided for those locations for which long-term hourly observations were available (at least 12 years of data). Consequently, many U.S. locations listed in previous versions (1993 and before) of this chapter are no longer listed because they lacked long-term data. The number of Canadian and international locations has increased significantly.

Warm-season temperature and humidity conditions correspond to annual percentile values of 0.4, 1.0, and 2.0. Cold-season conditions are based on annual percentiles of 99.6 and 99.0. The use of annual percentiles to define the design conditions ensures that they represent the same probability of occurrence anywhere, regardless of the seasonal distribution of extreme temperature and humidity. In the 1993 and earlier versions of this chapter, seasonal rather than annual percentiles were used to define the design conditions. As a result, the summer and winter months used for the calculation of design conditions varied depending on location. For instance, summer design conditions for the U.S. were calculated over the four-month period from June through September, whereas Canadian summer design conditions were based on only the month of July. The following sections describe how the annual percentiles were selected to yield design conditions that are similar to those previously calculated on a seasonal basis for most of the United States.

For a variety of reasons, such as seasonal variations in solar geometry, occupancy, or use of a building, there is sometimes a need for percentiles of temperature and humidity for specific months. New monthly tables for some U.S. locations have been added in this edition to meet this need.

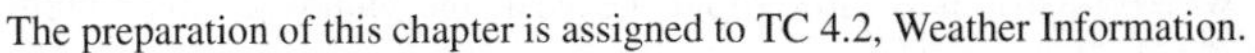

The preparation of this chapter is assigned to TC 4.2, Weather Information.

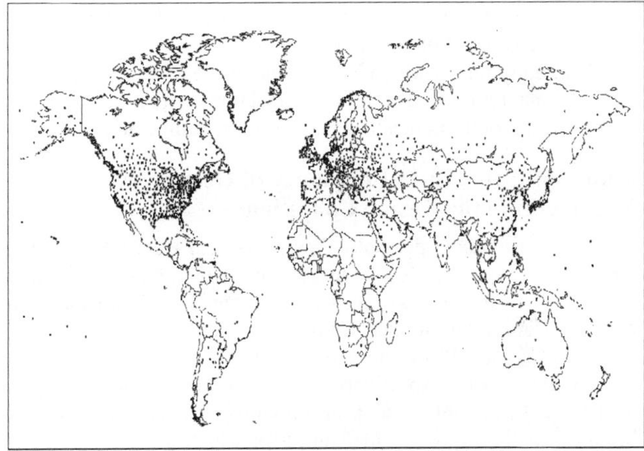

Fig. 1 Location of Weather Stations

CLIMATIC DESIGN CONDITIONS

Annual design conditions for the United States appear in Tables 1A and 1B, for Canada in Tables 2A and 2B, and for international locations in Tables 3A and 3B. Figure 1 is a world map showing locations.

Information on station location, period analyzed, heating design conditions, wind, mean annual extreme and standard deviation of minimum and maximum dry-bulb temperature, and mean daily temperature range is listed in Tables 1A, 2A, and 3A. Information on the design conditions for cooling and humidity control is provided in Tables 1B, 2B, and 3B.

The information provided in Tables 1A, 2A, and 3A in the indicated column numbers is

1. Name of the observing station as it appears in the data set from which it was abstracted, the World Meteorological Organization station number, latitude, longitude, elevation, standard pressure at elevation (see Chapter 6 for the equations used to calculate standard pressure), and the period analyzed
2. Dry-bulb temperature corresponding to 99.6% and 99.0% annual cumulative frequency of occurrence (cold)
3. Wind speed corresponding to 1.0%, 2.5% and 5.0% annual cumulative frequency of occurrence
4. Wind speed corresponding to the 0.4% and 1.0% cumulative frequency of occurrence for the coldest month (lowest average dry-bulb temperature), and the mean coincident dry-bulb temperature
5. Mean wind speed coincident with the 99.6% dry-bulb temperature in Column 2 and 0.4% dry-bulb temperature from Column 2 of Tables 1B, 2B, and 3B, and the wind direction most frequently occurring with the 99.6% and 0.4% dry-bulb

temperatures (direction is reported in degrees true: 360 represents a north wind, 90 east, etc.)

6. Average of annual extreme maximum and minimum dry bulb temperatures and standard deviations

Information provided in Tables 1B, 2B, and 3B includes

1. Station name
2. Dry-bulb temperature corresponding to 0.4%, 1.0%, and 2.0% annual cumulative frequency of occurrence and the mean coincident wet-bulb temperature (warm)
3. Wet-bulb temperature corresponding to 0.4%, 1.0%, and 2.0% annual cumulative frequency of occurrence and the mean coincident dry bulb temperature
4. Dew-point temperature corresponding to 0.4%, 1.0%, and 2.0% annual cumulative frequency of occurrence and the mean coincident dry-bulb temperature and the humidity ratio (calculated for the dew-point temperature at the standard atmospheric pressure at the elevation of the station)
5. Mean daily range, which is the mean of the difference between daily maximum and minimum dry-bulb temperatures for the warmest month (highest average dry-bulb temperature)

Values of Cumulative Frequency of Occurrence Representing the Design Conditions

Values of ambient dry-bulb, dew-point, and wet-bulb temperature and wind speed corresponding to the various annual percentiles represent the value that is exceeded on average by the indicated percentage of the total number of hours in a year (8760). The 0.4%, 1.0%, 2.0%, and 5.0% values are exceeded on average 35, 88, 175, and 438 h per year, respectively, for the period of record.

The 99.0% and 99.6% (cold) values are defined in the same way but are usually viewed as the values for which the corresponding weather element is less than the design condition for 88 and 35 h, respectively.

Mean coincident values are the average of the indicated weather element occurring concurrently with the corresponding design value.

These design conditions were calculated from the frequency distribution analyzed from data sets observed over several years. Gaps due to missing data were filled as needed, as explained in the section on Calculation of Design Conditions. The design values occur more frequently than the corresponding nominal percentile in some years and less frequently in others.

Data Sources

The following three primary sources of observational data sets were used for the calculation of design values:

1. Hourly weather observations from Surface Airways Meteorological and Solar Observing Network (SAMSON) data from NCDC (National Climatic Data Center) for 239 United States observing locations from 1961 through 1990 (NCDC 1991)
2. Hourly observational records in the DATSAV2 format (ASHRAE 1995b, Plantico and Lott 1995) for 538 United States locations for the period 1982 through 1993, including the 239 SAMSON locations, and for 860 international locations (most data for international locations consist of observations taken every 3 h)
3. Hourly weather records for the period 1953 through 1993 for 145 Canadian locations from the Canadian Weather Energy and Engineering Data Sets (CWEEDS) produced by Environment Canada (1993b)

Two primary periods of record were used in the calculations. The values for the United States SAMSON and Canadian CWEEDS locations are generally based on the period from 1961 through 1993. DATSAV2 data were used for the 1991 through 1993 period for the SAMSON sites. The values for international locations and the rest of the United States, whose data were analyzed from the DATSAV2 files, are based on the period 1982 through 1993. DATSAV2 is a comprehensive database containing hourly observations for locations around the world collected from global telecommunications circuits. It is quality-controlled and archived by the Air Force Combat Climatology Center at Asheville, NC. Tables 1A, 2A, and 3A indicate the period of record used for each location.

In summary, the data source for United States locations with the period identified as "6193" is SAMSON data supplemented with DATSAV2 data for the last 3 years. The source for United States locations with the period "8293" is DATSAV2.

Calculation of Design Conditions

Details of the analysis procedures are available in ASHRAE (1997c) and Colliver et al. (2000), including the measures used to ensure that the number and distribution of missing data, both by month and by hour of the day, did not introduce significant biases into the analysis. Generally, the annual cumulative frequency distribution was constructed from the relative frequency distributions compiled for each month. Each individual month's data were included if they met screening criteria for completeness and unbiased distribution of missing data. Although the minimum period of record selected for this analysis was 12 years (1982 through 1993), some variation and gaps in observing programs meant that some months' data were unusable due to incompleteness.

A station's design conditions were included in this chapter only if there were data from at least 8 months that met the screening criteria from the period of record for each month of the year. For instance, there had to be 8 months each for January, February, March, etc. whose data met the completeness screening criteria.

Gaps of up to 5 h were filled. A month's data were included if the month was at least 85% complete after filling and the difference between the number of day and nighttime observations was less than 60.

Relationship Between Design Conditions and Previously Published Design Temperatures

The design conditions in this chapter are calculated on a different basis from the design conditions published in the 1993 and previous editions of this Handbook. Previous design conditions were based on a 4-month summer period and a 3-month winter period in the United States, on the months of July and January in Canada, and on the warmest 4-month period and coldest 3-month period in international locations. Although generally suitable as design values, the different periods resulted in design temperatures representing different annual probabilities of occurrence, depending on the country; and within countries, on the distribution of temperature and humidity conditions throughout the year typical of regional climatic zones. The design conditions in this chapter explicitly represent the same annual probability of occurrence in any location, regardless of country or general climatic conditions.

The annual cumulative frequencies of occurrence representing the design dry-bulb temperatures generally correspond to the seasonal design temperatures in the following fashion for locations in the mid-latitude, continental locations (characterized by a hot summer and cold winter). The 0.4% annual value is about the same as the 1.0% summer design temperature in the 1993 *ASHRAE Handbook*. The 1.0% annual value is about 1°F lower than the 2.5% summer design temperature in the 1993 *ASHRAE Handbook*, and the 2.0% annual condition corresponds approximately to the 5.0% summer design temperature in the 1993 *ASHRAE Handbook*.

In Canadian continental locations, the 0.4% annual condition is about the same as the 2.5% July design temperature in the 1993 *ASHRAE Handbook*. In the United States for the Pacific region and southern coastal locations, where the extremes are generally more widely distributed throughout the year, the values in this chapter represent more extreme conditions than the design temperatures in the 1993 *ASHRAE Handbook*.

Annual 99.6% and 99.0% design conditions represent a slightly colder condition than the previous cold season design temperatures, although there is considerable variability in this relationship from location to location.

Further details concerning differences between the design conditions in this chapter and the 1993 and previous versions are described in ASHRAE (1997c) and Colliver et al. (2000).

Applicability and Characteristics of the Design Conditions

The sets of design values in this chapter represent different psychrometric conditions. Design data based on dry-bulb temperature represent peak occurrences of the sensible component of ambient outdoor conditions. Design values based on wet-bulb temperature are related to the enthalpy of the outdoor air. Conditions based on dew point relate to the peaks of the humidity ratio. The designer, engineer, or other user must decide which set(s) of conditions and probability of occurrence apply to the design situation under consideration. The addition of the new psychrometric design conditions allows for several viewpoints of operational peak loads. Additional sources of information on the frequency and duration of extremes of temperature and humidity are provided later in the section on Other Sources of Climatic Information. Further information is available from Harriman et al. (1999).

Heating Design Conditions. The 99.6% and 99.0% design conditions in Column 2 of Tables 1A, 2A, and 3A are often used in the sizing of heating equipment. In cold spells, dry-bulb temperatures below the design conditions can last for a week or more.

Columns 4 and 5 of Tables 1A, 2A, and 3A provide information useful for estimating peak loads accounting for infiltration. Column 4 provides extreme wind speeds only for the coldest month, with the mean coincident dry-bulb temperature. Column 5 provides the mean wind speed and direction coincident to the corresponding percentile design dry-bulb temperature.

Cooling and Dehumidification Design Conditions. The 0.4%, 1.0%, and 2.0% dry-bulb temperatures and mean coincident wet-bulb temperatures in Column 2 of Tables 1B, 2B, and 3B often represent conditions on hot, mostly sunny days. These are useful for cooling applications, especially air conditioning.

Design conditions based on wet-bulb temperature in Column 3 represent extremes of the total sensible plus latent heat of outdoor air. This information is useful for cooling towers, evaporative coolers, and fresh air ventilation system design.

The design conditions based on dew-point temperatures in Column 4 of Tables 1B, 2B, and 3B are directly related to extremes of humidity ratio, which represent peak moisture loads from the weather. Extreme dew-point conditions may occur on days with moderate dry-bulb temperatures resulting in high relative humidity. These values are especially useful for applications involving humidity control, such as desiccant cooling and dehumidification, cooling-based dehumidification, and fresh air ventilation systems. The values are also used as a check point when analyzing the behavior of cooling systems at part-load conditions, particularly when such systems are used for humidity control as a secondary function.

The humidity ratio values in Column 2 correspond to the combination of dew-point temperature and the mean coincident dry-bulb temperature calculated at the standard pressure at the elevation of the location.

Wind. Design wind speeds in Column 3 of Tables 1A, 2A, and 3A have been produced as a result of the requirement for the design of smoke management systems. Annual percentiles of 1.0, 2.5, and 5.0 have been determined appropriate for this application. The values for the United States SAMSON sites and Canadian locations have been taken from ASHRAE (1994), in which adjustments to the standard 33 ft anemometer height were made. Wind speed values for other locations are taken from ASHRAE (1997c), in which no adjustment for nonstandard anemometer height is made.

Annual Extreme Temperatures. Column 6 of Tables 1A, 2A, and 3A provides the mean and standard deviation of the annual extreme maximum and minimum dry-bulb temperatures. The probability of occurrence of very extreme conditions can be required for the operational design of equipment to ensure continuous operation and serviceability (regardless of whether the heating or cooling loads are being met). These values were calculated from the extremes of the hourly temperature observations. The true maximum and minimum temperatures for any day generally occur between hourly readings. Thus, the mean maximum and minimum temperatures calculated in this way will be about 1°F less extreme than the mean daily extreme temperatures observed with maximum and minimum thermometers.

Return period (or recurrence interval) is defined as the reciprocal of the annual probability of occurrence. For instance, the 50-year return period maximum dry-bulb temperature has a probability of occurring or being exceeded of 2.0% (i.e., 1/50) each year. This statistic does not indicate how often the condition will occur in terms of the number of hours each year (as in the design conditions based on percentiles) but describes the probability of the condition occurring at all in any year. The following method can be used to estimate the return period (recurrence interval) of extreme temperatures.

$$T_n = M + IFs \tag{1}$$

where

T_n = n-year return period value of extreme dry-bulb temperature to be estimated, years
M = mean of the annual extreme maximum or minimum dry-bulb temperatures, °F
s = standard deviation of the annual extreme maximum or minimum dry-bulb temperatures, °F
I = 1, if maximum dry-bulb temperatures are being considered
= –1 if minimum extremes are being considered

$$F = -\frac{\sqrt{6}}{\pi}\left\{0.5772 + \ln\left[\ln\left(\frac{n}{n-1}\right)\right]\right\}$$

For example, the 50-year return period extreme maximum dry-bulb temperature estimated for Terre Haute, Indiana is 104°F (M = 96°F, s = 3.2, n = 50, I = 1). Similarly, the 100-year return period extreme minimum dry-bulb temperature for Winnipeg, Manitoba, is –48°F (M = –33°F, s = 4.7, n = 100, I = –1).

This calculation is based on the assumptions that the annual maxima and minima are distributed according to the Gumbel (Type 1 Extreme Value) distribution and are fitted with the method of moments (Lowery and Nash 1970). The uncertainty or standard error using this method increases with increasing standard deviation, increasing value of return period, and decreasing length of the period of record. It can be significant. For instance, the standard error in the 50-year return period maximum dry-bulb temperature estimated at a location with a 12-year period of record can be 5°F or more. Thus, the uncertainty of return-period values estimated in this way will be greater for locations from the DATSAV2 data sets than those from the longer SAMSON and CWEEDS data sets.

Mean Daily Range. The mean daily range is the mean difference between the daily maximum and minimum temperatures during the hottest month. These values are calculated from the extremes of the hourly temperature observations. The true daily temperature range is generally about 2°F greater, for the same reason as explained in the previous section.

Monthly Tables

Selected monthly percentiles of temperature and humidity are provided in Tables 4A and 4B for 239 U.S. locations from the SAMSON data set. These monthly values are derived from the same analysis that results in the design conditions in Tables 1A and 1B. These monthly summaries provide additional information when seasonal variations in solar geometry and intensity, building or facility occupancy, or building use patterns require consideration. In particular, these values can be used when determining air-conditioning loads during periods of maximum solar radiation.

Table 4A contains the location name and WMO station number and the 0.4%, 1.0%, and 2.0% value of the wet-bulb temperature and mean coincident dry-bulb temperature for the indicated month and annual period. The percentiles for the annual period are the same as those in Table 1B for the same location and are listed in Table 4A for convenience.

Table 4B contains the location name and WMO station number and the 0.4%, 1.0%, and 2.0% value of the dry-bulb temperature and mean coincident wet-bulb temperature for the indicated month and annual period. The percentiles for the annual period are the same as those in Table 1B for the same location and are listed in Table 4B for convenience.

For a 30-day month, the 0.4%, 1.0%, and 2.0% values of occurrence represent the value that occurs or is exceeded for a total of 3, 7, or 14 h, respectively, per month on average over the period of record.

Monthly percentile values of dry-bulb or wet-bulb temperature may be higher or lower than the design conditions corresponding to the same nominal percentile, depending on the month and the seasonal distribution of the parameter at that location. Generally, for the hottest or most humid months of the year, the monthly percentile value will exceed the design condition for the same element corresponding to the same nominal percentile. For instance, Table 1B shows that the 0.4% design dry-bulb temperature at Mobile, Alabama, is 94°F. Table 4B shows that the 0.4 monthly percentile dry-bulb temperature exceeds 94°F for the months of June, July, and August, with values of 95°F, 96°F, and 95°F, respectively.

It is essential to remember that the design conditions are based on an annual analysis. The 0.4% annual value represents a value occurring or exceeded 35 h on average every year over the period of record, whereas the 0.4% monthly value occurs or is exceeded 3 h on average for the month that it represents, over the period of record.

Representativeness of Data and Sources of Uncertainty

The information in the tables was obtained by direct analysis of the observations from the indicated locations. The design values are provided and used as an estimate of the annual cumulative frequency of occurrence of the weather conditions at the recording station for several years into the future. Several sources of uncertainty affect the accuracy of using the design conditions to represent other locations or periods.

The most important of these factors is spatial representativeness. Most of the observed data for which design conditions were calculated were collected from airport observing sites, most of which are flat, grassy, open areas, away from buildings and trees or other local influences. Data representing the psychrometric conditions are generally properties of air masses, rather than local features, and tend to vary on regional scales. As a result, a particular value often may reasonably represent an area extending several miles. However, significant variations can occur with changes in local elevation, across large metropolitan areas, or in the vicinity of large bodies of water. Judgment must always be exercised in assessing the representativeness of the design conditions. It is especially important to note the elevation of locations because design conditions vary significantly for locations whose elevations differ by several hundred feet or more. An applied climatologist should be consulted in estimating design conditions for locations not explicitly listed in this chapter.

Weather conditions vary from year to year and, to some extent, from decade to decade due to the inherent variability of climate. Similarly, values representing design conditions vary depending on the period of record used in the analysis. Thus, due to short-term climatic variability, there is always some uncertainty in using the design conditions from one period to represent another period. Typically, the values of design dry-bulb temperature vary less than 2°F from decade to decade, but larger variations can occur. Differing periods used in the analysis can lead to differences in design conditions between nearby locations at similar elevations. Design conditions may show trends in areas of increasing urbanization or other regions experiencing extensive changes to land use. Longer term climatic change due to human or natural causes may also introduce trends into design conditions, but no conclusive evidence or consensus of opinion is available on the rapidity or nature of such changes.

Wind speed is very sensitive to local exposure features such as terrain and surface cover. Wind speed values in Columns 3, 4, and 5 of Tables 1A, 2A, and 3A are often representative of a flat, open exposure, such as at airports. Wind engineering methods, as described in Chapter 16, can be used to account for exposure differences between airport and building sites. This is a complex procedure, best undertaken by an experienced applied climatologist or wind engineer having knowledge of the exposure of the observing and building sites and surrounding regions.

OTHER SOURCES OF CLIMATIC INFORMATION

Joint Frequency Tables of Psychrometric Conditions

Most of the design values in this chapter were developed by a research project (ASHRAE 1997c, Colliver et al. 2000). The joint frequency tables from this project provide the annual and monthly joint frequency of occurrence of various psychrometric combinations for each location in the tables. The complete set of joint frequency tables is available on CD (ASHRAE 1999)

The Engineering Weather Data (NCDC 1999) CD, an update of Air Force *Manual* 88-29, was compiled by the U.S. Air Force Combat Climatology Center. This CD contains several tabular and graphical summaries of temperature, humidity, and wind speed information for hundreds of locations in the United States and around the world. In particular, it contains detailed joint frequency tables of temperature and humidity for each month, binned at 1°F and 3 h local time-of-day intervals.

The International Station Meteorological Climate Summary (ISMCS) is a CD-ROM containing several tables of climatic summary information for over 7000 locations around the world (NCDC 1996). Only several hundred of these locations have observed hourly weather data. A table providing the joint frequency of dry-bulb temperature and wet-bulb temperature depression is provided for the locations with hourly observations. It can be used as an aid in estimating design conditions for locations for which no other information is available.

The monthly frequency distribution of dry-bulb temperatures and mean coincident wet-bulb temperatures is available for 134 Canadian locations from Environment Canada (1983-1987).

Degree-Days and Climatic Normals

Degree-day summary information and climatic normals for the United States are available on CD-ROM in The Climatography of the U.S. (NCDC 1992a, 1992b, 1994); and, for Canada, in the Canadian Climate Normals, 1961 to 1990 (Environment Canada 1993a).

Typical Year Data Sets

Software exists to simulate the annual energy performance of buildings requiring a 1-year data set (8760 h) of weather conditions. Many data sets in different record formats have been developed to meet this requirement. The data represent a typical year from the viewpoint of weather-induced energy loads on a building. No explicit effort was made to represent extreme conditions, so these files do not represent design conditions.

Weather Year for Energy Calculations Version 2 (WYEC2) for 52 U.S. and 5 Canadian locations is available from ASHRAE (ASHRAE 1997d). A user manual and software toolkit explains the development, format, and characteristics of this data and provides access and toolkit software.

Typical Meteorological Year Version 2 (TMY2) files are available for 239 U.S. locations from the National Renewable Energy Laboratory (Marion and Urban 1995). These were produced using an objective statistical algorithm to select the most typical month from the long-term record. These files were originally intended for the design of solar energy systems, so solar radiation values were weighted heavily.

Canadian Weather Year for Energy Calculation (CWEC) files for 47 Canadian locations were developed for use with the Canadian National Energy Code, using the TMY algorithm and software (Environment Canada 1993b).

Examples of the use of these files for energy calculations in both residential and commercial buildings, including the differences among the files, are available in Crawley (1998) and Huang (1998).

Seasonal Percentiles of Dew-Point Temperature

Seasonal percentiles of dew-point temperature are available for 381 U.S. and Canadian locations in ASHRAE (1995a), with their development summarized in Colliver et al. (1995).

Sequences of Extreme Temperature and Humidity Durations

Colliver et al. (1998) and ASHRAE (1997b) compiled extreme sequences of 1-, 3-, 5-, and 7-day duration for 239 U.S. (SAMSON data) and 144 Canadian (CWEEDS data) locations based independently on the following five criteria: high dry-bulb temperature, high dew-point temperature, high enthalpy, low dry-bulb temperature, and low wet-bulb depression. For the criteria associated with high values, the sequences are selected according to annual percentiles of 0.4, 1.0, and 2.0. For the criteria corresponding to low values, annual percentiles of 99.6, 99.0, and 98.0 are reported. The data included for each hour of a sequence are solar radiation, dry-bulb and dew-point temperature, atmospheric pressure, and wind speed and direction. Accompanying information allows the user to go back to the source SAMSON and CWEEDS data and obtain sequences with different characteristics (i.e., different probability of occurrence, windy conditions, low or high solar radiation, etc.). These extreme sequences are available on CD (ASHRAE 1997a).

These sequences were developed primarily to assist the design of heating or cooling systems having a finite capacity before regeneration is required or systems that rely on thermal mass to limit loads. The information is also useful where information on the hourly weather sequence during extreme episodes is required for design.

Observational Data Sets

For detailed designs, custom analysis of the most appropriate long-term weather record is best. National weather services are generally the best source of long-term observational data. The WMO World Data Center A at the National Climate Data Center in Asheville, NC, collects and makes available a significant volume of archived weather observations from around the world.

The SAMSON and CWEEDS data sets provide long-term hourly data, including solar radiation values for the United States and Canada.

Increasingly, information about weather and climate services and data sets, as well as the data sets themselves, is becoming available through the Internet and World Wide Web. Information supplementary to this chapter may be posted from time to time on the ASHRAE Technical Committee 4.2 web site, the link to which is available from the ASHRAE web site (www.ashrae.org).

REFERENCES

ASHRAE. 1994. Wind data for design of smoke control systems. *Research Report* RP-816.

ASHRAE. 1995a. Design data for the 1%, 2½%, and 5% occurrences of extreme dew point temperature, with mean coincident dry-bulb temperature. *Research Report* RP-754.

ASHRAE. 1995b. Weather data sets for ASHRAE research. *Research Report* RP-889.

ASHRAE. 1997a. *Design weather sequence viewer* 2.1 (CD-ROM).

ASHRAE. 1997b. Sequences of extreme temperature and humidity for design calculations. *Research Report RP-828.*

ASHRAE. 1997c. Updating the tables of design weather conditions in the *ASHRAE Handbook of Fundamentals. Research Report RP-890.*

ASHRAE. 1997d. WYEC2 data and toolkit (CD-ROM).

ASHRAE. 1999. Weather data viewer 2.1 (CD-ROM).

Colliver, D.G., H. Zhang, R.S. Gates, and K.T. Priddy. 1995. Determination of the 1%, 2.5%, and 5% occurrences of extreme dew-point temperatures and mean coincident dry-bulb temperatures. *ASHRAE Transactions* 101(2):265-86.

Colliver D.G., R.S. Gates, H. Zhang, and K.T. Priddy. 1998. Sequences of extreme temperature and humidity for design calculations. *ASHRAE Transactions* 104(1a):133-44.

Colliver, D.G., R.S. Gates, T.F. Burkes, and H. Zhang. 2000. Development of the design climatic data for the 1997 *ASHRAE Handbook—Fundamentals. ASHRAE Transactions* 106(1).

Crawley, D.B. 1998. Which weather data should you use for energy simulations of commercial buildings? *ASHRAE Transactions* 104(2):498-515.

Environment Canada. 1983-1987. Principal station data. PSD 1 to 134. Atmospheric Environment Service, Downsview, Ontario.

Environment Canada. 1993A. Canadian climate normals (1961-1990).

Environment Canada. 1993b. Canadian weather energy and engineering data sets (CWEEDS files) and Canadian weather for energy calculations (CWEC files) user's manual.

Harriman, L.G., D.G. Colliver, and H.K. Quinn. 1999. New weather data for energy calculations. *ASHRAE Journal* 41(3):31-38.

Huang, J. 1998. The impact of different weather data on simulated residential heating and cooling loads. *ASHRAE Transactions* 104(2):516-27.

Lamming S.D. and J.R. Salmon. 1998. Wind data for design of smoke control systems *ASHRAE Transactions* 104(1A):742-51.

Lowery, M.D. and J.E. Nash. 1970. A comparison of methods of fitting the double exponential distribution. *Journal of Hydrology* 10(3):259-75.

Marion, W. and K. Urban. 1995. User's Manual for TMY2s, typical meteorological years, derived from the 1961-1990 National Solar Radiation Data Base. NREL/SP-463-7668, E95004064. National Renewable Energy Laboratory, Golden, CO.

NCDC. 1991. Surface airways meteorological and solar observing network (SAMSON) data set. National Climatic Data Center, Asheville, NC.

NCDC. 1992A. Monthly normals of temperature, precipitation, and heating and cooling degree-days. In *Climatography of the U.S.* #81.

NCDC. 1992B. Annual degree-days to selected bases (1961-1990). In *Climatography of the U.S.* #81.

NCDC. 1994. U.S. division and station climatic data and normals.

NCDC. 1996. International station meteorological climate summary (ISMCS).

NCDC. 1999. Engineering weather data.

Plantico, M.S. and J.N. Lott. 1995. Foreign weather data servicing at NCDC. *ASHRAE Transactions* 101(1):484-90.

Table 1A Heating and Wind Design Conditions—United States

Station	WMO#	Lat.	Long.	Elev., ft	StdP, psia	Dates	Heating Dry Bulb 99.6%	Heating Dry Bulb 99%	Extreme Wind Speed, mph 1%	2.5%	5%	Coldest Month 0.4% WS	0.4% MDB	1% WS	1% MDB	MWS/PWD to DB 99.6% MWS	99.6% PWD	0.4% MWS	0.4% PWD	Extr. Annual Daily Mean DB Max.	Mean DB Min.	StdD DB Max.	StdD DB Min.
1a	1b	1c	1d	1e	1f	1g	2a	2b	3a	3b	3c	4a	4b	4c	4d	5a	5b	5c	5d	6a	6b	6c	6d
ALABAMA																							
Anniston	722287	33.58	85.85	610	14.374	8293	19	24	16	14	13	18	47	15	46	6	300	7	240	98	10	3.2	7.4
Birmingham	722280	33.57	86.75	630	14.364	6193	18	23	19	17	15	20	41	18	42	7	340	9	320	98	9	3.3	6.4
Dothan	722268	31.32	85.45	400	14.484	8293	28	32	18	17	15	19	45	17	47	9	320	8	320	99	16	1.6	7.2
Huntsville	723230	34.65	86.77	643	14.357	6193	15	20	23	20	18	23	40	21	40	10	340	10	270	97	7	3.0	7.5
Mobile	722230	30.68	88.25	220	14.579	6193	26	30	22	19	17	23	48	21	48	10	360	9	320	97	18	1.9	6.3
Montgomery	722260	32.30	86.40	203	14.588	6193	24	27	20	17	15	20	45	18	45	7	360	8	270	98	15	2.9	6.3
Muscle Shoals/Florence	723235	34.75	87.62	551	14.405	8293	16	21	18	16	14	19	42	17	42	9	360	7	290	98	7	3.1	9.2
Ozark, Ft. Rucker	722269	31.28	85.72	299	14.538	8293	28	31	16	13	12	17	49	15	47	5	340	5	300	99	18	2.3	5.9
Tuscaloosa	722286	33.22	87.62	171	14.605	8293	20	24	17	14	13	18	47	16	51	5	360	7	240	99	11	1.8	6.8
ALASKA																							
Adak, NAS	704540	51.88	176.65	13	14.688	8293	19	23	34	30	27	40	34	34	35	4	210	10	170	67	11	3.4	2.9
Anchorage, Elementorf AFB	702720	61.25	149.80	213	14.583	8293	−13	−8	17	14	12	18	26	15	26	3	50	7	260	77	−18	3.2	6.5
Anchorage, Ft. Richardson	702700	61.27	149.65	377	14.496	8293	−19	−13	19	14	11	20	35	15	36	3	50	5	270	80	−23	2.2	6.3
Anchorage, Intl Airport	702730	61.17	150.02	131	14.626	6193	−14	−9	22	19	17	23	18	19	18	4	10	8	290	77	−18	2.9	7.2
Annette	703980	55.03	131.57	112	14.636	6193	13	17	31	27	23	31	41	28	40	10	40	8	320	81	10	3.8	5.4
Barrow	700260	71.30	156.78	13	14.688	6193	−41	−36	28	25	22	30	3	26	−1	7	140	12	90	65	−45	4.7	4.4
Bethel	702190	60.78	161.80	151	14.615	6193	−28	−24	31	27	24	34	8	30	5	13	20	12	360	78	−32	3.3	6.6
Bettles	701740	66.92	151.52	643	14.357	6193	−49	−44	18	16	14	19	11	16	7	2	340	8	190	85	−55	4.0	5.8
Big Delta, Ft. Greely	702670	64.00	145.73	1283	14.027	6193	−45	−39	34	29	25	38	0	33	3	3	180	9	180	84	−48	3.3	7.5
Cold Bay	703160	55.20	162.73	102	14.642	6193	6	10	38	34	30	46	34	40	34	15	340	16	140	67	2	4.0	5.3
Cordova	702960	60.50	145.50	43	14.673	8293	−4	1	22	19	16	24	40	22	38	1	340	8	240	79	−9	5.0	5.4
Deadhorse	700637	70.20	148.47	56	14.666	8293	−36	−34	32	28	25	34	−1	30	−7	12	240	12	60	78	−51	14.2	5.2
Dillingham	703210	59.05	158.52	95	14.645	8293	−20	−13	25	22	20	28	20	24	21	5	40	10	180	74	−27	3.1	9.4
Fairbanks, Eielson AFB	702650	64.67	147.10	548	14.407	8293	−33	−31	14	12	10	14	21	11	16	0	150	5	290	87	−46	3.8	7.7
Fairbanks, Intl Airport	702610	64.82	147.87	453	14.457	6193	−47	−41	18	15	13	16	11	12	11	2	10	8	220	87	−48	3.8	7.8
Galena	702220	64.73	156.93	151	14.615	8293	−33	−31	18	15	13	19	14	16	15	0	270	5	320	84	−50	2.5	10.4
Gulkana	702710	62.15	145.45	1578	13.877	6193	−44	−39	26	24	21	22	17	19	18	3	360	7	180	82	−46	3.2	7.4
Homer	703410	59.63	151.50	72	14.657	8293	0	4	22	20	18	23	24	21	27	9	30	10	270	70	−5	4.0	6.8
Juneau	703810	58.37	134.58	23	14.683	8293	4	7	27	23	20	29	39	25	38	5	360	9	230	81	−1	2.5	4.9
Kenai	702590	60.57	151.25	95	14.645	8293	−22	−14	23	20	18	25	25	22	24	2	30	9	270	75	−27	3.4	7.4
Ketchikan	703950	55.35	131.70	95	14.645	8293	13	20	25	22	19	29	42	24	42	5	280	11	320	78	7	1.8	5.2
King Salmon	703260	58.68	156.65	49	14.669	6193	−24	−19	32	28	24	33	36	28	36	7	360	12	270	78	−31	3.5	7.2
Kodiak, State USCG Base	703500	57.75	152.50	112	14.636	6193	7	12	34	30	26	34	28	30	30	18	300	11	320	76	1	3.6	6.1
Kotzebue	701330	66.87	162.63	16	14.687	6193	−36	−31	35	31	28	38	14	32	14	7	70	12	300	75	−39	4.8	6.5
McGrath	702310	62.97	155.62	338	14.517	6193	−47	−42	18	16	14	18	23	14	12	1	310	7	340	83	−52	3.3	7.0
Middleton Island	703430	59.43	146.33	46	14.671	8293	18	21	40	34	30	42	35	37	36	18	330	8	260	66	15	4.9	6.8
Nenana	702600	64.55	149.08	361	14.505	8293	−51	−44	16	14	12	18	10	15	8	2	250	7	60	87	−52	4.1	7.2
Nome	702000	64.50	165.43	23	14.683	6193	−31	−26	30	26	23	31	17	28	18	4	20	12	260	76	−35	4.2	6.3
Northway	702910	62.97	141.93	1722	13.803	8293	−34	−32	15	13	12	14	−13	12	−6	0	300	7	290	83	−54	2.7	5.9
Port Heiden	703330	56.95	158.62	95	14.645	8293	−6	−2	38	32	28	38	36	32	35	17	60	15	160	74	−11	4.3	7.4
Saint Paul Island	703080	57.15	170.22	30	14.680	6193	−2	3	41	37	33	47	24	41	21	19	350	14	240	58	−3	5.2	6.9
Sitka	703710	57.07	135.35	66	14.661	8293	16	21	23	21	19	24	40	22	41	8	70	9	230	76	11	6.1	5.0
Talkeetna	702510	62.30	150.10	358	14.507	6193	−28	−21	17	16	14	19	13	17	15	4	50	8	200	82	−35	2.8	8.0
Valdez	702750	61.13	146.35	33	14.678	8293	4	7	24	19	16	28	13	22	15	15	70	10	240	76	1	3.6	6.1
Yakutat	703610	59.52	139.67	30	14.680	6193	−3	2	24	19	16	25	36	21	33	2	100	9	320	75	−8	4.0	7.0
ARIZONA																							
Flagstaff	723755	35.13	111.67	7011	11.335	6193	1	8	21	18	17	21	29	18	30	3	20	9	220	89	−10	2.5	7.3
Kingman	723700	35.27	113.95	3389	12.983	8293	22	27	26	23	20	24	46	21	43	5	90	13	240	103	15	1.8	6.8
Page	723710	36.93	111.45	4278	12.561	8293	20	24	19	16	13	16	42	12	40	4	300	7	360	104	8	3.6	12.2
Phoenix, Intl Airport	722780	33.43	112.02	1106	14.118	6193	34	37	19	16	14	17	59	14	58	5	90	9	270	114	30	2.2	4.6
Phoenix, Luke AFB	722785	33.53	112.38	1089	14.126	8293	35	38	19	15	13	16	58	13	55	4	340	9	210	115	30	2.2	3.8
Prescott	723723	34.65	112.42	5043	12.208	6193	15	20	22	19	17	21	42	18	42	6	190	11	230	98	7	2.2	6.2
Safford, Agri Center	722747	32.82	109.68	3117	13.114	8293	21	26	17	14	12	15	50	13	48	4	110	7	310	106	11	3.8	11.5
Tucson	722740	32.12	110.93	2556	13.388	6193	31	34	24	21	18	24	56	21	56	7	140	12	300	108	25	2.8	4.0
Winslow	723740	35.02	110.73	4882	12.281	8293	10	14	26	22	19	24	46	19	45	5	140	9	250	100	3	4.9	6.3
Yuma	722800	32.65	114.60	207	14.586	8293	40	44	19	17	15	20	59	17	58	4	30	7	280	116	29	1.8	11.9
ARKANSAS																							
Blytheville, Eaker AFB	723408	35.97	89.95	256	14.560	8293	12	18	22	19	17	23	36	21	38	10	10	6	240	99	6	5.0	9.0
Fayetteville	723445	36.00	94.17	1250	14.044	8293	6	13	21	19	18	21	44	19	44	9	350	10	190	100	−1	3.2	9.2
Ft. Smith	723440	35.33	94.37	463	14.451	6193	13	19	20	18	16	21	46	18	41	9	320	9	270	102	6	3.9	6.6
Little Rock, AFB	723405	34.92	92.15	312	14.531	6193	16	21	20	18	16	20	42	18	42	9	360	9	200	101	10	3.8	6.2
Texarkana	723418	33.45	93.98	390	14.489	8293	20	25	19	17	15	20	47	18	48	9	50	9	190	101	13	3.1	7.6
CALIFORNIA																							
Alameda, NAS	745060	37.78	122.32	13	14.688	8293	40	42	21	18	16	20	51	17	52	6	120	8	300	93	25	4.9	14.2
Arcata/Eureka	725945	40.98	124.10	217	14.581	6193	30	32	21	19	17	21	53	18	51	5	90	10	320	82	26	4.5	3.2
Bakersfield	723840	35.43	119.05	492	14.436	6193	32	35	19	16	14	19	56	14	54	5	90	12	310	108	28	2.3	3.4
Barstow/Daggett	723815	34.85	116.78	1926	13.701	6193	28	32	30	27	23	30	58	25	54	6	270	12	290	111	22	2.5	4.8
Blue Canyon	725845	39.28	120.72	5285	12.097	8293	21	24	15	12	11	16	35	14	35	5	70	6	290	89	11	3.8	18.4
Burbank/Glendale	722880	34.20	118.35	774	14.289	8293	39	41	18	14	12	20	56	17	57	2	330	8	180	106	33	3.1	4.5

WMO# = World Meteorological Organization number
AFB = Air Force Base NAS = Naval Air Station
WSO = Weather Service Office
AAF = Army Air Field
MCAS = Marine Corps Air Station
DB = dry-bulb temperature, °F
Lat. = North latitude °
Long. = West longitude °
Elev. = elevation, ft
StdP = standard pressure at station elevation, psia
WS = wind speed, mph
PWD = prevailing wind direction, ° True

Table 1B Cooling and Dehumidification Design Conditions—United States

Station	Cooling DB/MWB						Evaporation WB/MDB						Dehumidification DP/MDB and HR									Range of DB
	0.4%		1%		2%		0.4%		1%		2%		0.4%			1%			2%			
	DB	MWB	DB	MWB	DB	MWB	WB	MDB	WB	MDB	WB	MDB	DP	HR	MDB	DP	HR	MDB	DP	HR	MDB	
1	2a	2b	2c	2d	2e	2f	3a	3b	3c	3d	3e	3f	4a	4b	4c	4d	4e	4f	4g	4h	4i	5
ALABAMA																						
Anniston	95	76	93	76	90	75	79	90	78	88	77	86	77	143	84	76	137	82	75	133	81	19.6
Birmingham	94	75	92	75	90	74	78	89	77	88	76	87	75	135	83	74	131	82	73	127	81	18.7
Dothan	95	76	93	76	92	76	80	90	79	88	78	87	77	144	83	76	139	82	76	136	82	17.5
Huntsville	94	75	92	74	90	74	78	89	77	88	76	86	75	135	83	74	130	82	73	126	81	18.5
Mobile	94	77	92	76	91	76	79	89	79	88	78	87	77	142	83	76	139	83	76	135	82	16.5
Montgomery	95	76	93	76	91	76	79	91	78	89	78	88	76	139	85	75	134	84	75	130	83	18.7
Muscle Shoals/Florence	96	76	94	75	92	74	78	90	78	89	77	87	76	137	82	75	133	82	74	130	81	20.0
Ozark, Ft. Rucker	95	77	94	77	92	76	81	90	79	89	78	88	78	146	85	77	142	84	76	138	83	18.0
Tuscaloosa	95	77	94	77	92	76	80	90	79	89	78	88	77	142	84	76	137	83	75	134	82	19.6
ALASKA																						
Adak, NAS	59	55	57	53	55	51	55	59	53	57	51	54	53	59	58	51	55	56	49	51	53	9.7
Anchorage, Elemendorf AFB	71	58	69	57	66	56	60	69	58	66	57	64	57	69	62	55	65	61	53	61	60	12.6
Anchorage, Ft. Richardson	74	60	71	58	68	57	61	72	59	69	58	66	56	69	64	54	63	62	53	61	61	15.5
Anchorage, Intl Airport	71	59	68	57	65	56	60	69	58	66	57	63	56	68	62	55	64	61	53	61	60	12.6
Annette	74	61	70	59	66	57	62	72	60	68	58	65	58	71	65	56	68	63	55	65	61	10.5
Barrow	57	51	52	49	48	46	52	56	49	52	46	48	49	53	54	46	46	51	44	42	48	10.6
Bethel	72	59	68	57	64	55	60	69	58	66	56	63	56	68	62	55	64	60	53	60	58	13.4
Bettles	79	61	75	59	72	58	63	76	61	73	59	70	58	72	66	56	67	64	54	63	63	19.4
Big Delta, Ft. Greely	78	59	75	58	71	56	61	74	59	72	58	69	56	70	65	54	65	63	52	61	61	17.3
Cold Bay	60	54	57	53	55	52	55	59	54	56	53	55	54	62	56	53	59	55	51	56	54	7.4
Cordova	70	59	67	57	63	56	60	69	58	65	56	62	56	67	63	54	62	60	53	60	59	13.5
Deadhorse	66	57	61	54	58	53	58	64	55	62	53	58	54	61	62	51	56	59	49	51	56	13.7
Dillingham	69	57	66	56	62	54	59	67	57	64	55	61	56	67	62	53	61	59	52	57	57	13.1
Fairbanks, Eielson AFB	81	61	78	60	75	59	64	78	62	75	60	72	58	74	66	56	69	66	54	63	66	19.4
Fairbanks, Intl Airport	81	61	77	59	74	58	63	77	61	74	59	71	58	72	65	56	68	64	54	64	63	18.6
Galena	78	61	74	59	71	58	63	74	61	71	59	69	58	73	66	56	69	65	54	63	64	15.3
Gulkana	77	58	73	56	69	55	59	73	57	70	56	67	53	63	62	51	60	60	50	55	59	20.3
Homer	65	56	62	55	60	54	57	63	56	61	55	59	54	64	59	53	60	58	52	57	57	11.9
Juneau	74	60	69	58	66	57	61	71	59	68	58	64	57	70	63	56	67	61	55	64	60	13.9
Kenai	68	56	65	55	62	54	58	65	56	63	55	61	55	64	59	53	60	58	52	58	57	13.3
Ketchikan	71	60	68	59	66	58	62	69	60	67	59	64	59	74	64	57	71	62	56	68	61	10.3
King Salmon	71	58	67	56	64	55	59	68	57	65	56	62	56	66	61	54	62	60	52	58	58	15.5
Kodiak, State USCG Base	68	58	65	56	62	55	59	66	57	63	56	61	56	67	61	55	64	59	53	61	57	11.2
Kotzebue	68	59	64	58	61	56	60	67	58	64	56	61	57	70	64	55	65	61	54	62	59	8.8
McGrath	77	60	73	58	70	56	61	74	59	70	58	67	57	69	63	55	65	62	54	62	61	17.4
Middleton Island	62	54	60	51	59	51	55	61	54	59	53	57	52	57	56	51	56	56	50	54	55	5.8
Nenana	80	60	76	59	73	57	62	75	60	73	59	70	57	69	65	55	65	65	53	60	63	21.2
Nome	69	57	65	55	61	54	58	66	56	63	55	60	55	64	61	53	60	59	51	56	57	10.9
Northway	78	58	74	57	71	56	60	76	58	71	57	69	54	66	62	53	62	61	51	60	59	20.0
Port Heiden	64	54	61	52	59	51	56	62	54	60	52	58	51	57	59	50	54	57	49	51	55	9.7
Saint Paul Island	54	51	52	50	51	49	51	53	50	52	49	50	50	55	52	49	52	51	48	50	50	5.4
Sitka	66	59	64	58	61	57	60	65	59	62	58	60	58	74	62	57	71	60	56	68	59	9.2
Talkeetna	77	60	73	58	70	57	62	74	60	70	58	67	57	71	64	56	67	62	54	64	61	16.4
Valdez	69	56	66	55	62	54	58	67	56	64	55	61	53	60	59	53	59	57	52	57	56	12.2
Yakutat	66	56	63	55	60	54	58	62	57	60	56	59	56	67	58	55	64	57	54	62	57	12.0
ARIZONA																						
Flagstaff	85	56	83	55	80	55	61	74	60	73	59	72	58	93	65	56	88	64	55	83	63	27.6
Kingman	99	64	97	63	95	62	71	82	67	85	66	86	67	112	77	62	92	75	59	85	76	24.8
Page	99	62	97	62	95	61	66	85	65	86	64	86	60	92	74	58	85	74	56	80	74	23.8
Phoenix, Intl Airport	110	70	108	70	106	70	76	97	75	96	74	95	71	118	82	69	111	84	67	104	85	23.0
Phoenix, Luke AFB	110	71	107	71	105	71	78	97	76	97	75	96	74	130	85	71	118	85	69	111	86	25.2
Prescott	94	60	91	60	89	60	67	81	66	80	64	79	63	104	71	61	98	71	60	93	70	25.4
Safford, Agri Center	102	66	99	66	97	65	71	89	71	89	69	88	67	111	77	66	106	76	64	102	77	34.7
Tucson	104	65	102	65	100	65	72	88	71	87	70	86	69	116	76	67	111	76	66	106	77	29.4
Winslow	95	60	93	60	91	59	65	80	64	81	63	80	61	95	71	59	91	69	58	85	69	27.4
Yuma	111	72	109	72	106	72	80	96	78	95	77	95	76	136	87	74	127	88	71	117	89	23.8
ARKANSAS																						
Blytheville, Eaker AFB	97	78	95	77	93	77	82	92	80	91	78	89	78	149	88	77	142	86	76	135	85	18.7
Fayetteville	95	75	93	75	90	74	78	90	77	89	76	87	75	136	85	74	132	84	72	124	81	21.4
Ft. Smith	99	76	96	76	93	75	79	92	78	91	77	90	75	134	85	74	130	84	73	126	83	21.5
Little Rock, AFB	97	77	95	77	92	76	80	92	79	91	78	89	77	141	86	76	137	85	75	133	84	19.5
Texarkana	97	77	95	77	93	76	80	91	79	90	78	89	77	143	85	76	139	85	75	135	84	20.5
CALIFORNIA																						
Alameda, NAS	83	65	79	64	76	63	67	79	65	76	64	73	62	85	70	61	80	69	60	78	67	14.8
Arcata/Eureka	70	60	67	59	65	58	62	67	61	65	60	64	60	78	64	59	75	63	58	71	62	15.5
Bakersfield	104	70	101	69	99	69	73	98	71	96	70	95	64	92	84	62	85	83	60	79	83	26.5
Barstow/Daggett	107	68	105	67	102	67	72	95	71	95	69	95	66	103	81	63	91	85	59	81	85	27.8
Blue Canyon	84	59	81	57	79	56	62	80	60	78	58	75	54	74	70	52	69	70	50	64	68	16.6
Burbank/Glendale	98	69	95	69	92	68	74	90	72	89	71	86	69	108	80	67	103	78	66	98	77	23.4

DP = dew-point temperature, °F
MDB = mean coincident dry-bulb temp., °F
StdD = standard deviation, °F
MWB = mean coincident wet-bulb temp., °F
MWS = mean coincident wind speed, mph
HR = humidity ratio, grains of moisture per lb of dry air
A = airport
ANGB = Air National Guard Base
MCAF = Marine Corps Air Facility
NAF = Naval Air Facility
NAWS = Naval Air Weapons Station
RAF = Royal Air Force

Table 1A Heating and Wind Design Conditions—United States

Station	WMO#	Lat.	Long.	Elev., ft	StdP, psia	Dates	Heating Dry Bulb 99.6%	Heating Dry Bulb 99%	Extreme Wind Speed, mph 1%	Extreme Wind Speed, mph 2.5%	Extreme Wind Speed, mph 5%	Coldest Month 0.4% WS	Coldest Month 0.4% MDB	Coldest Month 1% WS	Coldest Month 1% MDB	MWS/PWD to DB 99.6% MWS	MWS/PWD to DB 99.6% PWD	MWS/PWD to DB 0.4% MWS	MWS/PWD to DB 0.4% PWD	Extr. Annual Daily Mean DB Max.	Extr. Annual Daily Mean DB Min.	Extr. Annual Daily StdD DB Max.	Extr. Annual Daily StdD DB Min.
1a	1b	1c	1d	1e	1f	1g	2a	2b	3a	3b	3c	4a	4b	4c	4d	5a	5b	5c	5d	6a	6b	6c	6d
Fairfield, Travis AFB	745160	38.27	121.93	62	14.662	8293	31	34	28	24	22	26	53	22	52	4	20	9	240	105	26	3.4	4.0
Fresno	723890	36.77	119.72	328	14.522	6193	30	32	17	15	13	17	53	14	52	4	90	9	290	107	26	2.2	3.8
Lancaster/Palmdale	723816	34.73	118.22	2346	13.492	8293	22	24	30	28	25	29	48	26	49	2	260	14	240	107	15	2.0	5.9
Lemoore, Reeves NAS	747020	36.33	119.95	236	14.570	8293	30	32	19	16	14	20	49	16	51	4	150	7	360	110	17	4.1	10.4
Long Beach	722970	33.82	118.15	39	14.675	6193	40	43	19	16	14	19	58	16	58	4	300	10	270	102	35	4.5	2.8
Los Angeles	722950	33.93	118.40	105	14.640	6193	43	45	21	18	16	20	56	17	56	6	70	10	250	97	38	5.1	3.0
Marysville, Beale AFB	724837	39.13	121.43	112	14.636	8293	31	34	20	17	14	23	53	19	53	3	20	5	200	106	26	3.2	4.1
Merced, Castle AFB	724810	37.38	120.57	187	14.596	8293	30	32	18	15	12	21	51	17	49	2	110	9	320	104	26	2.7	3.6
Mount Shasta	725957	41.32	122.32	3543	12.909	8293	16	21	14	12	10	14	36	12	37	4	60	4	180	95	10	2.7	6.7
Mountain View, Moffet NAS	745090	37.42	122.05	39	14.675	8293	36	39	19	17	15	19	54	16	52	4	60	4	180	95	10	2.7	6.7
Ontario	722865	34.05	117.60	942	14.202	8293	35	38	22	19	17	28	62	21	57	1	140	9	330	98	23	2.5	12.2
Oxnard, Pt. Mugu NAWS	723910	34.12	119.12	7	14.692	8293	39	41	22	19	16	25	57	21	58	4	10	13	240	108	29	3.4	2.5
Paso Robles	723965	35.67	120.63	837	14.257	8293	26	29	22	20	18	21	52	18	51	5	20	12	50	93	24	5.0	10.6
Red Bluff	725910	40.15	122.25	354	14.508	8293	29	32	23	21	19	26	53	23	50	3	110	11	300	108	21	2.2	4.9
Riverside, March AFB	722860	33.88	117.27	1539	13.896	8293	34	36	18	15	13	22	51	18	55	6	340	9	160	111	25	3.2	3.8
Sacramento, Mather Field	724835	38.55	121.30	95	14.645	8293	30	32	20	17	14	24	53	20	51	1	210	9	300	107	29	2.3	3.2
Sacramento, McClellan AFB	724836	38.67	121.40	75	14.655	8293	31	34	20	16	14	23	53	19	52	2	120	6	310	105	26	5.0	4.3
Sacramento, Metro	724839	38.70	121.58	23	14.683	6193	31	33	22	19	17	23	51	20	50	2	340	5	220	107	27	2.5	4.9
Salinas	724917	36.67	121.60	85	14.650	8293	33	35	21	19	18	23	51	21	51	3	340	8	220	107	27	6.6	3.0
San Bernardino, Norton AFB	722866	34.10	117.23	1158	14.091	8293	34	36	17	13	11	21	56	16	55	6	130	11	310	95	29	4.7	2.2
San Diego, Intl Airport	722900	32.73	117.17	30	14.680	6193	44	46	18	16	15	20	59	16	60	2	50	8	250	109	29	2.5	2.7
San Diego, Miramar NAS	722930	32.85	117.12	420	14.474	8293	39	42	13	11	9	15	59	12	59	3	70	10	310	95	39	6.4	4.4
San Francisco	724940	37.62	122.38	16	14.687	6193	37	39	29	26	23	27	53	22	52	3	90	6	310	102	27	4.0	13.7
San Jose Intl Airport	724945	37.37	121.93	56	14.666	8293	35	38	20	18	17	20	56	17	56	5	160	13	300	94	33	4.3	3.0
Santa Barbara	723925	34.43	119.83	10	14.690	8293	34	37	20	17	14	19	58	16	58	1	160	10	320	101	27	3.1	9.0
Santa Maria	723940	34.90	120.45	240	14.569	6193	32	35	23	21	19	21	59	18	59	1	40	10	260	97	28	6.7	6.5
Stockton	724920	37.90	121.25	26	14.681	8293	30	32	22	19	17	24	52	21	49	4	110	11	300	95	27	5.0	2.8
Victorville, George AFB	723825	34.58	117.38	2874	13.232	8293	27	30	22	19	16	22	49	18	47	4	110	11	280	106	26	3.1	3.4
COLORADO																3	160	9	180	106	21	3.1	5.6
Alamosa	724620	37.45	105.87	7543	11.108	6193	−17	−11	26	23	21	23	33	20	30	3	190	12	240	88	−27	2.0	7.9
Colorado Springs	724660	38.82	104.72	6171	11.701	6193	−2	4	29	25	21	28	35	23	33	7	20	12	160	95	−9	2.0	6.9
Craig	725700	40.50	107.53	6283	11.652	8293	−20	−12	26	20	17	22	33	17	27	2	270	9	250	93	−31	2.0	10.6
Denver	724699	39.75	104.87	5331	12.076	8293	−3	3	24	21	18	25	39	21	40	6	180	9	160	97	−11	2.3	7.0
Eagle	724675	39.65	106.92	6539	11.539	6193	−13	−7	22	19	17	20	33	18	32	3	90	11	230	93	−23	3.2	7.8
Grand Junction	724760	39.12	108.53	4839	12.301	6193	2	7	22	19	17	17	33	14	30	5	70	11	290	100	−3	2.0	8.5
Limon	724665	39.27	103.67	5364	12.062	8293	−6	1	27	23	21	27	29	22	25	9	160	12	200	96	−13	2.2	6.5
Pueblo	724640	38.28	104.52	4721	12.355	6193	−1	5	32	27	24	30	44	26	43	5	270	12	140	102	−12	1.9	7.7
Trinidad	724645	37.27	104.33	5761	11.883	8293	−2	6	25	22	19	24	41	21	42	5	290	10	210	98	−10	2.0	6.8
CONNECTICUT																							
Bridgeport	725040	41.17	73.13	16	14.687	6193	8	12	27	23	21	34	29	30	29	14	320	14	230	93	2	2.8	4.9
Hartford, Brainard Field	725087	41.73	72.65	20	14.685	6193	2	6	23	20	18	23	25	20	26	7	320	11	250	97	−6	2.4	5.7
Windsor Locks, Bradley Fld	725080	41.93	72.68	180	14.600	8293	3	8	21	19	17	22	30	20	29	7	360	11	240	97	−5	2.0	5.8
DELAWARE																							
Dover, AFB	724088	39.13	75.47	30	14.680	8293	14	18	22	19	17	23	36	21	35	8	340	9	240	97	6	3.2	6.1
Wilmington	724089	39.68	75.60	79	14.654	6193	10	14	25	22	19	27	29	23	30	11	290	11	240	96	3	2.7	6.8
FLORIDA																							
Apalachicola	722200	29.73	85.03	20	14.685	8293	31	35	19	17	15	19	51	17	51	6	360	9	220	93	23	6.7	7.4
Cape Canaveral, NASA	747946	28.62	80.72	10	14.690	8293	38	42	19	17	15	21	60	19	60	8	320	8	220	96	29	1.4	6.1
Daytona Beach	722056	29.18	81.05	36	14.676	6193	34	37	21	19	17	22	61	19	61	7	310	11	240	96	27	1.9	4.4
Ft. Lauderdale/Hollywood	722025	26.07	80.15	23	14.683	8293	46	50	22	20	18	22	69	20	71	9	330	11	120	97	39	1.1	6.1
Ft. Myers	722106	26.58	81.87	16	14.687	8293	42	47	19	18	16	20	64	18	66	6	30	9	70	97	34	1.3	4.7
Gainesville	722146	29.68	82.27	151	14.615	8293	30	33	19	17	14	19	65	17	62	4	300	9	270	97	21	1.8	7.2
Homestead, AFB	722026	25.48	80.38	7	14.692	8293	48	52	17	15	13	17	70	15	70	6	360	7	120	95	41	2.2	5.6
Jacksonville, Cecil Fld NAS	722067	30.22	81.88	82	14.652	8293	31	34	18	16	14	19	62	17	62	3	290	7	270	100	20	2.0	8.8
Jacksonville, Intl Airport	722060	30.50	81.70	30	14.680	6193	29	32	21	18	17	21	54	19	55	6	310	9	230	98	22	2.1	5.1
Jacksonville, Mayport Naval	722066	30.40	81.42	16	14.687	8293	34	39	19	17	14	21	54	18	55	6	310	7	270	99	20	2.2	13.1
Key West	722010	24.55	81.75	20	14.685	6193	55	58	22	20	18	24	65	21	66	12	50	9	140	91	51	1.2	4.0
Melbourne	722040	28.10	80.65	36	14.676	8293	38	43	21	19	18	22	62	20	62	9	320	11	120	97	30	1.8	6.7
Miami, Intl Airport	722020	25.82	80.28	13	14.688	6193	46	50	23	20	18	22	68	20	69	10	340	11	150	94	39	2.1	5.1
Miami, New Tamiami A	722029	25.65	80.43	10	14.690	8293	45	49	21	19	18	21	72	19	72	8	360	11	130	95	39	2.0	6.3
Milton, Whiting Field NAS	722226	30.72	87.02	200	14.589	8293	28	31	18	16	14	19	50	17	52	6	340	6	330	99	17	1.8	8.6
Orlando	722050	28.43	81.32	105	14.640	8293	37	42	20	18	16	21	66	19	65	8	330	9	290	96	29	1.6	6.5
Panama City, Tyndall AFB	747750	30.07	85.58	16	14.687	8293	33	37	18	16	14	19	52	17	52	9	360	7	240	94	24	2.3	6.3
Pensacola, Sherman AFB	722225	30.35	87.32	30	14.680	8293	28	32	23	20	18	25	43	22	48	9	360	10	200	100	15	6.3	7.6
Saint Petersburg	722116	27.92	82.68	10	14.690	8293	43	47	21	19	17	22	65	20	63	11	10	10	230	97	35	1.8	4.7
Sarasota/Bradenton	722115	27.40	82.55	30	14.680	8293	39	43	22	19	17	23	67	20	67	5	40	9	270	97	29	1.6	12.1
Tallahassee	722140	30.38	84.37	69	14.659	6193	25	28	18	16	14	19	52	17	54	3	350	8	360	98	17	2.0	4.6
Tampa, Intl Airport	722110	27.97	82.53	10	14.690	6193	36	40	19	17	15	21	59	19	59	8	20	10	270	95	29	1.2	4.8
Valparaiso, Eglin AFB	722210	30.48	86.53	85	14.650	8293	30	33	19	16	14	18	49	16	51	6	360	7	210	97	19	2.0	6.1
Vero Beach	722045	27.65	80.42	26	14.681	8293	39	43	20	19	17	21	67	19	67	8	310	11	240	96	31	2.0	6.5

WMO# = World Meteorological Organization number
AFB = Air Force Base NAS = Naval Air Station
WSO = Weather Service Office
AAF = Army Air Field
MCAS = Marine Corps Air Station
DB = dry-bulb temperature, °F
Lat. = North latitude °
Long. = West longitude °
Elev. = elevation, ft
StdP = standard pressure at station elevation, psia
WS = wind speed, mph
PWD = prevailing wind direction, ° True

Table 1B Cooling and Dehumidification Design Conditions—United States

Station	Cooling DB/MWB						Evaporation WB/MDB						Dehumidification DP/MDB and HR									Range of DB
	0.4%		1%		2%		0.4%		1%		2%		0.4%			1%			2%			
	DB	MWB	DB	MWB	DB	MWB	WB	MDB	WB	MDB	WB	MDB	DP	HR	MDB	DP	HR	MDB	DP	HR	MDB	
1	2a	2b	2c	2d	2e	2f	3a	3b	3c	3d	3e	3f	4a	4b	4c	4d	4e	4f	4g	4h	4i	5
Fairfield, Travis AFB	98	67	94	67	91	66	70	92	69	90	67	88	62	85	76	61	79	74	59	75	73	29.0
Fresno	103	71	101	70	98	69	73	98	71	96	70	94	64	92	85	62	85	84	61	80	82	30.9
Lancaster/Palmdale	101	66	98	65	96	64	70	94	69	91	67	90	62	92	80	60	84	81	58	78	81	27.9
Lemoore, Reeves NAS	103	72	101	71	98	70	75	97	73	96	71	94	67	101	89	65	94	87	62	85	86	32.9
Long Beach	92	67	88	67	84	66	71	85	70	82	69	80	67	101	76	66	96	75	65	92	75	16.7
Los Angeles	85	64	81	64	78	64	70	78	69	76	68	75	67	99	75	66	95	73	65	92	72	10.9
Marysville, Beale AFB	101	70	98	69	95	68	72	97	71	95	69	92	63	86	85	61	80	82	60	76	81	29.9
Merced, Castle AFB	99	69	97	69	94	68	72	96	71	93	69	92	64	90	81	62	82	84	60	78	81	30.2
Mount Shasta	91	62	88	61	85	60	64	87	63	84	61	82	56	76	74	53	69	73	51	64	71	32.0
Mountain View, Moffet NAS	88	65	84	65	80	64	68	82	67	80	65	78	62	83	74	61	80	73	60	76	72	18.0
Ontario	102	71	98	70	95	69	75	94	73	92	72	90	70	113	80	68	106	80	66	101	78	27.7
Oxnard, Pt. Mugu NAWS	83	62	79	64	77	64	70	77	69	75	67	74	68	103	74	66	97	73	65	93	72	14.6
Paso Robles	102	68	98	67	95	65	70	97	68	94	67	91	61	81	76	58	75	73	57	71	71	37.8
Red Bluff	105	70	102	69	98	67	72	98	71	95	69	93	65	94	82	62	85	80	61	80	78	29.5
Riverside, March AFB	101	68	98	68	95	67	72	92	71	91	70	90	67	104	79	65	97	80	62	90	79	29.0
Sacramento, Mather Field	101	69	97	68	95	67	71	97	69	94	68	92	61	80	79	60	77	77	58	74	76	33.7
Sacramento, McClellan AFB	102	70	98	69	95	68	72	97	71	95	69	92	63	85	84	61	80	81	60	77	79	29.7
Sacramento, Metro	100	69	97	69	94	68	72	96	70	94	69	91	62	84	82	61	79	80	59	76	78	33.3
Salinas	83	63	78	62	75	61	66	78	65	75	63	72	62	82	69	60	78	68	59	76	67	18.7
San Bernardino, Norton AFB	103	70	101	70	97	69	74	94	73	94	71	92	68	107	83	66	101	83	65	95	82	31.5
San Diego, Intl Airport	85	67	81	67	79	67	73	79	71	78	70	77	70	111	77	68	104	76	67	99	74	8.9
San Diego, Miramar NAS	92	69	88	67	85	67	72	85	71	83	69	81	68	104	78	67	99	77	65	96	75	17.5
San Francisco	83	63	78	62	74	61	64	79	63	75	62	72	59	76	67	58	73	66	57	71	65	16.7
San Jose Intl Airport	93	67	89	66	86	65	70	88	68	85	67	83	63	85	77	61	81	76	60	78	74	22.3
Santa Barbara	83	64	80	64	77	64	69	77	67	76	66	74	66	96	74	65	91	71	63	85	70	18.0
Santa Maria	86	63	82	62	78	61	66	81	65	78	64	75	61	80	70	60	77	69	59	74	68	19.4
Stockton	100	69	97	68	94	67	71	96	70	94	68	92	62	83	78	60	78	78	59	75	77	30.4
Victorville, George AFB	101	65	98	65	96	64	69	88	68	88	67	88	65	102	78	61	90	79	59	83	78	28.3
COLORADO																						
Alamosa	84	55	82	55	80	54	60	75	59	74	58	73	55	87	62	54	81	62	52	77	62	31.2
Colorado Springs	90	58	87	58	84	58	63	78	62	77	61	76	59	92	66	57	88	66	56	83	65	24.9
Craig	88	57	85	56	83	55	60	79	59	78	57	77	53	77	66	52	72	65	50	68	64	36.4
Denver	93	60	90	59	87	59	65	81	63	80	62	79	60	96	69	58	90	68	57	85	68	26.9
Eagle	88	58	86	57	83	57	62	80	60	78	59	76	57	88	64	55	82	65	53	76	65	36.1
Grand Junction	96	61	94	60	92	60	65	84	64	83	63	82	60	93	70	58	87	71	56	79	71	26.6
Limon	90	60	88	60	85	59	64	79	63	78	62	77	60	96	67	59	92	66	58	88	66	26.8
Pueblo	97	62	94	62	92	62	67	84	66	83	65	83	63	104	71	62	98	71	60	92	71	29.4
Trinidad	93	61	90	60	87	60	65	84	64	83	63	81	60	96	71	58	91	69	57	86	69	28.3
CONNECTICUT																						
Bridgeport	86	73	84	72	82	71	76	83	74	81	73	79	74	126	79	72	120	78	71	115	77	14.1
Hartford, Brainard Field	91	73	88	72	85	70	76	87	74	84	73	82	73	121	81	71	116	79	70	110	78	20.9
Windsor Locks, Bradley Fld	92	73	88	71	85	70	76	87	74	84	72	82	72	119	81	71	114	79	69	109	77	20.9
DELAWARE																						
Dover, AFB	93	76	89	75	87	74	79	88	78	86	76	84	77	141	84	76	135	82	74	129	81	16.2
Wilmington	91	75	89	74	86	73	78	87	76	85	75	83	75	132	82	74	125	81	72	120	80	17.0
FLORIDA																						
Apalachicola	92	79	90	78	89	78	81	88	80	87	79	87	79	148	85	78	145	84	77	141	84	13.3
Cape Canaveral, NASA	92	78	90	78	89	77	80	88	79	87	79	87	78	145	84	77	141	84	76	138	83	16.0
Daytona Beach	92	77	90	77	88	77	79	88	79	87	78	86	77	141	84	76	137	84	76	134	83	15.4
Ft. Lauderdale/Hollywood	92	78	90	78	89	78	81	88	80	87	79	87	78	147	85	78	145	84	77	141	84	11.3
Ft. Myers	94	77	93	77	92	77	80	89	80	88	79	87	78	147	84	78	144	83	77	140	83	16.9
Gainesville	94	77	92	77	90	76	80	89	79	88	78	87	77	143	84	76	139	83	76	136	82	18.7
Homestead, AFB	92	79	90	79	89	78	81	89	80	88	80	87	79	150	87	78	145	86	77	141	85	11.7
Jacksonville, Cecil Fld NAS	96	76	95	76	93	76	79	91	78	90	77	89	76	138	84	75	134	83	75	130	82	20.0
Jacksonville, Intl Airport	94	77	93	77	91	77	80	90	79	89	78	88	77	142	85	76	138	84	76	134	83	17.8
Jacksonville, Mayport Naval	95	78	92	78	90	77	81	89	80	89	79	88	78	147	86	77	142	85	77	139	85	15.3
Key West	90	79	89	79	89	79	81	87	80	87	80	87	79	149	85	78	146	85	77	143	85	8.1
Melbourne	93	79	91	79	89	79	82	89	81	88	80	87	80	155	86	79	150	85	78	146	85	15.3
Miami, Intl Airport	91	77	90	77	89	77	80	87	79	87	78	86	78	144	83	77	141	83	76	138	82	11.4
Miami, New Tamiami A	92	78	91	78	90	77	80	89	79	88	79	87	78	145	83	77	141	83	76	138	83	15.5
Milton, Whiting Field NAS	95	78	93	77	92	76	81	90	80	89	79	88	78	148	86	77	143	85	76	138	84	18.5
Orlando	94	76	93	76	92	76	79	88	79	88	78	87	77	142	83	76	139	82	76	136	81	16.6
Panama City, Tyndall AFB	91	79	89	79	88	79	83	88	82	87	81	86	81	160	86	80	154	85	79	150	84	12.2
Pensacola, Sherman AFB	93	78	92	78	90	78	81	89	80	88	79	88	79	150	85	78	144	85	76	138	85	15.3
Saint Petersburg	94	80	93	79	92	79	82	90	82	89	81	88	80	156	86	80	153	85	79	150	85	13.5
Sarasota/Bradenton	93	80	92	79	90	79	82	90	81	89	80	88	79	153	87	79	148	86	78	146	86	15.8
Tallahassee	95	77	93	76	91	76	80	89	79	88	78	87	77	142	83	76	138	82	76	135	82	18.5
Tampa, Intl Airport	92	77	91	77	90	77	80	88	79	88	78	87	78	144	85	77	140	84	76	137	83	15.0
Valparaiso, Eglin AFB	92	78	90	78	89	77	81	88	80	87	79	86	79	149	85	78	144	84	77	141	83	13.9
Vero Beach	92	77	90	78	89	77	80	88	79	88	79	87	77	141	85	77	139	84	76	137	84	15.7

DP = dew-point temperature, °F
MDB = mean coincident dry-bulb temp., °F
StdD = standard deviation, °F
MWB = mean coincident wet-bulb temp., °F
MWS = mean coincident wind speed, mph
HR = humidity ratio, grains of moisture per lb of dry air
A = airport
ANGB = Air National Guard Base
MCAF = Marine Corps Air Facility
NAF = Naval Air Facility
NAWS = Naval Air Weapons Station
RAF = Royal Air Force

Table 1A Heating and Wind Design Conditions—United States

Station	WMO#	Lat.	Long.	Elev., ft	StdP, psia	Dates	Heating Dry Bulb		Extreme Wind Speed, mph			Coldest Month				MWS/PWD to DB				Extr. Annual Daily			
												0.4%		1%		99.6%		0.4%		Mean DB		StdD DB	
							99.6%	99%	1%	2.5%	5%	WS	MDB	WS	MDB	MWS	PWD	MWS	PWD	Max.	Min.	Max.	Min.
1a	1b	1c	1d	1e	1f	1g	2a	2b	3a	3b	3c	4a	4b	4c	4d	5a	5b	5c	5d	6a	6b	6c	6d
West Palm Beach	722030	26.68	80.12	20	14.685	6193	43	47	24	21	19	24	69	21	70	9	320	12	110	94	35	2.0	5.0
GEORGIA																							
Albany	722160	31.53	84.18	194	14.593	8293	27	30	19	17	15	19	50	18	50	4	360	9	250	100	17	2.2	7.2
Athens	723110	33.95	83.32	810	14.270	6193	20	25	19	17	15	20	40	18	40	10	290	9	270	98	11	3.5	6.6
Atlanta	722190	33.65	84.42	1033	14.155	6193	18	23	22	19	17	23	37	21	36	12	320	9	300	96	9	3.5	7.3
Augusta	722180	33.37	81.97	148	14.617	6193	21	25	20	18	15	21	45	19	46	5	290	9	250	100	13	3.7	5.6
Brunswick	722137	31.15	81.38	20	14.685	8293	30	34	18	17	16	19	49	18	49	8	350	10	250	98	22	2.5	7.7
Columbus, Ft. Benning	722250	32.33	85.00	233	14.572	8293	23	27	16	13	11	17	46	15	46	3	320	5	240	100	14	2.9	6.7
Columbus, Metro Airport	722255	32.52	84.93	397	14.486	6193	23	27	17	15	14	18	44	16	46	7	310	9	310	99	14	2.3	6.1
Macon	722170	32.70	83.65	361	14.505	6193	23	27	19	17	15	20	46	18	45	7	320	9	270	100	14	2.7	6.4
Marietta, Dobbins AFB	722270	33.92	84.52	1070	14.136	8293	21	26	18	16	13	20	35	18	38	9	340	6	300	97	12	3.6	6.7
Rome	723200	34.35	85.17	643	14.357	8293	15	21	14	12	10	14	42	13	42	5	340	6	270	98	4	3.8	7.0
Savannah	722070	32.13	81.20	49	14.669	6193	26	29	20	17	15	21	49	19	49	7	270	9	270	98	18	3.0	5.4
Valdosta, Moody AFB	747810	30.97	83.20	233	14.572	8293	30	34	15	13	12	16	53	14	52	4	360	5	300	99	21	2.5	7.6
Valdosta, Regional Airport	722166	30.78	83.28	203	14.588	8293	28	31	17	15	14	18	55	16	56	4	340	8	300	99	17	3.2	7.7
Waycross	722130	31.25	82.40	151	14.615	8293	29	32	16	14	12	16	52	14	52	4	250	7	240	98	21	7.0	7.6
HAWAII																							
Ewa, Barbers Point NAS	911780	21.32	158.07	49	14.669	8293	59	61	20	18	16	22	73	19	75	5	40	11	60	93	35	1.6	21.4
Hilo	912850	19.72	155.07	36	14.676	6193	61	63	19	16	14	21	76	18	76	7	230	12	110	88	58	1.6	1.8
Honolulu	911820	21.35	157.93	16	14.687	6193	61	63	23	21	20	23	74	21	75	5	320	15	60	91	58	1.9	2.2
Kahului	911900	20.90	156.43	66	14.661	6193	59	61	27	25	24	32	76	28	76	6	160	19	50	92	54	1.5	4.4
Kaneohe, MCAS	911760	21.45	157.77	10	14.690	8293	67	68	20	18	17	21	74	19	74	7	190	10	70	88	40	1.4	29.0
Lihue	911650	21.98	159.35	148	14.617	6193	60	62	26	24	21	25	73	23	73	8	270	14	60	87	57	1.4	3.0
Molokai	911860	21.15	157.10	449	14.458	8293	60	61	24	22	21	22	74	21	74	4	70	13	60	92	43	4.0	22.0
IDAHO																							
Boise	726810	43.57	116.22	2867	13.235	6193	2	9	24	21	18	22	37	19	37	6	130	11	320	103	−4	2.7	9.1
Burley	725867	42.55	113.77	4150	12.621	8293	−5	2	23	21	19	23	30	22	28	7	60	8	280	98	−11	4.0	8.5
Idaho Falls	725785	43.52	112.07	4741	12.346	8293	−12	−6	27	23	21	28	32	23	29	7	360	12	180	96	−20	3.6	9.0
Lewiston	727830	46.38	117.02	1437	13.948	8293	6	15	20	17	14	24	38	20	40	5	280	7	310	103	3	2.7	9.9
Mountain Home, AFB	726815	43.05	115.87	2995	13.173	8293	0	5	23	21	18	23	33	21	31	2	90	8	350	105	−6	3.2	8.5
Mullan	727836	47.47	115.80	3317	13.017	8293	−1	7	10	10	9	11	18	9	21	2	10	4	10	92	−7	2.0	7.9
Pocatello	725780	42.92	112.60	4478	12.468	6193	−7	0	29	25	23	30	36	27	36	6	50	11	250	98	−15	2.3	9.1
ILLINOIS																							
Belleville, Scott AFB	724338	38.55	89.85	453	14.457	8293	3	10	21	18	15	23	32	20	31	7	360	7	190	100	−3	3.1	7.2
Chicago, Meigs Field	725340	41.78	87.75	623	14.367	8293	−4	3	23	22	19	26	17	23	30	12	240	13	220	97	−10	3.2	8.1
Chicago, O'Hare Intl A	725300	41.98	87.90	673	14.342	6193	−6	−1	26	23	21	27	24	23	23	10	270	12	230	96	−12	2.8	6.5
Decatur	725316	39.83	88.87	682	14.337	8293	−2	3	24	22	20	27	24	24	27	13	310	12	210	99	−10	5.8	7.2
Glenview, NAS	725306	42.08	87.82	653	14.352	8293	−3	4	22	19	17	23	17	20	25	11	250	10	240	98	−10	3.1	7.7
Marseilles	744600	41.37	88.68	738	14.308	8293	−5	1	26	22	20	28	18	25	21	12	290	10	250	96	−11	4.0	5.9
Moline/Davenport IA	725440	41.45	90.52	594	14.383	6193	−8	−3	26	23	20	28	16	24	18	9	290	12	200	97	−14	2.7	6.0
Peoria	725320	40.67	89.68	663	14.347	6193	−6	−1	25	22	20	26	16	23	19	9	290	11	180	96	−12	3.3	6.1
Quincy	724396	39.95	91.20	768	14.292	8293	−4	2	26	23	20	28	23	24	22	12	330	12	210	97	−10	3.6	8.1
Rockford	725430	42.20	89.10	741	14.306	6193	−10	−4	26	23	21	26	18	23	20	9	290	13	200	95	−16	3.1	5.5
Springfield	724390	39.85	89.67	614	14.373	6193	−4	2	25	23	21	27	25	24	27	10	270	12	230	97	−11	2.8	5.5
West Chicago	725305	41.92	88.25	758	14.297	8293	−7	0	23	21	19	25	13	23	20	11	290	11	240	96	−14	3.2	7.7
INDIANA																							
Evansville	724320	38.05	87.53	387	14.491	6193	3	9	22	19	17	22	33	20	34	7	320	9	240	97	−4	2.7	8.5
Ft. Wayne	725330	41.00	85.20	827	14.262	6193	−4	2	25	23	20	27	19	24	22	10	250	12	230	95	−11	3.6	5.2
Indianapolis	724380	39.73	86.27	807	14.272	6193	−3	3	24	21	19	25	26	22	27	8	230	11	230	94	−10	2.8	6.8
Lafayette, Purdue Univ	724386	40.42	86.93	607	14.376	8293	−5	3	22	20	18	24	26	22	27	9	270	12	220	97	−11	3.8	7.7
Peru, Grissom AFB	725335	40.65	86.15	810	14.270	8293	−3	4	24	21	18	29	20	24	22	11	270	9	210	96	−8	3.8	7.4
South Bend	725350	41.70	86.32	774	14.289	6193	−2	3	25	23	20	26	22	23	23	13	230	12	230	95	−10	3.3	5.8
Terre Haute	724373	39.45	87.32	584	14.388	8293	−3	5	23	20	18	23	31	21	32	8	150	11	230	96	−10	3.2	7.9
IOWA																							
Burlington	725455	40.78	91.13	699	14.328	8293	−4	1	21	19	17	24	12	21	18	9	310	11	200	98	−10	4.0	6.8
Cedar Rapids	725450	41.88	91.70	869	14.240	8293	−11	−5	25	22	20	29	12	26	14	10	300	11	180	96	−15	3.6	5.4
Des Moines	725460	41.53	93.65	965	14.190	6193	−9	−4	27	24	21	28	14	24	19	11	320	12	180	98	−15	3.4	5.1
Ft. Dodge	725490	42.55	94.18	1165	14.087	8293	−13	−7	27	23	21	29	10	26	10	11	340	11	190	96	−17	4.9	4.9
Lamoni	725466	40.62	93.95	1122	14.109	8293	−6	0	19	17	15	21	23	19	20	7	320	9	210	99	−12	4.3	6.8
Mason City	725485	43.15	93.33	1214	14.062	6193	−15	−10	27	23	22	30	9	27	12	12	300	14	200	97	−23	3.6	11.4
Ottumwa	725465	41.10	92.45	846	14.251	8293	−5	0	29	26	23	31	20	28	24	13	320	15	200	98	−12	4.0	6.8
Sioux City	725570	42.40	96.38	1102	14.119	6193	−11	−6	29	25	22	31	14	28	16	11	320	14	180	99	−18	3.6	4.7
Spencer	726500	43.17	95.15	1339	13.998	8293	−16	−11	24	22	20	25	13	23	13	10	300	12	180	99	−20	6.3	4.0
Waterloo	725480	42.55	92.40	879	14.234	6193	−14	−9	27	24	22	29	10	25	13	9	300	13	180	96	−20	3.5	5.9
KANSAS																							
Concordia	724580	39.55	97.65	1483	13.925	8293	−4	3	28	25	22	28	32	25	32	13	360	16	10	104	−8	4.0	9.4
Dodge City	724510	37.77	99.97	2592	13.370	6193	0	6	30	27	24	31	31	27	32	13	10	17	200	104	−6	2.8	5.6
Ft Riley, Marshall AAF	724550	39.05	96.77	1066	14.138	8293	−2	5	21	18	16	20	39	18	37	5	350	9	180	104	−5	3.1	9.0
Garden City	724515	37.93	100.72	2890	13.224	8293	−3	4	30	26	23	29	32	25	34	12	360	16	190	104	−9	2.7	6.5
Goodland	724650	39.37	101.70	3688	12.840	6193	−3	2	32	28	24	31	27	27	30	12	270	13	180	102	−11	2.9	6.6

WMO# = World Meteorological Organization number
AFB = Air Force Base
NAS = Naval Air Station
WSO = Weather Service Office
AAF = Army Air Field
MCAS = Marine Corps Air Station
DB = dry-bulb temperature, °F
Lat. = North latitude °
Long. = West longitude °
Elev. = elevation, ft
StdP = standard pressure at station elevation, psia
WS = wind speed, mph
PWD = prevailing wind direction, ° True

Table 1B Cooling and Dehumidification Design Conditions—United States

Station	Cooling DB/MWB						Evaporation WB/MDB						Dehumidification DP/MDB and HR									Range of DB
	0.4%		1%		2%		0.4%		1%		2%		0.4%			1%			2%			
	DB	MWB	DB	MWB	DB	MWB	WB	MDB	WB	MDB	WB	MDB	DP	HR	MDB	DP	HR	MDB	DP	HR	MDB	
1	2a	2b	2c	2d	2e	2f	3a	3b	3c	3d	3e	3f	4a	4b	4c	4d	4e	4f	4g	4h	4i	5
West Palm Beach	91	78	90	78	89	77	80	88	79	88	78	87	77	143	84	77	139	84	76	137	83	13.1
GEORGIA																						
Albany	96	76	95	76	93	75	79	90	78	89	78	88	77	141	83	76	136	82	75	133	81	19.8
Athens	94	75	92	75	90	74	78	89	77	87	76	86	75	133	82	74	129	81	73	125	80	18.4
Atlanta	93	75	91	74	88	73	77	88	76	87	75	85	74	133	82	73	128	81	72	124	80	17.3
Augusta	96	76	94	76	92	75	79	91	78	89	77	88	76	135	84	75	130	83	74	127	82	20.2
Brunswick	93	78	91	79	88	78	81	89	80	88	79	87	78	147	86	78	144	85	77	141	84	14.4
Columbus, Ft. Benning	97	76	94	76	92	76	80	91	79	89	78	88	77	142	85	76	136	83	75	133	82	20.5
Columbus, Metro Airport	95	76	93	75	91	75	79	89	78	88	77	87	76	139	82	75	134	82	74	130	81	18.0
Macon	96	76	94	75	92	75	79	91	78	89	77	88	76	136	83	75	132	82	74	129	82	19.3
Marietta, Dobbins AFB	94	74	91	74	89	74	77	88	76	87	75	86	74	134	82	73	130	81	72	123	79	17.1
Rome	96	74	94	74	91	74	78	90	77	89	76	88	75	134	83	74	130	83	73	127	83	20.7
Savannah	95	77	93	76	91	76	79	90	78	89	78	87	77	139	84	76	135	83	75	132	82	17.5
Valdosta, Moody AFB	95	77	94	77	92	76	80	91	79	89	78	88	77	142	85	76	139	84	76	135	83	17.8
Valdosta, Regional Airport	95	77	94	76	92	76	80	90	79	89	78	88	77	144	83	76	139	82	76	136	82	19.4
Waycross	96	76	94	76	93	75	78	91	78	90	77	89	75	134	84	75	130	83	74	127	83	20.3
HAWAII																						
Ewa, Barbers Point NAS	92	73	90	72	89	72	76	86	75	86	75	85	74	126	83	72	118	82	71	113	82	15.8
Hilo	85	74	84	74	83	73	76	82	76	81	75	81	75	130	79	74	127	79	73	123	78	13.3
Honolulu	89	73	88	73	87	73	76	84	75	84	74	84	74	125	80	72	120	80	71	116	79	12.2
Kahului	89	74	88	74	87	73	76	85	76	85	75	84	74	127	80	73	122	80	72	118	80	15.6
Kaneohe, MCAS	86	75	85	74	84	74	78	82	77	82	76	82	76	138	81	75	133	81	74	128	80	7.4
Lihue	85	75	85	74	84	74	77	83	76	82	75	82	75	132	80	74	128	80	73	125	79	9.6
Molokai	88	73	87	73	86	72	76	85	75	83	74	83	74	128	80	73	124	79	71	118	79	13.3
IDAHO																						
Boise	96	63	94	63	91	62	66	90	64	89	63	87	58	79	72	55	72	71	53	67	71	30.3
Burley	94	63	90	62	87	61	67	86	65	84	63	83	60	90	75	58	84	72	56	78	72	29.0
Idaho Falls	92	61	89	60	86	60	64	84	63	82	61	81	58	88	71	56	81	69	54	73	68	34.0
Lewiston	97	65	93	64	90	63	67	91	65	89	64	87	58	76	72	56	71	71	54	65	71	26.5
Mountain Home, AFB	99	63	96	62	93	61	66	91	64	91	63	89	58	79	71	54	70	69	52	64	71	32.8
Mullan	87	62	84	61	80	60	65	81	63	79	62	77	60	86	69	58	80	68	56	75	66	28.1
Pocatello	93	61	90	60	87	59	64	84	62	83	61	82	57	83	70	55	76	70	53	70	69	32.1
ILLINOIS																						
Belleville, Scott AFB	95	78	93	77	90	76	80	92	78	90	77	88	77	141	87	76	136	85	74	131	84	19.8
Chicago, Meigs Field	92	74	89	73	86	71	77	88	76	85	74	83	74	132	84	72	121	80	71	115	80	16.0
Chicago, O'Hare Intl A	91	74	88	73	86	71	77	88	75	85	74	83	74	130	84	72	123	82	71	115	80	19.6
Decatur	94	76	91	75	88	74	79	90	78	89	76	86	76	140	86	75	133	84	73	127	83	20.0
Glenview, NAS	93	75	89	73	87	71	78	90	76	87	74	84	74	130	85	72	120	82	70	113	81	17.6
Marseilles	93	74	89	73	86	71	78	89	76	86	74	84	75	135	85	73	126	82	71	117	81	19.4
Moline/Davenport IA	93	76	90	74	87	73	78	90	77	87	75	85	75	134	85	73	127	83	72	120	82	20.0
Peoria	92	76	89	74	86	73	78	89	77	86	75	84	75	137	85	74	130	83	72	123	81	19.5
Quincy	94	76	91	75	88	74	78	89	77	88	76	85	76	138	84	74	132	82	73	126	82	18.9
Rockford	91	74	88	73	85	71	77	87	75	85	74	82	74	132	84	73	124	81	71	116	79	19.8
Springfield	93	76	91	75	88	74	79	89	77	88	76	85	76	139	86	75	132	84	73	125	82	19.4
West Chicago	91	75	88	74	86	72	78	88	76	85	74	83	76	138	85	74	130	83	71	119	80	19.8
INDIANA																						
Evansville	94	77	92	76	90	75	79	90	78	89	77	87	76	137	86	75	132	84	73	126	83	19.8
Ft. Wayne	90	74	88	73	85	71	77	86	75	84	74	82	74	131	83	72	124	81	71	117	79	19.9
Indianapolis	91	75	88	74	86	73	78	88	77	86	75	83	75	137	84	74	131	82	73	125	81	18.9
Lafayette, Purdue Univ	93	75	90	75	88	73	79	89	77	86	75	84	76	139	85	74	132	83	73	125	82	20.9
Peru, Grissom AFB	93	75	89	75	87	73	79	89	77	86	75	83	76	142	85	75	134	83	73	127	81	18.5
South Bend	90	73	87	72	85	71	77	86	75	84	73	81	74	130	83	72	123	80	71	116	78	18.6
Terre Haute	93	76	90	76	88	75	80	89	78	87	76	85	77	144	86	76	136	84	74	131	82	19.6
IOWA																						
Burlington	94	76	91	76	88	73	78	89	77	88	76	85	75	136	85	74	131	83	72	124	82	18.7
Cedar Rapids	93	75	89	74	86	72	78	89	76	86	74	84	75	136	84	74	129	83	71	120	80	20.0
Des Moines	93	76	90	74	87	73	78	89	76	87	75	85	74	133	85	73	126	83	71	120	81	18.5
Ft. Dodge	92	75	88	73	86	71	77	88	75	86	74	83	74	133	84	72	123	82	70	116	79	18.5
Lamoni	96	74	92	74	89	72	77	89	76	87	75	85	74	134	83	73	127	82	71	120	80	18.9
Mason City	91	74	88	73	85	71	77	87	75	85	74	82	75	135	84	73	126	82	71	118	80	20.8
Ottumwa	95	75	92	75	88	73	78	90	76	88	75	86	75	136	84	74	130	83	72	121	81	18.7
Sioux City	94	75	90	74	88	72	78	89	76	87	75	85	75	135	86	73	127	84	71	120	82	20.4
Spencer	91	75	88	73	85	71	77	88	75	86	73	82	74	134	84	72	123	82	70	117	79	20.2
Waterloo	91	75	88	73	85	71	77	87	75	85	74	83	74	132	84	72	124	82	71	117	80	20.0
KANSAS																						
Concordia	100	73	96	72	93	72	77	90	76	89	74	88	74	133	84	72	123	82	70	118	81	22.5
Dodge City	100	70	97	70	94	69	74	90	73	89	71	88	70	120	79	68	114	78	67	109	77	24.3
Ft Riley, Marshall AAF	99	75	96	74	93	74	78	90	77	90	75	88	75	136	86	73	130	83	71	120	82	22.7
Garden City	100	69	97	69	94	69	73	89	72	89	71	88	69	118	79	67	113	78	66	108	77	27.5
Goodland	97	66	94	66	91	65	70	86	69	84	68	84	66	111	74	65	106	73	63	100	73	26.5

DP = dew-point temperature, °F
MDB = mean coincident dry-bulb temp., °F
StdD = standard deviation, °F
MWB = mean coincident wet-bulb temp., °F
MWS = mean coincident wind speed, mph
HR = humidity ratio, grains of moisture per lb of dry air
A = airport
ANGB = Air National Guard Base
MCAF = Marine Corps Air Facility
NAF = Naval Air Facility
NAWS = Naval Air Weapons Station
RAF = Royal Air Force

Table 1A Heating and Wind Design Conditions—United States

Station	WMO#	Lat.	Long.	Elev., ft	StdP, psia	Dates	Heating Dry Bulb		Extreme Wind Speed, mph			Coldest Month				MWS/PWD to DB				Extr. Annual Daily			
												0.4%		1%		99.6%		0.4%		Mean DB		StdD DB	
							99.6%	99%	1%	2.5%	5%	WS	MDB	WS	MDB	MWS	PWD	MWS	PWD	Max.	Min.	Max.	Min.
1a	1b	1c	1d	1e	1f	1g	2a	2b	3a	3b	3c	4a	4b	4c	4d	5a	5b	5c	5d	6a	6b	6c	6d
Russell	724585	38.87	98.82	1864	13.732	8293	−4	3	29	26	23	29	33	25	35	11	10	16	190	105	−8	3.6	8.5
Salina	724586	38.80	97.65	1273	14.032	8293	−3	4	27	23	22	28	33	24	34	11	360	15	180	106	−7	2.3	9.9
Topeka	724560	39.07	95.62	886	14.231	6193	−2	4	25	22	20	25	28	22	29	9	320	12	180	100	−8	3.8	7.4
Wichita, Airport	724500	37.65	97.43	1339	13.998	6193	2	8	29	25	23	28	30	26	31	13	360	16	200	105	−4	2.9	6.3
Wichita, McConnell AFB	724505	37.62	97.27	1371	13.982	8293	2	10	25	23	20	25	38	23	36	11	360	12	190	105	−1	2.7	7.7
KENTUCKY																							
Bowling Green	746716	36.97	86.42	548	14.407	8293	7	14	20	19	17	21	40	19	40	6	220	9	230	97	−2	3.2	10.3
Covington/Cincinnati A	724210	39.05	84.67	876	14.236	6193	1	7	22	20	18	25	30	22	33	9	250	10	230	95	−7	3.1	8.5
Ft. Campbell, AAF	746710	36.67	87.50	571	14.395	8293	9	15	19	16	14	20	40	17	43	4	330	6	240	98	0	3.1	9.5
Ft. Knox, Godman AAF	724240	37.90	85.97	755	14.299	8293	9	15	17	15	13	18	42	16	39	4	290	6	270	97	0	4.1	7.9
Jackson	724236	37.60	83.32	1381	13.977	8293	7	14	17	14	13	18	44	16	40	7	230	6	230	94	−3	2.7	9.4
Lexington	724220	38.03	84.60	988	14.179	6193	4	10	21	19	17	23	38	20	38	8	270	9	240	94	−4	3.5	8.3
Louisville	724230	38.18	85.73	489	14.438	6193	6	12	22	19	17	22	40	20	34	10	290	10	250	96	−1	3.1	7.9
Paducah	724350	37.07	88.77	413	14.477	8293	7	13	22	19	17	22	45	19	42	8	40	9	180	98	−1	2.9	9.4
LOUISIANA																							
Alexandria, England AFB	747540	31.33	92.55	89	14.648	8293	27	30	16	13	12	17	53	15	49	7	360	3	180	98	20	2.2	6.3
Baton Rouge	722317	30.53	91.15	69	14.659	6193	27	30	20	18	16	21	48	19	49	8	360	8	270	97	20	2.2	5.4
Bossier City, Barksdale AFB	722485	32.50	93.67	167	14.607	8293	22	27	18	16	14	19	49	16	51	7	360	5	180	99	15	2.3	6.7
Lafayette	722405	30.20	91.98	43	14.673	8293	28	32	21	18	16	21	54	19	53	9	10	8	200	97	19	1.6	8.1
Lake Charles	722400	30.12	93.22	33	14.678	6193	29	32	22	19	17	24	50	21	49	10	20	8	230	96	23	2.3	4.7
Leesville, Ft. Polk	722390	31.05	93.20	328	14.522	8293	27	30	16	13	12	16	51	14	52	4	20	4	180	98	20	2.0	5.9
Monroe	722486	32.52	92.03	79	14.654	8293	22	27	19	17	15	20	50	18	47	9	10	7	230	99	17	1.8	8.5
New Orleans, Intl Airport	722310	29.98	90.25	30	14.680	6193	30	34	21	19	17	21	48	19	49	7	340	8	360	96	23	2.0	5.3
New Orleans, Lakefront A	722315	30.05	90.03	10	14.690	8293	35	39	22	19	18	21	49	20	50	14	360	9	300	94	21	8.1	12.4
Shreveport	722480	32.47	93.82	259	14.558	6193	22	26	20	18	16	22	46	19	48	9	360	8	180	99	16	3.1	5.6
MAINE																							
Augusta	726185	44.32	69.80	351	14.510	8293	−3	1	23	21	19	25	20	22	22	10	320	11	210	93	−10	3.1	3.4
Bangor	726088	44.80	68.83	194	14.593	8293	−7	−2	22	19	17	24	18	21	20	6	300	10	240	94	−16	2.9	5.9
Brunswick, NAS	743920	43.88	69.93	75	14.655	8293	−2	2	20	17	15	21	27	19	25	4	340	9	190	96	−12	7.9	6.1
Caribou	727120	46.87	68.02	623	14.367	6193	−14	−10	28	24	22	30	13	27	11	10	270	13	250	90	−23	2.8	4.5
Limestone, Loring AFB	727125	46.95	67.88	745	14.304	8293	−13	−9	23	20	18	25	12	22	11	7	300	9	260	91	−20	2.3	2.9
Portland	726060	43.65	70.32	62	14.662	6193	−3	2	24	21	18	24	26	21	25	7	320	12	270	93	−13	3.6	5.5
MARYLAND																							
Camp Springs, Andrews AFB	745940	38.82	76.87	282	14.546	8293	13	18	21	18	16	23	30	21	32	7	350	9	230	98	4	2.9	6.7
Baltimore, BWI Airport	724060	39.18	76.67	154	14.614	6193	11	15	24	21	19	25	31	22	31	10	290	11	280	97	4	2.9	5.8
Lex Park, Patuxent River NAS	724040	38.28	76.40	39	14.675	8293	16	21	20	17	15	22	30	19	35	9	340	9	270	98	8	2.3	6.1
Salisbury	724045	38.33	75.52	52	14.668	8293	13	18	20	18	16	20	35	19	37	6	10	9	240	97	4	2.7	5.8
MASSACHUSETTS																							
Boston	725090	42.37	71.03	30	14.680	6193	7	12	29	25	23	30	30	27	28	17	320	14	270	96	0	2.7	4.7
East Falmouth, Otis ANGB	725060	41.65	70.52	131	14.626	8293	11	14	26	22	20	26	34	23	33	9	300	10	240	90	5	2.5	3.8
Weymouth, S Weymth NAS	725097	42.15	70.93	161	14.610	8293	6	11	19	16	14	18	29	16	29	7	320	9	260	97	−2	3.8	3.8
Worcester	725095	42.27	71.88	1010	14.167	6193	0	5	27	23	20	29	22	26	21	14	270	10	270	90	−6	1.9	4.1
MICHIGAN																							
Alpena	726390	45.07	83.57	692	14.332	6193	−7	−1	21	19	17	22	20	19	20	5	270	11	240	93	−17	3.4	5.9
Detroit, Metro	725370	42.23	83.33	663	14.347	6193	0	5	27	23	21	28	28	24	27	11	240	13	230	95	−7	3.0	5.4
Flint	726370	42.97	83.75	764	14.294	6193	−2	3	25	22	20	27	24	23	23	8	230	13	230	93	−10	3.1	5.0
Grand Rapids	726350	42.88	85.52	804	14.274	6193	0	5	25	22	20	26	25	23	24	8	180	13	240	93	−9	2.1	5.3
Hancock	727440	47.17	88.50	1079	14.131	6193	−9	−4	21	19	18	23	18	20	16	8	270	10	250	90	−16	2.9	5.6
Harbor Beach	725386	44.02	82.80	600	14.379	8293	9	12	26	22	19	26	27	23	27	10	220	9	230	94	2	2.9	4.1
Jackson	725395	42.27	84.47	1001	14.172	8293	−3	4	20	19	17	23	22	20	23	9	240	11	210	93	−11	2.5	5.6
Lansing	725390	42.77	84.60	873	14.238	6193	−3	2	26	23	20	28	23	25	24	8	290	13	250	94	−13	2.8	5.9
Marquette, Sawyer AFB	727435	46.35	87.40	1220	14.059	8293	−11	−6	24	21	18	26	18	23	17	6	280	10	210	91	−18	4.7	4.7
Marquette/Ishpeming A	727430	46.53	87.55	1424	13.955	8293	−13	−8	22	19	18	22	20	20	16	8	270	11	230	90	−22	4.5	4.5
Mount Clemens, ANGB	725377	42.62	82.83	581	14.390	8293	3	7	21	18	16	25	21	21	24	7	280	9	230	95	−3	4.0	2.7
Muskegon	726360	43.17	86.25	633	14.362	6193	3	7	27	24	22	28	25	25	26	10	290	12	200	90	−5	2.7	5.0
Oscoda, Wurtsmith AFB	726395	44.45	83.40	633	14.362	8293	0	3	21	19	17	23	26	21	24	6	220	11	200	95	−7	4.1	4.7
Pellston	727347	45.57	84.80	719	14.318	8293	−9	−3	26	23	20	28	22	24	22	4	300	14	250	92	−21	3.1	4.9
Saginaw	726379	43.53	84.08	669	14.343	8293	0	4	23	21	19	25	22	22	23	10	260	13	240	96	−6	5.8	4.5
Sault Ste. Marie	727340	46.47	84.37	725	14.314	6193	−12	−7	23	20	18	24	19	21	18	7	90	10	230	89	−22	3.5	5.4
Seul Choix Point	726399	45.92	85.92	591	14.385	8293	0	4	28	24	22	30	27	26	27	9	300	8	200	82	−5	2.3	6.3
Traverse City	726387	44.73	85.58	623	14.367	6193	−3	2	21	19	18	23	23	21	23	7	180	13	230	94	−13	2.8	7.3
MINNESOTA																							
Alexandria	726557	45.87	95.40	1424	13.955	8293	−20	−15	25	22	20	28	12	24	8	10	300	14	180	96	−26	3.6	4.5
Brainerd, Pequot Lakes	727500	46.60	94.32	1280	14.029	8293	−24	−17	11	10	9	11	8	10	11	3	320	5	190	95	−30	7.9	6.8
Duluth	727450	46.83	92.18	1417	13.958	6193	−21	−16	25	22	20	25	12	22	11	10	310	12	230	90	−28	2.8	4.7
Hibbing	727455	47.38	92.83	1352	13.992	8293	−25	−20	20	19	17	20	13	19	13	6	330	11	200	92	−34	2.5	4.7
International Falls	727470	48.57	93.38	1184	14.077	6193	−29	−23	22	20	18	22	10	20	8	6	270	11	180	92	−37	3.4	3.8
Minneapolis-St. Paul	726580	44.88	93.22	837	14.257	6193	−16	−11	25	22	20	25	12	22	14	9	300	14	180	97	−22	3.5	5.4
Redwood Falls	726556	44.55	95.08	1024	14.160	8293	−17	−12	26	22	20	28	14	24	15	11	280	14	180	99	−22	4.1	5.2
Rochester	726440	43.92	92.50	1319	14.008	6193	−17	−12	29	26	24	32	12	28	12	13	300	15	200	94	−23	3.7	5.2

WMO# = World Meteorological Organization number
AFB = Air Force Base NAS = Naval Air Station
WSO = Weather Service Office
AAF = Army Air Field
MCAS = Marine Corps Air Station
DB = dry-bulb temperature, °F
Lat. = North latitude °
Long. = West longitude °
Elev. = elevation, ft
StdP = standard pressure at station elevation, psia
WS = wind speed, mph
PWD = prevailing wind direction, ° True

Table 1B Cooling and Dehumidification Design Conditions—United States

Station	Cooling DB/MWB						Evaporation WB/MDB						Dehumidification DP/MDB and HR									Range of DB
	0.4%		1%		2%		0.4%		1%		2%		0.4%			1%			2%			
	DB	MWB	DB	MWB	DB	MWB	WB	MDB	WB	MDB	WB	MDB	DP	HR	MDB	DP	HR	MDB	DP	HR	MDB	
1	2a	2b	2c	2d	2e	2f	3a	3b	3c	3d	3e	3f	4a	4b	4c	4d	4e	4f	4g	4h	4i	5
Russell	100	72	96	72	94	72	76	91	75	90	73	88	72	126	83	71	120	82	69	116	80	24.1
Salina	101	74	97	73	94	73	77	92	76	90	75	89	74	132	85	72	123	83	71	118	82	23.0
Topeka	96	75	93	75	90	75	79	90	78	89	76	88	76	139	87	74	132	85	73	126	83	20.3
Wichita, Airport	100	73	97	73	94	73	77	91	76	90	74	89	73	129	83	72	123	82	71	118	81	22.2
Wichita, McConnell AFB	100	73	97	73	94	73	77	92	76	90	75	89	74	133	84	72	124	83	71	119	82	21.8
KENTUCKY																						
Bowling Green	94	76	91	75	88	74	78	89	77	87	76	86	76	136	84	75	132	82	74	127	81	20.0
Covington/Cincinnati A	91	74	89	73	86	72	77	87	76	86	74	83	74	132	84	73	126	81	72	120	80	18.9
Ft. Campbell, AAF	95	77	93	76	90	76	80	90	78	89	77	87	77	143	85	76	136	84	74	132	83	19.4
Ft. Knox, Godman AAF	94	76	92	74	89	74	78	90	77	88	76	86	76	138	85	74	132	83	73	126	82	19.4
Jackson	90	74	87	73	85	72	77	87	76	85	74	83	74	135	83	73	130	81	71	122	79	18.2
Lexington	91	74	89	73	86	72	77	87	75	86	74	83	74	130	83	72	124	81	71	120	80	18.4
Louisville	93	76	90	75	88	74	78	90	77	88	76	86	75	134	85	74	129	84	73	125	82	18.2
Paducah	96	77	93	76	92	75	80	91	79	90	78	88	77	143	86	76	138	85	75	132	83	20.2
LOUISIANA																						
Alexandria, England AFB	95	78	94	78	92	77	81	90	80	90	79	89	78	147	86	77	142	85	76	138	85	18.4
Baton Rouge	94	78	92	77	91	77	80	89	79	88	78	87	78	145	84	77	141	84	76	137	83	16.7
Bossier City, Barksdale AFB	96	77	94	77	93	77	80	90	79	90	78	89	77	144	84	76	139	83	76	134	83	20.0
Lafayette	94	78	93	78	91	77	80	89	80	89	79	88	78	146	84	77	143	83	77	140	83	17.1
Lake Charles	93	78	91	78	90	77	80	88	80	88	79	87	78	148	84	78	145	84	77	141	83	16.2
Leesville, Ft. Polk	95	77	94	76	92	76	79	89	79	88	78	87	77	144	83	76	140	82	76	136	82	18.2
Monroe	96	78	94	78	93	77	81	91	80	90	79	89	78	147	86	77	143	85	77	139	84	19.3
New Orleans, Intl Airport	93	79	92	78	90	78	81	90	80	88	80	87	79	151	86	78	146	85	77	142	84	15.5
New Orleans, Lakefront A	93	78	92	78	90	77	81	88	80	87	79	87	79	150	85	78	145	84	77	141	83	11.9
Shreveport	97	77	95	77	93	76	79	91	79	90	78	89	76	139	84	76	135	83	75	132	83	19.1
MAINE																						
Augusta	87	71	84	69	80	67	73	83	71	80	69	77	70	113	77	68	106	75	67	100	74	18.4
Bangor	87	71	84	69	81	67	73	83	71	81	69	77	70	111	78	68	104	75	67	99	73	20.5
Brunswick, NAS	87	71	84	69	80	67	73	83	71	80	70	77	70	111	78	69	105	76	67	100	74	19.1
Caribou	85	69	82	67	79	66	72	81	70	77	68	76	70	112	76	68	104	75	66	97	72	19.5
Limestone, Loring AFB	84	68	80	66	78	64	71	79	69	76	67	74	68	107	75	67	101	72	65	94	71	18.7
Portland	86	71	83	70	80	68	74	83	72	80	70	77	71	114	79	69	107	76	67	101	74	18.7
MARYLAND																						
Camp Springs, Andrews AFB	94	75	91	74	88	73	78	88	77	87	75	85	75	134	83	74	129	82	73	124	80	18.7
Baltimore, BWI Airport	93	75	91	74	88	73	78	88	76	86	75	85	75	132	83	74	125	81	72	120	80	18.8
Lex Park, Patuxent River NAS	93	76	90	75	87	74	79	88	77	87	76	85	76	136	84	75	131	83	74	125	82	15.8
Salisbury	93	77	90	76	88	75	80	88	78	86	77	85	78	144	84	76	137	82	75	132	81	18.7
MASSACHUSETTS																						
Boston	91	73	87	71	84	70	75	87	74	83	72	81	72	119	80	71	113	79	69	108	78	15.3
East Falmouth, Otis ANGB	85	72	82	72	79	69	75	81	74	78	72	76	74	125	78	72	118	76	71	113	75	14.6
S. Weymouth NAS	92	73	87	72	85	71	77	87	75	84	73	81	74	129	82	72	118	79	70	111	78	19.6
Worcester	85	71	83	69	80	68	74	82	72	80	70	77	71	119	78	69	112	76	68	105	75	16.6
MICHIGAN																						
Alpena	87	71	84	69	81	67	74	83	72	81	70	78	71	116	79	69	107	76	67	100	74	22.9
Detroit, Metro	90	73	87	72	84	70	76	86	74	84	73	81	73	125	83	71	118	80	70	111	78	20.4
Flint	88	73	86	71	83	70	75	84	74	82	72	80	73	125	81	71	116	78	69	110	77	20.6
Grand Rapids	89	73	86	71	84	70	76	85	74	83	72	81	73	126	81	71	118	79	70	112	77	20.7
Hancock	86	71	83	69	80	67	73	82	71	80	70	77	70	116	79	69	109	76	67	103	74	20.6
Harbor Beach	90	71	86	69	83	68	74	86	72	83	70	80	70	113	82	68	106	80	67	100	78	14.4
Jackson	88	74	86	73	84	71	77	86	75	83	73	81	74	134	83	72	123	81	71	117	78	20.3
Lansing	89	73	86	72	84	70	76	85	74	83	73	81	73	127	81	72	120	79	70	114	78	21.7
Marquette, Sawyer AFB	86	69	83	68	79	65	72	83	70	79	68	75	69	113	77	67	106	74	66	99	73	22.1
Marquette/Ishpeming, A	85	69	82	67	78	65	72	82	70	78	68	75	69	111	77	67	104	75	65	98	72	22.1
Mount Clemens, ANGB	90	74	87	72	84	71	77	87	75	83	73	80	74	131	83	72	120	81	70	113	78	19.6
Muskegon	85	71	83	70	81	69	75	82	73	80	71	78	72	122	80	70	115	77	69	109	76	18.1
Oscoda, Wurtsmith AFB	89	72	86	71	83	69	75	86	73	83	71	79	72	120	80	70	112	79	68	106	77	21.4
Pellston	87	71	85	69	81	68	74	83	72	81	70	78	70	115	78	69	108	76	67	103	75	23.9
Saginaw	90	74	87	72	84	70	77	86	75	84	73	81	74	132	83	72	120	80	70	112	78	21.2
Sault Ste. Marie	83	69	80	68	77	66	72	80	70	77	68	74	69	111	76	67	103	74	65	95	72	21.9
Seul Choix Point	78	66	76	65	74	64	70	76	68	72	66	71	68	106	74	67	101	72	65	94	70	13.9
Traverse City	89	71	86	70	83	68	74	84	72	82	70	80	71	117	80	69	109	78	67	103	76	22.0
MINNESOTA																						
Alexandria	89	72	86	70	83	69	75	86	73	82	71	80	72	123	82	70	116	79	68	109	77	19.3
Brainerd, Pequot Lakes	88	70	85	68	81	66	72	85	70	82	68	78	68	108	81	66	102	77	65	96	75	21.6
Duluth	84	69	81	67	78	65	72	81	69	78	67	75	68	110	77	66	102	75	64	94	72	20.2
Hibbing	85	70	81	68	78	66	73	82	71	78	68	75	70	116	78	68	108	76	66	101	73	23.2
International Falls	86	69	83	67	80	66	72	82	70	79	68	77	69	112	78	67	103	75	65	96	73	21.8
Minneapolis-St. Paul	91	73	88	71	85	70	76	88	74	84	72	82	73	124	83	71	116	81	69	109	79	19.1
Redwood Falls	92	74	88	72	86	70	77	89	75	85	73	82	75	135	83	72	123	81	70	116	80	20.7
Rochester	88	72	85	71	82	70	76	85	74	82	72	80	73	128	81	71	120	79	69	111	77	19.7

DP = dew-point temperature, °F
MDB = mean coincident dry-bulb temp., °F
StdD = standard deviation, °F
MWB = mean coincident wet-bulb temp., °F
MWS = mean coincident wind speed, mph
HR = humidity ratio, grains of moisture per lb of dry air
A = airport
ANGB = Air National Guard Base
MCAF = Marine Corps Air Facility
NAF = Naval Air Facility
NAWS = Naval Air Weapons Station
RAF = Royal Air Force

Table 1A Heating and Wind Design Conditions—United States

Station	WMO#	Lat.	Long.	Elev., ft	StdP, psia	Dates	Heating Dry Bulb 99.6%	Heating Dry Bulb 99%	Extreme Wind Speed, mph 1%	Extreme Wind Speed, mph 2.5%	Extreme Wind Speed, mph 5%	Coldest Month 0.4% WS	Coldest Month 0.4% MDB	Coldest Month 1% WS	Coldest Month 1% MDB	MWS/PWD to DB 99.6% MWS	MWS/PWD to DB 99.6% PWD	MWS/PWD to DB 0.4% MWS	MWS/PWD to DB 0.4% PWD	Extr. Annual Daily Mean DB Max.	Extr. Annual Daily Mean DB Min.	Extr. Annual Daily StdD DB Max.	Extr. Annual Daily StdD DB Min.
1a	1b	1c	1d	1e	1f	1g	2a	2b	3a	3b	3c	4a	4b	4c	4d	5a	5b	5c	5d	6a	6b	6c	6d
Saint Cloud	726550	45.55	94.07	1024	14.160	6193	−20	−14	22	20	18	23	11	20	10	8	300	12	200	95	−27	3.0	5.6
Tofte	727554	47.58	90.83	791	14.280	8293	−10	−6	24	20	17	25	16	22	18	8	260	8	330	86	−19	4.5	4.9
MISSISSIPPI																							
Biloxi, Keesler AFB	747686	30.42	88.92	33	14.678	8293	31	35	17	14	13	18	49	16	50	8	360	7	210	97	23	2.0	7.4
Columbus, AFB	723306	33.65	88.45	220	14.579	8293	20	25	18	15	13	19	43	16	46	6	360	6	240	100	12	2.7	6.8
Greenwood	722359	33.50	90.08	154	14.614	8293	20	24	19	17	14	19	46	18	47	6	360	6	180	99	13	2.3	7.6
Jackson	722350	32.32	90.08	331	14.520	6193	21	25	20	18	16	21	45	19	46	7	340	8	270	98	14	2.7	5.8
McComb	722358	31.18	90.47	413	14.477	8293	23	28	17	14	13	17	49	15	49	6	350	7	230	98	15	2.0	7.2
Meridian	722340	32.33	88.75	308	14.532	6193	21	25	19	17	15	19	43	17	46	6	360	8	360	99	13	2.9	5.9
Tupelo	723320	34.27	88.77	361	14.505	8293	18	22	19	17	15	20	44	17	44	7	10	7	260	99	10	2.9	8.5
MISSOURI																							
Cape Girardeau	723489	37.23	89.57	341	14.515	8293	6	13	21	19	18	22	35	20	36	9	360	10	200	100	−1	2.9	9.2
Columbia	724450	38.82	92.22	899	14.224	6193	−1	5	25	22	20	25	27	22	28	11	310	11	200	99	−8	4.4	6.2
Joplin	723495	37.15	94.50	981	14.182	8293	3	11	23	21	19	24	50	21	47	10	10	11	220	100	−2	4.0	9.4
Kansas City	724460	39.32	94.72	1024	14.160	6193	−1	4	26	23	20	26	34	23	33	10	320	13	190	100	−7	4.1	6.6
Poplar Bluff	723300	36.77	90.47	479	14.443	8293	8	13	18	15	13	17	40	15	38	7	360	7	200	101	2	6.8	9.4
Spickard/Trenton	725400	40.25	93.72	886	14.231	8293	1	6	23	20	18	25	29	22	31	8	360	11	200	100	−5	5.2	7.4
Springfield	724400	37.23	93.38	1270	14.034	6193	3	9	24	21	19	23	35	21	35	10	340	10	230	98	−4	3.5	6.4
St. Louis, Intl Airport	724340	38.75	90.37	564	14.398	6193	2	8	26	23	20	26	26	23	27	12	290	11	240	99	−5	3.5	6.2
Warrensburg, Whiteman AFB	724467	38.73	93.55	869	14.240	8293	1	7	22	19	17	23	34	21	34	9	360	9	190	101	−5	4.0	7.7
MONTANA																							
Billings	726770	45.80	108.53	3570	12.896	6193	−13	−7	28	24	22	30	25	27	30	10	230	10	240	99	−19	2.8	6.2
Bozeman	726797	45.78	111.15	4475	12.469	8293	−20	−12	21	18	15	20	36	17	34	4	140	9	360	96	−29	2.9	7.7
Butte	726785	45.95	112.50	5545	11.980	8293	−22	−14	23	21	18	21	29	19	30	4	150	13	120	92	−34	2.5	7.9
Cut Bank	727796	48.60	112.37	3855	12.760	6193	−21	−16	34	30	27	40	36	34	36	7	320	13	270	93	−28	4.0	5.7
Glasgow	727680	48.22	106.62	2297	13.516	6193	−22	−17	29	26	23	28	18	25	15	8	330	13	160	99	−29	3.2	6.6
Great Falls, Intl Airport	727750	47.48	111.37	3658	12.854	6193	−19	−13	33	29	26	34	38	31	38	7	240	12	230	98	−25	3.2	7.4
Great Falls, Malmstrom AFB	727755	47.50	111.18	3527	12.917	8293	−17	−11	28	24	21	33	38	29	38	4	240	8	260	99	−22	3.2	7.9
Havre	727770	48.55	109.77	2598	13.367	8293	−25	−19	24	21	19	26	35	23	33	6	240	9	270	102	−33	5.0	8.1
Helena	727720	46.60	112.00	3898	12.740	6193	−18	−10	25	22	19	25	40	22	35	5	290	12	280	96	−24	3.3	7.2
Kalispell	727790	48.30	114.27	2972	13.184	6193	−12	−3	24	20	17	25	12	21	18	7	20	9	170	95	−19	2.9	8.6
Lewistown	726776	47.05	109.47	4167	12.613	6193	−18	−12	26	23	20	29	35	25	35	7	250	11	90	95	−25	3.5	7.3
Miles City	742300	46.43	105.87	2628	13.352	6193	−19	−13	27	23	20	28	25	23	27	8	290	11	140	102	−25	2.7	6.5
Missoula	727730	46.92	114.08	3189	13.079	6193	−9	−1	22	19	17	22	17	19	21	7	120	10	290	97	−15	2.9	8.2
NEBRASKA																							
Bellevue, Offutt AFB	725540	41.12	95.92	1047	14.148	8293	−5	1	22	19	17	26	23	22	23	8	330	10	190	100	−9	4.5	6.3
Grand Island	725520	40.97	98.32	1857	13.736	6193	−8	−2	30	26	23	29	21	26	19	11	270	15	180	102	−14	3.2	5.2
Lincoln	725510	40.85	96.75	1188	14.076	8293	−7	−2	27	23	21	28	25	24	27	9	350	15	180	103	−11	6.5	8.1
Norfolk	725560	41.98	97.43	1552	13.890	6193	−11	−5	29	25	22	33	20	28	21	11	340	15	190	101	−18	3.0	5.4
North Platte	725620	41.13	100.68	2785	13.275	6193	−10	−4	29	25	22	28	24	24	26	7	320	12	180	101	−16	2.9	6.6
Omaha, Eppley Airfield	725500	41.30	95.90	981	14.182	6193	−7	−2	26	23	20	27	21	23	17	10	340	12	180	100	−14	3.3	4.8
Omaha, WSO	725530	41.37	96.02	1332	14.002	8293	−8	−2	22	20	18	25	23	22	25	10	310	11	170	98	−14	4.0	6.5
Scottsbluff	725660	41.87	103.60	3957	12.712	6193	−11	−3	30	26	22	32	35	27	35	8	300	11	300	101	−19	2.9	8.0
Sidney	725610	41.10	102.98	4304	12.549	8293	−8	−1	29	24	22	31	32	26	35	9	290	12	160	101	−18	4.5	8.6
Valentine	725670	42.87	100.55	2598	13.367	8293	−16	−8	27	23	21	26	25	23	28	9	250	15	180	104	−22	4.3	8.5
NEVADA																							
Elko	725825	40.83	115.78	5135	12.166	6193	−5	1	21	18	16	20	36	16	37	4	70	10	230	98	−13	3.2	8.0
Ely	724860	39.28	114.85	6263	11.660	6193	−6	0	28	24	21	26	33	22	30	11	190	13	230	93	−15	2.3	7.3
Las Vegas, Intl Airport	723860	36.08	115.17	2178	13.575	6193	27	30	30	26	23	25	48	22	49	7	250	12	230	111	21	2.2	4.7
Mercury	723870	36.62	116.02	3310	13.021	8293	24	28	25	22	19	25	44	21	42	8	50	12	230	102	19	17.6	6.3
N. Las Vegas, Nellis AFB	723865	36.23	115.03	1870	13.729	8293	28	31	24	21	18	23	52	19	49	2	20	9	210	112	21	2.0	4.5
Reno	724880	39.50	119.78	4400	12.505	6193	8	13	26	22	19	26	46	21	44	3	160	10	290	99	1	2.2	8.4
Tonopah	724855	38.05	117.08	5427	12.033	6193	7	13	25	22	20	24	37	22	36	9	340	12	180	98	1	2.0	6.5
Winnemucca	725830	40.90	117.80	4314	12.544	6193	1	7	23	19	17	21	39	18	38	5	160	11	250	101	−9	2.3	10.2
NEW HAMPSHIRE																							
Concord	726050	43.20	71.50	344	14.513	6193	−8	−2	23	20	17	23	20	20	21	4	320	10	230	95	−18	2.9	5.5
Lebanon	726116	43.63	72.30	597	14.381	8293	−7	−3	18	15	14	18	25	16	26	2	360	9	220	94	−17	2.0	5.2
Mount Washington	726130	44.27	71.30	6266	11.659	8293	−23	−19	88	81	73	99	−14	92	−15	73	280	21	270	65	−33	2.3	4.1
Portsmouth, Pease AFB	726055	43.08	70.82	102	14.642	8293	4	9	21	18	16	22	26	20	27	8	280	8	270	94	−2	2.2	3.2
NEW JERSEY																							
Atlantic City	724070	39.45	74.57	66	14.661	6193	8	13	27	23	20	29	36	25	34	9	310	11	250	96	0	2.9	5.7
Millville	724075	39.37	75.07	82	14.652	8293	10	15	19	18	17	20	35	19	35	7	300	11	240	96	0	2.3	7.4
Newark	725020	40.70	74.17	30	14.680	6193	10	14	26	23	20	27	28	23	29	13	260	13	230	98	4	2.6	4.8
Teterboro	725025	40.85	74.07	10	14.690	8293	10	14	21	19	17	21	29	19	30	11	280	12	240	97	2	2.5	5.6
Trenton, McGuire AFB	724096	40.02	74.60	135	14.624	8293	11	15	22	19	17	23	31	21	31	8	330	8	240	97	2	2.2	5.2
NEW MEXICO																							
Alamogordo, Holloman AFB	747320	32.85	106.10	4094	12.647	8293	20	23	20	17	14	18	50	15	48	3	10	8	250	102	13	2.9	3.6
Albuquerque	723650	35.05	106.62	5315	12.084	6193	13	18	29	25	22	26	34	22	37	8	360	10	240	100	6	2.6	7.3
Carlsbad	722687	32.33	104.27	3294	13.028	8293	19	23	25	22	19	25	57	21	54	8	340	12	150	104	9	3.6	7.0
Clayton	723600	36.45	103.15	4970	12.241	8293	1	9	30	27	24	30	40	26	39	10	40	13	200	98	−5	2.5	7.4

WMO# = World Meteorological Organization number
AFB = Air Force Base
NAS = Naval Air Station
WSO = Weather Service Office
AAF = Army Air Field
MCAS = Marine Corps Air Station
DB = dry-bulb temperature, °F
Lat. = North latitude °
Long. = West longitude °
Elev. = elevation, ft
StdP = standard pressure at station elevation, psia
WS = wind speed, mph
PWD = prevailing wind direction, ° True

Table 1B Cooling and Dehumidification Design Conditions—United States

Station	Cooling DB/MWB						Evaporation WB/MDB						Dehumidification DP/MDB and HR									Range of DB
	0.4%		1%		2%		0.4%		1%		2%		0.4%			1%			2%			
	DB	MWB	DB	MWB	DB	MWB	WB	MDB	WB	MDB	WB	MDB	DP	HR	MDB	DP	HR	MDB	DP	HR	MDB	
1	2a	2b	2c	2d	2e	2f	3a	3b	3c	3d	3e	3f	4a	4b	4c	4d	4e	4f	4g	4h	4i	5
Saint Cloud	91	72	88	71	85	70	76	87	74	84	72	82	72	125	83	70	116	80	68	109	78	21.5
Tofte	79	64	75	62	71	61	66	74	64	72	63	70	64	92	70	61	83	69	59	77	68	13.0
MISSISSIPPI																						
Biloxi, Keesler AFB	92	79	91	78	89	78	81	89	80	88	80	87	79	151	86	78	147	85	78	144	84	13.0
Columbus, AFB	96	77	94	76	92	76	80	91	78	89	78	88	77	141	85	76	136	83	75	132	82	19.3
Greenwood	96	78	94	78	93	77	81	91	80	90	79	89	78	148	86	77	143	85	76	139	84	19.1
Jackson	95	77	93	76	92	76	80	90	79	89	78	88	77	142	84	76	138	83	75	134	82	19.2
McComb	94	76	92	76	91	76	79	89	78	88	78	87	77	141	83	76	138	82	75	135	81	19.8
Meridian	96	77	94	76	92	76	79	91	78	90	77	88	76	139	84	75	134	83	74	130	83	20.3
Tupelo	96	76	94	76	92	75	79	89	78	89	77	88	76	137	83	75	134	83	74	131	82	18.9
MISSOURI																						
Cape Girardeau	96	77	94	77	91	76	80	92	78	90	78	88	77	141	86	76	136	85	75	132	83	19.8
Columbia	95	75	92	75	89	74	78	89	77	88	75	86	75	137	85	74	130	83	72	124	82	20.3
Joplin	96	75	94	75	91	74	78	90	77	89	76	88	75	137	85	74	132	85	72	125	83	20.0
Kansas City	96	75	93	75	90	74	78	90	77	89	76	87	75	137	86	74	130	84	73	125	83	18.8
Poplar Bluff	95	77	92	76	90	76	80	90	78	88	77	87	77	144	85	76	138	83	75	133	82	20.0
Spickard/Trenton	96	74	93	73	89	72	78	88	76	88	75	86	76	139	83	73	128	83	71	118	81	19.6
Springfield	95	74	92	74	89	74	78	89	76	88	75	86	74	134	84	73	128	83	72	124	81	20.8
St. Louis, Intl Airport	95	76	93	75	90	74	79	90	78	88	76	87	76	138	85	75	132	83	73	127	82	18.3
Warrensburg, Whiteman AFB	96	76	93	76	90	75	79	91	78	90	76	88	76	139	86	75	134	85	73	128	83	19.3
MONTANA																						
Billings	93	63	90	62	87	61	65	86	64	84	62	83	59	83	71	57	78	71	55	74	70	25.8
Bozeman	91	61	87	60	85	59	64	83	62	82	61	81	58	83	69	56	78	67	53	71	66	31.7
Butte	86	57	84	56	80	55	60	76	58	76	57	76	54	76	61	52	70	63	50	66	62	31.5
Cut Bank	87	60	84	59	80	58	62	81	60	79	59	77	56	77	67	54	70	65	51	65	63	26.1
Glasgow	94	64	90	63	86	62	68	85	66	83	64	82	62	91	74	60	83	71	58	77	69	25.3
Great Falls, Intl Airport	92	61	88	60	85	59	64	84	62	82	61	81	57	81	69	55	74	67	53	69	66	27.2
Great Falls, Malmstrom AFB	93	62	89	61	86	60	65	85	63	83	62	81	59	84	71	57	78	69	54	71	68	26.3
Havre	94	63	90	62	86	61	66	87	64	84	62	82	60	84	72	58	78	69	56	74	68	27.9
Helena	90	60	87	59	84	59	63	82	61	81	60	80	57	80	68	55	73	66	52	68	66	28.0
Kalispell	89	62	86	61	82	60	65	83	63	81	61	79	59	82	69	57	76	67	55	71	67	29.9
Lewistown	89	61	86	60	82	59	64	81	63	80	61	78	58	85	71	56	79	69	54	74	67	28.3
Miles City	97	66	93	65	90	64	69	89	67	86	66	84	63	95	76	61	88	75	59	82	73	25.9
Missoula	91	62	88	61	85	60	65	84	63	82	62	81	58	82	68	56	76	68	55	71	66	31.3
NEBRASKA																						
Bellevue, Offutt AFB	95	76	91	75	88	74	79	89	77	88	76	86	76	141	85	74	134	83	73	127	82	18.4
Grand Island	97	72	93	72	90	70	76	89	74	88	73	86	72	127	82	70	120	81	69	113	79	22.4
Lincoln	97	74	94	74	91	73	78	90	76	89	75	87	75	136	84	73	130	83	71	121	82	22.3
Norfolk	95	74	92	72	89	72	76	90	75	88	73	86	73	129	83	71	121	82	70	115	81	20.8
North Platte	95	69	92	69	89	68	73	87	72	86	70	85	69	118	80	67	111	78	66	105	77	25.5
Omaha, Eppley Airfield	95	75	92	75	89	73	78	90	77	88	75	86	75	136	85	73	128	84	72	121	82	19.9
Omaha, WSO	94	75	90	75	87	73	77	89	76	87	74	85	74	134	84	72	126	83	71	120	82	17.6
Scottsbluff	95	65	92	64	89	64	69	87	68	85	66	84	64	102	76	62	97	74	60	91	73	28.9
Sidney	95	63	92	63	89	63	67	84	66	84	64	84	62	97	73	60	91	72	58	86	71	27.9
Valentine	97	68	94	67	90	67	72	90	71	89	69	87	67	110	79	65	103	78	63	94	77	26.5
NEVADA																						
Elko	95	60	92	59	90	58	63	85	61	84	60	84	57	84	68	54	75	66	51	67	67	38.4
Ely	89	56	87	56	85	55	60	78	59	78	58	78	55	82	64	53	75	64	50	68	65	34.6
Las Vegas, Intl Airport	108	66	106	66	103	65	71	95	70	93	69	93	65	102	79	63	92	81	60	84	85	24.8
Mercury	102	65	100	64	98	63	69	88	67	89	66	89	64	102	72	60	89	77	58	80	80	25.9
N. Las Vegas, Nellis AFB	108	68	106	67	104	66	72	94	71	94	70	94	67	106	79	64	97	82	61	86	84	26.3
Reno	95	61	92	60	90	59	63	87	62	86	60	85	56	77	69	53	69	69	50	63	68	37.3
Tonopah	94	58	92	57	89	57	62	83	61	82	60	81	56	83	67	53	74	68	50	67	69	31.1
Winnemucca	97	61	94	60	92	59	63	88	62	87	60	86	56	79	68	53	69	67	50	62	68	37.4
NEW HAMPSHIRE																						
Concord	90	71	87	70	84	68	74	85	73	82	71	79	71	118	79	70	111	77	68	105	76	24.1
Lebanon	88	71	86	69	83	68	74	84	72	82	70	79	70	113	79	69	108	77	67	103	75	23.0
Mount Washington	60	56	58	54	56	54	58	59	56	57	54	56	58	90	58	56	84	57	54	78	55	8.5
Portsmouth, Pease AFB	89	72	85	70	83	70	75	84	73	82	72	79	73	123	85	71	113	77	69	106	76	18.2
NEW JERSEY																						
Atlantic City	91	74	88	73	86	72	77	87	76	84	75	82	75	131	81	74	125	80	72	120	79	18.1
Millville	92	75	89	74	87	73	78	87	76	86	75	83	75	134	81	74	129	80	73	125	80	18.7
Newark	93	74	90	73	87	71	77	88	76	85	74	83	74	127	81	73	121	80	71	116	80	15.9
Teterboro	92	76	89	74	87	73	78	88	77	87	75	83	76	134	84	74	128	82	72	119	81	18.4
Trenton, McGuire AFB	93	75	90	74	87	73	78	89	76	87	75	84	75	132	83	74	127	82	72	118	80	18.9
NEW MEXICO																						
Alamogordo, Holloman AFB	98	63	96	63	93	63	68	87	67	85	67	85	65	106	72	62	98	72	61	92	73	30.2
Albuquerque	96	60	93	60	91	60	65	83	64	82	64	81	61	98	68	60	93	69	58	89	69	25.4
Carlsbad	101	65	98	66	96	66	72	88	71	87	70	85	69	121	76	68	116	76	67	111	75	25.4
Clayton	94	62	91	62	88	62	67	84	65	84	65	82	61	98	72	60	94	71	59	90	70	26.1

DP = dew-point temperature, °F
MDB = mean coincident dry-bulb temp., °F
StdD = standard deviation, °F
MWB = mean coincident wet-bulb temp., °F
MWS = mean coincident wind speed, mph
HR = humidity ratio, grains of moisture per lb of dry air
A = airport
ANGB = Air National Guard Base
MCAF = Marine Corps Air Facility
NAF = Naval Air Facility
NAWS = Naval Air Weapons Station
RAF = Royal Air Force

Table 1A Heating and Wind Design Conditions—United States

							Heating Dry Bulb		Extreme Wind Speed, mph			Coldest Month				MWS/PWD to DB				Extr. Annual Daily			
												0.4%		1%		99.6%		0.4%		Mean DB		StdD DB	
Station	WMO#	Lat.	Long.	Elev., ft	StdP, psia	Dates	99.6%	99%	1%	2.5%	5%	WS	MDB	WS	MDB	MWS	PWD	MWS	PWD	Max.	Min.	Max.	Min.
1a	1b	1c	1d	1e	1f	1g	2a	2b	3a	3b	3c	4a	4b	4c	4d	5a	5b	5c	5d	6a	6b	6c	6d
Clovis, Cannon AFB	722686	34.38	103.32	4295	12.554	8293	10	15	26	23	20	26	40	23	39	8	50	11	220	101	5	2.3	4.0
Farmington	723658	36.75	108.23	5502	11.999	8293	8	13	23	21	18	22	35	19	34	6	60	10	240	99	−1	3.8	7.2
Gallup	723627	35.52	108.78	6470	11.570	8293	−1	5	23	20	18	19	39	18	37	1	140	11	270	94	−12	2.3	7.9
Roswell	722680	33.30	104.53	3668	12.849	8293	14	20	22	19	17	20	51	18	48	8	360	11	140	105	6	4.7	6.5
Truth Or Consequences	722710	33.23	107.27	4859	12.292	8293	22	26	25	21	18	24	43	21	41	8	350	10	170	102	6	2.9	23.8
Tucumcari	723676	35.18	103.60	4065	12.661	6193	9	15	25	22	20	28	50	23	45	8	50	12	230	102	1	3.1	7.2
NEW YORK																							
Albany	725180	42.75	73.80	292	14.541	6193	−7	−2	24	22	19	23	20	20	22	5	300	10	230	95	−18	3.0	6.3
Binghamton	725150	42.22	75.98	1631	13.850	6193	−2	2	24	21	19	24	20	22	19	13	270	11	220	89	−9	3.1	4.4
Buffalo	725280	42.93	78.73	705	14.325	6193	2	5	29	26	23	34	25	30	24	12	270	13	240	91	−6	2.3	5.3
Central Islip	725035	40.80	73.10	98	14.643	8293	11	15	22	20	18	23	32	21	31	10	340	11	210	94	2	3.2	5.4
Elmira/Corning	725156	42.17	76.90	955	14.195	8293	−2	3	21	19	17	23	20	20	27	5	240	11	210	95	−10	3.6	6.1
Glens Falls	725185	43.33	73.62	328	14.522	8293	−10	−4	18	16	14	19	22	17	22	2	350	10	190	93	−20	2.7	4.7
Massena	726223	44.93	74.85	213	14.583	6193	−15	−10	21	18	17	23	22	21	22	4	270	10	230	92	−27	3.0	6.0
New York, JFK Airport	744860	40.65	73.78	23	14.683	6193	11	15	27	24	21	30	29	27	28	17	320	13	230	96	6	2.6	4.6
New York, La Guardia A	725030	40.77	73.90	30	14.680	8293	13	17	28	25	22	30	29	27	28	18	310	12	280	97	6	2.2	4.3
Newburgh	725038	41.50	74.10	492	14.436	8293	6	10	23	20	18	26	17	23	26	8	260	10	230	92	−4	3.1	6.5
Niagara Falls	725287	43.10	78.95	591	14.385	8293	4	7	26	22	20	30	24	27	23	11	240	13	230	91	−4	3.2	6.1
Plattsburgh, AFB	726225	44.65	73.47	236	14.570	8293	−9	−4	21	18	16	22	27	19	24	2	350	8	260	93	−17	2.7	4.7
Poughkeepsie	725036	41.63	73.88	167	14.607	8293	2	6	18	16	14	19	25	17	25	3	250	9	250	96	−8	3.1	5.9
Rochester	725290	43.12	77.67	554	14.403	6193	1	5	27	23	21	29	22	26	21	10	230	12	250	93	−7	2.8	4.9
Rome, Griffiss AFB	725196	43.23	75.40	505	14.429	8293	−5	1	22	19	16	23	22	20	22	3	330	8	260	93	−15	2.5	3.8
Syracuse	725190	43.12	76.12	407	14.481	6193	−3	2	26	22	20	28	20	25	21	7	90	11	250	93	−13	3.0	6.0
Watertown	726227	44.00	76.02	325	14.524	8293	−12	−6	21	19	18	24	24	21	25	5	80	11	240	90	−25	2.9	7.0
White Plains	725037	41.07	73.70	440	14.463	8293	7	12	19	17	15	19	29	18	29	13	310	9	260	95	0	2.9	4.5
NORTH CAROLINA																							
Asheville	723150	35.43	82.55	2169	13.580	6193	11	16	25	22	19	26	26	23	28	11	340	9	340	91	3	2.6	6.8
Cape Hatteras	723040	35.27	75.55	10	14.690	6193	26	29	26	22	20	27	47	23	47	11	340	11	230	91	20	2.0	4.9
Charlotte	723140	35.22	80.93	768	14.292	6193	18	23	20	17	15	20	44	18	45	6	50	9	240	97	10	2.9	6.0
Cherry Point, MCAS	723090	34.90	76.88	30	14.680	8293	24	28	19	16	15	19	43	17	48	5	10	7	240	100	12	2.5	8.5
Fayetteville, Ft. Bragg	746930	35.13	78.93	243	14.567	8293	22	27	17	14	12	19	42	16	44	4	10	6	240	100	15	3.8	6.8
Goldsboro, Johnson AFB	723066	35.33	77.97	108	14.638	8293	22	27	17	14	12	18	46	15	44	4	270	8	260	100	14	3.1	7.4
Greensboro	723170	36.08	79.95	886	14.231	6193	15	19	19	17	15	20	40	18	40	7	290	8	230	96	7	2.7	5.0
Hickory	723145	35.73	81.38	1188	14.076	8293	18	23	17	15	13	18	41	16	41	4	320	9	240	97	8	3.2	6.8
Jacksonville, New River MCAF	723096	34.72	77.45	26	14.681	8293	23	27	18	16	14	19	49	17	47	5	350	7	240	99	13	2.0	8.8
New Bern	723095	35.07	77.05	20	14.685	8293	22	27	18	16	14	19	49	17	47	6	10	8	240	99	13	1.4	8.3
Raleigh/Durham	723060	35.87	78.78	440	14.463	6193	16	20	21	18	16	21	42	19	43	8	360	9	240	96	9	2.9	5.3
Wilmington	723013	34.27	77.90	33	14.678	6193	23	27	21	19	17	22	51	20	48	7	320	10	220	97	17	2.2	5.7
Winston-Salem	723193	36.13	80.22	971	14.187	8293	18	23	19	17	15	21	38	19	38	7	290	8	240	96	8	2.7	5.8
NORTH DAKOTA																							
Bismarck	727640	46.77	100.75	1660	13.835	6193	−21	−16	29	25	22	29	13	25	16	7	290	13	180	100	−30	3.6	6.4
Devils Lake	727580	48.10	98.87	1453	13.940	8293	−23	−19	26	22	20	27	12	24	10	9	300	11	10	98	−27	7.0	5.0
Fargo	727530	46.90	96.80	899	14.224	6193	−22	−17	31	27	24	32	7	28	7	8	180	14	160	98	−27	3.4	4.4
Grand Forks, AFB	727575	47.97	97.40	912	14.217	8293	−20	−16	27	24	21	30	9	26	13	7	290	13	180	98	−25	4.7	4.9
Minot, AFB	727675	48.42	101.35	1667	13.832	8293	−21	−16	28	24	21	30	18	27	16	10	310	12	150	101	−25	3.8	7.4
Minot, Intl Airport	727676	48.27	101.28	1716	13.807	6193	−20	−16	28	24	22	30	14	27	14	12	290	13	200	98	−25	3.0	4.8
Williston	727670	48.18	103.63	1906	13.711	8293	−24	−18	27	23	21	28	25	24	20	8	220	14	150	101	−30	4.5	8.3
OHIO																							
Akron/Canton	725210	40.92	81.43	1237	14.050	6193	0	5	24	21	19	25	26	22	26	11	270	10	230	92	−7	2.9	7.0
Cincinnati, Lunken Field	724297	39.10	84.42	482	14.441	8293	5	12	21	19	17	22	35	19	33	9	260	10	210	96	−3	3.2	9.4
Cleveland	725240	41.42	81.87	804	14.274	6193	1	6	26	23	20	27	28	24	28	12	230	12	230	93	−6	2.8	6.3
Columbus, Intl Airport	724280	40.00	82.88	817	14.267	6193	1	6	23	20	18	24	30	21	25	9	270	11	270	94	−6	2.6	7.1
Columbus, Rickenbckr AFB	724285	39.82	82.93	745	14.304	8293	3	10	21	18	16	23	26	20	27	7	210	8	270	96	−4	4.5	8.6
Dayton, Intl Airport	724290	39.90	84.20	1004	14.170	6193	−1	5	24	21	19	25	26	22	28	11	270	11	240	95	−8	2.9	7.0
Dayton, Wright-Patterson	745700	39.83	84.05	823	14.263	8293	1	8	21	18	16	23	28	21	30	7	270	9	240	96	−7	3.2	7.7
Findlay	725366	41.02	83.67	810	14.270	8293	−2	4	23	20	19	25	34	22	29	11	250	12	210	94	−9	3.8	7.9
Mansfield	725246	40.82	82.52	1296	14.020	6193	−1	4	25	22	20	28	28	25	26	13	240	12	240	91	−8	2.8	6.0
Toledo	725360	41.60	83.80	692	14.332	6193	−2	3	23	20	18	25	25	22	21	10	250	11	230	95	−10	3.0	5.4
Youngstown	725250	41.27	80.67	1184	14.077	6193	−1	4	23	21	19	24	22	22	21	10	230	10	230	91	−8	2.5	5.8
Zanesville	724286	39.95	81.90	899	14.224	8293	2	9	19	18	16	21	32	19	31	7	240	9	220	94	−7	3.6	8.5
OKLAHOMA																							
Altus, AFB	723520	34.67	99.27	1378	13.978	8293	13	19	23	21	19	24	40	21	42	9	20	10	190	107	7	3.4	7.7
Enid, Vance AFB	723535	36.33	97.92	1306	14.015	8293	5	12	26	23	20	27	38	23	38	12	10	11	190	105	1	3.6	6.7
Lawton, Ft. Sill/Post Field	723550	34.65	98.40	1188	14.076	8293	12	19	24	21	19	26	35	22	36	11	10	11	170	103	8	2.5	7.4
McAlester	723566	34.88	95.78	771	14.291	8293	10	17	20	18	16	21	47	19	45	9	360	9	190	102	4	4.0	8.3
Oklahoma City, Tinker AFB	723540	35.42	97.38	1293	14.022	8293	10	17	24	22	19	25	42	22	42	10	10	11	190	103	6	3.2	6.1
Oklahoma City, W. Rogers A	723530	35.40	97.60	1302	14.017	6193	9	15	29	25	23	29	33	26	37	15	360	13	180	103	4	3.4	4.9
Tulsa	723560	36.20	95.90	676	14.340	6193	9	14	25	23	21	24	46	22	40	11	360	12	180	103	3	3.6	5.6
OREGON																							
Astoria	727910	46.15	123.88	23	14.683	6193	25	29	25	22	19	29	51	24	49	8	90	12	320	87	20	4.5	6.1

WMO# = World Meteorological Organization number
AFB = Air Force Base
NAS = Naval Air Station
WSO = Weather Service Office
AAF = Army Air Field
MCAS = Marine Corps Air Station
DB = dry-bulb temperature, °F
Lat. = North latitude °
Long. = West longitude °
Elev. = elevation, ft
StdP = standard pressure at station elevation, psia
WS = wind speed, mph
PWD = prevailing wind direction, ° True

Table 1B Cooling and Dehumidification Design Conditions—United States

Station	Cooling DB/MWB						Evaporation WB/MDB						Dehumidification DP/MDB and HR									Range of DB
	0.4%		1%		2%		0.4%		1%		2%		0.4%			1%			2%			
	DB	MWB	DB	MWB	DB	MWB	WB	MDB	WB	MDB	WB	MDB	DP	HR	MDB	DP	HR	MDB	DP	HR	MDB	
1	2a	2b	2c	2d	2e	2f	3a	3b	3c	3d	3e	3f	4a	4b	4c	4d	4e	4f	4g	4h	4i	5
Clovis, Cannon AFB	96	64	93	64	91	65	70	84	69	83	68	83	66	114	75	65	109	74	64	105	73	24.5
Farmington	94	60	92	60	89	60	65	83	64	83	63	82	60	94	69	58	90	68	57	85	69	28.8
Gallup	89	57	87	56	85	56	62	76	61	76	60	75	59	94	65	57	90	64	56	85	64	30.6
Roswell	98	65	96	65	94	65	70	87	69	86	68	85	66	111	73	65	108	73	64	104	73	24.8
Truth Or Consequences	97	61	95	61	93	61	66	85	65	85	64	84	60	94	71	59	90	71	58	87	72	25.0
Tucumcari	98	64	95	65	93	64	69	87	68	85	67	84	65	109	73	64	104	73	63	100	72	24.9
NEW YORK																						
Albany	90	71	86	70	84	69	74	85	73	82	71	79	72	118	79	70	111	77	68	106	76	23.7
Binghamton	85	70	82	69	80	67	73	81	71	79	70	77	70	118	77	69	111	75	67	106	74	17.5
Buffalo	86	70	84	69	81	68	74	82	72	80	71	78	71	118	78	70	113	77	68	106	75	17.7
Central Islip	88	73	85	72	83	70	76	83	75	81	74	79	74	129	79	73	124	78	71	116	77	15.1
Elmira/Corning	90	72	87	71	84	69	75	86	73	82	72	80	72	122	81	70	116	78	69	110	76	24.1
Glens Falls	88	73	85	71	83	70	76	85	74	82	72	80	74	127	81	71	116	79	69	108	76	22.1
Massena	87	72	84	71	82	69	75	84	73	81	71	79	72	118	80	70	111	78	68	105	76	21.8
New York, JFK Airport	91	74	88	72	85	71	76	86	75	84	74	82	74	125	80	72	120	80	71	114	79	13.9
New York, La Guardia A	92	74	89	73	86	72	77	87	76	85	74	83	74	129	81	73	125	80	71	116	80	14.6
Newburgh	88	74	85	72	83	70	76	85	74	83	73	80	74	130	82	72	119	80	70	111	78	17.1
Niagara Falls	87	72	85	71	83	69	75	84	74	81	72	79	73	125	81	71	116	78	69	111	76	18.9
Plattsburgh, AFB	86	71	83	69	80	68	74	82	72	80	70	78	71	115	79	69	108	76	67	102	75	19.6
Poughkeepsie	92	75	88	72	85	71	76	87	75	85	73	82	74	126	82	71	116	80	70	111	78	23.0
Rochester	89	73	86	71	83	70	75	85	74	82	72	80	73	123	81	71	116	79	69	109	77	20.1
Rome, Griffiss AFB	88	71	86	70	83	69	74	84	73	82	71	79	71	117	80	70	111	78	68	105	76	22.9
Syracuse	88	72	85	71	83	70	75	85	73	82	72	80	72	120	80	70	113	78	69	107	77	20.3
Watertown	85	71	83	70	80	69	74	82	72	80	71	77	71	118	78	70	111	77	69	106	75	20.5
White Plains	89	74	87	72	84	71	76	86	75	83	73	80	74	128	80	72	120	80	71	114	78	18.0
NORTH CAROLINA																						
Asheville	88	72	85	71	83	70	75	84	73	82	72	80	72	128	79	71	123	78	70	118	76	19.4
Cape Hatteras	88	78	86	77	85	77	80	86	79	84	78	83	78	147	83	77	142	83	76	138	82	11.4
Charlotte	94	74	91	74	89	73	77	88	76	87	75	86	74	130	82	73	125	80	72	122	80	17.8
Cherry Point, MCAS	95	79	92	78	90	77	81	91	80	90	79	88	78	146	87	77	141	86	76	136	85	16.6
Fayetteville, Ft. Bragg	96	77	94	76	92	75	79	91	78	89	77	88	76	139	84	76	135	83	75	131	83	18.2
Goldsboro, Johnson AFB	96	77	94	76	91	76	80	91	78	89	77	87	76	139	84	76	135	83	75	132	82	18.4
Greensboro	92	75	90	74	88	73	77	88	76	86	75	85	74	132	82	73	127	81	72	123	80	18.5
Hickory	94	73	91	72	88	72	76	87	75	85	74	84	74	133	80	73	128	80	71	120	78	19.6
Jacksonville, New River MCAF	94	79	92	78	89	77	81	90	79	89	78	87	78	145	86	77	140	85	76	136	84	17.1
New Bern	94	78	92	78	90	76	81	91	79	89	78	87	78	144	86	77	139	84	76	134	83	17.1
Raleigh/Durham	93	76	90	75	88	74	78	88	77	87	76	85	75	134	82	74	130	81	73	125	80	18.8
Wilmington	93	79	91	78	89	77	80	89	79	88	78	86	78	146	85	77	141	83	76	137	83	15.7
Winston-Salem	92	74	89	74	87	73	77	86	76	86	75	85	74	134	81	73	129	80	72	121	79	17.6
NORTH DAKOTA																						
Bismarck	93	68	90	67	86	66	72	86	70	84	68	82	68	109	79	65	100	77	63	92	75	26.5
Devils Lake	91	69	87	67	84	66	72	86	70	83	68	80	68	108	78	66	100	77	63	90	75	21.1
Fargo	91	71	88	70	85	69	75	86	73	84	71	81	72	122	82	69	112	80	67	104	77	22.3
Grand Forks, AFB	91	71	88	69	85	68	75	86	72	83	70	80	71	118	81	69	109	78	67	101	76	22.9
Minot, AFB	94	68	90	67	86	66	72	87	70	85	68	82	68	109	80	65	100	78	63	90	75	24.7
Minot, Intl Airport	92	67	88	66	84	65	71	85	69	83	67	80	67	106	78	64	96	75	62	89	73	22.9
Williston	96	67	92	66	87	65	71	87	69	86	67	83	66	103	78	63	92	76	61	85	73	25.7
OHIO																						
Akron/Canton	88	72	85	71	83	70	75	84	73	82	72	80	72	125	80	71	118	78	69	113	77	18.8
Cincinnati, Lunken Field	93	74	90	75	88	73	77	89	76	87	75	84	75	132	82	74	128	81	72	120	80	20.0
Cleveland	89	73	86	72	84	71	76	85	74	83	72	81	73	125	82	71	118	80	70	112	78	18.6
Columbus, Intl Airport	90	74	88	73	86	71	77	87	75	85	74	82	73	128	82	72	123	81	71	117	79	19.3
Columbus, Rickenbckr AFB	92	74	89	73	87	72	77	88	75	86	74	84	74	130	83	72	120	82	70	115	79	19.8
Dayton, Intl Airport	90	74	88	73	86	71	76	87	75	85	74	82	73	129	82	72	123	80	71	117	79	19.2
Dayton, Wright-Paterson	92	74	89	74	87	73	78	88	76	86	74	84	75	136	84	73	127	83	71	119	81	19.8
Findlay	90	74	87	72	85	71	76	86	75	83	73	81	74	132	81	72	121	80	71	116	78	18.9
Mansfield	88	73	85	72	83	71	76	85	74	83	73	80	73	128	81	71	122	79	70	116	78	17.8
Toledo	90	73	87	72	85	71	77	86	75	84	73	81	74	129	82	72	122	80	70	115	78	20.9
Youngstown	88	72	85	70	83	69	74	84	73	82	71	79	72	122	80	70	116	78	69	110	76	20.6
Zanesville	90	74	88	73	86	71	76	87	75	85	74	82	74	130	82	72	120	80	71	116	78	20.7
OKLAHOMA																						
Altus, AFB	102	73	100	73	97	73	77	93	76	92	75	91	74	132	84	72	124	83	71	119	82	23.6
Enid, Vance AFB	101	74	98	74	95	73	77	92	76	91	75	90	73	130	85	71	121	83	70	116	82	21.8
Lawton, Ft. Sill/Post Field	99	73	97	73	95	73	77	90	76	90	75	89	74	135	83	73	129	82	71	121	81	20.7
McAlester	98	76	96	76	93	76	79	92	78	91	77	89	76	141	85	75	137	83	74	133	83	21.8
Oklahoma City, Tinker AFB	98	74	96	75	94	74	78	92	77	91	76	89	75	138	87	74	132	85	72	123	83	19.4
Oklahoma City, W. Rogers A	99	74	96	74	94	73	77	91	76	90	75	89	73	129	83	72	125	82	71	120	81	21.0
Tulsa	100	76	97	76	94	75	79	92	78	92	77	90	76	137	87	74	132	85	73	127	84	19.5
OREGON																						
Astoria	76	64	72	62	69	61	65	75	63	71	62	68	61	81	69	60	76	66	59	74	65	14.2

DP = dew-point temperature, °F
MDB = mean coincident dry-bulb temp., °F
StdD = standard deviation, °F
MWB = mean coincident wet-bulb temp., °F
MWS = mean coincident wind speed, mph
HR = humidity ratio, grains of moisture per lb of dry air
A = airport
ANGB = Air National Guard Base
MCAF = Marine Corps Air Facility
NAF = Naval Air Facility
NAWS = Naval Air Weapons Station
RAF = Royal Air Force

Table 1A Heating and Wind Design Conditions—United States

Station	WMO#	Lat.	Long.	Elev., ft	StdP, psia	Dates	Heating Dry Bulb		Extreme Wind Speed, mph			Coldest Month				MWS/PWD to DB				Extr. Annual Daily			
												0.4%		1%		99.6%		0.4%		Mean DB		StdD DB	
							99.6%	99%	1%	2.5%	5%	WS	MDB	WS	MDB	MWS	PWD	MWS	PWD	Max.	Min.	Max.	Min.
1a	1b	1c	1d	1e	1f	1g	2a	2b	3a	3b	3c	4a	4b	4c	4d	5a	5b	5c	5d	6a	6b	6c	6d
Eugene	726930	44.12	123.22	374	14.498	6193	21	26	20	18	16	22	46	19	45	8	360	12	360	99	16	3.7	7.9
Hillsboro	726986	45.53	122.95	203	14.588	8293	19	24	19	17	15	23	26	19	34	8	60	9	360	100	15	3.8	6.3
Klamath Falls	725895	42.15	121.73	4091	12.649	8293	4	10	25	22	19	28	39	23	33	6	320	9	320	97	−4	4.1	8.6
Meacham	726885	45.52	118.40	4055	12.666	8293	−9	0	12	10	9	13	33	11	33	1	130	5	360	93	−21	4.7	12.2
Medford	725970	42.37	122.87	1329	14.003	6193	21	24	19	16	13	20	51	15	50	3	130	9	290	104	15	3.4	6.4
North Bend	726917	43.42	124.25	13	14.688	6193	30	32	25	23	20	23	51	20	50	7	140	14	340	82	24	4.0	5.5
Pendleton	726880	45.68	118.85	1496	13.918	6193	3	11	28	24	20	27	44	23	42	6	140	9	310	102	−1	3.6	11.2
Portland	726980	45.60	122.60	39	14.675	6193	22	27	25	21	18	28	37	25	39	13	120	11	340	99	18	4.4	6.0
Redmond	726835	44.25	121.15	3077	13.133	6193	1	9	20	17	16	20	42	18	41	6	320	11	340	98	−7	3.2	10.2
Salem	726940	44.92	123.00	200	14.589	6193	20	25	23	19	17	25	46	22	46	6	350	10	360	100	14	3.3	6.6
Sexton Summit	725975	42.62	123.37	3842	12.767	8293	21	24	24	22	19	27	37	24	38	9	120	6	340	89	16	4.9	8.5
PENNSYLVANIA																							
Allentown	725170	40.65	75.43	384	14.493	6193	5	10	27	23	21	28	26	25	24	9	270	11	240	95	−2	2.8	5.2
Altoona	725126	40.30	78.32	1503	13.915	8293	5	10	20	18	17	23	20	20	22	9	270	8	250	92	−5	3.6	7.9
Bradford	725266	41.80	78.63	2142	13.593	6193	−6	−1	19	18	16	22	22	19	21	7	270	9	240	87	−15	2.8	5.0
Du Bois	725125	41.18	78.90	1818	13.756	8293	0	5	21	19	17	23	20	21	20	11	280	10	270	90	−9	3.1	7.0
Erie	725260	42.08	80.18	738	14.308	6193	2	7	27	24	22	29	28	26	28	14	200	12	250	90	−4	3.1	6.4
Harrisburg	725115	40.20	76.77	308	14.532	6193	9	13	22	20	18	24	29	22	29	8	270	10	250	97	2	3.3	5.7
Philadelphia, Intl Airport	724080	39.88	75.25	30	14.680	6193	11	15	24	21	19	26	31	23	30	12	290	11	230	96	5	2.8	5.6
Philadelphia, Northeast A	724085	40.08	75.02	121	14.631	8293	11	15	21	19	17	22	30	19	29	10	300	10	260	97	3	2.5	6.1
Philadelphia, Willow Gr NAS	724086	40.20	75.15	361	14.505	8293	10	14	18	15	13	19	30	16	30	5	300	6	250	99	2	5.4	5.8
Pittsburgh, Allegheny Co. A	725205	40.35	79.93	1253	14.042	8293	4	11	21	19	17	23	24	21	24	11	250	11	240	94	−4	3.1	9.4
Pittsburgh, Intl Airport	725200	40.50	80.22	1224	14.057	6193	2	7	25	21	19	26	24	23	25	10	260	11	230	93	−6	3.1	6.7
Wilkes-Barre/Scranton	725130	41.33	75.73	948	14.199	6193	2	7	20	18	16	21	26	19	25	8	230	11	220	92	−5	2.8	4.9
Williamsport	725140	41.25	76.92	525	14.419	6193	2	7	23	20	18	24	23	21	25	8	270	10	250	94	−6	3.1	5.9
RHODE ISLAND																							
Providence	725070	41.73	71.43	62	14.662	6193	5	10	27	23	21	27	31	23	32	12	340	13	230	95	−2	3.7	5.0
SOUTH CAROLINA																							
Beaufort, MCAS	722085	32.48	80.72	39	14.675	8293	28	31	18	15	13	19	46	17	45	4	300	7	270	101	13	2.9	8.3
Charleston	722080	32.90	80.03	49	14.669	6193	25	28	22	19	17	22	52	19	51	7	20	10	230	98	18	2.3	5.6
Columbia	723100	33.95	81.12	226	14.576	6193	21	24	20	17	15	20	48	18	49	5	220	9	240	100	13	3.1	5.6
Florence	723106	34.18	79.72	148	14.617	8293	23	27	19	17	15	20	51	18	50	7	360	10	240	100	14	2.7	7.6
Greer/Greenville	723120	34.90	82.22	971	14.187	6193	19	23	20	18	16	21	45	18	44	6	50	9	230	97	11	2.6	5.5
Myrtle Beach, AFB	747910	33.68	78.93	26	14.681	8293	25	29	18	15	13	18	49	15	47	4	360	7	290	98	17	2.9	7.4
Sumter, Shaw AFB	747900	33.97	80.47	243	14.567	8293	24	29	18	16	14	19	48	17	48	5	10	8	240	100	17	3.1	6.1
SOUTH DAKOTA																							
Chamberlain	726530	43.80	99.32	1739	13.795	8293	−13	−7	27	24	21	28	18	25	20	11	270	13	190	106	−12	8.1	18.4
Huron	726540	44.38	98.22	1289	14.024	6193	−17	−12	29	25	22	29	14	25	15	9	290	14	180	102	−25	4.6	5.9
Pierre	726686	44.38	100.28	1742	13.794	6193	−14	−9	29	25	22	32	15	27	20	11	320	14	180	106	−20	3.8	5.7
Rapid City	726620	44.05	103.07	3169	13.089	6193	−11	−5	36	31	27	37	26	32	26	9	350	13	160	102	−17	3.4	5.4
Sioux Falls	726510	43.58	96.73	1427	13.953	6193	−16	−11	28	25	22	30	15	26	17	8	310	15	180	100	−23	4.1	4.9
TENNESSEE																							
Bristol	723183	36.48	82.40	1519	13.906	6193	9	14	20	17	15	21	35	19	36	6	270	8	250	92	−1	3.0	7.5
Chattanooga	723240	35.03	85.20	689	14.333	6193	15	20	19	17	15	20	37	18	38	7	360	8	280	97	7	3.6	7.0
Crossville	723265	35.95	85.08	1880	13.724	8293	7	15	16	14	13	18	33	16	36	4	310	8	270	93	−3	4.0	8.6
Jackson	723346	35.60	88.92	433	14.467	8293	12	18	20	18	16	21	46	19	44	9	360	8	240	98	4	2.3	8.8
Knoxville	723260	35.82	83.98	981	14.182	6193	13	19	21	18	15	21	48	19	45	7	50	8	250	95	4	3.0	7.8
Memphis	723340	35.05	90.00	285	14.544	6193	16	21	22	19	17	22	42	20	42	10	20	9	240	99	9	2.8	7.2
Nashville	723270	36.13	86.68	591	14.385	6193	10	16	22	19	17	22	46	20	42	8	340	9	230	97	1	3.3	7.8
TEXAS																							
Abilene	722660	32.42	99.68	1791	13.769	6193	16	22	27	24	22	26	48	23	46	12	0	11	140	102	10	2.8	5.9
Amarillo	723630	35.23	101.70	3606	12.879	6193	6	12	30	27	24	30	40	27	38	14	20	15	200	100	−1	2.8	5.5
Austin	722540	30.30	97.70	620	14.369	6193	25	30	23	20	18	25	41	22	43	12	10	10	180	101	20	2.4	5.9
Beaumont/Port Arthur	722410	29.95	94.02	23	14.683	6193	29	32	22	20	18	23	51	21	51	10	340	9	200	97	22	2.5	4.5
Beeville, Chase Field NAS	722556	28.37	97.67	190	14.595	8293	28	33	22	20	18	23	58	20	53	13	350	9	150	104	22	2.5	8.3
Brownsville	722500	25.90	97.43	20	14.685	6193	36	40	27	24	22	26	64	23	62	13	330	16	160	98	31	2.4	5.2
College Station/Bryan	722445	30.58	96.37	322	14.525	8293	22	29	21	19	17	21	47	19	49	12	350	9	170	101	17	2.3	8.3
Corpus Christi	722510	27.77	97.50	43	14.673	6193	32	36	28	25	23	27	59	24	58	13	360	15	140	98	25	1.9	5.2
Dallas/Ft. Worth, Intl A	722590	32.90	97.03	597	14.381	8293	17	24	26	23	21	26	46	24	47	13	350	10	170	103	14	3.1	8.3
Del Rio, Laughlin AFB	722615	29.37	100.78	1083	14.130	8293	28	32	22	19	17	22	47	19	50	7	10	9	140	105	22	3.1	5.8
El Paso	722700	31.80	106.40	3917	12.731	6193	21	25	25	21	18	24	51	21	49	5	20	8	180	104	14	3.3	6.1
Ft. Worth, Carswell AFB	722595	32.77	97.45	650	14.354	8293	18	24	22	20	18	23	43	20	45	11	10	8	10	103	15	2.2	8.1
Ft. Worth, Meacham Field	722596	32.82	97.37	709	14.323	6193	19	24	27	24	21	27	40	24	44	13	350	10	180	103	14	3.0	5.8
Guadalupe Pass	722620	31.83	104.80	5453	12.022	8293	13	19	51	45	41	50	39	46	37	19	70	13	250	98	10	2.9	6.8
Houston, Hobby Airport	722435	29.65	95.28	46	14.671	8293	29	34	22	20	18	23	52	21	52	13	350	7	190	98	24	2.0	8.1
Houston, Inter Airport	722430	29.97	95.35	108	14.638	6193	27	31	20	18	16	22	47	20	52	8	340	10	180	98	22	3.1	5.4
Junction	747400	30.50	99.77	1713	13.808	8293	19	23	19	16	15	19	53	16	53	6	360	9	150	104	12	2.3	6.7
Killeen, Ft. Hood	722576	31.07	97.83	1014	14.165	8293	20	27	22	19	17	22	48	19	53	11	360	9	160	102	15	2.0	8.6
Kingsville, NAS	722516	27.50	97.82	49	14.669	8293	31	36	23	21	19	22	61	20	60	11	360	11	150	102	18	2.0	10.1
Laredo	722520	27.55	99.47	509	14.427	8293	32	36	24	22	20	22	59	20	62	9	320	13	140	106	28	2.2	6.7

WMO# = World Meteorological Organization number
AFB = Air Force Base NAS = Naval Air Station
WSO = Weather Service Office
AAF = Army Air Field
MCAS = Marine Corps Air Station
DB = dry-bulb temperature, °F
Lat. = North latitude °
Long. = West longitude °
Elev. = elevation, ft
StdP = standard pressure at station elevation, psia
WS = wind speed, mph
PWD = prevailing wind direction, ° True

Table 1B Cooling and Dehumidification Design Conditions—United States

Station	Cooling DB/MWB						Evaporation WB/MDB						Dehumidification DP/MDB and HR									Range of DB
	0.4%		1%		2%		0.4%		1%		2%		0.4%			1%			2%			
	DB	MWB	DB	MWB	DB	MWB	WB	MDB	WB	MDB	WB	MDB	DP	HR	MDB	DP	HR	MDB	DP	HR	MDB	
1	2a	2b	2c	2d	2e	2f	3a	3b	3c	3d	3e	3f	4a	4b	4c	4d	4e	4f	4g	4h	4i	5
Eugene	91	67	87	65	83	64	69	87	67	84	65	81	62	83	74	60	78	73	59	74	71	27.6
Hillsboro	92	69	88	67	84	65	71	89	68	86	66	82	64	90	79	61	82	75	60	78	72	26.6
Klamath Falls	91	64	87	62	85	61	66	87	64	84	63	81	58	85	74	57	80	73	55	75	71	34.2
Meacham	87	59	84	58	80	57	61	82	59	80	58	78	52	67	66	50	63	64	49	59	64	37.1
Medford	98	67	95	66	91	65	69	94	67	91	66	88	60	81	75	58	76	74	56	71	73	33.7
North Bend	71	60	69	60	67	59	62	69	61	67	60	66	60	76	65	58	73	64	57	70	63	12.8
Pendleton	97	64	93	63	90	62	66	92	64	90	63	87	57	74	71	55	68	69	53	62	70	27.2
Portland	90	67	86	66	83	64	69	87	67	84	65	80	62	83	75	60	78	72	59	75	71	21.6
Redmond	93	62	89	61	86	59	63	88	62	86	60	83	55	71	68	52	65	67	50	60	66	35.0
Salem	92	67	87	66	83	64	68	89	67	85	65	81	61	81	75	59	76	73	58	72	71	27.9
Sexton Summit	83	60	80	59	77	58	62	80	61	77	59	74	55	76	70	53	69	68	52	66	66	18.9
PENNSYLVANIA																						
Allentown	90	73	88	72	85	71	76	86	74	84	73	82	73	123	81	71	117	79	70	111	78	19.4
Altoona	89	72	86	70	83	69	74	85	72	83	71	80	71	119	79	69	113	77	68	109	76	19.4
Bradford	83	69	80	68	78	66	72	79	70	77	68	75	69	116	75	68	111	73	66	105	72	21.2
Du Bois	86	70	84	69	81	67	72	81	71	79	70	78	70	116	76	69	112	74	67	108	73	19.4
Erie	85	72	83	70	80	70	74	82	73	80	71	78	72	122	79	70	115	77	69	109	76	15.6
Harrisburg	92	74	89	73	86	72	77	87	76	85	74	83	74	130	82	73	123	80	72	118	79	18.8
Philadelphia, Intl Airport	92	75	89	74	87	73	78	88	77	86	75	84	75	132	83	74	126	81	73	121	80	17.7
Philadelphia, Northeast A	93	76	90	74	88	73	78	88	77	87	75	84	76	135	83	74	129	82	72	121	82	19.1
Philadelphia, Willow Gr NAS	93	75	90	74	88	72	78	89	76	87	75	85	74	131	83	73	125	82	71	116	81	19.4
Pittsburgh, Allegheny Co. A	90	72	87	71	85	70	75	85	74	84	72	81	71	122	79	70	117	78	69	113	77	18.0
Pittsburgh, Intl Airport	89	72	86	70	84	69	74	85	73	82	71	80	71	121	80	70	115	78	68	109	77	19.5
Wilkes-Barre/Scranton	88	71	85	70	83	69	74	83	73	81	71	79	71	120	79	70	115	77	69	109	76	18.8
Williamsport	90	73	87	71	84	70	76	85	74	83	73	80	73	125	80	72	118	78	70	113	77	20.3
RHODE ISLAND																						
Providence	89	73	86	71	83	70	76	85	74	82	73	80	73	124	80	72	118	78	70	112	77	17.4
SOUTH CAROLINA																						
Beaufort, MCAS	95	78	93	78	92	77	80	90	80	89	79	88	78	145	85	77	141	85	76	137	84	16.7
Charleston	94	78	92	77	90	77	80	90	79	88	78	87	78	145	84	77	139	83	76	134	83	16.2
Columbia	96	76	94	75	92	74	78	90	77	89	77	87	75	134	82	75	130	81	74	127	81	19.9
Florence	96	76	94	76	92	76	80	90	78	89	78	88	77	142	85	76	136	83	75	132	82	19.8
Greer/Greenville	93	74	91	74	88	73	77	88	76	87	75	85	74	130	81	73	126	80	72	122	80	18.2
Myrtle Beach, AFB	93	79	90	78	88	78	81	89	80	88	79	87	79	150	87	78	144	86	77	140	84	14.4
Sumter, Shaw AFB	95	76	93	75	90	75	78	89	77	88	76	86	76	136	83	75	132	82	74	129	81	18.5
SOUTH DAKOTA																						
Chamberlain	98	72	94	71	90	70	76	91	74	89	72	87	71	124	84	70	116	82	68	109	80	23.8
Huron	95	72	91	71	88	70	76	89	74	87	72	84	72	126	84	70	117	81	69	110	79	24.1
Pierre	99	70	95	69	91	68	74	90	72	89	71	86	70	116	81	68	109	80	66	102	78	25.6
Rapid City	95	65	91	65	88	64	70	85	68	84	67	82	65	104	76	63	98	75	61	92	73	25.3
Sioux Falls	94	73	90	72	87	71	76	89	75	87	73	84	73	127	84	71	119	82	69	112	80	22.1
TENNESSEE																						
Bristol	89	72	87	72	85	71	75	85	74	84	73	82	72	125	81	71	120	79	70	116	77	19.2
Chattanooga	94	75	92	75	89	74	78	89	77	88	76	86	75	134	82	74	130	82	73	125	81	19.5
Crossville	89	73	87	72	85	72	76	85	74	83	73	82	74	134	80	72	125	79	71	121	78	19.8
Jackson	95	77	93	76	91	76	80	91	78	90	78	88	76	140	85	75	135	85	75	132	84	19.8
Knoxville	92	74	90	74	87	73	77	88	76	86	75	85	74	131	82	73	127	81	72	123	80	18.1
Memphis	96	78	94	77	92	77	80	92	79	91	78	89	77	143	87	76	137	86	75	133	84	16.8
Nashville	94	76	92	75	90	74	78	89	77	88	76	86	75	134	83	74	130	82	73	126	81	19.1
TEXAS																						
Abilene	99	71	97	71	95	71	75	89	74	89	73	88	71	123	81	70	119	80	69	115	79	20.5
Amarillo	96	67	94	66	92	66	71	86	70	86	69	85	67	112	76	65	107	75	64	104	74	23.3
Austin	98	74	96	74	94	74	78	89	77	88	76	87	76	137	81	75	134	80	74	130	80	20.1
Beaumont/Port Arthur	94	79	92	79	91	78	81	90	81	89	80	88	79	152	86	79	148	85	78	145	84	15.9
Beeville, Chase Field NAS	101	77	98	77	96	77	82	91	81	91	80	90	80	155	86	78	148	85	78	144	84	21.6
Brownsville	95	78	94	77	93	77	80	89	79	88	79	88	78	146	83	77	142	83	77	140	82	16.5
College Station/Bryan	98	75	96	75	94	75	79	89	78	89	78	88	77	141	82	76	138	81	75	134	81	21.4
Corpus Christi	95	78	94	78	92	78	81	90	80	89	79	88	79	148	84	78	146	83	77	143	83	16.5
Dallas/Ft. Worth, Intl A	100	74	98	74	96	74	78	92	77	91	76	90	75	132	82	74	130	82	73	126	81	20.3
Del Rio, Laughlin AFB	101	72	98	73	96	72	78	91	77	90	76	89	75	136	82	74	131	82	72	124	81	20.9
El Paso	101	64	98	64	96	64	70	85	69	84	68	84	67	114	73	65	109	73	64	103	74	28.0
Ft. Worth, Carswell AFB	100	75	97	75	96	75	79	92	78	91	77	90	76	141	85	75	135	84	74	130	84	19.3
Ft. Worth, Meacham Field	100	75	98	74	96	74	78	91	77	90	76	89	75	135	83	74	131	82	73	127	82	20.0
Guadalupe Pass	92	61	89	60	87	60	66	82	65	80	64	79	62	102	71	60	96	71	59	91	71	20.9
Houston, Hobby Airport	94	77	93	77	92	77	80	89	80	88	79	87	78	147	84	78	144	83	77	141	82	16.6
Houston, Inter Airport	96	77	94	77	92	77	80	90	79	89	79	88	78	144	83	77	141	83	76	137	83	18.2
Junction	100	72	98	71	96	71	76	89	75	88	74	87	73	130	80	71	121	79	70	118	79	24.8
Killeen, Ft. Hood	98	74	96	73	95	74	78	90	77	89	76	88	75	137	81	74	132	81	73	128	80	21.4
Kingsville, NAS	97	77	96	78	95	78	81	91	80	91	80	90	79	149	85	78	144	84	77	141	84	19.8
Laredo	102	73	101	74	98	74	79	92	78	91	77	89	76	138	82	75	136	81	75	132	81	21.2

DP = dew-point temperature, °F
MDB = mean coincident dry-bulb temp., °F
StdD = standard deviation, °F
MWB = mean coincident wet-bulb temp., °F
MWS = mean coincident wind speed, mph
HR = humidity ratio, grains of moisture per lb of dry air
A = airport
ANGB = Air National Guard Base
MCAF = Marine Corps Air Facility
NAF = Naval Air Facility
NAWS = Naval Air Weapons Station
RAF = Royal Air Force

Table 1A Heating and Wind Design Conditions—United States

Station	WMO#	Lat.	Long.	Elev., ft	StdP, psia	Dates	Heating Dry Bulb 99.6%	Heating Dry Bulb 99%	Extreme Wind Speed, mph 1%	2.5%	5%	Coldest Month 0.4% WS	0.4% MDB	1% WS	1% MDB	MWS/PWD to DB 99.6% MWS	99.6% PWD	0.4% MWS	0.4% PWD	Extr. Annual Daily Mean DB Max.	Mean DB Min.	StdD DB Max.	StdD DB Min.
1a	1b	1c	1d	1e	1f	1g	2a	2b	3a	3b	3c	4a	4b	4c	4d	5a	5b	5c	5d	6a	6b	6c	6d
Lubbock, Intl Airport	722670	33.65	101.82	3241	13.054	6193	11	17	30	26	23	30	43	27	44	12	0	14	160	102	4	2.6	5.6
Lubbock, Reese AFB	722675	33.60	102.05	3337	13.008	8293	11	18	25	22	19	25	48	22	44	10	20	11	170	102	6	3.1	4.9
Lufkin	722446	31.23	94.75	289	14.543	6193	23	27	18	16	14	18	44	17	46	6	330	8	230	99	17	3.2	5.3
Marfa	722640	30.37	104.02	4859	12.292	8293	15	19	24	21	18	25	44	22	45	5	360	9	220	97	5	2.3	5.0
McAllen	722506	26.18	98.23	108	14.638	8293	34	40	24	22	20	23	68	21	68	11	350	14	130	106	27	4.3	8.1
Midland/Odessa	722650	31.95	102.18	2861	13.238	6193	17	22	28	25	22	27	50	23	48	9	20	13	180	103	9	2.6	6.9
San Angelo	722630	31.37	100.50	1909	13.709	6193	20	24	26	23	21	25	52	22	51	10	20	11	160	103	13	2.8	6.1
San Antonio, Intl Airport	722530	29.53	98.47	794	14.279	6193	26	30	22	19	17	23	43	20	45	10	350	10	160	100	19	2.9	5.2
San Antonio, Kelly AFB	722535	29.38	98.58	689	14.333	8293	27	32	19	17	15	21	51	18	52	8	360	8	160	103	22	2.9	6.5
San Antonio, Randolph AFB	722536	29.53	98.28	761	14.296	8293	27	31	19	17	15	20	45	17	48	7	340	7	150	101	20	2.2	6.7
Sanderson	747300	30.17	102.42	2838	13.250	8293	23	28	19	16	13	20	44	17	48	6	360	7	120	102	9	2.9	8.3
Victoria	722550	28.85	96.92	118	14.633	6193	29	33	26	23	21	26	50	23	51	12	360	12	180	99	23	2.5	5.2
Waco	722560	31.62	97.22	509	14.427	6193	22	26	26	23	21	29	38	25	42	13	360	12	180	104	16	2.8	6.4
Wichita Falls, Sheppard AFB	723510	33.98	98.50	1030	14.157	6193	14	19	29	25	23	28	42	25	43	12	360	13	180	107	7	3.4	6.6
UTAH																							
Cedar City	724755	37.70	113.10	5623	11.945	6193	2	8	26	22	20	24	38	21	39	4	140	12	200	97	−6	2.3	8.3
Ogden, Hill AFB	725755	41.12	111.97	4787	12.325	8293	6	11	22	19	17	22	27	19	28	9	110	6	190	96	1	2.9	6.3
Salt Lake City	725720	40.78	111.97	4226	12.586	6193	6	11	27	23	20	27	42	22	40	7	160	11	340	100	−3	1.9	6.7
VERMONT																							
Burlington	726170	44.47	73.15	341	14.515	6193	−11	−6	23	21	18	24	30	21	27	6	70	11	180	93	−19	2.7	5.6
Montpelier/Barre	726145	44.20	72.57	1165	14.087	8293	−10	−6	21	19	17	22	20	20	20	4	320	9	220	91	−18	3.6	5.9
VIRGINIA																							
Ft. Belvoir	724037	38.72	77.18	69	14.659	8293	12	18	18	14	12	19	35	17	34	2	320	6	160	100	2	2.3	7.6
Hampton, Langley AFB	745980	37.08	76.37	10	14.690	8293	21	24	22	19	17	22	41	20	40	10	330	9	240	97	13	3.2	6.1
Lynchburg	724100	37.33	79.20	938	14.204	6193	12	17	19	17	15	21	35	18	35	8	360	9	230	95	5	2.9	5.8
Newport News	723086	37.13	76.50	43	14.673	8293	18	22	19	18	16	20	40	18	41	8	350	10	220	99	11	2.3	4.7
Norfolk	723080	36.90	76.20	30	14.680	6193	20	24	25	22	20	26	40	23	40	12	340	12	230	97	14	2.8	5.4
Oceana, NAS	723075	36.82	76.03	23	14.683	8293	22	25	21	19	17	21	42	19	42	8	310	9	220	98	14	1.8	6.8
Quantico, MCAS	724035	38.50	77.30	13	14.688	8293	16	21	17	14	12	19	36	15	38	6	340	5	230	100	8	3.6	5.9
Richmond	724010	37.50	77.33	177	14.602	6193	14	18	20	18	16	21	40	18	39	7	340	10	230	98	6	2.6	5.8
Roanoke	724110	37.32	79.97	1175	14.082	6193	12	17	23	20	17	27	31	23	32	10	320	10	290	96	4	3.3	5.6
Sterling	724030	38.95	77.45	322	14.525	6193	9	14	22	19	16	25	32	21	31	7	340	10	250	97	−1	3.3	7.0
Washington, R. Reagan A	724050	38.85	77.03	66	14.661	8293	15	20	23	20	18	24	34	21	35	11	340	11	170	99	8	2.5	6.8
WASHINGTON																							
Bellingham	727976	48.80	122.53	157	14.612	8293	15	21	23	20	18	28	33	23	34	17	40	9	290	87	11	3.1	7.4
Hanford	727840	46.57	119.60	732	14.311	8293	5	12	25	21	18	24	44	19	44	6	20	8	20	105	2	3.1	9.0
Olympia	727920	46.97	122.90	200	14.589	6193	18	23	21	18	16	21	45	19	45	5	180	8	50	94	10	4.0	8.1
Quillayute	727970	47.95	124.55	203	14.588	6193	23	27	33	27	21	41	45	35	45	7	60	9	240	87	19	8.4	6.4
Seattle, Intl Airport	727930	47.45	122.30	449	14.458	6193	23	28	22	19	17	24	44	21	44	10	10	10	350	92	19	3.6	6.8
Spokane, Fairchild AFB	727855	47.62	117.65	2461	13.435	6193	1	7	27	23	20	28	39	25	38	7	50	9	240	98	−7	3.2	8.7
Stampede Pass	727815	47.28	121.33	3967	12.708	8293	3	10	21	19	16	27	19	22	25	13	90	7	100	84	2	3.2	7.2
Tacoma, McChord AFB	742060	47.13	122.48	322	14.525	8293	18	24	18	15	13	22	45	18	46	2	180	7	20	94	12	2.7	6.8
Walla Walla	727846	46.10	118.28	1204	14.067	8293	4	12	22	19	17	24	49	22	47	6	180	9	300	105	1	3.2	11.7
Wenatchee	727825	47.40	120.20	1243	14.047	8293	3	9	22	19	17	17	36	12	31	3	100	9	280	101	−2	2.5	7.2
Yakima	727810	46.57	120.53	1066	14.138	6193	4	11	24	20	17	23	47	19	43	7	250	7	90	101	−2	3.2	8.5
WEST VIRGINIA																							
Bluefield	724125	37.30	81.20	2858	13.240	8293	5	12	15	13	12	18	34	15	33	6	270	6	290	88	−6	4.0	8.5
Charleston	724140	38.37	81.60	981	14.182	6193	6	11	18	16	14	20	38	18	34	7	250	8	240	94	−2	2.8	6.7
Elkins	724170	38.88	79.85	1998	13.665	6193	−2	5	20	18	16	22	30	19	30	4	280	8	290	88	−12	2.8	5.4
Huntington	724250	38.37	82.55	837	14.257	6193	6	11	19	16	14	20	32	17	32	8	270	8	270	94	−2	5.0	7.6
Martinsburg	724177	39.40	77.98	558	14.402	8293	8	14	21	18	16	23	33	20	34	7	270	9	290	99	−3	4.0	8.3
Morgantown	724176	39.65	79.92	1247	14.045	8293	4	11	18	15	13	19	32	17	33	6	210	8	240	93	−4	3.6	8.6
Parkersburg	724273	39.35	81.43	860	14.245	8293	4	11	18	16	14	20	32	18	29	7	240	8	270	95	−4	3.1	9.2
WISCONSIN																							
Eau Claire	726435	44.87	91.48	906	14.221	6193	−18	−13	22	19	17	21	14	20	13	7	250	13	220	95	−25	3.2	5.7
Green Bay	726450	44.48	88.13	702	14.326	6193	−13	−8	25	22	20	25	19	22	18	10	270	12	200	93	−19	2.8	5.6
La Crosse	726430	43.87	91.25	663	14.347	6193	−14	−8	23	20	18	23	13	21	13	7	310	12	180	97	−21	3.2	6.2
Madison	726410	43.13	89.33	866	14.241	6193	−11	−6	24	21	19	25	16	22	17	8	300	12	230	94	−18	3.2	6.0
Milwaukee	726400	42.95	87.90	692	14.332	6193	−7	−2	28	24	22	28	19	24	20	13	290	15	220	95	−12	3.2	6.7
Wausau	726463	44.93	89.63	1201	14.069	8293	−15	−9	19	17	15	19	16	17	17	7	300	10	200	93	−22	3.1	4.7
WYOMING																							
Big Piney	726710	42.57	110.10	6969	11.353	8293	−22	−15	24	20	17	22	25	19	21	3	60	11	260	87	−33	2.7	8.5
Casper	725690	42.92	106.47	5289	12.096	6193	−13	−5	34	30	27	35	35	32	32	9	260	13	240	97	−22	2.2	8.4
Cheyenne, Warren AFB	725640	41.15	104.82	6142	11.714	6193	−7	0	34	29	26	38	36	33	34	10	290	13	290	92	−15	2.2	7.5
Cody	726700	44.52	109.02	5095	12.184	8293	−14	−7	34	28	23	35	35	30	35	6	40	11	70	95	−20	4.1	9.4
Gillette	726650	44.35	105.53	4035	12.675	8293	−16	−7	28	25	22	30	34	27	33	8	260	11	140	101	−20	5.9	10.1
Lander	725760	42.82	108.73	5558	11.974	6193	−14	−7	23	19	16	25	38	19	37	3	120	10	270	95	−20	2.5	7.8
Rock Springs	725744	41.60	109.07	6759	11.444	6193	−9	−2	28	25	23	32	25	29	24	7	70	13	280	90	−17	2.0	8.0
Sheridan	726660	44.77	106.97	3967	12.708	6193	−14	−8	28	24	20	29	32	23	27	5	280	9	120	99	−22	3.0	6.4
Worland	726665	43.97	107.95	4245	12.577	8293	−22	−13	22	19	16	20	28	17	28	3	210	9	220	103	−30	2.2	10.4

WMO# = World Meteorological Organization number
AFB = Air Force Base
NAS = Naval Air Station
WSO = Weather Service Office
AAF = Army Air Field
MCAS = Marine Corps Air Station
DB = dry-bulb temperature, °F
Lat. = North latitude °
Long. = West longitude °
Elev. = elevation, ft
StdP = standard pressure at station elevation, psia
WS = wind speed, mph
PWD = prevailing wind direction, ° True

Table 1B Cooling and Dehumidification Design Conditions—United States

Station	Cooling DB/MWB						Evaporation WB/MDB						Dehumidification DP/MDB and HR									Range of DB
	0.4%		1%		2%		0.4%		1%		2%		0.4%			1%			2%			
	DB	MWB	DB	MWB	DB	MWB	WB	MDB	WB	MDB	WB	MDB	DP	HR	MDB	DP	HR	MDB	DP	HR	MDB	
1	2a	2b	2c	2d	2e	2f	3a	3b	3c	3d	3e	3f	4a	4b	4c	4d	4e	4f	4g	4h	4i	5
Lubbock, Intl Airport	97	67	95	67	93	67	73	87	72	86	71	85	69	120	77	68	115	76	67	111	76	22.1
Lubbock, Reese AFB	98	67	95	67	93	67	73	87	72	86	71	85	69	122	78	68	115	77	66	110	77	23.8
Lufkin	97	76	95	77	93	76	79	90	79	89	78	89	77	143	83	76	139	83	75	134	82	20.9
Marfa	94	62	92	61	89	62	68	82	67	81	66	80	65	110	72	63	103	71	62	98	71	31.3
McAllen	100	76	98	76	97	76	80	91	80	90	79	89	78	146	83	77	143	82	77	140	82	20.7
Midland/Odessa	99	67	97	67	95	67	73	87	72	86	71	86	69	120	76	68	115	75	67	111	75	23.7
San Angelo	100	70	97	70	95	70	75	90	74	89	73	88	71	123	80	70	118	79	69	116	78	22.3
San Antonio, Intl Airport	98	73	96	73	94	74	78	87	77	87	76	86	76	139	81	75	135	81	74	132	80	19.1
San Antonio, Kelly AFB	99	74	97	74	96	74	79	89	78	88	77	88	77	145	83	76	140	82	75	136	81	20.5
San Antonio, Randolph AFB	98	74	96	74	94	74	78	90	77	89	76	88	76	138	82	75	134	81	74	132	81	22.3
Sanderson	97	67	95	68	94	68	74	86	73	86	72	86	70	123	79	69	119	78	68	114	77	20.7
Victoria	95	76	94	76	92	77	80	88	79	88	78	87	78	145	83	77	141	82	76	139	82	17.4
Waco	101	75	99	75	97	75	79	93	78	92	77	91	75	135	83	74	131	82	74	127	82	21.6
Wichita Falls, Sheppard AFB	103	74	100	73	98	73	77	93	76	92	75	91	73	129	82	72	124	82	71	120	81	23.9
UTAH																						
Cedar City	93	59	91	59	88	58	64	80	62	80	61	79	59	93	68	57	85	68	55	78	68	28.5
Ogden, Hill AFB	93	61	90	60	87	60	65	83	64	81	62	81	60	91	72	57	83	73	55	77	73	22.0
Salt Lake City	96	62	94	62	92	61	66	85	65	85	64	85	60	92	73	58	84	73	56	77	73	27.7
VERMONT																						
Burlington	87	71	84	69	82	68	74	83	72	81	70	78	71	115	79	69	109	77	67	102	75	20.4
Montpelier/Barre	85	70	83	68	80	67	72	82	70	80	69	77	69	111	78	67	106	75	66	99	73	21.1
VIRGINIA																						
Ft. Belvoir	95	78	93	76	89	75	80	92	78	89	77	87	77	139	86	75	133	85	74	127	83	20.9
Hampton, Langley AFB	94	78	91	77	88	76	80	90	79	89	78	86	77	141	85	76	136	84	75	132	83	14.9
Lynchburg	93	74	90	74	88	73	77	88	76	87	75	85	74	129	81	73	125	80	72	120	79	18.2
Newport News	95	78	92	77	89	76	80	91	78	89	77	87	77	139	84	76	135	83	75	132	82	18.2
Norfolk	93	77	91	76	88	75	79	89	77	87	77	85	76	135	83	75	130	82	74	126	81	15.3
Oceana, NAS	94	77	91	76	88	75	79	89	78	87	77	86	77	139	85	76	134	83	74	129	82	15.7
Quantico, MCAS	94	77	92	76	89	74	79	91	78	89	76	87	76	136	87	75	130	85	73	125	83	18.5
Richmond	94	76	92	75	89	74	79	90	78	88	76	86	76	137	84	75	131	82	74	126	81	19.1
Roanoke	92	73	89	72	87	71	75	88	74	86	73	84	72	123	80	71	118	79	70	115	78	19.6
Sterling	93	75	90	74	88	73	77	88	76	87	75	85	74	130	83	73	125	81	72	120	80	21.0
Washington, National A	95	76	92	76	89	74	79	89	78	88	76	86	76	137	83	75	132	83	74	127	81	16.6
WASHINGTON																						
Bellingham	79	65	76	64	74	62	67	78	65	75	63	72	61	81	73	60	78	70	59	74	67	16.7
Hanford	100	67	96	65	93	64	68	96	66	94	65	90	58	73	72	56	68	75	53	62	74	26.5
Olympia	87	67	83	65	79	64	68	85	66	81	64	78	61	81	73	60	76	71	58	73	69	25.2
Quillayute	80	62	74	61	70	59	64	76	62	72	60	67	60	76	65	58	74	63	57	71	62	15.4
Seattle, Intl Airport	85	65	81	64	78	62	66	83	65	79	63	76	60	78	71	59	74	69	57	71	68	18.3
Spokane, Fairchild AFB	92	62	89	61	85	60	65	86	63	84	61	82	57	77	68	55	71	68	53	67	67	26.1
Stampede Pass	78	57	74	56	71	54	59	74	57	71	56	69	53	70	63	51	65	61	50	62	58	16.0
Tacoma, McChord AFB	86	65	82	63	78	62	67	83	65	80	63	76	60	79	71	59	76	70	58	72	68	22.5
Walla Walla	98	66	95	65	92	64	68	92	67	91	65	88	60	82	74	58	76	72	57	71	72	27.0
Wenatchee	95	67	92	65	88	63	67	91	66	89	64	85	59	78	75	57	73	75	55	68	74	25.2
Yakima	95	65	92	64	88	63	67	90	66	89	64	86	59	78	75	57	71	74	55	67	72	31.1
WEST VIRGINIA																						
Bluefield	85	69	83	69	80	67	72	81	71	79	70	77	69	120	75	68	116	75	67	111	73	16.4
Charleston	91	73	88	73	86	71	76	86	75	85	74	82	73	129	81	72	123	80	71	118	78	19.1
Elkins	85	71	83	70	81	69	73	82	72	80	71	78	71	121	78	69	116	77	68	111	75	21.1
Huntington	91	74	89	73	86	72	77	87	76	85	74	83	74	132	82	73	127	81	72	121	79	19.1
Martinsburg	94	74	91	73	88	72	77	87	75	86	74	85	74	130	81	72	120	80	71	116	79	21.8
Morgantown	89	72	87	71	85	70	75	85	74	83	73	82	72	124	79	71	119	78	70	115	76	20.3
Parkersburg	91	74	88	72	86	72	76	87	75	85	74	82	74	132	82	72	122	80	71	118	78	19.6
WISCONSIN																						
Eau Claire	90	73	87	71	84	70	76	86	74	83	72	81	73	125	82	71	116	80	69	109	78	20.6
Green Bay	88	73	85	72	82	70	76	85	74	82	72	80	73	124	82	71	116	79	69	109	77	20.7
La Crosse	91	74	88	73	85	71	77	87	75	84	74	82	75	132	83	73	125	81	71	117	78	20.1
Madison	90	73	87	72	84	70	76	86	74	84	72	82	73	126	83	71	118	80	69	111	78	21.9
Milwaukee	89	74	86	72	83	70	76	86	74	83	72	81	73	127	83	71	119	80	70	111	78	16.6
Wausau	88	71	85	70	82	69	74	83	72	82	71	78	71	120	79	69	113	77	68	108	75	19.6
WYOMING																						
Big Piney	83	54	80	53	78	52	56	75	55	74	53	74	50	69	60	48	64	60	45	57	59	32.8
Casper	92	59	89	58	86	58	62	81	61	80	60	79	57	85	66	55	78	66	53	73	65	30.4
Cheyenne, Warren AFB	87	58	85	57	82	57	62	77	61	76	60	75	58	90	66	56	85	65	55	80	64	25.7
Cody	91	59	87	58	84	57	61	83	60	81	58	80	54	76	70	52	69	66	50	64	65	25.4
Gillette	94	61	91	61	87	60	65	84	63	83	62	82	59	88	73	57	80	69	54	73	68	28.6
Lander	90	59	87	58	85	57	62	81	61	80	59	80	56	81	69	53	74	68	51	69	67	26.7
Rock Springs	86	54	84	54	82	53	58	75	57	74	56	74	54	78	62	51	71	61	49	66	61	27.7
Sheridan	93	62	90	61	86	61	66	85	64	83	63	81	60	88	71	58	82	71	56	76	69	29.1
Worland	96	63	93	63	90	62	67	88	66	86	64	84	61	94	75	59	86	75	57	80	73	31.0

DP = dew-point temperature, °F
MDB = mean coincident dry-bulb temp., °F
StdD = standard deviation, °F
MWB = mean coincident wet-bulb temp., °F
MWS = mean coincident wind speed, mph
HR = humidity ratio, grains of moisture per lb of dry air
A = airport
ANGB = Air National Guard Base
MCAF = Marine Corps Air Facility
NAF = Naval Air Facility
NAWS = Naval Air Weapons Station
RAF = Royal Air Force

Table 2A Heating and Wind Design Conditions—Canada

Station	WMO#	Lat.	Long.	Elev., ft	StdP, psia	Dates	Heating Dry Bulb 99.6%	Heating Dry Bulb 99%	Extreme Wind Speed, mph 1%	2.5%	5%	Coldest Month 0.4% WS	0.4% MDB	1% WS	1% MDB	MWS/PWD to DB 99.6% MWS	99.6% PWD	0.4% MWS	0.4% PWD	Extr. Annual Daily Mean DB Max.	Mean DB Min.	StdD DB Max.	StdD DB Min.
1a	1b	1c	1d	1e	1f	1g	2a	2b	3a	3b	3c	4a	4b	4c	4d	5a	5b	5c	5d	6a	6b	6c	6d
ALBERTA																							
Calgary Intl A	718770	51.12	114.02	3556	12.902	6193	−22	−17	28	24	21	32	29	28	27	7	0	11	160	89	−28	2.7	5.6
Cold Lake A	711200	54.42	110.28	1785	13.772	6193	−31	−26	21	18	16	21	18	18	11	3	270	9	180	88	−40	3.2	6.3
Coronation	718730	52.07	111.45	2595	13.369	6193	−27	−23	25	21	19	28	12	23	13	9	320	11	160	92	−35	3.1	6.3
Edmonton Intl A	711230	53.30	113.58	2372	13.479	6193	−28	−23	24	21	18	24	12	21	11	6	180	9	180	87	−36	3.1	8.1
Fort McMurray A	719320	56.65	111.22	1211	14.064	6193	−32	−29	17	15	13	18	16	15	11	3	90	9	250	90	−42	3.6	4.9
Grande Prairie A	719400	55.18	118.88	2195	13.567	6193	−32	−27	27	23	19	29	32	24	28	3	320	8	270	87	−41	2.7	6.7
Lethbridge A	718740	49.63	112.80	3048	13.147	6193	−22	−16	36	32	28	45	39	39	38	5	250	13	270	94	−30	3.2	6.5
Medicine Hat A	718720	50.02	110.72	2349	13.490	6193	−24	−19	26	22	20	29	36	25	33	5	230	11	220	97	−32	3.6	7.2
Peace River A	710680	56.23	117.43	1873	13.727	6193	−32	−27	21	18	17	22	30	19	24	4	0	9	270	87	−42	2.9	6.7
Red Deer A	718780	52.18	113.90	2969	13.186	6193	−27	−21	22	19	17	27	13	22	13	6	200	10	180	88	−35	3.1	6.5
Rocky Mtn. House	719280	52.43	114.92	3245	13.052	6193	−25	−20	19	16	13	19	26	16	20	3	340	8	160	87	−36	2.7	5.0
Vermilion A		53.35	110.83	2028	13.650	6193	−30	−25	22	19	17	21	13	19	11	3	270	11	180	90	−43	3.6	6.7
Whitecourt	719300	54.15	115.78	2566	13.383	6193	−30	−24	17	15	14	19	22	17	18	4	270	7	90	87	−41	2.0	5.4
BRITISH COLUMBIA																							
Abbotsford A	711080	49.03	122.37	190	14.595	6193	15	20	20	17	15	29	33	25	34	12	90	7	220	92	10	4.0	6.7
Cape St. James	710310	51.93	131.02	302	14.536	6193	25	29	50	46	40	60	40	54	42	22	50	11	300	69	22	3.4	5.6
Castlegar A	718840	49.30	117.63	1624	13.853	6693	5	9	18	15	14	21	18	19	21	8	0	7	180	98	−3	2.9	7.0
Comox A	718930	49.72	124.90	79	14.654	6193	21	25	29	25	21	31	43	28	42	7	290	7	340	87	17	3.8	5.2
Cranbrook A	718800	49.60	115.78	3081	13.131	7093	−15	−8	20	18	16	20	33	18	33	2	200	10	210	94	−22	2.7	6.7
Fort Nelson A	719450	58.83	122.58	1253	14.042	6193	−33	−30	16	14	12	15	8	12	2	1	220	5	120	88	−42	3.2	6.3
Fort St. John A	719430	56.23	120.73	2280	13.524	6193	−30	−25	25	22	19	29	23	25	22	7	0	9	230	86	−36	2.9	6.3
Kamloops A	718870	50.70	120.45	1135	14.103	6693	−8	−1	23	20	18	25	26	22	27	4	90	8	270	98	−14	2.7	8.8
Penticton A	718890	49.47	119.60	1129	14.106	6193	5	10	23	20	17	28	34	25	35	8	340	9	180	96	1	2.7	7.2
Port Hardy A	711090	50.68	127.37	72	14.657	6193	22	26	28	24	21	33	38	29	39	8	110	9	340	76	18	3.4	4.9
Prince George A	718960	53.88	122.68	2267	13.531	6193	−25	−18	21	18	15	27	32	23	24	2	0	6	180	85	−37	9.2	5.9
Prince Rupert A	718980	54.30	130.43	112	14.636	6393	7	13	28	23	20	30	44	26	43	6	70	8	270	75	2	4.5	6.8
Quesnel A	711030	53.03	122.52	1788	13.770	6193	−22	−14	17	15	14	19	18	17	20	1	340	5	340	92	−31	4.0	8.3
Sandspit A	711010	53.25	131.82	20	14.685	6193	21	25	38	32	27	42	44	37	42	18	320	9	270	72	18	3.4	5.0
Smithers A	719500	54.82	127.18	1716	13.807	6193	−19	−12	17	15	13	18	23	16	19	3	140	6	320	88	−26	4.0	7.0
Spring Island	714790	50.12	127.93	322	14.525	6193	29	31	41	35	29	44	46	40	45	6	50	6	320	78	25	6.1	4.5
Terrace A	719510	54.47	128.58	712	14.321	6193	−2	2	26	23	20	32	10	29	14	19	0	8	270	89	−5	4.0	5.8
Tofino A	711060	49.08	125.77	79	14.654	6193	25	29	24	20	18	29	46	24	45	5	70	7	290	81	21	4.0	5.6
Vancouver Intl A	718920	49.18	123.17	7	14.692	6193	18	24	22	19	16	25	41	21	42	6	90	7	290	82	14	2.9	6.3
Victoria Intl A	717990	48.65	123.43	62	14.662	6193	23	26	20	16	14	24	37	20	38	10	50	6	90	87	18	3.2	5.8
Williams Lake A	711040	52.18	122.05	3084	13.130	6193	−20	−14	22	19	17	24	29	21	30	3	320	6	140	88	−29	4.0	7.9
MANITOBA																							
Brandon A	711400	49.92	99.95	1342	13.997	6193	−29	−24	27	23	20	28	2	24	2	9	270	12	160	94	−36	3.1	4.5
Churchill A	719130	58.75	94.07	95	14.645	6193	−36	−33	34	30	26	36	−11	30	−14	15	270	13	230	86	−41	4.5	4.0
Dauphin A	718550	51.10	100.05	1001	14.172	6193	−28	−23	28	25	22	31	2	28	5	9	250	13	200	93	−36	3.4	4.3
Portage La Prairie A	718510	49.90	98.27	883	14.233	6193	−25	−21	26	23	20	29	1	25	1	9	250	12	180	95	−31	3.4	4.1
The Pas A	718670	53.97	101.10	889	14.229	6193	−32	−28	24	21	19	25	−6	22	−2	6	290	11	160	89	−40	3.4	4.1
Thompson A	710790	55.80	97.87	715	14.320	6893	−38	−34	20	18	16	19	−5	17	−7	3	270	10	180	89	−48	3.8	4.3
Winnipeg Int'l A	718520	49.90	97.23	784	14.284	6193	−27	−23	29	25	23	30	5	26	5	7	320	13	180	94	−33	3.4	4.7
NEW BRUNSWICK																							
Charlo A	717110	47.98	66.33	125	14.629	6793	−14	−10	24	21	19	27	3	24	8	11	250	11	250	89	−21	2.7	4.5
Chatham A	717170	47.00	65.45	102	14.642	6193	−12	−7	24	21	18	27	16	24	16	7	270	11	230	93	−20	2.2	4.3
Fredericton A	717000	45.87	66.53	66	14.661	6193	−12	−7	23	20	17	25	17	22	18	5	270	11	230	92	−21	2.5	5.4
Moncton A	717050	46.12	64.68	233	14.572	6193	−10	−5	26	23	20	30	19	26	19	13	270	13	250	89	−17	2.0	4.5
Saint John A	716090	45.32	65.88	358	14.507	6193	−9	−4	26	23	20	32	24	28	23	9	340	11	230	84	−18	4.0	4.7
NEWFOUNDLAND																							
Battle Harbour	718170	52.30	55.83	26	14.681	6193	−14	−10	40	35	32	48	17	42	15	18	270	17	230	78	−18	4.7	5.9
Bonavista	711960	48.67	53.12	89	14.648	6193	3	7	43	38	34	48	21	42	23	24	280	17	230	81	0	2.7	5.8
Cartwright	718180	53.70	57.03	46	14.671	6493	−18	−15	36	30	27	40	18	35	19	12	220	12	210	84	−23	4.1	4.5
Daniels Harbour	711850	50.23	57.58	62	14.662	6693	−7	−3	40	35	31	45	16	39	21	12	270	15	230	75	−12	3.2	6.5
Deer Lake A	718090	49.22	57.40	72	14.657	6693	−13	−7	24	22	18	26	21	22	19	3	240	14	220	86	−23	2.3	6.7
Gander Intl A	718030	48.95	54.57	495	14.434	6193	−4	0	32	28	25	37	20	32	22	16	270	13	230	85	−8	2.9	6.3
Goose A	718160	53.32	60.37	160	14.362	6193	−23	−20	26	22	20	30	3	26	5	11	250	12	250	89	−29	4.0	3.6
Hopedale	719000	55.45	60.23	26	14.681	6493	−21	−18	36	30	27	40	13	35	11	12	250	13	250	79	−25	4.5	5.8
St. John's A	718010	47.62	52.73	459	14.453	6193	3	7	37	33	29	41	24	37	25	17	290	17	250	82	−2	2.5	4.9
Stephenville A	718150	48.53	58.55	85	14.650	6193	−2	3	33	28	23	36	21	30	21	11	50	9	250	79	−7	2.5	6.3
Wabush Lake A	718250	52.93	66.87	1808	13.760	6193	−33	−30	23	20	17	25	−5	21	−4	5	270	12	240	82	−43	3.8	4.3
NORTHWEST TERRITORIES																							
Cape Parry A	719480	70.17	124.68	56	14.666	6193	−34	−33	31	28	25	34	−12	29	−12	7	270	9	110	66	−42	4.9	4.3
Fort Smith A	719340	60.02	111.95	666	14.345	6193	−34	−32	18	17	15	20	−3	18	−5	3	150	10	180	88	−48	3.4	5.6
Inuvik UA	719570	68.30	133.48	223	14.577	7393	−43	−40	17	15	14	19	−7	17	−7	1	70	8	180	84	−52	2.5	5.0
Norman Wells A	710430	65.28	126.80	243	14.567	6193	−40	−36	24	21	18	29	−4	24	−7	2	170	8	140	87	−49	3.1	5.4
Yellowknife A	719360	62.47	114.45	676	14.340	6193	−39	−36	22	19	17	23	−8	20	−9	5	50	10	160	82	−47	3.6	4.5
NUNAVUT																							
Baker Lake	719260	64.30	96.08	59	14.664	6393	−41	−39	36	32	28	42	−26	37	−26	12	0	10	270	77	−50	4.9	4.3
Cambridge Bay A	719250	69.10	105.12	89	14.648	6193	−38	−35	35	31	27	36	−19	31	−19	9	320	11	140	68	−50	4.9	3.8
Chesterfield	719164	63.33	90.72	36	14.676	6393	−35	−34	33	29	26	35	−26	31	−26	14	320	13	320	73	−49	13.9	4.9
Coral Harbour A	719150	64.20	83.37	210	14.584	6193	−40	−38	37	32	27	39	−5	32	−7	9	340	13	270	71	−49	3.8	5.2
Hall Beach A	710810	68.78	81.25	26	14.681	6193	−42	−38	33	29	25	34	−20	30	−21	10	320	11	180	64	−53	5.2	5.4

WMO# = World Meteorological Organization number Elev. = elevation, ft PWD = prevailing wind direction, ° True DB = dry-bulb temperature, °F
Lat. = North latitude, ° Long. = West longitude, ° StdP = standard pressure at station elevation, psia WS = wind speed, mph

Table 2B Cooling and Dehumidification Design Conditions—Canada

	Cooling DB/MWB						Evaporation WB/MDB						Dehumidification DP/MDB and HR									
	0.4%		1%		2%		0.4%		1%		2%		0.4%			1%			2%			Range
Station	DB	MWB	DB	MWB	DB	MWB	WB	MDB	WB	MDB	WB	MDB	DP	HR	MDB	DP	HR	MDB	DP	HR	MDB	of DB
1	2a	2b	2c	2d	2e	2f	3a	3b	3c	3d	3e	3f	4a	4b	4c	4d	4e	4f	4g	4h	4i	5
ALBERTA																						
Calgary Intl A	83	60	80	59	77	57	62	78	61	75	59	73	57	79	67	55	74	65	53	69	64	22.0
Cold Lake A	82	64	78	62	75	60	66	78	64	76	62	72	62	88	71	60	82	69	58	76	67	20.0
Coronation	85	62	82	60	78	59	65	80	63	78	61	75	60	84	69	58	78	67	56	73	66	22.1
Edmonton Intl A	82	63	78	62	75	60	66	77	64	75	62	73	62	89	73	59	83	69	57	77	66	21.8
Fort McMurray A	84	64	80	62	76	60	66	79	64	76	62	73	61	84	71	59	78	68	57	74	66	22.0
Grande Prairie A	81	62	78	60	75	58	64	77	62	74	60	71	59	81	68	57	76	65	55	71	63	20.9
Lethbridge A	88	61	84	61	81	60	65	81	63	79	61	77	59	84	70	57	78	69	55	73	67	24.8
Medicine Hat A	90	63	87	62	84	61	66	84	64	83	62	80	60	83	70	58	78	69	56	72	68	25.0
Peace River A	81	62	78	60	75	59	65	77	62	75	60	71	60	82	69	58	76	67	56	71	65	21.4
Red Deer A	82	62	79	61	76	59	65	78	63	75	61	73	60	85	71	58	80	69	56	74	66	22.9
Rocky Mtn. House	80	62	78	61	75	59	64	78	63	75	61	72	60	87	70	58	81	68	56	76	66	22.5
Vermilion A	83	64	80	62	77	61	66	78	64	77	62	74	62	88	72	60	83	69	58	77	67	22.0
Whitecourt	80	61	77	60	74	59	65	77	62	74	60	71	60	86	69	58	80	67	57	75	64	23.4
BRITISH COLUMBIA																						
Abbotsford A	85	67	80	66	77	64	69	83	66	79	64	76	62	85	77	61	80	73	59	76	70	21.4
Cape St. James	64	59	62	58	60	57	60	63	59	61	58	60	59	74	61	58	71	59	57	69	59	7.6
Castlegar A	92	64	88	63	84	62	67	85	65	83	63	81	60	83	71	59	78	70	57	74	68	27.9
Comox A	80	63	76	62	73	61	65	76	63	74	62	71	60	78	68	59	75	66	58	71	65	16.4
Cranbrook A	88	61	85	60	81	59	63	83	62	80	60	78	57	77	66	55	72	66	53	67	65	24.8
Fort Nelson A	82	62	78	60	75	59	64	77	62	74	60	72	60	80	68	58	75	66	56	70	64	21.1
Fort St. John A	79	61	76	59	73	58	63	76	61	72	59	70	59	80	67	57	74	65	55	69	63	18.7
Kamloops A	93	65	88	63	85	62	66	88	65	85	63	81	59	78	70	57	74	69	56	69	68	24.7
Penticton A	90	65	87	64	83	63	67	85	65	83	64	81	60	81	73	59	76	72	57	72	71	26.5
Port Hardy A	68	59	65	58	63	57	60	66	59	64	58	62	58	72	62	57	69	61	56	67	60	12.4
Prince George A	81	60	78	59	74	58	63	78	61	74	59	71	58	77	66	56	73	64	54	67	62	23.2
Prince Rupert A	66	58	63	57	61	56	60	64	58	62	57	60	58	71	61	57	69	60	56	66	59	10.4
Quesnel A	85	62	81	60	77	59	64	80	62	77	61	74	59	81	67	57	75	65	56	71	63	25.4
Sandspit A	68	60	65	59	63	58	61	66	60	64	59	62	59	75	62	58	71	61	57	69	60	8.6
Smithers A	81	61	77	59	73	58	62	78	61	74	59	71	57	74	65	56	70	64	54	65	62	22.0
Spring Island	68	60	66	59	63	58	61	66	60	64	59	62	59	76	62	58	74	61	57	71	60	8.8
Terrace A	83	62	78	60	74	59	64	79	62	76	60	72	58	73	66	56	69	64	55	65	64	17.1
Tofino A	72	62	68	60	66	58	62	70	60	67	59	64	59	76	63	58	73	62	57	71	60	12.2
Vancouver Intl A	76	65	74	64	71	62	66	75	64	72	63	70	62	83	71	61	79	70	60	76	68	14.0
Victoria Intl A	79	63	75	62	72	61	64	77	63	74	61	71	59	75	69	58	71	67	57	69	65	18.4
Williams Lake A	83	59	79	57	75	56	60	77	59	75	57	72	55	73	62	53	67	61	51	62	60	22.0
MANITOBA																						
Brandon A	87	67	84	66	80	65	71	82	69	80	66	77	67	105	77	65	97	74	62	88	72	23.6
Churchill A	77	62	72	60	67	58	64	74	61	70	58	66	59	76	68	56	68	65	53	60	62	16.7
Dauphin A	87	67	84	66	80	64	70	82	68	80	66	77	67	102	77	64	94	74	62	86	72	22.1
Portage La Prairie A	88	68	85	67	81	65	72	83	69	80	68	78	68	108	78	66	99	75	64	92	73	20.5
The Pas A	83	66	79	64	76	62	68	79	66	76	64	73	64	93	73	62	85	70	60	80	68	18.4
Thompson A	83	64	79	62	76	60	66	78	64	75	62	73	62	85	71	60	78	69	58	73	66	23.0
Winnipeg Int'l A	87	68	84	67	81	66	72	82	70	80	68	78	68	107	78	66	99	75	64	92	73	20.5
NEW BRUNSWICK																						
Charlo A	83	68	79	66	75	65	70	78	68	76	66	73	68	102	74	66	97	72	64	90	70	18.4
Chatham A	86	69	83	67	79	65	71	81	69	78	68	75	68	104	75	67	98	73	65	92	71	20.3
Fredericton A	86	69	83	68	79	66	72	82	70	79	68	76	68	104	77	67	99	75	65	93	72	20.7
Moncton A	83	68	80	67	77	65	71	79	69	77	68	74	68	104	75	67	99	73	65	93	71	19.4
Saint John A	78	65	75	64	72	62	68	75	66	72	64	69	66	96	71	64	90	69	62	84	66	16.9
NEWFOUNDLAND																						
Battle Harbour	65	58	60	55	58	53	59	63	56	60	54	57	57	69	61	54	62	58	52	57	55	10.4
Bonavista	74	65	71	63	68	62	67	72	65	69	63	67	65	92	70	63	85	68	61	80	66	11.7
Cartwright	75	62	70	59	67	58	63	72	61	68	59	66	59	76	67	57	70	65	55	64	63	17.5
Daniels Harbour	69	63	66	62	65	61	65	67	63	66	62	64	64	90	67	62	83	65	60	78	63	9.7
Deer Lake A	81	66	77	64	74	62	68	77	66	74	65	71	66	95	73	64	88	71	61	81	69	21.4
Gander Intl A	79	65	76	63	72	62	68	75	66	72	64	70	65	95	71	63	88	70	61	83	68	17.8
Goose A	81	63	77	61	73	60	66	77	63	73	61	71	61	82	70	59	76	68	57	71	65	18.2
Hopedale	70	59	65	57	61	55	60	68	58	64	55	61	57	69	64	54	62	61	52	57	59	12.6
St. John's A	76	65	73	64	70	63	68	73	66	71	64	68	66	98	71	64	92	69	62	85	67	15.7
Stephenville A	74	64	71	64	69	63	67	71	65	69	64	68	66	95	70	64	88	68	62	83	66	12.4
Wabush Lake A	76	60	72	58	68	57	63	71	61	69	59	66	60	83	66	58	76	63	56	70	62	16.9
NORTHWEST TERRITORIES																						
Cape Parry A	58	53	54	50	50	47	53	58	50	53	47	50	50	53	56	47	48	52	45	43	50	9.7
Fort Smith A	82	63	78	61	75	60	65	78	63	75	61	72	61	81	69	59	76	67	57	71	66	21.4
Inuvik UA	78	60	75	59	71	57	62	75	60	72	58	69	56	67	67	53	61	65	52	57	64	18.4
Norman Wells A	80	62	77	60	74	59	64	77	62	74	60	71	59	75	68	57	71	66	56	67	65	18.5
Yellowknife A	77	60	74	59	70	57	62	73	60	71	59	68	58	73	66	56	68	65	53	62	63	14.4
NUNAVUT																						
Baker Lake	69	57	65	55	61	53	59	67	56	63	54	60	55	63	64	52	57	59	50	53	57	16.4
Cambridge Bay A	60	53	57	51	53	48	54	59	51	56	49	53	50	55	56	48	50	53	46	46	51	12.4
Chesterfield	66	54	60	52	56	50	55	65	52	60	50	56	50	53	60	48	49	55	46	46	52	13.7
Coral Harbour A	64	53	60	51	56	50	55	63	52	59	50	56	50	54	58	48	50	55	46	46	52	14.8
Hall Beach A	56	50	52	47	49	45	50	55	48	52	45	49	47	48	53	44	42	50	42	39	47	9.9

MDB = mean coincident dry-bulb temperature, °F
MWB = mean coincident wet-bulb temperature, °F
MWS = mean coincident wind speed, mph
StdD = standard deviation, °F
HR = humidity ratio, grains of moisture per lb of dry air
A = airport
DP = dew-point temperature, °F

Table 2A Heating and Wind Design Conditions—Canada

Station	WMO#	Lat.	Long.	Elev., ft	StdP, psia	Dates	Heating Dry Bulb 99.6%	Heating Dry Bulb 99%	Extreme Wind Speed, mph 1%	Extreme Wind Speed, mph 2.5%	Extreme Wind Speed, mph 5%	Coldest Month 0.4% WS	Coldest Month 0.4% MDB	Coldest Month 1% WS	Coldest Month 1% MDB	MWS/PWD to DB 99.6% MWS	MWS/PWD to DB 99.6% PWD	MWS/PWD to DB 0.4% MWS	MWS/PWD to DB 0.4% PWD	Extr. Annual Daily Mean DB Max.	Extr. Annual Daily Mean DB Min.	Extr. Annual Daily StdD DB Max.	Extr. Annual Daily StdD DB Min.
1a	1b	1c	1d	1e	1f	1g	2a	2b	3a	3b	3c	4a	4b	4c	4d	5a	5b	5c	5d	6a	6b	6c	6d
Iqaluit A (Frobisher)	719090	63.75	68.55	108	14.638	6193	−39	−36	32	28	25	40	−12	34	−12	4	320	12	320	68	−43	4.3	4.7
Resolute A	719240	74.72	94.98	220	14.579	6393	−42	−40	39	34	30	42	−15	37	−18	9	320	12	110	55	−49	3.8	4.9
NOVA SCOTIA																							
Greenwood A	713970	44.98	64.92	92	14.647	6193	−2	3	29	25	22	35	25	30	23	7	300	15	250	89	−11	2.3	5.4
Halifax Intl A	713950	44.88	63.50	476	14.445	6993	−2	2	27	23	20	30	26	27	26	11	320	12	200	87	−8	2.7	4.3
Sable Island	716000	43.93	60.02	13	14.688	6193	14	17	38	34	23	43	30	39	29	24	290	13	200	73	10	2.5	4.0
Shearwater A	716010	44.63	63.50	167	14.607	6193	1	5	29	25	22	33	24	28	24	12	340	10	230	85	−5	3.2	4.1
Sydney A	717070	46.17	60.05	203	14.588	6193	−1	3	30	26	23	35	23	30	24	13	270	14	230	87	−7	2.5	4.5
Truro	713980	45.37	63.27	131	14.626	6193	−9	−4	24	21	18	30	25	26	25	5	0	11	270	86	−16	3.2	5.8
Yarmouth A	716030	43.83	66.08	141	14.621	6193	7	10	28	25	22	30	26	28	27	12	320	10	190	79	1	2.7	3.6
ONTARIO																							
Armstrong A	718410	50.30	89.03	1152	14.094	6193	−33	−30	22	19	17	21	−2	18	−3	3	270	12	0	88	−49	3.2	4.9
Atikokan	717480	48.75	91.62	1289	14.024	6793	−31	−27	16	14	12	16	6	14	6	1	270	8	230	89	−42	3.6	3.8
Big Trout Lake	718480	53.83	89.87	735	14.309	6793	−32	−30	23	21	18	24	−1	21	−2	6	290	9	200	86	−42	3.2	4.9
Earlton A	717350	47.70	79.85	797	14.277	6193	−27	−21	21	19	17	24	14	21	12	4	320	12	200	91	−38	3.8	5.0
Geraldton	718340	49.78	86.93	1152	14.094	6893	−32	−28	21	18	16	22	3	19	4	2	270	11	0	87	−45	4.1	6.7
Gore Bay A	717330	45.88	82.57	633	14.362	6193	−12	−7	27	23	21	29	23	26	22	7	0	11	180	86	−21	4.5	5.9
Kapuskasing A	718310	49.42	82.47	745	14.304	6193	−30	−25	20	17	16	21	5	19	5	4	270	10	230	90	−39	2.7	5.0
Kenora A	718500	49.78	94.37	1348	13.993	6193	−27	−22	21	18	17	21	7	19	6	8	320	11	180	89	−32	3.4	5.0
London A	716230	43.03	81.15	912	14.217	6193	−3	2	25	22	20	30	21	26	21	9	250	11	250	90	−11	3.4	5.6
Mount Forest	716310	43.98	80.75	1362	13.987	6293	−7	−3	25	21	19	28	19	25	18	6	90	10	250	87	−16	2.2	4.5
Muskoka A	716300	44.97	79.30	925	14.211	6193	−17	−11	21	19	18	23	21	20	21	7	320	9	270	87	−30	2.2	5.6
North Bay A	717310	46.35	79.43	1217	14.060	6193	−18	−13	20	17	16	23	16	20	16	6	0	10	230	86	−26	3.2	5.4
Ottawa Int'l A	716280	45.32	75.67	374	14.498	6193	−13	−8	23	20	17	27	17	23	15	9	290	10	250	91	−19	2.7	5.0
Sault Ste. Marie A	712600	46.48	84.52	630	14.364	6293	−13	−8	27	23	20	28	18	24	18	4	90	9	220	88	−25	3.2	5.9
Simcoe	715270	42.85	80.27	791	14.280	6293	−2	3	24	20	18	28	23	24	24	10	270	11	230	91	−10	2.9	4.7
Sioux Lookout A	718420	50.12	91.90	1280	14.029	6193	−30	−25	17	15	14	19	2	17	2	4	270	9	200	89	−39	3.2	4.9
Sudbury A	717300	46.62	80.80	1142	14.099	6193	−19	−14	30	26	23	30	13	27	13	10	0	14	230	89	−27	4.0	5.2
Thunder Bay A	717490	48.37	89.32	653	14.352	6193	−22	−18	24	21	18	25	9	22	6	9	250	11	200	90	−31	3.8	4.7
Timmins A	717390	48.57	81.37	968	14.189	6193	−28	−23	21	19	17	22	6	20	7	5	180	10	250	91	−39	3.6	4.7
Toronto Int'l A	716240	43.67	79.63	568	14.397	6593	−4	1	26	22	20	29	22	26	23	9	340	12	270	92	−11	2.9	5.6
Trenton A	716210	44.12	77.53	282	14.546	6193	−8	−3	26	22	19	30	23	26	24	6	50	12	230	89	−16	3.1	5.6
Wiarton A	716330	44.75	81.10	728	14.313	6193	−5	0	25	22	19	27	27	24	25	7	340	12	230	88	−15	2.3	7.0
Windsor A	715380	42.27	82.97	623	14.367	6193	2	6	28	24	21	30	22	26	22	12	230	12	250	94	−5	2.7	5.4
PRINCE EDWARD ISLAND																							
Charlottetown A	717060	46.28	63.13	177	14.602	6193	−6	−2	24	22	20	35	17	29	18	13	270	12	230	85	−12	2.3	4.7
Summerside A	717020	46.43	63.83	79	14.654	6193	−5	−1	32	28	25	40	18	35	18	15	270	13	200	85	−11	2.7	4.1
QUEBEC																							
Bagotville A	717270	48.33	71.00	522	14.421	6193	−23	−19	26	23	20	30	3	26	5	6	270	10	270	91	−30	2.5	5.0
Baie Comeau A	711870	49.13	68.20	72	14.657	6593	−19	−14	27	24	21	30	14	25	14	11	270	13	230	82	−28	3.2	6.8
Grindstone Island		47.38	61.87	194	14.593	6193	−1	3	49	43	39	55	20	49	21	24	290	18	250	79	−5	2.0	5.6
Kuujjuarapik A	719050	55.28	77.77	39	14.675	6193	−33	−30	29	25	22	26	1	24	−2	8	120	13	180	85	−43	4.0	4.7
Kuujuaq A	719060	58.10	68.42	121	14.631	6193	−34	−31	29	25	21	31	2	27	−2	5	230	11	180	82	−41	3.4	3.6
La Grande Riviere A	718270	53.63	77.70	640	14.359	7793	−33	−30	22	20	17	23	0	20	−2	6	270	13	240	85	−39	3.6	3.4
Lake Eon A	714210	51.87	63.28	1841	13.744	6193	−31	−27	23	21	18	23	4	21	0	5	270	10	230	80	−42	2.5	4.0
Mont Joli A	717180	48.60	68.22	171	14.605	6193	−12	−8	28	25	22	35	8	30	9	15	290	14	230	87	−18	3.1	4.5
Montreal Intl A	716270	45.47	73.75	118	14.633	6193	−12	−7	23	20	18	30	19	26	18	7	250	11	230	90	−19	2.3	4.5
Montreal Mirabel A	716278	45.68	74.03	269	14.553	7693	−16	−11	21	18	16	25	13	21	11	6	240	9	240	88	−25	1.8	4.3
Nitchequon		53.20	70.90	1759	13.785	6193	−33	−31	25	22	20	29	−6	25	−5	6	270	11	230	78	−46	3.6	4.7
Quebec A	717140	46.80	71.38	240	14.569	6193	−16	−11	24	21	19	30	13	26	11	10	250	12	250	89	−23	2.3	4.9
Riviere Du Loup	717150	47.80	69.55	486	14.439	6693	−13	−10	21	19	17	24	14	21	13	10	180	11	230	85	−18	2.0	4.7
Roberval A	717280	48.52	72.27	587	14.386	6193	−23	−19	23	20	18	27	8	23	10	7	270	12	220	90	−30	3.1	4.3
Schefferville A	718280	54.80	66.82	1709	13.810	6193	−33	−31	26	23	21	30	−10	26	−9	7	320	12	270	81	−43	4.0	5.0
Sept–Iles A	718110	50.22	66.27	180	14.600	6893	−20	−15	27	23	20	30	16	25	13	8	300	11	220	82	−26	4.3	4.3
Sherbrooke A	716100	45.43	71.68	791	14.280	6393	−20	−14	19	17	15	22	15	19	14	5	110	9	250	88	−30	2.2	5.0
St. Hubert A	713710	45.52	73.42	89	14.648	6193	−12	−7	27	23	20	30	18	27	19	7	20	13	250	91	−20	2.5	4.9
Ste. Agathe Des Monts	717200	46.05	74.28	1296	14.020	6693	−19	−15	19	17	15	22	12	19	11	4	290	9	270	86	−28	2.9	4.0
Val d'Or A	717250	48.07	77.78	1106	14.118	6193	−27	−22	21	18	16	22	12	20	10	5	310	11	230	88	−37	2.9	4.9
SASKATCHEWAN																							
Broadview	718610	50.38	102.68	1975	13.676	6693	−30	−25	26	23	20	28	9	25	9	8	290	12	160	94	−37	3.6	5.8
Estevan A	718620	49.22	102.97	1906	13.711	6193	−25	−21	30	26	23	32	12	28	12	10	290	14	180	97	−32	3.2	5.4
Moose Jaw A	718640	50.33	105.55	1893	13.718	6193	−26	−21	30	26	23	35	21	30	19	9	290	13	180	97	−33	3.2	5.4
North Battleford A	718760	52.77	108.25	1798	13.765	6193	−31	−26	25	22	20	26	8	22	7	4	320	11	140	93	−38	3.1	5.4
Prince Albert A	718690	53.22	105.68	1404	13.965	6193	−34	−29	23	20	18	24	2	21	3	3	270	11	180	91	−43	3.8	5.9
Regina A	718630	50.43	104.67	1893	13.718	6193	−29	−24	30	27	23	34	8	29	10	9	270	14	180	96	−36	3.6	5.0
Saskatoon	718660	52.17	106.68	1654	13.838	6193	−31	−26	27	23	20	27	12	23	6	7	290	13	180	94	−37	3.8	5.9
Swift Current A	718700	50.28	107.68	2684	13.325	6193	−25	−21	33	28	25	36	17	30	14	14	270	13	180	95	−33	3.1	5.8
Uranium City A	710760	59.57	108.48	1043	14.150	6393	−38	−35	20	18	16	18	−7	16	−9	4	70	7	230	86	−49	4.0	3.4
Wynyard	718650	51.77	104.20	1841	13.744	6593	−29	−25	26	23	21	27	13	24	11	8	270	13	180	92	−35	3.4	6.1
Yorkton A	711380	51.27	102.47	1634	13.848	6193	−30	−26	25	22	20	27	6	23	2	7	290	12	180	93	−37	3.2	5.6
YUKON TERRITORY																							
Burwash A	719670	61.37	139.05	2644	13.344	6793	−34	−32	28	24	21	29	31	25	31	0	290	9	110	79	−55	3.4	6.5
Whitehorse A	719640	60.72	135.07	2306	13.511	6193	−34	−31	23	20	18	28	17	24	15	2	340	8	140	83	−47	4.0	5.9

WMO# = World Meteorological Organization number Elev. = elevation, ft PWD = prevailing wind direction, ° True DB = dry-bulb temperature, °F
Lat. = North latitude, ° Long. = West longitude, ° StdP = standard pressure at station elevation, psia WS = wind speed, mph

Table 2B Cooling and Dehumidification Design Conditions—Canada

Station	Cooling DB/MWB						Evaporation WB/MDB						Dehumidification DP/MDB and HR									Range of DB
	0.4%		1%		2%		0.4%		1%		2%		0.4%			1%			2%			
	DB	MWB	DB	MWB	DB	MWB	WB	MDB	WB	MDB	WB	MDB	DP	HR	MDB	DP	HR	MDB	DP	HR	MDB	
1	2a	2b	2c	2d	2e	2f	3a	3b	3c	3d	3e	3f	4a	4b	4c	4d	4e	4f	4g	4h	4i	5
Iqualuit A (Frobisher)	60	50	57	48	53	46	51	59	49	55	47	53	46	46	53	44	43	52	42	40	50	12.4
Resolute A	50	45	48	43	44	41	46	50	43	47	41	44	42	39	47	40	36	45	38	34	43	8.5
NOVA SCOTIA																						
Greenwood A	84	69	80	67	78	66	72	79	70	77	69	75	69	108	76	68	102	74	66	96	72	20.0
Halifax Intl A	80	68	78	66	75	64	70	77	69	74	67	72	68	106	73	67	99	71	65	95	69	16.7
Sable Island	70	67	69	66	67	65	68	69	67	68	66	67	68	101	69	66	97	68	65	92	66	8.3
Shearwater A	78	66	75	64	72	63	69	74	67	72	66	69	68	102	71	66	96	69	65	92	68	13.1
Sydney A	81	68	78	67	75	65	71	78	69	75	67	72	68	105	74	67	99	72	65	93	70	17.3
Truro	79	69	77	67	75	66	71	77	69	75	68	72	69	106	75	68	102	73	66	97	70	19.3
Yarmouth A	73	66	71	64	69	63	68	71	66	69	64	67	67	98	69	65	93	68	63	86	66	13.1
ONTARIO																						
Armstrong A	81	66	78	65	75	63	69	78	67	75	65	73	67	102	73	64	94	71	61	85	69	24.7
Atikokan	84	67	80	66	77	64	71	80	69	77	67	74	68	109	75	66	102	73	64	95	71	23.0
Big Trout Lake	79	64	75	62	72	61	67	75	65	72	63	70	64	91	70	62	84	68	60	78	67	16.4
Earlton A	85	69	81	67	78	65	71	80	69	77	67	75	69	108	76	67	101	74	65	95	72	21.6
Geraldton	81	66	78	65	75	63	69	77	67	75	65	73	67	103	74	65	97	72	62	87	69	22.1
Gore Bay A	80	68	78	67	75	65	71	77	69	75	68	73	69	108	74	67	102	72	66	97	71	16.4
Kapuskasing A	84	67	80	65	77	64	70	79	68	77	66	74	67	102	75	65	95	72	62	87	70	22.5
Kenora A	84	67	81	65	78	64	70	79	68	77	66	75	68	106	75	66	99	73	63	92	72	16.4
London A	85	71	83	70	80	69	74	82	72	80	71	77	71	119	79	70	113	77	68	108	75	19.8
Mount Forest	83	70	80	68	77	67	72	80	70	77	69	75	70	115	77	68	109	74	67	103	73	20.3
Muskoka A	84	69	80	68	78	66	72	80	70	77	68	75	70	113	76	68	106	74	66	100	72	20.7
North Bay A	81	67	78	66	76	64	70	77	68	75	67	73	68	108	74	66	102	72	65	96	70	17.1
Ottawa Int'l A	86	70	83	69	80	67	73	82	71	80	69	78	70	112	78	68	106	76	67	99	75	18.5
Sault Ste. Marie A	83	69	79	67	76	65	71	79	69	76	67	74	69	108	76	67	102	73	65	95	71	20.9
Simcoe	85	72	83	70	80	69	74	83	72	80	71	77	71	118	79	70	112	77	68	107	75	19.3
Sioux Lookout A	84	67	80	65	78	63	70	80	68	76	66	74	67	104	74	65	97	71	63	90	70	18.9
Sudbury A	84	67	81	66	78	64	70	80	68	77	66	74	68	105	74	66	99	72	64	92	71	19.1
Thunder Bay A	84	68	80	66	77	64	70	80	68	77	66	73	67	101	76	65	93	72	62	86	70	21.8
Timmins A	84	67	81	65	78	64	70	80	68	77	66	74	67	102	74	65	95	72	63	88	70	23.0
Toronto Int'l A	87	71	84	70	81	68	74	83	72	80	70	78	71	116	79	69	110	77	68	104	75	20.2
Trenton A	84	71	81	70	78	68	74	81	72	78	70	76	71	116	78	70	110	76	68	104	74	18.0
Wiarton A	82	70	80	69	77	67	72	80	70	77	69	75	70	113	77	68	106	75	67	101	73	18.0
Windsor A	89	73	86	71	83	70	76	85	74	83	73	80	73	125	82	71	118	79	70	112	77	17.5
PRINCE EDWARD ISLAND																						
Charlottetown A	79	69	77	67	74	65	71	77	69	74	67	72	68	105	75	67	99	72	65	94	71	15.1
Summerside A	79	68	77	66	74	65	71	76	69	74	67	72	68	104	74	67	99	72	65	94	70	14.4
QUEBEC																						
Bagotville A	84	67	80	65	77	63	69	79	68	76	66	74	66	99	74	65	93	72	62	85	70	19.8
Baie Comeau A	75	63	71	61	69	60	65	71	63	69	61	67	62	85	68	61	79	66	59	75	64	17.1
Grindstone Island	73	66	70	65	68	64	68	71	67	69	65	68	67	100	70	66	95	69	64	90	68	8.6
Kuujjuarapik A	75	61	70	59	66	57	63	72	61	68	58	65	60	76	67	57	69	64	54	62	62	15.8
Kuujuaq A	74	60	69	57	65	55	62	69	59	66	56	64	58	72	65	55	65	63	52	58	60	18.7
La Grande Riviere A	78	62	75	60	71	58	65	73	62	70	60	68	62	83	68	59	77	65	57	70	63	21.2
Lake Eon A	74	60	70	58	67	57	63	70	61	67	59	64	60	83	65	58	78	63	57	74	62	16.4
Mont Joli A	80	67	76	65	74	64	69	77	67	75	65	72	66	97	75	64	90	72	62	83	70	16.4
Montreal Intl A	85	71	83	70	80	68	74	82	72	80	70	77	70	112	79	69	106	77	67	101	75	17.6
Montreal Mirabel A	84	71	81	69	78	67	73	81	71	78	69	76	70	110	78	68	103	75	66	98	73	20.2
Nitchequon	72	60	69	58	66	57	63	68	61	65	59	64	61	85	65	59	79	63	57	74	62	14.0
Quebec A	84	70	80	68	78	66	73	80	70	77	68	75	70	111	77	68	103	75	66	95	73	19.1
Riviere Du Loup	79	68	76	67	74	65	70	76	69	74	67	72	68	106	74	67	99	72	65	93	70	15.5
Roberval A	83	68	79	66	76	65	71	79	69	77	67	74	68	105	76	66	98	74	64	92	72	17.8
Schefferville A	74	58	69	57	66	55	61	69	59	66	57	64	58	76	64	56	71	62	53	64	60	16.0
Sept-Iles A	72	60	69	59	67	58	64	69	62	67	60	65	61	82	66	60	77	64	58	73	62	13.9
Sherbrooke A	84	70	80	68	78	66	72	81	70	78	68	75	69	109	77	67	103	75	66	97	73	22.3
St. Hubert A	86	71	83	69	80	68	74	83	72	80	70	78	71	113	78	69	106	77	67	101	75	19.1
Ste. Agathe Des Monts	81	68	78	66	76	65	71	77	69	75	67	73	69	110	75	67	103	72	65	97	70	19.4
Val d'Or A	83	67	80	65	77	63	69	79	68	76	66	73	67	102	73	65	95	72	62	88	70	20.7
SASKATCHEWAN																						
Broadview	87	65	83	64	79	63	69	81	66	79	64	76	64	97	74	62	89	71	60	83	70	23.8
Estevan A	90	66	86	65	83	64	70	83	68	81	66	79	66	103	76	63	92	73	61	86	71	23.8
Moose Jaw A	90	64	87	64	83	62	68	83	66	81	64	79	63	92	73	61	85	71	59	79	69	23.8
North Battleford A	86	64	82	63	78	61	67	80	65	78	63	76	62	88	71	60	83	69	58	77	68	21.1
Prince Albert A	84	65	81	64	78	62	68	80	65	77	64	75	63	90	74	61	84	71	59	78	69	21.8
Regina A	89	64	85	64	82	62	68	82	66	80	64	78	63	92	75	61	85	71	59	79	69	23.6
Saskatoon	87	64	84	63	80	62	67	81	65	79	63	77	62	88	73	60	82	70	58	76	68	22.7
Swift Current A	88	63	84	62	80	61	66	81	64	79	62	77	61	88	71	59	82	68	57	76	67	23.0
Uranium City A	79	62	76	60	72	58	64	76	62	72	60	69	59	78	68	58	74	66	56	69	64	17.1
Wynyard	85	64	81	63	78	62	68	79	65	77	63	75	64	94	74	61	85	70	59	81	68	20.7
Yorkton A	86	65	82	64	79	62	68	80	66	78	64	75	64	96	74	62	87	71	60	81	69	22.0
YUKON TERRITORY																						
Burwash A	73	57	69	55	66	53	57	71	56	68	54	64	52	62	60	50	58	58	48	55	57	21.4
Whitehorse A	77	57	73	55	69	53	58	73	56	70	55	67	52	62	61	50	58	59	48	55	58	20.9

MDB = mean coincident dry-bulb temperature, °F
MWB = mean coincident wet-bulb temperature, °F
MWS = mean coincident wind speed, mph
StdD = standard deviation, °F
HR = humidity ratio, grains of moisture per lb of dry air
A = airport
DP = dew-point temperature, °F

Table 3A Heating and Wind Design Conditions—World Locations

Station	WMO#	Lat.	Long.	Elev., ft	StdP, psia	Dates	Heating Dry Bulb 99.6%	Heating Dry Bulb 99%	Extreme Wind Speed, mph 1%	Extreme Wind Speed, mph 2.5%	Extreme Wind Speed, mph 5%	Coldest Month 0.4% WS	Coldest Month 0.4% MDB	Coldest Month 1% WS	Coldest Month 1% MDB	MWS/PWD to DB 99.6% MWS	MWS/PWD to DB 99.6% PWD	MWS/PWD to DB 0.4% MWS	MWS/PWD to DB 0.4% PWD	Extr. Annual Daily Mean DB Max.	Extr. Annual Daily Mean DB Min.	Extr. Annual Daily StdD DB Max.	Extr. Annual Daily StdD DB Min.
1a	1b	1c	1d	1e	1f	1g	2a	2b	3a	3b	3c	4a	4b	4c	4d	5a	5b	5c	5d	6a	6b	6c	6d
ALGERIA																							
Algiers	603900	36.72N	3.25E	82	14.656	8293	36	38	25	21	19	26	55	22	55	2	200	12	60	105	32	4.5	2.5
Annaba	603600	36.83N	7.82E	13	14.693	8293	39	41	25	22	20	27	54	22	53	5	220	11	240	106	34	5.8	2.2
Biskra	605250	34.80N	5.73E	285	14.549	8293	41	43	34	30	26	33	56	29	58	7	10	11	180	113	36	2.7	2.3
Constantine	604190	36.28N	6.62E	2276	13.530	8293	31	33	24	20	17	26	46	22	46	2	320	12	240	103	27	2.5	1.4
El Golea	605900	30.57N	2.87E	1302	14.021	8293	33	35	26	22	19	26	58	22	57	4	320	8	180	113	28	2.3	2.7
Oran	604900	35.63N	0.60W	295	14.544	8293	35	38	29	24	21	33	57	26	58	2	200	14	240	104	28	8.8	12.4
Tebessa	604750	35.48N	8.13E	2667	13.337	8293	29	31	26	22	19	30	44	23	49	1	280	11	220	103	25	3.1	2.0
ARGENTINA																							
Buenos Aires	875760	34.82S	58.53W	65	14.665	8293	31	34	23	20	18	22	53	19	53	4	270	11	270	98	27	2.5	2.0
Comodoro Rivadavia	878600	45.78S	67.50W	150	14.620	8293	29	31	43	37	33	43	51	36	50	11	270	22	290	97	26	7.2	2.0
Cordoba	873440	31.32S	64.22W	1555	13.892	8293	31	34	27	24	21	28	66	24	60	3	270	16	20	100	25	3.1	3.2
Junin Airport	875480	34.55S	60.95W	266	14.559	8293	30	32	26	23	20	26	55	23	55	1	270	12	360	99	26	5.9	2.0
Formosa	871620	26.20S	58.23W	196	14.596	8293	40	44	30	26	23	28	67	26	64	6	270	17	360	104	34	1.8	3.2
Marcos Juarez	874670	32.70S	62.15W	374	14.502	8293	29	32	28	24	22	27	57	24	55	3	50	11	360	101	24	3.4	3.8
Mendoza	874180	32.83S	68.78W	2309	13.514	8293	30	33	22	18	15	19	56	15	53	3	230	11	50	102	26	2.7	1.8
Paso De Los Libres	872890	29.68S	57.15W	229	14.578	8293	36	39	31	26	24	31	53	27	57	1	180	13	360	101	33	2.0	2.3
Posadas	871780	27.37S	55.97W	410	14.483	8293	39	43	22	19	17	23	71	21	66	6	180	11	360	101	34	2.2	4.7
Reconquista	872700	29.18S	59.67W	173	14.608	8293	37	40	27	21	20	27	60	21	58	2	200	12	50	103	32	4.9	2.5
Resistencia	871550	27.45S	59.05W	170	14.610	8293	35	39	20	17	15	19	69	16	64	2	50	11	20	103	31	2.3	3.4
Rio Gallegos	879250	51.62S	69.28W	62	14.667	8293	17	22	52	46	40	40	37	33	39	6	270	20	320	90	12	12.4	7.9
Rosario	874800	32.92S	60.78W	82	14.656	8293	30	33	30	26	23	28	53	25	53	2	180	13	360	98	26	2.9	2.7
Salta Airport	870470	24.85S	65.48W	3990	12.701	8293	30	33	19	16	14	22	78	17	63	2	270	10	70	97	25	3.1	2.5
San Juan	873110	31.57S	68.42W	1961	13.687	8293	28	31	32	28	24	28	52	24	52	1	360	11	180	107	23	2.5	2.9
San Miguel De Tucuman	871210	26.85S	65.10W	1476	13.932	8293	37	40	22	18	14	18	58	14	56	5	360	12	90	102	33	3.4	2.7
ARMENIA																							
Yerevan	377890	40.13N	44.47E	2919	13.214	8293	7	11	22	16	14	14	41	10	32	1	180	6	210	101	3	6.3	6.8
ASCENSION ISLAND																							
Georgetown	619020	7.97S	14.40W	259	14.563	8293	69	70	26	24	23	25	76	24	76	16	90	19	120	87	65	1.8	2.9
AUSTRALIA																							
Adelaide	946720	34.93S	138.52E	13	14.693	8293	39	41	27	24	22	26	53	23	54	2	50	13	310	104	35	3.1	2.2
Alice Springs	943260	23.80S	133.90E	1774	13.781	8293	34	36	20	17	16	17	64	15	62	2	270	8	100	108	29	1.8	3.6
Brisbane	945780	27.38S	153.10E	16	14.691	8293	44	46	22	19	17	22	60	19	61	4	220	11	20	95	39	3.8	2.0
Cairns	942870	16.88S	145.75E	22	14.688	8293	56	59	19	17	15	18	73	16	72	8	170	8	120	97	50	2.5	8.8
Canberra	949260	35.30S	149.18E	1893	13.722	8293	26	29	24	21	18	24	46	22	48	0	310	12	310	97	20	4.0	8.1
Darwin	941200	12.40S	130.87E	98	14.648	8293	64	66	19	17	15	18	80	17	80	7	140	12	290	98	60	2.9	2.5
Kalgoorlie/Boulder	946370	30.77S	121.45E	1181	14.083	8293	36	38	22	19	17	24	59	21	58	1	220	9	320	108	31	3.4	1.3
Learmonth	943020	22.23S	114.08E	19	14.690	8293	49	51	25	23	21	22	68	20	67	5	210	14	210	112	45	2.7	2.9
Perth	946100	31.93S	115.95E	95	14.650	8293	41	43	24	21	19	23	58	19	58	1	50	10	270	107	36	3.4	2.2
Port Hedland	943120	20.37S	118.62E	19	14.690	8293	51	54	23	20	19	24	68	22	70	5	160	12	120	111	46	2.5	2.2
Sydney Intl Airport	947670	33.95S	151.18E	10	14.695	8293	42	44	25	22	20	25	58	20	56	2	320	12	300	103	38	5.2	3.4
Townsville	942940	19.25S	146.75E	19	14.690	8293	48	52	21	19	17	20	72	18	72	0	190	9	50	101	43	3.8	2.7
AUSTRIA																							
Aigen/Ennstal (Mil)	111570	47.53N	14.13E	2129	13.604	8293	2	7	19	16	14	22	38	19	34	1	60	7	60	89	−6	2.7	6.3
Graz	112400	47.00N	15.43E	1138	14.105	8293	5	12	17	14	11	16	35	11	33	1	180	7	140	90	1	3.4	8.1
Innsbruck	111200	47.27N	11.35E	1945	13.695	8293	10	14	19	15	12	18	42	15	39	1	260	8	70	91	5	2.7	6.8
Klagenfurt	112310	46.65N	14.33E	1482	13.929	8293	4	9	13	11	9	10	31	7	27	3	310	6	100	90	−1	3.8	7.0
Linz	110100	48.23N	14.20E	1026	14.163	8293	5	12	24	20	17	28	39	25	38	6	90	7	110	91	1	3.8	9.9
Salzburg	111500	47.80N	13.00E	1476	13.932	8293	7	12	17	15	13	20	42	17	38	3	130	7	330	92	2	4.1	8.3
Vienna, Hohe Warte	110350	48.25N	16.37E	656	14.355	8293	12	17	23	19	17	27	43	23	42	6	240	9	140	91	9	3.2	6.7
Vienna, Schwechat	110360	48.12N	16.57E	623	14.372	8293	9	14	27	24	21	31	44	27	38	6	320	12	150	92	5	3.1	7.2
Zeltweg	111650	47.20N	14.75E	2237	13.550	8293	0	5	18	15	13	19	37	16	36	1	250	7	190	88	−6	3.4	7.6
AZORES																							
Lajes	85090	38.77N	27.10W	180	14.604	8293	46	48	28	23	21	30	56	26	57	3	300	8	250	83	41	1.8	3.8
BAHAMAS																							
Nassau	780730	25.05N	77.47W	22	14.688	8293	57	60	21	19	17	21	72	19	71	3	300	10	130	93	52	1.3	3.1
BAHRAIN																							
Al-Manamah	411500	26.27N	50.65E	6	14.697	8293	52	54	25	23	21	26	56	24	58	13	290	11	340	109	47	2.5	5.6
BELARUS																							
Babruysk (Bobruysk)	269610	53.12N	29.25E	541	14.415	8293	−9	−2	20	18	16	20	27	18	30	4	210	8	200	87	−16	4.0	21.1
Homyel (Gomel')	330410	52.45N	31.00E	416	14.480	8293	−6	0	18	15	14	17	28	15	28	4	330	7	150	87	−8	4.5	7.9
Hrodna (Grodno)	268250	53.68N	23.83E	442	14.466	8293	−5	1	26	22	19	25	33	22	32	5	270	9	180	87	−5	3.8	9.2
Mahilyow (Mogilev)	268630	53.90N	30.32E	633	14.367	8293	−9	−3	23	21	18	23	31	21	28	7	30	9	200	85	−11	3.4	7.6
Minsk	268500	53.87N	27.53E	767	14.297	8293	−5	0	17	14	13	17	23	15	24	5	300	9	70	85	−8	3.8	7.2
Vitsyebsk (Vitebsk)	266660	55.17N	30.13E	577	14.396	8293	−8	−2	19	17	15	21	26	17	28	3	30	8	210	84	−11	3.4	6.3
BELGIUM/LUXEMBOURG																							
Antwerp	64500	51.20N	4.47E	45	14.676	8293	16	21	24	21	18	28	47	24	44	7	50	7	90	89	15	4.1	7.2

WMO# = World Meteorological Organization number
Lat. = latitude, °
Elev. = elevation, ft
Long. = longitude, °
PWD = prevailing wind direction, ° True
StdP = standard pressure at station elevation, psia
DB = dry-bulb temperature, °F
WS = wind speed, mph

Table 3B Cooling and Dehumidification Design Conditions—World Locations

Station	Cooling DB/MWB						Evaporation WB/MDB						Dehumidification DP/MDB and HR									Range of DB
	0.4%		1%		2%		0.4%		1%		2%		0.4%			1%			2%			
	DB	MWB	DB	MWB	DB	MWB	WB	MDB	WB	MDB	WB	MDB	DP	HR	MDB	DP	HR	MDB	DP	HR	MDB	
1	2a	2b	2c	2d	2e	2f	3a	3b	3c	3d	3e	3f	4a	4b	4c	4d	4e	4f	4g	4h	4i	5
ALGERIA																						
Algiers	95	71	92	71	89	72	77	87	76	85	75	83	74	129	82	73	125	81	72	118	80	20.9
Annaba	94	71	90	72	87	73	78	84	77	83	76	83	76	134	82	75	130	81	73	125	80	17.1
Biskra	108	69	106	69	103	69	72	97	71	97	71	96	65	92	83	63	86	82	61	82	83	20.0
Constantine	100	66	96	67	93	66	71	88	70	87	69	86	66	106	76	65	100	76	63	94	75	27.0
El Golea	109	69	107	68	104	68	72	101	71	101	69	99	62	87	82	60	80	84	58	76	85	25.7
Oran	92	69	89	69	86	69	75	84	74	83	73	81	73	123	80	72	119	79	71	115	79	18.4
Tebessa	99	65	97	64	94	64	69	89	68	87	67	86	64	97	73	62	92	72	61	88	72	27.2
ARGENTINA																						
Buenos Aires	93	73	90	72	87	71	76	86	75	84	74	83	74	126	80	72	120	79	71	115	78	21.6
Comodoro Rivadavia	87	61	84	60	80	58	63	82	62	78	60	76	57	70	66	55	65	64	53	60	65	18.7
Cordoba	94	73	91	72	89	71	77	88	76	86	74	84	74	136	84	72	127	82	71	120	80	21.1
Junin Airport	92	72	89	71	87	71	76	85	74	84	73	83	73	125	80	72	118	78	70	111	77	21.6
Formosa	98	76	96	76	94	76	80	91	80	90	79	89	78	145	85	77	141	84	76	136	83	18.5
Marcos Juarez	95	74	92	74	90	73	78	88	77	87	75	85	76	136	85	74	129	82	73	123	80	22.1
Mendoza	96	68	93	67	91	67	73	89	71	87	70	85	68	112	81	66	106	80	65	100	79	22.1
Paso De Los Libres	96	74	94	75	92	73	79	89	78	88	76	86	76	137	84	75	132	83	74	126	82	20.2
Posadas	97	76	95	76	93	75	80	91	79	90	78	89	77	142	87	76	136	85	75	132	84	19.1
Reconquista	96	78	94	77	91	76	81	91	80	89	79	87	79	150	87	77	144	86	76	138	84	17.8
Resistencia	98	76	95	76	93	76	80	91	79	90	78	88	77	144	86	77	140	85	76	134	83	20.2
Rio Gallegos	76	57	72	55	70	54	59	73	57	70	55	67	53	60	63	51	56	61	49	52	60	19.4
Rosario	93	74	90	73	88	72	78	87	76	85	75	83	75	132	83	74	125	81	72	119	79	20.5
Salta Airport	89	65	87	66	84	66	72	82	71	80	70	79	69	123	76	68	119	76	67	115	74	19.1
San Juan	100	69	97	68	94	68	73	93	72	91	71	89	68	110	82	66	102	82	64	97	81	23.8
San Miguel De Tucuman	96	74	93	74	91	74	79	90	78	88	76	87	76	144	85	75	138	84	74	132	83	17.5
ARMENIA																						
Yerevan	96	69	94	69	91	67	72	91	70	90	69	88	65	104	85	63	95	82	61	90	81	24.5
ASCENSION ISLAND																						
Georgetown	86	75	85	75	85	75	77	84	76	83	76	82	75	132	81	74	130	80	74	126	80	7.7
AUSTRALIA																						
Adelaide	95	64	92	64	88	63	70	83	68	81	66	80	66	97	74	64	89	73	61	81	72	19.4
Alice Springs	104	65	102	64	100	64	73	83	72	83	71	83	71	123	77	69	115	77	67	104	77	24.7
Brisbane	88	72	86	72	84	72	77	84	76	82	75	81	75	133	81	74	126	80	73	121	79	13.7
Cairns	91	78	90	77	88	77	81	87	80	87	79	86	79	150	85	78	146	84	77	141	83	13.1
Canberra	90	63	87	63	83	62	67	79	66	78	65	77	64	96	71	62	90	70	60	84	68	23.9
Darwin	93	75	92	76	91	76	82	89	81	88	81	87	81	160	86	79	152	85	79	151	85	13.0
Kalgoorlie/Boulder	102	65	99	64	95	63	70	87	68	85	67	86	65	98	74	63	90	71	61	84	71	24.5
Learmonth	105	70	102	70	99	70	79	88	78	87	77	87	77	140	84	76	134	83	74	127	83	23.6
Perth	99	67	95	66	91	65	72	87	70	85	69	83	67	101	75	66	95	74	64	91	74	22.5
Port Hedland	105	71	102	71	100	71	82	92	81	91	81	90	80	156	87	79	151	86	78	147	86	19.3
Sydney Intl Airport	90	68	85	67	82	68	73	82	72	79	71	78	71	115	77	70	111	76	69	107	75	12.1
Townsville	92	76	91	76	89	76	81	88	80	87	79	85	79	150	84	78	147	84	77	141	83	11.7
AUSTRIA																						
Aigen/Ennstal (Mil)	82	66	79	64	76	63	67	79	66	76	64	74	63	93	74	61	88	72	60	83	69	20.0
Graz	85	69	82	68	79	65	71	81	69	78	67	77	67	105	75	66	99	73	64	94	71	20.5
Innsbruck	85	65	82	64	79	62	66	81	65	78	64	74	62	91	69	61	86	67	59	81	67	20.5
Klagenfurt	85	66	82	65	79	63	68	81	66	79	65	76	64	92	72	62	88	71	61	84	68	23.0
Linz	85	66	82	65	79	63	69	81	67	77	65	74	65	95	70	63	90	69	62	88	68	19.6
Salzburg	86	67	82	65	79	64	68	82	66	80	65	77	63	91	72	61	85	71	60	83	70	18.9
Vienna, Hohe Warte	86	68	83	67	80	66	70	83	69	80	67	77	66	98	74	65	94	73	64	90	72	16.7
Vienna, Schwechat	86	67	83	66	80	65	69	83	67	79	66	78	64	92	73	63	88	72	62	83	71	19.1
Zeltweg	82	66	79	65	76	63	67	80	66	76	64	74	63	94	74	62	89	72	60	85	69	21.1
AZORES																						
Lajes	80	71	78	71	77	69	72	78	72	77	71	75	71	115	75	70	111	75	69	106	74	11.3
BAHAMAS																						
Nassau	91	79	90	78	89	78	81	87	80	87	80	87	79	151	83	79	150	83	78	144	83	12.4
BAHRAIN																						
Al-Manamah	103	77	101	78	99	79	87	94	86	94	85	93	86	189	93	84	181	92	84	176	92	12.4
BELARUS																						
Babruysk (Bobruysk)	82	65	79	65	76	63	68	78	67	76	65	74	65	94	73	63	88	71	61	83	69	20.7
Homyel (Gomel')	83	66	80	65	77	64	69	78	67	76	65	74	65	95	72	64	89	71	62	84	70	16.2
Hrodna (Grodno)	82	66	78	64	75	63	68	79	66	75	64	73	65	92	72	63	87	71	61	81	69	19.4
Mahilyow (Mogilev)	80	66	77	64	75	63	68	77	66	75	64	72	65	94	72	63	88	70	61	83	68	16.9
Minsk	81	65	78	64	75	62	67	78	66	75	64	73	64	91	71	62	85	69	60	80	69	17.3
Vitsyebsk (Vitebsk)	80	65	77	64	74	63	68	77	66	74	64	72	65	94	72	63	88	70	61	83	68	15.1
BELGIUM/LUXEMBOURG																						
Antwerp	82	68	79	66	76	65	70	80	68	77	66	74	66	97	75	65	91	73	63	85	71	16.2

MDB = mean coincident dry-bulb temp., °F
MWB = mean coincident wet-bulb temp., °F
MWS = mean coincident wind speed, mph
StdD = standard deviation, °F
HR = humidity ratio, grains of moisture per lb of dry air
A = airport
DP = dew-point temperature, °F

Table 3A Heating and Wind Design Conditions—World Locations

Station	WMO#	Lat.	Long.	Elev., ft	StdP, psia	Dates	Heating Dry Bulb		Extreme Wind Speed, mph			Coldest Month				MWS/PWD to DB				Extr. Annual Daily			
												0.4%		1%		99.6%		0.4%		Mean DB		StdD DB	
							99.6%	99%	1%	2.5%	5%	WS	MDB	WS	MDB	MWS	PWD	MWS	PWD	Max.	Min.	Max.	Min.
1a	1b	1c	1d	1e	1f	1g	2a	2b	3a	3b	3c	4a	4b	4c	4d	5a	5b	5c	5d	6a	6b	6c	6d
Brussels	64510	50.90N	4.53E	190	14.599	8293	15	21	27	23	21	31	46	26	45	7	50	8	60	89	15	3.8	8.6
Charleroi	64490	50.47N	4.45E	629	14.368	8293	15	21	25	22	20	29	43	24	43	10	50	8	50	89	14	3.8	7.9
Florennes	64560	50.23N	4.65E	980	14.186	8293	13	19	24	21	19	29	44	24	41	9	70	8	170	87	13	3.6	7.9
Koksijde	64000	51.08N	2.65E	29	14.684	8293	17	22	30	26	23	33	46	29	45	10	90	10	100	87	14	3.2	7.7
Luxembourg	65900	49.62N	6.22E	1243	14.051	8293	13	18	24	21	18	27	41	23	34	11	50	9	80	88	12	3.6	7.2
Oostende	64070	51.20N	2.87E	16	14.691	8293	18	22	34	30	26	40	46	33	45	11	70	10	100	86	15	3.2	6.5
Saint Hubert	64760	50.03N	5.40E	1827	13.755	8293	11	16	22	20	18	26	35	22	31	11	90	8	30	82	10	2.9	8.1
BENIN																							
Cotonou	653440	6.35N	2.38E	29	14.684	8293	71	72	19	17	16	20	80	18	80	5	20	12	200	98	65	4.0	6.7
Parakou	653300	9.35N	2.62E	1289	14.028	8293	65	67	14	12	11	12	76	11	76	4	40	5	40	103	55	5.0	9.2
BERMUDA																							
Hamilton, Bermuda	780160	32.37N	64.68W	9	14.695	8293	55	57	29	26	23	30	65	28	64	16	310	11	190	90	47	1.3	7.6
BOLIVIA																							
Cochabamba	852230	17.45S	66.10W	8303	10.793	8293	35	37	22	18	13	22	66	17	66	0	180	6	360	89	30	2.0	2.3
La Paz	852010	16.52S	68.18W	13169	8.926	8293	25	27	19	17	14	22	52	19	52	2	330	8	60	69	21	4.7	2.0
BOSNIA-HERZEGOVINA																							
Banja Luka	132420	44.78N	17.22E	511	14.430	8293	10	16	13	10	8	14	52	10	50	2	320	4	360	98	6	4.0	7.6
BRAZIL																							
Arch De Fernando	824000	3.85S	32.42W	183	14.603	8293	73	74	18	16	15	19	79	17	79	7	130	10	150	98	63	6.5	15.5
Belem	821930	1.38S	48.48W	52	14.672	8293	72	73	17	14	12	15	83	13	83	2	90	6	90	96	63	3.2	14.0
Brasilia	833780	15.87S	47.93W	3480	12.943	8293	48	51	17	14	12	18	71	15	72	0	90	7	90	94	44	3.2	5.8
Campinas	837210	23.00S	47.13W	2168	13.584	8293	47	50	24	22	21	24	65	23	63	8	150	7	330	96	42	2.9	4.5
Campo Grande	836120	20.47S	54.67W	1824	13.756	8293	46	50	25	22	21	27	60	24	62	16	180	11	360	98	41	3.2	3.8
Caravelas	834970	17.63S	39.25W	13	14.693	8293	61	63	23	20	18	19	75	17	75	1	240	12	60	95	57	4.0	3.2
Curitiba	838400	25.52S	49.17W	2979	13.185	8293	36	40	18	16	13	20	68	18	65	2	270	9	300	92	26	3.4	11.5
Fortaleza	823980	3.78S	38.53W	82	14.656	8293	72	73	20	18	17	19	84	18	84	5	180	13	120	97	65	4.0	9.7
Goiania	834240	16.63S	49.22W	2450	13.444	8293	53	55	18	15	12	16	77	13	78	1	180	4	360	98	42	1.6	10.8
Maceio	829930	9.52S	35.78W	377	14.501	8293	67	68	18	15	13	17	78	14	78	0	180	10	100	95	58	4.0	10.6
Manaus, E. Gomes A	821110	3.03S	60.05W	6	14.697	8293	70	71	12	11	10	12	85	11	85	1	120	5	60	100	59	1.6	14.0
Manaus, Ponta Pelada	823320	3.15S	59.98W	275	14.554	8293	71	73	15	13	12	15	83	13	83	4	30	10	90	98	64	2.0	13.1
Natal	825990	5.92S	35.25W	170	14.610	8293	70	71	21	19	18	22	83	20	82	6	150	11	100	95	62	4.7	9.7
Porto Alegre	839710	30.00S	51.18W	9	14.695	8293	40	43	19	17	14	20	53	17	55	2	240	7	290	100	37	3.1	5.2
Recife	828990	8.07S	34.85W	62	14.667	8293	70	71	17	15	13	19	79	17	79	5	240	12	120	97	66	3.1	4.0
Rio De Janeiro, Galeao	837460	22.82S	43.25W	19	14.690	8293	59	61	19	17	15	17	74	15	73	3	320	7	50	107	50	2.5	9.4
Salvador	832480	12.90S	38.33W	19	14.690	8293	68	70	21	19	17	21	76	20	76	2	180	12	90	95	59	3.1	12.8
Santarem	822440	2.43S	54.72W	236	14.575	8293	72	73	19	18	16	19	81	17	81	6	210	12	90	96	64	3.2	16.0
Sao Paulo	837800	23.62S	46.65W	2634	13.353	8293	48	50	15	13	12	14	66	12	64	4	160	6	330	94	43	2.7	3.8
Vitoria	836490	20.27S	40.28W	13	14.693	8293	61	63	22	19	17	18	77	16	76	1	210	11	30	98	54	2.0	10.3
BULGARIA																							
Botevgrad	156270	42.67N	24.83E	7837	10.987	8293	−7	−2	76	62	53	77	15	64	11	30	320	9	270	66	−8	3.1	6.1
Burgas	156550	42.48N	27.48E	91	14.651	8293	17	21	31	24	21	31	37	26	33	9	270	11	110	94	13	3.1	5.2
Lom	155110	43.82N	23.25E	108	14.643	8293	14	18	28	23	18	31	35	27	40	2	270	4	50	96	9	4.7	7.7
Musala	156150	42.18N	23.58E	9603	10.266	8293	−10	−5	63	45	36	77	19	63	16	17	20	7	320	65	−13	6.8	5.8
Plovdiv	156250	42.13N	24.75E	606	14.380	8293	14	19	30	26	23	32	41	27	41	2	40	6	90	99	6	4.1	7.2
Ruse	155350	43.85N	25.95E	147	14.622	8293	12	16	32	27	21	32	39	27	33	5	50	8	270	101	7	5.4	4.9
Sofia	156140	42.65N	23.38E	1952	13.692	8293	10	14	21	17	14	22	33	18	32	2	360	5	110	94	3	4.9	8.5
Varna	155520	43.20N	27.92E	141	14.625	8293	17	20	32	26	21	40	29	30	33	15	360	9	90	92	13	4.0	4.9
BRUNEI																							
Brunei Intl Airport	963150	4.93N	114.93E	49	14.674	8293	71	72	16	14	12	18	82	16	82	0	220	8	320	96	67	2.5	4.7
CANARY ISLANDS																							
Las Palmas	600300	27.93N	15.38W	82	14.656	8293	55	57	32	30	28	27	65	24	65	4	320	19	30	94	51	4.3	0.9
Santa Cruz De Tenerife	600250	28.05N	16.57W	236	14.575	8293	57	57	29	27	24	28	68	25	68	8	360	19	60	101	51	5.4	6.5
CAPE VERDE																							
Sal Island	85940	16.73N	22.95W	180	14.604	8293	63	64	27	25	24	28	72	26	72	10	30	15	60	91	58	1.8	4.3
CHILE																							
Antofagasta	854420	23.43S	70.43W	393	14.492	8293	50	52	22	20	19	23	60	20	59	6	90	15	190	80	44	2.7	3.6
Arica	854060	18.33S	70.33W	193	14.597	8293	52	55	20	18	16	17	64	16	64	4	90	15	210	85	46	2.9	4.3
Concepcion	856820	36.77S	73.05W	52	14.672	8293	35	37	30	26	23	36	54	29	53	4	140	19	240	92	30	15.5	1.4
Iquique	854180	20.53S	70.18W	170	14.610	8293	54	55	22	20	18	20	64	18	63	6	160	14	210	84	50	2.5	2.3
La Serena	854880	29.92S	71.20W	479	14.447	8293	42	44	18	16	15	19	55	15	56	6	120	13	270	81	35	11.5	6.3
Puerto Montt	857990	41.42S	73.08W	282	14.551	8293	28	30	27	23	20	29	49	26	48	2	280	10	150	80	24	3.8	2.3
Punta Arenas	859340	53.00S	70.85W	121	14.636	8293	23	25	45	39	32	41	43	30	38	6	290	18	270	76	17	6.5	4.3
Santiago	855740	33.38S	70.78W	1561	13.889	8293	29	32	19	16	14	16	49	12	50	2	20	12	210	101	25	9.2	2.3
Temuco	857430	38.75S	72.63W	393	14.492	8293	29	31	22	19	15	29	51	23	52	2	70	10	240	91	25	8.5	1.8
CHINA																							
Anda	508540	46.38N	125.32E	492	14.440	8293	−19	−16	24	21	17	18	7	15	6	3	250	12	250	92	−23	3.1	4.3
Andirlangar	518480	37.93N	83.65E	4146	12.627	8293	−1	2	15	12	11	10	32	8	29	1	200	6	40	103	−5	2.3	3.2

WMO# = World Meteorological Organization number
Lat. = latitude, °
Elev. = elevation, ft
Long. = longitude, °
PWD = prevailing wind direction, ° True
StdP = standard pressure at station elevation, psia
DB = dry-bulb temperature, °F
WS = wind speed, mph

Table 3B Cooling and Dehumidification Design Conditions—World Locations

Station	Cooling DB/MWB						Evaporation WB/MDB						Dehumidification DP/MDB and HR									Range of DB
	0.4%		1%		2%		0.4%		1%		2%		0.4%			1%			2%			
	DB	MWB	DB	MWB	DB	MWB	WB	MDB	WB	MDB	WB	MDB	DP	HR	MDB	DP	HR	MDB	DP	HR	MDB	
1	2a	2b	2c	2d	2e	2f	3a	3b	3c	3d	3e	3f	4a	4b	4c	4d	4e	4f	4g	4h	4i	5
Brussels	82	67	79	66	76	65	69	80	68	77	66	74	66	95	74	64	90	72	62	85	71	16.9
Charleroi	82	67	79	65	76	64	69	79	67	76	65	73	65	94	74	63	89	71	62	85	69	16.7
Florennes	80	66	77	65	74	63	68	78	66	75	65	72	65	97	73	63	90	70	62	85	68	16.2
Koksijde	79	67	75	65	72	64	68	76	66	73	65	71	65	93	72	64	88	70	62	83	68	15.1
Luxembourg	82	65	79	63	76	62	67	78	65	76	64	72	64	92	71	62	87	68	60	82	67	17.1
Oostende	77	66	73	65	70	63	67	74	66	72	64	70	65	92	71	64	88	69	62	83	67	14.0
Saint Hubert	77	64	74	62	72	61	66	74	64	72	62	69	63	91	69	61	85	67	59	80	65	14.6
BENIN																						
Cotonou	90	79	89	80	89	80	82	88	81	88	81	87	80	156	85	80	153	85	79	151	85	8.6
Parakou	98	71	97	71	95	71	78	91	78	90	77	89	75	138	84	74	135	84	74	132	83	20.7
BERMUDA																						
Hamilton, Bermuda	88	78	88	78	86	77	80	85	80	85	79	84	79	150	84	78	144	83	77	141	83	8.3
BOLIVIA																						
Cochabamba	84	56	83	56	81	56	61	76	60	76	59	74	57	97	64	56	91	62	55	90	62	27.0
La Paz	63	44	62	44	61	43	49	58	48	57	47	56	45	73	51	44	71	50	43	68	49	22.7
BOSNIA-HERZEGOVINA																						
Banja Luka	92	69	88	69	85	67	72	85	71	83	69	81	68	106	79	66	99	76	65	94	73	22.9
BRAZIL																						
Arch De Fernando	88	78	87	78	86	77	80	85	79	85	79	85	78	147	84	77	144	83	77	141	83	8.1
Belem	92	78	90	78	90	78	81	87	80	87	80	87	79	150	84	79	149	84	77	143	83	14.8
Brasilia	89	65	88	65	86	65	71	80	70	79	70	79	69	120	73	68	118	72	68	116	72	23.4
Campinas	92	73	90	71	88	72	76	85	75	84	74	83	74	136	79	73	134	78	72	128	77	17.8
Campo Grande	95	71	93	71	92	71	77	87	76	86	76	86	75	141	80	74	134	80	73	132	79	18.4
Caravelas	89	77	88	77	87	77	79	87	78	85	78	85	77	142	81	77	140	81	76	137	80	13.7
Curitiba	87	69	85	68	83	68	73	81	72	80	71	78	71	128	75	70	123	74	69	119	74	17.5
Fortaleza	90	78	90	78	89	78	80	87	80	87	79	86	79	151	81	79	149	81	77	143	81	11.2
Goiania	93	67	92	68	91	68	75	85	74	84	73	84	72	129	78	71	125	77	70	123	76	23.0
Maceio	90	75	89	75	88	75	78	85	78	84	77	84	77	141	81	76	136	80	75	134	80	14.2
Manaus, E. Gomes A	97	78	95	78	94	78	83	90	82	89	81	88	81	161	86	80	158	85	79	151	83	20.3
Manaus, Ponta Pelada	94	77	93	78	92	78	81	89	80	88	80	88	79	153	84	79	151	84	78	146	83	14.2
Natal	90	78	90	77	89	77	80	87	79	86	79	86	78	148	82	77	142	82	77	141	82	12.2
Porto Alegre	95	76	92	75	90	74	79	89	77	87	76	86	76	136	83	75	132	82	74	126	80	17.1
Recife	92	78	91	78	90	78	80	89	79	88	79	88	77	143	83	77	142	83	77	140	83	11.3
Rio De Janeiro, Galeao	102	79	99	77	97	77	82	95	81	94	80	91	79	150	86	77	142	84	77	141	84	19.3
Salvador	90	78	88	78	88	78	80	87	80	86	79	85	79	150	85	77	142	84	77	141	84	10.8
Santarem	93	77	92	77	91	78	80	89	80	88	79	88	78	146	83	77	144	82	77	143	82	14.0
Sao Paulo	89	69	88	69	86	69	73	81	72	81	71	80	71	127	77	70	122	75	69	120	75	14.9
Vitoria	93	77	92	77	90	77	80	88	79	87	78	86	78	148	83	77	141	81	76	137	82	14.6
BULGARIA																						
Botevgrad	60	49	57	48	55	47	52	57	50	54	49	53	50	72	53	49	68	52	47	64	50	7.7
Burgas	87	72	84	71	82	69	74	82	73	81	71	79	72	118	79	70	111	77	68	104	76	20.0
Lom	90	74	87	72	85	71	75	88	73	85	72	83	71	114	83	69	108	81	68	102	79	19.1
Musala	56	44	53	43	51	42	47	53	45	51	44	49	44	62	48	43	58	47	41	55	46	9.2
Plovdiv	93	70	90	70	87	69	73	89	72	87	70	84	68	105	83	66	99	80	65	94	78	21.4
Ruse	94	73	91	71	88	70	74	90	72	88	71	86	69	107	85	67	100	80	66	95	78	20.7
Sofia	88	66	85	65	82	64	68	83	67	82	65	80	63	92	74	61	87	72	60	83	72	21.8
Varna	85	72	83	71	81	70	74	82	73	80	71	79	72	120	80	70	112	78	69	106	77	17.3
BRUNEI																						
Brunei Intl Airport	92	79	92	79	91	79	82	89	82	88	81	88	80	158	86	80	155	85	79	152	85	14.0
CANARY ISLANDS																						
Las Palmas	86	68	84	68	81	69	76	80	75	78	73	78	75	132	79	73	125	77	72	118	76	9.9
Santa Cruz De Tenerife	91	68	87	68	84	69	75	82	74	81	73	80	72	119	80	71	117	79	70	111	78	12.4
CAPE VERDE																						
Sal Island	86	75	85	75	84	75	78	82	77	82	77	81	77	141	81	76	136	80	75	133	80	8.8
CHILE																						
Antofagasta	77	68	75	67	74	65	69	75	68	74	67	73	67	100	73	66	97	73	65	92	71	10.4
Arica	83	71	81	69	79	68	73	79	72	78	70	78	71	114	78	69	107	78	67	100	76	11.9
Concepcion	76	63	74	62	72	61	65	72	64	71	63	70	63	85	67	61	80	66	60	78	65	20.2
Iquique	80	68	79	67	78	67	69	78	68	78	67	77	66	95	75	65	92	75	64	89	74	11.7
La Serena	72	63	71	63	70	62	64	71	64	70	63	69	62	84	67	61	81	67	60	78	66	11.9
Puerto Montt	73	62	70	61	68	60	64	71	62	68	61	67	61	81	66	60	77	65	58	72	64	17.6
Punta Arenas	64	54	61	53	59	51	56	62	54	60	53	58	52	59	57	51	55	56	49	52	55	13.0
Santiago	89	65	88	65	86	64	68	85	67	84	66	82	61	85	76	60	81	75	59	78	74	31.5
Temuco	81	64	78	63	75	61	66	77	64	76	62	73	61	82	71	59	76	70	58	72	67	24.7
CHINA																						
Anda	87	69	85	68	82	67	75	82	73	80	71	78	72	122	79	71	115	77	69	108	75	14.9
Andirlangar	98	63	96	62	93	61	66	89	65	88	63	86	60	90	71	58	83	71	55	76	72	26.6

MDB = mean coincident dry-bulb temp., °F
MWB = mean coincident wet-bulb temp., °F
MWS = mean coincident wind speed, mph
StdD = standard deviation, °F
HR = humidity ratio, grains of moisture per lb of dry air
A = airport
DP = dew-point temperature, °F

Table 3A Heating and Wind Design Conditions—World Locations

Station	WMO#	Lat.	Long.	Elev., ft	StdP, psia	Dates	Heating Dry Bulb 99.6%	Heating Dry Bulb 99%	Extreme Wind Speed, mph 1%	2.5%	5%	Coldest Month 0.4% WS	0.4% MDB	1% WS	1% MDB	MWS/PWD to DB 99.6% MWS	99.6% PWD	0.4% MWS	0.4% PWD	Extr. Annual Daily Mean DB Max.	Mean DB Min.	StdD DB Max.	StdD DB Min.
1a	1b	1c	1d	1e	1f	1g	2a	2b	3a	3b	3c	4a	4b	4c	4d	5a	5b	5c	5d	6a	6b	6c	6d
Anyang	538980	36.12N	114.37E	249	14.568	8293	18	21	16	14	12	14	36	12	35	1	290	7	180	100	13	2.2	3.6
Baoding	546020	38.85N	115.57E	62	14.667	8293	13	16	15	12	11	14	32	11	28	2	220	7	200	100	9	3.4	4.3
Bayan Mod	524950	40.75N	104.50E	4360	12.527	8293	−5	−1	28	23	20	26	20	21	22	6	270	8	180	96	−10	2.0	4.1
Beijing	545110	39.93N	116.28E	180	14.604	8293	13	15	21	17	14	21	30	18	28	4	340	7	200	101	8	3.4	3.1
Bengbu	582210	32.95N	117.37E	72	14.662	8293	22	24	16	14	12	15	38	13	37	3	20	8	180	99	16	2.3	4.1
Changchun	541610	43.90N	125.22E	780	14.290	8293	−12	−9	29	23	19	26	18	21	19	6	270	9	250	90	−17	2.5	4.9
Changsha (576790)	576870	28.23N	112.87E	223	14.582	8293	29	31	17	15	12	17	42	14	40	3	320	7	200	101	25	3.8	6.1
Chengdu	562940	30.67N	104.02E	1666	13.836	8293	32	34	11	9	8	9	44	7	44	0	20	4	200	93	27	2.7	2.5
Dalian	546620	38.90N	121.63E	318	14.532	8293	10	13	28	24	22	29	14	26	19	15	360	9	180	91	6	4.0	4.1
Dandong	544970	40.05N	124.33E	45	14.676	8293	1	5	21	17	15	23	16	19	18	5	340	6	220	89	−3	2.2	5.2
Datong	534870	40.10N	113.33E	3507	12.930	8293	−6	−3	23	19	17	20	15	18	17	5	360	9	160	92	−11	1.3	3.4
Deqen	564440	28.50N	98.90E	11443	9.556	8293	18	20	26	22	19	32	29	26	31	2	320	12	180	75	14	5.4	4.7
Dinghai	584770	30.03N	122.12E	121	14.636	8293	31	33	21	17	15	20	39	17	41	7	340	6	140	95	27	1.6	2.5
Erenhot	530680	43.65N	112.00E	3169	13.093	8293	−20	−16	29	25	21	25	8	21	9	5	70	12	140	96	−26	2.0	4.9
Fuzhou	588470	26.08N	119.28E	278	14.552	8293	40	42	18	16	14	17	54	14	52	7	320	10	140	100	35	1.6	3.1
Golmud	528180	36.42N	94.90E	9215	10.421	8293	0	4	20	17	15	17	20	14	19	3	250	6	70	86	−6	2.2	4.3
Guangzhou	592870	23.13N	113.32E	26	14.686	8293	42	44	15	13	11	15	53	13	53	6	360	5	270	98	37	1.4	2.7
Guilin	579570	25.33N	110.30E	544	14.413	8293	34	36	21	18	15	22	43	19	43	11	20	6	20	97	30	1.6	4.5
Guiyang	578160	26.58N	106.72E	3523	12.922	8293	28	30	15	12	11	13	48	11	45	4	40	8	160	91	26	3.4	1.6
Hami	522030	42.82N	93.52E	2424	13.457	8293	0	3	16	13	10	9	20	8	17	3	40	2	250	102	−4	3.1	4.1
Hangzhou	584570	30.23N	120.17E	141	14.625	8293	28	30	17	14	12	17	40	14	40	4	340	8	160	100	22	1.8	3.8
Harbin	509530	45.75N	126.77E	469	14.452	8293	−21	−16	23	19	16	18	13	16	8	2	200	11	180	91	−27	2.3	4.1
Hefei	583210	31.87N	117.23E	118	14.637	8293	24	27	19	16	14	17	39	15	39	4	340	8	180	98	19	2.3	5.4
Hohhot	534630	40.82N	111.68E	3494	12.936	8293	−4	−1	20	17	14	17	11	15	11	2	360	8	180	92	−8	2.5	3.1
Jinan	548230	36.68N	116.98E	190	14.599	8293	18	20	20	17	15	18	33	16	36	5	70	10	200	98	13	2.0	4.1
Jingdezhen	585270	29.30N	117.20E	196	14.596	8293	28	31	14	12	10	14	43	12	42	4	40	7	250	101	22	3.6	11.0
Jinzhou	543370	41.13N	121.12E	229	14.578	8293	2	5	25	21	18	23	18	20	20	6	360	9	200	94	−1	3.1	4.5
Jixi	509780	45.28N	130.95E	767	14.297	8293	−14	−10	23	20	17	23	3	21	5	6	250	8	270	91	−17	2.9	4.1
Kashi	517090	39.47N	75.98E	4235	12.585	8293	8	12	15	11	9	8	29	6	27	1	320	4	180	96	5	3.2	6.3
Korla	516560	41.75N	86.13E	3061	13.145	8293	6	10	21	17	14	14	24	12	24	1	70	4	40	100	2	2.5	7.6
Kowloon	450070	22.33N	114.18E	78	14.658	8293	48	51	22	20	18	20	62	19	63	8	330	10	250	96	45	2.2	2.5
Kunming	567780	25.02N	102.68E	6207	11.688	8293	32	34	19	16	14	19	60	17	59	2	140	11	250	86	27	5.9	4.3
Lanzhou	528890	36.05N	103.88E	4980	12.240	8293	10	13	10	8	7	6	28	5	26	0	90	5	70	94	6	2.0	3.4
Lhasa	555910	29.67N	91.13E	11975	9.358	8293	13	16	16	13	11	16	39	13	40	4	90	6	290	85	2	7.7	19.4
Liuzhou	590460	24.35N	109.40E	318	14.532	8293	38	40	14	12	10	12	50	11	48	6	340	6	200	99	35	1.8	1.8
Longzhou	594170	22.37N	106.75E	423	14.477	8293	43	45	10	9	7	10	61	8	59	1	90	5	200	100	38	2.0	3.4
Macau	450110	22.20N	113.53E	193	14.597	8293	44	47	19	17	15	19	50	17	51	13	360	7	200	94	42	1.6	2.7
Mudanjiang	540940	44.57N	129.60E	242	14.571	8293	−15	−12	22	18	15	20	5	17	7	3	180	7	200	91	−20	2.5	4.3
Nanchang	586060	28.60N	115.92E	164	14.613	8293	30	32	16	14	12	17	42	14	41	6	20	6	220	100	26	2.2	4.0
Nanjing	582380	32.00N	118.80E	39	14.679	8293	23	26	18	15	13	17	38	15	37	2	340	9	220	98	18	2.5	3.6
Nanning	594310	22.82N	108.35E	239	14.573	8293	42	44	12	10	8	10	62	9	61	2	90	4	160	100	38	2.2	2.2
Nenjiang	505570	49.17N	125.23E	797	14.281	8293	−32	−27	27	23	19	20	−1	16	−1	2	20	13	160	91	−36	4.3	5.2
Qingdao	548570	36.07N	120.33E	252	14.566	8293	19	21	31	27	23	34	22	29	25	17	340	11	160	90	15	2.2	3.1
Qiqihar	507450	47.38N	123.92E	485	14.444	8293	−18	−14	23	19	17	18	4	15	4	4	290	10	180	92	−22	3.8	4.9
Shanghai	583670	31.17N	121.43E	22	14.688	8293	26	29	19	17	15	19	38	17	39	6	290	8	200	98	21	2.0	2.7
Shantou	593160	23.40N	116.68E	9	14.695	8293	44	47	18	15	13	16	57	14	57	7	20	8	200	95	40	1.8	3.2
Shaoguan	590820	24.80N	113.58E	223	14.582	8293	36	39	14	12	10	13	52	11	52	2	360	5	180	101	33	2.2	2.3
Shenyang	543420	41.77N	123.43E	141	14.625	8293	−6	−2	22	18	15	19	22	16	20	3	70	9	200	92	−12	2.3	5.0
Shijiazhuang	536980	38.03N	114.42E	265	14.559	8293	14	17	16	13	11	16	29	11	34	2	160	6	160	101	10	2.5	4.3
Taiyuan	537720	37.78N	112.55E	2555	13.392	8293	4	8	21	17	14	20	27	16	29	2	90	6	180	94	−4	2.0	4.3
Tangshan	545340	39.67N	118.15E	95	14.650	8293	7	10	20	17	14	20	27	17	28	2	290	7	180	94	2	1.8	5.0
Tianjin	545270	39.10N	117.17E	16	14.691	8293	14	17	18	14	12	19	25	16	25	4	340	6	250	98	8	2.3	4.1
Urumqi	514630	43.78N	87.62E	3015	13.167	8293	−10	−5	18	13	11	10	19	8	15	2	180	8	320	98	−11	3.1	6.3
Weifang	548430	36.70N	119.08E	167	14.611	8293	12	15	24	21	18	23	30	20	29	6	290	10	200	99	8	2.7	2.5
Wenzhou	586590	28.02N	120.67E	22	14.688	8293	33	36	15	12	11	14	46	12	45	4	290	7	90	97	29	2.3	1.8
Wuhan	574940	30.62N	114.13E	75	14.660	8293	27	29	15	12	11	13	40	12	39	2	20	8	220	98	21	2.5	5.8
Xi'an	570360	34.30N	108.93E	1305	14.019	8293	20	23	14	12	10	12	36	10	35	1	40	5	40	99	14	2.2	4.5
Xiamen	591340	24.48N	118.08E	456	14.459	8293	43	45	21	18	16	19	52	18	52	10	70	8	160	95	40	1.3	2.5
Xining	528660	36.62N	101.77E	7421	11.163	8293	4	7	15	13	11	14	25	12	25	1	290	6	140	85	−1	2.0	3.1
Xuzhou	580270	34.28N	117.15E	137	14.627	8293	19	22	16	14	12	15	36	13	36	2	270	8	160	97	14	2.0	4.5
Yaxian	599480	18.23N	109.52E	22	14.688	8293	59	62	17	14	12	18	70	15	71	6	20	8	180	96	56	3.8	3.2
Yichang	574610	30.70N	111.30E	439	14.468	8293	30	32	11	9	8	10	46	8	45	2	110	7	140	100	28	2.2	2.3
Yichun	507740	47.72N	128.90E	761	14.300	8293	−28	−24	19	16	13	16	1	13	1	1	220	7	220	90	−33	3.1	4.0
Yinchuan	536140	38.48N	106.22E	3648	12.863	8293	−1	4	19	15	12	19	27	11	26	2	20	6	20	92	−5	2.0	5.0
Yingkou	544710	40.67N	122.20E	13	14.693	8293	1	4	26	22	19	24	26	20	23	6	20	10	200	90	−4	1.6	6.1
Yining	514310	43.95N	81.33E	2175	13.581	8293	−10	−3	16	13	10	12	26	9	24	1	70	3	270	97	−13	2.9	6.7
Yueyang	575840	29.38N	113.08E	170	14.610	8293	30	32	17	14	13	16	39	14	39	7	20	8	250	98	27	2.2	3.4

WMO# = World Meteorological Organization number
Lat. = latitude, °
Elev. = elevation, ft
Long. = longitude, °
PWD = prevailing wind direction, ° True
StdP = standard pressure at station elevation, psia
DB = dry-bulb temperature, °F
WS = wind speed, mph

Table 3B Cooling and Dehumidification Design Conditions—World Locations

Station	Cooling DB/MWB						Evaporation WB/MDB						Dehumidification DP/MDB and HR									Range of DB
	0.4%		1%		2%		0.4%		1%		2%		0.4%			1%			2%			
	DB	MWB	DB	MWB	DB	MWB	WB	MDB	WB	MDB	WB	MDB	DP	HR	MDB	DP	HR	MDB	DP	HR	MDB	
1	2a	2b	2c	2d	2e	2f	3a	3b	3c	3d	3e	3f	4a	4b	4c	4d	4e	4f	4g	4h	4i	5
Anyang	94	73	92	74	90	74	81	88	79	86	78	84	79	151	86	78	144	84	76	138	83	14.4
Baoding	94	72	91	72	89	73	80	87	78	85	77	84	78	146	84	76	139	83	75	132	82	14.8
Bayan Mod	91	59	89	58	86	57	64	80	62	79	60	78	59	88	69	56	80	68	54	74	69	20.7
Beijing	94	71	91	71	89	71	79	87	78	84	76	83	77	142	83	76	134	82	74	130	81	15.7
Bengbu	95	80	93	79	90	77	82	92	82	90	80	88	80	156	87	79	152	87	78	147	86	11.9
Changchun	86	70	84	70	82	69	76	83	74	80	72	78	74	130	81	72	123	78	71	116	76	13.5
Changsha (576790)	96	80	94	80	92	79	82	91	81	90	81	90	80	158	87	79	153	86	78	148	85	12.6
Chengdu	89	76	87	76	85	75	80	86	78	84	77	83	78	155	84	77	148	82	76	144	81	12.2
Dalian	86	74	84	72	82	72	77	82	76	80	75	78	76	135	80	75	132	78	73	126	78	9.9
Dandong	85	75	82	73	80	72	77	82	76	80	75	78	76	135	80	75	131	78	74	126	77	12.2
Datong	87	63	85	62	82	62	69	79	68	77	66	76	67	111	73	65	104	72	63	98	72	19.3
Deqen	67	52	65	52	63	52	55	62	54	61	54	60	53	93	58	52	90	57	51	87	56	12.1
Dinghai	90	81	88	80	86	79	82	89	81	87	80	85	80	156	87	79	151	85	78	148	84	9.7
Erenhot	91	61	88	60	85	59	66	79	64	78	63	77	62	94	70	60	88	69	58	81	69	21.1
Fuzhou	96	81	94	80	92	80	82	93	81	92	80	90	79	154	89	79	150	88	78	148	86	13.9
Golmud	80	52	78	51	75	50	55	75	53	72	51	71	47	69	60	45	62	60	43	57	58	18.9
Guangzhou	94	80	93	79	91	79	82	89	81	89	81	88	80	156	85	80	153	84	79	150	84	12.4
Guilin	94	78	92	78	91	78	81	88	80	87	80	87	79	155	84	78	150	83	78	147	83	13.3
Guiyang	87	70	85	70	83	70	73	82	72	81	72	80	71	130	77	70	126	77	69	123	76	12.6
Hami	97	66	94	65	92	64	70	90	68	89	66	87	63	95	80	61	87	77	58	80	75	23.9
Hangzhou	96	81	94	80	92	80	83	92	82	91	81	89	81	160	89	79	152	86	79	149	85	13.5
Harbin	87	69	84	68	82	68	75	82	74	79	72	78	73	127	80	72	119	78	70	111	76	15.3
Hefei	95	81	92	80	90	79	82	92	82	90	81	89	80	157	88	79	152	87	78	148	86	11.0
Hohhot	88	64	85	63	82	62	70	80	68	77	66	76	67	113	75	65	106	73	63	99	72	18.9
Jinan	95	73	92	74	90	73	80	89	79	87	78	85	78	145	85	76	139	84	75	134	83	13.1
Jingdezhen	97	80	94	79	93	79	82	92	81	91	80	89	79	153	85	78	149	85	78	145	85	14.9
Jinzhou	88	72	85	71	83	70	77	83	76	81	75	79	76	136	81	74	130	79	73	124	78	13.5
Jixi	87	70	83	69	81	68	74	83	72	79	71	77	72	121	79	70	113	77	68	106	76	15.1
Kashi	92	66	90	65	88	64	69	85	68	84	66	83	65	108	78	62	99	76	60	92	75	22.1
Korla	95	67	93	65	90	65	70	90	68	87	66	86	63	97	81	61	89	78	59	83	76	19.1
Kowloon	92	79	91	79	90	79	82	88	81	87	81	87	81	160	85	79	152	85	79	150	85	8.1
Kunming	80	63	78	63	77	63	68	76	67	75	67	74	66	120	72	65	118	71	65	116	71	15.3
Lanzhou	89	64	86	63	84	62	67	81	66	80	65	78	64	106	74	62	99	73	60	92	72	20.0
Lhasa	77	51	74	51	72	51	56	69	55	68	54	67	51	89	59	50	85	58	50	83	58	21.1
Liuzhou	95	78	94	78	92	78	81	90	80	89	80	89	79	151	85	78	148	85	78	145	84	12.6
Longzhou	96	80	95	80	93	79	82	92	82	91	81	90	80	158	86	79	155	86	79	152	85	13.7
Macau	92	81	90	81	89	80	83	90	82	88	81	88	81	161	87	80	157	86	80	156	86	7.6
Mudanjiang	87	71	84	69	81	68	74	84	72	80	71	78	72	120	80	70	113	78	68	106	76	16.0
Nanchang	96	80	94	80	92	80	83	91	82	90	81	89	81	160	87	80	155	87	79	151	86	12.4
Nanjing	94	81	92	80	89	79	83	90	82	89	81	88	81	160	87	80	154	87	79	150	86	11.2
Nanning	95	79	93	79	92	79	82	90	81	89	81	88	80	158	86	80	155	85	79	152	85	11.9
Nenjiang	86	67	82	66	80	66	72	79	70	77	69	75	70	113	76	68	106	74	66	100	73	17.1
Qingdao	85	74	83	74	81	74	78	82	78	80	77	80	78	145	81	77	140	80	76	136	79	8.1
Qiqihar	87	69	84	69	82	68	75	82	73	80	71	78	72	123	80	70	113	77	69	106	75	14.2
Shanghai	94	81	92	81	89	80	83	91	82	89	81	88	81	161	88	80	157	87	79	150	85	11.5
Shantou	91	80	90	80	89	80	82	89	81	87	81	87	80	156	85	79	153	85	79	150	84	9.4
Shaoguan	96	78	95	78	93	78	81	91	80	90	80	89	78	149	85	78	146	84	77	144	84	13.7
Shenyang	88	74	86	73	84	72	78	85	76	83	75	81	75	133	82	74	128	81	73	122	79	14.4
Shijiazhuang	95	72	92	72	90	73	80	87	79	85	77	84	78	149	85	77	141	83	75	134	82	15.5
Taiyuan	89	69	87	68	85	68	75	83	73	81	72	80	72	132	79	71	125	78	69	117	77	19.6
Tangshan	90	74	88	73	86	72	79	85	78	83	76	82	77	142	83	76	136	82	74	130	80	13.9
Tianjin	92	74	90	73	88	73	80	86	78	85	77	84	78	145	84	77	139	83	75	133	82	13.0
Urumqi	91	61	89	60	86	60	64	83	63	82	62	81	58	79	66	56	74	67	54	70	68	18.4
Weifang	93	73	90	73	88	73	81	88	79	85	78	83	79	150	85	77	142	83	76	136	82	14.9
Wenzhou	93	82	91	81	89	80	83	91	82	89	81	88	81	160	88	80	156	87	79	153	85	11.7
Wuhan	95	81	93	80	91	80	83	91	82	90	81	89	81	161	87	80	156	87	79	152	87	11.0
Xi'an	95	73	92	73	90	72	78	89	77	87	76	85	76	141	84	74	134	83	73	128	81	16.0
Xiamen	91	79	90	79	88	79	81	89	80	87	80	86	79	153	85	78	150	84	78	148	83	10.6
Xining	81	57	78	56	75	55	61	72	59	72	58	70	57	91	65	55	86	63	54	81	62	19.6
Xuzhou	94	77	91	76	89	75	82	90	80	88	79	86	80	155	86	78	148	85	77	143	84	12.2
Yaxian	91	80	90	80	89	80	83	89	82	88	82	87	81	160	87	80	157	86	80	155	86	9.5
Yichang	96	79	93	79	91	78	82	92	81	90	80	88	80	158	88	79	152	87	78	148	85	13.0
Yichun	86	69	83	67	80	67	73	81	71	79	69	76	70	113	78	68	106	75	67	101	73	18.2
Yinchuan	88	67	86	66	84	65	71	82	70	81	68	79	68	118	78	66	111	76	64	104	75	19.1
Yingkou	86	76	84	74	83	73	78	84	76	82	75	81	76	136	82	75	130	80	74	125	80	12.1
Yining	91	67	89	66	86	65	69	87	68	85	66	83	63	94	76	62	89	74	60	85	73	23.0
Yueyang	94	81	92	80	91	80	83	91	82	90	81	89	81	161	89	80	155	88	79	150	86	9.0

MDB = mean coincident dry-bulb temp., °F
MWB = mean coincident wet-bulb temp., °F
MWS = mean coincident wind speed, mph
StdD = standard deviation, °F
HR = humidity ratio, grains of moisture per lb of dry air
A = airport
DP = dew-point temperature, °F

Table 3A Heating and Wind Design Conditions—World Locations

							Heating Dry Bulb		Extreme Wind Speed, mph			Coldest Month				MWS/PWD to DB				Extr. Annual Daily			
												0.4%		1%		99.6%		0.4%		Mean DB		StdD DB	
Station	WMO#	Lat.	Long.	Elev., ft	StdP, psia	Dates	99.6%	99%	1%	2.5%	5%	WS	MDB	WS	MDB	MWS	PWD	MWS	PWD	Max.	Min.	Max.	Min.
1a	1b	1c	1d	1e	1f	1g	2a	2b	3a	3b	3c	4a	4b	4c	4d	5a	5b	5c	5d	6a	6b	6c	6d
Zhangjiakou	544010	40.78N	114.88E	2381	13.478	8293	1	4	17	15	13	16	15	14	16	7	340	6	140	95	−3	1.1	3.6
Zhanjiang	596580	21.22N	110.40E	91	14.651	8293	46	48	16	13	12	14	54	12	56	8	340	8	250	97	42	3.1	4.0
Zhengzhou	570830	34.72N	113.65E	364	14.508	8293	19	21	21	17	14	24	40	19	38	3	180	8	160	100	13	2.5	4.1
COLOMBIA																							
Bogota	802220	4.70N	74.13W	8359	10.770	8293	36	39	21	18	14	23	64	19	63	0	320	10	90	82	30	8.5	2.7
COOK ISLANDS																							
Rarotonga Island	918430	21.20S	159.82W	22	14.688	8293	62	64	25	22	20	26	71	23	72	1	150	11	80	89	58	3.1	2.5
CROATIA																							
Pula	132090	44.90N	13.92E	206	14.591	8293	25	27	26	21	17	26	35	22	38	7	20	6	270	92	21	2.0	3.6
Split	133330	43.53N	16.30E	68	14.663	8293	29	32	24	19	16	23	41	19	44	9	340	8	230	94	19	7.0	16.7
Zagreb	131310	45.73N	16.07E	351	14.514	8293	8	14	19	16	13	17	39	14	39	2	240	6	230	92	2	5.8	8.3
CUBA																							
Guantanamo	783670	19.90N	75.15W	55	14.670	8293	67	68	22	20	18	21	85	19	84	8	360	12	130	100	61	4.7	8.3
CYPRUS																							
Akrotiri	176010	34.58N	32.98E	75	14.660	8293	40	43	25	22	20	29	53	26	54	5	350	10	260	95	36	3.1	4.3
Larnaca	176090	34.88N	33.63E	6	14.697	8293	37	40	27	23	20	28	54	24	54	7	310	12	200	98	33	2.2	3.4
Paphos	176000	34.72N	32.48E	26	14.686	8293	39	42	24	21	18	30	55	25	55	9	30	9	280	92	36	3.2	3.4
CZECH REPUBLIC																							
Brno	117230	49.15N	16.70E	807	14.276	8293	6	12	24	21	18	26	30	21	31	8	60	10	180	91	4	2.9	7.2
Cheb	114060	50.08N	12.40E	1545	13.897	8293	4	10	16	14	12	17	37	14	36	2	40	5	220	90	1	3.8	6.3
Ostrava	117820	49.68N	18.12E	839	14.259	8293	1	9	23	20	19	26	32	23	33	5	20	10	190	90	−3	3.1	9.9
Plzen	114480	49.65N	13.27E	1194	14.077	8293	2	9	21	19	17	24	41	20	38	2	20	8	120	92	−1	4.0	9.2
Praded Mountain	117350	50.07N	17.23E	4895	12.279	8293	−2	2	47	41	36	51	20	43	22	19	20	12	180	72	−4	3.1	7.7
Prague	115180	50.10N	14.28E	1200	14.073	8293	3	10	28	23	20	31	39	27	36	4	10	8	160	91	0	3.6	8.8
Pribyslav	116590	49.58N	15.77E	1758	13.790	8293	3	9	29	25	22	30	34	27	31	5	360	9	130	87	−2	4.9	7.2
DENMARK																							
Alborg	60300	57.10N	9.87E	9	14.695	8293	8	15	29	26	23	32	45	28	42	6	220	11	100	82	7	4.0	12.4
Copenhagen	61800	55.63N	12.67E	16	14.691	8293	12	18	29	26	24	30	40	27	38	11	360	11	160	82	13	3.2	8.1
Hammerodde	61930	55.30N	14.78E	36	14.681	8293	20	22	44	37	34	45	34	41	34	20	70	12	230	80	22	3.4	5.9
Mon Island	61790	54.95N	12.55E	49	14.674	8293	18	22	43	35	32	46	37	41	35	14	320	9	70	78	19	4.0	7.7
Odense	61200	55.47N	10.33E	55	14.670	8293	14	18	29	26	23	30	42	27	40	8	40	11	120	84	9	4.1	9.5
Skagen	60410	57.77N	10.65E	22	14.688	8293	15	20	41	36	32	41	36	36	38	17	40	10	360	76	16	3.4	7.9
Tirstrup	60700	56.30N	10.62E	82	14.656	8293	9	16	27	24	21	28	40	24	39	5	20	11	280	82	7	3.4	11.0
ECUADOR																							
Guayaquil	842030	2.15S	79.88W	29	14.684	8293	67	68	16	15	13	17	74	16	74	8	210	7	40	95	51	2.3	11.3
Quito	840710	0.15S	78.48W	9225	10.417	8293	45	46	17	15	13	15	64	13	64	1	350	9	150	84	40	7.7	3.2
EGYPT																							
Alexandria	623180	31.20N	29.95E	22	14.688	8293	44	46	24	21	18	29	56	25	58	5	190	10	340	102	37	3.2	3.8
Cairo	623660	30.13N	31.40E	242	14.571	8293	45	46	21	19	16	23	58	19	62	6	210	13	350	108	38	2.9	4.9
Luxor	624050	25.67N	32.70E	288	14.547	8293	40	42	16	14	12	15	64	13	64	2	180	6	330	115	34	3.1	1.8
ESTONIA																							
Kopu/Cape Ristna	261150	58.92N	22.07E	29	14.684	8293	5	11	30	25	21	29	38	24	37	5	80	6	70	80	7	4.1	12.1
Tallinn	260380	59.35N	24.80E	144	14.623	8293	−4	3	21	18	16	22	34	19	32	6	140	8	40	82	−3	4.3	8.6
FAEROE ISLANDS																							
Torshavn	60110	62.02N	6.77W	127	14.632	8293	26	28	41	34	31	48	42	43	43	13	320	11	210	65	22	3.4	2.5
FIJI																							
Nadi	916800	17.75S	177.45E	59	14.669	8293	61	63	20	17	16	19	78	17	79	4	120	13	350	95	56	3.6	5.9
Nausori	916830	18.05S	178.57E	22	14.688	8293	62	64	20	18	16	20	75	18	74	1	320	11	60	91	59	2.0	1.8
FINLAND																							
Helsinki	29740	60.32N	24.97E	183	14.603	8293	−11	−3	22	20	18	24	35	22	32	5	340	11	210	83	−12	3.1	9.5
Jyvaskyla	29350	62.40N	25.68E	475	14.449	8293	−21	−13	20	17	15	23	28	19	26	2	330	9	180	83	−22	4.3	7.6
Kauhava	29130	63.10N	23.03E	144	14.623	8293	−20	−14	21	19	17	23	31	21	31	2	80	9	230	82	−21	2.3	8.1
Kuopio	29170	63.02N	27.80E	334	14.523	8293	−21	−14	19	17	15	21	30	19	28	1	140	7	170	82	−21	2.5	7.9
Lahti	29650	60.97N	25.63E	275	14.554	8293	−15	−7	14	12	11	15	33	13	31	1	350	6	150	83	−19	2.7	7.2
Pello	28440	66.80N	24.00E	275	14.554	8293	−25	−20	14	13	11	14	25	12	24	1	300	7	340	82	−30	4.9	5.4
Pori	29520	61.47N	21.80E	55	14.670	8293	−12	−4	25	22	19	30	37	25	35	5	90	11	140	82	−13	2.5	7.9
Suomussalmi	28790	64.90N	29.02E	734	14.314	8293	−21	−17	17	14	13	18	30	15	27	1	360	7	270	83	−26	10.1	6.7
Tampere	29440	61.42N	23.58E	367	14.506	8293	−15	−8	19	17	15	21	34	19	32	2	10	9	10	84	−17	2.5	8.6
Turku	29720	60.52N	22.27E	193	14.597	8293	−10	−3	21	19	17	25	33	21	32	6	40	9	230	83	−11	2.2	9.7
FRANCE																							
Bordeaux	75100	44.83N	0.70W	200	14.594	8293	22	27	22	19	16	24	51	20	51	4	40	7	80	97	19	2.7	7.4
Clermont-Ferrand	74600	45.78N	3.17E	1082	14.134	8293	16	20	24	20	17	26	49	22	49	3	360	8	20	97	11	2.9	7.0
Dijon	72800	47.27N	5.08E	744	14.309	8293	14	20	23	19	17	25	45	22	43	7	20	10	170	92	12	3.4	8.1
Brest	71100	48.45N	4.42W	337	14.521	8293	27	30	26	23	21	28	48	26	46	8	120	9	40	85	24	4.0	4.3
Lyon	74810	45.73N	5.08E	787	14.287	8293	17	23	26	22	18	26	44	22	45	3	20	12	180	95	15	2.7	7.4
Marseille	76500	43.45N	5.23E	118	14.637	8293	25	28	38	32	28	38	44	32	44	9	360	13	280	95	22	3.2	5.6
Montpellier	76430	43.58N	3.97E	19	14.690	8293	25	29	28	25	22	27	49	24	50	7	340	13	180	96	21	2.5	5.4

WMO# = World Meteorological Organization number
Lat. = latitude, °
Elev. = elevation, ft
Long. = longitude, °
PWD = prevailing wind direction, ° True
StdP = standard pressure at station elevation, psia
DB = dry-bulb temperature, °F
WS = wind speed, mph

Table 3B Cooling and Dehumidification Design Conditions—World Locations

Station	Cooling DB/MWB 0.4% DB	0.4% MWB	1% DB	1% MWB	2% DB	2% MWB	Evaporation WB/MDB 0.4% WB	0.4% MDB	1% WB	1% MDB	2% WB	2% MDB	Dehumidification DP/MDB and HR 0.4% DP	0.4% HR	0.4% MDB	1% DP	1% HR	1% MDB	2% DP	2% HR	2% MDB	Range of DB
1	2a	2b	2c	2d	2e	2f	3a	3b	3c	3d	3e	3f	4a	4b	4c	4d	4e	4f	4g	4h	4i	5
Zhangjiakou	89	65	87	65	84	64	72	81	71	80	69	79	70	119	78	67	111	76	66	104	75	17.1
Zhanjiang	93	80	92	80	90	80	82	89	82	88	81	87	81	162	86	80	158	85	80	155	85	8.5
Zhengzhou	94	74	92	74	90	74	81	89	80	87	78	85	79	154	86	78	147	84	77	141	83	14.6
COLOMBIA																						
Bogota	70	56	68	56	68	56	60	66	59	65	58	65	57	97	64	57	94	63	56	90	62	20.7
COOK ISLANDS																						
Rarotonga Island	85	78	85	78	84	77	80	84	79	83	79	83	79	148	82	78	146	82	77	142	82	8.3
CROATIA																						
Pula	89	71	86	69	84	68	74	82	72	82	71	81	71	117	79	69	109	77	68	103	75	19.1
Split	91	70	89	69	86	68	72	86	71	85	70	84	68	104	79	66	97	78	65	92	76	18.5
Zagreb	88	70	85	70	83	68	72	85	71	83	69	80	68	106	78	67	99	77	65	94	74	22.1
CUBA																						
Guantanamo	94	78	93	78	92	78	82	91	81	91	80	90	79	150	89	78	146	88	77	141	87	15.3
CYPRUS																						
Akrotiri	91	71	89	72	87	73	78	85	77	84	76	83	76	137	82	75	132	82	74	127	81	13.0
Larnaca	93	71	91	72	89	72	78	86	77	85	76	84	76	134	83	75	131	83	74	125	82	17.8
Paphos	88	76	86	76	85	76	79	86	78	85	77	83	77	141	85	76	137	84	75	132	83	15.5
CZECH REPUBLIC																						
Brno	85	66	82	65	79	64	68	82	67	79	65	76	64	92	74	63	88	71	61	83	70	19.4
Cheb	83	65	80	64	77	62	67	80	65	76	63	74	62	89	71	61	84	68	59	80	67	19.4
Ostrava	85	67	82	65	78	64	69	82	67	79	65	76	64	93	73	63	88	71	61	84	70	20.7
Plzen	84	67	81	65	78	64	69	82	67	78	65	75	64	94	74	63	89	73	61	84	69	20.3
Praded Mountain	65	55	63	54	60	53	57	63	56	60	54	58	56	79	59	54	74	57	52	70	56	9.7
Prague	84	65	80	64	77	63	67	79	66	76	64	74	63	90	72	62	85	69	60	82	69	20.0
Pribyslav	81	65	77	64	74	62	66	78	65	74	63	72	62	90	71	61	85	69	59	80	67	18.5
DENMARK																						
Alborg	77	63	74	61	71	60	65	75	63	71	61	69	61	80	68	59	75	67	58	71	65	15.1
Copenhagen	77	63	74	62	71	60	65	74	63	71	62	69	61	80	68	60	77	67	59	74	66	14.6
Hammerodde	73	64	70	63	68	62	66	70	64	69	63	67	64	90	68	63	85	67	61	80	66	7.0
Mon Island	74	65	71	63	69	62	66	71	65	70	63	67	64	90	69	62	84	68	61	80	66	10.1
Odense	78	64	75	63	72	61	66	75	64	73	62	70	62	85	70	61	79	68	59	75	66	17.3
Skagen	72	66	69	64	67	63	67	70	65	69	64	66	66	94	69	64	88	67	62	84	65	9.5
Tirstrup	77	64	75	62	72	61	65	74	64	71	62	69	63	85	68	61	80	66	59	75	65	17.8
ECUADOR																						
Guayaquil	92	76	91	76	90	76	80	88	79	87	79	85	78	148	85	77	143	83	77	141	82	13.3
Quito	72	54	70	54	69	54	58	67	57	66	57	65	55	93	61	54	88	59	54	88	59	18.4
EGYPT																						
Alexandria	90	71	88	73	86	74	77	85	76	84	75	83	74	128	82	73	125	82	72	120	81	11.3
Cairo	100	69	97	69	95	69	75	89	74	87	74	86	72	119	79	71	117	78	70	112	78	23.9
Luxor	110	72	108	71	106	71	75	103	74	102	73	101	66	97	92	64	90	91	62	85	90	30.6
ESTONIA																						
Kopu/Cape Ristna	73	64	70	63	68	61	66	71	64	68	62	67	64	88	69	62	83	67	60	78	65	8.6
Tallinn	77	64	74	62	71	61	66	73	64	71	62	69	63	88	69	61	82	68	59	76	66	14.8
FAEROE ISLANDS																						
Torshavn	58	54	56	53	55	53	55	57	54	56	53	54	55	64	56	54	62	55	53	60	54	5.4
FIJI																						
Nadi	90	77	89	77	88	77	80	87	79	86	79	85	78	146	83	77	143	83	77	140	82	14.2
Nausori	88	78	87	78	86	78	80	86	79	85	79	84	79	148	83	78	146	82	77	142	82	11.0
FINLAND																						
Helsinki	79	64	75	61	73	61	66	74	64	72	62	69	63	85	67	61	80	67	59	75	65	17.6
Jyvaskyla	78	62	75	61	71	60	65	73	63	72	61	69	62	84	69	60	78	65	58	72	64	17.6
Kauhava	77	62	74	61	70	59	65	73	63	70	61	68	62	83	68	60	76	65	58	71	64	18.0
Kuopio	78	62	75	61	71	60	65	74	63	71	62	69	62	84	69	60	79	66	58	74	65	12.8
Lahti	79	64	76	63	73	61	66	76	64	74	62	71	63	86	70	60	79	68	59	75	66	19.1
Pello	76	61	72	60	69	58	63	72	61	69	59	66	60	78	66	58	72	64	56	67	62	15.8
Pori	77	62	74	61	71	59	65	73	63	71	61	68	62	83	67	60	76	65	58	72	64	16.2
Suomussalmi	76	61	72	60	69	58	63	73	61	70	59	66	60	78	66	58	73	64	56	69	62	15.1
Tampere	79	62	76	61	73	60	65	75	63	72	61	70	62	83	67	59	76	66	58	71	64	18.7
Turku	79	63	75	61	72	60	65	74	63	72	61	69	62	84	69	60	77	66	58	73	65	16.4
FRANCE																						
Bordeaux	90	70	86	69	83	68	73	85	71	82	70	79	70	111	79	68	104	75	66	98	74	20.3
Clermont-Ferrand	89	70	86	68	82	67	72	85	70	83	68	80	67	104	77	65	98	76	64	93	74	22.7
Dijon	87	68	84	68	81	67	71	82	70	80	68	78	68	106	77	66	99	74	64	93	72	20.9
Brest	78	65	74	65	71	63	67	76	65	72	64	68	65	93	69	64	89	67	62	85	66	13.7
Lyon	90	69	86	69	83	67	72	84	70	82	68	80	68	105	78	66	99	75	64	93	73	21.4
Marseille	90	71	87	70	85	69	75	84	73	82	71	81	72	118	79	70	111	78	68	104	77	18.7
Montpellier	90	71	87	70	84	69	76	82	75	81	73	80	75	131	79	73	123	78	71	116	76	18.5

MDB = mean coincident dry-bulb temp., °F
MWB = mean coincident wet-bulb temp., °F
MWS = mean coincident wind speed, mph
StdD = standard deviation, °F
HR = humidity ratio, grains of moisture per lb of dry air
A = airport
DP = dew-point temperature, °F

Table 3A Heating and Wind Design Conditions—World Locations

Station	WMO#	Lat.	Long.	Elev., ft	StdP, psia	Dates	Heating Dry Bulb 99.6%	Heating Dry Bulb 99%	Extreme Wind Speed, mph 1%	Extreme Wind Speed, mph 2.5%	Extreme Wind Speed, mph 5%	Coldest Month 0.4% WS	Coldest Month 0.4% MDB	Coldest Month 1% WS	Coldest Month 1% MDB	MWS/PWD to DB 99.6% MWS	MWS/PWD to DB 99.6% PWD	MWS/PWD to DB 0.4% MWS	MWS/PWD to DB 0.4% PWD	Extr. Annual Daily Mean DB Max.	Extr. Annual Daily Mean DB Min.	Extr. Annual Daily StdD DB Max.	Extr. Annual Daily StdD DB Min.
1a	1b	1c	1d	1e	1f	1g	2a	2b	3a	3b	3c	4a	4b	4c	4d	5a	5b	5c	5d	6a	6b	6c	6d
Nancy	71800	48.68N	6.22E	711	14.326	8293	14	17	21	18	16	23	44	21	39	7	60	8	220	92	10	4.0	6.7
Nantes	72220	47.17N	1.60W	88	14.653	8293	23	27	24	21	18	27	52	24	50	7	60	9	60	93	20	3.2	6.7
Nice	76900	43.65N	7.20E	32	14.683	8293	35	37	25	21	17	24	53	19	52	11	340	8	160	90	31	2.7	4.9
Nimes	76450	43.87N	4.40E	203	14.592	8293	26	30	23	20	18	23	39	20	43	10	20	9	40	97	24	2.0	6.3
Orleans	72490	47.98N	1.75E	410	14.483	8293	17	22	26	23	20	30	50	26	48	8	60	9	80	93	10	3.1	20.3
Paris, De Gaulle	71570	49.02N	2.53E	357	14.511	8293	18	23	27	23	20	32	49	27	46	10	60	9	60	92	16	4.0	8.3
Paris, Orly	71490	48.73N	2.40E	314	14.533	8293	19	23	25	22	19	29	48	24	47	8	20	8	100	92	17	3.6	7.6
St.-Quentin	70610	49.82N	3.20E	331	14.525	8293	17	22	27	23	20	32	46	28	46	11	60	9	120	88	13	4.1	7.7
Strasbourg	71900	48.55N	7.63E	505	14.434	8293	12	17	22	19	16	26	48	21	41	6	340	8	20	93	10	2.5	8.1
Toulouse	76300	43.63N	1.37E	501	14.435	8293	22	27	22	19	17	22	48	19	48	5	280	7	140	99	19	3.6	8.1
FRENCH POLYNESIA																							
Moruroa Island	919520	21.82S	138.80W	9	14.695	8293	67	68	29	26	24	31	72	28	71	16	140	11	60	90	65	3.8	0.9
Papeete, Tahiti	919380	17.55S	149.62W	6	14.697	8293	68	69	22	19	16	22	78	19	79	3	120	7	260	91	65	0.7	2.0
GERMANY																							
Aachen	105010	50.78N	6.10E	672	14.346	8293	14	19	23	20	18	26	46	23	44	4	50	6	210	90	14	3.4	8.5
Ahlhorn (Ger-AFB)	102180	52.88N	8.23E	184	14.603	8293	11	16	25	22	19	30	45	25	42	7	90	9	10	90	9	4.0	9.5
Berlin	103840	52.47N	13.40E	160	14.615	8293	11	15	23	20	18	26	44	21	41	8	80	8	150	93	10	3.8	8.8
Bitburg	106100	49.95N	6.57E	1227	14.060	8293	12	18	23	20	18	27	42	23	38	11	60	7	10	90	11	3.6	6.8
Bremen	102240	53.05N	8.80E	16	14.691	8293	12	16	25	22	20	28	44	24	42	8	70	10	100	90	9	6.1	9.5
Bremerhaven	101290	53.53N	8.58E	36	14.681	8293	15	19	31	27	24	34	43	30	43	8	60	10	130	88	16	4.0	7.4
Dresden	104880	51.13N	13.78E	741	14.310	8293	8	13	22	19	16	23	42	20	41	4	320	7	260	92	6	2.9	11.2
Dusseldorf	104000	51.28N	6.78E	144	14.623	8293	14	20	23	21	18	26	45	23	44	6	60	9	130	92	13	2.7	8.8
Eggebek (Ger-Navy)	100340	54.63N	9.35E	72	14.662	8293	11	17	28	25	22	32	40	28	38	7	30	11	90	86	7	3.1	9.5
Ehrenberg	105440	50.50N	9.95E	3034	13.158	8293	5	10	34	30	27	37	27	33	25	15	100	11	190	83	5	3.1	8.5
Frankfurt Am Main	106370	50.05N	8.60E	370	14.504	8293	12	17	23	20	17	25	45	21	42	7	30	9	40	93	10	3.1	7.7
Grafenwohr	106870	49.70N	11.95E	1361	13.991	8293	−2	5	15	12	11	16	39	13	36	1	10	5	10	92	−7	4.1	9.5
Greifswald	101840	54.10N	13.40E	19	14.690	8293	9	15	23	20	18	25	41	21	40	4	250	8	50	89	8	4.1	11.0
Hamburg	101470	53.63N	10.00E	52	14.672	8293	11	16	23	20	18	24	42	21	40	6	60	11	90	89	10	4.5	9.0
Hannover	103380	52.47N	9.70E	177	14.606	8293	9	14	23	20	18	25	43	22	41	6	80	9	110	90	8	4.0	10.1
Heidelberg	107340	49.40N	8.65E	357	14.511	8293	14	19	17	14	12	18	44	15	43	4	170	6	10	96	12	2.9	8.6
Hof	106850	50.32N	11.88E	1863	13.737	8293	3	9	22	19	17	23	36	21	34	6	140	7	150	87	0	2.9	7.0
Husum (Ger-AFB)	100260	54.52N	9.15E	121	14.636	8293	12	17	29	26	23	32	43	29	42	9	50	9	90	86	8	4.9	9.0
Kap Arkona	100910	54.68N	13.43E	134	14.629	8293	17	21	43	38	34	46	37	42	37	15	360	12	70	81	17	3.6	8.3
Kiel/Holtenau(Ger-Navy)	100460	54.38N	10.15E	102	14.646	8293	14	19	25	22	19	28	41	24	38	9	40	9	160	85	10	4.9	10.6
Koln	105130	50.87N	7.17E	324	14.528	8293	11	17	21	18	16	25	45	21	43	4	110	8	130	92	8	3.1	10.6
Lahr	108050	48.37N	7.83E	511	14.430	8293	11	17	19	16	14	22	47	20	45	4	20	5	120	94	8	2.7	10.1
Landsberg (Ger-AFB)	108570	48.07N	10.90E	2060	13.638	8293	5	10	26	22	19	30	41	26	37	3	70	6	260	90	0	3.2	7.7
Leck (Ger-AFB)	100220	54.80N	8.95E	43	14.677	8293	11	17	28	25	22	32	42	28	40	5	80	10	110	85	4	3.8	11.0
Leipzig	104690	51.42N	12.23E	436	14.470	8293	8	13	28	24	21	30	42	25	41	6	70	9	190	92	6	3.2	12.2
Memmingen (Ger-AFB)	109470	47.98N	10.23E	2113	13.612	8293	5	10	25	21	18	29	38	26	38	5	50	9	220	90	−1	4.3	8.1
Munich	108660	48.13N	11.70E	1735	13.801	8293	4	10	27	22	18	29	42	24	40	4	80	8	30	90	−1	3.6	8.1
Neuburg (Ger-AFB)	108530	48.72N	11.22E	1270	14.038	8293	3	10	21	18	15	23	40	20	40	4	60	4	200	92	−1	4.1	11.9
Nordholz (Ger-Navy)	101360	53.77N	8.67E	102	14.646	8293	12	17	30	26	23	34	42	29	41	9	80	11	120	88	12	4.5	8.3
Ramstein (US-AFB)	106140	49.43N	7.60E	780	14.290	8293	11	16	19	17	14	21	44	18	43	2	10	5	240	93	7	3.8	7.9
Sollingen (Can-AFB)	107220	48.77N	8.10E	419	14.478	8293	13	17	22	19	17	26	45	23	44	6	30	6	10	94	10	3.2	8.1
Stuttgart	107380	48.68N	9.22E	1374	13.984	8293	9	14	21	18	15	23	41	20	40	4	90	7	90	92	4	4.0	10.3
GEORGIA																							
Batumi	374840	41.65N	41.63E	19	14.690	8293	29	32	30	27	24	31	49	28	50	13	130	9	300	91	21	9.0	12.2
K'ut'aisi (Kutaisi)	373950	42.27N	42.63E	380	14.499	8293	28	30	49	41	36	49	45	40	47	9	90	22	90	97	23	3.2	4.1
Sokhumi (Sukhumi)	372600	42.87N	41.13E	42	14.677	8293	29	31	18	15	13	19	44	16	42	6	50	9	320	90	25	3.2	4.0
Tbilisi	375490	41.68N	44.95E	1532	13.904	8293	21	24	49	42	37	51	36	45	36	6	320	10	180	97	16	3.1	4.0
GIBRALTAR																							
North Front	84950	36.15N	5.35W	16	14.691	8293	46	48	33	28	25	36	57	32	57	9	270	14	200	97	40	4.7	4.3
GREECE																							
Andravida	166820	37.92N	21.28E	39	14.679	8293	32	34	22	19	17	27	56	22	55	2	130	12	350	97	27	2.7	2.7
Athens	167160	37.90N	23.73E	49	14.674	8293	34	37	23	21	19	25	47	22	47	8	360	14	30	99	31	4.3	3.1
Elefsis (Hel-AFB)	167180	38.07N	23.55E	102	14.646	8293	33	36	23	21	19	24	49	22	46	4	360	12	10	104	30	5.2	3.2
Iraklion	167540	35.33N	25.18E	127	14.632	8293	41	44	32	29	25	40	53	33	51	14	340	11	320	97	37	3.6	3.6
Larisa	166480	39.63N	22.42E	242	14.571	8293	23	26	19	16	13	20	46	17	46	1	360	7	270	105	16	4.5	5.4
Preveza	166430	38.95N	20.77E	13	14.693	8293	36	38	26	22	19	31	51	24	50	10	40	10	250	95	33	4.1	2.7
Rodhos	167490	36.40N	28.08E	36	14.681	8293	41	44	24	22	20	28	49	23	49	13	360	14	270	96	38	3.6	3.6
Soudha	167460	35.48N	24.12E	495	14.439	8293	39	41	26	22	20	30	52	26	49	10	20	10	300	100	36	2.7	3.2
Thessaloniki	166220	40.52N	22.97E	13	14.693	8293	25	28	29	24	20	32	42	28	43	8	110	10	180	99	19	4.5	3.2
GREENLAND																							
Dundas, Thule Ab	42020	76.53N	68.50W	193	14.597	8293	−35	−33	30	24	19	30	−2	24	0	7	80	9	310	63	−37	7.6	3.2
Godthab	42500	64.17N	51.75W	229	14.578	8293	−12	−8	45	36	30	41	19	33	19	13	350	10	20	64	−8	5.2	6.1
Kangerlussuaq	42310	67.00N	50.80W	173	14.608	8293	−42	−38	22	19	16	23	22	19	20	9	60	11	70	69	−41	2.5	8.1

WMO# = World Meteorological Organization number
Lat. = latitude, °
Elev. = elevation, ft
Long. = longitude, °
PWD = prevailing wind direction, ° True
StdP = standard pressure at station elevation, psia
DB = dry-bulb temperature, °F
WS = wind speed, mph

Table 3B Cooling and Dehumidification Design Conditions—World Locations

Station	Cooling DB/MWB 0.4% DB	0.4% MWB	1% DB	1% MWB	2% DB	2% MWB	Evaporation WB/MDB 0.4% WB	0.4% MDB	1% WB	1% MDB	2% WB	2% MDB	Dehumidification DP/MDB and HR 0.4% DP	0.4% HR	0.4% MDB	1% DP	1% HR	1% MDB	2% DP	2% HR	2% MDB	Range of DB
1	2a	2b	2c	2d	2e	2f	3a	3b	3c	3d	3e	3f	4a	4b	4c	4d	4e	4f	4g	4h	4i	5
Nancy	86	69	82	67	79	65	71	82	69	79	67	76	67	100	76	65	94	73	63	90	71	20.9
Nantes	86	68	83	68	79	66	72	82	70	79	68	76	68	104	76	66	97	73	65	92	70	18.2
Nice	84	74	83	73	81	72	78	82	76	80	75	79	76	136	81	75	131	80	73	123	79	12.2
Nimes	92	69	89	69	87	68	73	84	72	83	71	81	71	113	77	69	107	76	67	100	75	19.6
Orleans	86	68	83	66	80	65	71	82	69	79	67	76	67	102	78	65	93	73	63	88	71	21.2
Paris, De Gaulle	86	69	82	68	79	66	71	81	69	79	67	76	68	104	76	66	97	74	64	91	72	18.7
Paris, Orly	86	69	82	67	79	65	71	82	69	79	67	76	67	99	75	65	93	73	63	88	71	18.4
St.-Quentin	82	68	79	67	76	65	70	79	68	77	66	74	67	102	75	65	94	72	63	88	70	18.9
Strasbourg	87	70	84	68	81	67	71	82	70	80	68	78	68	104	76	66	98	75	64	92	72	20.7
Toulouse	91	70	88	69	84	68	73	86	71	82	70	81	70	111	79	68	104	77	66	98	75	21.4
FRENCH POLYNESIA																						
Moruroa Island	87	79	86	78	85	78	80	85	80	84	79	83	79	151	83	78	147	82	78	144	82	6.8
Papeete, Tahiti	89	79	89	79	88	78	81	88	80	87	79	86	78	148	86	78	145	85	77	141	85	11.0
GERMANY																						
Aachen	83	66	80	65	77	64	68	80	67	78	65	75	64	93	73	63	88	71	61	83	70	15.5
Ahlhorn (Ger-AFB)	84	65	80	65	77	64	68	80	66	78	64	74	64	90	71	62	85	70	61	80	69	18.2
Berlin	86	66	82	65	79	64	68	81	67	79	65	75	64	90	72	62	85	70	61	80	69	16.7
Bitburg	84	66	80	65	77	64	68	80	67	78	65	75	64	94	72	63	89	71	61	83	70	18.5
Bremen	82	66	79	65	76	63	68	78	66	76	65	73	65	91	71	63	85	71	61	80	68	18.0
Bremerhaven	81	66	77	64	73	63	68	77	66	74	65	72	65	94	71	64	88	70	62	83	69	11.7
Dresden	85	66	82	65	78	63	68	81	66	78	65	76	64	91	72	62	86	71	61	81	69	17.6
Dusseldorf	85	67	82	65	79	64	69	81	67	79	66	76	65	92	73	63	88	72	62	83	71	17.5
Eggebek (Ger-Navy)	80	65	77	63	73	62	66	77	65	74	63	71	63	85	70	61	80	68	59	75	67	17.3
Ehrenberg	75	62	72	60	69	59	64	72	62	70	60	67	60	87	68	58	81	65	57	77	63	13.5
Frankfurt Am Main	87	67	83	66	80	64	69	82	67	80	66	77	65	93	73	63	88	71	62	83	70	19.8
Grafenwohr	85	66	82	66	79	64	68	81	66	79	65	76	63	90	69	62	88	71	61	83	69	25.0
Greifswald	81	66	77	65	74	63	68	78	66	75	64	72	65	91	72	63	85	70	61	80	68	16.4
Hamburg	82	66	79	64	75	63	68	78	66	76	64	73	64	90	72	62	85	70	61	80	68	16.7
Hannover	84	67	80	65	77	64	69	80	67	78	65	74	65	93	72	63	87	71	61	82	69	18.7
Heidelberg	90	69	86	67	83	66	71	86	69	84	67	80	66	98	77	64	92	75	63	87	71	20.0
Hof	81	64	77	62	74	61	66	77	64	74	62	71	62	89	70	60	83	67	59	79	66	18.5
Husum (Ger-AFB)	79	64	76	64	72	62	67	76	65	73	63	70	64	90	70	62	84	68	61	79	66	15.5
Kap Arkona	74	65	71	64	69	62	66	72	65	70	63	68	64	90	70	62	85	68	61	80	67	9.2
Kiel/Holtenau(Ger-Navy)	78	64	75	63	72	62	66	76	64	73	62	70	62	84	71	61	79	68	59	75	67	15.5
Koln	85	67	82	65	79	64	69	81	67	79	65	76	65	92	73	63	87	71	62	83	69	19.8
Lahr	86	69	84	68	80	66	71	83	69	81	68	78	67	99	77	65	94	74	64	91	73	20.7
Landsberg (Ger-AFB)	83	66	79	64	77	63	67	81	65	77	64	74	62	90	72	61	85	70	59	80	69	20.2
Leck (Ger-AFB)	79	65	76	63	72	62	67	77	64	75	63	71	63	85	71	61	80	68	59	75	67	16.4
Leipzig	85	66	82	65	78	64	68	81	67	78	65	76	64	91	73	63	86	71	61	81	70	18.5
Memmingen (Ger-AFB)	83	66	80	65	77	63	67	81	65	78	64	74	61	88	73	59	82	70	59	80	69	20.0
Munich	84	66	81	64	78	63	67	80	66	78	65	76	63	91	72	62	87	71	60	83	69	20.2
Neuburg (Ger-AFB)	85	66	81	65	79	64	68	82	66	80	65	76	63	90	73	61	85	71	59	79	69	22.7
Nordholz (Ger-Navy)	81	65	77	64	74	62	67	77	65	74	64	71	64	88	70	62	83	69	60	79	67	14.8
Ramstein (US-AFB)	86	68	83	66	80	65	70	83	67	79	66	78	65	94	72	63	88	72	61	83	71	22.3
Sollingen (Can-AFB)	87	69	84	68	81	66	71	83	69	81	68	78	67	99	75	66	97	74	64	92	72	19.8
Stuttgart	84	66	81	65	78	63	68	81	66	78	65	76	63	91	74	62	87	71	60	83	70	19.4
GEORGIA																						
Batumi	82	73	80	72	79	71	75	80	74	79	72	77	73	123	79	72	118	78	71	113	76	10.4
K'ut'aisi (Kutaisi)	90	71	87	70	84	70	76	83	74	81	73	79	73	125	79	72	120	78	71	116	77	14.9
Sokhumi (Sukhumi)	84	73	82	73	80	72	76	81	75	80	74	79	74	127	79	73	122	78	72	118	77	13.1
Tbilisi	92	70	89	70	87	69	73	88	72	86	70	84	68	111	81	67	105	79	65	99	78	18.4
GIBRALTAR																						
North Front	88	69	85	68	82	68	74	79	73	78	72	77	72	120	76	72	117	76	71	113	75	12.6
GREECE																						
Andravida	91	70	89	71	87	71	75	84	74	83	73	82	73	124	80	72	117	79	70	111	78	21.2
Athens	93	69	91	68	89	68	75	85	73	85	72	83	71	116	83	69	108	82	68	102	80	16.9
Elefsis (Hel-AFB)	97	70	95	68	92	68	74	88	73	88	71	86	70	110	83	68	103	82	66	96	80	18.2
Iraklion	88	66	86	67	84	68	74	82	73	81	72	80	71	116	80	70	110	79	68	104	79	10.6
Larisa	97	69	93	68	91	68	71	91	70	88	69	86	66	98	74	65	92	74	64	89	74	25.2
Preveza	88	71	86	71	84	72	76	83	75	82	74	81	74	126	80	73	122	79	72	118	79	14.4
Rodhos	90	71	87	71	86	70	76	82	75	82	74	81	74	126	79	73	122	79	72	117	79	10.1
Soudha	93	66	90	67	88	66	72	82	71	81	70	80	70	111	75	68	105	75	67	99	75	15.3
Thessaloniki	92	70	90	69	88	69	73	87	72	84	71	84	69	108	81	68	102	79	66	97	78	20.9
GREENLAND																						
Dundas, Thule Ab	54	44	51	43	48	41	45	53	43	50	42	48	39	36	47	37	33	45	36	31	44	8.3
Godthab	57	49	54	47	51	46	50	55	48	53	47	50	48	49	50	46	46	49	44	43	47	10.4
Kangerlussuaq	65	51	63	50	60	49	52	64	51	62	49	59	45	45	52	44	43	52	43	41	52	18.2

MDB = mean coincident dry-bulb temp., °F
MWB = mean coincident wet-bulb temp., °F
MWS = mean coincident wind speed, mphHR = humidity ratio, grains of moisture per lb of dry air
StdD = standard deviation, °F
A = airportDP = dew-point temperature, °F

Table 3A Heating and Wind Design Conditions—World Locations

Station	WMO#	Lat.	Long.	Elev., ft	StdP, psia	Dates	Heating Dry Bulb		Extreme Wind Speed, mph			Coldest Month				MWS/PWD to DB				Extr. Annual Daily			
												0.4%		1%		99.6%		0.4%		Mean DB		StdD DB	
							99.6%	99%	1%	2.5%	5%	WS	MDB	WS	MDB	MWS	PWD	MWS	PWD	Max.	Min.	Max.	Min.
1a	1b	1c	1d	1e	1f	1g	2a	2b	3a	3b	3c	4a	4b	4c	4d	5a	5b	5c	5d	6a	6b	6c	6d
Narsarsuaq	42700	61.18N	45.42W	85	14.655	8293	−18	−12	47	39	31	53	34	46	35	2	60	17	70	68	−15	2.5	10.8
GUAM																							
Andersen AFB (Guam)	912180	13.58N	144.93E	606	14.380	8293	74	75	20	18	16	19	79	18	79	9	70	9	90	91	72	2.9	1.1
HUNGARY																							
Budapest	128390	47.43N	19.27E	606	14.380	8293	8	14	36	29	24	35	40	27	40	2	170	10	200	94	2	2.2	7.9
Debrecen	128820	47.48N	21.63E	367	14.506	8293	6	11	22	18	15	22	35	19	35	4	50	6	90	92	2	2.7	6.7
Nagykanizsa	129250	46.45N	16.98E	462	14.456	8293	8	14	19	16	14	19	33	16	35	5	360	6	230	91	1	2.7	8.3
Pecs	129420	46.00N	18.23E	666	14.350	8293	12	16	22	18	15	23	36	20	33	6	320	7	270	93	8	2.3	5.8
Siofok	129350	46.92N	18.03E	354	14.513	8293	12	17	30	26	21	30	31	25	39	4	320	5	270	91	7	3.1	8.3
Szombathely	128120	47.27N	16.63E	725	14.319	8293	10	15	29	24	20	27	25	21	32	8	270	7	180	91	7	3.1	7.0
ICELAND																							
Akureyri	40630	65.68N	18.08W	88	14.653	8293	8	11	30	26	22	34	33	30	37	7	160	11	180	72	2	3.1	2.2
Keflavik	40180	63.97N	22.60W	177	14.606	8293	17	20	41	34	30	47	33	41	34	14	20	11	350	66	12	6.5	2.0
Raufarhofn	40770	66.45N	15.95W	32	14.683	8293	10	13	37	32	28	43	31	38	31	14	230	14	320	68	4	3.8	3.1
Reykjavik	40300	64.13N	21.90W	200	14.594	8293	14	17	41	34	31	47	36	42	36	10	90	12	360	65	10	3.4	2.5
INDIA																							
Ahmadabad	426470	23.07N	72.63E	180	14.604	8293	52	55	16	13	11	14	74	12	74	2	360	7	270	111	43	3.1	6.8
Bangalore	432950	12.97N	77.58E	3021	13.164	8293	59	60	13	12	10	11	71	10	72	3	90	4	90	99	54	2.0	3.4
Bombay	430030	19.12N	72.85E	45	14.676	8293	62	64	15	13	12	12	79	11	80	0	360	7	320	101	56	2.3	2.9
Calcutta	428090	22.65N	88.45E	19	14.690	8293	54	56	13	11	9	7	73	7	73	0	360	4	180	103	50	2.0	1.4
Cuddalore	433290	11.77N	79.77E	39	14.679	8293	68	69	14	12	11	13	79	12	80	2	320	6	250	105	63	2.7	3.1
Goa/Panaji	431920	15.48N	73.82E	196	14.596	8293	67	69	17	14	12	12	83	10	83	5	50	6	320	99	62	2.7	5.4
Hyderabad	431280	17.45N	78.47E	1788	13.775	8293	58	60	21	19	17	13	76	11	77	1	50	8	320	107	53	2.0	3.6
Jaipur	423480	26.82N	75.80E	1279	14.033	8293	44	47	16	13	11	12	64	10	64	0	90	9	320	110	39	2.2	3.1
Madras	432790	13.00N	80.18E	52	14.672	8293	68	69	17	14	13	13	80	11	80	2	290	8	270	106	65	2.2	1.8
Nagpur	428670	21.10N	79.05E	1017	14.168	8293	53	55	17	14	12	11	75	8	74	2	360	6	320	113	49	2.3	3.2
New Delhi	421820	28.58N	77.20E	708	14.327	8293	44	46	17	14	12	14	66	13	66	2	270	7	320	110	41	2.2	2.0
Poona	430630	18.53N	73.85E	1833	13.752	8293	50	52	12	10	8	8	79	6	78	0	70	3	270	105	45	1.3	2.7
Sholapur	431170	17.67N	75.90E	1571	13.884	8293	61	63	8	7	6	6	74	5	75	1	90	2	320	108	56	3.1	3.1
Trivandrum	433710	8.48N	76.95E	209	14.589	8293	72	73	14	13	11	17	83	15	83	2	360	6	320	99	65	3.1	4.7
INDIAN OCEAN ISLANDS																							
Diego Garcia Isl.	619670	7.30S	72.40E	9	14.695	8293	73	75	21	19	17	21	80	20	80	11	110	7	90	95	68	4.0	10.8
IRELAND																							
Belmullet	39760	54.23N	10.00W	32	14.683	8293	30	32	39	34	30	45	48	40	47	8	90	11	180	76	26	4.1	3.6
Birr	39650	53.08N	7.88W	236	14.575	8293	24	28	23	20	18	28	45	25	45	1	90	7	150	77	20	8.1	5.0
Claremorris	39700	53.72N	8.98W	226	14.580	8293	26	28	29	26	22	34	46	30	45	6	70	8	90	77	21	4.3	4.0
Clones	39740	54.18N	7.23W	291	14.546	8293	25	28	28	24	21	30	45	28	46	4	60	7	120	78	20	4.1	4.7
Cork	39550	51.85N	8.48W	531	14.420	8293	29	32	34	30	27	40	44	34	45	13	40	9	330	75	26	3.6	3.4
Dublin	39690	53.43N	6.25W	278	14.552	8293	29	31	31	28	24	35	44	30	44	9	250	11	230	77	26	3.2	3.2
Kilkenny	39600	52.67N	7.27W	209	14.589	8293	25	28	27	22	19	29	47	26	47	2	360	7	180	79	20	4.0	4.1
Malin	39800	55.37N	7.33W	82	14.656	8293	31	33	45	41	36	50	43	45	43	14	170	16	200	72	28	2.2	3.1
Mullingar	39710	53.53N	7.37W	341	14.520	8293	25	28	26	22	20	29	43	26	46	3	70	9	100	77	20	3.8	3.4
Rosslare	39570	52.25N	6.33W	82	14.656	8293	32	34	32	29	26	36	41	31	43	15	90	11	220	73	30	3.4	2.3
Shannon	39620	52.70N	8.92W	65	14.665	8293	28	31	31	27	23	37	45	31	46	6	70	9	110	78	24	4.3	3.1
Valentia Observatory	39530	51.93N	10.25W	45	14.676	8293	31	33	33	30	26	38	47	34	48	7	60	10	270	77	27	4.0	2.7
ISRAEL																							
Jerusalem	401840	31.78N	35.22E	2473	13.433	8293	33	35	24	21	19	28	41	24	42	6	270	10	290	98	31	5.4	2.3
Lod	401800	32.00N	34.90E	160	14.615	8293	40	42	23	19	17	26	56	22	54	4	150	11	320	103	36	2.9	2.9
Ovda (Isr-AFB/Civ)	401980	30.00N	34.83E	1417	13.963	8293	36	39	23	19	17	25	49	21	54	5	210	10	40	105	32	1.3	2.3
Tel Aviv-Yafo	401760	32.10N	34.78E	13	14.693	8293	44	46	28	22	19	30	56	24	56	7	120	9	310	99	41	5.4	2.5
ITALY																							
Bologna/Borgo (AFB)	161400	44.53N	11.30E	138	14.627	8293	22	25	16	13	11	14	41	11	37	1	220	5	80	97	18	3.4	5.9
Brindisi	163200	40.65N	17.95E	32	14.683	8293	36	39	26	22	19	30	51	26	51	9	360	10	180	99	31	4.3	2.3
Catania	164600	37.47N	15.05E	55	14.670	8293	35	37	23	19	16	26	54	22	55	5	230	10	90	103	31	5.4	2.2
Genova	161200	44.42N	8.85E	9	14.695	8293	32	36	27	24	22	28	43	26	44	15	40	7	50	92	29	2.3	4.1
Messina	164200	38.20N	15.55E	167	14.611	8293	43	45	19	16	14	20	56	17	57	4	310	6	60	97	38	4.9	4.0
Milan, Linate	160800	45.43N	9.28E	337	14.521	8293	21	25	16	12	10	20	46	14	46	1	90	5	220	95	18	5.9	5.0
Milan, Malpensa	160660	45.62N	8.73E	692	14.336	8293	15	18	18	12	9	19	43	11	40	0	360	4	210	94	9	3.1	4.7
Naples	162890	40.85N	14.30E	236	14.575	8293	32	34	23	18	15	26	49	21	49	4	340	8	200	98	28	3.8	2.9
Palermo	164050	38.18N	13.10E	111	14.641	8293	44	46	30	26	22	32	57	28	55	11	210	11	250	100	38	5.2	4.7
Perugia	161810	43.08N	12.50E	672	14.346	8293	24	27	21	18	16	22	46	19	45	1	360	5	270	96	17	2.9	7.2
Pisa	161580	43.68N	10.38E	3	14.698	8293	26	29	22	19	16	22	49	19	47	4	90	9	270	96	21	2.0	5.0
Rome	162420	41.80N	12.23E	9	14.695	8293	30	32	28	24	21	28	48	23	50	9	60	12	270	94	26	5.2	3.1
Ronchi Legionari Ab	161080	45.82N	13.48E	39	14.679	8293	21	25	22	18	14	28	38	20	38	2	10	5	220	96	17	2.3	4.1
Torino	160590	45.22N	7.65E	941	14.207	8293	20	23	13	9	7	16	48	10	41	0	260	2	70	92	16	2.9	4.0
Venice	161050	45.50N	12.33E	19	14.690	8293	23	26	22	17	13	28	36	21	33	4	60	6	160	92	19	4.3	4.0

WMO# = World Meteorological Organization number
Lat. = latitude, °
Elev. = elevation, ft
Long. = longitude, °
PWD = prevailing wind direction, ° True
StdP = standard pressure at station elevation, psia
DB = dry-bulb temperature, °F
WS = wind speed, mph

Table 3B Cooling and Dehumidification Design Conditions—World Locations

Station	Cooling DB/MWB						Evaporation WB/MDB						Dehumidification DP/MDB and HR									Range of DB
	0.4%		1%		2%		0.4%		1%		2%		0.4%			1%			2%			
	DB	MWB	DB	MWB	DB	MWB	WB	MDB	WB	MDB	WB	MDB	DP	HR	MDB	DP	HR	MDB	DP	HR	MDB	
1	2a	2b	2c	2d	2e	2f	3a	3b	3c	3d	3e	3f	4a	4b	4c	4d	4e	4f	4g	4h	4i	5
Narsarsuaq	65	50	62	49	60	47	52	60	51	58	50	57	49	52	51	48	49	50	46	46	49	12.2
GUAM																						
Andersen AFB (Guam)	88	79	87	79	87	78	81	86	80	85	80	85	80	158	84	79	154	83	78	150	83	7.6
HUNGARY																						
Budapest	90	69	86	68	84	66	71	87	69	85	67	80	65	94	71	64	92	74	63	87	74	22.0
Debrecen	88	71	85	70	82	68	72	86	71	83	69	81	67	102	80	66	97	78	64	92	76	20.3
Nagykanizsa	87	70	84	69	81	68	72	84	70	82	69	80	68	104	78	66	97	75	65	92	73	22.7
Pecs	88	70	86	69	83	68	72	86	70	84	69	81	67	102	80	65	96	79	64	90	76	18.9
Siofok	86	71	83	70	80	69	73	83	71	81	70	79	69	110	81	68	103	79	66	97	76	14.6
Szombathely	86	69	83	68	80	67	71	83	69	80	68	78	67	103	76	65	97	75	64	92	72	19.8
ICELAND																						
Akureyri	66	56	63	54	61	53	57	65	55	62	53	60	54	62	61	51	57	58	49	52	56	9.5
Keflavik	59	52	57	51	55	51	54	56	52	55	51	54	52	58	54	51	57	53	50	54	52	7.6
Raufarhofn	60	53	56	51	54	50	54	58	52	56	50	53	51	55	57	49	52	54	48	50	52	7.4
Reykjavik	60	53	58	52	56	51	54	58	53	56	52	55	53	59	56	51	56	54	50	54	53	8.5
INDIA																						
Ahmadabad	108	74	106	74	103	75	84	94	83	92	82	91	82	166	88	81	161	87	80	157	86	22.9
Bangalore	94	67	92	67	91	67	74	84	73	82	72	81	72	132	77	71	127	76	70	125	76	19.3
Bombay	95	73	93	74	92	75	82	89	81	88	81	88	80	156	86	80	153	86	79	150	85	9.4
Calcutta	99	78	97	79	95	79	85	94	84	92	83	91	83	171	90	82	167	89	82	164	88	18.0
Cuddalore	99	78	98	78	96	78	84	91	83	90	82	90	82	166	88	81	162	88	81	159	87	14.8
Goa/Panaji	93	77	92	77	91	77	83	88	82	88	81	87	81	163	87	80	158	86	79	153	85	10.4
Hyderabad	105	71	103	71	101	71	77	90	76	88	76	87	75	139	81	74	135	80	73	133	80	18.9
Jaipur	108	69	105	69	103	69	80	88	80	87	79	87	79	157	84	78	153	83	77	149	82	22.3
Madras	101	77	99	77	97	77	83	91	82	90	82	89	81	162	87	81	159	86	80	155	86	14.6
Nagpur	110	71	108	71	106	70	80	90	79	88	79	87	78	151	83	77	148	83	77	145	82	22.9
New Delhi	107	72	105	72	103	73	82	92	82	91	81	90	80	162	87	80	158	86	79	155	86	21.6
Poona	100	67	99	67	97	67	76	86	75	84	74	83	74	134	80	73	131	79	72	127	78	29.0
Sholapur	105	71	104	71	102	71	80	93	78	92	77	90	77	148	86	75	141	83	74	136	82	21.1
Trivandrum	92	78	91	78	90	78	81	88	81	88	80	87	79	153	85	79	150	84	78	148	84	11.7
INDIAN OCEAN ISLANDS																						
Diego Garcia Isl.	90	80	89	79	88	79	82	87	81	86	80	86	81	161	85	80	157	84	79	151	84	9.4
IRELAND																						
Belmullet	70	62	66	61	64	60	64	68	62	65	61	63	62	83	65	61	79	64	59	76	62	8.8
Birr	76	64	72	63	69	61	65	72	64	70	62	67	63	87	69	61	81	66	60	78	65	14.9
Claremorris	73	64	70	62	67	61	65	71	63	68	62	65	63	86	67	61	82	65	60	78	64	13.9
Clones	74	64	71	62	68	61	65	72	63	69	62	67	62	85	67	61	80	66	59	76	65	13.7
Cork	71	62	68	61	66	60	64	69	63	67	61	65	62	86	65	61	82	64	60	78	63	12.1
Dublin	72	63	69	61	67	60	64	69	63	67	61	66	62	85	67	61	80	65	59	76	64	12.6
Kilkenny	76	64	72	62	70	61	65	72	64	70	62	68	63	87	68	61	82	66	60	78	65	15.8
Malin	67	61	65	60	63	59	62	65	61	64	59	62	60	79	64	59	76	62	58	72	61	7.6
Mullingar	74	64	70	62	67	61	65	71	63	69	62	66	63	86	67	61	81	66	60	78	64	14.4
Rosslare	68	62	66	61	64	60	63	66	62	65	61	64	62	83	65	61	79	64	59	76	62	8.8
Shannon	75	64	71	63	68	61	65	72	64	70	62	67	63	85	68	61	81	67	60	78	65	12.1
Valentia Observatory	71	63	68	62	66	61	65	69	64	67	62	65	63	88	66	62	83	65	61	80	64	9.4
ISRAEL																						
Jerusalem	89	65	86	64	84	63	70	81	69	79	68	78	67	110	74	66	104	72	65	100	71	18.4
Lod	94	69	90	72	88	72	78	87	76	85	76	85	75	132	83	74	126	82	73	122	82	17.1
Ovda (Isr-AFB/Civ)	100	65	97	65	95	65	73	89	71	87	69	85	68	109	80	66	102	78	64	95	76	25.0
Tel Aviv-Yafo	88	69	86	74	85	74	78	84	77	83	76	83	76	137	83	75	132	82	74	127	82	9.9
ITALY																						
Bologna/Borgo (AFB)	93	75	90	73	88	72	77	89	75	87	74	85	73	125	83	72	118	81	70	111	81	20.3
Brindisi	90	73	86	74	84	75	80	84	79	83	77	82	79	149	83	77	141	82	75	133	81	13.0
Catania	95	72	91	73	89	72	79	85	78	84	76	83	77	141	82	75	133	81	74	126	80	20.9
Genova	86	72	84	72	82	72	76	83	75	81	74	80	74	129	82	73	123	81	72	117	79	10.4
Messina	89	72	88	73	86	74	79	84	78	83	77	83	77	142	83	76	136	83	75	132	82	9.4
Milan, Linate	89	73	87	72	85	71	76	85	74	84	73	82	73	123	81	71	117	80	70	111	78	18.2
Milan, Malpensa	90	74	87	73	85	72	77	86	75	85	74	83	74	131	83	72	121	81	70	115	79	23.0
Naples	92	73	89	73	87	73	79	85	77	84	76	83	77	142	83	75	133	82	73	125	80	19.8
Palermo	92	71	88	73	86	75	80	85	79	84	78	83	79	149	85	77	142	83	76	137	82	9.5
Perugia	92	70	90	69	86	69	73	87	72	84	70	83	69	110	79	68	104	77	66	99	76	25.0
Pisa	89	72	87	71	85	71	76	84	75	83	73	82	74	125	80	72	118	79	71	113	78	21.2
Rome	87	74	86	74	84	74	79	83	78	82	76	81	77	142	83	76	136	81	75	131	80	17.8
Ronchi Legionari Ab	91	72	88	71	86	71	76	83	74	83	72	82	74	125	80	71	116	79	70	109	78	21.2
Torino	87	72	85	71	83	70	75	84	74	82	72	80	72	125	78	71	120	79	70	113	77	18.9
Venice	87	74	85	73	83	71	77	83	75	82	74	81	75	132	81	73	123	80	71	116	78	16.4

MDB = mean coincident dry-bulb temp., °F
MWB = mean coincident wet-bulb temp., °F
MWS = mean coincident wind speed, mph
StdD = standard deviation, °F
HR = humidity ratio, grains of moisture per lb of dry air
A = airport
DP = dew-point temperature, °F

Table 3A Heating and Wind Design Conditions—World Locations

Station	WMO#	Lat.	Long.	Elev., ft	StdP, psia	Dates	Heating Dry Bulb 99.6%	Heating Dry Bulb 99%	Extreme Wind Speed, mph 1%	Extreme Wind Speed, mph 2.5%	Extreme Wind Speed, mph 5%	Coldest Month 0.4% WS	Coldest Month 0.4% MDB	Coldest Month 1% WS	Coldest Month 1% MDB	MWS/PWD to DB 99.6% MWS	MWS/PWD to DB 99.6% PWD	MWS/PWD to DB 0.4% MWS	MWS/PWD to DB 0.4% PWD	Extr. Annual Daily Mean DB Max.	Extr. Annual Daily Mean DB Min.	Extr. Annual Daily StdD DB Max.	Extr. Annual Daily StdD DB Min.
1a	1b	1c	1d	1e	1f	1g	2a	2b	3a	3b	3c	4a	4b	4c	4d	5a	5b	5c	5d	6a	6b	6c	6d
JAMAICA																							
Kingston	783970	17.93N	76.78W	29	14.684	8293	71	72	33	31	29	33	83	31	84	6	330	25	110	96	68	2.3	1.4
Montego Bay	783880	18.50N	77.92W	9	14.695	8293	70	71	29	26	23	28	80	26	80	5	140	21	70	95	61	3.6	15.1
JAPAN																							
Aomori	475750	40.82N	140.77E	9	14.695	8293	18	20	21	19	17	22	31	19	31	8	230	10	220	92	15	2.0	3.1
Asahikawa	474070	43.77N	142.37E	380	14.499	8293	−2	2	13	11	10	12	27	10	25	2	80	6	270	90	−8	2.5	4.9
Atsugi	476790	35.45N	139.45E	213	14.587	8293	29	30	23	20	18	21	47	19	46	4	360	10	180	95	24	2.7	2.0
Fukuoka	478080	33.58N	130.45E	39	14.679	8293	30	32	21	18	16	21	40	19	41	8	10	11	10	96	26	2.0	2.3
Hakodate	474300	41.82N	140.75E	118	14.637	8293	13	16	20	18	15	21	32	18	30	5	290	7	220	86	9	2.2	2.7
Hamamatsu	476810	34.75N	137.70E	157	14.617	8293	30	32	22	20	18	22	44	20	44	8	10	13	270	94	26	2.7	1.8
Hiroshima	477650	34.40N	132.47E	173	14.608	8293	30	31	21	18	16	20	45	17	43	6	20	10	220	95	27	3.2	2.7
Hyakuri (Jasdf)	477150	36.18N	140.42E	114	14.639	8293	19	23	22	19	16	19	43	17	42	2	250	11	130	92	14	2.0	2.2
Kadena	479310	26.35N	127.77E	147	14.622	8293	50	52	24	21	18	22	61	20	62	10	70	12	240	94	45	2.0	2.5
Kagoshima	478270	31.57N	130.55E	16	14.691	8293	33	35	17	14	13	15	49	13	49	6	300	9	270	94	29	2.0	2.0
Kumamoto	478190	32.82N	130.72E	127	14.632	8293	28	30	15	13	11	14	42	12	44	3	360	8	240	96	24	2.2	2.0
Maebashi	476240	36.40N	139.07E	370	14.504	8293	26	28	18	15	13	18	43	16	44	7	330	8	110	98	22	2.9	3.1
Maizuru	477500	35.45N	135.32E	72	14.662	8293	28	30	19	16	13	18	42	15	41	4	240	7	20	95	25	1.8	2.7
Matsumoto	476180	36.25N	137.97E	2004	13.666	8293	16	18	18	16	14	17	40	16	40	3	30	9	170	93	11	2.5	2.5
Matsuyama	478870	33.83N	132.78E	111	14.641	8293	31	33	14	12	10	14	42	12	43	4	110	8	270	93	28	1.6	2.2
Miho (Civ/Jasdf)	477430	35.48N	133.25E	29	14.684	8293	30	31	24	21	19	26	35	23	37	14	260	13	260	93	26	2.9	2.2
Miyako Jima Island	479270	24.78N	125.28E	134	14.629	8293	54	56	26	22	20	22	64	21	64	13	360	13	210	92	51	1.1	2.2
Morioka	475840	39.70N	141.17E	515	14.428	8293	14	18	19	17	15	18	31	16	33	4	140	9	190	92	10	1.8	3.2
Nagasaki	478170	32.73N	129.87E	114	14.639	8293	33	35	17	15	13	16	49	14	49	5	300	6	230	94	30	1.8	1.6
Nagoya	476350	35.25N	136.93E	55	14.670	8293	27	29	22	18	16	22	45	19	44	3	350	9	10	96	23	2.7	2.3
Naha	479300	26.18N	127.65E	26	14.686	8293	53	55	29	26	23	30	58	27	59	16	10	13	200	91	51	1.1	2.0
Naze	479090	28.38N	129.50E	22	14.688	8293	49	50	16	14	12	16	56	14	56	7	190	7	210	94	45	3.8	1.3
New Tokyo Intl A	476860	35.77N	140.38E	144	14.623	8293	23	25	22	19	16	21	46	18	43	5	330	11	10	93	17	2.3	4.0
Niigata	476040	37.92N	139.05E	22	14.688	8293	27	29	24	21	18	27	38	23	36	11	200	10	140	97	24	4.0	3.8
Nyutabaru (Jasdf)	478540	32.08N	131.45E	269	14.558	8293	28	31	21	18	16	22	46	19	48	7	270	12	230	95	24	2.2	2.0
Oita	478150	33.23N	131.62E	42	14.677	8293	30	32	17	14	13	16	44	15	43	6	170	9	10	94	26	1.6	3.2
Osaka	477710	34.78N	135.45E	49	14.674	8293	28	30	19	17	15	18	44	16	44	5	10	9	10	96	26	2.0	2.2
Owase	476630	34.07N	136.20E	88	14.653	8293	30	32	17	14	12	17	46	15	47	3	280	9	70	95	27	2.5	2.7
Sapporo	474120	43.05N	141.33E	62	14.667	8293	12	15	16	14	12	15	32	13	29	3	130	7	150	89	8	2.7	3.6
Sendai	475900	38.27N	140.90E	141	14.625	8293	24	26	23	20	17	24	37	21	37	6	350	9	130	91	21	3.6	2.3
Shimonoseki	477620	33.95N	130.93E	62	14.667	8293	33	35	24	20	17	21	44	19	44	10	330	9	100	91	30	1.4	2.3
Shizuhama (Jasdf)	476580	34.82N	138.30E	32	14.683	8293	30	32	24	22	19	25	47	23	47	9	10	13	260	95	27	2.7	2.0
Tokyo, Intl Airport	476710	35.55N	139.78E	26	14.686	8293	31	32	28	24	21	27	50	24	45	7	280	14	10	94	27	2.5	3.1
Tosashimizu	478980	32.72N	133.02E	108	14.643	8293	34	36	23	19	17	21	55	17	52	8	350	8	250	88	31	1.8	2.2
Wakkanai	474010	45.42N	141.68E	36	14.681	8293	11	13	27	23	21	29	26	25	25	9	190	10	240	81	9	3.6	4.0
JORDAN																							
Amman	402700	31.98N	35.98E	2536	13.402	8293	33	35	23	20	17	27	43	22	43	6	90	8	290	101	26	2.9	12.1
KAZAKHSTAN																							
Almaty (Alma Ata)	368700	43.23N	76.93E	2778	13.283	8293	−3	3	10	8	6	8	29	6	26	1	360	3	330	98	−4	3.1	5.8
Aqmola (Tselinograd)	351880	51.13N	71.37E	1141	14.104	8293	−21	−17	24	21	18	26	20	22	15	5	270	6	230	98	−26	5.8	5.8
Aqtobe (Aktyubinsk)	352290	50.30N	57.23E	744	14.309	8293	−19	−14	25	22	19	27	19	23	20	2	190	8	60	99	−26	4.0	3.6
Atyrau (Gur'yev)	357000	47.02N	51.85E	−49	14.726	8293	−7	−3	30	26	22	35	18	28	22	4	90	10	140	103	−12	2.5	5.0
Oral (Ural'sk)	351080	51.25N	51.40E	118	14.637	8293	−18	−13	28	23	21	31	17	28	16	6	360	11	140	100	−24	4.1	4.9
Pavlodar	360030	52.28N	76.95E	403	14.487	8293	−24	−19	22	19	17	23	21	20	20	5	60	7	230	97	−27	3.1	6.5
Qaraghandy (Karaganda)	353940	49.80N	73.13E	1820	13.758	8293	−18	−13	22	19	17	26	20	22	19	5	120	9	20	96	−22	3.2	6.8
Qostanay (Kustanay)	289520	53.22N	63.62E	511	14.430	8293	−21	−17	25	21	19	27	14	23	19	6	180	9	90	97	−26	3.1	5.2
Semey (Semipalatinsk)	361770	50.35N	80.25E	643	14.362	8293	−25	−20	20	17	14	22	27	18	23	1	90	6	90	99	−30	3.4	7.0
Zhambyl (Dzhambul)	383410	42.85N	71.38E	2142	13.597	8293	−5	1	27	22	16	27	36	22	36	2	180	9	20	102	−7	3.1	7.0
KENYA																							
Arissa	637230	0.47S	39.63E	482	14.446	8293	70	71	29	26	22	30	83	28	83	7	180	8	180	105	59	7.6	8.3
Kisumu	637080	0.10S	34.75E	3759	12.810	8293	60	62	22	19	17	19	76	16	79	4	90	13	230	100	52	11.2	7.0
Lodwar	636120	3.12N	35.62E	1689	13.824	8293	69	71	23	20	19	21	83	19	83	4	270	13	90	106	58	7.9	9.7
Nairobi	637400	1.32S	36.92E	5328	12.082	8293	49	51	23	21	19	17	70	15	69	6	240	14	60	90	41	4.5	4.9
Nakuru	637140	0.27S	36.10E	6236	11.675	8293	47	48	19	16	13	17	70	14	70	2	350	9	360	95	40	9.9	5.2
KOREA, NORTH (Democratic People's Republic of Korea)																							
Anju	470500	39.62N	125.65E	88	14.653	8293	−1	4	19	16	14	17	20	14	19	3	140	5	230	90	−7	2.7	7.4
Ch'ongjin	470080	41.78N	129.82E	141	14.625	8293	7	10	15	12	10	15	17	13	17	6	320	3	90	88	3	3.1	5.0
Changjin	470310	40.37N	127.25E	3546	12.911	8293	−19	−15	20	18	16	20	4	18	4	1	320	6	320	85	−25	4.7	3.4
Haeju	470690	38.03N	125.70E	265	14.559	8293	10	13	21	18	15	19	22	17	23	7	320	7	180	91	7	3.6	4.5
Hamhung	470410	39.93N	127.55E	124	14.634	8293	7	10	18	15	13	21	20	18	22	7	360	9	230	93	2	2.9	4.3
Namp'o	470600	38.72N	125.37E	154	14.618	8293	8	11	23	19	17	21	21	18	21	7	320	6	270	91	6	5.8	5.0
P'yongyang	470580	39.03N	125.78E	124	14.634	8293	3	7	14	12	10	15	19	13	19	2	110	4	270	91	−1	2.3	6.5
Sinuiju	470350	40.10N	124.38E	22	14.688	8293	3	7	17	15	13	19	19	16	18	6	50	5	230	92	−4	3.8	5.0
Wonsan	470550	39.18N	127.43E	118	14.637	8293	12	15	16	13	11	16	27	14	25	6	270	4	250	93	7	2.9	4.7

WMO# = World Meteorological Organization number
Lat. = latitude, °
Elev. = elevation, ft
Long. = longitude, °
PWD = prevailing wind direction, ° True
StdP = standard pressure at station elevation, psia
DB = dry-bulb temperature, °F
WS = wind speed, mph

Table 3B Cooling and Dehumidification Design Conditions—World Locations

	Cooling DB/MWB						Evaporation WB/MDB						Dehumidification DP/MDB and HR									
	0.4%		1%		2%		0.4%		1%		2%		0.4%			1%			2%			Range
Station	DB	MWB	DB	MWB	DB	MWB	WB	MDB	WB	MDB	WB	MDB	DP	HR	MDB	DP	HR	MDB	DP	HR	MDB	of DB
1	2a	2b	2c	2d	2e	2f	3a	3b	3c	3d	3e	3f	4a	4b	4c	4d	4e	4f	4g	4h	4i	5
JAMAICA																						
Kingston	92	78	91	78	90	78	81	89	80	89	79	88	79	150	85	77	143	85	77	141	85	11.7
Montego Bay	90	79	90	79	89	78	80	88	80	88	79	87	78	148	86	77	142	85	77	141	86	11.3
JAPAN																						
Aomori	86	74	83	73	80	71	76	84	74	81	73	79	74	125	80	72	120	80	71	115	78	13.0
Asahikawa	86	73	82	70	80	68	74	84	73	81	71	77	72	118	82	70	112	78	68	106	76	15.7
Atsugi	90	77	88	76	86	76	79	87	78	84	77	84	77	141	82	76	135	83	75	132	81	11.7
Fukuoka	93	77	91	78	88	77	79	89	79	87	78	86	77	141	83	76	134	82	75	133	82	13.1
Hakodate	82	73	80	72	77	71	75	81	73	78	72	76	73	123	79	72	117	77	70	111	76	11.0
Hamamatsu	90	76	88	76	86	76	79	86	79	84	78	82	78	148	81	77	142	81	77	140	81	11.9
Hiroshima	91	78	89	77	87	77	79	88	78	87	78	85	77	142	84	76	138	83	76	134	82	11.7
Hyakuri (Jasdf)	89	78	86	77	84	76	79	87	78	84	77	83	77	142	83	77	139	82	75	134	81	13.3
Kadena	92	81	91	81	90	80	83	90	82	89	82	88	81	162	87	81	161	87	80	159	86	9.7
Kagoshima	91	78	90	78	88	77	80	87	79	86	79	85	78	146	84	77	142	83	77	140	83	11.2
Kumamoto	92	78	91	77	89	77	80	88	79	87	78	85	78	146	83	77	142	82	76	139	82	14.0
Maebashi	92	77	90	76	87	75	78	89	77	87	76	85	75	134	82	74	130	83	73	125	82	13.5
Maizuru	91	77	89	77	87	76	79	88	78	87	77	85	76	137	83	75	132	82	74	130	81	13.7
Matsumoto	89	72	87	72	84	71	74	85	73	83	72	82	71	123	79	70	118	78	69	113	77	16.9
Matsuyama	91	77	89	76	87	76	78	87	77	86	77	85	76	136	82	75	133	82	74	129	82	12.1
Miho (Civ/Jasdf)	90	78	88	77	86	76	79	87	78	85	77	84	77	141	84	76	134	82	75	132	81	11.5
Miyako Jima Island	90	80	89	80	88	80	82	87	81	87	81	86	81	160	85	80	156	84	79	152	84	8.5
Morioka	87	75	84	73	82	71	76	84	75	82	74	79	74	130	80	73	125	79	72	120	78	13.9
Nagasaki	90	77	88	78	86	77	80	85	79	84	78	84	79	150	84	78	146	83	77	141	82	10.1
Nagoya	93	77	90	76	88	75	79	88	78	86	77	85	77	141	82	75	133	81	75	131	82	14.0
Naha	90	80	88	80	88	80	82	87	81	87	81	86	81	159	86	79	151	85	79	151	85	6.8
Naze	90	79	89	79	88	79	81	88	80	87	80	87	79	150	85	78	147	85	78	145	84	9.7
New Tokyo Intl Airport	89	78	87	78	85	76	79	87	78	85	77	83	77	143	82	77	141	82	75	134	81	13.5
Niigata	90	77	88	76	85	75	78	87	77	85	76	84	76	135	83	75	130	82	74	125	82	10.8
Nyutabaru (Jasdf)	90	78	88	78	86	77	79	86	79	85	78	83	78	149	82	77	143	82	77	141	82	11.0
Oita	91	78	89	77	87	77	79	87	78	86	78	85	77	141	83	76	136	82	75	133	82	12.2
Osaka	93	77	91	76	89	75	79	88	78	87	78	86	77	141	83	76	134	81	75	132	82	14.4
Owase	90	76	87	76	85	76	78	86	78	85	77	84	76	139	83	76	134	82	75	130	81	11.0
Sapporo	84	73	81	71	78	69	74	82	73	80	71	77	72	118	80	70	111	78	69	105	76	11.7
Sendai	86	75	84	74	81	73	77	83	76	82	75	80	75	134	81	74	129	80	73	124	78	9.5
Shimonoseki	88	77	87	77	85	76	79	86	78	85	77	84	77	139	83	76	136	82	75	133	82	7.9
Shizuhama (Jasdf)	91	79	89	78	86	77	80	88	79	86	79	84	79	150	83	77	143	83	77	140	82	11.9
Tokyo, Intl Airport	91	78	88	77	86	77	80	88	79	86	78	84	77	143	83	77	141	83	76	134	82	11.2
Tosashimizu	86	78	85	78	83	78	80	84	80	83	79	83	79	151	83	78	147	83	78	145	82	5.9
Wakkanai	77	71	74	69	72	68	72	76	70	74	69	72	71	113	75	69	106	73	67	100	72	7.6
JORDAN																						
Amman	95	65	92	65	89	64	71	83	70	82	68	81	68	115	76	66	106	74	65	100	72	20.3
KAZAKHSTAN																						
Almaty (Alma Ata)	91	65	89	64	86	63	67	85	66	84	65	83	62	92	76	60	84	73	58	79	73	19.8
Aqmola (Tselinograd)	89	64	85	62	82	62	67	81	65	79	64	78	63	89	71	61	83	69	59	77	69	19.4
Aqtobe (Aktyubinsk)	93	67	90	65	86	64	69	87	67	85	66	82	64	90	73	62	85	72	60	80	72	23.0
Atyrau (Gur'yev)	97	68	94	67	91	66	72	87	70	85	69	84	68	104	78	66	96	76	64	89	76	20.2
Oral (Ural'sk)	93	67	89	66	86	65	70	86	68	84	67	82	65	94	75	64	88	74	62	83	72	22.5
Pavlodar	90	65	87	65	84	63	68	82	67	80	66	79	65	92	72	63	86	71	61	80	71	20.2
Qaraghandy (Karaganda)	89	61	85	61	82	60	65	79	63	78	62	77	61	85	68	59	78	67	57	74	66	20.3
Qostanay (Kustanay)	90	66	86	65	83	64	69	83	67	81	66	79	65	93	73	63	87	72	61	83	71	18.7
Semey (Semipalatinsk)	91	66	87	64	84	64	68	84	67	82	66	80	64	90	72	62	86	72	61	81	71	22.5
Zhambyl (Dzhambul)	96	64	93	64	90	63	67	89	65	88	64	86	60	83	72	58	78	72	56	73	72	25.0
KENYA																						
Arissa	99	74	98	74	97	74	79	90	78	89	77	88	76	139	83	75	135	81	75	132	80	19.1
Kisumu	90	66	89	66	87	67	72	82	71	82	71	81	69	123	76	68	119	76	67	116	75	19.8
Lodwar	100	69	99	69	98	69	75	87	74	86	74	87	73	129	79	72	125	79	71	120	79	19.8
Nairobi	84	60	83	60	81	60	66	75	65	74	64	73	63	106	67	63	104	66	62	102	66	24.3
Nakuru	84	57	83	57	82	57	64	75	63	73	62	72	61	101	66	60	97	65	59	95	65	27.5
KOREA, NORTH (Democratic People's Republic of Korea)																						
Anju	86	75	83	74	82	73	77	83	76	81	75	80	76	136	80	75	130	79	73	125	78	13.5
Ch'ongjin	81	71	78	70	76	69	74	79	73	76	71	74	73	123	77	71	116	75	70	110	74	9.4
Changjin	77	65	74	63	72	62	69	74	67	71	65	70	67	115	71	66	108	70	64	102	68	16.2
Haeju	86	75	84	73	82	72	78	84	76	81	75	79	76	136	80	75	132	79	74	128	78	10.8
Hamhung	87	73	84	72	81	71	77	84	75	82	74	78	74	130	81	73	124	79	72	119	77	12.1
Namp'o	85	76	83	75	82	73	78	83	77	81	76	80	76	139	80	75	134	79	74	130	78	11.3
P'yongyang	87	75	85	74	83	72	78	84	76	82	75	80	76	135	81	75	130	80	74	125	79	13.7
Sinuiju	87	75	84	73	82	72	77	83	76	81	75	79	76	135	80	75	130	78	74	125	78	12.8
Wonsan	88	75	85	72	82	71	78	85	76	82	74	80	75	134	82	74	126	80	72	120	79	9.4

MDB = mean coincident dry-bulb temp., °F
MWB = mean coincident wet-bulb temp., °F
MWS = mean coincident wind speed, mph
StdD = standard deviation, °F
HR = humidity ratio, grains of moisture per lb of dry air
A = airport
DP = dew-point temperature, °F

Table 3A Heating and Wind Design Conditions—World Locations

Station	WMO#	Lat.	Long.	Elev., ft	StdP, psia	Dates	Heating Dry Bulb 99.6%	Heating Dry Bulb 99%	Extreme Wind Speed, mph 1%	Extreme Wind Speed, mph 2.5%	Extreme Wind Speed, mph 5%	Coldest Month 0.4% WS	Coldest Month 0.4% MDB	Coldest Month 1% WS	Coldest Month 1% MDB	MWS/PWD to DB 99.6% MWS	MWS/PWD to DB 99.6% PWD	MWS/PWD to DB 0.4% MWS	MWS/PWD to DB 0.4% PWD	Extr. Annual Daily Mean DB Max.	Extr. Annual Daily Mean DB Min.	Extr. Annual Daily StdD DB Max.	Extr. Annual Daily StdD DB Min.
1a	1b	1c	1d	1e	1f	1g	2a	2b	3a	3b	3c	4a	4b	4c	4d	5a	5b	5c	5d	6a	6b	6c	6d
KOREA, SOUTH (Republic of Korea)																							
Cheju	471820	33.50N	126.55E	87	14.653	8293	30	32	27	24	22	28	37	25	40	14	40	14	230	93	26	2.2	2.5
Inch'on	471120	37.48N	126.63E	229	14.578	8293	12	15	22	19	16	23	24	20	22	11	320	7	230	92	9	6.8	4.7
Kangnung	471050	37.75N	128.90E	88	14.653	8293	16	20	18	16	14	19	29	17	29	12	250	6	90	95	12	3.1	4.1
Kwangju	471560	35.13N	126.92E	236	14.575	8293	19	22	17	15	13	17	32	15	33	4	20	7	250	93	15	2.7	3.2
Osan	471220	37.08N	127.03E	39	14.679	8293	7	11	17	14	12	16	28	14	29	3	10	6	10	95	2	2.2	5.4
Seoul	471100	37.55N	126.80E	62	14.667	8293	7	10	19	17	15	19	24	16	25	3	10	9	160	92	2	1.6	6.1
Taegu	471430	35.88N	128.62E	200	14.594	8293	17	20	20	18	16	22	31	19	32	8	290	9	270	96	12	2.5	3.1
Taejon	471330	36.30N	127.40E	255	14.565	8293	12	15	15	13	11	12	37	11	34	1	110	6	270	95	8	3.4	2.7
Ulsan	471520	35.55N	129.32E	108	14.643	8293	20	22	16	14	12	18	31	16	31	7	320	8	140	95	15	3.1	3.4
KUWAIT																							
Kuwait	405820	29.22N	47.98E	180	14.604	8293	38	41	26	23	21	24	61	21	60	4	300	14	340	121	33	2.3	2.3
KYRGYZSTAN																							
Bishkek (Frunze)	383530	42.85N	74.53E	2083	13.627	8293	−8	−2	21	17	15	19	32	15	33	3	150	8	220	101	−11	2.2	7.6
Tianshan (Mtn Stn)	369820	41.92N	78.23E	11856	9.402	8293	−27	−23	22	19	17	20	4	17	1	1	360	11	210	68	−32	7.7	4.0
LATVIA																							
Liepaja	264060	56.55N	21.02E	26	14.686	8293	1	9	28	24	21	27	39	23	38	7	30	9	120	83	3	2.9	11.0
Riga	264220	56.97N	24.07E	9	14.695	8293	−3	4	24	21	18	23	37	21	36	4	40	9	150	85	−3	3.6	13.1
LIBYA																							
Banghazi	620530	32.08N	20.27E	433	14.471	8293	44	46	30	27	23	29	55	23	57	5	90	15	350	106	39	1.8	2.9
Tripoli	620100	32.67N	13.15E	265	14.559	8293	39	41	23	21	19	22	59	19	58	4	240	13	60	114	35	3.1	2.0
LIECHTENSTEIN																							
Vaduz	69900	47.13N	9.53E	1519	13.911	8293	12	17	22	17	13	22	50	18	48	3	180	10	320	89	8	2.0	6.7
LITHUANIA																							
Kaunas	266290	54.88N	23.88E	246	14.570	8293	−4	3	23	20	18	23	31	21	32	6	70	8	180	86	−2	3.6	8.5
Klaipeda	265090	55.70N	21.15E	32	14.683	8293	1	8	31	26	22	29	40	24	39	8	70	8	140	83	4	3.1	10.3
Vilnius	267300	54.63N	25.28E	511	14.430	8293	−5	2	25	23	20	25	29	22	29	5	70	11	140	86	−5	2.9	7.7
MACEDONIA																							
Skopje	135860	41.97N	21.65E	784	14.288	8293	10	15	20	17	14	19	36	15	34	1	50	4	270	100	4	4.5	9.4
MADEIRA ISLANDS																							
Funchal	85210	32.68N	16.77W	180	14.604	8293	53	55	30	27	23	34	61	29	62	8	310	11	30	87	50	5.2	1.8
MALAYSIA																							
George Town	486010	5.30N	100.27E	13	14.693	8293	73	73	15	13	11	13	82	12	83	2	350	8	270	96	70	4.0	1.3
Kota Baharu	486150	6.17N	102.28E	16	14.691	8293	71	72	17	15	14	18	81	17	81	1	190	9	90	95	68	2.9	1.4
Kuala Lumpur	486470	3.12N	101.55E	72	14.662	8293	71	72	16	14	12	13	85	11	85	1	340	8	270	98	68	3.1	3.4
Kuantan	486570	3.78N	103.22E	52	14.672	8293	70	71	16	14	12	16	83	15	82	5	350	8	230	99	58	5.2	23.4
Malacca	486650	2.27N	102.25E	29	14.684	8293	72	72	16	13	12	17	84	15	84	3	10	8	20	97	66	3.4	5.6
Sitiawan	486200	4.22N	100.70E	26	14.686	8293	71	72	13	12	10	11	84	10	85	1	60	7	180	99	66	5.9	4.9
Kuching	964130	1.48N	110.33E	88	14.653	8293	71	72	12	11	9	13	82	11	82	2	260	5	360	99	67	4.1	7.4
Miri	964490	4.33N	113.98E	59	14.669	8293	72	73	18	15	13	18	82	16	83	2	120	9	270	99	66	8.3	10.3
MALI																							
Bamako	612910	12.53N	7.95W	1250	14.048	8293	59	62	20	17	15	18	77	16	77	7	40	9	80	110	50	6.1	6.7
MALTA																							
Luqa	165970	35.85N	14.48E	298	14.542	8293	44	46	26	23	20	29	56	26	56	6	270	9	310	99	38	4.1	3.1
MARSHALL ISLANDS																							
Kwajalein Atoll	913660	8.73N	167.73E	26	14.686	8293	76	77	25	23	22	28	81	26	82	12	70	11	70	95	60	7.0	23.6
MAURITANIA																							
Nouadhibou	614150	20.93N	17.03W	9	14.695	8293	55	57	32	30	28	30	63	28	63	14	360	14	20	101	48	2.9	6.3
Nouakchott	614420	18.10N	15.95W	9	14.695	8293	55	57	23	21	19	26	75	24	75	9	60	14	80	113	44	1.3	6.7
MEXICO																							
Acapulco	768056	16.77N	99.75W	16	14.691	8293	68	70	23	19	17	17	84	14	84	2	320	17	200	97	60	2.7	8.6
Merida	766440	20.98N	89.65W	32	14.683	8293	57	60	34	23	19	22	77	19	77	4	90	15	140	103	47	2.2	2.2
Mexico City	766790	19.43N	99.08W	7329	11.202	8293	39	42	51	22	18	51	52	22	66	5	90	11	360	88	32	2.2	4.1
Puerto Vallarta (766010)	766014	20.68N	105.25W	19	14.690	8293	59	60	18	14	12	12	79	12	78	0	10	17	330	94	54	1.6	1.4
Tampico (765491)	765494	22.28N	97.87W	78	14.658	8293	50	53	32	24	21	34	59	28	62	9	270	11	90	97	43	4.3	6.5
Veracruz	766910	19.20N	96.13W	45	14.676	8293	57	59	46	34	29	47	70	35	68	4	330	22	90	101	50	4.0	4.5
MICRONESIA																							
Chuuk Intl/Moen	913340	7.47N	151.85E	7	14.697	8293	75	76	21	18	17	21	81	19	82	9	100	9	40	102	56	7.9	25.7
MIDWAY ISLAND																							
Midway Island NAF	910660	28.22N	177.37W	13	14.693	8293	59	60	24	22	20	29	67	26	67	10	360	9	110	89	46	1.8	20.2
MOLDOVA																							
Chisinau (Kishinev)	338150	47.02N	28.87E	590	14.389	8293	6	10	15	13	12	17	31	14	29	5	300	6	200	91	4	3.8	5.8
MONGOLIA																							
Ulaanbataar	442920	47.93N	106.98E	4317	12.547	8293	−23	−19	23	21	17	19	0	15	1	2	320	8	270	89	−27	5.2	4.7
Ulaangom	442120	49.97N	92.08E	3070	13.140	8293	−40	−37	18	13	11	9	−29	7	−29	1	180	5	50	89	−43	5.0	4.0
MOROCCO																							
Al Hoceima	601070	35.18N	3.85W	45	14.676	8293	44	46	24	21	18	24	58	19	58	3	180	12	360	98	39	9.0	4.1

WMO# = World Meteorological Organization number
Lat. = latitude, °
Elev. = elevation, ft
Long. = longitude, °
PWD = prevailing wind direction, ° True
StdP = standard pressure at station elevation, psia
DB = dry-bulb temperature, °F
WS = wind speed, mph

Table 3B Cooling and Dehumidification Design Conditions—World Locations

Station	Cooling DB/MWB						Evaporation WB/MDB						Dehumidification DP/MDB and HR									Range of DB
	0.4%		1%		2%		0.4%		1%		2%		0.4%			1%			2%			
	DB	MWB	DB	MWB	DB	MWB	WB	MDB	WB	MDB	WB	MDB	DP	HR	MDB	DP	HR	MDB	DP	HR	MDB	
1	2a	2b	2c	2d	2e	2f	3a	3b	3c	3d	3e	3f	4a	4b	4c	4d	4e	4f	4g	4h	4i	5
KOREA, SOUTH (Republic of Korea)																						
Cheju	89	78	86	79	85	78	82	86	80	85	79	84	81	160	84	79	151	83	78	148	83	9.7
Inch'on	87	76	84	74	82	73	77	84	76	82	75	80	76	136	81	75	131	80	74	127	79	10.4
Kangnung	90	75	87	74	85	72	78	87	76	85	75	82	75	132	82	74	126	82	73	122	80	10.4
Kwangju	90	78	88	76	85	75	80	87	78	85	77	83	78	144	83	76	139	82	76	135	81	12.6
Osan	90	78	88	76	86	75	80	88	78	85	77	83	78	148	86	77	139	83	75	132	82	14.4
Seoul	89	77	86	75	84	74	80	86	78	83	77	80	79	150	82	77	142	81	76	134	79	14.4
Taegu	92	78	90	76	87	74	79	89	78	87	77	85	77	141	85	76	135	83	74	130	82	13.1
Taejon	90	76	88	75	86	74	79	85	77	84	76	83	77	141	82	76	136	81	75	131	80	14.4
Ulsan	91	78	89	77	86	76	80	89	78	86	78	84	77	142	84	76	138	83	76	134	82	11.5
KUWAIT																						
Kuwait	117	69	115	69	113	68	82	94	78	91	75	92	79	153	92	75	132	87	70	112	85	27.7
KYRGYZSTAN																						
Bishkek (Frunze)	95	67	93	65	90	65	69	90	68	87	66	85	63	92	77	61	87	74	60	83	74	25.6
Tianshan (Mtn Stn)	57	42	54	41	51	39	44	54	42	51	41	49	40	57	46	38	53	44	36	50	43	20.9
LATVIA																						
Liepaja	76	64	73	62	70	62	67	72	65	70	63	68	64	90	70	62	85	68	61	80	66	10.3
Riga	79	65	76	64	73	62	67	75	66	73	64	71	65	91	71	63	85	69	61	80	68	14.2
LIBYA																						
Banghazi	99	72	95	71	92	70	78	89	76	86	75	85	75	134	83	74	127	81	72	123	80	16.7
Tripoli	107	76	103	74	100	73	81	99	78	94	76	90	76	139	88	74	130	84	73	123	83	24.8
LIECHTENSTEIN																						
Vaduz	83	67	80	65	78	64	68	80	67	78	65	75	64	94	74	63	90	72	61	85	71	16.6
LITHUANIA																						
Kaunas	80	67	77	65	74	63	69	78	67	75	65	72	65	94	73	63	88	71	62	83	68	16.6
Klaipeda	77	65	73	64	70	62	67	74	65	71	64	69	65	91	71	63	85	69	61	80	67	9.7
Vilnius	81	65	78	64	75	62	68	78	66	74	64	72	64	91	71	62	85	70	61	80	68	16.2
MACEDONIA																						
Skopje	95	68	92	68	89	67	71	90	70	88	68	86	65	94	78	63	88	76	62	86	75	27.4
MADEIRA ISLANDS																						
Funchal	81	69	79	69	77	68	72	78	71	76	70	76	70	111	76	68	105	75	68	102	74	8.5
MALAYSIA																						
George Town	91	79	90	78	90	78	82	88	81	87	81	87	80	158	85	79	151	84	79	150	84	13.3
Kota Baharu	91	79	90	79	90	79	81	88	80	88	80	87	79	150	85	78	147	84	78	145	84	12.8
Kuala Lumpur	94	78	93	78	92	78	81	90	80	89	80	89	79	152	85	79	149	84	78	146	84	16.2
Kuantan	92	79	91	79	90	79	81	89	80	88	80	88	79	150	85	78	147	84	78	146	84	15.3
Malacca	92	78	91	78	90	78	81	88	81	88	80	87	79	151	84	79	150	84	78	147	83	15.3
Sitiawan	92	79	91	79	90	79	81	90	81	89	80	88	79	151	85	79	150	85	78	147	85	14.8
Kuching	93	79	92	78	91	78	81	90	80	89	80	88	79	151	86	78	148	85	77	143	83	15.8
Miri	90	79	89	79	89	79	82	88	81	87	81	87	80	155	86	79	152	85	79	150	84	11.9
MALI																						
Bamako	104	69	103	69	101	69	79	91	78	89	78	88	77	147	83	76	140	82	75	139	82	22.1
MALTA																						
Luqa	92	71	88	72	86	72	77	84	76	82	75	82	75	134	80	74	127	80	73	125	79	14.4
MARSHALL ISLANDS																						
Kwajalein Atoll	89	79	88	79	88	79	81	87	80	87	80	86	79	150	86	78	147	85	78	146	85	7.6
MAURITANIA																						
Nouadhibou	92	69	88	69	86	69	76	83	74	81	73	81	74	126	79	72	118	78	70	112	77	15.8
Nouakchott	107	70	103	69	100	69	81	88	80	87	79	86	79	150	84	78	148	84	77	141	83	23.0
MEXICO																						
Acapulco	92	80	92	80	91	80	82	90	81	89	81	89	79	151	87	79	150	86	79	150	86	13.0
Merida	100	76	98	76	95	76	81	91	80	90	79	89	78	147	86	77	143	84	77	141	83	22.5
Mexico City	84	57	82	57	80	56	62	74	61	73	60	72	59	98	65	57	92	64	57	92	63	24.8
Puerto Vallarta (766010)	92	81	91	81	90	80	83	90	82	90	81	89	81	160	88	79	153	87	79	150	86	14.2
Tampico (765491)	92	80	90	80	90	80	83	89	82	88	81	87	82	168	88	80	158	86	79	152	84	11.3
Veracruz	94	80	92	80	91	80	82	91	81	90	80	89	79	152	85	79	150	85	78	148	85	14.9
MICRONESIA																						
Chuuk Intl/Moen	88	80	88	80	87	79	81	87	81	87	80	86	79	151	86	79	148	85	78	147	85	7.4
MIDWAY ISLAND																						
Midway Island NAF	87	75	86	75	86	75	77	84	76	84	76	84	75	132	82	74	129	82	74	126	82	8.1
MOLDOVA																						
Chisinau (Kishinev)	86	67	84	66	81	65	70	81	68	80	67	78	66	98	76	64	92	74	63	87	72	16.4
MONGOLIA																						
Ulaanbataar	82	60	78	59	75	58	63	75	61	73	59	71	58	86	67	56	80	66	55	75	64	17.6
Ulaangom	82	61	79	60	77	59	63	78	62	76	60	73	58	80	68	56	75	68	54	70	66	19.3
MOROCCO																						
Al Hoceima	87	73	84	72	82	72	77	82	76	81	75	80	76	134	80	74	130	79	73	125	78	11.2

MDB = mean coincident dry-bulb temp., °F
MWB = mean coincident wet-bulb temp., °F
MWS = mean coincident wind speed, mph
StdD = standard deviation, °F
HR = humidity ratio, grains of moisture per lb of dry air
A = airport
DP = dew-point temperature, °F

Table 3A Heating and Wind Design Conditions—World Locations

Station	WMO#	Lat.	Long.	Elev., ft	StdP, psia	Dates	Heating Dry Bulb 99.6%	Heating Dry Bulb 99%	Extreme Wind Speed, mph 1%	Extreme Wind Speed, mph 2.5%	Extreme Wind Speed, mph 5%	Coldest Month 0.4% WS	Coldest Month 0.4% MDB	Coldest Month 1% WS	Coldest Month 1% MDB	MWS/PWD to DB 99.6% MWS	MWS/PWD to DB 99.6% PWD	MWS/PWD to DB 0.4% MWS	MWS/PWD to DB 0.4% PWD	Extr. Annual Daily Mean DB Max.	Extr. Annual Daily Mean DB Min.	Extr. Annual Daily StdD DB Max.	Extr. Annual Daily StdD DB Min.
1a	1b	1c	1d	1e	1f	1g	2a	2b	3a	3b	3c	4a	4b	4c	4d	5a	5b	5c	5d	6a	6b	6c	6d
Casablanca	601550	33.57N	7.67W	203	14.592	8293	42	44	21	18	16	23	58	19	58	5	180	8	360	95	37	5.9	2.3
Casablanca/Nouasser	601560	33.37N	7.58W	675	14.345	8293	38	40	23	20	18	26	57	20	57	2	160	13	340	107	34	4.0	1.3
Midelt	601950	32.68N	4.73W	4970	12.245	8293	29	31	32	27	23	41	46	28	45	6	260	9	200	96	25	1.3	2.2
Ouarzazate	602650	30.93N	6.90W	3740	12.819	8293	33	35	32	27	22	29	54	22	51	2	320	12	240	102	29	1.1	1.6
Oujda	601150	34.78N	1.93W	1541	13.899	8293	34	36	30	26	22	31	54	27	55	4	240	15	360	106	29	2.7	2.2
Safi	601850	32.28N	9.23W	147	14.622	8293	42	44	21	19	17	20	59	18	58	7	60	12	20	106	37	5.2	2.3
Tanger	601010	35.73N	5.90W	68	14.663	8293	41	43	43	37	32	41	56	34	57	4	100	24	80	99	36	3.8	3.2
NETHERLANDS																							
Amsterdam	62400	52.30N	4.77E	−6	14.703	8293	17	21	31	27	24	35	47	31	44	11	70	11	70	86	16	3.4	8.3
Beek	63800	50.92N	5.78E	380	14.499	8293	14	19	27	24	21	30	45	26	43	10	60	8	40	90	13	3.8	8.5
De Bilt	62600	52.10N	5.18E	13	14.693	8293	16	20	19	17	15	21	44	19	42	7	50	8	70	87	14	3.6	8.1
Eindhoven	63700	51.45N	5.42E	72	14.662	8293	16	21	25	21	19	27	44	23	44	7	40	9	50	89	13	3.6	7.9
Gilze/Rijen	63500	51.57N	4.93E	42	14.677	8293	15	20	23	20	18	27	46	23	42	9	20	10	70	89	13	3.2	7.4
Groningen	62800	53.13N	6.58E	13	14.693	8293	14	18	28	24	22	31	45	27	44	7	50	9	100	87	11	3.2	7.7
Leeuwarden	62700	53.22N	5.75E	6	14.697	8293	16	20	29	26	22	32	43	29	43	8	80	11	80	85	13	2.3	7.4
Rotterdam	63440	51.95N	4.45E	−13	14.707	8293	17	21	30	27	24	33	44	29	44	9	50	10	90	86	15	3.4	6.7
NETHERLANDS ANTILLES																							
Willemstad	789880	12.20N	68.97W	219	14.584	8293	74	75	23	22	21	23	82	22	81	10	100	18	80	96	71	3.2	1.3
NEW CALEDONIA																							
Noumea	915920	22.27S	166.45E	236	14.575	8293	61	62	27	24	22	26	68	23	68	6	60	12	80	93	58	2.9	2.0
NEW ZEALAND																							
Auckland	931190	37.02S	174.80E	19	14.690	8293	35	37	31	28	25	32	53	28	53	10	240	13	20	85	35	13.7	2.2
Christchurch	937800	43.48S	172.55E	111	14.641	8293	28	30	27	23	21	24	48	21	48	1	280	16	300	92	25	10.8	1.3
Taiaroa Head	938960	45.77S	170.73E	249	14.568	8293	38	39	52	46	40	52	44	46	44	19	240	17	320	78	35	3.2	1.4
Wellington (934340)	934360	41.33S	174.80E	22	14.688	8293	35	36	42	38	33	40	50	34	50	14	10	17	360	84	35	14.4	2.2
NIGER																							
Agadez	610240	16.97N	7.98E	1646	13.846	8293	51	53	32	27	23	35	71	32	71	7	100	11	120	113	40	4.3	7.9
Niamey	610520	13.48N	2.17E	744	14.309	8293	60	62	23	19	17	24	73	22	74	6	40	8	40	112	52	3.1	4.5
NORWAY																							
Bergen	13110	60.30N	5.22E	164	14.613	8293	16	20	26	23	20	30	41	27	40	3	60	8	240	78	12	2.7	7.2
Bodo	11520	67.27N	14.37E	42	14.677	8293	9	13	38	33	29	44	32	39	28	19	80	11	100	76	8	4.0	4.7
Oslo/Fornebu	14880	59.90N	10.62E	55	14.670	8293	0	5	19	17	15	22	40	19	39	2	360	7	180	85	−1	5.6	8.6
Oslo/Gardermoen	13840	60.20N	11.08E	669	14.348	8293	−8	−2	20	17	15	22	36	19	34	2	30	7	180	82	−10	4.5	10.3
Stavanger	14150	58.88N	5.63E	29	14.684	8293	13	18	30	26	23	30	38	27	39	3	150	12	320	79	11	3.4	7.4
Svinoy (Lgt-H)	12050	62.33N	5.27E	134	14.629	8293	28	29	53	47	42	58	46	50	42	14	140	13	150	70	24	4.1	4.9
Tromso	10250	69.68N	18.92E	32	14.683	8293	6	10	30	27	23	34	36	30	35	2	170	8	180	75	4	2.5	3.6
Trondheim	12710	63.47N	10.93E	55	14.670	8293	−1	6	28	23	20	32	36	27	38	9	120	10	260	83	−2	3.2	8.1
Utsira	14030	59.30N	4.88E	183	14.603	8293	23	27	48	43	39	49	39	45	37	14	90	11	120	72	17	3.4	18.2
OMAN																							
Masqat	412560	23.58N	58.28E	49	14.674	8293	61	63	20	17	15	18	74	16	73	5	200	11	340	116	52	2.3	8.3
Salalah	413160	17.03N	54.08E	65	14.665	8293	63	65	21	19	16	28	70	23	73	10	20	12	200	101	52	5.0	10.1
Thamarit	413140	17.67N	54.03E	1459	13.941	8293	48	51	32	30	27	21	72	19	71	7	160	11	340	111	42	2.7	2.7
Tur'at Masirah	412880	20.67N	58.90E	62	14.667	8293	63	65	27	24	23	25	66	22	68	14	300	13	210	106	53	2.7	9.5
PANAMA																							
Panama	788060	8.92N	79.60W	52	14.672	8293	73	73	17	15	13	15	81	12	82	2	10	12	10	99	67	4.5	9.7
Tocumen	787920	9.05N	79.37W	36	14.681	8293	68	68	16	14	12	13	80	12	81	0	300	10	30	96	58	3.2	11.3
PARAGUAY																							
Asuncion	862180	25.27S	57.63W	331	14.525	8293	41	44	23	20	18	25	71	22	69	2	180	14	360	103	35	4.3	2.7
PERU																							
Arequipa	847520	16.32S	71.55W	8267	10.808	8293	42	43	26	21	18	31	54	26	54	6	30	14	240	79	34	3.4	3.4
Cuzco	846860	13.55S	71.98W	10659	9.853	8293	32	34	24	20	15	22	62	17	63	0	90	4	330	77	28	3.6	2.5
Iquitos	843770	3.75S	73.25W	413	14.482	8293	66	68	19	13	11	16	77	12	78	2	170	4	330	98	53	3.2	18.2
Lima	846280	12.00S	77.12W	42	14.677	8293	57	58	24	20	18	21	62	18	63	4	170	13	170	87	50	2.2	6.1
Pisco	846910	13.75S	76.28W	22	14.688	8293	53	55	25	21	19	23	65	19	65	1	90	11	210	89	47	4.0	6.3
Talara	843900	4.57S	81.25W	295	14.544	8293	60	61	46	42	33	46	64	42	65	22	150	15	190	93	55	3.2	7.7
PHILIPPINES																							
Angeles, Clark AFB	983270	15.18N	120.55E	643	14.362	8293	68	69	15	12	11	13	83	12	82	5	10	7	120	99	65	2.0	2.2
Baguio	983280	16.42N	120.60E	4924	12.266	8293	52	54	22	15	12	14	65	12	65	3	90	4	140	93	49	5.4	1.6
Cebu/Mandaue	986460	10.30N	123.97E	78	14.658	8293	73	74	18	15	14	19	82	17	82	5	40	9	40	98	67	3.2	10.3
Olongapo	984260	14.80N	120.27E	56	14.670	8293	70	71	21	18	16	21	85	19	84	4	70	11	70	100	67	4.3	2.5
Manila, Aquino Airport	984290	14.52N	121.00E	68	14.663	8293	69	71	41	36	32	40	83	36	83	2	90	21	90	99	43	1.8	14.2
POLAND																							
Bialystok	122950	53.10N	23.17E	495	14.439	8293	−4	3	18	16	14	18	28	16	30	2	310	6	180	87	−5	3.6	11.2
Gdansk	121500	54.38N	18.47E	452	14.461	8293	1	9	31	26	23	32	41	27	36	2	130	11	10	86	−2	5.2	11.0
Katowice	125600	50.23N	19.03E	931	14.212	8293	3	9	19	17	15	22	40	19	38	2	20	8	250	89	−2	2.9	9.5
Kielce	125700	50.82N	20.70E	856	14.251	8293	−1	6	20	18	16	21	31	19	33	4	60	7	190	88	−5	2.9	10.6
Kolobrzeg	121000	54.18N	15.58E	16	14.691	8293	10	16	18	16	14	19	38	17	36	5	110	6	140	90	11	4.1	8.1

WMO# = World Meteorological Organization number
Lat. = latitude, °
Elev. = elevation, ft
Long. = longitude, °
PWD = prevailing wind direction, ° True
StdP = standard pressure at station elevation, psia
DB = dry-bulb temperature, °F
WS = wind speed, mph

Table 3B Cooling and Dehumidification Design Conditions—World Locations

Station	Cooling DB/MWB						Evaporation WB/MDB						Dehumidification DP/MDB and HR									Range of DB
	0.4%		1%		2%		0.4%		1%		2%		0.4%			1%			2%			
	DB	MWB	DB	MWB	DB	MWB	WB	MDB	WB	MDB	WB	MDB	DP	HR	MDB	DP	HR	MDB	DP	HR	MDB	
1	2a	2b	2c	2d	2e	2f	3a	3b	3c	3d	3e	3f	4a	4b	4c	4d	4e	4f	4g	4h	4i	5
Casablanca	85	72	81	72	79	72	75	80	74	79	73	77	74	126	78	73	122	77	72	118	76	9.2
Casablanca/Nouasser	96	72	91	71	87	70	75	90	73	86	72	84	70	114	80	69	108	77	68	106	77	19.8
Midelt	92	58	90	58	89	58	62	83	61	81	60	80	56	81	66	55	76	67	53	72	67	24.5
Ouarzazate	100	62	98	61	97	61	66	90	64	89	63	89	58	84	72	56	77	71	54	71	71	24.7
Oujda	98	70	94	69	91	69	74	90	72	87	71	85	70	116	80	68	111	78	67	106	77	24.7
Safi	94	71	89	70	85	70	74	86	73	83	72	81	71	116	78	70	111	77	69	107	75	14.8
Tanger	92	71	89	71	86	70	74	86	73	84	72	81	71	114	79	70	110	78	68	104	77	16.7
NETHERLANDS																						
Amsterdam	80	66	77	65	74	64	69	77	67	74	65	72	66	94	72	64	90	69	63	85	68	14.8
Beek	83	67	79	65	76	64	69	79	67	76	66	74	66	97	74	64	92	71	63	87	70	16.4
De Bilt	82	66	79	65	75	64	69	79	67	75	65	73	65	92	73	64	88	70	62	83	69	16.0
Eindhoven	83	67	80	65	77	64	69	80	67	77	65	74	65	92	73	63	87	70	62	83	69	17.8
Gilze/Rijen	82	66	79	65	76	63	68	79	67	76	65	73	65	92	72	63	87	69	62	83	68	17.3
Groningen	81	67	77	65	74	64	69	77	67	74	65	71	66	95	74	64	90	70	63	85	68	17.5
Leeuwarden	79	66	75	64	71	63	67	76	65	72	64	70	64	90	70	63	85	68	61	81	67	13.7
Rotterdam	80	67	77	65	74	64	69	78	67	75	65	72	66	96	73	64	90	71	63	85	68	14.6
NETHERLANDS ANTILLES																						
Willemstad	91	80	90	80	90	79	82	89	81	88	81	87	79	153	86	79	153	86	79	150	86	9.5
NEW CALEDONIA																						
Noumea	88	76	86	76	85	75	79	85	78	84	77	82	77	143	83	76	139	81	75	134	80	9.4
NEW ZEALAND																						
Auckland	77	66	76	66	74	66	70	75	69	73	67	72	68	104	72	67	99	71	66	94	70	11.3
Christchurch	83	62	79	61	76	60	65	77	64	74	62	71	62	83	67	60	78	67	59	74	65	17.5
Taiaroa Head	69	57	66	57	64	56	61	65	60	63	59	62	59	76	62	58	74	61	57	71	60	8.6
Wellington (934340)	74	64	71	63	70	62	66	71	65	69	64	68	64	90	69	63	86	67	62	83	67	9.7
NIGER																						
Agadez	108	67	107	67	105	66	75	92	74	92	73	91	71	122	81	70	116	82	68	111	82	22.5
Niamey	108	71	106	71	105	70	80	95	79	94	78	93	76	142	85	76	138	84	75	135	84	23.8
NORWAY																						
Bergen	73	59	68	57	65	55	61	68	59	64	58	63	59	74	61	57	70	60	56	66	59	11.5
Bodo	70	59	66	57	63	55	60	67	58	64	56	62	57	69	61	55	65	59	54	62	58	9.0
Oslo/Fornebu	80	63	77	62	73	60	65	75	63	71	62	69	62	83	67	60	78	66	59	74	65	15.8
Oslo/Gardermoen	78	60	75	58	71	57	62	73	60	70	59	67	59	76	63	57	71	62	55	66	61	18.0
Stavanger	73	59	70	58	66	57	62	69	60	66	59	64	60	77	64	58	74	62	57	69	61	11.3
Svinoy (Lgt-H)	64	57	61	56	60	56	59	62	58	60	56	59	58	71	60	56	68	59	55	65	58	4.1
Tromso	68	57	64	55	61	54	59	65	56	63	55	60	56	66	61	54	62	59	52	57	57	10.8
Trondheim	75	60	71	59	68	58	63	70	61	68	60	65	61	80	66	59	75	64	57	70	61	12.4
Utsira	67	58	64	58	61	57	60	64	59	62	58	60	59	76	62	58	73	60	57	70	59	5.2
OMAN																						
Masqat	109	73	107	73	105	73	86	93	85	93	84	92	84	181	91	83	174	91	82	169	90	14.9
Salalah	92	71	91	76	90	76	82	88	82	87	81	87	81	160	86	80	158	86	79	152	85	9.7
Thamarit	108	69	106	68	104	68	79	94	78	91	76	90	76	142	86	74	134	84	73	130	82	25.2
Tur'at Masirah	99	74	96	76	94	77	84	90	82	89	82	88	82	168	88	81	160	86	80	156	86	15.5
PANAMA																						
Panama	95	76	93	77	92	77	82	89	81	89	81	88	80	158	86	79	152	85	79	150	85	15.8
Tocumen	93	78	92	77	91	77	81	89	80	88	80	88	79	150	85	78	148	85	77	143	84	17.5
PARAGUAY																						
Asuncion	98	75	95	75	94	75	80	91	79	90	78	89	77	144	86	76	136	83	75	134	83	18.5
PERU																						
Arequipa	75	55	74	54	73	53	59	71	58	69	57	69	55	90	64	54	85	62	53	83	61	23.4
Cuzco	72	52	71	52	70	51	55	67	54	67	53	66	50	80	61	49	76	60	48	74	59	23.9
Iquitos	93	80	92	80	91	80	82	91	81	90	81	90	79	153	87	79	151	87	78	147	87	17.1
Lima	86	75	84	74	82	73	76	83	75	81	74	80	74	126	80	73	124	80	72	118	79	11.5
Pisco	86	75	83	73	82	72	76	84	74	82	73	80	73	123	83	72	117	81	70	111	79	12.4
Talara	90	76	88	75	87	74	79	86	78	84	77	83	77	143	82	76	139	82	75	134	81	14.2
PHILIPPINES																						
Angeles, Clark AFB	97	78	95	77	93	77	82	89	82	89	81	87	81	164	86	80	161	86	79	154	85	17.6
Baguio	82	71	79	70	77	69	74	78	72	77	71	76	72	145	77	71	135	75	69	130	74	14.8
Cebu/Mandaue	93	81	92	81	91	80	82	90	82	90	81	89	80	153	87	79	152	87	79	150	87	12.4
Olongapo	98	77	96	77	95	78	83	91	82	90	81	89	81	160	88	79	152	86	79	150	86	17.1
Manila, Aquino Airport	95	81	93	80	92	79	83	91	82	90	82	89	81	161	89	80	158	88	79	152	87	15.8
POLAND																						
Bialystok	81	66	78	65	75	64	69	78	67	75	65	73	66	97	74	64	90	71	62	84	69	19.1
Gdansk	80	65	77	63	73	62	67	77	65	73	63	71	64	91	70	61	82	68	59	77	67	17.5
Katowice	83	67	80	65	77	64	68	80	67	77	65	74	64	94	72	63	88	71	62	85	69	18.4
Kielce	83	67	80	65	76	64	68	79	67	76	65	74	65	94	73	63	90	71	62	84	69	20.2
Kolobrzeg	80	65	75	63	71	63	67	74	65	72	64	70	64	90	70	63	85	68	61	80	67	12.1

MDB = mean coincident dry-bulb temp., °F
MWB = mean coincident wet-bulb temp., °F
MWS = mean coincident wind speed, mph
StdD = standard deviation, °F
HR = humidity ratio, grains of moisture per lb of dry air
A = airport
DP = dew-point temperature, °F

Table 3A Heating and Wind Design Conditions—World Locations

Station	WMO#	Lat.	Long.	Elev., ft	StdP, psia	Dates	Heating Dry Bulb 99.6%	99%	Extreme Wind Speed, mph 1%	2.5%	5%	Coldest Month 0.4% WS	MDB	1% WS	MDB	MWS/PWD to DB 99.6% MWS	PWD	0.4% MWS	PWD	Extr. Annual Daily Mean DB Max.	Min.	StdD DB Max.	Min.
1a	1b	1c	1d	1e	1f	1g	2a	2b	3a	3b	3c	4a	4b	4c	4d	5a	5b	5c	5d	6a	6b	6c	6d
Krakow	125660	50.08N	19.80E	777	14.292	8293	−1	6	20	18	16	24	41	20	36	3	60	6	240	89	−3	3.2	9.4
Lodz	124650	51.73N	19.40E	616	14.375	8293	2	9	22	20	17	24	38	20	34	5	90	8	130	92	0	4.0	11.2
Lublin	124950	51.22N	22.40E	787	14.287	8293	−2	5	21	18	16	21	29	19	30	5	180	5	220	88	−2	2.5	10.6
Poznan	123300	52.42N	16.83E	301	14.540	8293	3	11	22	19	17	23	35	20	35	3	90	8	220	92	3	3.8	11.2
Przemysl	126950	49.80N	22.77E	918	14.219	8293	1	7	23	21	19	27	34	23	33	5	270	6	250	86	−2	1.8	9.9
Snezka	125100	50.73N	15.73E	5291	12.098	8293	−3	2	80	68	62	85	20	78	20	36	340	13	200	70	−3	3.4	9.5
Suwalki	121950	54.13N	22.95E	610	14.379	8293	−5	2	26	21	18	27	36	22	30	3	20	7	300	87	−5	4.0	8.8
Szczecin	122050	53.40N	14.62E	9	14.695	8293	7	13	22	19	17	23	42	20	38	4	40	9	220	90	7	4.3	11.0
Torun	122500	53.03N	18.58E	236	14.575	8293	1	9	17	15	13	16	36	15	34	4	30	7	110	91	0	2.9	11.5
Warsaw	123750	52.17N	20.97E	351	14.514	8293	0	8	24	21	19	25	31	22	34	5	90	9	150	91	0	3.8	11.2
Wroclaw	124240	51.10N	16.88E	396	14.490	8293	2	10	21	18	16	22	42	19	39	4	110	8	170	92	0	3.4	10.8
PORTUGAL																							
Beja	85620	38.02N	7.87W	810	14.275	8293	36	38	22	19	17	23	55	19	53	7	90	10	180	104	32	2.2	3.1
Braganca	85750	41.80N	6.73W	2270	13.533	8293	26	28	22	19	16	24	38	20	43	2	180	8	240	97	22	2.5	3.6
Coimbra	85490	40.20N	8.42W	459	14.458	8293	35	38	22	17	14	23	54	19	53	4	180	6	310	101	32	2.2	2.2
Evora	85570	38.57N	7.90W	1053	14.149	8293	37	39	23	20	18	23	50	20	48	11	320	8	300	100	33	2.5	3.4
Faro	85540	37.02N	7.97W	13	14.693	8293	41	43	24	21	19	25	57	22	57	5	20	11	110	97	36	2.7	3.8
Lisbon	85360	38.78N	9.13W	403	14.487	8293	39	41	23	20	18	22	55	19	55	4	50	11	330	101	35	3.1	2.5
Portalegre	85710	39.28N	7.42W	1935	13.700	8293	34	37	24	20	18	24	48	22	47	11	290	8	240	98	30	2.7	3.2
Porto	85450	41.23N	8.68W	239	14.573	8293	35	37	24	21	18	26	54	22	54	7	90	9	330	95	30	2.3	2.5
Viana Do Castelo	85430	41.70N	8.80W	59	14.669	8293	33	35	19	16	13	20	55	17	56	1	50	6	160	97	29	2.3	1.8
PUERTO RICO																							
Ceiba, Roosevelt Rds	785350	18.25N	65.63W	39	14.679	8293	68	70	17	16	14	18	79	17	79	2	330	11	80	94	66	4.1	1.1
San Juan	785260	18.43N	66.00W	62	14.667	8293	69	69	19	17	16	19	81	17	81	3	190	12	170	94	56	2.2	22.3
QATAR																							
Ad Dawhah	411700	25.25N	51.57E	32	14.683	8293	51	53	25	22	20	21	65	19	64	7	290	15	350	115	43	1.8	7.0
ROMANIA																							
Bucharest	154200	44.50N	26.13E	298	14.542	8293	8	14	20	17	15	20	28	18	29	3	250	5	230	97	2	3.8	6.5
Cluj-Napoca	151200	46.78N	23.57E	1354	13.994	8293	4	8	20	17	14	19	32	15	30	2	270	6	140	90	−4	3.4	5.9
Constanta	154800	44.22N	28.63E	45	14.676	8293	15	19	31	26	23	35	29	31	34	12	360	8	180	91	11	5.4	6.1
Craiova	154500	44.23N	23.87E	639	14.363	8293	10	15	32	23	19	31	30	23	38	5	270	4	180	97	5	3.8	8.5
Galati	153100	45.50N	28.02E	236	14.575	8293	7	12	27	21	19	28	25	22	27	9	20	8	230	94	4	3.6	5.4
Omul Mountain	152800	45.45N	25.45E	8231	10.823	8293	−13	−7	89	75	54	90	8	89	8	33	230	6	230	73	−18	13.7	10.4
Satu Mare	150100	47.78N	22.88E	406	14.485	8293	0	6	22	19	16	23	33	19	33	2	90	5	230	93	−4	4.0	7.7
Timisoara	152470	45.77N	21.25E	288	14.547	8293	9	14	19	17	14	18	33	15	34	4	360	6	200	97	2	2.9	7.7
RUSSIA																							
Abakan	298650	53.75N	91.40E	803	14.278	8293	−29	−24	23	20	17	22	18	19	18	1	350	5	50	91	−32	3.6	6.3
Aldan	310040	58.62N	125.37E	2237	13.550	8293	−41	−37	14	12	11	14	−2	12	−1	2	200	5	180	86	−45	1.8	5.9
Aleksandrovsk-Sahal	320610	50.90N	142.17E	102	14.646	8293	−17	−13	31	27	22	35	6	29	18	7	130	12	220	80	−22	4.5	4.9
Anadyr’	255630	64.78N	177.57E	203	14.592	8293	−37	−34	48	40	34	52	10	47	14	14	320	9	140	73	−38	5.4	4.9
Apuka	259560	60.45N	169.58E	26	14.686	8293	−18	−14	39	33	30	44	12	39	10	15	60	12	270	68	−24	3.8	4.9
Arkhangel’sk	225500	64.53N	40.47E	42	14.677	8293	−29	−23	17	15	13	18	22	16	23	2	130	6	140	85	−32	4.9	7.9
Armavir	370310	44.98N	41.12E	524	14.423	8293	4	10	23	19	15	24	31	21	32	2	140	6	90	97	−5	4.5	5.8
Astrakhan’	348800	46.27N	48.03E	59	14.669	8293	−1	4	26	21	19	27	22	22	22	7	270	11	90	100	−7	4.5	9.2
Barnaul	298380	53.40N	83.70E	826	14.266	8293	−21	−16	29	23	20	32	17	26	19	7	170	9	60	91	−25	7.0	9.2
Blagoveshchensk	315100	50.25N	127.50E	449	14.463	8293	−27	−23	19	17	14	18	4	14	1	2	310	7	180	91	−31	3.8	5.0
Borzya	309650	50.38N	116.52E	2244	13.546	8293	−37	−32	26	22	19	19	−1	16	−6	3	80	10	150	89	−42	3.2	4.9
Bratsk	303090	56.07N	101.83E	1604	13.868	8293	−32	−27	22	17	15	19	7	15	7	3	280	7	130	87	−36	4.0	7.9
Bryansk	268980	53.33N	34.23E	711	14.326	8293	−8	−3	23	20	18	25	29	21	24	7	110	10	220	86	−12	3.1	6.8
Chelyabinsk	286420	55.30N	61.53E	744	14.309	8293	−19	−15	29	24	21	29	17	25	15	7	340	12	190	90	−23	4.3	5.9
Cherepovets	271130	59.12N	37.93E	429	14.473	8293	−25	−17	22	18	15	23	26	20	23	2	20	6	170	84	−32	3.1	9.2
Chita	307580	52.02N	113.33E	2247	13.545	8293	−33	−29	24	21	19	21	10	17	9	0	310	8	210	90	−36	3.4	3.4
Dudinka	230740	69.40N	86.17E	62	14.667	8293	−50	−44	32	27	23	29	−1	24	−2	6	100	10	60	82	−55	4.9	4.5
Egvekinot	253780	66.35N	179.12W	85	14.655	8293	−34	−29	35	30	26	41	1	36	1	2	190	8	160	75	−36	9.4	6.7
Groznyy	372350	43.35N	45.68E	531	14.420	8293	5	10	24	21	18	25	28	22	29	3	270	9	90	96	−2	4.1	4.7
Habarovsk/Novy	317350	48.52N	135.17E	236	14.575	8293	−22	−18	23	20	18	21	2	19	−1	4	200	9	250	90	−26	3.1	2.9
Irkutsk	307100	52.27N	104.35E	1683	13.828	8293	−29	−24	23	21	18	21	2	18	2	4	80	8	190	87	−35	2.7	6.8
Izhevsk	284110	56.82N	53.27E	518	14.427	8293	−21	−16	24	21	18	26	18	23	19	5	100	11	160	89	−27	4.7	4.3
Juzno-Kurilsk	321650	44.02N	145.87E	131	14.630	8293	10	13	34	29	25	36	21	32	22	13	320	8	320	76	9	4.7	4.9
Juzno-Sahalinsk	321500	46.92N	142.73E	102	14.646	8293	−11	−8	19	16	14	21	11	18	13	3	360	7	180	83	−16	2.5	4.1
Kaliningrad	267020	54.70N	20.62E	88	14.653	8293	−3	6	16	14	12	16	36	14	34	4	350	5	120	88	1	3.8	9.7
Kaluga	277030	54.57N	36.37E	659	14.353	8293	−12	−7	21	18	16	23	20	18	21	3	340	9	130	85	−23	2.5	17.6
Kazan’	275950	55.78N	49.18E	380	14.499	8293	−18	−12	28	24	21	30	14	27	14	9	330	10	170	88	−23	3.4	5.4
Kirov	271960	58.65N	49.62E	482	14.446	8293	−27	−18	22	19	17	23	22	20	18	7	250	9	90	87	−31	3.4	6.8
Kolpashevo	292310	58.30N	82.88E	249	14.568	8293	−37	−30	20	17	15	19	14	16	13	3	290	7	160	86	−41	3.1	8.3
Krasnodar	349290	45.03N	39.15E	108	14.643	8293	3	9	23	20	18	25	41	21	30	7	50	8	90	94	−4	2.3	8.3
Krasnoyarsk	295740	56.00N	92.88E	908	14.224	8293	−29	−24	23	19	16	26	14	22	18	3	290	6	50	90	−31	6.3	7.0

WMO# = World Meteorological Organization number
Lat. = latitude, °
Elev. = elevation, ft
Long. = longitude, °
PWD = prevailing wind direction, ° True
StdP = standard pressure at station elevation, psia
DB = dry-bulb temperature, °F
WS = wind speed, mph

Table 3B Cooling and Dehumidification Design Conditions—World Locations

	Cooling DB/MWB						Evaporation WB/MDB						Dehumidification DP/MDB and HR									
	0.4%		1%		2%		0.4%		1%		2%		0.4%			1%			2%			Range
Station	DB	MWB	DB	MWB	DB	MWB	WB	MDB	WB	MDB	WB	MDB	DP	HR	MDB	DP	HR	MDB	DP	HR	MDB	of DB
1	2a	2b	2c	2d	2e	2f	3a	3b	3c	3d	3e	3f	4a	4b	4c	4d	4e	4f	4g	4h	4i	5
Krakow	85	69	81	67	77	65	70	82	68	79	67	76	66	99	76	64	93	73	63	88	71	19.6
Lodz	84	66	80	65	77	63	68	80	66	77	65	74	64	92	72	62	86	71	61	82	70	18.7
Lublin	82	67	79	66	76	64	69	80	67	77	65	74	65	96	74	64	90	72	62	85	70	18.0
Poznan	85	66	81	64	78	63	68	80	67	78	65	75	64	92	72	63	86	71	61	80	69	19.6
Przemysl	82	67	79	66	76	64	69	79	67	76	66	74	66	98	74	64	92	72	63	88	70	14.9
Snezka	64	55	60	54	58	52	57	61	55	58	53	57	55	79	58	53	74	57	52	69	55	7.9
Suwalki	80	66	77	65	74	63	68	77	66	74	64	72	65	94	73	63	88	70	61	83	68	18.5
Szczecin	83	67	80	66	77	64	69	80	67	77	66	74	66	96	74	64	90	72	62	84	70	16.9
Torun	84	67	80	65	77	64	69	80	67	77	65	75	65	92	72	63	86	71	61	80	70	18.4
Warsaw	84	67	81	66	77	64	70	82	68	78	66	76	66	97	76	64	91	72	63	86	71	19.8
Wroclaw	84	67	81	66	78	64	69	81	67	78	66	76	65	94	73	63	88	71	62	83	71	19.1
PORTUGAL																						
Beja	99	70	95	68	92	67	71	94	70	92	69	89	65	94	74	63	89	74	62	85	73	29.5
Braganca	92	65	88	64	85	63	67	88	66	85	64	83	60	85	72	59	81	70	58	78	69	24.3
Coimbra	93	70	89	69	85	68	72	88	71	85	69	83	67	100	79	65	96	76	64	92	74	21.4
Evora	96	68	93	66	89	66	69	90	68	88	67	85	64	93	70	63	89	71	62	85	69	23.6
Faro	89	69	86	68	84	69	73	82	72	80	71	80	71	113	77	69	108	77	68	103	76	17.1
Lisbon	93	69	90	68	86	68	73	87	71	83	70	81	68	106	76	68	103	76	66	97	75	18.9
Portalegre	94	66	91	65	88	64	68	89	67	87	66	85	62	90	69	61	85	69	59	81	69	19.4
Porto	86	67	82	66	79	65	69	81	68	78	67	75	66	98	72	65	93	70	64	92	69	17.3
Viana Do Castelo	90	70	86	69	82	67	72	87	70	83	69	80	67	100	76	66	96	74	65	92	72	18.7
PUERTO RICO																						
Ceiba, Roosevelt Rds	90	78	89	78	88	77	80	87	80	87	79	86	79	150	85	77	143	84	77	141	84	10.1
San Juan	92	77	90	78	89	78	81	88	80	87	79	87	78	147	85	78	146	84	77	143	84	12.2
QATAR																						
Ad Dawhah	109	71	107	72	105	72	87	94	86	93	85	93	85	184	92	84	180	91	83	171	92	19.4
ROMANIA																						
Bucharest	91	72	88	70	86	69	74	87	73	85	71	83	71	118	78	69	109	77	68	104	75	23.9
Cluj-Napoca	85	68	82	67	79	65	71	80	69	78	67	76	68	107	75	65	99	73	64	93	71	20.5
Constanta	83	72	81	72	79	71	75	80	74	79	72	78	74	126	78	72	118	77	71	113	76	12.2
Craiova	92	74	89	73	86	71	77	89	74	86	73	83	73	127	83	71	118	80	69	110	79	22.0
Galati	89	72	86	70	84	69	74	84	72	83	71	80	71	116	80	69	108	77	67	102	75	20.2
Omul Mountain	58	50	54	48	52	47	52	56	50	53	48	51	50	74	54	49	69	52	47	64	49	11.2
Satu Mare	88	71	85	69	82	68	72	85	71	83	69	80	68	106	78	67	100	76	65	94	74	23.2
Timisoara	92	70	88	69	85	67	71	87	70	85	69	82	67	100	75	66	96	74	64	91	73	23.0
RUSSIA																						
Abakan	85	64	82	63	78	62	67	81	66	77	64	75	63	90	71	62	85	70	60	80	69	18.9
Aldan	81	61	77	60	74	59	64	77	62	74	60	71	60	83	67	58	77	66	56	72	65	18.4
Aleksandrovsk-Sahal	74	64	71	63	68	61	67	72	65	69	63	67	65	92	69	63	85	67	61	80	65	11.3
Anadyr'	65	56	62	55	59	53	57	64	55	61	53	58	54	62	60	52	58	58	50	55	56	9.7
Apuka	61	55	58	53	56	52	56	60	54	57	53	55	54	62	57	53	60	56	51	56	54	8.5
Arkhangel'sk	79	65	75	63	72	62	67	76	65	73	63	70	64	90	72	62	83	69	60	77	67	16.9
Armavir	90	69	87	69	84	68	73	83	71	82	70	81	70	111	79	68	104	77	66	98	75	21.8
Astrakhan'	94	71	91	69	88	68	74	87	72	85	71	83	71	113	79	69	105	78	67	99	76	19.4
Barnaul	85	66	81	64	78	63	68	80	67	78	65	75	64	93	73	63	88	72	61	83	70	16.9
Blagoveshchensk	86	69	83	69	80	67	73	81	72	78	70	77	71	116	77	69	109	76	68	104	74	16.6
Borzya	83	63	80	62	77	62	67	76	66	75	64	73	64	97	72	62	92	70	61	87	69	19.1
Bratsk	81	63	77	62	74	61	66	76	64	74	63	71	62	90	70	61	84	69	59	78	67	16.6
Bryansk	81	66	78	65	75	64	68	76	67	75	65	73	65	97	72	64	90	71	62	85	69	14.6
Chelyabinsk	85	66	82	65	79	64	69	80	67	78	66	76	65	94	74	63	90	72	62	85	70	16.6
Cherepovets	79	67	76	65	73	63	69	77	66	74	65	72	66	97	74	64	90	70	62	83	68	18.2
Chita	84	65	80	63	77	61	68	79	66	76	64	73	64	98	74	62	91	70	60	85	68	20.5
Dudinka	76	62	72	60	67	58	64	73	61	70	59	67	60	77	69	57	69	66	55	64	63	14.4
Egvekinot	64	53	60	52	57	50	55	62	53	60	51	56	51	56	57	50	53	54	48	50	52	9.2
Groznyy	91	71	88	69	86	68	73	87	71	84	70	82	69	108	79	67	102	77	66	99	75	18.5
Habarovsk/Novy	86	70	83	69	80	68	73	81	72	79	70	78	71	115	76	69	108	76	67	102	73	16.2
Irkutsk	81	63	78	62	74	61	66	76	65	74	63	72	63	92	70	61	86	68	59	80	67	20.0
Izhevsk	85	66	81	65	78	64	69	80	67	78	65	75	65	94	74	63	89	72	62	84	70	17.3
Juzno-Kurilsk	68	65	66	64	65	63	66	67	65	66	63	64	65	94	67	64	90	65	63	86	64	5.6
Juzno-Sahalinsk	78	68	75	67	72	65	70	75	68	72	66	70	69	106	73	67	99	71	64	91	69	14.0
Kaliningrad	80	66	77	64	74	62	68	78	66	74	64	72	64	91	72	62	85	70	61	79	68	15.3
Kaluga	80	67	77	66	74	63	69	78	67	75	65	73	65	97	74	64	91	72	62	85	69	16.6
Kazan'	85	67	81	66	78	64	70	81	68	77	66	75	66	98	75	64	91	73	62	85	71	16.6
Kirov	82	67	79	65	75	64	69	79	67	76	65	73	66	97	75	64	90	72	62	85	70	17.3
Kolpashevo	82	66	79	65	76	63	69	78	67	76	65	73	66	97	73	64	90	72	62	84	69	18.0
Krasnodar	90	71	87	70	84	69	74	85	72	83	71	81	70	111	79	69	106	78	67	100	77	21.1
Krasnoyarsk	83	65	80	63	77	62	68	79	66	76	64	73	64	93	72	62	88	70	61	82	69	19.1

MDB = mean coincident dry-bulb temp., °F
MWB = mean coincident wet-bulb temp., °F
MWS = mean coincident wind speed, mph
StdD = standard deviation, °F
HR = humidity ratio, grains of moisture per lb of dry air
A = airport
DP = dew-point temperature, °F

Table 3A Heating and Wind Design Conditions—World Locations

Station	WMO#	Lat.	Long.	Elev., ft	StdP, psia	Dates	Heating Dry Bulb 99.6%	Heating Dry Bulb 99%	Extreme Wind Speed, mph 1%	Extreme Wind Speed, mph 2.5%	Extreme Wind Speed, mph 5%	Coldest Month 0.4% WS	Coldest Month 0.4% MDB	Coldest Month 1% WS	Coldest Month 1% MDB	MWS/PWD to DB 99.6% MWS	MWS/PWD to DB 99.6% PWD	MWS/PWD to DB 0.4% MWS	MWS/PWD to DB 0.4% PWD	Extr. Annual Daily Mean DB Max.	Extr. Annual Daily Mean DB Min.	Extr. Annual Daily StdD DB Max.	Extr. Annual Daily StdD DB Min.
1a	1b	1c	1d	1e	1f	1g	2a	2b	3a	3b	3c	4a	4b	4c	4d	5a	5b	5c	5d	6a	6b	6c	6d
Kurgan	286610	55.47N	65.40E	259	14.563	8293	−25	−19	28	23	19	30	19	27	20	5	30	10	200	92	−31	3.8	7.6
Kursk	340090	51.73N	36.27E	688	14.338	8293	−9	−3	27	22	19	29	18	25	23	5	360	9	90	88	−14	2.3	10.1
Kyakhta	309250	50.37N	106.45E	2627	13.357	8293	−21	−18	19	14	12	12	6	9	6	0	150	6	180	91	−25	3.2	3.4
Magadan	259130	59.58N	150.78E	387	14.496	8293	−20	−15	22	20	17	24	18	21	15	1	50	9	270	73	−21	5.9	8.1
Magnitogorsk	288380	53.35N	59.08E	1253	14.046	8293	−19	−15	23	21	17	25	6	23	10	1	30	9	170	93	−25	5.4	3.6
Markovo	255510	64.68N	170.42E	108	14.643	8293	−54	−49	18	15	12	18	10	15	8	0	200	5	180	81	−59	2.9	6.3
Moscow	276120	55.75N	37.63E	511	14.430	8293	−10	−4	17	14	13	19	26	16	21	3	20	4	210	86	−14	3.2	5.4
Moscow, Vnukovo	275185	55.65N	37.27E	666	14.350	8293	−11	−6	23	21	18	27	27	23	27	6	10	9	240	86	−16	3.4	5.6
Murmansk	221130	68.97N	33.05E	167	14.611	8293	−20	−12	27	24	22	30	27	27	26	10	210	10	150	80	−19	3.6	10.3
Nikolayevsk	313690	53.15N	140.70E	223	14.582	8293	−28	−23	22	19	17	27	17	19	2	4	290	11	120	85	−31	3.8	4.5
Nikolskoe/Beringa	326180	55.20N	165.98E	20	14.690	8293	12	15	40	34	31	43	27	39	28	14	360	10	170	63	9	4.5	4.0
Nizhniy Novgorod	275530	56.22N	43.82E	269	14.558	8293	−17	−11	21	19	16	22	18	19	24	4	350	9	150	89	−23	3.2	5.8
Nizhniy Tagil	282400	57.88N	60.07E	846	14.256	8293	−26	−20	17	15	13	19	3	16	1	5	20	8	250	88	−30	3.4	7.6
Novokuznetsk	298460	53.73N	87.18E	1010	14.171	8293	−23	−17	29	24	21	33	20	28	18	4	340	7	150	88	−26	3.2	7.7
Novosibirsk	296340	55.03N	82.90E	580	14.394	8293	−25	−19	27	23	19	28	19	25	18	7	220	10	180	89	−29	4.5	8.6
Nyurba	246390	63.28N	118.33E	423	14.477	8293	−63	−58	17	14	12	13	−9	12	−13	2	20	6	20	92	−66	9.0	5.9
Olekminsk	249440	60.40N	120.42E	741	14.310	8293	−54	−48	17	14	12	15	−3	12	−2	2	90	6	90	91	−56	4.7	6.7
Omsk	286980	54.93N	73.40E	403	14.487	8293	−24	−18	25	21	18	26	17	22	16	7	210	10	120	93	−28	4.3	6.7
Orel	279060	53.00N	36.03E	666	14.350	8293	−10	−4	26	22	20	28	23	23	25	7	360	10	280	87	−14	2.7	7.4
Orenburg	351210	51.78N	55.22E	357	14.511	8293	−18	−13	27	23	21	29	19	26	19	5	260	11	160	98	−24	4.3	4.5
Ozernaja	325940	51.48N	156.48E	95	14.650	8293	4	7	44	36	31	45	28	38	26	11	80	16	90	67	−2	5.0	3.1
Penza	279620	53.13N	45.02E	570	14.399	8293	−14	−9	27	23	20	30	18	26	23	6	220	11	130	92	−21	4.9	5.6
Perm'	282250	58.02N	56.30E	564	14.403	8293	−24	−18	21	18	16	22	16	18	17	6	280	10	200	90	−28	4.1	6.7
Petropavlovsk-Kamca	325400	52.97N	158.75E	79	14.658	8293	5	8	30	25	21	30	24	27	22	8	360	8	330	76	0	3.6	3.1
Petrozavodsk	228200	61.82N	34.27E	367	14.506	8293	−19	−11	17	15	13	18	27	16	24	4	90	6	290	83	−19	2.5	10.4
Pskov	262580	57.80N	28.42E	137	14.627	8293	−13	−4	20	17	15	22	31	19	30	2	150	6	130	85	−14	4.0	9.0
Rostov-Na-Donu	347310	47.25N	39.82E	252	14.566	8293	2	5	31	27	23	35	28	31	20	11	80	11	110	93	−3	3.4	4.9
Rubtsovsk	360340	51.50N	81.22E	705	14.329	8293	−26	−20	31	27	23	33	20	30	23	4	360	8	30	96	−29	3.6	7.7
Ryazan'	277310	54.62N	39.72E	557	14.406	8293	−10	−6	22	18	15	23	24	20	24	6	310	7	140	88	−15	2.5	6.1
Rybinsk	272250	58.00N	38.83E	374	14.502	8293	−19	−12	20	18	16	21	20	19	23	5	320	8	210	85	−24	2.9	7.9
Samara (Kuybyshev)	289000	53.25N	50.45E	144	14.623	8293	−17	−12	25	22	18	24	21	22	20	2	320	8	110	93	−23	4.1	4.1
Saratov	341720	51.57N	46.03E	511	14.430	8293	−8	−5	27	23	20	29	28	25	21	13	300	11	70	91	−13	3.8	3.6
Smolensk	267810	54.75N	32.07E	790	14.285	8293	−9	−4	18	15	13	17	33	15	29	6	280	6	230	84	−12	3.2	6.3
Sochi	371710	43.45N	39.90E	52	14.672	8293	28	30	19	16	14	23	43	19	44	8	70	8	270	88	22	2.7	3.8
St Petersburg	260630	59.97N	30.30E	13	14.693	8293	−9	−2	19	16	14	23	32	19	31	3	40	6	180	84	−11	3.6	8.5
Svobodnyy	314450	51.45N	128.12E	646	14.360	8293	−35	−30	17	15	13	15	−1	14	−3	1	290	7	180	90	−41	3.6	5.0
Syktyvkar	238040	61.72N	50.83E	390	14.494	8293	−32	−25	19	17	15	20	19	18	18	4	10	7	140	87	−36	3.2	7.2
Tambov	279470	52.73N	41.47E	456	14.459	8293	−13	−8	26	23	20	29	25	23	27	5	360	9	140	92	−20	5.0	7.0
Tayshet	295940	55.95N	98.00E	990	14.181	8293	−34	−29	19	16	14	18	16	15	14	2	330	7	40	89	−39	3.1	6.8
Ufa	287220	54.75N	56.00E	344	14.518	8293	−24	−18	23	20	18	26	23	22	19	3	140	8	50	93	−30	6.5	6.5
Ulan Ude	308230	51.80N	107.43E	1673	13.833	8293	−33	−29	32	27	23	23	4	17	6	0	50	8	20	91	−37	4.0	4.1
Urup Island	321860	46.20N	150.50E	230	14.578	8293	12	14	46	40	35	50	25	45	24	24	320	11	180	71	11	3.8	3.2
Ust'ilimsk	301170	58.03N	102.73E	1318	14.013	8293	−40	−35	20	17	15	19	10	17	10	9	300	7	170	87	−46	2.5	11.0
Ust-Kamcatsk	324080	56.22N	162.47E	89	14.653	8293	−23	−18	29	25	22	34	27	29	25	6	360	9	180	76	−30	8.1	4.5
Vladimir	275320	56.13N	40.38E	557	14.406	8293	−16	−10	22	19	17	23	21	21	21	7	10	9	210	86	−21	3.1	7.6
Vladivostok	319600	43.12N	131.90E	604	14.382	8293	−8	−4	35	31	27	33	4	30	4	21	360	10	240	86	−12	2.3	5.6
Volgograd	345600	48.68N	44.35E	475	14.449	8293	−6	−2	32	28	25	34	21	29	19	9	340	15	110	96	−11	3.8	5.0
Vologda	270370	59.23N	39.87E	429	14.473	8293	−27	−18	20	17	15	22	17	18	21	4	330	7	190	84	−32	3.4	7.9
Voronezh	341220	51.70N	39.17E	505	14.434	8293	−9	−5	25	21	18	28	24	24	25	6	20	10	160	90	−14	3.4	7.6
Yakutsk	249590	62.08N	129.75E	337	14.521	8293	−61	−58	15	13	12	11	−35	9	−37	2	340	7	110	90	−62	4.3	4.7
Yekaterinburg	284400	56.80N	60.63E	777	14.292	8293	−22	−17	20	17	15	19	17	17	14	5	280	10	300	89	−27	4.0	6.5
Yelets	279280	52.63N	38.52E	551	14.410	8293	−11	−6	17	15	13	18	27	16	25	4	10	6	140	91	−16	6.8	7.7
Zyryanka	254000	65.73N	150.90E	141	14.625	8293	−56	−52	18	15	13	14	−19	12	−21	1	150	6	160	89	−59	3.1	4.1
SAMOA																							
Pago Pago	917650	14.33S	170.72W	9	14.695	8293	72	73	25	23	22	26	78	24	78	5	310	11	80	92	58	3.2	16.0
SAUDI ARABIA																							
Abha	411120	18.23N	42.65E	6837	11.413	8293	41	44	23	21	19	25	59	22	60	2	180	13	20	92	35	5.6	2.9
Al Jawf	403610	29.78N	40.10E	2244	13.546	8293	32	35	26	22	19	26	49	22	52	5	50	9	320	111	28	4.5	3.6
Al Madinah	404300	24.55N	39.70E	2070	13.633	8293	48	50	20	18	16	19	65	17	67	7	60	9	300	114	42	2.2	4.0
Al Wajh	404000	26.20N	36.47E	52	14.672	8293	53	55	28	25	23	27	68	24	69	7	20	13	270	105	47	3.4	4.7
Ar'ar	403570	30.90N	41.13E	1811	13.763	8293	32	34	22	19	17	21	51	18	52	4	270	7	240	111	28	2.0	3.1
At Ta'if	410360	21.48N	40.55E	4753	12.344	8293	42	45	23	21	19	24	65	22	66	5	90	10	50	101	35	3.2	2.9
Az Zahran	404160	26.27N	50.15E	55	14.670	8293	45	47	26	23	20	22	61	19	61	9	290	13	360	116	40	2.3	3.8
Ha'il	403940	27.43N	41.68E	3323	13.018	8293	31	34	23	20	18	23	57	20	60	4	180	8	180	108	25	3.6	4.9
Hafar Al Batin	403730	28.33N	46.17E	1165	14.092	8293	36	39	27	23	21	24	56	22	56	6	270	10	240	117	31	2.3	1.8
Jiddah	410240	21.67N	39.15E	39	14.679	8293	59	61	23	21	19	23	78	21	77	6	30	13	330	114	54	4.5	3.4
Jizan	411400	16.90N	42.58E	9	14.695	8293	68	70	20	18	16	17	82	16	82	5	100	10	230	107	58	4.5	10.3

WMO# = World Meteorological Organization number
Lat. = latitude, °
Elev. = elevation, ft
Long. = longitude, °
PWD = prevailing wind direction, ° True
StdP = standard pressure at station elevation, psia
DB = dry-bulb temperature, °F
WS = wind speed, mph

Table 3B Cooling and Dehumidification Design Conditions—World Locations

	Cooling DB/MWB						Evaporation WB/MDB						Dehumidification DP/MDB and HR									
	0.4%		1%		2%		0.4%		1%		2%		0.4%			1%			2%			Range
Station	DB	MWB	DB	MWB	DB	MWB	WB	MDB	WB	MDB	WB	MDB	DP	HR	MDB	DP	HR	MDB	DP	HR	MDB	of DB
1	2a	2b	2c	2d	2e	2f	3a	3b	3c	3d	3e	3f	4a	4b	4c	4d	4e	4f	4g	4h	4i	5
Kurgan	87	67	83	66	80	65	70	82	68	80	67	77	66	96	75	64	90	73	63	85	72	18.7
Kursk	83	67	80	66	78	65	69	78	68	77	66	75	66	99	74	64	92	72	62	87	71	17.1
Kyakhta	83	62	80	61	76	60	65	78	64	76	62	73	61	89	69	59	83	68	57	78	67	18.0
Magadan	65	55	61	53	59	52	57	62	55	59	54	57	54	64	58	53	61	56	52	58	55	9.7
Magnitogorsk	85	65	82	64	79	62	68	80	66	78	64	76	64	92	73	62	86	72	60	80	70	18.9
Markovo	76	61	72	59	69	57	62	73	60	70	58	67	58	71	67	55	66	63	54	62	62	18.4
Moscow	82	67	79	65	76	64	69	78	67	76	65	73	65	96	73	64	90	72	62	85	70	14.8
Moscow, Vnukovo	81	66	77	65	75	63	68	77	67	76	65	73	65	94	74	63	88	69	61	83	68	16.4
Murmansk	74	59	70	57	66	56	61	71	59	68	57	65	57	70	64	55	64	63	53	60	60	12.2
Nikolayevsk	78	67	75	65	73	64	69	76	67	73	65	71	66	95	73	64	90	71	62	85	68	15.5
Nikolskoe/Beringa	57	54	55	53	54	52	55	56	53	55	52	54	54	62	56	53	59	54	51	56	53	4.5
Nizhniy Novgorod	83	67	80	66	77	64	70	80	68	77	66	74	66	98	75	64	92	72	63	85	71	17.5
Nizhniy Tagil	83	65	80	64	76	63	68	78	66	76	64	74	64	93	72	63	88	71	61	83	68	18.4
Novokuznetsk	83	64	80	64	77	63	68	79	66	77	65	74	64	93	72	62	88	71	61	83	70	17.8
Novosibirsk	83	66	80	64	77	63	69	79	67	77	65	74	66	97	72	64	91	71	62	85	70	16.9
Nyurba	84	66	80	64	76	61	68	80	66	77	63	73	64	92	73	62	84	71	59	76	68	22.9
Olekminsk	85	65	81	64	77	62	67	81	65	78	64	74	63	88	71	61	82	72	59	76	68	20.2
Omsk	87	66	84	64	81	64	69	82	67	80	65	78	64	90	74	62	85	72	61	80	70	19.3
Orel	82	66	80	66	77	64	69	79	67	76	65	74	65	96	73	64	90	72	62	85	70	16.6
Orenburg	92	66	88	65	85	64	69	85	67	83	66	80	64	90	72	62	85	72	61	80	71	20.9
Ozernaja	60	56	58	54	56	53	57	60	55	57	54	55	56	66	58	54	62	56	53	59	55	5.9
Penza	85	65	82	64	79	63	69	81	67	78	65	75	65	94	74	64	90	72	62	85	70	18.7
Perm'	84	67	81	65	77	64	69	80	67	78	65	75	65	95	75	63	89	72	62	84	70	16.0
Petropavlovsk-Kamca	69	60	66	58	63	56	61	68	58	65	57	62	58	71	64	55	65	61	54	62	59	9.5
Petrozavodsk	77	64	73	62	70	60	66	74	64	71	62	69	64	90	70	61	82	67	59	76	65	13.7
Pskov	80	66	77	64	74	63	69	76	66	74	65	72	65	94	73	63	88	70	61	81	69	16.4
Rostov-Na-Donu	89	69	86	68	83	67	72	85	71	82	69	80	68	104	78	67	99	76	65	95	75	18.5
Rubtsovsk	87	67	84	66	81	64	69	81	68	79	66	77	66	98	73	64	92	71	62	87	71	20.2
Ryazan'	83	67	80	66	77	64	69	80	68	77	66	74	66	97	74	64	92	72	63	87	71	14.9
Rybinsk	80	68	77	65	74	64	69	78	67	74	65	72	66	96	73	64	90	72	63	86	69	13.0
Samara (Kuybyshev)	88	68	85	67	82	66	72	84	70	81	68	78	67	101	77	65	94	76	63	88	73	20.5
Saratov	87	66	84	65	81	64	69	82	68	80	66	78	65	94	74	63	89	73	62	84	72	15.1
Smolensk	79	66	76	65	74	63	68	76	66	74	65	72	65	95	72	63	90	71	62	84	69	15.1
Sochi	83	73	81	73	80	72	75	81	74	80	73	78	74	126	80	72	120	78	71	116	77	13.5
St Petersburg	79	65	76	64	73	62	67	76	66	74	64	71	64	90	72	62	84	69	61	79	68	13.5
Svobodnyy	85	67	82	67	79	66	72	80	70	77	68	76	69	109	75	67	103	73	66	97	72	18.7
Syktyvkar	83	66	79	65	75	62	69	80	67	76	65	73	65	94	74	63	88	71	61	82	68	17.1
Tambov	86	67	83	66	80	64	70	82	68	78	66	77	66	97	75	64	92	73	63	86	71	18.5
Tayshet	83	64	79	63	76	62	67	78	66	76	64	73	64	92	72	62	86	70	60	81	68	19.3
Ufa	87	67	83	66	80	65	70	82	68	80	67	78	66	98	76	64	91	73	63	86	72	19.1
Ulan Ude	85	63	82	62	78	61	67	78	65	76	63	74	62	90	72	61	84	70	59	79	68	20.7
Urup Island	63	59	60	57	57	56	60	62	58	59	56	57	60	78	61	57	71	58	55	65	56	9.2
Ust'ilimsk	82	63	78	62	75	61	66	76	64	74	62	73	63	90	70	61	83	69	58	77	67	20.2
Ust-Kamcatsk	67	57	63	56	60	54	59	64	57	62	55	59	56	66	60	54	63	59	53	60	57	10.6
Vladimir	82	67	78	66	75	64	69	79	67	76	65	73	66	97	74	64	91	71	62	85	70	15.3
Vladivostok	78	69	75	67	72	66	72	75	70	73	69	71	71	116	73	69	110	72	68	104	70	8.6
Volgograd	90	65	88	65	85	64	69	84	68	82	66	80	65	93	73	63	89	72	62	84	72	18.9
Vologda	80	66	76	64	74	63	69	78	67	73	65	71	65	96	74	64	90	71	62	83	68	16.9
Voronezh	85	66	82	65	79	63	69	80	67	78	66	75	65	94	73	64	90	72	62	85	71	17.5
Yakutsk	85	66	81	64	78	62	68	82	66	78	64	75	64	89	74	61	82	71	59	76	72	21.2
Yekaterinburg	84	66	81	65	77	64	69	80	67	77	65	74	65	96	75	64	90	71	62	85	70	17.1
Yelets	84	67	81	66	78	64	69	80	68	77	66	76	66	97	74	64	92	72	63	88	70	17.1
Zyryanka	83	63	79	62	74	60	66	79	64	76	61	72	60	79	72	58	73	70	56	66	68	17.3
SAMOA																						
Pago Pago	88	80	88	80	87	80	82	87	81	86	81	86	80	155	86	79	153	85	79	150	85	9.4
SAUDI ARABIA																						
Abha	87	56	86	56	84	56	67	76	66	75	65	74	65	119	72	63	111	71	62	108	71	21.4
Al Jawf	105	63	103	62	101	61	65	98	64	95	63	96	57	76	67	55	70	66	52	63	67	26.3
Al Madinah	113	65	110	65	109	64	69	97	68	97	67	98	62	91	74	60	82	75	57	76	76	23.8
Al Wajh	95	72	93	76	91	78	83	90	82	89	82	89	82	167	88	81	159	88	79	152	87	13.1
Ar'ar	107	68	105	68	103	67	73	103	71	100	69	98	62	90	92	60	81	90	57	75	86	25.6
At Ta'if	97	65	95	65	94	65	72	89	70	87	69	87	67	117	80	65	109	80	63	102	79	20.9
Az Zahran	111	71	109	71	107	72	86	94	84	94	83	92	84	180	91	82	168	90	80	158	90	23.9
Ha'il	105	65	104	64	102	64	68	98	67	97	65	97	59	85	71	57	80	72	55	74	71	28.3
Hafar Al Batin	113	67	112	66	110	66	71	99	69	100	67	100	64	94	74	61	85	72	59	78	71	27.5
Jiddah	104	72	102	73	100	74	83	94	82	94	81	92	81	159	89	79	151	89	79	150	88	22.0
Jizan	102	83	100	83	99	83	87	98	86	98	85	97	84	178	96	83	171	96	82	168	95	12.6

MDB = mean coincident dry-bulb temp., °F
MWB = mean coincident wet-bulb temp., °F
MWS = mean coincident wind speed, mph
StdD = standard deviation, °F
HR = humidity ratio, grains of moisture per lb of dry air
A = airport
DP = dew-point temperature, °F

Table 3A Heating and Wind Design Conditions—World Locations

Station	WMO#	Lat.	Long.	Elev., ft	StdP, psia	Dates	Heating Dry Bulb		Extreme Wind Speed, mph			Coldest Month				MWS/PWD to DB				Extr. Annual Daily			
												0.4%		1%		99.6%		0.4%		Mean DB		StdD DB	
							99.6%	99%	1%	2.5%	5%	WS	MDB	WS	MDB	MWS	PWD	MWS	PWD	Max.	Min.	Max.	Min.
1a	1b	1c	1d	1e	1f	1g	2a	2b	3a	3b	3c	4a	4b	4c	4d	5a	5b	5c	5d	6a	6b	6c	6d
Khamis Mushayt	411140	18.30N	42.80E	6738	11.456	8293	40	43	21	18	16	22	61	20	61	2	150	10	30	97	35	6.8	4.1
Makkah	410300	21.48N	39.83E	1017	14.168	8293	59	62	14	12	11	15	77	13	78	4	20	8	300	118	53	2.0	4.3
Qasim	404050	26.30N	43.77E	2132	13.602	8293	37	39	21	18	16	20	60	17	57	2	30	8	90	115	32	3.8	3.2
Rafha	403620	29.63N	43.48E	1466	13.937	8293	33	35	25	22	20	24	54	22	56	4	270	9	300	115	27	2.7	2.7
Riyadh	404380	24.72N	46.72E	2007	13.664	8293	41	44	22	19	17	21	60	18	60	4	320	11	360	115	35	1.4	2.7
Tabuk	403750	28.37N	36.63E	2526	13.407	8293	34	37	25	20	17	25	60	20	60	2	110	10	270	107	30	2.2	2.3
Turayf	403560	31.68N	38.67E	2667	13.337	8293	29	32	25	22	20	26	46	22	46	6	270	9	270	106	25	2.3	3.4
Yanbu'al Bahr	404390	24.15N	38.07E	3	14.698	8293	52	54	26	23	21	25	72	22	72	3	10	17	270	114	47	1.8	2.0
SENEGAL																							
Dakar	616410	14.73N	17.50W	78	14.658	8293	61	62	23	21	19	23	69	21	69	10	360	10	360	100	54	4.0	7.7
Saint Louis	616000	16.05N	16.45W	13	14.693	8293	60	61	23	20	18	23	76	20	76	7	40	11	80	108	53	2.9	3.6
Tambacounda	616870	13.77N	13.68W	164	14.613	8293	63	65	17	15	13	17	82	16	81	3	80	6	100	110	53	3.2	6.8
Ziguinchor	616950	12.55N	16.27W	75	14.660	8293	61	63	14	12	10	15	82	12	81	1	40	6	60	106	54	0.9	7.4
SINGAPORE																							
Singapore	486980	1.37N	103.98E	52	14.672	8293	73	74	18	16	14	18	83	17	84	4	330	11	30	93	65	2.0	12.1
SLOVAKIA																							
Bratislava	118160	48.20N	17.20E	427	14.475	8293	9	14	21	18	16	23	35	19	36	3	50	8	160	94	5	2.9	8.8
Chopok Mountain	119160	48.93N	19.58E	6601	11.516	8293	−6	−3	52	46	42	61	6	55	6	29	330	11	180	63	−8	2.0	6.5
Kosice	119680	48.70N	21.27E	761	14.300	8293	8	12	29	25	21	30	21	26	25	9	350	8	180	89	4	2.5	6.7
Lomnicky Stit (Peak)	119300	49.20N	20.22E	8645	10.652	8293	−12	−8	52	44	38	59	−1	51	2	23	310	6	180	58	−13	2.5	5.9
Zilina	118410	49.23N	18.62E	1033	14.159	8293	2	8	18	15	13	19	37	15	31	4	70	7	250	89	−3	2.2	7.0
SLOVENIA																							
Ljubljana	130140	46.22N	14.48E	1263	14.041	8293	9	13	14	11	9	12	34	10	35	1	290	7	130	93	3	4.5	5.8
SOUTH AFRICA																							
Bloemfontein	684420	29.10S	26.30E	4422	12.498	8293	26	28	24	21	19	21	55	18	58	1	220	12	270	98	22	2.5	3.1
Cape Town	688160	33.98S	18.60E	137	14.627	8293	38	41	32	29	26	31	57	28	58	0	40	12	170	94	34	2.9	1.4
Durban	685880	29.97S	30.95E	26	14.686	8293	50	52	27	23	21	24	70	21	69	1	340	14	30	93	46	2.2	2.0
Johannesburg	683680	26.13S	28.23E	5577	11.969	8293	34	37	22	19	17	19	55	18	54	9	210	9	300	89	29	1.8	3.1
Marion Island	689940	46.88S	37.87E	72	14.662	8293	30	32	60	53	47	59	38	52	43	17	200	22	290	68	24	9.0	4.3
Port Elizabeth	688420	33.98S	25.60E	196	14.596	8293	43	46	33	29	26	31	58	28	60	2	270	10	290	97	38	3.8	2.2
Pretoria	682620	25.73S	28.18E	4337	12.538	8293	39	41	14	12	11	13	61	11	59	1	220	4	270	94	35	2.9	2.0
SPAIN																							
Barcelona	81810	41.28N	2.07E	19	14.690	8293	32	35	21	17	15	21	50	18	48	8	350	9	210	90	28	2.9	3.4
Granada	84190	37.18N	3.78W	1833	13.752	8293	25	28	21	18	16	20	49	17	49	0	230	12	180	103	19	1.6	5.0
La Coruna	80010	43.37N	8.42W	219	14.584	8293	39	41	27	23	20	29	53	26	53	6	140	7	60	86	35	3.1	2.7
Madrid	82210	40.45N	3.55W	1909	13.714	8293	24	26	22	19	17	23	47	19	47	0	360	8	240	102	20	1.8	3.1
Malaga	84820	36.67N	4.48W	22	14.688	8293	39	41	27	23	20	32	55	28	56	10	320	13	320	103	33	3.2	3.1
Palma	83060	39.55N	2.73E	26	14.686	8293	31	33	23	20	18	24	54	21	54	0	60	10	60	99	26	3.2	2.2
Salamanca	82020	40.95N	5.50W	2608	13.366	8293	23	25	27	22	19	28	46	24	45	1	80	7	300	98	18	2.2	4.5
Santander	80230	43.47N	3.82W	213	14.587	8293	36	39	24	19	16	28	51	23	51	5	110	7	40	91	34	4.0	2.5
Santiago De Compostela	80420	42.90N	8.43W	1204	14.072	8293	30	32	22	19	16	23	51	21	49	3	90	6	280	97	24	4.1	4.0
Sevilla	83910	37.42N	5.90W	101	14.646	8293	34	37	20	18	15	20	55	18	55	2	30	8	240	109	30	2.7	3.1
Valencia	82840	39.50N	0.47W	203	14.592	8293	34	36	27	23	19	33	57	27	56	4	280	12	120	100	29	4.0	2.7
Zaragoza	81605	41.67N	1.05W	862	14.247	8293	28	30	28	24	22	29	46	26	49	5	10	7	90	101	25	10.3	4.7
SWEDEN																							
Goteborg, Landvetter	25260	57.67N	12.30E	554	14.408	8293	3	10	26	23	20	27	38	24	37	9	40	9	310	83	2	4.1	9.7
Goteborg, Save	25120	57.78N	11.88E	173	14.608	8293	3	10	27	24	21	28	41	25	39	5	50	9	290	82	3	3.1	9.2
Jonkoping	25500	57.77N	14.08E	761	14.300	8293	−4	5	25	22	20	27	41	24	38	7	30	10	50	83	−7	4.7	10.3
Kalmar	26720	56.73N	16.30E	52	14.672	8293	5	10	27	23	21	28	41	26	41	6	270	11	270	84	3	4.5	7.0
Karlsborg	25440	58.52N	14.53E	334	14.523	8293	2	9	27	23	20	30	38	26	38	11	50	6	190	81	3	3.8	10.1
Karlstad	24180	59.37N	13.47E	180	14.604	8293	−5	1	22	20	18	26	38	23	38	4	350	9	200	81	−5	3.6	9.9
Kiruna	20440	67.82N	20.33E	1482	13.929	8293	−22	−17	26	23	20	30	29	26	28	4	210	10	190	76	−26	2.3	5.4
Malmo	26360	55.55N	13.37E	347	14.516	8293	7	14	30	27	24	31	37	28	36	9	340	12	140	82	8	3.4	9.7
Ostersund/Froso	22260	63.18N	14.50E	1213	14.066	8293	−14	−7	27	23	20	35	34	28	32	3	320	7	280	80	−17	3.1	9.4
Soderhamn	23760	61.27N	17.10E	118	14.637	8293	−7	0	22	19	17	24	34	21	29	6	290	10	130	84	−7	2.9	7.6
Stockholm, Arlanda	24600	59.65N	17.95E	200	14.594	8293	−2	5	24	21	18	28	37	24	36	4	350	8	180	84	−1	3.4	10.3
Stockholm, Bromma	24640	59.35N	17.95E	36	14.681	8293	−1	5	21	19	17	20	37	18	36	4	320	9	200	84	−1	3.8	9.7
Sundsvall	23660	62.53N	17.45E	32	14.683	8293	−14	−8	24	20	17	28	39	23	32	3	310	10	140	82	−14	3.1	9.4
Ungskar	26660	56.03N	15.80E	9	14.695	8293	11	16	41	37	34	43	34	39	37	11	20	12	250	74	13	4.3	9.4
Uppsala	24580	59.88N	17.60E	134	14.629	8293	−4	3	24	21	19	28	35	25	37	6	330	9	230	82	−4	2.7	11.9
Visby	25900	57.67N	18.35E	154	14.618	8293	12	16	31	28	25	33	33	30	34	12	20	12	210	81	8	3.8	8.5
SWITZERLAND																							
Geneva	67000	46.25N	6.13E	1364	13.989	8293	18	23	20	17	15	21	38	18	38	6	230	8	210	92	14	2.0	6.8
Interlaken	67340	46.67N	7.88E	1902	13.717	8293	15	19	16	13	11	15	35	12	32	6	190	8	280	88	11	3.4	7.4
Jungfrau Mountain	67300	46.55N	7.98E	11732	9.448	8293	−15	−11	48	41	33	48	3	43	6	16	310	11	140	48	−19	6.5	6.8
La Chaux-De-Fonds	66120	47.08N	6.80E	3343	13.009	8293	6	12	20	17	15	22	35	19	33	3	230	7	250	86	−1	8.6	11.0
Locarno	67620	46.17N	8.88E	649	14.358	8293	21	24	16	13	11	16	43	13	42	4	90	6	240	89	16	3.6	5.2

WMO# = World Meteorological Organization number
Lat. = latitude, °
Elev. = elevation, ft
Long. = longitude, °
PWD = prevailing wind direction, ° True
StdP = standard pressure at station elevation, psia
DB = dry-bulb temperature, °F
WS = wind speed, mph

Table 3B Cooling and Dehumidification Design Conditions—World Locations

Station	Cooling DB/MWB						Evaporation WB/MDB						Dehumidification DP/MDB and HR									Range of DB
	0.4%		1%		2%		0.4%		1%		2%		0.4%			1%			2%			
	DB	MWB	DB	MWB	DB	MWB	WB	MDB	WB	MDB	WB	MDB	DP	HR	MDB	DP	HR	MDB	DP	HR	MDB	
1	2a	2b	2c	2d	2e	2f	3a	3b	3c	3d	3e	3f	4a	4b	4c	4d	4e	4f	4g	4h	4i	5
Khamis Mushayt	88	57	87	57	86	56	66	75	65	74	64	74	63	111	71	62	106	71	61	102	70	22.1
Makkah	113	76	111	76	109	75	82	104	81	102	80	101	77	147	93	76	139	94	74	133	94	27.2
Qasim	110	67	109	65	107	64	74	97	70	96	68	98	68	111	84	63	92	80	59	81	75	29.3
Rafha	111	69	109	68	107	67	72	104	70	104	69	103	63	90	75	61	83	72	58	75	75	29.7
Riyadh	111	64	110	64	108	64	69	96	67	97	66	96	62	91	73	60	83	72	57	76	72	25.2
Tabuk	104	64	102	63	100	63	68	95	67	95	65	94	59	80	77	56	74	77	55	71	76	26.6
Turayf	102	64	99	63	97	62	68	92	66	91	65	90	60	87	78	58	78	75	56	73	74	27.4
Yanbu'al Bahr	109	76	106	75	104	76	83	96	82	95	81	94	81	159	90	79	150	89	77	142	88	25.7
SENEGAL																						
Dakar	89	74	88	77	86	77	81	86	80	85	79	84	79	152	84	79	150	83	78	146	83	9.7
Saint Louis	101	69	97	69	94	69	82	88	82	87	81	86	81	160	85	80	156	85	79	153	84	16.2
Tambacounda	106	70	104	70	103	70	81	90	80	89	79	88	79	150	85	78	146	84	77	142	83	22.7
Ziguinchor	101	72	98	71	96	72	83	90	82	90	82	89	81	161	88	80	156	87	80	153	86	27.7
SINGAPORE																						
Singapore	91	79	90	79	90	79	81	88	81	87	80	87	79	152	84	79	152	84	79	150	84	11.3
SLOVAKIA																						
Bratislava	89	69	86	68	83	67	71	85	69	83	68	80	66	98	77	65	92	75	63	88	74	22.1
Chopok Mountain	59	52	56	51	54	50	54	57	52	55	51	53	52	75	54	51	71	53	49	67	52	7.9
Kosice	85	68	82	67	79	65	70	83	68	80	67	77	65	95	76	64	91	74	62	86	72	19.3
Lomnicky Stit (Peak)	53	47	51	45	49	43	49	52	47	50	45	48	47	66	50	45	62	48	43	57	46	8.5
Zilina	85	67	82	65	78	64	68	82	67	79	65	76	64	92	73	62	88	71	61	84	70	21.8
SLOVENIA																						
Ljubljana	86	68	83	67	80	65	70	83	68	80	67	78	65	96	74	64	92	72	62	88	72	22.3
SOUTH AFRICA																						
Bloemfontein	93	60	91	60	89	60	67	79	66	79	65	78	64	104	69	62	99	69	61	94	68	26.3
Cape Town	87	67	83	67	81	65	70	82	69	80	68	77	67	99	73	65	94	72	65	92	72	15.8
Durban	87	75	85	75	83	74	78	84	77	83	76	81	76	136	81	75	132	81	74	128	80	9.9
Johannesburg	84	60	82	60	80	60	65	77	65	76	64	75	62	104	69	61	99	68	60	96	67	18.7
Marion Island	57	54	55	52	53	51	55	56	53	54	52	53	54	62	55	53	59	54	51	56	52	8.1
Port Elizabeth	85	66	81	68	79	68	73	78	72	77	71	76	71	117	76	70	112	74	69	108	74	12.1
Pretoria	89	64	88	63	86	63	69	80	68	80	67	79	66	111	73	64	106	72	63	102	72	17.6
SPAIN																						
Barcelona	85	74	84	74	82	73	77	83	76	82	74	80	75	133	82	74	126	80	73	121	79	15.1
Granada	99	67	96	67	93	66	70	91	69	89	68	88	64	97	76	63	92	75	61	85	74	33.7
La Coruna	77	65	74	65	72	64	67	75	66	72	65	71	65	92	70	64	90	69	63	87	68	9.4
Madrid	97	69	94	68	92	67	71	94	70	91	68	89	64	96	82	62	90	80	60	84	79	29.2
Malaga	93	68	90	68	86	68	75	82	74	81	73	80	73	123	79	72	118	78	70	112	77	16.4
Palma	91	74	89	73	86	73	78	85	77	84	76	83	77	139	83	75	132	82	73	125	81	22.3
Salamanca	93	65	90	64	86	63	67	88	65	86	64	83	61	87	73	59	82	69	57	78	68	28.6
Santander	80	67	76	67	75	66	71	76	69	74	68	73	69	107	73	67	102	72	66	97	71	9.4
Santiago De Compostela	88	69	84	67	80	66	71	85	69	82	67	77	66	101	76	65	95	73	63	90	70	21.2
Sevilla	104	75	100	72	97	71	76	98	74	96	72	92	70	110	84	68	104	80	66	97	78	30.1
Valencia	90	71	88	72	86	72	76	85	75	83	74	82	74	127	81	73	125	80	72	118	80	16.6
Zaragoza	97	69	93	69	90	68	72	89	71	89	70	85	67	101	77	66	99	76	64	93	78	24.1
SWEDEN																						
Goteborg, Landvetter	78	62	75	60	72	59	64	74	62	71	60	68	61	80	66	59	76	64	57	71	62	14.9
Goteborg, Save	78	62	74	61	71	60	65	73	64	70	62	68	63	85	68	61	80	66	59	76	65	13.7
Jonkoping	79	61	75	60	72	58	64	74	62	71	60	68	61	81	65	59	76	64	57	72	62	19.6
Kalmar	79	63	75	62	72	60	66	75	64	72	62	70	63	85	69	61	79	67	59	74	65	18.7
Karlsborg	76	63	73	61	70	60	65	74	63	70	61	68	61	82	68	60	77	66	58	73	65	14.2
Karlstad	77	63	74	62	71	60	65	73	64	70	62	68	63	85	67	61	80	65	59	76	64	15.7
Kiruna	70	56	66	54	63	52	58	66	56	63	54	61	54	66	60	52	62	59	50	57	57	13.7
Malmo	77	62	74	61	71	60	66	72	64	70	62	68	64	90	68	62	85	66	60	80	65	14.2
Ostersund/Froso	74	58	70	57	67	55	61	71	58	67	57	65	56	70	63	54	66	61	53	62	60	13.5
Soderhamn	77	62	73	60	70	59	64	73	62	71	60	68	61	80	67	58	74	65	57	69	64	16.2
Stockholm, Arlanda	80	63	77	61	73	59	65	74	63	71	62	69	63	85	67	61	80	65	59	75	64	16.2
Stockholm, Bromma	79	63	76	61	72	60	66	74	64	72	62	69	63	86	68	61	80	66	59	75	65	15.8
Sundsvall	75	62	72	59	69	58	64	71	62	68	60	66	62	83	66	59	76	64	58	71	62	15.8
Ungskar	71	65	68	63	67	62	66	69	64	68	63	66	64	90	69	63	85	67	61	80	65	6.8
Uppsala	78	62	75	61	71	59	65	73	63	71	61	69	62	82	69	59	76	67	58	71	65	16.7
Visby	75	63	72	62	70	61	66	71	64	70	63	68	64	90	68	62	84	66	61	80	65	13.1
SWITZERLAND																						
Geneva	86	66	83	65	80	64	68	82	67	80	65	77	64	93	72	62	89	71	61	84	70	22.1
Interlaken	82	65	79	65	76	63	67	80	65	77	64	74	63	91	72	61	87	70	60	83	69	17.8
Jungfrau Mountain	43	33	41	32	39	32	37	40	35	38	34	37	35	47	38	33	43	36	32	41	35	6.8
La Chaux-De-Fonds	78	62	74	61	71	59	63	75	62	72	60	70	59	85	67	58	80	66	56	76	65	17.6
Locarno	84	70	82	69	80	67	72	82	71	80	69	78	69	108	78	67	102	76	66	97	74	17.8

MDB = mean coincident dry-bulb temp., °F
MWB = mean coincident wet-bulb temp., °F
MWS = mean coincident wind speed, mph
StdD = standard deviation, °F
HR = humidity ratio, grains of moisture per lb of dry air
A = airport
DP = dew-point temperature, °F

Table 3A Heating and Wind Design Conditions—World Locations

Station	WMO#	Lat.	Long.	Elev., ft	StdP, psia	Dates	Heating Dry Bulb		Extreme Wind Speed, mph			Coldest Month				MWS/PWD to DB				Extr. Annual Daily			
												0.4%		1%		99.6%		0.4%		Mean DB		StdD DB	
							99.6%	99%	1%	2.5%	5%	WS	MDB	WS	MDB	MWS	PWD	MWS	PWD	Max.	Min.	Max.	Min.
1a	1b	1c	1d	1e	1f	1g	2a	2b	3a	3b	3c	4a	4b	4c	4d	5a	5b	5c	5d	6a	6b	6c	6d
Lugano	67700	46.00N	8.97E	905	14.225	8293	25	28	18	14	11	17	45	13	45	3	360	6	190	90	17	3.1	9.9
Payerne	66100	46.82N	6.95E	1610	13.864	8293	14	19	18	15	13	20	47	18	42	4	40	7	230	90	11	3.1	7.6
Saentis (Aut)	66800	47.25N	9.35E	8202	10.835	8293	−3	1	41	36	32	44	16	40	17	14	260	6	210	63	−8	3.2	8.3
San Bernardino	67830	46.47N	9.18E	5374	12.061	8293	6	10	23	21	19	26	22	23	22	4	310	7	120	73	1	2.5	8.5
Zurich	66600	47.38N	8.57E	1866	13.735	8293	13	18	20	16	13	23	44	20	42	6	60	5	230	89	11	2.7	7.9
SYRIA																							
Damascus	400800	33.42N	36.52E	1984	13.676	8293	25	28	26	23	20	25	50	22	49	3	30	7	210	105	19	2.9	3.8
TAIWAN																							
Hsinchu	467570	24.82N	120.93E	88	14.653	8293	47	50	22	18	16	19	57	17	55	9	40	9	270	99	44	4.3	3.1
Hualien	466990	23.98N	121.60E	62	14.667	8293	53	55	19	15	13	19	63	16	62	6	250	8	140	96	48	6.1	2.5
Kaohsiung	467400	22.58N	120.35E	29	14.684	8293	52	55	21	17	15	19	66	17	65	8	360	13	280	97	48	3.6	2.5
T'aichung	467510	24.18N	120.65E	367	14.506	8293	46	48	20	17	16	20	62	18	61	5	30	10	240	98	37	1.6	4.1
T'ainan (593580)	467410	23.00N	120.22E	45	14.676	8293	51	53	20	17	15	19	62	18	61	11	20	11	200	98	47	5.2	2.9
Taipei	466960	25.07N	121.55E	19	14.690	8293	48	50	20	18	16	19	63	18	63	4	110	11	290	98	42	2.7	3.8
Taipei Intl Airport	466860	25.08N	121.22E	108	14.643	8293	48	50	29	27	25	28	56	26	57	16	40	16	260	97	43	1.4	3.2
TAJIKISTAN																							
Dushanbe	388360	38.55N	68.78E	2634	13.353	8293	19	23	13	10	8	13	41	9	42	2	60	3	270	105	13	6.3	5.8
Khujand (Leninabad)	385990	40.22N	69.73E	1404	13.969	8293	17	21	29	26	23	30	31	27	33	14	240	11	240	105	10	5.8	8.6
THAILAND																							
Bangkok	484560	13.92N	100.60E	39	14.679	8293	65	68	19	16	14	14	82	12	81	4	40	11	180	102	61	3.1	2.7
Chiang Mai	483270	18.78N	98.98E	1030	14.161	8293	53	56	17	13	11	13	72	10	75	1	360	7	190	103	48	1.6	1.8
Chiang Rai	483030	19.92N	99.83E	1295	14.025	8293	49	52	10	8	7	9	69	7	71	0	20	5	180	102	44	1.6	2.5
Chumphon	485170	10.48N	99.18E	16	14.691	8293	66	68	18	15	13	19	83	17	83	0	30	9	120	100	63	2.0	2.3
Hat Yai	485690	6.92N	100.43E	114	14.639	8293	70	71	18	16	14	19	83	16	83	0	330	7	240	100	67	2.0	1.4
Phetchabun	483790	16.43N	101.15E	380	14.499	8293	56	59	10	9	8	10	79	9	79	0	360	6	180	104	46	1.3	19.1
Phrae	483300	18.17N	100.17E	531	14.420	8293	55	58	10	8	7	7	76	7	79	0	30	5	240	105	48	2.0	5.8
Tak	483760	16.88N	99.15E	406	14.485	8293	57	60	18	14	12	8	80	7	80	0	270	5	270	106	52	1.3	2.9
TRINIDAD & TOBAGO																							
Port of Spain	789700	10.62N	61.35W	49	14.674	8293	68	70	19	17	16	20	84	18	84	0	90	12	90	94	60	2.0	11.7
TUNISIA																							
Bizerte	607140	37.25N	9.80E	9	14.695	8293	38	40	29	25	22	32	54	27	54	1	320	12	100	106	33	4.9	0.9
Gabes	607650	33.88N	10.10E	16	14.691	8293	42	44	20	17	15	22	59	19	57	4	230	10	250	107	36	5.0	3.4
Gafsa	607450	34.42N	8.82E	1030	14.161	8293	36	38	26	23	20	26	52	21	53	5	60	9	240	109	31	2.7	1.6
Kelibia	607200	36.85N	11.08E	98	14.648	8293	42	44	23	19	17	25	55	22	55	5	300	9	300	97	38	6.3	2.0
Qairouan (Kairouan)	607350	35.67N	10.10E	223	14.582	8293	40	42	17	14	12	16	54	13	57	2	240	7	180	111	35	3.1	2.9
Tunis	607150	36.83N	10.23E	13	14.693	8293	41	43	26	23	21	28	55	24	56	6	240	11	180	107	35	5.2	1.3
TURKEY																							
Adana	173500	37.00N	35.42E	216	14.585	8293	32	34	19	17	14	21	48	19	49	6	30	8	210	103	26	1.8	3.4
Ankara	171280	40.12N	32.98E	3113	13.120	8293	2	8	21	18	15	19	33	16	33	1	20	8	270	95	−2	4.9	9.5
Erzurum	170960	39.92N	41.27E	5767	11.884	8293	−23	−17	24	22	20	26	24	22	22	0	310	10	90	88	−28	3.2	7.0
Eskisehir	171240	39.78N	30.57E	2575	13.382	8293	12	16	19	17	15	19	32	16	31	3	120	9	320	96	6	2.5	6.3
Istanbul	170600	40.97N	28.82E	121	14.636	8293	26	29	23	21	20	26	32	23	37	14	360	13	60	95	21	4.1	6.3
Izmir/Cigli(Cv/AFB)	172180	38.50N	27.02E	16	14.691	8293	28	31	26	23	21	31	57	26	53	6	360	14	350	101	24	4.0	1.8
Malatya	172000	38.43N	38.08E	2785	13.279	8293	10	16	23	21	18	25	31	21	32	4	210	7	60	102	3	2.5	6.5
Van	171700	38.45N	43.32E	5449	12.027	8293	6	9	16	13	11	15	32	12	31	4	90	3	300	90	2	4.7	5.8
TURKMENISTAN																							
Ashgabat (Ashkhabad)	388800	37.97N	58.33E	688	14.338	8293	20	23	22	19	17	19	42	16	39	4	110	10	90	110	15	2.5	4.0
Dashhowuz (Tashauz)	383920	41.83N	59.98E	288	14.547	8293	5	10	22	19	18	21	33	19	33	6	200	10	360	108	−1	1.8	5.9
UNITED KINGDOM & NORTHERN IRELAND																							
Aberdeen/Dyce	30910	57.20N	2.22W	213	14.587	8293	22	27	29	25	22	33	41	29	42	3	360	11	170	77	13	2.9	8.6
Aberporth	35020	52.13N	4.57W	439	14.468	8293	26	29	41	36	32	47	44	42	46	14	90	13	130	80	24	4.9	4.1
Aughton	33220	53.55N	2.92W	183	14.603	8293	26	28	26	23	21	29	44	25	44	8	130	9	130	82	24	4.0	4.1
Aviemore	30630	57.20N	3.83W	721	14.321	8293	15	21	29	25	22	33	40	30	41	1	360	9	200	77	9	9.5	8.6
Belfast	39170	54.65N	6.22W	265	14.559	8293	27	29	28	25	22	32	43	29	44	4	180	9	110	77	22	3.8	3.4
Birmingham	35340	52.45N	1.73W	324	14.528	8293	21	24	23	20	18	26	44	23	43	4	70	9	100	84	15	4.5	7.7
Bournemouth	38620	50.78N	1.83W	36	14.681	8293	22	25	27	23	21	30	49	26	47	4	20	10	90	84	18	4.1	4.0
Bristol	37260	51.47N	2.60W	36	14.681	8293	26	29	24	21	18	27	47	23	47	8	70	9	90	85	23	4.0	5.4
Camborne	38080	50.22N	5.32W	288	14.547	8293	30	32	34	30	28	37	44	34	45	13	50	13	100	76	27	3.8	5.4
Cardiff	37150	51.40N	3.35W	219	14.584	8293	25	28	31	28	24	38	44	33	44	14	60	9	60	83	22	4.0	4.5
Edinburgh	31600	55.95N	3.35W	134	14.629	8293	21	25	28	25	22	33	47	30	44	2	250	9	250	79	16	3.8	5.9
Exeter	38390	50.73N	3.42W	98	14.648	8293	24	27	27	24	21	33	47	28	46	7	40	10	150	82	22	4.5	3.4
Finningley	33600	53.48N	1.00W	55	14.670	8293	23	26	28	24	22	31	45	27	47	4	170	10	150	84	18	4.3	3.8
Glasgow	31400	55.87N	4.43W	26	14.686	8293	21	24	30	26	23	39	46	34	45	2	270	9	230	80	15	3.4	5.6
Hemsby	34960	52.68N	1.68E	45	14.676	8293	27	29	30	26	23	34	42	31	40	15	70	12	230	81	23	3.4	5.2
Herstmonceux	38840	50.87N	0.33E	55	14.670	8293	24	27	31	27	23	34	47	30	47	9	40	9	130	80	18	9.7	9.0
Jersey/Channel Islands	38950	49.22N	2.20W	275	14.554	8293	28	31	33	29	26	39	47	33	46	13	100	10	70	83	24	4.0	5.9
Kirkwall	30170	58.95N	2.90W	68	14.663	8293	29	31	41	34	30	41	41	36	39	11	270	11	120	70	26	2.7	2.0

WMO# = World Meteorological Organization number
Lat. = latitude, °
Elev. = elevation, ft
Long. = longitude, °
PWD = prevailing wind direction, ° True
StdP = standard pressure at station elevation, psia
DB = dry-bulb temperature, °F
WS = wind speed, mph

Table 3B Cooling and Dehumidification Design Conditions—World Locations

	Cooling DB/MWB						Evaporation WB/MDB						Dehumidification DP/MDB and HR									
	0.4%		1%		2%		0.4%		1%		2%		0.4%			1%			2%			Range
Station	DB	MWB	DB	MWB	DB	MWB	WB	MDB	WB	MDB	WB	MDB	DP	HR	MDB	DP	HR	MDB	DP	HR	MDB	of DB
1	2a	2b	2c	2d	2e	2f	3a	3b	3c	3d	3e	3f	4a	4b	4c	4d	4e	4f	4g	4h	4i	5
Lugano	85	71	83	69	81	68	73	82	71	80	70	79	69	112	79	68	106	77	66	100	76	17.6
Payerne	84	67	81	66	78	64	69	81	67	78	65	76	64	95	74	63	91	73	61	86	70	19.8
Saentis (Aut)	57	47	55	46	53	44	50	54	48	52	47	51	48	67	51	46	63	49	45	60	48	7.7
San Bernardino	69	55	67	54	64	53	58	65	57	64	55	62	56	81	62	54	76	60	52	71	58	14.9
Zurich	83	66	80	65	77	63	67	80	66	77	65	74	64	94	71	62	90	70	61	86	69	16.0
SYRIA																						
Damascus	101	64	98	64	96	64	69	85	68	85	67	84	66	103	72	64	97	72	62	91	71	33.8
TAIWAN																						
Hsinchu	93	81	92	81	91	81	83	91	82	90	81	90	80	158	89	80	154	88	79	150	87	12.4
Hualien	90	80	89	80	88	80	82	88	81	87	81	87	80	156	87	79	153	86	79	149	86	9.7
Kaohsiung	92	79	90	79	90	79	82	88	81	87	80	87	80	158	85	79	151	84	79	150	84	11.5
T'aichung	94	82	93	82	91	81	84	92	83	91	82	90	82	169	90	81	162	89	80	159	89	15.1
T'ainan (593580)	92	81	91	81	90	80	83	89	82	88	81	88	81	162	86	80	158	85	80	155	85	9.9
Taipei	94	80	93	80	92	80	82	91	81	90	81	90	80	153	86	79	150	86	79	148	86	13.3
Taipei Intl Airport	93	80	92	80	91	80	82	90	82	90	81	89	80	159	88	79	152	86	79	149	86	13.1
TAJIKISTAN																						
Dushanbe	99	67	97	67	95	66	71	93	69	91	68	90	64	98	83	62	91	81	60	85	80	25.6
Khujand (Leninabad)	99	67	96	66	94	66	70	92	69	91	67	90	63	90	79	61	84	79	59	79	78	23.0
THAILAND																						
Bangkok	99	80	97	79	95	78	84	94	83	91	82	89	82	164	88	81	160	87	80	156	87	16.7
Chiang Mai	100	72	98	72	96	73	79	89	78	88	78	87	77	146	83	76	139	82	75	138	81	24.5
Chiang Rai	98	72	96	72	94	73	79	89	79	88	78	87	77	147	84	76	144	83	76	141	82	25.0
Chumphon	95	79	94	79	92	79	82	92	81	91	80	90	79	150	87	78	147	87	78	145	86	16.7
Hat Yai	95	77	94	77	93	77	80	89	80	89	79	88	79	149	83	78	144	82	77	143	82	18.0
Phetchabun	101	78	99	78	97	78	82	92	81	91	81	90	79	153	87	79	150	87	78	148	86	20.9
Phrae	101	76	99	77	97	76	81	92	81	90	80	90	79	153	86	78	149	85	78	146	85	22.1
Tak	102	74	101	74	99	74	80	90	79	89	79	88	78	146	83	77	144	83	76	140	83	18.7
TRINIDAD & TOBAGO																						
Port of Spain	91	77	90	77	90	77	80	87	79	87	79	86	78	147	83	77	143	82	77	141	82	14.2
TUNISIA																						
Bizerte	97	72	92	72	89	72	77	86	76	84	75	83	74	129	81	73	123	81	72	118	80	18.2
Gabes	96	71	92	73	89	73	80	87	78	86	77	85	77	142	85	76	136	85	75	130	84	11.7
Gafsa	105	68	101	69	98	68	73	92	72	90	71	89	69	110	79	67	104	78	66	99	78	23.8
Kelibia	89	73	86	73	85	73	78	83	77	82	76	81	76	138	81	75	133	81	74	127	80	13.1
Qairouan (Kairouan)	105	71	100	71	97	71	76	90	75	88	74	88	73	125	81	72	118	80	70	112	80	25.4
Tunis	98	73	94	73	91	73	78	88	77	86	76	85	76	134	82	74	128	82	73	123	81	21.8
TURKEY																						
Adana	97	71	94	71	92	72	79	89	78	87	77	86	76	137	82	75	133	82	74	128	82	19.8
Ankara	90	63	86	63	84	62	65	84	64	83	63	81	59	83	73	57	78	72	55	74	70	28.4
Erzurum	84	61	82	60	79	59	64	80	62	78	61	76	58	88	74	56	82	72	54	76	70	29.9
Eskisehir	90	68	87	67	85	66	71	85	69	83	67	83	66	105	80	64	97	78	62	92	75	25.9
Istanbul	86	70	84	69	83	69	74	82	72	80	71	78	72	117	78	70	111	76	68	105	76	15.3
Izmir/Cigli(Cv/AFB)	96	72	93	71	91	70	74	92	73	90	72	89	68	104	83	67	98	82	66	95	81	23.0
Malatya	97	68	95	67	93	66	70	94	68	93	67	91	61	88	87	59	82	85	57	77	83	27.4
Van	84	66	82	66	80	65	70	81	68	80	66	78	66	116	80	64	108	78	62	101	77	19.4
TURKMENISTAN																						
Ashgabat (Ashkhabad)	104	67	102	67	99	67	73	94	72	92	70	91	66	99	85	64	92	85	63	87	84	24.1
Dashhowuz (Tashauz)	103	74	99	73	96	71	77	98	75	95	74	93	70	112	92	68	105	90	66	98	88	24.3
UNITED KINGDOM & NORTHERN IRELAND																						
Aberdeen/Dyce	71	62	68	60	65	58	64	70	61	67	60	64	60	79	67	59	75	64	57	70	62	13.0
Aberporth	72	62	68	61	65	60	64	69	62	66	61	64	62	84	65	61	80	64	59	77	62	9.4
Aughton	75	64	72	62	69	60	65	73	63	69	62	67	62	84	68	61	80	66	59	76	64	10.8
Aviemore	75	61	71	60	67	58	63	72	61	69	59	65	59	76	65	57	71	65	56	67	62	15.5
Belfast	72	62	69	61	67	60	64	70	62	67	61	65	61	82	66	60	78	65	58	74	64	12.8
Birmingham	78	64	75	62	72	61	65	75	64	72	62	70	62	84	68	61	80	67	59	76	65	16.9
Bournemouth	78	65	75	63	72	62	66	76	65	72	63	69	63	86	69	62	83	67	61	80	66	18.0
Bristol	79	65	76	63	73	62	67	76	65	73	63	70	63	85	71	62	82	68	60	78	66	12.8
Camborne	71	62	68	61	66	60	64	67	63	66	62	65	63	86	65	62	83	64	60	79	63	8.8
Cardiff	77	64	74	63	71	62	66	74	64	71	63	68	63	87	69	62	83	66	60	79	65	14.8
Edinburgh	72	61	69	60	66	59	63	69	61	67	60	65	60	78	66	59	74	64	57	71	63	14.6
Exeter	78	65	75	64	72	62	67	76	65	73	64	70	64	88	70	62	84	68	61	81	67	15.8
Finningley	78	64	75	63	72	61	65	75	64	72	62	70	62	83	69	60	78	67	59	75	65	17.3
Glasgow	75	63	71	61	67	59	64	72	62	69	60	66	61	79	67	59	75	65	58	71	63	14.6
Hemsby	74	65	71	63	69	62	66	72	64	70	63	67	63	87	69	62	83	67	61	79	65	13.9
Herstmonceux	76	65	74	63	71	62	66	75	65	72	63	69	63	88	70	62	83	67	61	80	66	15.3
Jersey/Channel Islands	76	65	73	63	70	62	66	74	64	71	63	68	63	87	67	62	83	65	61	81	65	11.0
Kirkwall	64	59	62	57	60	56	60	63	58	61	57	59	58	73	62	57	69	59	56	66	58	9.2

MDB = mean coincident dry-bulb temp., °F MWS = mean coincident wind speed, mph HR = humidity ratio, grains of moisture per lb of dry air
MWB = mean coincident wet-bulb temp., °F StdD = standard deviation, °F A = airport DP = dew-point temperature, °F

Table 3A Heating and Wind Design Conditions—World Locations

							Heating Dry Bulb		Extreme Wind Speed, mph			Coldest Month				MWS/PWD to DB				Extr. Annual Daily			
												0.4%		1%		99.6%		0.4%		Mean DB		StdD DB	
Station	WMO#	Lat.	Long.	Elev., ft	StdP, psia	Dates	99.6%	99%	1%	2.5%	5%	WS	MDB	WS	MDB	MWS	PWD	MWS	PWD	Max.	Min.	Max.	Min.
1a	1b	1c	1d	1e	1f	1g	2a	2b	3a	3b	3c	4a	4b	4c	4d	5a	5b	5c	5d	6a	6b	6c	6d
Lerwick	30050	60.13N	1.18W	275	14.554	8293	28	30	43	38	34	45	42	41	42	14	350	12	180	66	24	3.4	2.0
Leuchars	31710	56.38N	2.87W	39	14.679	8293	24	27	32	28	24	38	44	33	43	5	250	13	240	78	19	4.7	4.9
London, Gatwick	37760	51.15N	0.18W	203	14.592	8293	22	25	23	20	18	26	45	23	43	3	80	8	70	85	16	4.5	6.1
London, Heathrow	37720	51.48N	0.45W	78	14.658	8293	25	28	22	20	18	26	47	22	44	6	20	10	90	87	21	4.1	4.1
Lyneham	37400	51.50N	1.98W	511	14.430	8293	22	26	26	23	20	29	43	26	41	11	30	10	70	83	18	4.1	6.3
Lynemouth	32620	55.02N	1.42W	98	14.648	8293	28	31	45	38	33	47	42	44	41	16	190	15	260	76	24	5.0	4.3
Manchester	33340	53.35N	2.27W	255	14.565	8293	24	27	25	22	20	28	42	25	43	6	90	9	130	83	20	4.3	3.8
Nottingham	33540	53.00N	1.25W	383	14.497	8293	23	26	24	21	19	28	41	25	43	8	20	8	210	84	19	4.7	5.8
Oban	31140	56.42N	5.47W	13	14.693	8293	26	29	30	24	21	34	42	30	44	3	180	6	180	78	22	3.2	4.0
Plymouth	38270	50.35N	4.12W	88	14.653	8293	29	31	34	30	26	39	49	33	48	8	80	10	80	81	26	4.0	3.8
Stansted Airport	36830	51.88N	0.23E	347	14.516	8293	23	26	25	22	19	28	44	24	43	7	30	10	130	83	18	4.1	6.3
Stornoway	30260	58.22N	6.32W	42	14.677	8293	29	31	37	32	29	42	45	37	43	6	300	10	160	70	24	3.2	3.1
Valley	33020	53.25N	4.53W	36	14.681	8293	27	30	40	34	31	42	46	37	47	10	80	9	50	80	25	4.1	3.8
Wyton Raf	35660	52.35N	0.12W	134	14.629	8293	22	26	27	23	21	30	44	26	44	8	40	10	100	85	18	4.3	7.4
UKRAINE																							
Chernihiv (Chernigov)	331350	51.48N	31.28E	449	14.463	8293	−7	−1	22	19	17	22	25	20	23	5	110	9	160	89	−10	4.3	9.4
Chernivtsi (Chernovtsky)	336580	48.27N	25.97E	787	14.287	8293	2	7	27	22	19	28	26	24	26	8	320	8	110	89	−2	3.4	6.1
Dnipropetrovs'k	345040	48.37N	35.08E	465	14.454	8293	0	4	27	23	21	29	27	25	28	9	50	11	90	93	−3	2.5	6.3
Donets'k	345190	48.07N	37.77E	741	14.310	8293	−1	3	30	26	22	37	21	32	23	6	70	11	100	91	−5	3.8	5.4
Kerch	339830	45.40N	36.42E	160	14.615	8293	11	15	28	24	21	34	26	28	26	11	30	11	40	90	9	3.8	5.6
Kharkiv (Khar'kov)	343000	49.87N	36.13E	498	14.437	8293	−3	2	23	20	18	27	28	23	22	6	30	11	110	90	−5	2.7	7.7
Kherson	339020	46.67N	32.62E	157	14.617	8293	4	8	23	20	18	28	27	24	26	7	270	8	80	93	2	2.2	5.2
Kirovohrad (Kirovograd)	337110	48.48N	32.25E	564	14.403	8293	−2	2	22	19	17	23	28	21	29	5	310	11	100	90	−5	2.5	5.2
Kryvyy Rih (Krivoy Rog)	337910	47.93N	33.33E	410	14.483	8293	0	4	27	23	20	32	24	26	26	9	50	10	90	92	−3	2.3	6.3
Kyyiv (Kiev)	333450	50.40N	30.45E	551	14.410	8293	−2	3	22	18	16	22	17	18	19	7	270	6	180	87	−3	3.4	9.2
Luhans'k	345230	48.60N	39.27E	203	14.592	8293	−4	0	30	25	21	32	17	28	22	6	90	10	90	93	−9	3.4	5.8
Mariupol' (Zdanov)	347120	47.07N	37.50E	229	14.578	8293	4	8	35	31	27	38	26	36	23	12	70	10	90	89	1	2.3	4.3
Odesa	338370	46.48N	30.63E	114	14.639	8293	7	12	27	23	21	30	27	26	27	11	360	10	180	92	4	3.1	6.3
Poltava	335060	49.60N	34.55E	521	14.425	8293	−3	2	24	21	18	28	20	23	19	6	360	7	70	89	−6	2.9	8.3
Rivne (Rovno)	333010	50.58N	26.13E	767	14.297	8293	−3	3	27	23	19	27	28	24	29	6	270	9	130	87	−5	2.7	10.1
Simferopol'	339460	45.02N	33.98E	593	14.387	8293	8	13	27	23	20	28	30	25	28	9	50	11	50	92	5	2.9	5.8
Sumy	332750	50.88N	34.78E	570	14.399	8293	−7	−1	23	21	18	23	29	21	28	6	330	8	130	89	−10	4.0	9.4
Uzhhorod (Uzhgorod)	336310	48.63N	22.27E	387	14.496	8293	6	10	19	16	14	21	30	16	32	3	100	7	170	90	0	3.2	6.8
Vinnytsya (Vinnitsa)	335620	49.23N	28.47E	977	14.188	8293	−2	3	28	23	20	28	25	25	24	8	340	9	180	87	−5	2.9	8.1
Zaporizhzhya	346010	47.80N	35.25E	282	14.551	8293	0	5	24	21	17	28	28	24	29	8	360	9	220	92	−2	2.2	5.4
Zhytomyr (Zhitomir)	333250	50.27N	28.63E	744	14.309	8293	−4	2	24	21	18	24	33	21	30	6	90	9	190	87	−5	3.1	8.6
UNITED ARAB EMIRATES																							
Abu Dhabi	412170	24.43N	54.65E	88	14.653	8293	52	54	22	19	17	21	69	18	69	4	200	9	320	116	46	0.9	2.5
Dubai	411940	25.25N	55.33E	16	14.691	8293	54	55	21	19	17	22	66	19	68	4	170	11	270	114	49	2.5	1.8
Ra's Al Khaymah	411840	25.62N	55.93E	101	14.646	8293	49	52	18	15	14	17	69	15	71	2	210	10	320	115	43	1.4	2.3
Sharjah	411960	25.33N	55.52E	108	14.643	8293	49	51	19	17	15	20	68	17	69	5	120	10	270	115	42	2.2	4.1
URUGUAY																							
Colonia Del Sacramento	865600	34.45S	57.83W	75	14.660	8293	39	41	32	27	23	30	51	26	49	9	50	9	360	95	34	5.9	3.1
Montevideo	865800	34.83S	56.00W	104	14.644	8293	35	38	32	28	24	30	53	27	55	8	330	14	360	97	31	3.1	2.2
Paso De Los Toros	864600	32.80S	56.52W	246	14.570	8293	34	36	26	23	19	25	55	22	55	2	280	10	330	100	30	2.3	2.2
Rocha	865650	34.48S	54.30W	59	14.669	8293	34	36	24	20	18	23	53	20	53	1	310	9	360	96	29	4.3	2.5
Salto	863600	31.38S	57.95W	111	14.641	8293	34	37	19	17	14	19	57	17	59	1	120	9	20	103	30	2.5	2.7
Treinta Y Tres	865000	33.22S	54.38W	151	14.620	8293	33	35	19	15	12	22	52	17	53	2	270	6	290	99	29	2.3	1.8
UZBEKISTAN																							
Samarqand (Samarkand)	386960	39.70N	67.00E	2375	13.481	8293	12	17	26	22	19	22	38	19	36	8	140	10	50	101	9	2.3	6.3
Tashkent	384570	41.27N	69.27E	1604	13.868	8293	13	18	13	11	10	14	45	11	42	2	90	4	300	105	10	2.2	5.4
VANUATU																							
Luganville	915540	15.52S	167.22E	144	14.623	8293	66	68	19	17	14	18	77	16	77	2	290	9	100	89	62	1.6	3.1
VENEZUELA																							
Caracas	804150	10.60N	66.98W	157	14.617	8293	70	71	12	11	9	12	81	11	81	1	140	6	340	97	59	2.2	14.0
VIETNAM																							
Ho Chi Minh City	489000	10.82N	106.67E	62	14.667	8293	68	70	39	26	17	25	83	16	82	4	360	9	160	101	56	5.0	20.5
WAKE ISLAND																							
Wake Island	912450	19.28N	166.65E	13	14.693	8293	71	72	29	26	24	30	74	28	75	15	40	14	80	94	67	3.1	2.5
WALLIS & FUTUNA ISLAND																							
Wallis Islands	917530	13.23S	176.17W	88	14.653	8293	72	73	21	18	17	21	80	19	80	2	160	9	100	91	68	3.1	2.3
YUGOSLAVIA																							
Belgrade	132720	44.82N	20.28E	324	14.528	8293	11	16	25	20	17	23	31	20	32	6	10	6	120	97	6	4.0	8.3
Palic	130670	46.10N	19.77E	344	14.518	8293	10	15	17	15	12	17	38	15	36	4	50	5	180	95	4	5.2	7.6
Podgorica (Titograd)	134620	42.37N	19.25E	108	14.643	8293	25	27	24	21	17	24	43	21	42	7	360	7	180	99	20	2.2	4.3
ZIMBABWE																							
Harare	677750	17.92S	31.13E	4931	12.263	8293	45	46	20	18	16	19	60	17	60	6	120	10	60	90	40	3.2	3.1

WMO# = World Meteorological Organization number
Lat. = latitude, °
Elev. = elevation, ft
Long. = longitude, °
PWD = prevailing wind direction, ° True
StdP = standard pressure at station elevation, psia
DB = dry-bulb temperature, °F
WS = wind speed, mph

Table 3B Cooling and Dehumidification Design Conditions—World Locations

	Cooling DB/MWB						Evaporation WB/MDB						Dehumidification DP/MDB and HR									
	0.4%		1%		2%		0.4%		1%		2%		0.4%			1%			2%			Range
Station	DB	MWB	DB	MWB	DB	MWB	WB	MDB	WB	MDB	WB	MDB	DP	HR	MDB	DP	HR	MDB	DP	HR	MDB	of DB
1	2a	2b	2c	2d	2e	2f	3a	3b	3c	3d	3e	3f	4a	4b	4c	4d	4e	4f	4g	4h	4i	5
Lerwick	60	56	58	55	57	54	57	59	56	58	55	56	56	69	58	55	66	57	54	64	56	6.8
Leuchars	72	61	69	59	66	58	62	69	61	67	59	64	60	76	65	58	73	63	57	69	62	14.0
London, Gatwick	80	65	76	63	74	62	67	77	65	73	64	71	63	88	70	62	83	68	60	79	67	17.6
London, Heathrow	81	66	78	64	75	63	67	79	66	75	64	72	63	88	70	62	83	69	61	80	68	16.6
Lyneham	78	64	75	62	72	61	66	75	64	72	62	69	62	85	68	61	81	66	59	77	65	15.8
Lynemouth	69	61	67	60	65	58	62	67	61	65	60	64	61	80	64	59	76	63	58	73	62	8.8
Manchester	77	63	74	62	71	60	65	74	63	71	62	69	62	83	68	60	79	67	59	74	65	13.7
Nottingham	78	64	74	63	71	61	66	75	64	72	62	69	63	87	70	61	81	68	60	77	65	16.0
Oban	73	61	69	60	66	58	63	71	61	67	60	64	60	77	65	58	73	64	57	70	62	10.4
Plymouth	75	63	72	62	69	61	65	72	64	69	63	67	63	85	67	62	83	66	61	80	64	11.0
Stansted Airport	79	64	76	63	73	62	66	76	64	72	63	70	63	87	70	61	82	67	60	78	66	16.7
Stornoway	65	59	62	58	60	56	60	64	58	61	57	60	58	74	62	57	71	60	56	67	59	8.6
Valley	74	63	70	61	67	60	64	72	62	68	61	65	62	82	66	60	78	64	59	75	63	10.6
Wyton Raf	79	64	76	63	73	62	67	76	65	73	63	71	63	87	70	62	82	68	60	78	66	16.7
UKRAINE																						
Chernihiv (Chernigov)	84	68	81	67	78	65	70	80	68	78	67	75	66	99	75	65	93	73	63	89	71	18.2
Chernivtsi (Chernovtsky)	83	68	80	66	78	65	69	80	68	78	66	76	66	98	74	64	92	73	63	88	72	16.2
Dnipropetrovs'k	87	67	85	66	82	65	70	82	69	80	67	78	67	99	75	65	94	74	63	89	72	18.7
Donets'k	86	66	83	66	80	65	70	82	68	80	67	76	65	97	74	64	92	73	63	88	72	19.6
Kerch	85	69	83	69	81	68	73	81	71	80	70	78	70	111	78	68	105	77	67	99	75	14.6
Kharkiv (Khar'kov)	85	67	83	65	80	65	70	80	68	78	67	76	66	98	74	65	93	73	63	88	71	16.4
Kherson	89	68	86	68	83	66	71	83	70	81	68	79	68	102	75	66	97	75	65	92	73	21.2
Kirovohrad (Kirovograd)	86	65	83	65	80	64	69	79	67	78	66	76	65	96	73	64	90	72	62	85	70	20.5
Kryvyy Rih (Krivoy Rog)	87	67	85	66	82	65	70	82	68	80	67	78	66	98	75	64	92	73	63	87	72	21.1
Kyyiv (Kiev)	83	67	80	66	77	65	69	79	68	77	66	75	66	99	74	65	93	73	63	88	71	16.6
Luhans'k	88	67	85	66	82	64	70	83	68	81	67	78	66	96	76	64	90	72	63	86	72	19.1
Mariupol' (Zdanov)	84	71	82	70	80	69	74	81	72	79	71	78	72	118	78	69	109	77	68	103	75	15.1
Odesa	86	67	84	67	81	66	72	79	70	79	69	77	70	109	76	67	102	74	66	96	73	18.4
Poltava	85	66	82	66	80	65	69	79	68	78	66	76	66	98	75	64	92	73	63	87	71	17.6
Rivne (Rovno)	82	67	79	66	76	64	69	79	67	77	66	74	66	97	75	64	92	72	62	86	70	18.5
Simferopol'	87	67	85	66	82	65	70	81	69	79	67	78	67	100	74	65	94	73	63	89	72	20.3
Sumy	84	67	81	66	78	64	70	79	68	77	66	75	66	99	74	64	92	72	63	88	71	17.1
Uzhhorod (Uzhgorod)	86	69	83	68	81	66	71	83	69	81	67	79	66	97	77	64	92	75	63	88	73	18.7
Vinnytsya (Vinnitsa)	82	66	79	65	77	64	68	78	67	76	65	74	65	97	73	64	91	71	62	86	70	18.2
Zaporizhzhya	88	67	85	66	82	65	71	82	69	80	67	78	67	100	76	65	94	74	64	89	72	20.2
Zhytomyr (Zhitomir)	82	67	79	65	77	64	69	79	67	77	65	74	65	96	74	64	91	71	62	86	70	18.9
UNITED ARAB EMIRATES																						
Abu Dhabi	111	74	108	74	106	75	87	95	85	94	85	93	85	182	91	84	178	91	82	169	90	23.0
Dubai	107	75	105	75	103	76	86	94	85	94	85	93	85	181	91	84	177	91	82	169	91	17.5
Ra's Al Khaymah	111	76	109	77	107	77	86	100	85	99	84	98	83	174	94	82	169	93	81	160	92	22.7
Sharjah	110	77	107	77	106	77	86	99	85	97	84	96	84	178	92	82	169	91	81	160	91	23.9
URUGUAY																						
Colonia Del Sacramento	88	74	86	73	84	72	76	85	75	83	74	81	74	127	82	73	122	80	72	117	79	14.9
Montevideo	89	72	86	71	83	70	76	83	74	81	73	79	74	125	79	72	118	77	71	116	76	16.7
Paso De Los Toros	95	72	91	72	88	71	77	87	75	85	74	83	74	127	82	72	122	80	71	117	78	20.3
Rocha	89	73	86	72	83	71	76	84	75	82	74	80	74	130	80	73	123	78	72	118	77	19.1
Salto	98	74	95	74	92	74	79	91	77	89	76	87	76	134	85	74	129	83	73	123	81	22.0
Treinta Y Tres	92	73	89	72	86	71	77	86	75	84	74	81	74	130	80	73	123	79	72	118	78	20.9
UZBEKISTAN																						
Samarqand (Samarkand)	96	67	94	66	92	66	70	92	68	89	67	89	62	91	81	60	85	77	58	80	76	24.8
Tashkent	100	71	98	68	95	68	75	95	72	92	70	90	68	111	89	65	99	84	63	91	82	26.8
VANUATU																						
Luganville	87	78	86	78	85	77	80	85	79	84	79	84	78	148	83	78	146	82	77	143	82	10.4
VENEZUELA																						
Caracas	92	84	91	83	90	83	86	90	85	89	84	88	86	190	89	84	181	88	84	177	87	12.6
VIETNAM																						
Ho Chi Minh City	95	77	94	77	93	77	81	90	80	89	80	89	79	150	86	78	147	85	77	143	84	14.8
WAKE ISLAND																						
Wake Island	89	79	89	79	88	78	81	86	80	86	80	85	80	155	85	79	150	84	78	146	84	8.1
WALLIS & FUTUNA ISLAND																						
Wallis Islands	88	80	87	80	87	80	82	87	81	86	81	86	80	157	85	79	153	85	79	151	84	8.5
YUGOSLAVIA																						
Belgrade	92	71	89	70	86	69	73	87	71	85	70	83	68	105	80	67	99	78	65	94	76	22.1
Palic	90	70	87	69	84	67	71	87	70	84	69	82	67	99	77	65	94	76	64	90	75	20.3
Podgorica (Titograd)	95	71	93	71	90	70	73	90	72	89	71	87	68	104	79	67	99	80	66	96	78	21.1
ZIMBABWE																						
Harare	86	62	84	61	83	61	68	76	67	76	67	74	66	116	70	65	112	69	65	110	69	21.1

MDB = mean coincident dry-bulb temp., °F
MWB = mean coincident wet-bulb temp., °F
MWS = mean coincident wind speed, mph
StdD = standard deviation, °F
HR = humidity ratio, grains of moisture per lb of dry air
A = airport
DP = dew-point temperature, °F

Table 4A Selected Monthly Percentiles of Wet-Bulb Temperature with Mean Dry-Bulb Temperature for U.S. Locations

Location and WMO#	%	Jan		Feb		Mar		Apr		May		Jun		Jul		Aug		Sep		Oct		Nov		Dec		Annual	
		WB	MDB	WB	MDB	WB	MDB	WB	MDB	WB	MDB	WB	MDB	WB	MDB	WB	MDB	WB	MDB	WB	MDB	WB	MDB	WB	MDB	WB	MDB
1a	1b	2a	2b	3a	3b	4a	4b	5a	5b	6a	6b	7a	7b	8a	8b	9a	9b	10a	10b	11a	11b	12a	12c	13a	13b	14a	14b
ALABAMA																											
Birmingham	0.4	64.3	67.7	66.1	69.9	68.9	76.8	71.6	79.6	75.2	85.1	78.3	90.0	79.7	91.2	79.2	89.4	77.3	86.8	73.4	80.3	69.6	73.6	67.6	72.1	78.2	89.3
722280	1	62.8	65.9	64.4	68.7	67.6	74.5	70.5	78.2	74.3	84.0	77.5	89.0	79.0	90.4	78.4	89.1	76.6	86.4	72.1	79.1	68.1	72.2	66.1	69.5	77.2	88.2
	2	61.3	64.0	62.7	66.6	66.4	73.2	69.3	77.3	73.4	82.9	76.7	88.0	78.4	89.8	77.9	88.8	76.0	86.0	71.1	78.0	67.0	71.8	64.3	67.8	76.3	86.6
Huntsville	0.4	62.7	65.8	64.3	68.2	67.4	75.2	71.5	79.0	74.7	84.0	77.6	88.9	79.6	91.8	79.1	88.8	77.2	87.5	72.7	79.2	68.2	72.4	65.7	69.4	78.0	88.9
723230	1	60.9	64.2	62.5	66.3	66.0	73.1	70.0	78.7	73.6	82.6	77.0	88.4	79.0	90.9	78.3	88.6	76.4	86.4	71.4	78.3	66.6	71.3	63.8	67.0	77.0	87.6
	2	59.3	62.3	60.9	64.8	64.6	71.1	68.6	76.7	72.7	81.6	76.2	87.3	78.4	89.8	77.6	88.2	75.7	85.6	70.3	77.0	65.3	69.8	62.4	65.9	76.0	86.0
Mobile	0.4	69.8	72.5	70.6	73.6	71.9	76.3	74.5	81.0	76.8	84.4	80.0	89.8	80.4	90.6	80.3	89.9	79.3	87.2	76.5	82.5	73.9	77.9	72.4	74.8	79.4	88.9
722230	1	68.8	71.8	69.5	72.3	71.2	75.6	73.8	80.0	76.0	83.8	79.2	89.0	79.9	90.0	79.7	89.0	78.7	86.8	75.7	81.4	72.9	76.2	71.5	74.1	78.5	87.8
	2	67.6	70.5	68.4	71.2	70.5	74.5	72.9	78.8	75.3	83.5	78.4	88.2	79.4	89.4	79.3	88.5	78.2	86.3	74.8	80.7	71.8	75.9	70.6	73.2	77.9	87.0
Montgomery	0.4	66.9	71.9	68.2	73.2	70.2	77.3	73.3	81.4	76.3	86.8	79.4	91.9	80.6	92.9	80.4	92.0	78.7	89.1	75.2	83.3	71.7	78.1	70.2	75.3	79.3	90.6
722260	1	65.3	69.5	66.8	71.3	69.2	77.0	72.2	80.1	75.4	85.6	78.5	90.4	80.0	91.6	79.9	91.3	78.1	88.5	74.0	80.8	70.4	75.6	68.9	73.8	78.4	89.3
	2	63.8	67.6	65.4	70.0	68.2	74.8	71.1	79.3	74.6	84.5	78.0	89.7	79.4	90.8	79.3	90.2	77.4	87.8	73.0	81.0	69.1	74.4	67.3	72.0	77.5	88.0
ARIZONA																											
Flagstaff	0.4	41.3	54.0	43.4	56.8	45.2	60.7	49.3	68.3	53.2	68.4	58.8	76.7	63.2	74.3	63.2	75.2	60.3	71.4	53.4	62.9	46.9	59.4	42.3	52.9	61.4	73.5
723755	1	39.8	50.1	41.8	54.4	43.8	58.3	47.8	66.3	52.3	69.3	57.6	78.9	62.3	73.9	62.3	74.5	59.1	70.8	51.7	63.7	45.4	58.9	40.6	51.3	60.1	72.6
	2	38.5	49.0	40.3	52.0	42.7	57.5	46.7	65.2	51.3	68.5	56.4	78.8	61.6	73.6	61.5	73.8	58.1	71.1	50.6	64.7	44.1	57.8	39.1	49.5	58.9	71.7
Phoenix,	0.4	58.3	65.2	59.6	75.6	61.2	82.2	64.2	93.3	67.4	97.3	73.6	97.8	76.5	97.9	77.2	98.3	75.9	96.7	70.1	84.2	63.1	77.3	58.2	65.6	75.6	96.5
Intl Airport	1	56.9	64.4	58.4	73.0	60.0	81.1	63.1	91.8	66.4	95.3	72.0	99.4	76.0	97.8	76.5	96.7	75.1	96.3	69.2	83.1	61.5	73.1	56.8	63.4	74.7	95.7
722780	2	55.8	64.8	57.4	70.0	59.0	80.1	62.1	90.3	65.7	94.5	70.6	100.8	75.4	96.8	75.9	95.8	74.2	94.1	68.2	82.8	60.3	73.0	55.7	64.4	73.8	95.0
Prescott	0.4	47.5	58.4	49.5	64.9	50.6	68.1	53.5	75.9	58.5	78.0	63.3	85.3	68.5	80.6	69.0	83.2	65.8	81.9	59.2	70.1	52.3	66.5	48.0	56.1	66.7	81.0
723723	1	46.1	56.8	48.0	61.4	49.5	66.4	52.4	74.3	57.4	76.9	62.2	86.2	67.3	80.6	68.0	81.4	64.8	80.3	57.9	69.1	50.8	63.9	46.4	56.1	65.5	80.0
	2	44.6	55.5	46.5	60.1	48.5	65.6	51.5	73.4	56.4	76.5	61.2	85.8	66.5	80.9	67.1	81.6	63.9	79.9	56.7	70.6	49.6	63.3	45.0	54.7	64.4	78.8
Tucson	0.4	55.1	65.0	56.2	71.8	57.6	82.3	60.6	87.7	64.4	91.2	69.8	91.7	72.8	87.8	73.4	90.5	72.0	88.0	66.8	78.4	59.7	73.9	56.1	64.1	72.0	87.8
722740	1	54.0	64.4	55.1	70.5	56.4	79.5	59.4	84.8	63.5	89.5	68.9	92.2	72.2	87.8	72.8	90.5	71.2	87.9	65.9	76.6	58.5	71.8	54.4	63.5	71.1	86.7
	2	52.8	64.3	54.1	68.4	55.4	78.1	58.3	84.0	62.5	87.5	68.0	92.9	71.6	87.5	72.3	88.7	70.5	87.0	64.8	78.1	57.4	71.0	53.4	63.0	70.3	86.2
ARKANSAS																											
Fort Smith	0.4	61.1	67.7	61.7	70.6	67.4	78.1	71.5	82.4	76.1	87.4	79.4	91.9	80.1	94.9	80.2	94.8	78.1	89.9	73.4	82.2	67.8	74.6	64.0	71.1	78.8	92.4
723440	1	58.9	64.2	60.1	67.8	66.0	76.4	70.3	79.8	74.9	86.2	78.6	91.2	79.4	93.5	79.5	93.8	77.2	88.8	72.1	81.4	66.3	73.9	62.4	68.3	77.8	91.1
	2	56.3	60.7	58.7	66.3	64.5	74.7	69.1	79.5	73.8	84.1	77.9	90.1	78.7	92.5	78.8	92.7	76.4	88.2	70.7	79.3	64.9	72.0	60.4	65.6	76.9	89.8
Little Rock,	0.4	64.2	69.5	64.3	68.9	69.3	76.9	73.0	81.2	76.6	86.6	80.5	92.2	81.3	93.7	81.3	92.7	79.3	90.2	74.9	82.5	69.8	75.6	66.7	70.8	80.0	92.0
AFB	1	62.8	66.3	62.9	68.2	68.0	75.3	71.8	79.1	75.8	85.8	79.7	90.6	80.6	93.3	80.4	91.8	78.5	88.5	73.4	81.0	68.4	73.6	64.9	68.7	79.0	90.8
723405	2	60.7	64.5	61.1	66.6	66.8	74.1	70.7	78.2	75.0	84.9	78.9	89.2	80.1	93.0	79.8	91.1	77.7	87.2	72.2	79.5	67.2	71.9	63.1	66.7	78.0	89.3
CALIFORNIA																											
Arcata/Eureka	0.4	57.8	60.4	58.9	63.2	56.5	60.1	58.1	62.7	60.1	65.7	61.4	66.8	62.2	66.2	64.1	69.2	64.6	72.5	62.1	70.8	60.2	63.8	58.3	60.3	62.2	67.4
725945	1	56.5	59.2	57.4	60.9	55.4	59.1	56.4	60.9	58.9	63.8	60.1	65.2	61.2	65.1	62.9	67.1	63.3	69.5	60.8	66.6	59.0	62.4	56.9	59.3	60.7	65.4
	2	55.4	58.1	56.1	59.3	54.4	58.1	55.3	59.6	57.7	61.9	59.2	64.0	60.3	64.6	61.9	65.8	62.1	67.6	59.7	65.4	58.0	61.3	55.7	58.2	59.6	63.9
Bakersfield	0.4	61.1	67.8	62.0	71.8	63.8	77.1	66.3	88.9	69.4	97.1	72.5	101.1	74.4	102.6	75.0	97.7	72.8	96.3	68.7	92.7	62.8	73.7	59.4	65.1	72.7	97.8
723840	1	58.7	64.9	60.7	70.0	62.4	74.7	64.9	86.5	68.2	95.3	71.2	100.9	73.5	100.9	73.9	96.0	71.8	91.5	67.6	90.8	61.4	74.0	57.8	63.3	71.3	96.0
	2	57.0	63.3	59.4	68.4	61.2	73.3	63.5	84.0	67.2	92.5	70.2	98.7	72.6	99.2	72.9	96.4	70.9	92.7	66.6	87.8	60.2	72.3	56.2	62.9	70.1	95.1
Barstow/	0.4	56.0	66.4	56.5	68.8	57.9	78.1	60.5	88.9	65.3	94.8	69.1	104.3	73.9	96.7	74.4	94.3	72.4	88.2	65.0	91.8	59.1	79.6	57.2	68.6	72.3	95.2
Daggett	1	54.0	63.2	55.2	69.2	56.9	77.2	59.4	87.6	64.1	93.3	68.1	103.4	73.1	96.2	73.4	94.5	71.2	90.8	63.8	90.5	57.5	74.7	54.1	65.8	70.8	95.2
723815	2	52.4	62.0	53.9	67.9	55.9	76.1	58.4	85.6	63.1	91.3	67.2	102.8	72.3	96.0	72.4	95.6	70.0	89.6	62.8	88.4	56.3	72.8	52.3	62.4	69.3	95.0
Fresno	0.4	59.4	63.5	61.4	68.6	64.2	75.1	66.2	86.6	70.0	95.6	72.4	99.7	74.7	101.9	75.3	100.2	72.4	97.5	68.3	92.4	61.8	72.1	59.2	62.1	72.9	98.2
723890	1	57.1	62.4	60.2	67.7	62.5	74.1	64.9	83.4	68.7	93.2	71.4	98.2	73.9	100.4	74.0	98.0	71.3	96.5	67.3	90.0	60.7	71.1	56.7	60.6	71.4	96.0
	2	55.4	60.0	59.0	66.9	61.2	72.1	63.5	81.5	67.4	91.2	70.4	96.9	73.0	99.0	73.0	97.4	70.3	93.6	66.2	87.5	59.7	70.6	55.0	59.0	70.1	94.2
Long Beach	0.4	60.4	68.5	61.6	70.8	62.6	74.8	65.0	80.7	67.3	82.2	69.4	83.8	71.7	86.4	73.5	88.9	73.4	87.9	69.4	83.1	65.1	74.6	61.1	66.0	71.4	84.9
722970	1	59.5	67.1	60.7	68.4	61.5	71.9	63.6	77.1	66.0	78.7	68.4	82.3	71.0	84.2	72.4	87.0	72.3	85.2	68.4	80.0	64.1	71.8	60.1	66.7	70.2	82.4
	2	58.8	65.7	59.8	66.8	60.5	70.0	62.5	74.5	64.8	75.1	67.4	79.8	70.2	82.8	71.4	85.1	71.4	84.1	67.5	78.4	63.1	70.9	59.1	65.5	69.0	80.3
Los Angeles	0.4	60.4	65.7	61.2	67.2	61.4	68.9	63.8	73.9	65.4	74.8	67.9	78.6	70.3	78.2	71.7	80.4	71.3	82.6	68.3	77.0	64.2	71.7	60.9	65.1	69.8	78.4
722950	1	59.5	64.5	60.2	66.1	60.5	67.7	62.6	71.5	64.3	72.3	66.8	76.0	69.4	77.0	70.9	80.0	70.3	80.5	67.3	75.3	63.3	69.9	59.8	64.4	68.6	76.4
	2	58.6	63.5	59.3	64.7	59.5	65.9	61.7	70.0	63.4	70.5	65.9	74.4	68.7	76.1	70.1	78.2	69.4	78.9	66.4	73.7	62.4	68.5	58.9	63.5	67.6	74.8
Sacramento,	0.4	58.2	60.5	61.0	66.5	63.8	73.1	67.4	82.4	70.1	90.6	72.9	98.5	73.9	100.3	75.0	97.9	70.8	95.4	67.3	89.3	62.7	70.8	58.3	61.2	71.6	96.2
Metro	1	56.4	59.1	59.3	64.7	62.0	71.6	65.4	80.0	68.6	89.1	71.5	97.1	72.5	98.1	72.4	96.4	69.8	93.5	66.2	86.5	61.3	69.9	56.5	59.8	70.0	93.7
724839	2	55.1	57.9	58.1	63.3	60.4	69.4	63.6	77.5	67.3	87.3	70.2	94.9	71.5	96.3	71.1	95.3	68.9	91.3	65.2	84.3	59.9	67.7	55.0	58.4	68.5	91.1
San Diego,	0.4	61.9	65.6	62.4	67.4	61.9	69.4	64.7	73.0	66.2	72.9	69.6	77.8	73.2	80.0	73.9	81.0	75.1	82.8	69.4	77.0	65.4	71.0	62.3	65.1	72.5	79.4
Intl Airport	1	61.0	64.4	61.4	66.5	61.1	67.8	63.7	71.3	65.1	72.4	68.4	76.0	72.1	79.0	73.1	80.1	73.9	80.7	68.6	75.6	64.4	70.4	61.0	65.1	71.0	78.0
722900	2	60.1	63.9	60.6	65.9	60.3	66.6	62.5	70.3	64.2	71.1	67.6	75.0	71.1	77.9	72.3	79.2	72.8	79.8	68.0	74.3	63.5	69.2	60.0	64.6	69.6	76.5
San Francisco	0.4	58.7	61.1	59.3	63.9	59.7	67.3	62.5	76.5	63.4	78.1	65.5	83.4	65.7	83.1	65.5	80.8	66.4	83.5	64.5	79.1	60.6	65.9	59.0	60.8	64.4	78.6
724940	1	57.3	59.4	58.1	61.7	58.4	64.6	60.3	71.2	61.9	74.8	63.8	79.7	64.4	79.1	64.3	77.0	65.0	78.6	63.4	74.8	59.6	64.8	57.9	59.9	63.0	74.9
	2	56.1	58.1	57.0	61.0	57.3	63.1	58.9	67.5	60.7	72.6	62.4	76.0	63.3	76.0	63.4	74.5	63.8	76.1	62.4	72.2	59.0	64.0	56.6	58.9	61.8	71.8
Santa Maria	0.4	60.2	65.4	61.9	70.5	61.7	73.1	63.7	81.6	63.2	75.4	64.3	79.8	68.8	74.1	67.5	82.8	69.5	86.5	65.4	82.9	62.1	73.3	59.8	66.7	66.3	80.9
723940	1	59.1	64.9	60.4	69.8	60.7	70.4	62.5	79.1	62.1	73.6	63.4	77.6	66.9	78.4	66.6	80.6	68.0	83.0	64.5	78.6	61.0	69.3	58.5	66.1	64.7	77.6
	2	58.1	64.2	59.4	66.8	59.4	68.0	61.2	76.7	61.1	71.5	62.5	75.5	65.5	78.6	65.6	78.7	66.3	80.3	63.7	76.7	60.2	68.8	57.3	66.7	63.5	75.1
COLORADO																											
Alamosa	0.4	37.7	47.9	40.2	53.4	43.4	62.3	47.2	67.0	52.1	70.3	58.6	76.7	61.7	77.2	63.2	75.9	58.2	72.8	51.7	65.2	43.3	57.9	37.5	46.9	59.8	75.2
724620	1	35.6	46.0	38.1	50.9	41.9	60.6	46.1	66.0	51.0	70.1	57.4	75.6	60.5	76.4	61.0	75.3	56.7	72.4	50.3	63.4	42.1	56.7	35.7	44.7	58.6	73.8
	2	34.0	44.0	36.8	48.6	40.5	58.5	45.1	63.5	50.2	68.6	56.6	76.0	59.9	75.8	60.1	74.8	55.5	70.7	48.8	64.1	40.9	54.2	34.2	43.7	57.5	72.8
Boulder	0.4	44.0	60.9	46.3	65.9	48.0	70.4	53.2	74.9	58.8	74.3	64.4	82.2	66.9	81.6	66.1	80.4	62.5	79.8	54.4	75.3	48.7	69.1	45.0	64.6	64.5	80.9
724699	1	42.4	58.7	44.2	61.7	46.4	68.0	52.0	72.5	57.6	73.3	63.4	81.3	65.8	81.9	65.2	80.3	61.1	77.0	53.3	74.5	47.4	67.1	43.3	61.1	63.2	79.8
	2	40.9	56.5	42.4	59.7	45.2	65.6	50.9	70.5	56.6	73.1	62.4	80.2	65.0	82.0	64.3	79.4	60.0	76.6	52.4	73.3	46.1	65.4	41.7	58.4	62.0	78.5
Colorado Springs	0.4	42.2	59.8	44.3	64.7	46.0	67.9	50.8	71.0	56.5	71.8	61.9	79.4	64.8	79.4	64.4	77.7	60.9	76.7	53.1	75.4	46.0	66.9	42.3	61.0	62.8	77.5
724660	1	40.4	58.1	42.1	60.8	44.2	64.8	49.4	68.8	55.4	70.2	61.0	78.1	63.9	79.1	63.5	77.1	59.7	75.4	51.8	73.0	44.8	64.5	40.9	58.5	61.5	77.2
	2	38.9	55.2	40.1	57.7	42.9	63.1	48.4	67.9	54.4	69.6	60.2	77.2	63.1	78.2	62.9	76.7	58.7	74.9	50.7	71.2	43.7	62.8	39.4	55.4	60.5	76.3
Eagle	0.4	37.7	43.5	40.5	51.0	45.3	59.3	49.8	70.4	55.1	76.3	61.4	84.3	64.5	83.8	63.7	83.7	59.3	77.8	53.3	71.9	45.4	60.0	39.0	46.7	61.8	80.3
724675	1	36.4	41.6	38.9	48.3	43.6	58.3	48.5	67.8	54.0	75.2	59.8	81.5	63.2	82.6	62.5	80.6	58.0	77.3	51.6	70.4	43.6	56.8	37.4	44.4	60.3	78.1
	2	34.9	40.5	37.5	46.3	42.1	56.7	47.1	65.6	53.0	73.1	58.6	81.1	62.2	81.3	61.6	78.4	57.0	76.3	50.3	67.6	42.1	54.3	35.7	41.9	59.0	76.2
Grand Junction	0.4	40.7	48.1	45.6	55.6	49.8	67.6	53.4	76.0	58.8	79.0	63.6	86.4	67.0	86.7	66.8	86.1	63.0	80.8	56.5	67.6	48.8	62.5	42.9	50.3	65.1	84.1
724760	1	39.2	46.8	44.2	55.6	48.0	63.6	52.3	73.6	57.7	78.3	62.6	87.2	66.2	84.9	65.9	84.7	61.9	78.8	55.3	68.6	47.5	59.2	41.1	48.1	63.9	82.9
	2	37.6	44.0	42.6	52.4	46.6	62.7	51.2	72.4	56.7	77.9	61.6	87.2	65.4	84.6	65.1	83.2	61.0	79.0	54.1	69.7	46.2	57.1	39.2	45.7	62.8	82.2

WMO# = World Meteorological Organization number WB = wet-bulb temperature, °F MDB = mean coincident dry-bulb temperature, °F

Table 4B Selected Monthly Percentiles of Dry-Bulb Temperature with Mean Wet-Bulb Temperature for U.S. Locations

Location and WMO#	%	Jan		Feb		Mar		Apr		May		Jun		Jul		Aug		Sep		Oct		Nov		Dec		Annual	
		DB	MWB	DB	MWB	DB	MWB	DB	MWB	DB	MWB	DB	MWB	DB	MWB	DB	MWB	DB	MWB	DB	MWB	DB	MWB	DB	MWB	DB	MWB
1a	1b	2a	2b	3a	3b	4a	4b	5a	5b	6a	6b	7a	7b	8a	8b	9a	9b	10a	10b	11a	11b	12a	12c	13a	13b	14a	14b
ALABAMA																													
Birmingham	0.4	69.8	60.8	75.5	61.3	81.9	63.2	85.8	66.9	91.9	70.8	95.5	75.2	98.8	76.2	96.5	75.7	94.5	73.9	86.2	68.8	78.3	64.8	73.2	64.8	94.4	75.1
722280	1	67.3	59.6	73.1	60.1	79.8	62.8	84.3	66.2	89.6	70.5	93.9	74.5	96.6	75.8	94.6	74.9	92.6	73.4	84.6	68.3	76.5	64.3	71.4	64.2	92.1	74.9
	2	65.5	59.3	70.4	58.8	77.5	62.2	82.9	65.2	87.5	70.7	92.3	74.2	94.8	76.0	93.2	75.2	90.9	73.4	82.8	67.8	74.7	63.2	69.4	63.3	90.1	74.3
Huntsville	0.4	67.7	59.5	72.8	60.6	80.1	62.3	85.9	67.6	90.5	69.5	95.1	74.2	97.7	76.3	96.4	75.4	93.5	74.0	85.2	68.9	77.1	63.5	71.4	64.2	94.1	74.8
723230	1	65.6	59.0	70.3	58.3	77.8	62.3	84.0	66.3	88.5	70.1	93.4	73.7	96.1	75.7	94.8	74.7	91.4	73.9	83.5	67.4	75.0	62.9	69.0	63.1	91.7	74.4
	2	63.6	57.6	67.8	57.4	75.6	61.3	82.2	64.9	86.9	70.3	91.9	73.7	94.5	75.9	93.2	74.7	89.8	72.9	81.8	66.4	72.9	62.4	67.0	60.7	89.5	73.9
Mobile	0.4	75.0	67.3	77.2	65.1	82.0	67.0	86.4	70.3	91.7	72.4	95.4	76.5	96.3	77.2	95.3	77.6	93.4	75.6	88.1	73.2	81.6	69.9	76.8	70.4	93.8	76.5
722230	1	73.4	66.8	75.1	64.9	80.2	65.1	84.7	69.5	90.0	72.5	94.1	76.1	95.0	77.5	94.1	77.1	92.2	75.6	86.7	71.4	79.9	68.9	75.4	69.2	92.1	76.3
	2	71.6	65.4	73.4	65.1	78.7	65.3	83.3	68.7	88.5	71.7	92.9	75.6	93.9	77.0	92.9	76.8	91.0	75.5	85.4	70.6	78.4	68.3	74.1	68.5	90.5	75.9
Montgomery	0.4	74.0	63.9	77.0	64.7	83.0	66.4	86.1	69.1	92.1	72.4	97.3	75.9	97.9	78.0	97.9	76.5	95.6	74.8	88.9	72.2	81.8	68.2	78.0	67.8	95.2	76.3
722260	1	71.6	62.3	74.9	62.8	81.2	65.2	84.7	68.6	90.3	72.4	95.5	75.7	96.2	77.2	96.0	76.7	94.0	74.6	87.1	70.7	79.8	67.1	75.6	66.5	93.2	76.0
	2	69.2	61.6	72.7	61.8	79.5	64.3	83.4	67.9	88.8	72.3	94.0	75.2	94.9	77.0	94.4	76.9	92.4	74.6	85.5	69.4	77.9	65.6	73.4	65.8	91.3	75.8
ARIZONA																													
Flagstaff	0.4	57.5	40.5	61.2	42.4	64.0	43.5	72.4	47.2	79.0	50.2	89.3	56.0	88.4	56.6	85.6	57.0	82.0	55.6	77.4	49.4	66.7	44.2	58.3	40.4	85.1	55.5
723755	1	54.0	38.5	57.8	40.1	62.2	42.3	70.0	46.0	77.1	49.9	87.1	54.9	86.8	56.4	83.8	57.5	80.0	54.7	75.4	48.8	64.4	43.3	55.9	38.9	82.6	55.2
	2	51.2	37.3	55.2	38.5	60.4	41.3	68.0	45.2	75.4	49.1	85.2	54.2	85.3	56.0	82.3	57.2	78.6	54.2	73.3	48.7	61.8	42.6	53.2	37.7	80.1	55.0
Phoenix,	0.4	78.0	53.6	84.3	55.4	90.7	58.8	98.7	62.0	105.5	64.7	113.4	68.3	112.4	70.4	110.5	72.3	107.9	70.3	100.9	64.6	87.6	58.1	77.6	53.2	109.8	69.7
Intl Airport	1	75.6	52.5	81.8	54.8	88.3	57.7	96.5	61.5	103.4	63.9	111.6	68.0	110.9	70.6	109.0	72.2	106.1	69.9	98.8	64.4	85.6	58.1	75.5	52.7	107.6	69.6
722780	2	73.5	52.0	79.5	54.3	86.0	57.0	94.4	60.7	101.3	63.4	109.9	67.5	109.7	70.5	107.7	71.9	104.6	70.1	96.9	64.4	83.7	57.1	73.4	51.7	105.6	69.5
Prescott	0.4	64.7	45.4	70.2	46.8	74.5	48.3	81.6	51.1	87.4	55.1	96.9	59.5	97.3	60.8	94.0	62.5	90.9	60.6	85.5	54.4	73.4	48.9	64.2	44.4	93.5	60.2
723723	1	62.0	44.2	67.6	46.0	72.0	47.5	79.3	50.9	85.3	54.2	95.3	59.3	95.5	61.0	92.2	62.7	89.1	60.3	83.6	53.9	71.2	48.8	62.1	43.6	90.9	60.1
	2	59.6	42.8	65.1	44.7	69.8	47.2	77.1	50.2	83.5	53.8	93.5	58.3	93.9	61.0	90.6	62.2	87.4	59.6	81.4	53.6	69.2	47.7	60.1	42.7	88.6	59.8
Tucson	0.4	77.5	51.2	82.5	53.8	87.8	56.0	95.4	58.1	101.3	61.4	108.8	65.9	107.3	66.3	103.5	68.4	101.5	66.0	96.5	61.2	85.0	56.0	77.0	51.1	104.2	65.3
722740	1	74.6	50.4	79.5	52.3	85.1	54.8	92.8	56.8	99.3	60.9	106.6	65.3	105.4	66.4	102.1	68.6	100.0	66.0	94.3	60.8	82.9	54.8	75.1	50.6	101.9	65.1
	2	72.5	49.9	77.4	51.5	82.7	53.7	90.4	56.5	97.1	60.0	105.0	64.7	103.9	66.1	100.6	68.0	98.5	65.5	92.1	59.9	81.1	53.9	72.9	50.1	99.6	65.1
ARKANSAS																													
Fort Smith	0.4	70.0	57.6	75.5	57.1	82.5	63.3	87.8	66.9	90.5	72.5	96.9	76.5	102.7	74.9	102.4	75.9	96.9	73.7	89.8	68.6	79.7	63.5	72.2	62.1	98.6	75.8
723440	1	66.6	55.6	72.4	56.9	80.1	61.9	85.4	66.6	89.2	72.0	95.0	75.9	100.9	75.4	100.4	75.6	94.8	74.2	87.6	68.3	77.5	63.1	69.8	60.9	95.8	75.7
	2	63.7	54.2	69.2	56.1	77.7	62.0	83.5	65.6	87.7	71.5	93.5	75.7	99.2	76.0	98.6	75.7	93.0	74.1	85.2	68.0	75.2	62.4	67.0	58.3	93.2	75.3
Little Rock,	0.4	71.3	60.6	76.3	58.0	81.9	65.1	85.9	68.5	91.2	73.0	97.1	76.0	102.2	76.9	100.9	76.4	95.9	75.9	88.7	69.6	79.5	65.9	72.4	64.7	97.4	77.1
Afb	1	68.4	59.3	72.5	58.3	79.5	64.0	83.9	67.3	89.8	72.4	95.4	76.2	100.0	78.2	99.0	76.3	93.7	75.0	86.4	69.1	76.9	65.1	70.1	62.8	94.7	76.9
723405	2	65.9	58.2	69.7	57.9	77.2	63.5	82.2	67.2	88.3	71.9	94.0	75.5	98.0	78.0	97.1	77.2	92.1	74.9	84.6	68.9	74.7	64.0	67.6	61.7	92.4	76.3
CALIFORNIA																													
Arcata/eureka	0.4	65.3	52.0	67.4	55.3	65.5	53.3	66.1	54.9	68.3	58.9	69.7	60.2	69.1	59.7	71.2	62.1	77.5	60.6	75.7	59.3	67.3	56.4	63.2	53.1	70.4	59.9
725945	1	62.6	53.1	64.3	54.1	62.3	53.7	63.3	54.1	65.1	58.0	67.0	59.0	67.3	59.4	69.0	61.8	72.9	60.6	71.4	58.5	65.1	56.2	61.5	54.2	67.4	59.0
	2	60.6	52.9	62.2	53.4	60.2	52.7	61.3	53.9	63.0	56.8	65.2	58.3	66.2	59.3	67.5	60.6	69.9	60.8	67.8	58.1	63.3	56.5	60.1	53.5	65.4	58.3
Bakersfield	0.4	72.0	56.8	77.5	57.7	82.5	59.8	93.1	63.6	100.4	67.4	106.4	69.8	107.0	72.2	106.1	72.0	102.9	70.1	96.8	67.3	81.9	58.9	70.8	56.0	103.9	70.4
723840	1	68.9	56.3	74.8	56.6	79.9	58.7	89.7	63.0	97.9	66.9	104.3	70.0	105.4	70.9	104.4	71.0	100.9	70.1	93.9	66.3	79.2	58.5	68.2	54.6	101.3	69.4
	2	66.3	55.0	72.6	56.7	77.5	58.6	87.2	61.5	95.6	65.8	102.2	69.3	104.0	70.6	102.8	70.3	99.0	69.1	91.8	65.3	76.7	57.9	66.2	53.5	98.8	68.6
Barstow/	0.4	72.5	51.2	79.2	54.1	85.5	55.4	93.9	59.5	99.5	62.9	108.5	67.3	110.4	70.0	109.4	69.4	104.4	67.1	98.1	63.5	83.5	57.6	73.5	53.3	107.3	68.3
Daggett	1	70.3	50.7	76.5	52.4	82.8	54.6	91.2	58.2	97.5	62.6	107.1	66.7	108.9	69.1	107.9	69.5	102.7	67.1	95.7	62.6	80.7	55.4	71.0	51.8	104.9	67.4
723815	2	68.2	49.0	74.5	51.4	80.4	54.0	88.9	57.4	95.6	61.6	105.5	66.0	107.7	68.4	106.3	69.2	101.0	66.0	93.3	61.5	78.2	54.4	68.4	50.2	102.4	66.5
Fresno	0.4	66.9	55.2	74.2	58.4	80.1	61.3	91.0	62.8	99.1	67.7	104.9	70.3	106.5	71.8	105.4	72.1	101.9	70.5	95.6	67.5	78.8	58.5	66.5	54.8	103.2	70.5
723890	1	64.3	54.8	72.0	57.6	77.3	59.8	88.6	62.6	96.9	66.9	103.3	69.6	105.2	71.1	103.8	70.9	99.9	69.7	92.8	66.1	76.2	58.3	64.1	53.7	100.5	69.6
	2	62.2	54.2	69.8	56.8	75.2	58.6	85.8	61.6	94.8	66.1	101.4	68.9	103.4	70.9	102.0	70.7	97.9	68.6	90.0	64.6	74.0	57.7	61.9	53.7	97.9	68.7
Long Beach	0.4	82.0	56.3	81.7	57.4	83.1	58.7	88.8	59.6	88.7	63.1	92.3	67.1	92.0	68.9	94.0	70.8	97.6	67.5	95.8	63.2	87.0	58.9	80.6	54.6	91.7	67.1
722970	1	79.2	55.5	79.1	56.3	78.6	58.2	84.3	60.0	84.0	63.9	87.1	66.6	89.0	69.2	90.9	70.5	93.5	68.5	91.2	63.0	83.5	58.3	77.7	54.7	87.6	66.5
	2	75.8	54.9	76.3	56.3	75.6	57.0	80.6	59.5	79.8	62.4	83.8	66.2	86.7	68.4	88.3	69.8	90.4	68.5	87.4	62.2	80.5	57.0	74.8	54.1	84.3	66.2
Los Angeles	0.4	79.6	55.3	79.4	56.2	77.7	55.6	82.1	59.4	80.8	62.5	85.1	65.6	82.4	67.7	84.0	69.3	91.5	65.9	90.4	61.6	83.9	58.1	78.9	54.3	84.6	63.8
722950	1	76.6	54.6	75.8	55.0	73.7	56.3	77.2	58.6	75.9	62.7	78.9	65.3	80.0	67.8	81.9	69.6	87.1	66.4	85.9	60.8	80.5	56.6	75.8	53.9	80.6	63.7
	2	73.4	53.5	73.0	54.8	71.0	54.8	74.2	58.4	73.0	62.0	76.0	64.6	78.3	67.8	80.0	69.0	83.1	66.9	82.0	59.7	77.7	56.8	73.0	53.0	77.8	64.3
Sacramento, Metro	0.4	64.2	53.7	71.5	57.5	77.0	60.3	86.5	64.8	96.2	66.4	103.0	69.8	104.3	71.4	103.0	71.0	100.2	69.0	93.7	66.0	77.3	59.1	64.0	54.5	100.0	69.4
724839	1	61.7	53.4	69.3	57.1	74.5	59.5	84.1	62.5	93.5	66.0	100.4	69.2	101.9	70.6	100.9	70.2	97.7	68.2	91.0	64.9	74.8	58.6	62.1	54.5	96.8	68.6
	2	59.9	53.5	67.0	56.5	72.2	58.7	81.5	61.8	91.2	65.5	98.1	68.4	99.7	69.9	98.8	69.5	95.2	67.1	87.9	63.8	72.0	57.8	60.2	53.3	93.6	67.5
San Diego,	0.4	78.5	54.8	79.0	56.6	78.0	56.9	82.0	58.2	80.3	60.3	85.1	68.0	84.8	70.7	85.2	71.0	90.9	68.8	90.0	62.5	82.6	57.4	76.4	53.5	85.0	67.2
Intl Airport	1	75.8	54.3	76.3	55.9	74.8	56.1	78.5	59.2	76.2	62.7	80.1	67.0	82.6	70.1	83.5	71.3	87.4	68.7	85.4	62.1	79.2	56.8	74.2	53.5	81.4	66.9
722900	2	72.9	53.8	73.1	55.1	72.2	55.7	75.1	58.6	73.5	62.6	77.4	66.9	80.5	69.9	81.5	70.8	84.4	69.3	81.7	62.0	76.4	56.1	71.9	53.2	78.9	66.5
San Francisco	0.4	64.1	53.4	68.7	56.1	71.5	56.8	79.1	61.7	82.8	61.7	88.6	64.1	85.9	65.0	83.0	64.0	88.5	63.8	85.1	61.4	72.9	56.9	63.5	54.1	83.1	62.6
724940	1	62.1	53.4	66.0	53.6	68.2	55.5	74.6	58.2	78.7	60.6	83.2	62.7	81.1	63.5	79.5	63.5	84.9	63.0	81.4	60.3	70.1	56.2	61.7	54.3	78.1	61.6
	2	60.5	54.4	63.8	54.2	65.8	54.6	71.3	57.1	74.3	59.5	78.4	61.6	77.3	62.6	76.1	62.6	81.1	62.1	77.8	60.5	67.6	55.8	60.3	54.6	73.9	60.5
Santa Maria	0.4	77.9	55.8	78.9	57.2	78.4	57.6	87.1	62.2	80.0	59.9	86.2	63.1	88.2	64.7	84.7	66.4	91.6	65.0	91.4	63.0	83.1	57.0	78.8	55.6	85.9	63.0
723940	1	75.2	55.1	76.1	56.5	75.3	56.6	83.0	60.9	76.2	60.5	80.0	61.9	83.3	65.1	81.8	65.7	87.4	65.5	87.1	60.0	80.5	56.9	76.9	54.2	81.5	62.1
	2	73.0	53.6	73.4	55.9	72.6	56.8	79.1	59.0	73.2	59.8	76.8	61.6	79.9	64.4	79.6	65.3	83.5	65.3	83.5	60.8	78.0	57.0	74.4	53.5	78.1	61.0
COLORADO																													
Alamosa	0.4	50.7	36.3	56.1	37.9	65.1	42.1	71.2	45.4	79.1	49.5	86.5	55.0	87.4	54.9	85.1	57.1	81.2	53.7	73.5	47.8	61.9	41.7	50.4	35.6	84.2	55.1
724620	1	47.7	34.6	53.7	36.6	63.1	40.8	69.2	44.3	77.1	49.0	85.1	53.9	86.1	56.0	83.7	56.4	79.7	53.0	72.0	47.2	59.7	41.1	47.8	34.6	82.1	54.8
	2	44.7	33.4	50.9	36.2	60.5	39.6	67.5	43.6	75.2	47.9	83.7	53.7	84.7	55.8	82.3	56.2	78.1	52.9	70.3	46.8	57.3	39.9	45.4	33.2	79.8	54.3
Boulder	0.4	63.2	42.1	67.6	45.2	73.5	46.6	80.2	51.0	85.8	54.7	94.4	58.9	96.4	60.0	94.7	60.1	90.5	58.2	82.0	51.5	72.5	47.2	66.3	44.0	92.8	59.6
724699	1	60.1	41.8	64.3	43.0	70.5	45.8	77.8	49.9	83.8	54.3	92.5	58.8	95.1	59.7	92.8	59.3	88.3	57.6	80.1	51.4	69.8	46.4	63.0	42.4	90.2	59.4
	2	57.3	40.4	61.2	42.1	67.9	44.2	75.4	48.9	81.7	53.5	91.0	58.8	93.7	60.1	91.2	59.2	86.2	56.8	78.1	50.9	67.2	45.4	59.7	41.3	87.4	59.2
Colorado Springs	0.4	61.8	40.9	66.0	43.4	70.7	45.1	77.5	48.6	83.5	51.9	92.3	56.9	93.8	58.0	91.1	58.3	86.5	56.4	80.2	51.1	70.4	44.9	62.7	41.6	89.8	58.0
724660	1	59.3	40.0	62.5	41.8	67.8	43.4	75.2	47.8	81.4	51.4	90.3	56.7	92.1	58.2	89.2	57.8	85.0	55.9	78.3	50.2	66.6	43.7	59.8	40.3	87.0	57.6
	2	56.2	38.2	59.1	39.2	64.7	42.2	73.0	47.0	79.2	51.4	88.2	56.3	90.5	58.5	87.5	58.0	83.2	55.7	76.0	49.3	64.3	43.0	57.1	39.0	84.2	57.5
Eagle	0.4	46.0	36.3	53.1	39.6	64.0	42.8	73.2	48.0	81.3	52.2	89.8	58.0	91.8	59.3	90.4	58.8	86.1	55.5	77.0	51.1	62.3	44.3	48.8	37.6	88.2	58.4
724675	1	43.5	34.9	50.5	38.2	61.1	42.9	70.7	46.8	79.5	51.8	88.0	57.2	90.2	58.9	88.6	58.9	84.0	55.8	74.9	49.5	59.2	42.9	46.0	35.6	85.8	57.3
	2	41.5	34.0	48.0	36.7	58.2	41.6	68.6	45.8	77.7	51.2	86.4	56.4	88.9	58.9	87.1	58.7	81.7	55.1	72.6	49.1	56.4	41.0	43.3	35.0	83.2	56.6
Grand Junction	0.4	50.8	38.8	60.6	43.5	71.4	47.4	81.1	51.4	88.3	55.6	98.5	60.6	99.4	62.0	97.6	61.6	91.8	59.3	81.3	52.9	66.1	47.3	52.5	40.9	96.2	60.9
724760	1	48.1	37.7	58.3	41.9	69.1	46.8	78.9	50.4	86.6	54.8	96.7	60.4	97.9	61.7	96.2	61.8	90.0	59.0	79.7	52.2	63.4	45.8	50.1	39.7	94.0	60.2
	2	45.8	36.6	55.6	41.0	66.5	45.4	76.6	49.9	84.9	54.2	95.0	58.9	96.7	61.0	94.8	61.1	88.2	57.8	77.7	51.7	60.9	44.5	47.6	37.8	91.6	59.7

WMO# = World Meteorological Organization number DB = dry-bulb temperature, °F MWB = mean coincident wet-bulb temperature, °F

Table 4A Selected Monthly Percentiles of Wet-Bulb Temperature with Mean Dry-Bulb Temperature for U.S. Locations

Location and WMO#	%	Jan		Feb		Mar		Apr		May		Jun		Jul		Aug		Sep		Oct		Nov		Dec		Annual	
		WB	MDB	WB	MDB	WB	MDB	WB	MDB	WB	MDB	WB	MDB	WB	MDB	WB	MDB	WB	MDB	WB	MDB	WB	MDB	WB	MDB	WB	MDB
1a	1b	2a	2b	3a	3b	4a	4b	5a	5b	6a	6b	7a	7b	8a	8b	9a	9b	10a	10b	11a	11b	12a	12c	13a	13b	14a	14b
Pueblo	0.4	46.5	64.0	48.5	70.8	49.7	71.7	55.3	72.7	61.2	78.5	66.9	85.2	68.9	85.6	69.4	84.9	65.1	81.8	57.0	76.1	49.5	72.6	46.6	65.1	67.4	83.6
724640	1	44.5	61.2	46.7	66.5	48.4	70.4	53.7	73.8	60.1	79.1	65.7	85.3	68.2	84.5	68.4	84.2	64.0	80.2	56.0	79.1	48.4	69.4	44.9	62.4	66.2	83.2
	2	42.9	59.5	44.8	63.7	47.4	68.8	52.6	72.3	59.0	78.0	64.8	84.0	67.4	84.2	67.7	83.2	63.1	79.8	55.1	78.3	47.4	67.4	43.2	59.9	65.1	82.7
CONNECTICUT																											
Bridgeport	0.4	50.0	51.7	48.3	51.7	52.2	57.4	60.9	70.8	69.2	76.9	73.7	83.8	77.4	86.1	77.9	84.9	75.9	82.6	68.8	72.1	63.1	66.4	54.4	56.4	75.8	82.6
725040	1	47.4	49.8	45.6	48.6	50.2	54.7	58.8	67.6	67.2	74.1	72.5	81.4	76.4	85.1	77.0	82.7	75.0	80.7	67.7	71.7	61.4	63.9	52.0	53.9	74.3	80.7
	2	44.4	46.2	43.4	46.0	48.6	52.8	56.3	62.8	65.3	72.1	71.5	79.4	75.3	83.8	76.2	82.0	74.0	78.8	66.3	70.2	59.9	62.1	50.2	52.2	73.1	79.2
Hartford,	0.4	51.3	54.5	52.8	56.5	58.9	66.2	64.6	79.3	71.6	85.5	75.5	88.2	78.0	91.8	77.7	89.1	75.8	87.5	68.5	74.3	64.4	68.6	56.8	59.2	75.8	87.4
Brainard Field	1	46.8	51.0	49.6	54.6	55.9	62.6	62.1	73.6	69.8	83.6	74.3	87.2	77.2	90.1	76.5	87.4	74.4	84.5	66.4	73.9	61.8	66.4	53.0	56.5	74.2	84.1
725087	2	42.7	46.3	45.6	50.4	53.0	60.6	60.0	69.9	68.3	79.4	73.3	85.4	76.2	88.7	75.4	85.3	73.1	81.7	65.0	71.7	59.3	63.2	49.4	52.1	72.7	81.5
DELAWARE																											
Wilmington	0.4	55.3	59.0	57.4	62.1	63.5	72.9	66.8	80.1	74.1	84.5	77.3	87.9	79.7	89.9	79.0	89.0	77.4	86.7	70.7	76.3	66.2	70.9	60.6	64.3	77.5	87.0
724089	1	52.4	55.4	54.8	58.4	60.6	69.3	65.1	77.3	72.1	82.2	76.2	86.6	78.6	89.3	78.1	86.8	76.3	85.0	69.4	74.7	63.5	68.2	58.1	61.5	76.2	84.7
	2	49.4	51.6	51.8	56.1	57.9	63.8	63.3	72.8	70.7	80.0	75.2	85.1	77.8	88.1	77.2	85.6	75.3	83.8	68.0	72.7	61.8	65.4	55.2	57.8	74.9	83.3
FLORIDA																											
Daytona Beach	0.4	71.0	77.6	71.1	80.0	72.3	81.4	74.5	83.1	77.1	86.2	79.4	88.9	80.4	89.6	80.3	88.3	79.5	86.6	77.6	83.9	74.8	79.9	72.7	79.1	79.4	87.9
722056	1	70.1	76.7	70.2	78.3	71.5	80.4	73.3	82.3	76.2	84.8	78.7	88.3	79.9	88.8	79.9	87.8	78.9	86.3	77.0	83.6	74.0	79.3	71.6	77.7	78.5	87.1
	2	69.3	76.0	69.2	76.9	70.8	79.4	72.5	81.3	75.4	83.8	78.2	87.5	79.3	88.5	79.3	87.7	78.4	85.8	76.3	82.6	73.2	78.7	70.5	76.6	78.0	86.4
Jacksonville,	0.4	69.9	75.9	70.4	77.8	72.1	80.9	74.6	83.7	77.5	86.8	80.1	92.2	81.4	92.7	81.0	91.1	79.8	88.3	77.4	84.3	74.2	79.6	71.6	77.5	79.9	90.0
Intl Airport	1	68.9	74.6	69.4	76.0	71.2	80.3	73.4	82.3	76.5	85.4	79.3	90.6	80.5	91.2	80.3	90.1	79.2	88.0	76.7	83.4	73.3	78.4	70.7	75.8	79.1	89.1
722060	2	67.9	73.7	68.3	74.6	70.3	78.9	72.4	81.2	75.5	84.3	78.6	89.4	80.0	90.3	79.7	89.5	78.6	87.2	76.0	81.7	72.4	77.4	69.6	74.6	78.3	87.8
Key West	0.4	75.7	80.3	76.0	80.7	77.0	81.8	78.3	84.0	79.9	86.0	81.3	88.0	81.3	88.8	81.5	87.9	80.9	87.2	80.2	86.4	78.7	82.5	76.6	80.6	80.6	87.4
722010	1	75.1	79.5	75.3	79.9	76.3	80.8	77.5	83.1	79.2	84.9	80.7	87.4	80.6	88.1	81.1	87.4	80.4	87.0	79.5	85.6	78.1	81.7	76.1	79.7	80.1	87.0
	2	74.4	78.5	74.7	79.2	75.7	80.2	77.0	82.5	78.4	84.4	80.2	86.9	80.3	87.7	80.5	87.2	80.1	86.8	79.0	85.2	77.4	81.6	75.5	79.1	79.5	86.7
Miami,	0.4	74.3	79.6	74.2	80.2	75.3	81.9	76.9	84.0	78.5	84.3	79.9	87.2	80.2	87.7	80.3	87.5	80.0	86.2	79.2	84.7	77.4	82.3	75.8	80.7	79.5	86.7
Intl Airport	1	73.7	79.0	73.4	79.4	74.5	80.9	76.0	82.7	77.5	83.9	79.2	86.5	79.5	87.6	80.0	87.0	79.4	86.2	78.4	84.5	76.9	82.1	74.9	79.7	79.0	86.5
722020	2	73.1	78.5	72.9	78.9	74.1	80.1	75.2	82.5	77.0	83.5	78.5	85.8	79.1	87.5	79.5	86.7	79.1	86.2	78.0	84.5	76.3	81.1	74.1	78.4	78.4	85.9
Tallahassee	0.4	70.7	74.1	70.5	73.9	72.4	78.2	74.4	81.4	77.2	85.2	79.9	90.6	80.9	90.9	80.4	90.0	79.4	87.1	76.8	83.3	73.3	77.5	72.6	75.5	79.5	88.9
722140	1	69.5	72.7	69.5	72.9	71.5	76.6	73.5	80.4	76.3	84.2	79.2	89.7	80.2	89.6	79.9	89.4	78.8	86.8	76.0	82.1	72.5	76.6	71.6	74.5	78.7	88.1
	2	68.2	71.1	68.3	71.8	70.6	75.2	72.6	79.4	75.5	83.8	78.5	88.4	79.5	88.9	79.4	89.0	78.2	86.5	75.2	82.1	71.9	76.0	70.3	73.1	78.1	87.1
Tampa,	0.4	72.3	78.3	72.4	79.2	74.2	80.0	76.1	82.2	78.2	86.9	80.1	87.6	81.3	89.1	81.1	89.1	79.9	87.7	78.3	85.8	75.7	81.8	74.2	79.4	80.0	88.2
Intl Airport	1	71.2	76.4	71.5	77.6	73.4	80.0	75.2	82.1	77.3	86.0	79.5	87.8	80.5	88.4	80.4	88.7	79.3	87.9	77.7	85.3	74.9	80.9	73.1	78.4	79.2	88.1
722110	2	70.3	75.2	70.7	76.2	72.5	79.6	74.3	81.1	76.5	84.8	79.0	88.1	80.0	88.2	79.9	88.5	78.8	87.8	77.1	84.7	74.1	79.8	72.1	77.0	78.4	87.2
West Palm Beach	0.4	73.6	79.9	74.1	80.5	75.0	82.7	76.3	83.2	78.3	85.1	80.0	87.4	80.4	88.7	80.4	88.8	80.2	87.4	78.7	85.1	77.0	82.6	75.0	81.0	79.6	87.9
722030	1	73.0	79.4	73.3	80.2	74.2	81.3	75.2	82.5	77.6	84.4	79.3	87.3	79.9	88.5	80.1	88.4	79.5	87.1	78.2	84.6	76.3	81.6	74.1	80.0	79.0	87.6
	2	72.3	78.1	72.5	79.3	73.4	80.2	74.3	82.0	76.9	83.7	78.7	86.9	79.4	88.3	79.6	88.1	79.1	87.0	77.5	84.3	75.5	80.8	73.3	79.2	78.4	86.7
GEORGIA																											
Athens	0.4	62.4	66.2	64.2	68.4	67.3	74.1	70.1	78.3	75.0	84.3	77.1	89.7	79.2	91.7	78.5	90.1	77.2	87.6	73.3	79.5	69.0	72.0	67.0	69.3	77.5	88.9
723110	1	60.9	64.3	62.5	66.7	65.9	72.0	68.8	77.5	73.8	82.9	76.4	88.9	78.4	90.5	78.0	89.8	76.3	86.2	71.7	77.3	67.6	71.8	65.3	68.3	76.5	87.4
	2	59.2	61.9	61.0	65.0	64.5	71.5	67.8	76.4	72.7	82.0	75.7	87.9	77.7	89.5	77.4	88.6	75.5	84.9	70.5	76.4	66.4	70.0	63.2	65.9	75.7	86.1
Atlanta	0.4	62.5	65.7	64.1	68.3	66.0	74.9	69.8	77.5	73.9	84.0	76.9	89.1	79.7	92.5	78.4	89.0	76.5	86.1	72.2	78.0	68.1	72.0	66.1	69.7	77.2	88.1
722190	1	61.1	64.2	62.8	66.6	64.9	71.9	68.2	76.8	72.7	82.8	76.0	87.6	78.4	89.6	77.5	88.5	75.5	85.1	70.8	76.9	67.0	71.4	64.3	67.2	76.0	86.6
	2	59.4	62.4	61.1	65.1	63.7	71.3	67.2	75.1	71.8	81.6	75.3	86.5	77.5	88.2	76.8	88.0	74.7	84.8	69.5	76.0	65.9	69.8	62.5	65.6	75.1	85.4
Augusta	0.4	65.1	69.1	66.7	73.6	68.5	78.6	71.9	81.4	75.9	87.4	78.6	91.5	80.4	92.8	80.1	91.3	78.1	89.5	74.5	82.2	71.1	76.7	68.4	73.3	78.7	91.0
722180	1	63.6	68.9	65.0	71.2	67.4	76.6	70.4	80.2	74.3	84.8	77.9	90.2	79.6	92.6	79.3	91.0	77.3	88.4	73.4	80.2	69.9	75.5	66.7	71.0	77.8	89.4
	2	62.0	65.8	63.4	69.1	66.2	74.3	69.2	79.0	73.3	84.2	77.2	89.3	79.1	91.8	78.4	90.0	76.6	87.4	72.3	79.8	68.7	73.9	65.1	70.0	76.9	88.0
Columbus,	0.4	66.3	70.1	66.9	70.6	69.8	77.1	72.4	81.1	75.9	84.8	79.3	90.4	80.3	93.0	79.7	90.4	78.1	87.0	74.8	82.6	70.7	75.8	69.8	73.4	78.6	89.4
Metro Airport	1	64.8	68.1	65.5	70.3	68.4	75.8	71.2	80.5	74.7	83.9	78.4	89.5	79.4	91.5	79.0	89.4	77.4	86.0	73.5	80.8	69.5	73.9	68.3	72.4	77.8	88.2
722255	2	63.5	66.6	64.3	69.0	67.4	74.3	70.2	78.7	73.9	83.0	77.6	88.5	78.8	90.4	78.4	88.7	76.8	85.2	72.5	79.8	68.6	72.7	66.8	70.6	77.0	87.0
Macon	0.4	65.8	69.5	67.2	72.5	69.4	77.2	72.6	81.8	76.0	86.4	78.8	93.1	80.3	93.5	80.1	91.6	78.1	88.2	74.5	82.5	71.0	75.3	69.5	73.5	78.7	90.9
722170	1	64.4	68.3	65.8	71.0	68.3	75.8	71.2	79.8	75.0	85.2	78.1	91.4	79.4	92.3	79.3	91.0	77.4	87.5	73.6	80.9	69.7	74.6	67.9	71.8	77.8	89.4
	2	63.1	67.0	64.5	69.2	67.4	74.3	70.2	79.2	74.0	82.7	77.3	89.5	78.8	91.6	78.7	90.3	76.9	87.0	72.6	79.9	68.7	73.4	66.4	70.4	77.0	88.0
Savannah	0.4	67.3	72.6	68.2	75.9	70.1	79.2	73.2	81.8	76.5	85.1	79.2	90.7	81.0	91.4	80.4	91.0	78.5	87.5	76.4	82.7	72.6	77.9	69.8	74.8	79.3	90.0
722070	1	66.3	71.9	67.0	74.1	68.9	77.9	71.9	80.6	75.6	84.6	78.4	89.7	80.3	90.6	79.7	90.2	78.0	87.2	75.3	81.7	71.4	76.3	68.5	73.7	78.4	88.7
	2	65.2	70.5	65.8	71.4	67.9	75.5	70.7	79.6	74.7	83.8	77.9	88.8	79.5	90.4	79.2	89.5	77.4	86.1	74.2	79.9	70.6	75.1	67.2	72.1	77.6	87.3
HAWAII																											
Hilo	0.4	73.4	80.3	73.6	80.7	73.5	79.9	74.4	80.8	75.2	82.0	75.0	81.7	76.4	81.7	77.3	82.8	77.3	83.7	77.4	83.7	76.2	81.8	74.4	80.5	76.4	82.1
912850	1	72.8	79.2	72.9	80.7	72.7	79.0	73.4	79.8	74.4	81.4	74.4	81.1	75.7	81.1	76.5	81.9	76.6	82.9	76.5	82.4	75.5	81.0	73.9	80.0	75.6	81.4
	2	72.2	78.6	72.2	79.5	72.1	78.2	72.8	78.9	73.8	80.8	73.9	80.7	75.2	81.0	76.1	81.4	76.1	82.2	76.0	81.6	74.9	80.4	73.2	79.3	74.9	81.0
Honolulu	0.4	74.1	79.0	73.4	80.2	73.3	80.3	73.3	80.9	74.4	83.0	74.5	84.4	76.1	85.1	76.7	85.9	77.2	85.1	77.2	84.3	75.5	82.6	74.8	80.7	76.0	84.4
911820	1	73.4	78.8	72.6	79.3	72.4	79.7	72.6	80.6	73.8	82.3	73.9	84.3	75.3	84.6	75.8	85.6	76.3	85.2	76.4	83.7	74.9	82.3	74.1	80.0	75.1	84.1
	2	72.8	78.5	72.0	78.6	71.8	79.3	72.1	80.4	73.2	82.2	73.3	83.9	74.5	84.1	75.2	85.4	75.7	85.0	75.7	83.4	74.3	81.8	73.4	79.5	74.3	83.5
Kahului	0.4	73.9	80.5	73.4	81.4	73.4	81.9	73.5	81.8	75.8	83.8	75.5	85.8	76.5	85.7	77.3	86.1	77.3	86.2	77.3	86.3	75.8	83.1	75.1	82.1	76.4	85.2
911900	1	73.1	79.8	72.6	80.3	72.6	80.8	73.0	81.8	74.8	82.8	75.0	85.2	75.9	85.2	76.5	85.1	76.5	86.0	76.6	85.5	75.1	83.1	74.2	81.3	75.5	84.5
	2	72.4	78.9	72.1	79.7	72.0	80.4	72.4	81.3	74.0	81.9	74.3	84.6	75.3	84.8	75.9	84.5	76.0	85.4	76.0	84.5	74.4	82.4	73.4	80.5	74.8	84.0
Lihue	0.4	73.8	78.0	73.8	79.0	73.4	79.5	74.1	79.8	74.9	81.5	76.0	83.5	77.4	83.2	77.9	83.6	77.5	83.9	77.4	82.7	76.0	79.7	74.5	78.5	76.9	82.9
911650	1	73.1	78.1	72.8	77.7	72.8	79.1	73.3	79.2	74.3	80.7	75.2	82.7	76.5	82.2	77.2	83.3	77.1	83.5	76.6	82.1	75.3	79.8	74.0	78.1	76.1	82.0
	2	72.4	77.5	72.1	77.4	72.1	78.2	72.8	78.7	73.9	80.1	74.5	81.8	75.9	81.7	76.6	82.7	76.5	83.0	76.1	81.6	74.7	79.6	73.4	77.9	75.3	81.6
IDAHO																											
Boise	0.4	45.0	49.6	49.2	58.1	51.8	67.1	57.4	75.7	62.4	84.4	66.2	91.2	68.3	90.9	67.9	93.4	63.9	85.6	57.4	78.3	51.4	59.9	46.1	52.1	65.8	89.9
726810	1	43.9	49.1	47.2	54.8	50.3	64.4	55.8	73.0	60.6	82.8	65.0	89.3	67.1	91.3	66.5	91.2	62.6	84.0	56.3	76.5	50.2	58.5	44.4	49.7	64.3	88.7
	2	42.8	47.6	45.5	52.7	49.0	61.0	54.1	72.0	59.4	80.8	63.8	87.5	66.1	90.6	65.6	89.3	61.4	82.4	55.2	73.6	48.8	56.2	43.2	49.0	62.9	86.8
Pocatello	0.4	41.2	47.2	45.8	53.8	48.3	62.5	53.7	72.5	59.4	79.0	64.7	86.3	66.3	85.4	65.9	83.9	61.3	79.4	55.0	76.0	48.7	58.1	42.7	48.7	63.9	84.2
725780	1	39.7	45.3	43.6	51.3	46.9	59.9	52.1	70.6	58.0	77.9	63.2	85.0	65.3	85.7	64.4	83.9	60.1	79.3	53.9	72.6	47.1	56.1	40.8	46.4	62.4	83.2
	2	38.4	43.2	41.8	48.6	45.3	56.8	50.7	68.4	56.6	75.4	62.0	84.2	64.3	85.0	63.5	83.3	59.0	78.7	52.7	71.3	45.7	54.6	39.3	44.4	61.0	82.0
ILLINOIS																											
Chicago, O'Hare	0.4	51.3	54.1	51.1	55.9	62.2	70.3	66.8	77.7	72.9	82.9	76.5	88.0	79.8	91.5	79.3	90.4	76.1	85.0	68.8	79.6	61.9	66.5	57.7	60.5	77.2	87.8
Intl Airport	1	46.5	49.1	47.9	52.1	59.8	66.6	65.2	74.7	71.4	81.4	75.4	87.3	78.5	90.0	78.3	89.2	74.8	84.3	67.0	76.2	60.3	65.2	54.7	57.4	75.3	85.1
725300	2	42.5	45.2	44.8	48.8	57.5	63.8	63.6	72.2	70.1	81.0	74.4	85.5	77.5	88.5	77.2	87.3	73.5	83.0	65.3	73.8	58.7	62.5	50.1	53.3	73.6	82.6

WMO# = World Meteorological Organization number WB = wet-bulb temperature, °F MDB = mean coincident dry-bulb temperature, °F

Table 4B Selected Monthly Percentiles of Dry-Bulb Temperature with Mean Wet-Bulb Temperature for U.S. Locations

Location and WMO#	%	Jan		Feb		Mar		Apr		May		Jun		Jul		Aug		Sep		Oct		Nov		Dec		Annual	
		DB	MWB	DB	MWB	DB	MWB	DB	MWB	DB	MWB	DB	MWB	DB	MWB	DB	MWB	DB	MWB	DB	MWB	DB	MWB	DB	MWB	DB	MWB
1a	1b	2a	2b	3a	3b	4a	4b	5a	5b	6a	6b	7a	7b	8a	8b	9a	9b	10a	10b	11a	11b	12a	12c	13a	13b	14a	14b
Pueblo	0.4	66.4	45.4	72.0	48.2	77.0	48.6	83.1	52.8	91.1	57.1	100.6	61.7	101.1	63.3	98.4	63.2	94.0	60.6	86.4	54.3	74.8	48.8	67.6	45.3	97.2	62.2
724640	1	63.5	43.2	68.7	46.0	74.3	47.4	81.0	51.6	89.3	56.1	98.4	61.0	99.4	63.4	96.7	62.8	92.2	59.7	84.7	53.9	72.0	47.8	64.3	43.9	94.3	62.1
	2	60.9	42.2	65.4	44.1	71.7	46.6	78.9	50.4	87.4	55.7	96.2	60.3	97.8	62.9	95.0	62.5	90.2	59.7	82.5	53.4	69.4	46.6	61.5	42.6	91.6	61.9
CONNECTICUT																											
Bridgeport	0.4	52.7	49.3	52.9	44.8	59.9	49.5	74.8	58.4	80.4	66.9	87.9	71.3	91.6	75.1	89.0	74.2	86.1	74.4	75.8	66.1	67.1	60.9	57.6	53.4	86.4	73.1
725040	1	49.8	46.2	49.9	44.5	57.7	49.0	70.4	56.8	77.7	63.4	85.1	70.4	89.4	73.4	87.2	73.9	83.6	72.1	73.3	64.2	65.1	59.5	54.9	51.1	83.9	71.8
	2	47.5	43.5	47.1	42.6	55.0	46.8	66.4	53.7	75.4	62.7	82.8	69.3	87.0	72.8	85.7	73.4	81.4	71.6	71.7	63.8	63.2	58.7	52.8	49.3	81.6	70.7
Hartford, Brainard	0.4	55.5	51.0	59.4	52.0	71.4	55.8	83.3	62.8	90.2	68.8	93.1	72.9	95.2	75.9	93.0	75.3	90.4	74.6	80.0	64.7	71.2	60.5	60.3	54.9	91.2	73.4
Field	1	51.1	45.4	54.7	48.5	66.4	53.6	78.1	59.7	86.5	67.3	91.2	71.5	93.3	74.9	91.1	74.2	87.2	72.8	77.2	63.7	67.9	59.6	57.0	51.5	88.2	71.7
725087	2	47.0	41.6	51.3	45.5	62.1	51.8	73.7	57.0	83.4	66.0	88.9	71.6	91.5	74.4	89.2	73.0	84.2	71.0	74.4	62.3	64.6	58.1	53.1	48.2	85.4	70.2
DELAWARE																											
Wilmington	0.4	59.9	52.2	64.7	53.4	75.4	60.7	83.6	64.4	89.3	70.9	92.8	73.9	95.5	76.4	93.1	75.1	91.1	75.3	80.6	67.1	73.4	63.6	65.1	59.5	91.2	74.8
724089	1	56.2	50.9	61.1	53.1	71.8	58.6	80.1	64.0	86.6	69.9	90.8	73.8	93.5	75.7	91.2	75.4	88.8	74.3	78.0	66.0	70.1	62.8	61.9	57.4	88.7	73.6
	2	53.0	48.7	57.5	50.8	67.1	55.7	76.0	61.4	84.4	67.9	89.0	72.9	91.9	75.3	89.6	74.7	86.4	73.1	75.8	65.5	67.2	59.8	58.6	54.2	86.2	72.7
FLORIDA																											
Daytona Beach	0.4	81.0	68.9	83.2	69.4	85.7	69.9	88.9	70.9	91.5	73.1	93.9	77.0	94.5	77.7	93.4	77.4	90.5	76.7	87.5	74.8	83.6	72.6	81.5	70.4	91.8	77.0
722056	1	79.6	68.8	81.3	68.6	84.2	69.3	87.1	70.2	89.8	72.7	92.3	76.9	93.0	77.6	91.9	78.0	89.3	76.9	86.2	74.4	82.3	72.0	80.2	69.7	90.0	76.7
	2	78.0	67.9	79.0	67.8	82.4	69.1	85.4	70.3	88.1	72.2	90.9	76.7	91.7	77.8	90.6	77.8	88.4	76.9	85.1	74.3	81.1	71.5	78.8	69.3	88.4	76.6
Jacksonville,	0.4	80.0	67.7	82.2	68.3	86.1	68.8	89.6	69.6	93.9	72.3	96.5	77.5	97.2	78.6	95.4	78.0	93.1	77.2	89.3	74.8	82.8	71.6	80.8	69.5	94.4	77.3
Intl Airport	1	78.2	67.5	80.3	66.8	84.4	68.8	88.2	69.4	91.8	72.6	95.0	76.9	95.8	78.4	94.3	78.1	91.8	77.0	87.4	73.5	81.5	70.7	79.1	68.4	92.6	77.1
722060	2	76.2	66.2	78.3	65.8	82.6	68.0	86.7	69.5	90.1	72.0	93.5	76.5	94.5	77.9	93.2	78.1	90.6	76.9	85.8	73.2	80.2	69.9	77.6	67.6	90.8	76.6
Key West	0.4	82.0	74.5	82.3	74.4	84.0	75.2	86.4	76.7	88.2	77.6	90.4	79.1	91.4	79.1	91.5	78.9	90.5	79.0	88.3	78.7	85.1	77.2	82.4	74.8	90.3	78.9
722010	1	81.0	73.8	81.5	74.1	83.2	74.9	85.4	75.9	87.4	77.3	89.6	78.8	90.8	79.0	91.0	78.9	89.8	78.8	87.5	78.2	84.3	76.7	81.6	74.2	89.4	78.8
	2	80.2	73.1	80.7	73.7	82.3	74.4	84.4	75.1	86.8	77.1	89.1	78.7	90.2	78.8	90.3	78.9	89.2	78.7	87.0	77.9	83.5	76.1	80.7	73.8	88.6	78.5
Miami,	0.4	83.1	71.8	84.1	71.7	86.3	71.3	89.2	72.8	90.1	73.5	92.0	76.9	93.1	77.3	92.5	77.8	91.2	77.8	89.4	76.4	85.6	75.3	83.3	73.4	91.1	77.3
Intl Airport	1	82.0	71.8	82.7	71.3	85.0	72.3	87.5	72.8	88.8	73.7	90.7	77.1	91.8	77.6	91.5	77.7	90.3	77.7	88.3	76.3	84.7	74.6	82.2	72.5	89.9	77.1
722020	2	80.8	71.3	81.6	70.9	83.8	71.7	86.1	72.3	87.7	73.6	89.6	76.8	90.8	77.6	90.7	77.6	89.5	77.5	87.3	76.1	83.7	73.9	81.2	71.8	88.9	76.9
Tallahassee	0.4	76.7	67.5	79.2	64.6	84.4	66.3	89.1	68.4	93.2	72.1	96.8	76.8	97.0	77.9	96.3	77.4	94.3	75.8	89.3	72.5	82.5	69.8	78.8	68.5	94.6	76.5
722140	1	75.1	66.6	77.3	64.9	82.6	66.2	87.3	68.7	91.6	72.0	95.1	76.0	95.6	77.6	95.0	77.5	93.1	75.5	87.9	72.1	81.1	69.1	77.1	68.3	92.8	76.1
	2	73.3	65.7	75.3	64.5	80.9	65.0	85.6	67.8	90.2	71.2	93.8	75.2	94.4	77.2	93.7	77.2	91.9	75.8	86.5	71.5	79.7	68.1	75.5	67.3	91.2	75.6
Tampa,	0.4	81.2	68.8	82.4	69.6	84.7	70.5	87.9	70.7	92.2	73.7	93.8	76.4	93.3	77.9	93.7	77.9	92.8	76.7	89.9	75.5	86.0	72.7	82.9	71.4	92.4	77.1
Intl Airport	1	80.0	69.5	81.2	68.3	83.6	70.2	86.7	71.1	90.8	73.7	92.5	76.3	92.5	77.9	92.9	77.9	91.8	76.8	88.7	75.1	84.5	72.1	81.3	69.6	91.2	77.1
722110	2	78.7	68.6	79.8	68.6	82.6	69.9	85.6	70.5	89.7	73.0	91.5	76.3	91.9	77.8	92.1	77.8	91.1	76.8	87.6	74.5	83.3	71.6	80.0	69.4	90.1	76.8
West Palm Beach	0.4	82.9	71.8	84.1	71.7	86.5	72.2	89.2	71.7	90.1	74.5	91.9	76.9	93.6	77.6	92.7	78.2	91.2	77.6	89.0	76.3	85.5	74.5	83.3	72.5	91.2	77.5
722030	1	81.5	71.3	82.7	71.3	85.0	71.8	87.3	71.9	88.6	73.7	90.7	77.1	92.1	77.5	91.6	78.2	90.3	77.6	88.1	75.9	84.3	74.0	82.2	72.2	90.0	77.6
	2	80.3	70.7	81.4	71.1	83.5	71.7	85.5	71.9	87.5	73.7	89.7	77.4	91.0	77.8	90.9	78.3	89.5	77.5	87.1	75.8	83.1	73.7	81.1	72.0	88.9	77.3
GEORGIA																											
Athens	0.4	69.2	57.7	74.0	58.1	80.4	62.3	85.9	65.3	90.9	70.9	96.4	73.9	98.6	75.5	96.0	75.6	92.2	74.7	85.2	69.0	77.7	64.0	72.6	63.6	94.3	74.5
723110	1	66.4	56.6	71.4	58.5	78.4	62.1	84.2	64.7	89.0	69.8	94.4	73.5	97.0	75.1	94.1	75.1	90.6	73.9	83.2	68.1	75.9	63.9	69.9	63.4	91.9	74.6
	2	64.2	55.5	69.0	56.9	76.2	60.6	82.4	63.9	87.4	69.8	92.7	73.5	95.3	75.1	92.6	75.4	89.1	73.2	81.3	66.8	73.7	62.1	67.6	61.6	89.6	73.8
Atlanta	0.4	68.5	58.0	73.3	58.2	80.2	62.0	85.1	65.0	88.8	70.8	94.4	73.8	97.5	75.7	95.2	75.3	91.9	74.2	83.8	68.4	77.1	64.4	71.5	63.3	93.1	74.8
722190	1	66.3	58.4	71.0	57.2	78.1	61.7	83.3	63.7	87.3	69.3	92.7	73.1	95.5	75.2	93.4	75.2	89.9	73.4	82.0	67.1	75.0	63.1	69.4	62.3	90.7	74.1
	2	64.1	56.8	68.5	57.0	76.0	60.0	81.6	63.1	86.0	68.5	91.2	73.0	93.8	75.3	91.8	74.9	88.4	72.7	80.3	66.3	73.1	62.1	67.3	60.4	88.4	73.1
Augusta	0.4	73.6	61.1	77.8	63.3	83.3	64.7	88.5	66.3	93.1	71.8	98.7	75.1	100.3	76.3	98.0	76.4	94.9	75.3	88.0	70.5	81.9	67.1	76.8	66.4	96.2	76.0
722180	1	70.9	60.8	75.3	61.3	81.7	63.8	86.9	65.8	91.3	71.3	96.5	74.7	98.4	76.1	96.2	76.5	93.2	75.2	86.3	70.3	79.2	65.9	74.4	64.8	93.8	75.7
	2	68.2	59.2	72.9	59.7	79.7	62.7	85.2	65.5	89.6	70.4	94.7	74.5	96.7	76.5	94.9	76.5	91.6	74.7	84.5	68.9	77.2	64.7	71.8	62.7	91.7	75.1
Columbus,	0.4	72.9	62.3	76.2	61.3	83.1	65.1	87.5	66.5	91.3	71.3	97.1	74.4	99.0	76.7	97.2	76.0	94.5	73.8	88.3	70.8	80.3	66.0	76.4	66.2	95.3	75.6
Metro Airport	1	70.8	61.7	74.2	61.2	81.1	63.9	86.1	66.9	89.8	70.5	95.5	74.8	97.4	76.3	95.4	75.7	93.0	73.9	86.5	69.7	78.5	65.2	74.3	66.0	93.2	75.1
722255	2	68.3	60.8	72.0	60.2	79.4	63.0	84.5	66.1	88.6	70.4	94.2	74.2	95.8	76.1	94.2	75.6	91.5	73.7	84.5	69.1	76.8	64.7	72.4	65.1	91.4	74.7
Macon	0.4	73.2	61.8	77.4	62.5	83.3	65.2	88.5	67.7	93.6	70.7	98.5	75.3	100.1	77.0	98.2	76.6	95.0	74.9	88.9	71.1	81.5	67.9	76.2	66.2	96.3	75.8
722170	1	70.8	60.9	75.0	61.8	81.4	64.0	86.7	66.3	91.9	71.4	96.8	74.9	98.3	76.3	96.3	76.6	93.4	74.7	86.9	70.0	79.2	65.8	74.2	65.0	94.0	75.2
	2	68.4	60.3	72.6	60.3	79.4	63.4	85.1	66.0	90.3	70.9	95.1	74.5	96.6	76.3	94.8	76.1	91.9	74.1	85.1	69.0	77.3	65.1	72.2	64.6	91.9	74.8
Savannah	0.4	76.6	65.1	79.5	65.7	84.3	66.5	89.3	67.4	93.2	72.2	97.4	76.1	99.0	77.2	96.2	77.9	93.2	76.0	87.5	72.0	81.9	69.3	78.4	67.8	95.0	76.8
722070	1	74.3	64.0	77.0	63.5	82.1	65.4	87.3	67.0	91.2	72.0	95.6	75.7	97.2	77.5	94.8	77.6	91.7	76.0	86.0	71.6	80.0	69.0	76.6	66.1	92.8	76.3
	2	72.0	62.9	74.7	63.0	80.1	65.2	85.3	67.1	89.5	71.2	93.8	75.3	95.3	77.5	93.5	77.5	90.3	75.6	84.4	71.1	78.3	67.4	74.2	65.0	90.7	75.9
HAWAII																											
Hilo	0.4	84.3	70.6	84.7	71.2	83.9	71.0	83.3	72.0	85.0	73.6	85.1	72.9	85.3	73.9	85.7	75.1	86.4	75.1	86.2	75.0	84.5	74.1	84.1	72.6	85.3	74.0
912850	1	82.9	70.6	83.2	70.7	82.7	70.4	82.2	71.6	83.8	72.8	84.2	72.5	84.3	73.7	84.8	74.7	85.4	74.8	85.3	74.6	83.6	73.5	82.8	72.0	84.3	73.6
	2	81.7	70.3	82.0	70.4	81.5	69.9	81.2	71.1	82.7	72.2	83.4	72.1	83.5	73.4	84.2	74.2	84.6	74.5	84.4	74.3	82.8	73.1	81.8	71.4	83.3	73.2
Honolulu	0.4	83.2	70.7	83.3	70.2	84.9	70.1	85.3	70.4	87.2	71.6	88.5	72.1	89.5	73.1	90.4	73.5	90.6	74.1	89.4	73.7	87.4	72.9	85.1	72.2	89.2	73.2
911820	1	82.3	70.3	82.5	70.0	84.0	70.0	84.5	70.0	86.3	71.3	87.8	71.7	88.9	72.8	89.6	73.3	89.7	73.7	88.6	73.4	86.4	72.7	83.7	71.2	88.3	72.9
	2	81.5	69.9	81.8	69.8	83.1	69.5	83.8	69.8	85.6	71.0	87.2	71.6	88.2	72.5	89.1	73.1	89.0	73.4	87.8	73.1	85.6	72.5	82.9	70.8	87.3	72.6
Kahului	0.4	83.7	71.9	84.2	70.9	86.2	71.1	86.2	71.3	88.0	72.2	89.2	72.9	90.0	74.2	90.3	75.0	90.3	74.9	89.3	74.4	87.5	73.2	85.1	72.8	89.0	74.0
911900	1	82.8	70.9	83.3	70.6	84.8	70.6	85.2	71.3	87.0	72.4	88.2	72.4	88.8	74.1	89.2	74.3	89.4	74.5	88.4	74.2	86.6	73.2	84.1	72.1	87.9	73.5
	2	81.9	70.3	82.4	70.2	83.6	70.4	84.2	70.9	86.1	71.9	87.3	72.4	87.8	73.8	88.3	73.9	88.5	74.0	87.7	73.9	85.6	72.7	83.3	71.6	86.8	73.1
Lihue	0.4	81.9	70.8	82.7	70.4	82.2	71.1	82.8	72.3	83.6	72.8	85.3	75.0	85.5	75.4	87.0	75.8	86.4	75.6	85.5	74.9	83.3	72.8	82.1	71.8	85.4	74.9
911650	1	80.8	70.3	81.3	69.9	81.2	70.8	81.7	71.5	82.8	72.5	84.3	74.1	85.0	74.9	86.0	75.5	85.8	75.2	84.9	74.4	82.5	72.7	81.0	71.6	84.5	74.4
	2	79.8	70.2	80.2	70.1	80.3	70.5	80.9	71.2	82.2	72.1	83.5	73.3	84.4	74.3	85.3	74.8	85.2	74.9	84.2	74.1	82.0	72.6	80.2	71.5	83.9	74.0
IDAHO																											
Boise	0.4	51.8	44.6	60.3	47.0	70.1	50.2	80.7	55.4	90.9	59.9	98.6	62.9	100.4	64.9	100.1	64.7	92.7	61.1	82.2	56.0	64.5	48.4	54.6	44.5	96.4	63.4
726810	1	49.8	43.2	57.4	45.8	67.0	49.6	77.7	53.6	88.0	59.7	96.0	62.6	98.4	64.2	97.8	63.8	90.2	60.7	79.3	55.0	61.7	47.8	51.9	43.6	93.5	62.5
	2	48.0	42.1	54.8	44.4	63.7	47.7	74.6	52.8	84.8	58.2	93.4	62.1	96.7	63.9	95.8	63.3	88.1	59.8	76.3	54.1	59.1	47.1	49.7	42.6	90.5	61.5
Pocatello	0.4	49.3	39.7	56.4	44.8	65.3	46.7	77.5	51.2	84.5	56.3	94.2	61.3	96.1	61.4	96.2	60.7	89.6	58.0	80.4	53.5	62.9	46.7	50.9	39.7	93.0	60.6
725780	1	46.0	38.8	53.3	42.0	62.5	45.6	75.0	50.4	82.5	55.6	92.0	60.6	94.7	61.3	94.3	60.5	87.5	57.5	77.6	52.2	60.3	45.6	48.1	39.7	90.1	59.9
	2	44.1	37.7	50.5	40.5	59.4	44.5	71.7	49.8	80.4	54.8	89.4	60.2	93.3	61.0	92.8	60.2	85.3	57.1	74.7	51.5	57.7	43.9	45.7	38.2	87.2	59.2
ILLINOIS																											
Chicago, O'Hare	0.4	53.8	50.8	57.4	49.6	73.7	59.1	83.1	63.9	88.7	69.4	93.0	71.2	95.6	76.1	94.0	76.9	90.3	73.8	82.7	65.9	69.9	59.4	61.3	57.2	91.1	74.4
Intl Airport	1	50.0	46.1	53.4	47.5	69.7	58.1	79.7	62.6	86.3	68.9	90.7	71.8	93.5	76.0	92.0	75.6	87.6	72.6	79.7	65.5	66.7	58.6	57.9	55.4	88.4	73.0
725300	2	45.8	42.2	49.6	44.3	65.8	55.6	75.9	61.3	84.1	68.0	89.1	71.6	91.5	75.9	89.9	74.7	85.1	70.9	76.5	62.9	63.9	56.9	53.7	49.4	85.5	71.4

WMO# = World Meteorological Organization number DB = dry-bulb temperature, °F MWB = mean coincident wet-bulb temperature, °F

Table 4A Selected Monthly Percentiles of Wet-Bulb Temperature with Mean Dry-Bulb Temperature for U.S. Locations

Location and WMO#	%	Jan		Feb		Mar		Apr		May		Jun		Jul		Aug		Sep		Oct		Nov		Dec		Annual	
		WB	MDB	WB	MDB	WB	MDB	WB	MDB	WB	MDB	WB	MDB	WB	MDB	WB	MDB	WB	MDB	WB	MDB	WB	MDB	WB	MDB	WB	MDB
1a	1b	2a	2b	3a	3b	4a	4b	5a	5b	6a	6b	7a	7b	8a	8b	9a	9b	10a	10b	11a	11b	12a	12c	13a	13b	14a	14b
Moline/Davenport	0.4	49.7	52.3	50.4	57.5	62.1	71.2	67.6	80.5	73.4	84.5	77.9	89.5	80.5	92.5	80.4	91.9	77.0	87.8	70.0	80.8	61.6	67.2	56.7	60.4	78.2	89.6
Intl Airport	1	45.1	49.1	47.9	54.6	59.9	67.6	65.9	76.9	72.1	83.0	76.6	87.8	79.5	91.4	79.3	90.5	75.5	85.5	68.2	76.2	60.0	65.2	53.6	56.2	76.5	86.9
725440	2	41.6	45.2	45.2	50.2	57.5	64.3	64.4	74.6	70.9	81.6	75.6	87.2	78.6	90.0	78.2	89.0	74.3	83.5	66.6	74.4	58.6	62.8	48.9	52.1	75.0	84.8
Peoria	0.4	52.0	54.0	52.6	57.2	63.3	71.5	68.8	78.1	74.2	82.6	78.3	87.8	81.1	91.5	80.2	90.1	77.3	87.5	69.5	77.3	62.3	66.9	58.5	60.8	78.4	88.9
725320	1	48.0	49.5	50.2	54.9	61.0	69.0	66.7	75.3	72.7	81.4	77.1	86.5	79.8	89.9	79.1	89.8	75.7	85.0	68.0	75.6	60.9	66.4	55.2	57.1	76.8	86.2
	2	44.2	46.3	47.9	51.6	58.9	63.8	65.0	73.2	71.3	80.6	76.0	85.4	78.8	89.6	78.1	88.2	74.4	82.8	66.5	73.1	59.4	63.6	51.3	53.4	75.1	84.2
Rockford	0.4	46.7	48.3	47.9	53.2	60.8	69.4	66.7	78.6	72.8	82.2	77.0	88.9	79.7	91.4	79.4	89.5	75.6	85.7	68.8	78.3	60.4	65.5	55.4	57.5	77.2	87.3
725430	1	42.6	44.6	45.2	49.0	58.2	64.6	64.9	76.8	71.1	81.3	75.5	86.7	78.7	89.7	78.2	87.7	74.4	83.9	67.2	74.7	59.0	62.6	51.9	53.8	75.3	84.6
	2	39.2	41.5	42.4	45.6	55.7	61.4	63.0	72.0	69.8	80.8	74.4	84.9	77.7	88.0	77.0	86.4	73.3	81.8	65.5	72.4	57.5	60.9	46.3	48.3	73.6	82.0
Springfield	0.4	55.1	57.4	55.8	59.5	64.5	72.9	69.9	78.9	75.3	84.1	78.5	89.0	81.8	90.9	80.8	91.4	77.8	88.0	70.4	78.2	64.2	69.4	60.1	62.7	79.0	89.4
724390	1	51.3	53.8	53.5	58.5	62.3	69.5	67.9	76.9	73.7	81.8	77.4	87.8	80.4	90.3	79.8	89.9	76.5	86.6	68.7	77.0	62.6	67.5	57.3	60.0	77.4	87.7
	2	47.8	50.9	50.7	55.2	60.6	67.5	66.3	74.9	72.4	81.3	76.4	86.7	79.5	90.1	78.8	89.1	75.3	85.1	67.5	75.1	61.2	65.9	54.2	56.2	75.9	85.1
INDIANA																											
Evansville	0.4	59.0	62.9	59.8	65.6	66.2	74.6	70.3	79.9	75.7	85.9	79.5	90.8	80.8	92.5	80.4	91.2	78.1	88.1	71.4	79.0	65.8	71.1	61.7	66.0	79.0	90.2
724320	1	56.7	60.3	57.9	61.9	64.4	71.5	68.9	78.8	74.5	84.6	78.4	89.8	80.0	92.4	79.5	90.1	77.1	87.4	70.0	77.3	64.4	69.6	59.6	63.7	77.7	88.5
	2	54.2	57.3	55.9	59.7	62.4	69.3	67.4	76.1	73.4	82.6	77.4	88.7	79.3	91.0	78.6	89.5	76.0	86.3	68.4	76.4	63.1	67.9	57.8	61.4	76.5	86.7
Fort Wayne	0.4	52.7	54.9	52.2	56.0	62.0	67.6	66.3	75.8	73.3	81.8	76.6	87.4	79.2	89.5	79.1	89.4	76.3	85.2	68.1	76.1	62.1	67.5	58.8	61.7	76.9	86.3
725330	1	49.0	50.7	49.7	54.3	60.0	66.9	64.9	75.3	71.6	81.3	75.5	86.3	78.1	88.7	78.0	87.8	74.9	84.1	66.5	74.0	60.5	64.7	56.3	58.3	75.2	84.4
	2	44.7	47.1	47.0	50.9	58.0	64.3	63.5	72.8	70.3	80.4	74.5	84.8	77.1	87.2	76.7	85.1	73.7	82.6	65.3	72.8	59.4	63.8	53.1	54.9	73.6	81.9
Indianapolis	0.4	57.4	60.6	56.1	59.4	63.3	69.5	67.5	76.4	74.0	83.1	78.3	87.4	80.2	90.1	80.1	89.7	77.0	86.8	69.9	76.7	64.4	68.9	59.8	62.3	78.1	87.9
724380	1	53.0	54.9	53.6	57.3	61.5	68.0	66.0	73.8	72.7	81.9	77.0	86.2	79.2	89.6	79.0	88.6	75.9	85.1	68.4	74.2	63.1	67.1	58.2	61.0	76.7	85.5
	2	49.4	51.7	51.1	55.3	59.9	66.8	64.8	73.1	71.4	81.0	75.8	85.3	78.4	88.7	78.0	87.2	74.8	83.8	67.1	73.0	61.6	65.4	55.9	58.7	75.2	83.4
South Bend	0.4	52.5	54.4	51.5	56.4	61.7	69.3	66.1	76.6	73.2	81.6	76.5	86.5	79.1	90.6	78.7	88.4	75.3	84.6	68.4	76.4	62.3	66.3	58.6	61.2	76.7	86.4
725350	1	49.0	51.3	48.6	53.1	59.5	67.5	64.6	73.8	71.3	79.8	75.3	85.9	78.0	88.3	77.7	87.5	74.3	82.4	66.9	74.6	60.8	65.0	56.6	58.8	74.9	84.0
	2	44.2	46.6	45.8	50.0	57.4	64.1	63.3	71.7	70.0	79.5	74.3	84.1	77.0	87.0	76.5	85.9	73.4	80.8	65.2	71.9	59.2	63.4	52.8	55.4	73.3	81.3
IOWA																											
Des Moines	0.4	47.1	50.9	50.6	57.8	60.9	71.2	67.3	80.2	73.2	83.2	78.1	89.3	80.2	92.3	78.9	91.2	76.6	86.7	69.4	78.6	61.0	66.4	56.3	59.1	77.6	89.2
725460	1	43.2	47.3	47.1	56.2	58.7	66.2	65.5	76.9	71.7	81.4	77.0	87.8	79.2	90.7	78.0	90.0	75.4	86.1	67.6	76.9	59.3	63.3	51.6	54.6	76.2	87.4
	2	40.4	46.1	44.6	52.2	56.2	65.3	63.9	75.1	70.1	80.5	75.9	86.7	78.2	89.7	77.1	89.0	74.3	84.1	65.8	73.2	57.5	61.7	45.9	51.2	74.7	85.3
Mason City	0.4	38.6	42.8	44.7	52.4	58.6	67.1	65.4	79.3	71.9	82.0	77.7	88.6	80.1	89.9	79.4	88.9	76.1	85.7	68.7	74.8	58.9	62.3	50.8	52.5	77.3	87.4
725485	1	36.8	39.8	41.6	46.7	55.4	61.8	63.6	75.4	70.2	79.9	76.2	87.1	78.9	88.8	78.1	87.9	74.2	82.6	66.4	72.9	56.9	60.3	42.7	44.9	75.4	85.0
	2	35.4	38.0	38.8	42.5	52.2	58.2	61.4	71.9	68.7	78.7	75.0	85.7	77.7	87.5	76.8	86.9	72.8	80.8	64.4	71.3	54.4	58.4	38.8	41.6	73.6	82.0
Sioux City	0.4	42.6	50.4	48.6	58.2	59.6	74.0	66.1	81.8	72.5	84.0	77.6	89.6	81.0	91.4	79.4	91.0	76.2	87.8	68.2	77.2	58.1	64.2	48.3	53.2	78.0	89.1
725570	1	40.4	47.3	45.8	54.7	56.5	66.1	64.3	77.5	71.0	82.0	76.6	88.5	79.6	90.1	78.4	90.2	74.9	85.9	66.5	76.1	56.0	61.8	43.6	49.7	76.3	87.3
	2	38.3	44.2	43.3	50.1	54.1	64.8	62.5	74.9	69.6	81.0	75.6	87.9	78.6	89.5	77.3	88.8	73.7	82.9	64.3	72.7	53.5	59.5	40.9	46.0	74.6	85.2
Waterloo	0.4	41.4	45.6	46.1	53.2	59.8	69.1	65.8	78.4	72.8	82.2	77.4	89.0	79.9	91.0	79.2	89.1	76.3	86.1	68.8	76.6	59.4	63.0	53.8	56.5	77.2	87.3
725480	1	38.7	41.7	42.5	48.6	56.7	64.8	63.9	75.1	71.1	80.7	76.1	87.7	78.8	89.5	78.0	87.9	74.7	83.1	66.8	75.9	57.5	61.6	48.3	50.6	75.4	85.3
	2	36.9	40.0	40.1	44.5	53.7	60.9	62.1	71.7	69.6	80.3	75.0	85.7	77.6	87.8	76.9	86.7	73.3	81.5	64.8	73.4	55.1	59.3	42.6	45.3	73.7	82.7
KANSAS																											
Dodge City	0.4	49.2	64.0	53.3	70.3	58.9	71.2	65.1	78.3	70.8	84.8	74.8	90.1	75.4	91.1	75.2	89.7	72.3	87.4	66.2	77.8	59.1	66.5	50.8	62.2	73.7	90.3
724510	1	46.8	60.9	50.9	65.6	56.9	72.3	63.2	75.3	69.5	82.0	73.8	90.3	74.4	91.2	74.2	90.0	71.2	86.5	65.0	76.3	56.9	64.3	48.3	59.7	72.5	89.4
	2	44.7	57.5	48.7	63.0	55.2	69.7	61.7	74.6	68.1	80.7	72.9	89.6	73.7	90.8	73.4	90.0	70.4	85.7	63.7	74.6	55.0	63.2	46.2	57.0	71.4	87.8
Goodland	0.4	45.7	62.4	49.6	68.3	53.2	71.6	59.5	75.1	66.0	78.1	70.5	85.2	72.3	86.4	71.4	84.9	68.3	83.1	60.2	72.7	52.4	66.6	47.4	63.4	70.1	85.5
724650	1	43.8	59.2	47.2	64.1	51.3	68.6	58.3	74.1	64.7	76.9	69.6	84.4	71.3	85.6	70.4	85.2	67.2	81.9	58.8	73.9	50.6	66.1	45.3	58.6	69.0	84.2
	2	42.1	55.9	45.6	61.4	49.6	66.6	56.9	73.4	63.6	76.3	68.7	83.9	70.5	85.7	69.8	85.0	66.2	80.7	57.4	73.3	49.1	63.7	43.2	57.2	67.9	83.8
Topeka	0.4	54.5	59.7	56.7	64.3	63.8	73.0	70.1	81.7	74.6	84.3	79.4	90.3	80.9	91.9	80.1	91.8	77.8	88.5	71.6	78.9	63.9	69.9	58.1	62.1	78.8	90.4
724560	1	50.1	55.3	53.7	62.1	62.5	71.6	68.7	79.5	73.3	83.5	78.4	88.8	80.0	91.6	79.2	90.9	76.7	88.3	69.7	77.8	62.1	67.8	55.9	59.9	77.5	88.9
	2	47.3	53.9	50.8	61.2	60.4	69.6	67.1	76.6	72.0	81.9	77.4	87.5	79.1	90.8	78.4	89.8	75.7	86.9	68.1	76.1	60.4	66.2	52.1	56.7	76.2	87.6
Wichita, Airport	0.4	55.0	58.4	56.5	64.9	64.4	74.2	69.2	80.2	74.2	85.1	77.4	92.4	78.2	91.5	78.3	91.1	76.4	88.5	71.0	80.3	63.6	69.3	58.2	61.5	76.7	90.7
724500	1	50.3	55.5	54.6	62.0	62.6	71.1	67.9	77.9	72.8	83.6	76.4	90.7	77.3	91.1	77.3	90.9	75.3	87.2	69.2	77.3	62.0	67.2	55.4	59.2	75.5	89.8
	2	47.7	54.1	52.1	62.5	60.6	69.5	66.7	76.2	71.7	81.9	75.6	89.3	76.6	90.9	76.4	90.5	74.3	85.7	67.7	75.7	60.3	65.6	52.3	55.8	74.4	88.6
KENTUCKY																											
Covington/	0.4	57.5	61.0	57.8	61.7	63.9	71.7	67.4	76.8	73.5	82.7	77.9	86.9	79.8	91.1	78.6	88.4	75.7	86.0	69.3	76.5	64.3	69.7	60.1	64.9	77.2	87.2
Cincinnati OH,	1	55.5	58.6	55.9	60.1	61.9	69.7	66.1	75.8	72.3	81.8	76.7	86.3	78.8	89.6	77.5	87.7	74.9	85.0	68.0	75.7	62.8	66.5	58.6	62.8	75.7	85.5
Intl Airport	2	52.3	55.1	53.6	57.3	60.1	67.2	64.8	74.5	71.2	80.6	75.5	85.2	77.7	88.0	76.7	86.7	74.1	83.3	66.8	73.9	61.3	65.3	56.5	60.0	74.3	83.2
724210																											
Lexington	0.4	57.7	62.1	58.3	62.6	64.1	73.3	67.8	78.0	73.7	82.5	76.5	86.5	78.9	89.7	77.9	88.6	75.9	86.1	69.9	77.2	65.0	72.0	60.5	65.0	76.6	87.2
724220	1	55.8	59.5	57.1	61.6	62.5	70.0	66.3	76.0	72.4	81.9	75.7	85.9	78.0	88.6	77.1	88.4	75.0	84.9	68.5	74.6	63.4	68.6	59.2	63.4	75.4	85.5
	2	53.6	57.0	55.4	59.2	60.8	68.3	65.0	74.8	71.4	80.9	74.9	85.1	77.1	87.8	76.2	87.3	74.1	83.1	67.3	74.4	61.9	67.2	57.9	61.7	74.2	83.3
Louisville	0.4	59.0	63.8	59.8	64.4	65.0	74.2	68.8	78.9	74.8	84.5	78.4	89.0	80.5	92.9	79.5	90.9	77.4	87.5	71.3	78.3	66.0	71.1	61.4	66.0	78.3	89.5
724230	1	57.0	60.2	58.1	63.5	63.5	71.6	67.4	77.6	73.8	83.4	77.5	88.1	79.5	91.8	78.9	90.4	76.4	86.5	70.0	76.8	64.6	69.1	59.8	64.6	77.1	87.6
	2	54.5	58.2	56.2	59.9	61.8	69.9	66.3	76.2	72.7	81.8	76.4	87.1	78.6	90.0	78.1	89.8	75.5	85.4	68.5	76.3	63.2	68.0	58.3	62.2	75.8	85.6
LOUISANA																											
Baton Rouge	0.4	70.0	74.5	71.1	75.6	73.0	79.9	74.7	82.5	78.3	87.2	80.3	90.7	81.2	91.3	81.3	90.7	80.0	87.6	77.4	84.1	74.1	78.4	73.0	76.0	80.1	89.1
722317	1	68.9	73.1	69.9	74.4	71.9	77.5	74.0	81.9	77.4	85.9	79.8	89.8	80.4	90.0	80.5	89.3	79.3	87.3	76.3	82.9	73.2	77.6	71.5	75.3	79.3	88.3
	2	67.8	72.3	68.5	72.6	71.0	76.9	73.3	80.8	76.5	84.4	79.2	88.6	80.0	89.2	80.1	88.6	78.5	86.8	75.2	82.3	72.2	77.0	70.5	74.0	78.4	87.3
Lake Charles	0.4	70.2	73.5	70.9	74.1	72.3	76.8	75.3	80.8	78.2	85.6	81.3	88.4	81.4	89.6	81.3	89.7	80.4	87.5	78.4	84.3	75.2	78.9	73.1	75.7	80.4	88.2
722400	1	69.3	72.1	69.8	72.2	71.6	75.8	74.4	80.1	77.5	84.5	80.4	87.7	80.8	88.8	80.7	88.9	79.9	87.1	77.5	83.5	74.2	77.9	72.1	74.1	79.8	87.7
	2	68.3	70.3	68.8	70.6	71.0	74.6	73.9	79.3	76.8	83.5	79.9	87.5	80.3	88.4	80.3	88.4	79.4	86.8	76.6	82.2	73.3	76.9	71.1	73.4	79.2	87.0
New Orleans,	0.4	71.0	74.8	72.3	76.1	74.4	78.7	76.4	82.1	78.4	86.0	81.3	90.4	82.5	92.5	82.3	91.4	81.3	89.1	78.3	84.9	75.3	79.4	73.5	77.0	81.2	89.5
Intl Airport	1	69.8	74.1	71.2	74.7	73.4	77.9	75.3	81.6	77.7	85.6	80.5	89.3	81.9	91.2	81.5	89.8	80.4	87.8	77.2	84.1	74.4	78.5	72.7	75.9	80.3	88.3
722310	2	68.7	72.6	70.1	73.3	72.5	77.1	74.5	80.5	77.2	85.0	80.0	88.6	81.2	89.9	81.0	88.8	79.9	87.0	76.2	82.2	73.4	77.2	71.6	75.1	79.5	87.3
Shreveport	0.4	66.9	72.6	67.7	72.7	70.4	78.6	74.3	82.1	77.5	87.2	80.4	91.4	80.2	93.0	80.3	92.4	79.3	90.0	76.4	84.2	72.1	77.7	69.7	74.1	79.3	91.2
722480	1	65.7	69.9	66.3	71.0	69.3	76.9	73.2	81.4	76.7	86.3	79.4	90.1	79.5	92.3	79.8	92.0	78.6	89.3	75.3	82.6	70.8	75.8	68.2	72.6	78.5	90.2
	2	64.7	68.9	64.8	70.3	68.4	75.0	72.3	79.7	75.8	85.1	78.6	89.2	79.2	91.9	79.3	91.3	78.0	89.0	74.2	81.0	69.7	75.0	66.8	71.0	77.9	89.3

WMO# = World Meteorological Organization number WB = wet-bulb temperature, °F MDB = mean coincident dry-bulb temperature, °F

Table 4B Selected Monthly Percentiles of Dry-Bulb Temperature with Mean Wet-Bulb Temperature for U.S. Locations

Location and WMO#	%	Jan		Feb		Mar		Apr		May		Jun		Jul		Aug		Sep		Oct		Nov		Dec		Annual	
		DB	MWB	DB	MWB	DB	MWB	DB	MWB	DB	MWB	DB	MWB	DB	MWB	DB	MWB	DB	MWB	DB	MWB	DB	MWB	DB	MWB	DB	MWB
1a	1b	2a	2b	3a	3b	4a	4b	5a	5b	6a	6b	7a	7b	8a	8b	9a	9b	10a	10b	11a	11b	12a	12c	13a	13b	14a	14b
Moline/Davenport	0.4	53.5	49.0	59.3	50.3	74.9	59.6	84.7	64.6	89.6	69.8	94.7	74.4	97.0	77.9	96.7	77.6	92.0	73.7	83.7	67.9	70.3	58.5	60.8	56.8	93.0	75.8
Intl Airport	1	49.9	44.7	54.7	46.2	70.9	56.4	81.5	64.0	87.6	69.4	92.8	73.8	95.2	77.0	93.6	77.5	89.0	73.5	80.9	66.1	67.0	58.6	56.5	51.8	90.0	74.2
725440	2	46.1	41.3	50.9	44.7	66.9	55.4	78.2	62.2	85.5	67.9	90.8	73.0	93.4	76.5	91.6	76.8	86.6	72.3	78.0	64.8	64.4	57.0	53.0	48.0	87.2	72.8
Peoria	0.4	54.5	49.9	59.6	50.1	75.0	59.6	83.2	65.5	87.9	70.2	93.2	74.0	97.0	77.4	96.2	76.9	90.4	74.0	82.2	66.8	70.6	60.0	61.2	57.4	91.9	75.7
725320	1	51.0	47.2	55.5	48.8	71.8	58.2	80.4	63.8	86.3	68.9	91.3	73.6	94.6	77.3	92.8	76.8	88.3	73.2	79.8	65.5	67.6	59.3	58.0	54.2	89.1	74.2
	2	47.4	43.2	52.3	47.1	67.8	56.7	77.5	62.8	84.4	68.6	89.6	72.9	92.7	76.6	90.7	76.6	86.1	73.1	77.0	63.9	64.6	58.0	54.5	51.2	86.4	73.0
Rockford	0.4	48.9	47.1	54.1	47.2	72.0	58.6	83.5	64.5	88.0	68.9	92.9	72.7	95.1	75.7	94.0	75.9	89.2	72.6	82.1	67.8	68.0	58.6	58.1	55.0	90.6	74.3
725430	1	45.2	42.1	49.7	44.1	68.1	55.1	79.9	62.7	86.3	68.3	90.7	72.3	93.0	76.2	91.3	75.4	87.0	72.6	78.9	64.9	64.3	57.7	54.0	51.8	87.8	72.7
	2	42.2	38.6	46.2	41.4	63.6	54.3	75.9	62.0	84.4	67.5	88.8	72.2	91.0	75.2	89.0	74.9	84.5	71.4	76.1	63.5	61.5	56.0	49.3	45.4	85.3	71.1
Springfield	0.4	58.6	53.8	62.4	54.7	76.2	61.5	84.2	65.5	90.2	71.0	95.3	74.6	97.8	78.0	96.2	77.6	92.4	74.3	85.2	67.4	73.1	60.6	63.2	57.5	93.4	76.1
724390	1	54.6	50.8	59.0	51.5	73.2	60.4	81.8	65.1	88.3	70.1	93.2	73.9	95.6	78.1	94.0	77.1	90.1	73.6	82.6	66.3	70.4	60.5	60.7	56.9	90.7	75.2
	2	51.3	46.9	55.9	50.7	69.9	58.4	79.4	64.3	86.4	69.4	91.6	73.6	93.8	77.0	91.9	77.3	88.0	73.4	79.9	64.5	67.4	59.4	57.0	52.7	88.1	73.7
INDIANA																											
Evansville	0.4	64.2	57.5	67.9	57.1	78.4	62.5	85.3	66.6	90.2	71.6	95.9	75.9	97.8	78.0	97.1	77.7	92.6	73.8	85.1	68.0	75.5	62.2	66.7	61.0	94.0	76.5
724320	1	60.9	56.2	64.3	55.5	75.4	61.3	83.4	65.5	88.5	70.7	94.0	75.4	95.5	77.6	94.9	76.8	90.9	74.0	83.0	66.4	73.2	61.5	64.3	58.8	91.6	75.5
	2	58.1	53.2	61.8	53.9	72.8	60.8	81.2	64.9	87.0	70.4	92.4	74.3	94.1	77.2	93.1	76.4	89.3	74.6	80.8	65.7	70.7	60.5	62.1	57.2	89.5	74.6
Fort Wayne	0.4	55.4	52.4	57.6	50.6	72.7	59.7	82.5	63.5	88.2	69.5	92.3	70.7	95.0	74.6	92.5	75.6	89.6	72.7	81.7	64.3	70.8	59.9	61.7	58.5	90.3	73.7
725330	1	51.6	48.3	53.6	48.8	69.7	58.6	79.6	63.3	86.2	68.5	90.5	71.5	92.5	75.0	90.3	75.2	87.1	72.5	79.2	64.7	67.5	59.1	58.7	55.6	87.7	72.7
	2	47.4	44.4	50.8	46.6	66.2	56.4	75.5	61.6	84.2	68.0	88.9	71.7	90.7	75.0	88.6	74.4	84.9	71.3	76.1	63.0	64.5	58.2	55.6	52.6	85.3	71.2
Indianapolis	0.4	60.5	57.0	62.6	53.2	75.2	61.0	82.0	64.6	87.9	70.3	92.1	73.0	95.5	76.6	93.1	76.3	90.3	73.4	82.2	65.6	72.5	61.2	63.6	58.3	90.8	75.3
724380	1	56.0	52.3	59.2	51.3	72.2	58.9	79.7	63.6	86.3	69.9	90.4	72.4	93.3	76.9	91.1	77.2	88.4	73.8	79.8	65.4	70.2	60.8	61.2	57.4	88.4	74.2
	2	52.3	48.6	56.2	50.8	69.2	57.6	76.9	63.0	84.6	68.7	88.8	73.2	91.4	76.6	89.4	76.3	86.4	72.8	77.3	64.5	67.6	59.6	58.8	55.6	86.2	72.9
South Bend	0.4	54.5	52.3	57.7	50.0	73.2	59.0	82.1	62.7	88.1	68.5	92.5	71.2	94.9	75.9	92.8	76.6	89.4	72.4	80.9	64.8	70.3	60.1	61.7	58.7	90.0	73.4
725350	1	51.3	48.4	53.6	49.1	69.5	56.9	78.6	62.7	86.1	68.5	90.3	71.5	92.4	75.4	90.3	74.9	86.2	71.7	78.4	64.4	66.9	58.6	58.9	56.0	87.4	72.3
	2	47.3	43.4	50.2	45.2	65.8	55.7	74.7	60.9	84.0	67.2	88.6	70.8	90.3	74.7	88.4	74.2	84.0	70.9	75.5	63.3	64.2	58.2	55.2	52.5	84.7	70.8
IOWA																											
Des Moines	0.4	52.8	44.2	60.9	47.0	76.1	57.9	84.7	65.0	87.5	68.9	94.9	74.9	98.5	76.1	97.8	75.7	91.6	74.0	83.1	66.6	68.8	58.0	59.9	54.3	93.4	75.5
725460	1	49.3	43.4	56.3	46.2	71.2	56.3	81.7	63.1	85.9	68.6	92.0	73.7	95.9	76.5	95.0	76.3	88.7	73.7	80.3	65.1	65.6	56.8	55.7	49.7	90.1	74.2
	2	45.9	39.9	52.4	44.5	67.1	54.2	78.1	61.7	84.1	68.1	90.0	73.4	93.9	76.2	92.4	75.4	86.3	72.0	77.5	63.5	63.1	55.9	51.6	45.2	87.2	72.8
Mason City	0.4	43.9	37.9	52.2	43.1	70.3	57.0	83.9	62.3	88.0	67.2	94.3	74.9	95.1	76.6	94.0	76.4	89.6	74.0	82.0	64.6	64.9	56.5	52.6	50.8	91.1	74.2
725485	1	40.4	36.1	46.9	40.6	63.9	52.9	79.9	62.0	85.7	65.8	91.7	72.4	93.0	75.9	91.4	75.7	86.5	71.2	78.5	62.8	61.8	54.4	46.1	42.2	88.0	72.7
	2	38.1	35.0	43.0	38.2	59.2	51.5	75.7	59.2	83.2	65.3	89.5	71.5	91.2	75.1	89.2	74.8	83.9	70.2	75.1	61.3	59.2	53.5	41.9	38.5	85.1	71.4
Sioux City	0.4	52.1	41.7	59.8	46.5	76.2	56.3	87.6	62.9	89.4	67.3	96.7	73.4	98.3	75.9	96.2	76.2	91.9	71.8	83.9	64.5	67.2	53.6	54.9	46.5	93.6	74.9
725570	1	48.1	39.5	55.2	45.4	71.5	55.4	83.7	62.5	87.3	68.0	94.0	73.7	95.8	76.1	94.1	75.6	89.1	72.1	80.8	63.2	64.3	53.6	50.9	43.9	90.4	73.9
	2	44.7	38.1	51.3	43.1	66.8	52.6	79.9	60.9	85.2	67.1	91.8	73.3	93.8	76.2	91.9	75.2	86.6	70.5	77.9	61.6	61.5	51.4	47.0	40.1	87.5	72.3
Waterloo	0.4	46.2	40.3	53.1	45.3	72.6	56.2	83.0	62.8	87.9	69.3	93.2	74.0	95.7	76.0	95.1	77.4	89.8	72.6	81.9	66.2	66.0	56.5	56.7	53.9	91.2	74.5
725480	1	42.8	38.0	48.8	43.2	67.0	54.3	79.7	61.6	85.8	68.3	91.1	73.0	93.5	75.5	91.4	75.3	86.9	73.1	78.6	64.6	62.8	55.3	51.0	47.9	88.2	72.9
	2	40.5	36.5	45.1	39.3	62.3	53.3	75.6	59.9	83.6	66.7	89.3	72.2	91.7	75.4	89.5	75.0	84.4	70.7	75.5	62.3	60.2	54.1	46.4	41.9	85.4	71.4
KANSAS																											
Dodge City	0.4	66.1	47.6	74.9	51.7	82.3	55.4	87.9	59.9	93.4	64.0	101.3	69.3	104.2	70.4	102.7	69.3	97.7	67.9	89.7	60.3	76.2	54.5	66.3	48.1	100.1	70.1
724510	1	62.4	45.4	70.0	48.7	79.0	54.7	84.9	58.3	90.8	64.7	98.7	69.9	102.4	69.8	100.8	70.3	95.6	67.2	87.3	60.1	72.9	53.2	62.8	46.2	97.1	69.5
	2	58.6	44.0	66.0	47.3	75.4	52.6	82.3	58.3	87.9	64.0	96.7	68.9	100.9	70.1	99.1	70.1	93.4	67.5	84.4	59.5	69.5	51.3	59.1	45.1	94.0	69.1
Goodland	0.4	65.2	44.7	71.1	48.9	78.7	50.0	85.3	55.0	89.7	59.7	99.7	64.4	101.1	65.8	99.2	66.1	94.6	63.1	86.9	55.7	72.3	49.3	65.6	46.1	96.9	65.9
724650	1	60.8	42.8	66.4	46.1	74.2	49.4	82.0	54.5	86.9	59.3	97.0	65.3	99.2	66.3	97.3	65.9	92.2	63.2	84.2	55.7	69.8	49.4	61.9	44.1	93.8	65.6
	2	56.6	41.4	62.9	45.0	71.0	48.2	78.9	54.2	84.5	59.1	94.5	64.6	97.4	66.6	95.4	65.1	90.0	62.4	81.4	55.1	66.8	47.8	58.0	42.1	90.7	65.3
Topeka	0.4	62.3	52.4	69.4	52.2	80.1	60.4	86.4	65.9	88.8	70.5	95.5	75.3	101.3	75.5	100.2	74.8	95.5	73.0	87.3	65.1	74.1	60.3	64.4	54.5	95.9	75.4
724560	1	58.1	48.2	65.8	51.7	76.2	58.5	83.5	65.4	87.1	70.1	93.6	75.4	98.5	76.0	97.6	75.3	93.3	73.4	84.4	66.0	71.1	58.8	61.5	52.7	92.9	75.2
	2	54.8	46.3	62.0	50.1	73.0	57.8	81.1	64.3	85.2	69.4	91.8	75.4	96.4	76.1	95.4	75.5	91.0	73.0	81.7	65.5	68.4	57.6	58.8	51.7	90.3	74.7
Wichita, Airport	0.4	62.7	50.0	71.6	53.1	79.8	59.9	86.2	64.2	90.9	70.8	100.3	72.2	106.5	72.2	103.4	72.7	98.5	71.5	88.7	64.6	74.0	59.0	63.9	52.8	100.3	72.7
724500	1	59.2	48.6	67.9	51.3	76.4	59.0	83.1	64.8	88.5	69.6	98.1	72.9	103.7	72.7	101.3	72.6	95.7	71.8	85.6	65.3	71.0	58.2	61.2	53.1	97.3	72.7
	2	55.7	46.3	64.4	49.9	73.3	57.9	80.4	63.2	86.4	68.6	95.9	72.9	101.4	72.8	99.5	72.8	93.2	71.2	82.9	64.5	68.6	57.6	58.8	50.6	94.1	72.5
KENTUCKY																											
Covington/	0.4	61.8	55.9	64.9	54.2	76.4	60.3	82.9	64.9	87.4	70.2	92.1	72.4	95.4	75.9	93.7	75.1	90.2	73.4	81.9	66.6	73.1	61.9	66.0	58.8	91.0	74.4
Cincinnati OH,	1	59.1	54.6	62.1	54.1	73.6	59.4	80.7	63.6	85.8	70.0	90.5	72.6	93.2	76.0	91.5	74.6	88.4	72.4	80.0	65.0	70.2	59.4	63.3	57.2	88.6	73.4
Intl Airport	2	55.7	51.0	59.4	52.6	70.7	58.6	78.3	62.4	84.1	68.5	89.0	73.0	91.5	75.8	89.9	74.2	86.2	71.9	77.7	63.9	67.7	58.8	60.8	56.1	86.2	72.3
724210																											
Lexington	0.4	63.8	56.3	66.4	53.9	76.7	60.4	82.4	64.8	87.3	70.1	91.3	71.6	94.8	75.2	93.9	74.7	90.1	72.8	81.8	64.9	74.6	63.5	67.0	59.1	90.7	73.8
724220	1	60.4	53.7	64.1	54.0	74.4	59.7	80.7	63.6	85.7	69.5	89.7	72.2	92.4	74.6	91.7	74.2	88.2	72.1	80.1	65.0	72.1	61.2	64.4	57.4	88.5	73.2
	2	57.7	53.0	61.7	53.1	71.7	58.5	78.7	63.3	84.1	68.9	88.5	72.2	91.1	74.9	90.1	73.8	86.6	72.0	78.1	64.3	69.5	59.4	62.2	57.2	86.3	72.3
Louisville	0.4	64.7	56.8	68.5	55.5	78.7	62.4	84.5	65.8	87.9	72.4	93.0	74.9	96.8	77.4	95.8	77.1	91.7	74.6	83.5	68.4	75.4	63.1	67.6	59.3	92.6	76.2
724230	1	61.9	55.5	65.5	55.1	75.9	59.8	82.8	65.0	86.5	71.0	91.4	74.6	94.5	77.0	93.7	77.0	89.9	74.5	81.4	66.7	72.8	61.3	65.2	58.6	90.4	75.4
	2	59.2	53.7	62.7	54.1	73.3	59.7	80.8	64.3	85.1	70.8	90.3	74.5	93.1	76.9	92.1	76.3	88.2	73.8	79.7	66.6	70.5	60.6	63.3	57.7	88.2	74.2
LOUISANA																											
Baton Rouge	0.4	77.1	68.1	78.7	67.6	83.7	68.2	86.8	71.2	91.7	74.7	95.2	77.8	95.3	78.0	95.4	77.7	93.3	77.1	89.3	73.4	82.7	70.4	78.8	70.5	93.6	77.5
722317	1	75.1	67.4	76.9	66.0	82.0	68.7	85.4	71.0	90.3	73.9	94.0	77.2	94.3	77.9	94.2	77.9	92.2	76.8	87.6	72.9	81.3	70.0	77.0	69.2	92.2	77.2
	2	73.1	66.0	74.9	64.9	80.4	67.5	84.1	70.6	89.1	73.4	93.0	76.4	93.3	77.7	93.1	78.1	91.1	76.4	86.1	72.4	79.8	69.5	75.5	68.6	90.8	76.8
Lake Charles	0.4	74.9	68.2	76.2	65.8	80.1	69.0	85.4	69.3	89.9	73.6	93.2	77.5	94.8	77.7	95.4	78.3	93.0	76.0	88.2	74.8	82.2	70.8	77.2	70.9	92.9	77.7
722400	1	73.4	67.6	74.6	66.4	78.6	68.1	83.6	70.6	88.4	73.1	92.2	77.2	93.7	77.9	94.1	78.0	91.8	76.3	87.0	73.9	80.8	71.1	75.8	70.1	91.4	77.6
	2	71.6	66.1	73.0	65.9	77.3	67.3	82.2	70.5	87.3	73.5	91.2	77.3	92.7	78.0	93.0	78.1	90.5	76.8	85.9	73.2	79.4	70.5	74.4	69.6	90.0	77.4
New Orleans,	0.4	77.4	69.5	78.5	69.1	82.4	69.6	85.7	72.5	90.0	74.7	94.1	79.2	95.8	80.1	95.1	79.4	92.4	78.1	88.4	75.9	82.4	72.4	79.0	71.7	93.1	79.0
Intl Airport	1	75.4	68.3	76.8	67.7	80.7	69.9	84.4	73.1	88.8	74.5	93.0	78.4	94.4	79.8	93.8	78.8	91.3	77.7	86.6	74.8	81.0	72.1	77.4	70.8	91.5	78.3
722310	2	73.4	67.9	75.1	67.5	79.3	69.7	83.3	72.1	87.7	74.4	92.0	77.8	93.3	79.4	92.6	78.7	90.3	77.7	85.3	73.5	79.6	71.3	76.1	70.0	90.1	78.0
Shreveport	0.4	76.0	64.4	79.2	62.6	84.0	65.6	86.5	68.9	90.6	74.6	96.2	77.8	99.7	77.4	99.6	76.6	96.6	76.2	90.2	71.3	81.6	67.7	76.4	67.0	96.7	76.9
722480	1	73.3	63.3	76.2	62.2	81.8	65.2	84.9	70.2	89.3	74.1	94.9	76.9	98.1	77.1	98.2	76.8	94.7	76.0	88.4	70.8	79.9	67.2	74.4	66.0	94.6	76.7
	2	70.6	61.8	73.9	60.6	79.7	64.8	83.6	68.8	88.1	73.2	93.5	76.6	96.6	77.1	96.9	76.7	93.2	75.8	86.5	70.8	78.2	67.1	72.5	65.2	92.6	76.3

WMO# = World Meteorological Organization number   DB = dry-bulb temperature, °F   MWB = mean coincident wet-bulb temperature, °F

Table 4A Selected Monthly Percentiles of Wet-Bulb Temperature with Mean Dry-Bulb Temperature for U.S. Locations

Location and WMO#	%	Jan		Feb		Mar		Apr		May		Jun		Jul		Aug		Sep		Oct		Nov		Dec		Annual	
		WB	MDB	WB	MDB	WB	MDB	WB	MDB	WB	MDB	WB	MDB	WB	MDB	WB	MDB	WB	MDB	WB	MDB	WB	MDB	WB	MDB	WB	MDB
1a	1b	2a	2b	3a	3b	4a	4b	5a	5b	6a	6b	7a	7b	8a	8b	9a	9b	10a	10b	11a	11b	12a	12c	13a	13b	14a	14b
MAINE																											
Caribou	0.4	43.2	43.9	43.2	45.1	46.6	53.4	57.3	64.6	67.2	81.6	72.8	80.9	75.3	83.8	74.9	85.4	70.6	76.9	64.3	70.3	57.3	58.8	46.0	46.3	72.3	80.5
727120	1	38.1	39.2	40.5	44.2	44.4	48.9	54.8	60.3	65.0	74.8	70.8	79.3	74.2	83.5	72.6	79.7	68.7	75.3	62.0	68.1	54.5	56.1	41.6	43.4	70.2	77.3
	2	35.6	37.3	37.4	39.7	42.3	46.6	51.9	58.6	63.4	73.4	69.2	77.2	73.1	82.0	71.3	77.5	67.1	71.6	59.6	64.4	51.6	53.6	38.7	40.0	68.2	75.6
Portland	0.4	47.0	49.0	46.2	48.8	51.9	58.0	58.9	67.2	68.5	79.2	73.3	84.3	76.4	87.8	76.2	86.1	73.7	83.7	64.3	71.2	58.1	61.0	51.8	53.4	73.7	82.9
726060	1	43.6	45.0	43.9	45.9	48.7	54.2	56.1	65.2	66.3	76.2	71.8	82.4	75.4	86.4	74.6	83.5	72.0	80.2	62.7	68.7	56.4	58.8	49.2	51.3	71.7	79.8
	2	40.3	42.5	41.8	44.5	46.2	50.8	53.7	62.0	63.9	73.2	70.4	81.0	74.1	83.8	73.4	81.5	70.1	76.3	61.3	66.8	54.6	57.1	46.6	48.8	69.8	77.1
MARYLAND																											
Glen Burnie/	0.4	56.4	61.5	59.0	64.8	64.2	75.3	67.3	80.5	74.2	85.0	77.8	89.2	79.8	91.5	79.3	88.8	77.3	86.9	71.1	77.3	66.8	72.9	61.5	65.4	77.7	88.2
Baltimore,	1	53.8	58.1	56.3	63.2	61.6	70.9	65.8	77.6	72.8	82.8	76.6	87.5	78.8	90.1	78.2	88.4	76.3	85.9	69.9	75.3	64.4	69.1	59.0	63.3	76.3	86.4
BWI Airport 724060	2	50.6	54.5	53.3	58.5	59.3	67.9	64.2	74.8	71.3	81.0	75.7	86.4	78.0	89.0	77.3	87.3	75.3	85.1	68.4	74.0	63.0	66.8	56.2	59.9	75.1	84.7
MASSACHUSETTS																											
Boston	0.4	54.1	55.4	53.4	57.7	58.1	64.1	63.4	78.8	70.6	84.9	75.3	87.9	77.3	90.7	77.5	88.1	75.3	87.2	68.2	74.1	64.0	68.7	58.1	60.6	75.4	86.5
725090	1	50.5	54.0	50.1	53.6	55.2	62.8	61.2	72.5	68.7	80.4	73.8	86.3	76.4	89.7	76.4	86.1	74.0	83.5	66.3	72.2	61.6	65.3	55.7	57.9	73.7	83.1
	2	46.5	50.3	46.8	50.8	52.5	57.7	58.9	67.5	67.0	77.1	72.5	84.7	75.5	87.7	75.2	84.3	72.8	80.4	64.8	69.8	59.5	64.1	52.6	56.0	72.2	80.5
Worcester	0.4	50.8	52.7	50.9	53.2	55.6	62.6	61.8	74.5	69.1	79.7	73.2	83.5	75.9	85.1	76.6	84.4	74.0	83.0	65.3	70.2	61.2	65.4	54.8	57.6	73.7	81.7
725095	1	46.6	50.6	47.5	50.3	52.9	59.7	59.2	70.1	67.1	77.7	71.6	81.9	74.8	84.0	74.9	82.1	72.3	80.2	63.6	69.0	59.0	61.7	52.5	55.1	71.8	79.9
	2	42.7	45.3	44.0	47.6	50.0	56.0	57.3	66.1	65.2	75.2	70.4	80.0	73.8	82.8	73.5	79.9	70.7	76.8	62.2	68.5	56.8	60.1	48.7	51.6	70.1	77.2
MICHIGAN																											
Alpena	0.4	42.3	44.5	43.0	47.6	55.1	61.1	63.7	78.9	69.9	79.8	73.5	85.9	76.9	87.3	76.0	84.8	73.0	83.1	65.7	74.4	58.1	62.4	51.0	53.0	73.6	83.2
726390	1	39.2	41.6	40.0	43.4	51.2	58.5	61.7	72.8	68.0	77.3	71.9	83.8	75.5	86.6	74.5	83.8	71.5	79.1	63.6	70.0	55.4	58.5	45.3	47.2	71.5	80.7
	2	36.9	39.4	37.6	41.2	47.4	54.6	58.8	67.5	66.4	75.9	70.5	81.5	74.1	84.2	73.4	81.5	69.8	76.4	61.4	67.3	52.5	55.9	41.0	43.2	69.6	78.0
Detroit, Metro	0.4	50.7	53.0	49.3	53.5	60.2	67.6	66.0	75.5	73.3	82.9	76.2	87.2	78.5	90.9	78.4	89.5	75.3	84.6	67.9	76.7	61.0	65.1	57.0	60.0	75.9	85.9
725370	1	45.9	48.3	46.5	50.7	57.4	64.4	64.2	74.2	71.7	81.8	75.0	86.0	77.4	88.8	77.0	87.2	74.0	83.2	65.8	73.4	59.2	63.3	53.8	56.5	74.2	83.6
	2	42.4	44.6	43.8	46.9	54.9	62.2	62.5	72.5	70.1	80.9	73.8	84.5	76.1	86.5	75.5	84.8	72.7	81.3	64.0	71.0	57.5	62.1	50.0	52.7	72.5	81.4
Flint	0.4	48.3	50.4	47.1	50.6	59.3	65.4	65.5	76.2	73.0	82.7	74.9	85.1	78.6	88.8	78.1	87.2	74.7	84.2	67.2	74.7	60.8	65.3	56.6	59.3	75.3	84.2
726370	1	44.9	47.4	44.7	47.9	56.6	62.9	63.8	73.6	71.3	80.8	73.9	85.1	77.1	86.4	76.5	85.3	73.2	81.2	65.7	72.6	59.0	62.5	53.0	54.7	73.5	82.2
	2	40.9	44.0	42.1	45.5	53.6	59.9	62.1	71.4	69.5	78.3	72.9	83.6	75.7	84.9	75.1	83.0	72.0	78.6	63.9	70.5	57.3	61.2	48.8	51.4	71.8	80.1
Grand Rapids	0.4	48.6	50.5	48.3	52.2	60.0	68.2	66.3	77.1	72.2	81.4	75.8	85.6	78.2	88.3	78.0	88.3	74.8	84.2	68.3	76.6	60.5	65.5	56.1	58.6	75.8	85.2
726350	1	44.7	46.6	44.6	48.1	57.5	64.7	64.6	74.3	70.9	81.0	74.6	85.3	77.1	87.2	76.9	86.7	73.6	81.2	66.2	73.6	58.7	62.1	52.7	54.6	74.0	82.5
	2	40.2	42.5	41.9	45.0	54.4	60.7	62.7	70.6	69.5	80.3	73.5	84.0	76.0	85.8	75.7	84.7	72.4	79.7	64.4	71.0	57.3	60.8	47.6	50.8	72.3	80.6
Hancock	0.4	41.9	43.1	41.9	45.7	54.3	58.5	63.2	73.1	68.6	79.5	73.5	84.3	75.8	86.2	75.9	85.6	72.3	79.7	65.0	71.5	57.0	60.5	47.4	49.1	73.3	82.1
727440	1	38.1	40.3	39.4	41.9	50.9	54.6	60.6	70.7	67.2	78.7	72.2	81.0	74.7	84.3	74.5	83.2	70.8	76.9	63.2	69.2	55.1	58.6	43.1	44.8	71.3	79.5
	2	35.5	37.0	36.9	39.4	47.3	51.4	58.4	65.1	65.7	75.7	70.9	79.8	73.5	82.2	73.4	81.5	69.5	74.6	61.3	66.9	53.0	55.9	38.9	40.3	69.5	77.0
Lansing	0.4	49.2	50.9	48.7	51.4	60.1	66.0	66.7	76.9	73.2	81.2	76.0	85.6	78.5	89.1	77.6	88.0	75.3	83.6	67.7	76.0	60.9	65.3	56.5	58.8	75.9	85.2
725390	1	45.5	48.1	45.6	49.3	57.6	63.7	64.5	73.9	71.3	80.6	74.7	84.7	77.4	87.4	76.4	86.5	74.1	81.7	66.1	73.1	59.0	63.8	52.9	54.8	74.2	82.8
	2	41.4	44.0	42.8	45.6	54.9	60.6	62.8	70.8	69.8	79.4	73.6	83.9	76.3	85.5	75.5	85.0	73.0	80.2	64.4	71.0	57.5	61.2	48.3	50.4	72.6	80.6
Muskegon	0.4	45.9	47.5	44.5	48.2	58.1	65.2	64.2	73.6	70.3	78.3	73.3	83.0	76.4	84.7	77.2	85.7	73.8	81.2	66.5	73.1	59.2	62.2	53.1	55.0	74.5	82.0
726360	1	42.5	44.1	42.1	44.9	55.1	60.5	62.3	70.4	68.5	77.4	72.3	81.7	75.4	83.6	76.1	83.9	72.7	79.4	65.0	71.0	57.7	60.7	50.2	51.8	72.8	80.0
	2	39.2	40.6	39.8	42.8	52.4	58.7	60.3	67.6	67.2	76.5	71.2	80.0	74.5	82.3	75.0	82.2	71.5	77.3	63.5	69.2	56.2	59.1	45.9	47.9	71.2	77.8
Sault Ste. Marie	0.4	37.2	38.2	38.1	39.6	44.5	48.0	60.1	69.1	69.1	79.9	71.6	81.5	75.5	84.3	74.5	83.0	72.3	80.2	63.1	68.8	54.2	55.1	43.9	44.8	72.0	80.3
727340	1	35.2	36.1	36.1	37.4	42.5	45.9	57.2	65.0	66.8	75.9	69.8	80.0	74.0	82.4	73.0	80.4	70.1	75.0	61.0	66.3	51.6	53.9	39.6	40.5	69.8	76.9
	2	33.8	34.7	34.8	36.2	40.6	43.4	54.2	59.8	64.5	74.0	68.2	77.1	72.3	81.3	71.4	78.4	68.1	72.3	59.0	63.6	49.0	51.2	37.0	37.9	67.7	74.4
Traverse City	0.4	41.4	43.2	42.5	46.7	56.1	64.0	64.8	77.0	71.0	82.3	74.6	87.3	76.8	89.3	76.5	86.1	74.1	83.5	67.7	75.6	58.8	62.6	52.6	56.2	74.2	84.1
726387	1	38.6	41.5	39.7	43.2	53.2	59.3	63.0	74.2	69.2	81.5	73.1	85.4	75.6	87.5	75.0	84.3	72.6	81.1	65.7	72.9	56.9	61.0	47.1	50.2	72.3	81.9
	2	36.8	39.6	37.6	40.8	50.0	56.6	60.7	70.7	67.4	78.1	71.5	83.3	74.4	84.9	73.9	82.3	71.1	77.8	63.7	70.2	54.7	58.5	42.2	43.9	70.4	79.5
MINNESTOA																											
Duluth	0.4	34.7	37.9	37.8	44.4	46.2	53.3	57.5	67.8	66.4	78.2	71.8	80.4	74.6	85.8	73.6	83.9	70.6	78.9	61.8	68.0	52.0	54.2	38.6	39.9	71.5	80.9
727450	1	32.9	35.4	35.4	39.6	43.1	48.3	54.7	65.4	64.4	76.1	69.6	79.9	73.4	83.9	72.2	82.1	68.5	74.5	59.8	65.4	48.7	51.1	34.7	36.3	69.3	77.8
	2	31.5	33.6	33.6	37.2	40.3	46.5	51.8	62.3	62.7	72.5	67.8	78.1	72.1	82.3	70.9	79.6	66.8	72.5	57.7	64.4	45.2	49.6	33.2	34.7	67.1	75.3
International Falls	0.4	34.1	37.6	37.5	42.5	46.0	53.9	58.8	72.1	68.0	82.2	72.1	83.1	75.6	85.6	74.5	84.1	71.4	77.7	61.5	70.0	50.0	52.1	35.5	37.2	72.2	82.1
727470	1	32.1	34.3	35.2	39.4	43.5	49.1	56.4	68.7	66.4	80.2	70.3	80.2	74.0	84.4	73.1	82.3	68.9	75.1	59.1	67.9	46.8	53.1	33.1	34.7	69.9	79.0
	2	30.6	32.4	33.5	36.6	40.8	46.8	53.9	65.8	64.5	75.8	68.9	78.9	72.6	83.2	71.7	80.9	66.8	73.7	57.3	64.6	44.3	48.0	31.7	33.4	67.8	76.5
Minneapolis-	0.4	37.4	41.6	43.3	50.5	56.7	64.2	64.4	77.0	71.0	82.1	76.4	88.7	78.8	90.0	78.1	89.7	75.2	86.1	67.2	73.6	57.2	60.9	46.3	48.6	76.0	87.5
St. Paul	1	35.5	38.7	39.8	44.4	53.1	60.9	62.4	73.8	69.4	80.1	75.2	86.5	77.6	89.2	76.8	88.4	73.3	82.9	65.5	72.6	54.3	59.1	39.4	41.5	74.1	84.4
726580	2	34.2	37.0	37.5	40.9	50.0	56.6	60.3	71.1	67.9	78.2	74.0	84.9	76.4	88.0	75.4	86.3	71.6	80.0	63.2	71.5	51.7	56.1	37.0	39.7	72.2	82.0
Rochester	0.4	37.3	40.4	43.3	47.3	57.2	64.4	64.5	77.3	70.9	79.9	76.0	85.9	79.3	88.2	77.4	87.1	74.4	81.9	66.5	72.9	57.2	61.1	48.2	50.6	75.5	84.6
726440	1	35.7	38.3	39.1	42.8	53.9	58.7	62.3	72.7	69.3	78.9	74.8	84.4	77.4	86.3	76.2	85.4	72.5	80.0	64.6	71.4	55.1	58.5	41.8	43.8	73.6	81.8
	2	34.3	36.5	36.7	39.2	50.2	55.4	60.3	69.7	67.9	76.3	73.4	82.9	76.0	84.6	75.0	84.2	71.0	77.9	62.5	69.7	52.5	57.0	37.5	39.5	71.7	79.7
Saint Cloud	0.4	36.6	41.2	42.0	47.1	54.4	61.9	64.1	75.6	70.3	81.2	75.7	88.0	78.6	89.2	77.5	88.6	75.0	84.8	66.4	70.8	54.9	57.7	40.0	43.7	75.7	86.6
726550	1	35.1	38.6	39.1	43.5	51.1	59.6	61.8	71.7	68.9	81.6	74.5	86.4	77.3	88.1	76.4	87.0	73.0	81.8	64.7	71.5	52.0	55.3	37.0	39.5	73.8	84.3
	2	33.6	36.3	36.9	41.1	47.9	54.3	58.9	69.9	67.4	79.6	73.3	83.8	76.2	87.2	75.2	85.4	71.2	79.5	62.4	70.0	49.2	54.8	35.1	37.1	71.8	81.6
MISSISSIPPI																											
Jackson	0.4	67.9	73.2	69.1	74.8	71.2	78.8	74.1	82.2	77.0	86.1	80.1	91.0	80.8	93.3	80.5	90.5	79.2	89.2	76.1	83.2	72.5	77.8	70.6	74.9	79.6	90.4
722350	1	66.5	70.9	67.8	72.8	70.2	77.2	73.2	80.6	76.1	84.9	79.1	90.3	80.2	92.3	80.1	90.3	78.5	88.1	74.8	81.4	71.2	75.6	69.3	73.2	78.7	89.2
	2	65.2	69.5	66.4	70.9	69.2	75.9	72.2	79.0	75.1	84.4	78.4	89.2	79.7	91.3	79.5	89.6	78.0	87.5	73.6	80.1	70.0	75.3	68.1	71.9	78.0	88.1
Meridian	0.4	67.2	71.2	68.5	73.3	70.5	77.7	73.6	81.2	76.5	86.8	79.5	89.9	80.8	93.3	80.3	90.3	78.7	88.5	75.2	83.0	71.2	75.5	70.2	74.6	79.3	90.6
722340	1	65.8	69.5	67.2	71.2	69.5	76.1	72.5	79.8	75.5	85.4	78.5	89.6	80.1	92.3	79.5	90.4	78.1	87.9	74.0	80.9	70.1	75.0	69.2	73.5	78.3	89.5
	2	64.5	68.4	65.7	70.3	68.6	74.8	71.5	78.7	74.7	84.1	77.9	89.6	79.4	91.5	79.1	90.4	77.3	87.6	72.9	78.9	69.1	74.7	67.7	72.0	77.4	88.3
MISSOURI																											
Columbia	0.4	56.3	61.4	56.9	64.4	64.0	73.9	68.9	80.3	75.0	83.6	78.7	89.2	80.1	91.4	79.5	90.6	76.8	88.0	70.3	79.4	63.8	70.4	60.2	65.0	78.1	89.2
724450	1	53.2	57.4	54.8	61.5	62.3	71.7	67.4	78.7	73.3	81.6	77.7	87.8	79.3	90.7	78.4	89.8	75.7	87.2	68.8	75.9	62.2	69.3	57.7	62.8	76.8	87.8
	2	50.0	54.0	52.5	59.9	60.5	70.0	66.1	78.1	71.8	80.4	76.6	86.9	78.5	89.7	77.6	88.9	74.7	85.9	67.5	75.2	60.6	66.8	54.9	59.0	75.4	86.4
Kansas City	0.4	55.8	61.4	56.0	64.1	63.9	72.4	69.9	81.6	74.3	83.2	78.4	89.5	80.0	91.3	79.9	90.5	77.3	89.6	71.8	79.2	63.7	70.7	59.1	63.3	78.3	90.1
724460	1	51.7	56.5	53.3	60.9	62.1	71.4	68.5	79.9	72.7	82.2	77.6	88.6	79.3	91.2	79.0	90.4	76.3	88.0	69.4	78.9	62.4	69.0	56.5	61.1	77.1	89.0
	2	48.2	53.3	50.7	59.0	60.4	69.1	66.8	77.1	71.4	81.6	76.8	88.1	78.5	90.6	78.1	89.9	75.4	86.5	68.1	75.7	60.8	66.9	53.3	57.8	75.9	87.4

WMO# = World Meteorological Organization number WB = wet-bulb temperature, °F MDB = mean coincident dry-bulb temperature, °F

Table 4B Selected Monthly Percentiles of Dry-Bulb Temperature with Mean Wet-Bulb Temperature for U.S. Locations

Location and WMO#	%	Jan		Feb		Mar		Apr		May		Jun		Jul		Aug		Sep		Oct		Nov		Dec		Annual	
		DB	MWB	DB	MWB	DB	MWB	DB	MWB	DB	MWB	DB	MWB	DB	MWB	DB	MWB	DB	MWB	DB	MWB	DB	MWB	DB	MWB	DB	MWB
1a	1b	2a	2b	3a	3b	4a	4b	5a	5b	6a	6b	7a	7b	8a	8b	9a	9b	10a	10b	11a	11b	12a	12c	13a	13b	14a	14b
MAINE																											
Caribou	0.4	43.8	43.4	47.1	42.4	55.5	45.7	69.8	55.4	84.0	64.7	87.4	67.4	89.3	72.0	87.0	72.5	81.3	68.2	72.9	63.2	59.4	56.4	47.1	45.3	84.9	68.9
727120	1	39.9	37.1	43.3	40.9	50.6	41.6	65.3	52.3	80.3	62.6	85.3	66.8	87.3	71.2	84.8	69.7	78.3	66.3	68.9	61.2	56.7	53.8	43.8	41.5	81.7	66.9
	2	37.6	35.2	40.1	36.7	47.6	40.8	61.2	50.1	77.2	59.5	83.2	66.3	85.2	70.5	82.8	68.0	75.9	64.8	65.4	58.2	54.3	51.0	40.3	38.1	78.8	65.8
Portland	0.4	49.1	45.8	50.6	44.9	60.8	49.8	72.6	56.2	84.3	65.8	89.6	70.5	91.2	74.7	89.9	72.6	85.7	72.7	75.4	62.1	63.9	55.2	54.3	49.6	86.4	71.2
726060	1	46.2	42.6	47.2	42.4	56.1	47.5	67.1	54.4	80.2	63.9	86.2	70.2	88.8	73.3	87.3	72.3	82.5	70.6	71.8	60.7	60.2	54.3	51.6	48.7	83.1	69.5
	2	43.3	39.1	44.7	40.7	52.2	44.4	63.1	51.2	75.9	61.2	83.4	68.4	86.5	71.2	85.1	71.3	79.2	68.4	68.6	59.4	58.0	52.6	48.9	46.2	80.0	67.7
MARYLAND																											
Glen Burnie/	0.4	62.5	56.3	68.9	55.0	78.3	61.7	86.1	64.9	91.0	70.9	94.3	74.2	97.5	76.1	95.4	75.9	93.1	73.7	82.7	67.0	75.5	64.9	67.7	59.9	93.2	74.6
Baltimore,	1	58.4	51.3	64.2	53.1	74.5	58.6	82.6	63.3	88.3	68.8	92.6	73.8	95.5	75.7	93.3	75.2	90.6	73.9	80.0	66.6	72.2	61.4	64.3	57.2	90.6	73.8
BWI Airport	2	55.3	49.5	60.9	52.5	70.2	57.7	78.9	61.8	85.9	68.5	90.9	72.9	93.7	75.5	91.5	74.7	88.4	72.8	77.5	65.6	69.5	60.1	61.0	55.1	88.0	72.5
724060																											
MASSACHUSETTS																											
Boston	0.4	57.6	53.0	58.9	53.9	69.3	54.9	80.8	62.2	88.1	68.0	93.2	72.2	95.6	75.7	92.4	74.4	89.8	74.0	78.9	65.0	70.7	61.9	61.9	56.6	90.5	72.7
725090	1	54.0	49.7	55.0	49.5	63.4	53.4	75.0	59.8	84.4	65.9	90.8	71.7	93.2	74.8	90.7	73.9	86.2	72.1	75.9	63.3	67.9	59.2	58.9	55.2	87.2	71.4
	2	50.8	46.2	51.6	45.9	59.2	51.0	70.7	56.3	81.1	64.1	88.1	70.7	90.9	73.8	88.5	72.7	83.2	70.3	73.3	62.3	64.7	58.3	56.3	52.2	84.1	69.9
Worcester	0.4	53.7	49.8	55.4	48.0	66.0	51.6	79.0	59.0	84.4	65.9	87.2	71.2	89.4	73.0	87.6	73.6	84.5	72.8	76.2	61.8	67.0	58.8	58.3	54.1	85.4	70.9
725095	1	50.4	46.0	51.7	46.9	62.1	50.7	74.0	58.0	81.5	64.4	85.3	69.7	87.7	72.9	85.3	71.8	81.8	71.4	73.5	60.7	63.9	56.8	55.7	51.9	82.7	69.4
	2	46.5	41.9	48.1	43.4	58.0	48.5	69.7	54.6	78.9	63.2	83.3	68.4	85.9	72.0	83.6	71.1	79.1	68.8	70.8	60.7	61.2	55.8	52.0	48.5	80.1	67.9
MICHIGAN																											
Alpena	0.4	45.3	42.3	47.7	40.9	63.1	53.8	81.6	62.2	85.0	65.5	90.9	70.3	93.1	73.5	89.5	73.5	84.8	71.3	76.1	64.1	63.7	55.5	54.7	50.3	87.4	71.1
726390	1	42.5	38.8	44.3	39.6	58.9	49.9	75.1	59.9	82.4	65.3	88.1	69.3	90.4	73.2	87.2	72.5	82.0	69.4	72.2	60.9	60.1	54.6	47.9	44.6	84.1	68.8
	2	39.7	36.6	41.5	36.8	54.5	47.0	70.2	56.8	79.4	64.2	85.3	68.8	88.0	71.8	84.9	71.1	79.2	67.7	69.0	59.1	57.0	52.4	43.4	40.5	80.8	67.3
Detroit, Metro	0.4	52.8	50.4	55.2	48.5	71.6	58.1	81.7	63.5	88.2	69.7	91.8	72.1	94.6	74.6	92.0	76.3	88.4	72.7	80.5	64.1	68.4	58.8	60.5	56.6	89.8	73.0
725370	1	48.9	45.8	51.0	45.8	67.3	56.3	78.2	62.2	85.6	68.9	89.9	71.6	92.2	74.9	89.8	74.6	86.0	71.9	77.1	64.1	65.2	58.1	57.1	53.4	87.0	71.8
	2	44.9	42.0	47.5	43.1	62.8	52.8	73.9	60.6	83.3	67.8	88.1	71.5	90.2	73.9	87.9	72.9	83.6	71.3	74.1	61.6	62.6	56.8	52.9	49.3	84.4	70.4
Flint	0.4	51.1	48.8	51.7	45.3	69.5	55.3	80.8	63.4	86.2	69.7	91.1	70.7	93.0	73.3	90.3	75.5	87.6	72.2	78.9	64.9	67.9	58.6	60.1	55.7	88.3	72.7
726370	1	48.0	44.1	48.7	44.3	66.0	55.2	77.2	62.1	84.0	68.6	88.6	70.9	90.7	74.4	88.2	74.0	84.4	71.3	75.8	63.3	64.8	57.7	56.2	52.5	85.6	71.2
	2	44.1	41.0	45.6	41.7	61.7	52.8	73.2	60.6	81.9	66.7	86.7	70.6	88.8	73.8	86.4	72.4	81.9	69.7	73.0	62.2	62.1	56.5	51.6	48.4	83.1	69.5
Grand Rapids	0.4	50.5	48.2	52.4	46.6	69.9	57.5	80.7	64.0	87.6	68.9	91.3	71.2	92.5	74.2	91.9	73.8	87.2	72.1	79.2	65.3	67.6	57.7	59.0	55.3	89.2	73.0
726350	1	46.9	44.1	48.8	44.6	66.4	55.1	77.8	62.8	85.1	68.4	89.3	70.8	90.9	73.8	89.8	74.1	84.5	71.7	76.6	64.4	64.1	56.9	55.2	52.0	86.4	71.3
	2	43.1	39.8	45.3	41.4	62.4	54.6	73.7	60.3	82.7	67.3	87.7	70.5	89.4	74.2	87.5	72.7	82.1	69.8	73.9	63.2	61.4	55.9	50.6	47.6	83.8	69.7
Hancock	0.4	43.4	41.5	45.6	41.1	61.0	51.2	76.6	61.8	85.1	66.1	87.9	69.9	90.9	72.7	88.4	73.7	82.3	70.2	76.2	62.3	62.7	55.0	50.0	47.8	85.5	70.7
727440	1	40.6	37.7	43.0	38.9	56.1	49.1	71.9	58.6	82.4	64.2	85.5	69.2	88.1	71.9	86.1	72.6	79.5	68.3	73.0	60.7	59.6	54.2	45.5	43.0	82.6	68.7
	2	37.5	35.0	40.0	35.9	52.2	45.6	68.4	56.4	79.9	63.0	83.4	68.3	85.8	71.5	84.2	71.6	77.3	66.3	69.6	59.7	56.7	52.0	40.5	38.2	80.0	67.2
Lansing	0.4	52.2	47.9	52.3	47.5	70.1	55.8	80.5	64.3	87.2	69.2	91.7	70.8	93.3	74.3	92.5	75.4	87.9	72.6	79.6	64.5	68.6	58.9	59.0	55.8	89.4	73.3
725390	1	48.6	45.5	49.6	45.4	66.7	55.2	77.8	63.3	84.9	70.1	89.5	71.4	91.5	74.3	89.9	74.7	84.9	71.8	76.8	64.4	64.8	58.1	55.4	52.3	86.4	71.6
	2	44.4	41.3	45.9	42.3	62.6	53.7	73.6	60.9	82.4	67.6	87.5	71.1	89.8	74.1	87.6	72.9	82.4	70.5	73.6	62.4	61.8	56.4	50.8	47.7	83.7	70.3
Muskegon	0.4	47.7	45.4	48.9	43.9	66.7	55.4	77.0	63.5	83.9	66.3	87.3	70.0	89.1	73.4	88.5	74.6	83.6	71.2	75.8	64.7	64.1	57.5	55.2	52.6	85.4	71.3
726360	1	44.1	42.0	45.3	41.0	63.1	52.4	73.2	60.0	81.9	65.0	85.4	69.2	87.1	72.7	86.1	73.6	81.5	69.9	73.6	63.4	61.6	56.4	52.1	50.4	83.1	69.9
	2	41.4	38.8	42.6	39.4	59.6	51.4	70.1	57.3	79.8	65.1	83.8	68.5	85.6	71.9	84.2	72.6	79.8	70.0	71.2	61.7	59.7	55.4	48.3	45.7	80.9	68.8
Sault Ste. Marie	0.4	38.7	36.8	40.3	36.0	50.3	43.0	72.2	58.1	82.7	67.0	85.4	69.2	88.3	72.1	85.7	71.4	81.0	71.6	72.6	62.0	56.4	51.7	45.5	43.5	83.2	69.4
727340	1	36.4	34.7	38.6	35.6	47.1	40.9	67.3	54.0	79.8	64.2	82.8	67.1	85.9	72.1	83.6	70.2	77.7	68.0	68.2	59.3	54.4	50.8	41.1	38.6	79.9	67.6
	2	35.0	33.5	37.0	34.1	44.3	39.2	63.4	51.4	76.9	61.4	80.2	66.5	83.5	70.3	81.6	69.8	75.0	66.0	64.9	57.9	52.0	48.4	38.2	36.7	76.8	65.7
Traverse City	0.4	44.7	40.0	47.0	41.8	66.6	54.3	82.0	63.9	87.7	68.4	92.8	70.5	93.6	73.3	91.8	73.1	85.9	71.9	78.5	65.4	65.6	56.0	55.7	52.9	89.0	71.4
726387	1	42.3	38.1	44.4	39.0	62.3	51.0	77.3	61.1	84.7	67.2	89.7	70.5	91.5	73.2	88.7	72.4	83.4	70.6	75.3	63.6	61.9	55.0	50.0	46.0	85.6	69.6
	2	40.0	36.3	41.4	36.6	57.6	49.6	72.3	57.6	81.4	64.6	87.0	68.7	89.5	72.1	86.4	70.9	80.9	68.9	72.3	61.7	59.5	54.4	45.0	41.1	82.6	68.1
MINNESTOA																											
Duluth	0.4	39.2	34.2	44.5	37.2	55.1	44.0	73.6	54.5	82.7	63.2	85.7	68.0	89.3	72.8	88.0	70.6	81.6	67.8	72.5	57.9	57.6	47.7	41.0	37.5	84.3	68.8
727450	1	36.2	32.1	40.5	35.0	50.4	41.9	68.8	52.0	79.6	61.2	83.3	67.0	86.8	70.8	85.3	70.7	78.1	66.7	69.1	57.0	53.3	46.5	37.1	33.2	80.9	66.7
	2	33.9	31.2	37.8	32.9	46.7	40.0	64.8	50.3	77.0	58.7	80.9	65.7	84.6	70.1	82.9	68.6	75.8	63.6	65.9	55.2	50.2	45.0	35.1	33.0	78.0	65.0
International Falls	0.4	37.5	33.7	42.8	36.6	55.5	44.8	76.4	56.3	86.8	65.2	88.6	68.1	90.9	73.4	89.4	71.1	82.1	67.7	74.5	58.5	57.2	48.9	38.6	34.7	86.2	69.0
727470	1	34.7	31.7	40.3	34.7	51.5	42.1	72.0	53.4	83.6	63.8	86.2	67.7	88.4	71.6	86.5	70.3	79.1	66.7	70.6	57.6	52.6	44.8	35.4	32.7	82.8	67.0
	2	32.5	30.4	37.4	32.8	47.8	39.7	67.8	52.9	80.5	61.7	83.7	65.9	86.1	70.2	84.5	68.8	76.2	64.0	66.7	55.6	49.5	44.2	33.7	31.4	79.8	65.6
Minneapolis-	0.4	42.0	36.5	50.1	43.1	67.0	53.7	82.1	61.2	88.4	66.1	93.6	72.6	96.1	74.6	94.0	75.0	89.0	72.4	80.0	63.2	63.9	54.5	49.3	45.5	91.0	72.9
St. Paul	1	39.6	34.9	45.4	39.4	61.8	51.6	77.6	59.9	85.7	66.0	91.0	72.3	93.7	74.5	91.3	73.9	85.7	70.6	76.2	62.2	60.4	52.8	43.0	38.2	87.9	71.4
726580	2	37.6	33.8	42.1	36.6	57.8	48.0	73.8	58.1	83.0	64.7	88.7	70.8	91.5	73.3	89.2	73.4	83.0	70.0	73.1	60.8	57.4	50.6	40.2	36.6	85.0	69.9
Rochester	0.4	41.4	37.4	48.2	43.2	65.9	54.3	80.3	61.5	85.4	66.4	90.7	71.3	92.9	73.8	91.1	74.2	85.4	72.7	78.3	62.7	62.4	54.6	50.6	48.2	88.0	72.4
726440	1	38.4	35.1	43.3	38.5	60.4	51.9	76.1	62.1	83.2	66.1	88.3	70.9	90.7	74.1	88.2	74.5	82.9	70.2	74.9	61.5	59.8	53.7	43.8	41.3	85.1	71.2
	2	36.7	34.1	40.0	36.0	56.6	49.2	72.1	58.3	80.6	65.2	86.3	70.1	88.5	73.7	86.2	73.3	80.3	67.9	72.1	60.5	57.3	51.9	39.8	37.2	82.3	69.6
Saint Cloud	0.4	41.5	36.2	47.9	40.7	66.1	54.8	81.2	59.4	88.1	66.7	93.5	71.1	95.8	73.5	93.2	73.7	88.4	72.9	78.1	61.9	62.7	51.1	44.2	38.9	90.5	72.2
726550	1	39.3	34.6	44.4	38.6	60.2	49.8	77.2	59.4	85.1	67.2	90.5	70.9	93.3	73.7	91.0	72.7	85.5	70.2	75.4	60.1	58.6	48.8	40.6	36.5	87.5	71.1
	2	37.0	32.8	41.7	36.1	55.9	46.4	73.3	58.0	82.5	64.3	88.5	69.7	90.9	73.5	89.1	72.5	82.7	68.5	72.5	59.7	55.3	48.4	37.6	34.6	84.7	69.5
MISSISSIPPI																											
Jackson	0.4	75.4	64.5	78.6	65.0	83.6	67.4	86.4	69.9	91.4	72.8	96.9	75.4	98.5	78.5	97.6	77.2	95.9	75.8	88.8	70.9	81.8	68.0	77.2	68.2	95.3	76.6
722350	1	72.8	64.4	76.4	64.0	81.7	65.9	85.1	69.7	89.9	71.9	95.2	75.5	96.8	77.6	96.0	76.9	93.8	75.4	87.2	70.4	80.1	67.6	75.3	67.6	93.4	76.4
	2	70.3	62.8	73.9	62.3	79.9	64.8	83.6	68.3	88.7	72.0	93.9	75.8	95.2	77.3	94.8	76.8	92.3	75.4	85.6	69.7	78.4	66.8	73.4	66.3	91.8	76.2
Meridian	0.4	74.5	63.9	78.5	63.9	83.9	66.6	86.8	69.0	91.7	72.7	97.3	75.5	99.4	78.0	97.5	77.0	96.2	74.2	88.8	70.6	81.1	67.1	77.0	67.6	95.6	76.5
722340	1	72.1	62.2	75.8	62.4	81.9	65.1	85.1	68.5	90.2	72.4	95.5	75.5	97.3	77.1	96.0	76.9	94.0	74.0	86.7	69.5	79.4	66.0	75.0	66.6	93.5	76.1
	2	69.8	62.2	73.4	62.1	79.7	63.9	83.8	67.7	88.8	72.4	94.0	75.2	95.7	77.4	94.7	76.7	92.5	74.5	85.0	68.9	77.6	65.5	72.9	65.4	91.6	75.7
MISSOURI																											
Columbia	0.4	63.8	54.8	68.4	54.0	79.2	60.6	84.8	65.2	87.5	69.5	94.6	74.0	101.7	74.3	98.9	74.9	93.3	73.6	85.8	66.0	74.9	61.1	66.1	58.6	94.7	74.9
724450	1	59.0	50.6	64.9	53.0	76.1	59.6	82.7	64.8	86.0	69.5	92.0	74.8	97.4	75.8	96.5	74.9	91.1	73.6	82.8	65.9	72.1	59.9	63.5	56.8	91.6	74.7
	2	55.8	48.5	61.1	50.7	73.1	58.5	80.6	64.1	84.4	69.2	90.2	74.1	95.2	75.7	94.6	74.5	89.0	73.2	80.0	65.0	69.4	58.8	60.1	53.3	89.1	73.7
Kansas City	0.4	61.7	54.3	67.9	52.6	78.6	60.4	86.3	65.8	88.4	69.6	95.9	74.9	101.0	75.5	100.0	76.0	94.2	74.0	86.9	66.2	74.8	60.6	64.5	56.7	95.9	75.3
724460	1	58.6	49.7	64.1	51.3	75.5	58.7	83.1	65.4	86.9	68.6	93.4	74.6	98.7	75.4	97.7	74.9	92.0	73.7	84.2	66.4	71.6	59.0	62.2	54.8	92.8	75.0
	2	55.0	46.9	60.7	49.8	72.5	57.7	80.5	64.5	85.2	69.0	91.6	74.4	96.4	75.7	95.6	74.9	89.9	73.5	81.4	65.2	68.9	58.4	59.3	51.8	90.1	74.2

WMO# = World Meteorological Organization number DB = dry-bulb temperature, °F MWB = mean coincident wet-bulb temperature, °F

Table 4A Selected Monthly Percentiles of Wet-Bulb Temperature with Mean Dry-Bulb Temperature for U.S. Locations

Location and WMO#	%	Jan		Feb		Mar		Apr		May		Jun		Jul		Aug		Sep		Oct		Nov		Dec		Annual	
		WB	MDB	WB	MDB	WB	MDB	WB	MDB	WB	MDB	WB	MDB	WB	MDB	WB	MDB	WB	MDB	WB	MDB	WB	MDB	WB	MDB	WB	MDB
1a	1b	2a	2b	3a	3b	4a	4b	5a	5b	6a	6b	7a	7b	8a	8b	9a	9b	10a	10b	11a	11b	12a	12c	13a	13b	14a	14b
Springfield	0.4	56.4	62.1	58.8	66.1	64.6	73.8	69.4	79.8	74.1	83.1	78.2	88.6	79.5	90.6	78.5	90.0	76.5	87.9	70.9	78.9	64.5	70.2	60.7	64.9	77.5	89.0
724400	1	54.3	58.0	56.9	63.2	63.2	71.3	68.0	78.0	72.9	82.3	77.2	87.1	78.7	89.9	77.7	89.8	75.5	86.2	69.4	77.5	63.0	69.3	59.0	63.8	76.3	87.9
	2	52.3	55.9	54.8	61.6	61.7	69.6	66.9	76.3	71.7	80.7	76.2	85.9	77.9	89.3	77.0	89.8	74.7	85.1	67.9	74.2	61.6	67.5	56.7	61.2	75.2	86.4
St. Louis,	0.4	58.1	64.1	59.0	66.1	65.4	75.8	70.3	80.5	74.7	85.1	78.8	90.0	81.1	93.0	79.7	92.1	78.4	88.6	71.5	78.2	65.4	70.7	62.0	66.2	78.9	90.2
Intl Airport	1	54.8	58.0	56.7	61.9	64.0	73.4	68.4	78.9	73.5	83.2	77.8	88.7	80.2	92.3	79.0	91.2	77.1	87.5	70.3	77.0	64.2	69.8	59.7	62.5	77.5	88.4
724340	2	52.0	56.2	54.2	60.2	62.4	70.2	67.1	76.9	72.4	82.2	77.0	87.5	79.3	90.7	78.3	90.3	76.0	85.6	69.0	75.5	62.8	67.3	57.0	60.0	76.3	86.7
MONTANA																											
Billings	0.4	42.4	54.1	45.7	61.3	48.3	66.7	54.9	78.0	60.6	80.3	66.5	87.5	68.4	86.4	66.8	87.9	62.8	84.6	55.6	77.6	48.7	64.8	43.6	56.4	65.4	86.2
726770	1	40.4	51.0	43.2	55.8	46.7	64.3	53.3	74.7	59.4	78.6	65.2	85.7	67.2	86.5	65.5	88.3	61.3	82.2	54.5	76.4	46.9	61.5	41.6	53.3	63.9	84.3
	2	38.9	48.7	41.8	54.3	45.3	61.7	51.8	71.2	58.3	77.5	64.0	84.1	66.1	86.5	64.4	86.4	60.1	80.7	53.2	72.6	45.2	58.4	40.0	50.2	62.4	82.8
Cut Bank	0.4	43.4	52.3	43.8	54.1	45.3	59.8	51.6	71.8	57.5	74.9	61.6	80.1	66.3	79.8	63.7	85.9	59.8	80.3	53.7	75.5	47.2	60.1	43.3	51.6	62.2	81.2
727796	1	41.2	49.9	41.5	51.2	43.9	56.2	49.9	69.1	56.1	72.8	60.6	79.9	64.7	78.4	62.9	83.6	58.2	78.4	52.4	73.0	45.1	57.7	41.2	49.6	60.4	79.3
	2	39.3	47.2	39.7	48.7	42.1	52.7	48.2	65.5	54.6	71.3	59.6	78.6	63.2	78.8	61.9	83.3	56.9	78.7	51.3	70.2	43.3	53.7	39.5	47.4	58.9	77.4
Glasgow	0.4	39.0	44.5	43.3	52.7	49.0	63.4	56.2	76.3	63.3	83.3	68.9	88.9	72.3	84.1	68.5	86.4	63.3	84.2	56.6	76.4	48.0	59.0	41.5	48.9	67.7	84.9
727680	1	37.4	42.7	41.0	49.2	46.7	59.7	54.2	72.6	61.6	80.4	67.3	85.7	70.6	83.8	67.2	86.2	62.0	79.7	55.1	73.7	46.1	57.3	39.1	46.0	65.5	83.4
	2	35.9	40.5	39.1	45.3	44.6	56.4	52.6	70.4	60.1	77.4	66.0	84.0	68.9	83.6	66.0	85.9	60.8	79.4	53.5	70.5	43.9	53.4	37.0	42.4	63.8	81.5
Great Falls,	0.4	43.6	54.3	45.4	58.6	48.0	65.0	54.3	75.5	60.3	78.7	65.1	85.1	67.3	84.7	65.1	86.6	61.0	84.6	55.1	76.9	47.9	61.0	44.5	55.2	63.9	83.8
Intl Airport	1	41.8	51.2	43.4	55.5	46.3	61.8	52.7	72.5	58.6	76.8	63.7	83.6	65.6	84.3	63.9	84.4	59.6	81.9	53.9	74.7	46.4	59.6	42.7	52.9	62.2	82.2
727750	2	40.2	49.7	41.8	52.9	44.8	58.5	51.1	69.2	57.1	75.8	62.5	82.1	64.4	83.8	62.9	83.6	58.5	80.4	52.5	73.0	45.1	57.6	41.0	50.8	60.7	80.5
Helena	0.4	41.4	50.0	44.3	55.5	47.3	64.9	53.7	72.5	59.0	77.7	64.7	81.1	65.5	83.1	64.3	82.6	61.4	80.8	54.2	72.2	47.0	59.5	42.5	52.4	63.0	82.2
727720	1	39.9	48.1	42.3	51.9	45.6	60.9	52.0	69.6	57.7	77.2	63.0	81.8	64.3	83.1	63.2	82.4	60.0	79.1	53.0	70.5	45.2	57.0	40.4	49.3	61.4	81.3
	2	38.5	45.9	40.7	50.1	44.0	57.5	50.4	67.9	56.5	75.0	61.7	80.7	63.3	82.6	62.2	82.2	58.7	78.1	51.7	68.8	43.5	53.9	39.0	47.0	60.1	79.8
Kalispell	0.4	40.9	44.1	42.1	47.9	47.4	58.7	55.2	71.8	61.3	79.4	65.4	83.0	67.6	85.7	66.6	84.6	61.8	79.7	55.2	69.7	48.0	53.1	42.0	44.9	64.6	83.0
727790	1	38.9	41.9	40.5	45.8	45.4	55.6	53.2	68.6	59.4	75.5	64.3	81.7	66.2	85.3	65.0	83.4	60.5	78.0	53.7	67.4	45.6	50.5	40.0	44.0	63.0	81.1
	2	37.4	40.6	39.1	43.5	43.5	52.6	51.4	64.8	57.9	73.3	63.1	80.1	65.1	83.9	64.0	83.1	59.4	76.5	52.3	64.3	43.8	48.1	38.5	41.3	61.4	78.7
Lewistown	0.4	42.7	51.2	44.6	56.0	46.9	63.1	52.9	73.7	59.6	77.5	65.4	81.0	68.1	82.2	66.1	85.8	62.3	80.6	56.3	72.7	48.1	64.2	43.4	54.6	64.4	81.4
726776	1	40.6	49.5	42.4	52.1	45.1	58.7	51.5	69.2	58.0	76.2	63.9	80.0	66.5	81.7	64.7	84.0	60.0	78.2	54.3	73.2	46.0	58.8	41.4	51.4	62.5	79.9
	2	38.9	46.8	40.4	49.3	43.6	56.3	50.0	67.0	56.5	72.9	62.7	78.2	65.1	81.5	63.4	82.8	58.5	76.8	52.7	70.6	44.2	56.3	39.8	48.1	60.9	78.2
Miles City	0.4	41.5	49.1	45.6	57.4	51.2	66.3	57.7	77.7	64.9	83.5	70.7	89.7	72.6	90.3	70.2	89.9	65.1	84.3	57.4	77.8	49.7	63.1	42.4	51.3	69.3	88.8
742300	1	39.7	47.2	43.4	53.4	49.3	63.5	55.7	75.0	63.2	82.1	69.3	88.8	71.3	90.5	68.9	88.2	63.8	84.6	56.2	75.4	47.5	59.2	40.4	48.8	67.3	86.4
	2	37.9	43.7	41.7	51.0	47.4	60.7	54.3	72.0	61.7	80.3	67.8	86.6	70.1	89.4	67.7	88.0	62.6	82.3	54.9	73.1	45.5	56.2	38.5	45.6	65.6	84.4
Missoula	0.4	40.8	44.5	42.8	50.7	48.4	61.0	55.8	71.7	62.1	80.2	66.1	85.0	67.1	87.1	66.5	85.3	61.9	84.5	55.3	71.4	47.6	54.3	41.7	46.6	64.6	83.5
727730	1	38.8	42.2	41.3	47.5	46.9	58.4	53.9	69.8	60.2	76.7	64.7	83.2	65.9	86.5	65.0	82.9	60.8	80.9	54.1	69.8	45.6	51.7	39.7	43.6	63.0	82.2
	2	37.5	40.6	40.0	45.1	45.2	55.2	52.4	67.8	58.6	75.0	63.6	81.8	64.8	83.9	63.9	83.6	59.6	77.7	52.6	66.5	44.0	49.7	38.2	41.4	61.5	80.6
NEBRASKA																											
Grand Island	0.4	44.9	57.7	49.9	63.1	58.4	73.2	65.4	79.1	71.2	83.6	76.3	88.7	78.0	91.0	77.4	89.3	73.8	84.4	66.1	77.7	57.1	63.0	47.4	59.2	75.6	89.1
725520	1	42.8	53.4	47.3	59.9	56.3	68.4	63.7	78.1	69.5	81.2	74.9	90.0	76.8	90.7	76.2	88.8	72.6	85.4	64.3	73.6	54.5	61.9	45.0	54.3	74.1	87.8
	2	40.4	49.3	45.4	56.5	53.9	66.1	61.8	75.2	68.2	80.1	73.7	88.1	76.0	89.7	75.2	88.1	71.4	84.7	62.5	72.9	52.6	59.7	42.6	52.7	72.6	85.8
Norfolk	0.4	44.7	54.3	48.7	60.4	58.3	69.8	66.3	84.7	71.1	84.4	76.8	89.7	78.3	92.0	78.1	91.5	74.3	87.2	64.2	75.9	57.6	62.8	46.2	57.1	76.4	89.8
725560	1	42.7	53.7	46.7	57.4	56.1	67.4	64.3	80.4	69.7	81.5	75.9	89.7	77.4	91.0	77.2	90.6	73.3	86.7	62.1	72.9	55.1	60.0	43.6	51.8	75.0	88.2
	2	40.1	48.7	44.4	53.5	54.1	63.8	62.3	75.0	68.3	79.5	74.8	88.8	76.5	90.0	76.3	89.6	72.2	84.9	60.6	72.1	52.6	58.9	41.4	49.3	73.4	86.0
North Platte	0.4	43.5	57.7	48.1	64.3	54.0	73.8	61.6	78.6	67.9	80.1	74.2	89.8	75.5	87.9	74.5	88.7	70.7	84.6	61.8	74.6	53.1	66.3	45.8	60.1	73.1	87.4
725620	1	41.6	53.7	46.3	61.3	52.0	68.2	60.0	75.1	66.5	79.3	72.5	87.9	74.4	87.9	73.4	87.9	69.3	83.9	60.0	74.5	51.1	63.6	43.4	56.3	71.5	86.1
	2	39.6	50.1	44.3	58.3	50.2	66.7	58.3	74.3	65.2	77.7	71.4	86.3	73.6	87.4	72.4	86.9	68.1	82.6	58.6	74.2	49.0	61.0	41.1	53.5	70.1	84.7
Omaha, Eppley	0.4	45.5	52.1	51.0	59.6	61.3	73.0	67.7	82.0	73.6	83.4	78.7	90.8	81.2	91.9	79.9	90.5	76.5	87.3	69.7	77.2	60.7	66.4	53.9	56.1	78.2	90.1
Airfield	1	43.2	50.4	47.9	58.8	59.1	69.8	66.0	78.9	72.1	83.6	77.5	89.5	79.8	91.1	78.8	90.6	75.4	85.8	68.0	77.2	58.8	64.0	48.9	54.8	76.6	88.3
725500	2	40.9	46.9	45.8	53.8	56.6	66.6	64.3	75.5	70.6	81.5	76.3	88.6	78.7	90.8	77.7	89.9	74.3	84.4	66.2	75.7	56.5	62.2	44.5	51.2	75.1	86.4
Scottsbluff	0.4	43.3	56.7	47.2	62.9	50.3	71.3	56.7	76.9	63.2	78.6	68.7	86.5	71.5	87.8	70.7	87.9	66.1	86.4	57.5	76.4	50.4	68.6	46.2	61.5	69.0	86.6
725660	1	41.7	53.5	45.1	60.2	48.6	68.3	55.4	75.7	62.0	78.1	67.6	86.4	70.4	87.5	69.4	87.1	64.9	82.9	56.0	76.6	48.6	65.3	43.7	57.8	67.5	85.1
	2	40.1	51.3	43.3	57.7	47.2	65.3	54.1	72.0	60.8	77.4	66.7	84.8	69.5	87.2	68.6	86.2	63.6	81.6	54.7	74.8	47.0	63.2	41.6	55.0	66.2	83.6
NEVADA																											
Elko	0.4	43.1	50.0	46.4	57.5	48.3	64.8	52.4	73.2	58.1	80.6	62.6	88.9	65.3	85.9	65.1	84.8	61.5	83.5	55.0	75.4	49.5	60.2	42.8	51.5	63.0	84.7
725825	1	41.3	47.8	44.7	53.7	46.9	62.1	51.0	70.6	56.6	77.9	61.3	86.9	64.2	85.7	64.1	83.5	60.0	81.6	53.4	75.9	47.5	58.4	41.4	48.6	61.4	84.3
	2	39.6	45.8	43.1	52.4	45.7	59.2	49.8	68.5	55.3	75.8	60.1	85.3	63.3	85.8	63.0	82.1	58.9	78.4	52.2	73.5	46.0	55.8	39.9	45.9	59.9	83.5
Ely	0.4	40.9	50.1	44.0	55.6	45.5	60.4	50.1	70.6	54.8	77.3	59.4	81.6	62.2	80.9	62.4	79.4	58.5	76.3	52.0	72.2	47.0	59.6	41.7	52.5	60.3	78.1
724860	1	39.4	48.5	42.0	52.5	44.1	59.6	48.5	69.1	53.5	75.0	58.1	82.8	61.2	79.3	61.3	78.1	57.4	76.1	50.8	72.3	45.0	58.2	39.7	50.4	59.0	78.0
	2	37.8	45.8	40.5	50.6	42.9	57.9	47.1	67.0	52.5	72.5	57.0	81.8	60.4	79.0	60.4	77.2	56.3	75.4	49.9	71.5	43.4	57.0	38.0	48.2	57.6	77.8
Las Vegas,	0.4	52.4	62.2	55.0	69.8	56.8	78.9	60.2	88.2	64.2	91.0	68.6	99.5	73.1	96.6	73.4	94.3	70.2	89.0	62.9	80.8	56.8	71.9	52.1	62.5	71.4	94.6
Intl Airport	1	50.6	60.6	53.5	66.4	55.4	76.3	58.9	86.8	63.2	89.3	67.3	100.1	72.2	96.3	72.5	94.5	69.1	88.6	61.7	81.5	55.5	70.0	50.4	59.5	70.1	93.1
723860	2	49.0	59.5	52.1	65.9	54.4	74.7	57.7	83.7	62.2	89.9	66.2	99.7	71.4	96.2	71.8	94.2	68.1	89.3	60.5	81.9	54.3	70.0	48.9	58.5	69.0	92.6
Reno	0.4	46.9	56.5	49.4	62.6	49.8	68.7	54.2	75.9	59.2	81.6	63.3	88.3	65.1	87.6	65.0	87.0	60.8	84.7	57.0	80.4	50.9	63.7	47.5	56.1	63.0	87.2
724880	1	44.7	53.6	47.5	59.6	48.3	66.1	52.5	74.2	58.0	80.3	62.1	87.6	64.1	87.2	63.9	86.1	59.9	83.9	55.7	79.9	49.6	63.4	45.7	55.4	61.6	86.4
	2	43.1	52.2	46.0	58.1	47.1	63.7	51.3	72.3	56.9	79.9	60.9	86.3	63.2	87.9	62.9	85.5	59.1	84.3	54.5	77.7	48.4	61.9	44.1	53.2	60.3	85.0
Tonopah	0.4	44.6	53.6	45.6	57.3	48.0	66.8	51.1	76.4	56.2	80.9	61.2	88.5	64.2	84.1	64.2	81.5	60.3	78.3	54.4	77.6	48.4	63.2	43.2	54.1	62.2	83.0
724855	1	42.5	52.7	44.3	56.6	46.6	64.6	49.7	74.0	55.0	79.3	59.8	87.6	63.2	84.2	63.3	82.0	59.3	78.3	53.1	76.2	46.9	61.7	41.4	50.8	60.8	82.3
	2	40.9	49.4	43.0	55.5	45.2	61.5	48.6	71.9	53.9	77.1	58.4	86.5	62.3	83.7	62.4	82.0	58.3	78.7	52.1	74.8	45.5	59.4	40.2	48.7	59.5	81.3
Winnemucca	0.4	47.0	54.0	49.0	60.8	50.6	68.7	53.8	74.4	59.2	85.5	63.5	91.8	66.2	88.7	65.3	90.3	60.8	81.3	55.9	80.8	50.2	64.0	46.7	55.2	63.4	87.9
725830	1	44.5	53.2	47.2	58.5	48.9	65.8	52.3	73.5	57.9	81.0	62.3	90.7	65.0	88.9	64.1	86.8	59.9	80.8	54.8	79.6	48.9	62.3	45.0	53.6	61.8	87.3
	2	43.0	50.7	45.5	56.1	47.4	62.3	51.2	71.8	56.8	80.6	61.0	89.2	63.9	87.4	63.1	85.8	59.0	83.2	53.7	76.8	47.6	60.1	43.3	51.5	60.3	86.1
NEW HAMPSHIRE																											
Concord	0.4	48.3	51.1	48.1	52.0	55.7	63.5	63.5	77.5	70.1	85.0	74.5	85.8	76.7	87.8	76.4	87.4	74.1	84.5	66.3	72.5	61.7	65.8	54.1	56.2	74.4	85.0
726050	1	42.7	46.4	45.0	49.7	51.8	60.0	60.7	71.9	68.3	81.6	73.0	83.7	75.7	87.3	75.1	85.6	72.7	82.3	64.4	71.5	58.8	62.9	49.6	53.4	72.7	82.0
	2	39.1	42.5	42.1	45.5	49.2	56.0	58.5	67.7	66.4	78.9	71.9	82.8	74.9	86.1	73.8	82.5	71.3	79.2	62.9	69.3	56.3	60.6	45.0	47.5	71.1	79.4
NEW JERSEY																											
Atlantic City	0.4	55.9	59.2	57.0	61.5	62.1	72.1	66.3	79.8	74.2	85.2	77.0	87.5	79.3	89.8	78.9	88.6	77.1	85.7	70.8	76.3	65.9	70.5	60.1	62.6	77.2	86.6
724070	1	53.9	56.2	54.5	58.1	59.4	66.3	64.5	76.3	72.0	79.4	75.9	87.1	78.3	88.6	77.9	86.8	76.1	84.3	69.4	74.2	64.1	67.2	58.3	60.6	75.9	84.1
	2	51.9	54.1	52.3	55.8	56.7	62.6	62.3	71.5	70.3	79.6	74.9	85.9	77.5	87.2	77.0	85.9	75.0	83.0	68.2	72.8	62.4	64.8	56.3	58.4	74.7	82.3
Newark	0.4	55.8	61.1	55.7	59.8	62.6	71.4	65.6	80.5	72.9	86.8	76.4	90.2	79.0	91.2	78.2	90.3	77.1	88.5	70.1	74.9	65.4	71.1	59.8	63.0	76.7	87.7
725020	1	51.8	54.8	52.7	57.9	59.2	66.2	63.9	76.3	71.5	83.6	75.2	87.5	77.9	90.8	77.3	88.0	76.1	85.4	68.8	74.0	63.2	67.0	57.2	60.0	75.5	85.3
	2	48.4	52.2	50.1	54.7	56.1	62.8	62.2	72.9	70.0	81.0	74.1	85.9	76.9	88.5	76.5	86.3	75.1	83.9	67.5	72.9	61.6	65.3	54.0	56.5	74.2	83.0

WMO# = World Meteorological Organization number WB = wet-bulb temperature, °F MDB = mean coincident dry-bulb temperature, °F

Table 4B Selected Monthly Percentiles of Dry-Bulb Temperature with Mean Wet-Bulb Temperature for U.S. Locations

Location and WMO#	%	Jan		Feb		Mar		Apr		May		Jun		Jul		Aug		Sep		Oct		Nov		Dec		Annual	
		DB	MWB	DB	MWB	DB	MWB	DB	MWB	DB	MWB	DB	MWB	DB	MWB	DB	MWB	DB	MWB	DB	MWB	DB	MWB	DB	MWB	DB	MWB
1a	1b	2a	2b	3a	3b	4a	4b	5a	5b	6a	6b	7a	7b	8a	8b	9a	9b	10a	10b	11a	11b	12a	12c	13a	13b	14a	14b
Springfield	0.4	65.1	53.5	70.7	55.6	78.4	61.3	84.2	66.5	86.4	70.5	94.0	74.3	99.5	73.6	98.6	74.0	93.1	73.0	86.4	65.1	74.9	62.3	67.5	56.8	94.7	74.3
724400	1	61.4	51.8	67.2	53.1	75.9	60.2	81.3	65.4	85.0	70.4	91.7	73.9	96.8	74.5	96.8	74.2	90.4	72.3	83.3	65.7	72.5	59.8	65.1	56.6	91.9	74.3
	2	58.5	49.2	64.2	52.4	73.2	58.8	79.4	64.4	83.6	69.5	89.9	74.1	94.8	75.0	95.0	73.9	88.5	72.8	80.8	65.5	70.2	59.5	62.5	55.5	89.3	73.6
St. Louis,	0.4	65.9	55.1	71.0	55.2	80.9	61.0	86.8	65.5	89.4	70.9	94.8	75.2	100.2	77.6	99.7	76.9	94.1	75.3	85.9	67.2	76.0	61.0	67.8	60.0	95.2	76.2
Intl Airport	1	61.0	51.2	66.8	53.9	77.5	60.6	84.2	65.2	87.9	69.9	93.2	75.1	97.7	77.1	96.8	76.4	92.1	74.8	83.2	67.3	73.3	60.2	64.5	56.2	92.5	75.4
724340	2	57.4	50.2	62.9	51.5	74.3	59.6	81.8	64.3	86.2	69.4	91.9	74.5	95.8	76.7	94.8	76.0	90.0	73.7	80.4	65.6	70.6	59.8	61.5	55.5	90.0	74.4
MONTANA																											
Billings	0.4	55.7	41.5	61.6	44.8	69.0	47.0	80.2	53.4	86.8	58.8	95.3	62.7	97.6	63.6	97.4	63.3	90.6	60.0	81.3	54.6	66.0	47.9	57.9	43.2	93.3	62.5
726770	1	52.6	39.9	58.0	42.4	65.6	46.2	76.5	52.6	83.7	57.8	92.5	62.5	95.6	63.1	94.9	62.6	88.1	59.1	77.6	53.5	62.8	46.2	54.5	41.0	90.2	61.9
	2	49.9	37.9	55.2	41.1	62.5	45.0	73.3	51.0	81.1	56.5	89.9	62.0	93.4	62.8	93.1	62.5	85.4	58.6	74.4	52.2	59.7	44.5	51.8	39.0	86.8	61.2
Cut Bank	0.4	54.5	41.2	56.2	42.6	61.7	44.4	74.4	50.6	80.3	55.2	86.7	59.9	89.8	60.8	93.2	61.1	87.1	57.6	77.9	52.7	61.8	46.1	53.8	42.2	87.3	59.5
727796	1	51.2	39.9	52.6	40.4	58.0	43.2	70.4	48.9	77.1	54.5	84.4	59.0	87.6	59.9	90.7	60.4	84.0	57.3	74.4	51.6	58.7	44.8	51.1	40.3	83.9	58.6
	2	47.9	38.6	50.0	39.5	54.4	41.1	66.7	47.4	74.5	53.8	82.2	58.3	85.8	60.4	88.6	59.9	80.9	55.7	71.4	50.8	55.1	42.4	48.0	38.7	80.3	57.5
Glasgow	0.4	45.8	38.5	54.0	42.8	65.0	47.5	80.1	55.0	88.5	61.7	96.2	65.6	97.4	65.5	98.5	64.7	90.4	60.6	79.8	54.5	61.5	47.2	50.0	40.6	93.5	63.7
727680	1	43.3	36.9	49.5	40.3	61.6	45.9	76.4	53.0	85.0	59.7	92.8	64.5	95.2	64.7	96.0	63.7	87.3	60.1	75.9	54.2	57.9	45.1	46.8	38.7	90.0	62.9
	2	41.1	35.7	46.1	38.5	57.6	44.0	72.9	51.8	82.1	58.6	89.8	63.5	93.0	64.0	93.6	63.1	84.4	59.2	72.4	53.0	54.3	43.1	43.4	36.6	86.3	62.1
Great Falls,	0.4	56.0	41.3	60.5	44.0	66.6	47.1	78.7	53.2	84.4	57.4	92.2	61.0	95.4	61.6	96.6	61.1	89.8	58.4	80.2	54.0	64.6	46.2	57.1	43.0	91.8	60.8
Intl Airport	1	53.1	41.1	57.4	42.6	63.1	45.2	75.1	51.6	82.2	56.8	89.7	60.9	93.5	61.2	94.3	60.8	87.2	57.6	77.0	52.8	61.4	45.4	54.8	41.4	88.3	59.9
727750	2	50.7	39.1	54.8	41.2	60.1	43.7	71.7	50.3	79.6	56.0	86.9	60.9	91.6	61.4	92.3	60.4	84.7	57.2	74.2	52.2	58.7	44.3	52.1	40.4	84.9	59.2
Helena	0.4	52.6	40.7	56.7	42.8	66.8	46.8	76.6	51.9	84.0	57.0	91.8	59.9	95.1	60.5	94.6	60.4	88.2	59.0	76.9	52.3	60.9	45.3	54.2	41.7	90.4	60.1
727720	1	48.9	39.5	53.8	41.8	62.4	44.8	73.2	51.0	81.5	55.7	89.2	60.3	92.8	60.9	92.2	60.2	85.5	57.8	73.5	52.0	58.2	44.8	50.7	39.9	87.2	59.4
	2	46.5	38.0	51.2	39.9	59.1	43.5	70.2	49.4	78.8	55.5	86.4	59.5	90.0	60.2	90.0	59.8	82.4	56.8	70.5	50.8	55.3	42.9	47.8	38.4	83.8	58.6
Kalispell	0.4	44.6	39.7	49.3	40.9	61.0	46.5	75.1	54.3	83.6	58.7	88.1	63.3	92.5	63.6	93.5	63.2	85.4	59.4	72.5	54.0	54.4	46.0	47.0	41.2	89.1	62.1
727790	1	42.8	38.7	46.8	39.5	57.1	44.2	71.2	52.1	80.0	57.6	86.0	62.0	90.8	62.8	91.5	62.2	82.5	58.8	69.6	52.6	51.8	44.7	44.3	39.2	85.7	61.2
	2	41.2	36.9	44.5	38.5	54.0	42.5	67.5	50.3	77.0	56.9	83.7	61.1	89.1	62.5	89.5	61.6	79.8	58.1	66.3	51.2	49.5	43.1	42.2	38.2	82.2	60.2
Lewistown	0.4	53.7	41.9	57.2	43.5	64.6	46.4	75.4	51.6	81.4	57.1	87.1	61.2	93.1	62.2	94.7	62.5	88.4	58.0	79.5	54.8	64.9	46.6	56.2	42.0	89.4	60.6
726776	1	50.6	39.3	53.1	41.5	60.9	44.2	72.2	50.1	79.0	55.8	84.3	61.2	90.9	62.1	92.6	61.3	85.7	57.2	76.4	53.2	61.4	45.1	52.9	40.5	85.5	60.0
	2	48.1	38.2	50.5	39.9	57.3	42.8	69.1	49.3	76.2	55.7	81.8	60.6	88.5	61.4	90.3	60.7	83.0	56.9	73.4	51.5	58.1	43.5	50.1	38.7	82.1	59.3
Miles City	0.4	51.0	40.2	59.2	44.1	69.2	49.6	82.6	55.7	89.7	61.6	99.5	66.1	101.2	66.6	100.3	64.9	94.0	62.0	81.5	55.9	64.8	48.8	53.6	41.0	96.7	65.5
742300	1	47.5	39.2	55.1	43.0	65.3	48.3	78.4	54.6	87.0	60.8	95.4	65.9	98.9	67.3	98.2	65.1	90.9	61.5	77.7	55.6	61.0	46.4	49.5	39.7	93.4	64.7
	2	44.5	37.7	51.7	41.4	61.8	46.8	75.3	53.3	84.4	60.2	92.8	65.0	96.8	66.4	96.5	64.7	87.7	60.6	74.7	54.1	57.7	44.8	46.5	38.2	89.8	63.8
Missoula	0.4	45.2	40.0	52.3	41.8	63.6	47.6	77.0	54.6	84.5	59.9	91.6	63.0	95.4	63.0	95.8	62.6	87.6	60.1	75.0	54.2	56.4	45.7	47.9	41.4	91.4	62.1
727730	1	43.0	38.4	48.7	40.2	60.3	45.4	73.4	52.9	81.5	58.5	89.4	63.1	93.4	62.5	93.4	61.7	84.9	59.2	72.0	52.9	53.2	44.3	44.8	39.2	88.2	61.2
	2	41.4	37.2	46.2	39.2	57.4	44.6	70.0	51.4	78.6	57.3	86.7	62.0	91.7	62.4	91.5	61.5	82.1	58.4	68.5	51.7	50.6	43.7	41.9	37.4	84.6	60.2
NEBRASKA																											
Grand Island	0.4	58.8	44.3	66.1	48.1	79.0	56.4	87.8	62.4	91.0	66.4	99.2	72.1	101.6	72.5	99.2	72.4	94.3	69.0	85.8	60.8	70.2	53.1	61.1	45.6	96.6	72.1
725520	1	54.4	41.8	61.8	46.8	73.7	53.8	83.8	60.8	88.1	64.9	96.7	71.4	99.5	72.6	96.7	72.0	91.5	69.7	82.6	61.1	66.8	51.7	56.6	44.0	93.1	71.5
	2	50.1	39.7	57.4	44.7	69.5	51.9	80.0	59.7	85.1	65.3	94.3	71.4	97.4	73.1	94.6	72.3	89.1	69.1	79.8	59.4	63.9	50.4	52.7	42.3	89.7	70.4
Norfolk	0.4	58.1	43.5	63.6	47.8	76.7	56.7	90.1	63.2	88.5	66.0	97.7	72.9	99.5	75.0	97.5	74.1	93.9	70.9	83.5	60.7	68.0	53.8	57.2	47.8	95.0	74.1
725560	1	53.8	42.0	58.9	45.7	71.2	53.5	86.4	63.1	86.3	66.7	95.6	73.5	97.3	74.4	95.4	75.1	91.7	69.9	80.3	59.6	64.5	51.3	53.2	43.0	92.0	72.2
	2	49.2	39.4	54.9	43.9	66.7	51.7	82.4	60.5	83.7	65.9	93.1	72.7	95.2	74.6	93.6	73.5	89.5	69.3	77.2	58.1	61.5	49.1	49.5	40.8	89.0	71.5
North Platte	0.4	59.3	42.9	66.9	46.6	76.1	51.6	86.1	57.2	88.6	62.5	97.1	70.6	99.8	69.4	98.3	68.3	93.8	66.7	85.6	58.2	71.4	50.6	61.8	45.1	95.0	68.9
725620	1	54.8	40.8	62.9	44.9	72.6	50.2	82.5	58.3	85.9	62.3	94.1	69.4	97.7	69.7	95.9	69.2	91.4	65.9	82.7	57.1	68.0	48.6	57.9	42.8	91.7	68.9
	2	51.1	39.3	59.4	44.3	69.2	49.5	78.6	55.8	83.4	62.1	91.6	69.1	95.6	69.7	93.5	69.0	88.9	64.7	79.9	56.5	64.1	47.8	53.8	41.1	88.6	67.9
Omaha, Eppley	0.4	54.9	43.7	63.2	49.1	78.9	58.4	86.8	63.5	89.8	69.7	97.1	74.7	100.2	75.4	98.3	76.3	93.2	72.6	84.7	66.4	70.0	56.7	59.2	51.1	95.0	75.2
Airfield	1	51.2	41.7	58.6	47.4	74.2	55.9	83.4	63.0	87.5	68.7	94.7	74.5	97.7	76.3	95.8	75.8	90.6	72.5	81.9	64.3	67.1	55.8	55.7	47.6	91.8	74.5
725500	2	47.8	40.4	54.8	44.9	70.1	55.2	80.2	61.9	85.4	68.1	92.2	74.0	95.5	76.2	93.6	75.6	88.2	71.8	79.1	63.5	64.5	54.4	52.2	44.9	88.9	73.4
Scottsbluff	0.4	58.9	41.6	65.8	46.6	74.0	49.3	83.3	54.0	88.8	58.4	97.5	64.3	100.8	65.5	97.4	64.6	94.0	62.9	84.0	55.8	71.3	49.5	62.7	44.4	95.3	64.7
725660	1	55.5	40.7	61.4	43.8	70.3	47.5	80.5	53.4	86.3	58.5	94.9	63.9	98.6	65.1	95.2	65.1	91.3	61.5	81.2	54.2	68.1	47.5	59.7	42.9	92.2	64.3
	2	52.6	39.2	59.1	42.9	67.1	46.2	77.4	52.8	84.0	58.3	92.5	63.1	96.5	65.3	93.3	64.7	88.6	60.4	78.6	53.6	64.5	46.4	55.9	41.1	89.0	64.3
NEVADA																											
Elko	0.4	52.9	41.9	61.2	44.5	67.8	47.1	77.6	50.7	85.5	55.6	95.7	59.8	98.6	62.2	97.6	60.9	91.9	58.7	82.8	52.8	66.4	46.7	53.5	40.8	94.8	59.9
725825	1	49.8	39.9	57.5	43.4	65.0	45.0	75.2	49.7	83.0	54.4	93.5	59.0	96.8	61.2	95.5	59.7	89.9	58.1	80.0	51.9	63.7	45.2	50.8	40.0	92.3	58.8
	2	47.1	38.5	54.5	41.9	61.9	44.4	72.3	48.6	80.8	53.6	91.4	58.5	95.4	60.7	94.1	59.4	88.0	56.7	77.6	51.1	60.2	44.3	48.3	38.7	89.6	58.0
Ely	0.4	53.5	38.9	59.5	41.9	64.8	43.7	73.9	48.3	80.8	53.6	90.0	55.9	92.4	57.5	91.1	56.5	86.3	55.1	77.9	50.2	65.9	44.5	56.0	39.9	89.2	56.4
724860	1	50.2	37.7	56.3	40.0	62.1	43.0	71.7	47.4	78.7	52.1	88.5	55.7	91.0	57.2	89.7	56.6	84.5	54.8	76.1	49.6	62.8	43.4	53.2	38.7	87.0	55.9
	2	47.4	36.4	53.3	39.5	59.8	41.9	69.3	46.4	76.7	50.6	86.9	55.1	89.8	56.8	88.4	56.4	82.7	54.2	73.9	48.7	60.2	41.8	49.9	37.3	84.7	55.3
Las Vegas,	0.4	68.8	48.9	76.6	51.7	84.0	54.4	92.9	58.1	101.0	61.6	109.5	65.2	111.2	66.5	109.8	67.3	103.9	64.4	95.7	60.0	79.5	54.5	68.1	47.9	107.9	66.1
Intl Airport	1	66.3	47.6	73.8	50.5	81.7	54.0	90.8	57.4	98.7	60.6	108.0	64.8	109.8	66.3	108.0	67.0	102.5	64.6	93.7	59.1	77.4	53.1	65.6	47.6	105.6	65.5
723860	2	64.1	47.0	71.4	49.8	79.5	53.0	88.6	56.3	96.4	59.8	106.4	63.8	108.5	66.3	106.6	67.0	100.9	64.2	91.3	58.3	75.1	52.4	63.6	46.7	103.2	64.7
Reno	0.4	59.5	43.9	66.5	47.5	71.6	48.5	80.2	52.3	88.2	57.6	94.9	60.7	98.5	61.8	97.4	61.1	92.7	58.8	84.9	55.6	69.7	49.2	60.6	44.1	94.6	60.5
724880	1	56.8	43.3	63.6	45.6	68.8	47.2	77.9	51.5	85.4	56.4	93.0	60.1	96.8	61.3	95.7	60.8	90.4	58.0	82.6	55.0	67.2	47.7	57.9	44.0	92.1	59.7
	2	54.3	41.6	61.1	44.1	66.2	46.1	75.2	50.7	83.1	55.3	91.2	58.9	95.2	60.9	94.3	60.2	88.6	57.5	80.2	53.7	64.5	46.4	55.2	43.0	89.5	58.8
Tonopah	0.4	58.1	42.5	63.2	43.1	69.8	46.4	79.1	50.0	85.7	54.8	94.9	58.2	97.0	59.0	96.3	58.4	90.1	56.6	83.1	52.6	68.6	45.9	56.6	40.4	93.9	58.0
724855	1	55.0	40.5	61.2	42.5	67.7	45.0	76.3	48.9	83.6	53.4	93.2	56.5	95.7	58.7	94.7	58.7	88.3	56.3	81.0	51.8	66.1	44.7	54.8	39.2	91.6	57.3
	2	52.6	39.1	59.1	41.5	65.2	44.2	73.7	47.9	81.8	52.2	91.4	55.9	94.5	58.3	93.1	58.3	86.8	55.6	78.8	50.8	63.7	44.3	52.8	38.8	89.4	56.7
Winnemucca	0.4	57.7	44.9	64.5	47.2	71.7	47.4	81.0	51.7	90.1	58.2	97.7	60.3	100.5	62.2	100.1	61.6	93.4	57.9	85.2	54.6	69.3	48.2	58.7	44.8	96.8	60.5
725830	1	54.7	43.2	61.7	45.1	68.6	47.6	78.3	51.2	87.1	56.4	95.9	60.6	98.9	61.4	98.0	61.0	91.6	57.9	82.5	53.7	66.1	47.2	55.8	43.1	94.4	59.7
	2	52.3	41.5	59.2	44.5	65.8	46.2	75.6	50.0	84.7	55.1	93.9	59.8	97.3	61.0	96.3	60.0	89.6	57.1	80.0	52.9	63.5	46.3	53.1	42.1	91.8	58.6
NEW HAMPSHIRE																											
Concord	0.4	52.1	48.6	54.1	46.9	67.5	52.8	81.4	60.8	89.7	66.0	92.0	71.9	94.4	73.7	91.7	73.9	87.7	72.8	78.9	62.0	68.5	59.8	57.6	53.2	89.7	71.0
726050	1	47.4	42.0	50.0	44.4	62.1	50.8	76.0	57.9	86.4	66.1	89.7	70.6	92.2	73.0	89.5	72.2	85.1	70.6	75.4	62.0	64.8	56.7	53.3	48.4	86.5	69.9
	2	43.7	38.5	46.3	41.1	57.4	47.5	72.0	54.8	83.2	64.6	87.2	69.5	90.1	71.8	87.6	71.2	82.4	69.0	72.2	60.1	61.5	54.3	48.8	43.7	83.8	68.4
NEW JERSEY																											
Atlantic City	0.4	61.2	52.1	65.4	54.5	75.2	59.1	83.0	64.1	90.1	70.6	93.9	73.5	95.4	75.6	93.0	75.4	90.7	74.9	80.4	66.8	72.9	63.2	64.5	57.2	91.3	74.2
724070	1	57.2	52.0	60.7	52.0	69.8	56.6	79.3	61.9	87.1	68.0	91.4	72.8	93.6	75.7	90.8	75.0	88.0	74.1	78.1	66.6	69.7	61.1	61.8	56.5	88.3	73.4
	2	54.8	51.4	57.2	50.9	65.1	55.1	74.8	59.9	84.0	67.9	89.2	73.2	91.7	75.0	89.0	74.4	85.5	72.5	75.7	65.5	67.0	60.2	59.6	55.8	85.6	72.3
Newark	0.4	60.2	53.0	63.3	52.6	75.2	58.9	84.4	63.2	90.9	70.3	94.9	73.1	97.9	75.4	94.9	76.1	92.7	75.5	80.9	66.4	73.1	62.0	64.4	57.4	93.0	74.1
725020	1	56.3	50.6	59.5	51.8	70.6	56.8	80.2	61.4	88.4	68.2	92.9	72.5	95.6	75.4	92.4	75.0	89.9	74.3	78.5	64.5	69.9	61.1	61.1	56.1	90.0	72.6
	2	52.6	47.2	55.5	48.5	66.1	54.7	76.1	59.5	85.6	67.7	90.9	71.3	93.5	74.9	90.5	74.1	86.9	72.3	76.4	63.8	67.0	59.0	57.8	53.1	87.3	71.4

WMO# = World Meteorological Organization number DB = dry-bulb temperature, °F MWB = mean coincident wet-bulb temperature, °F

Table 4A Selected Monthly Percentiles of Wet-Bulb Temperature with Mean Dry-Bulb Temperature for U.S. Locations

Location and WMO#	%	Jan WB	Jan MDB	Feb WB	Feb MDB	Mar WB	Mar MDB	Apr WB	Apr MDB	May WB	May MDB	Jun WB	Jun MDB	Jul WB	Jul MDB	Aug WB	Aug MDB	Sep WB	Sep MDB	Oct WB	Oct MDB	Nov WB	Nov MDB	Dec WB	Dec MDB	Annual WB	Annual MDB
1a	1b	2a	2b	3a	3b	4a	4b	5a	5b	6a	6b	7a	7b	8a	8b	9a	9b	10a	10b	11a	11b	12a	12c	13a	13b	14a	14b
NEW MEXICO																											
Albuquerque	0.4	45.1	55.1	47.2	63.6	50.1	69.8	54.0	77.2	59.2	78.7	64.2	82.9	66.4	84.9	66.8	84.0	64.5	81.1	58.6	70.6	51.4	61.2	45.1	54.9	65.3	83.4
723650	1	43.3	54.2	45.8	60.6	48.6	67.8	52.7	75.4	57.9	78.2	63.3	82.7	65.8	84.9	66.2	83.3	63.6	79.5	57.4	68.8	49.8	60.1	43.4	53.4	64.4	82.4
	2	41.9	52.4	44.5	58.6	47.4	67.1	51.6	73.7	56.9	77.5	62.4	82.4	65.3	84.7	65.5	82.7	62.9	77.9	56.2	69.0	48.3	60.1	42.2	52.0	63.5	81.1
Tucumcari	0.4	49.5	65.6	52.2	68.9	54.8	74.6	59.9	77.7	65.1	79.6	69.3	88.6	71.2	87.0	70.9	87.3	68.7	82.8	62.8	77.4	55.6	69.2	50.4	63.0	69.4	87.0
723676	1	47.8	64.9	50.3	69.2	52.7	72.7	58.4	74.2	63.9	79.9	68.3	86.4	70.3	87.2	70.1	86.8	67.8	82.8	61.4	74.6	54.2	67.1	49.0	63.4	68.4	85.4
	2	46.3	61.5	48.9	66.6	51.2	72.6	57.0	72.7	62.8	78.7	67.4	85.4	69.7	87.6	69.3	86.0	67.1	82.1	60.3	73.7	52.5	68.1	47.5	62.7	67.4	83.9
NEW YORK																											
Albany	0.4	48.6	51.4	48.0	51.5	55.6	63.5	63.5	77.9	70.6	83.7	74.5	86.0	76.8	87.6	76.4	87.6	74.2	85.1	66.4	72.1	61.7	65.6	54.1	56.1	74.4	84.8
725180	1	42.7	46.7	45.0	49.7	51.8	59.7	60.8	72.0	68.9	81.9	73.2	83.8	75.8	87.0	75.1	85.5	72.7	82.3	64.7	72.1	58.9	63.1	49.5	53.1	72.7	82.0
	2	39.0	42.2	42.3	45.4	49.3	55.6	58.5	68.0	66.9	80.2	72.0	82.7	74.9	85.7	73.9	82.8	71.3	79.2	63.1	69.5	56.4	60.6	45.1	47.7	71.2	79.4
Binghamton	0.4	49.1	51.0	49.5	51.7	57.5	67.2	62.3	76.0	69.6	79.5	72.8	81.8	75.1	84.1	74.3	83.3	73.0	80.4	64.3	70.1	59.8	63.7	54.2	56.9	72.6	81.1
725150	1	45.4	47.0	47.1	50.1	54.5	62.2	60.9	72.2	68.2	77.5	71.3	80.3	74.0	82.7	73.2	82.2	71.6	78.8	63.0	68.3	58.0	62.1	51.5	53.9	71.1	78.8
	2	41.3	43.8	44.0	46.9	52.0	58.5	59.1	69.1	66.8	75.5	70.2	79.3	72.9	81.7	72.1	79.8	70.5	76.8	61.7	67.3	55.7	58.8	47.7	49.6	69.5	76.7
Buffalo	0.4	50.3	53.9	50.5	54.3	57.7	67.0	63.7	75.8	70.4	79.8	73.9	83.5	75.6	84.8	75.5	83.5	74.2	81.4	65.8	73.2	60.2	65.5	56.4	59.3	73.8	81.6
725280	1	45.5	49.6	47.7	51.0	54.9	63.2	61.9	74.0	69.1	79.0	72.4	82.3	74.9	83.8	74.3	81.7	72.8	80.3	64.4	71.0	58.7	62.8	53.7	57.0	72.2	79.8
	2	42.7	45.8	44.7	48.8	52.8	59.7	59.9	69.7	67.5	76.1	71.2	80.5	74.1	82.1	73.3	80.7	71.6	77.9	63.1	69.0	56.6	61.2	50.4	53.8	70.8	77.9
Massena	0.4	45.0	48.0	46.8	50.0	54.7	63.2	63.6	73.0	71.2	82.8	75.4	84.8	77.5	88.1	76.8	85.1	74.7	82.8	65.9	74.5	59.5	65.6	51.9	53.9	74.7	84.0
726223	1	41.4	44.7	43.7	46.7	51.5	60.3	61.1	70.8	69.7	80.7	74.1	83.5	76.2	86.2	75.2	84.3	73.2	80.2	64.4	71.2	57.1	60.8	48.0	50.5	72.8	81.3
	2	38.9	41.5	41.1	44.2	48.6	54.5	58.7	67.7	67.9	77.4	72.7	81.8	75.0	84.9	73.9	82.1	71.5	77.0	62.6	68.9	54.6	58.6	43.8	45.9	71.0	78.7
New York, John F	0.4	53.2	56.2	54.3	57.2	59.1	68.5	64.6	77.5	72.4	83.5	75.7	89.4	78.0	89.2	78.2	88.2	76.3	86.0	69.6	74.7	64.1	69.1	57.5	60.1	76.3	85.9
Kennedy Airport	1	50.6	53.9	51.7	56.0	56.2	62.7	62.6	72.8	70.8	81.4	74.4	86.7	77.2	87.5	77.2	86.4	75.3	84.3	68.3	72.8	61.9	65.3	55.4	57.8	74.9	83.9
744860	2	47.9	51.2	49.1	53.2	53.5	59.6	60.5	70.3	69.1	79.2	73.3	84.6	76.3	86.3	76.3	84.8	74.3	81.8	66.9	71.8	60.2	63.8	53.0	55.4	73.6	81.6
Rochester	0.4	49.5	53.0	50.6	54.4	59.0	68.9	65.2	76.5	71.8	82.0	76.1	85.9	77.9	88.3	77.3	87.0	75.2	84.1	66.8	75.1	61.3	66.0	57.0	61.3	75.4	84.6
725290	1	45.0	48.9	47.7	51.2	56.2	63.5	63.4	74.6	70.5	80.9	74.6	84.4	76.8	86.6	75.7	84.1	73.9	82.2	65.3	72.2	59.1	63.6	53.0	56.8	73.5	81.9
	2	42.5	46.7	44.4	47.8	53.6	61.9	61.5	71.9	69.1	79.4	73.3	82.7	75.8	85.4	74.4	82.0	72.6	79.8	63.9	71.1	57.3	62.2	49.6	53.4	71.8	80.0
Syracuse	0.4	49.5	54.5	50.3	54.7	58.9	69.8	65.1	77.8	71.9	82.4	74.9	85.8	77.5	87.8	77.2	86.4	75.0	85.1	66.2	74.8	61.4	67.1	56.2	59.3	74.9	84.5
725190	1	44.8	48.4	46.9	51.7	55.5	64.5	63.1	74.9	70.3	81.3	73.7	84.7	76.3	86.1	75.8	83.5	73.7	82.1	64.9	72.5	59.5	64.5	53.1	56.7	73.2	82.4
	2	41.7	45.8	44.0	47.7	52.7	60.9	60.8	70.7	68.5	78.1	72.6	82.9	75.2	85.3	74.4	82.4	72.4	79.9	63.5	69.9	57.1	61.9	48.6	53.1	71.6	79.8
NORTH CAROLINA																											
Asheville	0.4	57.8	62.2	59.1	63.9	62.1	72.0	66.2	75.6	71.2	78.2	74.7	85.1	76.5	86.4	75.6	85.3	73.4	82.2	68.4	74.7	64.4	67.4	62.2	64.6	74.5	83.8
723150	1	55.7	58.9	57.0	61.4	60.9	67.4	64.4	73.2	69.7	77.4	73.6	83.2	75.7	85.3	74.9	84.7	72.6	81.3	67.2	71.9	63.1	66.8	60.3	62.1	73.3	82.1
	2	53.9	56.5	55.2	59.8	59.7	65.7	63.2	71.4	68.8	76.8	72.7	82.1	74.9	84.4	74.2	83.3	71.9	80.0	66.1	71.1	61.6	65.7	58.3	61.2	72.2	80.2
Cape Hatteras	0.4	66.2	68.5	65.6	69.4	67.5	71.6	69.4	74.9	75.7	80.5	79.2	84.8	81.3	87.7	81.1	86.7	79.0	83.9	76.1	80.6	71.5	75.2	68.7	71.9	79.9	85.5
723040	1	64.9	67.0	64.0	67.3	66.2	69.6	68.4	73.8	74.4	79.3	78.4	83.8	80.7	86.9	80.4	85.9	78.3	83.7	74.8	78.8	70.5	73.5	67.5	70.3	78.9	83.8
	2	63.3	65.5	62.7	65.4	65.1	68.6	67.5	72.5	73.5	78.3	77.7	83.0	80.2	86.1	79.8	85.2	77.6	83.2	73.7	77.2	69.5	72.4	66.2	68.6	78.0	83.1
Charlotte	0.4	61.5	64.7	63.0	68.8	65.4	74.4	69.0	80.4	74.4	84.7	77.1	88.0	78.1	89.8	77.8	89.2	76.2	86.5	72.2	79.7	68.2	72.9	66.2	69.1	76.8	88.2
723140	1	59.9	63.8	61.4	66.1	64.2	72.1	67.6	77.3	73.0	82.8	76.1	87.4	77.4	89.1	77.2	88.5	75.4	85.9	70.9	76.7	67.1	71.8	64.0	66.3	75.8	87.0
	2	57.9	61.4	59.5	63.4	62.8	69.6	66.4	76.2	71.9	82.2	75.3	86.8	76.9	88.7	76.6	87.8	74.7	84.9	69.6	75.8	65.7	69.5	61.9	64.8	74.9	85.7
Greensboro	0.4	61.0	65.2	62.0	68.0	65.4	73.7	68.4	77.0	74.3	83.6	77.5	87.2	78.9	90.2	78.2	88.1	75.7	85.8	71.1	77.5	67.9	72.1	65.0	67.9	77.1	87.5
723170	1	58.6	62.4	60.1	64.3	64.0	71.2	67.2	75.4	72.7	81.8	76.3	86.7	78.1	88.8	77.4	87.7	74.9	84.6	69.8	76.5	66.4	69.8	62.7	65.0	76.0	86.1
	2	56.3	59.5	58.4	61.9	62.5	69.4	66.0	74.2	71.7	80.8	75.4	86.5	77.3	87.9	76.8	87.2	74.2	83.4	68.5	74.6	64.8	69.4	60.8	63.0	74.9	85.0
Raleigh/Durham	0.4	63.1	66.2	64.0	69.9	66.3	76.3	69.2	78.1	74.8	84.6	78.2	88.0	79.8	90.5	79.2	89.1	76.9	86.4	72.7	78.7	68.5	72.7	66.6	70.3	77.9	88.4
723060	1	61.4	64.7	62.2	67.9	65.1	73.9	67.9	78.5	73.6	83.2	77.2	87.5	78.9	90.0	78.4	88.8	76.1	85.5	71.3	77.3	67.5	71.2	64.5	68.0	76.8	86.8
	2	59.5	63.6	60.4	64.8	63.7	71.1	66.7	76.6	72.5	82.0	76.2	86.9	78.1	88.7	77.8	88.2	75.3	84.3	70.3	75.9	66.1	70.0	62.4	66.2	75.7	85.1
Wilmington	0.4	67.5	70.1	68.0	73.1	69.5	75.8	72.8	80.5	76.9	86.3	80.1	90.5	82.1	92.3	81.4	90.8	79.4	87.5	76.5	82.3	73.2	77.2	70.0	73.0	80.2	89.4
723013	1	66.1	68.6	66.4	71.4	68.4	74.5	71.5	78.9	75.7	84.1	79.2	89.1	81.2	91.4	80.8	89.6	78.5	86.1	75.4	80.5	71.9	75.4	68.1	71.3	79.2	87.8
	2	64.7	67.7	64.9	69.1	67.2	73.2	70.1	77.6	74.6	82.2	78.4	87.8	80.4	90.0	80.2	88.9	78.1	85.4	74.2	78.8	70.5	74.2	66.6	69.7	78.3	86.2
NORTH DAKOTA																											
Bismarck	0.4	38.7	45.1	43.6	52.3	50.5	65.8	59.1	75.3	68.3	82.1	73.0	86.3	76.2	88.1	73.2	88.9	68.9	84.3	59.5	74.3	50.1	59.9	40.1	47.5	72.1	85.8
727640	1	37.2	42.5	40.9	48.6	47.7	60.3	57.1	73.8	66.2	79.4	71.3	85.8	74.7	86.6	72.0	88.0	67.1	82.8	57.8	72.3	47.3	57.3	37.9	43.1	70.0	84.4
	2	35.8	40.6	38.6	44.4	45.2	56.1	55.1	69.8	64.2	77.3	70.0	84.2	73.1	86.4	70.9	87.3	65.4	79.4	56.3	69.8	45.1	54.0	36.2	41.0	68.1	82.1
Fargo	0.4	35.9	39.2	40.6	45.6	51.5	60.3	62.4	77.5	70.2	84.9	76.4	85.7	78.4	88.7	77.1	89.8	73.5	85.0	63.7	72.1	52.3	57.1	39.2	43.7	75.1	86.4
727530	1	34.3	36.9	37.7	42.0	48.0	57.0	59.8	74.4	68.4	80.6	74.5	85.2	77.0	87.2	75.5	87.8	71.6	82.1	61.6	69.8	49.2	55.1	35.6	39.2	73.0	83.7
	2	33.0	35.0	35.7	39.1	45.3	52.4	57.3	69.8	66.8	78.2	72.8	83.3	75.6	86.2	74.2	85.8	69.5	78.3	59.7	69.5	47.0	52.1	33.7	36.0	70.9	81.4
Minot,	0.4	38.1	43.0	41.5	49.5	48.7	61.9	57.9	78.1	66.6	81.2	71.3	84.1	75.3	87.1	73.4	88.7	68.3	86.7	59.0	72.5	49.1	58.8	38.9	44.9	71.1	84.6
Intl Airport	1	36.8	41.6	39.3	45.5	46.3	57.8	56.1	72.9	64.5	81.5	69.9	83.7	73.6	85.7	72.0	86.3	65.5	80.0	57.3	71.1	46.3	54.9	37.1	42.5	68.9	82.8
727676	2	35.4	39.8	37.3	42.1	43.9	53.7	54.1	70.2	62.7	78.1	68.4	82.4	71.9	82.9	70.4	84.4	63.7	79.3	55.8	68.4	44.3	51.6	35.4	40.3	66.8	79.5
OHIO																											
Akron/Canton	0.4	53.1	56.0	53.5	56.9	60.0	67.7	65.0	74.6	72.4	79.2	75.2	85.5	77.2	86.7	76.8	86.6	74.0	83.3	66.8	74.3	61.7	66.8	57.6	61.1	74.9	84.2
725210	1	50.6	53.1	51.1	54.4	58.1	65.1	63.2	74.1	70.4	78.6	74.1	83.7	76.1	86.3	75.6	84.6	73.0	82.3	65.3	71.0	60.0	64.5	55.7	58.6	73.3	81.7
	2	47.3	50.4	49.1	52.4	56.3	63.8	61.8	72.4	68.6	78.0	72.9	82.0	75.2	85.0	74.4	82.7	72.0	80.5	63.9	70.3	58.6	62.6	53.2	56.3	71.8	79.8
Cleveland	0.4	54.1	58.0	54.2	57.8	61.9	70.1	66.3	77.1	73.1	81.2	76.0	86.5	77.9	89.0	77.5	88.0	75.2	84.8	67.8	76.1	63.0	68.2	57.7	62.4	75.6	85.4
725240	1	50.9	53.9	51.6	55.9	59.3	67.6	64.5	75.6	71.3	81.1	74.8	85.1	77.0	87.6	76.3	85.6	74.1	83.5	66.1	73.4	60.9	65.0	55.8	59.6	74.0	82.9
	2	47.0	50.4	48.8	52.6	57.1	65.1	63.0	73.5	69.8	79.9	73.6	83.1	76.0	86.4	75.1	83.8	73.0	81.6	64.5	71.6	59.3	63.4	53.3	56.9	72.4	80.7
Columbus,	0.4	55.4	58.9	55.5	59.7	62.1	71.7	66.4	76.5	74.2	84.2	77.2	87.2	78.6	88.8	78.1	88.8	75.7	86.6	68.6	76.3	63.0	68.9	58.7	62.3	76.5	86.9
Intl Airport	1	52.8	55.8	53.7	57.1	60.1	69.0	64.9	75.9	72.4	80.9	76.1	86.2	77.4	88.1	77.2	88.2	74.4	84.0	67.1	74.0	61.4	66.8	56.9	60.5	75.0	84.5
724280	2	49.1	52.1	51.2	55.6	58.4	66.1	63.4	74.1	70.8	79.7	74.9	84.7	76.6	87.2	76.1	86.1	73.3	82.7	65.7	73.0	59.8	64.9	54.7	58.1	73.5	82.1
Dayton,	0.4	55.5	58.4	55.8	59.5	62.4	70.8	66.9	75.3	73.6	83.1	76.9	87.0	78.5	89.5	78.4	88.7	75.2	86.0	68.4	75.1	63.7	68.9	59.0	62.1	76.4	86.8
Intl Airport	1	53.2	55.6	53.4	56.3	60.3	67.9	65.3	74.0	72.1	80.5	75.5	85.6	77.6	88.8	77.2	87.5	74.3	84.1	67.0	73.9	62.1	66.7	57.2	59.6	74.9	84.6
724290	2	49.4	51.9	51.2	55.2	58.7	64.5	63.9	73.1	70.7	79.9	74.5	84.6	76.6	87.6	76.2	85.9	73.2	82.3	65.6	72.8	60.4	64.7	54.6	57.3	73.5	82.2
Mansfield	0.4	54.3	59.3	53.6	56.3	60.9	69.2	64.5	75.0	72.2	80.2	75.7	85.0	78.0	88.0	77.5	85.8	74.5	82.8	67.3	73.8	62.6	68.3	58.3	61.5	75.5	84.5
725246	1	51.9	54.7	51.4	54.8	58.8	65.7	63.3	74.6	70.2	78.4	74.6	83.9	76.7	86.6	76.4	84.6	73.4	82.4	66.0	72.3	61.0	64.4	56.3	59.3	74.1	82.5
	2	48.4	52.3	49.2	52.9	56.7	64.4	62.1	72.4	68.5	76.0	73.7	82.3	75.8	85.1	75.3	83.9	72.4	80.8	64.7	71.6	59.3	63.8	54.0	56.6	72.6	80.2
Toledo	0.4	52.2	54.6	51.6	55.1	61.0	68.3	66.0	76.1	73.3	82.8	76.0	87.5	78.9	90.1	79.4	89.5	75.9	84.9	67.7	76.9	61.8	66.7	57.7	60.9	76.5	85.7
725360	1	48.8	51.2	48.7	53.2	58.3	67.4	64.3	75.2	71.7	81.4	75.0	86.6	77.9	88.1	77.9	87.7	74.6	82.9	66.0	73.8	60.0	63.8	55.1	57.5	74.8	83.7
	2	44.3	46.8	45.8	49.5	56.3	63.5	62.9	73.9	70.2	80.7	74.0	83.9	76.7	86.4	76.5	84.8	73.4	80.8	64.6	71.7	58.3	62.5	52.2	54.5	73.2	81.2
Youngstown	0.4	52.6	55.1	53.2	58.1	60.1	67.4	64.4	75.5	72.0	80.4	74.6	85.1	77.1	86.7	76.1	86.3	74.1	83.0	66.6	74.5	61.9	68.0	57.4	61.3	74.4	84.0
725250	1	49.6	52.3	50.6	54.5	57.9	65.5	62.9	75.4	70.0	79.3	73.7	83.1	76.0	85.8	74.9	84.5	73.0	81.6	65.1	71.0	60.1	65.2	55.5	58.5	72.9	81.5
	2	46.1	49.5	48.2	52.3	55.9	63.5	61.5	73.3	68.4	77.9	72.4	81.6	74.9	84.9	73.8	83.0	71.9	79.7	63.7	69.5	58.2	62.1	53.0	55.5	71.3	79.2

WMO# = World Meteorological Organization number WB = wet-bulb temperature, °F MDB = mean coincident dry-bulb temperature, °F

Table 4B Selected Monthly Percentiles of Dry-Bulb Temperature with Mean Wet-Bulb Temperature for U.S. Locations

Location and WMO#	%	Jan		Feb		Mar		Apr		May		Jun		Jul		Aug		Sep		Oct		Nov		Dec		Annual	
		DB	MWB	DB	MWB	DB	MWB	DB	MWB	DB	MWB	DB	MWB	DB	MWB	DB	MWB	DB	MWB	DB	MWB	DB	MWB	DB	MWB	DB	MWB
1a	1b	2a	2b	3a	3b	4a	4b	5a	5b	6a	6b	7a	7b	8a	8b	9a	9b	10a	10b	11a	11b	12a	12c	13a	13b	14a	14b
NEW MEXICO																											
Albuquerque	0.4	60.5	43.3	68.1	44.6	75.5	47.9	83.1	51.7	90.3	55.1	99.6	60.0	99.5	60.7	95.1	61.7	91.9	59.1	83.9	53.5	70.6	47.5	60.9	42.7	95.6	60.2
723650	1	58.0	41.1	65.7	44.2	73.4	47.1	80.9	50.6	88.2	54.2	97.6	58.7	97.6	61.6	93.9	61.7	89.9	59.9	81.5	52.9	68.4	47.1	58.3	41.7	93.0	60.0
	2	55.5	40.4	63.1	43.1	71.2	46.3	79.1	49.7	86.4	53.7	95.7	58.2	96.0	61.3	92.5	61.7	87.9	59.6	79.5	53.1	66.1	46.0	55.8	40.3	90.5	59.9
Tucumcari	0.4	69.9	47.0	76.5	50.2	81.5	52.3	87.3	55.5	94.3	59.4	102.3	63.2	100.7	65.9	98.3	64.9	94.1	63.8	88.4	57.9	78.4	52.8	70.2	48.0	97.6	64.3
723676	1	67.3	46.8	73.1	49.0	78.4	50.8	85.6	54.8	92.3	58.6	99.8	62.6	99.0	66.1	96.7	65.6	92.3	63.0	86.3	57.0	75.3	51.5	67.7	47.2	95.1	64.6
	2	64.6	45.4	69.9	47.2	76.1	49.7	83.5	53.8	90.5	58.2	97.8	62.1	97.6	65.8	95.4	65.3	90.4	63.0	84.1	56.7	72.6	50.5	65.1	46.1	92.6	64.2
NEW YORK																											
Albany	0.4	52.3	48.4	53.7	47.0	67.2	53.5	81.4	60.8	89.2	66.6	92.0	71.9	94.0	74.1	91.8	74.0	87.7	73.0	78.9	62.0	68.4	59.7	57.4	53.1	89.6	71.2
725180	1	47.5	42.1	50.0	44.4	62.1	50.8	76.0	58.5	86.1	66.3	89.6	70.8	91.8	73.0	89.4	72.2	85.0	70.6	75.5	62.2	64.7	56.8	53.2	48.4	86.4	69.9
	2	43.6	38.3	46.3	41.2	57.3	47.5	72.1	55.2	83.0	65.2	87.0	69.5	89.9	71.8	87.5	71.4	82.3	69.1	72.2	60.5	61.4	54.4	48.9	44.1	83.6	68.6
Binghamton	0.4	51.8	47.3	54.3	47.4	69.3	56.9	79.5	61.1	83.5	65.5	86.1	69.9	88.9	71.1	86.9	71.3	83.7	70.2	74.8	61.7	65.3	58.0	57.4	53.5	84.9	70.1
725150	1	48.1	45.4	50.7	46.8	64.6	53.4	75.6	58.9	81.3	65.5	84.2	69.2	87.0	71.4	85.0	70.9	81.5	70.2	72.2	60.2	62.9	55.0	54.2	50.1	82.2	68.7
	2	44.5	40.4	47.7	43.0	60.4	51.2	71.7	56.6	79.2	64.2	82.3	68.1	85.4	70.9	83.1	69.7	79.2	68.6	70.0	59.6	60.5	54.5	50.9	46.5	79.7	67.2
Buffalo	0.4	53.9	49.6	55.5	47.9	70.2	55.8	78.8	62.5	84.1	67.3	87.9	69.6	89.9	71.3	88.2	73.2	85.2	72.5	76.9	63.9	67.4	57.6	60.3	55.2	86.0	70.2
725280	1	50.1	45.3	52.6	46.8	65.8	54.5	75.5	60.5	82.0	66.0	86.1	69.3	87.9	70.9	86.2	71.1	83.1	70.5	74.7	61.8	65.0	55.8	57.5	53.4	83.6	69.2
	2	46.5	41.4	49.4	44.9	61.2	51.2	72.1	57.7	80.0	64.7	84.4	68.6	86.2	70.5	84.6	70.1	80.8	68.8	72.4	60.6	62.8	54.9	54.2	50.1	81.3	68.0
Massena	0.4	49.8	45.0	51.0	45.8	64.5	52.7	78.1	61.6	86.8	68.8	88.7	73.0	91.2	74.9	89.2	74.6	84.6	73.0	76.1	63.3	66.3	57.8	55.5	51.2	87.2	72.3
726223	1	45.0	41.1	47.4	42.7	59.9	51.3	73.8	59.3	83.6	67.4	86.8	72.1	89.6	74.0	87.1	72.8	82.0	71.8	73.8	62.9	62.8	56.2	51.1	48.1	84.4	70.5
	2	42.0	38.2	44.7	41.1	55.9	47.5	69.9	55.5	80.6	65.4	84.8	70.5	87.9	72.7	85.3	71.7	79.7	69.5	71.1	60.3	59.1	54.2	46.7	43.4	81.7	68.9
New York, John F	0.4	57.7	51.0	60.9	51.1	70.6	57.8	80.3	61.7	88.4	70.2	92.7	73.1	95.4	75.4	92.1	76.2	90.2	74.3	79.1	66.4	71.6	61.4	61.1	54.3	90.5	73.5
Kennedy Airport	1	54.4	49.6	57.5	50.6	66.3	54.6	76.1	61.1	85.3	68.0	90.4	72.3	93.2	74.7	90.0	74.2	87.8	73.6	76.6	64.7	67.9	59.2	58.9	54.3	87.7	72.3
744860	2	51.8	47.3	54.4	48.6	62.5	51.8	72.7	58.4	82.6	66.9	88.4	71.1	91.2	74.1	88.4	73.7	85.1	71.4	74.2	64.0	65.2	59.0	56.4	52.5	85.1	71.0
Rochester	0.4	53.9	47.6	55.5	49.1	72.4	57.1	81.1	63.6	87.1	69.0	89.9	72.2	93.2	74.6	90.5	74.1	87.6	72.8	78.9	64.5	68.3	58.6	60.9	56.4	88.5	72.6
725290	1	50.1	44.9	52.1	47.0	66.4	54.6	77.7	61.5	84.4	68.1	88.1	71.8	91.1	73.9	88.5	73.0	85.3	72.6	76.5	63.2	65.5	57.0	57.4	51.8	85.8	71.1
	2	46.7	41.7	49.0	43.9	62.1	52.4	73.9	59.8	82.2	66.2	86.5	71.0	89.2	73.5	86.4	71.7	82.6	71.0	73.8	61.8	63.1	56.1	53.6	49.0	83.1	69.5
Syracuse	0.4	55.1	49.9	57.2	48.6	72.2	57.0	81.4	62.8	86.8	69.5	90.0	72.7	92.0	74.7	90.5	74.2	87.2	72.9	78.8	63.0	69.2	58.3	60.3	53.9	88.1	72.4
725190	1	49.6	43.4	52.5	46.4	67.0	53.5	77.4	61.6	84.5	67.3	87.9	71.3	90.3	74.0	88.3	72.9	84.7	72.6	75.6	63.1	66.0	58.0	57.2	52.4	85.4	70.7
	2	45.9	41.4	48.9	43.4	62.2	51.8	73.7	58.8	81.8	66.0	86.0	70.2	88.5	73.2	86.2	71.8	82.3	71.1	72.8	61.3	63.0	56.2	53.6	47.7	82.9	69.6
NORTH CAROLINA																											
Asheville	0.4	64.6	53.1	69.7	54.1	76.3	58.8	82.7	61.6	84.6	67.3	89.5	71.6	91.7	73.2	90.0	73.3	86.4	71.3	79.5	64.1	73.2	58.9	67.8	58.5	87.8	72.0
723150	1	61.6	52.2	66.1	52.4	73.9	56.5	80.5	60.6	82.5	66.3	87.4	70.8	90.2	72.9	88.1	72.5	84.4	70.6	77.5	63.4	71.1	57.9	64.9	57.9	85.3	71.0
	2	59.2	50.9	63.2	52.2	71.4	55.8	78.5	59.8	81.0	65.6	85.7	70.1	88.4	72.4	86.5	72.2	82.9	69.9	75.6	62.1	68.8	57.8	62.4	55.7	83.2	70.3
Cape Hatteras	0.4	69.4	64.6	70.6	63.9	73.7	64.9	78.4	67.5	83.7	73.6	87.4	77.1	90.3	79.8	89.4	78.9	87.3	77.1	82.9	73.2	77.1	69.8	72.3	67.4	87.5	78.1
723040	1	67.8	63.5	68.3	62.7	72.0	64.4	76.5	66.5	81.9	71.7	85.9	76.6	88.9	79.2	88.3	78.3	86.1	76.6	81.0	72.4	75.6	68.4	70.9	66.5	86.2	77.4
	2	66.3	62.7	66.4	61.4	70.3	63.1	74.9	65.6	80.3	71.5	84.8	76.0	87.8	78.7	87.3	78.0	85.1	76.0	79.4	71.8	74.0	67.7	69.5	65.2	84.8	76.6
Charlotte	0.4	68.2	57.1	73.2	59.1	80.5	61.0	86.4	64.2	90.2	69.6	94.2	73.6	97.9	74.3	95.7	74.2	91.8	73.0	84.2	67.8	77.8	64.1	70.9	63.8	93.5	74.2
723140	1	65.3	56.2	69.9	56.5	77.7	60.4	84.5	63.8	88.3	70.0	92.5	73.4	96.1	74.4	94.0	74.4	90.3	73.0	82.4	66.6	75.3	63.0	68.7	62.2	91.1	73.8
	2	63.1	55.8	67.3	55.5	75.1	59.4	82.5	62.9	86.4	69.2	90.9	72.8	94.4	74.3	92.5	74.4	88.8	72.9	80.5	66.6	73.1	62.0	66.2	59.0	88.9	73.1
Greensboro	0.4	66.7	57.0	71.8	56.4	80.3	62.2	85.8	63.7	88.9	69.7	93.5	74.6	96.1	75.9	94.7	75.3	91.1	72.9	83.1	67.8	76.3	63.6	70.0	63.0	92.2	74.8
723170	1	64.2	56.5	68.0	55.7	76.9	60.9	83.9	62.8	87.1	69.4	91.7	74.4	94.6	75.1	92.7	75.3	89.3	72.6	81.1	66.3	73.8	61.6	67.6	61.2	89.9	73.9
	2	61.6	54.0	65.4	54.3	74.0	58.7	81.6	62.4	85.4	68.8	90.1	73.3	92.8	75.3	91.1	75.2	87.5	72.2	79.2	65.0	71.7	60.3	64.8	58.4	87.6	73.0
Raleigh/Durham	0.4	69.5	60.0	74.0	58.9	81.4	63.3	87.3	64.1	89.7	70.9	94.1	75.4	96.6	75.7	94.8	76.4	91.6	74.4	84.4	68.5	78.2	64.2	72.4	63.7	92.8	75.9
723060	1	66.9	58.9	70.7	58.5	78.6	61.9	85.0	64.4	87.9	70.2	92.2	74.8	94.9	76.5	93.0	76.3	89.7	74.0	82.1	68.0	75.6	63.4	70.0	62.4	90.3	74.7
	2	64.4	57.6	67.7	57.1	76.0	60.3	82.8	63.4	86.2	69.6	90.5	74.1	93.4	76.3	91.4	75.8	88.0	73.2	80.1	66.5	73.6	62.6	67.6	61.3	88.2	73.8
Wilmington	0.4	73.4	64.3	76.1	65.0	81.7	65.4	87.5	68.0	90.9	72.3	94.8	77.8	96.6	78.8	95.0	79.3	92.0	77.6	85.7	72.7	80.1	70.0	75.8	66.8	93.1	78.5
723013	1	70.9	63.0	73.8	64.0	78.9	65.4	85.2	67.1	88.7	72.2	92.7	77.1	95.0	79.3	93.4	79.4	90.1	76.8	83.7	71.8	78.2	69.6	73.7	65.7	90.8	77.6
	2	68.7	62.7	71.5	61.7	76.4	64.3	82.8	67.1	86.8	71.2	91.0	76.9	93.4	79.1	92.1	78.9	88.5	76.2	82.1	71.0	76.6	67.3	71.7	65.0	88.7	76.7
NORTH DAKOTA																											
Bismarck	0.4	45.9	38.1	53.1	43.1	67.3	49.8	81.7	56.4	88.2	64.3	95.2	68.2	98.7	69.5	98.1	68.5	92.5	65.6	80.4	57.1	64.0	49.1	48.9	38.5	93.4	67.9
727640	1	43.0	36.7	49.0	40.4	61.5	47.1	77.8	55.0	85.1	62.8	92.3	67.8	95.7	69.9	95.7	67.4	88.8	63.7	76.6	56.6	59.1	46.0	44.4	37.3	89.8	67.0
	2	40.8	35.6	45.0	38.3	57.0	44.0	73.7	53.4	82.1	61.8	89.0	66.8	92.9	68.7	93.3	67.5	85.5	62.6	73.1	54.8	55.1	44.1	41.5	35.8	86.2	65.9
Fargo	0.4	39.4	35.3	46.0	40.0	61.5	49.8	82.1	59.2	88.7	66.0	92.5	71.7	95.5	73.0	96.0	72.6	90.0	68.2	79.0	60.0	60.3	49.3	44.4	39.5	91.2	71.2
727530	1	37.3	34.0	42.3	37.7	57.1	48.1	77.6	57.3	86.3	64.7	90.1	70.1	92.8	73.3	93.5	71.8	86.8	68.1	74.9	58.9	56.8	47.9	39.5	35.3	87.8	70.2
	2	35.3	32.8	39.1	35.6	52.8	44.6	73.0	55.6	83.6	63.8	87.6	69.7	90.4	71.5	91.1	71.7	83.8	66.7	71.4	57.5	53.4	46.0	36.5	33.2	84.8	68.6
Minot,	0.4	43.5	37.6	50.0	41.6	64.1	46.9	80.8	56.0	88.5	63.2	94.0	65.7	96.1	70.0	96.4	66.8	90.7	63.9	78.1	57.0	60.7	47.5	46.3	38.3	91.9	67.2
Intl Airport	1	41.9	36.5	46.1	38.7	58.8	45.8	76.8	54.8	85.0	62.2	90.4	65.6	93.5	68.2	94.2	66.8	87.0	63.2	74.5	55.7	56.7	45.4	43.1	37.0	87.9	65.8
727676	2	39.9	35.2	43.1	36.9	54.4	43.3	72.3	53.0	81.8	61.0	87.5	64.9	91.0	67.8	91.9	67.5	83.5	61.6	71.1	54.1	53.5	43.0	40.5	35.2	84.3	64.7
OHIO																											
Akron/Canton	0.4	56.6	52.3	58.8	50.4	73.1	57.3	80.3	61.3	85.9	66.9	89.8	70.7	92.0	72.4	90.6	74.0	86.7	71.8	77.9	62.7	69.5	59.5	61.9	56.1	88.0	71.9
725210	1	53.8	50.0	56.1	50.5	69.7	56.0	77.8	61.0	83.6	67.3	87.8	71.0	89.8	73.2	88.2	72.8	84.7	71.5	75.9	62.6	66.8	57.4	59.4	54.4	85.3	70.8
	2	50.7	46.8	53.6	48.3	66.2	54.3	75.1	60.7	81.7	66.3	86.1	69.8	88.3	72.8	86.2	71.8	82.7	70.4	73.8	61.6	64.5	56.7	57.1	52.7	82.9	69.6
Cleveland	0.4	58.5	52.9	59.3	51.3	74.4	59.0	81.2	64.2	86.6	69.6	90.4	72.0	92.6	74.8	90.5	75.4	88.1	72.9	79.5	64.1	70.7	60.0	63.5	57.0	88.6	73.3
725240	1	54.6	49.9	56.7	51.0	70.6	57.3	78.7	63.2	84.5	68.8	88.4	71.8	90.8	74.8	88.6	74.2	85.8	72.5	76.8	64.3	68.0	58.8	60.2	54.2	86.0	71.8
	2	51.3	46.4	53.6	48.6	67.2	56.0	75.5	61.5	82.7	67.6	86.9	71.5	89.2	74.2	86.6	72.9	83.8	71.6	74.5	62.2	65.1	57.5	57.4	53.0	83.5	70.5
Columbus,	0.4	59.4	54.6	61.9	52.6	75.8	59.0	82.3	63.3	88.3	69.8	91.7	71.6	94.2	75.3	92.6	74.8	89.8	73.0	80.8	65.5	72.0	60.9	63.3	57.2	90.3	73.6
Intl Airport	1	56.2	52.2	59.1	52.7	72.9	59.5	80.3	62.9	86.4	69.2	90.1	72.0	92.1	75.0	90.8	75.1	87.7	72.5	78.9	64.4	69.4	58.9	61.1	55.9	87.9	72.8
724280	2	52.8	48.2	56.4	50.7	69.3	56.6	77.4	62.4	84.3	68.3	88.6	72.0	90.6	74.4	89.0	74.1	85.8	71.7	76.7	62.6	66.5	58.3	58.6	54.2	85.5	71.4
Dayton,	0.4	59.1	55.0	61.9	52.6	74.6	59.9	81.6	63.1	87.5	69.7	92.1	70.3	95.2	74.6	93.1	76.3	89.4	73.0	80.6	64.7	72.0	61.3	62.6	56.2	90.3	73.7
Intl Airport	1	55.9	52.9	58.7	52.8	71.7	57.9	79.3	62.6	85.9	68.8	90.2	72.1	92.7	74.9	90.8	74.7	87.3	71.9	78.4	64.4	69.3	60.4	60.5	55.9	87.9	72.6
724290	2	52.4	48.2	55.5	49.8	68.6	56.4	76.4	62.3	84.0	68.7	88.6	71.6	90.7	74.7	89.0	74.1	85.4	71.2	76.1	62.5	66.6	58.3	58.2	54.1	85.6	71.2
Mansfield	0.4	58.3	54.2	59.9	49.6	73.2	57.4	80.5	62.5	84.0	68.3	89.8	72.8	92.1	74.9	89.5	74.7	86.7	72.7	78.3	64.8	69.9	60.5	62.3	57.2	87.8	73.0
725246	1	55.6	51.8	56.4	49.3	70.2	57.1	78.0	61.3	82.2	66.6	87.7	71.1	90.2	74.3	87.7	73.8	84.8	71.6	75.8	64.0	68.0	58.5	59.7	55.8	85.3	71.9
	2	52.2	47.8	53.9	48.7	66.9	55.1	75.3	60.4	80.4	66.1	85.7	71.0	88.4	73.9	86.2	72.7	82.9	70.7	73.8	62.6	65.3	57.8	57.2	53.0	83.0	70.6
Toledo	0.4	55.1	52.0	56.9	49.9	73.2	58.4	82.8	63.4	89.3	69.5	92.8	71.4	94.5	74.7	92.7	77.1	89.6	72.3	81.1	64.4	69.9	58.7	61.0	57.4	90.3	73.3
725360	1	51.6	48.0	53.3	48.1	69.6	56.7	79.7	62.8	86.7	69.3	90.7	71.1	92.4	74.8	90.3	75.3	87.0	71.5	78.3	63.9	66.8	58.4	58.2	54.3	87.4	72.0
	2	47.7	44.2	50.1	44.9	65.5	54.8	75.5	60.9	84.3	67.1	88.8	71.2	90.3	74.8	88.2	73.9	84.6	71.2	75.4	62.1	63.8	57.0	54.9	52.3	84.8	70.7
Youngstown	0.4	55.7	51.4	59.3	50.2	73.4	57.7	81.5	62.1	85.4	67.9	89.1	70.4	91.7	72.9	90.3	73.7	87.0	70.6	78.1	63.5	70.2	59.2	62.3	56.4	87.5	71.9
725250	1	52.6	48.9	56.0	50.0	69.4	55.5	78.6	60.9	83.6	67.0	87.2	70.6	89.6	73.4	87.7	72.8	84.6	70.9	76.0	61.9	67.1	56.9	59.5	53.6	85.0	70.4
	2	49.5	45.6	52.7	47.5	65.9	54.6	75.4	59.9	81.8	65.8	85.6	70.2	88.0	73.0	85.8	71.0	82.6	70.2	73.5	60.9	64.3	56.4	56.7	52.7	82.6	69.0

WMO# = World Meteorological Organization number DB = dry-bulb temperature, °F MWB = mean coincident wet-bulb temperature, °F

Table 4A Selected Monthly Percentiles of Wet-Bulb Temperature with Mean Dry-Bulb Temperature for U.S. Locations

Location and WMO#	%	Jan		Feb		Mar		Apr		May		Jun		Jul		Aug		Sep		Oct		Nov		Dec		Annual	
		WB	MDB	WB	MDB	WB	MDB	WB	MDB	WB	MDB	WB	MDB	WB	MDB	WB	MDB	WB	MDB	WB	MDB	WB	MDB	WB	MDB	WB	MDB
1a	1b	2a	2b	3a	3b	4a	4b	5a	5b	6a	6b	7a	7b	8a	8b	9a	9b	10a	10b	11a	11b	12a	12c	13a	13b	14a	14b
OKLAHOMA																											
Oklahoma City,	0.4	58.8	62.1	60.3	68.6	65.8	73.7	70.3	81.1	75.3	87.0	78.4	91.1	78.1	91.9	77.6	92.6	76.6	89.8	72.9	82.6	66.4	71.8	61.7	65.3	76.9	91.0
Will Rogers Airport	1	55.6	60.0	58.4	66.0	64.3	73.2	69.4	79.1	74.2	85.1	77.4	90.5	77.4	91.9	77.0	91.6	75.7	88.8	71.4	79.2	64.5	70.3	60.0	64.1	75.9	90.3
723530	2	52.6	58.5	56.7	65.8	62.9	71.7	68.2	77.2	73.1	83.1	76.7	89.5	76.8	91.7	76.3	91.2	75.0	87.8	69.6	77.9	63.0	69.4	57.9	62.1	75.1	89.3
Tulsa	0.4	59.0	63.4	61.2	68.2	67.4	76.0	72.0	81.6	76.6	87.4	80.5	92.6	81.0	94.8	79.9	93.8	78.7	91.7	73.6	82.7	67.6	71.9	63.5	67.9	79.2	92.3
723560	1	57.3	62.3	59.7	66.9	65.6	74.6	70.6	80.5	75.4	85.9	79.5	90.5	80.2	94.0	79.2	93.6	77.8	90.4	72.3	81.2	66.1	71.9	61.3	65.7	78.1	91.6
	2	54.5	58.5	57.9	66.7	64.0	71.8	69.4	79.1	74.4	84.0	78.6	89.5	79.4	93.0	78.4	93.2	76.9	89.6	70.9	79.1	64.6	71.8	59.2	63.5	77.1	90.4
OREGON																											
Astoria	0.4	53.3	55.1	55.6	58.7	54.6	59.6	57.2	65.1	63.2	74.7	65.5	75.7	66.8	78.0	68.8	80.3	66.6	79.0	62.5	72.4	57.1	59.9	55.4	57.6	65.0	74.6
727910	1	52.3	53.9	54.1	57.7	53.2	57.2	55.5	61.9	61.3	70.7	63.4	72.0	65.2	74.8	66.6	76.0	65.2	76.1	60.7	67.8	55.9	58.1	53.8	55.9	63.0	70.7
	2	51.5	53.2	52.6	56.1	52.2	55.7	54.3	59.8	59.5	67.0	61.9	69.2	63.8	72.2	64.7	73.0	63.8	72.8	59.5	64.7	55.1	57.4	52.8	54.7	61.5	67.9
Eugene	0.4	55.6	58.5	56.5	60.0	57.9	64.4	62.3	73.2	68.7	81.2	69.9	88.1	70.8	90.2	70.5	92.9	67.8	84.9	62.9	76.3	58.9	61.6	56.2	58.0	68.5	87.0
726930	1	54.1	56.1	55.2	58.3	56.3	62.7	60.3	70.8	66.4	79.4	68.3	84.6	69.4	89.5	69.3	90.4	66.4	83.7	61.7	73.4	57.3	60.5	54.3	56.4	66.8	84.4
	2	52.8	55.2	53.7	56.9	54.8	60.8	58.7	68.3	64.6	77.1	67.0	82.7	68.2	88.0	68.1	87.5	65.4	82.1	60.4	70.9	56.0	58.8	52.8	54.8	65.2	81.4
Medford	0.4	51.4	55.9	54.7	63.0	57.4	70.7	62.0	80.1	67.2	88.4	69.4	94.1	70.9	96.8	70.2	97.8	67.5	91.6	62.5	84.1	55.8	62.6	52.6	57.4	68.6	94.1
725970	1	49.9	55.0	52.9	62.0	55.4	66.8	60.0	76.6	65.1	84.2	68.3	92.0	69.7	95.2	69.3	96.2	66.3	90.6	61.3	81.6	54.2	59.6	50.9	54.0	67.1	91.3
	2	48.8	53.4	51.4	59.0	54.0	63.6	58.2	73.5	63.3	82.4	67.1	90.0	68.8	94.7	68.2	93.6	65.1	88.3	60.1	78.9	53.0	58.3	48.9	52.8	65.5	88.3
North Bend	0.4	57.0	59.9	58.0	62.4	56.3	61.0	57.4	64.1	61.0	69.6	62.2	68.9	62.7	68.8	64.2	69.3	63.9	71.8	62.0	70.9	59.5	62.5	57.5	60.0	62.3	68.6
726917	1	55.8	58.8	56.4	60.7	55.0	59.1	56.0	62.6	59.3	64.8	61.0	67.7	61.9	68.0	63.1	68.2	62.6	69.9	60.7	67.6	58.4	61.8	56.3	59.0	61.1	67.2
	2	54.5	57.2	55.1	58.4	54.0	58.3	54.9	60.2	58.1	62.9	59.9	65.9	61.1	67.0	62.3	67.4	61.6	68.3	59.7	65.7	57.4	60.5	55.2	57.9	60.0	65.8
Pendleton	0.4	51.4	58.2	52.7	59.6	54.2	64.4	59.1	73.9	65.4	84.8	67.0	90.1	69.4	92.2	67.0	94.0	64.1	87.8	59.4	77.8	53.8	61.8	51.4	57.9	65.9	91.9
726880	1	49.0	55.3	50.8	57.4	52.6	62.6	57.3	73.0	63.5	81.6	65.7	88.0	67.6	94.7	66.0	93.2	62.9	85.9	57.9	74.4	51.9	59.8	49.4	55.6	64.3	89.6
	2	47.3	53.5	49.1	55.7	51.2	60.2	56.0	70.9	61.6	80.2	64.4	88.0	66.3	93.9	65.2	92.3	61.9	84.1	56.7	72.6	50.3	57.0	47.9	54.3	62.8	86.7
Portland	0.4	54.1	57.1	55.8	59.2	56.3	65.0	61.4	72.4	67.2	84.3	69.2	87.8	70.9	91.2	71.2	92.3	68.3	85.5	63.1	76.7	57.8	61.9	54.9	57.3	68.8	87.4
726980	1	52.4	55.3	54.2	58.5	55.0	63.3	59.4	70.0	64.9	79.4	67.7	85.4	69.6	89.3	69.9	90.0	67.1	83.8	61.6	72.2	56.1	59.5	53.3	56.2	67.1	83.9
	2	51.2	54.1	52.6	57.1	53.6	60.9	57.9	67.8	63.2	76.9	66.4	83.0	68.5	87.9	68.8	87.0	65.8	80.5	60.5	69.5	54.7	58.2	51.7	54.5	65.4	80.4
Redmond	0.4	47.8	55.2	49.7	58.9	51.0	65.2	55.8	73.1	61.6	83.0	64.6	86.6	65.5	88.1	65.3	88.4	61.8	86.2	57.5	80.0	51.7	61.4	48.5	54.9	63.4	88.2
726835	1	45.6	52.5	47.9	56.2	49.1	63.5	54.1	73.0	59.7	80.2	63.3	86.7	64.4	89.1	64.3	88.9	60.7	84.7	56.1	76.9	50.0	60.6	46.4	53.4	61.8	86.0
	2	44.1	50.5	46.5	54.8	47.6	60.0	52.5	69.4	57.9	76.8	61.8	84.6	63.4	88.6	63.3	88.4	59.7	82.7	54.9	73.9	48.3	57.0	44.6	52.0	60.3	82.9
Salem	0.4	54.5	57.4	56.3	60.0	56.9	63.4	61.4	71.3	67.1	83.8	69.9	88.1	71.0	92.0	70.5	93.0	67.4	85.4	63.2	76.1	58.0	61.2	55.3	57.7	68.4	88.5
726940	1	53.3	55.9	54.8	58.2	55.4	62.2	59.2	70.3	65.2	80.9	68.1	86.1	69.6	90.6	69.2	91.5	66.3	84.3	61.5	73.8	56.5	59.7	53.4	56.0	66.7	84.9
	2	52.0	54.6	53.5	57.0	54.1	60.8	57.7	68.2	63.5	77.8	66.7	83.6	68.4	88.9	68.1	88.9	65.3	82.3	60.1	70.5	55.1	58.2	52.3	55.0	65.0	81.4
PENNSYLVANIA																											
Allentown	0.4	53.9	56.8	54.3	57.0	61.9	70.8	65.4	79.7	72.0	83.0	75.4	87.3	77.8	89.8	77.4	86.6	75.5	85.7	68.4	73.4	64.5	69.0	58.5	61.7	75.5	85.9
725170	1	49.1	52.5	51.3	55.5	59.0	65.9	63.6	75.4	70.7	81.8	74.2	85.4	76.8	89.1	76.2	85.7	74.5	84.2	67.3	72.7	62.6	66.3	54.4	56.5	74.2	83.8
	2	44.6	47.6	47.4	52.0	56.0	62.7	62.0	73.4	69.2	80.3	73.2	83.9	75.8	87.2	75.2	84.3	73.4	82.1	65.9	72.0	60.5	64.2	50.1	52.9	72.8	81.7
Bradford	0.4	48.8	50.4	48.7	50.7	57.6	64.6	61.5	72.8	68.8	76.3	71.3	80.1	74.0	82.1	73.7	81.3	72.0	79.1	64.1	70.4	59.3	62.3	54.1	55.8	71.6	79.4
725266	1	44.6	45.9	46.0	48.8	55.2	61.2	59.9	70.4	66.9	75.6	70.3	78.9	72.9	81.6	72.2	79.8	70.3	77.1	62.5	67.8	57.7	61.6	51.4	53.1	70.0	77.2
	2	41.2	43.6	43.5	46.2	52.7	59.3	58.3	68.5	65.4	74.7	69.4	77.8	72.0	80.1	71.1	78.0	69.0	74.4	61.3	66.1	55.7	58.5	48.7	50.2	68.4	74.5
Erie	0.4	52.4	55.6	52.2	55.7	59.9	70.2	64.5	74.2	72.1	79.9	74.2	83.7	77.1	85.8	76.1	84.3	74.3	82.1	66.2	74.3	61.2	66.3	57.6	62.5	74.3	81.8
725260	1	49.0	52.6	49.7	53.4	57.2	64.7	62.7	72.1	70.0	78.7	73.0	82.0	75.8	83.0	74.8	82.4	73.0	80.9	64.6	71.7	59.5	64.2	55.3	58.6	72.8	80.1
	2	44.7	48.6	46.9	50.8	54.9	61.9	61.1	69.7	68.2	76.7	71.9	80.3	74.7	82.0	73.9	80.9	71.8	78.4	63.3	70.0	57.8	62.0	52.2	56.2	71.3	78.0
Harrisburg	0.4	53.6	57.6	53.7	57.7	62.5	72.0	66.7	80.8	74.4	84.4	76.9	87.0	79.6	89.8	79.2	88.4	77.2	86.2	69.4	75.3	65.4	69.8	58.8	62.2	77.2	87.3
725115	1	48.9	52.5	50.6	55.2	59.6	67.6	65.1	77.4	72.6	81.4	75.5	86.0	78.5	88.8	78.0	87.9	76.0	84.4	68.2	73.3	63.4	67.4	54.6	58.8	75.5	84.9
	2	44.7	48.2	48.1	53.8	57.0	63.8	63.3	74.9	70.9	79.4	74.6	85.1	77.4	88.0	76.9	86.5	74.7	83.2	66.8	71.8	60.9	65.6	50.5	53.4	74.2	82.7
Philadelphia,	0.4	56.3	58.6	57.3	61.3	63.3	71.8	67.1	81.8	74.0	85.1	77.5	88.4	79.9	91.9	79.4	89.9	77.7	87.4	71.2	76.3	66.7	71.6	60.6	62.5	77.8	87.9
Intl Airport	1	53.9	56.7	55.0	58.5	60.9	69.1	65.3	78.1	72.4	82.8	76.5	87.4	78.9	90.0	78.4	88.3	76.7	85.8	69.9	74.6	64.5	68.2	58.5	60.8	76.5	85.5
724080	2	50.3	53.2	52.2	56.2	58.3	64.8	63.5	74.1	71.0	81.9	75.5	86.3	78.1	88.6	77.5	86.8	75.6	83.8	68.4	73.8	62.6	66.3	55.5	58.0	75.1	83.7
Pittsburgh,	0.4	54.1	58.4	54.7	58.7	60.6	69.3	64.8	75.9	71.4	81.3	74.4	85.4	76.8	87.3	76.3	87.5	74.3	84.0	66.7	74.0	61.4	68.0	57.5	62.6	74.4	84.5
Intl Airport	1	51.1	54.3	51.9	55.5	58.6	66.9	63.3	74.6	70.1	80.2	73.4	84.2	75.7	86.4	75.1	85.3	73.0	81.9	65.2	72.5	60.0	65.6	56.1	60.6	72.9	82.3
725200	2	47.9	51.2	49.5	55.1	56.7	66.1	61.8	73.4	68.4	79.2	72.3	83.1	74.7	85.0	74.0	83.4	71.8	80.8	63.7	71.1	58.4	64.0	53.8	57.5	71.4	80.3
Wilkes-Barre/	0.4	52.2	55.3	53.1	55.9	59.8	69.4	63.9	77.9	71.0	81.3	74.3	84.0	76.4	86.9	75.7	85.6	74.2	83.1	66.9	72.8	62.4	66.5	56.6	59.3	74.2	83.4
Scranton	1	48.6	51.0	50.3	53.9	57.2	64.7	62.2	74.0	69.3	79.2	73.1	82.9	75.4	85.8	74.7	83.8	73.2	81.1	65.4	70.4	60.5	64.2	53.2	56.0	72.8	81.3
725130	2	43.8	47.1	47.2	50.6	54.5	62.0	60.4	71.2	67.8	77.7	72.1	81.7	74.5	84.1	73.9	82.7	72.0	79.7	64.0	69.7	58.4	62.9	49.6	52.1	71.4	79.2
Williamsport	0.4	49.9	53.3	52.2	55.3	60.4	69.8	65.1	78.5	72.1	82.0	75.3	84.6	77.8	88.5	77.2	87.4	75.1	84.8	67.7	72.3	62.7	66.4	54.7	57.6	75.5	84.8
725140	1	45.3	47.9	48.8	54.2	57.4	64.6	63.7	74.8	70.3	80.5	74.2	83.8	76.9	87.3	76.2	85.9	73.9	82.4	66.5	71.6	60.7	64.5	50.7	52.4	74.0	82.5
	2	42.0	45.2	45.3	49.1	54.3	61.8	62.0	71.4	68.9	79.4	73.2	82.3	75.9	85.6	75.2	84.0	72.8	80.3	65.0	70.0	58.4	61.8	46.9	49.8	72.7	80.2
RHODE ISLAND																											
Providence	0.4	54.8	56.0	53.4	56.0	57.4	62.9	63.2	77.3	70.8	83.1	75.4	87.3	78.1	89.8	78.1	88.2	75.7	85.5	68.9	72.6	64.0	68.5	58.1	60.2	75.7	85.1
725070	1	51.2	54.4	50.5	53.8	54.8	60.6	60.6	71.3	68.8	80.4	73.7	84.2	76.9	88.1	76.6	84.9	74.4	81.7	67.1	71.7	61.8	65.4	55.8	58.0	74.1	81.5
	2	47.5	49.9	47.8	51.6	52.5	58.0	58.7	67.4	66.9	76.9	72.4	82.7	75.8	86.6	75.6	83.3	73.3	78.9	65.7	70.9	59.9	63.9	53.2	56.0	72.7	79.5
SOUTH CAROLINA																											
Charleston	0.4	66.9	72.3	68.2	75.4	69.4	77.8	72.8	82.6	76.8	84.9	80.3	90.3	82.2	92.7	81.0	91.2	79.4	86.7	76.3	81.7	72.4	77.6	69.8	74.6	80.0	89.5
722080	1	65.7	70.0	66.6	72.7	68.4	75.8	71.3	80.4	75.8	84.7	79.4	89.1	81.2	90.9	80.3	90.4	78.6	86.2	75.4	81.0	71.4	75.8	68.4	72.6	79.0	88.2
	2	64.4	68.7	65.2	70.0	67.3	74.5	70.2	78.7	74.8	83.7	78.5	87.8	80.4	89.6	79.6	89.5	78.0	86.0	74.4	80.0	70.5	74.7	67.0	70.8	78.1	86.9
Columbia	0.4	66.5	71.7	67.0	73.2	68.4	78.2	71.9	82.6	75.8	85.8	78.3	91.2	79.8	91.3	79.4	91.5	78.1	89.1	74.4	82.1	70.8	76.3	68.6	73.1	78.3	90.1
723100	1	64.7	68.1	65.1	70.6	67.4	76.5	70.5	80.6	74.5	84.5	77.4	89.6	79.0	91.0	78.6	90.8	77.4	87.6	73.4	80.4	69.5	74.6	67.1	71.9	77.4	88.6
	2	62.9	66.9	63.5	68.2	66.3	75.5	69.3	78.9	73.4	83.4	76.7	88.1	78.4	90.1	78.2	90.0	76.7	86.4	72.4	79.0	68.4	73.6	65.3	69.8	76.5	87.1
Greer/Greenville	0.4	61.0	65.3	62.6	66.6	65.1	74.5	69.5	79.5	74.3	83.7	76.7	88.8	78.4	90.5	77.7	88.6	75.9	85.9	71.6	79.2	67.7	72.0	65.2	67.5	76.6	88.0
723120	1	59.2	63.4	60.4	64.3	64.0	71.6	67.7	77.9	73.1	82.6	76.0	88.4	77.6	89.4	77.0	88.1	75.2	84.9	70.2	77.2	66.3	70.6	62.9	65.5	75.6	86.6
	2	57.3	61.2	58.6	63.1	62.6	68.8	66.4	76.3	72.0	81.9	75.3	86.8	77.0	88.6	76.3	87.1	74.5	83.7	69.1	75.6	64.9	68.6	61.2	64.2	74.8	85.0
SOUTH DAKOTA																											
Huron	0.4	40.6	48.0	45.0	55.3	56.2	70.8	63.7	79.3	71.0	82.8	76.9	90.5	78.8	90.6	77.5	90.0	73.6	87.6	64.4	74.6	53.9	60.3	42.9	50.3	76.1	89.3
726540	1	38.5	44.1	42.7	50.4	52.8	63.3	62.1	74.8	69.7	81.4	75.4	87.9	77.5	90.4	76.4	88.8	72.2	85.3	62.5	71.3	51.7	57.6	40.2	46.7	74.2	86.6
	2	36.7	41.3	40.6	46.4	50.0	59.7	60.2	73.5	68.0	79.2	74.1	86.0	76.7	89.8	75.3	87.3	70.8	82.9	60.7	70.5	49.1	57.1	38.0	43.5	72.4	84.2
Pierre	0.4	42.3	52.2	46.2	59.8	53.7	69.2	61.5	78.7	69.2	84.0	75.1	88.8	77.1	94.2	74.6	92.1	71.7	89.1	61.5	77.2	52.4	61.1	43.3	53.0	74.0	90.3
726686	1	40.1	48.0	44.1	55.6	51.4	66.5	59.9	77.0	67.5	81.5	73.8	88.4	75.8	91.8	73.8	90.5	70.0	85.5	60.0	75.1	49.9	60.9	41.1	49.8	72.3	88.6
	2	38.2	44.6	42.1	51.8	49.0	61.6	58.1	73.9	66.1	79.7	72.5	87.2	74.7	91.0	72.9	90.0	68.4	84.1	58.5	73.2	47.7	59.0	39.2	46.6	70.8	86.2
Rapid City	0.4	44.2	57.6	46.3	61.7	51.1	70.4	57.4	77.3	65.0	79.7	71.2	85.2	72.1	87.2	71.2	86.0	66.1	85.0	57.3	75.8	49.9	66.1	45.3	59.9	69.8	85.3
726620	1	42.2	54.9	44.3	58.2	49.1	67.0	55.7	74.6	63.4	78.1	69.8	83.5	71.0	87.4	70.0	85.5	64.7	82.9	56.0	76.7	48.4	64.8	42.9	56.2	68.1	83.8
	2	40.4	51.3	42.7	55.5	47.4	63.2	54.3	73.0	62.1	76.6	68.5	82.5	70.2	86.3	68.9	85.4	63.3	81.3	54.8	75.1	46.5	61.0	41.1	52.9	66.5	81.9

WMO# = World Meteorological Organization number WB = wet-bulb temperature, °F MDB = mean coincident dry-bulb temperature, °F

Table 4B Selected Monthly Percentiles of Dry-Bulb Temperature with Mean Wet-Bulb Temperature for U.S. Locations

Location and WMO#	%	Jan		Feb		Mar		Apr		May		Jun		Jul		Aug		Sep		Oct		Nov		Dec		Annual	
		DB	MWB	DB	MWB	DB	MWB	DB	MWB	DB	MWB	DB	MWB	DB	MWB	DB	MWB	DB	MWB	DB	MWB	DB	MWB	DB	MWB	DB	MWB
1a	1b	2a	2b	3a	3b	4a	4b	5a	5b	6a	6b	7a	7b	8a	8b	9a	9b	10a	10b	11a	11b	12a	12c	13a	13b	14a	14b
OKLAHOMA																											
Oklahoma City,	0.4	67.6	52.0	76.2	56.0	81.8	60.0	87.0	66.3	90.5	70.5	97.0	74.9	103.8	73.0	102.1	73.5	97.3	73.5	89.7	65.6	77.4	60.9	69.0	57.1	99.2	73.8
Will Rogers Airport	1	64.3	50.8	72.8	54.9	79.1	60.0	84.6	65.9	88.6	71.0	95.0	74.1	101.9	73.2	100.4	73.6	95.4	72.8	86.8	67.3	75.1	59.8	66.4	55.1	96.2	73.5
723530	2	61.9	51.0	69.6	53.2	76.5	59.2	82.4	64.0	86.8	70.3	93.3	74.3	100.0	73.7	99.0	73.6	93.3	72.5	84.3	66.5	72.7	60.0	64.2	55.3	93.5	73.3
Tulsa	0.4	68.3	53.7	75.5	56.4	82.4	61.6	87.8	66.8	89.7	74.3	96.8	76.2	104.1	75.9	102.9	75.8	97.7	74.9	90.5	67.5	79.0	61.4	70.2	59.0	99.6	76.1
723560	1	65.0	53.6	72.2	55.1	79.5	61.6	85.4	66.5	88.3	72.7	95.0	76.3	101.9	76.2	101.3	75.5	95.2	74.3	87.4	69.5	76.4	61.8	67.7	58.0	96.7	75.7
	2	62.2	52.2	69.0	54.2	76.8	60.8	83.3	65.5	86.9	71.1	93.5	75.9	100.1	76.6	99.8	75.8	93.3	74.9	85.1	68.4	74.0	61.9	65.1	57.3	94.0	75.3
OREGON																											
Astoria	0.4	57.8	50.7	62.6	51.4	64.2	51.6	68.7	54.8	75.5	62.9	77.0	64.6	79.8	66.7	81.6	67.4	82.5	64.7	74.2	61.1	62.2	54.4	58.4	54.6	76.4	63.6
727910	1	55.8	50.3	59.6	51.4	61.1	50.7	65.1	54.1	71.5	61.1	72.9	62.8	75.7	64.4	77.5	65.5	79.5	63.8	70.7	59.1	60.1	54.3	56.5	53.1	72.0	62.3
	2	54.4	50.4	57.5	51.5	58.5	50.2	61.9	53.2	67.8	59.2	69.7	61.4	72.8	63.5	73.8	64.5	75.5	62.4	67.7	58.0	58.5	54.0	55.1	52.1	68.9	60.8
Eugene	0.4	59.1	54.8	62.7	53.9	67.8	55.0	75.1	61.4	83.9	67.2	91.1	68.3	96.0	68.8	96.7	68.5	91.9	65.1	81.6	60.9	64.1	57.0	59.0	54.9	91.2	66.8
726930	1	57.1	53.3	60.5	53.2	65.3	54.5	72.0	59.7	81.0	65.1	87.4	66.7	93.3	67.6	93.7	67.0	89.0	64.1	77.6	59.7	61.5	56.3	57.2	53.9	87.1	65.4
	2	55.4	52.3	58.3	52.7	63.1	53.6	69.2	57.7	78.0	63.7	84.8	65.8	91.0	66.9	91.1	67.0	86.0	63.9	73.9	59.0	59.7	55.1	55.3	52.6	83.4	64.2
Medford	0.4	60.5	48.1	68.0	52.5	73.4	55.4	82.8	60.1	92.1	65.3	98.5	67.7	102.2	67.7	103.5	68.4	98.3	65.4	88.7	61.3	66.3	53.3	59.2	50.8	98.3	66.8
725970	1	57.9	47.6	64.4	50.6	70.6	54.2	79.6	58.5	88.5	64.0	95.6	66.7	100.4	67.9	100.6	67.3	95.6	64.1	84.7	59.9	63.4	52.1	56.4	48.5	94.8	65.8
	2	55.3	47.2	61.6	49.8	67.4	52.4	76.6	57.5	85.3	62.2	93.1	65.6	98.3	67.1	98.2	66.5	92.9	63.2	81.2	59.1	60.9	50.8	54.2	47.7	91.2	64.5
North Bend	0.4	61.8	55.6	65.5	55.8	64.4	53.3	68.0	56.4	71.1	60.4	70.9	61.1	71.3	61.1	72.3	61.5	78.2	60.3	74.9	60.4	65.4	56.8	62.0	56.2	71.2	60.4
726917	1	59.9	53.8	62.8	54.2	61.7	53.0	64.0	54.8	66.6	57.9	68.4	59.8	69.9	60.8	70.6	61.4	73.3	60.8	70.4	59.2	63.4	56.3	60.1	55.0	69.0	60.0
	2	58.2	53.7	60.4	54.1	59.8	52.6	61.6	54.1	64.3	56.8	66.8	58.9	68.7	60.1	69.7	61.0	70.8	60.3	67.6	58.1	61.8	56.2	58.6	54.4	67.2	59.0
Pendleton	0.4	59.4	49.7	62.4	50.5	67.7	52.5	78.6	57.4	89.3	63.0	96.5	64.5	102.0	67.4	101.0	65.3	92.0	62.5	80.7	57.7	64.7	50.8	60.2	49.4	96.6	64.4
726880	1	57.4	48.1	59.5	49.4	65.1	51.3	75.1	56.8	86.1	62.3	94.2	64.0	99.7	65.8	98.3	64.7	89.7	61.6	77.5	56.8	62.0	50.3	57.3	47.8	93.0	63.2
	2	55.1	46.6	57.3	48.0	62.4	49.7	71.9	54.9	83.0	60.4	91.7	62.9	97.4	64.8	96.1	64.2	87.3	60.8	74.9	55.7	59.3	49.1	55.1	47.1	89.6	61.9
Portland	0.4	57.9	53.0	62.8	52.9	68.5	54.0	76.4	59.7	86.5	64.6	91.3	68.0	96.4	68.6	97.3	69.4	91.2	64.7	80.0	60.9	63.3	54.6	58.4	54.6	90.3	67.1
726980	1	56.2	51.5	60.3	51.2	66.2	53.6	73.3	58.2	83.5	63.8	87.7	66.2	92.5	68.5	93.3	68.3	88.6	64.6	76.3	59.4	61.2	54.7	56.3	52.9	86.2	65.8
	2	54.6	50.4	58.4	51.3	63.9	51.9	70.4	56.5	80.1	62.5	84.9	65.5	89.5	67.8	90.1	67.0	85.5	63.8	73.1	58.7	59.5	53.4	54.6	51.6	82.6	64.4
Redmond	0.4	57.9	46.7	61.9	47.3	69.3	49.1	78.9	54.1	86.3	59.4	92.5	62.2	97.0	63.4	97.3	63.0	91.0	60.3	82.7	55.8	65.8	49.2	57.7	45.6	92.6	61.7
726835	1	54.5	44.5	59.5	45.9	65.7	47.4	75.4	53.0	83.3	57.8	90.1	61.8	94.9	62.4	94.8	62.2	88.4	59.1	80.0	55.5	62.2	48.1	55.1	45.1	89.2	60.5
	2	51.8	42.9	56.9	44.7	62.7	46.3	71.6	51.2	80.1	56.8	87.7	60.5	93.0	61.8	92.5	61.5	85.9	58.2	76.5	53.6	58.8	46.9	52.9	43.3	85.8	59.2
Salem	0.4	58.7	52.6	63.7	52.6	67.9	54.0	75.5	59.9	85.6	66.0	93.0	67.9	96.8	68.7	97.8	68.5	92.7	64.9	81.7	60.1	63.1	55.8	58.8	55.1	91.7	67.2
726940	1	56.8	52.9	60.5	52.6	65.6	53.6	72.4	57.9	82.6	64.0	89.1	66.7	93.9	68.2	94.5	67.4	89.5	64.0	77.4	58.9	60.8	55.5	56.7	53.0	87.4	65.5
	2	55.1	51.6	58.6	52.2	63.2	52.5	69.5	56.7	79.5	62.7	86.0	65.6	91.2	67.6	91.3	66.6	86.4	63.3	74.0	58.9	59.2	54.4	55.1	51.8	83.4	64.2
PENNSYLVANIA																											
Allentown	0.4	57.0	52.2	61.3	51.8	74.2	58.6	84.0	63.0	88.6	69.6	92.2	72.9	94.8	75.1	92.3	73.9	90.2	73.0	79.2	65.4	71.5	62.4	61.8	57.7	90.4	72.7
725170	1	52.9	48.3	56.5	49.8	69.8	56.5	79.6	62.1	86.0	67.4	90.2	71.8	92.9	73.9	90.2	72.9	87.6	72.5	76.5	64.5	68.2	60.2	57.7	53.4	87.7	71.8
	2	48.8	44.0	52.9	47.6	65.7	54.6	75.3	59.5	83.6	66.4	88.2	71.0	91.0	73.9	88.5	72.9	85.2	70.7	74.2	63.8	65.3	58.9	53.6	49.7	85.2	70.5
Bradford	0.4	50.9	48.7	53.2	46.2	68.0	55.6	78.0	59.0	82.0	64.9	84.0	68.4	87.2	70.8	85.2	70.2	80.9	70.5	74.4	60.9	65.1	56.6	56.3	53.3	82.8	68.7
725266	1	47.5	43.5	50.0	44.8	63.9	53.0	74.8	57.8	80.2	64.0	82.3	67.7	85.1	70.6	83.0	69.6	79.1	69.1	72.0	59.6	62.5	55.4	53.8	50.6	80.4	67.6
	2	44.2	40.4	47.4	42.5	60.3	51.6	71.5	56.8	78.1	62.4	81.0	67.4	83.4	69.8	81.2	68.9	77.1	67.3	69.8	58.3	60.1	54.2	50.9	47.7	78.0	65.9
Erie	0.4	56.7	51.6	57.7	51.2	72.7	57.1	77.7	61.8	83.3	67.9	87.5	70.6	88.9	73.0	87.6	73.9	85.7	72.1	78.2	63.7	69.5	59.0	63.1	57.1	85.3	71.6
725260	1	53.3	49.1	54.6	48.5	68.2	55.4	75.2	60.9	81.4	67.0	85.6	70.1	86.5	73.0	85.8	72.9	83.0	72.1	75.1	63.0	66.4	57.2	59.4	54.3	82.8	70.4
	2	49.6	44.4	51.4	46.2	64.6	53.3	72.7	59.8	79.3	66.3	83.9	69.5	85.1	72.2	83.9	71.5	80.7	69.9	72.5	61.0	63.7	56.2	56.3	51.4	80.4	69.5
Harrisburg	0.4	59.6	52.4	62.3	50.8	75.6	59.6	84.7	64.0	89.1	72.2	93.5	73.1	97.1	74.4	94.0	74.9	91.2	74.3	80.2	67.3	72.2	61.6	63.2	57.6	91.9	74.3
725115	1	53.7	48.2	57.5	49.2	71.4	55.8	81.5	63.6	86.7	68.3	91.3	72.7	94.6	74.9	91.9	75.7	88.8	73.0	77.5	65.3	69.5	61.6	59.1	53.3	89.0	73.1
	2	49.6	43.5	53.8	47.3	67.3	55.4	77.4	61.4	84.3	67.8	89.3	72.5	92.6	74.5	89.9	74.7	85.9	72.7	75.0	63.9	66.7	60.3	54.4	48.4	86.4	71.9
Philadelphia,	0.4	60.0	55.0	63.8	53.9	75.6	60.1	85.5	65.0	90.7	71.4	93.8	73.7	96.6	77.3	94.2	76.3	91.3	75.9	81.1	68.9	74.1	64.6	64.0	59.8	92.1	75.1
Intl Airport	1	56.8	52.8	60.4	54.0	71.6	58.4	80.8	63.5	87.4	69.7	91.9	74.0	94.3	76.0	92.2	75.8	88.9	73.9	78.5	66.9	70.4	62.6	61.3	57.6	89.4	74.0
724080	2	53.7	49.9	57.1	52.0	67.3	55.7	76.6	60.9	84.9	68.6	90.0	73.1	92.5	76.0	90.5	75.1	86.8	73.2	76.2	65.6	67.4	60.6	58.7	54.3	87.0	72.7
Pittsburgh,	0.4	59.0	52.7	61.8	51.4	75.3	57.7	81.9	62.0	86.4	67.7	90.5	71.3	93.0	73.6	91.7	74.3	87.8	72.5	78.6	63.0	71.2	59.4	63.6	56.4	88.6	71.8
Intl Airport	1	55.5	49.7	58.3	49.8	72.0	56.1	79.6	60.9	84.4	67.5	88.3	70.3	90.8	72.8	89.3	72.8	85.6	71.3	76.9	62.3	68.4	57.5	61.1	54.0	85.9	70.4
725200	2	52.3	47.2	55.4	48.4	68.5	54.6	76.8	60.1	82.4	65.7	86.6	70.0	89.0	72.7	87.1	71.7	83.5	69.8	74.8	61.1	65.8	55.7	58.2	53.1	83.6	69.0
Wilkes-Barre/	0.4	56.2	52.0	57.9	51.2	72.1	56.8	82.5	61.8	86.5	67.5	89.3	70.9	92.6	72.1	89.5	73.0	86.3	71.9	77.2	63.4	68.9	60.5	60.3	54.6	87.8	71.1
Scranton	1	51.8	47.9	54.4	49.8	67.5	54.3	78.5	60.1	84.2	65.8	87.5	70.5	90.4	72.6	87.7	72.0	84.2	70.8	74.8	62.6	65.9	59.0	56.3	52.7	85.1	70.2
725130	2	47.7	43.5	51.5	46.4	64.1	53.8	74.9	58.4	82.0	64.8	85.6	69.7	88.3	72.1	85.9	71.7	82.2	70.0	72.4	61.4	63.5	57.2	52.7	48.8	82.6	68.7
Williamsport	0.4	53.6	49.0	57.1	49.3	73.4	57.6	83.8	63.0	89.2	68.0	91.3	72.0	95.0	74.6	91.3	74.7	87.9	72.6	77.8	64.7	68.6	59.9	57.2	54.3	89.5	72.5
725140	1	49.2	44.3	54.1	48.6	69.1	54.8	79.5	61.6	86.1	67.3	89.0	72.0	92.3	73.3	89.3	73.6	85.5	71.9	75.2	62.9	65.9	59.0	53.4	49.7	86.5	71.2
	2	45.4	41.3	50.8	44.1	64.5	52.8	76.1	59.7	83.7	66.1	86.9	70.2	90.1	73.4	87.3	72.7	83.1	70.6	72.5	61.8	63.3	57.2	50.4	45.7	83.8	69.8
RHODE ISLAND																											
Providence	0.4	57.2	54.2	58.5	50.8	67.5	54.3	79.6	61.2	87.3	68.5	91.2	72.4	94.4	76.1	91.2	74.9	89.4	74.0	78.0	64.3	70.1	61.2	61.5	55.6	89.2	73.0
725070	1	54.3	50.8	54.8	50.5	63.2	52.8	74.3	58.9	83.7	67.1	88.8	71.4	92.0	74.6	89.2	73.9	85.7	72.1	75.4	64.4	67.1	60.6	59.1	54.8	86.0	71.3
	2	50.9	46.7	52.2	46.5	59.6	50.4	70.6	55.8	80.3	64.9	86.5	70.3	89.5	73.6	87.2	73.4	82.5	70.4	73.1	63.0	64.3	58.9	56.3	52.8	83.2	69.9
SOUTH CAROLINA																											
Charleston	0.4	75.1	65.0	78.0	64.9	83.2	65.2	88.3	68.1	92.0	72.3	95.8	77.0	97.2	78.6	95.2	78.9	92.3	77.0	87.1	72.5	81.3	69.6	76.9	67.4	93.7	78.1
722080	1	72.9	63.0	75.5	64.0	80.8	65.3	86.3	67.2	90.2	72.4	93.9	76.9	95.5	79.1	93.8	78.7	90.9	76.9	85.1	71.2	79.4	68.7	75.0	66.3	91.6	77.2
	2	70.3	62.5	73.1	61.7	78.6	64.7	84.2	67.4	88.2	72.2	92.1	76.5	94.1	78.9	92.6	78.4	89.4	76.3	83.4	70.9	78.0	67.3	73.0	65.0	89.6	76.5
Columbia	0.4	74.6	62.9	78.2	62.7	84.4	64.6	89.5	66.6	93.0	71.2	97.8	74.3	100.8	75.0	98.1	76.1	94.3	75.0	87.6	71.0	81.4	68.7	76.6	65.6	96.1	75.5
723100	1	71.3	60.5	75.4	61.5	82.1	63.3	87.9	65.6	91.4	70.2	96.1	74.2	98.8	75.0	96.3	76.2	92.6	74.9	85.7	69.4	79.3	66.1	74.5	64.5	93.7	74.9
	2	68.5	59.9	72.6	58.1	79.7	62.4	86.1	65.6	90.0	70.1	94.4	74.0	96.8	75.5	94.9	75.7	91.3	74.7	84.0	68.6	77.3	64.4	72.0	64.1	91.5	74.4
Greer/Greenville	0.4	68.8	57.0	72.5	57.9	80.0	61.3	85.8	65.2	90.4	69.5	94.8	73.7	97.4	74.2	94.8	74.7	90.8	73.1	83.6	67.0	77.1	63.5	69.9	60.3	93.1	73.9
723120	1	65.2	55.7	69.9	56.6	77.5	60.4	84.1	63.8	88.1	69.5	92.9	73.2	95.7	74.3	93.0	74.0	89.1	73.4	82.1	66.7	74.4	61.8	67.9	60.3	90.7	73.6
	2	62.7	54.3	66.9	53.7	75.0	59.9	82.3	63.4	86.3	69.1	91.3	72.9	93.9	74.6	91.5	74.4	87.4	72.8	80.2	65.4	72.4	60.8	65.5	58.6	88.3	73.1
SOUTH DAKOTA																											
Huron	0.4	49.0	39.9	56.5	44.1	72.6	54.4	84.5	60.5	87.3	67.2	96.3	72.4	100.4	74.0	99.5	72.0	94.8	68.9	82.1	60.2	65.3	50.5	51.8	43.0	95.1	72.2
726540	1	44.9	38.1	51.9	42.3	66.9	50.7	80.2	59.5	84.7	67.0	93.2	72.8	98.4	73.2	96.6	71.7	91.1	68.8	78.4	59.5	61.3	48.3	47.5	39.9	91.4	71.2
	2	41.8	36.3	47.5	39.8	61.4	48.4	76.0	57.7	82.5	66.2	90.8	70.9	95.7	73.2	93.9	71.4	87.9	68.1	74.8	58.2	58.0	48.0	43.9	38.0	88.0	70.2
Pierre	0.4	53.0	41.5	60.6	46.1	72.5	53.3	85.4	58.6	90.3	65.1	99.4	70.5	105.3	71.7	103.1	70.7	97.6	66.4	85.3	58.5	67.6	49.5	54.5	42.5	98.8	69.9
726686	1	48.8	39.8	56.6	43.7	68.0	49.9	81.6	57.6	87.4	64.2	95.9	70.2	102.1	71.7	100.3	69.5	94.2	66.7	81.0	57.9	63.6	48.2	50.5	40.5	94.7	69.4
	2	45.4	37.6	52.2	41.7	63.2	47.9	77.6	56.1	84.6	63.6	92.8	69.3	99.6	71.2	97.9	69.4	90.7	65.4	77.4	56.7	59.7	46.2	47.2	38.7	90.9	68.3
Rapid City	0.4	59.4	43.2	63.7	44.7	72.6	50.2	82.3	55.0	86.9	60.4	97.7	64.3	100.9	66.5	98.8	64.9	94.3	61.9	84.1	55.3	69.6	48.7	60.9	44.0	95.1	64.8
726620	1	56.0	41.5	60.2	43.5	68.4	48.2	79.4	54.4	84.2	60.3	93.1	64.4	98.1	66.0	96.6	64.3	91.2	61.2	80.7	54.3	66.0	47.6	57.2	42.8	91.2	64.5
	2	52.6	39.9	56.8	42.3	64.7	46.6	75.8	52.9	81.5	59.5	89.6	65.4	95.4	66.3	94.3	64.5	88.5	60.4	77.7	53.4	62.4	46.0	53.6	40.3	87.6	63.9

WMO# = World Meteorological Organization number DB = dry-bulb temperature, °F MWB = mean coincident wet-bulb temperature, °F

Table 4A Selected Monthly Percentiles of Wet-Bulb Temperature with Mean Dry-Bulb Temperature for U.S. Locations

Location and WMO#	%	Jan WB	Jan MDB	Feb WB	Feb MDB	Mar WB	Mar MDB	Apr WB	Apr MDB	May WB	May MDB	Jun WB	Jun MDB	Jul WB	Jul MDB	Aug WB	Aug MDB	Sep WB	Sep MDB	Oct WB	Oct MDB	Nov WB	Nov MDB	Dec WB	Dec MDB	Annual WB	Annual MDB
1a	1b	2a	2b	3a	3b	4a	4b	5a	5b	6a	6b	7a	7b	8a	8b	9a	9b	10a	10b	11a	11b	12a	12c	13a	13b	14a	14b
Sioux Falls	0.4	40.0	46.8	45.3	55.2	57.6	68.7	64.1	77.9	70.6	82.5	76.5	88.8	78.5	91.2	77.9	90.6	74.6	86.9	66.3	75.9	55.9	60.6	44.8	52.5	76.2	88.5
726510	1	37.8	43.8	42.9	50.1	54.5	65.0	62.4	75.4	69.1	81.3	75.3	87.4	77.5	90.3	76.9	88.8	73.0	85.3	64.1	72.4	53.4	58.7	40.5	46.7	74.5	86.9
	2	36.0	40.6	40.4	46.9	51.7	61.5	60.5	72.9	67.9	79.6	74.1	86.4	76.6	89.3	75.8	87.2	71.5	82.4	62.0	70.5	50.9	57.0	38.4	43.4	72.7	84.4
TENNESSEE																											
Bristol	0.4	57.0	62.1	59.0	63.5	62.6	72.7	67.0	75.8	72.1	80.9	75.2	86.1	76.6	87.0	76.3	86.8	73.9	84.0	68.8	76.4	64.2	70.5	60.6	66.8	75.0	85.3
723183	1	55.2	59.8	57.1	62.1	61.1	70.2	65.4	75.0	70.8	79.8	74.2	84.7	75.9	86.0	75.5	86.0	73.0	83.7	67.2	73.6	62.8	68.7	58.9	63.9	73.9	83.5
	2	53.5	57.8	55.2	60.0	59.7	67.6	64.0	74.0	69.8	80.0	73.3	83.7	75.2	85.4	74.9	85.1	72.2	82.5	66.0	72.5	61.3	67.2	57.1	61.8	72.7	82.1
Chattanooga	0.4	60.7	64.0	62.3	66.5	66.2	75.3	69.3	80.0	74.8	84.2	77.7	88.5	80.0	93.2	79.3	90.7	76.9	87.1	72.4	79.4	67.4	71.9	64.5	68.5	77.8	89.0
723240	1	59.0	62.3	60.5	64.7	64.6	72.8	68.1	77.7	73.8	83.4	77.0	88.4	78.8	90.9	78.3	89.4	76.1	85.9	70.9	76.7	65.9	70.1	62.8	66.1	76.7	87.6
	2	57.1	60.2	59.0	63.4	63.2	70.0	67.0	76.2	72.8	82.5	76.2	86.9	78.1	89.8	77.5	88.5	75.3	85.0	69.7	75.6	64.6	68.6	61.0	64.0	75.7	86.0
Knoxville	0.4	59.7	65.0	61.2	66.2	65.7	73.1	69.2	77.8	74.3	83.0	77.3	87.0	79.0	90.7	77.9	88.7	75.9	86.2	71.2	79.3	66.6	72.2	64.3	69.1	76.9	87.8
723260	1	58.1	62.9	59.5	64.6	64.0	71.8	67.9	76.4	73.2	81.6	76.3	86.4	78.1	89.1	77.2	88.2	75.2	85.2	69.7	76.2	65.1	70.1	61.9	65.6	75.8	86.3
	2	56.0	59.6	58.0	63.1	62.5	69.5	66.6	75.7	72.1	80.7	75.5	85.5	77.3	88.4	76.5	87.3	74.4	84.1	68.3	74.6	63.7	68.5	59.8	64.5	74.8	84.6
Memphis	0.4	63.8	67.5	65.2	70.1	69.1	75.6	72.4	80.8	76.9	86.4	80.3	90.7	81.8	93.8	81.4	92.1	79.3	89.8	74.8	82.0	69.4	73.9	67.3	70.5	80.1	91.6
723340	1	62.2	65.6	63.2	68.2	67.9	73.8	71.3	79.5	76.0	85.7	79.3	90.1	81.1	93.6	80.5	91.5	78.4	88.8	73.2	79.8	68.2	73.3	65.3	68.5	79.1	90.5
	2	60.5	64.9	61.4	66.3	66.4	72.5	70.3	77.8	75.0	84.6	78.5	89.4	80.4	92.6	79.8	91.0	77.5	87.2	71.7	79.1	67.1	71.8	63.4	66.9	78.1	88.9
Nashville	0.4	62.2	66.0	63.0	67.5	66.9	75.3	70.5	77.8	75.4	84.1	78.2	89.8	79.5	90.9	79.0	90.0	77.2	87.6	72.5	79.2	67.9	73.7	64.7	67.7	78.0	89.4
723270	1	60.1	63.2	61.3	66.1	65.4	72.5	69.1	78.0	74.1	83.7	77.3	88.7	78.9	90.3	78.3	89.4	76.3	86.2	71.0	78.4	66.4	70.8	62.7	66.4	77.0	88.0
	2	58.1	61.0	59.7	63.7	64.0	70.8	67.9	75.8	73.2	82.7	76.6	87.7	78.2	90.0	77.7	88.8	75.5	85.4	69.8	76.8	65.1	69.2	61.1	64.4	76.0	86.3
TEXAS																											
Abilene	0.4	60.7	67.3	61.2	71.4	66.9	78.4	71.2	83.5	74.7	87.7	77.2	93.1	76.1	90.6	75.4	90.6	75.1	88.4	72.4	82.6	66.9	73.4	62.3	68.3	75.1	89.4
722660	1	58.5	65.0	59.7	69.7	65.3	75.8	69.9	81.8	73.5	86.8	76.1	91.4	75.2	90.2	74.8	89.9	74.2	87.4	71.3	80.9	65.7	72.8	60.9	67.4	74.1	88.8
	2	56.3	63.8	58.2	68.2	64.0	73.6	68.8	79.4	72.5	85.1	75.3	89.4	74.5	89.9	74.2	89.7	73.5	86.6	70.3	80.3	64.4	71.8	59.6	65.5	73.2	87.9
Amarillo	0.4	50.4	65.2	52.9	70.3	58.0	69.5	62.4	76.6	67.7	79.8	71.4	87.3	72.3	88.3	72.1	86.2	70.1	82.7	64.9	78.4	58.4	66.8	52.5	61.8	70.8	86.3
723630	1	48.3	61.7	50.8	66.7	55.9	70.7	61.2	74.2	66.7	79.7	70.6	86.3	71.5	87.5	71.3	86.0	69.2	83.2	63.6	75.5	56.2	65.5	50.2	61.3	69.7	85.6
	2	46.2	61.1	49.1	64.7	54.1	70.2	59.9	72.9	65.6	78.4	69.9	85.9	70.9	87.1	70.5	85.5	68.4	82.6	62.4	74.6	54.7	63.7	48.2	60.9	68.9	85.0
Austin	0.4	68.1	71.6	67.8	72.1	71.4	79.1	74.7	84.9	78.3	87.9	79.0	91.3	78.5	90.5	78.3	90.0	78.3	87.7	76.6	84.6	72.7	78.1	69.3	73.3	78.0	89.1
722540	1	66.8	70.7	66.6	71.9	70.4	76.9	73.6	82.7	77.3	87.1	78.3	90.3	78.0	89.6	77.6	89.4	77.5	86.6	76.0	82.4	71.6	76.6	68.2	72.0	77.2	88.0
	2	65.3	69.5	65.5	71.5	69.5	75.1	72.8	81.0	76.3	85.5	77.8	89.5	77.4	88.9	77.2	89.1	77.1	86.1	75.2	82.7	70.8	75.7	67.2	70.7	76.4	86.7
Beaumont/	0.4	70.4	73.6	70.5	73.7	72.6	77.7	76.4	81.3	79.1	85.3	82.0	90.1	82.4	90.9	82.3	90.9	81.4	89.1	79.0	85.8	75.6	79.5	73.5	75.6	81.4	89.6
Port Arthur	1	69.3	72.3	69.4	72.4	71.8	76.5	75.6	80.2	78.3	84.7	81.3	89.1	81.8	90.4	81.6	90.3	80.8	88.5	78.2	84.4	74.7	78.9	72.1	74.8	80.5	88.7
722410	2	68.3	70.6	68.5	71.3	71.1	75.2	74.8	79.5	77.6	84.1	80.5	88.2	81.3	89.6	81.2	89.8	80.3	87.8	77.4	83.2	73.8	77.6	71.0	73.6	79.9	88.1
Brownsville	0.4	72.4	78.9	73.0	81.2	74.8	84.7	77.8	87.5	80.0	89.1	81.0	90.4	80.2	89.2	80.2	88.7	80.5	88.5	79.3	87.5	76.6	82.8	74.5	79.9	80.0	88.6
722500	1	71.6	77.7	71.9	77.7	74.0	83.0	77.0	86.5	79.3	88.3	80.4	89.4	79.5	88.8	79.7	88.6	80.0	88.3	78.8	87.2	76.1	82.0	73.8	78.7	79.4	88.3
	2	71.0	76.5	71.1	76.6	73.3	81.3	76.2	85.0	78.7	87.2	80.0	88.8	79.3	88.8	79.4	88.8	79.4	88.0	78.3	86.4	75.4	81.5	73.1	77.6	78.9	88.0
Corpus Christi	0.4	72.4	78.1	72.2	79.1	75.3	83.1	77.9	86.2	80.5	88.3	81.6	90.7	81.0	91.9	81.2	91.4	81.4	89.2	80.1	87.3	77.1	81.8	73.7	78.3	80.7	89.5
722510	1	71.4	75.7	71.1	76.3	74.0	81.1	77.1	84.5	79.7	87.0	81.1	89.6	80.4	90.9	80.5	90.3	80.8	88.6	79.3	86.1	76.3	81.1	72.7	77.6	80.1	88.7
	2	70.4	74.1	70.1	74.5	73.1	78.9	76.3	83.1	79.1	86.1	80.5	88.5	80.1	90.4	80.3	89.7	80.3	88.0	78.6	85.2	75.4	80.6	71.9	75.9	79.4	87.8
El Paso	0.4	51.6	61.0	53.8	67.5	55.4	76.1	61.0	84.0	64.5	86.0	69.3	86.5	71.3	85.0	71.3	86.1	69.8	82.5	65.1	77.9	58.1	69.7	53.2	63.6	69.9	84.6
722700	1	50.1	61.1	52.3	68.3	54.3	76.4	59.2	79.3	63.3	83.7	68.4	87.0	70.5	85.2	70.6	85.0	69.1	82.0	63.8	76.8	56.4	67.7	51.5	60.7	69.0	84.3
	2	48.8	60.3	51.1	66.8	53.2	75.7	57.7	78.7	62.3	82.7	67.6	86.5	69.9	85.2	70.0	84.4	68.4	81.6	62.7	75.4	55.1	66.4	50.1	60.4	68.1	84.3
Fort Worth,	0.4	64.4	68.7	65.1	72.0	69.8	77.9	73.8	81.4	77.5	88.4	79.4	92.9	79.4	91.6	78.5	93.0	78.3	89.4	75.2	84.8	70.4	75.8	67.0	71.5	78.2	91.1
Meacham Field	1	63.1	67.5	63.8	70.6	68.5	76.6	72.7	81.0	76.3	86.5	78.6	91.7	78.5	90.9	78.0	92.8	77.5	88.1	74.3	82.1	69.4	75.3	65.6	70.3	77.3	90.2
722596	2	61.5	65.8	62.4	69.4	67.6	74.7	71.7	80.0	75.4	85.5	77.9	91.2	78.0	90.7	77.4	91.9	77.0	87.5	73.4	80.9	68.4	74.2	64.4	68.7	76.4	89.4
Houston, Inter-	0.4	70.3	74.7	70.8	74.9	72.6	77.9	76.3	83.2	78.3	86.9	80.5	90.6	80.5	91.1	80.4	90.6	80.3	89.4	78.2	85.8	75.1	80.3	72.2	76.3	79.9	89.6
continental Airport	1	69.4	73.7	69.7	73.4	71.7	77.0	75.2	82.9	77.7	86.4	80.0	90.3	80.1	90.4	80.0	89.7	79.7	88.3	77.3	84.5	74.2	78.9	71.2	74.5	79.2	88.8
722430	2	68.6	72.7	68.5	72.3	71.0	75.9	74.3	81.1	77.1	85.8	79.4	89.4	79.6	89.7	79.4	89.4	79.2	87.3	76.6	83.3	73.3	77.9	70.2	74.0	78.5	87.9
Lubbock,	0.4	53.8	62.0	55.8	70.7	60.5	68.9	64.7	77.1	69.9	83.7	73.8	88.2	74.1	87.9	73.4	88.7	72.5	84.9	67.6	77.2	61.3	68.3	56.2	62.9	72.5	87.0
Intl Airport	1	51.8	63.2	54.1	69.5	58.9	69.0	63.6	75.6	68.9	81.6	72.7	87.2	73.2	87.5	72.8	88.0	71.5	84.6	66.3	76.2	59.5	65.9	53.6	61.7	71.5	86.2
722670	2	49.4	63.9	52.4	65.8	57.3	68.8	62.7	74.2	67.9	80.3	71.9	86.0	72.5	87.1	72.2	87.4	70.7	83.6	65.2	75.3	58.1	64.4	51.7	62.0	70.5	85.4
Lufkin	0.4	69.1	74.2	69.0	71.7	72.0	78.7	75.7	83.1	78.1	87.6	80.2	90.6	80.4	91.1	80.3	90.4	80.3	89.7	77.6	84.9	72.8	78.1	71.4	74.9	79.4	89.9
722446	1	67.8	72.4	67.9	71.4	71.0	77.4	74.5	82.3	77.2	86.2	79.5	89.8	79.8	90.9	79.5	90.2	79.4	88.5	76.5	83.4	71.8	77.3	69.9	73.6	78.8	89.4
	2	66.7	70.9	66.6	71.4	70.0	75.5	73.7	81.3	76.4	85.0	79.1	89.3	79.3	90.3	79.2	90.2	78.7	88.0	75.4	82.4	71.0	76.1	68.5	72.5	78.1	88.8
Midland/Odessa	0.4	56.7	63.5	58.5	71.1	62.8	72.3	67.1	77.1	70.7	85.3	74.2	86.7	74.0	87.7	73.4	86.0	72.3	85.2	69.7	79.6	64.3	70.3	58.2	64.3	72.6	86.6
722650	1	54.7	61.3	56.6	68.1	61.4	72.3	66.1	76.7	69.9	83.9	73.3	86.3	73.2	88.3	72.6	86.2	71.5	85.0	68.5	79.1	62.5	69.5	56.7	63.7	71.6	86.3
	2	52.8	62.6	55.0	67.3	60.0	72.7	65.1	76.5	68.9	81.7	72.4	86.0	72.4	87.5	72.1	86.6	71.0	84.6	67.5	77.2	61.0	67.8	54.8	61.3	70.7	86.0
San Angelo	0.4	61.5	68.2	61.5	70.8	66.6	77.9	70.6	83.8	74.2	88.2	76.2	91.6	75.4	90.6	75.2	91.2	75.0	88.0	72.8	84.1	67.9	73.7	62.8	69.3	74.7	89.6
722630	1	60.0	66.1	60.3	69.7	65.4	75.2	69.5	82.0	73.3	87.3	75.4	90.8	74.8	90.3	74.5	90.3	74.2	87.6	71.8	81.8	66.5	74.4	61.6	68.1	73.8	88.6
	2	57.9	64.4	59.1	67.9	64.3	74.9	68.6	80.5	72.4	85.6	74.7	89.6	74.2	89.6	74.0	89.8	73.5	86.6	70.8	80.8	65.3	73.3	60.5	66.5	73.0	87.6
San Antonio,	0.4	67.4	71.4	67.9	72.1	71.1	79.2	74.7	84.2	78.0	87.5	79.5	89.6	78.9	88.3	78.1	88.5	78.2	86.5	76.4	83.7	72.5	77.6	69.4	73.3	78.0	87.3
Intl Airport	1	66.4	70.8	66.8	72.8	70.2	77.2	74.1	83.3	77.0	86.4	78.6	88.6	78.2	87.1	77.4	87.7	77.4	86.0	75.8	83.0	71.5	76.6	68.3	72.2	77.1	86.7
722530	2	65.4	69.2	65.7	71.7	69.4	75.4	73.3	81.4	76.2	85.3	78.0	87.8	77.5	86.9	77.0	87.1	77.0	85.8	75.1	82.6	70.8	76.0	67.4	71.2	76.4	85.9
Victoria	0.4	70.4	75.1	70.5	74.5	72.8	79.6	76.2	82.1	79.1	86.8	80.4	90.1	80.2	89.8	80.1	88.4	80.2	86.9	78.4	85.7	76.0	80.3	72.8	75.6	79.5	88.1
722550	1	69.5	73.6	69.4	73.1	72.0	77.6	75.3	82.0	78.2	86.5	80.0	88.8	79.6	88.9	79.5	88.5	79.5	86.7	77.8	84.9	74.9	80.2	71.6	75.0	79.1	87.9
	2	68.8	72.7	68.5	72.4	71.3	76.4	74.5	81.0	77.5	85.4	79.4	88.0	79.3	88.8	79.2	88.5	79.1	86.8	77.2	83.9	74.1	78.7	70.8	74.2	78.4	87.2
Waco	0.4	65.6	70.3	66.3	73.5	70.6	79.7	75.0	83.1	77.9	90.2	80.1	94.2	79.3	94.8	79.1	93.9	79.1	91.0	76.4	86.1	71.6	78.8	67.7	72.8	78.5	92.7
722560	1	64.4	69.0	65.1	72.7	69.6	78.5	73.9	82.4	77.1	88.6	79.2	93.1	78.5	94.0	78.4	93.8	78.3	90.3	75.5	84.4	70.7	76.9	66.4	70.5	77.7	91.8
	2	63.0	67.2	63.9	70.3	68.4	75.6	72.8	81.1	76.2	87.2	78.4	92.1	78.1	93.6	78.0	93.7	77.6	89.6	74.7	83.2	69.7	75.2	65.3	69.4	77.0	90.6
Wichita Falls,	0.4	60.6	65.7	62.7	73.8	67.8	78.0	72.5	84.1	76.4	89.9	79.0	94.7	78.4	94.7	77.9	94.1	77.2	91.6	74.4	85.1	67.7	74.9	63.0	69.1	77.2	93.2
Sheppard AFB	1	58.0	63.0	60.7	70.5	66.5	77.3	71.4	82.3	75.2	88.7	78.0	93.9	77.5	94.3	77.2	93.7	76.3	90.1	73.0	81.8	66.4	73.3	61.5	67.0	76.3	92.1
723510	2	55.4	63.4	58.8	68.4	65.0	75.4	70.2	81.0	74.2	86.7	77.2	92.5	76.9	93.9	76.5	93.2	75.7	89.2	71.7	81.4	65.2	72.8	59.7	64.9	75.4	91.2
UTAH																											
Cedar City	0.4	44.8	53.9	46.5	58.7	48.3	64.3	51.9	72.8	56.9	76.7	61.9	85.2	66.1	79.3	65.4	81.0	61.5	78.9	55.1	69.3	49.0	63.2	44.6	55.6	63.6	79.7
724755	1	42.3	52.5	45.2	55.8	47.1	63.0	50.6	70.8	55.9	75.4	60.4	83.8	64.9	79.8	64.4	80.0	60.6	78.0	53.8	69.3	47.6	61.2	42.7	53.2	62.3	79.5
	2	40.6	49.5	43.6	55.6	45.9	61.9	49.5	69.5	54.9	75.2	59.2	86.0	64.0	79.5	63.5	79.8	59.8	77.0	52.7	70.0	46.3	59.7	41.2	50.6	61.1	79.4
Salt Lake City	0.4	44.1	51.6	47.3	58.0	51.0	66.9	55.9	76.2	61.7	80.0	65.7	88.8	68.2	86.7	68.5	84.3	64.1	82.9	57.2	75.4	51.2	61.5	44.3	53.6	66.1	85.3
725720	1	42.0	48.7	46.0	55.3	49.5	64.3	54.4	74.0	60.2	78.4	64.2	87.8	67.1	86.9	67.3	83.4	63.0	79.8	56.1	74.7	49.5	59.9	43.0	50.9	64.7	85.3
	2	40.4	46.3	44.7	54.1	48.3	62.3	53.4	72.0	59.0	77.6	63.0	86.5	66.2	86.3	66.2	83.8	62.1	81.2	55.1	72.4	48.2	59.1	41.4	48.7	63.5	84.5

WMO# = World Meteorological Organization number WB = wet-bulb temperature, °F MDB = mean coincident dry-bulb temperature, °F

Table 4B Selected Monthly Percentiles of Dry-Bulb Temperature with Mean Wet-Bulb Temperature for U.S. Locations

Location and WMO#	%	Jan DB	Jan MWB	Feb DB	Feb MWB	Mar DB	Mar MWB	Apr DB	Apr MWB	May DB	May MWB	Jun DB	Jun MWB	Jul DB	Jul MWB	Aug DB	Aug MWB	Sep DB	Sep MWB	Oct DB	Oct MWB	Nov DB	Nov MWB	Dec DB	Dec MWB	Annual DB	Annual MWB
1a	1b	2a	2b	3a	3b	4a	4b	5a	5b	6a	6b	7a	7b	8a	8b	9a	9b	10a	10b	11a	11b	12a	12c	13a	13b	14a	14b
Sioux Falls	0.4	48.5	39.2	56.1	44.6	73.3	55.2	85.3	60.9	87.9	67.3	95.6	72.4	99.4	74.5	96.7	73.0	91.6	71.0	81.5	61.3	64.7	51.3	52.3	45.3	93.9	73.1
726510	1	44.3	37.5	51.3	41.9	68.1	52.4	81.1	60.5	85.4	66.5	93.1	71.9	97.0	74.3	94.5	73.0	88.8	69.8	78.1	60.7	61.8	50.2	47.4	40.2	90.3	71.9
	2	41.0	35.4	47.6	40.1	63.1	50.3	76.5	58.9	82.8	65.5	90.7	71.0	94.8	73.9	92.4	73.2	86.2	69.3	74.9	59.1	58.9	49.1	43.6	37.9	87.2	70.5
TENNESSEE																											
Bristol	0.4	64.4	54.9	68.6	57.5	76.6	58.9	82.4	63.4	86.2	69.0	90.3	72.0	92.6	72.4	92.3	73.0	88.5	71.0	81.2	63.8	74.8	61.5	69.2	59.9	89.0	72.2
723183	1	61.4	52.7	65.6	53.7	74.0	58.6	80.5	62.0	84.3	67.9	88.6	71.3	90.9	72.9	90.0	73.5	87.1	70.4	79.4	64.1	72.1	59.7	65.4	57.5	86.8	71.6
	2	58.9	52.0	62.8	52.1	71.8	57.2	78.7	61.1	82.8	67.4	87.2	70.6	89.3	72.9	88.0	72.9	85.6	70.5	77.7	63.2	69.7	58.4	62.7	55.5	84.8	70.9
Chattanooga	0.4	66.4	58.1	72.4	58.5	80.2	62.9	86.2	65.5	90.4	71.4	94.7	74.0	98.8	76.9	96.6	76.2	93.0	74.3	84.5	68.4	76.4	62.6	70.3	63.3	94.1	75.1
723240	1	63.8	56.3	69.4	57.4	77.7	61.2	84.2	64.2	88.4	71.2	93.0	73.6	96.7	76.2	94.9	75.6	91.0	73.6	82.4	66.9	73.9	62.2	67.6	60.6	91.6	74.7
	2	61.7	54.5	66.6	55.3	75.2	59.9	82.5	63.5	86.5	70.2	91.5	73.9	94.9	75.8	93.2	75.4	89.2	72.9	80.8	66.4	71.7	61.3	65.3	60.0	89.4	73.9
Knoxville	0.4	66.4	58.4	71.0	60.3	78.1	61.4	84.0	64.4	88.1	69.5	92.3	71.9	95.7	75.3	94.5	74.9	91.1	72.7	82.4	68.3	75.9	63.5	70.7	62.1	91.9	74.1
723260	1	63.9	56.3	68.2	56.0	75.8	60.8	82.2	64.3	86.3	69.7	90.5	72.9	94.2	75.3	92.5	74.5	89.2	72.5	80.8	67.0	73.3	62.1	67.8	61.2	89.5	73.7
	2	60.9	54.1	65.6	55.0	73.6	59.8	80.5	63.9	84.9	69.2	89.3	72.6	92.6	75.3	91.1	74.8	87.6	72.4	79.2	65.8	71.1	61.4	64.6	58.3	87.3	73.0
Memphis	0.4	69.5	61.3	74.5	61.1	80.6	63.9	86.1	67.4	91.1	73.4	96.2	76.1	100.1	79.1	97.8	77.7	95.7	75.7	87.1	70.4	78.9	64.4	72.1	65.3	95.7	77.6
723340	1	67.3	60.2	71.6	59.8	78.6	63.4	84.2	66.8	89.5	72.5	94.7	76.2	98.0	78.8	96.1	77.9	92.9	76.1	85.3	69.2	76.7	65.1	70.0	63.8	93.5	77.1
	2	65.2	58.8	69.2	58.6	76.5	62.9	82.6	66.7	88.2	72.4	93.4	75.9	96.3	78.5	94.7	77.8	91.2	75.7	83.7	68.2	74.5	64.0	67.7	62.0	91.7	76.5
Nashville	0.4	67.3	58.9	71.9	56.8	80.2	62.9	85.5	65.4	89.5	71.9	95.3	74.3	97.9	76.2	97.3	75.8	93.5	73.3	84.7	68.2	77.5	64.1	70.2	62.4	94.2	75.5
723270	1	65.0	58.5	69.5	57.9	77.8	62.1	83.6	65.4	87.9	71.4	93.4	74.5	96.2	76.1	95.4	75.6	91.5	74.0	83.3	67.7	74.9	63.0	68.0	60.4	91.7	74.8
	2	62.8	56.2	67.3	56.3	75.2	61.1	82.0	65.0	86.4	70.6	91.9	73.5	94.5	76.2	93.5	75.4	90.0	73.4	81.7	66.7	72.7	61.0	66.0	59.8	89.6	74.2
TEXAS																											
Abilene	0.4	75.7	55.3	80.9	56.0	87.3	60.4	92.2	62.8	97.0	67.4	100.3	70.6	102.0	70.6	100.8	70.8	97.6	69.7	91.3	64.1	82.0	60.5	74.7	55.0	98.9	70.6
722660	1	72.6	53.2	77.8	54.7	84.4	60.2	89.9	63.5	94.7	67.1	98.1	71.3	100.4	70.5	99.4	71.2	95.6	69.9	89.1	65.8	79.4	61.0	72.3	56.2	96.8	70.7
	2	69.9	52.6	74.7	54.0	81.6	59.4	87.5	63.0	92.5	67.5	96.2	71.4	99.1	70.6	98.2	70.8	94.0	69.9	87.1	66.5	77.2	60.2	69.9	55.8	94.6	70.5
Amarillo	0.4	70.1	48.3	76.2	50.5	82.0	52.6	88.5	56.7	94.1	59.7	100.4	64.7	99.1	66.8	97.5	67.2	94.2	64.5	88.5	59.3	77.4	52.9	70.2	48.7	96.2	66.6
723630	1	66.5	46.5	72.3	48.7	79.6	52.4	85.7	56.7	92.1	58.9	97.7	65.5	97.5	66.8	96.0	67.5	92.3	64.5	86.2	58.5	74.9	52.6	67.3	47.8	93.8	66.2
	2	63.4	44.8	68.8	47.4	76.5	51.4	83.4	54.3	90.0	59.5	95.5	65.4	96.3	67.2	94.4	67.2	90.6	64.6	84.1	58.5	72.5	51.7	64.4	46.0	91.5	65.9
Austin	0.4	77.7	60.3	82.7	60.7	87.1	63.3	89.9	67.0	93.1	73.5	97.8	75.6	100.3	73.2	101.1	73.9	97.9	73.8	91.7	72.4	84.1	68.2	78.3	64.5	98.3	74.1
722540	1	75.3	61.6	79.3	62.3	84.0	65.2	87.7	68.4	91.4	73.2	96.0	75.2	99.1	73.7	99.8	73.9	96.2	73.6	90.1	71.2	81.8	67.3	76.0	63.6	96.4	74.1
	2	73.2	60.9	76.6	60.9	81.5	64.1	85.9	68.9	89.9	73.5	94.4	74.8	98.0	74.1	98.6	73.9	94.7	73.9	88.5	71.1	80.2	67.0	74.1	63.9	94.4	74.1
Beaumont/	0.4	75.0	68.4	77.3	65.0	80.8	67.9	86.0	69.2	90.0	75.4	94.2	78.7	96.1	79.3	96.9	79.0	94.3	77.4	89.3	75.7	83.0	71.2	77.8	71.2	94.0	78.8
Port Arthur	1	73.6	66.7	75.4	65.8	79.2	67.6	84.2	71.3	88.7	75.3	93.0	78.2	94.9	79.6	95.2	78.9	93.0	78.0	87.9	75.5	81.3	72.3	76.3	70.3	92.4	78.6
722410	2	72.2	66.4	73.7	65.8	77.8	67.1	82.6	71.2	87.5	75.3	92.0	78.2	93.9	79.1	94.0	79.1	91.7	78.2	86.4	74.5	80.0	71.5	74.9	70.0	90.9	78.4
Brownsville	0.4	82.1	70.5	84.3	66.8	88.8	70.3	91.3	75.4	92.4	77.5	95.5	78.6	96.6	77.3	96.5	77.2	95.2	77.3	91.7	77.6	87.3	73.9	83.8	71.1	95.1	77.5
722500	1	80.4	70.1	82.2	69.3	86.5	71.3	89.4	74.7	91.4	77.4	94.4	78.3	95.7	77.2	95.8	77.4	94.3	77.3	90.6	77.2	85.9	74.0	82.3	71.5	93.8	77.4
	2	79.0	69.2	80.4	68.7	84.3	70.9	87.9	74.3	90.5	77.3	93.4	78.0	94.8	77.4	95.0	77.4	93.2	77.4	89.4	76.4	84.8	73.6	81.0	70.8	92.5	77.3
Corpus Christi	0.4	80.8	68.8	84.1	66.4	88.0	68.1	89.9	71.0	91.4	77.9	94.4	78.9	96.3	78.3	96.8	77.5	95.2	78.3	91.5	76.7	86.2	73.6	82.7	69.3	94.9	77.9
722510	1	78.7	68.9	81.0	65.9	84.9	66.9	87.9	72.6	90.1	77.6	93.4	78.7	95.5	78.0	95.8	77.8	94.0	77.8	90.0	77.2	84.6	73.6	80.8	70.3	93.5	77.8
	2	76.7	68.3	78.8	67.1	82.8	68.2	86.1	73.5	88.8	77.1	92.4	78.5	94.7	77.9	95.0	77.8	92.8	78.0	88.8	76.4	83.4	72.8	79.2	69.6	92.2	77.8
El Paso	0.4	70.7	48.2	77.0	51.0	82.8	52.6	90.4	57.4	96.4	60.6	105.6	64.2	103.2	65.9	99.6	65.4	96.5	63.8	89.8	59.5	78.5	54.2	70.4	49.0	100.5	64.3
722700	1	68.0	47.3	74.5	49.8	80.7	52.7	88.2	56.1	94.9	59.4	103.5	63.8	101.4	65.6	98.3	65.6	95.1	64.2	87.9	59.3	76.5	53.1	68.7	48.1	97.9	64.0
	2	65.8	46.4	72.3	48.9	79.0	52.0	86.4	55.2	93.3	58.9	101.9	63.8	99.9	65.4	97.0	65.7	93.5	63.9	86.0	59.3	74.6	52.1	66.6	47.2	95.6	63.8
Fort Worth,	0.4	74.7	57.7	79.4	59.3	84.5	64.7	89.1	68.4	92.8	74.9	100.1	76.2	103.7	74.0	102.5	74.2	98.2	74.5	92.7	68.5	82.8	64.8	75.7	62.6	100.0	74.5
Meacham Field	1	71.8	57.5	76.7	59.0	81.9	64.1	86.2	67.9	91.0	73.5	97.3	75.4	101.6	74.1	101.2	74.2	96.7	73.8	90.0	69.5	80.4	64.9	73.2	61.7	97.8	74.4
722596	2	69.2	56.8	74.1	58.1	79.5	62.8	84.2	67.4	89.3	72.2	95.9	75.1	100.1	74.3	100.1	74.5	95.0	73.9	87.9	69.4	78.3	63.8	70.7	61.8	95.6	74.4
Houston, Inter-	0.4	78.3	68.4	81.0	64.8	83.7	67.1	88.0	72.1	91.4	75.2	96.3	77.6	98.4	77.2	98.9	76.2	95.3	76.2	90.5	74.5	84.8	70.7	79.1	69.1	95.9	76.9
continental Airport	1	76.3	66.5	78.4	65.6	81.8	68.3	86.0	71.3	90.2	75.3	94.9	77.3	96.9	76.7	97.5	76.6	93.9	76.4	89.1	73.9	83.2	71.2	77.7	68.7	94.0	76.8
722430	2	74.5	65.5	76.2	64.3	80.2	66.3	84.4	71.4	89.0	74.8	93.6	76.8	95.7	76.8	96.2	76.7	92.6	76.2	87.6	73.1	81.6	70.7	76.2	68.1	92.3	76.8
Lubbock,	0.4	72.9	49.5	79.0	52.6	84.4	54.0	90.3	58.3	97.0	62.2	101.7	67.0	100.1	68.1	97.9	68.6	94.5	66.7	89.0	61.0	79.6	54.4	72.4	50.3	97.3	67.2
Intl Airport	1	69.7	48.1	75.3	50.8	82.1	53.9	87.8	58.1	94.6	61.9	99.5	65.7	98.2	68.3	96.5	69.3	92.9	66.6	86.9	60.3	77.2	53.6	69.6	49.1	94.9	67.4
722670	2	66.7	47.5	72.5	50.1	79.4	53.1	85.9	56.3	92.4	61.3	97.6	65.8	96.8	68.8	95.0	69.1	91.5	66.3	85.0	60.2	74.8	53.3	67.0	48.1	92.6	67.3
Lufkin	0.4	77.1	64.9	78.7	62.8	84.9	67.3	87.0	72.4	90.9	74.2	95.5	76.9	99.0	76.9	100.5	76.1	96.3	75.5	91.1	72.7	82.9	68.4	77.3	67.8	96.5	76.3
722446	1	74.8	65.8	76.4	62.9	82.3	66.7	85.6	71.8	89.7	74.0	94.3	77.1	97.7	76.9	98.4	76.6	94.9	76.2	89.3	72.3	80.8	68.7	75.8	67.4	94.5	76.5
	2	72.7	64.0	74.7	61.7	80.1	65.6	84.3	70.5	88.7	74.0	93.2	76.6	96.2	76.7	97.0	76.3	93.5	76.0	87.7	72.0	79.2	67.9	74.3	67.1	92.7	76.3
Midland/Odessa	0.4	75.1	51.2	81.3	54.2	87.1	55.4	92.9	59.0	99.2	63.5	102.4	65.3	101.8	68.2	101.1	68.5	97.8	67.4	91.1	62.3	82.2	55.7	75.5	51.9	99.3	67.3
722650	1	72.2	49.6	78.0	52.1	84.4	54.4	90.8	57.6	96.9	61.1	100.4	66.2	100.0	68.4	99.6	68.3	95.4	67.5	88.9	62.4	79.8	56.6	73.0	51.3	96.9	67.0
	2	69.7	49.0	75.3	51.5	82.2	54.0	88.6	58.0	95.0	60.5	98.8	66.6	98.4	68.2	98.1	68.2	93.6	66.9	87.1	62.0	77.4	55.6	70.5	50.5	94.7	67.1
San Angelo	0.4	77.5	53.1	82.3	56.6	88.5	60.1	93.8	62.7	99.3	66.9	101.1	71.1	102.1	70.0	102.4	70.4	98.0	70.2	91.5	68.0	83.3	59.6	76.6	55.5	99.5	70.2
722630	1	74.6	53.7	79.4	55.4	85.5	59.7	91.7	62.3	96.6	67.3	99.0	71.1	100.6	70.3	100.4	70.5	96.2	70.2	89.2	65.6	80.6	61.4	73.9	55.6	97.3	70.2
	2	71.7	53.3	76.5	55.1	83.1	59.7	89.7	62.5	94.2	67.2	97.1	70.6	99.5	70.3	99.0	70.4	94.5	70.1	87.3	66.4	78.7	61.4	71.8	55.2	95.2	70.2
San Antonio,	0.4	78.4	60.8	83.0	61.5	87.9	64.6	91.3	68.1	95.4	71.1	98.3	74.4	99.5	72.5	100.2	73.7	96.8	72.6	91.9	72.1	84.3	67.1	78.2	64.5	97.5	73.4
Intl Airport	1	75.6	60.6	80.4	60.7	85.0	63.8	88.6	68.5	92.9	72.8	96.7	74.7	98.2	73.1	98.7	73.3	95.4	73.0	90.2	71.0	82.2	67.4	76.4	63.5	95.8	73.4
722530	2	73.5	60.2	77.8	60.3	82.5	63.8	86.8	67.7	90.9	72.7	95.2	74.2	97.2	73.4	97.6	73.4	94.2	73.5	88.6	70.8	80.4	67.0	74.6	63.1	94.0	73.6
Victoria	0.4	78.4	66.7	81.1	63.4	85.0	67.1	87.9	67.3	91.4	75.9	95.3	77.3	97.2	75.8	98.1	75.8	95.6	75.4	90.9	74.1	85.1	72.0	79.8	69.2	95.4	76.3
722550	1	76.6	66.5	78.4	63.2	82.6	66.0	86.3	70.8	89.7	75.9	94.2	77.2	96.1	76.0	96.7	76.0	94.2	76.0	89.3	74.6	83.4	72.0	78.3	68.8	93.9	76.4
	2	74.9	66.0	76.5	64.1	80.9	65.9	85.0	70.8	88.5	75.1	93.1	77.1	95.2	76.5	95.5	76.3	92.8	76.1	88.0	74.4	81.9	71.1	77.0	68.4	92.4	76.6
Waco	0.4	76.3	60.5	80.5	61.3	85.7	65.0	90.1	68.2	93.2	75.2	99.6	76.5	103.0	75.1	104.0	75.3	99.2	74.9	93.1	69.2	84.2	66.9	77.0	62.6	100.7	75.1
722560	1	73.6	59.1	77.6	60.9	83.0	65.2	87.4	69.5	91.8	74.5	98.1	76.6	101.8	74.9	102.2	75.1	97.4	74.8	91.2	70.0	81.8	66.8	74.4	62.6	98.7	75.0
	2	70.9	59.1	75.1	59.6	80.6	64.2	85.4	68.3	90.4	73.4	96.5	76.0	100.8	74.9	101.0	75.2	96.1	74.6	89.1	70.9	79.8	66.2	72.2	61.8	96.6	74.9
Wichita Falls,	0.4	73.5	54.7	80.4	58.2	88.0	61.9	91.6	67.0	96.7	70.7	103.3	73.7	106.7	72.8	105.2	73.1	100.1	72.8	92.7	66.8	81.6	62.8	73.7	58.1	102.6	73.5
Sheppard AFB	1	70.6	54.2	76.9	56.7	84.7	61.4	89.1	66.5	94.3	71.4	100.2	73.9	105.1	73.3	103.4	73.4	98.3	72.8	90.0	68.5	79.2	62.5	71.1	57.2	99.9	73.4
723510	2	67.5	52.4	73.6	55.1	81.2	62.2	86.6	65.6	92.0	71.0	98.4	74.2	103.4	73.5	102.0	73.5	96.5	72.7	87.9	68.0	76.9	61.9	68.5	55.5	97.5	73.4
UTAH																											
Cedar City	0.4	58.0	42.5	63.7	44.8	69.5	46.4	77.0	50.2	85.1	54.6	95.1	58.6	96.4	59.7	94.5	59.4	89.7	57.7	81.4	52.0	68.2	47.1	59.0	42.3	93.0	58.9
724755	1	54.4	41.5	60.7	43.0	67.0	45.4	74.9	49.2	82.5	53.1	93.4	58.4	95.0	59.7	93.1	59.3	88.1	57.8	79.6	51.5	66.0	45.7	56.3	41.0	90.6	58.5
	2	51.0	39.4	58.1	41.7	64.7	44.7	72.7	48.5	80.6	52.5	91.4	57.5	93.5	59.0	91.6	59.2	86.4	56.8	77.6	50.4	63.7	45.0	53.7	39.8	88.2	57.8
Salt Lake City	0.4	52.8	42.3	60.7	45.6	69.7	49.3	80.5	53.9	87.2	56.6	98.2	61.3	99.7	62.9	98.1	62.5	92.7	60.2	82.0	55.2	67.0	49.2	56.1	43.2	96.4	62.0
725720	1	50.3	40.9	58.1	44.0	67.2	48.3	77.7	53.2	85.3	56.5	96.5	61.1	98.4	62.8	96.5	61.9	90.5	59.8	79.5	54.2	64.1	47.8	53.3	41.8	94.2	61.7
	2	47.5	39.8	55.5	43.7	64.7	47.2	75.2	52.0	83.7	56.5	94.5	61.0	97.1	62.6	95.3	62.1	88.7	59.2	77.0	53.5	61.5	46.7	50.0	40.5	91.8	61.1

WMO# = World Meteorological Organization number   DB = dry-bulb temperature, °F   MWB = mean coincident wet-bulb temperature, °F

Table 4A Selected Monthly Percentiles of Wet-Bulb Temperature with Mean Dry-Bulb Temperature for U.S. Locations

Location and WMO#	%	Jan		Feb		Mar		Apr		May		Jun		Jul		Aug		Sep		Oct		Nov		Dec		Annual	
		WB	MDB	WB	MDB	WB	MDB	WB	MDB	WB	MDB	WB	MDB	WB	MDB	WB	MDB	WB	MDB	WB	MDB	WB	MDB	WB	MDB	WB	MDB
1a	1b	2a	2b	3a	3b	4a	4b	5a	5b	6a	6b	7a	7b	8a	8b	9a	9b	10a	10b	11a	11b	12a	12c	13a	13b	14a	14b
VERMONT																											
Burlington	0.4	45.4	47.8	48.4	51.0	56.0	62.9	63.1	75.2	70.4	83.2	74.2	84.9	76.2	87.7	75.9	86.3	74.6	81.6	64.8	72.0	59.7	64.7	53.6	55.9	73.9	83.4
726170	1	41.8	44.8	43.4	46.2	52.3	59.3	60.8	71.4	68.4	78.3	72.9	82.8	75.1	85.8	74.4	83.7	72.9	79.8	63.1	69.7	57.5	62.8	49.0	51.6	72.0	80.5
	2	38.2	41.1	40.7	44.0	49.3	55.9	58.2	66.7	66.5	76.9	71.5	81.5	74.1	84.4	73.1	81.0	71.1	76.9	61.7	67.4	54.9	59.2	44.4	47.6	70.3	78.4
VIRGINIA																											
Lynchburg	0.4	58.4	63.0	60.7	66.7	64.8	73.9	66.9	75.5	74.7	84.2	76.6	87.3	78.3	90.1	77.5	90.1	75.8	87.5	70.8	78.3	66.5	70.0	63.1	67.1	76.5	88.1
724100	1	56.1	63.5	58.3	63.4	63.1	74.2	65.7	76.8	73.1	81.7	75.8	87.1	77.5	89.3	76.8	89.1	75.1	87.3	69.6	77.3	65.2	69.9	61.3	65.4	75.5	86.5
	2	53.6	59.0	56.0	61.6	61.3	70.2	64.6	76.2	71.7	80.2	75.0	85.9	76.9	88.6	76.1	87.7	74.3	85.2	68.2	75.2	63.9	68.5	59.0	63.0	74.5	84.8
Norfolk	0.4	63.5	68.1	65.3	70.4	67.1	75.9	69.1	80.2	75.3	86.2	78.4	90.9	80.3	91.5	79.7	89.7	77.9	87.3	73.1	79.5	69.3	75.2	66.6	71.3	78.5	89.2
723080	1	61.7	66.1	63.3	69.0	65.7	74.8	67.8	78.4	73.7	83.0	77.4	89.4	79.5	91.0	79.1	89.4	77.0	87.2	71.9	79.2	68.2	73.1	64.6	68.4	77.4	87.4
	2	59.5	64.2	61.0	65.5	64.1	71.4	66.6	76.9	72.4	81.7	76.7	87.6	78.8	90.2	78.3	88.2	76.2	85.6	70.9	76.6	66.8	71.7	62.9	67.0	76.5	85.4
Richmond	0.4	62.1	65.7	62.7	68.7	66.7	76.4	68.9	80.4	75.8	85.0	78.8	91.0	80.7	92.7	80.0	91.2	77.9	88.9	72.3	79.1	69.1	72.9	65.4	69.2	78.8	89.8
724010	1	59.6	64.1	60.8	66.9	65.1	74.1	67.8	79.2	74.4	84.3	77.8	90.0	80.0	91.2	79.2	90.2	76.9	87.7	71.3	77.8	67.5	71.1	63.1	66.5	77.6	88.0
	2	56.8	60.6	58.6	63.7	63.3	71.1	66.7	77.9	73.2	82.8	77.0	88.6	79.2	90.1	78.4	89.2	76.1	85.7	70.1	76.6	66.0	70.3	61.0	64.5	76.4	86.0
Roanoke	0.4	56.3	62.5	58.9	65.2	63.4	74.9	65.9	79.2	72.7	83.5	75.2	87.7	77.1	89.7	76.5	88.2	74.9	87.2	69.3	76.8	65.5	69.9	60.8	64.8	75.2	87.5
724110	1	54.0	61.6	56.5	62.8	61.4	70.8	64.8	77.4	71.2	82.0	74.3	86.4	76.2	88.5	75.7	87.8	74.0	85.5	67.9	74.5	63.7	68.7	59.0	64.0	74.1	85.6
	2	51.9	57.3	54.2	61.0	59.7	68.5	63.7	74.7	70.0	79.3	73.4	85.3	75.4	87.8	75.1	87.0	73.1	83.8	66.7	73.9	61.9	66.6	56.6	61.0	73.1	83.7
Sterling	0.4	57.0	63.8	59.1	63.1	64.0	74.3	67.3	78.6	74.6	85.4	77.6	88.1	79.5	91.7	78.6	89.9	77.9	85.3	70.7	77.3	66.7	72.5	62.2	66.2	77.4	88.2
724030	1	53.3	57.8	57.0	61.5	61.9	71.3	66.0	78.4	73.2	83.1	76.6	87.5	78.5	90.4	77.8	88.5	76.4	85.8	69.5	75.9	64.9	70.0	59.8	64.1	76.1	86.5
	2	50.5	54.1	54.0	58.9	60.0	68.1	64.6	76.1	71.6	81.0	75.6	86.2	77.6	89.2	77.0	87.3	75.2	84.9	68.1	73.3	63.0	67.7	56.9	60.5	74.9	84.5
WASHINGTON																											
Olympia	0.4	53.4	54.6	54.3	57.9	55.0	59.7	59.4	70.6	64.6	80.0	68.1	86.0	70.0	88.8	70.4	89.5	66.8	84.1	61.6	72.2	55.8	58.6	53.3	55.4	67.7	85.1
727920	1	51.8	53.1	53.1	55.5	53.7	59.0	57.5	67.6	63.1	79.0	66.7	83.2	68.8	86.4	69.0	88.2	65.3	79.4	60.3	68.4	54.5	56.8	51.6	53.2	65.9	81.2
	2	50.5	52.0	51.9	54.5	52.3	57.6	55.9	65.6	61.5	75.3	65.5	80.6	67.6	85.2	67.8	85.0	64.3	77.4	59.0	64.6	53.3	55.3	50.5	52.0	64.1	77.7
Quillayute	0.4	52.4	54.2	53.8	58.0	52.7	57.2	55.5	64.3	61.5	75.7	63.6	78.0	65.7	82.3	67.0	83.9	65.1	79.0	60.5	68.4	55.7	57.3	53.1	54.0	63.8	75.9
727970	1	50.9	51.9	52.3	55.4	51.5	55.1	54.3	62.2	59.6	72.4	62.1	74.7	64.3	77.9	65.2	78.6	63.6	75.6	59.2	64.4	54.9	56.0	52.2	53.0	61.8	71.5
	2	50.1	51.1	50.8	53.3	50.5	53.7	53.1	60.5	58.0	66.5	60.2	70.8	62.7	74.1	63.9	74.7	62.3	72.6	58.2	62.3	54.0	55.2	51.3	52.1	60.2	67.4
Seattle,	0.4	52.0	54.2	53.6	58.1	53.7	58.8	58.5	68.6	63.7	80.4	66.3	84.0	68.1	86.7	69.1	87.5	66.0	80.6	60.8	70.9	55.5	58.5	52.8	55.0	66.3	83.0
Intl Airport	1	50.6	53.0	52.4	56.1	52.4	58.3	56.7	68.3	62.1	77.4	65.2	81.8	67.0	85.1	67.8	84.9	64.7	78.7	59.5	67.5	54.3	57.1	51.2	53.4	64.6	79.3
727930	2	49.7	52.1	51.2	55.0	51.3	57.1	55.0	65.1	60.5	73.8	63.9	79.3	66.0	83.2	66.6	82.8	63.5	76.3	58.5	64.9	53.1	55.6	50.1	52.1	63.0	76.1
Spokane,	0.4	44.6	47.8	47.5	51.3	50.7	60.3	57.5	73.0	62.2	82.7	65.4	85.9	66.9	89.1	66.3	88.5	62.5	84.1	57.8	74.2	50.1	55.0	46.0	48.7	64.5	86.1
Fairchild AFB	1	42.5	45.2	45.8	49.9	49.1	56.5	55.0	69.2	60.0	78.6	64.0	83.7	65.6	87.9	65.3	87.7	61.3	82.3	56.0	71.0	48.3	51.6	43.7	45.9	62.9	83.8
727855	2	41.0	43.7	44.3	48.5	47.6	54.6	53.1	64.5	58.6	75.7	62.9	81.9	64.6	86.5	64.3	86.3	60.1	80.3	54.5	67.2	47.0	49.8	42.1	44.3	61.4	81.5
Yakima	0.4	47.0	53.3	49.9	59.3	53.3	65.1	59.1	74.8	64.3	85.3	67.1	91.8	70.7	92.0	69.7	91.4	66.0	86.0	60.4	77.8	52.5	60.5	48.2	54.0	67.4	90.4
727810	1	45.3	51.7	48.4	56.6	51.8	62.7	57.0	71.5	62.5	83.2	65.8	89.0	69.0	90.6	68.4	90.7	64.5	84.9	58.7	75.1	51.1	57.4	45.9	50.9	65.6	88.5
	2	43.5	49.3	46.9	54.7	50.1	60.1	55.2	70.0	60.8	80.1	64.7	86.6	67.5	90.9	67.4	90.1	63.4	83.5	57.1	71.8	49.6	55.7	43.9	48.2	64.1	85.8
WEST VIRGINIA																											
Charleston	0.4	57.3	63.9	58.3	65.9	62.9	74.3	66.2	77.3	73.8	83.2	76.3	86.5	78.4	89.8	77.6	87.5	75.7	85.1	69.5	76.4	64.3	72.7	60.4	67.4	76.2	86.2
724140	1	55.6	62.2	56.6	63.0	61.4	71.6	65.0	78.0	72.4	81.0	75.3	85.9	77.4	88.1	76.7	86.8	74.7	83.9	68.1	75.8	62.8	71.0	58.9	65.2	74.9	84.5
	2	53.8	59.1	54.9	61.5	59.8	69.3	63.9	77.0	70.9	80.1	74.4	84.5	76.5	86.6	75.8	85.5	73.8	82.6	66.8	74.3	61.3	68.6	57.4	63.5	73.7	82.4
Elkins	0.4	54.3	59.0	55.7	61.5	60.4	69.9	64.2	74.8	69.7	78.1	73.4	81.4	75.4	84.1	74.8	83.5	73.1	81.8	67.0	73.2	62.2	69.3	58.9	64.6	73.3	81.9
724170	1	52.7	56.8	54.2	59.3	58.5	67.1	62.6	73.9	68.7	77.2	72.4	80.9	74.6	83.3	73.8	82.1	72.1	80.1	65.4	71.8	60.3	67.2	56.9	62.8	71.9	80.0
	2	50.6	55.4	52.4	57.5	56.9	65.4	61.4	72.1	67.7	75.6	71.4	80.2	73.6	82.3	72.9	81.4	71.2	78.8	64.1	70.8	58.8	63.9	54.9	59.7	70.6	78.4
Huntington	0.4	57.9	64.7	58.5	63.9	63.3	73.7	67.4	79.0	74.5	83.6	77.4	87.0	79.1	89.6	78.8	89.2	75.4	86.0	69.4	76.4	64.8	72.0	60.8	67.4	77.0	86.6
724250	1	56.2	61.7	56.8	63.0	61.8	71.8	65.8	77.6	73.1	81.9	76.3	85.9	78.1	88.5	77.6	87.3	74.7	85.0	68.2	75.5	63.2	69.7	59.4	65.3	75.7	85.2
	2	54.2	58.4	55.0	60.7	60.1	69.2	64.6	76.6	71.7	81.1	75.3	85.1	77.3	87.2	76.7	86.3	73.9	83.4	67.0	74.2	61.7	68.1	57.6	62.5	74.4	83.1
WISCONSIN																											
Eau Claire	0.4	38.0	41.0	42.6	47.6	56.3	65.7	64.6	76.5	71.7	83.3	76.1	88.5	78.8	90.0	78.1	90.0	74.4	83.8	67.1	73.3	57.2	60.1	47.8	51.2	75.7	86.2
726435	1	35.8	38.8	39.4	43.3	52.7	60.2	62.7	73.5	70.1	81.3	74.8	86.4	77.4	89.0	76.6	87.3	72.8	81.2	65.4	71.5	54.8	58.2	40.2	41.4	73.8	83.3
	2	34.4	37.1	37.4	40.9	49.3	56.2	60.8	70.6	68.5	78.9	73.4	84.7	76.1	87.0	75.2	84.9	71.4	78.5	63.3	70.7	51.8	56.2	37.2	39.2	71.9	81.0
Green Bay	0.4	39.4	41.4	40.9	43.2	56.8	62.5	65.5	76.6	72.5	82.5	76.2	88.2	78.4	88.4	78.3	88.1	74.5	82.8	67.5	74.5	57.9	61.4	48.4	50.1	75.5	84.9
726450	1	37.2	39.2	38.2	41.0	52.8	58.2	63.2	72.9	70.9	80.4	74.9	85.4	77.1	86.4	76.5	85.0	73.1	80.3	65.2	71.9	55.7	58.5	43.9	45.5	73.6	82.2
	2	35.5	37.5	36.6	39.2	49.2	54.2	60.9	68.6	69.0	78.0	73.6	83.1	75.9	84.7	75.0	84.3	71.3	77.7	62.9	68.1	53.3	56.3	39.3	40.5	71.7	79.8
La Crosse	0.4	40.9	43.8	44.3	49.4	58.0	67.5	65.8	78.3	73.4	84.3	77.6	88.8	80.8	90.4	79.6	90.2	76.4	86.1	68.3	75.1	58.7	61.5	50.4	52.9	77.3	86.8
726430	1	38.0	41.3	41.1	45.4	55.1	62.0	63.8	75.1	71.1	79.6	76.2	86.5	79.3	88.7	78.0	89.2	74.9	83.1	66.6	73.3	57.0	60.2	44.2	46.5	75.3	84.3
	2	35.9	38.5	38.8	42.4	51.9	58.7	61.9	71.1	69.5	78.7	74.8	85.0	77.7	87.0	76.8	85.7	73.3	80.6	64.6	71.2	54.2	58.1	39.5	41.5	73.6	81.8
Madison	0.4	43.6	46.6	44.9	50.0	59.2	67.1	65.2	77.7	72.0	81.7	76.6	87.6	78.8	90.0	78.3	88.6	75.2	84.2	67.6	76.8	59.1	62.9	52.5	55.2	76.0	86.4
726410	1	40.0	42.9	41.9	46.2	56.4	63.1	63.4	74.3	70.5	81.0	75.0	86.2	77.4	88.0	76.8	86.5	73.5	82.1	66.0	74.0	57.4	61.0	48.0	49.5	74.0	83.5
	2	37.2	39.9	39.4	43.1	53.6	60.6	61.5	71.0	69.1	79.6	73.7	84.4	76.3	86.9	75.5	85.6	72.0	79.5	64.2	70.3	55.6	58.8	43.0	45.2	72.2	81.5
Milwaukee	0.4	45.7	47.9	45.5	49.6	59.8	66.0	65.3	76.4	72.1	81.5	76.0	86.7	79.0	90.0	78.5	88.7	75.9	83.8	68.3	78.0	60.2	64.0	54.3	57.0	76.3	85.8
726400	1	41.5	43.7	42.5	45.9	57.3	63.9	63.5	74.2	70.4	80.6	74.8	85.6	77.7	88.1	77.6	87.0	74.2	82.8	66.1	73.6	58.5	62.1	49.3	51.8	74.3	83.3
	2	38.6	40.9	39.9	43.5	53.6	60.4	61.4	70.0	68.8	77.9	73.6	84.4	76.6	86.3	76.2	85.3	72.6	79.2	64.1	70.3	56.5	60.1	45.4	47.5	72.4	80.6
WYOMING																											
Casper	0.4	39.3	47.8	42.6	55.3	46.7	63.3	51.4	71.8	57.2	75.6	62.8	82.0	64.4	82.4	64.3	84.3	59.5	80.4	52.4	72.8	45.9	60.8	40.8	51.4	62.4	81.1
725690	1	37.9	45.7	40.7	51.4	44.6	60.2	50.1	69.5	56.1	73.1	61.8	81.1	63.5	82.2	63.2	82.4	58.3	78.0	51.3	71.7	44.2	59.0	39.3	49.1	61.0	80.0
	2	36.6	43.9	39.2	48.9	43.3	58.8	48.9	67.7	55.0	72.7	60.8	80.1	62.6	81.2	62.2	81.0	57.3	77.4	50.3	69.7	42.6	56.6	37.9	46.8	59.8	79.0
Cheyenne,	0.4	39.9	53.9	42.4	57.3	44.8	64.3	50.3	68.2	56.4	69.8	62.3	76.9	64.4	78.2	64.3	77.2	60.1	74.0	51.8	71.8	45.3	61.9	41.2	56.0	62.3	76.8
Warren AFB	1	38.4	51.7	40.5	54.5	43.3	61.4	49.2	67.6	55.2	70.1	61.3	77.3	63.4	78.4	63.0	76.7	58.6	72.8	50.4	69.9	44.0	60.6	39.5	53.3	60.9	75.8
725640	2	37.1	49.4	38.9	52.8	41.9	58.8	48.1	65.5	54.1	68.9	60.3	76.6	62.6	77.6	62.0	75.8	57.3	72.2	49.3	68.6	42.5	58.5	38.1	51.0	59.6	74.7
Lander	0.4	38.6	48.5	41.8	51.8	45.3	62.8	51.6	72.3	56.4	74.6	62.2	82.0	64.4	83.4	64.3	82.9	59.4	78.7	52.9	71.0	45.5	59.0	40.2	50.6	62.1	81.4
725760	1	37.1	46.4	39.9	50.1	43.6	59.8	49.9	70.3	55.2	73.9	61.1	81.6	63.4	82.8	63.0	82.3	58.2	79.0	51.4	70.9	43.8	57.1	38.4	49.1	60.5	80.2
	2	35.8	44.7	38.2	48.4	42.4	57.6	48.5	67.1	54.2	72.6	60.0	79.6	62.5	81.9	61.8	80.9	57.0	77.5	50.3	69.6	42.2	55.2	36.9	47.0	59.2	79.5
Rock Springs	0.4	37.3	43.6	39.7	47.5	42.9	58.0	47.8	68.5	52.2	69.1	57.4	81.0	60.9	72.9	59.6	72.4	57.4	72.0	49.2	64.5	43.3	54.5	38.5	45.6	58.2	74.6
725744	1	35.9	41.1	37.8	43.8	41.4	55.9	46.5	65.5	51.1	69.6	56.3	78.6	59.6	74.8	58.7	72.1	55.4	69.6	48.1	65.1	41.7	52.5	36.8	42.8	56.8	74.1
	2	34.5	39.2	36.4	43.1	40.0	53.1	45.3	63.3	50.2	68.9	55.2	76.9	58.7	75.3	57.9	72.3	54.1	69.6	47.0	64.8	40.2	51.0	35.2	40.8	55.5	73.8
Sheridan	0.4	41.4	52.6	44.9	59.8	48.7	66.7	55.9	75.1	61.4	79.9	67.4	86.6	68.6	83.8	66.4	87.6	62.8	83.3	55.7	77.4	48.7	65.9	43.6	57.7	65.6	84.6
726660	1	39.9	49.7	42.9	55.2	47.2	63.8	53.6	72.4	60.3	77.8	66.1	83.9	67.1	83.9	65.4	86.1	61.0	81.7	54.3	74.9	46.8	62.9	41.7	54.8	64.1	83.0
	2	38.7	48.1	41.2	52.0	45.7	61.3	52.1	70.0	59.2	75.6	65.0	82.9	66.1	85.0	64.5	85.4	59.9	81.3	53.0	72.8	45.2	60.3	39.8	50.7	62.6	81.4

WMO# = World Meteorological Organization number WB = wet-bulb temperature, °F MDB = mean coincident dry-bulb temperature, °F

Table 4B Selected Monthly Percentiles of Dry-Bulb Temperature with Mean Wet-Bulb Temperature for U.S. Locations

Location and WMO#	%	Jan		Feb		Mar		Apr		May		Jun		Jul		Aug		Sep		Oct		Nov		Dec		Annual	
		DB	MWB	DB	MWB	DB	MWB	DB	MWB	DB	MWB	DB	MWB	DB	MWB	DB	MWB	DB	MWB	DB	MWB	DB	MWB	DB	MWB	DB	MWB
1a	1b	2a	2b	3a	3b	4a	4b	5a	5b	6a	6b	7a	7b	8a	8b	9a	9b	10a	10b	11a	11b	12a	12c	13a	13b	14a	14b
VERMONT																											
Burlington	0.4	49.0	45.1	52.3	46.9	66.0	56.4	78.8	60.9	86.8	67.5	89.6	70.6	92.2	73.6	89.3	72.7	84.1	72.7	74.7	63.5	65.8	59.3	56.2	52.7	87.4	70.9
726170	1	45.0	41.7	47.2	41.4	61.0	50.7	73.4	57.2	83.5	65.2	87.3	69.8	90.2	73.0	87.0	71.5	81.6	71.4	72.2	61.7	63.0	57.0	53.1	48.2	84.3	69.3
	2	41.8	37.5	44.4	40.5	56.5	48.5	69.8	56.1	80.4	63.9	85.3	69.4	88.0	72.4	85.2	70.6	79.2	69.1	69.6	60.1	60.4	54.3	48.5	44.2	81.5	68.0
VIRGINA																											
Lynchburg	0.4	67.6	56.6	72.2	56.5	80.2	62.6	87.7	63.7	88.1	69.8	92.7	74.4	96.6	74.5	95.2	74.2	92.6	73.8	82.7	68.0	76.7	60.8	70.3	61.9	92.7	74.4
724100	1	63.2	53.5	67.7	54.0	76.8	60.7	84.7	62.1	86.6	69.5	91.1	73.6	94.8	74.6	93.4	74.2	90.2	73.5	80.9	66.1	74.2	59.9	67.1	59.4	90.1	73.6
	2	59.9	52.0	64.8	52.6	73.4	58.5	82.3	61.6	84.8	68.5	89.5	72.9	93.3	74.6	91.7	73.9	88.3	72.4	79.1	65.2	71.5	58.9	63.5	57.4	87.9	72.6
Norfolk	0.4	70.5	61.7	74.4	62.0	81.0	65.2	86.5	65.5	90.1	71.8	94.7	76.3	96.7	77.6	95.7	77.1	93.2	75.9	84.6	70.2	78.8	66.7	72.7	63.9	93.2	76.6
723080	1	67.2	59.9	70.9	59.5	78.0	62.3	84.1	65.0	88.0	70.5	92.5	75.6	95.1	77.2	93.5	77.1	90.7	75.4	82.2	69.7	76.4	64.9	70.4	62.9	90.7	75.6
	2	64.6	58.6	67.8	59.0	74.8	61.2	81.7	64.3	86.2	69.7	91.0	75.0	93.5	77.1	91.7	76.6	88.3	74.5	79.7	68.8	74.1	64.1	68.1	62.1	88.2	74.7
Richmond	0.4	68.1	59.0	73.0	56.7	81.8	64.4	89.1	65.4	91.6	72.2	95.6	77.0	97.6	77.7	96.3	77.2	93.9	75.2	85.1	68.9	79.0	64.5	71.5	62.8	94.1	76.3
724010	1	64.8	56.9	69.7	57.4	78.7	62.5	86.1	65.0	89.5	70.5	93.6	75.7	96.0	77.1	94.3	77.3	91.7	75.3	82.5	68.3	76.1	63.6	69.2	61.3	91.6	75.4
	2	62.2	54.4	66.4	56.3	75.2	59.6	83.5	64.3	87.5	69.8	91.9	74.6	94.6	76.8	92.6	76.7	89.4	73.9	80.3	66.9	73.5	62.1	66.0	58.2	89.2	74.3
Roanoke	0.4	65.9	54.3	70.8	55.6	79.6	60.5	86.0	62.8	88.5	68.9	92.5	72.4	96.6	73.6	94.8	73.0	91.3	72.9	82.4	65.4	75.5	60.0	68.2	58.2	91.7	72.8
724110	1	62.0	51.6	67.1	53.2	76.3	58.4	83.2	61.4	86.5	68.0	90.9	72.1	94.4	73.0	92.3	73.1	89.2	72.3	80.3	64.1	72.7	59.7	65.3	56.7	89.3	72.0
	2	59.0	49.6	63.8	51.7	73.1	56.7	80.6	60.6	84.4	66.8	89.3	71.3	92.5	73.4	90.4	72.9	87.0	71.2	78.4	63.5	70.1	58.1	62.5	54.6	86.8	70.9
Sterling	0.4	63.0	53.9	68.5	54.3	78.4	61.0	85.5	63.7	89.3	71.2	93.5	74.5	96.9	75.8	94.9	75.6	92.8	75.2	82.8	66.6	75.6	63.5	68.8	61.5	92.6	75.0
724030	1	59.3	52.0	64.2	53.0	74.9	59.5	82.6	63.7	87.2	69.9	91.8	73.6	94.9	75.7	92.9	75.1	90.4	73.9	80.4	66.3	72.7	61.3	64.8	58.4	90.0	73.8
	2	55.2	48.8	61.0	54.0	71.0	58.3	79.6	62.8	85.2	68.8	89.9	73.5	93.2	75.6	91.3	74.5	88.4	73.2	77.9	65.7	70.0	60.1	61.5	56.1	87.6	72.7
WASHINGTON																											
Olympia	0.4	55.2	53.0	61.4	50.6	66.2	51.9	74.8	57.1	84.1	63.0	88.7	65.9	92.2	68.0	92.8	68.6	86.7	64.9	74.4	59.6	59.9	54.3	55.7	52.7	87.1	66.8
727920	1	53.8	51.1	58.4	50.6	63.4	50.6	71.3	56.0	80.4	62.3	85.2	65.6	89.2	67.6	90.0	67.8	83.1	63.5	71.0	58.8	57.8	53.3	53.5	51.1	83.0	64.8
	2	52.4	49.9	56.0	50.4	60.9	49.9	68.0	55.1	77.1	60.4	82.5	64.4	86.4	66.9	87.1	67.0	80.7	63.1	67.8	57.6	56.1	52.3	52.2	50.2	79.4	63.5
Quillayute	0.4	56.0	50.1	61.8	50.5	62.5	50.1	69.9	54.1	79.6	60.1	82.2	61.7	84.5	63.3	85.3	66.6	82.9	63.2	72.7	59.6	59.4	51.7	54.6	51.9	79.5	62.3
727970	1	53.4	49.2	58.3	49.6	59.6	49.2	66.2	52.2	74.8	57.9	76.6	61.5	80.6	62.8	80.9	63.9	79.8	62.0	68.5	56.1	57.2	53.3	53.4	51.6	73.9	60.7
	2	51.8	48.7	55.4	49.2	57.1	48.5	62.7	51.8	69.8	56.6	72.0	60.2	76.4	62.6	76.9	62.6	76.0	60.7	65.3	56.3	55.7	53.1	52.4	50.8	69.5	59.0
Seattle,	0.4	56.2	49.5	62.9	50.6	64.7	50.8	74.0	54.6	82.6	62.4	86.8	65.2	89.9	66.5	90.6	67.6	84.8	63.5	73.5	59.5	61.4	51.3	55.8	52.4	85.0	64.9
Intl Airport	1	54.4	49.2	59.8	49.6	62.4	50.1	70.4	55.6	78.9	60.7	83.7	64.2	86.7	66.2	87.4	66.5	81.8	62.9	70.4	58.5	58.9	52.1	54.1	50.5	81.4	63.6
727930	2	52.8	48.6	57.2	48.9	60.1	49.2	67.2	53.3	75.5	60.0	80.9	62.8	84.3	65.3	84.6	65.4	79.1	62.2	67.4	57.1	57.2	51.9	52.6	49.2	77.7	62.3
Spokane,	0.4	48.4	44.5	53.5	46.3	62.5	49.5	75.8	56.2	85.7	60.1	91.5	62.5	96.3	64.0	96.5	62.7	88.8	61.0	76.3	55.9	55.7	47.9	49.3	44.9	92.1	62.3
Fairchild AFB	1	46.0	42.0	51.2	44.6	59.6	47.7	71.3	53.5	82.5	58.9	89.3	62.0	94.2	63.0	94.2	62.7	85.7	59.7	73.1	54.9	53.0	47.5	46.8	42.7	88.7	61.3
727855	2	43.8	40.7	49.2	43.2	57.0	46.0	68.0	51.9	79.4	56.8	86.7	61.4	92.4	62.3	92.2	62.2	83.4	59.0	70.2	53.3	51.1	45.9	44.9	41.6	85.2	60.2
Yakima	0.4	55.7	45.7	60.7	48.7	68.1	51.1	78.5	57.3	90.0	62.9	95.6	65.3	99.7	67.1	99.8	67.0	90.7	64.1	79.8	59.1	62.1	51.2	55.7	47.0	95.2	65.4
727810	1	52.4	44.6	58.5	47.3	65.4	49.9	75.9	55.2	86.8	60.8	93.0	63.6	97.7	66.7	97.2	66.5	88.3	62.9	76.8	57.5	59.7	49.7	52.1	44.7	91.7	64.3
	2	50.2	43.1	56.0	45.9	63.0	48.7	73.1	53.9	83.9	60.0	90.7	63.2	95.8	65.5	95.0	65.5	86.2	62.1	73.6	56.4	57.6	48.2	49.1	43.0	88.1	63.0
WEST VIRGINIA																											
Charleston	0.4	66.6	55.5	70.1	55.0	79.7	59.2	86.3	63.4	88.3	68.5	91.4	72.0	94.1	74.8	93.7	73.3	90.2	71.9	81.9	65.8	76.6	61.7	70.1	58.6	90.5	73.0
724140	1	62.9	53.1	67.0	54.1	77.1	58.3	83.9	62.7	86.6	67.2	90.0	71.6	92.2	74.0	91.6	73.1	88.0	72.6	80.2	65.7	73.9	59.8	66.8	56.8	88.1	72.5
	2	60.3	52.1	64.1	52.7	74.6	57.0	82.0	61.6	85.3	67.6	88.6	72.0	90.6	74.6	89.6	73.1	86.1	71.7	78.2	64.2	71.6	58.5	64.4	55.9	86.0	71.3
Elkins	0.4	61.3	50.9	65.1	54.0	75.5	56.9	81.1	59.5	83.3	66.2	85.9	69.7	88.9	71.0	88.2	70.4	84.5	70.0	77.9	62.7	72.4	58.5	66.4	57.1	85.3	70.5
724170	1	58.6	50.4	62.1	52.5	72.4	54.7	79.1	59.9	81.7	65.3	84.4	69.1	87.1	71.4	86.4	70.5	83.1	70.5	76.1	62.2	69.8	56.9	63.4	55.4	83.1	69.6
	2	56.0	49.8	59.8	49.8	69.9	54.6	77.1	58.8	80.3	64.5	83.2	68.9	85.5	71.6	84.5	70.6	81.5	69.6	74.4	60.9	67.4	56.0	60.4	53.6	81.2	68.6
Huntington	0.4	65.9	57.2	69.7	55.3	80.3	60.3	86.1	63.3	87.9	70.3	91.6	72.6	94.8	75.9	93.9	74.9	90.5	72.6	82.1	66.7	76.4	62.7	69.7	59.7	90.9	74.3
724250	1	62.7	53.6	66.9	53.6	77.2	58.3	83.9	63.2	86.4	69.1	90.2	72.5	92.7	75.0	92.0	74.8	88.2	72.9	80.3	65.2	73.6	60.5	66.7	57.6	88.5	73.4
	2	59.9	52.2	63.8	52.6	74.5	57.7	81.8	62.3	85.0	68.0	88.8	72.7	91.2	75.1	90.3	74.7	86.4	72.2	78.4	64.5	71.1	58.6	64.0	56.2	86.3	72.2
WISCONSIN																											
Eau Claire	0.4	41.8	37.2	47.9	41.6	66.7	54.7	82.4	63.4	87.6	68.1	92.6	73.6	95.5	75.4	92.9	75.0	87.2	72.6	78.3	63.7	62.6	53.1	50.2	48.3	90.0	73.3
726435	1	39.5	35.5	44.2	39.0	61.7	51.6	77.6	59.3	85.5	67.2	89.9	72.3	93.1	74.9	90.4	74.6	84.4	70.3	75.2	62.8	59.4	53.7	42.8	38.9	86.9	71.1
	2	37.3	33.9	41.4	36.5	57.4	48.2	74.0	58.1	83.3	65.0	87.9	70.8	90.7	73.7	88.3	73.1	81.7	69.0	72.2	60.9	57.1	51.1	39.7	36.5	84.0	69.6
Green Bay	0.4	41.9	38.1	43.9	39.8	64.7	55.8	80.0	64.5	85.5	69.2	91.9	71.7	92.5	74.9	90.8	75.9	85.3	72.5	76.9	65.4	62.7	55.8	50.3	48.1	88.2	73.2
726450	1	39.5	36.8	41.8	37.9	59.1	51.6	75.5	61.2	83.4	68.5	89.4	71.9	90.3	74.5	88.4	74.7	82.6	71.1	73.6	63.0	59.5	54.2	45.8	43.8	85.1	71.5
	2	37.8	35.2	39.8	36.0	55.0	48.5	71.0	58.6	81.1	66.9	86.9	70.9	88.3	73.7	86.1	72.7	80.2	69.1	70.9	61.4	56.4	52.7	41.6	38.4	82.3	69.9
La Crosse	0.4	44.4	39.6	50.3	43.2	70.2	57.3	83.2	62.9	88.1	69.5	93.8	73.7	96.0	77.3	94.2	77.1	89.2	73.9	79.3	65.0	64.4	55.6	52.5	50.0	91.1	74.4
726430	1	41.5	37.5	46.6	40.2	64.4	53.2	79.3	61.8	86.0	67.9	91.3	72.9	93.5	76.0	91.5	75.8	86.2	71.5	76.5	63.7	61.2	54.6	47.0	43.3	88.0	72.7
	2	39.4	35.6	43.6	38.2	60.5	50.1	75.4	59.7	83.5	66.6	89.2	72.3	91.6	75.2	89.3	74.2	83.3	70.7	74.1	62.3	59.0	53.7	42.5	38.4	85.1	71.3
Madison	0.4	47.4	43.5	51.1	43.3	70.0	55.4	81.4	63.2	86.7	67.8	91.8	72.8	94.4	74.6	92.5	75.5	87.4	72.5	79.3	66.3	65.9	56.2	55.4	52.7	89.7	73.3
726410	1	43.3	39.6	46.9	41.1	65.7	54.3	78.0	61.5	84.7	68.0	89.9	72.2	92.2	74.5	90.2	74.3	85.0	70.7	76.4	63.7	62.6	56.0	50.0	46.9	86.9	71.8
	2	40.5	36.8	43.9	39.0	61.6	52.2	74.1	59.6	82.7	66.9	88.3	71.7	90.1	74.1	88.1	73.8	82.7	70.4	73.6	62.0	59.4	54.3	46.0	42.3	84.1	70.4
Milwaukee	0.4	48.7	44.9	51.0	44.5	70.3	57.1	80.9	63.3	85.5	68.2	92.1	72.8	93.8	75.3	93.0	76.6	87.8	73.0	79.7	66.7	66.6	58.5	57.0	53.7	89.4	74.0
726400	1	44.4	41.1	46.7	41.6	65.6	56.3	76.3	62.3	83.3	67.4	89.8	72.0	91.5	75.5	90.3	75.2	84.8	72.7	75.9	64.6	63.2	57.6	52.1	49.2	85.9	72.1
	2	41.4	38.1	44.3	39.4	60.9	53.3	72.2	58.3	80.9	67.0	87.2	71.1	89.5	74.9	87.7	73.9	82.2	70.4	72.7	61.8	60.3	56.3	47.9	45.2	83.0	70.4
WYOMING																											
Casper	0.4	50.0	38.3	57.1	42.4	66.1	45.4	76.7	49.9	83.2	53.7	93.5	59.0	95.8	59.6	93.7	59.6	89.2	56.4	78.7	50.5	63.3	44.9	53.6	39.7	91.7	58.6
725690	1	47.3	36.5	53.1	39.5	62.7	43.5	73.7	48.6	81.2	53.3	91.2	57.9	94.2	59.0	92.1	58.8	86.9	56.0	76.2	49.9	60.6	43.3	50.4	38.2	89.1	57.8
	2	45.3	35.2	50.5	38.5	60.1	42.6	71.0	48.1	79.1	53.1	88.8	57.3	92.5	58.8	90.6	58.3	84.7	54.7	73.4	48.9	58.0	41.8	48.2	37.1	86.4	57.5
Cheyenne,	0.4	56.0	38.7	60.2	40.6	66.9	44.3	74.3	48.2	80.9	52.2	88.7	57.2	91.7	58.8	89.6	57.7	85.2	54.6	77.1	49.8	65.3	44.2	58.2	39.5	87.4	57.5
Warren AFB	1	53.3	37.2	57.0	39.3	63.7	42.5	72.2	47.6	78.6	51.5	86.6	57.0	90.0	58.3	87.8	57.3	83.3	54.8	75.2	49.1	63.0	42.8	55.7	38.7	84.7	57.0
725640	2	50.9	36.4	54.1	38.1	60.3	41.0	69.5	46.5	76.3	51.7	84.6	56.8	88.4	57.9	86.1	56.9	81.3	54.2	72.8	48.0	60.4	41.8	52.8	37.3	82.0	56.9
Lander	0.4	50.4	37.4	55.2	40.4	65.0	44.7	75.4	49.2	81.7	52.7	92.0	58.1	94.3	59.5	92.2	59.0	86.7	56.3	77.4	50.9	61.4	44.0	52.6	39.3	90.2	58.7
725760	1	47.6	36.5	52.0	38.9	61.9	42.4	72.8	49.6	79.9	52.6	89.5	57.9	92.4	59.2	90.5	58.8	84.6	55.0	74.4	50.1	58.8	43.2	49.8	37.8	87.4	57.8
	2	45.3	35.2	49.4	37.3	58.9	41.9	69.9	47.6	77.7	52.0	87.3	57.4	90.9	58.8	89.0	58.2	82.6	54.3	71.9	49.0	56.0	41.8	47.3	36.6	84.6	57.2
Rock Springs	0.4	45.1	36.6	50.7	37.7	60.7	41.7	71.2	46.9	77.9	49.6	88.0	55.1	90.2	55.4	88.2	54.4	83.1	53.3	72.8	47.0	57.8	41.3	47.4	36.8	86.3	54.3
725744	1	42.5	34.9	47.7	36.7	57.5	40.5	69.2	45.2	76.0	49.2	86.0	54.2	88.4	55.1	86.8	54.3	81.4	51.9	70.8	46.3	55.3	40.4	44.5	35.2	84.0	53.8
	2	40.2	33.3	45.2	35.4	54.8	39.4	66.8	44.1	74.0	48.2	84.2	53.4	86.9	54.4	85.3	53.9	79.3	51.3	68.7	46.0	53.0	39.2	42.2	33.6	81.6	53.3
Sheridan	0.4	55.0	40.7	59.8	44.1	69.6	47.5	78.6	53.8	85.2	58.4	95.1	63.9	98.1	63.5	97.0	63.0	92.0	59.5	81.7	53.9	68.4	47.9	59.2	42.9	93.4	62.2
726660	1	51.4	39.3	56.1	42.5	66.3	46.6	75.5	52.5	82.6	57.9	91.4	63.2	95.9	62.7	95.1	62.2	89.1	58.2	78.4	52.8	64.7	46.4	55.6	41.1	89.9	61.1
	2	48.8	37.6	53.4	40.5	62.5	45.2	72.7	50.8	80.0	57.2	88.4	61.9	93.6	62.9	93.1	61.6	86.3	58.1	75.2	51.9	60.6	44.9	51.9	39.2	86.4	60.8

WMO# = World Meteorological Organization number DB = dry-bulb temperature, °F MWB = mean coincident wet-bulb temperature, °F

CHAPTER 28

RESIDENTIAL COOLING AND HEATING LOAD CALCULATIONS

THIS CHAPTER covers the engineering basis of modified residential load calculation procedures for the nonengineer. The procedures described in Chapter 29 may be used to calculate a heating or cooling load for residential buildings.

RESIDENTIAL FEATURES

With respect to heating and cooling load calculation and equipment sizing, the unique features distinguishing residences from other types of buildings are the following:

Unlike many other structures, residences are usually occupied and conditioned 24 h per day, virtually every day of the cooling and heating seasons.

- Residential system loads are primarily imposed by heat loss or gain through structural components and by air leakage or ventilation. Internal loads, particularly those from occupants and lights, are small in comparison to those in commercial or industrial structures.
- Most residences are conditioned as a single zone. Unit capacity cannot be redistributed from one area to another as loads change from hour to hour; however, exceptions do occur.
- Most residential cooling systems use units of relatively small capacity (about 18,000 to 60,000 Btu/h cooling, 60,000 to 110,000 Btu/h heating). Because loads are largely affected by outside conditions, and few days each season are design days, the unit operates at only partial load during most of the season; thus, an oversized unit is detrimental to good system performance, especially for cooling in areas of high wet-bulb temperature.
- Dehumidification occurs during cooling unit operation only, and space condition control is usually limited to use of room thermostats (sensible heat-actuated devices).
- Multifamily living units are similar to single-family detached houses, but the living units may not all have surfaces exposed in all directions. This affects load calculation.

Categories of Residences

Single-Family Detached. A house in this category usually has exposed walls in four directions, often more than one story, and a roof. The cooling system is a single-zone, unitary system with a single thermostat. Two-story houses may have a separate cooling system for each floor. The rooms are reasonably open and generally have a centralized air return. In this configuration, both air and load from rooms are mixed, and a load-leveling effect, which requires a distribution of air to each room that is different from a pure commercial system, results. Because the amount of air supplied to each room is based on the load for that room, proper load calculation procedures must be used.

Multifamily Buildings. Unlike single-family detached units, multifamily units by definition do not have exposed surfaces facing in all directions. Rather, each unit has only one or two exposed surfaces and possibly a roof. Two exposed walls will be at right angles, and both east and west walls will not be exposed in a given living unit. Each living unit has a single unitary cooling system or a single fan-coil unit, and the rooms are relatively open to one another. This configuration does not have the same load-leveling effect as a single-family detached house, but it is not a commercial building. Therefore, a specific load calculation procedure is required.

Other Categories. Many buildings do not fall into either of the above categories. Critical to the designation of a single-family detached building is the exposure of both east and west walls. Therefore, some multifamily structures should be treated as single-family detached when the exposed surfaces are oriented in a particular way. Examples include duplexes or apartments with either exposed east, west, and south walls or exposed east, west, and north walls, with or without a roof; and apartments, town houses, or condominiums with only east and west or north and south exposed walls.

COOLING LOAD

LOAD COMPONENTS

A cooling load calculation determines total sensible cooling load due to heat gain (1) through structural components (walls, floors, and ceilings); (2) through windows; (3) caused by infiltration and ventilation; and (4) due to occupancy. The latent portion of the cooling load is evaluated separately. While the entire structure may be considered a single zone, equipment selection and system design should be based on a room-by-room calculation. For proper design of the distribution system, the amount of conditioned air required by each room must be known.

Peak Load Computation

To select a properly sized cooling unit, the peak or maximum load (**block load**) for each zone must be computed. Because this procedure may vary considerably for different types of buildings, each building type has to be considered; the block load for a single-family detached house with one central system is the sum of all the room loads. If the house has a separate system for each zone, each zone block load (i.e., the sum of the loads for all rooms in each zone) is required. When a house is zoned with one central cooling system, the block load must be computed for the complete house as if it were one zone. In multifamily structures, each living unit has a zone load that equals the sum of the room loads. For apartments with separate systems, the block load for each unit establishes the system size. Apartment buildings with a central cooling system (i.e., a hydronic system with fan-coils in each apartment) require a block load calculation for the complete structure to size the central system; each unit

The preparation of this chapter is assigned to TC 4.1, Load Calculation Data and Procedures.

load establishes the size of the fan-coil and air distribution system for each apartment. One of the methods discussed in Chapter 29 may be used to calculate the block load.

Indoor Temperature Swing

For hour-by-hour load calculations, allowing for a swing in indoor temperature results in lower peak loads. Because the indoor temperature does swing, such an allowance gives a more reasonable equipment capacity. The tables in this section are based on an assumed indoor temperature swing of no more than 3°F on a design day, when the residence is conditioned 24 h per day and the thermostat is set at 75°F.

Cooling Load Due to Heat Gain Through Structure

The sensible cooling load due to heat gains through the walls, floor, and ceiling of each room is calculated using appropriate **cooling load temperature differences** (CLTDs) (Tables 1 and 2) and U-factors for summer conditions. For ceilings under naturally vented attics or beneath vented flat roofs, the combined U-factor for the roof, vented space, and ceiling should be used. The mass of the walls is a variable in Table 2 and is important in calculating energy use, but it is not used in Table 1 because of the averaging technique required to develop the CLTDs. Values in Tables 1 and 2 assume a dark color because color is an unpredictable variable in any residence.

Daily range (outdoor temperature swing on a design day) significantly affects the equivalent temperature difference. Tables 1 and 2 list daily temperature ranges classified as high, medium, and low. Tables 1, 2, and 3 in Chapter 27 list outdoor daily ranges of dry-bulb temperature for different locations.

Cooling Load Due to Heat Gain Through Windows

Direct application of procedures for calculating cooling load due to heat gain for flat glass (discussed in Chapters 29 and 30) results in unrealistically high cooling loads for residential installations. Window **glass load factors** (GLFs), modified for single- and multifamily residential cooling load calculations and including solar heat load plus air-to-air conduction, are given in Tables 3 and 4. Table 5 lists the **shading coefficients** (SCs) and U-factors used to compile Tables 3 and 4.

Table 1 CLTD Values for Single-Family Detached Residences[a]

Daily Temperature Range[b]	Design Temperature, °F											
	85		90			95			100		105	110
	L	M	L	M	H	L	M	H	M	H	M	H
All walls and doors												
North	8	3	13	8	3	18	13	8	18	13	18	23
NE and NW	14	9	19	14	9	24	19	14	24	19	24	29
East and West	18	13	23	18	13	28	23	18	28	23	28	33
SE and SW	16	11	21	16	11	26	21	16	26	21	26	31
South	11	6	16	11	6	21	16	11	21	16	21	26
Roofs and ceilings												
Attic or flat built-up	42	37	47	42	37	51	47	42	51	47	51	56
Floors and ceilings												
Under conditioned space, over unconditioned room, or over crawl space	9	4	12	9	4	14	12	9	14	12	14	19
Partitions												
Inside or shaded	9	4	12	9	4	14	12	9	14	12	14	19

[a]Cooling load temperature differences (CLTDs) for single-family detached houses, duplexes, or multifamily, with both east and west exposed walls or only north and south exposed walls, °F.

[b]L denotes low daily range, less than 16°F; M denotes medium daily range, 16 to 25°F; and H denotes high daily range, greater than 25°F.

In application, the area of each window is multiplied by the appropriate GLF. The effects of permanent outside shading devices should be considered separately in determining the cooling load. Shaded glass is considered the same as north-facing glass. The **shade line factor** (SLF) is the ratio of the distance a shadow falls beneath the edge of an overhang to the width of the overhang (Table 6). Therefore, assuming the overhang is at the top of the window, the shade line equals the SLF times the overhang width. The shaded and sunlit glass areas may then be computed separately. The tabulated values are the average of the shade line values for 5 h of maximum solar intensity on each wall orientation shown. Northeast- and northwest-facing windows are not effectively protected by roof overhangs; in most cases, they should not be considered shaded.

Infiltration

Natural air leakage in residential structures is less in summer than in winter, largely because wind velocities are lower in most localities. The data in Tables 7 and 8 showing space air changes per hour (ACH) apply to both single- and multifamily housing,

Table 2 CLTD Values for Multifamily Residences[a]

Daily Temperature Range[b]		Design Temperature, °F											
		85		90			95			100		105	110
		L	M	L	M	H	L	M	H	M	H	M	H
Walls and doors[c]													
N	Light	14	11	19	16	12	24	21	17	26	22	27	32
	Medium	13	10	18	15	11	23	20	16	25	21	26	31
	Heavy	9	6	15	11	7	20	16	12	21	17	22	27
NE	Light	23	17	28	22	17	33	27	22	32	26	31	36
	Medium	20	15	25	20	16	30	25	21	29	25	29	34
	Heavy	16	12	21	17	13	26	22	18	26	22	26	31
E	Light	32	27	37	32	27	43	38	32	42	37	42	47
	Medium	30	24	34	29	24	40	34	29	39	33	39	44
	Heavy	23	18	28	23	18	34	29	23	33	28	33	38
SE	Light	31	27	35	31	26	41	37	31	42	37	42	47
	Medium	28	22	32	27	22	37	32	27	37	33	38	43
	Heavy	21	16	26	22	17	32	27	22	31	27	32	37
S	Light	25	22	29	26	22	35	31	26	36	32	37	43
	Medium	22	18	26	22	18	31	26	22	31	27	32	38
	Heavy	16	11	20	16	12	26	21	17	26	21	27	33
SW	Light	39	36	44	40	35	50	46	40	51	47	52	58
	Medium	33	29	37	34	29	44	40	35	45	40	46	52
	Heavy	23	18	28	24	19	36	31	25	35	30	36	42
W	Light	44	41	48	45	40	54	51	46	56	52	57	63
	Medium	37	33	41	38	33	46	42	38	48	43	49	55
	Heavy	26	22	31	27	23	37	32	27	37	32	38	44
NW	Light	33	30	37	34	30	43	39	34	44	40	45	50
	Medium	28	25	32	29	24	37	33	29	39	35	40	45
	Heavy	20	16	25	20	16	31	26	21	31	26	32	37
Roof and ceiling													
Attic or flat built-up	Light	58	53	65	60	55	70	65	60	70	65	72	77
Flat built-up	Medium or heavy	21	18	23	21	18	25	23	21	25	23	25	28
Floors and ceiling													
Under or over unconditioned space, crawl space		9	4	12	9	4	14	12	9	14	12	14	19
Partitions													
Inside or shaded		9	4	12	9	4	14	12	9	14	12	14	19

[a]Cooling load temperature differences (CLTDs) for multifamily low-rise or single-family detached if zoned with separate temperature control for each zone, °F.

[b]L denotes low daily range, less than 16°F; M denotes medium daily range, 16 to 25°F; and H denotes high daily range, greater than 25°F.

[c]Light denotes lightweight; medium denotes medium-weight; and heavy denotes heavyweight construction.

Table 3 Window Glass Load Factors (GLFs) for Single-Family Detached Residences[a]

Design Temperature, °F	Regular Single Glass						Regular Double Glass						Heat-Absorbing Double Glass						Clear Triple Glass		
	85	90	95	100	105	110	85	90	95	100	105	110	85	90	95	100	105	110	85	90	95
No inside shading																					
North	34	36	41	47	48	50	30	30	34	37	38	41	20	20	23	25	26	28	27	27	30
NE and NW	63	65	70	75	77	83	55	56	59	62	63	66	36	37	39	42	44	44	50	50	53
East and West	88	90	95	100	102	107	77	78	81	84	85	88	51	51	54	56	59	59	70	70	73
SE and SW[b]	79	81	86	91	92	98	69	70	73	76	77	80	45	46	49	51	54	54	62	63	65
South[b]	53	55	60	65	67	72	46	47	50	53	54	57	31	31	34	36	39	39	42	42	45
Horizontal skylight	156	156	161	166	167	171	137	138	140	143	144	147	90	91	93	95	96	98	124	125	127
Draperies, venetian blinds, translucent roller shades, fully drawn																					
North	18	19	23	27	29	33	16	16	19	22	23	26	13	14	16	18	19	21	15	16	18
NE and NW	32	33	38	42	43	47	29	30	32	35	36	39	24	24	27	29	29	32	28	28	30
East and West	45	46	50	54	55	59	40	41	44	46	47	50	33	33	36	38	38	41	39	39	41
SE and SW[b]	40	41	46	49	51	55	36	37	39	42	43	46	29	30	32	34	35	37	35	36	38
South[b]	27	28	33	37	38	42	24	25	28	31	31	34	20	21	23	25	26	28	23	24	26
Horizontal skylight	78	79	83	86	87	90	71	71	74	76	77	79	58	59	61	63	63	65	69	69	71
Opaque roller shades, fully drawn																					
North	14	15	20	23	25	29	13	14	17	19	20	23	12	12	15	17	17	20	13	13	15
NE and NW	25	26	31	34	36	40	23	24	27	30	30	33	21	22	24	26	27	29	23	23	26
East and West	34	36	40	44	45	49	32	33	36	38	39	42	29	30	32	34	35	37	32	32	35
SE and SW[b]	31	32	36	40	42	46	29	30	33	35	36	39	26	27	29	31	32	34	29	29	31
South[b]	21	22	27	30	32	36	20	20	23	26	27	30	18	19	21	23	24	26	19	20	22
Horizontal skylight	60	61	64	68	69	72	57	57	60	62	63	65	52	52	55	57	57	59	56	57	59

[a]Glass load factors (GLFs) for single-family detached houses, duplexes, or multifamily residences, with both east and west exposed walls or only north and south exposed walls, Btu/h·ft².

[b]Correct by +30% for latitude of 48° and by −30% for latitude of 32°. Use linear interpolation for latitude from 40 to 48 and from 40 to 32°.

To obtain GLF for other combinations of glass and/or inside shading: $GLF_a = (SC_a/SC_t)(GLF_t - U_tD_t) + U_aD_t$, where the subscripts a and t refer to the alternate and table values, respectively. SC_t and U_t are given in Table 5. $D_t = (t_a - 75)$, where $t_a = t_o - (DR/2)$; t_o is the outdoor design temperature and DR is the daily range.

Table 4 Window Glass Load Factors (GLFs) for Multifamily Residences[a]

Design Temperature, °F	Regular Single Glass						Regular Double Glass						Heat-Absorbing Double Glass						Clear Triple Glass		
	85	90	95	100	105	110	85	90	95	100	105	110	85	90	95	100	105	110	85	90	95
No inside shading																					
North	40	44	49	54	58	64	34	36	39	42	44	47	23	24	26	29	30	33	30	32	34
NE	88	89	91	95	97	100	78	79	80	83	84	85	52	52	53	55	55	57	71	71	73
East	136	137	139	142	144	147	120	121	122	125	126	127	79	79	81	83	83	84	109	109	111
SE	129	130	134	139	141	144	109	113	116	119	120	122	72	75	77	79	79	81	99	103	105
South[b]	88	91	96	101	105	110	76	78	81	84	86	89	50	52	54	56	58	60	68	70	72
SW	154	159	164	169	174	179	134	137	140	143	145	148	89	91	93	95	97	99	121	123	125
West	174	178	183	188	192	197	151	154	157	160	162	165	100	102	104	106	108	110	137	139	141
NW	123	127	132	137	141	147	107	109	112	115	117	121	71	72	75	77	79	81	96	98	100
Horizontal	249	252	256	261	264	268	218	220	223	226	228	230	144	146	148	150	152	154	198	200	202
Draperies, venetian blinds, translucent roller shades, fully drawn																					
North	21	25	29	33	36	40	18	21	23	26	28	31	15	17	19	21	23	25	17	19	21
NE	43	44	46	50	51	52	39	40	41	44	45	46	33	33	34	36	36	37	39	39	40
East	67	68	70	74	75	76	61	62	63	65	66	67	50	50	51	54	54	55	60	60	61
SE	64	65	69	73	74	77	58	59	61	63	64	66	48	48	50	52	52	54	57	57	59
South[b]	45	48	52	56	59	63	40	42	44	47	49	52	33	34	36	39	40	42	38	40	42
SW	79	83	87	91	94	98	70	72	75	78	80	83	57	59	62	64	66	68	68	69	71
West	89	92	96	100	103	107	79	81	84	86	88	91	65	66	69	71	72	75	76	78	80
NW	63	66	70	74	77	81	56	58	61	63	66	68	46	48	50	52	54	56	54	55	57
Horizontal	126	128	132	135	137	141	113	115	117	120	121	124	93	94	96	98	100	102	110	111	113
Opaque roller shades, fully drawn																					
North	17	21	25	29	32	36	15	17	20	23	25	28	14	15	18	20	22	24	15	16	18
NE	33	34	35	39	40	42	31	32	33	36	35	37	29	28	30	32	32	34	32	31	33
East	51	52	53	57	61	65	48	49	50	53	52	55	45	45	46	48	48	49	49	49	50
SE	49	50	53	57	58	61	46	47	49	52	52	55	42	43	45	47	47	49	46	46	48
South[b]	35	38	42	46	49	53	32	34	37	40	42	42	29	31	33	35	37	39	32	33	35
SW	61	65	69	73	77	81	57	59	62	65	67	70	52	54	56	58	60	62	56	58	60
West	68	71	75	80	83	87	64	66	68	71	73	76	58	60	62	64	66	68	63	64	66
NW	49	52	56	60	63	67	45	47	50	53	55	58	41	43	45	47	49	51	45	46	48
Horizontal	97	99	102	106	108	111	91	93	95	97	99	102	83	85	87	89	90	92	90	92	93

[a]Glass load factors (GLFs) for multifamily low-rise or single-family detached residences if zoned with separate temperature control for each zone, Btu/h·ft².

[b]Correct by +30% for latitude of 48° and by −30% for latitude of 32°. Use linear interpolation for latitude from 40 to 48 and from 40 to 32°.

To obtain GLF for other combinations of glass and/or inside shading: $GLF_a = (SC_a/SC_t)(GLF_t - U_tD_t) + U_aD_t$, where the subscripts a and t refer to the alternate and table values, respectively. SC_t and U_t are given in Table 5. $D_t = (t_a - 75)$, where $t_a = t_o - (DR/2)$; t_o is the outdoor design temperature and DR is the daily range.

although most of the raw data were for single-family structures (McQuiston 1984). Construction may be defined as follows:

Tight. Good multifamily construction with close-fitting doors, windows, and framing is considered tight. New houses with full vapor retarder, no fireplace, well-fitted windows, weather-stripped doors, one story, and less than 1500 ft^2 floor area fall into this category.

Medium. Medium structures include new, two-story frame houses or one-story houses more than 10 years old with average maintenance, a floor area greater than 1500 ft^2, average fit windows and doors, and a fireplace with damper and glass closure. Below-average multifamily construction falls in this category.

Loose. Loose structures are poorly constructed single- and multifamily residences with poorly fitted windows and doors. Examples include houses more than 20 years old, of average maintenance, having a fireplace without damper or glass closure, or having more than an average number of vented appliances. Average manufactured homes are in this category.

Ventilation

Residential air-conditioning systems may introduce outdoor air, although it is not a code requirement in most localities. Positive ventilation should be considered, however, if the anticipated infiltration is less than about 0.5 ACH. When positive means of introducing outdoor air are used, controls, either manual or automatic, should be provided, and an energy recovery device should be considered.

Occupancy

Even though occupant density is low, occupancy loads should be estimated. Sensible heat gain per sedentary occupant is assumed to be 230 Btu/h. To prevent gross oversizing, the number of occupants should not be overestimated. Recent census studies recommend that the total number of occupants be based on two persons for the first bedroom, plus one person for each additional bedroom. The occupancy load should then be distributed equally among the living areas because the maximum load occurs when most of the residents occupy these areas.

Table 5 Shading Coefficients and U-Factors for Residential Windows

	Inside Shade					
	None		Drapery, Venetian Blind, or Translucent Roller Shade		Opaque Roller Shade	
Glass Type	SC	*U*	SC	*U*	SC	*U*
Single	1.00	1.04	0.50	0.81	0.38	0.81
Double	0.88	0.61	0.45	0.55	0.36	0.55
Heat-absorbing	0.58	0.45	0.37	0.44	0.33	0.44
Triple	0.80	0.44	0.44	0.40	0.36	0.40

Note: U is in $Btu/h \cdot ft^2 \cdot °F$.

Table 6 Shade Line Factors (SLFs)

Direction Window Faces	Latitude, Degrees N						
	24	32	36	40	44	48	52
East	0.8	0.8	0.8	0.8	0.8	0.8	0.8
SE	1.8	1.6	1.4	1.3	1.1	1.0	0.9
South	9.2	5.0	3.4	2.6	2.1	1.8	1.5
SW	1.8	1.6	1.4	1.3	1.1	1.0	0.9
West	0.8	0.8	0.8	0.8	0.8	0.8	0.8

Note: Shadow length below the overhang equals the shade line factor times the overhang width. Values are averages for the 5 h of greatest solar intensity on August 1.

Household Appliances

Appliance loads are concentrated mainly in the kitchen and laundry areas. Based on contemporary living conditions in single-family houses, a sensible load of 1600 Btu/h should be divided between the kitchen and/or laundry and the adjoining room or rooms. A sensible utility load of 1600 Btu/h may be added if the laundry room contains continuously operating appliances such as refrigerators and/or freezers. For multifamily units, the sensible heat gain values should be about 1200 Btu/h. These values assume that the cooking range and clothes dryer are vented. Further allowances should be considered when unusual lighting intensities, computers, or other equipment is present.

Air Distribution System—Heat Loss/Gain

Whenever the air distribution system is outside the conditioned space (i.e., in attics, crawl spaces, or other unconditioned spaces) heat loss or gains to the ducts or pipes must be included in the calculated load and should be considered in equipment selection.

Latent Heat Sources

The latent cooling load has three main sources: outdoor air, occupants, and miscellaneous sources, such as cooking, laundry, and

Table 7 Winter Air Exchange Rates (ACH) as Function of Airtightness

	Outdoor Design Temperature, °F									
Class	50	40	30	20	10	0	−10	−20	−30	−40
Tight	0.41	0.43	0.45	0.47	0.49	0.51	0.53	0.55	0.57	0.59
Medium	0.69	0.73	0.77	0.81	0.85	0.89	0.93	0.97	1.00	1.05
Loose	1.11	1.15	1.20	1.23	1.27	1.30	1.35	1.40	1.43	1.47

Note: Values are for 15 mph wind and indoor temperature of 68°F.

Table 8 Summer Air Exchange Rates (ACH) as Function of Airtightness

	Outdoor Design Temperature, °F					
Class	85	90	95	100	105	110
Tight	0.33	0.34	0.35	0.36	0.37	0.38
Medium	0.46	0.48	0.50	0.52	0.54	0.56
Loose	0.68	0.70	0.72	0.74	0.76	0.78

Note: Values are for 7.5 mph wind and indoor temperature of 75°F.

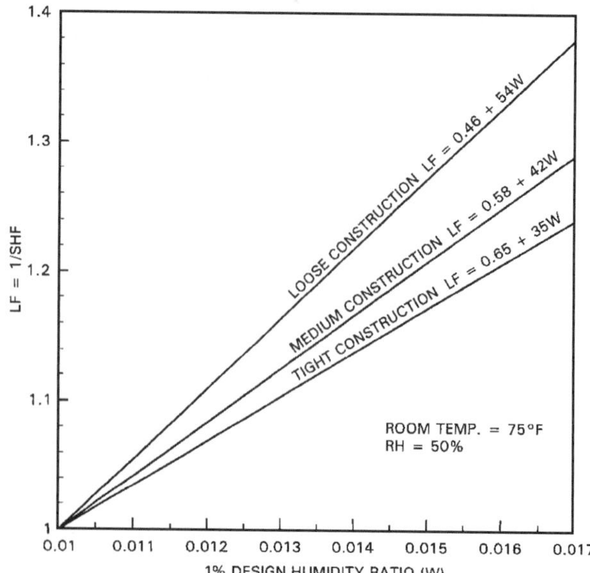

Fig. 1 Effect of Infiltration on Latent Load Factor

bathing. The miscellaneous latent loads are largely covered by outdoor air because most residences have exhaust fans and clothes dryers that vent most of the moisture from these sources. This vent air is accounted for in the infiltration calculation. McQuiston (1984) estimated latent load factors for typical houses located in geographic regions ranging from very dry to very wet using the transfer function method (Figure 1). A **latent factor** LF (LF = 1/SHF) of 1.3 or a **sensible heat factor** SHF (SHF = sensible load/total load) of 0.77 matches the performance of typical residential vapor compression cooling systems. Homes in almost all other regions of North America have cooling loads with an SHF greater than 0.77 and latent factors less than 1.3. Figure 1 may be used to estimate the total cooling load by reading LF as a function of the design humidity ratio and airtightness. Then $q_{total} = (\text{LF})q_{sensible}$. If the humidity ratio is less than 0.01, set LF = 1.0.

LOAD CALCULATION

The cooling load calculation procedures are summarized in Table 9.

Example 1. A single-family detached house (Figure 2) is located in the south central United States at 36°N latitude.

Roof construction. Conventional roof-attic-ceiling combination, vented to remove moisture with 6 in. of fibrous batt insulation and vapor retarder ($U = 0.05$ Btu/h·ft²·°F).

Wall construction. Frame with 4 in. face brick, 3.5 in. fibrous batt insulation, 0.75 in. polystyrene sheathing, and 0.5 in. gypsum wallboard ($U = 0.06$ Btu/h·ft²·°F). Ceiling height is 8 ft throughout.

Floor construction. 4 in. concrete slab on grade.

Fenestration. Clear double glass, 0.125 in. thick, in and out. Assume closed, medium-color venetian blinds. The window glass has a 2 ft overhang at the top.

Doors. Solid core flush with all-glass storm doors ($U = 0.32$ Btu/h·ft²·°F).

Outdoor design conditions. Temperature of 96°F dry bulb with a 24°F daily range and a humidity ratio of 0.0136 lb vapor/lb dry air (74.6°F wet bulb).

U-factors for all external surfaces are based on a 7.5 mph wind velocity.

Indoor design conditions. Temperature of 75°F dry bulb and 50% rh.

Occupancy. Four persons, based on two for the master bedroom and one for each additional bedroom. Assign to the living room.

Appliances and lights. Assume 1600 Btu/h for the kitchen, and assign 50% to the living room. Assume 1600 Btu/h for the utility room, and assign 25% to the kitchen and 25% to the storage room.

The conditioning equipment is located in the garage, and the construction of the house is considered medium.

Find the sensible, latent, and total cooling load; size the cooling unit; and compute the air quantity for each room.

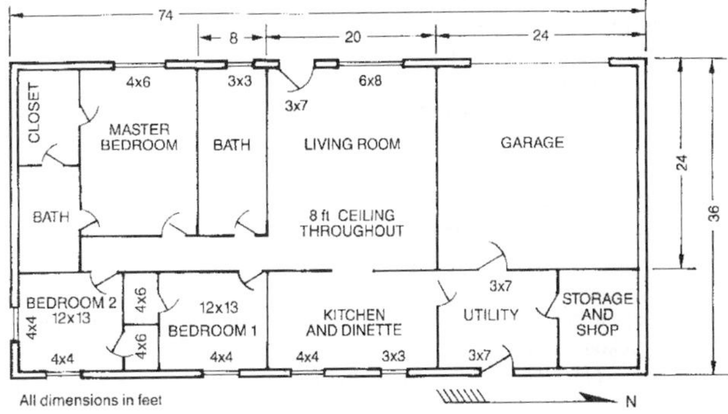

Fig. 2 Floor Plan of Single-Family Detached House

Table 9 Summary of Procedures for Residential Cooling Load Calculations

Load Source	Equation	Tables and Notes
Glass and window areas	$q = (\text{GLF})A$	Glass load factors may be found in Tables 3 and 4 according to window orientation, type of glass, type of interior shading, and outdoor design temperature. The GLF includes effects of both transmission and solar radiation. Glass shaded by overhangs is treated as north glass. Table 6 gives shade line factors.
Doors	$q = U_d A(\text{CLTD})$	Door CLTD values are in Tables 1 and 2 according to orientation, outdoor design temperature, and design daily temperature range.
Above-grade exterior walls	$q = U_w A(\text{CLTD})$	Wall CLTD values are in Tables 1 and 2 based on the outdoor design temperature, daily range, and orientation.
Partitions to unconditioned space	$q = U_p A \Delta t$	Where Δt is the temperature difference across the partition.
Ceilings and roofs	$q = U_r A(\text{CLTD})$	Tables 1 and 2 for CLTD, based on outdoor design temperature and daily range.
Exposed floors	$q = U_f A(\text{CLTD})$	Tables 1 and 2 for CLTD, based on outdoor design temperature and daily range.
Infiltration	$q = 1.1Q\Delta t$ $Q = \text{ACH} \times (\text{room volume})/60$	Air exchange rates are given in Tables 7 and 8.
Internal loads— People, appliances, lights	Plan 230 Btu/h per person.	Divide occupants evenly among rooms not used as bedrooms. If number of occupants is not known, assume two people for first bedroom and one person for each additional bedroom. The appliance and light load of 1600 Btu/h is divided between the kitchen and adjoining room and the laundry and adjoining room. Use 1200 Btu/h for multifamily units.
Total loads	Total cooling load = LF × (Sum of individual sensible cooling load components)	Load factors are from Figure 1 according to outdoor design humidity ratio and airtightness classification.

q = sensible cooling load, Btu/h
Δt = design temperature difference between outside and inside air, °F
A = area of applicable surface, ft²
U = U-factors for appropriate construction, Btu/h·ft²·°F
Q = volumetric airflow rate, cfm
ACH = air changes per hour, 1/h
GLF = glass load factor, Btu/h·ft²
CLTD = cooling load temperature difference, °F
LF = latent load multiplier

Solution: The cooling load must be made on a room-by-room basis to determine the proper distribution of air. The calculations follow the procedure outlined in the section on Load Components.

Walls, roof, windows, and doors. The calculations for the living room and the kitchen, where $q = UA(\text{CLTD})$ for the walls, roof, and door and $q = A(\text{GLF})$ for the windows, are outlined in Table 10. The glass shaded by the overhang is treated as north-facing glass, with the shaded area computed using Table 6.

Internal and infiltration sensible cooling loads. Compute as follows.

For the living room:

Infiltration. Using Table 8,

Q = ACH (room volume)/60

Q = 0.5 × 3840/60 = 32

$q = 1.1Q(t_o - t_i) = 1.1Q\Delta t$

$q = 1.1 \times 32(96 - 75) = 740$ Btu/h

Occupants. Assuming 230 Btu/h per person,

$q = 230 \times$ (persons)

$q = 230 \times 4 = 920$ Btu/h

Appliances. Assuming that 50% of the kitchen appliance load is picked up in the living room,

$q = 0.5 \times$ (kitchen appliance load)

$q = 0.5 \times 1600 = 800$ Btu/h

For the kitchen:

Infiltration.

$Q = 0.5 \times 1920/60 = 16$

$q = 1.1 \times 16(96 - 75) = 370$ Btu/h

No occupants.

$q = 0$

Appliances. Assuming that 25% of the utility appliance load is picked up in the kitchen,

$q = (1600/2) + (1600/4) = 1200$ Btu/h

Table 10 Transmission Cooling Load for Example 1

Item	Net Area, ft²	GLF, Btu/h·ft²	U-Factor, Btu/h·ft²·°F	CLTD, °F	Cooling Load, Btu/h	Reference
Living Room						
West wall	91		0.06	24	131	Table 1
Partition (garage)	192		0.07	12	161	Table 1
Roof	480		0.05	48	1152	Table 1
West door	21		0.32	24	161	Table 1
West glass	35	44			1540	Table 3
Shaded glass	13	19			247	Table 3
Kitchen						
East wall	135		0.06	24	194	Table 1
Roof	240		0.05	48	576	Table 1
East glass	14	44			616	Table 3
Shaded glass	11	19			209	Table 3

Table 11 Summary of Sensible Cooling Load Estimate for Example 1

Room	Roof, Walls, and Doors	Glass	People	Appliances	Infiltration	Total Room Btu/h (q_{rm})	Total Room cfm (Q_{rm})
Living room	1,605	1,787	920	800	740	5,852	296
Kitchen	770	825		1,200	370	3,165	160
Utility and storage	1,404			1,200	443	3,047	154
Bedroom No. 1	559	544			278	1,381	70
Bedroom No. 2	686	848			278	1,812	91
Master bedroom and bath	1,682	816			813	3,311	167
Bath	540	276			295	1,111	56
Total	7,246	5,096	920	3,200	3,217	19,679	994
Duct loss (10%)						1,968	
Outdoor ventilation air						1,600	
					Total	23,247 Btu/h	

For the total sensible cooling load for these two rooms and the cooling load for the remaining rooms, see Table 11. At this point, the sensible cooling load for the house is 19,679 Btu/h. Depending on the design of the air distribution system, heat losses from the supply and return ducts may add to the cooling load. These may be more accurately estimated after designing the system; however, to size the cooling unit, duct losses should be included initially. If all ducts are in the attic space, a duct loss of 10% of the space sensible cooling load is reasonable. For a counterflow system, with ducts below the slab, a 5% loss is more reasonable.

An infiltration rate of 0.5 ACH may not be adequate for good indoor air quality, so some outdoor air should be introduced. This additional cooling load may be estimated in the same way as the infiltration load.

Assume that the entire duct system is in the attic; that is, the total sensible cooling load with a 10% duct loss is 1.1 × 19,679 = 21,647 Btu/h. Also, assume that additional outdoor air is needed to assure good indoor air quality, so the total infiltration and outdoor ventilation air is 0.75 ACH. This increases the infiltration rate by 50%, or about 1600 Btu/h. The total sensible cooling load is then increased to 23,247 Btu/h (Table 11).

The total cooling load (sensible plus latent) may be estimated by applying the latent factor (LF) from Figure 1. For a design humidity ratio of 0.0136 lb vapor per lb dry air, LF = 1.15 for a house of medium construction. Hence, the total cooling load equals 1.15 × 23,247 = 26,734 Btu/h.

The load raises the temperature of the cooling air 18 to 21°F as it leaves the rooms. The total design flow from the air conditioner can be estimated by the following equation:

$$Q_{tot} = \frac{q}{1.1\Delta t} \tag{1}$$

where

Q_{tot} = total airflow, cfm

q = total sensible load, Btu/h

1.1 = density times specific heat of cooling air times 60 min/h

Δt = temperature difference of air entering and leaving room, °F

For a temperature difference of 18°F, the total airflow is estimated from Equation (1) as

$$Q_{tot} = \frac{19{,}679}{1.1 \times 18} = 994 \text{ cfm}$$

The exact design flow can be determined only after the cooling unit has been selected. Then, the supply air quantities can be computed. Air should be supplied to each room on the basis of the room sensible cooling load:

$$Q_{rm} = Q_{tot}(q_{rm}/q)$$

where

Q_{rm} = airflow to each room, cfm

q_{rm} = room sensible cooling load, Btu/h

Thus, for the example,

$$Q_{rm} = (994/19{,}679)q_{rm}$$

If the living space in Example 1 were a multifamily unit (assume that the north, south, and east walls are not exposed surfaces), the calculation procedure would be the same, except that Table 2 would have been used for the CLTDs and Table 4 for the GLFs. Assumptions regarding infiltration, ventilation, and appliance loads are different for smaller multifamily units.

HEATING LOAD

Calculating a residential heating load involves estimating the maximum (block) heat loss of each room or space to be heated and the simultaneous maximum (block) heat loss for the building, while maintaining a selected indoor air temperature during periods of design outdoor weather conditions. Heat losses are mainly

- Transmission losses or heat transferred through the confining walls, glass, ceiling, floor, or other surfaces
- Infiltration losses or energy required to warm outdoor air leaking in through cracks and crevices around doors and windows,

through open doors and windows, and through porous building materials

GENERAL PROCEDURE

To calculate a design heating load, prepare the following information about building design and weather data at design conditions.

1. Select outdoor design weather conditions: temperature, wind direction, and wind speed. Winter climatic data can be found in Chapter 27, or selected weather conditions and temperatures appropriate for the application may be used. Weather station data may differ significantly from values in Chapter 27.
2. Select the indoor air temperature to be maintained in each space during design weather conditions.
3. Temperatures in adjacent unheated spaces, attached garages, and attics can be estimated at the outdoor ambient temperature.
4. Select or compute heat transfer coefficients for outside walls and glass; for inside walls, nonbasement floors, and ceilings if these are next to unheated spaces; and for the roof if it is next to heated spaces.
5. Determine the net area of outside wall, glass, and roof next to heated spaces, as well as any cold walls, floors, or ceilings next to unheated spaces. These determinations can be made from building plans or from the actual building, using inside dimensions.
6. Compute transmission heat losses for each kind of wall, glass, floor, ceiling, and roof in the building by multiplying the heat transfer coefficient in each case by the area of the surface and the temperature difference between indoor air and outdoor air or adjacent lower temperature spaces.
7. Compute heat losses from basement or grade-level slab floors using the methods in this chapter.
8. Select unit values, and compute the energy associated with infiltration of cold air around outside doors, windows, porous building materials, and other openings. These unit values depend on the kind or width of crack, wind speed, and the temperature difference between indoor and outdoor air. An alternative method is to use air changes (see Chapter 26).
9. When positive ventilation using outdoor air is provided by an air-heating or air-conditioning unit, the energy required to warm the outdoor air to the space temperature must be provided by the unit. The principle for calculation of this load component is identical to that for infiltration. If mechanical exhaust from the space is provided in an amount equal to the outdoor air drawn in by the unit, the unit must also provide for natural infiltration losses. If no mechanical exhaust is used and the outdoor air supply equals or exceeds the amount of natural infiltration that can occur without ventilation, some reduction in infiltration may occur.
10. The sum of the coincidental transmission losses or heat transmitted through the confining walls, floor, ceiling, glass, and other surfaces, plus the energy associated with cold air entering by infiltration or the ventilation air required to replace mechanical exhaust, represents the total heating load.
11. Include the pickup loads that may be required in intermittently heated buildings using night thermostat setback. Pickup loads frequently require an increase in heating equipment capacity to bring the temperature of structure, air, and material contents to the specified temperature. See Figure 9.
12. Use materials and data in Chapters 25, 26, 27, and others as appropriate to the calculations. See Table 12.

SELECTING HEATING DESIGN CONDITIONS

The ideal solution to a basic heating system design is a plant with a maximum output capacity equal to the heating load that develops with the most severe local weather conditions. However, this solution is usually uneconomical. Weather records show that severe weather conditions do not repeat annually. If heating systems were designed for maximum weather conditions, excess capacity would exist during most of the system's operating life. In many cases, an occasional failure of a heating plant to maintain a preselected indoor design temperature during brief periods of severe weather is not critical.

Table 12 Summary of Loads, Equations, and References for Calculating Design Heating Loads

Heating Load	Equation	Reference, Table, Description
Roofs, ceilings, walls, glass	$q = U A \Delta t$	U → Chapter 25, Tables 1, 2, and 4 Δt → Temperature difference between inside and outside design dry bulbs, Chapter 27. For temperatures in unheated spaces, see Equation (2); for attic temperatures, see Equation (3). A → Area calculated from plans
Walls below grade	$q = U A \Delta t$	U → See Table 14. Δt → Use Figure 6 to assist in determining Δt.
Floors Above grade	$q = U A \Delta t$	Δt → For crawl space temperatures, see Equation (4).
On grade	$q = F_2 P \Delta t$	F_2 → See Table 16. Δt → See Equation (6). P → Perimeter of slab
Below grade	$q = U A \Delta t$	Δt → Use Figure 6 to assist in determining Δt. U → See Table 15.
Infiltration and ventilation air		
Sensible	$q_s = 0.018\, Q\, \Delta t$	Q → Volume of outdoor air entering building. See Chapter 26 for estimating methods for infiltration.
Latent	$q_l = 80.7\, Q\, \Delta W$	ΔW → Humidity ratio difference, if humidification is to be added

Outdoor Design Temperature

Before selecting an outdoor design temperature from Chapter 27, the designer should consider the following for residential buildings:

- Is the structure heavy, medium, or light?
- Is the structure insulated?
- Is the structure exposed to high wind?
- Is the load from infiltration or ventilation high?
- Is there more glass area than normal?
- During what part of the day will the structure be used?
- What is the nature of occupancy?
- Will there be long periods of operation at reduced indoor temperature?
- What is the amplitude between local maximum and minimum daily temperatures?
- Are there local conditions that cause significant variation from temperatures reported by the weather service?
- What auxiliary heating devices will be in the building?

Before selecting an outdoor design temperature, the designer must keep in mind that, if the outdoor to indoor design temperature difference is exceeded, the indoor temperature may fall, depending on (1) the thermal mass of the structure and its contents, (2) whether the internal load was included in calculations, (3) the duration of the cold period, and (4) internal heat generated by appliances, etc.

The effect of wind on the heating requirements of any building should be considered because

- Wind movement increases the heat transmission of walls, glass, and roof, affecting poorly insulated walls to a much greater extent than well-insulated walls.
- Wind materially increases the infiltration of cold air through cracks around doors and windows and even through building materials themselves (see Chapter 26).

Theoretically, on a design basis, the most unfavorable combination of temperature and wind speed should be chosen. A building may require more heat on a windy day with a moderately low outdoor temperature than on a quiet day with a much lower outdoor temperature. The worst combination of wind and temperature varies by building because wind speed has a greater effect on buildings with relatively high infiltration rates. The building heating load may be calculated for several combinations of temperature and wind speed on record, and the worst combination may be selected; however, except for critical applications, designers generally find such a degree of refinement unnecessary. No correlation has been shown between the design temperatures in Chapter 27 and the simultaneous maximum wind speed. If a designer prefers the air change method for computing infiltration rates, such correlation is not important. Designers who use the crack method can use a leakage rate at a wind speed of 15 mph, unless local experience has established that another speed is more appropriate. Abnormally high wind speeds may have an effect on infiltration and the U-factor of the building components (see Chapter 23).

Indoor Design Temperature

The indoor temperature for comfort heating may vary depending on building use, type of occupancy, or code requirements. Chapter 8 and ASHRAE *Standards* 55 and 55a define the relationship between temperature and comfort.

ESTIMATING TEMPERATURES IN ADJACENT UNHEATED SPACES

Heat loss from heated rooms to unheated rooms or spaces must be based on the estimated or assumed temperature in such unheated spaces. This temperature will be in between the indoor and outdoor temperatures. If the surface area adjacent to the heated room and that exposed to the outdoors are equal and if the heat transfer coefficients are equal, the temperature in the unheated space may be assumed equal to the mean of the indoor and outdoor design temperatures. If, however, the surface areas and coefficients are unequal, the temperature in the unheated space should be estimated by

$$\begin{aligned} t_u = {} & [t_i(A_1U_1 + A_2U_2 + A_3U_3 + \text{etc.}) \\ & + t_o(60\rho c_p Q_o + A_aU_a + A_bU_b + A_cU_c + \text{etc.})] \\ & \div (A_1U_1 + A_2U_2 + A_3U_3 + \text{etc.} \\ & + 60\rho c_p Q_o + A_aU_a + A_bU_b + A_cU_c + \text{etc.}) \end{aligned} \quad (2)$$

where

ρc_p = density times specific heat of air = 0.018 Btu/ft³·°F for standard air
t_u = temperature in unheated space, °F
t_i = indoor design temperature of heated room, °F
t_o = outdoor design temperature, °F
A_1, A_2, A_3, etc. = areas of surfaces of unheated space adjacent to heated spaces, ft²
A_a, A_b, A_c, etc. = areas of surfaces of unheated space exposed to outdoors, ft²
U_1, U_2, U_3, etc. = heat transfer coefficients of surfaces of A_1, A_2, A_3, etc., Btu/h·ft²·°F
U_a, U_b, U_c, etc. = heat transfer coefficients of surfaces of A_a, A_b, A_c, etc., Btu/h·ft²·°F
Q_o = rate of introduction of outside air into unheated space by infiltration and/or ventilation, cfm

Example 2. Calculate the temperature in an unheated space adjacent to a heated room with surface areas (A_1, A_2, and A_3) of 100, 120, and 140 ft² and overall heat transfer coefficients (U_1, U_2, and U_3) of 0.15, 0.20, and 0.25 Btu/h·ft²·°F, respectively. The surface areas of the unheated space exposed to the outdoors (A_a and A_b) are 100 and 140 ft², respectively, and the corresponding overall heat transfer coefficients are 0.10 and 0.30 Btu/h·ft²·°F. The sixth surface is on the ground and can be neglected for this example, as can the effect of introduction of outdoor air into the unheated space. Assume t_i = 70°F and t_o = −10°F.

Solution: Substituting into Equation (2),

$$\begin{aligned} t_u = {} & [70(100 \times 0.15 + 120 \times 0.20 + 140 \times 0.25) \\ & + (-10)(100 \times 0.10 + 140 \times 0.30)] \\ & \div (100 \times 0.15 + 120 \times 0.20 + 140 \times 0.25 \\ & + 100 \times 0.10 + 140 \times 0.30) \\ t_u = {} & 4660/126 = 37°\text{F} \end{aligned}$$

Temperatures in unheated spaces with large glass areas and two or more surfaces exposed to the outdoors (e.g., sleeping porches and sun parlors) are generally assumed to be the same as that of the outdoors.

Attic Temperature

An attic is a space having an average distance of 1 ft or more between a ceiling and the underside of the roof. Estimating attic temperature is a special case of estimating temperature in an adjacent unheated space and can be done using

$$t_a = \frac{A_cU_ct_c + t_o(60\rho c_pA_cV_c + A_rU_r + A_wU_w + A_gU_g)}{A_c(U_c + 60\rho c_pV_c) + A_rU_r + A_wU_w + A_gU_g} \quad (3)$$

where

ρc_p = air density times specific heat = 0.018 Btu/ft³·°F for standard air
t_a = attic temperature, °F
t_c = indoor temperature near top floor ceiling, °F
t_o = outdoor temperature, °F
A_c = area of ceiling, ft²
A_r = area of roof, ft²
A_w = area of net vertical attic wall surface, ft²
A_g = area of attic glass, ft²
U_c = heat transfer coefficient of ceiling, Btu/h·ft²·°F, based on surface conductance of 2.2 Btu/h·ft²·°F (upper surface, see Table 2 in Chapter 25); 2.2 = reciprocal of one-half the air space resistance
U_r = heat transfer coefficient of roof, Btu/h·ft²·°F, based on surface conductance of 2.2 Btu/h·ft²·°F (upper surface, see Table 2 in Chapter 25); 2.2 = reciprocal of one-half the air space resistance
U_w = heat transfer coefficient of vertical wall surface, Btu/h·ft²·°F
U_g = heat transfer coefficient of glass, Btu/h·ft²·°F
V_c = rate of introduction of outside air into the attic space by ventilation per square foot of ceiling area, cfm/ft²

Example 3. Calculate the temperature in an unheated attic assuming t_c = 70°F; t_o = 10°F; A_c = 1000 ft²; A_r = 1200 ft²; A_w = 100 ft²; A_g = 10 ft²; U_r = 0.50 Btu/h·ft²·°F; U_c = 0.40 Btu/h·ft²·°F; U_w = 0.30 Btu/h·ft²·°F; U_g = 1.13 Btu/h·ft²·°F; and V_c = 0.5 cfm/ft².

Solution: Substituting these values into Equation (3),

$$\begin{aligned} t_a = {} & [(1000 \times 0.40 \times 70) + 10(60 \times 0.018 \times 1000 \times 0.5 + 1200 \times 0.50 \\ & + 100 \times 0.30 + 10 \times 1.13)] \\ & \div [1000(0.40 + 60 \times 0.018 \times 0.5) + 1200 \times 0.50 + 100 \times 0.30 \\ & + 10 \times 1.13] \\ t_a = {} & 39{,}813/1581 = 25.2°\text{F} \end{aligned}$$

Equation (3) includes the effect of air interchange that would take place through attic vents or louvers intended to preclude attic condensation. Test data from Joy et al. (1956), Joy (1958), and Rowley et al. (1940) indicate that a reduction in the temperature difference between attic air and outside air is linear as attic ventilation rates increase from 0 to 0.5 cfm/ft^2 of the ceiling area. When attic ventilation meets the requirements in Chapter 24, 0.5 cfm/ft^2 is the approximate ventilation rate for design conditions. This reduction in temperature difference affects the overall heat loss of a residence with an insulated ceiling by only 1 or 2%.

Equation (3) does not consider factors such as heat exchange between chimney and attic or solar radiation to and from the roof. Because of these effects, attic temperatures are frequently higher than values calculated using Equation (3). However, Equation (3) can be used to calculate attic temperature because the resulting error is generally less than that introduced by neglecting the roof and assuming that the attic temperature is equal to the outdoor air temperature.

When relatively large louvers are installed (customary in southern regions of the United States), the attic temperature is often assumed to be the average of the indoor and outdoor air temperatures.

For an approximate method of calculating heat losses through attics, the combined ceiling and roof coefficient may be used (see Table 5 in Chapter 25).

CALCULATING HEAT LOSS FROM CRAWL SPACES

A crawl space can be considered a half basement. To prevent ground moisture from evaporating and causing a condensation problem, sheets of vapor retarder (e.g., polyethylene film) are used to cover the ground surface (see Chapter 24). Most codes require crawl spaces to be adequately vented all year round. However, venting the crawl space in the heating season causes substantial heat loss through the floor.

The space may be insulated in several ways: the crawl space ceiling (floor above the crawl space) can be insulated, or the perimeter wall can be insulated either on the outside or on the inside. If the floor above is insulated, the crawl space vents should be kept open because the temperature of the crawl space is likely to be below the dew point of the indoor space. If the perimeter wall is insulated, the vents should be kept closed in the heating season and open the remainder of the year.

Crawl Space Temperature

The crawl space temperature depends on such factors as venting, heating ducts, and the heating plant. When the crawl space is well ventilated, its temperature is close to that of the ambient air temperature. When the crawl space vent is closed for the heating season, or if the space is used as a plenum (i.e., part of the forced-air heating system), the crawl space temperature approaches that of the indoor conditioned space. In the former case, the floor above the crawl space, the heating ductwork, and the utility pipes should be insulated similarly to the walls and ceiling of a house.

The following steady-state equation can be used to estimate the temperature of a crawl space.

$$q_f = q_p + q_g + q_a$$

where

q_f = heat loss through floor into crawl space, Btu/h
q_p = heat loss from crawl space through foundation walls and sill box, Btu/h
q_g = heat loss into ground, Btu/h
q_a = heat loss due to ventilation of crawl space, Btu/h

Latta and Boileau (1969) estimated the air exchange rate for an uninsulated basement at 0.67 ACH under winter conditions. In more detail, the above equation can be repeated as

$$U_f A_f (t_i - t_c) = U_p A_p (t_c - t_o) + U_g A_g (t_c - t_g) + 0.67 \rho c_p V_c (t_c - t_o) \quad (4)$$

where

t_i = indoor air temperature (i.e., air above ceiling of crawl space), °F
t_o = outdoor air temperature, °F
t_g = ground temperature (constant), °F
t_c = crawl space temperature, °F
A_f = area of floor above, ft^2
A_p = area of perimeter, exposed foundation wall plus sill box, ft^2
A_g = area of ground below ($A_f = A_g$), ft^2
U_f = average heat transfer coefficient through floor, Btu/h·ft^2·°F
U_g = average heat transfer coefficient through ground (horizontal air film and 10 ft of soil), Btu/h·ft^2·°F
U_p = combined heat transfer coefficient of sill box and foundation wall (both above and below grade), Btu/h·ft^2·°F
V_c = volume of crawl space, ft^3
ρc_p = volumetric heat capacity of air = 0.018 Btu/ft^3·°F
0.67 = assumed air exchange rate, volumes/hour

Example 4. A crawl space of 1200 ft^2 with a 140 ft perimeter is considered. The construction of the perimeter wall is shown in Figure 3. The indoor, outdoor, and the deep-down ground temperatures are 70, 10, and 50°F, respectively. Estimate the heat loss and crawl space temperature with and without insulation. The heat transmission coefficient (U-factor) for each component is indicated in Table 13.

Solution: Three cases are examined.

Case A. This base case is a vented and uninsulated crawl space. The crawl space temperature approaches that of the outdoors, 10°F, and the heat loss is 0.25 × 1200(70 – 10) = 18,000 Btu/h.

Case B. The crawl space is vented. The floor above is insulated with an R-11 blanket; no insulation on the perimeter. The temperature of the crawl space approaches that of the outdoors, 10°F. The heat loss is calculated as

$$q_f = 1200 \times 0.076(70 - 10) = 5470 \text{ Btu/h}$$

Case C. The crawl space is not vented during the heating season. The floor above is not insulated, but the perimeter wall is insulated with R-5.4 down to 3 ft below grade.

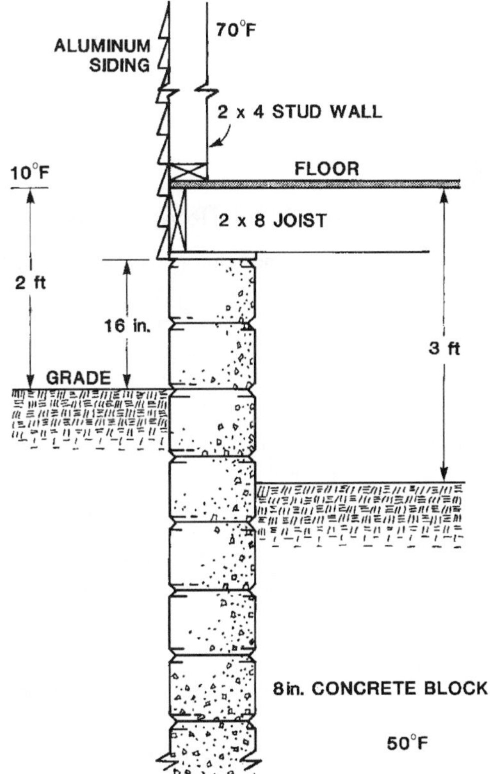

Fig. 3 Uninsulated Crawl Space

Table 13 Estimated U-Factors for Insulated and Uninsulated Crawl Spaces

Component	Uninsulated Btu/h·°F per ft of Perimeter	Insulated[a] Btu/h·°F per ft of Perimeter
16 in. exposed concrete blocks	0.7	0.18
7.5 in. sill box	0.188	0.071
1st 12 in. block wall below grade	0.355	0.127
2nd 12 in. block wall below grade	0.22	0.14
3rd 12 in. block wall below grade	0.133	0.1
Total for perimeter wall	1.6	0.62
	Btu/h·ft²·°F	Btu/h·ft²·°F
Ground	0.077	0.077
Floor above crawl space	0.25	0.076[a]

[a]Perimeter walls are insulated with R-5.4; the floor is insulated with R-11 blanket or batts.

$$q_f = 1200 \times 0.25(70 - t_c)$$
$$q_p = 140 \times 0.62(t_c - 10)$$
$$q_g = 1200 \times 0.077(t_c - 50)$$
$$q_a = 1200 \times 3 \times 0.67 \times 0.018(t_c - 10)$$

The crawl space temperature is solved using Equation (4): t_c = 51.5°F. The heat loss is 5550 Btu/h.

The results show that base case A can potentially lose the most heat. However, when the floor above is insulated, the crawl space must be vented to eliminate any condensation potential, and the heating ductwork and utility pipeline in the crawl space must be adequately insulated. When the perimeter is insulated, the vents must be closed during the heating season and opened for the rest of the year; the heating ductwork and utility pipeline do not need insulation.

Case	Venting	Insulation	Heat Loss Through Floor Above, Btu/h	Temperature of Crawl Space, °F
A	Yes	None	18,000	10
B	Yes	R-11 on floor above	5,470	10
C	No	R-5.4 on perimeter wall	5,550	51.5

CALCULATING TRANSMISSION HEAT LOSS

Steady-state heat loss by conduction and convection heat transfer through any surface is

$$q = UA(t_i - t_o) \quad (5)$$

where

q = heat transfer through wall, glass, roof, ceiling, floor, or other exposed surface, Btu/h
A = area of surface, ft²
U = air-to-air heat transfer coefficient, Btu/h·ft²·°F
t_i = indoor air temperature near surface involved, °F
t_o = outdoor air temperature or temperature of adjacent unheated space, °F

Example 5. Calculate the transmission loss through an 8 in. brick wall having an area of 150 ft², if the indoor temperature t_i is 70°F, and the outdoor temperature t_o is −10°F.

Solution: The overall heat transfer coefficient U of a plain 8 in. brick wall is 0.41 Btu/h·ft²·°F. Substituting into Equation (5),

$$q = 150 \times 0.41[70 - (-10)] = 4920 \text{ Btu/h}$$

Through Ceiling and Roof

Transmission heat loss through top floor ceilings, attics, and roofs may be estimated by either of two methods:

1. Substitute in Equation (5) the ceiling area A, the indoor/outdoor temperature difference $(t_i - t_o)$, and the proper U-factor:

   Flat roofs. Use appropriate coefficients in Equation (3) if side walls extend appreciably above the ceiling or the floor below.

   Pitched roofs. Calculate the combined roof and ceiling coefficient as outlined in Chapter 25.

2. For *pitched roofs*, estimate the attic temperature (based on the indoor and outdoor design temperatures) using Equation (3), and substitute for t_o in Equation (5), obtaining the value of t_a, together with the ceiling area A and the ceiling U-factor. Attic temperatures do not need to be calculated for *flat roofs*, as the ceiling-roof heat loss can be determined as suggested in Method 1 above.

From the Basement

The basement interior is considered conditioned space if a minimum temperature of 10°F below indoor design air temperature is maintained over the heating season. In many instances, the house heating plant, water heater, and heating ducts are in the basement, so it remains at or above 50°F.

Heat transmission from the below-grade portion of the basement wall to the ambient air cannot be estimated by simple, one-dimensional heat conduction. In fact, field measurement of an uninsulated basement by Latta and Boileau (1969) showed that the isotherms near the wall are not parallel lines but closer to radial lines centered at the intersection of the grade line and the wall. Therefore, heat flow paths approximately follow a concentric circular pattern (Figure 4).

Such heat flow paths are altered when insulation is added to the wall or floor. An extreme case would be no heat loss from the basement wall and floor (i.e., infinite insulation applied to the wall and floor). In this case, the isotherms would be horizontal lines parallel to the grade line, and the heat flow would be vertical. When finite insulation or partial insulation is applied to the wall and floor, the heat flow paths take shapes somewhere between the circular and vertical lines (Figure 5).

Ground Temperature. Ground temperatures assumed for estimating basement heat losses will differ for basement floors and walls. The temperatures under floors are generally higher than those adjacent to walls. This is discussed further in the section on Basement Design Temperatures.

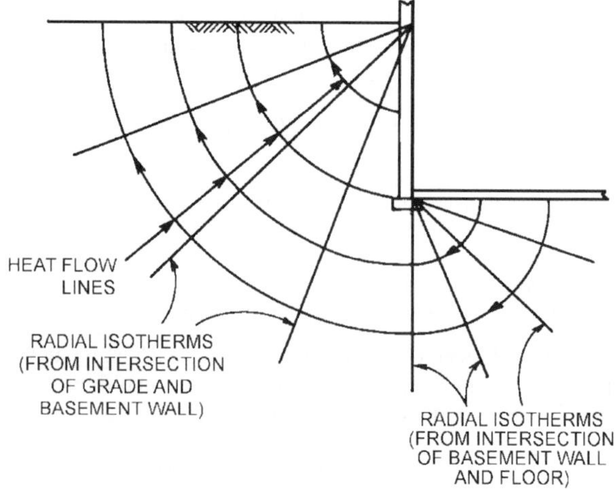

Fig. 4 Heat Flow from Basement

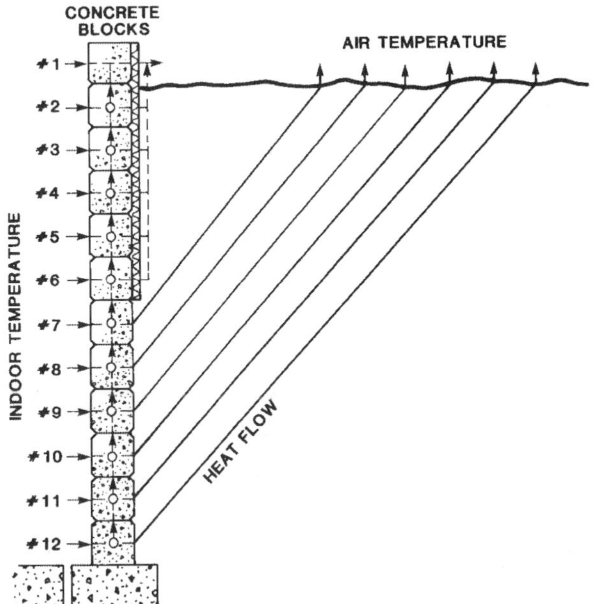

Fig. 5 Heat Flow Path for Partially Insulated Basement Wall

Through Basement Walls

Houghten et al. (1942) observed nonuniform heat flux across the basement wall with respect to the depth of the wall because each heat flow path contains a different thermal resistance. For a basement wall that has its top portion exposed to ambient air, heat may be conducted vertically through the concrete wall and dissipated to the ambient from the top portion of the wall (Wang 1979, Bligh et al. 1978). Under certain conditions, this vertical heat flux becomes significant and should not be ignored.

Once the heat paths are known or assumed, a steady-state analysis can calculate the overall heat transmission coefficient for each segment of the basement wall. Referring to Figures 4 and 5, the total thermal resistance for each depth increment of the basement wall can be found by summing the thermal resistances along each heat flow path. Based on these resistances, the heat loss at each depth increment can be estimated for a unit temperature difference between the basement and the average mean winter temperature. Table 14 lists such heat loss values at different depths for an uninsulated and an insulated concrete wall (Latta and Boileau 1969). Also listed are the lengths of the heat flow path through the soil (circular path).

Table 14 Heat Loss Below Grade in Basement Walls

Depth, ft	Path Length Through Soil, ft	Heat Loss Coefficient, Btu/h·ft²·°F[a]							
		Uninsulated		R-4.17		R-8.34		R-12.5	
0 to 1	0.68	0.410	Σ[b]	0.152	Σ[b]	0.093	Σ[b]	0.067	Σ[b]
1 to 2	2.27	0.222	0.632	0.116	0.268	0.079	0.172	0.059	0.126
2 to 3	3.88	0.155	0.787	0.094	0.362	0.068	0.240	0.053	0.179
3 to 4	5.52	0.119	0.906	0.079	0.441	0.060	0.300	0.048	0.227
4 to 5	7.05	0.096	1.002	0.069	0.510	0.053	0.353	0.044	0.271
5 to 6	8.65	0.079	1.081	0.060	0.570	0.048	0.401	0.040	0.311
6 to 7	10.28	0.069	1.150	0.054	0.624	0.044	0.445	0.037	0.348

Source: Latta and Boileau (1969).
[a]Soil conductivity was assumed to be 9.6 Btu·in/h·ft²·°F.
[b]Σ = heat loss to current depth.

Table 15 Heat Loss Through Basement Floors

Depth of Foundation Wall below Grade, ft	Heat Loss Coefficient, Btu/h·ft²·°F — Shortest Width of House, ft			
	20	24	28	32
5	0.032	0.029	0.026	0.023
6	0.030	0.027	0.025	0.022
7	0.029	0.026	0.023	0.021

Note: $\Delta t = (t_a - A)$

Through Basement Floors

The same steady-state design used for the basement wall can be applied to the basement floor, except that the length of the heat flow path is longer (see Figure 4). Thus, the heat loss through the basement floor is much smaller than that through the wall. An average value for the heat loss through the basement floor can be multiplied by the floor area to give total heat loss from the floor. Table 15 lists typical values.

Basement Design Temperatures

Although internal design temperature is given by basement air temperature, none of the usual external design air temperatures apply because of the heat capacity of the soil. However, ground surface temperature fluctuates about a mean value by an amplitude A, which varies with geographic location and surface cover. Therefore, suitable external design temperatures can be obtained by subtracting A for the location from the mean winter air temperature t_a. Values for t_a can be obtained from meteorological records, and A can be estimated from the map in Figure 6. This map is part of one prepared by Chang (1958) giving annual ranges in ground temperature at a depth of 4 in.

Example 6. Consider a basement 28 ft wide by 30 ft long sunk 6 ft below grade, with R-8.34 insulation applied to the top 2 ft of the wall below grade. Assume an internal air temperature of 70°F and an external design temperature $(t_a - A)$ of 20°F.

Solution:

Wall (using Table 14)

First foot below grade	0.093 Btu/h·ft·°F
Second foot below grade	0.079 Btu/h·ft·°F
Third foot below grade	0.155 Btu/h·ft·°F
Fourth foot below grade	0.119 Btu/h·ft·°F
Fifth foot below grade	0.096 Btu/h·ft·°F
Sixth foot below grade	0.079 Btu/h·ft·°F
Total per foot length of wall	0.621 Btu/h·ft·°F

Basement perimeter .... 2(28 + 30) = 116 ft
Total wall heat loss .... 0.62 × 116 = 72 Btu/h·°F

Floor (using Table 15)

Average heat loss per ft² .... 0.025 Btu/h·ft²·°F
Floor area 28 × 30 .... 840 ft²
Total floor heat loss .... 0.025 × 840 = 21 Btu/h·°F

Total

Total basement heat loss below grade ... 72 + 21 = 93 Btu/h·°F
Design temperature difference .... 70 – 20 = 50°F
Maximum rate of heat loss from below-grade basement .... 93 × 50 = 4650 Btu/h

If a basement is completely below grade and unheated, its temperature ranges between that in the rooms above and that of the ground. Basement windows lower the basement temperature when it is cold outdoors, and heat given off by the heating plant increases the basement temperature. The exact basement temperature is indeterminate if the basement is not heated. In general, heat from the heating plant sufficiently warms the air near the basement ceiling to make unnecessary an allowance for floor heat loss from rooms located over the basement.

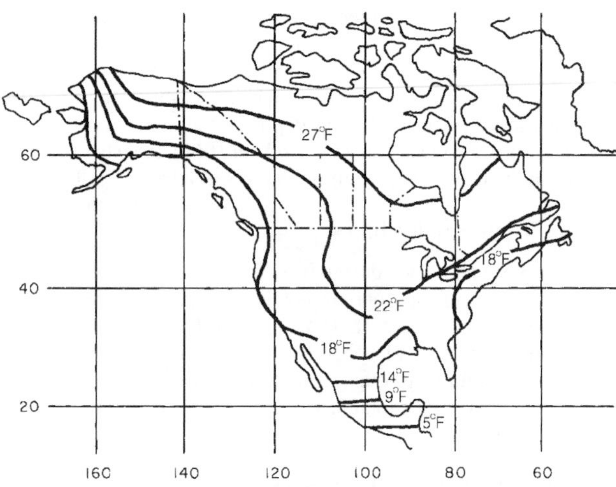

Fig. 6 Lines of Constant Amplitude

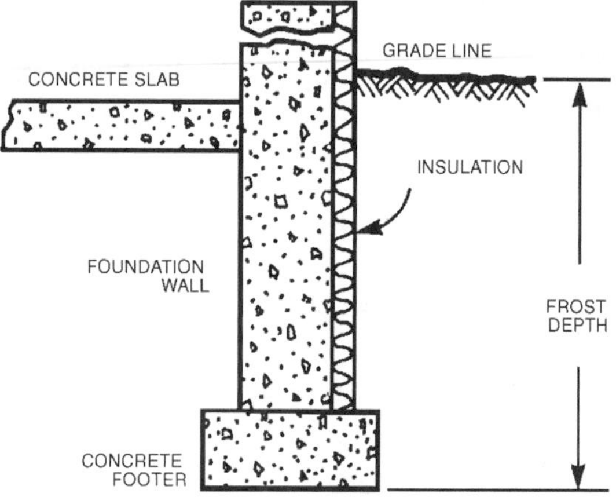

Fig. 7 "I"-Shaped or Vertical Insulation System

Transient Calculations for Basement Walls

The heat loss from basement walls can be estimated more accurately with a finite element or finite difference computer program by transient simulations (Wang 1979, Bligh et al. 1978). The solution is in the form of heat loss over time, which can be converted to an average U-factor. This approach also offers the possibility for estimating the depth below grade to which insulation is economical. Direct and indirect evidence of hollow concrete block walls shows that a convective path exists within the blocks vertically along the wall (Harrje et al. 1979). Therefore, insulation should be arranged to reduce this convective heat transfer.

Peony et al. (1979) showed that the dynamic thermal performance of a masonry wall is better when insulation is placed on the exterior. Moreover, transient simulation showed that insulation is more effective when it is placed on the exterior side of the basement wall. Depending on the exposed portion of the block wall and the temperature difference between indoor and outdoor air, exterior application can be 10 to 20% more efficient than a corresponding interior application. However, such exterior insulation must be installed properly to maintain its integrity.

Calculating Transmission Heat Loss from Floor Slabs

Concrete slab floors may be (1) unheated, relying for warmth on heat delivered above floor level by the heating system, or (2) heated, containing heated pipes or ducts that constitute a radiant slab or portion of it for complete or partial heating of the house.

The perimeter insulation of a slab-on-grade floor is quite important for comfort and energy conservation. In unheated slab floors, the floor edge must be insulated in order to keep the floor warm. Downdrafts from windows or exposed walls can create pools of chilly air over considerable areas of the floor. In heated slab floors, the floor edge must be insulated to prevent excessive heat loss from the heating pipe or duct embedded in the floor or from the baseboard heater.

Wang (1979) and Bligh et al. (1978) found that heat loss from an unheated concrete slab floor is mostly through the perimeter rather than through the floor and into the ground. Total heat loss is more nearly proportional to the length of the perimeter than to the area of the floor, and it can be estimated by the following equation for both unheated and heated slab floors:

$$q = F_2 P(t_i - t_o) \tag{6}$$

where

q = heat loss through perimeter, Btu/h

F_2 = heat loss coefficient per foot of perimeter (see Table 16), Btu/h·ft·°F

P = perimeter or exposed edge of floor, ft

t_i = indoor temperature, °F (For the heated slab, t_i is the weighted average heating duct or pipe temperature.)

t_o = outdoor temperature, °F

Vertical "I"-shaped systems are used to insulate slab floor perimeters. In the "I" system, the insulation is placed vertically next to the exposed slab edge, extending downward below grade, as shown in Figure 7.

Breaks or joints must be avoided when the insulation is installed; otherwise, local thermal bridges can be formed, and the overall efficiency of the insulation is reduced.

Transient Calculations for Floor Slabs

Figure 8 shows four basic slab-on-grade constructions analyzed with a finite element computer program by Wang (1979). Figures 8A-C represent unheated slabs; Figure 8D can be considered a heated slab. Each was investigated with and without insulation of R-5.4 under three climatic conditions (7433, 5350, and 2950 degree-days). Table 16 lists the results in terms of heat loss coefficient F_2, based on degree-days.

Table 16 shows that the heat loss coefficient F_2 is sensitive to both construction and insulation. The reverse loss, or heat loss into the ground and outward through the edges of the slab and foundation wall, is significant when heating pipes, heating ducts, or baseboard heaters are placed near the slab perimeters. To prevent reverse loss, the designer may find it advantageous to use perimeter insulation even in warmer climates. For severe winter regions (above 6000 degree-days), the insulation value should be increased to $R > 10$ ft²·°F·h/Btu.

Figure 8A shows that this construction benefits from the wall insulation between block and brick; the insulation is extended roughly 16 in. below the slab floor. Without this wall insulation, the heat loss coefficient F_2 would be close to that of the 4 in. block wall construction (Figure 8B). Table 16 can be used to estimate F_2 under different degree-days of heating season weather.

CALCULATING INFILTRATION HEAT LOSS

Infiltration of outside air causes both **sensible** and **latent** heat loss. The energy required to raise the temperature of outdoor infiltrating air to indoor air temperature is the sensible component. The energy associated with net loss of moisture from the space is the latent component. Infiltration is discussed in detail in Chapter 26.

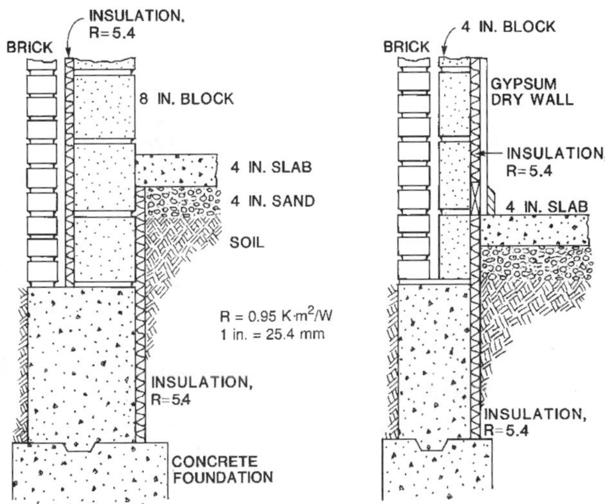

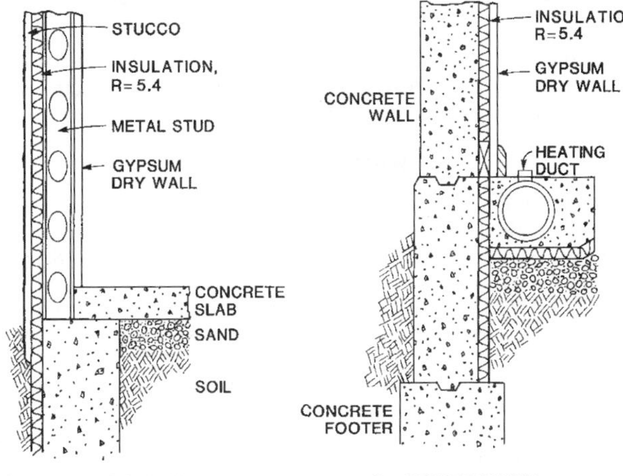

Fig. 8 Slab-on-Grade Foundation Insulation

Sensible Heat Loss

The energy required to warm outdoor air entering by infiltration to the temperature of the room is given by

$$q_s = c_p Q \rho (t_i - t_o) \tag{7}$$

where

q_s = heat flow required to raise temperature of air leaking into building from t_o to t_i, Btu/h
c_p = specific heat of air, Btu/lb·°F
Q = volumetric flow of outdoor air entering building, ft³/h
ρ = density of air at temperature t_o, lb/ft³

Using standard air (ρ = 0.075 lb/ft³ and c_p = 0.24 Btu/lb·°F), Equation (7) reduces to

$$q_s = 0.018 Q (t_i - t_o) \tag{8}$$

The volumetric flow Q of outdoor air entering depends on wind speed and direction, width of cracks or size of openings, type of openings, and other factors explained in Chapter 26. Two methods used to obtain the quantity of infiltration air are the **crack length** and the **air change**. Louvers and doors and the direction they face, as well as any other factors affecting infiltration, may need to be considered.

Table 16 Heat Loss Coefficient F_2 of Slab Floor Construction, Btu/h·°F per ft of Perimeter

Construction	Insulation	Degree-Days (65°F Base) 2950	5350	7433
8 in. block wall, brick facing	Uninsulated	0.62	0.68	0.72
	R-5.4 from edge to footer	0.48	0.50	0.56
4 in. block wall, brick facing	Uninsulated	0.80	0.84	0.93
	R-5.4 from edge to footer	0.47	0.49	0.54
Metal stud wall, stucco	Uninsulated	1.15	1.20	1.34
	R-5.4 from edge to footer	0.51	0.53	0.58
Poured concrete wall with duct near perimeter[a]	Uninsulated	1.84	2.12	2.73
	R-5.4 from edge to footer, 3 ft under floor	0.64	0.72	0.90

[a]Weighted average temperature of the heating duct was assumed at 110°F during the heating season (outdoor air temperature less than 65°F).

Latent Heat Loss

When moisture must be added to the indoor air to maintain winter comfort conditions, the energy needed to evaporate an amount of water equivalent to what is lost by infiltration (latent component of infiltration heat loss) must be determined. This energy may be calculated by

$$q_l = Q \rho (W_i - W_o) h_{fg} \tag{9}$$

where

q_l = heat flow required to increase moisture content of air leakage into building from W_o to W_i, Btu/h
Q = volumetric flow of outdoor air entering building, ft³/h
ρ = density of air at temperature t_i, lb/ft³
W_i = humidity ratio of indoor air, $\text{lb/lb}_{\text{dry air}}$
W_o = humidity ratio of outdoor air, $\text{lb/lb}_{\text{dry air}}$
h_{fg} = latent heat of vapor at t_i, Btu/lb

If the latent heat of vapor h_{fg} is 1076 Btu/lb, and the air density is 0.075 lb/ft³, Equation (7) reduces to

$$q_l = 80.7 Q (W_i - W_o) \tag{10}$$

Crack Length Method

The basis of calculation for the crack method is that the amount of crack used for computing the infiltration heat loss should not be less than one-half the total length of crack in the outside walls of the room. In a building without partitions, air entering through cracks on the windward side must leave through cracks on the leeward side. Therefore, one-half the total crack for each side and end of the building is used for calculation. In a room with one exposed wall, all the crack is used. With two, three, or four exposed walls, either the wall with the crack that will result in the greatest air leakage or at least one-half the total crack is used, whichever is greater.

In residences, total infiltration loss of the house is generally considered equal to the sum of infiltration losses of the various rooms. But, at any given time, infiltration takes place only on the windward side or sides and not on the leeward. Therefore, for determining total heat requirements of larger buildings, it is more accurate to base total infiltration loss on the wall with the most total crack or on at least half the total crack in the building, whichever is greater. When the crack method rather than Equations (8) and (10) is used for estimating leakage, the heat loss in terms of the crack length may be expressed as

$$q_s = 0.018BL(t_i - t_o) \quad (11)$$

and

$$q_l = 80.7BL(W_i - W_o) \quad (12)$$

where

B = air leakage for wind velocity and type of window or door crack involved, ft³/h per foot of crack

L = length of window or door crack to be considered, ft

Air Change Method

Some designers base infiltration on an estimated number of air changes rather than the length of window cracks. The number of air changes given in Chapter 26 should be considered only as a guide. When calculating infiltration losses by the air change method, Equations (8) and (10) can be used by substituting for Q the volume of the room multiplied by the number of air changes.

Exposure Factors

Some designers use empirical exposure factors to increase calculated heat loss of rooms or spaces on the side(s) of the building exposed to prevailing winds. However, exposure factors are not needed with the method of calculating heat loss described in this chapter. Instead, they may be (1) regarded as safety factors to allow for additional capacity for rooms or spaces exposed to prevailing winds or (2) used to account for the effects of radiation loss, particularly in the case of multistory buildings. Tall buildings may have severe infiltration heat losses induced by stack effect that require special analysis. Although a 15% exposure allowance is often assumed, the actual allowance, if any, is largely a matter of experience and judgment; no test data are available from which to develop rules for the many conditions encountered.

PICKUP LOAD

For intermittently heated buildings and night thermostat setback, additional heat is required to raise the temperature of air, building materials, and material contents of a building to the specified temperature. The pickup load, which is the rate at which this additional heat must be supplied, depends on the heat capacity of the structure, its material contents, and the time in which these are to be heated.

Relatively little information on pickup load exists; however, Smith (1941, 1942) addressed pickup loads for buildings heated only occasionally, such as auditoriums and churches. Nelson and MacArthur (1978) studied the relationship between thermostat setback, furnace capacity, and recovery time. Based on this limited information, the following design guidelines are offered.

Because design outdoor temperatures generally provide a substantial margin for outdoor temperatures typically experienced during operating hours, many engineers make no allowance for this additional heat in most buildings. However, the additional heat should be computed and allowed for, as conditions require. In the case of intermittently heated buildings, an additional 10% capacity should be provided.

In buildings with setback-type thermostats, the furnace must be oversized to reestablish the space temperature in an acceptable time. The amount of oversizing depends on such factors as the amount of setback, inside-to-outside temperature difference, building construction, and acceptable pickup time. Figure 9 shows this relationship for a particular residence. As a rule for residences, a 10°F night setback requires 40% oversizing for acceptable pickup time and minimum energy requirements (Nelson and MacArthur 1978). For smaller setback, the oversizing can be proportionally less. If daytime as well as night setback is practiced, oversizing of up to 60% is warranted.

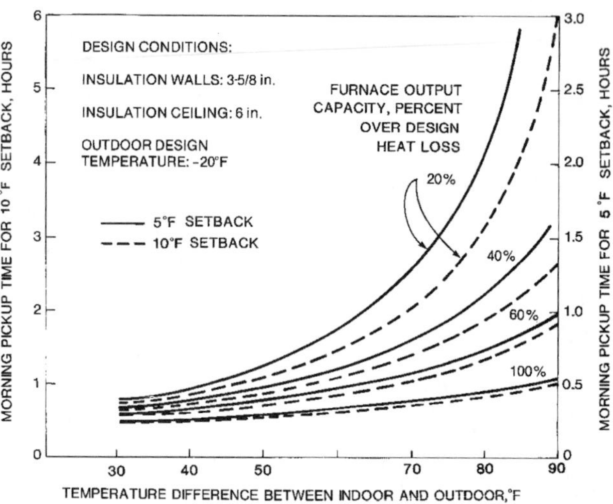

Fig. 9 Furnace Operating Times Required to Pick Up Space Temperature Following 5 and 10°F Night Setback

REFERENCES

ASHRAE. 1992. Thermal environmental conditions for human occupancy. ANSI/ASHRAE *Standard* 55-1992.

ASHRAE. 1995. Addendum to ANSI/ASHRAE 55-1992. ANSI/ASHRAE *Standard* 55a-1995.

Bligh, T.P., P. Shipp, and G. Meixel. 1978. Energy comparisons and where to insulate earth sheltered buildings and basements. Earth covered settlements, U.S. Department of Energy Conference, Fort Worth, TX.

Chang, J.H. 1958. *Ground temperature.* Bluehill Meteorological Observatory, Harvard University, Cambridge, MA.

Harrje, D.T., G.S. Dutt, and J. Beyea. 1979. Locating and eliminating obscure but major energy losses in residential housing. *ASHRAE Transactions* 85(2).

Houghten, F.C., S.I. Taimuty, C. Gutberlet, and C.J. Brown. 1942. Heat loss through basement walls and floors. *ASHVE Transactions* 48:369.

Joy, F.A. 1958. Improving attic space insulating values. *Heating, Piping and Air Conditioning* 30(1):223.

Joy, F.A., J.J. Zabrony, and S. Bhaduri. 1956. Insulating value of reflective elements in an attic under winter conditions. Pennsylvania State University, University Park, PA.

Latta, J.K. and G.G. Boileau. 1969. Heat losses from house basements. *Canadian Building* 19(10):39.

McQuiston, F.C. 1984. A study and review of existing data to develop a standard methodology for residential heating and cooling load calculations. *ASHRAE Transactions* 90(2A):102-36.

McQuiston, F.C. and J.D. Spitler. 1992. *Cooling and heating load calculation manual*, 2nd ed. ASHRAE, Atlanta.

Nelson, L.W. and J.W. MacArthur. 1978. Energy savings through thermostat setback. *ASHRAE Transactions* 84(2):319-34.

Peony, B.A., F.J. Powell, and D.M. Burch. 1979. Dynamic thermal performance of an experimental masonry building. NBS *Report* 10 664, National Institute of Standards and Technology, Gaithersburg, MD.

Rowley, F.B., A.B. Algren, and C.E. Lund. 1940. Methods of moisture control and their application to building construction. *Bulletin* No. 17 XLIII(4):28. University of Minnesota Engineering Experiment Station.

Smith, E.G. 1941. Heat requirement of intermittently heated buildings. Texas A&M *Engineering Experiment Station Series* No. 62 (November). College Station, TX.

Smith, E.G. 1942. A method of compiling tables for intermittent heating. *Heating, Piping, and Air Conditioning* 14(6):386.

Wang, F.S. 1979. Mathematical modeling and computer simulation of insulation systems in below grade applications. ASHRAE/DOE Conference on Thermal Performance of the Exterior Envelopes of Buildings, Orlando, FL.

CHAPTER 29

NONRESIDENTIAL COOLING AND HEATING LOAD CALCULATION PROCEDURES

THIS CHAPTER presents two load calculation methods that represent a significant departure from those in common use. The technology involved, however—the principle of calculating a heat balance for a given space—is not new. The first of the two methods is the **heat balance** (HB) **method**. The calculation procedures and scientific principles are explained in equation format. These equations are coded in a generic computer program named Hbfort, released with *Cooling and Heating Load Calculation Principles* (Pedersen et al. 1998), and linked to a user interface program to allow input and output in either inch-pound or SI units. The source code for these programs has been refined and enhanced under ASHRAE *Research Project* 987 and can be found in the *ASHRAE Load Calculation Toolkit*.

The second method is called the **radiant time series** (RTS) **method**, which is a simplified method directly related to and derived from the HB calculation procedure. This chapter presents the principles of both procedures rather than a working tool for the cooling load practitioner. The load prediction of an actual, multiple-room building requires a complex computer program.

While the procedures described in this chapter are the most reliable means for estimating cooling load for a defined building space, the methods described in earlier editions of the Handbook series are valid for many applications. Each of these earlier procedures is, however, a simplification of the heat balance principles, and their use requires experience to deal with atypical situations or unusual circumstances. In fact, any cooling or heating load estimate is no better than the assumptions used to define conditions and parameters such as physical makeup of the various envelope surfaces, conditions of occupancy and use, and ambient weather conditions outside the building. The experience of the practitioner can never be ignored.

The procedures described in this chapter are concerned with a given space or zone in a building. When estimating the loads for a group of spaces such as might be handled by a single air-handling system, the assembled zones must be analyzed to consider (1) the simultaneous effects taking place; (2) any diversification of heat gains for occupants, lighting, or other internal load sources; (3) ventilation; and/or (4) any other unique circumstances. With large buildings that involve more than a single HVAC system, simultaneous loads and any additional diversities also must be considered. The methods presented in this chapter are expressed as hourly load summaries, reflecting 24 h input schedules and profiles of the individual load variables. Specific systems and applications may require different profiles.

In comparing the HB and RTS methods against the methods described in earlier versions of this chapter, the primary difference is the directness of approach as opposed to the simplifying techniques necessitated by the limited computer capability available in earlier days. The **transfer function method** (TFM), for example, required many calculation steps. Also, this method was originally designed for energy analysis with emphasis on daily, monthly, and annual energy use and, thus, was more oriented to average hourly cooling loads than peak design loads.

The **total equivalent temperature differential method with time averaging** (TETD/TA) has been a highly reliable—if subjective—method of load estimating since its initial presentation in the 1967 *ASHRAE Handbook—Fundamentals*. Originally conceived as a manual method of calculation, it proved suitable only as a computer application because of the need to calculate an extended profile of hourly heat gain values from which the radiant components had to be averaged over a time perceived to represent the general mass of the building involved. Because this perception of thermal storage characteristics of a given building was almost entirely subjective, with little specific information for the user to judge variations, the TETD/TA method's primary usefulness has always been to the experienced engineer.

The **cooling load temperature differential method with solar cooling load factors** (CLTD/CLF) was an attempt to simplify the two-step TFM and TETD/TA methods into a single-step technique that allowed proceeding directly from raw data to cooling load without the intermediate conversion of radiant heat gain to cooling load. A series of factors were taken from cooling load calculation results (produced by more sophisticated methods) as "equivalent temperature differences" for use in traditional conduction ($q = UA\Delta t$) equa-

The preparation of this chapter is assigned to TC 4.1, Load Calculation Data and Procedures.

tions. The results, however, are approximate cooling load values rather than simple heat gain values. The simplifications required for this process limit the applicability of this method. In fact, past versions of this chapter listed several areas of potential use that should be avoided as beyond the range of applicability.

COOLING LOAD PRINCIPLES

The variables affecting cooling load calculations are numerous, often difficult to define precisely, and always intricately interrelated. Many cooling load components vary in magnitude over a wide range during a 24 h period. Because these cyclic changes in load components often are not in phase with each other, each must be analyzed to establish the maximum cooling load for a building or zone. A zoned system (a system of conditioning equipment serving several independent areas, each with its own temperature control) need provide no greater total cooling load capacity than the largest hourly summary of simultaneous zone loads throughout a design day; however, it must handle the peak cooling load for each zone at its individual peak hour. At certain times of the day during the heating or intermediate seasons, some zones may require heating while others require cooling.

Calculation Accuracy

A realistic cooling load calculation gives values adequate for acceptable system performance. Variation in the heat transmission coefficients of typical building materials and composite assemblies, the differing motivations and skills of those who construct the building, and the manner in which the building is actually operated are some of the variables that make a precise calculation impossible. Even if the designer uses reasonable procedures to account for these factors, the calculation can never be more than a good estimate of the actual cooling load.

Heat Flow Rates

In air-conditioning design, four related heat flow rates, each of which varies with time, must be differentiated: (1) space heat gain, (2) space cooling load, (3) space heat extraction rate, and (4) cooling coil load.

Space Heat Gain. This instantaneous rate of heat gain is the rate at which heat enters into and/or is generated within a space. Heat gain is classified by (1) the mode in which it enters the space and (2) whether it is a sensible or latent gain.

Mode of entry. The mode of entry includes (1) solar radiation through transparent surfaces; (2) heat conduction through exterior walls and roofs; (3) heat conduction through ceilings, floors, and interior partitions; (4) heat generated in the space by occupants, lights, and appliances; (5) energy transfer as a result of ventilation and infiltration of outdoor air; and (6) miscellaneous heat gains.

Sensible or latent heat. Sensible heat is added directly to the conditioned space by conduction, convection, and/or radiation. Latent heat gain occurs when moisture is added to the space (e.g., from vapor emitted by occupants and equipment). To maintain a constant humidity ratio, water vapor must condense on the cooling apparatus and be removed at a rate equal to the rate it is added to the space. The amount of energy required to offset the latent heat gain essentially equals the product of the rate of condensation and the latent heat of condensation. In selecting cooling apparatus, it is necessary to distinguish between sensible and latent heat gain. Every cooling apparatus has a maximum sensible heat removal capacity and a maximum latent heat removal capacity for particular operating conditions.

Space Cooling Load. This is the rate at which heat must be removed from the space to maintain a constant space air temperature. The sum of all space instantaneous heat gains at any given time does not necessarily (or even frequently) equal the cooling load for the space at that same time.

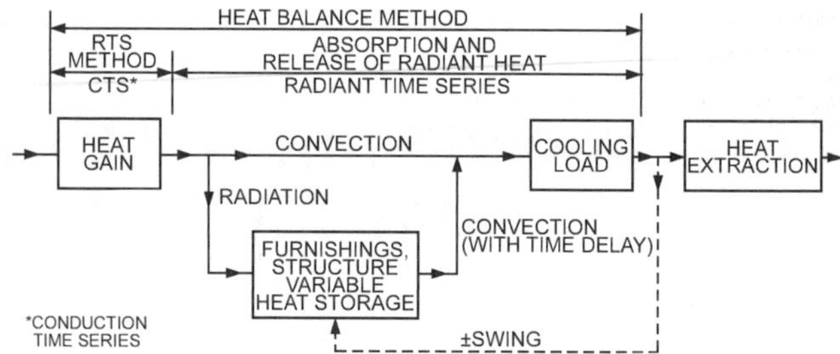

Fig. 1 Origin of Difference Between Magnitude of Instantaneous Heat Gain and Instantaneous Cooling Load

Radiant heat gain. Radiant energy must first be absorbed by the surfaces that enclose the space (walls, floor, and ceiling) and the objects in the space (furniture, etc.). When these surfaces and objects become warmer than the surrounding air, some of their heat is transferred to the air by convection. The composite heat storage capacity of these surfaces and objects determines the rate at which their respective surface temperatures increase for a given radiant input and thus governs the relationship between the radiant portion of heat gain and its corresponding part of the space cooling load (Figure 1). The thermal storage effect is critically important in differentiating between instantaneous heat gain for a given space and its cooling load at that moment. Predicting the nature and magnitude of this phenomenon in order to estimate a realistic cooling load for a particular combination of circumstances has long been a subject of interest to design engineers. The section on Bibliography lists some of the early work on the subject.

Space Heat Extraction Rate. The rate at which heat is removed from the conditioned space equals the space cooling load only if the room air temperature is held constant. Along with the intermittent operation of the cooling equipment, the control system characteristics usually permit a minor cyclic variation or swing in room temperature. Therefore, a proper simulation of the control system gives a more realistic value of energy removal over a fixed period than using the values of the space cooling load. However, this concept is primarily important for estimating energy use over time; it is not needed to calculate design peak cooling load for equipment selection.

Cooling Coil Load. The rate at which energy is removed at the cooling coil that serves one or more conditioned spaces equals the sum of the instantaneous space cooling loads (or space heat extraction rate if it is assumed that the space temperature does not vary) for all the spaces served by the coil, plus any external loads. Such external loads include fan heat gain, duct heat gain, and outdoor air heat and moisture brought into the cooling equipment to satisfy the ventilation requirement.

Cooling Load Estimation in Practice

Frequently, a cooling load must be calculated before every parameter in the conditioned space can be properly or completely defined. An example is a cooling load estimate for a new building with many floors of unleased spaces where detailed partition requirements, furnishings, lighting selection, and layout cannot be predefined. Potential tenant modifications once the building is occupied also must be considered. The load estimating process requires proper engineering judgment that includes a thorough understanding of heat balance fundamentals.

INITIAL DESIGN CONSIDERATIONS

To calculate a space cooling load, detailed building design information and weather data at selected design conditions are required. Generally, the following steps should be followed.

Data Assembly

Building Characteristics. Obtain characteristics of the building. Building materials, component size, external surface colors, and shape are usually determined from building plans and specifications.

Configuration. Determine building location, orientation, and external shading from building plans and specifications. Shading from adjacent buildings can be determined by a site plan or by visiting the proposed site but should be carefully evaluated as to its probable permanence before it is included in the calculation. The possibility of abnormally high ground-reflected solar radiation (i.e., from adjacent water, sand, or parking lots) or solar load from adjacent reflective buildings should not be overlooked.

Outdoor Design Conditions. Obtain appropriate weather data, and select outdoor design conditions. For outdoor design conditions for a large number of weather stations, see Chapter 27. Note, however, that these values for the design dry-bulb and mean coincident wet-bulb temperatures may vary considerably from data traditionally used in various areas. Use judgment to ensure that results are consistent with expectations. Also, consider prevailing wind velocity and the relationship of a project site to the selected weather station.

In recent years, several research projects have greatly expanded the amount of available weather data (Colliver et al. 1995, 1998, 2000). In addition to the conventional dry-bulb with mean coincident wet-bulb, data are now available for wet-bulb and dew-point with mean coincident dry-bulb. The peak load for a space that requires both large quantities of outside air and close control of moisture may occur at peak wet-bulb or peak dew-point conditions when the corresponding dry-bulb temperature is significantly lower than normal design conditions.

Indoor Design Conditions. Select indoor design conditions, such as indoor dry-bulb temperature, indoor wet-bulb temperature, and ventilation rate. Include permissible variations and control limits.

Operating Schedules. Obtain a proposed schedule of lighting, occupancy, internal equipment, appliances, and processes that contribute to the internal thermal load. Determine the probability that the cooling equipment will be operated continuously or shut off during unoccupied periods (e.g., nights and/or weekends).

Date and Time. Select the time of day and month to do the cooling load calculation. Frequently, several different times of day and several different months must be analyzed to determine the peak load time. The particular day and month are often dictated by peak solar conditions. For southern exposures in north latitudes above 32° having large fenestration areas, the peak space cooling load usually occurs in December or January. To calculate a space cooling load under these conditions, the warmest temperature for the winter months must be known. These data can be found for the United States in Chapter 27, Table 4B.

Additional Considerations

The proper design and sizing of all-air or air-and-water central air-conditioning systems require more than calculation of the cooling load in the space to be conditioned. The type of air-conditioning system, fan energy, fan location, duct heat loss and gain, duct leakage, heat extraction lighting systems, and type of return air system all affect system load and component sizing. Adequate system design and component sizing require that system performance be analyzed as a series of psychrometric processes.

HEAT SOURCES AND HEAT GAIN CALCULATION CONCEPTS

TIME DELAY EFFECT

The energy absorbed by walls, floor, furniture, etc., contributes to space cooling load only after a time lag, with some part of this energy still present and reradiating after the heat sources have been switched off or are no longer present (Figure 2).

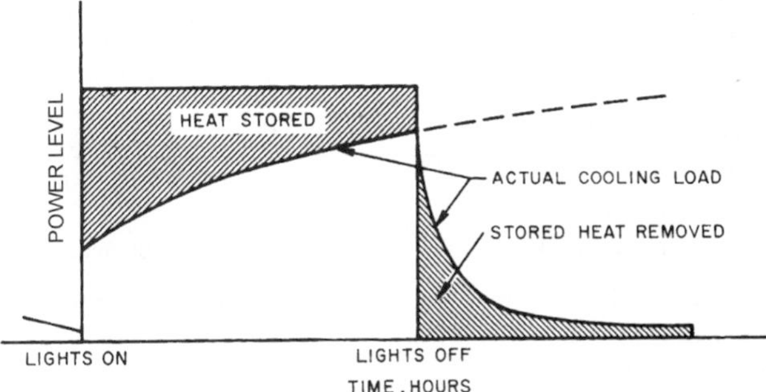

Fig. 2 Thermal Storage Effect in Cooling Load from Lights

There is always significant delay between the time of switching on or otherwise activating a heat source and the point when reradiated energy equals that being instantaneously stored. This time lag must be considered when calculating cooling load because the load felt by the space can be much lower than the instantaneous heat gain being generated, and the peak load for the space may be affected significantly.

PEOPLE

Table 1 gives representative rates at which heat and moisture are given off by human beings in different states of activity. Often these sensible and latent heat gains constitute a large fraction of the total load. Even for short-term occupancy, the extra heat and moisture brought in by people may be significant. Chapter 8 should be consulted for detailed information; however, Table 1 summarizes design data representing conditions commonly encountered.

The conversion of sensible heat gain from people to space cooling load is affected by the thermal storage characteristics of that space, since some percentage of the sensible load is radiant energy. Latent heat gains are considered instantaneous.

LIGHTING

Because lighting is often the major space cooling load component, an accurate estimate of the space heat gain it imposes is needed. Calculation of this load component is not straightforward; the rate of cooling load due to lighting at any given moment can be quite different from the heat equivalent of power supplied instantaneously to those lights.

Instantaneous Heat Gain from Lighting

The primary source of heat from lighting comes from light-emitting elements, or lamps, although significant additional heat may be generated from associated appurtenances in the light fixtures that house such lamps. Generally, the instantaneous rate of heat gain from electric lighting may be calculated from

$$q_{el} = 3.41 W F_{ul} F_{sa} \tag{1}$$

where

q_{el} = heat gain, Btu/h
W = total light wattage
F_{ul} = lighting use factor
F_{sa} = lighting special allowance factor

The **total light wattage** is obtained from the ratings of all lamps installed, both for general illumination and for display use.

The **lighting use factor** is the ratio of the wattage in use, for the conditions under which the load estimate is being made, to the total

Table 1 Representative Rates at Which Heat and Moisture Are Given Off by Human Beings in Different States of Activity

		Total Heat, Btu/h		Sensible	Latent	% Sensible Heat that is Radiant[b]	
Degree of Activity	Location	Adult Male	Adjusted, M/F[a]	Heat, Btu/h	Heat, Btu/h	Low *V*	High *V*
Seated at theater	Theater, matinee	390	330	225	105		
Seated at theater, night	Theater, night	390	350	245	105	60	27
Seated, very light work	Offices, hotels, apartments	450	400	245	155		
Moderately active office work	Offices, hotels, apartments	475	450	250	200		
Standing, light work; walking	Department store; retail store	550	450	250	200	58	38
Walking, standing	Drug store, bank	550	500	250	250		
Sedentary work	Restaurant[c]	490	550	275	275		
Light bench work	Factory	800	750	275	475		
Moderate dancing	Dance hall	900	850	305	545	49	35
Walking 3 mph; light machine work	Factory	1000	1000	375	625		
Bowling[d]	Bowling alley	1500	1450	580	870		
Heavy work	Factory	1500	1450	580	870	54	19
Heavy machine work; lifting	Factory	1600	1600	635	965		
Athletics	Gymnasium	2000	1800	710	1090		

Notes:
1. Tabulated values are based on 75°F room dry-bulb temperature. For 80°F room dry bulb, the total heat remains the same, but the sensible heat values should be decreased by approximately 20%, and the latent heat values increased accordingly.
2. Also refer to Table 4, Chapter 8, for additional rates of metabolic heat generation.
3. All values are rounded to nearest 5 Btu/h.

[a] Adjusted heat gain is based on normal percentage of men, women, and children for the application listed, with the postulate that the gain from an adult female is 85% of that for an adult male, and that the gain from a child is 75% of that for an adult male.

[b] Values approximated from data in Table 6, Chapter 8, where *V* is air velocity with limits shown in that table.

[c] Adjusted heat gain includes 60 Btu/h for food per individual (30 Btu/h sensible and 30 Btu/h latent).

[d] Figure one person per alley actually bowling, and all others as sitting (400 Btu/h) or standing or walking slowly (550 Btu/h).

installed wattage. For commercial applications such as stores, the use factor would generally be unity.

The **special allowance factor** is for fluorescent fixtures and/or fixtures that are either ventilated or installed so that only part of their heat goes to the conditioned space. For fluorescent or high-intensity discharge fixtures, the special allowance factor accounts primarily for ballast losses. Table 2 shows that the special allowance factor for a two-lamp fluorescent fixture ranges from 0.94 for T8 lamps with an electronic ballast to 1.21 for energy-saver T12 lamps with a standard electromagnetic ballast. High-intensity discharge fixtures, such as metal halide, may have special allowance factors varying from 1.07 to 1.44, depending on the lamp wattage and quantity of lamps per fixture, and should be dealt with individually. A wide variety of lamp and ballast combinations is available, and ballast catalog data provide the overall fixture wattage.

For ventilated or recessed fixtures, manufacturers' or other data must be sought to establish the fraction of the total wattage that may be expected to enter the conditioned space directly (and subject to time lag effect) versus that which must be picked up by return air or in some other appropriate manner.

Light Heat Components

Cooling load caused by lights recessed into ceiling cavities is made up of two components: one part (known as the **heat-to-space** load) comes from the light heat directly contributing to the space heat gain, and the other is the light heat released into the above-ceiling cavity, which (if used as a return air plenum) is mostly picked up by the return air that passes over or through the light fixtures. In such a ceiling return air plenum, this second part of the load (sometimes referred to as **heat-to-return**) never enters the conditioned space. It does, however, add to the overall load and significantly influences the load calculation.

Even though the total cooling load imposed on the cooling coil from these two components remains the same, the larger the fraction of heat output picked up by the return air, the more the space cooling load is reduced. The minimum required airflow rate for the conditioned space is decreased as the space cooling load decreases. Supply fan power decreases accordingly, which ultimately results in reduced energy consumption for the system and possibly reduced equipment size as well.

For ordinary design load estimation, the heat gain for each component may be calculated simply as a fraction of the total lighting load by using judgment to estimate heat-to-space and heat-to-return percentages (Mitalas and Kimura 1971).

Return Air Light Fixtures

Two generic types of return air light fixture are available—those that allow and those that do not allow return air to flow through the lamp chamber. The first type is sometimes called a heat-of-light fixture. The percentage of light heat released through the plenum side of various ventilated fixtures can be obtained from lighting fixture manufacturers. For representative data, see Nevins et al. (1971). Even unventilated fixtures lose some heat to plenum spaces; however, most of the heat ultimately enters the conditioned space from a dead-air plenum or is picked up by return air via ceiling return air openings. The percentage of heat to return air ranges from 40 to 60% for heat-to-return ventilated fixtures or 15 to 25% for unventilated fixtures.

Plenum Temperatures

As heat from lighting is picked up by the return air, the temperature differential between the ceiling cavity and the conditioned space causes part of that heat to flow from the ceiling back to the conditioned space. Return air from the conditioned space can be ducted to capture light heat without passing through a ceiling plenum as such, or the ceiling space can be used as a return air plenum, causing the distribution of light heat to be handled in distinctly different ways. Most plenum temperatures do not rise more than 1 to 3°F above space temperature, thus generating only a relatively small thermal gradient for heat transfer through plenum surfaces but

Table 2 Typical Nonincandescent Light Fixtures

Description	Ballast	Watts/Lamp	Lamps/Fixture	Lamp Watts	Fixture Watts	Special Allowance Factor
Compact Fluorescent Fixtures						
Twin, (1) 5 W lamp	Mag-Std	5	1	5	9	1.80
Twin, (1) 7 W lamp	Mag-Std	7	1	7	10	1.43
Twin, (1) 9 W lamp	Mag-Std	9	1	9	11	1.22
Quad, (1) 13 W lamp	Mag-Std	13	1	13	17	1.31
Quad, (2) 18 W lamp	Mag-Std	18	2	36	45	1.25
Quad, (2) 22 W lamp	Mag-Std	22	2	44	48	1.09
Quad, (2) 26 W lamp	Mag-Std	26	2	52	66	1.27
Twin, (2) 40 W lamp	Mag-Std	40	2	80	85	1.06
Quad, (1) 13 W lamp	Electronic	13	1	13	15	1.15
Quad, (1) 26 W lamp	Electronic	26	1	26	27	1.04
Quad, (2) 18 W lamp	Electronic	18	2	36	38	1.06
Quad, (2) 26 W lamp	Electronic	26	2	52	50	0.96
Twin or multi, (2) 32 W lamp	Electronic	32	2	64	62	0.97
Fluorescent Fixtures						
(1) 18 in., T8 lamp	Mag-Std	15	1	15	19	1.27
(1) 18 in., T12 lamp	Mag-Std	15	1	15	19	1.27
(2) 18 in., T8 lamp	Mag-Std	15	2	30	36	1.20
(2) 18 in., T12 lamp	Mag-Std	15	2	30	36	1.20
(1) 24 in., T8 lamp	Mag-Std	17	1	17	24	1.41
(1) 24 in., T12 lamp	Mag-Std	20	1	20	28	1.40
(2) 24 in., T12 lamp	Mag-Std	20	2	40	56	1.40
(1) 24 in., T12 HO lamp	Mag-Std	35	1	35	62	1.77
(2) 24 in., T12 HO lamp	Mag-Std	35	2	70	90	1.29
(1) 24 in., T8 lamp	Electronic	17	1	17	16	0.94
(2) 24 in., T8 lamp	Electronic	17	2	34	31	0.91
(1) 36 in., T12 lamp	Mag-Std	30	1	30	46	1.53
(2) 36 in., T12 lamp	Mag-Std	30	2	60	81	1.35
(1) 36 in., T12 ES lamp	Mag-Std	25	1	25	42	1.68
(2) 36 in., T12 ES lamp	Mag-Std	25	2	50	73	1.46
(1) 36 in., T12 HO lamp	Mag-Std	50	1	50	70	1.40
(2) 36 in., T12 HO lamp	Mag-Std	50	2	100	114	1.14
(2) 36 in., T12 lamp	Mag-ES	30	2	60	74	1.23
(2) 36 in., T12 ES lamp	Mag-ES	25	2	50	66	1.32
(1) 36 in., T12 lamp	Electronic	30	1	30	31	1.03
(1) 36 in., T12 ES lamp	Electronic	25	1	25	26	1.04
(1) 36 in., T8 lamp	Electronic	25	1	25	24	0.96
(2) 36 in., T12 lamp	Electronic	30	2	60	58	0.97
(2) 36 in., T12 ES lamp	Electronic	25	2	50	50	1.00
(2) 36 in., T8 lamp	Electronic	25	2	50	46	0.92
(2) 36 in., T8 HO lamp	Electronic	25	2	50	50	1.00
(2) 36 in., T8 VHO lamp	Electronic	25	2	50	70	1.40
(1) 48 in., T12 lamp	Mag-Std	40	1	40	55	1.38
(2) 48 in., T12 lamp	Mag-Std	40	2	80	92	1.15
(3) 48 in., T12 lamp	Mag-Std	40	3	120	140	1.17
(4) 48 in., T12 lamp	Mag-Std	40	4	160	184	1.15
(1) 48 in., T12 ES lamp	Mag-Std	34	1	34	48	1.41
(2) 48 in., T12 ES lamp	Mag-Std	34	2	68	82	1.21
(3) 48 in., T12 ES lamp	Mag-Std	34	3	102	100	0.98
(4) 48 in., T12 ES lamp	Mag-Std	34	4	136	164	1.21
(1) 48 in., T12 ES lamp	Mag-ES	34	1	34	43	1.26
(2) 48 in., T12 ES lamp	Mag-ES	34	2	68	72	1.06
(3) 48 in., T12 ES lamp	Mag-ES	34	3	102	115	1.13
(4) 48 in., T12 ES lamp	Mag-ES	34	4	136	144	1.06
(1) 48 in., T8 lamp	Mag-ES	32	1	32	35	1.09
(2) 48 in., T8 lamp	Mag-ES	32	2	64	71	1.11
(3) 48 in., T8 lamp	Mag-ES	32	3	96	110	1.15
(4) 48 in., T8 lamp	Mag-ES	32	4	128	142	1.11
(1) 48 in., T12 ES lamp	Electronic	34	1	34	32	0.94
(2) 48 in., T12 ES lamp	Electronic	34	2	68	60	0.88
(3) 48 in., T12 ES lamp	Electronic	34	3	102	92	0.90
(4) 48 in., T12 ES lamp	Electronic	34	4	136	120	0.88
(1) 48 in., T8 lamp	Electronic	32	1	32	32	1.00
(2) 48 in., T8 lamp	Electronic	32	2	64	60	0.94
(3) 48 in., T8 lamp	Electronic	32	3	96	93	0.97
(4) 48 in., T8 lamp	Electronic	32	4	128	120	0.94
(1) 60 in., T12 lamp	Mag-Std	50	1	50	63	1.26
(2) 60 in., T12 lamp	Mag-Std	50	2	100	128	1.28
(1) 60 in., T12 HO lamp	Mag-Std	75	1	75	92	1.23
(2) 60 in., T12 HO lamp	Mag-Std	75	2	150	168	1.12
(1) 60 in., T12 ES VHO lamp	Mag-Std	135	1	135	165	1.22
(2) 60 in., T12 ES VHO lamp	Mag-Std	135	2	270	310	1.15
(1) 60 in., T12 HO lamp	Mag-ES	75	1	75	88	1.17
(2) 60 in., T12 HO lamp	Mag-ES	75	2	150	176	1.17
(1) 60 in., T12 lamp	Electronic	50	1	50	44	0.88
(2) 60 in., T12 lamp	Electronic	50	2	100	88	0.88
(1) 60 in., T12 HO lamp	Electronic	75	1	75	69	0.92
(2) 60 in., T12 HO lamp	Electronic	75	2	150	138	0.92
(1) 60 in., T8 lamp	Electronic	40	1	40	36	0.90
(2) 60 in., T8 lamp	Electronic	40	2	80	72	0.90
(3) 60 in., T8 lamp	Electronic	40	3	120	106	0.88
(4) 60 in., T8 lamp	Electronic	40	4	160	134	0.84
(1) 72 in., T12 lamp	Mag-Std	55	1	55	76	1.38
(2) 72 in., T12 lamp	Mag-Std	55	2	110	122	1.11
(3) 72 in., T12 lamp	Mag-Std	55	3	165	202	1.22
(4) 72 in., T12 lamp	Mag-Std	55	4	220	244	1.11
(1) 72 in., T12 HO lamp	Mag-Std	85	1	85	120	1.41
(2) 72 in., T12 HO lamp	Mag-Std	85	2	170	220	1.29
(1) 72 in., T12 VHO lamp	Mag-Std	160	1	160	180	1.13
(2) 72 in., T12 VHO lamp	Mag-Std	160	2	320	330	1.03
(2) 72 in., T12 lamp	Mag-ES	55	2	110	122	1.11
(4) 72 in., T12 lamp	Mag-ES	55	4	220	244	1.11
(2) 72 in., T12 HO lamp	Mag-ES	85	2	170	194	1.14
(4) 72 in., T12 HO lamp	Mag-ES	85	4	340	388	1.14
(1) 72 in., T12 lamp	Electronic	55	1	55	68	1.24
(2) 72 in., T12 lamp	Electronic	55	2	110	108	0.98
(3) 72 in., T12 lamp	Electronic	55	3	165	176	1.07
(4) 72 in., T12 lamp	Electronic	55	4	220	216	0.98
(1) 96 in., T12 ES lamp	Mag-Std	60	1	60	75	1.25
(2) 96 in., T12 ES lamp	Mag-Std	60	2	120	128	1.07
(3) 96 in., T12 ES lamp	Mag-Std	60	3	180	203	1.13
(4) 96 in., T12 ES lamp	Mag-Std	60	4	240	256	1.07
(1) 96 in., T12 ES HO lamp	Mag-Std	95	1	95	112	1.18
(2) 96 in., T12 ES HO lamp	Mag-Std	95	2	190	227	1.19
(3) 96 in., T12 ES HO lamp	Mag-Std	95	3	285	380	1.33
(4) 96 in., T12 ES HO lamp	Mag-Std	95	4	380	454	1.19
(1) 96 in., T12 ES VHO lamp	Mag-Std	185	1	185	205	1.11
(2) 96 in., T12 ES VHO lamp	Mag-Std	185	2	370	380	1.03
(3) 96 in., T12 ES VHO lamp	Mag-Std	185	3	555	585	1.05
(4) 96 in., T12 ES VHO lamp	Mag-Std	185	4	740	760	1.03
(2) 96 in., T12 ES lamp	Mag-ES	60	2	120	123	1.03
(3) 96 in., T12 ES lamp	Mag-ES	60	3	180	210	1.17
(4) 96 in., T12 ES lamp	Mag-ES	60	4	240	246	1.03
(2) 96 in., T12 ES HO lamp	Mag-ES	95	2	190	207	1.09
(4) 96 in., T12 ES HO lamp	Mag-ES	95	4	380	414	1.09

Table 2 Typical Nonincandescent Light Fixtures (*Concluded*)

Description	Ballast	Watts/Lamp	Lamps/Fixture	Lamp Watts	Fixture Watts	Special Allowance Factor	Description	Ballast	Watts/Lamp	Lamps/Fixture	Lamp Watts	Fixture Watts	Special Allowance Factor
(1) 96 in., T12 ES lamp	Electronic	60	1	60	69	1.15	(1) 96 in., T8 HO lamp	Electronic	59	1	59	68	1.15
(2) 96 in., T12 ES lamp	Electronic	60	2	120	110	0.92	(1) 96 in., T8 VHO lamp	Electronic	59	1	59	71	1.20
(3) 96 in., T12 ES lamp	Electronic	60	3	180	179	0.99	(2) 96 in., T8 lamp	Electronic	59	2	118	109	0.92
(4) 96 in., T12 ES lamp	Electronic	60	4	240	220	0.92	(3) 96 in., T8 lamp	Electronic	59	3	177	167	0.94
(1) 96 in., T12 ES HO lamp	Electronic	95	1	95	80	0.84	(4) 96 in., T8 lamp	Electronic	59	4	236	219	0.93
(2) 96 in., T12 ES HO lamp	Electronic	95	2	190	173	0.91	(2) 96 in., T8 HO lamp	Electronic	86	2	172	160	0.93
(4) 96 in., T12 ES HO lamp	Electronic	95	4	380	346	0.91	(4) 96 in., T8 HO lamp	Electronic	86	4	344	320	0.93
(1) 96 in., T8 lamp	Electronic	59	1	59	58	0.98							
Circular Fluorescent Fixtures													
Circlite, (1) 20 W lamp	Mag-PH	20	1	20	20	1.00	(2) 8 in. circular lamp	Mag-RS	22	2	44	52	1.18
Circlite, (1) 22 W lamp	Mag-PH	22	1	22	20	0.91	(1) 12 in. circular lamp	Mag-RS	32	1	32	31	0.97
Circline, (1) 32 W lamp	Mag-PH	32	1	32	40	1.25	(2) 12 in. circular lamp	Mag-RS	32	2	64	62	0.97
(1) 6 in. circular lamp	Mag-RS	20	1	20	25	1.25	(1) 16 in. circular lamp	Mag-Std	40	1	40	35	0.88
(1) 8 in. circular lamp	Mag-RS	22	1	22	26	1.18							
High-Pressure Sodium Fixtures													
(1) 35 W lamp	HID	35	1	35	46	1.31	(1) 250 W lamp	HID	250	1	250	295	1.18
(1) 50 W lamp	HID	50	1	50	66	1.32	(1) 310 W lamp	HID	310	1	310	365	1.18
(1) 70 W lamp	HID	70	1	70	95	1.36	(1) 360 W lamp	HID	360	1	360	414	1.15
(1) 100 W lamp	HID	100	1	100	138	1.38	(1) 400 W lamp	HID	400	1	400	465	1.16
(1) 150 W lamp	HID	150	1	150	188	1.25	(1) 1000 W lamp	HID	1000	1	1000	1100	1.10
(1) 200 W lamp	HID	200	1	200	250	1.25							
Metal Halide Fixtures													
(1) 32 W lamp	HID	32	1	32	43	1.34	(1) 250 W lamp	HID	250	1	250	295	1.18
(1) 50 W lamp	HID	50	1	50	72	1.44	(1) 400 W lamp	HID	400	1	400	458	1.15
(1) 70 W lamp	HID	70	1	70	95	1.36	(2) 400 W lamp	HID	400	2	800	916	1.15
(1) 100 W lamp	HID	100	1	100	128	1.28	(1) 750 W lamp	HID	750	1	750	850	1.13
(1) 150 W lamp	HID	150	1	150	190	1.27	(1) 1000 W lamp	HID	1000	1	1000	1080	1.08
(1) 175 W lamp	HID	175	1	175	215	1.23	(1) 1500 W lamp	HID	1500	1	1500	1610	1.07
Mercury Vapor Fixtures													
(1) 40 W lamp	HID	40	1	40	50	1.25	(1) 250 W lamp	HID	250	1	250	290	1.16
(1) 50 W lamp	HID	50	1	50	74	1.48	(1) 400 W lamp	HID	400	1	400	455	1.14
(1) 75 W lamp	HID	75	1	75	93	1.24	(2) 400 W lamp	HID	400	2	800	910	1.14
(1) 100 W lamp	HID	100	1	100	125	1.25	(1) 700 W lamp	HID	700	1	700	780	1.11
(1) 175 W lamp	HID	175	1	175	205	1.17	(1) 1000 W lamp	HID	1000	1	1000	1075	1.08

Abbreviations: Mag = electromagnetic; ES = energy saver; Std = standard; HID = high-intensity discharge; HO = high output; VHO = very high output; PH = preheat; RS = rapid start

a relatively large percentage reduction in space cooling load. (Many engineers believe that a major reason for plenum temperatures not becoming more elevated is due to leakage into the plenum from supply air ducts normally concealed there.)

Energy Balance

Where the ceiling space is used as a return air plenum, an energy balance requires that the heat picked up from the lights into the return air (1) become a part of the cooling load to the return air (represented by a temperature rise of the return air as it passes through the ceiling space), (2) be partially transferred back into the conditioned space through the ceiling material below, and/or (3) may be partially "lost" (from the space) through the floor surfaces above the plenum. In a multistory building, the conditioned space frequently gains heat through its floor from a similar plenum below, offsetting the loss just mentioned. The radiant component of heat leaving the ceiling or floor surface of a plenum is normally so small that all such heat transfer is considered convective for calculation purposes.

Figure 3 shows a schematic diagram of a typical return air plenum. The following equations, using the heat flow directions shown

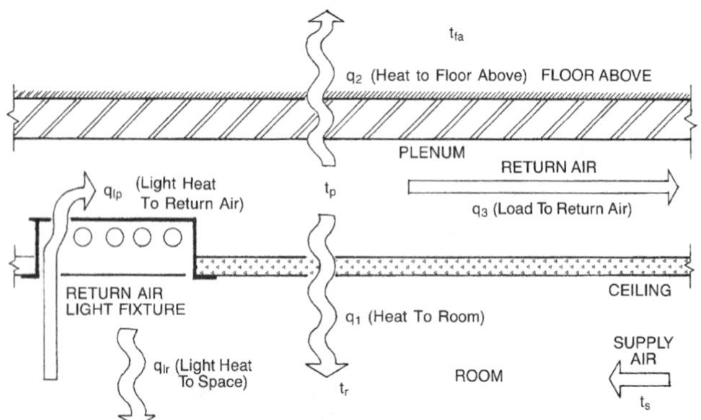

Fig. 3 Schematic Diagram of Typical Return Air Plenum

in Figure 3, represent the heat balance of a return air plenum design for a typical interior room in a multifloor building:

$$q_1 = U_c A_c (t_p - t_r) \tag{2}$$

$$q_2 = U_f A_f (t_p - t_{fa}) \quad (3)$$

$$q_3 = 1.1Q(t_p - t_r) \quad (4)$$

$$q_{lp} - q_2 - q_1 - q_3 = 0 \quad (5)$$

$$Q = \frac{q_r + q_1}{1.1(t_r - t_s)} \quad (6)$$

where

q_1 = heat gain to space from plenum through ceiling, Btu/h
q_2 = heat loss from plenum through floor above, Btu/h
q_3 = heat gain "pickup" by return air, Btu/h
Q = return airflow, cfm
q_{lp} = light heat gain to plenum via return air, Btu/h
q_{lr} = light heat gain to space, Btu/h
q_f = heat gain from plenum below, through floor, Btu/h
q_w = heat gain from exterior wall, Btu/h
q_r = space cooling load, including appropriate treatment of q_{lr}, q_f, and/or q_w, Btu/h
t_p = plenum temperature, °F
t_r = space temperature, °F
t_{fa} = space temperature of floor above, °F
t_s = supply air temperature, °F

From heat balance Equation (5), t_p can be found as the resultant return air temperature or plenum temperature. The results, although rigorous and best solved by computer, are important in determining the cooling load, which affects equipment size selection, future energy consumption, and other factors.

Equations (2) through (6) are simplified to illustrate the heat balance relationship. Heat gain into a return air plenum is not limited to the heat of lights alone. Exterior walls directly exposed to the ceiling space will transfer heat directly to or from the return air. For single-story buildings or the top floor of a multistory building, the roof heat gain or loss enters or leaves the ceiling plenum rather than entering or leaving the conditioned space directly. The supply air quantity calculated by Equation (6) is only for the conditioned space under consideration and is assumed equal to the return air quantity.

The amount of airflow through a return plenum above a conditioned space may not be limited to that supplied into the space under consideration; it will, however, have no noticeable effect on plenum temperature if the surplus comes from an adjacent plenum operating under similar conditions. Where special conditions exist, heat balance Equations (2) through (6) must be modified appropriately. Finally, even though the building's thermal storage has some effect, the amount of heat entering the return air is small and may be considered as convective for calculation purposes.

ELECTRIC MOTORS

Instantaneous heat gain from equipment operated by electric motors within a conditioned space is calculated as

$$q_{em} = 2545(P/E_M)F_{UM}F_{LM} \quad (7)$$

where

q_{em} = heat equivalent of equipment operation, Btu/h
P = motor power rating, hp
E_M = motor efficiency, decimal fraction < 1.0
F_{UM} = motor use factor, 1.0 or decimal fraction < 1.0
F_{LM} = motor load factor, 1.0 or decimal fraction < 1.0

The motor use factor may be applied when motor use is known to be intermittent with significant nonuse during all hours of operation (e.g., overhead door operator). For conventional applications, its value would be 1.0.

The motor load factor is the fraction of the rated load being delivered under the conditions of the cooling load estimate. In Equation (7), it is assumed that both the motor and the driven equipment are in the conditioned space. If the motor is outside the space or airstream,

$$q_{em} = 2545PF_{UM}F_{LM} \quad (8)$$

When the motor is inside the conditioned space or airstream but the driven machine is outside,

$$q_{em} = 2545P\left(\frac{1.0 - E_M}{E_M}\right)F_{UM}F_{LM} \quad (9)$$

Equation (9) also applies to a fan or pump in the conditioned space that exhausts air or pumps fluid outside that space.

Tables 3A and 3B give average efficiencies and related data representative of typical electric motors, generally derived from the lower efficiencies reported by several manufacturers of open, drip-proof motors. These reports indicate that totally enclosed

Table 3A Average Efficiencies and Related Data Representative of Typical Electric Motors

				Location of Motor and Driven Equipment with Respect to Conditioned Space or Airstream		
				A	B	C
Motor Nameplate or Rated Horsepower	Motor Type	Nominal rpm	Full Load Motor Efficiency, %	Motor in, Driven Equipment in, Btu/h	Motor out, Driven Equipment in, Btu/h	Motor in, Driven Equipment out, Btu/h
0.05	Shaded pole	1500	35	360	130	240
0.08	Shaded pole	1500	35	580	200	380
0.125	Shaded pole	1500	35	900	320	590
0.16	Shaded pole	1500	35	1,160	400	760
0.25	Split phase	1750	54	1,180	640	540
0.33	Split phase	1750	56	1,500	840	660
0.50	Split phase	1750	60	2,120	1,270	850
0.75	3-phase	1750	72	2,650	1,900	740
1	3-phase	1750	75	3,390	2,550	850
1.5	3-phase	1750	77	4,960	3,820	1,140
2	3-phase	1750	79	6,440	5,090	1,350
3	3-phase	1750	81	9,430	7,640	1,790
5	3-phase	1750	82	15,500	12,700	2,790
7.5	3-phase	1750	84	22,700	19,100	3,640
10	3-phase	1750	85	29,900	24,500	4,490
15	3-phase	1750	86	44,400	38,200	6,210
20	3-phase	1750	87	58,500	50,900	7,610
25	3-phase	1750	88	72,300	63,600	8,680
30	3-phase	1750	89	85,700	76,300	9,440
40	3-phase	1750	89	114,000	102,000	12,600
50	3-phase	1750	89	143,000	127,000	15,700
60	3-phase	1750	89	172,000	153,000	18,900
75	3-phase	1750	90	212,000	191,000	21,200
100	3-phase	1750	90	283,000	255,000	28,300
125	3-phase	1750	90	353,000	318,000	35,300
150	3-phase	1750	91	420,000	382,000	37,800
200	3-phase	1750	91	569,000	509,000	50,300
250	3-phase	1750	91	699,000	636,000	62,900

Table 3B Typical Overload Limits with Standard Motors

	Horsepower			
Motor Type	0.05 to 0.25	0.16 to 0.33	0.67 to 0.75	1 and up
AC open	1.4	1.35	1.25	1.15
AC TEFC[a] and DC	—	1.0	1.0	1.0

Note: Some shaded pole, capacitor start, and special purpose motors have a service factor varying from 1.0 up to 1.75.

[a]Some totally enclosed fan-cooled (TEFC) motors have a service factor above 1.0.

fan-cooled (TEFC) motors are slightly more efficient. For speeds lower or higher than those listed, efficiencies may be 1 to 3% lower or higher, depending on the manufacturer. Should actual voltages at motors be appreciably higher or lower than rated nameplate voltage, efficiencies in either case will be lower. If electric motor load is an appreciable portion of cooling load, the motor efficiency should be obtained from the manufacturer. Also, depending on design, the maximum efficiency might occur anywhere between 75 to 110% of full load; if underloaded or overloaded, the efficiency could vary from the manufacturer's listing.

Overloading or Underloading

Heat output of a motor is generally proportional to the motor load, within the overload limits. Because of typically high no-load motor current, fixed losses, and other reasons, F_{LM} is generally assumed to be unity, and no adjustment should be made for underloading or overloading unless the situation is fixed, can be accurately established, and the reduced load efficiency data can be obtained from the motor manufacturer.

Radiation and Convection

Unless the manufacturer's technical literature indicates otherwise, the heat gain normally should be equally divided between radiant and convective components for the subsequent cooling load calculations.

APPLIANCES

In a cooling load estimate, heat gain from all appliances—electrical, gas, or steam—should be taken into account. Because of the variety of appliances, applications, schedules, use, and installations, estimates can be very subjective. Often, the only information available about heat gain from equipment is that on its nameplate.

Cooking Appliances

These appliances include common heat-producing cooking equipment found in conditioned commercial kitchens. Marn (1962) concluded that appliance surfaces contributed most of the heat to commercial kitchens and that when applicances were installed under an effective hood, the cooling load was independent of the fuel or energy used for similar equipment performing the same operations.

Gordon et al. (1994) and Smith et al. (1995) found that gas appliances may exhibit slightly higher heat gains than their electric counterparts under wall-canopy hoods operated at typical ventilation rates. This is due to the fact that the heat contained in the combustion products exhausted from a gas appliance may increase the temperatures of the appliance and surrounding surfaces, as well as the hood above the appliance, more than the heat produced by its electric counterpart. These higher temperature surfaces radiate heat to the kitchen, adding moderately to the radiant gain directly associated with the appliance cooking surface.

Marn (1962) confirmed that where the appliances are installed under an effective hood, only radiant gain adds to the cooling load; convected and latent heat from the cooking process and combustion products are exhausted and do not enter the kitchen. Gordon et al. (1994) and Smith et al. (1995) substantiated these findings.

Sensible Heat Gain for Hooded Cooking Appliances. To establish a heat gain value, nameplate energy input ratings may be used with appropriate usage and radiation factors. Where specific rating data are not available (nameplate missing, equipment not yet purchased, etc.) or as an alternative approach, recommended heat gains listed in Table 5 for a wide variety of commonly encountered equipment items may be used. In estimating the appliance load, probabilities of simultaneous use and operation for different appliances located in the same space must be considered.

The radiant heat gain from hooded cooking equipment can range from 15 to 45% of the actual appliance energy consumption (Talbert et al. 1973, Gordon et al. 1994, Smith et al. 1995). This ratio of heat gain to appliance energy consumption may be expressed as a radiation factor. It is a function of both appliance type and fuel source. The radiation factor F_R is applied to the average rate of appliance energy consumption, determined by applying usage factor F_U to the nameplate or rated energy input. Marn (1962) found that radiant heat temperature rise can be substantially reduced by shielding the fronts of cooking appliances. Although this approach may not always be practical in a commercial kitchen, radiant gains can also be reduced by adding side panels or partial enclosures that are integrated with the exhaust hood.

Heat Gain from Meals. For each meal served, the heat transferred to the dining space is approximately 50 Btu/h, of which 75% is sensible and 25% is latent.

Heat Gain for Electric and Steam Appliances. The average rate of appliance energy consumption can be estimated from the nameplate or rated energy input q_{input} by applying a duty cycle or usage factor F_U. Thus, the sensible heat gain $q_{sensible}$ for generic types of electric, steam, and gas appliances installed under a hood can be estimated using one of the following equations:

$$q_{sensible} = q_{input} F_U F_R \tag{10}$$

or

$$q_{sensible} = q_{input} F_L \tag{11}$$

where F_L is defined as the ratio of sensible heat gain to the manufacturer's rated energy input.

Table 4 lists usage factors, radiation factors, and load factors based on appliance energy consumption rate for typical electrical, steam, and gas appliances under standby or idle conditions.

Unhooded Equipment. For all cooking appliances not installed under an exhaust hood or directly vent-connected and located in the conditioned area, the heat gain may be estimated as 50% ($F_U = 0.50$)

Table 4A Hooded Electric Appliance Usage Factors, Radiation Factors, and Load Factors

Appliance	Usage Factor F_U	Radiation Factor F_R	Load Factor $F_L = F_U F_R$ Elec/Steam
Griddle	0.16	0.45	0.07
Fryer	0.06	0.43	0.03
Convection oven	0.42	0.17	0.07
Charbroiler	0.83	0.29	0.24
Open-top range without oven	0.34	0.46	0.16
Hot-top range			
without oven	0.79	0.47	0.37
with oven	0.59	0.48	0.28
Steam cooker	0.13	0.30	0.04

Sources: Alereza and Breen (1984), Fisher (1998).

Table 4B Hooded Gas Appliance Usage Factors, Radiation Factors, and Load Factors

Appliance	Usage Factor F_U	Radiation Factor F_R	Load Factor $F_L = F_U F_R$ Gas
Griddle	0.25	0.25	0.06
Fryer	0.07	0.35	0.02
Convection oven	0.42	0.20	0.08
Charbroiler	0.62	0.18	0.11
Open-top range without oven	0.34	0.17	0.06

Sources: Alereza and Breen (1984), Fisher (1998).

or the rated hourly input, regardless of the type of energy or fuel used. On average, 34% of the heat may be assumed to be latent and the remaining 66% sensible. Note that cooking appliances ventilated by "ductless" hoods should be treated as unhooded appliances from the perspective of estimating heat gain. In other words, all energy consumed by the appliance and all moisture produced by the cooking process is introduced to the kitchen as a sensible or latent cooling load.

Recommended Heat Gain Values. As an alternative procedure, Table 5 lists recommended rates of heat gain from typical commercial cooking appliances. The data in the "with hood" columns assume installation under a properly designed exhaust hood connected to a mechanical fan exhaust system.

Hospital and Laboratory Equipment

Hospital and laboratory equipment items are major sources of heat gain in conditioned spaces. Care must be taken in evaluating the probability and duration of simultaneous usage when many components are concentrated in one area, such as a laboratory, an operating room, etc. Commonly, heat gain from equipment in a laboratory ranges from 15 to 70 Btu/h·ft^2 or, in laboratories with outdoor exposure, as much as four times the heat gain from all other sources combined.

Medical Equipment. It is more difficult to provide generalized heat gain recommendations for medical equipment than for general office equipment because medical equipment is much more varied in type and in application. Some heat gain testing has been done and can be presented, but the equipment included represents only a small sample of the type of equipment that may be encountered.

The data presented for medical equipment in Table 6 are relevant for portable and bench-top equipment. Medical equipment is very specific and can vary greatly from application to application. The data are presented to provide guidance in only the most general sense. For large equipment, such as MRI, engineers must obtain heat gain from the manufacturer.

Laboratory Equipment. Equipment in laboratories is similar to medical equipment in that it will vary significantly from space to space. Chapter 13 of the 1999 *ASHRAE Handbook—Applications* discusses heat gain from equipment, stating that it may range from 5 to 25 W/ft^2 in highly automated laboratories. Table 7 lists some values for laboratory equipment, but, as is the case for medical equipment, it is for general guidance only. Wilkins and Cook (1999) also examined laboratory equipment heat gains.

Office Equipment

Computers, printers, copiers, calculators, checkwriters, posting machines, etc., can generate 3 to 4 Btu/h·ft^2 for general offices or 6 to 7 Btu/h·ft^2 for purchasing and accounting departments. ASHRAE *Research Project* 822 developed a method to measure the actual heat gain from equipment in buildings and the radiant/convective percentages (Hosni et al. 1998; Jones et al. 1998). This methodology was then incorporated into ASHRAE *Research Project* 1055 and applied to a wide range of equipment (Hosni et al. 1999) as a follow-up to independent research by Wilkins et al. (1991) and Wilkins and McGaffin (1994). Komor (1997) found similar results. Analysis of measured data showed that results for office equipment could be generalized, but results from laboratory and hospital equipment proved too diverse. The following general guidelines for office equipment are a result of these studies.

Nameplate Versus Measured Energy Use. Nameplate data rarely reflect the actual power consumption of office equipment. Actual power consumption of such equipment is assumed equal to the total (radiant plus convective) heat gain, but the ratio of such energy to the nameplate value varies widely. ASHRAE *Research Project* 1055 (Hosni et al. 1999) found that for general office equipment with nameplate power consumption of less than 1000 W, the actual ratio of total heat gain to nameplate ranged from 25% to 50%, but when all tested equipment is considered, the range is broader. Generally, if the nameplate value is the only information known and no actual heat gain data are available for similar equipment, it would be conservative to use 50% of nameplate as heat gain and more nearly correct if 25% of nameplate were used. Much better results can be obtained, however, by considering the heat gain as being predictable based on the type of equipment.

Office equipment is grouped into categories such as computers, monitors, printers, facsimile machines, and copiers, with heat gain results within each group analyzed to establish patterns.

Computers. Based on tests by Hosni et al. (1999) and Wilkins and McGaffin (1994), nameplate values on computers should be ignored when performing cooling load calculations. Table 8 presents typical heat gain values for computers with varying degrees of safety factor.

Monitors. Based on monitors tested by Hosni et al. (1999), heat gain correlates approximately with screen size as

$$q_{mon} = 5S - 20 \qquad (12)$$

where

q_{mon} = heat gain from monitor, W
S = nominal screen size, in.

Wilkins and McGaffin tested ten monitors (13 to 19 in.), finding the average heat gain value to be 60 W. This testing was done in 1992 when DOS was prevalent and the Windows™ operating system was just being introduced. Monitors displaying Windows consumed more power than those displaying DOS. Table 8 tabulates typical values.

Laser Printers. Hosni et al. (1999) found that the power consumed by laser printers, and therefore the heat gain, depended largely on the level of throughput for which the printer was designed. It was observed that smaller printers are used more intermittently and that larger printers may run continuously for longer periods. Table 9 presents data on laser printers.

These data can be applied by taking the value for continuous operation and then applying an appropriate diversity factor. This would likely be most appropriate for larger open office areas. Another approach could be to take the value that most closely matches the expected operation of the printer with no diversity. This may be appropriate when considering a single room or small area.

Copiers. Hosni et al. (1999) also tested five copy machines considered to be of two types, desktop and office (freestanding high-volume copiers). Larger machines used in production environments were not addressed. Table 9 summarizes of the results. It was observed that desktop copiers rarely operated continuously but that office copiers frequently operated continuously for periods of an hour or more.

Miscellaneous Office Equipment. Table 10 presents data on miscellaneous office equipment such as vending machines and mailing equipment.

Diversity. The ratio of the measured peak electrical load at the equipment panels to the sum of the maximum electrical load of each individual item of equipment is the usage diversity. A small, one- or two-person office containing equipment listed in Tables 8 through 10 can be expected to contribute heat gain to the space at the sum of the appropriate listed values. Progressively larger areas with many equipment items will always experience some degree of usage diversity resulting from whatever percentage of such equipment is not in operation at any given time.

Wilkins and McGaffin (1994) measured diversity in 23 areas within five different buildings totaling over 275,000 ft^2. Diversity was found to range between 37 and 78%, with the average (normalized based on area) being 46%. Figure 4 illustrates the relationship between nameplate, the sum of the peaks, and the actual electical load with diversity accounted for, based on the average of the total

Table 5 Recommended Rates of Heat Gain From Typical Commercial Cooking Appliances

Appliance	Size	Energy Rate, Btu/h		Recommended Rate of Heat Gain,[a] Btu/h			
				Without Hood			With Hood
		Rated	Standby	Sensible	Latent	Total	Sensible
Electric, No Hood Required							
Barbeque (pit), per pound of food capacity	80 to 300 lb	136	—	86	50	136	42
Barbeque (pressurized), per pound of food capacity	44 lb	327	—	109	54	163	50
Blender, per quart of capacity	1 to 4 qt	1,550	—	1,000	520	1,520	480
Braising pan, per quart of capacity	108 to 140 qt	360	—	180	95	275	132
Cabinet (large hot holding)	16.2 to 17.3 ft^3	7,100	—	610	340	960	290
Cabinet (large hot serving)	37.4 to 406 ft^3	6,820	—	610	310	920	280
Cabinet (large proofing)	16 to 17 ft^3	693	—	610	310	920	280
Cabinet (small hot holding)	3.2 to 6.4 ft^3	3,070	—	270	140	410	130
Cabinet (very hot holding)	17.3 ft^3	21,000	—	1,880	960	2,830	850
Can opener		580	—	580	—	580	0
Coffee brewer	12 cup/2 burners	5,660	—	3,750	1,910	5,660	1,810
Coffee heater, per boiling burner	1 to 2 burners	2,290	—	1,500	790	2,290	720
Coffee heater, per warming burner	1 to 2 burners	340	—	230	110	340	110
Coffee/hot water boiling urn, per quart of capacity	11.6 qt	390	—	256	132	388	123
Coffee brewing urn (large), per quart of capacity	23 to 40 qt	2,130	—	1,420	710	2,130	680
Coffee brewing urn (small), per quart of capacity	10.6 qt	1,350	—	908	445	1,353	416
Cutter (large)	18 in. bowl	2,560	—	2,560	—	2,560	0
Cutter (small)	14 in. bowl	1,260	—	1,260	—	1,260	0
Cutter and mixer (large)	30 to 48 qt	12,730	—	12,730	—	12730	0
Dishwasher (hood type, chemical sanitizing), per 100 dishes/h	950 to 2000 dishes/h	1,300	—	170	370	540	170
Dishwasher (hood type, water sanitizing), per 100 dishes/h	950 to 2000 dishes/h	1,300	—	190	420	610	190
Dishwasher (conveyor type, chemical sanitizing), per 100 dishes/h	5000 to 9000 dishes/h	1,160	—	140	330	470	150
Dishwasher (conveyor type, water sanitizing), per 100 dishes/h	5000 to 9000 dishes/h	1,160	—	150	370	520	170
Display case (refrigerated), per 10 ft^3 of interior	6 to 67 ft^3	1,540	—	617	0	617	0
Dough roller (large)	2 rollers	5,490	—	5,490	—	5,490	0
Dough roller (small)	1 roller	1,570	—	140	—	140	0
Egg cooker	12 eggs	6,140	—	2,900	1,940	4,850	1,570
Food processor	2.4 qt	1,770	—	1,770	—	1,770	0
Food warmer (infrared bulb), per lamp	1 to 6 bulbs	850	—	850	—	850	850
Food warmer (shelf type), per square foot of surface	3 to 9 ft^2	930	—	740	190	930	260
Food warmer (infrared tube), per foot of length	39 to 53 in.	990	—	990	—	990	990
Food warmer (well type), per cubic foot of well	0.7 to 2.5 ft^3	3,620	—	1,200	610	1,810	580
Freezer (large)	73	4,570	—	1,840	—	1,840	0
Freezer (small)	18	2,760	—	1,090	—	1,090	0
Griddle/grill (large), per square foot of cooking surface	4.6 to 11.8 ft^2	9,200	—	615	343	958	343
Griddle/grill (small), per square foot of cooking surface	2.2 to 4.5 ft^2	8,300	—	545	308	853	298
Hot dog broiler	48 to 56 hot dogs	3,960	—	340	170	510	160
Hot plate (double burner, high speed)		16,720	—	7,810	5,430	13,240	6,240
Hot plate (double burner, stockpot)		13,650	—	6,380	4,440	10,820	5,080
Hot plate (single burner, high speed)		9,550	—	4,470	3,110	7,580	3,550
Hot water urn (large), per quart of capacity	56 qt	416	—	161	52	213	68
Hot water urn (small), per quart of capacity	8 qt	738	—	285	95	380	123
Ice maker (large)	220 lb/day	3,720	—	9,320	—	9,320	0
Ice maker (small)	110 lb/day	2,560	—	6,410	—	6,410	0
Microwave oven (heavy duty, commercial)	0.7 ft^3	8,970	—	8,970	—	8,970	0
Microwave oven (residential type)	1 ft^3	2,050 to 4,780	—	2,050 to 4,780	—	2,050 to 4,780	0
Mixer (large), per quart of capacity	81 qt	94	—	94	—	94	0
Mixer (small), per quart of capacity	12 to 76 qt	48	—	48	—	48	0
Press cooker (hamburger)	300 patties/h	7,510	—	4,950	2,560	7510	2,390
Refrigerator (large), per 10 ft^3 of interior space	25 to 74 ft^3	753	—	300	—	300	0
Refrigerator (small), per 10 ft^3 of interior space	6 to 25 ft^3	1,670	—	665	—	665	0
Rotisserie	300 hamburgers/h	10,920	—	7,200	3,720	10,920	3,480
Serving cart (hot), per cubic foot of well	1.8 to 3.2 ft^3	2,050	—	680	340	1,020	328
Serving drawer (large)	252 to 336 dinner rolls	3,750	—	480	34	510	150
Serving drawer (small)	84 to 168 dinner rolls	2,730	—	340	34	380	110
Skillet (tilting), per quart of capacity	48 to 132 qt	580	—	293	161	454	218
Slicer, per square foot of slicing carriage	0.65 to 0.97 ft^2	680	—	682	—	682	216
Soup cooker, per quart of well	7.4 to 11.6 qt	416	—	142	78	220	68
Steam cooker, per cubic foot of compartment	32 to 64 qt	20,700	—	1,640	1,050	2,690	784
Steam kettle (large), per quart of capacity	80 to 320 qt	300	—	23	16	39	13
Steam kettle (small), per quart of capacity	24 to 48 qt	840	—	68	45	113	32
Syrup warmer, per quart of capacity	11.6 qt	284	—	94	52	146	45

Table 5 Recommended Rates of Heat Gain From Typical Commercial Cooking Appliances (*Concluded*)

		Energy Rate, Btu/h		Recommended Rate of Heat Gain,[a] Btu/h			
				Without Hood			With Hood
Appliance	Size	Rated	Standby	Sensible	Latent	Total	Sensible
Toaster (bun toasts on one side only)	1400 buns/h	5,120	—	2,730	2,420	5,150	1,640
Toaster (large conveyor)	720 slices/h	10,920	—	2,900	2,560	5,460	1,740
Toaster (small conveyor)	360 slices/h	7,170	—	1,910	1,670	3,580	1,160
Toaster (large pop-up)	10 slice	18,080	—	9,590	8,500	18,080	5,800
Toaster (small pop-up)	4 slice	8,430	—	4,470	3,960	8,430	2,700
Waffle iron	75 in^2	5,600	—	2,390	3,210	5,600	1,770
Electric, Exhaust Hood Required							
Broiler (conveyor infrared), per square foot of cooking area	2 to 102 ft^2	19,230	—	—	—	—	3,840
Broiler (single deck infrared), per square foot of broiling area	2.6 to 9.8 ft^2	10,870	—	—	—	—	2,150
Charbroiler, per linear foot of cooking surface	2 to 8 linear ft	11,000	9,300	—	—	—	2,800
Fryer (deep fat)	35 to 50 lb oil	48,000	2,900	—	—	—	1,200
Fryer (pressurized), per pound of fat capacity	13 to 33 lb	1565	—	—	—	—	59
Oven (full-size convection)		41,000	4,600	—	—	—	2,900
Oven (large deck baking with 537 ft^3 decks), per cubic foot of oven space	15 to 46 ft^3	1,670	—	—	—	—	69
Oven (roasting), per cubic foot of oven space	7.8 to 23 ft^3	27,350	—	—	—	—	113
Oven (small convection), per cubic foot of oven space	1.4 to 5.3 ft^3	10,340	—	—	—	—	147
Oven (small deck baking with 272 ft^3 decks), per cubic foot of oven space	7.8 to 23 ft^3	2,760	—	—	—	—	113
Open range top, per 2 element section	2 to 6 elements	14,000	4,600	—	—	—	2,100
Range (hot top/fry top), per square foot of cooking surface	4 to 8 ft^2	7,260	—	—	—	—	2,690
Range (oven section), per cubic foot of oven space	4.2 to 11.3 ft^3	3,940	—	—	—	—	160
Griddle, per linear foot of cooking surface	2 to 8 linear ft	19,500	3,100	—	—	—	1,400
Gas, No Hood Required							
Broiler, per square foot of broiling area	2.7 ft^2	14,800	660[b]	5,310	2,860	8,170	1,220
Cheese melter, per square foot of cooking surface	2.5 to 5.1 ft^2	10,300	660[b]	3,690	1,980	5,670	850
Dishwasher (hood type, chemical sanitizing), per 100 dishes/h	950 to 2,000 dishes/h	1,740	660[b]	510	200	710	230
Dishwasher (hood type, water sanitizing), per 100 dishes/h	950 to 2,000 dishes/h	1,740	660[b]	570	220	790	250
Dishwasher (conveyor type, chemical sanitizing), per 100 dishes/h	5,000 to 9,000 dishes/h	1,370	660[b]	330	70	400	130
Dishwasher (conveyor type, water sanitizing), per 100 dishes/h	5,000 to 9,000 dishes/h	1,370	660[b]	370	80	450	140
Griddle/grill (large), per square foot of cooking surface	4.6 to 11.8 ft^2	17,000	330	1,140	610	1,750	460
Griddle/grill (small), per square foot of cooking surface	2.5 to 4.5 ft^2	14,400	330	970	510	1,480	400
Hot plate	2 burners	19,200	1,325[b]	11,700	3,470	15,200	3,410
Oven (pizza), per square foot of hearth	6.4 to 12.9 ft^2	4,740	660[b]	623	220	843	85
Gas, Exhaust Hood Required							
Braising pan, per quart of capacity	105 to 140 qt	9,840	660[b]	—	—	—	2,430
Broiler, per square foot of broiling area	3.7 to 3.9 ft^2	21,800	530	—	—	—	1,800
Broiler (large conveyor, infrared), per square foot of cooking area/minute	2 to 102 ft^2	51,300	1,990	—	—	—	5,340
Broiler (standard infrared), per square foot of broiling area	2.4 to 9.4 ft^2	1,940	530	—	—	—	1,600
Charbroiler (large), per linear foot of cooking area	2 to 8 linear ft	36,000	22,000	—	—	—	3,800
Fryer (deep fat)	35 to 50 oil cap.	80,000	5,600	—	—	—	1,900
Oven (bake deck), per cubic foot of oven space	5.3 to 16.2 ft^3	7,670	660[b]	—	—	—	140
Oven (convection), full size		70,000	29,400	—	—	—	5,700
Oven (pizza), per square foot of oven hearth	9.3 to 25.8 ft^2	7,240	660[b]	—	—	—	130
Oven (roasting), per cubic foot of oven space	9 to 28 ft^3	4,300	660[b]	—	—	—	77
Oven (twin bake deck), per cubic foot of oven space	11 to 22 ft^3	4,390	660[b]	—	—	—	78
Range (burners), per 2 burner section	2 to 10 burners	33,600	1,325	—	—	—	6,590
Range (hot top or fry top), per square foot of cooking surface	3 to 8 ft^2	11,800	330	—	—	—	3,390
Range (large stock pot)	3 burners	100,000	1,990	—	—	—	19,600
Range (small stock pot)	2 burners	40,000	1,330	—	—	—	7,830
Griddle, per linear foot of cooking surface	2 to 8 linear ft	25,000	6,300				1,600
Range top, open burner (per 2 burner section)	2 to 6 elements	40,000	13,600				2,200
Steam							
Compartment steamer, per pound of food capacity/h	46 to 450 lb	280	—	22	14	36	11
Dishwasher (hood type, chemical sanitizing), per 100 dishes/h	950 to 2,000 dishes/h	3,150	—	880	380	1,260	410
Dishwasher (hood type, water sanitizing), per 100 dishes/h	950 to 2,000 dishes/h	3,150	—	980	420	1,400	450
Dishwasher (conveyor, chemical sanitizing), per 100 dishes/h	5,000 to 9,000 dishes/h	1,180	—	140	330	470	150
Dishwasher (conveyor, water sanitizing), per 100 dishes/h	5,000 to 9,000 dishes/h	1,180	—	150	370	520	170
Steam kettle, per quart of capacity	13 to 32 qt	500	—	39	25	64	19

Sources: Alereza and Breen (1984), Fisher (1998).

[a]In some cases, heat gain data are given per unit of capacity. In those cases, the heat gain is calculated by: q = (recommended heat gain per unit of capacity) × (capacity)

[b]Standby input rating is given for entire appliance regardless of size.

Table 6 Recommended Heat Gain from Typical Medical Equipment

Equipment	Nameplate, W	Peak, W	Average, W
Anesthesia system	250	177	166
Blanket warmer	500	504	221
Blood pressure meter	180	33	29
Blood warmer	360	204	114
ECG/RESP	1440	54	50
Electrosurgery	1000	147	109
Endoscope	1688	605	596
Harmonical scalpel	230	60	59
Hysteroscopic pump	180	35	34
Laser sonics	1200	256	229
Optical microscope	330	65	63
Pulse oximeter	72	21	20
Stress treadmill	N/A	198	173
Ultrasound system	1800	1063	1050
Vacuum suction	621	337	302
X-ray system	968		82
X-ray system	1725	534	480
X-ray system	2070		18

Source: Hosni et al. (1999).

Table 7 Recommended Heat Gain from Typical Laboratory Equipment

Equipment	Nameplate, W	Peak, W	Average, W
Analytical balance	7	7	7
Centrifuge	138	89	87
Centrifuge	288	136	132
Centrifuge	5500	1176	730
Electrochemical analyzer	50	45	44
Electrochemical analyzer	100	85	84
Flame photometer	180	107	105
Fluorescent microscope	150	144	143
Fluorescent microscope	200	205	178
Function generator	58	29	29
Incubator	515	461	451
Incubator	600	479	264
Incubator	3125	1335	1222
Orbital shaker	100	16	16
Oscilloscope	72	38	38
Oscilloscope	345	99	97
Rotary evaporator	75	74	73
Rotary evaporator	94	29	28
Spectronics	36	31	31
Spectrophotometer	575	106	104
Spectrophotometer	200	122	121
Spectrophotometer	N/A	127	125
Spectro fluorometer	340	405	395
Thermocycler	1840	965	641
Thermocycler	N/A	233	198
Tissue culture	475	132	46
Tissue culture	2346	1178	1146

Source: Hosni et al. (1999).

area tested. Data on actual diversity can be used as a guide, but diversity varies significantly with occupancy. The proper diversity factor for an office of mail order catalog telephone operators is different from that for an office of sales representatives who travel regularly.

Heat Gain per Unit Area. Wilkins (1998) and Wilkins and Hosni (2000) summarized the recent research on a heat gain per unit area basis. The diversity testing showed that the actual heat gain per unit area, or load factor, ranged from 0.44 to 1.08 W/ft^2, with an average (normalized based on area) of 0.81 W/ft^2. Spaces tested

Table 8 Recommended Heat Gain from Typical Computer Equipment

	Continuous, W	Energy Saver Mode, W
Computers[a]		
Average value	55	20
Conservative value	65	25
Highly conservative value	75	30
Monitors[b]		
Small monitor (13 to 15 in.)	55	0
Medium monitor (16 to 18 in.)	70	0
Large monitor (19 to 20 in.)	80	0

Sources: Hosni et al. (1999), Wilkins and McGaffin (1994).
[a]Based on 386, 486, and Pentium grade.
[b]Typical values for monitors displaying Windows environment.

Table 9 Recommended Heat Gain from Typical Laser Printers and Copiers

	Continuous, W	1 page per min., W	Idle, W
Laser Printers			
Small desktop	130	75	10
Desktop	215	100	35
Small office	320	160	70
Large office	550	275	125
Copiers			
Desktop copier	400	85	20
Office copier	1,100	400	300

Source: Hosni et al. (1999).

Table 10 Recommended Heat Gain from Miscellaneous Office Equipment

Appliance	Maximum Input Rating, W	Recommended Rate of Heat Gain, W
Mail-processing equipment		
Folding machine	125	80
Inserting machine, 3,600 to 6,800 pieces/h	600 to 3,300	390 to 2,150
Labeling machine, 1,500 to 30,000 pieces/h	600 to 6,600	390 to 4,300
Postage meter	230	150
Vending machines		
Cigarette	72	72
Cold food/beverage	1,150 to 1,920	575 to 960
Hot beverage	1,725	862
Snack	240 to 275	240 to 275
Other		
Bar code printer	440	370
Cash registers	60	48
Check processing workstation, 12 pockets	4,800	2,470
Coffee maker, 10 cups	1,500	1,050 W sens., 1,540 Btu/h latent
Microfiche reader	85	85
Microfilm reader	520	520
Microfilm reader/printer	1,150	1,150
Microwave oven, 1 ft^3	600	400
Paper shredder	250 to 3,000	200 to 2,420
Water cooler, 32 qt/h	700	350

Table 11 Recommended Load Factors for Various Types of Offices

Load Density of Office	Load Factor, W/ft^2	Description
Light	0.5	Assumes 167 ft^2/workstation (6 workstations per 1000 ft^2) with computer and monitor at each plus printer and fax. Computer, monitor, and fax diversity 0.67, printer diversity 0.33.
Medium	1	Assumes 125 ft^2/workstation (8 workstations per 1000 ft^2) with computer and monitor at each plus printer and fax. Computer, monitor, and fax diversity 0.75, printer diversity 0.50.
Medium/ Heavy	1.5	Assumes 100 ft^2/workstation (10 workstations per 1000 ft^2) with computer and monitor at each plus printer and fax. Computer and monitor diversity 0.75, printer and fax diversity 0.50.
Heavy	2	Assumes 83 ft^2/workstation (12 workstations per 1000 ft^2) with computer and monitor at each plus printer and fax. Computer and monitor diversity 1.0, printer and fax diversity 0.50.

Source: Wilkins and McGaffin (1994).

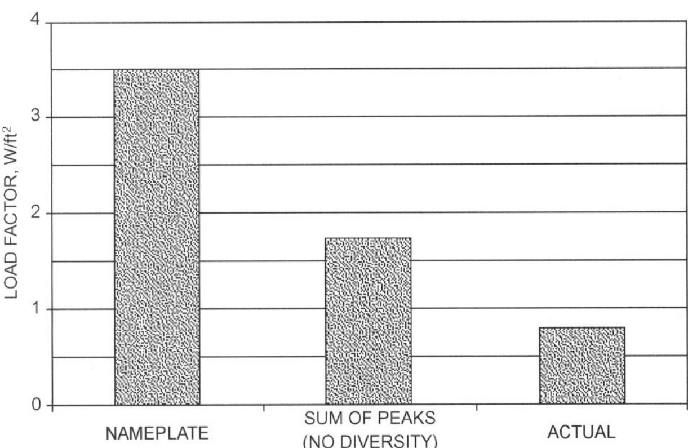

Fig. 4 Office Equipment Load Factor Comparison
(Wilkins and McGaffin 1994)

were fully occupied and highly automated, comprising 21 unique areas in five buildings, with a computer and monitor at every workstation. Table 11 presents a range of load factors with a subjective description of the type of space to which they would apply. Table 12 presents more specific data that can be used to better quantify the amount of equipment in a space and the expected load factor. The medium load density is likely to be appropriate for most standard office spaces. Medium/heavy or heavy load densities may be encountered but can be considered extremely conservative estimates even for densely populated and highly automated spaces.

Radiant Convective Split. Hosni et al. (1999) found that the radiant-convective split for equipment was fairly uniform, the most important differentiating feature being whether or not the equipment had a cooling fan. Table 13 is a summary of those results.

Table 12 Cooling Load Estimates for Various Office Load Densities

	Number	Each, W	Total, W	Diversity	Load, W
Light Load Density[a]					
Computers	6	55	330	0.67	220
Monitors	6	55	330	0.67	220
Laser printer—small desk top	1	130	130	0.33	43
Fax machine	1	15	15	0.67	10
Total Area Load					494
Recommended equipment load factor = 0.5 W/ft^2					
Medium Load Density[a]					
Computers	8	65	520	0.75	390
Monitors	8	70	560	0.75	420
Laser printer—desk	1	215	215	0.5	108
Fax machine	1	15	15	0.75	11
Total Area Load					929
Recommended equipment load factor = 1.0 W/ft^2					
Medium/Heavy Load Density[a]					
Computers	10	65	650	1	650
Monitors	10	70	700	1	700
Laser printer—small office	1	320	320	0.5	160
Facsimile machine	1	30	30	0.5	15
Total Area Load					1525
Recommended equipment load factor = 1.5 W/ft^2					
Heavy Load Density[a]					
Computers	12	75	900	1	900
Monitors	12	80	960	1	960
Laser printer-small office	1	320	320	0.5	160
Facsimile machine	1	30	30	0.5	15
Total Area Load					2035
Recommended equipment load factor = 2.0 W/ft^2					

Source: Wilkins and McGaffin (1994).
[a] See Table 11 for descriptions of load densities.

Table 13 Summary of Radiant-Convective Split for Office Equipment

Device	Fan	Radiant	Convective
Computer	Yes	10 to 15%	85 to 90%
Monitor	No	35 to 40%	60 to 65%
Computer and monitor	–	20 to 30%	70 to 80%
Laser printer	Yes	10 to 20%	80 to 90%
Copier	Yes	20 to 25%	75 to 80%
Fax machine	No	30 to 35%	65 to 70%

Source: Hosni et al. (1999).

HEAT GAIN THROUGH FENESTRATION AREAS

The primary weather-related variable influencing the cooling load for a building is solar radiation. The effect of solar radiation is more pronounced and immediate in its impact on exposed nonopaque surfaces. The calculation of solar heat gain and conductive heat transfer through various glazing materials and associated mounting frames, with or without interior and/or exterior shading devices, is discussed in Chapter 30. This chapter covers the application of such data to the overall heat gain evaluation and the conversion of the calculated heat gain into a composite cooling load for the conditioned space. Table 14 includes some useful solar equations.

Fenestration Direct Solar, Diffuse Solar, and Conductive Heat Gains

For fenestration heat gain, use the following equations:

Direct beam solar heat gain q_b:

$$q_b = AE_D \, \text{IAC}(\theta) \quad (13)$$

Diffuse solar heat gain q_d:

$$q_d = A(E_d + E_r)\langle \text{SHGC}\rangle_D \, \text{IAC} \quad (14)$$

Conductive heat gain q_c:

$$q_c = UA(T_{out} - T_{in}) \quad (15)$$

Total fenestration heat gain Q:

$$Q = q_b + q_d + q_c \quad (16)$$

where

A = window area, ft^2
E_D, E_d, and E_r = direct, diffuse, and ground-reflected irradiance, calculated using the equations in Table 14
SHGC(θ) = direct solar heat gain coefficient as a function of incident angle θ; may be interpolated between values in Table 13 of Chapter 30
$\langle \text{SHGC}\rangle_D$ = diffuse solar heat gain coefficient (also referred to as hemispherical SHGC); from Table 13 of Chapter 30
T_{in} = inside temperature, °F
T_{out} = outside temperature, °F
U = overall U-factor, including frame and mounting orientation from Table 4 of Chapter 30, $\text{Btu/h·ft}^2\text{·°F}$
IAC = inside shading attenuation coefficient, = 1.0 if no inside shading device

If specific window manufacturer's SHGC and U-factor data are available, those should be used. For fenestration equipped with inside shading (blinds, drapes or shades), IAC is listed in Tables 19, 20, and 22 of Chapter 30. The inside shading attentuation coefficients given are used to calculate both direct and diffuse solar heat gains.

Exterior Shading

Nonuniform exterior shading, caused by roof overhangs, side fins, or building projections, requires separate hourly calculations for the externally shaded and unshaded areas of the window in question, with the inside shading SHGC still used to account for any internal shading devices. The areas, shaded and unshaded, depend on the location of the shadow line on a surface in the plane of the glass. Sun (1968) developed fundamental algorithms for analysis of shade patterns. McQuiston and Spitler (1992) provide graphical data to facilitate shadow line calculation.

An alternative, more accurate method described by Todorovic et al. (1993) first calculates cooling loads as if the external shading were absent, then adjusts (reduces) the result to account for the shading effect. This correction applies a "negative cooling load factor," calculated in much the same way as a conventional cooling load but using the time-varying area of the shaded portion of the glass as the heat gain element. Todorovic (1987) describes the solution of the moving shade line problem in the context of consequent cooling load.

The equations for calculating shade angles [Chapter 30, Equations (107) to (110)] can be used to determine the shape and area of moving shadow falling across a given window from external shading elements during the course of a design day. Thus, a subprofile of heat gain for that window can be created by separating its sunlit and shaded areas for each hour.

Table 14 Solar Equations

Solar Angles

All angles are in degrees. The solar azimuth ϕ and the surface azimuth ψ are measured in degrees from south; angles to the east of south are negative, and angles to the west of south are positive. Calculate solar altitude, azimuth, and surface incident angles as follows:

Apparent solar time AST, in decimal hours:

$$\text{AST} = \text{LST} + \text{ET}/60 + (\text{LSM} - \text{LON})/15$$

Hour angle H, degrees:

$$H = 15(\text{hours of time from local solar noon}) = 15(\text{AST} - 12)$$

Solar altitude β:

$$\sin\beta = \cos L \cos\delta \cos H + \sin L \sin\delta$$

Solar azimuth ϕ:

$$\cos\phi = (\sin\beta \sin L - \sin\delta)/(\cos\beta \cos L)$$

Surface-solar azimuth γ:

$$\gamma = \phi - \psi$$

Incident angle θ:

$$\cos\phi = \cos\beta \cos\gamma \sin\Sigma + \sin\beta \cos\Sigma$$

where

ET = equation of time, decimal minutes
L = latitude
LON = local longitude, decimal degrees of arc
LSM = local standard time meridian, decimal degrees of arc
= 60° for Atlantic Standard Time
= 75° for Eastern Standard Time
= 90° for Central Standard Time
= 105° for Mountain Standard Time
= 120° for Pacific Standard Time
= 135° for Alaska Standard Time
= 150° for Hawaii-Aleutian Standard Time
LST = local standard time, decimal hours
δ = solar declination, °
ψ = surface azimuth, °
Σ = surface tilt from horizontal, horizontal = 0°

Values of ET and δ are given in Table 7 of Chapter 30 for the 21st day of each month.

Direct, Diffuse, and Total Solar Irradiance

Direct normal irradiance E_{DN}

If $\beta > 0$ $\quad E_{DN} = \left[\dfrac{A}{\exp(B/\sin\beta)}\right]\text{CN}$

Otherwise, $\quad E_{DN} = 0$

Surface direct irradiance E_D

If $\cos\theta > 0$ $\quad E_D = E_{DN}\cos\theta$

Otherwise, $\quad E_D = 0$

Ratio Y of sky diffuse on vertical surface to sky diffuse on horizontal surface

If $\cos\theta > -0.2$ $\quad Y = 0.55 + 0.437\cos\theta + 0.313\cos^2\theta$

Otherwise, $\quad Y = 0.45$

Diffuse irradiance E_d

Vertical surfaces	$E_d = CYE_{DN}$
Surfaces other than vertical	$E_d = CE_{DN}(1 + \cos\Sigma)/2$
Ground-reflected irradiance	$E_r = E_{DN}(C + \sin\beta)\rho_g(1 - \cos\Sigma)/2$
Total surface irradiance	$E_t = E_D + E_d + E_r$

where

A = apparent solar constant
B = atmospheric extinction coefficient
C = sky diffuse factor
CN = clearness number multiplier for clear/dry or hazy/humid locations. See Figure 5 in Chapter 32 of the 1999 *ASHRAE Handbook—Applications* for CN values.
E_d = diffuse sky irradiance
E_r = diffuse ground-reflected irradiance
ρ_g = ground reflectivity

Values of A, B, and C are given in Table 7 of Chapter 30 for the 21st day of each month. Values of ground reflectivity ρ_g are given in Table 10 of Chapter 30.

Temperature Considerations

To estimate the conductive heat gain through fenestration at any time, applicable values of the outdoor and indoor dry-bulb temperatures must be used. Chapter 27 gives monthly cooling load design values of outdoor dry-bulb temperatures for many locations. These are generally midafternoon temperatures; for other times, the method described below in the section on Outdoor Air Temperatures can be used to estimate temperatures. Where local microclimatic conditions prevail or data are not included in Chapter 27, local weather stations or the National Oceanic and Atmospheric Administration can supply temperature data. Winter design temperatures should not be used because such data are for heating design rather than coincident conductive heat gain with sunlit glass during the winter months.

HEAT GAIN THROUGH EXTERIOR SURFACES

Heat gain through exterior opaque surfaces is derived from the same elements of solar radiation and thermal gradient as that for fenestration areas. It differs primarily as a function of the mass and nature of the wall or roof construction, since those elements affect the rate of conductive heat transfer through the composite assembly to the interior surface.

SOL-AIR TEMPERATURE

Sol-air temperature is the temperature of the outdoor air that in the absence of all radiation changes gives the same rate of heat entry into the surface as would the combination of incident solar radiation, radiant energy exchange with the sky and other outdoor surroundings, and convective heat exchange with the outdoor air.

Heat Flux into Exterior Sunlit Surfaces

The heat balance at a sunlit surface gives the heat flux into the surface q/A as

$$\frac{q}{A} = \alpha E_t + h_o(t_o - t_s) - \varepsilon \Delta R \tag{17}$$

where

α = absorptance of surface for solar radiation
E_t = total solar radiation incident on surface, Btu/h·ft²
h_o = coefficient of heat transfer by long-wave radiation and convection at outer surface, Btu/h·ft²·°F
t_o = outdoor air temperature, °F
t_s = surface temperature, °F
ε = hemispherical emittance of surface
ΔR = difference between long-wave radiation incident on surface from sky and surroundings and radiation emitted by blackbody at outdoor air temperature, Btu/h·ft²

Assuming the rate of heat transfer can be expressed in terms of the sol-air temperature t_e,

$$\frac{q}{A} = h_o(t_e - t_s) \tag{18}$$

and from Equations (17) and (18),

$$t_e = t_o + \frac{\alpha E_t}{h_o} - \frac{\varepsilon \Delta R}{h_o} \tag{19}$$

Horizontal Surfaces. For horizontal surfaces that receive long-wave radiation from the sky only, an appropriate value of ΔR is about 20 Btu/h·ft², so that if $\varepsilon = 1$ and $h_o = 3.0$ Btu/h·ft²·°F, the long-wave correction term is about 7°F (Bliss 1961) and the correction itself thus –7°F.

Vertical Surfaces. Because vertical surfaces receive long-wave radiation from the ground and surrounding buildings as well as from the sky, accurate ΔR values are difficult to determine. When solar radiation intensity is high, surfaces of terrestrial objects usually have a higher temperature than the outdoor air; thus, their long-wave radiation compensates to some extent for the sky's low emittance. Therefore, it is common practice to assume $\varepsilon \Delta R = 0$ for vertical surfaces.

Tabulated Temperature Values

The sol-air temperatures in Table 15 have been calculated based on $\varepsilon \Delta R/h_o$ values of 7°F for horizontal surfaces and 0°F for vertical surfaces; total solar intensity values used for the calculations were calculated using the equations included in Table 14.

Surface Colors

Sol-air temperature values are given for two values of the parameter α/h_o (Table 15); the value of 0.15 is appropriate for a light-colored surface, while 0.30 represents the usual maximum value for this parameter (i.e., for a dark-colored surface or any surface for which the permanent lightness can not reliably be anticipated).

Example 1. Calculating sol-air temperature using solar equations. Calculate the sol-air temperature at 3 P.M., Central Daylight Time on July 21 for a vertical light-colored wall surface, facing southwest, located at 40° North latitude and 90° West longitude, with an outdoor temperature of 94°F. The clearness number CN is assumed to be 1.0 and ground reflectivity $\rho_g = 0.2$.

Solution: Sol-air temperature is calculated using Equation (19). For a light-colored wall, $\alpha/h_o = 0.15$, and for vertical surfaces, $\varepsilon \Delta R/h_o = 0$. The solar irradiance E_t must be determined using the equations in Table 14.

Solar Angles:

ψ = southwest orientation = +45°. Σ = surface tilt from horizontal (where horizontal = 0°) = 90° for vertical wall surface. Local solar time (LST) is one hour earlier than Daylight Time, so 3 P.M. DST = 2 P.M. LST = hour 14. Calculate solar altitude, solar azimuth, surface solar azimuth, and incident angle as follows.

From Table 7 in Chapter 30, solar position data and constants for July 21 are

$$\begin{aligned} ET &= -6.2 \text{ min} \\ \delta &= 20.6° \\ A &= 344 \text{ Btu/h·ft}^2 \\ B &= 0.207 \\ C &= 0.136 \end{aligned}$$

Local standard meridian (LSM) for Central Time Zone = 90°. Apparent solar time AST

$$\begin{aligned} AST &= LST + ET/60 + (LSM - LON)/15 \\ &= 14 + (-6.2/60) + [(90 - 90)/15] \\ &= 13.897 \end{aligned}$$

Hour angle H, degrees:

$$\begin{aligned} H &= 15(AST - 12) \\ &= 15(13.897 - 12) \\ &= 28.45° \end{aligned}$$

Solar altitude β:

$$\begin{aligned} \sin\beta &= \cos L \cos\delta \cos H + \sin L \sin\delta \\ &= \cos(40)\cos(20.6)\cos(28.45) + \sin(40)\sin(20.6) \\ &= 0.8566 \\ \beta &= \sin^{-1}(0.8566) = 58.9° \end{aligned}$$

Solar azimuth ϕ:

$$\begin{aligned} \cos\phi &= (\sin\beta \sin L - \sin\delta)/(\cos\beta \cos L) \\ &= [(\sin(58.9)\sin(40) - \sin(20.6)]/[\cos(58.9)\cos(40)] \\ &= 0.502 \\ \phi &= \cos^{-1}(0.502) = 59.9° \end{aligned}$$

Surface-solar azimuth γ:

$$\begin{aligned} \gamma &= \phi - \psi \\ &= 59.9 - 45 \\ &= 14.9° \end{aligned}$$

Table 15 Sol-Air Temperatures for July 21, 40° North Latitude

Local Standard Hour	Air Temp. t_o, °F	Light-Colored Surface, α/h_o = 0.15: N	NE	E	SE	S	SW	W	NW	Horiz.
1	76	76	76	76	76	76	76	76	76	69
2	76	76	76	76	76	76	76	76	76	69
3	75	75	75	75	75	75	75	75	75	68
4	74	74	74	74	74	74	74	74	74	67
5	74	74	74	74	74	74	74	74	74	67
6	74	82	94	96	85	76	76	76	76	73
7	75	83	103	109	97	79	78	78	78	84
8	77	82	104	114	105	83	82	82	82	95
9	80	86	102	115	111	92	86	86	86	106
10	83	89	97	111	113	100	90	89	89	115
11	87	94	94	106	112	108	96	94	94	123
12	90	97	97	98	107	112	109	99	97	128
13	93	100	100	100	103	114	117	110	100	130
14	94	101	101	101	101	112	123	121	107	127
15	95	101	101	101	101	108	126	129	116	123
16	94	99	99	99	99	101	123	131	120	114
17	93	101	97	97	97	97	117	129	122	104
18	91	100	93	93	93	93	105	116	114	92
19	87	88	87	87	87	87	88	89	89	80
20	85	85	85	85	85	85	85	85	85	78
21	83	83	83	83	83	83	83	83	83	76
22	81	81	81	81	81	81	81	81	81	74
23	79	79	79	79	79	79	79	79	79	72
24	77	77	77	77	77	77	77	77	77	70

Local Standard Hour	Air Temp. t_o, °F	Dark-Colored Surface, α/h_o = 0.30: N	NE	E	SE	S	SW	W	NW	Horiz.
1	76	76	76	76	76	76	76	76	76	69
2	76	76	76	76	76	76	76	76	76	69
3	75	75	75	75	75	75	75	75	75	68
4	74	74	74	74	74	74	74	74	74	67
5	74	74	74	74	74	74	74	74	74	67
6	74	90	114	117	97	78	78	78	78	80
7	75	91	132	144	119	82	82	82	82	100
8	77	87	132	152	134	89	86	86	86	120
9	80	92	124	150	141	104	91	91	91	139
10	83	96	112	140	142	118	96	96	96	154
11	87	101	102	125	138	129	104	101	101	166
12	90	104	104	105	124	135	128	107	104	172
13	93	107	107	107	114	136	141	127	108	173
14	94	107	107	107	108	131	152	147	119	167
15	95	107	107	107	107	122	156	162	136	157
16	94	105	104	104	104	108	152	169	147	141
17	93	108	101	101	101	101	140	164	150	122
18	91	108	95	95	95	95	119	142	137	101
19	87	89	87	87	87	87	89	90	91	81
20	85	85	85	85	85	85	85	85	85	78
21	83	83	83	83	83	83	83	83	83	76
22	81	81	81	81	81	81	81	81	81	74
23	79	79	79	79	79	79	79	79	79	72
24	77	77	77	77	77	77	77	77	77	70

Incident angle θ:

$$\begin{aligned}\cos\theta &= \cos\beta\cos\gamma\sin\Sigma + \sin\beta\cos\Sigma \\ &= \cos(58.9)\cos(14.9)\sin(90) + \sin(58.9)\cos(90) \\ &= 0.499 \\ \theta &= \cos^{-1}(0.499) = 60.1°\end{aligned}$$

Direct, Diffuse, and Total Solar Irradiance:

Direct normal irradiance E_{DN}

$$\begin{aligned}E_{DN} &= \left[\frac{A}{\exp(B/\sin\beta)}\right]CN \\ &= \frac{344}{\exp[0.207/\sin(58.9)]}(1.0) \\ &= 270 \text{ Btu/h·ft}^2\end{aligned}$$

Surface direct irradiance E_D

$$\begin{aligned}E_D &= E_{DN}\cos\theta \\ &= (270)\cos(60.1) \\ &= 135 \text{ Btu/h·ft}^2\end{aligned}$$

Ratio Y of sky diffuse on vertical surface to sky diffuse on horizontal surface

$$\begin{aligned}Y &= 0.55 + 0.437\cos\theta + 0.313\cos^2\theta \\ &= 0.55 + 0.437\cos(60.1) + 0.313\cos^2(60.1) \\ &= 0.846\end{aligned}$$

Diffuse irradiance E_d

Vertical surfaces

$$\begin{aligned}E_d &= CYE_{DN} \\ &= (0.136)(0.846)(270) \\ &= 31.1 \text{ Btu/h·ft}^2\end{aligned}$$

Ground-reflected irradiance E_r

$$\begin{aligned}E_r &= E_{DN}(C + \sin\beta)\rho_g(1 - \cos\Sigma)/2 \\ &= (270)[0.136 + \sin(58.9)](0.2)[1 - \cos(90)]/2 \\ &= 26.8 \text{ Btu/h·ft}^2\end{aligned}$$

Total surface irradiance E_t

$$\begin{aligned}E_t &= E_D + E_d + E_r \\ &= 135 + 31.1 + 26.8 \\ &= 193 \text{ Btu/h·ft}^2\end{aligned}$$

Sol-air temperature [from Equation (19)]:

$$\begin{aligned}t_e &= t_o + \alpha E_t/h_o - \varepsilon\Delta R/h_o \\ &= 94 + (0.15)(193) - 0 \\ &= 123°\text{F}\end{aligned}$$

This procedure was used to calculate the sol-air temperatures included in Table 15. Due to the tedious solar angle and intensity calculations, use of a simple computer spreadsheet or other computer software implementing these calculations can reduce the effort involved. A spreadsheet is illustrated in Table 16, calculating a 24 h sol-air temperature profile for the data of this example.

OUTDOOR AIR TEMPERATURES

The hourly air temperatures in Column 2, Table 16, are for a location with a design temperature of 95°F and a range of 21°F. To compute corresponding temperatures for other locations, select a suitable design temperature from Table 4B of Chapter 27 for the month being calculated and note the outdoor daily range from Table 1B of Chapter 27. For each hour, take the percentage of the daily range indicated in Table 17 of this chapter and subtract it from the design temperature.

Example 2. Air temperature calculation. Calculate the July dry-bulb temperature at 1200 h for Reno, Nevada.

Solution: From Table 1B, Chapter 27, the daily range is 37.3°F, and from Table 4B, Chapter 27, the 1% design dry-bulb temperature for July is 96.8°F. From Table 17, the percentage of the daily range at 1200 h is 23%. Thus, the dry-bulb temperature at 1200 is 96.8 – (0.23 × 37.3) = 88.2°F.

Table 16 Solar Calculations—Sol-Air Temperature for Example 1

GENERAL INPUT:	
Time Zone	= 3
Month	= 7
Longitude	= 90
Latitude	= 40
Clearness	= 1
Ground Refl.	= 0.2

SURFACE INPUT:	
Azimuth	= 45
Tilt	= 90

SOL-AIR TEMP. INPUT:	
Absorptance	= 0.45
h (outside)	= 3
Emittance	= 1
ΔR	= 0

SELECTED DATA:	
For Month	= 7
Equation of Time	= −6.2
Declination	= 20.6
A	= 344
B	= 0.207
C	= 0.136
Local Std Time Meridian	= 90

SOLAR CONSTANTS[a]

Month	Equation of Time, min.	Declination, degrees	A Btu/h·sf	B (Dimensionless Ratios)	C (Dimensionless Ratios)
1	−11.2	−20	390	0.142	0.058
2	−13.9	−10.8	385	0.144	0.060
3	−7.5	0	376	0.156	0.071
4	1.1	11.6	360	0.180	0.097
5	3.3	20	350	0.196	0.121
6	−1.4	23.45	345	0.205	0.134
7	−6.2	20.6	344	0.207	0.136
8	−2.4	12.3	351	0.201	0.122
9	7.5	0	365	0.177	0.092
10	15.4	−10.5	378	0.160	0.073
11	13.8	−19.8	387	0.149	0.063
12	1.6	−23.45	391	0.142	0.057

TIME ZONE MERIDIANS

Time Zone		Standard Meridian
1	Atlantic	60
2	Eastern	75
3	Central	90
4	Mountain	105
5	Pacific	120
6	Alaska	135
7	Hawaii	150

					DIRECT BEAM SOLAR HEAT GAIN			DIFFUSE SOLAR HEAT GAIN						
Local Standard Hour	Apparent Solar Time, hours	Hour Angle	Solar Altitude	Solar Azimuth	Direct Normal Irradiance	Surface Incident Angle	Surface Direct Irradiance	Ground Diffuse	Y Ratio	Sky Diffuse	Total Diffuse Irradiance	Total Surface Irradiance	Outside Temp.	Sol-Air Temp.
1	0.90	−166.55	−28.1	−165.7	—	139.3	0.0	0.0	0.45	—	0.0	0.0	76	76
2	1.90	−151.55	−23.8	−150.8	—	151.6	0.0	0.0	0.45	—	0.0	0.0	76	76
3	2.90	−136.55	−17.1	−137.7	—	162.7	0.0	0.0	0.45	—	0.0	0.0	75	75
4	3.90	−121.55	−8.6	−126.2	—	167.8	0.0	0.0	0.45	—	0.0	0.0	74	74
5	4.90	−106.55	1.3	−116.2	0	161.1	0.0	0.0	0.45	0.0	0.0	0.0	74	74
6	5.90	−91.55	11.9	−107.0	126	149.7	0.0	4.3	0.45	7.7	12.1	12.1	74	76
7	6.90	−76.55	23.1	−98.1	203	137.3	0.0	10.7	0.45	12.4	23.2	23.2	75	78
8	7.90	−61.55	34.6	−88.8	239	124.7	0.0	16.8	0.45	14.6	31.4	31.4	77	82
9	8.90	−46.55	46.0	−78.0	258	112.2	0.0	22.1	0.45	15.8	37.9	37.9	80	86
10	9.90	−31.55	56.8	−63.6	269	100.0	0.0	26.1	0.48	17.7	43.8	43.8	83	90
11	10.90	−16.55	66.0	−41.0	274	88.4	7.9	28.8	0.56	21.0	49.8	57.6	87	96
12	11.90	−1.55	70.6	4.4	276	75.4	69.8	29.8	0.68	25.6	55.4	125.1	90	109
13	12.90	13.45	67.5	34.6	275	67.8	103.7	29.1	0.76	28.4	57.5	161.2	93	117
14	13.90	28.45	58.9	59.8	270	60.1	134.8	26.8	0.85	31.1	57.9	192.7	94	123
15	14.90	43.45	48.3	75.4	261	55.0	149.5	23.0	0.90	32.0	55.1	204.6	95	126
16	15.90	58.45	37.0	86.7	244	53.4	145.3	18.0	0.92	30.6	48.5	193.9	94	123
17	16.90	73.45	25.5	96.2	213	55.6	120.2	12.0	0.90	25.9	38.0	158.2	93	117
18	17.90	88.45	14.2	105.1	148	61.1	71.5	5.6	0.83	16.8	22.4	93.9	91	105
19	18.90	103.45	3.4	114.2	11	69.3	3.7	0.2	0.74	1.1	1.3	5.0	87	88
20	19.90	118.45	−6.6	124.0	—	79.1	0.0	0.0	0.64	—	0.0	0.0	85	85
21	20.90	133.45	−15.5	135.2	—	90.2	0.0	0.0	0.55	—	0.0	0.0	83	83
22	21.90	148.45	−22.6	147.9	—	101.9	0.0	0.0	0.45	—	0.0	0.0	81	81
23	22.90	163.45	−27.5	162.5	—	114.2	0.0	0.0	0.45	—	0.0	0.0	79	79
24	23.90	178.45	−29.4	178.3	—	126.7	0.0	0.0	0.45	—	0.0	0.0	77	77

[a] See Table 7 in Chapter 30.

Table 17 Percentage of Daily Temperature Range

Time, h	%	Time, h	%	Time, h	%
1	87	9	71	17	10
2	92	10	56	18	21
3	96	11	39	19	34
4	99	12	23	20	47
5	100	13	11	21	58
6	98	14	3	22	68
7	93	15	0	23	76
8	84	16	3	24	82

Data Limitations

The outdoor daily range is the difference between the average daily maximum and average daily minimum temperatures during the warmest month. More reliable results could be obtained by determining or estimating the shape of the temperature curve for typical hot days at the building site and considering each month separately. Peak cooling load is often determined by solar heat gain through fenestration; this peak may occur in winter months and/or at a time of day when outside air temperature is not at its peak.

HEAT GAIN THROUGH INTERIOR SURFACES

Whenever a conditioned space is adjacent to a space with a different temperature, transfer of heat through the separating physical section must be considered. The heat transfer rate is given by

$$q = UA(t_b - t_i) \tag{20}$$

where

q = heat transfer rate, Btu/h
U = coefficient of overall heat transfer between adjacent and conditioned space, Btu/h·ft²·°F
A = area of separating section concerned, ft²
t_b = average air temperature in adjacent space, °F
t_i = air temperature in conditioned space, °F

Values of U can be obtained from Chapter 25. Temperature t_b may differ greatly from t_i. The temperature in a kitchen or boiler room, for example, may be as much as 15 to 50°F above the outdoor air temperature. Actual temperatures in adjoining spaces should be measured when possible. Where nothing is known except that the adjacent space is of conventional construction, contains no heat sources, and itself receives no significant solar heat gain, $t_b - t_i$ may be considered to be the difference between the outdoor air and conditioned space design dry-bulb temperatures minus 5°F. In some cases, the air temperature in the adjacent space will correspond to the outdoor air temperature or higher.

Floors

For floors directly in contact with the ground or over an underground basement that is neither ventilated nor conditioned, heat transfer may be neglected for cooling load estimates.

INFILTRATION AND VENTILATION HEAT GAIN

Ventilation

Outdoor air must be introduced to ventilate conditioned spaces. Chapter 26 suggests minimum outdoor air requirements for representative applications, but the minimum levels are not necessarily adequate for all psychological attitudes and physiological responses. Where maximum economy in space and load is essential, as in submarines or other restricted spaces, as little as 1 cfm of outdoor air per person can be sufficient, provided that recirculated air is adequately decontaminated (Consolazio and Pecora 1947).

Local codes and ordinances frequently specify ventilation requirements for public places and for industrial installations. For example, minimum requirements for safe practice in hospital operating rooms are given in NFPA *Standard* 99. Although 100% outdoor air is sometimes used in operating rooms, this standard does not require it, and limiting the outdoor air to 6 to 8 changes per hour is finding increasing acceptance.

ASHRAE *Standard* 62 recommends minimum ventilation rates for most common applications. For general applications, such as offices, 20 cfm per person is suggested.

Ventilation air is normally introduced at the air-conditioning apparatus rather than directly into the conditioned space and thus becomes a cooling coil load component instead of a space load component. Calculations for estimating this heat gain are discussed later in the section on Heat Gain Calculations Using Standard Air Values.

Reducing heat gain from outdoor air by using filtered recirculated air in combination with outdoor air should be considered. Recirculated air can also be treated to control odor (see Chapter 13 of this volume and Chapter 44 of the 1999 *ASHRAE Handbook—Applications*).

Infiltration

The principles of estimating infiltration in buildings, with emphasis on the heating season, are discussed in Chapter 26. For the cooling season, infiltration calculations are usually limited to doors and windows. Air leakage through doors can be estimated using the information in Chapter 26. Table 3 in Chapter 26, adjusted for the average wind velocity in the locality, may be used to compute infiltration for windows. In calculating window infiltration for an entire structure, the total window area on all sides of the building is not involved since wind does not act on all sides simultaneously. In any case, infiltration from all windows in any two adjacent wall exposures should be included. A knowledge of the prevailing wind direction and velocity is helpful in selecting exposures.

When economically feasible, sufficient outdoor air should be introduced as ventilation air through the air-conditioning equipment to maintain a constant outward escape of air and thus eliminate the infiltration portion of the gain. The pressure maintained must overcome wind pressure through cracks and door openings. When the quantity of outside air introduced through the cooling equipment is not sufficient to maintain the required pressure to eliminate infiltration, the entire infiltration load should be included in the space heat gain calculations.

Standard Air Volumes

Because the specific volume of air varies appreciably, calculations are more accurate when made on the basis of air mass instead of volume. However, volume values are often required for selection of coils, fans, ducts, etc., in which cases volume values based on measurement at standard conditions may be used for accurate results. One standard value is 0.075 lb (dry air)/ft³ (13.33 ft³/lb). This density corresponds to about 60°F at saturation and 69°F dry air (at 14.7 psia). Because air usually passes through the coils, fans, ducts, etc., at a density close to standard, the accuracy desired normally requires no correction. When airflow is to be measured at a particular condition or point, such as at a coil entrance or exit, the corresponding specific volume can be read from the psychrometric chart.

Example 3. Standard air calculation. Assume outdoor air at standard conditions is flowing at 1000 cfm. What is the flow rate when the outdoor air is at 95°F dry-bulb and 75°F wet-bulb (14.3 ft³/lb)?

Solution: The measured rate at that condition should be 1000(14.3/13.33) = 1070 cfm.

Heat Gain Calculations Using Standard Air Values

Air-conditioning design often requires calculation of the following.

1. Total heat

Total heat gain q_t corresponding to the change of a given standard flow rate Q_s through an enthalpy difference Δh is

$$q_t = 60 \times 0.075 Q_s \Delta h = 4.5 Q_s \Delta h \tag{21}$$

where 60 = min/h and 0.075 = lb (dry air)/ft³.

2. Sensible heat

Sensible heat gain q_s corresponding to the change of dry-bulb temperature Δt for given airflow (standard conditions) Q_s is

$$q_s = 60 \times 0.075(0.24 + 0.45W)Q_s \Delta t \tag{22}$$

where

0.24 = specific heat of dry air, Btu/lb·°F
W = humidity ratio, lb (water)/lb (dry air)
0.45 = specific heat of water vapor, Btu/lb·°F

The specific heats are for a range from about –100 to 200°F. When $W = 0$, the value of $60 \times 0.075 (0.24 + 0.45W) = 1.08$; when $W = 0.01$, the value is 1.10; when $W = 0.02$, the value is 1.12; and when $W = 0.03$, the value is 1.14. Because a value of W = 0.01 approximates conditions found in many air-conditioning problems, the sensible heat change (in Btu/h) can normally be found as

$$q_s = 1.10 Q_s \Delta t \quad (23)$$

3. Latent heat

Latent heat gain q_l corresponding to the change of humidity ratio ΔW for given airflow (standard conditions) Q_s is

$$q_l = 60 \times 0.075 \times 1076 Q_s \Delta W = 4840 Q_s \Delta W \quad (24)$$

where 1076 is the approximate heat content of 50% rh vapor at 75°F less the heat content of water at 50°F. 50% rh at 75°F is a common design condition for the space, and 50°F is normal condensate temperature from cooling and dehumidifying coils.

The constants 4.5, 1.10, and 4840 are useful in air-conditioning calculations at sea level (14.7 psia) and for normal temperatures and moisture ratios. For other conditions, more precise values should be used. For an altitude of 5000 ft (12.2 psia), appropriate values are 3.73, 0.92, and 4020.

LATENT HEAT GAIN FROM MOISTURE THROUGH PERMEABLE BUILDING MATERIALS

The diffusion of moisture through all common building materials is a natural phenomenon that is always present. Chapters 23 and 24 cover the principles and specific methods used to control moisture. Moisture transfer through walls is often neglected in the usual comfort air-conditioning application because the actual rate is quite small and the corresponding latent heat gain is insignificant. The permeability and permeance values for various building materials are given in Table 9, Chapter 25. Vapor retarders are frequently installed to keep moisture transfer to a minimum.

Certain industrial applications call for a low moisture content to be maintained in a conditioned space. In such cases, the latent heat gain accompanying moisture transfer through walls may be greater than any other latent heat gain. This gain is computed by

$$q_m = (M/7000) A \Delta p_v (h_g - h_f) \quad (25)$$

where

q_m = latent heat gain, Btu/h
M = permeance of wall assembly, perms or grains/(ft^2·h·in. Hg)
7000 = grains/lb
A = area of wall surface, ft^2
Δp_v = vapor pressure difference, in. Hg
h_g = enthalpy at room conditions, Btu/lb
h_f = enthalpy of water condensed at cooling coil, Btu/lb
= 1076 Btu/lb when room temperature is 75°F and condensate off coil is 50°F

HEAT GAIN FROM MISCELLANEOUS SOURCES

The calculation of the cooling load is affected by such factors as (1) type of HVAC system, (2) effectiveness of heat exchange surfaces, (3) fan location, (4) duct heat gain or loss, (5) duct leakage, (6) heat-extraction lighting systems, (7) type of return air system, and (8) sequence of controls. System performance needs to be analyzed as a sequence of individual psychrometric processes. The most straightforward method first defines all known (or desired) state points on a psychrometric chart. Next, the actual entering and leaving dry- and wet-bulb conditions are calculated for such components as the cooling and/or heating coils (based on zone or space load), the amount of outside air introduced into the system through the equipment, and the amount of heat gain or loss at various points.

This overall process must verify that the space conditions originally sought can actually be met by the designed system by considering all sensible and latent heat changes to the air as it travels from the space conditions through the return air system and equipment back to the conditioned space. If the design is successful (i.e., within the degree of correctness of the various design assumptions), appropriate equipment components can safely be selected. If not, the designer must judge if the results will be "close enough" to satisfy the needs of the project, or if one or more assumptions and/or design criteria must first be modified and the calculations rerun.

Heat Gain from Fans

Fans that circulate air through HVAC systems add energy to the system by one or all of the following processes:

- Temperature rise in the airstream from fan inefficiency. Depending on the equipment, fan efficiencies generally range between 50 and 70%, with an average value of 65%. Thus, some 35% of the energy required by the fan appears as instantaneous heat gain to the air being transported.
- Temperature rise in the airstream as a consequence of air static and velocity pressure. The "useful" 65% of the total fan energy that creates pressure to move air spreads out throughout the entire air transport system in the process of conversion to sensible heat. Designers commonly assume that the temperature change equivalent of this heat occurs at a single point in the system, depending on fan location as noted below.
- Temperature rise from heat generated by motor and drive inefficiencies. The relatively small gains from fan motors and drives are normally disregarded unless the motor and/or drive are physically located within the conditioned airstream. Equations (7), (8), and (9) may be used to estimate heat gains from typical motors. Belt drive losses are often estimated as 3% of the motor power rating. Conversion to temperature rise is calculated by Equation (26).

The location of each fan relative to other elements (primarily the cooling coil) and the type of system (e.g., single zone, multizone, double-duct, terminal reheat, VAV), along with the concept of equipment control (space temperature alone, space temperature and relative humidity, etc.), must be known before the analysis can be completed. A fan located upstream of the cooling coil (blowthrough supply fan, return air fan, outside air fan) adds the heat equivalent of its inefficiency to the airstream at that point; thus, a slightly elevated entering dry-bulb temperature to the cooling coil results. A fan located downstream of the cooling coil raises the dry-bulb temperature of air leaving the cooling coil. This rise can be offset by reducing the cooling coil temperature or, alternatively, by increasing airflow across the cooling coil as long as its impact on space conditions is considered.

Duct Heat Gain and Leakage

Unless return air duct systems are extensive or subjected to rigorous conditions, only the heat gained or lost by supply duct systems is significant; it is normally estimated as a percentage of space sensible cooling load (usually about 1%) and applied to the dry-bulb temperature of the air leaving the coil in the form of an equivalent temperature reduction.

Air leakage out of (or into) ductwork can have much greater impact than conventional duct heat gain or loss, but it is normally about the same or less. Outward leakage from supply ducts is a direct loss of cooling and/or dehumidifying capacity and must be offset by increased airflow (sometimes reduced supply air temperatures) unless it enters the conditioned space directly. Inward leakage to return ducts causes temperature and/or humidity variations, but these are often ignored under ordinary circumstances due to the low temperature and pressure differentials involved. Chapter 34 has further details on duct sealing and leakage.

A well-designed and installed duct system should not leak more than 1 to 3% of the total system airflow. All HVAC equipment and volume control units connected into a duct system are usually delivered from manufacturers with allowable leakage not exceeding 1 or 2% of maximum airflow rating. Where duct systems are specified to be sealed and leak tested, both low- and medium-pressure types can be constructed and required to fall within this range. Designers normally assume this loss to approximate 1% of the space load, handled in a similar manner to that for duct heat gain. Latent heat considerations are frequently ignored.

Poorly designed or installed duct systems can have leakage rates of 10 to 30%. Leakage from low-pressure lighting troffer connections lacking proper taping and sealing can be 35% or more of the terminal air supply. Improperly sealed high-pressure systems can leak as much as 10% or more from the high-pressure side alone. Such extremes destroy the validity of any load calculation. Although leaks do not always affect overall system loads enough to cause problems, they will always adversely impact required supply air quantities. Also, uninsulated supply ductwork running through return air plenums results in high thermal leakage, which reduces the space-cooling capability of the supply air and may cause condensation during a warm startup.

HEAT BALANCE METHOD OF COOLING LOAD CALCULATION

The estimation of cooling load for a space involves calculating a surface-by-surface conductive, convective, and radiative heat balance for each room surface and a convective heat balance for the room air. Sometimes called the **exact solution**, these principles form the foundation for all methods described in this chapter.

Some of the computations required by this rigorous approach to calculating space cooling load make the use of modern digital computers essential. The heat balance procedure is not new. Many energy calculation programs have used it in some form for many years. The first implementation that incorporated all the elements to form a complete method was NBSLD (Kusuda 1967). The heat balance procedure is also implemented in both the BLAST and TARP energy analysis programs (Walton 1983). Prior to the implementation of ASHRAE *Research Project* 875, the method had never been described completely or in a form applicable to cooling load calculations. The papers resulting from RP-875 describe the heat balance procedure in detail (Pedersen et al. 1997, Liesen and Pedersen 1997, McClellan and Pedersen 1997).

The HB method is codified in the software called Hbfort that accompanies *Cooling and Heating Load Calculation Principles* (Pedersen et al. 1998).

HEAT BALANCE MODEL ASSUMPTIONS

All calculation procedures involve some kind of model. All models require simplifying assumptions and therefore are approximate. The most fundamental assumption is that the air in the thermal zone can be modeled as **well mixed**, meaning it has a uniform temperature throughout the zone. ASHRAE *Research Project* 664 (Fisher et al. 1997) established that this assumption is valid over a wide range of conditions.

The next major assumption is that the surfaces of the room (walls, windows, floor, etc.) can be treated as having

- Uniform surface temperatures
- Uniform long-wave (LW) and short-wave (SW) irradiation
- Diffuse radiating surfaces
- One-dimensional heat conduction within

The resulting formulation is called the **heat balance model**. It is important to note that the foregoing assumptions, although common, are quite restrictive and set certain limits on the information that can be obtained from the model.

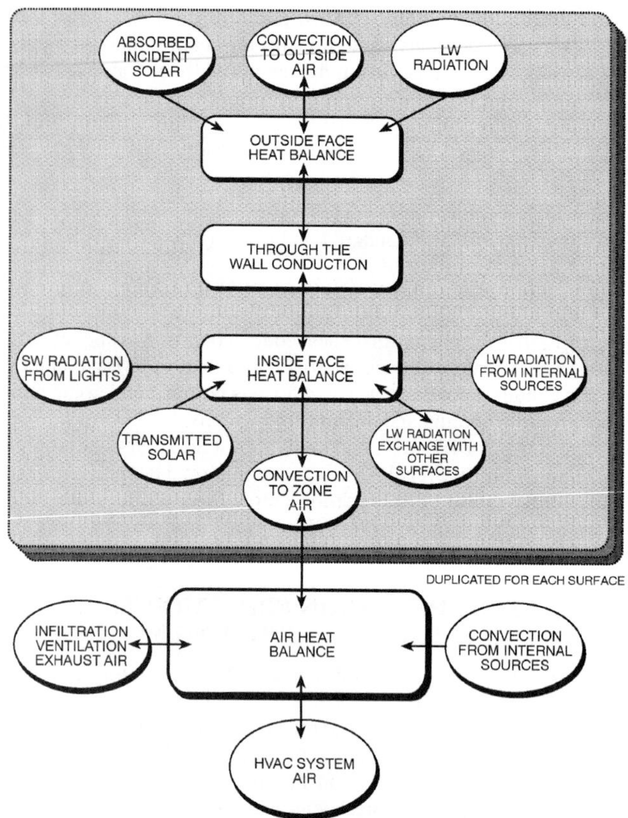

Fig. 5 Schematic of Heat Balance Processes in a Zone

ELEMENTS OF HEAT BALANCE MODEL

Within the framework of the foregoing assumptions, the heat balance model can be viewed as four distinct processes:

1. Outside face heat balance
2. Wall conduction process
3. Inside face heat balance
4. Air heat balance

Figure 5 shows the relationship between these processes for a single opaque surface. The top part of the figure, inside the shaded box, is repeated for each of the surfaces enclosing the zone. The process for transparent surfaces would be similar but would have the absorbed solar component appear in the conduction process block instead of at the outside face. Also, the absorbed component would split into an inward-flowing fraction and an outward-flowing fraction. These components would participate in the surface heat balances.

Outside Face Heat Balance

The heat balance on the outside face of each surface is

$$q''_{\alpha sol} + q''_{LWR} + q''_{conv} - q''_{ko} = 0 \tag{26}$$

where

$q''_{\alpha sol}$ = absorbed direct and diffuse solar radiation flux (q/A), Btu/h·ft²
q''_{LWR} = net long-wave radiation flux exchange with air and surroundings, Btu/h·ft²
q''_{conv} = convective exchange flux with outside air, Btu/h·ft²
q''_{ko} = conductive flux (q/A) into wall, Btu/h·ft²

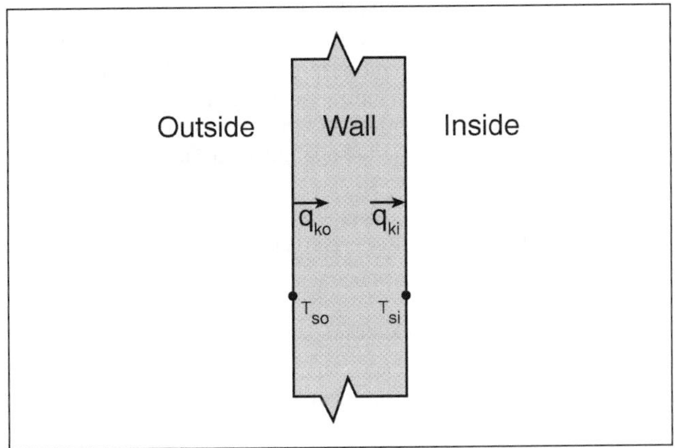

Fig. 6 Schematic of Wall Conduction Process

All terms are positive for net flux to the face except the conduction term, which is traditionally taken to be positive in the direction from outside to inside the wall.

Each of the heat flux terms in Equation (26) has been modeled in several ways, and in some formulations the first three terms are combined by using the sol-air temperature.

Wall Conduction Process

The wall conduction process has been formulated in more ways than any of the other processes. Among the possible ways to model this process are

1. Numerical finite difference
2. Numerical finite element
3. Transform methods
4. Time series methods

This process introduces part of the time dependence inherent in the load calculation process. Figure 6 shows schematically the surface temperatures on the inside and outside faces of the wall element and corresponding conductive heat fluxes away from the outside face and toward the inside face. All four quantities are functions of time. The direct formulation of the process has the two temperature functions as input or known quantities and the two heat fluxes as outputs or resultant quantities.

In some models, the surface heat transfer coefficients are included as part of the wall element. Then the temperatures in question are the inside and outside air temperatures. This is not an acceptable formulation because it hides the heat transfer coefficients and prohibits changing them as airflow conditions change. Also, it prohibits treating the internal long-wave radiation exchange appropriately.

Since the heat balances on both sides of the element induce both the temperature and heat flux, the solution technique must deal with this simultaneous condition. From a computational standpoint, two methods that have been used widely are a finite difference procedure and a method using conduction transfer functions. Because of the computational time advantage, the conduction transfer function formulation has been selected for the heat balance procedure presented here.

Inside Face Heat Balance

The heart of the heat balance method is the internal heat balance involving the inside faces of the zone surfaces. This heat balance has many heat transfer components, and they are all coupled. Both long-wave (LW) and short-wave (SW) radiation are important, as well as wall conduction and convection to the air. The inside face heat balance for each surface can be written as follows:

$$q''_{LWX} + q''_{SW} + q''_{LWS} + q''_{ki} + q''_{sol} + q''_{conv} = 0 \qquad (27)$$

where

q''_{LWX} = net long-wave radiant flux exchange between zone surfaces, Btu/h·ft²
q''_{SW} = net short-wave radiation flux to surface from lights, Btu/h·ft²
q''_{LWS} = long-wave radiation flux from equipment in zone, Btu/h·ft²
q''_{ki} = conductive flux through the wall, Btu/h·ft²
q''_{sol} = transmitted solar radiative flux absorbed at surface, Btu/h·ft²
q''_{conv} = convective heat flux to zone air, Btu/h·ft²

These terms are explained in the following paragraphs.

LW Radiation Exchange among Zone Surfaces. The two limiting cases for modeling internal LW radiation exchange are

1. Zone air is completely transparent to LW radiation
2. Zone air completely absorbs LW radiation from surfaces in the zone

Most heat balance models treat air as completely transparent, and then it does not participate in the LW radiation exchange among the surfaces in the zone. The second model is attractive because it can be formulated simply using a combined radiative and convective heat transfer coefficient from each surface to the zone air and in that way decouples the radiant exchange among surfaces in the zone. However, because the transparent air model allows the radiant exchange and is more realistic, the second model is inferior.

Furniture in a zone increases the amount of surface area that can participate in the radiative and convective heat exchanges. It also adds thermal mass to the zone. These two changes can affect the time response of the zone cooling load.

SW Radiation from Lights. The short-wavelength radiation from lights is usually assumed to be distributed over the surfaces in the zone in some manner. The HB procedure retains this approach but allows the distribution function to be changed.

LW Radiation from Internal Sources. The traditional model for this source defines a radiative/convective split for the heat introduced into a zone from equipment. The radiative part is then distributed over the surfaces in the zone in some manner. This is not a completely realistic model, and it departs from the heat balance principles. If it were handled in a true heat balance model, the equipment surfaces would be treated just as other LW radiant sources within the zone. However, since information about the surface temperature of equipment is rarely known, it is reasonable to keep the radiative/convective split concept even though it ignores the true nature of the radiant exchange. ASHRAE *Research Project* 1055 (Hosni et al. 1999) determined radiative/convective splits for many additional equipment types, as listed in Table 13.

Transmitted Solar Heat Gain. The calculation procedure for determining transmitted solar energy through fenestration as described in Chapter 30 uses the solar heat gain coefficient (SHGC) directly rather than relating it to double-strength glass as is done when using a shading coefficient (SC). The difficulty with this plan is that the SHGC includes both the transmitted solar and the inward-flowing fraction of the solar radiation absorbed in the window. With the heat balance method, this latter part should be added to the conduction component so that it can be included in the inside face heat balance.

Transmitted solar radiation is also distributed over surfaces in the zone in a prescribed manner. It is possible to calculate the actual position of beam solar radiation, but this involves partial surface irradiation, which is inconsistent with the rest of the zone model that assumes uniform conditions over an entire surface.

Using SHGC to Calculate Solar Heat Gain

The total solar heat gain through fenestration consists of the directly transmitted solar radiation plus the inward-flowing fraction of the solar radiation that is absorbed in the glazing system.

Both parts contain beam and diffuse contributions. The transmitted radiation goes directly onto surfaces within the zone and is accounted for in the surface inside heat balance. The zone heat balance model accommodates the resulting heat fluxes without difficulty. The second part, the inward-flowing fraction of the absorbed solar radiation, gets involved in an interaction with the other surfaces of the enclosure through long-wave radiant exchange and with the zone air through convective heat transfer. As such, it is dependent both on the geometric and radiative properties of the zone enclosure and the convection characteristics inside and outside the zone. The solar heat gain coefficient (SHGC) combines the transmitted solar radiation and the inward-flowing fraction of the absorbed radiation. The SHGC is defined as

$$\text{SHGC} = \tau + \sum_{k=1}^{n} N_k \alpha_k \tag{28}$$

where

τ = solar transmittance of glazing
α_k = solar absorptance of the *k*th layer of the glazing system
n = number of layers
N_K = inward-flowing fraction of absorbed radiation in the *k*th layer

Note that Equation (28) is written in a generic way. It can be written for a specific incidence angle and/or radiation wavelength and integrated over the wavelength and/or angle, but the principle is the same in each case. Refer to Chapter 30 for the specific expressions.

Unfortunately, the inward-flowing fraction N interacts with the zone in many ways. This interaction can be expressed as

$N = f$(inside convection coefficient, outside convection coefficient, glazing system overall heat transfer coefficient, zone geometry, zone radiation properties)

The only way to model these interactions correctly is to combine the window model with the zone heat balance model and solve the two simultaneously. This combination has been done recently in some energy analysis programs but is not generally available in load calculation procedures. In addition, the SHGC used for rating glazing systems is based on specific values of the inside, outside, and overall heat transfer coefficients and does not include any zonal long-wavelength radiation considerations. So, the challenge is to devise a way to use SHGC values within the framework of a heat balance calculation in the most accurate way possible. This will be done in the following paragraphs.

Using SHGC Data. The normal incidence SHGC used to rate and characterize glazing systems is not sufficient for determining solar heat gain for load calculations. These calculations require solar heat gain as a function of the incident solar angle in order to determine the hour-by-hour gain profile. Thus, it is necessary to use angular SHGC values and also diffuse SHGC values. These can be obtained from the Window 4.1 program (LBL 1994). This program does a detailed optical and thermal simulation of a glazing system and, when applied to a single clear layer, produces the information shown in Table 18.

Table 18 shows the parameters as a function of incident solar angle and also the diffuse values. The specific parameters shown are

V_{tc} = transmittance in the visible spectrum
R_f and R_b = front and back surface visible reflectances
T_{sol} = solar transmittance [symbol τ in Equations (28), (29), and (30)]
R_f and R_b = front and back surface solar reflectances
A_{bs1} = solar absorptance for layer 1, which is the only layer in this case [symbol α in Equations (28), (29), and (30)]
SHGC = solar heat gain coefficient at the center of the glazing

The parameters used for heat gain calculations are T_{sol}, A_{bs}, and SHGC. For the specific convective conditions assumed in the Window 4.1 program, the inward-flowing fraction of the absorbed solar can be obtained by rearranging Equation (28) to give:

$$N_k \alpha_k = \text{SHGC} - \tau \tag{29}$$

This quantity, when multiplied by the appropriate incident solar intensity, will provide the amount of absorbed solar radiation that flows inward. In the heat balance formulation for zone loads, this heat flux is combined with that caused by conduction through the glazing and included in the surface heat balance.

The outward-flowing fraction of the absorbed solar radiation is used in the heat balance on the outside face of the glazing and is determined from

$$(1 - N_k)\,\alpha_k = \alpha_k - N_k \alpha_k = \alpha_k - (\text{SHGC} - \tau) \tag{30}$$

If there is more than one layer, the appropriate summation of absorptances must be done.

There is some potential inaccuracy in using the Window 4.1 SHGC values because the inward-flowing fraction part was determined under specific conditions for the inside and outside heat transfer coefficients. However, it is possible to rerun the program with inside and outside coefficients of one's own choosing. Normally, however, this is not a large effect, and only in highly absorptive glazing systems might it cause significant error.

Table 18 Single Layer Glazing Data Produced by Window 4.1

	Incident Angle										
Parameter	**0**	**10**	**20**	**30**	**40**	**50**	**60**	**70**	**80**	**90**	**Diffuse (Hemis.)**
Vt_c	0.901	0.901	0.9	0.897	0.89	0.871	0.824	0.706	0.441	0	0.823
R_f	0.081	0.081	0.082	0.083	0.09	0.108	0.155	0.271	0.536	1	0.146
R_b	0.081	0.081	0.082	0.083	0.09	0.108	0.155	0.271	0.536	1	0.146
T_{sol}	0.85	0.85	0.848	0.844	0.835	0.814	0.766	0.652	0.399	0	0.77
R_f	0.075	0.074	0.075	0.076	0.082	0.099	0.144	0.255	0.509	1	0.136
R_b	0.075	0.074	0.075	0.076	0.082	0.099	0.144	0.255	0.509	1	0.136
A_{bs1}	0.075	0.076	0.077	0.08	0.083	0.087	0.091	0.093	0.092	0	0.084
SHGC	0.87	0.87	0.868	0.865	0.857	0.837	0.79	0.677	0.423	0	0.792

Source: LBL (1994).

For solar heat gain calculations, then, it seems reasonable to use the generic window property data that comes from Window 4.1. Considering Table 18, the procedure is as follows:

1. Determine the angle of incidence for the glazing.
2. Determine the corresponding SHGC.
3. Evaluate $N_k\alpha_k$ using Equation (28).
4. Multiply T_{sol} by the incident beam radiation intensity to get the transmitted beam solar radiation.
5. Multiply $N_k\alpha_k$ by the incident beam radiation intensity to get inward-flowing absorbed heat.
6. Repeat steps 2 through 5 with the diffuse parameters and diffuse radiation.
7. Add the beam and diffuse components of transmitted and inward-flowing absorbed heat.

Table 13 in Chapter 30 contains SHGC information for many additional glazing systems. That table is similar to Table 18 but is slightly abbreviated. Again, the information needed for heat gain calculations is T_{sol}, SHGC, and A_{bs}.

The same caution regarding the inside and outside heat transfer coefficients applies to the information in Table 13 in Chapter 30. Those values were also obtained with specific inside and outside heat transfer coefficients, and the inward-flowing fraction N is dependent upon those values.

Convection to Zone Air. The inside convection coefficients presented in past editions of this chapter and used in most load calculation procedures and energy programs are based on very old, natural convection experiments and do not accurately describe the heat transfer coefficients in a mechanically ventilated zone. In previous load calculation procedures, these coefficients were buried in the procedures and could not be changed. A heat balance formulation keeps them as working parameters. In this way, research results such as those from ASHRAE *Research Project* 664 (Fisher 1998) can be incorporated into the procedures. It also permits determining the sensitivity of the load calculation to these parameters.

Air Heat Balance

In heat balance formulations aimed at determining cooling loads, the capacitance of the air in the zone is neglected and the air heat balance is done as a quasi-steady balance in each time period. Four factors contribute to the air heat balance:

$$q_{conv} + q_{CE} + q_{IV} + q_{sys} = 0 \tag{31}$$

where

q_{conv} = convective heat transfer from surfaces, Btu/h
q_{CE} = convective parts of the internal loads, Btu/h
q_{IV} = sensible load due to infiltration and ventilation air, Btu/h
q_{sys} = heat transfer to/from the HVAC system, Btu/h

The **convection from zone surfaces** q_{conv} is the sum of all the convective heat transfer quantities from the inside surface heat balance. This comes to the air via the convective heat transfer coefficient on the surfaces.

The **convective parts of the internal loads** q_{CE} is the companion to q''_{LWS}, the radiant contribution from internal loads described previously [Equation (27)]. It is added directly to the air heat balance. Such a treatment also violates the tenets of the heat balance approach because surfaces producing the internal loads exchange heat with the zone air through normal convective processes. However, once again, this level of detail is generally not included in the heat balance, so it is included directly into the air heat balance instead.

In keeping with the well-mixed model for the zone air, any air that enters by way of **infiltration or ventilation** (q_{IV}) is immediately mixed with the zone air. The amount of infiltration air is uncertain. Sometimes it is related to the indoor-outdoor temperature difference and wind speed; however it is determined, it is added directly to the air heat balance.

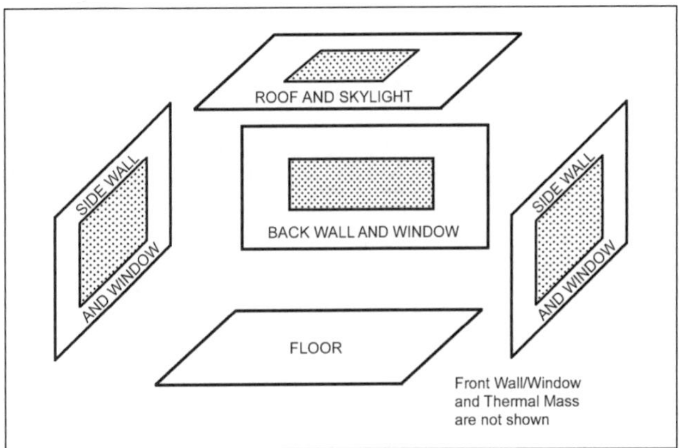

Fig. 7 Schematic View of General Heat Balance Zone

The conditioned air that enters the zone from the **HVAC system** and provides q_{sys} is also mixed directly with the zone air.

GENERAL ZONE FOR LOAD CALCULATION

The heat balance procedure is tailored to a single thermal zone, shown in Figure 7. The definition of a thermal zone depends on how the fixed temperature is going to be controlled. If air is circulated through an entire building or an entire floor in such a way that it is uniformly well stirred, the entire building or floor could be considered a thermal zone. On the other hand, if each room has a different control scheme, each room may need to be considered as a separate thermal zone. The framework needs to be flexible enough to accommodate any zone arrangement, but the heat balance aspect of the procedure also requires that a complete zone be described. This zone consists of four walls, a roof or ceiling, a floor, and a "thermal mass surface" (described in the section on Input Required for Heat Balance Procedure). Each wall and the roof can include a window (or skylight in the case of the roof). This makes a total of 12 surfaces, any of which may have zero area if it is not present in the zone to be modeled.

The heat balance processes for this general zone are formulated for a 24 h steady-periodic condition. The variables of the problem are the inside and outside temperatures of the 12 surfaces plus either the HVAC system energy required to maintain a specified air temperature or the air temperature, if the system capacity is specified. This makes a total of 25×24 or 600 variables. While it is possible to set up the problem for a simultaneous solution of these variables, the relatively weak coupling of the problem from one hour to the next permits a double iterative approach. One iteration is through all the surfaces in each hour, and the other iteration is through the 24 h of a day. This procedure automatically reconciles the nonlinear aspects of the surface radiative exchange and the other heat flux terms.

MATHEMATICAL DESCRIPTION OF HEAT BALANCE PROCEDURE

Conduction Process

Because it links the outside and inside heat balances, the wall conduction process regulates the time dependence of the cooling load. For the heat balance procedure presented here, the wall conduction process is formulated using **conduction transfer functions** (CTFs), which relate conductive heat fluxes to the current and past surface temperatures and the past heat fluxes. The general form for the inside heat flux is

$$q''_{ki}(t) = -Z_o T_{si,\theta} - \sum_{j=1}^{nz} Z_j T_{si,\theta-j\delta} + Y_o T_{so,\theta} + \sum_{j=1}^{nz} Y_j T_{so,\theta-j\delta} + \sum_{j=1}^{nq} \Phi_j q''_{ki,\theta-j\delta} \quad (32)$$

For the outside heat flux, the form is

$$q''_{ko}(t) = -Y_o T_{si,\theta} - \sum_{j=1}^{nz} Y_j T_{si,\theta-j\delta} + X_o T_{so,\theta} + \sum_{j=1}^{nz} X_j T_{so,\theta-j\delta} + \sum_{j=1}^{nq} \Phi_j q''_{ko,\theta-j\delta} \quad (33)$$

where

X_j = outside CTF, $j = 0,1,\ldots nz$
Y_j = cross CTF, $j = 0,1,\ldots nz$
Z_j = inside CTF, $j = 0,1,\ldots nz$
Φ_j = flux CTF, $j = 1,2,\ldots nq$
θ = time
δ = time step
T_{si} = inside face temperature, °F
T_{so} = outside face temperature, °F
q''_{ki} = conductive heat flux on inside face, Btu/h·ft²
q''_{ko} = conductive heat flux on outside face, Btu/h·ft²

The subscript following the comma indicates the time period for the quantity in terms of the time step δ. Also, the first terms in the series have been separated from the rest in order to facilitate solving for the current temperature in the solution scheme.

The two summation limits nz and nq depend on wall construction and depend somewhat on the scheme used for calculating the CTFs. If $nq = 0$, the CTFs are generally referred to as **response factors**, but then theoretically nz is infinite. The values for nz and nq are generally set to minimize the amount of computation. A development of CTFs can be found in Hittle (1981).

Heat Balance Equations

The primary variables in the heat balance for the general zone are the 12 inside face temperatures and the 12 outside face temperatures at each of the 24 h, assigning i as the surface index and j as the hour index, or, in the case of CTFs, the sequence index. Then, the primary variables are:

$T_{so_{i,j}}$ = outside face temperature, $i = 1,2,\ldots 12$; $j = 1,2,\ldots 24$

$T_{si_{i,j}}$ = inside face temperature, $i = 1,2,\ldots 12$; $j = 1,2,\ldots 24$

In addition, q_{sys_j} = cooling load, $j = 1,2,\ldots 24$

Equations (26) and (33) are combined and solved for T_{so} to produce 12 equations applicable in each time step:

$$T_{so_{i,j}} = \left(\sum_{k=1}^{nz} T_{si_{i,j-k}} Y_{i,k} - \sum_{k=1}^{nz} T_{so_{i,j-k}} Z_{i,k} - \sum_{k=1}^{nq} \Phi_{i,k} q''_{ko_{i,j-k}} + q''_{\alpha sol_{i,j}} + q''_{LWR_{i,j}} + T_{si_{i,j}} Y_{i,0} + T_{o_j} h_{co_{i,j}} \right) / (Z_{i,0} + h_{co_{i,j}}) \quad (34)$$

where

T_o = outside air temperature
h_{co} = outside convection coefficient, introduced by using $q''_{conv} = h_{co}(T_o - T_{so})$

Equation (34) shows the need for separating $Z_{i,0}$ because the contribution of the current surface temperature to the conductive flux can be collected with the other terms involving that temperature.

Equations (27) and (32) are combined and solved for T_{si} to produce the next 12 equations:

$$T_{si_{i,j}} = \left(T_{si_{i,j}} Y_{i,0} + \sum_{k=1}^{nz} T_{so_{i,j-k}} Y_{i,k} - \sum_{k=1}^{nz} T_{si_{i,j-k}} Z_{i,k} + \sum_{k=1}^{nq} \Phi_{i,k} q''_{ki_{i,j-k}} + T_{a_j} h_{ci_j} + q''_{LWS} + q''_{LWX} + q''_{SW} + q''_{sol} e \right) / (Z_{i,0} + h_{ci_{i,j}}) \quad (35)$$

where

T_a = zone air temperature
h_{ci} = convective heat transfer coefficient on the inside, obtained from $q''_{conv} = h_{ci}(T_a - T_{si})$

Note that in Equations (34) and (35), the opposite surface temperature at the current time appears on the right-hand side. The two equations could be solved simultaneously to eliminate those variables. Depending on the order of updating the other terms in the equations, this can have a beneficial effect on solution stability.

The remaining equation comes from the air heat balance, Equation (31). This provides the cooling load q_{sys} at each time step:

$$q_{sys_j} = \sum_{i=1}^{12} A_i h_{ci}(T_{si_{i,j}} - T_{a_j}) + q_{CE} + q_{IV} \quad (36)$$

In Equation (36), the convective heat transfer term is expanded to show the interconnection between the surface temperatures and the cooling load.

Overall HB Iterative Solution Procedure

The iterative HB procedure consists of a series of initial calculations that proceed sequentially, followed by a double iteration loop as shown in the following steps:

1. Initialize areas, properties, and face temperatures for all surfaces, 24 h.
2. Calculate incident and transmitted solar flux for all surfaces and hours.
3. Distribute transmitted solar energy to all inside faces, 24 h.
4. Calculate internal load quantities for all 24 h.
5. Distribute LW, SW, and convective energy from internal loads to all surfaces for all hours.
6. Calculate infiltration and ventilation loads for all hours.
7. Iterate the heat balance according to the following scheme:

```
For Day = 1 to Maxdays
    For j = 1 to 24                {hours in the day}
        For SurfaceIter = 1 to MaxIter
            For i = 1 to 12            {The twelve zone surfaces}
                Evaluate Equations (34) and (35)
            Next i
        Next SurfaceIter
        Evaluate Equation (36)
    Next j
If not converged, Next Day
```

8. Display results.

Generally, four or six surface iterations are sufficient to provide convergence. The convergence check on the day iteration should be based on the difference between the inside and the outside conductive heat flux terms q_k. A limit such as requiring the difference between all inside and outside flux terms to be less than 1% of either flux works well.

INPUT REQUIRED FOR HEAT BALANCE PROCEDURE

Previous methods for calculating cooling loads attempted to simplify the procedure by precalculating representative cases and grouping the results with various correlating parameters. This generally tended to reduce the amount of information required to apply the procedure. In the case of the heat balance procedure, no precalculations are made, so the procedure requires a fairly complete description of the zone.

Global Information. Because the procedure incorporates a solar calculation, some global information is required, including latitude, longitude, time zone, month, day of month, north axis of the zone, and the zone height (floor to floor). Additionally, to take full advantage of the flexibility of the method to incorporate, for example, variable outside heat transfer coefficients, such things as wind speed, wind direction, and terrain roughness may be specified. Normally, these variables and others default to some reasonable set of values, but the flexibility remains.

Wall Information (Each Wall). Because the walls are involved in three of the fundamental processes (external and internal heat balance and wall conduction), each wall of the zone requires a fairly large set of variables. They include

- Facing angle with respect to building north
- Tilt (degrees from horizontal)
- Area
- Solar absorptivity outside
- Long-wave emissivity outside
- Short-wave absorptivity inside
- Long-wave emissivity inside
- Exterior boundary temperature condition (solar vs. nonsolar)
- External roughness
- Layer-by-layer construction information

Again, some of these parameters can be defaulted, but they are changeable, and they indicate the more fundamental character of the heat balance method since they are related to true heat transfer processes.

Window Information (Each Window). The situation for windows is similar to that for walls, but the windows require some additional information because of their role in the solar load. The necessary parameters include:

- Area
- Normal solar transmissivity
- Normal SHGC
- Normal total absorptivity
- Long-wave emissivity outside
- Long-wave emissivity inside
- Surface-to-surface thermal conductance
- Reveal (for solar shading)
- Overhang width (for solar shading)
- Distance from overhang to window (for solar shading)

Roof and Floor Details. The roof and floor surfaces are specified similarly to walls. The main difference is that the ground outside boundary condition will probably be specified more often for a floor.

Thermal Mass Surface Details. An "extra" surface, called a thermal mass surface, can serve several functions. It is included in the radiant heat exchange with the other surfaces in the space but is only exposed to the inside air convective boundary condition. As an example, this surface would be used to account for the movable partitions in a space. The construction of the partitions is specified layer by layer, similar to specification for walls, and those layers store and release heat via the same conduction mechanism as walls. As a general definition, the extra thermal mass surface should be sized to represent all of the surfaces in the space that are exposed to the air mass, except the walls, roof, floor, and windows. In the formulation, both sides of the thermal mass participate in the exchange.

Internal Heat Gain Details. The space can be subjected to several internal heat sources: people, lights, electrical equipment, and infiltration. In the case of infiltration, the energy is assumed to go immediately into the air heat balance, so it is the least complicated of the heat gains. For the others, several parameters must be specified. These include the following fractions:

- Fraction of heat gain that is sensible energy
- Fraction of heat gain that is latent energy
- Fraction of energy that enters as short-wave radiation
- Fraction of energy that enters as long-wave radiation
- Fraction of energy that enters the air immediately as convection
- Activity level of people
- Fraction of energy of lighting heat gain that goes directly to the return air

Radiant Distribution Functions. As mentioned previously, the generally accepted assumptions for the heat balance method include specifying the distribution of radiant energy from several sources to the surfaces that enclose the space. This requires a distribution function that specifies the fraction of the total radiant input that is absorbed by each surface. The types of radiation that require distribution functions are

- Long-wave radiation from equipment and lights
- Short-wave radiation from lights
- Transmitted solar radiation

Other Required Information. Additional flexibility is included in the model so that results of research can be incorporated easily. This includes the capability to specify such things as

- Heat transfer coefficients/convection models
- Solar coefficients
- Sky models

The amount of input information required may seem extensive, but many of the parameters can be set to default values in most routine applications. However, all of the parameters listed can be changed when necessary to fit unusual circumstances or when additional information is obtained.

RADIANT TIME SERIES (RTS) METHOD

The radiant time series (RTS) method is a new simplified method for performing design cooling load calculations that is derived from the heat balance (HB) method described above. It effectively replaces all other simplified (non-heat-balance) methods, such as the transfer function method (TFM), the cooling load temperature difference/cooling load factor (CLTD/CLF) method, and the total equivalent temperature difference/time averaging (TETD/TA) method.

The casual observer might well ask why yet another load calculation method is necessary. This method was developed in response to the desire to offer a method that is rigorous, yet does not require iterative calculations, and that quantifies each component contribution to the total cooling load. In addition, it is desirable for the user to be able to inspect and compare the coefficients for different construction and zone types in a form illustrating their relative impact on the result. These characteristics of the RTS method make it easier to apply engineering judgment during the cooling load calculation process.

The RTS method is suitable for peak design load calculations, but it should not be used for annual energy simulations due to its inherent limiting assumptions. The RTS method, while simple in concept, involves too many calculations to be used practically as a manual method, although it can easily be implemented in a simple computerized spreadsheet, as illustrated in the examples. For a manual cooling load calculation method, refer to the CLTD/CLF method included in the 1997 *ASHRAE Handbook—Fundamentals*.

RTS COOLING LOAD ASSUMPTIONS AND PRINCIPLES

Design cooling loads are based on the assumption of **steady-periodic conditions** (i.e., the design day's weather, occupancy, and heat gain conditions are identical to those for preceding days such that the loads repeat on an identical 24 h cyclical basis). Thus, the heat gain for a particular component at a particular hour is the same as 24 h prior, which is the same as 48 h prior, etc. This assumption is the basis for the RTS derivation from the HB method.

Cooling load calculations must address two time-delay effects inherent in building heat transfer processes: (1) delay of conductive heat gain through opaque massive exterior surfaces (walls, roofs, or floors) and (2) delay of radiative heat gain conversion to cooling loads.

Exterior walls and roofs conduct heat due to temperature differences between outdoor and indoor air. In addition, solar energy on exterior surfaces is absorbed, then transferred by conduction to the building interior. Due to the mass and thermal capacity of the wall or roof construction materials, there is a substantial time delay in heat input at the exterior surface becoming heat gain at the interior surface.

As described earlier in the section on Cooling Load Principles, most heat sources transfer energy to a room by a combination of convection and radiation. The convection part of heat gain immediately becomes cooling load. The radiation part must first be absorbed by the finishes and mass of the interior room surfaces and becomes cooling load only when it is later transferred by convection from those surfaces to the room air. Thus, radiant heat gains become cooling loads over a delayed period of time.

OVERVIEW OF THE RADIANT TIME SERIES METHOD

Figure 8 gives an overview of the radiant time series method. In the calculation of solar radiation, transmitted solar heat gain through windows, sol-air temperature, and infiltration, the RTS method is exactly the same as previous simplified methods (TFM and TETD/TA). Important areas that are different include the computation of conductive heat gain, the splitting of all heat gains into radiant and convective portions, and the conversion of radiant heat gains into cooling loads.

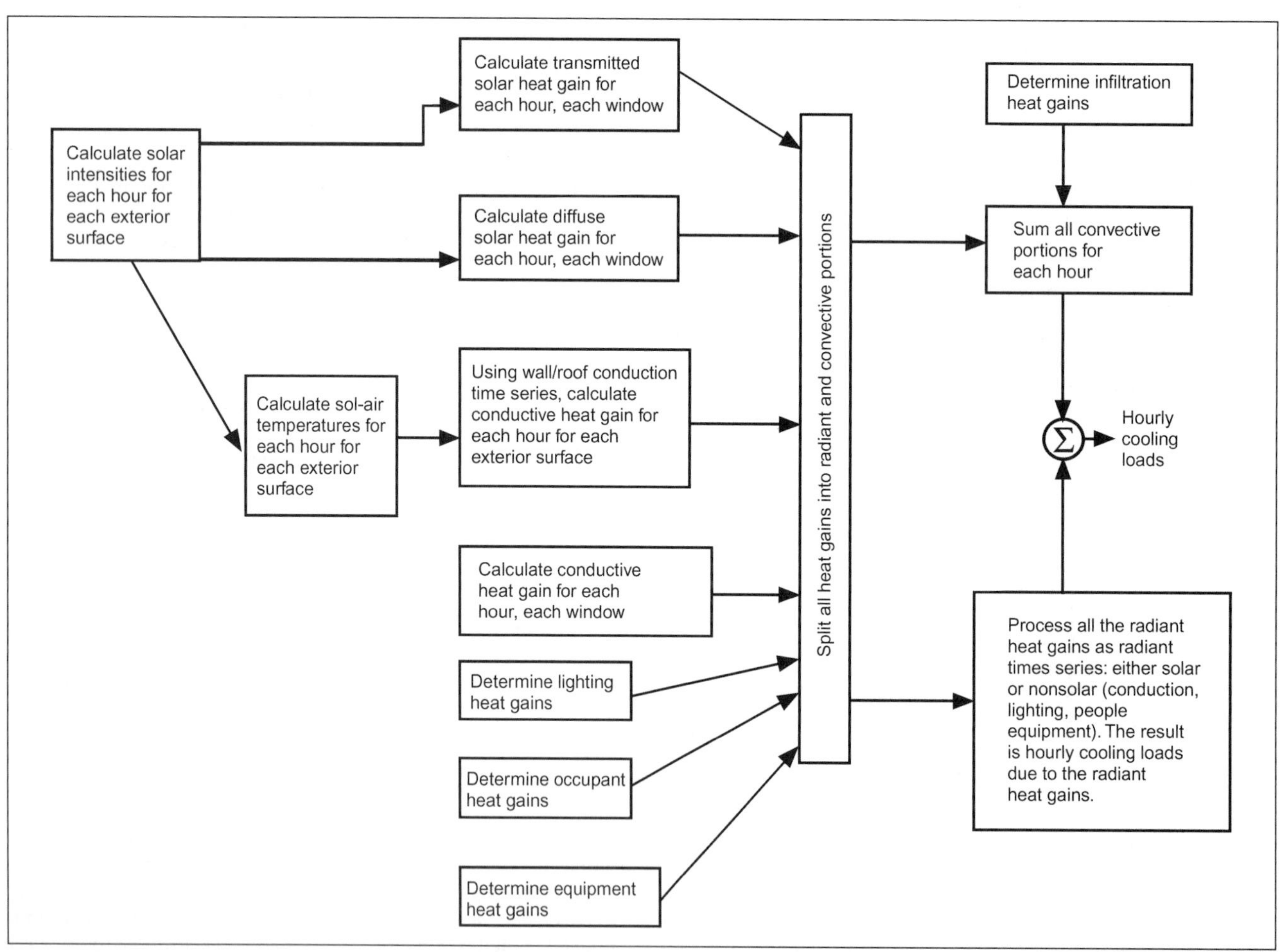

Fig. 8 Overview of Radiant Time Series Method

The RTS method accounts for both conduction time delay and radiant time delay effects by multiplying hourly heat gains by 24 h time series. The time series multiplication, in effect, distributes heat gains over time. Series coefficients, which are called **radiant time factors** and **conduction time factors**, are derived using the heat balance method. Radiant time factors reflect the percentage of an earlier radiant heat gain that becomes cooling load during the current hour. Likewise, conduction time factors reflect the percentage of an earlier heat gain at the exterior of a wall or roof that becomes heat gain at the inside during the current hour. By definition, each radiant or conduction time series must total 100%.

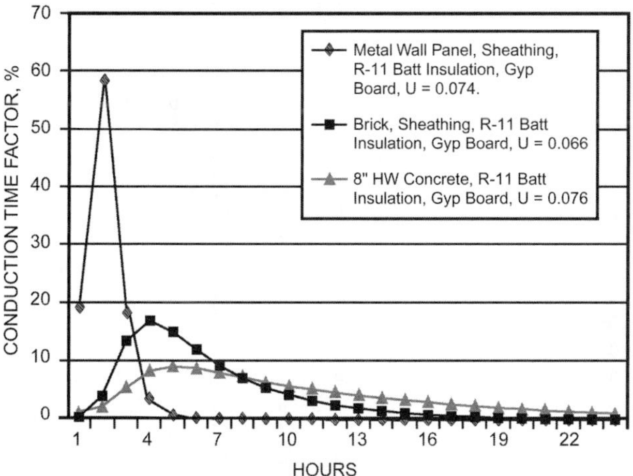

Fig. 9 CTS for Light to Heavy Walls

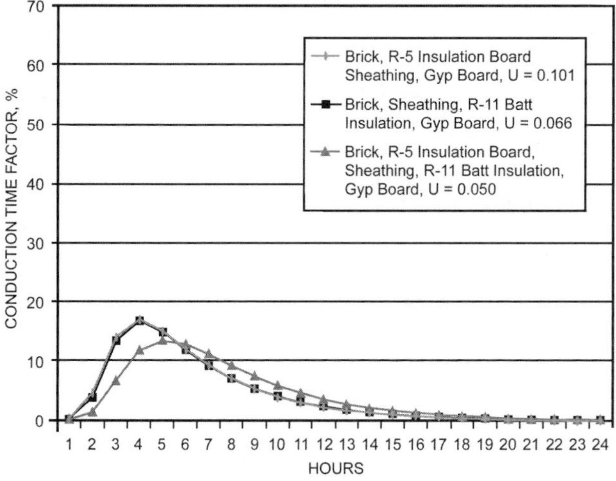

Fig. 10 CTS for Walls with Similar Mass and Increasing Insulation

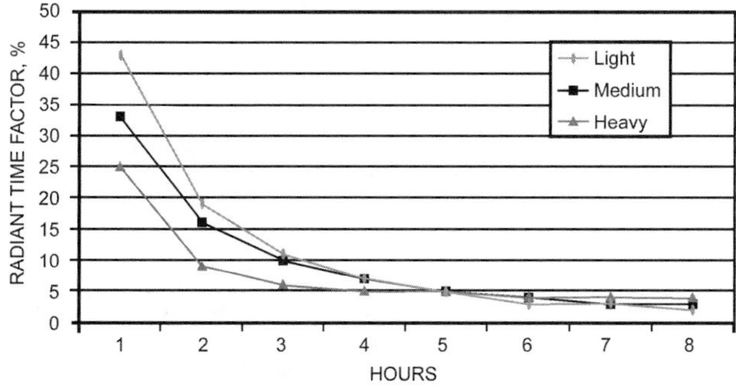

Fig. 11 RTS for Light to Heavy Construction

These series can be used to easily compare the time-delay impact of one construction versus another. This ability to compare choices is of particular benefit in the design process, when all construction details may not have been decided. Comparison can illustrate the magnitude of difference between the choices, allowing the engineer to apply judgment and make more informed assumptions in estimating the load.

Figure 9 illustrates CTS values for three walls with similar U-factors but with light to heavy construction. Figure 10 illustrates CTS for three walls with similar construction but with different amounts of insulation, thus with significantly different U-factors. Figure 11 illustrates RTS values for zones varying from light to heavy construction.

RADIANT TIME SERIES PROCEDURE

The general procedure for calculating cooling load for each load component (lights, people, walls, roofs, windows, appliances, etc.) with RTS is as follows:

1. Calculate 24 h profile of component heat gain for design day (for conduction, first account for conduction time delay by applying conduction time series).
2. Split heat gains into radiant and convective parts (see Table 19 for radiant and convective fractions).
3. Apply appropriate radiant time series to radiant part of heat gains to account for time delay in conversion to cooling load.
4. Sum convective part of heat gain and delayed radiant part of heat gain to determine cooling load for each hour for each cooling load component.

After calculating cooling loads for each component for each hour, sum those to determine the total cooling load for each hour and select the hour with the peak load for design of the air-conditioning system. This process should be repeated for multiple design months to determine the month when the peak load occurs, especially with windows on southern exposures (northern exposure in southern latitudes), which can result in higher peak room cooling loads in winter months than in summer.

Calculating Conductive Heat Gain Using Conduction Time Series

In the RTS method, conduction through exterior walls and roofs is calculated using conduction time series (CTS). Wall and roof conductive heat input at the exterior is defined by the familiar conduction equation as

Table 19 Convective and Radiant Percentages of Total Sensible Heat Gain

Heat Gain Source	Radiant Heat, %	Convective Heat, %
Transmitted solar, no inside shade	100	0
Window solar, with inside shade	63	37
Absorbed (by fenestration) solar	63	37
Fluorescent lights, suspended, unvented	67	33
Fluorescent lights, recessed, vented to return air	59	41
Fluorescent lights, recessed, vented to return air and supply air	19	81
Incandescent lights	80	20
People	See Table 1	
Conduction, exterior walls	63	37
Conduction, exterior roofs	84	16
Infiltration and ventilation	0	100
Machinery and appliances (see Table 13)	20 to 80	80 to 20

Sources: Pedersen et al. (1998), Hosni et al. (1999).

Table 20 Wall Conduction Time Series (CTS)

	CURTAIN WALLS			STUD WALLS				EIFS			BRICK WALLS									
Wall Number =	**1**	**2**	**3**	**4**	**5**	**6**	**7**	**8**	**9**	**10**	**11**	**12**	**13**	**14**	**15**	**16**	**17**	**18**	**19**	**20**
U-factor, Btu/h·ft²·°F	0.075	0.076	0.075	0.074	0.074	0.071	0.073	0.118	0.054	0.092	0.101	0.066	0.050	0.102	0.061	0.111	0.124	0.091	0.102	0.068
Total *R*	13.3	13.2	13.3	13.6	13.6	14.0	13.8	8.5	18.6	10.8	9.9	15.1	20.1	9.8	16.3	9.0	8.1	11.0	9.8	14.6
Mass, lb/ft²	6.3	4.3	16.4	5.2	17.3	5.2	13.7	7.5	7.8	26.8	42.9	44.0	44.2	59.6	62.3	76.2	80.2	96.2	182.8	136.3
Thermal Capacity, Btu/ft²·°F	1.5	1.0	3.3	1.2	3.6	1.6	3.0	1.8	1.9	5.9	8.7	8.7	8.7	11.7	12.4	15.7	15.3	19.0	38.4	28.4
Hour	**Conduction Time Factors, %**																			
0	18	25	8	19	6	7	5	11	2	1	0	0	0	1	2	2	1	3	4	3
1	58	57	45	59	42	44	41	50	25	2	5	4	1	1	2	2	1	3	4	3
2	20	15	32	18	33	32	34	26	31	6	14	13	7	2	2	2	3	3	4	3
3	4	3	11	3	13	12	13	9	20	9	17	17	12	5	3	4	6	3	4	4
4	0	0	3	1	4	4	4	3	11	9	15	15	13	8	5	5	7	3	4	4
5	0	0	1	0	1	1	2	1	5	9	12	12	13	9	6	6	8	4	4	4
6	0	0	0	0	1	0	1	0	3	8	9	9	11	9	7	6	8	4	4	5
7	0	0	0	0	0	0	0	0	2	7	7	7	9	9	7	7	8	5	4	5
8	0	0	0	0	0	0	0	0	1	6	5	5	7	8	7	7	8	5	4	5
9	0	0	0	0	0	0	0	0	0	6	4	4	6	7	7	6	7	5	4	5
10	0	0	0	0	0	0	0	0	0	5	3	3	5	7	6	6	6	5	4	5
11	0	0	0	0	0	0	0	0	0	5	2	2	4	6	6	6	6	5	5	5
12	0	0	0	0	0	0	0	0	0	4	2	2	3	5	5	5	5	5	5	5
13	0	0	0	0	0	0	0	0	0	4	1	2	2	4	5	5	4	5	5	5
14	0	0	0	0	0	0	0	0	0	3	1	2	2	4	5	5	4	5	5	5
15	0	0	0	0	0	0	0	0	0	3	1	1	1	3	4	4	3	5	4	4
16	0	0	0	0	0	0	0	0	0	3	1	1	1	3	4	4	3	5	4	4
17	0	0	0	0	0	0	0	0	0	2	1	1	1	2	3	4	3	4	4	4
18	0	0	0	0	0	0	0	0	0	2	0	0	1	2	3	3	2	4	4	4
19	0	0	0	0	0	0	0	0	0	2	0	0	1	2	3	3	2	4	4	4
20	0	0	0	0	0	0	0	0	0	2	0	0	0	1	3	3	2	4	4	4
21	0	0	0	0	0	0	0	0	0	1	0	0	0	1	2	2	1	4	4	4
22	0	0	0	0	0	0	0	0	0	1	0	0	0	1	2	2	1	4	4	3
23	0	0	0	0	0	0	0	0	0	0	0	0	0	0	1	1	1	3	4	3
Total Percentage	100	100	100	100	100	100	100	100	100	100	100	100	100	100	100	100	100	100	100	100
Layer ID from outside to inside (see Table 22)	F01	F01	F01	F01	F01	F01	F01	F01	F01	F01	F01	F01	F01	F01	F01	F01	F01	F01	F01	F01
	F09	F08	F10	F08	F10	F11	F07	F06	F06	F06	M01	M01	M01	M01	M01	M01	M01	M01	M01	M01
	F04	F04	F04	G03	G03	G02	G03	I01	I01	I01	F04	F04	F04	F04	F04	F04	F04	F04	F04	F04
	I02	I02	I02	I04	I04	I04	I04	G03	G03	G03	I01	G03	I01	I01	M03	I01	I01	I01	I01	M15
	F04	F04	F04	G01	G01	G04	G01	F04	I04	M03	G03	I04	G03	M03	I04	M05	M01	M13	M16	I04
	G01	G01	G01	F02	F02	F02	F02	G01	G01	F04	F04	G01	I04	F02	G01	G01	F02	F04	F04	G01
	F02	F02	F02	—	—	—	—	F02	F02	G01	G01	F02	G01	—	F02	F02	—	G01	G01	F02
	—	—	—	—	—	—	—	—	—	F02	F02	—	F02	—	—	—	—	F02	F02	—

Wall Number Descriptions

1. Spandrel glass, R-10 insulation board, gyp board
2. Metal wall panel, R-10 insulation board, gyp board
3. 1 in. stone, R-10 insulation board, gyp board
4. Metal wall panel, sheathing, R-11 batt insulation, gyp board
5. 1 in. stone, sheathing, R-11 batt insulation, gyp board
6. Wood siding, sheathing, R-11 batt insulation, 1/2 in. wood
7. 1 in. stucco, sheathing, R-11 batt insulation, gyp board
8. EIFS finish, R-5 insulation board, sheathing, gyp board
9. EIFS finish, R-5 insulation board, sheathing, R-11 batt insulation, gyp board
10. EIFS finish, R-5 insulation board, sheathing, 8 in. LW CMU, gyp board
11. Brick, R-5 insulation board, sheathing, gyp board
12. Brick, sheathing, R-11 batt insulation, gyp board
13. Brick, R-5 insulation board, sheathing, R-11 batt insulation, gyp board
14. Brick, R-5 insulation board, 8 in. LW CMU
15. Brick, 8 in. LW CMU, R-11 batt insulation, gyp board
16. Brick, R-5 insulation board, 8 in. HW CMU, gyp board
17. Brick, R-5 insulation board, brick
18. Brick, R-5 insulation board, 8 in. LW concrete, gyp board
19. Brick, R-5 insulation board, 12 in. HW concrete, gyp board
20. Brick, 8 in. HW concrete, R-11 batt insulation, gyp board

$$q_{i,\theta-n} = UA(t_{e,\theta-n} - t_{rc}) \qquad (37)$$

where

$q_{i,\theta-n}$ = conductive heat input for the surface n hours ago, Btu/h
U = overall heat transfer coefficient for the surface, Btu/h·ft²·°F
A = surface area, ft²
$t_{e,\theta-n}$ = sol-air temperature n hours ago, °F
t_{rc} = presumed constant room air temperature, °F

Conductive heat gain through walls or roofs can be calculated using conductive heat inputs for the current hours and past 23 h and conduction time series:

$$q_\theta = c_0 q_{i,\theta} + c_1 q_{i,\theta-1} + c_2 q_{i,\theta-2} + c_3 q_{i,\theta-3} + \ldots + c_{23} q_{i,\theta-23} \qquad (38)$$

where

q_θ = hourly conductive heat gain for the surface, Btu/h
$q_{i,\theta}$ = heat input for the current hour
$q_{i,\theta-n}$ = heat input n hours ago
c_0, c_1, etc. = conduction time factors

Conduction time factors for representative wall and roof types are included in Tables 20 and 21. Those values were derived by first calculating conduction transfer functions for each example wall and roof construction. The assumption of steady-periodic heat input conditions for design load calculations allowed the conduction transfer functions to be reformulated into periodic response factors

Table 20 Wall Conduction Time Series (CTS) (*Concluded*)

	CONCRETE BLOCK WALL						PRECAST AND CAST-IN-PLACE CONCRETE WALLS								
Wall Number =	**21**	**22**	**23**	**24**	**25**	**26**	**27**	**28**	**29**	**30**	**31**	**32**	**33**	**34**	**35**
U-factor, Btu/h·ft²·°F	0.067	0.059	0.073	0.186	0.147	0.121	0.118	0.074	0.076	0.115	0.068	0.082	0.076	0.047	0.550
Total *R*	14.8	16.9	13.7	5.4	6.8	8.2	8.4	13.6	13.1	8.7	14.7	12.2	13.1	21.4	1.8
Mass, lb/ft²	22.3	22.3	46.0	19.3	21.9	34.6	29.5	29.6	53.8	59.8	56.3	100.0	96.3	143.2	140.0
Thermal Capacity, Btu/ft²·°F	4.8	4.8	10.0	4.1	4.7	7.4	6.1	6.1	10.8	12.1	11.4	21.6	20.8	30.9	30.1
Hour	**Conduction Time Factors, %**														
0	0	1	0	1	1	0	1	2	1	3	1	2	1	0	1
1	2	11	3	1	10	8	1	2	2	3	2	2	2	2	11
2	8	21	12	2	20	18	3	3	3	4	5	3	4	8	21
3	12	20	16	5	18	18	6	5	6	5	8	3	7	12	20
4	12	15	15	7	14	14	8	6	7	6	9	5	8	12	15
5	11	10	12	9	10	11	9	6	8	6	9	5	8	11	10
6	9	7	10	9	7	8	9	6	8	6	8	6	8	9	7
7	8	5	8	8	5	6	9	6	7	5	7	6	8	8	5
8	7	3	6	8	4	4	8	6	7	5	6	6	7	7	3
9	6	2	4	7	3	3	7	6	6	5	6	6	6	6	2
10	5	2	3	6	2	2	7	5	6	5	5	6	6	5	2
11	4	1	3	6	2	2	6	5	5	5	5	5	5	4	1
12	3	1	2	5	1	2	5	5	5	4	4	5	4	3	1
13	2	1	2	4	1	1	4	5	4	4	4	5	4	2	1
14	2	0	1	4	1	1	4	4	4	4	3	4	4	2	0
15	2	0	1	3	1	1	3	4	3	4	3	4	3	2	0
16	1	0	1	3	0	1	2	4	3	4	3	4	3	1	0
17	1	0	1	2	0	0	2	3	3	4	2	4	3	1	0
18	1	0	0	2	0	0	1	3	2	4	2	4	2	1	0
19	1	0	0	2	0	0	1	3	2	3	2	3	2	1	0
20	1	0	0	2	0	0	1	3	2	3	2	3	2	1	0
21	1	0	0	2	0	0	1	3	2	3	2	3	1	1	0
22	1	0	0	1	0	0	1	3	2	3	1	3	1	1	0
23	0	0	0	1	0	0	1	2	2	2	1	3	1	0	0
Total Percentage	100	100	100	100	100	100	100	100	100	100	100	100	100	100	100
Layer ID from outside to inside (see Table 22)	F01	F01	F01	F01	F01	F01	F01	F01	F01	F01	F01	F01	F01	F01	F01
	F07	M08	M08	M09	M11	M11	M11	F06	M13	F06	M15	M16	M16	F07	M08
	M05	F02	F04	F04	I01	I04	I02	I01	I04	I02	I04	I05	F02	M05	F02
	I04	—	G01	G01	F04	G01	M11	M13	G01	M15	G01	G01	—	I04	—
	G01	—	F02	F02	G01	F02	F02	G01	F02	G01	F02	F02	—	G01	—
	F02	—	—	—	F02	—	—	F02	—	F02	—	—	—	F02	—

Wall Number Descriptions

21. 8 in. LW CMU, R-11 batt insulation, gyp board
22. 8 in. LW CMU with fill insulation, R-11 batt insulation, gyp board
23. 1 in. stucco, 8 in. HW CMU, R-11 batt insulation, gyp board
24. 8 in. LW CMU with fill insulation
25. 8 in. LW CMU with fill insulation, gyp board
26. 12 in. LW CMU with fill insulation, gyp board
27. 4 in. LW concrete, R-5 board insulation, gyp board
28. 4 in. LW concrete, R-11 batt insulation, gyp board
29. 4 in. LW concrete, R-10 board insulation, 4 in. LW concrete
30. EIFS finish, R-5 insulation board, 8 in. LW concrete, gyp board
31. 8 in. LW concrete, R-11 batt insulation, gyp board
32. EIFS finish, R-10 insulation board, 8 in. HW concrete, gyp board
33. 8 in. HW concrete, R-11 batt insulation, gyp board
34. 12 in. HW concrete, R-19 batt insulation, gyp board
35. 12 in. HW concrete

as demonstrated by Spitler and Fisher (1999a). The periodic response factors were further simplified by dividing the 24 periodic response factors by the respective overall wall or roof U-factor to form the conduction time series (CTS). The conduction time factors can then be used in Equation (38) and provide a means for comparison of time delay characteristics between different wall and roof constructions. Construction material data used in the calculations for walls and roofs included in Tables 20 and 21 are listed in Table 22.

Heat gains calculated for walls or roofs using periodic response factors (and thus CTS) are identical to those calculated using conduction transfer functions for the steady periodic conditions assumed in design cooling load calculations. The methodology for calculating periodic response factors from conduction transfer functions was originally developed as part of ASHRAE *Research Project* 875 (Spitler et al. 1997, Spitler and Fisher 1999b).

Example 4. Wall heat gain using conduction time series. Using the data from Example 1 and Table 15, calculate the heat gain at 3 P.M. Central Daylight Time on July 21 through 100 ft² of a wall composed of light-colored 4 in. brick, 2 in. of insulation (R = 10) and 8 in. lightweight concrete block. An air space is included between the brick and insulation. Overall U-factor of the wall is 0.068 Btu/h·ft²·°F. Inside room temperature is 75°F.

Solution: Conductive heat gain is calculated using Equations (37) and (38). First, calculate the 24 h heat input profile using Equation (37) and the sol-air temperatures from Table 15 for a southwest-facing wall with light exterior color:

$$q_{i,1} = (0.068)(100)(76-75) = 7 \text{ Btu/h}$$
$$q_{i,2} = (0.068)(100)(76-75) = 7$$
$$q_{i,3} = (0.068)(100)(75-75) = 0$$
$$q_{i,4} = (0.068)(100)(74-75) = -7$$
$$q_{i,5} = (0.068)(100)(74-75) = -7$$
$$q_{i,6} = (0.068)(100)(76-75) = 7$$

Table 21 Roof Conduction Time Series (CTS)

	SLOPED FRAME ROOFS						WOOD DECK		METAL DECK ROOFS					CONCRETE ROOFS					
Roof Number	1	2	3	4	5	6	7	8	9	10	11	12	13	14	15	16	17	18	19
U-factor, Btu/h·ft²·°F	0.044	0.040	0.045	0.041	0.042	0.041	0.69	0.058	0.080	0.065	0.057	0.036	0.052	0.054	0.052	0.051	0.056	0.055	0.042
Total *R*	22.8	25.0	22.2	24.1	23.7	24.6	14.5	17.2	12.6	15.4	17.6	27.6	19.1	18.6	19.2	19.7	18.0	18.2	23.7
Mass, lb/ft²	5.5	4.3	2.9	7.1	11.4	7.1	10.0	11.5	4.9	6.3	5.1	5.6	11.8	30.6	43.9	57.2	73.9	97.2	74.2
Thermal Capacity, Btu/ft·°F	1.3	0.8	0.6	2.3	3.6	2.3	3.7	3.9	1.4	1.6	1.4	1.6	2.8	6.6	9.3	12.0	16.3	21.4	16.2
Hour	**Conduction Time Factors, %**																		
0	6	10	27	1	1	1	0	1	18	4	8	1	0	1	2	2	2	3	1
1	45	57	62	17	17	12	7	3	61	41	53	23	10	2	2	2	2	3	2
2	33	27	10	31	34	25	18	8	18	35	30	38	22	8	3	3	5	3	6
3	11	5	1	24	25	22	18	10	3	14	7	22	20	11	6	4	6	5	8
4	3	1	0	14	13	15	15	10	0	4	2	10	14	11	7	5	7	6	8
5	1	0	0	7	6	10	11	9	0	1	0	4	10	10	8	6	7	6	8
6	1	0	0	4	3	6	8	8	0	1	0	2	7	9	8	6	6	6	7
7	0	0	0	2	1	4	6	7	0	0	0	0	5	7	7	6	6	6	7
8	0	0	0	0	0	2	5	6	0	0	0	0	4	6	7	6	6	6	6
9	0	0	0	0	0	1	3	5	0	0	0	0	3	5	6	6	5	5	5
10	0	0	0	0	0	1	3	5	0	0	0	0	2	5	5	6	5	5	5
11	0	0	0	0	0	1	2	4	0	0	0	0	1	4	5	5	5	5	5
12	0	0	0	0	0	0	1	4	0	0	0	0	1	3	5	5	4	5	4
13	0	0	0	0	0	0	1	3	0	0	0	0	1	3	4	5	4	4	4
14	0	0	0	0	0	0	1	3	0	0	0	0	0	3	4	4	4	4	3
15	0	0	0	0	0	0	1	3	0	0	0	0	0	2	3	4	4	4	3
16	0	0	0	0	0	0	0	2	0	0	0	0	0	2	3	4	3	4	3
17	0	0	0	0	0	0	0	2	0	0	0	0	0	2	3	4	3	4	3
18	0	0	0	0	0	0	0	2	0	0	0	0	0	1	3	3	3	3	2
19	0	0	0	0	0	0	0	2	0	0	0	0	0	1	2	3	3	3	2
20	0	0	0	0	0	0	0	1	0	0	0	0	0	1	2	3	3	3	2
21	0	0	0	0	0	0	0	1	0	0	0	0	0	1	2	3	3	3	2
22	0	0	0	0	0	0	0	1	0	0	0	0	0	1	2	3	2	2	2
23	0	0	0	0	0	0	0	0	0	0	0	0	0	1	1	2	2	2	2
	100	100	100	100	100	100	100	100	100	100	100	100	100	100	100	100	100	100	100
Layer ID from outside to inside (see Table 22)	F01	F01	F01	F01	F01	F01	F01	F01	F01	F01	F01	F01	F01	F01	F01	F01	F01	F01	F01
	F08	F08	F08	F12	F14	F15	F13	F13	F13	F13	F13	F13	M17	F13	F13	F13	F13	F13	F13
	G03	G03	G03	G05	G05	G05	G03	G03	G03	G03	G03	G03	F13	G03	G03	G03	G03	G03	M14
	F05	F05	F05	F05	F05	F05	I02	I02	I02	I02	I03	I02	G03	I03	I03	I03	I03	I03	F05
	I05	I05	I05	I05	I05	I05	G06	G06	F08	F08	F08	I03	I03	M11	M12	M13	M14	M15	I05
	G01	F05	F03	F05	F05	F05	F03	F05	F03	F05	F03	F08	F08	F03	F03	F03	F03	F03	F16
	F03	F16	—	G01	G01	G01	—	F16	—	F16	—	—	F03	—	—	—	—	—	F03
	—	F03	—	F03	F03	F03	—	F03	—	F03	—	—	—	—	—	—	—	—	—

Roof Number Descriptions

1. Metal roof, R-19 batt insulation, gyp board
2. Metal roof, R-19 batt insulation, suspended acoustical ceiling
3. Metal roof, R-19 batt insulation
4. Asphalt shingles, wood sheathing, R-19 batt insulation, gyp board
5. Slate or tile, wood sheathing, R-19 batt insulation, gyp board
6. Wood shingles, wood sheathing, R-19 batt insulation, gyp board
7. Membrane, sheathing, R-10 insulation board, wood deck
8. Membrane, sheathing, R-10 insulation board, wood deck, suspended acoustical ceiling
9. Membrane, sheathing, R-10 insulation board, metal deck
10. Membrane, sheathing, R-10 insulation board, metal deck, suspended acoustical ceiling
11. Membrane, sheathing, R-15 insulation board, metal deck
12. Membrane, sheathing, R-10 plus R-15 insulation boards, metal deck
13. 2-in. concrete roof ballast, membrane, sheathing, R-15 insulation board, metal deck
14. Membrane, sheathing, R-15 insulation board, 4-in. LW concrete
15. Membrane, sheathing, R-15 insulation board, 6-in. LW concrete
16. Membrane, sheathing, R-15 insulation board, 8-in. LW concrete
17. Membrane, sheathing, R-15 insulation board, 6-in. HW concrete
18. Membrane, sheathing, R-15 insulation board, 8-in. HW concrete
19. Membrane, 6-in HW concrete, R-19 batt insulation, suspended acoustical ceiling

$$q_{i,7} = (0.068)(100)(78-75) = 20$$
$$q_{i,8} = (0.068)(100)(82-75) = 48$$
$$q_{i,9} = (0.068)(100)(86-75) = 75$$
$$q_{i,10} = (0.068)(100)(90-75) = 102$$
$$q_{i,11} = (0.068)(100)(96-75) = 143$$
$$q_{i,12} = (0.068)(100)(109-75) = 231$$
$$q_{i,13} = (0.068)(100)(117-75) = 286$$
$$q_{i,14} = (0.068)(100)(123-75) = 326$$
$$q_{i,15} = (0.068)(100)(126-75) = 347$$
$$q_{i,16} = (0.068)(100)(123-75) = 326$$
$$q_{i,17} = (0.068)(100)(117-75) = 286$$
$$q_{i,18} = (0.068)(100)(105-75) = 204$$
$$q_{i,19} = (0.068)(100)(88-75) = 88$$
$$q_{i,20} = (0.068)(100)(85-75) = 68$$
$$q_{i,21} = (0.068)(100)(83-75) = 54$$
$$q_{i,22} = (0.068)(100)(81-75) = 41$$
$$q_{i,23} = (0.068)(100)(79-75) = 27$$
$$q_{i,24} = (0.068)(100)(77-75) = 14$$
$$\text{Total 24 h heat input} = 2693 \text{ Btu/h}$$

These data are used with conduction time series to calculate the wall heat gain. From Table 20, the most similar wall construction is wall number 14. This is a brick and block wall that has similar mass and thermal capacity. Using Equation (38), the conduction time factors for wall 14 can be used in conjunction with the 24 h heat input profile calculated above to determine the wall heat gain at 3 P.M. Central Daylight Time (which is actually 2 P.M. or hour 14 local standard time):

Table 22 Thermal Properties and Code Numbers of Layers Used in Wall and Roof Descriptions for Tables 20 and 21

Layer ID	Description	Thickness, in.	Conductivity, Btu·in/h·ft²·°F	Density, lb/ft³	Specific Heat, Btu/lb·°F	Resistance, ft²·°F·h/Btu	*R*	Mass, lb/ft²	Thermal Capacity, Btu/ft²·°F	Notes
F01	Outside surface resistance	—	—	—	—	0.25	0.25	—	—	1
F02	Inside vertical surface resistance	—	—	—	—	0.68	0.68	—	—	2
F03	Inside horizontal surface resistance	—	—	—	—	0.92	0.92	—	—	3
F04	Wall air space resistance	—	—	—	—	0.87	0.87	—	—	4
F05	Ceiling air space resistance	—	—	—	—	1.00	1.00	—	—	5
F06	EIFS finish	0.375	5.00	116.0	0.20	—	0.08	3.63	0.73	6
F07	1 in. stucco	1.000	5.00	116.0	0.20	—	0.20	9.67	1.93	6
F08	Metal surface	0.030	314.00	489.0	0.12	—	0.00	1.22	0.15	7
F09	Opaque spandrel glass	0.250	6.90	158.0	0.21	—	0.04	3.29	0.69	8
F10	1 in. stone	1.000	22.00	160.0	0.19	—	0.05	13.33	2.53	9
F11	Wood siding	0.500	0.62	37.0	0.28	—	0.81	1.54	0.43	10
F12	Asphalt shingles	0.125	0.28	70.0	0.30	—	0.44	0.73	0.22	
F13	Built-up roofing	0.375	1.13	70.0	0.35	—	0.33	2.19	0.77	
F14	Slate or tile	0.500	11.00	120.0	0.30	—	0.05	5.00	1.50	
F15	Wood shingles	0.250	0.27	37.0	0.31	—	0.94	0.77	0.24	
F16	Acoustic tile	0.750	0.42	23.0	0.14	—	1.79	1.44	0.20	11
F17	Carpet	0.500	0.41	18.0	0.33	—	1.23	0.75	0.25	12
F18	Terrazzo	1.000	12.50	160.0	0.19	—	0.08	13.33	2.53	13
G01	5/8 in. gyp board	0.625	1.11	50.0	0.26	—	0.56	2.60	0.68	
G02	5/8 in. plywood	0.625	0.80	34.0	0.29	—	0.78	1.77	0.51	
G03	1/2 in. fiberboard sheathing	0.500	0.47	25.0	0.31	—	1.06	1.04	0.32	14
G04	1/2 in. wood	0.500	1.06	38.0	0.39	—	0.47	1.58	0.62	15
G05	1 in. wood	1.000	1.06	38.0	0.39	—	0.94	3.17	1.24	15
G06	2 in. wood	2.000	1.06	38.0	0.39	—	1.89	6.33	2.47	15
G07	4 in. wood	4.000	1.06	38.0	0.39	—	3.77	12.67	4.94	15
I01	R-5, 1 in. insulation board	1.000	0.20	2.7	0.29	—	5.00	0.23	0.07	16
I02	R-10, 2 in. insulation board	2.000	0.20	2.7	0.29	—	10.00	0.45	0.13	16
I03	R-15, 3 in. insulation board	3.000	0.20	2.7	0.29	—	15.00	0.68	0.20	16
I04	R-11, 3-1/2 in. batt insulation	3.520	0.32	1.2	0.23	—	11.00	0.35	0.08	17
I05	R-19, 6-1/4 in. batt insulation	6.080	0.32	1.2	0.23	—	19.00	0.61	0.14	17
I06	R-30, 9-1/2 in. batt insulation	9.600	0.32	1.2	0.23	—	30.00	0.96	0.22	17
M01	4 in. brick	4.000	6.20	120.0	0.19	—	0.65	40.00	7.60	18
M02	6 in. LW concrete block	6.000	3.39	32.0	0.21	—	1.77	16.00	3.36	19
M03	8 in. LW concrete block	8.000	3.44	29.0	0.21	—	2.33	19.33	4.06	20
M04	12 in. LW concrete block	12.000	4.92	32.0	0.21	—	2.44	32.00	6.72	21
M05	8 in. concrete block	8.000	7.72	50.0	0.22	—	1.04	33.33	7.33	22
M06	12 in. concrete block	12.000	9.72	50.0	0.22	—	1.23	50.00	11.00	23
M07	6 in. LW concrete block (filled)	6.000	1.98	32.0	0.21	—	3.03	16.00	3.36	24
M08	8 in. LW concrete block (filled)	8.000	1.80	29.0	0.21	—	4.44	19.33	4.06	25
M09	12 in. LW concrete block (filled)	12.000	2.04	32.0	0.21	—	5.88	32.00	6.72	26
M10	8 in. concrete block (filled)	8.000	5.00	50.0	0.22	—	1.60	33.33	7.33	27
M11	4 in. lightweight concrete	4.000	3.70	80.0	0.20	—	1.08	26.67	5.33	
M12	6 in. lightweight concrete	6.000	3.70	80.0	0.20	—	1.62	40.00	8.00	
M13	8 in. lightweight concrete	8.000	3.70	80.0	0.20	—	2.16	53.33	10.67	
M14	6 in. heavyweight concrete	6.000	13.50	140.0	0.22	—	0.44	70.00	15.05	
M15	8 in. heavyweight concrete	8.000	13.50	140.0	0.22	—	0.59	93.33	20.07	
M16	12 in. heavyweight concrete	12.000	13.50	140.0	0.22	—	0.89	140.0	30.10	
M17	2 in. LW concrete roof ballast	2.000	1.30	40	0.20	—	1.54	6.7	1.33	28

Notes: The following notes give sources for the data in this table.

1. Chapter 25, Table 1 for 7.5 mph wind
2. Chapter 25, Table 1 for still air, horizontal heat flow
3. Chapter 25, Table 1 for still air, downward heat flow
4. Chapter 25, Table 3 for 1.5 in. space, 90°F, horizontal heat flow, 0.82 emittance
5. Chapter 25, Table 3 for 3.5 in. space, 90°F, downward heat flow, 0.82 emittance
6. EIFS finish layers approximated by Chapter 25, Table 4 for 3/8 in. cement plaster, sand aggregate
7. Chapter 38, Table 3 for steel (mild)
8. Chapter 25, Table 4 for architectural glass
9. Chapter 25, Table 4 for marble and granite
10. Chapter 25, Table 4, density assumed same as Southern pine
11. Chapter 25, Table 4 for mineral fiberboard, wet molded, acoustical tile
12. Chapter 25, Table 4 for carpet and rubber pad, density assumed same as fiberboard
13. Chapter 25, Table 4, density assumed same as stone
14. Chapter 25, Table 4 for nail-base sheathing
15. Chapter 25, Table 4 for Southern pine
16. Chapter 25, Table 4 for expanded polystyrene
17. Chapter 25, Table 4 for glass fiber batt, specific heat per glass fiber board
18. Chapter 25, Table 4 for clay fired brick
19. Chapter 25, Table 4, 16 lb block, 8 in. × 16 in. face
20. Chapter 25, Table 4, 19 lb block, 8 in. × 16 in. face
21. Chapter 25, Table 4, 32 lb block, 8 in. × 16 in. face
22. Chapter 25, Table 4, 33 lb normal weight block, 8 in. × 16 in. face
23. Chapter 25, Table 4, 50 lb normal weight block, 8 in. × 16 in. face
24. Chapter 25, Table 4, 16 lb block, vermiculite fill
25. Chapter 25, Table 4, 19 lb block, 8 in. × 16 in. face, vermiculite fill
26. Chapter 25, Table 4, 32 lb block, 8 in. × 16 in. face, vermiculite fill
27. Chapter 25, Table 4, 33 lb normal weight block, 8 in. × 16 in. face, vermiculite fill
28. Chapter 25, Table 4 for 40 lb/ft³ LW concrete

$$q_{14} = c_0 q_{i,14} + c_1 q_{i,13} + c_2 q_{i,12} + c_3 q_{i,11} + \ldots + c_{23} q_{i,15}$$
$$= (0.01)(326) + (0.01)(286) + (0.02)(231) + (0.05)(143) + (0.08)(102) + (0.09)(75) + (0.09)(48) + (0.09)(20) + (0.08)(7) + (0.07)(-7) + (0.07)(-7) + (0.06)(0) + (0.05)(7) + (0.04)(7) + (0.04)(14) + (0.03)(27) + (0.03)(41) + (0.02)(54) + (0.02)(68) + (0.02)(88) + (0.01)(204) + (0.01)(286) + (0.01)(326) + (0.0)(347)$$
$$= 54 \text{ Btu/h}$$

Table 23 Wall Heat Gain for Example 4

Wall Area = 100 ft^2
Inside Temperature = 75°F
Wall U-Factor = 0.068
CTS Table 20
Wall Number: 14

Hour	Sol-Air Temp., °F	Inside Temp., °F	Heat Input, Btu/h	CTS	Heat Gain, Btu/h
1	76	75	7	1%	165
2	76	75	7	1%	151
3	75	75	—	2%	136
4	74	75	−7	5%	121
5	74	75	−7	8%	106
6	76	75	7	9%	92
7	78	75	20	9%	79
8	82	75	48	9%	68
9	86	75	75	8%	58
10	90	75	102	7%	50
11	96	75	143	7%	45
12	109	75	231	6%	44
13	117	75	286	5%	47
14	123	75	326	4%	54
15	126	75	347	4%	68
16	123	75	326	3%	87
17	117	75	286	3%	110
18	105	75	204	2%	135
19	88	75	88	2%	158
20	85	75	68	2%	177
21	83	75	54	1%	188
22	81	75	41	1%	191
23	79	75	27	1%	186
24	77	75	14	0%	177
			2693	100%	2693

Due to the tedious calculations involved, use of a simple computer spreadsheet or other computer software implementing these calculations can reduce the effort involved. A spreadsheet is illustrated in Table 23, calculating a 24 h heat gain profile for the data of this example.

Calculating Cooling Load Using RTS

The instantaneous cooling load is defined as the rate at which heat energy is convected to the zone air at a given point in time. The computation of cooling load is complicated by the radiant exchange between surfaces, furniture, partitions, and other mass in the zone. Most heat gain sources transfer energy by both convection and radiation. Radiative heat transfer introduces to the process a time dependency that is not easily quantified. Radiation is absorbed by the thermal masses in the zone and then later transferred by convection into the space. This process creates a time lag and dampening effect. The convection portion of heat gains, on the other hand, is assumed to immediately become cooling load in the hour in which that heat gain occurs.

Heat balance procedures calculate the radiant exchange between surfaces based on their surface temperatures and emissivities, but they typically rely on estimated "radiative-convective splits" to determine the contribution of internal loads, including people, lighting, appliances, and equipment, to the radiant exchange. The radiant time series procedure further simplifies the heat balance procedure by also relying on an estimated radiative-convective split of wall and roof conductive heat gain instead of simultaneously solving for the instantaneous convective and radiative heat transfer from each surface, as is done in the heat balance procedure.

Thus, the cooling load for each load component (lights, people, walls, roofs, windows, appliances, etc.) for a particular hour is the sum of the convective portion of the heat gain for that hour plus the time-delayed portion of radiant heat gains for that hour and the previous 23 h. Table 19 contains recommendations for splitting each of the heat gain components into convective and radiant portions.

The radiant time series method converts the radiant portion of hourly heat gains to hourly cooling loads using radiant time factors, the coefficients of the radiant time series. Radiant time factors are used to calculate the cooling load for the current hour on the basis of current and past heat gains. The radiant time series for a particular zone gives the time-dependent response of the zone to a single pulse of radiant energy. The series shows the portion of the radiant pulse that is convected to the zone air for each hour. Thus, r_0 represents the fraction of the radiant pulse convected to the zone air in the current hour r_1 in the previous hour, and so on. The radiant time series thus generated is used to convert the radiant portion of hourly heat gains to hourly cooling loads according to the following equation:

$$Q_{r,\theta} = r_0 q_{r,\theta} + r_1 q_{r,\theta-1} + r_2 q_{r,\theta-2} + r_3 q_{r,\theta-3} + \ldots + r_{23} q_{r,\theta-23} \quad (39)$$

where

$Q_{r,\theta}$ = radiant cooling load (Q_r) for the current hour (θ), Btu/h
$q_{r,\theta}$ = radiant heat gain for the current hour, Btu/h
$q_{r,\theta-n}$ = radiant heat gain n hours ago, Btu/h
r_0, r_1, etc. = radiant time factors

The radiant cooling load for the current hour, which is calculated using RTS and Equation (39), is added to the convective portion to determine the total cooling load for that component for that hour.

Radiant time factors are generated by a heat balance based procedure. A separate series of radiant time factors is theoretically required for each unique zone and for each unique radiant energy distribution function assumption. For most common design applications, RTS variation depends primarily on the overall massiveness of the construction and the thermal responsiveness of the surfaces the radiant heat gains strike.

One of the goals in developing RTS was to provide a simplified method that was based directly on the heat balance method; thus, it was deemed desirable to generate the RTS coefficients directly from a heat balance. To this end, a heat balance computer program was developed. This program, called Hbfort, which may be used to generate RTS coefficients, is included as part of *Cooling and Heating Load Calculation Principles* (Pedersen et al. 1998). The RTS procedure is described by Spitler et al. (1997). The procedure for generating RTS coefficients may be thought of as analogous to the custom weighting factor generation procedure used by DOE 2.1 (Kerrisk et al. 1981; Sowell 1988a, 1988b). In both cases, a zone model is pulsed with a heat gain. In the case of DOE 2.1, the resulting loads are used to estimate the best values of the transfer function method weighting factors to most closely match the load profile. In the case of the procedure described here, a unit periodic heat gain pulse is used to generate loads for a 24 h period. As long as the heat gain pulse is a unit pulse, the resulting loads are equivalent to the RTS coefficients.

Two different series of radiant time factors are utilized—one for direct transmitted solar heat gain (radiant energy assumed to be distributed to the floor and furnishings only) and one for all other types of heat gains (radiant energy assumed to be uniformly distributed on all internal surfaces). Nonsolar RTS apply to radiant heat gains from people, lights, appliances, walls, roofs, and floors. Also, for diffuse

Table 24 Representative Nonsolar RTS Values for Light to Heavy Construction

	Light						Medium						Heavy						Interior Zones					
																			Light		Medium		Heavy	
	With Carpet			No Carpet			With Carpet			No Carpet			With Carpet			No Carpet								
% Glass	10%	50%	90%	10%	50%	90%	10%	50%	90%	10%	50%	90%	10%	50%	90%	10%	50%	90%	With Carpet	No Carpet	With Carpet	No Carpet	With Carpet	No Carpet
Hour	Radiant Time Factor, %																							
0	47	50	53	41	43	46	46	49	52	31	33	35	34	38	42	22	25	28	46	40	46	31	33	21
1	19	18	17	20	19	19	18	17	16	17	16	15	9	9	9	10	9	9	19	20	18	17	9	9
2	11	10	9	12	11	11	10	9	8	11	10	10	6	6	5	6	6	6	11	12	10	11	6	6
3	6	6	5	8	7	7	6	5	5	8	7	7	4	4	4	5	5	5	6	8	6	8	5	5
4	4	4	3	5	5	5	4	3	3	6	5	5	4	4	4	5	5	4	4	5	3	6	4	5
5	3	3	2	4	3	3	2	2	2	4	4	4	4	3	3	4	4	4	3	4	2	4	4	4
6	2	2	2	3	3	2	2	2	2	4	3	3	3	3	3	4	4	4	2	3	2	4	3	4
7	2	1	1	2	2	2	1	1	1	3	3	3	3	3	3	4	4	4	2	2	1	3	3	4
8	1	1	1	1	1	1	1	1	1	3	2	2	3	3	3	4	3	3	1	1	1	3	3	4
9	1	1	1	1	1	1	1	1	1	2	2	2	3	3	2	3	3	3	1	1	1	2	3	3
10	1	1	1	1	1	1	1	1	1	2	2	2	3	2	2	3	3	3	1	1	1	2	3	3
11	1	1	1	1	1	1	1	1	1	2	2	2	2	2	2	3	3	3	1	1	1	2	2	3
12	1	1	1	1	1	1	1	1	1	1	1	1	2	2	2	3	3	3	1	1	1	1	2	3
13	1	1	1	0	1	0	1	1	1	1	1	1	2	2	2	3	3	2	1	1	1	1	2	3
14	0	0	1	0	1	0	1	1	1	1	1	1	2	2	2	3	2	2	1	0	1	1	2	3
15	0	0	1	0	0	0	1	1	1	1	1	1	2	2	2	2	2	2	0	0	1	1	2	3
16	0	0	0	0	0	0	1	1	1	1	1	1	2	2	2	2	2	2	0	0	1	1	2	3
17	0	0	0	0	0	0	1	1	1	1	1	1	2	2	2	2	2	2	0	0	1	1	2	2
18	0	0	0	0	0	0	1	1	1	1	1	1	2	2	1	2	2	2	0	0	1	1	2	2
19	0	0	0	0	0	0	0	1	0	0	1	1	2	2	1	2	2	2	0	0	1	0	2	2
20	0	0	0	0	0	0	0	0	0	0	1	1	2	1	1	2	2	2	0	0	0	0	2	2
21	0	0	0	0	0	0	0	0	0	0	1	1	2	1	1	2	2	2	0	0	0	0	2	2
22	0	0	0	0	0	0	0	0	0	0	1	0	1	1	1	2	2	2	0	0	0	0	1	2
23	0	0	0	0	0	0	0	0	0	0	0	0	1	1	1	2	2	1	0	0	0	0	1	2
	100	100	100	100	100	100	100	100	100	100	100	100	100	100	100	100	100	100	100	100	100	100	100	100

Table 25 Representative Solar RTS Values for Light to Heavy Construction

	Light						Medium						Heavy					
	With Carpet			No Carpet			With Carpet			No Carpet			With Carpet			No Carpet		
% Glass	10%	50%	90%	10%	50%	90%	10%	50%	90%	10%	50%	90%	10%	50%	90%	10%	50%	90%
Hour	Radiant Time Factor, %																	
0	53	55	56	44	45	46	52	54	55	28	29	29	47	49	51	26	27	28
1	17	17	17	19	20	20	16	16	15	15	15	15	11	12	12	12	13	13
2	9	9	9	11	11	11	8	8	8	10	10	10	6	6	6	7	7	7
3	5	5	5	7	7	7	5	4	4	7	7	7	4	4	3	5	5	5
4	3	3	3	5	5	5	3	3	3	6	6	6	3	3	3	4	4	4
5	2	2	2	3	3	3	2	2	2	5	5	5	2	2	2	4	4	4
6	2	2	2	3	2	2	2	1	1	4	4	4	2	2	2	3	3	3
7	1	1	1	2	2	2	1	1	1	4	3	3	2	2	2	3	3	3
8	1	1	1	1	1	1	1	1	1	3	3	3	2	2	2	3	3	3
9	1	1	1	1	1	1	1	1	1	3	3	3	2	2	2	3	3	3
10	1	1	1	1	1	1	1	1	1	2	2	2	2	2	2	3	3	3
11	1	1	1	1	1	1	1	1	1	2	2	2	2	2	1	3	3	2
12	1	1	1	1	1	0	1	1	1	2	2	2	2	1	1	2	2	2
13	1	1	0	1	0	0	1	1	1	2	2	2	2	1	1	2	2	2
14	1	0	0	0	0	0	1	1	1	1	1	1	2	1	1	2	2	2
15	1	0	0	0	0	0	1	1	1	1	1	1	1	1	1	2	2	2
16	0	0	0	0	0	0	1	1	1	1	1	1	1	1	1	2	2	2
17	0	0	0	0	0	0	1	1	1	1	1	1	1	1	1	2	2	2
18	0	0	0	0	0	0	1	1	1	1	1	1	1	1	1	2	2	2
19	0	0	0	0	0	0	0	0	0	1	1	1	1	1	1	2	2	2
20	0	0	0	0	0	0	0	0	0	1	1	1	1	1	1	2	2	2
21	0	0	0	0	0	0	0	0	0	0	0	0	1	1	1	2	2	2
22	0	0	0	0	0	0	0	0	0	0	0	0	1	1	1	2	1	1
23	0	0	0	0	0	0	0	0	0	0	0	0	1	1	1	2	1	1
	100	100	100	100	100	100	100	100	100	100	100	100	100	100	100	100	100	100

Table 26 RTS Representative Zone Construction for Tables 24 and 25

Construction Class	Exterior Wall	Roof/Ceiling	Partitions	Floor	Furnishings
Light	steel siding, 2 in. insulation, air space, 3/4 in. gyp	4 in. LW concrete, ceiling air space, acoustic tile	3/4 in. gyp, air space, 3/4 in. gyp	acoustic tile, ceiling air space, 4 in. LW concrete	1 in. wood @ 50% of floor area
Medium	4 in. face brick, 2 in. insulation, air space, 3/4 in. gyp	4 in. HW concrete, ceiling air space, acoustic tile	3/4 in. gyp, air space, 3/4 in. gyp	acoustic tile, ceiling air space, 4 in. HW concrete	1 in. wood @ 50% of floor area
Heavy	4 in. face brick, 8 in. HW concrete air space, 2 in. insulation, 3/4 in. gyp	8 in. HW concrete, ceiling air space, acoustic tile	3/4 in. gyp, 8 in. HW concrete block, 3/4 in. gyp	acoustic tile, ceiling air space, 8 in. HW concrete	1 in. wood @ 50% of floor area

Table 27 Wall Heat Gain and Cooling Load for Example 5

Wall Area = 100 ft^2
Inside Temp. = 75°F
Wall U-factor = 0.068

CTS Table 20
Wall Number: 14

RTS medium-weight construction, 50% glass, with carpet

Hour	Sol-Air Temp	Inside Temp	Heat Input	CTS	Heat Gain	Split Heat Gain: Convective % 37%	Split Heat Gain: Radiant % 63%	Nonsolar RTS	Radiant Cooling Load	Total Cooling Load
1	76	75	7	1%	165	61	104	49%	100	161
2	76	75	7	1%	151	56	95	17%	95	151
3	75	75	—	2%	136	50	86	9%	88	139
4	74	75	–7	5%	121	45	77	5%	82	127
5	74	75	–7	8%	106	39	67	3%	74	114
6	76	75	7	9%	92	34	58	2%	68	102
7	78	75	20	9%	79	29	50	2%	61	90
8	82	75	48	9%	68	25	43	1%	55	80
9	86	75	75	8%	58	21	36	1%	49	71
10	90	75	102	7%	50	18	31	1%	45	63
11	96	75	143	7%	45	17	28	1%	41	58
12	109	75	231	6%	44	16	28	1%	39	55
13	117	75	286	5%	47	17	30	1%	39	56
14	123	75	326	4%	54	20	34	1%	40	60
15	126	75	347	4%	68	25	43	1%	45	70
16	123	75	326	3%	87	32	55	1%	51	83
17	117	75	286	3%	110	41	69	1%	61	101
18	105	75	204	2%	135	50	85	1%	72	121
19	88	75	88	2%	158	58	99	1%	83	141
20	85	75	68	2%	177	65	111	1%	93	158
21	83	75	54	1%	188	70	118	0%	101	170
22	81	75	41	1%	191	71	120	0%	105	176
23	79	75	27	1%	186	69	117	0%	106	175
24	77	75	14	0%	177	66	112	0%	104	170
			2693	100%	2693	996	1696	100%	1696	2693

solar heat gain and direct solar heat gain from fenestration with inside shading (blinds, drapes, etc.), the nonsolar RTS should be used. Radiation from those sources is assumed to be more uniformly distributed onto all room surfaces. Effect of beam solar radiation distribution assumptions is addressed by Hittle (1999).

Representative RTS data for light, medium, and heavyweight constructions are provided in Tables 24 and 25. Those were calculated using the Hbfort computer program (Pedersen et al. 1998) with zone characteristics listed in Table 26. Customized RTS values may be calculated using the HB method where the zone is not reasonably similar to these typical zones or where more precision is desired.

ASHRAE *Research Project* 942 compared HB and RTS results over a wide range of zone types and input variables (Spitler et al. 1998, Rees et al. 2000). In general, total cooling loads calculated using the RTS method closely agreed with or were slightly higher than those of the HB method with the same inputs. The project examined more than 5000 test cases of varying zone parameters. The dominating variable was overall thermal mass, and results were grouped into lightweight, U.S. medium-weight, U.K. medium-weight, and heavyweight construction. Best agreement between RTS and HB results was obtained for light- and medium-weight construction. Greater differences occurred in heavyweight cases, with RTS generally predicting slightly higher peak cooling loads than HB. Greater differences also were observed in zones with extremely high internal radiant loads and large glazing areas or with a very lightweight exterior envelope. In this case, heat balance calculations predict that some of the internal radiant load will be transmitted to the outdoor environment and never becomes cooling load within the space. The RTS methodology does not account for energy being transferred out of the space to the environment and thus predicted higher cooling loads.

Example 5. Wall cooling load using radiant time series. Using the data from Example 4, calculate the cooling load at 3 P.M. Central Daylight Time on July 21 through 100 ft^2 of a wall composed of 4 in. brick, 2 in. of insulation (R = 10), and 8 in. lightweight concrete block in an office building of average commercial construction.

Solution: Total cooling load for the wall is calculated by summing the convective and radiant portions. The convective portion is simply the wall heat gain for the hour being calculated, from Example 4, times the convective fraction for walls from Table 19:

$$Q_c = (54)(37\%) = 20 \text{ Btu/h}$$

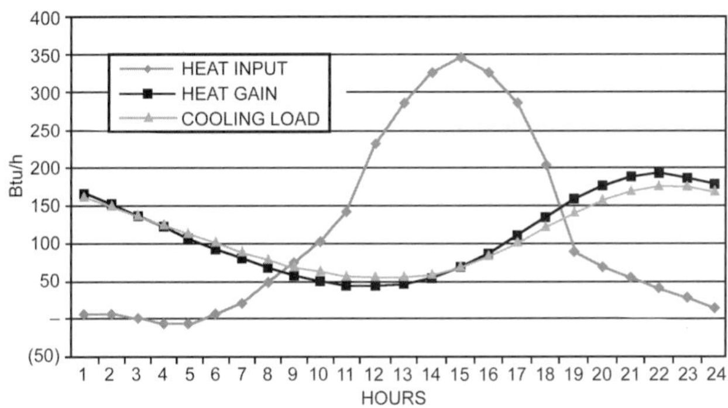

Fig. 12 Wall Heat Input, Heat Gain for Example 5 and Cooling Load Using CTS and RTS

The radiant portion of the cooling load is calculated using conductive heat gains for the current and past 23 h, the radiant fraction for walls from Table 19 (63%), and radiant time series from Table 24, in accordance with Equation (39). From Table 24, select the RTS for medium-weight construction, assuming 50% glass and carpeted floors as is often found in modern office construction. Take the wall heat gains from Table 23. Thus, the radiant cooling load for the wall is

$$\begin{aligned}Q_{r,14} &= r_0(0.63)q_{i,14} + r_1(0.63)q_{i,13} + r_2(0.63)q_{i,12} + r_3(0.63)q_{i,11} + \ldots \\ &\quad + r_{23}(0.63)q_{i,15} \\ &= (0.49)(0.63)(54) + (0.17)(0.63)(47) + (0.09)(0.63)(44) \\ &\quad + (0.05)(0.63)(45) + (0.03)(0.63)(50) + (0.02)(0.63)(58) \\ &\quad + (0.02)(0.63)(68) + (0.01)(0.63)(79) + (0.01)(0.63)(92) \\ &\quad + (0.01)(0.63)(106) + (0.01)(0.63)(121) + (0.01)(0.63)(136) \\ &\quad + (0.01)(0.63)(151) + (0.01)(0.63)(165) + (0.01)(0.63)(177) \\ &\quad + (0.01)(0.63)(186) + (0.01)(0.63)(191) + (0.01)(0.63)(188) \\ &\quad + (0.01)(0.63)(177) + (0.01)(0.63)(158) + (0.00)(0.63)(135) \\ &\quad + (0.00)(0.63)(110) + (0.00)(0.63)(87) + (0.00)(0.63)(68) \\ &= 40 \text{ Btu/h}\end{aligned}$$

The total wall cooling load at the designated hour is thus

$$Q_{wall} = Q_c + Q_{r14} = 20 + 40 = 60 \text{ Btu/h}$$

Again, due to the tedious calculations involved, use of a simple computer spreadsheet or other computer software implementing these calculations can reduce the effort involved. The spreadsheet illustrated in Table 23 is expanded in Table 27 to include splitting the heat gain into convective and radiant portions, applying RTS to the radiant portion, and totaling the convective and radiant loads to determine a 24 h cooling load profile for the data of this example. Figure 12 shows wall heat input, heat gain, and total cooling load versus time for the 24 h calculated.

Example 6. Window cooling load using radiant time series. Calculate the cooling load for a 20 ft^2 double-glazed bronze low-e window at 3 P.M. Central Daylight Time on July 21, at the location and for the conditions defined in Example 1, in an office building of average commercial construction. Use SHGC data for glass type 17f from Table 13 in Chapter 30. Use window U-factor = 0.56.

Solution: To determine the window cooling load, first calculate the 24 h heat gain profile for the window, then split those heat gains into radiant and convective portions, apply the appropriate RTS to the radiant portion, then sum the convective and radiant cooling load components to determine total window cooling load at the designated time. The window heat gain components are calculated using Equations (13) through (15).

From Example 1, at hour 14 standard time (3 P.M. Central Daylight Time):

$$E_D = 135 \text{ Btu/h·ft}^2$$
$$E_d = 31.1 \text{ Btu/h·ft}^2$$
$$E_r = 26.8 \text{ Btu/h·ft}^2$$
$$\theta = 60.1°$$

From Chapter 30, Table 13, for glass type 17f,

$$\text{SHGC}(\theta) = \text{SHGC}(60.1) = 0.349 \text{ (interpolated)}$$
$$\langle \text{SHGC} \rangle_D = 0.38$$

Therefore, window heat gain components for hour 14 are

$$q_{b14} = AE_D\,\text{SHGC}(\theta) = (20)(135)(0.349) = 942 \text{ Btu/h}$$
$$q_{d14} = A(E_d + E_r)\langle\text{SHGC}\rangle_D = (20)(31.1 + 26.8)(0.38) = 440 \text{ Btu/h}$$
$$q_{c14} = UA(t_{out} - t_{in}) = (0.56)(20)(94 - 75) = 213 \text{ Btu/h}$$

This procedure is repeated to determine these values for a 24 h heat gain profile, as illustrated in Table 28.

Total cooling load for the window is calculated by summing the convective and radiant portions. For windows with inside shading (blinds, drapes, etc.), the direct beam, diffuse, and conductive heat gains may be summed and treated together in calculating cooling loads. However, in this example, the window does not have inside shading, and the direct beam solar heat gain should be treated separately from the diffuse and conductive heat gains. The direct beam heat gain, without inside shading, is treated as 100% radiant, and solar RTS factors from Table 25 are used to convert the beam heat gains to cooling loads. The diffuse and conductive heat gains can be totaled and split into radiant and convection portions according to Table 19 percentages, and nonsolar RTS factors from Table 24 are used to convert the radiant portion to cooling load.

The solar beam cooling load is calculated using heat gains for the current hour and past 23 h and radiant time series from Table 25, in accordance with Equation (39). From Table 25, select the solar RTS for medium-weight construction, assuming 50% glass and carpeted floors as is often found in modern office construction. Using Table 28 values for direct solar heat gain, the radiant cooling load for the window direct beam solar component is

$$\begin{aligned}Q_{b,14} &= r_0q_{b,14} + r_1q_{b,13} + r_2q_{b,12} + r_3q_{b,11} + \ldots + r_{23}q_{b,15} \\ &= (0.54)(942) + (0.16)(596) + (0.08)(279) + (0.04)(4) \\ &\quad + (0.03)(0) + (0.02)(0) + (0.01)(0) + (0.01)(0) + (0.01)(0) \\ &\quad + (0.01)(0) + (0.01)(0) + (0.01)(0) + (0.01)(0) + (0.01)(0) \\ &\quad + (0.01)(0) + (0.01)(0) + (0.01)(0) + (0.01)(0) + (0.01)(0) \\ &\quad + (0.00)(21) + (0.00)(487) + (0.00)(894) + (0.00)(1113) \\ &\quad + (0.00)(1121) \\ &= 626 \text{ Btu/h}\end{aligned}$$

This process is repeated for other hours; results are listed in Table 29.

For diffuse and conductive heat gains, the radiant fraction according to Table 19 is 63%. The radiant portion is processed using nonsolar RTS coefficients from Table 24. The results are listed in Table29.

The total window cooling load at the designated hour is thus

$$Q_{window} = Q_b + Q_{diff+cond} = 626 + 581 = 1207 \text{ Btu/h}$$

Again, due to the tedious calculations involved, use of a simple computer spreadsheet or other computer software implementing these calculations can reduce the effort involved. The spreadsheet illustrated in Table 28 is expanded in Table 29 to include splitting the heat gain into convective and radiant portions, applying RTS to the radiant portion, and totaling the convective and radiant loads to determine a 24 h cooling load profile for the data of this example.

Example 7. Internal cooling load using radiant time series. Calculate the cooling load at 10 A.M. for a 1000 ft^2 office space due to suspended fluorescent lights with a total load of 2000 W, in an office building of average commercial construction. The lights are 50% on at 7 A.M., 100% on from 8 A.M. to 6 P.M., then off the remainder of the time.

Solution: To determine the lighting cooling load, first calculate the 24 h heat gain profile for the lighting load, then split those heat gains into radiant and convective portions, apply the appropriate RTS to the radiant portion, and sum the convective and radiant cooling load components to determine total cooling load at the designated time. The lighting heat gain profile, based on the schedule indicated and converting watts to Btu/h, is

$$q_1 = (2000 \text{ W})3.41(0\%) = 0$$
$$q_2 = (2000 \text{ W})3.41(0\%) = 0$$
$$q_3 = (2000 \text{ W})3.41(0\%) = 0$$
$$q_4 = (2000 \text{ W})3.41(0\%) = 0$$
$$q_5 = (2000 \text{ W})3.41(0\%) = 0$$
$$q_6 = (2000 \text{ W})3.41(0\%) = 0$$

Table 28 Solar Calculations—Solar Heat Gain Using SHGC—for Example 6

Time Zone =	3
Month =	7
Longitude =	90
Latitude =	40
Clearness =	1
Ground Refl. =	0.2
Room Temp. =	75
For Month =	7
Equation of Time =	−6.2
Declination =	20.6
A =	344
B =	0.207
C =	0.136
Local Standard Time Meridian =	90

Surface:

Azimuth 45

Tilt 90

Fenestration

Area	20 ft^2
U	0.56
IAC	1.0

Angle	SHGC:
0	0.450
40	0.420
50	0.400
60	0.350
70	0.270
80	0.140
90	—
Hemis:	0.380

Month	Equation of Time, min.	Declination degrees	A: Btu/h·ft^2	B (Dimensionless Ratios)	C (Dimensionless Ratios)
1	−11.2	−20	390	0.142	0.058
2	−13.9	−10.8	385	0.144	0.060
3	−7.5	0	376	0.156	0.071
4	1.1	11.6	360	0.180	0.097
5	3.3	20	350	0.196	0.121
6	−1.4	23.45	345	0.205	0.134
7	−6.2	20.6	344	0.207	0.136
8	−2.4	12.3	351	0.201	0.122
9	7.5	0	365	0.177	0.092
10	15.4	−10.5	378	0.160	0.073
11	13.8	−19.8	387	0.149	0.063
12	1.6	−23.45	391	0.142	0.057

Time Zone	Std. Meridian
1 Atlantic	60
2 Eastern	75
3 Central	90
4 Mountain	105
5 Pacific	120
6 Alaska	135
7 Hawaii	150

					Direct Beam Solar Heat Gain					Diffuse Solar Heat Gain						Conduction		Total
Local Standard Hour	Apparent Solar Time, hours	Hour Angle	Solar Altitude	Solar Azimuth	Direct Normal Irradiance	Surface Incident Angle	Surface Direct, Btu/h/ft^2	Direct SHGC	Direct Solar Heat Gain, Btu/h	Ground Diffuse	Y Ratio	Sky Diffuse	Total Diffuse, Btu/h/ft^2	Hemis. SHGC	Diffuse Solar Heat Gain, Btu/h	Outside Temp.	Conductive Heat Gain, Btu/h	Window Heat Gain, Btu/h
1	0.90	−166.55	−28.1	−165.7	—	139.3	0.0	—	0.0	0.0	0.45	0.00	0.0	0.380	0.0	76	11.2	11
2	1.90	−151.55	−23.8	−150.8	—	151.6	0.0	—	0.0	0.0	0.45	0.00	0.0	0.380	0.0	76	11.2	11
3	2.90	−136.55	−17.1	−137.7	—	162.7	0.0	—	0.0	0.0	0.45	0.00	0.0	0.380	0.0	75	—	—
4	3.90	−121.55	−8.6	−126.2	—	167.8	0.0	—	0.0	0.0	0.45	0.00	0.0	0.380	0.0	74	−11.2	−11
5	4.90	−106.55	1.3	−116.2	0	161.1	0.0	—	0.0	0.0	0.45	0.00	0.0	0.380	0.0	74	−11.2	−11
6	5.90	−91.55	11.9	−107.0	126	149.7	0.0	—	0.0	4.3	0.45	7.74	12.1	0.380	91.7	74	−11.2	81
7	6.90	−76.55	23.1	−98.1	203	137.3	0.0	—	0.0	10.7	0.45	12.43	23.2	0.380	176.1	75	—	176
8	7.90	−61.55	34.6	−88.8	239	124.7	0.0	—	0.0	16.8	0.45	14.62	31.4	0.380	238.9	77	22.4	261
9	8.90	−46.55	46.0	−78.0	258	112.2	0.0	—	0.0	22.1	0.45	15.79	37.9	0.380	287.7	80	56.0	344
10	9.90	−31.55	56.8	−63.6	269	100.0	0.0	—	0.0	26.1	0.48	17.66	43.8	0.380	332.9	83	89.6	423
11	10.90	−16.55	66.0	−41.0	274	88.4	7.9	0.023	3.6	28.8	0.56	20.99	49.8	0.380	378.3	87	134.4	516
12	11.90	−1.55	70.6	4.4	276	75.4	69.8	0.200	279.4	29.8	0.68	25.56	55.4	0.380	420.7	90	168.0	868
13	12.90	13.45	67.5	34.6	275	67.8	103.7	0.287	595.7	29.1	0.76	28.39	57.5	0.380	437.2	93	201.6	1,235
14	13.90	28.45	58.9	59.8	270	60.1	134.8	0.349	941.6	26.8	0.85	31.08	57.9	0.380	440.0	94	212.8	1,594
15	14.90	43.45	48.3	75.4	261	55.0	149.5	0.375	1121.3	23.0	0.90	32.04	55.1	0.380	418.4	95	224.0	1,764
16	15.90	58.45	37.0	86.7	244	53.4	145.3	0.383	1113.1	18.0	0.92	30.56	48.5	0.380	368.9	94	212.8	1,695
17	16.90	73.45	25.5	96.2	213	55.6	120.2	0.372	894.2	12.0	0.90	25.94	38.0	0.380	288.7	93	201.6	1,384
18	17.90	88.45	14.2	105.1	148	61.1	71.5	0.341	487.1	5.6	0.83	16.79	22.4	0.380	170.5	91	179.2	837
19	18.90	103.45	3.4	114.2	11	69.3	3.7	0.276	20.6	0.2	0.74	1.07	1.3	0.380	9.7	87	134.4	165
20	19.90	118.45	−6.6	124.0	—	79.1	0.0	0.151	0.0	0.0	0.64	0.00	0.0	0.380	0.0	85	112.0	112
21	20.90	133.45	−15.5	135.2	—	90.2	0.0	—	0.0	0.0	0.55	0.00	0.0	0.380	0.0	83	89.6	90
22	21.90	148.45	−22.6	147.9	—	101.9	0.0	—	0.0	0.0	0.45	0.00	0.0	0.380	0.0	81	67.2	67
23	22.90	163.45	−27.5	162.5	—	114.2	0.0	—	0.0	0.0	0.45	0.00	0.0	0.380	0.0	79	44.8	45
24	23.90	178.45	−29.4	178.3	—	126.7	0.0	—	0.0	0.0	0.45	0.00	0.0	0.380	0.0	77	22.4	22

$$q_7 = (2000\text{ W})3.41(50\%) = 3410$$
$$q_8 = (2000\text{ W})3.41(100\%) = 6820$$
$$q_9 = (2000\text{ W})3.41(100\%) = 6820$$
$$q_{10} = (2000\text{ W})3.41(100\%) = 6820$$
$$q_{11} = (2000\text{ W})3.41(100\%) = 6820$$
$$q_{12} = (2000\text{ W})3.41(100\%) = 6820$$
$$q_{13} = (2000\text{ W})3.41(100\%) = 6820$$
$$q_{14} = (2000\text{ W})3.41(100\%) = 6820$$
$$q_{15} = (2000\text{ W})3.41(100\%) = 6820$$
$$q_{16} = (2000\text{ W})3.41(100\%) = 6820$$
$$q_{17} = (2000\text{ W})3.41(100\%) = 6820$$
$$q_{18} = (2000\text{ W})3.41(100\%) = 6820$$
$$q_{19} = (2000\text{ W})3.41(0\%) = 0$$
$$q_{20} = (2000\text{ W})3.41(0\%) = 0$$
$$q_{21} = (2000\text{ W})3.41(0\%) = 0$$
$$q_{22} = (2000\text{ W})3.41(0\%) = 0$$
$$q_{23} = (2000\text{ W})3.41(0\%) = 0$$
$$q_{24} = (2000\text{ W})3.41(0\%) = 0$$

The convective portion is simply the lighting heat gain for the hour being calculated times the convective fraction for unvented fluorescent lighting from Table 19:

$$Q_{c,10} = (6820)(33\%) = 2251 \text{ Btu/h}$$

Table 29 Fenestration Cooling Load for Example 6

	Direct Beam Solar Cooling Load						Diffuse Solar and Conduction Cooling Load								Total
Hour	Direct Solar Heat Gain, Btu/h	Convect., %, 0%	Rad., %, 100%	Solar RTS Coefficients from Table 25	Direct Solar Radiant Cooling Load	Direct Solar Total Cooling Load	Diffuse Solar Heat Gain	Cond. Heat Gain, Btu/h	Tot Diff. & Cond., Btu/h	Convect., %, 37%	Rad., %, 63%	Nonsolar RTS Coefficients from Table 24	Diff. & Cond. Radiant Cooling Load	Diff. & Cond. Total Cooling Load	Total Window Cooling Load
1	0	—	—	54%	55	55	—	11	11	4.14	7.06	49%	52	56	110
2	0	—	—	16%	55	55	—	11	11	4.14	7.06	17%	48	52	106
3	0	—	—	8%	55	55	—	0	—	—	—	9%	42	42	96
4	0	—	—	4%	55	55	—	–11	–11	–4.14	–7.06	5%	34	30	85
5	0	—	—	3%	55	55	0	–11	–11	–4.14	–7.05	3%	30	26	81
6	0	—	—	2%	55	55	92	–11	81	29.79	50.73	2%	55	84	139
7	0	—	—	1%	52	52	176	0	176	65.17	110.97	2%	90	155	207
8	0	—	—	1%	46	46	239	22	261	96.68	164.62	1%	128	225	270
9	0	—	—	1%	36	36	288	56	344	127.16	216.52	1%	167	294	330
10	0	—	—	1%	25	25	333	90	423	156.33	266.18	1%	205	361	386
11	4	—	4	1%	16	16	378	134	513	189.69	322.99	1%	247	437	453
12	279	—	279	1%	157	157	421	168	589	217.83	370.89	1%	288	506	662
13	596	—	596	1%	367	367	437	202	639	236.34	402.42	1%	320	557	924
14	942	—	942	1%	626	626	440	213	653	241.53	411.26	1%	340	581	1,207
15	1121	—	1,121	1%	815	815	418	224	642	237.69	404.71	1%	348	585	1,400
16	1113	—	1,113	1%	888	888	369	213	582	215.23	366.48	1%	335	550	1,439
17	894	—	894	1%	812	812	289	202	490	181.41	308.88	1%	306	487	1,299
18	487	—	487	1%	583	583	171	179	350	129.40	220.34	1%	253	383	966
19	21	—	21	1%	266	266	10	134	144	53.30	90.76	1%	172	225	491
20	0	—	—	0%	152	152	—	112	112	41.44	70.56	1%	132	173	325
21	0	—	—	0%	100	100	—	90	90	33.15	56.45	0%	107	140	239
22	0	—	—	0%	74	74	—	67	67	24.86	42.34	0%	88	113	187
23	0	—	—	0%	60	60	—	45	45	16.58	28.22	0%	74	90	150
24	0	—	—	0%	55	55	—	22	22	8.29	14.11	0%	61	69	124
	5,457	—	5,457	100%	5,457	5,457	4,060	2,162	6,221	2,302	3,920	100%	3,919	6,221	11,678

The radiant portion of the cooling load is calculated using lighting heat gains for the current hour and past 23 h, the radiant fraction from Table 19 (67%), and radiant time series from Table 24, in accordance with Equation (39). From Table 24, select the RTS for medium-weight construction, assuming 50% glass and carpeted floors as is often found in modern office construction. Thus, the radiant cooling load for lighting is

$$
\begin{aligned}
Q_{r,10} &= r_0(0.67)q_{10} + r_1(0.67)q_9 + r_2(0.67)q_8 + r_3(0.67)q_7 + \ldots + r_{23}(0.67)q_{11} \\
&= (0.49)(0.67)(6820) + (0.17)(0.67)(6820) \\
&+ (0.09)(0.67)(6820) + (0.05)(0.67)(3410) + (0.03)(0.67)(0) \\
&+ (0.02)(0.67)(0) + (0.02)(0.67)(0) + (0.01)(0.67)(0) \\
&+ (0.01)(0.67)(0) + (0.01)(0.67)(0) + (0.01)(0.67)(0) \\
&+ (0.01)(0.67)(0) + (0.01)(0.67)(0) + (0.01)(0.67)(0) \\
&+ (0.01)(0.67)(0) + (0.01)(0.67)(0) + (0.01)(0.67)(6820) \\
&+ (0.01)(0.67)(6820) + (0.01)(0.67)(6820) \\
&+ (0.01)(0.67)(6820) + (0.00)(0.67)(6820) \\
&+ (0.00)(0.67)(6820) + (0.00)(0.67)(6820) \\
&+ (0.00)(0.67)(6820) \\
&= 3724 \text{ Btu/h}
\end{aligned}
$$

The total lighting cooling load at the designated hour is thus

$$Q_{light} = Q_{c,10} + Q_{r,10} = 2251 + 3724 = 5975 \text{ Btu/h}$$

COMPARISON WITH PREVIOUS METHODS

Some users of this chapter during the past eight editions may have been disappointed that it did not contain (nor was it accompanied by) fully developed multiple-room, multiple-system calculation computer programs for immediate use by HVAC designers. Nor have the several attempts to create and present hand-calculation procedures been acceptable in practice, as they were simply incapable of dealing realistically with contemporary demands. As ASHRAE evolves in the 21st century, the issue of practical load calculation procedures clearly demands computer operation.

Who should provide the computer applications for this purpose? While it could be argued that ASHRAE itself is the best source of experience and talent to generate such tools, the nature of ASHRAE as a technical society prevents this type of activity. Instead, ASHRAE has provided technically reliable mechanisms for both the heat balance and the radiant time series procedures to be accurately implemented, as the most scientifically correct methodologies available, anticipating that various qualified private enterprises will take these frameworks and complete their development into practical applications for HVAC designers.

The user may question what benefits may be expected now that the TFM, TETD/TA, and CLTD/CLF procedures presented in earlier chapter versions have been superseded (*not* invalidated or discredited). The primary benefit will be improved accuracy, with reduced dependency upon purely subjective input (such as determining a proper time-averaging period for TETD/TA, or ascertaining appropriate safety factors to add to the "rounded off" TFM results). As a generic example, the space sensible cooling load for the traditional little ASHRAE store building (used for example purposes since the 1940s) was calculated by means of the heat balance procedure and independently calculated by application of the radiant time series procedure, with each set of results plotted as one of the load profile curves of Figure 13. Also plotted on this chart are the corresponding curves produced by

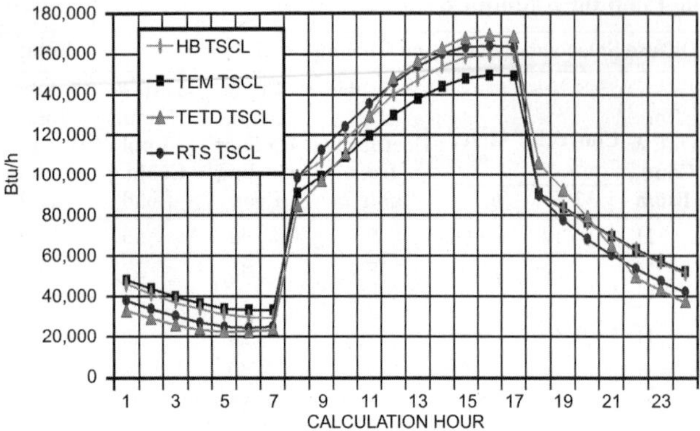

Fig. 13 Cooling Load Profile Comparison

the TFM and TETD/TA methodologies in the 1997 edition of this chapter. The user may draw his or her own conclusions from this chart.

HEATING LOAD PRINCIPLES

Techniques for estimating design heating load for commercial, institutional, and industrial applications are essentially the same as for those estimating design cooling loads for such uses, except that (1) temperatures outside the conditioned spaces are generally lower than the space temperatures maintained; (2) credit for solar heat gains or for internal heat gains is not included; and (3) the thermal storage effect of building structure or content is ignored. Heat losses (negative heat gains) are thus considered to be instantaneous, heat transfer essentially conductive, and latent heat treated only as a function of replacing space humidity lost to the exterior environment.

This simplified approach is justified because it evaluates "worst case" conditions that can reasonably occur during a heating season. The worst case is the load that must be met under design interior and exterior conditions, including infiltration and/or ventilation, but with no solar effect (at night or on cloudy winter days) and before the periodic presence of people, lights, and appliances has an offsetting effect.

Safety Factors and Load Allowances. Before mechanical cooling became a usual procedure, buildings included much less insulation, large operable windows, and generally more infiltration-prone assemblies than the energy-efficient and much tighter buildings typical of today. Allowances of 10 to 20% of the net calculated heating load for piping losses to unheated spaces and 10 to 20% more for a warm-up load were common practice, along with occasional other safety factors reflecting the experience and/or concern of the individual designer. Such measures are rarely used today, with the uncompensated net heating load normally considered as having an adequate margin for error. Armstrong et al. (1992a, 1992b) provide a design method to deal with warm-up and cool-down load.

Cooling Needs During Noncooling Months. Perimeter spaces exposed to high solar heat gain often justify mechanical cooling during sunlit portions of traditional heating months, as do completely interior spaces with significant internal heat gain. Such spaces can also have significant heating loads during nonsunlit hours or after periods of nonoccupancy when adjacent spaces have cooled below interior design temperatures. The loads involved can be estimated conventionally to offset or to compensate for them and prevent overheating, but they have no direct relationship to design heating loads for the spaces in question.

Other Considerations. Calculation of design heating load estimates for this general category of applications has essentially become a subset of the more involved and complex estimation of cooling loads for such spaces. Chapter 31 discusses using the heating load estimate to predict or analyze energy consumption over time. Special provisions to deal with special problems are covered in the 1999 *ASHRAE Handbook—Applications* and the 2000 *ASHRAE Handbook—Systems and Equipment.*

REFERENCES

Alereza, T. and J.P. Breen, III. 1984. Estimates of recommended heat gain due to commercial appliances and equipment. *ASHRAE Transactions* 90(2A):25-58.

ASHRAE. 1999. Ventilation for acceptable indoor air quality. ASHRAE *Standard* 62-1999.

Armstrong, P.R., C.E. Hancock III, and J.E. Seem. 1992a. Commercial building temperature recovery—Part I: Design procedure based on a step response model. *ASHRAE Transactions* 98(1):381-96.

Armstrong, P.R., C.E. Hancock III, and J.E. Seem. 1992b. Commercial building temperature recovery—Part II: Experiments to verify the step response model. *ASHRAE Transactions* 98(1):397-410.

Bliss, R.J.V. 1961. Atmospheric radiation near the surface of the ground. Solar Energy 5(3):103.

Consolazio, W. and L.J. Pecora. 1947. Minimal replenishment air required for living spaces. ASHVE *Standard* 53:127.

Colliver, D.G., H. Zhang, R.S. Gates, and K.T. Priddy. 1995. Determination of the 1%, 2.5%, and 5% occurences of extreme dew-point temperatures and mean coincident dry-bulb temperatures. *ASHRAE Transactions* 101(2):265-86.

Colliver, D.G., R.S. Gates, H. Zhang, and K.T. Priddy. 1998. Sequences of extreme temperature and humidity for design calculations. *ASHRAE Transactions* 104(1A):133-44.

Colliver, D.G., R.S. Gates, T.F. Burke, and H. Zhang. 2000. Development of the design climatic data for the 1997 ASHRAE Handbook—Fundamentals. *ASHRAE Transactions* 106(1):3.14.

Fisher, D.E. and C.O. Pedersen. 1997. Convective heat transfer in building energy and thermal load calculations. *ASHRAE Transactions* 103(2): 137-48.

Fisher, D.R. 1998. New recommended heat gains for commercial cooking equipment. *ASHRAE Transactions* 104(2):953-60.

Gordon, E.B., D.J. Horton, and F.A. Parvin. 1994. Development and application of a standard test method for the performance of exhaust hoods with commercial cooking appliances. *ASHRAE Transactions* 100(2): 988-99.

Hittle, D.C. and C.O. Pedersen. 1981. Calculating building heating loads using the frequency of multi-layered slabs. *ASHRAE Transactions* 87(2):545-68.

Hittle, D.C. 1999. The effect of beam solar radiation on peak cooling loads. *ASHRAE Transactions* 105(2):510-13.

Hosni, M.H., B.W. Jones, and J.M. Snipes. 1998. Total heat gain and the split between radiant and convective heat gain from office and laboratory equipment in buildings. *ASHRAE Transactions* 104(1A):356-65.

Hosni, M.H., B.W. Jones, and H. Xu. 1999. Experimental results for heat gain and radiant/convective split from equipment in buildings. *ASHRAE Transactions* 105(2):527-39.

Jones, B.W., M.H. Hosni, and J.M. Snipes. 1998. Measurements of radiant heat gain from office equipment using a scanning radiometer. *ASHRAE Transactions* 104(1B):1775-83.

Kerrisk, J.F., N.M. Schnurr, J.E. Moore, and B.D. Hunn. 1981. The custom weighting-factor method for thermal load calculations in the DOE-2 computer program. *ASHRAE Transactions* 87(2):569-84.

Komor, P. 1997. Space cooling demands from office plug loads. *ASHRAE Journal* 39(12):41-44.

Kusuda, T. 1967. NBSLD, The computer program for heating and cooling loads for buildings. Bss 69 and NBSIR 74-574. National Bureau of Standards.

LBL. 1994. Window 4.1: Program description. Lawrence Berkeley Laboratory.

Liesen, R.J. and C.O. Pedersen. 1997. An evaluation of inside surface heat balance models for cooling load calculations. *ASHRAE Transactions* 103(2):485-502.

McClellan, T.M. and C.O. Pedersen. 1997. The sensitivity of cooling load calculations to outside heat balance models. *ASHRAE Transactions* 103(2):469-84.

McQuiston, F.C. and J.D. Spitler. 1992. *Cooling and heating load calculation manual*. 2nd ed. ASHRAE.

Marn, W.L. 1962. Commercial gas kitchen ventilation studies. *Research Bulletin* No. 90 (March). Gas Association Laboratories, Cleveland, OH.

Mitalas, G.P. 1972. Transfer function method of calculating cooling loads, heat extraction rate, and space temperature. *ASHRAE Transactions* 14(12):52.

Mitalas, G.P. and K. Kimura. 1971. A calorimeter to determine cooling load caused by lights. *ASHRAE Transactions* 77(2)65.

Nevins, R.G., H.E. Straub, and H.D. Ball. 1971. Thermal analysis of heat removal troffers. *ASHRAE Transactions* 77(2):58-72.

NFPA. 1999. Standard for health care facilities. *Standard* 99-99. National Fire Protection Association, Quincy, MA.

Pedersen, C.O., D.E. Fisher, and R.J. Liesen. 1997. A heat balance based cooling load calculation procedure. *ASHRAE Transactions* 103(2):459-68.

Pedersen, C.O., D.E. Fisher, J.D. Spitler, and R.J. Liesen. 1998. *Cooling and heating load calculation principles*. ASHRAE, Atlanta.

Rees, S.J., J.D. Spitler, M.G. Davies, and P. Haves. 2000. Qualitative comparison of North American and UK cooling load calculation models. *International Journal of Heating, Ventilating, Air-Conditioning and Refrigerating Research* 6(1):75-99.

Smith, V.A., R.T. Swierczyna, C.N. Claar. 1995. Application and enhancement of the standard test method for the performance of commercial kitchen ventilation systems. *ASHRAE Transactions* 101(2).

Sowell, E.F. 1988a. Cross-check and modification of the DOE-2 program for calculation of zone weighting factors. *ASHRAE Transactions* 94(2):737-53.

Sowell, E.F. 1988b. Load calculations for 200,640 zones. *ASHRAE Transactions* 94(2):716-36.

Spitler, J.D. and D.E. Fisher. 1999a. Development of periodic response factors for use with the radiant time series method. *ASHRAE Transactions* 105(2).

Spitler, J.D. and D.E. Fisher. 1999b. On the relationship between the radiant time series and transfer function methods for design cooling load calculations. *International Journal of Heating, Ventilating, Air-Conditioning and Refrigerating Research* 5(2).

Spitler, J.D., D.E. Fisher, and C.O. Pedersen. 1997. The radiant time series cooling load calculation procedure. *ASHRAE Transactions* 103(2).

Spitler, J.D., S.J. Rees, and P. Haves. 1998. Quantitive comparison of North American and UK cooling load calculation procedures—Part 1: methodology, Part II: results. *ASHRAE Transactions* 104(2):36-46, 47-61.

Sun, T.-Y. 1968. Shadow area equations for window overhangs and side-fins and their application in computer calculation. *ASHRAE Transactions* 74(1):I-1.1 to I-1.9.

Talbert, S.G., L.J. Canigan, and J.A. Eibling. 1973. An experimental study of ventilation requirements of commercial electric kitchens. *ASHRAE Transactions* 79(1):34.

Todorovic, B., L. Marjanovic, and D. Kovacevic. 1993. Comparison of different calculation procedures for cooling load from solar radiation through a window. *ASHRAE Transactions* 99(2):559-64.

Walton, G. 1983. Thermal analysis research program reference manual. National Bureau of Standards.

Wilkins, C.K. 1998. Electronic equipment heat gains in buildings. *ASHRAE Transactions* 104(1B):1784-89.

Wilkins, C.K., R. Kosonen, and T. Laine. 1991. An analysis of office equipment load factors. *ASHRAE Journal* 33(9):38-44.

Wilkins, C.K. and N. McGaffin. 1994. Measuring computer equipment loads in office buildings. *ASHRAE Journal* 36(8):21-24.

Wilkins, CK. and M.R. Cook. 1999. Cooling loads in laboratories. *ASHRAE Transactions* 105(1):744-49.

Wilkins, C.K. and M.H. Hosni. 2000. Heat gain from office equipment. *ASHRAE Journal* 42(6):33-44.

BIBLIOGRAPHY

Alford, J.S., J.E. Ryan, and F.O. Urban. 1939. Effect of heat storage and variation in outdoor temperature and solar intensity on heat transfer through walls. ASHVE Transactions 45:387.

American Gas Association. 1948. A comparison of gas and electric use for commercial cooking. Cleveland, OH.

American Gas Association. 1950. Gas and electric consumption in two college cafeterias. Cleveland, OH.

ASHRAE. 1975. Procedure for determining heating and cooling loads for computerized energy calculations, algorithms for building heat transfer subroutines.

ASHRAE. 1979. Cooling and heating load calculation manual.

BLAST Support Office. 1991. *BLAST User Reference*. Urbana-Champaign. University of Illinois.

Brisken, W.R. and G.E. Reque. 1956. Thermal circuit and analog computer methods, thermal response. *ASHAE Transactions* 62:391.

Buchberg, H. 1955. Electric analog prediction of the thermal behavior of an inhabitable enclosure. *ASHAE Transactions* 61:339-386.

Buchberg, H. 1958. Cooling load from thermal network solutions ASHAE *Standard* 64:111.

Buffington, D.E. 1975. Heat gain by conduction through exterior walls and roofs—transmission matrix method. *ASHRAE Transactions* 81(2):89.

Burch, D.M., B.A. Peavy, and F.J. Powell. 1974. Experimental validation of the NBS load and indoor temperature prediction model. *ASHRAE Transactions* 80(2):291.

Burch, D.M., J.E. Seem, G.N. Walton, and B.A. Licitra. 1992. Dynamic evaluation of thermal bridges in a typical office building. *ASHRAE Transactions* 98:291-304.

Butler, R. 1984. The computation of heat flows through multi-layer slabs. *Building and Environment* 19(3):197-206.

Ceylan, H.T., and G.E. Myers. 1985. Application of response-coefficient method to heat-conduction transients. *ASHRAE Transactions* 91:30-39.

Chiles, D.C. and E.F. Sowell. 1984. A counter-intuitive effect of mass on zone cooling load response. *ASHRAE Transactions* 91(2A):201-208.

Chorpening, B.T. 1997. The sensitivity of cooling load calculations to window solar transmission models. *ASHRAE Transactions* 103(1).

Clarke, J.A. 1985. *Energy simulation in building design*. Adam Hilger Ltd., Boston.

Davies, M.G. 1996. A time-domain estimation of wall conduction transfer function coefficients. *ASHRAE Transactions* 102(1).

DeAlbuquerque, A.J. 1972. Equipment loads in laboratories. *ASHRAE Journal* 14(10):59.

Falconer, D.R., E.F. Sowell, J.D. Spitler, and B.B. Todorovic. 1993. Electronic tables for the ASHRAE load calculation manual. *ASHRAE Transactions* 99(1):193-200.

Harris, S. M. and F.C. McQuiston. 1988. A study to categorize walls and roofs on the basis of thermal response. *ASHRAE Transactions* 94(2):688-714.

Headrick, J.B. and D.P. Jordan. 1969. Analog computer simulation of heat gain through a flat composite roof section. *ASHRAE Transactions* 75(2):21.

Hittle, D.C. 1981. Calculating building heating and cooling loads using the frequency response of multilayered slabs, Ph.D. thesis, Department of Mechanical and Industrial Engineering, University of Illinois, Urbana, IL.

Hittle, D.C. and R. Bishop. 1983. An improved root-finding procedure for use in calculating transient heat flow through multilayered slabs. *International Journal of Heat and Mass Transfer* 26:1685-93.

Houghton, D.G., C. Gutherlet, and A.J. Wahl. 1935. ASHVE *Research Report* No. 1001—Cooling requirements of single rooms in a modern office building. *ASHVE Transactions* 41:53.

Kimura and Stephenson. 1968. Theoretical study of cooling loads caused by lights. *ASHRAE Transactions* 74(2):189-97.

Kusuda, T. 1969. Thermal response factors for multilayer structures of various heat conduction systems. *ASHRAE Transactions* 75(1):246

Leopold, C.S. 1947. The mechanism of heat transfer, panel cooling, heat storage. *Refrigerating Engineering* 7:33.

Leopold, C.S. 1948. Hydraulic analogue for the solution of problems of thermal storage, radiation, convection, and conduction. *ASHVE Transactions* 54:3-9.

Livermore, J.N. 1943. Study of actual vs predicted cooling load on an air conditioning system. *ASHVE Transactions* 49:287.

Mackey, C.O. and N.R. Gay, 1949. Heat gains are not cooling loads. *ASHVE Transactions* 55:413.

Mackey, C.O. and N.R. Gay. 1952. Cooling load from sunlit glass. *ASHVE Transactions* 58:321.

Mackey, C.O. and N.R. Gay. 1954. Cooling load from sunlit glass and wall. *ASHVE Transactions* 60:469.

Mackey, C.O. and L.T. Wright, Jr. 1944. Periodic heat flow—homogeneous walls or roofs. *ASHVE Transactions* 50:293.

Mackey, C.O. and L.T. Wright, Jr. 1946. Periodic heat flow—composite walls or roofs. *ASHVE Transactions* 52:283.

Mast, W.D. 1972. Comparison between measured and calculated hour heating and cooling loads for an instrumented building. *ASHRAE Symposium Bulletin* No. 72-2.

McBridge, M.F., C.D. Jones, W.D. Mast, and C.F. Sepsey. 1975. Field validation test of the hourly load program developed from the ASHRAE algorithms. *ASHRAE Transactions* 1(1):291.

Mitalas, G.P. 1968. Calculations of transient heat flow through walls and roofs. *ASHRAE Transactions* 74(2):182-88.

Mitalas, G.P. 1969. An experimental check on the weighting factor method of calculating room cooling load. *ASHRAE Transactions* 75(2):22.

Mitalas, G.P. 1973. Calculating cooling load caused by lights. *ASHRAE Transactions* 75(6):7.

Mitalas, G.P. and D.G. Stephenson. 1967. Room thermal response factors. *ASHRAE Transactions* 73(2): III.2.1.

Mitalas, G.P. and J.G. Arsenault. 1970. Fortran IV program to calculate Z-transfer functions for the calculation of transient heat transfer through walls and roofs. *Use of Computers For Environmental Engineering Related to Buildings*, 633-68. National Bureau of Standards, Gaithersburg, MD.

Mitalas, G.P. 1978. Comments on the Z-transfer function method for calculating heat transfer in buildings. *ASHRAE Transactions*, 84(1):667-74.

Nottage, H.B. and G.V. Parmelee. 1954. Circuit analysis applied to load estimating. *ASHVE Transactions* 60:59.

Nottage, H.B. and G.V. Parmelee. 1955. Circuit analysis applied to load estimating. *ASHAE Transactions* 61(2):125.

Ouyang, K. and F. Haghighat. 1991. A procedure for calculating thermal response factors of multi-layer walls-state space method. *Building and Environment* 26(2):173-77.

Parmelee, G.V., P. Vance, and A.N. Cherny. 1957. Analysis of an air conditioning thermal circuit by an electronic differential analyzer. *ASHAE Transactions* 63:129.

Paschkis, V. 1942. Periodic heat flow in building walls determined by electric analog method. *ASHVE Transactions* 48:75.

Peavy, B.A. 1978. Determination and verification of thermal response factors for thermal conduction applications.

Peavy, B.A. 1978. A note on response factors and conduction transfer functions. *ASHRAE Transactions* 84(1):688-90.

Peavy, B.A., F.J. Powell, and D.M. Burch. 1975. Dynamic thermal performance of an experimental masonry building. *NBS Building Science Series* 45 (July).

Romine, T.B., Jr. 1992. Cooling load calculation, Art or science? *ASHRAE Journal*, 34(1), p. 14.

Rudoy, W. 1979. Don't turn the tables. *ASHRAE Journal* 21(7):62.

Rudoy, W. and F. Duran. 1975. Development of an improved cooling load calculation method. *ASHRAE Transactions* 81(2):19-69.

Seem, J.E., S.A. Klein, W.A. Beckman, and J.W. Mitchell. 1989. Transfer functions for efficient calculation of multidimensional transient heat transfer. *Journal of Heat Transfer* 111:5-12.

Sowell, E.F. and D.C. Chiles. 1984a. Characterization of zone dynamic response for CLF/CLTD tables. *ASHRAE Transactions* 91(2A):162-78.

Sowell, E.F. and D.C. Chiles. 1984b. Zone descriptions and response characterization for CLF/CLTD calculations. *ASHRAE Transactions* 91(2A): 179-200.

Spitler, J.D. 1996. Annotated guide to building load calculation models and algorithms. ASHRAE, Atlanta.

Spitler, J.D., F.C. McQuiston, and K.L. Lindsey. 1993. The CLTD/SCL/CLF cooling load calculation method. *ASHRAE Transactions* 99(1).

Spitler, J.D. and F.C. McQuiston. 1993. Development of a revised cooling and heating calculation manual. *ASHRAE Transactions* 99(1).

Stephenson, D.G. 1962. Method of determining non-steady-state heat flow through walls and roofs at buildings. The *Journal of the Institution of Heating and Ventilating Engineers* 30:5.

Stephenson, D.G. and G.P. Mitalas. 1967. Cooling load calculation by thermal response factor method. *ASHRAE Transactions* 73(2):III.1.1.

Stephenson, D.G. and G.P. Mitalas. 1971. Calculation of heat transfer functions for multi-layer slabs. *ASHRAE Transactions* 77(2):117-26.

Stewart, J.P. 1948. Solar heat gain through walls and roofs for cooling load calculations. *ASHVE Transactions* 54:361.

Sun, T.-Y. 1968. Computer evaluation of the shadow area on a window cast by the adjacent building, *ASHRAE Journal*, September 1968.

Todorovic, B. 1987. The effect of the changing shade line on the cooling load calculations. ASHRAE videotape "Practical applications for cooling load calculations."

Todorovic B. 1982. Cooling load from solar radiation through partially shaded windows, taking heat storage effect into account. *ASHRAE Transactions* 88(2):924-937.

Todorovic, B. 1984. Distribution of solar energy following its transmittal through window panes. *ASHRAE Transactions* 90(1B):806-15.

Todorovic, B. 1989. Heat storage in building structure and its effect on cooling load; Heat and mass transfer in building materials and structure. Hemisphere publishing, New York, 603-14.

Todorovic, B. and D. Curcija. 1984. Calculative procedure for estimating cooling loads influenced by window shading, using negative cooling load method. *ASHRAE Transactions* 2:662.

Vild, D.J. 1964. Solar heat gain factors and shading coefficients. *ASHRAE Journal* 6(10):47.

Wilkins, C.K. and N. McGaffin. 1994. Measuring computer equipment loads in office buildings. *ASHRAE Journal* 36(8):21-24.

York, D.A. and C.C. Cappiello. 1981. DOE-2 engineers manual (Version 2.1A), Lawrence Berkeley Laboratory and Los Alamos National Laboratory.

CHAPTER 30

FENESTRATION

FENESTRATION is an architectural term that refers to the arrangement, proportion, and design of window, skylight, and door systems within a building. Fenestration components include glazing material, either glass or plastic; framing, mullions, muntins, dividers, and opaque door slabs; external shading devices; internal shading devices; and integral (between-glass) shading systems. For our purposes, **fenestration** and **fenestration systems** will refer to the basic assemblies and components of exterior window, skylight, and door systems within the building envelope. Fenestration can serve as a physical and/or visual connection to the outdoors, as well as a means to admit solar radiation. The solar radiation provides natural lighting, referred to as **daylighting**, and heat gain to a space. Fenestration can be fixed or operable, and operable units can allow natural ventilation to a space and egress in low-rise buildings.

Fenestration affects building energy use through four basic mechanisms—thermal heat transfer, solar heat gain, air leakage, and daylighting. The energy impacts of fenestration can be minimized by (1) using daylight to offset lighting requirements, (2) using appropriate glazings and shading strategies to control solar heat gain to supplement heating through passive solar gain and minimize cooling requirements, (3) using appropriate glazing to minimize conductive heat loss, and (4) specifying low air leakage fenestration products. In addition, natural ventilation strategies can reduce energy use for cooling and fresh air requirements.

Today designers, builders, energy codes, and energy-efficiency incentive programs [such as Energy Star (www.energystar.gov) and the LEED Green Building Program (www.usgbc.org)] are asking more and more from fenestration systems. Window, skylight, and door manufacturers are responding with new and improved products to meet those demands. With the advent of computer simulation software, designing to improve thermal performance of fenestration products has become much easier. Through participation in rating and certification programs (such as those of the National Fenestration Rating Council) that require the use of this software, fenestration manufacturers can take credit for these improvements through certified ratings that are credible to designers, builders, and code officials. A designer should consider architectural requirements, thermal performance, economic criteria, and human comfort when selecting fenestration. Typically, a wide range of fenestration products are available that meet the specifications for a project. Refining the specifications to improve the energy performance and enhance a living or work space can result in lower energy costs, increased productivity, and improved thermal and visual comfort. Carmody et al. (1996) and CEA (1995) provide guidance for carrying out these requirements.

FENESTRATION COMPONENTS

Fenestration consists of glazing, framing, and in some cases shading devices and insect screens.

Glazing

The glazing unit may have single glazing or multiple glazing. The most common glazing material is glass, although plastic is also used. The glass or plastic may be clear, tinted, coated, laminated, patterned, or obscured. **Clear** glass transmits more than 80% of the incident solar radiation and more than 75% of the visible light. **Tinted** glass is available in many colors, all of which differ in the amount of solar radiation and visible light they transmit and absorb. **Coatings** on glass affect the transmission of solar radiation, and visible light may affect the absorptance of room temperature radiation. Some coatings are highly reflective (such as mirrors), while others are designed to have a very low reflectance. Some coatings result in a visible light transmittance that is as much as 1.4 times higher than the solar heat gain coefficient (desirable for good daylighting while minimizing cooling loads). **Laminated** glass is made of two panes of glass adhered together. The interlayer between the two panes of glass is typically plastic and may be clear, tinted, or coated. **Patterned** glass is a durable ceramic frit applied to a glass surface in a decorative pattern. **Obscured** glass is translucent and is typically used in privacy applications.

Insulating Glazing Units

Insulating glazing units (IGUs) are hermetically sealed, multiple-pane assemblies consisting of two or more glazing layers held and bonded at their perimeter by a spacer bar typically containing a desiccant material. The desiccated spacer is surrounded on at least two sides by a sealant that adheres the glass to the spacer. Figure 1 shows the construction of a typical IGU.

The preparation of this chapter is assigned to TC 4.5, Fenestration.

Glazing. Common types of glass used in IGUs are clear, tinted, and low emissivity (low-e). Due to its energy efficiency, daylighting, and comfort benefits, low-e coated glass is now used in more than 30% of all the fenestration products installed in the United States. Tinted and reflective glazing can also be used to reduce solar heat gain through fenestration products. Low-e coatings can also be applied to thin plastic films for use in IGUs. There are two types of low-e coating: **high-solar-gain** and **low-solar-gain**. The first of these primarily reduces heat conduction through the glazing system and is intended for cold climates. The second, for hot climates, reduces solar heat gain by blocking admission of the infrared portion of the solar spectrum. There are two ways of achieving low-solar-gain low-e performance. The first is with a special multilayer solar infrared reflecting coating. The second is with a solar infrared absorbing outer glass. To protect the inner glazing and the building interior from the absorbed heat from this outer glass, a cold-climate-type low-e coating is also used to reduce conduction of heat from the outer pane to the inner one. In addition, argon and krypton gas are used in lieu of air in the gap between the panes in combination with low-e glazing to further reduce energy transfer. Some manufacturers construct IGUs with one or more suspended, low-e coated plastic films between the glass panes and with a spacer that has better insulating properties and a dual sealant that improves the seal around the gas spaces.

Spacer. The spacer serves to separate the panes of glass and to provide the surface for primary and secondary sealant adhesion. Several types of spacers are used in IGU construction today. Each type provides different heat transfer properties depending on the spacer material and geometry.

Heat transfer at the edge of the IGU is greater than at the center of the IGU due to greater heat flow through the spacer system. Spacer systems have been developed to minimize the heat flow at the edge of the IGU. These spacer systems are referred to as **warm edge spacers**. In IGU construction, warm edge spacer designs reduce edge heat transfer by substituting materials that have lower thermal conductivity than aluminum (e.g., stainless steel, galvanized steel, tin plated steel, polymers, or foamed silicone). Traditional spacers are often made of aluminum.

Fusing or bending the corners of the spacer minimizes moisture and hydrocarbon vapor transmission into the air space through the corners. Desiccants such as molecular sieve or silica gel are also used to absorb moisture that was initially trapped in the IGU during assembly or gradually diffuses through the seals after construction.

Sealant(s). Several different sealant configurations are being used successfully in modern IGU construction. In all sealant configurations, the primary seal minimizes moisture and hydrocarbon transmission. In dual-seal construction, the secondary seal provides structural integrity between the lites of the IGU. A secondary seal ensures long-term adhesion and greater resistance to solvents, oils, and short-term water immersion. In typical dual-seal construction, the primary seal is made of compressed polyisobutylene (PIB), and the secondary seal is made of silicone, polysulfide, or polyurethane. Single-seal construction depends on a single sealant to provide adhesion of the glass to the spacer as well as minimizing moisture and hydrocarbon transmission. Single-seal construction is generally more cost-efficient than dual-seal systems. A third type of sealant used in IGU construction takes advantage of advanced cross-linking polymers that provide both low moisture transmission and equivalent structural properties to dual-seal systems. These sealants are typically referred to as dual seal equivalent (DSE) materials.

Desiccants. Typical desiccants include molecular sieve, silica gel, or a matrix of both materials. Desiccants are used to absorb moisture that was initially trapped in the IGU during assembly or that gradually diffused through the seals after construction.

Gas Fill. The hermetically sealed space between glass panes in an IGU is most often filled with air. In some cases, argon and krypton gas are used in lieu of air in the space between the panes to further reduce the energy transfer.

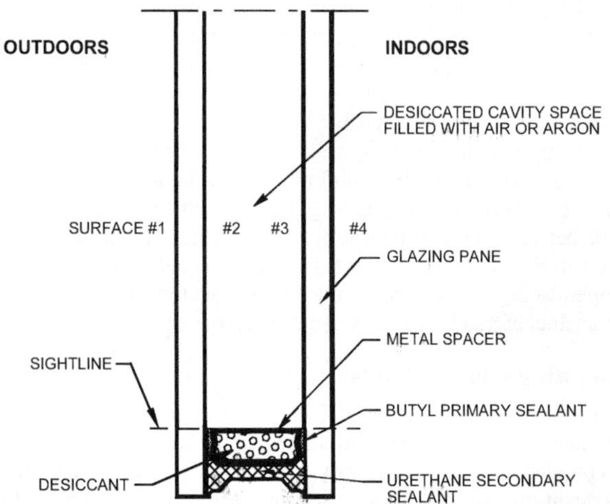

Fig. 1 Insulating Glazing Unit (IGU) Construction Detail

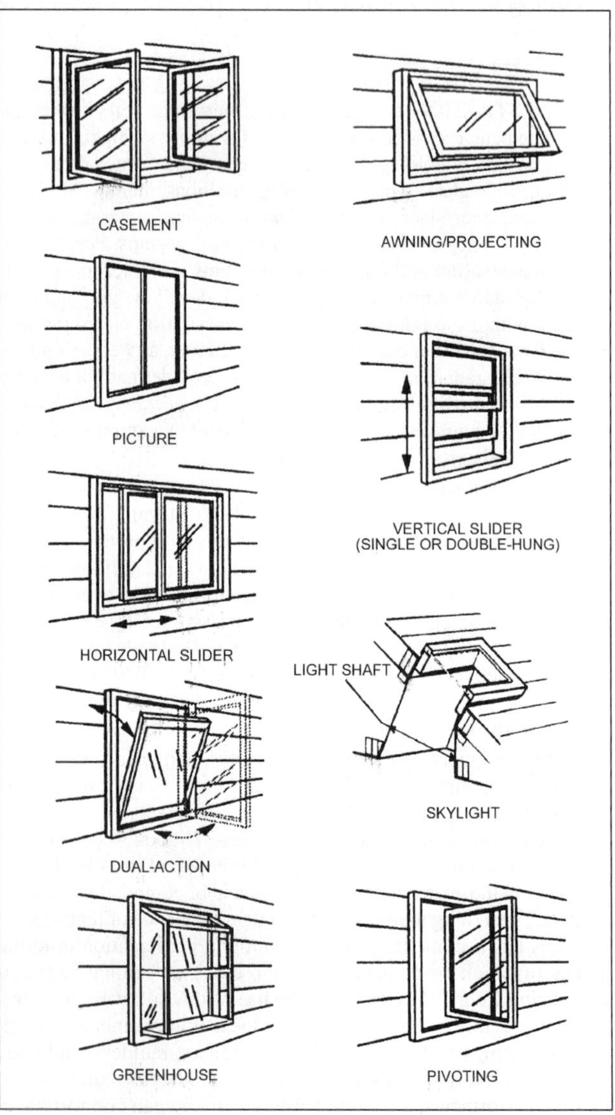

Fig. 2 Types of Residential Windows

Framing

The three main categories of window framing materials are wood, metal, and polymers. **Wood** has good structural integrity and insulating value but low resistance to weather, moisture, warpage, and organic degradation (from mold and insects). **Metal** is durable and has excellent structural characteristics, but it has very poor thermal performance. The metal of choice in windows is almost exclusively aluminum, due to ease of manufacture, low cost, and light weight, but aluminum has a thermal conductivity roughly 1000 times that of wood or polymers. The poor thermal performance of metal-frame windows can be improved with a thermal break (a nonmetal component that separates the metal frame exposed to the outside from the surfaces exposed to the inside). **Polymer** frames are made of extruded vinyl or poltruded fiberglass (glass-reinforced polyester). Their thermal and structural performance is similar to that of wood, although vinyl frames for large windows must be reinforced.

Manufacturers sometimes combine these materials as clad units (e.g., vinyl-clad aluminum, aluminum-clad wood, vinyl-clad wood) to increase durability, improve thermal performance, or improve aesthetics. In addition, curtain wall systems for commercial buildings may be structurally glazed, and the exterior "framing" is simply rubber gaskets or silicone.

Residential windows can be categorized by operator type, as shown in Figure 2. Traditionally there are several basic window types: casements, fixed picture windows, horizontal and vertical sliders, pivoting, awning, or projecting windows, dual acting windows, and special applications such as skylights and greenhouse or garden window inserts. The glazing system can be mounted either directly in the frame (a direct-glazed or direct-set window, which is not operable) or in a sash that moves in the frame (for an operating window). In operable windows, a weather-sealing system between the frame and sash reduces air and water leakage.

Shading

Shading devices are available in a wide range of products that differ greatly in their appearance and energy performance. Shading devices include interior and exterior blinds, integral blinds, interior and exterior screens, shutters, draperies, and roller shades. Shading devices on the exterior of the glazing reduce solar heat gain more effectively than interior devices. However, interior devices are easier to operate and adjust. Some products help insulate the indoors from the outdoors, while others redirect incoming solar radiation to minimize visual and thermal discomfort. Overhangs and vegetation can be effective shading too.

DETERMINING FENESTRATION ENERGY FLOW

Energy flows through fenestration via (1) conductive and convective heat transfer caused by the temperature difference between outdoor and indoor air, (2) net long-wave (above 2500 nm) radiative exchange between the fenestration and its surrounding and between glazing layers, and (3) short-wave (below 2500 nm) solar radiation incident on the fenestration product, either directly from the sun or reflected from the ground or adjacent objects. Simplified calculations are based on the observation that the temperatures of the sky, ground, and surrounding objects (and hence their radiant emission) correlate with the exterior air temperature. The radiative interchanges are then approximated by assuming that all the radiating surfaces (including the sky) are at the same temperature as the outdoor air. With this assumption, the basic equation for the instantaneous energy flow Q through a fenestration is

$$Q = UA_{pf}(t_{out} - t_{in}) + (\text{SHGC})A_{pf}E_t \tag{1}$$

where

Q = instantaneous energy flow, Btu/h
U = overall coefficient of heat transfer (U-factor), Btu/h·ft²·°F
t_{in} = interior air temperature, °F
t_{out} = exterior air temperature, °F
A_{pf} = total projected area of fenestration, ft²
SHGC = solar heat gain coefficient, nondimensional
E_t = incident total irradiance, Btu/h·ft²

The quantities U and SHGC are instantaneous performance indices. The principal justification for Equation (1) is its simplicity, achieved by collecting all the linked radiative, conductive, and convective energy transfer processes into U and SHGC. These quantities vary because (1) convective heat transfer rates vary as fractional powers of temperature differences or free-stream speeds, (2) variations in temperature due to the weather or climate are small on the absolute temperature scale (°R) that controls radiative heat transfer rates, (3) fenestration systems always involve at least two thermal resistances in series, and (4) solar heat gain coefficients depend on solar incident angle and spectral distribution.

In the discussion of this chapter, Q is divided into two parts:

$$Q = Q_{th} + Q_{sol} \tag{2}$$

where

Q_{th} = instantaneous energy flow due to indoor-outdoor temperature difference (thermal energy flow)
Q_{sol} = instantaneous energy flow due to solar radiation (solar energy flow)

The section on U-Factor (Thermal Transmittance) deals with Q_{th}, while the section on Solar Heat Gain and Visible Transmittance discusses Q_{sol}. In the latter section, both the effects of direct solar radiation and those of solar radiation scattered by the sky or ground are included.

Equation (1) presents a fenestration as it might appear on a building plan: a featureless, planar object filling an opening in the building envelope. Real fenestrations, however, are composite three-dimensional objects that may consist of frames, sashes, mullions, and other structural elements, as well as glazing systems. The latter in turn may contain structural spacers as well as glazing layers. There may in addition be shading elements, either as separate attachments or integrated into the glazing system.

The heat transfer through such an assembly of elements is calculated by dividing the fenestration area into parts, each of which has an energy flow that is more simply calculated than the total:

$$Q = \sum_{v} A_v q_v \tag{3}$$

where

q_v = energy flux (energy flow per unit area) of the vth part
A_v = area of the vth part

This subdivision is applied to each of the terms in Equation (2) separately; for example, the thermal heat transfer through glazings and frames is frequently different, so that it is useful to make the following separation:

$$Q_{th} = A_f q_f + A_g q_g \tag{4}$$

where the subscript f refers to the frame, and g refers to the glazing (both for thermal energy flow). Similarly, solar radiation will have a different effect on the frame and the glazed area of a fenestration (since the former is generally opaque), so that

$$Q_{sol} = A_{op} q_{op} + A_s q_s \tag{5}$$

where the subscript *op* refers to the (opaque) frame (for solar energy flow), and *s* refers to the (solar-transmitting) glazing. This division into frame and glazing areas can be and usually is different for the solar and thermal energy flows. Subdivisions of this sort, when Equation (3) is compared with Equation (1), effectively make the overall U-factor and solar heat gain coefficient area-averaged quantities. This area averaging is described explicitly in the appropriate sections below. Note that in more complicated fenestrations, where the glazing portion may contain opaque shading elements, the opaque portion is that part that can never under any conditions admit solar radiation in any form other than heat. A window with a closed, perfectly opaque blind would not be considered an opaque element because sometimes the blind may be open. A section of curtain wall consisting of wall or frame elements with an exterior cover of glass (for uniform appearance) would be an opaque element in spite of its transparent covering.

A second type of subdivision occurs when, for a given part of the fenestration system, energy flow is driven by physical processes that are more complicated than those assumed in Equation (1). For example, the heat transfer through a glazing consists of a "contact" (i.e., glass-to-air) part and a radiative part, and the latter (q_R) may depend on radiant temperatures that are different from the air temperatures in Equation (1):

$$q = q_C + q_R \tag{6}$$

U-FACTOR (THERMAL TRANSMITTANCE)

In the absence of sunlight, air infiltration, and moisture condensation, the first term in Equation (1) represents the rate of thermal heat transfer through a fenestration system. Most fenestration systems consist of transparent multipane glazing units and opaque elements comprising the sash and frame (hereafter called **frame**). The glazing unit's heat transfer paths include a one-dimensional center-of-glass contribution and a two-dimensional edge contribution. The frame contribution is primarily two-dimensional.

Consequently, the total rate of heat transfer through a fenestration system can be calculated knowing the separate heat transfer contributions of the center glass, edge glass, and frame. (When present, glazing dividers, such as decorative grilles and muntins, also affect heat transfer, and their contribution must be considered.) The overall U-factor is estimated using area-weighted U-factors for each contribution by

$$U_o = \frac{U_{cg}A_{cg} + U_{eg}A_{eg} + U_f A_f}{A_{pf}} \tag{7}$$

where the subscripts *cg*, *eg*, and *f* refer to the center-of-glass, edge-of-glass, and frame, respectively. A_{pf} is the area of the fenestration product's rough opening in the wall or roof less installation clearances. When a fenestration product has glazed surfaces in only one direction (typical windows), the sum of the areas equals the projected area. Skylights, greenhouse/garden windows, bay/bow windows, etc., because they extend beyond the plane of the wall/roof, have greater surface area for heat loss than a window with a similar glazing option and frame material; consequently, U-factors for such products are expected to be greater.

DETERMINING FENESTRATION U-FACTORS

Center-of-Glass U-Factor

Heat flow across the central glazed portion of a multipane unit must consider both convective and radiative transfer in the gas space. Convective heat transfer is estimated based on high-aspect-ratio, natural convection correlations for vertical and inclined air layers (El Sherbiny et al. 1982, Shewen 1986, Wright 1996). Radiative heat transfer (ignoring gas absorption) is quantified using a more fundamental approach. Computational methods solving the combined heat transfer problem have been devised (Rubin 1982a, 1982b, Hollands and Wright 1982).

Especially for single glass, U-factors depend strongly on indoor and outdoor film coefficients. The U-factor for single glass is

$$U = \frac{1}{1/h_o + 1/h_i + L/k} \tag{8}$$

where

h_o, h_i = outdoor and indoor respective glass surface heat transfer coefficients, Btu/h·ft²·°F
L = glass thickness, in.
k = thermal conductivity, Btu·in/h·ft²·°F

Values for U_{cg} at standard indoor and outdoor conditions depend on such glazing construction features as the number of glazing lights, the gas-space dimensions, the orientation relative to vertical, the emissivity of each surface, and the composition of the fill gas. Several computer programs can be used to estimate glazing unit heat transfer for a wide range of glazing construction (Arasteh et al. 1994, Finlayson and Arasteh 1993, Wright 1995c). The National Fenestration Rating Council calls for WINDOW 4.1 (LBL 1994) as a standard calculation method for the center glazing. In Canada, the VISION program (Wright 1995b) is used to determine center-glazing properties for the Canadian Standards Association (CSA *Standard* A440.2).

Figure 3 shows the effect of gas space width on U_{cg} for vertical double- and triple-paned glazing units. U-factors are plotted for air, argon, and krypton fill gases and for high (uncoated) and low (coated) values of surface emissivity. Gas space widths greater than 0.5 in. have no significant effect on U_{cg}, but greater glazing unit thicknesses decrease U_o since the length of the shortest heat flow path through the frame increases. A low-emissivity coating combined with krypton gas fill offers significant potential for reducing heat transfer in narrow gap-width glazing units.

Edge-of-Glass U-Factor

Insulating glazing units usually have continuous spacer members around the glass perimeter to separate the glazing and provide an edge seal. Aluminum spacers greatly increase conductive heat transfer between the contacted inner and outer glazing, thereby degrading the thermal performance of the glazing unit locally. The edge-of-glass area is typically taken to be a band 2.5 in. wide around the sightline. The width of this area is determined from the extent of two-dimensional heat transfer effects in current computer models, which are based on conduction-only analysis. In reality, due to convective and radiative effects, this area may extend beyond 2.5 in. (Beck et al. 1995, Curcija and Goss 1994, Wright and Sullivan 1995b), and it will depend on the type of insulating glazing unit and its thickness.

In low-conductivity frames, the heat flow at the edge-of-glass and frame area is through the spacer, and so the type of spacer has a greater impact on the edge-of-glass and frame U-factor. In metal frames, the edge-of-glass and frame U-factor varies little with the type of spacer (metal or insulating) because there is a significant heat flow through the highly conductive frame near the edge-of-glass area.

Frame U-Factor

Fenestration frame elements consist of all structural members exclusive of the glazing units and include sash, jamb, head, and sill members; meeting rails and stiles; mullions; and other glazing dividers. Estimating the rate of heat transfer through the frame is

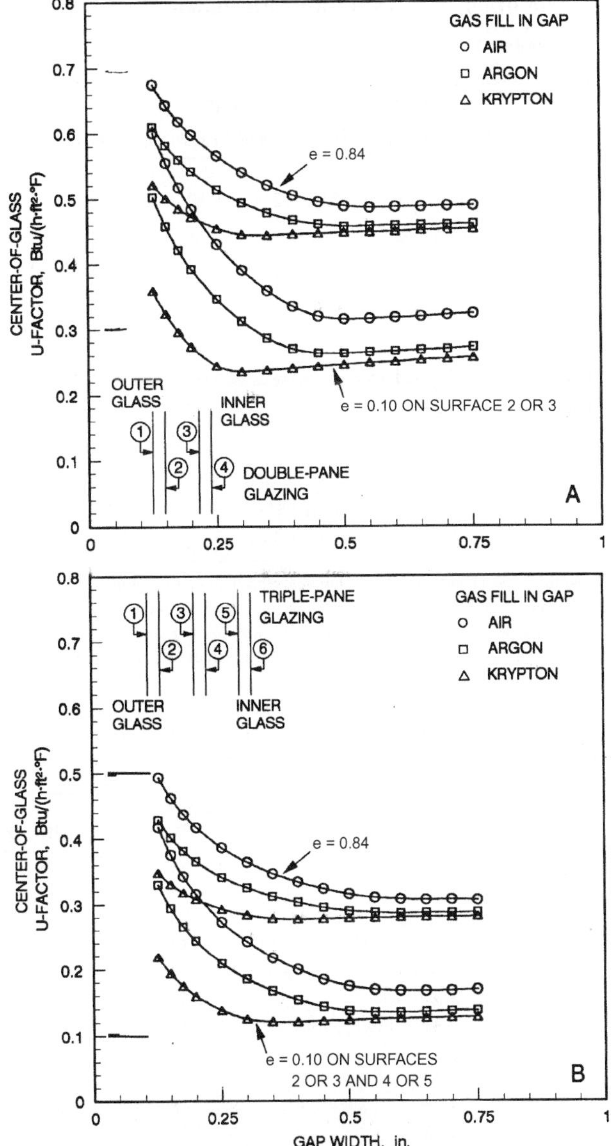

Fig. 3 Center-of-Glass U-Factor for Vertical Double- and Triple-Pane Glazing Units

complicated by (1) the variety of fenestration products and frame configurations, (2) the different combinations of materials used for frames, (3) the different sizes available, and, to a lesser extent, (4) the glazing unit width and spacer type. Internal dividers or grilles have little impact on the fenestration U-factor, provided there is at least a 1/8 in. gap between the divider and each panel of glass.

Computer simulations found that frame heat loss in most fenestration is controlled by a single component or controlling resistance, and only changes in this component significantly affect frame heat loss (EEL 1990). For example, the frame U-factor for thermally broken aluminum fenestration products is largely controlled by the depth of the thermal break material in the heat flow direction. For aluminum frames without a thermal break, the inside film coefficient provides most of the resistance to heat flow. For vinyl- and wood-framed fenestrations, the controlling resistance is the shortest distance between the inside and outside surfaces, which usually depends on the thickness of the sealed glazing unit.

Carpenter and McGowan (1993) experimentally validated frame U-factors for a variety of fixed and operable fenestration product types, sizes, and materials using computer modeling techniques. Table 1 lists frame U-factors for a variety of frame and spacer materials and glazing unit thicknesses. Frame and edge U-factors are normally determined by two-dimensional computer simulation. The National Fenestration Rating Council requires that frame and edge U-factors be determined using the THERM (Arasteh et al. 2000) and FRAME (EEL 1995) computer programs. The Canadian Standards Association requires that frame and edge U-factors be determined using FRAME.

Curtain Wall Construction

A curtain wall is an exterior building wall that carries no roof or floor loads and consists entirely or principally of glass and other surfacing materials supported by a framework. A curtain wall typically has a metal frame. To improve the thermal performance of standard metal frames, manufacturers provide both traditional thermal breaks as well as thermally improved products. The traditional thermal break type is poured and debridged, where urethane is poured into a metal U-channel in the frame and then the bottom of the channel is removed by machine. For this system to work well, there must be a thermal break between the interior and the exterior for all of the frame components, including those in any operable sash. Skip debridging (incomplete pour and debridging used for increased structural strength) can significantly degrade the U-factor. Bolts that penetrate the thermal break also degrade performance, but to a lesser degree. Griffith et al. (1998) showed that stainless steel bolts spaced 12 in. on center increased the frame U-factor by 18%. The paper also concluded that, in general, the isothermal planes method referenced in Chapter 25 provides a conservative approach to determining U-factors.

Thermally improved curtain wall products are a more recent development. In these products, most of the metal frame tends to be located on the interior with only a metal cap exposed on the exterior. Plastic spacers isolate the glazing assembly from both the metal cap on the exterior and the metal frame on the interior. These products can have significantly better thermal performance than standard metal frames, but it is important to minimize the number and area of the bolts that penetrate from exterior to interior.

INDOOR AND OUTDOOR SURFACE HEAT TRANSFER COEFFICIENTS

Part of the overall thermal resistance of a fenestration system is due to the convective and radiative heat transfer between the exposed surfaces and the environment. Surface heat transfer coefficients h_o and h_i at the outer and inner glazing surfaces, respectively, combine the effects of radiation and convection.

The wind speed and orientation of the building are important in determining h_o. This relationship has long been studied, and many correlations have been proposed for h_o as a function of wind speed. However, no universal relationship has been accepted, and limited field measurements at low wind speeds by Klems (1989) show significant difference with values used by others.

Convective heat transfer coefficients are usually determined at standard temperature and air velocity conditions on each side. Wind speed can vary from less than 0.5 mph for calm weather, free convection conditions, to over 65 mph for storm conditions. A standard value of 5.1 Btu/h·ft^2·°F corresponding to a 15 mph wind is often used to represent winter design conditions. At near-zero wind speed, h_o varies with outside air and surface temperature, orientation to vertical, and air moisture content. At low wind speeds, the overall surface heat transfer coefficient can be as low as 1.2 Btu/h·ft^2·°F (Yazdanian and Klems 1993).

For natural convection at the inner surface of a vertical fenestration product, the inner surface coefficient h_i depends on the indoor air and glass surface temperatures and on the emissivity of the glass inner surface. Table 2 shows the variation of h_i for winter (t_i = 70°F)

Table 1 Representative Fenestration Frame U-Factors in Btu/h·ft²·°F—Vertical Orientation

		Product Type/Number of Glazing Layers																
		Operable			Fixed			Garden Window		Plant-Assembled Skylight			Curtain Wall[e]			Sloped/Overhead Glazing[e]		
Frame Material	Type of Spacer	Single[b]	Double[c]	Triple[d]	Single[b]	Double[c]	Triple[d]	Single[b]	Double[c]	Single[b]	Double[c]	Triple[d]	Single[f]	Double[g]	Triple[h]	Single[f]	Double[g]	Triple[h]
Aluminum without thermal break	All	2.38	2.27	2.20	1.92	1.80	1.74	1.88	1.83	7.85	7.02	6.87	3.01	2.96	2.83	3.05	3.00	2.87
Aluminum with thermal break[a]	Metal	1.20	0.92	0.83	1.32	1.13	1.11			6.95	5.05	4.58	1.80	1.75	1.65	1.82	1.76	1.66
	Insulated	n/a	0.88	0.77	n/a	1.04	1.02			n/a	4.75	4.12	n/a	1.63	1.51	n/a	1.64	1.52
Aluminum-clad wood/ reinforced vinyl	Metal	0.60	0.58	0.51	0.55	0.51	0.48			4.86	3.93	3.66						
	Insulated	n/a	0.55	0.48	n/a	0.48	0.44			n/a	3.75	3.43						
Wood/vinyl	Metal	0.55	0.51	0.48	0.55	0.48	0.42	0.90	0.85	2.50	2.08	1.78						
	Insulated	n/a	0.49	0.40	n/a	0.42	0.35	n/a	0.83	n/a	2.02	1.71						
Insulated fiberglass/ vinyl	Metal	0.37	0.33	0.32	0.37	0.33	0.32											
	Insulated	n/a	0.32	0.26	n/a	0.32	0.26											
Structural glazing	Metal												1.80	1.27	1.04	1.82	1.28	1.05
	Insulated												n/a	1.02	0.75	n/a	1.02	0.75

Note: This table should only be used as an estimating tool for the early phases of design.
[a]Depends strongly on width of thermal break. Value given is for 3/8 in.
[b]Single glazing corresponds to individual glazing unit thickness of 1/8 in. (nominal).
[c]Double glazing corresponds to individual glazing unit thickness of 3/4 in. (nominal).
[d]Triple glazing corresponds to individual glazing unit thickness of 1-3/8 in. (nominal).
[e]Glass thickness in curtainwall and sloped/overhead glazing is 1/4 in.
[f]Single glazing corresponds to individual glazing unit thickness of 1/4 in. (nominal).
[g]Double glazing corresponds to individual glazing unit thickness of 1 in. (nominal).
[h]Triple glazing corresponds to individual glazing unit thickness of 1-3/4 in. (nominal).
n/a Not applicable

Table 2 Indoor Surface Heat Transfer Coefficient h_i in Btu/h·ft²·°F—Vertical Orientation (Still Air Conditions)

			Winter Conditions			Summer Conditions		
Glazing ID	Glazing Type	Glazing Height, ft	Glass Temp., °F	Temp. Diff., °F	h_i, Btu/h·ft²·°F	Glass Temp., °F	Temp. Diff., °F	h_i, Btu/h·ft²·°F
1	Single glazing	2	17	53	1.41	89	14	1.41
		4	17	53	1.31	89	14	1.33
		6	17	53	1.25	89	14	1.29
5	Double glazing with 1/2 in. air space	2	45	25	1.36	89	14	1.41
		4	45	25	1.27	89	14	1.33
		6	45	25	1.22	89	14	1.29
23	Double glazing with e = 0.1 on surface 2 and 1/2 in. argon space	2	56	14	1.31	87	12	1.38
		4	56	14	1.23	87	12	1.31
		6	56	14	1.19	87	12	1.27
43	Triple glazing with e = 0.1 on surfaces 2 and 5 and 1/2 in. argon spaces	2	63	7	1.25	93	18	1.45
		4	63	7	1.18	93	18	1.36
		6	63	7	1.15	93	18	1.32

Notes:
Glazing ID refers to fenestration assemblies in Table 4.
Winter conditions: room air temperature t_i = 70°F, outdoor air temperature t_o = 0°F, no solar radiation
Summer conditions: room air temperature t_i = 75°F, outdoor air temperature t_o = 89°F, direct solar irradiance E_D = 248 Btu/h·ft²
$h_i = h_{ic} + h_{iR} = 0.30(\Delta T/L)^{0.25} + e\sigma(T_g^4 - T_i^4)/\Delta T$
where $\Delta T = T_g - T_i$; T_g = glass temperature; L = glazing height

and summer (t_i = 75°F) design conditions, for a range of glass types and heights. Designers often use h_i = 1.46 Btu/h·ft²·°F, which corresponds to t_i = 70°F, glass temperature of 15°F, and uncoated glass with e_g = 0.84. For summer conditions, the same value (h_i = 1.46 Btu/h·ft²·°F) is normally used, and it corresponds approximately to glass temperature of 95°F, t_i = 75°F, and e_g = 0.84. For winter conditions, this most closely approximates single glazing with clear glass that is 2 ft tall, but it overestimates the value as the glazing unit conductance decreases and height increases. For summer conditions, this value approximates all types of glass that are 2 ft tall but, again, is less accurate as the glass height increases. If the room surface of the glass has a low-e coating, the h_i values are about halved at both winter and summer conditions.

Heat transfer between the glazing surface and its environment is driven not only by the local air temperatures but also by the radiant temperatures to which the surface is exposed. The radiant temperature of the indoor environment is generally assumed to be equal to the indoor air temperature. While this is a safe assumption where a small fenestration product is exposed to a large room with surface temperatures equal to the air temperature, it is not valid in rooms where the fenestration product is exposed to other large areas of glazing surfaces (e.g., greenhouse, atrium) or to other cooled or heated surfaces (Parmelee and Huebscher 1947).

The radiant temperature of the outdoor environment is frequently assumed to be equal to the outdoor air temperature. This assumption may be in error, since additional radiative heat loss occurs between a fenestration and the clear sky (Duffie and Beckman 1980). Therefore, for clear-sky conditions, some effective outdoor temperature $t_{o,e}$ should replace t_o in Equation (1). For methods for determining $t_{o,e}$, see, for example, work by AGSL (1992). Note that a fully cloudy sky is assumed in ASHRAE design conditions.

The air space in an insulating glass panel made up of glass with no reflective coating on the air space surfaces has a coefficient h_s of 1.3 Btu/h·ft²·°F. When a reflective coating is applied to an air space surface, h_s can be selected from Table 3 by first calculating the effective air space emissivity $e_{s,e}$ by Equation (9):

Table 3 Air Space Coefficients for Horizontal Heat Flow

Air Space Thickness, in.	Air Space Temp., °F	Air Temp. Diff., °F	Air Space Coefficient h_s, Btu/h·ft²·°F — Effective Emissivity $e_{s,e}$ 0.82	0.72	0.40	0.20	0.10	0.05
0.5	5	10	0.88	0.82	0.60	0.46	0.39	0.35
		25	0.90	0.83	0.61	0.48	0.41	0.37
		55	1.00	0.93	0.71	0.57	0.50	0.47
		70	1.05	0.98	0.76	0.62	0.55	0.51
		90	1.10	1.03	0.81	0.67	0.60	0.57
	32	10	1.00	0.92	0.66	0.50	0.42	0.38
		25	1.01	0.93	0.67	0.51	0.43	0.39
		55	1.08	1.00	0.74	0.57	0.49	0.45
		70	1.12	1.04	0.78	0.62	0.53	0.49
		90	1.17	1.09	0.83	0.67	0.58	0.54
	50	10	1.09	1.00	0.71	0.53	0.44	0.39
		25	1.10	1.01	0.72	0.54	0.44	0.40
		55	1.14	1.05	0.76	0.58	0.49	0.44
		70	1.18	1.09	0.80	0.62	0.53	0.48
		90	1.23	1.14	0.85	0.67	0.57	0.53
	85	10	1.28	1.16	0.81	0.59	0.48	0.42
		25	1.28	1.17	0.81	0.59	0.48	0.43
		55	1.30	1.19	0.84	0.62	0.51	0.45
		70	1.33	1.21	0.86	0.64	0.53	0.47
		90	1.36	1.25	0.90	0.67	0.56	0.51
	120	10	1.48	1.35	0.92	0.66	0.52	0.46
		25	1.49	1.35	0.92	0.66	0.52	0.46
		55	1.50	1.37	0.94	0.67	0.54	0.47
		70	1.51	1.38	0.95	0.68	0.55	0.48
		90	1.53	1.40	0.97	0.70	0.57	0.50
0.4	5	10	0.96	0.89	0.67	0.54	0.47	0.43
		55	1.00	0.93	0.71	0.57	0.50	0.47
		90	1.07	1.01	0.78	0.64	0.58	0.54
	32	10	1.09	1.00	0.74	0.58	0.50	0.46
		55	1.11	1.03	0.76	0.60	0.52	0.48
		90	1.15	1.07	0.81	0.64	0.56	0.52
	50	10	1.18	1.09	0.79	0.61	0.52	0.48
		55	1.19	1.10	0.81	0.63	0.54	0.49
		90	1.22	1.13	0.84	0.66	0.57	0.52
	85	10	1.37	1.26	0.90	0.68	0.57	0.51
		55	1.38	1.26	0.91	0.69	0.58	0.52
		90	1.40	1.26	0.93	0.70	0.59	0.54
	120	10	1.58	1.45	1.02	0.75	0.62	0.55
		55	1.59	1.45	1.02	0.76	0.62	0.56
		90	1.60	1.46	1.03	0.77	0.63	0.57
0.3	5	<90	1.10	1.03	0.81	0.68	0.61	0.57
	32	<90	1.23	1.15	0.89	0.72	0.64	0.60
	50	<90	1.32	1.23	0.94	0.76	0.67	0.62
	85	<90	1.52	1.41	1.06	0.84	0.72	0.67
	120	<90	1.74	1.61	1.18	0.92	0.78	0.72
0.25	5	<90	1.20	1.13	0.91	0.77	0.70	0.67
	32	<90	1.34	1.26	0.99	0.83	0.75	0.71
	50	<90	1.43	1.34	1.05	0.87	0.78	0.74
	85	<90	1.64	1.53	1.18	0.96	0.84	0.79
	120	<90	1.87	1.74	1.31	1.04	0.91	0.84
0.2	5	<90	1.36	1.29	1.07	0.93	0.86	0.83
	32	<90	1.50	1.42	1.16	1.00	0.92	0.88
	50	<90	1.61	1.52	1.23	1.05	0.95	0.91
	85	<90	1.83	1.71	1.36	1.14	1.03	0.97
	120	<90	2.07	1.93	1.51	1.24	1.10	1.04

$$e_{s,e} = \frac{1}{1/e_o + 1/e_i - 1} \qquad (9)$$

where e_o and e_i are the hemispherical emissivities of the two air space surfaces. Hemispherical emissivity of ordinary uncoated glass is 0.84 over a wavelength range of 0.4 to 40 μm.

REPRESENTATIVE U-FACTORS FOR FENESTRATION PRODUCTS

Table 4 lists computed U-factors for a variety of generic fenestration products. The table is based on ASHRAE-sponsored research involving laboratory testing and computer simulation of various fenestration products. In the past, test data were used to provide more accurate results for specific products (Hogan 1988). Computer simulations (with validation by testing) are now accepted as the standard method for accurate product-specific U-factor determination. The simulation methodologies are specified in the National Fenestration Rating Council's NFRC 100 or Canadian Standards Association (CSA) *Standard* A440.2 and are based on algorithms published in ISO *Standard* 15099. The *International Energy Conservation Code* (ICC 2000) and various state energy codes in the United States, the *National Energy Code* in Canada, and ASHRAE *Standards* 90.1 and 90.2 all cite these standards. Fenestration needs to be rated in accordance with the NFRC or CSA standards for code compliance. The use of Table 4 should be limited to that of an estimating tool for the early phases of design.

Values in Table 4 are listed at winter design conditions for vertical installation and for skylights and other sloped installations with glazing surfaces that are sloped 20° from the horizontal. Data are based on center-of-glass and edge-of-glass component U-factors and assume that there are no dividers. However, they apply only to the specific design conditions described in the footnotes in the table, and they are typically used only to determine peak load conditions for sizing heating equipment. While these U-factors have been determined for winter conditions, they can also be used to estimate heat gain during peak cooling conditions, since conductive gain, which is one of several variables, is usually a small portion of the total heat gain for fenestration in direct sunlight. Glazing designs and framing materials may be compared in choosing a fenestration system that needs a specific winter design U-factor.

Table 4 lists 48 glazing types. (A subset of these types is included in Table 13, which lists solar heat gain coefficients and visible light transmittance.) The multiple glazing categories are appropriate for sealed glazing units and the addition of storm sash to other glazing units. No distinction is made between flat and domed units such as skylights. For acrylic domes, use an average gas-space width to determine the U-factor. Note that garden window and sloped/pyramid/barrel vault skylight U-factors are approximately twice those of other similar products. While this is partially due to the difference in slope in the case of the sloped/pyramid/barrel vault skylights, it is largely because these products project out from the surface of the wall or roof. For instance, the skylight surface area, which includes the curb, can vary from 13 to 240% greater than the rough opening area, depending on the size and mounting method. Unless otherwise noted, all multiple-glazed units are filled with dry air. Argon units are assumed to be filled with 90% argon (Elmahdy and Yusuf 1995). U-factors for CO_2-filled units are similar to argon fills. For spaces up to 1/2 in., argon/SF_6 (sulfur hexafluoride) mixtures up to 70% SF_6 are generally the same as argon fills. The use of krypton gas can provide U-factors lower than those for argon for glazing spaces less than 1/2 in.

Table 4 provides data for six values of hemispherical emissivity and for 1/4 and 1/2 in. gas space widths. The emissivity of various low-e glasses varies considerably between manufacturers and processes. When the emissivity is between the listed values, interpolation may be used. When manufacturers' data are not available for

Table 4 U-Factors for Various Fenestration Products in Btu/h·ft²·°F

Product Type		Glass Only		Vertical Installation: Operable (including sliding and swinging glass doors)					Vertical Installation: Fixed				
ID	**Frame Type / Glazing Type**	**Center of Glass**	**Edge of Glass**	**Aluminum Without Thermal Break**	**Aluminum with Thermal Break**	**Reinforced Vinyl/ Aluminum Clad Wood**	**Wood/ Vinyl**	**Insulated Fiberglass/ Vinyl**	**Aluminum Without Thermal Break**	**Aluminum with Thermal Break**	**Reinforced Vinyl/ Aluminum Clad Wood**	**Wood/ Vinyl**	**Insulated Fiberglass/ Vinyl**
	Single Glazing												
1	1/8 in. glass	*1.04*	*1.04*	1.27	1.08	0.90	0.89	0.81	1.13	1.07	0.98	0.98	0.94
2	1/4 in. acrylic/polycarbonate	*0.88*	*0.88*	1.14	0.96	0.79	0.78	0.71	0.99	0.92	0.84	0.84	0.81
3	1/8 in. acrylic/polycarbonate	*0.96*	*0.96*	1.21	1.02	0.85	0.83	0.76	1.06	1.00	0.91	0.91	0.87
	Double Glazing												
4	1/4 in. air space	*0.55*	*0.64*	0.87	0.65	0.57	0.55	0.49	0.69	0.63	0.56	0.56	0.53
5	1/2 in. air space	*0.48*	*0.59*	0.81	0.60	0.53	0.51	0.44	0.64	0.57	0.50	0.50	0.48
6	1/4 in. argon space	*0.51*	*0.61*	0.84	0.62	0.55	0.53	0.46	0.66	0.59	0.53	0.52	0.50
7	1/2 in. argon space	*0.45*	*0.57*	0.79	0.58	0.51	0.49	0.43	0.61	0.54	0.48	0.48	0.45
	Double Glazing, *e* = 0.60 on surface 2 or 3												
8	1/4 in. air space	*0.52*	*0.62*	0.84	0.63	0.55	0.53	0.47	0.67	0.60	0.54	0.53	0.51
9	1/2 in. air space	*0.44*	*0.56*	0.78	0.57	0.50	0.48	0.42	0.60	0.53	0.47	0.47	0.45
10	1/4 in. argon space	*0.47*	*0.58*	0.81	0.59	0.52	0.50	0.44	0.63	0.56	0.50	0.49	0.47
11	1/2 in. argon space	*0.41*	*0.54*	0.76	0.55	0.48	0.46	0.40	0.58	0.51	0.45	0.44	0.42
	Double Glazing, *e* = 0.40 on surface 2 or 3												
12	1/4 in. air space	*0.49*	*0.60*	0.82	0.61	0.53	0.51	0.45	0.64	0.58	0.51	0.51	0.49
13	1/2 in. air space	*0.40*	*0.54*	0.75	0.54	0.48	0.45	0.40	0.57	0.50	0.44	0.44	0.41
14	1/4 in. argon space	*0.43*	*0.56*	0.78	0.57	0.50	0.47	0.41	0.59	0.53	0.46	0.46	0.44
15	1/2 in. argon space	*0.36*	*0.51*	0.72	0.52	0.45	0.43	0.37	0.53	0.47	0.41	0.40	0.38
	Double Glazing, *e* = 0.20 on surface 2 or 3												
16	1/4 in. air space	*0.45*	*0.57*	0.79	0.58	0.51	0.49	0.43	0.61	0.54	0.48	0.48	0.45
17	1/2 in. air space	*0.35*	*0.50*	0.71	0.51	0.44	0.42	0.36	0.53	0.46	0.40	0.39	0.37
18	1/4 in. argon space	*0.38*	*0.52*	0.74	0.53	0.46	0.44	0.38	0.55	0.48	0.42	0.42	0.40
19	1/2 in. argon space	*0.30*	*0.46*	0.67	0.47	0.41	0.39	0.33	0.48	0.41	0.36	0.35	0.33
	Double Glazing, *e* = 0.10 on surface 2 or 3												
20	1/4 in. air space	*0.42*	*0.55*	0.77	0.56	0.49	0.47	0.41	0.59	0.52	0.46	0.45	0.43
21	1/2 in. air space	*0.32*	*0.48*	0.69	0.49	0.42	0.40	0.35	0.50	0.43	0.37	0.37	0.35
22	1/4 in. argon space	*0.35*	*0.50*	0.71	0.51	0.44	0.42	0.36	0.53	0.46	0.40	0.39	0.37
23	1/2 in. argon space	*0.27*	*0.44*	0.65	0.45	0.39	0.37	0.31	0.46	0.39	0.33	0.33	0.31
	Double Glazing, *e* = 0.05 on surface 2 or 3												
24	1/4 in. air space	*0.41*	*0.54*	0.76	0.55	0.48	0.46	0.40	0.58	0.51	0.45	0.44	0.42
25	1/2 in. air space	*0.30*	*0.46*	0.67	0.47	0.41	0.39	0.33	0.48	0.41	0.36	0.35	0.33
26	1/4 in. argon space	*0.33*	*0.48*	0.70	0.49	0.43	0.41	0.35	0.51	0.44	0.38	0.38	0.36
27	1/2 in. argon space	*0.25*	*0.42*	0.63	0.44	0.38	0.36	0.30	0.44	0.37	0.32	0.31	0.29
	Triple Glazing												
28	1/4 in. air spaces	*0.38*	*0.52*	0.72	0.51	0.44	0.43	0.38	0.55	0.48	0.42	0.41	0.40
29	1/2 in. air spaces	*0.31*	*0.47*	0.67	0.46	0.40	0.39	0.34	0.49	0.42	0.36	0.35	0.34
30	1/4 in. argon spaces	*0.34*	*0.49*	0.69	0.48	0.42	0.41	0.35	0.51	0.45	0.39	0.38	0.36
31	1/2 in. argon spaces	*0.29*	*0.45*	0.65	0.44	0.38	0.37	0.32	0.47	0.40	0.34	0.34	0.32
	Triple Glazing, *e* = 0.20 on surface 2, 3, 4, or 5												
32	1/4 in. air spaces	*0.33*	*0.48*	0.69	0.47	0.41	0.40	0.35	0.50	0.44	0.38	0.37	0.36
33	1/2 in. air spaces	*0.25*	*0.42*	0.62	0.41	0.36	0.35	0.30	0.43	0.37	0.31	0.30	0.29
34	1/4 in. argon spaces	*0.28*	*0.45*	0.65	0.44	0.38	0.37	0.32	0.46	0.40	0.34	0.33	0.32
35	1/2 in. argon spaces	*0.22*	*0.40*	0.60	0.39	0.34	0.33	0.28	0.41	0.34	0.29	0.28	0.27
	Triple Glazing, *e* = 0.20 on surfaces 2 or 3 and 4 or 5												
36	1/4 in. air spaces	*0.29*	*0.45*	0.65	0.44	0.38	0.37	0.32	0.47	0.40	0.34	0.34	0.32
37	1/2 in. air spaces	*0.20*	*0.39*	0.58	0.38	0.32	0.31	0.27	0.39	0.33	0.27	0.26	0.25
38	1/4 in. argon spaces	*0.23*	*0.41*	0.61	0.40	0.34	0.33	0.29	0.42	0.35	0.30	0.29	0.28
39	1/2 in. argon spaces	*0.17*	*0.36*	0.56	0.36	0.30	0.29	0.25	0.37	0.30	0.25	0.24	0.23
	Triple Glazing, *e* = 0.10 on surfaces 2 or 3 and 4 or 5												
40	1/4 in. air spaces	*0.27*	*0.44*	0.64	0.43	0.37	0.36	0.31	0.45	0.39	0.33	0.32	0.31
41	1/2 in. air spaces	*0.18*	*0.37*	0.57	0.36	0.31	0.30	0.25	0.37	0.31	0.25	0.25	0.23
42	1/4 in. argon spaces	*0.21*	*0.39*	0.59	0.39	0.33	0.32	0.27	0.40	0.34	0.28	0.27	0.26
43	1/2 in. argon spaces	*0.14*	*0.34*	0.54	0.33	0.28	0.27	0.23	0.34	0.28	0.22	0.21	0.20
	Quadruple Glazing, *e* = 0.10 on surfaces 2 or 3 and 4 or 5												
44	1/4 in. air spaces	*0.22*	*0.40*	0.60	0.39	0.34	0.33	0.28	0.41	0.34	0.29	0.28	0.27
45	1/2 in. air spaces	*0.15*	*0.35*	0.54	0.34	0.29	0.28	0.24	0.35	0.28	0.23	0.22	0.21
46	1/4 in. argon spaces	*0.17*	*0.36*	0.56	0.36	0.30	0.29	0.25	0.37	0.30	0.25	0.24	0.23
47	1/2 in. argon spaces	*0.12*	*0.32*	0.52	0.32	0.27	0.26	0.22	0.32	0.26	0.20	0.20	0.19
48	1/4 in. krypton spaces	*0.12*	*0.32*	0.52	0.32	0.27	0.26	0.22	0.32	0.26	0.20	0.20	0.19

Notes:

1. All heat transmission coefficients in this table include film resistances and are based on winter conditions of 0°F outdoor air temperature and 70°F indoor air temperature, with 15 mph outdoor air velocity and zero solar flux. With the exception of single glazing, small changes in the indoor and outdoor temperatures will not significantly affect overall U-factors. The coefficients are for vertical position except skylight and sloped glazing values, which are for 20° from horizontal with heat flow up.

2. Glazing layer surfaces are numbered from the outdoor to the indoor. *Double*, *triple*, and *quadruple* refer to the number of glazing panels. All data are based on 1/8 in. glass, unless otherwise noted. Thermal conductivities are 0.53 Btu/h·ft·°F for glass and 0.11 Btu/h·ft·°F for acrylic and polycarbonate.

3. Standard spacers are metal. Edge-of-glass effects assumed to extend over the 2-1/2 in. band around perimeter of each glazing unit.

Table 4 U-Factors for Various Fenestration Products in Btu/h·ft^2·°F (*Concluded*)

Vertical Installation					Sloped Installation										
Garden Windows		Curtain Wall			Glass Only (Skylights)		Manufactured Skylight				Site-Assembled Sloped/Overhead Glazing			
Aluminum Without Thermal Break	Wood/ Vinyl	Aluminum Without Thermal Break	Aluminum with Thermal Break	Structural Glazing	Center of Glass	Edge of Glass	Aluminum Without Thermal Break	Aluminum with Thermal Break	Reinforced Vinyl/ Aluminum Clad Wood	Wood/ Vinyl	Aluminum Without Thermal Break	Aluminum with Thermal Break	Structural Glazing	ID
2.60	2.31	1.22	1.11	1.11	*1.19*	*1.19*	1.98	1.89	1.75	1.47	1.36	1.25	1.25	1
2.33	2.06	1.08	0.96	0.96	*1.03*	*1.03*	1.82	1.73	1.60	1.31	1.21	1.10	1.10	2
2.46	2.19	1.15	1.04	1.04	*1.11*	*1.11*	1.90	1.81	1.68	1.39	1.29	1.18	1.18	3
1.81	1.61	0.79	0.68	0.63	*0.58*	*0.66*	1.31	1.11	1.05	0.84	0.82	0.70	0.66	4
1.71	1.53	0.73	0.62	0.57	*0.57*	*0.65*	1.30	1.10	1.04	0.84	0.81	0.69	0.65	5
1.76	1.56	0.75	0.64	0.60	*0.53*	*0.63*	1.27	1.07	1.00	0.80	0.77	0.66	0.62	6
1.67	1.49	0.70	0.59	0.55	*0.53*	*0.63*	1.27	1.07	1.00	0.80	0.77	0.66	0.62	7
1.77	1.58	0.76	0.65	0.61	*0.54*	*0.63*	1.27	1.08	1.01	0.81	0.78	0.67	0.63	8
1.65	1.48	0.69	0.58	0.54	*0.53*	*0.63*	1.27	1.07	1.00	0.80	0.77	0.66	0.62	9
1.70	1.52	0.72	0.61	0.56	*0.49*	*0.60*	1.23	1.03	0.97	0.76	0.74	0.63	0.58	10
1.61	1.44	0.67	0.56	0.51	*0.49*	*0.60*	1.23	1.03	0.97	0.76	0.74	0.63	0.58	11
1.73	1.54	0.74	0.63	0.58	*0.51*	*0.61*	1.25	1.05	0.99	0.78	0.76	0.64	0.60	12
1.59	1.43	0.66	0.55	0.51	*0.50*	*0.61*	1.24	1.04	0.98	0.77	0.75	0.64	0.59	13
1.64	1.47	0.69	0.57	0.53	*0.44*	*0.56*	1.18	0.99	0.92	0.72	0.70	0.58	0.54	14
1.53	1.38	0.63	0.51	0.47	*0.46*	*0.58*	1.20	1.00	0.94	0.74	0.71	0.60	0.56	15
1.67	1.49	0.70	0.59	0.55	*0.46*	*0.58*	1.20	1.00	0.94	0.74	0.71	0.60	0.56	16
1.52	1.37	0.62	0.51	0.46	*0.46*	*0.58*	1.20	1.00	0.94	0.74	0.71	0.60	0.56	17
1.56	1.40	0.64	0.53	0.49	*0.39*	*0.53*	1.14	0.94	0.88	0.68	0.65	0.54	0.50	18
1.44	1.30	0.57	0.46	0.42	*0.40*	*0.54*	1.15	0.95	0.89	0.68	0.66	0.55	0.51	19
1.62	1.45	0.68	0.57	0.52	*0.44*	*0.56*	1.18	0.99	0.92	0.72	0.70	0.58	0.54	20
1.47	1.33	0.59	0.48	0.44	*0.44*	*0.56*	1.18	0.99	0.92	0.72	0.70	0.58	0.54	21
1.52	1.37	0.62	0.51	0.46	*0.36*	*0.51*	1.11	0.91	0.85	0.65	0.63	0.52	0.47	22
1.40	1.26	0.55	0.44	0.39	*0.38*	*0.52*	1.13	0.93	0.87	0.67	0.65	0.53	0.49	23
1.61	1.44	0.67	0.56	0.51	*0.42*	*0.55*	1.17	0.97	0.91	0.70	0.68	0.57	0.52	24
1.44	1.30	0.57	0.46	0.42	*0.43*	*0.56*	1.17	0.98	0.91	0.71	0.69	0.58	0.53	25
1.49	1.34	0.60	0.49	0.44	*0.34*	*0.49*	1.09	0.89	0.83	0.63	0.61	0.50	0.45	26
1.37	1.24	0.53	0.42	0.38	*0.36*	*0.51*	1.11	0.91	0.85	0.65	0.63	0.52	0.47	27
see note 7	see note 7	0.63	0.52	0.47	*0.39*	*0.53*	1.12	0.89	0.84	0.64	0.64	0.53	0.48	28
		0.57	0.46	0.41	*0.36*	*0.51*	1.10	0.87	0.81	0.61	0.62	0.51	0.45	29
		0.60	0.49	0.43	*0.35*	*0.50*	1.09	0.86	0.80	0.60	0.61	0.50	0.44	30
		0.55	0.45	0.39	*0.33*	*0.48*	1.07	0.84	0.79	0.59	0.59	0.48	0.42	31
see note 7	see note 7	0.59	0.48	0.42	*0.34*	*0.49*	1.08	0.85	0.79	0.59	0.60	0.49	0.43	32
		0.52	0.41	0.35	*0.31*	*0.47*	1.05	0.82	0.77	0.57	0.57	0.46	0.41	33
		0.54	0.44	0.38	*0.28*	*0.45*	1.02	0.79	0.74	0.54	0.55	0.44	0.38	34
		0.49	0.38	0.33	*0.27*	*0.44*	1.01	0.78	0.73	0.53	0.54	0.43	0.37	35
see note 7	see note 7	0.55	0.45	0.39	*0.29*	*0.45*	1.03	0.80	0.75	0.55	0.56	0.45	0.39	36
		0.48	0.37	0.31	*0.27*	*0.44*	1.01	0.78	0.73	0.53	0.54	0.43	0.37	37
		0.50	0.39	0.34	*0.24*	*0.42*	0.99	0.75	0.70	0.50	0.51	0.40	0.35	38
		0.45	0.34	0.29	*0.22*	*0.40*	0.97	0.74	0.69	0.49	0.50	0.39	0.33	39
see note 7	see note 7	0.54	0.43	0.37	*0.27*	*0.44*	1.01	0.78	0.73	0.53	0.54	0.43	0.37	40
		0.46	0.35	0.29	*0.25*	*0.42*	0.99	0.76	0.71	0.51	0.52	0.41	0.36	41
		0.48	0.38	0.32	*0.21*	*0.39*	0.96	0.73	0.68	0.48	0.49	0.38	0.32	42
		0.42	0.32	0.26	*0.20*	*0.39*	0.95	0.72	0.67	0.47	0.48	0.37	0.31	43
see note 7	see note 7	0.49	0.38	0.33	*0.22*	*0.40*	0.97	0.74	0.69	0.49	0.50	0.39	0.33	44
		0.43	0.32	0.27	*0.19*	*0.38*	0.94	0.71	0.66	0.46	0.47	0.36	0.30	45
		0.45	0.34	0.29	*0.18*	*0.37*	0.93	0.70	0.65	0.45	0.46	0.35	0.30	46
		0.41	0.30	0.24	*0.16*	*0.35*	0.91	0.68	0.63	0.43	0.44	0.33	0.28	47
		0.41	0.30	0.24	*0.13*	*0.33*	0.88	0.65	0.60	0.40	0.42	0.31	0.25	48

4. Product sizes are described in Figure 4, and frame U-factors are from Table 1.
5. Use U = 0.60 Btu/h·ft^2·°F for glass block with mortar but without reinforcing or framing.
6. The use of this table should be limited to that of an estimating tool for the early phases of design.
7. Values for triple- and quadruple-glazed garden windows are not listed as these are not common products.
8. Minor differences exist between the data in this table and U-factors determined using NFRC 100-91 because the data in this table are generated using modified heat transfer correlations for glazing cavities (Wright 1996) and indoor fenestration surfaces (Curcija and Goss 1995b).

low-e glass, assume that glass with a pyrolytic (hard) coating has an emissivity of 0.40 and that glass with a sputtered (soft) coating has an emissivity as low as 0.36. Tinted glass does not change the winter U-factor. Also, some reflective glass may have an emissivity less than 0.84. Values listed are for insulating glass units using aluminum edge spacers. If an insulated or nonmetallic spacer is used, the U-factors are approximately 0.03 Btu/h·ft^2·°F lower.

Fenestration product types are subdivided first by vertical versus sloped installation and then into two general categories—manufactured and site-assembled. "Manufactured" is intended to represent products delivered as a complete unit to the site. These products are typically installed in low-rise residential and small commercial/institutional/industrial buildings. For vertical sliders, horizontal sliders, casement, awning, pivoted, and dual-action windows, and for sliding and swinging glass doors, use the operable category. For picture windows, use the fixed category. For products that project out from the surface of the wall, use the garden window category. For skylights, use the sloped skylight category.

"Site-assembled" is intended to represent products where frame extrusions are assembled on site into a fenestration product and then glazing is added on site. These products are typically installed in high-rise residential and larger commercial/institutional/industrial buildings. Curtain walls are typically made up of vision (transparent portion) and spandrel (opaque portion) panels. Table 4 contains representative U-factors for the vision panel (including mullions) for these assemblies. The spandrel portion of curtain walls usually consists of a metal pan filled with insulation and covered with a sheet of glass or other weatherproof covering. Although the U-factor in the center of the spandrel panel can be quite low, the metal pan is a thermal bridge, significantly increasing the U-factor of the assembly. Two-dimensional simulation validated by testing of a curtain wall having an aluminum frame with a thermal break found that the U-factor for the edge of the spandrel panel (the 2-1/2 in. band around the perimeter adjacent to the frame) was 40% of the way toward the U-factor of the frame. The U-factor was 0.06 for the center of the spandrel, 0.45 for the edge of the spandrel, and 1.06 for the frame (Carpenter and Elmahdy 1994). Two-dimensional heat transfer analysis or physical testing is recommended to determine the U-factor of spandrel panels. Use the sloped/overhead glazing category for sloped glazing panels comparable to curtain walls.

Physical testing of double-glazed units showed U-factors of 1.0 Btu/(h·ft^2·°F for a thermally broken aluminum pyramidal skylight and 1.3 Btu/(h·ft^2·°F) for an aluminum-frame half-round barrel vault (both normalized to a rough opening of 8 ft by 8 ft). Until more conclusive results are available, U-factors for these systems can be estimated by multiplying the "site-assembled sloped/overhead glazing" values in Table 4 by the ratio of total product surface area (including curbs) to rough opening area. These ratios range from 1.2 to 2.0 for low-slope skylights, 1.4 to 2.1 for pyramid assemblies sloped at 45°, and 1.7 to 2.9 for semicircular barrel vault assemblies. An example calculation is provided in Example 4.

The U-factors in Table 4 are based on the definitions of the six product types, frame sizes, and proportion of frame to glass area shown in Figure 4. Four of the products are manufactured fenestration products. The operable category glazing units are 15 ft^2 in area, and the overall size corresponds to a 3 ft by 5 ft fenestration product. The fixed (nonoperable) category is about 16 ft^2 in area, and the overall size corresponds to a 4 ft by 4 ft window. The garden window category is 15 ft^2 in projected area (35 ft^2 in surface area) and 5 ft wide by 3 ft high by 15 in. deep. The manufactured skylight category is a nominal 8 ft^2 in area, corresponding to a 2 ft by 4 ft skylight. The nominal dimensions of a roof-mounted skylight correspond to centerline spacing of roof framing members; consequently, the rough opening dimensions are 22.5 in. by 46.5 in. The curtain wall and sloped/overhead glazing categories are a nominal 16 ft^2 in area, representing repeating 4 ft by 4 ft panels. The nominal dimensions correspond to centerline spacing of the head and sill and vertical mullions.

Six frame types are listed (although not all for any one category) in order of improving thermal performance. The most conservative assumption is to use the frame category of aluminum frame without a thermal break (although there are products on the market that have

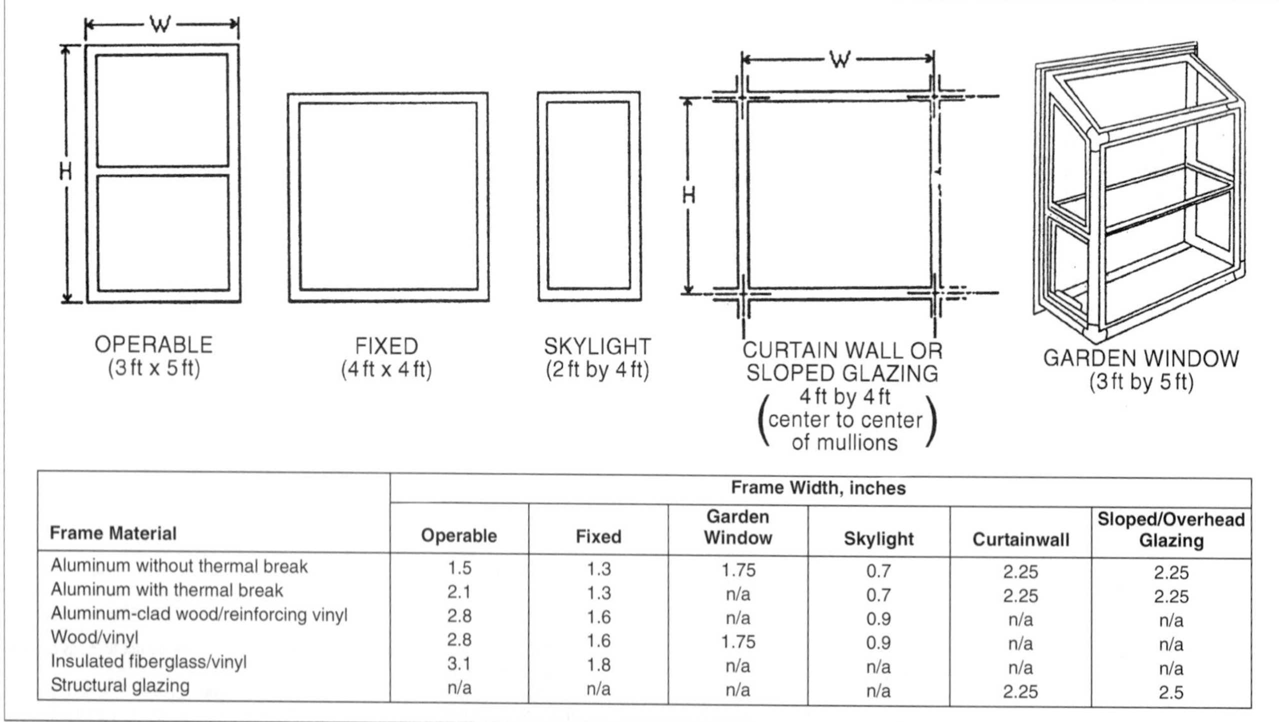

Frame Material	Frame Width, inches					
	Operable	Fixed	Garden Window	Skylight	Curtainwall	Sloped/Overhead Glazing
Aluminum without thermal break	1.5	1.3	1.75	0.7	2.25	2.25
Aluminum with thermal break	2.1	1.3	n/a	0.7	2.25	2.25
Aluminum-clad wood/reinforcing vinyl	2.8	1.6	n/a	0.9	n/a	n/a
Wood/vinyl	2.8	1.6	1.75	0.9	n/a	n/a
Insulated fiberglass/vinyl	3.1	1.8	n/a	n/a	n/a	n/a
Structural glazing	n/a	n/a	n/a	n/a	2.25	2.5

Fig. 4 Standard Fenestration Units

Table 5 Glazing U-Factors for Various Wind Speeds

Wind Speed, mph		
15	7.5	0
U-Factor, Btu/h·ft²·°F		
0.10	0.10	0.10
0.20	0.20	0.19
0.30	0.29	0.28
0.40	0.38	0.37
0.50	0.47	0.45
0.60	0.56	0.53
0.70	0.65	0.61
0.80	0.74	0.69
0.90	0.83	0.78
1.00	0.92	0.86
1.10	1.01	0.94
1.20	1.10	1.02
1.30	1.19	1.10

higher U-factors). The aluminum frame with a thermal break is for frames having at least a 3/8 in. thermal break between the inside and outside for all members including both the frame and the operable sash, if applicable. (Products are available with significantly wider thermal breaks, which achieve considerable improvement.) The reinforced vinyl/aluminum clad wood category represents vinyl-frame products, such as sliding glass doors or large windows that have extensive metal reinforcing within the frame and wood products with extensive metal, usually on the exterior surface of the frame. Both of these factors provide short circuits, which degrade the thermal performance of the frame material. The wood/vinyl frame is meant to represent the improved thermal performance that is possible if the thermal short circuits from the previous frame category do not exist. Insulated fiberglass/vinyl represents fiberglass or vinyl frames that do not have metal reinforcing and whose frame cavities are filled with insulation. For several site-assembled product types, there is a structural glazing frame category that is intended to represent products where sheets of glass are butt-glazed to each other using a sealant only, and none of the framing members is exposed to the exterior. For glazing with a steel frame, use aluminum frame values. For aluminum window with wood trim or vinyl cladding, use the values for aluminum. Frame type refers to the primary unit. Thus, when storm sash is added over another fenestration product, use the values given for the nonstorm product.

To estimate the overall U-factor of a fenestration product that differs significantly from the assumptions given in Table 4 and/or Figure 4, first determine the area that is frame/sash, center-of-glass, and edge-of-glass (based on a 2-1/2 in. band around the perimeter of each glazing unit). Next, determine the appropriate component U-factors. These can be taken either from the standard values listed in italics in Table 4 for glass, from the values in Table 1 for frames, or from some other source such as test data or computed factors. Finally, multiply the area and the component U-factors, sum these products, and then divide by the rough opening in the building envelope where this product will fit to obtain the overall U-factor U_o.

Table 5 provides approximate data to convert the overall U-factor at one wind condition to a U-factor at another.

REPRESENTATIVE U-FACTORS FOR DOORS

Doors are often an overlooked component in the thermal integrity of the building envelope. Although swinging and revolving doors represent a small portion of the shell in residential, commercial, and institutional buildings, their U-factor is usually many times higher than that of the walls or ceilings. In some storage and industrial buildings, loading bay doors (overhead doors) represent a significant area of high heat loss. Table 6 contains representative U-factors for swinging, overhead, and revolving doors determined through computer simulation (Carpenter and Hogan 1996). These are generic values, and product-specific values determined in accordance with standards should be used whenever available. NFRC 100, Section B, and CSA *Standard* A453 give procedures for evaluating the performance of swinging doors. Overhead doors are often evaluated in accordance with National Association of Garage Door Manufacturers (NAGDM) *Standard* 105. Where these standards are cited in codes, they must be used for compliance.

Table 6 U-Factors of Doors in Btu/h·ft²·°F

Door Type	No Glazing	Single Glazing	Double Glazing with 1/2 in. Air Space	Double Glazing with $e = 0.10$, 1/2 in. Argon
SWINGING DOORS (Rough Opening—38 in. × 82 in.)				
Slab Doors				
Wood slab in wood frame[a]	0.46			
6% glazing (22 in. × 8 in. lite)	—	0.48	0.46	0.44
25% glazing (22 in. × 36 in. lite)	—	0.58	0.46	0.42
45% glazing (22 in. × 64 in. lite)	—	0.69	0.46	0.39
More than 50% glazing		Use Table 4 (operable)		
Insulated steel slab with wood edge in wood frame[a]	0.16			
6% glazing (22 in. × 8 in. lite)	—	0.21	0.19	0.18
25% glazing (22 in. × 36 in. lite)	—	0.39	0.26	0.23
45% glazing (22 in. × 64 in. lite)	—	0.58	0.35	0.26
More than 50% glazing		Use Table 4 (operable)		
Foam insulated steel slab with metal edge in steel frame[b]	0.37			
6% glazing (22 in. × 8 in. lite)	—	0.44	0.41	0.39
25% glazing (22 in. × 36 in. lite)	—	0.55	0.48	0.44
45% glazing (22 in. × 64 in. lite)	—	0.71	0.56	0.48
More than 50% glazing		Use Table 4 (operable)		
Cardboard honeycomb slab with metal edge in steel frame	0.61			
Stile- and -Rail Doors				
Sliding glass doors/ French doors		Use Table 4 (operable)		
Site-Assembled Stile- and -Rail Doors				
Aluminum in aluminum frame	—	1.32	0.93	0.79
Aluminum in aluminum frame with thermal break	—	1.13	0.74	0.63
REVOLVING DOORS (Rough Opening—82 in. × 84 in.)				
Aluminum in aluminum frame				
Open	—	1.32	—	—
Closed	—	0.65	—	—
SECTIONAL OVERHEAD DOORS (Nominal—10 ft × 10 ft)				
Annunciated steel (nominal U = 1.15)	1.15	—	—	—
Insulated steel (nominal U = 0.11)	0.24	—	—	—
Insulated steel with thermal break (nominal U = 0.08)	0.13	—	—	—

[a] Thermally broken sill (add 0.03 Btu/h·ft²·°F for non-thermally broken sill)
[b] Non-thermally broken sill
[c] Nominal U-factors are through the center of the insulated panel before consideration of thermal bridges around the edges of the door sections and due to the frame.

Swinging doors can be divided into two categories: slab and stile-and-rail. A stile-and-rail door is a swinging door with a full-glass insert supported by horizontal rails and vertical stiles. The stiles and rails are typically either solid wood members or extruded aluminum or vinyl, as shown in Figure 5. Most residential doors are slab type with either solid wood, steel, or fiberglass skin over foam insulation in a wood frame with aluminum sill. The edges of the steel skin door are normally wood to provide a thermal break. In

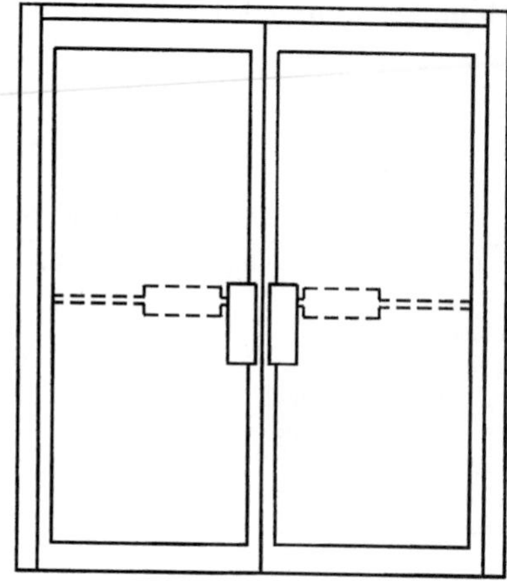

Fig. 5 Details of Stile-and-Rail Door

commercial construction, doors are either steel skin over foam insulation in a steel frame (i.e., utility doors) or a full glass door made up of aluminum stiles, rails, and frame (i.e., entrance doors). The most important factors affecting door U-factor are material construction, glass size, and glass type. Frame depth, slab width, and number of panels have a minor effect on door performance. Sidelites and double doors have U-factors similar to a single door of the same construction. For wood slab doors in a wood frame, the glazing area does not have much effect on the U-factor. For the insulated steel slab in a wood frame, however, glazing area has a strong effect on U-factor. Typical commercial insulated slab doors have a U-factor approximately twice that of residential insulated doors, the prime reason being thermal bridging of the slab edge and the steel frame. Stile-and-rail doors, even if thermally broken, have U-factors 50% higher than a full-glass commercial steel slab door.

There are three generic types of overhead doors: roll-up, uninsulated sectional, and insulated sectional. Metal roll-up doors consist of small metal plates of approximately 2.5 in. in height that “roll up” around a metal rod to open. Sectional doors consist of a series of 2 ft high sections that travel in a track to open. There is a wide range in the design of insulated overhead doors. Factors affecting heat transfer include width of insulation, thermal break design (if any), and design of interior skin. For the uninsulated sectional door, there is very little difference between the center value and the total value: essentially the value of single glazing. The center of the insulated door has low U-factors, but thermal bridging at the door and section edges significantly increases the total U-factor. For doors without thermally broken edges, the total value is 2.5 to 3.3 times greater than the center value. The addition of a good thermal break design reduces this increase to a 1.6 multiplier.

Many commercial buildings use revolving entrance doors. Most of these doors are of similar design: single glazing in an aluminum frame without thermal break. The door, however, can be in two positions: closed (X shaped as viewed from above) or open (+ shaped). At nighttime, these doors are locked in the X position, effectively creating a double-glazed system. During the daytime, the door revolves and is often left positioned so that there is only one glazing between the inside and outside (+ position). U-factors are given in Table 6 for both positions.

EXAMPLES

Example 1. Estimate the U-factor for a manufactured fixed fenestration product with a reinforced vinyl frame and double-glazing with a sputter-type low-e coating (e = 0.10). The gap is 0.5 in. wide and argon-filled, and the spacer is metal.

Solution: Locate the glazing system type in the first column of Table 4 (ID = 23), then find the appropriate product type (fixed) and frame type (reinforced vinyl). The U-factor listed (in the tenth column of U-factors) is 0.33 Btu/h·ft²·°F.

Example 2. Estimate a representative U-factor for a wood-framed, 38 in. by 82 in. swinging French door with eight 11 in. by 16 in. panes (true divided panels), each consisting of clear double-glazing with a 0.25 in. air space and a metal spacer.

Solution: Without more detailed information, assume that the dividers have the same U-factor as the frame and that the divider edge has the same U-factor as the edge-of-glass. Calculate the center-of-glass, edge-of-glass, and frame areas:

$$A_{cg} = 8[(11-5)(16-5)] = 528 \text{ in}^2$$

$$A_{eg} = 8(11 \times 16) - 528 = 880 \text{ in}^2$$

$$A_f = (38 \times 82) - 8(11 \times 16) = 1708 \text{ in}^2$$

Select the center-of-glass, edge-of-glass, and frame U-factors. These component U-factors are 0.55 and 0.64 (from Table 4, glazing ID = 4, U-factor columns 1 and 2) and 0.51 Btu/h·ft²·°F (from Table 1, wood frame, metal spacer, operable, double-glazing), respectively. From Equation (7),

$$U_o = \frac{(0.55 \times 528) + (0.64 \times 880) + (0.51 \times 1708)}{(38 \times 82)}$$

$$= 0.55 \text{ Btu/h} \cdot \text{ft}^2 \cdot {}^\circ\text{F}$$

Example 3. Estimate the overall average U-factor for a multifloor curtain wall assembly that is part vision glass and part opaque spandrel. The typical floor-to-floor height is 12 ft, and the building module is 4 ft as reflected in the spacing of the mullions both horizontally and vertically. For a representative section 4 ft wide and 12 ft tall, one of the modules is glazed and the other two are opaque. The mullions are aluminum frame with a thermal break 3 in. wide and centered on the module. The IGU is double glazing with a pyrolytic low-e coating (e = 0.40) and has a 1/2 in. gap filled with air and a metal spacer. The spandrel panel has a metal pan backed by R-20 insulation and no intermediate reinforcing members.

Solution: It is necessary to calculate the U-factor for the glazed module and for the opaque spandrel modules and then to do an area-weighted average to determine the average U-factor for the overall curtain wall assembly.

First, calculate the overall U-factor for the glazed module. Calculate the center-of-glass, edge-of-glass, and frame areas. The glazed area is 45 in. by 45 in. (48 in. module, 1.5 in. of mullions on each edge).

$$A_{cg} = (45-5)(45-5) = 1600 \text{ in}^2$$

$$A_{eg} = (45 \times 45) - 1600 = 425 \text{ in}^2$$

$$A_f = (48 \times 48) - (45 \times 45) = 279 \text{ in}^2$$

Select the center-of-glass, edge-of-glass, and frame U-factors. These component U-factors are 0.40 and 0.54 (from Table 4, ID = 13, columns 1 and 2) and 1.75 Btu/h·ft²·°F (from Table 1, aluminum frame with a thermal break, metal spacer, curtain wall, double glazing), respectively. From Equation (7),

$$U_{glazing\ module} = \frac{(0.40 \times 1600) + (0.54 \times 425) + (1.75 \times 279)}{(48 \times 48)}$$

$$= 0.59 \text{ Btu/h} \cdot \text{ft}^2 \cdot {}^\circ\text{F}$$

Then, calculate the overall U-factor for the two opaque spandrel modules. The center-of-spandrel, edge-of spandrel, and frame areas are the same as the glazed module. The frame U-factor is the same. Calculate the center-of-spandrel U-factor. In this particular case, the R-value of the insulation does not need to be rated as there are no intermediate framing members penetrating it and providing thermal short circuits. When the resistance of the insulation is added to the exterior air film resistance of 0.17 and the interior air film resistance of 0.68 $ft^2 \cdot °F \cdot h/Btu$ (from Table 1, Chapter 25), the total resistance is 20.85 $ft^2 \cdot °F \cdot h/Btu$ and the U-factor is 1/20.85 = 0.05 $Btu/h \cdot ft^2 \cdot °F$. The edge-of-spandrel U-factor is 40% of the way to the frame U-factor, which is 0.05 + [0.40 × (1.75 − 0.05)] = 0.73 $Btu/h \cdot ft^2 \cdot °F$.

$$U_{opaque\ spandrel\ module} = \frac{(0.05 \times 1600) + (0.73 \times 425) + (1.75 \times 279)}{(48 \times 48)}$$

$$= 0.38\ Btu/(h \cdot ft^2 \cdot °F)$$

Finally, calculate the overall average U-factor for the curtain wall assembly, including the one module of vision glass and the two modules of opaque spandrel.

$$U_{curtain\ wall} = \frac{[0.59 \times (48 \times 48)] + [0.38 \times 2 \times (48 \times 48)]}{3 \times (48 \times 48)}$$

$$= 0.45\ Btu/h \cdot ft^2 \cdot °F$$

Note that even with double glazing having a low-e coating and with R-20 in the opaque areas, this curtain wall with metal pans only has an overall R-value of approximately 2.

Example 4. Estimate the U-factor for a semicircular barrel vault that is 18 ft wide (9 ft tall) and 30 ft long mounted on a 6 in. curb. The barrel vault has an aluminum frame without a thermal break. The glazing is double with a 1/2 in. gap width filled with air and a low-e coating (e = 0.20).

Solution: An approximation can be made by multiplying the U-factor for a site-assembled sloped/overhead glazing product having the same frame and glazing features by the ratio of the surface area (including the curb) of the barrel vault to the rough opening area in the roof that the barrel vault fits over. First, determine the surface area (including the curb) of the barrel vault:

Area of the curved portion of the barrel vault
= (π × diameter/2) × length
= (3.14 × 18/2) × 30 = 848 ft^2

Area of the two ends of the barrel vault
= 2 × (π × $radius^2$)/2 = πr^2
= (3.14 × 9^2) = 254 ft^2

Area of the curb
= perimeter × curb height
= (18 + 30 + 18 + 30) × 6/12 = 48 ft^2

Total surface area of the barrel vault
= 848 + 254 + 48 = 1150 ft^2

Second, determine the rough opening area in the roof that the barrel vault fits over:
= length × width
= 18 × 30 = 540 ft^2

Third, determine the ratio of the surface area to the rough opening area:
= 1150/540 = 2.13

Fourth, determine the U-factor from Table 4 of a site-assembled sloped/overhead glazing product having the same frame and glazing features. The U-factor is 0.71 $Btu/(h \cdot ft^2 \cdot °F)$ (ID = 17, 12th column on the second page of Table 4).

Fifth, determine the estimated U-factor of the barrel vault.

$U_{barrel\ vault}$
= $U_{sloped\ overhead\ glazing}$ × surface area/rough opening for the barrel vault
= 0.71 × 2.13 = 1.51 $Btu/h \cdot ft^2 \cdot °F$

SOLAR HEAT GAIN AND VISIBLE TRANSMITTANCE

Fenestration solar heat gain has two components. First is directly transmitted solar radiation. The quantity of radiation entering the fenestration directly is governed by the solar transmittance of the glazing system. Multiplying the incident irradiance by the glazing area and its solar transmittance yields the solar heat entering the fenestration directly. The second component is the absorbed solar radiation, radiation that is removed from the main beam and absorbed in the glazing and framing materials of the window, some of which is subsequently conducted to the interior of the building.

DETERMINING INCIDENT SOLAR FLUX

Solar Radiation

The flux of solar radiation on a surface normal (perpendicular) to the sun's rays above the earth's atmosphere at the mean earth-sun distance of 92.9 × 10^6 miles (Allen 1973) is defined as the solar constant E_{sc}. The currently accepted value is 433 $Btu/h \cdot ft^2$ (Iqbal 1983). Because the earth's orbit is slightly elliptical, the extraterrestrial radiant flux E_o varies from a maximum of 448 $Btu/h \cdot ft^2$ on January 3, when the earth is closest to the sun (aphelion), to a minimum of 419 $Btu/h \cdot ft^2$ on July 4, when the earth-sun distance reaches its maximum (perihelion).

The earth's orbital velocity also varies throughout the year, so **apparent solar time**, as determined by a solar time sundial, varies somewhat from the **mean time** kept by a clock running at a uniform rate. This variation, called the **equation of time**, is given in Table 7. The conversion between local standard time and solar time involves two steps. First the equation of time is added to the local standard time, and then a longitude correction is added. This longitude correction is four minutes of time per degree difference between the local (site) longitude and the longitude of the **local standard meridian** for that time zone. Standard meridians are found every 15° from 0° at Greenwich, England (Greenwich Meridian). In the United States and Canada, these values are 60° for Atlantic Standard Time, 75° for Eastern Standard Time, 90° for Central Standard Time, 105° for Mountain Standard Time, 120° for Pacific Standard Time, 135° for Alaska Standard Time, and 150° for Hawaii-Aleutian Standard Time.

Equation (10) relates apparent solar time (AST) to local standard time (LST) as follows:

$$AST = LST + ET/60 + (LSM - LON)/15 \tag{10}$$

where

AST = apparent solar time, decimal hours
LST = local solar time, decimal hours
ET = equation of time, decimal minutes
LSM = local standard time meridian, decimal ° of arc
LON = local longitude, decimal ° of arc

Because the earth's equatorial plane is tilted at an angle of 23.45° to the orbital plane, the **solar declination** δ (the angle between the

Table 7 Extraterrestrial Solar Irradiance and Related Data

	E_o, $Btu/h \cdot ft^2$	Equation of Time, min	Declination δ, degrees	A $Btu/h \cdot ft^2$	B (Dimensionless Ratios)	C
Jan	448.8	−11.2	−20.0	390	0.142	0.058
Feb	444.2	−13.9	−10.8	385	0.144	0.060
Mar	437.7	−7.5	0.0	376	0.156	0.071
Apr	429.9	1.1	11.6	360	0.180	0.097
May	423.6	3.3	20.0	350	0.196	0.121
June	420.2	−1.4	23.45	345	0.205	0.134
July	420.3	−6.2	20.6	344	0.207	0.136
Aug	424.1	−2.4	12.3	351	0.201	0.122
Sep	430.7	7.5	0.0	365	0.177	0.092
Oct	437.3	15.4	−10.5	378	0.160	0.073
Nov	445.3	13.8	−19.8	387	0.149	0.063
Dec	449.1	1.6	−23.45	391	0.142	0.057

Note: Data are for 21st day of each month during the base year of 1964.

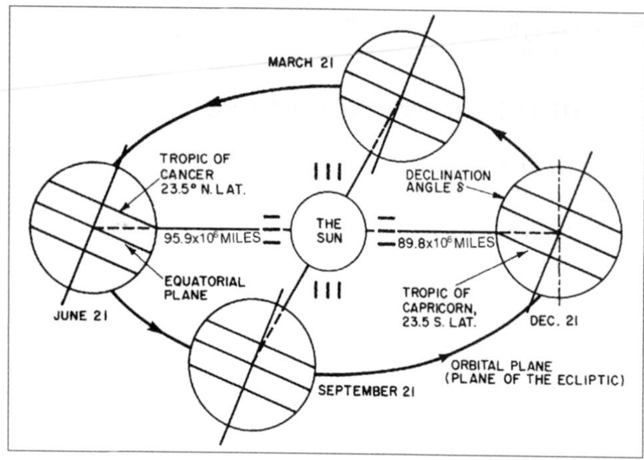

Fig. 6 Motion of Earth around Sun

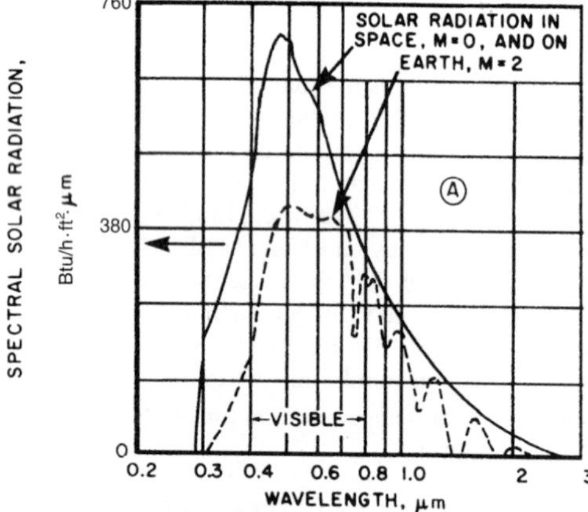

Fig. 7 Terrestrial and Extraterrestrial Solar Spectral Irradiances

earth-sun line and the equatorial plane) varies through out the year, as shown in Figure 6, Table 7, and Equation (11). This variation causes the changing seasons with their unequal periods of daylight and darkness. The following equation can be used to estimate the declination from the day of year η, but it is more accurate to look up the actual declination in an astronomical or nautical almanac for the actual year and date in question.

$$\delta = 23.45 \sin \{[360(284 + \eta)]/365\} \tag{11}$$

The spectral distribution of solar radiation beyond the earth's atmosphere (Figure 7) resembles the radiant energy emitted by a blackbody at about 11,000°R. The peak solar spectral irradiance of 675 Btu/h·ft²·µm is reached at 0.451 µm (451 nm) in the green portion of the visible spectrum.

In passing through the earth's atmosphere, the sun's radiation is reflected, scattered, and absorbed by dust, gas molecules, ozone, water vapor, and water droplets (fog and clouds). The extent of this depletion at any given time is determined by atmospheric composition and length of the atmospheric path traversed by the sun's rays. This length is expressed in terms of the air mass m, which is the ratio of the mass of atmosphere in the actual earth-sun path to the mass that would exist if the sun were directly overhead at sea level (m = 1.0). For most purposes, the air mass at any time equals the cosecant of the solar altitude multiplied by the ratio of the existing barometric pressure to standard pressure. Beyond the atmosphere, m = 0.

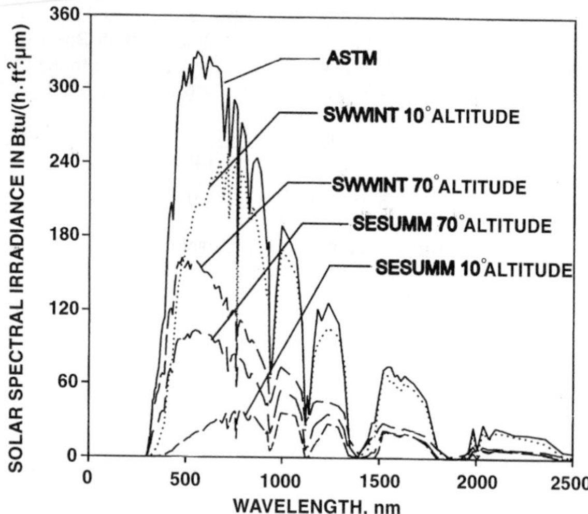

Fig. 8 Comparison of Standard Air Mass m = 1.5 Solar Spectrum with Direct Beam Spectra Through Atmospheres
Characteristic of southwest in winter (SWWINT) and southeastern U.S. in summer (SESUMM) for two solar altitude angles (McCluney 1996)

Table 8 Portions of Total Solar Spectral Irradiance Contained in Portions of Visible Spectrum

Wavelength, nm		Percent Irradiance	Percent Illuminance
Start	End		
370	770	54.4	100.0
380	760	52.2	100.0
390	750	50.2	99.9
400	740	47.4	99.9
410	730	44.9	99.8
420	720	41.9	99.8
430	710	39.5	99.8
440	700	36.7	99.8
450	690	35.3	99.5
460	680	31.1	99.1

Note: The integrated total irradiance = 301 Btu/h·ft² and illuminance = 100 klx.

Most ultraviolet solar radiation is absorbed by the ozone in the upper atmosphere, while part of the radiation in the short-wave portion of the spectrum is scattered by air molecules, imparting the blue color to the sky. Water vapor in the lower atmosphere causes the characteristic absorption bands observed in the solar spectrum at sea level (Figure 7). For a solar altitude β of 41.8° (air mass m = 1.5), the total solar direct beam flux on a clear day at sea level can be divided into spectral regions as follows. Less than 3% of the total is in the ultraviolet, 47% is in the visible region, and the remaining 50% is in the infrared (ASTM *Standard* E 891). The maximum spectral irradiance occurs at 0.61 µm, and little solar energy (less than 5% of the spectrum) exists at wavelengths beyond 2.1 µm.

It is interesting to see what fraction of the total solar irradiance lies in the visible part of the spectrum. Since the limits of the visible portion vary from observer to observer (and because the eye is not very sensitive to radiation at the spectral limits of vision), the fractions of total irradiance and illuminance found between different spectral limits at the edge of the visible portion of the spectrum can be calculated. The results are shown in Table 8 for the ASTM air mass m = 1.5 terrestrial spectrum shown in Figure 8.

The solar spectral distribution shown in Figure 7 for m = 0 is the World Radiation Center's 1985 standard extraterrestrial spectrum for a solar constant of 433 Btu/h·ft² (Wehrli 1985). The one for m = 1.5 in Figure 8 is from ASTM *Standard* E 891. This latter takes no account of monthly variations in irradiance caused by changes in the

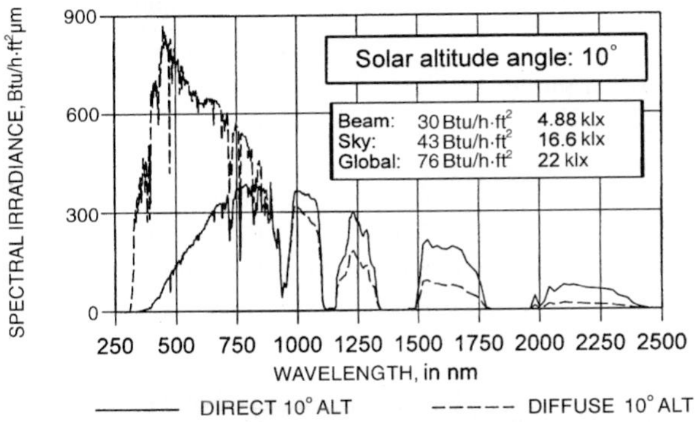

Fig. 9 Comparison of Direct and Diffuse Solar Spectra for Low Solar Altitude Angle

earth-sun distance and by variations in the atmosphere's constituent particulates and gases.

When variations in atmospheric constituents and air mass are considered, the solar spectral distribution is seen to vary, as illustrated in Figure 8, for two different atmospheric conditions and for two solar altitude angles, and in Figure 9 for both direct and diffuse radiation components and a low sun angle. It is clear that the spectral distribution for low sun angle beam radiation is significantly shifted toward longer wavelengths. This shift can be seen visually as a reddening of the sun near to the horizon. Clear sky diffuse radiation is generally shifted toward the blue end of the spectrum.

Upon passage through the atmosphere, extraterrestrial solar radiation is reduced in magnitude due to absorption by atmospheric gases and particulates. The strength of this absorption varies with wavelength, and the terrestrial solar spectrum exhibits definite "dips" in regions of strong absorption, called **absorption bands**. The most prominent atmospheric gases contributing to this effect are listed below:

- **Ozone.** Strongest absorption in the ultraviolet, some in the visible. Concentration variable.
- **H_2O.** Strongest absorption in near and far IR. Highly variable.
- **CO_2.** Strongest absorption in near and far IR. Slightly variable.
- **O_2, CH_4, N_2O, CFCs.** Strongest absorption mostly in the IR. Concentration almost constant.
- **NO_2.** Strongest absorption in the visible. Highly variable in polluted areas.

The effect of aerosols and other particulates on terrestrial solar radiation can be significant. Diffuse sky radiation is solar beam radiation that has been multiply scattered out of the direct beam and downward through the atmosphere to the earth's surface. This scattering is produced by 30 different atmospheric molecules (of which the above are the most significant optically) and by larger particles of different types, including aerosols of water, dust, smoke, and particulates of other kinds.

More information on atmospheric optics can be found in Chapter 44 of the Optical Society of America's *Handbook of Optics* (Bass 1995) and in Iqbal (1983).

Glazing systems exhibiting strong spectral selectivity (strong changes in their optical properties over the solar spectrum) will selectively pass more or less radiation in different parts of the spectrum. This effect can cause substantial changes in the solar heat gain coefficient of the glazing system when the shape of the solar spectrum shifts appreciably. This in turn can cause errors in solar heat gain predictions when the actual solar radiation on a fenestration system has a spectrum that is different from the standard spectrum used to determine the solar heat gain coefficient of that system (McCluney 1996). These errors are typically 5 to 10% but can be substantially greater in special cases.

Some short-wavelength radiation scattered by air molecules, dust, and other particulates in the atmosphere reaches the earth in the form of diffuse sky radiation E_d. Since this diffuse radiation comes from all parts of the sky, its irradiance is difficult to predict and varies as moisture and particulate content and sun angle change throughout any given day. For completely overcast conditions, the diffuse component accounts for all solar radiant heat gain of fenestrations.

The total short-wavelength irradiance E_t reaching a terrestrial surface is the sum of the direct solar radiation E_D, the diffuse sky radiation E_d, and the solar radiation E_r reflected from surrounding surfaces. The irradiance on the fenestration aperture of the direct beam component E_D is the product of the direct normal irradiation E_{DN} and the cosine of the angle of incidence θ between the incoming solar rays and a line normal (perpendicular) to the surface:

$$E_t = E_{DN} \cos\theta + E_d + E_r \tag{12}$$

A method for computing all the factors on the right side of Equation (12) is presented in the sections on Direct Normal Irradiance and Diffuse and Ground-Reflected Radiation. Perez et al. (1986), Gueymard (1987), *Solar Energy* (1988), and Gueymard (1993) give more detailed models, which separate the diffuse sky radiation into different components. Gueymard (1995) provides a comprehensive spectrally based model for calculating the spectral and broadband totals of all three terms in Equation (12), for cloudless sky conditions. The Gueymard model allows user input of the concentrations of a variety of atmospheric constituents, including particulates.

The importance of the diffuse component is illustrated in Figure 9, which shows that at low sun angles the diffuse component contains more radiant flux than the direct beam component, even on a clear day, and that the spectral distributions of the two components are quite different. Although the total irradiances are relatively modest for both of these components, they are not insignificant for annual energy performance calculations.

Vertical windows receive considerable quantities of diffuse sky radiation over the course of a year. The diffuse component is an important part of solar radiant heat gain.

Determining Solar Angle

The sun's position in the sky is conveniently expressed in terms of the solar altitude β above the horizontal and the solar azimuth ϕ measured from the south (Figure 10). These angles, in turn, depend on the local latitude L; the solar declination δ, which is a function of the date [Table 7 or Equation (11)]; and the apparent solar time, expressed as the hour angle H, where

$$H = 15(\text{AST} - 12) \tag{13}$$

Equations (14) and (15) relate β and ϕ to the three angles just mentioned:

$$\sin\beta = \cos L \cos\delta \cos H + \sin L \sin\delta \tag{14}$$

$$\cos\phi = \frac{\sin\beta \sin L - \sin\delta}{\cos\beta \cos L} \tag{15}$$

Figure 10 shows the solar position angles and incident angles for horizontal and vertical surfaces. Line OQ leads to the sun, the north-south line is NOS, and the east-west line is EOW. Line OV is perpendicular to the horizontal plane in which the solar azimuth ϕ (angle HOS) and the surface azimuth Ψ (angle POS) are located. Angle HOP is the surface solar azimuth γ, defined as

$$\gamma = \phi - \psi \tag{16}$$

The solar azimuth ϕ is positive for afternoon hours and negative for morning hours. Likewise, surfaces that face west have a positive surface azimuth ψ; those facing east have a negative surface azimuth (Table 9). If γ is greater than 90° or less than −90°, the surface

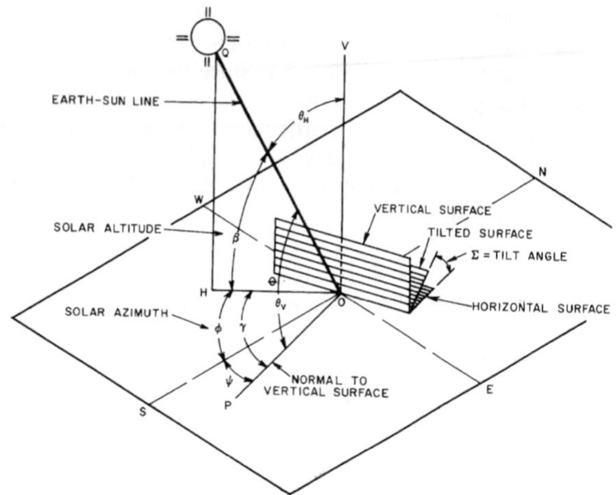

Fig. 10 Solar Angles for Vertical and Horizontal Surfaces

Table 9 Surface Orientations and Azimuths, Measured from South

Orientation	N	NE	E	SE	S	SW	W	NW
Surface azimuth ψ	180°	−135°	−90°	−45°	0	45°	90°	135°

is in the shade. Table 9 gives values in degrees for the surface azimuth ψ, applicable to the orientations of interest.

The angle of incidence θ for any surface is defined as the angle between the incoming solar rays and a line normal to that surface. For the horizontal surface shown in Figure 10, the incident angle θ_H is QOV; for the vertical surface, the incident angle θ_V is QOP.

For any surface, the incident angle θ is related to β, γ, and the tilt angle of the surface Σ by

$$\cos\theta = \cos\beta\,\cos\gamma\sin\Sigma + \sin\beta\cos\Sigma \tag{17}$$

where Σ = tilt angle of surface from horizontal.

When the surface is horizontal, Σ = 0° and

$$\cos\theta_H = \sin\beta \tag{18}$$

For a vertical surface, Σ = 90° and

$$\cos\theta_V = \cos\beta\cos\gamma \tag{19}$$

Direct Normal Irradiance

At the earth's surface on a clear day, direct normal irradiation, or solar irradiance E_{DN}, is represented by

$$E_{DN} = \frac{A}{\exp(B/\sin\beta)} \tag{20}$$

where

A = apparent solar irradiation at air mass $m = 0$ (Table 7)
B = atmospheric extinction coefficient (Table 7)

Values of A and B vary during the year because of seasonal changes in the dust and water vapor content of the atmosphere and because of the changing earth-sun distance. Equation (20) does not give the maximum value of E_{DN} that can occur in each month but yields values that are representative of conditions on cloudless days for a relatively dry and clear atmosphere. For very clear atmospheres, E_{DN} can be 15% higher than indicated by Equation (20), using values of A and B in Table 7.

For locations where clear, dry skies predominate (e.g., at high elevations) or, conversely, where hazy and humid conditions are frequent, values found by using Equation (20) and Table 7 should be multiplied by the Clearness Numbers in Threlkeld and Jordan (1958), reproduced as Figure 5 in Chapter 32 of the 1999 *ASHRAE Handbook—Applications*. This broadband model should only be used for determining fenestration solar gain when the glazing system is not strongly spectrally selective and when its angular selectivity closely matches that of single-pane glass.

Table 10 Solar Reflectances of Foreground Surfaces

Foreground Surface	Incident Angle 20°	30°	40°	50°	60°	70°
New concrete	0.31	0.31	0.32	0.32	0.33	0.34
Old concrete	0.22	0.22	0.22	0.23	0.23	0.25
Bright green grass	0.21	0.22	0.23	0.25	0.28	0.31
Crushed rock	0.20	0.20	0.20	0.20	0.20	0.20
Bitumen and gravel roof	0.14	0.14	0.14	0.14	0.14	0.14
Bituminous parking lot	0.09	0.09	0.10	0.10	0.11	0.12

Adapted from Threlkeld (1962).

Diffuse and Ground-Reflected Radiation

The following equations can be used to generate E_d and E_r, where all angles are in degrees. The solar azimuth ϕ and the surface azimuth ψ are measured in degrees from south; angles to the east of south are negative, and angles to the west of south are positive. Values of A, B, and C are given in Table 7 for the 21st day of each month. Values for other dates can be obtained by interpolation.

The ratio Y of sky diffuse radiation on a vertical surface to sky diffuse radiation on a horizontal surface is given by

$$\begin{aligned} Y &= 0.55 + 0.437\cos\theta + 0.313\cos^2\theta \quad &\text{for } \cos\theta > -0.2 \\ Y &= 0.45 &\text{for } \cos\theta \le -0.2 \end{aligned} \tag{21}$$

Diffuse irradiance E_d is given by

$$E_d = CYE_{DN} \quad \text{for vertical surfaces} \tag{22}$$

$$E_d = CE_{DN}\frac{1+\cos\Sigma}{2} \quad \text{for surfaces other than vertical} \tag{23}$$

Ground-reflected irradiance E_r is given by

$$E_r = E_{DN}(C + \sin\beta)\rho_g\frac{1-\cos\Sigma}{2} \quad \text{for surfaces at all orientations} \tag{24}$$

where ρ_g is ground reflectivity, often taken to be 0.2 for typical mixture of ground surfaces. For other surfaces, Table 10 lists angle-dependent solar reflectances.

Example 5. Find the solar azimuth and altitude at 3:00 P.M. central daylight savings time on June 21 in St. Louis, MO.

Solution: Central daylight savings time of 3:00 P.M. re-expressed in decimal hours at local standard time is LST = 14.0. The latitude and longitude of St. Louis, MO, are 38.6°N and 90.2°W, respectively. The local standard meridian of the central time zone is 90°W. The equation of time (Table 7) is –1.4 min. From Equation (10), apparent solar time (AST) = 14.0 – 1.4/60 + (90 – 90.2)/15 = 13.9633. From Equation (13), $H = 15 \times (13.9633 - 12) = 29.45°$. Table 7 gives the solar declination δ on June 21 as 23.45°.

Thus, by Equation (14),

$$\sin\beta = \cos(38.6°)\cos(23.45°)\cos(29.45°) + \sin(38.6°)\sin(23.45°) = 0.873$$
$$\beta = 60.76°$$

Using Equation (15),

$$\cos\phi = \frac{\sin(60.76)\sin(38.6) - \sin(23.45)}{\cos(60.76)\cos(38.6)} = 0.384$$

$$\phi = 67.4°$$

Example 6. For the conditions of Example 5, find the incident angle at a window facing west.

Solution: From Example 5, ϕ = 67.4°. From Table 9, ψ = 90°. From Equation (16), γ = 67.4° − 90° = −22.6°.

A negative surface solar azimuth γ indicates that the sun is south of the normal to the surface. Thus, using Equation (19),

$$\cos\theta = \cos(60.76°)\cos(-22.6°) = 0.451$$
$$\theta = 63.2°$$

Example 7. Find the direct, diffuse, and ground-reflected components of the solar irradiation on the window in Example 6.

Solution: From Example 5, sin β = 0.873, and from Table 7, A = 345 Btu/h·ft² and B = 0.205. Therefore, from Equation (20),

$$E_{DN} = 345/[\exp(0.205/0.873)] = 272.8 \text{ Btu/h}\cdot\text{ft}^2$$

$$E_{DN}\cos\theta = 123 \text{ Btu/h}\cdot\text{ft}^2$$

From Table 7, C = 0.134. From Equation (21), Y = 0.55 + 0.437 × 0.451 + 0.313 × $(0.451)^2$ = 0.811. From Equation (22),

$$E_d = (0.134)(0.811)(272.8) = 29.6 \text{ Btu/h}\cdot\text{ft}^2$$

From Equation (24), assuming a ground reflectivity ρ_g of 0.2,

$$E_r = (272.8)(0.134 + 0.873)(0.2)(1)/2 = 27.5 \text{ Btu/h}\cdot\text{ft}^2$$

Thermal Infrared Radiation

Any material above a temperature of absolute zero emits electromagnetic radiation. The rate of emission depends upon the temperature of the material and can be expressed in a simple equation, called the Stefan-Boltzmann law:

$$E_b = \sigma T^4 \quad (25)$$

where

E_b = hemispherical total emissive power of black body, Btu/h·ft²
T = temperature, °R
σ = Stefan-Boltzmann constant, 0.1714 × 10^{-8} Btu/h·ft²·°R⁴

The **emissivity** e of a surface is the ratio of the emission of thermal radiant flux from the surface to the flux that would be emitted by a blackbody emitter at the same temperature. Given the temperature and emissivity of a surface, the emitted irradiance spectrum can be computed from

$$E = e\sigma T^4$$

where E is the emitted irradiance in units of flux per unit area, e is the emissivity of the surface, σ is the Stefan-Boltzmann constant, and T is the absolute temperature of the surface.

The maximum value of **hemispherical emissivity** e for any material is 1.0, in which case the surface emits the theoretical maximum amount of radiation possible. In this case, the surface is called a **blackbody**, and the radiation emitted by the surface is called **blackbody radiation**.

Occasionally, a related quantity called **normal emissivity** is used. The relationship between them is illustrated in Figure 11. The spectral distribution of blackbody radiation is illustrated in Figure 12 for temperatures ranging from 540°R (room temperature) to 36,000°R.

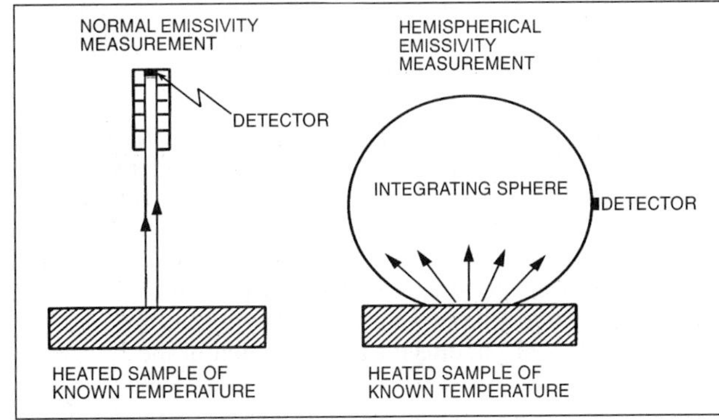

Fig. 11 Illustration of Difference Between Normal and Hemispherical Emissivity

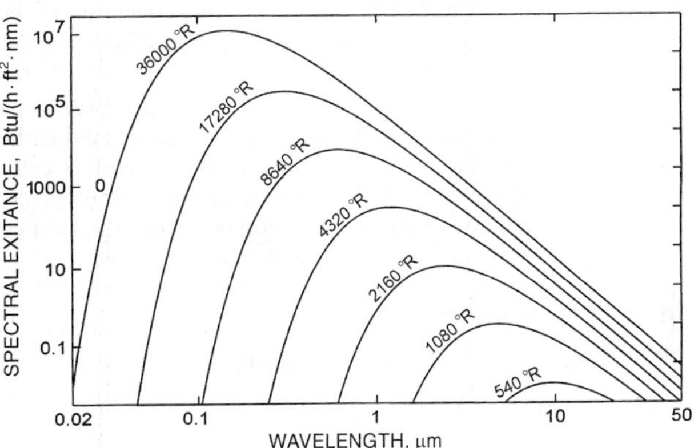

Fig. 12 Spectral Distributions of Blackbody Radiation at Different Source Temperatures

OPTICAL PROPERTIES

Solar radiation (including both direct rays from the sun and diffuse rays from the sky, clouds, and surrounding objects) incident on a fenestration system is partly transmitted and partly reflected by the glazings of that system. An additional fraction is absorbed within the glazings and/or the coatings on their surfaces. The fraction of incident flux that is reflected is called the **reflectance** R, the fraction absorbed is called the **absorptance** $\mathcal{A}$, and the fraction transmitted is the **transmittance** T. The sum of the transmittance T, absorptance $\mathcal{A}$, and reflectance R of a glazing layer is unity:

$$T + R + \mathcal{A} = 1 \quad (26)$$

However, this is complicated by the fact that radiation incident on a surface can have nonconstant distributions over the directions of incidence and over the wavelength (or frequency) scale. Thus, when measuring one or more of the optical properties, the wavelength distribution and direction of incident (and emerging) radiation must be specified.

Angular Dependence

The concept of solid angle is needed to understand angular dependence. A solid angle is defined and enclosed by all rays joining a point to a closed curve. For a closed curve on a sphere of radius R, the solid angle ϖ is the ratio of the projected area A on the sphere to the square of R. A sphere has a solid angle of 4π steradians (4π sr); a hemisphere has a 2π sr solid angle.

Radiation incident on a point in a surface comes to that point from many directions in some solid angle. For a cone of half angle α, the solid angle defined by the circular top and point bottom of that cone is given by

$$\varpi = 2\pi(1 - \cos\alpha) \quad (27)$$

In measuring transmittance or reflectance, a sample is illuminated over a specified solid angle. The reflected or transmitted flux is then collected within another solid angle. The size of a conical solid angle and the direction of its axis need to be specified to obtain meaningful results. A conical solid angle is bounded by a right circular cone.

ASTM *Standards* E 903, E 1084, E 971, and E 972, as well as NFRC 300, which refers to ASTM *Standard* E 971, refer to **conical-hemispherical** measurements of optical properties. The reason is that for most thermal or HVAC design calculations, only the total flux transmitted into a **hemispherical** solid angle, due to the direct solar beam incident in a small **conical** solid angle, is of interest. All transmitted solar irradiance is considered heat gain in these applications, regardless of the directional distribution of the transmitted radiation.

For complex fenestrations (those with nonspecular components), for many daylighting applications, and for some passive solar space-heating applications, the directional distribution of transmitted radiation is of interest. In such cases, it is important to know what is called the **biconical transmittance and reflectance** of fenestration systems, defined for conical solid angles of incidence and emergence (see Figure 13).

Biconical optical properties are needed (1) to treat diffuse sky as well as direct beam radiation, (2) to handle the directional distribution of the flux entering a room through a window, and (3) to calculate the angle-dependent optical and solar gain properties of multiple-pane window systems and complex glazing systems, including those with integral or attached shading devices.

Spectral Dependence

Frequency and wavelength are related through

$$\lambda f = c \tag{28}$$

where

λ = wavelength, m
f = frequency, Hz
c = speed of light = 3.0×10^8 m/s in air at standard atmospheric pressure

The wavelength dependence of radiometric quantities is denoted with a subscript λ attached to the optical quantity, thus ϕ_λ, E_λ, and L_λ. The wavelength dependency of optical properties is denoted by the functional notation, thus $\alpha(\lambda)$, $\rho(\lambda)$, $\tau(\lambda)$, and $e(\lambda)$.

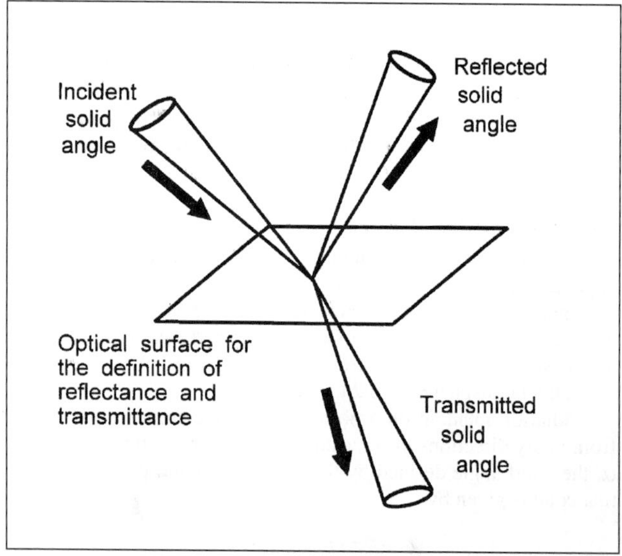

Fig. 13 Geometry for Definition of Biconical Transmittance and Reflectance

All solar radiant flux becomes heat when it is absorbed by materials such as glazings, window frames, and room surfaces. This includes radiation in the UV, visible, and IR portions of the spectrum. Recall from Equation (26) that the sum of the transmittance, reflectance, and absorptance is 1.0.

There is an additional relationship among the optical properties that is of interest and importance. It is called Kirchhoff's law (McCluney 1994a):

$$\mathcal{A}(\lambda, \theta, \phi) = e(\lambda, \theta, \phi) \tag{29}$$

where θ and ϕ are angles defining the directional dependence of the spectral absorptance $\mathcal{A}(\lambda)$ and the spectral emissivity $e(\lambda)$. A consequence of Equations (26) and (29) is that for opaque materials a good absorber is a good emitter and a poor reflector, and vice versa, but only on a wavelength-by-wavelength basis or over a defined wavelength interval. A surface appearing to be an excellent reflector in the visible portion of the spectrum may have a high emissivity (and low reflectance) over most of the infrared spectrum, or vice versa.

Source Spectra. Radiation incident on a surface has a distribution not only over direction within some solid angle but also over a range of wavelengths. The latter distribution is called a **spectrum.** For terrestrial applications, it is only after the extraterrestrial solar spectrum has been modified by passage through the atmosphere that it is of interest (Figures 8 and 9).

SOLAR-OPTICAL PROPERTIES OF GLAZING

The solar-optical properties of a glazing are the wavelength-integrated (or total) transmittance, reflectance, and absorptance of the glazing to incident solar radiation. If the spectral optical properties $[T(\lambda), R(\lambda), \mathcal{A}(\lambda)]$ of the glazing and the spectral irradiance $E(\lambda)$ incident on the glazing are known, the solar optical properties can be calculated from

$$p = \frac{\int_{\lambda_{min}}^{\lambda_{max}} p(\lambda)E(\lambda)d\lambda}{\int_{\lambda_{min}}^{\lambda_{max}} E(\lambda)d\lambda} \tag{30}$$

where p stands for transmittance, reflectance, or absorptance, and the limits of integration are those values of wavelength outside of which the solar spectral irradiance is negligibly small.

If $E(\lambda)$ in Equation (30) is a standard solar spectral irradiance distribution, then the optical property resulting from this equation is called the **solar optical property.** If the spectral properties $p(\lambda)$ are available only at a set of discrete wavelengths λ_k (e.g., measured data), then the following discrete sum approximation of Equation (30) can be calculated:

$$p = \sum_{k=1}^{M} w_k p(\lambda_k) \tag{31}$$

where $w_k = \dfrac{E(\lambda_k)}{\sum_{k=1}^{M} E(\lambda_k)\Delta\lambda_k}$ is the spectral weight for the wavelength λ_k and where $\Delta\lambda_k = 1/2(\lambda_{k+1} - \lambda_{k-1})$.

Unless explicitly stated otherwise, solar-optical properties used in this chapter are calculated using the tabulated ASTM *Standard* E 891 solar spectrum and Equation (31).

Many window glazings do not have strong spectral selectivity over the solar spectrum, so their spectral optical properties can be considered constant, even if the source spectrum changes substantially. In these cases, the transmitted spectral irradiance can be determined by multiplying the incident irradiance by the solar transmittance.

Figure 14 shows the spectral transmittance at normal incidence of typical architectural glasses. The approximate transmittance of total incident solar radiation through clear float glass at an incident angle of 0° ranges from 86% for 3/32 in. thick glass to 84% for 1/8 in. thick glass to 78% for 1/4 in. thick glass. Actual transmittance varies with the amount of iron or other absorbers in the glass. Low iron content glass has a relatively constant spectral transmittance over the entire solar spectrum.

Figure 15 shows the normal incidence spectral transmittances of several common commercially available glazings. Figure 16 shows the normal incidence spectral transmittances and exterior reflectances of a variety of additional coated and tinted glasses, indicating the strong spectral selectivity that is now available from some glass and window manufacturers.

Angular Dependence of Glazing Optical Properties

As Figure 17 shows, the optical properties of a single sheet of clear glass depend on the angle of incidence. This variation is small for incident angles below 40° but becomes significant at larger angles.

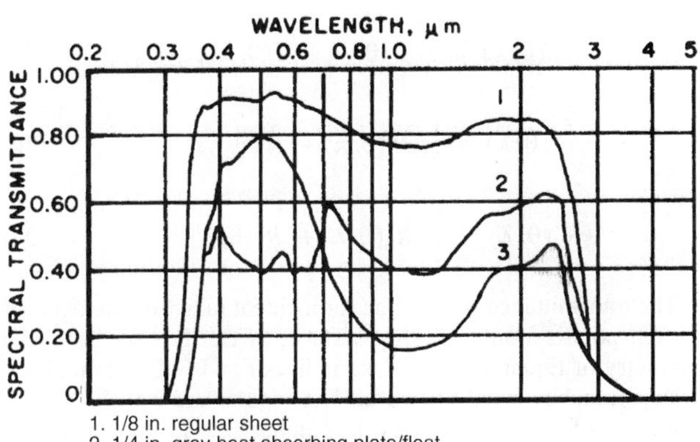

Fig. 14 Spectral Transmittance for Typical Architectural Glass

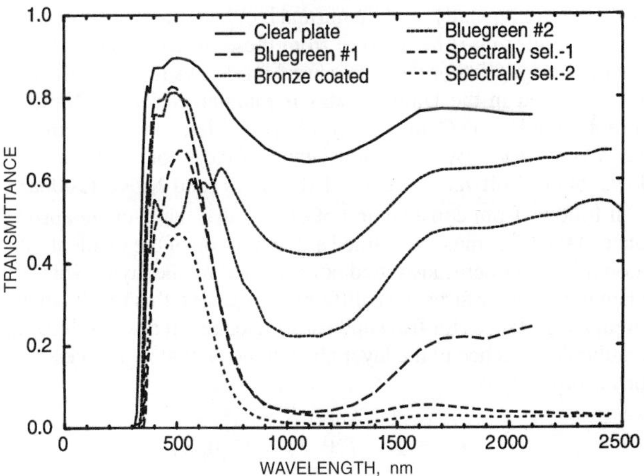

Fig. 15 Spectral Transmittances of Commercially Available Glazings
(McCluney 1993)

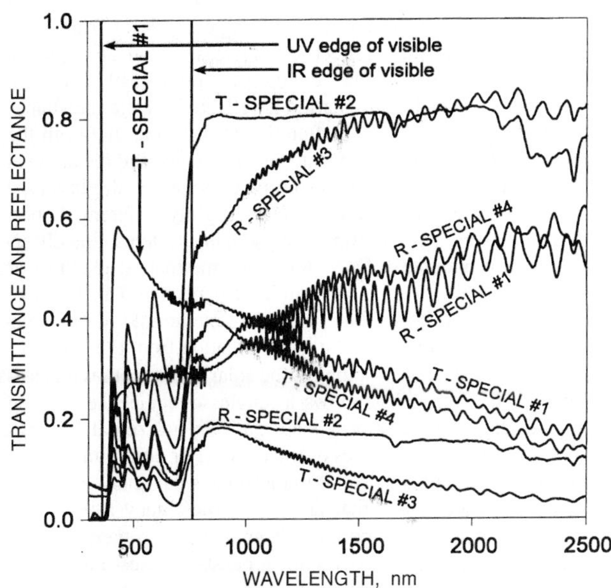

Fig. 16 Spectral Transmittances and Reflectances of Strongly Spectrally Selective Commercially Available Glazings
(McCluney 1996)

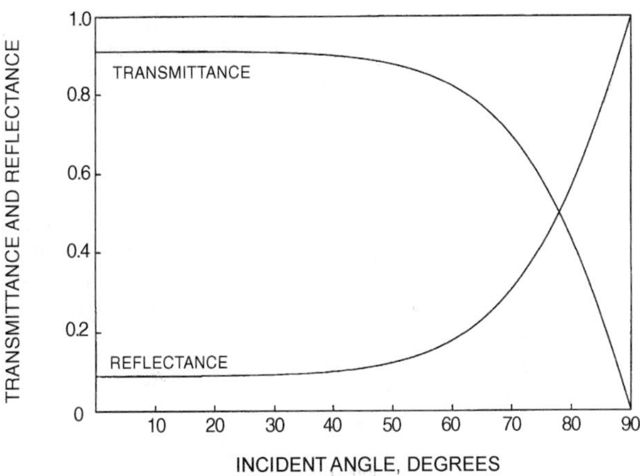

Fig. 17 Transmittance and Reflectance of Plane, Parallel, Glass Plate
refractive index $n = 1.55$, thickness $t = 1/8$ in., absorptivity $\alpha = 0.0003$ in^{-1}

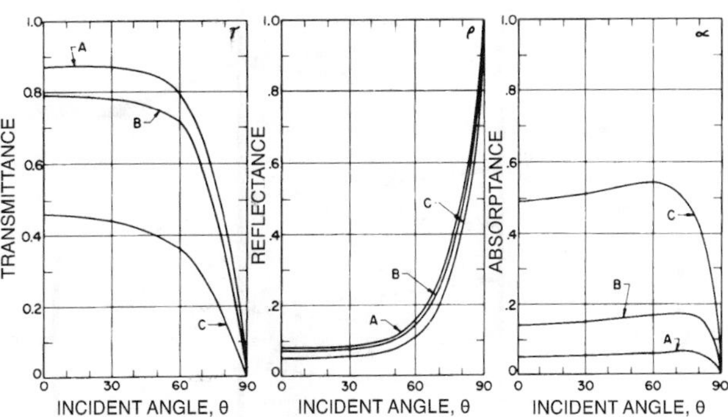

Fig. 18 Variations with Incident Angle of Solar-Optical Properties for (A) Double-Strength Sheet Glass, (B) Clear Plate Glass, and (C) Heat-Absorbing Plate Glass

Figure 18 compares the properties of glasses of different thickness and composition. As the incident angle increases from zero, transmittance diminishes, reflectance increases, and absorptance first increases because of the lengthened optical path and then decreases as more incident radiation is reflected. While the shapes of the property curves are superficially similar, note that both the magnitude of the transmittance at normal incidence and the angle at which the transmittance changes significantly vary with glass type and thickness. The three curves all have slightly different shapes. For coated glasses or for multiple-pane glazing systems, this difference is more pronounced. One cannot assume that all glazings or glazing systems have a universal angular dependence. This is one of the inadequacies of the shading coefficient methodology for determining solar heat gain that led to its elimination from this edition.

In North America, peak summertime solar gains occur with east- and west-facing vertical windows at angles of incidence ranging from about 25 to 55°. The peak solar gain for horizontal glazings occurs typically at small angles of incidence. For north- and south-facing vertical glazings, peak summertime solar gains occur at angles of incidence greater than about 40° (McCluney 1994b).

Angles of incidence important for annual energy performance calculations range from 5° to over 80° for east- and west-facing vertical windows and for horizontal glazings. This range is only slightly diminished for south-facing windows. For north-facing windows, the direct beam solar gains are small and their angles of incidence range from 62 to 86° (McCluney 1994b).

Optical Properties of Single Glazing Layers

The optical properties of a single layer of glazing material are outlined in Figure 19. The layer has a thickness d and is characterized by a surface reflectivity and transmissivity, ρ and τ, for each of the two surfaces (denoted f and b in the figure) and an absorptivity, α, which is a volumetric property of the material (assumed of uniform composition). In general, τ and ρ are characteristics of the interface between the material and the adjacent medium; they may in principle be different for the two surfaces (e.g., for a coated surface, or where a material layer is adjacent to another material rather than air). All three properties, transmissivity, reflectivity, and absorptivity, depend on the wavelength of the incident radiation, and τ and ρ also depend on the incident angle θ of the radiation incident on the layer.

The transmittance T and the front reflectance R^f of a **layer** (as opposed to a **surface**) contain the effects of multiple reflections between the two surfaces of the layer, as indicated in Figure 19, as well as the effects of absorption during the passage through the layer (one or more times), due to the volume absorptivity α. The same is true of the back reflectance R^b, which is the reflectance of the layer for radiation incident on back side b and which is not illustrated in the figure. For non-normal incidence, surface reflectances are in general different for the two possible polarizations of light, conventionally denoted s (TE) and p (TM). We distinguish these below by a subscript μ (= s,p) on the surface reflectance. [A pre-subscript is used where there is a possibility of confusion with later notational additions.] The transmittance and reflectances are given by

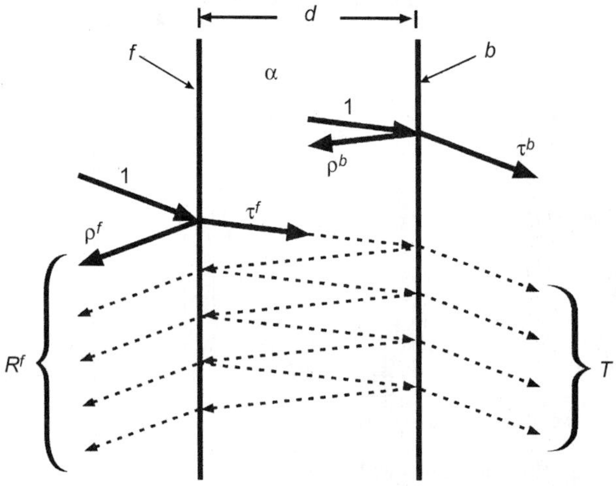

Fig. 19 Optical Properties of a Single Glazing Layer

$$ {}_{\mu}T(\theta, \lambda) = \frac{\tau_{\mu}^{f}(\theta, \lambda)\tau_{\mu}^{b}(\theta, \lambda)e^{-\frac{\alpha(\lambda)d}{\cos\zeta}}}{1 - \rho_{\mu}^{f}(\theta, \lambda)\rho_{\mu}^{b}(\theta, \lambda)e^{-\frac{2\alpha(\lambda)d}{\cos\zeta}}} \quad (32) $$

$$ {}_{\mu}R^{f}(\theta, \lambda) = \rho_{\mu}^{f}(\theta, \lambda) + \rho_{\mu}^{b}(\theta, \lambda)\,{}_{\mu}T(\theta, \lambda)e^{-\frac{\alpha(\lambda)d}{\cos\zeta}} \quad (33) $$

$$ {}_{\mu}R^{b}(\theta, \lambda) = \rho_{\mu}^{b}(\theta, \lambda) + \rho_{\mu}^{f}(\theta, \lambda)\,{}_{\mu}T(\theta, \lambda)e^{-\frac{\alpha(\lambda)d}{\cos\zeta}} \quad (34) $$

where ζ is the angle at which radiation incident at angle θ propagates within the glazing layer (the refracted angle). Since sunlight is unpolarized, the transmittance and reflectances of an isolated glazing layer are then calculated from

$$ T(\theta, \lambda) = \frac{1}{2}[{}_{s}T(\theta, \lambda) + {}_{p}T(\theta, \lambda)] \quad (35) $$

$$ R^{f}(\theta, \lambda) = \frac{1}{2}[{}_{s}R^{f}(\theta, \lambda) + {}_{p}R^{f}(\theta, \lambda)] \quad (36) $$

$$ R^{b}(\theta, \lambda) = \frac{1}{2}[{}_{s}R^{b}(\theta, \lambda) + {}_{p}R^{b}(\theta, \lambda)] \quad (37) $$

The transmittance is the same for incident radiation incident (of a given polarization) on either surface, as can be seen from the symmetry of Equation (32) in the indices f and b. Front and back reflectances, however, may differ. The angular and wavelength dependence of these quantities is emphasized in the equations through explicit function reference [e.g., $T(\theta,\lambda)$]. This dependence will not always be made explicit (in the interest of brevity of equations) but should not be forgotten.

The transmittance and reflectances are the basic measurable quantities for an isolated glazing layer in air. Measurements on glazing layers are typically made at normal incidence, and the properties at other angles must be inferred from these measurements. A systematic compilation of these measured properties for most glazings manufactured in the United States is maintained by the National Fenestration Rating Council, Silver Spring, MD, and is available on the World Wide Web at http://www.nfrc.org or at http://windows.lbl.gov/software/ [see also LBL (1994) and NFRC (2000a)].

It follows from conservation of energy that the average absorptance of the layer must be defined as the fraction of the incident radiation that is neither transmitted nor reflected by the layer. Note that when the layer surfaces have different properties, this results in different absorptances for front and back incidence. It also produces an angular dependence in the layer absorptance a that is not present in the absorptivity α:

$$ {}_{\mu}a^{f}(\theta, \lambda) = 1 - {}_{\mu}T(\theta, \lambda) - {}_{\mu}R^{f}(\theta, \lambda) \quad (38a) $$

$$ {}_{\mu}a^{b}(\theta, \lambda) = 1 - {}_{\mu}T(\theta, \lambda) - {}_{\mu}R^{b}(\theta, \lambda) \quad (38b) $$

and, since these are linear relations,

$$a^f(\theta, \lambda) = 1 - T(\theta, \lambda) - R^f(\theta, \lambda) \quad (38c)$$

$$a^b(\theta, \lambda) = 1 - T(\theta, \lambda) - R^b(\theta, \lambda) \quad (38d)$$

for the unpolarized quantities. A lowercase symbol is used here in anticipation of the discussion of multilayer glazing systems below.

Uncoated Glazings

For uncoated glazings, the interface reflectivities ρ of the two surfaces are the same and may be determined from the Fresnel equations:

$$\rho_s(\theta, \lambda) = \left(\frac{\sin(\theta - \zeta)}{\sin(\theta + \zeta)}\right)^2 \quad (39a)$$

$$\rho_p(\theta, \lambda) = \left(\frac{\tan(\theta - \zeta)}{\tan(\theta + \zeta)}\right)^2 \quad (39b)$$

where ζ is the refracted angle and may be calculated from Snell's law and the real part of the refractive index n relative to air:

$$\sin\theta = n\sin\zeta \qquad \text{and} \qquad \zeta = \arcsin\left(\frac{\sin\theta}{n}\right) \quad (40)$$

At normal incidence, the two polarizations are indistinguishable, and Equation (39) reduces to

$$\rho(0, \lambda) = \frac{[n(\lambda) - 1]^2}{[n(\lambda) + 1]^2} \qquad \text{and} \qquad n(\lambda) = \frac{1 + \sqrt{\rho(0, \lambda)}}{1 - \sqrt{\rho(0, \lambda)}} \quad (41)$$

and since

$$\tau_\mu(\theta, \lambda) + \rho_\mu(\theta, \lambda) = 1 \quad (42)$$

for any surface, a measurement of the spectral transmittance and reflectance at normal incidence may be used with Equations (32) and (33) to determine the refractive index and absorptance as a function of wavelength. Note that n is also wavelength-dependent, although for most glazing materials the dependence is weak over the solar spectrum. Once these quantities are known, the equations may be used to calculate the properties at all angles.

ASHRAE "Standard" Glass

In the discussion of single glazing, and historically in the context of the solar heat gain factor and shading coefficient methodology, ASHRAE has used as a calculation standard the properties of "one-eighth-inch, clear double-strength glass" (DSG). The wavelength-averaged properties of this standard glazing are calculated from the following equations:

$$T_{DSG}(\theta) = \sum_{n=0}^{5} (ts)_n \cos^n\theta \quad (43)$$

$$a^f_{DSG}(\theta) = a^b_{DSG}(\theta) = \sum_{n=0}^{5} (as)_n \cos^n\theta \quad (44)$$

where the coefficients $(ts)_n$ and $(as)_n$ are given in Table 11. The hemispherical average quantities may be calculated by averaging Equations (43) and (44), which yields

$$\langle T_{DSG}\rangle_D = 2\sum_{n=0}^{5} \frac{(ts)_n}{n+2} \quad (45)$$

Table 11 Coefficients for Double-Strength Glass (DSG) for Calculation of Transmittance and Absorptance

n	$(as)_n$	$(ts)_n$
0	0.01154	−0.00885
1	0.77674	2.71235
2	−3.94657	−0.62062
3	8.57881	−7.07329
4	−8.38135	9.75995
5	3.01188	−3.89922

$$\langle a^f_{DSG}\rangle_D = \langle a^b_{DSG}\rangle_D = 2\sum_{n=0}^{5} \frac{(as)_n}{n+2} \quad (46)$$

Determining the Properties of Uncoated Glazing Layers from Normal Incidence Measurements

For uncoated glazings, the front and back transmissivities, reflectivities, transmittances, reflectances, and isolated-layer absorptances are equal, the two polarizations are indistinguishable, and at normal incidence Equations (32) and (33) become

$$T(0, \lambda) = \frac{[\tau(0, \lambda)]^2 e^{-\alpha(\lambda)d}}{1 - [\rho(0, \lambda)]^2 e^{-2\alpha(\lambda)d}} \quad (47)$$

$$R(0, \lambda) = \rho(0, \lambda)[1 + T(0, \lambda)e^{-\alpha(\lambda)d}] \quad (48)$$

while Equation (42) becomes

$$\tau(0, \lambda) + \rho(0, \lambda) = 1 \quad (49)$$

These three equations can be solved to yield $\rho(0, \lambda)$ and $\alpha(\lambda)$:

$$\rho(0, \lambda) = \frac{P - \sqrt{P^2 - 4[2 - R(0, \lambda)]R(0, \lambda)}}{2[2 - R(0, \lambda)]} \quad (50)$$

where

$$P = [T(0, \lambda)]^2 - [R(0, \lambda)]^2 + 2R(0, \lambda) \quad (51)$$

$$\alpha(\lambda) = -\frac{1}{d}\ln\left[\frac{R(0, \lambda) - \rho(0, \lambda)}{\rho(0, \lambda)T(0, \lambda)}\right] \quad (52)$$

The real part of the refractive index (relative to air) is then calculated by solving Equation (41):

$$n(\lambda) = \frac{1 + \sqrt{\rho(0, \lambda)}}{1 - \sqrt{\rho(0, \lambda)}} \quad (53)$$

Thus, given spectroscopic measurements of the transmittance and reflectance at normal incidence of an uncoated glazing layer, one can determine the basic parameters needed for a complete calculation of the optical properties of that layer.

Example 8. Construct an approximate model of the optical properties of a single layer of uncoated 1/8 in. clear glass, suitable for use in calculations involving selective glazings that have different properties in the visible and NIR regions. Use this model to calculate the properties under the conditions of Example 5.

Solution: The spectral transmittance and reflectance of clear 1/8 in. glass are shown in Figure 20. [The source of these data is "generic" clear glass in the NFRC (2000b) spectral data library.] It can be seen that the variation with wavelength is not very great, and although a single average transmittance and reflectance would be adequate, for use with spectrally selective glazings it is better to construct the "2-band" model also shown in the figure, which characterizes the properties separately for the visible region (320 to 780 nm wavelength) and the NIR region (780 to 2500 nm). About 99% of the ASTM solar spectrum lies within these two regions. The corresponding normal incidence properties are $T(0,\text{vis}) = 0.876$, $R(0,\text{vis}) = 0.081$, $T(0,\text{NIR}) = 0.791$, and $R(0,\text{NIR}) = 0.069$.

From Equation (51), one then calculates $P(\text{vis}) = 1.922$ and $P(\text{NIR}) = 1.758$. Equation (50) yields surface reflectances of $\rho(0,\text{vis}) = 0.44$ and $\rho(0,\text{NIR}) = 0.041$, and Equation (53) gives refractive indices for the two wavelength regions of $n(\text{vis}) = 1.530$ and $n(\text{NIR}) = 1.507$. From Equation (52), we obtain the values $\alpha(\text{vis}) = 0.0149$ and $\alpha(\text{NIR}) = 0.0508$. These are the parameters necessary for calculating the optical properties.

For the conditions of Example 5, the incident angle is found in Example 6 to be 63.2°. At this incident angle, Equation (40) gives refracted angles of $\zeta(\text{vis}) = 35.7°$ and $\zeta(\text{NIR}) = 36.3°$. For polarization *s*, Equation (39a) gives $\rho_s(63.2°,\text{vis}) = 0.218$, $\rho_s(63.2°,\text{NIR}) = 0.210$, and for polarization *p*, Equation (39b) gives $\rho_p(63.2°,\text{vis}) = 0.0066$, and $\rho_p(63.2°,\text{NIR}) = 0.0072$. Putting these values into Equations (32) through (34) gives for the transmittances and reflectances for each polarization: ${}_sT(63.2°,\text{vis}) = 0.604$, ${}_sR(63.2°,\text{vis}) = 0.343$, ${}_sT(63.2°,\text{NIR}) = 0.533$, ${}_sR(63.2°,\text{NIR}) = 0.303$, ${}_pT(63.2°,\text{vis}) = 0.934$, ${}_pR(63.2°,\text{vis}) = 0.012$, ${}_pT(63.2°,\text{NIR}) = 0.816$, and ${}_pR(63.2°,\text{NIR}) = 0.012$.

Note that the reflected radiation is almost completely polarized. The unpolarized quantities are then calculated from Equations (35) and (36): $T(63.2°,\text{vis}) = 1/2(0.604 + 0.934) = 0.769$ and $R(63.2°,\text{vis}) = 1/2\,(0.343 + 0.012) = 0.178$, and, similarly, $T(63.2°,\text{NIR}) = 0.674$ and $R(63.2°,\text{NIR}) = 0.157$.

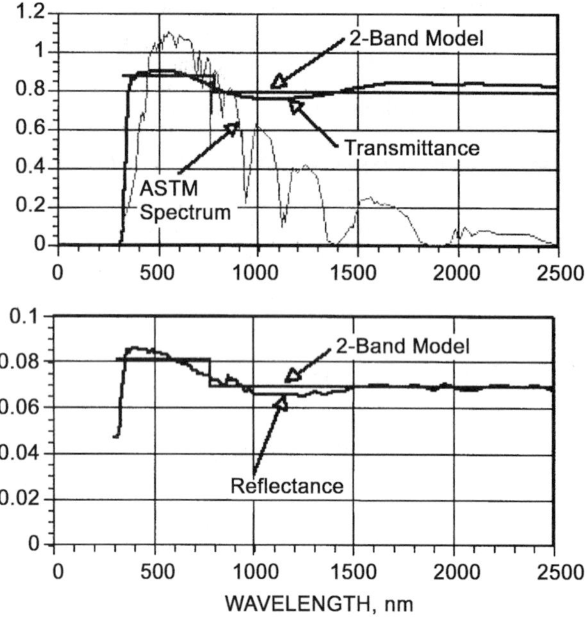

Fig. 20 Transmittance and Reflectance of Clear 1/8 in. Glass Approximated by a 2-Band Model of Spectrally Weighted Transmittance and Reflectance

Table 12 Polynomial Coefficients for Calculation of Reference Angular Functions for Coated Glazings

Condition		*m* 0	1	2	3	4
$T(0) > 0.645$	TA	−0.0015	3.355	−3.840	1.460	0.0288
	RA	0.999	−0.563	2.043	−2.532	1.054
$T(0) \le 0.645$	TA	−0.002	2.813	−2.341	−0.05725	0.599
	RA	0.997	−1.868	6.513	−7.862	3.225

Coated Glazings

While in principle the equations in the preceding sections could be used to calculate the properties of coated glazings, this is not currently practical. To obtain the necessary basic information about the structure of complex coatings would require spectrophotometric measurements at angles other than normal incidence. The instruments for making such measurements are not widely available nor are there yet standardized procedures for making the measurements and extracting coating properties from them.

Until such capabilities are available, the following approximate procedure should be used to model coated glazings (Finlayson and Arasteh 1993). Coated glazing properties should vary from these estimates by no more than ±20% at 60° incidence (Rubin et al. 1999).

The spectral transmittance and reflectances at any incident angle are approximated from those at normal incidence by

$$T(\theta, \lambda) = T(0, \lambda) T_{REF}(\theta) \tag{54}$$

$$R^i(\theta, \lambda) = R^i(0, \lambda)[1 - R_{REF}(\theta)] + R_{REF}(\theta) \quad (i = f, b) \tag{55}$$

where

$$T_{REF}(\theta) = \sum_{m=0}^{4} (TA)_m \cos^m\theta \tag{56}$$

and

$$R_{REF}(\theta) = \sum_{m=0}^{4} (RA)_m \cos^m\theta - T_{REF}(\theta) \tag{57}$$

The constants in Equations (56) and (57) are selected from the entries in Table 12 appropriate to the value of the spectrally averaged transmittance at normal incidence, $T(0)$. Since Equations (54) and (55) make wavelength-independent modifications to the properties at normal incidence, the spectral averaging may be done first, yielding

$$T(\theta) = T(0)[T_{REF}(\theta)] \tag{58}$$

$$R^i(\theta) = R^i(0)[1 - R_{REF}(\theta)] + R_{REF}(\theta) \quad (i = f, b) \tag{59}$$

Example 9. Construct an approximate model of the optical properties of a single layer of 1/8 in. clear glass with a selective low-emissivity coating on surface 2. Use this model to calculate the properties under the conditions of Example 5.

Solution: The spectral transmittance and reflectances of selective 1/8 in. glass at normal incidence are shown in Figure 21. [The source of these data is "EE-170-3.CIG" glass in the NFRC (2000b) spectral data library.] The 2-band model shown in the figure has the following normal-incidence spectral properties: $T(0,\text{vis}) = 0.657$, $R^f(0,\text{vis}) = 0.132$, $R^b(0,\text{vis}) = 0.122$, $T(0,\text{NIR}) = 0.088$, $R^f(0,\text{NIR}) = 0.614$, and $R^b(0,\text{NIR}) = 0.830$. It can be seen from Figure 21 that this model crudely represents the difference in glazing optical properties for the two spectral regions.

For the crude optical model available for coated glass, the angular dependence enters only through T_{REF} and R_{REF}. Following the method of Example 11 to average the spectral properties, one finds that the spectral average transmittance at normal incidence for this glazing, $T(0)$, is 0.379. Since this is less than 0.645, constants for computing these two functions using Equations (56) and (57) are taken from the lower two rows of Table 12. At an incident angle of 63.2° (from Example 6), these functions are $T_{REF}(63.2°) = 0.810$ and $R_{REF}(63.2°) = 0.081$. The transmittance and reflectances are then calculated for each wavelength region using Equations (58) and (59):

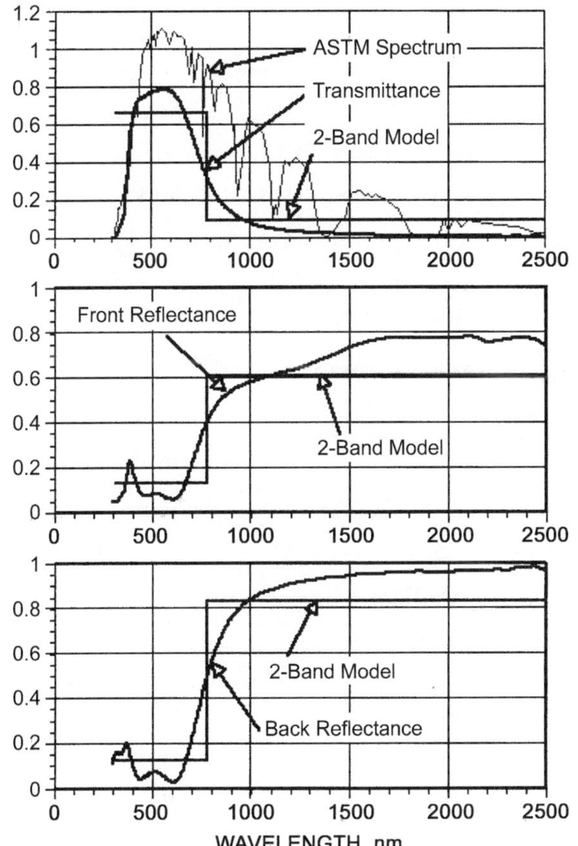

Fig. 21 Transmittance and Reflectance at Normal Incidence of a Selective Low-e Glass Approximated by a 2-Band Model of Spectrally Weighted Transmittance and Reflectance

$T(63.2°,\text{vis}) = 0.532$, $R^f(63.2°,\text{vis}) = 0.202$, $R^b(63.2°,\text{vis}) = 0.193$, $T(63.2°,\text{NIR}) = 0.071$, $R^f(63.2°,\text{NIR}) = 0.646$, and $R^b(63.2°,\text{NIR}) = 0.844$

Optical Properties of Multiple-Layer Glazing Systems

For the optical properties of glazing systems consisting of multiple glazing layers, interreflections may occur between layers, which means that the effect of a particular layer on the overall properties may depend on its position within the assembly as well as on the transmittance and reflectances of the particular layer. We must therefore expand the glazing layer considerations above to apply to the overall properties of systems and subsystems of glazing layers. The notation for doing this is illustrated in Figure 22.

In the following section, polarization indices are omitted from the equations. In principle, these equations should be used to make separate calculations of the optical properties of the glazing system for each of the two polarizations, *s* and *p*, and the results should be averaged using Equations (35) through (37). This should be done if a highly accurate result is desired (and data are available). However, since accurate data on coated glazings are lacking and polarization effects are relatively small, the polarization-averaged quantities will be used for each layer. This is the approximation used in commonly available computer calculations such as Wright (1995b) and LBL (1994).

The position of each layer in a multilayer glazing system consisting of *L* layers is characterized by its layer number *n* as shown in Figure 22. (By convention, layer 1 is the layer closest to the sun.) The individual layer transmittances, reflectances, and absorptances for the *n*th layer (the properties of the layer when isolated in air) are

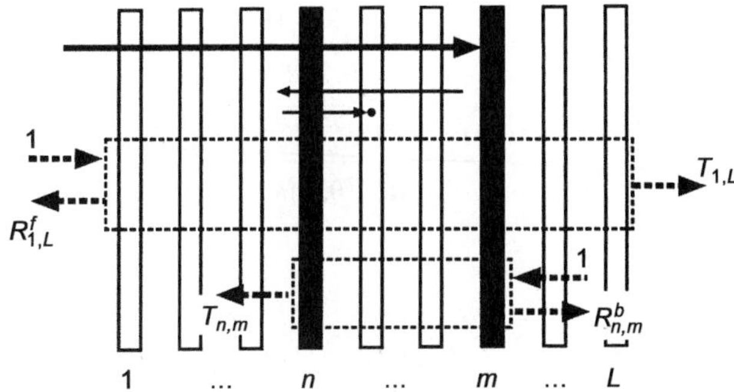

Fig. 22 Multilayer Glazings Considered as Systems and Subsystems

denoted by adding a single subscript to the property symbol: T_n, R^f_n, R^b_n, a^f_n and a^b_n. The properties of a subsystem consisting of the layers *n* through *m*, inclusive (*n* and *m* may be written as capital letters *N* and *M* to avoid confusion and emphasize their role in specifying a subsystem), are denoted by symbols with two indices for transmittances and reflectances and by script capital letters for absorptances ($\mathcal{A}$):

$T_n(\theta,\lambda)$ = isolated-layer transmittance of the *n*th layer (in an *L*-layer system)

$T_{N,M}(\theta,\lambda)$ = transmittances of the subsystem consisting of layers *N* through *M* (in an *L*-layer system)

and similarly for reflectances, while

$a^f_n(\theta,\lambda)$ = isolated-layer front absorptance for the *n*th layer (in an *L*-layer system)

$\mathcal{A}^f_{n:(N,M)}(\theta,\lambda)$ = actual front absorptance for the *n*th layer in the subsystem consisting of layers *N* through *M* (in an *L*-layer system); the fraction of the radiation incident on layer *N* that is absorbed in layer *n*, including effects of multiple reflections from layers *N* through *M*

$\mathcal{A}^b_{n:(N,M)}(\theta,\lambda)$ = actual back absorptance for the *n*th layer in the subsystem consisting of layers *N* through *M* (in an *L*-layer system); the fraction of the (backward-going) radiation incident on layer *M* that is absorbed in layer *n*, including effects of multiple reflections from layers *N* through *M*.

A quantity such as $T_{1,L}(\theta,\lambda)$ refers to the total overall system property. Note that $\mathcal{A}^f_{n:(n,n)}(\theta,\lambda) = a^f_n$ and $\mathcal{A}^b_{n:(n,n)}(\theta,\lambda) = a^b_n$. (A subsystem consisting of one layer is the same as an isolated layer.)

The properties of any subsystem can be calculated by use of the following recursion relations and proceeding from left to right in Figure 22 (Finlayson and Arasteh 1993):

$$T_{n,m+1}(\theta,\lambda) = \frac{T_{n,m}(\theta,\lambda)T_{m+1}(\theta,\lambda)}{1 - R^b_{n,m}(\theta,\lambda)R^f_{m+1}(\theta,\lambda)} \tag{60}$$

$$R^f_{n,m+1}(\theta,\lambda) = R^f_{n,m}(\theta,\lambda) + \frac{[T_{n,m}(\theta,\lambda)]^2 R^b_{m+1}(\theta,\lambda)}{1 - R^b_{n,m}(\theta,\lambda)R^f_{m+1}(\theta,\lambda)} \tag{61}$$

$$R^b_{n,m+1}(\theta,\lambda) = R^b_{m+1}(\theta,\lambda) + \frac{[T_{m+1}(\theta,\lambda)]^2 R^b_{n,m}(\theta,\lambda)}{1 - R^b_{n,m}(\theta,\lambda)R^f_{m+1}(\theta,\lambda)} \tag{62}$$

where it is always the case that $m \geq n$, and a subsystem consisting of one layer is the same as an isolated layer [e.g., $T_{n,n}(\theta,\lambda) \equiv T_n(\theta,\lambda)$]. These equations allow one to build up the properties of the *L*-layer system by beginning with the isolated properties of the first layer and successively adding additional layers. The absorptance of the *n*th layer in the system is then calculated from

$$\mathcal{A}^f_{n:(1,L)}(\theta,\lambda) = \frac{T_{1,n-1}(\theta,\lambda)a^f_n(\theta,\lambda)}{1-R^b_{1,n-1}(\theta,\lambda)R^f_{n,L}(\theta,\lambda)} + \frac{T_{1,n}(\theta,\lambda)R^f_{n+1,L}(\theta,\lambda)a^b_n(\theta,\lambda)}{1-R^b_{1,n}(\theta,\lambda)R^f_{n+1,L}(\theta,\lambda)} \quad (63)$$

or

$$\mathcal{A}^b_{n:(1,L)}(\theta,\lambda) = \frac{T_{n+1,L}(\theta,\lambda)a^b_n(\theta,\lambda)}{1-R^b_{1,n}(\theta,\lambda)R^f_{n+1,L}(\theta,\lambda)} + \frac{T_{n,L}(\theta,\lambda)R^f_{1,n-1}(\theta,\lambda)a^f_n(\theta,\lambda)}{1-R^b_{1,n}(\theta,\lambda)R^b_{n+1,L}(\theta,\lambda)} \quad (64)$$

Note that in Equations (63) and (64), combinations of subscripts will arise for the first and last layer absorptances that refer to nonexistent layers (the layer before the first or after the last). Rather than write separate equations for these special cases, we define the nonexistent layers to have transmittance of one and reflectance of zero:

$$T_{1,0} = T_{L+1,L} = 1$$
$$R^f_{L+1,L} = R^b_{L+1,L} = R^f_{1,0} = R^b_{1,0} = 0 \quad (65)$$

Example 10. Calculate the properties of a double-glazed unit having the coated glass of Example 9 as an outer glass and the clear glass of Example 8 as an inner glass, under the conditions of Example 5.

Solution: From Example 9, the optical properties of the outer glazing, which is layer 1, at 63.2° (Example 6) are $T_1(63.2°,\text{vis}) = 0.532$, $R^f_1(63.2°,\text{vis}) = 0.202$, and $R^b_1(63.2°,\text{vis}) = 0.193$ for the visible region and $T_1(63.2°,\text{NIR}) = 0.071$, $R^f_1(63.2°,\text{NIR}) = 0.646$, and $R^b_1(63.2°,\text{NIR}) = 0.844$ for the NIR region. From Example 8, the corresponding (unpolarized) values for the inner glazing (layer 2) are $T_2(63.2°,\text{vis}) = 0.769$ and $R^f_2(63.2°,\text{vis}) = R^b_2(63.2°,\text{vis}) = 0.178$ for the visible region, and $T_2(63.2°,\text{NIR}) = 0.674$ and $R^f_2(63.2°,\text{NIR}) = R^b_2(63.2°,\text{NIR}) = 0.157$ for the NIR region.

Inserting these values into Equation (60) gives the following for the glazing system's overall transmittance:

$$T_{1,2}(63.2°,\text{vis}) = \frac{T_1(63.2°,\text{vis})T_2(63.2°,\text{vis})}{1-R^b_1(63.2°,\text{vis})R^f_2(63.2°,\text{vis})} = \frac{(0.532)(0.769)}{1-(0.193)(0.178)} = 0.424$$

and

$$T_{1,2}(63.2°,\text{NIR}) = \frac{T_1(63.2°,\text{NIR})T_2(63.2°,\text{NIR})}{1-R^b_1(63.2°,\text{NIR})R^f_2(63.2°,\text{NIR})} = \frac{(0.071)(0.674)}{1-(0.844)(0.157)} = 0.055$$

Similarly, Equation (61) gives $R^f_{1,2}(63.2°,\text{vis}) = 0.202 + (0.532)^2 \times (0.178)/[1-(0.193)(0.178)] = 0.254$ and $R^f_{1,2}(63.2°,\text{NIR}) = 0.446$ for the front reflectances in the two spectral regions, and Equation (62) gives the back reflectances, $R^b_{1,2}(63.2°,\text{vis}) = 0.178 + (0.769)^2 \times (0.193)/[1-(0.193)(0.178)] = 0.296$ and $R^b_{1,2}(63.2°,\text{NIR}) = 0.600$. In order to calculate the layer absorptances in the glazing system, we must first calculate the isolated-layer absorptances for both glazings using Equations (38c) and (38d):

$$a^f_1(63.2°,\text{vis}) = 1 - 0.532 - 0.202 = 0.266$$
$$a^b_1(63.2°,\text{vis}) = 0.275$$
$$a^f_1(63.2°,\text{NIR}) = 0.283$$
$$a^b_1(63.2°,\text{NIR}) = 0.085$$
$$a^f_2(63.2°,\text{vis}) = a^b_2(63.2°,\text{vis}) = 0.053$$
$$a^f_2(63.2°,\text{NIR}) = a^b_2(63.2°,\text{NIR}) = 0.165$$

We then use Equations (63) and (65) to calculate the layer front absorptances in the glazing system:

$$\mathcal{A}^f_{1:(1,2)}(63.2°,\text{vis}) = a^f_1(63.2°,\text{vis}) + \frac{T_1(63.2°,\text{vis})R^f_2(63.2°,\text{vis})a^b_1(63.2°,\text{vis})}{1-R^b_1(63.2°,\text{vis})R^f_2(63.2°,\text{vis})} = 0.266 + \frac{(0.532)(0.178)(0.275)}{1-(0.193)(0.178)} = 0.293$$

$$\mathcal{A}^f_{2:(1,2)}(63.2°,\text{vis}) = \frac{T_1(63.2°,\text{vis})a^f_2(63.2°,\text{vis})}{1-R^b_1(63.2°,\text{vis})R^f_2(63.2°,\text{vis})} = \frac{(0.532)(0.053)}{1-(0.193)(0.178)} = 0.029$$

$$\mathcal{A}^f_{1:(1,2)}(63.2°,\text{NIR}) = 0.285$$

$$\mathcal{A}^f_{2:1,2}(63.2°,\text{NIR}) = 0.014$$

and Equations (64) and (65) to calculate the back absorptances: $\mathcal{A}^b_{1:(1,2)}(63.2°,\text{vis}) = 0.219$, $\mathcal{A}^b_{2:(1,2)}(63.2°,\text{vis}) = 0.061$, $\mathcal{A}^b_{1:(1,2)}(63.2°,\text{NIR}) = 0.066$, and $\mathcal{A}^b_{2:(1,2)}(63.2°,\text{NIR}) = 0.279$.

Spectral Averaging of Glazing Properties

For calculating fenestration heat transfer, it is generally sufficient to use spectrally averaged glazing system properties as defined in Equation (30) or (31). While for special combinations of climate and location it may be desirable to carry out this averaging using variant solar spectra as weighting functions in these equations, as discussed in the section on Solar Radiation under Determining Incident Solar Flux, it is seldom either feasible or necessary to carry out heat transfer calculations using a detailed, time-dependent solar spectrum and the spectral glazing properties.

For multiple-layer glazing systems, the spectral averaging of Equation (30) or (31) should in general be applied to the system spectral properties at each angle [Equations (60) through (64)]. Since all glazing layer properties are to some extent both angle and wavelength dependent, and since these equations are nonlinear in the glazing properties, this is the only procedure that is valid in principle. In general, when a glazing property is used without a specific function reference to wavelength, it assumes that this averaging procedure has been carried out. For example, for a multiple-layer glazing,

$$T_{n,m}(\theta) = \frac{\int_{\lambda_{min}}^{\lambda_{max}} E_{STD}(\lambda)T_{n,m}(\theta,\lambda)d\lambda}{\int_{\lambda_{min}}^{\lambda_{max}} E_{STD}(\lambda)d\lambda} \quad (66)$$

where $T_{n,m}(\theta,\lambda)$ has been obtained from the individual layer properties and the application of Equation (60). The analogous definitions apply for $R^f_{n,m}(\theta)$, $R^b_{n,m}(\theta)$, $\mathcal{A}^f_{n:(1,L)}(\theta)$, and $\mathcal{A}^b_{n:(1,L)}(\theta)$

and the other wavelength-dependent quantities appearing in the section on Optical Properties of Multiple-Layer Glazing Systems. Unless otherwise stated, the standard spectrum, $E_{STD}(\lambda)$, used in the calculation is the ASTM *Standard* E 891 spectrum.

However, as explained in this section, many glazings have properties that vary little over the wavelength region spanned by the solar spectrum. For systems containing only these glazings, much computational labor can be saved by calculating the wavelength-averaged properties (as a function of incident angle) for each isolated glazing layer and using these properties in Equations (60) through (64). While this procedure is less accurate, it may prove adequate for systems involving clear and nearly clear glazings. In using it, one should, however, examine the spectral properties of the glazings to ensure that the potential errors in the approximation are understood. One should particularly note Figures 15, 16, 23, and 24. While clear architectural glazings are generally engineered to have constant spectral properties in the visible region to avoid objectionable coloration, this may not apply to the substantial part of the solar spectrum outside the visible region.

The values of $T_{1,m}(\theta)$, $R^f_{1,\ m}(\theta)$, and $R^b_{1,\ m}(\theta)$ are tabulated for a number of common single ($m = 1$), double ($m = 2$), and triple ($m = 3$) glazing systems in Table 13, and these may be used in combination with the wavelength-averaged versions of Equations (63), (64), (38c), and (38d). One obtains the wavelength-averaged equation by simply dropping the references to wavelength in the equations. At the level of accuracy implied by the use of Table 13, it will be sufficient to use the wavelength-averaged properties listed in the table for layers or subsystems in Equations (63) and (64) to calculate the layer absorptances in the system. For example, if one were interested in a double-glazed system with a tinted and a clear glass, the properties listed in Table 13 for single-glazed clear and single-glazed tinted glass could be used for the layer properties in calculating the in-system layer absorptances. A problem arises for some low-e coated glazings, which are not listed in Table 13 as single glazings because they are never utilized that way (due to the fragility of the coating). In this case, one must use the formulas or consult the NFRC (2000b) data library and its associated (free) computer software to obtain the layer properties.

Even when some glazings have a significant spectral dependence, it may be possible to simplify the calculation by evaluating Equation (31) over only a small number of selected wavelengths. For example, a common situation is a glazing system consisting primarily of clear glazing layers but containing one "selective" or "hot climate" low-e glazing layer (see Figure 24 and associated text). This glazing layer will have very different properties in the visible and near-infrared parts of the solar spectrum. One could, therefore, calculate its spectral average separately over the two spectral regions and utilize the center wavelengths of the two regions in Equation (31)—a "two-band" model. If the other glazings in the system have properties that can be considered constant within each of the two spectral regions, a model of this type will generally produce adequate results.

Example 11. Calculate the spectral average properties of the glazings in Examples 8, 9, and 10. How do the results using a 2-band model compare with those for a single-band model? How do the results obtained from the formulas compare with those obtained by linearly interpolating the values in Table 13?

Solution: One must first calculate the spectral weights to be applied to the 2-band model. This is done by integrating the ASTM spectrum (numerically) to determine the fraction of energy contained in the visible (320 to 780 nm) and NIR (780 to 2500 nm) bands. Energy outside of either band is neglected (about 1.4% of the solar spectrum). These weights are $w_{vis} = 0.512$ and $w_{NIR} = 0.488$. The spectral average properties are then calculated using Equation (31). For the selective glazing of Example 9, the spectral average properties at 63.2° are $T_1(63.2°) = (0.512)(0.532) + (0.488)(0.071) = 0.307$, $R^f_1(63.2°) = 0.419$, $R^b_1(63.2°) = 0.511$, $a^f_1(63.2°) = 0.274$, and $a^b_1(63.2°) = 0.182$. For the clear glazing of Example 8, the spectral average properties are $T_2(63.2°) = 0.723$, $R^f_2(63.2°) = R^b_2(63.2°) = 0.168$, and $a^f_2(63.2°) = a^b_2(63.2°) = 0.107$. For the double-glazed system of Example 10, the spectral average properties are $T_{1,2}(63.2°) = 0.244$, $R^f_{1,2}(63.2°) = 0.446$, $R^b_{1,2}(63.2°) = 0.444$, $\mathcal{A}^f_{1:(1,2)}(63.2°) = 0.289$, $\mathcal{A}^b_{1:(1,2)}(63.2°) = 0.144$, $\mathcal{A}^f_{2:(1,2)}(63.2°) = 0.022$, and $\mathcal{A}^b_{2:(1,2)}(63.2°) = 0.167$.

To obtain the results for a single-band model, we repeat the calculations of Example 10 using the spectral average quantities for the two glazing layers instead of the separate values for the visible and NIR bands. When we do this, we obtain the spectral average properties $T_{1,2}(63.2°) = 0.243$, $R^f_{1,2}(63.2°) = 0.436$, $R^b_{1,2}(63.2°) = 0.460$, $\mathcal{A}^f_{1:(1,2)}(63.2°) = 0.285$, $\mathcal{A}^b_{1:(1,2)}(63.2°) = 0.144$, $\mathcal{A}^f_{2:(1,2)}(63.2°) = 0.037$, and $\mathcal{A}^b_{2:(1,2)}(63.2°) = 0.153$. As can be seen, the results for the two methods are not identical but are relatively close together for this glazing system.

The exact glazing system used in this example is not listed in Table 13 (as will frequently be the case in practice), so we must locate the glazing system that is closest in properties. The emissivity of the selective coating on surface 2 of the outer glazing is 0.05, and the transmittance and reflectances for the insulating glass unit at normal incidence are $T(0°) = 0.331$, $R^f(0°) = 0.388$, and $R^b(0°) = 0.396$. The most similar unit listed in Table 13 is ID 25a, which has $T(0°) = 0.37$, $R^f(0°) = 0.35$, and $R^b(0°) = 0.39$ (i.e., slightly lower front reflectance and therefore higher transmittance).

Near the required angle of incidence, the table values listed are $T(60°) = 0.29$, $R^f(60°) = 0.40$, $R^b(60°) = 0.43$ and $T(70°) = 0.22$, $R^f(70°) = 0.47$, and $R^b(70°) = 0.50$. (The subscripts 1,2 are understood from the system description.) Linear interpolation in angle yields the values $[T_{1,2}(63.2°)]_{25a} = 0.27$, $[R^f_{1,2}(63.2°)]_{25a} = 0.42$, and $[R^b_{1,2}(63.2°)]_{25a} = 0.45$.

Since the normal incidence transmittance of ID 25a in the table was about 0.04 higher than that of the glazing system in which we are interested (and the front reflectance correspondingly 0.04 lower), we might correct the above calculation by multiplying by the ratio of the normal incidence value for the desired glazing to that of table entry 25a. When we do this, we obtain the values $[T_{1,2}(63.2°)]_{table} = 0.27 \times 0.331/0.37 = 0.24$, $[R^f_{1,2}(63.2°)]_{table} = 0.42 \times 0.388/0.35 = 0.47$, and $[R^b_{1,2}(63.2°)]_{table} = 0.45 \times 0.396/0.39 = 0.46$. These values are quite close to the ones obtained above from use of the exact formulas.

Angular Averaging of Glazing Properties

As stressed in the previous sections, the solar-optical properties of glazings depend on the incident angle of the radiation passing through the glazing. It is relatively simple to account for this dependence in the case of beam solar radiation, since at any particular time the radiation can be considered to be incident from a single direction, and from this direction the incident angle can be calculated as described in the section on Determining Solar Angle under Determining Solar Incident Flux. However, for the diffuse solar and ground-reflected radiation, the situation is more complicated. In principle, the energy flow through the glazing should be a sum over the individual energy flows resulting from incident radiation from each direction; for example, the radiant flux directly transmitted through an L-layer glazing due to sky and ground radiation would be

$$q_{trans} = \int_{\lambda} \int \int_{hem} E_{sky,\ ground}(\theta, \phi, \lambda) T_{1,\ L}(\theta, \lambda)(\cos\theta) d\varpi d\lambda \quad (67)$$

where λ denotes solar wavelengths and *hem* stands for hemispherical. $E_{sky,\ ground}(\theta, \phi, \lambda)$ is the spectral radiance of the portion of the sky viewed in the direction (θ, ϕ) from the glazing, and $d\varpi$ is the infinitesimal element of solid angle corresponding to this direction. While such calculations can be carried out for specific sky conditions using detailed sky data or models, such as those cited in the text following Equation (12) in the section on Solar Radiation, the labor of such a calculation will be worthwhile only for very specific purposes. More generally, a drastic simplifying assumption, alluded to in Equation (12), is made. Both the sky and ground radiation are

Table 13 Visible Transmittance (T_v), Solar Heat Gain Coefficient (SHGC), Solar Transmittance (T), Front Reflectance (R^f), Back Reflectance (R^b), and Layer Absorptances ($\mathcal{A}_n^f$) for Glazing and Window Systems

ID	Glass Thick., in.	Glazing System	Center Glazing T_v		Center-of-Glazing Properties: Incidence Angles Normal 0.00	40.00	50.00	60.00	70.00	80.00	Hemis., Diffuse	Total Window SHGC at Normal Incidence: Aluminum Operable	Aluminum Fixed	Other Frames Operable	Other Frames Fixed	Total Window T_v at Normal Incidence: Aluminum Operable	Aluminum Fixed	Other Frames Operable	Other Frames Fixed
Uncoated Single Glazing																			
1a	1/8	CLR	0.90	SHGC	0.86	0.84	0.82	0.78	0.67	0.42	0.78	0.75	0.78	0.64	0.75	0.77	0.80	0.66	0.78
				T	0.83	0.82	0.80	0.75	0.64	0.39	0.75								
				R^f	0.08	0.08	0.10	0.14	0.25	0.51	0.14								
				R^b	0.08	0.08	0.10	0.14	0.25	0.51	0.14								
				$\mathcal{A}_1^f$	0.09	0.10	0.10	0.11	0.11	0.11	0.10								
1b	1/4	CLR	0.88	SHGC	0.81	0.80	0.78	0.73	0.62	0.39	0.73	0.71	0.74	0.60	0.71	0.75	0.79	0.64	0.77
				T	0.88	0.87	0.85	0.80	0.69	0.43	0.80								
				R^f	0.08	0.09	0.11	0.15	0.27	0.53	0.14								
				R^b	0.08	0.09	0.11	0.15	0.27	0.53	0.14								
				$\mathcal{A}_1^f$	0.16	0.17	0.18	0.19	0.19	0.17	0.17								
1c	1/8	BRZ	0.68	SHGC	0.73	0.71	0.68	0.64	0.55	0.34	0.65	0.64	0.67	0.54	0.64	0.58	0.61	0.50	0.59
				T	0.65	0.62	0.59	0.55	0.46	0.27	0.56								
				R^f	0.06	0.07	0.08	0.12	0.22	0.45	0.12								
				R^b	0.06	0.07	0.08	0.12	0.22	0.45	0.12								
				$\mathcal{A}_1^f$	0.29	0.31	0.32	0.33	0.33	0.29	0.31								
1d	1/4	BRZ	0.54	SHGC	0.62	0.59	0.57	0.53	0.45	0.29	0.54	0.54	0.56	0.46	0.54	0.45	0.48	0.39	0.47
				T	0.49	0.45	0.43	0.39	0.32	0.18	0.41								
				R^f	0.05	0.06	0.07	0.11	0.19	0.42	0.10								
				R^b	0.05	0.68	0.66	0.62	0.53	0.33	0.10								
				$\mathcal{A}_1^f$	0.46	0.49	0.50	0.51	0.49	0.41	0.48								
1e	1/8	GRN	0.82	SHGC	0.70	0.68	0.66	0.62	0.53	0.33	0.63	0.62	0.64	0.52	0.61	0.70	0.73	0.60	0.71
				T	0.61	0.58	0.56	0.52	0.43	0.25	0.53								
				R^f	0.06	0.07	0.08	0.12	0.21	0.45	0.11								
				R^b	0.06	0.07	0.08	0.12	0.21	0.45	0.11								
				$\mathcal{A}_1^f$	0.33	0.35	0.36	0.37	0.36	0.31	0.35								
1f	1/4	GRN	0.76	SHGC	0.60	0.58	0.56	0.52	0.45	0.29	0.54	0.53	0.55	0.45	0.53	0.65	0.68	0.55	0.66
				T	0.47	0.44	0.42	0.38	0.32	0.18	0.40								
				R^f	0.05	0.06	0.07	0.11	0.20	0.42	0.10								
				R^b	0.05	0.06	0.07	0.11	0.20	0.42	0.10								
				$\mathcal{A}_1^f$	0.47	0.50	0.51	0.51	0.49	0.40	0.49								
1g	1/8	GRY	0.62	SHGC	0.70	0.68	0.66	0.61	0.53	0.33	0.63	0.62	0.64	0.52	0.61	0.52	0.55	0.45	0.54
				T	0.61	0.58	0.56	0.51	0.42	0.24	0.53								
				R^f	0.06	0.07	0.08	0.12	0.21	0.44	0.11								
				R^b	0.06	0.07	0.08	0.12	0.21	0.44	0.11								
				$\mathcal{A}_1^f$	0.33	0.36	0.37	0.37	0.37	0.32	0.35								
1h	1/4	GRY	0.46	SHGC	0.59	0.57	0.55	0.51	0.44	0.28	0.52	0.53	0.54	0.44	0.52	0.39	0.41	0.34	0.40
				T	0.46	0.42	0.40	0.36	0.29	0.16	0.38								
				R^f	0.05	0.06	0.07	0.10	0.19	0.41	0.10								
				R^b	0.05	0.06	0.07	0.10	0.19	0.41	0.10								
				$\mathcal{A}_1^f$	0.49	0.52	0.54	0.54	0.52	0.43	0.51								
1i	1/4	BLUGRN	0.75	SHGC	0.62	0.59	0.57	0.54	0.46	0.30	0.55	0.55	0.57	0.46	0.54	0.64	0.67	0.55	0.65
				T	0.49	0.46	0.44	0.40	0.33	0.19	0.42								
				R^f	0.06	0.06	0.07	0.11	0.20	0.43	0.11								
				R^b	0.06	0.06	0.07	0.11	0.20	0.43	0.11								
				$\mathcal{A}_1^f$	0.45	0.48	0.49	0.49	0.47	0.38	0.48								
Reflective Single Glazing																			
1j	1/4	SS on CLR 8%	0.08	SHGC	0.19	0.19	0.19	0.18	0.16	0.10	0.18	0.18	0.18	0.15	0.17	0.07	0.07	0.06	0.07
				T	0.06	0.06	0.06	0.05	0.04	0.03	0.05								
				R^f	0.33	0.34	0.35	0.37	0.44	0.61	0.36								
				R^b	0.50	0.50	0.51	0.53	0.58	0.71	0.52								
				$\mathcal{A}_1^f$	0.61	0.61	0.60	0.58	0.52	0.37	0.57								
1k	1/4	SS on CLR 14%	0.14	SHGC	0.25	0.25	0.24	0.23	0.20	0.13	0.23	0.23	0.24	0.19	0.22	0.12	0.12	0.10	0.12
				T	0.11	0.10	0.10	0.09	0.07	0.04	0.09								
				R^f	0.26	0.27	0.28	0.31	0.38	0.57	0.30								
				R^b	0.44	0.44	0.45	0.47	0.52	0.67	0.46								
				$\mathcal{A}_1^f$	0.63	0.63	0.62	0.60	0.55	0.39	0.60								

Table 13 Visible Transmittance (T_v), Solar Heat Gain Coefficient (SHGC), Solar Transmittance (T), Front Reflectance (R^f), Back Reflectance (R^b), and Layer Absorptances ($\mathcal{A}^f_n$) for Glazing and Window Systems (*Continued*)

ID	Glass Thick., in.	Glazing System	Center Glazing T_v		Center-of-Glazing Properties, Incidence Angles: Normal 0.00	40.00	50.00	60.00	70.00	80.00	Hemis., Diffuse	Total Window SHGC at Normal Incidence: Aluminum Operable	Aluminum Fixed	Other Frames Operable	Other Frames Fixed	Total Window T_v at Normal Incidence: Aluminum Operable	Aluminum Fixed	Other Frames Operable	Other Frames Fixed
1l	1/4	SS on CLR 20%	0.20	SHGC	0.31	0.30	0.30	0.28	0.24	0.16	0.28	0.28	0.29	0.24	0.27	0.17	0.18	0.15	0.17
				T	0.15	0.15	0.14	0.13	0.11	0.06	0.13								
				R^f	0.21	0.22	0.23	0.26	0.34	0.54	0.25								
				R^b	0.38	0.38	0.39	0.41	0.48	0.64	0.41								
				$\mathcal{A}^f_1$	0.64	0.64	0.63	0.61	0.56	0.40	0.60								
1m	1/4	SS on GRN 14%	0.12	SHGC	0.25	0.25	0.24	0.23	0.21	0.14	0.23	0.23	0.24	0.19	0.22	0.10	0.11	0.09	0.10
				T	0.06	0.06	0.06	0.06	0.04	0.03	0.06								
				R^f	0.14	0.14	0.16	0.19	0.27	0.49	0.18								
				R^b	0.44	0.44	0.45	0.47	0.52	0.67	0.46								
				$\mathcal{A}^f_1$	0.80	0.80	0.78	0.76	0.68	0.48	0.75								
1n	1/4	TI on CLR 20%	0.20	SHGC	0.29	0.29	0.28	0.27	0.23	0.15	0.27	0.27	0.27	0.22	0.26	0.17	0.18	0.15	0.17
				T	0.14	0.13	0.13	0.12	0.09	0.06	0.12								
				R^f	0.22	0.22	0.24	0.26	0.34	0.54	0.26								
				R^b	0.40	0.40	0.42	0.44	0.50	0.65	0.43								
				$\mathcal{A}^f_1$	0.65	0.65	0.64	0.62	0.57	0.40	0.62								
1o	1/4	TI on CLR 30%	0.30	SHGC	0.39	0.38	0.37	0.35	0.30	0.20	0.35	0.35	0.36	0.30	0.34	0.26	0.27	0.22	0.26
				T	0.23	0.22	0.21	0.19	0.16	0.09	0.20								
				R^f	0.15	0.15	0.17	0.20	0.28	0.50	0.19								
				R^b	0.32	0.33	0.34	0.36	0.43	0.60	0.36								
				$\mathcal{A}^f_1$	0.63	0.65	0.64	0.62	0.57	0.40	0.62								
Uncoated Double Glazing																			
5a	1/8	CLR CLR	0.81	SHGC	0.76	0.74	0.71	0.64	0.50	0.26	0.66	0.67	0.69	0.56	0.66	0.69	0.72	0.59	0.70
				T	0.70	0.68	0.65	0.58	0.44	0.21	0.60								
				R^f	0.13	0.14	0.16	0.23	0.36	0.61	0.21								
				R^b	0.13	0.14	0.16	0.23	0.36	0.61	0.21								
				$\mathcal{A}^f_1$	0.10	0.11	0.11	0.12	0.13	0.13	0.11								
				$\mathcal{A}^f_2$	0.07	0.08	0.08	0.08	0.07	0.05	0.07								
5b	1/4	CLR CLR	0.78	SHGC	0.70	0.67	0.64	0.58	0.45	0.23	0.60	0.61	0.63	0.52	0.61	0.66	0.69	0.57	0.68
				T	0.61	0.58	0.55	0.48	0.36	0.17	0.51								
				R^f	0.11	0.12	0.15	0.20	0.33	0.57	0.18								
				R^b	0.11	0.12	0.15	0.20	0.33	0.57	0.18								
				$\mathcal{A}^f_1$	0.17	0.18	0.19	0.20	0.21	0.20	0.19								
				$\mathcal{A}^f_2$	0.11	0.12	0.12	0.12	0.10	0.07	0.11								
5c	1/8	BRZ CLR	0.62	SHGC	0.62	0.60	0.57	0.51	0.39	0.20	0.53	0.55	0.57	0.46	0.54	0.53	0.55	0.45	0.54
				T	0.55	0.51	0.48	0.42	0.31	0.14	0.45								
				R^f	0.09	0.10	0.12	0.16	0.27	0.49	0.15								
				R^b	0.12	0.13	0.15	0.21	0.35	0.59	0.19								
				$\mathcal{A}^f_1$	0.30	0.33	0.34	0.36	0.37	0.34	0.33								
				$\mathcal{A}^f_2$	0.06	0.06	0.06	0.06	0.05	0.03	0.06								
5d	1/4	BRZ CLR	0.47	SHGC	0.49	0.46	0.44	0.39	0.31	0.17	0.41	0.44	0.46	0.37	0.43	0.40	0.42	0.35	0.41
				T	0.38	0.35	0.32	0.27	0.20	0.08	0.30								
				R^f	0.07	0.08	0.09	0.13	0.22	0.44	0.12								
				R^b	0.10	0.11	0.13	0.19	0.31	0.55	0.17								
				$\mathcal{A}^f_1$	0.48	0.51	0.52	0.53	0.53	0.45	0.50								
				$\mathcal{A}^f_2$	0.07	0.07	0.07	0.07	0.06	0.04	0.07								
5e	1/8	GRN CLR	0.75	SHGC	0.60	0.57	0.54	0.49	0.38	0.20	0.51	0.53	0.55	0.45	0.53	0.63	0.66	0.54	0.65
				T	0.52	0.49	0.46	0.40	0.30	0.13	0.43								
				R^f	0.09	0.10	0.12	0.16	0.27	0.50	0.15								
				R^b	0.12	0.13	0.15	0.21	0.35	0.60	0.19								
				$\mathcal{A}^f_1$	0.34	0.37	0.38	0.39	0.39	0.35	0.37								
				$\mathcal{A}^f_2$	0.05	0.05	0.05	0.04	0.04	0.03	0.04								
5f	1/4	GRN CLR	0.68	SHGC	0.49	0.46	0.44	0.39	0.31	0.17	0.41	0.43	0.45	0.37	0.43	0.57	0.60	0.49	0.59
				T	0.39	0.36	0.33	0.29	0.21	0.09	0.31								
				R^f	0.08	0.08	0.10	0.14	0.23	0.45	0.13								
				R^b	0.10	0.11	0.13	0.19	0.31	0.55	0.17								
				$\mathcal{A}^f_1$	0.49	0.51	0.05	0.53	0.52	0.43	0.50								
				$\mathcal{A}^f_2$	0.05	0.05	0.05	0.05	0.04	0.03	0.05								

Table 13 Visible Transmittance (T_v), Solar Heat Gain Coefficient (SHGC), Solar Transmittance (T), Front Reflectance (R^f), Back Reflectance (R^b), and Layer Absorptances ($\mathcal{A}_n^f$) for Glazing and Window Systems (*Continued*)

	Glazing System				Center-of-Glazing Properties							Total Window SHGC at Normal Incidence				Total Window T_v at Normal Incidence			
					Incidence Angles							Aluminum		Other Frames		Aluminum		Other Frames	
ID	Glass Thick., in.		Center Glazing T_v		Normal 0.00	40.00	50.00	60.00	70.00	80.00	Hemis., Diffuse	Operable	Fixed	Operable	Fixed	Operable	Fixed	Operable	Fixed
5g	1/8	GRY CLR	0.56	SHGC	0.60	0.57	0.54	0.48	0.37	0.20	0.51	0.53	0.55	0.45	0.52	0.48	0.50	0.41	0.49
				T	0.51	0.48	0.45	0.39	0.29	0.12	0.42								
				R^f	0.09	0.09	0.11	0.16	0.26	0.48	0.14								
				R^b	0.12	0.13	0.15	0.21	0.34	0.59	0.19								
				$\mathcal{A}_1^f$	0.34	0.37	0.39	0.40	0.41	0.37	0.37								
				$\mathcal{A}_2^f$	0.05	0.06	0.06	0.05	0.05	0.03	0.05								
5h	1/4	GRY CLR	0.41	SHGC	0.47	0.44	0.42	0.37	0.29	0.16	0.39	0.42	0.43	0.35	0.41	0.35	0.37	0.30	0.36
				T	0.36	0.32	0.29	0.25	0.18	0.07	0.28								
				R^f	0.07	0.07	0.08	0.12	0.21	0.43	0.12								
				R^b	0.10	0.11	0.13	0.18	0.31	0.55	0.17								
				$\mathcal{A}_1^f$	0.51	0.54	0.56	0.57	0.56	0.47	0.53								
				$\mathcal{A}_2^f$	0.07	0.07	0.07	0.06	0.05	0.03	0.06								
5i	1/4	BLUGRN CLR	0.67	SHGC	0.50	0.47	0.45	0.40	0.32	0.17	0.43	0.45	0.46	0.38	0.44	0.57	0.60	0.49	0.58
				T	0.40	0.37	0.34	0.30	0.22	0.10	0.32								
				R^f	0.08	0.08	0.10	0.14	0.24	0.46	0.13								
				R^b	0.11	0.11	0.14	0.19	0.31	0.55	0.17								
				$\mathcal{A}_1^f$	0.47	0.49	0.50	0.51	0.50	0.42	0.48								
				$\mathcal{A}_2^f$	0.06	0.06	0.06	0.05	0.04	0.03	0.05								
5j	1/4	HI-P GRN CLR	0.59	SHGC	0.39	0.37	0.35	0.31	0.25	0.14	0.33	0.35	0.36	0.30	0.34	0.50	0.53	0.43	0.51
				T	0.28	0.26	0.24	0.20	0.15	0.06	0.22								
				R^f	0.06	0.07	0.08	0.12	0.21	0.43	0.11								
				R^b	0.10	0.11	0.13	0.19	0.31	0.55	0.17								
				$\mathcal{A}_1^f$	0.62	0.65	0.65	0.65	0.62	0.50	0.63								
				$\mathcal{A}_2^f$	0.03	0.03	0.03	0.03	0.02	0.01	0.03								
Reflective Double Glazing																			
5k	1/4	SS on CLR 8%, CLR	0.07	SHGC	0.13	0.12	0.12	0.11	0.10	0.06	0.11	0.13	0.13	0.11	0.12	0.06	0.06	0.05	0.06
				T	0.05	0.05	0.04	0.04	0.03	0.01	0.04								
				R^f	0.33	0.34	0.35	0.37	0.44	0.61	0.37								
				R^b	0.38	0.37	0.38	0.40	0.46	0.61	0.40								
				$\mathcal{A}_1^f$	0.61	0.61	0.60	0.58	0.53	0.37	0.56								
				$\mathcal{A}_2^f$	0.01	0.01	0.01	0.01	0.01	0.01	0.01								
5l	1/4	SS on CLR 14%, CLR	0.13	SHGC	0.17	0.17	0.16	0.15	0.13	0.08	0.16	0.17	0.17	0.13	0.15	0.11	0.12	0.09	0.11
				T	0.08	0.08	0.08	0.07	0.05	0.02	0.07								
				R^f	0.26	0.27	0.28	0.31	0.38	0.57	0.30								
				R^b	0.34	0.33	0.34	0.37	0.44	0.60	0.36								
				$\mathcal{A}_1^f$	0.63	0.64	0.64	0.63	0.61	0.56	0.60								
				$\mathcal{A}_2^f$	0.02	0.02	0.02	0.02	0.02	0.02	0.02								
5m	1/4	SS on CLR 20%, CLR	0.18	SHGC	0.22	0.21	0.21	0.19	0.16	0.09	0.20	0.21	0.21	0.17	0.20	0.15	0.16	0.13	0.16
				T	0.12	0.11	0.11	0.09	0.07	0.03	0.10								
				R^f	0.21	0.22	0.23	0.26	0.34	0.54	0.25								
				R^b	0.30	0.30	0.31	0.34	0.41	0.59	0.33								
				$\mathcal{A}_1^f$	0.64	0.64	0.63	0.62	0.57	0.41	0.61								
				$\mathcal{A}_2^f$	0.03	0.03	0.03	0.03	0.02	0.02	0.03								
5n	1/4	SS on GRN 14%, CLR	0.11	SHGC	0.16	0.16	0.15	0.14	0.12	0.08	0.14	0.16	0.16	0.13	0.14	0.09	0.10	0.08	0.10
				T	0.05	0.05	0.05	0.04	0.03	0.01	0.04								
				R^f	0.14	0.14	0.16	0.19	0.27	0.49	0.18								
				R^b	0.34	0.33	0.34	0.37	0.44	0.60	0.36								
				$\mathcal{A}_1^f$	0.80	0.80	0.79	0.76	0.69	0.49	0.76								
				$\mathcal{A}_2^f$	0.01	0.01	0.01	0.01	0.01	0.01	0.01								
5o	1/4	TI on CLR 20%, CLR	0.18	SHGC	0.21	0.20	0.19	0.18	0.15	0.09	0.18	0.20	0.20	0.16	0.19	0.15	0.16	0.13	0.16
				T	0.11	0.10	0.10	0.08	0.06	0.03	0.09								
				R^f	0.22	0.22	0.24	0.27	0.34	0.54	0.26								
				R^b	0.32	0.31	0.32	0.35	0.42	0.59	0.35								
				$\mathcal{A}_1^f$	0.65	0.66	0.65	0.63	0.58	0.41	0.62								
				$\mathcal{A}_2^f$	0.02	0.02	0.02	0.02	0.02	0.01	0.02								
5p	1/4	TI on CLR 30%, CLR	0.27	SHGC	0.29	0.28	0.27	0.25	0.20	0.12	0.25	0.27	0.27	0.22	0.26	0.23	0.24	0.20	0.23
				T	0.18	0.17	0.16	0.14	0.10	0.05	0.15								
				R^f	0.15	0.15	0.17	0.20	0.29	0.51	0.19								
				R^b	0.27	0.27	0.28	0.31	0.40	0.58	0.31								
				$\mathcal{A}_1^f$	0.64	0.64	0.63	0.62	0.58	0.43	0.61								
				$\mathcal{A}_2^f$	0.04	0.04	0.04	0.04	0.03	0.02	0.04								

Table 13 Visible Transmittance (T_v), Solar Heat Gain Coefficient (SHGC), Solar Transmittance (T), Front Reflectance (R^f), Back Reflectance (R^b), and Layer Absorptances ($\mathcal{A}^f_n$) for Glazing and Window Systems (*Continued*)

	Glazing System				Center-of-Glazing Properties							Total Window SHGC at Normal Incidence				Total Window T_v at Normal Incidence			
					Incidence Angles							Aluminum		Other Frames		Aluminum		Other Frames	
ID	Glass Thick., in.		Center Glazing T_v		Normal 0.00	40.00	50.00	60.00	70.00	80.00	Hemis., Diffuse	Operable	Fixed	Operable	Fixed	Operable	Fixed	Operable	Fixed
Low-e Double Glazing, e = 0.2 on surface 2																			
17a	1/8	LE CLR	0.76	SHGC	0.65	0.64	0.61	0.56	0.43	0.23	0.57	0.57	0.59	0.49	0.57	0.65	0.68	0.55	0.66
				T	0.59	0.56	0.54	0.48	0.36	0.18	0.50								
				R^f	0.15	0.16	0.18	0.24	0.37	0.61	0.22								
				R^b	0.17	0.18	0.20	0.26	0.38	0.61	0.24								
				$\mathcal{A}^f_1$	0.20	0.21	0.21	0.21	0.20	0.16	0.20								
				$\mathcal{A}^f_2$	0.07	0.07	0.08	0.08	0.07	0.05	0.07								
17b	1/4	LE CLR	0.73	SHGC	0.60	0.59	0.57	0.51	0.40	0.21	0.53	0.53	0.55	0.45	0.53	0.62	0.65	0.53	0.64
				T	0.51	0.48	0.46	0.41	0.30	0.14	0.43								
				R^f	0.14	0.15	0.17	0.22	0.35	0.59	0.21								
				R^b	0.15	0.16	0.18	0.23	0.35	0.57	0.22								
				$\mathcal{A}^f_1$	0.26	0.26	0.26	0.26	0.25	0.19	0.25								
				$\mathcal{A}^f_2$	0.10	0.11	0.11	0.11	0.10	0.07	0.10								
Low-e Double Glazing, e = 0.2 on surface 3																			
17c	1/8	CLR LE	0.76	SHGC	0.70	0.68	0.65	0.59	0.46	0.24	0.61	0.62	0.64	0.52	0.61	0.65	0.68	0.55	0.66
				T	0.59	0.56	0.54	0.48	0.36	0.18	0.50								
				R^f	0.17	0.18	0.20	0.26	0.38	0.61	0.24								
				R^b	0.15	0.16	0.18	0.24	0.37	0.61	0.22								
				$\mathcal{A}^f_1$	0.11	0.12	0.13	0.13	0.14	0.15	0.12								
				$\mathcal{A}^f_2$	0.14	0.14	0.14	0.13	0.11	0.07	0.13								
17d	1/4	CLR LE	0.73	SHGC	0.65	0.63	0.60	0.54	0.42	0.21	0.56	0.57	0.59	0.49	0.57	0.62	0.65	0.53	0.64
				T	0.51	0.48	0.46	0.41	0.30	0.14	0.43								
				R^f	0.15	0.16	0.18	0.23	0.35	0.57	0.22								
				R^b	0.14	0.15	0.17	0.22	0.35	0.59	0.21								
				$\mathcal{A}^f_1$	0.17	0.19	0.20	0.21	0.22	0.22	0.19								
				$\mathcal{A}^f_2$	0.17	0.17	0.17	0.15	0.13	0.07	0.16								
17e	1/8	BRZ LE	0.58	SHGC	0.57	0.54	0.51	0.46	0.35	0.18	0.48	0.51	0.52	0.43	0.50	0.49	0.52	0.42	0.50
				T	0.46	0.43	0.41	0.36	0.26	0.12	0.38								
				R^f	0.12	0.12	0.14	0.18	0.28	0.50	0.17								
				R^b	0.14	0.15	0.17	0.23	0.35	0.60	0.21								
				$\mathcal{A}^f_1$	0.31	0.34	0.35	0.37	0.38	0.35	0.34								
				$\mathcal{A}^f_2$	0.11	0.11	0.10	0.10	0.08	0.04	0.10								
17f	1/4	BRZ LE	0.45	SHGC	0.45	0.42	0.40	0.35	0.27	0.14	0.38	0.40	0.42	0.34	0.40	0.38	0.40	0.33	0.39
				T	0.33	0.30	0.28	0.24	0.17	0.07	0.26								
				R^f	0.09	0.09	0.10	0.14	0.23	0.44	0.13								
				R^b	0.13	0.14	0.16	0.21	0.34	0.58	0.20								
				$\mathcal{A}^f_1$	0.48	0.51	0.52	0.54	0.53	0.45	0.50								
				$\mathcal{A}^f_2$	0.11	0.11	0.10	0.09	0.07	0.04	0.09								
17g	1/8	GRN LE	0.70	SHGC	0.55	0.52	0.50	0.44	0.34	0.17	0.46	0.49	0.50	0.41	0.48	0.60	0.62	0.51	0.61
				T	0.44	0.41	0.38	0.33	0.24	0.11	0.36								
				R^f	0.11	0.11	0.13	0.17	0.27	0.48	0.16								
				R^b	0.14	0.15	0.17	0.23	0.35	0.60	0.21								
				$\mathcal{A}^f_1$	0.35	0.38	0.39	0.41	0.42	0.37	0.38								
				$\mathcal{A}^f_2$	0.11	0.10	0.10	0.09	0.07	0.04	0.09								
17h	1/4	GRN LE	0.61	SHGC	0.41	0.39	0.36	0.32	0.25	0.13	0.34	0.38	0.39	0.32	0.37	0.53	0.55	0.45	0.54
				T	0.29	0.26	0.24	0.21	0.15	0.06	0.23								
				R^f	0.08	0.08	0.09	0.13	0.22	0.43	0.13								
				R^b	0.13	0.14	0.16	0.21	0.34	0.58	0.20								
				$\mathcal{A}^f_1$	0.53	0.57	0.58	0.59	0.58	0.48	0.56								
				$\mathcal{A}^f_2$	0.10	0.09	0.09	0.08	0.06	0.03	0.08								
17i	1/8	GRY LE	0.53	SHGC	0.54	0.51	0.49	0.44	0.33	0.17	0.46	0.47	0.49	0.40	0.47	0.43	0.45	0.37	0.44
				T	0.43	0.40	0.38	0.33	0.24	0.11	0.35								
				R^f	0.11	0.11	0.13	0.17	0.27	0.48	0.16								
				R^b	0.14	0.15	0.17	0.22	0.35	0.60	0.21								
				$\mathcal{A}^f_1$	0.36	0.39	0.40	0.42	0.42	0.38	0.39								
				$\mathcal{A}^f_2$	0.10	0.10	0.10	0.09	0.07	0.04	0.09								

Table 13 Visible Transmittance (T_v), Solar Heat Gain Coefficient (SHGC), Solar Transmittance (T), Front Reflectance (R^f), Back Reflectance (R^b), and Layer Absorptances ($\mathcal{A}_n^f$) for Glazing and Window Systems (*Continued*)

	Glazing System				Center-of-Glazing Properties							Total Window SHGC at Normal Incidence				Total Window T_v at Normal Incidence			
					Incidence Angles							Aluminum		Other Frames		Aluminum		Other Frames	
ID	Glass Thick., in.		Center Glazing T_v		Normal 0.00	40.00	50.00	60.00	70.00	80.00	Hemis., Diffuse	Operable	Fixed	Operable	Fixed	Operable	Fixed	Operable	Fixed
17j	1/4	GRY LE	0.37	SHGC	0.39	0.37	0.35	0.31	0.24	0.13	0.33	0.35	0.36	0.30	0.34	0.31	0.33	0.27	0.32
				T	0.27	0.25	0.23	0.20	0.14	0.06	0.21								
				R^f	0.09	0.09	0.11	0.14	0.23	0.44	0.14								
				R^b	0.13	0.14	0.16	0.22	0.34	0.58	0.20								
				$\mathcal{A}_1^f$	0.55	0.58	0.59	0.59	0.58	0.48	0.56								
				$\mathcal{A}_2^f$	0.09	0.09	0.08	0.07	0.06	0.03	0.08								
17k	1/4	BLUGRN LE	0.62	SHGC	0.45	0.42	0.40	0.35	0.27	0.14	0.37	0.40	0.42	0.34	0.40	0.53	0.55	0.45	0.54
				T	0.32	0.29	0.27	0.23	0.17	0.07	0.26								
				R^f	0.09	0.09	0.10	0.14	0.23	0.44	0.13								
				R^b	0.13	0.14	0.16	0.21	0.34	0.58	0.20								
				$\mathcal{A}_1^f$	0.48	0.51	0.53	0.54	0.54	0.45	0.51								
				$\mathcal{A}_2^f$	0.11	0.10	0.10	0.09	0.07	0.03	0.09								
17l	1/4	HI-P GRN LE	0.55	0.241	0.34	0.31	0.30	0.26	0.20	0.11	0.28	0.31	0.32	0.26	0.30	0.47	0.49	0.40	0.48
				T	0.22	0.19	0.18	0.15	0.10	0.04	0.17								
				R^f	0.07	0.07	0.08	0.11	0.20	0.41	0.11								
				R^b	0.13	0.14	0.16	0.21	0.33	0.58	0.20								
				$\mathcal{A}_1^f$	0.64	0.67	0.68	0.68	0.66	0.53	0.65								
				$\mathcal{A}_2^f$	0.08	0.07	0.06	0.06	0.04	0.02	0.06								
Low-e Double Glazing, e = 0.1 on surface 2																			
21a	1/8	LE CLR	0.76	SHGC	0.65	0.64	0.62	0.56	0.43	0.23	0.57	0.48	0.50	0.41	0.47	0.64	0.67	0.55	0.65
				T	0.59	0.56	0.54	0.48	0.36	0.18	0.50								
				R^f	0.15	0.16	0.18	0.24	0.37	0.61	0.22								
				R^b	0.17	0.18	0.20	0.26	0.38	0.61	0.24								
				$\mathcal{A}_1^f$	0.20	0.21	0.21	0.21	0.20	0.16	0.20								
				$\mathcal{A}_2^f$	0.07	0.07	0.08	0.08	0.07	0.05	0.07								
21b	1/4	LE CLR	0.72	SHGC	0.60	0.59	0.57	0.51	0.40	0.21	0.53	0.45	0.47	0.38	0.45	0.61	0.64	0.53	0.63
				T	0.51	0.48	0.46	0.41	0.30	0.14	0.43								
				R^f	0.14	0.15	0.17	0.22	0.35	0.59	0.21								
				R^b	0.15	0.16	0.18	0.23	0.35	0.57	0.22								
				$\mathcal{A}_1^f$	0.26	0.26	0.26	0.26	0.25	0.19	0.25								
				$\mathcal{A}_2^f$	0.10	0.11	0.11	0.11	0.10	0.07	0.10								
Low-e Double Glazing, e = 0.1 on surface 3																			
21c	1/8	CLR LE	0.75	SHGC	0.60	0.58	0.56	0.51	0.40	0.22	0.52	0.53	0.55	0.45	0.53	0.64	0.67	0.55	0.65
				T	0.48	0.45	0.43	0.37	0.27	0.13	0.40								
				R^f	0.26	0.27	0.28	0.32	0.42	0.62	0.31								
				R^b	0.24	0.24	0.26	0.29	0.38	0.58	0.28								
				$\mathcal{A}_1^f$	0.12	0.13	0.14	0.14	0.15	0.15	0.13								
				$\mathcal{A}_2^f$	0.14	0.15	0.15	0.16	0.16	0.10	0.15								
21d	1/4	CLR LE	0.72	SHGC	0.56	0.55	0.52	0.48	0.38	0.20	0.49	0.50	0.51	0.42	0.49	0.61	0.64	0.53	0.63
				T	0.42	0.40	0.37	0.32	0.24	0.11	0.35								
				R^f	0.24	0.24	0.25	0.29	0.38	0.58	0.28								
				R^b	0.20	0.20	0.22	0.26	0.34	0.55	0.25								
				$\mathcal{A}_1^f$	0.19	0.20	0.21	0.22	0.23	0.22	0.21								
				$\mathcal{A}_2^f$	0.16	0.17	0.17	0.17	0.16	0.10	0.16								
21e	1/8	BRZ LE	0.57	SHGC	0.48	0.46	0.44	0.40	0.31	0.17	0.42	0.43	0.44	0.36	0.42	0.48	0.51	0.42	0.50
				T	0.37	0.34	0.32	0.27	0.20	0.08	0.30								
				R^f	0.18	0.17	0.19	0.22	0.30	0.50	0.21								
				R^b	0.23	0.23	0.25	0.29	0.37	0.57	0.28								
				$\mathcal{A}_1^f$	0.34	0.37	0.38	0.39	0.39	0.35	0.37								
				$\mathcal{A}_2^f$	0.11	0.12	0.12	0.12	0.11	0.07	0.11								
21f	1/4	BRZ LE	0.45	SHGC	0.39	0.37	0.35	0.31	0.24	0.13	0.33	0.35	0.36	0.30	0.34	0.38	0.40	0.33	0.39
				T	0.27	0.24	0.22	0.19	0.13	0.05	0.21								
				R^f	0.12	0.12	0.13	0.16	0.24	0.44	0.16								
				R^b	0.19	0.20	0.22	0.25	0.34	0.55	0.24								
				$\mathcal{A}_1^f$	0.51	0.54	0.55	0.56	0.55	0.46	0.53								
				$\mathcal{A}_2^f$	0.10	0.10	0.10	0.10	0.09	0.05	0.10								

Table 13 Visible Transmittance (T_v), Solar Heat Gain Coefficient (SHGC), Solar Transmittance (T), Front Reflectance (R^f), Back Reflectance (R^b), and Layer Absorptances ($\mathcal{A}^f_n$) for Glazing and Window Systems (*Continued*)

ID	Glass Thick., in.	Glazing System	Center Glazing T_v		Center-of-Glazing Properties, Incidence Angles: Normal 0.00	40.00	50.00	60.00	70.00	80.00	Hemis., Diffuse	Total Window SHGC at Normal Incidence: Aluminum Operable	Aluminum Fixed	Other Frames Operable	Other Frames Fixed	Total Window T_v at Normal Incidence: Aluminum Operable	Aluminum Fixed	Other Frames Operable	Other Frames Fixed
21g	1/8	GRN LE	0.68	SHGC	0.46	0.44	0.42	0.38	0.30	0.16	0.40	0.41	0.42	0.34	0.40	0.58	0.61	0.50	0.59
				T	0.36	0.32	0.30	0.26	0.18	0.08	0.28								
				R^f	0.17	0.16	0.17	0.20	0.29	0.48	0.20								
				R^b	0.23	0.23	0.25	0.29	0.37	0.57	0.27								
				$\mathcal{A}^f_1$	0.38	0.41	0.42	0.43	0.43	0.38	0.40								
				$\mathcal{A}^f_2$	0.10	0.11	0.11	0.11	0.10	0.06	0.10								
21h	1/4	GRN LE	0.61	SHGC	0.36	0.33	0.31	0.28	0.22	0.12	0.30	0.32	0.33	0.27	0.31	0.52	0.54	0.44	0.53
				T	0.24	0.21	0.19	0.16	0.11	0.05	0.18								
				R^f	0.11	0.10	0.11	0.14	0.22	0.43	0.14								
				R^b	0.19	0.20	0.22	0.25	0.34	0.55	0.24								
				$\mathcal{A}^f_1$	0.56	0.59	0.61	0.61	0.59	0.48	0.58								
				$\mathcal{A}^f_2$	0.09	0.09	0.09	0.08	0.08	0.04	0.08								
21i	1/8	GRY LE	0.52	SHGC	0.46	0.44	0.42	0.38	0.30	0.16	0.39	0.41	0.42	0.35	0.40	0.44	0.46	0.38	0.45
				T	0.35	0.32	0.30	0.25	0.18	0.08	0.28								
				R^f	0.16	0.16	0.17	0.20	0.28	0.48	0.20								
				R^b	0.23	0.23	0.25	0.29	0.37	0.57	0.27								
				$\mathcal{A}^f_1$	0.39	0.42	0.43	0.44	0.44	0.38	0.41								
				$\mathcal{A}^f_2$	0.10	0.11	0.11	0.11	0.10	0.06	0.10								
21j	1/4	GRY LE	0.37	SHGC	0.34	0.32	0.30	0.27	0.21	0.12	0.28	0.31	0.32	0.26	0.30	0.31	0.33	0.27	0.32
				T	0.23	0.20	0.18	0.15	0.11	0.04	0.17								
				R^f	0.11	0.11	0.12	0.15	0.23	0.44	0.15								
				R^b	0.20	0.20	0.22	0.25	0.34	0.55	0.24								
				$\mathcal{A}^f_1$	0.58	0.60	0.61	0.61	0.59	0.48	0.59								
				$\mathcal{A}^f_2$	0.08	0.08	0.08	0.08	0.07	0.04	0.08								
21k	1/4	BLUGRN LE	0.62	SHGC	0.39	0.37	0.34	0.31	0.24	0.13	0.33	0.35	0.36	0.30	0.34	0.53	0.55	0.45	0.54
				T	0.28	0.25	0.23	0.20	0.14	0.06	0.22								
				R^f	0.12	0.12	0.13	0.16	0.24	0.44	0.16								
				R^b	0.23	0.23	0.25	0.28	0.37	0.57	0.27								
				$\mathcal{A}^f_1$	0.51	0.54	0.56	0.56	0.55	0.46	0.53								
				$\mathcal{A}^f_2$	0.08	0.09	0.08	0.08	0.08	0.05	0.08								
21l	1/4	HI-P GRN W/LE CLR	0.57	SHGC	0.31	0.30	0.29	0.26	0.21	0.12	0.27	0.28	0.29	0.24	0.27	0.48	0.51	0.42	0.50
				T	0.22	0.21	0.19	0.17	0.12	0.06	0.18								
				R^f	0.07	0.07	0.09	0.13	0.22	0.46	0.12								
				R^b	0.23	0.23	0.24	0.28	0.37	0.57	0.27								
				$\mathcal{A}^f_1$	0.67	0.68	0.67	0.66	0.62	0.46	0.65								
				$\mathcal{A}^f_2$	0.04	0.05	0.05	0.05	0.04	0.03	0.04								
Low-e Double Glazing, e = 0.05 on surface 2																			
25a	1/8	LE CLR	0.72	SHGC	0.41	0.40	0.38	0.34	0.27	0.14	0.36	0.37	0.38	0.31	0.36	0.61	0.64	0.53	0.63
				T	0.37	0.35	0.33	0.29	0.22	0.11	0.31								
				R^f	0.35	0.36	0.37	0.40	0.47	0.64	0.39								
				R^b	0.39	0.39	0.40	0.43	0.50	0.66	0.42								
				$\mathcal{A}^f_1$	0.24	0.26	0.26	0.27	0.28	0.23	0.26								
				$\mathcal{A}^f_2$	0.04	0.04	0.04	0.04	0.03	0.03	0.04								
25b	1/4	LE CLR	0.70	SHGC	0.37	0.36	0.34	0.31	0.24	0.13	0.32	0.34	0.34	0.28	0.33	0.60	0.62	0.51	0.61
				T	0.30	0.28	0.27	0.23	0.17	0.08	0.25								
				R^f	0.30	0.30	0.32	0.35	0.42	0.60	0.34								
				R^b	0.35	0.35	0.35	0.38	0.44	0.60	0.37								
				$\mathcal{A}^f_1$	0.34	0.35	0.35	0.36	0.35	0.28	0.34								
				$\mathcal{A}^f_2$	0.06	0.07	0.07	0.06	0.06	0.04	0.06								
25c	1/4	BRZ W/LE CLR	0.42	SHGC	0.26	0.25	0.24	0.22	0.18	0.10	0.23	0.24	0.25	0.20	0.23	0.36	0.37	0.31	0.37
				T	0.18	0.17	0.16	0.14	0.10	0.05	0.15								
				R^f	0.15	0.16	0.17	0.21	0.29	0.51	0.20								
				R^b	0.34	0.34	0.35	0.37	0.44	0.60	0.37								
				$\mathcal{A}^f_1$	0.63	0.63	0.63	0.61	0.57	0.42	0.60								
				$\mathcal{A}^f_2$	0.04	0.04	0.04	0.04	0.03	0.03	0.04								

Table 13 Visible Transmittance (T_v), Solar Heat Gain Coefficient (SHGC), Solar Transmittance (T), Front Reflectance (R^f), Back Reflectance (R^b), and Layer Absorptances ($\mathcal{A}_n^f$) for Glazing and Window Systems (*Continued*)

	Glazing System				Center-of-Glazing Properties							Total Window SHGC at Normal Incidence				Total Window T_v at Normal Incidence			
					Incidence Angles							Aluminum		Other Frames		Aluminum		Other Frames	
ID	Glass Thick., in.		Center Glazing T_v		Normal 0.00	40.00	50.00	60.00	70.00	80.00	Hemis., Diffuse	Operable	Fixed	Operable	Fixed	Operable	Fixed	Operable	Fixed
25d	1/4	GRN W/LE CLR	0.60	SHGC	0.31	0.30	0.28	0.26	0.21	0.12	0.27	0.28	0.29	0.23	0.27	0.51	0.53	0.44	0.52
				T	0.22	0.21	0.20	0.17	0.13	0.06	0.18								
				R^f	0.10	0.10	0.12	0.16	0.25	0.48	0.15								
				R^b	0.35	0.34	0.35	0.37	0.44	0.60	0.37								
				$\mathcal{A}_1^f$	0.64	0.64	0.64	0.63	0.59	0.43	0.62								
				$\mathcal{A}_2^f$	0.05	0.05	0.05	0.05	0.04	0.03	0.05								
25e	1/4	GRY W/LE CLR	0.35	SHGC	0.24	0.23	0.22	0.20	0.16	0.09	0.21	0.23	0.23	0.19	0.21	0.30	0.31	0.26	0.30
				T	0.16	0.15	0.14	0.12	0.09	0.04	0.13								
				R^f	0.12	0.13	0.15	0.18	0.26	0.49	0.17								
				R^b	0.34	0.34	0.35	0.37	0.44	0.60	0.37								
				$\mathcal{A}_1^f$	0.69	0.69	0.68	0.67	0.62	0.45	0.66								
				$\mathcal{A}_2^f$	0.03	0.03	0.03	0.03	0.03	0.02	0.03								
25f	1/4	BLUE W/LE CLR	0.45	SHGC	0.27	0.26	0.25	0.23	0.18	0.11	0.24	0.25	0.26	0.21	0.24	0.38	0.40	0.33	0.39
				T	0.19	0.18	0.17	0.15	0.11	0.05	0.16								
				R^f	0.12	0.12	0.14	0.17	0.26	0.49	0.16								
				R^b	0.34	0.34	0.35	0.37	0.44	0.60	0.37								
				$\mathcal{A}_1^f$	0.66	0.66	0.65	0.64	0.60	0.44	0.63								
				$\mathcal{A}_2^f$	0.04	0.04	0.04	0.04	0.04	0.03	0.04								
25g	1/4	HI-P GRN W/LE CLR	0.53	SHGC	0.27	0.26	0.25	0.23	0.18	0.11	0.23	0.25	0.26	0.21	0.24	0.45	0.47	0.39	0.46
				T	0.18	0.17	0.16	0.14	0.10	0.05	0.15								
				R^f	0.07	0.07	0.09	0.13	0.22	0.46	0.12								
				R^b	0.35	0.34	0.35	0.38	0.44	0.60	0.37								
				$\mathcal{A}_1^f$	0.71	0.72	0.71	0.69	0.64	0.47	0.68								
				$\mathcal{A}_2^f$	0.04	0.04	0.04	0.04	0.03	0.02	0.04								
Triple Glazing																			
29a	1/8	CLR CLR CLR	0.74	SHGC	0.68	0.65	0.62	0.54	0.39	0.18	0.57	0.60	0.62	0.51	0.59	0.63	0.66	0.54	0.64
				T	0.60	0.57	0.53	0.45	0.31	0.12	0.49								
				R^f	0.17	0.18	0.21	0.28	0.42	0.65	0.25								
				R^b	0.17	0.18	0.21	0.28	0.42	0.65	0.25								
				$\mathcal{A}_1^f$	0.10	0.11	0.12	0.13	0.14	0.14	0.12								
				$\mathcal{A}_2^f$	0.08	0.08	0.09	0.09	0.08	0.07	0.08								
				$\mathcal{A}_3^f$	0.06	0.06	0.06	0.06	0.05	0.03	0.06								
29b	1/4	CLR CLR CLR	0.70	SHGC	0.61	0.58	0.55	0.48	0.35	0.16	0.51	0.54	0.56	0.46	0.53	0.60	0.62	0.51	0.61
				T	0.49	0.45	0.42	0.35	0.24	0.09	0.39								
				R^f	0.14	0.15	0.18	0.24	0.37	0.59	0.22								
				R^b	0.14	0.15	0.18	0.24	0.37	0.59	0.22								
				$\mathcal{A}_1^f$	0.17	0.19	0.20	0.21	0.22	0.21	0.19								
				$\mathcal{A}_2^f$	0.12	0.13	0.13	0.13	0.12	0.08	0.12								
				$\mathcal{A}_3^f$	0.08	0.08	0.08	0.08	0.06	0.03	0.08								
29c	1/4	HI-P GRN CLR CLR	0.53	SHGC	0.32	0.29	0.27	0.24	0.18	0.10	0.26	0.31	0.32	0.26	0.30	0.45	0.47	0.39	0.46
				T	0.20	0.17	0.15	0.12	0.07	0.02	0.15								
				R^f	0.06	0.07	0.08	0.11	0.20	0.41	0.11								
				R^b	0.13	0.14	0.16	0.22	0.35	0.57	0.20								
				$\mathcal{A}_1^f$	0.64	0.67	0.68	0.68	0.66	0.53	0.65								
				$\mathcal{A}_2^f$	0.06	0.06	0.05	0.05	0.05	0.03	0.05								
				$\mathcal{A}_3^f$	0.04	0.04	0.04	0.03	0.02	0.01	0.04								
Triple Glazing, e = 0.2 on surface 2																			
32a	1/8	LE CLR CLR	0.68	SHGC	0.60	0.58	0.55	0.48	0.35	0.17	0.51	0.53	0.55	0.45	0.53	0.58	0.61	0.50	0.59
				T	0.50	0.47	0.44	0.38	0.26	0.10	0.41								
				R^f	0.17	0.19	0.21	0.27	0.41	0.64	0.25								
				R^b	0.19	0.20	0.22	0.29	0.42	0.63	0.26								
				$\mathcal{A}_1^f$	0.20	0.20	0.20	0.21	0.21	0.17	0.20								
				$\mathcal{A}_2^f$	0.08	0.08	0.08	0.09	0.08	0.07	0.08								
				$\mathcal{A}_3^f$	0.06	0.06	0.06	0.06	0.05	0.03	0.06								

Table 13 Visible Transmittance (T_v), Solar Heat Gain Coefficient (SHGC), Solar Transmittance (T), Front Reflectance (R^f), Back Reflectance (R^b), and Layer Absorptances ($\mathcal{A}_n^f$) for Glazing and Window Systems (*Continued*)

	Glazing System				Center-of-Glazing Properties							Total Window SHGC at Normal Incidence				Total Window T_v at Normal Incidence			
					Incidence Angles							Aluminum		Other Frames		Aluminum		Other Frames	
ID	Glass Thick., in.		Center Glazing T_v		Normal 0.00	40.00	50.00	60.00	70.00	80.00	Hemis., Diffuse	Operable	Fixed	Operable	Fixed	Operable	Fixed	Operable	Fixed
32b	1/4	LE CLR CLR	0.64	SHGC	0.53	0.50	0.47	0.41	0.29	0.14	0.44	0.47	0.49	0.40	0.47	0.54	0.57	0.47	0.56
				T	0.39	0.36	0.33	0.27	0.17	0.06	0.30								
				R^f	0.14	0.15	0.17	0.21	0.31	0.53	0.20								
				R^b	0.16	0.16	0.19	0.24	0.36	0.57	0.22								
				$\mathcal{A}_1^f$	0.28	0.31	0.31	0.34	0.37	0.31	0.31								
				$\mathcal{A}_2^f$	0.11	0.11	0.11	0.11	0.10	0.08	0.11								
				$\mathcal{A}_3^f$	0.08	0.08	0.08	0.07	0.05	0.03	0.07								
Triple Glazing, e = 0.2 on surface 5																			
32c	1/8	CLR CLR LE	0.68	SHGC	0.62	0.60	0.57	0.49	0.36	0.16	0.52	0.55	0.57	0.46	0.54	0.58	0.61	0.50	0.59
				T	0.50	0.47	0.44	0.38	0.26	0.10	0.41								
				R^f	0.19	0.20	0.22	0.29	0.42	0.63	0.26								
				R^b	0.18	0.19	0.21	0.27	0.41	0.64	0.25								
				$\mathcal{A}_1^f$	0.11	0.12	0.13	0.14	0.15	0.15	0.13								
				$\mathcal{A}_2^f$	0.09	0.10	0.10	0.10	0.10	0.08	0.10								
				$\mathcal{A}_3^f$	0.11	0.11	0.11	0.10	0.08	0.04	0.10								
32d	1/4	CLR CLR LE	0.64	SHGC	0.56	0.53	0.50	0.44	0.32	0.15	0.47	0.50	0.51	0.42	0.49	0.54	0.57	0.47	0.56
				T	0.39	0.36	0.33	0.27	0.17	0.06	0.30								
				R^f	0.16	0.16	0.19	0.24	0.36	0.57	0.22								
				R^b	0.14	0.15	0.17	0.21	0.31	0.53	0.20								
				$\mathcal{A}_1^f$	0.17	0.19	0.20	0.21	0.22	0.22	0.19								
				$\mathcal{A}_2^f$	0.13	0.14	0.14	0.14	0.13	0.10	0.13								
				$\mathcal{A}_3^f$	0.15	0.16	0.15	0.14	0.12	0.05	0.14								
Triple Glazing, e = 0.1 on surface 2 and 5																			
40a	1/8	LE CLR LE	0.62	SHGC	0.41	0.39	0.37	0.32	0.24	0.12	0.34	0.37	0.38	0.31	0.36	0.53	0.55	0.45	0.54
				T	0.29	0.26	0.24	0.20	0.13	0.05	0.23								
				R^f	0.30	0.30	0.31	0.34	0.41	0.59	0.33								
				R^b	0.30	0.30	0.31	0.34	0.41	0.59	0.33								
				$\mathcal{A}_1^f$	0.25	0.27	0.28	0.30	0.32	0.27	0.28								
				$\mathcal{A}_2^f$	0.07	0.08	0.08	0.08	0.07	0.06	0.07								
				$\mathcal{A}_3^f$	0.08	0.09	0.09	0.09	0.07	0.04	0.08								
40b	1/4	LE CLR LE	0.59	SHGC	0.36	0.34	0.32	0.28	0.21	0.10	0.30	0.33	0.34	0.27	0.32	0.50	0.53	0.43	0.51
				T	0.24	0.21	0.19	0.16	0.10	0.03	0.18								
				R^f	0.34	0.34	0.35	0.38	0.44	0.61	0.37								
				R^b	0.23	0.23	0.25	0.28	0.36	0.56	0.27								
				$\mathcal{A}_1^f$	0.24	0.25	0.26	0.28	0.30	0.25	0.26								
				$\mathcal{A}_2^f$	0.10	0.11	0.11	0.11	0.10	0.07	0.10								
				$\mathcal{A}_3^f$	0.09	0.09	0.09	0.08	0.07	0.03	0.08								
Triple Glazing, e = 0.05 on surface 2 and 4																			
40c	1/8	LE LE CLR	0.58	SHGC	0.27	0.25	0.24	0.21	0.16	0.08	0.23	0.25	0.25	0.21	0.24	0.49	0.52	0.42	0.50
				T	0.18	0.17	0.16	0.13	0.08	0.03	0.14								
				R^f	0.41	0.41	0.42	0.44	0.50	0.65	0.44								
				R^b	0.46	0.45	0.46	0.48	0.53	0.68	0.47								
				$\mathcal{A}_1^f$	0.27	0.28	0.28	0.29	0.30	0.24	0.28								
				$\mathcal{A}_2^f$	0.12	0.12	0.12	0.12	0.11	0.07	0.12								
				$\mathcal{A}_3^f$	0.02	0.02	0.02	0.02	0.01	0.01	0.02								
40d	1/4	LE LE CLR	0.55	SHGC	0.26	0.25	0.23	0.21	0.16	0.08	0.22	0.24	0.25	0.20	0.23	0.47	0.49	0.40	0.48
				T	0.15	0.14	0.12	0.10	0.07	0.02	0.12								
				R^f	0.33	0.33	0.34	0.37	0.43	0.60	0.36								
				R^b	0.39	0.38	0.38	0.40	0.46	0.61	0.40								
				$\mathcal{A}_1^f$	0.34	0.36	0.36	0.37	0.36	0.28	0.35								
				$\mathcal{A}_2^f$	0.15	0.15	0.15	0.14	0.12	0.08	0.14								
				$\mathcal{A}_3^f$	0.03	0.03	0.03	0.03	0.02	0.01	0.03								

KEY:
CLR = clear, BRZ = bronze, GRN = green, GRY = gray, BLUGRN = bluegreen, SS = stainless steel reflective coating, TI = titanium reflective coating
Reflective coating descriptors include percent visible transmittance as *x*%.
HI-P GRN = high performance green tinted glass, LE = low-emissivity coating
T_v = visible transmittance, T = solar transmittance, SHGC = solar heat gain coefficient, and H. = hemispherical SHGC

ID #'s refer to U-factors in Table 5.

assumed to be **ideally diffuse** (i.e., to have a sky radiance that is independent of direction). In addition, the spectral dependence is assumed to be the same as for beam solar radiation. These simplifications are implicit in the use of the quantities E_d and E_r in Equation (12). To calculate these quantities from one of the detailed sky or ground models mentioned in that section (or from detailed data), rather than from the simpler equations given elsewhere in this chapter, one would use the following equations:

$$E_d = \int_{\lambda} \int\int_{sky} E_{sky}(\theta, \phi, \lambda)(\cos\theta)d\varpi d\lambda \qquad (68)$$

$$E_r = \int_{\lambda} \int\int_{ground} E_{ground}(\theta, \phi, \lambda)(\cos\theta)d\varpi d\lambda \qquad (69)$$

Since these are assumed to be direction-independent quantities, and since for vertical glazings the sky and ground subtend equal solid angles, in most cases the fact that these quantities originate from distinct solid angle regions is ignored. More careful consideration must be given to these quantities for tilted glazings or for direction-dependent shading (such as overhangs and venetian blinds) when high accuracy in the results is desired.

The optical properties to be used with these quantities are then the **hemispherical averages** of the corresponding angle-dependent properties. For example, for an L-layer unshaded glazing, the average transmittance, front reflectance, and layer-specific overall front absorptance would be

$$\langle T_{1,L}\rangle_D = \frac{\int\int_{hem} T_{1,L}(\theta)(\cos\theta)d\varpi}{\int\int_{hem}(\cos\theta)d\varpi} = 2\int_0^{\pi/2} T_{1,L}(\theta)(\cos\theta)(\sin\theta)d\theta \qquad (70)$$

$$\langle R^f_{1,L}\rangle_D = \frac{\int\int_{hem} R^f_{1,L}(\theta)(\cos\theta)d\varpi}{\int\int_{hem}(\cos\theta)d\varpi} = 2\int_0^{\pi/2} R^f_{1,L}(\theta)(\cos\theta)(\sin\theta)d\theta \qquad (71)$$

$$\langle \mathcal{A}^f_{n:(1,L)}\rangle_D = \frac{\int\int_{hem} \mathcal{A}^f_{n:(1,L)}(\theta)(\cos\theta)d\varpi}{\int\int_{hem}(\cos\theta)d\varpi} = 2\int_0^{\pi/2} \mathcal{A}^f_{n:(1,L)}(\theta)(\cos\theta)(\sin\theta)d\theta \qquad (72)$$

Table 14 Angular Weighting Function and Double Glazing System (Example 10) Transmittance Calculated for the Angular Mesh Used to Calculate Hemispherical Average Transmittance

θ°	Angular Weight	$T_{1,2}(\theta)$
0	0.000	0.331
10	0.060	0.329
20	0.113	0.325
30	0.153	0.318
40	0.174	0.309
50	0.174	0.294
60	0.153	0.260
70	0.113	0.194
80	0.060	0.094
90	0.000	0.000

and similarly for the back reflectance and layer-specific absorptances. The simplified expression at the end of each equation follows from the fact that the quantities being averaged are independent of the angle ϕ.

Example 12. Calculate the hemispherical average properties for the two glazings and the glazing system in Examples 8, 9, and 10, and compare the results with Table 13.

Solution: Equations (70) through (72) are used, and the integration is done numerically, using an angular grid of 10° intervals extending from 0° to 90°. To carry out this calculation, all of the properties in the three examples must be recalculated at each angle. A single-band model rather than a 2-band is used to reduce the calculational labor. Values of $T_{1,2}(\theta)$ used in calculating $\langle T_{1,2}\rangle_D$ are shown in Table 14, together with the angular weighting functions used in the numerical integration. The angular weighting functions are the values, at each angle, of 2(cosθ)(sinθ)Δθ at that angle, where Δθ is the grid interval (10°) expressed in radians.

The resulting values are $\langle T_1\rangle_D = 0.335$, $\langle R^f_1\rangle_D = 0.401$, $\langle R^b_1\rangle_D = 0.496$, $\langle a^f_1\rangle_D = 0.264$, and $\langle a^b_1\rangle_D = 0.169$ for the selective glazing of Example 9; $\langle T_2\rangle_D = 0.761$, $\langle R^f_2\rangle_D = \langle R^b_2\rangle_D = 0.137$, and $\langle a^f_2\rangle_D = \langle a^b_2\rangle_D = 0.101$ for the clear glazing of Example 8; $\langle T_{1,2}\rangle_D = 0.277$, $\langle R^f_{1,2}\rangle_D = 0.405$, $\langle R^b_{1,2}\rangle_D = 0.442$, and $\langle \mathcal{A}^f_{1:(1,2)}\rangle_D = 0.271$, $\langle \mathcal{A}^b_{1:(1,2)}\rangle_D = 0.138$, $\langle \mathcal{A}^f_{2:(1,2)}\rangle_D = 0.036$, and $\langle \mathcal{A}^b_{2:(1,2)}\rangle_D = 0.141$ for the double-glazing system of Example 10.

The corresponding values listed for system 25a in Table 13 are $[\langle T_{1,2}\rangle_D]_{25a} = 0.31$, $[\langle R^f_{1,2}\rangle_D]_{25a} = 0.39$, and $[\langle R^b_{1,2}\rangle_D]_{25a} = 0.42$. When we correct them as in Example 11 for the differences between the glazing system of interest and ID 25a at normal incidence, we obtain $[\langle T_{1,2}\rangle_D]_{table} = 0.31 \times 0.331/0.37 = 0.28$, $[\langle R^f_{1,2}\rangle_D]_{table} = 0.39 \times 0.388/0.35 = 0.43$, and $[\langle R^b_{1,2}\rangle_D]_{table} = 0.42 \times 0.396/0.39 = 0.43$. These are reasonably close to the values we obtained from the calculation using the more detailed equations.

Spectrally Selective Glazing

The spectral range from 350 nm to over 50 μm contains radiation from both the sun and sky incident upon fenestration systems as well as the longer wavelength **thermal radiation**. Thermal radiation is emitted by warm bodies both outside and inside the building. Figure 23 shows the human eye spectral response, the solar spectrum for an air mass m = 1.5, and a room temperature blackbody radiation spectrum. The blackbody radiation is scaled to compare with the solar spectrum. Spectral selectivity was defined previously as strong changes in the optical properties of a glazing system over the spectrum. Figure 23 illustrates the basic concept in terms of the spectral reflectance and Figure 24, the spectral transmittance of ideal low solar gain glazings. One can see in this figure the human eye spectral response (called the human photopic visibility function), an air mass 1.5 solar spectrum, and a room temperature (75°F) blackbody radiation spectrum. The latter has been scaled up to better compare it with the shape of the solar spectrum. What is clear from this diagram is the separation of the solar spectrum from the emission spectrum characteristic of an interior room of a multiple-pane glazing system.

Almost all architectural glass is opaque to the long-wave radiation emitted by surfaces at temperatures below about 2200°F. This characteristic produces the **greenhouse effect**, by which solar radiation passing through a window is partially retained inside by the following mechanism. Radiation absorbed by surfaces within the room is emitted as long-wavelength radiation, and it cannot escape directly through the glass since it is opaque to all radiation beyond 4.5 μm. Instead, the radiation from the room surfaces falling on the glass is absorbed and re-emitted to both sides as determined by several parameters, such as the inside and outside film heat transfer coefficients, the surface emissivities, and other glazing properties.

A good reflector in the long-wavelength infrared portion of the spectrum can be a poor reflector and a good transmitter in the solar portion. Because of the conservation of energy ($T + R + A = 1.0$), a high reflectance in the long-wavelength infrared portion of the spectrum means a low transmittance and absorptance. Because of Kirchhoff's law [Equation (29)], a low absorptance means a low emissivity as well. This is the principle of operation of the high-solar-gain (or cold-climate) **low-e coating** on window glass. Such a coating has high transmittance over the entire solar spectrum, producing high solar heat gain while being highly reflective to the long-wavelength infrared radiation emitted by the interior surfaces, reflecting this radiation back inward. The *low-e* in low-e coating refers to a low emissivity over the long-wavelength portion of the spectrum.

Another characteristic of glazing is that at a given wavelength, or over a defined range of wavelengths, the transmittance is the same in both directions. A glazing with a coating or consisting of multiple glazing layers will have one solar transmittance value, but it will have two reflectance values—one for radiation approaching each side of the system.

Figure 24 shows a hypothetical glazing system with improved performance for hot climates. In this case, the sharp **reflectance edge** that the ideal cold-climate low-e coating exhibited just past the end of the solar spectrum in Figure 23 is shifted closer to the edge of the visible portion of the spectrum, thereby increasing the solar near-infrared (NIR) reflectance of the glazing. This results, as seen in Figure 24, in a drop in the hot-climate transmittance to the right of the visible portion of the spectrum. The effect is to reflect the near-infrared portion of the solar spectrum outside, reducing solar gain, while still admitting visible light in the wavelength region below about 800 nm. This hot-climate, solar-gain-rejecting coating also exhibits low emissivity over the long-wavelength spectrum, and is therefore also properly termed a low-e coating. To distinguish the cold- from the hot-climate version, a glazing with this type of spectral response is often termed **selective low-e**. This is somewhat of a misnomer, because both hot- and cold-climate glazings are spectrally selective. Another terminology is **high-solar-gain low-e** glazing system for cold climates, contrasted with **low-solar-gain low-e** glazing system for hot climates.

The reduced infrared transmittance for the hot-climate glazing in Figure 24 is ideally achieved by high reflectance and low absorptance (meaning also low emissivity). It can also be accomplished with high infrared absorptance, if the flow of the absorbed solar radiation to the interior of the building can be reduced, introducing a second approach to the construction of a hot-climate, low-solar-gain glazing system. In this case, the outer pane of a multiple-pane glazing system is made to have good visible transmittance but high absorptance over the solar infrared spectrum. To protect the interior of the building from the heat of this absorbed radiation, additional glazings, gas spaces, and cold-climate type, low-e coatings are added.

By this means, radiation, conduction, and convection of heat from the hot outer pane to the interior ones and to the interior of the building are reduced because of the coating, the insulating gas space, and the additional panes. Such a glazing system for hot climates is insulated primarily *not* to protect the building from conductive heat losses in winter but to protect the interior from the solar radiant heat absorbed by the hot outer pane in summer. Several manufacturers offer this kind of nonreflecting spectrally selective glazing system for commercial buildings having large cooling loads. (If this glazing system is placed in a flip window, one that rotates open for cleaning and can be closed with either side facing outward, then flipping it over in winter can make it an effective trap for passive solar heating of the interior (McCluney and Jindra 2000). Such a system will have two different SHGC values, depending upon which way the window is flipped.

Figure 24 shows that glazings intended for hot climates should have (1) high transmittance over the visible portion of the spectrum to let daylight in for both illumination and view and (2) low transmittance over all other portions of the spectrum to reduce solar heat gain. In contrast, glazings intended for very cold climates should have high transmittance over the whole solar spectrum, from 380 nm

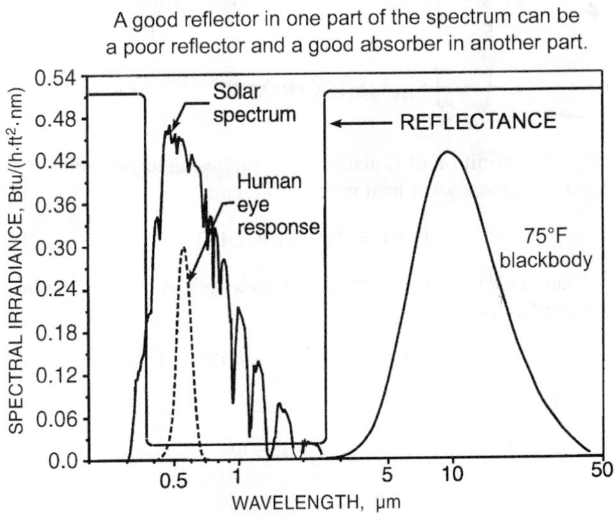

A good reflector in one part of the spectrum can be a poor reflector and a good absorber in another part.

Fig. 23 Solar Spectrum, Human Eye Response Spectrum, Scaled Blackbody Radiation Spectrum, and Idealized Glazing Reflectance Spectrum

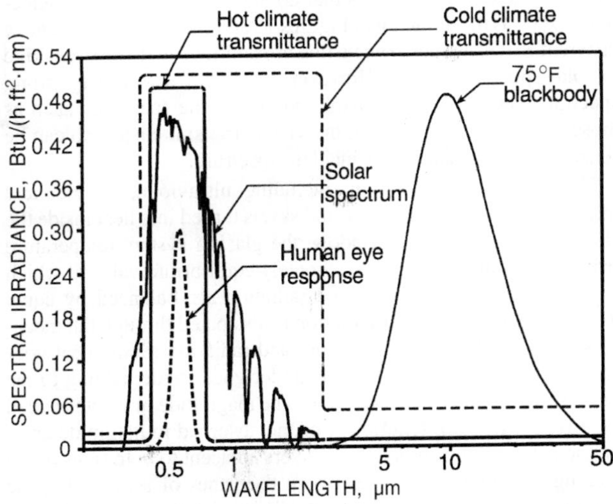

Fig. 24 Demonstration of Two Spectrally Selective Glazing Concepts, Showing Ideal Spectral Transmittances for Glazings Intended for Hot and Cold Climates

to over 3500 nm, for maximum admission of solar radiant heat gain and light. In addition, glazings for cold climates should have low transmittance over the long-wavelength portion of the spectrum in order to block the radiant heat emitted by the relatively warm interior surfaces of buildings, preventing its escape to the outside.

Extreme spectral selectivity in glazing systems in the visible portion of the spectrum can produce an unwanted color shift in the transmitted light. The color of the transmitted light and its color-rendering properties should be considered in the design.

SOLAR HEAT GAIN COEFFICIENT

Fenestration solar heat gain has two components. First is the directly transmitted solar radiation. The quantity of radiation entering the fenestration directly is governed by the solar transmittance of the glazing system. Multiplying the incident irradiance by the glazing area and its solar transmittance yields the solar heat entering the fenestration directly. The second component is the inward-flowing portion of the absorbed solar radiation, radiation that is absorbed in the glazing and framing materials of the window and is subsequently conducted, convected, and radiated to the interior of the building.

The solar heat gain coefficient values presented in this chapter are based on a standard spectral irradiance distribution for air mass 1.5. This spectrum is recommended by the National Fenestration Rating Council (NFRC) for the purpose of rating fenestrations for instantaneous energy performance using defined environmental and incident irradiance conditions. The NFRC standard spectrum can be found in ASTM *Standard* E 891. For real solar gain calculation situations, and with glazing systems exhibiting spectral selectivity over the solar spectrum, the incident solar spectrum used should be altered from the representative one contained in this standard in order to be more realistic for the atmospheric conditions at the location and time of the calculation. Gueymard (1987a, 1987b, 1993a, 1993b, and 1995) has developed an algorithm to guide a computerized methodology for calculating realistic solar spectra that are sensitive to changes in the atmospheric constituents.

The ASTM standard spectrum is different from Parry Moon's (Moon 1940) air mass 2 spectrum used by the glazing industry in the past. Both are different from solar spectral distributions incident upon fenestrations with different atmospheric conditions and for different sun angles. These differences will have little impact on the SHGC of glazing systems containing only glazings with relatively flat spectral transmittances, that is, glazings that are not strongly spectrally selective. Glazings that do exhibit strong spectral selectivity (such as those shown in Figures 15 and 16), however, can have different SHGC values. The visible transmittances are less sensitive to solar spectral changes. However, for glazings with very strong spectral selectivity in the visible portion of the spectrum, such as those exhibiting strong color, the visible transmittance also can be sensitive to the shape of the incident spectrum.

Absorbed solar radiation, including ultraviolet, visible, and infrared radiation from the sun and sky, is turned into heat inside the absorbing material. In a window, the glazing system temperature rises as a result to some approximately equilibrium value at which the energy gains from absorbed radiation are balanced by equal losses. The absorbed solar radiation is dissipated through the mechanisms of conduction, convection, and radiation. Some heat goes outside the building, and the remainder goes inside, adding to the directly transmitted solar radiation. The magnitude of what is called the **inward-flowing fraction** N of the absorbed radiation depends on the nature of the air boundary layers adjacent to both sides of the glazing, including any gas between the panes of a multiple-pane glazing system (N_i is often used to distinguish the inward-flowing fraction from the outward-flowing fraction, N_o. However, since only the inward-flowing fraction is used here, the subscript i is dropped for clarity.).

The concept of the solar heat gain coefficient is best illustrated for the case of a single glass pane in direct sunlight. Let E_D be the direct solar irradiance incident upon a single pane of glass, T be its solar transmittance, $\mathcal{A}$ be the solar absorptance, and N be the inward-flowing fraction of the absorbed radiation. In this case, the total solar gain (heat flow per unit area) q_b that enters the space due to the incident solar radiation is

$$q_b = E_D(T + N\mathcal{A}) \tag{73}$$

in units of energy flux per unit area, Btu/(h·ft^2). Multiple glazings are discussed in the section on Calculation of Solar Heat Gain Coefficient.

The quantity in parenthesis in Equation (73) is called the **solar heat gain coefficient** or **SHGC**. It is the fraction of incident irradiance that enters through the glazing and becomes heat gain. It includes both the directly transmitted portion and the absorbed and re-emitted portion:

$$\text{SHGC} = T + N\mathcal{A} \tag{74}$$

The SHGC is needed to determine the solar radiant heat gain through a window's glazing system. The SHGC for certain defined conditions of spectrum and incident angle θ should be included along with U-factor and other instantaneous performance properties in any manufacturer's description of a window's energy performance. Since the optical properties $\mathcal{A}$ and T vary with the angle of incidence [defined as the angle between the rays incident on the glazing and the normal (perpendicular) to the glazing], according to Equation (74), the solar heat gain coefficient is also a function of angle of incidence. Once the incident irradiance and SHGC are known for a given angle of incidence, the solar gain (from direct beam radiation) can be computed with the following equation:

$$q_b = E_D\text{SHGC} \tag{75}$$

Optical properties also vary with wavelength. The quantities $\mathcal{A}$ and T are spectral averages, as described by Equation (30).

Calculation of Solar Heat Gain Coefficient

In the most general way, the solar heat gain q and the solar heat gain coefficient SHGC are defined as angle-dependent and spectrally dependent properties:

$$q(\theta) = \int_\lambda E_D(\lambda)[T(\theta,\lambda) + N\mathcal{A}(\theta,\lambda)]d\lambda$$

$$= \int_\lambda E_D(\lambda)\text{SHGC}(\theta,\lambda)d\lambda \tag{76}$$

where the definition of Equation (74) implies an angle- and wavelength-dependent solar heat gain coefficient:

$$\text{SHGC}(\theta,\lambda) = T(\theta,\lambda) + N\mathcal{A}(\theta,\lambda) \tag{77}$$

This can in turn be used to define a wavelength-averaged solar heat gain coefficient:

$$\text{SHGC}(\theta) = \frac{\int_\lambda E_D(\lambda)[T(\theta,\lambda) + N1\mathcal{A}(\theta,\lambda)]d\lambda}{\int_\lambda E_D(\lambda)d\lambda} \tag{78}$$

where

$E_D(\lambda)$ = incident solar spectral irradiance
$T(\theta,\lambda)$ = spectral transmittance of the glazing system
$\mathcal{A}(\theta,\lambda)$ = total spectral absorptance of the glazing system

Starting with this edition of the Handbook, Equations (76) through (78) indicate the preferred way of determining the solar gain of glazing systems and calculating the solar heat gain coefficient. At least two computer programs are available to assist in the calculation (Arasteh et al. 1994, AGSL 1992). This approach has been adopted by the National Fenestration Rating Council in NFRC 200 for the rating, certification, and labeling of windows for energy performance and by the Canadian Standards Association (CSA *Standard* A440.2). The method is valid for strongly spectrally selective glazing systems as well as for nonselective ones. In these programs, the overall system optical properties at a given incident angle are calculated for each wavelength and the results averaged following Equation (31). The ASTM *Standard* E 891 spectrum is used in the averaging. The wavelength-averaged properties (at a given incident angle) can then be used in Equation (76).

When a glazing system is not strongly spectrally selective, the solar-weighted spectral broadband values of the optical properties can be used, and the integral over wavelength shown in Equations (76) and (78) is not needed. In this case, and for a general unshaded glazing system, which may consist of several glazing layers, each glazing layer will have its own individual inward-flowing fraction of the absorbed radiation for that layer. Let the glazings be numbered from the outside inward, and k be the glazing index. Then, the SHGC is given by

$$\text{SHGC}(\theta) = T^f_{1,L}(\theta) + \sum_{k=1}^{L} N_k \mathcal{A}^f_{k:(1,L)}(\theta) \tag{79}$$

where

$T^f_{1,L}(\theta)$ = front transmittance of the glazing system
L = number of glazing layers
$\mathcal{A}^f_{k:(1,L)}$ = absorptance of layer k
N_k = inward-flowing fraction for layer k

The wavelength-independent quantities in this equation are obtained by the averaging procedure of Equation (66). Conversely, the wavelength-dependent form of Equation (79) is obtained by adding the wavelength dependence to the indicated angle dependence; the N_k and the layer decomposition are independent of wavelength.

The solar heat gain coefficient is a combination of the solar-optical properties discussed in the section on Solar Optical Properties of Glazings and a new set of quantities N_k, which account for the fact that of the energy absorbed in a given layer, the fraction that reaches the interior space will depend on the location of the layer. The N_k are thermal in origin; they depend on the heat transfer properties of the assembly rather than on its optical properties.

The inward-flowing fractions can be calculated from simplified heat transfer models, using the following equation:

$$N_k = \frac{U}{h_{o,k}} \tag{80}$$

where

U = U-factor of the glazing
$h_{o,k}$ = effective heat transfer coefficient between the exterior environment and the kth glazing layer

The effective heat transfer coefficient can be calculated in a one-dimensional model as the reciprocal of the sum of the thermal resistances of all elements between layer k and the exterior (see Chapter 3). For example, for single glazing,

$$N = \frac{U}{h_o} \tag{81}$$

where U is the U-factor of the glazing and h_o is the exterior heat transfer coefficient (see Chapter 3). For double glazing, the two inward-flowing fractions are

$$N_1 = \frac{U}{h_o}, \qquad N_2 = \frac{h_o + U}{h_o} \tag{82}$$

where the numeric subscripts obey the usual layer numbering convention (1 is the exterior glazing).

In more complicated multilayer glazing systems, it is advisable to perform a detailed heat transfer analysis of the system to determine the values of N_k, since the effective heat transfer coefficients and U depend (weakly) on the glazing layer temperatures and other environmental conditions (e.g., Finlayson and Arasteh 1993, LBL 1994, and Wright 1995c).

Diffuse Radiation

For incident radiation that is diffuse, the hemispherical average solar heat gain coefficient must be used. This may be calculated either by a direct hemispherical averaging of Equation (79) using the analog of Equation (70) as follows:

$$\langle \text{SHGC}(\theta) \rangle_D = \frac{\iint_{hem} \text{SHGC}(\theta)(\cos\theta)d\varpi}{\iint_{hem} (\cos\theta)d\varpi} = 2\int_0^{\pi/2} \text{SHGC}(\theta)(\cos\theta)d\theta \tag{83}$$

or equivalently by simply using the hemispherically averaged solar-optical properties in Equation (79):

$$\langle \text{SHGC} \rangle_D = \langle T^f_{1,L} \rangle_D + \sum_{k=1}^{L} N_k \langle \mathcal{A}^f_{k:(1,L)} \rangle_D \tag{84}$$

In any case, N_k is unaffected in the averaging process, since it does not depend on the incident angle or the wavelength. Considerations of the solar spectrum discussed previously in the sections on Spectral Averaging of Glazing Properties and Angular Averaging of Glazing Properties also apply to the solar heat gain coefficient.

Solar Gain Through Frame and Other Opaque Elements

Figure 25 illustrates the mechanisms by which a window provides solar gain. It is assumed that all of the directly transmitted solar radiation is absorbed at indoor surfaces, where it is converted to heat. Solar gain also enters a building through opaque elements such as the frame and any mullion or dividers that are part of the fenestration system because a portion of the solar energy absorbed at the surfaces of these elements is redirected to the indoor side by heat transfer.

The solar heat gain coefficient of the fenestration system can be calculated while accounting for solar gain through the opaque elements by area-weighting the solar heat gain coefficients of the glazing, frame, and M divider elements. Thus,

$$\text{SHGC} = \frac{\text{SHGC}_g A_g + \text{SHGC}_f A_f + \sum_{i=1}^{M} A_i \text{SHGC}_i}{A_g + A_f + \sum_{i=1}^{M} A_i} \tag{85}$$

where $SHGC_g$, $SHGC_f$, and $SHGC_i$ are the solar heat gain coefficients of the glazed area, frame, and ith divider, respectively. A_g, A_f, and A_i are the corresponding projected areas.

In some cases, it is useful to have an overall SHGC for the opaque elements only, which is defined by

$$SHGC_{op} = \frac{SHGC_f A_f + \sum_{i=1}^{M} A_i SHGC_i}{A_{op}} \tag{86}$$

where

$$A_{op} = A_f + \sum_{i=1}^{M} A_i$$

$SHGC_f$ can be estimated (Wright 1995a) using

$$SHGC_f = \alpha_f^s \left(\frac{U_f}{h_f}\right)\left(\frac{A_f}{A_{surf}}\right) \tag{87}$$

where α_f^s is the solar absorptivity of the outdoor surface of the frame, U_f is the frame U-factor, and h_f is the heat transfer coefficient (radiative plus convective) between the frame and the outdoor environment. The projected-to-surface area ratio (A_f/A_{surf}) corrects for the fact that U_f is based on projected area A_f and h_f is based on the exposed outdoor frame surface area A_{surf}. $SHGC_i$ can be calculated in the same way:

$$SHGC_i = \alpha_i^s \left(\frac{U_i}{h_i}\right)\left(\frac{A_i}{A_{surf,\,i}}\right) \tag{88}$$

The outdoor side heat transfer coefficients h_f and h_i can be estimated using ASHRAE (1996):

$$h_f \text{ or } h_i = h_{co} + 4\sigma e_f T_{out}^3 \tag{89}$$

where h_{co} is the convective heat transfer coefficient between the frame (or divider) surface and the outdoor environment, e_f is the emissivity (long-wave) of the outdoor frame (or divider) surface, T_{out} is the outdoor absolute temperature, and σ is the Stefan-Boltzmann constant.

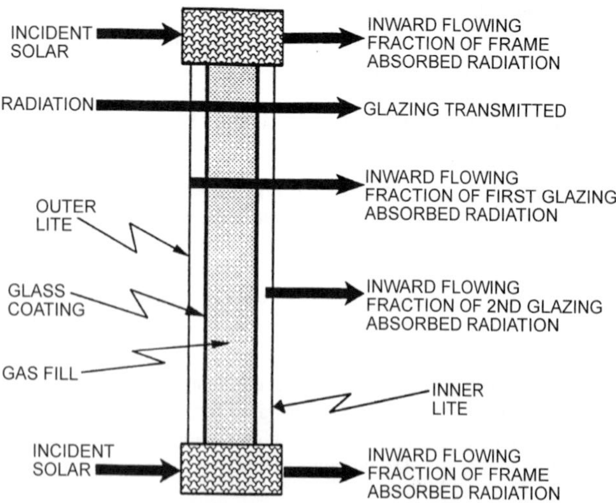

Fig. 25 Components of Solar Radiant Heat Gain with Double-Pane Window, Including Both Frame and Glazing Contributions

Shading Coefficient—A Historical Perspective

Before modern complex windows were marketed in quantity, the determination of fenestration solar gain was substantially simpler. Frame and edge effects were largely ignored, and attention focused on the glazing, which was typically made up of single-pane clear or tinted glass. ASHRAE provided a method for calculating the incident solar beam irradiance for any direction of incidence and tabulated the resulting solar gains through single-pane clear glass in units of flux per unit area in tables of what were called **solar heat gain factors** (SHGFs). These factors, having units of Btu/(h·ft²), are to be distinguished from the solar heat gain coefficient, which is dimensionless. Solar heat gain factors provide the total solar radiant heat gain through a standard single-glazing system, including both the directly transmitted radiant component and the inward flowing fraction of the radiation absorbed in the glazing system.

The engineer's job is relatively easy with this method. One first figures out the angle of direct beam incidence on the glass for a typical peak solar gain date and time. This is done by (1) using equations provided in the section on Determining Solar Angle under Determining Incident Solar Flux or (2) by looking up the data for a latitude close to that of the building being designed for a chosen glazing orientation and for a given time of day expected to produce peak solar gain, in Tables 16 through 22 in Chapter 29 of the 1997 *ASHRAE Handbook—Fundamentals*. These tables provide the solar heat gain factor for clear single-pane glass (the so-called **standard reference glazing**, having $T = 0.86$, $R = 0.08$, and $a = 0.06$ at normal incidence) under these conditions.

The next step is simply a matter of multiplying the solar heat gain factor by the area of the glazing, producing the solar gain expected, in Btu/h. This provides directly the solar gain for single-pane clear glass.

In order to deal with windows having shades or tinted glass, the concept of a **shading coefficient** was introduced. The idea behind the shading coefficient was to find a multiplicative factor for the shaded glass that allowed the engineer to correct the previously determined solar gain number through clear glass to the proper value for the window with a shade or tint. This could be done because the commonly used shading systems (interior shades and drapes) were assumed not to have strongly varying angular properties, leaving the single glazing as the only determinant of angular dependence. Spectral variation was not considered. This concept was later extended to tinted glazings, which often do not violate the basic underlying assumptions, and to glass blocks and single-glazed windows with interior venetian blinds, for which the application is more dubious.

The shading coefficient was defined to be the ratio of the solar heat gain coefficient of a glazing system at a particular angle of incidence and incident solar spectrum to that for standard reference glazing at the same angle of incidence and spectral distribution:

$$SC = \frac{SHGC(\theta)_{test}}{SHGC(\theta)_{ref}} \tag{90}$$

For single-pane clear and many single-pane tinted glazings, this ratio remains constant as the solar spectral shape varies and as the angle of incidence varies. Thus, a single number can be used to convert from the reference SHGC to the SHGC for the tinted glazing at the angle of incidence selected. Due to its lack of sensitivity to angle of incidence, the same SC value works for beam radiation at any angle of incidence, as well as for diffuse radiation. This meant that the complicated calculation of solar heat gain transmission through glass could be done once and tabulated, a quantity appearing in previous editions of this Handbook as the SHGF. For other systems, the SHGF obtained from the table could simply be multiplied by the SC, a great simplification. The value of the SC for standard reference glass is 1.0, but the SHGC for this glass is 0.87 at normal incidence, using ASHRAE standard summer conditions, and for the standard ASTM solar spectrum. This was useful for dealing with glazings of different thicknesses (and was later extended to tinted glazings). Because the principal property causing angular dependence in single glazing is reflectance, angular properties do not depend strongly on thickness, so that the SC defined by Equation (90) remains approximately constant. While tinted glazings, having different spectral transmission from the reference glazing, have a very different SHGC, this difference does not depend strongly on angle, so again SC is approximately constant. Since it is easiest to determine the transmittance of a glazing at normal incidence, it became standard practice to calculate the SC for clear and tinted single glazing from the following relation:

$$SC = \frac{SHGC}{0.87} \tag{91}$$

This equation applies only to the glazing portion of single-pane tinted or clear windows. It does not include frame effects. It may still be used to determine the SC of commercially available, single, uncoated glazing products from the solar heat gain coefficient (or vice versa) published by the manufacturer.

Solar Heat Gain Coefficient, Visible Transmittance, and Spectrally Averaged Solar-Optical Property Values

Table 13 lists the visible transmittance, solar transmittance, front and back reflectance, and solar heat gain coefficients for common glazing and window systems. The window systems include windows with aluminum or metal frames and windows with other frames that have a lower conductivity (e.g., thermally broken aluminum, wood, vinyl, and fiberglass). As can be seen in Table 13, the total window solar heat gain coefficient varies with the type of operator, size of the fenestration product, and type of frame.

The glazing T_v, T_{sol}, R^f, R^b, and SHGC values have been calculated using manufacturers' spectral data following methods described in the section on Solar-Optical Properties of Glazing (Finlayson et al. 1994, Wright 1995a). The glazing values are given for 1/8 in. and 1/4 in. glass and will vary with glass thickness and glass manufacturer. The values shown are average values and may vary by ±0.05. It is recommended that actual values be determined using the methods described in this chapter with detailed spectral data from NFRC (1994). The front reflectance is the reflectance of the unit to the outside, and the back reflectance is the reflectance to the room side.

The visible transmittances are center-glazing values at normal incidence. A rule of thumb is to select a glazing unit whose visible transmittance is greater than its solar heat gain coefficient, especially if daylighting strategies will be used in the building. For maximum light with minimum solar gain, there are fenestration products available having visible transmittance that is 1.4 times their SHGC.

The solar heat gain coefficients are center-glazing values and total window values. The center-glazing solar heat gain coefficients are given at normal incidence (0°) and at 40°, 50°, 60°, 70°, and 80° incidence angles. For angles other than those listed, straight-line interpolation can be used between the two closest angles for which values are shown.

The solar transmittances and front and back reflectances are also center-glazing values and are given at normal incidence (0°) and at 40°, 50°, 60°, 70°, and 80° incidence angles. The solar absorptances can be calculated from these values using Equation (26). The effective inward-flowing fraction of absorbed radiation for the entire system (not layer-specific values) can be determined from Equation (74) by inserting the solar transmittance and corresponding SHGC.

The total window solar heat gain coefficients in Table 13 assume normal incidence. The operable and fixed window sizes in Table 4 were used. To calculate the frame area, the frame heights shown in Figure 4 for aluminum and aluminum-clad wood/wood/vinyl were used. The frame area for the aluminum windows is 15% for the operable size and 11% for the fixed size. The frame area for other frames is 27% for the operable size and 13% for the fixed size. The ratio of projected frame area to frame surface area is assumed to be 1.0, based on Wright (1995a).

The frame solar heat gain coefficients used to determine the total window solar heat gain coefficients are calculated according to the section on solar heat gain coefficients for frames and other nonglazing elements in this chapter. The frame U-factors are taken from Table 1. The frame absorptance is assumed to be 0.5. The outside film coefficient is 3.9 Btu/h·ft^2·°F, corresponding to a wind speed of 7.5 mph. For the aluminum window, the frame solar heat gain coefficient is 0.14 for the operable window and 0.11 for the fixed window. For the other frames, the frame solar heat gain coefficient was found to vary between 0.02 and 0.07 for the various lower conductivity frame types. A frame solar heat gain coefficient of 0.04 is used for the operable window, and 0.03 is used for the fixed window. These values correspond directly to the aluminum-clad wood/reinforced vinyl frames.

For energy calculations on a daylit building, the visible transmittance for the entire window should be used. The visible transmittance of a window can be calculated by multiplying the fraction of glazing area by the center-glazing visible transmittance (see Example 13).

The U-factor of a window listed in Table 13 can be found in Table 4. The ID number for each entry in Table 13 refers to an ID number in Table 4. For a particular glazing system in Table 13, the corresponding glazing system should be the glazing system with that ID number or following that ID number in Table 4. Remember that while the gap width and gas fill have a negligible impact on the solar heat gain coefficient and other optical properties, they are important factors when determining U-factors.

Example 13. Estimate the overall visible light transmittance for an operable wood casement window with clear, uncoated 1/4 in. double glazing.

Solution: From Table 13, ID 5b, the center-glazing visible light transmittance is 0.78. The operable window has 27% frame area with a wood frame. The overall visible light transmittance is

$$T_v = 0.27(0) + 0.73(0.78) = 0.57$$

Passive Solar Gain

Energy analysis of a fenestration product should include the value of passive solar gain through the product in winter. As described in Chapter 32 of the 1999 *ASHRAE Handbook—Applications*, the magnitude of this energy gain depends on such variables as latitude and orientation. In some cases, properly designed and operated fenestration allows more energy into the building over a heating season than it loses, thus making it energy-contributing rather than energy-consuming. Excessive solar gain during the cooling season must be controlled, however.

Direct beam admission to occupied spaces can often produce severe localized glare and overheating conditions. Judicious use of shades and other fenestration control strategies, as well as placement and orientation of workstations and furniture, can alleviate these problems in most cases.

Solar Gain Rejection and Internal Load Dominated Buildings

For some buildings in certain climates, preventing solar gain is more important from an energy perspective than improving thermal insulation using multiple panes of glazing. For example, internal load dominated buildings in cool, clear climates can have substantial daytime solar and internal heat gains. These gains can be rejected by conduction through the building envelope and/or forced ventilation through the HVAC system. Preventing excessive solar gain through the fenestration systems of such buildings is very important.

Airflow Windows

If properly managed, airflow between panes of a double-glazed window can improve fenestration performance. In normal use, a venetian blind is located between the glazing layers. Ventilation air from the room enters the double-glazed cavity, flows over the blind, and is, in some designs, exhausted from the building or returned through the ducts to the central HVAC system. In cold weather, the window acts as a heat exchanger when sunlit so that the inner glass temperature nearly equals the room air temperature and improves thermal comfort.

The apparent conductance across the inner glazing is very low, but this is misleading since additional heat is lost to the outdoors from the moving airstream in the window cavity. During sunny winter days, the blind acts as a solar air collector; heat removed by the moving air can be used elsewhere in the building. In the summer, the window can have a very low shading coefficient if the blinds are appropriately placed since the majority of solar gains are removed from the window. These systems can control window heat transfer under many different operating conditions. Sodergren and Bostrom (1971) and Brandle and Boehm (1982) give details on airflow or exhaust windows.

Skylights

Skylight solar heat gain strongly depends on the configuration of the space below or adjacent to (i.e., in sloped applications) the skylight formed by the skylight curb and any associated light well.

Five aspects must be considered (1) the transmittance and absorptance of the skylight unit, (2) the transmitted solar flux that reaches the aperture of the light well, (3) whether that aperture is covered by a diffuser, (4) the transmitted solar flux that strikes the walls of the light well, and (5) the reflectance of the walls of the light well. Data for flat skylights, which may be considered as sloped glazings, are found in Tables 4 and 13.

Domed Skylights. Solar and total heat gains for domed skylights can be determined by the same procedure used for windows. Table 15 gives SHGCs for plastic domed skylights at normal incidence. Manufacturers' literature has further details. Given the poorly defined incident angle conditions for domed skylights, it is best to use these values without correction for incident angle, together with the correct (angle-dependent) value of incident solar irradiance. Results should be considered approximate. In the absence of other data, these values may also be used to make estimates for skylights on slanted roofs.

Glass Block Walls

Glass block can be used for light transmission through exterior walls when optical clarity for view is not needed or wanted. Table 16 describes a variety of glass block patterns and gives solar heat gain coefficients to be applied to the solar irradiances so that approximate instantaneous solar heat gains can be calculated.

Convection and low-temperature radiative heat gain for all hollow glass block panels fall within a narrow range. Differences in SHGCs are largely the result of differences in the transmittance of the glass blocks for solar radiation. Solar heat gain coefficients for any particular glass block pattern vary depending on orientation and time of day. The SHGC for western exposures in the morning (in the

Table 15 Solar Heat Gain Coefficients for Domed Horizontal Skylights

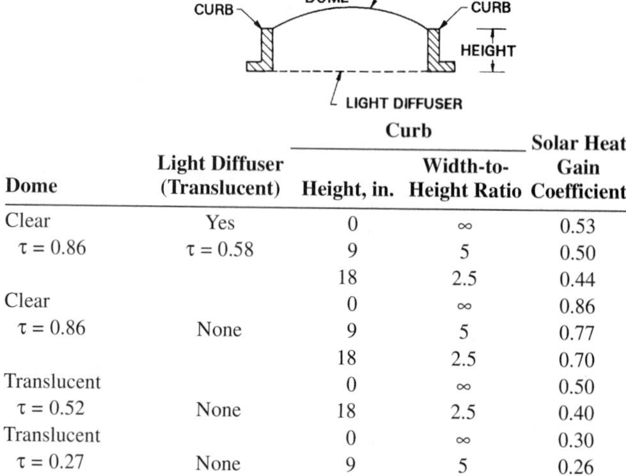

Dome	Light Diffuser (Translucent)	Curb Height, in.	Curb Width-to-Height Ratio	Solar Heat Gain Coefficient
Clear	Yes	0	∞	0.53
τ = 0.86	τ = 0.58	9	5	0.50
		18	2.5	0.44
Clear		0	∞	0.86
τ = 0.86	None	9	5	0.77
		18	2.5	0.70
Translucent		0	∞	0.50
τ = 0.52	None	18	2.5	0.40
Translucent		0	∞	0.30
τ = 0.27	None	9	5	0.26
		18	2.5	0.24

Table 16 Solar Heat Gain Coefficients and U-Factors for Standard Hollow Glass Block Wall Panels

A OUTDOOR SURFACE
D INDOOR SURFACE
B C
E CAVITY PARTITION

Type of Glass Block[a]	Description of Glass Block	Solar Heat Gain Coefficient In Sun	Solar Heat Gain Coefficient In Shade[b]	U-Factor,[c] Btu/(h·ft²·°F)
Type I	Glass colorless or aqua A, D: Smooth B, C: Smooth or wide ribs, or flutes horizontal or vertical, or shallow configuration E: None	0.57	0.35	0.51
Type IA	Same as Type I except ceramic enamel on A	0.23	0.17	0.51
Type II	Same as Type I except glass fiber screen partition E	0.38	0.30	0.48
Type III	Glass colorless or aqua A, D: Narrow vertical ribs or flutes. B, C: Horizontal light-diffusing prisms, or horizontal light-directing prisms E: Glass fiber screen	0.29	0.23	0.48
Type IIIA	Same as Type III except E: Glass fiber screen with green ceramic spray coating or glass fiber screen and gray glass or glass fiber screen with light-selecting prisms	0.22	0.42	0.48
Type IV	Same as Type I except reflective oxide coating on A	0.14	0.44	0.51

[a]All values are for 7-3/4 by 7-3/4 by 3-7/8 in. block, set in light-colored mortar. For 11-3/4 by 11-3/4 by 3-7/8 in. block, increase coefficients by 15%, and for 5-3/4 by 5-3/4 by 3-7/8 in. block reduce coefficients by 15%.

[b]For NE, E, and SE panels in shade, add 50% to the values listed for panels in the shade.

[c]Values shown are the same for all size block.

shade) is depressed because of the heat storage within the block, whereas the SHGC for eastern exposures in the afternoon (in the shade) is elevated as the stored heat is dissipated. Time lag effects from heat storage are estimated by using solar gains and air-to-air temperature differences for one hour earlier than the time for which the load calculation is made.

Calorimeter tests of Type 1A glass block showed little difference in solar heat gains between glass block with either black or white ceramic enamel on the exterior of the block. Because white and black ceramic enamel surfaces represent the two extremes for reflecting or absorbing solar energy, glass block with enamel surfaces of other colors should have solar heat gain coefficients between these values. Since glass blocks are good examples of strongly angularly selective fenestrations, all cautions in the section on Angular Dependence of Glazing Optical Properties apply here.

Plastic Materials for Glazing

Generally, the factors outlined for glass apply also to glazing materials such as acrylic, polycarbonate, polystyrene, or other plastic panels. If the solar transmittance, absorptance, and reflectance are known, an SHGC and a shading coefficient can be calculated in the same way as for glass. These properties can be obtained from the manufacturer or be determined by simple laboratory tests. The National Fenestration Rating Council has developed standards for testing the optical properties of glazing (NFRC 301, NFRC 300).

In selecting plastic panels for glazing, possible deterioration from the sun, expansion and contraction because of temperature extremes, and possible damage from abrasion are concerns.

CALCULATION OF SOLAR HEAT GAIN

As indicated in the section on Determining Fenestration Energy Flow, the solar energy flow through a fenestration may be divided into two parts, opaque and glazing portions, q_{op} and q_s, respectively, as given in Equation (5).

The glazing solar energy flux can be split into that due to incident beam radiation (b) and that due to incident diffuse radiation (d), which includes both diffuse sky radiation and radiation scattered (reflected) from the ground:

$$q_s = q_b + q_d \tag{92}$$

The net heat balance that would occur for a sunlit glazing if there were no diffuse radiation is shown in Figure 26. This net heat balance does not include any of the heat flows contained in Q_{th} in Equation (1) (i.e., those resulting from inside-outside temperature differences). The heat balance due to sunlight is pictured as an effect superimposed on the thermal effect, with, for example, glazing temperatures somewhat elevated over their value without the sunlight. This superposition picture cannot be carried too far, however, because the heat flows indicated in Figure 26 as resulting from convection and radiation depend in part on processes that are nonlinear with respect to temperature [e.g., Equation (25)], so that in reality the two effects cannot be separated. To calculate them, one would need the actual glazing (and other) temperatures, not simply the incremental temperature rise due to the sunlight.

One can see from Figure 26 that the glazing solar energy flow consists of two parts:

$$q_b = q_{bt} + q_{ba} \tag{93}$$

where

q_{bt} = glazing solar energy flux due to transmitted incident beam radiation

q_{ba} = glazing solar energy flux due to inward heat flow of absorbed beam radiation by convection and radiation

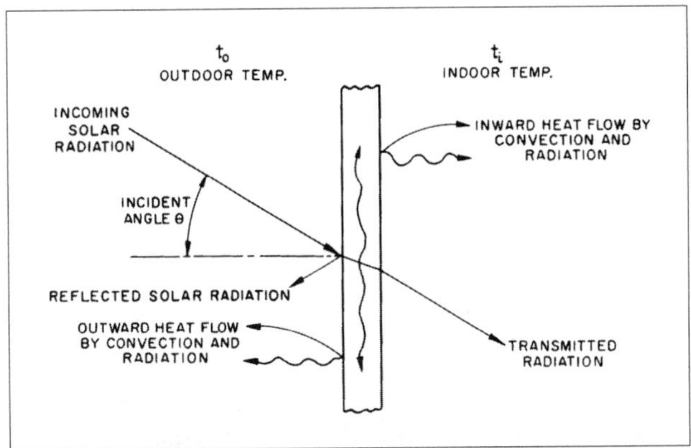

Fig. 26 Instantaneous Heat Balance for Sunlit Glazing Material

However, there is always also diffuse incident radiation, so that the glazing heat flux is in reality given by

$$q_s = q_t + q_a \tag{94}$$

where

q_t = glazing solar energy flux due to all transmitted incident solar radiation

q_a = glazing solar energy flux due to inward flow of absorbed incident solar radiation (by heat transfer processes)

It is convenient to separate the solar energy fluxes due to the two types of incident radiation for purposes of calculation. We can do this by making a definition analogous to Equation (93) for the diffuse solar heat flux:

$$q_d = q_{dt} + q_{da} \tag{95}$$

In order to calculate solar energy fluxes, one must first calculate the incident angle θ from the local standard time and the longitude using Equations (10) and (13) through (17). The direct normal solar irradiance E_{DN} is then calculated from Equation (20), the diffuse sky irradiance E_d from Equation (22) or (23), and the ground-reflected radiation E_r from Equation (24). Note that the latter two are assumed to be ideally diffuse radiation. The total incident irradiance E_t may be calculated from Equation (12).

Because of the distinctions between beam and diffuse incident radiation, one must determine the overall solar heat gain coefficient $SHGC_o$, if desired, by first calculating Q_{sol} and then solving Equations (2) and (1).

Opaque Fenestration Elements

The opaque portion solar energy flux is calculated from

$$q_{op} = (E_{DN}\cos\theta + E_d + E_r)\mathrm{SHGC}_{op} \tag{96}$$

where $SHGC_{op}$ is obtained from Equation (86).

Glazing Systems

The following discussion of glazing system solar energy fluxes applies to unshaded glazings. For shaded glazings, see the section on Shading Devices and Fenestration Attachments.

The glazing solar energy flux due to incident beam radiation is calculated from

$$q_b = E_{DN}\cos\theta\,\mathrm{SHGC}(\theta) \tag{97}$$

where the beam solar heat gain coefficient is given by Equation (79). If instead one desires the solar radiant and heat fluxes separately, one can calculate the glazing transmitted solar flux (which is solar radiation traveling in the incident direction) from

$$q_{bt} = E_{DN}\cos\theta T^f_{1,L}(\theta) \tag{98}$$

and the inward-flowing absorbed solar flux (which is heat) from

$$q_{ba} = E_{DN}\cos\theta \sum_{k=1}^{L} N_k \mathcal{A}^f_{k:(1,L)}(\theta) \tag{99}$$

Calculation of $T^f_{1,L}(\theta)$ and $\mathcal{A}^f_{k:(1,L)}(\theta)$ in these equations is described in the section on Optical Properties of Multiple-Layer Glazing Systems, and determination of N_k is discussed in the section on Calculation of Solar Heat Gain Coefficient.

The glazing solar energy flux due to diffuse incident radiation is calculated from

$$q_d = (E_d + E_r)\langle \text{SHGC}\rangle_D \tag{100}$$

where the hemispherically averaged solar heat gain coefficient is calculated from Equation (83). The solar radiant and heat fluxes can be separately calculated from

$$q_{dt} = (E_d + E_r)\langle T_{1,L}(\theta)\rangle_D \tag{101}$$

which is diffusely distributed solar radiation (note that effects of finite glazing size and thickness are neglected), and

$$q_{da} = (E_d + E_r)\sum_{n=1}^{L} N_n \langle \mathcal{A}^f_{n:1,L}\rangle_D \tag{102}$$

The hemispherically averaged optical properties are calculated using Equations (70) and (72).

Example 14. Calculate the solar energy flux through the glazing system of Example 10 under the conditions of Example 5 using (a) Table 13 and (b) the equations for optical properties.

Solution:

(a) The exact glazing system used in this example is not listed in Table 13 (as will frequently be the case in practice), so we must locate the glazing system that is closest in properties. The emissivity of the selective coating on surface 2 of the outer glazing is 0.05, and the transmission and reflectances for the insulating glass unit at normal incidence are $T(0°) = 0.331$, $R^f(0°) = 0.388$, and $R^b(0°) = 0.396$. The most similar unit is ID 25a, which has $T(0°) = 0.37$, $R^f(0°) = 0.35$, and $R^b(0°) = 0.39$ (i.e., slightly lower front reflectance and therefore higher transmittance). Near the required angle of incidence, the table values listed for the solar heat gain coefficient are $[\text{SHGC}(60°)]_{25a} = 0.34$ and $[\text{SHGC}(70°)]_{25a} = 0.27$, and from the diffuse column of the table we obtain $[\langle\text{SHGC}\rangle_D]_{25a} = 0.36$. Linear interpolation in angle yields the value $[\text{SHGC}(63.2°)]_{25a} = 0.34 + (0.27 - 0.34)/(70° - 60°) \times (63.2° - 60°) = 0.32$. This value can then be approximately corrected for the (small) difference between the glazing of interest and ID 25a in Table 13 as follows. We first note that entry 25a has a transmittance that is 0.04 higher than the glazing in question and a front reflectance that is 0.04 lower. This means that the absorptance of the two glazings is the same to within the accuracy of Table 13. Referring to Equation (79), we see that with the absorptance held constant, a change in the transmittance of the glazing produces an additive correction to the SHGC that is equal to the difference in transmittance. Referring to Examples 11 and 12 where the transmittances for this case are calculated, we make the correction:

$$\begin{aligned}\text{SHGC}(63.2°) &= [\text{SHGC}(63.2°)]_{25a} \\ &\quad + \{T_{1,2}(0°) - [T_{1,2}(0°)]_{25a}\} \times \frac{T_{1,2}(63.2°)}{T_{1,2}(0°)} \\ &= 0.32 + (0.33 - 0.37) \times \frac{0.24}{0.33} = 0.32 - 0.3 = 0.29\end{aligned}$$

and

$$\begin{aligned}\langle\text{SHGC}\rangle_D &= [\langle\text{SHGC}\rangle_D]_{25a} \\ &\quad + \{T_{1,2}(0°) - [T_{1,2}(0°)]_{25a}\} \times \frac{\langle T_{1,2}\rangle_D}{T_{1,2}(0°)} \\ &= 0.36 + (0.33 - 0.37) \times \frac{0.25}{0.33} = 0.36 - 0.3 = 0.33\end{aligned}$$

Note that these values are very close to those calculated below in part (b) of the solution. The solar energy flux is then calculated from Equations (92), (97), and (100), using the incident solar fluxes calculated in Example 7:

$$\begin{aligned}q_s = q_b + q_d &= E_{DN}\cos(63.2°)\text{SHGC}(63.2°) + (E_d + E_r)\langle\text{SHGC}\rangle_D \\ &= (123)(0.29) + (29.6 + 27.5)(0.33) = 54.5 \text{ Btu/h}\cdot\text{ft}^2\end{aligned}$$

(b) We use the previously calculated optical properties of the glazing to determine the SHGC. To do this we need the inward-flowing fractions N_1 and N_2 for the two glazing layers. The simplest way to obtain these is to do a heat transfer analysis of the glazing and determine the glazing temperatures under the appropriate conditions. We assume an exterior temperature t_o of 89°F, a wind speed of 7.5 mph, and an indoor temperature t_i of 75°F. Using methods presented in the section on Determining Fenestration U-Factors, the heat transfer analysis yields a central glazing U of 0.215 Btu/h·ft²·°F and glazing layer temperatures of t_1 = 88.2°F and t_2 = 77.5°F. Since the temperature differences will be proportional to the thermal resistances (which are inversely proportional to the effective heat transfer coefficients), $N_1 = (t_1 - t_o)/(t_i - t_o) = 0.064$ and $N_2 = (t_2 - t_o)/(t_i - t_o) = 0.821$. We can then calculate the solar heat gain coefficient for beam radiation incident at 63.2° (from Examples 5 and 6), using the results from Example 11, from Equation (79):

$$\begin{aligned}\text{SHGC}(63.2°) &= T_{1,2}(63.2°) + N_1\mathcal{A}^f_{1:(1,2)}(63.2°) + N_2\mathcal{A}^f_{2:(1,2)}(63.2°) \\ &= 0.244 + (0.064)(0.289) + (0.821)(0.022) = 0.281\end{aligned}$$

and for diffuse radiation, using the results from Example 12, from Equation (84):

$$\begin{aligned}\langle\text{SHGC}\rangle_D &= \langle T_{1,2}\rangle_D + N_1\langle\mathcal{A}^f_{1:(1,2)}\rangle_D + N_2\langle\mathcal{A}^f_{2:(1,2)}\rangle_D \\ &= 0.277 + (0.064)(0.271) + (0.821)(0.036) = 0.324\end{aligned}$$

The solar energy flux is then calculated from Equations (92), (97), and (100), using the incident solar fluxes calculated in Example 7:

$$\begin{aligned}q_s = q_b + q_d &= E_{DN}\cos(63.2°)\text{SHGC}(63.2°) + (E_d + E_r)\langle\text{SHGC}\rangle_D \\ &= (123)(0.281) + (29.6 + 27.5)(0.324) = 53.1 \text{ Btu/h}\cdot\text{ft}^2\end{aligned}$$

SHADING DEVICES AND FENESTRATION ATTACHMENTS

Fenestrations with shading devices and attachments have a degree of optical and geometric complexity far greater than that of fenestrations with unshaded glazings. ASHRAE has sponsored the development of a method for determining the SHGC of glazing systems of arbitrary complexity. However, there are as yet insufficient data to make this method easy to use as a design tool. For the large number of shading systems for which there are no adequate data, it is still necessary to use the simplified solar heat gain coefficient data to estimate solar gain. In the following section, a detailed calculation method is described, and it should be used for those cases where the necessary data are available.

Table 17 Measured Layer Inward-Flowing Fractions N_k for Typical Fenestration

System	Bind Angle Below Horizontal	Inner Shading Layer	Inner Glass	Between-Pane Shading	Outer Glass	Exterior Shading Layer
Single glazing with interior shade		0.80 ± 0.08			0.08 ± 0.06	
Single glazing with interior venetian blind	−45°	0.69 ± 0.05			0.24 ± 0.09	
	30°	0.83 ± 0.08			0.21 ± 0.07	
	Closed	0.72 ± 0.07			0.14 ± 0.05	
Single glazing with exterior venetian blind	45°				0.46 ± 0.12	0.04 ± 0.01
Double glazing with interior shade		0.85 ± 0.10	0.52 ± 0.12		0.28 ± 0.06	
Double glazing with interior venetian blind	45°	0.86 ± 0.06	0.69 ± 0.14		0.21 ± 0.09	
Double glazing with between-pane blind	45°		0.69 ± 0.14	0.45 ± 0.06	0.34 ± 0.10	
	−45°		0.76 ± 0.10	0.40 ± 0.07	0.27 ± 0.14	
Low-e double glazing with between-pane blind	35°		0.46 ± 0.12	0.38 ± 0.05	0.32 ± 0.11	
Double glazing with exterior venetian blind	45°		0.73 ± 0.13		0.28 ± 0.12	0.03 ± 0.02

Generalized Calculation Procedure for Shading Devices and Fenestration Attachments

A complex fenestration system is one that contains one or more nonspecular optical elements in the glazed area of the window. A **nonspecular optical element** is one for which light (or short-wave infrared radiation) incident on the element from a single spatial direction does not emerge traveling in a single transmitted direction and/or a single reflected direction. Examples of nonspecular elements are shades, drapes, blinds, honeycombs, figured glass, ground glass, and other diffusers, lenses, prisms, and holographic glazings.

These systems may have a more complicated angle dependence than do the specular glazings discussed previously. The optical properties and solar heat gain coefficient may now depend on both angles defining the incident direction rather than simply on the angle θ between the incident direction and the surface normal. In addition, radiation incident from a given direction is not necessarily reflected or transmitted in a unique direction but may be distributed over a variety of directions. One must therefore introduce the concept of the **directional-hemispherical** transmittance (reflectance) [i.e., the fraction of radiation incident from a given direction that is transmitted (reflected) into the complete outgoing hemisphere].

With these extensions, the equation for the solar heat gain coefficient of a system with L layers becomes

$$\mathrm{SHGC}(\theta, \phi) = T^{fH}_{1,L}(\theta, \phi) + \sum_{k=1}^{L} N_k \mathcal{A}^f_{k:(1,L)}(\theta, \phi) \qquad (103)$$

where

θ = incident angle relative to normal layer
ϕ = azimuthal angle (in plane of layer, about normal)
$T^{fH}_{1,L}(\theta, \phi)$ = directional-hemispherical front transmittance of system
$\mathcal{A}^f_{i:(1,L)}(\theta, \phi)$ = directional absorptance of ith layer in system
N_k = inward-flowing fraction of absorbed energy for ith layer in system

The usual approach to complex fenestration solar heat gain has been either direct measurement or calculation of SHGC(θ, ϕ) at some specified incident direction, usually either normal incidence or θ = 30°, ϕ = 0°. Such values, with SHGC re-expressed as shading coefficient, are given for a variety of systems in Tables 18 through 20 and 22. Although in the past it has been assumed that the result would be the same for all incident directions, this is not generally a true assumption, and users of these tables should be aware that they apply only to incident directions of 30° or less and may not include important azimuthal angle dependence. While the values in these tables may give an approximate "first guess" in the absence of better information, for accurate and reliable data, one should rely on a measurement or calculation for the incident directions of interest unless there is independent reason for believing that incident angle or azimuthal dependencies are unimportant.

Solar-Thermal Separation

It is possible to use Equation (103) to calculate the solar heat gain coefficient for a system from separate determinations of the system transmittance $T^{fH}_{1,L}(\theta, \phi)$, the layer absorptances $\mathcal{A}^f_{k:(1,L)}(\theta, \phi)$, and the inward-flowing fractions N_k. This has been termed **solar-thermal separation**, since the processes determining the N_k are thermal in nature, while those determining $T^{fH}_{1,L}(\theta, \phi)$ and $\mathcal{A}^f_{k:(1,L)}(\theta,\phi)$ are solar-optical. Table 17 gives calorimetrically determined values of N_k for a number of generic glazing/shading systems (Klems and Kelley 1996). These are independent of the solar-optical properties of the particular system.

The transmittances and layer absorptances in Equation (103) may be determined by a variety of different methods, ranging from calculation to overall system measurement. The computation procedure has advantages, for example, when one wishes to compare the performance of differing glazings or shading system colors in the same general configuration. Optical data are sometimes more readily available or more economically obtained than overall calorimetric measurements of SHGC. For example, $T^{fH}_{1,L}(\theta, \phi)$ is rather simply measurable by an optical technique using an integrating sphere. The layer absorptances $\mathcal{A}^f_{k:(1,L)}(\theta, \phi)$ present a more difficult problem of determination, since the data easily available are likely to be the directional absorptance of an **isolated** layer, $a^f_k(\theta, \phi)$, whereas $\mathcal{A}^f_{k:(1,L)}(\theta, \phi)$ is the **in-system** layer absorptance (i.e., it includes the contributions of absorbed radiation multiply reflected back to the ith layer from all the other layers in the system). Isolated layer absorptances $a^f_k(\theta, \phi)$ must be corrected for this effect.

Calculating System Transmittance and Absorptances from Layer Properties

The key feature of nonspecular elements is that they produce distributions of outgoing radiation (in the solar-optical spectral region) in the transmitted and/or reflected hemisphere, even for incident radiation from a single direction. This means that they are characterized by bidirectional transmittance (BTDF) and reflectance (BRDF) distribution functions that give the outgoing radiance (energy flux per unit area per unit solid angle) as a fraction of the incident irradiance (energy flux per unit area):

$$\begin{aligned} I(\theta_o, \phi_o; \theta_i, \phi_i) &= \tau(\theta_o, \phi_o; \theta_i, \phi_i)E \\ J(\theta_o, \phi_o; \theta_i, \phi_i) &= \rho(\theta_o, \phi_o; \theta_i, \phi_i)E \end{aligned} \qquad (104)$$

where

θ_o, ϕ_o = angles specifying outgoing direction
θ_i, ϕ_i = angles specifying incident direction

E = incident irradiance
I, J = transmitted, reflected radiance

In a real nonspecular element, these quantities are also functions of the location in space at which the radiation is incident on the fenestration, as can be seen by visualizing a venetian blind. However, this level of detail is only useful if one wishes to find the detailed spatial images of the outgoing radiation patterns. For determining the solar heat gain, it is sufficient to consider the process as spatially averaged over the nonspecular device so that it can be considered as a thin uniform layer with only angular dependence (Klems 1994a). By dividing the (transmission or reflection) hemisphere into a grid of solid angle sections, the bidirectional property functions can be approximated as matrices (Klems 1994b).

There are many standard commercial computer programs executable on personal computers that can perform the matrix calculations, including many popular spreadsheet programs. Use of this bi-angular grid to characterize a nonspecular layer with azimuthal dependence requires handling matrices that are 145 × 145 elements, and for this level of complexity a special-purpose computer program for handling the large amount of data involved is probably desirable.

Figure 27 shows the results of such a calculation for an interior buff-colored blind (slat reflectance 62%) in combination with sealed double glazing (1/8 in. glass panes). The calculation used bidirectional transmittance and reflectance measurements averaged over an 8 in. square section of the blind, with 1 in. slats. These measurements were used to construct layer property matrices with the clear glazing properties taken from published literature (Rubin 1985). The SHGC was calculated using Equation (103), solar-optical transmittances and absorptances calculated by the matrix method of Klems (1994b), and inward-flowing fractions from Table 17 (Klems and Kelley 1996, Klems and Warner 1995).

Simplified Calculation Procedure

The detailed methodology has been shown to agree with measurement (Klems et al. 1996) and should be used whenever accurate calculations are necessary and the data needed are available. Once the necessary data are available for a shading layer, Klems and Warner (1997) have shown how to calculate the performance for any combination of that shading layer and unshaded glazings. However, the data necessary to use this exact calculation method are not readily available for many shading systems and are difficult to obtain by measurement for some (e.g., venetian blinds). It will therefore be necessary to use a simpler and less accurate method for most design purposes.

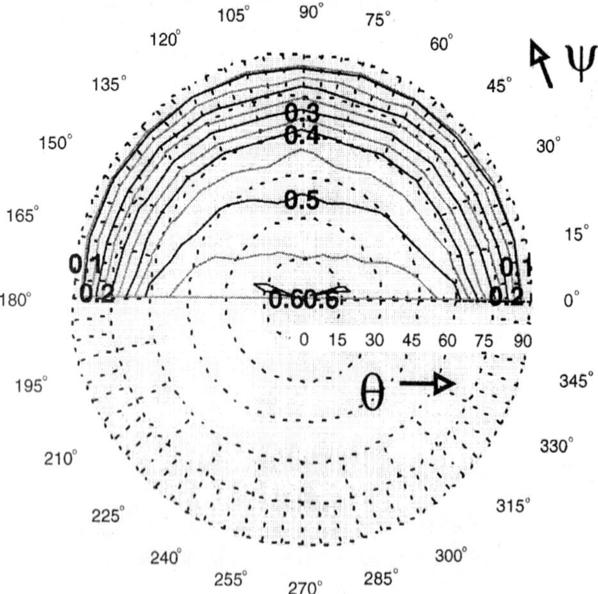

Fig. 27 Contour Plot of Beam SHGC for Double Glazed Window with Interior Venetian Blind with Slats Tilted at 45°

Two simplified approaches are described below. For some simple glazing systems (single glazing and clear double glazing) at solar incident angles below 30°, measurements of the solar heat gain coefficient exist for a number of shading configurations. These are presented in tables, and the first approach, described in the sections on Exterior Shading and Indoor and Between Glass Shading Devices on Simple Fenestrations, may be used. For situations not covered by these tables, the second approach, consisting of the following simplifying approximation together with the section on Completely Shaded Glazings, should be used.

First, one must determine the fraction (if any) of the incident radiation that passes through the fenestration essentially without encountering the shading element. This is termed the **unshaded fraction** F_u. For example, an overhang might shadow only part of a fenestration, a venetian blind or louver might under some conditions permit radiation to pass through it without encountering any of the slats, and for a drapery of open weave, some radiation passes through the gaps between threads and is effectively specularly transmitted (as opposed to radiation that is diffusely scattered by the threads). For this fraction of the incident radiation, the fenestration is essentially an unshaded one. The unshaded fraction is determined from geometric considerations, either those described below for exterior shading or those described for draperies. Once the unshaded fraction has been determined, the glazed area A_G is divided into two equivalent glazings, an unshaded one of area $F_u A_G$ and a completely shaded one of area $(1 - F_u)A_G$. Heat gain through the unshaded equivalent glazing is calculated using the methods given in the section on Calculation of Solar Heat Gain, in the subsection on Glazing Systems, for unshaded glazings. Heat gain through the completely shaded equivalent glazing is calculated by the methods given in the section on Completely Shaded Glazings, which assume that the shading layer is a uniform diffuse reflector or transmitter. A **uniform diffuse reflector** or **transmitter** is one for which a given incident irradiance produces an outgoing reflected or transmitted radiance that is the same in all directions and is independent of the incident direction.

EXTERIOR SHADING

The most effective way to reduce the solar load on fenestration is to intercept direct radiation from the sun before it reaches the glass. Fenestration products fully shaded from the outside reduce solar heat gain as much as 80%. In one way or another, fenestration can be shaded by roof overhangs, vertical and horizontal architectural projections, awnings, heavily proportioned exterior louvers, insect or shading screens, patterned screens having a weave designed for sunlight interception, sun screens of narrow fixed louvers, or a variety of vegetative shades including trees, hedges, and trellis vines. In all exterior shading structures, the air must move freely to carry away heat absorbed by the shading and glazing materials. See manufacturers' instructions regarding proper installation for achieving the expected performance by providing suitable free convection ventilation between the shading and glazing. Also, consider the geometry of the structures relative to changing sun position to determine the times and quantities of direct sunlight penetration. Detailed discussions of the effectiveness of various outside shading devices are given in Pennington (1968), Yellott (1972), and Ewing and Yellott (1976).

The general effect of exterior shading is to attenuate the solar radiation. Some of the beam radiation may pass through or around the shading, and this is accounted for by the unshaded fraction F_u

Table 18 Unshaded Fractions (F_u) and Exterior Solar Attenuation Coefficients (EAC) for Louvered Sun Screens

Profile Angle	Group 1			Group 2			Group 3			Group 4		
	Transmittance	F_u	EAC	Transmittance	F_u	EAC	Transmittance	F_u	EAC	Transmittance	F_u	EAC
10°	0.23	0.20	0.15	0.25	0.13	0.02	0.4	0.33	0.18	0.48	0.29	0.3
20°	0.06	0.02	0.15	0.14	0.03	0.02	0.32	0.24	0.18	0.39	0.2	0.3
30°	0.04	0.00	0.15	0.12	0.01	0.02	0.21	0.13	0.18	0.28	0.08	0.3
≥ 40°	0.04	0.00	0.15	0.11	0.00	0.02	0.07	0.00	0.18	0.2	0.00	0.3

Group 1: Black, width over spacing ratio 1.15/1; 23 louvers/in. Group 2: Light color; high reflectance, otherwise same as Group 1. Group 3: Black or dark color; w/s ratio 0.85/1; 17 louvers/in. Group 4: Light color or unpainted aluminum; high reflectance; otherwise same as Group 3. U-factor = 0.85 Btu/(h·ft²·°F) for all groups when used with single glazing.

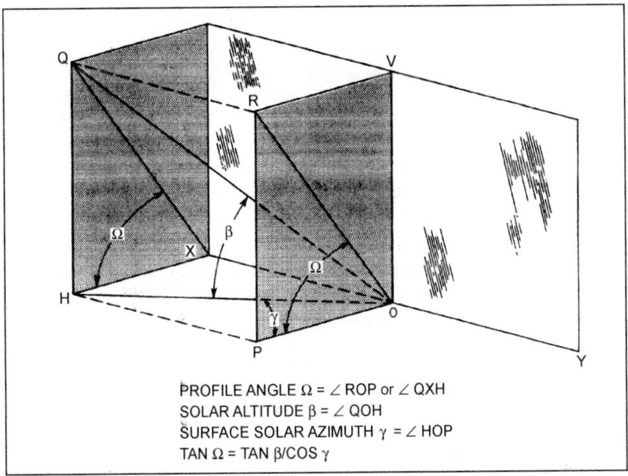

Fig. 28 Profile Angle for South-Facing Slat-Type Sunshades

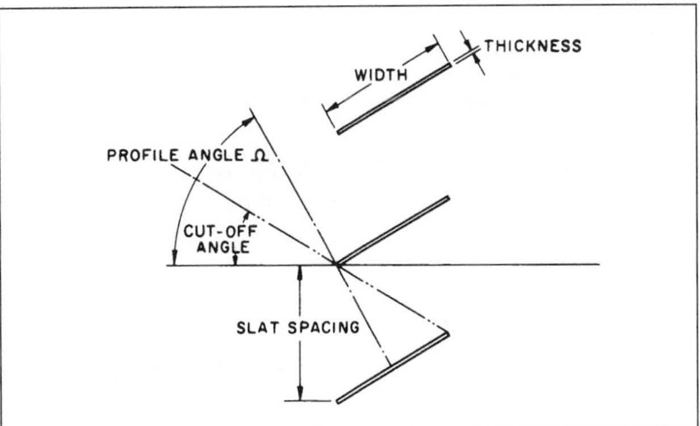

Fig. 29 Geometry of Slat-Type Sunshades

defined above. Some fraction of the remainder of the solar radiation will be transmitted by the shading system and will be incident on the glazing, but in the form of diffuse radiation. The actual angular distribution of this radiation may be quite complex, but for simplicity it is treated here as uniformly diffuse radiation, and it is considered to compose a fraction EAC, the exterior solar attenuation coefficient, of the total incident radiation interacting with the exterior shading system. The solar energy flux through the glazing with exterior shading is then given by

$$q = F_u E_{DN} \cos\theta \text{SHGC}(\theta) + [(1 - F_u)E_{DN}\cos\theta + E_d + E_r]\text{EAC}\langle\text{SHGC}\rangle_D \quad (105)$$

The quantities SHGC(θ) and $\langle\text{SHGC}\rangle_D$ in Equation (105) are the beam and diffuse solar heat gain coefficients of the unshaded glazing system. The second of the two terms on the right-hand side of the equation is the flux through the *completely shaded* glazing and is discussed in more detail in Equation (114) and the associated text, which makes clear the physical basis of the EAC.

Louvers and Sunshades

The ability of horizontal panels or louvers to intercept the direct component of solar radiation depends on their geometry and the profile or shadow-line angle Ω (Figure 28), defined as the angular difference between a horizontal plane and a plane tilted about a horizontal axis in the plane of the fenestration until it includes the sun. The profile angle can be calculated by

$$\tan \Omega = \tan \beta / \cos \gamma \quad (106)$$

For slat-type sunshades, the transmitted solar radiation consists of straight-through and transmitted-through components. When the profile angle Ω is above the cutoff angle (see Figure 29), straight-through transmission of direct radiation is completely eliminated, but the transmitted diffuse and the reflected-through components remain. Their magnitude depends largely on the reflectance of the sunshade surfaces and of exterior objects.

Narrow horizontal louvers, fabricated in conventional width-to-spacing ratios and framed as window screens, retain their shading characteristics, while gaining in effective transparency (view) by eliminating the coarse striation pattern of wide louvers. Table 18 gives values of F_u and EAC for several types of louvered sun screens. Commercially available sun screens completely exclude direct solar radiation when the profile angle exceeds approximately 26° (Groups 1 and 2) or 40° (Groups 3 and 4). Group designations are defined in the footnote to Table 18.

Roof Overhangs: Horizontal and Vertical Projections

In the northern hemisphere, horizontal projections can considerably reduce solar heat gain on south, southeast, and southwest exposures during late spring, summer, and early fall. On east and west exposures during the entire year, and on south exposures in winter, the solar altitude is generally so low that to be effective, horizontal projections must be excessively long.

The shadow width S_W and shadow height S_H (Figure 30) produced by the vertical and horizontal projections (P_V and P_H), respectively, can be calculated using the surface solar azimuth γ and the horizontal profile angle Ω determined by Equation (106).

$$S_W = P_V \, |\tan \gamma| \quad (107)$$

$$S_H = P_H \tan \Omega \quad (108)$$

Note: When the surface solar azimuth γ is greater than 90° and less than 270°, the fenestration product is completely in the shade; thus, $S_w = W + R_w$ and $A_{SL} = 0$.

The sunlit A_{SL} and shaded A_{SH} areas of the fenestration product are variable during the day and can be calculated for each moment using the following relations (see Figure 30):

$$A_{SL} = [W - (S_W - R_W)][H - (S_H - R_H)] \quad (109)$$

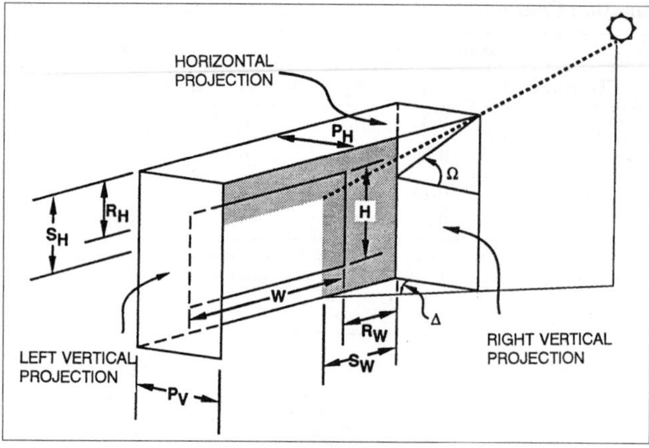

Fig. 30 Vertical and Horizontal Projections and Related Profile Angles for Vertical Surface Containing Fenestration

$$A_{SH} = A - A_{SL} \tag{110}$$

where A is total fenestration product area. McCluney (1990) described an algorithm for computer use to calculate the unshaded fraction of a window equipped with overhangs, awnings, or side fins.

Example 15. A window in the southwest wall of a building at 40°N latitude is 34.5 in. wide and 58.5 in. high. The depth of the horizontal and vertical projections is 6 in., and they are located 3 in. beyond the edges of the window.

(a) Find the sunlit and shaded area of the window at 3:00 P.M. on July 21.

(b) Find the depth of the projections necessary to fully shade the window just described.

Solution:

(a) The wall azimuth ψ for a southwest wall is +45° (Table 9). The solar azimuth ϕ can be calculated using Equation (15). At 3:00 P.M., $H = 0.25 \times 180 = 45°$; from Table 7, for July 21, $\delta = 20.6°$.

Find the solar altitude β using Equation (14):

$$\sin\beta = \cos(40)\cos(20.6)\cos(45) + \sin(40)\sin(20.6)$$

$$\beta = 47.2°$$

Find the solar azimuth ϕ using Equation (15):

$$\cos\phi = [\sin(47.2)\sin(40) - \sin(20.6)]/[\cos(47.2)\cos(40)]$$

$$\phi = 76.7°$$

Thus, from Equation (16), $\gamma = 76.7 - 45 = 31.7°$.

Using Equation (107), the width of the vertical projection shadow is

$$S_W = 6\,|\tan(31.7)| = 3.71 \text{ in.}$$

Using Equation (106), the profile angle for the horizontal projection is

$$\tan\Omega = \tan(47.2)/\cos(31.7)$$

$$\Omega = 51.8°$$

Using Equation (108), the height of the horizontal projection shadow is

$$S_H = 6\tan(51.8) = 7.62 \text{ in.}$$

Using Equations (109) and (110), the sunlit and shaded area of the window are now

$$A_{SL} = [34.5 - (3.71 - 3)]\,[58.5 - (7.62 - 3)]/144 = 12.6 \text{ ft}^2$$

$$A_{SH} = (34.5 \times 58.5)/144 - 12.6 = 1.4 \text{ ft}^2$$

(b) The shadow length necessary to fully shade the given window $S_{H(fs)}$ and $S_{W(fs)}$ from the horizontal and vertical projection are given by (see Figure 30)

$$S_{H(fs)} = 58.5 + 3 = 61.5 \text{ in.}$$

$$S_{W(fs)} = 34.5 + 3 = 37.5 \text{ in.}$$

Thus, using Equations (107) and (108),

$$P_{V(fs)} = 37.5\phi\cot(31.7)\phi = 60.7 \text{ in.}$$

$$P_{H(fs)} = 61.5\cot(51.8) = 48.4 \text{ in.}$$

For this example, because both horizontal and vertical projections do not need to fully shade the window, a horizontal projection of 48.4 in. is satisfactory. Also, to accurately analyze the influence of external projections, an hour-by-hour calculation must be performed over the periods of the year for which shadowing is desired.

Example 16. Suppose that the glazing of Examples 10 and 14 has a black exterior louver shade with a width-to-spacing ratio of 1.15/1. Find the solar energy flux through the system if it is on a west-facing orientation under the conditions of Example 5.

Solution: In Example 5, the solar altitude was found to be $\beta = 60.76°$ and the solar azimuth was found to be $\phi = 67.4°$. The incident angle was found in Example 6 to be $\theta = 63.2°$. Referring to Table 9, we see that $\psi = 90°$ for a west-facing window, and so $\gamma = \phi - \psi = 67.4° - 90° = -22.6°$. We can discard the sign of γ as irrelevant to the calculation of profile angle, since it merely determines whether the sun lies to the south or the north of the normal to the glazing. From Equation (106), we find that the profile angle is $\Omega = \arctan(\tan\beta/\tan\gamma) = \arctan[\tan(60.76°)/\tan(22.6°)] = 76.9°$.

From the footnote to Table 18, we see that the louver system corresponds to Group 1, and from the table, we see that for profile angles greater than or equal to 40° we should use $F_u = 0$ and EAC = 0.15. In Example 14, the (unshaded) glazing system was determined to have SHGC(63.2°) = 0.29 and $\langle SHGC\rangle_D = 0.33$, respectively, using Table 13. In Example 7, the incident solar fluxes for these conditions were calculated to be $E_{DN}\cos(\theta) = 123$ Btu/(h·ft²), $E_d = 29.6$ Btu/(h·ft²), and $E_r = 27.5$ Btu/(h·ft²). Using Equation (105), we calculate

$$q = (0)(123)(0.29) + [(1-0)(123) + 29.6 + 27.5](0.15)(0.33) = 8.9 \text{ Btu/h·ft}^2$$

Equations for Computer Calculations of External Shadowing of Inclined Surfaces

Incident angle: $\theta = \cos^{-1}(\cos\beta\cos\gamma\sin\Sigma + \sin\beta\cos\Sigma)$

Vertical surface: $\theta_V = \cos^{-1}(\cos\beta\cos\gamma)$

Horizontal surface: $\theta_H = \cos^{-1}(\sin\beta)$

Vertical projection profile angle:

$$\Delta = \tan^{-1}\left(\frac{\sin\gamma\cos\beta}{\cos\theta}\right);\ |\theta| < 90°$$

For $\theta > 90°$ $A_{SL} = 0$ and $A_{SH} = A$

Vertical surface: $\Delta_V = \tan^{-1}(\gamma)$ for $90° > |\gamma| > 270°$

Horizontal surface: $\Delta_H = \tan^{-1}\left(\frac{\sin\gamma}{\tan\beta}\right)$ for all γ

Horizontal projection profile:

$$\text{Angle: } \Omega = \tan^{-1}\left(\frac{\sin\beta\sin\Sigma - \cos\beta\cos\gamma\cos\Sigma}{\cos\theta}\right) \text{for } |\theta| < 90°$$

$$\text{Vertical surface: } \Omega_V = \tan^{-1}\left(\frac{\tan\beta}{\cos\gamma}\right) \text{ for } 90° > |\gamma| > 270°$$

$$\text{Horizontal surface: } \Omega_H = \tan^{-1}\left(\frac{\cos\gamma}{\tan\beta}\right) \text{ for } 90° < |\gamma| < 270°$$

Length of shadow from vertical projection: $S_W = P_V \mid \tan\Delta \mid$

Length of shadow from horizontal projection: $S_H = P_H \mid \tan\Omega \mid$

Sunlit and shaded areas of the fenestration product are determined using Equations (109) and (110).

where

ϕ = solar azimuth
β = solar altitude
γ = surface solar azimuth
Σ = surface tilt angle
P_V = vertical projection depth
P_H = horizontal projection depth
W = fenestration product width
H = fenestration product height
R_W = width of opaque surface between fenestration product and vertical projection
R_H = height of opaque surface between fenestration product and horizontal projection
A = total projected area of the fenestration product
θ = angle of incidence
Ω = horizontal projection profile angle
Δ = vertical projection profile angle

INDOOR AND BETWEEN-GLASS SHADING DEVICES ON SIMPLE FENESTRATIONS

Venetian Blinds and Roller Shades

Most fenestration has some type of internal shading to provide privacy and aesthetic effects, as well as to give varying degrees of sunshine control (Ozisik and Schutrum 1960). This section provides a method of calculating the approximate SHGC for such fenestrations for a selection of simple and somewhat common shading elements and glazing systems. The information presented here is based on measurements that predate many modern developments in fenestration systems. For this reason, desired systems may not be represented in the tables, or it may be difficult to determine whether the desired conditions match those of the table. If information is not available in this section or greater accuracy or detail is desired, the reader should refer to the subsequent section on Completely Shaded Glazings, where calculation formulas are presented. However, the present state of data availability on shading devices does not permit great accuracy in the calculation of heat flow through shaded fenestrations. The user should proceed with caution.

As discussed in the text and equations associated with Equation (129), the details of solar transmission through a glazing with interior shading are quite complex. However, if one is interested only in the approximate total heat flux through the fenestration, rather than in the detailed separation into transmitted and absorbed (and thermally retransmitted) fluxes and the distribution of the absorbed energy among the fenestration layers, it will frequently be the case that measurements made on a fenestration under one set of conditions can be extrapolated to other fenestrations and conditions to give an adequate answer. In this case, the heat flux is represented by

$$q = [E_{DN}\cos\theta \text{SHGC}(\theta) + (E_d + E_r)\langle\text{SHGC}\rangle_D]\text{IAC} \quad (111)$$

where the quantity in the brackets represents the heat flow through the unshaded glazing system [compare Equations (97) and (100)] and the constant interior solar attenuation coefficient IAC represents the fraction of that heat flow that enters the room, some energy having been excluded by the shading. The solar heat gain coefficients in Equation (111) may be obtained from Table 13 or calculated by Equations (79) and (84) for unshaded glazing. Table 19 gives values of IAC (derived from measurements) for a variety of glazing and shading combinations.

It will be noticed that the IAC bears a certain similarity to the shading coefficient. There is, however, an important difference: we must calculate the solar heat flux through the unshaded glazing at the appropriate angle before applying the IAC. With the shading coefficient, only the angular dependence of single glazing was included (through the now-discarded SHGF). The effectiveness of any internal shading device depends on its ability to reflect incoming solar radiation back through the fenestration before it can be absorbed and converted into heat within the building. Table 21 lists approximate values of solar-optical properties for the typical indoor shading devices described in Tables 19 and 20.

For shading between the panes of a double-glazing system, an equation similar to Equation (111) can be used, but in this case one defines the unshaded glazing system to be the portion of the glazing that is exterior to the shading and defines a new coefficient, the between-pane solar attenuation coefficient (BAC), to describe the effect of the shading system together with the portion of the glazing that is interior to it:

$$q = [E_{DN}\cos\theta \text{SHGC}(\theta) + (E_d + E_r)\langle\text{SHGC}\rangle_D]_{ext}\text{BAC} \quad (112)$$

Values of BAC for those systems for which measurements are available are given in Table 20. For other systems, one should use the detailed equations of the section on Completely Shaded Glazings.

Example 17. Find the approximate SHGC and solar heat flux for a heat-absorbing double glazing with a light-colored interior venetian blind under the conditions of Example 5. The heat-absorbing double glazing has a visible transmittance of 0.41.

Solution: Referring to Table 19, we see that an interior, light venetian blind, heat-absorbing double glazing has an IAC value of 0.66. For the conditions of Example 5, the incident angle was found in Example 6 to be $\theta = 63.2°$, and in Example 7 the incident solar fluxes for these conditions were calculated to be $E_{DN}\cos\theta$ = 123 Btu/h·ft², E_d = 29.6 Btu/h·ft², and E_r = 27.5 Btu/(h·ft²). Referring to Table 13, we see that the closest glazing system corresponds to ID 5h, 1/4 in. gray glass exterior, clear interior, which has SHGC(60°) = 0.37, SHGC(70°) = 0.29, and $\langle\text{SHGC}\rangle_D$ = 0.39. Interpolating in incident angle between the two table values gives SHGC(63.2°) = 0.34. Using Equation (111), we find the solar heat flux q = [123 × 0.34 + (29.6 + 27.5) × 0.39]0.66 = 42 Btu/h·ft².

There is some ambiguity about the meaning of the SHGC in this case. One could pick out the terms in Equation (111) corresponding to either the effective beam or diffuse SHGC for the overall system, but, given the fact that the IAC values in Table 19 were measured in outdoor calorimeters so that the incident radiation consisted of both beam and diffuse radiation, it is most reasonable to calculate an effective SHGC for these conditions by dividing the estimated solar heat flux by the total incident intensity:

$$\text{SHGC}_{eff} = \frac{42}{123 + 29.6 + 27.5} = 0.23$$

Example 18. Estimate the solar heat flux for the glazing and conditions of Example 17, but assume that the venetian blind is between the glazing panes.

Solution: Referring to Table 20, we see that a light venetian blind between the panes of a heat-absorbing double-glazing system has a BAC value of 0.28. We use this value in Equation (112), but in contrast to Example 17, we need the SHGC values of the exterior glazing layer rather than the full glazing system. In Example 17, we used ID 5h,

Table 19 Interior Solar Attenuation Coefficients (IAC) for Single or Double Glazings Shaded by Interior Venetian Blinds or Roller Shades

					IAC				
	Nominal Thickness[b]	Glazing Solar Transmittance[b]			Venetian Blinds		Roller Shades		
Glazing System[a]	Each Pane, in.	Outer Pane	Single or Inner Pane	Glazing SHGC[b]	Medium	Light	Opaque Dark	Opaque White	Translucent Light
Single Glazing Systems									
Clear, residential	1/8[c]		0.87 to 0.80	0.86	0.75[d]	0.68[d]	0.82	0.40	0.45
Clear, commercial	1/4 to 1/2		0.80 to 0.71	0.82					
Clear, pattern	1/8 to 1/2		0.87 to 0.79						
Heat absorbing, pattern	1/8			0.59					
Tinted	3/16, 7/32		0.74, 0.71						
Above glazings, automated blinds[e]				0.86	0.64	0.59			
Above glazings, tightly closed vertical blinds				0.85	0.30	0.26			
Heat absorbing[f]	1/4		0.46	0.59	0.84	0.78	0.66	0.44	0.47
Heat absorbing, pattern	1/4								
Tinted	1/8, 1/4		0.59, 0.45						
Heat absorbing or pattern			0.44 to 0.30	0.59	0.79	0.76	0.59	0.41	0.47
Heat absorbing	3/8		0.34						
Heat absorbing or pattern			0.29 to 0.15						
			0.24	0.37	0.99	0.94	0.85	0.66	0.73
Reflective coated glass				0.26 to 0.52	0.83	0.75			
Double Glazing Systems [g]									
Clear double, residential	1/8	0.87	0.87	0.76	0.71[d]	0.66[d]	0.81	0.40	0.46
Clear double, commercial	1/4	0.80	0.80	0.70					
Heat absorbing double[f]	1/4	0.46	0.8	0.47	0.72	0.66	0.74	0.41	0.55
Reflective double				0.17 to 0.35	0.90	0.86			
Other Glazings (Approximate)					0.83	0.77	0.74	0.45	0.52
± Range of Variation [h]					0.15	0.17	0.16	0.21	0.21

[a] Systems listed in the same table block have the same IAC.
[b] Values or ranges given for identification of appropriate IAC value; where paired, solar transmittances and thicknesses correspond. SHGC is for unshaded glazing at normal incidence.
[c] Typical thickness for residential glass.
[d] From measurements by Van Dyke and Konen (1982) for 45° open venetian blinds, 35° solar incidence, and 35° profile angle.
[e] Use these values only when operation is automated for exclusion of beam solar (as opposed to daylight maximization). Also applies to tightly closed horizontal blinds.
[f] Refers to gray, bronze, and green tinted heat-absorbing glass (on exterior pane in double glazing)
[g] Applies either to factory-fabricated insulating glazing units or to prime windows plus storm windows.
[h] The listed approximate IAC value may be higher or lower by this amount, due to glazing/shading interactions and variations in the shading properties (e.g., manufacturing tolerances).

Table 20 Between-Glass Solar Attenuation Coefficients (BAC) for Double Glazing with Between-Glass Shading

					Type of Shading		
	Nominal Thickness,	Solar Transmittance[a]			Venetian Blinds		Louvered
Type of Glass	Each Pane	Outer Pane	Inner Pane	Description of Air Space	Light	Medium	Sun Screen
Clear out, Clear in	3/32, 1/8 in.	0.87	0.87	Shade in contact with glass or shade separated from glass by air space.	0.33	0.36	0.43
Clear out, Clear in	1/4 in.	0.80	0.80	Shade in contact with glass-voids filled with plastic.	—	—	0.49
Heat-absorbing[b] out, Clear in				Shade in contact with glass or shade separated from glass by air space.	0.28	0.30	0.37
	1/4 in.	0.46	0.80	Shade in contact with glass-voids filled with plastic.	—	—	0.41

[a] Refer to manufacturers' literature for exact values.
[b] Refers to gray, bronze and green tinted heat-absorbing glass.

Table 21 Properties of Representative Indoor Shading Devices Shown in Tables 19 and 20

	Solar-Optical Properties (Normal Incidence)		
Indoor Shade	Transmittance	Reflectance	Absorptance
Venetian blinds[a] (ratio of slat width to slat spacing 1.2, slat angle 45°)			
Light colored slat	0.05	0.55	0.40
Medium colored slat	0.05	0.35	0.60
Vertical blinds			
White louvers	0.00	0.77	0.23
Roller shades			
Light shades (translucent)	0.25	0.60	0.15
White shade (opaque)	0.00	0.65	0.35
Dark colored shade (opaque)	0.00	0.20	0.80

[a] Values in this table and Tables 19 and 20 are based on horizontal venetian blinds. However, tests show that these values can be used for vertical blinds with good accuracy.

which Table 13 shows had an exterior glazing that is 1/4 in. gray glass. We find that ID 1h is the entry for clear 1/4 in. gray glass, for which SHGC(60°) = 0.51, SHGC(70°) = 0.44, and $\langle SHGC \rangle_D = 0.52$. Interpolating the beam SHGC values in incident angle gives SHGC(63.2°) = 0.49. We then use Equation (112) to calculate

$$q = [123 \times 0.49 + (29.6 + 27.5) \times 0.52]0.41 = 37 \text{ Btu/h·ft}^2$$

This indicates that having the blind between the glazings yields slightly better solar heat gain rejection than an interior placement.

Draperies

Draperies reduce heating and cooling loads, depending on the type and the use by the occupant. Rudoy and Duran (1975) found annual reductions between 5 and 20%. An approximate model for determining the SHGC of free-standing vertical interior shades was developed by McCluney and Mills (1993). Solar heat gain may be estimated by using Equation (111) and the IAC values listed in Table 22.

The solar optical properties of drapery fabrics can be determined accurately by laboratory tests (Yellott 1963), and manufacturers can usually supply solar transmittance and reflectance values for their products. In addition to these properties, the **openness factor** (ratio of the open area between the fibers to the total area of the fabric) is a useful property that can be measured exactly (Keyes 1967, Pennington and Moore 1967). It can also be estimated by inspection, since the human eye can readily distinguish between tightly woven fabrics that permit little direct radiation to pass between the fibers and loosely woven fabrics that allow the sun's rays to pass freely.

Drapery fabrics can be classified in terms of their solar-optical properties as having specific values of fabric transmittance and reflectance. Fabric reflectance is the major factor in determining the ability of a fabric to reduce solar heat gain. Based on their appearance, draperies can also be classified by yarn color as dark, medium, and light and by weave as closed, semiopen, and open. The apparent color of a fabric is determined by the reflectance of the yarn itself. The figure in Table 22 shows yarn reflectance. Figure 31 classifies drapery fabrics into nine types, rated by openness and yarn reflectances.

Figure 31 guides in the use of Table 22 for a fabric-glass combination when the solar-optical properties are unknown. Whenever possible, fabric reflectance and transmittance values should be obtained from the manufacturer, which permits more accurate solar heat flux estimates to be made. Visual estimations of openness and yarn reflectance, interpreted through Table 22, are valuable in judging the effectiveness of drapes for (1) protection from excessive radiant energy from either sunlight or sun-heated glass, (2) brightness control, (3) providing either outward view or privacy, and (4) sound control.

Table 22 applies to glass and a single drape hung with 100% fullness (drapery width is twice the width of the draped area). If the drapery is hung flat, like a fenestration product shade, a different IAC applies; with a low transmittance and high reflectance, the IAC is appreciably lower. As an extreme example, a flat opaque drapery having an aluminized or similar coating with reflectance of 0.80, in combination with 1/4 in. clear glass, has an SHGC of 0.18, compared with 0.32 for this material in draped form. Pennington and Moore (1967) explain the effect of folding drapery materials to provide 100% fullness and describe a method for calculating SHGC when materials are used flat.

Example 19. A drape with 100% fullness, having a fabric transmittance of 0.20 and a fabric reflectance of 0.40, is used with 1/4 in. glass. What IAC should be used?

Solution: Locating the point on the ideogram in Table 22 corresponding to fabric reflectance 0.40 and transmittance 0.20, we find that it is close to the diagonal line labeled "F". In column F of the table, we find a value of 0.58 on the line corresponding to 1/4 in. clear glass. This is the desired IAC value.

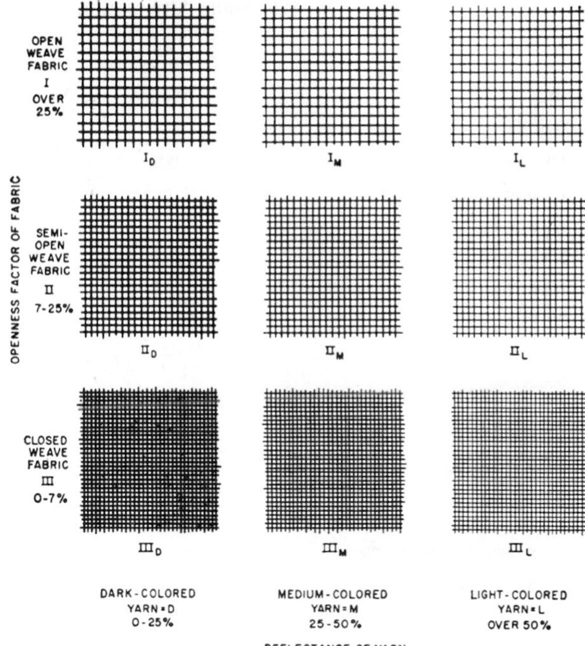

Note: Classes may be approximated by eye. With closed fabrics, no objects are visible through the material, but large light or dark areas may show. Semiopen fabrics do not permit details to be seen, and large objects are clearly defined. Open fabrics allow details to be seen, and the general view is relatively clear with no confusion of vision. The yarn color or shade of light or dark may be observed to determine whether the fabric is light, medium, or dark.

Fig. 31 Designation of Drapery Fabrics

Example 20. Determine the fabric designator for a fabric having an openness factor of 0.10 and a yarn reflectance of 0.60.

Solution: On the figure in Table 22, these lines intersect in the area of Designator II_L. Refer also to Figure 30. Fabric is semiopen and light in color. Additional information: probable fabric reflectance is 0.50, and fabric transmittance is 0.35.

COMPLETELY SHADED GLAZINGS

This section presents approximate models (Klems 2001) for calculating the heat flow through glazing systems with shading for all cases not covered by the sections on Exterior Shading and Indoor and Between-Glass Shading Devices on Simple Fenestrations (i.e., glazing or shading systems not covered by the tables or with incident angles above 30°). With the exception of exterior fins and overhangs, all shading is modeled as a **planar, ideally diffuse layer** parallel to the glass layers, and the models do not apply if there is more than one shading layer. The effects of specular transmission or partial shading are assumed to have been separated out using the method described above under Simplified Calculation Procedure.

The shading layer is denoted by the subscript *S*. To apply these models, one or more of the following quantities must be known for the shading (the number depends on the position of the shading in the glazing system):

T_S^{fH} = front directional-hemispherical transmittance of the shading layer
T_S^{bH} = back directional-hemispherical transmittance of the shading layer
R_S^{fH} = front directional-hemispherical reflectance of the shading layer
R_S^{bH} = back directional-hemispherical reflectance of the shading layer

Table 22 Interior Solar Attenuation Coefficients for Single and Insulating Glass with Draperies

Glazing	Glass Trans-mission	Glazing SHGC (No Drapes)	IAC A	B	C	D	E	F	G	H	I	J
Single glass												
1/8 in. clear	0.86	0.87	0.87	0.82	0.74	0.69	0.64	0.59	0.53	0.48	0.42	0.37
1/4 in. clear	0.8	0.83	0.84	0.79	0.74	0.68	0.63	0.58	0.53	0.47	0.42	0.37
1/2 in. clear	0.71	0.77	0.84	0.80	0.75	0.69	0.64	0.59	0.55	0.49	0.44	0.40
1/4 in. heat absorbing	0.46	0.58	0.85	0.81	0.78	0.73	0.69	0.66	0.61	0.57	0.54	0.49
1/2 in. heat absorbing	0.24	0.44	0.86	0.84	0.80	0.78	0.76	0.72	0.68	0.66	0.64	0.60
Reflective coated	—	0.52	0.95	0.90	0.85	0.82	0.77	0.72	0.68	0.63	0.60	0.55
	—	0.44	0.92	0.88	0.84	0.82	0.78	0.76	0.72	0.68	0.66	0.62
	—	0.35	0.90	0.88	0.85	0.83	0.80	0.75	0.73	0.70	0.68	0.65
	—	0.26	0.83	0.80	0.80	0.77	0.77	0.77	0.73	0.70	0.70	0.67
Insulating glass, 1/4-in. air space												
(1/8 in. out and 1/8 in. in)	0.76	0.77	0.84	0.80	0.73	0.71	0.64	0.60	0.54	0.51	0.43	0.40
Insulating glass 1/2-in. air space												
Clear out and clear in	0.64	0.72	0.80	0.75	0.70	0.67	0.63	0.58	0.54	0.51	0.45	0.42
Heat absorbing out and clear in	0.37	0.48	0.89	0.85	0.82	0.78	0.75	0.71	0.67	0.64	0.60	0.58
Reflective coated	—	0.35	0.95	0.93	0.93	0.90	0.85	0.80	0.78	0.73	0.70	0.70
	—	0.26	0.97	0.93	0.90	0.90	0.87	0.87	0.83	0.83	0.80	0.80
	—	0.17	0.95	0.95	0.90	0.90	0.85	0.85	0.80	0.80	0.75	0.75

Interior Solar Attenuation (IAC)

Notes:

1. Interior attenuation coefficients are for draped fabrics.
2. Other properties are for fabrics in flat orientation.
3. Use fabric reflectance and transmittance to obtain accurate IAC values.
4. Use openness and yarn reflectance or openness and fabric reflectance to obtain the various environmental characteristics, or to obtain approximate IAC values.

Classification of Fabrics

I = Open weave

II = Semiopen weave

III = Closed weave

D = Dark color

M = Medium color

L = Light color

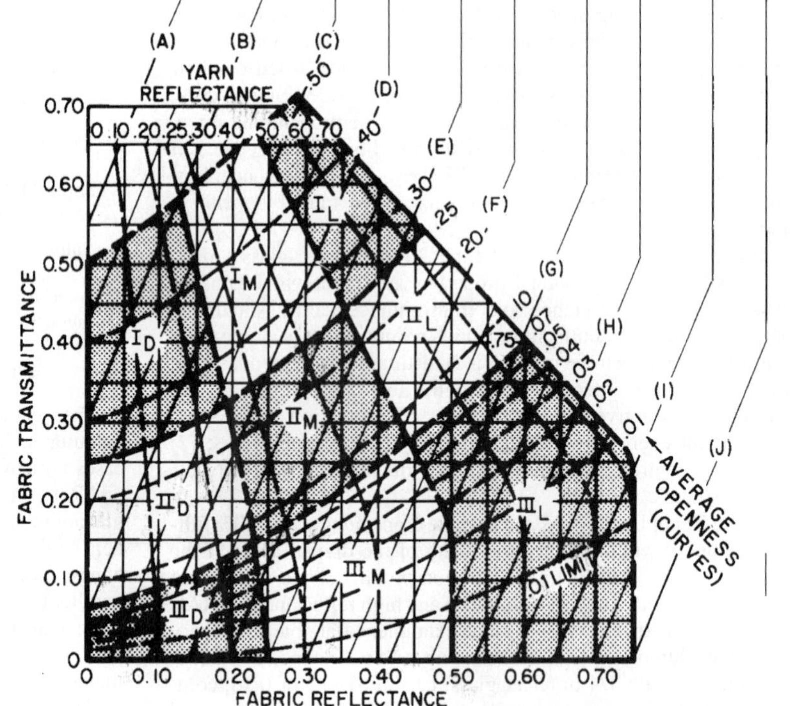

To obtain fabric designator (III_L, I_M, etc.). Using either (1) fabric transmittance and fabric reflectance coordinates, or (2) openness and yarn reflectance coordinates, find a point on the chart and note the designator for that area. If properties are not known, the classification may be approximated by eye as described in the note in Figure 31.

To obtain interior attenuation (IAC). (1) Locate drapery fabric as a point using its known properties, or approximate using its fabric classification designator. For accuracy, use fabric transmittance and fabric reflectance; (2) follow diagonal IAC lines to lettered columns in the table. Find IAC value in selected column on line corresponding to glazing used. For example, IAC is 0.4 for 1/4-in. clear single glass with III_L drapery (Column H).

Note: Unlike specular transmittance, front and back directional-hemispherical transmittances are not necessarily equal.

To match the assumptions of the model, the specular component should be *excluded* from the transmittance but *included* in the reflectances. Since the shading layer is assumed to be an ideal diffuser, the above quantities are the same for all incident directions. In addition, the notation $\langle\ \rangle_D$ will be applied to various optical properties of specular glazings. This notation means the hemispherical average of the quantity over all incident directions and is the quantity labeled "diffuse" in Table 13. For example, for a particular glass layer j, $\langle T_j \rangle_D$ would be the hemispherical average transmittance of the glass layer, which is also the transmittance for ideally diffuse incident radiation.

Another addition to the notation is needed for discussing absorption in this section. In the treatment of unshaded multiple glazings, the notation $\mathcal{A}_n^f(\theta)$ was used to denote the actual front absorption of the nth layer of an L-layer system. Since only one system was under consideration, no ambiguity could arise from the fact that the notation does not specifically reference the number of layers in the system. Here it will be necessary to discuss multi-layer systems that are subdivided into one or more subsystems. Since the actual absorption depends on the other layers in the system because of reflections from those layers and because it is referenced to the incident intensity on the front (or back) surface of the system, rather than on the specific layer, it is necessary to include the particular subsystem in the notation. The following notation will be used. We assume that the overall system has L layers that are numbered sequentially from outside to inside, as before. Now $\mathcal{A}_{n:(M,N)}^f(\theta)$ is the actual front absorption of layer n in subsystem of layers M through N, and similarly for back absorption. In this notation, the front absorptance of layer n in the overall system, which previously was called $\mathcal{A}_n^f(\theta)$, would now be denoted $\mathcal{A}_{n:(1,L)}^f(\theta)$. Note that for the subsystem (M, N) the front absorptance $\mathcal{A}_{n:(M,N)}^f$ is the fraction of the radiation incident on the front of layer M that is absorbed in layer n, while the back absorptance $\mathcal{A}_{n:(M,N)}^b$ is the fraction of the radiation incident on the back side of layer N that is absorbed in layer n.

In all cases, we assume that there is a specular glazing of L layers, labeled as described in the section on Optical Properties of Multiple-Layer Glazing Systems in the discussion of unshaded glazing optical properties. The shading layer, denoted S, is not counted in this labeling. Differing locations of the shading layer are discussed in the following sections.

Exterior Shading by Overhangs, Fins, etc.

In this case, we neglect (1) reduction of the sky view by the shading and (2) reflection of diffuse radiation by the face of the shading seen by the glazing. It will usually be the case that this results in an overestimate of the heat flow, but exceptions are possible (e.g., highly reflective ground and light-colored overhang). Here the effect of the shading is simply to eliminate the direct beam radiation, and one is left with an unshaded glazing with diffuse radiation incident; the relevant optical properties are the hemispherical averages of those of the unshaded glazing system.

$$q = (E_d + E_r)\langle \mathrm{SHGC} \rangle_D \tag{113}$$

The designer should be aware that because the diffuse solar radiation is small compared to that of the direct beam, reflection of direct beam radiation from other surfaces visible to the glazing can result in significant additional heat flow (e.g., beam radiation falling on the "wrong" side of a fin or overhang intended to shade an adjacent window).

Exterior Shading Layer

Examples of planar exterior shading layers include exterior louvers or venetian blinds. It is assumed that air has free flow between the exterior shading system and the glazing system. Under these conditions, as Table 17 shows, the fraction of the energy absorbed in the shading that flows inward is small. If this energy flow is neglected, then the whole effect of the shading layer is to modify the intensity of the (diffuse) radiation incident on the unshaded glazing, and the heat flow is

$$q = [(1 - F_u)E_{DN}\cos\theta + E_d + E_r] \frac{T_S^{fH}}{1 - R_S^{bH}\langle R_{1,L}^f \rangle_D} \langle \mathrm{SHGC} \rangle_D \tag{114}$$

where $\langle \mathrm{SHGC} \rangle_D$ is for the unshaded glazing system. To compute the amount of energy absorbed in the shading layer, one must estimate the isolated-layer absorptances of the shading layer from

$$\begin{aligned} a_S^f &= 1 - T_S^{fH} - R_S^{fH} \\ a_S^b &= 1 - T_S^{bH} - R_S^{bH} \end{aligned} \tag{115}$$

When attached to the glazing, and using $S = 0$ to fit into the labeling scheme, the shading layer front absorptance is then

$$\mathcal{A}_{S:(S,L)}^f = a_S^f + \frac{a_S^b T_S^{fH} \langle R_{1,L}^f \rangle_D}{1 - R_S^{bH} \langle R_{1,L}^f \rangle_D} \tag{116}$$

where $R_{1,L}^f$ refers to the unshaded glazing system. $R_{1,L}^f$ may either be obtained from Table 13 or calculated as described previously in the section on Solar-Optical Properties of Glazings. The addition to the heat flow in Equation (114) is then

$$\Delta q = [(1 - F_u)E_{DN}\cos\theta + E_d + E_r]\mathcal{A}_S^f N_S \tag{117}$$

where the inward-flowing fraction N_S for the shading layer is taken from the appropriate line of Table 17.

Between-Glass Shading Layer

One must first identify the position of the shading layer within the L-layer glazing system, remembering that the labeling of glazing layers did not count the shading layer S. We count the layers, beginning with the one in contact with the exterior air, until we reach the layer next to the shading layer. Let the number of that layer be K. We now relabel the shading layer by setting $S = K + 1$ and relabel all subsequent glazing layers, making the one to the interior of S layer $K + 2$ (or $S + 1$), and continuing until we reach glazing layer L, which becomes $L + 1$ in the new labeling scheme. We now have an $L + 1$ layer system that consists of three parts: (1) an exterior glazing subsystem of $S - 1$ layers, (2) the shading layer S, and (3) an interior glazing subsystem, consisting of layers $S + 1$ to $L + 1$, a total of $(L + 1) - S$ or $L - K$ layers in all. Both the exterior glazing and interior glazing subsystems are assemblies of unshaded, specular glazing layers, and their optical properties are calculated by the methods described in the section on Optical Properties of Multiple-Layer Glazing Systems for unshaded glazings. The exterior and interior glazing subsystems will often be either a single or double glazing, and, if they are sufficiently common, their properties may then be found in Table 13.

For calculational purposes, it is convenient to consider two other subsystems, one consisting of the shading layer together with the exterior glazing, an S-layer subsystem consisting of layers 1 through S, and the other consisting of the shading layer together with the interior glazing, consisting of layers S through $L + 1$, a total of $L - S + 2$ or $L - K + 1$ layers. For the former subsystem, we will need the directional-hemispherical transmittance:

$$T_{1,S}^{fH}(\theta) = \frac{T_{1,S-1}(\theta)T_S^{fH}}{1 - R_S^{fH}\langle R_{1,S-1}^b\rangle_D} \tag{118}$$

and the hemispherical back reflectance (which is the same for all incident directions):

$$R_{1,S}^{bH} = R_S^{bH} + \frac{\langle R_{1,S-1}^b(\theta)\rangle_D T_S^{bH} T_S^{fH}}{1 - R_S^{fH}\langle R_{1,S-1}^b\rangle_D} \tag{119}$$

For the latter subsystem, we need the hemispherical front reflectance:

$$R_{S,L+1}^{fH} = R_S^{fH} + \frac{\langle R_{S+1,L+1}^f\rangle_D T_S^{bH} T_S^{fH}}{1 - R_S^{bH}\langle R_{S+1,L+1}^f\rangle_D} \tag{120}$$

The directional-hemispherical transmittance of the total system is then

$$T_{1,L+1}^{fH}(\theta) = \frac{T_{1,S}^{fH}(\theta)T_{S+1,L+1}^{fH}}{1 - R_{1,S}^{bH}\langle R_{S+1,L+1}^f\rangle_D} = \frac{T_{1,S-1}(\theta)T_S^{fH}T_{S+1,L+1}^{fH}}{[1 - R_S^{fH}\langle R_{1,S-1}^b\rangle_D][1 - R_{1,S}^{bH}\langle R_{S+1,L+1}^f\rangle_D]} \tag{121}$$

The absorptances for the kth layer are
for $k < S$

$$\mathcal{A}_{k:(1,L+1)}^f(\theta) = \mathcal{A}_{k:(1,S-1)}^f(\theta) + \langle\mathcal{A}_{k:(1,S-1)}^b\rangle_D \frac{T_{1,S}^{fH}(\theta)}{1 - R_{S,L+1}^{fH}\langle R_{1,S-1}^b\rangle_D} \tag{122}$$

for $k = S$

$$\mathcal{A}_{k:(1,L+1)}^f(\theta) = \frac{a_S^f T_{1,S-1}(\theta)}{1 - R_{S,L+1}^{fH}\langle R_{1,S-1}^b\rangle_D} + \frac{a_S^b T_{1,S}^{fH}(\theta)\langle R_{S+1,L+1}^{fH}\rangle_D}{1 - R_{1,S}^{bH}\langle R_{S+1,L+1}^{fH}\rangle_D} \tag{123}$$

for $k > S$

$$\mathcal{A}_{k:(1,L+1)}^f(\theta) = \langle\mathcal{A}_{k:(S+1,L+1)}^f\rangle_D \frac{T_{1,S}^{fH}(\theta)}{1 - R_{1,S}^{bH}\langle R_{S+1,L+1}^{fH}\rangle_D} \tag{124}$$

The hemispherical average of the appropriate above equation then gives $\langle\mathcal{A}_{k:(1,L+1)}^f\rangle_D$.

For convenience, we split the heat flow up into three parts: that due to transmitted direct beam radiation (subscript B), that due to transmitted diffuse beam and ground-reflected radiation (subscript d), and that due to energy absorbed in the various layers and flowing inward by the normal mechanisms of heat transfer (subscript a):

$$q = q_B + q_d + q_a \tag{125}$$

We note that q_B and q_d consist of diffusely distributed radiation in the solar wavelength range, while q_a is a heat flux. The quantities in Equation (125) are given by

$$q_B = E_{DN}\cos(\theta)T_{1,L+1}^{fH}(\theta) \tag{126}$$

$$q_d = (E_d + E_r)\langle T_{1,L+1}^{fH}\rangle_D \tag{127}$$

$$q_a = E_{DN}\cos(\theta)\sum_{k=1}^{L+1}\mathcal{A}_{k:(1,L+1)D}^f(\theta)N_k + (E_d + E_r)\sum_{k=1}^{L+1}\langle\mathcal{A}_{k:(1,L+1)}^f\rangle_D N_k \tag{128}$$

The values of the layer-specific inward-flowing fractions N_k are chosen from the appropriate lines of Table 17. All types of shading should be considered equivalent when using Table 17 with this simplified method; accordingly, corresponding entries in the lines for differing blind tilts should be averaged in the case of double glazing.

For systems with more than two glazing layers, for which no information exists in Table 17, the following estimate should be used. If the glazing system contains no low-e coating, then information for double glazing with between-pane shading should be averaged; if the system contains one or more low-e coatings, then the line corresponding to low-e double glazing should be chosen. For $k < S$, use for N_k the value in Table 17 corresponding to the outer glass; for $k = S$, use the value for the shading layer; and for $k > S$, use the value corresponding to the inner glass.

Interior Shading Layer

For interior shading, we can consider the shading layer S to form the $(L + 1)$th layer of an $(L + 1)$ layer system. The total directional-hemispherical transmittance of this system is given by

$$T_{1,L+1}^{fH}(\theta) = \frac{T_{1,L}(\theta)T_S^{fH}}{1 - R_S^{fH}\langle R_{1,L}^b\rangle_D} \tag{129}$$

and the total diffuse transmittance of the system is the hemispherical average of this quantity is given by

$$\langle T_{1,L+1}^{fH}\rangle_D = \frac{\langle T_{1,L}(\theta)\rangle_D T_S^{fH}}{1 - R_S^{fH}\langle R_{1,L}^b\rangle_D} \tag{130}$$

In discussing layer absorptance, we need to consider a layer both in its role in the L-layer unshaded system and in the $(L + 1)$-layer system formed by adding the shading layer. The L-layer system is an unshaded one, so all quantities pertaining to that system can be calculated from Equations (60) through (65), (79), and (84) for the optical properties of multilayer, unshaded glazings above. If it is a sufficiently common system, its properties may be found in Table 13, and the table values may be linearly interpolated to the desired incident angle; see also the previous section on Spectral Averaging of Glazing Properties and Examples 11, 12, and 13. The addition of the shading layer causes an additional contribution to the absorptance in each layer of the L-layer system due to the backward incidence of diffuse radiation reflected by the shade:

$$\mathcal{A}^f_{k:(1,L+1)}(\theta) = \mathcal{A}^f_{k:(1,L)}(\theta) + \langle \mathcal{A}^b_{k:(1,L)} \rangle_D \frac{T_{1,L}(\theta) R_S^{fH}}{1 - R_S^{fH} \langle R^b_{1,L} \rangle_D} \quad (131)$$

and the hemispherical average of this quantity over incident direction is given by

$$\langle \mathcal{A}^f_{k:(1,L+1)} \rangle_D = \langle \mathcal{A}^f_{k:(1,L)} \rangle_D + \langle \mathcal{A}^b_{k:(1,L)} \rangle_D \frac{\langle T_{1,L} \rangle_D R_S^{fH}}{1 - R_S^{fH} \langle R^b_{1,L} \rangle_D} \quad (132)$$

The absorptance in the shade layer is

$$\mathcal{A}^f_{S:(1,L+1)}(\theta) \equiv \mathcal{A}^f_{L+1:(1,L+1)}(\theta) = \frac{T_{1,L}(\theta) a_S^f}{1 - R_S^{fH} \langle R^b_{1,L} \rangle_D} \quad (133)$$

with, again, a hemispherical average given by

$$\langle \mathcal{A}^f_{L+1:(1,L+1)} \rangle_D = \frac{\langle T_{1,L} \rangle_D a_S^f}{1 - R_S^{fH} \langle R^b_{1,L} \rangle_D} \quad (134)$$

The heat flow is calculated from Equations (125) through (128).

The inward-flowing fractions N_k in Equation (128) are chosen from Table 17 when possible. The most thermally resistive glazing system with interior shading in this table is double glazing. For higher resistance systems (low-e double glazing, triple glazing, or higher, etc.), there are three possible procedures:

1. Calculate the N_k ($k < S$) for the unshaded glazing, and for the shading layer ($k = S$), take the value from the Inner Shading Layer column of Table 17 for the system that is closest to the desired one.
2. Treat the shading layer as an additional (uncoated, unless the shading layer has low emissivity on the exterior side) glazing and calculate the N_k for the resulting ($L + 1$)-layer glazing.
3. Take N_1, N_2, and the value for the interior shading layer from the table entries for double glazing with interior shade, and take $N_k = N_2$ for all subsequent glazing layers. Using as an alternative assumption $N_L = N_2$ and $N_k = N_1$ for all $k < L$ will give an estimate of the range of error to be expected.

Method (1) will produce values of N_k for $k \le L$ that tend to be too high, and a value of N_S ($S = L + 1$) that may be too low. Method (2) will produce values of N_k for $k \le L$ that tend to be too low and a value of N_S that also tends to be too low (since some heat that transfers outward by convection from layer S would be carried to the interior with an unsealed air space). By comparing the resulting solar energy fluxes using each of the three methods, one can obtain an estimate of the sensitivity of the results to uncertainty in calculating N_k.

Example 21. Calculate the solar energy flux through the glazing system of Example 10 with a closed interior translucent light shade, under the conditions of Example 5.

Solution: From Table 21, we take the properties of a translucent light shade to be $T_S^{fH} = 0.25$, $R_S^{fH} = 0.60$, and $a_S^f = 0.15$. We assume that the reflectance is diffuse and that the transmittance is independent of incident angle. By using method (1) above for determining the inward-flowing fractions, we can take the values of $N_1 = 0.064$ and $N_2 = 0.821$ calculated in Example 14, and we take $N_S = 0.85$ from the entry in Table 17 for double glazing with interior shade. We use the property values for an incident angle θ of 63.2° (from Example 6) that were calculated for the unshaded glazing in Example 11: $T_{1,2}(63.2°) = 0.244$, $R^f_{1,2}(63.2°) = 0.446$, $R^b_{1,2}(63.2°) = 0.444$, $\mathcal{A}^f_{1:(1,2)}(63.2°) = 0.289$, $\mathcal{A}^b_{1:(1,2)}(63.2°) = 0.144$, $\mathcal{A}^f_{2:(1,2)}(63.2°) = 0.022$, $\mathcal{A}^b_{2:(1,2)}(63.2°) = 0.167$. We use the hemispherical average properties calculated in Example 12: $\langle T_{1,2} \rangle_D = 0.277$, $\langle R^f_{1,2} \rangle_D = 0.405$, $\langle R^b_{1,2} \rangle_D = 0.442$, $\langle \mathcal{A}^f_{1:(1,2)} \rangle_D = 0.271$, $\langle \mathcal{A}^b_{1:(1,2)} \rangle_D = 0.138$, $\langle \mathcal{A}^f_{2:(1,2)} \rangle_D = 0.036$, and $\langle \mathcal{A}^b_{2:(1,2)} \rangle_D = 0.141$. In the shaded fenestration, the shade comprises layer 3 (i.e., $S = 3$); the total transmittance at an incident angle of 63.2° is calculated from Equation (129).

$$T^{fH}_{1,S}(63.2°) = \frac{T_{1,2}(63.2°) T_S^{fH}}{1 - R_S^{fH} \langle R^b_{1,2} \rangle_D} = \frac{(0.244)(0.25)}{1 - (0.60)(0.442)} = 0.083$$

The layer absorptances are calculated from Equation (131):

$$\mathcal{A}^f_{1:(1,S)}(63.2°) = \mathcal{A}^f_{1:(1,2)}(63.2°) + \langle \mathcal{A}^b_{1:(1,2)} \rangle_D \frac{T_{1,2}(63.2°) R_S^{fH}}{1 - R_S^{fH} \langle R^b_{1,2} \rangle_D}$$

$$= 0.289 + (0.138)\frac{(0.244)(0.60)}{1 - (0.60)(0.442)} = 0.316$$

$$\mathcal{A}^f_{2:(1,S)}(63.2°) = \mathcal{A}^f_{2:(1,2)}(63.2°) + \langle \mathcal{A}^b_{2:(1,2)} \rangle_D \frac{T_{1,2}(63.2°) R_S^{fH}}{1 - R_S^{fH} \langle R^b_{1,2} \rangle_D}$$

$$= 0.022 + (0.141)\frac{(0.244)(0.60)}{1 - (0.60)(0.442)} = 0.050$$

and Equation (133):

$$\mathcal{A}^f_{S:(1,S)}(63.2°) = \frac{T_{1,2}(63.2°) a_S^f}{1 - R_S^{fH} \langle R^b_{1,2} \rangle_D} = \frac{(0.244)(0.15)}{1 - (0.60)(0.442)} = 0.050$$

From this we calculate the solar heat gain coefficient, using Equation (79),

$$\text{SHGC}(63.2°) = T^f_{1,S}(63.2°) + \sum_{k=1}^{S(=3)} N_k \mathcal{A}^f_{k:(1,2)}(63.2°)$$

$$= T^f_{1,S}(63.2°) + N_1 \mathcal{A}^f_{1:(1,S)}(63.2°)$$

$$+ N_2 \mathcal{A}^f_{2:(1,S)}(63.2°) + N_S \mathcal{A}^f_{S:(1,S)}(63.2°) = (0.083)$$

$$+ (0.064)(0.316) + (0.821)(0.050) + (0.85)(0.050) = 0.187$$

The corresponding hemispherical average properties are calculated from Equations (130), (131), and (134):

$$\langle T^{fH}_{1,S} \rangle_D = \frac{\langle T_{1,2} \rangle_D T^{fH}_{1,S}}{1 - R_S^{fH} \langle R^b_{1,2} \rangle_D} = \frac{(0.277)(0.25)}{1 - (0.60)(0.442)} = 0.094$$

$$\langle \mathcal{A}^f_{1:(1,S)} \rangle_D = \langle \mathcal{A}^f_{1:(1,2)} \rangle_D + \langle \mathcal{A}^b_{1:(1,2)} \rangle_D \frac{\langle T_{1,2} \rangle_D R_S^{fH}}{1 - R_S^{fH} \langle R^b_{1,2} \rangle_D}$$

$$= 0.271 + (0.138)\frac{(0.277)(0.25)}{1 - (0.60)(0.442)} = 0.302$$

$$\langle \mathcal{A}^f_{2:(1,S)} \rangle_D = \langle \mathcal{A}^f_{2:(1,2)} \rangle_D + \langle \mathcal{A}^b_{2:(1,2)} \rangle_D \frac{\langle T_{1,2} \rangle_D R_S^{fH}}{1 - R_S^{fH} \langle R^b_{1,2} \rangle_D}$$

$$= 0.036 + (0.141)\frac{(0.277)(0.60)}{1 - (0.60)(0.442)} = 0.068$$

$$\langle \mathcal{A}^f_{S:(1,S)} \rangle_D = \frac{\langle T_{1,2} \rangle_D a_S^f}{1 - R_S^{fH} \langle R^b_{1,2} \rangle_D} = \frac{(0.277)(0.15)}{1 - (0.60)(0.442)} = 0.057$$

Using these values, one calculates the diffuse solar heat gain coefficient from Equation (84):

$$\langle \text{SHGC} \rangle_D = \langle T^f_{1,S} \rangle_D + \sum_{k=1}^{S(=3)} N_k \langle \mathcal{A}^f_{k:(1,2)} \rangle_D$$

$$= \langle T^f_{1,S} \rangle_D + N_1 \langle \mathcal{A}^f_{1:(1,S)} \rangle_D + N_2 \langle \mathcal{A}^f_{2:(1,S)} \rangle_D + N_S \langle \mathcal{A}^f_{S:(1,S)} \rangle_D$$

$$= 0.094 + (0.064)(0.302) + (0.821)(0.068) + (0.85)(0.057)$$

$$= 0.218$$

In Example 7, the incident solar irradiances are calculated: $E_{DN}\cos(63.2°) = 123$ Btu/h·ft², $E_d\cos(63.2°) = 29.6$ Btu/h·ft², and $E_r\cos(63.2°) = 27.4$ Btu/h·ft². From these values, the solar energy flux is calculated from Equations (92), (97), and (100):

$$q_s = q_b + q_d = [E_{DN}\cos(63.2°)]\text{SHGC}(63.2°) + (E_d + E_r)\langle \text{SHGC} \rangle_D$$

$$= (123)(0.187) + (29.6 + 27.5)(0.218) = 35.4 \text{ Btu/h·ft}^2$$

VISUAL AND THERMAL CONTROLS

The ideal fenestration system permits optimum light, heat, ventilation, and visibility; minimizes moisture and sound transfer between the exterior and the interior; and produces a satisfactory physiological and psychological environment. The controls of an optimum system will react to varying climatological and occupant demands. Fixed controls may have operations or cost advantages or both but do not react to physical and psychological variations. Variable controls are, therefore, more effective in energy conservation and environmental satisfaction.

Operational Effectiveness of Shading Devices

Shading devices vary in their operational effectiveness. Some devices such as overhangs, light shelves, and tinted glazings do not require operation, have long life expectancies, and do not degrade significantly over their effective life. Other types of shading devices, especially operable interior shades, may have reduced effectiveness due to less than optimal operation and degradation of effectiveness over time. It is important to evaluate operational effectiveness when considering the actual heat rejection potential of shading devices.

The performance of shading devices for the reduction of peak cooling loads and annual energy use should account for operational effectiveness or reliability in actual operation. Passive devices, such as architectural elements and glazing tinting, are considered 100% effective in operation. Glazing coatings and adherent films may degrade over time. Shade screens are removable and may be assumed to be operated seasonally, but in any given population of users, some will remain in place all year long and some will not be installed or removed at optimum times. Automated shading devices controlled for optimum thermal operation are considered more effective than manual devices, but controls require ongoing maintenance, and some occupants may object to the lack of personal control with totally automated devices. Automated shading devices may also be operated for nonthermal purposes such as glare and daylighting optimization, and this may reduce thermal effectiveness. Manually operated devices will be subject to wide variation in use effectiveness, and this diversity in effective use should be considered when evaluating performance.

Indoor Shading Devices

While the thermal comfort of occupants within the glazed space may be paramount to the HVAC designer, other factors (see Table 22 and Figure 31) that should be considered, some of which may be more important to the user, include the following:

Radiant Energy Protection. Unshaded fenestration products become sources of radiant heat by transmitting short-wave solar radiation and by emitting long-wave radiation to dissipate some of the absorbed solar energy. In winter, glass temperatures usually fall below room air temperature, which may produce thermal discomfort to occupants near the fenestration. In summer, individuals seated near the unshaded fenestration product may experience discomfort from both direct solar rays and long-wave radiation emitted by sun-heated glass. In winter, loss of heat by radiation to cold glass can also cause discomfort. Tightly woven, highly reflective drapes minimize such discomfort; drapes with high openness factors are less effective because they permit short-wave and long-wave radiation to pass more freely. Light-colored shading devices with maximum total surface usually provide the best protection since they absorb less heat and tend to lose heat readily by convection to the conditioned air.

Outward Vision. Outward vision is normally desirable in both business and living spaces. Open-weave, dark-colored fabrics of uniform pattern permit maximum outward vision, while uneven pattern weaves reduce the ability to see out. A semiopen weave modifies the view without completely obscuring the outdoors. Tightly woven fabrics block off outward vision completely.

Privacy. Venetian blinds, either vertical or horizontal, can be adjusted and, when completely closed, afford full privacy. When draperies are closed, the degree of privacy is determined by their color and tightness of weave and the source of the principal illumination. To obscure the view so completely that not even shadows or silhouettes can be detected, fully opaque materials are used. Generally, the more brightly lit side of a partially shaded glazing is the most visible from the opposite side, making the interior fairly private in daytime, but not at night.

Brightness Control. Visual comfort is essential in many occupied areas, and freedom from glare is an important factor in performing tasks. *Discomfort glare* is produced by uneven brightnesses in occupied spaces, with areas or spots that are much brighter than

Table 23 Summary of Environmental Control Capabilities of Draperies

Item	Designator (Table 22 and Figure 31)								
	I_D	I_M	I_L	II_D	II_M	II_L	III_D	III_M	III_L
1. Protection from direct solar radiation and long-wave radiation to or from window areas	Fair	Fair	Fair	Fair	Good	Good	Fair	Good	Good
2. Effectiveness in allowing outward vision through fenestration	Good	Good	Fair	Fair	Fair	Some	None	None	None
3. Effectiveness in attaining privacy (limiting inward vision from outside)	None	None	Poor[a] Good[a]	Poor	Fair	Fair[a] Good[a]	Good[b]	Good[b]	Good[b]
4. Protection against excessive brightness and glare from sunshine and external objects	Mild	Mild	Mild[c] Poor[c]	Good	Good	Good[c] Poor[c]	Good	Good	Good[c] Poor[c]
5. Effectiveness in modifying unattractive or distracting view out of window	Little	Little	Some	Some	Good	Good	Blocks	Blocks	Blocks

[a]Good when bright illumination is on the viewing side.
[b]To obscure view completely, material must be completely opaque.
[c]Poor rating applies to white fabric in direct sunlight. Use off-white color to avoid excessive transmitted light.

surrounding surfaces. Windows themselves, when they look out onto bright skies or brightly reflecting surfaces, can be glare sources if care is not taken to keep surround brightnesses comparable. A maximum brightness ratio of about 3 to 1 is sometimes quoted. Moderation of this ratio can be achieved through the use of interior furnishings and wall coverings, which on average have moderately high diffuse reflectances and access to admitted daylight. Conversely, dark interior surfaces, and those shaded from daylight illumination, will accentuate the brightness difference between the window and its surroundings. Interior surface brightness can also be elevated by ample use of interior electric lighting, but this can have adverse consequences for the building's energy use. In general, larger window apertures admit more sunlight, increasing interior brightnesses without affecting the perceived brightness of the window, all other things being equal.

An important guideline is the dictum that direct sunlight must not strike the eye, and reflected sunlight from bright or shiny surfaces is equally disturbing and even disabling. A tightly woven white fabric with high solar transmittance attains such brilliance when illuminated by direct sunshine that, by contrast with its surroundings, it creates excessive glare. Off-white colors should be used so their surface brightness is not too great. Venetian blinds permit considerable light to enter by interreflection between slats. When two shading devices are used, the one on the inside (away from the fenestration product) should be darker and more open. With this arrangement, the inside device can be used to control brightness for the other shading devices and, when used alone, to reduce brightness while still permitting some view of the outside.

View Modification. When the view is unattractive or distracting, draperies modify the view to some degree, depending on the fabric weave and color (summarized in Table 23), but the fenestration product remains as an effective connection to the outside.

Sound Control. Indoor shading devices, particularly draperies, can absorb some of the sounds originating within the room but have little or no effect in preventing outdoor sounds from entering. For excessive internally generated sound, the usual remedy is to apply acoustical treatment to the ceiling and other room surfaces. While these materials can be effective in controlling sound, they are often located on the two horizontal surfaces (ceiling and floor) and leave the opposing vertical surfaces of glass and bare wall to reflect sound. The noise reduction coefficient (NRC = average absorptance coefficient at four frequencies) for venetian blinds is about 0.10, compared to 0.02 for glass and 0.03 for plaster. For drapery fabrics at 100% fullness, NRC ranges from 0.10 to 0.65, depending on the tightness of weave. Class III (tightly woven) fabrics have NRC values of 0.35 to 0.65. Figure 32 shows the relationship between NRC and openness factor for fabrics of normal weight.

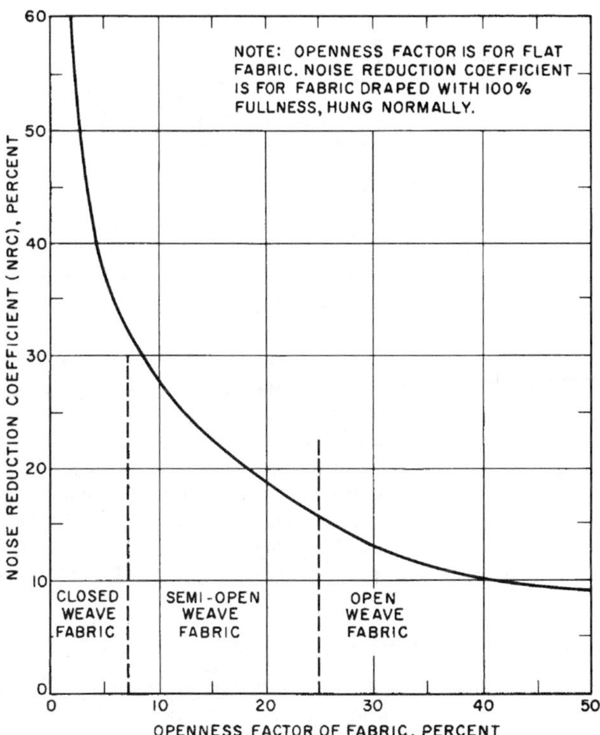

Fig. 32 Noise Reduction Coefficient Versus Openness Factor for Draperies

Example 22. To select a drapery fabric, consider the five environmental factors listed in Table 23. Choose a fabric designator that has suitable performance for all the factors important to the case being considered. If this is not possible, make compromises resulting in an acceptable designator. Determine from Table 23 if the IAC for the chosen designator is satisfactory. Specific cases follow:

1. Where modification of a distracting view is necessary but a degree of outward vision is needed and an IAC of 0.50 with 1/4 in. gray glass is satisfactory (see Table 23, Item 5), select II_M or II_L; from Item 2, select II_M. The IAC for II_M on Table 23 is approximately 0.46, therefore satisfactory.
2. Where protection from radiation is paramount and minimum IAC is necessary (see Table 23, Item 1), select a closed weave, III_M or III_L. Since the IAC for III_L is lowest (see Table 23), choose III_L.
3. When good outward vision is desired, together with some reduction in brightness, choose I_M or I_D (Table 23).

Double Drapery

Double draperies (two sets of drapery covering the same area) have a light, open weave on the fenestration product side for outward vision and daylight when desired and a heavy, closed weave or opaque drapery on the room side to block out sunlight and provide privacy when desired. When properly selected and used, double draperies can provide a reduced U-factor and a lowered IAC.

The reduced U-factor results principally from adding a semi-closed air space to the barrier. A U-factor of about 0.57 Btu/ h·ft^2·°F is achieved using double draperies with single glass, and about 0.37 Btu/h·ft^2·°F with insulating glass.

To most effectively reduce solar heat gain, the drapery exposed to sunlight should have high reflectance and low transmittance. The light, open-weave drapery should be opened when the heavy drapery is closed to prevent entry of sunlight.

Properly used double draperies give (1) extreme flexibility of vision and light intensity, (2) a lowered U-factor and IAC, and (3) an improved comfort condition, since the room-side drapery is more nearly at room temperature. Table 23 gives characteristics of individual draperies. For large areas, the IAC should be calculated in detail to determine the cooling load.

AIR LEAKAGE

Infiltration Through Fenestration

Air infiltration through fenestration products affects occupant comfort and energy consumption. Infiltration should not be confused with ventilation. Infiltration is the uncontrolled inward leakage of air caused by pressure effects of wind or differences in air density, such as the stack effect. While fenestration products can be operated to intentionally provide natural ventilation and increase comfort, infiltration should be reasonably minimized to avoid unpleasant accompanying problems. If additional air is required, controlled ventilation is preferable to infiltration. Mechanical ventilation provides air in a comfortable manner and when desired. For infiltration, however, the peak supply is more likely to occur as an uncomfortable draft and when least desired, such as during a storm or the coldest weather.

ASHRAE/IESNA *Standard* 90.1, ASHRAE's energy standard for all buildings other than low-rise residential buildings, establishes an air leakage maximum of 0.4 cfm per square foot of gross fenestration product area (1.0 cfm/ft^2 for swinging entrance doors and revolving doors). This air leakage is as determined in accordance with NFRC 400 and ASTM *Standard* E 283 and allows direct comparison of all fenestration products: operable and fixed, windows and doors.

Most manufactured fenestration products achieve these reasonable standards of maximum air infiltration. However, products that do not completely seal, such as jalousie windows or doors, are not likely to do so and are most appropriate for installation in unconditioned spaces.

For products achieving this infiltration standard, the energy consumption due to infiltration is likely to be significantly less than the energy associated with U-factor and solar heat gain coefficient. Also, while overall air infiltration is a significant component in determining a building's heating and cooling loads, the infiltration through fenestration products meeting the above standard is generally likely to be a small portion of that total. Chapter 26 presents calculation procedures for air infiltration.

Indoor Air Movement

Because supply air grilles are frequently located directly below fenestration products, air sweeps the interior glass surface. Heated supply air should be directed away from the glass to prevent large temperature differences between the center and edges of the glass. These thermal effects must be considered, particularly when annealed glass is used and air is forced over the glass surface during the heating season. Direct flow of heated air over the glass surface can increase the heat transfer coefficient and the temperature difference, causing a substantial increase in heat loss; it may also lead to thermally induced stress and risk of glass breakage.

Systems designed predominantly for cooling lower the glass temperature and rapidly pick up the cooling load. Both tend to improve the comfort condition of the space. However, the air-conditioned space has an increased net heat gain caused by (1) increase in the solar heat gain coefficient (SHGC) due to delivery of a larger portion of the absorbed heat to the indoor space, (2) increase in the fenestration U-factor because of the greater convection effect at the indoor surface, and (3) increase in the air-to-air temperature difference since supply air rather than room air is in contact with the indoor glass surface. The principal increase in heat gain with clear glass is the result of the higher U-factor and the greater air-to-air temperature difference.

DAYLIGHTING

DAYLIGHT PREDICTION

Daylighting is the illumination of building interiors with sunlight and sky light and is known to affect visual performance, lighting quality, health, human performance, and energy efficiency. In many European countries with predominantly cloudy skies, there are codes regulating minimum window size, minimum daylight factor, and window position in order to provide views to all occupants and to create a minimum interior brightness level. For practical reasons, daylighting provides backup interior illumination in the event of power outages. Daylighting may have some positive or negative health effects on the skin, eyes, hormone secretion, and mood. Its temporal variation, intensity, spectral content, and diurnal and temporal variation may be used to combat jet lag, sick building syndrome, and other health problems.

In terms of energy efficiency, daylighting can provide substantial whole-building energy reductions in nonresidential buildings through the use of electric lighting controls. Daylight admission can displace the need for electric lighting at the perimeter zone with vertical windows and at the core zone with skylights. Lighting and its associated cooling energy use constitute 30 to 40% of a nonresidential building's energy use. Energy use reductions can be achieved, perhaps less reliably, in residential buildings with manual or automated switching of electric lights on and off to match space occupancy. For internal load-dominated buildings, daylight admission must be balanced against solar heat admission to achieve optimum energy efficiency. Since the heat gains from solar radiation typically define peak load conditions, daylighting is also a very effective method of decreasing peak demand. The use of daylighting can not only decrease annual operating costs through energy efficiency but may also reduce capital cost due to mechanical downsizing.

For conventional sidelit nonresidential buildings, three fundamental relationships for daylight optimization are given as a function of (1) glazing properties and (2) window area or the **window-to-wall area ratio** (WWR), which is defined as the ratio of the transparent glazing area to the exterior floor-to-floor wall area:

1. Annual cooling energy use (including fan energy use) increases linearly with solar radiation admission, as indicated by the product of SHGC and WWR, but is affected by decreases in electric lighting heat gains.
2. Annual lighting energy use decreases exponentially/asymptotically with daylight admission, as indicated by the product of T_v and WWR.
3. Annual heating energy use (including fan energy use) increases linearly with decreased lighting heat gains.

Figure 33 illustrates the first two relationships for a prototypical nonresidential building. A similar relationship can be demonstrated with skylights.

To determine the fenestration design that achieves an optimum balance between daylight admission and solar rejection requires iterative calculations where the glazing area and/or glazing solar-optical properties are varied parametrically. For each case, the following general steps should be taken for each hour over a year:

1. **Interior Daylight Illuminance.** Determine the building characteristics, configuration, outdoor design conditions, and operating schedules as described in the section on Initial Design Considerations in Chapter 29. These include building orientation, exterior obstructions, ground reflectance, etc. Determine the depth from the window wall for each electric lighting zone. Typical sidelighting windows can effectively daylight the perimeter zone to a depth of 1.5 times the head height of the window. In private offices, one dimming zone is typically cost-effective, while in open plan offices, two zones are cost-effective.

   Select a typical task location within each of the lighting zones. Determine interior daylight illuminance due to all window and skylight sources at these locations. Interior illuminance may be determined using computer simulation tools or physical scale models. Comprehensive explanations of simple and computer-based tools are available (IEA 1999). The majority of these tools can model simple box geometry with noncomplex fenestration systems. Some advanced simulation tools, such as Radiance (Ward 1990) and Adeline (Erhorn and Dirksmöller 2000), are capable of modeling complex geometry and fenestration systems with adequate bidirectional solar-optical data, but this capability is not routine.
2. **Lighting Energy Use.** Determine the type of lamps, ballasts, and control system to be used in the perimeter zones. Determine whether the lamp is capable of being dimmed or switched. For example, fluorescent lamps can be dimmed, while metal halides cannot be switched or dimmed. Cold, outdoor applications of some lamps may prevent switching. For electronic dimming bal-

lasts, obtain dimming power and light output characteristics. Obtain control specifications to determine how the system will respond to available light; dead-band ranges, response times, and commissioning will affect the sensitivity and accuracy of the system. The type of switching (on-off, bilevel, multilevel, and continuous dimming controls) will be dictated by both the type of lamp and space use.

Determine the task illuminance design set point for each zone. Determine the percentage electric lighting power reduction $F_{daylight}$ [see Equation (135)] that will result with automatic daylight controls, and apply to the installed wattage. Simplified methods for calculating lighting power reductions based on task illuminance levels are given in Robbins (1986). More sophisticated programs (Choi and Mistrick 1999) model commercially available photosensor dimming control systems (typically located in the ceiling above the work plane task) more rigorously; the spectral and bidirectional response of the photosensor to incident flux is used to determine voltage output, which is then used by the ballast controller algorithm to determine the lighting power reduction. Response delays and commissioning set points will further affect this predicted output. Lights may also be switched manually, but there are no modeling prediction tools for manual switching. Field tests (Jennings et al. 1999) indicate that with bilevel switching, 45% of the lighting zone-hours were at less than full power lighting, with 28% at only one-third of full lighting output levels. Manual switching occurred less in public spaces. Occupancy and other types of switching may occur as well and should be accounted for as a confounding effect with any daylighting controls.

3. **Mechanical Energy Use.** Determine mechanical energy use due to the fenestration loads and reduced electric lighting heat gains.

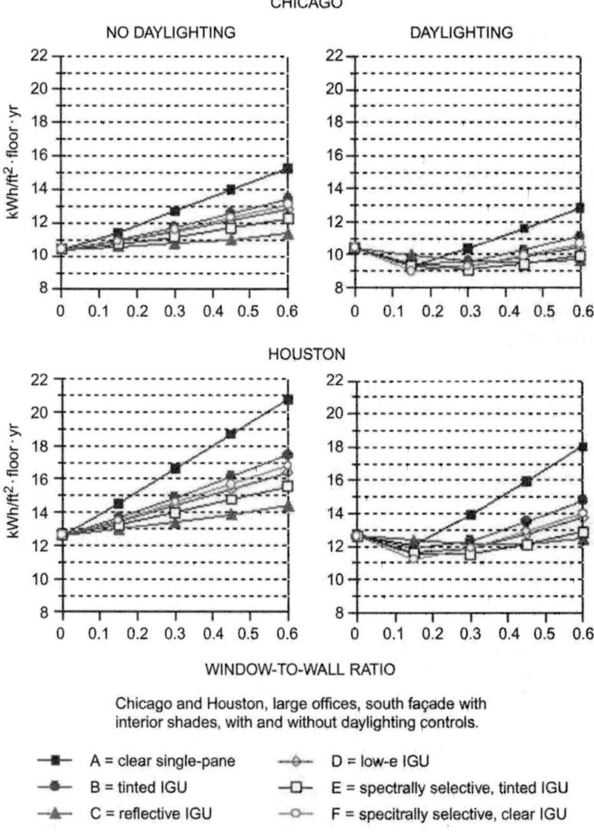

Fig. 33 Window-to-Wall Ratio Versus Annual Electricity Use (kWh/ft^2·floor·year)

Fenestration heat gains and losses may be computed using the section on Determining Fenestration Energy Flow. Instantaneous lighting heat gains q_{el}, described by Equation (1) in Chapter 29, must be multiplied by the power reduction factor $F_{daylight}$.

Mechanical loads and energy use may then be determined as described in Chapter 29. There have been many studies investigating the magnitude of change in heating and cooling energy use associated with reductions of lighting energy use in nonresidential buildings, as will be realized with daylighting controls. In a DOE-2.1E simulation study (Sezgen and Koomey 2000), the greatest savings were generated in hospitals, large offices, and large hotels; for every $1.00 saved through lighting energy efficiency, additional savings as a result of reduced HVAC were $0.26, $0.16, and $0.14, respectively. These results emphasize the need to include HVAC effects when assessing the impact of daylighting. Simplified design tools are available that enable one to conduct such parametric runs for preliminary analysis. For vertical windows, the BEEM program can be used to determine the relative impacts of daylighting on energy use, peak demand, and costs for a given window and lighting control system (Rundquist 1991). Skylighting tools based on regressions using DOE-2 data or simplified DOE-2 procedures are also available (AAMA 1988, Heschong et al. 1998). More comprehensive building energy prediction tools combined with daylighting algorithms, such as DOE-2.1E (Winkelmann 1983), implement hour-by-hour calculations using existing weather data and enable one to evaluate glare, visual comfort, and quality of light as well.

In the United States, a general rule has been that the fenestration area should be at least 20% of the floor area. In Europe, a similar rule was based on a minimum illumination value on the normal work plane from a standard overcast sky condition. In general, it is more energy-efficient to use larger window areas to elevate interior surface brightnesses as a glare reduction strategy than to increase interior electric lighting levels. As window area increases, interior brightness increases while window brightness remains the same. Of course there are mitigating considerations, such as the higher cost of larger windows and increased heat transfers through larger windows. The latter problem can be mitigated through the use of insulating multiple-pane windows and special coatings to reduce solar gain without serious loss of light transmission, as discussed in the section on Selecting Fenestration.

The secondary visual benefit of fenestration is the amount and quality of light it produces in the work environment. One general rule determined the need for auxiliary electric light by assuming that daylight was adequate for a depth of two-and-one-half times the height of the fenestration product into the room based on a normal sill height. To prevent excessive glare, all fenestration should have sun controls. Variable and removable controls are often more effective in daylight than fixed controls.

For more accurate evaluation of daylight distribution within a space, several prediction tools, such as the *Recommended Practice of Daylighting* (IESNA 1999), are available. This practice shows a simple way of calculating the daylight distribution on the work plane from windows and skylights with and without controls. Many other daylight prediction tools calculate illuminance from radiant flux transfer or ray tracing.

Any or all of the various daylight prediction tools can be used to compare the relative value of daylight distribution from alternative fenestration systems, but ultimately the designer must evaluate costs and benefits to choose between alternative designs. This may be based on energy use or, more properly, on overall costs and benefits to the client. Also, the negative possibility of total loss of productivity from an electric brown-out in a space with no natural ventilation or daylight may be as important as the benefits of many energy-saving schemes.

LIGHT TRANSMITTANCE AND DAYLIGHT USE

When daylight is to be the primary lighting system, the minimum expected daylight in the building must be calculated for the building performance cycle and integrated into the lighting calculations. IESNA (1999) gives daylight design and calculation procedures. In some glazing applications, such as artists' studios and showrooms, maximum transmittance may be required for adequate daylighting of the interior. Regular clear glass, produced by float, plate, or sheet process, may be the logical choice.

When daylight is a supplementary light source, the electric lighting can be designed independently of the daylight system. However, adequate switching must be included in the electric distribution to substitute available daylight for electric lighting by automatic or prescribed manual control whenever possible and practical. Photosensitive controls automatically adjust shading devices to provide uniform illumination and reduce energy consumption. Manual control is less effective.

Buildings with large areas of glass usually have insulating glass units with clear, tinted, or reflective coatings. The tinted and reflecting units reduce the brightness contrast between fenestration products and other room surfaces and provide a relatively glare-free environment for most daylight conditions.

Tables 13 and 24 list typical solar energy transmittances and daylight transmittances for various glass types. Manufacturers' literature has more appropriate values.

The color of glass chosen for a building depends largely on where and how it is used. For commercial building lobbies, showroom fenestration products, and other areas where maximum visibility from exterior to interior is required, regular clear glass is generally best. Clear glass with a low-e coating is also suitable for these locations, including for retail storefronts, as it only decreases the light transmittance by 10% or so. For other glass areas, a tinted glass may best complement the interior colors. Bronze, gray, and reflective-film glasses also give some privacy to building occupants during daylight hours. Patterned, etched, or sandblasted glass that diffuses lighting is available. In warm climates, a tinted outer glass in an insulated double-pane system can have solar heat gain rejection benefits, while providing good color rendering illumination of the interior without apparent color.

The primary purpose of a fenestration product is not just to save energy but to provide a view of the exterior. One sees out of a fenestration product by virtue of the light from the outside that comes through that fenestration product into the occupant's eyes. The light from outside is valuable not only for views of the outdoors but for providing daylight illumination of the interior.

Table 24 Daylight Transmittance for Various Types of Glass

Type of Glass	Visible Transmittance, T_v
1/8 in. regular sheet or float glass	0.86 to 0.91
1/8 in. gray sheet	0.31 to 0.71
3/16 in. gray sheet	0.61
7/32 in. gray sheet	0.14 to 0.56
1/4 in. gray sheet	0.52
1/4 in. green/float glass	0.75
1/4 in. gray plate glass	0.44
1/4 in. bronze plate glass	0.49
1/2 in. gray plate glass	0.21
1/2 in. bronze plate glass	0.25
Coated glasses (single, laminated, insulating)	0.07 to 0.50

Some buildings are designed especially to use the daylight coming through fenestration systems in displacing electric lighting and its attendant energy costs. Using daylighting to displace electric lighting benefits the energy bill directly through reduced direct consumption by the lighting systems involved and indirectly through reduced electrically produced heat gain that may have to be removed by the air-conditioning system.

The light-transmitting properties of fenestration systems are therefore of great importance, not only for permitting views of the outdoors but also for admitting daylight to reduce electric lighting. It is conceivable that one could design a fenestration product with excellent solar heat gain performance for hot climates (meaning a very low solar heat gain coefficient) but very poor view and daylight illumination performance. If this problem is bad enough, it can cause occupants to turn on electric lights indoors during the daytime, which adds to the electric bill and possibly causes problems of thermal discomfort as well.

The light-transmitting property of a fenestration product is called the visible **transmittance** T_v. It is similar to the solar-weighted solar transmittance, except that an additional weighting function is needed, in this case to account for the spectral response of the human eye.

In most applications, it is important to have a high visible transmittance. In northern climates, a good solar heat gain is also important for offsetting wintertime heating costs. In southern climates, a low solar heat gain is good for offsetting summertime cooling costs. In the latter situation, it is difficult to have both a high visible transmittance and a low solar heat gain coefficient. Figures 34 and 35 show a plot of visible transmittance versus SHGC for a number of glazing systems covering a range of spectral selectivities. The data are for normal incidence and a single, ASTM standard solar spectral distribution.

A rule of thumb is to select a glazing unit having a visible transmittance greater than its shading solar heat gain coefficient, especially if daylighting strategies will be used in the building. For maximum light with minimum solar gain, there are fenestration products available having a visible transmittance 1.4 times the SHGC.

Three different zones are delineated on Figure 35. In the **neutral zone**, it is possible to have colorless glazing systems, meaning glazings with approximately uniform transmittance over the visible spectrum. Of course, one can have glazings in this zone with color, but this is not necessary. In the **color zone**, the only way to achieve higher visible transmittance for a given level of solar heat gain coefficient is by stripping off some of the red and blue wavelengths at the edges of the V-λ function with a spectrally selective glazing transmittance, imparting color to the transmitted radiation (or by

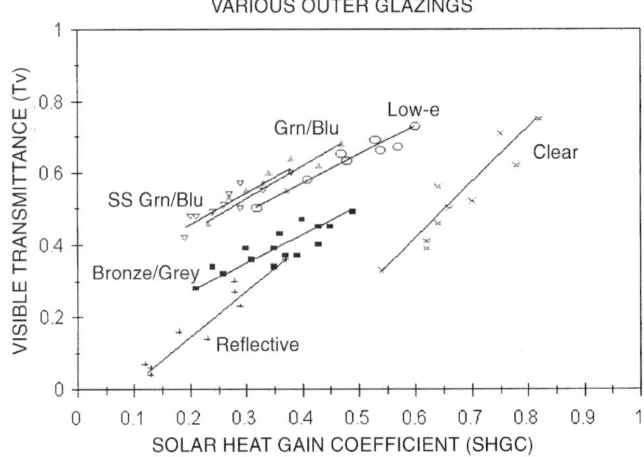

Fig. 34 Visible Transmittance Versus SHGC for Several Glazings with Different Spectral Selectivities

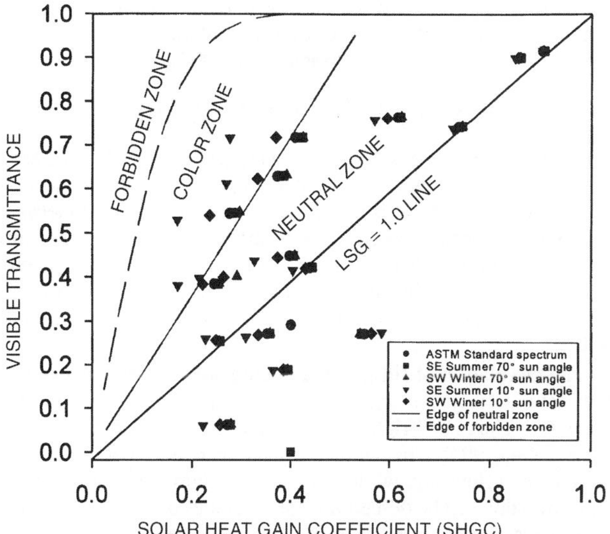

Fig. 35 Visible Transmittance Versus SHGC at Various Spectral Selectivities (McCluney 1996)

otherwise altering the spectral transmittance and hence the color over the visible portion of the spectrum). In the **forbidden zone**, no combination of visible transmittance and solar heat gain coefficient is possible for normal incidence and for the solar spectral distribution used. (Changing the solar spectral distribution used to calculate T_v and SHGC will shift the transition curves somewhat. A low solar altitude angle, direct-beam spectrum will move the curves to the left on the plot in Figure 35.) It can be seen that the glazings that transmit more solar radiant heat than light cluster on the lower portion of the plot.

The T_v versus SHGC chart can be a useful tool for illustrating the degree of spectral selectivity attained by a glazing system. These concepts lead to an index of spectral selectivity that can be useful. It is called the **light-to-solar-gain ratio**, or LSG, defined as

$$\text{LSG} = \frac{T_v}{\text{SHGC}} \tag{135}$$

Some characteristic values for T_v, SHGC, and LSG are given in Table 25 for several different glazings, using the ASTM standard spectral distribution at normal incidence to calculate the values.

The LSG can be useful in spotting errors in the calculation of the SHGC. Values of SHGC that lie outside reasonable ranges can be spotted fairly quickly and used to identify possible problems in calculations or measurements. In general, it is very difficult and therefore unlikely to have a useful glazing system for buildings with an LSG value greater than 2.0. Values near 0 should be particularly suspect, since they indicate a glazing that transmits considerably more heat than light and would be unlikely candidates for general use. Generally, a high value of LSG is desired for residential buildings in hot climates, to maximize daylight admission with minimal solar heat gain. This is also true for internal load-dominated nonresidential buildings in many climates, since solar gain rejection is often desired for such buildings, even in cool or cold climates. An LSG value somewhat below 1.0 would be appropriate in cold climates for residential buildings and nonresidential buildings without strong internal cooling loads.

Table 25 Spectral Selectivity of Several Glazings

Glazing	T_v	SHGC	LSG
Reflective blue-green	0.33	0.38	0.87
Film on clear glass	0.19	0.22	0.86
Green tinted, medium	0.75	0.69	1.09
Green low-e	0.71	0.49	1.45
Sun-control low-e + green	0.36	0.23	1.56
Super low-e + clear	0.71	0.40	1.77
Super low-e + green	0.60	0.30	2.00

SELECTING FENESTRATION

Since fenestration systems provide so many functions and because environmental conditions and user needs vary widely, it is difficult to make a completely optimal selection of a fenestration system. Aesthetic and cost considerations are perhaps the most important to residential users, with visual and comfort performance also being of interest. Considering annual energy costs, peak load consequences, and acoustic characteristics, the choice is seldom optimal. The HVAC system designer, fortunately, has a more restricted range of interests, mainly dealing with the energy consequences of a particular fenestration selection. This section therefore focuses on fenestration energy performance determination.

ANNUAL ENERGY PERFORMANCE

Instantaneous energy performance indices (U-factor, solar heat gain coefficient, air leakage, etc.) are typically used to compare fenestration systems under a fixed set of conditions. However, the absolute and relative effect of these indices on a building's heating and cooling load can fluctuate as environmental conditions change. As a result, these indices alone are not good indicators of the annual energy performance (energy savings/costs) attributable to the fenestration. Furthermore, fenestration annual energy performance is difficult to quantify in and of itself because of the many dynamic responses that occur between the fenestration system and the total environment in which it is installed. The four basic mechanisms of fenestration energy performance that were each addressed previously in the chapter—thermal heat transfer, solar heat gains, air leakage, and daylighting—should all be taken into account but are not independent of many other parameters that influence fenestration annual energy performance. As a result, the annual energy performance of fenestration systems can be accurately determined only when a large number of variables are considered. Building type and orientation, climate (weather, temperature, wind speed), microclimate (shading from adjacent buildings, trees, terrain), occupant usage patterns, and certain HVAC parameters can significantly affect the annual energy impacts of fenestration systems.

For these reasons, the most effective means of establishing fenestration annual energy performance is through detailed, dynamic, hourly computer simulations for the specific building and climate of interest. Since the instantaneous performance of the fenestration will often vary by differing magnitudes as climatic conditions change, the most accurate simulation results are obtained when these variances are accounted for in a building energy simulation computer program. After constructing the building energy simulation model following the procedures defined in Chapters 28 and 29 (for residential and commercial construction, respectively), specific changes to the fenestration system can be modeled, and the annual energy performance changes attributable to fenestration can be quantified. These analytical techniques do not consider issues of performance durability for the various instantaneous indices and should only be used as an initial annual energy performance indicator (Mathis and Garries 1995).

Simplified Techniques for Rough Estimates of Fenestration Annual Energy Performance

While dynamic hourly modeling is certainly the most accurate technique for determining fenestration annual energy performance, it is not readily available to many decision makers and end users of fenestration products simply because it may not be practical or cost-effective. Under these circumstances, it may be useful to assess the relative importance of, or balance the trade-off between, the known instantaneous performance indices of U-factor, SHGC, air leakage, and T_v for any given fenestration system when considering heating, cooling, and lighting loads for many different building types and climates. Mitchell et al. (1999) and Huang et al. (1999) describe personal computer programs that are being developed to run this simplified analysis for residential windows.

Broad generalizations can be made for some classifications of building types and climates. For instance, with large commercial buildings, which require substantial cooling energy use because of high internal loads, significant thermal mass, or high orientation dependency, the primary objective may be to place the most emphasis on low SHGC to reduce the cooling load. Also, an evaluation of commercial fenestration annual energy use can take into account the trade-off between artificial lighting and the natural daylighting benefits associated with a particular fenestration system. Contrary to this, electric lighting loads in low-rise detached residential buildings are typically very small in comparison to the heating and cooling loads because of high envelope-dependent energy use, egress requirements, and occupant usage patterns, and therefore the energy influence of daylighting may be neglected altogether. Yet, despite these generalizations, the problem still exists of balancing and assessing the impact of each of the remaining parameters to establish the seasonal or annual energy performance for cases in which detailed computer modeling is not performed.

Realizing the need for characterization of fenestration annual energy performance, scientists in many different countries have been working over the last several years to develop simplified annual energy performance indices for fenestration. These simplified techniques typically involve using the instantaneous fenestration performance indices to quantify building- and climate-independent scalars of annual or seasonal energy performance for rating purposes. Many of these performance indices have value in that they can be relatively independent of building type, climate, distribution of products, orientation, and other items needed for hourly dynamic building energy analyses. These normalized, scalar-based approaches are also limited in accuracy for the same reasons. A further limitation with the simplified techniques is that they do not have broad applicability to varied building types (commercial versus residential buildings, for example). The usefulness of these scalar-based approaches can be increased when limiting the comparison to a single building type. Currently, the simplified techniques for characterizing fenestration annual energy performance are applicable only to fenestration systems for detached residential buildings and are not appropriate for use with multifamily residential or commercial building fenestration systems.

Simplified Residential Annual Energy Performance Ratings

Annual energy performance ratings can provide a simple means of product comparisons for consumers. Such ratings have been derived with many assumptions, usually to suit local climatic conditions.

The Canadian Standards Association (CSA *Standard* A440.2) developed a simplified energy rating applicable to residential heating in the Canadian climate, which has been adopted in the 1995 *National Energy Code for Houses*. The standard also provides for specific energy ratings to compare products by orientation and climate.

In the United States, where heating and cooling are both significant, the NFRC is developing a rating system that includes both effects (Crooks et al. 1995, Arasteh et al. 2000).

CONDENSATION RESISTANCE

Water vapor condenses in a film on fenestration surfaces that are at temperatures below the dew-point temperature of the inside air. If the surface temperature is below freezing, frost forms. Sometimes, condensation occurs first, and ice from the condensed water forms when temperatures drop below freezing. Condensation frequently occurs on single glazing and on aluminum frames without a thermal break. The edge-seal creates a thermal bridge at the perimeter of the IGU.

The circulation of fill gas due to temperature differences in the IGU cavity contributes to the condensation problem at the bottom of the indoor glazing (Wright and Sullivan 1995a, 1995b; Curcija and Goss 1994, 1995a). In winter, fill gas near the indoor glazing is warmed and flows up, while gas near the outdoor glazing is cooled and flows down. The descending gas becomes progressively colder until it reaches the bottom of the cavity. There, the gas turns and flows to the indoor glazing, resulting in higher heat transfer rates at the bottom. Thus, the bottom edge of the indoor glazing is cooled both by edge-seal conduction and by fill-gas convection. The combined effect of these two heat transfer mechanisms is shown in Figure 36. The surface isotherms show a wider band of cold glass at the bottom of the window. Typical condensation patterns match these isotherms. The vertical indoor surface temperature profile also shows the effect of edge-seal conduction and that the minimum indoor surface temperature is near the bottom edge of the glass.

Condensation to the fenestration and surrounding structures can cause extensive structural, aesthetic, and health problems. Specific examples include peeling of paint, rotting of wood, saturation of insulation, and mold growth. Ice can render doors and windows inoperable and prevent egress during an emergency.

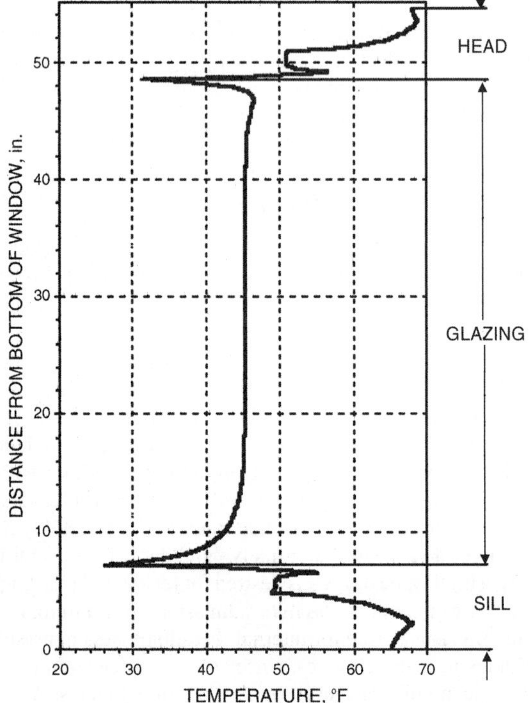

Fig. 36 Temperature Distribution on Indoor Surfaces of Insulating Glazing Unit

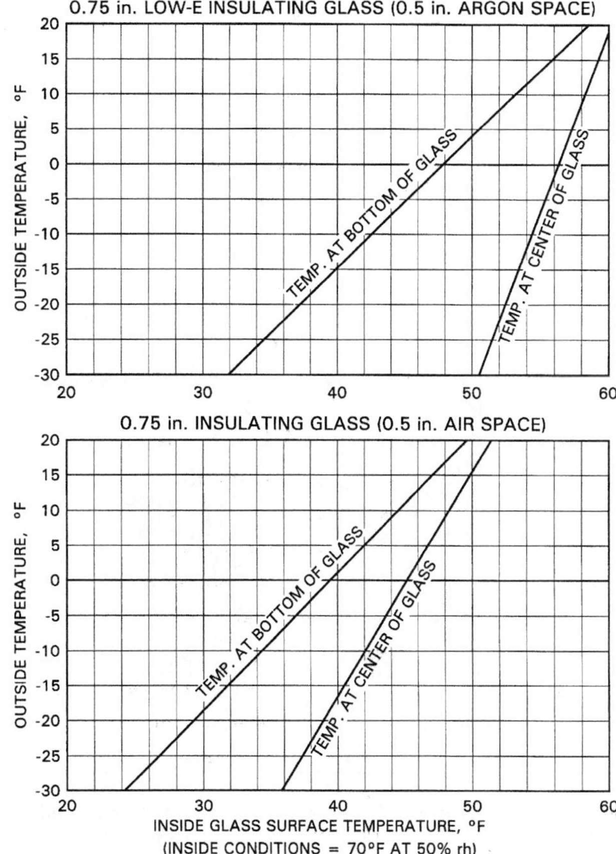

Fig. 37 Minimum Indoor Surface Temperatures Before Condensation Occurs

Energy-efficient housing has been accompanied by reduced ventilation. The resulting increase in indoor humidity has contributed to the condensation problem. However, the solution does not lie in the reduction of humidity levels to a minimum. Relative humidity below 20% and above 70% can increase health risks and reduce comfort. Generally, a minimum of 30% rh should be maintained, and 40% to 50% is more desirable (Sterling et al. 1985).

Minimum indoor surface temperatures can be quantified in a variety of ways. Sullivan et al. (1996), Griffith et al. (1996), Elmahdy (1996), Zhao et al. (1996), and de Abreu et al. (1996) demonstrated good agreement between detailed two-dimensional numerical simulation and surface temperature measurements using thermographs. Wright and Sullivan (1995c), and Curcija et al. (1996) developed simplified simulation models to predict condensation resistance. Estimates of center-glass and bottom-edge surface temperatures that can be expected for two different glazing systems exposed to a range of outdoor temperature are shown in Figure 36. Both glazing systems include insulating foam edge seals. High-performance glazing systems (e.g., low-e/argon and insulated spacers) permit significantly higher indoor humidity levels.

Current measures of condensation resistance of a fenestration system are the **condensation index** (CI) as defined by NFRC (2000a), the **condensation resistance factor** (CRF) as defined by AAMA (1988), or the **temperature index** (I), as defined in CSA *Standards* A440 and A440.1.

The condensation index is a measure of condensation potential that is based on both area and temperature weighting and is expressed as a minimum of center-of-glazing, edge-of-glazing, and frame CIs. The novelty of this index lies in the fact that it is determined using computer simulation tools unless the overall thermal performance cannot be validated with testing. In the case that thermal performance cannot be validated, a testing option for determining CI is used.

Computer simulation is done for characteristic two-dimensional cross sections in much the same way that U-factors are determined. The basic difference between U-factor and CI simulations is that more advanced models are used for CI calculations. This is necessary because temperatures are intrinsically local quantities, as opposed to U-factors, which are average quantities, and it is necessary to provide better models for convective heat transfer in glazing cavities and convective and radiative heat transfer on indoor fenestration boundaries. The most general expression of the formula for calculating frame, center-of-glazing, and edge-of-glazing CI is given by the following equation:

$$\mathrm{CI} = \left\{ 1 - \left\{ \left[\frac{1}{3} \sum_{j=1}^{3} \left[\frac{\sum_i (t_{dpp,j} - t_i)^+ A_i}{(t_{dpp,j} - t_o) A} \right] \right]^{1/3} \right\} \right\} \times 100 \qquad (136)$$

where

i = frame, center-of-glazing, or edge-of-glazing section
j = 30%, 50%, and 70% relative humidity
$t_{dpp} = t_{dp} + 0.5$°F
t_{dp} = dew-point temperature, °F
$+$ = positive values only

The other two standards define the values by a single dimensionless number as

$$\mathrm{CRF\ or\ } I = \frac{t - t_c}{t_h - t_c} \qquad (137)$$

where t_h and t_c are the warm and cold side temperatures, respectively. Figure 38 can be used to determine the acceptable range of CRF/*I* for a specific climatic zone.

The two standards differ in the methods used to determine temperature. The CSA test procedure is based on thermocouple measurements at the coldest location on the frame plus three locations on the glass, each 3/8 in. above the bottom sightline. The AAMA procedure specifies two separate factors: one for the frame (CRF_F), which uses weighted frame temperature obtained from surface temperature measurements at predetermined and roving locations on the frame, and one for the IGU (CRF_G), which uses the average of six temperatures measured at predetermined locations near the top, middle, and bottom of the glazed area.

Inside details can significantly alter the potential for condensation on window surfaces. Items such as venetian blinds, roll blinds, insect screens, and drapes increase the thermal resistance between the indoor space and the window and lower the temperature of the window surfaces. These window treatments do not prevent migration of moisture, so they can cause increased condensation. Figure 39 shows different situations that affect the potential for condensation. Note that window reveal plays an important role. If the window is placed near the outside of the wall, the increase in the outdoor film coefficient and decrease in the indoor film coefficient cause colder window surfaces. This effect is more pronounced near the corners of the recess where the indoor film coefficient is locally suppressed because air movement is restricted. Also, blinds should be placed at least 4 in. from the plane of the wall to allow some natural convection between the window and the blind.

Air leakage, especially in operable sections of fenestration, is another important cause of low surface temperature. Leakage near the edge-of-glass sections can further increase the potential for

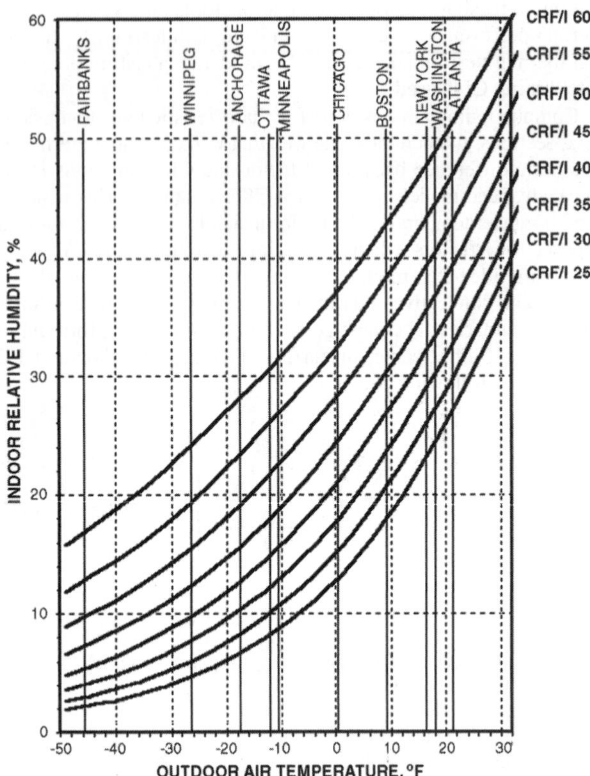

Fig. 38 Minimum Condensation Resistance Requirements (t_h = 68°F)

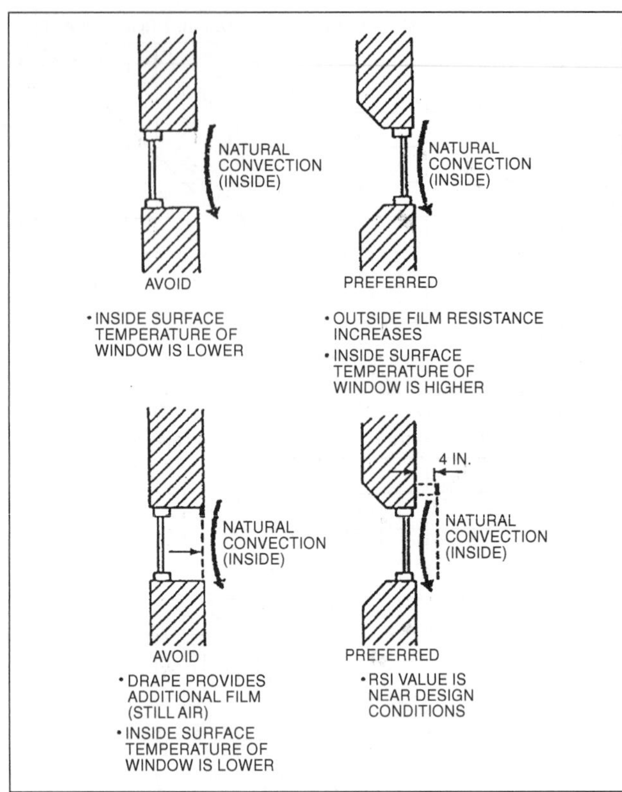

Fig. 39 Location of Fenestration Product Reveals and Blinds/Drapes and Their Effect on Condensation Resistance

condensation. However, the drier outdoor air decreases the relative humidity near the leakage sites and, in some cases, offsets the undesirable effect of the lower surface temperatures. The net effect of air leakage cannot readily be determined experimentally or with simulation.

OCCUPANT COMFORT AND ACCEPTANCE

Human thermal comfort is an immediate sensation that reflects building occupants' perceived response to many physical factors. Unlike much building design that is based primarily on long-term energy and economic considerations, comfort-related design focuses on, and must take heed of, short-term responses of the body's physiology to its surroundings.

Windows influence thermal comfort through a combination of three mechanisms: long-wave radiation exchange, absorption of solar radiation, and convective draft effects (Figure 40). An understanding of these phenomena is important to help designers evaluate the benefits of improved windows and create comfortable buildings. Although it is well understood that high-performance windows can reduce building energy consumption, a better understanding of their impact on comfort might lead to further savings. For example, Hawthorne and Reilly (2000) suggest that significant energy consumption is caused by the standard practice of using perimeter duct distribution in houses to mitigate potential discomfort caused by windows. They found that perimeter heating is often not necessary when high-performance windows are installed and that heating energy savings of 10 to 15% could result from installing a simpler, less expensive duct system. Better windows can allow thermostat settings to be lowered with no loss of comfort. Another simulation study (Lyons et al. 2000) examined the relative magnitudes of a residential window's physical influences under a wide variety of winter and summer climates, glazing parameters, and clothing levels. They found that

- Long-wave, thermal radiation influences of the window dominate unless direct sun strikes the occupant
- Direct solar load has a major influence on perceptions of comfort
- For most residential-size windows, draft effects are generally small

With all but highly insulating windows, the inside surface temperature of the window is heavily influenced by exterior conditions, and this temperature can significantly affect the radiant heat exchange between an occupant and the environment. If this heat exchange moves outside the acceptable range, discomfort will result. Mean radiant temperature (MRT) is commonly used to simplify the characterization of the radiant environment. On a cold day, the inside surface temperature can easily drop below 15°F for a clear single-pane window and below 40°F for a clear, double-pane window. If the occupant is sitting sufficiently near the window, MRT could drop to 55°F for the single-pane case and 62°F for the double-pane case. Based on ASHRAE *Standard* 55, even the use of the clear double-pane window could result in discomfort. [This example assumes an outdoor air temperature of 0°F, indoor air temperature of 72°F, non-window surface temperatures of 72°F, occupant-window view factor of 0.3, 0.9 clo (standard winter indoor clothing), and activity level of 1 met.] In addition to the MRT effect, a cold inside glass surface can induce a downward draft that increases air movement, contributing to further discomfort. If direct solar radiation strikes the glazing or occupant, the situation is much more complex.

In winter, the warming effect of sunlight on skin and clothing is often welcome, depending on the compounding effect of other factors such as air temperature. Windows also absorb and transmit a significant amount of solar radiation. Because of such absorption, a solar-heated window may improve MRT for a nearby person. The premise of passive solar design is that occupants will welcome, or at least tolerate, solar gain in exchange for savings on heating energy. However, it is desirable that the onset of discomfort be able to be

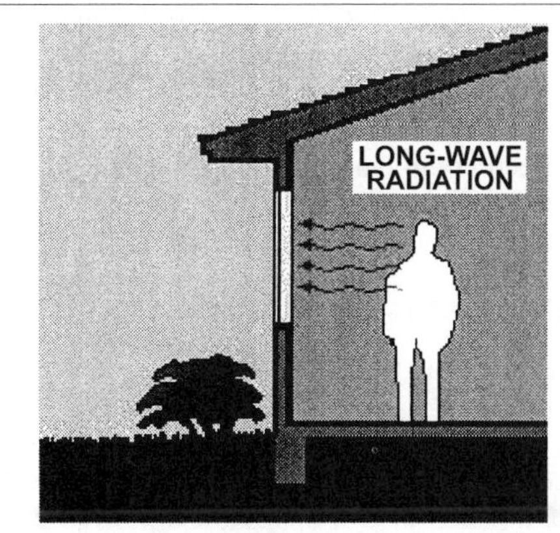

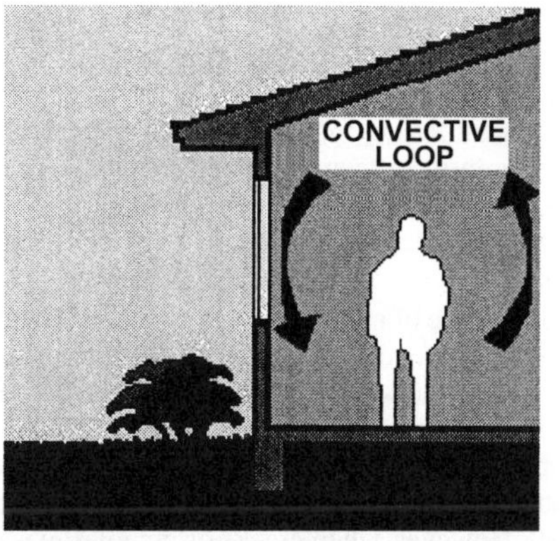

Fig. 40 Fenestration Impacts on Thermal Comfort: Long-Wave Radiation, Solar Radiation, Convective Draft

predicted; otherwise, the energy-saving design may be defeated if occupants draw shades to prevent overheating.

In summer, solar-heated glass may become uncomfortably hot and, in commercial premises, actually devalue rented space near windows. The inside surface of body-tinted, heat-absorbing glass can routinely reach temperatures above 120°F in summer conditions, raising MRT by as much as 15°F. This can be ameliorated with the addition of a second pane of glass on the inside. Transmitted radiation often causes discomfort if it falls directly on the occupant. A person sitting near a window in direct solar radiation can experience heat gain equivalent to a 20°F rise in MRT [Arens et al. (1986)]. Similarly, in residential applications, the perceived need for solar control is affected both by the contribution of window surfaces to MRT and by overheating due to direct solar load.

Advances in window technology, especially high-performance glazings, mean that the designer has a choice of potential glazing systems. On the basis of annual energy performance for heating, cooling, and lighting, these alternatives may give similar outcomes. However, because they represent different combinations of U-factor, SHGC, and inside glass surface temperature, their comfort outcomes may differ considerably. Research continues to develop tools that will help designers evaluate such difficult trade-offs. In the meantime, several general rules of thumb may be followed:

- In heating-dominated climates, windows with the lowest U-factor tend to give the best comfort outcomes. However, there is likely to be a trade-off between the twin goals of maximizing instantaneous comfort and minimizing annual energy consumption.
- In cooling-dominated climates or for orientations where cooling loads are of concern, windows with the lowest rise in surface temperature for a given SHGCtend to give the best comfort outcomes.

Sound Reduction

Proper acoustical treatment of exterior walls can decrease noise levels in certain areas. The airtightness of a wall is the primary factor to consider in reducing sound transmission from the exterior. Once walls and fenestration products are tight, the choice of glass and draperies becomes important. Draperies do not prevent sound from coming through the fenestration; they act as an absorber for sound that does penetrate. Table 26 lists average sound transmission losses for various types of glass. These averages apply for the frequency range of 125 to 4000 Hz and were determined by tests based on ASTM *Standard* E 90.

Table 26 Sound Transmittance Loss for Various Types of Glass

Type of Glass	Sound Transmittance Loss, dB
1/8 in. double-strength sheet glass	24
1/4 in. plate or float glass	27
1/2 in. plate glass	32
3/4 in. plate glass	35
1 in. plate glass	36
1/4 in. laminated glass (9/20 in. plastic interlayer)	30
1 in. insulating glass	32
1/2 in. laminated glass (9/20 in. plastic interlayer)	34
Insulating glass, 6 in. air space, 1/4 in. plate or float glass	40

Strength and Safety

In addition to its thermal, visual, and aesthetic functions, glass for building exteriors must also perform well structurally. Wind loads are specified in most building codes, and these requirements may be adequate for many structures. However, detailed wind tunnel tests should be run for tall or unusually shaped buildings and for buildings where the surroundings create unusual wind patterns. The strength of annealed, heat-strengthened, tempered, laminated, and insulated glass is given in ASTM *Standard* E 1300.

Thermal expansion and contraction of glass can result in breakage of ordinary annealed glass. This expansion and contraction can be caused by solar radiation onto partly shaded glass, by heat traps from drop ceilings and tight-fitting drapes, or by HVAC ducts incorrectly directed toward the glazing. High-performance tinted and reflective glasses with low-e coatings are usually more vulnerable to thermal stress breakage than clear glass. Heat treating (heat strengthening or fully tempering) the glass resists thermal stress breakage. Heat-strengthened glass, although not a safety glass, is usually preferred to tempered (safety) glass because it typically has less distortion and is much less likely to have spontaneous breakage. Spontaneous breakage can occur on very rare occasions in tempered glass. The glass manufacturer or fabricator should be consulted for information on thermal stress performance.

Building codes may require glass in certain positions to perform with certain breakage characteristics, which can be satisfied by tempered, laminated, or wired glass. In this case, glass should meet *Federal Standard* 16 CFR 1201 or other appropriate breakage performance requirements.

Life-Cycle Costs

Alternative building shells should be compared to ensure satisfactory energy use and total energy budget compliance, if required. ASHRAE *Standards* 90.1 and 90.2 should be used as a starting point. A life-cycle cost model should be developed for each system considered. See Chapter 35 of the 1999 *ASHRAE Handbook—Applications*.

DURABILITY

The service life and long-term performance of fenestration systems depend on the durability of all the components that make up the system. Representative samples of IGUs are usually tested (for seal durability) according to test methods to ensure the integrity of the seal. Failure of IGUs is usually indicated by loss of adhesion of sealant to the glass; as a result, fogging occurs inside the glazing cavity.

In the case of argon-filled units, the seal failure means the loss of argon and, hence, degradation in the thermal characteristics of the unit. Extensive work was done at the National Research Council of Canada to study the durability of IGUs filled with argon gas (Elmahdy and Yusuf 1995). The results indicated that, under normal conditions, argon loss due to diffusion through the sealant is very small. However, when cracks or pinholes exist in the sealant, most of the argon gas escapes from the unit, which implies that the implementation of stringent quality control procedures is essential for the production of durable IGUs.

The degradation of organic materials and other chemical components in the IGUs, as a result of exposure to ultraviolet radiation, is also among the factors affecting the durability and service life of fenestration systems. The use of low-e coating on glass tends to enhance the appearance of chemical deposits on the glass surface. Also, the insertion of muntin bars in the glazing cavities may result in excessive rate of unit failure during the ultraviolet volatile (fogging) test unless strict quality assurance processes are implemented. The current ASTM (United States) and CGSB (Canada) durability standards are being reviewed to reflect the emergence of new technologies in the fenestration industry.

Insulating glass products have been studied in a 15-year correlation study by the Sealed Insulating Glass Manufacturers Association (SIGMA). During this study, it was found that long-term performance and durability of insulating glass correlated well with the test level to which such a unit's construction had been manufactured with regard to the ASTM *Standard* E 773 test method and ASTM *Standard* E 774 specification for sealed insulating glass. The units showing the highest percentage of resistance to seal failure were those that were tested in conformance with the ASTM *Standard* E 774 Class CBA standard. Units that did not qualify to the A level showed a definite correlation to a higher percentage of failure. During the field correlation studies, it was found that units glazed in compliance with the SIGMA recommendations perform for longer periods than units not constructed properly, having deficiencies in the glazing system, or not meeting the ASTM requirements.

The durability of fenestration systems is also dependent on the durability of other system components such as the weatherstripping, gaskets, glazing tapes, air seals, and hardware. The wear and tear of these elements with time and use may result in excessive air and water leakage, which will affect the overall performance and the service life of the system. Excessive water leakage may result in damage to the fenestration product, especially the edge seal, as well as the wall section where the product is mounted. Excessive air leakage may lead to frost buildup and condensation on the fenestration surfaces.

Studies conducted at the National Research Council of Canada (Elmahdy 1995) and elsewhere (Patenaude 1995) showed that when windows are tested at high pressure and temperature differentials, they experience air leakage rates which exceed those determined at 75 Pa and zero temperature differential (these conditions are used in rating the window air leakage in U.S. and Canadian standards). In other studies (CANMET 1991, 1993), the effect of pressure and motion cycling on windows resulted in excessive degradation in almost all the window performance factors, particularly the condensation resistance, ease of operation, air leakage, and water leakage.

In order to predict long-term performance, the unit construction for insulating glass should be subjected to a test and certification program such as ASTM *Standard* E 774 Class CBA level and the requirements of SIGMA or CGSB *Standard* 12.8 certified by the Insulating Glass Manufacturers Association of Canada (IGMAC) or equivalent.

In addition to affecting the fenestration performance factors mentioned above, durability may also affect long-term energy performance.

CODES AND STANDARDS

National Fenestration Rating Council (NFRC)

The National Fenestration Rating Council (NFRC) was formed in 1989 to respond to a need for fair, accurate, and credible ratings for fenestration products. NFRC has adopted rating procedures for U-factor (NFRC 100), solar heat gain coefficient and visible transmittance (NFRC 200), optical properties (NFRC 300), emissivities (NFRC 301), and air leakage (NFRC 400). To provide certified ratings, manufacturers follow the requirements in the NFRC *Product Certification Program* (PCP) which involves working with laboratories accredited to the NFRC *Laboratory Accreditation Program* (LAP) and independent certification and inspection agencies accredited through the NFRC *Certification Agency Program* (CAP).

NFRC 100 was the first of the NFRC rating procedures approved and thus the first NFRC procedure adopted into energy codes in the United States. NFRC 100 requires the use of a combination of state-of-the-art computer simulations and improved thermal testing to determine U-factors for the whole product. The next step is product certification. NFRC has a series of checks and balances to ensure that the rating system is accurately and uniformly employed. Products and their ratings are authorized for certification by an NFRC-licensed independent certification and inspection agency (IA). Finally, two labels are required: the temporary label, which contains the product ratings, and a permanent label, which allows tracking back to the IA and information in the NFRC *Product Directory*. In addition to informing the buyer, the temporary label provides the building inspector with the information necessary to verify energy code compliance. The permanent label provides access to energy rating information for a future owner, property manager, building inspector, lending agency, or building energy rating organization.

This process has a number of noteworthy features that make it superior to previous fenestration energy rating systems and correct past problems:

- The procedures provide a means for manufacturers to take credit for all the nuances and refinement in their product design and a common basis for others to compare product claims.
- The involvement of independent laboratories and the IA provides architects, engineers, designers, contractors, consumers, building officials, and utility representatives with greater confidence that the information is unbiased.
- Requiring simulation and testing provides an automatic check on accuracy. This also remedies a shortcoming of previous energy code requirements that relied on testing alone, which allowed manufacturers to perform several tests and then use the best one for code purposes.
- The certification process indicates that the manufacturer is consistently producing the product that was rated. This corrects a past concern that manufacturers were able to make an exceptionally high quality sample and obtain a good rating in a test but not consistently produce that product.
- There is now a readily visible temporary label that can be used by the building inspector to quickly verify compliance with the energy code.
- There is now a permanent label that enables future access to energy rating information.

While the NFRC program is similar for other fenestration characteristics, there are differences worth pointing out. The solar heat gain coefficient and visible transmittance ratings (NFRC 200), which have been referenced in several codes, are based on simulation alone. Optical properties (NFRC 300) and emissivity (NFRC 301) are based on measurements by the manufacturer, with independent verification. The air leakage ratings (NFRC 400) are based on testing alone. For site-assembled fenestration products (such as curtain walls and window walls), there is an NFRC label certificate that fulfills the labeling requirements and serves the certification purpose. There must be a separate NFRC label certificate for each "individual product" in a particular project.

United States Energy Policy Act (EPAct)

In the United States, the 1992 Energy Policy Act (EPAct) required the development of national fenestration energy rating systems and specified NFRC as the preferred developer. (The U.S. Department of Energy was to establish procedures if the NFRC did not.) While this recognition provided an impetus for NFRC to develop the desired procedures and programs, the EPAct sections on energy codes have been a key factor in their implementation.

EPAct set energy code baselines for state energy codes. The ICC 2000 *International Energy Conservation Code* (IECC) and ASHRAE/IESNA *Standard* 90.1-1999, *Energy Standard for Buildings Except Low-Rise Residential Buildings* are the current successors to the versions cited in the 1992 legislation. The majority of states have adopted the predecessors to the 2000 *IECC* (including the 1998 IECC and the CABO 1995 *Model Energy Code*) and to ASHRAE/IESNA *Standard* 90.1-1999 (i.e., ASHRAE/IESNA *Standard* 90.1-1989) into their codes either directly or by reference when adopting a building code published by one of the three national code organizations in the United States. The ICC 2000 *International Building Code* (the U.S. model building code jointly developed by ICBO, BOCA, and SBCCI) references the 2000 *International Energy Conservation Code.*

The ICC 2000 International Energy Conservation Code

The ICC *2000 International Energy Conservation Code (IECC)* references NFRC 100 for U-factor (as did the 1998 *IECC* and the 1995 *Model Energy Code*) and NFRC 200 for solar heat gain coefficient (SHGC) (as did the 1998 *IECC*). Section 102.3, which applies to all occupancies, requires U-factors of fenestration products (windows, doors, and skylights) to be determined in accordance with NFRC 100 by an accredited independent laboratory and labeled and certified by the manufacturer. While the language does not specify NFRC accreditation, it both requires the use of the NFRC rating procedure by an independent entity and requires labeling and certification.

ASHRAE/IESNA *Standard* 90.1-1999

In 1999, ASHRAE and IESNA published a comprehensive update to the 1989 version of *Standard* 90.1. The fenestration rating, labeling, and certification criteria are in Sections 5.2.2 and 5.2.3. U-factors are to be determined in accordance with NFRC 100, solar heat gain coefficient in accordance with NFRC 200, visible transmittance in accordance with NFRC 300, and air leakage in accordance with NFRC 400.

For further information on U.S. energy codes, the Building Codes Assistance Project (BCAP) publishes a bimonthly summary entitled "Status of State Energy Codes," which provides information on current codes and pending legislation. For additional information, contact BCAP at 1200 18th Street NW, Suite 900, Washington DC 20036; voice: 202-530-2200; fax: 202-331-9588; e-mail: bcap@ase.org; website: http://solstice.crest.org/efficiency/bcap/update. html.

Canadian Standards Association (CSA)

In Canada, the Canadian Standards Association (CSA) promulgates fenestration energy rating standards. CSA *Standard* A440.2 addresses most fenestration products, and CSA *Standard* A453 addresses doors. These are companion standards to NFRC 100. NFRC and CSA have established a Thermal Harmonization Task Force to attempt to harmonize their fenestration energy rating standards.

SYMBOLS

a = absorptance in a layer, considered as an isolated layer
A = apparent solar constant
A = total projected area of the fenestration product
AST = apparent solar time
B = atmospheric extinction coefficient
C = sky diffuse factor
e = hemispherical emissivity
E_{DN} = direct normal irradiance
E_D = direct irradiance
E_d = diffuse sky irradiance
E_r = diffuse ground reflected irradiance
E_t = total irradiance
ET = equation of time
h = surface heat transfer coefficient
H = fenestration product height
H = hour angle
k = thermal conductivity
L = latitude
LON = longitude
LSM = local standard meridian
LST = local standard time
n = refractive index
P_V = vertical projection depth
P_H = horizontal projection depth
q = instantaneous energy flux
Q = instantaneous energy flow
R = reflectance of a layer or collection of layers (system or subsystem)
R_H = height of opaque surface between fenestration product and horizontal projection
R_W = width of opaque surface between fenestration product and vertical projection
SHGC = solar heat gain coefficient
t = relative temperature
T = absolute temperature
T = transmittance of a layer or collection of layers (system or subsystem)
U = overall coefficient of heat transfer
W = fenestration product width
Y = ratio of vertical/horizontal sky diffuse
α = material absorptivity
β = solar altitude
δ = declination
Δ = vertical projection profile angle
ϕ = solar azimuth
γ = surface solar azimuth
η = day of year
λ = wavelength
θ = incident angle
ρ_g = ground reflectance
ϖ = solid angle
Ω = horizontal projection profile angle
ξ = refractive angle
ψ = surface azimuth
Σ = surface tilt
$\mathcal{A}$ = absorptance in a layer or a collection of layers (system or subsystem)

REFERENCES

AAMA. 1987. *Skylight handbook: Design guidelines*. American Architectural Manufacturers Association, Schamberg, IL.

AAMA. 1988. Voluntary test method for thermal transmittance and condensation resistance of windows, doors and glazed wall sections. Publication No. AAMA 1503.1-88.

AGSL. 1992. Vision3, Glazing system thermal analysis—User manual, Department of Mechanical Engineering, University of Waterloo, Waterloo, Ontario, Canada, August.

Allen, C.W. 1973. *Astrophysical quantities*. The Athlone Press, University of London, London.

Arasteh, D. 1989. An analysis of edge heat transfer in residential windows. *Proceedings of ASHRAE/DOE/BTECC Conference, Thermal Performance of the Exterior Envelopes of Buildings IV*, Orlando, FL, 376-87.

Arasteh, D. et al. 1994. WINDOW 4.1: A PC program for analyzing window thermal performance in accordance with standard NFRC procedures. *Publication* LBL-35298, Lawrence Berkeley Laboratory, Energy & Environment Division, Berkeley, CA, 1993.

Arasteh, D. et al. 1995. Recent technical improvements to the WINDOW computer program. *Proceedings of Window Innovations Conference '95*, Toronto, Canada.

Arasteh, D., J. Huang, R. Mitchell, B. Clear, and C. Kohler. 2000. A database of window annual energy use in typical North American single family houses. *ASHRAE Transactions* 106(2).

Arens, E.A., R. Gonzalez, and L. Berglund. 1986. Thermal comfort under an extended range of environmental conditions. *ASHRAE Transactions* 92(1).

ASHRAE. 1988. Method of measuring solar-optical properties of materials. ANSI/ASHRAE *Standard* 74-1988.

ASHRAE. 1996. Standard method for determining and expressing the heat transfer and total optical properties of fenestration products. Draft ASHRAE *Standard*, February.

ASHRAE. 1999. Energy standard for buildings except low-rise residential buildings. ASHRAE/IESNA *Standard* 90.1-1999.

ASTM E. 1981. Recommended practice for laboratory measurements of air borne sound transmission loss of building partitions. *Standard* E 90-81. American Society for Testing and Materials, West Conshohocken, PA.

ASTM. 1982. Standard test method for solar absorptance, reflectance, and transmittance of materials using integrating spheres. *Standard* E 90-82.

ASTM. 1986. Standard test method for solar transmittance (terrestrial) of sheet materials using sunlight. *Standard* E 1084-86.

ASTM. 1987. Standard tables for terrestrial direct normal solar spectral irradiance for air mass 1.5. *Standard* E 891-87 (now part of ASTM *Standard* G 159-98).

ASTM. 1987. Standard tables for terrestrial solar spectral irradiance at air mass 1.5 for a 37° tilted surface. *Standard* E 892-87 (now part of ASTM *Standard* G 159-98).

ASTM. 1988. Standard practice for calculation of photometric transmittance and reflectance of materials to solar radiation. *Standard* E 971-88.

ASTM. 1988. Standard test method for solar photometric transmittance of sheet materials using sunlight. *Standard* E 972-88.

ASTM. 1988. Standard test method for seal durability of sealed insulated glass units. *Standard* E 773-88.

ASTM. 1991. Standard test method for determining rate of air leakage through exterior windows, curtain walls, and doors under specified pressure differences across the specimen. *Standard* E 283-91.

ASTM. 1992. Standard specification for sealed insulated glass units. *Standard* E 774-92.

ASTM. 1994. Standard practice for determining the minimum thickness and type of glass required to resist a specific load. *Standard* E 1300-94.

Bass, M., ed. 1995. *Handbook of optics, Vol. I—Fundamentals, techniques, and design*, 2nd ed. Optical Society of America, McGraw-Hill, New York.

Beck, F.A., B.T. Griffith, D, Turler, and D. Arasteh. 1995. Using infrared thermography for the creation of a window surface temperature database to validate computer heat transfer models. *Proceedings of Window Innovations Conference '95*, Toronto, Canada.

Bird, R.E. and R.L. Hulstrom. 1982. Terrestrial solar spectral data sets. SERI/TR-642-1149. Solar Energy Research Institute.

Bliss, R.W. 1961. Atmospheric radiation near the surface of the ground. *Solar Energy* 5(3):103.

Brandle, K. and R.F. Boehm. 1982. Air flow windows: Performance and applications. *Proceedings of Thermal Performance of the Exterior Envelopes of Buildings II*, ASHRAE/DOE Conference, ASHRAE SP38, December.

CANMET. 1991. A study of the long term performance of operating and fixed windows subjected to pressure cycling. Catalogue No. M91-7/214-1993E. Efficiency and Alternative Energy Technology Branch, CANMET, Ottawa, Ontario, Canada.

CANMET. 1993. Long term performance of operating windows subjected to motion cycling. Catalogue No. M91-7/235-1993E. Efficiency and Alternative Energy Technology Branch, CANMET, Ottawa, Ontario, Canada.

Carpenter, S. and A. Elmahdy. 1994. Thermal performance of complex fenestration systems. *ASHRAE Transactions*.

Carpenter, S. and J. Hogan. 1996. Recommended U-factors for swinging, overhead and revolving doors. *ASHRAE Transactions*.

Carpenter, S. and A. McGowan. 1993. Effect of framing systems on the thermal performance of windows. *ASHRAE Transactions* 99(1).

CEA. 1995. *Energy-efficient residential and commercial windows reference guide*. Canadian Electricity Association, Montreal, PQ.

Choi, A.-S. and R.G. Mistrick. 1999. Analysis of daylight responsive dimming system performance. *Building and Environment* 34:231-43.

Crooks, B.P. et al. 1995. NFRC efforts to develop a residential fenestration annual energy rating methodology. *Proceedings of Window Innovations Conference '95*, Toronto, Canada.

CSA. 1990. Windows. *Publication* CAN/CSA-A440-90. Canadian Standards Association, Etobicoke, Ontario.

CSA. 1991. Windows/User selection guide to CSA standards. CAN/CSA A440.1-M90.

CSA. 1993. Energy performance evaluation of windows and sliding glass doors. CAN/CSA A440.2-93.

CSA. 1995. Energy performance evaluation of swinging doors. A453-95.

Curcija, D. and W.P. Goss. 1994. Two-dimensional finite element model of heat transfer in complete fenestration systems. *ASHRAE Transactions* 100(2).

Curcija, D. and W.P. Goss. 1995a. Three-dimensional finite element model of heat transfer in complete fenestration systems. *Window Innovations Conference '95*, Toronto, Canada, June.

Curcija, D. and W.P. Goss. 1995b. New correlations for convective heat transfer coefficient on indoor fenestration surfaces—Compilation of more recent work. *Thermal Performance of the Exterior Envelopes of Buildings VI*. ASHRAE.

Curcija, D., W.P. Goss, J.P. Power, and Y. Zhao. 1996. "Variable-h" model for improved prediction of surface temperatures in fenestration systems. *Technical Report*, University of Massachusetts at Amherst.

de Abreu, P., R.A. Fraser, H.F. Sullivan, and J.L. Wright. 1996. A study of insulated glazing unit surface temperature profiles using two-dimensional computer simulation. *ASHRAE Transactions* 102(2).

Duffie, J.A. and W.A. Beckman. 1980. Solar engineering of thermal processes. John Wiley & Sons, New York.

Elmahdy, H. 1996. Surface temperature measurement of insulating glass units using infrared thermography. *ASHRAE Transactions* 102(2).

Elmahdy, A.H. and S.A. Yusuf. 1995. Determination of argon concentration and assessment of the durability of high performance insulating glass units filled with argon gas. *ASHRAE Transactions* 101(2).

Elmahdy, A.H. 1995. Air leakage characteristics of windows subjected to simultaneous temperature and pressure differentials. *Proceedings Window Innovation Conference '95*, CANMET.

El Sherbiny, S.M., et al. 1982. Heat transfer by natural convection across vertical and inclined air layers. *Journal of Heat Transfer* 104:96-102.

EEL. 1990. FRAME/VISION window performance modelling and sensitivity analysis. Institute for Research in Construction, National Research Council of Canada, Ottawa.

EEL. 1995. The FRAMEplus Toolkit for heat transfer assessment of building components. Enermodal Engineering, Kitchener, ON, and Denver, CO.

Solar Energy. 1988. Erratum 40:175.

Erhorn, H. and M. Dirksmöller, eds. 2000. Documentation of the software package ADELINE 3. Fraunhofer-Institut für Bauphysik, Stuttgart.

Ewing, W.B. and J.I. Yellott. 1976. Energy conservation through the use of exterior shading of fenestration. *ASHRAE Transactions* 82(1):703-33.

Finlayson, E.U. and D. Arasteh. 1993. WINDOW 4.0: Documentation of calculation procedures. *Publication* LBL-33943/UC-350, Lawrence Berkeley Laboratory, Energy & Environment Division, Berkeley, CA.

Federal Standard 16 CFR 1201. Safety standard for architectural glazing materials.

Galanis, N. and R. Chatiguy. 1986. A critical review of the ASHRAE solar radiation model. *ASHRAE Transactions* 92(1).

Gates, D.M. 1966. Spectral distribution of solar radiation at the earth's surface. *Science* 151(2):3710.

Griffith, B.T., D. Turler, and D. Arasteh. 1996. Surface temperature of insulated glazing units: Infrared thermography laboratory measurements. *ASHRAE Transactions* 102(2).

Griffith, B., E. Finlayson, M. Yazdanian, and D. Arasteh. 1998. The significance of bolts in the thermal performance of curtain-wall frames for glazed facades. *ASHRAE Transactions* 105(1).

Gueymard, C.A. 1987a. An anisotropic solar irradiance model for tilted surfaces and its comparison with selected engineering algorithms. *Solar Energy* 38:367-86. Erratum, *Solar Energy* 40:175 (1988).

Gueymard, C. 1987b. *Solar Energy* 38:367-86.

Gueymard, C.A. 1993a. Critical analysis and performance assessment of clear sky solar irradiance models using theoretical and measured data. *Solar Energy.*

Gueymard, C. 1993b. Development and performance assessment of a clear sky spectral radiation model, *Proceedings, 22nd Annual Conference*, Solar '93, Washington, D.C., American Solar Energy Society, 433-38.

Gueymard, C. 1995. A simple model of the atmospheric radiative transfer of sunshine: Algorithms and performance assessment. FSEC-PF-270-95, Florida Solar Energy Center, Cocoa, FL, November.

Hawthorne W. and M.S. Reilly. 2000. The impact of glazing selection on residential duct design and comfort. *ASHRAE Transactions* 106(1).

Heschong, L., D. Mahone, F. Rubinstein, and J. McHugh. 1998. *Skylighting guidelines.* The Heschong Mahone Group, Sacramento, CA.

Hickey, J.R. et al. 1981. Observations of the solar constant and its variations: Emphasis on Nimbus 7 results. *Symposium on the Solar Constant and the Special Distribution of Solar Irradiance*, IAMAP 3rd Scientific Assembly, Hamburg, Germany.

Hogan, J.F. 1988. A summary of tested glazing U-values and the case for an industry wide testing program. *ASHRAE Transactions* 94(2).

Hollands, K.G.T. and J.L. Wright. 1982. Heat loss coefficients and effective $\tau\alpha$ products for flat plate collectors with diathermous covers. *Solar Energy* 30:211-16.

Huang, Y.J., J.C. Bull, and R.L. Ritschard. 1987. Technical documentation for a residential energy use database developed in support of ASHRAE Special Project 53. Lawrence Berkeley Laboratory *Report* LBL-24306.

Huang, J., R. Mitchell, D. Arasteh, and S. Selkowitz. 1999. Residential fenestration performance analysis using RESFEN 3.1. *Thermal Performance of the Exterior Envelopes of Building VII*, ASHRAE.

ICC. 2000. *International energy conservation code.* International Code Council, Falls Church, VA.

IEA. 1999. Daylighting simulation: Methods, algorithms, and resources. A Report of IEA SHC Task 21/ECBCS Annex 29, December. Available at http://www.iea-shc.org or LBNL *Report* 44296, Lawrence Berkeley National Laboratory.

IESNA. 1999. IESNA recommended practice of daylighting. IES RP-5-99. Illuminating Engineering Society of North America, New York.

Iqbal, M. 1983. *An introduction to solar radiation.* Academic Press, Toronto.

ISO. 2000. Thermal performance of windows, doors and shading devices—Detailed calculations. ISO 15099: Final Draft International Standard (FDIS). International Organization for Standardization, Geneva.

Jennings, J.D., F.M. Rubinstein, D. DiBartolomeo, and S. Blanc. 1999. Comparison of control options in private offices in an advanced lighting controls testbed. LBNL-43096, Lawrence Berkeley National Laboratory.

Keyes, M.W. 1967. Analysis and rating of drapery materials used for indoor shading. *ASHRAE Transactions* 73(1):8.4.1.

Klems, J.H. 1989. U-values, solar heat gain, and thermal performance: Recent studies using the MoWitt. *ASHRAE Transactions* 95(1).

Klems, J.H. 1994a. A new method for predicting the solar heat gain of complex fenestration systems: I. Overview and derivation of the matrix layer calculation. *ASHRAE Transactions* 100(1):1065-72.

Klems, J.H. 1994b. A new method for predicting the solar heat gain of complex fenestration systems: II. Detailed description of the matrix layer calculation. *ASHRAE Transactions* 100(1):1073-86.

Klems, J.H. 2001. Solar heat gain through fenestration systems containing shading: Procedures for estimating performance from minimal data. LBNL-46682. Windows and Daylighting Group, Lawrence Berkeley National Laboratory, Berkeley, CA.

Klems, J.H. and G.O. Kelley. 1996. Calorimetric measurements of inward-flowing fraction for complex glazing and shading systems. *ASHRAE Transactions* 102(1):947-54.

Klems, J.H. and J.L. Warner. 1995. Measurement of bi-directional optical properties of complex shading devices. *ASHRAE Transactions* 101(1): 791-801.

Klems, J.H. and J.L. Warner. 1997. Solar heat gain coefficient of complex fenestrations with a venetian blind for differing slat tilt angles. *ASHRAE Transactions* 103(1):1026-34.

Klems, J.H., J.L. Warner, and G.O. Kelley. 1996. A comparison between calculated and measured SHGC for complex fenestration systems. *ASHRAE Transactions* 102(1):931-39.

LBL. 1994. WINDOW 4.1: A PC program for analyzing window thermal performance of fenestration products. LBL-35298. Windows and Daylighting Group, Lawrence Berkeley Laboratory, Berkeley, CA.

LBL. 1996. THERM 1.0: A PC program for analyzing the two-dimensional heat transfer through building products. LBL-37371. Windows and Daylighting Group.

Lyons, P.R.A., D.K. Arasteh, and C. Huizenga. 2000. Window performance for human thermal comfort. *ASHRAE Transactions* 106(1).

Mathis, R.C. and R. Garries. 1995. Instant, annual life: A discussion on the current practice and evolution of fenestration energy performance rating. *Proceedings of Window Innovations Conference '95*, Toronto, Canada.

McCabe, M.E., et al. 1986. U-value measurements for windows and movable insulations from hot box tests in two commercial laboratories. *ASHRAE Transactions* 92(1).

McCluney, R. 1987. Determining solar radiant heat gain of fenestration systems. *Passive Solar Journal* 4(4):439-87.

McCluney, R. 1990. Awning shading algorithm update. *ASHRAE Transactions* 96(1).

McCluney, R. 1991. The death of the shading coefficient? *ASHRAE Journal* (3):36-45.

McCluney, R. 1993. Sensitivity of optical properties and solar gain of spectrally selective glazing systems to changes in solar spectrum. *Solar '93*, 22nd American Solar Energy Society Conference, Washington, D.C.

McCluney, R. 1994a. *Introduction to radiometry and photometry.* Artech House, Boston.

McCluney, R. 1994b. Angle of incidence and diffuse radiation influences on glazing system solar gain. *Proceedings Solar '94 Conference*, American Solar Energy Society, San Jose, CA.

McCluney, R. 1996. Sensitivity of fenestration solar gain to source spectrum and angle of incidence. *ASHRAE Transactions* 102(2).

McCluney, R. and C. Gueymard. 1992. SUNSPEC 1.0, Operating manual. FSEC-SW-3-92. Florida Solar Energy Center, Cocoa, FL.

McCluney, R. and L.R. Mills. 1993. Effect of interior shade on window solar gain. *ASHRAE Transactions* 99(2).

Mitchell, R., J. Huang, D. Arasteh, R. Sullivan, and S. Phillip. 1999. RESFEN 3.1: A PC program for calculating the heating and cooling energy use of windows in residential buildings—Program description. LBNL-40682 Rev. BS-371. Lawrence Berkeley National Laboratory.

Moon, P. 1940. Proposed standard solar radiation curves for engineering use. *Journal of the Franklin Institute* 11:583.

NAGDM. 1992. Test method for thermal transmittance and air infiltration of garage doors. *Standard* 105-1992. National Association of Garage Door Manufacturers, Chicago.

NFRC. 1993. NFRC 301-93: Standard test method for emissivity of specular surfaces using spectrometric measurements. National Fenestration Rating Council, Silver Spring, MD.

NFRC. 1994. NFRC 300-94: Procedures for determining solar optical properties of simple fenestration products.

NFRC. 1995. NFRC 200-95: Procedure for determining fenestration product solar heat gain coefficients at normal incidence.

NFRC. 1995. NFRC 400-95: Procedure for determining fenestration product air leakage.

NFRC. 1997. NFRC 100-97: Procedure for determining fenestration product U-factors.

NFRC. 1997. NFRC 100-97 Attachment A: NFRC test procedure for measuring the steady-state thermal transmittance of fenestration systems.

NFRC. 1999. NFRC 100-99 Section B: Procedure for determining door system product thermal properties (Currently limited to U-values).

NFRC. 2000a. *Draft* NFRC 500: Calculational method and test procedure to determine the condensation index of fenestration products.

NFRC. 2000b. Spectral data library available through www.nfrc.org/software.html.

Ozisik, N. and L.F. Schutrum. 1960. Solar heat gain factors for windows with drapes. *ASHRAE Transactions* 66:228.

Parmelee, G.V. and W.W. Aubele. 1952. Radiant energy emission of atmosphere and ground. *ASHRAE Transactions* 58:85.

Parmelee, G.V. and R.G. Huebscher. 1947. Forced convection heat transfer from flat surfaces. *ASHVE Transactions* 245-84.

Patenaude, A. 1995. Air infiltration rate of windows under temperature and pressure differentials. *Proceedings Window Innovation Conference '95.* CANMET.

Pennington, C.W. 1968. How louvered sun screens cut cooling, heating loads. *Heating, Piping, and Air Conditioning* (December).

Pennington, C.W. et al. 1964. Experimental analysis of solar heat gain through insulating glass with indoor shading. *ASHRAE Journal* 2:27.

Pennington, C.W. and G.L. Moore. 1967. Measurement and application of solar properties of drapery shading materials. *ASHRAE Transactions* 73(1):8.3.1.

Pennington, C.W., C. Morrison, and R. Pena. 1973. Effect of inner surface air velocity and temperatures upon heat loss and gain through insulating glass. *ASHRAE Transactions* 79(2).

Perez, R. et al. 1986. An anisotropic hourly diffuse radiation model for sloping surfaces—Description, performance validation, and site dependency evaluation. *Solar Energy* 36:481-98.

Reilly, M.S. 1994. Spacer effects on edge-of-glass and frame heat transfer. *ASHRAE Transactions* 94:31-33.

Reilly, M.S., F.C. Winkelmann, D.K. Arasteh, and W.L. Carroll. 1992. Modeling windows in DOE-2.1E. *Energy and Buildings* 22(1995):59-66.

Robbins, C.L. 1986. *Daylighting: Design and analysis.* Van Nostrand Reinhold, New York.

Rubin, M. 1982a. Solar optical properties of windows. *Energy Research* 6:122-33.

Rubin, M. 1982b. Calculating heat transfer through windows. *Energy Research* 6:341-49.

Rubin, M. 1984. Optical constants and bulk optical properties of soda lime silica glasses for windows. *Report* No. 13572, Applied Sciences Division, Lawrence Berkeley Laboratory, Berkeley, CA.

Rubin, M. 1985. Optical properties of soda lime silica glasses. *Solar Energy Materials* 12:275-88.

Rubin, M., R. Powles, and K. von Rottkay. 1999. Models for the angle-dependent optical properties of coated glazing materials. *Solar Energy* 66(4):267-76.

Rudoy, W. and F. Duran. 1975. Effect of building envelope parameters on annual heating/cooling load. *ASHRAE Journal* (7):19.

Rundquist, R.A. 1991. Calculation procedure for daylighting and fenestration effects on energy and peak demand. *ASHRAE Transactions* 97(2).

Selkowitz, S.E. 1979. Thermal performance of insulating windows. *ASHRAE Transactions* 85(2):669-85.

Sezgen, O. and J.G. Koomey. 1998. Interactions between lighting and space conditioning energy use in U.S. commercial buildings. LBNL-39795, Lawrence Berkeley National Laboratory.

Shewen, E.C. 1986. A Peltier-effect technique for natural convection heat flux measurement applied to the rectangular open cavity. *Ph.D. thesis*, Department of Mechanical Engineering, University of Waterloo, Ontario.

Smith, W.A. and C.W. Pennington. 1964. Shading coefficients for glass block panels. *ASHRAE Journal* 5(12):31.

Sodergren, D. and T. Bostrom. 1971. Ventilating with the exhaust air window. *ASHRAE Journal* 13(4):51.

Sterling, E.M., A. Arundel, and T.D. Sterling. 1985. Criteria for human exposure in occupied buildings. *ASHRAE Transactions* 91(1).

Sullivan, H.F., J.L. Wright, and R.A. Fraser. 1996. Overview of a project to determine the surface temperatures of insulated glazing units: Thermographic measurement and 2-D simulation. *ASHRAE Transactions* 102(2).

Sullivan, H.F. and J.L. Wright. 1987. Recent improvements and sensitivity of the VISION glazing system thermal analysis program. *Proceedings of the 12th Passive Solar Conference*, ASES/SESCI, Portland, OR, 145-49.

TSC. 1996. WinSARC: Solar angles and radiation calculation for MS Windows. *User's Manual.* Tait Solar Co., Tempe, AZ.

Threlkeld, J.L. and R.C. Jordan. 1958. Direct solar radiation available on clear days. *ASHRAE Transactions* 64:45.

Vild, D.J. 1964. Solar heat gain factors and shading coefficients. *ASHRAE Journal* 10:47.

Ward, G.W. 1990. Visualization. *Lighting Design + Application* 20(6):4-20.

Wehrli. 1985. Extraterrestrial solar spectrum. *Publication* No. 615, Physikalisch-Metrologisches Observatorium and World Radiation Data Center, Davos, Switzerland.

Winkelmann, F.C. 1983. Daylighting calculation in DOE-2. LBNL-11353. Lawrence Berkeley National Laboratory.

Wright, J.L. 1995a. Summary and comparison of methods to calculate solar heat gain. *ASHRAE Transactions* 101(1).

Wright, J.L. 1995b. VISION4 glazing system thermal analysis: User manual. Advanced Glazing System Laboratory, University of Waterloo, August.

Wright, J.L. 1995c. VISION4 glazing system thermal analysis: Reference manual. Advanced Glazing System Laboratory, University of Waterloo, May.

Wright, J.L. 1996. A correlation to quantify convective heat transfer between window glazings. *ASHRAE Transactions* 102(1):940-46.

Wright, J.L. and H.F. Sullivan. 1995a. A 2-D numerical model for natural convection in a vertical, rectangular window cavity. *ASHRAE Transactions* 100(2).

Wright, J.L. and H.F. Sullivan. 1995b. A 2-D numerical model for glazing system thermal analysis. *ASHRAE Transactions* 101(1).

Wright, J.L. and H.F. Sullivan. 1995c. A simplified method for the numerical condensation resistance analysis of windows. *Window Innovations '95*, Toronto, Canada, June.

Yazdanian, M. and J.H. Klems. 1993. Measurement of the exterior convective film coefficient for windows in low-rise buildings. *ASHRAE Transactions* 100(1).

Yellott, J.I. 1963. Selective reflectance—A new approach to solar heat control. *ASHRAE Transactions* 69:418.

Yellott, J.I. 1966. Shading coefficients and sun-control capability of single glazing. *ASHRAE Transactions* 72(1):72.

Yellott, J.I. 1972. Effect of louvered sun screens upon fenestration heat loss. *ASHRAE Transactions* 78(l):199-204.

Zhao, Y., D. Curcija, and W.P. Goss. 1996. Condensation resistance validation project—Detailed computer simulations using finite element methods. *ASHRAE Transactions* 102(2).

BIBLIOGRAPHY

Brambley, M.R. and S.S. Penner. 1979. Fenestration devices for energy conservation I. Energy savings during the cooling season. *Energy* (February).

Burkhardt, W.C. 1975. Solar optical properties of gray and brown solar control series transparent acrylic sheet. *ASHRAE Transactions* 81(1): 384-97.

Burkhardt, W.C. 1976. Acrylic plastic glazing; properties, characteristics and engineering data. *ASHRAE Transactions* 82(1):683.

Collins, B.L. 1975. Windows and people: A literature survey, psychological reaction with and without windows. National Bureau of Standards, *Building Science Series* 70.

Energy, Mines and Resources Canada. 1987. Enermodal Engineering Ltd.—The effect of frame design on window heat loss phase 1. Report prepared for Renewable Energy Branch, Ottawa.

Johnson, B. 1985. Heat transfer through windows. Swedish Council for Building Research, Stockholm.

Lawrence Berkeley Laboratory. 1987. *WINDOW 3.0 users guide.* Windows and Daylighting Group, Lawrence Berkeley Laboratory, Berkeley, CA.

Iqbal, M. 1983. *An introduction to solar radiation*, p. 389. Academic Press, New York.

McCluney, W.R. 1968. Radiometry and photometry. *American Journal of Physics* 36:977-79.

Meyer-Arendt, J.R. 1968. Radiometry and photometry: Units and conversion factors. *Applied Optics* 7:2081-84.

Moore, G.L. and C.W. Pennington. 1967. Measurement and application of solar properties of drapery shading materials. *ASHRAE Transactions* 73(1):3.1-15.

Nicodemus, F.E. 1963. Radiance. *American Journal of Physics* 31:368.

Nicodemus, F.E. et al. 1976-84. Self-study manual on optical radiation measurements, Part I—Concepts. NBS *Technical Notes* 910-1, 910-2, 910-3, and 910-4. National Bureau of Standards (now National Institute of Standards and Technology, Gaithersburg, MD).

Nicodemus, F.E. et al. 1977. Geometrical considerations and nomenclature for reflectance. NBS *Monograph* 160, NBS (now NIST), October.

Pennington, C.W. and D.E. McDuffie, Jr. 1970. Effect of inner surface air velocity and temperature upon heat gain and loss through glass fenestration. *ASHRAE Transactions* 76:190.

Pierpoint, W. and J. Hopkins. 1982. The derivation of a new area source equation. Presented at the Annual Conference of the Illuminating Engineering Society, Atlanta, August.

Threlkeld, J.L. 1962. *Thermal environmental engineering*, p. 321. Prentice Hall, New York.

Van Dyke, R.L. and T.P. Konen. 1982. Energy conservation through interior shading of windows: An analysis, test and evaluation of reflective venetian blinds. LBL-14369. Lawrence Berkeley Laboratory, March.

Yellott, J.I. 1965. Drapery fabrics and their effectiveness in sun control. *ASHRAE Transactions* 71(l):260-72.

CHAPTER 31

ENERGY ESTIMATING AND MODELING METHODS

GENERAL CONSIDERATIONS

THE ENERGY requirements and fuel consumption of HVAC systems have a direct impact on the cost of operating a building and an indirect impact on the environment. This chapter discusses methods for estimating energy use for two purposes: modeling for the design of buildings and HVAC systems and associated design optimization (**forward modeling**); and modeling the energy use of existing buildings for establishing baselines and calculating retrofit savings (**inverse modeling**).

FORWARD AND INVERSE MODELS

A mathematical **model** is a description of the behavior of a system. It is made up of three components (Beck and Arnold 1977):

1. **Input variables** (statisticians call these regressor variables while physicists refer to them as forcing variables), which act on the system. Note that there are two types of such variables: controllable by the experimenter, and such uncontrollable variables as climate.
2. **System structure and parameters/properties**, which provide the necessary physical description of the system (for example, thermal mass or mechanical properties of the elements).
3. **Output** (or response or dependent) variables, which describe the reaction of the system to the input variables. Energy use is often a response variable.

The science of mathematical modeling as applied to physical systems involves determining the third component of a system when the other two components are given or specified. We can broadly differentiate between two distinct categories of modeling, the choice of which is dictated essentially by the objective or purpose behind the investigation (Rabl 1988).

Forward or Classical Approach. The objective is to predict the output variables of a specified model with known structure and known parameters when subject to specified input variables. In order to ensure accuracy of prediction, the models have tended to become increasingly complex, especially with the advent of cheap and powerful computing power. This approach presumes detailed knowledge not only of the various natural phenomena affecting system behavior but also of the magnitude of various interactions (e.g., effective thermal mass, heat and mass transfer coefficients, etc.). The main advantage of this approach is that the system need not be physically built in order to predict its behavior. Thus, this approach is ideal in the preliminary design and analysis stage and is most often employed as such.

Forward modeling as applied to building energy use begins with a physical description of the building system or component of interest. For example, we define the building geometry, geographical location, physical characteristics (such as wall material and thickness), type of equipment and operating schedules, type of HVAC system, building operating schedules, plant equipment, etc. The peak and average energy use of such a building can then be predicted or simulated by the forward simulation model. The primary benefits of this method are that it is based on sound engineering principles usually taught in colleges and universities and consequently has gained widespread acceptance by the design and professional community. Major government-developed simulation codes, such as BLAST, DOE-2, and EnergyPlus, are based on forward simulation models. Figure 1 is a flow chart that illustrates the ordering of the analysis that is typically performed by a building energy simulation program.

Inverse or Data-Driven Approach. In this case, the input and output variables are known and measured, and the objective is to determine a mathematical description of the system and to estimate the system parameters. In contrast to the forward approach, the inverse approach is relevant to the case when the system has already been built and actual performance data are available for model development and/or identification. Two types of performance data can be used: nonintrusive and intrusive. **Intrusive data** are gathered under conditions of certain predetermined or planned experiments on the system in order to elicit system response under a wider range of system performance than would have occurred under normal system operation. Such performance data allow for more accurate model specification and identification. When constraints on system operation do not permit such tests to be performed, the model must be identified from **nonintrusive data** obtained under normal operation.

The inverse modeling approach often allows identification of system models that are not only simpler to use but that are more accurate predictors of future system performance than forward models. The inverse approach arises in many fields, such as physics, biology, engineering, and economics. Although several monographs, textbooks, and even specialized technical journals are available in this area, the approach has not been widely adopted in energy-related curricula and has yet to diffuse in a significant and pervasive manner (as has the forward approach) into the building professional community.

The preparation of this chapter is assigned to TC 4.7, Energy Calculations.

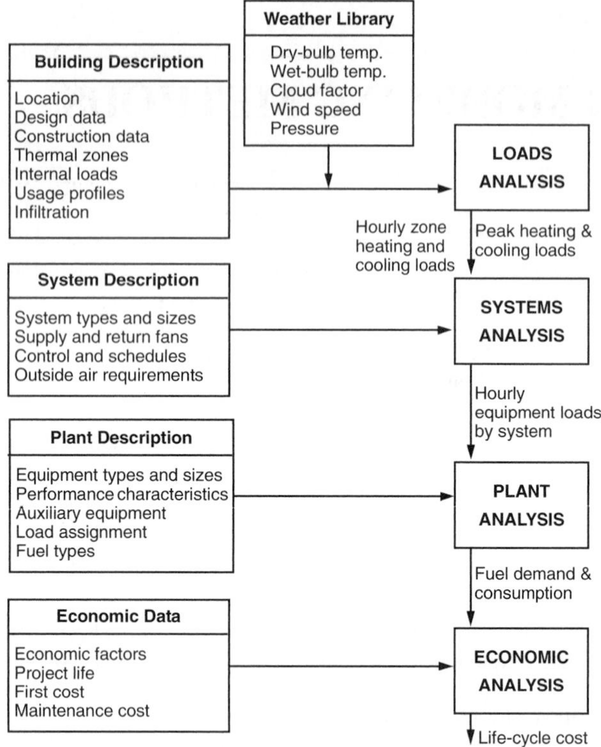

Fig. 1 Flow Chart for Building Energy Simulation Program
(Ayres and Stamper 1995)

CHARACTERISTICS OF MODELS

Forward Models

Although the procedures for estimating energy requirements vary considerably in their degree of complexity, they all have three common elements: the calculation of (1) space load, (2) secondary equipment load, and (3) primary equipment energy requirements. Here, secondary refers to equipment that distributes the heating, cooling, or ventilating medium to conditioned spaces, while primary refers to central plant equipment that converts fuel or electric energy to heating or cooling effect. A major distinction is made between **steady-state** methods (based on degree-days or temperature bins) and **dynamic** methods (e.g., based on transfer functions).

The first step in calculating energy requirements is to determine the **space load,** which is the amount of energy that must be added to or extracted from a space to maintain thermal comfort. The simplest procedures assume that the energy required to maintain comfort is only a function of the outdoor dry-bulb temperature. More detailed methods consider solar effects, internal gains, heat storage in the walls and interiors, and the effects of wind on both building envelope heat transfer and infiltration. Chapters 28 and 29 discuss load calculation in detail.

While energy calculations are similar to the heating and cooling load calculations used to size equipment, they are not the same. Energy calculations are based on average use and typical weather conditions rather than on maximum use and worst-case weather. Currently, the most sophisticated procedures are based on hourly profiles for climatic conditions and operational characteristics for a number of typical days of the year or on 8760 h of operation per year.

The second step translates the space load to a **load on the secondary equipment**. This can be a simple estimate of duct or piping losses or gains or a complex hour-by-hour simulation of an air system, such as variable air volume with outdoor-air cooling. This step must include the calculation of all forms of energy required by the secondary system (i.e., electrical energy to operate fans and/or pumps, as well as energy in the form of heated or chilled water).

The third step calculates the fuel and **energy required by the primary equipment** to meet these loads and the peak demand on the utility system. It considers equipment efficiencies and part-load characteristics. It is often necessary to keep track of the different forms of energy, such as electrical, natural gas, or oil. In some cases, where calculations are required to assure compliance with codes or standards, these energies must be converted to source energy or resource consumed, as opposed to energy delivered to the building boundary.

Often, energy calculations lead to an economic analysis to establish the cost-effectiveness of conservation measures (ASHRAE *Standard* 90.1). Thus, thorough energy analysis provides intermediate data, such as time of energy usage and maximum demand, so that utility charges can be accurately estimated. Although not part of the energy calculations, estimated capital equipment costs should be included in such an analysis.

Complex and often unexpected interactions can occur between the systems or between various modes of heat transfer. For example, radiant heating panels impact the space loads by raising the mean radiant temperature in the space (Howell and Suryanarayana 1990). As a result, the air temperature can be lowered while maintaining comfort. Compared to a conventional heated air system, radiant panels create a greater temperature difference from the inside surface to the outside air. Thus, conduction losses through the walls and roof increase because the inside surface temperatures are greater. At the same time, the heating load due to infiltration or ventilation decreases because of the reduced indoor air to outdoor air temperature difference. The infiltration rate may also decrease because the reduced air temperature difference reduces the stack effect.

Inverse Models

The inverse model has to meet requirements very different from the forward model. The inverse model can only contain a relatively small number of parameters because of the limited and often repetitive information contained in the performance data. (For example, building operation from one day to the next is fairly repetitive.) The inverse model is thus a much simpler model that contains fewer terms representative of aggregated or macroscopic parameters (e.g., overall building heat loss coefficient and time constants). Since the model parameters are deduced from actual building performance, it is much more likely to accurately capture the as-built system performance, thus allowing more accurate prediction of future system behavior under certain specific circumstances. Performance data collection and model formulation need to be appropriately tailored for the specific circumstance, which often requires a higher level of skill and expertise of the user. In general, inverse models are less flexible than forward models in evaluating energy implications of different design and operational alternatives, and so they are not a substitute in this regard.

To better understand the uses of inverse models, consider some of the questions that a building professional may ask about an existing building whose energy consumption is known (Rabl 1988):

- How does the consumption compare with design predictions (and, in case of discrepancies, are they due to anomalous weather, to unintended building operation, to improper operation, or to other causes)?
- How would the consumption change if the thermostat settings, ventilation rates, or indoor lighting levels were changed?
- How much energy could be saved by retrofits to the building shell, changes to air handler operation from CV to VAV, or changes in the various control settings?
- If the retrofits are implemented, can one verify that the savings are due to the retrofit and not to other causes (e.g., the weather)?
- How can one detect faults in HVAC equipment and optimize control and operation?

All the above questions are better addressed by the inverse approach. The forward approach could also be used, for example, by going back to the blueprints of the building and of the HVAC system, and repeating the analysis performed at the design stage while using actual building schedules and operating modes. This, however, is tedious and labor-intensive, and materials and equipment often perform differently under as-built conditions than as specified. Tuning the forward simulation model is often awkward and labor intensive, although it is still an option (as adopted in the calibrated inverse approach).

CHOOSING AN ANALYSIS METHOD

The most important step in selecting an energy analysis method is to match the method capabilities with project requirements. The method must be capable of evaluating all design options with sufficient accuracy to make correct choices. The following factors apply generally (Sonderegger 1985):

- **Accuracy.** The method should be sufficiently accurate to allow correct choices. Because of the many parameters involved in energy estimation, absolutely accurate energy prediction is not possible (Waltz 1992).
- **Sensitivity.** The method should be sensitive to the design options being considered. The difference in energy use between two choices should be accurate.
- **Versatility.** The method should allow the analysis of all options under consideration. When different methods must be used to consider different options, an accurate estimate of the differential energy use cannot be made.
- **Speed and cost.** The total time (gathering data, preparing input, calculations, and analysis of output) to make an analysis should be appropriate to the potential benefits gained. With greater speed, more options can be considered in a given time. The cost of analysis is largely determined by the total time of analysis.
- **Reproducibility.** The method should not allow so many vaguely defined choices that different analysts would get completely different results (Corson 1992).
- **Ease of use.** This impacts both on the economics of analysis (speed) and the reproducibility of results.

ASHRAE *Standard* 140, Method of Test for the Evaluation of Building Energy Analysis Computer Programs, has been developed to identify and diagnose differences in predictions that may possibly be caused by algorithmic differences, modeling limitations, or coding errors.

Selecting Energy Analysis Computer Programs

The selection of a building energy analysis program depends on its application, the number of times it will be used, the experience of the user, and the hardware available to run it. The first criterion is the capability of the program to deal with the application. For example, if the effect of a shading device is to be analyzed on a building that will also be shaded by other buildings part of the time, the capability of analyzing detached shading is an absolute requirement, regardless of any other factors.

Because almost all manual methods are now implemented on a computer, the selection of an energy analysis method is the selection of a computer program. Today, all well-known programs run on microcomputers. However, the cost of the computer facilities and the software itself are typically a small part of running a building energy analysis. The major costs are the cost of learning to use the program and the cost of using it. Major issues that influence the cost of learning a program include (1) complexity of the input procedures, (2) quality of the user's manual, and (3) availability of a good support system to answer questions. As the user becomes more experienced, the cost of learning becomes less important, but the need to obtain and enter a complex set of input data will continue to consume the time of even an experienced user until such data are readily available in electronic form compatible with simulation programs.

The **complexity of input** is largely influenced by the availability of default values for the input variables. Default values can be used as a simple set of input data when detail is not needed or when the building design is very conventional, but additional complexity can be supplied when needed. Secondary defaults, which can be supplied by the user, are also useful in the same way. Some programs allow the user to specify a level of detail. Then the program requests only the information appropriate to that level of detail, using default values for all others.

The **quality of the output** is another factor to consider. Reports should be easy to read and uncluttered. The titles and headings should be unambiguous. Units should be stated explicitly. The user's manual should explain the meanings of the data presented. Graphic output can be very helpful. In most cases, simple summaries of overall results are the most useful, but very detailed output is needed for certain studies and also for debugging program input during the early stages of an analysis.

Before a final decision is made, manuals for the most suitable programs should be obtained and reviewed, and, if possible, demonstration versions of the programs should be obtained and run. During this last part of the selection process, support from the software supplier should be tested. The availability of training should be considered when choosing a more complex program.

The **availability of weather data** and the availability of a weather data processing subroutine or program are major features of a program. Some programs include subroutine or supplementary programs that allow the user to create a weather file for any site for which weather data is available. Programs that do not have this capability must have weather files for various sites created by the program supplier. In that case, the available weather data and the terms on which the supplier will create new weather data files must be checked.

Auxiliary capabilities, such as economic analysis and design calculations, are a final concern in selecting a program. An economic analysis may include only the ability to calculate annual energy bills from utility rates, or it might extend to calculations or even to life-cycle cost optimization. An integrated program may save time because some input data will have been entered already for other purposes.

The results of computer calculations should be accepted with caution, as the software vendor does not accept responsibility for the correctness of calculations or the use of the program. A manual calculation should be run to develop a good understanding of the underlying physical processes and building behavior. In addition, the user should (1) review the computer program documentation to determine what calculation procedures are used, (2) compare the results with manual calculations and measured data, and (3) conduct sample tests to confirm that the program delivers acceptable results.

Tools for Energy Analysis

The most accurate methods for calculating building energy consumption are the most costly because of their intense computational requirements and the needed expertise on the part of the designer or analyst. Simulation programs that assemble component models into system models and then exercise those models with weather and occupancy data are preferred by experts for determining energy use in buildings. This chapter provides descriptions of methods to model heating and cooling loads, including heat flow through building foundations, and methods to model the performance of secondary and primary HVAC equipment. The chapter continues with information about system-level modeling.

Often, energy consumption at a system or whole-building level must be estimated quickly to study trends, compare systems, or study building effects such as envelope characteristics. For these purposes, simpler methods may be used. Degree-day and bin methods are two

Table 1 Classification of Analysis Methods For Building Energy Use

Method	Forward	Inverse: Empirical or Black-Box	Inverse: Calibrated Simulation	Inverse: Physical or Gray-Box	Comments
Steady-State Methods					
Simple linear regression (Kissock et al. 1998, Ruch and Claridge 1991)	—	X	—	—	One dependent parameter, one independent parameter. May have slope and *y*-intercept.
Multiple linear regression (Dhar 1995, Dhar et al. 1998, 1999a,b, Katipamula et al. 1998, Sonderegger 1998)	—	X	—	—	One dependent parameter, multiple independent parameters.
Modified degree-day method	X	—	—	—	Based on fixed reference temperature of 65°F.
Variable-base degree-day method, or 3-P change point models (Fels 1986, Reddy et al. 1997, Sonderegger 1998)	X	X	—	X	Variable base reference temperatures.
Change point models: 4-P, 5-P (Fels 1986, Kissock et al. 1998)	—	X	—	X	Uses daily or monthly utility billing data and average period temperatures.
ASHRAE bin method and inverse bin method (Thamilseran and Haberl 1995)	X	X	—	—	Hours in temperature bin times load for that bin.
ASHRAE TC 4.7 modified bin method (Knebel 1983)	X	—	—	—	Modified bin method with cooling load factors.
Multistep parameter identification (Reddy et al. 1999)	—	—	—	X	Uses daily data to determine overall heat loss and ventilation of large buildings.
Dynamic Methods					
Thermal network (Sonderegger 1977, Rabl 1988, Reddy 1989)	X	—	—	X	Uses equivalent thermal parameters (inverse mode).
Response factors (Stephenson and Mitalas 1967, Mitalas and Stephenson 1967, Mitalas 1968, Kusuda 1969)	X	—	—	—	Tabulated or as used in simulation programs.
Fourier analysis (Shurcliff 1984, Subbarao 1988)	X	—	X	X	Frequency domain analysis convertible to time domain.
ARMA model (Subbarao 1986, Rabl 1988, Reddy 1989)	—	—	—	X	Autoregressive Moving Average model.
PSTAR (Subbarao 1988)	X	—	X	X	Combination of ARMA and Fourier series, includes loads in time domain.
Modal analysis (Bacot et al. 1984, Rabl 1988)	X	—	—	X	Building described by diagonalized differential equation using nodes.
Differential equation (Rabl 1988)	—	—	—	X	Analytical linear differential equation.
Computer simulation: DOE-2, BLAST (Norford et al. 1994, Haberl and Bou-Saada 1998, Manke et al. 1996)	X	—	X	—	Hourly simulation programs with system models.
Computer emulation (HVACSIM+, TRNSYS) (Clark 1985, Klein et al. 1994)	X	—	—	—	Subhourly simulation programs.
Artificial neural networks (Kreider and Wang 1991, Kreider and Haberl 1994)	—	X	—	—	Connectionist models.

simpler approaches described in the chapter. Finally, the chapter addresses inverse models in detail.

Table 1 classifies methods for analyzing building energy use as forward or inverse and steady-state or dynamic. The U.S. Department of Energy maintains an up-to-date listing of building energy software with links to other sites that describe energy modeling tools: www.eren.doe.gov/buildings/tools_directory/

COMPONENT MODELING AND LOADS

CALCULATING SPACE SENSIBLE LOADS

Calculating instantaneous space sensible load is a key step in any building energy simulation. The **heat balance method** and the **weighting factor method** are two methods used for these calculations. A third method, the **thermal network method**, while not widely used, shows promise.

The **instantaneous space sensible load** is the rate of heat flow into the space air mass. This quantity, sometimes called the cooling load, differs from heat gain in that heat gain usually contains a radiative component that passes through the air and is absorbed by other bounding surfaces. Instantaneous space sensible load is entirely convective; even loads from internal equipment, lights, and occupants enter the air by convection from the surface of such objects or by convection from room surfaces that have absorbed the radiant component of energy emitted from these sources. However, some adjustment must be made when radiant cooling and heating systems are evaluated because some of the space load is offset directly by radiant transfer without convective transfer to the air mass.

For equilibrium, the instantaneous space sensible load must match the heat removal rate of the conditioning equipment. Any imbalance in these rates changes the energy stored in the air mass. Customarily, however, the thermal mass (heat capacity) of the air itself is ignored in an analysis, so that the air is always assumed to be in thermal equilibrium. Under these assumptions, the instantaneous space sensible load and the rate of heat removal are equal in magnitude and opposite in sign.

The weighting factor method and the heat balance method use conduction transfer functions (or their equivalents) to calculate transmission heat gain or loss. The principal difference is in the methods used to calculate the subsequent internal heat transfers to the room. Experience with both methods has indicated largely the same results, provided the weighting factors are determined for the specific building under analysis.

Heat Balance Method

The heat balance method for calculating net space sensible loads is more fundamental than the weighting factor method. Its development relies on the first law of thermodynamics (conservation of energy) and the principles of matrix algebra. Because it requires fewer assumptions than the weighting factor method, it is also more flexible. However, the heat balance method requires more calculations at each point in the simulation process, using more computer time. The weighting factors used in the weighting factor method are determined with a heat balance procedure. Although not necessary, linearization is commonly used to simplify the radiative transfer formulation.

The heat balance method allows the net instantaneous sensible heating and/or cooling load to be calculated on the space air mass. Generally, a heat balance equation is written for each enclosing surface, plus one equation for room air. This set of equations can then be solved for the unknown surface and air temperatures. Once these temperatures are known, they can be used to calculate the convective heat flow to or from the space air mass. The heat balance method is developed in Chapter 29 for use in design cooling load calculations, so a fuller description is omitted here. However, one fundamental difference is that the heat balance procedure described in Chapter 29 is aimed at obtaining the design cooling load for a fixed zone air temperature. For building energy analysis purposes, it is preferable to know the actual heat extraction rate.

This may be determined by recasting Equation (36) of Chapter 29 so that the system heat transfer is determined simultaneously with the zone air temperature. The system heat transfer is the rate at which heat is transferred to the space by the system. Although this can be done by simultaneously modeling the zone and the system (Taylor et al. 1990, 1991), it is convenient to make a simple, piece wise-linear representation of the system known as a control profile. This usually takes the form

$$q_{sys_j} = a + bt_{a_j} \tag{1}$$

where

q_{sys_j} = system heat transfer at time step j, Btu/h
a, b = coefficients that apply over a certain range of zone air temperatures
t_{a_j} = zone air temperature at time step j, °F

The system heat transfer q_{sys_j} may be considered positive when heating is provided to the space and negative when cooling is provided. It is equal in magnitude but opposite in sign to the zone cooling load, as defined in Chapter 29, when the zone air temperature is fixed.

Substituting Equation (1) into Equation (36) of Chapter 29 and solving for the zone air temperature,

$$t_{a_j} = \frac{a + \sum_{i=1}^{N} A_i h_{ci} t_{si_{i,j}} + \rho c V_{infil_j} t_{o_j} + \rho c V_{vent_j} t_{v_j} + q_{c,int_j}}{-b + \sum_{i=1}^{N} A_i h_{ci} + \rho c V_{infil_j} + \rho c V_{vent_j}} \tag{2}$$

where

t_{a_j} = zone air temperature at time step j, °F
N = number of zone surfaces
A_i = area of ith surface, ft²
h_{ci} = convection coefficient for ith surface
$t_{si_{i,j}}$ = surface temperature for ith surface at time step j, °F
ρ = density, lb_m/ft^3
c = specific heat of air, $Btu/lb_m \cdot °F$
V = volumetric flow rate of air, ft^3/h
t_{o_j} = outdoor air temperature at time step j, °F
t_{v_j} = ventilation air temperature at time step j, °F
q_{c,int_j} = sum of the convective portions of all internal heat gains at time step j, Btu/h

The zone air heat balance equation [Equation (2)] must be solved simultaneously with the interior and exterior surface heat balance equations [Equations (35) and (34)] described in Chapter 29. Also, it is necessary to determine the correct temperature range to use the proper set of a and b coefficients. This may be done iteratively. Once the zone air temperature is found, the actual system heat transfer rate may be found directly from Equation (1).

Beyond the treatment of the system heat transfer, there are a few other considerations that may be important in building energy analysis programs. These include simulations over periods as long as a year, treatment of radiant cooling and heating systems, treatment of interzone heat transfer, modeling of convection heat transfer, and modeling of radiation heat transfer.

The heat balance method presented in Chapter 29 assumes the use of a single design day. When utilized in a building energy analysis program, it is most commonly used with a year's worth of design weather data. In this case, the first day of the year is usually simulated several times until a steady-periodic response is obtained. Then, each day is simulated sequentially, and, where needed, historical data for surface temperatures and heat fluxes from the previous day are utilized.

When radiant cooling and heating systems are evaluated, the radiant source should be identified as a room surface. The calculation procedure considers the radiant source in the heat balance analysis. Therefore, the heat balance method is preferred over the weighting factor method for evaluating radiant systems. Strand and Pedersen (1997) have described the implementation of heat-source conduction transfer functions, which may be used for modeling of radiant panels, into a heat balance-based building simulation program.

In principle, the heat balance method described above extends directly to multiple spaces, with heat transfer between zones. In this case, some surface temperatures will appear in the surface heat balance equations for two different zones. In practice, however, the size of the coefficient array required for solving the simultaneous equations becomes prohibitively large, and the solution time excessive. For this reason, many programs solve only one space at a time and assume that the adjacent space temperatures are either the same as the space in question or some assigned, constant value. Other approaches may remove this limitation (Walton 1980).

Relatively simple exterior and interior convection models may be used for design cooling load calculation procedures. However, it may be desirable to use more sophisticated exterior convection models (Cooper and Tree 1973, Fracastoro et al. 1982, Melo and Hammond 1991, Walton 1983, Yazdanian and Klems 1993) that incorporate the effects of wind speed, wind direction, surface orientation, etc. Likewise, a number of more detailed interior convection correlations have been published for use in buildings (Alamdari and Hammond 1982, 1983; Altmayer et al. 1983; Bauman et al. 1983; Bohn et al. 1984; Chandra and Kerestecioglu 1984; Khalifa and Marshall 1990; Spitler et al. 1991; Walton 1983).

Also, more detailed models of exterior long-wave radiation transfer [e.g., Cole (1976) and Walton (1983)] and interior long-wave radiation transfer have been implemented in detailed building simulation programs. See Carroll (1980), Davies (1988), Kamal and Novak (1991), Steinman et al. (1989) and Walton (1980) for further discussion of interior radiation heat transfer models.

Weighting Factor Method

The weighting factor method of calculating instantaneous space sensible load represents a compromise between simpler methods, such as a steady-state calculation, that ignore the ability of building mass to store energy, and more complex methods, such as complete energy balance calculations. With this method, space heat gains at constant space temperature are determined from a physical description of the building, ambient weather conditions, and internal load profiles. Along with the characteristics and availability of heating and cooling systems for the building, space heat gains are used to calculate air temperatures and heat extraction rates. This discussion is in terms of heat gains, cooling loads, and heat extraction rates. Heat losses, heating loads, and heat addition rates are merely different terms for the same quantities, depending on the direction of the heat flow.

The weighting factors represent *Z*-transfer functions (York and Cappiello 1982, Kerrisk et al. 1981). The *Z*-transform is a method for solving differential equations with discrete data. Two groups of weighting factors are used: heat gain and air temperature.

Heat gain weighting factors represent transfer functions that relate space cooling load to instantaneous heat gains. A set of weighting factors is calculated for each group of heat sources that differ significantly in (1) the relative amounts of energy appearing as convection to the air versus radiation, and (2) in the distribution of radiant energy intensities on different surfaces.

Air temperature weighting factors represent a transfer function that relates room air temperature to the net energy load of the room. The weighting factors for a particular heat source are determined by introducing a unit pulse of energy from that source into the room's network. The network is a set of equations that represents a heat balance for the room. At each time step (1 h intervals), including the initial introduction, the energy flow to the room air represents the amount of the pulse that becomes a cooling load. Thus, a long sequence of cooling loads can be generated from which the weighting factors are calculated. Similarly, a unit pulse change in room air temperature can be used to produce a sequence of cooling loads.

A two-step process is used to determine the air temperature and heat extraction rate of a room or building zone for a given set of conditions. First, the room air temperature is assumed to be fixed at some reference value. This reference temperature is usually chosen as the average air temperature expected for the room over the simulation period. Instantaneous heat gains are calculated based on this constant air temperature. Various types of heat gains are considered. Some, such as solar energy entering through windows or energy from lighting, people, or equipment, are independent of the reference temperature. Others, such as conduction through walls, depend directly on the reference temperature.

A space sensible cooling load for the room, defined as the rate at which energy must be removed from the room to maintain the reference value of the air temperature, is calculated for each type of instantaneous heat gain. The cooling load generally differs from the instantaneous heat gain because some energy from the heat gain is absorbed by walls or furniture and stored for later release to the air. At an hour θ, the calculation uses present and past values of the instantaneous heat gain (q_θ, $q_{\theta-1}$), past values of the cooling load ($Q_{\theta-1}$, $Q_{\theta-2}$, ...) and the **heat gain weighting factors** (v_0, v_1, v_2, ..., w_1, w_2, ...) for the type of heat gain under consideration. Thus, for each type of heat gain q_θ, the cooling load Q_θ is calculated as

$$Q_\theta = v_0 q_\theta + v_1 q_{\theta-1} + \ldots - w_1 Q_{\theta-1} - w_2 Q_{\theta-2\ldots} \quad (3)$$

The heat gain weighting factors are a set of parameters that quantitatively determine how much of the energy entering a room is stored and how rapidly the stored energy is released during later hours. Mathematically, the weighting factors are parameters in a *Z*-transfer function relating the heat gain to the cooling load.

These weighting factors differ for different heat gain sources because the relative amounts of convective and radiative energy leaving various sources differ and because the distribution of radiative energy can differ. The heat gain weighting factors also differ for different rooms because room construction influences the amount of incoming energy stored by walls or furniture and the rate at which it is released. Sowell (1988) showed the effects of 14 zone design parameters on zone dynamic response. After the first step, the cooling loads from various heat gains are added to give a total cooling load for the room.

In the second step, the total cooling load is used—along with information on the HVAC system attached to the room and a set of **air temperature weighting factors**—to calculate the actual heat extraction rate and air temperature. The actual heat extraction rate differs from the cooling load (1) because, in practice, the air temperature can vary from the reference value used to calculate the cooling load or (2) because of HVAC system characteristics. The deviation of the air temperature t_θ from the reference value at hour θ is calculated as

$$t_\theta = 1/g_0 + [(Q_\theta - \mathrm{ER}_\theta) + P_1(Q_{\theta-1} - \mathrm{ER}_{\theta-1}) + P_2(Q_{\theta-2} - \mathrm{ER}_{\theta-2}) + \ldots - g_1 t_{\theta-1} - g_2 t_{\theta-2} - \ldots] \quad (4)$$

where ER_θ is the energy removal rate of the HVAC system at hour θ, and g_0, g_1, g_2, ..., P_1, P_2, ... are the air temperature weighting factors, which incorporate information about the room, particularly the thermal coupling between the air and the storage capacity of the massive elements.

Tables of values of weighting factors for typical building rooms are presented in the table below. One of the three groups of weighting factors, for light, medium, and heavy construction rooms, can be used to approximate the behavior of any room. Some automated simulation techniques allow weighting factors to be calculated specifically for the building under consideration. This option improves the accuracy of the calculated results, particularly for a building with an unconventional design. McQuiston and Spitler (1992) provided electronic tables of weighting factors for a large number of parametrically defined zones.

Normalized Coefficients of Space Air Transfer Functions

Room Envelope Construction	g_0^*	g_1^*	g_2^*	p_0	p_1
	Btu/h·ft·°F			Dimensionless	
Light	1.68	−1.73	0.05	1.0	−0.82
Medium	1.81	−1.89	0.08	1.0	−0.87
Heavy	1.85	−1.95	0.10	1.0	−0.93

Two assumptions are made in the weighting factor method. First, the processes modeled are linear. This assumption is necessary because heat gains from various sources are calculated independently and summed to obtain the overall result (i.e., the superposition principle is used). Therefore, nonlinear processes such as radiation or natural convection must be approximated linearly. This assumption does not represent a significant limitation because these processes can be linearly approximated with sufficient accuracy for most calculations. The second assumption is that system properties influencing the weighting factors are constant (i.e., they are not functions of time). This assumption is necessary because only one set of weighting factors is used during the entire simulation period. This assumption can limit the use of weighting factors in situations where important room properties vary during the calculation. Two examples are the distribution of solar radiation incident on the interior walls of a room, which can vary hourly, and inside surface heat transfer coefficients.

When the weighting factor method is used, a combined radiative-convective heat transfer coefficient is used as the inside surface heat transfer coefficient. This value is assumed constant even though in a real room (1) the radiant heat transferred from a surface depends on the temperature of other room surfaces (not on the room air temperature) and (2) the combined heat transfer coefficient is not constant. Under these circumstances, an average value of the property must be used to determine the weighting factors. Cumali et al. (1979) have investigated extensions to the weighting factor method to eliminate this limitation.

Thermal Network Methods

Although implementations of the thermal network method vary, they all have in common the discretization of the building into a network of nodes, with interconnecting paths through which energy flows. In many respects, thermal network models may be considered a refinement of the heat balance method. Where the heat balance model generally uses one node for zone air, the thermal network method might use multiple nodes. For each heat transfer element (wall, roof, floor, etc.), the heat balance model generally has one interior surface node and one exterior surface node; the thermal network model may include additional nodes. Heat balance models generally use simple methods for distributing radiation from lights; thermal network models may model the lamp, ballast, and luminaire housing separately. Furthermore, thermal network models depend on a heat balance at each node to determine the node temperature and the energy flow between all connected nodes. The energy flows may include conduction, convection, short-wave radiation, and long-wave radiation.

For any mode of energy flow, a range of techniques may be used to model the energy flow between two nodes. Taking conduction heat transfer as an example, the simplest thermal network model would be a resistance-capacitance network (Sowell 1990). By refining the network discretization, the models become what are commonly thought of as finite difference or finite volume models (Clarke 1985, Lewis and Alexander 1990, Walton 1993).

Thermal network models generally use a set of algebraic and differential equations. In most implementations, the solution procedure is separated from the models so that, in theory, different solvers might be used to perform the simulation. In contrast, most heat balance programs and weighting factor programs interweave the solution technique with the models. Various solution techniques have been used in conjunction with thermal network models. Examples include graph theory combined with Newton-Raphson and predictor-corrector ordinary differential equation integration (Buhl et al. 1990) and the use of Euler explicit integration combined with sparse matrix techniques (Walton 1993).

Of the three zone models discussed, thermal network models are the most flexible and have the greatest potential for high accuracy. As a trade-off, they also require the most computation time, and, in current implementations, they require more user effort to take advantage of the flexibility.

GROUND HEAT TRANSFER

The thermal performance of building foundations, including guidelines for placement of insulation, is described in Chapter 24 of this volume and Chapter 42 of the 1999 *ASHRAE Handbook—Applications*. Chapter 28 provides information needed to calculate transmission heat losses through slab foundations and through basement walls and floors. These calculations are appropriate for design loads but are not intended for estimating annual energy usage. This section provides simplified calculation methods suitable for energy estimates over time periods of arbitrary length.

The thermal performance of building foundations has been largely ignored. It is estimated that in the early 1970s, only 10% of the total energy use of a typical U.S. home was attributed to the heat transfer from its foundation (Labs et al. 1988). Since then, the thermal performance of above-grade building elements has improved significantly, and the contribution of ground-coupled heat transfer to total energy use in a typical U.S. home has increased. Shipp and Broderick (1983) estimated that the heat transfer from an uninsulated basement in Columbus, Ohio, can represent up to 67% of the total building envelope heating load.

Earth-contact heat transfer, rated at 1 to 3 quadrillion Btu of energy annually in U.S. buildings, has an impact similar to infiltration on annual heating and cooling loads in residential buildings (Claridge 1988a). Adding insulation to building foundations is estimated to save up to 0.5 quadrillion Btu of annual energy use in the U.S. (Labs et al. 1988).

Simplified Calculation Method for Slab Foundations and for Basements

A simplified design tool for calculating heat loss for slabs and basements is presented by modifying the design tool for slab-on-grade floors developed by Krarti and Chuangchid (1999). The proposed design tool is easy to use and requires straightforward input parameters with continuously variable values, including foundation size, insulation R-values, soil thermal properties, and indoor and outdoor temperatures. The simplified method provides a set of equations that are suitable for estimating the design, seasonal, and annual total heat loss of a slab or a basement as a function of a wide range of variables.

When the indoor temperature of the building is maintained constant, the ground-coupled heat transfer $q(\theta)$ varies with time according to the following equation:

$$q(\theta) = q_{mean} + q_{amp}\sin(\omega\theta + \phi) \tag{5}$$

where

q_{mean} = annual-mean heat loss/gain, Btu/h
q_{amp} = heat loss/gain amplitude, Btu/h
ϕ = phase lag between total slab heat loss/gain and soil surface temperature, s
ω = annual angular frequency ($\omega = 1.992 \times 10^{-7}$ rad/s)
θ = time, s

Equation (5) is convenient and flexible because it can be used to calculate the foundation heat loss/gain not only at any time but also at design conditions and for any time period (such as a heating season or 1 year). In particular, the design heat loss/gain load q_{des} for a slab foundation is obtained as follows:

$$q_{des} = q_{mean} + q_{amp} \tag{6}$$

The parameters q_{mean} and q_{amp} are functions of such variables as building dimensions, soil properties, and insulation R-values. Expressions developed by nondimensional analysis allow the calculation of q_{mean} and q_{amp}.

The soil conductivity is normalized to form four parameters—U_o, G, H, and D:

$$U_o = \frac{k_s}{(A/P)_{eff,\,b}} \tag{7}$$

where

k_s = soil thermal conductivity, Btu·ft/h·ft²·°F
P = slab perimeter, ft
A = slab area, ft²

For mean calculations,

$$(A/P)_{eff,\,b,\,mean} = [1 + b_{eff}(-0.4 + e^{-H_b})](A/P)_b \tag{8}$$

For annual calculations,

$$(A/P)_{eff,\,b,\,amp} = (1 + b_{eff}e^{-H_b})(A/P)_b \tag{9}$$

where

$$H_b = \frac{(A/P)_b}{k_s R_{eq}} \tag{10}$$

$$b_{eff} = \frac{B}{(A/P)_b} \tag{11}$$

B = basement depth, ft (0 ft for slab)

$$G = k_s R_{eq}\sqrt{\frac{\omega}{\alpha_s}} \tag{12}$$

where

R_{eq} = equivalent thermal resistance for the entire slab, ft²·h·°F/Btu
α_s = soil thermal diffusivity, ft²/s

For uniform insulation configurations (placed horizontally beneath the slab floor),

$$R_{eq} = R_f + R_i \tag{13}$$

where

R_f = thermal resistance of the floor, ft²·h·°F/Btu
R_i = thermal resistance of insulation, ft²·h·°F/Btu

For partial insulation configurations (both horizontal and vertical),

$$R_{eq} = \frac{R_f}{\left[1 - \left(\frac{c}{A/P}\frac{R_i}{(R_i + R_f)}\right)\right]} \tag{14}$$

where c = insulation length of slab, ft.

$$H = \frac{(A/P)_{eff,\,b}}{k_s R_{eq}} \tag{15}$$

$$D = \ln\left[(1 + H)\left(1 + \frac{1}{H}\right)^H\right] \tag{16}$$

The effective heat-transfer coefficients for mean heat flow $U_{eff,mean}$ and heat-flow amplitude $U_{eff,amp}$, Btu/h·ft²·°F, are

Table 2 Coefficients *m* and *a* for Slab-Foundation Heat Transfer Calculations

Insulation Placement	*m*	*a*
Uniform—Horizontal	0.40	0.25
Partial—Horizontal	0.34	0.20
Vertical	0.28	0.13

$$U_{eff,\,mean} = mU_oD \tag{17}$$

$$U_{eff,\,amp} = aU_oD^{0.16}G^{-0.6} \tag{18}$$

where the dimensionless coefficients m and a depend on the insulation placement configurations and are provided in Table 2.

The annual-mean slab foundation and basement heat loss/gain can now be defined as

$$q_{mean} = U_{eff,\,mean}A(t_a - t_r) \tag{19}$$

where

t_a = annual average ambient dry-bulb temperature, °F
t_r = annual average indoor dry-bulb temperature, °F

The heat loss/gain amplitude for slab foundations and basements is

$$q_{amp} = U_{eff,\,amp}At_{amp} \tag{20}$$

where t_{amp} = annual amplitude ambient temperature, °F.

This simplified model for slab-foundation and basement heat flows provides accurate predictions when A/P is larger than 1.5 ft. To illustrate the use of the simplified models, two examples are presented: one for a slab-on-grade floor for a building insulated with uniform horizontal insulation, and one for a basement structure insulated with uniform insulation.

Example 1. Calculation for Slab Foundations. Determine the annual mean and annual amplitude of total slab heat loss for the slab foundation illustrated in Figure 2. The building is located in Denver, Colorado.

Solution:

Step 1. Provide the required input data.

Dimensions
Slab width = 32.81 ft
Slab length = 49.22 ft
Ratio of slab area to slab perimeter, A/P = 9.84 ft

4 in. thick reinforced concrete slab, thermal resistance R_f = 2.84 h·ft²·°F/Btu

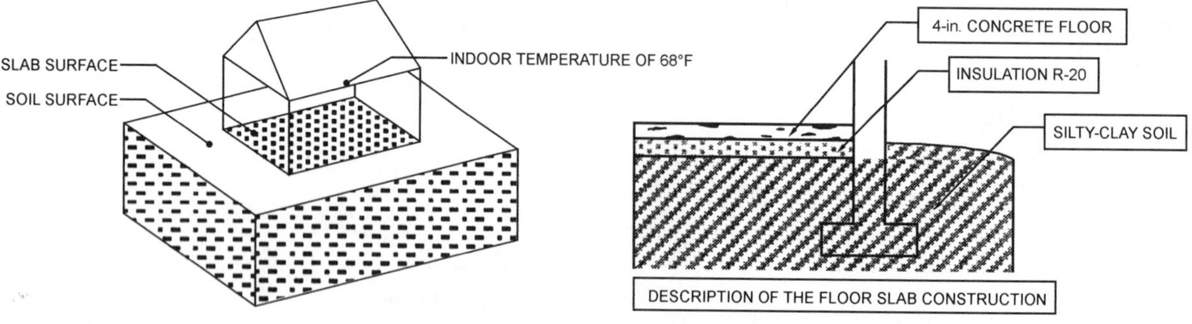

Fig. 2 Slab Foundation for Example 1

Soil Thermal Properties
Soil thermal conductivity k_s = 0.70 Btu/h·ft·°F
Soil density ρ = 168.56 lb_m/ft^3
Soil thermal diffusivity $\alpha_s = 6.43 \times 10^{-6}$ ft^2/s

Insulation
Uniform insulation R-value R_i = 20.0 h·ft²·°F/Btu

Temperatures
Indoor temperature t_r = 68°F
Annual average ambient temperature t_a = 43°F
Annual amplitude ambient temperature t_{amp} = 36°F
Annual angular frequency $\omega = 1.992 \times 10^{-7}$ rad/s

Step 2. Calculate q_{mean} and q_{amp} values.

The various normalized parameters are first calculated using Equations (7) through (18). Then, the annual mean and amplitude of the foundation slab heat loss/gain are determined using Equations (19) and (20).

$$U_o = \frac{k_s}{(A/P)} = \frac{0.70}{9.84} = 0.0711$$

$$H = \frac{(A/P)}{k_s R_{eq}} = \frac{9.84}{0.70(2.84 + 20)} = 0.6155$$

$$D = \ln\left[(1 + H)\left(1 + \frac{1}{H}\right)^H\right] = 1.0736$$

$$G = k_s R_{eq}\sqrt{\frac{\omega}{\alpha_s}} = 0.70(2.84 + 20)\sqrt{\frac{1.992\times10^{-7}}{6.43\times10^{-6}}} = 2.8141$$

Therefore,

$$q_{mean} = U_{eff,mean}A(t_r - t_a) = 0.40 \times 0.0711 \times 1.0736 \times 1615 \times (68 - 43) = 1233 \text{ Btu/h}$$

and

$$q_{amp} = U_{eff,amp}At_{amp} = 0.25 \times 0.0711 \times 1.0736^{0.16} \times 2.8141^{-0.6} \times 1615 \times 36 = 562 \text{ Btu/h}$$

Example 2. Calculation for Basements. Determine the annual mean and annual amplitude of total basement heat loss for a building located in Denver, Colorado.

Solution:

Step 1. Provide the required input data.

Dimensions
Basement width = 32.81 ft
Basement length = 49.22 ft
Basement wall height B = 4.92 ft
Basement slab and wall total area = 2422 ft²
Ratio of slab and wall area to slab and wall perimeter, $(A/P)_b$ = 11.91 ft
4 in. thick reinforced concrete slab, thermal resistance R_f = 2.84 h·ft²·°F/Btu

Soil Thermal Properties
Soil thermal conductivity k_s = 0.70 Btu·ft/h·ft²·°F
Soil thermal diffusivity $\alpha_s = 4.812 \times 10^{-6}$ ft^2/s

Insulation
Uniform insulation R-value R_i = 6.54 h·ft²·°F/Btu

Temperatures
Indoor temperature, t_r = 71.6°F
Annual average ambient temperature, t_a = 50°F
Annual amplitude ambient temperature, t_{amp} = 23°F
Annual angular frequency, $\omega = 1.992 \times 10^{-7}$ rad/s

Step 2. Calculate q_{mean} and q_{amp} values.

The normalized parameters are first calculated using Equations (7) through (18). Then, the annual mean and amplitude of the basement heat loss are determined using Equations (19) and (20).

$$H_b = \frac{(A/P)_b}{k_s R_{eq}} = \frac{11.91}{0.70(2.84 + 6.54)} = 1.8139$$

$$b_{eff} = \frac{B}{(A/P)_b} = \frac{4.92}{11.91} = 0.4131$$

$$(A/P)_{eff,b,mean} = [1 + 0.4131(-0.4 + e^{-1.8139})] \times 11.91 = 10.7440$$

$$(A/P)_{eff,b,amp} = (1 + 0.4131e^{-1.8139}) \times 11.91 = 12.7120$$

$$U_{o,mean} = \frac{k_s}{(A/P)_{eff,b,mean}} = \frac{0.70}{10.7440} = 0.0652$$

$$U_{o,amp} = \frac{k_s}{(A/P)_{eff,b,amp}} = \frac{0.70}{12.7120} = 0.0551$$

$$H_{mean} = \frac{(A/P)_{eff,b,mean}}{k_s R_{eq}} = \frac{10.7440}{0.70(2.84 + 6.54)} = 1.6363$$

$$H_{amp} = \frac{(A/P)_{eff,b,amp}}{k_s R_{eq}} = \frac{12.7120}{0.70(2.84 + 6.54)} = 1.9360$$

$$D_{mean} = \ln\left[(1 + H_{mean})\left(1 + \frac{1}{H_{mean}}\right)^{H_{mean}}\right] = 1.7498$$

$$D_{amp} = \ln\left[(1 + H_{amp})\left(1 + \frac{1}{H_{amp}}\right)^{H_{amp}}\right] = 1.8832$$

$$G = k_s R_{eq}\sqrt{\frac{\omega}{\alpha_s}} = 0.70(2.84 + 6.54)\sqrt{\frac{1.992\times10^{-7}}{48.12\times10^{-7}}} = 1.3359$$

Therefore,

$$q_{mean} = U_{eff,mean}A(t_a - t_r) = 0.4 \times 0.0652 \times 1.7498 \times 2422 \times (71.6 - 50) = 2387 \text{ Btu/h}$$

and

$$q_{amp} = U_{eff,amp}At_{amp} = 0.25 \times 0.0551 \times 1.8832^{0.16} \times 1.3359^{-0.6} \times 2422 \times 23 = 714 \text{ Btu/h}$$

Table 3 compares results of the simplified method presented here and the more exact interzone temperature profile estimation (ITPE) (Krarti 1994a, 1994b; Krarti et al. 1988a, 1988b).

Table 3 Example 2 Heat Loss per Unit Area for the Simplified and ITPE Methods

Method	Mean (q_{mean}), Btu/h	Amplitude (q_{amp}), Btu/h
Simplified	2387	714
ITPE solution	2245	724

SECONDARY SYSTEM COMPONENTS

Secondary HVAC systems generally include all elements of the overall building energy system between a central heating and cooling plant and the building zones. The precise definition depends heavily on the building design. A secondary system typically includes air-handling equipment, air distribution systems with the associated ductwork, dampers, fans, and heating, cooling, and humidity conditioning equipment. Secondary systems also include the liquid distribution systems between the central plant and the zone and air-handling equipment, including piping, valves, and pumps.

While the exact design of secondary systems varies dramatically among buildings, they are composed of a relatively small set of generic HVAC components. These components include distribution components (e.g., pumps/fans, pipes/ducts, valves/dampers, headers/plenums, fittings) and heat and mass transfer components (e.g., heating coils, cooling and dehumidifying coils, liquid heat exchangers, air heat exchangers, evaporative coolers, steam injectors). Most secondary systems can be described by simply connecting these components to form the complete system.

Energy estimation through computer simulation often mimics the modular construction of secondary systems by using modular simulation elements [e.g., the ASHRAE *HVAC 2 Toolkit* (Brandemuehl 1993, Brandemuehl and Gabel 1994), the simulation program TRNSYS (Klein et al. 1994), and Annex 10 activities of the International Energy Agency]. To the extent that the secondary system consumes energy and transfers energy between the building and central plant, an energy analysis can be performed by characterizing the energy consumption of the individual components and the energy transferred among system components. In fact, few of the secondary components consume energy directly, except fans, pumps, furnaces, direct expansion air-conditioning package units with gas-fired heaters, and inline heaters. In this chapter, secondary components are divided into two categories: distribution components and heat and mass transfer components.

Fans, Pumps, and Distribution Systems

The distribution system of an HVAC system affects energy consumption in two ways. First, fans and pumps consume electrical energy directly, based on the flow and pressures under which the device operates. Ducts and dampers, or pipes and valves, and the system control strategies affect the flow and pressures at the fan or pump. Second, thermal energy is often transferred to (or from) the fluid due to heat transfer through pipes and ducts and due to the electrical input to fans and pumps. The analysis of system components should, therefore, account for both direct electrical energy consumption and thermal energy transfer.

Fan and pump performance are discussed in Chapters 18 and 39 of the 2000 *ASHRAE Handbook—Systems and Equipment*. In addition, Chapter 34 of this handbook covers pressure loss calculations for airflow in ducts and duct fittings. Chapter 35 presents a similar discussion for fluid flow in pipes. While these chapters do not specifically focus on energy estimation, energy use is governed by the same performance characteristics and engineering relationships. Strictly speaking, performance calculations of the fan and air distribution systems in a building require a detailed pressure balance on the entire network. For example, in an air distribution system, the airflow through the fan depends on its physical characteristics, the operating speed, and the pressure differential across the fan. The pressure drop through the duct system depends on the duct design, the position of all dampers, and the airflow through the fan. The interaction between the fan and duct system results in a set of coupled, nonlinear algebraic equations. Models and subroutines for performing these calculations are available in the ASHRAE *HVAC 2 Toolkit* (Brandemuehl 1993).

While a detailed analysis of a distribution system requires flow and pressure balancing among the components, nearly all commercially available energy analysis methods approximate the effect of the interactions with part-load performance curves. This procedure eliminates the need to calculate pressure drop through the distribution system at off-design conditions. The part-load curves are often expressed in terms of a **power input ratio** as a function of the part-load ratio, defined as the ratio of part-load flow to design flow:

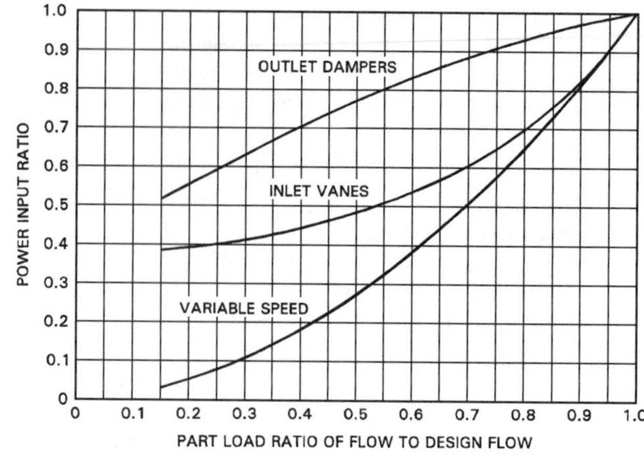

Fig. 3 Part-Load Curves for Typical Fan Operating Strategies

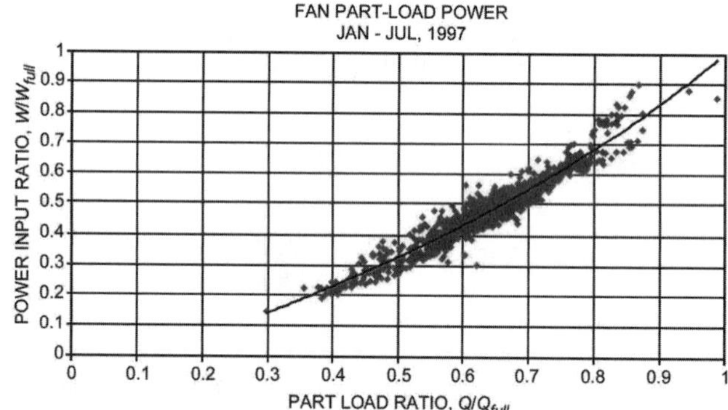

Fig. 4 Fan Part-Load Curve Obtained from Measured Field Data under ASHRAE 823-RP

$$\text{PIR} = \frac{W}{W_{full}} = f_{plr}\left(\frac{Q}{Q_{full}}\right) \qquad (21)$$

where

PIR = power input ratio
W = fan motor power at part load, W
W_{full} = fan motor power at full load or design, W
Q = fan airflow rate at part load, cfm
Q_{full} = fan airflow rate at full load or design, cfm
f_{plr} = regression function, typically polynomial

The exact shape of the part-load curve depends on the effect of flow control on the pressure and fan efficiency and may be calculated using a detailed analysis or measured field data. Figure 3 shows the relationship for three typical fan control strategies, as represented in a simulation program (York and Cappiello 1982). Within the simulation program, the curves are represented by polynomial regression equations. Models and subroutines for performing these calculations are also available in the ASHRAE *HVAC 2 Toolkit* (Brandemuehl 1993).

Figure 4 shows an example of a similar curve for the part-load operation of a fan system in a monitored building (Brandemuehl and Bradford 1999). In this particular case, the fan system represents ten separate air handlers, each with supply and return fans, operating with variable-speed fan control to maintain a set duct static pressure. Notice that, although the shape of the curve is similar to the variable-speed curve of Figure 3, the measured data for this particular system exhibit a more linear relationship between power and flow.

Heat transferred to the airstream due to fan operation increases the temperature of the air. While the shaft power of the fan has a direct effect on the heat transfer, motor inefficiencies also heat the air if the motor is mounted inside the airstream. For pumps, this contribution is typically assumed to be zero.

The following equation provides a convenient and general model to calculate the heat transferred to the fluid:

$$q_{fluid} = [\eta_m + (1 - \eta_m) f_{m,\,loss}]W \tag{22}$$

where

q_{fluid} = heat transferred to the fluid, Btu/h
$f_{m,loss}$ = fraction of the motor heat loss transferred to the fluid stream, dimensionless (= 1 if fan mounted in airstream, = 0 if fan mounted outside airstream)
W = fan motor power, Btu/h
η_m = motor efficiency

Heat and Mass Transfer Components

Secondary HVAC systems comprise such heat and mass transfer components as steam-based air-heating coils, chilled water cooling and dehumidifying coils, shell-and-tube liquid heat exchangers, air-to-air heat exchangers, evaporative coolers, and steam injectors. While these components do not consume energy directly, their thermal performance dictates the interactions between the building loads and the energy-consuming primary components (e.g., chillers, boilers). In particular, the performance of the secondary components determines the entering fluid conditions for primary components, which in turn determine the energy efficiencies of the primary equipment. Accurate energy calculations cannot be performed without appropriate models of the system heat and mass transfer components.

For example, the load on a chiller is typically described as the sum of zone sensible and latent loads, plus any heat gain from ducts, plenums, fans, pumps, and piping. However, the energy consumption of the chiller is determined not only by the load but also by the return chilled water temperature and flow rate. The return water condition is determined by the cooling coil performance and the part-load operating strategy of the air and water distribution system. The cooling coil might typically be controlled to maintain a constant leaving air temperature by modulating the water flow through the coil. In such a scenario, the cooling coil model must be able to calculate the leaving air humidity, leaving water temperature, and leaving water flow rate given the cooling coil design characteristics, entering air temperature and humidity, entering airflow, and entering water temperature.

Virtually all building energy simulation programs include, and require, models of heat and mass transfer components. In general, these models are relatively simple. While a coil designer might use a detailed tube-by-tube analysis of conduction and convection heat transfer and condensation on fin surfaces to develop an optimal combination of fin and tube geometry, an energy analyst is more interested in determining the changes in leaving fluid states as operating conditions vary during the year. In addition, the energy analyst is likely to have limited design data on the equipment and, therefore, requires a model with very few parameters that depend on equipment geometry and detailed design characteristics.

A typical approach to modeling heat and mass transfer components for energy calculations is based on an **effectiveness-NTU heat exchanger model** (Kays and London 1984). The effectiveness-NTU model is described in most heat transfer textbooks and is briefly discussed in Chapter 3. It is particularly appropriate for describing the leaving fluid conditions when the entering fluid conditions and equipment design characteristics are known. In addition, this model requires only a single parameter to describe the characteristics of the exchanger—the overall transfer coefficient *UA*—which can be determined from limited design performance data.

Because the classical effectiveness methods were developed for sensible heat exchangers, they are used to perform energy calculations for a variety of sensible heat exchangers in HVAC systems. For typical finned-tube air-heating coils, the crossflow configuration with both fluid streams unmixed is most appropriate. The same configuration typically applies to air-to-air heat exchangers. For liquid-to-liquid exchangers, tube-in-tube equipment can be modeled as parallel or counterflow, depending on flow directions; and shell-and-tube equipment can be modeled as either counter- or crossflow, depending on the extent of baffling and the number of tube passes.

The energy analyst must determine the *UA* to describe the operations of a specific heat exchanger. There are typically two approaches to determine this important parameter: direct calculation and from manufacturers' data. Given detailed information about the materials, geometry, and construction of the heat exchanger, it is possible to apply fundamental heat transfer principles to calculate the overall heat transfer coefficient. An alternative to direct calculation, and the method most appropriate for energy estimation, is to use manufacturers' performance data or direct measurements of installed performance. In reporting the design performance of a heat exchanger, a manufacturer typically gives the rate of heat transfer under various operating conditions, with the operating conditions described in terms of entering fluid flow rates and temperatures. The effectiveness and *UA* can be calculated from the given heat transfer rate and entering fluid conditions.

Example 3. An energy analyst seeks to perform an evaluation of a hot water heating system that includes a hot water heating coil. The energy analysis program uses an effectiveness-NTU model of the coil and requires the *UA* of the coil as an input parameter. While detailed information on the coil geometry and heat transfer surfaces is not available, the manufacturer states that the 1-row hot water heating coil will deliver 818,000 Btu/h of heat under the following design conditions:

Design Performance
Entering water temperature t_{hi} = 175°F
Water mass flow rate = 661 lb/min
Entering air temperature t_{ci} = 68°F
Air mass flow rate = 1058 lb/min
Design heat transfer q = 818,000 Btu/h

Solution: The solution proceeds by first determining the heat exchanger *UA* from the design data, and then using the *UA* to predict performance at the off-design conditions. The effectiveness-NTU relationships are used for both steps. The key assumption is that the *UA* is constant for both operating conditions. This example uses the nomenclature as described in Chapter 3 in the section on Overall Heat Transfer.

a) An examination of the flow rates and fluid specific heats allows calculation of the capacity rates C_h and C_c at design conditions and the capacity rate ratio Z.

$$C_h = (\dot{m}c_p)_h = (661)(1.00)(60) = 39{,}660 \text{ Btu/h} \cdot {}^\circ\text{F}$$

$$C_c = (\dot{m}c_p)_c = (1058)(0.24)(60) = 15{,}235 \text{ Btu/h} \cdot {}^\circ\text{F}$$

$$C_{max} = C_h \qquad C_{min} = C_c$$

$$Z = \frac{C_{min}}{C_{max}} = 0.384$$

b) The effectiveness can be directly calculated from the heat transfer definition.

$$\varepsilon = \frac{(t_{co} - t_{ci})}{(t_{hi} - t_{ci})} = \frac{q/C_c}{(t_{hi} - t_{ci})} = \frac{(818{,}000/15{,}235)}{(175 - 68)} = 0.502$$

c) The effectiveness-NTU relationships for a crossflow heat exchanger with both fluids unmixed allows calculation of the effectiveness in terms of the capacity rate ratio Z and the NTU [the relationships are available from most heat transfer textbooks and specifically can be found in Kays and London (1984)].

Given the effectiveness and capacity rate ratio, the NTU can be determined to be NTU = 0.804.

d) The heat transfer UA is then determined from the definition of the NTU.

$$UA = C_{min}\text{NTU} = (15{,}235)(0.804) = 12{,}250 \text{ Btu/h} \cdot °\text{F}$$

Application to Cooling and Dehumidifying Coils

The analysis of air-cooling and dehumidifying coils requires coupled, nonlinear heat and mass transfer relationships. These relationships form the basis for all HVAC components with moisture transfer, including cooling coils, cooling towers, air washers, and evaporative coolers. While the complex heat and mass transfer theory that is presented in many textbooks is often required for cooling coil design, simpler models based on effectiveness concepts are usually more appropriate for energy estimation. For example, the bypass factor is a form of effectiveness in the approach of the leaving air temperature to the apparatus dewpoint, or coil surface, temperature.

While the effectiveness-NTU method is typically developed and applied in the analysis of sensible heat exchangers, it can also be used to analyze other types of exchangers such as cooling and dehumidifying coils that couple heat and mass transfer. By redefining the state variables, the capacity rates, and the overall exchange coefficient of these enthalpy exchangers, the effectiveness concept may be used to calculate heat transfer rates and leaving fluid states. For sensible heat exchangers, the state variable is temperature, the capacity is the product of mass flow and fluid specific heat, and the overall transfer coefficient is the conventional overall heat transfer coefficient. For cooling and dehumidifying coils, the state variable becomes moist air enthalpy, the capacity has units of mass flow, and the overall heat transfer coefficient is modified to reflect enthalpy exchange. This approach forms the basis for models by Threlkeld (1970), Elmahdy and Mitalas (1977), Braun (1988), and Brandemuehl (1993). The same principles also underlie the coil model described in Chapter 21 of the 2000 *ASHRAE Handbook—Systems and Equipment*.

The effectiveness model is based on the observation that, for a given set of entering air and liquid conditions, the heat and mass transfer are bounded by thermodynamic maximum values. Figure 5 shows the limits for leaving air states on a psychrometric chart. Specifically, the leaving chilled water temperature cannot be warmer than the entering air temperature and the leaving air temperature and humidity cannot be lower than the conditions of saturated moist air at the temperature of the entering chilled water.

Figure 5 also shows that the performance of a cooling coil requires the evaluation of two different effectivenesses to identify the leaving air temperature and humidity. An overall effectiveness can be used to describe the approach of the leaving air enthalpy to the minimum possible value. An air-side effectiveness, related to the coil bypass factor, describes the approach of the leaving air temperature to the effective wet coil surface temperature.

The effectiveness analysis is accomplished for wet coils by establishing a common state variable for both the moist air and liquid streams. As implied by the lower limit of the entering chilled water temperature, this common state variable is the moist air enthalpy. In other words, all liquid and coil temperatures are transformed to the enthalpy of saturated moist air at the liquid or coil temperature. Changes in liquid temperature can similarly be expressed in terms of changes in saturated moist air enthalpy through a saturation specific heat $c_{p,sat}$ defined by the following:

$$c_{p,\,sat} = \frac{\Delta h_{l,\,sat}}{\Delta t_l} \tag{23}$$

Using the definition of Equation (23), the basic effectiveness relationships discussed in Chapter 3 can be written as

$$q = C_a(h_{a,\,ent} - h_{a,\,lvg}) = C_l(h_{l,\,sat,\,lvg} - h_{l,\,sat,\,ent}) \tag{24}$$

$$q = \varepsilon C_{min}(h_{a,\,ent} - h_{l,\,sat,\,ent}) \tag{25}$$

$$C_a = \dot{m}_a \tag{26}$$

$$C_l = \frac{(\dot{m}c_p)_l}{c_{p,\,sat}} \tag{27}$$

$$C_{min} = \min(C_a, C_l) \tag{28}$$

where

q = heat transfer from air to water, Btu/h
C = fluid capacity, lb/h
$\dot{m}_a$ = dry air mass flow rate, lb/h
$\dot{m}_l$ = liquid mass flow rate, lb/h
$c_{p,l}$ = liquid specific heat, Btu/lb · °F
$c_{p,sat}$ = saturation specific heat, defined by Equation (23), Btu/lb · °F
h_a = enthalpy of moist air, Btu/lb
$h_{l,sat}$ = enthalpy of saturated moist air at the temperature of the liquid, Btu/lb

The cooling coil effectiveness of Equation (25) is defined, then, as the ratio of moist air enthalpies in Figure 5. As in the case of sensible heat exchangers, the effectiveness is also a function of the physical coil characteristics and can be obtained by modeling the coil as a counterflow heat exchanger. However, since the heat transfer calculations are performed based on enthalpies, the overall transfer coefficient must be based on enthalpy potential rather than temperature potential. The enthalpy-based heat transfer coefficient UA_h is related to the conventional temperature-based coefficient by the specific heat:

$$q = UA\Delta t = UA_h\Delta h$$
$$UA_h = \frac{UA\Delta t}{\Delta h} = \frac{UA}{c_p} \tag{29}$$

A similar analysis can be performed to evaluate the air-side effectiveness, which identifies the leaving air temperature. While the overall enthalpy-based effectiveness is based on an overall heat transfer coefficient between the chilled water and the air, the air-side effectiveness is based on a heat transfer coefficient between the coil surface and the air.

As with sensible heat exchangers, the overall heat transfer coefficients UA can be determined either from direct calculation from coil properties or from manufacturers' performance data. While a sensible heat exchanger is modeled with a single effectiveness and

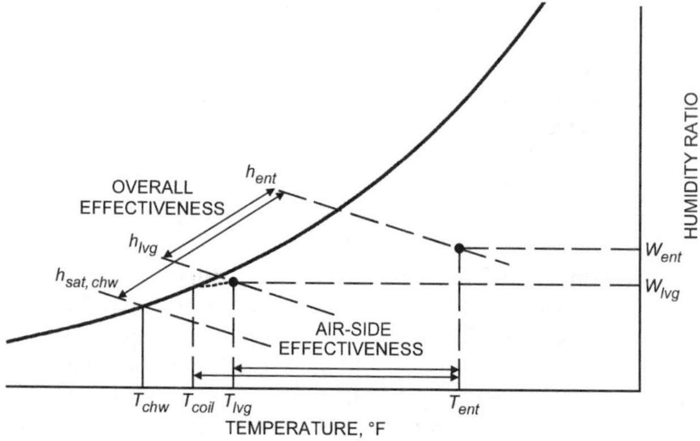

Fig. 5 Psychrometric Schematic of Cooling Coil Processes

can be described by a single parameter *UA*, a wet cooling and dehumidifying coil requires two parameters to describe the two effectivenesses shown in Figure 5. These two parameters are the internal and external *UA*s—one describes the heat transfer between the chilled water and the air-side surface through the pipe wall and the other between the surface and the moist air. These two parameters can be determined from the sensible and latent capacity of a cooling coil at a single rating condition. A significant advantage of the effectiveness-NTU method is that the component can be described with as little as one measured data point or one manufacturer's design calculation.

PRIMARY SYSTEM COMPONENTS

Primary HVAC systems consume energy and deliver heating and cooling to a building, usually through secondary systems. Primary equipment generally includes chillers, boilers, cooling towers, cogeneration equipment, and plant-level thermal storage equipment. In particular, primary equipment generally represents the major energy-consuming equipment of a building, so accurate characterization of building energy use relies on accurate modeling of primary equipment energy consumption.

Modeling Strategies

The energy consumption characteristics of primary equipment generally depend on equipment design, load conditions, environmental conditions, and equipment control strategies. For example, chiller performance depends on the basic equipment design features (e.g., heat exchange surfaces, compressor design), the temperatures and flow through the condenser and evaporator, and the methods for controlling the chiller at different loads and operating conditions (e.g., inlet guide vane control on centrifugal chillers to maintain leaving chilled water temperature set point). In general, these variables that dictate energy consumption vary constantly and require calculations on an hourly basis.

Regression Models. While many secondary components (e.g., heat exchangers, valves) are readily described by fundamental engineering principles, the complex nature of most primary equipment has discouraged the use of first-principle models for energy calculations. Instead, the energy consumption characteristics of primary equipment have traditionally been modeled using simple equations developed by regression analysis of manufacturers' published design data. Because published data are often available only for full-load design conditions, additional correction functions are used to correct the full-load data to part-load conditions. The functional form of the regression equations and correction functions takes many forms, including exponentials, Fourier series, and, most of the time, second- or third-order polynomials. The selection of an appropriate functional form depends on the behavior of the equipment. In some cases, energy consumption is calculated using direct interpolation from tables of data. However, this method often requires excessive data input and computer memory.

The typical approach to modeling primary equipment in energy simulation programs is to assume the following functional form for equipment power consumption:

$$P = \text{PIR} \times \text{Load}$$
$$\text{PIR} = \text{PIR}_{nom} f_1(t_a, t_b, \ldots) f_2(\text{PLR}) \qquad (30)$$

$$C_{avail} = C_{nom} f_3(t_a, t_b, \ldots)$$
$$\text{PLR} = \frac{\text{Load}}{C_{avail}} \qquad (31)$$

where

P = equipment power, kW
PIR = energy input ratio
PIR_{nom} = energy input ratio under nominal full-load conditions
Load = power delivered to the load, kW
C_{avail} = available equipment capacity, kW
C_{nom} = nominal equipment capacity, kW
f_1 = function relating full-load power at off-design conditions (t_a, t_b, ...) to full-load power at design conditions
f_2 = fraction full-load power function, relating part-load power to full-load power
f_3 = function relating available capacity at off-design conditions (t_a, t_b, ...) to nominal capacity
t_a, t_b = various operating temperatures that affect power
PLR = part-load ratio

The part-load ratio is defined as the ratio of the load to the available equipment capacity at given off-design operating conditions. Like the power, the available, or full-load, capacity will be a function of operating conditions.

The particular forms of the off-design functions f_1 and f_3 depend on the specific type of primary equipment. For example, for fossil-fuel boilers, full-load capacity and power (or fuel use) can be affected by the thermal losses to ambient temperature. However, these off-design functions are typically considered to be unity in most building simulation programs. For chillers, both capacity and power are affected by the condenser and evaporator temperatures. These two temperatures are often characterized in terms of their secondary fluids. For direct expansion air-cooled chillers, the operating temperatures are typically the wet-bulb temperature of the air entering the evaporator and the dry-bulb temperature of the air entering the condenser. For liquid chillers, the temperatures are usually the leaving chilled water temperature and the entering condenser water temperature.

As an example, consider the performance of a direct expansion (DX) packaged single-zone rooftop unit. The nominal rated performance of these units is typically given for an outdoor air temperature of 95°F and evaporator entering coil conditions of 80°F dry-bulb temperature and 67°F wet-bulb temperature. However, the performance changes as the outdoor temperature and entering coil conditions vary. To account for these effects, the DOE-2.1E simulation program expresses the off-design functions f_1 and f_3 with biquadratic functions of the outdoor dry-bulb temperature and the coil entering wet-bulb temperature.

$$f_1(t_{wb,\,ent}, t_{oa})$$
$$= a_0 + a_1 t_{wb,\,ent} + a_2 t_{wb,\,ent}^2 + a_3 t_{oa} + a_4 t_{oa}^2 + a_5 t_{wb,\,ent} t_{oa} \qquad (32)$$

$$f_3(t_{wb,\,ent}, t_{oa})$$
$$= c_0 + c_1 t_{wb,\,ent} + c_2 t_{wb,\,ent}^2 + c_3 t_{oa} + c_4 t_{oa}^2 + c_5 t_{wb,\,ent} t_{oa} \qquad (33)$$

The constants in Equations (32) and (33) are given in Table 4.

The fraction full-load power function f_2 represents the change in equipment efficiency at part-load conditions and depends heavily on the control strategies used to match load and capacity. Figure 6 shows several possible shapes of these functional relationships. (Notice that these curves are similar to the fan part-load curves of Figure 3.) Curve 1 represents equipment with constant efficiency, independent of load. Curve 2 represents equipment that is most efficient in the middle of its operating range. Curve 3 represents equipment that is most efficient at full load. Note that these types of curves apply to both boilers and chillers.

Table 4 Correlation Coefficients for Off-Design Relationships

Correlation	0	1	2	3	4	5
f_1	−1.063931	0.0306584	0.0001269	0.0154213	0.0000497	0.0002096
f_3	0.8740302	0.0011416	0.0001711	−0.002957	0.0000102	0.0000592

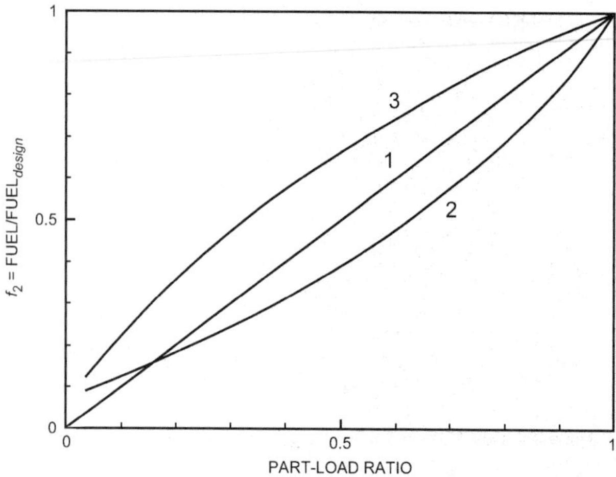

Fig. 6 Possible Part-Load Power Curves

First-Principle Models. As with the secondary components, engineering first principles can also be used to develop models of primary equipment. Lebrun et al. (1999), Gordon and Ng (1994, 1995), Gordon et al. (1995), and others have sought to develop such models in which unknown model parameters are extracted from measured or published manufacturers' data.

The energy analyst is often faced with choosing the appropriate model for the job. For example, a complex boiler model is not appropriate if the boiler in question operates at virtually constant efficiency. By similar arguments, a regression-based model might be appropriate when the user has a full dataset of reliable in-situ measurements of the plant. However, first-principle physical models generally have several advantages over pure regression models:

- Physical models allow confident extrapolation outside the range of available data.
- Regression is still required to obtain values for unknown physical parameters. However, the values of these parameters usually have physical significance. The engineer can capitalize on this significance to estimate default parameter values, diagnose errors in data analysis through checks for realistic parameter values, and even evaluate potential performance improvements.
- The number of unknown parameters is generally much smaller than the number of unknown coefficients in the typical regression model. For example, the standard ARI compressor model requires as many as 30 coefficients, 10 coefficients each for the regressions of capacity, power, and refrigerant flow. By comparison, a physical compressor model may have as few as four or five unknown parameters. As a result, the physical models require fewer measured data.
- Data on part-load operation of chillers and boilers are notoriously difficult to obtain. Part-load corrections often represent the greatest uncertainty in the regression models, while causing the greatest effect on annual energy predictions. By comparison, physical models of full-load operation often allow direct extension to part-load operation with little additional required data.

While physical models of primary HVAC equipment are generally based on fundamental engineering analysis and found in many HVAC textbooks, the models described here are specifically based on the work of Bourdouxhe et al. (1994a, 1994b, 1994c) in the development of the ASHRAE *HVAC 1 Toolkit* (Lebrun et al. 1999). The behavior of each elementary component is characterized by a limited number of physical parameters, such as heat exchanger heat transfer area or centrifugal compressor impeller blade angle. Values of these parameters are identified, or tuned, based on regression fits of overall performance compared to measured or published data.

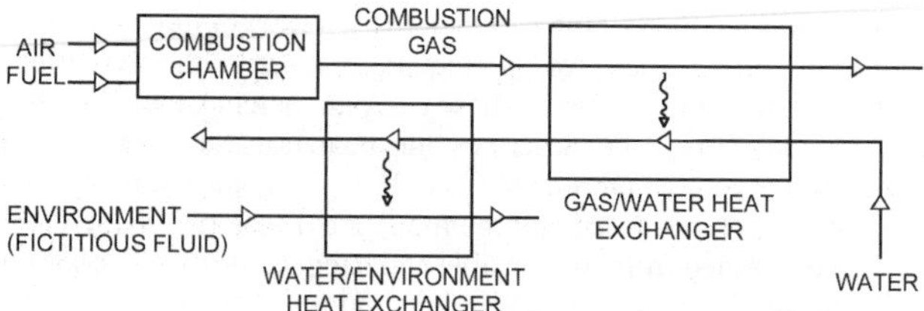

Fig. 7 Boiler Modeled with Elementary Components

While physical models are based on physical characteristics, the values obtained through a regression analysis of manufacturers' data are not necessarily representative of the actual measured values. Strictly speaking, the parameter values are regression coefficients with somewhat fictitious values, identified to minimize the error in overall system performance. In other words, errors in the fundamental models of the equipment are offset by over or underestimation of the parameter values.

Boiler Model

The thermal model of a water boiler operating in a steady-state regime is shown in Figure 7. It consists of an adiabatic combustion chamber and two heat exchangers. These three components interact through the following fluids: air, fuel, combustion gas, water, and a fictitious fluid representing the environment.

The adiabatic combustion chamber and the two heat exchangers can be modeled using classical thermodynamic principles. For example, the boiler heat exchangers can be described using effectiveness-NTU models. That is, knowing the flow rates and entering fluid temperatures, the heat transfer is calculated using a single overall heat transfer coefficient UA. For the water/environment heat exchanger, the environment is considered to be an isothermal reservoir and can be modeled as a fictitious fluid with infinite capacitance. The values of the overall heat transfer coefficients of the two heat exchangers are selected, or tuned to manufacturer's data, to represent the behavior of a particular boiler.

Most boilers are equipped with combustion control systems that vary fuel input to satisfy changing heating loads. The model of Figure 7 can reflect the physical effects of varying fuel, fluid flow, and stand-by losses to realistically show the degradation of performance at part-load operation. In addition, single-stage boilers with on-off control and two-stage boilers with two distinct firing rates can be modeled as "quasi-static" systems, where the cyclic operation is represented by a linear combination of two steady-state regimes.

Vapor Compression Chiller Models

Figure 8 shows a schematic of a vapor compression chiller. In this case, the components include two heat exchangers, an expansion valve, and a compressor with a motor and transmission. The components of a chiller are linked through the refrigerant. For energy estimating, a simplified approach is sufficient to represent the refrigerant as a "perfect" fluid with fictitious property values. That is, refrigerant liquid is modeled as incompressible, and vapor properties are described by ideal gas laws with effective average values of property parameters, such as specific heat.

Condenser and Evaporator Modeling. Both condensers and evaporators are modeled as classical heat exchangers. The two heat exchangers are each assumed to have a constant overall heat transfer coefficient. In addition, the models used in chiller systems suffer from one additional assumption—the refrigerant fluid is assumed to be isothermal for both heat exchangers, which effectively ignores the superheated and subcooled regions of the heat exchanger. The assumption of an isothermal refrigerant is particularly crude for the condenser, which sees very high refrigerant temperatures from the compressor discharge. The effect of the assumption is to signifi-

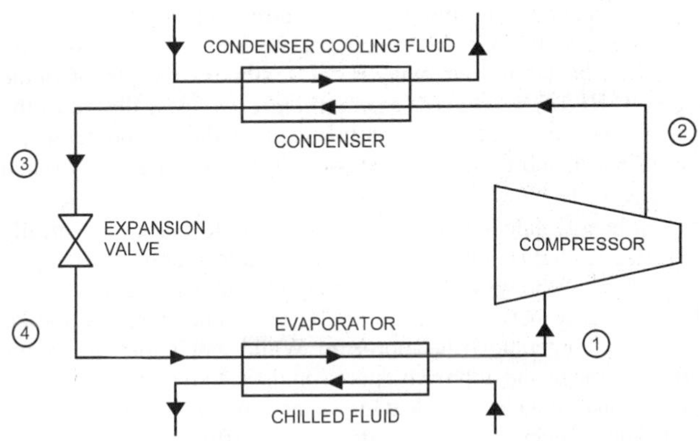

Fig. 8 Chiller Model Using Elementary Components

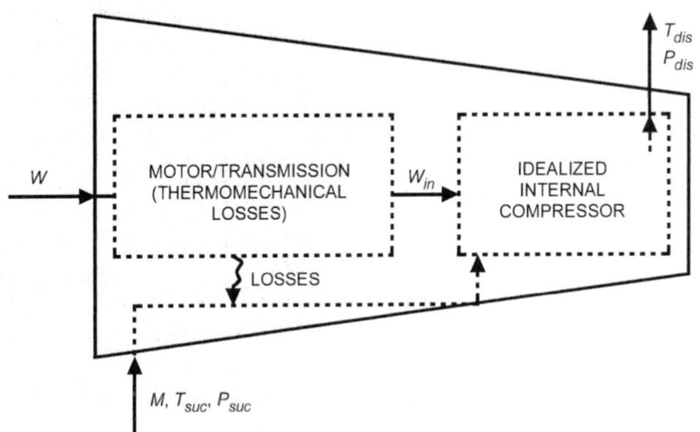

Fig. 9 General Schematic of Compressor

cantly underestimate the mean temperature difference between refrigerant and water in the heat exchanger. Fortunately, this systematic error is compensated by a significant overestimate of the corresponding heat transfer coefficient.

General Compressor Modeling. The modeling of real compressors requires the description of many thermomechanical losses within the compressor. Such losses could include heat loss, fluid friction, throttling losses in valves, or motor and transmission inefficiencies. While some of these losses can be modeled within the compressor, others are too complex or unknown to describe in a model for energy calculations.

The general approach used here for compressor modeling is described in the Figure 9. The compressor is described by two distinct internal elements: an idealized internal compressor and a motor-transmission element to account for unknown losses. Schematically, the motor-transmission subsystem represents an inefficiency of energy conversion. The losses from these inefficiencies are assumed to heat the fluid prior to compression. Mathematically, it can be modeled by the following linear relationship:

$$W = W_{lo} + (1 + \alpha)W_{in} \tag{34}$$

where

W = electrical power for a hermetic or semihermetic compressor, or shaft power for an open compressor
W_{in} = idealized internal compressor power (depends on type of compressor)
W_{lo} = constant electromechanical loss
α = proportional power loss factor

W_{lo} and α are empirical parameters determined by performing a regression analysis on manufacturers' data. Other parameters are also required to model W_{in}, the idealized internal compressor power, depending on the type of compressor.

The following sections describe different modeling techniques for reciprocating, screw, and centrifugal compressors. Detailed modeling techniques are available in the ASHRAE *HVAC 1 Toolkit* (Lebrun et al. 1999) and associated references.

Modeling the Reciprocating Compressor. The conceptual schematic for a reciprocating compressor, for use with the general model, is shown in Figure 10. The refrigerant enters the compressor at state 1 and is heated to state 1a by the thermomechanical losses of the motor-transmission model in Figure 9. The refrigerant undergoes an isentropic compression process to state 2s, followed by a throttling process to the compressor discharge at state 2. The throttling valve is a simplified approach to model the known losses within the compressor due to pressure drops across the suction and discharge valves. Perhaps a more accurate model would include pressure losses at both the inlet and outlet of the compressor, but analysis of compressor data reveals that this simpler model is

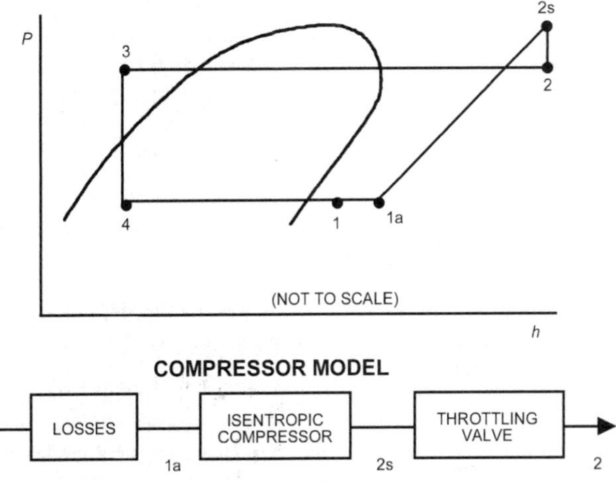

Fig. 10 Schematic of Reciprocating Compressor Model

adequate for modeling of typical reciprocating compressors. In fact, many compressors can be adequately modeled with no throttling valve at all.

The refrigerant flow rate through the system must be determined to predict chiller and compressor performance. In general, volumetric flow depends on the pressure difference across the compressor. The compressor refrigerant flow rate is a decreasing function of the pressure ratio due to the re-expansion of the vapor in the clearance volume. With the refrigerant vapor modeled as an ideal gas, the volumetric flow rate is given by the following:

$$V = V_s\left[1 + C_f - C_f\left(\frac{p_{ex}}{p_{suc}}\right)^{1/\gamma}\right] \tag{35}$$

where

V = volumetric flow rate
V_s = swept volumetric flow rate (geometric displacement of the compressor)
C_f = clearance factor = $V_{clearance}/V_s$
P_{ex}/P_{suc} = cylinder pressure ratio
γ = specific heat ratio

V_s and C_f must be identified using data for the actual reciprocating compressor.

While the models discussed apply to full-load operation, Equation (35) is also valid at part-load conditions. However, the internal power use can be different at part load depending on the particular strategy for capacity modulation, such as on-off cycling, cylinder unloading, hot-gas bypass, or variable-speed motor. In most cases, simple physical models can be developed to describe these methods, which generally vary the swept volumetric rate. Additional thermomechanical losses can also be modeled flow but often involve additional parameters. For example, the effect of cylinder unloading can be modeled by the following relationship:

$$W_{in} = W_{in,FL} + \left(1 - \frac{N_c}{N_{c,FL}}\right) W_{pump} \tag{36}$$

where

N_c = number of cylinders in use
$N_{c,FL}$ = number of cylinders in use in full-load regime
W_{pump} = internal power of the compressor when all the cylinders are unloaded ("pumping" power)
$W_{in,FL}$ = full-load power

The variable W_{pump} characterizes the part-load regime of the reciprocating compressor, and it is assumed to be constant throughout the entire part-load range.

In summary, a realistic physical model of a reciprocating compressor, covering both full-load and part-load operations, can be developed based on six parameters: the constant and proportional loss terms of the motor-transmission model W_{lo} and α, the swept volumetric flow rate V_s of the compressor cylinders, the cylinder clearance volume factor C_f, the fictitious exhaust valve flow area A_{ex}, and the zero-load pumping power of the unloaded compressor W_{pump}. The entire chiller can then be modeled with two additional parameters for the overall heat transfer coefficients of the condenser and evaporator.

Modeling of Other Compressors and Chillers. From a modeling perspective, the thermodynamic processes of a screw compressor are similar to those of a reciprocating compressor. Physically, the screw compressor transports an initial volumetric flow rate of refrigerant vapor to a higher pressure and density by squeezing it into a smaller space. A realistic physical model of a variable-volume-ratio, twin-screw compressor, covering both full-load and part-load operations, can be developed based on five parameters: the constant and proportional loss terms of the motor-transmission model of Equation (34), the swept volumetric flow rate of the compressor screw, the internal leakage area, and a pumped pressure differential for diverted flow at part-load (Lebrun et al. 1999). The entire chiller can then be modeled with two additional parameters for the overall heat transfer coefficients of the condenser and evaporator.

An idealized internal model of a centrifugal compressor, to be used in conjunction with Equation (34) and Figure 9, can be based on an ideal analysis of a single-stage compressor composed of an isentropic impeller and isentropic diffuser. In addition to the thermomechanical loss parameters of Equation (34), only three additional parameters are required by this centrifugal compressor model: the peripheral speed of the impeller, the inclination of the vanes at the impeller exhaust, and the impeller exhaust area.

The refrigerant cycle of an absorption chiller is the same as for a vapor compression cycle, except for the absorption-generation subsystem in place of the compressor. The absorption-generation subsystem includes an absorber, a steam-fired generator, a recovery heat exchanger, a pump, and a control valve. All components except the pump and control valve can be modeled as heat exchangers.

Cooling Tower Model

A cooling tower is used in primary systems to reject heat from the chiller condenser. The controls typically control the tower fans and pumps to maintain a desired water temperature entering the condenser. Like cooling and dehumidifying coils in secondary systems, the performance of a cooling tower has a strong influence on the energy consumption of the chiller. In addition, tower fans consume electrical energy directly.

Fundamentally, a cooling tower is a direct contact heat and mass exchanger. Equations describing the fundamental processes are given in Chapter 5 and in many HVAC textbooks. Chapter 36 of the 2000 *ASHRAE Handbook—Systems and Equipment* describes the specific performance of cooling towers. In addition, cooling tower performance subroutines are available in Lebrun et al. (1999) and Klein et al. (1994).

For energy calculations, cooling tower performance is typically described in terms of the outdoor wet-bulb temperature, the temperature drop of the water flowing through the tower (the range), and the difference between the leaving water temperature and the air wet-bulb temperature (the approach). While simple models assume constant range and approach, more sophisticated models use rating performance data to relate leaving water temperature to the outdoor wet-bulb temperature, water flow, and airflow. Simple cooling tower models, such as those based on a single overall transfer coefficient that can be directly inferred from a single tower rating point, are often appropriate for energy calculations.

SYSTEM MODELING

OVERALL MODELING STRATEGIES

In developing a simulation model for building energy prediction, two basic issues must be considered—(1) modeling of components or subsystems and (2) the overall modeling strategy. Modeling of components, which was discussed in the section on Component Modeling and Loads, results in sets of equations describing the individual components. The overall modeling strategy refers to the **sequence** and **procedures** used to solve these equations. The accuracy of results and the computer resources required to achieve these results depend on the modeling strategy.

In most building energy programs, the load models are executed for every space for every hour of the simulation period. (Practically all models use 1 h as the time step, which excludes any information on phenomena occurring in a shorter time span.) The load model is followed by running models for every secondary system, one at a time, for every hour of the simulation. Finally, the plant simulation model is executed again for the entire period. Each sequential execution processes the **fixed** output of the preceding step.

This procedure is illustrated in Figure 11. The solid lines represent data passed from one model to the next. The dashed lines represent information, usually provided by the user, about one model passed to the preceding model. For example, the system information consists of a piecewise-linear function of zone temperature that gives the system capacity.

Because of this loads-systems-plants sequence, certain phenomena cannot be modeled precisely. For example, if the heat balance method for computing loads is used, and some component in the system simulation model cannot meet the load, the program can only report the current load. In actuality, the space temperature should readjust until the load matches the equipment capacity, but this cannot be modeled because the loads have been precalculated and fixed. If the weighting factor method is used for loads, this problem is partially overcome, because loads are continually readjusted during the system simulation. However, the weighting factor

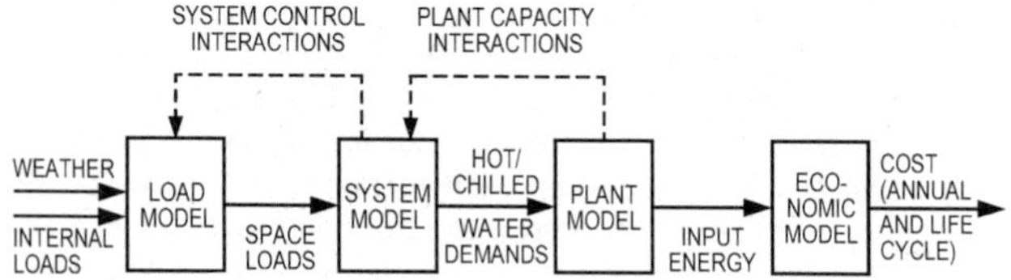

Fig. 11 Overall Modeling Strategy

technique is based on linear mathematics, and wide departures of room temperatures from those used during execution of the load program can introduce errors.

A similar problem arises in plant simulation. For example, in an actual building, as the load on the central plant varies the supply chilled water temperature also varies. This variation in turn affects the capacity of the secondary system equipment. In an actual building, when the central plant becomes overloaded, space temperatures should rise to reduce the load. However, in most energy estimating programs, this condition cannot occur; thus, only the overload condition can be reported. These are some of the penalties associated with decoupling of the load, system, and plant models.

An alternative strategy, in which all calculations are performed at each time step, is conceivable. Here the load, system, and plant equations are solved simultaneously at each time interval. With this strategy, unmet loads and imbalances cannot occur; conditions at the plant are immediately reflected to the secondary system and then to the load model, forcing them to readjust to the instantaneous conditions throughout the building. The results of this modeling strategy are superior to those currently available, although the magnitude and importance of the improvement are uncertain.

The principal disadvantage of the alternative approach, and the reason that it has not been widely used, is that it demands more computing resources. However, programs that, to one degree or another, implement simultaneous solution of the loads, system, and plant models have been developed by Clarke (1985), Park et al. (1985), Klein et al. (1994), and Metcalfe et al. (1995).

An economic model, as shown in Figure 11, calculates energy costs (and sometimes capital costs) based on the estimated required input energy. Thus, the simulation model calculates energy usage and cost for any given input weather and internal loads. By applying this model (i.e., determining output for given inputs) at each hour (or other suitable interval), the hour-by-hour energy consumption and cost can be determined. Maintaining running sums of these quantities yields monthly or annual energy usage and costs.

These models only compare design alternatives; a large number of uncontrolled and unknown factors usually rule out such models for accurate prediction of utility bills. For example, Miller (1980) found that the dynamics of control of components may have at least minor effects on predicted energy use. The section on Bibliography lists several models, which are also described in Walton (1983) and York and Cappiello (1981). Generally, the load models tend to be the most complex and time-consuming, while the central plant model is the least complex.

Because the detailed models are computationally intensive, several simplified methods have been developed. These methods include the degree-day method, the bin method, and correlation methods, and they are presented in the next two sections.

DEGREE-DAY AND BIN METHODS

Degree-day methods are the simplest methods for energy analysis and are appropriate if the building use and the efficiency of the HVAC equipment are constant. Where efficiency or conditions of use vary with outdoor temperature, the consumption can be calculated for different values of the outdoor temperature and multiplied by the corresponding number of hours; this approach is used in various **bin methods**. When the indoor temperature is allowed to fluctuate or when interior gains vary, models other than simple steady-state models must be used.

Even in an age when computers can easily calculate the energy consumption of a building, the concepts of degree-days and balance point temperature remain valuable tools. The severity of a climate can be characterized concisely in terms of degree-days. Also, the degree-day method and its generalizations can provide a simple estimate of annual loads, which can be accurate if the indoor temperature and internal gains are relatively constant and if the heating or cooling systems operate for a complete season. For these reasons, basic steady-state methods continue to be important.

Balance Point Temperature

The balance point temperature t_{bal} of a building is defined as that value of the outdoor temperature t_o at which, for the specified value of the interior temperature t_i, the total heat loss q_{gain} is equal to the heat gain from sun, occupants, lights, and so forth.

$$q_{gain} = K_{tot}(t_i - t_{bal}) \tag{37}$$

where K_{tot} is the total heat loss coefficient of the building in Btu/h·°F. For any of the steady-state methods described in this section, heat gains must be the average for the period in question, not for the peak values. In particular, solar radiation must be based on averages, not peak values. The balance point temperature is therefore obtained as

$$t_{bal} = t_i - \frac{q_{gain}}{K_{tot}} \tag{38}$$

Heating is needed only when t_o drops below t_{bal}. The rate of energy consumption of the heating system is

$$q_h = \frac{K_{tot}}{\eta_h}[t_{bal} - t_o(\theta)]^+ \tag{39}$$

where η_h is the efficiency of the heating system, also designated on an annual basis as the annual fuel use efficiency (AFUE), θ is time, and the plus sign above the bracket indicates that only positive values are to be counted. If t_{bal}, K_{tot}, and η_h are constant, the annual heating consumption can be written as an integral:

$$Q_{h,yr} = \frac{K_{tot}}{\eta_h}\int [t_{bal} - t_o(\theta)]^+ d\theta \tag{40}$$

This integral of the temperature difference conveniently summarizes the effect of outdoor temperatures on a building. In practice, it is approximated by summing averages over short time intervals (daily or hourly); the result is termed **degree-days** or **degree-hours**.

Annual Degree-Day Method

Annual Degree-Days. If daily average values of outdoor temperature are used for evaluating the integral, the degree-days for heating $DD_h(t_{bal})$ are obtained as

$$DD_h(t_{bal}) = (1\text{ day})\sum_{\text{days}} (t_{bal} - t_o)^+ \tag{41}$$

with dimensions of °F·days. Here the summation is to extend over the entire year or over the heating season. It is a function of t_{bal}, reflecting the roles of t_i, heat gain, and loss coefficient. The balance point temperature t_{bal} is also known as the **base** of the degree-days. In terms of degree-days, the annual heating consumption is

$$Q_{h,yr} = \frac{K_{tot}}{\eta_h}DD_h(t_{bal}) \tag{42}$$

Heating degree-days or degree-hours for a balance point temperature of 65°F have been widely tabulated based on the observation that this has represented average conditions in typical buildings in the past. The 65°F base is assumed whenever t_{bal} is not indicated explicitly. The extension of degree-day data to different bases is discussed later.

Cooling degree-days can be calculated using an equation analogous to Equation (41) for heating degree-days as

$$DD_c(t_{bal}) = (1\text{ day})\sum_{\text{days}}(t_o - t_{bal})^+ \quad (43)$$

While the definition of the balance point temperature is the same as that for heating, in a given building its numerical value for cooling is generally different from that for heating because q_i, K_{tot}, and t_i can be different. According to Claridge et al. (1987), t_{bal} can include both solar and internal gains as well as losses to the ground.

Calculating cooling energy consumption using degree-days is more difficult than heating. For cooling, the equation analogous to Equation (42) is

$$Q_{c,\,yr} = \frac{K_{tot}}{\eta_h} DD_c(t_{bal}) \quad (44)$$

for a building whose K_{tot} does not change. That assumption is generally acceptable during the heating season, when windows are closed and the air exchange rate is fairly constant. However, during the intermediate or cooling season, heat gains can be eliminated, and the onset of mechanical cooling can be postponed by opening windows or increasing the ventilation. (In buildings with mechanical ventilation, this is called the **economizer** mode.) Mechanical air conditioning is needed only when the outdoor temperature extends beyond the threshold t_{max}. This threshold is given by an equation analogous to Equation (38), with the replacement of the closed window heat transmission coefficient K_{tot} by its value K_{max} for open windows:

$$t_{max} = t_i - \frac{q_{gain}}{K_{max}} \quad (45)$$

K_{max} varies considerably with wind speed, but a constant value can be assumed for simple cases. The resulting sensible cooling load is shown schematically in Figure 12 as a function of t_o. The solid line is the load with open windows or increased ventilation; the dashed line shows the load if K_{tot} were kept constant. The annual cooling load for this mode can be calculated by breaking the area under the solid line into a rectangle and a triangle, or

$$Q_c = K_{tot}[DD_c(t_{max}) + (t_{max} - t_{bal})N_{max}] \quad (46)$$

where $DD_c(t_{max})$ are the cooling degree-days for base t_{max}, and N_{max} is the number of days during the season when t_o rises above t_{max}.

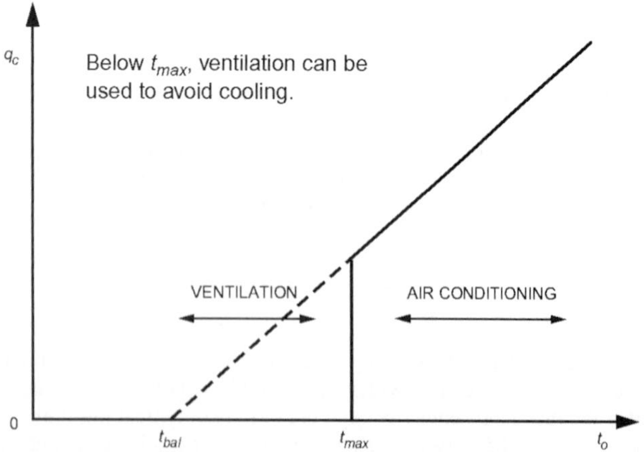

Fig. 12 Cooling Load as Function of Outdoor Temperature t_o

This is merely a schematic model of air conditioning. In practice, heat gains and ventilation rates vary, as does the occupant behavior in using the windows and the air conditioner. Also, in commercial buildings with economizers, the extra fan energy for increased ventilation must be added to the calculations. Finally, air-conditioning systems are often turned off during unoccupied periods. Therefore, cooling degree-hours better represent the period when equipment is operating than cooling degree-days because degree-days assume uninterrupted equipment operation as long as there is a cooling load.

Latent loads can form an appreciable part of a building's cooling load. The degree-day method can be used to estimate the latent load during the cooling season on a monthly basis by adding the following term to Equation (46):

$$q_{latent} = \dot{m}h_{fg}(W_o - W_i) \quad (47)$$

where

q_{latent} = monthly latent cooling load, Btu/h
$\dot{m}$ = monthly infiltration (total airflow), lb/h
h_{fg} = heat of vaporization of water, Btu/lb
W_o = outdoor humidity ratio (monthly averaged)
W_i = indoor humidity ratio (monthly averaged)

The degree-day method assumes that t_{bal} is constant, which is not well satisfied in practice. Solar gains are zero at night, and internal gains tend to be highest during the evening. The pattern for a typical house is shown in Figure 13. As long as t_o always stays below t_{bal}, the variations average out without changing the consumption. But for the situation in Figure 13, t_o rises above t_{bal} from shortly after 10:00 A.M. to 10:00 P.M.; the consequences for energy consumption depend on the thermal inertia and on the control of the HVAC system. If this building had low inertia and if temperature control were critical, heating would be needed at night and cooling during the day. In practice, this effect is reduced by thermal inertia and by the dead band of the thermostat, which allows t_i to float.

The closer t_o is to t_{bal}, the greater the uncertainty. If the occupants keep the windows closed during mild weather, t_i will rise above the set point. If they open the windows, the potential benefit of heat gains is reduced. In either case, the true values of t_{bal} become uncertain. Therefore, the degree-day method, like any steady-state method, is unreliable for estimating the consumption during mild weather. In fact, the consumption becomes most sensitive to occupant behavior and cannot be predicted with certainty.

Despite these problems, the degree-day method (using an appropriate base temperature) can give remarkably accurate results for the annual heating energy of single-zone buildings dominated by

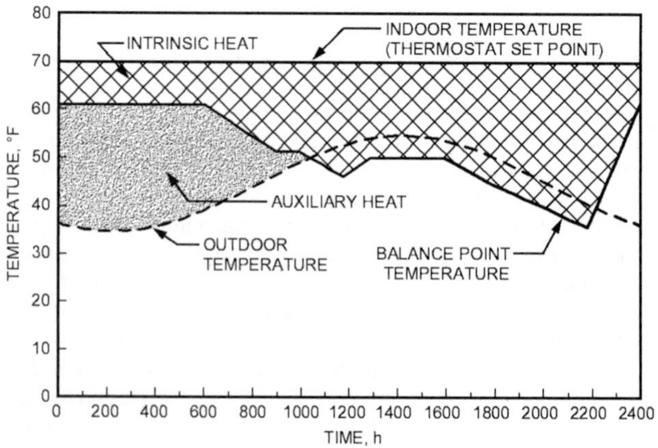

Fig. 13 Variation of Balance Point Temperature and Internal Gains for a Typical House
(Nisson and Dutt 1985)

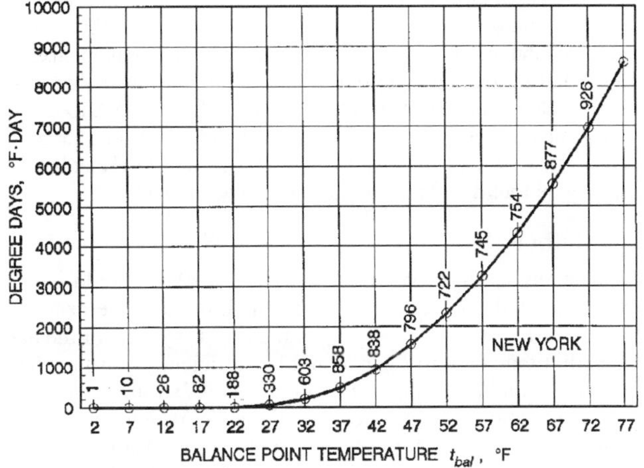

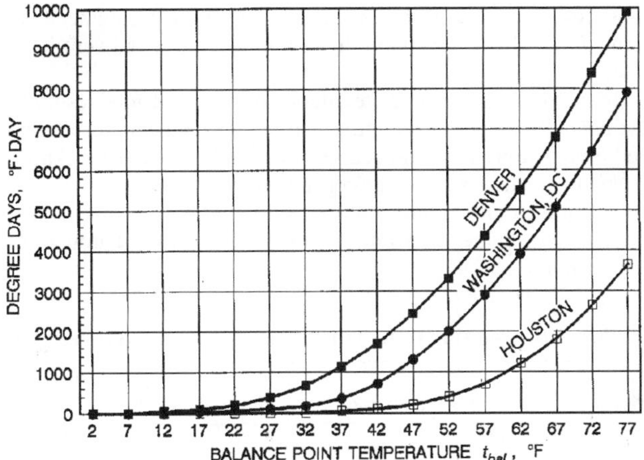

Fig. 14 Annual Heating Days $DD_h(t_{bal})$ as Function of Balance Temperature t_{bal}

losses through the walls and roof and/or ventilation. Typical buildings have time constants that are about 1 day, and a building's thermal inertia essentially averages over the diurnal variations, especially if t_i is allowed to float. Furthermore, the energy consumption in mild weather is small; hence, a relatively large error here has only a small effect on the total for the season.

Variable-Base Annual Degree-Days. The calculation of Q_h from degree-days $DD_h(t_{bal})$ depends on the value of t_{bal}. This value varies widely from one building to another because of widely differing personal preferences for thermostat settings and setbacks and because of different building characteristics. In response to the fuel crises of the 1970s, heat transmission coefficients have been reduced, and thermostat setback has become common. At the same time, the energy use by appliances has increased. These trends all reduce t_{bal} (Fels and Goldberg 1986). Hence, in general, degree-days with the traditional base 65°F are not to be used.

Figure 14A shows how the heating degree-days vary with t_{bal} for a particular site, in this case, New York. The plot is obtained by evaluating Equation (41) with data for the number of hours per year during which t_o is within 5°F temperature intervals centered at 77°F, 72°F, 67°F, 62°F, ..., 7°F. The data for the number of hours in each interval, or **bin**, are included as labels in this plot. Analogous curves, without these labels, are shown in Figure 14B for three other sites: Houston, Washington, and Denver. If the annual average of t_o is known, the cooling degree-days to any base below 72 ± 2.5°F can also be found.

Seasonal Efficiency. The seasonal efficiency η_h of heating equipment depends on such factors as steady-state efficiency, sizing, cycling effects, and energy conservation devices. Sometimes it is much lower than and other times it is comparable to steady-state efficiency. Alereza and Kusuda (1982) developed expressions to estimate the seasonal efficiency for a variety of furnaces, if information on rated input and output is available. These expressions correlate seasonal efficiency with variables determined by using the equipment simulation capabilities of a large hourly simulation program and typical equipment performance curves supplied by the National Institute of Standards and Technology (NIST):

$$\eta = \frac{\eta_{ss} CF_{pl}}{1 + \alpha_D} \tag{48}$$

where

η_{ss} = steady-state efficiency (rated output/input)
α_D = fraction of heat loss from ducts
CF_{pl} = part-load correction factor

The dimensionless term CF_{pl} is a characteristic of the part-load efficiency of the heating equipment, which may be calculated as follows:

Gas Forced-Air Furnaces

With pilot

$$CF_{pl} = 0.6328 + 0.5738(RLC) - 0.3323(RLC)^2$$

With intermittent ignition

$$CF_{pl} = 0.7791 + 0.1983(RLC) - 0.0711(RLC)^2$$

With intermittent ignition and loose stack damper

$$CF_{pl} = 0.9276 + 0.0732(RLC) - 0.0284(RLC)^2$$

Oil Furnaces Without Stack Damper

$$CF_{pl} = 0.7092 + 0.6515(RLC) - 0.4711(RLC)^2$$

Resistance Electric Furnaces

$$CF_{pl} = 1.0$$

These equations are based on many annual simulations for the equipment. The dimensionless ratio RLC of building design load to the capacity (rated output) of the equipment is defined as follows:

$$RLC = \frac{BLC}{CHT}(t_{bal} - t_{od})(1 + \alpha_D)$$

where

BLC = building loss coefficient, Btu/h·°F
t_{od} = outside design temperature, °F
CHT = capacity (rated output) of heating equipment, Btu/h

The building loss coefficient BLC can be defined as design-day heat loss/$(t_{bal} - t_{od})$. The design-day heat loss includes both infiltration and ground losses. Duct losses as a percentage of the design-day heat loss are added via the factor $(1 + \alpha_D)$. RLC assumes values in the range 0 to 1.0, appropriate for typical cases when the heating equipment is oversized. Seasonal efficiency is also discussed by Chi and Kelly (1978), Parker et al. (1980), and Mitchell (1983).

Monthly Degree-Days

Many formulas have been proposed for estimating the degree-days relative to an arbitrary base when detailed data are not available. The basic idea is to assume a typical probability distribution of temperature data, characterized by its average $\bar{t}_o$ and by its standard deviation σ. Erbs et al. (1983) developed a model that needs as input only the average $\bar{t}_o$ for each month of the year. The standard deviations σ_m for each month are then estimated from the correlation

$$\sigma_m = 3.54 - 0.0290\bar{t}_o + 0.0644\sigma_{yr} \tag{49}$$

This is a dimensional equation with t and σ in °F; σ_{yr} is the standard deviation of the monthly average temperatures about the annual average $\bar{t}_{o,yr}$:

$$\sigma_{yr} = \sqrt{\frac{1}{12}\sum_{1}^{12}(\bar{t}_o - \bar{t}_{o,yr})^2} \tag{50}$$

To obtain a simple expression for the degree-days, a normalized temperature variable ϕ is defined as

$$\phi = \frac{\bar{t}_{bal} - \bar{t}_o}{\sigma_m\sqrt{N}} \tag{51}$$

where N = number of days in the month (N has units of day/month and ϕ has units of $\sqrt{\text{month/day}}$). While temperature distributions can be different from month to month and location to location, most of this variability can be accounted for by the average and the standard deviation of $\bar{t}_o$. Being centered around $\bar{t}_o$ and scaled by σ_m, the quantity ϕ eliminates these effects. In terms of ϕ, the monthly heating degree-days for any location are well approximated by

$$DD_h(t_{bal}) = \sigma_m N^{1.5}\left[\frac{\phi}{2} + \frac{\ln(e^{-a\theta} + e^{a\theta})}{2a}\right] \tag{52}$$

where $a = 1.698\sqrt{\text{day/month}}$.

For nine locations spanning most climatic zones of the United States, Erbs et al. (1983) verified that the annual heating degree-days can be estimated with a maximum error of 315°F·days if Equation (52) is used for each month. For cooling degree-days, the largest error is 270°F·days. Such errors are quite acceptable, representing less than 5% of the total.

Table 5 lists monthly heating degree-days for New York City, using the model of Erbs et al. (1983), given monthly averages of t_o as reproduced in column 2 of Table 5. The degree-days are based on a balance temperature of 60°F. Column 2 lists the given values of monthly average outdoor temperature, and N is the number of days in the month. Intermediate quantities are shown in columns 4 and 5, and $t_{o,yr}$ and s_{yr} are shown at the bottom. Column 6 shows the monthly and annual results.

Table 6 contains degree-day data for several sites and monthly averaged outdoor temperatures needed for the algorithm. More complete tabulations of the latter are contained in Cinquemani et al. (1978) and in local climatological data summaries available from the National Climatic Data Center, Asheville, NC (NOAA 1973 and www.ncdc.noaa.gov). Monthly degree-day data at various bases, as well as other climatic information for 209 U.S. and 14 Canadian cities, may be found in Appendix 3 to Balcomb et al. (1982).

Bin Method

For many applications, the degree-day method should not be used, even with the variable-base method, because the heat loss coefficient K_{tot}, the efficiency η_h of the HVAC system, or the balance point temperature t_{bal} may not be sufficiently constant. The efficiency of a heat pump, for example, varies strongly with outdoor temperature; or the efficiency of the HVAC equipment may be affected indirectly by t_o when the efficiency varies with the load, a common situation for boilers and chillers. Furthermore, in most commercial buildings, the occupancy has a pronounced pattern, which affects heat gain, indoor temperature, and ventilation rate.

In such cases, a steady-state calculation can yield good results for the annual energy consumption if different temperature intervals and time periods are evaluated separately. This approach is known as the bin method because the consumption is calculated for several values of the outdoor temperature t_o and multiplied by the number of hours N_{bin} in the temperature interval (bin) centered around that temperature:

Table 5 Degree-Day Calculation from Monthly Averaged Data

Month	$\bar{t}_o$, °F	N, day/mo.	σ_m, °F	ϕ, $\sqrt{\text{mo./day}}$	$DD_h(t_{bal})$, °F·day
January	32.2	31	3.65	1.37	864
February	33.4	28	3.62	1.39	746
March	41.1	31	3.40	1.00	592
April	52.1	30	3.08	0.47	265
May	62.3	31	2.79	–0.15	67
June	71.6	30	2.52	–0.84	7
July	76.6	31	2.38	–1.26	2
August	74.9	31	2.41	–1.11	3
September	68.4	30	2.61	–0.59	16
October	58.7	31	2.88	0.08	123
November	46.4	30	3.22	0.72	391
December	35.5	31	3.56	1.24	762
$t_{o,yr}$	54.4			Sum	3837
σ_{yr}	15.8				

Note: Use Equation (52) to calculate $DD_h(t_{bal})$.

Table 6 Degree-Day and Monthly Average Temperatures for Various Locations

Site	Variable-Base Heating Degree-Day, °F·days[a]					Monthly Average Outdoor Temperature $\bar{t}_o$, °F[b]											
	65	60	55	50	45	Jan	Feb	Mar	Apr	May	Jun	Jul	Aug	Sep	Oct	Nov	Dec
Los Angeles, CA	1245	522	158	26	0	54.5	55.6	56.5	58.8	61.9	64.5	68.5	69.6	68.7	65.2	60.5	56.9
Denver, CO	6016	4723	3601	2653	1852	29.9	32.8	37.0	47.5	57.0	66.0	73.0	71.6	62.8	52.0	39.4	32.6
Miami, FL	206	54	8	0	0	67.2	67.8	71.3	75.0	78.0	81.0	82.3	82.9	81.7	77.8	72.2	68.3
Chicago, IL	6127	4952	3912	2998	2219	24.3	27.4	36.8	49.9	60.0	70.5	74.7	73.7	65.9	55.4	40.4	28.5
Albuquerque, NM	4292	3234	2330	1557	963	35.2	40.0	45.8	55.8	65.3	74.6	78.7	76.6	70.1	58.2	44.5	36.2
New York, NY	4909	3787	2806	1980	1311	32.2	33.4	41.1	52.1	62.3	71.6	76.6	74.9	68.4	58.7	47.4	35.5
Bismarck, ND	9044	7656	6425	5326	4374	8.2	13.5	25.1	43.0	54.4	63.8	70.8	69.2	57.5	46.8	28.9	15.6
Nashville, TN	3696	2758	1964	1338	852	38.3	41.0	48.7	60.1	68.5	76.6	79.6	78.5	72.0	60.9	48.4	40.4
Dallas/Ft. Worth, TX	2290	1544	949	526	250	45.4	49.4	55.8	66.4	73.8	81.6	85.7	85.8	78.2	68.0	55.9	48.2
Seattle, WA	4727	3269	2091	1194	602	39.7	43.5	45.5	50.4	56.5	61.3	65.7	64.9	60.6	54.2	45.7	42.0

[a]*Source*: NOAA (1973). [b]*Source*: Cinquemani et al. (1978).

Table 7 Sample Annual Bin Data

Site	100/ 104	95/ 99	90/ 94	85/ 89	80/ 84	75/ 79	70/ 74	65/ 69	60/ 64	55/ 59	50/ 54	45/ 49	40/ 44	35/ 39	30/ 34	25/ 29	20/ 24	15/ 19	10/ 14	5/ 9	0/ 4	–5/ –1
Chicago, IL			97	222	362	512	805	667	615	622	585	577	636	720	957	511	354	243	125	66	58	6
Dallas/Ft. Worth, TX	27	210	351	527	804	1100	947	705	826	761	615	615	523	364	289	57	29					
Denver, CO		3	118	235	348	390	472	697	699	762	783	718	665	758	713	565	399	164	106	65	80	22
Los Angeles, CA	8	8	9	17	53	194	632	1583	234	2055	1181	394	74	4								
Miami, FL			45	864	1900	2561	1605	871	442	222	105	77	36	12								
Nashville, TN		7	137	407	616	756	1100	866	706	692	650	670	720	582	342	280	107	71	29			
Seattle, WA				16	62	139	256	450	769	1353	1436	1461	1413	915	358	51	43	15	1			

Table 8 Calculation of Annual Heating Energy Consumption for Example 4

Climate			House	Heat Pump							Supplemental		
A	B	C	D	E	F	G	H	I	J	K	L	M	N
Temp. Bin, °F	Temp. Diff., $t_{bal} - t_{bin}$	Weather Data Bin, h	Heat Loss Rate, 1000 Btu/h	Heat Pump Integrated Heating Capacity, 1000 Btu/h	Cycling Capacity Adjustment Factor[a]	Adjusted Heat Pump Capacity, 1000 Btu/h[b]	Rated Electric Input, kW	Operating Time Fraction[c]	Heat Pump Supplied Heating, 10^6 Btu[d]	Seasonal Heat Pump Electric Consumption, KWh[e]	Space Load, 10^6 Btu[f]	Supplemental Heating Required, kWh[g]	Total Electric Energy Consumption[h]
62	2.3	740	1.8	44.3	0.760	33.7	3.77	0.05	1.30	146	1.30	—	146
57	7.3	673	5.5	41.8	0.783	32.7	3.67	0.17	3.72	417	3.72	—	417
52	12.3	690	9.3	39.3	0.809	31.8	3.56	0.29	6.42	719	6.42	—	719
47	17.3	684	13.1	36.8	0.839	30.9	3.46	0.42	8.95	1002	8.95	—	1002
42	22.3	790	16.9	29.9	0.891	26.6	3.23	0.63	13.31	1614	13.31	—	1614
37	27.3	744	20.6	28.3	0.932	26.4	3.15	0.78	15.35	1833	15.35	—	1833
32	32.3	542	24.4	26.6	0.979	26.0	3.07	0.94	13.22	1559	13.22	—	1559
27	37.3	254	28.2	25.0	1.000	25.0	3.00	1.00	6.35	762	7.16	236	998
22	42.3	138	31.9	23.4	1.000	23.4	2.92	1.00	3.23	403	4.41	345	748
17	47.3	54	35.7	21.8	1.000	21.8	2.84	1.00	1.18	153	1.93	220	373
12	52.3	17	39.5	19.3	1.000	19.3	2.74	1.00	0.33	47	0.67	101	147
7	57.3	2	43.3	16.8	1.000	16.8	2.63	1.00	0.03	5	0.09	16	21
2	62.3	0	47.0	14.3	1.000	—	—	—	—	—	—	—	—
								Totals:	73.39	8660	76.52	917	9578

[a]Cycling Capacity Adjustment Factor = 1 – C_d(1 – x), where C_d = degradation coefficient (default = 0.25 unless part load factor is known) and x = building heat loss per unit capacity at temperature bin. Cycling capacity = 1 at the balance point and below. The cycling capacity adjustment factor should be 1.0 at all temperature bins if the manufacturer includes cycling effects in the heat pump capacity (Column E) and associated electrical input (Column H).
[b]Column G = Column E × Column F
[c]Operating Time Factor equals smaller of 1 or Column D/Column G
[d]Column J = (Column I × Column G × Column C)/1000
[e]Column K = Column I × Column H × Column C
[f]Column L = Column C × Column D/1000
[g]Column M = (Column L – Column J) × 10^6/3413
[h]Column N = Column K + Column M

$$Q_{bin} = N_{bin} \frac{K_{tot}}{\eta_h} [t_{bal} - t_o]^+ \qquad (53)$$

The superscript plus sign indicates that only positive values are counted; no heating is needed when t_o is above t_{bal}. Equation (53) is evaluated for each bin, and the total consumption is the sum of the Q_{bin} over all bins.

In the United States, the necessary weather data are available in ASHRAE (1995) and USAF (1978). The bins are usually in 5°F increments and are often collected in three daily 8 h shifts. Mean coincident wet-bulb temperature data (for each dry-bulb bin) are used to calculate latent cooling loads from infiltration and ventilation. The bin method considers both occupied and unoccupied building conditions and gives credit for internal loads by adjusting the balance point. For example, a calculation could be performed for 42°F outdoors (representing all occurrences from 39.5 to 44.5°F) and with building operation during the midnight to 0800 shift. Because there are 23 5°F bins between –10 and 105°F and 3 8 h shifts, 69 separate operating points are calculated. For many applications, the number of calculations can be reduced. A residential heat pump (heating mode), for example, could be calculated for just the bins below 65°F without the three-shift breakdown.

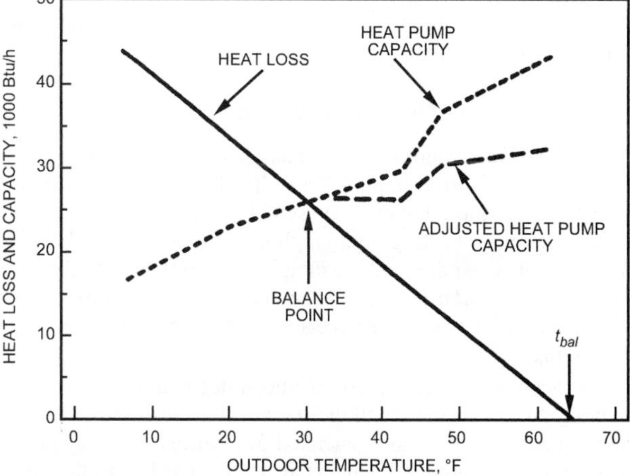

Fig. 15 Heat Pump Capacity and Building Load

The data included in Table 7 are samples of annual totals for a few sites, but ASHRAE (1995) and USAF (1978) include monthly data and data further separated into time intervals during the day.

Equipment performance may vary with load. For heat pumps, the U.S. Department of Energy has adopted test procedures to determine the effect of dynamic operations. The bin method uses these results for a specific heat pump to adjust the integrated capacity for the effect of part-load operation. Figure 15 compares the adjusted heat pump capacity to the building heat loss in Example 4 below. This type of curve must be developed for each model heat pump as applied to an individual profile. The heat pump cycles on and off above the balance point temperature to meet the house load, while supplemental heat is required at lower temperatures. This cycling can reduce performance, depending on the part-load factor at a given temperature. The cycling capacity adjustment factors used in this example to account for cycling degradation can be calculated from the equation in footnote a of Table 8.

Frosting and the necessary defrost cycle can reduce performance over steady-state conditions that do not include frosting. The effects of frosting and defrosting are already integrated into many (but not all) manufacturer's published performance data. Example 4 assumes that the manufacturer's data already accounts for the frosting/defrosting losses (as indicated by the characteristic notch of the capacity curve in Figure 15) and shows how to adjust an integrated performance curve for cycling losses.

Example 4. Estimate the energy requirements for a residence with a design heat loss of 40,000 Btu at 53°F design temperature difference. The inside design temperature is 70°F. It is estimated that internal heat gains average 4280 Btu/h. Assume the selection of a 3 ton heat pump having the characteristics given in Columns E and H of Table 8 and in Figure 15.

Solution: The design heat loss is based on no internal heat generation. The heat pump system energy input is the net heat requirement of the space (i.e., the envelope loss minus the internal heat generation). The net heat loss per degree and the heating/cooling balance temperature may be computed:

$$K_{tot} = \text{HL}/\Delta t = 40{,}000/53 = 755 \text{ Btu/h}\cdot{}^\circ\text{F}$$

From Equation (38),

$$t_{bal} = 70 - (4280/755) = 64.3^\circ\text{F}$$

Table 8 is then computed, resulting in 9578 kWh.

The **modified bin method** (Knebel 1983) extends the basic bin method to account for weekday/weekend and partial-day occupancy effects, to calculate net building loads (conduction, infiltration, internal loads, and solar loads) at four temperatures, rather than interpolate from design values, and to better describe secondary and primary equipment performance.

CORRELATION METHODS

One way to simplify energy analyses is to correlate energy requirements to various inputs. Typically, the result of a correlation is a simple equation that may be used in a calculator or small computer program or to develop a graph that provides a quick insight into the energy requirements. Examples of correlation methods are in ASHRAE *Standard* 90.1, which includes several empirical equations that may be used to predict energy consumption by many types of buildings.

The accuracy of correlation methods depends on the size and accuracy of the database and the statistical means used to develop the correlation. A database generated from measured data can lead to accurate correlations (Lachal et al. 1992). The key to the proper use of a correlation is that the case being studied matches the cases used in developing the database. Inputs to the correlation (the independent variables) indicate the factors that are considered to have significant impact on energy consumption. A correlation is invalid either when an input parameter is used beyond its valid range (corresponding to extrapolation rather than interpolation) or when some important feature of the building/system is not included in the available inputs to the correlation.

SIMULATING SECONDARY AND PRIMARY SYSTEMS

Traditionally, most energy analysis programs include a set of preprogrammed models that represent various systems, such as variable air volume, terminal reheat, multizone, etc. In this scheme, the equations for each system are arranged so they can be solved sequentially. If this is not possible, then the smallest number of equations that must be solved simultaneously is solved using an appropriate technique. Furthermore, individual equations may vary from hour to hour in the simulation, depending on controls and operating conditions. For example, a dry coil uses different equations than a wet coil.

The primary disadvantage of this scheme is that it is relatively inflexible—in order to modify a system, the program source code may have to be modified and recompiled. Alternative strategies (Klein et al. 1994, Park et al. 1985) have viewed the system as a series of components (e.g. fan, coil, pump, duct, pipe, damper, thermostat) that may be organized in a component library. Users of the program specify the connections between the components. The program then resolves the specification of components and connections into a set of simultaneous equations.

A refinement of component-based modeling is known as equation-based modeling (Sowell and Moshier 1995, Buhl et al. 1993). In this scheme, the models do not follow well-defined rules for a solution, and input and output variables are not predetermined.

MODELING OF SYSTEM CONTROLS

From a mathematical viewpoint, controls represent equations that must be satisfied at each point during the simulation. For example, the room thermostat can be represented as a function relating heating and cooling delivery to space temperature. Similarly, cooling coil reset controls can be modeled as a relationship between outside or zone temperature and coil discharge temperature. An accurate secondary system model must ensure that all controls are properly represented and that the governing equations are satisfied at each simulation time step. This often creates a need for iteration or, alternately, for use of values from an earlier solution point.

The controls on space temperature affect the interaction between loads calculations and the secondary system simulation. A realistic model might require a dead band in space temperature in which no heating or cooling is called for; within this range, the true space sensible load is zero, and the true space temperature must be adjusted accordingly. If the thermostat has proportional control between zero and full capacity, the space temperature will rise in proportion to the load during cooling and fall similarly during heating. Capacity to heat or cool also varies with space temperature after the control device has reached its maximum because capacity is proportional to the **difference** between supply and space temperatures. Failure to properly model these phenomena results in overestimating required energy.

INTEGRATION OF SYSTEM MODELS

Energy calculations for secondary systems involve construction of the complete system from the set of HVAC components. For example, a VAV system is a single-path system that controls zone temperature by modulating the airflow while maintaining a constant supply air temperature. VAV terminal units, located at each zone, adjust the quantity of air reaching each zone depending on its load requirements. Reheat coils may be included to provide required heating for perimeter zones.

This VAV system simulation consists of a central air-handling unit and a VAV terminal unit with reheat coil located at each zone,

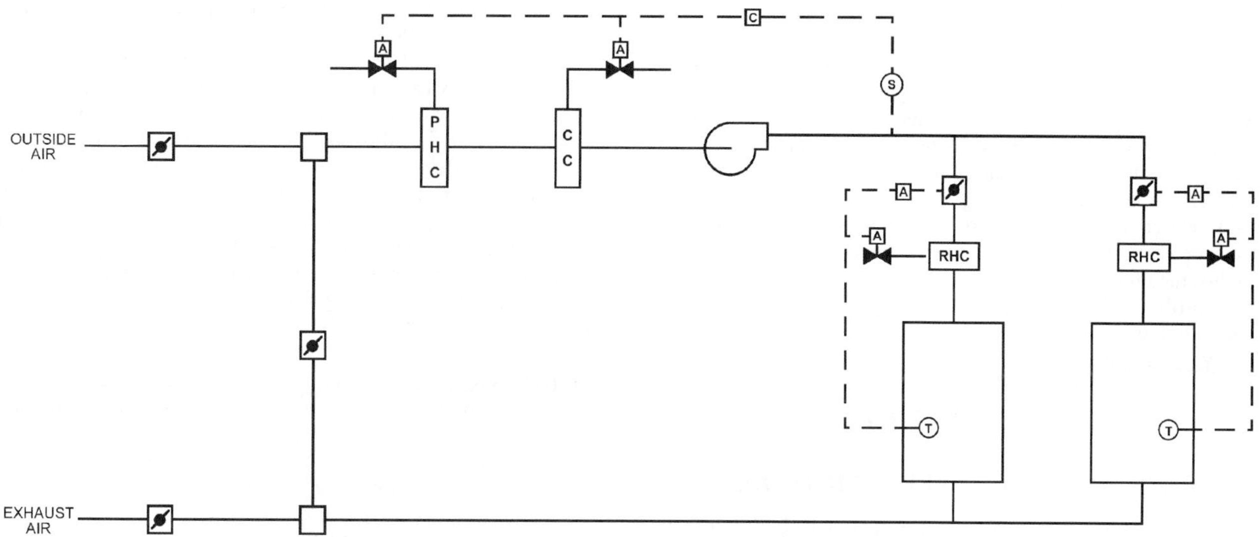

Fig. 16 Schematic of Variable Air Volume System with Reheat

as shown in Figure 16. The central air-handling unit includes a fan, a cooling coil, a preheat coil, and an outside air economizer. The supply air leaving the air-handling unit is controlled to a fixed set point. The VAV terminal unit at each zone varies the airflow to meet the cooling load. As the zone cooling load decreases, the VAV terminal unit decreases the zone airflow until the unit reaches its minimum position. If the cooling load continues to decrease, the reheat coil will be activated to meet the zone load. As the supply air volume leaving the unit decreases, the fan power consumption will also be reduced. A variable-speed drive is used to control the supply fan.

The simulation is based on system characteristics and zone design requirements. For each zone, the inputs include the sensible and latent loads, the zone set point temperature, and the minimum zone supply air mass flow. System characteristics include the supply air temperature set point, the entering water temperature of the reheat, preheat and cooling coils, the minimum mass flow of outside air, and the economizer temperature/enthalpy set point for minimum airflow.

The algorithm for performing the calculations for this VAV system is shown in Figure 17. The algorithm directs sequential calculations of system performance. Calculations proceed from the zones forward along the return air path to the cooling coil inlet and back through the supply air path to the cooling coil discharge.

Moving back along the supply air path, the fan entering air temperature is calculated by setting fan outlet air temperature to the system design supply air temperature. The known fan inlet air temperature is then used as both the cooling coil and preheat coil discharge air temperature set point. Moving forward along the return air path, the cooling coil entering air temperature can be determined by sequentially moving through the economizer cycle and the preheat coil.

Unlike temperature, the humidity ratio at any point in a system cannot be explicitly determined due to the dependence of the cooling coil performance on the mixed air humidity ratio. The latent load defines the difference between zone humidity and supply air humidity. However, the humidity ratio of the supply air depends on the humidity ratio entering the coil, which in turn depends on that of the return air. This calculation must be performed either by solving simultaneous equations or, as in this case, by an iterative process.

Assuming a trial value for the humidity ratio at the cooling coil discharge (e.g., 55°F, 90% rh), the humidity ratio at all other points throughout the system can be calculated. With known cooling coil inlet air conditions and a design discharge air temperature, the inverted cooling coil subroutine iterates on the coil fluid mass flow

```
BEGIN LOOP  Calculate zone related design requirements
    •  Calculate required supply airflow to meet zone load
    •  Sum actual zone mass airflow rate
    •  Sum zone latent loads
  IF zone equals last zone THEN Exit Loop
END LOOP
•  Calculate system return air temperature from zone temps
•  Assume an initial cooling coil leaving air humidity ratio

BEGIN LOOP  Iterate on cooling coil leaving air humidity ratio
    •  Calculate return air humidity ratio from latent loads
    •  Calculate supply fan power consumption and
       entering fan air temperature
    •  Calculate mixed air temperature and humidity
       ratio using an economizer cycle
  IF mixed air temperature is less than design
       supply air temperature THEN
         •  Calculate preheat coil load
  ELSE
         •  Calculate cooling coil load and leaving air
            humidity ratio
  ENDIF
  IF cooling coil leaving air humidity ratio converged
  THEN Exit Loop
END LOOP

BEGIN LOOP Calculate the zone reheat coil loads
  IF zone supply air temperature is greater than system
  design supply air temperature THEN
       •  Calculate reheat coil load
          (Subroutine: COILINV/HCDET)
  ENDIF
  •  Sum reheat coil loads for all zones
  IF zone equals last zone THEN Exit Loop
END LOOP
```

Fig. 17 Algorithm for Calculating Performance of VAV with System Reheat

to converge on the discharge air temperature with the discharge air humidity ratio as an output. The cooling coil discharge air humidity ratio is then compared to the previous discharge humidity ratio. This iterative process continues calculating through the loop several times until the values of the cooling coil discharge air humidity ratio stabilize within a specified tolerance.

This basic algorithm for simulation of a VAV system might be used in conjunction with a heat balance type of load calculation. For a weighting factor approach, it would have to be modified to allow zone temperatures to vary and consequently zone loads to be readjusted. It should also be enhanced to allow for possible limits on reheat temperature and/or cooling coil limits, zone humidity limits, outside air control (economizers), and/or heat-recovery devices, zone exhaust, return air fan, heat gain in the return air path because of lights, the presence of baseboard-type heaters, and more realistic control profiles. Most current building energy programs incorporate these and other features as user options, as well as algorithms for other types of systems.

INVERSE MODELING

CATEGORIES OF INVERSE METHODS

Inverse methods for energy-use estimation in buildings and related HVAC&R equipment can be classified into three broad categories. These approaches are widely disparate in data requirements, in time and effort needed to develop the associated models, in user skill, and in the sophistication and reliability that they provide.

Empirical or "Black-Box" Approach

With this approach, a simple or multi-variate regression model is identified between measured energy use and the various influential parameters (climatic variables, building occupancy). The form of the regression models can be either purely statistical or loosely based on some basic engineering formulation of energy use in the building. In any case, the identified model coefficients are such that no (or very little) physical meaning can be assigned to them. This approach can be used with any time scale—monthly, daily, hourly or subhourly—provided appropriate data are available. Single-variate, multivariate, change point, Fourier series, and artificial neural network (ANN) models fall under this category, as noted in Table 1.

Model identification is relatively straightforward, usually requires little effort, and can be used in several diverse types of circumstances. The empirical approach is thus the most widely used inverse approach. Although more sophisticated regression techniques such as maximum likelihood and two-stage regression schemes can be used for model identification, least-squares regression is most common. The purely statistical approach is usually adequate for evaluating demand-side management (DSM) programs to identify simple and conventional energy conservation measures in an actual building (lighting retrofits, air handler retrofits such as CV to VAV retrofits) and for baseline model development in energy conservation measurement and verification (M&V) projects (Fels 1986; MacDonald and Wasserman 1989; Ruch and Claridge 1991; Reddy et al. 1997; Claridge 1998b; Kissock et al. 1998; Katipamula et al. 1998; Dhar 1995; Dhar et al. 1998, 1999a, 1999b; Miller and Seem 1991; Kreider and Wang 1991; Krarti et al. 1998). It is also appropriate for modeling equipment such as pumps and fans, and even more elaborate equipment such as chillers and boilers, provided the necessary performance data are available (Phelan et al. 1996, Braun 1992, Englander and Norford 1992, Lorenzetti and Norford 1993). Although this approach allows detection or flagging of equipment or system faults, it is usually of limited value for diagnosis and on-line control (with ANN as a possible exception).

Calibrated Simulation Approach

With this approach, one uses an existing building simulation computer program and "tunes" or calibrates the various physical inputs to the program so that observed energy use matches closely with that predicted by the simulation program. Once that is achieved, more reliable predictions can be made than with the statistical approaches. The calibrated simulation approach is advocated in cases where only whole-building metering is available and M&V calls for estimating the energy savings of individual retrofits. Practitioners of this approach have largely tended to adopt existing and widely used forward simulation programs such as DOE-2 to subsequently perform the calibration with performance data. Hourly subaggregated monitored energy data (most compatible with the time step adopted by most building energy simulation programs) allow the development of the most accurate calibrated model, but analysts are usually forced to work with less data. Tuning can be done with monthly data or data that span only a few weeks or months over the year, but the resulting model is very likely to be increasingly less accurate with decrease in performance data.

The main reservations with the widespread use of calibrated simulation are that it is labor-intensive, requires a high level of user skill and knowledge in both simulation and practical building operation, is time-consuming, and is often dependent on the person doing the calibration. Several practical difficulties prevent achieving a calibrated simulation or a simulation that nearly reflects the actual building performance, including (1) the measurement and adaptation of weather data for use by the simulation programs (e.g., converting global horizontal solar into beam and diffuse solar radiation), (2) the choice of methods used to calibrate the model, and (3) the choice of methods used to measure the required input parameters for the simulation (i.e., the mass of the building, infiltration coefficients, and shading coefficients). Truly "calibrated" models have been achieved in only a few applications because they require a very large number of input parameters, a high degree of expertise, and enormous amounts of computing time, patience, and financial resources. Bronson et al. (1992), Haberl and Bou-Saada (1998), Kaplan et al. (1990), Corson (1992), Bou-Saada and Haberl (1995a, 1995b), Manke et al. (1996), and Norford et al. (1994) provide examples of different methods used to calibrate simulation models.

In recent years, Katipamula and Claridge (1993) and Liu and Claridge (1998) have suggested that models simpler than those used in the detailed simulation programs such as DOE-2 could also serve the intended purpose while allowing model calibration to be done much faster. Typically, the building is divided into two zones: an exterior or perimeter zone and an interior or a core zone. The core zone is assumed to be insulated from the envelope heat losses/gains, while the solar heat gains, infiltration heat loss/gain, and conduction gains/losses from the roof are taken to appear as loads on the external zone only. Given the internal load schedule, the building description, the type of HVAC system, and the climatic parameters, the HVAC system loads can be estimated for each hour of the day and for as many days of the year as needed by the simplified systems model. Since there are fewer parameters to vary, the calibration process is much faster. Therefore, these models have a significant advantage over general-purpose models in buildings where the HVAC systems can be adequately modeled. These studies, based on the ASHRAE Simplified Energy Analysis Procedure (Knebel 1983), illustrate the applicability of this method both to baseline model development for M&V purposes and as a diagnostic tool for identifying potential operational problems and for estimating potential savings from optimized operating parameters.

Gray-Box Approach

With this approach, a physical model is first formulated to represent the structure or physical configuration of the building or HVAC&R equipment or system, and then important parameters representative of certain key and aggregated physical parameters and characteristics are identified by a statistical analysis (Rabl and Riahle 1992). This approach requires a high level of user expertise both in setting up the appropriate modeling equations and in the estimation of these parameters. Often an intrusive experimental protocol is necessary for proper parameter estimation that also requires a certain amount of user expertise. This approach has great potential, especially for fault detection and diagnosis (FDD) and on-line control, but its applicability to whole-building energy use is limited. Examples of parameter estimation studies as applied to building energy use are Sonderegger (1977), Hammersten (1984), Subbarao (1988), Rabl (1988), Reddy (1989), Andersen and Brandemuehl (1992), Braun (1990), Reddy et al. (1999), Gordon and Ng (1995), and Guyon and Palomo (1999).

TYPES OF INVERSE MODELS

We distinguish between two types of models: steady-state and dynamic. **Steady-state models** are those that do not consider such effects such as thermal mass or capacitance that cause short-term temperature transients. Generally these models are appropriate for monthly, weekly, or daily data and are often used for baseline model development. **Dynamic models** capture effects such as building warm-up or cool-down periods and peak loads and are appropriate for building load control, FDD, and equipment control. A simple criterion to determine whether a model is steady-state or dynamic is to look for the presence of time-lagged variables, either in the response or regressor variables. Steady-state models do not contain time-lagged variables.

Steady-State Models

There are several types of steady-state models used for both building and equipment energy use. They are single-variate, multivariate, polynomial, and physical models.

Single-Variate Models. Single-variate models (i.e., models with one regressor variable only) are perhaps the most widely used. They formulate energy use in a building as a function of one driving force that impacts building energy use. An important aspect in identifying statistical models of baseline energy use is the choice of the functional form and the independent (or regressor) variables. Extensive studies (Fels 1986, Kissock et al. 1993, Katipamula et al. 1994, Reddy et al. 1997) have clearly indicated that the outdoor dry-bulb temperature is the most important regressor variable, especially at monthly time scales but also at daily time scales.

The simplest steady-state inverse model is one developed by regressing monthly utility consumption data against average billing-period temperatures. The model must identify the balance point temperatures (or change points) at which energy use switches from weather-dependent to weather-independent behavior. In its simplest form, the 65°F degree-day model is a change-point model that has a fixed change point at 65°F. Other examples include three- and five-parameter Princeton Scorekeeping Methods (PRISM) based on the variable-base degree-day concept (Fels 1986). An allied modeling approach for commercial buildings is the four-parameter (4-P) model developed by Ruch and Claridge (1991), which is based on the monthly mean temperature (and not degree-days). Table 9 shows the appropriate model functional forms. The three parameters are a weather-independent base-level use, a change point, and a temperature-dependent energy use, characterized as a slope of a line that is determined by regression. The four parameters include a change point, a slope above the change point, a slope below the change point, and the energy use associated with the change point. An inverse bin method has also been proposed to handle more than four change points (Thamilseran and Haberl 1995).

Figure 18 shows several types of steady-state, single-variate inverse models. Figure 18A shows a simple one-parameter, or constant, model, and Table 9 gives the equivalent notation for calculating the constant energy use using this model. Figure 18B shows a steady-state two-parameter (2-P) model where b_0 is the y-axis intercept and b_1 is the slope of the regression line for positive values of x, where x represents the ambient air temperature. The 2-P model represents cases when either heating or cooling is always required.

Figure 18C shows a three-parameter, change-point model. This model is typical of natural gas energy use in a single-family residence that uses gas for space heating and domestic water heating. In the notation of Table 9 for the three-parameter model, b_0 represents the baseline energy use and b_1 is the slope of the regression line for values of ambient temperature less than the change point b_2. In this type of notation, the superscript plus sign indicates that only positive values of the parenthetical expression are considered. Figure 18D shows a three-parameter model for cooling energy use, and Table 9 provides the appropriate analytic expression.

Figure 18E and Figure 18F illustrate four-parameter models for heating and cooling, respectively. The appropriate expressions for calculating the heating and cooling energy consumption using a four-parameter model are found in Table 9: b_0 represents the baseline energy exactly at the change point b_3, and b_1 and b_2 are the lower and upper region regression slopes for ambient air temperature below and above the change point b_3. Figure 18G illustrates a 5-P model (Fels 1986). Such a model is useful for modeling buildings that are electrically heated and cooled. The 5-P model has two change points and a base level consumption value.

The advantage of these steady-state inverse models is that their use can be easily automated and applied to large numbers of buildings where monthly utility billing data and average daily temperatures for the billing period are available. Steady-state single-variate inverse models have also been applied with success to daily data (Kissock et al. 1998). In such a case, the variable-base degree-day method and the monthly mean temperature models described earlier for utility billing data analysis become identical in their functional form. Single-variate models can also be applied to daily data to compensate for differences such as weekday and weekend use by separating the data accordingly and identifying models for each period separately.

Table 9 Single-Variate Models Applied to Utility Billing Data

Model Type	Independent Variable(s)	Form	Examples
One-parameter or constant (1-P)	None	$E = b_0$	Non-weather-sensitive demand
Two-parameter (2-P)	Temperature	$E = b_0 + b_1(T)$	
Three-parameter (3-P)	Degree-days/ Temperature	$E = b_0 + b_1(DD_{BT})$ $E = b_0 + b_1(b_2 - T)^+$ $E = b_0 + b_1(T - b_2)^+$	Seasonal weather-sensitive use (fuel in winter, electricity in summer for cooling)
Four-parameter change point (4-P)	Temperature	$E = b_0 + b_1(b_3 - T)^+ - b_2(T - b_3)^+$ $E = b_0 - b_1(b_3 - T)^+ + b_2(T - b_3)^+$	Energy use in commercial buildings
Five-parameter (5-P)	Degree-days/ Monthly mean temperature	$E = b_0 - b_1(DD_{TH}) + b_2(DD_{TC})$ $E = b_0 + b_1(b_3 - T)^+ + b_2(T - b_4)^+$	Heating and cooling supplied by same meter

Note: DD denotes degree-days and T is the monthly mean daily outdoor dry-bulb temperature.

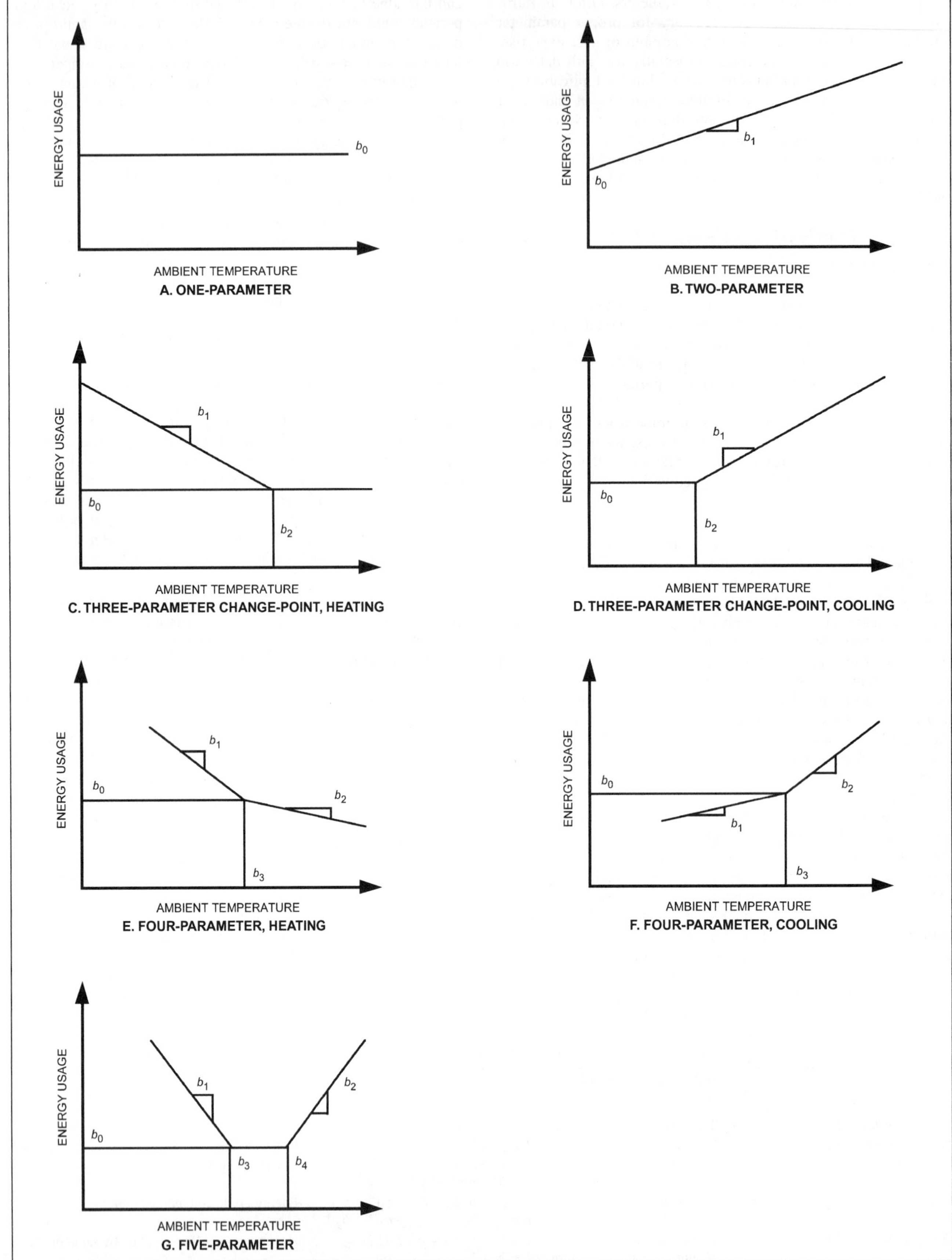

Fig. 18 Steady-State, Single-Variate Models Appropriate for Modeling Energy Use in Residential and Commercial Buildings

Disadvantages of steady-state single-variate inverse models include insensitivity to dynamic effects (e.g., thermal mass), insensitivity to variables other than temperature (e.g., humidity and solar gain), and inappropriateness for some buildings (e.g., buildings with strong on/off schedule-dependent loads or buildings with multiple change points). Moreover, the use of a single-variable, 3-P model such as the PRISM model (Fels 1986) has a physical basis only when energy use above a base level is linearly proportional to degree-days. This is a good approximation in the case of heating energy use in residential buildings where the heating load never exceeds the capacity of the heating system. However, commercial buildings, in general, have higher internal heat generation with simultaneous heating and cooling energy use and are strongly influenced by HVAC system type and control strategy. This makes energy use in commercial buildings less strongly influenced by outdoor air temperature alone. Therefore, it is not surprising that blind use of single-variate models has had mixed success at modeling energy use in commercial buildings (MacDonald and Wasserman 1989).

Change-point regression models work best with heating data from buildings with systems that have little or no part-load nonlinearities (i.e., systems that become less efficient as they begin to cycle on-off with part loads). In general, change-point regression models do not predict cooling loads as well because outdoor humidity has a large influence on latent loads on the cooling coil. Other factors that decrease the accuracy of change-point models include solar effects, thermal lags, and on-off HVAC schedules. Four-parameter models exhibit a better statistical fit than three-parameter models in buildings with continuous, year-round cooling or heating (e.g., grocery stores and office buildings with high internal loads). However, every model should be checked to ensure that the regression is not falsely indicating an unreasonable relationship.

A major advantage of using a steady-state inverse model to evaluate the effectiveness of energy conservation retrofits lies in its ability to factor out year-to-year weather variations by using a normalized annual consumption (NAC) (Fels 1986). Basically, the annual energy conservation savings can be calculated by comparing the difference obtained by multiplying the preretrofit and postretrofit parameters by the weather conditions for the average year. Typically, 10 to 20 years of average daily weather data from a nearby weather service site are used to calculate 365 days of average weather conditions, which are then used to calculate the average preretrofit and postretrofit conditions.

Utilities and government agencies have found it advantageous to prescreen many buildings against test regression models. Such inverse models can be used to develop comparative figures of merit for buildings in a similar standard industrial code (SIC) classification. In such applications, a minimum goodness of fit is usually established that determines whether the monthly utility billing data are well fitted by the one-, two-, three-, four-, or five-parameter model being tested. Comparative figures of merit can then be determined by dividing the parameters by the conditioned floor area to yield average daily energy use per unit area of conditioned space. For example, an area-normalized comparison of base-level parameters across residential buildings would be used to analyze weather-independent energy use. Such information can be used by energy auditors to focus their efforts on those systems needing assistance (Haberl and Komor 1990a, 1990b).

Multivariate Models. Two types of steady-state, multivariate models have been reported:

1. **Standard multiple-linear or change-point regression models,** where the set of data observations is treated without retaining the time series nature of the data (Katipamula et al. 1998).
2. **Fourier series models** that retain the time-series nature of the building energy use data and capture the diurnal and seasonal cycles according to which buildings are operated (Seem and Braun 1991; Dhar 1995; Dhar et al. 1998, 1999a, 1999b).

These models are a logical extension to single-variate models, provided that the choice of the variables to be included and their functional forms are based on the engineering principles on which HVAC systems and other systems in commercial buildings operate. The goal of modeling energy use by the multivariate approach is to characterize building energy use with a few readily available and reliable input variables. These input variables should be selected with care. The model should contain variables not affected by the retrofit and likely to change (for example, climatic variables) from preretrofit to postretrofit periods. Other less obvious variables, such as changes in operating hours, in base load, and in occupancy levels, should be included in the model if these are not energy conservation measures (ECMs) but variables that may change during the postretrofit period.

Environmental variables that meet the above criteria for modeling heating and cooling energy use include outdoor air dry-bulb temperature, solar radiation, and outdoor specific humidity. Some of these parameters are difficult to estimate or measure in an actual building and hence are not good candidates for regressor variables. Further, some of the variables vary little. Although their effect on energy use may be important, an inverse model will implicitly lump their effect into the parameter that represents constant load. In commercial buildings, internally generated loads, such as the heat given off by people, lights, and electrical equipment, also impact heating and cooling energy use. Such internal loads are difficult to measure in their entirety given the ambiguous nature of occupant loads and latent loads. However, monitored electricity used by internal lights and equipment is a good surrogate for total internal sensible loads (Reddy et al. 1999). For example, when the building is fully occupied, it is also likely to be experiencing high internal electric loads, and vice versa.

The effect of environmental variables is important for such buildings as offices but may be less so for mixed-use buildings (e.g., hotels and hospitals) and buildings such as retail buildings, schools, and assembly buildings. Differences in HVAC system behavior during occupied and unoccupied periods can be modeled by a dummy or indicator variable (Draper and Smith 1981). For some office buildings, there seems to be little need to include such a dummy variable, but its inclusion in the general functional form will provide added flexibility.

Several standard statistical tests exist for evaluating the goodness-of-fit of the model and the degree of influence that each of the independent variables exerts on the response variable (Draper and Smith 1981, Neter et al. 1989). Although energy use is in fact dependent on several variables, there are strong practical incentives for identifying the simplest model that results in acceptable accuracy. Multivariate models require more metering and are unusable if even one of the variables becomes unavailable. In addition, some of the regressor variables may be linearly correlated. This condition, called **multicollinearity**, can result in large uncertainty in the estimates of the regression coefficients (i.e., unintended error) and can also lead to poorer model prediction accuracy compared to a model where the regressors are not linearly correlated.

Several authors recommend using **principal component analysis** (PCA) to overcome multicollinearity effects. PCA was one of the strongest analysis methods in the ASHRAE Predictor Shootout I and II contests (Kreider and Haberl 1994, Haberl and Thamilseran 1996). Analysis of multiyear monitored daily energy use in a grocery store found a clear superiority of PCA over multivariate regression models (Ruch et al. 1993), but this conclusion is unproven for commercial building energy use in general. A more general evaluation by Reddy and Claridge (1994) of both analysis techniques using synthetic data from four different geographic locations in the U.S. found that injudicious use of PCA may exacerbate rather than overcome problems associated with multicollinearity. Draper and Smith (1981) also caution against indiscriminate use of PCA.

The functional basis of air-side heating and cooling use in various HVAC system types has been addressed by Reddy et al. (1995) and subsequently applied to monitored data in commercial buildings (Katipamula et al. 1994, 1998). Because the quadratic and cross-product terms of the engineering equations are not usually picked up by the multivariate models, one is often left with models for energy use that are strictly **linear**.

In addition to T_o, internal electric equipment and lighting load E_{int}, solar loads q_{sol}, and latent effects via the outdoor dew-point temperature T_{dp} are candidate regressor variables. In commercial buildings, a major portion of the latent load is due to fresh air ventilation. However, this load appears only when the outdoor air dew-point temperature exceeds the cooling coil temperature. Hence, the term $(T_{dp} - T_s)^+$ (where the + sign indicates that the term is to be set to zero if negative, and T_s is the mean surface temperature of the cooling coil, typically about 51 to 55°F) is a more realistic descriptor of the latent loads than is T_{dp} alone. The use of $(T_{dp} - T_s)^+$ as a regressor in the model is a simplification that seems to yield good accuracy.

Therefore, a multivariate linear regression model with an engineering basis has the following structure:

$$Q_{bldg} = \beta_0 + \beta_1(T_o - \beta_3)^- + \beta_2(T_o - \beta_3)^+ + \beta_4(T_{dp} - \beta_6)^- + \beta_5(T_{dp} - \beta_6)^+ + \beta_7 q_{sol} + \beta_8 E_{int} \quad (54)$$

Based on the above discussion, $\beta_4 = 0$. Introducing indicator variable terminology (Draper and Smith 1981), Equation (54) becomes identical to

$$Q_{bldg} = a + bT_o + cI + dIT_o + eT_{dp}^+ + fq_{sol} + gE_{int} \quad (55)$$

where the indicator variable I is introduced to handle the change in slope of the energy use due to T_o. The variable I is set equal to 1 for T_o values to the right of the change point (i.e., for high T_o range) and set equal to 0 for low T_o values. As with the single-variate segmented models (i.e., 3-P and 4-P models), a search method is used in order to determine the change point that minimizes the total sum of squares of residuals (Fels 1986, Kissock et al. 1993).

Another finding from the Katipamula et al. (1994) study was that the model given by Equation (55), appropriate for VAV systems, could be simplified for constant volume HVAC systems:

$$Q_{bldg} = a + bT_o + eT_{dp}^+ + fq_{sol} + gE_{int} \quad (56)$$

Note that instead of using $(T_{dp} - T_s)^+$, one could equally use the absolute humidity potential $(W_0 - W_s)^+$, where W_0 is the outdoor absolute humidity, and W_s, the absolute humidity level at the dew point of the cooling coil, is typically about 0.198 lb/lb. A final aspect to be kept in mind is that the term T_{dp}^+ should be omitted from the regressor variable set when regressing heating energy use because there are no latent loads on a heating coil.

The above multivariate models have been found to be very accurate for daily time scales and slightly less so for hourly time scales. This is because changes in the way the building is operated during the day and the night lead to different relative effects of the various regressors on energy use, which cannot be accurately modeled by one single hourly model. Breaking up the energy use data into hourly bins corresponding to each hour of the day and then identifying 24 individual hourly models leads to appreciably greater accuracy (Katipamula et al. 1994).

Polynomial Models. Historically, polynomial models have been widely used as pure statistical models to model the behavior of such equipment as pumps, fans, and chillers (Stoecker and Jones 1982). The theoretical aspects of calculating pump performance are well understood and documented. Pump capacity and efficiency are calculated from measurements of pump head, flow rate, and pump electrical power input. Phelan et al. (1996) have studied the predictive ability of linear and quadratic models for electricity consumed by pumps and water mass flow rate and concluded that quadratic models are superior to linear models. For fans, Phelan et al. (1996) have studied the predictive ability of linear and quadratic polynomial single-variate models of fan electricity consumption as a function of the supply air mass flow rate and concluded that, although quadratic models are superior in terms of predicting energy use, the linear model seems to be the better overall predictor of both energy and demand (i.e., maximum monthly power consumed by the fan). This is a noteworthy conclusion given that a third-order polynomial is warranted analytically as well as from monitored field data presented by previous authors (e.g., Englander and Norford 1992, Lorenzetti and Norford 1993).

Polynomial models have been used to correlate chiller (or evaporator) thermal cooling capacity or load Q_{evap} and the electrical power consumed by the chiller (or compressor) E_{comp} with the relevant number of influential physical parameters. For example, based on the functional form of the DOE-2 building simulation software (York and Cappiello 1982), models for part-load performance of energy equipment and plant, E_{comp}, can be modeled as the following triquadratic polynomial:

$$E_{comp} = a + bQ_{evap} + cT_{cond}^{in} + dT_{evap}^{out} + eQ_{evap}^2 + fT_{cond}^{in\,2} + gT_{evap}^{out\,2} + hQ_{evap}T_{cond}^{in} + iT_{evap}^{out}Q_{evap} + jT_{cond}^{in}T_{evap}^{out} + kQ_{evap}T_{cond}^{in}T_{evap}^{out} \quad (57)$$

In this model, there are 11 model parameters to identify. However, since all of them are unlikely to be statistically significant, a step-wise regression to the sample data set yields the optimal set of parameters to retain in a given model. Other authors, such as Braun (1992), have used slightly different polynomial forms.

Physical Models. In contrast to polynomial models, which have no physical basis (merely a convenient statistical one), physical models are based on fundamental thermodynamic or heat transfer considerations. These types of models are usually associated with the parameter estimation approach. Often such models are preferred because they generally have fewer parameters. Furthermore, their mathematical formulation can be traced to actual physical principles that govern the performance of the building or equipment. Hence the model coefficients tend to be more robust, leading to sounder model predictions. Only a few studies have used steady-state physical models for parameter estimation relating to commercial building energy use (for example, Reddy et al. 1999). Unlike in single-family residences, it is difficult to perform elaborately planned experiments in large buildings and obtain representative values of indoor fluctuations.

A physical model of a chiller has been proposed by Gordon and Ng (1994, 1995) and Gordon et al. (1995). It is a simple, analytical, universal model for chiller performance based on thermodynamic considerations and linearization of heat losses. The model predicts the dependence of chiller COP [defined as the ratio of chiller (or evaporator) thermal cooling capacity Q_{evap} divided by the electrical power consumed by the chiller (or compressor) E_{comp}] with certain key (and easily measurable) parameters such as the fluid (water or refrigerant) return temperature from the condenser T_{cond}^{in}, the fluid temperature leaving the evaporator (or the chilled water supply temperature to the building) T_{evap}^{out}, and the thermal cooling capacity of the evaporator. The complete Gordon-Ng model is a three-parameter model that takes the following form:

$$\left(\frac{1}{\text{COP}} + 1 - \frac{T_{cond}^{in}}{T_{evap}^{out}}\right)Q_{evap} = -A_0 + A_1 T_{cond}^{in} - A_2 \frac{T_{cond}^{in}}{T_{evap}^{out}} \quad (58)$$

The three parameters are identified by multiple linear regression. Results of applying such models to field monitored data are fully described by the Gordon et al. papers as well as by Phelan et al. (1996) and Haberl et al. (1997).

Dynamic Models

In general, steady-state inverse models are used with monthly and daily data containing one or more independent variables. Dynamic inverse models are usually used with hourly or subhourly data in cases where the thermal mass of a building is sufficiently significant to delay the heat gains or losses. Dynamic models traditionally required the solution of a set of differential equations. The disadvantages of dynamic inverse models include their complexity and the need for more detailed measurements to tune the model. Unlike steady-state inverse models, dynamic inverse models usually require a high degree of user interaction and knowledge of the building or system being modeled.

Several residential energy studies using dynamic inverse models based on parameter estimation approaches have been reported, most of them involving intrusive data gathering. Rabl (1988) has classified the various types of dynamic inverse models used for whole-building energy use and drawn attention to the common underlying features of these models. There are essentially four different types of model formulations: thermal network models, time series models, differential equation models, and modal models. They all qualify as parameter-estimation approaches. Table 1 lists several pertinent studies in each category. A few studies (Hammersten 1984, Rabl 1988, Reddy 1989) have evaluated these different approaches with the same data set. A number of papers have reported results of applying such different techniques such as thermal network models and ARMA models to residential and commercial building energy use (see Table 1). Examples of dynamic inverse models for commercial building are found in Rabl (1988), Andersen and Brandemuehl (1992), and Braun (1990).

Dynamic inverse models based on pure statistical approaches have also been reported. Two examples are machine learning (Miller and Seem 1991) and artificial neural networks (Miller and Seem 1991, Kreider and Wang 1991, Kreider and Haberl 1994).

Neural networks are considered to be intuitive because they learn by example rather than by following programmed rules. The ability to "learn" is one of the key aspects of neural networks. A neural network consists of one input layer (which can contain one or more inputs), one or more hidden layers, and an output or target layer. One challenge of this technology is to construct a net with sufficient complexity to learn accurately without imposing excessive computational time.

The weights of a net are initiated with small random numbers. Then, the weights are adjusted iteratively or "trained" so that the application of a set of inputs produces the desired set of outputs. Usually a network is trained with a training data set that consists of many input-output pairs. Artificial neural networks have been trained by a wide variety of methods (McClelland and Rumelhart 1988, Wasserman 1989). One such training method is called **back propagation**.

Neural networks have been useful in predicting energy use in commercial buildings for such reasons as

- Prediction of what a properly operating building should be doing compared to actual operation. If there is a difference, it can be used in an expert system to produce early diagnoses of building operation problems.
- Prediction of what a building, prior to an energy retrofit, would have consumed under present conditions. When compared to the measured consumption of the retrofitted building, the difference represents a good estimate of the energy savings due to the retrofit. This represents one of the few ways that actual energy savings can be determined after the preretrofit building configuration has ceased to exist.

EXAMPLES OF INVERSE METHODS

Modeling of Utility Bill Data

The following example (taken from Sonderegger 1998) illustrates a utility bill analysis. Assume that values of the utility bills over an entire year have been measured. To obtain the equation coefficients through regression, the utility bills must be normalized by the length of the time interval between utility bills. This is equivalent to expressing all utility bills, degree-days, and other independent variables by their daily averages.

An appropriate software program is used in which one assumes values for heating and cooling balance points; from these, the corresponding heating and cooling degree-days for each utility bill period are determined. Repeated regression is done till the regression equation represents the best fit to the meter data. The model coefficients are then assumed to be tuned. Some computer programs allow direct determination of these optimal model parameters without the user's manual tuning of the parameters.

A widely used statistic to gage the goodness-of-fit of the model is the **coefficient of determination** R^2. A value of $R^2 = 1$ indicates a perfect correlation between actual data and the regression equation; a value of $R^2 = 0$ indicates no correlation. For purposes of tuning for a performance contract, as a rule of thumb the value of R^2 should never be less than 0.75.

When more than one independent variable is included in the regression, the value of R^2 is no longer sufficient to determine the goodness-of-fit. The standard error of the estimate of the coefficients becomes the more important determinant. The smaller the standard error compared to the coefficient's magnitude, the more reliable the coefficient estimate. To facilitate the significance of individual coefficients, the so-called ***t*-statistics**, or simply ***t*-values**, are used. These are simply the ratio of the coefficient estimate divided by the standard error of the estimate.

The coefficient of each variable included in the regression has a *t*-statistic. For a coefficient to be statistically meaningful, the absolute value of its *t*-statistic must be at least 2.0. Another way of stating this is that under no circumstances should a variable be included in a regression if the standard error of its coefficient estimate is greater than half the magnitude of the coefficient. The latter is true even when including a variable that increases the R^2. Generally speaking, including more variables in a regression results in a higher R^2, but the significance of most individual coefficients will likely decrease.

Figure 19 illustrates how well a regression fit captures measured baseline energy use in a hospital building. Cooling degree-days are found to be a significant variable, with the best fit for a base temperature of 54°F.

There are good reasons why individual utility bills may be unsuitable to develop a baseline and should be excluded from the regression. For example, a bill may be atypically high because of a one-time equipment malfunction that was subsequently repaired. However, it is often tempting to look for reasons to exclude bills that fall far from "the line" and not question those that are close to it. For example, bills for periods containing vacations or production shutdowns may look anomalously low, but excluding them from the regression would result in a chronic overestimate of the future baseline during the same period.

Neural Network Models

Figure 20 shows results for a single neural network typical of several hundred networks constructed for an academic engineering center located in central Texas. The cooling load is created by solar gains, internal gains, outdoor air sensible heat, and outdoor

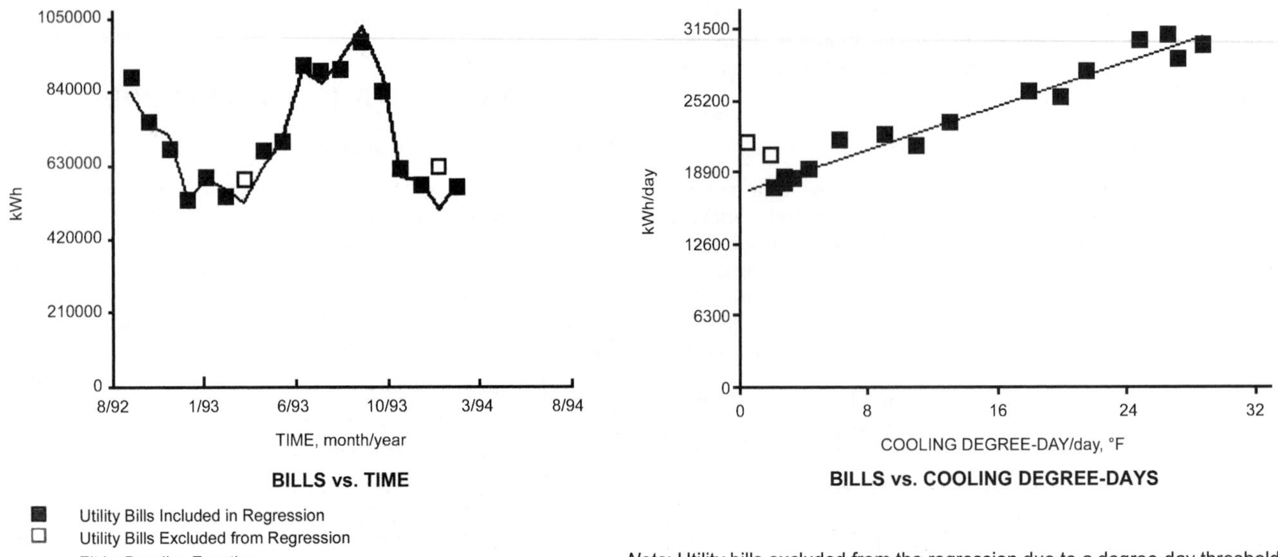

Fig. 19 Variable-Base Degree-Day Model Identification Using Electricity Utility Bills at a Hospital
(Sonderegger 1998)

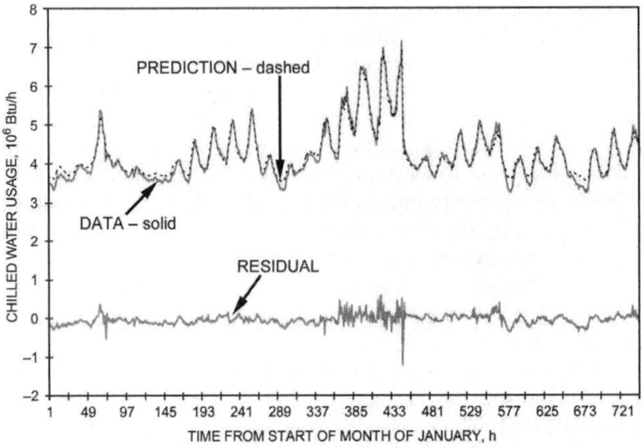

Fig. 20 Neural Network Prediction of Whole-Building, Hourly Chilled Water Consumption for a Commercial Building

air humidity loads. The neural network is used to predict the preretrofit energy consumption for comparison with measured consumption of the retrofitted building. Six months of preretrofit data were available with which to train a network. The solid lines show the known building consumption data while the dashed lines show the neural network predictions. This figure shows that a neural network trained for one period (September 1989) can predict energy consumption well into the future (here, January 1990).

The network used for this prediction had two hidden layers. The input layer contained eight neurons that receive eight different types of input data as listed below. The output layer consisted of one neuron that gave the output datum (chilled water consumption). Each training fact (i.e., training data set), therefore, contained eight input data (independent variables) and one pattern datum (dependent variable). The eight hourly input data used in each hour's data vector were selected on physical bases (Kreider and Rabl 1994) and were as follows:

- Hour number (0 to 2300)
- Ambient dry-bulb temperature
- Horizontal insolation
- Humidity ratio
- Wind speed
- Weekday/weekend binary flag (0, 1)
- Past hour's chilled water consumption
- Second past hour's chilled water consumption

These measured independent variables were able to predict the chilled water use to an RMS error of less than 4% (JCEM 1992).

The choice of an optimal network's configuration for a given problem remains an art. The number of hidden neurons and layers must be sufficient to meet the requirement of the given application. However, if too many neurons and layers are used, the network tends to memorize data rather than learning, that is, finding the underlying patterns in the data. Further, choosing an excessively large number of hidden layers significantly increases the required training time for certain learning algorithms. Anstett and Kreider (1993), Kreider and Wang (1991), Wang and Kreider (1992), and Krarti et al. (1998) report additional case studies for commercial buildings.

Closing Remarks

Steady-state and dynamic inverse models can be used with energy management and control systems to predict energy use (Kreider and Haberl 1994). Hourly or daily comparisons of measured energy use against predicted energy use can be used to determine whether systems are being left on unnecessarily or are in need of maintenance. Combinations of predicted energy use and a knowledge-based system can indicate above-normal energy use and diagnose the possible cause of the malfunction if sufficient historical information has been previously gathered (Haberl and Claridge 1987). Hourly systems that use artificial neural networks have also been constructed (Kreider and Wang 1991).

Table 10 presents a decision diagram for selecting a forward or inverse model where use of the model, degree of difficulty in understanding and applying the model, time scale for the data used by the model, calculation time, and input variables used by the models are the criteria used to choose a particular model.

Table 10 Capabilities of Different Forward and Inverse Modeling Methods

Methods	Usage[a]	Difficulty	Time Scale[b]	Calc. Time	Variables[c]	Accuracy
Simple linear regression	ES	Simple	D, M	Very fast	*T*	Low
Multiple linear regression	D, ES	Simple	D, M	Fast	*T, H, S, W, t*	Medium
ASHRAE bin method and inverse bin method	ES	Moderate	H	Fast	*T*	Medium
Change point models	D, ES	Simple	H, D, M	Fast	*T*	Medium
ASHRAE TC 4.7 modified bin method	ES, DE	Moderate	H	Medium	*T, S, tm*	Medium
Artificial Neural Networks	D, ES, C	Complex	S, H	Fast	*T, H, S, W, t, tm*	High
Thermal network	D, ES, C	Complex	S, H	Fast	*T, S, tm*	High
Fourier series analysis	D, ES, C	Moderate	S, H	Medium	*T, H, S, W, t, tm*	High
ARMA model	D, ES, C	Moderate	S, H	Medium	*T, H, S, W, t, tm*	High
Modal analysis	D, ES, C	Complex	S, H	Medium	*T, H, S, W, t, tm*	High
Differential equation	D, ES, C	Complex	S, H	Fast	*T, H, S, W, t, tm*	High
Computer simulation (component-based)	D, ES, C, DE	Very complex	S, H	Slow	*T, H, S, W, t, tm*	Medium
Computer simulation (fixed schematic)	D, ES, DE	Very complex	H	Slow	*T, H, S, W, t, tm*	Medium
Computer emulation	D, C	Very complex	S, H	Very slow	*T, H, S, W, t, tm*	High

Notes:
[a]Usage shown includes diagnostics (D), energy savings calculations (ES), design (DE), and control (C).
[b]Time scales shown are hourly (H), daily (D), monthly (M), and subhourly (S).
[c]Variables include temperature (*T*), humidity (*H*), solar (*S*), wind (*W*), time (*t*), and thermal mass (*tm*).

REFERENCES

Alamdari, F. and G.P. Hammond. 1982. Time-dependent convective heat transfer in warm-air heated rooms. Energy Conservation in the Built Environment: *Proceedings* CIB W67 Third International Symposium. pp. 209-20. Dublin, Ireland.

Alamdari, F. and G.P. Hammond. 1983. Improved data correlations for buoyancy-driven convection in rooms. *Building Services Engineering Research and Technology* 4(3):106-12.

Alereza, T. and T. Kusuda. 1982. Development of equipment seasonal performance models for simplified energy analysis methods. *ASHRAE Transactions* 88(2):249-62.

Altmayer, E.F., A.J. Gadgil, F.S. Bauman, and R.C. Kammerud. 1983. Correlations for convective heat transfer from room surfaces. *ASHRAE Transactions* 89(2A):61-77.

Andersen, I. and M.J. Brandemuehl. 1992. Heat storage in building thermal mass: A parametric study. *ASHRAE Transactions* 98(1):910-18.

Anstett, M. and J.F. Kreider. 1993. Application of artificial neural networks to commercial building energy use prediction. *ASHRAE Transactions* 99(1):505-17.

ASHRAE. 1995. *Bin and degree hour weather data for simplified energy calculations*.

ASHRAE. 1999. Energy standard for buildings except low-rise residential buildings. ANSI/ASHRAE/IESNA *Standard* 90.1-1999.

Ayres, M.J. and E. Stamper. 1995. Historical development of building energy calculations. *ASHRAE Transactions* 101(1):47-55.

Bacot, P., A. Neveu, and J. Sicard. 1984. Analyse modale des phenomenes thermiques en regime variable dans le batiment. *Revue Generale de Thermique* 267:189.

Balcomb, J.D., R.W. Jones, R.D. McFarland, and W.O. Wray. 1982. Expanding the SLR method. *Passive Solar Journal* 1(2).

Bauman, F., A. Gadgil, R. Kammerud, E. Altmayer, and M. Nansteel. 1983. Convective heat transfer in buildings: Recent research results. *ASHRAE Transactions* 89(1A):215-32.

Beck, J.V. and K.J. Arnold, 1977. *Parametric estimation in engineering and science*. John Wiley and Sons, New York.

Bohn, M.S., A.T. Kirkpatrick, and D.A. Olson. 1984. Experimental study of three-dimensional natural convection high-Rayleigh number. *Journal of Heat Transfer* 106:339-45.

Bou-Saada, T. and J. Haberl. 1995a. A weather-day typing procedure for disaggregating hourly end-use loads in an electrically heated and cooled building from whole-building hourly data. *Proceedings* 30th IECEC, pp. 349-56.

Bou-Saada, T. and J. Haberl. 1995b. An improved procedure for developing calibrated hourly simulation models. *Proceedings Building Simulation '95*. IBPSA. Madison, Wisc.

Bourdouxhe, J.P., M. Grodent, J. Lebrun, and C. Saavedra. 1994a. A toolkit for primary HVAC system energy calculation—Part 1: Boiler model. *ASHRAE Transactions* 100(2):759-73.

Bourdouxhe, J.P., M. Grodent, J. Lebrun, C. Saavedra, and K. Silva. 1994b. A toolkit for primary HVAC system energy calculation—Part 2: Reciprocating chiller models. *ASHRAE Transactions* 100(2):774-86.

Bourdouxhe, J.P., M. Grodent, and C. Silva. 1994c. Cooling tower model developed in a toolkit for primary HVAC system energy calculation—Part 1: Model description and validation using catalog data. *Proceedings* Fourth International Conference on System Simulation in Buildings.

Brandemuehl, M.J. 1993. *HVAC 2 toolkit: Algorithms and subroutines for secondary HVAC system energy calculations*. ASHRAE.

Brandemuehl, M.J. and S. Gabel. 1994. Development of a toolkit for secondary HVAC system energy calculations. *ASHRAE Transactions* 100(1):21-32.

Brandemuehl, M.J. and J.D. Bradford. 1999. Optimal supervisory control of cooling plants without storage. *Final Report* RP-823. ASHRAE.

Braun, J.E. 1988. Methodologies for the design and control of chilled water systems. *Ph.D. thesis*, University of Wisconsin-Madison.

Braun, J.E. 1990. Reducing energy costs and peak electrical demand through optimal control of building thermal mass. *ASHRAE Transactions* 96(2): 876-88.

Braun, J.E. 1992. A comparison of chiller-priority, storage-priority, and optimal control of an ice-storage system. *ASHRAE Transactions* 98(1): 893-902.

Bronson, D., S. Hinchey, J. Haberl, and D. O'Neal. 1992. A procedure for calibrating the DOE-2 simulation program to non-weather dependent loads. *ASHRAE Transactions* 98(1):636-52.

Buhl, W.F., A.E. Erdem, J.M. Nataf, F.C. Winkelmann, M.A. Moshier, and E.F. Sowell. 1990. The US EKS: Advances in the SPANK-based energy kernel system. *Proceedings* Third International Conference on System Simulation in Buildings, pp. 107-50.

Buhl, W.F., A.E. Erdem, F.C. Winkelmann, and E.F. Sowell. 1993. Recent improvements in SPARK: Strong component decomposition, multivalued objects and graphical interface. *Proceedings of Building Simulation '93*. IBPSA, pp. 283-390.

Carroll, J.A. 1980. An "MRT method" of computing radiant energy exchange in rooms. *Systems Simulation and Economic Analysis*, pp. 343-48. San Diego.

Chandra, S. and A.A. Kerestecioglu. 1984. Heat transfer in naturally ventilated rooms: Data from full-scale measurements. *ASHRAE Transactions* 90(1B):211-24.

Chi, J. and G.E. Kelly. 1978. A method for estimating the seasonal performance of residential gas and oil-fired heating systems. *ASHRAE Transactions* 84(1):405.

Cinquemani, V., J.R. Owenby, and R.G. Baldwin. 1978. Input data for solar systems. U.S. Department of Energy *Report* No. E(49-26)1041.

Clark, D.R. 1985. *HVACSIM+ building systems and equipment simulation program: Reference manual*. NBSIR 84-2996, U.S. Department of Commerce, Washington, D.C.

Clarke, J.A. 1985. *Energy simulation in building design*. Adam Hilger Ltd., Boston.

Claridge, D. 1988a. Design methods for earth-contact heat transfer. *Progress in Solar Energy*, K. Boer, ed. American Solar Energy Society, Boulder, CO.

Claridge, D. 1998b. A perspective on methods for analysis of measured energy data from commercial buildings, *ASME Journal of Solar Energy Engineering* 120:150.

Claridge, D.E., M. Krarti, and M. Bida. 1987. A validation study of variable-base degree-day cooling calculations. *ASHRAE Transactions* 93(2):90-104.

Cole, R.J. 1976. The longwave radiation incident upon the external surface of buildings. *The Building Services Engineer* 44:195-206.

Cooper, K.W. and D.R. Tree. 1973. A re-evaluation of the average convection coefficient for flow past a wall. *ASHRAE Transactions* 79:48-51.

Corson, G.C. 1992. Input-output sensitivity of building energy simulations. *ASHRAE Transactions* 98(1):618.

Cumali, Z., A.O. Sezgen, et al. 1979. Extensions of methods used in analyzing building thermal loads. *Proceedings* Thermal Performance of the Exterior Envelopes of Buildings, pp. 411-20.

Davies, M.G. 1988. Design models to handle radiative and convective exchange in a room. *ASHRAE Transactions* 94(2):173-95.

Dhar, A. 1995. Development of Fourier series and artificial neural network approaches to model hourly energy use in commercial buildings. *Ph.D. thesis*, ME Dept., Texas A&M University.

Dhar, A., T.A. Reddy, and D.E. Claridge. 1998. Modeling hourly energy use in commercial buildings with Fourier series functional forms. *Journal of Solar Energy Engineering* 120:217.

Dhar, A., T.A. Reddy, and D.E. Claridge. 1999a. A Fourier series model to predict hourly heating and cooling energy use in commercial buildings with outdoor temperature as the only weather variable. *Journal of Solar Energy Engineering* 121:47-53.

Dhar, A., T.A. Reddy, and D.E. Claridge. 1999b. Generalization of the Fourier series approach to model hourly energy use in commercial buildings. *Journal of Solar Energy Engineering* 121:54-62.

Draper, N. and H. Smith. 1981. *Applied regression analysis*, 2nd ed. John Wiley and Sons, New York.

Elmahdy, A.H. and G.P. Mitalas. 1977. A simple model for cooling and dehumidifying coils for use in calculating energy requirements for buildings. *ASHRAE Transactions* 83(2):103-17.

Englander, S.L. and L.K. Norford. 1992. Saving fan energy in VAV systems—Part 1: Analysis of a variable-speed-drive retrofit. *ASHRAE Transactions* 98(1):3-18.

Erbs, D.G., S.A. Klein, and W.A. Beckman. 1983. Estimation of degree-days and ambient temperature bin data from monthly-average temperatures. *ASHRAE Journal* 25(6):60.

Fels, M., ed. 1986. Measuring energy savings: The scorekeeping approach. *Energy and Buildings* 9.

Fels, M. and M. Goldberg, 1986. Refraction of PRISM results in components of saved energy. *Energy and Buildings* 9:169.

Fracastoro, G., M. Masoero, and M. Cali. 1982. Surface heat transfer in building components. *Proceedings* Thermal Performance of the Exterior Envelopes of Buildings II, pp. 180-203.

Gordon, J.M. and K.C. Ng. 1994. Thermodynamic modeling of reciprocating chillers. *Journal of Applied Physics* 75(6):2769-74.

Gordon, J.M and K.C. Ng. 1995. Predictive and diagnostic aspects of a universal thermodynamic model for chillers. *International Journal of Heat and Mass Transfer* 38(5):807-18.

Gordon, J.M., K.C. Ng, and H.T. Chua. 1995. Centrifugal chillers: Thermodynamic modeling and a case study. *International Journal of Refrigeration* 18(4):253-57.

Guyon, G. and E. Palomo. 1999. Validation of two French building energy programs—Part 2: Parameter estimation method applied to empirical validation. *ASHRAE Transactions* 105(2):709-20.

Haberl, J.S. and T.E. Bou-Saada. 1998. Procedures for calibrating hourly simulation models to measured building energy and environmental data. *ASME Journal of Solar Energy Engineering* 120(August):193.

Haberl, J.S. and D.E. Claridge. 1987. An expert system for building energy consumption analysis: Prototype results. *ASHRAE Transactions* 93(1):979-98.

Haberl, J. and P. Komor. 1990a. Improving commercial building energy audits: How annual and monthly consumption data can help. *ASHRAE Journal* 32(8):26-33.

Haberl, J. and P. Komor. 1990b. Improving commercial building energy audits: How daily and hourly data can help. *ASHRAE Journal* 32(9):26-36.

Haberl, J.S. and S. Thamilseran. 1996. The great energy predictor shootout II: Measuring retrofit savings and overview and discussion of results. *ASHRAE Transactions* 102(2):419-35.

Haberl, J.S., T.A. Reddy, I.E. Figuero, and M. Medina. 1997. Overview of LoanSTAR chiller monitoring—Analysis of in-situ chiller diagnostics using ASHRAE RP-827 test method. Paper presented at Cool Sense National Integrated Chiller Retrofit Forum, Presidio, San Francisco, September.

Hammersten, S. 1984. Estimation of energy balances for houses. National Swedish Institute for Building Research, M84.18, December.

Howell, R.H. and S. Suryanarayana. 1990. Sizing of radiant heating systems: Part I and Part II. *ASHRAE Transactions* 96(1):652-65.

JCEM. 1992. Final report: Artificial neural networks applied to LoanSTAR data. Joint Center for Energy Management *Report* TR/92/15.

Kamal, S. and P. Novak. 1991. Dynamic analysis of heat transfer in buildings with special emphasis on radiation. *Energy and Buildings* 17(3): 231-41.

Kaplan, M., J. McFerran, J. Jansen, and R. Pratt. 1990. Reconciliation of a DOE2.1C model with monitored end-use data from a small office building. *ASHRAE Transactions* 96(1):981.

Katipamula, S. and D.E. Claridge. 1993. Use of simplified systems model to measure retrofit energy savings. *Transactions of the ASME Journal of Solar Energy Engineering* 115(May):57-68.

Katipamula, S., T.A. Reddy, and D.E. Claridge. 1994. Development and application of regression models to predict cooling energy consumption in large commercial buildings. *Proceedings* of the 1994 ASME/JSME/JSES International Solar Energy Conference, San Francisco, March 27-30, p. 307.

Katipamula, S., T.A. Reddy, and D.E. Claridge. 1998. Multivariate regression modeling. *ASME Journal of Solar Energy Engineering* 120 (August):176.

Kays, W.M. and A.L. London. 1984. *Compact heat exchangers*, 3rd ed. McGraw-Hill, New York.

Kerrisk, J.F., N.M. Schnurr, et al. 1981. The custom weighting-factor method for thermal load calculation in the DOE-2 computer program. *ASHRAE Transactions* 87(2):569-84.

Khalifa, A.J.N. and R.H. Marshall. 1990. Validation of heat transfer coefficients on interior building surfaces using a real-sized indoor test cell. *International Journal of Heat and Mass Transfer* 33(10):2219-36.

Kissock, J.K., T.A. Reddy, J.S. Haberl, and D.E. Claridge. 1993. E-model: A new tool for analyzing building energy use data. *Proceedings* Intl. Indust. Energy Tech. Conf., Texas A&M University.

Kissock, J.K., T.A. Reddy, and D.E. Claridge. 1998. Ambient temperature regression analysis for estimating retrofit savings in commercial buildings. *ASME Journal of Solar Energy Engineering* 120:168.

Klein, S.A., W.A. Beckman, et al. 1994. *TRNSYS: A transient simulation program*. Engineering Experiment Station *Report* 38-14, University of Wisconsin-Madison.

Knebel, D.E. 1983. *Simplified energy analysis using the modified bin method*. ASHRAE.

Krarti, M. 1994a. Time varying heat transfer from slab-on-grade floors with vertical insulation. *Building and Environment* 29(1):55-61.

Krarti, M. 1994b. Time varying heat transfer from horizontally insulated slab-on-grade floors. *Building and Environment* 29(1):63-71.

Krarti, M. and P. Chuangchid. 1999. Cooler floor heat gain for refrigerated structures, *Final Report*, ASHRAE Project 953-TRP.

Krarti, M., D.E. Claridge, and J. Kreider. 1988a. The ITPE technique applied to steady-state ground-coupling problems. *International Journal of Heat and Mass Transfer* 31:1885-98.

Krarti, M., D.E. Claridge, and J. Kreider. 1988b. ITPE method applications to time-varying two-dimensional ground-coupling problems. *International Journal of Heat and Mass Transfer* 31:1899-1911.

Krarti, M., J.F. Kreider, D. Cohen, and P. Curtiss. 1998. Estimation of energy savings for building retrofits using neural networks. *ASME Journal of Solar Energy Engineering* 120:211.

Kreider, J.F. and J. Haberl. 1994. Predicting hourly building energy usage: The great predictor shootout—Overview and discussion of results. *ASHRAE Transactions* 100(2):1104-18.

Kreider, J.F. and A. Rabl. 1994. *Heating and cooling of buildings*. McGraw-Hill, New York.

Kreider, J.F. and X.A. Wang. 1991. Artificial neural networks demonstration for automated generation of energy use predictors for commercial buildings. *ASHRAE Transactions* 97(1):775-79.

Kusuda, T. 1969. Thermal response factors for multi-layer structures of various heat conduction systems. *ASHRAE Transactions* 75(1):246-71.

Labs, K., J. Carmody, R. Sterling, L. Shen, Y. Huang, and D. Parker. 1988. Building foundation design handbook, ORNL *Report* Sub/86-72143/1. Oak Ridge National Laboratory, Oak Ridge, TN.

Lachal, B., W.U. Weber, and O. Guisan. 1992. Simplified methods for the thermal analysis of multifamily and administrative buildings. *ASHRAE Transactions* 98.

Lebrun, J., J.-P. Bourdouxhe, and M. Grodent. 1999. *HVAC 1 toolkit: A toolkit for primary HVAC system energy calculation.* ASHRAE

Lewis, P.T. and D.K. Alexander. 1990. HTB2: A flexible model for dynamic building simulation. *Building and Environment*. pp. 7-16.

Liu, M. and D.E. Claridge. 1998. Use of calibrated HVAC system models to optimize system operation. *ASME Journal of Solar Energy Engineering* 120:131.

Lorenzetti, D.M. and L.K. Norford. 1993. Pressure reset control of variable air volume ventilation systems. *Proceedings* ASME International Solar Energy Conference, p. 445, April, Washington, D.C.

MacDonald, J.M. and D.M. Wasserman. 1989. Investigation of metered data analysis methods for commercial and related buildings. Oak Ridge National Laboratory *Report* ORNL/CON-279.

Manke, J.M., D.C. Hittle, and C.E. Hancock. 1996. Calibrating building energy analysis models using short-term data. *Proceedings* ASME International Solar Energy Conference, p.369, San Antonio, TX.

McClelland, J.L. and D.E. Rumelhart. 1988. *Exploration in parallel distributed processing*. MIT Press, Cambridge, MA.

McQuiston, F.C. and J. D. Spitler. 1992. *Cooling and Heating Load Calculation Manual*. ASHRAE

Melo, C. and G.P. Hammond. 1991. Modeling and assessing the sensitivity of external convection from building facades in heat and mass transfer in building materials and structures. pp. 683-95. Hemisphere Publishing, New York.

Metcalfe, R.R., R.D. Taylor, C.O. Pederson, R.J. Liesen, and D.E. Fisher. 1995. Incorporating a modular system simulation program into a large energy analysis program: The linking of IBLAST and HVACSIM+. *Proceedings* of Building Simulation '95.

Miller, D.E. 1980. The impact of HVAC process dynamics on energy use. *ASHRAE Transactions* 86(2):535-56.

Miller, R. and J. Seem. 1991. Comparison of artificial neural networks with traditional methods of predicting return from night setback. *ASHRAE Transactions* 97(2):500-08.

Mitalas, G.P. and D.G. Stephenson. 1967. Room thermal response factors. *ASHRAE Transactions* 73(1):III.2.1-III.2.10.

Mitalas, G.P. 1968. Calculations of transient heat flow through walls and roofs. *ASHRAE Transactions* 74(2):182-88.

Mitchell, J.W. 1983. *Energy engineering*. John Wiley and Sons, New York.

NOAA. 1973. Degree-days to selected bases. U.S. National Climatic Data Center, Asheville, NC.

Neter, J., W. Wasseran, and M. Kutner. 1989. *Applied linear regression models*, 2nd ed. Richard C. Irwin, Homewood, IL.

Nisson, J.D.N. and G. Dutt. 1985. *The superinsulated home book.* John Wiley and Sons, New York.

Norford, L.K., R.H. Socolow, E.S. Hsieh, and G.V. Spadaro. 1994. Two-to-one discrepancy between measured and predicted performance of a low-energy office building: Insights from a reconciliation based on the DOE-2 model. *Energy and Buildings* 21:121.

Park, C., D.R. Clark, and G.E. Kelly. 1985. An overview of HVACSIM+, a dynamic building/HVAC control systems simulation program. *Proceedings* First Building Energy Simulation Conference.

Parker, W.H., G.E. Kelly, and D. Didion. 1980. A method for testing, rating, and estimating the heating seasonal performance of heat pumps. National Bureau of Standards, NBSIR 80-2002, April.

Phelan, J., M.J. Brandemuehl, and M. Krarti. 1996. *Final Report* ASHRAE Project RP-827: Methodology development to measure in-situ chiller, fan, and pump performance. JCEM *Report* No. JCEM/TR/96-3, University of Colorado at Boulder.

Rabl, A. 1988. Parameter estimation in buildings: Methods for dynamic analysis of measured energy use. *Journal of Solar Energy Engineering* 110:52-66.

Rabl, A. and A. Riahle. 1992. Energy signature model for commercial buildings: Test with measured data and interpretation. *Energy and Buildings* 19:143-54.

Reddy, T. 1989. Application of dynamic building inverse models to three occupied residences monitored non-intrusively. *Proceedings* Thermal Performance of Exterior Envelopes of Buildings IV, ASHRAE/DOE/BTECC/CIBSE.

Reddy, T. and D. Claridge. 1994. Using synthetic data to evaluate multiple regression and principle component analyses for statistical modeling of daily building energy consumption. *Energy and Buildings* 24:35-44.

Reddy, T.A., S. Katipamula, J.K. Kissock, and D.E. Claridge. 1995. The functional basis of steady-state thermal energy use in air-side HVAC equipment. *Journal of Solar Energy Engineering* 117:31-39.

Reddy, T.A., N.F. Saman, D.E. Claridge, J.S. Haberl, W.D. Turner, and A. Chalifoux. 1997. Baselining methodology for facility level monthly energy use—Part 1: Theoretical aspects. *ASHRAE Transactions* 103(2): 336-47.

Reddy, T.A., S. Deng, and D.E. Claridge. 1999. Development of an inverse method to estimate overall building and ventilation parameters of large commercial buildings. *Journal of Solar Energy Engineering* 121:47.

Ruch, D. and D. Claridge. 1991. A four parameter change-point model for predicting energy consumption in commercial buildings. *Proceedings* ASME International Solar Energy Conference, 433-40.

Ruch, D., L. Chen, J. Haberl, and D. Claridge. 1993. A change-point principle component analysis (CP/CAP) method for predicting energy use in commercial buildings: The PCA model. *Journal of Solar Energy Engineering* 115:77-84.

Seem, J.E. and J.E. Braun. 1991. Adaptive methods for real-time forecasting of building electricity demand. *ASHRAE Transactions* 97(1):710.

Shipp, P.H. and T.B. Broderick. 1983. Analysis and comparison of annual heating loads for various basement wall insulation strategies using transient and steady-state models. *Thermal Insulation, Materials, and Systems for Energy Conservation in the 80's*. ASTM STP 789, F.A. Govan, D.M. Greason, and J.D. McAllister, eds. American Society for Testing and Materials, West Conshohocken, PA.

Shurcliff, W.A. 1984. *Frequency method of analyzing a building's dynamic thermal performance*. Cambridge, MA.

Sonderegger, R.C. 1977. Dynamic models of house heating based on equivalent thermal parameters. *Ph.D. thesis*, Center for Energy and Environmental Studies *Report* No. 57. Princeton University, Princeton, NJ.

Sonderegger, R.C. 1985. Thermal modeling of buildings as a design tool. *Proceedings* of CHMA 2000, Vol. 1.

Sonderegger, R.C. 1998. Baseline equation for utility bill analysis using both weather and non-weather related variables. *ASHRAE Transactions* 104(2):859-70.

Sowell, E.F. 1988. Classification of 200,640 parametric zones for cooling load calculations. *ASHRAE Transactions* 94(2):716-36.

Sowell, E.F. 1990. Lights: A numerical lighting/HVAC test cell. *ASHRAE Transactions* 96(2):780-86.

Sowell, E.F. and M.A. Moshier. 1995. HVAC component model libraries for equation-based solvers. *Proceedings of Building Simulation '95*. IBPSA, Madison, WI.

Spitler, J.D., C.O. Pedersen, and D.E. Fisher. 1991. Interior convective heat transfer in buildings with large ventilative flow rates. *ASHRAE Transactions* 97(1):505-15.

Steinman, M., L.N. Kalisperis, and L.H. Summers. 1989. The MRT-correction method: A new method of radiant heat exchange. *ASHRAE Transactions* 95(1):1015-27.

Stephenson, D.G. and G.P. Mitalas. 1967. Cooling load calculations by thermal response factor method. *ASHRAE Transactions* 73(1):III.1.1-III.1.7.

Stoecker, W.F. and J.W. Jones. 1982. *Refrigeration and air conditioning*, 2nd ed. McGraw-Hill, New York.

Strand, R.K. and C.O. Pedersen. 1997. Implementation of a radiant heating and cooling model into an integrated building energy analysis program. *ASHRAE Transactions* 103(1):949-58.

Subbarao, K. 1986. Thermal parameters for single and multi-zone buildings and their determination from performance data. SERI *Report* SERI/TR-253-2617. Solar Energy Research Institute, Golden, CO.

Subbarao, K. 1988. PSTAR—Primary and secondary terms analysis and renormalization: A unified approach to building energy simulations and short-term monitoring. SERI/TR-253-3175.

Taylor, R.D., C.O. Pedersen, and L. Lawrie. 1990. Simultaneous simulation of buildings and mechanical systems in heat balance based energy analysis programs. *Proceedings* Third International Conference on System Simulation in Buildings. Liege, Belgium.

Taylor, R.D. et al. 1991. Impact of simultaneous simulation of buildings and mechanical systems in heat balance based energy analysis programs on system response and control. *Proceedings* Building Simulation '91. Sophia Antipolis, Nice France: IBPSA.

Thamilseran, S. and J. Haberl. 1995. A bin method for calculating energy conservation retrofit savings in commercial buildings. *Proceedings* 1995 ASME/JSME/JSES Intl. Solar Energy Conference, pp. 111-24.

Threlkeld, J.L. 1970. *Thermal environmental engineering*, 2nd ed. Prentice-Hall, Englewood Cliffs, NJ.

USAF. 1978. *Engineering weather data*. Department of the Air Force *Manual* AFM 88-29. U.S. Government Printing Office, Washington, D.C.

Walton, G.N. 1980. A new algorithm for radiant interchange in room loads calculations. *ASHRAE Transactions* 86(2):190-208.

Walton, G.N. 1983. Thermal analysis research program reference manual. NBSIR 83-2655. National Institute of Standards and Technology, Gaithersburg, MD.

Walton, G.N. 1993. Computer programs for simulation of lighting/HVAC interactions. NISTIR 5322. NIST.

Waltz, J.P. 1992. Practical experience in achieving high levels of accuracy in energy simulations of existing buildings. *ASHRAE Transactions* 98(1): 606-17.

Wang, X.A. and J.F. Kreider.1992. Improved artificial neural networks for commercial building energy use prediction. *Journal of Solar Energy Engineering*.

Wasserman, P.D. 1989. *Neural computing, theory and practice*. Van Nostrand Reinhold, New York.

Yazdanian, M. and J. Klems. 1993. Measurement of the exterior convective film coefficient for windows in low-rise buildings. *ASHRAE Transactions* 100(1):1087-96.

York, D.A. and C.C. Cappiello, eds. 1982. *DOE-2 engineers manual*. Lawrence Berkeley Laboratory *Report* LBL-11353 (LA-8520-M, DE83004575). National Technical Information Services, Springfield, VA.

BIBLIOGRAPHY

NEMVP. 1996. *U.S. DOE North-America energy monitoring and verification protocol*. U.S. Department of Energy, Washington, D.C.

Shavit, G. 1995. Short-time-step analysis and simulation of homes and buildings during the last 100 years. *ASHRAE Transactions* 101(1):856-68.

Sowell, E.F. and G.N. Walton. 1980. Efficient computation of zone loads. *ASHRAE Transactions* 86(1):49-72.

Sowell, E.F. and D.C. Hittle. 1995. Evolution of building energy simulation methodology. *ASHRAE Transactions* 101(1):850-55.

Spitler, J.D. 1996. *Annotated guide to load calculation models and algorithms*. ASHRAE.

Subbarao, K., J. Burch, and C.E. Hancock. 1990. How to accurately measure the load coefficient of a residential building. *Journal of Solar Energy Engineering*.

U.S. Army. 1979. *BLAST, The building loads analysis and system thermodynamics program—Users manual*. U.S. Army Construction Engineering Research Laboratory *Report* E-153.

Yuill, G.K. 1990. *An annotated guide to models and algorithms for energy calculations relating to HVAC equipment*. ASHRAE.

CHAPTER 32

SPACE AIR DIFFUSION

TERMINOLOGY

ASPECT ratio. Ratio of length to width of an opening or core of a grille.

Axial flow jet. Stream of air whose motion is approximately symmetrical along a line, although some spreading and drop or rise can occur from diffusion and buoyancy effects.

Coefficient of discharge. Ratio of area at vena contracta to area of opening.

Cold air. General term used for supply air at 35 to 40°F.

Core area. Total plane area of that portion of a grille, included within lines tangent to the outer edges of the outer openings, through which air can pass.

Damper. Device used to vary the volume of air passing through a confined cross section by varying the cross-sectional area.

Diffuser. Outlet discharging supply air in various directions and planes.

Diffusion. Distribution of air within a space by an outlet discharging supply air in various directions and planes.

Draft. Undesired local cooling of a body caused by low temperature and movement of air.

Drop. Vertical distance that the lower edge of a horizontally projected airstream drops between the outlet and the end of its throw.

Effective area. Net area of an outlet or inlet device through which air can pass; equal to the free area times the coefficient of discharge.

Entrainment. Movement of room air into the jet caused by the airstream discharged from the outlet (secondary air motion).

Entrainment ratio. Total air divided by the air discharged from the outlet.

Envelope. Outer boundary of an airstream moving at a perceptible velocity.

Exhaust opening or inlet. Any opening through which air is removed from a space.

Free area. Total minimum area of the openings in the air outlet or inlet through which air can pass.

Grille. A covering for any area through which air passes.

Induction. See Entrainment.

Isothermal jet. Air jet with the same temperature as the surrounding air.

Lower zone. Room volume below the stratification level created by displacement ventilation.

Nonisothermal jet. Air jet with an initial temperature different from the surrounding air.

Outlet velocity. Average velocity of air emerging from the outlet, measured in the plane of the opening.

Primary air. Air delivered to the outlet by the supply duct.

Radius of diffusion. Horizontal axial distance an airstream travels after leaving an air outlet before the maximum stream velocity is reduced to a specified terminal level (e.g., 50, 100, 150, or 200 fpm).

Register. Grille equipped with a damper or control valve.

Spread. Divergence of the airstream in a horizontal and/or vertical plane after it leaves the outlet.

Stagnant zone. Area characterized by low air motion and stratification. This does not imply poor air quality.

Supply opening or outlet. Any opening through which supply air is delivered into a ventilated space being heated, cooled, humidified, or dehumidified. Supply outlets are classified according to their location in a room as sidewall, ceiling, baseboard, or floor outlets. However, because numerous designs exist, they are more accurately described by their construction features. (See Chapter 17 of the 2000 *ASHRAE Handbook—Systems and Equipment*.)

Temperature differential. Temperature difference between primary and room air.

Terminal velocity. Maximum airstream velocity at the end of the throw.

Throw. Horizontal or vertical axial distance an airstream travels after leaving an air outlet before the maximum stream velocity is reduced to a specified terminal velocity (e.g., 50, 100, 150, or 200 fpm), defined by ASHRAE *Standard* 70.

Total air. Mixture of discharged air and entrained air.

Upper zone. Room volume above the stratification level created by displacement ventilation.

Vane. Thin plate in the opening of a grille.

Vane ratio. Ratio of the depth of a vane to the space between two adjacent vanes.

Vena contracta. Smallest area of a fluid stream leaving an orifice.

PRINCIPLES OF JET BEHAVIOR

AIR JET CLASSIFICATION

As a rule, air supplied into rooms through the various types of outlets (e.g., grilles, ceiling diffusers, perforated panels) is distributed by turbulent air jets. These air jets are the primary factor affecting room air motion; Baturin (1972), Christianson (1989), and Murakami (1992) have further information on the relationship between the air jet and the occupied zone. If the air jet is not obstructed by walls, ceiling, or other obstructions, it is considered a **free jet**. If the air jet is attached to a surface, it is an **attached air jet**.

The preparation of this chapter is assigned to TC 5.3, Room Air Distribution.

Characteristics of the air jet in a room might be influenced by reverse flows created by the same jet entraining the ambient air. This air jet is called a **confined jet**. If the temperature of the supplied air is equal to the temperature of the ambient room air, the air jet is called an **isothermal jet**. A jet with an initial temperature different from the temperature of the ambient air is called a **nonisothermal jet**. The air temperature differential between supplied and ambient room air generates thermal forces in jets, affecting (1) the trajectory of the jet, (2) the location at which the jet attaches and separates from the ceiling/floor, and (3) the throw of the jet. The significance of such effects depends on the ratio between the thermal buoyancy of the air and inertial forces (characterized by the Archimedes number Ar).

Depending on diffuser type, air jets can be classified as follows:

- **Compact** air jets are formed by cylindrical tubes, nozzles, square or rectangular openings with a small aspect ratio (unshaded or shaded by perforated plates), grilles, etc. Compact air jets are three-dimensional and axisymmetric at least at some distance from the diffuser opening. The maximum velocity in the cross section of the compact jet is on the axis.
- **Linear** air jets are formed by slots or rectangular openings with a large aspect ratio. The jet flows are approximately two-dimensional. Air velocity is symmetric in the plane at which air velocities in the cross section are maximum. At some distance from the diffuser, linear air jets tend to transform into compact jets.
- **Radial** air jets are formed by ceiling cylindrical air diffusers with flat disks or multidiffusers that direct the air horizontally in all directions.
- **Conical** air jets are formed by cone-type or regulated multidiffuser ceiling-mounted air distribution devices. They have an axis of symmetry. The air flows parallel to the conical surface (the angle at the top of the cone is 120°) with the maximum velocities in the cross section perpendicular to the axis.
- **Incomplete radial** air jets are formed by outlets with grilles having diverging vanes and a forced angle of expansion. At a distance, this jet tends to transform into a compact one.
- **Swirling** air jets are formed by diffusers with vortex-forming devices. These devices create rotation, which has, in addition to the axial component of velocity vectors, tangential and radial ones. Depending on the type of air diffuser, swirling jets can be compact, conical, or radial.

Isothermal Jets

The shape of jets at a short distance from the outlet face is very similar whether the outlet is round, rectangular, grille-like, or a perforated panel. The jet discharged from a round opening forms an expanding cone; jets from rectangular outlets rapidly pass from a rectangular to an elliptical cross-sectional shape and then to a circular shape, at a rate depending primarily on the aspect ratio and jet width. Even for wide-angle grilles and annular outlets, the similarities permit the same performance analysis for both.

For many conditions of jet discharge, it is possible to analyze jet performance and determine (1) the angle of divergence of the jet boundary, (2) the velocity patterns along the jet axis, (3) the velocity profile at any cross section in the zone of maximum engineering importance, and (4) the entrainment ratios in the same zone (Tuve 1953).

Using the data in this section, the following must be considered:

1. Because the method of finding the jet velocities is based on several approximations, the two recommended equations [Equations (3) and (4)] must be used cautiously for extreme axial and radial distances.
2. The characteristics of the low-velocity regions of ventilating jets are not well understood, and the effects at various Reynolds numbers are not fully known for axial or radial jets.
3. The quantitative treatment of the forces governing room air diffusion problems is limited, and nonisothermal conditions involving buoyant forces are more difficult to predict.
4. Most investigations have addressed free jets, whereas airstreams in practical room air diffusion are not free streams but are influenced by walls, ceilings, floors, and other obstructions.

Angle of Divergence. The angle of divergence is well defined near the outlet face, but the boundary contours are billowy and easily affected by external influences. Near the outlet, as in the room, air movement has local eddies, vortices, and surges. The internal forces governing this air motion are extremely delicate (Nottage et al. 1952b).

Measured angles of divergence (spread) for discharge into large open spaces usually range from 20 to 24° with an average of 22°. Coalescing jets for closely spaced multiple outlets expand at smaller angles, averaging 18°, and jets discharging into relatively small spaces show even smaller angles of expansion (McElroy 1943). In cases where the outlet area is small compared to the dimensions of the space normal to the jet, the jet may be considered free as long as

$$X \le 1.5\sqrt{A_R} \tag{1}$$

where

X = distance from face of outlet, ft
A_R = cross-sectional area of confined space normal to jet, ft^2

Jet Expansion Zones. The full length of an air jet (compact, linear, radial, or conical), in terms of the maximum or centerline velocity and temperature differential at the cross section, can be divided into four zones:

Zone 1. A core zone; a short zone, extending about four diameters or widths from the outlet face, in which the maximum velocity (temperature) of the airstream remains practically unchanged.

Zone 2. A transition zone, the length of which depends on the type of outlet, aspect ratio of the outlet, initial airflow turbulence, and so forth.

Zone 3. A zone of fully established turbulent flow that may be 25 to 100 equivalent air outlet diameters (widths for slot-type air diffusers) long.

Zone 4. A zone of diffuser jet degradation, where the maximum air velocity and temperature decreases rapidly. The distance to this zone and its length depend on the velocities and turbulence characteristics of the ambient air. In a few diameters or widths, the air velocity becomes less than 50 fpm. The characteristics of this zone are still not well understood.

Zone 3 is of major engineering importance because, in most cases, the diffuser jet enters the occupied area within this zone.

Centerline Velocities in Zones 1 and 2. In Zone 1, the ratio V_x/V_o is constant and equal to the ratio of the center velocity of the jet at the start of expansion to the average velocity. The ratio V_x/V_o varies from approximately 1.0 for rounded entrance nozzles to about 1.2 for straight pipe discharges; it has much higher values for diverging discharge outlets.

Experimental evidence indicates that in Zone 2,

$$\frac{V_x}{V_o} = \sqrt{\frac{1.13KH_o}{X}} \tag{2}$$

where

V_x = centerline velocity at distance X from outlet, fpm
V_o = $V_c/C_d R_{fa}$ = average initial velocity at discharge from open-ended duct or across contracted stream at vena contracta of orifice or multiple-opening outlet, fpm
V_c = nominal velocity of discharge based on core area, fpm
C_d = discharge coefficient (usually between 0.65 and 0.90)
R_{fa} = ratio of free area to gross (core) area

Table 1 Recommended Values for Centerline Velocity Constant *K* for Commerical Supply Outlets

Outlet Type	Discharge Pattern	Area *A*	*K*[a]
High sidewall grilles (Figure 1A)	0° deflection [b]	Core	5.0
	Wide deflection	Core	3.7
High sidewall linear (Figure 1B)	Core less than 4 in. high [c]	Core	3.9
	Core more than 4 in. high	Core	4.4
Low sidewall (Figure 1C)	Up and on wall, no spread	Core	4.4
	Wide spread [c]	Core	2.6
Baseboard (Figure 1C)	Up and on wall, no spread	Duct	3.9
	Wide spread [e]	Duct	1.8
Floor (Figure 1C)	No spread [c]	Core	4.1
	Wide spread	Core	1.4
Ceiling circular directional (Fig. 1D)	360° horizontal [d]	Duct	1.0
	Four-way—little spread	Duct	3.3
Ceiling linear (Figure 1E)	One-way—horizontal along ceiling [c]	Core	4.8

[a] These values are representative for commercial outlets with discharge patterns as shown in Figure 1.
[b] Free area is about 80% of core area.
[c] Free area is about 50% of core area.
[d] Cone free area is greater than duct area.
[e] Face free area is greater than duct area.

H_o = width of jet at outlet or at vena contracta, ft
K = centerline velocity constant depending on outlet type and discharge pattern (see Table 1)
X = distance from outlet to measurement of centerline velocity V_x, ft

The aspect ratio (Tuve 1953) and turbulence (Nottage et al. 1952b) primarily affect the centerline velocities in Zones 1 and 2. Aspect ratio has little effect on the terminal zone of the jet when H_o is greater than 4 in. This is particularly true of nonisothermal jets. When H_o is very small, the induced air can penetrate the core of the jet, thus reducing the centerline velocities. The difference in performance between the radial outlet with a small H_o and the axial outlet with a large H_o shows the importance of the thickness of the jets.

When air is discharged from relatively large perforated panels, the constant velocity core formed by the coalescence of the individual jets extends a considerable distance from the panel face. In Zone 1, when the ratio $X/\sqrt{A_c}$ (Distance from Panel/$\sqrt{\text{Panel Area}}$) is less than 5, the following equation should be used for estimating centerline velocities (Koestel et al. 1949):

$$V_x = 1.2 V_o \sqrt{C_d R_{fa}} \tag{3}$$

Centerline Velocity in Zone 3. In Zone 3, maximum or centerline velocities of straight flow isothermal jets can be determined with accuracy from the following equations:

$$\frac{V_x}{V_o} = \frac{K D_o}{X} = \frac{1.13 K \sqrt{A_o}}{X} \tag{4}$$

$$V_x = \frac{1.13 K V_o \sqrt{A_o}}{X} = \frac{1.13 K Q}{X \sqrt{A_o}} \tag{5}$$

$$V_x = \frac{1.13 K Q}{X \sqrt{A_c C_d R_{fa}}} \tag{6}$$

where

K = proportionality constant
D_o = effective or equivalent diameter of stream at discharge from open-end duct or at contracted section, ft

Table 2 Recommended Values of Centerline Velocity Constant for Standard Openings

Type of Outlet	*K*: V_o = 500 to 1000 fpm	*K*: V_o = 2000 to 10,000 fpm
Free openings		
Round or square	5.0	6.2
Rectangular, large aspect ratio (<40)	4.3	5.3
Annular slots, axial or radial[a]	—	—
Grilles and grids		
Free area 40% or more	4.1	5.0
Perforated panels		
Free area 3 to 5%	2.7	3.3
Free area 10 to 20%	3.5	4.3

[a]For radial slots, use X/H instead of $X/\sqrt{A}$. H is height or width of slot.
Note: K is an index of loss in axial kinetic energy. Interpolate as required. Departures from maximum value indicate losses in Zones 1 and 2 when compared with the jet from a rounded-entrance, circular nozzle.

A_o = core area A_c or duct area, ft^2
A_c = measured gross (core) area of outlet, ft^2
Q = discharge from outlet, cfm

Because A_o equals the effective area of the stream, the flow area for commercial registers and diffusers, according to ASHRAE *Standard* 70, can be used in Equation (4) with the appropriate value of K.

Equation (4) is nondimensional and requires only that consistent units be used. Values of K are listed in Table 2 (Tuve 1953, Koestel et al. 1950).

In multiple-opening outlets and annular ring outlets, the streams coalesce into a solid jet before actual jet expansion takes place. This coalescence affects the proportionality constant K and accounts for some divergence in reported values for similar outlets.

For perforated panels of relatively large size, the values of K given in Table 2 apply only when the ratio $X/\sqrt{A_c}$ is larger than 5 (see the section on Centerline Velocities in Zones 1 and 2).

Low-velocity test results, in the range $V_x < 150$ fpm, indicate that normal values of K should be reduced about 20% for $V_x = 50$ fpm, as used later in Equation (9) for throw.

Determining Centerline Velocities. To correlate data from all four zones, centerline velocity ratios are plotted against distance from the outlet in Figure 2.

The airflow patterns of diffusers are related to the throw K-factors and to the throw distance. In general, diffusers that have a circular airflow pattern have a shorter throw than those with a directional or crossflow pattern. During cooling, the circular pattern has a tendency to curl back from the end of the throw toward the diffuser. This action reduces the drop and ensures that the cool air remains near the ceiling.

Cross-flow airflow patterns have a longer throw, and the individual side jets react in a manner similar to jets from sidewall grilles. The airflow jets with this type of pattern have a longer throw and the airflow does not roll back to the diffuser at the end of the throw. The airflow continues to move away from the diffuser at low velocities.

The following example illustrates the use of Table 1 and Figure 2.

Example 1. A 12 in. by 18 in. high sidewall grille with an 11.25 in. by 17.25 in. core area is selected. From Table 1, K = 5 for Zone 3. If the airflow is 600 cfm, what is the throw to 50 fpm, 100 fpm, and 150 fpm? The grille has 80% free area.

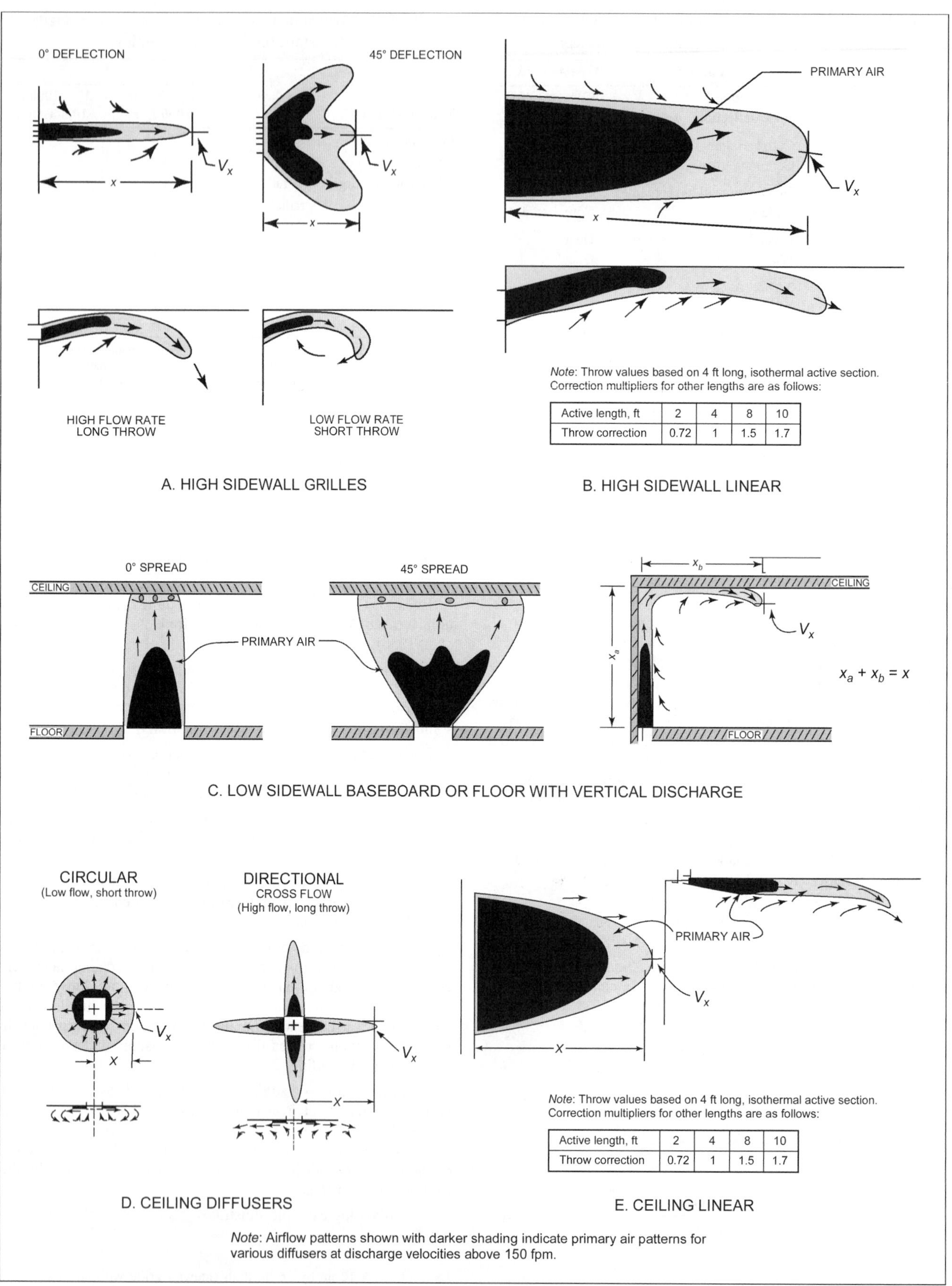

Active length, ft	2	4	8	10
Throw correction	0.72	1	1.5	1.7

Active length, ft	2	4	8	10
Throw correction	0.72	1	1.5	1.7

Fig. 1 Airflow Patterns of Different Diffusers

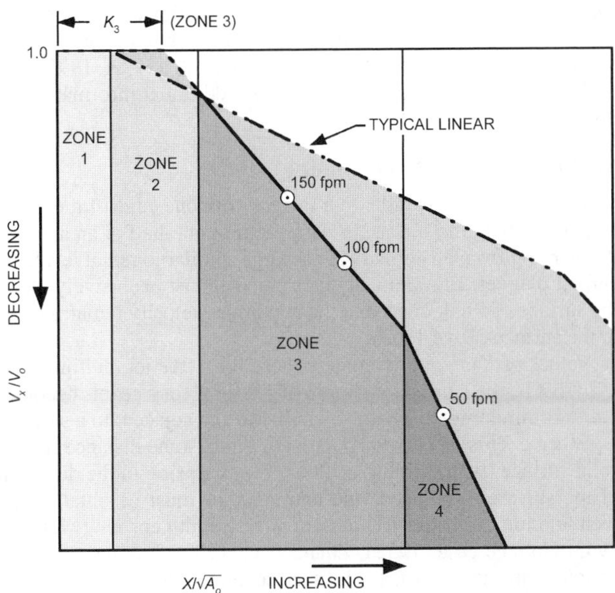

Fig. 2 Chart for Determining Centerline Velocities of Axial and Radial Jets

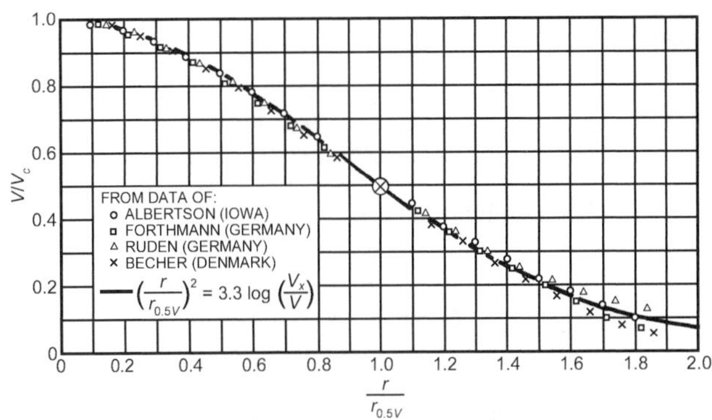

Fig. 3 Cross-Sectional Velocity Profiles for Straight-Flow Turbulent Jets

Solution:

From Equation (5):

$$X = \frac{1.13KQ}{V_x\sqrt{A_o}} = \frac{1.13 \times 5 \times 600}{V_x\sqrt{11.25 \times 17.25/144}} = \frac{2920}{V_x}$$

Solving for 50 fpm throw,

$$X = \frac{2920}{50} = 58 \text{ ft}$$

But according to Figure 2, 50 fpm is in Zone 4, which is typically 20% less than calculated in Equation (4), or

$$X = 58 \times 0.80 = 47 \text{ ft}$$

Solving for 100 fpm throw,

$$X = \frac{2920}{100} = 29 \text{ ft}$$

Solving for 150 fpm throw,

$$X = \frac{2920}{150} = 19 \text{ ft}$$

Throw. Equation (6) can be transposed to determine the throw X of an outlet if the discharge volume and the centerline velocity are known:

$$X = \frac{1.13K}{V_x}\frac{Q}{\sqrt{A_c C_d R_{fa}}} \tag{7}$$

Or, if $Z = \sqrt{C_d R_{fa}}$,

$$X = \frac{1.13K}{V_x}\frac{Q}{Z\sqrt{A_c}} \tag{8}$$

The maximum throw T_V is usually defined as the distance from the outlet face to where the centerline velocity is 50 fpm. Therefore, for V_T = 50 fpm,

$$T_V = X = \frac{1.13K}{50}\frac{Q}{Z\sqrt{A_c}} \tag{9}$$

Any other terminal centerline velocity could be inserted in Equation (9) for V_T.

Velocity Profiles of Jets. In Zone 3 of both axial and radial jets, the velocity distribution may be expressed by a single curve (Figure 3) in terms of dimensionless coordinates; this same curve can be used as a good approximation for adjacent portions of Zones 2 and 4. Temperature and density differences have little effect on cross-sectional velocity profiles.

Velocity distribution in Zone 3 can be expressed by the Gauss error function or probability curve, which is approximated by the following equation:

$$\left(\frac{r}{r_{0.5V}}\right)^2 = 3.3 \log\frac{V_x}{V} \tag{10}$$

where

r = radial distance of point under consideration from centerline of jet

$r_{0.5V}$ = radial distance in same cross-sectional plane from axis to point where velocity is one-half centerline velocity (i.e., $V = 0.5V_x$)

V_x = centerline velocity in same cross-sectional plane

V = actual velocity at point being considered

Experiments show that the conical angle for $r_{0.5V}$ is approximately one-half the total angle of divergence of a jet. The velocity profile curve for one-half of a straight-flow turbulent jet (the other half being a symmetrical duplicate) is shown in Figure 3. For multiple-opening outlets, such as grilles or perforated panels, the velocity profiles are similar, but the angles of divergence are smaller.

Entrainment Ratios. The following are equations for the entrainment of circular jets and of jets from long slots.

For third-zone expansion of circular jets,

$$\frac{Q_x}{Q_o} = \frac{2X}{1.13K\sqrt{A_o}} \tag{11}$$

By substituting from Equation (4),

$$\frac{Q_x}{Q_o} = 2\frac{V_o}{V_x} \tag{12}$$

For a continuous slot with active sections up to 10 ft and separated by 2 ft,

$$\frac{Q_x}{Q_o} = \sqrt{\frac{2}{1.13K}}\sqrt{\frac{X}{H_o}} \tag{13}$$

or, substituting from Equation (2),

$$\frac{Q_x}{Q_o} = \sqrt{2}\,\frac{V_o}{V_x} \tag{14}$$

where

Q_x = total volumetric flow rate at distance X from face of outlet, cfm
Q_o = discharge from outlet, cfm
X = distance from face of outlet, ft
K = proportionality constant
A_o = core area A_c or duct area, ft^2
H_o = width of slot, ft

The entrainment ratio Q_x/Q_o is important in determining total air movement at a given distance from an outlet. For a given outlet, the entrainment ratio is proportional to the distance X [Equation (11)] or to the square root of the distance X [Equation (13)] from the outlet. Equations (12) and (14) show that, for a fixed centerline velocity V_x, the entrainment ratio is proportional to the outlet velocity. Equations (12) and (14) also show that, at a given centerline and outlet velocity, a circular jet has greater entrainment and total air movement than a long slot. Comparing Equations (11) and (13), the long slot should have a greater rate of entrainment. The entrainment ratio at a given distance is less with a large K than with a small K.

Isothermal Radial Flow Jets

In a radial jet, as with an axial jet, the cross-sectional area at any distance from the outlet varies as the square of this distance. Centerline velocity gradients and cross-sectional velocity profiles are similar to those of Zone 3 of axial jets, and the angles of divergence are about the same.

A jet from a ceiling plaque has the same form as half of a free radial jet. The jet is wider and longer than a free jet, with the maximum velocity close to the surface. Koestel (1957) provides an equation for radial flow outlets.

Nonisothermal Jets

When the temperature of introduced air is different from the room air temperature, the behavior of the diffuser air jet is affected by the thermal buoyancy due to air density difference. The trajectory of a nonisothermal jet introduced horizontally is determined by the **Archimedes number** (Baturin 1972):

$$\mathrm{Ar} = \frac{gL_o(t_o - t_s)}{V_o^2 T_s} \tag{15}$$

where

g = gravitational acceleration rate, ft/min^2
L_o = length scale of diffuser outlet equal to hydraulic diameter of outlet, ft
t_o = initial temperature of jet, °F
t_s = temperature of surrounding air, °F
V_o = initial air velocity of jet, fpm
T_s = room air temperature, °R

The paths assumed by horizontally projected heated and chilled jets influenced by buoyant forces are significant in heating and cooling with wall outlets. Koestel's equation (1955) describes the behavior of these jets.

Helander and Jakowatz (1948), Helander et al. (1953, 1954, 1957), Yen et al. (1956), and Knaak (1957) developed equations for outlet characteristics that affect the downthrow of heated air. Koestel (1954, 1955) developed equations for temperatures and velocities in heated and chilled jets. Li et al. (1993, 1995) and Kirkpatrick and Elleson (1996) provide additional information on nonisothermal jets.

Surface Jets (Wall and Ceiling)

Jets discharging parallel to a surface with one edge of the outlet coinciding with the surface take the form of one-half of an axial jet discharging from an outlet twice as large, similar to radial jets from ceiling plaques. Entrainment takes place almost exclusively along the surface of a half cone, and the maximum velocity remains close to the surface (Tuve 1953).

Values of K are approximately those for a free jet multiplied by $\sqrt{2}$; that is, the normal maximum of 6.2 for K for free jets becomes 8.8 for a similar jet discharged parallel to and adjacent to a surface.

When a jet is discharged parallel to but at some distance from a solid surface (wall, ceiling, or floor), its expansion in the direction of the surface is reduced, and entrained air must be obtained by recirculation from the jet instead of from ambient air (McElroy 1943, Nottage et al. 1952a, Zhang et al. 1990). The restriction to entrainment caused by the solid surface induces the **Coanda effect**, which makes the jet attach to a surface a short distance after it leaves the diffuser outlet. The jet then remains attached to the surface for some distance before separating from the surface again.

In nonisothermal cases, the trajectory of the jet is determined by the balance between the thermal buoyancy and the Coanda effect, which depends on the jet momentum and the distance between the jet exit and the solid surface. The behavior of such nonisothermal surface jets has been studied by Kirkpatrick et al. (1991), Wilson et al. (1970), Oakes (1987), and Zhang et al. (1990), each addressing different factors. A more systematic study of these jets in room ventilation flows is needed to provide reliable guidelines for designing air diffusion systems.

MULTIPLE JETS

Twin parallel air jets act independently until they interfere. The point of interference and its distance from the outlets varies with the distance between the outlets. From the outlets to the point of interference, the maximum velocity, as for a single jet, is on the centerline of each jet. After interference, the velocity on a line midway between and parallel to the two jet centerlines increases until it equals the jet centerline velocity. From this point, a maximum velocity of the combined jet stream is on the midway line, and the profile seems to emanate from a single outlet of twice the area of one of the two outlets.

Koestel and Austin (1956) determined the spacing between outlets for noninterference between the jets. For a K value of 6.5, the outlets should be placed three to eight diameters apart, with V_o values from 500 to 1500 fpm.

AIRFLOW IN OCCUPIED ZONE

Mixing Systems. Laboratory experiments on jets usually involve recirculated air with negligible resistance to flow on the return path of the jet air. Experiments in mine tunnels of small cross-sectional areas, where the return flow of jet air to outlets meets considerable resistance, show that expansion of the jet terminates abruptly at a distance that is independent of discharge velocity and is only slightly affected by the size of the outlet. These distances are determined primarily by the size and length of the return path. In a long tunnel with a cross section of 5 ft by 6 ft, a jet may not travel more than 25 ft; in a tunnel with a relatively large section (25 ft by 60 ft), the jet may travel more than 250 ft. McElroy (1943) provides data on this phase of jet expansion.

Zhang et al. (1990) found that, for a given heat load and room air supply rate, air velocity in the occupied zone increases when the outlet discharge velocity increases. Therefore, the design supply air velocity should be high enough to maintain the jet traveling in the desired direction in order to ensure good mixing before it reaches the occupied zone. Excessively high outlet air velocity would induce high air velocity in the occupied zone and result in thermal discomfort.

Turbulence Production and Transport. The air turbulence within a room is mainly produced at the diffuser jet region by interaction of the supply air with the room air and with the solid surfaces (walls or ceiling) in the vicinity. It is then transported to other parts of the room, including the occupied zone (Zhang et al. 1992). Meanwhile, the turbulence is also damped by viscous effect. Air in the occupied zone usually contains very small amounts of turbulent kinetic energy compared to that in the jet region. Because turbulence may cause thermal discomfort (Fanger et al. 1989), air diffusion systems should be designed so that the primary mixing between the introduced air and the room air occurs away from occupied regions.

ROOM AIR DIFFUSION METHODS

Room air diffusion systems can be classified as mixing, displacement, unidirectional, underfloor, and task ambient conditioning.

MIXING SYSTEMS

In mixing systems, conditioned air is normally discharged from air outlets at velocities much greater than those acceptable in the occupied zone. Conditioned air temperature may be above, below, or equal to the air temperature in the occupied zone, depending on the heating/cooling load. The diffuser jets mix with the ambient room air by entrainment, which reduces the air velocity and equalizes the air temperature. The occupied zone is ventilated either by the decayed air jet directly or by the reverse flow created by the jets. Mixing air distribution creates relatively uniform air velocity, temperature, humidity, and air quality conditions in the occupied zone.

Outlet Classification and Performance

Straub et al. (1956) and Straub and Chen (1957) classified outlets into five groups:

Group A. Outlets mounted in or near the ceiling that discharge air horizontally.

Group B. Outlets mounted in or near the floor that discharge air vertically in a nonspreading jet.

Group C. Outlets mounted in or near the floor that discharge air vertically in a spreading jet.

Group D. Outlets mounted in or near the floor that discharge air horizontally.

Group E. Outlets mounted in or near the ceiling that project primary air vertically.

Analysis of outlet performance was based on primary air pattern, total air pattern, stagnant air layer, natural convection currents, return air pattern, and room air motion. Figures 4 through 8 show the room air motion characteristics of the five outlet groups; exterior walls are depicted by heavy lines. The principles of air diffusion emphasized by these figures are as follows:

1. The primary air (shown by dark envelopes in Figures 4 through 8) from the outlet down to a velocity of about 150 fpm can be treated analytically. The heating or cooling load has a strong effect on the characteristics of the primary air.
2. The total air, shown by gray envelopes in Figures 4 through 8, is influenced by the primary air and is of relatively high velocity (but less than 150 fpm). The total air is also influenced by the environment and drops during cooling or rises during heating; it is not subject to precise analytical treatment.
3. Natural convection currents form a stagnant zone from the ceiling down during cooling, and from the floor up during heating. This zone forms below the terminal point of the total air during heating and above the terminal point during cooling. Because this zone results from natural convection currents, the air velocities within it are usually low (approximately 20 fpm), and the air stratifies in layers of increasing temperatures. The concept of a stagnant zone is important in properly applying and selecting outlets because it considers the natural convection currents from warm and cold surfaces and internal loads.
4. A return inlet affects the room air motion only within its immediate vicinity. The intake should be located in the stagnant zone to return the warmest room air during cooling or the coolest room air during heating. The importance of the location depends on the relative size of the stagnant zone, which depends on the type of outlet.
5. The general room air motion (shown by white areas in Figures 4 through 8) is a gentle drift toward the total air. Room conditions are maintained by the entrainment of the room air into the total airstream. The room air motion between the stagnant zone and the total air is relatively slow and uniform. The highest air motion occurs in and near the total airstreams.

Group A Outlets. This group includes high sidewall grilles, sidewall diffusers, ceiling diffusers, linear ceiling diffusers, and similar outlets. High sidewall grilles and ceiling diffusers are illustrated in Figure 4.

The primary air envelopes (isovels) show a horizontal, two-jet pattern for the high sidewall and a 360° diffusion pattern for the ceiling outlet. Although variation of vane settings might cause a discharge in one, two, or three jets in the case of the sidewall outlet, or have a smaller diffusion angle for the ceiling outlet, the general effect in each is the same.

During cooling, the total air drops into the occupied zone at a distance from the outlet that depends on air quantity, supply velocity, temperature differential between supply and room air, deflection setting, ceiling effect, and type of loading within the space. Analytical methods of relating some of these factors are presented in the section on Principles of Jet Behavior.

The cooling diagram for the high sidewall outlet shows an overthrow condition, which causes the total air to drop along the opposite wall and flow slowly for some distance across the floor. Velocities of about 100 to 150 fpm may be found near the wall but will dissipate within about 4 in. of the wall.

The cooling diagram for the ceiling outlet shows that the total air movement is counteracted by the rising natural convection currents on the heated wall, and, therefore, drops before reaching the wall. On the other hand, the total air reaches the inside wall and descends for some distance along it. With this type of outlet, temperature variations in the room are minimized, with minimal stagnant volume. The maximum velocity and the maximum temperature variation occur in and near the total air envelope; therefore, the drop region becomes important because it is an area with high effective draft temperature θ [see Equation (16)]. Consequently, how far the air drops before velocities and temperatures reach acceptable limits must be known.

Because these outlets discharge horizontally near the ceiling, the warmest air in the room is mixed immediately with the cool primary air far above the occupied zone. Therefore, the outlets are capable of handling relatively large quantities of air at large temperature differentials.

During heating, warm supply air introduced at the ceiling can cause stratification in the space if there is insufficient induction of room air at the outlet. Selecting diffusers properly, limiting the room supply temperature differential, and maintaining air supply rates at a level high enough to ensure air mixing by induction provide adequate air diffusion and minimize stratification.

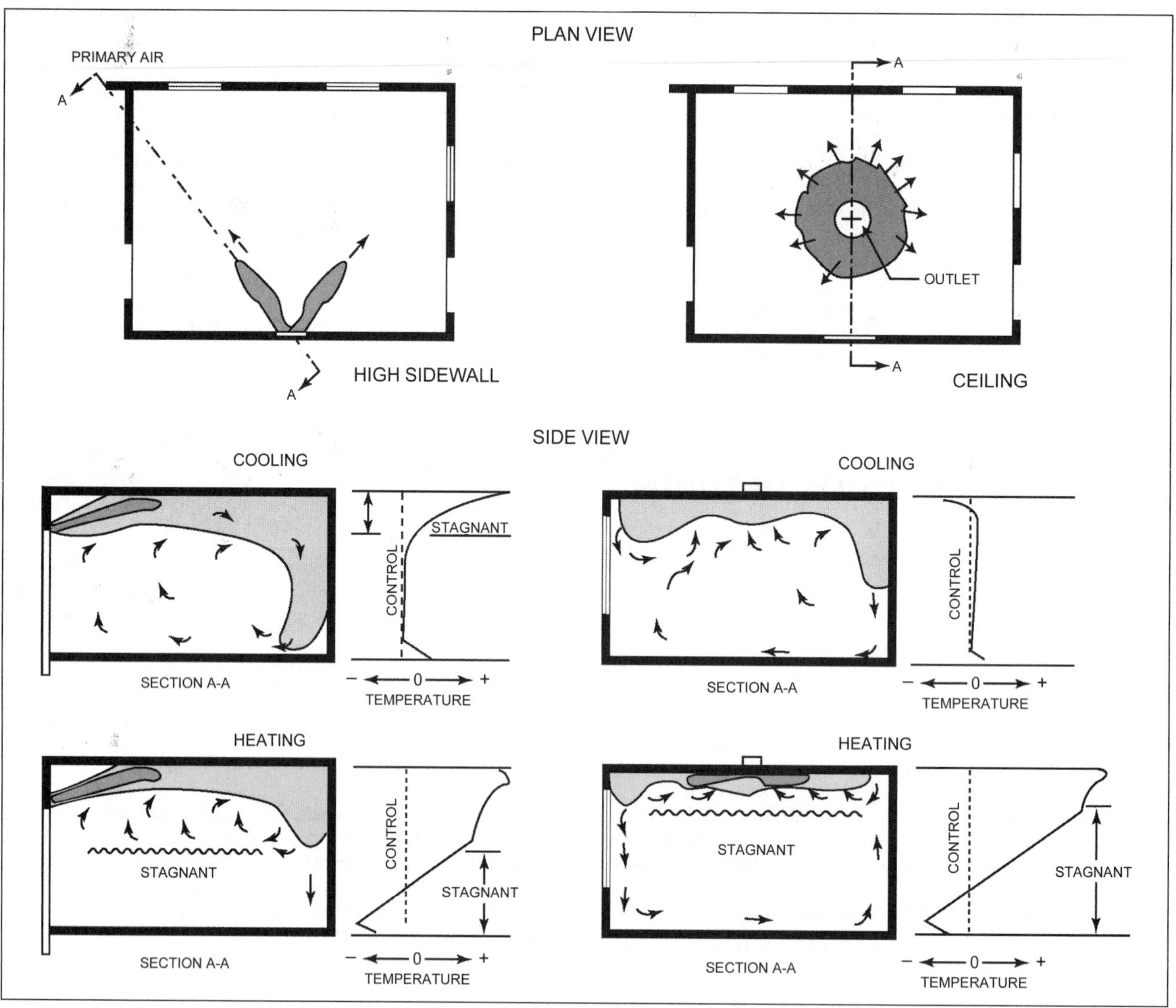

Fig. 4 Air Motion Characteristics of Group A Outlets
(Straub et al. 1956)

Several building codes and ASHRAE *Standard* 90.1 require sufficient insulation in exterior walls, so most perimeter spaces can be heated effectively by ceiling air diffusion systems. Interior spaces, which generally have only cooling demand conditions, seldom require long-term heating and are seldom a design problem.

Flow rate and velocity for both heating and cooling are the same for the outlets shown in Figure 4. The heating diagram for the sidewall unit shows that, under these conditions, the total air does not descend along the wall. Consequently, higher velocities might be beneficial in eliminating the stagnant zone, since high velocity causes some warm air to reach floor level and counteract stratification of the stagnant region.

The heating diagram for the ceiling outlet shows the effect of the natural convection currents that produce a larger throw toward the cold exposed wall. The velocity of the total air toward the exposed wall complements the natural convection currents. However, the warm total air loses its downward momentum at its terminal point, and buoyancy forces cause it to rise toward the primary air. Although these forces are complementary, the heating effect of the total air replaces the cool natural convection currents with warm total air.

Group B Outlets. This group includes floor registers, baseboard units, low sidewall units, linear-type grilles in the floor or windowsill, and similar outlets. Figure 5 illustrates a floor outlet adjacent to an inside wall.

Because these outlets have no deflecting vanes, the primary air is discharged in a single, vertical jet. When the total air strikes the ceiling, it fans out in all directions and, during cooling, follows the ceiling for some distance before dropping toward the occupied zone. During heating, the total airflow follows the ceiling across the room, then descends partway down the exterior wall.

The cooling diagram shows that a stagnant zone forms outside the total air region above its terminal point. Below the stagnant zone, air temperature is uniform, effecting complete cooling. Also, the space below the terminal point of the total air is cooled satisfactorily. For example, if total airflow is projected upward for 8 ft, the region from this level down to the floor will be cooled satisfactorily. This, however, does not apply to an extremely large space. Judgment to determine the acceptable size of the space outside the total air is needed. A distance of 15 to 20 ft between the drop region and the exposed wall is a conservative design value.

A comparison of Figures 4 and 5 for heating shows that the stagnant region is smaller for Group B outlets than for Group A outlets

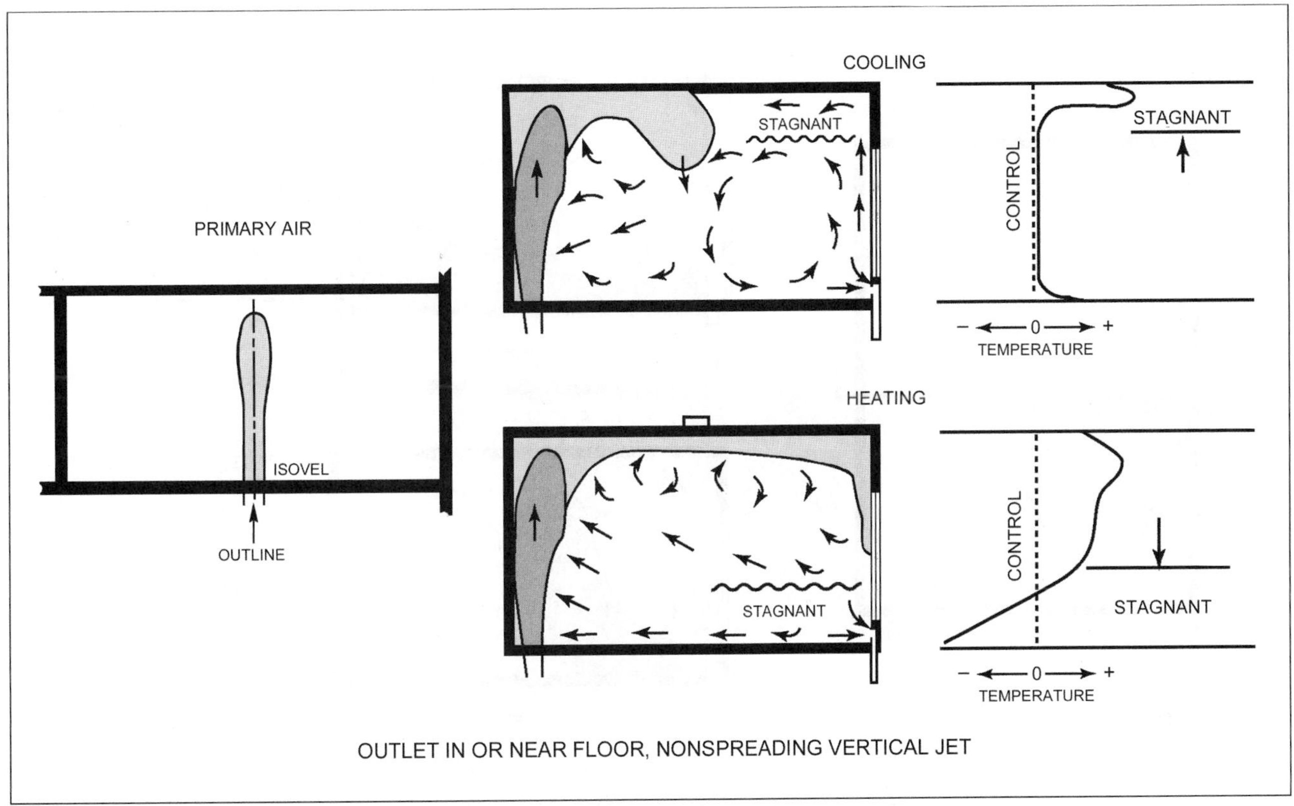

Fig. 5 Air Motion Characteristics of Group B Outlets
(Straub et al. 1956)

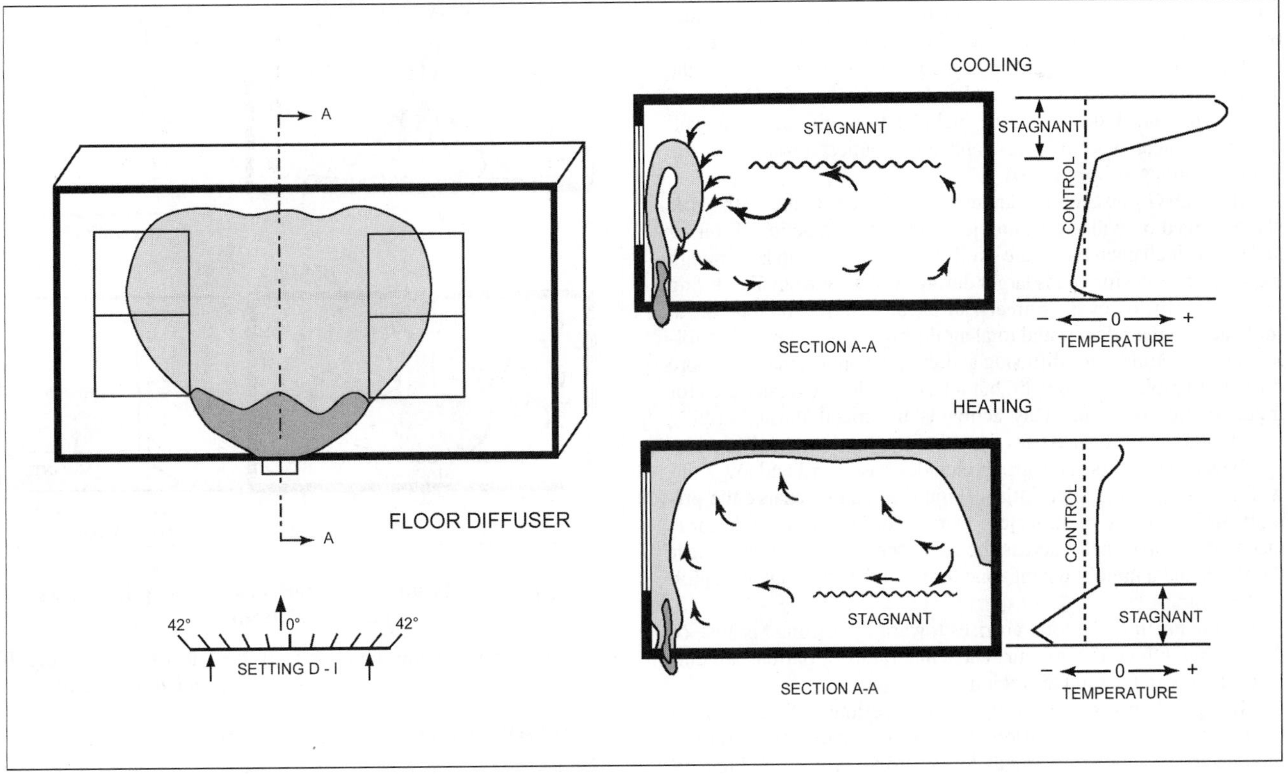

Fig. 6 Air Motion Characteristics of Group C Outlets
(Straub et al. 1956)

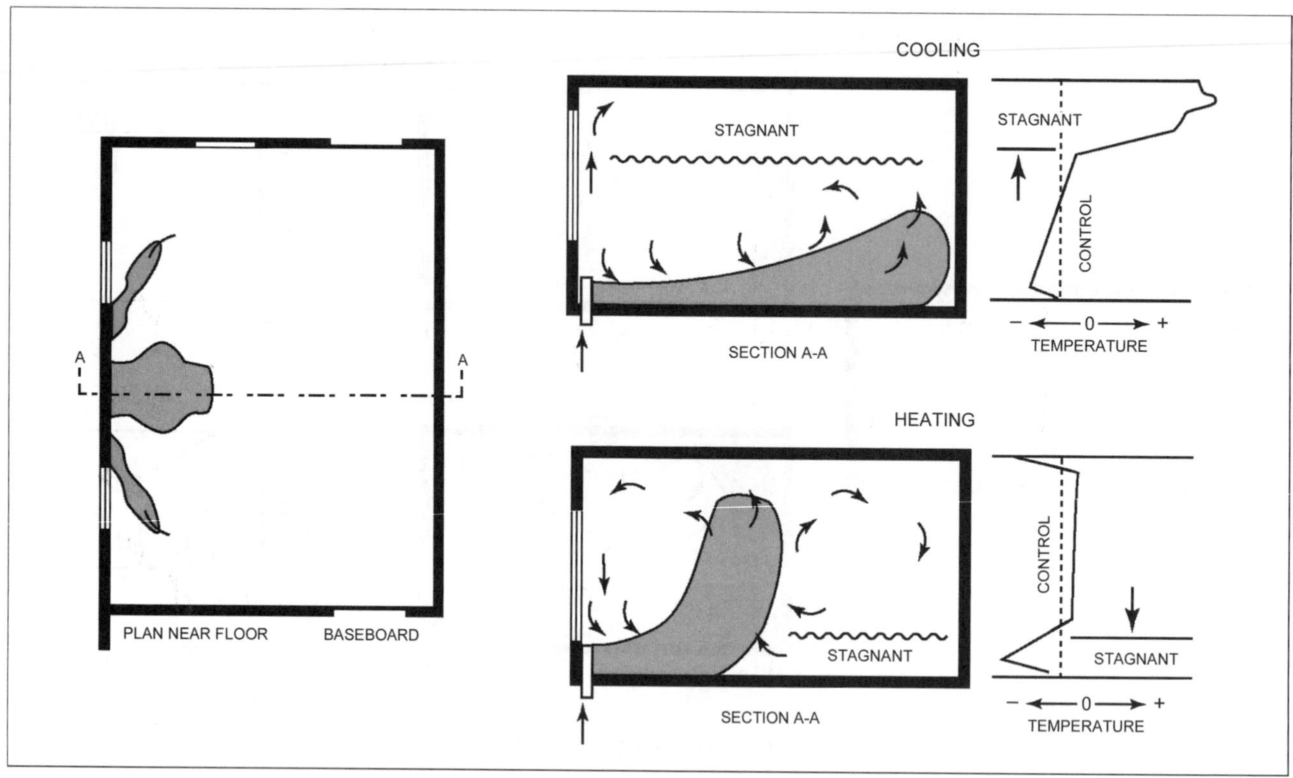

Fig. 7 Air Motion Characteristics of Group D Outlets
(Straub et al. 1956)

because the air entrained in the immediate vicinity of the outlet is taken mainly from the stagnant region, which is the coolest air in the room. This results in greater temperature equalization and less buoyancy in the total air than would occur with Group A outlets.

While the temperature gradients for both outlet groups are about the same, the stagnant layer for Group B is lower than that for Group A.

Group C Outlets. This group includes floor diffusers, sidewall diffusers, linear-type diffusers, and other outlets installed in the floor or windowsill (Figure 6).

Although Group C outlets are related to Group B outlets, they are characterized by wide-spreading jets and diffusing action. Total air and room air characteristics are similar to those of Group B, although the stagnant zone formed is larger during cooling and smaller during heating. Diffusion of the primary air usually causes the total air to fold back on the primary and total air during cooling, instead of following the ceiling. This diffusing action of the outlets makes it more difficult to project the cool air, but it also provides a greater area for induction of room air. This action is beneficial during heating because the induced air comes from the lower regions of the room.

Group D Outlets. This group includes baseboard and low sidewall registers and similar outlets (Figure 7) that discharge the primary air in single or multiple jets. During cooling, because the air is discharged horizontally across the floor, the total air remains near the floor, and a large stagnant zone forms in the entire upper region of the room.

During heating, the total air rises toward the ceiling because of the buoyant effect of warm air. The temperature variations are uniform, except in the total air region.

Group E Outlets. This group includes ceiling diffusers, linear-type grilles, sidewall diffusers and grilles, and similar outlets mounted or designed for vertical downward air projection. Figure 8 shows the heating and cooling diagrams for such a ceiling diffuser.

During cooling, the total air projects to and follows the floor, producing a stagnant region near the ceiling. During heating, the total airflow reaches the floor and folds back toward the ceiling. If projected air does not reach the floor, a stagnant zone results.

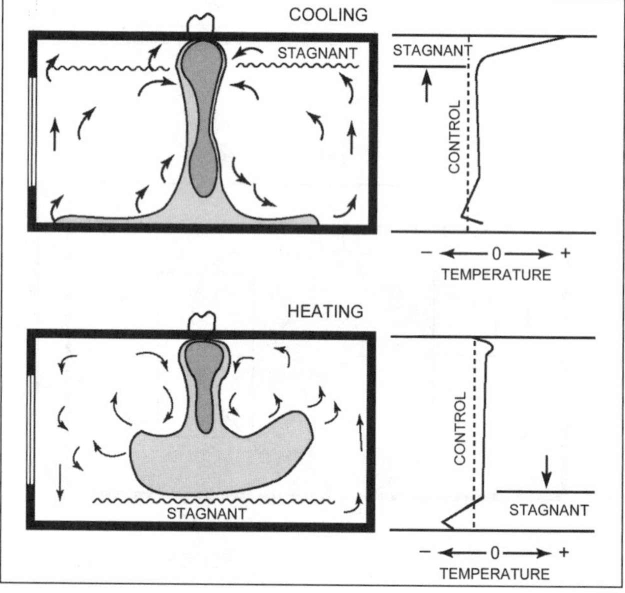

Fig. 8 Air Motion Characteristics of Group E Outlets
(Straub et al. 1956)

Factors Affecting Outlet Performance

Vanes. Vanes affect grille performance if their depth is at least equal to the distance between the vanes (vane ratio ≥ 1). If the vane ratio is less than unity, effective control by the vanes of the airstream discharged from the grille is impossible. Increasing the vane ratio above two has little or no effect, so vane ratios should be between one and two.

A grille discharging air uniformly forward (vanes in straight position) has a spread of 14 to 24°, depending on the type of outlet, the duct approach, and the discharge velocity. Turning the vanes influences the direction and throw of the discharged airstream.

A grille with diverging vanes (vertical vanes with uniformly increasing angular deflection from the centerline to a maximum at each end of 45°) has a spread of about 60° and reduces the throw considerably. With increasing divergence, the quantity of air discharged by a grille for a given upstream total pressure decreases.

A grille with converging vanes (vertical vanes with uniformly decreasing angular deflection from the centerline) has a slightly higher throw than a grille with straight vanes, but the spread is approximately the same for both settings. The airstream converges slightly for a short distance in front of the outlet and then spreads more rapidly than air discharged from a grille with straight vanes.

In addition to vertical vanes that normally spread the air horizontally, horizontal vanes may spread the air vertically. However, spreading the air vertically risks hitting beams or other obstructions or blowing primary air into the occupied zone at excessive velocities. On the other hand, vertical deflection may increase adherence to the ceiling and reduce the drop.

Beamed Ceilings and Obstructions. In spaces with exposed beams, the outlets should be located below the bottom of the lowest beam level, preferably low enough to employ an upward or arched air path. The air path should be arched sufficiently to miss the beams and prevent the primary or induced airstream from striking furniture and obstacles and producing objectionable drafts (Wilson 1970). Obstructions influence airflow patterns and can reduce air distribution efficiency. Obstructions can reduce jet throw, increase air velocities in portions of the occupied zone, and create stagnant zones.

Variable Air Volume (VAV) Systems. The design of air distribution systems is usually based on the full load (heating/cooling). When only a partial load exists, VAV systems reduce the supply airflow, which in turn reduces the air velocity at the outlet. Therefore, the different operation modes of the system (airflow and initial temperature difference) should be considered in designing a VAV system air distribution.

DISPLACEMENT VENTILATION

In displacement ventilation, conditioned air with a temperature slightly lower than the desired room air temperature in the occupied zone is supplied from air outlets at low air velocities (100 fpm or less). The outlets are located at or near the floor level, and the supply air is introduced directly to the occupied zone. Returns through which the warm room air is exhausted from the room are located at or close to the ceiling. The supply air is spread over the floor and then rises as it is heated by the heat sources in the occupied zone. Heat sources (e.g., person, computer) in the occupied zone create upward convective flows in the form of thermal plumes. These thermal plumes tend to remove heat and contaminants within the plume from the occupied zone (Figure 9).

The air volume in the plumes increases as they rise because the plumes entrain ambient air. A stratification level exists where the airflow rate in the plumes equals the supply airflow rate. Two distinct zones are thus formed within the room: one lower zone below the stratification level and with no recirculation flow (close to displacement flow), and one upper zone, with recirculation flow (Figure 9). The height of the lower zone depends on the supply airflow rate and the characteristics of heat sources and their distribution across the floor area. In a properly designed displacement ventilation system, the upper boundary of the lower zone is above the occupied zone so that the occupied zone can be ventilated effectively. For this type of system to function properly, a stable vertically stratified temperature field is essential.

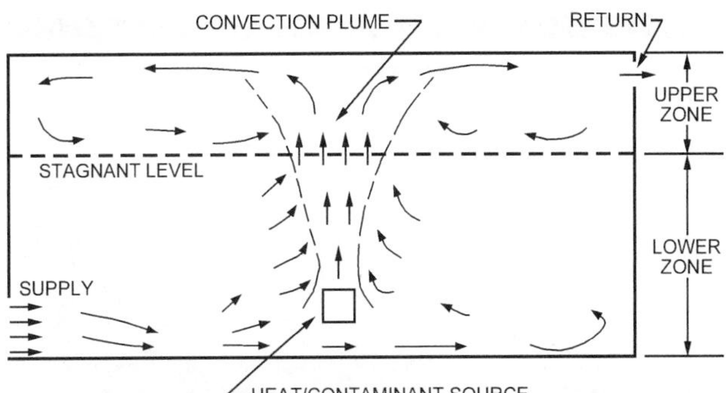

Fig. 9 Schematic of Displacement Ventilation

In contrast to mixing ventilation, displacement ventilation is designed to minimize mixing of air within the occupied zone. The objective of the displacement ventilation is to create conditions close to supply air conditions in the occupied zone. This type of ventilation was originally used in industrial buildings as an effective method for removing contaminants in the occupied zone. It is now also used for ventilating and cooling office buildings. However, local discomfort due to draft and vertical temperature gradient may be critical (Melikov and Nielsen 1989). Sandberg and Blomqvist (1989) suggest that the maximum convective cooling load in office buildings with displacement ventilation not exceed about 8 Btu/h·ft^2 so that the maximum vertical temperature gradient in the occupied zone will not be larger than 5°F. This is equivalent to 1 cfm/ft^2 at a maximum cooling differential of 7.5°F. Kegel and Schulz (1989) and Svensson (1989) suggest somewhat higher cooling load limits of 10 to 13 Btu/h·ft^2.

One way of increasing the cooling capacity of displacement ventilation systems is to recirculate some of the room air in the occupied zone through an induction circuit; that is, the room air is induced into the supply air and is mixed before discharge through the low-velocity air terminal device into the room. This reduces the room air temperature gradient for a given cooling load, thus allowing a cooling load limit of up to 16 Btu/h·ft^2 (Jackman 1991).

Air diffusers with a large outlet area are used to supply air at low velocity. Displacement ventilation has been compared with conventional mixing ventilation (Svensson 1989, Seppanen et al. 1989, Stymne et al. 1991). Design guidelines for displacement ventilation can be found in Scaret (1985), Jackman and Appleby (1990), Jackman (1991), and Shilkrot and Zhivov (1992).

UNIDIRECTIONAL AIRFLOW VENTILATION

In this type of ventilation, air is either (1) supplied from the ceiling and exhausted through the floor, or vice versa; or (2) supplied through the wall and exhausted through returns at the opposite wall. The outlets are uniformly distributed over the ceiling, floor, or wall to provide a low-turbulence "plug"-type flow across the entire room. This type of system is mainly used for clean room ventilation, in which the main objective is to remove contaminant particles from the room. Details about clean room ventilation are given in Chapter 15 of the 1999 *ASHRAE Handbook—Applications*. Unidirectional flow ventilation is also used in other areas, such as computer rooms and paint booths.

UNDERFLOOR AIR DISTRIBUTION AND TASK/AMBIENT CONDITIONING

Underfloor air distribution systems are installed with a raised floor through which conditioned air is delivered to the space through floor grilles or as part of the workstation furniture and partitions. Sometimes called localized ventilation, these systems

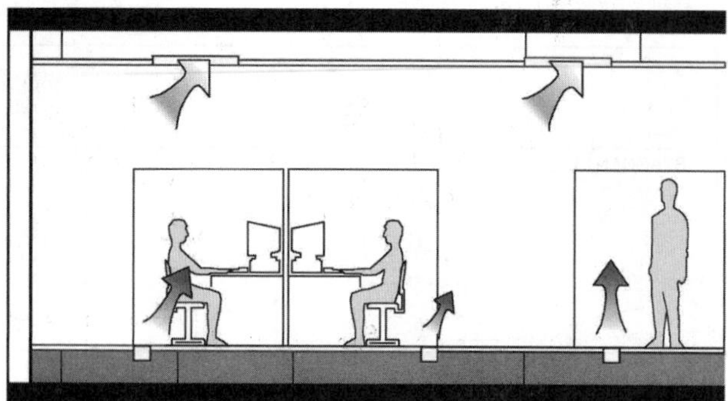

Fig. 10 Underfloor Air Distribution System

supply air to local areas that are often near building occupants or other specific locations in the space. In comparison to conventional ceiling-based air diffusion, underfloor air distribution systems generally have a larger number of supply diffusers directly in the occupied zone of the building (e.g., in floors, desks, workstation partitions, or theater seats). Air typically returns at or close to ceiling level, so that localized systems benefit from the same overall upward movement of air in the room as displacement ventilation systems. In cooling applications, this air movement efficiently removes heat and contaminant sources from the room.

Underfloor air distribution differs from displacement ventilation in that (1) it generally uses higher supply volumes, which enable higher cooling loads to be met; and (2) it supplies air at a higher velocity through smaller diffusers. Because air is delivered directly to the occupied zone, the supply air temperature is usually warmer (above 63 to 64°F) than that maintained for conventional ceiling distribution in order to avoid occupant discomfort due to drafts. Bauman et al. (1999), Hanzawa and Nagasawa (1990), Houghton (1995), Loudermilk (1999), McCarry (1995, 1998), Shute (1992, 1995), Sodec and Craig (1990), Spoormaker (1990), and Tanabe and Kimura (1996) describe the results of laboratory studies, case studies of actual installations, field experiments of system performance, and present design guidelines.

A well-designed underfloor air distribution system also requires less energy and is more flexible in providing and maintaining building services than traditional overhead systems. Extremely low operational static pressures in the underfloor air supply plenum can reduces central fan energy use. Thermal storage in the exposed structural mass in the underfloor plenum (e.g., concrete slab) can save energy and reduce peak cooling loads. The use of raised flooring provides maximum flexibility and significantly lowers costs associated with reconfiguring building services, particularly in buildings having high churn rates. (Churn rate is defined as the annual percentage of workers and their associated work spaces in a building that are reconfigured or undergo significant changes.) Figure 10 shows a schematic diagram of an underfloor air distribution system.

Task/Ambient Conditioning (TAC)

Task/ambient conditioning (TAC) is most commonly installed with underfloor air distribution (Arens et al. 1991; Bauman et al. 1991, 1993, 1995, 1998; Bauman and Arens 1996; Faulkner et al. 1993, 1999; Fisk et al. 1991; Matsunawa et al. 1995; Tsuzuki et al. 1999). TAC gives individuals some control over their local environment without adversely affecting that of nearby occupants. Typically, the occupant can control the speed, direction, and, in some cases, temperature of the incoming air supply. TAC systems have been most frequently installed in open-plan offices in which they provide supply air and, in some cases, radiant heating directly into workstations. Figure 11 shows an underfloor TAC system with a local (personal HVAC) diffuser located in the partition in front of the office worker (Matsunawa et al. 1995).

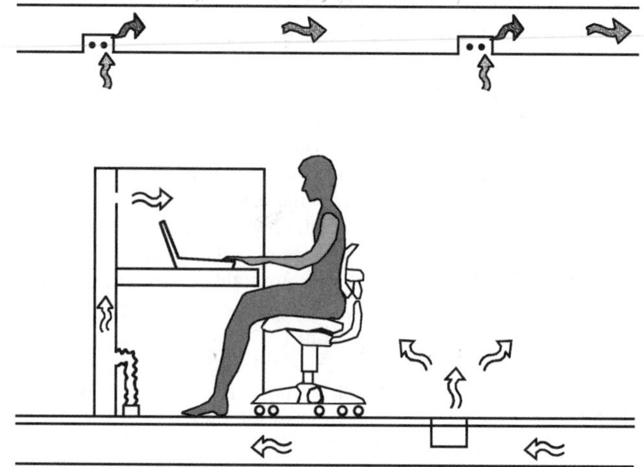

Fig. 11 Underfloor TAC and Personal HVAC System
(Matsunawa et al. 1995)

As further evidence of the benefits of providing personal control, field research has found that building occupants who have no individual control capabilities are twice as sensitive to changes in temperature as occupants who do have individual thermal control (de Dear and Brager 1999, Bauman et al. 1998).

METHODS OF EVALUATION

Standards for Satisfactory Conditions

The object of air diffusion in warm-air heating, ventilating, and air-conditioning is to create the proper combination of temperature, humidity, and air motion in the occupied zone of the conditioned room—from the floor to 6 ft above floor level (Miller 1989). The effective draft temperature combines the effects of air temperature, air motion, and relative humidity in terms of their physiological effects on a human body. Variation from accepted standard limits (ASHRAE *Standard* 55) may cause occupant discomfort. Lack of uniform conditions within the space or excessive fluctuation of conditions in the same part of the space also produces discomfort. Discomfort can arise due to any of the following conditions:

- Excessive air motion (draft)
- Excessive room air temperature variations (horizontal, vertical, or both)
- Failure to deliver or distribute air according to the load requirements at different locations
- Overly rapid fluctuation of room temperature

Draft. Koestel and Tuve (1955) and Reinmann et al. (1959) studied the effect of air motion on comfort and defined **draft** as any localized feeling of coolness or warmth of any portion of the body due to both air movement and air temperature, with humidity and radiation considered constant. The warmth or coolness of a draft was measured above or below a controlled room condition of 76°F dry-bulb at the center of the room, 30 in. above the floor, with air moving at about 30 fpm.

To define the **effective draft temperature** θ (the difference in temperature between any point in the occupied zone and the control condition), the investigators used the following equation proposed by Rydberg and Norback (1949) and modified by Straub in discussion of a paper by Koestel and Tuve (1955):

$$\theta = (t_x - t_c) - 0.07(V_x - 30) \quad (16)$$

where

θ = effective draft temperature, °F
t_x = local airstream dry-bulb temperature, °F
t_c = average (control) room dry-bulb temperature, °F
V_x = local airstream centerline velocity, fpm

Equation (16) accounts for the feeling of coolness produced by air motion and is used to establish the neutral line in Figure 12. In summer, the local airstream temperature t_x is below the control temperature t_c. Hence, both temperature and velocity terms are negative when velocity V_x is greater than 30 fpm, and they both add to the feeling of coolness. If, in winter, t_x is above t_c, any air velocity above 30 fpm subtracts from the feeling of warmth produced by t_x. Therefore, it is usually possible to have zero difference in effective temperature between location x and the control point in winter, but not in summer.

Houghten et al. (1938) presented data that make it possible to interpret statistically the percentage of room occupants that will object to a given draft condition. Figure 12 presents the data in the form used by Koestel and Tuve (1955). The data show that a person

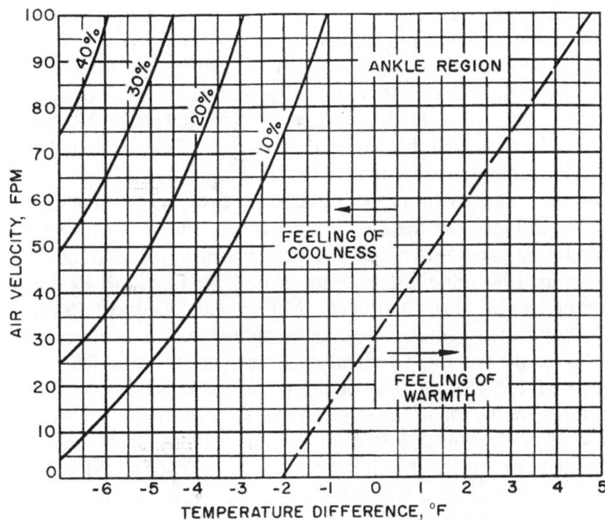

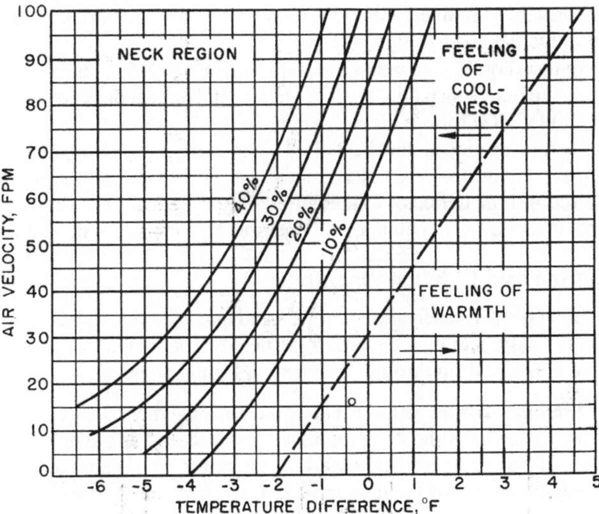

Fig. 12 Percentage of Occupants Objecting to Drafts in Air-Conditioned Room

tolerates higher velocities and lower temperatures at ankle level than at neck level. Because of this, conditions in the zone extending from approximately 30 to 60 in. above the floor are more critical than conditions nearer the floor.

Air Velocity. Room air velocities less than 50 fpm are generally preferred; however, Figure 12 shows that even higher velocities may be acceptable to some occupants. ASHRAE *Standard* 55 recommends elevated air speeds at elevated air temperatures. No minimum air speeds are recommended for comfort, although air speeds below 20 fpm are usually imperceptible.

Temperature Gradient. Figure 12 also shows that up to 20% of occupants will not accept an ankle-to-sitting-level gradient of about 4°F. Poorly designed or operated systems in a heating mode can create this condition, which emphasizes the importance of proper selection and operation of perimeter systems. The section on Outlet Classification and Performance describes possible regions of high room air velocities caused by various outlets; the section on Outlet Location and Selection describes how to evaluate acceptable air diffusion.

AIR DIFFUSION PERFORMANCE INDEX (ADPI)

A high percentage of people are comfortable in sedentary (office) occupations where the effective draft temperature θ, as defined in Equation (16), is between −3 and +2°F and the air velocity is less than 70 fpm. If several measurements of air velocity and air temperature are made throughout the occupied zone of an office, the ADPI is the percentage of locations where measurements were taken that meet these specifications for effective draft temperature and air velocity. If the ADPI is maximum (approaching 100%), the most desirable conditions are achieved (Miller and Nevins 1969, 1970, 1972, 1974; Miller 1971; Miller and Nash 1971; Nevins and Ward 1968; Nevins and Miller 1972).

The ADPI is based only on air velocity and the effective draft temperature (a combination of local temperature variations from the room average) and is not directly related to the dry-bulb temperature or relative humidity. These and similar effects, such as mean radiant temperature, must be accounted for separately according to ASHRAE *Standard* 55.

The ADPI is for cooling mode conditions; a measurement technique is specified in ASHRAE *Standard* 113. Heating conditions can be evaluated using ASHRAE *Standard* 55 guidelines or ISO *Standard* 7730. The ADPI technique uses isothermal throw data determined under ASHRAE *Standard* 70 (see Table 4) to predict what will happen under cooling conditions.

ADPI Selection Guide

Jet Throw. The throw of a jet is the distance from the outlet to a point where the maximum velocity in the stream cross section has been reduced to a selected terminal velocity. To estimate ADPI, **terminal velocity** V_T was selected for all diffusers as 50 fpm, except in the case of ceiling slot diffusers, where it was selected as 100 fpm. Each manufacturer gives data for the throw of a jet from various diffusers for isothermal conditions and without a boundary wall interfering with the jet.

The throw distance of a jet is denoted by T_V, where subscript V indicates the terminal velocity for which the throw is given. The **characteristic room length** L is the distance from the diffuser to the nearest boundary wall in the principle horizontal direction of the airflow. However, where air injected into the room does not impinge on a wall surface but collides with air from a neighboring diffuser, the characteristic length is one-half the distance between diffusers plus the distance the mixed jet travels downward to reach the occupied zone. Table 3 summarizes definitions of characteristic length for various diffusers.

The midplane between diffusers also can be considered the module line when diffusers serve equal modules throughout a

Table 3 Characteristic Room Length for Several Diffusers

Diffuser Type	Characteristic Length L
High sidewall grille	Distance to wall perpendicular to jet
Circular ceiling diffuser	Distance to closest wall or intersecting air jet
Sill grille	Length of room in direction of jet flow
Ceiling slot diffuser	Distance to wall or midplane between outlets
Light troffer diffusers	Distance to midplane between outlets plus distance from ceiling to top of occupied zone
Perforated, louvered ceiling diffusers	Distance to wall or midplane between outlets

Table 4 Air Diffusion Performance Index (ADPI) Selection Guide

Terminal Device	Room Load, Btu/h·ft^2	T_{50}/L for Maximum ADPI	Maximum ADPI	For ADPI Greater than	Range of T_{50}/L
High sidewall grilles	80	1.8	68	—	—
	60	1.8	72	70	1.5–2.2
	40	1.6	78	70	1.2–2.3
	20	1.5	85	80	1.0–1.9
Circular ceiling diffusers	80	0.8	76	70	0.7–1.3
	60	0.8	83	80	0.7–1.2
	40	0.8	88	80	0.5–1.5
	20	0.8	93	90	0.7–1.3
Sill grille, straight vanes	80	1.7	61	60	1.5–1.7
	60	1.7	72	70	1.4–1.7
	40	1.3	86	80	1.2–1.8
	20	0.9	95	90	0.8–1.3
Sill grille, spread vanes	80	0.7	94	90	0.6–1.5
	60	0.7	94	80	0.6–1.7
	40	0.7	94	—	—
	20	0.7	94	—	—
Ceiling slot diffusers (for T_{100}/L)	80	0.3	85	80	0.3–0.7
	60	0.3	88	80	0.3–0.8
	40	0.3	91	80	0.3–1.1
	20	0.3	92	80	0.3–1.5
Light troffer diffusers	60	2.5	86	80	<3.8
	40	1.0	92	90	<3.0
	20	1.0	95	90	<4.5
Perforated, louvered ceiling diffusers	11–50	2.0	96	90	1.4–2.7
				80	1.0–3.4

space, and a characteristic length consideration can be based on module dimension d.

Load Considerations. The recommendations in Table 4 cover cooling loads of up to 80 Btu/h per square foot of floor surface. The loading is distributed uniformly over the floor up to about 7 Btu/h·ft^2, lighting contributes about 10 Btu/h·ft^2, and the remainder is supplied by a concentrated load against one wall that simulates a business machine or a large sun-loaded window. Over this range of data, the maximum ADPI condition is lower for the highest loads; however, the optimum design condition changes only slightly with the load.

Design Conditions. The quantity of air must be known from other design specifications. If it is not known, the solution must be obtained by trial and error.

The devices for which data were obtained are (1) high sidewall grilles, (2) cone-type circular ceiling diffusers, (3) sill grilles, (4) two- and four-slot ceiling diffusers, (5) light troffer diffusers, and (6) square-faced perforated and louvered ceiling diffusers. Table 2 summarizes the results of the recommendations on values of T_V/L by giving the value of T_V/L at which the ADPI is a maximum for various loads, as well as a range of values of T_V/L for which ADPI is above a minimum specified value.

SYSTEM DESIGN

DESIGN CONSIDERATIONS

Noise

The noise generated by diffusers transmits to the occupied space directly and cannot be attenuated. Therefore, the diffusion system design should meet the sound level criteria specified in Chapter 46 of the 1999 *ASHRAE Handbook—Applications*.

Duct Approaches to Diffuser Outlets

The manner in which the airstream approaches the diffuser outlet is important. For correct air diffusion, the velocity of the airstream must be as uniform as possible over the entire cross-sectional area of the connecting duct and must be perpendicular to the outlet face. Effects of improper duct approach generally cannot be corrected by the diffuser.

If the system is designed carefully, a wall grille installed at the end of a horizontal duct and a ceiling outlet at the end of a vertical duct receive the air perpendicularly and at uniform velocity over the entire duct cross section. However, few outlets are installed in this way. Most sidewall outlets are installed either at the end of vertical ducts or in the side of horizontal ducts, and most ceiling outlets are attached either directly to the bottom of horizontal ducts or to special vertical takeoff ducts that connect the outlet with the horizontal duct. In all these cases, special devices for directing and equalizing the airflow are necessary for proper direction and diffusion of the air.

The influence of the duct approach on outlet performance has been investigated for vertical stack heads with plain openings (Nelson et al. 1940) or equipped with grilles (Nelson et al. 1942) and side outlets on horizontal ducts (Nelson and Smedberg 1943). In tests conducted with the stack heads, splitters or guide vanes in the elbows at the top of the vertical stacks are needed, regardless of the shape of the elbows (rounded, square, or expanding). Cushion chambers at the top of the stack heads are not beneficial. Figure 13 shows the direction of flow, diffusion, and velocity (measured 12 in. from the opening) of the air for various stack heads tested, expanding from a 14 in. by 6 in. stack to a 14 in. by 9 in. opening, without a grille. The air velocity for each was 500 fpm in the stack below the elbow, but the direction of flow and the diffusion pattern indicate performance obtained with nonexpanding elbows of similar shapes for velocities from 200 to 400 fpm.

In tests conducted with 3 in. by 10 in., 4 in. by 9 in., and 6 in. by 6 in. side outlets in a 6 in. by 20 in. horizontal duct at duct velocities of 200 to 1400 fpm in the horizontal duct section, multiple curved deflectors produced the best flow characteristics. Vertical guide strips in the outlet were not as effective as curved deflectors. A single scoop-type deflector at the outlet did not improve the flow pattern obtained from a plain outlet and, therefore, was not desirable.

Return and Exhaust Openings

Selection. The selection of return and exhaust openings depends on (1) velocity in the occupied zone near the openings, (2) permissible pressure drop through the openings, and (3) noise.

Velocity. Airflow patterns and room air movement are not influenced by the location of the return and exhaust outlets beyond a distance of one characteristic length of the return or exhaust opening (e.g., square root of the opening area). Air handled by the opening approaches the opening from all directions, and its velocity decreases rapidly as the distance from the opening increases. Therefore, drafty conditions rarely occur near return openings. Table 5 shows recommended return opening face velocities.

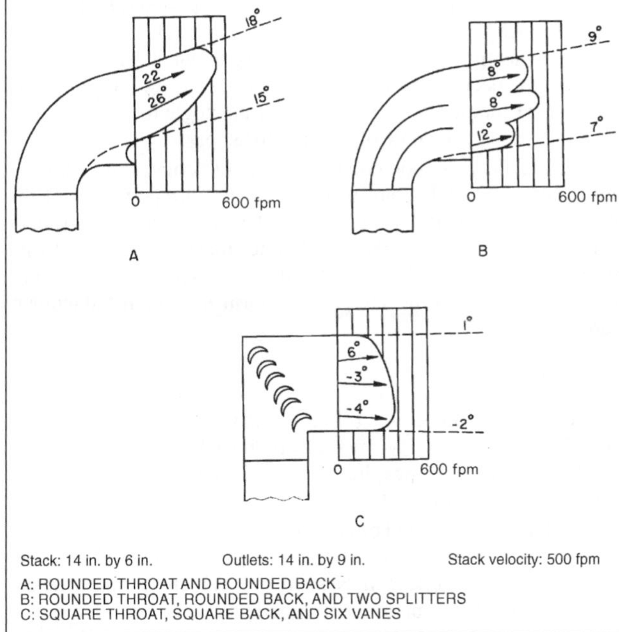

Fig. 13 Outlet Velocity and Air Direction Diagrams for Stack Heads with Expanding Outlets

Table 5 Recommended Return Inlet Face Velocities

Inlet Location	Velocity Across Gross Area, fpm
Above occupied zone	> 800
Within occupied zone, not near seats	600 to 800
Within occupied zone, near seats	400 to 600
Door or wall louvers	200 to 300
Through undercut area of doors	200 to 300

Permissible pressure drop. Permissible pressure drop depends on the choice of the designer. Proper pressure drop allowances should be made for control or directive devices.

Noise. The problem of noise in return openings is the same as that in supply outlets. In computing room noise levels resulting from the operation of an air-conditioning system, the return opening must be included as part of the total grille area.

Location. The openings should be located to minimize short-circuiting of supply air. If air is supplied by the jets attached to the ceiling, exhaust openings should be located between the jets or at the side of the room away from the supply air jets. In rooms with vertical temperature stratification, such as foundries, computer rooms, theaters, bars, kitchens, dining rooms, and club rooms, exhaust openings should be located near the ceiling to collect warm air, odors, and fumes.

For industrial rooms with gas release, selection of exhaust opening locations depends on the density of the released gases and their temperature; locations should be specified for each application.

Exhaust outlets located in walls and doors, depending on their elevation, have the characteristics of either floor or ceiling returns. In large buildings with many small rooms, return air may be brought through door grilles or door undercuts into the corridors and then to a common return or exhaust. If the pressure drop through door returns is excessive, the air diffusion to the room may be seriously unbalanced by opening or closing the doors. Outward leakage through doors or windows cannot be counted on for dependable results.

System Balancing

Ducts and diffusers in a system should be sized so that the supply of air is distributed properly. However, for flexibility, use standard sizes and allow for future redistribution; the system as designed may not be self-balancing. Chapter 36 of the 1999 *ASHRAE Handbook—Applications* describes the procedures used to balance air distribution systems.

DESIGN PROCEDURE

1. Determine the air volumetric flow requirements based on load and room size. For VAV systems, evaluation should include the range of flow rates from minimum occupied to design load.
2. Select the tentative diffuser type and location within room.
3. Determine the room's characteristic length L (Table 3).
4. Select the recommended T_V/L ratio from Table 4.
5. Calculate the throw distance T_V by multiplying the recommended T_V/L ratio from Table 4 by the room length L.
6. Locate appropriate outlet size from the manufacturer's catalog.
7. Ensure that this outlet meets other imposed specifications, such as for noise and for static pressure.

Example 2.

Specifications:

Room size	20 ft by 12 ft with 9 ft ceiling
Loading	Uniform, 10 Btu/h·ft^2 or 2400 Btu/h
Air volumetric flow	1 cfm/ft^2 or 240 cfm for the one outlet
Device	High sidewall grille, located at center of 12 ft endwall, 9 in. from ceiling

Calculations:

Characteristic length L	= 20 ft (length of room: Table 3)
Recommended T_V/L	= 1.5 (Table 4)
Throw to 50 fpm T_{50}	= 1.5 × 20 = 30 ft

Refer to the manufacturer's catalog for a size that gives this isothermal throw to 50 fpm. One manufacturer recommends the following sizes, when vanes are straight, discharging 240 cfm: 16 in. by 4 in., 12 in. by 5 in., or 10 in. by 6 in.

OUTLET LOCATION AND SELECTION

No criteria have been established for choosing among the six types of outlets to obtain an optimum ADPI. All outlets tested, when used according to these recommendations, can have ADPI values that are satisfactory (greater than 90% for loads less than 40 Btu/h·ft^2).

The design of an air distribution and air diffusion system is influenced by the same factors that influence the design of an air-conditioning plant—building use, size, and construction type. Location and selection of the supply outlets is further influenced by the interior design of the building, local sources of heat gain or loss, and outlet performance and design.

Local sources of heat gain or loss promote convection currents or cause stratification; they may, therefore, determine both the type and location of the supply outlets. Outlets should be located to neutralize any undesirable convection currents set up by a concentrated load. If a concentrated heat source is located at the occupancy level of the room, the heating effect can be counteracted (1) by directing cool air toward the heat source or (2) by locating an exhaust or return grille adjacent to the heat source. The second method is more economical for cooling applications, since heat is withdrawn at its source rather than dissipated into the conditioned space. Where lighting loads are heavy (5 W/ft^2) and ceilings relatively high (above 15 ft), outlets should be located below the lighting load, and the stratified warm air should be removed by an exhaust or return fan. An exhaust fan is recommended if the wet-bulb temperature of the air is above that of the outdoors; a return fan is recommended if the wet-bulb temperature is below this temperature. These methods reduce the requirements for supply air. Enclosed lights are more economical than exposed lights because a considerable portion of the energy is radiant.

Based on the analysis of the outlet performance tests conducted by Straub et al. (1956) and Straub and Chen (1957), the following are selection considerations for outlets in Groups A through E.

Group A Outlets

Outlets mounted in or near the ceiling with horizontal air discharge should not be used with temperature differentials exceeding 25°F during heating. Researchers have recommended that temperature differentials not exceed 15°F during heating (Hart and Int-Hout 1980, Lorch and Straub 1983). Consequently, such outlets should be used for heating buildings located in regions where winter heating is only a minor problem and, in northern latitudes, solely for interior spaces. However, these outlets are particularly suited for cooling and can be used with high airflow rates and large temperature differentials. They are usually selected for their cooling characteristics.

The performance of these outlets is affected by various factors. Vane deflection settings reduce throw and drop by changing air from a single straight jet to a wide-spreading or fanned-out jet. Accordingly, a sidewall outlet with 0° deflection has a longer throw and a greater drop than a ceiling diffuser with a single 360° angle of deflection. Sidewall grilles and similar outlets with other deflection settings may have performance characteristics between these two extremes.

Wide deflection settings also cause a ceiling effect, which increases the throw and decreases the drop. To prevent smudging, the total air should be directed away from the ceiling, but this is rarely practicable, except for very high ceilings. For optimum air diffusion in areas with normal ceilings, total air should scrub the ceiling surface.

Drop increases and throw decreases with larger temperature differentials. For constant temperature differential, airflow rate affects drop more than velocity. Therefore, to avoid drop, several small outlets may be better in a room than one large outlet.

With the data in the section on Principles of Jet Behavior, the throw may be selected for a portion of the distance between the outlet and wall or, preferably, for the entire distance. For outlets in opposite walls, the throw should be one-half the distance between the walls. Following these recommendations, the air drops before striking the opposite wall or the opposing airstream. To counteract specific sources of heat gain or to provide higher air motion in rooms with high ceilings, it may be necessary to select a longer throw. In no case should the drop exceed the distance from the outlet to the 6 ft level.

To maintain maximum ventilation effectiveness with ceiling diffusers, throws should be kept as long as possible. With VAV designs, some overthrow at maximum design volumes will be desirable—the highest induction can be maintained at reduced flows. Adequate induction by a ceiling-mounted diffuser prevents short-circuiting of unmixed supply air between supply outlet and ceiling-mounted returns.

Group B Outlets

In selecting these outlets, it is important to provide enough throw to project the air high enough for proper cooling in the occupied zone. An increase in supply air velocity improves air diffusion during both heating and cooling. Also, a terminal velocity of about 150 fpm is found at the same distance from the floor during both heating and cooling. Therefore, outlets should be selected from the data given in the section on Principles of Jet Behavior, with throw based on a terminal velocity of 150 fpm.

With outlets installed near the exposed wall, the primary air is drawn toward the wall, resulting in a wall effect similar to the ceiling effect for ceiling outlets. This scrubbing of the wall increases heat gain or loss. To reduce scrubbing, outlets should be installed some distance from the wall, or the supply air should be deflected at an angle away from the wall. However, to prevent the air from dropping into the occupied zone before it reaches maximum projection the distance should not be too large nor the angle too wide. A distance of 6 in. and an angle of 15° is satisfactory.

These outlets do not counteract natural convection currents unless sufficient outlets are installed around the perimeter of the space — preferably in locations of greatest heat gain or loss (under windows). The effect of drapes and blinds must be considered with outlets installed near windows. If installed correctly, outlets of this type handle large airflow rates with uniform air motion and temperatures.

Group C Outlets

These outlets can be used for heating, even with severe heat load conditions. Higher supply velocities produce better room air diffusion than lower velocities, but velocity is not critical in selecting these units for heating.

To achieve the required projection for cooling, the outlets should be used with temperature differentials of less than 15°F. With higher temperature differentials, supply air velocity is not sufficient to project the total air up to the desired level.

The outlets have been used successfully for residential heating, but they may also offer a solution for applications where heating requirements are severe and cooling requirements are moderate. For throw, refer to the section on Principles of Jet Behavior.

Group D Outlets

These outlets direct high-velocity total air into the occupied zone, and, therefore, are not recommended for comfort, particularly for summer cooling. For heating, outlet velocities should not be higher than 300 fpm, so that air velocities in the occupied zone will not be excessive. These outlets have been applied successfully to process installations where controlled air velocities are desired.

Group E Outlets

The different throws shown in the heating and cooling diagrams for these outlets become critical in selecting and applying the outlets. Because the total air enters the occupied zone for both cooling and heating, outlets are used for either cooling or heating, but seldom for both.

During cooling, temperature differential, supply air velocity, and airflow rate have considerable influence on projection. Therefore, low values of each should be selected.

During heating, it is important to select the correct supply air velocity to project the warm air into the occupied zone. Temperature differential is also critical because a small temperature differential reduces variation of the throw during the cyclic fluctuation of the supply air temperature. Vane setting for deflection is as important here as it is for Group B and C outlets.

Investigations by Nevins and Ward (1968) and Miller and Nevins (1969) in full-scale interior test rooms indicate that air temperatures and velocities throughout a room cooled by a ventilating ceiling are a linear function of room load (heat load per unit area) and are not affected significantly by variations in ceiling type, total air temperature differential, or air volumetric flow rate. Higher room loading produces wider room air temperature variations and higher velocities, which decrease performance.

These studies also found no appreciable difference in the performance of air-diffusing ceilings and circular ceiling diffusers for lower room loads (20 Btu/h·ft^2). For higher room loads (80 Btu/h·ft^2), an air-diffusing ceiling system has only slightly larger vertical temperature variations and slightly lower room air velocities than a ceiling diffuser system.

When the ventilating ceiling is used at exterior exposures, the additional load at the perimeter must be considered. During heating operation, the designer must provide for the cold wall effect, as with any ceiling supply diffusion system. The sound generated by the air supply device must also be considered in total system analysis to ensure that room sound levels do not exceed the design criteria.

RETURN AIR DESIGN FOR OPTIMUM PERFORMANCE

An HVAC system operating in the cooling mode performs best when generated heat is removed at its source rather than distributed throughout the conditioned space. Heat from solar and miscellaneous loads such as machinery and floor or desk-mounted lamps is difficult to remove at the source. However, return air flowing over ceiling-mounted lighting fixtures keeps most of that heat from being distributed into the conditioned space. In addition to increasing HVAC system efficiency, return air lighting fixtures improve light output and extend the life of the lamps. The manufacturers of fixtures, ceiling grids, and grilles give performance information (airflow rate, pressure drop, and heat removal rate) of their product. Ball et al. (1971) found that the heat removal performance of return air fixtures covers a narrow range.

With a suspended ceiling, low operating static pressure across the ceiling must be maintained. Failure to do so can result in return air being forced around the edges of the ceiling panels or, in some cases, through the ceiling panels. The result is often a soiled ceiling and a mechanical system that is choked for return air. To avoid this, the static pressure difference across the ceiling should be as low as possible. If necessary, slotted tees or grilles can be used with return air fixtures to obtain the specified pressure drop. A maximum pressure drop of 0.02 to 0.03 in. of water is acceptable under most conditions.

At the typical air supply rates found in office interior zone spaces (usually less than 1.5 cfm/ft^2) and with adequate induction at the supply diffusers, the location of the return diffuser has no effect on air patterns in the space. For most office spaces, it is only necessary that sufficient return outlets be provided to maintain inlet velocities within recommendations (see Table 5).

In spaces expected to operate in a cooling mode most of the time, returning the warmest air in the space can effectively reduce energy costs and increase circulation in the space. This is especially true in climates where economizer systems operate for long periods during the year. In spaces having very high ceilings, with atriums, skylights, or large vertical glass surfaces, and where the highest areas are unoccupied, air stratification may be used as an energy-saving measure by locating returns near the occupied zone.

CEILING-MOUNTED AIR DIFFUSER SYSTEMS

For the best thermal comfort conditions and highest ventilation effectiveness in an occupied space (i.e., office or retail store), the entire system performance of air diffusers should be considered. This is particularly true for open spaces, where airstreams from diffusers may interact with each other, and for perimeter spaces, where airstreams from diffusers interact with hot or cold perimeter walls. While throw data for individual diffusers are used in system design, an air diffuser system should maintain a high quality of air diffusion in the occupied space with low temperature variation, good air mixing, and no objectionable drafts in the occupied space (typically 6 in. to 6 ft above the floor).

Adequate ventilation requires that the selected diffusers effectively mix (by entrainment) the total air in the room with the supplied conditioned air, which is assumed to contain adequate ventilation air.

Interior Spaces

An interior space is conditioned exclusively for cooling loads, except after unoccupied periods when the space may have cooled to below a comfortable temperature. Tests by Miller and Nevins (1970), Miller and Nash (1971), Miller (1979), and Hart and Int-Hout (1981) suggest that the air diffusion performance index (ADPI) can be improved by moving diffusers closer together (i.e., specifying more diffusers for a given space and air quantity) and by limiting the value of the supply air/room air temperature difference. In a given system of diffusers, these studies found an optimum operating range of air volumetric flow rates at a given thermal load. The operating load varies with diffuser design, ceiling height, thermal load, and diffuser orientation. This information can be obtained by constructing a mock-up representing the proposed building space, with several alternatives tested for ADPI values, in accordance with ASHRAE *Standard* 113. Usually, the diffuser manufacturer has performed these tests and can provide the best choice of design options for a particular building. For a VAV system, the diffuser spacing selection should not be based on maximum or design air volumes but rather on the air volume range in which the system is expected to operate most of the time. For VAV applications, Miller (1979) recommends that the designer consider the expected variation in the outlet air volume to ensure that ADPI values remain above a specified minimum.

Perimeter Spaces

All-air mechanical systems that handle both heating and cooling thermal loads are commonly used in modern office buildings instead of baseboards for heating and forced air for cooling. State energy codes (most based on ASHRAE *Standard* 90 series) require that commercial buildings have exterior walls that meet minimum thermal performance criteria for a particular location. Typically, walls of new buildings have design heat losses as low as 200 to 300 Btu/h per linear foot of wall.

A successful all-air heating/cooling mechanical system requires the designer to consider several design variables that have been the subject of research by Hart and Int-Hout (1980), Lorch and Straub (1983), and Rousseau (1983). The most important design variables include

- Supply air/room air temperature difference
- Diffuser type and design
- Design heating and cooling loads
- Supply air volumetric flow rates
- Distance between diffusers and perimeter wall
- Direction of air throw (toward wall, away from wall, or both)
- Ceiling height
- Desired air diffusion performance criteria

The diffuser manufacturer is best able to recommend the use of equipment.

For an office environment in cooling mode, the design goal should be an ADPI greater than 80. The ADPI *should not* be used as a measure of performance for heating conditions. In both cases, ASHRAE *Standard* 55 recommends that the maximum temperature gradient (the difference in temperature between any two points) should not exceed 5°F.

Linear diffusers placed parallel to the perimeter wall perform well. For year-round operation, linear diffusers with two-way throw (i.e., both toward and away from the perimeter wall) work best. Lorch and Straub (1983) reported optimum performance with a diffuser that throws warm air toward the perimeter wall under heating load conditions and chilled air in both directions under cooling load conditions. All researchers found less than optimum performance with high discharge temperatures (greater than 15°F above ambient), both with one-way throw of air away from a cold wall and with one-way throw of chilled air toward the perimeter wall. Under heating load conditions, the supply air temperature must be limited to avoid excessive thermal stratification.

To resolve any uncertainty about performance, a mockup should be constructed with provisions for a cold wall; several variations of the design should be tested so that the best diffuser wall spacing and supply air volumes can be selected. The ADPI, room temperature gradients, or both, measured in accordance with ASHRAE *Standard* 113, can help gage system performance.

The following principles provide the best air diffusion quality and minimum energy use:

- For cooling load conditions, return air should exhaust from a location that takes advantage of any thermal stratification design. In many cases, this should be a high point in order to take advantage of rising warm air. Cooling supply air should be introduced as close to the heat sources as possible. Alternately, stratification designs may condition only part of the total space. In these cases, conditioned air is supplied and exhausted as close to the occupants as possible. In either case, comfort zone temperature gradients should be maintained within 5°F.
- For heating load conditions, thermal stratification should be discouraged. Heat should be introduced at points low in the large space. Ceiling-mounted fans may reduce stratification.

REFERENCES

Arens, E.A., F. Bauman, L. Johnston, and H. Zhang. 1991. Testing of localized ventilation systems in a new controlled environment chamber. *Indoor Air* 3:263-81.

ASHRAE. 1990. Method of testing for room air diffusion. ANSI/ASHRAE *Standard* 113-1990.

ASHRAE. 1991. Method of testing for rating the performance of air outlets and inlets. ANSI/ASHRAE *Standard* 70-1991.

ASHRAE. 1992. Thermal environmental conditions for human occupancy. ANSI/ASHRAE *Standard* 55-1992.

Ball, H.D., R.G. Nevins, and H.E. Straub. 1971. Thermal analysis of heat removal troffers. *ASHRAE Transactions* 77(2).

Baturin, V.V. 1972. *Fundamentals of industrial ventilation*, 3rd England ed. Translated by O.M. Blunn. Pergamon Press, New York.

Bauman, F. and E. Arens. 1996. Task/ambient conditioning systems: Engineering and application guidelines. Center for Environmental Design Research, University of California, Berkeley.

Bauman, F.S., E.A. Arens, S. Tanabe, H. Zhang, and A. Baharlo. 1995. Testing and optimizing the performance of a floor-based task conditioning system. *Energy and Buildings* 22(3):173-86.

Bauman, F.S., T.G. Carter, A.V. Baughman, and E.A. Arens. 1998. Field study of the impact of a desktop task/ambient conditioning system in office buildings. *ASHRAE Transactions* 104(1).

Bauman, F.S., L. Johnston, H. Zhang, and E. Arens. 1991. Performance testing of a floor-based, occupant-controlled office ventilation system. *ASHRAE Transactions* 97(1).

Bauman, F., P. Pecora, and T. Webster. 1999. How low can you go? Air flow performance of low-height underfloor plenums. Center for the Built Environment, University of California, Berkeley.

Bauman, F.S., H. Zhang, E. Arens, and C. Benton. 1993. Localized comfort control with a desktop task conditioning system: Laboratory and field measurements. *ASHRAE Transactions* 99(2).

Christianson, L.L., ed. 1989. *Building systems: Room air and air contaminant distribution*. ASHRAE, Atlanta.

de Dear, R. and G.S. Brager. 1999. Developing an adaptive model of thermal comfort and preference. *ASHRAE Transactions* 104 (1).

Fanger, P.O., A.K. Melikov, H. Hanzawa, and J. Ring. 1988. Air turbulence and sensation of draft. *Energy and Buildings* 12:21-39.

Faulkner, D., W.J. Fisk, and D.P. Sullivan. 1993. Indoor air flow and pollutant removal in a room with desktop ventilation. *ASHRAE Transactions* 99(2).

Faulkner, D., W.J. Fisk, D.P. Sullivan, and D.P. Wyon. 1999. Ventilation efficiencies of task/ambient conditioning systems with desk-mounted air supplies. *Proceedings Indoor Air '99*, Edinburgh, Scotland, 8-13 August.

Fisk, W.J., D. Faulkner, D. Pih, P. McNeel, F. Bauman, and E. Arens. 1991. Indoor air flow and pollutant removal in a room with task ventilation. *Indoor Air* 3:247-62.

Hanzawa, H., and Y. Nagasawa. 1990. Thermal comfort with underfloor air-conditioning systems. *ASHRAE Transactions* 96(2).

Hart, G.H. and D. Int-Hout. 1980. The performance of a continuous linear diffuser in the perimeter zone of an office environment. *ASHRAE Transactions* 86(2).

Hart, G.H. and D. Int-Hout. 1981. The performance of a continuous linear diffuser in the interior zone of an open office environment. *ASHRAE Transactions* 87(2).

Helander, L. and C.V. Jakowatz. 1948. Downward projection of heated air. *ASHVE Transactions* 54:71.

Helander, L., S.M. Yen, and R.E. Crank. 1953. Maximum downward travel of heated jets from standard long radius ASME nozzles. *ASHVE Transactions* 59:241.

Helander, L., S.M. Yen, and L.B. Knee. 1954. Characteristics of downward jets of heated air from a vertical delivery discharge unit heater. *ASHVE Transactions* 60:359.

Helander, L., S.M. Yen, and W. Tripp. 1957. Outlet characteristics that affect the downthrow of heated air jets. *ASHAE Transactions* 63:255.

Houghten, F.C., C. Gutberlet, and E. Witkowski. 1938. Draft temperatures and velocities in relation to skin temperatures and feelings of warmth. *ASHVE Transactions* 44:289.

Houghton, D. 1995. Turning air conditioning on its head: Underfloor air distribution offers flexibility, comfort, and efficiency. *E Source* TU-95-8. E Source, Inc., Boulder, CO.

ISO. 1994. Moderate thermal environments—Determination of the PMV and PPD indices and specification of the conditions for thermal comfort. ISO *Standard* 7730-1994. International Organization for Standardization, Geneva.

Jackman, P.J. 1991. Displacement ventilation. CIBSE National Conference. Chartered Institution of Building Services Engineers, London.

Jackman, P.J., and P.A. Appleby. 1990. Displacement flow ventilation. BSRIA Project Report. The Building Services Research and Information Association, Old Bracknell Lane, Berkshire, RG12 4AH, UK.

Kegel, B. and U.W. Schulz. 1989. Displacement ventilation for office buildings. Proceedings of the 10th AIVC Conference, Helsinki. Air Infiltration and Ventilation Center, University of Warwick Science Park, Barclays Venture Centre, Sir William Lyons Road, Coventry CV4 7EZ, UK.

Kirkpatrick, A. and J. Elleson. 1996. *Design guide for cold air distribution systems*. ASHRAE, Atlanta.

Kirkpatrick, A., T. Malmstrom, P. Miller, and V. Hassani. 1991. Use of low temperature air for cooling of buildings. Proceedings Building Simulation.

Knaak, R. 1957. Velocities and temperatures on axis of downward heated jet from 4-inch long-radius ASME nozzle. *ASHAE Transactions* 63:527.

Koestel, A. 1954. Computing temperatures and velocities in vertical jets of hot or cold air. *ASHVE Transactions* 60:385.

Koestel, A. 1955. Paths of horizontally projected heated and chilled air jets. *ASHAE Transactions* 61:213.

Koestel, A. 1957. Jet velocities from radial flow outlets. *ASHAE Transactions* 63:505.

Koestel, A. and J.B. Austin, Jr. 1956. Air velocities in two parallel ventilating jets. *ASHAE Transactions* 62:425.

Koestel, A. and G.L. Tuve. 1955. Performance and evaluation of room air distribution systems. *ASHRAE Transactions* 61:533.

Koestel, A., P. Hermann, and G.L. Tuve. 1949. Air streams from perforated panels. *ASHVE Transactions* 55:283.

Koestel, A., P. Hermann, and G.L. Tuve. 1950. Comparative study of ventilating jets from various types of outlets. *ASHVE Transactions* 56:459.

Li, Z., L.L. Christianson, and J.S. Zhang. 1995. Separation distances of nonisothermal air jets. Research Triangle Park, NC.

Li, Z., J.S. Zhang, A.M. Zhivov, and L.L. Christianson. 1993. Characteristics of diffuser air jets and airflow in the occupied regions of mechanically ventilated rooms: A literature review. *ASHRAE Transactions* 99(1):1119-27.

Lorch, F.A. and H.E. Straub. 1983. Performance of overhead slot diffusers with simulated heating and cooling conditions. *ASHRAE Transactions* 89(1).

Loudermilk, K. 1999. Underfloor air distribution solutions for open office applications. *ASHRAE Transactions* 105(1).

Matsunawa, K., H. Iizuka, and S. Tanabe. 1995. Development and application of an underfloor air-conditioning system with improved outlets for a "smart" building in Tokyo. *ASHRAE Transactions* 101(2):887.

McCarry, B.T. 1995. Underfloor air distribution systems: Benefits and when to use the system in building design. *ASHRAE Transactions* 101(2).

McCarry, B.T. 1998. Innovative underfloor system. *ASHRAE Journal* 40 (3).

McElroy, G.E. 1943. Air flow at discharge of fan-pipe lines in mines. U.S. Bureau of Mines *Report of Investigations*, 19.

Melikov, A.K. and J.B. Nielsen. 1989. Local thermal discomfort due to draft and vertical temperature difference in rooms with displacement ventilation. *ASHRAE Transactions* 95(2):1050-57.

Miller, P.L. 1971. Room air distribution performance of four selected outlets. *ASHRAE Transactions* 77(2):194.

Miller, P.L. 1979. Design of room air diffusion systems using the air diffusion performance index (ADPI). *ASHRAE Journal* 10:85.

Miller, P.L. 1989. Descriptive methods. In *Building systems: Room air and air contaminant distribution*, L.L. Christianson, ed. ASHRAE, Atlanta.

Miller, P.L. and R.T. Nash. 1971. A further analysis of room air distribution performance. *ASHRAE Transactions* 77(2):205.

Miller, P.L. and R.G. Nevins. 1969. Room air distribution with an air distributing ceiling—Part II. *ASHRAE Transactions* 75:118.

Miller, P.L. and R.G. Nevins. 1970. Room air distribution performance of ventilating ceilings and cone-type circular ceiling diffusers. *ASHRAE Transactions* 76(1):186.

Miller, P.L. and R.G. Nevins. 1972. An analysis of the performance of room air distribution systems. *ASHRAE Transactions* 78(2):191.

Murakami, S. 1992. New scales for ventilation efficiency and their application based on numerical simulation of room airflow. International Symposium on Room Air Convection and Ventilation Effectiveness.

Nelson, D.W. and G.E. Smedberg. 1943. Performance of side outlets on horizontal ducts. *ASHVE Transactions* 49:58.

Nelson, D.W., H. Krans, and A.F. Tuthill. 1940. The performance of stack heads. *ASHVE Transactions* 46:205.

Nelson, D.W., D.H. Lamb, and G.E. Smedberg. 1942. Performance of stack heads equipped with grilles. *ASHVE Transactions* 48:279.

Nevins, R.G. and P.L. Miller. 1972. Analysis, evaluation and comparison of room air distribution performance. *ASHRAE Transactions* 78(2):235.

Nevins, R.G. and E.D. Ward. 1968. Room air distribution with an air distributing ceiling. *ASHRAE Transactions* 74:VI.2.1.

Nottage, H.B., J.G. Slaby, and W.P. Gojsza. 1952a. Isothermal ventilation jet fundamentals. *ASHVE Transactions* 58:107.

Nottage, H.B., J.G. Slaby, and W.P. Gojsza. 1952b. Outlet turbulence intensity as a factor in isothermal-jet flow. *ASHVE Transactions* 58:343.

Oakes, W.C. 1987. Experimental investigation of Coanda jet. M.S. thesis, Michigan State University, East Lansing, MI.

Reinmann, J.J., A. Koestel, and G.L. Tuve. 1959. Evaluation of three room air distribution systems for summer cooling. *ASHRAE Transactions* 65:717.

Rousseau, W.H. 1983. Perimeter air diffusion performance index tests for heating with a ceiling slot diffuser. *ASHRAE Transactions* 89(1).

Rydberg, J. and P. Norback. 1949. Air distribution and draft. *ASHVE Transactions* 55:225.

Sandberg, M. and C. Blomqvist. 1989. Displacement ventilation in office rooms. *ASHRAE Transactions* 95(2):1041-49.

Scaret, E. 1985. *Ventilation by displacement Characterization and design implications.* Elsevier Science Publishers, New York.

Seppanen, O.A., W.J. Fisk., J. Eto, and D.T. Grimsrud. 1989. Comparison of conventional mixing and displacement air-conditioning and ventilating systems in U.S. commercial buildings. *ASHRAE Transactions* 95(2): 1028-40.

Shilkrot, E., and A. Zhivov. 1992. Room ventilation with designed vertical air temperature stratification. ROOMVENT-'92, Proceedings 3rd International Conference on Engineering Aero- and Thermodynamics of Ventilated Rooms.

Shute, R.W. 1992. Integrating access floor plenums for HVAC air distribution. *ASHRAE Journal* 34(10).

Shute, R.W. 1995. Integrated access floor HVAC: Lessons learned. *ASHRAE Transactions* 101(2).

Sodec, F., and R. Craig. 1990. The underfloor air supply system—The European experience. *ASHRAE Transactions* 96(2).

Spoormaker, H.J. 1990. Low-pressure underfloor HVAC system. *ASHRAE Transactions* 96(2).

Straub, H.E. and M.M. Chen. 1957. Distribution of air within a room for year-round air conditioning—Part II. University of Illinois Engineering Experiment Station *Bulletin* No. 442.

Straub, H.E., S.F. Gilman, and S. Konzo. 1956. Distribution of air within a room for year-round air conditioning—Part I. University of Illinois Engineering Experiment Station *Bulletin* No. 435.

Stymne, H., M. Sandberg, and M. Mattsson. 1991. Dispersion pattern of contaminants in a displacement ventilation room—Implications for demand control. *Proceedings*, 12th Air Movement and Ventilation Control Within Buildings, Ottawa. AIVC Conference.

Svensson, A.G.L. 1989. Nordic experiences of displacement ventilation systems. *ASHRAE Transactions* 95(2):1013-17.

Tanabe, S. and K. Kimura. 1996. Comparisons of ventilation performance and thermal comfort among displacement, underfloor and ceiling based air distribution systems by experiments in a real sized office chamber. *Proceedings*, 5th International Conference on Air Distribution in Rooms, ROOMVENT '96.

Tuve, G.L. 1953. Air velocities in ventilating jets. *ASHVE Transactions* 59:261.

Tsuzuki, K., E.A. Arens, F.S. Bauman, and D.P. Wyon. 1999. Individual thermal comfort control with desk-mounted and floor-mounted task/ambient conditioning (TAC) systems. *Proceedings Indoor Air '99*, Edinburgh, Scotland, 8-13 August.

Wilson, J.D., M.L. Esmay, and S. Persson. 1970. Wall-jet velocity and temperature profiles resulting from a ventilation inlet. *ASAE Transactions*.

Yen, S.M., L. Helander, and L.B. Knee. 1956. Characteristics of downward jets from a vertical discharge unit heater. *ASHAE Transactions* 62:123.

Zhang, J.S., L.L. Christianson, and G.L. Riskowski. 1990. Regional airflow characteristics in a mechanically ventilated room under nonisothermal conditions. *ASHRAE Transactions* 96(1):751-59.

Zhang, J.S., L.L. Christianson, G.J. Wu, and G.L. Riskowski. 1992. Detailed measurements of room air distribution for evaluating numerical simulation models. *ASHRAE Transactions* 98(1):58-65.

BIBLIOGRAPHY

Int-Hout, D. 1981. Measurement of room air diffusion in actual office environments to predict occupant thermal comfort. *ASHRAE Transactions* 87(2).

Poz, M.Y. 1991. Theoretical investigation and practical applications of nonisothrmal jets for the rooms ventilating. Current East/West HVAC Developments. IEI/CIBSE/ABOK Joint Conference.

Zhivov, A. 1990. Variable-air volume ventilation systems for industrial buildings. *ASHRAE Transactions* 96(2).

CHAPTER 33

HVAC COMPUTATIONAL FLUID DYNAMICS

THE APPLICATION of computational fluid dynamics (CFD) to room air motion, as summarized by Haghighat et al. (1992), began with investigations into room airflow (Nielsen 1974) and natural convection in enclosed cavities (Catton 1978, Ostrach 1982, Markatos and Pericleous 1984, Lin and Nansteel 1987, Hadjisophocleous et al. 1988, Gadgil et al. 1984, Chen et al. 1990b). Lemaire (1987) applied the CHAMPHxN code (Pun and Spalding 1976) coupled with radiation to predict air movement and heat transfer in a room heated by a radiator.

Convective heat and mass tranfer has been analyzed using a commercial code by Holmes (1982), Markatos (1983), Jones and Sullivan (1985), and Chen and Van der Kooi (1988). Murakami et al. (1988), Horstman (1988), and Chen et al. (1990a) developed numerical models of ventilation with contaminant transport.

Partitioned models (multiple zones) have also been developed to predict room air motion. Natural convection was investigated by Chang et al. (1982) and Kelkar and Patankar (1985). Contaminant transport models were used in the ventilation models of Haghighat et al. (1989, 1990).

Theory

The basis for ventilation computational fluid dynamics analyses are the incompressible Navier-Stokes equations. These equations describe the motion of a viscous Newtonian fluid. For example, the following represent these equations in two dimensions:

$$\frac{\partial u}{\partial t} + u\frac{\partial u}{\partial x} + v\frac{\partial u}{\partial y} = \frac{1}{\rho}X - \frac{1}{\rho}\frac{\partial p}{\partial x} + \nu\left(\frac{\partial^2 u}{\partial x^2} + \frac{\partial^2 u}{\partial y^2}\right) \quad (1)$$

$$\frac{\partial v}{\partial t} + u\frac{\partial v}{\partial x} + v\frac{\partial v}{\partial y} = \frac{1}{\rho}X - \frac{1}{\rho}\frac{\partial p}{\partial y} + \nu\left(\frac{\partial^2 v}{\partial x^2} + \frac{\partial^2 v}{\partial y^2}\right) \quad (2)$$

$$\frac{\partial u}{\partial x} + \frac{\partial v}{\partial y} = 0 \quad (3)$$

Because these equations are complex, only a few exact solutions have been obtained for very simple flow conditions. The typical ventilation system is well beyond the realm of an exact solution. Forms of the Navier-Stokes equations have been developed using numerical methods that divide the flow field into finite volumes or elements. The elements typically assume uniform properties throughout and exchange pressure, momentum, and viscous dissipation information with one another.

The solution is obtained iteratively either by time step or by a steady flow updating of each element. The new value of pressure or velocity of an element at each iteration may be incorporated into the overall solution gradually using relaxation methods, which help stabilize the solution, giving time for information to travel between the elements and allowing each to assert its influence on the entire flow field. An example of this method is the finite difference approximation of the **stream function-vorticity formulation** of the Navier-Stokes equations in two dimensions:

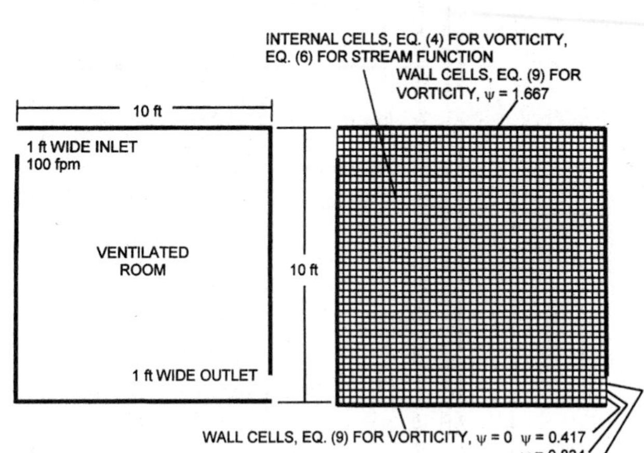

Fig. 1 Example Two-Dimensional Model

Example. A two-dimensional model of a ventilated room (Figure 1) is modeled using the stream function-vorticity formulation of the Navier-Stokes equations. The room is 10 ft by 10 ft. Air enters at 100 fpm through a 1 ft wide opening on the left wall near the ceiling and leaves through a 1 ft wide opening near the floor. The grid representing the room consists of 41 × 41 elements.

Solution:

The stream function-vorticity formulation of the Navier-Stokes equations begins with the **vorticity transport** equation:

$$u\frac{\partial \omega}{\partial x} + v\frac{\partial \omega}{\partial y} = \nu\left(\frac{\partial^2 \omega}{\partial x^2} + \frac{\partial^2 \omega}{\partial y^2}\right) \quad (4)$$

The **stream function** is related to the velocity:

$$u = \frac{\partial \psi}{\partial y}, \qquad v = -\frac{\partial \psi}{\partial x} \quad (5)$$

Vorticity is analogous to rotation of the fluid:

$$\omega = \left(\frac{\partial v}{\partial x} + \frac{\partial u}{\partial y}\right) = -\nabla^2 \psi \quad (6)$$

Most CFD models have some sort of turbulence model to account for eddy viscosity when the Reynolds number exceeds about 2000. The simplest of the turbulence models uses the Prandtl mixing length:

$$\nu_t = \kappa L C_\mu^{1/4} K^{1/2} \quad (7)$$

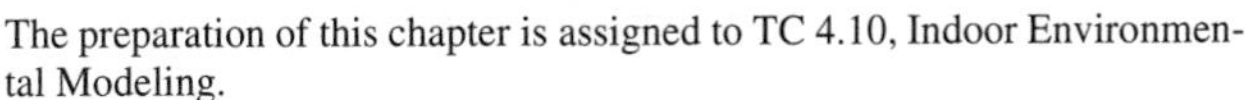

The preparation of this chapter is assigned to TC 4.10, Indoor Environmental Modeling.

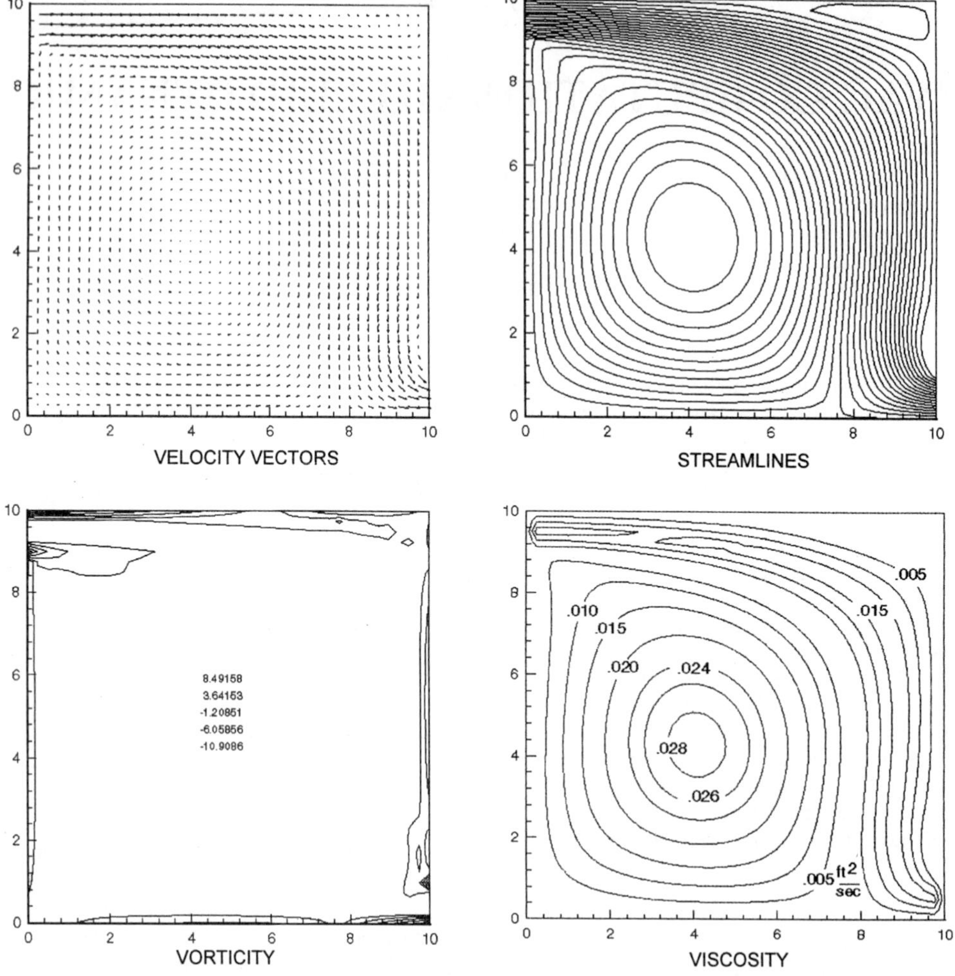

Fig. 2 Stream Function-Vorticity Solution for Ventilated Room (50,000 Iterations)

In recirculating flows, the length to the wall is difficult to measure and is flow-dependent. Equation (5) defines the length scale by the position of streamlines relative to the two boundary values of the stream function:

$$L = \left|\frac{\psi}{V}\right| \quad \text{or} \quad L = \frac{|\psi - \psi_B|}{|V|} \quad \text{whichever is smaller} \tag{8}$$

The turbulent kinetic energy K is a function of V', the magnitude of the fluctuating component of the velocity due to turbulence. V' can be assumed to be about 10% of V, the local average velocity, for typical room air models:

$$K = \frac{3}{2}(V')^2 = \frac{3}{2}(0.1V)^2 \tag{9}$$

Equations (7), (8), and (9) are combined to give a simple turbulence model based on the stream function alone:

$$v_t = \kappa|\psi|C_\mu^{1/4}\sqrt{\frac{3}{2}}(0.1)$$

$$\text{or} \quad v_t = \kappa|\psi - \psi_B|C_\mu^{1/4}\sqrt{\frac{3}{2}}(0.1) \quad \text{whichever is smaller} \tag{10}$$

Finite difference equations of the following form are used to approximate Equations (4) and (6):

$$\frac{\partial f}{\partial s} \approx \frac{f_{n+1} - f_{n-1}}{2\Delta S} \tag{11}$$

$$\frac{\partial^2 f}{\partial s^2} \approx \frac{f_{n+1} - 2f_n + f_{n-1}}{\Delta S^2} \tag{12}$$

The equations are rearranged so that the current value of the center cell or element is based on that of the surrounding cells. In addition, the vorticity transport Equation (4) is expanded for second-order accuracy as done by Dennis and Hudson (1978):

$$\omega_{i,j} = \frac{\begin{aligned}&\left(1 - \frac{\Delta x u_{i+1,j}}{2v_{i,j}} + \frac{\Delta x^2 u_{i+1,j}^2}{8v_{i,j}^2}\right)\omega_{i+1,j} + \left(1 + \frac{\Delta x u_{i-1,j}}{2v_{i,j}} + \frac{\Delta x^2 u_{i-1,j}^2}{8v_{i,j}^2}\right)\omega_{i-1,j} + \ldots \\ &\ldots + \left(1 - \frac{\Delta x v_{i,j+1}}{2v_{i,j}} + \frac{\Delta x^2 v_{i,j+1}^2}{8v_{i,j}^2}\right)\omega_{i,j+1} + \left(1 + \frac{\Delta x v_{i,j-1}}{2v_{i,j}} + \frac{\Delta x^2 v_{i,j-1}^2}{8v_{i,j}^2}\right)\omega_{i,j-1}\end{aligned}}{\left(4 + \frac{\Delta x^2(u_{i,j}^2 + v_{i,j}^2)}{4v_{i,j}^2}\right)} \tag{13}$$

$$\psi_{i,j} = \frac{\psi_{i+1,j} + \psi_{i-1,j} + \psi_{i,j+1} + \psi_{i,j-1} + \Delta x^2 \omega_{i,j}}{4} \tag{14}$$

The subscripted velocities shown in Equation (13) are obtained from Equation (5) in the form of Equation (11). For example, the v velocity at location $i,j-1$ is

$$v_{i,j-1} = -\frac{(\psi_{i+1,j-1}) - (\psi_{i-1,j-1})}{2\Delta x}$$

or

$$v_{i,j} = -\frac{(\psi_{i+1,j}) - (\psi_{i-1,j})}{2\Delta x} \quad (15)$$

Boundary conditions for vorticity are defined by the apparent rotation rate of the nearby fluid passing by:

$$\omega_{i,j}(\text{wall}) = -\frac{\psi_{adj} - \psi_{i,j(\text{wall})}}{\Delta x^2} \quad (16)$$

Relaxation parameters are applied to the vorticity and to the viscosity. The parameter δ for the vorticity is typically equal to 0.03, so that 3% of the new value and 97% of the old value is used. The viscosity is unstable, so a smaller relaxation parameter is required; 0.01 was used in this example. To illustrate this procedure, a new value of vorticity is calculated using Equation (13) or (16) (depending on the node type). The current value of vorticity is obtained by adding most of the old value to a fraction of the new value:

$$\omega_{i,j}(\text{current}) = \delta\omega_{i,j}(\text{new}) + (1-\delta)\omega_{i,j}(\text{old}) \quad (17)$$

The inlets and outlets have a fixed, uniform velocity. The stream function is set at each of the three nodes between the walls. Since the wall below the inlet has $\psi = 0$ and the wall above has $\psi = 1.667$, the stream function is divided evenly between the inlet nodes: $\psi_{1,38} = 0.25(1.667) = 0.417$; $\psi_{1,39} = 0.5(1.667) = 0.834$; $\psi_{1,40} = 0.75(1.667) = 1.25$. The same distribution is applied to the outlet nodes as shown in Figure 1. In a real room, the velocity would not be uniform across the outlet, but for an illustrative example, this simplification is appropriate.

The results of this analysis method are shown in Figure 2.

These results show the overall flow pattern of the room. Ventilation air travels along the ceiling, then attaches to the right wall before exiting. This generates a clockwise rotational flow where the eddy viscosity approaches two hundred times the molecular viscosity. At this point, the engineer would make a finer grid and run the model again to see if the solution has reached grid independence.

Finite Volume Formulations

Commercially available CFD programs are generally used to solve specific applications. Code development requires frequent validation and benchmarking and is best done by companies that specialize in that task. Most commercially available codes are based on finite volume formulations such as the SIMPLE algorithm (Patankar 1988). The finite volume method is distinguished from the finite element method in that the properties and flow conditions of the fluid are assumed uniform throughout the elemental volume and the exchange of momentum, energy, and pressure occurs at the faces.

Finite Element Formulations

Less common than the finite volume formulations, finite element formulations are usually available when the flow solver is packaged commercially with a stress analysis code. The finite element method may be differentiated from the finite volume method in the way the properties and flow conditions are point values and information travels between nodes, through the elements (Baker 1983). Until recently, the finite element method had the advantage of geometry adaptability.

Preprocessors

The first step in the CFD process is building the grid that mathematically represents the physical model. The preprocessor constructs the grid from such data as room dimensions, diffuser and duct dimensions, and internal features such as furniture or partitions that may affect flow. The preprocessor usually includes a computer aided design (CAD) package to allow the analyst to construct the required geometry. A good preprocessor usually includes a translator to other CAD packages so that geometry data is interchangeable.

The grid has an exact correspondence to the geometry; open areas, walls, diffusers, etc., are represented by the elements or volumes of computational fluid, boundaries, and blocked elements. The grid may be structured, where the element index or location is described by coordinates of $i(x)$, $j(y)$, and $k(z)$. Or, the grid may be unstructured, in which the elements are arbitrarily arranged and defined by associative identification.

Post Processors

After the flow solver has converged on the solution, the data is presented through the post processor. The typical post processor displays contour diagrams of equation variables such as pressure, temperature, velocity, turbulence, and contaminant concentration. Post processors are available that provide elaborate three-dimensional contour and vector plots with color scaling and animated output for time-dependent solutions.

Energy Equation

When the CFD application involves the exchange of heat, the energy equation is used. The energy equation is similar in structure to the Navier-Stokes equation:

$$\rho c\left(u\frac{\partial T}{\partial x} + v\frac{\partial T}{\partial y}\right) = k\left(\frac{\partial^2 T}{\partial x^2} + \frac{\partial^2 T}{\partial y^2}\right) \quad (18)$$

The energy equation accommodates several HVAC situations. For example, heat in a room may be lost or gained through walls, windows, heaters, air conditioners, or occupants. Natural convection may dominate the flow pattern during heating or cooling and affect the comfort of the occupants, causing drafts or reduced air change effectiveness.

The energy equation is also important in compressible flow (high velocity). When the Mach number exceeds 0.2, the effects of compressibility begin to appear. The fluid density and velocity are determined in part by the energy equation. The CFD applications that require these high velocities are usually components (valves, ducting, ejectors, etc.).

A special class of problems requires the solution of the **conduction equation** (Laplace's equation). This type of analysis is called **conjugate heat transfer**. An example would be a room with an insulated wall of known thermal conductivity, where the solver determines both the temperature profile across the wall and the flow in the room.

Multispecies Flows

Often, a system with a mixture of gases must be evaluated. For example, a vent hood in a laboratory may be used to evacuate toxic gases from the room. A CFD model can provide an accurate assessment of the hood performance under a wide range of conditions and help to optimize the design for energy consumption.

If the concentration of the second gas is low, its viscous and momentum effects may be ignored. These types of problems may be solved using the single-species equation with the second species being convected along. The diffusion term is handled separately. If the concentration of the second gas is high enough to affect the properties of the mixture, such as density or viscosity, then the fully coupled form of the transport equations must be used.

Another class of multispecies flow problems involves the change of phase of one or more components. The humidity of a room with an evaporative water source would be an example of this.

Chemically reacting flows are often analyzed with CFD. Combustion models have been used extensively. These flows are complicated by the interaction of the various species, the phase changes, and the changes in properties and heat of formation of the combustion products.

Viscosity Models

In the classical Navier-Stokes equations, the viscosity is assumed constant. For low Reynolds numbers (i.e., Re < 2500), the laminar form is used. When higher Reynolds numbers are encountered, a means to adjust the viscosity (the **eddy viscosity**) is required. These types of equations are frequently called **turbulence models** because the basis for eddy viscosity is the turbulence level. Several turbulence models are available, but the most accepted and benchmarked model is the k-ε model. Chen (1995) provides some background for the k-ε application to room airflow. The standard k-ε model is a two-equation model. The first equation describes the turbulent kinetic energy transport and generation rate:

$$\rho u\frac{\partial k}{\partial x} + \rho v\frac{\partial k}{\partial y} = \frac{\partial}{\partial x}\left[\left(\mu + \frac{\mu_t}{\sigma_k}\right)\frac{\partial k}{\partial x}\right] + \frac{\partial}{\partial y}\left[\left(\mu + \frac{\mu_t}{\sigma_k}\right)\frac{\partial k}{\partial y}\right] + \mu_t\left[2\left(\frac{\partial u}{\partial x}\right)^2 + 2\left(\frac{\partial v}{\partial y}\right)^2 + \left(\frac{\partial u}{\partial y} + \frac{\partial v}{\partial x}\right)^2\right] - \rho\varepsilon - 2\mu\left[\left(\frac{\partial\sqrt{k}}{\partial x}\right)^2 + \left(\frac{\partial\sqrt{k}}{\partial x}\right)^2\right] \quad (19)$$

Turbulent kinetic energy [Equation (9)] represents the fluctuating component of velocity.

The second equation in the k-ε model is the dissipation rate:

$$\rho u\frac{\partial \varepsilon}{\partial x} + \rho v\frac{\partial \varepsilon}{\partial y} = \frac{\partial}{\partial x}\left[\left(\mu + \frac{\mu_t}{\sigma_\varepsilon}\right)\frac{\partial \varepsilon}{\partial x}\right] + \frac{\partial}{\partial y}\left[\left(\mu + \frac{\mu_t}{\sigma_\varepsilon}\right)\frac{\partial \varepsilon}{\partial y}\right] + C_{\varepsilon 1}\mu_t\frac{\varepsilon}{k}\left[2\left(\frac{\partial u}{\partial x}\right)^2 + 2\left(\frac{\partial v}{\partial y}\right)^2 + \left(\frac{\partial u}{\partial y} + \frac{\partial v}{\partial x}\right)^2\right] - C_{\varepsilon 2}f_2\rho\frac{\varepsilon^2}{k} + E \quad (20)$$

where $$E \cong 2\frac{\mu_t}{\rho}\left[\left(\frac{\partial^2 u}{\partial x^2}\right)^2 + \left(\frac{\partial^2 v}{\partial x^2}\right)^2\right]$$

Equation (20) is solved in conjunction with the Equation (19) to obtain the local eddy viscosity:

$$\mu_t = C_\mu f_\mu \rho\frac{k^2}{\varepsilon} \quad (21)$$

This two-equation model has a counterpart for high Reynolds number flows, but most room air motion studies use a lower Reynolds number model.

The near wall conditions for the turbulence model are often defined with wall functions that are based on the classical turbulent boundary layer. They may have two layers (the laminar sublayer and the turbulent layer) or they may have three layers (a laminar sublayer, a buffer layer, and a fully turbulent outer layer). The two-layer wall function is shown as follows:

$$\frac{u}{u_\tau} = y^+ \quad \text{for} \quad 0 < y^+ \le 30 \quad (22)$$

$$\frac{u}{u_\tau} = \frac{1}{\kappa}\ln(Ey^+) \quad \text{for} \quad 30 < y^+ \le 300 \quad (23)$$

where $$y^+ = \frac{\rho u_\tau y}{\mu}$$

and $$\mu\frac{\partial u}{\partial y}\bigg|_{\text{wall}} = \tau_w = \rho u_\tau^2 \quad (24)$$

Direct numerical simulation (DNS) is an active area of study of turbulence modeling. The basis for DNS turbulence modeling is that there is actually no turbulence model at all. Rather, the motion of the individual eddies within the bulk flow are modeled themselves. This model requires an extremely small grid size in which an eddy cannot exist due to molecular viscosity. Due to the huge demand on computing resources; only small flow fields, about shoe box size at the most, have been simulated using DNS.

Since DNS is really not yet practical for room air simulations, a compromise between it and conventional turbulence models has been developed. This method, **large eddy simulation** (LES), uses an eddy viscosity for the subgrid scale turbulence, but fully captures the time-dependent behavior of the larger energy bearing eddies. The application to room airflow is just beginning (Emmerich and McGrattan 1998).

CFD Applications

Currently, CFD has been applied to room air motion. The room might represent an auditorium, an aircraft cabin, or an automobile interior. Useful parameters such as velocity and temperature distribution, air change effectiveness, and humidity are predicted for the comfort of the occupant.

External flow models have used CFD to predict building infiltration, heat transfer rates for heating and cooling loads, and wind loads for structural design.

Finally, CFD is a useful tool for other internal flows, especially when nonstandard components are analyzed. Complex manifolds can be flow-balanced in one step, and important data such as pressure drop, aero/hydrodynamic loads, and heat transfer rates are available with the currently obtainable tools.

SYMBOLS

adj = subscript denoting adjacent node
c = specific heat, J/(kg·K) based on units of ρ
$C_{\varepsilon 1}$ = constant = 1.45
$C_{\varepsilon 2}$ = constant = 1.92
C_μ = constant = 0.09
f = field variable such as velocity
f_2 = coefficient = $1 - 0.3\exp(-\mathrm{Re}_t^2)$, where Re_t is local Reynolds number = $\rho K^2/\mu\varepsilon$
f_μ = coefficient = $\exp[-3.4/(1 + \mathrm{Re}_t/50)^2]$
i = subscript index in x direction
j = subscript index in y direction
k = thermal conductivity, W/(m·K)
K = kinetic energy of turbulence, m^2/s^2
L = turbulence length scale, m
n = general subscript index
P = pressure, Pa
s = distance, m
T = temperature, K
u = fluid velocity in x direction, m/s
u_τ = friction velocity, m/s
v = fluid velocity in y direction, m/s
V = local velocity magnitude, m/s
V' = turbulent velocity fluctuation, m/s
x = horizontal distance, m
Δx = node spacing, m
y = vertical distance, m

y^+ = distance from wall in the universal velocity profile, ft
t = time, s
δ = relaxation parameter
ε = dissipation, ft^2/s^3
κ = von Karman constant = 0.41
μ = dynamic viscosity, slug/ft·s
μ_t = effective viscosity, slug/ft·s
ν = kinematic viscosity, ft^2/s
ν_t = effective viscosity, ft^2/s
ρ = density, slug/ft^3
σ_k = constant = 1.0
σ_ε = constant = 1.3
ψ = stream function, ft^2/s
ψ_B = boundary stream function, ft^2/s
ω = vorticity, 1/s

REFERENCES

Baker, A.J. 1983. *Finite element computational fluid mechanics*. Hemisphere Publishing, New York.

Catton, I. 1978. Natural convection in enclosures. *Heat Transfer*, 6.

Chang, L.C., J.R. Lloyd, and K.T. Yang. 1982. A finite difference study of natural convection in complex enclosures. *Proceedings of the 7th International Heat Transfer Conference* 2:183-88.

Chen, Q. and J. Van der Kooi. 1988. ACCURACY—A program for combined problems of energy analysis, indoor airflow, and air quality. *ASHRAE Transactions* 94(2):196-214.

Chen, Q. 1995. Comparison of different k-ε models for indoor air flow computations. *Numerical Heat Transfer*, Part B Fundamentals 28:353-69.

Chen, Q., A. Moser, and A. Huber. 1990a. Prediction of buoyant, turbulent flow by a low-Reynolds-number k-ε model. *ASHRAE Transactions* 96(1):564-73.

Chen, Q., A. Moser, and P. Suter. 1990b. Indoor air quality and thermal comfort under six kinds of air diffusion. *ASHRAE Transactions* 96(1).

Dennis, S.C.R. and J.D. Hudson. 1978. A difference method for solving the Navier-Stokes equations. *Numerical Methods in Laminar and Turbulent Flow*, pp. 69-80. Taylor Morgan Brebbia.

Emmerich, S.J. and K.B. McGrattan. 1998. Application of a large eddy simulation model to study room airflow. *ASHRAE Transactions* 104(1).

Gadgil, A., F. Bauman, E. Altmayer, and R.C. Kammerund. 1984. Verification of a natural simulation technique for natural convection. *ASME Journal of Solar Energy Engineering* 106:366-69.

Hadjisophocleous, G.V., A.C.M. Sousa, and J.E.S. Venart. 1988. Prediction of transient natural convection in enclosures of arbitrary geometry using a non orthogonal numerical model. *Numerical Heat Transfer* 13:373.

Haghighat, F., J.C.Y. Wang, and Z. Jiang. 1989. Natural convection and airflow pattern in a partitioned room with turbulent flow. *ASHRAE Transactions* 95(2):600-10.

Haghighat, F., J.C.Y. Wang, and Z. Jiang. 1990. Development of a three-dimensional numerical model to investigate the airflow and age distribution in a multizone enclosure. *Proceedings Fifth International Conference on Indoor Air Quality and Climate: Indoor Air '90* 4:183-88. International Conference on Indoor Air Quality and Climate, 2344 Haddinton Crescent, Ottawa, Ontario K1H 8J4, Canada.

Haghighat, F., J.C.Y. Wang, Z. Jiang, and F. Allard. 1992. Air movement in buildings using computational fluid dynamics. *Transactions of the ASME Journal of Solar Energy Engineering* 114:84-92.

Holmes, M.J. 1982. The application of fluid mechanics simulation program PHOENICS to a few typical HVAC problems. Ove Arup and Partners, London.

Horstman, R.H. 1988. Predicting velocity and contamination distribution in ventilated volumes using Navier-Stokes equations. *ASHRAE Conference, IAQ 88*, pp. 209-30.

Jones, P. and P.O. Sullivan. 1985. Modeling of air flow patterns in large single volume space. SERC Workshop: Development in Building Simulation Programs, Loughborough University.

Kelkar, K.M. and S.V. Patankar. 1985. Numerical prediction of natural convection in partitioned enclosures. *Numerical Heat Transfer*, pp. 63-71.

Lemaire, A.D. 1987. The numerical simulation of the air movement and heat transfer in a heated room resp. a ventilated atrium. *Proceedings of the International Conference on Air Distribution in Ventilated Space, ROOMVENT-87*. NORSKVVS, Oslo.

Lin, D.S. and N.W. Nansteel. 1987. Natural convection heat transfer in a square enclosure containing water near its density maximum. *International Journal of Heat and Mass Transfer* 30(11):2319-29.

Markatos, N.C. 1983. The computer analysis of building ventilation and heating problems. *Passive and Low Energy Architecture*, pp. 667-75.

Markatos, N.C. and K.A. Pericleous.1984. Laminar and turbulent natural convection in an enclosed cavity. *International Journal of Heat and Mass Transfer* 27(5):755-72.

Murakami, S., S. Kota, and Y. Suyama. 1988. Numerical and experimental study on turbulent diffusion fields in convection flow type clean rooms. *ASHRAE Transactions* 94(2):469-93.

Nielsen, P.V. 1974. Flow in air-conditioned rooms. *Ph.D. thesis*. Technical University of Denmark, Copenhagen.

Ostrach, S. 1982. Natural convection heat transfer in cavities and cells. *Proceedings of the International Heat Transfer Conference*. Hemisphere, Washington, D.C., pp. 365-79.

Patankar, S.V. 1988. Elliptic systems: Finite-difference method 1. *Handbook of numerical heat transfer*, Chap. 6, pp. 215-40. John Wiley and Sons, New York.

Pun, W.M., and D.B. Spalding. 1976. A general computer program for two-dimensional elliptic flows. HTS/76/2. Imperial College, London.

BIBLIOGRAPHY

Cebeci, T. 1988. Parabolic systems: Finite-difference method II. *Handbook of numerical heat transfer*. John Wiley and Sons, New York.

Chen, Q. and Z. Jiang. 1992. Significant questions in predicting room air motion. *ASHRAE Transactions* 98(1):929-39.

Lin, C.H., T. Han, and C.A. Koromilas. 1992. Effect of HVAC design parameters on passenger thermal comfort. SAE No. 920264. Society of Automotive Engineers, Warrendale, PA.

CHAPTER 34

DUCT DESIGN

COMMERCIAL, industrial, and residential air duct system design must consider (1) space availability, (2) space air diffusion, (3) noise levels, (4) duct leakage, (5) duct heat gains and losses, (6) balancing, (7) fire and smoke control, (8) initial investment cost, and (9) system operating cost.

Deficiencies in duct design can result in systems that operate incorrectly or are expensive to own and operate. Poor air distribution can cause discomfort, loss of productivity and even adverse health effects; lack of sound attenuators may permit objectionable noise levels. Poorly designed ductwork can result in unbalanced systems. Faulty duct construction or lack of duct sealing produces inadequate airflow rates at the terminals. Proper duct insulation eliminates the problem caused by excessive heat gain or loss.

In this chapter, system design and the calculation of a system's frictional and dynamic resistance to airflow are considered. Chapter 16 of the 2000 *ASHRAE Handbook—Systems and Equipment* examines duct construction and presents construction standards for residential, commercial, and industrial heating, ventilating, air-conditioning, and exhaust systems.

BERNOULLI EQUATION

The Bernoulli equation can be developed by equating the forces on an element of a stream tube in a frictionless fluid flow to the rate of momentum change. On integrating this relationship for steady flow, the following expression (Osborne 1966) results:

$$\frac{v^2}{2g_c} + \int \frac{dP}{\rho} + \frac{gz}{g_c} = \text{constant, ft} \cdot \text{lb}_f/\text{lb}_m \quad (1)$$

where

v = streamline (local) velocity, fps
g_c = dimensional constant, 32.2 $\text{lb}_m \cdot \text{ft/lb}_f \cdot \text{s}^2$
P = absolute pressure, lb_f/ft^2
ρ = density, lb_m/ft^3
g = acceleration due to gravity, ft/s^2
z = elevation, ft

Assuming constant fluid density within the system, Equation (1) reduces to

$$\frac{v^2}{2g_c} + \frac{P}{\rho} + \frac{gz}{g_c} = \text{constant, ft} \cdot \text{lb}_f/\text{lb}_m \quad (2)$$

Although Equation (2) was derived for steady, ideal frictionless flow along a stream tube, it can be extended to analyze flow through ducts in real systems. In terms of pressure, the relationship for fluid resistance between two sections is

$$\frac{\rho_1 V_1^2}{2g_c} + P_1 + \frac{g}{g_c}\rho_1 z_1 = \frac{\rho_2 V_2^2}{2g_c} + P_2 + \frac{g}{g_c}\rho_2 z_2 + \Delta p_{t,1\text{-}2} \quad (3)$$

where

V = average duct velocity, fps
$\Delta p_{t,1\text{-}2}$ = total pressure loss due to friction and dynamic losses between sections 1 and 2, lb_f/ft^2

In Equation (3), V (section average velocity) replaces v (streamline velocity) because experimentally determined loss coefficients allow for errors in calculating $\rho v^2/2g_c$ (velocity pressure) across streamlines.

On the left side of Equation (3), add and subtract p_{z1}; on the right side, add and subtract p_{z2}, where p_{z1} and p_{z2} are the values of atmospheric air at heights z_1 and z_2. Thus,

$$\frac{\rho_1 V_1^2}{2g_c} + P_1 + (p_{z1} - p_{z1}) + \frac{g}{g_c}\rho_1 z_1$$

$$= \frac{\rho_2 V_2^2}{2g_c} + P_2 + (p_{z2} - p_{z2}) + \frac{g}{g_c}\rho_2 z_2 + \Delta p_{t,1\text{-}2} \quad (4)$$

The atmospheric pressure at any elevation (p_{z1} and p_{z2}) expressed in terms of the atmospheric pressure p_a at the same datum elevation is given by

$$p_{z1} = p_a - \frac{g}{g_c}\rho_a z_1 \quad (5)$$

$$p_{z2} = p_a - \frac{g}{g_c}\rho_a z_2 \quad (6)$$

Substituting Equations (5) and (6) into Equation (4) and simplifying yields the total pressure change between sections 1 and 2. Assume no change in temperature between sections 1 and 2 (no heat exchanger within the section); therefore, $\rho_1 = \rho_2$. When a heat exchanger is located within the section, the average of the inlet and outlet temperatures is generally used. Let $\rho = \rho_1 = \rho_2$. $(P_1 - p_{z1})$ and $(P_2 - p_{z2})$ are gage pressures at elevations z_1 and z_2.

$$\Delta p_{t,1\text{-}2} = \left(p_{s,1} + \frac{\rho V_1^2}{2g_c}\right) - \left(p_{s,2} + \frac{\rho V_2^2}{2g_c}\right) + \frac{g}{g_c}(\rho_a - \rho)(z_2 - z_1) \quad (7a)$$

$$\Delta p_{t,1\text{-}2} = \Delta p_t + \Delta p_{se} \quad (7b)$$

The preparation of this chapter is assigned to TC 5.2, Duct Design.

$$\Delta p_t = \Delta p_{t,1\text{-}2} - \Delta p_{se} \qquad (7c)$$

where

$p_{s,1}$ = static pressure, gage at elevation z_1, lb_f/ft^2
$p_{s,2}$ = static pressure, gage at elevation z_2, lb_f/ft^2
V_1 = average velocity at section 1, fps
V_2 = average velocity at section 2, fps
ρ_a = density of ambient air, lb_m/ft^3
ρ = density of air or gas within duct, lb_m/ft^3
Δp_{se} = thermal gravity effect, lb_f/ft^2
Δp_t = total pressure change between sections 1 and 2, lb_f/ft^2
$\Delta p_{t,1-2}$ = total pressure loss due to friction and dynamic losses between sections 1 and 2, lb_f/ft^2

HEAD AND PRESSURE

The terms **head** and **pressure** are often used interchangeably; however, head is the height of a fluid column supported by fluid flow, while pressure is the normal force per unit area. For liquids, it is convenient to measure the head in terms of the flowing fluid. With a gas or air, however, it is customary to measure pressure on a column of liquid.

Static Pressure

The term $pg_c/\rho g$ is static head; p is static pressure.

Velocity Pressure

The term $V^2/2g$ refers to velocity head, and the term $\rho V^2/2g_c$ refers to velocity pressure. Although velocity head is independent of fluid density, velocity pressure, calculated by Equation (8), is not.

$$p_v = \rho\left(\frac{V}{1097}\right)^2 \qquad (8)$$

where

p_v = velocity pressure, in. of water
V = fluid mean velocity, fpm

For air at standard conditions (0.075 lb_m/ft^3), Equation (8) becomes

$$p_v = \left(\frac{V}{4005}\right)^2 \qquad (9)$$

Velocity is calculated by Equation (10) or (11).

$$V = 144Q/A \qquad (10)$$

where

Q = airflow rate, cfm
A = cross-sectional area of duct, in^2

$$V = Q/A \qquad (11)$$

where A = cross-sectional area of duct, ft^2.

Total Pressure

Total pressure is the sum of static pressure and velocity pressure:

$$p_t = p_s + \rho\left(\frac{V}{1097}\right)^2 \qquad (12)$$

or

$$p_t = p_s + p_v \qquad (13)$$

where

p_t = total pressure, in. of water
p_s = static pressure, in. of water

Pressure Measurement

The range, precision, and limitations of instruments for measuring pressure and velocity are discussed in Chapter 14. The manometer is a simple and useful means for measuring partial vacuum and low pressure. Static, velocity, and total pressures in a duct system relative to atmospheric pressure are measured with a pitot tube connected to a manometer. Pitot tube construction and locations for traversing round and rectangular ducts are presented in Chapter 14.

SYSTEM ANALYSIS

The total pressure change due to friction, fittings, equipment, and net **thermal gravity effect (stack effect)** for each section of a duct system is calculated by the following equation:

$$\Delta p_{t_i} = \Delta p_{f_i} + \sum_{j=1}^{m} \Delta p_{ij} + \sum_{k=1}^{n} \Delta p_{ik} - \sum_{r=1}^{\lambda} \Delta p_{se_{ir}} \qquad (14)$$

$$\text{for } i = 1, 2, \ldots, n_{up} + n_{dn}$$

where

Δp_{t_i} = net total pressure change for i-section, in. of water
Δp_{f_i} = pressure loss due to friction for i-section, in. of water
Δp_{ij} = total pressure loss due to j-fittings, including fan system effect (FSE), for i-section, in. of water
Δp_{ik} = pressure loss due to k-equipment for i-section, in. of water
$\Delta p_{se_{ir}}$ = thermal gravity effect due to r-stacks for i-section, in. of water
m = number of fittings within i-section
n = number of equipment within i-section
λ = number of stacks within i-section
n_{up} = number of duct sections upstream of fan (exhaust/return air subsystems)
n_{dn} = number of duct sections downstream of fan (supply air subsystems)

From Equation (7), the thermal gravity effect for each nonhorizontal duct with a density other than that of ambient air is determined by the following equation:

$$\Delta p_{se} = 0.192(\rho_a - \rho)(z_2 - z_1) \qquad (15)$$

where

Δp_{se} = thermal gravity effect, in. of water
z_1 and z_2 = elevation from datum in direction of airflow (Figure 1), ft
ρ_a = density of ambient air, lb_m/ft^3
ρ = density of air or gas within duct, lb_m/ft^3

Example 1. For Figure 1, calculate the thermal gravity effect for two cases: (a) air cooled to −30°F, and (b) air heated to 1000°F. The density of air at −30°F and 1000°F is 0.0924 lb_m/ft^3 and 0.0271 lb_m/ft^3, respectively. The density of the ambient air is 0.075 lb_m/ft^3. Stack height is 40 ft.

Solution:

$$\Delta p_{se} = 0.192(\rho_a - \rho)z$$

(a) For $\rho > \rho_a$ (Figure 1A),

$$\Delta p_{se} = 0.192(0.075 - 0.0924)40$$
$$= -0.13 \text{ in. of water}$$

(b) For $\rho < \rho_a$ (Figure 1B),

$$\Delta p_{se} = 0.192(0.075 - 0.0271)40$$
$$= +0.37 \text{ in. of water}$$

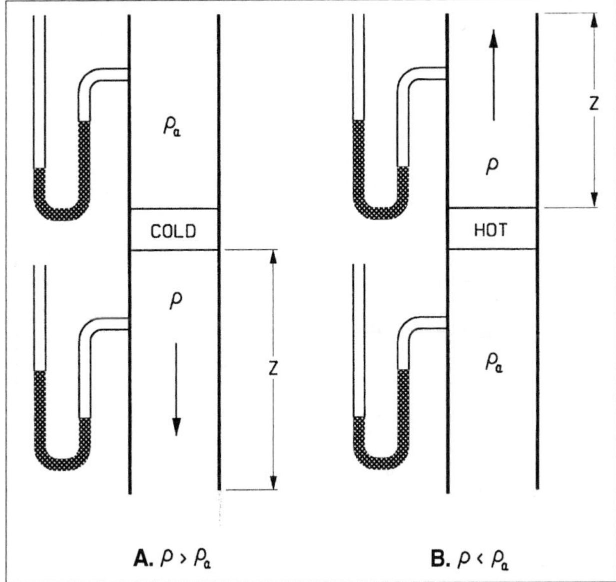

Fig. 1 Thermal Gravity Effect for Example 1

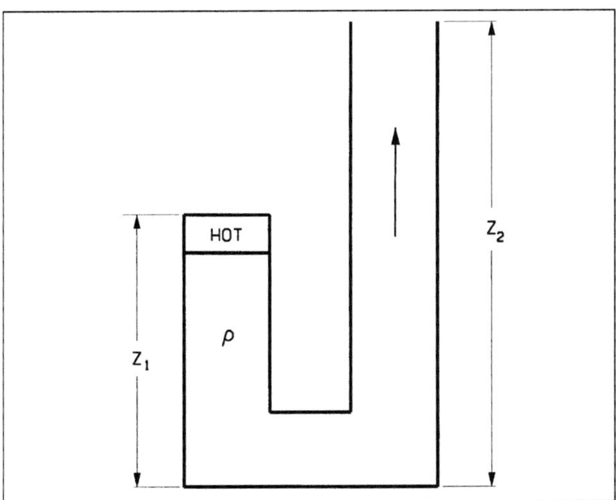

Fig. 2 Multiple Stacks for Example 2

Example 2. Calculate the thermal gravity effect for the two-stack system shown in Figure 2, where the air is 250°F and the stack heights are 50 and 100 ft. The density of 250°F air is 0.0558 lb_m/ft^3; ambient air is 0.075 lb_m/ft^3.

Solution:

$$\Delta p_{se} = 0.192(0.075 - 0.0558)(100 - 50)$$
$$= 0.18 \text{ in. of water}$$

For the system shown in Figure 3, the direction of air movement created by the thermal gravity effect depends on the initiating force. The initiating force could be fans, wind, opening and closing doors, and turning equipment on and off. If for any reason air starts to enter the left stack (Figure 3A), it creates a buoyancy effect in the right stack. On the other hand, if flow starts to enter the right stack (Figure 3B), it creates a buoyancy effect in the left stack. In both cases the produced thermal gravity effect is stable and depends on the stack height and magnitude of heating. The starting direction of flow is important when using natural convection for ventilation.

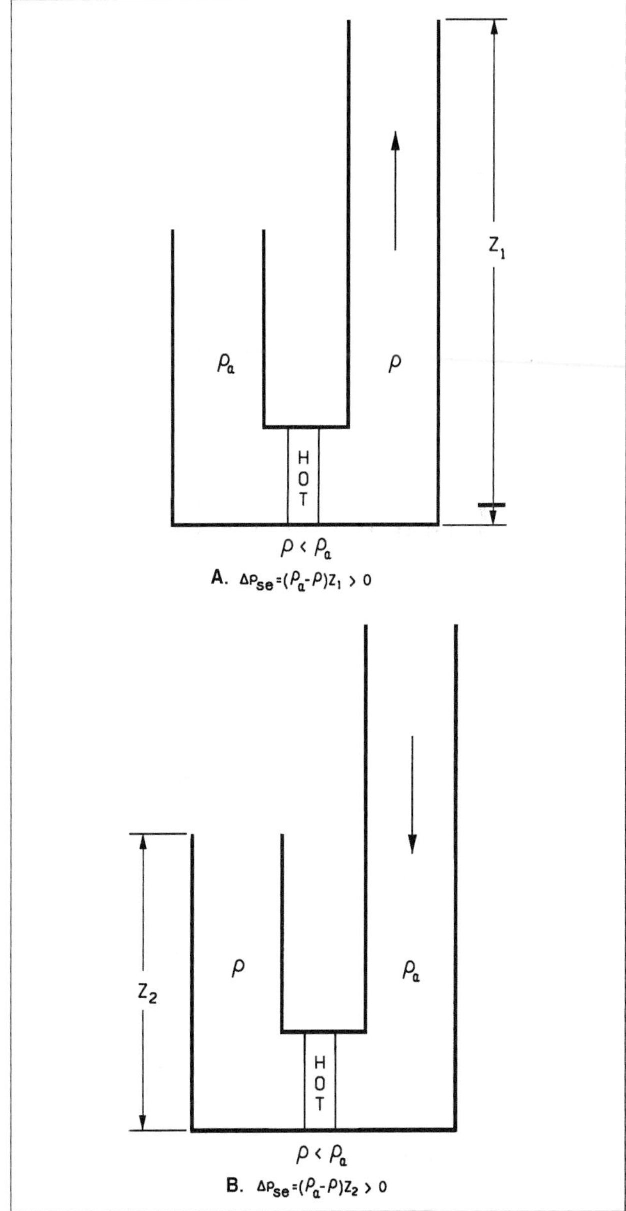

Fig. 3 Multiple Stack Analysis

To determine the fan total pressure requirement for a system, use the following equation:

$$P_t = \sum_{i \varepsilon F_{up}} \Delta p_{t_i} + \sum_{i \varepsilon F_{dn}} \Delta p_{t_i} \qquad \text{for } i = 1, 2, \ldots, n_{up} + n_{dn} \qquad (16)$$

where

F_{up} and F_{dn}=sets of duct sections upstream and downstream of a fan

P_t = fan total pressure, in. of water

ε = symbol that ties duct sections into system paths from the exhaust/return air terminals to the supply terminals

Figure 4 illustrates the use of Equation (16). This system has three supply and two return terminals consisting of nine sections connected in six paths: 1-3-4-9-7-5, 1-3-4-9-7-6, 1-3-4-9-8, 2-4-9-7-5, 2-4-9-7-6, and 2-4-9-8. Sections 1 and 3 are unequal area; thus, they are assigned separate numbers in accordance with the rules for identifying sections (see Step 4 in the section on HVAC Duct Design Procedures). To determine the fan pressure requirement, the following six equations, derived from Equation (16), are applied. These equations must be satisfied to attain pressure balancing for design airflow. Relying entirely on dampers is not economical and may create objectionable flow-generated noise.

Fig. 4 Illustrative 6-Path, 9-Section System

$$\left\{\begin{array}{l} P_t = \Delta p_1 + \Delta p_3 + \Delta p_4 + \Delta p_9 + \Delta p_7 + \Delta p_5 \\ P_t = \Delta p_1 + \Delta p_3 + \Delta p_4 + \Delta p_9 + \Delta p_7 + \Delta p_6 \\ P_t = \Delta p_1 + \Delta p_3 + \Delta p_4 + \Delta p_9 + \Delta p_8 \\ P_t = \Delta p_2 + \Delta p_4 + \Delta p_9 + \Delta p_7 + \Delta p_5 \\ P_t = \Delta p_2 + \Delta p_4 + \Delta p_9 + \Delta p_7 + \Delta p_6 \\ P_t = \Delta p_2 + \Delta p_4 + \Delta p_9 + \Delta p_8 \end{array}\right. \tag{17}$$

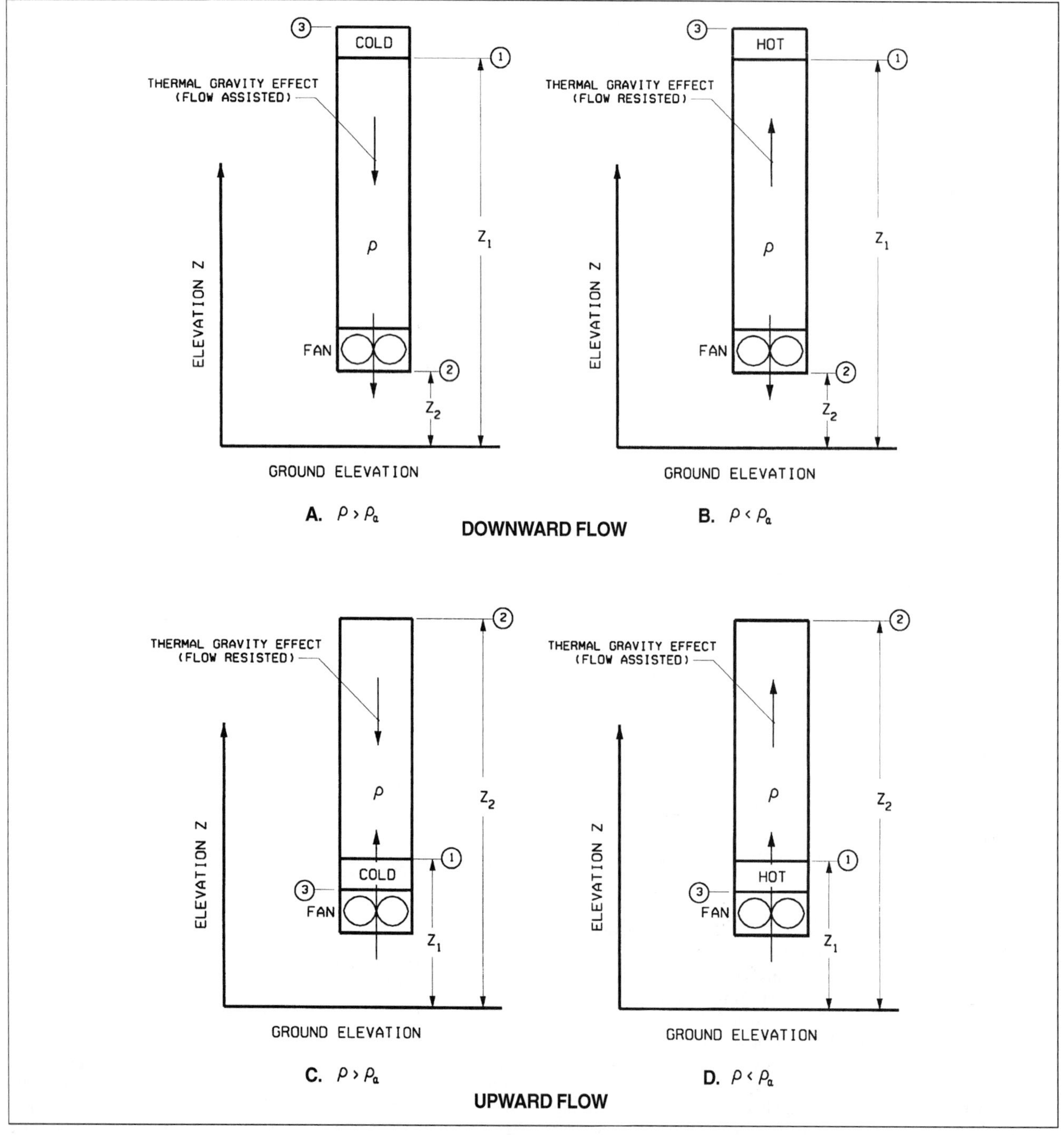

Fig. 5 Single Stack with Fan for Examples 3 and 4

Example 3. For Figures 5A and 5C, calculate the thermal gravity effect and fan total pressure required when the air is cooled to −30°F. The heat exchanger and ductwork (section 1 to 2) total pressure losses are 0.70 and 0.28 in. of water respectively. The density of −30°F air is 0.0924 lb_m/ft^3; ambient air is 0.075 lb_m/ft^3. Elevations are 70 ft and 10 ft as noted in the solutions below.

Solution:

(a) For Figure 5A (downward flow),

$$\begin{aligned}\Delta p_{se} &= 0.192(\rho_a - \rho)(z_2 - z_1)\\ &= 0.192(0.075 - 0.0924)(10 - 70)\\ &= 0.20 \text{ in. of water}\end{aligned}$$

$$\begin{aligned}P_t &= \Delta p_{t,3\text{-}2} - \Delta p_{se}\\ &= (0.70 + 0.28) - (0.20)\\ &= 0.78 \text{ in. of water}\end{aligned}$$

(b) For Figure 5C (upward flow),

$$\begin{aligned}\Delta p_{se} &= 0.192(\rho_a - \rho)(z_2 - z_1)\\ &= 0.192(0.075 - 0.0924)(70 - 10)\\ &= -0.20 \text{ in. of water}\end{aligned}$$

$$\begin{aligned}P_t &= \Delta p_{t,3\text{-}2} - \Delta p_{se}\\ &= (0.70 + 0.28) - (-0.20)\\ &= 1.18 \text{ in. of water}\end{aligned}$$

Example 4. For Figures 5B and 5D, calculate the thermal gravity effect and fan total pressure required when the air is heated to 250°F. The heat exchanger and ductwork (section 1 to 2) total pressure losses are 0.70 and 0.28 in. of water respectively. The density of 250°F air is 0.0558 lb_m/ft^3; ambient air is 0.075 lb_m/ft^3. Elevations are 70 ft and 10 ft as noted in the solutions below.

Solution:

(a) For Figure 5B (downward flow),

$$\begin{aligned}\Delta p_{se} &= 0.192(\rho_a - \rho)(z_2 - z_1)\\ &= 0.192(0.075 - 0.0558)(10 - 70)\\ &= -0.22 \text{ in. of water}\end{aligned}$$

$$\begin{aligned}P_t &= \Delta p_{t,3\text{-}2} - \Delta p_{se}\\ &= (0.70 + 0.28) - (-0.22)\\ &= 1.20 \text{ in. of water}\end{aligned}$$

(b) For Figure 5D (upward flow),

$$\begin{aligned}\Delta p_{se} &= 0.192(\rho_a - \rho)(z_2 - z_1)\\ &= 0.192(0.075 - 0.0558)(70 - 10)\\ &= 0.22 \text{ in. of water}\end{aligned}$$

$$\begin{aligned}P_t &= \Delta p_{t,3\text{-}2} - \Delta p_{se}\\ &= (0.70 + 0.28) - (0.22)\\ &= 0.76 \text{ in. of water}\end{aligned}$$

Example 5. Calculate the thermal gravity effect for each section of the system shown in Figure 6 and the net thermal gravity effect of the system. The density of ambient air is 0.075 lb_m/ft^3, and the lengths are as follows: z_1 = 50 ft, z_2 = 90 ft, z_4 = 95 ft, z_5 = 25 ft, and z_9 = 200 ft. The pressure required at section 3 is −0.1 in. of water. Write the equation to determine the fan total pressure requirement.

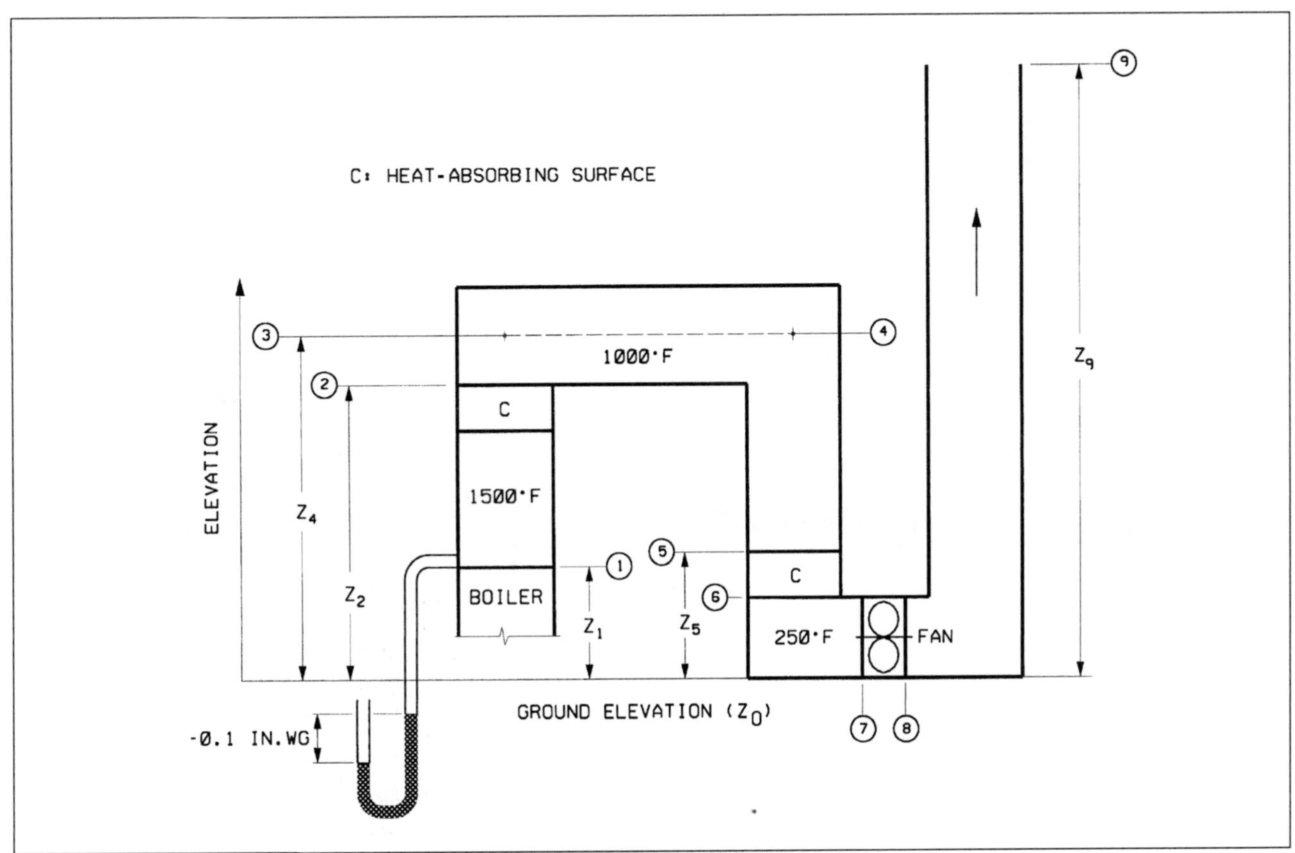

Fig. 6 Triple Stack System for Example 5

Solution: The following table summarizes the thermal gravity effect for each section of the system as calculated by Equation (15). The net thermal gravity effect for the system is 0.52 in. of water. To select a fan, use the following equation:

$$P_t = 0.1 + \Delta p_{t,1\text{-}7} + \Delta p_{t,8\text{-}9} - \Delta p_{se}$$
$$= 0.1 + \Delta p_{t,1\text{-}7} + \Delta p_{t,8\text{-}9} - 0.52$$
$$= \Delta p_{t,1\text{-}7} + \Delta p_{t,8\text{-}9} - 0.42$$

Path $(x\text{-}x')$	Temp., °F	ρ, lb_m/ft^3	Δz $(z_{x'} - z_x)$, ft	$\Delta\rho$ $(\rho_a - \rho_{x-x'})$, lb_m/ft^3	Δp_{se}, in. of water [Eq. (15)]
1-2	1500	0.0202	(90 – 50)	+0.0548	+0.42
3-4	1000	0.0271	0	+0.0479	0
4-5	1000	0.0271	(25 – 95)	+0.0479	–0.64
6-7	250	0.0558	0	+0.0192	0
8-9	250	0.0558	(200 – 0)	+0.0192	+0.74
Net Thermal Gravity Effect					0.52

PRESSURE CHANGES IN SYSTEM

Figure 7 shows total and static pressure changes in a fan/duct system consisting of a fan with both supply and return air ductwork. Also shown are the total and static pressure gradients referenced to atmospheric pressure.

For all constant-area sections, the total and static pressure losses are equal. At the diverging transitions, velocity pressure decreases, absolute total pressure decreases, and absolute static pressure can increase. The static pressure increase at these sections is known as **static regain**.

At the converging transitions, velocity pressure increases in the direction of airflow, and the absolute total and absolute static pressures decrease.

At the exit, the total pressure loss depends on the shape of the fitting and the flow characteristics. Exit loss coefficients C_o can be greater than, less than, or equal to one. The total and static pressure grade lines for the various coefficients are shown in Figure 3. Note that for a loss coefficient less than one, static pressure upstream of the exit is less than atmospheric pressure (negative). The static pressure just upstream of the discharge fitting can be calculated by subtracting the upstream velocity pressure from the upstream total pressure.

At section 1, the total pressure loss depends on the shape of the entry. The total pressure immediately downstream of the entrance equals the difference between the upstream pressure, which is zero (atmospheric pressure), and the loss through the fitting. The static pressure of the ambient air is zero; several diameters downstream, static pressure is negative, equal to the sum of the total pressure (negative) and the velocity pressure (always positive).

System resistance to airflow is noted by the total pressure grade line in Figure 7. Sections 3 and 4 include fan system effect pressure losses. To obtain the fan static pressure requirement for fan selection where the fan total pressure is known, use

$$P_s = P_t - p_{v,o} \tag{18}$$

where

P_s = fan static pressure, in. of water
P_t = fan total pressure, in. of water
$p_{v,o}$ = fan outlet velocity pressure, in. of water

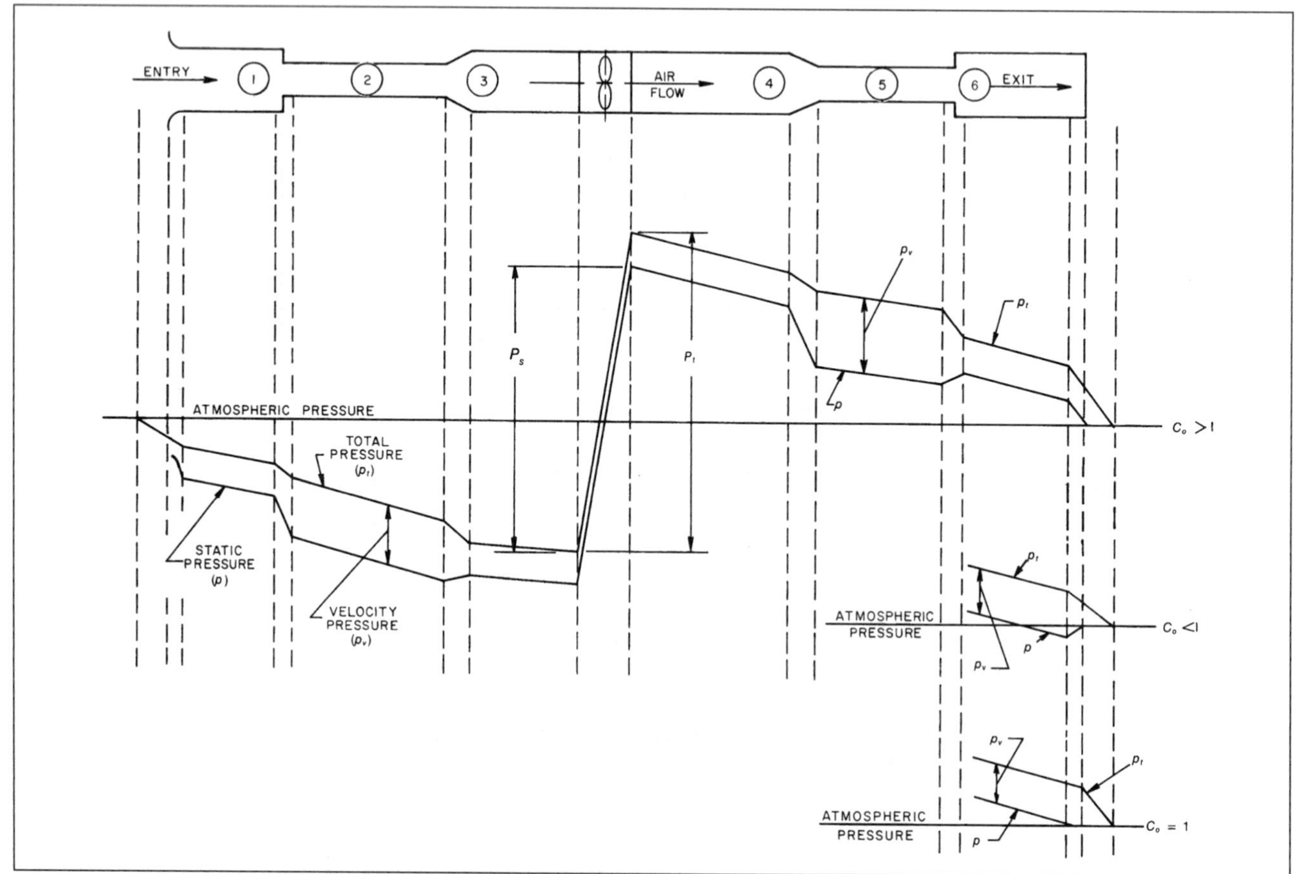

Fig. 7 Pressure Changes During Flow in Ducts

FLUID RESISTANCE

Duct system losses are the irreversible transformation of mechanical energy into heat. The two types of losses are (1) friction losses and (2) dynamic losses.

FRICTION LOSSES

Friction losses are due to fluid viscosity and are a result of momentum exchange between molecules in laminar flow and between individual particles of adjacent fluid layers moving at different velocities in turbulent flow. Friction losses occur along the entire duct length.

Darcy, Colebrook, and Altshul-Tsal Equations

For fluid flow in conduits, friction loss can be calculated by the Darcy equation:

$$\Delta p_f = \frac{12fL}{D_h}\rho\left(\frac{V}{1097}\right)^2 \quad (19)$$

where

Δp_f = friction losses in terms of total pressure, in. of water
f = friction factor, dimensionless
L = duct length, ft
D_h = hydraulic diameter [Equation (24)], in.
V = velocity, fpm
ρ = density, lb_m/ft^3

Within the region of laminar flow (Reynolds numbers less than 2000), the friction factor is a function of Reynolds number only.

For completely turbulent flow, the friction factor depends on Reynolds number, duct surface roughness, and internal protuberances such as joints. Between the bounding limits of hydraulically smooth behavior and fully rough behavior, is a transitional roughness zone where the friction factor depends on both roughness and Reynolds number. In this transitionally rough, turbulent zone the friction factor f is calculated by Colebrook's equation (Colebrook 1938-39). Colebrook's transition curve merges asymptotically into the curves representing laminar and completely turbulent flow. Because Colebrook's equation cannot be solved explicitly for f, use iterative techniques (Behls 1971).

$$\frac{1}{\sqrt{f}} = -2\log\left(\frac{12\varepsilon}{3.7D_h} + \frac{2.51}{\mathrm{Re}\sqrt{f}}\right) \quad (20)$$

where

ε = material absolute roughness factor, ft
Re = Reynolds number

A simplified formula for calculating friction factor, developed by Altshul (Altshul et al. 1975) and modified by Tsal, is

$$f' = 0.11\left(\frac{12\varepsilon}{D_h} + \frac{68}{\mathrm{Re}}\right)^{0.25}$$

$$\text{If } f' \geq 0.018: \quad f = f' \quad (21)$$

$$\text{If } f' < 0.018: \quad f = 0.85f' + 0.0028$$

Friction factors obtained from the Altshul-Tsal equation are within 1.6% of those obtained by Colebrook's equation.

Reynolds number (Re) may be calculated by using the following equation.

$$\mathrm{Re} = \frac{D_h V}{720\nu} \quad (22)$$

where ν = kinematic viscosity, ft^2/s.

For standard air, Re can be calculated by

$$\mathrm{Re} = 8.56 D_h V \quad (23)$$

Roughness Factors

The roughness factors ε listed in Table 1 are recommended for use with the Colebrook or Altshul-Tsal equation [Equations (20) and (21), respectively]. These values include not only material, but also duct construction, joint type, and joint spacing (Griggs and Khodabakhsh-Sharifabad 1992). Roughness factors for other materials are presented in Idelchik et al. (1994). Idelchik summarizes roughness factors for 80 materials including metal tubes; conduits made from concrete and cement; and wood, plywood, and glass tubes.

Swim (1978) conducted tests on duct liners of varying densities, surface treatments, transverse joints (workmanship), and methods of attachment to sheet metal ducts. As a result of these tests, Swim recommends for design 0.015 ft for spray-coated liners and 0.005 ft for liners with a facing material cemented onto the air side. In both cases, the roughness factor includes the resistance offered by mechanical fasteners and assumes good joints. Liners cut too short result in (1) loss of thermal performance, (2) possible condensation problems, (3) potential damage to the liner (erosion of the blanket or tearing away from the duct surface), and (4) the collection of dirt and debris and the initiation of biological problems. Liner density does not significantly influence flow resistance.

Manufacturers' data indicate that the absolute roughness for fully extended nonmetallic flexible ducts ranges from 0.0035 to 0.015 ft. For fully extended flexible metallic ducts, absolute roughness ranges from 0.0004 to 0.007 ft. This range covers flexible duct with the supporting wire exposed to flow or covered by the material. Figure 8 provides a pressure drop correction factor for straight flexible duct when less than fully extended.

Table 1 Duct Roughness Factors

Duct Material	Roughness Category	Absolute Roughness ε, ft
Uncoated carbon steel, clean (Moody 1944) (0.00015 ft) PVC plastic pipe (Swim 1982) (0.00003 to 0.00015 ft) Aluminum (Hutchinson 1953) (0.000015 to 0.0002 ft)	Smooth	0.0001
Galvanized steel, longitudinal seams, 4 ft joints (Griggs et al. 1987) (0.00016 to 0.00032 ft) Galvanized steel, continuously rolled, spiral seams, 10 ft joints (Jones 1979) (0.0002 to 0.0004 ft) Galvanized steel, spiral seam with 1, 2, and 3 ribs, 12 ft joints (Griggs et al. 1987) (0.00029 to 0.00038 ft)	Medium smooth	0.0003
Galvanized steel, longitudinal seams, 2.5 ft joints (Wright 1945) (0.0005 ft)	Average	0.0005
Fibrous glass duct, rigid Fibrous glass duct liner, air side with facing material (Swim 1978) (0.005 ft)	Medium rough	0.003
Fibrous glass duct liner, air side spray coated (Swim 1978) (0.015 ft) Flexible duct, metallic (0.004 to 0.007 ft when fully extended) Flexible duct, all types of fabric and wire (0.0035 to 0.015 ft when fully extended) Concrete (Moody 1944) (0.001 to 0.01 ft)	Rough	0.01

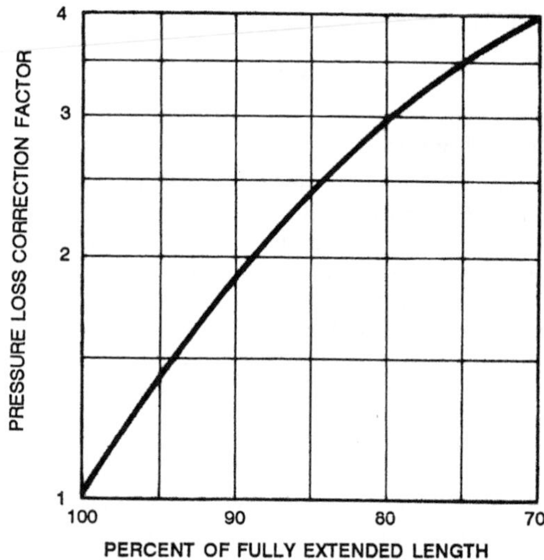

Fig. 8 Correction Factor for Unextended Flexible Duct

Friction Chart

Fluid resistance caused by friction in round ducts can be determined by the friction chart (Figure 9). This chart is based on standard air flowing through round galvanized ducts with beaded slip couplings on 48 in. centers, equivalent to an absolute roughness of 0.0003 ft.

Changes in barometric pressure, temperature, and humidity affect air density, air viscosity, and Reynolds number. No corrections to Figure 9 are needed for (1) duct materials with a medium smooth roughness factor, (2) temperature variations in the order of ±30°F from 70°F, (3) elevations to 1500 ft, and (4) duct pressures from −20 in. of water to +20 in. of water relative to the ambient pressure. These individual variations in temperature, elevation, and duct pressure result in duct losses within ±5% of the standard air friction chart.

For duct materials other than those categorized as medium smooth in Table 1, and for variations in temperature, barometric pressure (elevation), and duct pressures (outside the range listed), calculate the friction loss in a duct by the Altshul-Tsal and Darcy equations [Equations (21) and (19), respectively].

Noncircular Ducts

A momentum analysis can relate average wall shear stress to pressure drop per unit length for fully developed turbulent flow in a passage of arbitrary shape but uniform longitudinal cross-sectional area. This analysis leads to the definition of **hydraulic diameter**:

$$D_h = 4A/P \tag{24}$$

where

D_h = hydraulic diameter, in.
A = duct area, in^2
P = perimeter of cross section, in.

While the hydraulic diameter is often used to correlate noncircular data, exact solutions for laminar flow in noncircular passages show that such practice causes some inconsistencies. No exact solutions exist for turbulent flow. Tests over a limited range of turbulent flow indicated that fluid resistance is the same for equal lengths of duct for equal mean velocities of flow if the ducts have the same ratio of cross-sectional area to perimeter. From a series of experiments using round, square, and rectangular ducts having essentially the same hydraulic diameter, Huebscher (1948) found that each, for most purposes, had the same flow resistance at equal mean velocities. Tests by Griggs and Khodabakhsh-Sharifabad (1992) also indicated that experimental rectangular duct data for airflow over the range typical of HVAC systems can be correlated satisfactorily using Equation (20) together with hydraulic diameter, particularly when a realistic experimental uncertainty is accepted. These tests support using hydraulic diameter to correlate noncircular duct data.

Rectangular Ducts. Huebscher (1948) developed the relationship between rectangular and round ducts that is used to determine size equivalency based on equal flow, resistance, and length. This relationship, Equation (25), is the basis for Table 2.

$$D_e = \frac{1.30(ab)^{0.625}}{(a+b)^{0.250}} \tag{25}$$

where

D_e = circular equivalent of rectangular duct for equal length, fluid resistance, and airflow, in.
a = length one side of duct, in.
b = length adjacent side of duct, in.

To determine equivalent round duct diameter, use Table 2. Equations (21) or (20) and (19) must be used to determine pressure loss.

Flat Oval Ducts. To convert round ducts to spiral flat oval sizes, use Table 3. Table 3 is based on Equation (26) (Heyt and Diaz 1975), the circular equivalent of a flat oval duct for equal airflow, resistance, and length. Equations (21) or (20) and (19) must be used to determine friction loss.

$$D_e = \frac{1.55AR^{0.625}}{P^{0.250}} \tag{26}$$

where AR is the cross-sectional area of flat oval duct defined as

$$AR = (\pi a^2/4) + a(A-a) \tag{27}$$

and the perimeter P is calculated by

$$P = \pi a + 2(A-a) \tag{28}$$

where

P = perimeter of flat oval duct, in.
A = major axis of flat oval duct, in.
a = minor axis of flat oval duct, in.

DYNAMIC LOSSES

Dynamic losses result from flow disturbances caused by duct-mounted equipment and fittings that change the airflow path's direction and/or area. These fittings include entries, exits, elbows, transitions, and junctions. Idelchik et al. (1994) discuss parameters affecting fluid resistance of fittings and presents local loss coefficients in three forms: tables, curves, and equations.

Local Loss Coefficients

The dimensionless coefficient C is used for fluid resistance, because this coefficient has the same value in dynamically similar streams (i.e., streams with geometrically similar stretches, equal Reynolds numbers, and equal values of other criteria necessary for dynamic similarity). The fluid resistance coefficient represents the ratio of total pressure loss to velocity pressure at the referenced cross section:

$$C = \frac{\Delta p_j}{\rho(V/1097)^2} = \frac{\Delta p_j}{p_v} \tag{29}$$

where

C = local loss coefficient, dimensionless
Δp_j = total pressure loss, in. of water
ρ = density, lb$_m$/ft^3
V = velocity, fpm
p_v = velocity pressure, in. of water

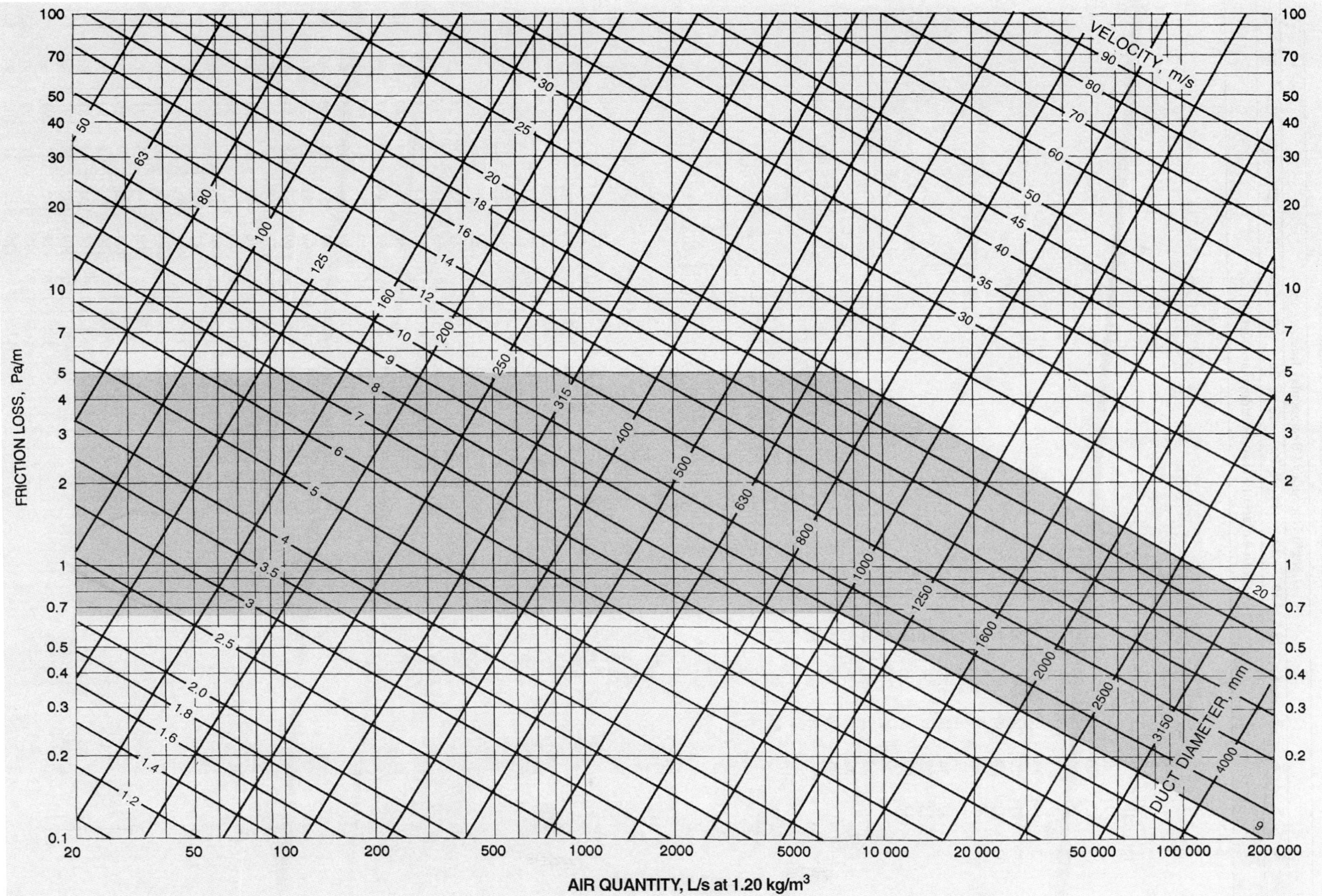

Fig. 9 Friction Chart for Round Duct ($\rho = 1.20\ kg/m^3$ and $\varepsilon = 0.09$ mm)

Table 2 Equivalent Rectangular Duct Dimensions

Circular Duct Diameter, in.	Length One Side of Rectangular Duct (*a*), in.																			
	4	5	6	7	8	9	10	12	14	16	18	20	22	24	26	28	30	32	34	36
	Length Adjacent Side of Rectangular Duct (*b*), in.																			
5	5																			
5.5	6	5																		
6	8	6																		
6.5	9	7	6																	
7	11	8	7																	
7.5	13	10	8	7																
8	15	11	9	8																
8.5	17	13	10	9																
9	20	15	12	10	8															
9.5	22	17	13	11	9															
10	25	19	15	12	10	9														
10.5	29	21	16	14	12	10														
11	32	23	18	15	13	11	10													
11.5		26	20	17	14	12	11													
12		29	22	18	15	13	12													
12.5		32	24	20	17	15	13													
13		35	27	22	18	16	14	12												
13.5		38	29	24	20	17	15	13												
14			32	26	22	19	17	14												
14.5			35	28	24	20	18	15												
15			38	30	25	22	19	16	14											
16			45	36	30	25	22	18	15											
17				41	34	29	25	20	17	16										
18				47	39	33	29	23	19	17										
19				54	44	38	33	26	22	19	18									
20					50	43	37	29	24	21	19									
21					57	48	41	33	27	23	20									
22					64	54	46	36	30	26	23	20								
23						60	51	40	33	28	25	22								
24						66	57	44	36	31	27	24	22							
25							63	49	40	34	29	26	24							
26							69	54	44	37	32	28	26	24						
27							76	59	48	40	35	31	28	25						
28								64	52	43	38	33	30	27	26					
29								70	56	47	41	36	32	29	27					
30								76	61	51	44	39	35	31	29	28				
31								82	66	55	47	41	37	34	31	29				
32								89	71	59	51	44	40	36	33	31				
33								96	76	64	54	48	42	38	35	33	30			
34									82	68	58	51	45	41	37	35	32			
35									88	73	62	54	48	44	40	37	34	32		
36									95	78	67	58	51	46	42	39	36	34		
37									101	83	71	62	55	49	45	41	38	36	34	
38									108	89	76	66	58	52	47	44	40	38	36	
39										95	80	70	62	55	50	46	43	40	37	36
40										101	85	74	65	58	53	49	45	42	39	37
41										107	91	78	69	62	56	51	47	44	41	39
42										114	96	83	73	65	59	54	50	46	44	41
43										120	102	88	77	69	62	57	53	49	46	43
44											107	93	81	73	66	60	55	51	48	45
45											113	98	86	76	69	63	58	54	50	47
46											120	103	90	80	72	66	61	56	53	49
47											126	108	95	84	76	69	64	59	55	52
48											133	114	100	89	80	73	67	62	58	54
49											140	120	105	93	84	76	70	65	60	56
50											147	126	110	98	88	80	73	68	63	59
51												132	115	102	92	83	76	71	66	61
52												139	121	107	96	87	80	74	69	64
53												145	127	112	100	91	83	77	71	67
54												152	133	117	105	95	87	80	74	70
55													139	123	110	99	91	84	78	72
56													145	128	114	104	95	87	81	75
57													151	134	119	108	98	91	84	78
58													158	139	124	112	102	94	87	81
59													165	145	130	117	107	98	91	85
60													172	151	135	122	111	102	94	88

Table 3 Equivalent Spiral Flat Oval Duct Dimensions

Circular Duct Diameter, in.	Minor Axis (a), in.											
	3	4	5	6	7	8	9	10	11	12	14	16
	Major Axis (A), in.											
5	8											
5.5	9	7										
6	11	9										
6.5	12	10	8									
7	15	12	10	8								
7.5	19	13	—	9								
8	22	15	11	—								
8.5		18	13	11	10							
9		20	14	12	—	10						
9.5		21	18	14	12	—						
10			19	15	13	11						
10.5			21	17	15	13	12					
11				19	16	14	—	12				
11.5				20	18	16	14	—				
12				23	20	17	15	13				
12.5				25	21	—	—	15	14			
13				28	23	19	17	16	—	14		
13.5				30	—	21	18	—	16	—		
14				33	—	22	20	18	17	15		
14.5				36	—	24	22	19	—	17		
15				39	—	27	23	21	19	18		
16				45	—	30	—	24	22	20	17	
17				52	—	35	—	27	24	21	19	
18				59	—	39	—	30	—	25	22	19

Circular Duct Diameter, in.	Minor Axis (a), in.										
	8	9	10	11	12	14	16	18	20	22	24
	Major Axis (A), in.										
19	46	—	34	—	28	23	21				
20	50	—	38	—	31	27	24	21			
21	58	—	43	—	34	28	25	23			
22	65	—	48	—	37	31	29	26			
23	71	—	52	—	42	34	30	27			
24	77	—	57	—	45	38	33	29	26		
25			63	—	50	41	36	32	29		
26			70	—	56	45	38	34	31		
27			76	—	59	49	41	37	34		
28					65	52	46	40	36		
29					72	58	49	43	39	35	
30					78	61	54	46	40	38	
31					81	67	57	49	44	39	37
32						71	60	53	47	42	40
33						77	66	56	51	46	41
34							69	59	55	47	44
35							76	65	58	50	46
36							79	68	61	53	49
37								71	64	57	52
38								78	67	60	55
40									77	69	62
42										75	68
44										82	74

Dynamic losses occur along a duct length and cannot be separated from friction losses. For ease of calculation, dynamic losses are assumed to be concentrated at a section (local) and to exclude friction. Frictional losses must be considered only for relatively long fittings. Generally, fitting friction losses are accounted for by measuring duct lengths from the centerline of one fitting to that of the next fitting. For fittings closely coupled (less than six hydraulic diameters apart), the flow pattern entering subsequent fittings differs from the flow pattern used to determine loss coefficients. Adequate data for these situations are unavailable.

For all fittings, except junctions, calculate the total pressure loss Δp_j at a section by

$$\Delta p_j = C_o p_{v,o} \tag{30}$$

where the subscript o is the cross section at which the velocity pressure is referenced. The dynamic loss is based on the actual velocity in the duct, not the velocity in an equivalent noncircular duct. For the cross section to reference a fitting loss coefficient, refer to Step 4 in the section on HVAC Duct Design Procedures. Where necessary (unequal area fittings), convert a loss coefficient from section o to section i using Equation (31), where V is the velocity at the respective sections.

$$C_i = \frac{C_o}{(V_i/V_o)^2} \tag{31}$$

For converging and diverging flow junctions, total pressure losses through the straight (main) section are calculated as

$$\Delta p_j = C_{c,s} p_{v,c} \tag{32}$$

For total pressure losses through the branch section,

$$\Delta p_j = C_{c,b} p_{v,c} \tag{33}$$

where $p_{v,c}$ is the velocity pressure at the common section c, and $C_{c,s}$ and $C_{c,b}$ are losses for the straight (main) and branch flow paths, respectively, each referenced to the velocity pressure at section c. To convert junction local loss coefficients referenced to straight and branch velocity pressures, use the following equation:

$$C_i = \frac{C_{c,i}}{(V_i/V_c)^2} \tag{34}$$

where

C_i = local loss coefficient referenced to section being calculated (see subscripts), dimensionless
$C_{c,i}$ = straight ($C_{c,s}$) or branch ($C_{c,b}$) local loss coefficient referenced to dynamic pressure at common section, dimensionless
V_i = velocity at section to which C_i is being referenced, fpm
V_c = velocity at common section, fpm

Subscripts:

b = branch
s = straight (main) section
c = common section

The junction of two parallel streams moving at different velocities is characterized by turbulent mixing of the streams, accompanied by pressure losses. In the course of this mixing, an exchange of momentum takes place between the particles moving at different velocities, finally resulting in the equalization of the velocity distributions in the common stream. The jet with higher velocity loses a part of its kinetic energy by transmitting it to the slower moving jet. The loss in total pressure before and after mixing is always large and positive for the higher velocity jet and increases with an increase in the amount of energy transmitted to the lower velocity jet. Consequently, the local loss coefficient, defined by Equation (29), will always be positive. The energy stored in the lower velocity jet increases as a result of mixing. The loss in total pressure and the local loss coefficient can, therefore, also have negative values for the lower velocity jet (Idelchik et al. 1994).

Table 4 Duct Fitting Codes

Fitting Function	Geometry	Category	Sequential Number
S: **S**upply	D: round (**D**iameter)	1. Entries 2. Exits	1,2,3...n
E: **E**xhaust/Return	R: **R**ectangular	3. Elbows 4. Transitions	
C: **C**ommon (supply and return)	F: **F**lat oval	5. Junctions 6. Obstructions 7. Fan and system interactions 8. Duct-mounted equipment 9. Dampers 10. Hoods	

Duct Fitting Database

A duct fitting database, developed by ASHRAE (1994), which includes 228 round and rectangular fittings with the provision to include flat oval fittings, is available from ASHRAE in electronic form with the capability to be linked to duct design programs.

The fittings are numbered (coded) as shown in Table 4. Entries and converging junctions are only in the exhaust/return portion of systems. Exits and diverging junctions are only in supply systems. Equal-area elbows, obstructions, and duct-mounted equipment are common to both supply and exhaust systems. Transitions and unequal-area elbows can be either supply or exhaust fittings. Fitting ED5-1 (see the section on Fitting Loss Coefficients) is an **E**xhaust fitting with a round shape (**D**iameter). The number 5 indicates that the fitting is a junction, and 1 is its sequential number. Fittings SR3-1 and ER3-1 are **S**upply and **E**xhaust fittings, respectively. The R indicates that the fitting is **R**ectangular, and the 3 identifies the fitting as an elbow. Note that the cross-sectional areas at sections 0 and 1 are not equal (see the section on Fitting Loss Coefficients). Otherwise, the elbow would be a **C**ommon fitting such as CR3-6. Additional fittings are reproduced in the section on Fitting Loss Coefficients to support the example design problems (see Table 12 for Example 8; see Table 14 for Example 9).

DUCTWORK SECTIONAL LOSSES

Darcy-Weisbach Equation

Total pressure loss in a duct section is calculated by combining Equations (19) and (29) in terms of Δp, where ΣC is the summation of local loss coefficients within the duct section. Each fitting loss coefficient must be referenced to that section's velocity pressure.

$$\Delta p = \left(\frac{12fL}{D_h} + \Sigma C\right)\rho\left(\frac{V}{1097}\right)^2 \qquad (35)$$

FAN-SYSTEM INTERFACE

Fan Inlet and Outlet Conditions

Fan performance data measured in the field may show lower performance capacity than manufacturers' ratings. The most common causes of deficient performance of the fan/system combination are improper outlet connections, nonuniform inlet flow, and swirl at the fan inlet. These conditions alter the aerodynamic characteristics of the fan so that its full flow potential is not realized. One bad connection can reduce fan performance far below its rating. No data have been published that account for the effects of fan inlet and outlet flexible vibration connectors.

Normally, a fan is tested with open inlets and a section of straight duct attached to the outlet (ASHRAE *Standard* 51). This setup results in uniform flow into the fan and efficient static pressure recovery on the fan outlet. If good inlet and outlet conditions are not provided in the actual installation, the performance of the fan suffers. To select and apply the fan properly, these effects must be considered, and the pressure requirements of the fan, as calculated by standard duct design procedures, must be increased.

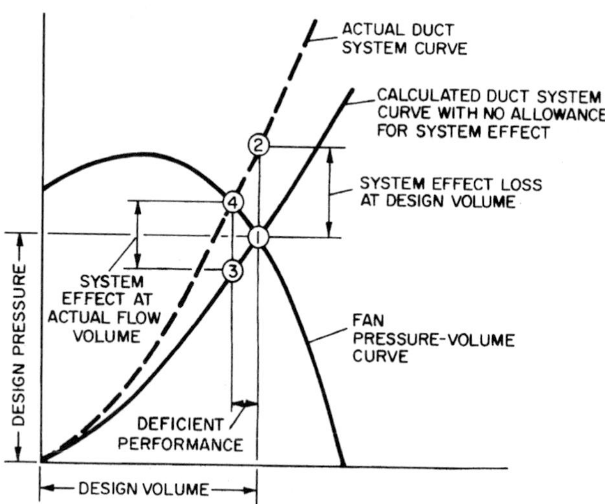

Fig. 10 Deficient System Performance with System Effect Ignored

Figure 10 illustrates deficient fan/system performance. The system pressure losses have been determined accurately, and a fan has been selected for operation at Point 1. However, no allowance has been made for the effect of system connections to the fan on fan performance. To compensate, a fan system effect must be added to the calculated system pressure losses to determine the actual system curve. The point of intersection between the fan performance curve and the actual system curve is Point 4. The actual flow volume is, therefore, deficient by the difference from 1 to 4. To achieve design flow volume, a fan system effect pressure loss equal to the pressure difference between Points 1 and 2 should be added to the calculated system pressure losses, and the fan should be selected to operate at Point 2.

Fan System Effect Coefficients

The system effect concept was formulated by Farquhar (1973) and Meyer (1973); the magnitudes of the system effect, called **system effect factors**, were determined experimentally in the laboratory of the Air Movement and Control Association (AMCA) (Brown 1973, Clarke et al. 1978) and published in their *Publication* 201 (AMCA 1990a). The system effect factors, converted to local loss coefficients, are in the *Duct Fitting Database* (ASHRAE 1994) for both centrifugal and axial fans. Fan system effect coefficients are only an approximation. Fans of different types and even fans of the same type, but supplied by different manufacturers, do not necessarily react to a system in the same way. Therefore, judgment based on experience must be applied to any design.

Fan Outlet Conditions. Fans intended primarily for duct systems are usually tested with an outlet duct in place (ASHRAE *Standard* 51). Figure 11 shows the changes in velocity profiles at various distances from the fan outlet. For 100% recovery, the duct, including transition, must meet the requirements for 100% effective duct length [L_e (Figure 11)], which is calculated as follows:

For $V_o > 2500$ fpm,

$$L_e = \frac{V_o\sqrt{A_o}}{10{,}600} \qquad (36)$$

For $V_o \leq 2500$ fpm,

$$L_e = \frac{\sqrt{A_o}}{4.3} \tag{37}$$

where

V_o = duct velocity, fpm
L_e = effective duct length, ft
A_o = duct area, in^2

As illustrated by Fitting SR7-1 in the section on Fitting Loss Coefficients, centrifugal fans should not abruptly discharge to the atmosphere. A diffuser design should be selected from Fitting SR7-2 (see the section on Fitting Loss Coefficients) or SR7-3 (see ASHRAE 1994).

Fan Inlet Conditions. For rated performance, air must enter the fan uniformly over the inlet area in an axial direction without prerotation. Nonuniform flow into the inlet is the most common cause of reduced fan performance. Such inlet conditions are not equivalent to a simple increase in the system resistance; therefore, they cannot be treated as a percentage decrease in the flow and pressure from the fan. A poor inlet condition results in an entirely new fan performance. An elbow at the fan inlet, for example Fitting ED7-2 (see the section on Fitting Loss Coefficients), causes turbulence and uneven flow into the fan impeller. The losses due to the fan system effect can be eliminated by including an adequate length of straight duct between the elbow and the fan inlet.

The ideal inlet condition allows air to enter axially and uniformly without spin. A spin in the same direction as the impeller rotation reduces the pressure-volume curve by an amount dependent on the intensity of the vortex. A counterrotating vortex at the inlet slightly increases the pressure-volume curve, but the power is increased substantially.

Inlet spin may arise from a great variety of approach conditions, and sometimes the cause is not obvious. Inlet spin can be avoided by providing an adequate length of straight duct between the elbow and the fan inlet. Figure 12 illustrates some common duct connections that cause inlet spin and includes recommendations for correcting spin.

Fans within plenums and cabinets or next to walls should be located so that air may flow unobstructed into the inlets. Fan performance is reduced if the space between the fan inlet and the enclosure is too restrictive. The system effect coefficients for fans in an enclosure or adjacent to walls are listed under Fitting ED7-1 (see the section on Fitting Loss Coefficients). The manner in which the airstream enters an enclosure in relation to the fan inlets also affects fan performance. Plenum or enclosure inlets or walls that are not symmetrical with the fan inlets cause uneven flow and/or inlet spin.

Testing, Adjusting, and Balancing Considerations

Fan system effects (FSEs) are not only to be used in conjunction with the system resistance characteristics in the fan selection process, but are also applied in the calculations of the results of testing, adjusting, and balancing (TAB) field tests to allow direct comparison to design calculations and/or fan performance data. Fan inlet swirl and the effect on system performance of poor fan inlet and outlet ductwork connections cannot be measured directly. Poor inlet flow patterns affect fan performance within the impeller wheel

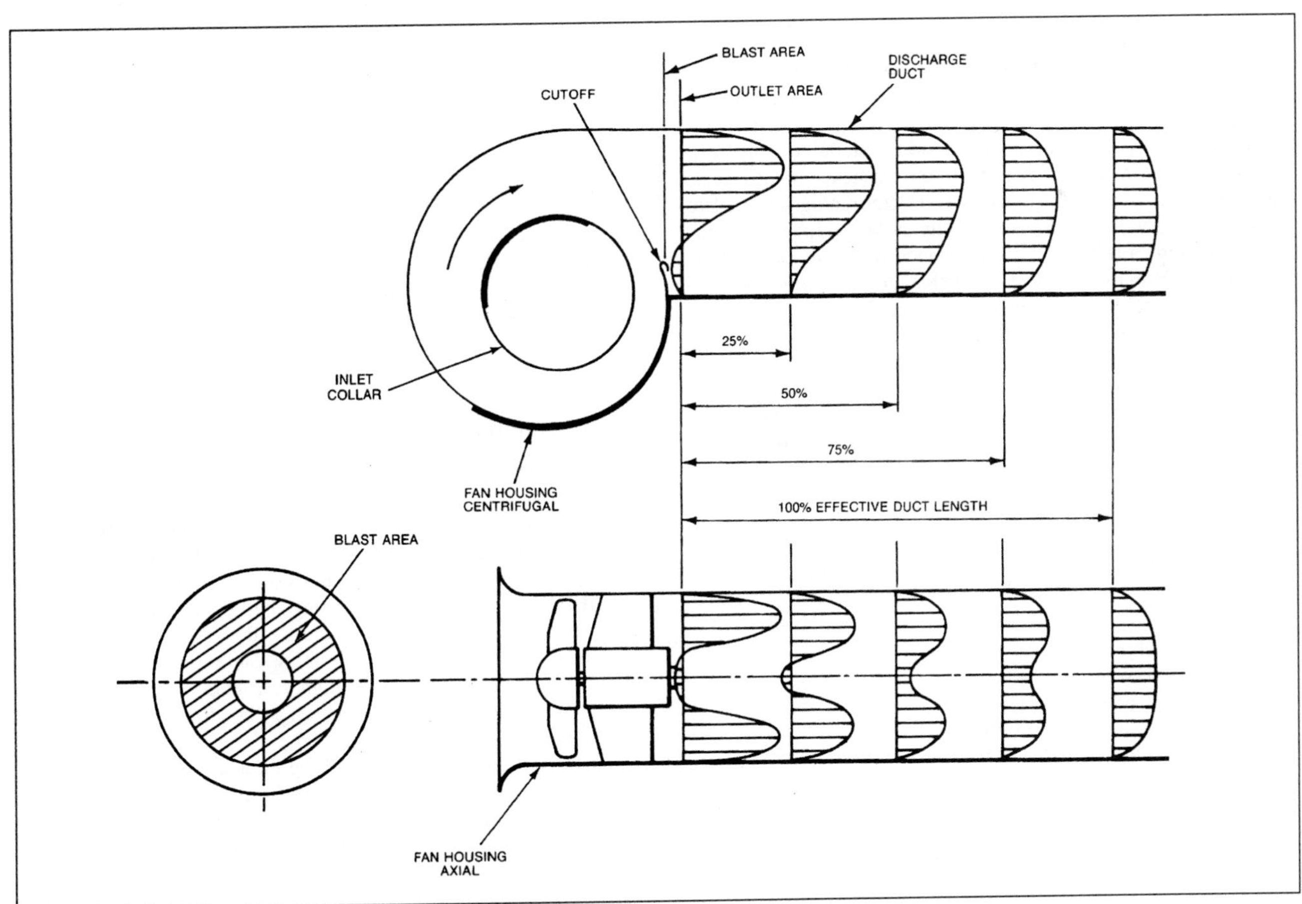

Fig. 11 Establishment of Uniform Velocity Profile in Straight Fan Outlet Duct
(Adapted by permission from AMCA *Publication* 201)

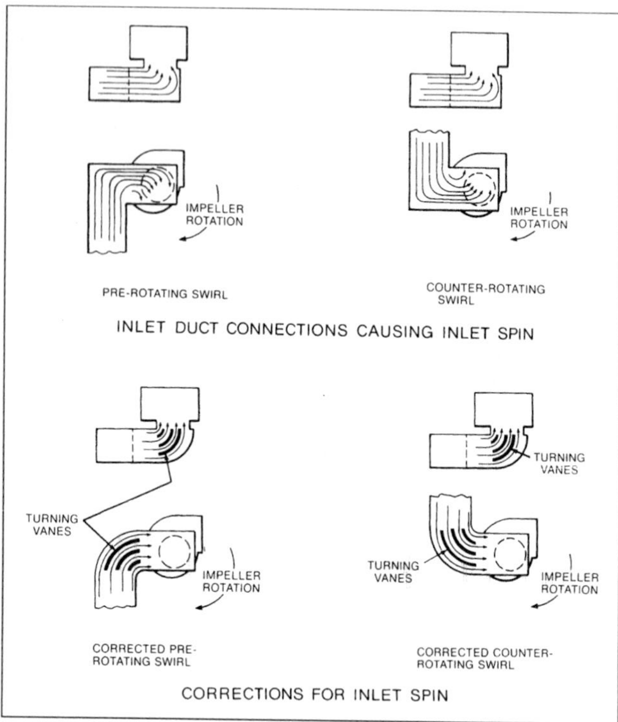

Fig. 12 Inlet Duct Connections Causing Inlet Spin and Corrections for Inlet Spin
(Adapted by permission from AMCA *Publication* 201)

(centrifugal fan) or wheel rotor impeller (axial fan), while the fan outlet system effect is flow instability and turbulence within the fan discharge ductwork.

The static pressure at the fan inlet and the static pressure at the fan outlet may be measured directly in some systems. In most cases, static pressure measurements for use in determining fan total (or static) pressure will not be made directly at the fan inlet and outlet, but at locations a relatively short distance from the fan inlet and downstream from the fan outlet. To calculate fan total pressure for this case from field measurements, use Equation (38), where $\Delta p_{x\text{-}y}$ is the summation of calculated total pressure losses between the fan inlet and outlet sections noted. Plane 3 is used to determine airflow rate. If necessary, use Equation (18) to calculate fan static pressure knowing fan total pressure. For locating measurement planes and calculation procedures, consult AMCA *Publication* 203 (AMCA 1990b).

$$P_t = (p_{s,5} + p_{v,5}) + \Delta p_{2\text{-}5} + \text{FSE}_2 + (p_{s,4} + p_{v,4}) + \Delta p_{4\text{-}1} + \text{FSE}_1 + \text{FSE}_{1,sw} \quad (38)$$

where

P_t = fan total pressure, in. of water
p_s = static pressure, in. of water
p_v = velocity pressure, in. of water
FSE = fan system effect, in. of water
$\Delta p_{x\text{-}y}$ = summarization of total pressure losses between planes x and y, in. of water

Subscripts (numerical subscripts same as used by AMCA *Publication* 203):

1 = fan inlet
2 = fan outlet
3 = plane of airflow measurement
4 = plane of static pressure measurement upstream of fan
5 = plane of static pressure measurement downstream of fan
sw = swirl

DUCT SYSTEM DESIGN

DESIGN CONSIDERATIONS

Space Pressure Relationships

Space pressure is determined by fan location and duct system arrangement. For example, a supply fan that pumps air into a space increases space pressure; an exhaust fan reduces space pressure. If both supply and exhaust fans are used, space pressure depends on the relative capacity of the fans. Space pressure is positive if supply exceeds exhaust and negative if exhaust exceeds supply (Osborne 1966). System pressure variations due to wind can be minimized or eliminated by careful selection of intake air and exhaust vent locations (Chapter 16).

Fire and Smoke Management

Because duct systems can convey smoke, hot gases, and fire from one area to another and can accelerate a fire within the system, fire protection is an essential part of air-conditioning and ventilation system design. Generally, fire safety codes require compliance with the standards of national organizations. NFPA *Standard* 90A examines fire safety requirements for (1) ducts, connectors, and appurtenances; (2) plenums and corridors; (3) air outlets, air inlets, and fresh air intakes; (4) air filters; (5) fans; (6) electric wiring and equipment; (7) air-cooling and -heating equipment; (8) building construction, including protection of penetrations; and (9) controls, including smoke control.

Fire safety codes often refer to the testing and labeling practices of nationally recognized laboratories, such as Factory Mutual and Underwriters Laboratories (UL). The *Building Materials Directory* compiled by UL lists fire and smoke dampers that have been tested and meet the requirements of UL *Standards* 555 and 555S. This directory also summarizes maximum allowable sizes for individual dampers and assemblies of these dampers. Fire dampers are 1.5 h or 3 h fire-rated. Smoke dampers are classified by (1) temperature degradation [ambient air or high temperature (250°F minimum)] and (2) leakage at 1 and 4 in. of water pressure difference (8 and 12 in. of water classification optional). Smoke dampers are tested under conditions of maximum airflow. UL's *Fire Resistance Directory* lists the fire resistance of floor/roof and ceiling assemblies with and without ceiling fire dampers.

For a more detailed presentation of fire protection, see Chapter 51 of the 1999 *ASHRAE Handbook—Applications* and the NFPA *Fire Protection Handbook*.

Duct Insulation

In all new construction (except low-rise residential buildings), air-handling ducts and plenums installed as part of an HVAC air distribution system should be thermally insulated in accordance with Section 6.2.4.2 of ASHRAE *Standard* 90.1. Duct insulation for new low-rise residential buildings should be in compliance with ASHRAE *Standard* 90.2. Existing buildings should meet the requirements of ASHRAE *Standard* 100. The insulation thicknesses in these standards are minimum values. Economic and thermal considerations may justify higher insulation levels. Additional insulation, vapor retarders, or both may be required to limit vapor transmission and condensation.

Duct heat gains or losses must be known for the calculation of supply air quantities, supply air temperatures, and coil loads (see Chapter 29 of this volume and Chapter 2 of the 2000 *ASHRAE Handbook—Systems and Equipment*). To estimate duct heat transfer and entering or leaving air temperatures, use the following equations:

$$q_l = \frac{UPL}{12}\left[\left(\frac{t_e + t_l}{2}\right) - t_a\right] \quad (39)$$

$$t_e = \frac{t_l(y+1) - 2t_a}{(y-1)} \quad (40)$$

$$t_l = \frac{t_e(y-1) + 2t_a}{(y+1)} \quad (41)$$

where

q_l = heat loss/gain through duct walls, Btu/h (negative for heat gain)
U = overall heat transfer coefficient of duct wall, Btu/h·ft²·°F
P = perimeter of bare or insulated duct, in.
L = duct length, ft
t_e = temperature of air entering duct, °F
t_l = temperature of air leaving duct, °F
t_a = temperature of air surrounding duct, °F
y = $10AV\rho c_p/UPL$ for rectangular ducts
= $2.5DV\rho c_p/UL$ for round ducts
A = cross-sectional area of duct, in²
V = average velocity, fpm
ρ = density of air, lb_m/ft^3
c_p = specific heat of air, Btu/lb_m·°F
D = diameter of duct, in.

Use Figure 13A to determine U-factors for insulated and uninsulated ducts. Lauvray (1978) has shown the effects of (1) compressing insulation wrapped externally on sheet metal ducts and (2) insulated flexible ducts with air-porous liners. For a 2 in. thick, 0.75 lb_m/ft^3 fibrous glass blanket compressed 50% during installation, the heat transfer rate increases approximately 20% (see Figure 13A). Pervious flexible duct liners also influence heat transfer significantly (see Figure 13B). At 2500 fpm, the pervious liner U-factor is 0.33 Btu/h·ft²·°F; for an impervious liner, U = 0.19 Btu/h·ft²·°F.

Example 6. A 65 ft length of 24 in. by 36 in. uninsulated sheet metal duct, freely suspended, conveys heated air through a space maintained above freezing at 40°F. Based on heat loss calculations for the heated zone, 17,200 cfm of standard air (c_p = 0.24 Btu/lb_m·°F) at a supply air temperature of 122°F is required. The duct is connected directly to the heated zone. Determine the temperature of the air entering the duct and the duct heat loss.

Solution: Calculate duct velocity using Equation (10):

$$V = \frac{(144)(17{,}200 \text{ cfm})}{(24 \text{ in.})(36 \text{ in.})} = 2900 \text{ fpm}$$

Calculate entering air temperature using Equation (40):

$$U = 0.73 \text{ Btu/h} \cdot \text{ft}^2 \cdot \text{°F (from Figure 13A)}$$

$$P = 2(24 \text{ in.} + 36 \text{ in.}) = 120 \text{ in.}$$

$$y = \frac{(10)(24 \text{ in.})(36 \text{ in.})(2900 \text{ fpm})(0.075 \text{ lb}_m/\text{ft}^3)(0.24)}{(0.73 \text{ Btu/h} \cdot \text{ft}^2 \cdot \text{°F})(120 \text{ in.})(65 \text{ ft})} = 79.2$$

$$t_e = \frac{122\text{°F}(79.2+1) - (2 \times 40\text{°F})}{(79.2-1)} = 124.1\text{°F}$$

Calculate duct heat loss using Equation (39):

$$q_l = \frac{(0.73 \text{ Btu/h} \cdot \text{ft}^2 \cdot \text{°F})(120 \text{ in.})(65 \text{ ft})}{12} \times \left[\frac{124.1\text{°F} + 122\text{°F}}{2} - 40\text{°F}\right]$$

$$= 39{,}400 \text{ Btu/h}$$

Example 7. Same as Example 6, except the duct is insulated externally with 2 in. thick fibrous glass with a density of 0.75 lb_m/ft^3. The insulation is wrapped with 0% compression.

Solution: All values except U remain the same as in Example 6. From Figure 13A, U = 0.15 Btu/h·ft²·°F at 2900 fpm. Therefore,

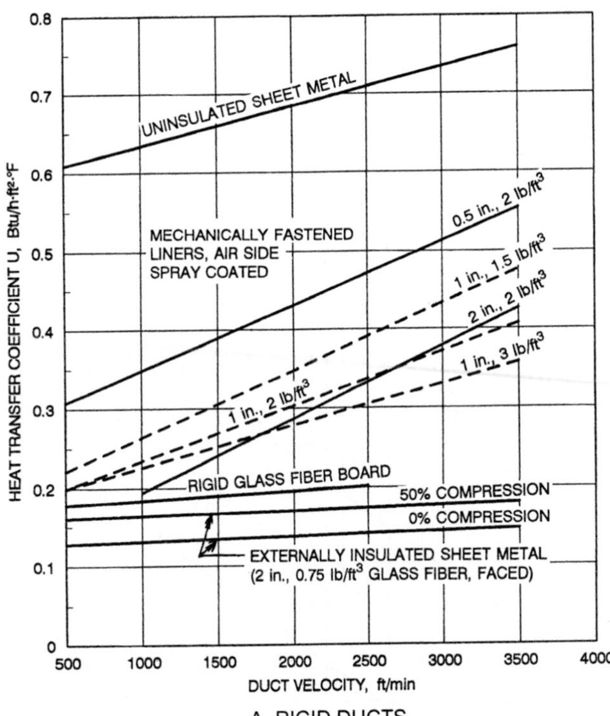

A. RIGID DUCTS

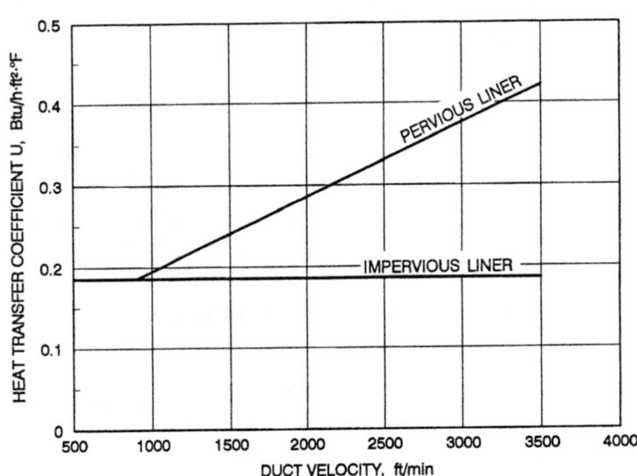

B. INSULATED FLEXIBLE DUCTS

Fig. 13 Duct Heat Transfer Coefficients

$$y = 385$$

$$t_e = 122.4\text{°F}$$

$$q_l = 8014 \text{ Btu/h}$$

Insulating this duct reduces heat loss to 20% of the uninsulated value.

Duct System Leakage

Leakage in all unsealed ducts varies considerably with the fabricating machinery used, the methods for assembly, and installation workmanship. For sealed ducts, a wide variety of sealing methods and products exists. Sealed and unsealed duct leakage tests (AISI/SMACNA 1972, ASHRAE/SMACNA/TIMA 1985, Swim and Griggs 1995) have confirmed that longitudinal seam, transverse joint, and assembled duct leakage can be represented by Equation (42) and that for the same construction, leakage is not significantly different in the negative and positive modes. A range of leakage rates

Table 5 Unsealed Longitudinal Seam Leakage, Metal Ducts

Type of Duct/Seam	Leakage, cfm per ft Seam Length[a] Range	Average
Rectangular		
Pittsburgh lock	0.01 to 0.56	0.16
Button punch snaplock	0.01 to 0.16	0.08
Round		
Snaplock	0.04 to 0.14	0.11
Grooved	0.11 to 0.18	0.12

[a]Leakage rate is at 1 in. of water static pressure.

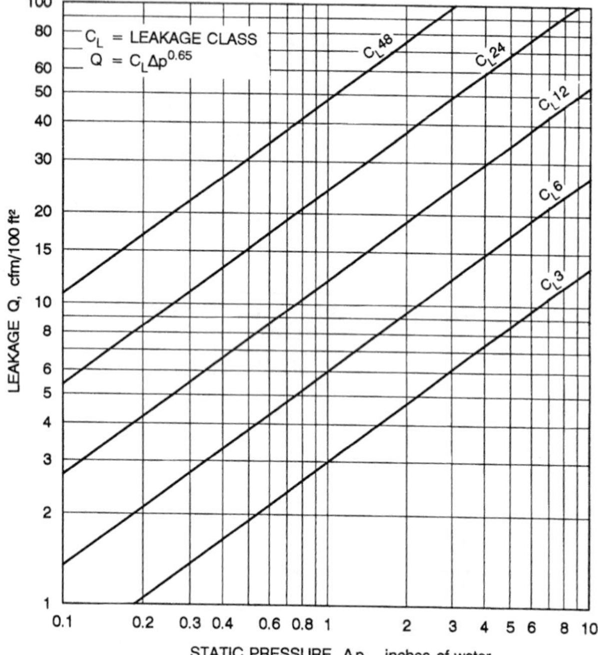

Fig. 14 Duct Leakage Classifications

for longitudinal seams commonly used in the construction of metal ducts is presented in Table 5. Longitudinal seam leakage for unsealed or unwelded metal ducts is about 10 to 15% of total duct leakage.

$$Q = C\Delta p_s^{\ N} \tag{42}$$

where

Q = duct leakage rate, cfm
C = constant reflecting area characteristics of leakage path
Δp_s = static pressure differential from duct interior to exterior, in. of water
N = exponent relating turbulent or laminar flow in leakage path

Analysis of the AISI/ASHRAE/SMACNA/TIMA data resulted in the categorization of duct systems into **leakage classes** C_L based on Equation (43), where the exponent N is assumed to be 0.65. A selected series of leakage classes based on Equation (43) is shown in Figure 14.

$$C_L = Q/\Delta p_s^{\ 0.65} \tag{43}$$

where

Q = leakage rate, cfm/100 ft^2 (surface area)
C_L = leakage class, cfm per 100 ft duct surface at 1 in. of water static pressure

Table 6 Duct Leakage Classification[a]

Duct Type	Predicted Leakage Class C_L [Eq. (43)] Sealed[b,c]	Unsealed[c]
Metal (flexible excluded)		
Round and flat oval	3	30 (6 to 70)
Rectangular		
≤ 2 in. of water (both positive and negative pressures)	12	48 (12 to 110)
> 2 and ≤ 10 in. of water (both positive and negative pressures)	6	48 (12 to 110)[c]
Flexible		
Metal, aluminum	8	30 (12 to 54)
Nonmetal	12	30 (4 to 54)
Fibrous glass		
Round	3	na
Rectangular	6	na

[a]The leakage classes listed in this table are averages based on tests conducted by AISI/SMACNA (1972), ASHRAE/SMACNA/TIMA (1985), and Swim and Griggs (1995).
[b]The leakage classes listed in the sealed category are based on the assumptions that for metal ducts, all transverse joints, seams, and openings in the duct wall are sealed at pressures over 3 in. of water, that transverse joints and longitudinal seams are sealed at 2 and 3 in. of water, and that transverse joints are sealed below 2 in. of water. Lower leakage classes are obtained by careful selection of joints and sealing methods.
[c]Leakage classes assigned anticipate about 25 joints per 100 linear feet of duct. For systems with a high fitting to straight duct ratio, greater leakage occurs in both the sealed and unsealed conditions.

Table 6 is a forecast of the leakage class attainable for commonly used duct construction and sealing practices. Connections of ducts to grilles, diffusers, and registers are not represented in the test data. Leakage classes listed are for a specific duct type, not a system with a variety of duct types, access doors, and other duct-mounted equipment. The designer is responsible for assigning acceptable system leakage rates. It is recommended that this be accomplished by using Table 7 as a guideline to specify a ductwork leakage class or by specifying a duct seal level as recommended by Table 8. The designer should take into account attainable leakage rates by duct type and the fact that casings of volume-controlling air terminal units may leak 1 to 2% of their maximum flow. The effects of such leakage should be anticipated, if allowed, and the ductwork should not be expected to compensate for equipment leakage. When a system leakage class is specified by a designer, it is a performance specification that should not be compromised by prescriptive sealing. A portion of a system may exceed its leakage class if the aggregate system leakage meets the allowable rate. Table 9 can be used to estimate the system percent leakage based on the system design leakage class and system duct surface area. Table 9 is predicated on assessment at an average of upstream and downstream pressures because use of the highest pressure alone could indicate an artificially high rate. When several duct pressure classifications occur in a system, ductwork in each pressure class should be evaluated independently to arrive at an aggregate leakage for the system.

Leakage tests should be conducted in compliance with SMACNA's *HVAC Air Duct Leakage Test Manual* (1985) to verify the intent of the designer and the workmanship of the the installing contractor. Leakage tests used to confirm leakage class should be conducted at the pressure class for which the duct is constructed. Leakage testing is also addressed in ASHRAE *Standard* 90.1.

Limited performance standards for metal duct sealants and tapes exist. For guidance in their selection and use refer to SMACNA's *HVAC Duct Construction Standards* (1995). Fibrous glass ducts and their closure systems are covered by UL *Standards* 181and 181A.

Table 7 Recommended Ductwork Leakage Class by Duct Type

Duct Type	Leakage Class, cfm/100 ft^2 at 1 in. of water
Metal (flexible excluded)	
Round	3
Flat oval	3
Rectangular	6
Flexible	6
Fibrous glass	
Round	3
Rectangular	6

Table 8A Recommended Duct Seal Levels[a]

Duct Location	Supply ≤ 2 in. of water	Supply > 2 in. of water	Exhaust	Return
Outdoors	A	A	A	A
Unconditioned spaces	B	A	B	B
Conditioned spaces (concealed ductwork)	C	B	B	C
Conditioned spaces (exposed ductwork)	A	A	B	B

[a]See Table 8B for definition of seal level.

Table 8B Duct Seal Levels

Seal Level	Sealing Requirements[a]
A	All transverse joints, longitudinal seams, and duct wall penetrations
B	All transverse joints and longitudinal seams
C	Transverse joints only

[a]Transverse joints are connections of two duct or fitting elements oriented perpendicular to flow. Longitudinal seams are joints oriented in the direction of airflow. Duct wall penetrations are openings made by screws, non-self-sealing fasteners, pipe, tubing, rods, and wire. Round and flat oval spiral lock seams need not be sealed prior to assembly, but may be coated after assembly to reduce leakage. All other connections are considered transverse joints, including but not limited to spin-ins, taps and other branch connections, access door frames, and duct connections to equipment.

For fibrous glass duct construction standards consult NAIMA (1997) and SMACNA (1992). Flexible duct performance and installation standards are covered by UL 181, UL 181B and ADC (1996). Soldered or welded duct construction is necessary where sealants are not suitable. Sealants used on exterior ducts must be resistant to weather, temperature cycles, sunlight, and ozone.

Shaft and compartment pressure changes affect duct leakage and are important to health and safety in the design and operation of contaminant and smoke control systems. Shafts should not be used for supply, return, and/or exhaust air without accounting for their leakage rates. Airflow around buildings, building component leakage, and the distribution of inside and outside pressures over the height of a building, including shafts, are discussed in Chapters 16 and 26. Smoke management system design is covered in Chapter 51 of the 1999 *ASHRAE Handbook—Applications* and in Klote and Milke (1992).

System Component Design Velocities

Table 10 summarizes face velocities for HVAC components in built-up systems. In most cases, the values are abstracted from pertinent chapters in the 2000 *ASHRAE Handbook—Systems and Equipment*; final selection of the components should be based on data in these chapters or from manufacturers.

Table 9 Leakage as Percentage of Airflow[a,b]

Leakage Class	System cfm per ft^2 Duct Surface[c]	Static Pressure, in. of water: 0.5	1	2	3	4	6
48	2	15	24	38	49	59	77
	2.5	12	19	30	39	47	62
	3	10	16	25	33	39	51
	4	7.7	12	19	25	30	38
	5	6.1	9.6	15	20	24	31
24	2	7.7	12	19	25	30	38
	2.5	6.1	9.6	15	20	24	31
	3	5.1	8.0	13	16	20	26
	4	3.8	6.0	9.4	12	15	19
	5	3.1	4.8	7.5	9.8	12	15
12	2	3.8	6	9.4	12	15	19
	2.5	3.1	4.8	7.5	9.8	12	15
	3	2.6	4.0	6.3	8.2	9.8	13
	4	1.9	3.0	4.7	6.1	7.4	9.6
	5	1.5	2.4	3.8	4.9	5.9	7.7
6	2	1.9	3	4.7	6.1	7.4	9.6
	2.5	1.5	2.4	3.8	4.9	5.9	7.7
	3	1.3	2.0	3.1	4.1	4.9	6.4
	4	1.0	1.5	2.4	3.1	3.7	4.8
	5	0.8	1.2	1.9	2.4	3.0	3.8
3	2	1.0	1.5	2.4	3.1	3.7	4.8
	2.5	0.8	1.2	1.9	2.4	3.0	3.8
	3	0.6	1.0	1.6	2.0	2.5	3.2
	4	0.5	0.8	1.3	1.6	2.0	2.6
	5	0.4	0.6	0.9	1.2	1.5	1.9

[a]Adapted with permission from HVAC *Air Duct Leakage Test Manual* (SMACNA 1985, Appendix A).
[b]Percentage applies to the airflow entering a section of duct operating at an assumed pressure equal to the average of the upstream and downstream pressures.
[c]The ratios in this column are typical of fan volumetric flow rate divided by total system surface. Portions of the systems may vary from these averages.

Louvers require special treatment since the blade shapes, angles, and spacing cause significant variations in louver-free area and performance (pressure drop and water penetration). Selection and analysis should be based on test data obtained in accordance with AMCA *Standard* 500-L (1999). This standard presents both pressure drop and water penetration test procedures and a uniform method for calculating the free area of a louver. Tests are conducted on a 48 in. square louver with the frame mounted flush in the wall. For the water penetration tests, the rainfall is 4 in/h, no wind, and the water flow down the wall is 0.25 gpm per linear foot of louver width.

Use Figure 15 for preliminary sizing of air intake and exhaust louvers. For air quantities greater than 7000 cfm per louver, the air intake gross louver openings are based on 400 fpm; for exhaust louvers, 500 fpm is used for air quantities of 5000 cfm per louver and greater. For air quantities less than these, refer to Figure 15. These criteria are presented on a per louver basis (i.e., each louver in a bank of louvers) to include each louver frame. Representative production-run louvers were used in establishing Figure 15, and all data used in that analysis are based on AMCA standard tests. For louvers larger than 16 ft^2, the free areas are greater than 45%, while for louvers less than 16 ft^2, the free areas are less than 45%. Unless specific louver data are analyzed, no louver should have a face area less than 4 ft^2. If debris collection on the screen of an intake louver is possible, or if louvers are located at grade with adjacent pedestrian traffic, louver face velocity should not exceed 100 fpm.

System and Duct Noise

The major sources of noise from air-conditioning systems are diffusers, grilles, fans, ducts, fittings, and vibrations. Chapter 46 of the 1999 *ASHRAE Handbook—Applications* discusses sound control for each of these sources. Sound control for terminal devices

Table 10 Typical Design Velocities for HVAC Components

Duct Element	Face Velocity, fpm
LOUVERS[a]	
Intake	
7000 cfm and greater	400
Less than 7000 cfm	See Figure 15
Exhaust	
5000 cfm and greater	500
Less than 5000 cfm	See Figure 15
FILTERS[b]	
Panel filters	
Viscous impingement	200 to 800
Dry-type, extended-surface	
Flat (low efficiency)	Duct velocity
Pleated media (intermediate efficiency)	Up to 750
HEPA	250
Renewable media filters	
Moving-curtain viscous impingement	500
Moving-curtain dry media	200
Electronic air cleaners	
Ionizing type	150 to 350
HEATING COILS[c]	
Steam and hot water	500 to 1000 200 min., 1500 max.
Electric	
Open wire	Refer to mfg. data
Finned tubular	Refer to mfg. data
DEHUMIDIFYING COILS[d]	400 to 500
AIR WASHERS[e]	
Spray type	300 to 600
Cell type	Refer to mfg. data
High-velocity spray type	1200 to 1800

[a]Based on assumptions presented in text.
[b]Abstracted from Chapter 24, 2000 *ASHRAE Handbook—Systems and Equipment*.
[c]Abstracted from Chapter 23, 2000 *ASHRAE Handbook—Systems and Equipment*.
[d]Abstracted from Chapter 21, 2000 *ASHRAE Handbook—Systems and Equipment*.
[e]Abstracted from Chapter 19, 2000 *ASHRAE Handbook—Systems and Equipment*.

consists of selecting devices that meet the design goal under all operating conditions and installing them properly so that no additional sound is generated. The sound power output of a fan is determined by the type of fan, airflow, and pressure. Sound control in the duct system requires proper duct layout, sizing, and provision for installing duct attenuators, if required. The noise generated by a system increases with both duct velocity and system pressure. Chapter 46 of the 1999 *ASHRAE Handbook—Applications* presents methods for calculating required sound attenuation.

Testing and Balancing

Each air duct system should be tested, adjusted, and balanced. Detailed procedures are given in Chapter 36 of the 1999 *ASHRAE Handbook—Applications*. To properly determine fan total (or static) pressure from field measurements taking into account fan system effect, refer to the section on Fan-System Interface. Equation (38) allows direct comparison of system resistance to design calculations and/or fan performance data. It is important that the system effect magnitudes be known prior to testing. If necessary, use Equation (18) to calculate fan static pressure knowing fan total pressure [Equation (38)]. For TAB calculation procedures of numerous fan/system configurations encountered in the field, refer to AMCA *Publication* 203 (AMCA 1990b).

DUCT DESIGN METHODS

Duct design methods for HVAC systems and for exhaust systems conveying vapors, gases, and smoke are the equal friction method, the static regain method, and the T-method. The section on Industrial Exhaust System Duct Design presents the design criteria and procedures for exhaust systems conveying particulates. Equal friction and static regain are nonoptimizing methods, while the T-method is a practical optimization method introduced by Tsal et al. (1988).

To ensure that system designs are acoustically acceptable, noise generation should be analyzed and sound attenuators and/or acoustically lined duct provided where necessary. Dampers must be installed throughout systems designed by equal friction, static regain, and the T-method because inaccuracies are introduced into these design methods by duct size round-off and the effect of close-coupled fittings on the total pressure loss calculations.

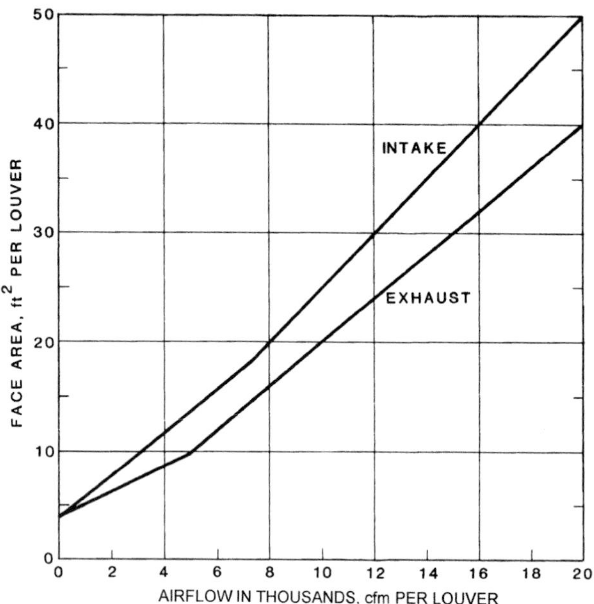

Parameters Used to Establish Figure	Intake Louver	Exhaust Louver
Minimum free area (48 in. square test section), %	45	45
Water penetration, oz/(ft²·0.25 h)	Negligible (less than 0.2)	na
Maximum static pressure drop, in. of water	0.15	0.25

Fig. 15 Criteria for Louver Sizing

Equal Friction Method

In the equal friction method, ducts are sized for a constant pressure loss per unit length. The shaded area of the friction chart (Figure 9) is the suggested range of friction rate and air velocity. When energy cost is high and installed ductwork cost is low, a low friction rate design is more economical. For low energy cost and high duct cost, a higher friction rate is more economical. After initial sizing, calculate the total pressure loss for all duct sections, and then resize sections to balance pressure losses at each junction.

Static Regain Method

The objective of the static regain method is to obtain the same static pressure at diverging flow junctions by changing downstream duct sizes. This design objective can be developed by rearranging Equation (7a) and setting $p_{s,2}$ equal to $p_{s,1}$ (neglecting thermal gravity effect term). Thus,

$$p_{s,1} - p_{s,2} = \Delta p_{t,1\text{-}2} - \left[\frac{\rho V_1^2}{2g_c} - \frac{\rho V_2^2}{2g_c}\right] \quad (44)$$

and

$$\Delta p_{t,1\text{-}2} = \frac{\rho V_1^2}{2g_c} - \frac{\rho V_2^2}{2g_c} \quad (45)$$

where $\Delta p_{t,1\text{-}2}$ is the total pressure loss from upstream of junction 1 to upstream of junction 2, or the terminal of section 2. The immediate downstream duct size that satisfies Equation (45) is determined by iteration.

To start the design of a system, a maximum velocity is selected for the root section (duct section upstream and/or downstream of a fan). In Figure 17, section 6 is the root for the return air subsystem. Section 19 is the root for the supply air subsystem. The shaded area on the friction chart (Figure 9) is the suggested range of air velocity. When energy cost is high and installed ductwork cost is low, a lower initial velocity is more economical. For low energy cost and high duct cost, a higher velocity is more economical. All other sections, except terminal sections, are sized iteratively by Equation (45). In Figure 17, terminal sections are 1, 2, 4, 7, 8, 11, 12, 15, and 16. Knowing the terminal static pressure requirements, Equation (45) is used to calculate the duct size of terminal sections. If the terminal is an exit fitting rather than a register, diffuser, or terminal box, the static pressure at the exit of the terminal section is zero.

The classical static regain method (Carrier Corporation 1960, Chun-Lun 1983) is based on Equation (46), where R is the static pressure regain factor, and Δp_r is the static pressure regain between junctions.

$$\Delta p_r = R\left(\frac{\rho V_1^2}{2g_c} - \frac{\rho V_2^2}{2g_c}\right) \quad (46)$$

Typically R-values ranging from 0.5 to 0.95 have been used. Tsal and Behls (1988) show that this uncertainty exists because the splitting of mass at junctions and the dynamic (fitting) losses between junctions are ignored. The classical static regain method using an R-value should not be used because R is not predictable.

T-Method Optimization

T-method optimization (Tsal et al. 1988) is a dynamic programming procedure based on the tee-staging idea used by Bellman (1957), except that phase level vector tracing is eliminated by optimizing locally at each stage. This modification reduces the number of calculations but requires iteration.

Optimization Basis. The objective function, Equation (47), includes both initial system cost and the present worth of energy. Hours of operation, annual escalation and interest rates, and amortization period are also required for optimization.

$$E = E_p(\text{PWEF}) + E_s \quad (47)$$

where

E = present worth owning and operating cost
E_p = first year energy cost
E_s = initial cost
PWEF = present worth escalation factor (Smith 1968), dimensionless

Energy cost is determined by

$$E_p = Q_f\left[\frac{1.176 \times 10^4(E_d + E_c T)}{\eta_f \eta_e}\right] P_t \quad (48)$$

where

Q_f = fan airflow rate, cfm
E_c = unit energy cost, cost/kWh
E_d = energy demand cost, cost/kW
T = system operating time, h/year
P_t = fan total pressure, in. of water
η_f = fan total efficiency, decimal
η_e = motor-drive efficiency, decimal

Energy cost depends on both applicable energy rates E_c and demand cost E_d. Since the difference in fan pressure between an optimized and a nonoptimized system is a small part of demand, it is usually neglected. Initial cost includes ducts and HVAC equipment, which is primarily the central handling unit. The cost of duct systems is given by the following equations:

Round $$E_s = S_d \pi DL/12 \quad (49)$$

Rectangular $$E_s = 2S_d(H + W)L/12 \quad (50)$$

where

S_d = unit ductwork cost/ft² (including material and labor)
H = duct height, in.
W = duct width, in.
L = duct length, ft

The cost of space required by ducts and equipment is another important factor of duct optimization. Including this cost reduces the size of ducts, thereby increasing energy consumption. Because the space available for ductwork is usually not used for anything else, its cost is ignored.

Both electrical energy rates and ductwork costs vary widely, by a factor of up to eight times for industrial users (DOE). Black iron rectangular ductwork can cost about 3.9 times that of spiral ductwork (Wendes 1989). Combining these ratios yields a factor of 30 to 1 based on locale and type of ductwork. Therefore, a great potential exists for reducing duct system life-cycle cost due to energy and ductwork cost variations.

The following constraints are necessary for duct optimization (Tsal and Adler 1987):

- *Continuity.* For each node, the flow in equals the flow out.
- *Pressure balancing.* The total pressure loss in each path must equal the fan total pressure; or, in effect, at any junction, the total pressure loss for all paths is the same.
- *Nominal duct size.* Ducts are constructed in discrete, nominal sizes. Each diameter of a round duct or height and width of a rectangular duct is rounded to the nearest increment, usually 1 or 2 in. If a lower nominal size is selected, the initial cost decreases, but the pressure loss increases and may exceed the fan pressure. If the higher nominal size is selected, the opposite is true—the initial cost increases, but the section pressure loss decreases. However, this lower pressure at one section may allow smaller ducts to be selected for sections that follow. Therefore, optimization must consider size rounding.
- *Air velocity restriction.* The maximum allowable velocity is an acoustic limitation (ductwork regenerated noise).
- *Construction restriction.* Architectural limits may restrict duct sizes. If air velocity or construction constraints are violated during an iteration, a duct size must be calculated. The pressure loss calculated for this preselected duct size is considered a fixed loss.

Calculation Procedure. The T-method comprises the following major procedures:

- *System condensing*. This procedure condenses a multiple-section duct system into a single imaginary duct section with identical hydraulic characteristics and the same owning cost as the entire system. By Equation (1.41) in Tsal et al. (1988), two or more converging or diverging sections and the common section at a junction can be replaced by one condensed section. By applying this equation from junction to junction in the direction to the root section (fan), the entire supply and return systems can be condensed into one section (a single resistance).
- *Fan selection*. From the condensed system, the ideal optimum fan total pressure P_t^{opt} is calculated and used to select a fan. If a fan with a different pressure is selected, its pressure P^{opt} is considered optimum.
- *System expansion*. The expansion process distributes the available fan pressure P^{opt} throughout the system. Unlike the condensing procedure, the expansion procedure starts at the root section and continues in the direction of the terminals.

Economic Analysis. Tsal et al. (1988) describe the calculation procedure and include an economic analysis of the T-method.

T-Method Simulation

T-method simulation, also developed by Tsal et al. (1990), determines the flow in each duct section of an existing system with a known operating fan performance curve. The simulation version of the T-method converges very efficiently. Usually three iterations are sufficient to obtain a solution with a high degree of accuracy.

Calculation procedure. The simulation version of the T-method includes the following major procedures:

- *System condensing*. This procedure condenses a branched tee system into a single imaginary duct section with identical hydraulic characteristics. Two or more converging or diverging sections and the common section at a junction can be replaced by one condensed section [by Equation (18) in Tsal et al. (1990)]. By applying this equation from junction to junction in the direction to the root section (fan), the entire system, including supply and return subsystems, can be condensed into one imaginary section (a single resistance).
- *Fan operating point*. This step determines the system flow and pressure by locating the intersection of the fan performance and system curves, where the system curve is represented by the imaginary section from the last step.
- *System expansion*. Knowing system flow and pressure, the previously condensed imaginary duct section is expanded into the original system with flow distributed in accordance with the ratio of pressure losses calculated in the system condensing step.

Simulation Applications. The need for duct system simulation appears in many HVAC problems. In addition to the following concerns that can be clarified by simulation, the T-method is an excellent design tool for simulating the flow distribution within a system with various modes of operation.

- Flow distribution in a variable air volume (VAV) system due to terminal box flow diversity
- Airflow redistribution due to HVAC system additions and/or modifications
- System airflow analysis for partially occupied buildings
- Necessity to replace fans and/or motors when retrofitting an air distribution system
- Multiple-fan system operating condition when one or more fans shut down
- Pressure differences between adjacent confined spaces within a nuclear facility when a design basis accident (DBA) occurs (Farajian et al. 1992)
- Smoke management system performance during a fire, when certain fire/smoke dampers close and others remain open

HVAC DUCT DESIGN PROCEDURES

The general procedure for HVAC system duct design is as follows:

1. Study the building plans, and arrange the supply and return outlets to provide proper distribution of air within each space. Adjust calculated air quantities for duct heat gains or losses and duct leakage. Also, adjust the supply, return, and/or exhaust air quantities to meet space pressurization requirements.
2. Select outlet sizes from manufacturers' data (see Chapter 32).
3. Sketch the duct system, connecting supply outlets and return intakes with the air-handling units/air conditioners. Space allocated for supply and return ducts often dictates system layout and ductwork shape. Use round ducts whenever feasible and avoid close-coupled fittings.
4. Divide the system into sections and number each section. A duct system should be divided at all points where flow, size, or shape changes. Assign fittings to the section toward the supply and return (or exhaust) terminals. The following examples are for the fittings identified for Example 6 (Figure 16), and system section numbers assigned (Figure 17). For converging flow fitting 3, assign the straight-through flow to section 1 (toward terminal 1), and the branch to section 2 (toward terminal 4). For diverging flow fitting 24, assign the straight-through flow to section 13 (toward terminals 26 and 29) and the branch to section 10 (toward terminals 43 and 44). For transition fitting 11, assign the fitting to upstream section 4 [toward terminal 9 (intake louver)]. For fitting 20, assign the unequal area elbow to downstream section 9 (toward diffusers 43 and 44). The fan outlet diffuser, fitting 42, is assigned to section 19 (again, toward the supply duct terminals).
5. Size ducts by the selected design method. Calculate system total pressure loss; then select the fan (refer to Chapter 18 of the 2000 *ASHRAE Handbook—Systems and Equipment*).
6. Lay out the system in detail. If duct routing and fittings vary significantly from the original design, recalculate the pressure losses. Reselect the fan if necessary.
7. Resize duct sections to approximately balance pressures at each junction.
8. Analyze the design for objectionable noise levels, and specify sound attenuators as necessary. Refer to the section on System and Duct Noise.

Example 8. For the system illustrated by Figures 16 and 17, size the ductwork by the equal friction method, and pressure balance the system by changing duct sizes (use 1 in. increments). Determine the system resistance and total pressure unbalance at the junctions. The airflow quantities are actual values adjusted for heat gains or losses, and ductwork is sealed (assume no leakage), galvanized steel ducts with transverse joints on 4 ft centers (ε = 0.0003 ft). Air is at standard conditions (0.075 lb_m/ft^3 density).

Because the primary purpose of Figure 16 is to illustrate calculation procedures, its duct layout is not typical of any real duct system. The layout includes fittings from the local loss coefficient tables, with emphasis on converging and diverging tees and various types of entries and discharges. The supply system is constructed of rectangular ductwork; the return system, round ductwork.

Solution: See Figure 17 for section numbers assigned to the system. The duct sections are sized within the suggested range of friction rate shown on the friction chart (Figure 9). Tables 11 and 12 give the total pressure loss calculations and the supporting summary of loss coefficients by sections. The straight duct friction factor and pressure loss were calculated by Equations (19) and (20). The fitting loss coefficients are from the *Duct Fitting Database* (ASHRAE 1994). Loss coefficients were calculated automatically by the database program (not by manual interpolation). The pressure loss values in Table 11 for the diffusers (fittings 43 and 44), the louver (fitting 9), and the air-measuring station (fitting 46) are manufacturers' data.

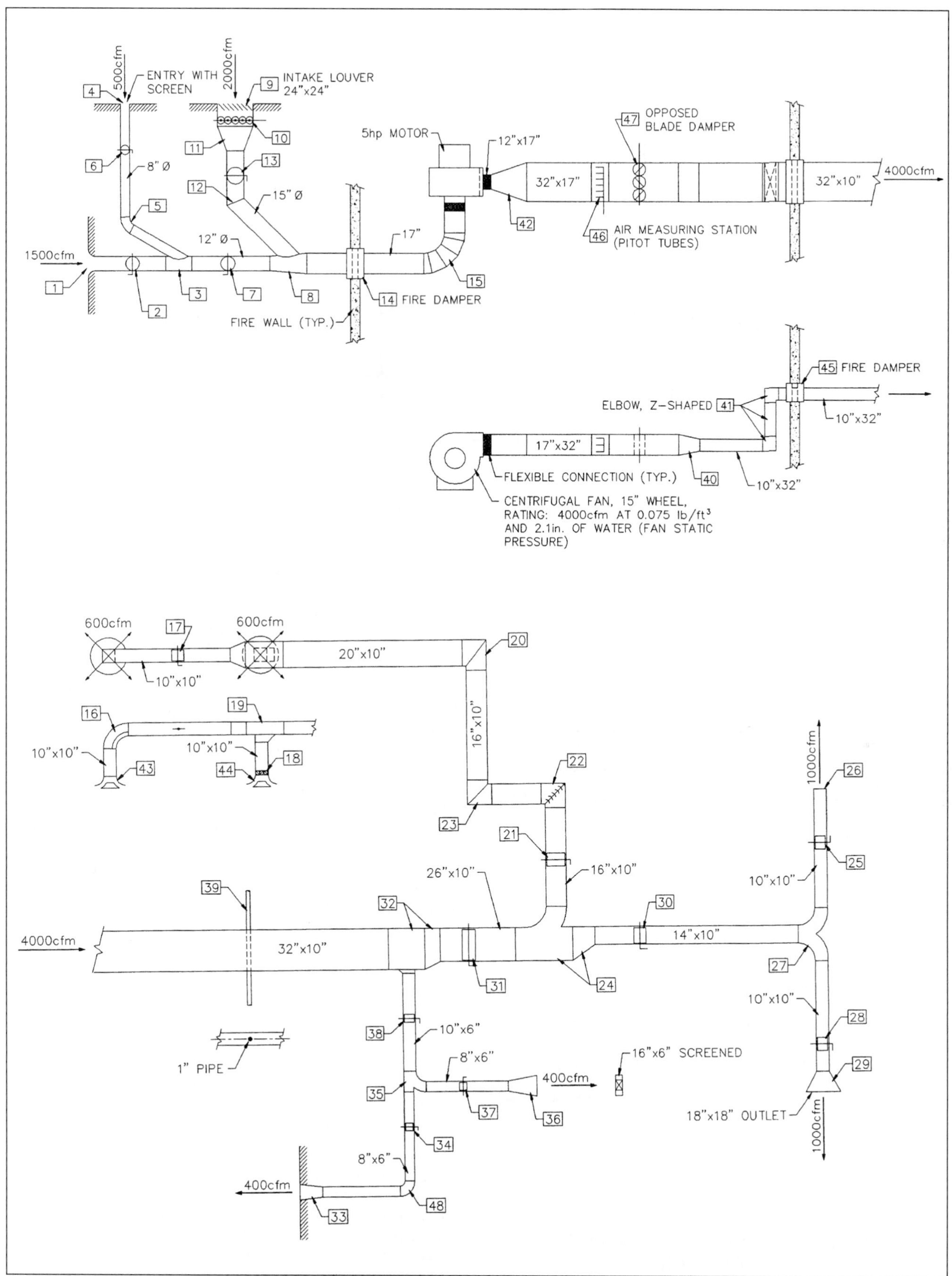

Fig. 16 Schematic for Example 8

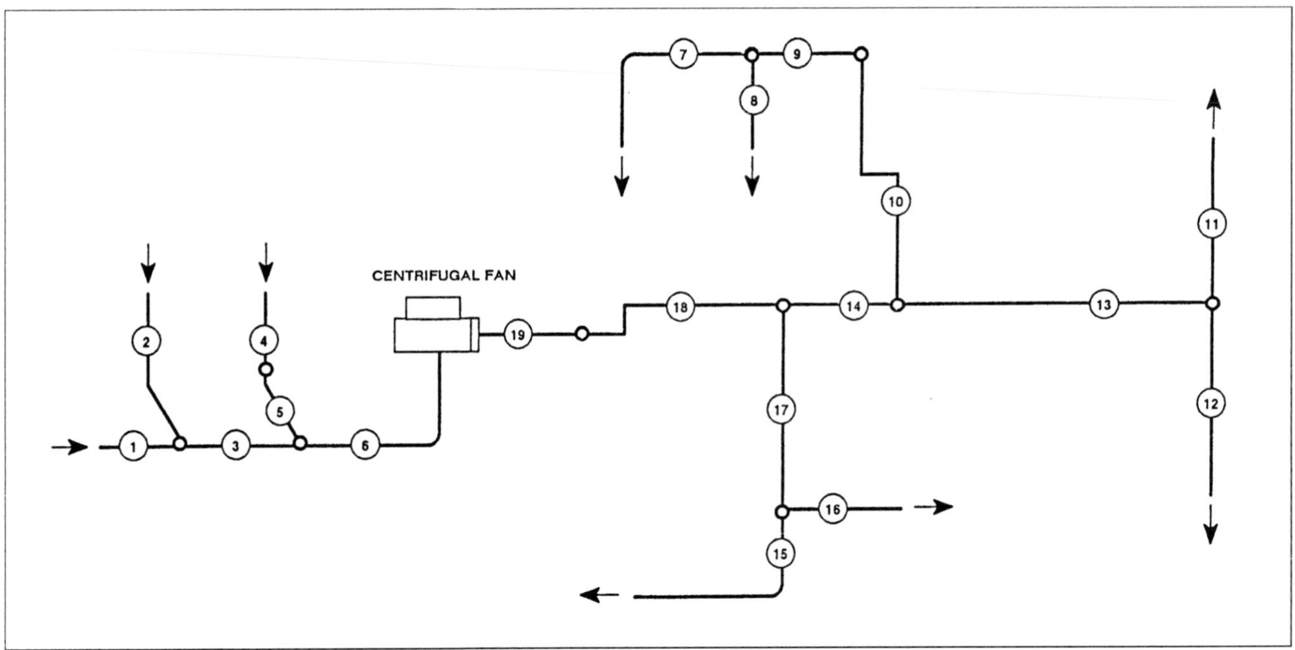

Fig. 17 System Schematic with Section Numbers for Example 8

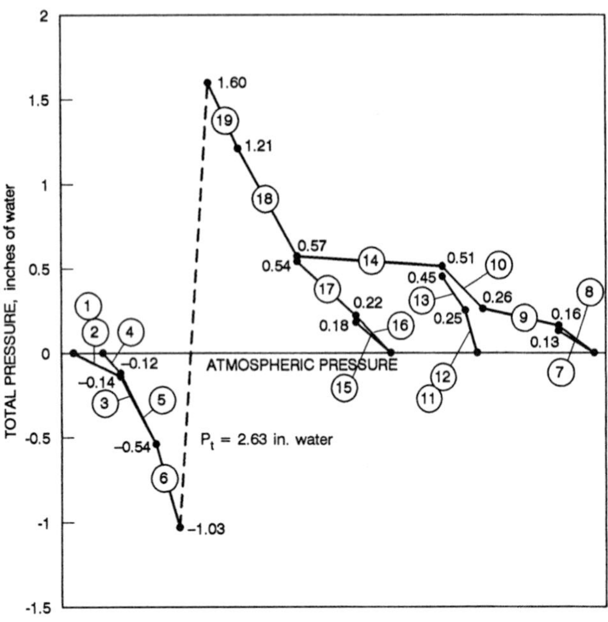

Fig. 18 Total Pressure Grade Line for Example 8

The pressure unbalance at the junctions may be noted by referring to Figure 18, the total pressure grade line for the system. The system resistance P_t is 2.63 in. of water. Noise levels and the need for duct silencers were not evaluated. To calculate the fan static pressure, use Equation (18):

$$P_s = 2.63 - 0.50 = 2.1 \text{ in. of water}$$

where 0.50 in. of water is the fan outlet velocity pressure.

INDUSTRIAL EXHAUST SYSTEM DUCT DESIGN

Chapter 29 of the 1999 *ASHRAE Handbook—Applications* discusses design criteria, including hood design, for industrial exhaust systems. Exhaust systems conveying vapors, gases, and smoke can be designed by equal friction, or T-method. Systems conveying particulates are designed by the constant velocity method at duct velocities adequate to convey particles to the system air cleaner. For contaminant transport velocities, see Table 5 in Chapter 29 of the 1999 *ASHRAE Handbook—Applications.*

Two pressure-balancing methods can be considered when designing industrial exhaust systems. One method uses balancing devices (e.g., dampers, blast gates) to obtain design airflow through each hood. The other approach balances systems by adding resistance to ductwork sections (i.e., changing duct size, selecting different fittings, and increasing airflow). This self-balancing method is preferred, especially for systems conveying abrasive materials. Where potentially explosive or radioactive materials are conveyed, the prebalanced system is mandatory because contaminants could accumulate at the balancing devices. To balance systems by increasing airflow, use Equation (51), which assumes that all ductwork has the same diameter and that fitting loss coefficients, including main and branch tee coefficients, are constant.

$$Q_c = Q_d(P_h/P_l)^{0.5} \tag{51}$$

where

Q_c = airflow rate required to increase P_l to P_h, cfm
Q_d = total airflow rate through low-resistance duct run, cfm
P_h = absolute value of pressure loss in high-resistance ductwork section(s), in. of water
P_l = absolute value of pressure loss in low-resistance ductwork section(s), in. of water

For systems conveying particulates, use elbows with a large centerline radius-to-diameter ratio (*r*/*D*), greater than 1.5 whenever possible. If *r*/*D* is 1.5 or less, abrasion in dust-handling systems can reduce the life of elbows. Elbows are often made of seven or more gores, especially in large diameters. For converging flow fittings, a 30° entry angle is recommended to minimize energy losses and abrasion in dust-handling systems. For the entry loss coefficients of hoods and equipment for specific operations, refer to Chapter 29 of the 1999 *ASHRAE Handbook—Applications* and to ACGIH (1998).

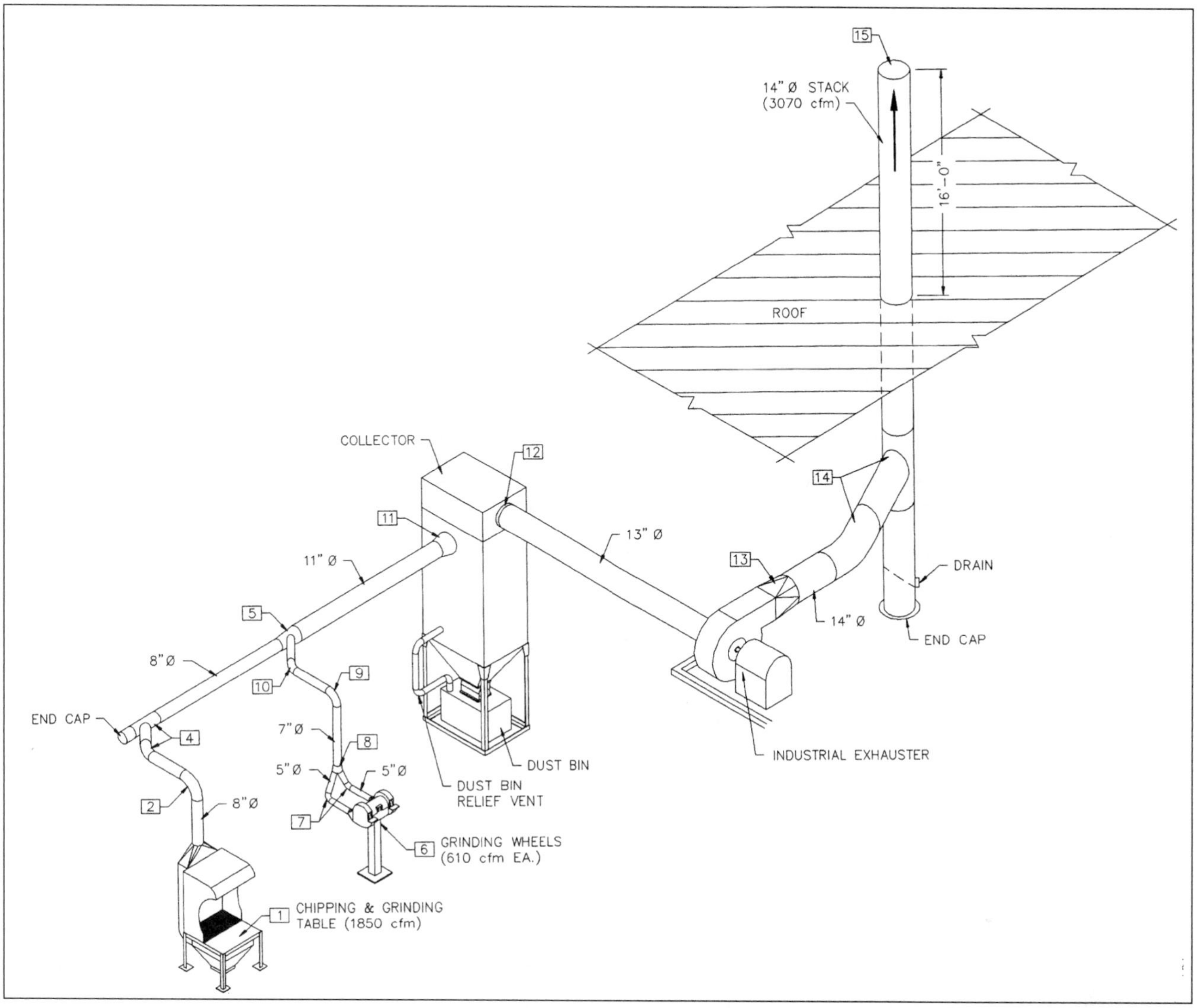

Fig. 19 Metalworking Exhaust System for Example 9

Example 9. For the metalworking exhaust system in Figures 19 and 20, size the ductwork and calculate the fan static pressure requirement for an industrial exhaust designed to convey granular materials. Pressure balance the system by changing duct sizes and adjusting airflow rates. The minimum particulate transport velocity for the chipping and grinding table ducts (sections 1 and 5, Figure 20) is 4000 fpm. For the ducts associated with the grinder wheels (sections 2, 3, 4, and 5), the minimum duct velocity is 4500 fpm. Ductwork is galvanized steel, with the absolute roughness being 0.0003 ft. Assume that air is standard and that duct and fittings are available in the following sizes: 3 in. through 9.5 in. diameters in 0.5 in. increments, 10 in. through 37 in. diameters in 1 in. diameter increments, and 38 in. through 90 in. diameters in 2 in. diameter increments.

The building is one story, and the design wind velocity is 20 mph. For the stack, use Design J shown in Figure 13 in Chapter 16 for complete rain protection. The stack height, determined by calculations from Chapter 16, is 16 ft above the roof. This stack height is based on minimized stack downwash; therefore, the stack discharge velocity must exceed 1.5 times the design wind velocity.

Solution: For the contaminated ducts upstream of the collector, initial duct sizes and transport velocities are summarized below. The 4474 fpm velocity in sections 2 and 3 is acceptable because the transport velocity is not significantly lower than 4500 fpm. For the next available duct size (4.5 in. diameter), the duct velocity is 5523 fpm, significantly higher than 4500 fpm.

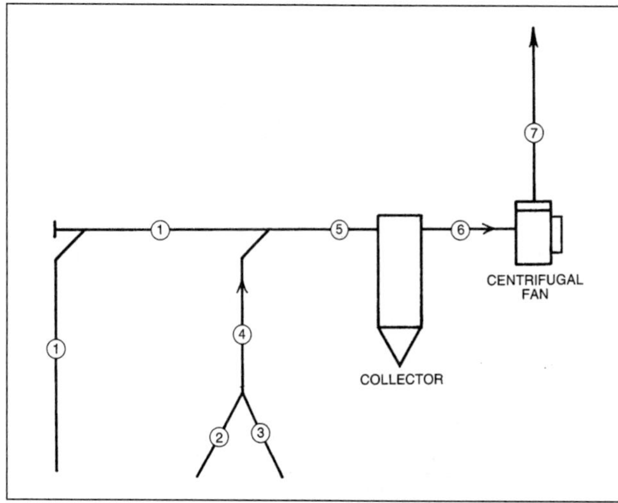

Fig. 20 System Schematic with Section Numbers for Example 9

Table 11 Total Pressure Loss Calculations by Sections for Example 8

Duct Section[a]	Fitting No.[b]	Duct Element	Airflow, cfm	Duct Size (Equivalent Round)	Velocity, fpm	Velocity Pressure, in. of water	Duct Length,[c] ft	Summary of Fitting Loss Coefficients[d]	Duct Pressure Loss/100 ft,[e] in. of water	Total Pressure Loss, in. of water	Section Pressure Loss, in. of water
1	—	Duct	1500	12 in. ϕ	1910	—	15	—	0.40	0.06	
	—	Fittings	1500	—	1910	0.23	—	0.33	—	0.08	0.14
2	—	Duct	500	8 in. ϕ	1432	—	60	—	0.39	0.23	
	—	Fittings	500	—	1432	0.13	—	−0.71	—	−0.09	0.14
3	—	Duct	2000	12 in. ϕ	2546	—	20	—	0.67	0.13	
	—	Fittings	2000	—	2546	0.40	—	0.67	—	0.27	0.40
4	—	Duct	2000	24 in. × 24 in. (26.2)	500	—	5	—	0.01	0.00	
	—	Fittings	2000	—	500	0.02	—	1.11	—	0.02	
	9	Louver	2000	24 in. × 24 in.	—	—	—	—	—	0.10[f]	0.12
5	—	Duct	2000	15 in. ϕ	1630	—	55	—	0.23	0.13	
	—	Fittings	2000	—	1630	0.17	—	1.73	—	0.29	0.42
6	—	Duct	4000	17 in. ϕ	2538	—	30	—	0.45	0.14	
	—	Fittings	4000	—	2538	0.40	—	0.87	—	0.35	0.49
7	—	Duct	600	10 in. × 10 in. (10.9)	864	—	14	—	0.11	0.02	
	—	Fittings	600	—	864	0.05	—	0.26	—	0.01	
	43	Diffuser	600	10 in. × 10 in.	—	—	—	—	—	0.10[f]	0.13
8	—	Duct	600	10 in. × 10 in. (10.9)	864	—	4	—	0.11	0.00	
	—	Fittings	600	—	864	0.05	—	1.25	—	0.06	
	44	Diffuser	600	10 in. × 10 in.	—	—	—	—	—	0.10[f]	0.16
9	—	Duct	1200	20 in. × 10 in. (15.2)	864	—	25	—	0.08	0.02	
	—	Fittings	1200	—	864	0.05	—	1.67	—	0.08	0.10
10	—	Duct	1200	16 in. × 10 in. (13.7)	1080	—	45	—	0.13	0.06	
	—	Fittings	1200	—	1080	0.07	—	2.66	—	0.19	0.25
11	—	Duc	1000	10 in. × 10 in. (10.9)	1440	—	10	—	0.29	0.03	
	—	Fittings	1000	—	1440	0.13	—	1.68	—	0.22	0.25
12	—	Duct	1000	10 in. × 10 in. (10.9)	1440	—	22	—	0.29	0.06	
	—	Fittings	1000	—	1440	0.13	—	1.45	—	0.19	0.25
13	—	Duct	2000	14 in. × 10 in. (12.9)	2057	—	35	—	0.47	0.16	
	—	Fittings	2000	—	2057	0.26	—	0.16	—	0.04	0.20
14	—	Duct	3200	26 in. × 10 in. (17.1)	1772	—	15	—	0.27	0.04	
	—	Fittings	3200	—	1772	0.20	—	0.12	—	0.02	0.06
15	—	Duct	400	8 in. × 6 in. (7.6)	1200	—	40	—	0.32	0.13	
	—	Fittings	400	—	1200	0.09		0.58	—	0.05	0.18
16	—	Duct	400	8 in. × 6 in. (7.6)	1200	—	20	—	0.32	0.06	
	—	Fittings	400	—	1200	0.09	—	1.74	—	0.16	0.22
17	—	Duct	800	10 in. × 6 in. (8.4)	1920	—	22	—	0.70	0.15	
	—	Fittings	800	—	1920	0.23	—	0.76	—	0.17	0.32
18	—	Duct	4000	32 in. × 10 in. (18.8)	1800	—	23	—	0.25	0.06	
	—	Fittings	4000	—	1800	0.20		2.91	—	0.58	0.64
19	—	Duct	4000	32 in. × 17 in. (25.2)	1059	—	12	—	0.06	0.01	
	—	Fittings	4000	—	1059	0.07	—	4.71	—	0.33	
	46	Air-measuring station	4000	—	—	—	—	—	—	0.05[f]	0.39

[a]See Figure 17.
[b]See Figure 16.
[c]Duct lengths are to fitting centerlines.
[d]See Table 12.
[e]Duct pressure based on a 0.0003 ft absolute roughness factor.
[f]Pressure drop based on manufacturers' data.

Table 12 Loss Coefficient Summary by Sections for Example 8

Duct Section	Fitting Number	Type of Fitting	ASHRAE Fitting No.[a]	Parameters	Loss Coefficient
1	1	Entry	ED1-3	$r/D = 0.2$	0.03
	2	Damper	CD9-1	$\theta = 0°$	0.19
	3	Wye (30°), main	ED5-1	$A_s/A_c = 1.0$, $A_b/A_c = 0.444$, $Q_s/Q_c = 0.75$	0.11 (C_s)
Summation of Section 1 loss coefficients					0.33
2	4	Entry	ED1-1	$L = 0$, $t = 0.064$ in. (16 gage)	0.50
	4	Screen	CD6-1	$n = 0.70$, $A_1/A_o = 1$	0.58
	5	Elbow	CD3-6	60°, $r/D = 1.5$, pleated	0.27
	6	Damper	CD9-1	$\theta = 0°$	0.19
	3	Wye (30°), branch	ED5-1	$A_s/A_c = 1.0$, $A_b/A_c = 0.444$, $Q_b/Q_c = 0.25$	−2.25 (C_b)
Summation of Section 2 loss coefficients					−0.71
3	7	Damper	CD9-1	$\theta = 0°$	0.19
	8	Wye (45°), main	ED5-2	$A_s/A_c = 0.498$, $A_b/A_c = 0.779$, $Q_s/Q_c = 0.5$	0.48 (C_s)
Summation of Section 3 loss coefficients					0.67

[a]*Duct Fitting Database* (ASHRAE 1994) data for fittings reprinted in the section on Fitting Loss Coefficients.

Table 12 Loss Coefficient Summary by Sections for Example 8 (*Concluded*)

Duct Section	Fitting Number	Type of Fitting	ASHRAE Fitting No.[a]	Parameters	Loss Coefficient
4	10	Damper	CR9-4	$\theta = 0°$, 5 blades (opposed), $L/R = 1.25$	0.52
	11	Transition	ER4-3	$L = 30$ in., $A_o/A_1 = 3.26$, $\theta = 17°$	0.59
Summation of Section 4 loss coefficients					1.11
5	12	Elbow	CD3-17	45°, mitered	0.34
	13	Damper	CD9-1	$\theta = 0°$	0.19
	8	Wye (45°), branch	ED5-2	$Q_b/Q_c = 0.5$, $A_s/A_c = 0.498$, $A_b/A_c = 0.779$	1.20 (C_b)
Summation of Section 5 loss coefficients					1.73
6	14	Fire damper	CD9-3	Curtain type, Type C	0.12
	15	Elbow	CD3-9	90°, 5 gore, $r/D = 1.5$	0.15
	—	Fan and system interaction	ED7-2	90° elbow, 5 gore, $r/D = 1.5$, $L = 34$ in.	0.60
Summation of Section 6 loss coefficients					0.87
7	16	Elbow	CR3-3	90°, $r/W = 0.70$, 1 splitter vane	0.14
	17	Damper	CR9-1	$\theta = 0°$, $H/W = 1.0$	0.08
	19	Tee, main	SR5-13	$Q_s/Q_c = 0.5$, $A_s/A_c = 0.50$	0.04 (C_s)
Summation of Section 7 loss coefficients					0.26
8	19	Tee, branch	SR5-13	$Q_b/Q_c = 0.5$, $A_b/A_c = 0.50$	0.73 (C_b)
	18	Damper	CR9-4	$\theta = 0°$, 3 blades (opposed), $L/R = 0.75$	0.52
Summation of Section 8 loss coefficients					1.25
9	20	Elbow	SR3-1	90°, mitered, $H/W_1 = 0.625$, $W_o/W_1 = 1.25$	1.67
Summation of Section 9 loss coefficients					1.67
10	21	Damper	CR9-1	$\theta = 0°$, $H/W = 0.625$	0.08
	22	Elbow	CR3-10	90°, single-thickness vanes, design 2	0.12
	23	Elbow	CR3-6	$\theta = 90°$, mitered, $H/W = 0.625$	1.25
	24	Tee, branch	SR5-1	$r/W_b = 1.0$, $Q_b/Q_c = 0.375$, $A_s/A_c = 0.538$, $A_b/A_c = 0.615$	1.21 (C_b)
Summation of Section 10 loss coefficients					2.66
11	25	Damper	CR9-1	$\theta = 0°$, $H/W = 1.0$	0.08
	26	Exit	SR2-1	$H/W = 1.0$, Re = 122,500	1.00
	27	Wye, dovetail	SR5-14	$r/W_c = 1.5$, $Q_{b1}/Q_c = 0.5$, $A_{b1}/A_c = 0.714$	0.60 (C_b)
Summation of Section 11 loss coefficients					1.68
12	28	Damper	CR9-1	$\theta = 0°$, $H/W = 1.0$	0.08
	29	Exit	SR2-5	$\theta = 19°$, $A_1/A_o = 3.24$, Re = 130,000	0.77
	27	Wye, dovetail	SR5-14	$r/W_c = 1.5$, $Q_{b2}/Q_c = 0.5$, $A_{b2}/A_c = 0.714$	0.60 (C_b)
Summation of Section 12 loss coefficients					1.45
13	30	Damper	CR9-1	$\theta = 0°$, $H/W = 0.71$	0.08
	24	Tee, main	SR5-1	$r/W_b = 1.0$, $Q_s/Q_c = 0.625$, $A_s/A_c = 0.538$, $A_b/A_c = 0.615$	0.08 (C_s)
Summation of Section 13 loss coefficients					0.16
14	31	Damper	CR9-1	$\theta = 0°$, $H/W = 0.38$	0.08
	32	Tee, main	SR5-13	$Q_s/Q_c = 0.8$, $A_s/A_c = 0.813$	0.04 (C_s)
Summation of Section 14 loss coefficients					0.12
15	48	Elbow	CR3-1	$\theta = 90°$, $r/W = 1.5$, $H/W = 0.75$	0.19
	33	Exit	SR2-6	$L = 18$ in., $D_h = 6.86$	0.28
	34	Damper	CR9-1	$\theta = 0°$, $H/W = 0.75$	0.08
	35	Tee, main	SR5-1	$r/W_b = 1.0$, $Q_s/Q_c = 0.5$, $A_s/A_c = 0.80$, $A_b/A_c = 0.80$	0.03 (C_s)
Summation of Section 15 loss coefficients					0.58
16	36	Exit	SR2-3	$\theta = 20°$, $A_1/A_o = 2.0$, Re = 70,000	0.63
	36	Screen	CR6-1	$n = 0.8$, $A_1/A_o = 2.0$	0.08
	37	Damper	CR9-1	$\theta = 0°$, $H/W = 0.75$	0.08
	35	Tee, branch	SR5-1	$r/W_b = 1.0$, $Q_b/Q_c = 0.5$, $A_s/A_c = 0.80$, $A_b/A_c = 0.80$	0.95 (C_b)
Summation of Section 16 loss coefficients					1.74
17	38	Damper	CR9-1	$\theta = 0°$, $H/W = 0.6$	0.08
	32	Tee, branch	SR5-13	$Q_b/Q_c = 0.2$, $A_b/A_c = 0.187$	0.68 (C_b)
Summation of Section 17 loss coefficients					0.76
18	39	Obstruction, pipe	CR6-4	Re = 15,000, $y = 0$, $d = 1$ in., $S_m/A_o = 0.1$, $y/H = 0$	0.17
	40	Transition	SR4-1	$\theta = 22°$, $A_o/A_1 = 0.588$, $L = 18$ in.	0.04
	41	Elbows, Z-shaped	CR3-17	$L = 42$ in., $L/W = 4.2$, $H/W = 3.2$, Re = 240,000	2.51
	45	Fire damper	CR9-6	Curtain type, Type B	0.19
Summation of Section 18 loss coefficients					2.91
19	42	Diffuser, fan	SR7-17	$\theta_1 = 28°$, $L = 40$ in., $A_o/A_1 = 2.67$, $C_1 = 0.59$	4.19 (C_o)
	47	Damper	CR9-4	$\theta = 0°$, 8 blades (opposed), $L/R = 1.39$	0.52
Summation of Section 19 loss coefficients					4.71

[a]*Duct Fitting Database* (ASHRAE 1994) data for fittings reprinted in the section on Fitting Loss Coefficients.

Duct Section	Design Airflow, cfm	Transport Velocity, fpm	Duct Diameter, in.	Duct Velocity, fpm
1	1800	4000	9	4074
2,3	610 each	4500	5	4474
4	1220	4500	7	4565
5	3020	4500	11	4576

The following tabulation summarizes design calculations up through the junction after sections 1 and 4.

Design No.	D_1, in.	Δp_1, in. of water	Δp_{2+4}, in. of water	Imbalance, $\Delta p_1 - \Delta p_{2+4}$
1	9	1.46	3.09	−1.63
2	8.5	2.00	3.08	−1.08
3	8	2.79	3.00	−0.21
4	7.5	3.92	2.88	+1.04

Q_1 = 1800 cfm
Q_2 = 610 cfm; D_2 = 5 in. dia.
Q_3 = 610 cfm; D_3 = 5 in. dia.
Q_4 = 1220 cfm; D_4 = 7 in. dia.

For the initial design, Design 1, the imbalance between section 1 and section 2 (or 3) is 1.63 in. of water, with section 1 requiring additional resistance. Decreasing section 1 duct diameter by 0.5 in. increments results in the least imbalance, 0.21 in. of water, when the duct diameter is 8 in. (Design 3). Because section 1 requires additional resistance, estimate the new airflow rate using Equation (51):

$$Q_{c,1} = (1800)(3.00/2.79)^{0.5} = 1870 \text{ cfm}$$

At 1870 cfm flow in section 1, 0.13 in. of water imbalance remains at the junction of sections 1 and 4. By trial-and-error solution, balance is attained when the flow in section 1 is 1850 cfm. The duct between the collector and the fan inlet is 13 in. round to match the fan inlet (12.75 in. diameter). To minimize downwash, the stack discharge velocity must exceed 2640 fpm, 1.5 times the design wind velocity (20 mph) as stated in the problem definition. Therefore, the stack is 14 in. round, and the stack discharge velocity is 2872 fpm.

Table 13 summarizes the system losses by sections. The straight duct friction factor and pressure loss were calculated by Equations (19) and (20). Table 14 lists fitting loss coefficients and input parameters necessary to determine the loss coefficients. The fitting loss coefficients are from the *Duct Fitting Database* (ASHRAE 1994). The fitting loss coefficient tables are included in the section on Fitting Loss Coefficients for illustration but can not be obtained exactly by manual interpolation since the coefficients were calculated by the duct fitting database algorithms (more significant figures). For a pressure grade line of the system, see Figure 21. The fan total pressure, calculated by Equation (16), is 7.89 in. of water. To calculate the fan static pressure, use Equation (18):

$$P_s = 7.89 - 0.81 = 7.1 \text{ in. of water}$$

where 0.81 in. of water is the fan outlet velocity pressure. The fan airflow rate is 3070 cfm, and its outlet area is 0.853 ft^3 (10.125 in. by 12.125 in.). Therefore, the fan outlet velocity is 3600 fpm.

The hood suction for the chipping and grinding table hood is 2.2 in. of water, calculated by Equation (18) from Chapter 29 of the 1999 *ASHRAE Handbook—Applications* [$HS = (1 + 0.25)(1.74) = 2.2$ in. of water, where 0.25 is the hood entry loss coefficient C_o, and 1.74 is the duct velocity pressure P_v a few diameters downstream from the hood]. Similarly, the hood suction for each of the grinder wheels is 1.7 in. of water:

$$HS_{2,3} = (1 + 0.4)(1.24) = 1.7 \text{ in. of water}$$

where 0.4 is the hood entry loss coefficient, and 1.24 in. of water is the duct velocity pressure.

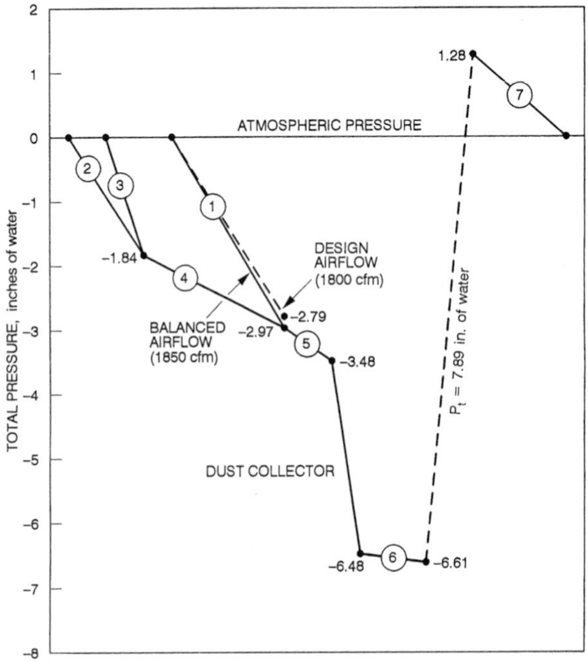

Fig. 21 Total Pressure Grade Line for Example 9

Table 13 Total Pressure Loss Calculations by Sections for Example 9

Duct Section[a]	Duct Element	Airflow, cfm	Duct Size	Velocity, fpm	Velocity Pressure, in. of water	Duct Length,[b] ft	Summary of Fitting Loss Coefficients[c]	Duct Pressure Loss/100 ft, in. of water[d]	Total Pressure Loss, in. of water	Section Pressure Loss, in. of water
1	Duct	1850	8 in. ϕ	5300	—	22.5	—	4.63	1.04	
	Fittings	1850		5300	1.75	—	1.10	—	1.93	2.97
2,3	Duct	610	5 in. ϕ	4474	—	9	—	5.94	0.53	
	Fittings	610		4474	1.24	—	1.06	—	1.31	1.84
4	Duct	1220	7 in. ϕ	4565	—	11.5	—	4.08	0.47	
	Fittings	1220		4565	1.30	—	0.51	—	0.66	1.13
5	Duct	3070	11 in. ϕ	4652	—	8.5	—	2.44	0.21	
	Fittings	3070		4652	1.35	—	0.22	—	0.30	0.51
—	Collector,[e] fabric	3070	—	—	—	—	—	—	3.0	3.0
6	Duct	3070	13 in. ϕ	3331	—	12	—	1.05	0.13	
	Fittings	3070		3331	0.69	—	0.00	—	0.00	0.13
7	Duct	3070	14 in. ϕ	2872	—	29	—	0.72	0.21	
	Fittings	3070		2872	0.51	—	2.09	—	1.07	1.28

[a]See Figure 20.
[b]Duct lengths are to fitting centerlines.
[c]See Table 14.
[d]Duct pressure based on a 0.0003 ft absolute roughness factor.
[e]Collector manufacturers set the fabric bag cleaning mechanism to actuate at a pressure difference of 3.0 in. of water between the inlet and outlet plenums. The pressure difference across the clean media is approximately 1.5 in. of water.

Table 14 Loss Coefficient Summary by Sections for Example 9

Duct Section	Fitting Number	Type of Fitting	ASHRAE Fitting No.[a]	Parameters	Loss Coefficient
1	1	Hood[b]	—	Hood face area: 3 ft by 4 ft	0.25
	2	Elbow	CD3-10	90°, 7 gore, $r/D = 2.5$	0.11
	4	Capped wye (45°), with 45° elbow	ED5-6	$A_b/A_c = 1$	0.64 (C_b)
	5	Wye (30°), main	ED5-1	$Q_s/Q_c = 0.60$, $A_s/A_c = 0.529$, $A_b/A_c = 0.405$	0.10 (C_s)
Summation of Section 1 loss coefficients					1.10
2,3	6	Hood[c]	—	Type hood: For double wheels, dia. = 22 in. each, wheel width = 4 in. each; type takeoff: tapered	0.40
	7	Elbow	CD3-12	90°, 3 gore, $r/D = 1.5$	0.34
	8	Symmetrical wye (60°)	ED5-9	$Q_b/Q_c = 0.5$, $A_b/A_c = 0.51$	0.32 (C_b)
Summation of Sections 2 and 3 loss coefficients					1.06
4	9	Elbow	CD3-10	90°, 7 gore, $r/D = 2.5$	0.11
	10	Elbow	CD3-13	60°, 3 gore, $r/D = 1.5$	0.19
	5	Wye (30°), branch	ED5-1	$Q_b/Q_c = 0.40$, $A_s/A_c = 0.529$, $A_b/A_c = 0.405$	0.21 (C_b)
Summation of Section 4 loss coefficients					0.51
5	11	Exit, conical diffuser to collector	ED2-1	$L = 24$ in., $L/D_o = 2.18$, $A_1/A_o \approx 16$	0.22
Summation of Section 5 loss coefficients					0.22
6	12	Entry, bellmouth from collector	ER2-1	$r/D_1 = 0.20$	0.00 (C_1)
Summation of Section 6 loss coefficients					0.00
7	13	Diffuser, fan outlet[d]	SR7-17	Fan outlet size: 10.125 in. by 12.125 in., $A_o/A_1 = 1.596$ (assume 14 in. by 14 in. outlet rather than 16 in. round), $L = 18$ in.	0.45 (C_o)
	14	Capped wye (45°), with 45° elbow	ED5-6	$A_b/A_c = 1$	0.64 (C_b)
	15	Stackhead	SD2-6	$D_e/D = 1$	1.0
Summation of Section 7 loss coefficients					2.09

[a]*Duct Fitting Database* (ASHRAE 1994) data for fittings reprinted in the section on Fitting Loss Coefficients.
[b]From *Industrial Ventilation* (ACGIH 1998, Figure VS-80-19).
[c]From *Industrial Ventilation* (ACGIH 1998, Figure VS-80-11).
[d]Fan specified: Industrial exhauster for granular materials: 21 in. wheel diameter, 12.75 in. inlet diameter, 10.125 in. by 12.125 in. outlet, 7.5 hp motor.

REFERENCES

ACGIH. 1998. *Industrial ventilation: A manual of recommended practice*, 23rd ed. American Conference of Governmental Industrial Hygienists, Lansing, MI.

ADC. 1996. Flexible duct performance and installation standards, 3rd ed. Air Diffusion Council.

AISI/SMACNA. 1972. Measurement and analysis of leakage rates from seams and joints of air handling systems.

Altshul, A.D., L.C. Zhivotovckiy, and L.P. Ivanov. 1987. *Hydraulics and aerodynamics*. Stroisdat Publishing House, Moscow.

AMCA. 1999. Laboratory method of testing louvers for rating. *Standard* 500-L. Air Movement and Control Association, Arlington Heights, IL.

AMCA. 1990a. Fans and systems. *Publication* 201.

AMCA. 1990b. Field performance measurement of fan systems. *Publication* 203.

ASHRAE. 1993. Energy-efficient design of new low-rise residential buildings. ASHRAE *Standard* 90.2-1993.

ASHRAE. 1994. *Duct fitting database*.

ASHRAE. 1995. Energy conservation in existing buildings. ASHRAE/IESNA *Standard* 100-1995. Addendum 1-1996.

ASHRAE. 1999. Energy standard for buildings except low-rise residential buildings. ASHRAE/IESNA *Standard* 90.1-1999.

ASHRAE. 1999. Laboratory methods for testing fans for aerodynamic performance rating. ANSI/ASHRAE *Standard* 51-1999. Also ANSI/AMCA 210-99.

ASHRAE/SMACNA/TIMA. 1985. Investigation of duct leakage. ASHRAE *Research Project* 308.

Behls, H.F. 1971. Computerized calculation of duct friction. *Building Science Series* 39, p. 363. National Institute of Standards and Technology, Gaithersburg, MD.

Bellman, R.E. 1957. *Dynamic programming*. Princeton University Press, New York.

Brown, R.B. 1973. Experimental determinations of fan system effect factors. In *Fans and systems*, ASHRAE Symposium Bulletin LO-73-1, Louisville, KY (June).

Carrier Corporation. 1960. Air duct design. Chapter 2 in *System design manual*, Part 2: Air distribution. pp.17-63. Syracuse, NY.

Chun-Lun, S. 1983. Simplified static-regain duct design procedure. *ASHRAE Transactions* 89(2A):78.

Clarke, M.S., J.T. Barnhart, F.J. Bubsey, and E. Neitzel. 1978. The effects of system connections on fan performance. *ASHRAE Transactions* 84(2): 227-63.

Colebrook, C.F. 1938-39. Turbulent flow in pipes, with particular reference to the transition region between the smooth and rough pipe laws. *Journal of the Institution of Civil Engineers* 11:133. London.

DOE. *Electrical sales and revenue*, latest edition. Department of Energy, Washington, D.C. (To purchase, call Energy Information Administration 202-512-1800.)

Farajian, T., G. Grewal, and R.J. Tsal. 1992. Post-accident air leakage analysis in a nuclear facility via T-method airflow simulation. 22nd DOE/NRC Nuclear Air Cleaning and Treatment Conference, Denver, CO, October.

Farquhar, H.F. 1973. System effect values for fans. In *Fans and systems*, ASHRAE Symposium Bulletin LO-73-1, Louisville, KY (June).

Griggs, E.I. and F. Khodabakhsh-Sharifabad. 1992. Flow characteristics in rectangular ducts. *ASHRAE Transactions* 98(1).

Griggs, E.I., W.B. Swim, and G.H. Henderson. 1987. Resistance to flow of round galvanized ducts. *ASHRAE Transactions* 93(1):3-16.

Heyt, J.W. and M.J. Diaz. 1975. Pressure drop in flat-oval spiral air duct. *ASHRAE Transactions* 81(2):221-32.

Huebscher, R.G. 1948. Friction equivalents for round, square and rectangular ducts. *ASHVE Transactions* 54:101-18.

Hutchinson, F.W. 1953. Friction losses in round aluminum ducts. *ASHVE Transactions* 59:127-38.

Idelchik, I.E., M.O. Steinberg, G.R. Malyavskaya, and O.G. Martynenko. 1994. *Handbook of hydraulic resistance*, 3rd ed. CRC Press/Begell House, Boca Raton, Ann Arbor, London, Tokyo.

Jones, C.D. 1979. Friction factor and roughness of United Sheet Metal Company spiral duct. United Sheet Metal, Division of United McGill Corp., Westerville, OH (August). Based on data contained in Friction loss tests, United Sheet Metal Company Spiral Duct, Ohio State University Engineering Experiment Station, File No. T-1011, September, 1958.

Klote, J.H. and J. Milke. 1992. *Design of smoke management systems.* ASHRAE, Atlanta.

Lauvray, T.L. 1978. Experimental heat transmission coefficients for operating air duct systems. *ASHRAE Journal* (June):69.

Meyer, M.L. 1973. A new concept: The fan system effect factor. In *Fans and systems*, ASHRAE Symposium Bulletin LO-73-1, Louisville, KY (June).

Moody, L.F. 1944. Friction factors for pipe flow. *ASME Transactions* 66:671.

NAIMA. 1997. Fibrous glass duct construction standards, 3rd ed. North American Insulation Manufacturers Association.

NFPA. *Fire protection handbook*, latest ed. National Fire Protection Association, Quincy, MA.

NFPA. Installation of air conditioning and ventilating systems. ANSI/NFPA *Standard* 90A, latest ed.

Osborne, W.C. 1966. *Fans*. Pergamon Press Ltd., London.

SMACNA. 1985. *HVAC air duct leakage manual.* Sheet Metal and Air Conditioning Contractors' National Association, Chantilly, VA.

SMACNA. 1992. Fibrous glass duct construction standards, 6th ed.

SMACNA. 1995. HVAC duct construction standards, metal and flexible, 2nd ed.

Smith, G.W. 1968. *Engineering economy: Analysis of capital expenditures.* The Iowa State University Press, Ames, IA.

Swim, W.B. 1978. Flow losses in rectangular ducts lined with fiberglass. *ASHRAE Transactions* 84(2):216.

Swim, W.B. 1982. Friction factor and roughness for airflow in plastic pipe. *ASHRAE Transactions* 88(1):269.

Swim, W.B. and E.I. Griggs. 1995. Duct leakage measurement and analysis. *ASHRAE Transactions* 101(1).

Tsal, R.J. and M.S. Adler. 1987. Evaluation of numerical methods for ductwork and pipeline optimization. *ASHRAE Transactions* 93(1):17-34.

Tsal, R.J. and H.F. Behls. 1988. Fallacy of the static regain duct design method. *ASHRAE Transactions* 94(2):76-89.

Tsal, R.J., H.F. Behls, and R. Mangel. 1988. T-method duct design, Part I: Optimization theory; Part II: Calculation procedure and economic analysis. *ASHRAE Transactions* 94(2):90-111.

Tsal, R.J., H.F. Behls, and R. Mangel. 1990. T-method duct design, Part III: Simulation. *ASHRAE Transactions* 96(2).

UL. Published annually. *Building materials directory.* Underwriters Laboratories, Northbrook, IL.

UL. Published annually. *Fire resistance directory.*

UL. Factory-made air ducts and air connectors. UL *Standard* 181, latest ed.

UL. Closure systems for use with rigid air ducts and air connectors. UL *Standard* 181A, latest ed.

UL. Closure systems for use with rigid air ducts and air connectors. UL *Standard* 181B, latest ed.

UL. Fire dampers. *Standard* UL 555, latest ed.

UL. Smoke dampers. *Standard* 555S, latest ed.

Wendes, H.C. 1989. *Sheet metal estimating.* Wendes Mechanical Consulting Services, Elk Grove Village, IL.

Wright, D.K., Jr. 1945. A new friction chart for round ducts. *ASHVE Transactions* 51:303-16.

BIBLIOGRAPHY

SMACNA. 1987. Duct research destroys design myths. Videotape (VHS). Sheet Metal and Air Conditioning Contractors' National Association, Chantilly, VA.

FITTING LOSS COEFFICIENTS

Fittings to support Examples 8 and 9 and some of the more common fittings are reprinted here.

For the complete fitting database see the *Duct Fitting Database* (ASHRAE 1994).

ROUND FITTINGS

CD3-1 Elbow, Die Stamped, 90 Degree, r/D = 1.5

D, in.	3	4	5	6	7	8	9	10
C_o	0.30	0.21	0.16	0.14	0.12	0.11	0.11	0.11

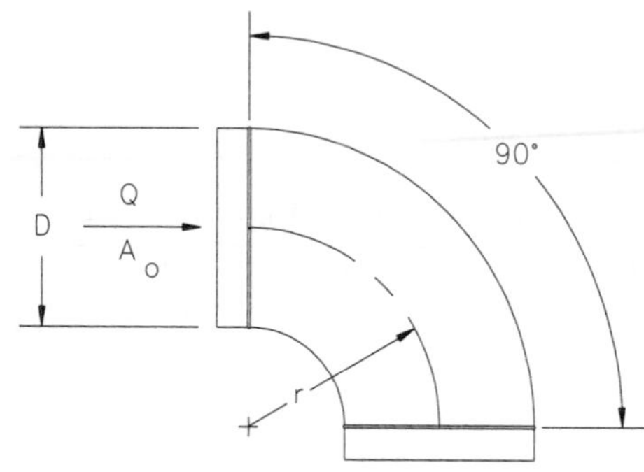

CD3-3 Elbow, Die Stamped, 45 Degree, r/D = 1.5

D, in.	3	4	5	6	7	8	9	10
C_o	0.18	0.13	0.10	0.08	0.07	0.07	0.07	0.07

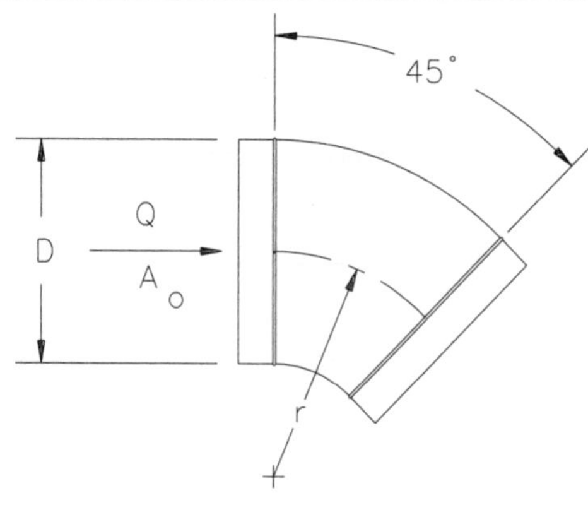

CD3-5 Elbow, Pleated, 90 Degree, r/D = 1.5

D, in.	4	6	8	10	12	14	16
C_o	0.57	0.43	0.34	0.28	0.26	0.25	0.25

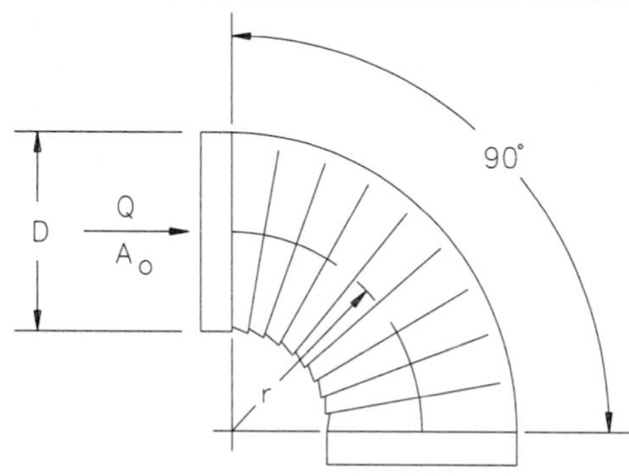

CD3-6 Elbow, Pleated, 60 Degree, r/D = 1.5

D, in.	4	6	8	10	12	14	16
C_o	0.45	0.34	0.27	0.23	0.20	0.19	0.19

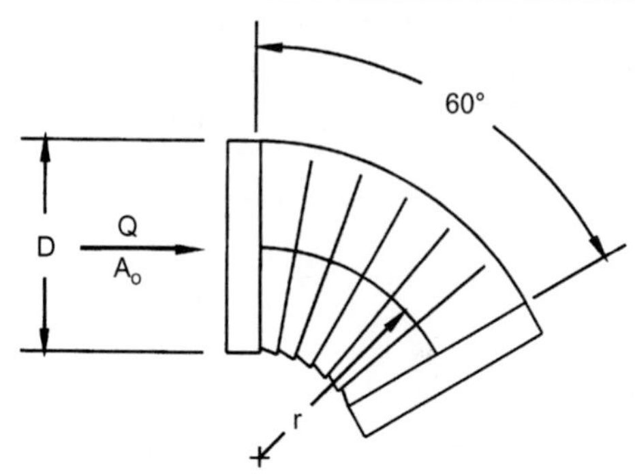

CD3-7 Elbow, Pleated, 45 Degree, r/D = 1.5

D, in.	4	6	8	10	12	14	16
C_o	0.34	0.26	0.21	0.17	0.16	0.15	0.15

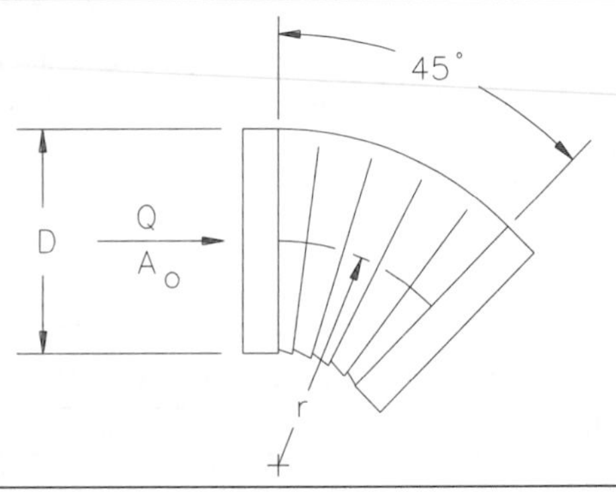

CD3-9 Elbow, 5 Gore, 90 Degree, r/D = 1.5

D, in.	3	6	9	12	15	18	21	24	27	30	60
C_o	0.51	0.28	0.21	0.18	0.16	0.15	0.14	0.13	0.12	0.12	0.12

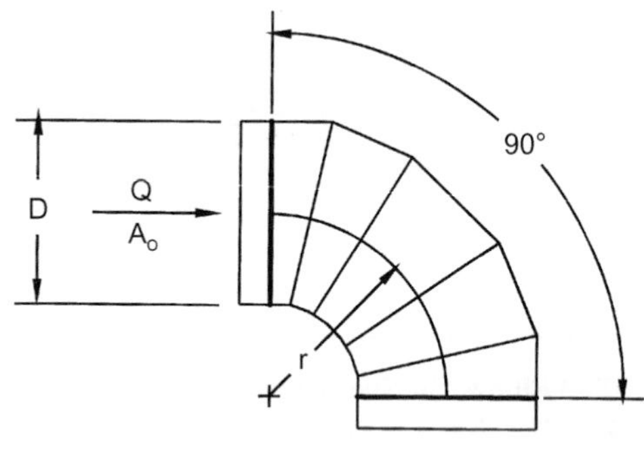

CD3-10 Elbow, 7 Gore, 90 Degree, r/D = 2.5

D, in.	3	6	9	12	15	18	27	60
C_o	0.16	0.12	0.10	0.08	0.07	0.06	0.05	0.03

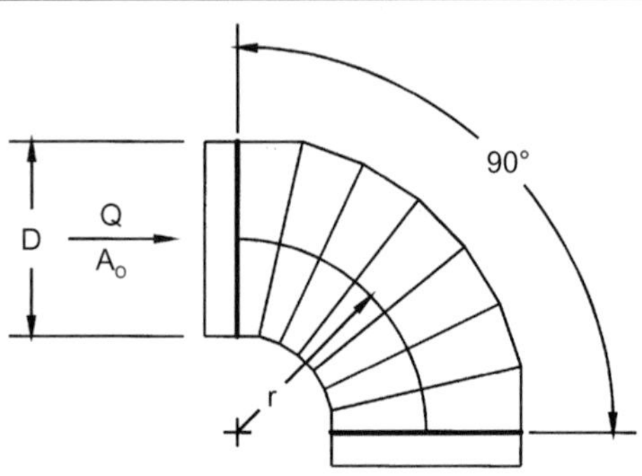

CD3-12 Elbow, 3 Gore, 90 Degree, r/D = 0.75 to 2.0

r/D	0.75	1.00	1.50	2.00
C_o	0.54	0.42	0.34	0.33

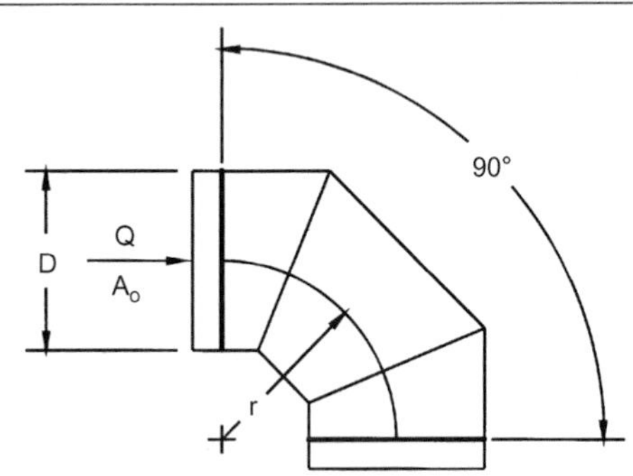

CD3-13 Elbow, 3 Gore, 60 Degree, r/D = 1.5

D, in.	3	6	9	12	15	18	21	24	27	30	60
C_o	0.40	0.21	0.16	0.14	0.12	0.12	0.11	0.10	0.09	0.09	0.09

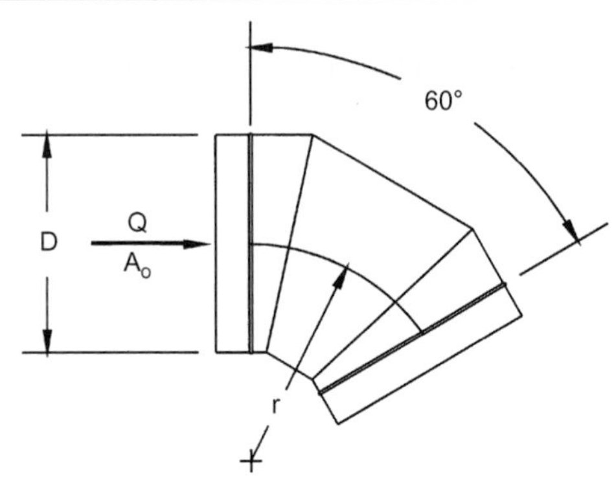

CD3-14 Elbow, 3 Gore, 45 Degree, r/D = 1.5

D, in.	3	6	9	12	15	18	21	24	27	30	60
C_o	0.31	0.17	0.13	0.11	0.11	0.09	0.08	0.08	0.07	0.07	0.07

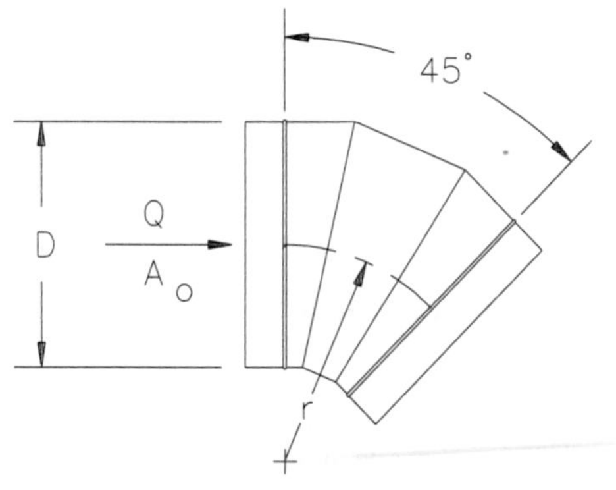

CD3-17 Elbow, Mitered, 45 Degree

D, in.	3	6	9	12	15	18	21	24	27	60
C_o	0.34	0.34	0.34	0.34	0.34	0.34	0.34	0.34	0.34	0.34

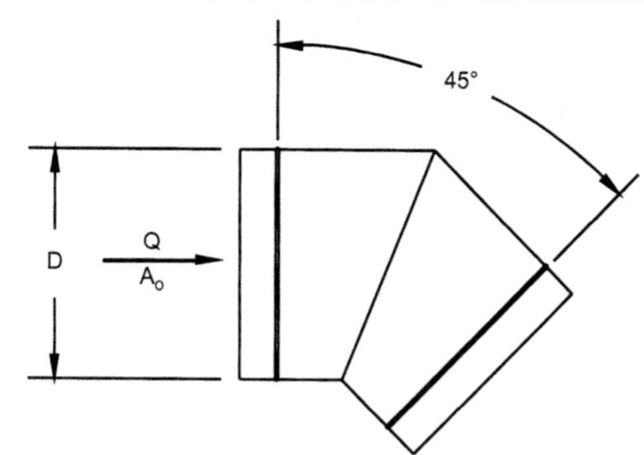

CD6-1 Screen (Only)

	C_o Values												
	n												
A_1/A_o	0.30	0.35	0.40	0.45	0.50	0.55	0.60	0.65	0.70	0.75	0.80	0.90	1.00
0.2	155.00	102.50	75.00	55.00	41.25	31.50	24.25	18.75	14.50	11.00	8.00	3.50	0.00
0.3	68.89	45.56	33.33	24.44	18.33	14.00	10.78	8.33	6.44	4.89	3.56	1.56	0.00
0.4	38.75	25.63	18.75	13.75	10.31	7.88	6.06	4.69	3.63	2.75	2.00	0.88	0.00
0.5	24.80	16.40	12.00	8.80	6.60	5.04	3.88	3.00	2.32	1.76	1.28	0.56	0.00
0.6	17.22	11.39	8.33	6.11	4.58	3.50	2.69	2.08	1.61	1.22	0.89	0.39	0.00
0.7	12.65	8.37	6.12	4.49	3.37	2.57	1.98	1.53	1.18	0.90	0.65	0.29	0.00
0.8	9.69	6.40	4.69	3.44	2.58	1.97	1.52	1.17	0.91	0.69	0.50	0.22	0.00
0.9	7.65	5.06	3.70	2.72	2.04	1.56	1.20	0.93	0.72	0.54	0.40	0.17	0.00
1.0	6.20	4.10	3.00	2.20	1.65	1.26	0.97	0.75	0.58	0.44	0.32	0.14	0.00
1.2	4.31	2.85	2.08	1.53	1.15	0.88	0.67	0.52	0.40	0.31	0.22	0.10	0.00
1.4	3.16	2.09	1.53	1.12	0.84	0.64	0.49	0.38	0.30	0.22	0.16	0.07	0.00
1.6	2.42	1.60	1.17	0.86	0.64	0.49	0.38	0.29	0.23	0.17	0.13	0.05	0.00
1.8	1.91	1.27	0.93	0.68	0.51	0.39	0.30	0.23	0.18	0.14	0.10	0.04	0.00
2.0	1.55	1.03	0.75	0.55	0.41	0.32	0.24	0.19	0.15	0.11	0.08	0.04	0.00
2.5	0.99	0.66	0.48	0.35	0.26	0.20	0.16	0.12	0.09	0.07	0.05	0.02	0.00
3.0	0.69	0.46	0.33	0.24	0.18	0.14	0.11	0.08	0.06	0.05	0.04	0.02	0.00
4.0	0.39	0.26	0.19	0.14	0.10	0.08	0.06	0.05	0.04	0.03	0.02	0.01	0.00
6.0	0.17	0.11	0.08	0.06	0.05	0.04	0.03	0.02	0.02	0.01	0.01	0.00	0.00

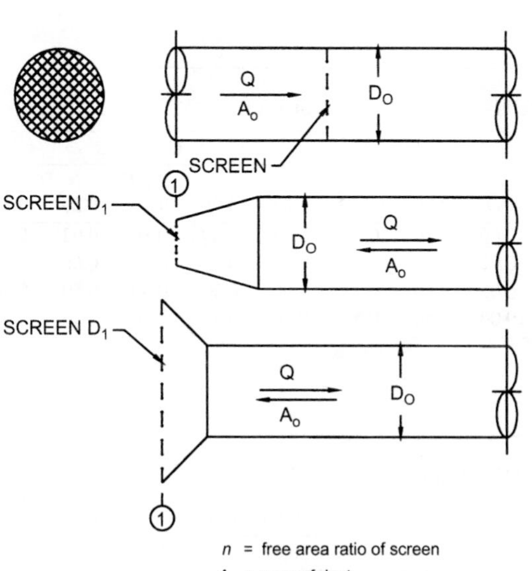

n = free area ratio of screen
A_o = area of duct
A_1 = cross-sectional area of duct or fitting where screen is located

CD6-4 Round Duct, Depressed to Avoid an Obstruction

$C_o = 0.24$

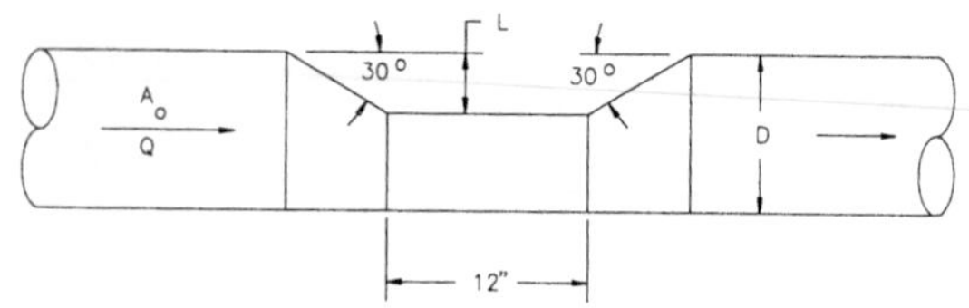

CD9-1 Damper, Butterfly

	C_o Values											
	θ											
D/D_o	0	10	20	30	40	50	60	70	75	80	85	90
0.5	0.19	0.27	0.37	0.49	0.61	0.74	0.86	0.96	0.99	1.02	1.04	1.04
0.6	0.19	0.32	0.48	0.69	0.94	1.21	1.48	1.72	1.82	1.89	1.93	2.00
0.7	0.19	0.37	0.64	1.01	1.51	2.12	2.81	3.46	3.73	3.94	4.08	6.00
0.8	0.19	0.45	0.87	1.55	2.60	4.13	6.14	8.38	9.40	10.30	10.80	15.00
0.9	0.19	0.54	1.22	2.51	4.97	9.57	17.80	30.50	38.00	45.00	50.10	100.00
1.0	0.19	0.67	1.76	4.38	11.20	32.00	113.00	619.00	2010.00	10350.00	99999.00	99999.00

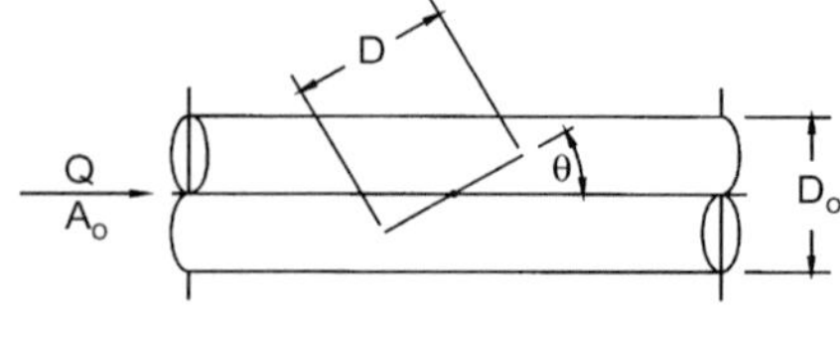

CD9-3 Fire Damper, Curtain Type, Type C

$C_o = 0.12$

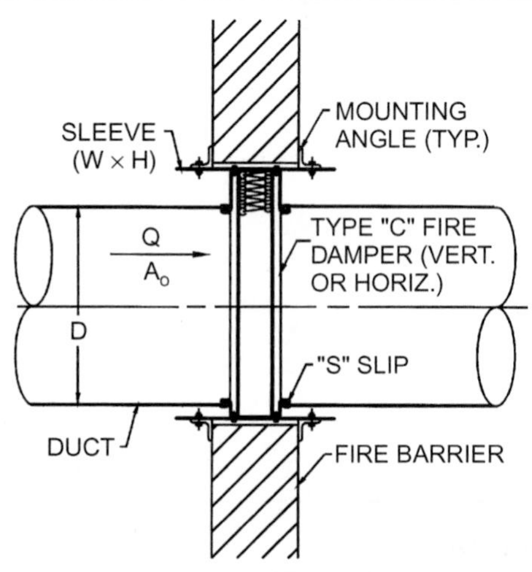

ED1-1 Duct Mounted in Wall

	C_o Values								
	L/D								
t/D	0.00	0.002	0.01	0.05	0.10	0.20	0.30	0.50	10.00
0.00	0.50	0.57	0.68	0.80	0.86	0.92	0.97	1.00	1.00
0.02	0.50	0.51	0.52	0.55	0.60	0.66	0.69	0.72	0.72
0.05	0.50	0.50	0.50	0.50	0.50	0.50	0.50	0.50	0.50
10.00	0.50	0.50	0.50	0.50	0.50	0.50	0.50	0.50	0.50

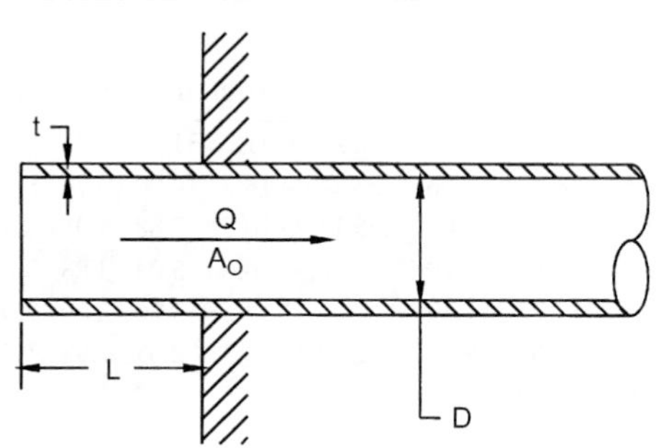

ED1-3 Bellmouth, with Wall

r/D	0.00	0.01	0.02	0.03	0.04	0.05	0.06	0.08	0.10	0.12	0.16	0.20	10.00
C_o	0.50	0.44	0.37	0.31	0.26	0.22	0.20	0.15	0.12	0.09	0.06	0.03	0.03

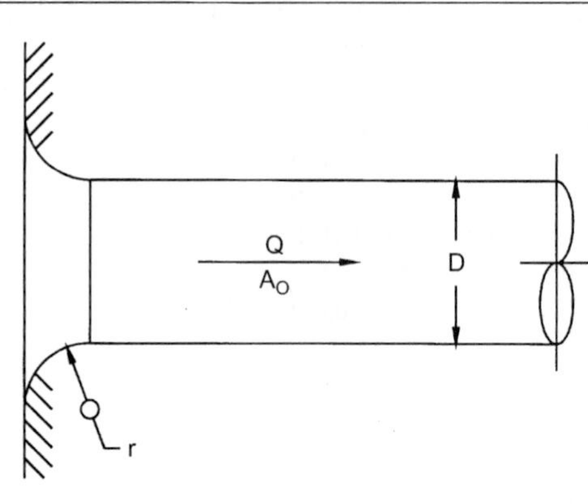

ED2-1 Conical Diffuser, Round to Plenum, Exhaust/Return Systems

	C_o Values										
	L/D_o										
A_1/A_o	0.5	1.0	2.0	3.0	4.0	5.0	6.0	8.0	10.0	12.0	14.0
1.5	0.03	0.02	0.03	0.03	0.04	0.05	0.06	0.08	0.10	0.11	0.13
2.0	0.08	0.06	0.04	0.04	0.04	0.05	0.05	0.06	0.08	0.09	0.10
2.5	0.13	0.09	0.06	0.06	0.06	0.06	0.06	0.06	0.07	0.08	0.09
3.0	0.17	0.12	0.09	0.07	0.07	0.06	0.06	0.07	0.07	0.08	0.08
4.0	0.23	0.17	0.12	0.10	0.09	0.08	0.08	0.08	0.08	0.08	0.08
6.0	0.30	0.22	0.16	0.13	0.12	0.10	0.10	0.09	0.09	0.09	0.08
8.0	0.34	0.26	0.18	0.15	0.13	0.12	0.11	0.10	0.09	0.09	0.09
10.0	0.36	0.28	0.20	0.16	0.14	0.13	0.12	0.11	0.10	0.09	0.09
14.0	0.39	0.30	0.22	0.18	0.16	0.14	0.13	0.12	0.10	0.10	0.10
20.0	0.41	0.32	0.24	0.20	0.17	0.15	0.14	0.12	0.11	0.11	0.10
	Optimum Angle θ										
A_1/A_o	0.5	1.0	2.0	3.0	4.0	5.0	6.0	8.0	10.0	12.0	14.0
1.5	34	20	13	9	7	6	4	3	2	2	2
2.0	42	28	17	12	10	9	8	6	5	4	3
2.5	50	32	20	15	12	11	10	8	7	6	5
3.0	54	34	22	17	14	12	11	10	8	8	6
4.0	58	40	26	20	16	14	13	12	10	10	9
6.0	62	42	28	22	19	16	15	12	11	10	9
8.0	64	44	30	24	20	18	16	13	12	11	10
10.0	66	46	30	24	22	19	17	14	12	11	10
14.0	66	48	32	26	22	19	17	14	13	11	11
20.0	68	48	32	26	22	20	18	15	13	12	11

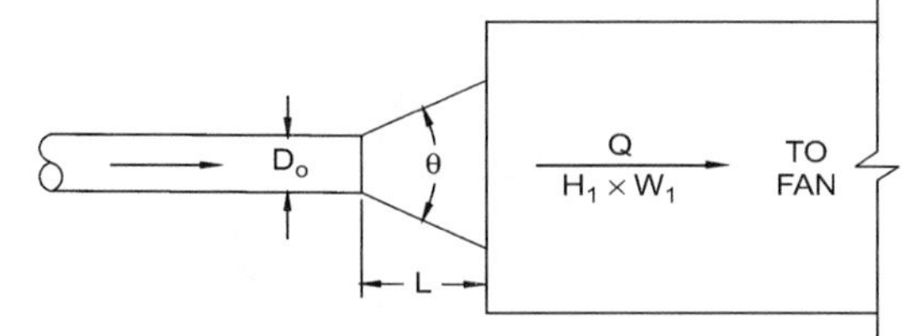

ED4-1 Transition, Round to Round, Exhaust/Return Systems

	C_o Values									
	θ									
A_o/A_1	10	15	20	30	45	60	90	120	150	180
0.06	0.21	0.29	0.38	0.60	0.84	0.88	0.88	0.88	0.88	0.88
0.10	0.21	0.28	0.38	0.59	0.76	0.80	0.83	0.84	0.83	0.83
0.25	0.16	0.22	0.30	0.46	0.61	0.68	0.64	0.63	0.62	0.62
0.50	0.11	0.13	0.19	0.32	0.33	0.33	0.32	0.31	0.30	0.30
1.00	0.00	0.00	0.00	0.00	0.00	0.00	0.00	0.00	0.00	0.00
2.00	0.20	0.20	0.20	0.20	0.22	0.24	0.48	0.72	0.96	1.04
4.00	0.80	0.64	0.64	0.64	0.88	1.12	2.72	4.32	5.60	6.56
6.00	1.80	1.44	1.44	1.44	1.98	2.52	6.48	10.10	13.00	15.10
10.00	5.00	5.00	5.00	5.00	6.50	8.00	19.00	29.00	37.00	43.00

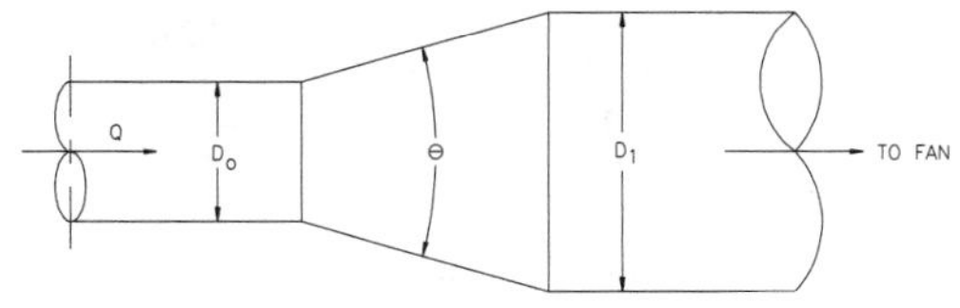

ED4-2 Transition, Round to Rectangular, Exhaust/Return Systems

	C_o Values									
	θ									
A_o/A_1	10	15	20	30	45	60	90	120	150	180
0.06	0.30	0.54	0.53	0.65	0.77	0.88	0.95	0.98	0.98	0.93
0.10	0.30	0.50	0.53	0.64	0.75	0.84	0.89	0.91	0.91	0.88
0.25	0.25	0.36	0.45	0.52	0.58	0.62	0.64	0.64	0.64	0.64
0.50	0.15	0.21	0.25	0.30	0.33	0.33	0.33	0.32	0.31	0.30
1.00	0.00	0.00	0.00	0.00	0.00	0.00	0.00	0.00	0.00	0.00
2.00	0.24	0.28	0.26	0.20	0.22	0.24	0.49	0.73	0.97	1.04
4.00	0.89	0.78	0.79	0.70	0.88	1.12	2.72	4.33	5.62	6.58
6.00	1.89	1.67	1.59	1.49	1.98	2.52	6.51	10.14	13.05	15.14
10.00	5.09	5.32	5.15	5.05	6.50	8.05	19.06	29.07	37.08	43.05

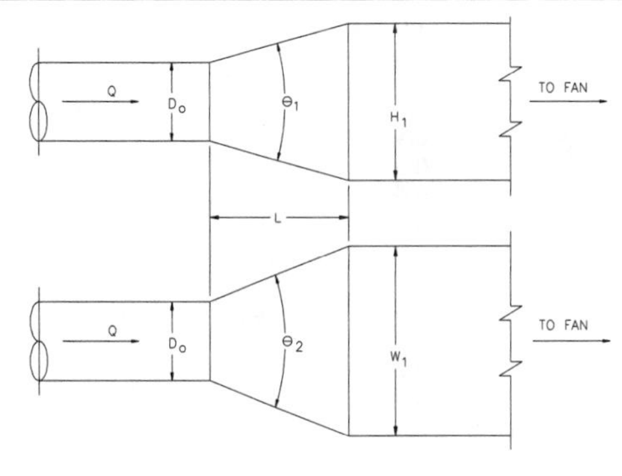

ED5-1 Wye, 30 Degree, Converging

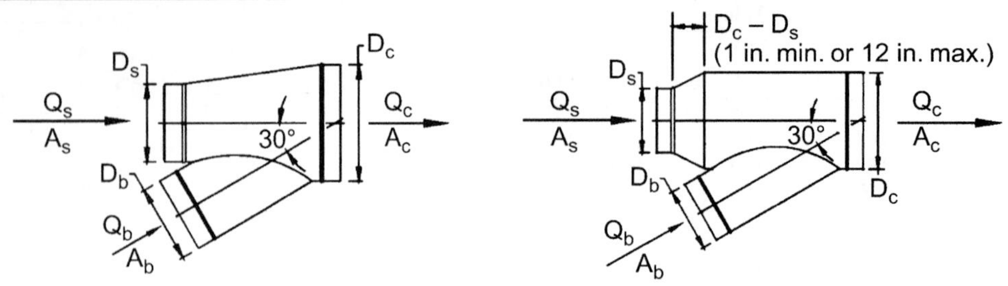

C_b Values

A_s/A_c	A_b/A_c	Q_b/Q_c 0.1	0.2	0.3	0.4	0.5	0.6	0.7	0.8	0.9
0.2	0.2	−24.17	−3.78	−0.60	0.30	0.64	0.77	0.83	0.88	0.98
	0.3	−55.88	−9.77	−2.57	−0.50	0.25	0.55	0.67	0.70	0.71
	0.4	−99.93	−17.94	−5.13	−1.45	−0.11	0.42	0.62	0.68	0.68
	0.5	−156.51	−28.40	−8.37	−2.62	−0.52	0.30	0.62	0.71	0.69
	0.6	−225.62	−41.13	−12.30	−4.01	−0.99	0.20	0.66	0.78	0.75
	0.7	−307.26	−56.14	−16.90	−5.61	−1.51	0.11	0.73	0.90	0.86
	0.8	−401.44	−73.44	−22.18	−7.44	−2.08	0.04	0.84	1.06	1.01
	0.9	−508.15	−93.02	−28.15	−9.49	−2.71	−0.03	0.99	1.27	1.20
	1.0	−627.39	−114.89	−34.80	−11.77	−3.39	−0.08	1.18	1.52	1.43
0.3	0.2	−13.97	−1.77	0.08	0.59	0.77	0.84	0.88	0.92	1.06
	0.3	−33.06	−5.33	−1.09	0.10	0.51	0.66	0.71	0.72	0.74
	0.4	−59.43	−10.08	−2.52	−0.41	0.32	0.59	0.67	0.68	0.66
	0.5	−93.24	−16.11	−4.30	−1.00	0.14	0.56	0.69	0.70	0.66
	0.6	−134.51	−23.45	−6.44	−1.68	−0.03	0.57	0.76	0.77	0.70
	0.7	−183.25	−32.08	−8.93	−2.45	−0.21	0.61	0.87	0.88	0.79
	0.8	−239.47	−42.01	−11.77	−3.32	−0.38	0.69	1.02	1.03	0.91
	0.9	−303.16	−53.25	−14.97	−4.27	−0.56	0.80	1.21	1.23	1.07
	1.0	−374.32	−65.79	−18.53	−5.32	−0.73	0.94	1.45	1.47	1.27
0.4	0.2	−9.20	−0.85	0.39	0.71	0.82	0.87	0.90	0.94	1.09
	0.3	−22.31	−3.24	−0.38	0.39	0.64	0.73	0.76	0.78	0.85
	0.4	−40.52	−6.48	−1.37	0.02	0.48	0.64	0.67	0.66	0.65
	0.5	−63.71	−10.50	−2.50	−0.33	0.40	0.63	0.69	0.67	0.63
	0.6	−92.00	−15.37	−3.84	−0.71	0.33	0.67	0.75	0.71	0.65
	0.7	−125.40	−21.08	−5.40	−1.13	0.28	0.75	0.85	0.80	0.70
	0.8	−163.90	−27.65	−7.16	−1.59	0.25	0.86	1.00	0.93	0.80
	0.9	−207.52	−35.07	−9.14	−2.09	0.25	1.02	1.18	1.10	0.93
	1.0	−256.25	−43.35	−11.33	−2.63	0.26	1.21	1.42	1.31	1.09
0.5	0.2	−6.62	−0.36	0.54	0.77	0.85	0.88	0.90	0.95	1.11
	0.3	−16.42	−2.11	−0.01	0.54	0.72	0.78	0.80	0.83	0.96
	0.4	−30.26	−4.59	−0.79	0.22	0.54	0.64	0.66	0.64	0.64
	0.5	−47.68	−7.55	−1.61	−0.02	0.48	0.63	0.65	0.62	0.59
	0.6	−68.93	−11.13	−2.56	−0.28	0.45	0.67	0.69	0.65	0.58
	0.7	−94.00	−15.31	−3.65	−0.55	0.44	0.74	0.77	0.71	0.61
	0.8	−122.90	−20.12	−4.88	−0.83	0.46	0.85	0.90	0.81	0.68
	0.9	−155.63	−25.54	−6.25	−1.12	0.51	1.00	1.06	0.94	0.77
	1.0	−192.18	−31.58	−7.77	−1.43	0.59	1.19	1.26	1.12	0.90
0.6	0.2	−5.12	−0.10	0.62	0.79	0.85	0.87	0.90	0.95	1.11
	0.3	−13.00	−1.49	0.18	0.61	0.75	0.79	0.82	0.86	1.02
	0.4	−24.31	−3.55	−0.50	0.30	0.55	0.62	0.63	0.62	0.63
	0.5	−38.41	−5.94	−1.16	0.09	0.48	0.59	0.60	0.57	0.55
	0.6	−55.58	−8.80	−1.92	−0.12	0.45	0.61	0.62	0.57	0.52
	0.7	−75.83	−12.16	−2.79	−0.33	0.44	0.66	0.67	0.60	0.52
	0.8	−99.17	−16.00	−3.76	−0.54	0.46	0.74	0.76	0.67	0.56
	0.9	−125.60	−20.33	−4.83	−0.76	0.51	0.86	0.88	0.77	0.62
	1.0	−155.12	−25.14	−6.02	−0.99	0.58	1.02	1.04	0.90	0.71
0.7	0.2	−4.24	0.05	0.65	0.80	0.85	0.87	0.89	0.94	1.12
	0.3	−11.00	−1.15	0.27	0.63	0.75	0.79	0.82	0.87	1.06
	0.4	−20.82	−3.00	−0.38	0.31	0.52	0.59	0.60	0.59	0.61
	0.5	−32.99	−5.09	−0.98	0.10	0.43	0.53	0.54	0.52	0.51
	0.6	−47.78	−7.58	−1.67	−0.11	0.38	0.52	0.53	0.49	0.45
	0.7	−65.22	−10.50	−2.44	−0.32	0.34	0.53	0.54	0.49	0.43
	0.8	−85.32	−13.83	−3.30	−0.53	0.33	0.58	0.59	0.52	0.43
	0.9	−108.07	−17.58	−4.26	−0.75	0.34	0.66	0.67	0.58	0.46
	1.0	−133.48	−21.76	−5.30	−0.97	0.38	0.76	0.78	0.67	0.51

ED5-1 Wye, 30 Degree, Converging (*Continued*)

		C_b Values (*Concluded*)								
		Q_b/Q_c								
A_s/A_c	A_b/A_c	0.1	0.2	0.3	0.4	0.5	0.6	0.7	0.8	0.9
0.8	0.2	−3.75	0.11	0.65	0.79	0.84	0.86	0.88	0.94	1.12
	0.3	−9.88	−0.99	0.29	0.63	0.74	0.78	0.81	0.87	1.09
	0.4	−18.88	−2.75	−0.36	0.28	0.48	0.55	0.56	0.57	0.61
	0.5	−29.98	−4.71	−0.96	0.04	0.36	0.46	0.47	0.46	0.47
	0.6	−43.46	−7.05	−1.64	−0.20	0.26	0.41	0.43	0.41	0.39
	0.7	−59.34	−9.77	−2.40	−0.44	0.19	0.38	0.41	0.38	0.34
	0.8	−77.64	−12.88	−3.26	−0.69	0.13	0.38	0.42	0.37	0.31
	0.9	−98.35	−16.38	−4.20	−0.95	0.09	0.40	0.45	0.39	0.30
	1.0	−121.48	−20.27	−5.24	−1.23	0.06	0.45	0.51	0.43	0.31
0.9	0.2	−3.52	0.12	0.64	0.78	0.82	0.85	0.88	0.93	1.12
	0.3	−9.34	−0.95	0.28	0.60	0.71	0.76	0.80	0.87	1.10
	0.4	−17.96	−2.70	−0.40	0.22	0.43	0.50	0.53	0.54	0.60
	0.5	−28.58	−4.65	−1.05	−0.07	0.26	0.37	0.40	0.41	0.42
	0.6	−41.45	−6.97	−1.77	−0.35	0.12	0.28	0.32	0.32	0.32
	0.7	−56.61	−9.66	−2.58	−0.65	0.00	0.21	0.27	0.26	0.24
	0.8	−74.08	−12.74	−3.49	−0.97	−0.12	0.16	0.23	0.22	0.18
	0.9	−93.84	−16.21	−4.50	−1.30	−0.23	0.13	0.21	0.19	0.14
	1.0	−115.92	−20.06	−5.61	−1.66	−0.34	0.11	0.21	0.18	0.11
1.0	0.2	−3.48	0.10	0.62	0.76	0.81	0.84	0.87	0.92	1.11
	0.3	−9.22	−1.00	0.23	0.56	0.68	0.74	0.78	0.86	1.11
	0.4	−17.76	−2.79	−0.50	0.14	0.37	0.45	0.49	0.52	0.60
	0.5	−28.31	−4.82	−1.21	−0.20	0.15	0.28	0.33	0.35	0.38
	0.6	−41.06	−7.21	−2.01	−0.55	−0.04	0.15	0.22	0.23	0.25
	0.7	−56.09	−9.99	−2.91	−0.92	−0.23	0.03	0.12	0.14	0.15
	0.8	−73.39	−13.17	−3.92	−1.32	−0.41	−0.07	0.04	0.06	0.06
	0.9	−92.98	−16.75	−5.04	−1.75	−0.60	−0.17	−0.03	−0.01	−0.02
	1.0	−114.85	−20.74	−6.28	−2.21	−0.79	−0.26	−0.09	−0.07	−0.09

		C_s Values								
		Q_s/Q_c								
A_s/A_c	A_b/A_c	0.1	0.2	0.3	0.4	0.5	0.6	0.7	0.8	0.9
0.2	0.2	−16.02	−3.15	−0.80	0.04	0.45	0.69	0.86	0.99	1.10
	0.3	−11.65	−1.94	−0.26	0.32	0.60	0.77	0.90	1.01	1.10
	0.4	−8.56	−1.20	0.05	0.47	0.68	0.82	0.92	1.02	1.11
	0.5	−6.41	−0.71	0.25	0.57	0.73	0.84	0.93	1.02	1.11
	0.6	−4.85	−0.36	0.38	0.63	0.76	0.86	0.94	1.02	1.11
	0.7	−3.68	−0.10	0.48	0.68	0.79	0.87	0.95	1.03	1.11
	0.8	−2.77	0.10	0.56	0.71	0.81	0.88	0.95	1.03	1.11
	0.9	−2.04	0.26	0.62	0.74	0.82	0.89	0.95	1.03	1.11
	1.0	−1.45	0.38	0.66	0.76	0.83	0.89	0.96	1.03	1.11
0.3	0.2	−36.37	−7.59	−2.48	−0.79	−0.06	0.29	0.47	0.57	0.61
	0.3	−26.79	−5.07	−1.42	−0.27	0.21	0.42	0.53	0.59	0.61
	0.4	−19.94	−3.49	−0.80	0.02	0.35	0.49	0.56	0.60	0.62
	0.5	−15.18	−2.44	−0.41	0.20	0.43	0.54	0.58	0.61	0.62
	0.6	−11.73	−1.70	−0.13	0.32	0.49	0.56	0.60	0.61	0.62
	0.7	−9.13	−1.14	0.07	0.41	0.53	0.58	0.60	0.61	0.62
	0.8	−7.11	−0.72	0.23	0.48	0.57	0.60	0.61	0.62	0.62
	0.9	−5.49	−0.38	0.35	0.53	0.59	0.61	0.62	0.62	0.62
	1.0	−4.17	−0.11	0.45	0.58	0.61	0.62	0.62	0.62	0.62
0.4	0.2	−64.82	−13.76	−4.74	−1.81	−0.59	−0.02	0.24	0.36	0.39
	0.3	−47.92	−9.38	−2.93	−0.94	−0.16	0.19	0.34	0.39	0.40
	0.4	−35.81	−6.62	−1.88	−0.46	0.07	0.30	0.38	0.41	0.40
	0.5	−27.39	−4.78	−1.20	−0.16	0.22	0.36	0.41	0.42	0.41
	0.6	−21.28	−3.48	−0.73	0.04	0.31	0.41	0.43	0.43	0.41
	0.7	−16.68	−2.51	−0.38	0.20	0.38	0.44	0.45	0.43	0.41
	0.8	−13.10	−1.77	−0.12	0.31	0.44	0.46	0.46	0.44	0.41
	0.9	−10.24	−1.18	0.09	0.40	0.48	0.48	0.46	0.44	0.41
	1.0	−7.90	−0.69	0.26	0.47	0.51	0.50	0.47	0.44	0.41

ED5-1 Wye, 30 Degree, Converging (*Concluded*)

		C_s Values (*Concluded*)								
		Q_s/Q_c								
A_s/A_c	A_b/A_c	0.1	0.2	0.3	0.4	0.5	0.6	0.7	0.8	0.9
0.5	0.2	−101.39	−21.64	−7.61	−3.07	−1.19	−0.34	0.05	0.22	0.26
	0.3	−75.05	−14.87	−4.83	−1.75	−0.54	−0.03	0.19	0.26	0.27
	0.4	−56.18	−10.59	−3.21	−1.02	−0.20	0.13	0.26	0.29	0.27
	0.5	−43.04	−7.74	−2.16	−0.56	0.02	0.23	0.30	0.30	0.27
	0.6	−33.51	−5.72	−1.43	−0.24	0.16	0.30	0.33	0.31	0.28
	0.7	−26.34	−4.22	−0.90	−0.01	0.27	0.35	0.35	0.32	0.28
	0.8	−20.75	−3.06	−0.49	0.16	0.35	0.39	0.37	0.33	0.28
	0.9	−16.29	−2.14	−0.17	0.30	0.41	0.41	0.38	0.33	0.28
	1.0	−12.64	−1.39	0.10	0.41	0.46	0.44	0.39	0.33	0.28
0.6	0.2	−146.06	−31.26	−11.09	−4.56	−1.89	−0.68	−0.12	0.10	0.16
	0.3	−108.19	−21.55	−7.12	−2.69	−0.97	−0.24	0.07	0.17	0.17
	0.4	−81.04	−15.40	−4.80	−1.65	−0.48	−0.01	0.17	0.20	0.18
	0.5	−62.13	−11.31	−3.30	−0.99	−0.17	0.13	0.22	0.22	0.18
	0.6	−48.43	−8.41	−2.25	−0.54	0.03	0.22	0.26	0.24	0.18
	0.7	−38.10	−6.25	−1.49	−0.22	0.18	0.29	0.29	0.25	0.19
	0.8	−30.07	−4.59	−0.90	0.03	0.30	0.34	0.31	0.25	0.19
	0.9	−23.64	−3.27	−0.44	0.23	0.39	0.38	0.33	0.26	0.19
	1.0	−18.39	−2.20	−0.06	0.39	0.46	0.42	0.34	0.27	0.19
0.7	0.2	−198.85	−42.62	−15.17	−6.31	−2.68	−1.04	−0.29	0.01	0.08
	0.3	−147.33	−29.41	−9.78	−3.77	−1.44	−0.45	−0.04	0.10	0.10
	0.4	−110.40	−21.07	−6.64	−2.36	−0.77	−0.14	0.09	0.15	0.11
	0.5	−84.67	−15.50	−4.60	−1.48	−0.36	0.05	0.17	0.17	0.11
	0.6	−66.02	−11.56	−3.19	−0.86	−0.08	0.18	0.23	0.19	0.12
	0.7	−51.97	−8.63	−2.15	−0.42	0.12	0.27	0.27	0.20	0.12
	0.8	−41.04	−6.37	−1.35	−0.08	0.27	0.34	0.29	0.21	0.12
	0.9	−32.30	−4.58	−0.72	0.19	0.39	0.39	0.32	0.22	0.12
	1.0	−25.16	−3.12	−0.21	0.40	0.49	0.43	0.33	0.23	0.13
0.8	0.2	−259.75	−55.70	−19.86	−8.29	−3.56	−1.43	−0.46	−0.06	0.03
	0.3	−192.48	−38.47	−12.84	−4.99	−1.95	−0.66	−0.12	0.05	0.05
	0.4	−144.25	−27.58	−8.74	−3.16	−1.09	−0.26	0.05	0.11	0.06
	0.5	−110.65	−20.32	−6.08	−2.00	−0.55	−0.01	0.15	0.15	0.07
	0.6	−86.30	−15.17	−4.24	−1.20	−0.19	0.15	0.22	0.17	0.08
	0.7	−67.95	−11.34	−2.88	−0.62	0.08	0.27	0.27	0.19	0.08
	0.8	−53.67	−8.40	−1.84	−0.18	0.28	0.36	0.30	0.20	0.08
	0.9	−42.26	−6.05	−1.02	0.16	0.44	0.43	0.33	0.21	0.08
	1.0	−32.93	−4.15	−0.35	0.44	0.56	0.49	0.36	0.22	0.09
0.9	0.2	−328.76	−70.51	−25.16	−10.53	−4.54	−1.84	−0.62	−0.12	0.00
	0.3	−243.63	−48.72	−16.28	−6.35	−2.50	−0.87	−0.20	0.03	0.03
	0.4	−182.60	−34.94	−11.09	−4.03	−1.41	−0.37	0.02	0.10	0.04
	0.5	−140.07	−25.75	−7.74	−2.57	−0.74	−0.06	0.15	0.14	0.05
	0.6	−109.25	−19.24	−5.40	−1.56	−0.28	0.15	0.23	0.17	0.05
	0.7	−86.04	−14.40	−3.68	−0.83	0.06	0.30	0.30	0.20	0.06
	0.8	−67.96	−10.66	−2.37	−0.27	0.31	0.41	0.34	0.21	0.06
	0.9	−53.52	−7.70	−1.33	0.17	0.51	0.50	0.38	0.22	0.06
	1.0	−41.71	−5.29	−0.49	0.52	0.67	0.57	0.41	0.23	0.07
1.0	0.2	−405.88	−87.06	−31.07	−13.01	−5.62	−2.29	−0.77	−0.16	−0.02
	0.3	−300.78	−60.15	−20.11	−7.85	−3.10	−1.09	−0.26	0.02	0.02
	0.4	−225.44	−43.14	−13.70	−4.99	−1.76	−0.47	0.01	0.11	0.04
	0.5	−172.93	−31.80	−9.56	−3.18	−0.92	−0.09	0.17	0.17	0.05
	0.6	−134.89	−23.76	−6.68	−1.94	−0.35	0.17	0.28	0.20	0.06
	0.7	−106.23	−17.78	−4.56	−1.04	0.06	0.36	0.35	0.23	0.06
	0.8	−83.92	−13.18	−2.93	−0.35	0.37	0.50	0.41	0.25	0.06
	0.9	−66.08	−9.52	−1.65	0.19	0.62	0.61	0.46	0.26	0.07
	1.0	−51.51	−6.54	−0.61	0.63	0.81	0.70	0.49	0.28	0.07

ED5-2 Wye, 45 Degree, Converging

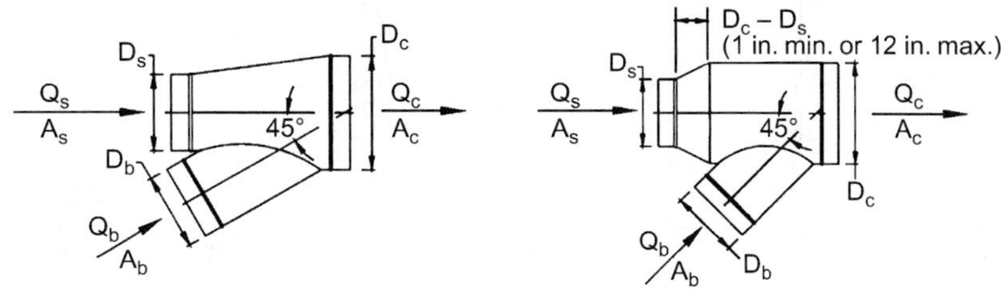

A_s/A_c	A_b/A_c	C_b Values, Q_b/Q_c 0.1	0.2	0.3	0.4	0.5	0.6	0.7	0.8	0.9
0.2	0.2	−25.19	−3.97	−0.64	0.32	0.67	0.82	0.90	0.96	1.08
	0.3	−58.03	−10.14	−2.63	−0.45	0.36	0.69	0.84	0.93	1.08
	0.4	−104.08	−18.80	−5.40	−1.51	−0.07	0.52	0.77	0.88	1.01
	0.5	−163.36	−29.97	−8.97	−2.87	−0.62	0.29	0.67	0.80	0.84
	0.6	−235.59	−43.47	−13.22	−4.44	−1.20	0.12	0.65	0.83	0.85
	0.7	−320.90	−59.38	−18.21	−6.25	−1.84	−0.04	0.68	0.91	0.93
	0.8	−419.32	−77.73	−23.95	−8.33	−2.56	−0.22	0.72	1.02	1.02
	0.9	−530.86	−98.50	−30.44	−10.66	−3.36	−0.40	0.79	1.16	1.14
	1.0	−655.51	−121.72	−37.68	−13.26	−4.25	−0.59	0.87	1.33	1.28
0.3	0.2	−14.27	−1.77	0.13	0.66	0.85	0.93	0.97	1.03	1.21
	0.3	−33.62	−5.28	−0.95	0.27	0.70	0.87	0.94	1.01	1.19
	0.4	−60.85	−10.26	−2.48	−0.30	0.47	0.77	0.88	0.93	1.04
	0.5	−95.87	−16.64	−4.44	−1.00	0.21	0.66	0.82	0.84	0.84
	0.6	−138.38	−24.26	−6.68	−1.73	0.01	0.66	0.88	0.91	0.88
	0.7	−188.60	−33.25	−9.32	−2.58	−0.20	0.68	0.98	1.02	0.95
	0.8	−246.54	−43.60	−12.34	−3.54	−0.43	0.72	1.11	1.15	1.03
	0.9	−312.21	−55.33	−15.76	−4.61	−0.68	0.78	1.26	1.31	1.13
	1.0	−385.59	−68.43	−19.56	−5.79	−0.94	0.86	1.45	1.49	1.24
0.4	0.2	−8.77	−0.64	0.54	0.85	0.95	0.99	1.03	1.09	1.31
	0.3	−21.41	−2.85	−0.10	0.63	0.87	0.96	1.00	1.06	1.26
	0.4	−39.30	−6.02	−1.05	0.28	0.72	0.87	0.91	0.92	1.00
	0.5	−62.10	−9.96	−2.16	−0.06	0.63	0.85	0.90	0.88	0.86
	0.6	−89.77	−14.65	−3.42	−0.38	0.61	0.93	0.99	0.95	0.90
	0.7	−122.46	−20.19	−4.88	−0.74	0.61	1.04	1.12	1.06	0.95
	0.8	−160.18	−26.56	−6.55	−1.15	0.62	1.18	1.29	1.19	1.01
	0.9	−202.93	−33.77	−8.44	−1.60	0.64	1.36	1.48	1.35	1.07
	1.0	−250.70	−41.83	−10.54	−2.09	0.68	1.56	1.71	1.53	1.15
0.5	0.2	−5.45	0.04	0.79	0.97	1.02	1.04	1.07	1.14	1.39
	0.3	−14.10	−1.39	0.40	0.84	0.97	1.00	1.02	1.07	1.28
	0.4	−26.48	−3.53	−0.24	0.59	0.83	0.89	0.88	0.85	0.86
	0.5	−41.84	−5.96	−0.80	0.51	0.88	0.97	0.95	0.90	0.87
	0.6	−60.61	−8.90	−1.46	0.43	0.97	1.09	1.06	0.97	0.90
	0.7	−82.80	−12.36	−2.22	0.35	1.09	1.25	1.20	1.08	0.93
	0.8	−108.39	−16.35	−3.09	0.27	1.24	1.45	1.38	1.20	0.96
	0.9	−137.41	−20.86	−4.07	0.19	1.42	1.68	1.59	1.35	0.99
	1.0	−169.84	−25.90	−5.15	0.11	1.63	1.95	1.83	1.52	1.02
0.6	0.2	−5.54	−0.08	0.70	0.91	0.98	1.01	1.05	1.14	1.42
	0.3	−14.48	−1.75	0.13	0.64	0.81	0.88	0.92	0.98	1.19
	0.4	−27.10	−4.14	−0.68	0.26	0.57	0.68	0.71	0.72	0.76
	0.5	−42.84	−6.91	−1.50	−0.02	0.47	0.64	0.68	0.69	0.70
	0.6	−62.07	−10.28	−2.48	−0.34	0.37	0.61	0.67	0.66	0.63
	0.7	−84.79	−14.26	−3.62	−0.71	0.27	0.59	0.67	0.63	0.54
	0.8	−111.02	−18.84	−4.92	−1.12	0.16	0.58	0.67	0.61	0.44
	0.9	−140.76	−24.03	−6.40	−1.57	0.04	0.58	0.68	0.59	0.31
	1.0	−174.01	−29.83	−8.04	−2.07	−0.08	0.58	0.70	0.56	0.15
0.7	0.2	−3.96	0.25	0.83	0.97	1.01	1.04	1.08	1.17	1.47
	0.3	−11.07	−1.10	0.34	0.71	0.83	0.87	0.90	0.95	1.13
	0.4	−20.92	−2.92	−0.27	0.43	0.65	0.72	0.73	0.73	0.77
	0.5	−33.20	−5.01	−0.85	0.24	0.59	0.69	0.71	0.69	0.70
	0.6	−48.21	−7.55	−1.55	0.03	0.53	0.68	0.69	0.65	0.61
	0.7	−65.95	−10.56	−2.37	−0.20	0.48	0.68	0.69	0.62	0.49
	0.8	−86.42	−14.01	−3.30	−0.46	0.43	0.68	0.69	0.58	0.35
	0.9	−109.65	−17.93	−4.35	−0.75	0.38	0.70	0.70	0.53	0.18
	1.0	−135.63	−22.32	−5.53	−1.07	0.33	0.72	0.71	0.48	−0.03

ED5-2 Wye, 45 Degree, Converging (*Continued*)

		C_b Values (Concluded)								
		Q_b/Q_c								
A_s/A_c	A_b/A_c	0.1	0.2	0.3	0.4	0.5	0.6	0.7	0.8	0.9
0.8	0.2	−2.78	0.50	0.91	1.01	1.03	1.05	1.09	1.18	1.49
	0.3	−8.58	−0.65	0.47	0.74	0.82	0.85	0.86	0.89	1.02
	0.4	−16.29	−2.00	0.05	0.56	0.71	0.75	0.74	0.74	0.78
	0.5	−25.98	−3.59	−0.37	0.44	0.68	0.73	0.72	0.69	0.69
	0.6	−37.82	−5.52	−0.87	0.31	0.65	0.72	0.70	0.64	0.58
	0.7	−51.83	−7.79	−1.44	0.17	0.63	0.73	0.69	0.59	0.43
	0.8	−68.01	−10.42	−2.10	0.01	0.62	0.75	0.69	0.53	0.25
	0.9	−86.37	−13.39	−2.84	−0.16	0.61	0.77	0.68	0.47	0.03
	1.0	−106.91	−16.73	−3.68	−0.35	0.61	0.79	0.68	0.38	−0.25
0.9	0.2	−1.87	0.68	0.98	1.03	1.05	1.06	1.09	1.18	1.49
	0.3	−6.70	−0.33	0.54	0.74	0.79	0.80	0.80	0.81	0.87
	0.4	−12.69	−1.29	0.29	0.66	0.76	0.77	0.75	0.74	0.78
	0.5	−20.37	−2.48	0.00	0.59	0.74	0.75	0.72	0.69	0.67
	0.6	−29.77	−3.94	−0.34	0.52	0.73	0.75	0.70	0.63	0.54
	0.7	−40.89	−5.66	−0.73	0.45	0.74	0.76	0.68	0.56	0.36
	0.8	−53.74	−7.64	−1.18	0.37	0.76	0.78	0.67	0.48	0.13
	0.9	−68.32	−9.89	−1.69	0.28	0.77	0.80	0.65	0.38	−0.15
	1.0	−84.66	−12.42	−2.27	0.18	0.80	0.83	0.62	0.26	−0.49
1.0	0.2	−1.17	0.81	1.02	1.05	1.05	1.06	1.09	1.18	1.48
	0.3	−5.09	−0.02	0.64	0.78	0.81	0.81	0.80	0.80	0.86
	0.4	−9.81	−0.72	0.48	0.74	0.79	0.78	0.76	0.74	0.77
	0.5	−15.89	−1.61	0.29	0.71	0.79	0.77	0.72	0.68	0.65
	0.6	−23.34	−2.69	0.07	0.68	0.80	0.77	0.69	0.60	0.49
	0.7	−32.15	−3.96	−0.18	0.66	0.82	0.78	0.67	0.51	0.27
	0.8	−42.35	−5.44	−0.47	0.64	0.85	0.79	0.63	0.41	0.00
	0.9	−53.94	−7.12	−0.80	0.61	0.88	0.81	0.60	0.28	−0.34
	1.0	−66.93	−9.01	−1.17	0.58	0.92	0.82	0.55	0.13	−0.75

		C_s Values								
		Q_s/Q_c								
A_s/A_c	A_b/A_c	0.1	0.2	0.3	0.4	0.5	0.6	0.7	0.8	0.9
0.2	0.2	−10.16	−2.08	−0.43	0.24	0.62	0.88	1.10	1.29	1.46
	0.3	−7.83	−1.20	0.03	0.50	0.77	0.97	1.14	1.30	1.46
	0.4	−5.62	−0.59	0.30	0.65	0.85	1.01	1.16	1.31	1.46
	0.5	−3.96	−0.18	0.48	0.74	0.90	1.04	1.18	1.32	1.47
	0.6	−2.71	0.12	0.60	0.80	0.94	1.06	1.19	1.32	1.47
	0.7	−1.75	0.34	0.70	0.85	0.96	1.07	1.19	1.32	1.47
	0.8	−0.99	0.52	0.77	0.88	0.98	1.08	1.20	1.32	1.47
	0.9	−0.38	0.66	0.82	0.91	0.99	1.09	1.20	1.33	1.47
	1.0	0.13	0.77	0.87	0.93	1.00	1.10	1.20	1.33	1.47
0.3	0.2	−23.33	−5.14	−1.67	−0.44	0.12	0.42	0.58	0.67	0.72
	0.3	−18.44	−3.44	−0.84	0.00	0.36	0.54	0.64	0.69	0.73
	0.4	−13.64	−2.22	−0.34	0.25	0.49	0.60	0.67	0.70	0.73
	0.5	−10.00	−1.37	0.00	0.41	0.57	0.64	0.69	0.71	0.73
	0.6	−7.26	−0.75	0.24	0.52	0.62	0.67	0.70	0.72	0.73
	0.7	−5.15	−0.29	0.41	0.60	0.66	0.69	0.71	0.72	0.73
	0.8	−3.48	0.07	0.55	0.66	0.69	0.70	0.71	0.72	0.73
	0.9	−2.14	0.36	0.65	0.71	0.72	0.72	0.72	0.72	0.73
	1.0	−1.03	0.60	0.74	0.75	0.73	0.73	0.72	0.72	0.73
0.4	0.2	−42.17	−9.48	−3.34	−1.23	−0.31	0.12	0.33	0.42	0.44
	0.3	−33.68	−6.60	−1.98	−0.53	0.05	0.31	0.41	0.45	0.45
	0.4	−25.24	−4.51	−1.13	−0.13	0.25	0.40	0.46	0.47	0.45
	0.5	−18.83	−3.04	−0.57	0.13	0.37	0.46	0.48	0.48	0.46
	0.6	−13.99	−1.97	−0.17	0.31	0.46	0.50	0.50	0.48	0.46
	0.7	−10.27	−1.17	0.12	0.44	0.52	0.53	0.51	0.49	0.46
	0.8	−7.32	−0.54	0.35	0.54	0.57	0.55	0.52	0.49	0.46
	0.9	−4.94	−0.04	0.53	0.62	0.61	0.57	0.53	0.49	0.46
	1.0	−2.98	0.37	0.68	0.68	0.64	0.58	0.54	0.50	0.46
0.5	0.2	−66.95	−15.18	−5.49	−2.21	−0.81	−0.16	0.14	0.26	0.28
	0.3	−53.80	−10.77	−3.45	−1.17	−0.27	0.11	0.26	0.30	0.29
	0.4	−40.66	−7.54	−2.16	−0.57	0.02	0.25	0.32	0.33	0.30
	0.5	−30.68	−5.27	−1.30	−0.18	0.21	0.33	0.36	0.34	0.30
	0.6	−23.15	−3.62	−0.69	0.09	0.33	0.39	0.38	0.35	0.30
	0.7	−17.34	−2.38	−0.24	0.29	0.42	0.43	0.40	0.35	0.30
	0.8	−12.75	−1.41	0.11	0.44	0.49	0.47	0.41	0.36	0.30
	0.9	−9.04	−0.64	0.39	0.56	0.55	0.49	0.43	0.36	0.30
	1.0	−5.99	0.00	0.61	0.65	0.59	0.51	0.43	0.36	0.30

ED5-2 Wye, 45 Degree, Converging (*Concluded*)

		C_s Values (*Concluded*)								
		Q_s/Q_c								
A_s/A_c	A_b/A_c	**0.1**	**0.2**	**0.3**	**0.4**	**0.5**	**0.6**	**0.7**	**0.8**	**0.9**
0.6	0.2	−97.90	−22.29	−8.18	−3.41	−1.39	−0.46	−0.03	0.13	0.16
	0.3	−79.03	−15.99	−5.28	−1.94	−0.64	−0.09	0.13	0.19	0.17
	0.4	−60.15	−11.37	−3.44	−1.09	−0.23	0.10	0.21	0.22	0.18
	0.5	−45.80	−8.13	−2.22	−0.55	0.03	0.22	0.26	0.24	0.18
	0.6	−34.97	−5.77	−1.35	−0.17	0.20	0.30	0.30	0.25	0.18
	0.7	−26.62	−3.98	−0.71	0.11	0.33	0.36	0.32	0.26	0.19
	0.8	−20.02	−2.59	−0.21	0.33	0.43	0.41	0.34	0.26	0.19
	0.9	−14.68	−1.48	0.18	0.49	0.51	0.44	0.35	0.27	0.19
	1.0	−10.29	−0.57	0.51	0.63	0.57	0.47	0.37	0.27	0.19
0.7	0.2	−135.28	−30.88	−11.42	−4.85	−2.08	−0.80	−0.21	0.02	0.06
	0.3	−109.64	−22.35	−7.50	−2.88	−1.07	−0.31	0.00	0.09	0.07
	0.4	−83.96	−16.08	−5.02	−1.73	−0.52	−0.05	0.11	0.13	0.08
	0.5	−64.44	−11.67	−3.36	−0.99	−0.17	0.11	0.18	0.15	0.09
	0.6	−49.71	−8.47	−2.19	−0.48	0.06	0.22	0.22	0.17	0.09
	0.7	−38.35	−6.04	−1.31	−0.10	0.24	0.30	0.26	0.18	0.09
	0.8	−29.37	−4.16	−0.64	0.18	0.37	0.36	0.28	0.19	0.09
	0.9	−22.12	−2.65	−0.10	0.41	0.47	0.40	0.30	0.19	0.09
	1.0	−16.14	−1.41	0.33	0.60	0.55	0.44	0.32	0.20	0.09
0.8	0.2	−179.32	−41.01	−15.25	−6.55	−2.88	−1.19	−0.41	−0.10	−0.04
	0.3	−145.86	−29.89	−10.14	−3.99	−1.58	−0.55	−0.13	0.00	−0.02
	0.4	−112.34	−21.71	−6.91	−2.50	−0.86	−0.22	0.01	0.05	−0.01
	0.5	−86.85	−15.96	−4.75	−1.54	−0.41	−0.01	0.10	0.08	0.00
	0.6	−67.62	−11.78	−3.22	−0.87	−0.10	0.13	0.16	0.10	0.00
	0.7	−52.79	−8.62	−2.08	−0.38	0.12	0.23	0.20	0.11	0.00
	0.8	−41.06	−6.16	−1.20	0.00	0.29	0.31	0.23	0.12	0.01
	0.9	−31.59	−4.19	−0.51	0.29	0.43	0.37	0.26	0.13	0.01
	1.0	−23.78	−2.58	0.06	0.53	0.54	0.42	0.28	0.14	0.01
0.9	0.2	−230.27	−52.75	−19.69	−8.53	−3.81	−1.63	−0.63	−0.22	−0.13
	0.3	−187.95	−38.69	−13.24	−5.29	−2.16	−0.83	−0.28	−0.10	−0.10
	0.4	−145.53	−28.34	−9.15	−3.41	−1.26	−0.41	−0.10	−0.04	−0.09
	0.5	−113.27	−21.07	−6.42	−2.19	−0.69	−0.15	0.01	0.00	−0.09
	0.6	−88.94	−15.78	−4.48	−1.35	−0.30	0.03	0.09	0.03	−0.08
	0.7	−70.16	−11.78	−3.04	−0.73	−0.02	0.16	0.14	0.04	−0.08
	0.8	−55.33	−8.67	−1.93	−0.25	0.20	0.26	0.18	0.06	−0.07
	0.9	−43.33	−6.18	−1.05	0.12	0.37	0.33	0.21	0.07	−0.07
	1.0	−33.46	−4.14	−0.34	0.42	0.50	0.39	0.24	0.08	−0.07
1.0	0.2	−288.39	−66.15	−24.77	−10.80	−4.88	−2.14	−0.87	−0.35	−0.22
	0.3	−236.14	−48.79	−16.81	−6.80	−2.85	−1.15	−0.44	−0.20	−0.19
	0.4	−183.77	−36.02	−11.76	−4.47	−1.73	−0.63	−0.22	−0.12	−0.18
	0.5	−143.95	−27.05	−8.39	−2.98	−1.03	−0.31	−0.08	−0.08	−0.17
	0.6	−113.91	−20.52	−6.00	−1.93	−0.55	−0.09	0.01	−0.04	−0.16
	0.7	−90.73	−15.58	−4.23	−1.17	−0.20	0.07	0.08	−0.02	−0.16
	0.8	−72.41	−11.74	−2.86	−0.58	0.06	0.19	0.13	−0.01	−0.16
	0.9	−57.61	−8.66	−1.77	−0.12	0.27	0.28	0.16	0.01	−0.15
	1.0	−45.42	−6.15	−0.88	0.25	0.44	0.36	0.20	0.02	−0.15

ED5-3 Tee, D_c < or = 10 in., Converging

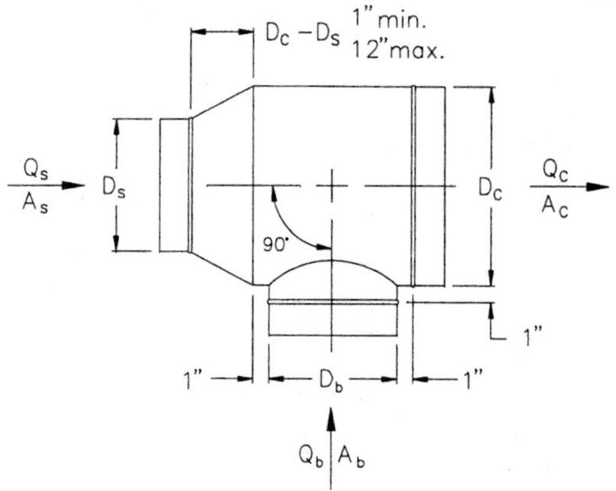

		C_b Values								
		Q_b/Q_c								
A_s/A_c	A_b/A_c	0.1	0.2	0.3	0.4	0.5	0.6	0.7	0.8	0.9
0.2	0.2	–24.56	–3.63	–0.36	0.59	0.93	1.08	1.14	1.19	1.27
	0.3	–56.72	–9.54	–2.15	–0.01	0.78	1.10	1.23	1.30	1.39
	0.4	–101.83	–17.86	–4.68	–0.87	0.52	1.09	1.32	1.41	1.48
	0.5	–159.91	–28.59	–7.98	–2.02	0.17	1.05	1.40	1.51	1.51
	0.6	–230.83	–41.68	–11.98	–3.39	–0.24	1.03	1.53	1.66	1.61
	0.7	–314.56	–57.10	–16.68	–4.98	–0.69	1.04	1.71	1.90	1.82
	0.8	–411.18	–74.90	–22.10	–6.82	–1.21	1.04	1.92	2.16	2.05
	0.9	–520.69	–95.08	–28.25	–8.90	–1.81	1.04	2.15	2.45	2.31
	1.0	–643.09	–117.63	–35.12	–11.24	–2.47	1.04	2.41	2.78	2.58
0.3	0.2	–14.05	–1.55	0.36	0.89	1.08	1.16	1.19	1.23	1.34
	0.3	–33.18	–4.91	–0.58	0.64	1.07	1.23	1.30	1.34	1.44
	0.4	–60.09	–9.68	–1.94	0.24	1.00	1.29	1.39	1.42	1.47
	0.5	–94.80	–15.89	–3.74	–0.33	0.87	1.32	1.46	1.46	1.38
	0.6	–136.97	–23.33	–5.84	–0.92	0.81	1.45	1.65	1.66	1.53
	0.7	–186.81	–32.14	–8.32	–1.62	0.74	1.61	1.88	1.88	1.70
	0.8	–244.33	–42.30	–11.19	–2.43	0.65	1.78	2.14	2.13	1.88
	0.9	–309.54	–53.82	–14.44	–3.35	0.54	1.98	2.42	2.41	2.08
	1.0	–382.43	–66.70	–18.08	–4.39	0.42	2.19	2.74	2.72	2.29
0.4	0.2	–8.95	–0.54	0.71	1.04	1.15	1.20	1.22	1.26	1.40
	0.3	–21.82	–2.70	0.16	0.94	1.19	1.29	1.32	1.35	1.47
	0.4	–39.99	–5.81	–0.67	0.73	1.19	1.35	1.39	1.39	1.41
	0.5	–63.37	–9.82	–1.75	0.45	1.18	1.42	1.47	1.43	1.32
	0.6	–91.72	–14.59	–2.97	0.20	1.26	1.60	1.67	1.60	1.43
	0.7	–125.23	–20.24	–4.41	–0.10	1.34	1.81	1.90	1.81	1.56
	0.8	–163.91	–26.77	–6.09	–0.45	1.43	2.04	2.16	2.03	1.69
	0.9	–207.76	–34.17	–7.99	–0.85	1.53	2.30	2.44	2.28	1.82
	1.0	–256.79	–42.45	–10.12	–1.30	1.63	2.58	2.75	2.54	1.95
0.5	0.2	–6.03	0.04	0.91	1.13	1.20	1.22	1.24	1.29	1.44
	0.3	–15.35	–1.46	0.56	1.09	1.25	1.30	1.32	1.35	1.46
	0.4	–28.59	–3.67	–0.01	0.96	1.26	1.34	1.35	1.32	1.29
	0.5	–45.45	–6.42	–0.66	0.85	1.33	1.45	1.45	1.38	1.24
	0.6	–65.92	–9.70	–1.41	0.78	1.46	1.64	1.63	1.53	1.32
	0.7	–90.12	–13.58	–2.29	0.69	1.61	1.86	1.85	1.70	1.39
	0.8	–118.07	–18.07	–3.32	0.57	1.78	2.11	2.09	1.89	1.46
	0.9	–149.75	–23.18	–4.49	0.43	1.96	2.38	2.35	2.09	1.53
	1.0	–185.19	–28.89	–5.81	0.27	2.16	2.67	2.63	2.31	1.57
0.6	0.2	–4.20	0.39	1.03	1.18	1.22	1.24	1.26	1.30	1.47
	0.3	–11.33	–0.72	0.79	1.16	1.27	1.30	1.31	1.33	1.43
	0.4	–21.57	–2.42	0.35	1.05	1.25	1.29	1.27	1.22	1.12
	0.5	–34.29	–4.35	–0.03	1.07	1.38	1.44	1.41	1.32	1.16
	0.6	–49.85	–6.73	–0.50	1.08	1.54	1.63	1.57	1.45	1.19
	0.7	–68.26	–9.55	–1.06	1.09	1.71	1.83	1.76	1.58	1.21
	0.8	–89.52	–12.81	–1.72	1.10	1.91	2.07	1.97	1.73	1.22
	0.9	–113.64	–16.52	–2.47	1.10	2.12	2.32	2.20	1.88	1.21
	1.0	–140.62	–20.68	–3.33	1.09	2.35	2.60	2.44	2.03	1.16

ED5-3 Tee, D_c < or = 10 in., Converging (*Continued*)

		C_b Values (*Concluded*)								
		Q_b/Q_c								
A_s/A_c	A_b/A_c	0.1	0.2	0.3	0.4	0.5	0.6	0.7	0.8	0.9
0.7	0.2	–3.00	0.62	1.10	1.21	1.23	1.24	1.26	1.31	1.49
	0.3	–8.74	–0.27	0.91	1.19	1.26	1.27	1.27	1.28	1.36
	0.4	–16.90	–1.59	0.58	1.11	1.25	1.27	1.24	1.18	1.06
	0.5	–26.99	–3.06	0.33	1.17	1.38	1.41	1.36	1.26	1.06
	0.6	–39.35	–4.86	0.02	1.22	1.54	1.57	1.50	1.35	1.05
	0.7	–53.97	–7.01	–0.35	1.29	1.72	1.76	1.65	1.45	1.02
	0.8	–70.87	–9.50	–0.79	1.35	1.91	1.97	1.82	1.54	0.96
	0.9	–90.04	–12.34	–1.31	1.41	2.12	2.19	2.00	1.64	0.86
	1.0	–111.50	–15.53	–1.89	1.46	2.34	2.43	2.19	1.73	0.72
0.8	0.2	–2.20	0.76	1.14	1.22	1.24	1.24	1.26	1.31	1.49
	0.3	–7.04	–0.01	0.95	1.18	1.23	1.23	1.23	1.22	1.27
	0.4	–13.77	–1.06	0.71	1.13	1.24	1.24	1.20	1.13	1.00
	0.5	–22.11	–2.24	0.54	1.20	1.36	1.36	1.30	1.19	0.97
	0.6	–32.33	–3.69	0.31	1.27	1.50	1.50	1.41	1.25	0.90
	0.7	–44.42	–5.41	0.04	1.34	1.66	1.65	1.53	1.30	0.81
	0.8	–58.40	–7.42	–0.29	1.42	1.83	1.83	1.65	1.35	0.67
	0.9	–74.28	–9.72	–0.67	1.49	2.01	2.01	1.78	1.38	0.49
	1.0	–92.06	–12.30	–1.12	1.56	2.21	2.20	1.92	1.40	0.24
0.9	0.2	–1.67	0.85	1.16	1.22	1.23	1.24	1.25	1.30	1.48
	0.3	–5.95	0.12	0.95	1.14	1.18	1.18	1.16	1.15	1.14
	0.4	–11.68	–0.74	0.77	1.12	1.20	1.20	1.16	1.08	0.93
	0.5	–18.85	–1.74	0.63	1.18	1.31	1.30	1.23	1.11	0.86
	0.6	–27.63	–2.98	0.44	1.24	1.42	1.41	1.31	1.13	0.75
	0.7	–38.04	–4.45	0.21	1.30	1.55	1.53	1.39	1.14	0.58
	0.8	–50.07	–6.17	–0.07	1.36	1.69	1.66	1.47	1.13	0.37
	0.9	–63.75	–8.14	–0.40	1.42	1.83	1.79	1.54	1.11	0.09
	1.0	–79.08	–10.36	–0.79	1.46	1.98	1.92	1.61	1.06	–0.26
1.0	0.2	–1.33	0.89	1.16	1.21	1.22	1.22	1.24	1.29	1.46
	0.3	–5.30	0.15	0.90	1.08	1.11	1.11	1.09	1.06	0.99
	0.4	–10.31	–0.57	0.78	1.09	1.16	1.15	1.11	1.03	0.86
	0.5	–16.71	–1.47	0.64	1.13	1.24	1.22	1.15	1.03	0.74
	0.6	–24.56	–2.59	0.46	1.17	1.32	1.30	1.20	1.01	0.57
	0.7	–33.87	–3.93	0.23	1.20	1.41	1.38	1.24	0.97	0.34
	0.8	–44.64	–5.49	–0.05	1.22	1.51	1.46	1.27	0.91	0.05
	0.9	–56.89	–7.29	–0.38	1.24	1.59	1.54	1.28	0.82	–0.33
	1.0	–70.62	–9.32	–0.77	1.24	1.68	1.61	1.28	0.69	–0.80

ED5–3 Tee, D_c < or = 10 in., Converging

		C_s Values								
		Q_s/Q_c								
A_s/A_c	A_b/A_c	0.1	0.2	0.3	0.4	0.5	0.6	0.7	0.8	0.9
0.2	0.2	18.11	3.42	1.62	1.11	0.90	0.80	0.74	0.70	0.68
	0.3	12.67	2.79	1.45	1.04	0.87	0.78	0.73	0.70	0.68
	0.4	9.98	2.47	1.36	1.01	0.85	0.77	0.72	0.69	0.67
	0.5	8.39	2.27	1.30	0.98	0.84	0.76	0.72	0.69	0.67
	0.6	7.34	2.13	1.26	0.96	0.83	0.76	0.72	0.69	0.67
	0.7	6.61	2.02	1.22	0.95	0.82	0.75	0.71	0.69	0.67
	0.8	6.08	1.94	1.19	0.93	0.81	0.75	0.71	0.68	0.67
	0.9	5.68	1.87	1.17	0.92	0.80	0.74	0.70	0.68	0.66
	1.0	4.55	1.61	1.05	0.86	0.76	0.71	0.68	0.66	0.65
0.3	0.2	44.33	7.19	2.80	1.57	1.08	0.84	0.71	0.63	0.57
	0.3	29.24	5.46	2.33	1.40	1.00	0.80	0.69	0.62	0.57
	0.4	21.88	4.59	2.09	1.30	0.96	0.78	0.67	0.61	0.56
	0.5	17.62	4.06	1.93	1.24	0.92	0.76	0.66	0.60	0.56
	0.6	14.90	3.71	1.82	1.19	0.90	0.74	0.65	0.59	0.55
	0.7	13.06	3.45	1.74	1.15	0.88	0.73	0.64	0.59	0.55
	0.8	11.78	3.26	1.67	1.12	0.86	0.72	0.63	0.58	0.54
	0.9	9.02	2.64	1.41	0.97	0.77	0.66	0.59	0.54	0.51
	1.0	8.36	2.52	1.36	0.95	0.75	0.65	0.58	0.54	0.51

ED5–3 Tee, D_c < or = 10 in., Converging (*Continued*)

		C_s Values (*Concluded*)								
		Q_s/Q_c								
A_s/A_c	A_b/A_c	0.1	0.2	0.3	0.4	0.5	0.6	0.7	0.8	0.9
0.4	0.2	78.99	12.25	4.42	2.26	1.39	0.97	0.74	0.60	0.50
	0.3	50.14	8.96	3.54	1.92	1.24	0.90	0.70	0.57	0.49
	0.4	36.26	7.32	3.08	1.74	1.16	0.85	0.67	0.56	0.48
	0.5	28.38	6.35	2.80	1.63	1.10	0.82	0.65	0.54	0.47
	0.6	23.50	5.72	2.61	1.54	1.05	0.79	0.63	0.53	0.46
	0.7	20.32	5.27	2.46	1.47	1.02	0.77	0.62	0.52	0.45
	0.8	14.94	4.13	1.98	1.21	0.85	0.65	0.53	0.46	0.40
	0.9	13.55	3.88	1.89	1.16	0.82	0.63	0.52	0.45	0.39
	1.0	12.66	3.69	1.80	1.12	0.79	0.62	0.51	0.44	0.39
0.5	0.2	114.73	17.76	6.27	3.07	1.79	1.16	0.81	0.60	0.46
	0.3	70.56	12.71	4.92	2.56	1.56	1.05	0.75	0.56	0.44
	0.4	49.68	10.24	4.23	2.29	1.43	0.98	0.71	0.54	0.42
	0.5	38.12	8.81	3.81	2.11	1.34	0.93	0.68	0.52	0.41
	0.6	31.23	7.90	3.53	1.99	1.27	0.88	0.65	0.50	0.39
	0.7	21.87	6.00	2.75	1.57	1.01	0.71	0.52	0.40	0.32
	0.8	19.30	5.57	2.59	1.49	0.96	0.67	0.50	0.38	0.30
	0.9	17.84	5.27	2.46	1.42	0.92	0.65	0.48	0.37	0.29
	1.0	17.16	5.05	2.36	1.36	0.88	0.62	0.46	0.35	0.28
0.6	0.2	142.32	22.64	8.06	3.91	2.23	1.39	0.92	0.63	0.44
	0.3	84.89	16.05	6.28	3.24	1.92	1.23	0.83	0.58	0.41
	0.4	58.43	12.90	5.39	2.88	1.75	1.14	0.78	0.55	0.39
	0.5	44.34	11.13	4.86	2.66	1.63	1.07	0.74	0.52	0.37
	0.6	29.06	8.20	3.69	2.04	1.25	0.81	0.55	0.38	0.26
	0.7	24.71	7.51	3.44	1.91	1.18	0.77	0.52	0.35	0.24
	0.8	22.56	7.06	3.26	1.81	1.11	0.72	0.48	0.33	0.22
	0.9	21.89	6.78	3.12	1.73	1.06	0.68	0.45	0.30	0.20
	1.0	22.24	6.61	3.00	1.65	1.00	0.65	0.43	0.28	0.18
0.7	0.2	152.32	25.82	9.48	4.66	2.65	1.63	1.04	0.68	0.44
	0.3	87.85	18.38	7.46	3.88	2.29	1.44	0.94	0.62	0.40
	0.4	59.34	14.92	6.47	3.48	2.09	1.33	0.87	0.58	0.37
	0.5	35.18	10.56	4.78	2.60	1.55	0.97	0.62	0.38	0.22
	0.6	28.26	9.51	4.41	2.42	1.45	0.90	0.57	0.35	0.19
	0.7	25.45	8.91	4.16	2.28	1.36	0.85	0.53	0.32	0.17
	0.8	25.21	8.60	3.99	2.18	1.29	0.79	0.49	0.28	0.14
	0.9	26.68	8.48	3.86	2.08	1.22	0.74	0.45	0.25	0.12
	1.0	29.34	8.49	3.77	2.01	1.16	0.70	0.41	0.22	0.10
0.8	0.2	136.74	26.38	10.30	5.22	3.01	1.85	1.17	0.74	0.45
	0.3	75.52	19.20	8.32	4.45	2.64	1.66	1.06	0.67	0.41
	0.4	37.55	12.79	5.92	3.23	1.91	1.17	0.72	0.42	0.21
	0.5	27.25	11.28	5.41	2.98	1.77	1.08	0.66	0.37	0.18
	0.6	24.23	10.57	5.10	2.81	1.66	1.01	0.60	0.33	0.14
	0.7	25.36	10.32	4.91	2.69	1.57	0.94	0.55	0.29	0.11
	0.8	29.09	10.37	4.80	2.59	1.50	0.88	0.50	0.25	0.08
	0.9	34.55	10.60	4.74	2.50	1.42	0.82	0.46	0.21	0.05
	1.0	41.23	10.98	4.71	2.43	1.36	0.77	0.41	0.18	0.01
0.9	0.2	90.70	23.73	10.34	5.54	3.28	2.05	1.30	0.81	0.47
	0.3	29.93	14.20	6.95	3.86	2.30	1.41	0.85	0.48	0.22
	0.4	16.27	12.21	6.28	3.55	2.12	1.29	0.77	0.42	0.18
	0.5	14.80	11.58	5.96	3.35	1.99	1.20	0.70	0.37	0.14
	0.6	19.43	11.62	5.81	3.23	1.89	1.13	0.64	0.32	0.10
	0.7	27.55	12.06	5.77	3.14	1.81	1.06	0.59	0.27	0.06
	0.8	37.84	12.73	5.79	3.07	1.74	0.99	0.53	0.23	0.02
	0.9	49.59	13.57	5.85	3.01	1.67	0.93	0.48	0.18	–0.02
	1.0	62.35	14.52	5.94	2.97	1.61	0.87	0.42	0.14	–0.06
1.0	0.2	–6.40	12.70	7.32	4.31	2.64	1.64	1.00	0.56	0.25
	0.3	–17.35	10.90	6.66	3.97	2.44	1.51	0.90	0.49	0.20
	0.4	–11.05	11.02	6.50	3.82	2.32	1.41	0.83	0.43	0.15
	0.5	2.15	11.91	6.54	3.74	2.23	1.33	0.76	0.38	0.10
	0.6	18.80	13.18	6.67	3.70	2.16	1.26	0.70	0.32	0.06
	0.7	37.42	14.67	6.86	3.68	2.09	1.19	0.63	0.26	0.01
	0.8	57.27	16.30	7.09	3.67	2.03	1.12	0.57	0.21	–0.04
	0.9	77.95	18.02	7.35	3.66	1.97	1.06	0.51	0.15	–0.09
	1.0	99.20	19.80	7.61	3.67	1.92	1.00	0.45	0.10	–0.14

ED5–3 Tee, D_c > 10 in., Converging *(Continued)*

		C_b Values								
		Q_b/Q_c								
A_s/A_c	A_b/A_c	0.1	0.2	0.3	0.4	0.5	0.6	0.7	0.8	0.9
0.2	0.2	−26.08	−4.19	−0.70	0.33	0.71	0.87	0.93	0.95	0.93
	0.3	−59.71	−10.53	−2.72	−0.43	0.43	0.78	0.91	0.95	0.91
	0.4	−106.78	−19.39	−5.53	−1.46	0.05	0.67	0.91	0.97	0.91
	0.5	−167.36	−30.77	−9.12	−2.78	−0.42	0.55	0.93	1.02	0.93
	0.6	−241.50	−44.68	−13.50	−4.37	−0.97	0.42	0.96	1.10	0.98
	0.7	−329.25	−61.15	−18.68	−6.25	−1.62	0.27	1.02	1.21	1.06
	0.8	−430.67	−80.18	−24.67	−8.42	−2.37	0.10	1.09	1.35	1.17
	0.9	−545.81	−101.78	−31.47	−10.89	−3.22	−0.08	1.17	1.52	1.31
	1.0	−674.72	−125.98	−39.08	−13.64	−4.17	−0.28	1.28	1.72	1.48
0.3	0.2	−15.50	−2.16	−0.04	0.58	0.81	0.90	0.93	0.94	0.91
	0.3	−35.76	−5.90	−1.20	0.16	0.66	0.85	0.92	0.92	0.88
	0.4	−64.09	−11.09	−2.78	−0.38	0.48	0.82	0.93	0.94	0.86
	0.5	−100.54	−17.73	−4.78	−1.06	0.29	0.80	0.97	0.98	0.87
	0.6	−145.16	−25.85	−7.21	−1.86	0.06	0.80	1.05	1.05	0.90
	0.7	−198.01	−35.46	−10.08	−2.81	−0.19	0.82	1.15	1.16	0.96
	0.8	−259.13	−46.56	−13.39	−3.89	−0.47	0.85	1.28	1.30	1.05
	0.9	−328.59	−59.18	−17.15	−5.11	−0.78	0.89	1.44	1.47	1.17
	1.0	−406.44	−73.33	−21.37	−6.48	−1.12	0.94	1.63	1.68	1.32
0.4	0.2	−10.31	−1.18	0.26	0.69	0.84	0.91	0.93	0.93	0.90
	0.3	−23.96	−3.65	−0.48	0.43	0.75	0.88	0.91	0.91	0.86
	0.4	−42.98	−7.03	−1.46	0.11	0.67	0.87	0.93	0.91	0.84
	0.5	−67.44	−11.35	−2.69	−0.26	0.59	0.90	0.97	0.94	0.84
	0.6	−97.39	−16.60	−4.17	−0.69	0.52	0.95	1.06	1.01	0.87
	0.7	−132.88	−22.81	−5.91	−1.17	0.46	1.03	1.17	1.11	0.92
	0.8	−173.96	−29.99	−7.90	−1.73	0.40	1.15	1.33	1.24	1.00
	0.9	−220.69	−38.15	−10.16	−2.35	0.35	1.29	1.51	1.40	1.11
	1.0	−273.12	−47.31	−12.70	−3.04	0.29	1.45	1.74	1.61	1.26
0.5	0.2	−7.26	−0.62	0.43	0.75	0.86	0.91	0.93	0.93	0.90
	0.3	−16.99	−2.35	−0.07	0.57	0.80	0.89	0.91	0.90	0.87
	0.4	−30.49	−4.67	−0.72	0.38	0.76	0.89	0.92	0.90	0.85
	0.5	−47.82	−7.61	−1.50	0.19	0.75	0.93	0.97	0.93	0.85
	0.6	−69.03	−11.17	−2.42	−0.03	0.76	1.01	1.05	0.98	0.88
	0.7	−94.17	−15.37	−3.49	−0.26	0.80	1.13	1.17	1.07	0.93
	0.8	−123.30	−20.22	−4.71	−0.50	0.87	1.29	1.33	1.20	1.02
	0.9	−156.48	−25.73	−6.09	−0.77	0.96	1.48	1.53	1.36	1.13
	1.0	−193.74	−31.92	−7.63	−1.07	1.06	1.71	1.77	1.56	1.28
0.6	0.2	−5.28	−0.27	0.54	0.78	0.88	0.91	0.93	0.93	0.91
	0.3	−12.43	−1.51	0.18	0.66	0.83	0.89	0.91	0.91	0.89
	0.4	−22.29	−3.15	−0.25	0.55	0.82	0.90	0.92	0.91	0.88
	0.5	−34.92	−5.19	−0.74	0.46	0.84	0.95	0.96	0.93	0.89
	0.6	−50.35	−7.64	−1.30	0.38	0.91	1.05	1.04	0.98	0.93
	0.7	−68.66	−10.52	−1.94	0.32	1.01	1.18	1.16	1.07	0.99
	0.8	−89.89	−13.83	−2.65	0.26	1.15	1.36	1.33	1.19	1.08
	0.9	−114.09	−17.61	−3.46	0.22	1.32	1.59	1.53	1.35	1.21
	1.0	−141.33	−21.84	−4.35	0.18	1.54	1.85	1.77	1.54	1.37
0.7	0.2	−3.90	−0.03	0.61	0.81	0.89	0.92	0.94	0.94	0.93
	0.3	−9.25	−0.94	0.35	0.72	0.85	0.90	0.92	0.92	0.92
	0.4	−16.54	−2.10	0.07	0.66	0.85	0.91	0.93	0.92	0.92
	0.5	−25.85	−3.51	−0.22	0.64	0.90	0.97	0.97	0.94	0.95
	0.6	−37.21	−5.18	−0.54	0.65	1.00	1.07	1.05	1.00	1.00
	0.7	−50.68	−7.13	−0.87	0.70	1.14	1.22	1.17	1.08	1.08
	0.8	−66.31	−9.37	−1.24	0.78	1.33	1.41	1.33	1.21	1.20
	0.9	−84.17	−11.92	−1.64	0.89	1.56	1.65	1.53	1.36	1.34
	1.0	−104.29	−14.78	−2.09	1.03	1.84	1.94	1.78	1.56	1.52
0.8	0.2	−2.90	0.15	0.67	0.83	0.90	0.93	0.94	0.95	0.96
	0.3	−6.91	−0.53	0.47	0.76	0.87	0.91	0.93	0.94	0.96
	0.4	−12.31	−1.34	0.30	0.74	0.88	0.93	0.94	0.95	0.98
	0.5	−19.16	−2.29	0.15	0.77	0.94	0.99	0.98	0.98	1.03
	0.6	−27.50	−3.39	0.01	0.84	1.06	1.09	1.06	1.03	1.11
	0.7	−37.38	−4.66	−0.11	0.97	1.23	1.24	1.18	1.12	1.21
	0.8	−48.87	−6.11	−0.22	1.15	1.46	1.45	1.35	1.25	1.35
	0.9	−62.01	−7.75	−0.33	1.37	1.73	1.70	1.55	1.41	1.52
	1.0	−76.85	−9.59	−0.44	1.63	2.06	2.00	1.80	1.61	1.73

ED5–3 Tee, D_c > 10 in., Converging *(Continued)*

		C_b Values *(Concluded)*								
		Q_b/Q_c								
A_s/A_c	A_b/A_c	0.1	0.2	0.3	0.4	0.5	0.6	0.7	0.8	0.9
0.9	0.2	–2.14	0.28	0.71	0.85	0.91	0.94	0.96	0.97	0.99
	0.3	–5.14	–0.21	0.57	0.80	0.88	0.92	0.95	0.97	1.02
	0.4	–9.09	–0.76	0.47	0.80	0.91	0.94	0.96	0.98	1.06
	0.5	–14.06	–1.36	0.42	0.86	0.98	1.01	1.01	1.02	1.14
	0.6	–20.08	–2.04	0.42	0.99	1.11	1.12	1.09	1.09	1.24
	0.7	–27.21	–2.79	0.47	1.17	1.30	1.27	1.21	1.19	1.38
	0.8	–35.50	–3.63	0.55	1.42	1.55	1.49	1.38	1.32	1.55
	0.9	–45.01	–4.57	0.66	1.72	1.86	1.75	1.59	1.49	1.75
	1.0	–55.79	–5.64	0.80	2.08	2.22	2.06	1.84	1.69	1.99
1.0	0.2	–1.54	0.39	0.74	0.87	0.92	0.95	0.97	0.99	1.03
	0.3	–3.75	0.03	0.64	0.83	0.90	0.94	0.97	1.00	1.08
	0.4	–6.57	–0.32	0.61	0.85	0.93	0.97	0.99	1.03	1.16
	0.5	–10.05	–0.65	0.64	0.94	1.02	1.03	1.04	1.08	1.26
	0.6	–14.24	–0.98	0.74	1.10	1.16	1.15	1.13	1.16	1.40
	0.7	–19.20	–1.32	0.91	1.33	1.37	1.31	1.26	1.27	1.57
	0.8	–24.98	–1.69	1.14	1.63	1.63	1.53	1.43	1.41	1.78
	0.9	–31.62	–2.10	1.42	2.00	1.96	1.80	1.64	1.59	2.02
	1.0	–39.19	–2.55	1.76	2.43	2.35	2.12	1.90	1.81	2.30

ED5–3 Tee, D_c > 10 in., Converging

		C_s Values								
		Q_s/Q_c								
A_s/A_c	A_b/A_c	0.1	0.2	0.3	0.4	0.5	0.6	0.7	0.8	0.9
0.2	0.2	20.43	3.28	1.45	0.98	0.81	0.73	0.69	0.66	0.64
	0.3	12.53	2.40	1.22	0.90	0.77	0.71	0.68	0.66	0.64
	0.4	8.78	1.98	1.12	0.86	0.76	0.70	0.67	0.66	0.64
	0.5	6.69	1.75	1.06	0.84	0.75	0.70	0.67	0.65	0.64
	0.6	5.43	1.61	1.02	0.83	0.74	0.70	0.67	0.65	0.64
	0.7	4.64	1.52	1.00	0.82	0.74	0.70	0.67	0.65	0.64
	0.8	4.15	1.47	0.98	0.81	0.74	0.69	0.67	0.65	0.64
	0.9	3.86	1.43	0.97	0.81	0.74	0.69	0.67	0.65	0.64
	1.0	3.71	1.42	0.97	0.81	0.73	0.69	0.67	0.65	0.64
0.3	0.2	51.24	7.11	2.49	1.33	0.90	0.70	0.60	0.54	0.50
	0.3	29.57	4.70	1.87	1.10	0.80	0.66	0.58	0.53	0.50
	0.4	19.40	3.57	1.58	1.00	0.76	0.64	0.57	0.52	0.50
	0.5	13.84	2.96	1.42	0.94	0.73	0.62	0.56	0.52	0.50
	0.6	10.58	2.59	1.32	0.90	0.72	0.62	0.56	0.52	0.49
	0.7	8.64	2.38	1.27	0.88	0.71	0.61	0.56	0.52	0.49
	0.8	7.52	2.25	1.23	0.87	0.70	0.61	0.56	0.52	0.49
	0.9	6.95	2.19	1.22	0.87	0.70	0.61	0.56	0.52	0.49
	1.0	6.76	2.17	1.21	0.86	0.70	0.61	0.55	0.52	0.49
0.4	0.2	90.30	12.10	3.91	1.85	1.08	0.74	0.55	0.45	0.38
	0.3	49.68	7.59	2.74	1.42	0.90	0.65	0.51	0.43	0.37
	0.4	30.96	5.51	2.21	1.23	0.82	0.61	0.49	0.42	0.37
	0.5	21.00	4.40	1.92	1.13	0.78	0.59	0.48	0.42	0.37
	0.6	15.43	3.78	1.76	1.07	0.75	0.58	0.48	0.41	0.37
	0.7	12.36	3.44	1.67	1.04	0.74	0.57	0.48	0.41	0.37
	0.8	10.86	3.27	1.63	1.02	0.73	0.57	0.47	0.41	0.37
	0.9	10.40	3.22	1.61	1.01	0.73	0.57	0.47	0.41	0.37
	1.0	10.67	3.25	1.62	1.02	0.73	0.57	0.47	0.41	0.37
0.5	0.2	126.36	16.99	5.39	2.42	1.32	0.81	0.54	0.38	0.28
	0.3	65.94	10.28	3.65	1.79	1.05	0.68	0.48	0.35	0.27
	0.4	38.84	7.27	2.87	1.51	0.93	0.63	0.45	0.34	0.27
	0.5	25.07	5.74	2.47	1.37	0.87	0.60	0.44	0.33	0.26
	0.6	17.98	4.95	2.27	1.29	0.84	0.58	0.43	0.33	0.26
	0.7	14.69	4.58	2.17	1.26	0.82	0.58	0.43	0.33	0.26
	0.8	13.78	4.48	2.15	1.25	0.82	0.57	0.43	0.33	0.26
	0.9	14.45	4.56	2.17	1.26	0.82	0.58	0.43	0.33	0.26
	1.0	16.24	4.76	2.22	1.28	0.83	0.58	0.43	0.33	0.26

ED5–3 Tee, D_c > 10 in., Converging (*Continued*)

		C_s Values (*Concluded*)								
		Q_s/Q_c								
A_s/A_c	A_b/A_c	0.1	0.2	0.3	0.4	0.5	0.6	0.7	0.8	0.9
0.6	0.2	146.22	20.32	6.54	2.92	1.54	0.89	0.54	0.33	0.20
	0.3	70.93	11.95	4.37	2.13	1.20	0.73	0.46	0.30	0.18
	0.4	38.66	8.37	3.44	1.80	1.06	0.67	0.43	0.28	0.18
	0.5	23.61	6.70	3.00	1.64	0.99	0.64	0.42	0.28	0.18
	0.6	17.17	5.98	2.82	1.57	0.97	0.62	0.41	0.27	0.18
	0.7	15.64	5.81	2.77	1.56	0.96	0.62	0.41	0.27	0.18
	0.8	17.19	5.98	2.82	1.57	0.97	0.62	0.41	0.27	0.18
	0.9	20.79	6.38	2.92	1.61	0.98	0.63	0.42	0.27	0.18
	1.0	25.82	6.94	3.07	1.66	1.00	0.64	0.42	0.28	0.18
0.7	0.2	137.78	20.74	7.01	3.21	1.70	0.96	0.54	0.29	0.13
	0.3	58.74	11.96	4.73	2.39	1.34	0.79	0.47	0.26	0.12
	0.4	27.78	8.52	3.84	2.06	1.21	0.73	0.44	0.24	0.11
	0.5	16.04	7.21	3.50	1.94	1.15	0.71	0.43	0.24	0.11
	0.6	13.91	6.97	3.44	1.92	1.14	0.70	0.42	0.24	0.11
	0.7	17.28	7.35	3.54	1.95	1.16	0.71	0.43	0.24	0.11
	0.8	24.08	8.10	3.73	2.02	1.19	0.72	0.43	0.24	0.11
	0.9	33.17	9.11	3.99	2.12	1.23	0.74	0.44	0.25	0.11
	1.0	43.86	10.30	4.30	2.23	1.28	0.76	0.45	0.25	0.11
0.8	0.2	92.97	17.35	6.57	3.21	1.75	0.99	0.55	0.27	0.08
	0.3	26.98	10.02	4.67	2.52	1.46	0.86	0.48	0.24	0.07
	0.4	6.75	7.77	4.09	2.31	1.37	0.81	0.46	0.23	0.06
	0.5	4.83	7.56	4.03	2.29	1.36	0.81	0.46	0.23	0.06
	0.6	12.05	8.36	4.24	2.37	1.39	0.83	0.47	0.23	0.07
	0.7	24.51	9.75	4.60	2.49	1.45	0.85	0.48	0.24	0.07
	0.8	40.23	11.49	5.05	2.66	1.52	0.88	0.50	0.24	0.07
	0.9	58.13	13.48	5.57	2.85	1.60	0.92	0.51	0.25	0.07
	1.0	77.56	15.64	6.13	3.05	1.68	0.96	0.53	0.26	0.08
0.9	0.2	10.77	10.05	5.20	2.91	1.70	0.99	0.55	0.25	0.04
	0.3	–21.27	6.49	4.28	2.57	1.56	0.93	0.52	0.24	0.04
	0.4	–19.11	6.73	4.34	2.60	1.57	0.93	0.52	0.24	0.04
	0.5	–3.28	8.49	4.80	2.76	1.64	0.97	0.54	0.24	0.04
	0.6	19.39	11.01	5.45	3.00	1.74	1.01	0.56	0.25	0.04
	0.7	45.97	13.96	6.21	3.27	1.86	1.07	0.58	0.27	0.05
	0.8	74.99	17.18	7.05	3.58	1.98	1.13	0.61	0.28	0.05
	0.9	105.64	20.59	7.93	3.89	2.12	1.19	0.64	0.29	0.06
	1.0	137.43	24.12	8.85	4.23	2.26	1.26	0.67	0.31	0.06
1.0	0.2	–99.78	–0.17	3.15	2.40	1.58	0.98	0.56	0.25	0.02
	0.3	–75.42	2.54	3.85	2.65	1.69	1.03	0.58	0.26	0.03
	0.4	–38.31	6.66	4.92	3.04	1.86	1.11	0.62	0.28	0.03
	0.5	3.90	11.35	6.14	3.48	2.04	1.20	0.66	0.29	0.04
	0.6	48.66	16.32	7.43	3.94	2.24	1.29	0.70	0.31	0.04
	0.7	94.88	21.46	8.76	4.43	2.45	1.38	0.75	0.33	0.05
	0.8	142.01	26.70	10.12	4.92	2.66	1.48	0.79	0.35	0.06
	0.9	189.74	32.00	11.49	5.41	2.87	1.58	0.84	0.37	0.07
	1.0	237.90	37.35	12.88	5.92	3.08	1.68	0.88	0.39	0.07

ED5-6 Capped Wye, Branch with 45-Degree Elbow, Branch 90 Degrees to Main, Converging, r/D_b = 1.5

A_b/A_c	0.1	0.2	0.3	0.4	0.5	0.6	0.7	0.8	0.9	1.0
C_b	1.26	1.07	0.94	0.86	0.81	0.76	0.71	0.67	0.64	0.64

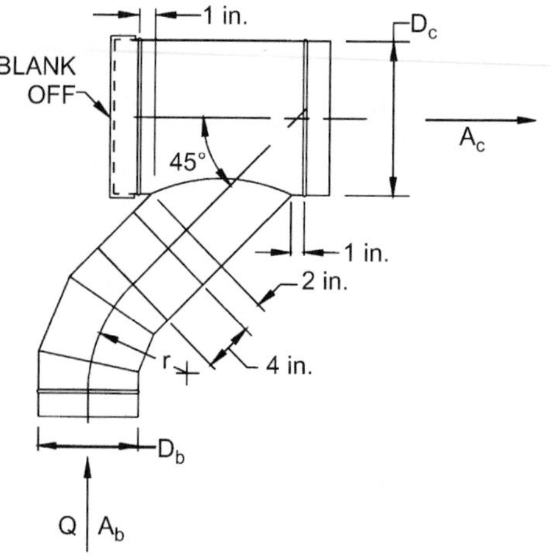

ED5-9 Symmetrical Wye, 60 Degree, $D_{b1} \geq D_{b2}$, Converging

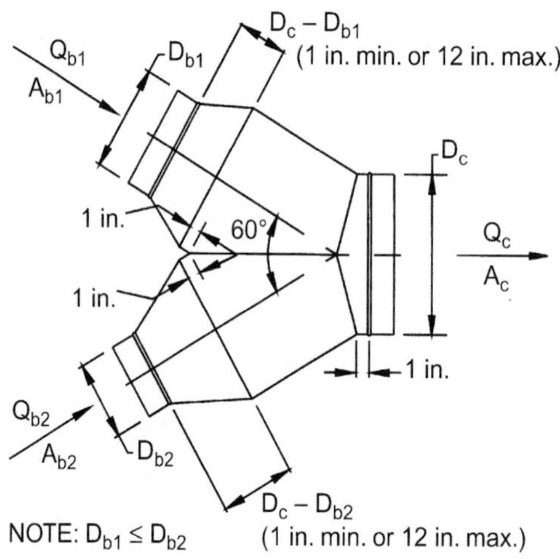

		C_{b1} Values								
		Q_{b1}/Q_c								
A_{b1}/A_c	A_{b2}/A_c	0.1	0.2	0.3	0.4	0.5	0.6	0.7	0.8	0.9
0.2	0.2	−11.95	−1.89	−0.09	0.41	0.62	0.74	0.80	0.80	0.79
	0.3	−11.95	−1.89	−0.09	0.41	0.62	0.74	0.80	0.80	0.79
0.3	0.2	−45.45	−9.39	−2.44	−0.41	0.33	0.68	0.89	1.03	1.13
	0.3	−16.88	−2.92	−0.09	0.59	0.86	1.02	1.09	1.10	1.08
0.4	0.2	−72.04	−14.00	−4.26	−1.24	−0.10	0.33	0.50	0.57	0.63
	0.3	−52.95	−9.91	−2.86	−0.69	0.07	0.30	0.40	0.49	0.62
	0.4	−28.86	−6.22	−2.15	−0.57	0.19	0.55	0.72	0.79	0.85
0.5	0.2	−126.04	−23.80	−7.44	−2.64	−0.85	−0.13	0.16	0.26	0.28
	0.3	−91.07	−16.91	−5.16	−1.73	−0.46	0.04	0.23	0.29	0.28
	0.4	−56.41	−10.07	−2.90	−0.82	−0.07	0.21	0.30	0.31	0.29
	0.5	−30.58	−5.23	−1.06	0.00	0.32	0.43	0.47	0.47	0.41
0.6	0.2	−209.81	−39.31	−12.13	−4.35	−1.54	−0.40	0.06	0.22	0.23
	0.3	−147.43	−27.69	−8.75	−3.20	−1.13	−0.29	0.05	0.17	0.18
	0.4	−85.06	−16.07	−5.38	−2.04	−0.71	−0.17	0.04	0.12	0.13
	0.5	−58.22	−11.03	−3.84	−1.49	−0.50	−0.09	0.07	0.11	0.12
	0.6	−40.57	−7.86	−2.60	−0.99	−0.26	0.00	0.14	0.21	0.25
0.7	0.2	−291.57	−54.52	−17.03	−6.21	−2.27	−0.68	−0.04	0.19	0.21
	0.3	−197.37	−38.02	−12.54	−4.92	−2.01	−0.76	−0.22	0.01	0.08
	0.4	−102.97	−21.41	−8.05	−3.64	−1.75	−0.84	−0.40	−0.17	−0.05
	0.5	−65.15	−14.75	−6.16	−3.07	−1.61	−0.85	−0.44	−0.22	−0.09
	0.6	−48.24	−11.70	−4.97	−2.59	−1.40	−0.76	−0.37	−0.15	−0.03
	0.7	−73.02	−16.68	−6.90	−3.29	−1.61	−0.80	−0.29	0.02	0.22
0.8	0.2	−373.33	−69.73	−21.93	−8.08	−3.00	−0.95	−0.13	0.15	0.20
	0.3	−247.31	−48.35	−16.32	−6.65	−2.89	−1.24	−0.49	−0.15	−0.02
	0.4	−120.88	−26.76	−10.71	−5.24	−2.78	−1.52	−0.84	−0.45	−0.24

ED5-9 Symmetrical Wye, 60 Degree, $D_{b1} \geq D_{b2}$, Converging (*Continued*)

		C_{b1} Values (*Concluded*)								
		Q_{b1}/Q_c								
A_{b1}/A_c	A_{b2}/A_c	0.1	0.2	0.3	0.4	0.5	0.6	0.7	0.8	0.9
	0.5	−72.08	−18.46	−8.48	−4.65	−2.71	−1.61	−0.95	−0.55	−0.31
	0.6	−55.91	−15.54	−7.35	−4.20	−2.54	−1.53	−0.89	−0.51	−0.30
	0.7	−80.68	−20.52	−9.27	−4.90	−2.75	−1.56	−0.80	−0.34	−0.06
	0.8	−105.46	−25.49	−11.19	−5.59	−2.96	−1.60	−0.72	−0.18	0.19
0.9	0.2	−479.24	−89.56	−28.39	−10.59	−4.04	−1.41	−0.36	0.01	0.09
	0.3	−305.31	−61.27	−21.50	−9.28	−4.39	−2.16	−1.07	−0.54	−0.29
	0.4	−131.17	−32.88	−14.60	−7.98	−4.74	−2.91	−1.79	−1.10	−0.68
	0.5	−67.90	−22.76	−12.17	−7.53	−4.89	−3.19	−2.05	−1.30	−0.81
	0.6	−68.95	−23.08	−12.11	−7.45	−4.84	−3.15	−2.01	−1.26	−0.79
	0.7	−90.48	−27.35	−13.58	−7.95	−4.97	−3.16	−1.96	−1.17	−0.65
	0.8	−112.02	−31.63	−15.05	−8.44	−5.11	−3.18	−1.90	−1.07	−0.51
	0.9	−130.32	−35.19	−16.07	−8.70	−5.18	−3.19	−1.88	−1.08	−0.53
1.0	0.2	−585.16	−109.39	−34.85	−13.11	−5.09	−1.86	−0.59	−0.13	−0.01
	0.3	−363.31	−74.20	−26.68	−11.91	−5.90	−3.08	−1.66	−0.94	−0.56
	0.4	−141.46	−39.00	−18.50	−10.71	−6.71	−4.29	−2.74	−1.74	−1.12
	0.5	−63.71	−27.06	−15.85	−10.41	−7.07	−4.77	−3.16	−2.05	−1.31
	0.6	−81.99	−30.62	−16.87	−10.70	−7.13	−4.77	−3.13	−2.02	−1.28
	0.7	−100.28	−34.19	−17.89	−11.00	−7.19	−4.76	−3.11	−1.99	−1.24
	0.8	−118.58	−37.76	−18.91	−11.29	−7.26	−4.76	−3.09	−1.96	−1.20
	0.9	−136.88	−41.32	−19.93	−11.55	−7.32	−4.77	−3.07	−1.98	−1.23
	1.0	−155.18	−44.89	−20.95	−11.80	−7.39	−4.78	−3.05	−1.99	−1.25

		C_{b2} Values								
		Q_{b2}/Q_c								
A_{b1}/A_c	A_{b2}/A_c	0.1	0.2	0.3	0.4	0.5	0.6	0.7	0.8	0.9
0.2	0.2	−11.95	−1.89	−0.09	0.41	0.62	0.74	0.80	0.80	0.79
	0.3	−11.95	−1.89	−0.09	0.41	0.62	0.74	0.80	0.80	0.79
0.3	0.2	−8.24	−1.18	0.05	0.42	0.61	0.73	0.78	0.77	0.76
	0.3	−16.88	−2.92	−0.09	0.59	0.86	1.02	1.09	1.10	1.08
0.4	0.2	−6.95	−1.00	0.16	0.53	0.67	0.71	0.72	0.72	0.71
	0.3	−16.21	−2.90	−0.44	0.40	0.79	0.98	1.05	1.06	1.05
	0.4	−28.86	−6.22	−2.15	−0.57	0.19	0.55	0.72	0.79	0.85
0.5	0.2	−4.82	−0.01	0.56	0.71	0.82	0.89	0.92	0.90	0.89
	0.3	−12.27	−1.17	0.44	0.88	1.11	1.25	1.29	1.25	1.23
	0.4	−20.76	−2.93	−0.21	0.48	0.73	0.84	0.88	0.87	0.82
	0.5	−30.58	−5.23	−1.06	0.00	0.32	0.43	0.47	0.47	0.41
0.6	0.2	−3.68	0.07	0.77	0.98	1.06	1.08	1.08	1.06	1.04
	0.3	−9.06	−0.55	0.86	1.27	1.42	1.48	1.49	1.46	1.42
	0.4	−17.62	−2.12	0.06	0.60	0.83	0.95	0.98	0.95	0.91
	0.5	−28.00	−4.26	−0.99	−0.16	0.20	0.39	0.45	0.41	0.38
	0.6	−40.57	−7.86	−2.60	−0.99	−0.26	0.00	0.14	0.21	0.25
0.7	0.2	−5.44	−0.40	0.55	0.86	0.98	1.02	1.04	1.03	1.02
	0.3	−9.36	−0.77	0.73	1.20	1.39	1.47	1.49	1.47	1.44
	0.4	−19.57	−3.09	−0.44	0.36	0.71	0.89	0.97	0.98	0.97
	0.5	−31.88	−6.02	−1.90	−0.63	−0.05	0.26	0.40	0.44	0.46
	0.6	−46.44	−9.82	−3.47	−1.41	−0.48	−0.04	0.21	0.36	0.45
	0.7	−73.02	−16.68	−6.90	−3.29	−1.61	−0.80	−0.29	0.02	0.22
0.8	0.2	−7.21	−0.87	0.33	0.73	0.90	0.97	1.00	1.00	0.99
	0.3	−9.67	−0.99	0.60	1.13	1.36	1.45	1.49	1.48	1.46
	0.4	−21.53	−4.06	−0.93	0.11	0.59	0.83	0.96	1.01	1.03
	0.5	−35.77	−7.77	−2.82	−1.09	−0.29	0.13	0.35	0.48	0.55
	0.6	−52.32	−11.78	−4.34	−1.83	−0.70	−0.09	0.28	0.51	0.65
	0.7	−78.89	−18.64	−7.76	−3.71	−1.83	−0.85	−0.22	0.16	0.42
	0.8	−105.46	−25.49	−11.19	−5.59	−2.96	−1.60	−0.72	−0.18	0.19
0.9	0.2	−4.98	−0.34	0.54	0.85	0.97	1.03	1.04	1.03	1.01
	0.3	−9.97	−1.21	0.48	1.06	1.32	1.44	1.49	1.49	1.48
	0.4	−23.54	−4.98	−1.39	−0.12	0.47	0.78	0.95	1.04	1.09
	0.5	−40.14	−9.57	−3.69	−1.56	−0.55	−0.01	0.31	0.51	0.63
	0.6	−58.25	−14.28	−5.64	−2.53	−1.08	−0.30	0.18	0.49	0.70
	0.7	−84.09	−21.02	−8.91	−4.38	−2.22	−1.04	−0.31	0.15	0.46
	0.8	−109.92	−27.77	−12.18	−6.22	−3.35	−1.79	−0.81	−0.19	0.23
	0.9	−130.32	−35.19	−16.07	−8.70	−5.18	−3.19	−1.88	−1.08	−0.53
1.0	0.2	−2.75	0.19	0.76	0.96	1.05	1.08	1.08	1.06	1.04
	0.3	−10.28	−1.43	0.35	0.99	1.29	1.43	1.49	1.50	1.50
	0.4	−25.56	−5.89	−1.86	−0.36	0.35	0.72	0.93	1.07	1.15
	0.5	−44.52	−11.37	−4.56	−2.02	−0.81	−0.14	0.27	0.54	0.72
	0.6	−64.19	−16.77	−6.94	−3.24	−1.47	−0.50	0.09	0.48	0.74
	0.7	−89.28	−23.41	−10.05	−5.05	−2.61	−1.24	−0.40	0.14	0.50
	0.8	−114.38	−30.04	−13.16	−6.86	−3.75	−1.97	−0.89	−0.20	0.27
	0.9	−134.78	−37.47	−17.06	−9.33	−5.57	−3.38	−1.97	−1.09	−0.49
	1.0	−155.18	−44.89	−20.95	−11.80	−7.39	−4.78	−3.05	−1.99	−1.25

ED7-1 Centrifugal Fan Located in Plenum or Cabinet

L/D_o	0.30	0.40	0.50	0.75
C_o	0.80	0.53	0.40	0.22

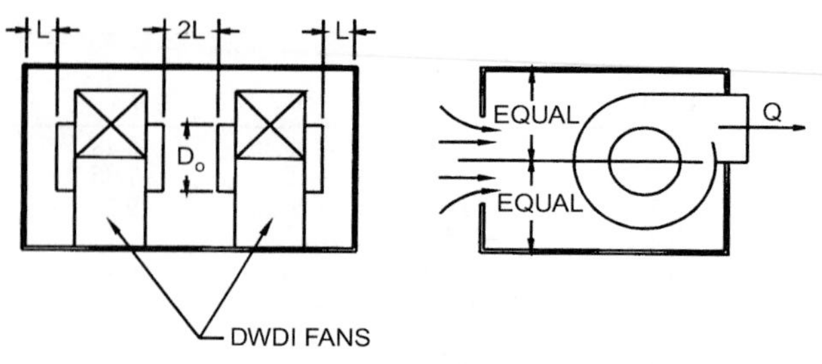

ED7-2 Fan Inlet, Centrifugal, SWSI, with 4 Gore Elbow

	C_o Values			
	L/D_o			
r/D_o	0.0	2.0	5.0	10.0
0.50	1.80	1.00	0.53	0.53
0.75	1.40	0.80	0.40	0.40
1.00	1.20	0.67	0.33	0.33
1.50	1.10	0.60	0.33	0.33
2.00	1.00	0.53	0.33	0.33
3.00	0.67	0.40	0.22	0.22

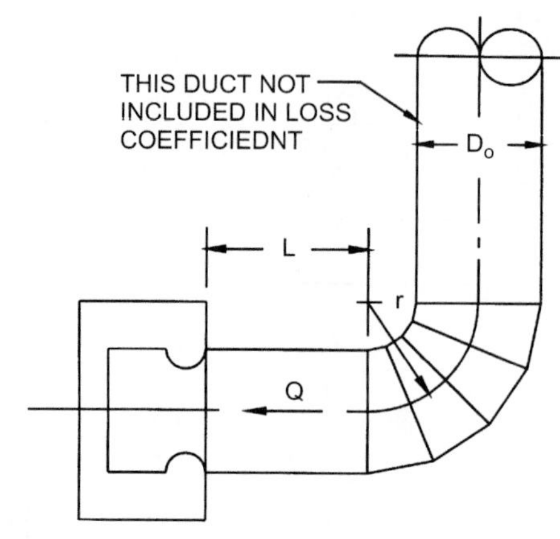

SD1-1 Bellmouth, Plenum to Round, Supply Air Systems

r/D_o	0.00	0.01	0.02	0.03	0.04	0.05	0.06	0.08	0.10	0.12	0.16	0.20	10.00
C_o	0.50	0.44	0.36	0.31	0.26	0.22	0.20	0.15	0.12	0.09	0.06	0.03	0.03

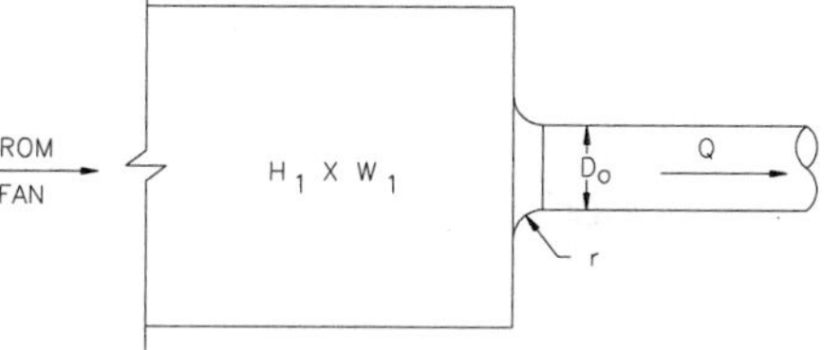

SD1-2 Conical Bellmouth/Sudden Contraction, Plenum to Round, Supply Air Systems

		C_o Values									
		θ									
A_o/A_1	L/D_o	0	10	20	30	45	60	90	120	150	180
0.10	0.025	0.46	0.43	0.42	0.40	0.38	0.37	0.38	0.40	0.43	0.46
	0.050	0.46	0.42	0.38	0.33	0.30	0.28	0.31	0.36	0.41	0.46
	0.075	0.46	0.39	0.32	0.28	0.23	0.21	0.26	0.32	0.39	0.46
	0.100	0.46	0.36	0.30	0.23	0.19	0.17	0.23	0.30	0.38	0.46
	0.150	0.46	0.34	0.25	0.18	0.15	0.14	0.21	0.29	0.37	0.46
	0.300	0.46	0.31	0.22	0.16	0.13	0.13	0.20	0.28	0.37	0.46
	0.600	0.46	0.25	0.17	0.12	0.10	0.11	0.19	0.27	0.36	0.46
0.20	0.025	0.42	0.40	0.38	0.36	0.34	0.34	0.35	0.37	0.39	0.42
	0.050	0.42	0.38	0.35	0.30	0.27	0.25	0.29	0.33	0.37	0.42
	0.075	0.42	0.36	0.30	0.25	0.21	0.19	0.24	0.30	0.36	0.42
	0.100	0.42	0.33	0.27	0.21	0.18	0.15	0.21	0.27	0.35	0.42
	0.150	0.42	0.31	0.23	0.17	0.13	0.13	0.19	0.26	0.34	0.42
	0.300	0.42	0.28	0.20	0.15	0.12	0.12	0.18	0.26	0.34	0.42
	0.600	0.42	0.23	0.15	0.11	0.10	0.10	0.17	0.25	0.33	0.42

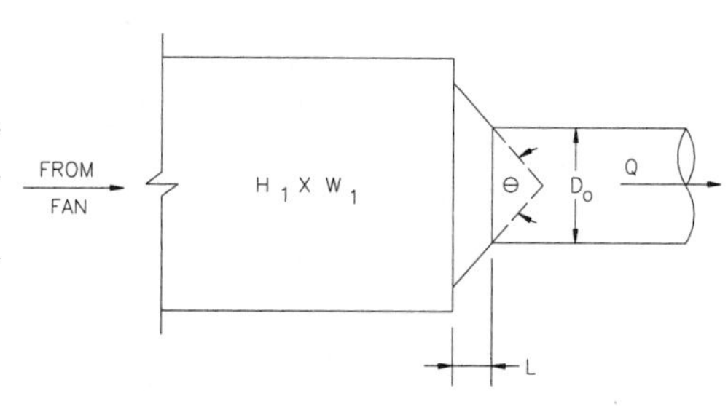

SD1-2 Conical Bellmouth/Sudden Contraction, Plenum to Round, Supply Air Systems (*Continued*)

		C_o Values (*Concluded*)									
		θ									
A_o/A_1	L/D_o	0	10	20	30	45	60	90	120	150	180
0.40	0.025	0.34	0.32	0.31	0.29	0.28	0.27	0.28	0.30	0.32	0.34
	0.050	0.34	0.31	0.28	0.25	0.22	0.20	0.23	0.26	0.30	0.34
	0.075	0.34	0.29	0.24	0.20	0.17	0.16	0.19	0.24	0.29	0.34
	0.100	0.34	0.27	0.22	0.17	0.14	0.12	0.17	0.22	0.28	0.34
	0.150	0.34	0.25	0.18	0.14	0.11	0.10	0.15	0.21	0.27	0.34
	0.300	0.34	0.23	0.16	0.12	0.10	0.10	0.15	0.21	0.27	0.34
	0.600	0.34	0.18	0.12	0.09	0.08	0.08	0.14	0.20	0.27	0.34
0.60	0.025	0.25	0.24	0.23	0.22	0.20	0.20	0.21	0.22	0.23	0.25
	0.050	0.25	0.23	0.21	0.18	0.16	0.15	0.17	0.19	0.22	0.25
	0.075	0.25	0.21	0.18	0.15	0.13	0.12	0.14	0.18	0.21	0.25
	0.100	0.25	0.20	0.16	0.13	0.11	0.09	0.12	0.16	0.21	0.25
	0.150	0.25	0.19	0.14	0.10	0.08	0.08	0.11	0.16	0.20	0.25
	0.300	0.25	0.17	0.12	0.09	0.07	0.07	0.11	0.15	0.20	0.25
	0.600	0.25	0.14	0.09	0.07	0.06	0.06	0.10	0.15	0.20	0.25
0.80	0.025	0.15	0.14	0.13	0.13	0.12	0.12	0.12	0.13	0.14	0.15
	0.050	0.15	0.13	0.12	0.11	0.10	0.09	0.10	0.12	0.13	0.15
	0.075	0.15	0.13	0.10	0.09	0.08	0.07	0.08	0.10	0.13	0.15
	0.100	0.15	0.12	0.10	0.07	0.06	0.05	0.07	0.10	0.12	0.15
	0.150	0.15	0.11	0.08	0.06	0.05	0.04	0.07	0.09	0.12	0.15
	0.300	0.15	0.10	0.07	0.05	0.04	0.04	0.07	0.09	0.12	0.15
	0.600	0.15	0.08	0.05	0.04	0.03	0.04	0.06	0.09	0.12	0.15
0.90	0.025	0.09	0.08	0.08	0.08	0.07	0.07	0.07	0.08	0.08	0.09
	0.050	0.09	0.08	0.07	0.06	0.06	0.05	0.06	0.07	0.08	0.09
	0.075	0.09	0.07	0.06	0.05	0.04	0.04	0.05	0.06	0.08	0.09
	0.100	0.09	0.07	0.06	0.04	0.04	0.03	0.04	0.06	0.07	0.09
	0.150	0.09	0.07	0.05	0.04	0.03	0.03	0.04	0.06	0.07	0.09
	0.300	0.09	0.06	0.04	0.03	0.03	0.02	0.04	0.05	0.07	0.09
	0.600	0.09	0.05	0.03	0.02	0.02	0.02	0.04	0.05	0.07	0.09

SD2-6 Stackhead

D_e/D	0.3	0.4	0.5	0.6	0.7	0.8	0.9	1.0
C_o	129	41.02	16.80	8.10	4.37	2.56	1.60	1.00

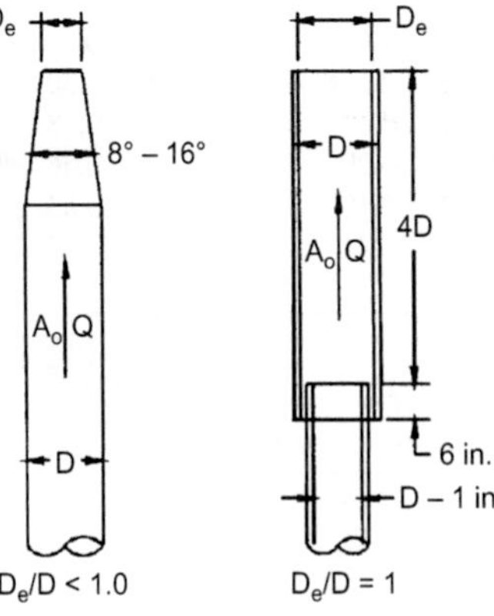

SD4-1 Transition, Round to Round, Supply Air Systems

	C_o Values									
	θ									
A_o/A_1	10	15	20	30	45	60	90	120	150	180
0.10	0.05	0.05	0.05	0.05	0.07	0.08	0.19	0.29	0.37	0.43
0.17	0.05	0.04	0.04	0.04	0.06	0.07	0.18	0.28	0.36	0.42
0.25	0.05	0.04	0.04	0.04	0.06	0.07	0.17	0.27	0.35	0.41
0.50	0.05	0.05	0.05	0.05	0.06	0.06	0.12	0.18	0.24	0.26
1.00	0.00	0.00	0.00	0.00	0.00	0.00	0.00	0.00	0.00	0.00
2.00	0.44	0.52	0.76	1.28	1.32	1.32	1.28	1.24	1.20	1.20
4.00	2.56	3.52	4.80	7.36	9.76	10.88	10.24	10.08	9.92	9.92
10.00	21.00	28.00	38.00	59.00	76.00	80.00	83.00	84.00	83.00	83.00
16.00	53.76	74.24	97.28	153.60	215.04	225.28	225.28	225.28	225.28	225.28

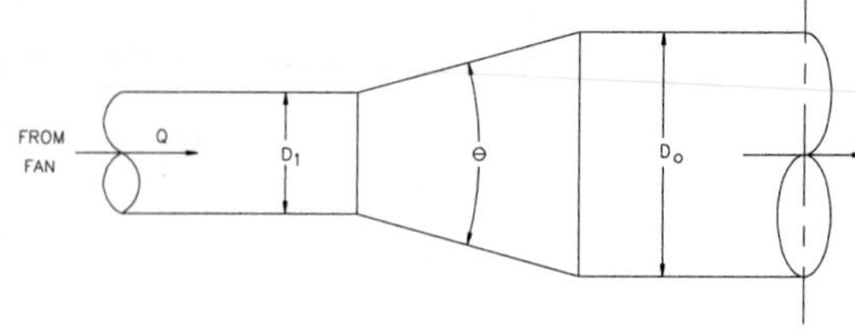

SD4-2 Transition, Rectangular to Round, Supply Air Systems

	C_o Values									
	θ									
A_o/A_1	10	15	20	30	45	60	90	120	150	180
0.10	0.05	0.05	0.05	0.05	0.07	0.08	0.19	0.29	0.37	0.43
0.17	0.05	0.05	0.04	0.04	0.06	0.07	0.18	0.28	0.36	0.42
0.25	0.06	0.05	0.05	0.04	0.06	0.07	0.17	0.27	0.35	0.41
0.50	0.06	0.07	0.07	0.05	0.06	0.06	0.12	0.18	0.24	0.26
1.00	0.00	0.00	0.00	0.00	0.00	0.00	0.00	0.00	0.00	0.00
2.00	0.60	0.84	1.00	1.20	1.32	1.32	1.32	1.28	1.24	1.20
4.00	4.00	5.76	7.20	8.32	9.28	9.92	10.24	10.24	10.24	10.24
10.00	30.00	50.00	53.00	64.00	75.00	84.00	89.00	91.00	91.00	88.00
16.00	76.80	138.24	135.68	166.40	197.12	225.28	243.20	250.88	250.88	238.08

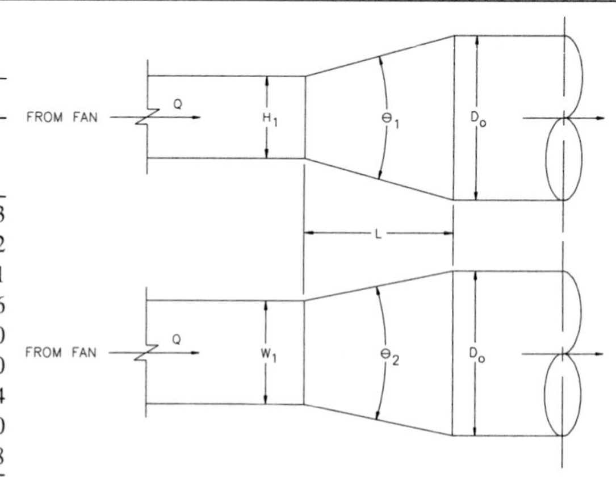

SD5-1 Wye, 45 Degree, Diverging

	C_b Values								
	Q_b/Q_c								
A_b/A_c	0.1	0.2	0.3	0.4	0.5	0.6	0.7	0.8	0.9
0.1	0.38	0.39	0.48						
0.2	2.25	0.38	0.31	0.39	0.46	0.48	0.45		
0.3	6.29	1.02	0.38	0.30	0.33	0.39	0.44	0.48	0.48
0.4	12.41	2.25	0.74	0.38	0.30	0.31	0.35	0.39	0.43
0.5	20.58	4.01	1.37	0.62	0.38	0.30	0.30	0.32	0.36
0.6	30.78	6.29	2.25	1.02	0.56	0.38	0.31	0.30	0.31
0.7	43.02	9.10	3.36	1.57	0.85	0.52	0.38	0.31	0.30
0.8	57.29	12.41	4.71	2.25	1.22	0.74	0.50	0.38	0.32
0.9	73.59	16.24	6.29	3.06	1.69	1.02	0.67	0.48	0.38

	C_s Values								
	Q_s/Q_c								
A_s/A_c	0.1	0.2	0.3	0.4	0.5	0.6	0.7	0.8	0.9
0.1	0.13	0.16							
0.2	0.20	0.13	0.15	0.16	0.28				
0.3	0.90	0.13	0.13	0.14	0.15	0.16	0.20		
0.4	2.88	0.20	0.14	0.13	0.14	0.15	0.15	0.16	0.34
0.5	6.25	0.37	0.17	0.14	0.13	0.14	0.14	0.15	0.15
0.6	11.88	0.90	0.20	0.13	0.14	0.13	0.14	0.14	0.15
0.7	18.62	1.71	0.33	0.18	0.16	0.14	0.13	0.15	0.14
0.8	26.88	2.88	0.50	0.20	0.15	0.14	0.13	0.13	0.14
0.9	36.45	4.46	0.90	0.30	0.19	0.16	0.15	0.14	0.13

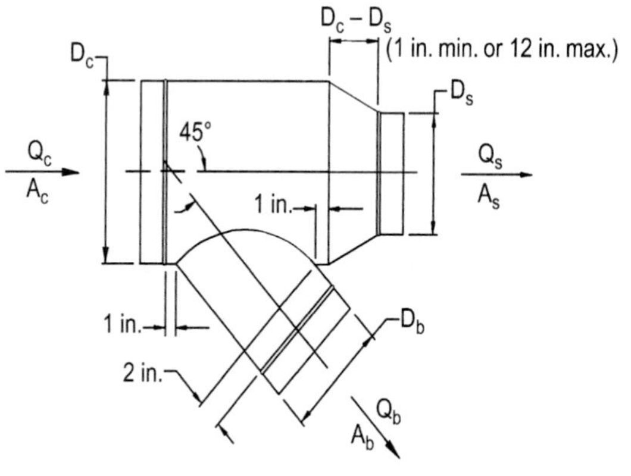

SD5-9 Tee, Diverging

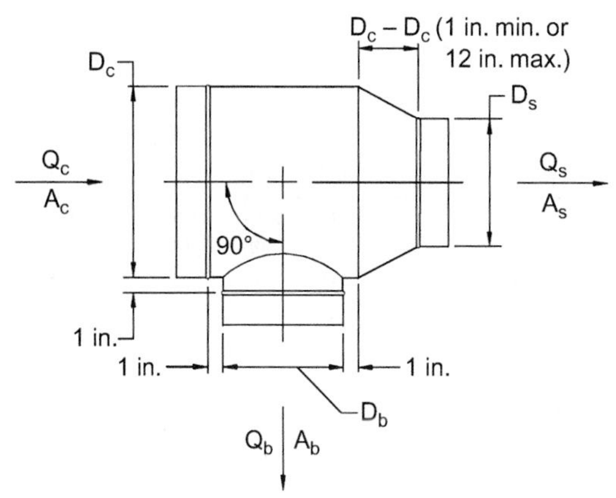

	C_b Values								
	Q_b/Q_c								
A_b/A_c	0.1	0.2	0.3	0.4	0.5	0.6	0.7	0.8	0.9
0.1	1.20	0.62	0.80	1.28	1.99	2.92	4.07	5.44	7.02
0.2	4.10	1.20	0.72	0.62	0.66	0.80	1.01	1.28	1.60
0.3	8.99	2.40	1.20	0.81	0.66	0.62	0.64	0.70	0.80
0.4	15.89	4.10	1.94	1.20	0.88	0.72	0.64	0.62	0.63
0.5	24.80	6.29	2.91	1.74	1.20	0.92	0.77	0.68	0.63
0.6	35.73	8.99	4.10	2.40	1.62	1.20	0.96	0.81	0.72
0.7	48.67	12.19	5.51	3.19	2.12	1.55	1.20	0.99	0.85
0.8	63.63	15.89	7.14	4.10	2.70	1.94	1.49	1.20	1.01
0.9	80.60	20.10	8.99	5.13	3.36	2.40	1.83	1.46	1.20

	C_s Values								
	Q_s/Q_c								
A_s/A_c	0.1	0.2	0.3	0.4	0.5	0.6	0.7	0.8	0.9
0.1	0.13	0.16							
0.2	0.20	0.13	0.15	0.16	0.28				
0.3	0.90	0.13	0.13	0.14	0.15	0.16	0.20		
0.4	2.88	0.20	0.14	0.13	0.14	0.15	0.15	0.16	0.34
0.5	6.25	0.37	0.17	0.14	0.13	0.14	0.14	0.15	0.15
0.6	11.88	0.90	0.20	0.13	0.14	0.13	0.14	0.14	0.15
0.7	18.62	1.71	0.33	0.18	0.16	0.14	0.13	0.15	0.14
0.8	26.88	2.88	0.50	0.20	0.15	0.14	0.13	0.13	0.14
0.9	36.45	4.46	0.90	0.30	0.19	0.16	0.15	0.14	0.13

SD5-10 Tee, Conical Branch Tapered into Body, Diverging

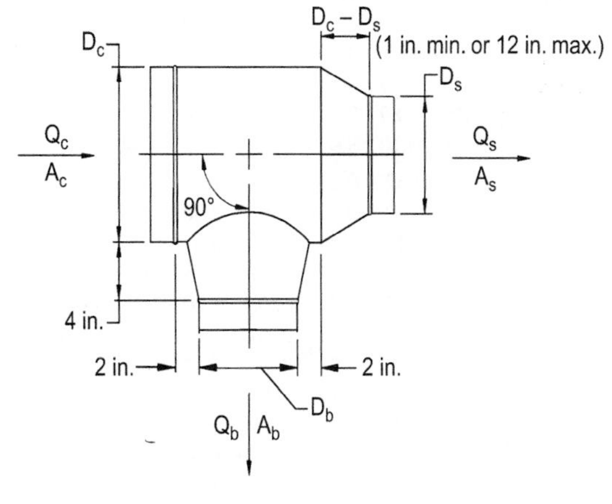

	C_b Values								
	Q_b/Q_c								
A_b/A_c	0.1	0.2	0.3	0.4	0.5	0.6	0.7	0.8	0.9
0.1	0.65	0.24							
0.2	2.98	0.65	0.33	0.24	0.18				
0.3	7.36	1.56	0.65	0.39	0.29	0.24	0.20		
0.4	13.78	2.98	1.20	0.65	0.43	0.33	0.27	0.24	0.21
0.5	22.24	4.92	1.98	1.04	0.65	0.47	0.36	0.30	0.26
0.6	32.73	7.36	2.98	1.56	0.96	0.65	0.49	0.39	0.33
0.7	45.26	10.32	4.21	2.21	1.34	0.90	0.65	0.51	0.42
0.8	59.82	13.78	5.67	2.98	1.80	1.20	0.86	0.65	0.52
0.9	76.41	17.75	7.36	3.88	2.35	1.56	1.11	0.83	0.65

	C_s Values								
	Q_s/Q_c								
A_s/A_c	0.1	0.2	0.3	0.4	0.5	0.6	0.7	0.8	0.9
0.1	0.13	0.16							
0.2	0.20	0.13	0.15	0.16	0.28				
0.3	0.90	0.13	0.13	0.14	0.15	0.16	0.20		
0.4	2.88	0.20	0.14	0.13	0.14	0.15	0.15	0.16	0.34
0.5	6.25	0.37	0.17	0.14	0.13	0.14	0.14	0.15	0.15
0.6	11.88	0.90	0.20	0.13	0.14	0.13	0.14	0.14	0.15
0.7	18.62	1.71	0.33	0.18	0.16	0.14	0.13	0.15	0.14
0.8	26.88	2.88	0.50	0.20	0.15	0.14	0.13	0.13	0.14
0.9	36.45	4.46	0.90	0.30	0.19	0.16	0.15	0.14	0.13

SD5-24 Cross, Diverging

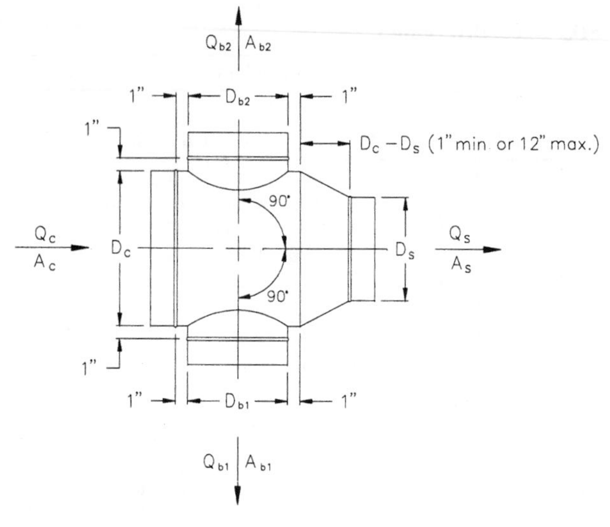

		C_{b1} Values								
		Q_{b1}/Q_c								
A_s/A_c	A_{b1}/A_c	0.1	0.2	0.3	0.4	0.5	0.6	0.7	0.8	0.9
0.20	0.1	2.07	2.08	1.62	1.30	1.08	0.93	0.81	0.72	0.64
	0.2		2.07	2.31	2.08	1.83	1.62	1.44	1.30	1.18
	0.3			2.07	2.34	2.24	2.08	1.91	1.76	1.62
	0.4			0.90	2.07	2.32	2.31	2.21	2.08	1.95
	0.5				1.28	2.07	2.30	2.33	2.27	2.18
	0.6					1.48	2.07	2.29	2.34	2.31
	0.7					0.55	1.60	2.07	2.27	2.33
	0.8						0.90	1.68	2.07	2.25
	0.9							1.12	1.74	2.07
0.35	0.1		3.25	3.11	2.69	2.32	2.03	1.80	1.61	1.46
	0.2			2.44	3.25	3.28	3.11	2.90	2.69	2.49
	0.3				1.69	2.88	3.25	3.31	3.23	3.11
	0.4					1.12	2.44	3.02	3.25	3.31
	0.5						0.69	2.04	2.73	3.09
	0.6							0.37	1.69	2.44
	0.7								0.11	1.38
	0.8									
	0.9									
0.55	0.1		1.50	1.56	1.38	1.20	1.06	0.94	0.84	0.77
	0.2			0.89	1.50	1.60	1.56	1.47	1.38	1.28
	0.3				0.38	1.20	1.50	1.59	1.59	1.56
	0.4					0.00	0.89	1.31	1.50	1.58
	0.5							0.61	1.09	1.36
	0.6								0.38	0.89
	0.7									0.17
	0.8									
	0.9									
0.80	0.1	1.20	0.62	0.80	1.28	1.99	2.92	4.07	5.44	7.02
	0.2	4.10	1.20	0.72	0.62	0.66	0.80	1.01	1.28	1.60
	0.3	8.99	2.40	1.20	0.81	0.66	0.62	0.64	0.70	0.80
	0.4	15.89	4.10	1.94	1.20	0.88	0.72	0.64	0.62	0.63
	0.5	24.80	6.29	2.91	1.74	1.20	0.92	0.77	0.68	0.63
	0.6	35.73	8.99	4.10	2.40	1.62	1.20	0.96	0.81	0.72
	0.7	48.67	12.19	5.51	3.19	2.12	1.55	1.20	0.99	0.85
	0.8	63.63	15.89	7.14	4.10	2.70	1.94	1.49	1.20	1.01
	0.9	80.60	20.10	8.99	5.13	3.36	2.40	1.83	1.46	1.20
1.00	0.1	1.20	0.62	0.80	1.28	1.99	2.92	4.07	5.44	7.02
	0.2	4.10	1.20	0.72	0.62	0.66	0.80	1.01	1.28	1.60
	0.3	8.99	2.40	1.20	0.81	0.66	0.62	0.64	0.70	0.80
	0.4	15.89	4.10	1.94	1.20	0.88	0.72	0.64	0.62	0.63
	0.5	24.80	6.29	2.91	1.74	1.20	0.92	0.77	0.68	0.63
	0.6	35.73	8.99	4.10	2.40	1.62	1.20	0.96	0.81	0.72
	0.7	48.67	12.19	5.51	3.19	2.12	1.55	1.20	0.99	0.85
	0.8	63.63	15.89	7.14	4.10	2.70	1.94	1.49	1.20	1.01
	0.9	80.60	20.10	8.99	5.13	3.36	2.40	1.83	1.46	1.20

	C_s Values								
	Q_s/Q_c								
A_s/A_c	0.1	0.2	0.3	0.4	0.5	0.6	0.7	0.8	0.9
0.1	0.13	0.16							
0.2	0.20	0.13	0.15	0.16	0.28				
0.3	0.90	0.13	0.13	0.14	0.15	0.16	0.20		
0.4	2.88	0.20	0.14	0.13	0.14	0.15	0.15	0.16	0.34
0.5	6.25	0.37	0.17	0.14	0.13	0.14	0.14	0.15	0.15
0.6	11.88	0.90	0.20	0.13	0.14	0.13	0.14	0.14	0.15
0.7	18.62	1.71	0.33	0.18	0.16	0.14	0.13	0.15	0.14
0.8	26.88	2.88	0.50	0.20	0.15	0.14	0.13	0.13	0.14
0.9	36.45	4.46	0.90	0.30	0.19	0.16	0.15	0.14	0.13

For the other branch, subscripts 1 and 2 change places.

SD5-25 Cross, Conical Branches Tapered into Body, Diverging

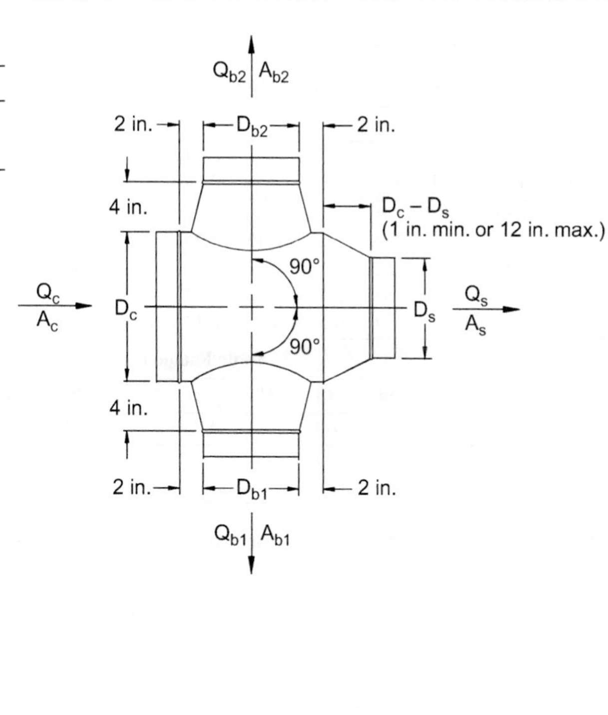

C_{b1} Values

		Q_{b1}/Q_c								
A_s/A_c	A_{b1}/A_c	0.1	0.2	0.3	0.4	0.5	0.6	0.7	0.8	0.9
0.20	0.1	2.07	2.08	1.62	1.30	1.08	0.93	0.81	0.72	0.64
	0.2		2.07	2.31	2.08	1.83	1.62	1.44	1.30	1.18
	0.3			2.07	2.34	2.24	2.08	1.91	1.76	1.62
	0.4			0.90	2.07	2.32	2.31	2.21	2.08	1.95
	0.5				1.28	2.07	2.30	2.33	2.27	2.18
	0.6					1.48	2.07	2.29	2.34	2.31
	0.7					0.55	1.60	2.07	2.27	2.33
	0.8						0.90	1.68	2.07	2.25
	0.9							1.12	1.74	2.07
0.35	0.1		3.25	3.11	2.69	2.32	2.03	1.80	1.61	1.46
	0.2			2.44	3.25	3.28	3.11	2.90	2.69	2.49
	0.3				1.69	2.88	3.25	3.31	3.23	3.11
	0.4					1.12	2.44	3.02	3.25	3.31
	0.5						0.69	2.04	2.73	3.09
	0.6							0.37	1.69	2.44
	0.7								0.11	1.38
	0.8									
	0.9									
0.55	0.1		1.50	1.56	1.38	1.20	1.06	0.94	0.84	0.77
	0.2			0.89	1.50	1.60	1.56	1.47	1.38	1.28
	0.3				0.38	1.20	1.50	1.59	1.59	1.56
	0.4					0.00	0.89	1.31	1.50	1.58
	0.5							0.61	1.09	1.36
	0.6								0.38	0.89
	0.7									0.17
	0.8									
	0.9									
0.80	0.1	0.65	0.24							
	0.2	2.98	0.65	0.33	0.24	0.18				
	0.3	7.36	1.56	0.65	0.39	0.29	0.24	0.20		
	0.4	13.78	2.98	1.20	0.65	0.43	0.33	0.27	0.24	0.21
	0.5	22.24	4.92	1.98	1.04	0.65	0.47	0.36	0.30	0.26
	0.6	32.73	7.36	2.98	1.56	0.96	0.65	0.49	0.39	0.33
	0.7	45.26	10.32	4.21	2.21	1.34	0.90	0.65	0.51	0.42
	0.8	59.82	13.78	5.67	2.98	1.80	1.20	0.86	0.65	0.52
	0.9	76.41	17.75	7.36	3.88	2.35	1.56	1.11	0.83	0.65
1.00	0.1	0.65	0.24							
	0.2	2.98	0.65	0.33	0.24	0.18				
	0.3	7.36	1.56	0.65	0.39	0.29	0.24	0.20		
	0.4	13.78	2.98	1.20	0.65	0.43	0.33	0.27	0.24	0.21
	0.5	22.24	4.92	1.98	1.04	0.65	0.47	0.36	0.30	0.26
	0.6	32.73	7.36	2.98	1.56	0.96	0.65	0.49	0.39	0.33
	0.7	45.26	10.32	4.21	2.21	1.34	0.90	0.65	0.51	0.42
	0.8	59.82	13.78	5.67	2.98	1.80	1.20	0.86	0.65	0.52
	0.9	76.41	17.75	7.36	3.88	2.35	1.56	1.11	0.83	0.65

C_s Values

	Q_s/Q_c								
A_s/A_c	0.1	0.2	0.3	0.4	0.5	0.6	0.7	0.8	0.9
0.1	0.13	0.16							
0.2	0.20	0.13	0.15	0.16	0.28				
0.3	0.90	0.13	0.13	0.14	0.15	0.16	0.20		
0.4	2.88	0.20	0.14	0.13	0.14	0.15	0.15	0.16	0.34
0.5	6.25	0.37	0.17	0.14	0.13	0.14	0.14	0.15	0.15
0.6	11.88	0.90	0.20	0.13	0.14	0.13	0.14	0.14	0.15
0.7	18.62	1.71	0.33	0.18	0.16	0.14	0.13	0.15	0.14
0.8	26.88	2.88	0.50	0.20	0.15	0.14	0.13	0.13	0.14
0.9	36.45	4.46	0.90	0.30	0.19	0.16	0.15	0.14	0.13

For the other branch, subscripts 1 and 2 change places

RECTANGULAR FITTINGS

CR3-1 Elbow, Smooth Radius, Without Vanes

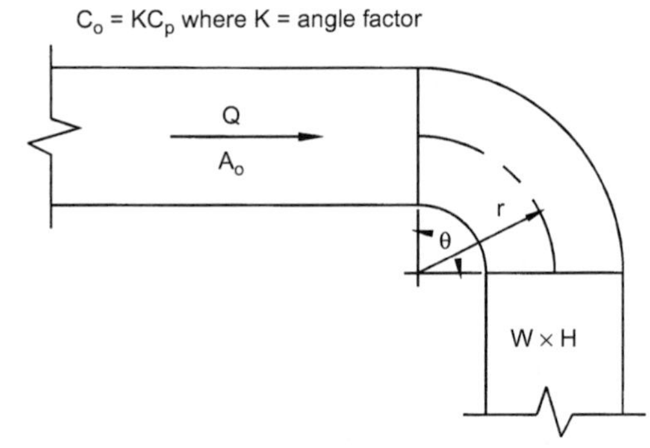

	C_p Values										
	H/W										
r/W	**0.25**	**0.50**	**0.75**	**1.00**	**1.50**	**2.00**	**3.00**	**4.00**	**5.00**	**6.00**	**8.00**
0.50	1.53	1.38	1.29	1.18	1.06	1.00	1.00	1.06	1.12	1.16	1.18
0.75	0.57	0.52	0.48	0.44	0.40	0.39	0.39	0.40	0.42	0.43	0.44
1.00	0.27	0.25	0.23	0.21	0.19	0.18	0.18	0.19	0.20	0.21	0.21
1.50	0.22	0.20	0.19	0.17	0.15	0.14	0.14	0.15	0.16	0.17	0.17
2.00	0.20	0.18	0.16	0.15	0.14	0.13	0.13	0.14	0.14	0.15	0.15

	Angle Factor *K*										
θ	**0**	**20**	**30**	**45**	**60**	**75**	**90**	**110**	**130**	**150**	**180**
K	0.00	0.31	0.45	0.60	0.78	0.90	1.00	1.13	1.20	1.28	1.40

CR3-3 Elbow, Smooth Radius, One Splitter Vane

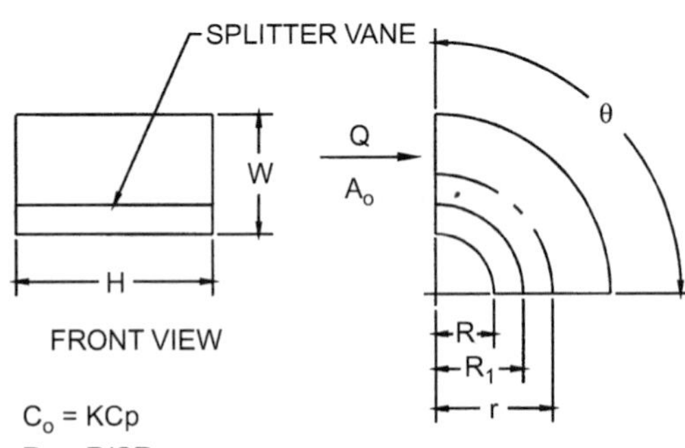

$C_o = KCp$

$R_1 = R/CR$

where

R = throat radius

R_1 = splitter vane radius

CR = curve ratio

K = angle factor

	C_p Values										
	H/W										
r/W	**0.25**	**0.50**	**1.00**	**1.50**	**2.00**	**3.00**	**4.00**	**5.00**	**6.00**	**7.00**	**8.00**
0.55	0.52	0.40	0.43	0.49	0.55	0.66	0.75	0.84	0.93	1.01	1.09
0.60	0.36	0.27	0.25	0.28	0.30	0.35	0.39	0.42	0.46	0.49	0.52
0.65	0.28	0.21	0.18	0.19	0.20	0.22	0.25	0.26	0.28	0.30	0.32
0.70	0.22	0.16	0.14	0.14	0.15	0.16	0.17	0.18	0.19	0.20	0.21
0.75	0.18	0.13	0.11	0.11	0.11	0.12	0.13	0.14	0.14	0.15	0.15
0.80	0.15	0.11	0.09	0.09	0.09	0.09	0.10	0.10	0.11	0.11	0.12
0.85	0.13	0.09	0.08	0.07	0.07	0.08	0.08	0.08	0.08	0.09	0.09
0.90	0.11	0.08	0.07	0.06	0.06	0.06	0.06	0.07	0.07	0.07	0.07
0.95	0.10	0.07	0.06	0.05	0.05	0.05	0.05	0.05	0.06	0.06	0.06
1.00	0.09	0.06	0.05	0.05	0.04	0.04	0.04	0.05	0.05	0.05	0.05

	Angle Factor *K*				
θ	**0**	**30**	**45**	**60**	**90**
K	0.00	0.45	0.60	0.78	1.00

	Curve Ratio CR									
r/W	**0.55**	**0.60**	**0.65**	**0.70**	**0.75**	**0.80**	**0.85**	**0.90**	**0.95**	**1.00**
CR	0.218	0.302	0.361	0.408	0.447	0.480	0.509	0.535	0.557	0.577

	Throat Radius/Width Ratio (*R/W*)									
r/W	**0.55**	**0.60**	**0.65**	**0.70**	**0.75**	**0.80**	**0.85**	**0.90**	**0.95**	**1.00**
R/W	0.05	0.10	0.15	0.20	0.25	0.30	0.35	0.40	0.45	0.50

CR3-6 Elbow, Mitered

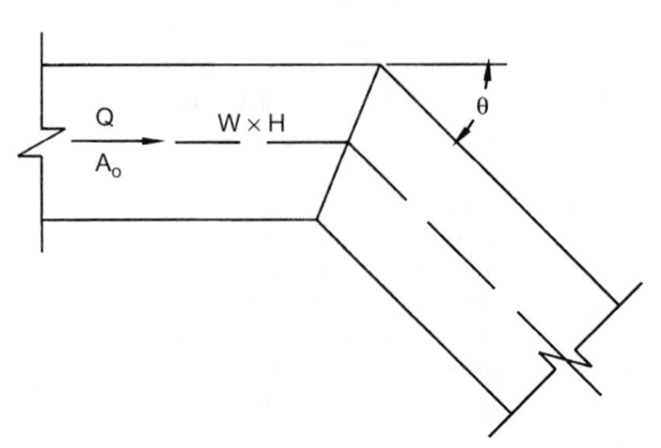

	C_o Values										
	H/W										
θ	**0.25**	**0.50**	**0.75**	**1.00**	**1.50**	**2.00**	**3.00**	**4.00**	**5.00**	**6.00**	**8.00**
20	0.08	0.08	0.08	0.07	0.07	0.07	0.06	0.06	0.05	0.05	0.05
30	0.18	0.17	0.17	0.16	0.15	0.15	0.13	0.13	0.12	0.12	0.11
45	0.38	0.37	0.36	0.34	0.33	0.31	0.28	0.27	0.26	0.25	0.24
60	0.60	0.59	0.57	0.55	0.52	0.49	0.46	0.43	0.41	0.39	0.38
75	0.89	0.87	0.84	0.81	0.77	0.73	0.67	0.63	0.61	0.58	0.57
90	1.30	1.27	1.23	1.18	1.13	1.07	0.98	0.92	0.89	0.85	0.83

CR3-9 Elbow, Mitered, 90 Degree, Single-Thickness Vanes (Design 1)

$C_o = 0.11$

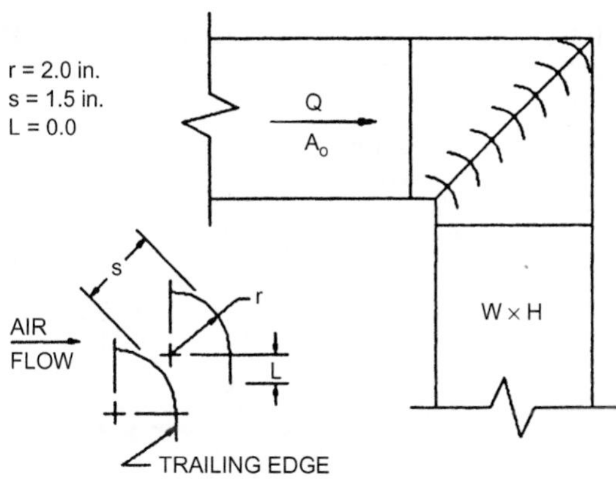

CR3-10 Elbow, Mitered, 90 Degree, Single-Thickness Vanes (Design 2)

$C_o = 0.12$

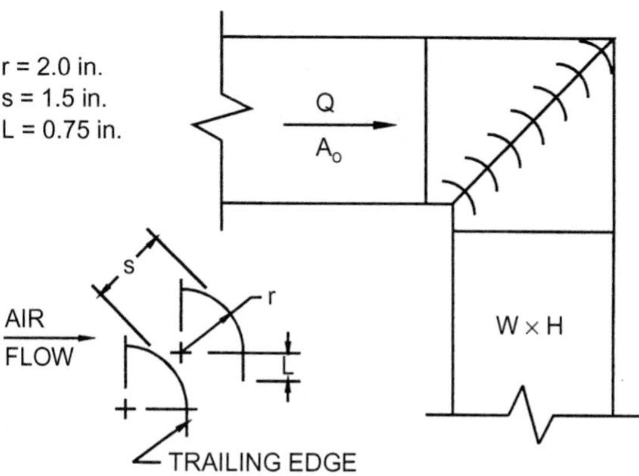

CR3-12 Elbow, Mitered, 90 Degree, Single-Thickness Vanes (Design 4)

$C_o = 0.33$

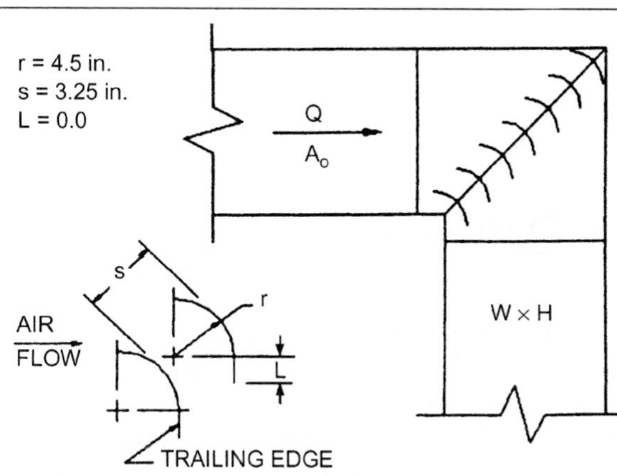

CR3-14 Elbow, Mitered, 90 Degree, Double-Thickness Vanes (Design 1)

$C_o = 0.38$

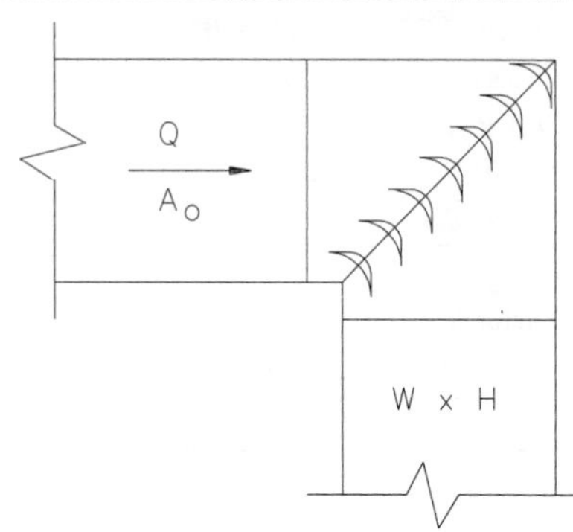

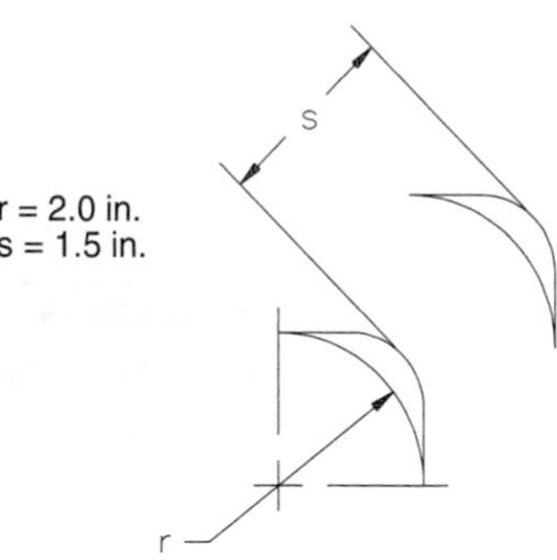

CR3-15 Elbow, Mitered, 90 Degree, Double-Thickness Vanes (Design 2)

$C_o = 0.25$

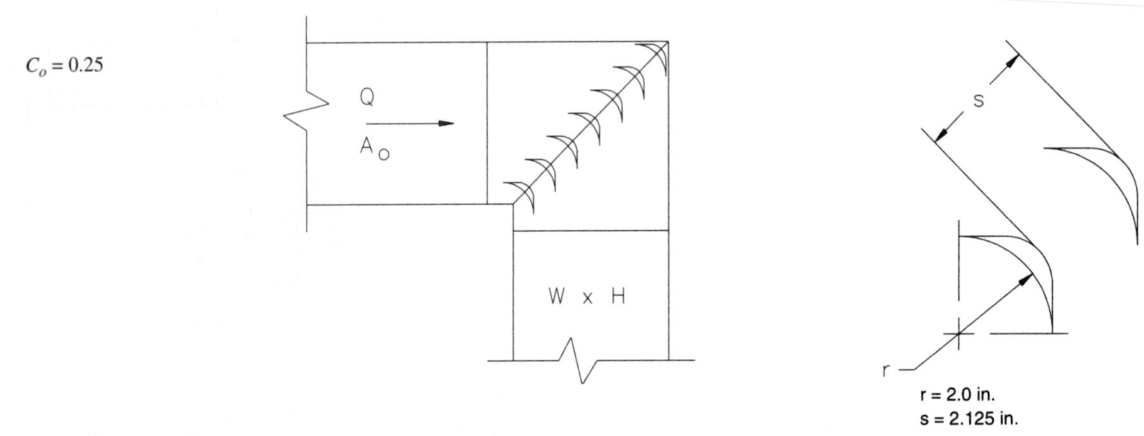

CR3-16 Elbow, Mitered, 90 Degree, Double-Thickness Vanes (Design 3)

$C_o = 0.41$

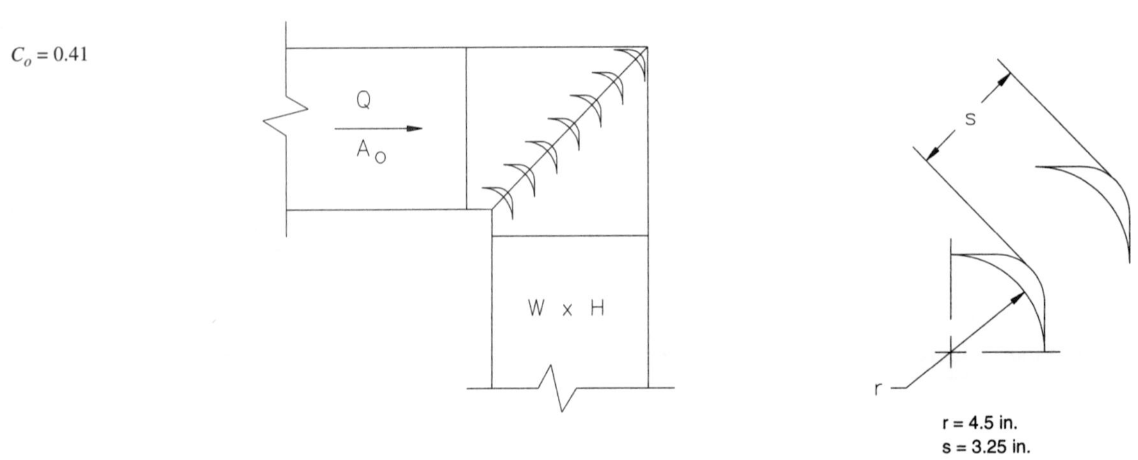

CR3-17 Elbow, Z-Shaped

	C_p Values													
	L/W													
H/W	0.0	0.4	0.6	0.8	1.0	1.2	1.4	1.6	1.8	2.0	4.0	8.0	10.0	100.0
0.25	0.00	0.68	0.99	1.77	2.89	3.97	4.41	4.60	4.64	4.60	3.39	3.03	2.70	2.53
0.50	0.00	0.66	0.96	1.72	2.81	3.86	4.29	4.47	4.52	4.47	3.30	2.94	2.62	2.46
0.75	0.00	0.64	0.94	1.67	2.74	3.75	4.17	4.35	4.39	4.35	3.20	2.86	2.55	2.39
1.00	0.00	0.62	0.90	1.61	2.63	3.61	4.01	4.18	4.22	4.18	3.08	2.75	2.45	2.30
1.50	0.00	0.59	0.86	1.53	2.50	3.43	3.81	3.97	4.01	3.97	2.93	2.61	2.33	2.19
2.00	0.00	0.56	0.81	1.45	2.37	3.25	3.61	3.76	3.80	3.76	2.77	2.48	2.21	2.07
3.00	0.00	0.51	0.75	1.34	2.18	3.00	3.33	3.47	3.50	3.47	2.56	2.28	2.03	1.91
4.00	0.00	0.48	0.70	1.26	2.05	2.82	3.13	3.26	3.29	3.26	2.40	2.15	1.91	1.79
6.00	0.00	0.45	0.65	1.16	1.89	2.60	2.89	3.01	3.04	3.01	2.22	1.98	1.76	1.66
8.00	0.00	0.43	0.63	1.13	1.84	2.53	2.81	2.93	2.95	2.93	2.16	1.93	1.72	1.61

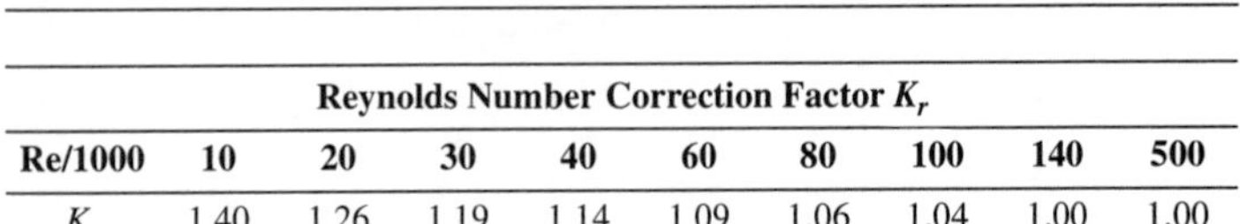

Reynolds Number Correction Factor K_r									
Re/1000	10	20	30	40	60	80	100	140	500
K_r	1.40	1.26	1.19	1.14	1.09	1.06	1.04	1.00	1.00

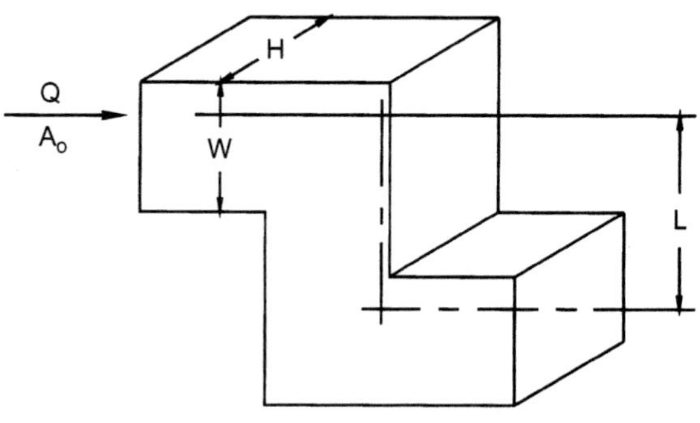

$C_o = K_r C_p$
where K_r = Reynolds number correction factor

CR6-1 Screen (Only)

A_1/A_o	C_o Values, n = 0.30	0.35	0.40	0.45	0.50	0.55	0.60	0.65	0.70	0.75	0.80	0.90	1.00
0.2	155.00	102.50	75.00	55.00	41.25	31.50	24.25	18.75	14.50	11.00	8.00	3.50	0.00
0.3	68.89	45.56	33.33	24.44	18.33	14.00	10.78	8.33	6.44	4.89	3.56	1.56	0.00
0.4	38.75	25.63	18.75	13.75	10.31	7.88	6.06	4.69	3.63	2.75	2.00	0.88	0.00
0.5	24.80	16.40	12.00	8.80	6.60	5.04	3.88	3.00	2.32	1.76	1.28	0.56	0.00
0.6	17.22	11.39	8.33	6.11	4.58	3.50	2.69	2.08	1.61	1.22	0.89	0.39	0.00
0.7	12.65	8.37	6.12	4.49	3.37	2.57	1.98	1.53	1.18	0.90	0.65	0.29	0.00
0.8	9.69	6.40	4.69	3.44	2.58	1.97	1.52	1.17	0.91	0.69	0.50	0.22	0.00
0.9	7.65	5.06	3.70	2.72	2.04	1.56	1.20	0.93	0.72	0.54	0.40	0.17	0.00
1.0	6.20	4.10	3.00	2.20	1.65	1.26	0.97	0.75	0.58	0.44	0.32	0.14	0.00
1.2	4.31	2.85	2.08	1.53	1.15	0.88	0.67	0.36	0.40	0.31	0.22	0.10	0.00
1.4	3.16	2.09	1.53	1.12	0.84	0.64	0.49	0.38	0.30	0.22	0.16	0.07	0.00
1.6	2.42	1.60	1.17	0.86	0.64	0.49	0.38	0.29	0.23	0.17	0.13	0.05	0.00
1.8	1.91	1.27	0.93	0.68	0.51	0.39	0.30	0.23	0.18	0.14	0.10	0.04	0.00
2.0	1.55	1.03	0.75	0.55	0.41	0.32	0.24	0.19	0.15	0.11	0.08	0.04	0.00
2.5	0.99	0.66	0.48	0.35	0.26	0.20	0.16	0.12	0.09	0.07	0.05	0.02	0.00
3.0	0.69	0.46	0.33	0.24	0.18	0.14	0.11	0.08	0.06	0.05	0.04	0.02	0.00
4.0	0.39	0.26	0.19	0.14	0.10	0.08	0.06	0.05	0.04	0.03	0.02	0.01	0.00
6.0	0.17	0.11	0.08	0.06	0.05	0.04	0.03	0.02	0.02	0.01	0.01	0.00	0.00

Q
A_o
$H_o \times W_o$
SCREEN
SCREEN $H_1 \times W_1$
①

n = free area ratio of screen
A_o = area of duct
A_1 = cross-sectional area of duct or fitting where screen is located

CR6-4 Obstruction, Smooth Cylinder in Rectangular Duct

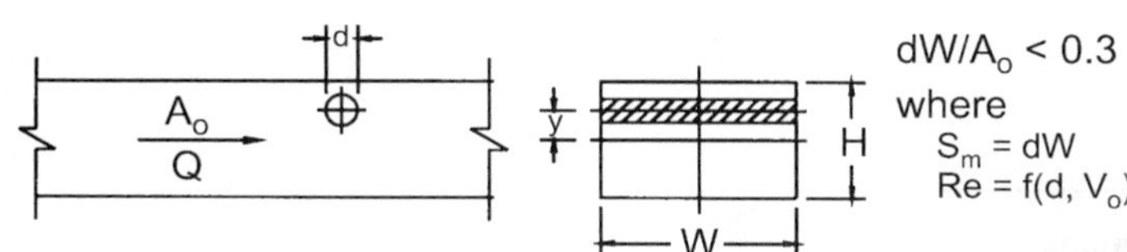

$dW/A_o < 0.3$
where
$S_m = dW$
$Re = f(d, V_o)$

y/H	Re/1000	C_o Values, S_m/A_o = 0.00	0.05	0.10	0.15	0.20
0.00	0.1	0.00	0.10	0.21	0.35	0.47
	0.5	0.00	0.08	0.17	0.28	0.38
	200	0.00	0.08	0.17	0.28	0.38
	300	0.00	0.07	0.16	0.26	0.35
	400	0.00	0.05	0.11	0.19	0.25
	500	0.00	0.04	0.09	0.14	0.19
	600	0.00	0.02	0.05	0.07	0.10
	1000	0.00	0.02	0.05	0.08	0.11
0.05	0.1	0.00	0.10	0.21	0.34	0.46
	0.5	0.00	0.08	0.17	0.27	0.37
	200	0.00	0.08	0.17	0.27	0.37
	300	0.00	0.07	0.15	0.25	0.34
	400	0.00	0.05	0.11	0.18	0.24
	500	0.00	0.04	0.08	0.13	0.18
	600	0.00	0.02	0.04	0.07	0.10
	1000	0.00	0.02	0.05	0.08	0.11
0.10	0.1	0.00	0.09	0.20	0.32	0.44
	0.5	0.00	0.07	0.16	0.26	0.35
	200	0.00	0.07	0.16	0.26	0.35
	300	0.00	0.07	0.15	0.24	0.32
	400	0.00	0.05	0.11	0.17	0.23
	500	0.00	0.04	0.08	0.13	0.18
	600	0.00	0.02	0.04	0.07	0.09
	1000	0.00	0.02	0.05	0.08	0.10
0.15	0.1	0.00	0.09	0.19	0.31	0.42
	0.5	0.00	0.07	0.15	0.25	0.34
	200	0.00	0.07	0.15	0.25	0.34
	300	0.00	0.06	0.14	0.23	0.31
	400	0.00	0.05	0.10	0.17	0.22
	500	0.00	0.04	0.08	0.12	0.17
	600	0.00	0.02	0.04	0.07	0.09
	1000	0.00	0.02	0.04	0.07	0.10
0.20	0.1	0.00	0.08	0.18	0.29	0.40
	0.5	0.00	0.07	0.14	0.24	0.32
	200	0.00	0.07	0.14	0.24	0.32
	300	0.00	0.06	0.13	0.22	0.29
	400	0.00	0.04	0.10	0.16	0.21
	500	0.00	0.03	0.07	0.12	0.16
	600	0.00	0.02	0.04	0.06	0.09
	1000	0.00	0.02	0.04	0.07	0.09
0.25	0.1	0.00	0.08	0.17	0.28	0.38
	0.5	0.00	0.06	0.14	0.22	0.30
	200	0.00	0.06	0.14	0.22	0.30
	300	0.00	0.06	0.12	0.20	0.28
	400	0.00	0.04	0.09	0.15	0.20
	500	0.00	0.03	0.07	0.11	0.15
	600	0.00	0.02	0.04	0.06	0.08
	1000	0.00	0.02	0.04	0.06	0.09
0.30	0.1	0.00	0.07	0.16	0.26	0.35
	0.5	0.00	0.06	0.13	0.21	0.28
	200	0.00	0.06	0.13	0.21	0.28
	300	0.00	0.05	0.12	0.19	0.26
	400	0.00	0.04	0.08	0.14	0.19
	500	0.00	0.03	0.06	0.10	0.14
	600	0.00	0.02	0.03	0.05	0.07
	1000	0.00	0.02	0.04	0.06	0.08
0.35	0.1	0.00	0.07	0.14	0.23	0.32
	0.5	0.00	0.05	0.11	0.19	0.25
	200	0.00	0.05	0.11	0.19	0.25
	300	0.00	0.05	0.11	0.17	0.23
	400	0.00	0.04	0.08	0.12	0.17
	500	0.00	0.03	0.06	0.09	0.13
	600	0.00	0.01	0.03	0.05	0.07
	1000	0.00	0.02	0.03	0.05	0.07
0.40	0.1	0.00	0.06	0.13	0.20	0.28
	0.5	0.00	0.05	0.10	0.16	0.22
	200	0.00	0.05	0.10	0.16	0.22
	300	0.00	0.04	0.09	0.15	0.20
	400	0.00	0.03	0.07	0.11	0.15
	500	0.00	0.02	0.05	0.08	0.11
	600	0.00	0.01	0.03	0.04	0.06
	1000	0.00	0.01	0.03	0.05	0.06

CR9-1 Damper, Butterfly

	C_o Values									
	θ									
H/W	0	10	20	30	40	50	60	65	70	90
0.12	0.04	0.30	1.10	3.00	8.00	23.00	60.00	100.00	190.00	99999
0.25	0.08	0.33	1.18	3.30	9.00	26.00	70.00	128.00	210.00	99999
1.00	0.08	0.33	1.18	3.30	9.00	26.00	70.00	128.00	210.00	99999
2.00	0.13	0.35	1.25	3.60	10.00	29.00	80.00	155.00	230.00	99999

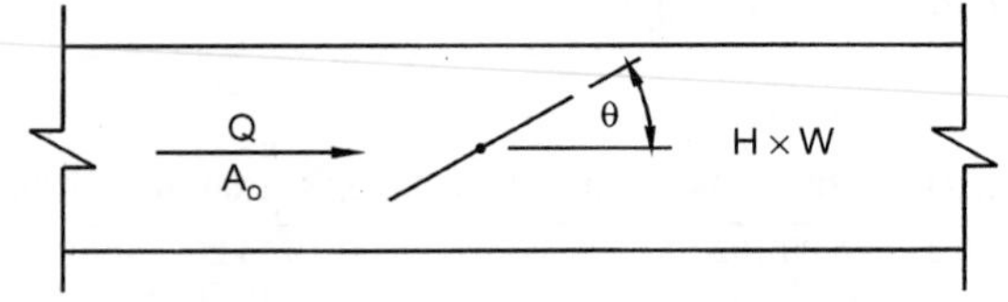

CR9-3 Damper, Parallel Blades

	C_o Values								
	θ								
L/R	0	10	20	30	40	50	60	70	80
0.3	0.52	0.79	1.49	2.20	4.95	8.73	14.15	32.11	122.06
0.4	0.52	0.84	1.56	2.25	5.03	9.00	16.00	37.73	156.58
0.5	0.52	0.88	1.62	2.35	5.11	9.52	18.88	44.79	187.85
0.6	0.52	0.92	1.66	2.45	5.20	9.77	21.75	53.78	288.89
0.8	0.52	0.96	1.69	2.55	5.30	10.03	22.80	65.46	295.22
1.0	0.52	1.00	1.76	2.66	5.40	10.53	23.84	73.23	361.00
1.5	0.52	1.08	1.83	2.78	5.44	11.21	27.56	97.41	495.31

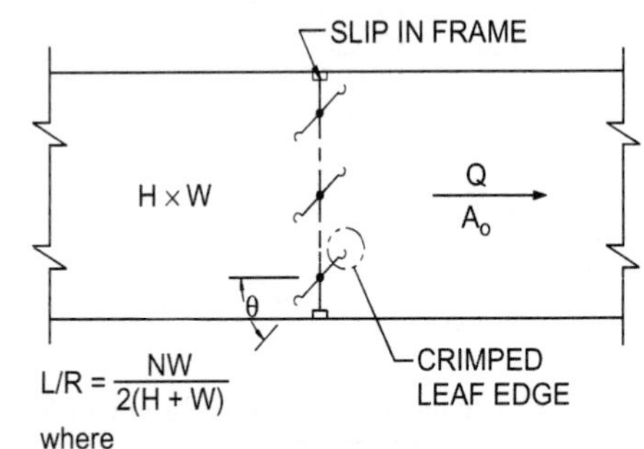

$$L/R = \frac{NW}{2(H + W)}$$

where
N = number of damper blades
W = duct dimension parallel to blade axis, in.
H = duct height, in.
L = sum of damper blade lengths, in.
R = perimeter of duct, in.

CR9-4 Damper, Opposed Blades

	C_o Values								
	θ								
L/R	0	10	20	30	40	50	60	70	80
0.3	0.52	0.79	1.91	3.77	8.55	19.46	70.12	295.21	807.23
0.4	0.52	0.85	2.07	4.61	10.42	26.73	92.90	346.25	926.34
0.5	0.52	0.93	2.25	5.44	12.29	33.99	118.91	393.36	1045.44
0.6	0.52	1.00	2.46	5.99	14.15	41.26	143.69	440.25	1163.09
0.8	0.52	1.08	2.66	6.96	18.18	56.47	193.92	520.27	1324.85
1.0	0.52	1.17	2.91	7.31	20.25	71.68	245.45	576.00	1521.00
1.5	0.52	1.38	3.16	9.51	27.56	107.41	361.00	717.05	1804.40

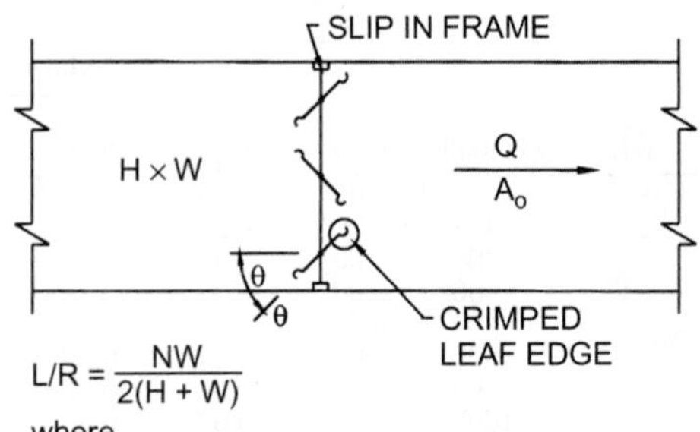

$$L/R = \frac{NW}{2(H + W)}$$

where
N = number of damper blades
W = duct dimension parallel to blade axis, in.
H = duct height, in.
L = sum of damper blade lengths, in.
R = perimeter of duct, in.

CR9-6 Fire Damper, Curtain Type, Type B

$C_o = 0.19$

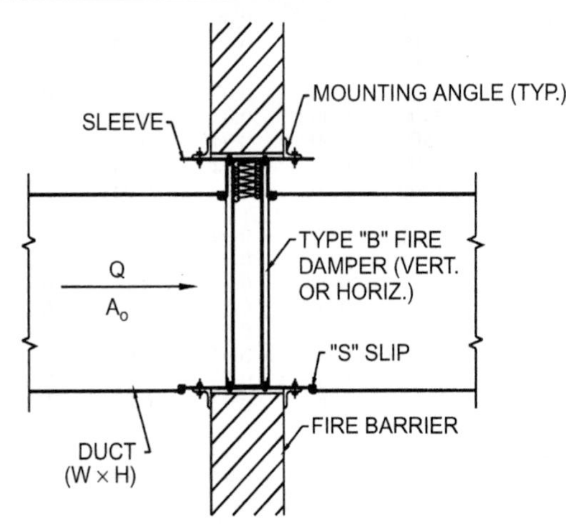

ER2-1 Bellmouth, Plenum to Round, Exhaust/Return Systems

	C_o Values												
	r/D_1												
A_o/A_1	0.00	0.01	0.02	0.03	0.04	0.05	0.06	0.08	0.10	0.12	0.16	0.20	10.00
1.5	0.22	0.20	0.15	0.14	0.12	0.10	0.09	0.07	0.05	0.04	0.03	0.01	0.01
2.0	0.13	0.11	0.08	0.08	0.07	0.06	0.05	0.04	0.03	0.02	0.02	0.01	0.01
2.5	0.08	0.07	0.05	0.05	0.04	0.04	0.03	0.02	0.02	0.01	0.01	0.00	0.00
3.0	0.06	0.05	0.04	0.03	0.03	0.02	0.02	0.02	0.01	0.01	0.01	0.00	0.00
4.0	0.03	0.03	0.02	0.02	0.02	0.01	0.01	0.01	0.01	0.01	0.00	0.00	0.00
8.0	0.01	0.01	0.01	0.00	0.00	0.00	0.00	0.00	0.00	0.00	0.00	0.00	0.00

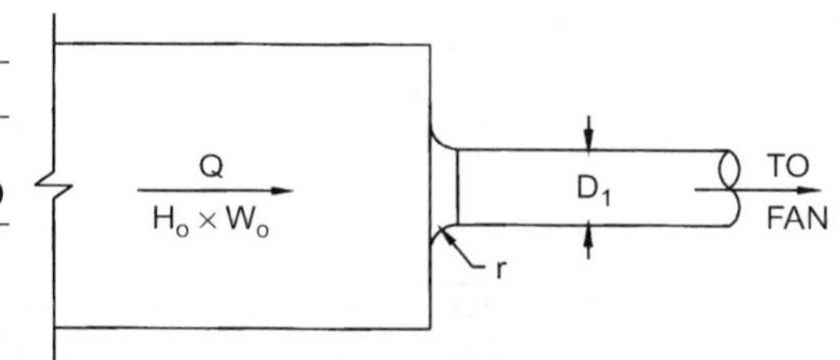

ER3-1 Elbow, 90 Degree, Variable Inlet/Outlet Areas, Exhaust/Return Systems

	C_o Values						
	W_1/W_o						
H/W_o	0.6	0.8	1.0	1.2	1.4	1.6	2.0
0.25	1.76	1.43	1.24	1.14	1.09	1.06	1.06
1.00	1.70	1.36	1.15	1.02	0.95	0.90	0.84
4.00	1.46	1.10	0.90	0.81	0.76	0.72	0.66
100.00	1.50	1.04	0.79	0.69	0.63	0.60	0.55

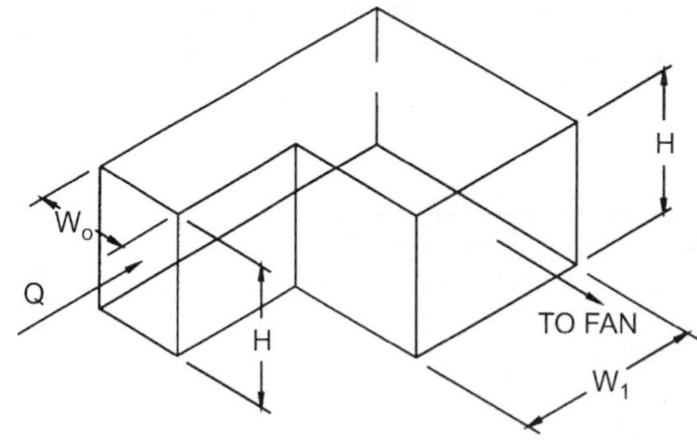

ER4-1 Transition, Rectangular, Two Sides Parallel, Symmetrical, Exhaust/Return Systems

	C_o Values									
	θ									
A_o/A_1	10	15	20	30	45	60	90	120	150	180
0.06	0.26	0.27	0.40	0.56	0.71	0.86	1.00	0.99	0.98	0.98
0.10	0.24	0.26	0.36	0.53	0.69	0.82	0.93	0.93	0.92	0.91
0.25	0.17	0.19	0.22	0.42	0.60	0.68	0.70	0.69	0.67	0.66
0.50	0.14	0.13	0.15	0.24	0.35	0.37	0.38	0.37	0.36	0.35
1.00	0.00	0.00	0.00	0.00	0.00	0.00	0.00	0.00	0.00	0.00
2.00	0.23	0.20	0.20	0.20	0.24	0.28	0.54	0.78	1.02	1.09
4.00	0.81	0.64	0.64	0.64	0.88	1.12	2.78	4.38	5.65	6.60
6.00	1.82	1.44	1.44	1.44	1.98	2.53	6.56	10.20	13.00	15.20
10.00	5.03	5.00	5.00	5.00	6.50	8.02	19.10	29.10	37.10	43.10

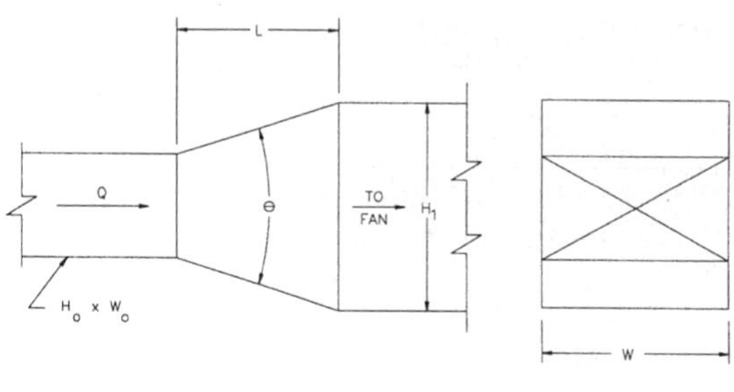

A_o/A_1 < or > 1

ER4-3 Transition, Rectangular to Round, Exhaust/Return Systems

	C_o Values									
	θ									
A_o/A_1	10	15	20	30	45	60	90	120	150	180
0.06	0.30	0.54	0.53	0.65	0.77	0.88	0.95	0.98	0.98	0.93
0.10	0.30	0.50	0.53	0.64	0.75	0.84	0.89	0.91	0.91	0.88
0.25	0.25	0.36	0.45	0.52	0.58	0.62	0.64	0.64	0.64	0.64
0.50	0.15	0.21	0.25	0.30	0.33	0.33	0.33	0.32	0.31	0.30
1.00	0.00	0.00	0.00	0.00	0.00	0.00	0.00	0.00	0.00	0.00
2.00	0.24	0.28	0.26	0.20	0.22	0.24	0.49	0.73	0.97	1.04
4.00	0.89	0.78	0.79	0.70	0.88	1.12	2.72	4.33	5.62	6.58
6.00	1.89	1.67	1.59	1.49	1.98	2.52	6.51	10.14	13.05	15.14
10.00	5.09	5.32	5.15	5.05	6.50	8.05	19.06	29.07	37.08	43.05

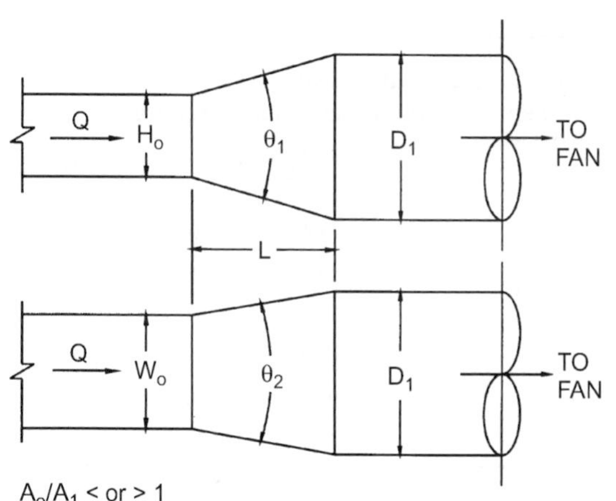

A_o/A_1 < or > 1
θ is larger of θ_1 and θ_2

ER5-2 Tee, Round Tap to Rectangular Main, Converging

Q_b/Q_c	0.1	0.2	0.3	0.4	0.5	0.6	0.7	0.8	0.9	1.0
C_b	−12.25	−1.31	0.64	0.94	1.27	1.43	1.40	1.45	1.52	1.49

Q_s/Q_c	0.1	0.2	0.3	0.4	0.5	0.6	0.7	0.8	0.9
C_s	2.15	11.91	6.54	3.74	2.23	1.33	0.76	0.38	0.10

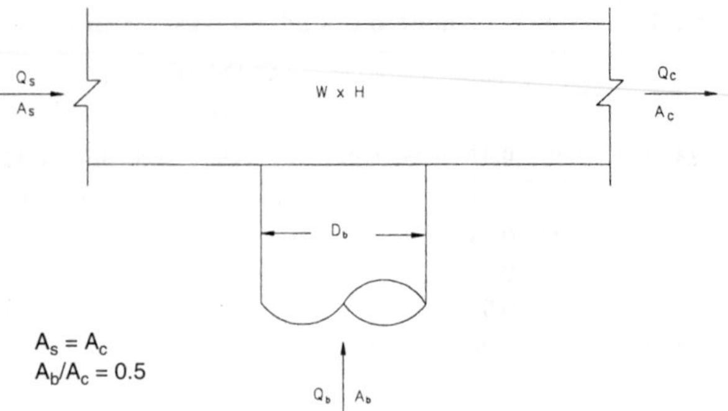

ER5-3 Tee, 45 Degree Entry Branch, Converging

Q_b/Q_c	0.1	0.2	0.3	0.4	0.5	0.6	0.7	0.8	0.9	1.0
C_b	-18.00	-3.25	-0.64	0.53	0.76	0.79	0.93	0.79	0.90	0.91

Q_s/Q_c	0.1	0.2	0.3	0.4	0.5	0.6	0.7	0.8	0.9
C_s	2.15	11.91	6.54	3.74	2.23	1.33	0.76	0.38	0.10

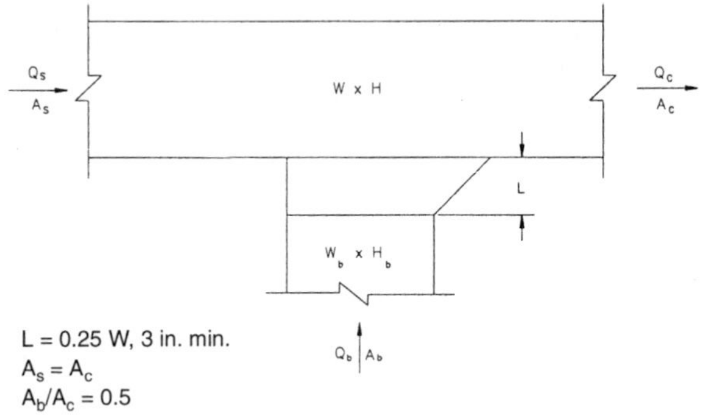

ER5-4 Wye, Symmetrical, Dovetail, $Q_b/Q_c = 0.5$, Converging

A_b/A_c	0.5	1.0
C_b	0.23	0.28

Branches are identical, $Q_{b1} = Q_{b2} = Q_b$, and $C_{b1} = C_{b2} = C_b$

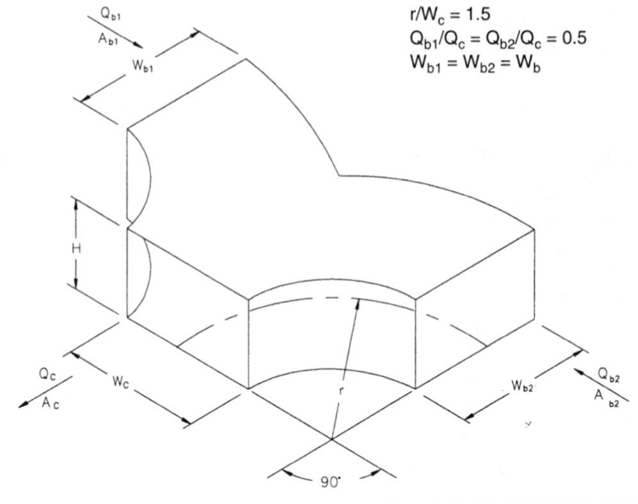

ER7-1 Fan Inlet, Centrifugal, SWSI, 90 Degree Smooth Radius Elbow (Square)

	C_o Values			
	L/H			
r/H	0.0	2.0	5.0	10.0
0.50	2.50	1.60	0.80	0.80
0.75	2.00	1.20	0.67	0.67
1.00	1.20	0.67	0.33	0.33
1.50	1.00	0.57	0.30	0.30
2.00	0.80	0.47	0.26	0.26

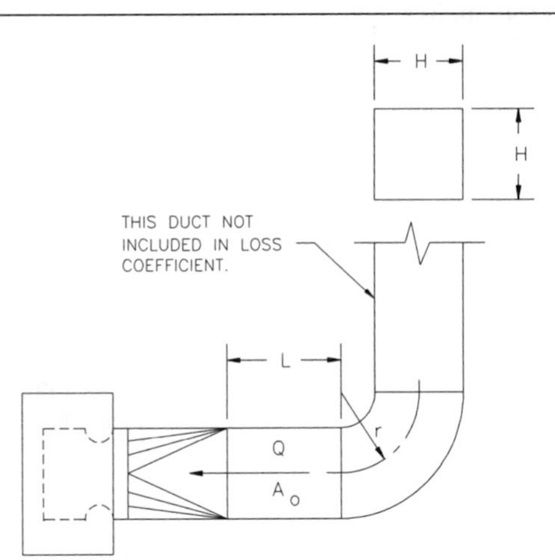

SR1-1 Conical Bellmouth/Sudden Contraction, Plenum to Rectangular, Supply Air Systems

		C_o Values									
		θ									
A_o/A_1	L/D_h	0	10	20	30	45	60	90	120	150	180
0.10	0.025	0.46	0.43	0.42	0.40	0.38	0.37	0.38	0.40	0.43	0.46
	0.050	0.46	0.42	0.38	0.33	0.30	0.28	0.31	0.36	0.41	0.46
	0.075	0.46	0.39	0.32	0.28	0.23	0.21	0.26	0.32	0.39	0.46
	0.100	0.46	0.36	0.30	0.23	0.19	0.17	0.23	0.30	0.38	0.46
	0.150	0.46	0.34	0.25	0.18	0.15	0.14	0.21	0.29	0.37	0.46
	0.300	0.46	0.31	0.22	0.16	0.13	0.13	0.20	0.28	0.37	0.46
	0.600	0.46	0.25	0.17	0.12	0.10	0.11	0.19	0.27	0.36	0.46
0.20	0.025	0.42	0.40	0.38	0.36	0.34	0.34	0.35	0.37	0.39	0.42
	0.050	0.42	0.38	0.35	0.30	0.27	0.25	0.29	0.33	0.37	0.42
	0.075	0.42	0.36	0.30	0.25	0.21	0.19	0.24	0.30	0.36	0.42
	0.100	0.42	0.33	0.27	0.21	0.18	0.15	0.21	0.27	0.35	0.42
	0.150	0.42	0.31	0.23	0.17	0.13	0.13	0.19	0.26	0.34	0.42
	0.300	0.42	0.28	0.20	0.15	0.12	0.12	0.18	0.26	0.34	0.42
	0.600	0.42	0.23	0.15	0.11	0.10	0.10	0.17	0.25	0.33	0.42
0.40	0.025	0.34	0.32	0.31	0.29	0.28	0.27	0.28	0.30	0.32	0.34
	0.050	0.34	0.31	0.28	0.25	0.22	0.20	0.23	0.26	0.30	0.34
	0.075	0.34	0.29	0.24	0.20	0.17	0.16	0.19	0.24	0.29	0.34
	0.100	0.34	0.27	0.22	0.17	0.14	0.12	0.17	0.22	0.28	0.34
	0,150	0.34	0.25	0.18	0.14	0.11	0.10	0.15	0.21	0.27	0.34
	0.300	0.34	0.23	0.16	0.12	0.10	0.10	0.15	0.21	0.27	0.34
	0.600	0.34	0.18	0.12	0.09	0.08	0.08	0.14	0.20	0.27	0.34
0.60	0.025	0.25	0.24	0.23	0.22	0.20	0.20	0.21	0.22	0.23	0.25
	0.050	0.25	0.23	0.21	0.18	0.16	0.15	0.17	0.19	0.22	0.25
	0.075	0.25	0.21	0.18	0.15	0.13	0.12	0.14	0.18	0.21	0.25
	0.100	0.25	0.20	0.16	0.13	0.11	0.09	0.12	0.16	0.21	0.25
	0.150	0.25	0.19	0.14	0.10	0.08	0.08	0.11	0.16	0.20	0.25
	0.300	0.25	0.17	0.12	0.09	0.07	0.07	0.11	0.15	0.20	0.25
	0.600	0.25	0.14	0.09	0.07	0.06	0.06	0.10	0.15	0.20	0.25
0.80	0.025	0.15	0.14	0.13	0.13	0.12	0.12	0.12	0.13	0.14	0.15
	0.050	0.15	0.13	0.12	0.11	0.10	0.09	0.10	0.12	0.13	0.15
	0.075	0.15	0.13	0.10	0.09	0.08	0.07	0.08	0.10	0.13	0.15
	0.100	0.15	0.12	0.10	0.07	0.06	0.05	0.07	0.10	0.12	0.15
	0.150	0.15	0.11	0.08	0.06	0.05	0.04	0.07	0.09	0.12	0.15
	0.300	0.15	0.10	0.07	0.05	0.04	0.04	0.07	0.09	0.12	0.15
	0.600	0.15	0.08	0.05	0.04	0.03	0.04	0.06	0.09	0.12	0.15
0.90	0.025	0.09	0.08	0.08	0.08	0.07	0.07	0.07	0.08	0.08	0.09
	0.050	0.09	0.08	0.07	0.06	0.06	0.05	0.06	0.07	0.08	0.09
	0.075	0.09	0.07	0.06	0.05	0.04	0.04	0.05	0.06	0.08	0.09
	0.100	0.09	0.07	0.06	0.04	0.04	0.03	0.04	0.06	0.07	0.09
	0.150	0.09	0.07	0.05	0.04	0.03	0.03	0.04	0.06	0.07	0.09
	0.300	0.09	0.06	0.04	0.03	0.03	0.02	0.04	0.05	0.07	0.09
	0.600	0.09	0.05	0.03	0.02	0.02	0.02	0.04	0.05	0.07	0.09

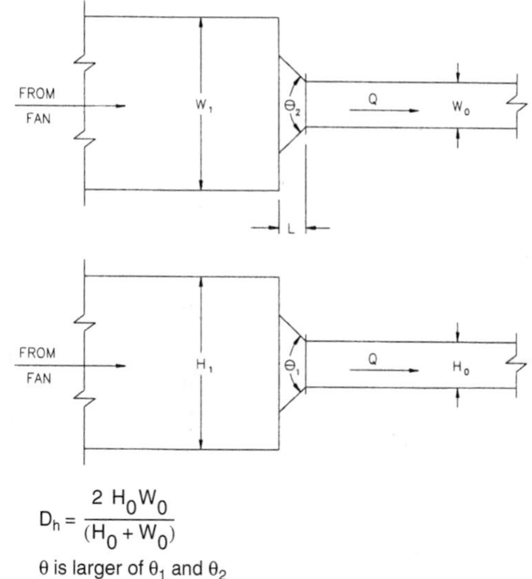

$$D_h = \frac{2\,H_0W_0}{(H_0 + W_0)}$$

θ is larger of θ_1 and θ_2

SR2-1 Abrupt Exit

H/W	0.1	0.2	0.9	1.0	1.1	4.0	5.0	10.0
C_o	1.55	1.55	1.55	2.00	1.55	1.55	1.55	1.55

$C_o = 1.0$

Note: Table is LAMINAR flow; $C_o = 1.0$ is TURBULENT flow.

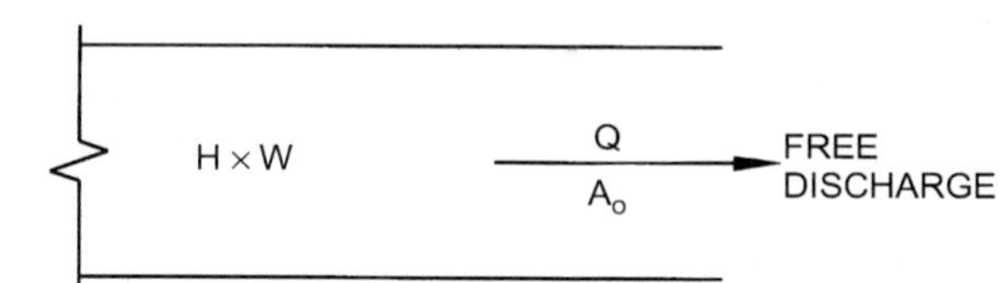

SR2-3 Plain Diffuser (Two Sides Parallel), Free Discharge

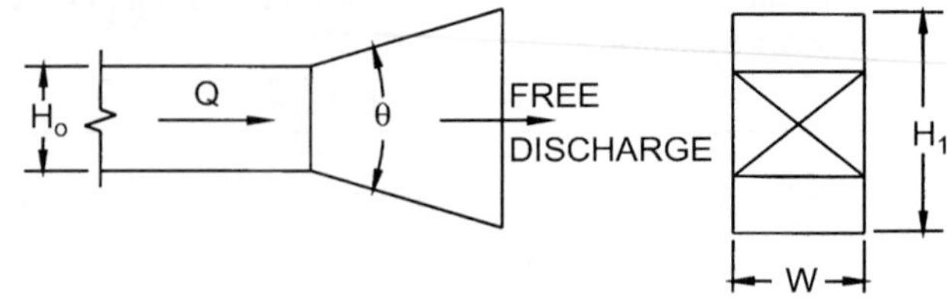

		C_o Values								
		θ								
A_1/A_o	Re/1000	8	10	14	20	30	45	60	90	120
1	50	0.00	0.00	0.00	0.00	0.00	0.00	0.00	0.00	0.00
	100	0.00	0.00	0.00	0.00	0.00	0.00	0.00	0.00	0.00
	200	0.00	0.00	0.00	0.00	0.00	0.00	0.00	0.00	0.00
	400	0.00	0.00	0.00	0.00	0.00	0.00	0.00	0.00	0.00
	2000	0.00	0.00	0.00	0.00	0.00	0.00	0.00	0.00	0.00
2	50	0.50	0.51	0.56	0.63	0.80	0.96	1.04	1.09	1.09
	100	0.48	0.50	0.56	0.63	0.80	0.96	1.04	1.09	1.09
	200	0.44	0.47	0.53	0.63	0.74	0.93	1.02	1.08	1.08
	400	0.40	0.42	0.50	0.62	0.74	0.93	1.02	1.08	1.08
	2000	0.40	0.42	0.50	0.62	0.74	0.93	1.02	1.08	1.08
4	50	0.34	0.38	0.48	0.63	0.76	0.91	1.03	1.07	1.07
	100	0.31	0.36	0.45	0.59	0.72	0.88	1.02	1.07	1.07
	200	0.26	0.31	0.41	0.53	0.67	0.83	0.96	1.06	1.06
	400	0.22	0.27	0.39	0.53	0.67	0.83	0.96	1.06	1.06
	2000	0.22	0.27	0.39	0.53	0.67	0.83	0.96	1.06	1.06
6	50	0.32	0.34	0.41	0.56	0.70	0.84	0.96	1.08	1.08
	100	0.27	0.30	0.41	0.56	0.70	0.84	0.96	1.08	1.08
	200	0.24	0.27	0.36	0.52	0.67	0.81	0.94	1.06	1.06
	400	0.20	0.24	0.36	0.52	0.67	0.81	0.94	1.06	1.06
	2000	0.18	0.24	0.34	0.50	0.67	0.81	0.94	1.05	1.05

SR2-5 Pyramidal Diffuser, Free Discharge

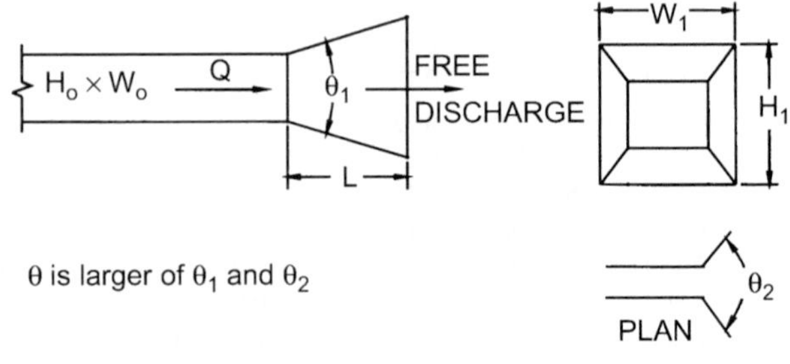

θ is larger of θ_1 and θ_2

		C_o Values								
		θ								
A_1/A_o	Re/1000	8	10	14	20	30	45	60	90	120
1	50	0.00	0.00	0.00	0.00	0.00	0.00	0.00	0.00	0.00
	100	0.00	0.00	0.00	0.00	0.00	0.00	0.00	0.00	0.00
	200	0.00	0.00	0.00	0.00	0.00	0.00	0.00	0.00	0.00
	400	0.00	0.00	0.00	0.00	0.00	0.00	0.00	0.00	0.00
	2000	0.00	0.00	0.00	0.00	0.00	0.00	0.00	0.00	0.00
2	50	0.65	0.68	0.74	0.82	0.92	1.05	1.10	1.08	1.08
	100	0.61	0.66	0.73	0.81	0.90	1.04	1.09	1.08	1.08
	200	0.57	0.61	0.70	0.79	0.89	1.04	1.09	1.08	1.08
	400	0.50	0.56	0.64	0.76	0.88	1.02	1.07	1.08	1.08
	2000	0.50	0.56	0.64	0.76	0.88	1.02	1.07	1.08	1.08
4	50	0.53	0.60	0.69	0.78	0.90	1.02	1.07	1.09	1.09
	100	0.49	0.55	0.66	0.78	0.90	1.02	1.07	1.09	1.09
	200	0.42	0.50	0.62	0.74	0.87	1.00	1.06	1.08	1.08
	400	0.36	0.44	0.56	0.70	0.84	0.99	1.06	1.08	1.08
	2000	0.36	0.44	0.56	0.70	0.84	0.99	1.06	1.08	1.08
6	50	0.50	0.57	0.66	0.77	0.91	1.02	1.07	1.08	1.08
	100	0.47	0.54	0.63	0.76	0.98	1.02	1.07	1.08	1.08
	200	0.42	0.48	0.60	0.73	0.88	1.00	1.06	1.08	1.08
	400	0.34	0.44	0.56	0.73	0.86	0.98	1.06	1.08	1.08
	2000	0.34	0.44	0.56	0.73	0.86	0.98	1.06	1.08	1.08
10	50	0.45	0.53	0.64	0.74	0.85	0.97	1.10	1.12	1.12
	100	0.40	0.48	0.62	0.73	0.85	0.97	1.10	1.12	1.12
	200	0.34	0.44	0.56	0.69	0.82	0.95	1.10	1.11	1.11
	400	0.28	0.40	0.55	0.67	0.80	0.93	1.09	1.11	1.11
	2000	0.28	0.40	0.55	0.67	0.80	0.93	1.09	1.11	1.11

SR2-6 Pyramidal Diffuser, with Wall

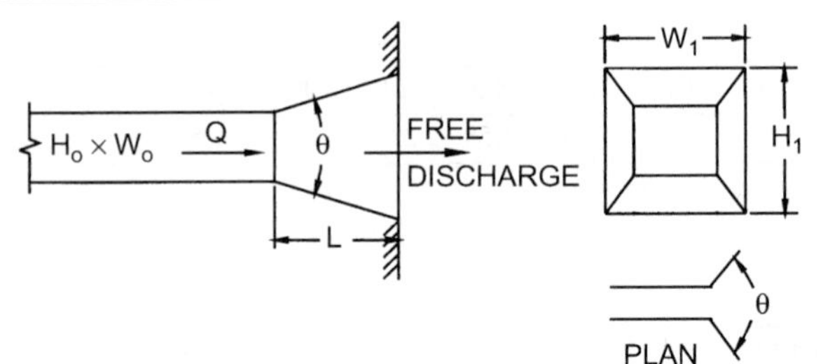

L/D_h	0.5	1.0	2.0	3.0	4.0	5.0	6.0	8.0	10.0	12.0	14.0
C_o	0.49	0.40	0.30	0.26	0.23	0.21	0.19	0.17	0.16	0.15	0.14
θ	26	19	13	11	9	8	7	6	6	5	5

θ is the optimum angle.

SR3-1 Elbow, 90 Degree, Variable Inlet/Outlet Areas, Supply Air Systems

	C_o Values						
	W_o/W_1						
H/W_1	0.6	0.8	1.0	1.2	1.4	1.6	2.0
0.25	0.63	0.92	1.24	1.64	2.14	2.71	4.24
1.00	0.61	0.87	1.15	1.47	1.86	2.30	3.36
4.00	0.53	0.70	0.90	1.17	1.49	1.84	2.64
100.00	0.54	0.67	0.79	0.99	1.23	1.54	2.20

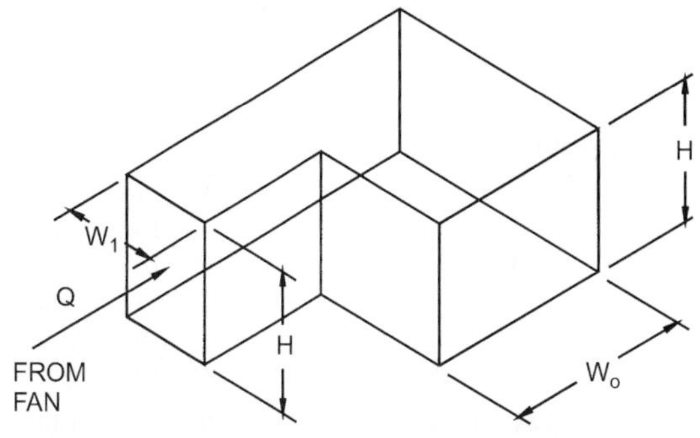

SR4-1 Transition, Rectangular, Two Sides Parallel, Symmetrical, Supply Air Systems

	C_o Values									
	θ									
A_o/A_1	10	15	20	30	45	60	90	120	150	180
0.10	0.05	0.05	0.05	0.05	0.07	0.08	0.19	0.29	0.37	0.43
0.17	0.05	0.04	0.04	0.04	0.05	0.07	0.18	0.28	0.36	0.42
0.25	0.05	0.04	0.04	0.04	0.06	0.07	0.17	0.27	0.35	0.41
0.50	0.06	0.05	0.05	0.05	0.06	0.07	0.14	0.20	0.26	0.27
1.00	0.00	0.00	0.00	0.00	0.00	0.00	0.00	0.00	0.00	1.00
2.00	0.56	0.52	0.60	0.96	1.40	1.48	1.52	1.48	1.44	1.40
4.00	2.72	3.04	3.52	6.72	9.60	10.88	11.20	11.04	10.72	10.56
10.00	24.00	26.00	36.00	53.00	69.00	82.00	93.00	93.00	92.00	91.00
16.00	66.56	69.12	102.40	143.36	181.76	220.16	256.00	253.44	250.88	250.88

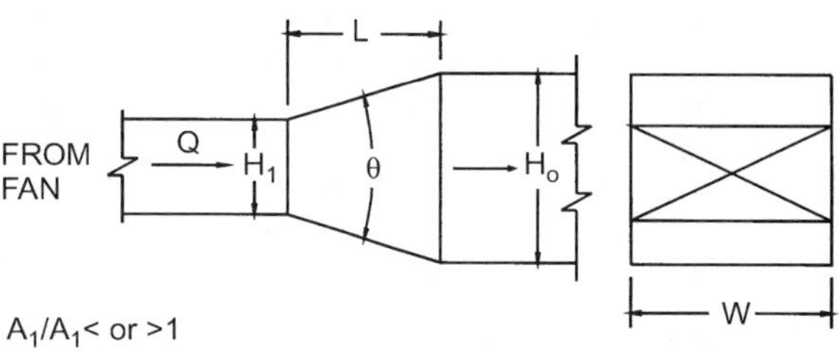

SR4-3 Transition, Round to Rectangular, Supply Air Systems

	C_o Values									
	θ									
A_o/A_1	10	15	20	30	45	60	90	120	150	180
0.10	0.05	0.05	0.05	0.05	0.07	0.08	0.19	0.29	0.37	0.43
0.17	0.05	0.05	0.05	0.04	0.06	0.07	0.18	0.28	0.36	0.42
0.25	0.06	0.05	0.05	0.04	0.06	0.07	0.17	0.27	0.35	0.41
0.50	0.06	0.07	0.07	0.05	0.06	0.06	0.12	0.18	0.24	0.26
1.00	0.00	0.00	0.00	0.00	0.00	0.00	0.00	0.00	0.00	0.00
2.00	0.60	0.84	1.00	1.20	1.32	1.32	1.32	1.28	1.24	1.20
4.00	4.00	5.76	7.20	8.32	9.28	9.92	10.24	10.24	10.24	10.24
10.00	30.00	50.00	53.00	64.00	75.00	84.00	89.00	91.00	91.00	88.00
16.00	76.80	138.24	135.68	166.40	197.12	225.28	243.20	250.88	250.88	238.08

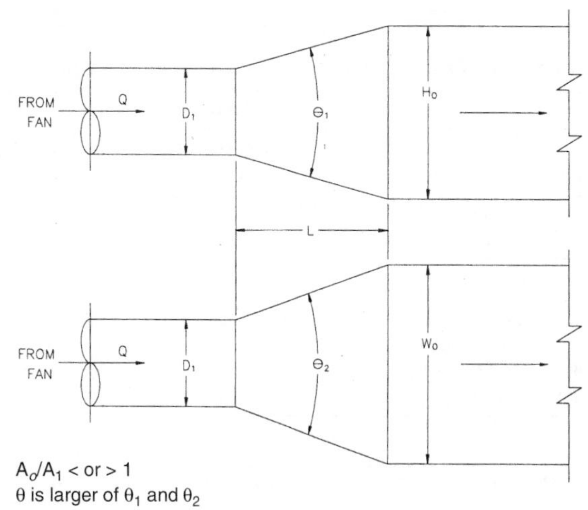

SR5-1 Smooth Wye of Type $A_s + A_b \geq A_c$, Branch 90° to Main, Diverging

		C_b Values								
		Q_b/Q_c								
A_s/A_c	A_b/A_c	0.1	0.2	0.3	0.4	0.5	0.6	0.7	0.8	0.9
0.50	0.25	3.44	0.70	0.30	0.20	0.17	0.16	0.16	0.17	0.18
	0.50	11.00	2.37	1.06	0.64	0.52	0.47	0.47	0.47	0.48
	1.00	60.00	13.00	4.78	2.06	0.96	0.47	0.31	0.27	0.26
0.75	0.25	2.19	0.55	0.35	0.31	0.33	0.35	0.36	0.37	0.39
	0.50	13.00	2.50	0.89	0.47	0.34	0.31	0.32	0.36	0.43
	1.00	70.00	15.00	5.67	2.62	1.36	0.78	0.53	0.41	0.36
1.00	0.25	3.44	0.78	0.42	0.33	0.30	0.31	0.40	0.42	0.46
	0.50	15.50	3.00	1.11	0.62	0.48	0.42	0.40	0.42	0.46
	1.00	67.00	13.75	5.11	2.31	1.28	0.81	0.59	0.47	0.46

		C_s Values								
		Q_s/Q_c								
A_s/A_c	A_b/A_c	0.1	0.2	0.3	0.4	0.5	0.6	0.7	0.8	0.9
0.50	0.25	8.75	1.62	0.50	0.17	0.05	0.00	−0.02	−0.02	0.00
	0.50	7.50	1.12	0.25	0.06	0.05	0.09	0.14	0.19	0.22
	1.00	5.00	0.62	0.17	0.08	0.08	0.09	0.12	0.15	0.19
0.75	0.25	19.13	3.38	1.00	0.28	0.05	−0.02	−0.02	0.00	0.06
	0.50	20.81	3.23	0.75	0.14	−0.02	−0.05	−0.05	−0.02	0.03
	1.00	16.88	2.81	0.63	0.11	−0.02	−0.05	0.01	0.00	0.07
1.00	0.25	46.00	9.50	3.22	1.31	0.52	0.14	−0.02	−0.05	−0.01
	0.50	35.00	6.75	2.11	0.75	0.24	0.00	−0.10	−0.09	−0.04
	1.00	38.00	7.50	2.44	0.81	0.24	−0.03	−0.08	−0.06	−0.02

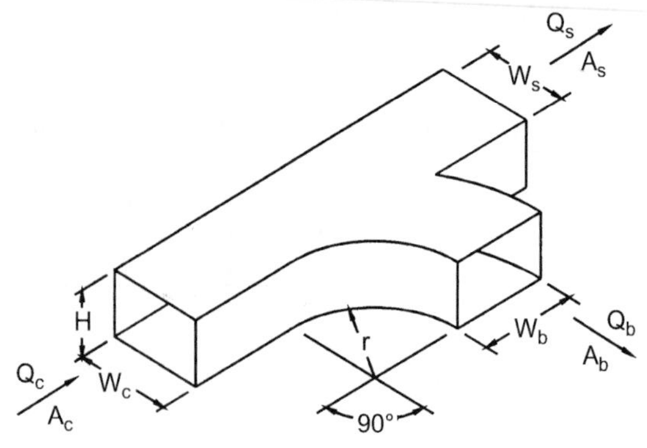

$r/W_b = 1.0$

$A_s = A_b \geq A_c$

SR5-3 Wye of the Type $A_s + A_b > A_c$, $A_s = A_c$, 45 Degree, Diverging

	C_b Values								
	Q_b/Q_c								
A_b/A_c	0.1	0.2	0.3	0.4	0.5	0.6	0.7	0.8	0.9
0.1	0.60	0.52	0.57	0.58	0.64	0.67	0.70	0.71	0.73
0.2	2.24	0.56	0.44	0.45	0.51	0.54	0.58	0.60	0.62
0.3	5.94	1.08	0.52	0.41	0.44	0.46	0.49	0.52	0.54
0.4	10.56	1.88	0.71	0.43	0.35	0.31	0.31	0.32	0.34
0.5	17.75	3.25	1.14	0.59	0.40	0.31	0.30	0.30	0.31
0.6	26.64	5.04	1.76	0.83	0.50	0.36	0.32	0.30	0.30
0.7	37.73	7.23	2.56	1.16	0.67	0.44	0.35	0.31	0.30
0.8	49.92	9.92	3.48	1.60	0.87	0.55	0.42	0.35	0.32

Q_s/Q_c	0.1	0.2	0.3	0.4	0.5	0.6	0.8	1.0
C_s	32.00	6.50	2.22	0.87	0.40	0.17	0.03	0.00

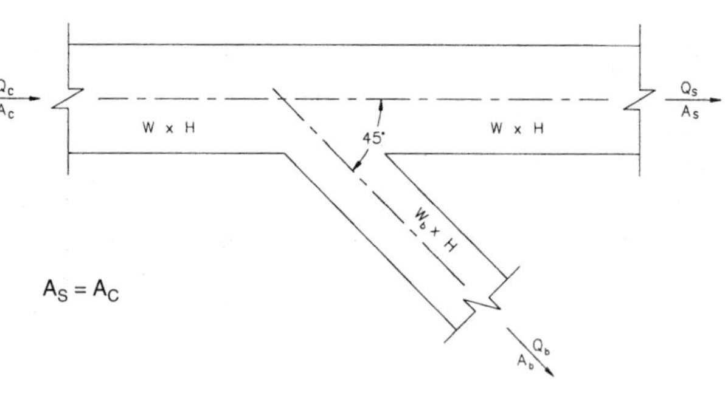

SR5-5 Tee of the Type $A_s + A_b > A_c$, $A_s = A_c$ Diverging

	C_b Values								
	Q_b/Q_c								
A_b/A_c	0.1	0.2	0.3	0.4	0.5	0.6	0.7	0.8	0.9
0.1	2.06	1.20	0.99	0.87	0.88	0.87	0.87	0.86	0.86
0.2	5.16	1.92	1.28	1.03	0.99	0.94	0.92	0.90	0.89
0.3	10.26	3.13	1.78	1.28	1.16	1.06	1.01	0.97	0.94
0.4	15.84	4.36	2.24	1.48	1.11	0.88	0.80	0.75	0.72
0.5	24.25	6.31	3.03	1.89	1.35	1.03	0.91	0.84	0.78
0.6	34.56	8.73	4.04	2.41	1.64	1.22	1.04	0.94	0.87
0.7	46.55	11.51	5.17	3.00	2.00	1.44	1.20	1.06	0.96
0.8	60.80	14.72	6.54	3.72	2.41	1.69	1.38	1.20	1.07

Q_s/Q_c	0.1	0.2	0.3	0.4	0.5	0.6	0.8	1.0
C_s	32.00	6.50	2.22	0.87	0.40	0.17	0.03	0.00

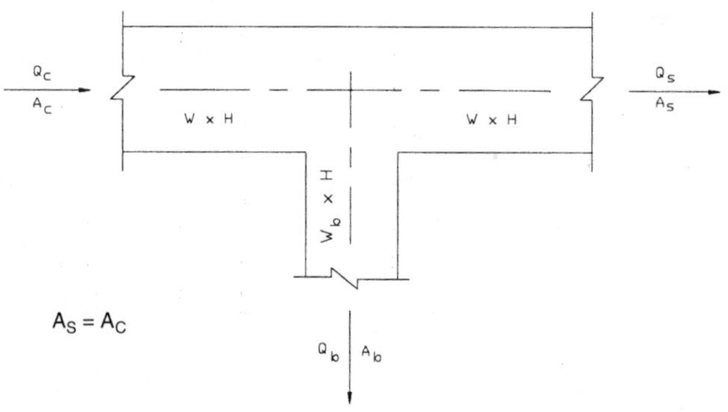

SR5-11 Tee, Rectangular Main to Round Tap, Diverging

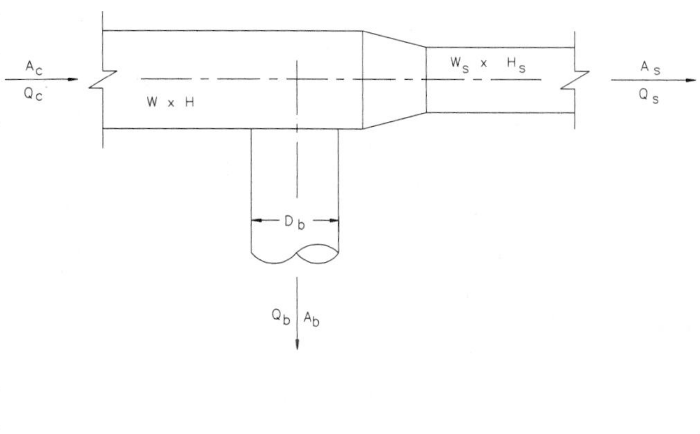

	C_b Values								
	Q_b/Q_c								
A_b/A_c	0.1	0.2	0.3	0.4	0.5	0.6	0.7	0.8	0.9
0.1	1.58	0.94	0.83	0.79	0.77	0.76	0.76	0.76	0.75
0.2	4.20	1.58	1.10	0.94	0.87	0.83	0.80	0.79	0.78
0.3	8.63	2.67	1.58	1.20	1.03	0.94	0.88	0.85	0.83
0.4	14.85	4.20	2.25	1.58	1.27	1.10	1.00	0.94	0.90
0.5	22.87	6.19	3.13	2.07	1.58	1.32	1.16	1.06	0.99
0.6	32.68	8.63	4.20	2.67	1.96	1.58	1.35	1.20	1.10
0.7	44.30	11.51	5.48	3.38	2.41	1.89	1.58	1.38	1.24
0.8	57.71	14.85	6.95	4.20	2.94	2.25	1.84	1.58	1.40
0.9	72.92	18.63	8.63	5.14	3.53	2.67	2.14	1.81	1.58

	C_s Values								
	Q_s/Q_c								
A_s/A_c	0.1	0.2	0.3	0.4	0.5	0.6	0.7	0.8	0.9
0.1	0.04								
0.2	0.98	0.04							
0.3	3.48	0.31	0.04						
0.4	7.55	0.98	0.18	0.04					
0.5	13.18	2.03	0.49	0.13	0.04				
0.6	20.38	3.48	0.98	0.31	0.10	0.04			
0.7	29.15	5.32	1.64	0.60	0.23	0.09	0.04		
0.8	39.48	7.55	2.47	0.98	0.42	0.18	0.08	0.04	
0.9	51.37	10.17	3.48	1.46	0.67	0.31	0.15	0.07	0.04

SR5-13 Tee, 45 Degree Entry Branch, Diverging

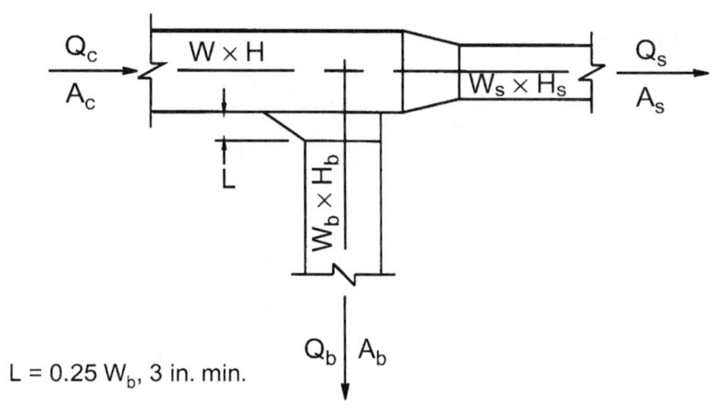

	C_b Values								
	Q_b/Q_c								
A_b/A_c	0.1	0.2	0.3	0.4	0.5	0.6	0.7	0.8	0.9
0.1	0.73	0.34	0.32	0.34	0.35	0.37	0.38	0.39	0.40
0.2	3.10	0.73	0.41	0.34	0.32	0.32	0.33	0.34	0.35
0.3	7.59	1.65	0.73	0.47	0.37	0.34	0.32	0.32	0.32
0.4	14.20	3.10	1.28	0.73	0.51	0.41	0.36	0.34	0.32
0.5	22.92	5.08	2.07	1.12	0.73	0.54	0.44	0.38	0.35
0.6	33.76	7.59	3.10	1.65	1.03	0.73	0.56	0.47	0.41
0.7	46.71	10.63	4.36	2.31	1.42	0.98	0.73	0.58	0.49
0.8	61.79	14.20	5.86	3.10	1.90	1.28	0.94	0.73	0.60
0.9	78.98	18.29	7.59	4.02	2.46	1.65	1.19	0.91	0.73

	C_s Values								
	Q_s/Q_c								
A_s/A_c	0.1	0.2	0.3	0.4	0.5	0.6	0.7	0.8	0.9
0.1	0.04								
0.2	0.98	0.04							
0.3	3.48	0.31	0.04						
0.4	7.55	0.98	0.18	0.04					
0.5	13.18	2.03	0.49	0.13	0.04				
0.6	20.38	3.48	0.98	0.31	0.10	0.04			
0.7	29.15	5.32	1.64	0.60	0.23	0.09	0.04		
0.8	39.48	7.55	2.47	0.98	0.42	0.18	0.08	0.04	
0.9	51.37	10.17	3.48	1.46	0.67	0.31	0.15	0.07	0.04

SR5-14 Wye, Symmetrical, Dovetail, Q_b/Q_c = 0.5, Diverging

A_b/A_c	0.5	1.0
C_b	0.30	1.00

Branches are identical: $Q_{b1} = Q_{b2} = Q_b$, and $C_{b1} = C_{b2} = C_b$

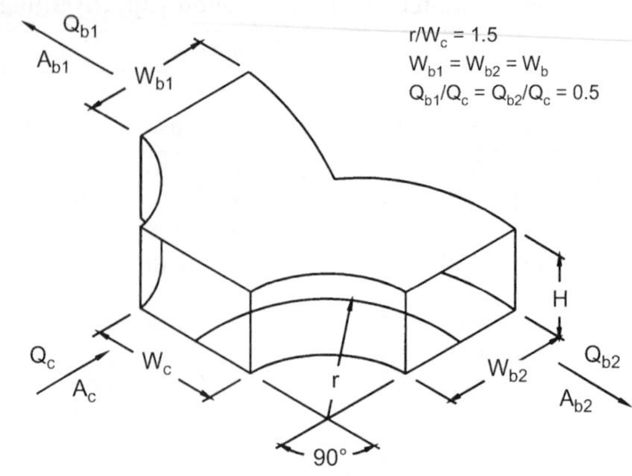

SR7-1 Fan, Centrifugal, Without Outlet Diffuser, Free Discharge

A_b/A_o	0.4	0.5	0.6	0.7	0.8	0.9	1.0
C_o	2.00	2.00	1.00	0.80	0.47	0.22	0.00

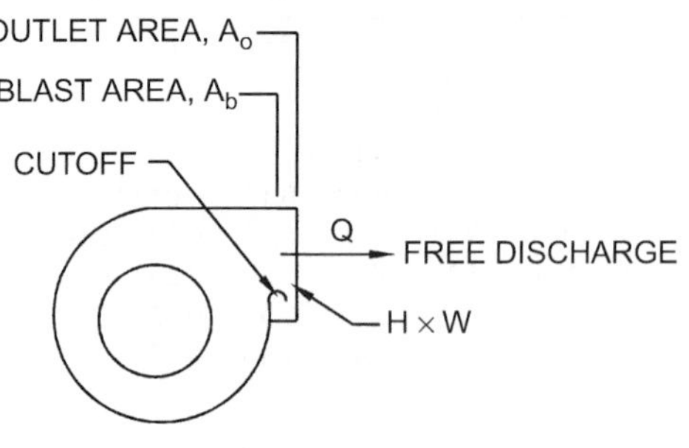

SR7-2 Plane Asymmetric Diffuser at Centrifugal Fan Outlet, Free Discharge

	C_o Values					
	A_1/A_o					
θ	1.5	2.0	2.5	3.0	3.5	4.0
10	0.51	0.34	0.25	0.21	0.18	0.17
15	0.54	0.36	0.27	0.24	0.22	0.20
20	0.55	0.38	0.31	0.27	0.25	0.24
25	0.59	0.43	0.37	0.35	0.33	0.33
30	0.63	0.50	0.46	0.44	0.43	0.42
35	0.65	0.56	0.53	0.52	0.51	0.50

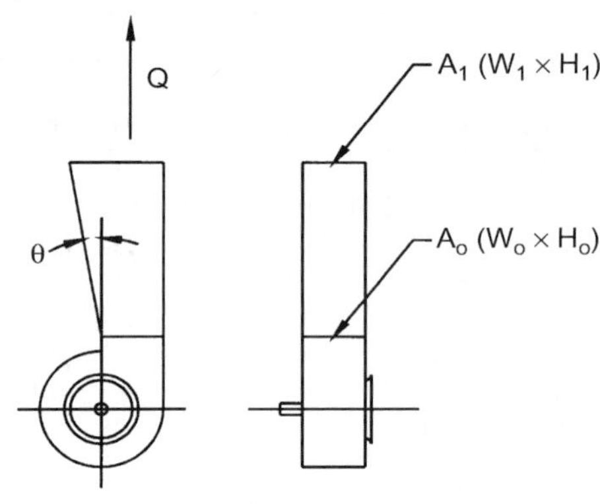

SR7-5 Fan Outlet, Centrifugal, SWSI, with Elbow (Position A)

	C_o Values					
	L/L_e					
A_b/A_o	0.00	0.12	0.25	0.50	1.00	10.00
0.4	3.20	2.50	1.80	0.80	0.00	0.00
0.5	2.20	1.80	1.20	0.53	0.00	0.00
0.6	1.60	1.40	0.80	0.40	0.00	0.00
0.7	1.00	0.80	0.53	0.26	0.00	0.00
0.8	0.80	0.67	0.47	0.18	0.00	0.00
0.9	0.53	0.47	0.33	0.18	0.00	0.00
1.0	0.53	0.47	0.33	0.18	0.00	0.00

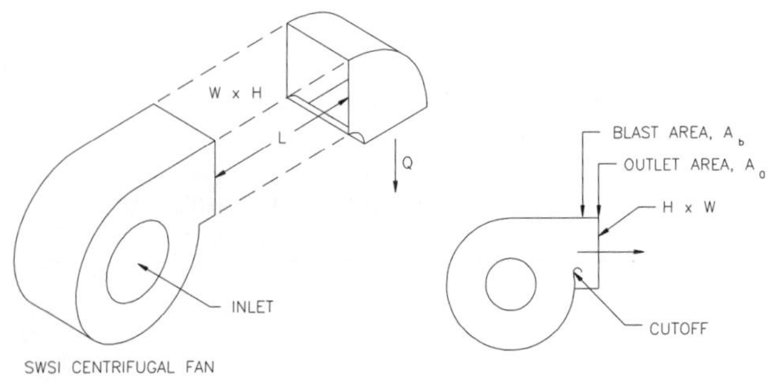

SR7-6 Fan Outlet, Centrifugal, SWSI, with Elbow (Position B)

	C_o Values					
	L/L_e					
A_b/A_o	0.00	0.12	0.25	0.50	1.00	10.00
0.4	3.80	3.20	2.20	1.00	0.00	0.00
0.5	2.90	2.20	1.60	0.67	0.00	0.00
0.6	2.00	1.60	1.20	0.53	0.00	0.00
0.7	1.40	1.00	0.67	0.33	0.00	0.00
0.8	1.00	0.80	0.53	0.26	0.00	0.00
0.9	0.80	0.67	0.47	0.18	0.00	0.00
1.0	0.67	0.53	0.40	0.18	0.00	0.00

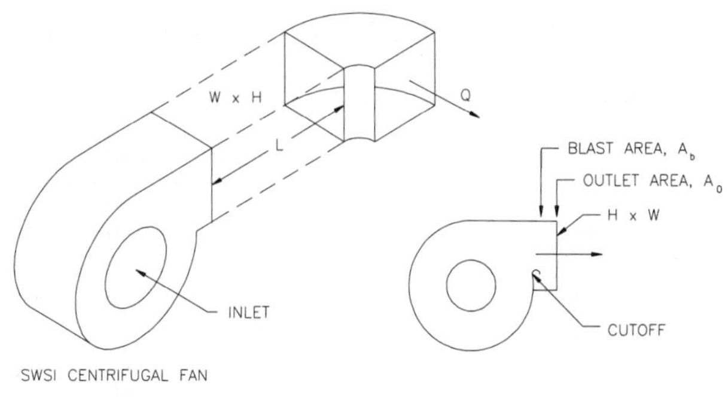

SR7-7 Fan Outlet, Centrifugal, SWSI, with Elbow (Position C)

	C_o Values					
	L/L_e					
A_b/A_o	0.00	0.12	0.25	0.50	1.00	10.00
0.4	5.50	4.50	3.20	1.60	0.00	0.00
0.5	3.80	3.20	2.20	1.00	0.00	0.00
0.6	2.90	2.50	1.60	0.80	0.00	0.00
0.7	2.00	1.60	1.00	0.53	0.00	0.00
0.8	1.40	1.20	0.80	0.33	0.00	0.00
0.9	1.20	0.80	0.67	0.26	0.00	0.00
1.0	1.00	0.80	0.53	0.26	0.00	0.00

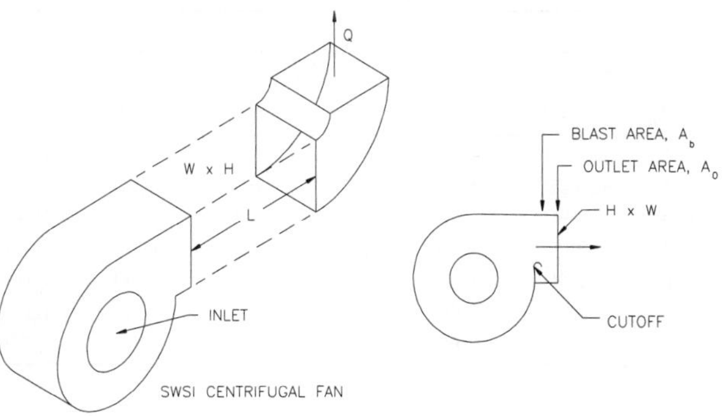

SR7-8 Fan Outlet, Centrifugal, SWSI, with Elbow (Position D)

	C_o Values					
	L/L_e					
A_b/A_o	0.00	0.12	0.25	0.50	1.00	10.00
0.4	5.50	4.50	3.20	1.60	0.00	0.00
0.5	3.80	3.20	2.20	1.00	0.00	0.00
0.6	2.90	2.50	1.60	0.80	0.00	0.00
0.7	2.00	1.60	1.00	0.53	0.00	0.00
0.8	1.40	1.20	0.80	0.33	0.00	0.00
0.9	1.20	0.80	0.67	0.26	0.00	0.00
1.0	1.00	0.80	0.53	0.26	0.00	0.00

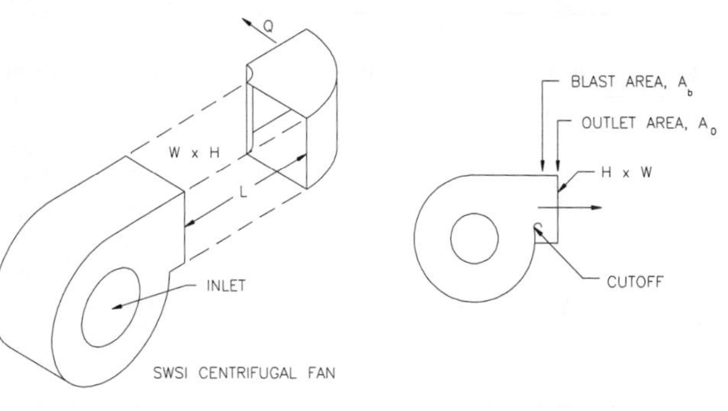

SR7-9 Fan Outlet, Centrifugal, DWDI, with Elbow (Position A)

	C_o					
	L/L_e					
A_b/A_o	0.00	0.12	0.25	0.50	1.00	10.00
0.4	3.20	2.50	1.80	0.80	0.00	0.00
0.5	2.20	1.80	1.20	0.53	0.00	0.00
0.6	1.60	1.40	0.80	0.40	0.00	0.00
0.7	1.00	0.80	0.53	0.26	0.00	0.00
0.8	0.80	0.67	0.47	0.18	0.00	0.00
0.9	0.53	0.47	0.33	0.18	0.00	0.00
1.0	0.53	0.47	0.33	0.18	0.00	0.00

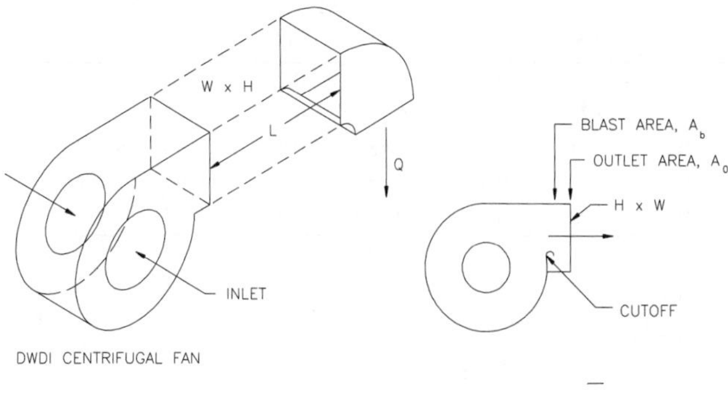

SR7-10 Fan Outlet, Centrifugal, DWDI, with Elbow (Position B)

	C_o					
	L/L_e					
A_b/A_o	0.00	0.12	0.25	0.50	1.00	10.00
0.4	4.80	4.00	2.90	1.30	0.00	0.00
0.5	3.60	2.90	2.00	0.84	0.00	0.00
0.6	2.50	2.00	1.50	0.66	0.00	0.00
0.7	1.80	1.30	0.84	0.41	0.00	0.00
0.8	1.25	1.00	0.66	0.33	0.00	0.00
0.9	1.00	0.84	0.59	0.23	0.00	0.00
1.0	0.84	0.66	0.50	0.23	0.00	0.00

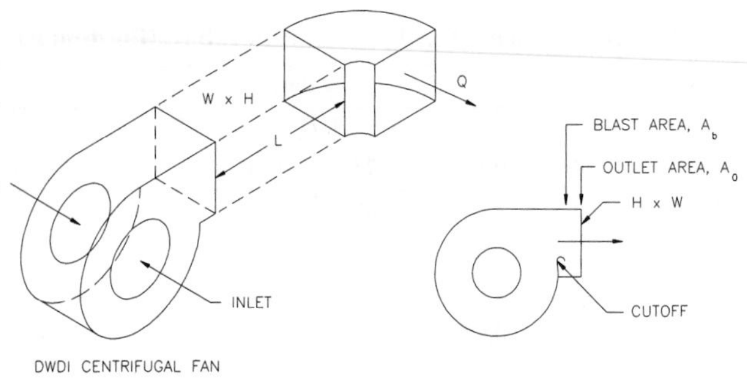

SR7-11 Fan Outlet, Centrifugal, DWDI, with Elbow (Position C)

	C_o					
	L/L_e					
A_b/A_o	0.00	0.12	0.25	0.50	1.00	10.00
0.4	5.50	4.50	3.20	1.60	0.00	0.00
0.5	3.80	3.20	2.20	1.00	0.00	0.00
0.6	2.90	2.50	1.60	0.80	0.00	0.00
0.7	2.00	1.60	1.00	0.53	0.00	0.00
0.8	1.40	1.20	0.80	0.33	0.00	0.00
0.9	1.20	0.80	0.67	0.26	0.00	0.00
1.0	1.00	0.80	0.53	0.26	0.00	0.00

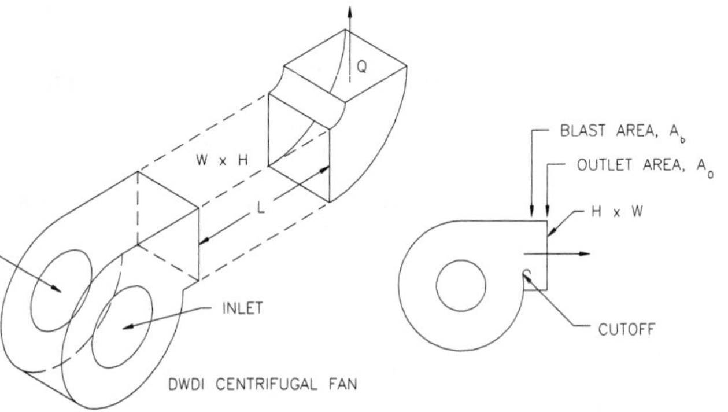

SR7-12 Fan Outlet, Centrifugal, DWDI, with Elbow (Position D)

	C_o					
	L/L_e					
A_b/A_o	0.00	0.12	0.25	0.50	1.00	10.00
0.4	4.70	3.80	2.70	1.40	0.00	0.00
0.5	3.20	2.70	1.90	0.85	0.00	0.00
0.6	2.50	2.10	1.40	0.68	0.00	0.00
0.7	1.70	1.40	0.85	0.45	0.00	0.00
0.8	1.20	1.00	0.68	0.26	0.00	0.00
0.9	1.00	0.68	0.57	0.22	0.00	0.00
1.0	0.85	0.68	0.45	0.22	0.00	0.00

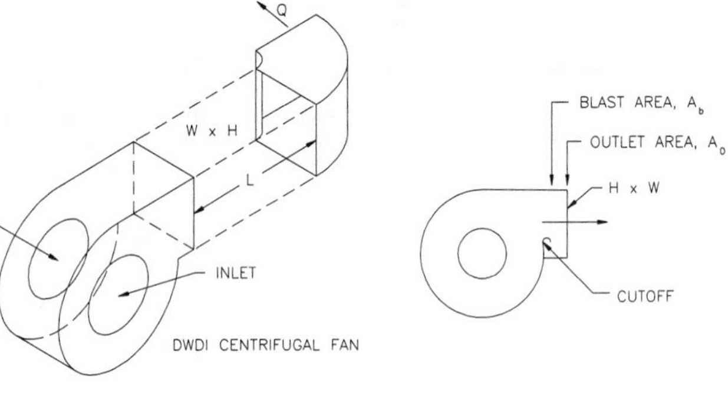

SR7-17 Pyramidal Diffuser at Centrifugal Fan Outlet with Ductwork

	C_1 Values					
	A_o/A_1					
θ	1.5	2.0	2.5	3.0	3.5	4.0
10	0.10	0.18	0.21	0.23	0.24	0.25
15	0.23	0.33	0.38	0.40	0.42	0.44
20	0.31	0.43	0.48	0.53	0.56	0.58
25	0.36	0.49	0.55	0.58	0.62	0.64
30	0.42	0.53	0.59	0.64	0.67	0.69

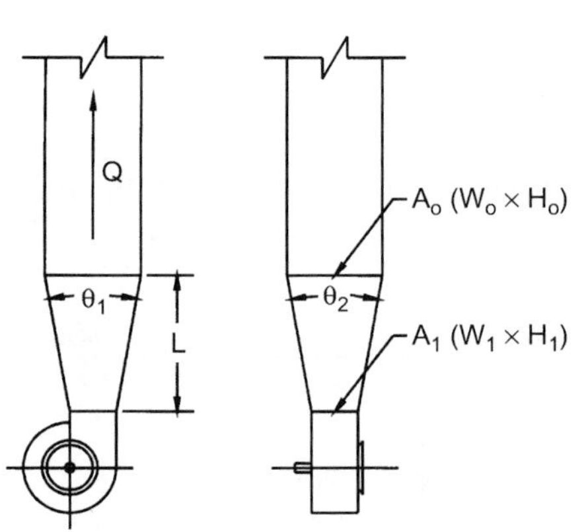

θ is larger of θ_1 and θ_2

CHAPTER 35

PIPE SIZING

THIS CHAPTER includes tables and charts to size piping for various fluid flow systems. Further details on specific piping systems can be found in appropriate chapters of the ASHRAE Handbook series.

Two related but distinct concerns emerge when designing a fluid flow system: sizing the pipe and determining the flow-pressure relationship. The two are often confused because they can use the same equations and design tools. Nevertheless, they should be determined separately.

The emphasis in this chapter is on the problem of sizing the pipe, and to this end design charts and tables for specific fluids are presented in addition to the equations that describe the flow of fluids in pipes. Once a system has been sized, it should be analyzed with more detailed methods of calculation to determine the pump head required to achieve the desired flow. Computerized methods are well suited to handling the details of calculating losses around an extensive system.

PRESSURE DROP EQUATIONS

Darcy-Weisbach Equation

Pressure drop caused by fluid friction in fully developed flows of all "well-behaved" (Newtonian) fluids is described by the Darcy-Weisbach equation:

$$\Delta p = f\left(\frac{L}{D}\right)\left(\frac{\rho}{g_c}\right)\left(\frac{V^2}{2}\right) \tag{1}$$

where

Δp = pressure drop, lb_f/ft^2
f = friction factor, dimensionless (from Moody chart, Figure 13 in Chapter 2)
L = length of pipe, ft
D = internal diameter of pipe, ft
ρ = fluid density at mean temperature, lb_m/ft^3
V = average velocity, fps
g_c = units conversion factor, 32.2 $ft \cdot lb_m/lb_f \cdot s^2$

This equation is often presented in head or specific energy form as

$$\Delta h = \left(\frac{\Delta p}{\rho}\right)\left(\frac{g_c}{g}\right) = f\left(\frac{L}{D}\right)\left(\frac{V^2}{2g}\right) \tag{2}$$

where

Δh = head loss, ft
g = acceleration of gravity, ft/s^2

In this form, the density of the fluid does not appear explicitly (although it is in the Reynolds number, which influences f).

The friction factor f is a function of pipe roughness ε, inside diameter D, and parameter Re, the Reynolds number:

$$Re = DV\rho/\mu \tag{3}$$

where

Re = Reynolds number, dimensionless
ε = absolute roughness of pipe wall, ft
μ = dynamic viscosity of fluid, $lb_m/ft \cdot s$

The friction factor is frequently presented on a Moody chart (Figure 13 in Chapter 2) giving f as a function of Re with ε/D as a parameter.

A useful fit of smooth and rough pipe data for the usual turbulent flow regime is the **Colebrook equation**:

$$\frac{1}{\sqrt{f}} = 1.74 - 2\log\left(\frac{2\varepsilon}{D} + \frac{18.7}{Re\sqrt{f}}\right) \tag{4}$$

Another form of Equation (4) appears in Chapter 2, but the two are equivalent. Equation (4) is more useful in showing behavior at limiting cases—as ε/D approaches 0 (smooth limit), the $18.7/Re\sqrt{f}$ term dominates; at high ε/D and Re (fully rough limit), the $2\varepsilon/D$ term dominates.

Equation (4) is implicit in f; that is, f appears on both sides, so a value for f is usually obtained iteratively.

Hazen-Williams Equation

A less widely used alternative to the Darcy-Weisbach formulation for calculating pressure drop is the Hazen-Williams equation, which is expressed as

$$\Delta p = 3.022L\left(\frac{V}{C}\right)^{1.852}\left(\frac{1}{D}\right)^{1.167}\left(\frac{\rho g}{g_c}\right) \tag{5}$$

or

$$\Delta h = 3.022L\left(\frac{V}{C}\right)^{1.852}\left(\frac{1}{D}\right)^{1.167} \tag{6}$$

where C = roughness factor.

Typical values of C are 150 for plastic pipe and copper tubing, 140 for new steel pipe, down to 100 and below for badly corroded or very rough pipe.

Valve and Fitting Losses

Valves and fittings cause pressure losses greater than those caused by the pipe alone. One formulation expresses losses as

$$\Delta p = K\left(\frac{\rho}{g_c}\right)\left(\frac{V^2}{2}\right) \quad \text{or} \quad \Delta h = K\left(\frac{V^2}{2g}\right) \tag{7}$$

where K = geometry- and size-dependent loss coefficient (Tables 1 through 5).

The preparation of this chapter is assigned to TC 6.1, Hydronic and Steam Equipment and Systems.

Table 1 K Factors—Threaded Pipe Fittings

Nominal Pipe Dia., in.	90° Standard Elbow	90° Long-Radius Elbow	45° Elbow	Return Bend	Tee-Line	Tee-Branch	Globe Valve	Gate Valve	Angle Valve	Swing Check Valve	Bell Mouth Inlet	Square Inlet	Projected Inlet
3/8	2.5	—	0.38	2.5	0.90	2.7	20	0.40	—	8.0	0.05	0.5	1.0
1/2	2.1	—	0.37	2.1	0.90	2.4	14	0.33	—	5.5	0.05	0.5	1.0
3/4	1.7	0.92	0.35	1.7	0.90	2.1	10	0.28	6.1	3.7	0.05	0.5	1.0
1	1.5	0.78	0.34	1.5	0.90	1.8	9	0.24	4.6	3.0	0.05	0.5	1.0
1-1/4	1.3	0.65	0.33	1.3	0.90	1.7	8.5	0.22	3.6	2.7	0.05	0.5	1.0
1-1/2	1.2	0.54	0.32	1.2	0.90	1.6	8	0.19	2.9	2.5	0.05	0.5	1.0
2	1.0	0.42	0.31	1.0	0.90	1.4	7	0.17	2.1	2.3	0.05	0.5	1.0
2-1/2	0.85	0.35	0.30	0.85	0.90	1.3	6.5	0.16	1.6	2.2	0.05	0.5	1.0
3	0.80	0.31	0.29	0.80	0.90	1.2	6	0.14	1.3	2.1	0.05	0.5	1.0
4	0.70	0.24	0.28	0.70	0.90	1.1	5.7	0.12	1.0	2.0	0.05	0.5	1.0

Source: *Engineering Data Book* (Hydraulic Institute 1979).

Table 2 K Factors—Flanged Welded Pipe Fittings

Nominal Pipe Dia., in.	90° Standard Elbow	90° Long-Radius Elbow	45° Long-Radius Elbow	Return Bend Standard	Return Bend Long-Radius	Tee-Line	Tee-Branch	Globe Valve	Gate Valve	Angle Valve	Swing Check Valve
1	0.43	0.41	0.22	0.43	0.43	0.26	1.0	13	—	4.8	2.0
1-1/4	0.41	0.37	0.22	0.41	0.38	0.25	0.95	12	—	3.7	2.0
1-1/2	0.40	0.35	0.21	0.40	0.35	0.23	0.90	10	—	3.0	2.0
2	0.38	0.30	0.20	0.38	0.30	0.20	0.84	9	0.34	2.5	2.0
2-1/2	0.35	0.28	0.19	0.35	0.27	0.18	0.79	8	0.27	2.3	2.0
3	0.34	0.25	0.18	0.34	0.25	0.17	0.76	7	0.22	2.2	2.0
4	0.31	0.22	0.18	0.31	0.22	0.15	0.70	6.5	0.16	2.1	2.0
6	0.29	0.18	0.17	0.29	0.18	0.12	0.62	6	0.10	2.1	2.0
8	0.27	0.16	0.17	0.27	0.15	0.10	0.58	5.7	0.08	2.1	2.0
10	0.25	0.14	0.16	0.25	0.14	0.09	0.53	5.7	0.06	2.1	2.0
12	0.24	0.13	0.16	0.24	0.13	0.08	0.50	5.7	0.05	2.1	2.0

Source: *Engineering Data Book* (Hydraulic Institute 1979).

Table 3 Approximate Range of Variation for K Factors

Fitting	Type	Variation	Fitting	Type	Variation
90° Elbow	Regular threaded	±20% above 2 in.	Tee	Threaded, line or branch	±25%
		±40% below 2 in.		Flanged, line or branch	±35%
	Long-radius threaded	±25%	Globe valve	Threaded	±25%
	Regular flanged	±35%		Flanged	±25%
	Long-radius flanged	±30%	Gate valve	Threaded	±25%
45° Elbow	Regular threaded	±10%		Flanged	±50%
	Long-radius flanged	±10%	Angle valve	Threaded	±20%
Return bend (180°)	Regular threaded	±25%		Flanged	±50%
	Regular flanged	±35%	Check valve	Threaded	±50%
	Long-radius flanged	±30%		Flanged	+200% −80%

Source: *Engineering Data Book* (Hydraulic Institute 1979).

Table 4 Comparison of K Factors from Rahmeyer (1999a) with Previous Reference Data for Elbows

	Previous Ref.[a] Values	Rahmeyer Values 2 fps	4 fps	8 fps
2 in. standard elbow (threaded)	0.60 to 1.0 (1.0)	0.54	0.6	0.68
4 in. long-radius elbow (welded)	0.22 to 0.24 (0.22)	0.28	0.26	0.24
4 in. standard elbow (welded)	0.31 to 0.34	0.40	0.37	0.34
2 in. reducing elbow (2 × 1.5) (threaded)	–	0.76	0.81	0.87
2 in. expanding elbow (1.5 × 2) (threaded)	–	0.65	0.59	0.54
4 in. reducing elbow (4 × 3) (welded)	–	0.72	0.57	0.45
4 in. expanding elbow (3 × 4) (welded)	–	0.28	0.28	0.27
2 in. reducer (2 × 1.5) (threaded)	–	0.99	0.53	0.28
2 in. expansion (1.5 × 2) (threaded)	–	0.20	0.16	0.13
4 in. reducer (4 × 3) (welded)	0.22	0.40	0.23	0.14
4 in. expansion (3 × 4) (welded)	–	0.13	0.11	0.11

[a] Previous references are Freeman (1941), Crane Co. (1976), and Hydraulic Institute (1979). Numbers in ( ) are from Table 1 and Table 2, above.

Table 5 Comparison of *K* Factors from Rahmeyer (1999b) with Previous Reference Data for Tees

	Previous Ref.[a] Values	Rahmeyer Values		
		2 fps	4 fps	8 fps
2 in. tee 100% branch (threaded)	1.20 to 1.80 (1.4)	–	0.93	–
2 in. tee 0% branch (flow through) (threaded)	0.50 to 0.90 (0.90)	–	0.19	–
2 in. tee 100% mix (threaded)	–	–	1.19	–
4 in. tee 100% branch (welded)	0.70 to 1.02 (0.70)	–	–	0.57
4 in. tee 0% branch (flow through) (welded)	0.15 to 0.34 (0.15)	–	–	0.06
4 in. tee 100% mix (welded)	–	–	–	0.49
2 in. reducing tee 100% branch (threaded)	–	–	3.75	–
2 in. reducing tee 0% branch (threaded)	–	–	2.55	–
2 in. reducing tee 100% mix (threaded)	–	–	3.36	–
4 in. reducing tee 100% branch (welded)	–	–	–	3.26
4 in. reducing tee 0% branch (welded)	–	–	–	0.04
4 in. reducing tee 100% mix (welded)	–	–	–	0.75

[a] Previous references are Freeman (1941), Crane Co. (1976), and Hydraulic Institute (1979). Numbers in ( ) are from Table 1 and Table 2, above.

Example 1. Determine the pressure drop for 60°F water flowing at 4 fps through a nominal 1 in., 90° threaded elbow.

Solution: From Table 1, the *K* for a 1 in., 90° threaded elbow is 1.5.

$\Delta p = 1.5 \times 62.4/32.2 \times 4^2/2 = 23.3$ lb/ft^2 or 0.16 psi

The loss coefficient for valves appears in another form as C_v, a dimensional coefficient expressing the flow through a valve at a specified pressure drop.

$$Q = C_v\sqrt{\Delta p} \tag{8}$$

where

Q = volumetric flow, gpm
C_v = valve coefficient, gpm at Δp = 1 psi
Δp = pressure drop, psi

See the section on Control Valve Sizing in Chapter 42 of the 2000 *ASHRAE Handbook—Systems and Equipment* for a more complete explanation of C_v.

Example 2. Determine the volumetric flow through a valve with C_v = 10 for an allowable pressure drop of 5 psi.

Solution: $Q = 10\sqrt{5} = 22.4$ gpm.

Alternative formulations express fitting losses in terms of equivalent lengths of straight pipe (Table 6, Table 7, and Figure 4). Pressure loss data for fittings are also presented in Idelchik (1986).

Equation (7) and the data in Table 1 and Table 2 are based on the assumption that separated flow in the fitting causes the *K* factors to be independent of Reynolds number. In reality, the *K* factor for most pipe fittings varies with Reynolds number. Tests by Rahmeyer (1999a,b) (sponsored as ASHRAE *Research Project* 968) on 2 in. threaded and 4 in. welded fittings demonstrate the variation and are shown in Table 4 and Table 5. The study also presents *K* factors of diverting and mixing flows that range between full through flow and full branch flow. It also examined the variation in *K* factors caused by variations in geometry among manufacturers and by surface defects in individual fittings.

Hegberg (1995) and Rahmeyer (1999a,b) discuss the origins of the data shown in Table 4 and Table 5. The Hydraulic Institute (1979) data appear to have come from Freeman (1941), work that was actually performed in 1895. The work of Giesecke (1926) and Giesecke and Badgett (1931, 1932a,b) may not be representative of present-day fittings.

Calculating Pressure Losses

The most common engineering design flow loss calculation selects a pipe size for the desired total flow rate and available or allowable pressure drop.

Because either formulation of fitting losses requires a known diameter, pipe size must be selected before calculating the detailed influence of fittings. A frequently used rule of thumb assumes that the design length of pipe is 50 to 100% longer than actual to account for fitting losses. After a pipe diameter has been selected on this basis, the influence of each fitting can be evaluated.

WATER PIPING

FLOW RATE LIMITATIONS

Stewart and Dona (1987) surveyed the literature relating to water flow rate limitations. Noise, erosion, and installation and operating costs all limit the maximum and minimum velocities in piping systems. If piping sizes are too small, noise levels, erosion levels, and pumping costs can be unfavorable; if piping sizes are too large, installation costs are excessive. Therefore, pipe sizes are chosen to minimize initial cost while avoiding the undesirable effects of high velocities.

Table 6 Water Velocities Based on Type of Service

Type of Service	Velocity, fps	Reference
General service	4 to 10	a, b, c
City water	3 to 7	a, b
	2 to 5	c
Boiler feed	6 to 15	a, c
Pump suction and drain lines	4 to 7	a, b

[a]Crane Co. (1976). [b]Carrier (1960). [c]Grinnell Company (1951).

Table 7 Maximum Water Velocity to Minimize Erosion

Normal Operation, h/yr	Water Velocity, fps
1500	15
2000	14
3000	13
4000	12
6000	10

Source: Carrier (1960).

A variety of upper limits of water velocity and/or pressure drop in piping and piping systems is used. One recommendation places a velocity limit of 4 fps for 2 in. pipe and smaller, and a pressure drop limit of 4 ft of water/100 ft for piping over 2 in. Other guidelines are based on the type of service (Table 6) or the annual operating hours (Table 7). These limitations are imposed either to control the levels of pipe and valve noise, erosion, and water hammer pressure or for economic reasons. Carrier (1960) recommends that the velocity not exceed 15 fps in any case.

Noise Generation

Velocity-dependent noise in piping and piping systems results from any or all of four sources: turbulence, cavitation, release of entrained air, and water hammer. In investigations of flow-related noise, Marseille (1965), Ball and Webster (1976), and Rogers (1953, 1954, 1956) reported that velocities on the order of 10 to 17 fps lie within the range of allowable noise levels for residential and commercial buildings. The experiments showed considerable variation in the noise levels obtained for a specified velocity. Generally, systems with longer pipe and with more numerous fittings and valves were noisier. In addition, sound measurements were taken under widely differing conditions; for example, some tests used plastic-covered pipe, while others did not. Thus, no detailed correlations relating sound level to flow velocity in generalized systems are available.

The noise generated by fluid flow in a pipe increases sharply if cavitation or the release of entrained air occurs. Usually the combination of a high water velocity with a change in flow direction or a decrease in the cross section of a pipe causing a sudden pressure drop is necessary to cause cavitation. Ball and Webster (1976) found that at their maximum velocity of 42 fps, cavitation did not occur in straight 3/8 and 1/2 in. pipe; using the apparatus with two elbows, cold water velocities up to 21 fps caused no cavitation. Cavitation did occur in orifices of 1:8 area ratio (orifice flow area is one-eighth of pipe flow area) at 5 fps and in 1:4 area ratio orifices at 10 fps (Rogers 1954).

Some data are available for predicting hydrodynamic (liquid) noise generated by control valves. The International Society for Measurement and Control compiled prediction correlations in an effort to develop control valves for reduced noise levels (ISA 1985). The correlation to predict hydrodynamic noise from control valves is

$$SL = 10\log C_v + 20\log\Delta p - 30\log t + 5 \quad (9)$$

where

SL = sound level, dB
C_v = valve coefficient, gpm/(psi)$^{0.5}$
Q = flow rate, gpm
Δp = pressure drop across valve, psi
t = downstream pipe wall thickness, in.

Air entrained in water usually has a higher partial pressure than the water. Even when flow rates are small enough to avoid cavitation, the release of entrained air may create noise. Every effort should be made to vent the piping system or otherwise remove entrained air.

Erosion

Erosion in piping systems is caused by water bubbles, sand, or other solid matter impinging on the inner surface of the pipe. Generally, at velocities lower than 100 fps, erosion is not significant as long as there is no cavitation. When solid matter is entrained in the fluid at high velocities, erosion occurs rapidly, especially in bends. Thus, high velocities should not be used in systems where sand or other solids are present or where slurries are transported.

Allowances for Aging

With age, the internal surfaces of pipes become increasingly rough, which reduces the available flow with a fixed pressure supply. However, designing with excessive age allowances may result in oversized piping. Age-related decreases in capacity depend on the type of water, type of pipe material, temperature of water, and type of system (open or closed) and include

- Sliming (biological growth or deposited soil on the pipe walls), which occurs mainly in unchlorinated, raw water systems.
- Caking of calcareous salts, which occurs in hard water (i.e., water bearing calcium salts) and increases with water temperature.
- Corrosion (incrustations of ferrous and ferric hydroxide on the pipe walls), which occurs in metal pipe in soft water. Because oxygen is necessary for corrosion to take place, significantly more corrosion takes place in open systems.

Allowances for expected decreases in capacity are sometimes treated as a specific amount (percentage). Dawson and Bowman (1933) added an allowance of 15% friction loss to new pipe (equivalent to an 8% decrease in capacity). The *HDR Design Guide* (1981) increased the friction loss by 15 to 20% for closed piping systems and 75 to 90% for open systems. Carrier (1960) indicates a factor of approximately 1.75 between friction factors for closed and open systems.

Obrecht and Pourbaix (1967) differentiated between the corrosive potential of different metals in potable water systems and concluded that iron is the most severely attacked, then galvanized steel, lead, copper, and finally copper alloys (i.e., brass). Hunter (1941) and Freeman (1941) showed the same trend. After four years of cold and hot water use, copper pipe had a capacity loss of 25 to 65%. Aged ferrous pipe has a capacity loss of 40 to 80%. Smith (1983) recommended increasing the design discharge by 1.55 for uncoated cast iron, 1.08 for iron and steel, and 1.06 for cement or concrete.

The Plastic Pipe Institute (1971) found that corrosion is not a problem in plastic pipe; the capacity of plastic pipe in Europe and the United States remains essentially the same after 30 years in use.

Extensive age-related flow data are available for use with the Hazen-Williams empirical equation. Difficulties arise in its application, however, because the original Hazen-Williams roughness coefficients are valid only for the specific pipe diameters, water velocities, and water viscosities used in the original experiments. Thus, when the *C*s are extended to different diameters, velocities, and/or water viscosities, errors of up to about 50% in pipe capacity can occur (Williams and Hazen 1933, Sanks 1978).

Water Hammer

When any moving fluid (not just water) is abruptly stopped, as when a valve closes suddenly, large pressures can develop. While detailed analysis requires knowledge of the elastic properties of the pipe and the flow-time history, the limiting case of rigid pipe and instantaneous closure is simple to calculate. Under these conditions,

$$\Delta p_h = \rho c_s V / g_c \quad (10)$$

where

Δp_h = pressure rise caused by water hammer, lb$_f$/ft^2
ρ = fluid density, lb$_m$/ft^3
c_s = velocity of sound in fluid, fps
V = fluid flow velocity, fps

The c_s for water is 4720 fps, although the elasticity of the pipe reduces the effective value.

Example 3. What is the maximum pressure rise if water flowing at 10 fps is stopped instantaneously?

Solution: $\Delta p_h = 62.4 \times 4720 \times 10/32.2 = 91{,}468 \text{ lb/ft}^2 = 635 \text{ psi}$

Other Considerations

Not discussed in detail in this chapter, but of potentially great importance, are a number of physical and chemical considerations: pipe and fitting design, materials, and joining methods must be appropriate for working pressures and temperatures encountered, as well as being suitably resistant to chemical attack by the fluid.

Other Piping Materials and Fluids

For fluids not included in this chapter or for piping materials of different dimensions, manufacturers' literature frequently supplies pressure drop charts. The Darcy-Weisbach equation, with the Moody chart or the Colebrook equation, can be used as an alternative to pressure drop charts or tables.

HYDRONIC SYSTEM PIPING

The Darcy-Weisbach equation with friction factors from the Moody chart or Colebrook equation (or, alternatively, the Hazen-Williams equation) is fundamental to calculating pressure drop in hot and chilled water piping; however, charts calculated from these equations (such as Figures 1, 2, and 3) provide easy determination of pressure drops for specific fluids and pipe standards. In addition, tables of pressure drops can be found in Hydraulic Institute (1979) and Crane Co. (1976).

The Reynolds numbers represented on the charts in Figures 1, 2, and 3 are all in the turbulent flow regime. For smaller pipes and/or lower velocities, the Reynolds number may fall into the laminar regime, in which the Colebrook friction factors are no longer valid.

Most tables and charts for water are calculated for properties at 60°F. Using these for hot water introduces some error, although the answers are conservative (i.e., cold water calculations overstate the pressure drop for hot water). Using 60°F water charts for 200°F water should not result in errors in Δp exceeding 20%.

Range of Usage of Pressure Drop Charts

General Design Range. The general range of pipe friction loss used for design of hydronic systems is between 1 and 4 ft of water per 100 ft of pipe. A value of 2.5 ft/100 ft represents the mean to which most systems are designed. Wider ranges may be used in specific designs if certain precautions are taken.

Piping Noise. Closed-loop hydronic system piping is generally sized below certain arbitrary upper limits, such as a velocity limit of 4 fps for 2 in. pipe and under, and a pressure drop limit of 4 ft per 100 ft for piping over 2 in. in diameter. Velocities in excess of 4 fps can be used in piping of larger size. This limitation is generally accepted, although it is based on relatively inconclusive experience with noise in piping. **Water velocity noise** is not caused by water but by free air, sharp pressure drops, turbulence, or a combination of these, which in turn cause cavitation or flashing of water into steam. Therefore, higher velocities may be used if proper precautions are taken to eliminate air and turbulence.

Air Separation

Air in hydronic systems is usually undesirable because it causes flow noise, allows oxygen to react with piping materials, and sometimes even prevents flow in parts of a system. Air may enter a system at an air-water interface in an open system or in an expansion tank in a closed system, or it may be brought in dissolved in makeup water. Most hydronic systems use air separation devices to remove air. The solubility of air in water increases with pressure and decreases with temperature; thus, separation of air from water is best achieved at the point of lowest pressure and/or highest temperature in a system. For more information, see Chapter 12 of the 2000 *ASHRAE Handbook—Systems and Equipment*.

In the absence of venting, air can be entrained in the water and carried to separation units at flow velocities of 1.5 to 2 fps or more in pipe 2 in. and under. Minimum velocities of 2 fps are therefore recommended. For pipe sizes 2 in. and over, minimum velocities corresponding to a head loss of 0.75 ft/100 ft are normally used. Maintenance of minimum velocities is particularly important in the upper floors of high-rise buildings where the air tends to come out of solution because of reduced pressures. Higher velocities should be used in **downcomer** return mains feeding into air separation units located in the basement.

Example 4. Determine the pipe size for a circuit requiring 20 gpm flow.

Solution: Enter Figure 1 at 20 gpm, read up to pipe size within normal design range (1 to 4 ft/100 ft), and select 1-1/2 in. Velocity is 3.1 fps, which is between 2 and 4. Pressure loss is 2.9 ft/100 ft.

Valve and Fitting Pressure Drop

Valves and fittings can be listed in elbow equivalents, with an elbow being equivalent to a length of straight pipe. Table 8 lists equivalent lengths of 90° elbows; Table 9 lists elbow equivalents for valves and fittings for iron and copper.

Example 5. Determine equivalent feet of pipe for a 4 in. open gate valve at a flow velocity of approximately 4 fps.

Solution: From Table 8, at 4 fps, each elbow is equivalent to 10.6 ft of 4 in. pipe. From Table 9, the gate valve is equivalent to 0.5 elbows. The actual equivalent pipe length (added to measured circuit length for pressure drop determination) will be 10.6 × 0.5, or 5.3 equivalent feet of 4 in. pipe.

Tee Fitting Pressure Drop. Pressure drop through pipe tees varies with flow through the branch. Figure 4 illustrates pressure drops for nominal 1 in. tees of equal inlet and outlet sizes and for the flow patterns illustrated. Idelchik (1986) also presents data for threaded tees.

Table 8 Equivalent Length in Feet of Pipe for 90° Elbows

Velocity,	Pipe Size														
fps	1/2	3/4	1	1-1/4	1-1/2	2	2-1/2	3	3-1/2	4	5	6	8	10	12
1	1.2	1.7	2.2	3.0	3.5	4.5	5.4	6.7	7.7	8.6	10.5	12.2	15.4	18.7	22.2
2	1.4	1.9	2.5	3.3	3.9	5.1	6.0	7.5	8.6	9.5	11.7	13.7	17.3	20.8	24.8
3	1.5	2.0	2.7	3.6	4.2	5.4	6.4	8.0	9.2	10.2	12.5	14.6	18.4	22.3	26.5
4	1.5	2.1	2.8	3.7	4.4	5.6	6.7	8.3	9.6	10.6	13.1	15.2	19.2	23.2	27.6
5	1.6	2.2	2.9	3.9	4.5	5.9	7.0	8.7	10.0	11.1	13.6	15.8	19.8	24.2	28.8
6	1.7	2.3	3.0	4.0	4.7	6.0	7.2	8.9	10.3	11.4	14.0	16.3	20.5	24.9	29.6
7	1.7	2.3	3.0	4.1	4.8	6.2	7.4	9.1	10.5	11.7	14.3	16.7	21.0	25.5	30.3
8	1.7	2.4	3.1	4.2	4.9	6.3	7.5	9.3	10.8	11.9	14.6	17.1	21.5	26.1	31.0
9	1.8	2.4	3.2	4.3	5.0	6.4	7.7	9.5	11.0	12.2	14.9	17.4	21.9	26.6	31.6
10	1.8	2.5	3.2	4.3	5.1	6.5	7.8	9.7	11.2	12.4	15.2	17.7	22.2	27.0	32.0

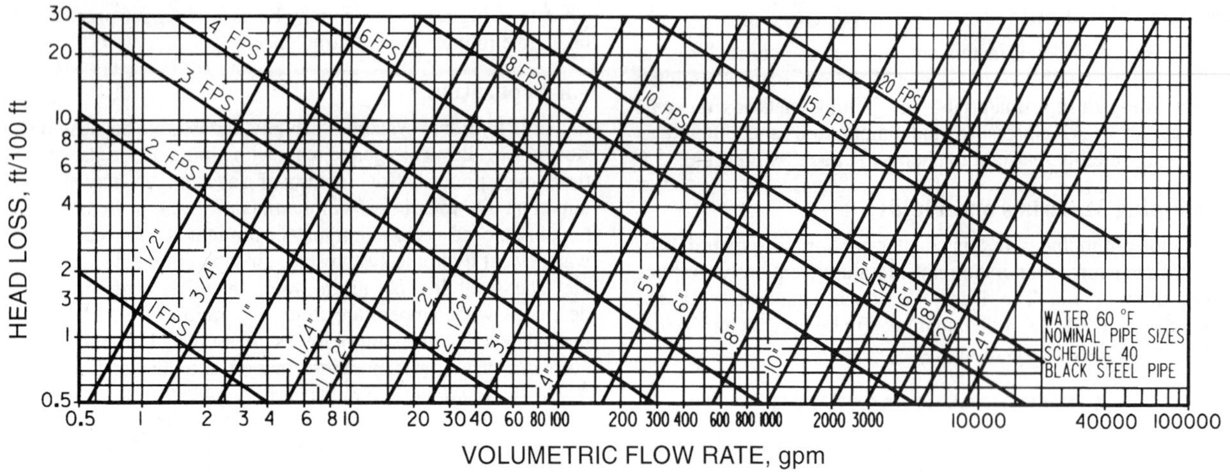

Fig. 1 Friction Loss for Water in Commercial Steel Pipe (Schedule 40)

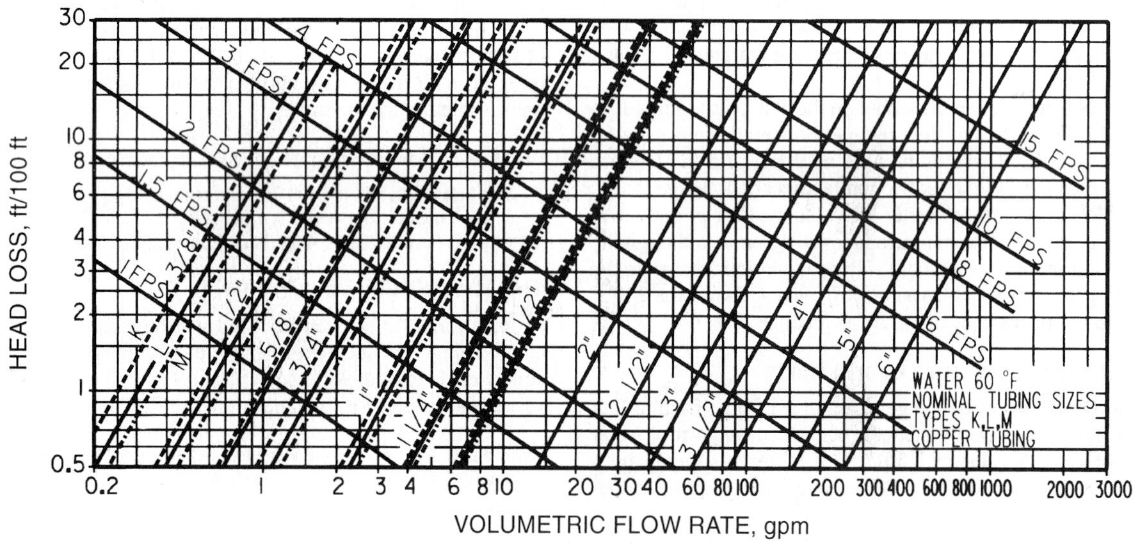

Fig. 2 Friction Loss for Water in Copper Tubing (Types K, L, M)

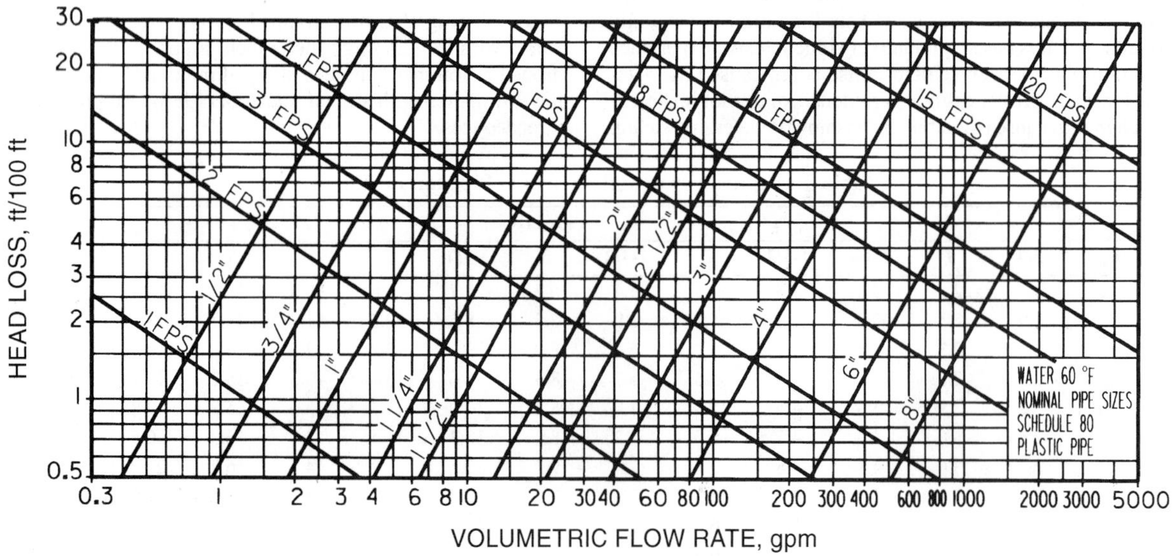

Fig. 3 Friction Loss for Water in Plastic Pipe (Schedule 80)

Table 9 Iron and Copper Elbow Equivalents[a]

Fitting	Iron Pipe	Copper Tubing
Elbow, 90°	1.0	1.0
Elbow, 45°	0.7	0.7
Elbow, 90° long-radius	0.5	0.5
Elbow, welded, 90°	0.5	0.5
Reduced coupling	0.4	0.4
Open return bend	1.0	1.0
Angle radiator valve	2.0	3.0
Radiator or convector	3.0	4.0
Boiler or heater	3.0	4.0
Open gate valve	0.5	0.7
Open globe valve	12.0	17.0

Source: Giesecke (1926) and Giesecke and Badgett (1931, 1932a).
[a]See Table 8 for equivalent length of one elbow.

Table 10 Proper Flow and Pressure Required During Flow for Different Fixtures

Fixture	Flow Pressure, psig[a]	Flow, gpm
Ordinary basin faucet	8	3.0
Self-closing basin faucet	12	2.5
Sink faucet—3/8 in.	10	4.5
Sink faucet—1/2 in.	5	4.5
Dishwasher	15-25	—[b]
Bathtub faucet	5	6.0
Laundry tube cock—1/4 in.	5	5.0
Shower	12	3-10
Ball cock for closet	15	3.0
Flush valve for closet	10-20	15-40[c]
Flush valve for urinal	15	15.0
Garden hose, 50 ft, and sill cock	30	5.0

[a]Flow pressure is the pressure in the pipe at the entrance to the particular fixture considered.
[b]Varies; see manufacturers' data.
[c]Wide range due to variation in design and type of flush valve closets.

Different investigators present tee loss data in different forms, and it is sometimes difficult to reconcile results from several sources. As an estimate of the upper limit to tee losses, a pressure or head loss coefficient of 1.0 may be assumed for entering and leaving flows (i.e., $\Delta p = 1.0\rho V_{in}^2/2 + 1.0\rho V_{out}^2/2$).

Example 6. Determine the pressure or head losses for a 1 in. (all openings) threaded pipe tee flowing 25% to the side branch, 75% through. The entering flow is 10 gpm (3.71 fps).

Solution: From Figure 4, bottom curve, the number of equivalent elbows for the through-flow is 0.15 elbows; the through-flow is 7.5 gpm (2.78 fps); and the pressure loss is based on the exit flow rate. Table 8 gives the equivalent length of a 1 in. elbow at 3 fps as 2.7 ft. Using Equations (1) and (2) with friction factor $f = 0.0290$ and diameter $D = 0.0874$ ft,

$$\Delta p = (0.15)(0.0290)(2.7/0.0874)(62.4/32.2)(2.78^2/2)$$
$$= 1.01 \text{ lb/ft}^2 = 0.00699 \text{ psi pressure drop, or}$$
$$\Delta h = (0.15)(0.0290)(2.7/0.0874)(2.78^2)/[(2)(32.2)]$$
$$= 0.0161 \text{ ft head loss}$$

From Figure 4, top curve, the number of equivalent elbows for the branch flow of 25% is 13 elbows; the branch flow is 2.5 gpm (0.93 fps); and the pressure loss is based on the exit flow rate. Table 8 gives the equivalent of a 1 in. elbow at 1 fps as 2.2 ft. Using Equations (1) and (2) with friction factor $f = 0.0350$ and diameter = 0.0874 ft,

$$\Delta p = (13)(0.0350)(2.2/0.0874)(62.4/32.2)(0.93^2/2)$$
$$= 9.60 \text{ lb/ft}^2 = 0.0667 \text{ psi pressure drop, or}$$
$$\Delta h = (13)(0.0350)(2.2/0.0874)(0.93^2)/[(2)(32.2)]$$
$$= 0.154 \text{ ft head loss}$$

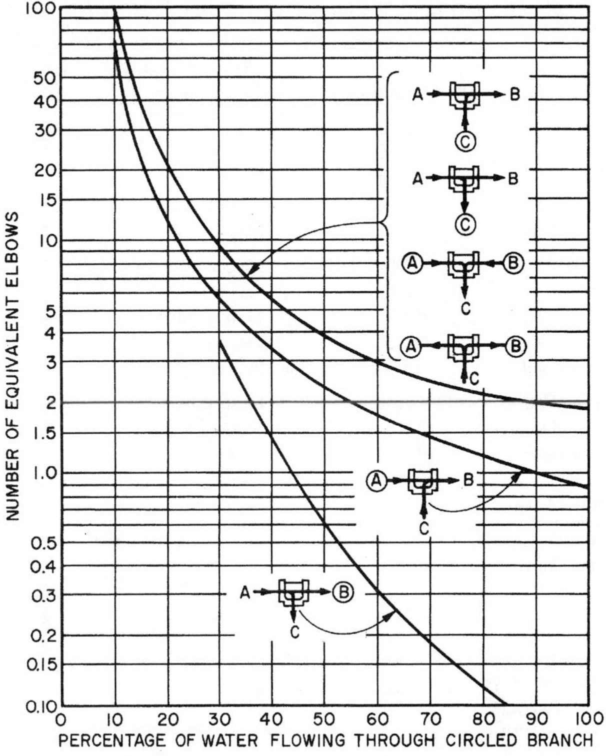

Notes: 1. Chart is based on straight tees (i.e., branches A, B, and C are the same size).
2. Pressure loss in desired circuit is obtained by selecting the proper curve according to illustrations, determining the flow at the circled branch, and multiplying the pressure loss for the same size elbow at the flow rate in the circled branch by the equivalent elbows indicated.
3. When the size of an outlet is reduced, the equivalent elbows shown in the chart do not apply. Therefore, the maximum loss for any circuit for any flow will not exceed 2 elbow equivalents at the maximum flow occurring in any branch of the tee.
4. Top curve is average of 4 curves, one for each circuit shown.

Fig. 4 Elbow Equivalents of Tees at Various Flow Conditions
(Giesecke and Badgett 1931, 1932b)

SERVICE WATER PIPING

Sizing of service water piping differs from sizing of process lines in that design flows in service water piping are determined by the probability of simultaneous operation of a multiplicity of individual loads such as water closets, urinals, lavatories, sinks, and showers. The full flow characteristics of each load device are readily obtained from manufacturers; however, service water piping sized to handle all load devices simultaneously would be seriously oversized. Thus, a major issue in sizing service water piping is to determine the diversity of the loads.

The procedure shown in this chapter uses the work of R.B. Hunter for estimating diversity (Hunter 1940, 1941). The present-day plumbing designer is usually constrained by building or plumbing codes, which specify the individual and collective loads to be used for pipe sizing. Frequently used codes (including the BOCA *National Plumbing Code*, *Standard Plumbing Code*, *Uniform Plumbing Code*, and *National Standard Plumbing Code*) contain procedures quite similar to those shown here. The designer must be aware of the applicable code for the location being considered.

Table 11 Demand Weights of Fixtures in Fixture Units[a]

Fixture or Group[b]	Occupancy	Type of Supply Control	Weight in Fixture Units[c]
Water closet	Public	Flush valve	10
Water closet	Public	Flush tank	5
Pedestal urinal	Public	Flush valve	10
Stall or wall urinal	Public	Flush valve	5
Stall or wall urinal	Public	Flush tank	3
Lavatory	Public	Faucet	2
Bathtub	Public	Faucet	4
Shower head	Public	Mixing valve	4
Service sink	Office, etc.	Faucet	3
Kitchen sink	Hotel or restaurant	Faucet	4
Water closet	Private	Flush valve	6
Water closet	Private	Flush tank	3
Lavatory	Private	Faucet	1
Bathtub	Private	Faucet	2
Shower head	Private	Mixing valve	2
Bathroom group	Private	Flush valve for closet	8
Bathroom group	Private	Flush tank for closet	6
Separate shower	Private	Mixing valve	2
Kitchen sink	Private	Faucet	2
Laundry trays (1 to 3)	Private	Faucet	3
Combination fixture	Private	Faucet	3

Source: Hunter (1941).

[a]For supply outlets likely to impose continuous demands, estimate continuous supply separately, and add to total demand for fixtures.

[b]For fixtures not listed, weights may be assumed by comparing the fixture to a listed one using water in similar quantities and at similar rates.

[c]The given weights are for total demand. For fixtures with both hot and cold water supplies, the weights for maximum separate demands can be assumed to be 75% of the listed demand for the supply.

Federal mandates are forcing plumbing fixture manufacturers to reduce design flows to many types of fixtures, but these may not yet be included in locally adopted codes. Also, the designer must be aware of special considerations; for example, toilet usage at sports arenas will probably have much less diversity than the codes allow and thus may require larger supply piping than the minimum specified by the codes.

Table 10 gives the rate of flow desirable for many common fixtures and the average pressure necessary to give this rate of flow. The pressure varies with fixture design.

In estimating the load, the rate of flow is frequently computed in **fixture units**, which are relative indicators of flow. Table 11 gives the demand weights in terms of fixture units for different plumbing fixtures under several conditions of service, and Figure 5 gives the estimated demand in gallons per minute corresponding to any total number of fixture units. Figures 6 and 7 provide more accurate estimates at the lower end of the scale.

The estimated demand load for fixtures used intermittently on any supply pipe can be obtained by multiplying the number of each kind of fixture supplied through that pipe by its weight from Table 11, adding the products, and then referring to the appropriate curve of Figure 5, 6, or 7 to find the demand corresponding to the total fixture units. In using this method, note that the demand for fixture or supply outlets other than those listed in the table of fixture units is not yet included in the estimate. The demands for outlets (e.g., hose connections and air-conditioning apparatus) that are likely to impose continuous demand during heavy use of the weighted fixtures should be estimated separately and added to demand for fixtures used intermittently to estimate total demand.

The Hunter curves in Figures 5, 6, and 7 are based on use patterns in residential buildings and can be erroneous for other usages such as sports arenas. Williams (1976) discusses the Hunter assumptions and presents an analysis using alternative assumptions.

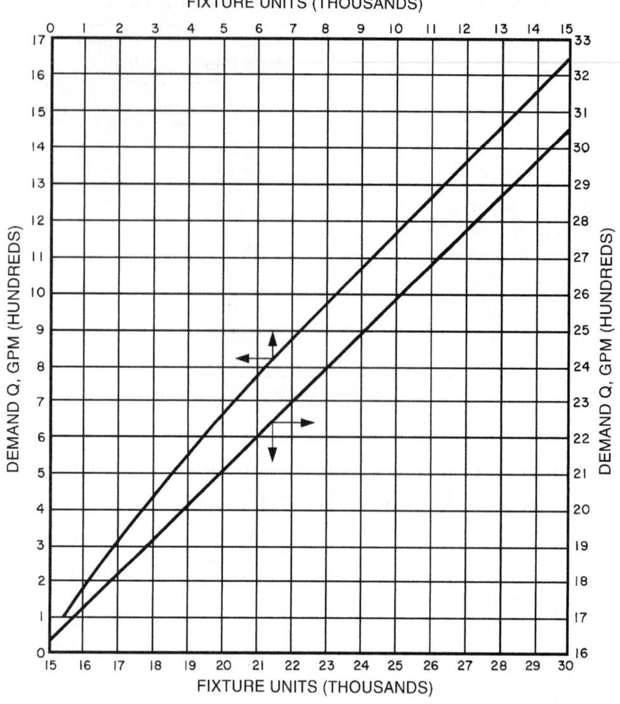

Fig. 5 Demand Versus Fixture Units, Mixed System, High Part of Curve
(Hunter 1941)

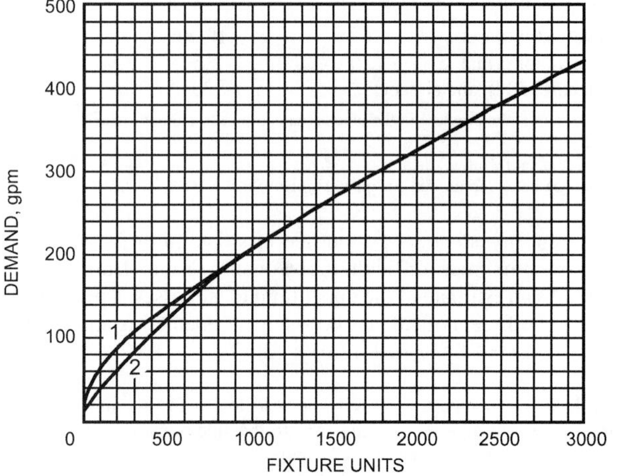

Fig. 6 Estimate Curves for Demand Load
(Hunter 1941)

So far, the information presented shows the *design rate of flow* to be determined in any particular section of piping. The next step is to determine the *size* of piping. As water flows through a pipe, the pressure continually decreases along the pipe due to loss of energy from friction. The problem is then to ascertain the minimum pressure in the street main and the minimum pressure required to operate the topmost fixture. (A pressure of 15 psig may be ample for most flush valves, but reference should be made to the manufacturers' requirements. Some fixtures require a pressure up to 25 psig. A minimum of 8 psig should be allowed for other fixtures.) The pressure differential overcomes pressure losses in the distributing system and the difference in elevation between the water main and the highest fixture.

The pressure loss (in psi) resulting from the difference in elevation between the street main and the highest fixture can be obtained

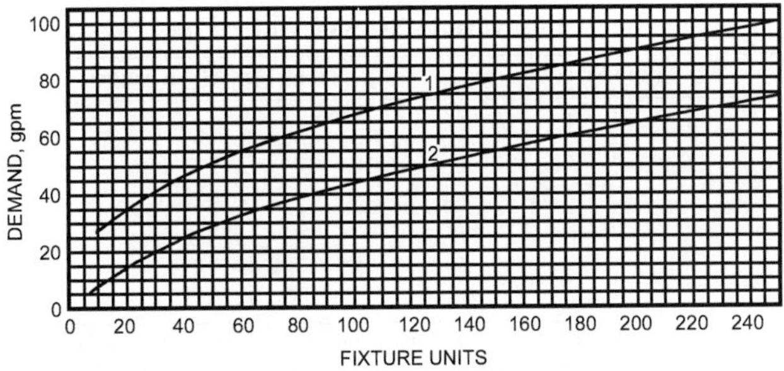

Fig. 7 Section of Figure 6 on Enlarged Scale

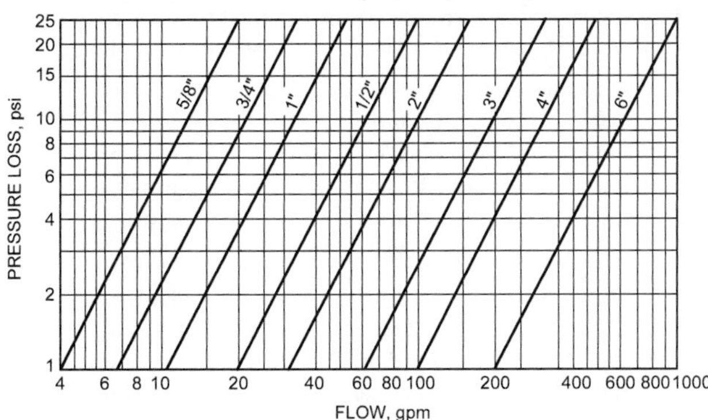

Fig. 8 Pressure Losses in Disk-Type Water Meters

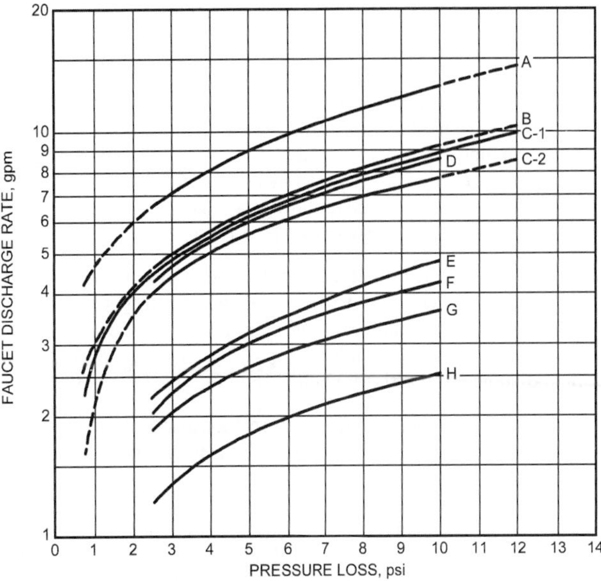

A. 1/2 in. laundry bibb (old style)
B. Laundry compression faucet
C-1. 1/2 in. compression sink faucet (mfr. 1)
C-2. 1/2 in. compression sink faucet (mfr. 2)
D. Combination compression bathtub faucets (both open)
E. Combination compression sink faucet
F. Basin faucet
G. Spring self-closing faucet
H. Slow self-closing faucet
(Dashed lines indicate recommended extrapolation)

Fig. 9 Variation of Pressure Loss with Flow Rate for Various Faucets and Cocks

by multiplying the difference in elevation in feet by the conversion factor 0.434.

Pressure losses in the distributing system consist of pressure losses in the piping itself, plus the pressure losses in the pipe fittings, valves, and the water meter, if any. Approximate design pressure losses and flow limits for disk-type meters for various rates of flow are given in Figure 8. Water authorities in many localities require compound meters for greater accuracy with varying flow; consult the local utility. Design data for compound meters differ from the data in Figure 8. Manufacturers give data on exact pressure losses and capacities.

Figure 9 shows the variation of pressure loss with rate of flow for various faucets and cocks. The water demand for hose bibbs or other large-demand fixtures taken off the building main frequently results in inadequate water supply to the upper floor of a building. This condition can be prevented by sizing the distribution system so that the pressure drops from the street main to all fixtures are the same. An ample building main (not less than 1 in. where possible) should be maintained until all branches to hose bibbs have been connected. Where the street main pressure is excessive and a pressure reducing valve is used to prevent water hammer or excessive pressure at the fixtures, the hose bibbs should be connected ahead of the reducing valve.

The principles involved in sizing upfeed and downfeed systems are the same. In the downfeed system, however, the difference in elevation between the overhead supply mains and the fixtures provides the pressure required to overcome pipe friction. Because friction pressure loss and height pressure loss are not additive, as in an upfeed system, smaller pipes may be used with a downfeed system.

Plastic Pipe

The maximum safe water velocity in a thermoplastic piping system under most operating conditions is typically 5 fps; however, higher velocities can be used in cases where the operating characteristics of valves and pumps are known so that sudden changes in flow velocity can be controlled. The total pressure in the system at any time (operating pressure plus surge of water hammer) should not exceed 150% of the pressure rating of the system.

Procedure for Sizing Cold Water Systems

The recommended procedure for sizing piping systems is outlined below.

1. Sketch the main lines, risers, and branches, and indicate the fixtures to be served. Indicate the rate of flow of each fixture.
2. Using Table 11, compute the demand weights of the fixtures in fixture units.
3. Determine the total demand in fixture units and, using Figure 5, 6, or 7, find the expected demand.
4. Determine the equivalent length of pipe in the main lines, risers, and branches. Because the sizes of the pipes are not known, the exact equivalent length of various fittings cannot be determined. Add the equivalent lengths, starting at the street main and proceeding along the service line, the main line of the building, and up the riser to the top fixture of the group served.
5. Determine the average minimum pressure in the street main and the minimum pressure required for the operation of the topmost fixture, which should be 8 to 25 psi.
6. Calculate the approximate design value of the average pressure drop per 100 ft of equivalent length of pipe determined in step 4.

$$\Delta p = (p_s - 0.434H - p_f - p_m)100/L \quad (11)$$

where

Δp = average pressure loss per 100 ft of equivalent length of pipe, psi
p_s = pressure in street main, psig
p_f = minimum pressure required to operate topmost fixture, psig
p_m = pressure drop through water meter, psi
H = height of highest fixture above street main, ft
L = equivalent length determined in step 4, ft

If the system is downfeed supply from a gravity tank, height of water in the tank, converted to psi by multiplying by 0.434, replaces the street main pressure, and the term 0.434H is added instead of subtracted in calculating Δp. In this case, H is the vertical distance of the fixture below the bottom of the tank.

7. From the expected rate of flow determined in step 3 and the value of Δp calculated in step 6, choose the sizes of pipe from Figure 1, 2, or 3.

Example 7. Assume a minimum street main pressure of 55 psig; a height of topmost fixture (a urinal with flush valve) above street main of 50 ft; an equivalent pipe length from water main to highest fixture of 100 ft; a total load on the system of 50 fixture units; and that the water closets are flush valve operated. Find the required size of supply main.

Solution: From Figure 7, the estimated peak demand is 51 gpm. From Table 10, the minimum pressure required to operate the topmost fixture is 15 psig. For a trial computation, choose the 1-1/2 in. meter. From Figure 8, the pressure drop through a 1-1/2 in. disk-type meter for a flow of 51 gpm is 6.5 psi.

The pressure drop available for overcoming friction in pipes and fittings is 55 – 0.434 × 50 – 15 – 6.5 = 12 psi.

At this point, estimate the equivalent pipe length of the fittings on the direct line from the street main to the highest fixture. The exact equivalent length of the various fittings cannot be determined since the pipe sizes of the building main, riser, and branch leading to the highest fixture are not yet known, but a first approximation is necessary to tentatively select pipe sizes. If the computed pipe sizes differ from those used in determining the equivalent length of pipe fittings, a recalculation using the computed pipe sizes for the fittings will be necessary. For this example, assume that the total equivalent length of the pipe fittings is 50 ft.

The permissible pressure loss per 100 ft of equivalent pipe is 12 × 100/(100 + 50) = 8 psi or 18 ft/100 ft. A 1-1/2 in. building main is adequate.

The sizing of the branches of the building main, the risers, and the fixture branches follows these principles. For example, assume that one of the branches of the building main carries the cold water supply for 3 water closets, 2 bathtubs, and 3 lavatories. Using the permissible pressure loss of 8 psi per 100 ft, the size of branch (determined from Table 11 and Figures 1 and 7) is found to be 1-1/2 in. Items included in the computation of pipe size are as follows:

Fixtures, No. and Type	Fixture Units (Table 11 and Note c)		Demand (Figure 7)	Pipe Size (Figure 1)
3 flush valves	3 × 6 =	18		
2 bathtubs	0.75 × 2 × 2 =	3		
3 lavatories	0.75 × 3 × 1 =	2.25		
Total	=	23.25	38 gpm	1-1/2 in.

Table 12 is a guide to minimum pipe sizing where flush valves are used.

Velocities exceeding 10 fps cause undesirable noise in the piping system. This usually governs the size of larger pipes in the system, while in small pipe sizes, the friction loss usually governs the selection because the velocity is low compared to friction loss. Velocity is the governing factor in downfeed systems, where friction loss is usually neglected. Velocity in branches leading to pump suctions should not exceed 5 fps.

Table 12 Allowable Number of 1 in. Flush Valves Served by Various Sizes of Water Pipe[a]

Pipe Size, in.	No. of 1 in. Flush Valves
1-1/4	1
1-1/2	2-4
2	5-12
2-1/2	13-25
3	26-40
4	41-100

[a]Two 3/4 in. flush valves are assumed equal to one 1 in. flush valve but can be served by a 1 in. pipe. Water pipe sizing must consider demand factor, available pressure, and length of run.

Table 13 Pressure Drops Used for Sizing Steam Pipe[a]

Initial Steam Pressure, psig	Pressure Drop per 100 ft	Total Pressure Drop in Steam Supply Piping
Vacuum return	2 to 4 oz/in^2	1 to 2 psi
0	0.5 oz/in^2	1 oz/in^2
1	2 oz/in^2	1 to 4 oz/in^2
2	2 oz/in^2	8 oz/in^2
5	4 oz/in^2	1.5 psi
10	8 oz/in^2	3 psi
15	1 psi	4 psi
30	2 psi	5 to 10 psi
50	2 to 5 psi	10 to 15 psi
100	2 to 5 psi	15 to 25 psi
150	2 to 10 psi	25 to 30 psi

[a]Equipment, control valves, and so forth must be selected based on delivered pressures.

If the street pressure is too low to adequately supply upper-floor fixtures, the pressure must be increased. Constant or variable speed booster pumps, alone or in conjunction with gravity supply tanks, or hydropneumatic systems may be used.

Flow control valves for individual fixtures under varying pressure conditions automatically adjust the flow at the fixture to a predetermined quantity. These valves allow the designer to (1) limit the flow at the individual outlet to the minimum suitable for the purpose, (2) hold the total demand for the system more closely to the required minimum, and (3) design the piping system as accurately as is practicable for the requirements.

STEAM PIPING

Pressure losses in steam piping for flows of dry or nearly dry steam are governed by Equations (1) through (7) in the section on Pressure Drop Equations. This section incorporates these principles with other information specific to steam systems.

Pipe Sizes

Required pipe sizes for a given load in steam heating depend on the following factors:

- The initial pressure and the total pressure drop that can be allowed between the source of supply and the end of the return system
- The maximum velocity of steam allowable for quiet and dependable operation of the system, taking into consideration the direction of condensate flow
- The equivalent length of the run from the boiler or source of steam supply to the farthest heating unit

Initial Pressure and Pressure Drop. Table 13 lists pressure drops commonly used with corresponding initial steam pressures for sizing steam piping.

Table 14 Comparative Capacity of Steam Lines at Various Pitches for Steam and Condensate Flowing in Opposite Directions

Pitch of Pipe, in/10 ft	Nominal Pipe Diameter, in.									
	3/4		1		1-1/4		1-1/2		2	
	Capacity	Maximum Velocity	Capacity	Maximum Velocity	Capacity	Maximum Velocity	Capacity	Maximum Velocity	Capacity	Maximum Velocity
1/4	3.2	8	6.8	9	11.8	11	19.8	12	42.9	15
1/2	4.1	11	9.0	12	15.9	14	25.9	16	54.0	18
1	5.7	13	11.7	15	19.9	17	33.0	19	68.8	24
1-1/2	6.4	14	12.8	17	24.6	20	37.4	22	83.3	27
2	7.1	16	14.8	19	27.0	22	42.0	24	92.9	30
3	8.3	17	17.3	22	31.3	25	46.8	26	99.6	32
4	9.9	22	19.2	24	33.4	26	50.8	28	102.4	32
5	10.5	22	20.5	25	38.5	31	59.2	33	115.0	33

Source: Laschober et al. (1966). Velocity in fps; capacity in lb/h.

Several factors, such as initial pressure and pressure required at the end of the line, should be considered, but it is most important that (1) the total pressure drop does not exceed the initial gage pressure of the system (and in practice it should never exceed one-half the initial gage pressure); (2) the pressure drop is not great enough to cause excessive velocities; (3) a constant initial pressure is maintained, except on systems specially designed for varying initial pressures (e.g., subatmospheric pressure), which normally operate under controlled partial vacuums; and (4) for gravity return systems, the pressure drop to the heating units does not exceed the water column available for removing condensate (i.e., the height above the boiler water line of the lowest point on the steam main, on the heating units, or on the dry return).

Maximum Velocity. For quiet operation, steam velocity should be 8000 to 12,000 fpm, with a maximum of 15,000 fpm. The lower the velocity, the quieter the system. When the condensate must flow against the steam, even in limited quantity, the velocity of the steam must not exceed limits above which the disturbance between the steam and the counterflowing water may (1) produce objectionable sound, such as water hammer, or (2) result in the retention of water in certain parts of the system until the steam flow is reduced sufficiently to permit the water to pass. The velocity at which these disturbances take place is a function of (1) pipe size; (2) the pitch of the pipe if it runs horizontally; (3) the quantity of condensate flowing against the steam; and (4) the freedom of the piping from water pockets that, under certain conditions, act as a restriction in pipe size. Table 14 lists maximum capacities for various size steam lines.

Table 15 Equivalent Length of Fittings to Be Added to Pipe Run

Nominal Pipe Diameter, in.	Length to Be Added to Run, ft				
	Standard Elbow	Side Outlet Tee[b]	Gate Valve[a]	Globe Valve[a]	Angle Valve[a]
1/2	1.3	3	0.3	14	7
3/4	1.8	4	0.4	18	10
1	2.2	5	0.5	23	12
1-1/4	3.0	6	0.6	29	15
1-1/2	3.5	7	0.8	34	18
2	4.3	8	1.0	46	22
2-1/2	5.0	11	1.1	54	27
3	6.5	13	1.4	66	34
3-1/2	8	15	1.6	80	40
4	9	18	1.9	92	45
5	11	22	2.2	112	56
6	13	27	2.8	136	67
8	17	35	3.7	180	92
10	21	45	4.6	230	112
12	27	53	5.5	270	132
14	30	63	6.4	310	152

[a]Valve in full-open position.
[b]Values apply only to a tee used to divert the flow in the main to the last riser.

Equivalent Length of Run. All tables for the flow of steam in pipes based on pressure drop must allow for pipe friction, as well as for the resistance of fittings and valves. These resistances are generally stated in terms of straight pipe; that is, a certain fitting produces a drop in pressure equivalent to the stated number of feet of straight run of the same size of pipe. Table 15 gives the number of feet of straight pipe usually allowed for the more common types of fittings and valves. In all pipe sizing tables in this chapter, the *length of run* refers to the *equivalent length of run* as distinguished from the *actual length* of pipe. A common sizing method is to assume the length of run and to check this assumption after pipes are sized. For this purpose, the length of run is usually assumed to be double the actual length of pipe.

Example 8. Using Table 15, determine the equivalent length in feet of pipe for the run illustrated.

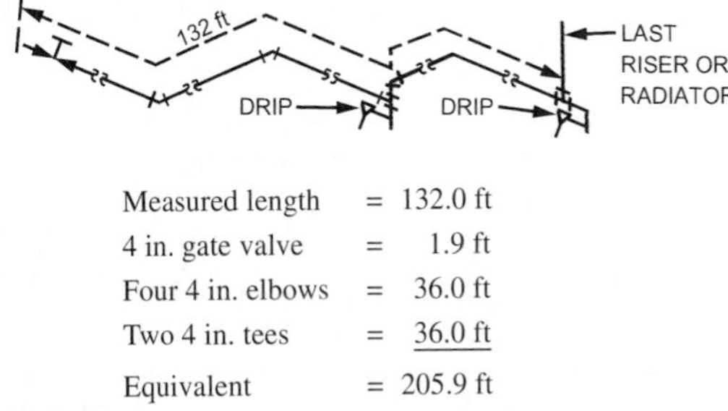

Measured length	=	132.0 ft
4 in. gate valve	=	1.9 ft
Four 4 in. elbows	=	36.0 ft
Two 4 in. tees	=	36.0 ft
Equivalent	=	205.9 ft

Sizing Charts

Figure 10 is the basic chart for determining the flow rate and velocity of steam in Schedule 40 pipe for various values of pressure drop per 100 ft, based on 0 psig saturated steam. Using the multiplier chart (Figure 11), Figure 10 can be used at all saturation pressures between 0 and 200 psig (see Example 10).

Figures 10A through 10D present charts for sizing steam piping for systems of 30, 50, 100, and 150 psig at various pressure drops. These charts are based on the Moody friction factor, which considers the Reynolds number and the roughness of the internal pipe surfaces; they contain the same information as the basic chart (Figure 10) but in a more convenient form.

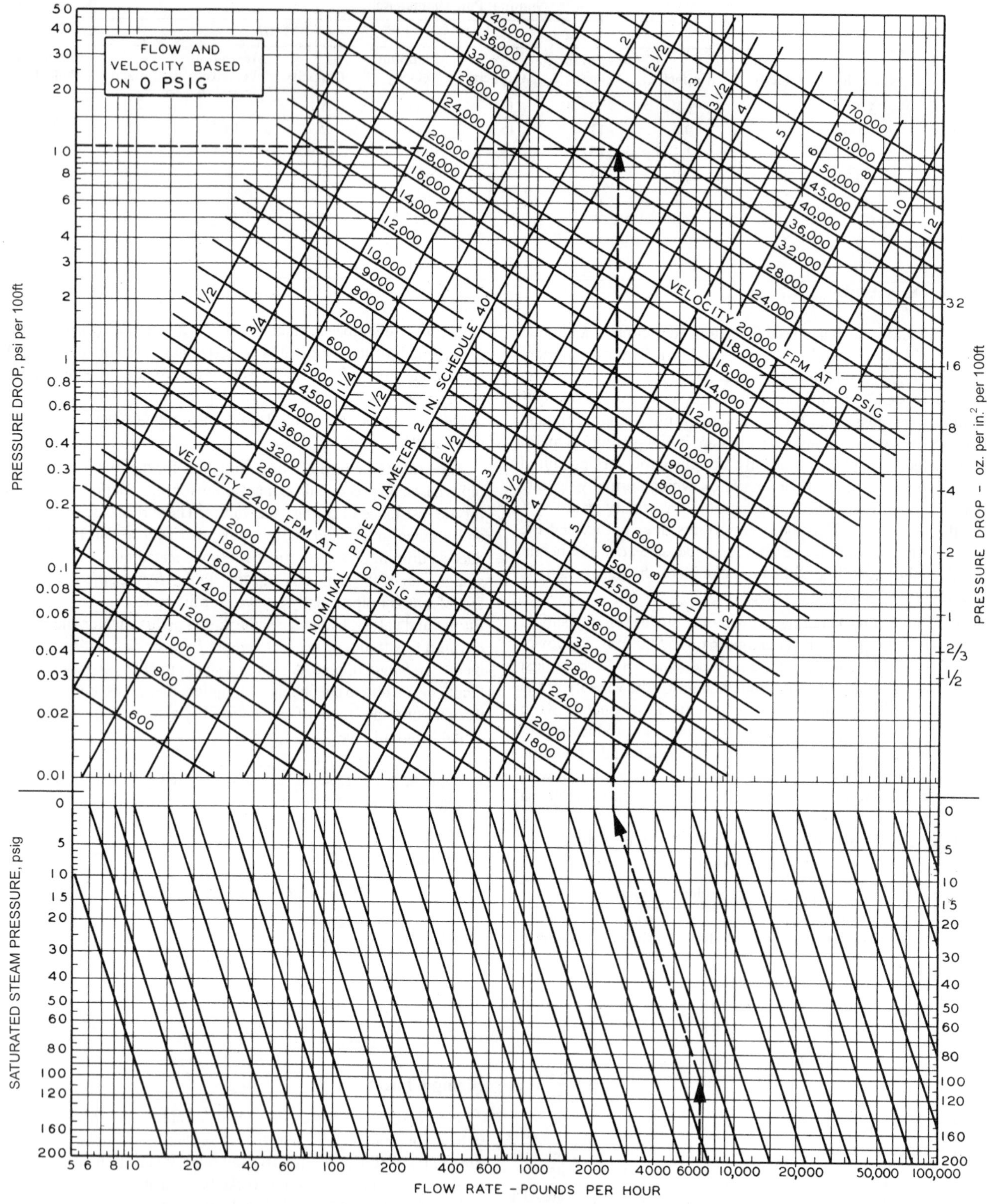

Notes: Based on Moody Friction Factor where flow of condensate does not inhibit the flow of steam.
See Figure 11 for obtaining flow rates and velocities of all saturation pressures between 0 and 200 psig; see also Examples 9 and 10.

Fig. 10 Flow Rate and Velocity of Steam in Schedule 40 Pipe at Saturation Pressure of 0 psig

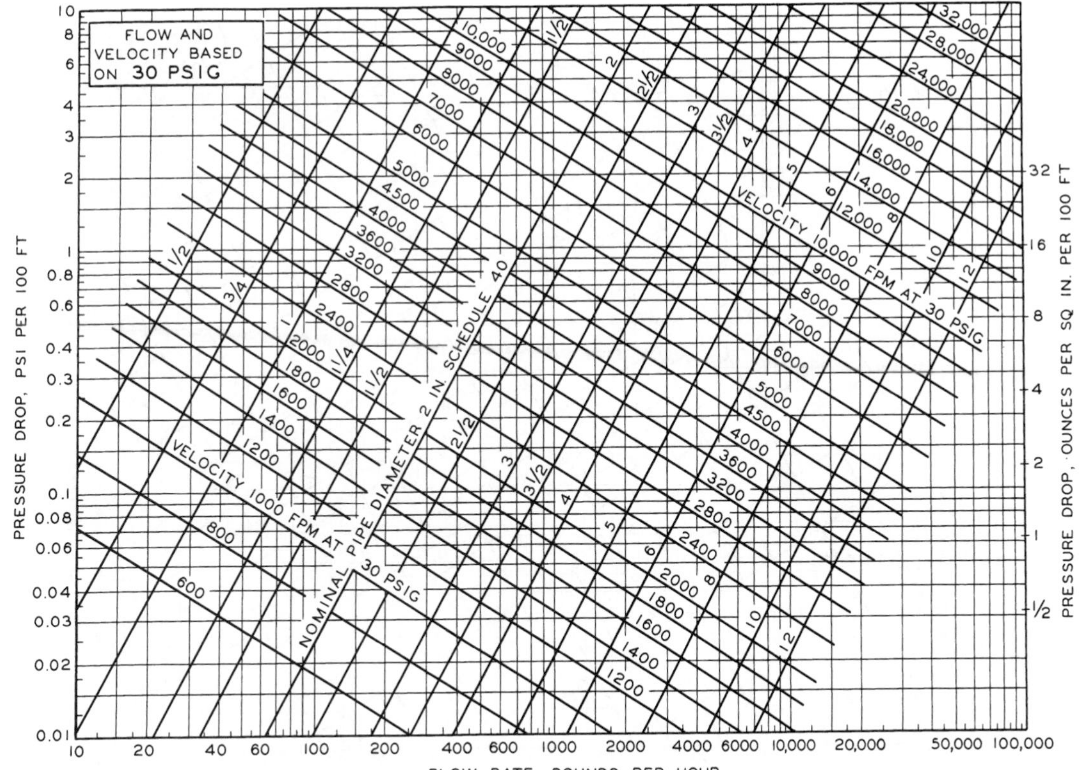

Notes: Based on Moody Friction Factor where flow of condensate does not inhibit the flow of steam. May be used for steam pressures from 23 to 37 psig with an error not exceeding 9%.

Fig. 10A Flow Rate and Velocity of Steam in Schedule 40 Pipe at Saturation Pressure of 30 psig

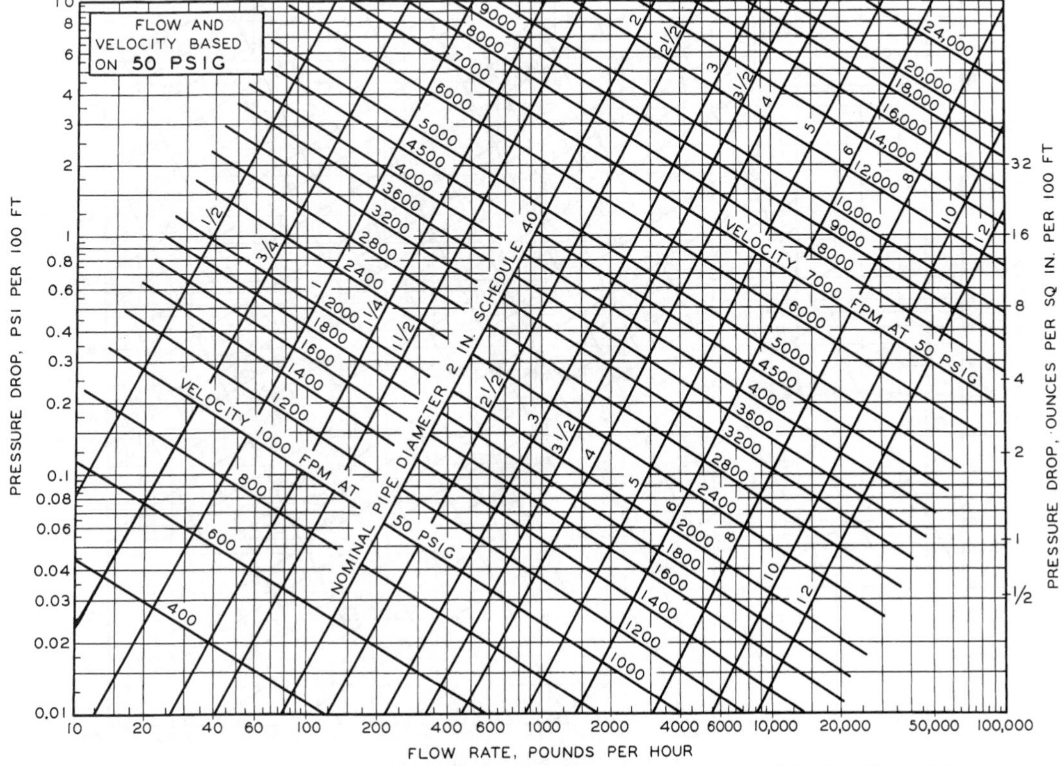

Notes: Based on Moody Friction Factor where flow of condensate does not inhibit the flow of steam. May be used for steam pressures from 40 to 60 psig with an error not exceeding 8%.

Fig. 10B Flow Rate and Velocity of Steam in Schedule 40 Pipe at Saturation Pressure of 50 psig

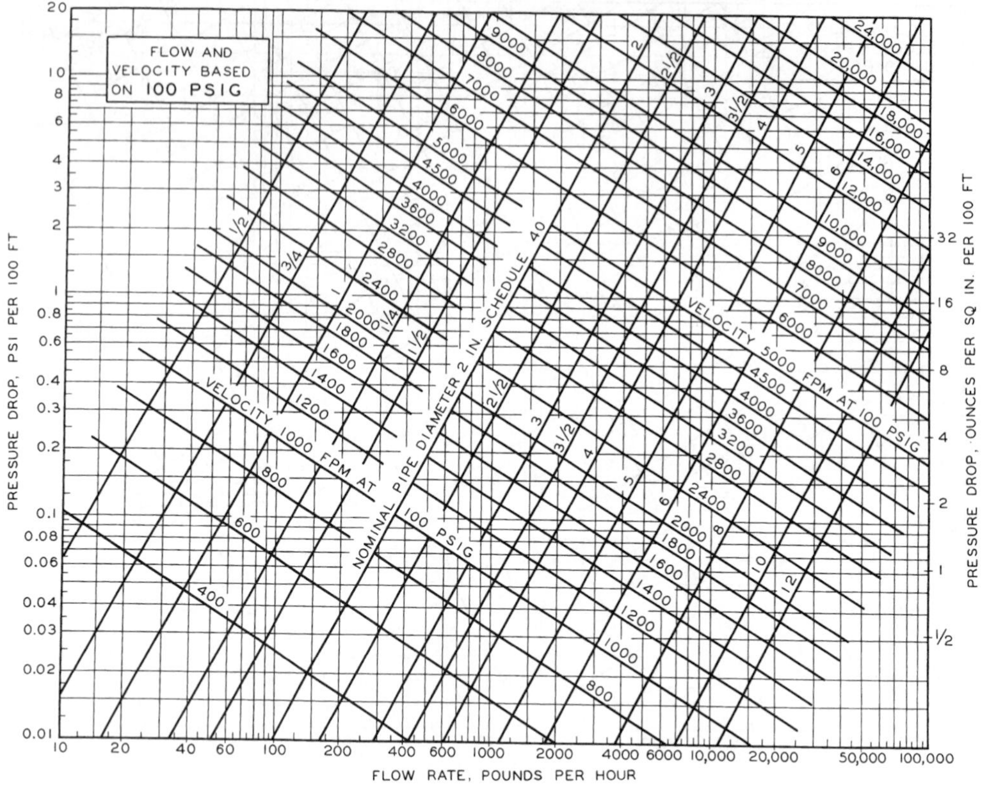

Notes: Based on Moody Friction Factor where flow of condensate does not inhibit the flow of steam.
May be used for steam pressures from 85 to 120 psig with an error not exceeding 8%.

Fig. 10C Flow Rate and Velocity of Steam in Schedule 40 Pipe at Saturation Pressure of 100 psig

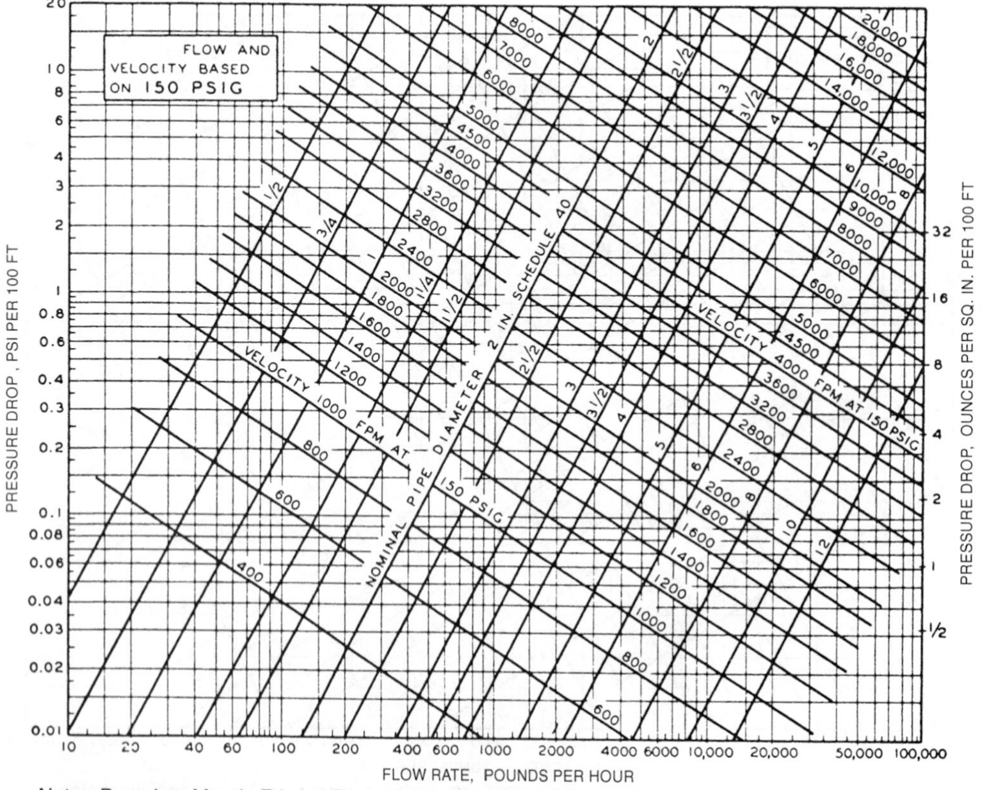

Notes: Based on Moody Friction Factor where flow of condensate does not inhibit the flow of steam.
May be used for steam pressures from 127 to 180 psig with an error not exceeding 8%.

Fig. 10D Flow Rate and Velocity of Steam in Schedule 40 Pipe at Saturation Pressure of 150 psig

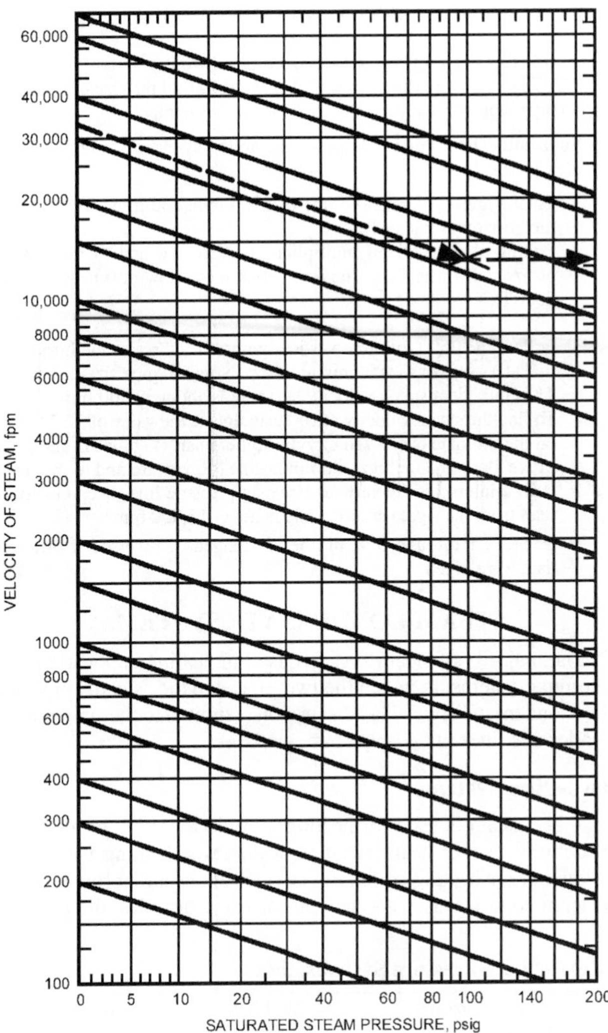

Fig. 11 Velocity Multiplier Chart for Figure 10

LOW-PRESSURE STEAM PIPING

Values in Table 16 (taken from Figure 10) provide a more rapid means of selecting pipe sizes for the various pressure drops listed and for systems operated at 3.5 and 12 psig. The flow rates shown for 3.5 psig can be used for saturated pressures from 1 to 6 psig, and those shown for 12 psig can be used for saturated pressures from 8 to 16 psig with an error not exceeding 8%.

Both Figure 10 and Table 16 can be used where the flow of condensate does not inhibit the flow of steam. Columns B and C of Table 17 are used in cases where steam and condensate flow in opposite directions, as in risers or runouts that are not dripped. Columns D, E, and F are for one-pipe systems and include risers, radiator valves and vertical connections, and radiator and riser runout sizes, all of which are based on the critical velocity of the steam to permit the counterflow of condensate without noise.

Return piping can be sized using Table 18, in which pipe capacities for wet, dry, and vacuum return lines are shown for several values of pressure drop per 100 ft of equivalent length.

Example 9. What pressure drop should be used for the steam piping of a system if the measured length of the longest run is 500 ft, and the initial pressure must not exceed 2 psig?

Solution: It is assumed, if the measured length of the longest run is 500 ft, that when the allowance for fittings is added, the equivalent length of run does not exceed 1000 ft. Then, with the pressure drop not over one-half of the initial pressure, the drop could be 1 psi or less. With a pressure drop of 1 psi and a length of run of 1000 ft, the drop per 100 ft would be 0.1 psi; if the total drop were 0.5 psi, the drop per 100 ft would be 0.05 psi. In both cases, the pipe could be sized for a desired capacity according to Figure 10.

On completion of the sizing, the drop could be checked by taking the longest line and actually calculating the equivalent length of run from the pipe sizes determined. If the calculated drop is less than that assumed, the pipe size is adequate; if it is more, an unusual number of fittings is probably involved, and either the lines must be straightened, or the next larger pipe size must be tried.

HIGH-PRESSURE STEAM PIPING

Many heating systems for large industrial buildings use high-pressure steam (15 to 150 psig). These systems usually have unit heaters or large built-up fan units with blast heating coils. Temperatures are controlled by a modulating or throttling thermostatic valve or by face or bypass dampers controlled by the room air temperature, fan inlet, or fan outlet.

Table 16 Flow Rate of Steam in Schedule 40 Pipe

Nominal Pipe Size, in.	Pressure Drop per 100 ft of Length: 1/16 psi (1 oz/in²), Sat. Press., psig 3.5	12	1/8 psi (2 oz/in²), Sat. Press., psig 3.5	12	1/4 psi (4 oz/in²), Sat. Press., psig 3.5	12	1/2 psi (8 oz/in²), Sat. Press., psig 3.5	12	3/4 psi (12 oz/in²), Sat. Press., psig 3.5	12	1 psi, Sat. Press., psig 3.5	12	2 psi, Sat. Press., psig 3.5	12
3/4	9	11	14	16	20	24	29	35	36	43	42	50	60	73
1	17	21	26	31	37	46	54	66	68	82	81	95	114	137
1-1/4	36	45	53	66	78	96	111	138	140	170	162	200	232	280
1-1/2	56	70	84	100	120	147	174	210	218	260	246	304	360	430
2	108	134	162	194	234	285	336	410	420	510	480	590	710	850
2-1/2	174	215	258	310	378	460	540	660	680	820	780	950	1,150	1,370
3	318	380	465	550	660	810	960	1,160	1,190	1,430	1,380	1,670	1,950	2,400
3-1/2	462	550	670	800	990	1,218	1,410	1,700	1,740	2,100	2,000	2,420	2,950	3,450
4	640	800	950	1,160	1,410	1,690	1,980	2,400	2,450	3,000	2,880	3,460	4,200	4,900
5	1,200	1,430	1,680	2,100	2,440	3,000	3,570	4,250	4,380	5,250	5,100	6,100	7,500	8,600
6	1,920	2,300	2,820	3,350	3,960	4,850	5,700	6,800	7,000	8,600	8,400	10,000	11,900	14,200
8	3,900	4,800	5,570	7,000	8,100	10,000	11,400	14,300	14,500	17,700	16,500	20,500	24,000	29,500
10	7,200	8,800	10,200	12,600	15,000	18,200	21,000	26,000	26,200	32,000	30,000	37,000	42,700	52,000
12	11,400	13,700	16,500	19,500	23,400	28,400	33,000	40,000	41,000	49,500	48,000	57,500	67,800	81,000

Notes:
1. Flow rate is in lb/h at initial saturation pressures of 3.5 and 12 psig. Flow is based on Moody friction factor, where the flow of condensate does not inhibit the flow of steam.
2. The flow rates at 3.5 psig cover saturated pressure from 1 to 6 psig, and the rates at 12 psig cover saturated pressure from 8 to 16 psig with an error not exceeding 8%.
3. The steam velocities corresponding to the flow rates given in this table can be found from Figures 10 and 11.

Table 17 Steam Pipe Capacities for Low-Pressure Systems

	Capacity, lb/h					
	Two-Pipe System		One-Pipe Systems			
Nominal Pipe Size, in.	Condensate Flowing Against Steam: Vertical	Condensate Flowing Against Steam: Horizontal	Supply Risers Upfeed	Radiator Valves and Vertical Connections	Radiator and Riser Runouts	
A	B[a]	C[b]	D[c]	E	F[b]	
3/4	8	7	6	—	7	
1	14	14	11	7	7	
1-1/4	31	27	20	16	16	
1-1/2	48	42	38	23	16	
2	97	93	72	42	23	
2-1/2	159	132	116	—	42	
3	282	200	200	—	65	
3-1/2	387	288	286	—	119	
4	511	425	380	—	186	
5	1,050	788	—	—	278	
6	1,800	1,400	—	—	545	
8	3,750	3,000	—	—	—	
10	7,000	5,700	—	—	—	
12	11,500	9,500	—	—	—	
16	22,000	19,000	—	—	—	

Notes:
1. For one- or two-pipe systems in which condensate flows against the steam flow.
2. Steam at an average pressure of 1 psig is used as a basis of calculating capacities.

[a]Do not use Column B for pressure drops of less than 1/16 psi per 100 ft of equivalent run. Use Figure 10 or Table 15 instead.

[b]Pitch of horizontal runouts to risers and radiators should be not less than 0.5 in/ft. Where this pitch cannot be obtained, runouts over 8 ft in length should be one pipe size larger than that called for in this table.

[c]Do not use Column D for pressure drops of less than 1/24 psi per 100 ft of equivalent run except on sizes 3 in. and over. Use Figure 10 or Table 15 instead.

Use of Basic and Velocity Multiplier Charts

Example 10. Given a flow rate of 6700 lb/h, an initial steam pressure of 100 psig, and a pressure drop of 11 psi/100 ft, find the size of Schedule 40 pipe required and the velocity of steam in the pipe.

Solution: The following steps are illustrated by the broken line on Figures 10 and 11.

1. Enter Figure 10 at a flow rate of 6700 lb/h, and move vertically to the horizontal line at 100 psig.
2. Follow along inclined multiplier line (upward and to the left) to horizontal 0 psig line. The equivalent mass flow at 0 psig is about 2500 lb/h.
3. Follow the 2500 lb/h line vertically until it intersects the horizontal line at 11 psi per 100 ft pressure drop. Nominal pipe size is 2-1/2 in. The equivalent steam velocity at 0 psig is about 32,700 fpm.
4. To find the steam velocity at 100 psig, locate the value of 32,700 fpm on the ordinate of the velocity multiplier chart (Figure 11) at 0 psig.
5. Move along the inclined multiplier line (downward and to the right) until it intersects the vertical 100 psig pressure line. The velocity as read from the right (or left) scale is about 13,000 fpm.

Note: Steps 1 through 5 would be rearranged or reversed if different data were given.

STEAM CONDENSATE SYSTEMS

The majority of steam systems used in heating applications are two-pipe systems, in which the two pipes are the "steam" pipe and the "condensate" pipe. This discussion is limited to the sizing of the condensate lines in two-pipe systems.

Two-Pipe Systems

When steam is used for heating a liquid to 215°F or less (e.g., in domestic water heat exchangers, domestic heating water converters, or air-heating coils), the devices are usually provided with a steam control valve. As the control valve throttles, the absolute

Table 18 Return Main and Riser Capacities for Low-Pressure Systems, lb/h

	Pipe Size, in.	1/32 psi (1/2 oz/in²) Drop per 100 ft			1/24 psi (2/3 oz/in²) Drop per 100 ft			1/16 psi (1 oz/in²) Drop per 100 ft			1/8 psi (2 oz/in²) Drop per 100 ft			1/4 psi (4 oz/in²) Drop per 100 ft			1/2 psi (8 oz/in²) Drop per 100 ft		
		Wet	Dry	Vac.	Wet	Dry	Vac.	Wet	Dry	Vac.	Wet	Dry	Vac.	Wet	Dry	Vac.	Wet	Dry	Vac.
	G	H	I	J	K	L	M	N	O	P	Q	R	S	T	U	V	W	X	Y
Return Main	3/4	—	—	—	—	—	42	—	—	100	—	—	142	—	—	200	—	—	283
Return Main	1	125	62	—	145	71	143	175	80	175	250	103	249	350	115	350	—	—	494
Return Main	1-1/4	213	130	—	248	149	244	300	168	300	425	217	426	600	241	600	—	—	848
Return Main	1-1/2	338	206	—	393	236	388	475	265	475	675	340	674	950	378	950	—	—	1,340
Return Main	2	700	470	—	810	535	815	1,000	575	1,000	1,400	740	1,420	2,000	825	2,000	—	—	2,830
Return Main	2-1/2	1,180	760	—	1,580	868	1,360	1,680	950	1,680	2,350	1,230	2,380	3,350	1,360	3,350	—	—	4,730
Return Main	3	1,880	1,460	—	2,130	1,560	2,180	2,680	1,750	2,680	3,750	2,250	3,800	5,350	2,500	5,350	—	—	7,560
Return Main	3-1/2	2,750	1,970	—	3,300	2,200	3,250	4,000	2,500	4,000	5,500	3,230	5,680	8,000	3,580	8,000	—	—	11,300
Return Main	4	3,880	2,930	—	4,580	3,350	4,500	5,500	3,750	5,500	7,750	4,830	7,810	11,000	5,380	11,000	—	—	15,500
Return Main	5	—	—	—	—	—	7,880	—	—	9680	—	—	13,700	—	—	19,400	—	—	27,300
Return Main	6	—	—	—	—	—	12,600	—	—	15,500	—	—	22,000	—	—	31,000	—	—	43,800
Riser	3/4	—	48	—	—	48	143	—	48	175	—	48	249	—	48	350	—	—	494
Riser	1	—	113	—	—	113	244	—	113	300	—	113	426	—	113	600	—	—	848
Riser	1-1/4	—	248	—	—	248	388	—	248	475	—	248	674	—	248	950	—	—	1,340
Riser	1-1/2	—	375	—	—	375	815	—	375	1,000	—	375	1,420	—	375	2,000	—	—	2,830
Riser	2	—	750	—	—	750	1,360	—	750	1,680	—	750	2,380	—	750	3,350	—	—	4,730
Riser	2-1/2	—	—	—	—	—	2,180	—	—	2,680	—	—	3,800	—	—	5,350	—	—	7,560
Riser	3	—	—	—	—	—	3,250	—	—	4,000	—	—	5,680	—	—	8,000	—	—	11,300
Riser	3-1/2	—	—	—	—	—	4,480	—	—	5,500	—	—	7,810	—	—	11,000	—	—	15,500
Riser	4	—	—	—	—	—	7,880	—	—	9680	—	—	13,700	—	—	19,400	—	—	27,300
Riser	5	—	—	—	—	—	12,600	—	—	15,500	—	—	22,000	—	—	31,000	—	—	43,800

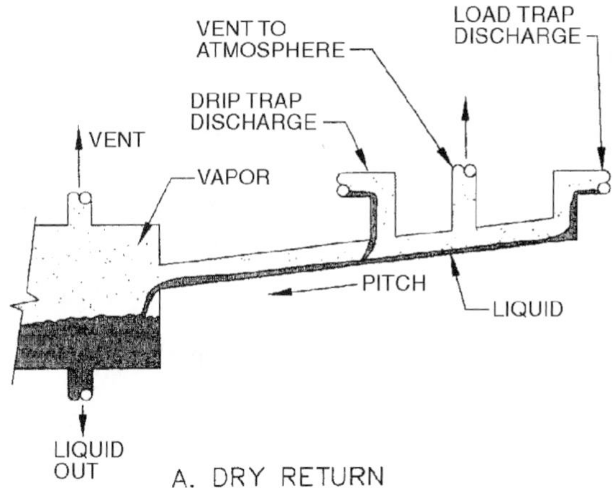

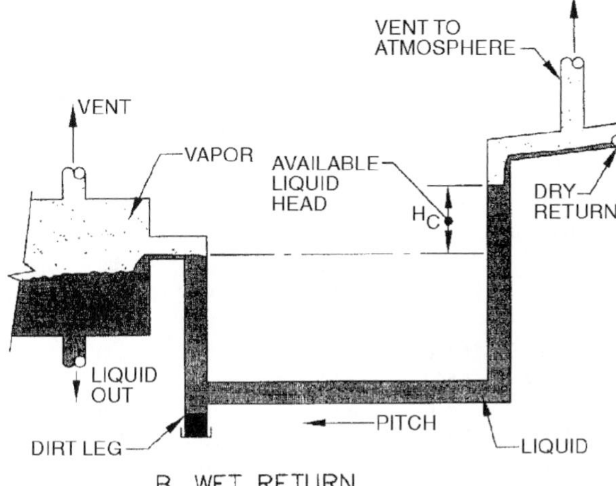

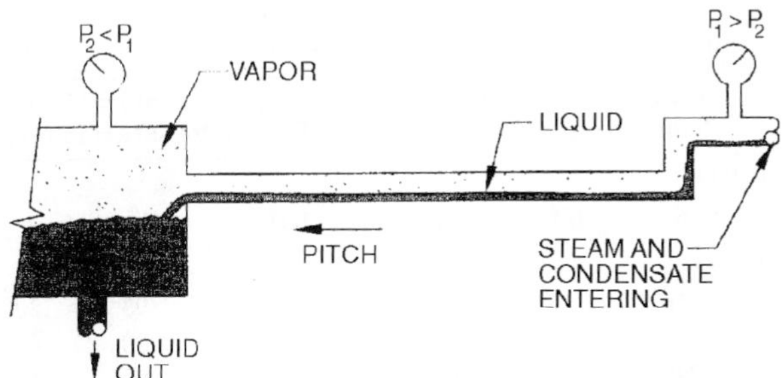

Fig. 12 Types of Condensate Return Systems

pressure in the load device decreases, removing all pressure motivation for flow in the condensate return system. In order to ensure the flow of steam condensate from the load device through the trap and into the return system, it is necessary to provide a vacuum breaker on the device ahead of the trap. This ensures a minimum pressure at the trap inlet of atmospheric pressure plus whatever liquid leg the designer has provided. Then, to ensure flow through the trap, it is necessary to design the condensate system so that it will never have a pressure above atmospheric in the condensate return line.

Vented (Open) Return Systems. To achieve this pressure requirement, the condensate return line is usually vented to the atmosphere (1) near the point of entrance of the flow streams from the load traps, (2) in proximity to all connections from drip traps, and (3) at transfer pumps or feedwater receivers.

Table 19 Vented Dry Condensate Return for Gravity Flow Based on Manning Equation

Nominal Diameter, in. IPS	Condensate Flow, lb/h[a,b] — Condensate Line Slope, in/ft: 1/16	1/8	1/4	1/2
1/2	38	54	76	107
3/4	80	114	161	227
1	153	216	306	432
1-1/4	318	449	635	898
1-1/2	479	677	958	1,360
2	932	1,320	1,860	2,640
2-1/2	1,500	2,120	3,000	4,240
3	2,670	3,780	5,350	7,560
4	5,520	7,800	11,000	15,600
5	10,100	14,300	20,200	28,500
6	16,500	23,300	32,900	46,500

[a] Flow is in lb/h of 180°F water for Schedule 40 steel pipes.
[b] Flow was calculated from Equation (12) and rounded.

With this design, the only motivation for flow in the return system is gravity. Return lines that are below the liquid level in the downstream receiver or boiler and are thus filled with liquid are called wet returns; those above the liquid level have both liquid and gas in the pipes and are called dry returns.

The dry return lines in a vented return system have flowing liquid in the bottom of the line and gas or vapor in the top (Figure 12A). The liquid is the condensate, and the gas may be steam, air, or a mixture of the two. The flow phenomenon for these dry return systems is open channel flow, which is best described by the **Manning equation**:

$$Q = \frac{1.49 A r^{2/3} S^{1/2}}{n} \qquad (12)$$

where

Q = volumetric flow rate, cfs
A = cross-sectional area of conduit, ft^2
r = hydraulic radius of conduit, ft
n = coefficient of roughness (usually 0.012)
S = slope of conduit, ft/ft

Table 19 is a solution to Equation (12) that shows pipe size capacities for steel pipes with various pitches. Recommended practice is to size vertical lines by the maximum pitch shown, although they would actually have a capacity far in excess of that shown. As the pitch increases, hydraulic jump that could fill the pipe and other transient effects that could cause water hammer should be avoided. Flow values in Table 19 are calculated for Schedule 40 steel pipe, with a factor of safety of 3.0, and can be used for copper pipes of the same nominal pipe size.

The flow characteristics of **wet return lines** (Figure 12B) are best described by the Darcy-Weisbach equation [Equation (1)]. The motivation for flow is the fluid head difference between the entering section of the flooded line and the leaving section. It is common practice, in addition to providing for the fluid head differential, to slope the return in the direction of flow to a collection point such as a dirt leg in order to clear the line of sediment or solids. Table 20 is a solution to Equation (1) that shows pipe size capacity for steel pipes with various available fluid heads. Table 20 can also be used for copper tubing of equal nominal pipe size.

Nonvented (Closed) Return Systems. For those systems in which there is a continual steam pressure difference between the point where the condensate enters the line and the point where it leaves (Figure 12C), Table 18 or Table 21, as applicable, can be used for sizing the condensate lines. Although these tables express condensate capacity without slope, common practice is to slope the

Table 20 Vented Wet Condensate Return for Gravity Flow Based on Darcy-Weisbach Equation

Nominal Diameter, in. IPS	Condensate Flow, lb/h[a,b]							
	Condensate Head, ft per 100 ft							
	0.5	1	1.5	2	2.5	3	3.5	4
1/2	105	154	192	224	252	278	302	324
3/4	225	328	408	476	536	590	640	687
1	432	628	779	908	1,020	1,120	1,220	1,310
1-1/4	901	1,310	1,620	1,890	2,120	2,330	2,530	2,710
1-1/2	1,360	1,970	2,440	2,840	3,190	3,510	3,800	4,080
2	2,650	3,830	4,740	5,510	6,180	6,800	7,360	7,890
2-1/2	4,260	6,140	7,580	8,810	9,890	10,900	11,800	12,600
3	7,570	10,900	13,500	15,600	17,500	19,300	20,900	22,300
4	15,500	22,300	27,600	32,000	35,900	39,400	42,600	45,600
5	28,200	40,500	49,900	57,900	64,900	71,300	77,100	82,600
6	45,800	65,600	80,900	93,800	105,000	115,000	125,000	134,000

[a] Flow is in lb/h of 180°F water for Schedule 40 steel pipes.
[b] Flow was calculated from Equation (1) and rounded.

Table 21 Flow Rate for Dry-Closed Returns

Pipe Dia. *D*, in.	Supply Pressure = 5 psig, Return Pressure = 0 psig			Supply Pressure = 15 psig, Return Pressure = 0 psig			Supply Pressure = 30 psig, Return Pressure = 0 psig			Supply Pressure = 50 psig, Return Pressure = 0 psig		
	$\Delta p/L$, psi/100 ft											
	1/16	1/4	1	1/16	1/4	1	1/16	1/4	1	1/16	1/4	1
	Flow Rate, lb/h											
1/2	240	520	1,100	95	210	450	60	130	274	42	92	200
3/4	510	1,120	2,400	210	450	950	130	280	590	91	200	420
1	1,000	2,150	4,540	400	860	1,820	250	530	1,120	180	380	800
1-1/4	2,100	4,500	9,500	840	1,800	3,800	520	1,110	2,340	370	800	1,680
1-1/2	3,170	6,780	14,200	1,270	2,720	5,700	780	1,670	3,510	560	1,200	2,520
2	6,240	13,300	a	2,500	5,320	a	1,540	3,270	a	1,110	2,350	a
2-1/2	10,000	21,300	a	4,030	8,520	a	2,480	5,250	a	1,780	3,780	a
3	18,000	38,000	a	7,200	15,200	a	4,440	9,360	a	3,190	6,730	a
4	37,200	78,000	a	14,900	31,300	a	9,180	19,200	a	6,660	13,800	a
6	110,500	a	a	44,300	a	a	27,300	a	a	19,600	a	a
8	228,600	a	a	91,700	a	a	56,400	a	a	40,500	a	a

Pipe Dia. *D*, in.	Supply Pressure = 100 psig, Return Pressure = 0 psig			Supply Pressure = 150 psig, Return Pressure = 0 psig			Supply Pressure = 100 psig, Return Pressure = 15 psig			Supply Pressure = 150 psig, Return Pressure = 15 psig		
	$\Delta p/L$, psi/100 ft											
	1/16	1/4	1	1/16	1/4	1	1/16	1/4	1	1/16	1/4	1
	Flow Rate, lb/h											
1/2	28	62	133	23	51	109	56	120	260	43	93	200
3/4	62	134	290	50	110	230	120	260	560	93	200	420
1	120	260	544	100	210	450	240	500	1,060	180	390	800
1-1/4	250	540	1,130	200	440	930	500	1,060	2,200	380	800	1,680
1-1/2	380	810	1,700	310	660	1,400	750	1,600	3,320	570	1,210	2,500
2	750	1,590	a	610	1,300	a	1,470	3,100	6,450	1,120	2,350	4,900
2-1/2	1,200	2,550	a	980	2,100	a	2,370	5,000	10,300	1,800	3,780	7,800
3	2,160	4,550	a	1,760	3,710	a	4,230	8,860	a	3,200	6,710	a
4	4,460	9,340	a	3,640	7,630	a	8,730	18,200	a	6,620	13,800	a
6	13,200	a	a	10,800	a	a	25,900	53,600	a	19,600	40,600	a
8	27,400	a	a	22,400	a	a	53,400	110,300	a	40,500	83,600	a

[a] For these sizes and pressure losses, the velocity is above 7000 fpm. Select another combination of size and pressure loss.

lines in the direction of flow to a collection point similar to wet returns to clear the lines of sediment or solids.

When saturated condensate at pressures above the return system pressure enters the return (condensate) mains, some of the liquid flashes to steam. This occurs typically at drip traps into a vented return system or at load traps leaving process load devices that are not valve-controlled and typically have no subcooling. If the return main is vented, the vent lines will relieve any excessive pressure and prevent a back pressure phenomenon that could restrict the flow through traps from valved loads; the pipe sizing would be as described above for vented dry returns. If the return line is not vented, the flash steam results in a pressure rise at that point and the piping could be sized as described above for closed returns, and in accordance with Table 18 or Table 21, as applicable.

The passage of the fluid through the steam trap is a throttling or constant enthalpy process. The resulting fluid on the downstream side of the trap can be a mixture of saturated liquid and vapor. Thus, in nonvented returns, it is important to understand the condition of the fluid when it enters the return line from the trap.

Table 22 Flash Steam from Steam Trap on Pressure Drop

Supply Pressure, psig	Return Pressure, psig	x, Fraction Vapor, Mass Basis	V_c, Fraction Vapor, Volume Basis
5	0	0.016	0.962
15	0	0.040	0.985
30	0	0.065	0.991
50	0	0.090	0.994
100	0	0.133	0.996
150	0	0.164	0.997
100	15	0.096	0.989
150	15	0.128	0.992

The condition of the condensate downstream of the trap can be expressed by the quality x, defined as

$$x = \frac{m_v}{m_l + m_v} \tag{13}$$

where

m_v = mass of saturated vapor in condensate
m_l = mass of saturated liquid in condensate

Likewise, the volume fraction V_c of the vapor in the condensate is expressed as

$$V_c = \frac{V_v}{V_l + V_v} \tag{14}$$

where

V_v = volume of saturated vapor in condensate
V_l = volume of saturated liquid in condensate

The quality and the volume fraction of the condensate downstream of the trap can be estimated from Equations (13) and (14), respectively.

$$x = \frac{h_1 - h_{f_2}}{h_{g_2} - h_{f_2}} \tag{15}$$

$$V_c = \frac{x v_{g_2}}{v_{f_2}(1 - x) + x v_{g_2}} \tag{16}$$

where

h_1 = enthalpy of liquid condensate entering trap evaluated at supply pressure for saturated condensate or at saturation pressure corresponding to temperature of subcooled liquid condensate
h_{f_2} = enthalpy of saturated liquid at return or downstream pressure of trap
h_{g_2} = enthalpy of saturated vapor at return or downstream pressure of trap
v_{f_2} = specific volume of saturated liquid at return or downstream pressure of trap
v_{g_2} = specific volume of saturated vapor at return or downstream pressure of trap

Table 22 presents some values for quality and volume fraction for typical supply and return pressures in heating and ventilating systems. Note that the percent of vapor on a mass basis x is small, while the percent of vapor on a volume basis V_c is very large. This indicates that the return pipe cross section is predominantly occupied by vapor. Figure 13 is a working chart to determine the quality of the condensate entering the return line from the trap for various combinations of supply and return pressures. If the liquid is subcooled entering the trap, the saturation pressure corresponding to the liquid temperature should be used for the supply or upstream pressure. Typical pressures in the return line are given in Table 23.

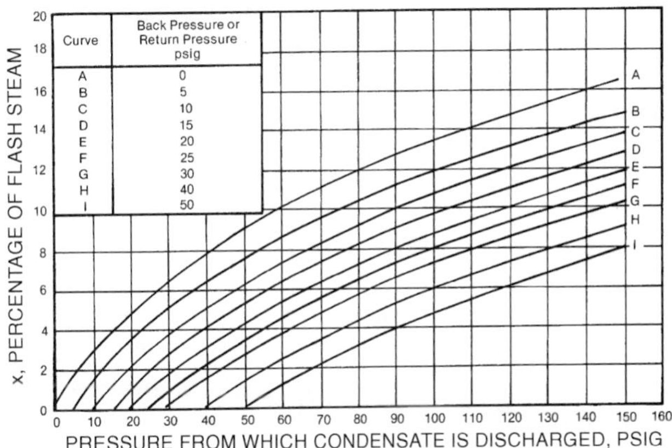

Fig. 13 Working Chart for Determining Percentage of Flash Steam (Quality)

Table 23 Estimated Return Line Pressures

Pressure Drop, psi/100 ft	Pressure in Return Line, psig	
	30 psig Supply	150 psig Supply
1/8	0.5	1.25
1/4	1	2.5
1/2	2	5
3/4	3	7.5
1	4	10
2	—	20

One-Pipe Systems

Gravity one-pipe air vent systems in which steam and condensate flow in the same pipe, frequently in opposite directions, are considered obsolete and are no longer being installed. Chapter 33 of the 1993 *ASHRAE Handbook—Fundamentals* or earlier ASHRAE Handbooks include descriptions of and design information for one-pipe systems.

GAS PIPING

Piping for gas appliances should be of adequate size and installed so that it provides a supply of gas sufficient to meet the maximum demand without undue loss of pressure between the point of supply (the meter) and the appliance. The size of gas pipe required depends on (1) maximum gas consumption to be provided, (2) length of pipe and number of fittings, (3) allowable pressure loss from the outlet of the meter to the appliance, and (4) specific gravity of the gas.

Gas consumption in ft^3/h is obtained by dividing the Btu input rate at which the appliance is operated by the average heating value of the gas in Btu/ft^3. Insufficient gas flow from excessive pressure losses in gas supply lines can cause inefficient operation of gas-fired appliances and sometimes create hazardous operations. Gas-fired appliances are normally equipped with a data plate giving information on maximum gas flow requirements or Btu input as well as inlet gas pressure requirements. The gas utility in the area of installation can give the gas pressure available at the utility's gas meter. Using the information, the required size of gas piping can be calculated for satisfactory operation of the appliance(s).

Table 24 gives pipe capacities for gas flow for up to 200 ft of pipe based on a specific gravity of 0.60. Capacities for pressures less than 1.5 psig may also be determined by the following equation from NFPA/IAS *National Fuel Gas Code*:

Table 24 Maximum Capacity of Gas Pipe in Cubic Feet per Hour

Nominal Iron Pipe Size, in.	Internal Diameter, in.	Length of Pipe, ft 10	20	30	40	50	60	70	80	90	100	125	150	175	200
1/4	0.364	32	22	18	15	14	12	11	11	10	9	8	8	7	6
3/8	0.493	72	49	40	34	30	27	25	23	22	21	18	17	15	14
1/2	0.622	132	92	73	63	56	50	46	43	40	38	34	31	28	26
3/4	0.824	278	190	152	130	115	105	96	90	84	79	72	64	59	55
1	1.049	520	350	285	245	215	195	180	170	160	150	130	120	110	100
1-1/4	1.380	1,050	730	590	500	440	400	370	350	320	305	275	250	225	210
1-1/2	1.610	1,600	1,100	890	760	670	610	560	530	490	460	410	380	350	320
2	2.067	3,050	2,100	1,650	1,450	1,270	1,150	1,050	990	930	870	780	710	650	610
2-1/2	2.469	4,800	3,300	2,700	2,300	2,000	1,850	1,700	1,600	1,500	1,400	1,250	1,130	1,050	980
3	3.068	8,500	5,900	4,700	4,100	3,600	3,250	3,000	2,800	2,600	2,500	2,200	2,000	1,850	1,700
4	4.026	17,500	12,000	9,700	8,300	7,400	6,800	6,200	5,800	5,400	5,100	4,500	4,100	3,800	3,500

Note: Capacity is in cubic feet per hour at gas pressures of 0.5 psig or less and a pressure drop of 0.3 in. of water; specific gravity = 0.60.

$$Q = 2313d^{2.623}\left(\frac{\Delta p}{CL}\right)^{0.541} \tag{17}$$

where

Q = flow rate at 60°F and 30 in. Hg, cfh
d = inside diameter of pipe, in.
Δp = pressure drop, in. of water
C = factor for viscosity, density, and temperature
= $0.00354(t + 460)s^{0.848}\mu^{0.152}$
t = temperature, °F
s = ratio of density of gas to density of air at 60°F and 30 in. Hg
μ = viscosity of gas, centipoise (0.012 for natural gas, 0.008 for propane)
L = pipe length, ft

Gas service in buildings is generally delivered in the "low-pressure" range of 7 in. of water. The maximum pressure drop allowable in piping systems at this pressure is generally 0.5 in. of water but is subject to regulation by local building, plumbing, and gas appliance codes (see also the NFPA/IAS *National Fuel Gas Code*).

Where large quantities of gas are required or where long lengths of pipe are used (e.g., in industrial buildings), low-pressure limitations result in large pipe sizes. Local codes may allow and local gas companies may deliver gas at higher pressures (e.g., 2, 5, or 10 psig). Under these conditions, an allowable pressure drop of 10% of the initial pressure is used, and pipe sizes can be reduced significantly. Gas pressure regulators at the appliance must be specified to accommodate higher inlet pressures. NFPA/IAS (1992) provides information on pipe sizing for various inlet pressures and pressure drops at higher pressures.

More complete information on gas piping can be found in the *Gas Engineers' Handbook* (1970).

FUEL OIL PIPING

The pipe used to convey fuel oil to oil-fired appliances must be large enough to maintain low pump suction pressure and, in the case of circulating loop systems, to prevent overpressure at the burner oil pump inlet. Pipe materials must be compatible with the fuel and must be carefully assembled to eliminate all leaks. Leaks in suction lines cause pumping problems that result in unreliable burner operation. Leaks in pressurized lines create fire hazards. Cast-iron or aluminum fittings and pipe are unacceptable. Pipe joint compounds must be selected carefully.

Oil pump suction lines should be sized so that at maximum suction line flow conditions, the maximum vacuum will not exceed 10 in. Hg for distillate grade fuels and 15 in. Hg for residual oils. Oil supply lines to burner oil pumps should not be pressurized by circulating loop systems or aboveground oil storage tanks to more than

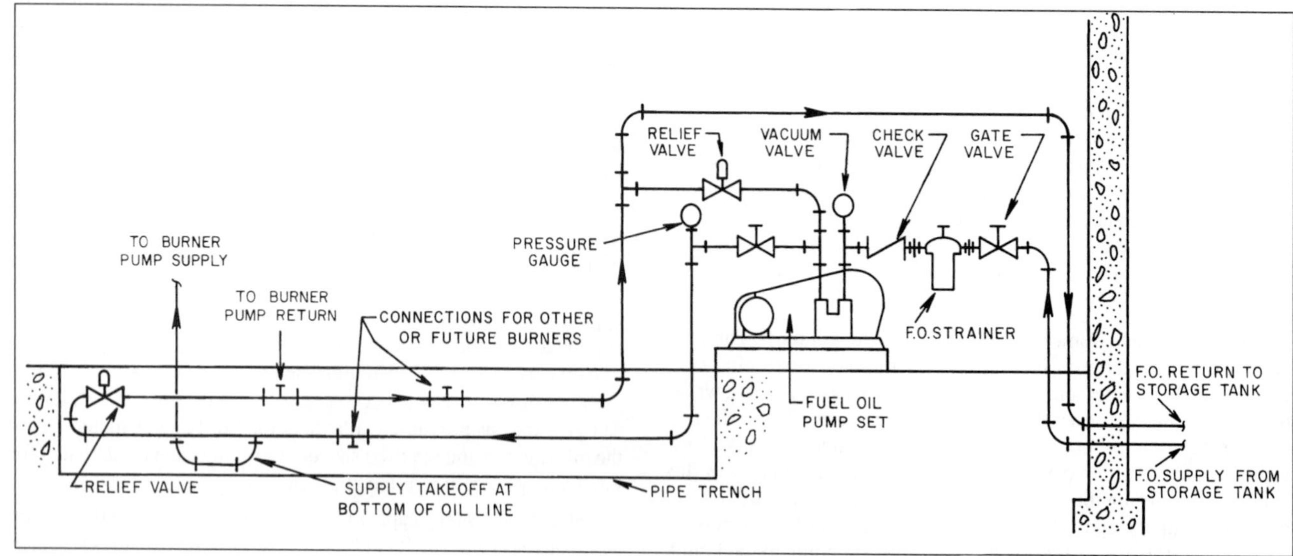

Fig. 14 Typical Oil Circulating Loop

Table 25 Recommended Nominal Size for Fuel Oil Suction Lines from Tank to Pump (Residual Grades No. 5 and No. 6)

Pumping Rate, gph	Length of Run in Feet at Maximum Suction Lift of 15 ft									
	25	50	75	100	125	150	175	200	250	300
10	1-1/2	1-1/2	1-1/2	1-1/2	1-1/2	1-1/2	2	2	2-1/2	2-1/2
40	1-1/2	1-1/2	1-1/2	2	2	2-1/2	2-1/2	2-1/2	2-1/2	3
70	1-1/2	2	2	2	2	2-1/2	2-1/2	2-1/2	3	3
100	2	2	2	2-1/2	2-1/2	3	3	3	3	3
130	2	2	2-1/2	2-1/2	2-1/2	3	3	3	3	4
160	2	2	2-1/2	2-1/2	2-1/2	3	3	3	4	4
190	2	2-1/2	2-1/2	2-1/2	3	3	3	4	4	4
220	2-1/2	2-1/2	2-1/2	3	3	3	4	4	4	4

Notes:
1. Pipe sizes smaller than 1 in. IPS are not recommended for use with residual grade fuel oils.
2. Lines conveying fuel oil from pump discharge port to burners and tank return may be reduced by one or two sizes, depending on piping length and pressure losses.

Table 26 Recommended Nominal Size for Fuel Oil Suction Lines from Tank to Pump (Distillate Grades No. 1 and No. 2)

Pumping Rate, gph	Length of Run in Feet at Maximum Suction Lift of 10 ft									
	25	50	75	100	125	150	175	200	250	300
10	1/2	1/2	1/2	1/2	1/2	1/2	1/2	3/4	3/4	1
40	1/2	1/2	1/2	1/2	1/2	3/4	3/4	3/4	3/4	1
70	1/2	1/2	3/4	3/4	3/4	3/4	3/4	1	1	1
100	1/2	3/4	3/4	3/4	3/4	1	1	1	1	1-1/4
130	1/2	3/4	3/4	1	1	1	1	1	1-1/4	1-1/4
160	3/4	3/4	3/4	1	1	1	1	1-1/4	1-1/4	1-1/4
190	3/4	3/4	1	1	1	1	1-1/4	1-1/4	1-1/4	2
220	3/4	1	1	1	1	1-1/4	1-1/4	1-1/4	1-1/4	2

5 psi, or pump shaft seals may fail. A typical oil circulating loop system is shown in Figure 14.

In assembling long fuel pipe lines, care should be taken to avoid air pockets. On overhead circulating loops, the line should vent air at all high points. Oil supply loops for one or more burners should be the continuous circulation type, with excess fuel returned to the storage tank. Dead-ended pressurized loops can be used, but air or vapor venting is more problematic.

Where valves are used, select ball or gate valves. Globe valves are not recommended because of their high pressure drop characteristics.

Oil lines should be tested after installation, particularly if they are buried, enclosed, or otherwise inaccessible. Failure to perform this test is a frequent cause of later operating difficulties. A suction line can be hydrostatically tested at 1.5 times its maximum operating pressure or at a vacuum of not less than 20 in. Hg. Pressure or vacuum tests should continue for at least 60 min. If there is no noticeable drop in the initial test pressure, the lines can be considered tight.

Pipe Sizes for Heavy Oil

Table 25 and Table 26 give recommended pipe sizes for handling No. 5 and No. 6 oils (residual grades) and No. 1 and No. 2 oils (distillate grades), respectively.

Storage tanks and piping and pumping facilities for delivering the oil from the tank to the burner are important considerations in the design of an industrial oil-burning system.

The construction and location of the tank and oil piping are usually subject to local regulations and National Fire Protection Association (NFPA) *Standards* 30 and 31.

REFERENCES

Ball, E.F. and C.J.D. Webster. 1976. Some measurements of water-flow noise in copper and ABS pipes with various flow velocities. *The Building Services Engineer* 44(2):33.

BOCA. 1992. *BOCA National plumbing code*, 9th ed. Building Officials and Code Administrators International, Country Club Hills, IL.

Carrier. 1960. Piping design. In *System design manual*. Carrier Air Conditioning Company, Syracuse, NY.

Crane Co. 1976. Flow of fluids through valves, fittings and pipe. *Technical Paper* No. 410. Crane Company, New York.

Dawson, F.M. and J.S. Bowman. 1933. Interior water supply piping for residential buildings. University of Wisconsin Experiment Station, No. 77.

Freeman, J.R. 1941. *Experiments upon the flow of water in pipes*. American Society of Mechanical Engineers, New York.

Gas engineers' handbook. 1970. The Industrial Press, New York.

Giesecke, F.E. 1926. Friction of water elbows. *ASHVE Transactions* 32:303.

Giesecke, F.E. and W.H. Badgett. 1931. Friction heads in one-inch standard cast-iron tees. *ASHVE Transactions* 37:395.

Giesecke, F.E. and W.H. Badgett. 1932a. Loss of head in copper pipe and fittings. *ASHVE Transactions* 38:529.

Giesecke, F.E. and W.H. Badgett. 1932b. Supplementary friction heads in one-inch cast-iron tees. *ASHVE Transactions* 38:111.

Grinnell Company. 1951. *Piping design and engineering*. Grinnell Company, Cranston, RI.

HDR design guide. 1981. Hennington, Durham and Richardson, Omaha, NE.

Hegberg, R.A. 1995. Where did the k-factors for pressure loss in fittings come from? *ASHRAE Transactions* 101(1): 1264-78.

Howell, R.H. 1985. Evaluation of sizing methods for steam condensate systems. *ASHRAE Transactions* 91(1).

Hunter, R.B. 1940. Methods of estimating loads in plumbing systems. NBS *Report* BMS 65. National Institute of Standards and Technology, Gaithersburg, MD.

Hunter, R.B. 1941. Water distributing systems for buildings. NBS *Report* BMS 79. National Institute of Standards and Technology.

Hydraulic Institute. 1979. *Engineering data book*. Hydraulic Institute, Parsippany, NJ.

IAPMO. 1994. *Uniform plumbing code*. International Association of Plumbing and Mechanical Officials, Walnut, CA.

Idelchik, I.E. 1986. *Handbook of hydraulic resistance*. Hemisphere Publishing Corporation, New York.

ISA. 1985. Flow equations for sizing control valves. ANSI/ISA *Standard* S75.01-85. International Society for Measurement and Control, Research Triangle Park, NC.

Laschober, R.R., G.Y. Anderson, and D.G. Barbee. 1966. Counterflow of steam and condensate in slightly pitched pipes. *ASHRAE Transactions* 72(1):157.

Marseille, B. 1965. Noise transmission in piping. *Heating and Ventilating Engineering* (June):674.

NAPHCC. 1996. *National standard plumbing code*. National Association of Plumbing-Heating-Cooling Contractors, Falls Church, VA.

NFPA. 1992. Installation of oil burning equipment. ANSI/NFPA *Standard* 31-92. National Fire Protection Association, Quincy, MA.

NFPA. 1993. Flammable and combustible liquids code. ANSI/NFPA *Standard* 30-93.

NFPA/IAS. 1992. *National fuel gas code*. ANSI/NFPA *Standard* 54-92. National Fire Protection Association, Quincy, MA. ANSI/IAS *Standard* Z223.1-92. American Gas Association, Arlington, VA.

Obrecht, M.F. and M. Pourbaix. 1967. Corrosion of metals in potable water systems. AWWA 59:977. American Water Works Association, Denver, CO.

Plastic Pipe Institute. 1971. *Water flow characteristics of thermoplastic pipe*. Plastic Pipe Institute, New York.

Rahmeyer, W.J. 1999a. Pressure loss coefficients of threaded and forged weld pipe fittings for ells, reducing ells, and pipe reducers. *ASHRAE Transactions* 105(2):334-54.

Rahmeyer, W.J. 1999b. Pressure loss coefficients of pipe fittings for threaded and forged weld pipe tees. *ASHRAE Transactions* 105(2):355-85.

Rogers, W.L. 1953. Experimental approaches to the study of noise and noise transmission in piping systems. *ASHVE Transactions* 59:347-60.

Rogers, W.L. 1954. Sound-pressure levels and frequencies produced by flow of water through pipe and fittings. *ASHRAE Transactions* 60:411-30.

Rogers, W.L. 1956. Noise production and damping in water piping. *ASHAE Transactions* 62:39.

Sanks, R.L. 1978. *Water treatment plant design for the practicing engineer.* Ann Arbor Science Publishers, Ann Arbor, MI.

SBCCI. 1994. *Standard plumbing code.* Southern Building Code Congress International, Birmingham, AL.

Smith, T. 1983. Reducing corrosion in heating plants with special reference to design considerations. *Anti-Corrosion Methods and Materials* 30 (October):4.

Stewart, W.E. and C.L. Dona. 1987. Water flow rate limitations. *ASHRAE Transactions* 93(2):811-25.

Williams, G.J. 1976. The Hunter curves revisited. *Heating/Piping/Air Conditioning* (November):67.

Williams, G.S. and A. Hazen. 1933. *Hydraulic tables.* John Wiley and Sons, New York.

CHAPTER 36

ABBREVIATIONS AND SYMBOLS

THIS CHAPTER contains information about abbreviations and symbols for heating, ventilating, air-conditioning, and refrigerating (HVAC&R) engineers.

Abbreviations are shortened forms of names and expressions used in text, drawings, and computer programs. This chapter discusses conventional English language abbreviations that may be different in other languages. A **letter symbol** represents a quantity or a unit, not its name, and is independent of language. Because of this, use of a letter symbol is preferred over abbreviations for unit or quantity terms. Letter symbols necessary for individual chapters are defined in the chapters where they occur.

Abbreviations are never used for mathematical signs, such as the equality sign (=) or division sign (/), except in computer programming, where the abbreviation functions as a letter symbol. Mathematical operations are performed only with symbols. Abbreviations should be used only where necessary to save time and space; avoid their usage in documents circulated in foreign countries.

Graphical symbols in this chapter of piping, ductwork, fittings, and in-line accessories can be used on scale drawings and diagrams.

Identifying piping by legend and color promotes greater safety and lessens the chance of error in emergencies. Piping identification is now required throughout the United States by the Occupational Safety and Health Administration (OSHA) for some industries and by many federal, state, and local codes.

ABBREVIATIONS FOR TEXT, DRAWINGS, AND COMPUTER PROGRAMS

Table 1 gives some abbreviations, as well as others commonly found on mechanical drawings and abbreviations (symbols) used in computer programming. Abbreviations specific to a single subject are defined in the chapters in which they appear. Additional abbreviations used on drawings can be found in the section on Graphical Symbols for Drawings.

The abbreviations (symbols) used for computer programming for the HVAC&R industries have been developed by ASHRAE Technical Committee 1.5, Computer Applications. These symbols identify computer variables, subprograms, subroutines, and functions commonly applied in the industry. Using these symbols enhances comprehension of the program listings and provides a clearly defined nomenclature in applicable computer programs.

Certain programming languages differentiate between real numbers (numbers with decimals) and integers (numbers without decimals) by reserving certain initial letters of a variable for integer numbers. Many of the symbols listed in this chapter begin with these letters and, in order to make them real numbers, must be prefixed with a noninteger letter.

Some symbols have two or more options listed. The longest abbreviation is preferred and should be used if possible. However, it is sometimes necessary to shorten the symbol to further identify the variable. For instance, the area of a wall cannot be defined as WALLAREA because some computer languages restrict the number of letters in a variable name. Therefore, a shorter variable symbol is applied, and WALLAREA becomes WALLA or WAREA.

Many advanced computer programming languages such as Basic, C, and C++ do not have the limitations of older computer language compilers. It is good programming practice to include the complete name of each variable and to define any abbreviations in the comments section at the beginning of each module of code. Abbreviations should be used to help clarify the variables in an equation and not to obscure the readability of the code.

In Table 1, the same symbol is sometimes used for different terms. This liberty is taken because it is unlikely that the two terms would be used in the same program. If such were the case, one of the terms would require a suffix or prefix to differentiate it from the other.

LETTER SYMBOLS

Letter symbols include symbols for physical quantities (quantity symbols) and symbols for the units in which these quantities are measured (unit symbols). **Quantity symbols**, such as *I* for electric current, are listed in this chapter and are printed in italic type. A **unit symbol** is a letter or group of letters such as ft for foot or a special sign such as ° for degrees and is printed in Roman type. Subscripts and superscripts are governed by the same principles. Letter symbols are restricted mainly to the English and Greek alphabets.

Quantity symbols may be used in mathematical expressions in any way consistent with good mathematical usage. The product of two quantities, *a* and *b*, is indicated by ab. The quotient is a/b, or ab^{-1}. To avoid misinterpretation, parentheses must be used if more than one slash (/) is employed in an algebraic term; for example, $(a/b)/c$ or $a/(b/c)$ is correct, but not $a/b/c$.

Subscripts and superscripts, or several of them separated by commas, may be attached to a single basic letter (kernel), but not to other subscripts or superscripts. A symbol that has been modified by a superscript should be enclosed in parentheses before an exponent is added $(X_a)^3$. Symbols can also have alphanumeric marks such as ′ (prime), + (plus), and * (asterisk).

More detailed information on the general principles of letter symbol standardization are in standards listed at the end of this chapter. The letter symbols, in general, follow these standards, which are out of print:

Y10.3M	Letter Symbols for Mechanics and Time-Related Phenomena
Y10.4-82	Letter Symbols for Heat and Thermodynamics

Other symbols chosen by an author for a physical magnitude not appearing in any standard list should be ones that do not already have different meanings in the field of the text.

The preparation of this chapter is assigned to TC 1.6, Terminology.

Table 1 Abbreviations for Text, Drawings, and Computer Programs

Term	Text	Drawings	Program
above finished floor	—	AFF	—
absolute	abs	ABS	ABS
accumulat(e, -or)	acc	ACCUM	ACCUM
air condition(-ing, -ed)	—	AIR COND	—
air-conditioning unit(s)	—	ACU	ACU
air-handling unit	—	AHU	AHU
air horsepower	ahp	AHP	AHP
alteration	altrn	ALTRN	—
alternating current	ac	AC	AC
altitude	alt	ALT	ALT
ambient	amb	AMB	AMB
American National Standards Institute[1]	ANSI	ANSI	—
American wire gage	AWG	AWG	—
ampere (amp, amps)	amp	AMP	AMP, AMPS
angle	—	—	ANG
angle of incidence	—	—	ANGI
apparatus dew point	adp	ADP	ADP
approximate	approx.	APPROX	—
area	—	—	A
atmosphere	atm	ATM	—
average	avg	AVG	AVG
azimuth	az	AZ	AZ
azimuth, solar	—	—	SAZ
azimuth, wall	—	—	WAZ
barometer(-tric)	baro	BARO	—
bill of material	b/m	BOM	—
boiling point	bp	BP	BP
brake horsepower	bhp	BHP	BHP
Brown & Sharpe wire gage	B&S	B&S	—
British thermal unit	Btu	BTU	BTU
Celsius	°C	°C	°C
center to center	c to c	C TO C	—
circuit	ckt	CKT	CKT
clockwise	cw	CW	—
coefficient	coeff.	COEF	COEF
coefficient, valve flow	C_v	C_v	CV
coil	—	—	COIL
compressor	cprsr	CMPR	CMPR
condens(-er, -ing, -ation)	cond	COND	COND
conductance	—	—	C
conductivity	cndct	CNDCT	K
conductors, number of (3)	3/c	3/c	—
contact factor	—	—	CF
cooling load	clg load	CLG LOAD	CLOAD
counterclockwise	ccw	CCW	—
cubic feet	ft^3	CU FT	CUFT, CFT
cubic inch	in^3	CU IN	CUIN, CIN
cubic feet per minute	cfm	CFM	CFM
cfm, standard conditions	scfm	SCFM	SCFM
cubic ft per sec, standard	scfs	SCFS	SCFS
decibel	dB	DB	DB
degree	deg. or °	DEG or °	DEG
density	dens	DENS	RHO
depth or deep	dp	DP	DPTH
dew-point temperature	dpt	DPT	DPT
diameter	dia.	DIA	DIA
diameter, inside	ID	ID	ID
diameter, outside	OD	OD	OD
difference or delta	diff., Δ	DIFF	D, DELTA
diffuse radiation	—		DFRAD
direct current	dc	DC	DC
direct radiation	dir radn	DIR RADN	DIRAD
dry	—		DRY
dry-bulb temperature	dbt	DBT	DB, DBT
effectiveness	—		EFT
effective temperature[2]	ET*	ET*	ET
efficiency	eff	EFF	EFF
efficiency, fin	—		FEFF
efficiency, surface	—		SEFF
electromotive force	emf	EMF	—
elevation	elev.	EL	ELEV
entering	entr	ENT	ENT
entering water temperature	EWT	EWT	EWT
entering air temperature	EAT	EAT	EAT
enthalpy	—	—	H
entropy	—	—	S
equivalent direct radiation	edr	EDR	—
equivalent feet	eqiv ft	EQIV FT	EQFT
equivalent inches	eqiv in	EQIV IN	EQIN
evaporat(-e, -ing, -ed, -or)	evap	EVAP	EVAP
expansion	exp	EXP	XPAN
face area	fa	FA	FA
face to face	f to f	F to F	—
face velocity	fvel	FVEL	FV
factor, correction	—	—	CFAC, CFACT
factor, friction	—	—	FFACT, FF
Fahrenheit	°F	°F	F
fan	—	—	FAN
feet per minute	fpm	FPM	FPM
feet per second	fps	FPS	FPS
film coefficient,[3] inside	—	—	FI, HI
film coefficient,[3] outside	—	—	FO, HO
flow rate, air	—	—	QAR, QAIR
flow rate, fluid	—	—	QFL
flow rate, gas	—	—	QGA, QGAS
foot or feet	ft	FT	FT
foot-pound	ft·lb	FT LB	—
freezing point	fp	FP	FP
frequency	Hz	HZ	—
gage or gauge	ga	GA	GA, GAGE
gallons	gal	GAL	GAL
gallons per hour	gph	GPH	GPH
gallons per minute	gpm	GPM	GPM
gallons per day	gpd	GPD	GPD
grains	gr	GR	GR
gravitational constant	g	G	G
greatest temp difference	GTD	GTD	GTD
head	hd	HD	HD
heat	—	—	HT
heater	—	—	HTR
heat gain	HG	HG	HG, HEATG
heat gain, latent	LHG	LHG	HGL
heat gain, sensible	SHG	SHG	HGS
heat loss	—	—	HL, HEATL
heat transfer	—	—	Q
heat transfer coefficient	U	U	U
height	hgt	HGT	HGT, HT
high-pressure steam	hps	HPS	HPS
high-temperature hot water	hthw	HTHW	HTHW
horsepower	hp	HP	HP
hour(s)	h	HR	HR
humidity, relative	rh	RH	RH
humidity ratio	W	W	W
incident angle	—	—	INANG
indicated horsepower	ihp	IHP	—
International Pipe Std	IPS	IPS	—
iron pipe size	ips	IPS	—
kelvin	K	K	K
kilowatt	kW	kW	KW
kilowatt hour	kWh	KWH	KWH
latent heat	LH	LH	LH, LHEAT
least mean temp. difference[4]	LMTD	LMTD	LMTD
least temp. difference[4]	LTD	LTD	LTD
leaving air temperature	lat	LAT	LAT
leaving water temperature	lwt	LWT	LWT
length	lg	LG	LG, L
linear feet	lin ft	LF	LF
liquid	liq	LIQ	LIQ
logarithm (natural)	ln	LN	LN
logarithm to base 10	log	LOG	LOG
low-pressure steam	lps	LPS	LPS
low-temp. hot water	lthw	LTHW	LTHW

Term	Text	Drawings	Program
Mach number	Mach	MACH	—
mass flow rate	mfr	MFR	MFR
maximum	max.	MAX	MAX
mean effective temp.	MET	MET	MET
mean temp. difference	MTD	MTD	MTD
medium-pressure steam	mps	MPS	MPS
medium-temp. hot water	mthw	MTHW	MTHW
mercury	Hg	HG	HG
miles per hour	mph	MPH	MPH
minimum	min.	MIN	MIN
minute	min	MIN	MIN
noise criteria	NC	NC	—
normally open	n o	N O	—
normally closed	n c	N C	—
not applicable	na	N/A	—
not in contract	n i c	N I C	—
not to scale	—	N T S	—
number	no.	NO	N, NO
number of circuits	—	—	NC
number of tubes	—	—	NT
ounce	oz	OZ	OZ
outside air	oa	OA	OA
parts per million	ppm	PPM	PPM
percent	%	%	PCT
phase (electrical)	ph	PH	—
pipe	—	—	PIPE
pounds	lb	LBS	LBS
pounds per square foot	psf	PSF	PSF
psf absolute	psfa	PSFA	PSFA
psf gage	psfg	PSFG	PSFG
pounds per square inch	psi	PSI	PSI
psi absolute	psia	PSIA	PSIA
psi gage	psig	PSIG	PSIG
pressure	—	PRESS	PRES, P
pressure, barometric	baro pr	BARO PR	BP
critical pressure	—	—	CRIP
pressure, dynamic (velocity)	vp	VP	VP
pressure drop or difference	PD	PD	PD, DELTP
pressure, static	sp	SP	SP
pressure, vapor	vap pr	VAP PR	VAP
primary	pri	PRI	PRIM
quart	qt	QT	QT
radian	—	—	RAD
radiat(-e, -or)	—	RAD	—
radiation	—	RADN	RAD
radius	—	—	R
Rankine	°R	°R	R
receiver	rcvr	RCVR	REC
recirculate	recirc.	RECIRC	RCIR, RECIR
refrigerant (12, 22, etc.)	R-12, R-22	R12, R22	R12, R22
relative humidity	rh	RH	RH
resist(-ance, -ivity, -or)	res	RES	RES, OHMS
return air	ra	RA	RA
revolutions	rev	REV	REV
revolutions per minute	rpm	RPM	RPM
revolutions per second	rps	RPS	RPS
roughness	rgh	RGH	RGH, E
safety factor	sf	SF	SF
saturation	sat.	SAT	SAT
Saybolt seconds Furol	ssf	SSF	SSF
Saybolt seconds Universal	ssu	SSU	SSU
sea level	sl	SL	SE
second	s	s	SEC
sensible heat	SH	SH	SH
sensible heat gain	SHG	SHG	SHG
sensible heat ratio	SHR	SHR	SHR
shading coefficient	—	—	SC
shaft horsepower	sft hp	SFT HP	SHP
solar	—	—	SOL
specification	spec	SPEC	—
specific gravity	SG	SG	—
specific heat	sp ht	SP HT	C

Term	Text	Drawings	Program
sp ht at constant pressure	c_p	c_p	CP
sp ht at constant volume	c_v	c_v	CV
specific volume	sp vol	SP VOL	V, CVOL
square	sq.	SQ	SQ
standard	std	STD	STD
standard time meridian	—	—	STM
static pressure	SP	SP	SP
suction	suct.	SUCT	SUCT, SUC
summ(-er, -ary, -ation)	—	—	SUM
supply	sply	SPLY	SUP, SPLY
supply air	sa	SA	SA
surface	—	—	SUR, S
surface, dry	—	—	SURD
surface, wet	—	—	SURW
system	—	—	SYS
tabulat(-e, -ion)	tab	TAB	TAB
tee	—	—	TEE
temperature	temp.	TEMP	T, TEMP
temperature difference	TD, Δt	TD	TD, TDIF
temperature entering	TE	TE	TE, TENT
temperature leaving	TL	TL	TL, TLEA
thermal conductivity	k	K	K
thermal expansion coeff.	—	—	TXPC
thermal resistance	R	R	RES, R
thermocouple	tc	TC	TC, TCPL
thermostat	T STAT	T STAT	T STAT
thick(-ness)	thkns	THKNS	THK
thousand circular mils	Mcm	MCM	MCM
thousand cubic feet	Mcf	MCF	MCF
thousand foot-pounds	kip ft	KIP FT	KIPFT
thousand pounds	kip	KIP	KIP
time	—	T	T
ton	—	—	TON
tons of refrigeration	tons	TONS	TONS
total	—	—	TOT
total heat	tot ht	TOT HT	—
transmissivity	—	—	TAU
U-factor	—	—	U
unit	—	—	UNIT
vacuum	vac	VAC	VAC
valve	v	V	VLV
vapor proof	vap prf	VAP PRF	—
variable	var	VAR	VAR
variable air volume	VAV	VAV	VAV
velocity	vel.	VEL	VEL, V
velocity, wind	w vel.	W VEL	W VEL
ventilation, vent	vent	VENT	VENT
vertical	vert.	VERT	VERT
viscosity	visc	VISC	MU, VISC
volt	V	V	E, VOLTS
volt ampere	VA	VA	VA
volume	vol.	VOL	VOL
volumetric flow rate	—	—	VFR
wall	—	—	W, WAL
water	—	—	WTR
watt	W	W	WAT, W
watt-hour	Wh	WH	WHR
weight	wt	WT	WT
wet bulb	wb	WB	WB
wet-bulb temperature	wbt	WBT	WBT
width	—	—	WI
wind	—	—	WD
wind direction	wdir	WDIR	WDIR
wind pressure	wpr	WPR	WP, WPRES
yard	yd	YD	YD
year	yr	YR	YR
zone	z	Z	Z, ZN

[1]Abbreviations of most proper names use capital letters in both text and drawings.

[2]The asterisk (*) is used with ET*, effective temperature, as in Chapter 8 of this volume.

[3]These are surface heat transfer coefficients.

[4]Letter L also used for *Logarithm of* these temperature differences in computer programming.

LETTER SYMBOLS

Symbol	Description of Item	Typical Units
a	acoustic velocity	fps or fpm
A	area	ft^2
b	breadth or width	ft
B	barometric pressure	psia or in. Hg
c	concentration	lb/ft^3, mol/ft^3
c	specific heat	Btu/lb·°F
c_p	specific heat at constant pressure	Btu/lb·°F
c_v	specific heat at constant volume	Btu/lb·°F
C	coefficient	—
C	fluid capacity rate	Btu/h·°F
C	thermal conductance	Btu/h·ft^2·°F
C_L	loss coefficient	—
C_P	coefficient of performance	—
d	prefix meaning differential	—
d or D	diameter	ft
D_e or D_h	equivalent or hydraulic diameter	ft
D_v	mass diffusivity	ft^2/s
e	base of natural logarithms	—
E	energy	Btu
E	electrical potential	V
f	film conductance (alternate for h)	Btu/h·ft^2·°F
f	frequency	Hz
f_D	friction factor, Darcy-Weisbach formulation	—
f_F	friction factor, Fanning formulation	—
F	force	lb$_f$
F_{ij}	angle factor (radiation)	—
g	gravitational acceleration	ft/s^2
G	mass velocity	lb/h·ft^2
h	heat transfer coefficient	Btu/h·ft^2·°F
h	hydraulic head	ft
h	specific enthalpy	Btu/lb
h_a	enthalpy of dry air	Btu/lb
h_D	mass transfer coefficient	lb/h·ft^2·lb per ft^3
h_s	enthalpy of moist air at saturation	Btu/lb
H	total enthalpy	Btu
I	electric current	A
J	mechanical equivalent of heat	ft·lb$_f$/Btu
k	thermal conductivity	Btu/h·ft·°F
k (or γ)	ratio of specific heats, c_p/c_v	—
K	proportionality constant	—
K_D	mass transfer coefficient	lb/h·ft^2
l or L	length	ft
L_p	sound pressure	dB
L_w	sound power	dB
m or M	mass	lb
M	molecular weight	lb/lb mol
n or N	number in general	—
N	rate of rotation	rpm
p or P	pressure	psi
p_a	partial pressure of dry air	psi
p_s	partial pressure of water vapor in moist air	psi
p_w	vapor pressure of water in saturated moist air	psi
P	power	hp, watts
q	time rate of heat transfer	Btu/h
Q	total heat transfer	Btu
Q	volumetric flow rate	cfm
r	radius	ft
r or R	thermal resistance	ft^2·h·°F/Btu
R	gas constant	ft·lb$_f$/lb$_m$·°R
s	specific entropy	Btu/lb·°R
S	total entropy	Btu/°R
t	temperature	°F
Δt_m or ΔT_m	mean temperature difference	°F
T	absolute temperature	°R
u	specific internal energy	Btu/lb
U	total internal energy	Btu
U	overall heat transfer coefficient	Btu/h·ft^2·°F
v	specific volume	ft^3/lb
V	total volume	ft^3
V	linear velocity	fps
w	mass rate of flow	lb/h
W	weight	lb$_f$
W	humidity ratio of moist air	lb (water)/lb (dry air)
W	work	ft·lb$_f$
W_s	humidity ratio of moist air at saturation	lb (water)/lb (dry air)
x	mole fraction	—
x	quality, mass fraction of vapor	—
x,y,z	lengths along principal coordinate axes	ft
Z	figure of merit	—
α	absolute Seebeck coefficient	V/°C
α	absorptivity, absorptance radiation	—
α	linear coefficient of thermal expansion	per °F
α	thermal diffusivity	ft^2/h
β	volume coefficient of thermal expansion	per °F
γ (or k)	ratio of specific heats, c_p/c_v	—
γ	specific weight	lb$_f$/ft^3
Δ	difference between values	—
ε	emissivity, emittance (radiation)	—
θ	time	s, h
η	efficiency or effectiveness	—
λ	wavelength	nm
μ	degree of saturation	—
μ	dynamic viscosity	lb/ft·h
ν	kinematic viscosity	ft^2/h
ρ	density	lb/ft^3
ρ	reflectivity, reflectance (radiation)	—
ρ	volume resistivity	Ω·cm
σ	Stefan-Boltzmann constant	Btu/h·ft^2·°R^4
σ	surface tension	lb$_f$/ft
τ	stress	lb$_f$/ft^2
τ	time	s, h
τ	transmissivity, transmittance (radiation)	—
ϕ	relative humidity	—

DIMENSIONLESS NUMBERS

Fo	Fourier number	$\alpha\tau/L^2$
Gr	Grashof number	$L^3\rho^2\beta g(\Delta t)/\mu^2$
Gz	Graetz number	wc_p/kL
j_D	Colburn mass transfer	$\mathrm{Sh}/\mathrm{Re}\,\mathrm{Sc}^{1/3}$
j_H	Colburn heat transfer	$\mathrm{Nu}/\mathrm{Re}\,\mathrm{Pr}^{1/3}$
Le	Lewis number	α/D_v
M	Mach number	V/a
Nu	Nusselt number	hD/k
Pe	Peclet number	GDc_p/k
Pr	Prandtl number	$c_p\mu/k$
Re	Reynolds number	$\rho VD/\mu$
Sc	Schmidt number	$\mu/\rho D_v$
Sh	Sherwood number	h_DL/D_v
St	Stanton number	h/Gc_p
Str	Strouhal number	fd/V

MATHEMATICAL SYMBOLS

equal to	$=$
not equal to	$\neq$
approximately equal to	$\approx$
greater than	$>$
less than	$<$
greater than or equal to	$\geq$
less than or equal to	$\leq$
plus	$+$
minus	$-$
plus or minus	$\pm$
a multiplied by b	$ab,\ a\cdot b,\ a\times b$
a divided by b	$\frac{a}{b},\ a/b,\ ab^{-1}$
ratio of the circumference of a circle to its diameter	π

a raised to the power n	a^n
square root of a	$\sqrt{a}$, $a^{0.5}$
infinity	∞
percent	%
summation of	Σ
natural log	ln
logarithm to base 10	log

SUBSCRIPTS

These are to be affixed to the appropriate symbols. Several subscripts may be used together to denote combinations of various states, points, or paths. Often the subscript indicates that a particular property is to be kept constant in a process.

a,b,...	referring to different phases, states or physical conditions of a substance, or to different substances
a	air
a	ambient
b	barometric (pressure)
c	referring to critical state or critical value
c	convection
db	dry bulb
dp	dew point
e	base of natural logarithms
f	referring to saturated liquid
f	film
fg	referring to evaporation or condensation
F	friction
g	referring to saturated vapor
h	referring to change of phase in evaporation
H	water vapor
i	referring to saturated solid
i	internal
if	referring to change of phase in melting
ig	referring to change of phase in sublimation
k	kinetic
L	latent
m	mean value
M	molar basis
o	referring to initial or standard states or conditions
p	referring to constant pressure conditions or processes
p	potential
r	refrigerant
r	radiant or radiation
s	referring to moist air at saturation
s	sensible
s	referring to isentropic conditions or processes
s	static (pressure)
s	surface
t	total (pressure)
T	referring to isothermal conditions or processes
v	referring to constant volume conditions or processes
v	vapor
v	velocity (pressure)
w	wall
w	water
wb	wet bulb
1,2,...	different points in a process, or different instants of time

GRAPHICAL SYMBOLS FOR DRAWINGS

Graphical symbols have been extracted from ASME *Standard* Y32.2.3 and ASME *Standard* Y32.2.4. Some of these symbols have been modified, and others have been added to reflect current practice. Symbols and quotations are used with permission of the publisher, the American Society of Mechanical Engineers.

Piping

Heating

High-pressure steam	——HPS——
Medium-pressure steam	——MPS——
Low-pressure steam	——LPS——
High-pressure condensate	——HPC——
Medium-pressure condensate	——MPC——
Low-pressure condensate	——LPC——
Boiler blowdown	——BBD——
Pumped condensate	——PC——
Vacuum pump discharge	——VPD——
Makeup water	——MU——
Atmospheric vent	——ATV——
Fuel oil discharge	——FOD——
Fuel oil gage	——FOG——
Fuel oil suction	——FOS——
Fuel oil return	——FOR——
Fuel oil tank vent	——FOV——
Low-temperature hot water supply	——HWS——
Medium-temperature hot water supply	——MTWS——
High-temperature hot water supply	——HTWS——
Low-temperature hot water return	——HWR——
Medium-temperature hot water return	——MTWR——
High-temperature hot water return	——HTWR——
Compressed air	——A——
Vacuum (air)	——VAC——
Existing piping	——(NAME)E——
Pipe to be removed	-XX- (NAME) -XX-

Air Conditioning and Refrigeration

Refrigerant discharge	——RD——
Refrigerant suction	——RS——
Brine supply	——B——
Brine return	——BR——
Condenser water supply	——C——
Condenser water return	——CR——
Chilled water supply	——CWS——
Chilled water return	——CWR——
Fill line	——FILL——
Humidification line	——H——
Drain	——D——
Hot/chilled water supply	——HCS——
Hot/chilled water return	——HCR——
Refrigerant liquid	——RL——
Heat pump water supply	——HPWS——
Heat pump water return	——HPWR——

Plumbing

Sanitary drain above floor or grade	——SAN——
Sanitary drain below floor or grade	----SAN----
Storm drain above floor or grade	——ST——
Storm drain below floor or grade	----ST----
Condensate drain above floor or grade	——CD——
Condensate drain below floor or grade	----CD----
Vent	----------
Cold water	-·-·-·-·-·-
Hot water	-··-··-··-
Hot water return	-···-···-···
Gas	—G————G
Acid waste	——ACID——
Drinking water supply	——DWS——
Drinking water return	——DWR——
Vacuum (air)	——VAC——
Compressed air	——A——
Chemical supply pipes[a]	——(NAME)——
Floor drain	——⊡D——
Funnel drain, open	Y

Fire Safety Devices[b]

Signal Initiating Detectors

Heat (thermal)

Smoke

Gas

Flame

Valves

Valves for Selective Actuators

Air line

Ball

Butterfly

Diaphragm

Gate

Gate, angle

Globe

Globe, angle

Plug valve

Three way

Valves Actuators

Manual

Non-rising sun

Outside stem & yoke

Lever

Gear

G

Electric

Motor

M

Solenoid

S

Pneumatic

Motor

M

Diaphragm

D

Valves, Special Duty

Check, swing gate

Check, spring

S

Control, electric-pneumatic

EP

Control, pneumatic-electric

PE

Hose end drain

Lock shield

Needle

Pressure reducing (number and specify)

PRV-1

Quick opening

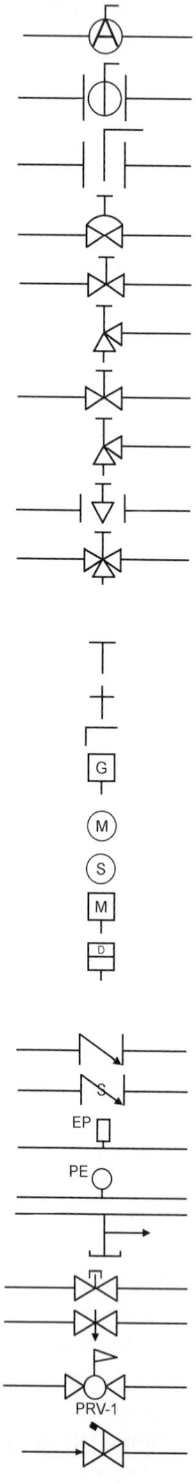

Quick closing, fusible link

Relief (R) or safety (S)

Solenoid

S

Square head cock

Unclassified (number and specify)

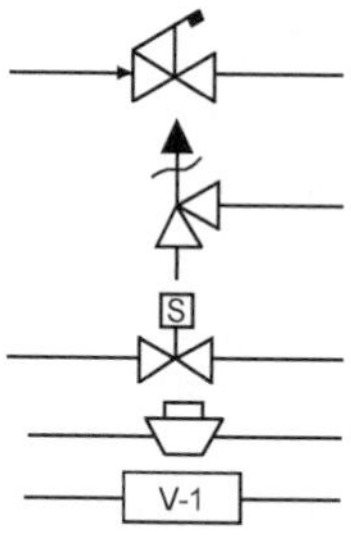

V-1

Fittings

The following fittings are shown with screwed connections. The symbol for the body of a fitting is the same for all types of connections, unless otherwise specified. The types of connections are often specified for a range of pipe sizes, but are shown with the fitting symbol where required. For example, an elbow would be:

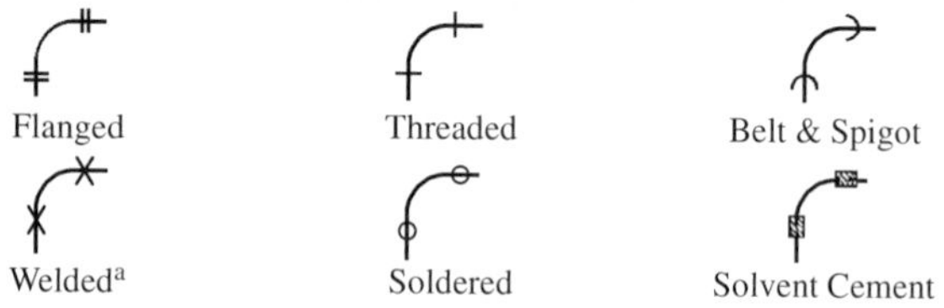

Flanged

Threaded

Belt & Spigot

Welded[a]

Soldered

Solvent Cement

Fitting	*Symbol*
Bushing	
Cap	
Connection, bottom	
Connection, top	
Coupling (joint)	
Cross	
Elbow, 90°	
Elbow, 45°	
Elbow, turned up	
Elbow, turned down	
Elbow, reducing (show sizes)	2, 4
Elbow, base	
Elbow, long radius	LR
Elbow, double branch	
Elbow, side outlet, outlet up	
Elbow, side outlet, outlet Down	
Lateral	
Reducer, concentric	
Reducer, eccentric straight invert	
Reducer, eccentric straight crown	
Tee	
Tee, outlet up	
Tee, outlet down	
Tee, reducing (show sizes)	2, 6, 4
Tee, side outlet, outlet up	

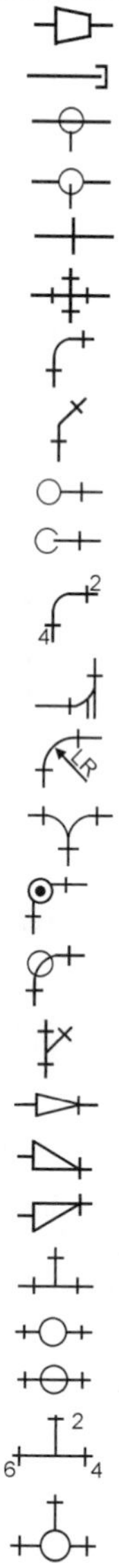

[a] See section on Piping Identification in this chapter.

[b] Refer to *Standard for Fire Safety Symbols*, 1999 Edition (NFPA *Standard* 170).

[a] Includes fusion, specify type.

Tee, side outlet, outlet down	
Tee, single sweep	
Union, screwed	
Union, flanged	

Piping Specialties

Air vent, automatic	AV
Air vent, manual	MV
Air separator	S
Alignment guide	
Anchor, intermediate	
Anchor, main	
Ball joint	
Expansion joint	EJ-1
Expansion loop	
Flexible connector	
Flowmeter, orifice	OFM-1
Flowmeter, venturi	VFM-1
Flow switch	FS
Hanger rod	H
Hanger spring	H
Heat exchanger, liquid	
Heat transfer surface (indicate type)	RAD-1
Pitch of pipe, rise (R) drop (D)	R
Pressure gage and cock	
Pressure switch	PS
Pump (indicate use)	CW-1
Pump suction diffuser	PSD
Spool piece, flanged	
Strainer	
Strainer, blow off	
Strainer, duplex	
Tank (indicate use)	FO
Thermometer	
Thermometer well, only	TW
Thermostat, electric	T
Thermostat, pneumatic	T
Thermostat, self-contained	T
Traps, steam (indicate type)	F&T

Unit heater (indicate type)	UH

Air Moving Devices and Components

***Fans* (indicate use)**[a]

Axial flow	R 1,2
Centrifugal	S 1,2
Propeller	E 1,2
Roof ventilator, intake	SRV-1
Roof ventilator, exhaust	ERV-1
Roof ventilator, louvered	

Ductwork[b]

Direction of flow	
Duct size, first figure is side down	12/20
Duct section, positive pressure, first figure is top	20/12
Duct section, negative pressure	20/12
Change of elevation rise (R) drop (D)	R
Access doors, vertical or horizontal	AD 10/10
Acoustical lining (insulation)	
Cowl, (gooseneck) and flashing	
Flexible connection	
Flexible duct	
Sound attenuator	SA
Terminal unit, mixing	H C TU M-1
Terminal unit, reheat	TU RH-1
Terminal unit, variable volume	TU VAV-1
Transition[a]	20/10 15/8
Turning vanes	
Detectors, fire and/or smoke	

Dampers

Backdraft damper	BDD
Pneumatic operated damper	POD

[a] Units of measurement are not shown herein, but should be shown on drawings. The first of the two dimensions on ducts indicates the side of the duct showing; on duct sections, the top; on grilles and registers, the horizontal edge.

[b] Adapted from SMACNA, Symbols for Ventilation and Air Conditioning Figure 4.2. *HVAC Duct System Design.*

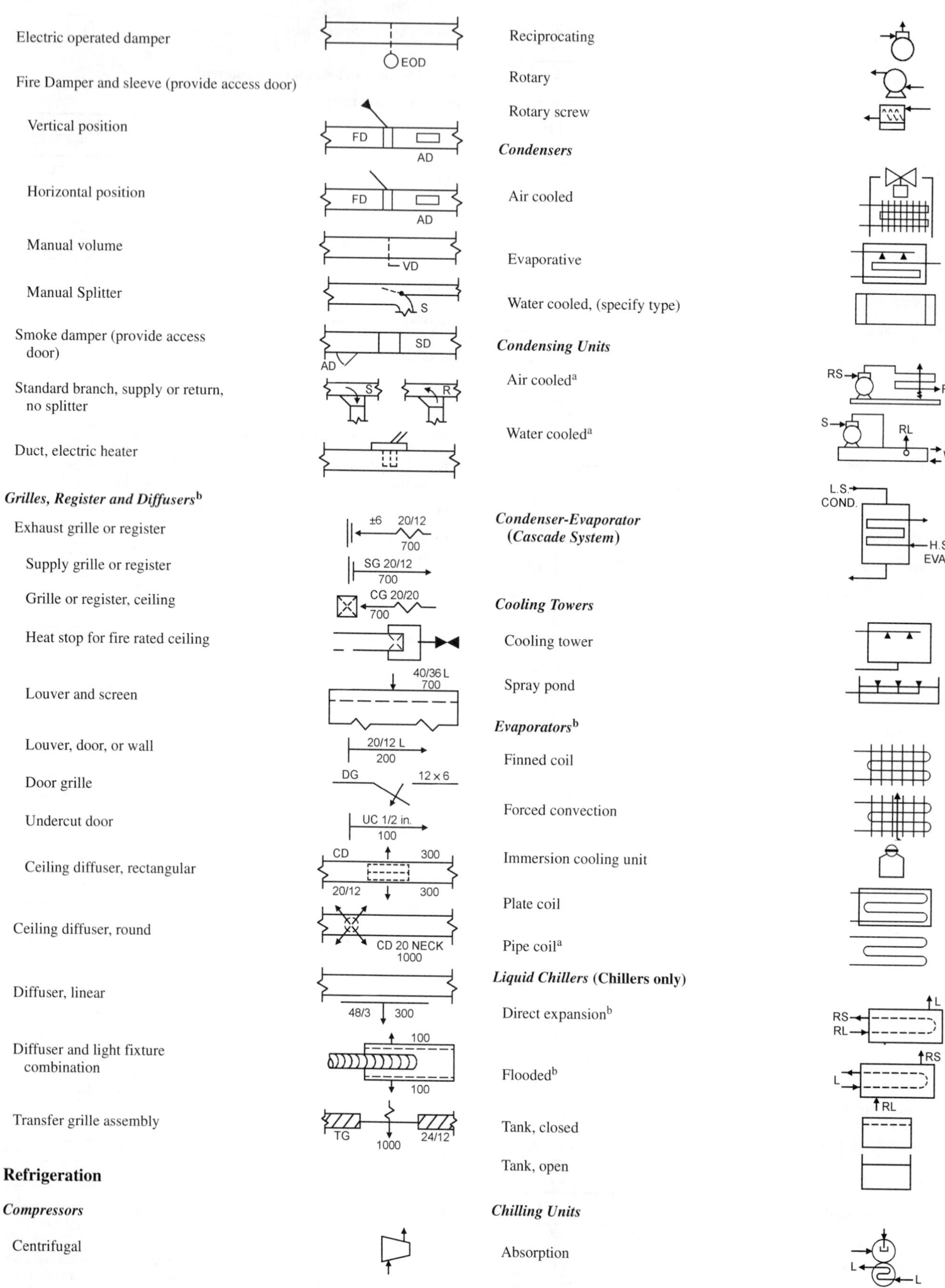

[a] Indicate flat on bottom or top (FOB or FOT), if applicable.
[b] Show volumetric flow rate at each device.

[a] L = Liquid being cooled, RL = Refrigerant liquid, RS = Refrigerant suction.
[b] Specify manifolding.

Centrifugal

Reciprocating

Rotary screw

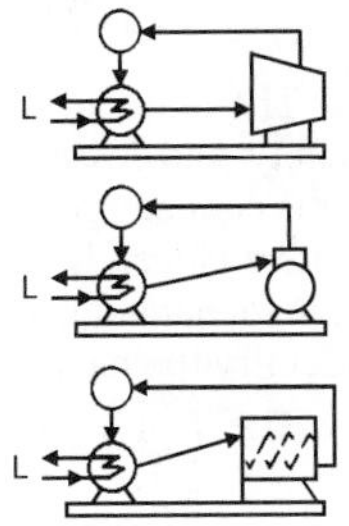

Controls

Refrigerant Controls

Capillary tube

Expansion valve, hand

Expansion valve, automatic

Expansion valve, thermostatic

Float valve, high side

Float valve, low side

Thermal bulb

Solenoid valve

Constant pressure valve, suction

Evaporator pressure regulating valve, thermostatic, throttling

Evaporator pressure regulating valve, thermostatic, snap-action

Evaporator pressure regulating valve, throttling-type, evaporator side

Compressor suction valve, pressure-limiting, throttling-type, compressor side

Thermosuction valve

Snap-action valve

Refrigerant reversing valve

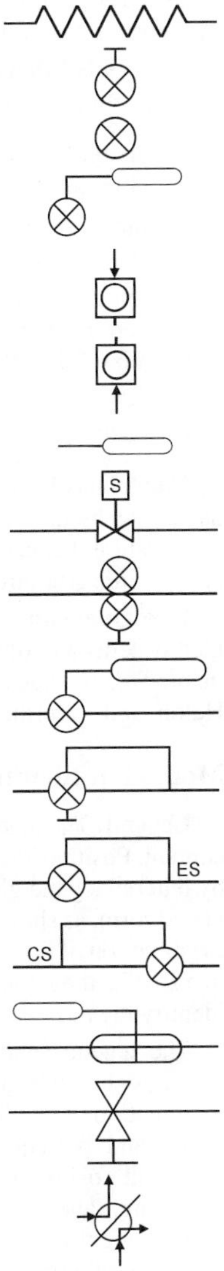

Temperature or Temperature-Actuated Electrical or Flow Controls

Thermostat, self-contained

Thermostat, Remote Bulb

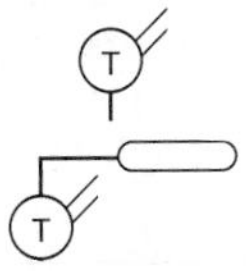

Pressure or Pressure-Actuated Electrical or Flow Controls

Pressure switch

Pressure switch, dual (high-low)

Pressure switch, differential oil pressure

Valve, automatic reducing

Valve, automatic bypass

Valve, pressure-reducing

Valve, condenser water regulating

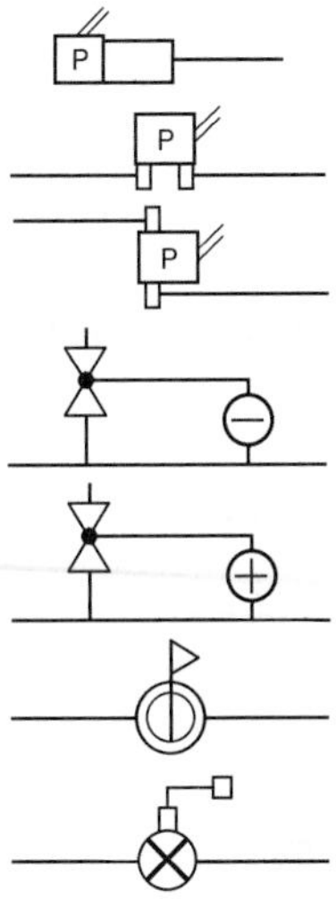

Auxiliary Equipment

Refrigerant

Filter

Strainer

Filter and drier

Scale trap

Drier

Vibration absorber

Heat exchanger

Oil separator

Sight glass

Fusible plug

Rupture disk

Receiver, high pressure, horizontal

Receiver, high-pressure, vertical

Receiver, low-pressure

Intercooler

Intercooler/desuperheater

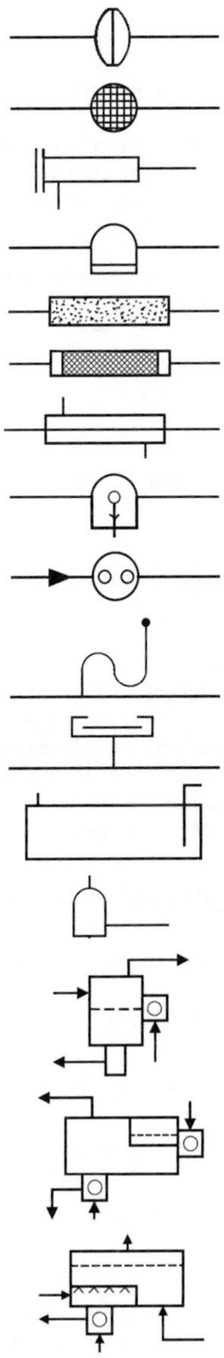

[a] Frequently used diagrammatically as evaporator and/or condenser with label indicating name and type.

[b] L = Liquid being cooled, RL = Refrigerant liquid, RS = Refrigerant suction.

Energy Recovery Equipment

Condenser, double bundle

Air to Air Energy Recovery

Rotary heat wheel — C, H

Coil loop — C, H

Heat pipe — C, H

Fixed plate — C, H

Plate fin, crossflow — C, H

Power Sources

Motor, electric (number for identification of description in specifications) — M-1

Engine (indicate fuel) — D

Gas turbine — G

Steam turbine — S

Steam turbine, condensing — S

Electrical Equipment[a]

Symbols for electrical equipment shown on mechanical drawings are usually geometric figures with an appropriate name or abbreviation, with details described in the specifications. The following are some common examples.[b]

Motor control — MC

Disconnect switch, unfused — U DS

Disconnect switch, fused — DSF

Time clock — TC

Automatic filter panel — AFP

Lighting panel — LP

Power panel — PP

[a] See ARI *Standard* 130 for preferred symbols of common electrical parts.

[b] Number each symbol if more than one; see ASME *Standard* Y32.4.

PIPING SYSTEM IDENTIFICATION

The material in piping systems is identified to promote greater safety and lessen the chances of error, confusion, or inaction in times of emergency. Primary identification should be by means of a lettered legend naming the material conveyed by the piping. In addition to, but not instead of lettered identification, color can be used to identify the hazards or use of the material.

The data in this section have been extracted from ASME *Standard* A13.1.

Definitions

Piping Systems. Piping systems include pipes of any kind, fittings, valves, and pipe coverings. Supports, brackets, and other accessories are not included. Pipes are defined as conduits for the transport of gases, liquids, semiliquids, or fine particulate dust.

Materials Inherently Hazardous to Life and Property. There are four categories of hazardous materials:

- Flammable or explosive materials that are easily ignited, including materials known as fire producers or explosives
- Chemically active or toxic materials that are corrosive or are in themselves toxic or productive of poisonous gases
- Materials at extreme temperatures or pressures that, when released from the piping, cause a sudden outburst with the potential for inflicting injury or property damage by burns, impingement, or flashing to vapor state
- Radioactive materials that emit ionizing radiation

Materials of Inherently Low Hazard. All materials that are not hazardous by nature, and are near enough to ambient pressure and temperature that people working on systems carrying these materials run little risk through their release.

Fire Quenching Materials. This classification includes sprinkler systems and other piped fire fighting or fire protection equipment. This includes water (for fire fighting), chemical foam, CO_2, Halon, and so forth.

Method of Identification

Legend. The legend is the primary and explicit identification of content. Positive identification of the content of the piping system is by lettered legend giving the name of the contents, in full or abbreviated form, as shown in Table 2. Arrows should be used to indicate the direction of flow. Use the legend to identify contents exactly and to provide temperature, pressure, and other details necessary to identify the hazard.

The legend shall be brief, informative, pointed, and simple. Legends should be applied close to valves and adjacent to changes in direction, branches, and where pipes pass through walls or floors, and as frequently as needed along straight runs to provide clear and positive identification. Identification may be applied by stenciling, tape, or markers (see Figure 1). The number and location of identification markers on a particular piping system is based on judgment.

Color. Colors listed in Table 3 are used to identify the characteristic properties of the contents. Color can be shown on or contiguous to the piping by any physical means, but it should be used in combination with a legend. Color can be used in continuous total length coverage or in intermittent displays.

Table 2 Examples of Legends

HOT WATER
AIR 100 PSIG
H.P. RETURN
STEAM 100 PSIG

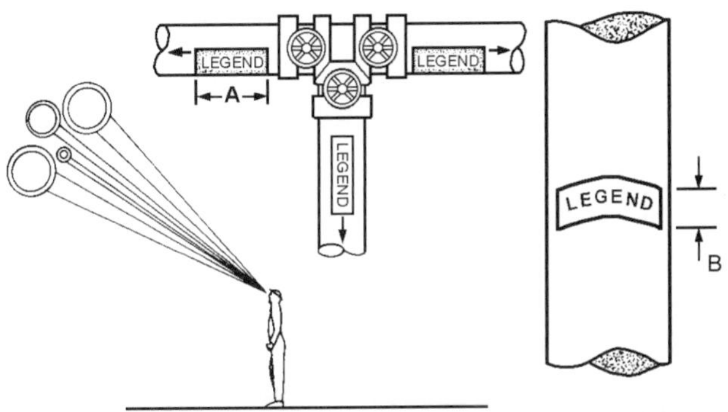

Fig. 1 Visibility of Pipe Markings

Visibility. Pipe markings should be highly visible. If pipe lines are above the normal line of vision, the lettering is placed below the horizontal centerline of the pipe (Figure 1).

Type and Size of Letters. Provide the maximum contrast between color field and legend (Table 3). Table 4 shows the size of letters recommended. Use of standard size letters of 1/2 in. or larger is recommended. For identifying materials in pipes of less than 3/4 in. in diameter and for valve and fitting identification, use a permanently legible tag.

Unusual or Extreme Situations. When the piping layout occurs in or creates an area of limited accessibility or is extremely complex, other identification techniques may be required. While a certain amount of imagination may be needed, the designer should always clearly identify the hazard and use the recommended color and legend guidelines.

Table 3 Classification of Hazardous Materials and Designation of Colors[a]

Classification	Color Field	Colors of Letters for Legend
Materials Inherently Hazardous		
Flammable or explosive	Yellow	Black
Chemically active or toxic	Yellow	Black
Extreme temperatures or pressures	Yellow	Black
Radioactive[b]	Purple	Yellow
Materials of Inherently Low Hazard		
Liquid or liquid admixture[c]	Green	Black
Gas or gaseous admixture	Blue	White
Fire Quenching Materials		
Water, foam, CO_2, Halon, etc.	Red	White

[a]When the color scheme above is used, the colors should be as recommended in the latest revision of NEMA *Standard* Z535.1.

[b]Previously specified radioactive markers using yellow or purple are acceptable if already installed and/or until existing supplies are depleted, subject to applicable federal regulations.

[c]Markers with black letters on a green color field are acceptable if already installed and/or until existing supplies are depleted.

Table 4 Size of Legend Letters

Outside Diameter of Pipe or Covering, in.	Length of Color Field A, in.	Size of Letters B, in..
3/4 to 1-1/4	8	1/2
1-1/2 to 2	8	3/4
2-1/2 to 6	12	1-1/4
8 to 10	24	2-1/2
over 10	32	3-1/2

CODES AND STANDARDS

ARI. 1982. Graphic electrical symbols for air-conditioning and refrigeration equipment. *Standard* 130.

ASME. 1996. Scheme for the identification of piping systems. *Standard* A13.1.

ASME. 1988. Glossary of terms concerning letter symbols. *Standard* Y10.1.

ASME. 1984. Letter symbols and abbreviations for quantities used in acoustics. *Standard* Y10.11.

ASME. 1987. Letter symbols for illuminating engineering. Y10.18.

ASME. 1999. Abbreviations and acronyms. *Standard* Y14.38-1999 (Revision and redesignation of ASME Y1.1-1989).

ASME. 1999. Graphical symbols for pipe fittings, valves, and piping. *Standard* Y32.2.3.

ASME. 1998. Graphical symbols for heating, ventilating, and air conditioning. *Standard* Y32.2.4.

ASME. 1999. Graphic symbols for plumbing fixtures for diagrams used in architecture and building construction. *Standard* Y32.4.

IEEE. 1998. American national standard letter symbols for units of measurement. IEEE *Standard* 260.1-1993.

IEEE. 1993. Mathematical signs and symbols for use in physical science and technology. *Standard* 260.3-1993.

IEEE. 1996. Letter symbols and abbreviations used in acoustics. *Standard* 260.4-1996.

NEMA. 1998. Safety color code. *Standard* Z535.1.

NFPA. 1999. Standard for fire safety symbols, 1999 Edition. *Standard* 170.

CHAPTER 37

UNITS AND CONVERSIONS

Table 1 Conversions to SI Units

Multiply	By	To Obtain
acre	0.4047	ha
atmosphere (standard)	*101.325	kPa
bar	*100	kPa
barrel (42 U.S. gal, petroleum)	159.0	L
	0.1590	m^3
Btu (International Table)	1.055	kJ
Btu/ft^2	11.36	kJ/m^2
Btu/ft^3	37.3	kJ/m^3
Btu/gal	279	kJ/m^3
Btu·ft/h·ft^2·°F	1.731	W/(m·K)
Btu·in/h·ft^2·°F (thermal conductivity, k)	0.1442	W/(m·K)
Btu/h	0.2931	W
Btu/h·ft^2	3.155	W/m^2
Btu/h·ft^2·°F (overall heat transfer coefficient, U)	5.678	W/(m^2·K)
Btu/lb	*2.326	kJ/kg
Btu/lb·°F (specific heat, c_p)	4.184	kJ/(kg·K)
bushel	0.03524	m^3
calorie, gram	4.184	J
calorie, kilogram (kilocalorie)	4.184	kJ
centipoise (dynamic viscosity, μ)	*1.00	mPa·s
centistokes (kinematic viscosity, ν)	*1.00	mm^2/s
clo	0.155	m^2·K/W
dyne/cm^2	*0.100	Pa
EDR hot water (150 Btu/h)	44.0	W
EDR steam (240 Btu/h)	70.3	W
EER	0.293	COP
ft	*0.3048	m
	*304.8	mm
ft/min, fpm	*0.00508	m/s
ft/s, fps	*0.3048	m/s
ft of water	2.99	kPa
ft of water per 100 ft pipe	0.0981	kPa/m
ft^2	0.09290	m^2
ft^2·h·°F/Btu (thermal resistance, R)	0.176	m^2·K/W
ft^2/s (kinematic viscosity, ν)	92,900	mm^2/s
ft^3	28.32	L
	0.02832	m^3
ft^3/min, cfm	0.4719	L/s
ft^3/s, cfs	28.32	L/s
ft·lb$_f$ (torque or moment)	1.356	N·m
ft·lb$_f$ (work)	1.356	J
ft·lb$_f$/lb (specific energy)	2.99	J/kg
ft·lb$_f$/min (power)	0.0226	W
footcandle	10.76	lx
gallon (U.S., *231 in^3)	3.7854	L
gph	1.05	mL/s
gpm	0.0631	L/s
gpm/ft^2	0.6791	L/(s·m^2)
gpm/ton refrigeration	0.0179	mL/J
grain (1/7000 lb)	0.0648	g
gr/gal	17.1	g/m^3
gr/lb	0.143	g/kg
horsepower (boiler) (33,470 Btu/h)	9.81	kW
horsepower (550 ft·lb$_f$/s)	0.7457	kW
inch	*25.4	mm
in. of mercury (60°F)	3.37	kPa
in. of water (60°F)	249	Pa
in/100 ft, thermal expansion	0.833	mm/m
in·lb$_f$ (torque or moment)	113	mN·m
To Obtain	**By**	**Divide**

Multiply	By	To Obtain
in^2	645.2	mm^2
in^3 (volume)	16.4	mL
in^3/min (SCIM)	0.273	mL/s
in^3 (section modulus)	16,390	mm^3
in^4 (section moment)	416,200	mm^4
km/h	0.2778	m/s
kWh	*3.60	MJ
kW/1000 cfm	2.12	kJ/m^3
kilopond (kg force)	9.81	N
kip (1000 lb$_f$)	4.45	kN
kip/in^2 (ksi)	6.895	MPa
litre	*0.001	m^3
met	58.15	W/m^2
micron (μm) of mercury (60°F)	133	mPa
mile	1.609	km
mile, nautical	*1.852	km
mph	1.609	km/h
	0.447	m/s
millibar	*0.100	kPa
mm of mercury (60°F)	0.133	kPa
mm of water (60°F)	9.80	Pa
ounce (mass, avoirdupois)	28.35	g
ounce (force or thrust)	0.278	N
ounce (liquid, U.S.)	29.6	mL
ounce inch (torque, moment)	7.06	mN·m
ounce (avoirdupois) per gallon	7.49	kg/m^3
perm (permeance)	57.45	ng/(s·m^2·Pa)
perm inch (permeability)	1.46	ng/(s·m·Pa)
pint (liquid, U.S.)	473	mL
pound		
lb (mass)	0.4536	kg
	453.6	g
lb$_f$ (force or thrust)	4.448	N
lb/ft (uniform load)	1.49	kg/m
lb$_m$/ft·h (dynamic viscosity, μ)	0.4134	mPa·s
lb$_m$/ft·s (dynamic viscosity, μ)	1490	mPa·s
lb$_f$·s/ft^2 (dynamic viscosity, μ)	47.88	Pa·s
lb/h	0.126	g/s
lb/min	0.00756	kg/s
lb/h [steam at 212°F (100°C)]	0.2843	kW
lb$_f$/ft^2	47.9	Pa
lb/ft^2	4.88	kg/m^2
lb/ft^3 (density, ρ)	16.0	kg/m^3
lb/gallon	120	kg/m^3
ppm (by mass)	*1.00	mg/kg
psi	6.895	kPa
quad (10^{15} Btu)	1.055	EJ
quart (liquid, U.S.)	0.9463	L
square (100 ft^2)	9.29	m^2
tablespoon (approximately)	15	mL
teaspoon (approximately)	5	mL
therm (U.S.)	105.5	MJ
ton, long (2240 lb)	1.016	Mg
ton, short (2000 lb)	0.907	Mg; t (tonne)
ton, refrigeration (12,000 Btu/h)	3.517	kW
torr (1 mm Hg at 0°C)	133	Pa
watt per square foot	10.76	W/m^2
yd	*0.9144	m
yd^2	0.8361	m^2
yd^3	0.7646	m^3
To Obtain	**By**	**Divide**

*Conversion factor is exact.

Notes: Units are U.S. values unless noted otherwise.

Litre is a special name for the cubic decimetre. 1 L = 1 dm^3 and 1 mL = 1 cm^3.

The preparation of this chapter is assigned to TC 1.6, Terminology.

Table 2 Conversion Factors

Pressure psi	in. of water (60°F)	in. Hg (32°F)	atmosphere	mm Hg (32°F)	bar	kgf/cm^2	pascal
1	= 27.708	= 2.0360	= 0.068046	= 51.715	= 0.068948	= 0.07030696	= 6894.8
0.036091	1	0.073483	2.4559×10^{-3}	1.8665	2.4884×10^{-3}	2.537×10^{-3}	248.84
0.491154	13.609	1	0.033421	25.400	0.033864	0.034532	3386.4
14.6960	407.19	29.921	1	760.0	1.01325*	1.03323	1.01325×10^5*
0.0193368	0.53578	0.03937	1.31579×10^{-3}	1	1.3332×10^{-3}	1.3595×10^{-3}	133.32
14.5038	401.86	29.530	0.98692	750.062	1	1.01972*	10^5*
14.223	394.1	28.959	0.96784	735.559	0.980665*	1	9.80665×10^4*
1.45038×10^{-4}	4.0186×10^{-3}	2.953×10^{-4}	9.8692×10^{-6}	7.50×10^{-3}	10^{-5}*	1.01972×10^{-5}*	1

Mass	lb (avoir.)	grain	ounce (avoir.)	kg
	1	= 7000*	= 16*	= 0.45359
	1.4286×10^{-4}	1	2.2857×10^{-3}	6.4800×10^{-5}
	0.06250	437.5*	1	0.028350
	2.20462	1.5432×10^4	35.274	1

Volume	cubic inch	cubic foot	gallon	litre	cubic metre (m^3)
	1	= 5.787×10^{-4}	= 4.329×10^{-3}	= 0.0163871	= 1.63871×10^{-5}
	1728*	1	7.48052	28.317	0.028317
	231.0*	0.13368	1	3.7854	0.0037854
	61.02374	0.035315	0.264173	1	0.001*
	6.102374×10^4	35.315	264.173	1000*	1

Energy	Btu	$ft \cdot lb_f$	calorie (cal)	joule (J) = watt-second (W·s)	watt-hour (W·h)
Note: MBtu, which is 1000 Btu, is confusing and is not used in the Handbook.	1	= 778.17	= 251.9958	= 1055.056	= 0.293071
	1.2851×10^{-3}	1	0.32383	1.355818	3.76616×10^{-4}
	3.9683×10^{-3}	3.08803	1	4.1868*	1.163×10^{-3}*
	9.4782×10^{-4}	0.73756	0.23885	1	2.7778×10^{-4}
	3.41214	2655.22	859.85	3600*	1

Density	lb/ft^3	lb/gal	g/cm^3	kg/m^3
	1	= 0.133680	= 0.016018	= 16.018463
	7.48055	1	0.119827	119.827
	62.4280	8.34538	1	1000*
	0.0624280	0.008345	0.001*	1

Specific Volume	ft^3/lb	gal/lb	cm^3/g	m^3/kg
	1	= 7.48055	= 62.4280	= 0.0624280
	0.133680	1	8.34538	0.008345
	0.016018	0.119827	1	0.001*
	16.018463	119.827	1000*	1

Viscosity (absolute) 1 poise = 1 dyne-sec/cm² = 0.1 Pa·s = 1 g/(cm·s)

Viscosity (absolute)	poise	$lb_f \cdot s/ft^2$	$lb_f \cdot h/ft^2$	$kg/(m \cdot s) = N \cdot s/m^2$	$lb_m/ft \cdot s$
	1	= 2.0885×10^{-3}	= 5.8014×10^{-7}	= 0.1*	= 0.0671955
	478.8026	1	2.7778×10^{-4}	47.88026	32.17405
	1.72369×10^6	3600*	1	1.72369×10^5	1.15827×10^5
	10*	0.020885	5.8014×10^{-6}	1	0.0671955
	14.8819	0.031081	8.6336×10^{-6}	1.4882	1

Temperature Scale		Temperature K	°C	°R	°F	Temperature Interval	K	°C	°R	°F
Kelvin	x K =	x	$x - 273.15$	$1.8x$	$1.8x - 459.67$	1 K =	1	1	9/5 = 1.8	9/5 = 1.8
Celsius	x°C =	$x + 273.15$	x	$1.8x + 491.67$	$1.8x + 32$	1°C =	1	1	9/5 = 1.8	9/5 = 1.8
Rankine	x°R =	$x/1.8$	$(x - 491.67)/1.8$	x	$x - 459.67$	1°R =	5/9	5/9	1	1
Fahrenheit	x°F =	$(x + 459.67)/1.8$	$(x - 32)/1.8$	$x + 459.67$	x	1°F =	5/9	5/9	1	1

Notes: Conversions with * are exact.
The Btu and calorie are based on the International Table.

All temperature conversions and factors are exact.
The term centigrade is obsolete and should not be used.

When making conversions, remember that a converted value is no more precise than the original value. For many applications, rounding off the converted value to the same number of significant figures as those in the original value provides sufficient accuracy.

Caution: The conversion values in Table 1 are rounded to three or four significant figures, which is sufficiently accurate for most applications. See ANSI *Standard* SI-10 (available from ASTM or IEEE) for additional conversions with more significant figures.

CHAPTER 38

PHYSICAL PROPERTIES OF MATERIALS

VALUES in the following tables are in consistent units to assist the engineer looking for approximate values. For data on refrigerants, see Chapter 19; for secondary coolants, see Chapter 21. Chapter 25 gives more information on the values for materials used in building construction and insulation. Many properties vary with temperature, material density, and composition. The references document the source of the values and provide more detail or values for materials not listed here.

Table 1 Properties of Vapor

Material	Molecular Mass	Normal Boiling Point, °F	Critical Temperature, °F	Critical Pressure, psia	Density, lb/ft^3	Specific Heat, Btu/lb·°F	Thermal Conductivity, Btu/h·ft·°F	Viscosity, lb/ft·h
Alcohol, Ethyl	46.07[a]	173.3[a]	469.6[b]	927.3[b]		0.362[j]	0.0073[a]	0.0343[j] (60)
Alcohol, Methyl	32.04[a]	148.9[a]	464.0[b]	1157[b]		0.322[j]	0.0174[r]	0.0358[j] (30)
Ammonia	17.03[a]	−28[a]	270.3[b]	1639[b]	0.0482[b]	0.525[aa]	0.0128[b]	0.0225[aa]
Argon	39.948[a]	−302.5*	−188.5*	704.9*	0.1114[b]	0.125[c]	0.0094[a]	0.0507[a]
Acetylene	26.04[a]	−118.5[a]	96.8[b]	911[b]	0.0732[b]	0.377[a]	0.0108[b]	0.0226[a]
Benzene	78.11[a]	176.2[a]	553.1[d]	714.2[d]	0.167[e] (176)	0.31[e] (176)	0.0041[e]	0.017[a]
Bromine	159.82[a]	137.8[a]	591.8[d]	1499[d]	0.38[f] (138)	0.055[f] (212)	0.0035[a]	0.041[a]
Butane	58.12[a]	31.1[a]	305.6[d]	550.7[d]	0.168[g]	0.377[aa]	0.0079[a]	0.017[a]
Carbon dioxide	44.01[a]	−109.3[a]	87.9[d]	1071[d]	0.123[g]	0.20[g]	0.0084[a]	0.033[h]
Carbon disulfide	76.13[h]	115.2[h]	534[h]	1046[h]		0.1431[p] (80)		
Carbon monoxide	28.01[a]	−312.7[a]	−220.4[d]	507[d]	0.078[d]	0.25[f]	0.0133[a]	0.040[a]
Carbon tetrachloride	153.84[g]	169.8[h]	541.8[h]	661[h]		0.206[q] (80)		0.0375[j]
Chlorine	70.91[a]	−30.3[a]	291.2[d]	1118[d]	0.201[d]	0.117[a]	0.0046[a]	0.030[a]
Chloroform	119.39[h]	143.1[h]	506.1[h]	794[h]		0.126[j]	0.0081[r]	0.038[j]
Ethyl chloride	64.52[h]	54.2[h]	369.0[h]	764[h]	0.1793[b]	0.426[r]	0.00504[j]	0.0378[q]
Ethylene	28.03[h]	−154.6[h]	49.9[h]	742[h]	0.0783[b]	0.352[aa]	0.0102[aa]	0.0231[aa]
Ethyl ether	74.12[h]	94.4[h]	378.8[h]	523[h]		0.589[h] (95)		0.0273[q]
Fluorine	38.00[h]	−304.5[h]	−200.5[h]	808[h]	0.1022[b]	0.194[j]	0.0147[j]	0.089[j]
Helium	4.0026[a]	−452.1[i]	−450.2[h]	33.21[i]	0.0111[i]	1.241[aa]	0.0823[aa]	0.0452[aa]
Hydrogen	2.0159[a]	−423.0[i]	−399.9[i]	190.8[i]	0.00562[i]	3.40[j]	0.0972[aa]	0.0203[aa]
Hydrogen chloride	36.461[a]	−120.8[a]	124.5[d]	1198[d]	0.1024[b]	0.191[j]	0.00757[j]	0.0321[j]
Hydrogen sulfide	34.080[a]	−77.3[a]	212.7[d]	1307[d]	0.0961[b]	0.238[j]	0.00751[j]	0.0281[j]
Heptane (m)	100.21[a]	209.2[a]	512.2[b]	394[b]	0.21[k]	0.476[j]	0.0107[j]	0.0168[j]
Hexane (m)	86.18[a]	154[a]	454.5[d]	440[d]	0.21[k]	0.449[j]	0.00971[j]	0.0182[j]
Isobutane	58.12[f]	−11.1*	275.0[j]	529.1[j]	0.154[s] (70)	0.376[aa]	0.0081[aa]	0.0168[aa]
Methane	16.04[a]	−263.2[a]	−115.18[j]	673.1[b]	0.0448[b]	0.520[aa]	0.0178[aa]	0.0250[aa]
Methyl chloride	50.49[a]	−11.6[a]	289.6[j]	968.5[b]	0.1440[b]	0.184[aa]	0.0054[aa]	0.0244[aa]
Naphthalene	128.19[a]	424.4*	876.2[j]	576.1[j]		0.313[q] (77)		
Neon	20.183[a]	−412.6[a]	−379.7[j]	391.3[j]		0.246[aa]	0.0268[aa]	0.0718[aa]
Nitric oxide	30.01[a]	−241.6[a]	−135.2[j]	949.4[j]		0.238[j]		0.0712[j]
Nitrogen	28.01[a]	−320.4[a]	−232.4[j]	492.3[b]		0.248[j]	0.0138[aa]	0.0402[aa]
Nitrous oxide	44.01[a]	−127.3[a]	97.5[j]	1049.3[j]		0.203[j]	0.01001[j] (80.3)	0.0543[j]
Nitrogen tetroxide	92.02[a]		316.8[j]	1469.6[j]		0.201[p] (80)	0.0232[r] (131)	
Oxygen	31.9977*	−297.3*	−181.5*	731.4*		0.218[j]	0.0141[aa]	0.0462[aa]
n-Pentane	72.53[a]	97.0*	385.9[j]	489.5[j]		0.400[a] (80)	0.00877[j] (80.3)	0.0282[j]
Phenol	74.11[b]	358.5[b]	786[b]	889[b]	0.16[k]	0.34[k]	0.0099[k]	0.029[k]
Propane	44.09[g]	−43.76*	206.1*	616.1*	0.126[g]	0.3753[j] (40)	0.0087[j]	0.0179[j]
Propylene	42.08[b]	−53.86[l]	197.2[l]	670.3[l]	0.120[l]	0.349[aa]	0.0081[aa]	0.0195[aa]
Sulfur dioxide	64.06[b]	14.0[b]	315[b]	1142[b]	0.183[b]	0.145[l]	0.0049[j]	0.0281[j]
Water vapor	18.02[b]	212.0[m]	705.18*	3200.0*	0.0373[m]	0.489[aa]	0.0143[m]	0.0293[aa]

*Data source unknown.

Notes: 1. Properties at 14.696 psia and 32°F, or the saturation temperature if higher than 32°F, unless otherwise noted in parentheses.

2. Superscript letters indicate data source from the section on References.

Table 2 Properties of Liquids

Name or Description	Normal Boiling Point, °F at 14.696 psia	Enthalpy of Vaporization, Btu/lb	Specific Heat, c_p Btu/lb·°F	Specific Heat Temp., °F	Viscosity lb/h·ft	Viscosity Temp., °F	Enthalpy of Fusion, Btu/lb	Density lb/ft³	Density Temp., °F	Thermal Conductivity Btu/h·ft·°F	Thermal Conductivity Temp., °F	Vapor Pressure mm of Hg	Vapor Pressure Temp., °F	Freezing Point, °F
Acetic acid	245.3[a]	174.1[b]	0.522[b]	79-203	2.956[f]	68	84.0[b]	65.49[a]	68	0.099[b]	68	400[a]	210	61.9[a]
Acetone	133.2[a]	228.9[b]	0.514[b]	37-73	0.801[f]	68	42.1[b]	49.4[a]	68	0.102[b]	86	400[a]	103	−139.6[a]
Allyl alcohol	206.6[a]	294.1[b]	0.655[b]	70-205	3.298[f]	68		53.31[a]	68	0.104[b]	77-86	400[a]	176	−200.2[a]
n-Amyl alcohol	280.6[i]	216.3[b]			9.686[f]	73.4	48.0[b]	51.06[f]	59	0.094[b]	86	100[a]	186	−110.2[a]
Ammonia	−28[a]	583.2[b]	1.099[b]	32	0.643[f]	−28.3	142.9[b]	43.50[b]	−50	0.29[b]	5-86	400[a]	−49.7	−107.9[a]
Alcohol, Ethyl	173.3[a]	367.5[b]	0.680[b]	32-208	2.889[f]	68	46.4[b]	49.27[a]	68	0.105[b]	68	100[a]	94.8	−179.1[a]
Alcohol, Methyl	148.9[a]	473.0[b]	0.601[b]	59-68	1.434[f]	68	42.7[a]	49.40[a]	68	0.124[b]	68	100[a]	70.2	−144.0[a]
Aniline	363.8[a]	186.6[b]	0.512[b]	46-180	10.806[f]	68	48.8[b]	63.77[a]	68	0.100[b]	32-68	10[a]	156.9	20.84[a]
Benzene	176.2[a]	169.4[h]	0.412[h]	68	1.58[a]	68	54.2[h]	54.9[d]	68	0.085[h]	68	75[d]	68	42[a]
Bromine	137.8[a]	79.4[d]	0.107[f]	68	2.39[a]	68	28.5[d]	194.7[f]	68	0.070[a]	77	165[d]	68	19[a]
n-Butyl alcohol	243.5[a]	254.3[h]	0.563[f]	68	7.13[f]	68	53.9[b]	50.6[a]	68	0.089[h]	68	5[d]	68	−130[a]
n-Butyric acid	326.3[a]	217.0[h]	0.515[f]	68	3.73[a]	68	54.1[a]	60.2[a]	68	0.094[h]	54	0.7[d]	68	20[a]
Calcium chloride brine (20% by mass)			0.744[i]	68	4.8[i]	68		73.8[i]	68	0.332[i]	68			2[i]
Carbon disulfide	115.3[a]	148.8[h]	0.240[i]	68	0.88[a]	68	24.8[d]	78.9[d]	68	0.093[b]	86	295[d]	68	−168[a]
Carbon tetrachloride	170.2[a]	83.7[h]	0.201[f]	68	2.34[a]	68	12.8[d]	99.5[d]	68	0.062[j]	68	87[d]	68	−9[a]
Chloroform	142.3[v]	106[v]	0.234[v]	68	1.36[v]	68		92.96[v]	68	0.075[v]	68	160[v]	68	−81.8[v]
n-Decane	345.2[b]		0.50[b]	68			86.9[b]	45.6[b]	68	0.086[b]	68	1.3[b]	68	−21.5[b]
Ethyl ether	94.06[v]	151[v]	0.541[v]	68	0.56[v]	68	42.4[v]	44.61[v]	68	0.081[b]	68	440[v]	68	−177.3[v]
Ethyl acetate	170.8[v]	183.8[v]	0.468[v]	68	1.09[v]	68	51.2[b]	52.3[v]	68	0.101[b]	68	72[b]	68	−116.3[v]
Ethyl chloride	54.2[j]	165.9[f] (68)	0.368[f]	32			29.68[a]	56.05[a]	68	0.179[f]	33.6	400[y]	53.1	−213.5[a]
Ethyl iodide	162.1[a]	82.1[f] (160)	0.368[f]	32	0.0239[f]	68		120.85[a]	68	0.214[f]	86	100[y]	64.4	−162.4*
Ethylene bromide	268.8[a]	99.2[f] (210)	0.174[f]	68	0.0694[f]	68	24.82[a]	136.05[a]	68			10[y]	65.5	49.2[a]
Ethylene chloride	182.3[a]	153.4[f] (308)	0.301[f]	68	0.0338[f]	68	38.02[a]	77.10[a]	68			60[y]	64.6	−31.64[a]
Ethylene glycol	388.4[a]	344.0[f] (651)					77.86[a]	69.22[a]	68	0.100[f]	68	1[y]	128	12.7[a]
Formic acid	213.3[a]	215.8[f] (420)	0.526[f]	68	0.0719[f]	68	118.89[a]	76.16[a]	68	0.104[a]	33	40[y]	75.2	47.1[a]
Glycerin (glycerol)	359* (20 mm)				43.1[f]	68		78.72[a]	68	0.113[a]	68	1[a]	125.5	68[a]
Heptane	209.2[a]	138[f]	0.532[j]	68	0.990[a]	68	60.4[b]	42.7[a]	68	0.0741[j]	68	35.5[y]	68	−132[a]
Hexane	154[a]	145[f]	0.538[j]	68	0.775[d]	68	65.0[b]	41.1[a]	68	0.0720[j]	68	120.0[y]	68	−139[a]
Hydrogen chloride	−120.8[a]	191[f]					23.6[f]	74.6[d]	b.p.					−174.6[a]
Isobutyl alcohol	226.4[a]	249[f]	0.116[f]	68	9.45[f]	68		50.0[f]	68	0.082[f]	68	9.7[y]	68	−162.4[a]
Kerosene	400-560[b]		0.50[n]	68	6.0[b]	68		51.2[a]	68	0.086[n]	68			
Linseed oil					104[b]	68		58[d]	68					−11†[a]
Methyl acetate	134.6[a]	177[f]	0.468[f]	68	0.940[f]	68		60.6[a]	68	0.093[f]	68	169.8[y]	68	−144.6[a]
Methyl iodide	108.5[a]	82.6[f]			1.21[f]	68		142[a]	68			320[y]	68	−87.7[a]
Naphthalene	411.4[a]	136[f]	0.402[f]	m.p.	2.18[b]	m.p.	64.9[b]	60.9[y]	m.p.			2.18[b]	68	176.4[a]
Nitric acid	186.8[v]	270[v]	0.42[v]	68	2.2[k]	68	71.5[v]	94.45[v]	68	0.16[v]	68	1.77[v]	68	−42.9[v]
Nitrobenzene	411.6[b]	142[b]	0.348[b]	68	5.20[b]	68	40.28[v]	75.2[b]	68	0.96[b]	68	< 0.01[b]	68	42.3[b]
Octane	258.3[b]	131.7[b]	0.51[b]	68	1.36[b]	68	77.70[b]	43.9[b]	68	0.084[b]	68	0.42[b]	68	−69.7[b]
Petroleum		98-165[w]	0.4-0.6[w]	68	19-2900[w]	68		40-66[w]						
n-Pentane	96.8[a]	153.6[h]	0.558[h]	68	0.546[d]	68	50.1[h]	39.1[a]	68	0.066[h]	68	425[d]	68	−201.5[a]
Propionic acid	286.0[a]	177.8[f]	0.473[h]	68	2.666[a]	68		61.9[a]	68	0.100*	54	3[d]	68	−5.4[a]
Sodium chloride brine														
20% by mass	220.8[a]		0.745[x]	68	3.80[x]	68		71.8[x]	68	0.337[x]	68	0.57[x]	68	2.6[x]
10% by mass	215.5[a]		0.865[x]	68	2.85[x]	68		66.9[x]	68	0.343[x]	68	0.65[x]	68	20.6[x]
Sodium hydroxide and water														
15% by mass	215.0[v]		0.864[b]	68				72.4[b]	68					−5.8[b]
Sulfuric acid and water														
100% by mass	550.0[v]		0.335[b]	68	53[b]	68		114.4[v]	68			< 0.01[b]	68	50.9[b]
95% by mass	575.0[v]		0.35[v]	68	52[v]	68		114.6[v]	68			< 0.01[v]	68	−18[v]
90% by mass	500.0[v]		0.39[v]	68	60[v]	68		113.4[v]	68	0.22[b]	68	< 0.01[v]	68	15.0[v]
Toluene ($C_6H_5CH_3$)	231[b]	156[b]	0.404[v]	68	1.42[v]	68	30.9[b]	54.1[b]	68	0.090[b]	68	0.88[b]	68	−139[b]
Turpentine	303[a]	123[v]	0.42[b]	68	1.32[b]	68		53.9[b]	68	0.073[b]	68			
Water	211.9*	970.3[m]	0.999[m]	68	2.39[m]	68	143.5[b]	62.32[m]	68	0.348[m]	68	17.59*	68	32.018[m]
Xylene [$C_6H_4(CH_3)_2$]														
Ortho	291[b]	149[b]	0.411[b]	68	2.01[b]	68	55.1[b]	55.0[b]	68	0.90[b]	68	0.196[b]	68	−13[b]
Meta	283[b]	147[b]	0.400[b]	68	1.52[b]	68	46.9[b]	54.1[b]	68	0.90[b]	68	0.218[b]	68	−53[b]
Para	281[b]	146[b]	0.393[b]	68	1.62[b]	68	69.3[b]	53.8[b]	68			0.227[b]	68	56[b]
Zinc sulfate and water														
10% by mass			0.90[b]	68	3.80[a]	68		69.2[r]	68	0.337[a]	68			29.7[a]
1% by mass			0.80[b]	68	2.54[a]	68		63.0[r]	68	0.346[a]	68			31.7[a]

*Data source unknown.
†Approximate solidification temperature.

Notes: Superscript letters indicate data source from the section on References.
m.p. = melting point b.p. = boiling point

Table 3 Properties of Solids

Material Description	Specific Heat, Btu/lb·°F	Density, lb/ft³	Thermal Conductivity, Btu/h·ft·°F	Emissivity: Ratio	Emissivity: Surface Condition
Aluminum (alloy 1100)	0.214[b]	171[u]	128[u]	0.09[n]	Commercial sheet
				0.20[n]	Heavily oxidized
Aluminum bronze (76% Cu, 22% Zn, 2% Al)	0.09[u]	517[u]	58[u]		
Asbestos: Fiber	0.25[b]	150[u]	0.097[u]		
Insulation	0.20[t]	36[b]	0.092[b]	0.93[b]	"Paper"
Ashes, wood	0.20[t]	40[b]	0.041[b] (122)		
Asphalt	0.22[b]	132[b]	0.43[b]		
Bakelite	0.35[b]	81[u]	9.7[u]		
Bell metal	0.086[t] (122)				
Bismuth tin	0.040*		37.6*		
Brick, building	0.2[b]	123[u]	0.4[b]	0.93*	
Brass: Red (85% Cu, 15% Zn)	0.09[u]	548[u]	87[u]	0.030[b]	Highly polished
Yellow (65% Cu, 35% Zn)	0.09[u]	519[u]	69[u]	0.033[b]	Highly polished
Bronze	0.104[t]	530[t]	17[d] (32)		
Cadmium	0.055[a]	540[f]	53.7[b]	0.02[d]	
Carbon (gas retort)	0.17[a]		0.20[b] (2)	0.81[a]	
Cardboard			0.04[b]		
Cellulose	0.32[b]	3.4[t]	0.033[t]		
Cement (portland clinker)	0.16[b]	120[i]	0.017[i]		
Chalk	0.215[t]	143[t]	0.48*	0.34*	About 250°F
Charcoal (wood)	0.20[t]	15[a]	0.03[a] (392)		
Chrome brick	0.17[b]	200[b]	0.67[b]		
Clay	0.22[b]	63[t]			
Coal	0.3[b]	90[t]	0.098[f] (32)		
Coal tars	0.35[b] (104)	75[b]	0.07[b]		
Coke (petroleum, powdered)	0.36[b] (752)	62[b]	0.55[b] (752)		
Concrete (stone)	0.156[b] (392)	144[b]	0.54[b]		
Copper (electrolytic)	0.092[u]	556[u]	227[u]	0.072[n]	Commercial, shiny
Cork (granulated)	0.485[t]	5.4[t]	0.028[t] (23)		
Cotton (fiber)	0.319[u]	95[u]	0.024[u]		
Cryolite ($AlF_3 \cdot 3NaF$)	0.253[b]	181[b]			
Diamond	0.147[b]	151[t]	27[t]		
Earth (dry and packed)		95[t]	0.037*	0.41*	
Felt		20.6[b]	0.03[b]		
Fireclay brick	0.198[b] (212)	112[t]	0.58[b] (392)	0.75[n]	At 1832°F
Fluorspar (CaF_2)	0.21[b]	199[v]	0.63[v]		
German silver (nickel silver)	0.09[u]	545[u]	19[u]	0.135[n]	Polished
Glass: Crown (soda-lime)	0.18[b]	154[u]	0.59[t] (200)	0.94[n]	Smooth
Flint (lead)	0.117[b]	267[u]	0.79[r]		
Heat-resistant	0.20[b]	139[t]	0.59[t] (200)		
"Wool"	0.157[b]	3.25[t]	0.022[t]		
Gold	0.0312[u]	1208[u]	172[t]	0.02[n]	Highly polished
Graphite: Powder	0.165*		0.106*		
Impervious	0.16[u]	117[u]	75[u]	0.75[n]	
Gypsum	0.259[b]	78[b]	0.25[b]	0.903[b]	On a smooth plate
Hemp (fiber)	0.323[u]	93[u]			
Ice: 32°F	0.487[t]	57.5[b]	1.3[b]	0.95*	
−4°F	0.465[t]		1.41*		
Iron: Cast	0.12[v] (212)	450[b]	27.6[b] (129)	0.435[b]	Freshly turned
Wrought		485[b]	34.9[b]	0.94[b]	Dull, oxidized
Lead	0.0309[u]	707[u]	20.1[u]	0.28[n]	Gray, oxidized
Leather (sole)		62.4[b]	0.092[b]		
Limestone	0.217[b]	103[b]	0.54[b]	0.36* to 0.90	At 145 to 380°F
Linen			0.05[b]		
Litharge (lead monoxide)	0.055[b]	490[b]			
Magnesia: Powdered	0.234[b] (212)	49.7[b]	0.35[b] (117)		
Light carbonate		13[b]	0.034[b]		
Magnesite brick	0.222[b] (212)	158[b]	2.2[b] (400)		
Magnesium	0.241[b]	108[u]	91[u]	0.55[n]	Oxidized
Marble	0.21[b]	162[b]	1.5[b]	0.931[b]	Light gray, polished
Nickel, polished	0.105[u]	555[u]	34.4[u]	0.045[n]	Electroplated
Paints: White lacquer				0.80[n]	
White enamel				0.91[n]	On rough plate
Black lacquer				0.80[n]	
Black shellac		63[u]	0.15[u]	0.91[n]	"Matte" finish
Flat black lacquer				0.96[n]	
Aluminum lacquer				0.39[n]	On rough plate

*Data source unknown.

Notes: 1. Values are for room temperature unless otherwise noted in parentheses.

2. Superscript letters indicate data source from the section on References.

Table 3 Properties of Solids (*Continued*)

Material Description	Specific Heat, Btu/lb·°F	Density, lb/ft^3	Thermal Conductivity, Btu/h·ft·°F	Emissivity	
				Ratio	Surface Condition
Paper	0.32*	58[b]	0.075[b]	0.92[b]	Pasted on tinned plate
Paraffin	0.4[bb]	47[bb]	0.14[b] (32)		
Plaster		132[b]	0.43[b] (167)	0.91[b]	Rough
Platinum	0.032[u]	1340[u]	39.9[u]	0.054[b]	Polished
Porcelain	0.18*	162[u]	1.3[u]	0.92[b]	Glazed
Pyrites (copper)	0.131[b]	262[b]			
Pyrites (iron)	0.136[b] (156)	310[v]			
Rock salt	0.219[u]	136[u]			
Rubber, vulcanized: Soft	0.48*	68.6[t]	0.08[t]	0.86[b]	Rough
Hard		74.3[t]	0.092[t]	0.95[b]	Glossy
Sand	0.191[b]	94.6[b]	0.19[b]		
Sawdust		12[b]	0.03[b]		
Silica	0.316[b]	140[v]	0.83[t] (200)		
Silver	0.0560[u]	654[u]	245[u]	0.02[n]	Polished and at 440°F
Snow: Freshly fallen		7[y]	0.34[t]		
At 32°F		31[t]	1.3[t]		
Steel (mild)	0.12[b]	489[b]	26.2[b]	0.12[n]	Cleaned
Stone (quarried)	0.2[b]	95[t]			
Tar: Pitch	0.59[v]	67[u]	0.51[v]		
Bituminous		75[t]	0.41[u]		
Tin	0.0556[u]	455[u]	37.5[u]	0.06[h]	Bright and at 122°F
Tungsten	0.032[u]	1210[u]	116[u]	0.032[n]	Filament at 80°F
Wood: Hardwoods—	0.45/0.65[b]	23/70[z]	0.065/0.148[z]		
Ash, white		43[z]	0.0992[z]		
Elm, American		36[z]	0.0884[z]		
Hickory		50[z]			
Mahogany		34[u]	0.075[u]		
Maple, sugar		45[z]	0.108[z]		
Oak, white	0.570[b]	47[z]	0.102[z]	0.90[n]	Planed
Walnut, black		39[z]			
Softwoods—	See Table 4,	22/46[z]	0.061/0.093[z]		
Fir, white	Chapter 24	27[z]	0.068[z]		
Pine, white		27[z]	0.063[z]		
Spruce		26[z]	0.065[z]		
Wool: Fiber	0.325[u]	82[u]			
Fabric		6.9/20.6[u]	0.021/0.037[u]		
Zinc: Cast	0.092[u]	445[u]	65[u]	0.05[n]	Polished
Hot-rolled	0.094[b]	445[b]	62[b]		
Galvanizing				0.23[n]	Fairly bright

*Data source unknown.

Notes: 1. Values are for room temperature unless otherwise noted in parentheses. 2. Superscript letters indicate data source from References.

REFERENCES

[a]*Handbook of chemistry and physics*, 63rd ed. 1982-83. Chemical Rubber Publishing Co., Cleveland, OH.

[b]Perry, R.H. *Chemical engineers' handbook*, 2nd ed., 1941, 5th ed., 1973. McGraw-Hill, New York.

[c]*Tables of thermodynamic and transport properties of air, argon, carbon dioxide, carbon monoxide, hydrogen, nitrogen, oxygen and steam.* 1960. Pergamon Press, Elmsford, NY.

[d]*American Institute of Physics handbook*, 3rd ed. 1972. McGraw-Hill, New York.

[e]Organick and Studhalter. 1948. *Thermodynamic properties of benzene. Chemical Engineering Progress* (November):847.

[f]Lange. 1972. *Handbook of chemistry*, rev. 12th ed. McGraw-Hill, New York.

[g]ASHRAE. 1969. *Thermodynamic properties of refrigerants.*

[h]Reid and Sherwood. 1969. *The properties of gases and liquids*, 2nd ed. McGraw-Hill, New York.

[i]Chapter 19, 1993 *ASHRAE Handbook—Fundamentals.*

[j]*T.P.R.C. data book*. 1966. Thermophysical Properties Research Center, W. Lafayette, IN.

[k]Estimated.

[l]Canjar, L.N., M. Goldman, and H. Marchman. 1951. Thermodynamic properties of propylene. *Industrial and Engineering Chemistry* (May):1183.

[m]*ASME steam tables*. 1967. American Society of Mechanical Engineers, New York.

[n]McAdams, W.H. 1954. *Heat transmission*, 3rd ed. McGraw-Hill, New York.

[o]Stull, D.R. 1947. Vapor pressure of pure substances (organic compounds). *Industrial and Engineering Chemistry* (April):517.

[p]*JANAF thermochemical tables*. 1965. PB 168 370. National Technical Information Service, Springfield, VA.

[q]*Physical properties of chemical compounds*. 1955-61. American Chemical Society, Washington, D.C.

[r]*International critical tables of numerical data*. 1928. National Research Council of USA, McGraw-Hill, New York.

[s]*Matheson gas data book*, 4th ed. 1966. Matheson Company, Inc., East Rutherford, NJ.

[t]Baumeister and Marks. 1967. *Standard handbook for mechanical engineers*. McGraw-Hill, New York.

[u]Miner and Seastone. *Handbook of engineering materials*. John Wiley and Sons, New York.

[v]Kirk and Othmer. 1966. *Encyclopedia of chemical technology*. Interscience Division, John Wiley and Sons, New York.

[w]Gouse and Stevens. 1960. *Chemical technology of petroleum*, 3rd ed. McGraw-Hill, New York.

[x]*Saline water conversion engineering data book*. 1955. M.W. Kellogg Co. for U.S. Department of Interior.

[y]Timmermans, J. *Physicochemical constants of pure organic compounds*, 2nd ed. American Elsevier, New York.

[z]*Wood handbook*. 1955. Handbook No. 72. Forest Products Laboratory, U.S. Department of Agriculture.

[aa]ASHRAE. 1976. *Thermophysical properties of refrigerants*.

[bb]Lane, G. ed. 1986. *Solar heat storage: Latent heat materials, Vol II—Technology*. CRC Press, Chicago.

ASHRAE HANDBOOK

Additions and Corrections

This report includes additional information and technical errors found between April 15, 1998 and April 20, 2001 in the inch-pound (I-P) edition of the 1998, 1999, and 2000 *ASHRAE Handbooks*. Occasional typographical errors and nonstandard symbol labels will be corrected in future volumes.

The authors and editor encourage you to notify them if you find other technical errors. Please send corrections to: Handbook Editor, ASHRAE, 1791 Tullie Circle NE, Atlanta, GA 30329, or e-mail ashrae@ashrae.org.

1998 Refrigeration

p. 8.2, Table 1 and Table 2. The equations in these tables were taken directly from the following source:

Choi, Y., and M.R. Okos. 1986. Effects of temperature and composition on the thermal properties of foods. In *Food Engineering and Process Applications*, Vol. 1, pp. 93-101. Elsevier Applied Science Publishers.

Analysis of the thermal diffusivity equations reported by Choi and Okos (1986) indicates that they are incorrect by a factor of 10^6. The thermal diffusivity equations should be as follows:

	Thermal Diffusivity, ft^2/h
Protein	$\alpha = 2.3170 \times 10^{-3} + 1.1364 \times 10^{-5}t - 1.7516 \times 10^{-8}t^2$
Fat	$\alpha = 3.9137 \times 10^{-3} - 2.6765 \times 10^{-6}t - 4.5790 \times 10^{-10}t^2$
Carbohydrate	$\alpha = 2.7387 \times 10^{-3} + 1.3198 \times 10^{-5}t - 2.7769 \times 10^{-8}t^2$
Fiber	$\alpha = 2.4818 \times 10^{-3} + 1.2873 \times 10^{-5}t - 2.6553 \times 10^{-8}t^2$
Ash	$\alpha = 4.5565 \times 10^{-3} + 8.9716 \times 10^{-6}t - 1.4644 \times 10^{-8}t^2$
Water	$\alpha = 4.6428 \times 10^{-3} + 1.5289 \times 10^{-5}t - 2.8730 \times 10^{-8}t^2$
Ice	$\alpha = 5.0909 \times 10^{-2} - 2.0371 \times 10^{-4}t + 1.1366 \times 10^{-6}t^2$

p. 8.5, Table 3. The specific heat above freezing for butter should be 0.52 Btu/lb·°F. Delete the value shown for the specific heat of butter below freezing.

p. 9.2, 1st line after Equation (4). Change t to θ.

p. 9.3, Table 2. The variable on the left side of the fifth equation (below the word *where*) should read v, not w.

p. 9.5, Equation (15). ω is in radians.

p. 9.5, Equation (20). Delete 2 from both the numerator and denominator in the first term on the right side. Then multiply the first term by 1.5 so the equation reads as follows:

$$E_o = 1.5\frac{\beta_1 + \beta_2 + \beta_1^2(1+\beta_2) + \beta_2^2(1+\beta_1)}{\beta_1\beta_2(1+\beta_1+\beta_2)} - \frac{[(\beta_1 - \beta_2)^2]^{0.4}}{15} \quad (20)$$

p. 9.6, Example 1. In Step 5, change the value of E_o to 2.06 to reflect the corrected Equation (20). As a result of this change, the value of E in the last equation in Step 5 is 1.45. Then in Step 8, θ = 3.38 h because E = 1.45.

p. 16.11, 1st column. Change the third sentence in the second paragraph in the section on Chilling and Freezing Variety Meats to read:

Quick Chilling. A better and more widely used method consists of quick ...

p. 18.4, 1st column. Change first heading to read "Irradiation of Freah Seafood."

p. 20.2, 1st column, 2nd para. Change second sentence to read, "Most of the shell's organic matter is protein."

p. 23.3, 1st column. Second heading should read, "Packaging, Loading, and Handling."

p. 23.3, Table 2. Beans, green should be in the third column. (Store at 50°F and 80 to 85% rh). Green beans suffer chilling injury if stored at 32 to 41°F.

p. 23.11, 2nd column, 7th line up. Change 45 cfm/lb to read 10 cfm/ton.

p. 24.2, 1st column. Insert the following paragraph to the top of the column before the paragraph on Flavor Fortification.

Evaporation-Water Removal. Removal of water follows juice extraction and preparation. Refer to the section on Concentration Methods for details.

p. 32.1, 1st para. Change the paragraph to read:

This chapter is a guide to specifying insulation systems for refrigeration piping, fittings, and vessels. It does not deal with HVAC systems or applications such as chilled water systems. Refer to Chapters 22, 23, 24, and 39 in the 1997 *ASHRAE Handbook—Fundamentals* for information about insulation and vapor barriers for these systems.

p. 32.3, Table 2. Change footnote a to read:

[a] Tested in accordance with ASTM E 84 for 1 in. thick insulation.

p. 34.2, 1st column, 7th line up. Change sentence to read,

The amount of control over each load source is indicated as an approximate percentage of the maximum reduction possible through effective design and operation.

p. 34.3, 2nd column, first equation on the page. The units of q_r should be:

Btu/h

not Btu/h-ft²

p. 34.4, 1st column. In the equation for q_r, the third value should be 0.1714, not 0.714.

p. 45.27, 2nd column. In line 2 of the second paragraph in the section on Short Tube Restrictors, the length-to-diameter ratio should read $3 < L/D < 20$.

p. 47.12, Figure 17. Relabel valve on upper left of heat reclaim coil to read, 3-WAY SOLENOID VALVE. Relabel valve on upper right of heat reclaim coil to read, CHECK VALVE.

1999 HVAC Applications

p. 15.1, Figure 1. Increments in the label for the horizontal scale should read: 0.1, 0.2, 0.5, 1, 2, 5, 10 μm.

p. 15.2, 1st column. FS 209 Class 10,000 particle count should not exceed 65 particles per cubic foot of a size 5 μm or larger.

p. 29.12, 1st column, 1st para. The last sentence should refer to Equation (25), not Equation (26).

p. 33.11, 1st column, last para. Change sentence to read, "Designers typically select a diffuser dimension to create an inlet Froude number of 1.0 or less."

p. 46.19, 2nd column. Replace text starting half-way down the column, Examples 3 and 4, and Figure 23 with the following.

For applications where the duct is exposed, the sound pressure level in an occupied space as a result of duct sound breakout can be obtained from

$$L_p = L_{w(out)} - 10\log(\pi rL) \quad (14)$$

where

L_p = sound pressure level at specified point in the space, dB
$L_{w(out)}$ = sound power level of sound radiated from outside surface of duct walls given by Equation (7), dB
r = distance between duct and position at which L_p is being calculated, ft
L = length of duct sound radiating surface, ft

Equation (14) does not apply when the duct is in a plenum, concealed behind a suspended ceiling. In such cases, simply subtract the attenuation for ceilings listed in Table 23 instead of $10\log(\pi rL)$ and assume the sound field in the room is uniform.

Example 3. A 12 in. by 48 in. by 15 ft long rectangular supply duct above a mineral fiber lay-in tile ceiling is constructed of 22 gage sheet metal. Given the sound power levels in the duct, what are the sound pressure levels at a listener 5 ft from the duct?

Solution:

	Octave Band Center Frequency, Hz						
	63	**125**	**250**	**500**	**1000**	**2000**	**4000**
$L_{w(in)}$	90	85	80	75	70	65	60
$-TL_{out}$ (Table 20)	−19	−22	−25	−28	−31	−37	−43
10 log (*S/A*)	16	16	16	16	16	16	16
$L_{w(out)}$	87	79	71	63	55	44	33
− Ceiling tile (Table 23)	−13	−16	−18	−20	−26	−31	−36
L_p, dB	74	63	53	43	29	13	−3

The RC level associated with the above L_p values is RC 28 (LFV*b*).

Example 4. Repeat Example 3 using a round duct of equivalent airflow area: 27 in. in diameter, 22 gage (0.853 mm), 15 ft long.

Solution:

	Octave Band Center Frequency, Hz						
	63	**125**	**250**	**500**	**1000**	**2000**	**4000**
$L_{w(in)}$	90	85	80	75	70	65	60
$-TL_{out}$ (Table 21)	−49	−50	−32	−30	−29	−25	−31
10 log (*S/A*)	14	14	14	14	14	14	14
$L_{w(out)}$	55	49	62	59	55	54	43
− Ceiling tile (Table 23)	−13	−16	−18	−20	−26	−31	−36
L_p, dB	42	33	44	39	29	23	7

The RC level associated with the above L_p values is RC 31(MF). The use of the round duct results in a slight increase in the RC level and eliminates the low-frequency rumble present in Example 3, but it introduces some mid-frequency roar that can be dealt with by another means. The two spectra are shown in Figure 23.

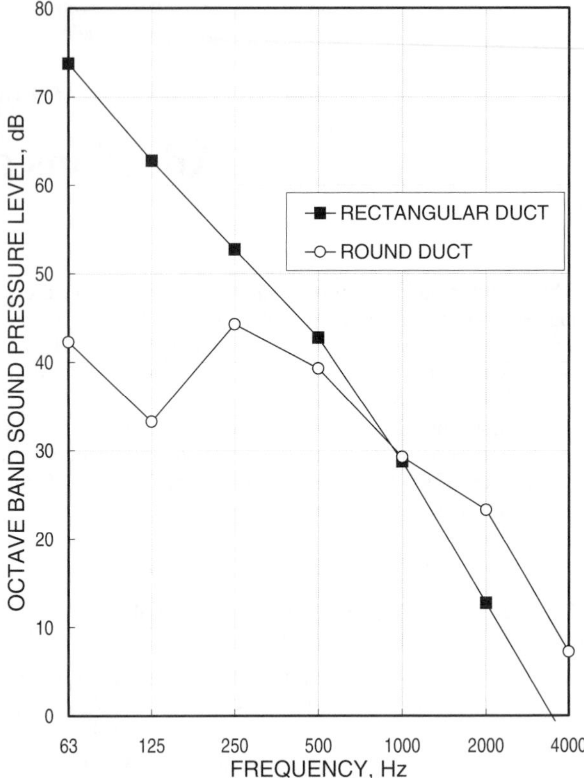

Fig. 23 Duct Breakout Sound Pressure Levels from Examples 3 and 4

p. 46.20, Table 23. Delete the footnote from the table.

p. 46.32, Equation (20). The last term in the equation should be deleted so that the equation reads as follows:

$$L_p = L_w + 10\log Q - 20\log d \quad (20)$$

p. 50.9, 1st column. In the section on Apples change the last two sentences to read,

"Generally, a cooler designed to exchange air every 3 minutes (20 air changes per hour) is the largest that can be installed. This capacity provides a complete air change every 1 to 1.5 minutes (40 to 60 air changes per hour) when the storage is loaded."

p. 50.9, 2nd column, 11th line up. Delete "per hour."

p. 51.10, Equation (17). The value of *B* is calculated for sea level standard pressure.

2000 HVAC Systems and Equipment

p. 1.3, 2nd column. In the Space requirements paragraph the first sentence should read:

"A decentralized system may or may not have an equipment room."

p. 7.11, 2nd column, second to last paragraph on the page. The Btu in the last sentence should be Btu/h.

p. 7.11, 2nd column, last paragraph on the page. The Btu in the first sentence should be Btu/h.

p. 7.33, 1st column, last paragraph on the page. The values lb/h should read lb:

7000 lb

7 lb

p. 7.33, 1st column, last line on the page. The "total energy requirement" should be "total power requirement".

p. 7.33, 2nd column, last sentence in the paragraph. The "more energy than the gas turbine" should be "more power than the gas turbine".

p. 7.33, 2nd column, last sentence in example 2. The "preceding energy costs" should be "preceding power costs".

p. 8.18, 1st column, 1st line of the column. Amplify should be raise. "...rejected heat, raise its temperature..."

p. 9.4, 2nd column, 1st paragraph, second to last sentence. Delete "by more than 25% of the sensible load."

p. 11.29, 2nd column, last line. Delete "visa" and insert "vice".

p. 11.30, 2nd column, first bulleted text. Capacity should be changed to power.

p. 18.7, 2nd column, Fig. 13. Should read:

"Curve shows performance of a fixed fan size running at a fixed speed."

p. 20.2, 1st column, last paragraph. Delete second sentence that reads, "Water vapor migrates from areas of high vapor pressure and lower temperature."

p. 30.31, 1st column, 1st reference. Change the year to 1992 and Customer Service to Handbook Editor.

CHAPTER I

COMPOSITE INDEX
ASHRAE HANDBOOK SERIES

This index covers the current Handbook series published by ASHRAE. The four volumes in the series are identified as follows:

R = 1998 Refrigeration

A = 1999 Applications

S = 2000 Systems and Equipment

F = 2001 Fundamentals

Alphabetization of the index is letter by letter; for example, **Heaters** precedes **Heat exchangers**, and **Floors** precedes **Floor slabs**.

The page reference for an index entry includes the book letter and the chapter number, which may be followed by a decimal point and the beginning page within the chapter. For example, the page number S31.4 means the information may be found in the 2000 Systems and Equipment volume, Chapter 31, beginning on page 4.

Each Handbook is revised and updated on a four-year cycle. Because technology and the interests of ASHRAE members change, some topics are not included in the current Handbook series but may be found in the earlier Handbook editions cited in the index.

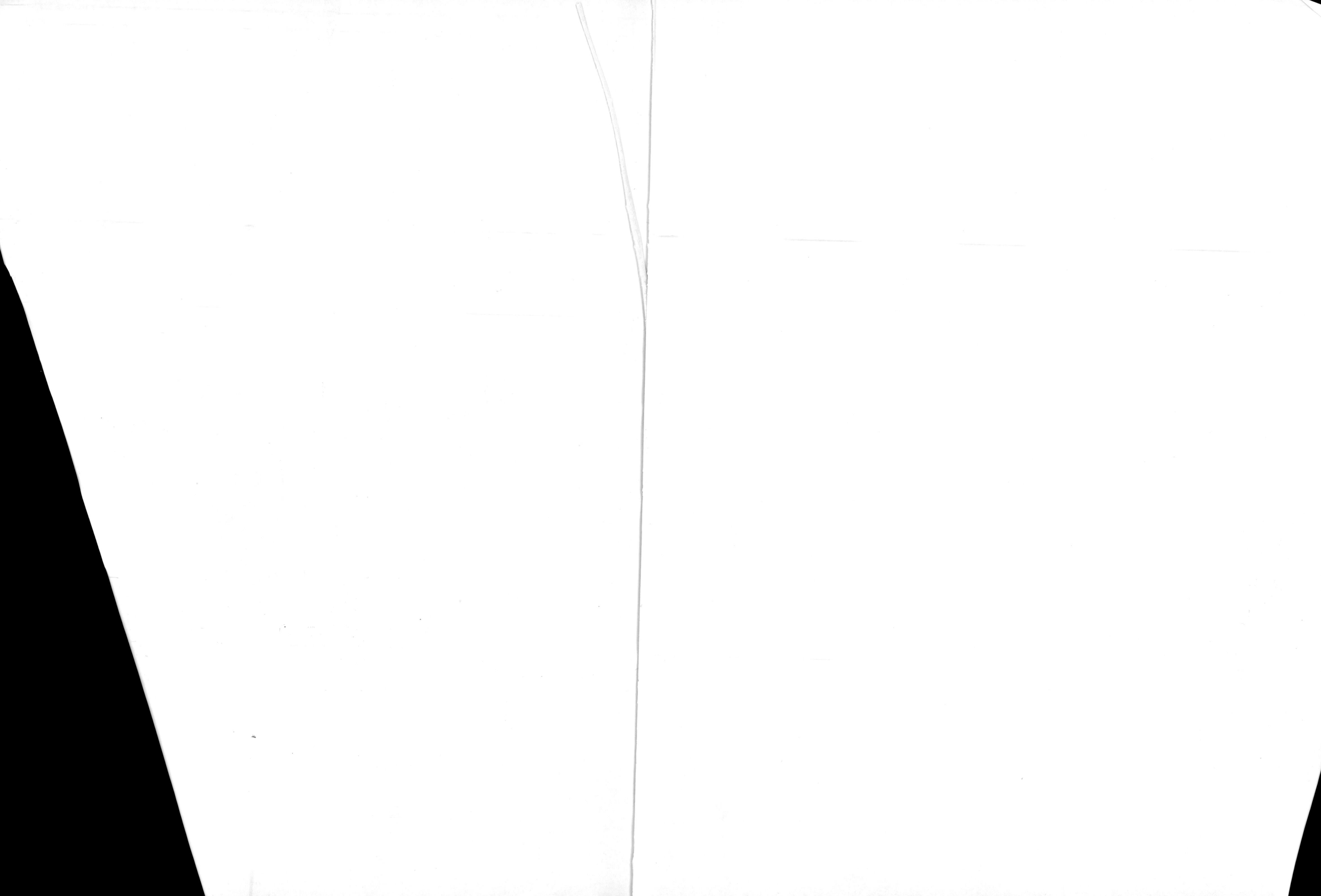